Encyclopedia of Cognitive Science

Encyclopedia of Cognitive Science

Volume 1

Editor-in-Chief

Lynn Nadel
University of Arizona

nature publishing group

London, New York and Tokyo

Published by
Nature Publishing Group, 2003
The Macmillan Building, 4 Crinan Street, London, N1 9XW, UK

Associated companies and representatives throughout the world

www.nature.com

ISBN: 0-333-792610

Distributed on behalf of the Nature Publishing Group in the United States and Canada by
Grove's Dictionaries, Inc.
345 Park Avenue South,
New York,
NY 10010-1707,
USA

British Library Cataloguing in Publication Data
Encyclopedia of cognitive science
1. Cognitive science – Encyclopedias
I. Nadel, Lynn
153′.03

Library of Congress Cataloging in Publication Data

A catalog for this record is available from the Library of Congress

Typeset by Kolam Information Services Pvt. Ltd., Pondicherry, India
Printed and bound by The Bath Press, England

Contents

Preface

The *Encyclopedia of Cognitive Science* (ECS) captures current thinking about the workings of the mind and brain, focusing on problems that are as old as recorded history, but reflecting new approaches and techniques that have emerged since the 1980s. Modern cognitive science emerged as a discipline on the 50s onwards (the history and pre-history of this field are traced in an overview article; see page XXX). It represents the convergence of ideas about mind and brain from a variety of disciplines, and as such is under constant revision.

The publishers and editor felt that it would be helpful, at this time in the evolution of cognitive science, to provide an in depth look at the best ideas and research, written by the individuals who have themselves been responsible for creating the current knowledge in the field. We aimed at a set of volumes that would provide food for thought for any interested and curious reader who wanted to know the best current answer to the kinds of questions a clever child might ask: where do my thoughts come from, why is the moon so large on the horizon, can animals talk, and so on. Our plan from the outset was to provide multiple levels of information so that readers at various levels could benefit from the articles. Some are written at an introductory level and provide an overview of a domain, others are written at some depth and with considerable technical detail. In all cases, however, we have attempted to make the material as accessible as possible to an intelligent reader by avoiding jargon wherever we could.

The project has taken us about five years from start to finish and reflects the efforts of many individuals who have devoted their time to planning, writing, reviewing, and re-writing the articles that ultimately appear in these volumes.

Ultimately, the success of a project of this sort completely depends upon the quality of the individuals who choose to write for it. We have been fortunate indeed to enlist the efforts of an outstanding group of section editors, who then recruited authors of distinction in every field to write the articles.

This is an exciting time in our field. The development of powerful computational devices has offered new metaphors for the mind, as well as a tool capable of modeling how minds are actually implemented. Within neuroscience, new techniques, including most prominently an array of neuroimaging methods that permit scientists to look at the human brain in action, have made it possible to ask and sometimes answer questions that have resisted analysis for centuries. Psychologists and philosophers are once again busily attacking the problem of consciousness – what it is, where it comes from, and why it exists. All this is shaped by a sharper interest in the cultural contexts within which minds are found, and a deep concern for how new knowledge about the workings of the mind can be brought to bear in the real world of education, engineering and more.

It is our hope that the Encyclopedia will convey this excitement to readers who approach it simply out of curiosity as well as those who look to it for detailed answers to specific questions. We started out hoping to appeal to high school students and undergraduates as well as our colleagues. We end up hoping that readers at both ends of this spectrum will benefit from our efforts and will take it upon themselves to contribute to the continuing evolution of cognitive science.

A project of this magnitude involves the efforts of many people over many years, and it is important to recognize the crucial roles they played. Gina Fullerlove and Andrew Diggle deserve the credit for initiating this project, for getting it started, and for nurturing it through its early years. When Andrew moved on, Gina continued, and was ably assisted by Roheena Anand, who took over the responsibility of day-to-day management of the project. In this effort she was lucky to have Christina Doyle and Fozia Khan at her side. Christina has done much of the hard work in the past year as the project took clear shape, and she alone of the initial group remains involved as it comes to its conclusion. As we moved from commissioning articles to other phases, the efforts of Daniel Price, Darren Smith and Jane Macmillan concerning production and marketing have been critical. Finally, Sean Pidgeon has stepped in near the end to make certain that the project's loose ends were tied. Without his help at this time we would not have been able to bring ECS to full fruition.

As Editor-in-Chief, I have been lucky to enlist the collaboration of an outstanding group of Section Editors: David Chalmers (Philosophy), Peter Culicover (Linguistics), Bob French (Computer

Science), Robert Goldstone (Psychology). They, along with Joel Levin (Education), Kevin McCabe (Economics) and John Odling-Smee (Anthropology), made the critical choices concerning topics to be covered and authors to write the articles. The quality of this project reflects their stellar work.

Finally, I would be remiss in not also thanking the people who gave up some part of their time with me so that I could work on ECS over the past five years: my wife Mary Peterson, and my children Yael, Leila, Misha, Ken and Melissa. I hope they agree that the end result was worth the sacrifices made.

Lynn Nadel
Tuscon, AZ, USA
October 2002

What is Cognitive Science?

Lynn Nadel and Massimo Piattelli-Palmarini[1]
Department of Psychology, University of Arizona, Tucson, AZ

Rich scientific disciplines defy simple definition, and cognitive science is no exception. For our present purposes, cognitive science can be defined broadly as the scientific study of minds and brains, be they real, artificial, human or animal. In practice, cognitive science has been more limited, largely restricting itself to domains in which there is reasonable hope of attaining real understanding. The richness and diversity of the contributions to this encyclopedia show that there are now many such domains.

By 'understanding' we mean going beyond common-sense intuitions, often to the point of radically subverting them. It is no longer surprising when a cognitive system is shown to work in highly unexpected ways. One of the insights of modern cognitive science is that the mind often works in counterintuitive ways.

Mature sciences owe much of their initial progress to the pursuit of phenomena and hypotheses within a few 'windows of opportunity', often opened by chance. Many of these seem, at their inception, to be quite far from the daily concerns of ordinary people, but they come to have great impact. Cognitive science has thrived on such opportunities. As an example, the analysis of language impairments following stroke or war injuries by a group of outstanding neurologists in the nineteenth century (Broca, 1878; Wernicke, 1874) led to important insights about both the organization of language and its instantiation in the brain (*See* Aphasia; Broca, Paul; Language Disorders; Wernicke–Geshwind Model). The use of rapid-succession photography by the French physiologist Marey (1830–1904; Braun, 1992) and by the expatriate Englishman Eadweard Muybridge in the U.S. (also, by a curious coincidence, born in 1830 and deceased in 1904) (Haas, 1976) made possible the analysis of the natural motion of people and horses and opened the way to our present understanding of the uniqueness of such motion in perception and planning (*See* Motion Perception, Neural Basis of; Motion Perception, Psychology of).

The development of highly selective neuron staining methods, which led to the emergence of the 'neuron doctrine', is yet another example. The ability to visualize fine details of the central nervous system (CNS) settled the debate between the Italian Camillo Golgi (1843–1926) and the Spaniard Santiago Ramón y Cajal (1852–1934) over the nature of connection and communication within the brain (*See* Golgi Staining; Neuron Doctrine). Golgi had argued that the nervous system was a meshwork of connected elements. Cajal, on the other hand, argued that there were discrete elements within the CNS, subsequently called neurons, that were not actually in contact, but instead communicated across a gap – the synapse. Stained sections supported Cajal, and the era of neurons and synapses began.[2,3] The subsequent study of neuronal transmission in the squid giant axon and the development of single neuron recording methods in vertebrates built upon this early work. Parallel developments in the study of logic and computation stimulated early attempts to apply mathematical principles to nervous system function (McCulloch and Pitts, 1943), and these in turn eventually led to modern approaches in computational neuroscience (*See* Computational Neuroscience: From Biology to Cognition).

Attempts to ameliorate diseases of the nervous system have played a significant role in the history of cognitive science. The treatment of recurrent epileptic seizures resistant to all drugs led to experimental surgical treatments that have produced fundamental knowledge about brain function (*See* Epilepsy). The use of radical resection of the temporal lobe had the unfortunate consequence of causing a massive amnesic syndrome, but the study of the patient known as H.M. (Scoville and Milner, 1957) and other such patients has contributed greatly to our understanding of memory. Similarly, the use of surgical section of the corpus callosum led to the discovery of the 'split brain', which has fascinated philosophers and the public at large ever since (Gazzaniga, 1970; Sperry, 1968; *See* Sperry, Roger; Split-brain Research).

In this general introduction we will briefly explore these and some other 'windows of opportunity' that helped create cognitive science as we know it today. We cannot be exhaustive but we hope in this short introduction to provide a broad roadmap to the field, where it has come from, and where it might be headed. We are aware that the names of protagonists are often introduced without proper biographical presentation, and that many concepts and terms are briefly characterized or sometimes just mentioned. It is our goal here to offer a general synopsis, partly historical, partly analytic, but hopefully sufficient to situate people and ideas in a vast web, trusting that the reader will be stimulated to locate related articles in this work. Our task is made somewhat easier by the existence of some excellent histories of cognitive science (e.g. Albertazzi, 2001; Baars, 1986; Bechtel *et al.*, 1999; Dupuy, 2000; Gardner, 1985), as well as numerous reflections on history by key participants (to which we shall refer below; also *see* History of Cognitive Science and Computational Modeling). Inevitably, our road map will reflect the journeys we ourselves have taken; we apologize in advance to those of our colleagues whose contributions to cognitive science have been overlooked in what follows.

SOME PREHISTORY

Interest in mind and brain is as old as recorded history, and any complete rendering of the prehistory of cognitive science would treat early philosophers at some length. That, however, is not our purpose. Rather, we will take it for granted that interest in fundamental questions about cognition and its physical bases has long existed, well before the term 'cognitive science' was coined, and pick up our story at the point where genuinely scientific analyses became possible in the nineteenth century. An interesting perspective on this gestational era is offered in Albertazzi (2001).

We can identify several strands of nineteenth century thought as clearly antecedent to modern cognitive science: the work of the neuropsychologists who studied the impact of brain damage on language and other cognitive abilities; the development by Darwin and others of the theory of evolution (later to be extended into theories of the evolution of the brain and mind; also *see* Evolutionary Psychology: Applications and Criticisms; Evolutionary Psychology: Theoretical Foundations); the creation of modern experimental psychology and psychophysics (Ebbinghaus, Helmholtz and Wundt deserve special mention here; also *see* Ebbinghaus, Hermann; Helmholtz, Hermann von; Wundt, Wilhelm); and the initial efforts of neurologists and psychiatrists to relate complex human conditions to underlying neuroanatomy and neuropathology.[4]

Several landmarks at the very end of the nineteenth century stand out: (1) the publication in 1890 by William James of *The Principles of Psychology* (James, 1890; also *see* James, William); (2) the aforementioned emergence of the neuron doctrine in the work of Cajal; and (3) the development of Freud's psychodynamic approach (*See* Freud, Sigmund). All three had profound and lasting effects, although many might argue that the influence of Freud's thinking has waned; time will tell. A fourth strand of intellectual development that has had a profound influence on cognitive science can also be traced back to the nineteenth century – the emergence of computational devices. Fascinating histories have been written about the role of key individuals such as Augusta Ada Byron (Stein, 1985) and Charles Babbage in this development, but real progress in this domain was not seen until the twentieth century. Finally, from philosophy and linguistics came the development of powerful systems of logic (Frege, 1879; Russell, 1900, 1919; Tarski, 1935, 1996; Whitehead and Russell, 1910; also *see* Frege, Gottlob).

These various advances began to provide the basis for formal treatments of many aspects of cognitive function, and it is these treatments that, in the fullness of time, combined with germane developments in experimental psychology, neuroscience, linguistics and anthropology. The convergence of these strands produced modern cognitive science.

THE BEHAVIORIST INTERREGNUM

Notwithstanding the explosion of possibilities offered by the developments noted above, the beginning of the twentieth century saw a turning away from many of the issues central to cognitive science, especially in North America. The behaviorist revolution, exemplified by John B. Watson (1924), can be viewed in retrospect as a reaction against the overly ambitious reach of early cognitive scientists.[5] Behaviorists rightly pointed out that not enough was known about what goes on inside the organism to ground any sort of meaningful theory. Instead, behaviorism argued that psychology should restrict itself to what could be observed, which ruled out consciousness and other products of introspection. Combined with an infectious enthusiasm for spreading 'rigorous' scientific methodology to all fields of inquiry, behaviorism effectively banished all appeals to internal states and representations (concepts, ideas, meanings, percepts, computations, etc.): better to focus on what could be observed and measured if one wanted to create a science.[6]

A strict behaviorist view, and its emphasis on general learning mechanisms, persisted for about

50 years in North America, during which time the center was occupied by narrowly conceived research programs, some of which bore considerable fruit. Clark Hull's (1943) early efforts at producing mathematical models of behavior were largely unsuccessful but they provided the foundation upon which a more modern and influential mathematical psychology was subsequently built (*See* Hull, Clark L.).[7]

Three major exceptions to the narrow behaviorist perspective of this era were Karl Lashley, Edward Chace Tolman and Egon Brunswik (*See* Lashley, Karl S.; Tolman, Edward C.). Lashley's (1929) work on neural mechanisms underlying intelligence and his thoughts about the localization of function in the nervous system provided a foundation for the subsequent efforts of many influential neuropsychologists of the 1940s and 1950s, including Donald Olding Hebb (Hebb, 1949; also *see* Hebb, Donald Olding). Tolman and Brunswik were strongly influenced by the Gestalt movement in Germany, which had focused on the role of higher-order factors such as form relations and organization in accounting for perception, insight and even learning (*See* Perception, Gestalt Principles of). They set the stage for a return of cognitive approaches in North America. Tolman's early book, *Purposive Behavior in Animals and Men* (1932), provided a roadmap for the pursuit of aspects of behavior that went beyond the observable. Brunswik, an expatriate Viennese, joined the psychology faculty of Berkeley in 1937 thanks to Tolman, but remained a maverick his entire life (committing suicide in 1955; see Bower, 2002). He introduced the idea of 'ecological validity' (usually associated with the much better known work of James Gibson and Ulric Neisser), defying the relevance of the narrow laboratory settings so dear to the behaviorists. Brunswik studied the role of sensory cues and subjective estimates in shaping perception and judgment. He advocated the view that knowledge is a probability-based process, developed a 'probabilistic functionalism', and was among the first to reveal subjective probability biases. His better known 'lens model' pictures systematic distortions at the interface between the external scene and the observer, whereby the structure of the environment is filtered by the structure of subjective perception and knowledge. This results in 'perceptual compromises' fit to serve the relevant purposes of the observer.[8]

Behaviorism was not as dominant in Europe as it was in North America, and elements of what was to become cognitive science proceeded in a variety of domains. The Gestalt psychologists, including Wertheimer, Kohler and Koffka, steadfastly retained a focus on the role of organization in perception and problem solving, with Kohler's work on 'insight' being a particularly important contribution. Bartlett's book *Remembering* (1932) explored the role of schemata in the formation of memory, and remains influential to the present day (*See* Bartlett, Frederic Charles). Piaget's (1930, 1954) comprehensive studies on development, which emphasized the internal models formed by children in comprehending the world, had an enormous impact in Europe, and subsequently in North America when they were made widely known after World War II (*See* Piaget, Jean).[9] Finally, there were the convergent contributions by Kurt Gödel, Alan Mathison Turing, Alonzo Church and Stephen C. Kleene on fundamental issues in computation, recursion and the theory of automata (Church, 1936; Kleene, 1936; Turing, 1936a,b; also *see* Computability and Computational Complexity; Turing, Alan).[10]

THE MODERN ERA BEGINS

In 1936, Turing published his crucial paper on 'Computability', in which he spelled out the design for a machine that could carry out any set of well-defined formal operations. Two years later, Shannon (1938) demonstrated that on–off electrical circuits could carry out basic mathematical procedures, an idea that ultimately led to the development of information theory (Shannon and Weaver, 1949/1998; also *see* Information Theory; Shannon, Claude). A critical early contribution was published by Kenneth Craik (1943), entitled *The Nature of Explanation*. Craik sought ways to link mental and mechanical operations, and settled on the notion of internal models that would become central to cognitive science in the future. He claimed (p. 85) that:

> thought is a term for the conscious working of a highly complex machine, built of parts having dimensions where the classical laws of mechanics are still very nearly true, and having dimensions where space is, to all intents and purposes, Euclidean. This mechanism, I have argued, has the power to represent, or parallel, certain phenomena in the external world as a calculating machine can parallel the development of strains in a bridge.

According to Craik, thought involved three critical steps: first, external processes were translated into words, numbers or symbols; second, these 'representations' were manipulated by processes such as reasoning to yield transformed symbols; and third, these transformed symbols were retranslated into external processes to yield a product, such as behavior. The critical assumption here is the idea that minds create internal models and then use these models to predict the future. Such a thought process allows an organism the luxury of trying out possible futures before settling on the one that would be most adaptive. For Craik, thoughts could not be separated from feelings, a perspective that early cognitive science ignored to its detriment. He died in a bicycling accident a few years after the publication of this work, and further development depended upon others.

One of the major historical forces propelling this development was World War II, which presented a set of military problems that required rapid computational solution. The breaking of military secret codes, the calculation of artillery fire trajectories, and several problems faced by the real-time reactions of airplane pilots soon occupied the best minds of the time, and enormous progress in understanding complex systems resulted. 'Cybernetics', arguably the most crucial of the disciplines that paved the way to cognitive science (Dupuy, 2000), was defined by one of its inventors (the MIT mathematician Norbert Wiener) as the science of communication between complex systems (natural or artificial) and of the control of such systems by intelligent agents. It derives its name from the Greek 'kybernetis' (the skipper of a boat and, by extension, the pilot of an airplane), betraying its origin in concrete problems posed by the war effort (Wiener, 1948).

Many of the participants in this fascinating era have provided at least some historical record, as noted already, and the interested reader is directed to these sources for a fascinating tour (e.g. McCulloch, 1988; Heims, 1991). Almost all agree that a few critical concerns fueled the enterprise: first, there was a strong desire to bring the insights of mathematical modeling (the solution of complex differential equations) and of mathematical logic into contact with both biology and engineering. Two of the key players were trained by giants in logic: Norbert Wiener under the guidance of Bertrand Russell, and Walter Pitts under Rudolph Carnap. Second, there was a strong commitment to the notion, outlined by Craik, that thought could be viewed as a computational process utilizing internal models, and hence cognition would ultimately be accounted for in terms of finite and specifiable procedures that could be performed by the 'computers' that were then being developed. In this context we must mention John von Neumann, who played a central role in the early days of cognitive science. His work on game theory (von Neumann and Morgenstern, 1944; also *see* Game Theory), and his contributions to the development of computers (von Neumann, 1951) were critical at the outset, and he was a major participant in the emergence of cybernetics until his early death in 1957 (*See* von Neumann, John). His posthumously published essay on *The Computer and the Brain* discusses the fundamental properties of computation in machine and brain, laying out 'an approach toward the understanding of the nervous system from the mathematician's point of view' (von Neumann, 1957, p. 1).

A powerful additional incentive, though initially mostly indirect, came from plans to put the new programmable computers and proto-robots (Ashby, 1960; Walter, 1953; *See* Walter, Grey) to good use in navigation and learnable self-steering, text translation between languages, selective and addressable archiving, automatic abstracting of documents, visual detection and discrimination, problem-solving and automated induction. The impact of a practically-oriented engineering perspective and of generous financial resources on what had previously been abstract and elusive domains of academic research generated some rather naive approaches and exaggerated expectations, but there were also fresh starts, original redefinitions of many problems and new models that were unencumbered by ancient paralyzing paradoxes and a stifling received wisdom.

Arguably the best example is to be found in linguistics, which had been mostly a literary, comparative and philological discipline. It suddenly received a new impulse to rethink its very foundations and to explore computer-assisted applications. These new approaches to the structure of language (Bar-Hillel, 1954; Chomsky, 1955, 1957, 1975; Harris, 1951, 1952a,b, 1957, 1986/1951) were in some measure a reaction to shotgun engineering attempts to build automatic translators. It is emblematic, for instance, that starting in the mid-1950s and for decades to come, the team of linguists that gave rise to generative grammar was assembled and then sustained in a school of electrical engineering.[11]

All in all, this early seminal period proved crucial in engendering the conviction that long-standing hard problems in the study of the mind and brain were open to radically new insights, and that an intense collaboration between different scientific fields (from logic to neurology, from text analysis to the mathematical theory of recursive functions) was not only possible, but mandatory. The end of World War II heralded an era of 'visiting' scientists, mobile young PhDs and ardent interdisciplinary exchanges. Not unlike the immense impact that high-level scientific meetings had previously had on physics in the 1920s and 1930s, the role that some meetings in the 1940s and 1950s had in shaping what later became cognitive science deserves special attention.

THE 1940s: SEMINAL MEETINGS FOR COGNITIVE SCIENCE

In reconstructing the history of the early years, one cannot escape the crucial role played by a number of critical meetings at which key participants from several fields were profitably brought together. The first such meeting, held in 1942 in New York City, focused on the topic of central inhibition in the nervous system. It was sponsored by the Josiah Macy Jr. Foundation, which would play an absolutely essential role in the birth of cognitive science over the next decade. At this meeting, Warren McCulloch and Arturo Rosenblueth presented material related to the papers they were about to publish (McCulloch and Pitts, 1943; Rosenblueth *et al.*, 1943; *See* McCulloch, Warren). These papers suggested, in rather different ways, that aspects of mental activity could be modeled in a formal way using idealized nervous system elements (an idea that was to be revamped many years later by the connectionists, with radically different mathematical models, as we shall see). McCulloch and Pitts showed that networks of on–off neurons could compute logical functions, while Rosenblueth *et al.* (1943), concerned with purpose, showed that goal-directed behavior could emerge in systems with feedback.

In the winter of 1943–44, another meeting was held at Princeton, attended by McCulloch and Pitts, by a prominent neuroanatomist, Lorente de No, and by two leading figures in the emerging computer paradigm, John von Neumann and Herman Goldstine. This line-up suggests the coming together of formal logic, neural networks, real neuroanatomy, and computation. Lorente de No (1938) had previously demonstrated, in elegant anatomical work, that conditions existed within the cerebral cortex for reverberatory circuits that could instantiate re-entrant loops. These loops were seen as critical to maintaining memory within the brain, a requirement that had been hinted at earlier by Kubie (1930) and was adopted in Hebb's (1949) neuropsychological theory of cell assemblies and phase sequences (*See* Hebbian Cell Assemblies). Hebb's book *The Organization of Behavior* stands out as a crucial contribution, merging psychology and physiology in pursuit of explanations of cognitive function.[12]

Ten Macy Foundation conferences on cybernetics were held between 1946 and 1953. The major moving force behind these meetings was Warren McCulloch, as portrayed in considerable detail by Heims (1991) and Dupuy (2000). No written record exists of the first five meetings, but proceedings of the final five were published. Over the course of these meetings, ideas central to cybernetics – such as control, feedback and communication – were imparted to a wide range of scientists. What distinguished these meetings was their interdisciplinarity. The anthropologists Gregory Bateson and Margaret Mead were as central to the proceedings as the neurophysiologist McCulloch, the mathematicians Wiener and von Neumann, the information theorist Shannon and the psychologists Kluver and Lewin (*See* Lewin, Kurt). Another distinguishing feature of these early meetings on cybernetics was the lack of concern with computation as a 'symbolic' activity. This perspective, which came to dominate cognitive science in the 1950s, was largely absent from these early formative discussions.

Some of the same individuals participated in the famous Hixon symposium, held at Caltech in 1948, and published a few years later (Jeffress, 1951).[13] It was at the Hixon meeting that von Neumann gave a paper on 'The general and logical theory of automata', and Lashley wowed the audience with his famous paper on 'Serial order' (Lashley, 1951), in which he took a strong stand against the reigning behaviorist stance, pointing out that the requirements of speech and language rendered stimulus–response theory implausible.

Another critical meeting was held September 10–12, 1956 at MIT – the second Symposium on Information Theory. George Miller has referred to the session held on September 11 as the actual birth day of cognitive science (but see Wildgen, 2001 for a different perspective). Speaking that day were Newell and Simon, who presented material on their logic theory machine; Rochester, who had been using a large digital computer in a failed effort to test Hebb's theory of cell assemblies (Rochester *et al.*, 1956); Chomsky, who showed how linguistics could produce results with considerable mathematical rigor; Miller, who talked about the limits of short-term memory; and Swets, who presented on signal detection theory and perceptual recognition. George Miller later claimed that he went away from this meeting with the feeling that a new science was emerging.

Some months before this germinal meeting, a research seminar on artificial intelligence was held at Dartmouth, attended by most of the individuals active in the area at the time (*See* Artifical Intelligence, Philosophy of).[14] Newell and Simon (1972) point out in the Historical Addendum to *Human Problem Solving* that Minsky's (1961) essay, 'Steps toward Artificial Intelligence' – first circulated in draft form at this summer meeting – captures the consensus views that existed at that time (available on Minsky's website at: http://web.media.mit.edu/~minsky/).

In meetings such as these one can see the mix of disciplines that would come to define much of modern cognitive science. One can also see the groping for methodologies that would permit scientists to ask meaningful questions about mental activity. The single-minded emphasis on behavior at the expense of cognition was clearly at an end. Within a few years the first of many centers of cognitive studies was started, at Harvard, by Jerome Bruner (focusing on spontaneous reasoning; Bruner *et al.*, 1956) and George Miller (focusing on language and memory; Chomsky and Miller, 1963; Miller and Chomsky, 1963). Roger Brown soon transferred there from MIT, enriching the research domain with his pioneering studies of first-language acquisition (Brown, 1958, 1973). The impact of this center was enormous: its interdisciplinary mix included faculty, research fellows, visitors and local-area researchers who over the years made substantial contributions to cognitive science.[15] Now, over 40 years later, many such programs exist.

In the 1950s, as clearly seen in the 1956 MIT meeting, the close connection between biology and cognitive science fell apart. Simon was perhaps the leading, but by no means the only, exponent of the view that to understand cognition one need not pay much if any attention to the underlying biology. Newell and Simon (1972), for example, characterized Hebb's position as 'confused' insofar as Hebb thought he was proposing a physiological account of behavior. In their view it was essential to interpose 'a specific layer of explanation lying between behavior, on the one side, and neurology on the other' (p. 876). This perspective foreshadowed a similar position, presently mainstream, adopted by Marr (1982) a decade later, in his classic treatise on vision (*See* Marr, David). The net result of this shift in emphasis was that the integrative perspective of the cyberneticists was sacrificed, and an era of symbolic modeling, of direct study of mental computations and representations, and of artificial intelligence research without much reference to real brains, blossomed.

It is of historical interest to consider why this happened. We would suggest a few possible reasons: (1) the early biologically driven approach to neural nets, as exemplified by the perceptron model (Rosenblatt, 1962; *See* Perceptron) apparently failed, the limitations of these early models being subsequently made clear by Minsky and Papert (1969/1990) – Hanson (1999) provides an interesting perspective on the competition between Rosenblatt and Minsky, and how the latter's views, tilted towards artificial intelligence and away from biology, carried the day – and initial attempts by Rochester and his colleagues (Rochester *et al.*, 1956) to simulate the neurobiology of Hebb and Peter Milner were largely unsuccessful, as noted already; (2) writers such as Chomsky argued persuasively that it was the formal properties of the mind–brain that mattered, not the underlying biology that allowed it to compute;[16] (3) one of the main representatives from neuroscience, Karl Lashley, was himself rather skeptical about the enterprise. At a symposium on 'The Brain and the Mind' at the American Neurological Association meeting in 1951, Lashley served as a discussant on several papers and had this to say about the enterprise of linking brains and computers:

> I suggest that we are more likely to find out how the brain works by studying the brain itself and the phenomena of behavior than by indulging in far-fetched physical analogies. The similarities in such comparisons are the product of an oversimplification of the problems of behavior (quoted in Beach *et al.*, 1960, p. xix).

What is more, Lashley's probabilistic views of nervous system function were very much at odds with the connectionist requirements of the cyberneticists. Lashley participated in many of the early critical meetings, and his views had a major impact. The rather more positive approach of Hebb, who disagreed with Lashley about the role of specific neural pathways, could not overcome Lashley's influence at that time.

Because of this schism, cognitive science and neuroscience developed separately after the 1950s. For the better part of 25 years, much of neuroscience was reductionist in scope and purpose, rarely speaking to questions of interest to cognitive scientists. On the other side, much if not all of cognitive science proceeded within a symbolic framework that required little or no contact with the brain. Neurons were relegated to the role of 'mere implementation'.

The then prevailing philosophy of mind, called 'functionalism' (Dennett, 1987; Fodor, 1975, 1981; Putnam, 1960, 1973) offered principled arguments as to why any physicobiological implementation of cognitive functions was secondary to a thorough abstract characterization of the logical structure of the mental representations and transformations involved in those functions (*See* Functionalism).[17]

EMPIRICAL RESULTS

The 1950s and beyond witnessed an impressive growth in empirical results in all the domains of cognitive science. Chomsky revolutionized the study of language; Broadbent (1958) and others focused on attention (*See* Attention); Bruner and his colleagues (1956) looked at thinking; Newell, Shaw and Simon (Newell *et al.*, 1958) produced the General Problem Solver; Hochberg (1956) studied the role of memory and other internal factors on perception; Sperling's (1960) work on brief visual presentations and partial report methods led to the notion of an iconic memory store; and there were various thrusts in artificial intelligence (e.g. Samuel's checkers program, 1959; also *see* Samuel's Checkers Player), including interesting work on mathematical neural networks (e.g. Selfridge, 1958; Rosenblatt, 1958). But there were problems. In many cases the successes were garnered in severely restricted systems, with no certainty that they would scale-up or generalize. Some domains were simply not part of the mix – the study of emotion, or consciousness, was ruled off limits (*See* Emotion; Consciousness). During this period, only those phenomena of which humans were somehow at least partially conscious qualified as 'cognitive' – implicit capacities did not make the grade, nor did any animals (*See* Implicit Cognition). Cognitive science defined its subject of study – the human mind – as an 'information processing' device, and models of such things as perception, memory, and the like were couched in terms of typically step-wise processes that led from an input through stages of transformation and representation to a possible output (*See* Information Processing).

Although cognitive science at this time paid little heed to the brain, neuroscience pushed ahead, making great strides in a number of areas. In the late 1940s, the discovery of the 'reticular activating system' (Moruzzi and Magoun, 1949) had a major impact (*See* Reticular Activating System). This landmark event shifted thinking about the brain in a fundamental way. It showed that, contrary to prior notions, the brain was not a passive organ waiting to respond to external stimulation. Instead, it was constantly active, and the critical question was no longer what brought it into activity but rather what kind of activity it engaged in. The selectivity of brain function was shown to reflect not just exogenous factors, but endogenous ones as well. The implications of this were enormous, as Hebb noted.[18]

Another critical research program centered around Penfield and the group of scientists at the Montreal Neurological Institute (MNI), largely focused on patients about to undergo surgery to control for epilepsy (*See* Penfield, Wilder). Penfield and Rasmussen (1949) pioneered the method of stimulating the exposed brain in areas adjacent to the presumed site of the focus as a means of determining which tissue should be excised and which spared. This method yielded several remarkable and widely reported results. First, this method generated the famous pictures of sensory and motor maps in the cortex, within which various body parts were represented in often unusual proportions. Second, punctate stimulation in the temporal lobe could apparently yield the retrieval of a highly specific and detailed memory. This finding strongly countered Lashley's nonlocalizationist perspective, and dramatically affected views about the organization of information in the brain. It seemed to promise that an approach depending upon specific neural connections might indeed have merit, much as Hebb (Penfield's colleague at McGill, and Lashley's former student) had proposed. At the same time, the study of H.M., another patient at the MNI – who had lost the capacity to memorize recent events, though he maintained some capacity to remember events that preceded the surgical bilateral section of part of the medial temporal lobe – generated enormously important information about the critical role of the hippocampal formation in memory (*See* Amnesia). The capacity of such patients to learn new procedural tasks, without any explicit memory of what they were doing or why, also raised considerable interest (Scoville and Milner, 1957; Milner, 1965).

A final critical discovery dependent on the study of epileptic patients followed upon the use of callosal section to prevent the spread of the epileptic focus from one brain hemisphere to the other. These 'split-brain' patients were quickly shown (by Roger Sperry and his colleagues)[19] to have remarkable psychological characteristics that have informed us for nearly 50 years about how cognitive functions are carried out in the brain.

Two very important meetings that were held in the 1950s involved many of the neuroscientists involved in the work just described. Both were sponsored by the Council for International Organizations of Medical Sciences (CIOMS). The first was held in August 1953 in the Laurentian Mountains near Montreal, and brought together researchers in various fields to discuss the implications of the reticular activating system – it was titled 'Brain Mechanisms and Consciousness' (Delafresnaye, 1954).[20] The second, held in August 1959 in Montevideo, Uruguay, brought together an even wider array of neuroscientists under the title 'Brain Mechanisms of Learning' (Delafresnaye, 1961).[21]

The inclusion of scientists from the USSR and Eastern Europe at the second meeting was particularly noteworthy, as the cold war had precluded such interactions for much of the period 1946–1957. Once again, the Macy Foundation had an important role to play, sponsoring (with the National Science Foundation) three yearly conferences beginning in 1958 on 'The Central Nervous System and Behavior' (Brazier, 1958, 1959, 1960; also *see* Brazier, Mary A. B.). The first meeting had a central goal of bringing Russian neurophysiology to the West; although no scientists from the USSR were present, the work of Sechenov, Pavlov (*See* Pavlov, Ivan Petrovich), Bechterev and others was the focus of the discussion.[22] The second conference broadened this base by including several prominent researchers from Eastern Europe – Bures (Prague, Czechoslovakia),[23] Grastyan (Pecs, Hungary) and Rusinov (Moscow, USSR). The third meeting included Luria (*See* Luria, Alexander R.) and Sokolov, both from Moscow. Yet another critical meeting held at this time (October, 1958) was the 'Moscow Colloquium on Electroencephalography of Higher Nervous Activity' (Jasper and Smirnov, 1960). This meeting ranged widely over many topics, and brought together the most prominent neuroscientists from both East and West.

The impact of these six meetings, focused on the brain, was immense. It is not an exaggeration to say that an entire generation of cognitive neuroscientists (although not yet called by that name) was weaned on the books from these meetings. While the meetings focused largely on arousal, memory, perception and the like, similar undercurrents were at play in the study of language.

By the mid 1950s, the view had emerged that the careful study of aphasia from a variety of perspectives could yield real gains in understanding language (*See* Aphasia). This feeling led to a six-week seminar, held in 1958 at the Boston VA Hospital.[24] The discussions at this seminar were captured in a book published some years later (Osgood and Miron, 1963). A few years after the Boston VA meeting, another meeting focused on aphasia was held in London, sponsored by the Ciba Foundation (De Reuck and O'Connor, 1964).[25] The spirit of both meetings was interdisciplinary, and though these efforts were not in the mainstream of cognitive science at the time, they were important in setting the agenda in cognitive neuropsychology.[26]

At a more neurophysiological level, tremendous discoveries were being made, largely in the wake of technical advances that made possible the recording of activity from individual neurons in response to carefully controlled inputs. Here, the work of Mountcastle (1957) in the somatosensory system, and Hubel and Wiesel (1962a,b) in the visual system, stand out as seminal, to be followed by a literal explosion of studies that continues to the present day. In this vein it is also important to mention the classic study by Lettvin *et al.* (1959) in the visual system of the frog. These authors claimed to have found the neuronal correlates of Kant's synthetic *a priori* knowledge (*See* Kant, Immanuel), propositions that were not derived from the external world, but that were nonetheless taken by the organism to be true (e.g. the axiom that a straight line is the shortest distance between two points). McCulloch (1988) later referred to this study as a first step towards experimental epistemology. All of these studies showed that the activity of neurons could be related in meaningful ways to certain properties of external stimulation without being a passive copy of the surrounding scene, and in so doing began the process of explaining how internal models of the world could be instantiated in the brain (*See* Neurons, Representation in).

Another key finding was the discovery by Olds and Milner (1954) of systems in the rat brain that subserve reward (*See* Brain Self-stimulation; Olds, James; Reward, Brain Mechanisms of). Although it has taken nearly half a century for the study of affect to be reintegrated with the study of thought, reasoning and 'pure' cognition, the basis for this synthesis was laid in these early studies. This finding was important at that time for another reason: it contributed to the demise of Hull's drive reduction theory. Along with contemporary studies demonstrating the power of curiosity and stimulation seeking in the perceptual domain, self-stimulation of the brain presented a form of behavior that simply could not be accounted for in terms of drives and their reduction. As the 1960s dawned, one could get the feeling of great progress, but in a compartmentalized way. The decade itself ended with the publication of a landmark book by Miller *et al.* (1960) that proposed a model for cognitive function applying the insights of Craik (internal representations) and cybernetics (feedback) to the problems identified by Lashley (serial order) and Chomsky (generative linguistics).

CONSOLIDATING THE GAINS

The 1960s were a period of consolidation of the gains made in the previous decade, but new problems were looming. Little progress was being made in connecting minds with brains, and the grand promises issued by proponents of the symbolic approach to artificial intelligence were beginning to look unattainable. It seemed that there were quite a few things that human brains could do much better, and even faster, than computers. The realization that visual recognition was actually a very complex phenomenon followed from repeated failures to get computers to solve even simple recognition problems (*See* Computer Vision; Pattern Recognition, Statistical). Language translation by machines, as we have said, also seemed a lot harder than was once imagined (*See* Machine Translation). During these years an ongoing program of informal meetings was organized through the 'Neurosciences Research Program' sponsored by MIT and several federal agencies. Francis O. Schmitt was the force behind this effort at the outset, which brought an interdisciplinary group of scientists together on a regular basis to discuss the latest developments. One such meeting, for example, held in 1964, was concerned with 'mathematical concepts of central nervous system function'. These meetings, and the reports they generated that were published as a Bulletin and informally circulated, had a significant impact on the emerging domain of neuroscience.

A signal event in the 1960s was the publication of Ulric Neisser's book, *Cognitive Psychology* (1967). This was the first textbook in the field, and it had a powerful didactic and organizational influence for many years. Sternberg's (1966) work on memory stages using reaction time studies (*See* Reaction Time) and Posner's (1969) approach to abstraction and recognition made important contributions to this new domain.

While many areas of cognitive science were in a consolidation phase in the 1960s, psycholinguistics blossomed into a highly influential discipline (*See* Psycholinguistics). The groundwork was laid in part by an ongoing seminar at MIT on language acquisition led by Roger Brown and attended regularly by Chomsky and others, which formulated the basic approach to the study of language acquisition that exists today. At the same time, Miller inspired a group of young psychologists to take as an object of experimental investigation Chomsky's theory of grammar as presented in his *Syntactic Structures* (1957). That model had rocked the linguistic world by proposing a 'deep' structure that represented the essential thematic relations between verbs and arguments, regardless of their surface order (harking back to the 'inner form' of sentences proposed by Wilhelm Wundt and others). Miller and his students created a set of experiments that seemed to demonstrate the 'psychological reality' of deep structure in memory for sentences. Thus was born an exciting period in which it seemed that results from the highly speculative and theoretical discipline of linguistics could be immediately tested in the laboratory: the rallying cry was 'one linguistic rule–one mental operation'. Extended and careful research as well as logical argument later disproved this idea. Fodor and Garrett (1966) noted that the relationship between the grammar and behavior must be 'abstract' to some degree: Bever (1970) proposed a direct 'strategies' model of comprehension that shortcuts linguistic rules. Their joint book (Fodor *et al.*, 1974) laid out systematically the implications of these ideas for the major areas of psycholinguistics: acquisition, perception and production.[27]

The fierce debate about the linguistic (or protolinguistic) capacities of higher primates started in earnest in these years (Gardner and Gardner, 1971; Gardner, 1989; Fouts, 1989), opposing the unshakable believers in the evolutionary continuity of all cognitive functions[28] to the *bona fide* linguists, led by Chomsky, who stressed that the unique central components of human languages (discrete infinity, boundless recursivity, constituency, generative power, compositionality) have no counterpart in the communicative systems found in animals, higher primates included (*See* Animal Language; Language Acquisition by Animals). The negative conclusions on the linguistic capabilities of chimpanzees by the prominent cognitive primatologist David Premack (Premack, 1972, 1986), and the thorough longitudinal study of the male chimp Nim Chimpsky at Columbia University (Seidenberg and Petitto, 1987; Terrace *et al.*, 1979) reset the debate for quite some time. Though remarkably intelligent and capable of sophisticated cognitive operations, chimpanzees are provably devoid of the most central components of human linguistic competence. While productive comparative studies between animals and humans in the domains of vision, motor control, brain development, acoustic perception and categorization were destined to blossom from the early 1960s to the present day, the comparative study of language dwindled as a result.[29,30]

In the period beginning in the late 1960s, neuroscience had pushed forward as well, but now the contributions were of a sort that clearly could connect with cognitive science. Work on the visual system began to make contact with perception as studied in cognitive laboratories; (Blakemore and Cooper, 1970; Gross *et al.*, 1972; Zeki, 1978), and a line of research on vision began that has yielded tremendous insights into how we see the world, and why we see it the way we do. Much of this research depended on neurophysiology carried out in primates, who in the early studies were anesthetized, then only stabilized, and in recent years even capable of moving about in the world. Studies carried out in ever more natural environments are producing increasingly sophisticated understandings of how the brain subserves vision.

Equally dramatic was a series of discoveries about the hippocampus, a brain structure implicated in memory since the pioneering work on H.M. in the 1950s. O'Keefe and Dostrovsky (1971), using new methods to record the activity of single neurons in freely moving animals, discovered 'place cells' in the hippocampus, neurons whose activity reflected the animal's location in the environment (*See* Place Cells). This discovery provided the basis for O'Keefe and Nadel's (1978, 1979) theory that the hippocampus instantiated 'cognitive maps' of the sort postulated by Tolman (1948) (*See* Animal Navigation and Cognitive Maps; Navigation; Navigation and Homing, Neural Basis of). This was one of the first neurophysiological research programs that made a direct connection between the activities of individual neurons and complex cognitive activities. Bliss and Lomo (1973) discovered a form of synaptic plasticity in the hippocampus (long-term potentiation, or LTP) that seemed to verify Hebb's early speculations about how learning might occur in the nervous system (*See* Hebb Synapses: Modeling of Neuronal Selectivity; Long-term Potentiation, Discovery of). Research on the hippocampus has continued at a furious pace ever since (*See* Hippocampus).

Another research program worth noting involved the efforts of a displaced psychiatrist, Eric Kandel, to uncover the neural mechanisms of learning and memory. Taking the bold step of shifting his research attention to an invertebrate (the sea slug, *Aplysia californica*), Kandel began the process of painstakingly working out the synaptic mechanisms underlying various forms of plasticity, laying the basis for an understanding of the cellular and molecular mechanisms of memory (*See* Learning in Simple Organisms).[31]

THE ROYAUMONT MEETING: A DEBATE BETWEEN (AND AROUND) JEAN PIAGET AND NOAM CHOMSKY, 1975

The Royaumont meeting was motivated largely by an attempt to 'reconcile' Chomsky's approach to language with Piaget's approach to cognition in general.[32] Both Chomsky and Piaget professed a deep link with biology, so a reconciliation seemed possible. Piaget opened the conference with a summary of basic assumptions that he believed would be received as innocent, obvious and hardly worth discussion. Much to his amazement, the whole meeting (three days) was dedicated to a multifaceted discussion of these very assumptions. The biologists questioned Piaget's reliance on autoregulation without specific pre-existing regulators[33] and his attempts to reintroduce in subtle ways the inheritance of acquired traits. Chomsky offered basic facts about language (mostly syntax) that, he claimed, could not be even remotely explained in terms of abstractions from motor schemas, nor by any general conceptual grasp of the world. Fodor argued that genuine conceptual novelty and any genuine potentiation of a pre-existing language could not be the result of learning. Other participants added their bit of specialized knowledge, some defending Piaget (notably his collaborators Cellerier and Inhelder, but also Papert, Bateson, Wilden and Toulmin), others siding with Chomsky and Fodor (Premack, Sperber, Mehler, Piattelli-Palmarini and, in indirect ways, Monod, Jacob and Changeux). The core issues, as they now appear more clearly in hindsight, were: (a) the modularity of mind and the autonomy of syntax; (b) the specificity of innate cognitive structures and the poverty of the stimulus; (c) reasons to keep rejecting (a) and (b) in spite of what Chomsky and Fodor presented as overwhelming evidence that one should accept them. Sequels of these issues and codas to point (c) are still very much alive as we write. While the arguments put forth by Chomsky and Fodor remain to many as strong as they were then, unshakable resistance to modularity, specificity and innateness survives in many quarters of cognitive science in various incarnations (Bates and Dick, 2000; Cowie, 1999, 2000; Elman *et al.*, 1996; Karmiloff-Smith, 1992, 1994; Karmiloff and Karmiloff-Smith, 2001).[34] In particular, debate rages over just how 'impoverished' the environment of the growing infant really is, and whether or not powerful abilities to extract statistical regularities from the environment might not make possible the ontogenetic, rather than phylogenetic, acquisition of various aspects of language including syntax.

As of the mid-1950s, Chomsky had argued for the 'autonomy of syntax', offering more or less incidentally the now famous example of the sentence 'Colorless green ideas sleep furiously', which every speaker of English identifies as syntactically well-formed even though it is utterly meaningless. Subsequent studies by Chomsky and his early collaborators (Chomsky, 1965) revealed basic syntactic principles of a very specific nature, common to many, and arguably to all, languages and dialects, as an integral part of the speaker–hearer's tacit 'knowledge of language'. These did not resemble in the least the then known basic principles of visual perception, motor control and generic reasoning, forming an integrated cluster of autonomous cognitive rules and representations. 'Knowledge of language' had to be kept separate from generic knowledge of the world. The metaphor of a 'language organ' defied all traditional conceptions of a small set of 'horizontal' multi-purpose mental faculties. These two separate strands of a modular conception of the mind–brain, one focusing on language and 'input systems' (Fodor, 1981, 1983) at the level of mental contents, representations and symbols, the other on central systems (memory, perception and planning) at the level of neuronal substrates in the animal and in humans, were to converge eventually, though emanating from different starting points.

On the neurobiological front, in the early 1970s Tulving (1972) suggested that there might be two rather different kinds of human memory – episodic and semantic. In 1974, three papers were published suggesting the same thing, but based instead on animal research (Hirsh, 1974; Gaffan, 1974; Nadel and O'Keefe, 1974). This notion applied to the brain an idea first promulgated by Tolman in a classic paper, 'There is more than one kind of learning' (1949). It was discussed by O'Keefe and Nadel (1978) for both animals and humans as part of their 'cognitive map' theory, and was also applied within the human amnesia literature by Kinsbourne and Wood (1975) and then Cohen and Squire (1980). The general notion that there are multiple neural modules concerned with different kinds of memory is now widely accepted (Schacter and Tulving, 1974). A somewhat similar history unfolded in the study of visual cognition, beginning with the seminal work of Ungerleider and Mishkin (1982) on 'two visual systems'. When a few years later Jerry Fodor published his landmark philosophical treatise *The Modularity of Mind* (1983), the strands were ready to be intertwined. It is safe to say that the idea that the brain comprises a large number of specialized modules is now the accepted wisdom – the challenges we face lie in figuring out how these semi-autonomous systems interact to generate cognition and behavior (*See* Modularity; Modularity in Neural Systems and Localization of Function).

In the 1970s and early 1980s, cognitive scientists made considerable strides, in particular in the study of spontaneous mental imagery (later assembled and expanded in Kosslyn, 1980; also *see* Imagery); mental rotation (Shepard, 1971; also *see* Mental Rotation); concept and category formation (Rosch, 1973, 1978; Smith and Medin, 1981; also *see* Concept Learning; Concept Learning and Categorization: Models); biases and heuristics in natural reasoning and decision making (Kahneman *et al.*, 1982; Kahneman and Tversky, 1972, 1973, 1979; also *see* Reasoning); memory (Tulving and Thompson, 1973; also *see* Memory; Memory Models; Memory: Implicit versus Explicit); abstraction (Posner, 1978); motion perception (Johansson, 1973); cognitive conceptual development in the child (Carey, 1985; Keil, 1979; Markman, 1989; Spelke, 1985, 1988; also *see* Categorization: Development of);[35] learnability theory (Osherson *et al.*, 1986; also *see* Learnability Theory); and the theory of Government and Binding (Chomsky, 1981; also *see* Binding Theory; Government–Binding Theory).

A major paradigm shift was in the offing, however, in the revival of biologically-inspired approaches to cognitive science. Beginning with the efforts of a group in San Diego, the 'connectionist' movement has made major inroads in a large number of fields previously dominated by the views first emphasized by Newell and Simon (*See* Connectionism). A critical focus of this approach was its assertion, sometimes explicit, sometimes implicit, that cognitive models should look more closely at biology. Instead of emphasizing the symbolic level that had been the bread and butter of cognitive science since the mid-1950s, this approach claimed to eschew symbols altogether, focusing instead on distributed representations and learning algorithms (Rumelhart *et al.*, 1985; McClelland *et al.*, 1985; also *see* Distributed Representations). Connectionism as a concept is an old idea – Hebb (1949) referred to it in his book. The label resurfaced in the mid-1980s as the name for a new approach to neural networks, one that has had a major impact on the domain of cognitive science in the past 20 years. Initially developed by John J. Hopfield in 1982, neural networks were abstract entities (later to be also implemented by physical hardware) that were explicitly inspired by real neuronal circuitry and were capable of automatic learning, rule extraction and generalization (Hopfield, 1982). Hopfield showed how the mathematical simplification of a neuron could allow an analysis of the behavior of large-scale neural networks, modeling progressive descents on an energy surface, thereby mimicking automated learning and automatic feature extraction from a corpus of different, but related, stimuli. Under an assortment of training procedures, with artificial equivalents of reinforcement and punishment, and with 'backpropagation' from one layer of terminal nodes back to layers of 'hidden' nodes (a powerful improvement that could solve some of the problems of older perceptrons), the remarkable potential of these artificial networks created a sensation (*See* Backpropagation). Cognitive scientists, in a number of places, paid very close attention and began to challenge the modularist–innatist theory of mind, and the very idea of dedicated mental rules and representations. The efficiency of such connectionist networks in extracting common features from certain families of inputs could equal, or even surpass, that of humans, as Stephen Grossberg of Boston University had noticed some years before (Grossberg, 1976; also *see* Adaptive Resonance Theory). But in spite of the adjective 'neural' and in spite of the liberal use of terms borrowed from biology (evolutionary landscape, fitness, adaptive behavior, etc.), the real proximity to 'wet' neurobiology remained questionable.[36]

The rise of connectionism in cognitive science goes hand in hand with neo-Piagetianism, touting the virtues of general intelligence and multi-purpose cognitive mechanisms powered by processes called stepwise abstractions, categorizations, thematizations and generalizations. Renewed invitations to go 'beyond' innatism and modularity (Bates and Elman, 1996; Elman, 1989; Elman *et al.*, 1996; Karmiloff-Smith, 1992, 1993, 1994) show how pertinent the arguments and counterarguments developed at the Royaumont conference remain. The often feisty debates between proponents of these two approaches to cognitive science continue to this day.

Since the early 1980s, a conception of learning that is radically different not only from the connectionist models but also from almost all previous models of learning has been developed in a new approach to the study of first-language acquisition (notice that even the word 'learning' has been expunged). Commandeering and reorganizing an array of studies on many languages and dialects in a series of lectures at the Scuola Normale in Pisa (Italy) in 1980, and then in a published book (Chomsky, 1981), Chomsky introduced the model called 'principles and parameters' (PP for short). In essence, this model reduces all the differences between human languages to a small universal set of syntactic nodes (parameters), for each of which there is a choice between only two admissible 'values'. The binary values for each parameter are labeled as '+' (the marked value) and '–' (the unmarked or default value). Under this idealization, the 'task' of the child learning his/her native language consists in appropriately 'fixing' the binary values of all the parameters in conformity with the set of values implicitly chosen by the surrounding community of speakers. James Higginbotham has summarized this idealization as the positioning of a set of 'switches' on a mental 'panel' (Higginbotham, 1982).[37]

THE EMERGENCE OF COGNITIVE NEUROSCIENCE: SOME EXCITING DEVELOPMENTS

Progress in cognitive neuroscience during the last decades has been nothing short of phenomenal, owing in large measure to the development of neuroimaging techniques that have made it possible to study the human brain during various cognitive activities (*See* Neuroimaging). The use of electroencephalographic methods with humans has a long history, but such methods involving surface recordings have inherent limitations, most specifically related to the spatial localization of the recorded signals. The use of many more recording sites and sophisticated analytic tools has engendered a new generation of such methods, most prominent in the domain of event-related potentials (ERPs), where recordings are synchronized to cognitive events of interest so that patterns of brain activation specifically related to those events can be identified. Such ERP methods have yielded considerable insights, especially in the temporal domain. However, spatial localization remains a problem. This is where the emergence of new neuroimaging techniques has been most productive. The critical insight here was the realization that methods could be devised to track metabolic and other consequences of neural activity in humans. The first widely used method, positron emission tomography (PET), depends upon the use of radioactive substances and the uptake of these materials by recently active neural tissue. More recently, a less invasive method, functional magnetic resonance imaging (fMRI), has been developed as an alternative. This method takes advantage of the fact that blood oxygenation levels change as a function of neural activity, and that oxygenated and deoxygenated blood (or, more precisely, hemoglobin) have different magnetic properties. This permits the detection, with powerful magnets, of those areas of the brain mobilized by some form of cognitive activity. Yet another method, magnetoencephalography (MEG), depends upon the very small, but measurable, magnetic fields engendered by neural activity. This method, though depending on considerable analysis to extract signals from noise, offers great promise given its ability to couple the real-time dynamic response (on a scale of milliseconds) with accurate spatial localization. Finally, transcranial stimulation (TMS) has emerged as a method to stimulate or, mostly, selectively inhibit areas on the cortical surface, and has been productively used to study in a very precise way the role of these surface structures in various cognitive functions.

Considered together, these methods have brought about a considerable explosion of research on the brain mechanisms of normal human cognitive function.[38] Where previous studies were limited to pathological cases and involved the problematic analysis of function from the deficits caused by pathology, these methods provide an entirely new window into the human mind. Not surprisingly, they have in the first instance confirmed many of our hard-won assumptions about which parts of the brain are engaged by what kinds of cognitive activity. Imaging studies have also shifted the focus of explanation away from reliance on discrete 'centers' of cognitive function towards the notion of an interaction between multiple brain areas. While some might argue, even at this stage, that neuroimaging is merely a modern-day version of phrenology (activations in the head, rather than bumps on the skull; Bates and Dick, 2000), clever researchers are beginning to develop new paradigms that offer the promise of real advances that would have been impossible with earlier techniques. Finally, the combination of neuroimaging with more traditional single and multiple neuron recording and with selective chemical labeling methods offers the promise of combining the insights that these approaches can offer.[39]

The productive use of more traditional methods has by no means ceased: consider for example the discovery in monkeys of the 'mirror neurons' (Rizzolatti and Craighero, 1998; also *see* Mirror Neurons). These neurons demonstrate a remarkable property: the same neuron is active when the animal either engages in an action or observes another animal engaging in the same action. The existence of such neurons raises questions of great import to philosophers of mind (Goldman and Galese, 1998; also *see* Simulation Theory). The possible role of a system of mirror neurons in the creation of internal mental models is obvious, and the implications of these findings for theories of the emergence of language is under active discussion (e.g. Rizzolatti and Arbib, 1998, 1999). Another major advance concerns the development of methods for simultaneously recording from many individual neurons, making it possible to study the activity of neural ensembles. These methods have been quite productively applied to the study of hippocampal 'place cells' (e.g. Wilson and McNaughton, 1993), where it has been possible to demonstrate that patterns of activity observed during a rat's daily activity have a high likelihood of recurring during a subsequent sleep episode.

New findings from developmental neuroscience and neuroanatomy have overturned the long-accepted view that nerve cells cannot be formed after the earliest stages of life (e.g. Gould and Gross, 2002; also *see* Neurogenesis). This, and other findings from the study of memory and perception, have reminded neuroscientists of the incredible dynamism of the brain. A major challenge for the future, at the very heart of the cognitive science enterprise, is to figure out how the stable world our minds construct, as pointed out by Craik, can be instantiated in a biological substrate that is constantly changing. Or, to put it as McCulloch (and Shakespeare) did: 'What's in a brain that ink might character' (McCulloch, 1964).

Thanks to the stunning discovery of systematic errors in spontaneous reasoning and decision making, notably by Amos Tversky and Daniel Kahneman, a progressive integration has begun between cognitive science and economics (for a pioneering survey, see Thaler, 1991, 1992). The recently explored neuronal bases of decision making, both in pathological cases (Adolphs *et al.*, 1996; Bechara *et al.*, 1994, 1999; Damasio, 1994; Damasio *et al.*, 1996) and in normal subjects (Breiter *et al.*, 2001) has suggested that a whole new domain, called neuroeconomics (McCabe *et al.*, 2001; Smith, in press; also *see* 'Neuroeconomics) may be just around the corner. The need to integrate standard economic analyses with what cognitive scientists have discovered about spontaneous heuristics and biases is now reported in the popular press.[40]

THE POSTMODERN ERA

Among the most striking changes in cognitive science since the 1980s years has been the shift in what is open to study. As we noted earlier, cognitive science started with the view that cognition is limited to those things humans can be conscious of. This position has been totally abandoned, and much of the domain is now concerned with phenomena that lie behind the veil of consciousness. Whether or not they are conscious, animals are very much a part of modern-day cognitive science. One prominent example concerns the study of emotion. Great strides have been made in linking emotion to traditional views of cognition, thereby returning the field to the point at which Craik left it more than 50 years ago. Finally, the grand-daddy of them all, consciousness itself, has become the focus of intense research interest within cognitive science (and beyond) in recent years (*See* Consciousness).[41]

Cognitive science has now reached the stage where one sees the production of integrated textbooks (Osherson, 1990/1995; Stillings *et al.*, 1987; Bechtel and Graham, 1999; Posner, 1989) and the publication of a concise encyclopedia (Wilson and Keil, 2000). Neuropsychology and cognitive neuroscience also now have their comprehensive sourcebooks (Posner, 2001; Gazzaniga, 1984, 2000; Kosslyn and Andersen, 1992).

Having grown into a rich and multifaceted domain, it is no surprise that cognitive science has witnessed, is witnessing, and will continue to witness disagreements, schisms, partial reconciliations and yet further splits in theories and methodological criteria. If we decide, with some simplification, to characterize as 'mainstream' or 'classical' cognitive science the individualist, largely innatist, modular and representational–mentalist (RTM) conception of the mind that characterized much of the 1980s and 1990s, there are clear signs that we may be entering a postclassical cognitive science (Piattelli-Palmarini, 2001). The innovative turn introduced by connectionist models in the mid-1980s has revamped an anti-modular and general-purpose conception of the mind–brain, soon contested by 'classic' cognitive scientists (Pinker and Mehler, 1988). The last few years have witnessed a partial (only partial, but not irrelevant) rapprochement between the two camps: many connectionists now countenance initial sets of prewired connections, and can explain the spontaneous tendency of parallel networks to locally cluster into modules (these modules are, however, conceived as emergent à la Piaget, not prewired; Elman *et al.*, 1996; Karmiloff-Smith, 1992, 1994). Symmetrically, on the other front, several linguists and developmental psychologists rooted in the generative tradition are presently searching for inductive (even statistical) components of early language acquisition, and report finding some that may play a crucial role (Nespor, 2001; Newport *et al.*, 2001; Ramus, in press; Ramus *et al.*, 1999; also *see* Language Acquisition).

Critics of modularity are also to be found in cognitive neuroscience, notably in the analysis of pathological deficits heretofore depicted as paradigms of modularity – prosopagnosia, for instance, with contrasting views, and contrasting observations, championed by Isabel Gauthier (Gauthier *et al.*, 1999, 2000a,b) on one side, and by Nancy Kanwisher and Morris Moscovitch (Kanwisher and Moscovitch, 2000) on the other (*See* Prosopagnosia). The nature and significance of earlier discoveries of specific language variants and deficits, from savants (Smith and Tsimpli, 1995) to Williams syndrome (Bellugi *et al.*, 1994, 1999, 2001; Bellugi and St. George, 2000; Stevens and Karmiloff-Smith, 1997; also *see* Williams Syndrome); to sign-languages (Kegl and McWorther, in press; Kegl *et al.*, 2000; Klima and Bellugi, 1979; Petitto, 1987; Petitto and Marentette, 1991; Emmorey and Lane, 2000; also *see* Sign Language), to SLI (Specific Language Impairment; Gopnik, 1990, 1994; Van der Lely, 1997; Van der Lely and Stollwerck, 1996; Wexler, 2002), is being questioned by researchers who conceptualize language as a specialization of general cognitive and communicative functions (Bates and Dick, 2000; Bates *et al.*, 1999; Karmiloff and Karmiloff-Smith, 2001; Karmiloff-Smith, 1998; Volterra *et al.*, 2002; Volterra and Erting, 2002). Even the legitimacy of combining data from language pathologies in the adult with data on developmental deficits in the child is being criticized in principle (Karmiloff-Smith *et al.*, in press).

Theories of language evolution that revise the approach of generative grammar (Pinker, 1994; Pinker and Bloom, 1990)[42], only pay lip service to it (Deacon, 1997), or fly in the face of decades of research in generative grammar (Dunbar, 1999; Lieberman, 2000; Tomasello, 1999), have recently been published. The age-old attempt to derive linguistic structures from motor control, considered moribund (cf. the exchange between Chomsky and Piaget on this point at the Royaumont debate described earlier), is being revamped under a different guise (Rizzolatti and Arbib, 1998, 1999). Cognitive innatism is being re-analyzed at its roots, and allegedly more promising alternatives are being offered (Cowie, 1999, 2000; Elman *et al.*, 1996).

The cognitive sciences today expand in every direction, as can be seen in the wide range of articles included in this encyclopedia. Neuroscience has been drawn back into the fold, and the area of cognitive neuroscience is one of the major growth industries in the field. We will let the articles in this work speak for themselves in filling in our history.

Among these articles are biographies of a variety of pioneers (now deceased) whose contributions were critical to the emergence of cognitive science. Also among these articles are a number of overarching reviews that attempt to address large domains within cognitive science (*See* Action; Cognitive Science: Experimental Methods; Cognitive Science: Philosophical Issues; Consciousness; Development; Perception).

It is impossible to anticipate the paths that cognitive science is going to take in the years to come. We are entering a postclassical era and there are reasons to believe that it will prove as productive and as innovative as the one that preceded it. This encyclopedia offers a complete and complex picture of the discipline up to the present. What will happen in the future will almost certainly surprise us.

Putting together the encyclopedia has been a challenge but also an opportunity, and the same can be said about looking into the history of the field. It is a fascinating history, peopled by intellectual giants and featuring ruminations about the big issues that have concerned thinkers for several millennia. A complete history of these times remains to be written; perhaps a reader of this encyclopedia will be sufficiently excited by its contents, and the genesis of these ideas, to take on that task.

NOTES

1. We gratefully acknowledge the help of several colleagues, Tom Bever, Howard Gardner and Morris Moscovitch, who read an early version of this document and whose suggestions vastly improved it. They, of course, are not responsible for its remaining shortcomings.
2. A century later, matters are not quite as settled as they appeared to be. There is now considerable evidence of direct contacts between at least some neurons in the CNS – so-called electrical synapses.
3. In 1906, Golgi and Cajal were jointly awarded the Nobel Prize 'in Physiology or Medicine'.
4. There were important 'schools' of neuropsychologists and neuropsychiatrists throughout Europe at this time. In Germany, Brodmann, Pick, Alzheimer, and Korsakoff are noteworthy. In France there were Janet and Charcot, among others. In England, Hughlings-Jackson, Head and Ferrier deserve note. In Russia, Sechenov was extremely influential, and had a major impact on Pavlov.
5. An interesting example is offered by Freud's *Project for a Scientific Psychology* (1895). This remarkable piece of work started out in a blaze of glory, defining in a very clear way what the issues were, and what kinds of answers would be necessary. Indeed, Karl Pribram (Pribram and Gill, 1976) resurrected this remarkable essay from near oblivion and justified the subsequent characterization of Freud, by Frank Sulloway, as a 'biologist of the mind' (Sulloway, 1979). However, Freud's approach went gloriously off the rails, and was ultimately abandoned by Freud altogether. This failure can now be seen as inevitable, given the tools that Freud had to work with at the time.
6. Somewhat later, spelling out in full a judgment that had been implicit all along, the pro-behaviorist analytic philosopher Quine referred to these mind-internal entities as 'creatures of darkness' (Quine, 1956; *See* Behaviorism, Philosophical). It was inevitable that the chief architects of an unrepentantly mentalistic, internalist, computational and representational theory of mind (usually referred to as RTM – Representational Theory of Mind), most notably Noam Chomsky and Jerry Fodor, found themselves fighting a long and sustained battle against behaviorism and its avatar in contemporary analytic philosophy (methodological behaviorism). Chomsky's destructive review of Skinner (Chomsky, 1959/1980) and Fodor's anti-behaviorist book-long essay *The Language of Thought* (Fodor, 1975) helped shape the turn away from behaviorism and much of modern mentalistic cognitive science for many years. Chomsky argued in favor of the specificity and universality of grammar. He defended an internalist, individualistic, intensional (as opposed to extensional) characterization of grammars (Chomsky, 1955, 1956, 1957, 1986) against a Wittgensteinian conception of language as a collective conventional public entity, individually mastered through the tuning up of a 'skill', eternally subject to the cumulative action of infinitesimally small variations, from one dialect to the next, from one generation to the next. (cf. Chomsky's and Fodor's writings in reaction to theses by Quine, Putnam, Davidson, Dummett, Kripke: Chomsky, 1980, 1988, 1995, 1998; Fodor, 1981, 1990; Fodor and Lepore, 1992).
7. Hull's attempt to generate mathematical treatments of learning, especially animal learning, was brought into the modern era by Rescorla and Wagner (1972) and Mackintosh (1974). Reacting to critical new findings such as the phenomenon of 'blocking' (e.g. Kamin, 1969), they produced learning rule equations that anticipated some of the more powerful algorithms to be developed within the connectionist framework 20 years later. (*See* Animal Learning)
8. Brunswik influenced Rosenblatt, who developed the perceptron model in the 1950s. He also had a significant influence on Julian Hochberg, who two decades after his student days at Berkeley talked about perception in terms of piecemeal perception, mental structures and the intentions of the viewer. Hochberg emphasized integration over inputs obtained from successive glances, hence the critical role of memory and other cognitive factors in perception, a line that has been continued by Hochberg's students to the present day (e.g. Peterson, 1999; also *see* Vision: Top-down Effects). An explicit revival of

Brunswikian models in decision making is presently advocated by Gerd Gigerenzer and his colleagues at the Max Planck Institute for Human Development in Berlin (Gigerenzer *et al.*, 1991); by A. J. Figueredo at the University of Arizona (Figueredo, 1992); and by Kenneth R. Hammond, now Emeritus Professor at the University of Colorado in Boulder, who has recently edited an extensive collection of reprints of Brunswik's original papers (Hammond and Stewart, 2001).

9. Piaget's work covered a variety of crucial domains, from the child's conception of numbers to moral judgment, from the development of the concept of causality to a cognitive approach to the historical development of physics and of science in general (*See* Piagetian Theory, Development of Conceptual Structure in). Many schoolteachers and avant-garde research teams in developmental psychology had tried with some success to apply Piaget's central ideas to their teaching in the classroom. The progressive spontaneous unfolding of higher cognitive capacities in the developing child had been modeled by Piaget and his collaborators at the University of Geneva through a series of successive 'stages', characterized as 'logically necessary' and universal, that the child attains one after the other at characteristic ages. Each stage was described in detail by means of novel characteristic mental operations that were either absent or only embryonic in the preceding stages. The internal mental engines of this stepwise process had been identified by Piaget as consisting of increasing auto-regulation, thematization, grouping, and 'reflective abstraction' (in French: abstraction réfléchissante). He pictured these processes as present already, at lower levels, in all living beings, making cognitive science continuous with biology. It hardly needs to be stressed how radically opposed to behaviorism this whole conception was, and how refreshing it appeared to many psychologists in the 1950s and 1960s. (see text and notes 31 and 33.)
10. These classical papers have been reprinted in van Heijenoort (1967) and in Davis (1965). For a modern systematic treatment see Lewis and Papadimitriou (1981); Rogers (1988). For the impact of these theories on early linguistic models see Bar-Hillel (1953a,b); Barton *et al.* (1987); Chomsky (1956).
11. The intellectual climate at the Research Laboratory of Electronics (RLE) of MIT in the early 1950s was one of excitement and imminent accomplishment. The injection of ideas and models from cybernetics and information theory into the study of language appeared capable of leading towards a full understanding of complex communication in humans, animals and machines. The modern scientific theory of language, and its momentous impact on cognitive science as a whole, largely originated in those laboratories, in the mid- and later 1950s, thanks to a disillusionment with both traditional classificatory linguistics and the statistical approaches to language that modeled it as a largely repetitive 'signal' interspersed with 'noise', in conformity with the then influential theory of information.
12. According to Hebb, perception could be accounted for in terms of organized sets of neurons (cell assemblies); thought would then follow from the concatenation of many of these assemblies into phase sequences. Memory involved changes in the efficacy of connections between neurons composing these ensembles. This deceptively simple approach had a dramatic impact on that subset of investigators willing to pay attention to the brain at the time, and over the past few decades has had an even wider effect. In Hebb's time his theory led to a variety of important research findings, including work on the effects of sensory deprivation, the stabilization of visual inputs to the retina, the impact of enriched or deprived rearing conditions, and a good deal more.
13. This meeting was presided over by Henry Brosin, a psychiatrist from Pittsburgh who played a significant facilitatory role throughout this early era. Brosin later retired to Tucson, Arizona, and a fair number of his books from these formative days, copiously annotated, have found their way into the hands of one of us (L.N.).
14. John McCarthy, Marvin Minsky, Nat Rochester, Claude Shannon, Oliver Selfridge, Herbert Gelernter, Alan Newell (*See* Newell, Alan) and Herb Simon (*See* Simon, Herbert A.), among others.
15. The list includes, in alphabetical order: Ursula Bellugi, Tom Bever, Roger Brown, Jerome Bruner, Susan Carey, Noam Chomsky, Jerry Fodor, Merrill Garrett, Janellen Huttenlocher, Roman Jakobson (*See* Jakobson, Roman), Dan Kahneman, Jerrold Katz, Paul Kolers, Pim Levelt, David McNeill, Jacques Mehler, George Miller, Don Norman, Eleanor Rosch, Dan Slobin, Amos Tversky (*See* Tversky, Amos), Peter Wason, Nancy Waugh.
16. Chomsky's position vis-à-vis the neurobiological foundations of language and mind deserves a word of clarification. Over many years he has insisted that all kinds of relevant data (the qualification 'relevant' is essential here) from any domain of science or even from everyday observation are of interest. Having constantly used the hyphenated expression brain–mind, and having always insisted that linguistics is part of the natural sciences (verbatim: part of biology 'at a suitably abstract level'), he is clearly aware of the potential power of the neurosciences to corroborate or refute abstract linguistic hypotheses. While encouraging a serious search for neuronal correlates, Chomsky maintains, however, that the neurosciences must cooperate with other relevant disciplines (e.g. linguistics, developmental psychology, the study of first and second language acquisition, genetics) and also must not forget the power of logic, of abstract arguments and even of physics and mathematics. His early participation in meetings with neurologists (especially on aphasia and other language deficits, as we have stated here), his close interaction with Eric Lenneberg, and his co-organization of a group

focused on bio-linguistics at MIT with the molecular biologist Salvador E. Luria and the neuropsychologist Hans-Lukas Teuber (*See* Teuber, Hans-Lukas) testify to his active interests in biology (for an insightful reconstruction of that initiative, see Jenkins (2000)). Some may interpret his attitude towards the brain sciences as tepid (at best) because of the quintessentially internalist–mentalist character of his theories in linguistics and a consistent refusal to give a privileged scientific status to data on brain structures and mechanisms, as compared to data on linguistic intuitions. Chomsky concurs with Fodor in cautioning against 'the intimidation by white blouses' (an expression used by Fodor in public discussions to refer to brain scientists) who present as truly scientific only 'wet' data, as opposed to cogent and rationally supported theories.

17. In recent times, Putnam (the acknowledged father of philosophical functionalism) has retreated from his earlier position, advocating a pivotal role for our intuitions about the material nature of cognitive systems (Putnam, 1987, 1988). Ned Block, a former student of Putnam's, has pointed out some serious 'troubles with functionalism', requiring a considerable expansion of this conception (Block, 1978). John Searle, over many years, has challenged the very consistency of functionalism, pleading for the centrality of a specific causal role attributed to the unique biological structure of the brain (Searle, 1980a,b, 1992, 1996; also *see* Chinese Room Argument, The). At the other extreme, John Haugeland has challenged the legitimacy of functionalism on more abstract grounds, espousing the holist doctrines of human cognition proposed by the German phenomenologists (Haugeland, 1981, 1997). As we write, nonetheless, some variant of functionalism still appears to constitute the spontaneous (and often implicit) philosophy of the mind/brain for most cognitive scientists. However, the recent development of refined brain imagery techniques and the growing number of publications dealing with specific brain correlates of higher cognitive functions (from language to numerical cognition, from decision making to categorization), might eventually modify somewhat the functional conceptual scheme of cognitive science. A direct match between abstract cognitive characterizations and real brain structures is now increasingly possible. Except for a minority of unrepentant symbolists, and, at the other extreme, of irreducible reductionists, the complex expression brain–mind is taking on a more insightful and richer unified meaning. It is interesting to realize that the core of this problem had already been identified, and vibrantly debated, in the early conferences on cybernetics.
18. More recently, the special epistemological importance of the spontaneous oscillatory activities of the brain has been stressed by Rodolfo Llinas and Jean-Pierre Changeux (2002) (*See* Neural Oscillations).
19. Awarded the Nobel Prize in 'Physiology or Medicine' in 1981, with David Hubel and Torsten Wiesel.
20. Papers were presented by Magoun, Moruzzi, Penfield, Hebb, Lashley and Kubie, among others.
21. Papers at this meeting were presented by Hebb, Olds, Magoun, Morrell, Hernandez-Peon, and a number of others. An important aspect of this meeting was the inclusion of a number of key researchers from Russia and both Western and Eastern Europe. Anokhin, Asratyan, Eibl-Eibesfeldt, Fessard, Grastyan, Jouvet, Konorski, Lissak, Naquet and Thorpe participated.
22. Participants included Magoun, Brazier, Doty, Olds, Pribram, Purpura, Galambos, John, Morrell, Sperry, and Teuber.
23. With whom L.N. subsequently was a postdoctoral fellow (1967–1970). The Prague laboratory of Jan Bures and his wife and scientific partner Olga Buresova became a mecca for neuroscientists from around the world, and remains so today. Their work on memory and more recently the spatial functions of the hippocampus has been influential for more than four decades.
24. Participants included Roger Brown, Noam Chomsky, Norman Geschwind, Kurt Goldstein, Harold Goodglass, Eric Lenneberg, Brenda Milner, Charles Osgood, Karl Pribram, and Hans-Lukas Teuber.
25. Participants at this meeting included Macdonald Critchley, Lord Brain, Roman Jakobson, Donald Broadbent, Alexander Romanovich Luria, Brenda Milner, Henri Hecaen, Oliver Zangwill, Hans-Lukas Teuber and Colin Cherry. There were a number of other important meetings held in the United Kingdom in the 1950s and 1960s, including a series of meetings held in London in the 1950s under the title 'London Symposium on Information Theory'. At the Fourth Symposium (Cherry, 1961), held in 1960, Averbach and Sperling reported on their methodology involving brief visual exposures to subjects who were asked to report only a subset of the presented material. Using this partial report method, the authors were able to show that immediate visual memory has available to it a good deal more information than subjects can retrieve when asked for a full report. The 'Mechanisation of Thought Processes' symposium held at the National Physical Laboratory in 1958 included presentations by Minsky, Mackay, McCarthy, Ashby, Uttley, Rosenblatt, Selfridge, McCulloch, Sutherland, Gregory, and Bar-Hillel. Other noteworthy attendees included Bartlett, Buerle, Cherry, Gabor, Wason, and J. Z. Young.
26. At about this time a small group of neuropsychologists, the International Neuropsychological Symposium, started to meet every year in Europe. According to Boller (1997), the group was launched in 1949 at a party that Henry Hecaen held at his home on the occasion of the International Congress of Psychiatry. After dinner, he outlined a proposal to found an international group to promote knowledge and understanding of brain functions and cognate issues on the borderland of neurology, psychology and psychiatry. This group, which continues to meet

yearly, strongly promoted the integrative, multidisciplinary perspective that became characteristic of cognitive science. Later on, in Italy, cognitive neuropsychology was to flourish beyond any other domain of cognitive science, gaining considerable international recognition. Individuals and groups were scattered in many different universities (Eduardo Bisiach in Turin, Anna Basso in Milan, Carlo Umiltà, Remo Job and Renzo de Renzi in Padua, Elisabetta Ladavas in Bologna, Gabriele Miceli in Rome, to name a few). Ever since the early 1980s, the annual international conferences held in Bressanone in January, under the auspices of the University of Padua, have regularly assembled the Italian 'contingent' of cognitive neuroscientists with colleagues from many other countries.

27. For a recent revisitation of the origins of psycholinguistics and for a development of the 'strategic' approach to language comprehension, see Townsend (2001).
28. For example, Allen and Beatrice Gardner, Duane and Sue Savage-Rumbaugh, Roger and Diane Fouts and, from a distant shore, Jean Piaget. (For a recent reappraisal, see Savage-Rumbaugh, in press.)
29. Except for a few unrepentant 'continuists', mainstream cognitive science in the domain of language followed different paths, unearthing deep and hard-to-detect similarities among distant languages, broaching the gap between syntax and semantics (*See* Semantics and Cognition), exploring subtle lexical structures (*See* Lexicon), modeling the 'logical' problem of language acquisition with formalized mathematical tools, developing computational models of linguistic competence and performance. The very issue of the biological evolution of language was to be tackled afresh at its roots by generative linguists, steering a course away from simplistic adaptationism and continuism, examining the possible conditions of the evolution of the very roots of linguistic competence (recursiveness, constituency, compositionality, infinite discreteness, generativity; Jenkins, 2000; Lightfoot, 1982; Nowak *et al.*, 2002; Piattelli-Palmarini, 1989; Uriagereka, 1998). For a recent reappraisal by Chomsky himself, see Hauser *et al.* (in press).
30. A special position in this debate between continuists and modularist–innatists was occupied by the influential biosemiotician Thomas Sebeok. He rejected wholesale all the experiments on the alleged linguistic abilities of apes, claiming a much deeper, more universal and more meaningful underlying substrate: the 'semiotic function'. He described incremental steps of complexification in this universal underlying substratum and insisted that a unified theory could range from the 'syntactic' (sic) nature of Mendeleeff's table of the chemical elements (Sebeok, 1995/2000) up to all systems of human communication, be they vocal, gestural, graphic or pictorial, passing through the genetic code, the immune code, the systems of communication between cells, between unicellular organisms (microsemiotics), plants (phytosemiotics) and the circuits of neurotransmitters in the nervous system (neurosemiotics). These incremental steps in the quality and complexity of signaling were analyzed as accruing to a common semiotic substrate, displaying a universal 'perfusion of signs' which, according to Sebeok, authorizes a unified conceptualization, a semiotic 'ecumenicalism' (Sebeok, 1977). Sebeok's conceptualization and his alleged semiotic 'theorems' and 'lemmas' have found attentive ears in some literary quarters and in some schools of communication (notably in Italy), but have remained, in the main, alien to cognitive science. The semantics of natural language has developed a radically different approach (for a textbook synthesis, see Larson and Segal (1995)).
31. See Kandel (1980) for a review of some of this research program, begun in the 1960s, presented at a symposium held in Texas in 1978 to commemorate the thirtieth anniversary of the Hixon Symposium and to honor Lloyd Jeffress, who edited the volume from that earlier historic meeting. Kandel was awarded the Nobel Prize in 2000, with Arvid Carlsson and Paul Greengard. His monumental textbook *Principles of Neural Science*, updated and translated into many languages, has been adopted in many countries.
32. The published volume (Piattelli-Palmarini, 1980), and the afterthoughts by its principal organizer (Piattelli-Palmarini, 1994) spare us from having to summarize this rich multidisciplinary conference (organized by Monod, Piattelli-Palmarini, Atran and Changeux). Besides Piaget and Chomsky, the main participants from cognitive science were Jerry Fodor, Barbel Inhelder, Guy Cellerier, David Premack, Seymour Papert, Gregory Bateson, Dan Sperber, Scott Atran and Jacques Mehler. Other disciplines were also represented: in attendance were the molecular biologists Jacques Monod and François Jacob, the neurobiologist Jean-Pierre Changeux, the philosopher of science Stephen Toulmin, the anthropologist Claude Lévy-Strauss, the ethologist Norbert Bischoff, and the mathematician Jean Petitot. Important additional contributions to the volume by invited participants who had been unable to attend came from the logician and philosopher Hilary Putnam and the mathematician René Thom. The post-conference exchange between Putnam, Chomsky and Fodor was included by Ned Block in his anthology of writings, *The Philosophy of Psychology*. In the preface to this book, Howard Gardner suggested that the meeting was one of the seminal events at the very origins of cognitive science.
33. François Jacob (the co-discoverer, with Jacques Monod, of the 'operon', a complex genetic unit of regulatory genes) stated that regulation can only take place as a result of pre-existing regulatory genes that actually and selectively kick in (or remain shut off) to regulate metabolic pathways. Piaget's conception of

progressive cascades of higher and higher auto-regulations was at odds with this very concrete finding. It quickly appeared to some participants that Piaget's concept was a metaphor, not a model. Piaget immediately retorted that Jacob's idea was exceedingly 'narrow', and that he knew other biologists who concurred with him in presenting a more general and more flexible picture of auto-regulation.

34. Over the years, readers of different scientific orientations have drawn strikingly different conclusions from the proceedings of this debate. One of us (M.P.P., the editor of the book) was told a few months after its publication, during a visit at MIT, that it was 'obvious' (sic, a qualification whose importance will be clear in a moment) that Chomsky and Fodor had won the debate. A few weeks later, in Geneva, he was told that it was 'obvious' (again, sic) that Piaget, Inhelder, Céllérier and Papert had won the debate. The self-imposed neutrality of MPP while editing the proceedings was thus powerfully, albeit indirectly, acknowledged. Many years later, this neutrality was abandoned in an afterthoughts piece in *Cognition* (1994), decidedly in favor of Chomsky and Fodor. But opposite conclusions and afterthoughts have been reached in other quarters. Those who have sided with Piaget typically accuse Chomsky and Fodor of having been quite ungracious in rejecting Piaget's many overtures and concessions to their positions, and of having countered his simple, untendentious and unassailable theses with a flurry of possibly (just possibly) relevant paradoxes and conundra, for which (so the story goes) they had no solutions to offer. Present and future readers will have to decide for themselves which conclusion is more correct.
35. The theories of conceptual development, categorization (Rosch, 1973, 1978, 1996; Smith *et al.*, 1988; Smith and Medin, 1981) and psychological similarity (Shepard, 1962, 1964, 1994; Tversky, 1977; Tversky and Gati, 1978; see the special issue of *Brain and Behavioral Sciences* on the work of Roger Shepard for a recent synthesis) combined with the theories of, and experiments on, lexical acquisition in the child, and with lexical semantics, opened the way to an integrated theory of conceptual and linguistic development at the interfaces between phonology and syntax, syntax and the lexicon, syntax and semantics (Bloom, 2000; Gopnik and Meltzoff, 1998; Jackendoff, 1983, 1990; Landau and Gleitman, 1985; Levin and Pinker, 1991; with a dissenting view by Fodor, 1994, 1998; also *see* Phonology; Phonological Processes; Syntax; Syntax, Aquisition of; Syntax and Semantics: Formal Approaches).
36. It is noteworthy that, since the mid-1980s, physicists in France (Marc Mézard and Gérard Toulouse), in Israel (Daniel Amit and Gabriele Veneziano) and in Italy and Argentina (Giorgio Parisi and Miguel Virasoro) have developed germane mathematical models for the so-called spin glasses (amorphous magnetic lattices in which each node is occupied by an element with a magnetic dipole – spin – that interacts mostly, but not exclusively, with its immediately adjacent neighbors, with a statistical tendency to propagate local dipole alignments to larger regions, in search of global minima (attractors) in the resulting energy surface; for a comprehensive synthesis, see Mézard *et al.*, 1987). Inspired by models of ferromagnetic lattices developed in the 1920s by the German physicist Ernest Ising (Brush, 1967) and temperature-sensitive probability decay functions initially proposed by Ludwig Boltzmann, the theories of spin glasses and Hopfield's models soon began to converge, jointly establishing many basic analogies between idealized magnetic lattices and idealized neural nets. The formal equivalence between spin glasses and neuronal networks, no matter how elegant and intellectually satisfactory at an abstract level, engendered in some neurobiologists doubts concerning the real applicability of these models to real brain circuits.
37. At the time (and until the mid-1990s) the nature of these parameters was conceived as quintessentially syntactic. Under the guidance of specific hypotheses about the parameters, researchers applied real acquisition data (and vast precompiled corpora) to their theoretical predictions of the subtle cascades of manifest consequences on linguistic expressions that the different possible parametric values were expected to produce. This kind of linguistico-developmental 'parametric' literature started to grow steadily (and still continues to the present date), but the early results were not always neat, and many intense discussions ensued. In 1995, the 'minimalist program' (Chomsky, 1995) shifted the localization of the parameters from the syntax proper to the morpho-lexical component of language. The problematic reinterpretation of older data and a new flurry of experiments carried out under this different model have blurred some of the simple and elegant contours of the initial experiments. This has stimulated some, but also disappointed other, researchers, who presently feel the need for a reappraisal of the very ideas behind the PP model. Whatever the ultimate destiny of the PP idealization, it has given a very productive impulse to the development of detailed non-inductivist models of learning. Novel kinds of fixation mechanisms and equally novel kinds of interactions between the linguistic external inputs and these mechanisms have been introduced. One quick and relevant measure is a steady increase in the number of scientific communications inspired by this approach presented each year at the Boston University Conference on Language Development. The Conferences started in 1976. Their proceedings from 1994 to the present have been published by Cascadilla Press (Somerville, Massachusetts).

An entire family of new mechanisms was proposed to account for non-inductive learning and the fixation and stabilization of rules and concepts in the child's developing mind–brain: the subset principle

(Berwick, 1985), greediness, conservativity, locking capacity, triggers, and default values (Wexler and Culicover, 1980; Wexler and Borer, 1986; Wexler and Manzini, 1986; Gibson and Wexler, 1994). A close constructive dialogue was established between formal theories of 'learnability' (Gold, 1967; Osherson *et al.*, 1986; Pinker, 1979) and empirical data on language acquisition in children in a variety of languages (for an early synthesis, see Wanner and Gleitman, 1982; for a recent one see Guasti, 2002). Arguably, the most remarkable property of these hypotheses, besides their non-inductivism, is that they constitute a cumulative attempt by many of these researchers to build a detailed selectional (as opposed to instructional) frame for language acquisition across languages. Selective models of the growth and development of the nervous system can be traced back to the paradigm of stabilization of developing synapses through selective activation, as initially established by Hubel and Wiesel's experiments on the development of the visual system in the selectively deprived kitten. Pasko Rakic and Patricia Goldman-Rakic (Goldman-Rakic, 1985; Rakic *et al.*, 1986) and Purves and Lichtman, among others, later extended and refined a model based on the overproduction of synapses in the developing cortex, followed by a massive trimming of the inactive connections. The wider implications of such selective models for neurobiology, cognitive science and beyond were soon highlighted by J.-P. Changeux and G. M. Edelman (Changeux *et al.*, 1984; Edelman and Mountcastle, 1978; Edelman and Reeke, 1982; for a recent reconstruction and wide philosophical consequences of selective theories, see Changeux, 2002).

It is to be expected that at least some of these new ideas will find a suitable place in future theories of learning and acquisition, possibly even beyond the domain of language. Over and beyond the interest of the PP model *per se*, this is a case study of a deep connection between the neurosciences and the study of the mind, one of many that have shaped modern cognitive science.

38. Many have contributed to this set of exciting developments, but the St. Louis group of Marcus Raichle, Peter Fox, Steve Peterson and Michael Posner, instrumental in getting things started, deserve special mention (Posner *et al.*, 1988).
39. This combination is currently being explored most productively by Nikos Logothetis and his group at the Max Planck Instiutute in Tübingen (Saleem *et al.*, 2002).
40. For instance, among many more, the following articles: *The New York Times*, March 30, 1997; *The New York Times Magazine*, February 11, 2001; *The Washington Post*, January 27, 2002.
41. This resurrection was initiated perhaps by researchers in cognitive psychology and neuropsychology (e.g. Moscovitch, 1995), but it has now been embraced by a much wider interdisciplinary network. A series of large multidisciplinary conferences on consciousness has been held at two-year intervals in Tucson, Arizona since 1994, each time attracting more than a thousand participants ranging from poets to physicists to physiologists.
42. The 1990 article in *Brain and Behavioral Sciences* by Pinker and Bloom has been received with warm, albeit sometimes cautious (Deacon, 1997), assent by many who have only general sympathy for generative grammar, or accept only parts of it, or remain prudently agnostic about the enterprise as a whole. The reconciliation between generative grammar and a neo-Darwinian adaptationist account of the evolution of language, so eagerly explored by Pinker and Bloom (for later developments and refinements see their subsequent books: Bloom, 2000; Pinker, 1994, 1997), reassures those who are reluctant to follow Chomsky all the way in his defense of radical discontinuity and the punctate appearance of the language faculty exclusively in humans. It assuages their fear that Chomsky's theses may involve an appeal to an evolutionary miracle, an exceedingly improbable 'hopeful monster'. The radical adversaries of generative grammar, predictably, have found no reason to be interested in this reconciliation, noting that the generative camp harbors embarrassing internal disagreements. Generative grammarians have, in the main, remained unabatedly critical of all extant adaptationist accounts of language evolution (see note 25), including the one offered by Pinker and Bloom (see the peer commentaries accompanying the *Brain and Behavioral Sciences* article), and some have brought into evidence the convergent critiques of adaptationist explanations independently and authoritatively developed by Stephen J. Gould and Richard C. Lewontin in evolutionary theory proper (one of the main targets of Pinker and Bloom was a paper by Piattelli-Palmarini (1989) in which this convergence between antecedently separate contributions was made explicit). At an international meeting in Venice (Italy), with Gould in the audience, Paul Bloom confessed with humor that he and Pinker would have felt uncomfortable challenging Chomsky on language and Gould on evolution, but felt reasonably comfortable challenging Chomsky on evolution and Gould on language.

REFERENCES

Adolphs R, Tranel D, Bechara A, Damasio H and Damasio AR (1996) Neuropsychological approaches to reasoning and decision-making. In: Damasio AR (ed.) *Neurobiology of Decision-making*. New York: Springer.

Albertazzi L (2001) *The Dawn of Cognitive Science: Early European Contributors*. Dordrecht: Kluwer.

Ashby WR (1960) *Design for a Brain: The Origins of Adaptive Behavior*. New York: Wiley.

Baars BJ (1986) *The Cognitive Revolution in Psychology*. New York: The Guilford Press.

Bar-Hillel Y (1953a) On recursive definitions in empirical sciences. In: *Proceedings of the 11th Congress of Philosophy, Brussels V*, pp. 160–165.

Bar-Hillel Y (1953b) A quasi-arithmetical notation for syntactic description. *Language* **29**: 47–58.

Bar-Hillel Y (1954) Logical syntax and semantics. *Language* **30**: 230–237.

Bartlett FC (1932) *Remembering: a Study in Experimental and Social Psychology*. Cambridge, UK: Cambridge University Press.

Barton EG, Berwick RC and Ristad ES (1987) *Computational Complexity and Natural Language*. Cambridge, MA: Bradford Books/MIT Press.

Bates E and Dick F (2000) Beyond phrenology: brain and language in the next millennium. *Brain and Language* **71**: 18–21.

Bates E and Elman J (1996) Learning rediscovered. *Science* **274**: 1849–1850.

Bates E, Elman J, Johnson M *et al.* (1999) Innateness and emergentism. In: Bechtel W and Graham G (eds) *A Companion to Cognitive Science*. Malden, MA and Oxford: Blackwell.

Beach FA, Hebb DO, Morgan CT and Nissen HW (eds) (1960) *The Neuropsychology of Lashley: Selected Papers of K. S. Lashley*. New York: McGraw-Hill.

Bechara A, Damasio AR and Damasio H (1994) Insensitivity to future consequences following damage to human prefrontal cortex. *Cognition* **50**: 7–15.

Bechara A, Damasio H, Damasio AR and Lee GP (1999) Different contribution of the human amygdala and ventromedial prefrontal cortex to decision-making. *The Journal of Neuroscience* **19**(13): 5473–5481.

Bechtel W, Abrahamsen A and Graham G (1999) The life of cognitive science. In: Bechtel W and Graham G (eds) *A Companion to Cognitive Science*, pp. 1–104. Malden, MA and Oxford: Blackwell.

Bechtel W and Graham G (eds) (1999) *A Companion to Cognitive Science*. Malden, MA and Oxford: Blackwell.

Bellugi U, Lichtenberger L, Mills D, Galaburda AM and Korenberg JR (1999) Bridging cognition, brain and molecular genetics: evidence from Williams syndrome. *Trends in Neurosciences* **22**(5): 197–207.

Bellugi U and St. George M (2000) Linking cognitive neuroscience and molecular genetics: new perspectives from Williams syndrome. *Journal of Cognitive Neuroscience* **12**(1): 1–107.

Bellugi U, St. George M and Galaburda AM (eds) (2001) *Journey from Cognition to Brain to Gene: Perspectives from Williams Syndrome*. Cambridge, MA: Bradford Books/MIT Press.

Bellugi U, Wang P and Jernigan TL (1994) Williams syndrome: An unusual neuropsychological profile. In: Broman S and Grafman J (eds) *Atypical Cognitive Deficits in Developmental Disorders: Implication for Brain Function*, pp. 23–56. Mahwah, NJ: Erlbaum.

Berwick RC (1985) *The Acquisition of Syntactic Knowledge*. Cambridge, MA: MIT Press.

Blakemore C and Cooper RM (1970) Development of the brain depends upon the visual environment. *Nature* **228**: 477–478.

Bliss TVP and Lomo T (1973) Long-lasting potentiation of synaptic transmission in the dentate area of the anaesthetised rabbit following stimulation of the perforant path. *Journal of Physiology* **232**: 331–356.

Block N (1978) Troubles with functionalism. In: Savage CW (ed.) *Perception and Cognition: Issues in the Foundations of Psychology: Minnesota Studies in the Philosophy of Science*, vol. 9. Minneapolis: University of Minnesota Press.

Block N (ed.) (1980) *Readings in the Philosophy of Psychology*, vol. 1. Cambridge, MA: Harvard University Press.

Block N (ed.) (1981) *Readings in the Philosophy of Psychology*, vol. 2. Cambridge, MA: Harvard University Press.

Bloom P (2000) *How Children Learn the Meanings of Words*. Cambridge, MA: Bradford Books/MIT Press.

Boller F (1998) History of the International Neuropsychological Symposium: a reflection of the evolution of a discipline. *Neuropsychologia* **37**: 17–26.

Bower B (2002) A maverick reclaimed: some psychologists say it's time that Egon Brunswik got his due. *Science News* **161**: 155–156.

Braun M (1992) *Picturing Time: The Work of Etienne-Jules Marey*. Chicago, IL: University of Chicago Press.

Brazier MAB (ed.) (1958) *The Central Nervous System and Behavior: Transactions of the First Conference*. New York: Josiah Macy Jr. Foundation.

Brazier MAB (ed.) (1959) *The Central Nervous System and Behavior: Transactions of the Second Conference*. New York: Josiah Macy Jr. Foundation.

Brazier MAB (ed.) (1960) *The Central Nervous System and Behavior: Transactions of the Third Conference*. New York: Josiah Macy Jr. Foundation.

Breiter HC, Aharon I, Kahneman D, Dale A and Shizgal P (2001) Functional imaging of neural responses to expectancy and experience of monetary gains and losses. *Neuron* **30**: 619–639.

Broadbent D (1958) *Perception and Communication*. New York: Pergamon Press.

Broca P (1878) Anatomie comparée des circomvolutions cérébrales. Le grand lobe limbique et la scissure limbique dans la série des mammifères. *Revue anthropologique* **1**: 385–498.

Brown R (1958) How shall a thing be called? *Psychological Review* **65**: 14–21.

Brown R (1973) *A First Language: The Early Stages*. Cambridge, MA: Harvard University Press.

Bruner JR, Goodnow JJ and Austin GA (1956) *A Study of Thinking*. New York: Wiley.

Brush SG (1967) History of the Lenz–Ising model. *Review of Modern Physics* **39**: 883–893.

Carey S (1985) *Conceptual Change in Childhood*. Cambridge, MA: Bradford Books/MIT Press.

Changeux J-P (2002) *L'homme de vérite*. Paris: Odile Jacob.

Changeux J-P (ed.) *L'unité de l'homme: invariants biologiques et universaux culturels*. Paris: Seuil.

Changeux J-P, Heidmann T and Patte P (1984) Learning by selection. In: Marler P and Terrace HS (eds) *The Biology of Learning*. New York: Springer Verlag.

Cherry C (ed.) (1961) *Information Theory: Fourth London Symposium*. London, UK: Butterworths.

Chomsky N (1955) *The Logical Structure of Linguistic Theory*. Cambridge, MA: MIT Press.

Chomsky N (1956) Three models for the description of language. In: Luce RD, Bush RR and Galanter E (eds) *Readings in Mathematical Psychology*, vol. 2, pp. 105–124. New York: John Wiley and Sons.

Chomsky N (1957) *Syntactic Structures*. The Hague: Mouton.

Chomsky N (1959/1980) A review of B. F. Skinner's 'Verbal Behavior'. *Language* **35**(1): 26–58.

Chomsky N (1965) *Aspects of the Theory of Syntax*. Cambridge, MA: MIT Press.

Chomsky N (1975) *The Logical Structure of Linguistic Theory*. Chicago, MA: University of Chicago Press.

Chomsky N (1980) *Rules and Representation*. Columbia University Press.

Chomsky N (1981) *Lectures on Government and Binding (The Pisa Lectures)*. Dordrecht: Foris.

Chomsky N (1986) *Knowledge of Language: Its Nature, Origin, and Use*. New York: Praeger Scientific.

Chomsky N (1995) Language and nature. *Mind* **104**: 1–61.

Chomsky N (1995) *The Minimalist Program*. Cambridge, MA: MIT Press.

Chomsky N (1998) *On Language*. New York: New Press.

Chomsky N (ed.) (1988) *Language and Interpretation: Philosophical Reflections and Empirical Inquiry*.

Chomsky N and Miller GA (1963) Introduction to the formal analysis of natural languages. In: Luce RD, Bush RR and Galanter E (eds) *Handbook of Mathematical Psychology*, vol. 2, pp. 269–321. New York: John Wiley and Sons.

Church A (1936) An unsolvable problem of elementary number theory. *The American Journal of Mathematics* **58**: 345–363.

Cohen NJ and Squire LR (1980) Preserved learning and retention of pattern-analyzing skill in amnesia: dissociation of knowing how and knowing that. *Science* **210**: 207–210.

Cowie F (1999) *What's Within? Nativism Reconsidered*. New York and Oxford, UK: Oxford University Press.

Cowie F (2000) Whistling 'Dixie': response to Fodor. In: Nani M and Marraffa M (eds) *E-Symposium on Fiona Cowie's 'What's Within?: Nativism Reconsidered' (SIFA E-Symposia)*. [http://host.uniroma3.it/progetti/kant/field/booksymp.html.]

Craik K (1943) *The Nature of Explanation*. Cambridge, UK: Cambridge University Press.

Damasio AR (1994) *Descartes' Error*. Grossett Putnam.

Damasio AR, Damasio H and Christen Y (1996) *Neurobiology of Decision-making*. Springer.

Davis M (ed.) (1965) *The Undecidable: Basic Papers on Undecidable Propositions, Unsolvable Problems and Computable Functions*. Hewlett, NY: Raven Press.

De Reuck AVS and O'Connor M (eds) (1964) *Disorders of Language*. Boston, MA: Little, Brown and Co.

Deacon TW (1997) *The Symbolic Species*. London: WW Norton and Company Inc.

Delafresnaye JF (ed.) (1954) *Brain Mechanims and Consciousness: A Symposium Organized by the Council for International Organizations of Medical Sciences (C.I.O.M.S.)*. Springfield, IL: Charles C Thomas.

Delafresnaye JF (ed.) (1961) *Brain Mechanims and Learning: A Symposium Organized by the Council for International Organizations of Medical Sciences (C.I.O.M.S.)*. Oxford, UK: Blackwell.

Dennett DC (1987) *The Intentional Stance*. Cambridge, MA: Bradford Books/MIT Press.

Dunbar R (1999) *Grooming, Gossip and the Evolution of Language*. Cambridge, MA: Harvard University Press.

Dupuy J-P (2000) *The Mechanization of the Mind: On the Origins of Cognitive Science*. Princeton, NJ: Princeton University Press.

Edelman GM (1987) *Neural Darwinism: The Theory of Neuronal Group Selection*. New York: Basic Books.

Edelman GM and Mountcastle VB (1978) *The Mindful Brain: Cortical Organization and the Group-selective Theory of Higher Brain Function*. Cambridge, MA: Bradford Books/MIT Press.

Edelman GM and Reeke GN (1982) Selective networks capable of representative transformations, limited generalizations, and associative memory. *Proceedings of the National Academy of Sciences of the USA* **79**: 2091–2095.

Elman JL (1989) Representation and Structure in Connectionist Models, *Center for Research in Language Technical Report 8903*. La Jolla, CA: University of California.

Elman JL, Bates EA, Johnson MH *et al.* (1996) *Rethinking Innateness: A Connectionist Perspective on Development*. Cambridge, MA: Bradford Books/MIT Press.

Emmorey K and Lane H (eds) (2000) *The Signs of Language Revisited: An Anthology in Honor of Ursula Bellugi and Edward Klima*. Mahwah, NJ: Lawrence Erlbaum.

Figueredo AJ (1992) Preparedness and plasticity: a stochastic optimality theory. *Human Behavior and Evolution Society: Fourth Annual Meeting*. Albuquerque, NM.

Fodor J and Garrett M (1966) Some reflections on competence and performance (with comments by N Stuart Sutherland and L Jonathan Cohen). In: Lyons J and Wales RJ (eds) *Psycholinguistics Papers (Proceedings of the Edinburgh Conference 1966)*, pp. 135–179. Edinburgh, UK and Chicago, IL: Edinburgh University Press/Aldine Publishing Company.

Fodor JA (1975) *The Language of Thought*. New York: Thomas Y Crowell.

Fodor JA (1981) *Representations*. Harvester Press.

Fodor JA (1983) *The Modularity of Mind: An Essay on Faculty Psychology*. Cambridge, MA: Bradford Books/MIT Press.

Fodor JA (1990) *A Theory of Content and Other Essays*. Cambridge, MA: Bradford Book/MIT Press.

Fodor JA (1994) Concepts: a potboiler. *Cognition* **50**: 95–113.

Fodor JA (1998) *Concepts: Where Cognitive Science Went Wrong*. New York and Oxford, UK: Oxford University Press.

Fodor JA, Bever TG and Garrett MF (1974) *The Psychology of Language: An Introduction to Psycholinguistics and Generative Grammar*. New York: McGraw-Hill.

Fodor JA and Lepore E (1992) *Holism*. New York and Oxford, UK: Blackwell.

Fouts RS and Fouts DH (1989) Loulis in conversation with cross-fostered chimpanzees. In: Gardner RA, Gardner BT and Van Cantfort TE (ed.) *Teaching Sign Language to Chimpanzee*. Albany, NY: CUNY Press.

Frege G (1879) *Begriffsschrift, eine der arithmetischen nachgebildete formelsprache des reinen denken*. Halle.

Gaffan D (1974) Recognition impaired and association intact in the memory of monkeys after transection of the fornix. *Journal of Comparative and Physiological Psychology* **86**: 1100–1109.

Gardner BT and Gardner RA (1971) *Behavior of Non-Human Primates*, vol. 4. New York: Academic Press.

Gardner H (1985) *The Mind's New Science: A History of the Cognitive Revolution*. New York: Basic Books.

Gardner RA, Gardner BT and Van Cantfort TE (ed.) (1989) *Teaching Sign Language to Chimpanzee*. Albany, NY: CUNY Press.

Gauthier I, Skudlarsky P, Gore JC and Anderson AW (2000a) Expertise for cars and birds recruits brain areas involved in face recognition. *Nature Neuroscience* **3**: 191–197.

Gauthier I, Tarr MJ, Anderson AW, Skudlarsky P and Gore JC (1999) Activation of the middle fusiform 'face area' increases with expertise in recognizing novel objects. *Nature Neuroscience* **2**: 568–573.

Gauthier I, Tarr MJ, Moylan J *et al.* (2000b) The fusiform 'face area' is part of a network that processes faces at the individual level. *Journal of Cognitive Neuroscience* **12**: 495–504.

Gazzaniga MS (1970) *The Bisected Brain*. New York: Appleton-Century-Crofts.

Gazzaniga MS (1984) *Handbook of Cognitive Neuroscience*. New York: Plenum Press.

Gazzaniga MS (ed.) (2000) *The New Cognitive Neurosciences*, 2nd edn. Cambridge, MA: Bradford Books/MIT Press.

Gibson E and Wexler K (1994) Triggers. *Linguistic Inquiry* **25**: 407–454.

Gigerenzer G, Hoffrage U and Kleinbölting H (1991) Probabilistic mental models: a Brunswikian theory of confidence. *Psychological Review* **98**(4): 506–528.

Gold EM (1967) Language identification in the limit. *Information and Control* **10**: 447–474.

Goldman AI and Gallese V (1998) Mirror neurons and the simulation theory of mind-reading. *Trends in Cognitive Science* **2**(12): 493–501.

Goldman-Rakic P (1985) Toward a neurobiology of cognitive development. In: Mehler RFJ (ed.) *Neonate Cognition: Beyond the Blooming Buzzing Confusion*. Hillsdale, NJ: Lawrence Erlbaum Associates.

Goldman-Rakic PS (1987) Circuitry of primate prefrontal cortex and the regulation of behavior by representation memory. In: Phern F (ed.) *Handbook of Physiology*, vol. 5.

Gopnik A and Meltzoff AN (eds) (1998) *Words, Thoughts, and Theories*. Cambridge, MA: MIT Press.

Gopnik M (1990) Feature-blind grammar and dysphasia **344**: 715.

Gopnik M (1994) Impairments of tense in a familial language disorder. *Journal of Neurolinguistics* **8**(2): 109–133.

Gould E and Gross CG (2002) Neurogenesis in adult mammals: some progress and problems. *The Journal of Neuroscience* **22**: 619–623.

Gross CG, Rocha-Miranda CE and Bender D (1972) Visual properties of neurons in inferotemporal cortex of the macaque. *Journal of Neurophysiology* **35**: 96–111.

Grossberg S (1976) Adaptive pattern classification and universal recoding: 1. parallel development and coding of neural feature detectors. *Biological Cybernetics* **23**: 121–134.

Guasti MT (2002) *Language Acquisition: The Growth of Grammar*. Cambridge, MA: MIT Press.

Haas RB (1976) *Muybridge: Man in Motion*. Berkeley, CA: University of California Press.

Hammond KR and Stewart TR (eds) (2001) *The Essential Brunswik: Beginnings, Explications, Applications*. New York and Oxford, UK: Oxford University Press.

Hanson SJ (1999) Connectionist neuroscience: representational and learning issues for neuroscience. In: Lepore E and Pylyshyn (eds) *What is Cognitive Science?*, pp. 401–427. Malden, MA: Blackwell.

Harris ZS (1951) *Methods of Structural Linguistics*. Chicago, IL: University of Chicago Press.

Harris ZS (1952a) Discourse analysis (Part 1). *Language* **28**: 18–23.

Harris ZS (1952b) Discourse analysis (Part 2): A sample text. *Language* **28**: 474–494.

Harris ZS (1957) Co-occurrence and transformation in linguistic structure. *Language* **33**: 293–340.

Harris ZS (1986/1951) *Structural Linguistics*. Chicago, IL: University of Chicago Press.

Haugeland J (1997) What is mind design. In: Haugeland J (ed.) *Mind Design II: Philosophy, Psychology, Artificial Intelligence*, pp. 1–28. Cambridge, MA: Bradford Books/MIT press. [revised and enlarged edition]

Haugeland J (ed.) (1981) *Mind Design: Philosophy, Psychology, Artificial Intelligence*. Cambridge, MA: Bradford Books/MIT press.

Hauser MD, Chomsky N and Fitch WT (in press) The faculty of language: what it is, who has it, and how did it evolve? *Science*.

Hebb DO (1949) *The Organization of Behavior: A Neuropsychological Theory*. New York: Wiley-Interscience.

Heims SJ (1991) *The Cybernetics Group*. Cambridge, MA: MIT Press.

Higginbotham JT (1982) Noam Chomsky's linguistic theory. *Social Research* **49**(1): 143–157.

Hirsh R (1974) The hippocampus and contextual retrieval of information from memory: a theory. *Behavioural Biology* **12**: 421–444.

Hochberg J (1956) Perception: toward the recovery of a definition. *Psychological Review* **63**: 400–405.

Hochberg J (1978) *Perception*. Englewood Cliffs, NJ: Prentice-Hall.

Hopfield JJ (1982) Neural networks and physical systems with emergent collective computational abilities. *Proceedings of the National Academy of Sciences of the U.S.A* **81**: 2554–2558.

Hubel DH and Wiesel TN (1962a) Receptive fields, binocular interaction and functional architecture in the cat's visual cortex. *Journal of Physiology* **160**: 106–154.

Hubel DH and Wiesel TN (1962b) Receptive fields of single neurons in the cat's striate cortex. *Journal of Physiology* **148**: 574–591.

Hull CL (1943) *Principles of Behavior*. New York: Appleton-Century-Crofts.

Jackendoff R (1983) *Semantics and Cognition*. Cambridge, MA: MIT Press.

Jackendoff R (1990) *Semantic Structures*. Cambridge, MA: MIT Press.

Jackendoff RP, Bloom P and Wynn K (eds) (1999) *Language, Logic, and Concepts: Essays in Honor of John MacNamara*. Cambridge, MA: MIT Press.

James WJ (1890) *The Principles of Psychology*. New York: Henry Holt.

Jasper HH and Smirmov GD (eds) (1960) The Moscow colloquium on the electroencephalography of higher nervous activity. *The International Journal of Electroencephalography and Clinical Neurophysiology* **13**(suppl.).

Jeffress LA (ed.) (1951) *Cerebral Mechanisms in Behavior: The Hixon Symposium*. New York: John Wiley.

Jenkins L (2000) *Biolinguistics: Exploring the Biology of Language*. Cambridge: Cambridge University Press.

Johansson G (1973) Visual perception of biological motion and a model for its analysis. *Perception and Psychophysics* **14**(2): 201–211.

Kahneman D, Slovic P and Tversky A (eds) (1982) *Judgment Under Uncertainty: Heuristics and Biases*. Cambridge, UK and New York: Cambridge University Press.

Kahneman D and Tversky A (1972) Subjective probability: a judgment of representativeness. *Cognitive Psychology* **3**: 43–454.

Kahneman D and Tversky A (1973) On the psychology of prediction. *Psychological Review* **80**: 237–251.

Kahneman D and Tversky A (1979) Prospect theory: an analysis of decision under risk. *Econometrica* **47**: 263–291.

Kamin LJ (1969) Predictability, surprise, attention and conditioning. In: Campbell BA and Church RM (eds) *Punishment and Aversive Behavior*, pp. 279–296. New York: Appleton-Century-Crofts.

Kandel ER (1980) Cellular insights into the multivariant nature of arousal. In: McFadden D (ed.) *Neural Mechanisms in Behavior*, pp. 260–291. New York: Springer-Verlag.

Kanwisher N and Moscovitch M (2000) The cognitive neuroscience of face processing: an introduction. *Cognitive Neuropsychology* **1/2/3**: 1–13.

Karmiloff K and Karmiloff-Smith A (2001) *Pathways to Language: From Fetus to Adolescent*. Cambridge, MA: Harvard University Press.

Karmiloff-Smith A (1992) *Beyond Modularity: A Developmental Perspective on Cognitive Science*. Cambridge, MA: MIT Press.

Karmiloff-Smith A (1993) Beyond Modularity: a developmental perspective on cognitive science. *European Journal of Disorders of Communication* **28**: 95–105.

Karmiloff-Smith A (1994) *Précis* of beyond modularity: a developmental perspective on cognitive science. *Behavioral and Brain Sciences* **17**: 693–745.

Karmiloff-Smith A (1998) Development itself is the key to understanding developmental disorders. *Trends in Cognitive Sciences* **2**(10): 389–398.

Kegl J and McWorther J (in press) Perspectives on an emerging language. *Proceedings of the 28th Annual Stanford Child Language Research Forum*. New York and Cambridge UK: Cambridge University Press.

Kegl J, Neidle C, MacLaughlin D, Bahan B and Lee RG (2000) *The Syntax of American Sign Language: Functional Categories and Hierarchical Structure*. Cambridge, MA: MIT Press.

Keil FC (1979) *Semantic and Conceptual Development: An Ontological Perspective*. Cambridge, MA: Harvard University Press.

Kinsbourne M and Wood F (1975) Short-term memory and pathological forgetting. In: Deutsch D and Deutsch AJ (eds) *Short-term Memory*. New York: Academic Press.

Kleene SC (1936) General recursive functions of natural numbers. *Mathematische Annalen* **112**: 727–742.

Klima ES and Bellugi U (1979) *The Signs of Language*. Cambridge, MA: Harvard University Press.

Kosslyn SM (1980) *Image and Mind*. Cambridge, MA: Harvard University Press.

Kosslyn SM and Andersen RA (eds) (1992) *Frontiers in Cognitive Neuroscience*. Cambridge, MA: Bradford Books/MIT Press.

Kubie L (1930) A theoretical application to some neurological problems of the propensities of excitation waves which move in closed circuits. *Brain* **53**: 166–177.

Landau B and Gleitman LR (1985) *Language and Experience: Evidence from the Blind Child*. Cambridge, MA: Harvard University Press.

Larson R and Segal G (1995) *Knowledge of Meaning: An Introduction to Semantic Theory*. Cambridge, MA: MIT Press.

Lashley KS (1929) *Brain Mechanisms and Intelligence*. Chicago, IL: University of Chicago Press.

Lashley KS (1951) The problem of serial order in behavior. In: Jeffress LA (ed.) *Cerebral Mechanisms in Behavior*, pp. 112–136. New York: John Wiley.

Lettvin JY, Maturana HR, McCulloch WS and Pitts WH (1959) What the frog's eye tells the frog's brain. *Proceedings of the IRE* **47**: 1940–1951.

Levin B and Pinker S (eds) (1991) *Lexical and Conceptual Semantics*. New York and Oxford: Blackwell.

Lewis HR and Papadimitriou CH (1981) *Elements of the Theory of Computation*. Prentice-Hall.

Lieberman P (2000) *Human Language and Our Reptilian Brain: The Subcortical Bases of Speech, Syntax, and Thought*. Cambridge, MA: Harvard University Press.

Lightfoot D (1982) *The Language Lottery: Toward a Biology of Grammars*. MIT Press.

Lorente de Nó R (1938) Analysis of the activity of the chains of internuncial neurons. *Journal of Neurophysiology* **1**: 207–244.

Mackintosh NJ (1975) A theory of attention: variations in the associability of stimuli with reinforcements. *Psychological Review* **82**: 276–298.

Markman EM (1989) *Categorization and Naming in Children: Problems of Induction*. Cambridge, MA: Bradford Books/MIT Press.

Marr D (1982) *Vision: A Computational Investigation into the Human Representation and Processing of Visual Information*. San Francisco, CA: WH Freeman.

McCabe KA, Houser D, Ryan L, Smith VL and Trouard T (2001) A functional imaging study of cooperation in two-person reciprocal exchange. *Proceedings of the National Academy of Sciences of the USA* **98**: 11832–11835.

McCulloch WS (1964) What's in the brain that ink may character? *The International Congress for Logic, Methodology and Philosophy of Science (Jerusalem, Israel, August 26)*.

McCulloch WS (1988) *Embodiments of Mind*. Cambridge, MA: MIT Press.

McCulloch WS and Pitts WH (1943) A logical calculus of the ideas immanent in nervous activity. *Bulletin of Mathematical Biophysics* **5**: 115–133.

Mézard M, Parisi G and Virasoro MA (1987) *Spin Glass Theory and Beyond*. Singapore and Teaneck, NJ: World Scientific.

Miller GA and Chomsky N (1963) Finitary models of language users. In: Luce RD, Bush RR and Galanter E (eds) *Handbook of Mathematical Psychology*, vol. 2, pp. 419–491. New York: John Wiley and Sons.

Miller GA, Galanter E and Pribram KH (1960) *Plans and the Structure of Behavior*. New York: Holt, Rinehart and Winston.

Milner B (1965) Memory disturbance after bilateral hippocampal lesions. In: Milner PM and Glickman SE (eds) *Cognitive Processes and the Brain*, pp. 97–111. Patterson, NJ: Van Nostrand.

Minsky M (1961) Steps toward Artificial Intelligence. *Proceedings of IRE* **49**: 8–30.

Minsky ML and Papert SA (1990) *Perceptrons*. MIT Press.

Moruzzi G and Magoun HW (1949) Brain stem reticular formation and activation of the EEG. *Electroencephalography and Clinical Neurophysiology* **1**: 455–473.

Moscovitch M (1995) Recovered consciousness: a hypothesis concerning modularity and episodic memory. *Journal of Clinical and Experimental Neuropsychology* **17**: 276–290.

Mountcastle VB (1957) Modality and topographic properties of single neurons of cat's somatic sensory cortex. *Journal of Neurophysiology* **20**: 408–434.

Nadel L and O'Keefe J (1974) The hippocampus in pieces and patches: an essay on modes of explanation in physiological psychology. In: Bellairs R and Gray EG (eds) *Essays on the Nervous System. A Festschrift for J.Z. Young*. Oxford, UK: Clarendon Press.

Neisser U (1967) *Cognitive Psychology*. Englewood Cliffs: Prentice-Hall.

Nespor M (2001) About parameters, prominence, and bootstrapping. In: Dupoux E (ed.) *Language, Brain, and Cognitive Development*, pp. 127–142. Cambridge, MA: Bradford Books/MIT Press.

Newell A, Shaw JC and Simon HA (1958) Elements of a theory of human problem-solving. *Psychological Review* **65**: 151–166.

Newell A and Simon HA (1972) *Human Problem Solving*. Englewood Cliffs, NJ: Prentice-Hall.

Newport EL, Bavelier D and Neville HJ (2001) Critical thinking about critical periods: perspectives on a critical period for language acquisition. In: Dupoux E (ed.) *Languge, Brain and Cognitive Development: Essays in Honour of Jacques Mehler*, pp. 481–502. Cambridge, MA: Bradford Books/MIT Press.

Nowak MA, Komarova NL and Niyogi P (2002) Computational and evolutionary aspects of language. *Nature* **417**: 611–617.

O'Keefe J and Dostrovsky J (1971) The hippocampus as a spatial map: preliminary evidence from unit activity in the freely-moving rat. *Brain Research* **34**: 171–175.

O'Keefe J and Nadel L (1978) *The Hippocampus as a Cognitive Map*. Oxford, UK: Clarendon Press.

O'Keefe J and Nadel L (1979) The hippocampus as a cognitive gap. *Behavioral and Brain Sciences* **2**(4).

Olds J and Milner PM (1954) Positive reinforcement produced by electrical stimulation of the septal area and other regions of the rat brain. *Journal of Comparative and Physiological Psychology* **47**: 419–427.

Osgood CE and Miron MS (eds) (1963) *Approaches to the Study of Aphasia. A Report of an Interdisciplinary Conference on Aphasia*. Urbana, IL: University of Illinois Press.

Osherson DN (ed.) (1990/1995) *An Invitation to Cognitive Science*. Cambridge, MA: Bradford Books/MIT Press. [1st edn (1990) in 3 volumes. 2nd edn (1995) in 4 volumes.]

Osherson DN, Stob M and Weinstein S (1986) *Systems that Learn*. Cambridge, MA: MIT Press.

Penfield W and Rasmussen T (1949) Vocalization and arrest of speech. *Archives of Neurology and Psychiatry* **61**: 21–27.

Peterson MA (1999) Top-down influences on distinctly perceptual processes. *Behavioral and Brain Sciences* **22**: 389–390.

Petitto LA (1987) On the autonomy of language and gesture: Evidence from the acquisition of personal pronouns in American Sign Language. *Cognition* **27**(1): 1–52.

Petitto LA and Marentette P (1991) Babbling in the manual mode: Evidence for the ontogeny of language. *Science* **251**: 1483–1496.

Piaget J (1930) *The Child's Conception of the World*. New York: Harcourt Brace.

Piaget J (1954) *The Construction of Reality in the Child*. Ballantine Books.

Piattelli-Palmarini M (1989) Evolution, selection and cognition: from 'learning' to parameter setting in biology and in the study of language. *Cognition* **31**: 1–44.

Piattelli-Palmarini M (1994) Ever since language and learning: after thoughts on the Piaget–Chomsky debate. *Cognition* **50**: 315–346.

Piattelli-Palmarini M (2001) Portrait of a 'classical' cognitive scientist: what I have learned from Jacques Mehler. In: Dupoux E (ed.) *Language, Brain, and Cognitive Development*, pp. 3–21. Cambridge: Bradford Books/MIT Press.

Piattelli-Palmarini M (ed.) (1980) *Language and Learning: The Debate between Jean Piaget and Noam Chomsky*. Cambridge, MA: Harvard University Press.

Pinker S (1979) Formal models of language learning. *Cognition* **7**: 217–283.

Pinker S (1990) Language acquisition. In: Osherson DN (ed.) *An Invitation to Cognitive Science. Volume 1: Language*. Cambridge, MA: Bradford Books/MIT Press.

Pinker S (1994) *The Language Instinct*. New York: William Morrow and Company Inc.

Pinker S (1997) *How the Mind Works*. New York: WW Norton and Company.

Pinker S and Bloom P (1990) Natural language and natural selection. *Behavioral and Brain Sciences* **12**: 707–784.

Pinker S and Mehler J (eds) (1988) *Connections and Symbols*. Cambridge, MA: MIT Press.

Posner MI (1969) Abstraction and the process of recognition. In: Bower GH and Spence JT (eds) *The Psychology of Learning and Motivation*, vol. 3, pp. 44–100. New York: Academic Press.

Posner MI (1978) *Chronometric Explorations of Mind*. Hillsdale: Lawrence Erlbaum Associates.

Posner MI (2001) Cognitive neuroscience: the synthesis of mind and brain. In: Dupoux E (ed.) *Languge, Brain, and Cognitive Development: Essays in Honour of Jacques Mehler*, pp. 403–416. Cambridge: Bradford Books/MIT Press.

Posner MI (ed.) (1989) *Foundations of Cognitive Science*. Cambridge, MA: Bradford Books/MIT Press.

Posner MI, Petersen SE, Fox PT and Raichle ME (1988) Localization of cognitive operations in the human brain. *Science* **240**.

Premack D (1972) Language in chimpanzees? *Science* **172**: 808–822.

Premack D (1986) *Gavagai! or the Future History of the Animal Language Controversy*. Cambridge, MA: Bradford Books/MIT Press.

Pribram KH and Gill MM (1976) *Freud's 'Project' Reassessed : Preface to Contemporary Cognitive Theory and Neuropsychology*. New York: Basic Books.

Putnam H (1960) Minds and Machines. In: *Collected Papers. Vol. 2: Mind, Language and Reality*, pp. 362–384. New York and Cambridge, UK: Cambridge University Press.

Putnam H (1973) Philosophy and our mental life. In: *Mind, Language and Reality. Philosophical Papers: Vol 2*, pp. 291–303. Cambridge: Cambridge University Press.

Putnam H (1987) *The Many Faces of Realism*. La Salle, IL: Open Court.

Putnam H (1988) *Representation and Reality*. Cambridge, MA: MIT Press/Bradford Books.

Quine WVO (1956) Quantifiers and propositional attitudes. *Journal of Philosophy* **53**(5): 177–187.

Rakic P, Bourgeois J-P, Eckenhoff MF, Zecevic N and Goldman-Rakic P (1986) Concurrent overproduction of synapses in diverse regions of the primate cerebral cortex. *Science* **232**: 232–234.

Ramus F (2002) Language discrimination by newborns: teasing apart phonotactic, rhythmic, and intonational cues. *Annual Review of Language Acquisition* **2**.

Ramus F, Nespor M and Mehler J (1999) Correlates of linguistic rhythm in the speech signal. *Cognition* **73**(3): 265–292.

Rescorla RA and Wagner AR (1972) A theory of Pavlovian conditioning: variations in the effectiveness of reinforcement and nonreinforcement. In: Black AH and Prokasy WF (eds) *Classical Conditioning II*, pp. 64–99. New York: Apple-Century-Crofts.

Rizzolatti G and Arbib MA (1998) Language within our grasp. *Trends in Neurosciences* **21**(5): 188–194. [Viewpoint.]

Rizzolatti G and Arbib MA (1999) From grasping to speech: imitation might provide a missing link. *Trends in Neurosciences* **22**(4): 152. [Reply.]

Rizzolatti G and Craighero L (1998) Spatial attention: mechanisms and theories. In: Sabourin M, Craik F and Robert M (eds) *Advances in Psychological Science. Volume 2: Biological and Cognitive Aspects*. Hove: Psychology Press.

Rochester N, Holland JH, Haibt LH and Duda WL (1956) Tests on a cell assembly theory of the action of the brain, using a large digital computer. *IRE Transactions on Information Theory* **2**: 80–93.

Rogers H (1967/1988) *Theory of Recursive Functions and Effective Computability*. MIT Press. [1st edn: 1967. Paperback edn: 1988]

Rosch E (1973) Natural categories. *Cognitive Psychology* **7**: 573–605.

Rosch E (1978) Principles of categorization. In: Rosch E and Lloyd BB (eds) *Cognition and Categorization*, pp. 27–48. Hillsdale, NJ: Lawrence Erlbaum.

Rosch E and Mervis CB (1996) Family resemblances: studies in the internal structure of categories. In: Geirsson H and Losonsky M (eds) *Readings in Language and Mind*, pp. 442–446. Cambridge: Blackwell.

Rosch E, Mervis CB, Gray WD and Johnson DM (1976) Basic objects in natural categories. *Cognitive Psychology* **8**: 382–439.

Rosenblatt F (1958) The perceptron: a probabilistic model for information storage and orgranization in the brain. *Psychological Review* **65**: 386–408.

Rosenblatt F (1962) *Principles of Neurodynamics: Perceptrons and the Theory of Brain Mechanisms*. Washington, DC: Spartan Books.

Rosenblueth A, Wiener N and Bigelow J (1943) Behavior, purpose and teleology. *Philosophy of Science* **10**: 18–24.

Rumelhart DE and McClelland JL (1986) *Parallel Distributed Processing: Explorations in the Microstructure of Cognition*. Cambridge: MIT Press.

Russell B (1900) *The Principles of Mathematics*, vol. 1. Cambridge, UK: Cambridge University Press.

Russell B (1919) *Introduction to Mathematical Philosophy*. London, UK: George Allen and Unwin Ltd.

Saleem KS, Pauls J, Augath MA *et al.* (2002) Magnetic resonance imaging of neuronal connections in the macaque monkey. *Neuron* **34**: 685–700.

Samuel AL (1959) Some studies in machine learning using the game of checkers. *IBM Journal of Research and Development* **3**: 210–229.

Savage-Rumbaugh ES, Murphy J, Sevcik RA *et al.* (in press) Language comprehension in ape and child. *Monographs of the Society for Research in Child Development*, vol. 58.

Schacter D and Tulving E (eds) (1994) *Memory Systems*, Cambridge, MA: MIT Press.

Scoville WB and Milner B (1957) Loss of recent memory after bilateral hippocampal lesion. *Journal of Neurology, Neurosurgery, and Psychiatry* **20**: 11–21.

Searle JR (1980a) Intrinsic intentionality. *Behavioral and Brain Sciences* **3**: 450–456.

Searle JR (1980b) Minds, brains and programs. *Behavioral and Brain Sciences* **3**: 417–424.

Searle JR (1992) *The Rediscovery of the Mind*. Cambridge, MA: MIT Press/Bradford Books.

Searle JR (1996) Is the brain's mind a computer program? In: Geirsson H and Losonsky M (eds) *Readings in Language and Mind*, pp. 264–271. New York and Oxford, UK: Blackwell.

Sebeok TA (1977) Ecumenicalism in semiotics. In: Sebeok TA (ed.) *A Perfusion of Signs*. Bloomington, IN: Indiana University Press.

Sebeok TA (1995/2000) Semiotics and the biological sciences: Initial conditions. [a lecture delivered at the Collegium Budapest.]

Seidenberg MS and Petitto LA (1979) Signing behavior in apes: a critical review. *Cognition* **7**: 177–215.

Seidenberg MS and Petitto LA (1987) Communication, symbolic communication, and language in child and chimpanzee. *Journal of Experimental Psychology (General)* **116**(3): 279–287, 1483–1496. [Comment on Savage-Rumbaugh, McDonald, Sevcik, Hopkins and Rupert.]

Selfridge OG (1958) Pandemonium: a paradigm for learning. *Mechanisation of Thought Processes: Proceedings of a Symposium held at the National Physical Laboratory, November 1958*, pp. 513–526. London: Her Majesty's Stationery Office.

Shannon CE (1938) A symbolic analysis of relay and switching circuits. *Transactions of the American Institute of Electrical Engineers* **57**: 1–11.

Shannon CE (1948) A mathematical theory of communication. *Bell System Technical Journal* **27**: 379–423, 623–656.

Shannon CE and Weaver W (1949/1998) *The Mathematical Theory of Communication*. Urbana, IL: University of Illinois Press.

Shepard RN (1962) The analysis of proximities: multidimensional scaling with an unknown distance function. *Psychometrika* **27**(2): 125–140.

Shepard RN (1964) On subjectively optimum selection among multiattribute alternatives. In: Bryan GL (eds) *Human Judgments and Optimality*, p. 166. New York: Wiley.

Shepard RN (1994) Perceptual-cognitive universals as reflections of the world. *Psychonomic Bulletin and Review* **1**: 2–28.

Shepard RN and Metzler J (1971) Mental rotation of three-dimensional objects. *Science* **171**: 701–703.

Smith E, Osherson D, Rips L and Keane M (1988) Combining prototypes: a selective modification model. *Cognitive Science* **12**: 485–527.

Smith EE and Medin DL (1981) *Categories and Concepts*. Cambridge, MA: Harvard University Press.

Smith N and Tsimpli I-M (1995) *The Mind of a Savant*. New York and Oxford, UK: Blackwell.

Smith VL (2003) Experimental methods in economics. *Encyclopedia of Cognitive Science*. London: Nature Publishing Group.

Spelke ES (1985) Perception of unity, persistence and identity: thoughts on infants' conception of objects. In: Mehler J and Fox R (eds) *Neonate Cognition: Beyond the Blooming Buzzing Confusion*. Hillsdale, NJ: Lawrence Erlbaum.

Spelke ES (1988) Where perceiving ends and thinking begins: the apprehension of objects in infancy. In: Yonas A (ed.) *Perceptual Development in Infancy*, vol. 20. Minneapolis, MI: University of Minnesota Press.

Sperling G (1960) The information available in brief visual presentations. *Psychological Monographs* **74**.

Sperry RW (1968) Hemisphere deconnection and unity in conscious awareness. *American Psychologist* **23**(10): 723–733.

Stein D (1985) *Ada: A Life and a Legacy*. Cambridge, MA: MIT Press.

Sternberg S (1966) High-speed memory scanning in human memory. *Science* **153**: 652–654.

Sternberg S (1969) The discovery of processing stages: extensions of Donders' method. *Acta Psychologica* **30**: 276–315.

Stevens T and Karmiloff-Smith A (1997) Word learning in a special population: do individuals with Williams syndrome obey lexical constraints? *Journal of Child Language* **24**: 737–765.

Stillings NA, Feinstein MH and Garfield JL (1987) *Cognitive Science: An Introduction*. Cambridge, MA: MIT Press.

Sulloway FJ (1979) *Freud, Biologist of the Mind : Beyond the Psychoanalytic Legend*. New York: Basic Books.

Tarski A (1935) Der Wahrheitsbegriff in den formalisierten sprachen. *Studia Philosophica* **1**: 261–405. [German translation of a book in Polish, 1933.]

Tarski A (1996) The semantic conception of truth and the foundations of semantics. In: Martinich AP (ed.) *The Philosophy of Language*, 3rd edn, pp. 61–84. New York, Oxford: Oxford University Press.

Terrace HS, Petitto LA, Sanders RJ and Bever TG (1979) Can an ape create a sentence? *Science* **206**: 891–902.

Thaler RH (1991) *Quasi-rational Economics*. New York: Russell Sage Foundation.

Thaler RH (1992) *The Winner's Curse: Paradoxes and Anomalies of Economic Life*. New York: Russell Sage Foundation/Free Press.

Tolman EC (1932) *Purposive Behavior in Animals and Men*. New York: Century.

Tolman EC (1948) Cognitive maps in rats and men. *Psychological Review* **55**: 189–208.

Tolman EC (1949) There is more than one kind of learning. *Psychological Review* **56**: 144–155.

Tomasello M (1999) *The Cultural Origins of Human Cognition*. Cambridge, MA: Harvard University Press.

Townsend DJ and Bever TG (2001) *Sentence Comprehension: The Integration of Habits and Rules*. Cambridge, MA: Bradford Books/MIT Press.

Tulving E (1972) Episodic and semantic memory. In: Tulving E and Donaldson W (eds) *Organization and Memory*. New York, Academic Press.

Tulving E and Thompson D (1973) Encoding specificity and retrieval processes in episodic memory. *Psychological Review* **80**: 352–373.

Turing AM (1936a) On computable numbers: with an application to the Entscheidungsproblem (Part 1). *Proceedings of the London Mathematical Society, Series 2* **42**: 230–265.

Turing AM (1936b) On computable numbers: with an application to the Entscheidungsproblem (Part 2). *Proceedings of the London Mathematical Society, Series 2* **43**: 544–546.

Tversky A (1977) Features of similarity. *Psychological Review* **84**: 327–352.

Tversky A and Gati I (1978) Studies of similarity. In: Rosch E and Lloyd B (eds) *Cognition and Categorization*, pp. 79–98. Hillsdale, NJ: Lawrence Erlbaum.

Ungerleider LG and Mishkin M (1982) Two cortical visual systems. In: Ingle DJ, Goodale MA and Mandfield RJW (eds) *Analysis of Visual Behavior*. Cambridge: MIT Press.

Uriagereka J (1998) *Rhyme and Reason: An Introduction to Minimalist Syntax*. Cambridge, MA: MIT Press.

Van der Lely HKJ (1997) Language and cognitive development in a grammatical SLI boy: modularity and innateness. *Journal of Neurolinguistics* **10**: 75–107.

Van der Lely HKJ and Stollwerck L (1996) A grammatical specific language impairment in children: an autosomal dominant inheritance? *Brain and Language* **52**: 484–504.

van Heijenoort J (ed.) (1967) *From Frege to Gödel: A Source Book of Mathematical Logic, 1897–1931*. Cambridge, MA: Harvard University Press.

Volterra V, Caselli MC, Capirci O, Vicari S and Tonucci F (2002) *Early Linguistic Abilities in Italian Children with Williams Syndrome*. [Paper presented at the Williams Syndrome and the Issue of Neurogenetic Developmental Disorders Conference, Budapest, Hungary.]

Volterra V and Erting CJ (eds) (2002) *From Gesture to Language in Hearing and Deaf Children*. Washington DC: Gallaudet University Press.

von Neumann J (1951) The general and logical theory of automata. In: Jeffress LA (ed.) *Cerebral Mechanisms in Behavior: The Hixon Symposium*, pp. 1–41. New York: John Wiley and Sons.

von Neumann J (1957) *The Computer and the Brain*. New Haven, CONN: Yale University Press.

von Neumann J and Morgenstern O (1944) *Theory of Games and Economic Behaviour*. Princeton, NJ: Princeton University Press. [1st edn: 1944. 2nd edn: 1947.]

Walter WG (1953) *The Living Brain*. New York: WW Norton.

Wanner E and Gleitman LR (eds) (1982) *Language Acquisition: The State of the Art*. Cambridge, UK: Cambridge University Press.

Watson JB (1924) *Behaviorism*. New York: Norton.

Wernicke C (1874) *Der aphasische symptomencomplex: eine psychologische studie auf anatomischer basis*. Breslau: Cohn und Weigert.

Wexler K (2002) Lenneberg's dream: learning, normal language development and specific language impairment. In: Schaffer J and Levy Y (eds) *Language Competence across Populations: Towards a Definition of Specific Language Impairment*, pp. 11–60. Mahwah, NJ: Erlbaum.

Wexler K and Borer H (1986) *The Maturation of Syntax*, vol. 39. School of Social Sciences, University of California, Irvine.

Wexler K and Culicover PW (1980) *Formal Principles of Language Acquisition*. Cambridge, MA: MIT Press.

Wexler K and Manzini MR (1986) Parameters and learnability in binding theory. In: Roeper T and Williams E (eds) *Parameters and Linguistic Theory*.

Whitehead AN and Russell B (1910) *Principia Mathematica*, vol. 1. Cambridge, UK: Cambridge University Press.

Wiener N (1948) *Cybernetics, or Control and Communication in the Animal and the Machine*. New York: John Wiley and Sons.

Wildgen W (2001) Kurt Lewin and the rise of cognitive science. In: Albertazzi L (ed.) *The Dawn of Cognitive Science: Early European Contributors*, pp. 299–332. Dordrecht, Holland: Kluwer Academic.

Wilson MA and McNaughton BL (1993) Dynamics of the hippocampal ensemble code for space. *Science* **261**: 1055–1058.
Wilson RA and Keil FC (eds) (2000) *The MIT Encyclopedia of the Cognitive Sciences*. Cambridge, MA: MIT Press.
Winograd T (1972) *Understanding Natural Language*. New York: Academic Press.
Zeki SM (1978) Functional specialization in the visual cortex of the rhesus monkey. *Nature* **274**: 423–428.

Academic Achievement

Introductory article

Harold W Stevenson, University of Michigan, Ann Arbor, Michigan, USA

CONTENTS

Worldwide studies of academic achievement in children and adolescents strengthen our understanding of the factors underlying differences in performance.

CONDUCTING CROSS-CULTURAL RESEARCH

This article describes the methodology and findings of several studies of academic achievement of children and adolescents in the US, Taiwan, China, and Japan. In the mid-1980s, when many of these studies were begun, large-scale studies of academic achievement in the US and East Asian countries were still rare, in part because China had not until then opened its doors to foreign researchers. Since then, there has been increasing interest in cross-cultural studies of academic achievement comparing East Asian children with their US peers.

The main purpose of most of these comparative studies is to identify factors that underlie cross-cultural differences in children's performance in academic subjects such as mathematics and science. Such research can generate ideas about what is possible in one's own culture, by drawing attention to those environmental conditions and strategies that have been found to be productive in others. East Asian societies have attracted the interest of many educators and researchers because of the excellence of their students' academic performance. International test results show that the academic performance of American children is consistently below that of their East Asian counterparts.

We begin with a consideration of several important methodological and interpretive issues, and then review some findings that illustrate the importance and relevance of cross-cultural research to educational policy and practice in the United States.

Methodological Issues in Developing Tests

Cross-cultural research is difficult to conduct, for it requires instruments that are culturally specific but are also capable of producing findings that can be applied meaningfully across cultures. Perhaps the greatest problem encountered in cross-cultural research in the social and behavioral sciences has been the use of materials that were initially developed for one culture and then applied inappropriately to other cultures. For example, a scale devised in New York for measuring depression would not necessarily be equally valid in East Asian cultures. Great care must be taken, therefore, in ensuring that research materials and procedures are culturally relevant, reliable, and valid for all cultures being studied.

Another problem often encountered in cross-cultural research is that sample sizes are too small to permit generalization to the larger population being studied. One must ensure that samples are representative of the larger population, and that study materials are prepared in the language of the culture under study. For example, an intelligence test such as the Stanford–Binet test developed in the US is not equally useful in China when translated into Chinese. Problems such as these make the research difficult to replicate and to apply.

Measuring Personality and Social Attributes

It is often difficult to translate psychological terms such as 'dependency', 'aggression', or 'anxiety', because of differences in the nuances or meanings of these concepts across different languages. Some researchers measure personality and social attributes through survey research, where large numbers of participants fill out scales designed to record subtle aspects of their reactions to, for example, mathematics instruction. Alternatively, researchers may conduct personal interviews with subjects, making it possible to probe and clarify subjects' responses through questions such as: 'How satisfied were you when you received your last semester's grade in math? Can you tell me more about why you felt this way?' Another technique for investigating differences in the meanings of terms for personality and social attributes across cultures involves the development of vignettes, prepared by residents of each country, to provide a detailed look at differences in the meanings of the terms across languages.

Studying Classroom Behavior

Studies of the classroom behavior of students and teachers require many hours of observation by researchers and thorough checks of the validity and reliability of their interpretations and conclusions. The advent of videotaping has significantly streamlined the process of observing children and teachers in action, as it has enabled researchers to gather substantial amounts of data without incurring the cost of observation on site. However, it is essential that videographers understand the purposes of the study, and the languages and cultures being studied, to ensure that the tapes are relevant to the research.

CULTURAL DIFFERENCES IN SCHOOLING

Here we briefly describe some results from a series of comparative studies concerned with differences in performance in mathematics and science of students in the US, Taiwan, and Japan.

Main Results

US students generally perform at lower levels at each grade level – from first grade through entrance into college – than do Chinese and Japanese students on a range of tests of cognitive development and academic achievement. Children's academic performance across countries bears little relation to the amount of money the countries spend per child on education. For example, the US spends a greater proportion of its gross national product on education than does China, yet consistently produces lower scores on tests and other indicators of children's cognitive development and academic achievement. Interestingly, the Chinese students who performed better than the American students were less satisfied with their performance than were the American students with theirs. This finding is consistent with the generally higher standards that Chinese and Japanese schools, parents, and students themselves, tend to set for students.

Thinking about Thinking

One explanation of the observed superiority among Chinese and Japanese students on academic tests concerns the manner in which problem-solving is presented. The East Asian teacher guides children through a problem presented in a lesson plan by encouraging them to define the problem in their own words, to determine an acceptable solution to the problem, and then to create additional methods for solving the problem. As the teacher walks around the classroom to observe students' efforts, he or she may offer hints to a confused child. After one acceptable solution is found, the child is asked if another one can be found. Depending on the child's reaction to this challenge, additional solutions may be requested of the student. The child quickly learns that problems can be approached in more than one manner, and that the way to success in solving problems is through close attention, careful thought, creative thinking, and discovery of alternative strategies.

In the US, students are typically expected to learn a prescribed strategy for solving problems and then to apply the strategy in new contexts. Less emphasis is placed on having children define the problems and then develop multiple approaches to their solution.

East Asian students commonly characterize mathematics as something they need to understand, not as something to memorize. The emphasis on problem-solving, rather than rote learning and memorization, is evident in the types of problems included in contemporary curricula in mathematics. Teachers are encouraged to present the questions in a meaningful context so that students become able to understand a rule, to provide a number of problems that are solvable by the same general rule, and to extend the discussion, opening it up for further comments.

Innate Ability

Children's own perceptions of the role that several factors play may also influence their level of academic achievement. These factors include the strength of the teacher, the level of effort students expend, their level of ability, their aspirations for educational achievement, their reasons for working hard, and their dissatisfaction with current performance. A common finding among East Asian participants is the belief that a child's behavior is highly malleable and can be readily changed, depending on the quality, interest, and relevance of the child's experiences. East Asian participants stressed the importance of the level of effort students apply to their studies, as an important predictor of their level of academic achievement – rather than their natural or innate ability, which US respondents favored as a stronger predictor. Thus, according to East Asian participants, students' test scores may rise or fall according to the quality of their experiences and the effort they apply to their studies.

Development and Expansion of Academies

What happens in the classroom constitutes only a part of the academic training that occurs in schools. A second group of educational institutions provides supplemental training to many students. These are called 'buxiban' in Taiwan, and 'juku' and 'yobiko' in Japan. These academies were set up to ensure that children would be taught aspects of the curriculum that could not be covered during the ordinary school hours. They may be especially relevant now that the school week in East Asia has been reduced from five and a half days to five days.

East Asian educators have developed a variety of extracurricular programs to accommodate differences in students' levels of academic, social, and personality development. Yobiko in Japan, for example, are academies designed specifically to help students who fail the college entrance examination to pass it in subsequent efforts. Buxiban in Taiwan, and yobiko in Japan, provide students with additional practice and review of the content of academic courses, and help them to prepare for college entrance examinations. Buxiban and juku may include activities for students with special abilities who wish to undertake projects of unusual complexity: for example, making their own communications equipment, developing their own pharmaceuticals, or presenting programs of folk dances from various regions of China. Children and young people who wish to become more involved in sports or the arts, or who are interested in filling a void of social experiences, are also often served by these institutions. They provide opportunities for athletics and art, and a meeting place for fostering social interactions among schoolmates in after-school activities.

In the past, admission to colleges and universities in Japan depended solely on students' scores on college entrance tests. The emphasis on test scores as the main basis for college admission guided many aspects of higher education in Japan, including the *Course of Study*, a guidebook that describes the content of the curriculum and what students should know in preparing for the examinations. While this practice may have served the purpose of selecting highly able college students in the past, many Japanese have responded to it unfavorably, claiming that its narrow focus on only one aspect of children's academic achievement is 'elitist'. Following the general practice in the United States, admission procedures have changed to permit grades in high school and letters of recommendation, as well as test results, to be considered as important factors in admission decisions. This 'recommendation' method has resulted in the introduction of personal and social aspects of achievement, such as extracurricular activities and other distinguishing characteristics of children, into the criteria for selecting high school and college students. A negative feature of these changes is that students have less time than was available earlier for traditional academic work. There is increasing interest in transforming the system of education so that lessons place less emphasis on memorization and more on thoughtful, creative problem-solving.

CONCLUSION

It has become increasingly clear that understanding a single culture is not sufficient to create new methods of educating students. Cross-cultural research on cognitive development, conducted with sound methodology, can give researchers and educators substantial insight into methods for improving educational practices and students' academic performance. Recent technological advances, such as fast computers and videography, as well as recent political developments, such as China opening its doors to foreign researchers, have enabled much data to be obtained and analyzed across a wide range of cultures. The findings from cross-cultural research on children's academic achievement have far-reaching implications for US education policy and practice. These findings identify mechanisms to help strengthen children's understanding of academic information and may help to narrow the gap between western and eastern children's performance in mathematics and science.

Further Reading

Cizek GJ (ed.) (1999) *Handbook of Educational Psychology*. San Diego, CA: Academic Press.

Johanek M (ed.) (2001) *A Faithful Mirror: Reflections on the College Board and Education in America*. New York, NY: College Board.

Munro DJ (1996) *The Imperial Style of Inquiry in Twentieth Century China*. Ann Arbor, MI: Center for Chinese Studies.

Paris S and Wellman H (eds) (1998) *Global Prospects for Education: Development, Culture, and Schooling*. Washington, DC: American Psychological Association.

Ravitch D (ed.) (1998) *Brookings Papers on Education Policy*. Washington, DC: Brookings Institution.

Rohlen T and LeTendre GL (eds) (1996) *Teaching and Learning in Japan*. New York, NY: Cambridge University Press.

Sing L (ed.) (1996) *Growing Up the Chinese Way: Chinese Child and Adolescent Development*. Hong Kong: Chinese University Press.

Stevenson HW (2000) *To Sum It Up: The TIMSS Case Studies of Education in Germany, Japan, and the United States*. Philadelphia, PA: Research for Better Schools.

Stevenson HW and Stigler JW (eds) (1992) *The Learning Gap*. New York, NY: Summit Books.

Stigler JW and Hiebert J (1999) *The Teaching Gap*. New York, NY: Free Press.

Acalculia

Intermediate article

John Whalen, University of Delaware, Newark, Delaware, USA

CONTENTS

Impairment of the ability to calculate; research into numerical impairments in patients with brain damage has revealed that human calculation abilities are composed of several distinct cognitive processes.

INTRODUCTION

What are the brain processes that underlie our fundamental numerical abilities such as calculation, estimation, and reading and writing numbers? A particularly successful line of research in answering this question has come from the study of impairments in numerical abilities as a result of brain injury.

While it was once thought that there was one calculation center in the brain that could be impaired (hence the term 'Acalculia' – the inability to calculate), it is now believed that there are many distinct numerical abilities. This change in our understanding is based on evidence that only some arithmetic abilities may be impaired as a result of brain injury, while others are spared. Individual brain-injured patients show some remarkably specific impairments. Note that these are not isolated cases: there are several documented reports of people with similar patterns of performance.

NUMERICAL IMPAIRMENTS AFTER BRAIN DAMAGE

Numeral Comprehension and Production Processes

A distinction has been drawn between the ability to perform calculations and the ability to comprehend and produce numbers. Patients such as the person known as D. R. C. reveal striking impairments after brain damage, for example the inability to remember simple arithmetic facts such as $2 \times 4 = 8$, even though other abilities such as comprehending and producing written and spoken numerals are unimpaired. This pattern implies the brain systems for comprehending and producing numerals are separate from those for calculation. Several cases with the same as well as the opposite pattern of impairment have been reported.

The ability to comprehend and produce numbers is also composed of several functional subcomponents. For example, some patients reveal highly specific impairments strictly limited to writing arabic numerals, such as writing down '100206' in response to hearing 'one hundred and twenty-six'. This example reveals that numerical processing has several distinct components, including numeral comprehension and numeral production (only production was impaired), verbal and arabic numeral processes (only arabic numeral production was impaired), and the ability to retrieve a single digit and the ability to put together several digits to make a larger number within one component (only number composition was impaired).

Dissociations between number and language processing

Skills such as reading 'number words' (e.g. 'five', 'sixty') were originally thought to be subsumed by language processes. However, at virtually all levels of processing, dissociations have been observed between numerical and linguistic abilities, suggesting a surprising degree of specificity in the human brain. There are several case reports of near-total impairment in one domain, with remarkable sparing of similar function in another. One brain-damaged patient reveals the ability to write numbers with remarkable ease, even performing complex arithmetic, while failing to be able to write simple words and even their own name. In contrast, other patients reveal word reading to be largely spared, while 'number word' reading was almost totally lost as a result of brain damage.

Localization of numeral comprehension and production processes

The brain regions involved in comprehending and producing written and spoken numerals generally mirror those involved in written and spoken language, despite the apparent functional uniqueness of number processing. Impairment in producing spoken and written numerals (both word and arabic forms) nearly always originates from damage to left hemisphere language production areas, while impairments in comprehending spoken numerals typically results from damage localized to left hemisphere language comprehension regions.

In contrast, arabic numeral comprehension is distributed across both brain halves (or hemispheres) more than other numerical and linguistic abilities. Evidence from patients with disconnected brain hemispheres (which permits the study of each brain hemisphere acting independently) reveal excellent comprehension abilities in both hemispheres.

Impairment of Calculation

Evidence from people with brain damage, as well as those with normal and impaired development, has revealed that several independent cognitive processes are required in order to perform calculations such as 274×59. First, there is a distinction between the ability to remember simple arithmetic facts (such as $6 \times 4 = 24$), and the ability to perform calculation procedures (such as those required for carrying numbers in multidigit multiplication). Several patients with brain damage have impairments that are specific to either fact retrieval or multidigit calculation procedures, indicating that the ability to retrieve arithmetic facts and the ability to perform multidigit calculations are represented by different neural substrates.

Within the simple process of remembering arithmetic facts like $6 \times 4 = 24$, there are multiple processes involved. Brain-damaged patients have also revealed selective impairment of the ability to recognize the arithmetic operator (e.g. + or −) and the ability to recognize the digits.

The independence of fact retrieval and numerical competence

Evidence from people with brain damage has revealed strong divisions between the ability to remember the answer to simple arithmetic problems (e.g. remembering $2+2=4$) and the ability to calculate an answer based on arithmetic principles. For example, one patient revealed a marked inability to retrieve previously known facts from memory (e.g. 8×7). However, given enough time the patient was able to answer a problem such as 8×7 by producing an elaborate strategy, such as using the answer to 8×10 and adjusting in accordance with the appropriate algebraic principle: $8 \times 7 = 8 \times 10 - 8 \times (10 - 7)$. Others patients are clearly able recall some facts, but are simply unable to use mathematical principles to derive other answers.

Representing arithmetic facts in memory

There is debate as to the form in which arithmetic facts are stored. One view is that arithmetic facts are stored in an abstract, meaning-based form that is not tied to a specific modality (e.g. spoken or written). In several cases of brain damage the patients have revealed calculation impairments that are consistent regardless of the form in which the problems are presented or answered. Perhaps more surprising is the fact that patients can successfully calculate in spite of severe impairments in representing the spoken form of numerals. For example, a patient can be presented with a problem such as '7 × 3', read the problem aloud as 'five times eight', say the answer as 'twenty-six' but nevertheless write the correct answer to the original problem: '21'. This pattern of performance appears incompatible with the notion that we store arithmetic facts purely in a sound-based form.

However, an alternative possibility is that arithmetic facts are stored and retrieved in a sound-based representation (like a nursery rhyme). According to this position, abstract magnitude representations are not related to arithmetic fact retrieval, but instead are involved in estimation and mathematical reasoning. Some people with brain damage are unable to determine exact arithmetic responses (e.g. $2+2=3$) but nevertheless can reject a highly implausible answer such as $2+2=9$, indicating that there is an approximate number representation which provides the meaning of numbers and may be separate from the process of retrieving exact arithmetic facts (Dehaene, 1997).

Localization of arithmetic processes

The ability to retrieve arithmetic facts from memory seems to be strongly localized to the left hemisphere. Patients with disconnected hemispheres perform at nearly normal levels in their left hemisphere, while having essentially no calculation abilities in the right hemisphere. A majority of fact-retrieval impairments are suffered as a result of left parietal lobe damage; more rarely they are the result of damaged subcortical structures and left language centers. In contrast, the ability to perform multidigit arithmetic appears to draw on the planning and working memory capacity of the frontal lobes (dorsolateral prefrontal cortex).

Exact and Approximate Calculation

Response latencies for determining the larger of two numerals (derived from healthy adults) suggests that we represent numerical quantities in terms of a magnitude representation similar to that used for light brightness, sound intensity and time duration. This magnitude representation is used to compare two or more numbers. Moyer and Landauer (1967) found that humans are faster at comparing numerals with a large difference (e.g. 1 and 9) than comparing two numerals with a small difference (e.g. 4 and 5). Further, when the differences between the numbers are equated (e.g. 2 and 3, versus 8 and 9), the smaller number pair (2 and 3) is compared more quickly than the larger number pair (8 and 9). These findings led to the conclusion that numerals are being translated into a magnitude representation along a mental number line with increasing imprecision the larger the quantity being represented (a psychophysical representation which conforms to Weber's law).

These magnitude representations are found in human adults, infants and children, and also in animals. This magnitude system allows us to meaningfully represent numerical quantities, estimate, and compare numerical quantities. It appears to be related to our representation of other magnitudes, such as time duration and distance (Whalen *et al.*, 1999). A crucial challenge for the development of numerical literacy is the formation of mappings between the exact number symbol representations that are learned in school, and the approximate numerical quantity representations that are present very early in life.

There is strong evidence that parietal regions of both brain hemispheres represent numerical magnitude. Patients with disconnected hemispheres have revealed the ability to perform number comparison in either hemisphere (Dehaene, 1997). Electrical brain signatures during number comparison reveal that approximately 0.1 s, after the presentation of two numbers there is bilateral activation of the parietal lobes, and that the signature varies according to the difficulty of the comparison (Dehaene, 1996).

INTEGRATION: THE TRIPLE CODE MODEL OF NUMERICAL PROCESSING

Stanislas Dehaene and colleagues were the first researchers to provide a detailed theory of number processing that includes both the different functional components and their localization in the brain (Dehaene, 1997). According to the 'triple code' model there are three separate number systems in the brain: verbal, visual, and magnitude systems (Figure 1).

The verbal system is located in the left hemisphere adjacent to language areas, and is responsible for comprehending and producing spoken numerals, as well as storing arithmetic table facts in memory. The visual number system takes input from primary visual centers, and is used to recognize and produce numerals in written word and arabic forms. The magnitude system is located in the parietal lobe of both hemispheres, and provides the ability to estimate and compare numbers.

According to the triple code model, even simple calculation (e.g. $8+9=?$) may require several processes and brain regions, including rote verbal memory (verbal system), elaboration (e.g. $8+9=7+10$, magnitude system), and strategy use (frontal lobes). This theory illustrates that even the simplest arithmetic (e.g. $2+2=?$) may require several processes and brain regions.

In summary, the cases of brain damage reported here reveal that acalculia is not a single impairment. Rather, we now know that many cognitive processes distributed across the brain together

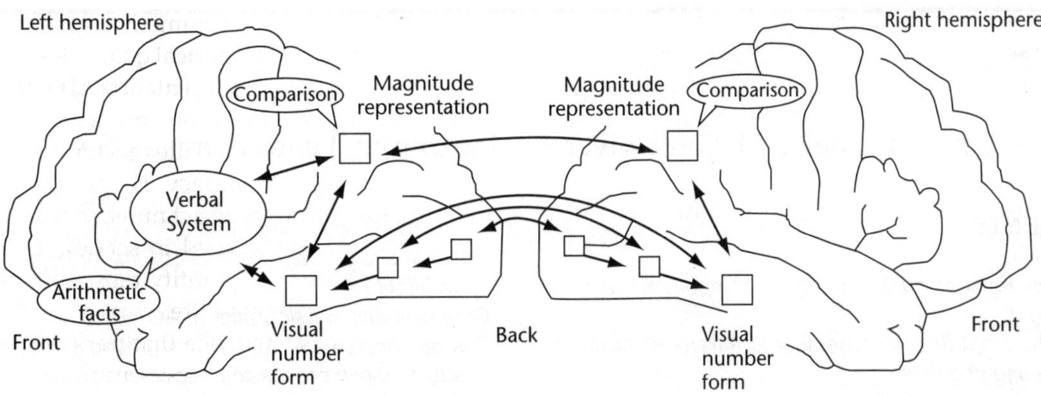

Figure 1. The triple code model of numerical processing is composed of three separate number systems. The verbal system is responsible for comprehending and producing spoken numbers, as well as storing arithmetic table facts such as $2 \times 2 = 4$ in memory. The visual system recognizes numbers written in arabic digit or written-word form. The magnitude system is responsible for deciding which of two numbers is the larger, estimating, and giving an approximate sense of how much a number represents.

provide our arithmetic and numerical competencies. The challenge for the future will be the precise identification of the brain regions involved in arithmetic and the elucidation of how these regions work together to produce our considerable arithmetic abilities.

References

Dehaene S (1996) The organization of brain activations in number comparison: event-related potentials and the additive-factors method. *Journal of Cognitive Neuroscience* **8**: 47–68.

Dehaene S (1997) *The Number Sense: How the Mind Creates Mathematics*. New York: Oxford University Press.

Moyer RS and Landauer TK (1967) Time required for judgments of numerical inequality. *Nature* **215**: 1519–1520.

Warrington EK (1982) The fractionation of arithmetical skills: a single case study. *Quarterly Journal of Experimental Psychology* **34A**: 31–51.

Whalen J, Gallistel CR and Gelman R (1999) Non-verbal counting in humans: the psychophysics of number representation. *Psychological Science* **10**: 130–137.

Further Reading

Butterworth B (1999) *What Counts: How Every Brain is Hardwired for Math*. New York: Free Press.

Dehaene S (1997) *The Number Sense: How the Mind Creates Mathematics*. New York: Oxford University Press.

Achievement

See **Academic Achievement**

ACT

Intermediate article

Christian Lebiere, Carnegie Mellon University, Pittsburgh, Pennsylvania, USA

CONTENTS

The ACT theory of cognition has been used to model a wide range of cognitive tasks. The latest version of the theory, ACT-R, is a hybrid architecture which combines a symbolic production system with a subsymbolic level of neural-like activation processes.

INTRODUCTION

In the twentieth century, cognitive psychology produced enormous amounts of data on all aspects of human behavior. The challenge of cognitive science in the twenty-first century is to provide a coherent, integrated and systematic scientific explanation of these phenomena. Computational modeling has become an increasingly popular tool in a wide range of sciences, from mathematics and physics to economics and the social sciences. By combining precision, lacking in many verbal theories, with the ability to deal with almost limitless complexity, a limitation of mathematical theories, computational modeling provides the most promising approach to a unified theory of cognition.

The 'adaptive control of thought' (ACT) theory of cognition attempts to provide such a unified framework by defining an integrated cognitive architecture that can be applied to a wide range of psychological tasks. The most recent version of the theory, ACT-R, is a hybrid architecture which combines elements of the symbolic and connectionist frameworks. By using a symbolic production system to provide the structure of behavior, it has a direct interpretive link to psychological data, and thus inherits a tractable level of analysis. By associating to each symbolic knowledge structure real-valued activation quantities that control its application, it enables the use of statistical learning mechanisms that provide neural-like adaptivity and generalization. The long-term goal of the ACT theory is to provide constrained computational models of a wide range of cognitive phenomena.

PRODUCTIONS AND CHUNKS AS ATOMIC COMPONENTS OF THOUGHT

How to represent knowledge is the first and most important question concerning the design of a cognitive system. As with data structures in a computer program, the knowledge structures assumed by a cognitive theory will fundamentally determine its characteristics: what knowledge can be represented, how it can be accessed and which learning processes can be used to acquire it. Since many cognitive systems, including ACT, are implemented as computer simulations, it is important to define which representational assumptions constitute theoretical claims and which are merely notational conventions. The ACT theory has from its inception made three theoretical assumptions regarding knowledge representation. The first assumption is known as the procedural–declarative distinction; that is, that there are two long-term repositories of knowledge, a procedural memory and a declarative memory. The second assumption is that 'chunks' are the basic units of knowledge in declarative memory. The third assumption is that production rules are the basic units of knowledge in procedural memory. These assumptions will be examined in detail in the rest of this section.

Unlike other cognitive theories and frameworks such as Soar or connectionism, ACT makes a fundamental distinction between declarative and procedural knowledge. This is not just a notational distinction, but a fundamental psychological claim about the existence of distinct memory systems with different properties. The best way to state the procedural–declarative distinction is in terms of a production system framework. Declarative memory holds factual knowledge, such as the knowledge that George Washington was the first president of the United States or that $3 + 4 = 7$, while procedural memory holds rules that access and modify declarative memory. This distinction corresponds closely to the common operational definition that declarative knowledge is verbalizable while procedural knowledge is not, with the caveat that some declarative knowledge might exist without being directly reportable because one might lack the necessary procedural knowledge to access and express it. These two memory systems have some common features, such as the build-up and decay of strength, but they also have distinct properties: for example, more flexible access to declarative than to procedural knowledge, and different acquisition and retention characteristics. Some recent results in cognitive neuroscience can be interpreted as supporting the procedural–declarative distinction: for example, results showing that damage to the hippocampus inhibited the creation of new declarative memories but not of new procedural memories (Squire, 1992).

The basic units of declarative memory are called chunks. The purpose of a chunk is to organize a set of elements (either chunks themselves or more basic perceptual components) into a long-term memory unit. Chunks can only contain a limited number of elements: as few as two, often three (which it has been argued is a theoretical optimum), and seldom more than five or six. Elements in a chunk assume specific relational roles: for example, in the chunk encoding '$3 + 4 = 7$', '3' is the first addend, '+' is the operator, '4' is the second addend and '7' is the sum. Finally, chunks can be organized hierarchically, since the elements of a chunk can be chunks themselves. For example, to memorize a long sequence, chunks of finite length can be used to store short pieces of the sequence and can themselves be aggregated into other chunks to encode the full sequence. Chunks have two possible origins: either as direct encodings of objects in the environment or as long-term encodings of particular internal elaborations, called goals. Thus when reading a sentence, every word read, together perhaps with some environmental characteristics such as its position in the sentence, constitutes a chunk. But the understanding of the sentence, which is the goal of the processing, is also available as one or more chunks holding an elaboration of its meaning as a result of the cognitive processing.

Production rules are the basic units of procedural memory. Production rules encode cognitive skills as condition–action pairs. A production rule tests the contents of the current goal and perhaps of declarative memory, then executes one or more actions, which can include modifying the current goal or changing the external environment. Just as for chunks, the size and complexity of productions are limited in that they only perform a limited number of retrievals from declarative memory (usually a single one), and those retrievals are performed sequentially. ACT makes four claims related to production rules. The first is 'modularity', i.e, that procedural knowledge takes the form of production rules that can be acquired and deployed independently. Complex skills can be decomposed into production rules that capture significant regularities in human behavior (Anderson, Conrad and Corbett, 1989). For example, performance in the learning of a programming language, which doesn't show any regularities when organized with respect to the number of problems, exhibits a very regular learning pattern when organized with respect to the production rules used in solving each problem.

The second claim is 'abstraction', i.e., that productions are general, applying across a wide range of problems. The third claim is 'goal factoring', which moderates the claim of abstraction by making each production specific to a particular goal type. Thus, a production that performs addition can apply to any addition problem (abstraction) but only to addition problems (goal factoring).

The fourth claim is 'condition–action asymmetry'. For example, while a chunk holding a multiplication fact can be used to solve either a multiplication or a division problem, productions used to compute the answer to a multiplication problem (e.g. by repeated addition) cannot be used to find the answer to a division problem.

According to the ACT theory, chunks and procedural units thus constitute the atomic components of thought.

FROM THE HAM THEORY OF MEMORY TO A FRAMEWORK FOR MODELING COGNITION

ACT has its roots in the 'human associative memory' (HAM) theory of human memory (Anderson and Bower, 1973), which represented declarative knowledge as a propositional network. HAM was implemented as a computer simulation in an attempt to handle complexity and to specify precisely how the model applied to the task, thus overcoming the major limitations of the mathematical theories of the 1950s and 1960s. Although it fell short of these goals, the theory was afterwards developed in significant ways.

The next step was the introduction of the first instance of ACT, ACTE (Anderson, 1976). It combined HAM's theory of declarative memory with a production system implementation of procedural memory, thus precisely specifying the process by which declarative knowledge was created and applied, and added basic activation processes to link procedural and declarative memory. While the distinction between procedural and declarative knowledge had little support at the time, it has found increasing popularity and support from recent neuroscientific evidence for a dissociation between declarative and procedural memories. The next major step in the evolution of the ACT theory was the ACT* system (Anderson, 1983), which added a more neural-like calculus of activation and a more plausible theory of production rule learning. ACT* was successfully applied to a wide range of psychological phenomena and had a profound influence on cognitive science, but, although some computer simulations were available, it existed primarily as a verbally specified mathematical theory.

The first computational implementation of the theory to be widely adopted was ACT-R (Anderson, 1993). ACT-R was introduced to capitalize on advances in skill acquisition (Anderson and Thompson, 1989) to improve production rule learning, and to tune the subsymbolic level to the structure of the environment to reflect the rational analysis of cognition (Anderson, 1990). Due to these theoretical advances as well as computational factors such as the standardization of the implementation language, Common LISP, and the exponentially increasing power of desktop computers, ACT-R has been adopted as a cognitive architecture by a growing group of researchers. The needs of that user community to apply ACT-R to an increasingly diverse set of tasks, and a growing need for neural plausibility, led to further advances which were embodied in a new version of the theory, ACT-R 4.0 (Anderson and Lebiere, 1998). As suggested by the use of version numbers, the changes in the theory were relatively minor and largely consisted of a reduction in the grain size of chunks and productions.

Through its widening range of applications and the increasing constraints on model development, ACT-R can be regarded as being well on the way to achieving HAM's long-term goal of providing a rigorous computational framework for modeling cognition. The technique of using computational simulation rather than mathematical analysis to provide tractability in complex domains has been widely adopted not only in cognitive modeling but in many other sciences as well. Because it tightly integrates a symbolic production system with a neural-like activation calculus, ACT can be termed a hybrid activation-based production system cognitive architecture. As such, it constitutes a general framework for modeling cognition (another example of such a framework being connectionism). While frameworks can often make general qualitative predictions, in order to obtain precise quantitative predictions for specific experiments they need to be instantiated into detailed theories, such as the various members of the ACT family, that exactly specify the system's mechanisms and equations. A serious potential problem resulting from the generality of frameworks is the possibility of instantiating them into competing theories and incorrectly citing those theories' accomplishments as support for the framework itself. To avoid that danger and to ensure cumulative progress in the development of the ACT theory, Anderson and Lebiere made available on the web every simulation described in Anderson and Lebiere (1998), and promised that the account provided by those models would remain valid in future instantiations of the theory (*The ACT Web*: see Further Reading list).

THE MECHANISMS OF ACT

While the specific mechanisms of the ACT theory have changed significantly over more than 20 years of evolution, the assumptions that underlie the theory have remained fairly constant. ACT operates in continuous time, predicting specific latencies for each step of cognition, such as production firing or declarative retrieval. At the symbolic level, the procedural–declarative distinction states that there are two memory systems, with a production rule component operating on a declarative component. Declarative knowledge is composed of chunks, each having a limited number of components, which can be described alternatively in terms of a propositional network. Procedural knowledge is composed of production rules, or condition–action pairs, which are the basic units of skills. Conditions apply to the state of declarative memory, while actions result in changes in declarative memory. Cognition is goal-directed and maintains at all times a current goal which a production must match in order to apply. At the subsymbolic level, real-valued quantities such as chunk activation and production utility control the application of that symbolic knowledge. Those subsymbolic parameters are learned, and reflect the past use of their respective symbolic structures. Activation computation is a dynamic process that involves the spreading of activation from a set of sources, usually the contents of the current goal, to related nodes, as well as time-related decay.

ACQUIRING PRODUCTIONS: ACT* AND PUPS

The learning of skills in the form of production rules is one of the more complex mechanisms of the ACT architecture, and has seen the most fundamental changes over the course of the development of the theory. In ACT*, the first version of the theory to provide a comprehensive account of the learning of production rules, procedural learning is accomplished by a set of mechanisms that compile into production rules the process of interpreting declarative knowledge. In other words, procedural skills are learned by doing. For example, if one dials a telephone number or enters one's password by explicitly remembering the number and then iteratively identifying and keying each digit (or letter), the knowledge compilation process would create a production that directly encodes and keys each character without retrieving it from declarative memory. Knowledge compilation is accomplished by two separate processes: composition, which takes a sequence of production firings and compiles them into a single equivalent production; and proceduralization, which takes an existing production and encodes the result of declarative retrievals (in our example, the phone number or password) directly into a new production. Unlike the learning of a declarative fact, which can take place in a single episode, the learning of a procedural skill in the form of production rules requires many iterations for the new knowledge to be refined into its final form and ready to be applied. ACT* has three mechanisms that take new production rules and tune them for optimal performance. The generalization process broadens the applicability of new production rules by making their conditions more general. The discrimination process performs the opposite task, narrowing the applicability of new rules by making their conditions more specific (in our example, the production would be specific to the person or account associated with the number or password). Finally, the strengthening process increases the production strength, allowing it to apply faster and more often.

Anderson and Thompson (1989) introduced a different conception of production learning in their 'penultimate production system' (PUPS). Following empirical studies indicating that analogical problem solving played a fundamental role in skill acquisition, they proposed a new mechanism of production creation based on an analogy process. Instead of automatically creating new productions as a function of an interpretive process, PUPS creates productions to encode the solving of a current problem by analogy with a solution to a previous problem encoded in declarative memory. The analogy process discovers a mapping in declarative memory from problem to solution, which is then encoded in a new production that can solve the problem directly without referring to previous examples. Analogy can therefore be regarded as a mechanism for generalizing from examples. Since it operates on explicit memory structures instead of an automatic goal-based trace of operations (as in ACT*), it allows for the addition to the example of chunks of conditions and heuristics to guide the generalization and discrimination processes that produce the final form of the new productions. This results in a more reliable and controlled mechanism.

RATIONAL ANALYSIS AND ITS IMPLICATIONS FOR COGNITIVE ARCHITECTURE

Inspired by Marr's theory of information-processing levels, Anderson (1990) introduced his 'rational analysis' of human cognition, based on the assumption of a rational level used to analyze the computations performed by human cognition. The general 'principle of rationality' states that a cognitive system operates at all times to optimize the adaptation of the behavior of the organism. This does not imply that human cognition is perfectly optimal, but it helps explain why cognition operates the way it does at the algorithmic level given its physical limitations at the biological level and the optimum defined by the rational level that it attempts to implement. Anderson's analysis provides strong guidance on theory development, because given a particular framework (say, an activation-based production system) it strongly constrains the set of possible mechanisms to those that satisfy the rational level.

The rational analysis can be applied to several aspects of human cognition, including memory, categorization, causal inference and problem solving. The task of human memory is analyzed in terms of a Bayesian estimation of the probabilities of needing a particular memory at a particular point in time. The analysis accounts for effects of recency, frequency and spacing; effects of context and word frequency; and priming and fan effects. It also provides an interpretation of the concept of activation in ACT-R as the logarithm of the odds that the corresponding chunk needs to be retrieved from memory. The rational analysis of categorization can be interpreted as supporting the creation of chunk types in ACT-R. Finally, the analysis of problem solving in terms of expected utilities of procedural operators led to the refinement of the conflict resolution process in ACT-R. In summary, the rational analysis of cognition provided strong guidance for the development of learning mechanisms in ACT-R (the R stands for 'rational') that automatically tune the subsymbolic level to the structure of the environment.

CONNECTIONIST CONSIDERATIONS: ACT-RN

Lebiere and Anderson (1993) describe ACT-RN, which is an attempt to implement ACT-R using standard connectionist constructs such as Hopfield networks and feedforward networks. Symbols are represented using distributed patterns of activation over pools of units. The current goal, or focus of attention, is located in a central memory that holds the components of the goal, which are the sources of activation in ACT-R. Separate declarative memories for each chunk type are implemented using associative memories in the form of simplified Hopfield networks with a separate pool of units for each component of the chunk. ACT-R's goal stack is implemented as a separate declarative memory that associates each goal to its parent goal. Procedural memory consists of pathways between central memory and declarative memories. Each production is represented as a single unit that tests the contents of the current goal in central memory, then, if successful, activates the proper connections to perform a retrieval from a single declarative memory (possibly the goal stack) and update the goal with the retrieval results. There is a rough correspondence between these constructs and neural locations. Goal memory can be associated with the prefrontal cortex, since damage in that area has been associated with loss of executive function. Declarative memory is distributed throughout the posterior cortex, with each type corresponding to hypercolumns or small cortical areas. Procedural memory might be located in the basal ganglia, which have extensive connections to all cortical areas and could therefore implement the functionality of production rules.

As a practical system, ACT-RN was found unsatisfactory in a number of ways. It only provided a partial implementation of ACT-R, with some features and mechanisms being too complex or difficult to map onto a connectionist substrate. While some models adapted well to the connectionist implementation, others, even simple ones, ran into computational hurdles. More fundamentally, it was an imperfect implementation of the ACT-R standard which only approximately reproduced the ACT-R mechanisms and equations. Nevertheless, ACT-RN provided the main impetus for further development of the theory. Some features of ACT-R that were too complex to be implemented in ACT-RN, such as complicated representational constructs in chunks and powerful pattern-matching primitives in productions, were abandoned in the later versions of the theory as being too computationally powerful for any neural implementation. This resulted in a welcome simplification of the theory. Conversely, a feature of ACT-RN which the connectionist implementation provided naturally – generalization based on distributed representations – was added to the theory in the form of similarity-based partial matching of production conditions (Lebiere *et al.*, 1994). Thus, although implementing ACT-R in a neural network did not directly result in a practical system, it did provide a functional theory of neural organization and imposed a strong direction on further developments of the ACT theory.

THE MECHANISMS OF ACT-R: INTEGRATING SYMBOLIC AND SUBSYMBOLIC PROCESSING AND LEARNING

The power of ACT-R as a hybrid architecture of cognition lies in its tight integration of the symbolic and subsymbolic levels. Because of this integration, it is able to combine the most desirable characteristics of symbolic systems, such as structured behavior and ease of analysis, with those of connectionist networks, such as generalization and fault-tolerance, while avoiding their most serious shortcomings, such as the overly deterministic behavior of symbolic systems and the intractability of learning characteristic of connectionist networks. At the symbolic level, ACT-R operates sequentially – only one production can fire in each cycle and only one chunk can be retrieved from memory at a time – corresponding to the basic sequential nature of human cognition. However, each of the basic steps of production selection and of declarative memory retrieval involves the parallel consideration of all relevant productions and memory chunks, reflecting the massively parallel nature of the human brain.

A typical production cycle in ACT-R works as follows. The conflict resolution process attempts to select the best production that matches the current goal. To that effect, the expected production utilities of all matching productions are computed in parallel and the best production is selected. Typically, that production attempts a retrieval of information from declarative memory. The activations of the relevant memory chunks are computed concurrently, as: the sum of the base-level activations, reflecting the history of use of each chunk; the activation spreads from the components of the goals, reflecting the specificity of that chunk to the current context; a partial matching penalty, allowing generalization to similar patterns; and a noise component providing stochasticity. Once the retrieval is complete, the activation parameters of the chunks involved are automatically adjusted by the subsymbolic learning mechanisms to reflect this experience. The production then executes its action, which typically consists of modifying the goal to incorporate the results of the declarative retrieval and perhaps performing an external action. If a goal is accomplished, the subsymbolic parameters controlling the utility of the productions involved in solving that goal will be automatically learned, and a chunk encoding the results of the goal will enter declarative memory.

Thus, the production system part of ACT-R provides the basic synchronization of a massively parallel system into a meaningful sequence of cognitive steps, while the subsymbolic part is continuously tuned to the statistical nature of the environment to provide the adaptivity characteristic of human cognition.

MATCHING HUMAN PERFORMANCE IN DIVERSE DOMAINS

The idea of using computational precision to eliminate the looseness of the merely verbal mapping between model and task is embodied in what Anderson and Lebiere (1998) call the 'no-magic doctrine', which consists of the following six tenets:

1. Theories must be experimentally grounded. To avoid unprincipled degrees of freedom in the mapping between task stimuli and model representations, ACT-R includes a 'perceptual motor' component called ACT-R/PM which interacts with the task through the same interface as human subjects.
2. Theories must provide a detailed and precise accounting for the data. Because of its experimental grounding, ACT-R makes precise predictions about every aspect of empirical data, including choice percentages, response latencies, etc.
3. Models must be learnable through experience. ACT-R has mechanisms capable of learning symbolic chunks and productions as well as their subsymbolic parameters.
4. Theories must be capable of dealing with complex cognitive phenomena. ACT-R is applicable to tasks ranging from sub-second psychology experiments to complex environments involving substantial knowledge and learning that may take hours.
5. Theories must have principled parameters. The parameters attached to symbolic knowledge structures are learned from experience. The modeler sets the architectural parameters, but variations are increasingly constrained and understood across tasks.
6. Theories must be neurally plausible. ACT-R is situated at a level of abstraction higher than actual brain structures, but the need to provide a plausible mapping between theoretical constructs and actual brain structures imposes useful and powerful constraints on theory development.

ACT-R has been applied to an increasing variety of cognitive tasks in domains as diverse as memory, categorization, problem solving, analogy, scientific discovery, human–computer interaction, decision theory and game theory. In each domain, ACT-R predicts a wide range of measurable aspects of human behavior at a very fine scale, including latency, errors, learning, eye movements and individual differences.

SUMMARY

ACT is a hybrid cognitive architecture that combines a symbolic production system with a subsymbolic level of neural-like activation processes that control the application of the symbolic structures. Learning mechanisms provide for the acquisition of symbolic knowledge and the statistical tuning of the subsymbolic layer to the structure of the environment, as specified by the rational analysis of cognition. The ACT architecture has been successfully applied to a wide variety of cognitive tasks to accurately predict many aspects of human behavior. (*See* **Computer Modeling of Cognition: Levels of Analysis**)

References

Anderson JR (1976) *Language, Memory, and Thought.* Hillsdale, NJ: Erlbaum.

Anderson JR (1983) *The Architecture of Cognition.* Cambridge, MA: Harvard University Press.

Anderson JR (1990) *The Adaptive Character of Thought.* Hillsdale, NJ: Erlbaum.

Anderson JR (1993) *Rules of the Mind.* Hillsdale, NJ: Erlbaum.

Anderson JR and Bower GH (1973) *Human Associative Memory*. Washington, DC: Winston and Sons.
Anderson JR, Conrad FG and Corbett AT (1989) Skill acquisition and the LISP Tutor. *Cognitive Science* **13:** 467–506.
Anderson JR and Lebiere C (1998) *The Atomic Components of Thought*. Mahwah, NJ: Erlbaum.
Anderson JR and Thompson R (1989) Use of analogy in a production system architecture. In: Ortony A *et al*. (eds) *Similarity and Analogy*, pp. 367–397. New York, NY: Cambridge University Press.
Lebiere C and Anderson JR (1993) A connectionist implementation of the ACT-R production system. In: *Proceedings of the Fifteenth Annual Meeting of the Cognitive Science Society*, pp. 635–640. Hillsdale, NJ: Erlbaum.
Lebiere C, Anderson JR and Reder LM (1994) Error modeling in the ACT-R production system. In: *Proceedings of the Sixteenth Annual Meeting of the Cognitive Science Society*, pp. 555–559. Hillsdale, NJ: Erlbaum.
Squire LR (1992) Memory and the hippocampus: a synthesis from findings with rats, monkeys and humans. *Psychological Review* **99**: 195–232.

Further Reading

The ACT Web. [http://act.psy.cmu.edu]
Anderson JR (1987) Skill acquisition: compilation of weak-method problem solutions. *Psychological Review* **94**: 192–210.
Anderson JR (1991) The adaptive nature of human categorization. *Psychological Review* **98**: 409–429.
Anderson JR, Bothell D, Lebiere C and Matessa M (1998) An integrated theory of list memory. *Journal of Memory and Language* **38**: 341–380.
Anderson JR, John BE, Just MA *et al*. (1995). Production system models of complex cognition. In: *Proceedings of the Seventeenth Annual Conference of the Cognitive Science Society*, pp. 9–12. Hillsdale, NJ: Erlbaum.
Anderson JR, Matessa M and Lebiere C (1997) ACT-R: A theory of higher level cognition and its relation to visual attention. *Human Computer Interaction* **12**: 439–462.
Anderson JR, Reder LM and Lebiere C (1996) Working memory: activation limitations on retrieval. *Cognitive Psychology* **30**: 221–256.
Anderson JR and Schooler LJ (1991) Reflections of the environment in memory. *Psychological Science* **2**: 396–408.
Corbett AT, Koedinger KR and Anderson JR (1997) Intelligent tutoring systems. In: Helander MG, Landauer TK and Prabhu P (eds) *Handbook of Human–Computer Interaction*, 2nd edn. Amsterdam: Elsevier.
Newell A (1973a) You can't play twenty questions with nature and win: projective comments on the nature of this symposium. In: Chase WD (ed.) *Visual Information Processing*, pp. 283–310. New York, NY: Academic Press.
Newell A (1973b) Production systems: models of control structures. In: Chase WG (ed.) *Visual Information Processing*, pp. 463–526. New York, NY: Academic Press.

Action

Intermediate article

Jos J Adam, Maastricht University, Maastricht, Netherlands
Martinus J Buekers, Katholieke Universiteit Leuven, Leuven, Belgium

CONTENTS

Action is the ability to move the body or body parts in a purposeful, coordinated manner in order to physically interact with the environment. It is based on the integration and cooperation of sensory and motor systems.

INTRODUCTION

A characteristic of humans is their ability to perform skilled motor actions. Some of these motor actions are carried out to perform mundane tasks such as walking, picking up a pencil, drinking a glass of water, or buttoning a shirt. Other motor actions accomplish more sophisticated tasks such as painting a picture, flying an airplane, and performing brain surgery. How does the brain generate and control motor actions? This is not a trivial question, as the body contains hundreds of muscle groups that act on scores of joints, thereby introducing a very large number of 'degrees of freedom' in movement. This article discusses some of the variables that shape human motor performance, and describes its main underlying control principles. (*See* **Motor Control and Learning**; **Performance**; **Reaction Time**)

STARTING, STOPPING, AND SEQUENCING ACTIONS

Starting Actions

To start an action, preparatory processes are first needed in order to formulate a movement plan. These preparatory processes include the processing of sensory information in order to perceive the state of the environment and the state of the actor. Based on this information, an action plan is formulated. For instance, when driving a car, one needs to visually scan the environment in order to detect significant environmental changes, such as a traffic light changing to amber. When reaching for a pencil, one first needs to determine its position in space. This kind of perceptual information is then related to the state of the actor (the effector), and a decision is made about what to do and how to do it. These preparatory processes take time, which can be measured by means of the reaction-time paradigm.

In a typical reaction-time task, the actor is first presented with a warning signal in order to increase alertness. Then, after a variable time interval (which can range between 0.2 s and 5 s), the imperative reaction signal (the stimulus) is presented, calling for a physical response. In the laboratory, the response is often a button-press or button-lift movement. The time that elapses between the presentation of the reaction signal and the start of the overt response defines the reaction time. Many variables influence reaction time. Some of the most influential are: the number of response alternatives; the complexity of the response; the spatial compatibility between the possible stimuli and responses; and the performer's subjective preference for speed or accuracy of responding.

Number of stimulus–response alternatives

In some situations, there is only one possible stimulus and only one correct response. The reaction time in this situation is called the 'simple' reaction time. For example, an athlete just before the start of a 100-meter race knows in advance what to expect and what to do, and the reaction time is typically short (less than 200 ms), representing the minimal time needed to perceive and act.

Other situations confront the performer with more uncertainty. For example, a boxer facing an opponent who can attack with the left or right fist must make a fast decision about what to do. In situations like this, the 'choice' reaction time is substantially longer, mainly reflecting the increased processing demands associated with selecting and programming the appropriate action. In general, as the number of stimulus–response alternatives increases, the choice reaction time increases. The exact relationship between reaction time and the number of stimulus–response alternatives is given by the Hick–Hyman Law, which states that reaction time increases linearly with the logarithm of the number of stimulus–response alternatives.

Response complexity

The nature of the response to be executed is an important determinant of reaction time. A complex action needs more programming time than a simple action, and therefore a longer reaction time. Increased response complexity can be defined in terms of additional movement parts, increased accuracy demands, and longer movement durations (e.g. Klapp, 1996).

Stimulus–response compatibility

The (spatial) relationship between the set of potential stimuli and the set of potential responses is one of the most important determinants of reaction time. For example, choice reaction time in a task involving left and right stimuli and left and right responses is slower when the stimulus and associated response are on opposite sides (i.e. a left stimulus requires a right response and a right stimulus requires a left response) than when they are on the same side (i.e. a left stimulus requires a left response and a right stimulus a right response).

In general, performance tends to be faster when there is a close, natural correspondence or association between the stimulus and its required response. According to recent theories of stimulus–response compatibility, when there is some kind of correspondence or similarity between the stimulus and response sets, each stimulus will automatically activate its natural – or habitually associated – response (e.g. Kornblum *et al.*, 1990). If this response is the correct one, it can be executed immediately and the reaction time is short: this situation represents a compatible or congruent stimulus–response assignment. If, however, the automatically activated response is incorrect – as is the case in a 'crossed' or incompatible stimulus–response assignment – then the automatically activated response must be aborted and inhibited, and consequently reaction time is lengthened.

The notion of automatic response activation has been confirmed by evidence from psychophysiological studies that recorded lateralized event-related brain potentials on the basis of electroencephalogram recordings (e.g. Eimer, 1995). These studies indicate that action control processes may proceed very fast and without conscious awareness.

Speed–accuracy bias

Human task performance can be error-prone. In fact, there is a trade-off between speed of responding and accuracy of responding: improvements in reaction time often co-occur with an increase in the number of errors, and improvements in accuracy with slower responses. Task constraints and subjective biases for speed or accuracy jointly determine the position on the speed–accuracy continuum.

Stopping Actions

Once the motor program has been formulated, it is implemented and executed. 'Movement time' is the time that elapses between the start of the movement and its termination. A movement stop may be incorporated in the motor program before execution. This is the case for extremely fast, ballistic movements that are executed to maximize speed (e.g. a boxer's punch). However, a movement stop may also be planned 'online', that is, during the execution of the movement. This is the case when accuracy is an important task requirement. Under such circumstances, feedback-based modifications of the movement are implemented during the latter parts of the movement – that is, when the movement approaches the target zone – in order to ensure that the movement terminates accurately.

Thus, for ballistic actions movement time is very short and mainly a function of the parameters specified by the motor program; while for accurate actions movement time is typically longer and a function of motor program specifications and feedback-based movement modifications. Note that this distinction describes two modes of motor control: open-loop and closed-loop control. Open-loop control is fast, but inflexible, because it just follows the instructions laid down in the motor program without the possibility of modification. Closed-loop control is more flexible, but slower, because of feedback processing and the implementation of movement corrections.

Research using the manual-aiming paradigm (whereby the hand is moved from a 'start' position to a 'target' position) has identified several important variables that influence movement time: movement distance; target size; the presence of non-targets (distractors); and the actor's subjective preference for speed or accuracy of moving.

Movement distance and target size

Increasing movement distance and decreasing target size both result in longer movement times. The exact relationship is expressed by Fitts' law. Fitts' law states that movement time increases linearly with the 'index of difficulty', defined as $\log_2 (2A/W)$ where A is the movement amplitude (or distance) and W is the width or size of the target (Fitts, 1954). According to the 'optimized submovement' model (Meyer *et al.*, 1988), Fitts' law is a consequence of a particular hybrid way of controlling movements: a preprogrammed, initial-impulse phase followed by a feedback-based control phase that allows the implementation of corrective submovements. This model allows actors either to make submovements in order to optimize spatial accuracy or to keep submovements to a minimum in order to maximize speed.

Presence of non-targets

Often, the target for an action does not appear in isolation but is surrounded by other objects (sometimes called non-targets or distractors). An example is the act of picking a particular apple from a basket full of fruit. Research has shown that the presence of additional objects slows down reaching performance (Tipper *et al.*, 1992). Moreover, there is an important asymmetry in the spatial nature of this interference effect: distractors close to the starting position of the hand interfere more than distractors further away (the 'proximity-to-hand' effect). According to the visuomotor account of distractor interference (Meegan and Tipper, 1999), both target and distractor automatically trigger the planning of movements towards their locations. Distractor interference, then, reflects the need to suppress or inhibit responses towards the distractor.

Speed–accuracy bias

It is not only external, physical constraints (e.g. size and distance of the target) that influence the control of movements; the intention or goal of the performer is important too. Individual performers may opt to emphasize speed or accuracy. It has been shown that the kinematics of the movements produced under these two strategies are different (Adam, 1992). Whereas a speed strategy results in a more or less symmetric movement profile (i.e. similar durations for the acceleration and deceleration phases), an accuracy strategy results in a much longer deceleration than acceleration phase. The longer deceleration phase associated with an accuracy strategy allows feedback information concerning endpoint accuracy to be monitored and used for movement adjustments.

Sequencing Actions

So far, we have considered discrete or one-element movements which have a single start and a single stop. Often, however, actions involve a series of consecutive movements that occur through time: for example, writing, speaking, piano playing, dialing a phone number, and entering a security code in a bank terminal.

The 'one-target advantage' phenomenon sheds some light on how the brain controls sequences of movements. A rapid aimed hand movement is executed faster when it is performed as a single, isolated movement than when it is followed by an additional movement. The one-target advantage phenomenon may be demonstrated by comparing performance in two conditions: participants are asked either to move as quickly as possible to the first target and stop (the one-tap condition) or to strike the first target and then immediately move on and hit a second target (the two-tap condition). Typically, movement time to the first target is about 20 ms shorter in the one-tap condition than in the two-tap condition, indicating that one of the effects of making a second movement is to slow down the first. This observation implies that the two movements are functionally interdependent.

According to the 'movement integration' account, the one-target advantage results from an anticipatory motor control strategy whereby the two movements in the two-tap condition are planned together in one overall response program before response initiation (Adam *et al.*, 2000). This overall response program specifies that the second movement should be implemented during the latter parts of the first movement so that a smooth and quick transition of the first movement into the second can occur. In other words, implementation of the second movement does not await termination of the first movement, but, rather, overlaps with and is superimposed on the execution of that first movement. The movement integration account of the one-target advantage is based on the notion that the serial ordering of movements is controlled by a motor program that represents an integrated series of movements.

Sequences of movements involving more than two elements may be represented hierarchically. This means that the mental representation of a sequence is not just a linear string of event-to-event associations; rather, elements are organized in groups, and superordinate relationships may exist among them, so that there are distinct (i.e., higher and lower) levels of control mediating the temporal structure of the movement sequence. Consistent with a hierarchical conception of motor control, interresponse times of individual elements in a movement sequence are not identical, but follow closely the hierarchical (or tree-like) structure of a sequence of keyboard responses (Rosenbaum *et al.*, 1983).

ACTION AND PRACTICE

As the proverb 'practice makes perfect' implies, high-level performance appears to be a function of repeated rehearsal of motor acts. Indeed, the elegance, speed, and accuracy of actions performed by experts (e.g. athletes, musicians, craftspeople) derive from many years of deliberate practice. The observation that repetition exerts a powerful influence on the quality of the motor act (e.g. Ericsson *et al.*, 1993; Helsen *et al.*, 1998) has inspired many researchers to concentrate on the mechanisms underlying skill acquisition. (*See* **Expertise**)

The Three-Stage Theory of Practice

Regarding actions as the overt consequences of the interactions of the human neuromuscular system with the environment, one can regard the learning of new motor skills as adaptations of the implemented input–output connections. Apparently the cerebellum plays an important role in this adaptation process, as the repeated soliciting of its circuits establishes, modifies, or strengthens the required input–output connections. The learning of motor skills has a strong biological foundation (this observation is supported by recent research using magnetic resonance imaging). Within these structural boundaries, the learning process can materialize in different ways. Regarding the mental representations of the action to be executed as the controlling and driving forces of movements, the observed practice effects appear to express the enhancement of these 'dynamical' representations. Actually, the adaptability of these representations (which illustrates well the plasticity of the brain functions) is a prerequisite for learning.

What form does this learning process take? The progress made by humans trying to learn new skills does not appear to follow a linear function. Ever since Fitts (1964) formulated his observation that learning follows a triphasic path, it has been clear that learning phases are fundamental features of long-term behavioral adaptations. Indeed, distinct learning phases have been described, each embodying specific characteristics and properties of motor behavior. The amount of mental effort put into the action by the performer is an important determinant of these phases. A strong cognitive involvement is observed in the initial ('cognitive') phase of the learning process, but the need for this cognitive activity gradually decreases as learning progresses. Numerous repetitions are required to pass from the second ('associative') phase, in which performers continually adjust and improve the movement pattern, to the last ('autonomous') phase in which real expertise appears. This final phase, with its smooth, elegant and parsimonious motor behavior, is the automated expression of years of deliberate practice, which has enabled the actor to eliminate cognitive interference and to use higher-order mental processes only to develop and implement strategic elements. A highly skilled performer transforms action into art.

The Importance of Feedback

Various mechanisms are available to facilitate and possibly shorten the route to expertise. Apart from verbal instructions and visual models (which are most effective during the cognitive phase), the most widely used strategy is to inform the actor about his or her performance after completing the action. This feedback can engender positive learning effects; however, neutral or even negative effects have also been reported, showing that the effectiveness of feedback strongly depends on the nature of the task. For example, when a player tries to score a goal, information about the result of his or her kick or throw is redundant as this information is intrinsically available in the task. Moreover, false or conflicting feedback information may engender incorrect behavioral adaptations (Buekers *et al.*, 1992). Nevertheless, the crucial element in learning appears to be the availability of (correct) information about the errors produced by the learner. On the basis of this error information the learner can adjust his or her movement in a subsequent trial and better meet the criteria of the task.

How should this feedback information be transmitted to the learner? Knowledge of results is an obvious form of feedback, but it is actually less efficient than one might expect. Knowledge of performance (information about the movement itself) appears to have a more pronounced facilitating effect on the learning of new motor skills. This form of feedback is primarily focused on movement characteristics (form, spatial, and temporal characteristics, etc.), and aims to change the movement errors that led to performance failure. The term 'transition feedback' has been proposed, referring to how specific elements of the movement must be changed to produce a successful outcome.

MOTOR IMAGERY

Many experiments have shown that performance can improve as a result of purely mental activity, in the absence of the actual action – for example, imagining oneself performing a perfect golf swing, lifting a heavy weight, or throwing a dart at the bullseye. This improvement has been observed for a variety of motor skills, such as tracking in a pursuit task, walking on a balance beam, hitting golf balls to a target, and muscular strength. However, although these effects are substantial, motor imagery cannot replace physical practice.

The observed performance benefits of motor imagery have sometimes been ascribed to motivational factors; however, it has been shown that motor imagery and overt practice stimulate – at least in part – common neural substrates. Apparently, motor imagery and overt practice rely to a large extent on the same cortical structures. For example, qualitatively similar event-related brain potentials have been reported when subjects imagined or executed specific hand movements. More recent studies using positron emission tomography or magnetic resonance imaging confirm this finding and support the hypothesis of functional equivalence, that movement preparation and motor imagery use the same brain processes. Thus, there seem to be relevant mental representations that are accessible to the actor even in the absence of external stimulation and overt action.

ACTION AND CONSCIOUSNESS

Consciousness is very difficult to define, but phenomenologically it refers to experience (Flanagan, 1998). Consciousness often, but not always, plays a role in the control of action. When learning a new motor skill, a substantial amount of conscious attention is typically needed to learn the basic procedures involved. In this early cognitive stage, verbal cues are often used. However, once actions are highly practiced they may no longer require conscious control; indeed, when automated, actions like writing or driving a car can be performed without conscious awareness at all. Thus, skilled actions that have progressed to Fitts' autonomous stage can be performed without consciousness; they are automatic in the sense that they can be carried out concurrent with other mental or motor activities.

The phenomenon of 'blindsight' also illustrates that consciousness is not always necessary for the control of action. Patients with cortical blindness in one hemifield can perform eye movements or pointing gestures towards visual information presented in their blind field even though they report that they are unaware of this information. Furthermore, the Ebbinghaus (or Titchener circles) illusion (a target circle surrounded by smaller circles appears to be larger than a target circle surrounded by larger circles) affects the conscious visual perception of the size of the target but not grasping movements towards it: the grip aperture of the grasping hand is determined by the actual and not by the subjectively perceived size of the target (Agliotti *et al.*, 1995). Observations like these indicate that actions are not always controlled by the conscious experience of objects. Indeed, Milner and Goodale (1995) argue that (conscious) visual perception and visuomotor control are separate functions, sensitive to different constraints, and mediated by different neural pathways. According to this view, visual information may affect action directly, without mediation by consciousness. (*See* **Blindsight**; **Visual Scene Perception**)

CONCLUSION

Motor actions are the principal means by which we interact with the world. Brain processes compute an abstract representation of the intended action. The time to prepare a motor representation or motor program is reflected in the reaction time, which is sensitive to factors such as the number of response choices, the complexity of the response, the compatibility between the potential stimuli and responses, and the performer's bias towards speed or accuracy. The time needed to complete a movement depends on variables such as movement distance, target size, the presence and spatial nature of non-targets, and, again, the actor's bias for speed or accuracy.

Movements can be fully preprogrammed and executed as such (open-loop control); the advantage is speed, though sometimes at the cost of accuracy. Accurate (and slower) movements are constrained by feedback processes that allow corrective submovements (closed-loop control).

Motor skill acquisition progresses via three phases: in the first, cognitive phase, conscious attention is required to understand and practice the basic procedures and movements involved; in the second, associative phase, feedback in the form of knowledge of performance is particularly effective in shaping successful movement patterns and eliminating errors; in the last, autonomous phase, motor performance is automated and freed from the involvement of conscious control, allowing the concurrent engagement in other tasks. Mental practice – especially in combination with physical practice – can improve motor performance. But even though consciousness is necessary for mental practice, and is very involved in the first two phases of skill acquisition, skilled motor action can, and often does, proceed unconsciously.

References

Adam JJ (1992) The effects of objectives and constraints on motor control strategy in reciprocal aiming movements. *Journal of Motor Behavior* **24**: 173–185.

Adam JJ, Nieuwenstein J, Huys R *et al.* (2000) Control of rapid aimed hand movements: the one-target advantage. *Journal of Experimental Psychology: Human Perception and Performance* **26**: 295–312.

Agliotti S, DeSouza JFX and Goodale MA (1995) Size-contrast illusions deceive the eye but not the hand. *Current Biology* **5**: 679–685.

Buekers MJ, Magill RA and Hall KG (1992) The effect of erroneous knowledge of results on skill acquisition when augmented information is redundant. *Quarterly Journal of Experimental Psychology* **44A**: 105–117.

Eimer M (1995) Stimulus–response compatibility and automatic response activation: evidence from psychophysiological studies. *Journal of Experimental Psychology: Human Perception and Performance* **21**: 837–854.

Ericsson KA, Krampe RT and Teschromer C (1993) The role of deliberate practice in the acquisition of expert performance. *Psychological Review* **100**: 363–406.

Fitts PM (1954) The information capacity of the human motor system in controlling the amplitude of movement. *Journal of Experimental Psychology* **47**: 381–391.

Fitts PM (1964) Perceptual–motor skill learning. In: Melton AW (ed.) *Categories of Human Learning*, pp. 243–285. New York, NY: Academic Press.

Flanagan O (1998) Consciousness. In: Bechtel W and Graham G (eds) *A Companion to Cognitive Science*, pp. 176–185. Oxford, UK: Blackwell.

Helsen WF, Starkes JL and Hodges NJ (1998) Team sports and the theory of deliberate practice. *Journal of Sport and Exercise Psychology* **20**: 12–34.

Klapp ST (1996) Reaction time analysis of central motor control. In: Zelaznik HN (ed.) *Advances in Motor Learning and Control*, pp. 13–35. Champaign, IL: Human Kinetics.

Kornblum ST, Hasbroucq T and Osman A (1990) Dimensional overlap: cognitive basis for stimulus–response compatibility – a model and taxonomy. *Psychological Review* **97**: 253–270.

Meegan DV and Tipper SP (1999) Visual search and target-directed action. *Journal of Experimental Psychology: Human Perception and Performance* **25**: 1347–1362.

Meyer DE, Abrams RA, Kornblum S, Wright CE and Smith JEK (1988) Optimality in human motor performance: ideal control of rapid aimed movements. *Psychological Review* **95**: 340–370.

Milner AD and Goodale MA (1995) *The Visual Brain in Action*. Oxford, UK: Oxford University Press.

Rosenbaum DA, Kenny S and Derr MA (1983) Hierarchical control of rapid movement sequences. *Journal of Experimental Psychology: Human Perception and Performance* **9**: 86–102.

Tipper SR, Lortie C and Baylis GC (1992) Selective reaching: evidence for action-centered attention. *Journal of Experimental Psychology: Human Perception and Performance* **18**: 891–905.

Further Reading

Gallistel CR (1999) Coordinate transformations in the genesis of directed action. In: Bly BM and Rumelhart DE (eds) *Cognitive Science*, pp. 1–42. San Diego, CA: Academic Press.

Gazzaniga MS, Ivry RB and Mangun GR (1998) Motor control. In: Gazzaniga MS, Ivry RB and Mangun GR (eds) *Cognitive Neuroscience*, pp. 371–422. New York, NY: Norton.

Jeannerod M (1997) *The Cognitive Neuroscience of Action*. Cambridge, MA: Blackwell.

Keele S (1986) Motor control. In: Boff JK, Kaufman L and Thomas JP (eds) *Handbook of Human Perception and Performance*, vol. II, pp. 1–60. New York, NY: John Wiley.

Rosenbaum DA (1991) *Human Motor Control*. San Diego, CA: Academic Press.

Schmidt RA and Lee TD (1999) *Motor Control and Learning*. Champaign, IL: Human Kinetics.

Shumway-Cook A and Woollacott MH (1995) *Motor Control: Theory and Practical Applications*. Baltimore, MD: Williams & Wilkins.

Action, Philosophical Issues about

Intermediate article

Alfred R Mele, Florida State University, Tallahassee, Florida, USA

CONTENTS

The philosophy of action concerns theories about what actions are, how they are to be explained, and the mental events and states associated with intentional action.

INTRODUCTION

In striving to analyze, understand, and explain actions, philosophers of action are concerned primarily with intentional actions. In discussions of freedom of action, intentional action also naturally occupies center stage. And although people are morally accountable for some unintentional actions, as in cases of negligence, moral assessment of actions is focused primarily on intentional actions. What is the nature of intentional action and how are intentional actions to be explained?

CENTRAL PHILOSOPHICAL ISSUES ABOUT ACTION

There are two main philosophical questions about actions: What is an action? How are actions to be explained? The first question directly raises two others: How are actions different from nonactions? (For example, how does an ordinary instance of running hard differ from an ordinary unintentional instance of breathing hard?) How do actions differ from one another? (For example, if I turn on my computer by pressing a button, are my turning it on and my pressing the button different actions or the same action?) If not all actions are intentional, the first question also raises another: What is it for an action to be intentional? The question about the explanation of actions is also a question for cognitive science. The challenge is to produce a theory of the springs of action in light of which we can, in principle, explain particular intentional actions – in light of which we can explain, for example, why you are reading this article, why I wrote it, and why the editor decided to publish it. If proper explanations of actions are causal explanations, part of what we would like to understand is how the events or states that explain actions help to produce actions.

PHILOSOPHICAL VIEWS AND THEORIES OF ACTION

According to a popular answer to the question how actions differ from nonactions (Brand, 1984; Davidson, 1980; Mele, 1992), actions are like sunburns in an important respect. The burn on Al's back is a sunburn partly in virtue of its having been caused by exposure to the sun's rays; a burn that looks and feels just the same is not a sunburn if it was caused by a heat lamp. Similarly, a certain event is Al's raising his left hand – an action – partly in virtue of its having been appropriately caused by mental items. An influential version of this view claims that reasons, understood as combinations of beliefs and desires, are causes of actions and that an event counts as an action partly in virtue of its having been suitably caused by a reason (Davidson, 1980). Alternative conceptions of action include an 'internalist' position according to which actions differ experientially from other events in a way that does not depend on how, or whether, they are caused; a conception of actions as composites of nonactional mental events or states (e.g. intentions) and pertinent nonactional effects (e.g. an arm's rising); and views identifying an action with the causing of a suitable nonactional product by an appropriate nonactional mental event or state – or, instead, by an agent (for discussion of these alternatives, see Davis, 1979; Ginet, 1990).

Promising theories about how actions differ from one another include a fine-grained theory (Goldman, 1970), a coarse-grained theory (Davidson, 1980), and componential theories (Ginet, 1990). According to a fine-grained theory of actions, *A* and *B* are different actions if, in performing them, the agent exemplifies different act-properties. For example, if Ann starts her car by turning a key, her starting the car and her turning the key are two different actions, since the act-properties at issue are distinct. A coarse-grained theory asserts that Ann's turning the key and her starting the car are the same action described in two different ways. A componential theory claims that Ann's starting her car is an action having various components, including her moving her arm, her turning the key, and the car's starting. Where the first two theories claim to find, alternatively, a collection of related actions, or a single action under different descriptions, component theories assert that there is a 'larger' action having 'smaller' actions among its parts.

Most philosophers agree that at least a sketch of an explanation of an intentional action can be provided by identifying the reasons for which the agent performed it. Whether reasons can have a place in *causal* explanations of actions is controversial. In 1963, Donald Davidson challenged anti-causalists about 'reasons-explanations' to provide an account of the reasons for which we act that does not treat (our having) those reasons as causes of relevant actions. Imagine that Ann has a pair of reasons for using her leaf blower this morning. First, she wants to blow the leaves off her lawn today and she regards this morning as a very convenient time. Second, she has a desire to repay her neighbor for awakening her yesterday with his leaf blower and she believes that blowing the leaves off her lawn this morning would do the trick. As it happens, Ann uses her leaf blower this morning only for one of these reasons. In virtue of what is it true that she uses it for this reason, and not for the other, if not that this reason (or her having it), and not the other, makes an appropriate causal contribution to her using it? Detailed attempts to meet this challenge have been revealingly problematic (for discussion, see Mele, 1992, chap. 13).

IMPACT OF COGNITIVE SCIENCE ON ISSUES ABOUT ACTION

Philosophers of action have explored the nature of psychological states and events thought to play important causal/explanatory roles in intentional action, including beliefs, desires, and intentions. Some philosophers appeal to intentions in an effort to avoid the problems 'deviant causal chains' pose for attempted analyses of intentional action featuring reasons as causes (Brand, 1984; Mele, 1992). The alleged problem is that whatever psychological causes are offered as necessary and sufficient for a resultant action's being intentional, scenarios can be constructed in which, owing to an atypical causal connection between the favored psychological antecedents and a pertinent resultant action, that action is not intentional. For example, Ann wants to awaken her husband and she believes that she may do so by making a loud noise. Motivated (causally) by this desire and belief, Ann may search in the dark for a suitable noise-maker. In her search, she may accidentally knock over a lamp, producing a loud crash. By so doing, she may awaken her husband, but her awakening him in this way is not an intentional action. The task for those who wish to analyze intentional action causally is to specify not only the psychological causes of actions associated with their being intentional but also the pertinent roles played by these causes. Some philosophers have sought help from cognitive science in undertaking this task (Brand, 1984; Mele, 1992).

Presumably, intentions play important roles in the production of intentional actions. Functions plausibly attributed to intention in the philosophical and psychological literature include initiating and motivationally sustaining intentional actions, guiding intentional behavior, helping to coordinate agents' behavior over time and their interaction with other agents, and prompting and appropriately terminating practical reasoning (see Brand, 1984; Bratman, 1987, 1999; Mele, 1992). The initiating, sustaining, and guiding roles are relevant to the problem of deviant causal chains.

Intentions, like many psychological states, have both a representational and an attitudinal dimension. The representational content of an intention may be understood as a *plan*. The intending *attitude* towards plans may be termed an *executive* attitude. Plans, on one conception, are purely representational and have no motivational power of their own (Brand, 1984; Mele, 1992). People may have any number of attitudes towards plans, in this sense. They may believe that a plan is elegant, admire it, hope that it is never executed, and so on.

To understand the executive dimension of intention, compare an intention to attend a concert with a *desire* to attend a concert. Both encompass motivation to attend a concert, and the content of each is or includes a representation of the prospective course of action. But although one can have a desire to attend a concert without being at all *settled* on doing so, intending to attend a concert is partially constituted by being settled on so doing. This is compatible with intentions' being revocable and revisable. Though Al is now settled on meeting a friend for dinner, he would cancel the arrangement were a pressing problem to arise at home.

An important motivational difference between desires and intentions may lie in their *access* to the mechanisms of intentional action. This difference coheres with the claim that intending to *A* entails being settled on *A*-ing while desiring to *A* does not. Whereas our becoming settled on *A*-ing straightaway is normally sufficient to initiate an *A*-ing at once, this is not true of the acquisition of desires to *A* straightaway. To be sure, someone's being settled now on *A*-ing later normally will not initiate an *A*-ing now. But if the intention is still present at the later time and the agent recognizes that the designated time has arrived, an attempt at *A*-ing will normally be immediately forthcoming. This is not true of someone who still has a mere desire at the later time to *A*; such a person may simply choose not to *A* and behave accordingly.

If intentions initiate actions, it is *proximal* intentions that do so – roughly, intentions to do something straightaway. (More precisely, it is the *acquisition* of a proximal intention that plays this role.) But why do proximal intentions initiate and sustain the actions that they do? Why, for example, does an intention to drive home tend to initiate and sustain one's driving home rather than one's cycling home or one's driving to a friend's house? Return to the representational side of intentions. An intention to *A* incorporates a plan for *A*-ing, and *which* intentional action or actions an intention generates is a partial function of the intention-embedded plan. In the limiting case, the plan is a simple representation of one's *A*-ing. Often, intention-embedded plans are more complex. For example, Al's proximal intention to check the oil in his car incorporates a plan that includes his first unlatching the hood, then opening the hood, then unscrewing the oil cap, and so on. An agent who successfully executes an intention is *guided* by the intention-embedded plan.

An intention-embedded plan identifies a goal and (in non-limiting cases) provides action-directions, as it were. Exactly how deep the representational content of intentions runs is a partly empirical and partly conceptual question. Even when what is intended is routine and very simple behavior for the agent – a doctor's signing a prescription, for example – a great deal is going on representationally. Some psychologists take the representational content of motor schemata to run quite deep, suggesting, for example, that motor schemata involved in handwriting include representations of the neuromuscular activity required to achieve the movement represented by their higher-level components. Standard philosophical conceptions of intention seem not to countenance such representations as parts of the representational content of a normal agent's intention to write her name, probably because of the apparent inaccessibility of these representations to consciousness. On standard conceptions, however, intentions guide behavior in a way that depends on their representational content. If plans embedded in standard writing intentions do not incorporate representations of low-level neuromuscular activity, they can provide guidance at a higher level, with the assistance of motor schemata that are external to intentions. Perhaps a solution to some problems posed by deviant causal chains can be produced by careful attention to the guiding function of proximal intentions.

RELEVANCE OF ACTION THEORY TO COGNITIVE SCIENCE

The philosophy of action has produced theories about what actions are and about how actions are to be explained. In both connections, philosophers have speculated about how mental events and states figure in the production of intentional actions. Theories of the former kind are useful to cognitive scientists concerned to explain human actions, insofar as these theories provide conceptions of what is to be explained. Some philosophical theories of the explanation of action are fertile ground for cognitive science. These theories have elements that are testable, or suggestive of empirical hypotheses, and the theories reveal interesting patterns in our own common-sense explanations of why we do what we do.

References

Brand M (1984) *Intending and Acting*. Cambridge, MA: MIT Press.

Bratman M (1987) *Intention, Plans, and Practical Reason*. Cambridge, MA: Harvard University Press.

Bratman M (1999) *Faces of Intention*. Cambridge, UK: Cambridge University Press.

Davidson D (1963) Actions, Reasons, and Causes. *Journal of Philosophy* **60**: 685–700.

Davidson D (1980) *Essays on Actions and Events*. Oxford, UK: Oxford University Press.

Davis L (1979) *Theory of Action*. Englewood Cliffs, NJ: Prentice-Hall.

Ginet C (1990) *On Action*. Cambridge, UK: Cambridge University Press.

Goldman A (1970) *A Theory of Human Action*. Englewood Cliffs, NJ: Prentice-Hall.

Mele A (1992) *Springs of Action*. New York, NY: Oxford University Press.

Further Reading

Audi R (1993) *Action, Intention, and Reason*. Ithaca, NY: Cornell University Press.

Bishop J (1989) *Natural Agency*. Cambridge, UK: Cambridge University Press.

Hornsby J (1980) *Actions*. London: Routledge & Kegan Paul.

McCann H (1998) *The Works of Agency*. Ithaca, NY: Cornell University Press.

Mele A (2003) *Motivation and Agency*. New York, NY: Oxford University Press.

Searle J (2001) *Rationality in Action*. Cambridge, MA: MIT Press.

Velleman JD (2000) *The Possibility of Practical Reason*. Oxford, UK: Oxford University Press.

Wilson G (1989) *The Intentionality of Human Action*. Stanford, CA: Stanford University Press.

Adaptive Resonance Theory

Advanced article

Stephen Grossberg, Boston University, Boston, Massachusetts, USA

CONTENTS

Adaptive resonance theory is a cognitive and neural theory about how the brain develops and learns to recognize and recall objects and events throughout life. It shows how processes of learning, categorization, expectation, attention, resonance, synchronization, and memory search interact to enable the brain to learn quickly and to retain its memories stably, while explaining many data about perception, cognition, learning, memory, and consciousness.

INTRODUCTION

The processes whereby our brains continue to learn about, recognize, and recall a changing world in a stable fashion throughout life are among the most important for understanding cognition. These processes include the learning of top-down expectations, the matching of these expectations against bottom-up data, the focusing of attention upon the expected clusters of information, and the development of resonant states between bottom-up and top-down processes as they reach an attentive consensus between what is expected and what is there in the outside world. It has been suggested that all conscious states in the brain are resonant states, and that these resonant states trigger learning of sensory and cognitive representations. The models which summarize these concepts are called 'adaptive resonance theory' (ART) models. ART was introduced by Grossberg in 1976 (see Carpenter and Grossberg, 1991), along with rules for competitive learning and self-organizing maps. Since then, psychophysical and neurobiological data in support of ART have been reported in experiments on vision, visual object recognition, auditory streaming, variable-rate speech perception, somatosensory perception, and cognitive–emotional interactions, among others (e.g. Carpenter and Grossberg, 1991, 1994; Grossberg, 1999a,b; and Grossberg and Merrill, 1996). In particular, ART mechanisms seem to be operative at all levels of the visual system, and these mechanisms may be realized by known laminar circuits of visual cortex. It is predicted that the same circuit realization of ART mechanisms will be found, suitably specialized, in the laminar circuits of all sensory and cognitive neocortex.

THE STABILITY–PLASTICITY DILEMMA

We experience the world as a whole. Although myriad signals relentlessly bombard our senses, we somehow integrate them into unified moments of conscious experience that cohere despite their diversity. Because of the apparent unity and coherence of our awareness, we can develop a sense of self that can gradually mature with our experiences of the world. This capacity lies at the heart of our ability to function as intelligent beings.

The apparent unity and coherence of our experiences is all the more remarkable when we consider several properties of how the brain copes with the environmental events that it processes. First and foremost, these events are highly context-sensitive. When we look at a complex picture or scene as a whole, we can often recognize its objects and its meaning at a glance, as in the picture of a familiar face. However, if we process the face piece by piece, as through a small aperture, then its significance may be greatly degraded. To cope with this context-sensitivity, the brain typically processes pictures and other sense data in parallel, as patterns of activation across a large number of feature-sensitive nerve cells, or neurons. The same is true for senses other than vision, such as audition. If the sound of the word 'go' is altered by clipping off the vowel 'o', then the consonant 'g' may sound like a chirp, quite unlike its sound as part of the word 'go'.

During vision, all the signals from a scene typically reach the photosensitive retinas of the eyes at virtually the same time, so parallel processing of all the scene's parts begins at the retina itself. During audition, successive sounds reach the ear at different times. Before an entire pattern of sounds, such as the word 'go', can be processed as a whole, it needs to be recoded, at a later processing stage, into a simultaneously available spatial pattern of activation. Such a processing stage is often called a working memory, and the activations that it stores are often called short-term memory (STM) traces. For example, when you hear an unfamiliar telephone number, you can temporarily store it in working memory while you walk over to the telephone and dial the number.

In order to determine which of these patterns represent familiar events and which do not, the brain matches the patterns against stored representations of previous experiences that have been acquired through learning. The learned experiences are stored in long-term memory (LTM) traces. One difference between STM and LTM traces concerns how they react to distractions. For example, if you are distracted by a loud noise before you dial an unfamiliar telephone number, its STM representation can be rapidly reset so that you forget it. On the other hand, you will not normally forget the LTM representation of your own name.

How does new information get stably stored in LTM? For example, after seeing a movie just once, we can tell our friends many details about it later on, even though the scenes flashed by very quickly. More generally, we can quickly learn about new environments, even if no one tells us how the rules of the environments differ. We can rapidly learn new facts, without being forced to just as rapidly forget what we already know. We do not need to avoid going out into the world for fear that, in learning to recognize a new friend's face, we will suddenly forget our parents' faces. This is sometimes called the problem of catastrophic forgetting.

Many learning algorithms can forget catastrophically. But the brain is capable of rapid yet stable autonomous learning of huge amounts of data in an ever-changing world. Discovering the brain's solution to this problem is as important for understanding ourselves as it is for developing new pattern recognition and prediction applications in technology.

The problem of learning quickly and stably without catastrophically forgetting past knowledge may be called the stability–plasticity dilemma. It must be solved by every brain system that needs to rapidly and adaptively respond to the flood of signals that subserves even the most ordinary experiences. If the brain's design is parsimonious, then we should expect to find similar design principles operating in all the brain systems that can stably learn an accumulating knowledge base in response to changing conditions throughout life. The discovery of such principles should also clarify how the brain unifies diverse sources of information into coherent moments of conscious experience.

LEARNING, EXPECTATION, ATTENTION, AND RESONANCE

Humans are intentional beings, who learn expectations about the world and make predictions about what is about to happen. Humans are also attentional beings, who focus their processing resources upon a restricted amount of incoming information at any time. Why are we both intentional and attentional beings, and are these two types of process related? The stability–plasticity dilemma, and its solution using resonant states, provides a unifying framework for understanding these questions.

Suppose you were asked to 'find the yellow ball within half a second, and you will win a $10 000 prize'. Activating an expectation of 'yellow balls' enables more rapid detection of a yellow ball, and with a more energetic neural response, than if you were not looking for one. Sensory and cognitive top-down expectations lead to 'excitatory matching' with confirmatory bottom-up data. On the other hand, mismatch between top-down expectations and bottom-up data can suppress the mismatched part of the bottom-up data, and thereby focus attention upon the matched, or expected, part of the bottom-up data.

This sort of excitatory matching and attentional focusing on bottom-up data using top-down expectations may generate resonant brain states: When there is a good enough match between bottom-up and top-down signal patterns, between two or more levels of processing, their positive feedback signals amplify and prolong their mutual activation, leading to a resonant state. The amplification and prolongation of the system's fast activations are sufficient to trigger learning in the more slowly varying adaptive weights that control the signal flow along pathways from cell to cell. Resonance thus provides a global context-sensitive indicator that the system is processing data worthy of learning.

ART proposes that there is an intimate connection between the mechanisms that enable us to learn quickly and stably about a changing world and the mechanisms that enable us to learn expectations about such a world, test hypotheses about it, and focus attention upon information that we find interesting. ART also proposes that, in order to solve the stability–plasticity dilemma, resonance must be the mechanism that drives rapid new learning.

Learning within the sensory and cognitive domains is often 'match learning'. Match learning occurs only if a good enough match occurs between bottom-up information and a learned top-down expectation that is specified by an active recognition category, or code. When such a match occurs, previously learned knowledge can be refined. If novel information cannot form a good enough match with the expectations that are specified by previously learned recognition categories, then a memory search, or hypothesis testing, is triggered, which leads to selection and learning of a new recognition category, rather than catastrophic forgetting of an old one. (Figure 1 illustrates how this happens in an ART model.) In contrast, learning within spatial and motor processes could be 'mismatch learning' that continuously updates sensory–motor maps or the gains of sensory–motor commands. As a result, we can stably learn what is happening in a changing world, thereby solving the stability–plasticity dilemma, while adaptively updating our representations of where objects are and how to act upon them using bodies whose parameters change continuously through time.

It has been mathematically proven that match learning within an ART model leads to stable memories in response to arbitrary lists of events to be learned (Carpenter and Grossberg, 1991). However, match learning has a serious potential weakness: if you can only learn when there is a good enough match between bottom-up data and learned top-down expectations, then how do you ever learn anything that you do not already know? ART proposes that this problem is solved by the brain by using a complementary interaction between processes of resonance and reset, which are proposed to control properties of attention and memory search, respectively. These complementary processes help our brains to balance the complementary demands of processing the familiar and the unfamiliar, the expected and the unexpected. One of these complementary processes is hypothesized to take place in the What cortical stream, notably in the visual, inferotemporal, and prefrontal cortex. It is here that top-down expectations are matched against bottom-up inputs (Chelazzi *et al.*, 1998; Miller *et al.*, 1996). When a top-down expectation achieves a good enough match with bottom-up data, this matching process focuses attention upon those feature clusters in the bottom-up input that are expected. If the expectation is close enough to the input pattern, then a state of resonance develops as the attentional focus is established.

Figure 1 illustrates these ideas in a simple two-level example. Here, a bottom-up input pattern, or vector, I activates a pattern X of activity across the feature detectors of the first level F_1. For example, a visual scene may be represented by the features comprising its boundary and surface representations. This feature pattern represents the relative importance of different features in I. In Figure 1(a), the pattern peaks represent more activated feature detector cells, the troughs less activated feature detectors. This feature pattern sends signals S through an adaptive filter to the second level F_2 at which a compressed representation Y (a recognition category, or symbol) is activated in response to the distributed input T. T is computed by multiplying the signal vector S by a matrix of adaptive weights, which can be altered through learning. The representation Y is compressed by competitive interactions across F_2 that allow only a small subset of its most strongly activated cells to remain active in response to T. The pattern Y in the figure indicates that a small number of category cells may be activated to different degrees. These category cells, in turn, send top-down signals U to F_1 (see Figure 1(b)). The vector U is converted into the top-down expectation V by being multiplied by another matrix of adaptive weights. When V is received by F_1, a matching process takes place between I and V, which selects that subset X^* of F_1 features that were 'expected' by the active F_2 category Y. The set of these selected features is the emerging 'attentional focus'.

RECONCILING DISTRIBUTED AND SYMBOLIC REPRESENTATIONS USING RESONANCE

If the top-down expectation is close enough to the bottom-up input pattern, then the pattern X^* of attended features reactivates the category Y which, in turn, reactivates X^*. The network thus enters a resonant state through a positive feedback loop that dynamically links, or binds, the attended features across X^* with their category, or symbol, Y.

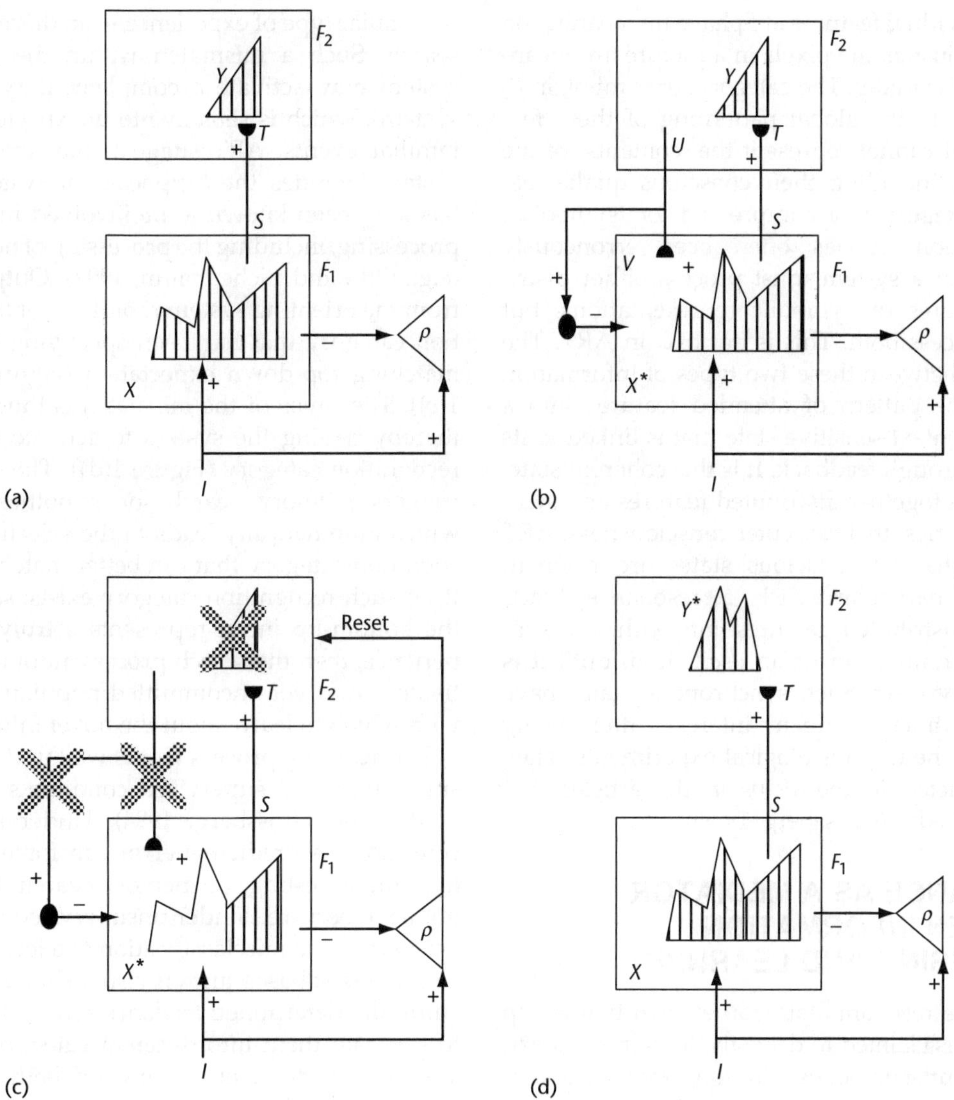

Figure 1. Search for a recognition code within an ART learning circuit. (a) The input pattern I is instated across the feature detectors at level F_1 as a short-term memory (STM) activity pattern X. I also nonspecifically activates the orienting system ρ; that is, all the input pathways converge on ρ and can activate it. X is represented by the hatched pattern across F_1. X both inhibits ρ and generates the output pattern S. S is multiplied by learned adaptive weights, which are long-term memory (LTM) traces. These LTM-gated signals are added at F_2 cells, or nodes, to form the input pattern T, which activates the STM pattern Y across the recognition categories coded at level F_2. (b) Pattern Y generates the top-down output pattern U, which is multiplied by top-down LTM traces and added at F_1 nodes to form a 'prototype' pattern V that encodes the learned expectation of the active F_2 nodes. Such a prototype represents the set of features shared by all the input patterns capable of activating Y. If V mismatches I at F_1, then a new STM activity pattern X^* is selected at F_1. X^* is represented by the hatched pattern. It consists of the features of I that are confirmed by V. Mismatched features are inhibited. The inactivated nodes, corresponding to unconfirmed features of X, are unhatched. The reduction in total STM activity which occurs when X is transformed into X^* causes a decrease in the total inhibition of ρ from F_1. (c) If inhibition decreases sufficiently, ρ releases a nonspecific arousal wave to F_2; that is, a wave of activation that activates all F_2 nodes equally. ('Novel events are arousing'.) This arousal wave resets the STM pattern Y at F_2 by inhibiting Y. (d) After Y is inhibited, its top-down prototype signal is eliminated, and X can be reinstated at F_1. The prior reset event maintains inhibition of Y during the search cycle. As a result, X can activate a different STM pattern Y^* at F_2. If the top-down prototype due to Y^* also mismatches I at F_1, then the search for an appropriate F_2 code continues, until an appropriate F_2 representation is selected. Such a search cycle represents a type of nonstationary hypothesis testing. When the search ends, an attentive resonance develops and learning of the attended data is initiated. (Adapted with permission from Grossberg, 1999b.)

The individual features at F_1 have no meaning on their own, just as the pixels in a picture are meaningless individually. The category, or symbol, in F_2 is sensitive to the global patterning of these features, but it cannot represent the 'contents' of the experience, including their conscious qualia, because a category is a compressed, or 'symbolic', representation. It has often been erroneously claimed that a system must process either distributed features or symbolic representations, but cannot process both. This is not true in ART. The resonance between these two types of information converts the pattern of attended features into a coherent context-sensitive state that is linked to its category through feedback. It is this coherent state, which joins together distributed features and symbolic categories, that can enter consciousness. ART proposes that all conscious states are resonant states. In particular, such a resonance binds spatially distributed features into either a synchronous equilibrium or an oscillation, until it is dynamically reset. Such synchronous states have recently attracted much interest after being reported in neurophysiological experiments. They were predicted in the 1970s in the articles that introduced ART (Grossberg, 1999b).

RESONANCE AS A MEDIATOR BETWEEN INFORMATION PROCESSING AND LEARNING

In ART, the resonant state, rather than bottom-up activation, is claimed to drive the learning process. The resonant state persists for long enough, and at a high enough activity level, to activate the slower learning processes in the adaptive weights that guide the flow of signals between bottom-up and top-down pathways between levels F_1 and F_2. This helps to explain how adaptive weights that were changed through previous learning can regulate the brain's present information processing, without learning about the signals that they are currently processing unless they can initiate a resonant state. Through resonance as a mediating event, one can see from a deeper viewpoint why humans are intentional beings who are continually predicting what may next occur, and why we tend to learn about the events to which we pay attention.

LEARNING AND HYPOTHESIS TESTING

A sufficiently strong mismatch between an active top-down expectation and a bottom-up input – for example, because the input represents an unfamiliar type of experience – can drive a memory search. Such a mismatch within the attentional system may activate a complementary 'orienting system', which is sensitive to unexpected and unfamiliar events. ART suggests that this orienting system includes the hippocampal system, which has long been known to be involved in mismatch processing, including the processing of novel events (e.g., Otto and Eichenbaum, 1992). Output signals from the orienting system rapidly reset the recognition category that has been specifying the poorly matching top-down expectation (Figure 1(b) and 1(c)). The cause of the mismatch is thus removed, thereby freeing the system to activate a different recognition category (Figure 1(d)). The reset event triggers memory search, or hypothesis testing, which automatically leads to the selection of a recognition category that can better match the input. If no such recognition category exists, say because the bottom-up input represents a truly novel experience, then the search process automatically activates an as-yet-uncommitted population of cells, with which to learn about the novel information.

This learning process works well under both unsupervised and supervised conditions (e.g., Carpenter and Grossberg, 1994). Under supervised conditions, a predictive error can force a cycle of hypothesis testing, or memory search, that might not have occurred under unsupervised conditions. For example, a misclassification of a letter F as an E could persist based just on visual similarity, unless culturally determined feedback forced the network to separate them into different categories. Such a search can discover a new or better-matching category with which to represent the novel data. Taken together, the interacting processes of attentive learning and orienting search achieve a type of error correction through hypothesis testing that can build an ever-growing, self-refining internal model of a changing world.

CONTROLLING THE GENERALITY OF KNOWLEDGE

What information is bound into object or event representations? Some scientists believe that exemplars, or individual experiences, can be learned and remembered, like familiar faces. But storing every exemplar requires huge amounts of memory, and leads to unwieldy memory retrieval. Others believe that we learn prototypes (Posner and Keele, 1970) that represent more general properties of the environment, for example, that everyone has a face. But then how do we learn specific episodic memories? ART provides an answer to this question.

ART systems learn prototypes whose generality is determined by a process of 'vigilance' control by environmental feedback or internal volition (Carpenter and Grossberg, 1991; Grossberg, 1999b). Low vigilance permits learning of general categories with abstract prototypes. High vigilance forces memory search to occur when even small mismatches exist between an exemplar and the category that it activates: for example, between letter exemplar F and letter category E. Given high enough vigilance, a category prototype may encode an individual exemplar. Vigilance is computed within the ART orienting system: see Figure 1. Here, bottom-up excitation from an input pattern I is balanced against inhibition from active features across level F_1. If a top-down expectation acts on F_1, then only the 'matched' features are active there. If the ratio of matched features in F_1 to all features in I is less than a vigilance parameter ρ (Figure 1(b)), then a reset, or 'novelty,' wave is activated (Figure 1(c)), which can trigger a search for another category.

The simplest rule for controlling vigilance during supervised learning is called match tracking. Here, a predictive error causes vigilance to increase until it is just higher than the ratio of active features in F_1 to total features in I. The error hereby forces vigilance to 'track' the degree of match between input exemplar and matched prototype. This is the minimal level of vigilance that can trigger a reset wave and thus a memory search for a new category. Match tracking realizes a minimax learning rule that maximizes category generality while minimizing predictive error. That is, it uses the least amount of memory resources that can prevent errors. ART models thus try to learn the most general category that is consistent with the data. This can lead to overgeneralization, like that seen in young children, until further learning causes category refinement. Benchmark studies of classifying complex databases have shown that the number of categories learned scales well with data complexity (e.g. Carpenter and Grossberg, 1994).

MEMORY CONSOLIDATION AND THE EMERGENCE OF RULES

As sequences of inputs are practiced over learning trials, the search process eventually converges upon stable categories. It has been mathematically proven (Carpenter and Grossberg, 1991) that familiar inputs directly access the category whose prototype provides the globally best match, while unfamiliar inputs engage the orienting subsystem to trigger memory searches for better categories, until they become familiar. This process continues until the available memory, which can be arbitrarily large, is fully utilized. The process whereby search is automatically disengaged is a form of memory consolidation that emerges from network interactions. Emergent consolidation does not preclude structural consolidation at individual cells, since the amplified and prolonged activities that subserve a resonance may be a trigger for learning-dependent cellular processes, such as protein synthesis and transmitter production. It has also been shown that the adaptive weights which are learned by some ART models can, at any stage of learning, be translated into if–then rules (e.g. Carpenter and Grossberg, 1994). Thus the ART model is a self-organizing rule-discovering production system as well as a neural network. These examples show that the claims of some cognitive scientists and AI practitioners that neural network models cannot learn rule-based behaviors are incorrect.

CORTICOHIPPOCAMPAL INTERACTIONS AND MEDIAL TEMPORAL AMNESIA

As noted above, the attentional subsystem of ART has been used to model aspects of inferotemporal cortex, while the orienting subsystem models part of the hippocampal system. The interpretation of ART dynamics in terms of inferotemporal cortex led Miller *et al.* (1991) to successfully test the prediction that cells in monkey inferotemporal cortex are reset after each trial in a working memory task. To illustrate the implications of an ART interpretation of inferotemporal-hippocampal interactions, we will review how a lesion of the ART model's orienting subsystem creates a formal memory disorder with symptoms much like the medial temporal amnesia that is caused in animals and human patients after hippocampal system lesions. In particular, such a lesion *in vivo* causes: unlimited anterograde amnesia; limited retrograde amnesia; failure of consolidation; tendency to learn the first event in a series; abnormal reactions to novelty, including perseverative reactions; normal priming; and normal information processing of familiar events. Unlimited anterograde amnesia occurs because the network cannot carry out the memory search to learn a new recognition code. Limited retrograde amnesia occurs because familiar events can directly access correct recognition codes. Before events become familiar, memory consolidation occurs, which utilizes the orienting subsystem (Figure 1(c)). This failure of consolidation would not necessarily prevent learning. Instead, it would learn coarser categories, because of the failure of vigilance control and memory search. For the same reason, learning may differentially influence the first recognition category activated by bottom-up processing, much as amnesics are particularly strongly bound to the first response they learn. Perseverative reactions can occur because the orienting subsystem cannot reset sensory representations or top-down expectations that may be persistently mismatched by bottom-up cues. The inability to search memory prevents ART from discovering more appropriate stimulus combinations to attend. Normal priming occurs because it is mediated by the attentional subsystem. Data supporting these predictions are summarized by Grossberg and Merrill (1996), who also note that these are not the only problems that can be caused by such a lesion: hippocampal structures can also play a role in learned spatial navigation and adaptive timing functions.

Knowlton and Squire (1993) have reported that amnesics can classify items as members of a large category even if they are impaired on remembering the individual items themselves. To account for these results, the authors propose that item and category memories are formed by distinct brain systems. Grossberg and Merrill (1996) suggest that their data could be explained by a single ART system in which the absence of vigilance control caused only coarse categories to form. Nosofsky and Zaki (2000) have quantitatively simulated the Knowlton and Squire data using a single-system model in which category sensitivity is low.

CORTICAL SUBSTRATES OF ART MATCHING

How are ART top-down matching rules implemented in the cerebral cortex of the brain? An answer to this question has been proposed as part of a rapidly developing theory of why the cerebral cortex is typically organized into six distinct layers of cells (Grossberg, 1999a). Earlier mathematical work had predicted that such a matching rule would be realized by a 'modulatory top-down on-center off-surround' network (e.g. Carpenter and Grossberg, 1991; Grossberg, 1999b). Figure 2 shows how such a matching circuit may be realized in the cortex. The top-down circuit generates outputs from cortical layer 6 of V2 that activate layer 6 of V1 via the vertical pathway between these layers that ends in an open triangle (which indicates an excitatory connection). Cells in layer 6 of V1, in turn, activate an 'on-center off-surround' circuit to

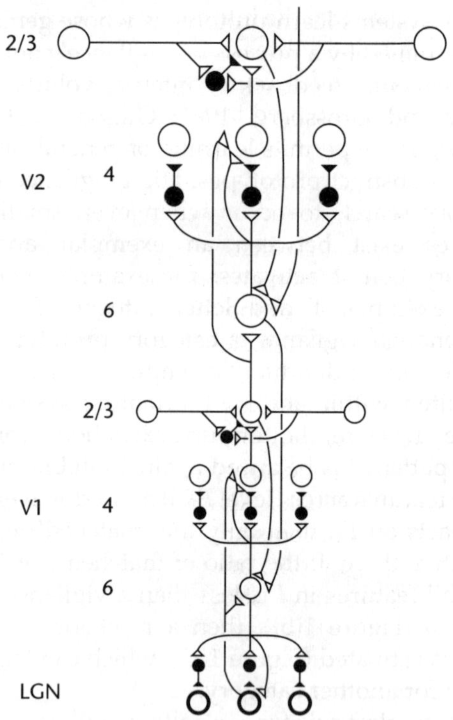

Figure 2. The LAMINART model. The model is a synthesis of feedforward (bottom-up), feedback (top-down), and horizontal interactions within and between the lateral geniculate nucleus (LGN) and visual cortical areas V1 and V2. Cells and connections with open symbols indicate excitatory interactions, and closed symbols indicate inhibitory interactions. The top-down connections from level 6 of V2 to level 6 of V1 indicate attentional feedback. (See Grossberg, 1999a and Grossberg and Raizada, 2000 for further discussion of how these circuits work.) (Adapted with permission from Grossberg and Raizada, 2000.)

layer 4 of V1. In this circuit, an excitatory cell (open circle) in layer 6 excites the excitatory cell immediately above it in layer 4 via the vertical pathway from layer 6 to layer 4 that ends in an open triangle. This excitatory interaction constitutes the 'on-center'. The same excitatory cell in layer 6 also excites nearby inhibitory cells (closed black circles) which, in turn, inhibit cells in layer 4. This spatially distributed inhibition constitutes the 'off-surround' of the layer 6 cell. The on-center is predicted to have a modulatory, or sensitizing, effect on layer 4, due to the balancing of excitatory and inhibitory inputs to layer 4 within the on-center. The inhibitory signals in the off-surround can strongly suppress unattended visual features. This arrangement shows how top-down attention can sensitize the brain to prepare for expected information that may or may not actually occur, without actively firing the sensitized target cells and thereby inadvertently creating hallucinations that the information is already there. When this balance breaks down, model 'hallucinations' may indeed occur, and these have many of the properties reported by schizophrenic patients.

CONCLUSION

Adaptive resonance theory is a neural and a cognitive theory of human and animal information processing. ART proposes how the processes whereby the brain can stably develop in the infant and learn throughout life constrain the form of perceptual and cognitive processes such as categorization, expectation, attention, synchronization, memory search, and consciousness in both normal and clinical patients. ART realizes a mechanistic unification of concepts about exemplar, prototype, distributed, symbolic, and rule-based processing. Recent models have shown how predicted ART matching properties may be realized in certain laminar circuits of visual cortex, and by extension in other sensory and cognitive neocortical areas.

Acknowledgments

Supported in part by the Defense Advanced Research Projects Agency and the Office of Naval Research (ONR N00014-95-1-0409), the National Science Foundation (NSF IRI-97-20333), and the Office of Naval Research (ONR N00014-95-1-0657).

References

Carpenter GA and Grossberg S (1991) *Pattern Recognition by Self-Organizing Neural Networks.* Cambridge, MA: MIT Press.

Carpenter GA and Grossberg S (1994) Integrating symbolic and neural processing in a self-organizing architecture for pattern recognition and prediction. In: Honavar V and Uhr L (eds) *Artificial Intelligence and Neural Networks: Steps Towards Principled Prediction,* pp. 387–421. San Diego, CA: Academic Press.

Chelazzi L, Duncan J, Miller EK and Desimone R (1998) Responses of neurons in inferior temporal cortex during memory-guided visual search. *Journal of Neurophysiology* **80**: 2918–2940.

Grossberg S (1999a) How does the cerebral cortex work? Learning, attention, and grouping by the laminar circuits of visual cortex. *Spatial Vision* **12**: 163–186.

Grossberg S (1999b) The link between brain learning, attention, and consciousness. *Consciousness and Cognition* **8**: 1–44.

Grossberg S and Merrill JWL (1996) The hippocampus and cerebellum in adaptively timed learning, recognition, and movement. *Journal of Cognitive Neuroscience* **8**: 257–277.

Grossberg S and Raizada RDS (2000) Contrast-sensitive perceptual grouping and object-based attention in the laminar circuits of primary visual cortex. *Vision Reseach* **40**: 1413–1432.

Knowlton BJ and Squire LR (1993) The learning of categories: parallel brain systems for item memory and category knowledge. *Science* **262**: 1747–1749.

Miller EK, Erickson CA and Desimone R (1996) Neural mechanisms of visual working memory in prefrontal cortex of the macaque. *Journal of Neuroscience* **16**: 5154–5167.

Miller EK, Li L and Desimone R (1991) A neural mechanism for working and recognition memory in inferior temporal cortex. *Science* **254**: 1377–1379.

Nosofsky RM and Zaki SR (2000) Category learning and amnesia: an exemplar model perspective. In: *Proceedings of the 2000 Memory Disorders Research Society Annual Meeting*. Toronto.

Posner MI and Keele SW (1970) Retention of abstract ideas. *Journal of Experimental Psychology* **83**: 304–308.

Otto T and Eichenbaum H (1992) Neuronal activity in the hippocampus during delayed non-match to sample performance in rats: evidence for hippocampal processing in recognition memory. *Hippocampus* **2**: 323–334.

Further Reading

Clark EV (1973) What's in a word? On the child's acquisition of semantics in his first language. In: Morre TE (ed.) *Cognitive Development and the Acquisition of Language*, pp. 65–110. New York, NY: Academic Press.

Goodale MA and Milner D (1992) Separate visual pathways for perception and action. *Trends in Neurosciences* **15**: 20–25.

Grossberg S (2000) How hallucinations may arise from brain mechanisms of learning, attention, and volition. *Journal of the International Neuropsychological Society* **6**: 583–592.

Lynch G, McGaugh JL and Weinberger NM (eds) (1984) *Neurobiology of Learning and Memory.* New York, NY: Guilford Press.

Mishkin M, Ungerleider LG and Macko KA (1983) Object vision and spatial vision: Two cortical pathways. *Trends in Neurosciences* **6**: 414–417.

Sokolov EN (1968) *Mechanisms of Memory.* Moscow: Moscow University Press.

Squire LR and Butters N (eds) (1984) *Neuropsychology of Memory.* New York, NY: Guilford Press.

Vinogradova OS (1975) Functional organization of the limbic system in the process of registration of information: facts and hypotheses. In: Isaacson RL and Pribram KH (eds) *The Hippocampus*, vol. II. pp. 3–69. New York, NY: Plenum Press.

Addiction

Introductory article

George V Rebec, Indiana University, Bloomington, Indiana, USA

CONTENTS

The essence of addiction is persistent performance of a behavior despite increasingly aversive consequences. This definition is most commonly used to describe the compulsive taking of drugs. An addict endures deteriorating health, social isolation, and other setbacks in maintaining a drug habit. In effect, addiction is the loss of control over drug use.

ADDICTIVE PROPERTIES OF DRUGS

Many substances, whether they occur in nature or are synthesized in laboratories, are potentially addictive. Intravenously administered heroin and the smokable forms of cocaine ('crack') and methamphetamine ('ice') stand out as drugs that pose a high risk of addiction. But the risk also extends to routinely available substances, such as nicotine and alcohol.

What makes drugs addictive? Part of the answer involves rapid entry into the brain. For a drug like heroin, rapid brain entry often elicits a brief, but intense, rush or 'high' that can drive the search for more of the drug. In fact, heroin has a higher potency than morphine even though heroin is converted to morphine once it enters the brain. The difference is that heroin has additional methyl groups that allow it to slip across membrane barriers more readily than morphine. The invention of the hypodermic syringe, which can deposit drugs directly into the bloodstream for ready access to the brain, became a key factor in the spread of heroin addiction at the end of the nineteenth century. Drugs that are inhaled or smoked also gain rapid access to the brain. Alcohol, because of its relatively simple molecular structure, is absorbed across mucus membranes and begins entering the blood and brain even before it reaches the stomach.

But rapid brain entry of a drug is not the only factor contributing to addiction, nor is the drug itself. Not all heroin and cocaine users become addicts. Many tobacco smokers and social drinkers also avoid compulsive use. Addiction develops only after regular, repeated exposure to a drug on a chronic basis. The factors that control such behavior are many, ranging from physiology to family history, and they interact in complex ways that thwart the search for a simple explanation.

There are certain features of addictive drugs, however, that make them more likely to be abused than other substances. One of these features is the ability to elicit pleasure or reward.

The Reward Model

That drug reward might play a role in addiction gained widespread acceptance after the demonstration that animals, equipped with appropriate catheters, would work to obtain drugs given intravenously. Rats, for example, will readily and repeatedly press a lever to self-administer many of the same drugs as humans. This was an important demonstration because it showed that there was nothing uniquely human about compulsive drug use. In fact, the animal experiments showed that drugs shared many similarities with naturally occurring rewards such as food, water, or sex. A hungry rat, for example, will work very hard for food just as it will go to great lengths for an intravenous injection of heroin or cocaine. Thus, drugs came to be viewed as positive reinforcers, subject to the same principles that govern the behavioral response to other positive reinforcers like food.

It should be noted that positive reinforcement is not necessarily the same as pleasure or reward. Reinforcement simply refers to a procedure that strengthens or increases the likelihood of a given behavioral response. It is impossible to know what the rat actually experiences. Although discussions of drug addiction often use 'reinforcement' and 'reward' interchangeably, these terms are not always equivalent.

When animal research revealed the powerful reinforcing properties of addictive drugs, the next step was to investigate the brain mechanisms underlying this behavior. Research revealed that the brain was equipped with a circuit that appeared to mediate the behavioral response to natural reinforcers as well as drugs. Although the investigation continues, now bolstered by powerful brain imaging techniques applied to human subjects, the reinforcing effects of drugs have been linked to specific neuronal and biochemical changes.

The Reward Circuit

The brain circuit that appeared to signal reinforcement, and that now has become known as the reward circuit, was identified in the 1950s when it was shown that rats would work to stimulate electrodes implanted in the medial forebrain bundle. Although this bundle contained fibers connecting a wide array of structures, attention centered on a group of axons extending from the ventral tegmental area in the midbrain to several forebrain structures in the limbic system and cerebral cortex. The axons were found to release dopamine as a transmitter, and it was dopamine that seemed to play a key role in virtually all forms of motivated behavior. In fact, the mesocorticolimbic dopamine pathway, as it came to be known, also appeared to be highly sensitive to drugs of abuse. Opiates, like heroin and morphine, and psychomotor stimulants, like cocaine and the amphetamines, all increased dopamine transmission. The same was found for nicotine and alcohol. Not only did all these drugs appear to elevate the synaptic level of dopamine, they were most likely to have this effect in the nucleus accumbens, an important target of the mesocorticolimbic pathway known to process autonomic, emotional, and cognitive signals. When blockade of the accumbal dopamine system was found to block drug self-administration behavior in rats, the link between drug reinforcement and accumbal dopamine transmission was firmly established.

Although all the major drugs of abuse increase dopamine transmission, they do so in different ways. Opiates promote the release of dopamine by stimulating receptors that normally respond to opiate peptides found naturally in the brain, the so-called endorphins. Although there are at least three different receptors that respond to endorphins, opiate drugs primarily activate the mu receptor. One effect of this activation is an increase in dopamine release. Nicotine and alcohol promote dopamine release by acting at different receptors: nicotine stimulates a group of receptors that respond to acetylcholine, and alcohol interacts with the major receptor for GABA, an amino acid found throughout the brain. Both acetylcholine and GABA, like many of the endorphins, are transmitters that play a role in dopamine release. When opiates, nicotine, or alcohol are introduced into this system, their receptor effects overwhelm any naturally occurring activity and dopamine is released in abnormally high amounts.

Psychomotor stimulants, on the other hand, increase dopamine transmission by interacting with a protein that transports newly released dopamine back into the neuron for re-release. Cocaine prevents the transporter from working and amphetamine forces it to operate in reverse. Thus, cocaine allows dopamine to accumulate outside the neuron and amphetamine causes dopamine release. The net effect of either drug is an increase in synaptic dopamine.

A drug-induced increase in dopamine transmission, however, is only the beginning of the reinforcement story. The mesocorticolimbic pathway is embedded in a complex network of structures, each of which processes information relevant to drug-induced behavior such as emotional state, environment, past experience, and many other key variables. Investigations at the membrane level, moreover, reveal that rather than exerting a powerful excitatory or inhibitory influence on individual neurons, dopamine modulates the effects of other transmitters, serving more as a filter or gain mechanism than as a conveyor of specific information. In addition, some neurons, such as the endorphin system, may respond to drugs of abuse independently of a change in dopamine. Thus, the drug-induced dopamine signal may be only one of many that contribute to what might become a sense of pleasure or reward. The reinforcing effects of most drugs are likely to involve both dopamine-dependent and dopamine-independent neuronal systems.

Whatever role dopamine, endorphins, and other transmitters play in drug reward, the reward model itself is simply a starting point for understanding addiction. Compulsive drug use is not driven by euphoria alone. Many addicts actually report a loss of pleasure after chronic drug use, but the habit persists. A case can be made that liking a drug, which may parallel drug-induced reward, is entirely different from wanting a drug, sometimes described as the craving that overwhelms an addict's behavior. Noteworthy in this regard is evidence that the transition to compulsive drug use may reside in the neuronal changes that occur when the brain reward circuit is exposed to drugs on a chronic basis.

Neuroadaptations Underlying Addiction

One effect of chronic administration of drugs of abuse is a change in the responsiveness of dopamine receptors. Although these receptors exist in multiple forms, they can be grouped into what are known as D1 or D2 families. Both families are *metabotropic* in that their activation leads to intracellular metabolic changes that regulate membrane excitability and gene expression. Whereas the D2 group appears to be critical for normal motor behavior, the D1 family may play a greater role in the neural adaptations that accompany learning. These same adaptations may also be involved in the behavioral changes that accompany repeated exposure to drugs of abuse. In fact, addiction itself is likely to involve some type of learning-related change in the behavioral response to a drug. By disrupting the normal flow of dopamine transmission, and thus changing D1 receptors, drugs of abuse may set in motion a series of neural events that can lead to addiction. A drug-induced change in this receptor, such as a down-regulation or loss of sensitivity, can occur within minutes of an amphetamine injection. This may explain why the first administration of psychomotor stimulants, which are typically taken in a series of administrations known as a run or binge, has the greatest euphoric effect. As the run continues, the drug-induced high loses its intensity. Over a series of runs, however, prolonged stimulant exposure appears to up-regulate or enhance the sensitivity of D1 receptors, and this effect may persist for weeks after the last drug administration.

Stimulation of the D1 receptor family also leads to a change in gene expression and ultimately the production of cellular proteins. Up-regulation of these receptors, therefore, could mean not only an increase in synaptic transmission but also long-term structural changes. In fact, changes in dopamine synaptic structure have been reported in rats after amphetamine exposure. By usurping the dopamine system and driving it to excess, addictive drugs may permanently alter the flow of information through limbic and cortical circuits in a way that increases the likelihood of further drug use.

CONTROL AND TREATMENT OF ADDICTION

By the time an addict appears for treatment, the drug habit is likely to have persisted for years. During this time, there has been ample opportunity for the problems that typically accompany addiction – family, legal, medical, occupational, and social troubles – to complicate and confuse the treatment process. The greater these accompanying problems are, the more difficult it will be for treatment to succeed.

The first step towards rehabilitation involves detoxification, the removal of the addictive drug from the patient's system. The next step, preventing a relapse, is far more difficult, but various behavioral approaches can be effective. These include teaching coping skills or attempting to desensitize an addict to the cues associated with drug-taking. But if chronic exposure to addictive drugs causes lasting changes in critical brain circuits, as ample evidence suggests, then medication may occupy an important place in the rehabilitation effort.

Detoxification

For heroin addicts, detoxification is best accomplished in conjunction with a drug such as methadone, which acts as a mild heroin substitute. Gradually decreasing the maintenance dose of methadone is part of the detoxification strategy. For nicotine addicts, detoxification can be achieved relatively simply by gradually reducing the dose of nicotine delivered by skin patch, chewing gum, or nasal spray. In some cases, a low, maintenance dose of nicotine may continue for several months to encourage abstinence. Detoxification is especially important for alcoholics because there is danger of death from overdose. In addition, an alcoholic suffers convulsions and other life-threatening reactions that increase in severity each time the drug is withdrawn. To minimize these effects, alcohol detoxification includes treatment with one or more benzodiazepines, which act as sedatives or minor tranquilizers. For cocaine or amphetamine addicts, detoxification becomes important in the event of an overdose, which can be lethal. Drugs can be used to counteract the racing heart, high blood pressure, and other dangerous effects associated with stimulant overdose.

Medication Strategies

Medication is used in one of three basic strategies to keep an addict off drugs: antagonizing or blocking the effect of the addictive drug; substituting another drug from the same class as the addictive drug but with less powerful effects; or the use of medication specifically designed to reduce craving. Medication is now available for patients dependent on opiates, nicotine, and alcohol.

In the treatment of opiate addiction, the antagonist strategy involves administering a drug that has a high affinity for the mu receptor but does not cause the same cellular reaction as heroin or morphine. One such drug is naltrexone. Rather than activating the mu receptor, naltrexone acts as a mu antagonist, binding to the mu receptor but preventing it from working. When given to heroin addicts, for example, naltrexone blocks the effects of a subsequent heroin injection. The problem with such treatment, however, is that all the subjective effects of heroin that an addict has come to expect are prevented. Another problem is that if heroin is still present in the body, naltrexone creates an immediate withdrawal syndrome. Thus, unless an addict is firmly committed to overcoming the habit, the antagonist strategy is rarely successful.

Most heroin addicts prefer the substitution strategy. In this case, the drug substituted for heroin acts like heroin itself in that it also stimulates the mu receptor. Unlike heroin, however, the substituted drug has a relatively slow onset of effects and stays attached to the receptor for prolonged periods of time. Thus, the addict experiences some heroin-like effects but in relatively mild form. Moreover, because the substituted drug stays attached to the receptor for a day or more, a subsequent injection of heroin during this time will not elicit the intense rush or high likely to trigger more craving. Methadone was introduced in the 1960s as the first such substitution drug for heroin addicts. Now, even longer-lasting substitutes, such as buprenorphine and L-acetylmethadol (LAAM), have been developed, which may attach to the mu receptor for up to three days. More than 100,000 heroin addicts in the USA are currently being treated with methadone or long-acting opiate substitutes.

Naltrexone is sometimes used in the treatment of alcoholism to reduce craving. This strategy emerged from evidence that endorphins play a role in reward and that blockade of mu opiate receptors decreased responding for alcohol in animals. It may be that the GABA system, which is sensitive to alcohol, interacts with endorphins in the forebrain reward circuit. That naltrexone can reduce craving in human alcoholics supports this view. Another anti-craving compound is acamprosate, which is used in Europe to reduce relapse in detoxified alcoholics. Although some of its brain actions are still obscure, acamprosate is known to modulate neuronal excitability in the nucleus accumbens by interacting with specific groups of glutamate and GABA receptors. Interestingly, nicotine craving can be reduced by bupropion, a drug commonly used to treat mild to moderate cases of clinical depression. Bupropion has a wide range of effects, but its ability to block nicotine receptors and modulate dopamine transmission may play key roles in reducing nicotine craving.

A fourth medication strategy, the use of nausea-inducing drugs, is available for treating alcoholics. They can obtain a prescription for disulfiram (antabuse), a drug that interferes with the normal metabolism of alcohol, creating an abundance of acetaldehyde. A high level of acetaldehyde causes headache, vomiting, disorientation, and other signs of nausea that are so repulsive that the impulse to drink disappears. The problem with this treatment strategy is compliance; an alcoholic who wants to drink again can simply stop taking disulfiram.

The development of medication for treating an addiction to cocaine or other stimulants is still in its infancy, but several possibilities are being explored. One is akin to the methadone strategy for heroin addiction and involves developing a drug that activates dopamine receptors to produce a partial stimulant-like effect with the hope that the addict could be gradually weaned away from the medication without a reinstatement of craving. Another approach is to interfere with the ability of cocaine or amphetamine to reach their main site of action, the dopamine transporter protein. In this case, the drugs would lose the dopamine-enhancing effect that may underlie addiction. No dramatic successes have emerged with either approach in clinical trials, but there is still much to be learned about the neuropharmacology of the dopamine system and how this system interacts with other transmitters. New medications continue to be developed, and further study of their mechanism of action is likely to lead to new strategies for treatment.

PHYSICAL AND PSYCHOLOGICAL ADDICTION

For most of the twentieth century, addiction was explained in terms of how the body responded when the drug was no longer available. The more severe the response, the stronger the addiction. Consider heroin, a drug that causes a wide range of effects, including analgesia, muscle relaxation, suppressed gag reflex, low core body temperature, and constipation. When the drug is stopped or withdrawn, an addict experiences exactly the opposite sensations: pain, tension, retching, fever, and diarrhea. These withdrawal symptoms, which could last for days or weeks depending on how much and for how long the drug was used, were considered a sign of addiction or, more appropriately, physical dependence. Without the drug, the body became physically sick. To avoid this condition, more of the drug was required, thus perpetuating addiction.

Physical dependence was also used to explain addiction to alcohol and nicotine because of the unpleasant physical reactions that occur when these drugs are withdrawn. In fact, a drug was not considered addicting unless there were clear and profound signs of physical dependence. The problem with this model is that it cannot explain addiction to many drugs, including stimulants like cocaine and the amphetamines. Apart from fatigue or depression, there are no profound withdrawal symptoms associated with these drugs, yet they are strongly addicting. To deal with this issue, some theorists proposed the concept of psychological dependence, the notion that drug withdrawal could lead to a mental or psychological state that triggered addiction. This model was even more difficult to accept because the concept of a mental or psychological addiction was impossible to define and thus impossible to study empirically. It never caught on as a useful model of addiction.

Remnants of the psychological dependence model, however, are apparent in the use of such terms as 'reward' or 'positive reinforcement' to explain a drug habit. Such terms are invoked when the compulsion to take drugs cannot be explained by physical dependence. In fact, even the physical dependence model cannot explain why former heroin addicts or alcoholics can relapse months or years later, long after the withdrawal syndrome has passed. Interesting in this regard is evidence that animals will work very hard to obtain addictive drugs, even at doses that fail to elicit signs of physical dependence. Thus, although physical dependence is a legitimate reason for why some addicts maintain a drug habit, the model has limited applications. As brain research has revealed, drug craving is a long-term process that most probably involves a dysfunction of the neural circuitry underlying motivational behavior. An overriding question is why only some persons exposed to addictive drugs become addicts.

PREDISPOSITION TO ADDICTION

There appear to be certain risk factors that make some individuals especially vulnerable to addiction. These include age, mental state, and personality type, as well as a variety of environmental and genetic conditions. No single factor by itself is a good predictor of addiction but as the number of risk factors increases, the likelihood of addiction also increases.

The highest rates of illicit drug use occur among 18- to 25-year-olds. Experimenting with drugs of abuse at earlier ages, including the pre-teen years, increases the chance of addiction. Some addicts also have co-existing psychiatric conditions, such as depression or schizophrenia, that may impair judgment. Although the psychiatric problem sometimes precedes the addiction, it is most often the case that the addiction develops first. Personality traits may also contribute to addiction, but the relationship is difficult to specify because, as with psychiatric problems, it is not clear if abnormalities in personality are a cause or a consequence of the addiction. In some cases, however, personality influences on drug-taking behavior have been studied in controlled laboratory settings, and the data suggest a problem for adventurous and antisocial personalities, the so-called novelty-seekers or risk-takers. Their lack of impulse control not only puts them in vulnerable situations with respect to drug use but may also complicate rehabilitation. The antisocial personality, for example, correlates with poor outcome in methadone maintenance programs.

Environmental influences on drug-taking behavior take many forms and range from parenting practices to the prevalence of public education efforts. In fact, efforts at informing people about the dangers of drugs are at their peak during times of peak drug abuse and then decline as drug abuse declines, which may help pave the way for the start of another peak period of drug abuse. Wide swings in drug use, such as a sixfold variation in alcohol consumption in the United States over the course of the twentieth century, largely reflect changes in social attitudes and public policy. The context in which drugs are taken is also important. This point was nicely illustrated during the Vietnam war in the 1960s and 1970s. Many United States servicemen became addicted to heroin while serving in Vietnam but either stopped the habit completely or dramatically lowered their heroin use when they returned home. In Vietnam, high-quality opiates were readily available at a cheap price, and any disapproving family members were likely to be thousands of miles away. These factors, combined with the high stress of battle conditions, made it easy to justify drug use. Many servicemen, moreover, disconnected their one-year tour of duty in Vietnam from the rest of their lives. Most, if not all, of these contributing factors disappeared when the tour of duty ended.

Genetics also may contribute to drug addiction, but genes alone are not destiny. In fact, only one in five of those genetically at risk for alcoholism actually become alcoholic. Another consideration is that even if the genetic influence to addiction is critical, there is likely to be more than one genetic factor involved. Genetics, for example, can contribute to the rate and efficiency at which a drug is metabolized as well as to how receptor proteins in the brain respond to a drug. Both of these factors will influence how a drug alters behavior. The strength of that influence, moreover, will depend on the influence of all the other risk factors that contribute to addiction.

Some attempts to specify the potential influence of genetic and nongenetic risk factors involve work with animal models. In this type of research, mice and rats are strategically bred to establish genomic correlations with drug-induced behavioral responses, to identify neurobiological mechanisms, and to localize the chromosome associated with specific drug-induced behavioral traits. In many cases, the identification of drug-response genes in mice indicates the appropriate gene location on human chromosomes. This work is being carried out for all major drugs of abuse and holds promise for assessing the genetic and environmental risk factors underlying addiction.

Further Reading

Berke JD and Hyman SE (2000) Addiction, dopamine, and the molecular mechanisms of memory. *Neuron* **25**: 515–532.

Berridge KC and Robinson TE (1998) What is the role of dopamine in reward: hedonic impact, reward learning, or incentive salience? *Brain Research Reviews* **28**: 309–369.

Koob GF and LeMoal M (2001) Drug addiction, dysregulation of reward, and allostasis. *Neuropsychopharmacology* **24**: 97–129.

Leshner AI (1997) Addiction is a brain disease, and it matters. *Science* **278**: 45–47.

Nestler EJ and Aghajanian GK (1997) Molecular and cellular basis of addiction. *Science* **278**: 58–63.

O'Brien CP (1997) A range of research-based pharmacotherapies for addiction. *Science* **278**: 66–70.

Pickens RW, Elmer GI, LaBuda MC and Uhl GR (1996) Genetic vulnerability to substance abuse. In: Schuster CR and Kuhar MJ (eds) *Pharmacological Aspects of Drug Dependence*, pp. 3–52. [*Handbook of Experimental Pharmacology*, vol. 118.] Berlin, Germany: Springer-Verlag.

Porrino LJ and Lyons D (2000) Orbital and medial prefrontal cortex and psychostimulant abuse: studies in animal models. *Cerebral Cortex* **10**: 326–333.

Tarter RE, Ammerman RT and Ott PJ (eds) (1998) *Handbook of Substance Abuse: Neurobehavioral Pharmacology*. New York, NY: Plenum.

White FJ and Kalivas PW (1998) Neuroadaptations involved in amphetamine and cocaine addiction. *Drug and Alcohol Dependence* **51**: 141–153.

Addiction, Neural Basis of

Intermediate article

Roy A Wise, National Institute on Drug Abuse, Baltimore, Maryland, USA

CONTENTS

Addictive drugs affect brain function by acting at specialized receptors for the endogenous chemical messengers that affect communication between nerve cells. The understanding of addiction depends on our understanding of these receptors, the brain circuits they are embedded in, and the functions that these circuits normally serve.

INTRODUCTION

Addiction is a term that is widely used but poorly defined, even by specialists (Maddux and Desmond, 2000). It is most commonly used to refer to compulsive drug-seeking and drug-taking behaviors, particularly when those behaviors are continued, despite repeated attempts to change, in the face of clearly harmful consequences. The term is also used increasingly to refer to compulsive eating, compulsive gambling, compulsive sexual behavior, and other compulsions that are maintained despite harmful consequences.

The 'official' definitions of the World Health Organization and the *Diagnostic and Statistical Manual of Mental Disorders* of the American Psychiatric Association continue to change as theoretical, legal, and political factors influence the committees in charge. The major problem is that the distinction between 'ordinary' habits and the habits termed 'addiction' is subjective and quantitative rather than based on principle. There was considerable debate in the 1970s as to whether cocaine and nicotine are addictive; by the 1990s it was widely accepted that they are. Part of the problem was theoretical; in the 1950s and 1960s physiological dependence was taken as the defining property of addiction. However, it has become clear that cocaine and nicotine can establish habits as compulsive and harmful as those associated with heroin and alcohol, despite their lack of dramatic dependence syndromes. As the conceptual strength of dependence theory eroded (see below), addiction became defined less by physiological criteria than by subjective ones. The criteria for the terms 'compulsive' and 'harmful', central to the contemporary definition of addiction, are largely matters of personal judgment.

VIEWS OF ADDICTION

Negative Reinforcement View

Lower animals can be trained to work compulsively for intravenous injections of such drugs as cocaine, nicotine, amphetamine, and heroin; thus these drugs can serve as reinforcers, stamping in response habits much as food does for a hungry animal. Despite the fact that lower animals are not subject to the peer pressures or the stresses of poverty or city life that are often blamed for human addiction, it appears that all mammals are at risk of addiction to such drugs as cocaine and heroin. Laboratory rats and monkeys learn to self-administer intravenous heroin and will do so to the point of physical dependence; they learn to self-administer intravenous cocaine, and will do so to the point of death. Thus the observation that addictive drugs are reinforcers has become the common denominator of contemporary addiction theory.

A drug can be viewed as reinforcing because it is an extra 'treat' in one's life – like an after-dinner mint – or because it is a 'treatment' for a need state – like the aspirin that alleviates headache or the meat and potatoes that alleviate hunger and restore energy balance. The once-dominant negative reinforcement view of addiction held that drug-taking becomes compulsive when the nervous or metabolic system has adapted to continued use of a drug and the drug has become necessary for normal bodily homeostasis. The obvious and objective physiological distress of an opiate addict in the early stages of drug withdrawal – the defining evidence of physical dependence – gave rise to this view (dependence theory), which dominated addiction theory until quite recently (Wise, 1987). In this view, initial drug-taking was attributed to peer pressure, thrill-seeking, or simple boredom, but subsequent drug-taking was seen as reflecting the need to self-medicate the withdrawal syndrome that soon developed. Because the dependence syndrome becomes stronger with continued drug use, apparent tolerance develops to dependence-producing drugs and progressively stronger doses of the drug are required to alleviate withdrawal distress. A variation of dependence theory, the self-medication hypothesis, raised the possibility that some individuals have preexisting conditions of stress or anxiety that, like withdrawal symptoms, are medicated by addictive drugs. This view was offered to explain why not all individuals who try addictive drugs come to take them compulsively. The view that all addiction reflects self-medication of preexisting distress syndromes has largely been discounted on evidence that happy, healthy animals, healthy suburban human adolescents – and, indeed, physicians – appear to be as readily addicted to cocaine or opiates as are the inner-city adolescents who were once the primary concern of addiction specialists.

Dependence theory, at least in its classic form, has largely been discredited as a sufficient explanation of addiction. Injections of heroin, the prototypical addictive drug, can establish compulsive drug-seeking and drug-taking even when it is available for too short a portion of each day to establish the classic opiate dependence syndrome (Deneau *et al.*, 1969). Indeed, it has been found that the classic opiate dependence syndrome is alleviated by drug injections into the periaqueductal gray matter, whereas the reinforcing effects of the drug are caused when the drug is injected into the nearby ventral tegmental area (Bozarth and Wise, 1984). People with alcoholism often forgo alcohol during periods of maximum withdrawal distress, only to begin working for alcohol when withdrawal stress has largely subsided. Thus the notion that drug-taking is compulsive only when needed to alleviate withdrawal distress does not fit with the basic facts of even the most classic addictions: those of opiates and alcohol.

Nor does classic dependence theory fit with the facts of compulsive use of nicotine, cannabis, or cocaine. Extensive use of these drugs does not produce the classic somatic dependence syndromes seen with the opiates, barbiturates, benzodiazepines, or alcohol. The withdrawal symptoms associated with barbiturates, benzodiazepines, and alcohol are similar for all three classes of drugs, and are alleviated by drugs from any of the other classes, suggesting a common mechanism for dependence. The withdrawal symptoms associated with opiates are similar, and are partially alleviated by these agents and thus at least partially mediated by the same mechanisms. However, withdrawal from cocaine, amphetamine, nicotine or cannabis produces mild symptoms by comparison, and in the case of cocaine and amphetamine, the somatic withdrawal symptoms are the converse of those of opiates. The general mood associated with opiate withdrawal is hyperexcitability and irritability, whereas cocaine or amphetamine withdrawal is associated with hypoexcitability and depression.

Thus, although there is a withdrawal syndrome associated with termination of regular use of many drugs, there is no common somatic withdrawal syndrome that can explain the compulsive use of the full range of addictive drugs. For these reasons, classic dependence theory is no longer accepted as an explanation or a defining property of addiction (Wise, 1987). As discussed below, however, a variant of negative reinforcement theory remains influential.

Positive Reinforcement View

An alternative view of addiction holds that the primary motivation for drug-taking is the seeking of euphoria or a drug 'high' (McAuliffe and Gordon, 1974). Rather than simply returning the addicted individual to a normal feeling state, this view holds that the drug is taken because it produces a better-than-normal feeling state. This view is strengthened by the fact that humans and lower animals will work for electrical stimulation of certain brain regions, stimulation that brings a state of pleasure which satisfies no biological need and was never experienced in mammalian evolutionary history. The view that addictive drugs produce elevated states of pleasure also explains, as dependence theory does not, why addictive drugs are strongly habit-forming prior to development of dependence, and why there is such a strong probability of relapse after detoxification.

The negative and the positive reinforcement views are not mutually exclusive. The alleviation of withdrawal distress certainly brings pleasure, and it is difficult to imagine that a larger dose than is needed to alleviate pain would not be desirable to someone who is self-medicating. This would explain the fact that people given methadone to medicate withdrawal distress still desire and often use street heroin when it is available. In the case of heroin addiction, it is reported that initial drug-taking results in drug euphoria, but that as chronic use progresses drug euphoria and positive reinforcement become progressively weaker while withdrawal-associated dysphoria and negative reinforcement become progressively more dominant factors in the maintenance of compulsive drug-seeking.

ADDICTION AND THE MESOLIMBIC DOPAMINE SYSTEM

The 1970s, when injected amphetamine and intranasal cocaine were becoming increasingly popular, brought the positive reinforcement view to the forefront of addiction theory. Cocaine users reported that cocaine caused euphoria and that no great distress was felt when the available supply of the drug had been used up. Nonetheless, the drug was taken compulsively for as long as it was available. It was widely accepted that this drug, when used by normal and successful individuals, was taken to 'get high' and not to self-medicate a withdrawal state or any preexisting abnormal distress state. As cocaine came to be taken by injection or by smoking freebase or 'crack' cocaine, it became much more obvious that this drug was strongly addictive. While many people became addicted to intranasal cocaine, intravenous cocaine or smoked freebase caused addiction much more rapidly, and 'graduation' from nasal use to smoking or injecting became a frequent consequence. Rats and monkeys given unlimited access to intravenous cocaine will take the drug to the point of death, losing a third or more of their body weight with a week or two of drug self-administration, punctuated by minimal sleep and food intake. This weight loss and sleep disturbance are not a withdrawal syndrome, however; they are exacerbated rather than alleviated by continued intoxication.

Cocaine is an inhibitor of monoamine neurotransmitter reuptake. The monoamine neurotransmitters are noradrenaline (norepinephrine), dopamine and serotonin; cocaine shares with amphetamine the ability to elevate the concentration of each of these substances in the junctions between communicating neurons. Whereas amphetamine causes the direct release of these transmitters, cocaine blocks their synaptic inactivation by blocking their reuptake by the cells that released them (thus prolonging and elevating their actions). The drug-induced elevation of extracellular dopamine levels accounts for the habit-forming effects of the cocaine and amphetamine; drugs that block the effects of dopamine block the ability of cocaine or amphetamine to establish drug-seeking response habits or preferences for the places in the environment where the drug has been experienced. Lesions of the mesolimbic branch of the forebrain dopamine projections also block the reinforcing effects of these drugs. Lesions or pharmacological blockade of the noradrenergic or serotonergic systems have no such effects. Dopamine-blocking drugs also eliminate the ability of normally rewarding brain stimulation to establish or maintain intracranial self-stimulation. Thus cocaine and amphetamine are habit-forming because they activate the reward system that was once termed a 'pleasure center in the brain' (see Wise and Bozarth, 1987).

Food, sex, and rewarding brain stimulation, as well as the addictive drugs amphetamine, cocaine, morphine, heroin, nicotine, cannabis, and alcohol, each elevate brain dopamine levels in the nucleus accumbens, where the mesolimbic dopamine system has its main termination. When cocaine, amphetamine, or heroin is self-administered, dopamine levels are elevated severalfold; animals allowed to self-administer these drugs do so whenever their dopamine levels fall to about 200% of normal. Food for hungry animals and sex for experienced males cause dopamine levels to increase, but the increases tend to peak at 150–200% of normal. Thus animals self-administering cocaine, amphetamine or heroin, at least, maintain their dopamine levels at higher values than are usually produced by the normal pleasures of life (Wise, 1998).

SITES OF THE REINFORCING ACTIONS OF ADDICTIVE DRUGS

The major classes of addictive drug act at receptors that are the normal targets of endogenous neurotransmitters or neuromodulators. Nicotine acts at the nicotinic class of receptors for the neurotransmitter acetylcholine. Cannabis acts at the receptors for the neuromodulator anandamide. Phencyclidine acts at the *N*-methyl-D-aspartate (NMDA) class of glutamate receptor. Opiates act at mu, delta, and kappa opioid receptors, receptors for the endogenous opioids enkephalin, β-endorphin, dynorphin, and endomorphin. Cocaine and amphetamine act at transporters for dopamine, noradrenaline, and serotonin, elevating the extracellular levels of these transmitters which then act at their own endogenous receptors.

Each of these drugs acts in multiple anatomical pathways, and thus is involved in multiple physiological functions. The sites of reinforcing actions of some addicting drugs have been localized, and each thus far identified is associated with the mesolimbic dopamine system or with the cells it targets in the nucleus accumbens. Nicotine's reinforcing action involves nicotinic cholinergic receptors localized to the mesolimbic dopamine neurons themselves, stimulating these cells to fire and to release dopamine. Morphine and heroin have reinforcing actions on cells containing γ-aminobutyric acid (GABA) that are found near to the dopamine cells and that normally inhibit dopaminergic cell firing. The opiates inhibit the GABA-containing cells, thereby disinhibiting the dopaminergic cells and increasing their firing rates. The opiates also inhibit, as does dopamine, the medium-sized spiny output neurons of nucleus accumbens.

The psychomotor stimulants amphetamine and cocaine have their reinforcing actions at the dopamine transporters in the nucleus accumbens. Amphetamine reverses the transporter causing it to expel rather than take up dopamine, and cocaine blocks the transporter, blocking the reuptake of dopamine released from cell firing. Dopamine, in turn, acts at its own receptors on nucleus accumbens neurons. Phencyclidine blocks NMDA-type receptors for the excitatory amino acid transmitter glutamate, thus reducing the excitatory input to the medium-sized spiny output neurons of nucleus accumbens. Thus the direct or indirect inhibition of medium spiny neuron output from nucleus accumbens appears to be the critical common consequence for the reinforcing actions of each of these addictive drugs. Phencyclidine is also habit-forming when injected into the medial prefrontal cortex. Again, the drug's ability to block NMDA receptors accounts for its rewarding action, but it is not yet known which population of NMDA receptors or which type of cortical neuron is involved (Wise, 1998).

The sites of reinforcing actions of cannabis, alcohol, barbiturates, benzodiazepines, and caffeine remain to be identified. Some but not all of these are expected to prove to have rewarding actions involving the mesolimbic dopamine system or its associated circuitry.

CONTEMPORARY THEORIES OF ADDICTION

Classic dependence theory, attributing the compulsive dimension of addiction to the need to alleviate the somatic, largely autonomic, symptoms of withdrawal distress, has been described above. While the conscious desire to alleviate somatic withdrawal distress doubtless contributes to drug-taking in people who are addicted, it is no longer seen as the defining property of addiction. Contemporary addiction theories stress the actions of addictive drugs in the reward circuitry of the brain: the circuitry that is activated by the natural pleasures of life as well as by the laboratory rewards of brain stimulation and addictive drugs.

The psychomotor stimulant theory of addiction (Wise and Bozarth, 1987) suggests that the dominant sedative effect of a number of addictive drugs is unrelated to the addictive liability of these drugs. Rather, it postulates that even the addictive depressants and sedatives (e.g. opiates, alcohol, cannabis) activate the brain circuitry of arousal and forward locomotion. Forward locomotion – the central behavioral component of approach behavior – has been postulated to be the unconditioned response to all positive reinforcers. In the cases of amphetamine and cocaine, the prototypic psychomotor stimulants, drug-induced locomotion and stereotyped orofacial movements associated with activation of the mesolimbic and adjacent nigrostriatal dopamine systems dominate the behavior of intoxicated animals. With opiates and alcohol this action is not obvious because it is masked by the dominant depressive actions of strong doses of these drugs. These depressants activate the mesolimbic dopamine system, however; they cause locomotion and act as stimulants at low doses or in the early stages of intoxication, when only low levels of the drug have reached the brain. The psychomotor stimulant theory attributes the habit-forming effects of addictive drugs to their ability to activate the brain mechanisms of reward and approach behaviors, and attributes the high-dose depressant effects of the drugs to actions in other parts of the brain. The psychomotor stimulant theory fits well with current data on cocaine, amphetamine, opiates, and nicotine. While alcohol and cannabis each activates the postulated reward circuitry, the sites and circuitry through which they do so are not yet known. Whether barbiturates or benzodiazepines activate this circuitry is controversial, and evidence suggests that the habit-forming effects of caffeine are likely to involve an independent mechanism. Thus the psychomotor stimulant theory offers a unified theory for some, but probably not all, addictive substances.

The psychomotor stimulant theory also introduced the notion of 'incentive motivation' to addiction theory. Incentive motivation is a construct from learning theory, designed to deal with the circularity of reinforcement theory. A reinforcer is an object or event which, when made contingent upon a given act by a given animal, increases the probability of recurrence of that act. To suggest that reinforcement explains the act it is defined by is circular; it implies the teleological view that the cause of an event can follow the event. It is the reinforcement history, not the coming reinforcer, that explains the probability of the act in question. Incentive motivational theory was an attempt to identify the precipitating precursor of the habitual act. It stressed the role of incentives that are present prior to the act – incentives that are attractive because of their past association with the drug – in eliciting and guiding the act. Incentive motivation is the principle by which addictive drugs are seen to establish conditioned place preferences: learned preferences for the portions of the environment where the drug has previously been experienced.

Advocates of the psychomotor stimulant theory pointed out – largely on the basis of observations from brain stimulation reward studies – that reinforcers not only have the ability to increase the probability of recurrence of acts that precede them, but that they (and their associated environmental predictors) have the ability to 'prime' a response habit, energizing the animal and focusing attention on the previously established habit. This proactive feature of reinforcers is the cornerstone of animal models of relapse to addiction. Because it encompasses the ability of a reward to precipitate anticipatory excitement and behavioral arousal as well as to consolidate the memory trace for recent acts, the psychomotor stimulant theory addresses drug reward (embracing both the proactive and the retroactive processes) rather than simply drug reinforcement.

Opponent process theory (Solomon and Corbit, 1973) is a negative reinforcement theory, a general view that subsumes classic dependence theory. However, whereas classic dependence theory attributed compulsion to the need to medicate somatic withdrawal distress involving the autonomic nervous system, contemporary opponent process theories (Dackis and Gold, 1985; Koob and Bloom, 1988) attribute the compulsive nature of addiction to neuroadaptations in the reward circuitry discussed above. The core postulate of classic opponent process theory is that homeostatic controls adjust in compensation for chronic intoxication, opposing the direct effects of the drug and desensitizing the nervous system to that drug. Such opponent processes are found in all the targets of a given drug, including the thermoregulatory system and neurotransmitter receptors in the gut. It is the unopposed effect of the postulated opponent-process neuroadaptations that explains the fact that the withdrawal symptoms associated with a given drug are the opposite of the acute effects of the drug. Thus, while one of the direct effects of opiates is to constrict the intestine and cause constipation, the compensatory relaxation, which is masked so long as the drug continues to oppose it, results in diarrhea when the drug wears off.

The contemporary versions of opponent process theory hold that the reward pathway itself becomes desensitized by chronic drug intoxication, and becomes progressively more difficult to activate by both normal rewards and by the addictive drug itself. This results in loss of responsiveness to (and, as a consequence, loss of interest in) the normal pleasures of life. It is also seen to result in the need to escalate the dosage in order to achieve the expected drug effect. Opponent process theory offers an explanation of tolerance and dependence, not only in the reward pathway but also in the autonomic nervous system and all the systems activated by a given drug. An early version of this theory that focused on the reward system was the dopamine depletion hypothesis (Dackis and Gold, 1985), which held that one consequence of such opponent processes is a decrease in extracellular levels of mesolimbic dopamine during psychomotor stimulant withdrawal. While there remains controversy as to its magnitude and significance, depletion of extracellular dopamine has been reported in animals withdrawn from cocaine, amphetamine, and opiates. Moreover, there is now evidence of development of multiple drug-induced opponent processes in the reward pathway.

First, animals undergoing withdrawal from amphetamine or cocaine have elevated brain stimulation reward thresholds; that is, it takes stronger stimulation of the reward system to motivate an animal undergoing withdrawal from psychomotor stimulants. While there is controversy as to whether there is tolerance or its opposite (sensitization) to the reward-specific effects of addictive drugs, under some circumstances animals have been shown to escalate drug intake as drug exposure is extended. Also, there is now considerable evidence for intracellular neuroadaptations within the dopamine system and within its target neurons in the nucleus accumbens following repeated cocaine and morphine treatments (Nestler and Aghajanian, 1997). Some of these neuroadaptations can be mimicked experimentally, and they reduce the effectiveness of cocaine in producing conditioned place preferences. Thus it is clear that opponent processes are called into play by chronic use of addictive drugs. It remains to be determined how important a role these neuroadaptations have in compulsive drug-taking. The neuroadaptations are clearly consequences of chronic drug intake, but it is not clear to what extent they become causes of subsequent intake.

Incentive salience theory (Robinson and Berridge, 1993), like contemporary opponent process theory, builds on the psychomotor stimulant theory, attributing the habit-forming actions of drugs of abuse to their ability to activate the reward circuitry associated with the mesolimbic dopamine system. Incentive salience theory, however, differs from opponent process theory in that it invokes proponent processes to explain the compulsive nature of addiction. It is well known that the psychomotor stimulant effects of amphetamine, cocaine, and opiates become progressively stronger with repeated intermittent intoxication. The incentive salience theory postulates that progressively increasing sensitivity to the drug gives drug-associated incentives progressively more control over behavior, and that it is this sensitization that makes drug-seeking compulsive. This sensitization has several known correlates, all involving the mesolimbic dopamine system and its associated afferents and efferents. Again, however, it is not yet clear what causal role psychomotor sensitization plays in the increasingly compulsive drug-seeking habits of addiction.

A major issue highlighted by incentive salience theory is the relative importance of the reinforcing effects of the drug (effects of the drug after it has been taken) and incentive motivational effects of the drug (effects related to the drug history and environmental associations with that history). Incentive salience theory calls attention to this under-appreciated incentive motivational postulate of traditional learning theory by relating the specialist terms 'incentive motivation' and 'reinforcement' to their subjective correlates 'wanting' and 'liking'. One wants the drug prior to having it: this is the cognitive correlate of incentive motivation and is presumably elicited by drug-associated stimuli that are present before the drug is. One likes the drug after having it: this is the presumed cognitive correlate of the reinforcement that stamps in the stimulus–stimulus associations between the drug actions and the environmental stimuli in the situation in which the drug is encountered. Opponents of this view point out that reinforcing events need not involve conscious pleasure: monkeys have been trained to lever-press for painful footshock, and we can learn to drink initially noxious solutions if they contain alcohol.

CONCLUSION

Much has been learned – though much remains to be learned – about the neural basis of addiction. Many addictions result from the ability of the addictive drug to activate primitive brain circuitry involved in the formation of simple and normal response habits such as feeding and sexual activity. Most addictive drugs activate this system and do so more strongly than do the normal pleasures of life. The activation of this system by addictive drugs seems to be associated, at least initially, with pleasure, though the nature of the pleasure is vaguely and variously described by human subjects.

Repeated drug use leads to adaptations within the central and peripheral nervous systems, and these adaptations are usually opposite in direction to the acute effects of the drugs that produce them. When a drug is taken often enough to build up significant neuroadaptations, withdrawal symptoms opposite to the effect of the drug are seen when drug use is terminated. Such withdrawal symptoms can be aversive, and the self-medication of these withdrawal symptoms is reported to be a significant factor in the inability of people with addiction to simply discontinue the use of opiates and barbiturates. Against the claim that self-medication is a sufficient explanation of continued heroin use, however, is the fact that people given methadone to medicate withdrawal distress often continue to use street heroin when it is readily available. Alleviation of withdrawal symptoms seems to play a minimal part in other addictions, and compulsive drug-seeking can be established with some drug regimens that cause minimal signs of addiction.

If repeated use is sufficiently intermittent, the stimulant effects of addictive drugs can become progressively stronger. It has been suggested that this might explain the progressively stronger control of behavior by addictive drugs. Drug tolerance and drug sensitization have each been demonstrated, but the optimal conditions for the two are different. Tolerance most clearly results from chronic intoxication, whereas sensitization results from intermittent intoxication. Since addiction involves periods of prolonged intoxication and also periods of intermittent drug withdrawal, each factor seems likely to play some part in addiction. Pavlovian conditioning – association of the drug with the environment in which the drug is experienced – is known to contribute both to tolerance and to sensitization, and the degree to which each results from simple pharmacological exposure or rather to the interaction of the drug with the environment remains to be clarified.

It is widely assumed that stressful life experiences predispose some individuals to addiction, and the role of stress in addiction is a topic of great contemporary interest. Some forms of stress can clearly reinstate drug-taking behavior in animal models, and a good deal of attention is turning from the factors that establish initial drug habits to the factors that can reinstate these habits once they have been broken. A priming 'taste' of the drug itself (such as just one drink or just one cigarette) is one of the strongest reinstating stimuli, and some stress stimuli are comparably effective. The mechanisms of interaction between stress systems and reward systems have not yet been identified.

The cognitive correlates of the various stages of addiction are complex and not well defined or identified. Pleasure, euphoria, craving, stress, wanting, liking, and satiety are terms frequently attributed to phases of addiction, but it remains to be determined which of these relate to causal factors and which to after-the-fact correlates.

References

Bozarth MA and Wise RA (1984) Anatomically distinct opiate receptor fields mediate reward and physical dependence. *Science* **224**: 516–518.

Dackis CA and Gold MS (1985) New concepts in cocaine addiction: the dopamine depletion hypothesis. *Neuroscience and Biobehavioral Reviews* **9**: 469–477.

Deneau G, Yanagita T and Seevers MH (1969) Self-administration of psychoactive substances by the monkey: a measure of psychological dependence. *Psychopharmacologia* **16**: 30–48.

Koob GF and Bloom FE (1988) Cellular and molecular mechanisms of drug dependence. *Science* **242**: 715–723.

Maddux JF and Desmond DP (2000) Addiction or dependence? *Addiction* **95**: 661–665.

McAuliffe WE and Gordon RA (1974) A test of Lindesmith's theory of addiction: the frequency of euphoria among long-term addicts. *American Journal of Sociology* **79**: 795–840.

Nestler EJ and Aghajanian GK (1997) Molecular and cellular basis of addiction. *Science* **278**: 58–63.

Robinson TE and Berridge KC (1993) The neural basis of drug craving: an incentive-sensitization theory of addiction. *Brain Research Reviews* **18**: 247–292.

Solomon RL and Corbit JD (1973) An opponent-process theory of motivation: II. Cigarette addiction. *Journal of Abnormal Psychology* **81**: 158–171.

Wise RA (1987) The role of reward pathways in the development of drug dependence. *Pharmacology and Therapeutics* **35**: 227–263.

Wise RA (1998) Drug-activation of brain reward pathways. *Drug and Alcohol Dependence* **51**: 13–22.

Wise RA and Bozarth MA (1987) A psychomotor stimulant theory of addiction. *Psychological Review* **94**: 469–492.

Further Reading

Berke JD and Hyman SE (2000) Addiction, dopamine, and the molecular mechanisms of memory. *Neuron* **25**: 515–532.

Carlezon WAJ, Thome J, Olson VG *et al.* (1998) Regulation of cocaine reward by CREB. *Science* **282**: 2272–2275.

Goldstein A (1994) *Addiction: From Biology to Drug Policy.* New York, NY: Freeman.

Nestler EJ (1997) Molecular mechanisms of opiate and cocaine addiction. *Current Opinion in Neurobiology* **7**: 713–719.

Robbins TW and Everitt BJ (1999) Drug addiction: bad habits add up. *Nature* **398**: 567–570.

Shaham Y and Stewart J (1995) Stress reinstates heroin-seeking in drug-free animals: an effect mimicking heroin, not withdrawal. *Psychopharmacology* **119**: 334–341.

Stewart J and Eikelboom R (1987) Conditioned drug effects. In: Iversen LL, Iversen SD and Snyder SH (eds) *Handbook of Psychopharmacology*, pp. 1–57. New York, NY: Plenum.

Wise RA (1996) Neurobiology of addiction. *Current Opinion in Neurobiology* **6**: 243–251.

Wise RA (2000) Addiction becomes a brain disease. *Neuron* **26**: 27–33.

Wise RA, Newton P, Leeb K *et al.* (1995) Fluctuations in nucleus accumbens dopamine concentration during intravenous cocaine self-administration in rats. *Psychopharmacology* **120**: 10–20.

Aesthetics

Intermediate article

Jerrold Levinson, University of Maryland, College Park, Maryland, USA

CONTENTS

Aesthetics is that branch of philosophy devoted to conceptual and theoretical inquiry into art and aesthetic experience.

THE DOMAIN OF AESTHETICS

We may usefully distinguish three conceptions of the domain of aesthetics, according to what is taken as the focus of attention:

- The practice of making and appreciating works of art.
- Aesthetic properties, features, or aspects of things.
- Aesthetic attitudes, perceptions, or experiences.

There are intimate relations among these three conceptions. Thus, art might be conceived as a practice in which people aim to make objects possessing valuable aesthetic properties, or that are apt to give subjects valuable aesthetic experiences; aesthetic properties might be conceived as those properties saliently possessed by works of art, or those on which aesthetic experience is centrally directed; and aesthetic perception might be conceived as the sort of perception that is central to the appreciation either of works of art, or of the aesthetic properties of things, whether natural or man-made. Finally, it can be argued that art, in its creative and receptive aspects, provides the richest and most varied arena for the exploration of aesthetic properties and the enjoyment of aesthetic experiences.

The aesthetics of nature may be included in the second or third of these conceptions, if it is understood as the study of certain distinctive properties of natural phenomena that can be classified as aesthetic (e.g. beauty, sublimity, grandeur), or of certain kinds of experience provoked by nature, or of certain kinds of attitudes to nature. The theory of criticism may be included in the first conception, if it is understood as the study of that part of the practice of art concerned with the reception of artworks, including their description, interpretation, and evaluation. Craft, too, can be understood as an art-related or quasi-artistic activity, and hence may be included in the first conception.

Art

One conception of art sees it as specially concerned with the exploration and contemplation of perceptible form for its own sake. This view has roots in the work of the eighteenth-century German philosopher Immanuel Kant, who thought that the beauty of objects and phenomena, whether natural or man-made, consisted in their ability to stimulate the free play of the cognitive faculties in virtue of their pure forms, both spatial and temporal, and without the mediation of concepts. In the early twentieth century, the English art theorist Clive Bell took a similar line, holding that spatial form was the only artistically relevant aspect of visual art, and that possessing 'significant form' was the necessary and sufficient condition of a work of art.

Another conception of art sees it as essentially a vehicle of expression or communication, especially of states of mind or nonpropositional contents. In the early twentieth century, the Italian philosopher Benedetto Croce claimed that the essence of art is in the expression of emotion. He emphasized the indissociability, even identity, of content and vehicle in art. The English philosopher R. G. Collingwood developed this line further, observing that making works of art was a way for the artist to articulate or make clear the nature of his or her emotional condition. The Russian novelist Leo Tolstoy identified art with emotional communication from one person to another by indirect means, namely, a structure of signs in an external medium.

A third conception of art sees it as concerned with the imitation or representation of the external world, perhaps in distinctive ways or by distinctive means. This conception can be found in the earliest works in the canon of aesthetics, the *Republic* of Plato and the *Poetics* of Aristotle. Modified so as to allow for representation of matters beyond the visible, it finds expression among later thinkers in the aesthetic theories of Lessing, Hegel, and

Schopenhauer. Some modern discussions of art as representation regard it broadly as semiotic or symbolic in nature.

Art has also been conceived as an activity aimed explicitly at the creation of beautiful objects, including representations of natural and human beauty; as an arena for the exhibition of skill, particularly in fashioning or manipulating objects capable of exciting admiration (Sparshott, 1982); as a development of play, stressing the structured and serious aspects of play (Gadamer, 1986); and as the sphere of experience per se, in which attention is drawn to the interplay of active (creative) and passive (receptive) phases in engagement with the external world (Dewey, 1934).

More recently, art has been conceived as the production of objects intended to afford aesthetic experience (Beardsley, 1981); as the investing of objects with 'aboutness' in the context of a specific cultural framework (Danto, 1981); as a particular social institution identified by its constituent rules and roles (Dickie, 1997; Davies, 1991); and as an activity identifiable only historically through a connection to earlier activities or objects whose art status is assumed (Wollheim, 1980; Levinson, 1990, 1996; Carroll, 2001).

Aesthetic Property

It is generally agreed that aesthetic properties are perceptual or observable, experienced in a fairly direct manner, and relevant to the aesthetic value of the objects that possess them. Various further characteristics of aesthetic properties have been proposed, including: having gestalt character; requiring a certain sensitivity for discernment; having an evaluative aspect; affording pleasure or displeasure in contemplation; not being governed by conditions, or applicable by rule; supervenience on lower-level perceptual properties; requiring imagination for attribution; requiring metaphorical thought for attribution; being notably revealed in aesthetic experience; and being notably present in works of art.

Although the relative status of the characteristics is debated, there is substantial intuitive agreement as to which perceivable properties of things are aesthetic. These include, for example: beauty, sublimity, grace, elegance, delicacy, harmony, balance, unity, power, anguish, sadness, tranquility, serenity, and melancholy. It is evident that expressive properties, which arguably belong only to works of art and not to natural objects, constitute a significant subset of aesthetic properties.

Aesthetic Experience

Among the characteristics that have been proposed as distinguishing aesthetic states of mind (whether attitudes, perceptions, emotions, or acts of attention) from others are: disinterestedness, or detachment from desires, needs and practical concerns; non-instrumentality, or being undertaken or sustained for their own sake; contemplation or absorbtion, with consequent effacement of the subject; focus on an object's form; focus on the relation between an object's form and its content or character; focus on the aesthetic features of an object; and centrality in the appreciation of works of art. It is still a matter of debate whether these criteria, either individually or in some combination, adequately define aesthetic experience.

PROBLEMS IN AESTHETICS

Evidently, among the problems aesthetics addresses are the interrelated characterizations of art, aesthetic properties, and aesthetic experience. These broad problems engender many more specific ones.

The issue of the definition of art leads naturally to many further issues: the ontology of art; the process of artistic creation; the demands of artistic appreciation; the concept of form in art; the role of media in art; the analysis of representation and expression in art; the nature of artistic style; the meaning of authenticity in art; and the principles of artistic interpretation and evaluation. The philosophy of art is, in fact, sometimes conceived of as metacriticism, or the theory of art criticism (Beardsley, 1981).

The ontology of art concerns the question of exactly what sort of object a work of art is, and how this might vary between different art forms. Philosophers have asked whether a work of art is physical or mental, abstract or concrete, singular or multiple, created or discovered, notationally definable or only culturally specifiable; and they have asked what authenticity of a work of art consists in.

Interest in the creative process in art concerns the question of whether the creative process can be characterized in any general way, and the relevance of knowledge of the creative process (and more generally of the historical context of creation) to appreciation of works of art.

Issues about artistic form include the status of formalism as a theory of art, the different kinds of form manifested in different art forms, and the relation of form to content and of form to medium.

Among the modes of meaning that inhere in works of art, perhaps the most important are representation and expression. Goodman (1976) argues that exemplification is an equally important mode.

Accounts of representation (usually with special reference to pictorial representation) have been proposed in terms of resemblance between object and representation; perceptual illusion (Gombrich, 1960); symbolic conventions (Goodman, 1976); 'seeing-in' (Wollheim, 1987); world-projection (Wolterstorff, 1980); make-believe (Walton, 1990); recognitional capacities (Schier, 1986); resemblance between experience of object and experience of representation (Budd, 1995; Hopkins, 1998); and information content (Lopes, 1996).

Accounts of artistic expression (usually with special reference to the expression of emotion) have been proposed in terms of personal expression by the artist; induced empathy with the artist; metaphorical exemplification; correspondence (Wollheim, 1987); evocation (Matravers, 1998); imaginative projection (Scruton, 1997); expressive appearance (Kivy, 1989; Davies, 1994); and imagined personal expression (Vermazen, 1986; Levinson, 1996).

Concerning artistic style, attention has focused on the distinction between individual and period style, on the psychological reality of style, on the interplay between style and representational objective, and on the role that cognizance of style plays in aesthetic appreciation.

Concerning the interpretation of art, attention has focused on the relevance of artists' intentions; on the diversity of interpretative aims; on the debate between critical monism and critical pluralism; on the similarities and differences between critical and performative interpretation; and on the relationship between interpretation and maximization of value.

Concerning the evaluation of art, attention has focused on the question of its objectivity or subjectivity; on the relation between artistic value and pleasurability; on the relation between the value of art as a whole and the value of individual works of art; on the existence of general criteria of value across art forms; and on the relevance of a work's historical influence, ethical import, emotional power, and cognitive reward to its evaluation as art.

Certain concepts are relevant to the understanding of many, if not all, works of art: for example, the concepts of intention, fiction, metaphor, genre, narrative, genius, forgery, performance, and tragedy.

There are other questions concerning the relationships between art and other domains or aspects of human life, such as emotion, knowledge, and morality. For example, there are the questions of how we can coherently have emotions for characters whom we know to be fictional; whether art can be a vehicle of knowledge; and whether art can contribute to moral education.

There are also questions relating to particular art forms: for example, whether photography is an inherently realistic medium; whether poetry can be usefully paraphrased; whether the basic form of music is local or global; and whether narration operates similarly in novels and films.

The question of the nature of aesthetic properties leads naturally to questions about realism in relation to such properties; the supervenience relation between aesthetic properties and the non-aesthetic properties on which they depend; the range of aesthetic properties to be found in the natural world; the special status of beauty among aesthetic properties; the difference between the beautiful and the sublime; the degree of objectivity of judgments of beauty; the relations between artistic, natural, and human beauty; and the relationship between the aesthetic properties of artworks and their artistic properties, such as originality or seminality (which, although appreciatively relevant, are not directly perceivable as are aesthetic properties).

Finally, discussions of aesthetic experience open into discussions of the nature of perception, reason, imagination, feeling, memory, and mood, in relation to art or nature.

UNDERSTANDING WORKS OF ART

Categorial Perception

Walton (1970) follows Beardsley and Sibley in taking aesthetic properties to be perceptual, gestalt-like, non-rule-governed, and dependent on an object's lower-level perceptual properties. But Walton insists that aesthetic properties are dependent as well on the artistic categories (for example, of style, genre or medium) into which works of art may be said to fall. The category of a work is partly a matter of art-historical context, including factors such as the artist's intention, the artist's previous work, the artistic traditions in which the artist worked, and the artistic problems to which the artist is responding. If, as Walton argues, a work's aesthetic properties do not reside in perceivable structure alone, it is even more evident that its artistic properties depend on non-perceivable factors.

Artistic Expression

For Goodman (1976), artistic expression is a matter of an artwork exemplifying or drawing attention to some property it metaphorically possesses, in virtue of its general symbolic functioning.

For Tormey (1971), artistic expression is a matter of an artwork's possessing expressive properties that are related to intentional states, and which are ambiguously constituted by the non-expressive structural features underlying them.

Wollheim (1987) suggests that expressiveness in painting is a matter of intuitive correspondence or fit between the appearances that works of art present and feeling states of the subject, which are then projected onto those works in complex ways.

Davies (1994) offers a theory of musical expressiveness in terms of resemblances between musical patterns and human emotional behavior, and explores the variety of responses that listeners have to perceived expressiveness.

Scruton (1997) locates the perception of musical expression in our ability to inhabit 'from the inside' the gestures that music appears to embody, and thus to adequately imagine the inner states correlative with such gestures.

Levinson (1996) accounts for musical expressiveness in terms of music's ready hearability as the personal expression of an indefinite agent or persona.

Pictorial Representation

There are various accounts of our capacity to see what pictures depict and then respond to those depictions in aesthetically relevant ways. Currently the two most influential theories are Wollheim's (1987) 'seeing-in' theory and Walton's (1990) 'make-believe' theory.

Wollheim's theory is a development of Wittgenstein's idea of aspect perception, or 'seeing-as', for example, seeing a gnarled tree as an old woman. 'Seeing-in' applies to the parts of a picture as well as to the picture as a whole; and it involves simultaneous ('twofold') awareness of the picture's surface and the depicted content.

For Wollheim, seeing-in is a primitive visual capacity, at first exercised on natural phenomena, like stained rock faces, and later deliberately harnessed for making images. A large part of the aesthetic interest in pictures is due to the twofoldness of seeing-in, whereby we appreciate what is depicted, in a virtual three-dimensional space, in relation to the real two-dimensional pattern of marks before us.

Walton understands pictures as props in visual games of guided imagining, or make-believe. The configuration of marks that constitutes a picture prompts us to imagine we are seeing an object, and we imagine that our seeing those marks is a seeing of the object. Pictures generate fictional worlds; and what it is correct to imagine seeing in a picture is determined by implicit rules and conventions. In addition, in interacting with a picture visually, the viewer generates transient fictional worlds specific to him or her.

It is not yet clear whether Wollheim's and Walton's proposals are reconcilable. For Walton, Wollheim's seeing-in is to be analyzed in terms of imagined seeing, whereas for Wollheim, seeing-in is an activity prior to and more fundamental than imagined seeing, however important such seeing is in later phases of pictorial appreciation.

AESTHETICS AND COGNITIVE SCIENCE

Underlying aesthetic experience are certain mental states and processes: those involved in creating, perceiving, understanding and appreciating works of art. Accordingly, much recent work in aesthetics has taken into account empirical research on the human mind.

Pictorial Perception

Schier (1986) appeals directly to facts about ordinary visual processing in support of a theory of pictures. He proposes that a representation is pictorial just insofar as it recruits the visual recognitional capacities subjects already possess for familiar objects, so that a picture represents an object *O* if it triggers in subjects who view it the same capacities for recognition that would be triggered by the sight of *O* in the world. Schier emphasizes that pictorial competence, unlike language learning, is characterized by 'natural generativity', whereby once a subject can decipher a few pictures of a given sort, he or she can generally decipher any number of such pictures, however novel.

Lopes (1996) maintains that the essence of pictorial representation is the furnishing of similar visual information by picture and object. He proposes an 'aspect-recognition' theory of depiction, according to which successful pictures embody nonconceptual aspectual information sufficient to trigger recognition of their objects in suitable perceivers. Developing a theme of Gombrich (1960), Lopes proposes that the essence of depiction as a mode of representation is its inevitable selectivity, so that a picture of whatever style (unlike a description) is explicitly noncommittal about certain represented properties of its object, precisely in virtue of being explicitly committal about others.

Lopes (1997) argues for the possibility of purely tactile pictures (though he overlooks certain experiential asymmetries between tactile and visual pictures (Hopkins, 2000)). And Lopes (1999) draws on color perception theory to demonstrate how pictures depict the colors of the world without actually replicating them.

Musical Comprehension

Raffman (1993) investigates aspects of the apparent ineffability of music in terms of facts about the mental processing of music. She sketches a cognitivist account of music perception that draws on the work of Jackendoff and Lerdahl, whereby an experienced listener unconsciously assigns a structural description to heard music in accordance with internalized rules governing musical parameters. Though a subject may become aware in an inarticulate way of how he or she is parsing the music, the representations involved elude verbal grasp. Raffman calls this 'structural ineffability'.

'Feeling ineffability' is a result of the fact that knowledge of music is sensory-perceptual: to know a piece is to know how it sounds. Knowledge of music depends on knowing what, say, a minor third actually sounds like.

Raffman also sketches a psychology of musical nuances – those highly specific values of pitch, rhythm and timbre that characterize any musical event – and shows how this underpins what she calls 'nuance ineffability'. Nuance ineffability arises from the fact that in aural experience we are conscious of differences, say, in pitch more subtle than we can inwardly label or classify: we are unable to remember and judge of such nuances. The basic problem is one of memory: reporting a perception requires retention of information in a manner that allows for stable association with verbal labels, but our ability to register musical nuances exceeds the mental schemata we seem to have available for storing them. Nuance ineffability in the reception of musical events poses a problem for those accounts of consciousness that identify it with sentential episodes: if we consciously experience aural nuances but cannot represent them propositionally, then there must be more to consciousness than sentential-type representation can allow.

DeBellis (1995) discusses statements of the form '*S* hears *x* as *F*'. He takes the meaning of such ascriptions to be given by the content of the mental states that ground or justify them; and he takes these mental states to be ones in which passages of music are represented, correctly or incorrectly, as having certain properties.

DeBellis argues that the music-hearing of an ordinary (experienced but untrained) listener is both weakly and strongly nonconceptual. Weak nonconceptuality is the claim that the ordinary listener's hearing of music does not involve those concepts in terms of which an analyst might describe that hearing. Strong nonconceptuality is the claim that the ordinary listener's hearing of music does not involve concepts of any sort, even narrowly perceptual ones. A consequence of both theses is that ordinary musical perception is not a process of acquiring beliefs, since beliefs presuppose concepts. Rather, the comprehending ordinary listener represents the music being heard as having some qualities or features, without thereby believing that it has those qualities or features.

DeBellis's argument for strong nonconceptuality is as follows. Current psychological theory suggests that ordinary listeners represent all heard sound events of a given kind *K* in the same way. Yet they generally prove unable to discriminate between *K* and non-*K* events, and generally fail to judge two *K* events to be similar. This suggests that such listeners lack even a perceptual concept of *K*, and that the conversion of ordinary listeners into expert listeners is in large part the acquisition of perceptual concepts for musical features which ordinary and expert listeners register alike.

Robinson (1994) explores the relevance of recent research on emotions to theories of musical expression.

Jackendoff (1991) proposes an explanation of musical affect in terms of discrepancies between conscious knowledge of musical progression and unconscious states of a postulated musical parser.

Levinson (1998a) questions the extent to which basic musical understanding requires a grasp of large-scale form.

Fictional Appreciation

Feagin (1996) uses simulation theory to help understand what responding appropriately to a work of literary fiction might involve. She proposes that appreciation of a literary text typically involves mental shifts in response to the flow of the text. Such mental shifting is a prerequisite to empathizing with fictional characters. Empathizing involves simulating another's mentality, in effect conforming one's own mind to that of one's target, by putting one's mind 'offline' and then 'inputting' what one takes to be the experiences of one's target, thus generating an affective 'output' in oneself. This account might explain how, in responding empathically to a work of fiction, one may thereby be acquiring real knowledge – knowledge of 'what it is like' to be a certain person in a certain situation – which is often said to be one of the rewards of reading imaginative literature.

Currie (1995a) criticizes meta-representational theories of pretence, according to which pretending involves decoupling inner symbols from their normal semantic implications or flagging such symbols with special 'pretence' markers. He suggests that such views confuse mental contents and psychological attitudes towards them, and make it hard to explain the specific character of individual pretendings.

Currie questions whether, as has been claimed, empirical studies of autism and related cognitive disorders support meta-representational theories of pretense. He argues instead that such studies support the identification of imaginative pretense with simulation. He suggests that imagination is an 'internal simulator', a part of our mental equipment that evolved for the purpose of strategy-testing. This hypothesis can explain some aspects of appreciation of literary fiction, in particular our capacity to be affected by the plights of fictional characters: empathizing with characters involves the same process as empathizing with real people, taking on their beliefs and desires in imagination. The only additional requirement is that we first imagine them to exist.

Currie (1995b) considers a number of central issues in film theory from a cognitivist perspective, such as how films represent, what cinematic content consists in, and how we interpret cinematic narratives. He argues that the essence of cinematic experience is not seeming to see, or even imagining seeing, the objects and events represented in a film, but rather 'impersonally visually imagining' those objects and events. Since imagining is construed as a form of simulation – whether of others' states or of one's own states on other occasions – the essence of cinematic experience is thus simulated perceptual belief.

Currie accounts for the special 'realism' of film in terms not of its capacity to induce perceptual illusion, but of its mode of representation, whereby temporal properties are represented by temporal ones and spatial properties by spatial ones, which makes cinematic experience of objects similar to experience of those objects in the world.

References

Beardsley M (1981) *Aesthetics: Problems in the Philosophy of Criticism*. Indianapolis, IN: Hackett. [First published 1958.]

Budd M (1995) *Values of Art*. London, UK: Penguin.

Carroll N (2001) *Beyond Aesthetics*. Cambridge, UK: Cambridge University Press.

Currie G (1995a) Imagination and simulation: aesthetics meets cognitive science. In: Davies M and Stone T (eds) *Mental Simulation*, pp. 151–169. Oxford, UK: Blackwell.

Currie G (1995b) *Image and Mind: Film, Philosophy, and Cognitive Science*. Cambridge, UK: Cambridge University Press.

Danto A (1981) *The Transfiguration of the Commonplace*. Cambridge, MA: Harvard University Press.

Davies S (1991) *The Definition of Art*. Ithaca, NY: Cornell University Press.

Davies S (1994) *Musical Meaning and Expression*. Ithaca, NY: Cornell University Press.

DeBellis M (1995) *Music and Conceptualization*. Cambridge, UK: Cambridge University Press.

Dewey J (1934) *Art as Experience*. New York, NY: G. P. Putnam.

Dickie G (1997) *The Art Circle*. Chicago, IL: Spectrum Press. [First published 1984.]

Feagin S (1996) *Reading With Feeling*. Ithaca, NY: Cornell University Press.

Gadamer H (1986) *The Relevance of the Beautiful and Other Essays*. Cambridge, UK: Cambridge University Press.

Goldman A (1995) *Aesthetic Value*. Boulder, CO: Westview Press.

Gombrich E (1960) *Art and Illusion*. Princeton, NJ: Princeton University Press.

Goodman N (1976) *Languages of Art*, 2nd edn. Indianapolis, IN: Hackett. [First published 1968.]

Hopkins R (1998) *Picture, Image and Experience*. Cambridge, UK: Cambridge University Press.

Hopkins R (2000) Touching pictures. *British Journal of Aesthetics* **40**: 149–167.

Jackendoff R (1991) Musical parsing and musical affect. *Music Perception* **9**: 199–230.

Kivy P (1989) *Sound Sentiment*. Philadelphia, PA: Temple University Press.

Levinson J (1990) *Music, Art, and Metaphysics*. Ithaca, NY: Cornell University Press.

Levinson J (1996) *The Pleasures of Aesthetics*. Ithaca, NY: Cornell University Press.

Levinson J (1998a) *Music in the Moment*. Ithaca, NY: Cornell University Press.

Lopes D (1996) *Understanding Pictures*. Oxford, UK: Oxford University Press.

Lopes D (1997) Art media and the sense modalities: tactile pictures. *Philosophical Quarterly* **47**: 425–440.

Lopes D (1999) Pictorial color: aesthetics and cognitive science. *Philosophical Psychology* **12**: 415–428.

Matravers D (1998) *Art and Emotion*. Oxford: Clarendon Press.

Raffman D (1993) *Language, Music, and Mind*. Cambridge, MA: MIT Press.

Robinson J (1994) The expression and arousal of emotion in music. *Journal of Aesthetics and Art Criticism* **52**: 13–22.
Schier F (1986) *Deeper Into Pictures*. Cambridge, UK: Cambridge University Press.
Scruton R (1997) *The Aesthetics of Music*. Oxford: Oxford University Press.
Sparshott F (1982) *Theory of the Arts*. Princeton, NJ: Princeton University Press.
Tormey A (1971) *The Concept of Expression*. Princeton, NJ: Princeton University Press.
Vermazen B (1986) Expression as expression. *Pacific Philosophical Quarterly* **67**: 196–224.
Walton K (1970) Categories of art. *Philosophical Review* **79**: 334–367.
Walton K (1990) *Mimesis as Make-Believe*. Cambridge, MA: Harvard University Press.
Wollheim R (1980) *Art and Its Objects*, 2nd edn. Cambridge, UK: Cambridge University Press. [First published 1968.]
Wollheim R (1987) *Painting as an Art*. Princeton, NJ: Princeton University Press.
Wolterstorff N (1980) *Worlds and Works of Art*. Oxford: Oxford University Press.

Further Reading

Beardsley M (1982) *The Aesthetic Point of View*. Ithaca, NY: Cornell University Press.
Currie G (1989) *An Ontology of Art*. London, UK: Macmillan.
Currie G (1990) *The Nature of Fiction*. Cambridge, UK: Cambridge University Press.
Kivy P (1990) *Music Alone*. Ithaca, NY: Cornell University Press.
Lamarque P (1996) *Fictional Points of View*. Ithaca, NY: Cornell University Press.
Levinson J (1998b) Wollheim on pictorial perception. *Journal of Aesthetics and Art Criticism* **56**: 227–233.
Scruton R (1974) *Art and Imagination*. London, UK: Methuen.
Sibley F (2001) *Approach to Aesthetics*. Oxford, UK: Oxford University Press.
Stecker R (1997) *ArtWorks: Definition, Meaning, Value*. University Park, PA: Penn State University Press.
Walton K (1987) Style and the products and processes of art. In: Lang B (ed.) *The Concept of Style*, pp. 72–103. Ithaca, NY: Cornell University Press.

Affective Disorders: Depression and Mania

Introductory article

Michael T Compton, Emory University School of Medicine, Atlanta, Georgia, USA
Charles L Raison, Emory University School of Medicine, Atlanta, Georgia, USA
Charles B Nemeroff, Emory University School of Medicine, Atlanta, Georgia, USA

CONTENTS

Affective disorders are psychiatric illnesses characterized by disturbances of affect or mood, which cause emotional, cognitive, and behavioral symptoms that significantly impair functioning. They are classified according to polarity (the degree of mood elevation or depression), and are treated with pharmacological and/or psychotherapeutic interventions accordingly.

INTRODUCTION

Affective disorders have been described and treated since the age of ancient Greco-Roman medicine. Hippocrates described melancholia (an excess of 'black bile') around 2400 years ago. Affective excitement, or mania, was also described by ancient physicians. Aretaeus of Cappadocia recognized that mania and depression were usually connected in the same individuals, indicating that the concept of bipolar disorder was anticipated in antiquity. Affective disorders continue to be subdivided according to polarity: unipolar disorders are characterized by depressive episodes only, whereas bipolar disorders are marked by the presence of hypomanic, manic or mixed episodes, with or without intervening depressive episodes.

CLASSES OF AFFECTIVE DISORDERS

Affective disorders are currently diagnosed using four specific mood episodes as building blocks: major depressive episode, manic episode, hypomanic episode, and mixed episode.

Mood Episodes

A major depressive episode consists of a period of at least 2 weeks during which five or more specific symptoms are experienced, representing a change from previous functioning. At least one of these symptoms must be either depressed mood, or loss of interest or pleasure (anhedonia). Other potential symptoms include significant change in appetite or weight, insomnia or excessive sleeping, psychomotor agitation or retardation, fatigue or loss of energy, feelings of worthlessness or inappropriate guilt, poor concentration or indecisiveness, and suicidal thoughts. Patients suffering from a major depressive episode often experience negative cognitive patterns, including helplessness, hopelessness, and preoccupation with inadequacy. Low self-esteem is commonly present. Psychotic symptoms accompany approximately 15% of depressive episodes.

A manic episode is a distinct period of abnormally elevated, expansive, or irritable mood lasting at least a week or requiring hospitalization. During this period, three or more specific symptoms (four or more if the mood is only irritable) typically occur: inflated self-esteem or grandiosity; decreased need for sleep; being more talkative than usual or feeling under pressure to keep talking; flight of ideas, or subjective experience that thoughts are racing; distractibility; increase in goal-directed activity or psychomotor agitation; and excessive involvement in pleasurable activities that have a high potential for painful consequences. Psychotic symptoms develop in approximately 25% of manic episodes and typically consist of grandiose delusions, hallucinations of deities, or paranoia born of a delusional sense of importance.

A hypomanic episode, which is less severe than mania, is a distinct period of elevated, expansive or irritable mood lasting for at least 4 days, during which time three or more symptoms of a manic episode are experienced (four if the mood is only irritable). The changes represent an unequivocal change in functioning, which is observable by others, but is not severe enough to cause marked impairment in functioning or to necessitate hospitalization, and there are no psychotic features present.

A mixed episode is a period of at least 1 week during which the above criteria for both a major depressive episode and a manic episode are experienced. Few patients have episodes that meet the full criteria for a mixed state, but dysphoric manias, in which a depressed and/or anxious mood coexists with manic symptoms, are frequent in patients with bipolar disorder.

For all four of the above mood episodes, the fourth edition of the *Diagnostic and Statistical Manual of Mental Disorders* (DSM-IV) specifies that the symptoms should not be due to the direct physiological effect of a substance (e.g. drug of abuse or medication) or a general medical condition.

Mood Disorders

Major depressive disorder is defined by the lifetime presence of at least one episode of major depression that is not precipitated by a medical illness, a medication, or a substance of abuse and is not better accounted for by bereavement. Dysthymic disorder is a mild form of depression that often has an early age of onset (in childhood, adolescence, or early adult life) and usually lasts for protracted periods (at least 2 years in adults, according to DSM-IV). Patients with dysthymia do not meet full criteria for major depression during the course of their illness. Dysthymia tends to be characterized more by emotional and cognitive symptoms, such as depressed mood, pessimism and poor self-esteem, than by the types of neurovegetative symptoms seen in major depression (i.e. changes in sleep, appetite, and physical activity level).

Bipolar I disorder is defined by the lifetime presence of a manic episode not judged to be caused by a medical illness or ingestion of recreational drugs or medications. Bipolar II disorder is characterized by the presence of major depressive episodes and periods of hypomania. Although hypomanic episodes tend, by definition, not to produce the degree of functional impairment caused by full manias, bipolar II disorder patients may actually have more overall impairment in their lives than many patients with bipolar I disorder, owing to the greater length and treatment resistance of depressive episodes in bipolar II disorder, as well as to the tendency of patients with bipolar II disorder to develop rapid cycling, a condition characterized by rapid progression from one mood episode to the next with little or no period of normal mood between episodes.

Cyclothymia, like dysthymia, is a chronic condition characterized by the presence of mood symptoms insufficient to meet the criteria for a major affective disorder. In the case of cyclothymia, these symptoms take the form of a repeated alternation of subsyndromal hypomanic and depressive symptoms. It appears that the presence of cyclothymia is a risk factor for the later development of both bipolar I and II disorders, although some patients never experience a full mania or major depression.

Affective disorders are distinguished from other psychiatric conditions not so much by specific symptoms as by the fact that alterations in mood and/or a loss in the ability to find pleasure in life are considered to be the primary derangements in the disease course of the individual. Beyond this central focus, however, patients with affective disorders can evince symptoms seen in a number of other psychiatric conditions. In addition, comorbidity is the rule rather than the exception in psychiatry. Other psychiatric conditions that are highly comorbid with affective disorders include anxiety disorders (generalized anxiety disorder, social phobia, panic disorder, and obsessive–compulsive disorder) and various alcohol and drug abuse and dependence problems.

ETIOLOGY OF AFFECTIVE DISORDERS

Genetics and Biological Factors

Affective disorders, especially bipolar disorder because of its greater degree of familiality (stronger evidence for a genetic basis), have been the subject of much research into their genetics and heredity. Despite methodological and interpretive limitations, several reproducible findings have emerged.

The risk of bipolar disorder in first-degree relatives of patients with bipolar disorder ranges between 3% and 8%, compared with a 1–1.5% rate in the general population. The risk of depressive disorders among first-degree relatives of patients with depression is two to three times that of the general population. If one parent has an affective disorder, a child's risk of an affective disorder is between 10% and 25%; if both parents are affected, the risk roughly doubles. Having more affected family members confers increased risk, especially when family members have bipolar disorder.

In addition to genetic factors it is becoming increasingly clear that, as a group, people with affective disorders differ physiologically from well-matched controls without histories of depression or mania. Evidence suggests that patients with affective disorders have consistent changes in brain structure and functioning. These changes include decreased volume in prefrontal and temporal cortex, as well as in the basal ganglia. Consistent with these structural changes, many patients with major depression demonstrate decreased functioning in these same areas when they are studied by modern imaging techniques. Postmortem studies suggest that changes in brain volume and function may reflect cell loss and/or atrophy as well as loss of functional connections (synapses) between nerve cells in these brain areas. Major depression is associated with changes in sleep architecture. Some of these changes remit when affective symptoms resolve, but others remain, and are even found in unaffected family members, suggesting that such sleep abnormalities may be a risk factor for the development of affective disorders.

Psychological and Social Factors

Major depression frequently develops in a psychological context of anxiety and/or neurosis, as well as in individuals with a long-term tendency toward excessive social inhibition. Many patients with bipolar disorders, especially bipolar II disorder, demonstrate premorbid psychological traits of moodiness, impulsivity, and acting-out behavior. The likelihood that a person will develop an affective disorder is also influenced by a number of environmental factors, many of which are psychosocial in nature. For example, it is known that childhood conditions such as the death of a parent, or exposure to neglect or to physical, sexual or emotional trauma, pose a significant risk for the development of affective disorders later in life. Although traumatic events early in life appear to be especially potent in fostering later depression, probably as a result – at least in part – of negatively influencing postnatal brain development, actual or perceived stress at any time in the life cycle is strongly associated with the development of depression.

Investigations into psychological and environmental contributions to the affective disorders tend to be empirical in nature, using biological and ethological approaches in an attempt to map out pathways through which psychological and environmental factors influence mood. However, many of these research interests have been anticipated by the more theoretically driven approaches to the issue that dominated the field of psychiatry until well into the last half of the twentieth century. Many such theories derive from psychoanalytic perspectives and tend to privilege early life experiences. Other schools of thought have advanced behavioral and cognitive factors as primary causes of affective disorders. Fewer psychological theories have tried to explain the experience of mania, and genetic vulnerability has been more clearly implicated. Nonetheless, psychological models have been proposed for manic reactions. For example, some theorists have described mania as a stance defensive against an underlying depression (for example, self-deprecation and excessive guilt of depression become replaced by grandiosity and elevated self-esteem characteristic of hypomania or mania). Key figures in the formulation of these models include Karl Abraham and Melanie Klein.

ONSET AND COURSE OF AFFECTIVE DISORDERS

Affective disorders are highly prevalent illnesses that significantly impair functioning, interpersonal relationships, and quality of life. In the USA, lifetime prevalence rates are 17.1% for major depressive episode, 6.4% for dysthymia, and 1.6% for manic episode. Epidemiological studies reveal that affective disorders are more prevalent among individuals under the age of 45 years (the average age of onset is 20–40 years). Women suffer from unipolar depressive disorders about twice as often as men. Certain points in the human life cycle confer increased risk for the development of depression: these include puberty, as well as childbirth in women and the passage through middle age in men (after middle age, rates of depression are equal between the genders). Menopause may be a risk factor for mood disturbance in certain women, but does not pose anything like the threat of the postpartum period, a time in which 20% of new mothers will develop a major depression. Rates of bipolar disorder do not appear to be affected by gender. The prevalence of affective disorders does not vary significantly by race or ethnicity.

It is increasingly recognized that both unipolar depressive disorder and bipolar disorder are frequently recurrent illnesses that lead to increased functional impairment with the passage of time and the accumulation of episodes. On average, people diagnosed with unipolar major depression in youth or early adulthood can expect five or six episodes over their life span, compared with an average of eight to nine major lifetime mood episodes in people with bipolar disorder. Twenty-five percent of patients with unipolar major depression demonstrate a chronic course. Similarly, 5% of patients with bipolar I disorder remain chronically manic, often providing a diagnostic conundrum to clinicians used to assigning a diagnosis of schizophrenia to anyone with a chronic psychotic illness. One-third of patients with unipolar depression will have only a single episode. These patients tend to be older at the time of onset and are less likely to have a history of affective disorders in their families.

Affective disorders underlie 50–70% of all cases of suicide, and individuals with serious depression (i.e. requiring hospitalization) have a 15% suicide rate. Rates of suicide in bipolar disorder are probably higher, approximately 20–25%.

TREATMENT

Pharmacologic and Somatic Treatments

Pharmacologic and somatic treatments for depression compare favorably with the pharmacologic treatment of other chronic medical disorders in terms of efficacy. A broad spectrum of effective antidepressant strategies are available, from electroconvulsive therapy through a wide array of antidepressant medications. In addition, psychotherapy and medications have a synergistic effect when used in combination.

The modern era of psychopharmacologic treatment was initiated in the mid-twentieth century with the discovery that lithium, monoamine oxidase inhibitors, and tricyclic antidepressants effectively treated affective disorders. Monoamine oxidase inhibitors (MAOIs) block the enzyme that metabolizes biogenic amines, increasing the availability of these neurotransmitters. Though especially efficacious for atypical depression, the use of MAOIs is greatly limited by their adverse effects, by their potential to induce possibly lethal hypertensive episodes in combination with foods that contain tyramine, and by a propensity to induce a frequently fatal condition known as 'serotonin syndrome' when taken in combination with medications that affect serotonin functioning. Tricyclic antidepressants (TCAs) are thought to work by blocking the reuptake of noradrenaline (norepinephrine) and serotonin into presynaptic nerve cells. However, in addition to these therapeutic actions at reuptake transporter sites, TCAs block other receptors leading to a host of unwanted side effects, including blurred vision, dry mouth, tachycardia, constipation, urinary retention, cognitive dysfunction, postural hypotension, dizziness, sedation, weight gain, and sexual effects. Of even more concern is that fact that TCAs are highly lethal in overdose as a result of direct effects on cardiac conduction.

As suggested by their name, selective serotonin reuptake inhibitors (SSRIs) work by blockade of the serotonin reuptake pump. Unlike the MAOIs or TCAs, these newer agents were specifically designed to diminish or abolish activity at other receptors known to mediate many side effects of the older drugs. As a result, SSRIs are associated with a decreased side-effect burden and significantly enhanced safety in overdose. However, these agents are not without their own limitations, including a high rate of sexual dysfunction, and a tendency, as a class, to block the metabolism of other medications, leading to potentially problematic interactions with other drugs. Antidepressants developed since the arrival of the first SSRIs show a trend toward reversing the specificity of action that was the goal in SSRI development. Examples of newer medications that use a combined action strategy include venlafaxine, which works by blocking the reuptake of both serotonin and noradrenaline, and mirtazapine, which blocks presynaptic noradrenaline autoreceptors and postsynaptic serotonin receptors.

Finally, although typically reserved for patients who have failed to respond to other treatments, electroconvulsive therapy (ECT) remains a highly effective therapy for depression. It is especially effective in melancholic and/or psychotic forms of depression and for patients whose illness has catatonic features. Nonetheless, ECT has long been hampered by the striking short-term memory loss associated with the treatment, as well as by the risks of undergoing anesthesia and the general difficulties inherent in implementing this typically hospital-based procedure.

Because bipolar disorder is a recurrent, severely impairing psychiatric disorder, the search for efficacious pharmacological treatments continues to be at the forefront of psychiatric research. Lithium is a naturally occurring salt that was found to have antimanic properties in 1949. In the experience of many experts, lithium remains the 'gold standard' for treatment of both the manic and depressive phases of bipolar disorder. In addition to its acute effects in curtailing mania and depression, lithium when taken chronically is known to protect against the advent of future mood episodes. Lithium treatment also decreases the lifetime risk of suicide in people with affective disorders, a finding that has been difficult to replicate with other pharmacological agents.

Several anticonvulsant medications have attracted attention as treatments for bipolar disorder. The use of valproic acid for the treatment of acute mania is supported by data suggesting that (like lithium) it is effective in protecting patients from future mood episodes. Valproic acid appears to be more effective than lithium in treating dysphoric manias and may give better results in patients who have a disease course characterized by manias that follow depressions, rather than the other way around. Lamotrigine may be effective for bipolar depression, with a minimal risk of the induction of mania or hypomania that accompanies the use of traditional antidepressants. Carbamazepine was the first anticonvulsant used in the treatment of bipolar disorder, but its use has largely been eclipsed by valproic acid, which has a more tolerable side-effect profile in many patients.

In addition to mood stabilizers such as lithium and the anticonvulsants, antipsychotic agents and benzodiazepines have been shown to be effective in the acute treatment of mania, both as additions to mood stabilizers and as single agents. Finally, ECT is an effective treatment in both the manic and depressed phases of bipolar disorder, and is especially effective when patients demonstrate catatonic symptoms.

Psychotherapeutic Treatments

Several forms of psychotherapy are roughly as effective as antidepressant drugs in the treatment of most cases of major depression. Current treatment of bipolar disorders favors the use of pharmacological and somatic interventions, but studies suggest that psychotherapy has a valuable adjunctive role in these conditions. Two commonly employed psychotherapeutic techniques that have received considerable empirical validation are interpersonal therapy and cognitive behavioral therapy.

Interpersonal therapy is descended from a long-held and valuable insight from the psychoanalytic tradition that interpersonal relationships are crucial to the maintenance of a normal mood state (euthymia) for most people, and that major depression is frequently associated with disturbances in the realm of interpersonal relations. The therapist directly explores and works with the patient's interpersonal difficulties, operating under the assumption that these difficulties are frequently causative of depression and that their resolution promotes a return to euthymia. Cognitive behavioral therapy is a highly structured, effective short-term psychotherapy that aims to correct negative thought patterns, specific distorted schemas, and cognitive errors. Unlike older psychoanalytic traditions, cognitive behavioral therapy specifically and directly confronts maladaptive cognitions through a process of gentle challenging, but also through the use of educational components and 'homework' aimed at encouraging the patient to practice more positive behavior in daily life. Interestingly, although these two psychotherapies are based on different – though not mutually exclusive – theoretical underpinnings, most studies suggest they are equally effective in the treatment of depression.

NEURAL CORRELATES AND THEORIES OF AFFECTIVE DISORDERS

No single physiological abnormality has been found to account for affective disorders, but many years of research have established a number of abnormalities that are present in many affectively ill patients. The current biological theories of affective disorders began to emerge in the 1950s coincident with the discovery of antidepressants and mood stabilizers. Based on the suspected mechanisms of action of antidepressants, early theories focused on putative deficiencies in single biogenic amine neurotransmitter systems, especially noradrenaline, and later, serotonin. While these early (and relatively simplistic) proposals have given way to far more complicated heuristic models, it remains true that affective disorders as a whole continue to be characterized by evidence of altered functioning in catecholamine (noradrenaline and dopamine) and indolamine (serotonin) neurotransmitter systems, as well as in other related systems such as the hypothalamic–pituitary–adrenal axis that, together with the sympathetic nervous system, constitutes the mammalian stress response system.

Biogenic Amines

Based on existing technologies, early psychobiological investigations into the role of biogenic amines in the pathogenesis of affective disorders focused on the concentration of these neurotransmitters (or their metabolites) in blood, urine, and cerebrospinal fluid. Taken as a whole these studies suggested a decrease in serotonergic activity in people with unmedicated depression. The relationship between noradrenaline production and depression appears more complex. Early studies reported decreased urinary levels of catecholamine metabolites in depression, and animal models of stress suggest that depleted noradrenaline in selected brain areas might contribute to the diminished energy, inability to feel pleasure, and decreased libido that frequently accompany depression. However, later research suggested that people with melancholic major depression have a markedly increased release of catecholamines into the cerebrospinal fluid.

Despite the initial attractiveness of these neurotransmitter models, significant evidence has accumulated countering the notion that affective disorders are caused by a simple neurotransmitter deficiency or excess. Current formulations of the role of biogenic amines in depression tend to focus on the potential restorative role that increased levels of these transmitters might have for 'downstream' neural functioning. Chronic alterations in aminergic availability fostered by antidepressant medication may stimulate intracellular second-messenger pathways to favor the production of neurotrophic factors that lead to enhanced functioning of neurons in brain areas believed to be centrally involved in mediating mood abnormalities.

Neuroendocrine Axes and Neuropeptide Systems

The hypothalamic–pituitary–adrenal (HPA) axis serves as the principal mammalian stress response system. Abnormalities of this axis are amongst the most replicable biological findings associated with affective disorders, especially major depression. These abnormalities in major depression include evidence for increased corticotrophin releasing hormone (CRH) production in the central nervous system, a blunted adrenocorticotrophin response to CRH stimulation, enlargement of pituitary and adrenal glands, increased blood levels of cortisol, and decreased tissue sensitivity to glucocorticoids. Evidence for heightened production of CRH in the brain is especially intriguing, given that this neuropeptide serves as a neurotransmitter in limbic brain regions linked to depression, in addition to its role in the hypothalamus as the primary activator of the HPA axis.

Hypothyroidism is frequently associated with a markedly depressed mood. Studies have documented alterations in thyroid axis activity in patients with depression, including increased thyrotrophin releasing hormone (TRH) concentrations in cerebrospinal fluid, altered thyrotrophin response to TRH stimulation, decreased nocturnal plasma thyrotrophin concentrations, and presence of antimicrosomal thyroid or antithyroglobulin antibodies. Similar abnormalities have been associated with bipolar disorder, the course of which is worsened by hypothyroidism.

CONCLUSION

Affective disorders, including unipolar depressive disorder and bipolar disorders, are prevalent psychiatric illnesses, affecting approximately 25% of individuals during the course of a lifetime. They are classified by the presence of discrete or chronic mood episodes characterized by abnormal depression or elevation of mood and mood-related functions below or above the individual's baseline. The causes of affective disorders are multifactorial, with both genetic and biological factors involved, and modified by psychosocial variables. Treatment consists of psychiatric management with pharmacological agents, ECT, and psychotherapy. Contemporary psychiatry and neuroscience are increasingly concerned with the neurobiological correlates of affective disorders, predictors of response to particular therapies, and achieving return to complete euthymia (remission).

Further Reading

American Psychiatric Association (1994) *Practice Guideline for the Treatment of Patients with Bipolar Disorder*. Washington, DC: American Psychiatric Association.

American Psychiatric Association (2000) *Practice Guideline for the Treatment of Patients with Major Depressive Disorder* (revision). Washington, DC: American Psychiatric Association.

American Psychiatric Association (2000) *Diagnostic and Statistical Manual of Mental Disorders*, 4th edn, revised. Washington, DC: American Psychiatric Association.

Goodwin F and Jamison K (eds) (1990) *Manic-Depressive Illness*. New York, NY: Oxford University Press.

Kaplan HI and Sadock BJ (eds) (2000) *Comprehensive Textbook of Psychiatry*, 7th edn. Baltimore, MD: Williams & Wilkins.

Aggression and Defense, Neurohormonal Mechanisms of

Introductory article

R J Blanchard, University of Hawaii, Honolulu, Hawaii, USA
C Markham, University of Hawaii, Honolulu, Hawaii, USA
D C Blanchard, University of Hawaii, Honolulu, Hawaii, USA

CONTENTS

Aggression and defense are important behavioral strategies throughout the animal kingdom, and are linked to activation of brain areas in the prefrontal cortex, limbic system, hypothalamus, and periaqueductal gray matter. While substantial progress has been made toward understanding the neural and neurotransmitter systems underlying defense, and the neurotransmitter and hormonal systems involved in aggression, the relationship between aggression and defense systems is poorly understood, as are the applicability of these findings to human aggression and violence.

INTRODUCTION

The medical, social, economic, and societal problems associated with human aggressive and defensive behaviors often create an incorrect view of these as essentially maladaptive or pathological. A comparative approach to analysis of aggression and defense is necessary to understand that these represent crucial life-crisis management strategies for all mammals and most inframammalian species. Research on situations eliciting aggression and defense and the hormonal mechanisms modulating them has provided information about the evolutionary functions of these biobehavioral systems, while work on their neural systems and pharmacological control may provide links to some specific psychopathological conditions.

In humans and in other mammalian species, agonistic interactions between members of the same species (conspecific) almost always consist of some mixture of aggressive and defensive behaviors. Both combatants are likely to be aggressive, as a nonaggressive participant could usually preclude or conclude the encounter by leaving. However, fights also typically involve pain and sometimes injury: that nonhuman animals always settle conspecific disputes by nonpainful or noninjurious displays is a myth. Thus self-defense is also typical of one or both participants in real-world encounters. This situation creates difficulties in making clear differentiations between aggression and defense, and has substantially hindered analysis of these two different patterns, and of their evolutionary functions.

STUDIES OF AGGRESSION AND DEFENSE IN ANIMALS

Laboratory studies can polarize aggressive and defensive tendencies of two combatants, making one extremely aggressive and the other exclusively defensive, in order to make the two behavior patterns easier to analyze. While such studies are not adequate for determining the range of conditions that will elicit aggression or defense, they do provide information on some focal antecedents for each of them, and enable relatively uncontaminated descriptions of the actual behaviors involved in the two biobehavioral systems. When findings from such laboratory research (typically involving rodents) are combined with ethological studies on a much broader range of mammalian and inframammalian species in their own natural habitats, a consistent view of the adaptive value of both aggression and defense begins to emerge.

CAUSES OF AGGRESSION AND DEFENSE

Offensive aggression is coming to be conceptualized as a behavior pattern elicited in the context of resource disputes, particularly in response to challenge (generally but not always from a member of the same species) to the individual's control over these resources or rights. While aggression may be adaptive in reducing conspecific challenge over crucial elements such as food, water, access to nesting sites or materials, and the like, in an evolutionary context a particularly valuable resource is access to a breeding partner. Thus male on male aggression is especially common during the breeding season, or in the presence of reproductive females. Similarly, for females offspring are a particularly important evolutionary resource, and many instances of female aggression occur in a maternal context. The common tendency to refer to such situations as involving 'defense' of mates, offspring or other resources produces an additional, but purely semantic, problem in conceptualizing the difference between the biobehavioral systems for aggression and defense. The defense system is a response to threat to bodily integrity. Aggression is a response to challenge to the individual's control of a resource. It does not aim at 'defending' the integrity of the resource, but in promoting the aggressor's claim to that resource. Successful aggression is thus an important means of social control.

Most mammalian species have evolved mechanisms that reduce or limit intraspecies aggression by limiting the circumstances under which one animal is likely to challenge another for crucial resources. Territoriality and dominance relationships are two such mechanisms. The first operates through avoidance by one animal or group of the territory of another animal or group, while the second involves recognition by animals within a group of the relative ranks of other animals, such that subordinates typically do not challenge those higher in rank to themselves. However, territorial encroachment and rank challenges do occur, and for most species acquisition of territory and rise in rank depend on successful aggression by the protagonist and sometimes its allies.

In contrast to these rather complex causes of aggression, defensive behaviors occur in response to bodily threat, from predators or from conspecific individuals. Antipredator situations provide an opportunity to describe and analyze these behaviors in situations that do not contain elements of aggression: predation is not aggression. Defensive behaviors represent attempts to avoid, evade, escape, conceal, bluff, threaten or otherwise neutralize threat from dangerous stimuli, thereby enhancing the animal's likelihood of coping successfully with the threat.

FACTORS CONTROLLING THE SUCCESS OF AGGRESSION AND DEFENSE

Successful aggression is adaptive, but aggression itself can be dangerous. This puts a premium on the ability of animals including humans to evaluate the probability of success prior to entering into an aggressive encounter. The individual's own history of victory or defeat is a particularly important determinant of future success in aggression, and a strong predictor of aggressive behavior. Members of most mammalian (and many other) species also pay particular attention to features of a potential opponent that indicate size, strength, health, and fighting ability, before entering a fight. Thus hyperexpression of such features becomes adaptive in discouraging challengers, a situation that has resulted in the evolution of ever-larger bodies (particularly for males of polygynous species, a factor in sexual dimorphism); more impressive weapon systems such as antlers; and enhancement of other body features that may be the focus of comparison, such as the mouth of the hippopotamus. Particular postures and movements that optimally present these features, sometimes eliciting retreat from a challenger, constitute many of the aggressive displays that have wrongly been believed to replace actual agonistic behavior in nonhuman animals.

Age and fighting ability are also factors in the success of aggression, and these are often evaluated in fights among pubertal and young adult males. However, play fights in prepubertal mammals appear to bear little direct relationship to adult fighting between the same animals: prepubertal males who later become dominant are less likely to initiate play fights than are those who later become subordinates, and the contact sites and behavior patterns seen in the fights of prepubertal mammals appear to be more similar to adult sexual behavior than to adult agonistic behaviors.

Successful defensive behaviors must counter the attacks of predators in addition to conspecific threats, and must be effective in a variety of habitats and situations. Thus few higher animals rely on a single type of defensive behavior. Most mammals appear to have a relatively consistent group of about half a dozen focal forms of defensive behavior, with some of these emphasized or rarely used in a particular species, but all represented to some degree. Given this range of defensive behavior, the major single factor in the success of defense is how well the particular defensive behavior fits the situation, i.e. the specific threat stimulus and features of the environment relevant to the success of a particular defense. Thus, although the success of both aggression and defense relies on utilization of information about both the opponent and the situation, the topography or form of defensive behavior is geared more to features of the environment (such as presence of an escape route, location of places of safety, and distance between predator and self), while the form of aggression largely reflects features and behaviors of the opponent.

AGGRESSION AND DEFENSE IN THE RAT

Conspecific aggressive and defensive behaviors have been described in greatest detail in laboratory rodents, particularly the rat. Conspecific attack in rats involves elements of approach and investigation, typically focused on olfactory assessment of the sex, age, and breeding condition of the opponent. The attack itself involves attempts to bite the back and flanks of the opponent, as well as a number of maneuvers that enable the attacker to reach this location. Conspecific defense in rats also relies heavily on the targeting of the attack towards the back of the opponent, in that if the defender can remove this site from its attacker, it is relatively safe from being bitten. One general method of removing a target site for attack is flight, for which the countering attack tactic is chasing. The defensive animal may also adopt an upright posture facing the attacker; this keeps its back – the attacker's target – out of reach. An experienced defender can also pivot smoothly to maintain frontal orientation if the attacker attempts to circle around to the defender's back, utilizing a lateral attack sequence. This consists of the attacker arching its back and moving laterally toward the upright defender, sometimes pushing it off balance or lunging in a forward circular motion toward the defender's back and flanks. In an even higher-level defense tactic the rat lies on its back, again concealing the target of conspecific attack. However, the attacker typically stands over the supine defender, sometimes pushing at it to induce movement and reveal the back target.

These defense tactics are adaptive because laboratory rats, as well as wild *Rattus norvegicus* and *R. rattus*, are reluctant to bite targets other than the back of a conspecific opponent. In particular they fail to bite an opponent's ventrum, even if this is the only body area exposed. While the strength of this prohibition on biting particular targets may vary somewhat from one species to another, some targeting of bites and blows appears to be common across mammalian species. The specific target of attack is typically one in which there is relatively little chance of lethal damage. Thus such bites or blows may produce pain, discouraging further resource or status challenge from the other, without killing a conspecific that may be a relative, or part of the attacker's social group. Some dedicated structures, such as antlers, provide weapons that are used in offensive attack, and also serve as the target of this attack. Typically such structures are not used in defense against predators.

Antipredator defensive behaviors are much less dependent on defense of particular body structures, since the attacks of predators are typically aimed at vulnerable body sites rather than away from them. Thus while defensive tactics like flight remain adaptive, some defenses that are effective against conspecifics (such as lying on the back) would not be useful against a predator. For most mammals, flight is a dominant antipredator defense when an escape route is available or a place of safety within reach. Immobility, often involving specific freezing postures, is also a common defensive behavior. This may be adaptive by helping the prey animal to avoid detection or because some predators selectively attack fleeing prey. In close encounters, defensive threat followed by defensive attack may be highly effective in discouraging the attacker: typically, an injured predator is at a severe disadvantage in future hunts. When there is a potential rather than a clear and obvious threat (e.g. novel stimuli or predator odors) defenses such as flight and defensive threat or attack may be useless or even counterproductive. To such stimuli the most prominent defensive behavior is risk assessment, a highly motivated information-gathering pattern involving orientation to the potential threat source, sensory (visual, auditory, olfactory) scanning, and approach with a low-back posture that minimizes detection of the defensive animal while allowing it to investigate the potential threat source.

NEURAL MECHANISMS OF DEFENSIVE BEHAVIOR

Neural systems underlying defensive behaviors are the focus of intense research interest. They have been investigated using a variety of research paradigms, including evaluation of behavioral effects of lesions or stimulation of particular brain sites, and analysis of regional intermediate early gene expression during confrontation with a threat source. These studies, in conjunction with tract tracing from cells in sites thus implicated, has yielded a detailed, and relatively consistent, view of the nervous control of particular defensive behaviors.

Studies of neuronal activation in association with aggression or defense have often used regional expression of *c-fos* messenger ribonucleic acid (mRNA) as a marker of activity in specific brain sites. A number of such studies, using a variety of different species (rat, mouse, hamster) and both conspecific defeat and predator exposure paradigms, have provided relatively consistent findings. Defensiveness is associated with activity in several limbic areas (e.g. amygdala, lateral septum), and in a ventral zone from the preoptic area through the hypothalamus to the midbrain periaqueductal gray (PAG) matter. Fos expression in both the septum and the central nucleus of the amygdala (CeA) following confrontation with a threat source (conspecific defeat) appears to vary from the initial to later exposures. Fos expression provides a good, though not necessarily comprehensive, overview of brain areas that respond to particular experiences or activities. However, Fos studies are somewhat difficult to interpret as they do not differentiate between areas that are directly involved in a particular function and those that respond to that function or to nonspecific events such as arousal, motor activity, or autonomic changes that may be associated with it. In addition, Fos does not indicate the organization of the systems that may be involved. Stimulation, lesion, and tract-tracing studies add significantly to systems research on defense, and such studies have shown that several of the areas in which Fos expression is seen during defense are indeed important in the elicitation and maintenance of defensiveness.

Stimulation of the PAG can elicit flight, freezing, and defensive threat responses in both rats and cats. These different responses are organized in terms of longitudinally coursing columns within the PAG. At more rostral levels, dorsal stimulation tends to elicit defensive vocalizations, while at more caudal levels such stimulation elicits flight and sometimes freezing. Stimulation of the ventrolateral column, in its caudal aspects, elicits an immobility response that has been variously interpreted as pain-related quiescence or as freezing. However, as lesions of this area sharply reduce kyphosis – an upright, crouched, posture shown by rat mothers during nursing – it is also possible that this area of the PAG is broadly involved with the maintenance of all active immobile reactions rather than with defense-related immobility only, or with specific types of defense-related immobility.

The dorsal premamillary nucleus of the hypothalamus has also received particular attention in a defense context. Lesions in this area appear to virtually abolish flight and freezing, and retrograde tracing studies indicate that it receives direct afferent connections from many of the telencephalic and diencephalic structures that show Fos expression after exposure to a predator. Other well-developed research efforts have been directed toward analysis of neural systems involved in specific defense-related behaviors. Siegel and his colleagues used combinations of lesions, stimulation, and local and systemic drug administration to characterize the control of defensive threat vocalizations (hissing) in the cat. They found that a number of sites in both limbic areas and in the ventral zone from preoptic area to PAG act to initiate and modulate this component of defense. Several such neurotransmitter-specific modulatory systems have been described, some involving direct connections from several areas of the amygdala to the PAG, while others consist of direct and indirect hypothalamic–PAG connections.

Similarly, Davis and his colleagues have concentrated on systems involved in potentiation of the startle reflex. Although the basic startle reflex is a simple three-synapse brain stem–spinal cord circuit, the amplitude of startle can be increased by conditioned cues (stimuli previously associated with shock) or unconditioned defense-enhancing manipulations such as testing in a brightly lit chamber. A particularly interesting finding is that the systems involved in conditioned and unconditioned potentiation of startle may be quite different: the CeA is strongly involved in conditioned potentiation, but is not crucial for unconditioned potentiation of startle, while damage to the bed nucleus of the stria terminalis abolishes unconditioned but not conditioned potentiation. These findings are consonant with the central role of the CeA in conditioning of freezing to shock cues, and with findings that CeA Fos expression changes with repeated threat exposure – all of which suggest that this nucleus may be more involved in plastic processes (learning, habituation) than in defense *per se*.

NEURAL MECHANISMS OF AGGRESSIVE BEHAVIOR

Research into the physiological mechanisms of aggression has tended to look at the pharmacological control of aggression rather than the specific neural systems involved in this behavior. Analysis of aggressive behaviors is additionally complicated by the fact that attack behaviors may involve offensive aggression, defensive attack, or predation; these may be difficult to differentiate under standard laboratory conditions. However, research programs have focused on sites in the hypothalamic area, with some differences in specific sites for different species. For example, in the hamster, vasopressin manipulations in the nucleus circularis, an area rich in vasopressin-expressing neurons, may profoundly affect aggressive behavior. However, in the rat, the nucleus circularis does not contain vasopressin-expressing neurons, suggesting that aggression changes associated with manipulations in this area must involve a different mechanism of action. In a number of hypothalamic sites, including the nucleus circularis (hamster), the intermediate hypothalamic area and part of the ventromedial nucleus (rat), electrical stimulation elicits an attack response, while lesions may reduce offensive attack. Neuronal tracing studies indicate connections to a number of areas similar to those implicated in defense, such as the prefrontal cortex, amygdala, septum and the PAG. Because stimulation and lesioning of some of these structures may impact both aggressive and defensive behaviors, it is not clear that the changes in one set of behaviors are independent of effects on the other. Similarly, because of the reliance of both aggressive behavior and sexual and maternal behaviors on pheromones, areas such as the medial nucleus of the amygdala that are strongly involved in the accessory olfactory system may influence both types of behavior. However, aggressive, defensive and reproductive behaviors appear to be well differentiated at the level of the hypothalamus. In addition, there are a number of hypothalamic sites in which stimulation elicits grooming, suggesting the existence of a number of roughly parallel systems in this area that are strongly involved in species-typical behaviors of great evolutionary importance.

Serotonin (5-hydroxytryptamine, 5-HT) is perhaps the most frequently investigated neurotransmitter in terms of effects on aggression. Deficiencies in serotonin have been implicated as potentially important in impulsivity and aggression in species ranging from lower mammals to humans. In the latter species, low levels of a serotonin metabolite in cerebrospinal fluid may signal reduced serotonin activity in the brain, which, interacting with other genetic and environmental factors, may be associated with instances of impulsive violence, or other impulsive actions. Two specific serotonin receptor subtypes, 5-HT_{1A} and 5-HT_{1B}, are of particular interest as the principal targets of a class of 'serenic' drugs that have proved to be extremely effective in reducing offensive aggression in a variety of animal models.

Vasopressin activity in the hypothalamus and related structures is associated with aggression enhancement, and serotonin may act at the level of the vasopressin receptor to modulate this behavior. In hamsters, pretreatment with fluoxetine, a serotonin reuptake inhibitor, increases serotonin levels in the anterior hypothalamus (an area known to influence aggressive behavior) and inhibits aggression. Systemic fluoxetine also reduces the increase in aggression seen when vasopressin is injected into the ventrolateral hypothalamus. This area contains both the 5-HT_{1A} and 5-HT_{1B} receptor subtypes, which have been particularly implicated in the control of aggression in animal models.

HORMONAL MECHANISMS OF AGGRESSION AND DEFENSE

The evidence for a relationship between hormones and aggressive behavior predates science: castration of male animals to reduce their aggressiveness has been common throughout human history. Testosterone generally enhances aggression across a range of mammalian species, including humans, but the relationship is highly variable, and may be more specific to times or situations involving reproductive behaviors, or dominance and territoriality related to reproductive advantage. One mechanism by which testosterone, like serotonin, may influence aggression is through hypothalamic vasopressin receptors. In the ventrolateral hypothalamus of male hamsters, vasopressin receptor binding disappears after castration, but can be maintained by treatment with testosterone. Findings that microinjections of vasopressin in this area enhance aggression in untreated male hamsters but fail to do so in castrates suggest that testosterone maintenance is important for the functioning of vasopressin receptors in this area.

Attempts to understand the mechanisms of the relationship between testosterone and aggression have emphasized that both testosterone and its metabolites may influence relevant neural systems and that the direction of effect may be different for some of these. In the brain, testosterone may be aromatized to estradiol, or metabolized by 5α-reductase to dihydrotestosterone. The functional pathways underlying male aggression may be modulated by either of these metabolites, being either estrogen-sensitive or androgen-sensitive, at various points. A finding of particular interest, in view of the centrality of 5-HT_{1A} and 5-HT_{1B} receptors in reducing aggression, is that in male mice treated with androgens, stimulation of either 5-HT_{1A} or 5-HT_{1B} receptors or combinations thereof reduces aggression. However, in estrogen-treated males, only combined, high-dose, 5-HT_{1A} plus 5-HT_{1B} receptor activation was effective. In animals given testosterone, which metabolizes into both androgens and estrogens, aggression fails to decrease with low doses of 5-HT_{1A} or 5-HT_{1B} alone or in combination, suggesting that estrogens may protect the neural systems involved in aggression from suppression by serotonin receptor activation. These relationships may also reflect action at particular sites along the brain pathways serving aggression, in that estrogen or androgen treatments differentially modulate effects of microinjections of 5-HT_{1A} or 5-HT_{1B} agonists into the lateral septum, or the medial preoptic nucleus of the hypothalamus. Such findings suggest the possibility of complex modulation of aggressive behavior, or of aggressive behavior as elicited in different sorts of situations, by reproduction-relevant steroids. These interactions are also interesting in the context of widespread environmental or food contamination by estrogenic chemicals.

Anabolic steroids are not uncommonly used illegally by athletes to enhance the development of muscle mass. Reports of enhanced aggressiveness or irritability accompanying such use have spurred experimental studies in animals. High-dose anabolic steroid treatment during early adolescence increases the aggressive response of male hamsters towards intruders, and these increases may be partly ameliorated by treatment with vasopressin receptor antagonists in the anterior hypothalamus. Treated hamsters showed enhanced vasopressin fiber density and peptide content in this area.

Gonadal and adrenal steroid hormones may interact in complex ways in association with aggression and defense. Both corticosterone and testosterone levels respond to acute and chronic defeat, with the former increasing and the latter decreasing. In fact, high circulating levels of free corticosterone may contribute to reductions in testosterone. However, corticosterone has recently been reported to enhance offensive aggression in male rats, while adrenalectomy produces a more rapid but 'deviant' attack aimed at the head of the opponent. The specificity of this enhancement is not yet entirely clear, as other behavioral consequences of corticosterone are poorly understood, but reports that blocking the mineralocorticoid receptor by spironolactone dramatically reduced territorial aggression are consistent with a view that adrenal hormones may have an important permissive role in aggression.

AGGRESSION AND DEFENSE IN HUMANS

Studies of aggression in humans have largely consisted of criminological data and articles on laboratory elicitation of aggression-like behaviors in response to some type of insult, irritation or other provocation. In general there has been little attempt to differentiate between offensive and defensive aggression, although a case has been made for anger or moral outrage as the human equivalent of offense, elicited by challenge to important rights or resources of the individual. Conversely, it has been suggested that human aggression represents defensive aggression. Because of the complexity of human verbal representations (to self and others) of the rationale or motivation for behavior (particularly behavior that is considered socially undesirable), the task of determining whether a particular instance of aggression is primarily offensive or defensive may be extremely difficult. However, this distinction is important in the treatment of some individual acts of violence by virtually every legal system yet devised, and enhanced attention to the differences in behavioral, as well as circumstantial, aspects of violent acts may yet shed light on these differences. As noted above, human aggression findings such as those involving low levels of serotonin metabolites, or effects of anabolic steroids, serve as important spurs to experimental research in animals. However, research attempting to link cognitive factors in human aggression to the circumstances that elicit aggression in lower animals has been sadly neglected. Similarly, normal human defensive behavior has received little attention. However, the potential involvement of particular defensive behaviors in specific psychopathological disorders, notably various types of anxiety disorders, is receiving increasing consideration. This is due in part to agreement between the effects of drugs on particular defensive behaviors and their efficacy against relevant anxiety disorders, and also reflects specific similarities of behavior between major symptoms of the anxiety disorder and those of the parallel defensive behavior.

Further Reading

Blanchard DC and Blanchard RJ (1984) Affect and aggression: an animal model applied to human behavior. In: Blanchard RJ and Blanchard DC (eds) *Advances in the Study of Aggression*, vol. 1, pp. 2–62. New York, NY: Academic Press.

Blanchard DC, Griebel G, Rodgers RJ and Blanchard RJ (1998) Benzodiazepine and serotonergic modulation of antipredator and conspecific defense. *Neuroscience and Biobehavioral Reviews* **22**(5): 597–612.

Canteras NS, Chiavegatto S, Valle LE and Swanson LW (1997) Severe reduction of rat defensive behavior to a predator by discrete hypothalamic chemical lesions. *Brain Research Bulletin* **44**(3): 297–305.

Delville Y, De Vries GJ and Ferris CF (2000) Neural connections of the anterior hypothalamus and agonistic behavior in golden hamsters. *Brain, Behavior and Evolution* **55**(2): 53–76.

Depaulis A, Keay KA and Bandler R (1992) Longitudinal neuronal organization of defensive reactions in the midbrain periaqueductal gray region of the rat. *Experimental Brain Research* **90**(2): 307–318.

Haller J, Halasz J, Mikics E, Kruk MR and Makara GB (2000) Ultradian corticosterone rhythm and the propensity to behave aggressively in male rats. *Journal of Neuroendocrinology* **12**(10): 937–940.

Kollack-Walker S, Don C, Watson SJ and Akil H (1999) Differential expression of c-fos mRNA within neurocircuits of male hamsters exposed to acute or chronic defeat. *Journal of Neuroendocrinology* **11**(7): 547–559.

Pope HG, Kouri EM and Hudson JI (2000) Effects of supraphysiologic doses of testosterone on mood and aggression in normal men: a randomized controlled trial. *Archives of General Psychiatry* **57**(2): 133–140.

Roeling TAP, Veening JG, Kruk MR *et al.* (1994) Efferent connections of the hypothalamic 'aggression area' in the rat. *Neuroscience* **59**(4): 1001–1024.

Siegel A, Roeling TA, Gregg TR and Kruk MR (1999) Neuropharmacology of brain-stimulation-evoked aggression. *Neuroscience and Biobehavioral Reviews* **23**(3): 359–389.

Simon NG, Cologer-Clifford A, Lu SF, McKenna SE and Hu S (1998) Testosterone and its metabolites modulate 5HT1A and 5HT1B agonist effects on intermale aggression. *Neuroscience and Biobehavioral Reviews* **23**(2): 325–336.

Aging and Cognition

Introductory article

Pat Rabbitt, University of Manchester, Manchester, UK

CONTENTS

The study of aging and cognition aims to identify the mental changes that occur in old age and relate them to corresponding neurophysiological processes.

INTRODUCTION

Cognitive aging cannot be equated with passage of calendar time because it is a complex of processes that proceed at different rates in different people and in different parts of the brains and central nervous systems (CNS) of individual persons. A main goal of cognitive gerontology is to identify the mental changes that occur in old age and to relate them to these neurophysiological changes. Demographics, socioeconomic status, and lifestyle are also important because factors such as gender, prolonged education, lengthy marriage to an intelligent spouse, complexity of workplace and social environments, level of income, and personality type all affect longevity and also maintenance of cognitive functioning in old age. Because the influences on cognitive aging are so numerous, and their interactions are so complex, individuals have strikingly different trajectories of aging. Consequently, as a population ages, the widening gap in competence between its most and least able members is more striking and informative than is the average decline in ability.

SENSORY CHANGES AFFECT, AND PREDICT, COGNITIVE ABILITY

Loss of efficiency of the sense organs is among the most obvious and widespread changes with age. Besides curtailing efficiency in obvious ways this increases the effort necessary to resolve degraded visual and auditory information, and this distracting difficulty makes it more difficult to remember, or to make useful associations to, material that has been correctly heard or seen. Losses of visual and auditory acuity, and also of balance, muscle strength, and lung capacity, are good markers for levels of cognitive efficiency in old age, probably because they reflect changes in the brain CNS resulting from cerebrovascular inefficiency and other causes.

Effects of 'Primary' and 'Secondary' Aging

For some investigators, the growing dependence of mental ability on physiological efficiency as old age advances begs the question of the relative extents to which cognitive changes are caused by 'primary' processes of 'normal' or 'usual' aging that may be genetically programmed to occur at different rates in different individuals, or by the 'secondary' effects of lifetime accumulations of pathologies and biological damage. Unsurprisingly there is clear evidence that pathologies that become common in later life, such as late onset diabetes, hypertension, or loss of cardiovascular and respiratory efficiency, can significantly reduce cognitive ability.

However, the boundaries between 'primary' and 'secondary' aging are seldom clear-cut or physiologically meaningful. For example, 'primary' heritable differences in immune system efficiency determine longevity and so also maintenance of cognitive efficiency, by protecting against the 'secondary' effects of pathology. Consequently, the effects of 'primary' aging have been defined by default after pathologies have been identified and taken into account. Alternatively, the relative impacts of primary and secondary effects have been estimated by taking individuals' calendar ages as a rough proxy for the progress of 'primary' aging and using multivariate statistics to estimate the relative amounts of variance in mental abilities between individuals that are associated with differences in their calendar ages and with their self-reported, or diagnosed, health status. Such analyses typically find that while differences in individuals' calendar ages account for only 15 to 20 percent of variance between them, differences in their self-reported health accounts for 2 percent or less.

These estimates can be misleading because while individuals' cognitive performance sharply declines with the number of different pathologies from which they suffer, people who volunteer for studies of cognition in old age are, typically, unusually fit, active, capable, and highly motivated members of their age groups. The variance associated with pathology is thus largely contributed by a few who, as individuals, have suffered severe losses. Another difficulty is that older people's self-reports of their health and cognitive efficiency must be cautiously interpreted because they cannot make absolute judgments of their status and so tend to compare themselves with their frail coevals. Perhaps for this reason, though the number of different pathologies that individuals report significantly increases with their ages, their subjective ratings of their general health may alter accordingly.

Pathologies that Directly Affect Cognitive Function

Apart from illnesses that indirectly affect mental ability by causing general physiological changes that have knock-on consequences for the central nervous system there are also pathologies, collectively known as 'dementias', that directly affect the brain. On its own the term 'dementia' is a label of convenience for changes in mental competence, much grosser than those observed in 'normal' aging, resulting from a diversity of different causes. For example, 'multi-infarct' dementia is a condition of diffuse brain damage often caused by small, but relatively numerous, cerebrovascular accidents (such as strokes, aneurisms, and minor narrowings and blockages of blood vessels) while others, such as Pick's disease, are more tightly defined in terms of causes and symptoms. The most common of these conditions, accounting for over 60 percent of all cases of dementias, is Alzheimer's disease (AD). This condition involves death of and changes in neurons resulting in the appearance of its main diagnostic criterion: 'senile plaques' and neurofibrillary tangles in brain cortex. Unfortunately these key signs can still only be ascertained on post mortem so that early diagnosis remains difficult. Other brain changes are neurotransmitter abnormalities, decreased brain volume, and, as is increasingly recognized, the incidence of many, and substantial, brain lesions that bring about marked loss of general cognitive ability and problems with memory, language, and general mental ability. Emotional problems such as anxiety, agitation, and depression, and marked personality changes may also occur. Any of these changes may also be encountered in dementias resulting from other causes.

The prevalence of AD sharply increases with age. 'Early onset' AD may occur in the 40s and 50s, the condition becomes more common in the 60s, and, it has been claimed, more than 30 percent of individuals aged 70 and above may be diagnosed as having cognitive changes consistent with a prognosis of AD. Early onset AD, occurring in middle age, progresses relatively rapidly, but when the condition occurs later in life it usually develops more slowly and patients may survive for a decade or longer.

For this article the issue is whether the marked and accelerating increase in incidence of AD with age means that it is the 'natural' and inevitable end state for all who survive long enough, or whether it should be regarded as a distinct pathology unrelated to 'normal' cognitive aging. The evidence is inconclusive. Although there are clearly genetic predisposing factors for AD, and there have been suggestions that viral or prion infection may be a causal or a triggering factor, all of the brain changes seen in AD, including the key diagnostic signs of neurofibrillary tangles and senile plaques, are found, though to a much less marked extent, in most healthy older individuals. In this respect the definition of 'normal' aging can, pragmatically, be treated as those changes in brain and behavior that are observed when a diagnosis of AD and of other dementias has been eliminated. Since, at the moment, this can only confidently be done at post mortem, the distinction is logically useful, but not always practically helpful.

Accounting for the Increased Variability between Individuals in Aging Populations

A main aim of cognitive gerontology is that by understanding the nature of declines in ability we may find means to delay them. It is therefore encouraging that trajectories of change vary markedly between individuals, because we may hope to discover what makes some more fortunate than others. For all mental skills, except those, such as vocabulary, that have been acquired early in life and practiced continuously ever since, plots of average levels of performance for successive age groups show a steadily accelerating decline after youth (Figure 1(a)). However, these averages are uninformative and might represent either of two, very different, limiting scenarios. One is that individuals' trajectories of change all have the form illustrated in Figure 1(a) but, because they accelerate at very different rates, they increasingly diverge

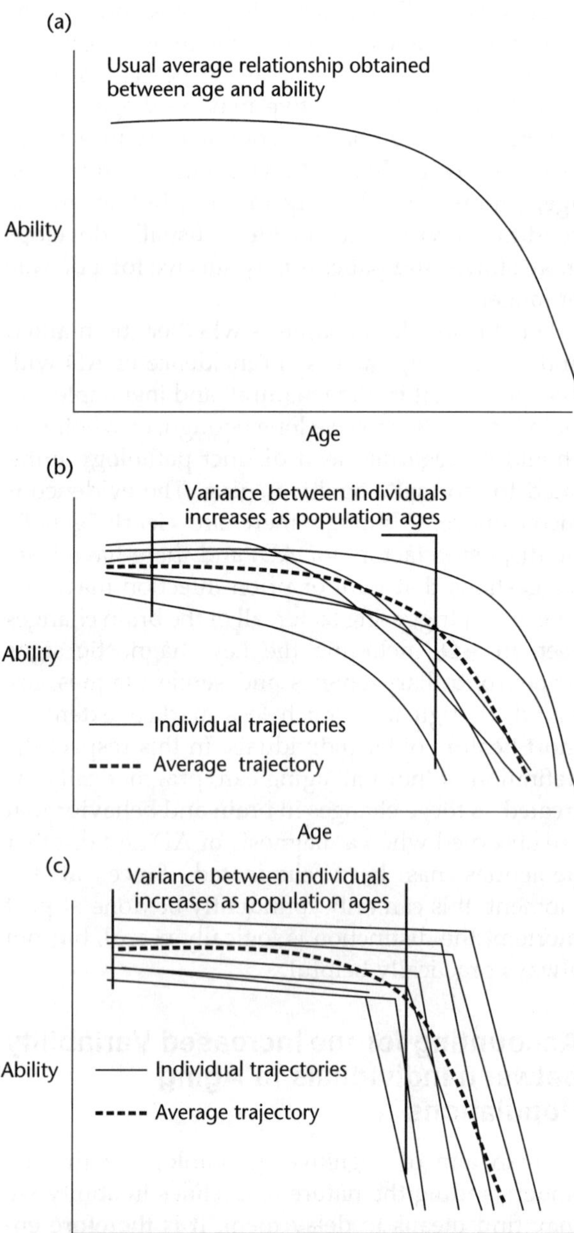

Figure 1. Relationship between age and ability. (a) Average performance levels show steadily accelerating decline after youth; (b) different rates of acceleration lead to increasing divergence in the population; (c) peak performance is maintained until terminal pathology causes abrupt decline.

as a population ages (Figure 1(b)). Another is that individuals attain peak performance early in life, and experience little loss until terminal pathology causes abrupt decline (Figure 1(c)). Either of these limiting cases can equally well account for both the form of observed average trajectories of change and the marked variance between individuals that increases as a population ages. Both indicate a degree variability between individual trajectories that encourages hopes for useful interventions.

In practice the best description will depend on the particular population examined. In socioeconomically disadvantaged populations, in which disease or accident sharply and abruptly curtails life expectancy, many individual trajectories of cognitive change will tend to show catastrophic 'terminal drops' rather than continuous declines. In affluent societies in which life expectancy is much longer, and medical interventions postpone death from pathology, gradual declines will become more common. Single cross-sectional comparisons between groups of different ages cannot distinguish true trajectories for individuals from average trajectories for populations. To do this we need longitudinal studies in which the same people are repeatedly assessed over many years.

DO DIFFERENT MENTAL ABILITIES CHANGE AT THE SAME OR AT DIFFERENT RATES?

'Global' Change Models

To relate behavioral to neurophysiological changes we must propose functional models. One limiting case is *global single factor models* which propose that changes in all mental abilities reflect changes in a unique 'master' functional characteristic, most typically 'information processing rate' or 'mental speed' or 'general fluid mental ability'. This would imply that all mental abilities 'age' at similar rates. Another limiting case is *modular aging models*, which are based on evidence that age affects some subsystems in the brain, and so the disparate cognitive abilities that they support, to different extents and at different rates.

If changes in performance on all cognitive tasks directly, and solely, reflect changes in a single, 'master' functional performance characteristic the effects of differences in age should completely disappear when the effects of individual differences in this characteristic have been directly measured and taken into consideration. So, for example, we may test whether age-related declines in 'general fluid intellectual ability' (gf), as assessed by unadjusted scores on intelligence tests (IT scores), causally determine declines in all other cognitive skills. On nearly all mental tasks individuals' levels of performance correlate negatively with their calendar ages and these correlations are, indeed, abolished or greatly reduced when differences in general mental ability, assessed by unadjusted scores on intelligence tests, are also taken into consideration. Another way to put this is that individual differences in decision speed or IT scores pick up all age-related changes in many simple laboratory tasks.

One explanation might be that IT scores directly reflect some single functional property of the central nervous system that determines levels of performance on all these tasks. A simpler, and more likely, explanation is that intelligence tests predict levels of performance over a very wide range of functionally distinct mental abilities because they include a correspondingly wide range of different problems that, collectively, make demands on all of them.

The same argument cannot dispose of a different theory, that the key performance characteristic on which all mental abilities rely is the maximum rate at which the brain can process information, and that this is directly reflected in the speed with which people can make very simple decisions. It early became apparent that age markedly slows average choice reaction time (CRT) and also that, as tasks become harder, so differences between the average decision times of older and younger groups markedly increase. Subsequent reanalyses of these data suggested that age apparently slows decisions of all kinds, and of all levels of difficulty, by the same proportional amount. That is, on any task, whatever particular demands it makes, decision times for older groups can accurately be estimated by multiplying decision times for younger groups by the same simple constant. This was taken as evidence that 'global slowing of information processing' affects all mental functions, so that changes in information-processing speed are the most sensitive and general indices of cognitive aging. Some authors go much further and suggest that individuals' IT scores, like their performance on all other tasks, directly reflect their maximum information-processing speeds because these, in turn, directly reflect the limiting efficiency of neurophysiological characteristics such as speed of synaptic conduction.

Recent work has questioned both the evidence and the methodological assumptions on which global single factor slowing models are based. It is increasingly recognized that changes in average decision times with age are better described as the consequences of greatly increased moment-to-moment variability in the speed of decisions rather than as the average slowing of all decisions. A direct consequence of this increased moment-to-moment variability is greater average variability from day to day and week to week and that, in both these respects, older people show much greater variability than do the young. When the same individuals are tested on a wide range of tasks, those who show greater moment-to-moment variability on one kind of task are also found to be consistently more variable on others. Findings that age increases variability within individuals at least partly account for findings that, when people are compared on any given occasion, age markedly increases variability between individuals.

There is also evidence that 'global slowing' does not affect all cognitive skills equally. While 'fluid intellectual abilities', indexed by IT scores, decision speed, and rate of learning of novel tasks, all markedly decline with age, language skills, that have become crystallized by practice throughout a lifetime, remain unchanged, or may even slightly improve into the late eighth or even the ninth decade of life. For example, age markedly affects the time taken to respond to simple signals such as lights or tones but has much less effect on speed of decisions about words. Age also does not reduce the speed with which people can carry out highly practised skills such as mental arithmetic. Sustained practice over a lifetime protects skills from age-related decline, and even tasks that have been relatively briefly practiced in the laboratory to the point when they become automatic are then less affected by age.

'Local' or 'Differential' Change Models

Post-mortem and brain imaging studies suggest that age affects frontal cortex more radically and earlier than other parts of the brain in terms of loss of volume, cell loss, reduced cerebral blood flow, and reduction in the concentration, synthesis, and number of receptor sites for dopamine and other neurotransmitters. Consistently, some investigators have found that clinical tests for frontal damage, and laboratory tasks that make demands on functions supported by the frontal and prefrontal cortex, are especially sensitive to age. These functions include the inhibition of unwanted information or of inappropriate responses, the ability rapidly and accurately to generate categories of words, and the ability to switch easily between different criteria for classification of signals. However, there have been inconsistencies of replication due to problems of measurement, problems of task familiarity, problems of construct validity, and, finally, and probably most basically, much neglected problems of participant selection. On balance, some frontal tasks do seem particularly sensitive to age-related changes, but it is not yet clear whether this is because the incidence of focal brain lesions increases in older populations or because all, or most, people suffer varying degrees of diffuse frontal changes.

MEMORY CHANGES IN OLD AGE

Both older people's subjective complaints and countless laboratory studies confirm that memory efficiency declines in old age and that, in general, the more difficult a memory task is, the greater will be the difference between the average numbers of errors made by young and by elderly adults. This general interaction between age and task difficulty is methodologically inconvenient, because to establish that age affects some functional processes more than others we must show that older people find it especially difficult to cope with a particular *kind* of task demand rather than simply a general increase in task difficulty. Neglect of this point makes brings into question much of the evidence that age affects some particular functional processes in memory, or some 'kinds' of memory, more than it affects others. Consequently, rather than reviewing evidence that age differentially affects speculatively different functional memory processes it may be more helpful to consider the effects of age in terms of the different uses that individuals make of their memories in their everyday lives.

WHAT IS MEMORY USED FOR?

Age changes in efficiency of immediate verbatim memory are slight but consistent, and seem to occur both because age speeds the rates at which memory traces decay and slows the rates at which they can be rehearsed. In everyday life a more common problem is to recover information selectively rather than completely, in some different order from that in which it is presented, or transformed in some way. The ability to meet such demands, particularly in order to schedule decisions and choices, has become known as 'working memory'.

Use of working memory allows people to transcend the limitations of their immediate verbatim recall for novel material ('memory span') by developing and using mnemonics to recode information about items or events. When sufficiently practiced, such recoding techniques seem to allow recall of almost limitless amounts of information but, because old age slows decision speeds, it also makes such recoding less efficient. The extent to which older persons' memories of events are limited by the amounts of information that they can immediately process is illustrated by findings that when older people are asked to recall accounts of actions performed by others, actions they have themselves performed, or actions they have imagined performing, they often correctly recall particular actions but have difficulty remembering whether they have performed, heard about, or imagined them.

Even when they do not deliberately and consciously use mnemonic techniques, people of all ages encode their experiences in terms of their expectations of what is most likely to have occurred. In this sense, video- or tape-recording that passively and unselectively registers information, subject only to failures to register or loss of information, provide poor metaphors for memory which is, rather, a dynamically selective and reconstructive process in which current motivations and lifetime knowledge of the world determine which aspects of new events are attended to, which are ignored, and how previous experience is used to shape interpretation. Because older people can process less information about events in unit time, and lose more of the information that they process, they must increasingly rely on their knowledge of the world to guide economical selection and to reconstruct events from sparse remembered detail. Increased reliance on interpretation and expectation can betray older people when they recreate events that contain unexpected and unfamiliar elements that cannot easily be inferred from context.

This dynamic view of memory highlights the counterintuitive point that humans and other animals typically use their memories to tell them what to do next rather than as archives of past experiences. This use of previously acquired information to anticipate future events and to formulate and execute appropriate plans to cope with them is termed 'prospective memory'. There is evidence that age does impair prospective memory but in everyday life these effects may not be noticeable because most of us, and perhaps especially the elderly, live predictable lives in structured environments that support us by providing contexts that cue us to perform appropriate actions. Such 'environmental support' is especially valuable to the elderly, and may serve them so well that the full extent of their memory impairment is not apparent until they are deprived of familiar routines and environments.

Once information has been successfully encoded there seems to be no limit on how much of it can be retained, or for how long it remains available. This uncertainty is partly due to the logical difficulty of demonstrating that any information has been permanently lost, rather than having become temporarily inaccessible. Nevertheless, older people's ubiquitous subjective complaints, and many laboratory studies, show that they find it more difficult to access information that they are sure they once knew. For example, when young and older adults are compared on recall of public events that they have both shared, the young do much better. This can be taken to show that the young have forgotten less, but it may also mean that they could encode the events better when they first experienced them so that their memory representations have always been correspondingly more detailed and elaborate.

The complexity of inferences necessary when making such comparisons is illustrated by studies of the relative frequency, and efficiency, with which people of different ages can spontaneously recall events from different times of their lives. Both elderly and young adults recall very recent events more often, and more vividly, than distant events. Apart from losses of information over time a likely explanation is that recent events may be relatively frequently recalled, pondered, and discussed because they tend to have implications for the immediate future. Older people recall relatively fewer events than do the young from intermediate periods of six months or a year previously, but also recall relatively more events from their adolescence and young adult lives than from their middle age. This may partly be because events are better and more elaborately encoded, and so longer remembered, in youth than in middle or in old age, but another factor is that because young adult experiences are often more engaging, and have more long-lasting and memorable consequences than those that occur later in life, they may have since been much more often recalled, reassessed, and discussed. For similar reasons, at whatever age they may occur, vivid experiences during dramatic historical periods, such as wars or marked social change, are better recalled than those in more humdrum epochs. These factors can explain the subjective contrast that older people experience between their vivid recall of remote events and tenuous grasp of the recent past. However, the need for a distinction between older people's subjective impressions of vividness, and the objective completeness and accuracy of their memories, is illustrated when documentation such as school records is available. Older individuals' recall of events in their early lives is often much less reliable than they suppose.

The idea that the accuracy and durability of memories depends on the period of life when they were laid down is supported by recent work showing that individuals who have attained similar standards in university degree examinations forget their hard-won, but seldom used, knowledge more rapidly if they acquired it in early middle age than as young adults.

In contrast to marked age decrements in accuracy of explicit, conscious, recall and recognition of items and events there is evidence that information that is held in memory, but cannot be overtly recalled, can nevertheless influence behavior. For example, people can solve anagrams of words that they have recently learned, but subsequently failed to remember or recognize, more easily than anagrams of words that they have not recently seen, and fail to register as recurring. Loss of conscious, 'explicit' memory, with preservation of unconscious, 'implicit' memory, is found not only in older people but also in patients suffering from amnesias resulting from local damage to the hippocampus and temporal lobes of their brains. This has been taken as evidence that explicit and implicit memories are retained by functionally separate systems that are differentially sensitive both to focal brain damage and to normal aging.

Investigations of the ways in which old age alters mental abilities, particularly memory, have produced extremely detailed descriptions of behavioral changes but, so far, much less information on how these changes relate to changes in the brain and central nervous system. One obvious reason is that our knowledge of relationships between brain and cognition has mainly depended on studies of rare patients with highly localized brain damage and with equally well-defined cognitive impairments. In contrast, old age brings about both diffuse and focal brain changes. Prior to recent advances in brain imaging, diffuse brain changes could not be assessed until death terminated investigations of behavior and post-mortem data became available. Brain imaging now allows detection of global and diffuse, as well as local and specific, brain changes while individuals' cognitive losses can still be investigated. Methodologies for acquisition and interpretation of behavioral data have also greatly improved. It seems likely that the first decade of the twenty-first century will see greater advances in our understanding of the effects of aging on cognition than have been possible during the entire twentieth century.

Further Reading

Craik FIM and Salthouse T (eds) (1992) *The Handbook of Aging and Cognition*. Hillsdale, NJ: Lawrence Erlbaum.

Kausler DH (1990) *Experimental Psychology, Cognition and Human Aging*. New York: Springer-Verlag.

Rabbitt PMA (2001) Methodology of cognitive gerontology. In: Wixtead J (ed.) *Stevens Handbook of Experimental Psychology*, vol. 4. New York, NY: John Wiley.

Salthouse TA (1991) *Theoretical Perspectives in Cognitive Aging*. Hillsdale, NJ: Lawrence Erlbaum.

Aging, Neural Changes in

Introductory article

Elizabeth A Kensinger, Massachusetts Institute of Technology, Cambridge, Massachusetts, USA
Suzanne Corkin, Massachusetts Institute of Technology, Cambridge, Massachusetts, USA

CONTENTS

Dementia is not an obligatory consequence of aging. Normal aging does, however, result in changes in cognition, caused by a combination of neurotransmitter abnormalities and alterations in the structure and function of the brain.

INTRODUCTION

The increasing proportion of older adults in the population of many countries has heightened interest in the cognitive and neural changes that accompany normal aging. The following sections elucidate the effect of aging on a range of cognitive capacities, and the neural changes that may explain the pattern of spared and impaired function. Particular attention is devoted to short-term and working memory (allowing the temporary storage and manipulation of information); declarative memory (requiring conscious awareness); and nondeclarative memory (formed without conscious awareness). The evidence presented here comes from behavioral studies in healthy volunteers (cognitive psychology), patients with brain lesions (neuropsychology), analyses of brain structure (volumetric magnetic resonance imaging) and observations of focal task-related changes in normal brain activation during performance of specific cognitive operations (functional neuroimaging using positron emission tomography or magnetic resonance imaging).

DECLARATIVE MEMORY

Long-term memory is broadly divided into two components: declarative and nondeclarative memory. Declarative (explicit) memory is formed with conscious awareness, and requires the participation of medial temporal lobe structures, including the hippocampus. Declarative memory is what we use to help us remember the items we need to pick up at the grocery store, or the name of a friend whom we haven't seen in years. In general, declarative memory is more affected by normal aging than is nondeclarative memory. As an example of declarative memory, read the following words, and try to remember them. After reading through the list, write down as many words as you can remember, without looking back at the list: table, orange, calendar, computer, paper, needle, napkin, chair, sneeze, movie, sleep, castle, build, lunch, flower, dragon, plant, cushion, dolphin, muscle.

SENSORY MEMORY (PERCEPTION)

The sensory systems are those through which we receive direct input from the world via receptors: touch, taste, smell, vision, and hearing. Our perception of that input, however, can be influenced by a number of factors, such as our internal state, or the context in which the sensation occurred. Imagine you hear a loud banging sound while you are walking by a construction site. Now imagine that you hear the identical sound while walking alone on a deserted street late at night. Although the sound (sensory information) is identical in the two instances, the interpretation of the sound (perception) may differ greatly because it occurs in different contexts.

Some sensory systems are degraded as part of the normal aging process. Most commonly, older adults experience hearing loss. By the age of 80 years the majority of older adults have significant hearing loss. They also have visual deficits, including poorer color and luminance contrast, and many have a loss of central vision due to macular degeneration. Sensory and perceptual deficits can hinder adults' performance on many tasks. For example, Murphy and colleagues found that older adults are more affected by background noise when trying to remember word pairs than are young adults. These researchers proposed that part of this increased effect is due to degraded sensory representations, though attentional reductions probably also contribute. Older adults also may have more difficulty discriminating isoluminant colors such as blue and green, and are slower and less accurate on tasks that require color discrimination, such as color-naming tests, the Stroop Test, and the Wisconsin Card Sorting Test.

Nevertheless, with modifications to testing procedures (such as louder stimuli) most older adults are successfully able to perceive information. Perceptual priming, requiring visual processing, is spared with normal aging. Older adults are also able to repeat word lists or digit strings (presented either aurally or visually), suggesting that their sensory deficits are not profound enough to affect immediate memory. It is, nonetheless, important to control for perceptual confounds when interpreting the performance of older adults, and to match (equate) the perceptual capabilities of young and older adults (either by matching individuals or, more feasibly, matching the stimuli so that young and older adults perceive them equally). Without taking these measures, it is unclear whether impaired performance in older adults stems from a purely cognitive deficit or from impaired perception. Particularly in memory studies, reduced perception may result in older adults having a degraded memory representation, subject to faster disruption over time.

SHORT-TERM MEMORY AND WORKING MEMORY

Short-term memory is a limited-capacity storage buffer for information to be remembered over a very short duration (a few seconds). The term 'short-term memory' is commonly used to mean recent memory, but that definition is not used by cognitive psychologists. Short-term memory consists of two components: a passive information store and an active rehearsal system. Working memory, in contrast, not only stores information but also updates and manipulates that information.

Read the following words, and try to keep them in mind for 10 s: hill, milk, goat, tool, foot, pie. This type of storage requires short-term memory. To succeed in repeating the words 10 s later, you might also have felt that you were 'rehearsing' those words (e.g. internally vocalizing them) to allow yourself to remember them. This phenomenon highlights the active rehearsal component of short-term memory. Now, read the words again, look away, and this time try to say them in alphabetical order. Simply rehearsing the words is insufficient to complete this task; rather, you also need to manipulate the words to place them in the proper sequence. This task, therefore, requires working memory.

The most widely accepted model of working memory, proposed by Baddeley and Hitch, defines working memory as consisting of three components: the central executive, the phonological or articulatory loop, and the visuospatial sketchpad. The central executive controls the allocation of attention, as well as the coordination and monitoring of activities, while the phonological loop and visuospatial sketchpad are slave systems of the central executive that temporarily maintain and manipulate verbal and nonverbal material, respectively.

Short-term memory is usually spared with aging, whereas working memory shows age-related decrements. This decline probably does not occur equally across all components of working memory, but rather targets only a subset of processes. Three components probably account for the majority of age-related working memory decline: processing speed, storage capacity, and inhibitory ability.

Processing Speed

Older adults are known to have a slowed speed of processing. Salthouse and colleagues proposed that decreased processing speed could account for some of the age-related declines in cognition. They suggested that cognitive performance suffers because (a) the slowed mental operations cannot be carried out within the necessary time frame, and (b) the increased time between mental operations makes it more difficult to access previously processed information. Processing speed can affect encoding because the quality and availability of perceptual information degrades over time, so information that is processed quickly will be encoded more effectively and, therefore, will have a more durable representation or memory trace.

The hypothesis of a relation between processing speed changes and cognitive decline has been confirmed in a number of studies. Longitudinal studies have shown that changes in speed of processing may predict longitudinal cognitive decline, and a number of researchers have found that controlling for speed eliminates age effects on various memory tasks.

Storage Capacity

Storage capacity is one component of short-term and working memory: the passive storage buffer that dictates how much information can be stored without rehearsal being used to 'refresh' that information. A reduction in storage capacity is likely to contribute to age-related working memory decline: older adults may be able to hold less information in mind. Reduced storage capacity could provide an alternate explanation to reduced processing speed. Thus, remembering what information was processed, or carrying out mental operations, would be restricted not by time pressure but by reduced storage capacity. Although storage capacity declines with age, it is not clear that this deficit is sufficient to explain the cognitive decrements in working memory that occur with aging.

Inhibitory Ability

Hasher and Zacks proposed the inhibitory deficit theory to account for changes in cognitive performance with age. 'Inhibition', in this theory, is the ability to ignore irrelevant information while focusing attention on pertinent information. The inability to filter out irrelevancies causes older participants' working memory to be filled with unneeded information, leaving less space for task-relevant memories. This explanation, therefore, is not completely dissociable from a storage capacity explanation for cognitive aging.

Researchers have found evidence for inhibitory deficits in older adults on a variety of tasks. Commonly used paradigms for assessing inhibition are task-switching or set-shifting. On these tasks, participants must first remember one set of rules or pay attention to one salient characteristic, and then must switch rules or attend to a different characteristic. Most investigators have found that these tasks are sensitive to aging effects, with older adults being less able to ignore the previously relevant information.

LONG-TERM MEMORY

Long-term memory can be divided into two categories: episodic and semantic. Episodic memory entails retrieving information from a particular episode, localized in space and time (e.g. remembering seeing the Eiffel Tower on your first trip to Paris), while semantic memory requires retrieving factual information independent of any specific episode (e.g. knowing that the Eiffel Tower is in Paris). Recall of the word list given at the beginning of this article required episodic memory. You had to bring the word to mind, and also correctly remember that the word was on the list you had just read. Accessing the meaning of the words, however, required semantic memory.

Episodic Memory

Episodic memory appears to be more affected by normal aging than other memory processes. All aspects of episodic memory are not affected uniformly, however.

Factual and source memory

Episodic memory can be subdivided into two components: factual memory and contextual or source memory. Normal aging results in a disproportionate impairment in source memory as compared with fact memory. Even when older adults remember a fact or event, they have more difficulty than younger adults pinpointing the specific contextual details, such as where and when they learned a fact. For example, Spencer and Raz tested young and older adults on a test requiring them to remember facts, some true and some fictitious (e.g. 'Angela Lansbury regularly consults with astrologists'). After a delay, participants were asked to complete the fact ('Angela Lansbury regularly consults with …'), and to say where they had learned the fact (experiment or elsewhere) and whether the fact had been presented on a blue or pink card. Older adults were disproportionately impaired on the source recall than on the fact recall.

Source memory is believed to rely on the brain's frontal lobes. Measures of frontal lobe function correlate with measures of source memory, and reductions in source memory have been shown to occur in amnesic patients with frontal lobe lesions. The frontal lobes are also critical for linking events together in time. Aging results in frontal lobe dysfunction, probably connected to the source memory deficits seen in older adults.

Recall and recognition

Older adults show poorer performance on recall tests ('What words were on the word list?') where no cue is provided, than on recognition tests ('Was "cloud" or "table" on the word list?') where retrieval cues are provided. In general, older adults show improved performance on episodic memory tests when cues are provided during encoding or retrieval phases.

The source memory decrement, and the benefit provided to older adults with cues, are probably related to the robustness of the memory trace encoded by the older adults. Aging seems to affect the quality of the representation, such that general gist-based information is more easily encoded and retrieved than richer, item-specific information that includes not only the to-be-remembered information, but also the context in which it was learned. This hypothesis is supported by the finding that on recognition tasks, older adults are more likely than young adults to say that an item is 'familiar' (they feel they have encountered the item before), but less likely to say that they 'recollect' the item (remember something specific about the item's presentation).

Semantic Memory

One of the most readily reported complaints by older adults is their declining ability to recall the names of people and objects. Word finding difficulties are among the most severe deficits in normal aging. Naming deficits result in slower speed of picture naming, a greater number of speech disfluencies, and an increased number of tip-of-the-tongue effects.

Picture naming

Older adults' naming deficit is particularly pronounced for proper names, though studies have also reported longer naming times for nonproper objects. The difficulty may be related in part to deficits in associative memory: the ability to form associations between a name and an object may be reduced in normal aging.

Tip-of-the-tongue effect

The tip-of-the-tongue effect occurs when a person has access to a word's meaning, but is unable to produce the phonological code. Older people report more tip-of-the-tongue experiences with everyday objects and with proper names than younger people. In addition, the accuracy of available information during a tip-of- the-tongue state is higher for young than old adults. For example, younger participants are more likely to state correctly the first letter of the word they are trying to remember than older participants. As with naming deficits, tip-of-the-tongue effects are more pronounced for proper names than for everyday objects. Better performance with everyday objects may be related to what Burke and colleagues refer to as 'summation of priming'. With everyday objects, connections from a variety of semantic associates converge on the correct name; but with proper names, older adults are handicapped without this type of summation.

NONDECLARATIVE MEMORY

Nondeclarative (implicit) memory is encoded and strengthened, across trials, without conscious awareness. It encompasses a heterogeneous group of processes and kinds of performance, including skill (motor) learning, repetition priming, and classical conditioning. These domains rely on distinct and separable neural substrates. Because of the task diversity, and range of necessary neural substrates, it is perhaps logical that nondeclarative memory is not uniformly impaired with aging.

Skill (Motor) Learning

In the 1960s, Milner demonstrated that the amnesic patient HM, while unable to form new declarative memories, could successfully learn a new motor skill. She asked HM to perform a mirror tracing task, in which he had to trace the outline of a star seen only in mirror-reversed view. Over 3 days of practice, his error scores decreased dramatically, and he maintained the learning from one day to the next, but he had no conscious recollection that he had done the task before. Corkin and colleagues administered additional skill-learning tasks to HM, confirming that he generally showed preserved learning. Other investigators have also reported that amnesic patients can learn and retain motor skill learning without awareness of prior exposure to the task.

These results indicate that the brain structures that support conscious, declarative memory and which are damaged in amnesia (the hippocampus and other medial temporal lobe structures) are not critical for skill learning. Skill learning is thought to rely on the motor cortex, supplementary motor area, cerebellum, basal ganglia, and posterior parietal cortex. Older adults have reductions in the amount of dopamine and acetylcholine in the basal ganglia; they also show cerebellar dysfunction. These changes may result in slower acquisition of some motor learning tasks.

No consensus exists as to how skill learning is affected by aging. Researchers have found every possible outcome: equal performance in young and older adults, better performance in older adults, and poorer performance in older adults.

Repetition Priming

Priming is broadly defined as a faster or biased response to a stimulus based on prior exposure to that stimulus, or a related stimulus. As an example of priming, try to complete these word stems with the first word that comes to mind: nap—, dol—, cas—, cus—, tab—, dra—. You may have responded with words from the list given at the beginning of this article, without being consciously aware that you had done so. This effect, based on prior exposure to a stimulus, is an example of repetition priming.

Priming is not a unitary construct; rather, multiple processes contribute to priming effects. For discussion purposes, we will divide priming into two categories: perceptual priming and conceptual priming. These types of priming are dissociable and rely on separate neural substrates.

Perceptual Priming

Perceptual priming is based on the sensory characteristics of a stimulus. For example, if participants are shown the pseudoword 'pabhan', they will later be more likely to recognize that pseudoword when it is flashed briefly, than another pseudoword flashed at the same rate. Keane and colleagues proposed that perceptual priming effects are mediated by a structural–perceptual memory system localized to the occipital lobe; this hypothesis has been supported by neuropsychological and functional imaging studies.

Older participants frequently perform as well as younger adults on perceptual priming tasks. For example, Schacter and colleagues presented young and older adults with black-and-white drawings of three-dimensional objects in either structurally possible or impossible configurations. When participants had to judge whether the briefly presented stimuli were possible or impossible objects, young and older adults showed the same magnitude and pattern of priming, with robust priming for possible objects and no priming for impossible objects.

The finding of spared perceptual priming with aging is consistent with its reliance on the occipital and temporoparietal cortex because aging is thought to spare primary cortices and modality-specific association areas, including the occipital lobe.

Conceptual Priming

In contrast to perceptual priming, conceptual priming relies primarily on the semantic representation of the stimulus. For example, if participants are first presented with the category word 'fruit', they will be faster at determining that the word 'apple' is a real word than if they were first presented with the category word 'furniture'. Keane and colleagues proposed that conceptual priming is mediated by a lexical–semantic memory system recruiting temporoparietal regions. This hypothesis has been supported also by neuropsychological and neuroimaging studies.

Some studies have reported age-related deficits in priming experiments that are conceptual in nature, including lexical priming and priming for new word associations. Other researchers, however, have reported spared performance in older adults. The discrepancy may have stemmed from different task designs, or individual variation within the older populations.

CLASSICAL CONDITIONING

One of the most commonly used forms of classical conditioning is the eyeblink response. In delay conditioning, a neutral stimulus (a tone) is followed repeatedly by a biologically relevant stimulus (an air puff to the eye), and the two stimuli coterminate. The measure of learning is the subsequent ability of the tone, by itself, to elicit a biologically relevant response (an eyeblink), the conditioned response. The strength of the conditioned response increases gradually with repetition, making it possible to document the number of trials needed to learn to a particular criterion. Older rabbits and older humans require significantly more trials than younger ones to acquire the association between the tone and the air puff, but considerable variability exists among older individuals. Results from neuroimaging and neuropsychology converge on the conclusion that the cerebellum is the critical neural substrate for delay conditioning. Because the cerebellum is affected by normal aging, the reduction in classical conditioning with normal aging is believed to result from less efficient cerebellar communication and output.

Trace conditioning differs from delay conditioning in that there is an unfilled interval between the offset of the neutral stimulus (the tone) and the onset of the biologically relevant stimulus (the air puff). The participant must therefore build up a representation, across trials, as to the relation between the tone and air puff. In addition to cerebellar recruitment, the hippocampus is critical for trace conditioning. The hippocampal contribution is likely to stem from the fact that delay conditioning is not purely a nondeclarative memory task; conscious awareness of the relation is mandatory for successful conditioning.

On the trace conditioning paradigm, young and middle-aged adults condition at a similar rate, but older animals and humans are impaired. These deficits may occur at an earlier age than deficits in delay conditioning, and may be more pronounced.

AGE OF COGNITIVE DECLINE

Methods of Assessment

The age of cognitive decline can be assessed using one of two designs: cross-sectional or longitudinal.

Cross-sectional studies use data collected from individuals considered to be representative of an entire population, and interpret differences among those individuals as indicative of differences across two or more populations. For example, a cross-sectional study of aging might examine the performance of adults in their twenties, fifties and eighties. If the 80-year-olds performed more poorly than the other groups, this difference would be attributed to age. This design requires that groups be equated (matched) on as many variables as possible (e.g. overall intelligence and perceptual ability, as well as lifestyle, psychological and medical factors) to assure that group differences are due to age and not to other differences.

A longitudinal study avoids many of these confounds by tracking the same group of individuals across time, and comparing their performance at different time points. For example, a group of adults might be tested every 5 years for 20 years. Because each individual serves as his or her own baseline, the investigator does not have to worry about confounds such as intelligence or education level. Changes in overall health or perceptual ability over time must still be considered, and longitudinal studies can also be confounded by nonrandom drop-out rates (e.g. in a memory study, individuals who believe their memory is failing might be more likely to drop out of the study than those who believe their memory is good).

Cognitive Performance

The worsening performance across an extensive age range has led many researchers to divide the older adult population into 'young-old' and 'old-old' subgroups. This dichotomy was first proposed by Neugarten, who noted that these groups were dissociable not only by chronological age, but also by lifestyle changes. The young-old have fewer health limitations than the old-old, and the old-old are more likely to be widows or widowers than the young-old.

Researchers have used this dissociation to examine the progression of cognitive changes into the later decades of life. Most studies have confirmed that memory loss does not reach a plateau in the sixth or seventh decades; rather, memory decline continues throughout the later decades. Adults over the age of 70 years perform significantly worse on a range of recognition tasks compared with individuals in their seventh decade of life. Deficits in semantic memory and conditioning can also become more pronounced in the old-old.

The age at onset of decline differs depending on the type of function assessed. Semantic memory, as measured by tip-of-the tongue effects, has been found to be altered between the fifth and sixth decades. Woodruff-Pak and colleagues, however, found that eyeblink classical conditioning decrements began almost a decade earlier, with 40-year-olds showing significant impairments. Episodic memory, in contrast, remains relatively stable until around the seventh decade.

ANATOMICAL CHANGES

Longitudinal studies have found decreases in overall grey and white matter volumes with age, as well as increases in volumes of ventricular cerebrospinal fluid. The changes are not uniform across all brain regions, however. For example, the prefrontal cortex and medial temporal areas are more affected than primary association cortices. The pattern of neural changes helps to clarify why some types of cognition are particularly affected by normal aging, while other cognitive processes are relatively spared.

Hippocampus and Other Medial Temporal Lobe Structures

Hippocampal function is impaired by normal aging. Functional neuroimaging studies have shown that the hippocampus and other medial temporal lobe structures are less activated by memory tests with aging, and these functional changes often correlate with memory performance. A quantitative imaging study assessing the volume of different brain regions also found that hippocampal volume is significantly correlated with performance on delayed recall tests. In fact, out of a variety of brain regions measured (including overall brain volume), hippocampal volume was the best predictor of delayed memory performance.

It is unclear whether there is substantial cell loss in this region, or whether the hippocampal dysfunction is related to neuropathological changes and cellular dysfunction affecting neuronal communication. On postmortem examination, adults over the age of 55 years typically show at least some entorhinal neurons that contain tangles, or where tangles are beginning to form. Brain neurochemistry also appears to be altered, with reductions in synaptic signaling. For example, long-term potentiation, thought to be a critical neural mechanism for learning and memory, is reduced with normal aging. Reductions in the number of NMDA (*N*-methyl-D-aspartate) receptors in the hippocampus, or reductions in the efficiency of the receptor, may mediate some of the age-related hippocampal dysfunction. Glucocorticoids, too, mediate hippocampal function, and increases in glucocorticoid levels may contribute to dysfunction.

Even studies that have found cell loss do not agree on which medial temporal lobe regions are most affected. While a number of studies found evidence for cell loss in the CA1 region of the hippocampus, not all studies have replicated this finding, using unbiased stereologic counting methods.

Cerebellum

Studies of humans, rats and rabbits suggest that the cerebellum, in particular the Purkinje (output) cells, is affected by aging. Older animals have fewer Purkinje cells, and those that remain have a reduced efficiency. Because Purkinje cells are the major output system of the cerebellum, damage to these cells results in dramatically reduced cerebellar output. Evidence for structural changes in humans comes from a magnetic resonance imaging (MRI) study showing significant negative correlations between age and grey matter volume in the cerebellar vermis and hemispheres.

Prefrontal Cortex

The function of the prefrontal cortex is affected by aging. Older adults perform more poorly than younger adults on tasks that measure frontal lobe capacities, including the Wisconsin Card Sorting Test and the Stroop Test. Neuroimaging studies have indicated changes in prefrontal activation, particularly in dorsolateral prefrontal cortex. Even on tasks where young and older adults perform at similar levels, prefrontal regions in older individuals show different patterns of activation, including recruitment of additional areas, and reduced activation in other regions relative to young adults.

As with the hippocampal region, it is unclear what neuropathological changes account for deficits in frontal lobe capacities. Researchers have proposed that neuronal shrinkage, or reductions in the number of presynaptic terminals, may be responsible for some of the age-related impairment. Axonal abnormalities may also underlie age-related deficits. In a volumetric MRI study, Double and colleagues found frontal lobe white-matter atrophy, suggestive of reductions in axonal processes. They suggested that slowed cognitive processing may occur because of a decrease in the speed of nerve conduction due to such axonal changes. These alterations may account for the working memory deficits with normal aging.

Neurotransmitter and Neuromodulator Abnormalities

A neurotransmitter is a chemical messenger that is released by one neuron, travels across a space between two neurons (a synapse), and binds to the second neuron. In this way, information is passed between neurons. A neuromodulator is a chemical that is not itself a transmitter, but affects the release of neurotransmitters.

Dopamine

Age-related changes in the dopaminergic system are well documented in humans, monkeys, and rodents. Levels of dopamine and tyrosine hydroxylase (an enzyme important for the production of dopamine) decrease with normal aging, and these reductions are particularly pronounced in the frontal lobes and basal ganglia. Postsynaptic alterations are also reported to occur with aging, including reductions in D2 dopamine receptors and some increases in D1 receptors.

Age-related reductions in dopamine levels may contribute to age-related working memory impairments. Dopamine depletion in the frontal lobes impairs performance on working memory tasks, and dopamine may be particularly important for inhibitory ability. Prefrontal cortex must sort out task-relevant information, and maintain that information in the face of other distractors. Dopaminergic systems may provide the basis for that allocation of attention. Dopamine may potentiate synapses associated with a reward (e.g. correct recall), thereby intensifying links between task-relevant computations, and weakening others.

Acetylcholine

Considerable evidence links acetylcholine to learning and memory. Acetylcholine is released when animals perform spatial memory tasks, and injection of cholinergic antagonists such as hyoscine (scopolamine) impairs memory acquisition in humans and nonhuman primates.

In aged animals, memory loss is correlated with hypoactive cholinergic neurons. For example, older rats show reduced excitatory postsynaptic potential amplitudes resulting from stimulation of the CA1 region of the hippocampus, suggesting that cholinergic neurons are less responsive in older animals. Older animals also show reductions in cholinergic receptor density that are particularly pronounced in the medial and caudal parts of the striatum, and in the frontal lobes.

Adrenal glucocorticoids

Stress hormones, too, are linked to the neural loss and dysfunction associated with normal aging. The adrenal cortex (in the adrenal glands, located near the kidneys) secretes glucocorticoids, which underlie our physical responses to threatening stimuli. In the short term, glucocorticoids are essential to our survival because under stressful conditions they increase the availability of energy substrates (blood glucose). Prolonged exposure to elevated glucocorticoid levels, however, can be detrimental, suppressing anabolic processes and depleting existing energy stores. With age, the stress response is not terminated as efficiently, causing glucocorticoid levels to remain elevated for significantly longer following stress.

The hippocampus, one of the main target sites for glucocorticoids, seems to be hardest hit by prolonged glucocorticoid exposure. When rats underwent experimental removal of the adrenal glands, disrupting glucocorticoid production, aged rats showed little or no evidence of hippocampal neuron loss as compared with control rats. These results link the production of glucocorticoids to the hippocampal atrophy that occurs with aging. Further evidence for this hypothesis comes from studies in Sapolsky's laboratory, showing that young rats treated with corticosterone show patterns of hippocampal cell loss similar to that in aged rats. Sapolsky and colleagues suggest that the effect of glucocorticoids on hippocampal neurons is probably related to metabolic changes stemming from the fact that glucocorticoids inhibit glucose uptake.

Rate of Decline

As discussed above, different cognitive processes decline at differing phases of the aging process. These differences are probably related to the times that neuropathological abnormalities appear in different brain regions. Cerebellar atrophy, thought to cause changes in the acquisition of a conditioned response, may start at an earlier age than most other brain changes, with significant atrophy present by the fifth decade of life. Other regions such as the medial temporal lobe or frontal lobe may not be altered until the seventh or eighth decades. Similarly, neurotransmitter changes, such as dopaminergic reductions, are thought to start around the seventh decade of life and to continue throughout the remaining adult years.

CONCLUSION

Aging does not affect all aspects of cognition uniformly. It does, however, affect a range of cognitive capacities. These changes are not static, but rather continue to intensify. Cognitive alterations are intimately linked to age-related changes in the neurotransmitter systems and in the structure and function of the brain.

Further Reading

Baddeley AD and Hitch GJ (1974) Working memory. In: Bower GH (ed.) *The Psychology of Learning and Motivation*. New York, NY: Academic Press.

Burke DM, MacKay DG, Worthley JS and Wade E (1991) On the tip of the tongue: what causes word finding failures in young and older adults? *Journal of Memory and Language* **30**: 542–579.

Craik FIM and Salthouse TA (1999) *The Handbook of Aging and Cognition*. Mahwah, NJ: Lawrence Erlbaum.

Double KL, Halliday GM, Kril JJ *et al.* (1996) Topography of brain atrophy during normal aging and Alzheimer's disease. *Neurobiology of Aging* **17**: 513–521.

Golman-Rakic PS and Brown RM (1981) Regional changes of monoamines in cerebral cortex and subcortical structures of aging rhesus monkeys. *Neuroscience* **6**: 177–187.

Hasher L and Zacks RT (1988) Working memory, comprehension, and aging: a review and a new view. In: Bower GH (ed.) *The Psychology of Learning and Motivation*, vol. 22, pp. 193–225. New York, NY: Academic Press.

Light LL and Burke DM (1993) *Language, Memory, and Aging*. New York, NY: Cambridge University Press.

Makman MH and Stefano GB (eds) (1993) *Neuroregulatory Mechanisms in Aging*. New York, NY: Pergamon Press.

Murphy DR, Craik FIM, Li KZ and Schneider BA (2000) Comparing the effects of aging and background noise on short-term memory performance. *Psychology and Aging* **15**: 323–334.

Park DC and Schwarz N (1999) *Cognitive Aging: A Primer*. Philadelphia, PA: Psychology Press.

Perfect TJ and Maylor EA (eds) (2000) *Models of Cognitive Aging*. New York, NY: Oxford University Press.

Salthouse TA (1996) The processing-speed theory of adult age differences in cognition. *Psychological Review* **103**: 403–428.

Schacter DL, Cooper LA and Valdisseri M (1992) Implicit and explicit memory for novel objects in older and younger adults. *Psychology and Aging* **7**: 299–308.

Schultz W, Dayan P and Montague PR (1997) A neural substrate of prediction and reward. *Science* **275**: 1593–1599.

Spencer WD and Raz N (1994) Memory for facts, source, and context: can frontal lobe dysfunction explain age-related differences? *Psychology and Aging* **9**: 149–159.

Sullivan EV, Desmond JE, Deshmukh A, Lim KO and Pfefferbaum A (2000) Cerebellar volume decline in normal aging, alcoholism, and Korsakoff's syndrome: relation to ataxia. *Neuropsychology* **14**: 341–352.

Woodruff-Pak DS (1997) *The Neuropsychology of Aging*. Malden, MA: Blackwell.

Agreement

Intermediate article

Marcel den Dikken, City University of New York, New York, USA

CONTENTS

Agreement in linguistics is a relationship of matching or systematic covariation of the features of constituents of a syntactic construct. All major syntactic categories and many minor categories can entertain agreement relationships of a variety of different kinds, typically involving subject–verb or modifier–head configurations.

AGREEMENT: GENERAL PROPERTIES

Agreement (or concord) is a relationship of matching or systematic covariation of the features of constituents of a syntactic construct. The constituents are said to agree in features: ϕ-features (where 'ϕ' is a cover for person, number, gender); case (e.g., Latin *illarum duarum bonarum feminarum* – 'of those two good women', with genitive feminine plural marked throughout); noun class (e.g., Bantu); or some other properties (e.g., categorial features, as in Chamorro complementizer agreement; or tense).

All major syntactic categories can entertain agreement relationships with other constituents. In many languages, finite verbs agree with their subjects ('subject agreement'), and there are also languages in which finite verbs can agree with one or more of their objects ('object agreement'), or with *wh*-extracted constituents ('*wh*-agreement' as in Bantu, Palauan, Chamorro); nonfinite verbs can also show agreement with their dependents (past participle agreement in Romance languages; inflected infinitives in Portuguese, Hungarian). Predicative adjectives can agree with their subjects, attributive adjectives with the head noun. Predicate nominals often agree with their subjects as well;

and possessed nouns may show agreement with their possessors. Finally, prepositions can agree with their objects (e.g., Celtic, Hungarian, Abkhaz).

Minor (or closed-class) syntactic categories may also show agreement. Determiners (articles, demonstratives) typically agree with the head noun of a complex noun phrase. Complementizers (subordinating conjunctions) may bear agreement inflection for either the subject of the clause they introduce (as in varieties of the Germanic languages) or a *wh*-phrase extracted out of that clause (as in Irish; cf. also relative clause constructions featuring inflected relative complementizers).

HEAD AGREEMENT AND DEPENDENT AGREEMENT

An important typological distinction between languages is that between head-marking and dependent-marking languages (Nichols, 1986). Head-marking languages are rich agreement languages; dependent-marking ones express relationships between heads and their dependents in other ways (typically with case-marking on the dependent). Abkhaz, a typical head-marking language, shows agreement inflection on the verb for several of its arguments, and possessive inflection on nouns and prepositions. Languages may combine a rich head-marking agreement system with a system of morphological case-marking on dependents (e.g., Hungarian); languages showing neither type of marking also exist (e.g., Chinese).

Agreement can be classified along head/dependent lines as well. We define head agreement as involving agreement marking realized on the head, not on the dependent. Interpretively, in a sentence with an overt subject and an inflected verb, the expression of ϕ-features on the subject is meaningful (the difference between singular and plural noun phrases is semantically significant) while that on the finite verb is not (it is merely the reflection of the agreement relationship with the subject). Thus, we may call the ϕ-features on the subject interpretable and those on the verb uninterpretable (cf. Chomsky (1995) for a theory in which this distinction plays a major role). In languages in which nominal arguments may remain unexpressed in the presence of ϕ-feature inflection on the head (so-called pro-drop languages), the inflection on the head may itself be taken to be meaningful – the inflection would be the argument of the head ('pronominal argument languages'; Jelinek, 1984). The pronominal agreement approach bears a strong relationship to analyses of clitic constructions; indeed, the dividing line between clitics and agreement is often difficult to draw and remains a contentious issue.

Agreement may also be marked on the dependent. The quintessential example of dependent agreement is the inflection of attributive adjectival modifiers of noun phrases, where the head noun determines the form of the modifying adjective. Another possible case of dependent agreement is that between anaphoric expressions and their antecedents (*They like themselves*); agreement here is not always strictly grammatical agreement, though (*Everybody thinks they are smart*). Both these dependent agreement cases can be reanalyzed as involving head agreement (Abney, 1986; Kayne, 2001); whether the theory needs to recognize two separate agreement types is not immediately obvious, therefore.

In addition to the above agreement patterns, in which one member is the dependent of the other member of the pair, we find agreement relationships between items where neither is a direct dependent of the other. We will encounter some of these in the next section; they may be assimilable to the head/dependent pattern in ways sketched in the last section.

INTRICACIES OF AGREEMENT

Agreement in ϕ-features exhibits a complicated distribution when it comes to the subset of features picked out. Number shows up cross-linguistically in all types of head agreement; person is frequently marked in finite verb agreement but not all languages having past participle or adjective agreement express person there (cf. Romance); gender, on the other hand, is much less commonly marked in finite verb agreement than in adjective agreement. Animacy and definiteness are two other major agreement features.

The question of whether we find agreement or not may be influenced by complicated syntactic factors, especially in the context of subject agreement and extraction. The position of a noun phrase *vis-à-vis* the agreeing verb may affect agreement possibilities: thus, in Arabic, prenominal subjects agree in all relevant features while postnominal subjects trigger person agreement only. In Berber and varieties of Celtic, *wh*-extracted subjects fail to agree with the verb except if the clause is negated, in which case subject agreement does show up (Ouhalla, 1993). And regular subject agreement can be suspended in sentences in which the finite verb agrees with a *wh*-constituent – as in Bantu (Kinyalolo, 1991) and varieties of American English (Kimball and Aissen, 1971: *the people who* *John* *think are in the garden*) – or with a subconstituent of the subject ('agreement attraction': *The identity of the participants are to remain a secret*).

Such 'overruling' tends to be skewed with respect to number: plurals can supplant regular singular agreement, but the opposite is much less common (cf. the frequently occurring *the key to the doors are missing* versus the much rarer *the keys to the door is missing*). This points towards number involving a privative opposition, with plural as the marked member. 'Overruling' of regular subject agreement also tends not to occur when the subject is pronominal. There is a robust cross-linguistic tendency for agreement between a head and a pronoun to be richer and more persistent than that between a head and a full noun phrase. Thus, in Welsh VSO sentences the verb does not show number agreement with full-nominal subjects, but subject pronouns must agree for number; *mutatis mutandis*, the same is found in Hungarian possessed noun phrases.

Within the realm of pronouns, first and second person pronouns often behave differently when it comes to agreement-related phenomena than do third person noun phrases (whether full-nominal or pronominal). Hungarian definiteness agreement between finite verbs and their objects yields straightforward results with third person objects, but first and second person object pronouns surprisingly trigger indefinite agreement on the verb. Splits between first and second person on the one hand, and third person on the other, characterize many so-called split ergative languages as well. The Mayan language Mocho, for instance, exhibits such a split. Other Mayan languages show split ergativity conditioned by tense, aspect, or clause type (main versus subordinate; Dixon, 1994, p. 201 and sections 4.3–4.4). Many morphologically ergative languages (e.g., Warlpiri) exhibit a nominative–accusative verb agreement pattern in tandem with an ergative–absolutive case system, showing that case and agreement patterns need not coincide.

Hybrid agreement patterns manifest themselves in a variety of forms. In French *Vous êtes loyal* [you-2PL are-2PL loyal-M.SG], the second person plural pronoun *vous* is used as a polite form with a singular referent, in which case it triggers second person plural agreement on the verb but singular agreement on the predicate; similarly, in Spanish *Su Majestad suprema está contento* [your supreme-F.SG Majesty-F.SG is happy-M.SG], *Majestad* triggers femine agreement on the attributive adjective regardless of the referent but has predicate agreement determined by the gender of the referent (here, masculine).

These kinds of hybrid agreement may also be classified as semantic agreement, with the ϕ-feature composition of the head being determined by the referent of the dependent rather than by the morphosyntactic features of the dependent *per se*. Semantic agreement seems to be confined to head agreement.

AGREEMENT IN CURRENT GRAMMATICAL THEORIES

Semantic agreement is the cornerstone of Dowty and Jacobson's (1988) theory of agreement, in which agreement relationships are given a semantic explanation. In Reed's (1991) functional approach to verb number in English, meaning is also the epicenter: for Reed, what is generally referred to as an agreement relationship between the subject and the finite verb is not a case of agreement at all; instead, the number specification of each is chosen independently of that of the other, with each contributing independently to the message the speaker seeks to convey. Naturally, the emphasis in this work is on lack of agreement.

A semantic theory of agreement faces difficulties wherever semantic factors fail to have the final say. Thus, while *the dog* can be pronominalized with either *it* or *he*, in a sentence like *That dog is so ferocious, it/he even tried to bite itself/himself*, the assignment of gender to the subject pronoun and the object anaphor has to be uniform: the combinations *it* + *himself* and *he* + *itself* are impossible. This uniformity is not semantically determined; instead, it cues the need for a morphosyntactic theory of agreement.

Pollard and Sag (1994), who contributed this argument against semantic approaches, offer a theory of agreement built on the feature-based formalism of head-driven phrase structure grammar but allowing agreement access to semantic and pragmatic information as well. In this theory, ϕ-features are taken to be part of the internal structure of referential indices, the latter being the key notion of their theory. Indices in this theory make both a semantic and a syntactic contribution; they are vital in the analysis of agreement phenomena and referential dependencies.

A unified approach to agreement and referential dependencies in terms of indices is found also in the early principles-and-parameters literature (Chomsky, 1981, 1986; Borer 1986). In more recent principles-and-parameters work (Chomsky 1995), however, indices are assumed not to play a theoretical role. Instead, agreement is represented in terms of a local structural configuration ('specifier–head agreement') or is established under a (potentially long-distance) Agree relationship (Chomsky, 2001). Of these two options, the former represents theories in which agreement is a combination of feature matching and a specific structural configuration under which such matching is 'checked' – the spec–head structure (see the next section for more discussion), or Chomsky's (1995) 'checking domain' (see Chung (1998) for a different approach, cast in terms of the Associate relationship). The more recent Agree approach reduces agreement strictly to feature matching, with specific structural configurations resulting not from the need to establish agreement but from other, unrelated requirements.

Agreement as feature checking is essentially a symmetrical relationship; the Agree approach, by contrast, conceives of agreement as an asymmetrical relationship between a 'probe' and a 'goal'. Early generative approaches to agreement were asymmetrical as well, with transformations copying ϕ-feature specifications from one member of the agreeing pair to the other. Asymmetrical agreement also characterizes Keenan's (1979) approach to agreement in terms of the function–argument relationship. In the framework of Generalized Phrase Structure Grammar (Gazdar *et al.*, 1985), agreement relations are likewise encoded asymmetrically, but in the more recent Head-Driven Phrase Structure Grammar (Pollard and Sag, 1994) agreement is treated in symmetrical terms.

AGREEMENT AS EVIDENCE FOR STRUCTURE

Agreement relationships are severely restricted: though there may be a variety of noun phrases present in the domain of a head, this head establishes agreement with only a narrow subset of those noun phrases. Thus, in *John ate Bill's cereal this morning*, there are four noun phrases surrounding the verb, but in no language will the verb agree with all four at the same time; at most, the verb agrees with the subject and the object. While the possessor can agree with the verb under special circumstances ('possessor ascension to direct object'), bare NP adverbs never agree.

There are structural reasons why agreement relationships are so restricted. Agreement can be established in specific structural configurations only, of which the subject or specifier relationship seems to be the canonical case. If all agreement relationships are taken to involve such a structure, the occurrence of agreement between any two constituents is evidence for a structure in which these constituents are in a specifier–head relationship. In the principles-and-parameters theory of generative grammar, this hypothesis has led to the introduction of agreement phrases (AgrPs) for objects of verbs and prepositions and in the complementizers system. Agreement thus plays a pivotal role in the establishment of syntactic structures in some theories.

Agreement between complementizer and *wh*-extracted constituents and between possessors and possessed nouns is readily recast in these structural terms. When the possessed object itself incorporates into the verb, the possessor in a sense becomes a derived specifier of the verb; similarly, when the finite verb is incorporated into the complementizer position, the subject becomes a derived specifier of the complementizers (Zwart, 1997). In this way 'possessor ascension' and complementizer–subject agreement may be captured. Kayne (1995) shows that a similar treatment is available for cases of agreement in which the finite verb of an embedded clause agrees not with the subject but with an extracted nonsubject (*the people who John think are in the garden*). More 'exotic' cases of agreement (like that between a subconstituent of a complex subject noun phrase and the finite verb in English 'agreement attraction' constructions like *The identity of the participants are to remain a secret*) may, when assimilated to specifier–head agreement, provide evidence for syntactic constituency or derivation as well (cf. Kayne, 1998; Den Dikken, 2000).

Often harder to recast as a specifier–head relationship, long-distance agreement between the matrix verb and an argument of the clause it embeds (as found in Daghestanian, Indic, and Finno-Ugric languages) is restricted in ways which likewise provide highly specific evidence for syntactic structure (see, e.g., Polinksy and Potsdam, 2001). Throughout, agreement is a key diagnostic in the syntactician's toolkit.

References

Abney S (1986) *The English Noun Phrase in Its Sentential Aspect*. Unpublished PhD dissertation, MIT.

Borer H (1986) I-subjects. *Linguistic Inquiry* **17**: 375–416.

Chomsky NA (1981) *Lectures on Government and Binding*. Dordrecht, Netherlands: Foris.

Chomsky NA (1986) *Barriers*. Cambridge, MA: MIT Press.

Chomsky NA (1995) *The Minimalist Program*. Cambridge, MA: MIT Press.

Chomsky NA (2001) Derivation by phase. In: Kenstowicz M (ed.) *Ken Hale: A Life in Language*. Cambridge, MA: MIT Press.

Chung S (1998) *The Design of Agreement: Evidence from Chamorro*. Chicago, IL: University of Chicago Press.
Dikken M den (2000) 'Pluringulars', pronouns and quirky agreement. *The Linguistic Review* **18**: 19–41.
Dixon RMW (1994) *Ergativity*. Cambridge, UK: Cambridge University Press.
Dowty D and Jacobson P (1988) Agreement as a semantic phenomenon. In: *Proceedings of the 5th Eastern States Conference on Linguistics*, pp. 1–17.
Gazdar G, Klein E, Pullum GK and Sag IA (1985) *Generalized Phrase Structure Grammar*. Cambridge, MA: Harvard University Press.
Jelinek E (1984) Empty categories, case and configurationality. *Natural Language and Linguistic Theory* **2**: 39–72.
Kayne RS (1998) A note on prepositions and complementizers. Posted on the MIT Press website celebrating Noam Chomsky's 70th birthday (http://addendum.mit.edu/chomskydisc/Kayne.html).
Kayne RS (1995) Agreement and verb morphology in three varieties of English. In Haider H, Olsen S and Vikner S (eds) *Studies in Comparative Germanic Syntax*, pp. 159–165. Dordrecht, Netherlands: Kluwer.
Kayne RS (2001) Pronouns and their antecedents. Ms. New York: New York University.
Keenan EL (1979) On surface form and logical form. *Studies in the Linguistic Sciences* **8**(2).
Kimball J and Aissen J (1971) I think, you think, he think. *Linguistic Inquiry* **2**: 241–246.
Kinyalolo K (1991) *Syntactic Dependencies and the Spec–Head Agreement Hypothesis in KiLega*. Unpublished PhD dissertation, UCLA.
Ouhalla J (1993) Subject-extraction, negation, and the anti-agreement effect. *Natural Language and Linguistic Theory* **11**: 477–518.
Pollard C and Sag IA (1994) *Head-Driven Phrase Structure Grammar*. Chicago, IL: University of Chicago Press.
Polinsky M and Potsdam E (2001) Long-distance greement and topic in Tsez. *Natural Language and Linguistic Theory* **19**: 583–646.
Reed W (1991) *Verb and Noun Number in English: A Functional Explanation*. London, UK: Longman.
Zwart CJW (1997) *Morphosyntax of Verb Movement. A Minimalist Approach to the Syntax of Dutch*. Dordrecht, Netherlands: Kluwer.

Further Reading

Barlow M and Ferguson CA (eds) (1988) *Agreement in Natural Language*. Stanford, CA: Center for the Study of Language and Information.
Corbett G (1991) *Gender*. Cambridge, UK: Cambridge University Press.
Corbett G (2000) *Number*. Cambridge, UK: Cambridge University Press.
Kathol A (1999) Agreement and the syntax–morphology interface in HPSG. In: Levine R and Green G (eds) *Studies in Contemporary Phrase Structure Grammar*, pp. 223–274, Cambridge, UK: Cambridge University Press.
Steele S (1990) *Agreement and Anti-Agreement: A Syntax of Luiseño*. Dordrecht, Netherlands: Kluwer.

Altruism

See **Human Altruism**

Alzheimer Disease

Intermediate article

Elizabeth A Kensinger, Massachusetts Institute of Technology, Cambridge, Massachusetts, USA
Suzanne Corkin, Massachusetts Institute of Technology, Cambridge, Massachusetts, USA

CONTENTS

Alzheimer disease is the leading cause of dementia among older adults. It results in neurochemical and neuroanatomical brain changes that cause increasing cognitive dysfunction.

INTRODUCTION

Alzheimer disease (AD) was first described by Alois Alzheimer in 1907. A German neuropathologist working in Emil Kraepelin's laboratory in Heidelberg, Alzheimer reported a case study of a 51-year-old woman who had psychiatric symptoms and memory problems. Upon her death, Alzheimer noted an abundance of neuritic plaques and neurofibrillary tangles in her brain. Although neuritic plaques had been described previously, Alzheimer was the first to recognize that their presence in large numbers was abnormal, and that the coexistence of the plaques and tangles signaled a previously unidentified disease of the cerebral cortex. This combination of neuritic ('senile') plaques and neurofibrillary tangles is now recognized as the neuropathologic signature of AD.

Alzheimer disease is the most common cause of dementia, accounting for approximately two-thirds of all cases. It is estimated that over 4 million people in the USA have AD, and by the eighth decade of life as many as one in two adults will develop the disease, which is characterized by neuronal changes, neurotransmitter abnormalities and decreased brain volume. These changes underlie cognitive deficits including memory loss, language dysfunction, and visuospatial and temporal disorientation.

DIAGNOSIS

Dementia is defined as an overall loss of intellectual function severe enough to impede daily activities. It consists of a group of behavioral symptoms that must occur together; these symptoms can have various etiologies. Alzheimer disease is defined as the presence of memory impairment plus one other area of cognitive dysfunction: language, motor, attention, executive function, personality, or object recognition, according to the fourth edition of the *Diagnostic and Statistical Manual of Mental Disorders* (DSM-IV). The deficits must have a gradual onset, and a continuous (and irreversible) progression. A definitive diagnosis of AD can be made only at postmortem examination by observing the hallmark neuritic plaques and neurofibrillary tangles. Antemortem, the diagnosis of 'probable' AD is given when an individual meets the criteria of the DSM-IV, the National Institute of Neurological Disorders and Stroke and the Alzheimer's Disease and Related Disorders Association (NINDS-ADRDA), and when all other causes of dementia have been eliminated. When made by a trained clinician, this exclusionary diagnosis is accurate in 80–90% of cases.

INCIDENCE AND PREVALENCE

The incidence of AD increases exponentially with advancing age. At age 70–75 years the incidence of AD is approximately 1% per year, but at age 80–85 years the annual incidence is over 6% per year (e.g. Hebert *et al.*, 1995). Overall, women remain at greater risk of developing AD than men. The prevalence of AD also increases with age (e.g. Fratiglioni *et al.*, 1991), with 30–50% of adults aged 70 years and above having a diagnosis of probable AD.

NEURONAL CHANGES

Alzheimer disease results in a range of neuronal changes, including cellular dysfunction and death.

Eventually, AD affects nearly all brain structures. Early in the disease, however, some brain regions (e.g. limbic structures) are affected to a much greater extent than others (e.g. the primary sensory cortices).

Neuropathologic Changes

The hallmarks of AD are neurofibrillary tangles (intracellular) and neuritic plaques (extracellular). These changes reduce the efficiency of neural communication. While normal aging is associated with these neuropathological changes as well, the number of plaques and tangles seen in the AD brain is far greater than in nondemented individuals.

Neurofibrillary tangles

Neurofibrillary tangles consist of small, paired, helical filaments (i.e. two fiber strands that are twisted around one another). They are found in the cell body and dendrites of neurons, and often appear flame-shaped, with a rounded cell body and a threadlike apical dendrite. They can also have a more spherical shape. Neurofibrillary tangles are composed of hyperphosphorylated tau protein. Typically, tau protein is a soluble component of the cell. When tau is overphosphorylated, however, it becomes insoluble and forms tangles. Because neurofibrils are frequently used for the transport of chemicals that will be made into neurotransmitters, the tangling of these fibrils renders them useless and can prevent neurotransmitter synthesis. Neurofibrillary tangles are not specific to AD, but also appear in other neurological disorders including Parkinson disease, Down syndrome, progressive supranuclear palsy and other forms of dementia.

Neuritic plaques

Neuritic plaques are dense deposits found outside the brain's nerve cells (extracellularly). They are spherical structures with a dense core of amyloid-β protein surrounded by a halo and a ring of abnormal (dystrophic) neurites. The halo component consists of other types of brain cells (astrocytes) and inflammatory cells (microglia). The dystrophic neurites represent dying nerve terminals and are small, threadlike structures consisting of abnormal neuronal dendrites. In addition to these 'typical' plaques, AD brains may also show 'diffuse' plaques, which have a loose accumulation of amyloid-β rather than a dense core, and no surrounding dystrophic neurites.

Patterns of deposition

Neurofibrillary tangles and neuritic plaques show different patterns of accumulation. Early in the course of AD, neurofibrillary tangles are confined primarily within the limbic structures. Neuritic plaques, however, appear throughout the cortex, even early in AD (Arriagada *et al.*, 1992).

Relation between neuropathology and disease

It is unknown whether these neuropathological changes cause AD, or whether they are epiphenomenal. Nonetheless, there does appear to be a link between the amount of tangles in the brain and the severity of AD. Researchers are now working on therapeutic approaches to reduce the formation of tangles and plaques in the brain, hoping that this reduction will halt, or reverse, disease progression.

Brain Atrophy

One of the most prominent features of AD is atrophy (shrinkage) of the medial temporal lobe. The entorhinal cortex (a gateway for information into the hippocampus) and the hippocampus are among the first regions affected. These regions lose about 50% of their neurons, a finding that accounts for the shrunken brain tissue. The volume of these regions, measured by neuroimaging techniques, can be used to identify people with early AD, and may even identify individuals with memory impairments who will later develop AD (e.g. Jack *et al.*, 1999). Another region of the brain that shows substantial cell loss early in AD is the nucleus basalis. This region of the ventral forebrain contains many of the brain's cholinergic neurons. Damage to this region reduces neurotransmission in pathways using the neurotransmitter acetylcholine.

As AD progresses the brain changes become more widespread. The ventricles of the brain expand as the surrounding tissue deteriorates. Sulci (the 'valleys' between the brain's folds) widen. Neocortical areas, including temporal and parietal neocortex, show increased atrophy. Eventually nearly all of the brain, including secondary and even primary sensory areas, is affected.

Neurotransmitter Abnormalities

The damage to the nucleus basalis results in reduced levels of choline acetyltransferase, the enzyme needed for acetylcholine formation. These reductions occur relatively early in the disease, but not all cholinergic pathways are affected equally:

the long-axon cholinergic neurons (e.g. those connecting the nucleus basalis and the cerebral cortex) are particularly vulnerable. Because of the dramatic reduction in cholinergic transmission, the first approved therapies for AD were aimed at increasing the amount of acetylcholine in the brain. Current treatment for AD is administration of acetylcholine esterase inhibitors. Acetylcholine esterase is the enzyme that breaks down acetylcholine into its constituent parts. Acetylcholine esterase inhibitors, therefore, enhance cholinergic neurotransmission by blocking the breakdown of the neurotransmitter. Acetylcholine levels remain higher and can have a longer-lasting effect before being broken down. This therapy, however, provides only a transient increase in memory performance, if any, and has no effect on disease progression.

The minimal effectiveness of acetylcholine esterase inhibitors suggests that acetylcholine deficiency is not the only cause of the cognitive dysfunction in AD. In fact, as the disease progresses, nearly all neurotransmitter systems become depleted. There appears to be much individual variation in the neurotransmitters most affected by AD and the absolute reductions. Estimates, however, suggest that levels of neurotransmitters including noradrenaline (norepinephrine), dopamine and serotonin can show reductions of up to 50% in the late stages of AD.

GENETICS

Alzheimer disease can be divided into two types: familial (inherited) and sporadic. Familial AD is relatively rare, representing less than 5% of total cases, and typically affects individuals at a younger age than sporadic AD (often before age 50 years, with cases reported of people developing the disease in their mid-20s). Sporadic AD usually has a later age at onset (after age 65 years). Some research suggests that familial and sporadic AD differ not only in terms of age at onset, but also in their cognitive profile. Familial AD may be associated with a more rapid cognitive decline and shorter time to death. It also may be linked to more verbal deficits and fewer visuospatial deficits than sporadic AD (Filley *et al.*, 1986).

Familial or Early-onset AD

Familial AD is linked to mutations in three genes: those coding for presenilin 1 (PS-1) on chromosome 14, presenilin 2 (PS-2) on chromosome 1, and amyloid precursor protein (APP) on chromosome 21 (Table 1). These mutations are causative: a person who has one of these genetic mutations will develop AD. Familial AD is inherited following an autosomal dominant pattern. This pattern means that if one parent has this form of AD, each offspring has a 50% chance of developing AD. Interestingly, all of these mutations appear to have a common effect: increased production of amyloid-β peptide (Aβ) 42, the main constituent of the amyloid plaques in the AD brain. The peptide is part of a larger precursor protein (APP), which can be cleaved (cut apart) in two different places, leading to the formation of two types of amyloid-β, one with 42 amino acids (Aβ42) and one with 40 (Aβ40). The first type appears to be more likely to form plaques in the brain than Aβ40; it may also be the more toxic form, and its presence may lead to neuronal death. Researchers are now working on ways to reduce the amount of Aβ42 formed from APP, in the hopes of stopping the formation of amyloid plaques, and perhaps also preventing further clinical decline.

Table 1. Molecular genetics of Alzheimer disease

AD Group	*Chromosome*	*Gene*	*Protein*
Familial AD, onset 50s	21	*APP*	Amyloid
Familial AD, onset 40s	14	*PS1*	Presenilin 1
Familial AD	1	*PS2*	Presenilin 2
Late-onset AD	19	*APOE*	ApoE ε2, ε3, ε4

AD, Alzheimer disease.

Sporadic or Late-onset AD

Sporadic AD probably has a multifactorial basis, including possible genetic and environmental influences. Although there are no causative mutations, there is a major genetic susceptibility factor. A gene that encodes apolipoprotein E (ApoE) is found on chromosome 19 (see Table 1). Everyone has this gene: it is essential for carrying cholesterol in our bloodstream. The gene has many alleles (or forms); the most common ones produce the ApoE variants ε2, ε3 and ε4. One allele is inherited from each parent. Being homozygous for the ε4 allele (i.e. having two ε4 alleles) is associated with an increased risk of developing AD (Saunders *et al.*, 1993). Being homozygous for the ε2 allele, in contrast, is associated with a reduced likelihood of developing AD. The allele is not predictive of AD, however; individuals without an ε4 allele can develop AD, and those who are homozygous for the ε4 allele can remain unaffected by AD. The ApoE alleles are thought to influence the development of late-onset AD in about one-third of the population. Dozens of other genetic risk factors have been suggested, but their roles have been researched less thoroughly than the role of ApoE.

Sporadic AD is also associated with other, nongenetic risk factors. The greatest risk factor is advanced age. One well-researched correlation is with decreased estrogen levels (e.g. Paganini-Hill and Henderson, 1994). It is believed that the reason women are at greater risk of developing AD than men is because of the postmenopausal drop in estrogen levels. Taking estrogen after menopause appears to decrease the likelihood of developing AD, though it does not seem to alter its progression in those who already have the disease. Head injury is another risk factor for AD. Particularly when unconsciousness has occurred, it increases the likelihood of developing AD, and more severe injury is associated with greater risk (e.g. Guo *et al.*, 2000). Head injury as far back as early childhood appears to influence the rate of AD development later in life.

ANIMAL MODELS

Animal models of AD can provide insight into the genetics, pathological progression and treatment of AD. Several transgenic mouse models have been created with the objective of clarifying the role of genetic factors. Researchers have engineered mice that express genes implicated in AD, such as those coding for APP, the presenilins and tau; for reviews see Janus and Westaway (2001), van Leuven (2000) and Duff and Rao (2001). The characteristic pathologic findings in these mice differ depending on the genetic alterations. Transgenic mice created by introduction of APP develop neuritic plaques, and show deficits in learning and memory; however, they do not develop neurofibrillary tangles, which are the other neuropathological hallmark of AD. Transgenic mice created by inserting human tau genes develop abnormal clumping of tau filaments (neurofibrillary tangles), as well as neuronal degeneration, but do not develop neuritic plaques. Mice engineered to express PS-1, in contrast, do not display abnormal pathology or cognitive impairment, but do show elevated levels of Aβ42 (the peptide associated with plaque formation). These models, therefore, provide insights into the contributions of genes implicated in the development of familial AD. Transgenic mice can also be used to examine whether and how genetic risk factors (e.g. expression of ApoE ε4) influence disease progression.

In addition, animal models can provide clues about the pathological progression of AD. Because AD appears to occur naturally only in humans, it has not been possible to examine the neuropathological changes in the brain at various stages of the disease. Rather, the samples available have by necessity come at the time of death. While much information can be garnered from analysis of endpoint data, animal models provide a means for systematic tracking of pathological changes. Hybridizing mice genetically altered to develop neuritic plaques with those engineered to manifest neurofibrillary tangles, has resulted in a strain of mice showing both of the neuropathological hallmarks of AD. This animal model may be particularly important in contributing to our understanding of how these two neuropathological features relate. Researchers are optimistic that use of such animal models will provide information about the relation between amyloid deposits and tau-containing tangles, and about the role they play in the development and progression of AD.

Animal models will also be important in testing potential treatments for AD. Once researchers have established an animal model that closely approximates the pathological and cognitive characteristics of AD, it will be possible to test the efficacy of treatments on these animals.

COGNITIVE FUNCTION

The signs of AD develop slowly, and it is often difficult to pinpoint the time of disease onset. Initial symptoms are mild, and can include forgetfulness, passivity, decreased work productivity, word-finding difficulties, and disorientation. As the disease progresses, nearly all aspects of function are affected, including memory, language, attention, vision, audition, and motor control.

Impaired Capacities

Episodic memory

Impairment of episodic memory – the recollection of events that occupy a specific spatial and temporal context – is typically the earliest and most prominent deficit in AD. Patients have difficulty forming new episodic memories (anterograde amnesia), and this impairment worsens with disease progression. Deficits in episodic memory, including delayed recall of verbal and nonverbal material, are the best way of distinguishing people with AD from healthy older adults (e.g. Locascio *et al.*, 1995). In contrast, however, people with AD remain capable of retrieving some long-term episodic memories. While remote memory is impaired, it does not show the pronounced decrements seen in the formation of new episodic memories. The loss of retrograde memory (retrograde amnesia) also appears to be temporally graded, with recent memories showing more degradation than remote memories. In fact, the capacity to recollect events from the remote past is often quite resilient in AD. Patients can even become preoccupied with the past, and can confuse their current environment with that of their youth. The degrees of anterograde and retrograde memory deficits are not significantly correlated (e.g. Greene and Hodges, 1996).

The episodic memory deficit is consistent with the neural changes early in AD: brain structures that support long-term memory (medial temporal lobe regions, including the hippocampus) are compromised in early AD, while other regions of the brain are less affected.

Emotional memory

Individuals are often better at remembering emotional compared with neutral information. This emotional memory enhancement effect appears to result from interactions between the amygdala and other regions of the medial temporal lobe, including the hippocampus. Alzheimer disease results in a substantial volumetric reduction in the amygdala, and this amygdaloid atrophy appears to reduce the emotional enhancement effect: people with AD do not show better memory for emotional information than for neutral information (Kensinger *et al.*, 2002) and their ability to remember emotional stimuli appears to correlate with amygdaloid volume.

Semantic memory

Semantic memory – general knowledge about the world – is relatively spared early in AD, but as the disease progresses significant deficits arise. Deficits occur on tasks requiring general knowledge retrieval, word definitions, word–picture matching, or picture naming. It is unclear whether the semantic deficit is related to a breakdown in the structure of semantic memory, to impaired access of semantic information, or to a combined deficit in structure and access.

The extent of the language deficits is useful for assessing the severity of AD (Locascio *et al.*, 1995). Initial deficits include increased 'tip of the tongue' effects, reduced fluency scores, and a difficulty with tests requiring confrontation naming. In later stages of the disease, deficits can include forgetting the names of spouse or children, and the inability to recall names of common objects. The progression of semantic memory deficits is associated with the expanding pathological changes of advancing AD: as perisylvian areas become affected, semantic memory deficits increase.

Visuospatial function

While early AD is associated with some disorientation, visuospatial dysfunction increases with disease progression. In the middle stages of the disease, it is common for patients to become lost while driving their car or on a walk, even when following a route that they have taken on many occasions.

Executive functions

People with AD show deficits in short-term memory and in on-line processing of information (Corkin, 1982). At least some of these deficits may be due to deficits in the 'central executive' in Baddeley's model of working memory. Becker (1988) suggested that AD might have two main deficits: one paralleling that of amnesia, and the other in the central executive. In support of the central executive hypothesis, people with AD have frequently been found to have poor dual-task performance while being capable of performing each component task at a normal level. Baddeley and colleagues also found that dual-task performance declined with disease progression, while single-task performance remained stable.

Classical conditioning

Most forms of nondeclarative memory are spared in AD. One notable exception, however, is seen with delay conditioning, in which the unconditioned stimulus occurs just before the offset of the conditioned stimulus. People with AD are impaired at acquiring a conditioned response (such as an eyeblink in response to a tone). They require more trials to learn this type of relation (e.g. Woodruff-Pak *et al.*, 1996). This deficit is probably not related to damage to the medial temporal lobe because amnesic patients with damage to these structures are capable of acquiring a conditioned response. The deficit is more likely to be related to cholinergic or cerebellar dysfunction.

Vision

Alzheimer disease results in changes in basic sensory and perceptual capabilities. People with AD often have more difficulty perceiving visual stimuli than do nondemented older adults. A significant correlation exists between the severity of perceptual deficits and the extent of cognitive dysfunction (e.g. Cronin-Golomb *et al.*, 1995). It is unclear whether this correlation is causal (e.g. visual deficits could cause poorer performance on a task of visual memory) or associative (e.g. patients with greater brain atrophy are likely to have both sensory deficits and memory dysfunction). The visual deficits are related to reductions in contrast sensitivity, color perception and discrimination, and visual acuity. The visual dysfunction is probably related to neuropathological changes in primary visual cortices and visual association areas because AD does not appear to affect the retina or optic nerve.

Preserved Capacities

Despite the range of capacities affected in AD, some domains remain relatively preserved until late in the course of the disease. Most prominently, many types of nondeclarative (implicit) memory are unaffected by early to moderate AD.

Priming

Priming can be broadly broken down into two subsets: conceptual and perceptual priming. People with AD show a dissociation in performance on these priming tasks: their performance on conceptual priming tasks is impaired, whereas their perceptual priming is normal. This dissociation probably reflects a disproportionate reliance on temporoparietal regions in conceptual priming. Conversely, perceptual priming appears to rely predominantly on occipital lobe regions that are less affected by AD.

Skill learning

Until the late stages of AD, patients are capable of learning new skills, ranging from motor learning to visual adaptation. The preservation of such learning is likely to be related to the relative preservation of brain regions important for nondeclarative learning, including the basal ganglia and frontal lobe.

TREATMENT

The treatment of AD consists predominantly of three types of approaches: (a) mitigating noncognitive disorders (psychiatric symptoms such as anxiety or paranoia, sleep disturbance), (b) restoring neurotransmitter function, and (c) protecting neurons from further death. The majority of drugs prescribed to restore neurotransmitter function have been cholinesterase inhibitors. These drugs, however, have had only minimal effectiveness in slowing the disease progression and have not been able to halt or reverse the disease's effects. There is some evidence that antioxidants (vitamin E), antiinflammatory drugs (ibuprofen), estrogen, and lipid-lowering agents (statins) may be neuroprotective, in as much as they slow the progression of AD. To date, however, there is no evidence that these treatments can slow disease progression in individuals already affected.

SIGNIFICANCE OF RESEARCH FOR UNDERSTANDING BRAIN FUNCTION

Research into AD has improved our understanding not only of the neurologic disorder but also of brain function. By observing the pattern of spared and impaired functions, researchers have learned that some types of memory (e.g. declarative) are affected by diffuse damage to the medial temporal lobe, while other types of memory (i.e. nondeclarative) remain relatively unaffected. Similar dissociations, such as between preserved perceptual priming and impaired conceptual priming, or between impaired anterograde memory and only moderately affected retrograde memory, have helped to uncover the dissociable neural mechanisms responsible for these cognitive functions. Similarly, the finding of reduced cholinergic function in AD sparked interest in the role of acetylcholine in long-term memory formation. Further research has demonstrated the necessity of the cholinergic system for successful episodic encoding.

By comparing performance in AD and amnesia, researchers have also been able to determine what memory dysfunction in AD is caused specifically by damage to the medial temporal lobe system, and what deficits may be related to neocortical damage (Corkin, 1982). This complementary interaction between neurology, neuropsychology, and neuroscience has allowed simultaneous advancements in the diagnosis and treatment of AD, and in our understanding of brain–behavior relations.

References

Arriagada PV, Growdon JH, Hedley-Whyte ET and Hyman BT (1992) Neurofibrillary tangles but not senile plaques parallel duration and severity of Alzheimer's disease. *Neurology* **42**: 631–639.

Becker JT (1988) Working memory and secondary memory deficits in Alzheimer's disease. *Journal of Clinical and Experimental Neuropsychology* **10**: 739–753.

Corkin S (1982) Some relationships between global amnesias and the memory impairments in Alzheimer's disease. In: Corkin S, Davis KL, Growdon JH, Usdin E and Wurtman RJ (eds) *Alzheimer's Disease*, vol. 19, A Report of Progress in Research, pp. 149–164. New York, NY: Raven Press.

Cronin-Golomb A, Corkin S and Growdon JH (1995) Visual dysfunction predicts cognitive deficits in

Alzheimer's disease. *Optomology and Visual Science* **72**: 168–176.

Duff K and Rao MV (2001) Progress in the modeling of neurodegenerative diseases in transgenic mice. *Current Opinion in Neurology* **14**: 441–447.

Filley CM, Kelly J and Heaton RK (1986) Neuropsychologic features of early- and late-onset Alzheimer's disease. *Archives of Neurology* **43**: 574–576.

Fratiglioni L, Grut M, Forsell Y *et al.* (1991) Prevalence of Alzheimer's disease and other dementias in an elderly urban population: relationship with age, sex, and education. *Neurology* **41**: 1886–1892.

Greene JDW and Hodges JR (1996) The fractionation of remote memory: evidence from the longitudinal study of dementia of Alzheimer's type. *Brain* **119**: 129–142.

Guo Z, Cupples LA, Kurz A *et al.* (2000) Head injury and the risk of AD in the MIRAGE study. *Neurology* **54**: 1316–1323.

Herbert LE, Scherr PA, Beckett LA *et al.* (1995) Age-specific incidence of Alzheimer's disease in a community population. *Journal of the American Medical Association* **273**: 1354–1359.

Jack CR, Peterson RC, Xy YC *et al.* (1999) Prediction of AD with MRI-based hippocampal volume in mild cognitive impairment. *Neurology* **52**: 1397–1403.

Janus C and Westaway D (2001) Transgenic mouse models of Alzheimer's disease. *Physiology and Behavior* **73**: 873–886.

Kensinger EA, Brierley B, Medford N, Growdon JH and Corkin S (2002) Effects of normal aging and Alzheimer's disease on emotional memory. *Emotion* **2**.

van Leuven F (2000) Single and multiple transgenic mice as models for Alzheimer's disease. *Progress in Neurobiology* **61**(3): 305–312.

Locascio JJ, Growdon JH and Corkin S (1995) Cognitive test performance in detecting, staging, and tracking Alzheimer's disease. *Archives of Neurology* **52**: 1087–1099.

Paganini-Hill A and Henderson VW (1994) Estrogen deficiency and risk of Alzheimer's disease in women. *American Journal of Epidemiology* **140**: 256–261.

Saunders A, Strittmater W, Schmechel D *et al.* (1993) Association of apolipoprotein E allele e4 with late-onset familial and sporadic Alzheimer's disease. *Neurology* **43**(8): 1467–1472.

Woodruff-Pak DA, Papka M, Romano S and Lo YT (1996) Eyeblink classical conditioning in Alzheimer's disease and cerebrovascular dementia. *Neurobiology of Aging* **17**: 505–512.

Further Reading

Baddeley AD, Bressi S, Della Sala S, Logie R and Spinnler H (1991) The decline of working memory in Alzheimer's disease: a longitudinal study. *Brain* **114**: 2521–2542.

Corkin S (1998) Functional MRI for studying episodic memory in aging and Alzheimer's disease. *Geriatrics* **53**: S13–S15.

Growdon JH, Wurtman RJ, Corkin S and Nitsch RM (eds) (2000) The molecular basis of dementia. *Annals of the New York Academy of Sciences*, vol. 920. New York, NY: New York Academy of Sciences.

Hodges JR (2000) Memory in the dementias. In: Tulving E and Craik FIM (eds) *The Oxford Handbook of Memory*. New York, NY: Oxford University Press.

Katzman R, Terry RP and Bick KL (eds) (1994) *Alzheimer's Disease*. New York: Raven Press.

Keane MM, Gabrieli JD, Fennema AC, Growdon JH and Corkin S (1991) Evidence for a dissociation between perceptual and conceptual priming in Alzheimer's disease. *Behavioral Neuroscience* **105**: 326–342.

Morris RG and Kopelman MD (1986) The memory deficits in Alzheimer-type dementia: a review. *Journal of Experimental Psychology* **A38**: 575–602.

Nebes RD (1989) Semantic memory in Alzheimer's disease. *Psychological Bulletin* **106**: 377–394.

Selkoe DJ (1999) Translating cell biology into therapeutic advances in Alzheimer's disease. *Nature* **399**: A23–A31.

Villareal DT and Morris JC (1998) The diagnosis of Alzheimer's disease. *Alzheimer's Disease Review* **3**: 142–152.

Amnesia

Intermediate article

AR Mayes, University of Liverpool, Liverpool, UK
NM Hunkin, University of Sheffield, Sheffield, UK

CONTENTS

Amnesia means loss of memory; more specifically, the term refers to the amnesic syndrome. In this syndrome, usually caused by regional brain damage, there is impaired recall and recognition of facts and episodes encountered both before and after the onset of the disorder, although intelligence and short-term or immediate memory are preserved.

INTRODUCTION

The term 'amnesia' is used in a general sense to mean any kind of loss of memory. The term also has a technical sense in which it refers to the amnesic syndrome. This memory disorder syndrome has four main features. The first is an impairment in the ability to recall or recognize facts or personally experienced episodes encountered following the brain injury; this is known as anterograde amnesia. The second feature is an impairment in the ability to recall or recognize facts and personally experienced episodes that were encountered and put into memory premorbidly; this is known as retrograde amnesia. In patients, the severity of anterograde amnesia is only weakly correlated with the severity of retrograde amnesia, and these disorders are often accompanied by two further features: preserved intelligence, and preserved short-term memory. Research has confirmed earlier, more informal observations, the first of which was made by Claparède in 1911, that people with amnesia can show good memory although they have no recall or recognition of how they acquired such memory. These 'implicit' memories are only shown by changed behavior, because patients are not aware that they are remembering.

Amnesia in the technical sense is, therefore, far from being a general disorder of all kinds of memory. It is referred to as a syndrome because some of its component symptoms probably have different causes, although it remains unclear to what extent the cause or causes of anterograde and retrograde amnesia differ. Most often these memory disorders are caused by damage to any one of several different brain regions, each of which plays a key role in certain kinds of memory (organic amnesia); but it is also known that amnesic symptoms can result from psychiatric causes that are not necessarily triggered by brain damage, and which typically only affect premorbid autobiographical memories (psychogenic amnesia).

Although Lawson was one of the first to describe a relatively selective memory disorder associated with chronic alcoholism in 1878, it was the Russian physician Korsakoff who, between 1887 and 1891, gave the classic description of the amnesic syndrome. Most of the cases he described were of chronic alcoholism, but he also described patients with neoplasm, carbon monoxide poisoning, diabetes and persistent vomiting. Korsakoff syndrome, the variant of amnesic syndrome that bears his name, is believed to be caused by thiamin deficiency. Most often thiamin deficiency is caused by chronic alcoholism, but in the twentieth century it became clear that other forms of nutritional disorder such as anorexia nervosa or persistent vomiting may cause thiamin deficiency and amnesia. Korsakoff also noted that patients with his syndrome usually showed lack of insight into their condition and often falsely recollected information (confabulated). These symptoms have subsequently been found not to be universal, or even particularly common, features of other kinds of amnesia.

Later work clearly showed that the amnesic syndrome could arise from a wide range of other causes including several kinds of vascular incident (such as infarctions of the posterior cerebral artery, and rupture and repair of anterior communicating artery aneurysms), herpes simplex encephalitis and some kinds of meningitis, as well as surgical treatment of temporal lobe epilepsy involving bilateral removal of the medial temporal lobes. The most famous amnesic patient, known by his initials as HM, had such an operation in 1953, and nearly fifty years later still showed a very severe anterograde amnesia and a less severe retrograde amnesia mainly affecting memories acquired in the years immediately preceding his surgery. It was first suggested by Ribot in 1882 that memories acquired further in the past before brain damage occurred are more resistant to impairment. In other words, he predicted that retrograde amnesia should be temporally graded such that older memories are less impaired.

The history of research on amnesia may be broken into three overlapping stages. The first stage, which began with Lawson, Korsakoff and Ribot, involved systematic clinical observation of patients. The second stage, which began after the Second World War, involved the application to amnesic patients of more formal experimental procedures drawn from mainstream cognitive psychology. In the third stage, modern *in vivo* brain imaging technologies made it possible to determine exactly what brain lesions cause which kinds of memory deficits. This work is complemented by the exploration of which brain regions are activated when normal people use the memory processes that are impaired in amnesic patients. For decades, work on human amnesic patients has been complemented by research on animal models of amnesia in which it is possible to localize damage more focally.

FORMS OF AMNESIA

Amnesia is divisible into organic and psychogenic forms. Psychogenic or functional amnesia, which includes fugue states and multiple personality disorders, typically has no apparent physical cause and is believed to have a psychiatric origin. People with psychogenic amnesia are believed to be neither merely pretending to be unable to remember, nor consciously trying not to remember or create new memories, although it is often possible to identify emotional benefits arising from such patients being unable to remember. This contrasts them with malingerers, who are pretending or consciously trying not to remember.

Fugue states consist of a sudden loss of all autobiographical memory together with the loss of personal identity. Typically, these states are short-lived, lasting only a few hours or days, although there are reports of fugues being maintained over long periods. Fugues tend to be precipitated by a traumatic event such as bereavement or marital strife, and the memory loss may be interpreted as a form of adaptive response in which patients dissociate themselves from their own identity as a means of escaping from the traumatic experience. In multiple personality disorder, a patient appears to have at least two different personalities, and sometimes as many as 20. The different personalities are used to compartmentalize the patient's experiences, and a particular personality needs to be active for particular experiences to be recalled. In addition to fugues and multiple personality states, there are cases when loss of autobiographical memory for the premorbid past is less complete. In these cases the person usually has a retrograde amnesia that is selective for autobiographical memory, but with no impairment of the ability to acquire new autobiographical memories. The ability to acquire new memories is only rarely disrupted in cases of psychogenic amnesia.

Organic and psychogenic amnesias can usually be distinguished from each other by examining the patient's medical history. If the patient has suffered a brain injury, and there is damage to brain areas known to be involved in memory functioning, an organic basis to the memory disorder is likely. In contrast, if there has been no brain trauma or evidence of brain disease, it is more likely that the memory disorder has a psychogenic basis. Some cases are more difficult to interpret; in these, there is evidence of brain damage that is either inconsistent with the pattern of memory impairment observed or does not seem extensive enough to explain the severity of the observed memory impairment. In such cases psychogenic factors may overlie an organic disorder or, conversely, organic factors such as epilepsy may exacerbate an underlying psychogenic disorder.

Like psychogenic amnesia, organic amnesia may be transient, although more often it is permanent. Transient organic amnesia may occur following electroconvulsive therapy and epileptic seizures. It also occurs in post-traumatic amnesia, during the early period following head injury, or emergence from coma in severely injured patients when patients have difficulty in storing and/or retrieving new information. Finally, best known is transient global amnesia, in which there is an abrupt onset of severe anterograde amnesia and a more variable, temporally graded retrograde amnesia. The attack typically lasts for a number of hours and then gradually resolves, leaving the patient with a dense loss of memory covering the period of the attack and a few hours preceding it, but with a more or less restored ability to lay down new memories. It has been shown that blood flow to the medial temporal lobes and sometimes the thalamic region is temporarily reduced during attacks. The cause of this reversible dysfunction of memory-related regions (see next section) is unknown, but is sometimes associated with migraine.

The severity of the memory deficits in permanent organic amnesia usually remains stable, although there may be some recovery following the initial brain injury. However, amnesia is also typically the major characteristic of the early stages of Alzheimer disease: in this neurodegenerative disease not only does the amnesia became worse as the disease progresses, but other cognitive and memory deficits also appear.

Psychogenic amnesia typically involves a retrograde amnesia for personal episodes and no anterograde amnesia. It is interesting, therefore, that it has recently been claimed that brain damage can also cause a retrograde amnesia in the context of minimal anterograde amnesia. Cases of such focal retrograde amnesia have been used to identify which brain structures are involved in remote memory function. However, such research is controversial because (a) it is difficult to explain how a patient can fail to recall a premorbidly formed memory while being able to store and retrieve new memories fairly normally, and (b) it is difficult to exclude the possibility that the amnesia has a psychogenic explanation.

NEUROANATOMY OF AMNESIA

Damage to any one of several distinct brain regions, which include the medial temporal lobes, the midline diencephalon, the basal forebrain and possibly parts of the prefrontal neocortex, is sufficient to cause the amnesic syndrome. Although these regions comprise distinct gray matter structures, they are all interconnected and there is evidence that damage to the fibre tracts, such as the fornix, which link the different regions can also cause the syndrome, or at least some of its characteristic memory impairments (Tranel and Damasio, 1995).

The medial temporal lobes include the perirhinal and parahippocampal cortices, which receive processed information from the posterior and anterior association neocortices. They provide nearly two-thirds of the cortical input to the hippocampus via an intermediate projection to the entorhinal cortex (Figure 1). The hippocampus, therefore, receives inputs of processed modality-specific and polysensory information, semantic information and possibly other kinds of information related to planning and movement activities from the association

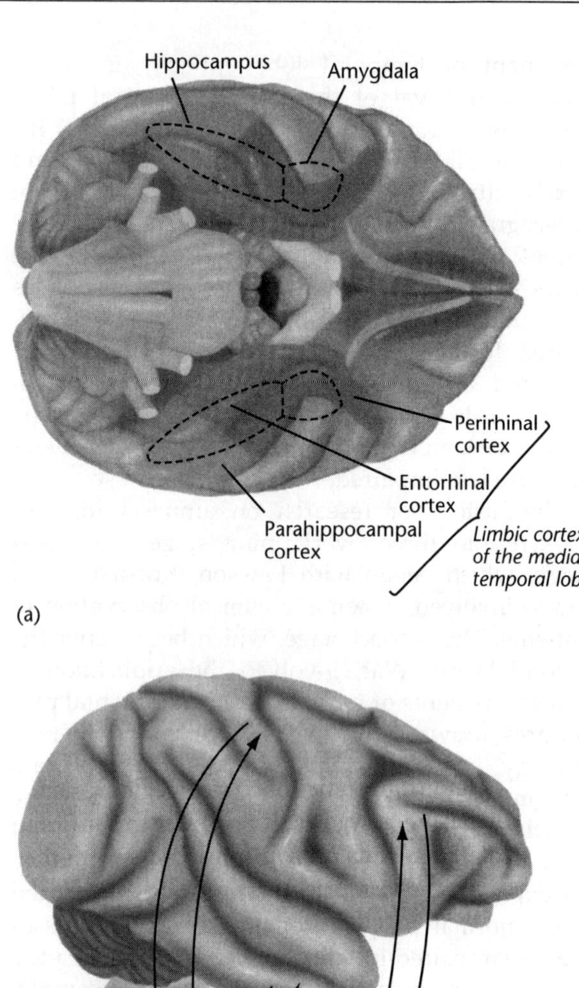

Figure 1. (a) Ventral view of a typical nonhuman primate (monkey) brain which illustrates the position of the medial temporal lobe cortices and the hippocampus and amygdala (which lie just above them). (b) Sagittal view of the same structure, that illustrates the two way connections which exist between the hippocampus and entorhinal cortex; the entorhinal, parahippocampal, and perirhinal cortices; and the parahippocampal and perirhinal cortices, and overlying neocortical structures.

cortices. Combined damage to all medial temporal lobe regions produces severe anterograde and retrograde amnesia in humans and monkeys (Zola-Morgan and Squire, 1993). Research with monkeys has shown that perirhinal cortex lesions alone cause severe anterograde amnesia with badly impaired recognition, whereas parahippocampal cortex lesions may primarily disrupt some kinds of spatial memory. Hippocampal lesions are known to produce mild anterograde and retrograde amnesia, although there is controversy about the nature of this impairment. In humans, this is because selective hippocampal damage is rare, and some of the patients reported have shown equivalent recognition and recall deficits, whereas others have shown moderately severe recall deficits but mild or undetectable recognition deficits. These selectively impaired patients show milder retrograde amnesias than are found in patients with large medial temporal lobe lesions. There is also evidence that severe (and perhaps focal) retrograde amnesia may depend on damage that includes parts of the anterior temporal neocortex and possibly parts of the prefrontal neocortex. The medial temporal lobes contain the amygdala, which is interconnected with the hippocampus and lies above the perirhinal cortex. Although studies with monkeys indicate that amygdala lesions do not usually impair recognition, there is growing evidence that this structure is essential to some aspects of emotional memory and that damage to it reduces the memory advantage of 'emotional' stimuli over neutral ones (Phelps *et al*, 1998).

Evidence from both animals and humans indicates that damage to either the mamillary bodies or the nuclei and fibre tracts in the midline thalamus also causes amnesia. It is not agreed, however, precisely which thalamic regions are implicated in amnesia, although there is good evidence that damage to either the anterior thalamic or the dorsomedial thalamic nuclei causes memory impairments. The memory disorder in Korsakoff syndrome is particularly associated with lesions of the mamillary bodies, and of the anterior and dorsomedial thalamic nuclei.

Structures in the basal forebrain modulate neocortical and medial temporal cortex activity, so it is not surprising that damage to these structures can produce amnesia. The septum and diagonal band of Broca project to the hippocampus via the fornix, and lesions to these basal forebrain structures have been found to cause amnesia. The same has also been claimed for lesions of the nucleus accumbens, a structure that also has hippocampal and other limbic system connections.

The prefrontal neocortex is a large region, and although lesions to some parts of it (e.g. the dorsolateral prefrontal cortex) have been found to disrupt particular kinds of memory in humans and animals, it is often argued that, in contrast to organic amnesia, these deficits are usually mild and are secondary to failures of planning that impair encoding and retrieval of complex information, and hence disrupt memory. It remains possible, however, that damage to those frontal regions, such as the orbitofrontal cortex, which receive the heaviest projections from relevant thalamic nuclei and the medial temporal lobes (Tranel and Damasio, 1995), causes a memory disturbance that is more similar to the amnesic syndrome.

Damage to the fornix, which connects the hippocampus directly (and indirectly via the mamillary bodies) to the anterior thalamus, has been shown to cause a mild amnesia. This includes a mild retrograde amnesia, although – as with hippocampal lesions – it is disputed to what extent the anterograde amnesia caused by the lesion involves recognition as well as recall. A similar disagreement extends to the mamillary bodies and the anterior thalamus.

The extent to which all the structures discussed above function as a unitary and integrated system subserving memory for facts and episodes, and to what degree damage to the different structures impairs these forms of memory in subtly different ways, is not yet clear. Figure 2 illustrates some of the connections between the structures implicated in amnesia. These connections may provide clues about the kinds of processing that may underlie fact and episode memory.

IMPLICIT AND EXPLICIT MEMORY

The terms 'explicit memory' and 'implicit memory' were introduced by Graf and Schacter (1985). Explicit memory is the form of memory required when people know that they are remembering (often a particular episode). It is believed to be essential for performing tasks involving free recall, cued recall or recognition of previously presented information. In contrast, implicit memory is the form of memory required to mediate performance on tasks in which memory is indicated by some behavioral change without the person being aware of remembering something. Implicit memory is functionally heterogeneous and includes skill learning, various forms of simple conditioning, and perceptual learning. It also includes an information-specific kind of implicit memory called priming, which mediates performance on tasks

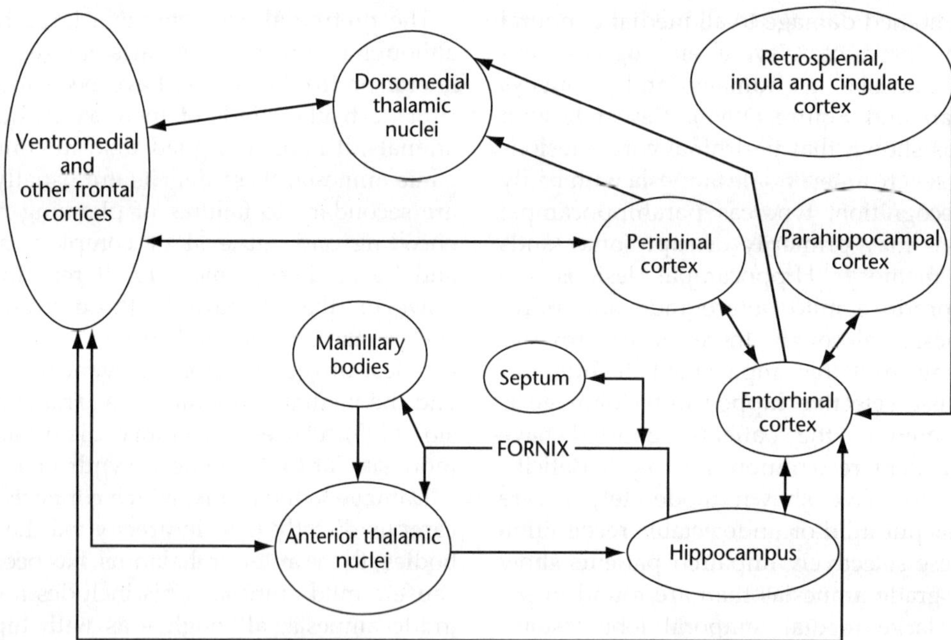

Figure 2. Flow chart illustrating the connections which exist in the primate brain between the components of the medial temporal lobes, the midline diencephalon, and the prefrontal neocortex. Damage to any of these structures is believed to cause at least some of the memory symptoms of the amnesic syndrome. The connections between the structures may provide a useful constraint on theories about what functional deficits underlie the amnesic syndrome.

such as fragment completion, perceptual identification and speeded reading. In experimental investigation into these tasks, participants first study specific information (e.g. words or pictures) and then, after a delay, are required to perform some operation related to that information (such as identifying a briefly displayed studied word). Techniques are used to minimize the degree to which participants use explicit memory to influence their performance. Priming is indicated to the extent that study influences subsequent behavior (e.g. identification) towards the studied information when the behavior is not affected by explicit memory. For example, the speed of reading studied words, compared with similar unstudied words, may be greater; or briefly displayed studied words may be better identified, even when the words are not recognized as having been studied.

Although people with organic amnesia are typically impaired at all tasks involving explicit memory, it has been suspected since the time of Claparède that implicit memory may be preserved. There is good evidence that such patients are relatively unimpaired at acquisition and retention of motor, perceptual and even cognitive skills. Impairments have only become apparent when normal participants' performance, unlike that of the amnesic patients, has been enhanced by their ability to use explicitly remembered strategies. Lacking explicit memory of the learning experiences, amnesic patients are often surprised by their own skills. Similar preservation of simple kinds of classical conditioning has been found. There is some evidence that patients show impaired conditioning when there is a short delay between the end of the conditioned stimulus and the start of the unconditioned stimulus, but it has been argued that this occurs because trace conditioning depends on explicit memory for the relationship between the stimuli. Patients also show preservation of several kinds of perceptual learning and memory.

Evidence about whether people with organic amnesia show preserved priming for all kinds of information is more conflicting. Most but not all researchers believe that patients show preserved priming when the studied information – whether perceptual or semantic in nature – is already familiar and in memory prior to study (e.g. words or famous faces). However, a meta-analysis by Gooding *et al* (2000) indicates that although this is true, amnesic patients' priming of novel items and novel associations is impaired. It might be argued that people with amnesia only show impairments on priming tasks when performance in normal people is facilitated by their superior explicit memory; but this explanation is somewhat implausible because the meta-analysis deliberately compared priming of similar kinds of familiar and novel information on which participants performed similar priming operations at test (for example, made word or non-word judgments as quickly as possible). Also, it cannot explain why priming of novel associations is impaired even when normal participants have no recognition of studied stimuli. There has been no study of whether people with amnesia show preservation of priming for premorbidly acquired memories, so it is unknown whether such priming might differentiate between organic and psychogenic amnesia. However, when both priming and explicit memory for the same information are impaired it is likely that the information never has been or is no longer in long-term storage.

WHAT CAUSES AMNESIA?

Theories about the causes of amnesia must specify what processes break down to cause the syndrome, what lesions are responsible for this, and why. Initially, theorists treated amnesia as a unitary deficit in which only one functional process was disrupted, and all the lesions that produce amnesia disrupted this process although perhaps to differing degrees. Most researchers now believe that the syndrome comprises several distinct functional deficits, each caused by a different lesion (or, more probably, set of lesions), because it is also believed that processes are mediated by systems of structures rather than individual structures.

Theories may differ with respect to: (a) whether encoding, storage or retrieval processes are disrupted in amnesia; (b) what kinds of fact and episode information the impaired processes directly affect; and (c) what specific structures are thought to be critically damaged and, thus, necessary for the execution of the process in question. It has been proposed that defective encoding (processing and representing) of semantic information causes amnesia, or at least the amnesia of Korsakoff syndrome (Butters and Cermak, 1980). The patients' relatively normal intelligence and ability to show normal memory benefits from directed semantic encoding argue against this view. Also, people with amnesia can answer questions normally about semantic (and some other kinds of) information when the questions are asked immediately following presentation. Amnesia is, therefore, unlikely to be caused by a semantic encoding deficit. However, some evidence suggests that perirhinal cortex lesions may impair integrative encoding of high-level sensory information. This possible visual object encoding deficit remains to be systematically tested in humans.

There have been few retrieval deficit theories of amnesia. The most popular such account suggests that people with amnesia may suffer excessively from interference during retrieval and so should show abnormal numbers of intrusion errors, and benefit abnormally from cues restricting the number of possible responses (Warrington and Weiskrantz, 1974). The evidence favoring these predictions is not good, and the theory also leaves unexplained why people with amnesia have this retrieval problem unless it is a secondary effect of a storage defect. Such a defect could cause a secondary problem with interference during retrieval if contextual associations to studied information were not stored. In such a case, contextual markers would not be available to identify which of several competing alternatives was the appropriate memory in a given situation, with the result that intrusion errors would be made unless appropriately restrictive cues were provided. In general, however, retrieval theories imply that people with amnesia store information normally (in anterograde amnesia) or retain it normally (in retrograde amnesia). This possibility is hard to refute decisively so it still needs to be proved convincingly that amnesic patients do not store fact and episode information normally. If such patients do store this information normally, then they should show preserved priming for the same fact or episode information for which their explicit memory is impaired.

Most theories have suggested that amnesia is caused by a failure of the consolidation processes that put all fact and episode information into long-term storage. More specifically, a failure to put the associations linking together the components of facts and episodes normally into memory storage in the minutes or hours following exposure is postulated to cause anterograde amnesia. Retrograde amnesia is explained in terms of a retention failure for the same kinds of association. However, many researchers believe that anterograde and retrograde amnesia are closely linked deficits because there is a slow consolidation process, perhaps involving rehearsal and repetition, through which long-term storage of memories is eventually achieved in the neocortex (Squire and Alvarez, 1995). If they are correct, then memories acquired closer in time to the brain injury should be more disrupted than memories acquired earlier. The alternative view proposes that such temporally graded retrograde amnesia does not occur because memories about episodic incidents (although perhaps not facts) continue to depend on the medial temporal lobes and perhaps other structures implicated in amnesia for as long as they last (Nadel and Moscovitch, 1997). The evidence is conflicting.

Aggleton and Brown (1999) have proposed one of the currently most influential theories. Their theory states that amnesia is functionally heterogeneous. They argue that lesions of the hippocampus, fornix, mamillary bodies or anterior thalamus disrupt rapid associative memory consolidation such that recall memory is greatly impaired, but item recognition – except when it relies heavily on recollection – is relatively intact. Another version of this view is that these lesions selectively impair recall because they impair the consolidation of contextual associations such as those involving spatial and temporal relations. In contrast, the theory postulates that lesions to the perirhinal cortex or some of its projections disrupt 'true recognition' or familiarity. Perirhinal cortex lesions also disrupt recollection because they prevent critical inputs reaching the hippocampus. This theory is controversial, as is the view that anterior temporal neocortex lesions disrupt long-term storage of facts and episodes so perhaps causing focal retrograde amnesia (see Kapur, 1992). However, both theories illustrate the increasing belief that the amnesic syndrome is functionally heterogeneous.

TREATMENT OR REMEDIATION

Organic amnesia is permanent and stable when it is caused by destruction of memory-relevant brain regions. In the future, the condition might possibly be treated effectively by transplantation of stem cells to replace some of the lost neurons and form appropriate connections; at present, however, only remediation using strategies to compensate for the permanently impaired memory processes is feasible. Psychogenic amnesia is typically transient, but when it is long-lasting it may be treated by hypnosis or by drugs such as amobarbital, both of which therapies can reduce patients' resistance to retrieving memories.

Remediation of permanent organic amnesia is important because severe memory impairment typically means that patients will be easily disoriented, have only a hazy sense of their past, cannot work, and may require careful supervision in the home. They may also be bored because social interactions and interesting pastimes that depend on memory, such as reading and watching television, can no longer be pursued.

Recent research has focused on alleviating the consequences of having a deficient memory rather than attempting to improve the impaired underlying memory processes. Four kinds of technique have been moderately successful. First, the use of cognitive strategies (such as imagery in teaching the name of a new acquaintance) has been effective, although patients rarely use these strategies unless specifically instructed to do so.

Second, Baddeley and Wilson (1994) have developed an 'errorless learning' technique, in which the possibility of making errors during training is eliminated by repeatedly providing the information to be learned and thereby preventing guessing. For example, patients learn a series of face–name associations by being repeatedly shown a picture of a face immediately followed by its name, which has to be read or copied. This method has been used successfully to teach a range of information. It is uncertain whether it relies on the patients' use of implicit memory or priming, or their residual explicit memory.

Third, Glisky and Schacter (1987) developed the method of 'vanishing cues' for teaching specific factual information to facilitate activities in a particular domain. They taught an amnesic patient a set of computer terms and procedures by gradually reducing the strength of cues given. The patient was able to return to employment where this knowledge could be used. Like the errorless learning technique, this method leads to very few errors being made, and evidence that the memory it produces is dependent on the cues matching those used during learning suggests that it may depend on priming.

Fourth, the ability of people with amnesia to cope may be improved by modifying their immediate environment – for example, by keeping to a regular routine or encouraging the use of a diary. The repetition produced by regularizing the local environment in this way may help patients by allowing them to form important long-term memories using their relatively intact, but slow to consolidate, neocortical storage sites. Like the methods of errorless learning and vanishing cues, the approach could also be effective because the patients' relatively good memory for local (and other) information may depend to a greater extent than does the memory of normal people on implicit memory or priming. There is evidence that this form of memory is much more sensitive than explicit memory to the effects of interference (see Mayes *et al*, 1987), so that the generation of errors would be much more likely to impede appropriate learning in people with amnesia than it would in normal people. Thus, the prevention of errors by making the local environment more regular and repetitive might facilitate learning in amnesic individuals as well as providing them with a less memory-demanding set of surroundings.

References

Aggleton JP and Brown MW (1999) Episodic memory, amnesia, and the hippocampal-anterior thalamic axis. *Behavioral and Brain Sciences* **22**: 425–443.

Baddeley A and Wilson BA (1994) When implicit learning fails: amnesia and the problem of error elimination. *Neuropsychologia* **32**: 53–68.

Butters N and Cermak LS (1980) *Alcoholic Korsakoff's Syndrome: An Information Processing Approach to Amnesia*. New York, NY: Academic Press.

Glisky EL and Schacter DL (1987) Acquisition of domain-specific knowledge in organic amnesia: training for computer-related work. *Neuropsychologia* **25**: 893–906.

Gooding PA, Mayes AR and van Eijk R (2000) A meta-analysis of indirect memory tests for novel material in organic amnesics. *Neuropsychologia* **38**: 666–676.

Graf P and Schacter DL (1985) Implicit and explicit memory for new associations in normal and amnesic subjects. *Journal of Experimental Psychology: Learning, Memory and Cognition* **11**: 501–518.

Kapur N (1992) Focal retrograde amnesia in neurological disease: a critical review. *Cortex* **29**: 217–234.

Mayes AR, Pickering A and Fairbairn A (1987) Amnesic sensitivity to proactive interference: its relationship to priming and the causes of amnesia. *Neuropsychologia* **25**: 211–220.

Nadel L and Moscovitch M (1997) Memory consolidation, retrograde amnesia and the hippocampal complex. *Current Opinion in Neurobiology* **7**: 217–227.

Phelps EA, LaBar KS, Anderson AK *et al.* (1998) Specifying the contributions of the human amygdala to emotional memory: a case study. *Neurocase* **4**: 527–540.

Squire LR and Alvarez P (1995) Retrograde amnesia and memory consolidation: a neurobiological perspective. *Current Opinion in Neurobiology* **5**: 169–177.

Tranel D and Damasio AR (1995) Neurobiological foundations of human memory. In: Baddeley AD, Wilson BA and Watts FN (eds) *Handbook of Memory Disorders*, pp. 27–50. New York, NY: Wiley.

Warrington EK and Weiskrantz L (1974) The effect of prior learning on subsequent retention in amnesic patients. *Neuropsychologia* **12**: 419–428.

Zola-Morgan S and Squire LR (1993) The neuroanatomy of amnesia. *Annual Review of Neuroscience* **16**: 547–563.

Further Reading

Cohen NJ and Eichenbaum H (1993) *Memory, Amnesia, and the Hippocampal System*. Cambridge, MA: MIT Press.

Mayes AR (1988) *Human Organic Memory Disorders*. Cambridge, UK: Cambridge University Press.

Mayes AR and Downes JJ (1997) *Theories of Organic Amnesia*. Hove, UK: Psychology Press.

Schacter DL and Glisky EL (1986) Memory remediation: restoration, alleviation and the acquisition of domain-specific knowledge. In: Uzzell BP and Gross Y (eds) *Clinical Neuropsychology of Intervention*. Dordrecht: Martinus Nijhoff.

Squire LR (1992) Memory and the hippocampus: a synthesis from findings with rats, monkeys and humans. *Psychological Review* **99**: 195–231.

Squire LR and Knowlton BJ (1995) Memory, hippocampus and brain systems. In: Gazzaniga M (ed.) *The Cognitive Neurosciences*. Cambridge, MA: MIT Press.

Parkin AJ and Leng NRC (1993) *Neuropsychology of the Amnesic Syndrome*. Hove, UK: LEA.

Amygdala

Intermediate article

Ralph Adolphs, University of Iowa, Iowa City, Iowa, USA

CONTENTS

The amygdala is a collection of nuclei in the telencephalon that connect with the sensory neocortex, frontal cortex and a variety of structures that regulate physiological responses. The amygdala participates in the regulation of emotion, attention, memory, and decision-making.

INTRODUCTION

The amygdala is an almond-shaped collection of nuclei which connect with disparate brain structures. In mammals, it is located in the anterior medial temporal lobe, just rostral to the hippocampal formation. The amygdala has a role in motivational, emotional, and cognitive processes that involve information of biological value to the organism. This function has been dissected at the finest grain in animal studies, whereas studies in humans have typically focused on the amygdala's contribution to social cognition.

NEUROANATOMY AND CONNECTIVITY

The amygdala's connections with other brain structures are enormously diverse: it connects with sensory and association neocortex, with hypothalamus, brainstem nuclei, basal forebrain, ventral striatum, thalamus, and hippocampus, and with other structures (Amaral *et al.*, 1992). This broad architecture permits the amygdala to link sensory representations of stimuli (in sensory and association neocortex) with modulation of both body state (via hypothalamus and brainstem) and cognition (via frontal cortex, basal forebrain, hippocampus, and other structures). The amygdala's complex connectivity with other brain regions is complemented by an equally complex internal connectivity among its various component nuclei.

Sensory Input to the Amygdala

The amygdala receives sensory information from all sensory modalities. Signals directly from the olfactory bulb are a prominent form of input in mammals other than primates. Projections from brainstem nuclei and insular cortex provide somatic visceral information, and projections from thalamus and temporal cortex provide auditory and visual information.

In primates, the visual inputs to the amygdala have been best studied. Visually responsive high-level neocortices in the anterior temporal lobe project to the lateral amygdala, and provide highly processed visual information. The amygdala in turn projects back to all those regions from which it receives inputs (via its basal nucleus), and also projects back to some earlier sensory cortices from which it does not receive direct inputs. In fact, the amygdala provides nonreciprocal feedback to all earlier stages of visual processing in the ventral visual stream (which subserves object recognition), including the striate cortex (Figure 1). The feedback from the amygdala terminates in layers 1 and 2 of the cortex and provides an interesting architecture by which the amygdala could, in principle, modulate perception even at the earliest stages of processing.

Connectivity with Brainstem and Hypothalamus

The amygdala connects with nuclei in the brainstem and hypothalamus via its central nucleus. These targets permit the amygdala to modulate the organism's somatic state, a function that constitutes an important component of emotion. For instance, emotionally arousing stimuli are known to trigger increases in sympathetic autonomic tone. The amygdala participates in such triggering via its projections to the paraventricular hypothalamus and periaqueductal gray matter, structures that in turn contain neurons that project to the intermediolateral cell column in the spinal cord, constituting the preganglionic sympathetic neurons. It is thus possible in many cases to trace a causal chain from

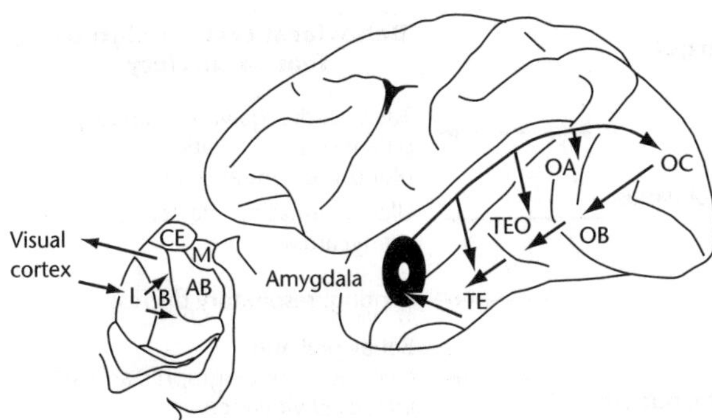

Figure 1. Connections of the amygdala with sensory neocortex. In the monkey, visual inputs to the amygdala originate in high-level visual cortices in the anterior temporal lobe. Feedback from the amygdala is both reciprocal and nonreciprocal, providing projections to all levels of visual processing, including V1. This architecture in principle permits the amygdala to participate in perceptual processing at even the earliest cortical stages. Inputs to, and outputs from, the amygdala rely on the lateral and the basal nucleus, respectively (inset at left). Reproduced with permission from Amaral *et al.* (1992), copyright John Wiley. **Key**: AB, accessory basal; B, basal; CE, central; L, lateral; M, medial; OA, OB, OC, sectors of occipital cortex; TE, TEO, sectors of temporal cortex.

the central nucleus of the amygdala to various effector structures in the brainstem and hypothalamus, all of which regulate autonomic, visceral, and endocrine states (Figure 2).

As with the sensory information, the amygdala's effector communications are not one-way. The amygdala receives information about the somatic changes it effects, via multiple channels. In addition to receiving visceral somatic information from the insular cortex, it also receives more direct information from brainstem nuclei such as the parabrachial nucleus and the nucleus of the solitary tract, as well as direct input from the spinal cord. The vagus nerve is an important route by which body-state information is conveyed to the brainstem and then to the amygdala.

Connectivity with Structures Mediating Cognitive Processes

The above two sets of connections – with sensory cortices on the one hand, and with autonomic and visceral effector structures on the other – would suffice to permit the amygdala to trigger emotional body-state changes in response to sensory stimuli. While this is certainly one function of the amygdala, it is only a small part of the whole story. Perception of emotionally salient stimuli triggers changes not only in the functioning of an animal's body, but also in the animal's cognition. Attention, memory, decision-making and other processes are all prominently modulated by emotion and motivation, and all involve the amygdala.

The amygdala's participation is by virtue of its connections with many other brain regions. For instance, the amygdala projects to cholinergic nuclei in the basal forebrain, which modulate attention; to the hippocampal formation, which is involved in memory; and to the prefrontal cortex, which is involved in planning, reasoning, and decision-making. All these structures, and all the cognitive processes they mediate, can be influenced by the amygdala. As a consequence, emotionally salient information from a sensory stimulus will lead to changes not only in an animal's body, but also in the way that its brain processes that information.

To put the amygdala's role in full perspective, we need to add yet another consideration. The amygdala links sensory percepts with both bodily and cognitive changes, but it does not do so on its own. Rather, it does so in conjunction with other structures that have similar roles. In particular, the amygdala functions in tandem with the orbitofrontal cortex and the ventral striatum, both of which are intimately involved in motivation and emotion.

Intrinsic Processing of the Amygdala

Inputs to the amygdala from the sensory neocortex terminate in the lateral nucleus. Outputs from the amygdala to brainstem and hypothalamus originate in the central nucleus, and outputs to many other structures, such as ventral striatum, cingulate cortex and orbitofrontal cortex, issue from the lateral and basal nuclei, among others. Connections between the lateral nucleus and the basal, central and other nuclei mediate considerable internal information processing (Pitkanen *et al.*, 1997). Put simply, there is a flow of information from the lateral to the central parts of the amygdala, either directly or via intermediate nuclei. Lesions of the lateral nucleus can therefore impair all aspects of an animal's conditioned response to a stimulus, whereas damage to other nuclei impairs only specific components of such a conditioned response (Amorapanth *et al.*, 2000).

AMYGDALA FUNCTION IN ANIMALS

The amygdala has been studied extensively in rats, cats and primates, with some convergent findings. Ever since the classical studies of Klüver and Bucy in the 1930s, the amygdala has been implicated in emotional and social behavior. This rather vague function has now been dissected by studies

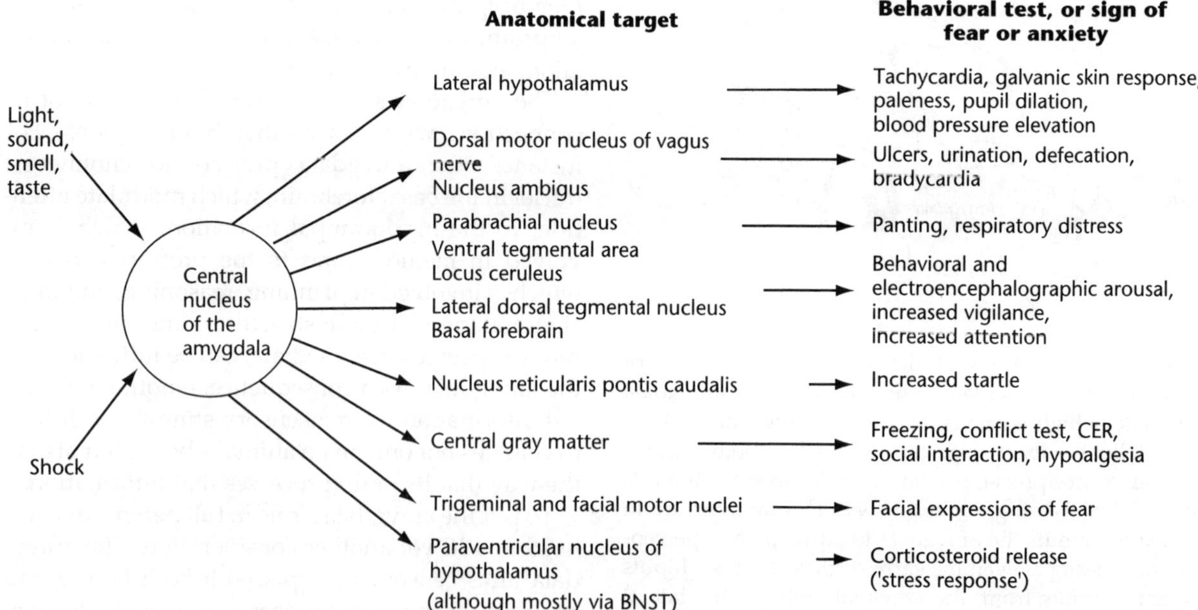

Figure 2. Outputs of the amygdala that mediate fear conditioning. Conditioned and unconditioned stimuli are associated in the lateral amygdala, which in turn projects to the central nucleus of the amygdala. The central nucleus can effect autonomic, endocrine, and visceromotor fear responses by virtue of its connections with structures in brainstem and hypothalamus. Reproduced with permission from Davis (1992), copyright John Wiley.

demonstrating a more specific role in associating sensory stimuli with their rewarding or punishing contingencies. Different nuclei within the amygdala contribute to different aspects of this function. It remains an open question whether the amygdala's basic function is related to reward and punishment in general, or whether it is more specifically related to processing the significance of social stimuli, although findings in animals are more consistent with the former possibility than with the latter.

Regulation of Physiological Responses

Consistent with its connections to hypothalamus and brainstem nuclei, the amygdala participates in regulating a wide array of physiological responses (Figure 2). Direct evidence for this role comes from experiments in which electrical stimulation of the amygdala by implanted electrodes produced changes in physiological measures such as blood pressure and heart rate, corroborating the amygdala's role as an autonomic control structure, and consistent with its known connectivity.

Fear Conditioning

The topic that has seen the most intense research in animal studies is fear conditioning. In fear conditioning an animal is presented with two different kinds of stimuli: an unconditioned stimulus (US) that is highly aversive, such as an electric shock, and a conditioned stimulus (CS) that is neutral to begin with, such as a light. At the beginning of the experiment the animal shows a physiological reaction to the electric shock, but not to the light. During the conditioning experiment, the electric shock is paired with the light stimulus. After several such CS–US pairings the animal learns that the light predicts the shock, a form of associative emotional memory. The final stage of the experiment is to present the CS without the US. The fear-conditioned animal will behave as though it is afraid of the light, a conditioned response that can be measured in a number of ways as indicated in the right-hand column in Figure 2. (*See* **Conditioning**)

Conditioned fear responses critically depend on the amygdala, and cells within the lateral amygdala increase their firing rates and the synchrony of their discharges following the CS. Importantly, the dependency of the acquisition and expression of fear-conditioned responses on the basolateral amygdala is not attributable solely to the amygdala's role in the behaviors used to assess fear conditioning, such as 'freezing' (Maren, 1999).

Current debates

Despite the popularity of the fear-conditioning paradigm, there are many issues still to resolve. Is the amygdala essential for all components of fear conditioning? Is the memory of the US–CS association actually stored in the amygdala? Affirmative answers to these questions have been given by some, but contested by others. One general issue is whether the amygdala is essential to the acquisition of fear-conditioned responses, or whether its function is better thought of as modulatory and important rather than essential (e.g. Wilensky *et al.*, 2000). Another issue concerns the role of fear conditioning during development, and its relation to emotional and social development. Fear conditioning is not seen at birth in all species; possibly its absence early in life permits an animal to form an attachment to its parent regardless of the circumstances (that is, even when the presence of the parent occurs together with aversive events).

Reconsolidation

If the amygdala is indeed the repository for the associations between conditioned and unconditioned stimuli in fear-conditioning experiments, it remains an open question precisely how such storage is implemented. One possibility is that a relatively permanent association is stored in a way that is insensitive to future perturbations. This appears unlikely at first, because it is known that fear-conditioned responses can be extinguished if the CS is presented for many times without the US. However, this extinction appears to rely not on an erasure of the memory trace within the amygdala, but rather on active inhibition of the amygdala's fear memory by the prefrontal cortex. It thus appears that some components of the fear memory may indeed be stored in the amygdala permanently, an aspect that may have implications for indelible fear responses in humans, for instance in posttraumatic stress disorder and phobias.

Although the memory of the fear conditioning in the amygdala may be relatively permanent, it is not static. Experimenters have found that reactivating the fear memory by presenting the CS rendered the memory trace plastic and susceptible to alteration. After reactivation of the memory, there was a period of 'reconsolidation' during which the memory could indeed be erased. Injections of drugs that inhibited protein synthesis, such as anisomycin, were found to erase the fear-conditioned memory, but only when given during such a period of reconsolidation (Nader *et al.*, 2000). This finding may have broad applicability to memory in general, including memory in humans, and emphasizes the dynamic nature of memory representations in the brain. (*See* **Memory Consolidation**)

Molecular mechanisms

What are the molecular mechanisms whereby the amygdala stores the associations learned during fear conditioning? There is evidence that such associative memory may rely on long-term potentiation (LTP) within the amygdala. The acquisition of fear-conditioned responses requires within the amygdala some of the molecular machinery known to be involved in the induction and maintenance of LTP, such as protein kinase A, protein synthesis, and perhaps even NMDA (*N*-methyl-D-aspastate) receptors (although this has not been conclusively shown). It thus appears likely that at least some components of the amygdala's role in associative memory are similar at the molecular level to those required for memory consolidation in the hippocampus.

Other Forms of Conditioning

The amygdala has also been implicated in other forms of conditioning besides the classical (Pavlovian) fear response. Conditioning to stimuli that are rewarding rather than aversive does not appear to rely on the amygdala to the same extent that conditioning to aversive stimuli does. However, the amygdala has been found to be important in reinforcer devaluation effects, in which the value of a rewarding US is changed: for instance, by using food reward as the US and then later feeding the animal to satiation on that food so that the US no longer has the same motivational value that it did at the time of conditioning. Normal animals are sensitive to such manipulations, but animals with amygdala damage are not. These studies suggest that a comprehensive function of the amygdala in associative emotional memory is the provision of access to current values of reinforcing stimuli, a function in which it participates intimately with the orbitofrontal cortex (Baxter *et al.*, 2000).

Modulation of Motivated Learning

Aversively motivated learning can be modulated by various neurotransmitter systems acting within the amygdala (Cahill and McGaugh, 1996; McGaugh, 2000). For instance, injection into the amygdala of one of a variety of drugs that modulate neurotransmission mediated by γ-aminobutyric acid (GABA), noradrenaline, or opiates immediately after training influences long-term memory for stimulus avoidance. Research suggests that the amygdala can influence such memory by modulating consolidation that actually takes place within other brain structures, such as the hippocampus and basal ganglia.

The amygdala thus participates in at least two distinct types of memory function: associative emotional memory (such as fear conditioning), which may be stored within the amygdala, and modulation of memory (including declarative memory in humans) that is itself dependent on other brain structures.

Influences on Attention

The amygdala's role in attentional processes has been investigated in a number of experiments (Holland and Gallagher, 1999). One component of attention, orienting behavior towards cues that have become associated with rewarding contingencies, has been found to rely on a circuit involving the central nucleus of the amygdala and its connections with the substantia nigra and the dorsal striatum. Another important component of attention, increased allocation of processing resources toward novel or surprising situations, appears to depend on the integrity of the central nucleus of the amygdala and its connections with cholinergic neurons in the substantia innominata and nucleus basalis, structures in the basal forebrain. Thus, the amygdala could modulate cholinergic neuromodulatory functions of the basal forebrain nuclei, and consequently modulate attention, vigilance, signal-to-noise and other aspects of information processing that depend on cholinergic modulation of cognition. Through circuits including components of amygdala, striatum, and basal forebrain, emotion may thus help to select particular aspects of the stimulus environment for disproportionate allocation of cognitive processing resources; in other words, the organism preferentially processes information about its environment that is most salient to its immediate survival and wellbeing. Recent studies in humans have demonstrated such a role (Anderson and Phelps, 2001).

Social Behavior

Although early experiments concerning the effects of amygdala lesions highlighted impairments in social behavior, this issue has received scant attention. In general, the findings of the effects of focal amygdala lesions have corroborated impairments in social behavior – typically an increase in tameness and placidity towards others (Weiskrantz, 1956; Emery *et al.*, 2001). However, how such an impairment plays out in a group of animals depends on the species and on the environmental circumstances in which the social behavior is assessed: monkeys whose amygdala has been removed can be the subjects of either increased affiliative or increased aggressive behavior from other monkeys in the troop. It remains an open question to what extent the deficits in social behavior that are seen following amygdala damage could be understood in terms of (or could be reducible to) impairments in more basic processing of reward and punishment.

AMYGDALA FUNCTION IN HUMANS

In contrast to animal studies, which implicate the amygdala in processing reward and punishment in a general sense, studies with humans point to a more restricted function in processing socially relevant stimuli. Furthermore, findings in humans indicate a role of disproportionate importance in processing stimuli related to negatively valenced stimuli that signal threat, ambiguity or distress. However, the apparent differences between animals and humans may be due more to the kinds of tasks and stimuli employed than to actual differences in amygdala function.

Conditioning

Damage to the human amygdala impairs the ability to acquire conditioned fear responses. One study showed that a patient with bilateral damage to the amygdala, but with an intact hippocampal memory system, was unable to acquire conditioned autonomic responses, but was nonetheless able to acquire declarative memory about the CS–US pairings: in essence, the patient could tell the experimenter that the light had been paired with the shock – so the patient knew, as a declarative fact, that light predicted shock – even though there was no fear conditioning (Bechara *et al.*, 1995). Functional imaging studies have likewise found evidence that the amygdala is activated in normal people during fear conditioning.

Recognition of Emotion

Both lesion and functional imaging studies of the brain have confirmed a role for the amygdala in perception and recognition of emotional stimuli. The evidence is strongest in the case of recognition of emotions from facial expressions, although there is some evidence for emotion perception from auditory, olfactory, and gustatory stimuli as well. A rare patient with bilateral amygdala lesions was found to have an impaired ability to recognize fear from faces, a finding subsequently replicated in several other studies, and corroborated by the finding that viewing faces showing expressions of fear activates the amygdala in normal people.

Despite these studies, it remains unclear precisely which emotions the amygdala is most involved in processing. While fear recognition is notably impaired in some people with amygdala damage, this is not always the case, and others with similar brain lesions are more impaired on recognition of anger, sadness or disgust (Adolphs *et al.*, 1999). Functional brain imaging of psychiatric patients has shown that some people with phobias or similar disorders have abnormal activation of the amygdala (for example, an abnormally high activation of the amygdala occurs in response to 'neutral' faces, in patients who are socially phobic). All these findings have led to proposals that the amygdala processes predominantly information about threat, about ambiguity or about distress. At this stage, all we can say for sure is that the amygdala participates in processing knowledge of emotion from a variety of social stimuli, such as facial expressions – but whether there is specificity regarding particular emotions, and what that specificity consists of, are issues requiring further research. Some functional imaging studies have found activation of the amygdala by pleasant stimuli also.

Another important point is that the amygdala does not appear essential for all knowledge regarding emotions. In particular, knowledge of emotions from explicit lexical stimuli is unaffected by amygdala damage. People with bilateral amygdala damage are quite able to use words such as 'fear' appropriately in conversation, and possess considerable knowledge about the kinds of situation that would normally make people feel afraid, or the kinds of behavior elicited by fear. However, the concept of fear that people with bilateral amygdala damage have is not entirely normal: it has gaps. It seems likely that those gaps are most prominent for emotional knowledge that cannot be explicitly encoded into language – such as knowing what a fearful face looks like.

Functional imaging studies have confirmed that the amygdala is activated by emotional facial expressions, especially ones showing fear. Such activation has been observed even when the face stimuli were presented so briefly that they could not be consciously recognized (Whalen *et al.*, 1998) There is evidence to suggest a differential role for the right amygdala in processing emotional visual stimuli below the level of conscious awareness, and for the left amygdala in processing emotional stimuli that are consciously perceived (Morris *et al.*, 1999).

It is less clear to what extent the amygdala is involved in recognition of emotional auditory stimuli; some lesion and functional imaging studies have suggested such a role, whereas others have not. The most robust activation of the amygdala to sensory input with emotional significance has been observed with olfactory stimuli, where unpleasant smells reliably activate the amygdala (Royet *et al.*, 2000).

Emotional Memory

As in animals, the amygdala in humans modulates memory for emotionally arousing stimuli. In humans, this has been studied in regard to declarative memory. When normal people are shown emotionally arousing pictures or film, they remember best those stimuli that they found the most emotionally arousing. Functional imaging studies have shown that activation of the amygdala at the time of encoding (when the stimuli were shown, while the subject was in the scanner) correlated with how well the person later remembered those same stimuli when questioned several weeks later (Canli *et al.*, 2000). Consonant with these findings, people with bilateral amygdala damage were less able to remember emotionally arousing pictures – they remembered them only as well as neutral pictures. Electrical stimulation of the vagus nerve – which provides visceral body-state information to brainstem nuclei and the amygdala – has been shown to enhance memory, consistent with the idea that emotional body states can modulate cognitive processes such as memory, at least in part via the amygdala.

Social Cognition

The human amygdala plays a more general role in social cognition as well. In one study, participants were asked to judge how trustworthy or approachable they found other people; those with bilateral amygdala damage gave abnormally positive ratings (Adolphs *et al.*, 1998). Other studies have found activation of the amygdala in response to viewing faces of people of another race: the amygdala habituated more rapidly when viewing faces of one's own race than of another race, and amygdala activation to unfamiliar faces of a different race correlated with implicit measures of race evaluation (Phelps *et al.*, 2000). Such racial outgroup responses may fit into the general scheme of threat detection and vigilance by the amygdala. Finally, there is evidence linking amygdala pathology to autism (Baron-Cohen *et al.*, 2000): some

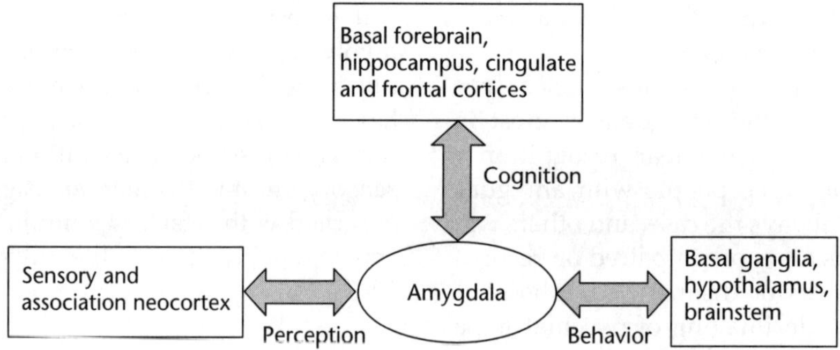

Figure 3. Connectivity of the amygdala at the system level. The amygdala's participation in multiple effector mechanisms permits mediation of a concerted, integrated emotional response in both body and brain.

structural and functional imaging studies suggest such a link, and people with autism are impaired on some of the same tasks as people with bilateral amygdala damage. Clearly, the amygdala is important in social cognition.

A SYSTEMS-LEVEL VIEW OF THE AMYGDALA

While it is clear that the amygdala broadly participates in emotion and social behavior, the mechanisms that underlie this function remain poorly understood. In general terms, the amygdala can participate in emotion in three different ways (Figure 3). First, it can link perception of stimuli to an emotional response, by virtue of its inputs from sensory cortices and its outputs to control structures such as the hypothalamus, brainstem nuclei, and periaqueductal gray matter; second, it can link perception of stimuli to modulation of cognition, by virtue of its connections with structures involved in decision-making, memory, and attention; and third, it can link early perceptual processing of stimuli with modulation of such perception via direct feedback. The amygdala's participation in all three of these mechanisms enables it to contribute globally to affective processing in a concerted manner, by modulating numerous processes simultaneously.

To which aspects of emotion could such mechanisms contribute? Certainly, the amygdala contributes to emotional responses to stimuli. It also contributes to the recognition of emotions, but this function is likely to be indirect: possibly the amygdala's modulation of perception by feedback, or its elicitation of components of an emotional response, can contribute to the brain's ability to reconstruct knowledge about the emotion signaled by a stimulus. Finally, the amygdala's role in feeling an emotion (in the conscious experience of emotion) is unclear, and may well be inessential.

References

Adolphs R, Tranel D and Damasio AR (1998) The human amygdala in social judgment. *Nature* **393**: 470–474.

Adolphs R, Tranel D, Hamann S *et al.* (1999) Recognition of facial emotion in nine subjects with bilateral amygdala damage. *Neuropsychologia* **37**: 1111–1117.

Amaral DG, Price JL, Pitkanen A and Carmichael ST (1992) Anatomical organization of the primate amygdaloid complex. In: Aggleton JP (ed.) *The Amygdala: Neurobiological Aspects of Emotion, Memory, and Mental Dysfunction*, pp. 1–66. New York, NY: Wiley-Liss.

Amorapanth P, LeDoux JE and Nader K (2000) Different lateral amygdala outputs mediate reactions and actions elicited by a fear-arousing stimulus. *Nature Neuroscience* **3**: 74–79.

Anderson AK and Phelps EA (2001) Lesions of the human amygdala impair enhanced perception of emotionally salient events. *Nature* **411**: 305–309.

Baron-Cohen S, Ring HA, Bullmore ET *et al.* (2000) The amygdala theory of autism. *Neuroscience and Biobehavioral Reviews* **24**: 355–364.

Baxter MG, Parker A, Lindner CCC, Izquierdo AD and Murray EA (2000) Control of response selection by reinforcer value requires interaction of amygdala and orbital prefrontal cortex. *Journal of Neuroscience* **20**: 4311–4319.

Bechara A, Tranel D, Damasio H *et al.* (1995) Double dissociation of conditioning and declarative knowledge relative to the amygdala and hippocampus in humans. *Science* **269**: 1115–1118.

Cahill L and McGaugh JL (1996) Modulation of memory storage. *Current Opinion in Neurobiology* **6**: 237–242.

Canli T, Zhao Z, Brewer J, Gabrieli JDE and Cahill L (2000) Event-related activation in the human amygdala associates with later memory for individual emotional experience. *Journal of Neuroscience* **20**: RC99 (91–95).

Emery NJ, Capitanio JP, Mason WA *et al.* (2001) The effects of bilateral lesions of the amygdala on dyadic social interactions in rhesus monkeys. *Behavioral Neuroscience* **115**: 515–544.

Holland PC and Gallagher M (1999) Amygdala circuitry in attentional and representational processes. *Trends in Cognitive Sciences* **3**: 65–73.

Maren S (1999) Neurotoxic basolateral amygdala lesions impair learning and memory but not the performance of conditional fear in rats. *Journal of Neuroscience* **19**: 8696–8703.

McGaugh JL (2000) Memory – a century of consolidation. *Science* **287**: 248–251.

Morris JS, Ohman A and Dolan RJ (1999) A subcortical pathway to the right amygdala mediating 'unseen' fear. *Proceedings of the National Academy of Sciences of the USA* **96**: 1680–1685.

Nader K, Schafe GE and LeDoux JE (2000) Fear memories require protein synthesis in the amygdala for reconsolidation after retrieval. *Nature* **406**: 722–726.

Phelps EA, O'Connor KJ, Cunningham WA *et al.* (2000) Performance on indirect measures of race evaluation predicts amygdala activation. *Journal of Cognitive Neuroscience* **12**: 729–738.

Pitkanen A, Savander V and LeDoux JE (1997) Organization of intra-amygdaloid circuitries in the rat: an emerging framework for understanding functions of the amygdala. *Trends in Neurosciences* **20**: 517–523.

Royet JP, Zald D, Versace R *et al.* (2000) Emotional responses to pleasant and unpleasant olfactory, visual, and auditory stimuli: a positron emission tomography study. *Journal of Neuroscience* **20**: 7752–7759.

Weiskrantz L (1956) Behavioral changes associated with ablation of the amygdaloid complex in monkeys. *Journal of Comparative Physiology and Psychology* **49**: 381–391.

Whalen PJ, Rauch SL, Etcoff NL *et al.* (1998) Masked presentations of emotional facial expressions modulate amygdala, activity without explicit knowledge. *Journal of Neuroscience* **18**: 411–418.

Wilensky AE, Schafe GE and LeDoux J (2000) The amygdala modulates memory consolidation of fear-motivated inhibitory avoidance learning but not classical fear conditioning. *Journal of Neuroscience* **20**: 7059–7066.

Further Reading

Adolphs R (1999) The human amygdala and emotion. *Neuroscientist* **5**: 125–137.

Aggleton J (ed.) (2000) *The Amygdala: A Functional Analysis*. New York, NY: Oxford University Press.

Davis M (1997) Neurobiology of fear responses: the role of the amygdala. *Journal of Neuropsychiatry and Clinical Neurosciences* **9**: 382–402.

Emery NJ and Amaral DG (1999) The role of the amygdala in primate social cognition. In: Lane RD and Nadel L (eds) *Cognitive Neuroscience of Emotion*, pp. 156–191. Oxford, UK: Oxford University Press.

LeDoux J (1996) *The Emotional Brain*. New York, NY: Simon & Schuster.

Rolls ET (1999) *The Brain and Emotion*. New York, NY: Oxford University Press.

Analgesia

See **Pain and Analgesia, Neural Basis of**

Analogical Reasoning, Psychology of

Intermediate article

Dedre Gentner, Northwestern University, Evanston, Illinois, USA

CONTENTS

Analogical reasoning is a kind of reasoning that applies between specific exemplars or cases, in which what is known about one exemplar is used to infer new information about another exemplar. The basic intuition behind analogical reasoning is that when there are substantial parallels across different situations, there are likely to be further parallels.

INTRODUCTION

Analogical thinking is ubiquitous in human cognition. First, analogies are used in explaining new concepts. Domains such as electricity or molecular motion, which cannot directly be perceived, are often taught by analogy to familiar concrete domains such as water flow or billiard-ball collisions. Within cognitive science, mental processes are likened to computer programs (e.g. neural networks; parallel or serial processes), or to searching within a space (e.g. mental distance; close or far associates). Such analogies can then serve as mental models to support reasoning in new domains. Another use of analogy is in making predictions within domains. When the stock market plunged in 2001 after the attack on the World Trade Center, many newswriters made an analogy to the 1929 Wall Street crash, and argued on this basis that the market would be higher after a few years (or that, because key causal conditions are different, the reverse would occur). Analogy is also important in creativity and scientific discovery, as discussed later. Finally, analogy is used in communication and persuasion. For example, environmentalists have compared the earth to Easter Island, where overpopulation and exploitation of the island's once-rich ecology led first to massive loss of species, and then to famine and societal collapse. Such persuasive analogies are meant to invite new inferences: for example, that continued population growth will lead to irreversible ecological decline.

Analogical processing involves several subprocesses. First, given a current topic, *analogical retrieval* is the process of being reminded of a past situation from long-term memory. Once two cases are present in working memory (either because of an analogical retrieval or simply through encountering two cases together), *analogical mapping* can occur. We will begin by discussing the mapping processes.

MAPPING AND USE

History

Important early work on analogical mapping came out of philosophy, notably Hesse's analysis of analogical models in science. Early psychological research on analogy focused on simple four-term analogies of the form a:b::c:d. In the 1970s and 1980s, artificial intelligence researchers introduced a new level of representational complexity and computational specificity. Patrick Winston explored computational algorithms for analogical matching and inference, and Jaime Carbonell modeled the transfer of solution methods from one problem to another. This kind of work inspired psychologists to lay out detailed process models of how analogies are represented and processed. The ensuing period has seen intense computational and psychological research, theory revision, and an expansion of the phenomena studied. The field of analogy continues to be characterized by extremely fruitful interchange between computational models and psychological research. (*See* **Analogy-making, Computational Models of**)

Analogical Mapping

Analogical mapping is the core process in analogy. In a typical instance of analogical mapping, a familiar situation – the *base* or *source* description – is matched with a less familiar situation – the *target* description. The familiar situation suggests ways of viewing the newer situation as well as further inferences about it. Analogical mapping requires *aligning* the two situations – that is, finding the correspondences between the two representations – and *projecting inferences* from the base to the target. Then the reasoner must *evaluate* the analogical match and its inferences. Two further processes that can occur are *re-representation* of one or both analogs to improve the match, and *abstraction* of the structure common to both analogs.

Structure-mapping theory (Gentner, 1983) aims to capture the psychological processes that carry out analogical mapping. According to this theory, the comparison process involves finding an alignment between the base and target representations that reveals common relational structure. On the basis of this alignment, further inferences are projected from base to target. People prefer to find an alignment that is *structurally consistent*: that is, there should be a *one-to-one correspondence* between elements in the base and elements in the target, and the arguments of corresponding predicates must also correspond (*parallel connectivity*). For example, in the analogy below, Timmy in (a) could be put in correspondence with Timmy in (b) (on the basis of a local entity match) or with Fluffy in (b) (on the basis of matching relational roles). People appear to entertain both possibilities during processing, but to settle on one or the other by the end of the process.

(a) Lassie rescued Timmy.
(b) Timmy rescued Fluffy.

Another important early theory was Holyoak's (1985) *pragmatic mapping theory*, which focused on the use of analogy in problem-solving and held that analogical mapping processes are oriented towards attaining goals (such as solutions to problems). According to pragmatic mapping theory, it is goal relevance that determines what is selected in analogy. Holyoak and Thagard (1989) later combined this pragmatic focus with structural factors in their multi-constraint approach to analogy.

Analogical inference projection is a crucial part of the mapping process. Once an alignment is achieved, further inferences can be made by projecting information from the base (or source) domain into the target domain. For example, in the above analogy, suppose we knew more about event (a), such as:

(a) Lassie rescued Timmy because she loves him. She has beatiful brown eyes.
(b) Timmy rescued Fluffy.

In this case, the likely inference in (b) is that Timmy rescued Fluffy because he loves Fluffy. This ability to invite new inferences is central to analogy's role in reasoning. Importantly, analogical inference is rather selective. For example, we are unlikely to make the inference here that Timmy has brown eyes (or that he has four legs, even if we also know this to be true of Lassie).

This illustrates the *selection problem* in theories of analogical inference. If people projected everything known about the base into the target, analogy would be useless in reasoning. Fortunately, people do not do this. Thus a central aim of theories of analogy is to characterize this selection process. At least three factors have been proposed.

Holyoak and his colleagues have emphasized *goal relevance*: the inferences projected are those that fit with the reasoner's current goals in problem-solving.

A second factor, proposed in structure-mapping theory, is relational connectivity – more specifically, *systematicity*: a preference for projecting from matching systems of relations connected by higher-order relations such as *cause*, rather than projecting local matches. In many cases, goal relevance and systematicity will make the same predictions, because problem-solving goals often involve a focus on causal systems.

A third factor in selecting inferences, proposed by Keane, is *adaptability*: the ease with which a possible inference can be modified to fit the target.

There is evidence for all three of these criteria. Spellman and Holyoak (1996) showed that when two possible mappings are available for a given analogy, people will select the mapping whose inferences are relevant to their goals. Evidence for systematicity comes from the finding that when people read analogous passages and make inferences from one to the other, they are more likely to import a fact from the base to the target when it is causally connected to other matching predicates (Clement and Gentner, 1991; Markman, 1997). Finally, Keane (1996) found evidence that the degree of adaptability predicts which inferences are made from an analogy.

There remain many open questions. For example, according to the structure-mapping account, many different higher-order relations can provide inferential selection – including causal relations, deontic relations such as permission and obligation, and spatial relations such as symmetry and transitive increase. The challenge then is to delineate the set of higher-order relations that can serve this purpose. Another open question is the time course of these constraints. For example, do goals have special priority *during* the analogical mapping process, or do the effects of goals occur through influencing the initial representations of the two analogs (*before* the mapping process) or through selecting among multiple possible interpretations (*after* the mapping process)?

Evaluation

Once the common alignment and the candidate inferences have been discovered, the analogy is evaluated. *Evaluating* an analogy involves at least three kinds of judgment: (1) *structural soundness*: whether the alignment and the projected inferences are structurally consistent; (2) *factual correctness*: whether the projected inferences are false, true, or indeterminate in the target; and (3) *relevance*: whether the analogical inferences are relevant to the current goals. In practice, the relative importance of these factors varies quite a bit. In domains where little is known, or where there is disagreement about the facts – for example, in politics – goal relevance may be more important than factual correctness.

Abstraction

In *analogical abstraction*, the common system that represents the interpretation of an analogy is extracted and stored. This kind of schema abstraction helps to promote transfer to new exemplars. When people are asked to compare two analogous passages, they are better able to later retrieve and use their common structure (given a relationally similar probe) than are people who were given only one of the stories (Gick and Holyoak, 1983). Further studies have shown that actively comparing two analogous passages leads to better subsequent retrieval than reading the two passages separately. These findings are consistent with the claim that analogical alignment promotes the common structure and makes it more available for later use.

Analogy in Real-world Reasoning

Analogy is often used in common-sense reasoning to provide plausible inferences. It must be noted that analogy is not a deductive process. There is no guarantee that the inferences from a given analogy will be true in the target, even if the analogy is carried out perfectly and all of the relevant statements are true in the base. However, the set of implicit constraints described above make analogy a relatively 'tight' form of inductive reasoning. This may be why analogy is heavily used in arenas such as law, where clear reasoning is important but formal principles are often not sufficient to decide issues.

The lack of deductive certainty in analogical reasoning has a positive side. It means that analogy can suggest genuinely new hypotheses, whose truth could not be deduced from current knowledge. One arena in which this kind of analogical inferencing has been extensively studied is scientific reasoning and discovery. Nancy Nersessian has examined the role of analogy and other model-based reasoning processes in the thought processes of Faraday and Maxwell. Paul Thagard has discussed analogy as a contributor to conceptual revolutions in science. Kevin Dunbar has observed scientists in working microbiology laboratories and has found that analogy plays a large role in the discovery process.

Analogy appears to be very important in children's thinking, as Halford, Goswami, and others have argued. Children often use analogies from known domains as a way to fill in gaps in their knowledge of other domains. For example, Inagaki and Hatano (1987) asked five-year-old children hypothetical questions like 'What would happen if a rabbit were continually given more water?' The children often answered by using an analogy to humans: for example, 'I would get sick if I kept drinking water, and I think the rabbit would too.' Interestingly, children's answers tended to be more accurate when they used such analogies than when they did not. Children were most likely to use analogies to humans when the target was somewhat similar to humans, suggesting that for children (as for adults) similar analogs are more likely to be retrieved and are easier to align with the target than dissimilar analogs.

FACTORS THAT INFLUENCE ANALOGICAL MAPPING AND USE

People's fluency in carrying out analogical mappings is influenced by three broad kinds of factors. First are factors internal to the analogical mapping itself, such as *systematicity* – whether the common relational system possesses higher-order connective structure – and *transparency* – the degree to which corresponding elements are similar. The second category includes characteristics of the reasoner, such as age and expertise. The third includes task factors such as processing load, time pressure, and context.

Transparency and systematicity have been found to be important in analogical problem-solving. The transparency of the mapping between two analogous algebra problems – that is, the similarity between corresponding objects – is a good predictor of people's ability to notice and apply solutions from one problem to the other. For example, Ross (1989) taught people algebra problems and later gave them new problems that followed the same principles. People were better able to map the solution from a prior problem to a current problem when the corresponding objects were highly similar between the two problems: for example, 'How many golf balls per golfer' → 'How many tennis balls per tennis player'. They performed worst in the *cross-mapped* condition, in which similar objects appeared in different roles across the two problems: for example, 'How many golf balls per golfer' → 'How many tennis players per tennis ball'.

The intrinsic factors of transparency and systematicity interact with characteristics of the reasoner, notably age and experience. Gentner and Toupin (1986) gave children a simple story and asked them to re-enact the story with new characters. Both six- and nine-year-olds performed far better when the corresponding characters were highly similar between the two stories than when they were different, and they performed worst when similar characters played different roles across the two stories (the *cross-mapped* condition). Thus both age groups were sensitive to the transparency of the correspondences. In addition, older children (but not younger children) benefited from systematicity – that is, from hearing a summary statement that provided an overarching social or causal moral.

The developmental change found here is an instance of the *relational shift*: a shift from focusing on object matches to focusing on relational matches. Some researchers have suggested that this shift is driven by gains in knowledge (Gentner and Rattermann, 1991), while others propose that it results from a developmental increase in processing capacity (Halford, 1993).

The third class of factors affecting analogical processing concerns task variables such as time pressure, processing load, and immediate context. One generalization that emerges from several studies is that making relational matches requires more time and processing resources than making object–attribute matches. For example, Goldstone and Medin (1994) found that when people are forced to terminate processing early, they are strongly influenced by local attribute matches (such as *A* with *A* in the example below), even in cases where they would choose a relational match (such as *A* with *P*) if given sufficient time:

A above M and *P above A*

Adult performance in mapping tasks is also influenced by immediately preceding experiences. For example, in the *one-shot mapping task* (Markman and Gentner, 1993) subjects are shown a pair of cross-mapped pictures, such as *a robot repairing a car* and *a man repairing a robot*. The experimenter points to the robot in the first picture and the subject indicates which object in the second picture 'goes with' it. Subjects often choose the object match (e.g. the other robot). However, if they have previously rated the similarity of the pair, they are likely to choose the relational correspondence (the repairman). These findings suggest that carrying out a similarity comparison induces a structural alignment.

Kubose, Holyoak, and Hummel used this one-shot mapping task to show that processing load influences analogical processing. The experimenter pointed to the cross-mapped object in the first picture (the robot) and subjects were instructed to point to the relational correspondence (the repairman) in the second picture. Subjects made more object-mapping errors when given an extra processing load, such as having to count backwards. Recent work by Holyoak and his colleagues also suggests that damage to the prefrontal cortex is associated with detriments in analogical tasks, although it is not clear whether this results from specific involvement of the prefrontal cortex in analogical processing or from more general factors such as inhibitory control.

Summary

Research on analogical mapping has revealed a set of basic phenomena that characterize human analogical processing (see Table 1). A striking feature of analogical mapping is the importance of systematic, structurally connected representations. Commonalities that are interconnected into a relational system are considered to be more central to a comparison than are those that are not. Connected systems are easier to map to a new domain than are unconnected sets, leading to better transfer in analogy and problem-solving. Systematicity also governs inferences: inferences are projected from interconnected systems in the base to fill out corresponding structure in the target. Even the

Table 1. Basic phenomena of analogy (adapted from Gentner and Markman, 1995, 1997; see also Hummel and Holyoak, 1997)

1 *Relational similarity*	Analogies involve relational commonalities; object commonalities are optional.
2 *Structural consistency*	Analogical mapping involves one-to-one correspondence and parallel connectivity.
3 *Systematicity*	In interpreting analogy, connected systems of relations are preferred over isolated relations.
4 *Candidate inferences*	Analogical inferences are generated via structural completion.
5 *Alignable differences*	Differences that are connected to the common system are rendered more salient by a comparison.
6 *Interactive mapping*	Analogy interpretation depends on both terms. The same term yields different interpretations in different pairings.
7 *Multiple interpretations*	Analogy allows multiple interpretations of a single comparison.
8 *Cross-mapping*	Analogies are more difficult to process when there are competing object matches.

differences associated with a similarity comparison are influenced by systematicity: the differences that are psychologically salient in a comparison are those that are connected to the common system. In addition, goal relevance may have effects over and above the effects of connected relational structure.

RETRIEVAL OF ANALOGS

So far, we have discussed the processing of an analogy once both analogs are present. When we turn to the issue of what leads people to think of analogies, we see a very different pattern of results. People often fail to retrieve potentially useful analogs, even when there is an excellent structural match, and even when they clearly have retained the material in memory. For example, Gick and Holyoak (1983) gave subjects a thought problem: how to cure an inoperable tumor without using a strong beam of radiation that would kill the surrounding flesh. Only about 10 percent of the participants came up with the ideal solution, which is to converge on the tumor with several weak beams of radiation. If given a prior analogous story in which soldiers converged on a fort, three times as many people (about 30 percent) produced convergence solutions. Yet the majority of participants still failed to think of the convergence solution. Surprisingly, when these people were simply told to think back to the story they had heard, the percentage of convergence solutions again tripled, to 80–90 percent. Thus, the fortress story had been retained in memory, but it was not retrieved by the analogous tumor problem. The implication is clear. Even when a prior experience has been successfully stored in memory, it might not be retrieved when a person encounters a new analogous situation where it would be useful.

When we ask what does facilitate analogical retrieval, one major factor emerges: the similarity between the analogs. As noted earlier, similarity is one of the factors that facilitates analogical mapping; but it has a much larger effect on retrieval. For example, Gentner *et al.* (1993) gave subjects a set of stories to remember and later showed them probe stories that were either surface-similar to their memory item (e.g. similar objects and characters) or structurally similar (i.e. analogous, with similar higher-order causal structure). Surface similarity was the best predictor of whether people would be reminded of the prior stories; people were two to five times more likely to retrieve prior stories with only surface commonalities than with only structural commonalities. However, their judgments of the goodness of the match were completely different. They rated the surface-similar pairs (including their own remindings) as low in inferential value and in similarity, and preferred the structurally similar pairs. A similar dissociation between reminding and use has been found in problem-solving tasks: remindings of prior problems are strongly influenced by surface similarity, even though structural similarity better predicts success in solving current problems (Ross, 1989). This failure to access potentially useful analogs (unless they are highly similar to the target) is an instance of the *inert knowledge* problem in education. One piece of good news is that it appears that domain expertise may improve matters somewhat. For example, Novick (1988) found that people with mathematics training retrieved fewer surface-similar lures in a problem-solving task than did novice mathematicians. Moreover, experts were quicker to reject these false matches than were novices.

One factor that may contribute to experts' success in analogical retrieval is *representational uniformity*: the extent to which the relations in the memory trace are represented similarly to those in the probe. Clement *et al.* (1994) explored the effect of relational predicate similarity on analogical access and mapping between stories. They found that retrieval was more likely when the probe and target contained synonymous terms (*manifest* similarity) than when they contained loosely similar predicates such as 'munched' and 'consumed' (*latent* similarity). However, unlike retrieval, analogical mapping when both situations were present was relatively unaffected by the latent–manifest distinction. In analogical reminding, with only the current situation in working memory, success depends on the degree of match of the pre-existing representations; whereas during mapping, with both situations present in working memory, there is opportunity for re-representation.

ANALOGICAL LEARNING

Analogical comparison can lead to new learning in at least four ways: analogical abstraction, inference projection, difference detection, and re-representation. The first two we have already discussed. In *analogical abstraction*, the structure common to base and target is noticed and extracted. Sometimes the common system is stored as a separate representation; this is referred to as *schema abstraction* (Gick and Holyoak, 1983). In *inference projection*, a proposition from the base is mapped to the target. If it is retained as part of the target structure, then learning has occurred. In *difference detection*, carrying out a comparison process leads people to notice certain differences – namely, those connected to the common structure. In *re-representation*, two non-identical predicates are aligned and decomposed (or abstracted) to find their commonalities, resulting in a re-representation that contains a common predicate: for example, comparing *chase*(dog, cat) and *follow*(detective, suspect) might result in *pursue* (entity1, entity2). A further kind of knowledge change, hypothesized to take place in scientific discovery, is *restructuring*, in which the target undergoes a radical change in structure.

COMPUTATIONAL MODELS

The interplay between computational models and psychological studies has been extremely productive in analogical research. Current models include Boicho Kokinov's AMBR model, which integrates retrieval and mapping; Keane's IAM model, which utilizes an incremental mapping algorithm; Halford's tensor product model; and the systems of Doug Hofstadter and his colleagues Melanie Mitchell and Robert French, which integrate perceptual processing with analogical matching. (*See* **Analogy-making, Computational Models of**)

Analogical modeling has made great strides, but there are still open questions. At present no model fully captures human analogy processing. Two challenges for analogical models are (1) the *selection problem* discussed above – namely, how to avoid indiscriminate inferencing; and (2) the problem of *representational flexibility* – that is, how to achieve a matching process that does not require absolute identity matches. Falkenhainer *et al.*'s (1989) SME, which uses a local-to-global alignment and inference process over structured symbolic representations, meets the benchmarks in Table 1 and can capture selective matching and inference, as well as schema abstraction. But it is not yet sufficiently flexible in its match process. Another leading model, Hummel and Holyoak's (1997) LISA model, uses a combination of distributed representations of concepts and structured representations of the relational connections (necessary for achieving structural consistency in mapping). It uses a connectionist temporal-binding algorithm and makes its matches in a serial order partly guided by the experimenter. LISA's use of distributed representations allows for flexible matching, and unlike most models of analogy, it attempts to capture working memory limitations. However, it has yet to solve the selectivity problem in inferencing.

THE FUTURE

Of the many research questions that remain, four stand out as particularly interesting and timely. First is the role of analogy in cognitive development: how much of children's rapid learning can be attributed to the processes of comparing and drawing inferences between partially similar situations? Second is tracing the neuropsychology of analogical processes: what areas of the brain are implicated, and what is the course of processing? Third is the exploration of analogy in animal cognition. Comparative research so far indicates that humans excel in analogical ability, yet this ability exists in certain other species as well – for example, in chimpanzees and dolphins. Cross-species comparisons may help us decompose the cognitive components of analogical ability. A final important research frontier is the integration of analogy into larger models of cognition.

References

Clement CA and Gentner D (1991) Systematicity as a selection constraint in analogical mapping. *Cognitive Science* **15**: 89–132.

Clement CA, Mawby R and Giles DE (1994) The effects of manifest relational similarity on analog retrieval. *Journal of Memory and Language* **33**: 396–420.

Falkenhainer B, Forbus KD and Gentner D (1989) The structure-mapping engine: algorithm and examples. *Artificial Intelligence* **41**: 1–63.

Gentner D (1983) Structure-mapping: a theoretical framework for analogy. *Cognitive Science* **7**(2): 155–170.

Gentner D and Markman AB (1995) Analogy-based reasoning in connectionism. In: Arbib MA (ed.) *The Handbook of Brain Theory and Neural Networks*, pp. 91–93. Cambridge, MA: MIT Press.

Gentner D and Markman AB (1997) Structure-mapping in analogy and similarity. *American Psychologist* **52**: 45–56.

Gentner D and Rattermann MJ (1991) Language and the career of similarity. In: Gelman SA and Byrnes JP (eds) *Perspectives on Thought and Language: Interrelations in Development*, pp. 225–277. London, UK: Cambridge University Press.

Gentner D, Rattermann MJ and Forbus KD (1993) The roles of similarity in transfer: separating retrievability from inferential soundness. *Cognitive Psychology* **25**: 524–575.

Gentner D and Toupin C (1986) Systematicity and surface similarity in the development of analogy. *Cognitive Science* **10**: 277–300.

Gick ML and Holyoak KJ (1983) Schema induction and analogical transfer. *Cognitive Psychology* **15**(1): 1–38.

Goldstone RL and Medin DL (1994) Time course of comparison. *Journal of Experimental Psychology: Learning, Memory and Cognition* **20**(1): 29–50.

Halford GS (1993) *Children's Understanding: The Development of Mental Models*. Hillsdale, NJ: Lawrence Erlbaum.

Holyoak KJ (1985) The pragmatics of analogical transfer. In: Bower GH (ed.) *The Psychology of Learning and Motivation: Advances in Research and Theory*, vol. 19, pp. 59–87. New York, NY: Academic Press.

Holyoak KJ and Thagard PR (1989) Analogical mapping by constraint satisfaction. *Cognitive Science* **13**(3): 295–355.

Hummel JE and Holyoak KJ (1997) Distributed representations of structure: a theory of analogical access and mapping. *Psychological Review* **104**(3): 427–466.

Inagaki K and Hatano G (1987) Young children's spontaneous personification as analogy. *Child Development* **58**: 1013–1020.

Keane MT (1996) On adaptation in analogy: tests of pragmatic importance and adaptability in analogical problem solving. *Quarterly Journal of Experimental Psychology* **49**: 1062–1085.

Markman AB (1997) Constraints on analogical inference. *Cognitive Science* **21**(4): 373–418.

Markman AB and Gentner D (1993) Structural alignment during similarity comparisons. *Cognitive Psychology* **25**: 431–467.

Novick LR (1988) Analogical transfer, problem similarity, and expertise. *Journal of Experimental Psychology: Learning, Memory and Cognition* **14**: 510–520.

Ross BH (1989) Distinguishing types of superficial similarities: different effects on the access and use of earlier problems. *Journal of Experimental Psychology: Learning, Memory and Cognition* **15**: 456–468.

Spellman BA and Holyoak KJ (1996) Pragmatics in analogical mapping. *Cognitive Psychology* **31**: 307–346.

Further Reading

Carbonell JG (1983) Learning by analogy: formulating and generalizing plans from past experience. In: Michalski RS, Carbonell JG and Mitchell TM (eds) *Machine Learning: An Artificial Intelligence Approach*, vol. 1, pp. 137–161. Palo Alto, CA: Tioga.

Dunbar K (1995) How scientists really reason: scientific reasoning in real-world laboratories. In: Sternberg RJ and Davidson JE (eds) *The Nature of Insight*, pp. 365–395. Cambridge, MA: MIT Press.

French R (1995) *The Subtlety of Sameness: A Theory and Computer Model of Analogy-making*. Cambridge, MA: MIT Press.

Gentner D, Holyoak KJ and Kokinov BN (eds) (2001) *The Analogical Mind: Perspectives from Cognitive Science*. Cambridge, MA: MIT Press.

Goswami U (1992) *Analogical Reasoning in Children*. Hillsdale, NJ: Lawrence Erlbaum.

Halford GS, Wilson WH and Phillips S (1998) Processing capacity defined by relational complexity: implications for comparative, developmental, and cognitive psychology. *Behavioral and Brain Sciences* **21**: 803–864.

Hesse MB (1966) *Models and Analogies in Science*. Notre Dame, IN: University of Notre Dame Press.

Hofstadter DR and Mitchell M (1994) The Copycat project: a model of mental fluidity and analogy-making. In: Holyoak KJ and Barnden JA (eds) *Advances in Connectionist and Neural Computation Theory*, vol. 2: *Analogical Connections*, pp. 31–112. Norwood, NJ: Ablex.

Holyoak KJ and Thagard PR (1995) *Mental Leaps: Analogy in Creative Thought*. Cambridge, MA: MIT Press.

Keane MT (1990) Incremental analogising: theory and model. In: Gilhooly KJ, Keane MTG, Logie RH and Erdos G (eds) *Lines of Thinking*, vol. 1. Chichester, UK: John Wiley.

Kokinov BN and Petrov AA (2001) Integrating memory and reasoning in analogy-making: the AMBR model. In: Gentner D, Holyoak KJ and Kokinov BN (eds) *The Analogical Mind: Perspectives from Cognitive Science*, pp. 161–196. Cambridge, MA: MIT Press.

Nersessian NJ (1984) *Faraday to Einstein: Constructing Meaning in Scientific Theories*. Dordrecht, Netherlands: Nijhoff.

Thagard P (1992) *Conceptual Revolutions*. Princeton, NJ: Princeton University Press.

Winston PH (1982) Learning new principles from precedents and exercises. *Artificial Intelligence* **19**: 321–350.

Analogy-making, Computational Models of

Intermediate article

Boicho Kokinov, New Bulgarian University, Sofia, Bulgaria
Robert M French, University of Liège, Liège, Belgium

CONTENTS

Analogy-making is the process of finding or constructing a common relational structure in the descriptions of two situations or domains and making inferences by transferring knowledge from the familiar domain (the 'base' or 'source') to the unfamiliar domain (the 'target'), thus enriching our knowledge about the latter.

INTRODUCTION

Analogy-making is crucial for human cognition. Many cognitive processes involve analogy-making in one way or another: perceiving a stone as a human face, solving a problem in a way similar to another problem previously solved, arguing in court for a case based on its common structure with another case, understanding metaphors, communicating emotions, learning, or translating poetry from one language to another (Gentner *et al*, 2001). All these applications require an abstract mapping to be established between two cases or domains based on their common structure (common systems of relations). This may require re-representation of one (or both) of the domains in terms of the other one (or in terms of a third domain). The first domain is called the base, or source, and the second is called the target.

Analogy-making is a basic cognitive ability. It appears to be present in humans from a very early age, and develops over time. It starts with the simple ability of babies to imitate adults and to recognize when adults are imitating them, progresses to children's being able to recognize an analogy between a picture and the corresponding real object, and culminates in the adult ability to make complex analogies between various situations. This seems to suggest that analogy-making serves as the basis for numerous other kinds of human thinking; hence the importance of developing computational models of analogy-making.

Analogy-making involves at least the following sub-processes: building a representation, retrieving a 'base' for the analogy, mapping this base to the 'target', transferring unmapped elements from the base to the target, thereby making inferences, evaluating the validity and applicability of these inferences, and learning from the experience – which includes generalizing from specific cases and, possibly, developing general mental schemata. There are, at present, no models that incorporate all of these sub-processes. Individual models focus on one or more of them.

Representation-building

The process of representation-building is absent from most models of analogy-making. Typically, representations are fed into the model. However, there are some models (e.g., ANALOGY, Copycat, Tabletop, Metacat) that do produce their own high-level representations based on essentially unprocessed input. These models (Mitchell, 1993; Hofstadter *et al*, 1995; French, 1995) attempt to build flexible, context-sensitive representations during the course of the mapping phase. Other models, such as AMBR (Kokinov and Petrov, 2001), perform re-representation of old episodes.

Retrieval

The retrieval process has been extensively studied experimentally. Superficial similarity is the most important factor in analogical retrieval: the retrieval of a base for analogy is easier if it shares similar objects, similar properties, and similar general themes with the target. Structural similarity, the familiarity of the domain from which the analogy is drawn, the richness of its representations, and the presence of generalized schemata, also facilitate retrieval. Most models of retrieval are based on exhaustive search of long-term memory (LTM) and on the assumption that old episodes have context-independent, encapsulated representations. There are, however, exceptions (e.g., AMBR) that rely on context-sensitive reconstruction of old episodes performed in interaction with the mapping process.

Mapping

Mapping is the core of analogy-making. All computational models of analogy-making include mapping mechanisms, i.e., means of discovering which elements of the base correspond to which elements of the target. The difficulty is that one situation can be mapped to a second situation in many different ways. We might, for example, make a mapping based on the color of the objects in both the base and target (the red-shirted person in the base domain would be mapped to the red-shirted person in the target domain). This would, in general, be a very superficial mapping (but might, none the less, be appropriate on occasion). We could also map the objects in the two domains based on their relational structures. For example, we could decide that it was important to map the giver–receiver relationship in the first domain to the same relationship in the target domain, ignoring the fact that in the base domain the giver had a red shirt and in the target domain the giver was wearing a blue shirt.

Experimental work has demonstrated that finding this type of structural isomorphism between base and target domains is crucial for mapping (Gentner, 1983). Object similarity also plays a role in mapping, although generally a secondary one. A third factor is the pragmatic importance of various elements in the target: we want to find mappings that involve the most important elements in the target. Searching for the appropriate correspondences between the base and the target is a computationally complex task that can become infeasibly time-consuming if the search is unconstrained.

Transfer

New knowledge then has to be inserted into the target domain based on the mapping. For example, suppose a new brand of car appears on the market, and that this car maps well onto another brand of car that is small, fast, and handles well on tight curves. But you also know that this latter brand of car is frequently in need of repair. You then wonder whether the new brand of car will also be in frequent need of repair.

Transfer is present in some form in most models of analogy-making, and is typically integrated with mapping. Transfer is considered by some researchers as an extension of the mapping already established, adding new elements to the target.

Evaluation

Evaluation is the process of establishing the likelihood that the transferred knowledge will be applicable to the target domain. In the example above, the evaluation process would have to assign the degree of confidence we would have that the new car would also be in frequent need of repair. Evaluation is often implicit in the mechanisms of mapping and transfer.

Learning

Only a few models of analogy-making have incorporated learning mechanisms. This is somewhat surprising since analogy-making is clearly a driving force behind much learning. However, some models are capable of generalization from the base and target, or from multiple exemplars, to form an abstract schema, as in LISA (Hummel and Holyoak, 1997) and the SEQL model based on SME (Falkenhainer *et al*, 1989).

Below we will review a number of important computational models of analogy-making belonging to different classes and following different approaches. First the 'symbolic' models will be presented. These models employ separate local representations of objects, relations, propositions and episodes (e.g., 'John', 'chair', 'run', 'greater than', 'John ate fish', 'my birthday party last year'). Then, 'connectionist' models will be presented. Here the objects, relations, and episodes are represented as overlapping patterns of activation in a neural network. Finally, a third, hybrid class of models will be presented. These models combine symbolic representations with connectionist activations. They are based on the idea that cognition is an emergent property of the collective behavior of many simple agents.

SYMBOLIC MODELS

ANALOGY

The earliest computational model of analogy-making, ANALOGY, was developed by Thomas

Evans (1964). This program solves multiple-choice geometric analogy problems of the form '*A* is to *B* as *C* is to what?' taken from intelligence tests and college entrance examinations.

An important feature of this program is that the input is not a high-level description of the problem, but a low-level description of each component of the figure – dots, simple closed curves or polygons, and sets of closed curves or polygons. The program builds its own high-level representation describing the figures in *A*, *B*, *C*, and all given alternatives for the answer, with their properties and relationships – for example `((P1 P2 P3) . ((INSIDE P1 P2) (LEFT P1 P3) (LEFT P2 P3)))`. Then the program represents the relationship between *A* and *B* as a set of possible rules describing how figure *A* is transformed into figure *B* – for example, `((MATCH P2 P4) (MATCH P1 P5) (REMOVE P3))` which means that figure P_2 from *A* corresponds to figure P_4 from *B*, P_1 corresponds to P_5, and the figure P_3 does not have a corresponding figure and is therefore deleted. Then each such rule is applied to *C* in order to get one of the alternative answers. In fact, each such rule would be generalized in such a way as to allow *C* to be applied to *D*. Finally, the most specific successful rule would be selected as an outcome. Arguably, one of the most significant features of the program is its ability to represent the target problem on its own – a feature that has been dropped in most recent models.

Structure Mapping Theory

The most influential family of computational models of analogy-making have been those based on Dedre Gentner's (1983) 'structure mapping theory' (SMT). This theory was the first to explicitly emphasize the importance of structural similarity between base and target domains, defined by common systems of relations between objects in the respective domains. Numerous psychological experiments have confirmed the crucial role of relational mappings in producing convincing and sound analogies. There are several important assumptions underlying the computational implementation of SMT called SME (Falkenhainer *et al*, 1989): (1) mapping is largely isolated from other analogy-making sub-processes (such as representation, retrieval and evaluation) and is based on independent mechanisms; (2) relational matches are preferred over property matches; (3) only relations that are identical in the two domains can be put into correspondence; (4) relations that are arguments of higher-order relations that can also be mapped have priority, following the 'systematicity principle' that favors systems of relations over isolated relations; and (5) two or three interpretations are constructed by a 'greedy merge' algorithm that generally finds the 'best' structurally coherent mapping. Early versions of SME mapped only identical relations and relied solely on relational structure. This purely structural approach was intended to ensure the domain independence of the mapping process. Recent versions of SME have made some limited use of pragmatic aspects of the situation, as well as re-representation techniques that allow initially non-matching predicates to match.

The MAC/FAC model of analogical retrieval (Forbus *et al*, 1995) was intended to be coupled with SME. This model assumes that episodes are encapsulated representations of past events; they have a dual encoding in LTM: a detailed predicate-calculus representation of all the properties and relations of the objects in an episode and a shorter summary (a vector representation indicating the relative frequencies of the predicates that are used in the detailed representation). The retrieval process has two stages. The first stage uses the vector representations to perform a superficial search for episodes that share predicates with the target problem. The episode vectors in LTM that are close to the target vector are selected for processing by the second stage. The second stage uses the detailed predicate-calculus representations of the episodes to select the one that best matches the target. These two stages reflect the dominance of superficial similarity as well as the influence of structural similarity.

Gentner's ideas – in particular, their emphasis on the structural aspects of analogical mappings – have been very influential in the area of computational analogy-making and have been applied in contexts ranging from child development to folk physics. Various improvements and variants of SME have been developed, and it has been included as a module in various practical applications.

Other Symbolic Models

A number of other symbolic models have helped to advance our understanding of analogy-making. Jaime Carbonell proposed the concept of derivational analogy, where the analogy is drawn not with the final solution of the old problem, but with its derivation, i.e., with the way of reaching the solution, an approach developed further by Manuela Veloso. Smadar Kedar-Cabelli developed a model of purpose-directed analogy-making in concept learning. Mark Burstein developed a model called CARL which learned from multiple analogies combining several bases. Mark Keane and his colleagues developed an incremental model of mapping, IAM, which helps explain the effects of order of presentation of the material observed in humans.

CONNECTIONIST MODELS

Research in the field of analogy-making has, until recently, been largely dominated by the symbolic approach, for an obvious reason: symbolic models are well equipped to process and compare the complex structures required for analogy-making. In the early years of the new connectionist paradigm, these structures were very difficult to represent in a connectionist network. However, advances in connectionist representation techniques have allowed distributed connectionist models of analogy to be developed. Most importantly, distributed representations provide a natural internal measure of similarity, thereby allowing the system to handle the problem of similar but not identical relations in a relatively straightforward manner. This ability is essential to analogy-making and has proved hard for symbolic models to implement.

Multiple Constraints Theory

The earliest attempt to design an architecture in which analogy-making was an emergent process of activation states of neuron-like objects was proposed by Keith Holyoak and Paul Thagard (1989) and implemented in a model called ACME. In this model, structural similarity, semantic similarity, and pragmatic importance determine a set of constraints to be simultaneously satisfied. The model is supplied with representations of the target and of the base, and proceeds to build a localist constraint-satisfaction connectionist network where each node corresponds to a possible pairing hypothesis for an element of the base and an element of the target. For example, if the base is 'train' and the target is 'car' then all elements of trains will be mapped to all elements of cars; there will therefore be hypothesis nodes created not only for 'locomotive → motor' but also for 'locomotive → license plate', 'locomotive → seat-belt buckle', etc. The excitatory and inhibitory links between these nodes implement the structural constraints. In this way, contradictory hypothesis nodes compete and do not become simultaneously active, while consistent ones mutually support each other. The network gradually moves towards an equilibrium state, and the best set of consistent mapping hypotheses (e.g., 'locomotive → motor', 'rails → road', etc.) wins. The relaxation of the network provides a parallel evaluation of all possible mappings and finds the best one, which is represented by the set of most active hypothesis nodes.

ARCS is another related model of retrieval. It is coupled with ACME and operates in a similar fashion. However, while mapping is dominated by structural similarity, retrieval is dominated by semantic similarity.

STAR

STAR-1 was the first distributed connectionist model of analogy-making (Halford *et al.*, 1994). It is based on the tensor product connectionist models developed by Smolensky. A proposition like `MOTHER-OF (CAT, KITTEN)` is represented by the tensor product of the three vectors corresponding to `MOTHER-OF`, `CAT`, and `KITTEN`: `MOTHER-OF` ⊗ `CAT` ⊗ `KITTEN`. The tensor product in this case is a three-dimensional array of numbers where the number in each cell is the product of the three corresponding coordinates. This representation allows any of the arguments, or the relational symbol, to be extracted by a generalized dot product operation: (`MOTHER-OF` ⊗ `CAT`) • (`MOTHER-OF` ⊗ `CAT` ⊗ `KITTEN`) = `KITTEN`. The LTM of the system is represented by a tensor that is the sum of all tensor products representing the individual statements (the main restriction being that the propositions are simple and have the same number of arguments). Using this type of representation, STAR-1 solves proportional analogy problems like 'cat is to kitten as mare is to what?'

STAR-2 (Wilson *et al*, 2001) maps complex analogies by sequentially focusing on various parts of the domains – simple propositions with no more than four dimensions – and finding the best map for the arguments of these propositions by parallel processing in the constraint satisfaction network (similarly to ACME). The fact that the number of units required for a tensor product representation increases exponentially with the number of arguments of a predicate implies processing constraints in the model. Wilson *et al* claim that humans are subject to similar processing constraints: specifically, they can, in general, handle a maximum of four dimensions of a situation concurrently. The primary interest of the modelers is in exploring and explaining capacity limitations of human beings and achieving a better understanding of the development of analogy-making capabilities in children.

LISA

John Hummel and Keith Holyoak (1997) proposed an alternative computational model of analogy-making using distributed representations of structure relying on dynamic binding. The idea is to introduce an explicit time axis so that patterns of activation can oscillate over time (thus the timing of activation becomes an additional parameter independent of the level of activation). Patterns of activation oscillating in synchrony are considered to be bound together, while those oscillating out of synchrony are not. For example, 'John hired Mary' requires synchronous oscillation of the patterns for 'John' and 'Employer' alternating with synchronous oscillation of the patterns for 'Mary' and 'Employee'. Periodic alternation of the activation of the two pairs represents the whole statement. However, if the statement is too complex there will be too many pairs that need to fire in synchrony. Based on research on single-cell recordings, Hummel and Holyoak believe that the number of such pairs of synchronously firing concepts cannot exceed six. Representations in LISA's working memory are distributed over the network of semantic primitives, but representations in long-term memory are localist – there are separate units representing the episode, the propositions, their components, and the predicates, arguments, and bindings. Retrieval is performed by spreading activation, while mapping is performed by learning new connections between the most active nodes. LISA successfully integrates retrieval of a base with the mapping of the base and target, even though retrieval and mapping are still performed sequentially (mapping starts only after one episode is retrieved).

HYBRID MODELS

Two groups of researchers have independently produced similar models of analogy-making based on the idea that high-level cognition emerges as a result of the continual interaction of relatively simple, low-level processing units, capable of doing only local computations. These models are a combination of the symbolic and connectionist approaches. Semantic knowledge is incorporated in order to compute the similarity between elements of the two domains in a context-sensitive way.

Copycat and Related Architectures

The family of Copycat and Tabletop architectures (Mitchell, 1993; Hofstadter *et al*, 1995; French, 1995) was explicitly designed to integrate top-down semantic information with bottom-up emergent processing. Copycat solves letter-string analogies of the form '*ABC* is to *ABD* as *KLM* is to what?' and gives plausible answers such as *KLN* or *KLD*. The architecture of Copycat involves a working memory, a semantic network (simulating long-term memory) defining the concepts used in the system and their relationships, and the 'coderack' – the procedural memory of the system – a store for small, nondeterministic computational agents ('codelets') working on the structures in the working memory and continually interacting with the semantic network. Codelets can build new structures or destroy old structures in working memory. The system gradually settles towards a consistent set of structures that will determine the mapping between the base and the target.

The most important feature of these models of analogy-making is their ability to build up their own representations of the problem, in contrast with most other models which receive the representations of the base and target as input. Thus these models abandon traditional sequential processing and allow representation-building and mapping to run in parallel and continually influence each other. The partial mapping can influence further representation-building, thus allowing the gradual construction by the program of context-sensitive representations. In this way, the mapping may force us to see a situation from an unusual perspective in terms of another situation, and this is an essential aspect of creative analogy-making.

AMBR

AMBR (Kokinov, 1994) solves problems by analogy. For example, 'how can you heat some water in a wooden vessel, being in the forest?' The solution, heating a knife in a fire and immersing it in the water, is found by analogy with boiling water in a glass using an immersion heater.

The AMBR model is based on DUAL, a general cognitive architecture. The LTM of DUAL consist of many micro-agents, each of which represents a small piece of knowledge. Thus concepts, instances and episodes are represented by (possibly overlapping) coalitions of micro-agents. Each micro-agent is hybrid: its symbolic part encodes the declarative or procedural knowledge it is representing, while its connectionist part computes the agent's activation level, which represents the relevance of this knowledge to the current context. The symbolic processors run at a speed proportional to their computed relevance, so the behavior of the system is highly context-sensitive. The AMBR model implements the interactive parallel work of recollection, mapping and transfer that emerge from the collective behavior of the agents and which produces the analogy. Recollection in AMBR-2 (Kokinov and Petrov, 2001) is reconstruction of the base episode in working memory by activating relevant aspects of event information, of general knowledge, and of other episodes, and forming a coherent representation which will correspond to the target problem. The model exhibits illusory memories, including insertions from general knowledge and blending with other episodes, and context and priming effects. Some of these phenomena have been experimentally confirmed in humans.

CONCLUSIONS

The field of computational modeling of analogy-making has moved from the early models, which were intended mainly to demonstrate that computers could, in fact, be programmed to do analogy-making, to complex models that make nontrivial predictions of human behavior. Researchers have come to appreciate the need for structural mapping of the base and target domains, for integration of and interaction between representation-building, retrieval, mapping and learning, and for systems that can potentially scale up to the real world. Computational models of analogy-making have now been applied to a large number of cognitive domains (Gentner *et al*, 2001). However, many issues remain to be explored in the endeavor to model the human capacity for analogy-making, one of our most important cognitive abilities.

References

Evans TG (1964) A heuristic program to solve geometric-analogy problems. In: *Proceedings of the Spring Joint Computer Conference* **25**: 327–338 [Reprinted in: Fischler M and Firschein O (eds) (1987) *Readings in Computer Vision*. Los Altos, CA: Morgan Kaufmann.]

Falkenhainer B, Forbus KD and Gentner D (1989) The structure-mapping engine: algorithm and examples. *Artificial Intelligence* **41**: 1–63.

Forbus K, Gentner D and Law K (1995) MAC/FAC: a model of similarity-based retrieval. *Cognitive Science* **19**(2): 141–205.

French R (1995) *The Subtlety of Sameness: A Theory and Computer Model of Analogy-Making*. Cambridge, MA: MIT Press.

Gentner D (1983) Structure-mapping: a theoretical framework for analogy. *Cognitive Science* **7**(2): 155–170.

Gentner D, Holyoak K and Kokinov B (eds) (2001) *The Analogical Mind: Perspectives From Cognitive Science*. Cambridge, MA: MIT Press.

Halford G, Wilson W, Guo J *et al* (1994) Connectionist implications for processing capacity limitations in analogies. In: Holyoak K and Barnden J (eds) *Advances in Connectionist and Neural Computation Theory*, vol. II 'Analogical Connections', pp. 363–415. Norwood, NJ: Ablex.

Hofstadter D and the Fluid Analogies Research Group (1995) *Fluid Concepts and Creative Analogies: Computer Models of the Fundamental Mechanisms of Thought*. New York, NY: Basic Books.

Holyoak K and Thagard P (1989) Analogical mapping by constraint satisfaction. *Cognitive Science* **13**: 295–355.

Hummel J and Holyoak K (1997) Distributed representations of structure: a theory of analogical access and mapping. *Psychological Review* **104**: 427–466.

Kokinov B (1994) A hybrid model of analogical reasoning. In: Holyoak K and Barnden J (eds) *Advances in Connectionist and Neural Computation Theory*, vol. II 'Analogical Connections', pp. 247–318. Norwood, NJ: Ablex.

Kokinov B and Petrov A (2001) Integration of memory and reasoning in analogy-making: the AMBR model. In: Gentner *et al* (2001), pp. 59–124.

Mitchell M (1993) *Analogy-Making as Perception: A Computer Model*. Cambridge, MA: MIT Press.

Wilson W, Halford G, Gray B and Phillips S (2001) The STAR-2 model for mapping hierarchically structured analogs. In: Gentner *et al* (2001), pp. 125–159.

Further Reading

Barnden J and Holyoak K (eds) (1994) *Advances in Connectionist and Neural Computation Theory*, vol. III 'Analogy, Metaphor, and Reminding'. Norwood, NJ: Ablex.

Gentner D, Holyoak K and Kokinov B (eds) (2001) *The Analogical Mind: Perspectives From Cognitive Science*. Cambridge, MA: MIT Press.

Hall R (1989) Computational approaches to analogical reasoning: a comparative analysis. *Artificial Intelligence* **39**: 39–120.

Holyoak K and Barnden J (eds) (1994) *Advances in Connectionist and Neural Computation Theory*, vol. II 'Analogical Connections'. Norwood, NJ: Ablex.

Holyoak K, Gentner D and Kokinov B (eds) (1998) *Advances in Analogy Research: Integration of Theory and Data from the Cognitive, Computational, and Neural Sciences*. Sofia: New Bulgarian University Press.

Holyoak K and Thagard P (1995) *Mental Leaps*. Cambridge, MA: MIT Press.

Vosniadou S and Ortony A (eds) (1989) *Similarity and Analogical Reasoning*. New York, NY: Cambridge University Press.

Anaphora

Advanced article

Gregory Carlson, University of Rochester, Rochester, New York, USA

CONTENTS

Anaphora refers to referentially dependent expressions in natural language which contribute their meaning by identifying another expression to give them their semantic value.

INTRODUCTION

Anaphora, in its primary instances, is the establishment of a referential dependency between two (or more) expressions. The pronoun 'him' in (1) below is one such instance of anaphora:

Mark felt that there was someone watching *him*. (1)

On the understanding that 'him' refers to Mark, the pronoun is the *anaphor* and the expression 'Mark' is the *antecedent*. Both expressions refer to the same individual. The relationship between these expressions is not an equal one, however, since the reference of the pronoun is dependent upon the reference of its antecedent, whereas the reference of the antecedent is established by virtue of its meaning alone. The term 'co-reference' is often used to describe this referential connection between anaphor and antecedent. But anaphor–antecedent relations must be distinguished from the phenomenon of *accidental co-reference*. This occurs when two independently referring expressions happen to refer to the same individual. So, for instance, in (2) the two italicized expressions will be co-referential, 'accidentally', only when the president of the company is also the company's best employee:

The president of the company rewarded *the best employee*. (2)

This requires an understanding where the company has a self-rewarding president, but there is no anaphoric connection established between the expressions. Thus, anaphora is a matter of co-reference, and something more.

ANAPHOR–ANTECEDENT RELATIONS

Anaphors depend upon their antecedents to determine their referential content. One reflection of this referential dependency is that in many instances an anaphor cannot be interpreted as co-referential with another noun phrase (NP). For instance, in the following examples, the pronouns cannot be construed as non-accidentally having the same reference as the italicized NPs:

Bob was nominated by him. (him $\neq$ Bob) (3)

She hoped that *Mary* would win the contest. (she $\neq$ Mary) (4)

This is because an anaphor cannot receive its reference from another NP if that NP does not have an appropriate syntactically defined relationship to the anaphor. This relationship is not simply one of linear precedence, as in many instances an anaphor may precede its antecedent (a phenomenon which is occasionally called *cataphora,* though more commonly *backward anaphora*):

Near *her*, *Jill* saw a snake. (5)

If *he* wins the race today, *Bret* will be a hero. (6)

Much research has focused on the question of the precise nature of this syntactic relationship. The research is detailed and extensive (for example, research on *Binding Theory*). Most agree that the notion of *c-command* is crucial (Langacker, 1966; Lasnik, 1976). In general, an anaphor cannot c-command its antecedent, and in examples such as (3) and (4) above where the two designated NPs cannot be interpreted co-referentially, the pronoun would c-command its antecedent, and a referential connection cannot be established. The reference for the pronoun in these instances needs to be determined by other means, such as finding another, appropriate antecedent for it, or by providing it with a *deictic* interpretation (discussed further below). (*See* **Binding Theory**)

One class of pronouns that has also received extensive attention is that of *reflexive pronouns*, exemplified below:

We found *ourselves* with too much to do. (7)

The professor taught *herself* French. (8)

These differ from the other personal pronouns in important respects. Primarily, the syntactic relations to their antecedents are much more limited. In general, reflexive pronouns may only have antecedents within the same clause, though the precise conditions remain a topic of detailed investigation. In the following examples, the reflexive pronoun may not be construed as co-referential with the italicized NPs:

We thought that [$_S$Jim liked ourselves] (9)

The professor remembered when [$_S$herself lived in Paris] (10)

As there is no appropriate antecedent for the reflexive pronoun within the same clause in these instances, the sentences are not grammatical.

Pronouns not only may find their reference by identifying an antecedent and using the reference of the antecedent as its own value, but they may function as *bound variables* as well. In such instances, the 'reference' of the pronoun is not determined by the reference of its antecedent NP, but rather by the assignment of values to variables that is determined by the quantifier, as in first-order logic. A representation of a sentence such as (11), with 'Every man' construed as the antecedent of 'he', would be as indicated:

Every man thinks that [*he* deserves a raise]
$\forall x$ [man(x) $\Rightarrow$ x thinks that [x deserves a raise]] (11)

Bound variable pronouns and their antecedents are syntactically more constrained than identity of reference pronouns and theirs. In the following examples, (12a) precludes any bound variable reading; this is despite the fact that a natural identity of reference reading is available when the antecedent NP has a clear referential value, as with proper names (12b):

a. The dean who placed *no student* on probation told *her* to check back in the fall.
b. The dean who placed *Hillary* on probation told *her* to check back in the fall. (12)

(12a) has no bound variable interpretation, because the antecedent is in a syntactic position which precludes this possibility. The relation that must hold, in the case of bound variable readings, is for the antecedent NP to c-command the pronoun.

The phenomenon of anaphora is much broader than the personal pronouns discussed thus far. One form of anaphora that has received much attention is *temporal anaphora* (Partee, 1984). This applies not only to pronouns referring back to time NPs,

The mail arrived *this morning*. I was at home *then* (= this morning) (13)

but also to the time introduced by the *tense* of a sentence:

Ali woke up. It was cold *then*. (14)

The study of temporal anaphora includes a wide variety of forms which are used to coordinate the time in one sentence with that of another. Beyond 'then', expressions such as 'when, before/after, until, as, while, since, immediately thereafter', and many others fall within this domain. Perhaps most significantly, tenses themselves appear to function anaphorically. In the examples below, the tense in the second sentence is understood as coordinated with the time reference in the first:

Samantha opened the door. She *had been* repairing the car. (15)

Daryl fell down. He *was* drunk. (16)

Our understanding that the repairing occurred prior to the door opening, or that the falling occurred while Daryl was drunk, is often attributed to temporal anaphora.

A wide variety of other anaphoric forms, beyond personal pronouns and temporal anaphora, make reference to an extensive array of other types of things. These include the demonstratives 'this' and 'that' (with or without a following noun), and epithetics such as 'the fool' or 'the bastard'. Other forms take as antecedents phrases that are not NPs. 'So' may take a verb phrase as an antecedent; 'such' takes a modifier; 'there' may take a locative prepositional phrase; 'one' may take a noun:

a. Sam tried to *win the race* before Al could do *so*.
b. If *intelligent* students attend college, *such* students usually do very well.
c. Everyone who was *at the party* had a good time *there*.
d. I own a big *car*, and you own a small *one*. (17)

In many cases the anaphor is expressed as null: that is, the anaphor is indicated by having some constituent missing from the sentence. The following sentence is missing a noun plus its modifying adjective at the point indicated by the blank:

Jack owns three *large dogs,* and I own two____. (18)

In this case, the 'blank' takes as its antecedent the portion of the NP italicized. It functions as an anaphor in the same way as a pronoun.

Null anaphora extends well beyond nouns and NPs. Verb phrases (VPs) can function as antecedents for null VPs (known as *VP ellipsis*):

If you want to____, we can *take a break.* (19)

Verbs, and verb complexes, can serve as antecedents in the *gapping* construction:

Joseph *ate* a bagel, and Samuel, ____ a grapefruit. (20)

Null complement anaphora takes complement sentences as antecedents:

Kevin claimed *that our television was broken, but I'm not so sure*____. (21)

This by no means exhausts the range of anaphora expressed by null expressions.

DISCOURSE ANAPHORA

'Discourse' is the normal mode of communication: the use of more than one independent sentence or utterance put together in a way that 'makes sense'. The discussion above was limited to those instances of anaphora that take place within the boundaries of a sentence. Anaphora takes place across sentence boundaries as well. Many instances of anaphora that appear within sentence boundaries take place as well in discourse:

Several team members were suspended. Reportedly, *they* had missed a practice. (22)

Most people want to *win a million dollars.* Doris doesn't____. (23)

Certain cases of anaphora that occur within the boundaries of a sentence do not function as discourse anaphors. For instance, the phenomena of reflexive pronouns, gapping, relative pronouns, and bound variable anaphora do not appear to be able to function this way.

One treatment of discourse anaphora is to treat all such pronouns as *free variables,* which are assigned a reference independently by an *assignment function,* which designates a referential value for any free variables within its domain (e.g. Cooper and Parsons, 1976). It becomes co-referential with a NP in a previous sentence by virtue of being assigned the same reference. (*See* **Semantics, Dynamic**)

So, in (24), if a function assigns the same referential value as the proper name 'Leonard' has, to the pronoun 'he' in the following sentence, then a co-referential reading arises:

Leonard is a famous composer. *He* writes operas. (24)

On the other hand, if 'he' is assigned a different value (e.g. Fred), then the discourse will be understood as saying that Fred writes operas, and no co-referential reading will occur. All phrases with which the pronoun is co-referential must have a reference value in the first place, if this is to be the appropriate analysis. The case of indefinite NPs in discourse raises questions, though. Indefinite NPs are those which appear with a number of different determiners, most prominently the indefinite article 'a(n)'. Such NPs can be 'referred back to' by anaphors in discourse:

A man walked into the room. *He* sat down. (25)

Most researchers, however, question whether an indefinite NP should be properly assigned a reference value (Kamp, 1981). This is because the reference value determines the truth-conditions of the sentence, and if one assigns a certain individual as the reference of 'a man' in a sentence such as (25), then it would be true if that *particular* man walked into the room, and false if he did not (regardless of whether any other man walked in). However, these are not the truth-conditions for such a sentence, since (an utterance of) the sentence will be true if any man whatsoever walked into the room (and false only if no man at all did). If one assigned a reference for 'a man' as some particular man, one could not characterize these truth-conditions. It appears that the truth-conditions of the sentence are best represented quantificationally, with an existential quantifier binding a variable:

$\exists x$ [man(x) & x walked into the room] (26)

This treatment raises some problems, however, when we turn to discourse anaphora. The representation of the two-sentence discourse in (25) would have the existential quantifier binding the instance of 'he' in the subsequent sentence:

$\exists x$ [man(x) & x walked into the room & x sat down] (27)

Since anaphors referring back to indefinites are both very common and natural, unlike those functioning as bound variables with their antecedent quantified expressions in another sentence, one would need to make a separation between classes of quantifiers, some of which could bind variables in other sentences, and others which could not. This has proven an unsatisfactory analysis of the phenomenon, however, for both syntactic and semantic reasons. One of the main issues has centered around the treatment of *donkey sentences* (or, *donkey anaphora*). Such examples are so called because of the example below, commonly cited in the literature:

Every farmer who owns *a donkey* beats *it*. (28)

These sentences pose a problem of logical representation that has been known since medieval times. The problem is this. If one were to take 'it' in this sentence to be a free variable assigned the same reference as 'a donkey', there is, very clearly, no particular donkey which this sentence is in any way 'about'. The other, more attractive, possibility is that the pronoun is functioning as a bound variable, bound by an existential quantifier that is taken to be the meaning of the indefinite article. However, the only consistent representation available is essentially as follows:

$\exists x$ [donkey(x) & Every farmer who owns x beats x] (29)

The truth-conditions of this (which are directly reflective of the meaning), however, are very different from the truth-conditions of the sentence itself. This formula is true only when there is some donkey or other that every owner of it beats, which is far from the meaning of the sentence itself. There is no consistent way of representing the meaning by treating the indefinite article as an existential quantifier which binds the pronoun in the predicate of the sentence.

Replacing the quantificational analysis is one where indefinites are treated as contributing no existential meaning on their own, but only a free variable and a property ascription; so indefinites have neither inherent reference nor inherent quantificational force. The free variable implicitly contributed by the indefinite is bound by an operation of *text closure*. This is where the existential force arises. So, a single sentence such as the following has the representation given immediately below it, and then text closure operates to bind the variable as indicated:

A man walked into the room.
man(x) & x walked into the room
$\Rightarrow$ Text closure
$\exists x$ [man(x) & x walked into the room] (30)

Text closure is formulated in such a way that it operates over stretches of discourse, as more sentences are added. So a two-sentence discourse would be represented and operated on by text closure as follows:

A man walked into the room. He sat down.
man(x) & x walked into the room & x sat down
$\Rightarrow$ Text closure
$\exists x$ [man(x) & x walked into the room & x sat down] (31)

This analysis allows us to distinguish quantified NPs from indefinites, on the one hand, and allows us to treat the pronouns as free variables at the same time. Also, though not presented here, it offers a solution to the donkey sentences problem.

This approach raises issues of its own, as illustrated in the following sentence:

There is a man in the garden. The dog is barking at *him*. (32)

The 'There is…' construction in English quite plausibly introduces an existential quantifier of its own, rendering the variable contributed by 'a man' unavailable for binding by text closure. But the pronoun in the second sentence could be bound by text closure. If this is so, then the text would have the meaning 'Some man is in the garden. The dog is barking at someone'. Another problem with text closure is that the representation

$\exists x$ [man(x) & x is in the garden & the dog is barking at x] (33)

will be true also in cases where there are more men in the garden than just one. However, the original text means – or possibly strongly implies – that there is one and only one man in the garden.

Evans (1980) has argued that there is a need for still another category of pronoun, which he calls *E-type pronouns*. These pronouns, like the bound variable and co-referential pronouns, share all the same forms, but function differently: they are disguised definite descriptions, picking out a unique individual given the information present in the context. Informally, the analysis of the pronoun 'him' in (32) would be:

There is a man in the garden. The dog is barking at him (= *the man that is in the garden*). (34)

Since these descriptions can contain pronouns, or variables of their own, one can obtain solutions to problems like the following (an example that is often called a *pronoun of laziness*, a term coined by Peter Geach since it was a 'lazy' way to avoid repeating an entire NP):

The woman who deposited *her paycheck* in the bank was wiser than the woman who gave *it* to her teenage son. (35)

In this case, analyzing 'it' as meaning 'her paycheck', with 'her' in this instance assigned the same value as the second woman rather than the first, will give the right reading. However, on a co-referential reading (or a bound variable reading) the paychecks would have to be one and the same.

IDENTITY OF SENSE AND IDENTITY OF REFERENCE ANAPHORA

A traditional distinction is made between what are called 'identity of sense' and 'identity of reference' anaphora. The distinction between sense and reference goes back to the writings of the philosopher Gottlob Frege. In the case of NP meanings, this distinction concerns whether the individuals designated by the antecedent and the anaphor must be interpreted as identical. So, in (36a) the cars driven by Lyle and Maria must have been the same; however, in (36b) they need not:

a. Lyle drove *a car*. Maria drove *it*, too.
b. Lyle drove *a car*. Maria drove *one*, too. (36)

The difference between the anaphors 'it' and 'one' (the latter taking a noun meaning as its antecedent) would seem to suggest that anaphors themselves fall into these two classes. While this is so to a certain extent, many instances of anaphora can be identified in which the same form can play both roles. Consider the following:

a. *The President* (of the United States) walked off the plane. *He* waved to the crowd.
b. *The President* is elected every four years. *He* has been from a southern state ten times. (37)

The *reference* of the phrase 'The President' is whoever happens to be in that office at the time – currently it is George W. Bush; the *sense*, on the other hand, is that which picks out the president at the time, whoever it may be. In (37a), 'he' refers to a certain individual – Bush, for instance. The sentence 'he' appears in would be synonymous with saying 'Bush waved at the crowd'. In (37b), 'he' is anaphoric to the sense of the term, not its reference. It does not follow that any particular president, such as the current one, has been from a southern state ten times. Rather, this instance of the pronoun is talking about the presidents of the USA across time, regardless of who that individual may currently be. That is, it refers to the sense of the antecedent, not its reference.

When we turn to other types of anaphora, it is more difficult to make this sense/reference distinction. Consider null VP anaphora:

Zelda will *get up early* if Harry does____. (38)

The question that arises in this case is whether VPs themselves have a sense/reference distinction in their meanings to begin with. If, for instance, VPs have individual events as their reference, and have classes of events as their sense, then VP anaphora would be sense anaphora (as presumably Harry waking and Zelda waking would be distinct events). One can make reference to individual events by using the pronoun 'it', as exemplified below, but VP anaphora does not appear to ever make reference to events in this particular way:

The train blew its whistle. *It* (= the blowing of the whistle) was heard for miles. (39)

A similar situation holds for anaphora to sentence meanings. In a Fregean analysis, the sense of a sentence is a proposition, and its reference is a truth-value (T or F). However, with Null Complement Anaphora, which takes sentences as antecedents, the proposition rather than the truth-value is clearly the value assigned to the anaphor, as any other proposition with the same truth value (e.g. 'Grass is green') does not yield a synonymous sentence:

Bruno was selling drugs. When the FBI found out____, he was arrested. (40)

Thus, the sense/reference distinction is most useful in describing anaphora to NP meanings.

PRAGMATIC ANAPHORA

Pragmatics, that is knowledge of how the world works and what it contains, the circumstances under which a sentence is uttered, and of how language is used, is crucial for the study of anaphora. In most instances, a pronoun or other anaphor could, in principle, find more than one unique antecedent in the sentence or discourse, as in the following:

Mary told the *woman* talking to *her sister* that *Lesley* was sick today. *She* then turned and walked away. (41)

While 'she' could, in principle, find any of the previous NPs as its antecedent, in practice only one 'makes sense' and so is the one that is readily understood (in this instance, Mary). *Centering Theory* is one proposal that attempts to deal with this phenomenon (Grosz *et al.*, 1995).

Another area requiring pragmatic knowledge to resolve reference of anaphora is *bridging inferences* (Clark, 1975). The listener or reader must make use of real-world knowledge to interpret a definite NP appropriately. For example:

John bought *a new car*. *The engine* was painted bright red. (42)

Here, one knows that the engine is the engine in the car that John bought, making use of real-world knowledge that cars have engines.

Much work in pragmatic anaphora focuses not on the process of selecting an appropriate antecedent from candidates given in the text or discourse, but on instances where the sentence or discourse itself provides no possible antecedents for an anaphor. For instance, imagine I was standing on the street with someone else when a man walks by very abruptly. Under these circumstances I can say:

He appears very upset. (43)

In so doing, I refer to the man that just walked by even though there is no expression within the previous discourse to serve as an antecedent for the pronoun. The man himself, in some sense, provides the 'antecedent' for the pronoun. When elements themselves in the real-world context of use provide the values for anaphoric expressions, they are said to be *pragmatically controlled*.

The example above might suggest that perceptual evidence establishes possible antecedents for *deictic* uses of pronouns. However, having the referent perceptually available is not always necessary. Consider the case where I am walking down the hallway at work, and a student is knocking on the door of the office of another faculty member. I can say, under the circumstances:

I haven't seen *her* today. (44)

even though the professor is not perceptually available for reference at the time. Thus, some contexts are 'rich' enough to support pragmatic control even in the absence of who or what is being referred to.

Most (but not all) instances of anaphora may be pragmatically controlled, including certain instances of reflexive pronouns and *logophoric* pronouns. These are pronouns, indicated by specialized forms in some languages, which are canonically used in indirect discourse to make reference to the person whose speech is reported (e.g. 'Ariel said that *he*[logophoric] was going to write a paper'). Below are instances of other types of anaphora that may be controlled pragmatically:

a. [Picking up a coat from the coat-check attendant] '*This* is torn!'
b. [Pointing through the glass at the candy counter] 'A green *one* and a red *one*, please.'
c. [Sally hides cigarettes in her room. Her sister, seeing this, says:] 'Better hope our parents don't find out____.'
d. [Trying on suits, and the salesman says:] '*Which* appeals to you most?' (45)

Certain instances of anaphora cannot be pragmatically controlled (Hankamer and Sag, 1976). These are the instances of *surface anaphora*, which, unlike *deep anaphora*, require a specifically linguistic antecedent. The distinction between deep and surface anaphors hinges not only on whether they may be pragmatically controlled, but also on whether, when there is a linguistic antecedent, the syntactic details of the antecedent determine the possibility of it serving as an antecedent, regardless of what meaning is expressed (which applies to surface anaphora but not deep).

Gapping, null anaphora to a verb or verb complex, requires an explicitly (surface) linguistic antecedent, even in very clear contexts, and cannot be pragmatically controlled; the example below is not felicitous:

[Bob throws a baseball] '. . . and Cary, a basketball.'
(Contrast with: 'Bob threw a basketball, and Cary, a baseball.') (46)

Sluicing likewise requires a linguistically introduced 'surface' antecedent, so the following too sounds strange:

[From outside, a scream; a shot is fired; and a thud] 'I wonder who____?' (47)

Similarly, the phenomenon of bound variable pronouns is not amenable to pragmatic control.

References

Clark H (1975) Bridging. In: Schank R and Nash-Webber B (eds) *Theoretical Issues in Natural Language Processing*. Cambridge, MA: MIT Press.

Cooper R and Parsons T (1976) Montague Grammar, generative semantics, and interpretive semantics. In:

Partee B (ed.) *Montague Grammar*, pp. 311–362. New York, NY: Academic Press.
Evans G (1980) Pronouns. *Linguistic Inquiry* **11**: 337–362.
Grosz B, Joshi A and Weinstein S (1995) Centering: a framework for modeling the local coherence of discourse. *Computational Linguistics* **21**: 2.
Hankamer J and Sag I (1976) Deep and surface anaphora. *Linguistic Inquiry* **7**: 391–428.
Kamp H (1981) A theory of truth and representation. In: Groenendijk J, Janssen T and Stokhof M (eds) *Formal Methods in the Study of Language*, pp. 277–322. Amsterdam, Netherlands: Mathematisch Centrum.
Langacker R (1966) On pronominalization and the chain of command. In: Reibel W and Schane S (eds) *Modern Studies in English*, pp. 160–186. Englewood Cliffs, NJ: Prentice-Hall.
Lasnik H (1976) Remarks on coreference. *Linguistic Analysis* **2**: 1–22.
Partee B (1984) Nominal and temporal anaphora. *Linguistics and Philosophy* **7**: 243–286.

Further Reading

Chierchia G (1992) Anaphora and dynamic binding. *Linguistics and Philosophy* **15**(2): 111–183.
Chomsky N (1981) *Lectures on Government and Binding*. Dordrecht, Netherlands: Foris.
Jacobson P (2000) Paycheck pronouns, Bach–Peters sentences, and variable-free semantics. *Natural Language Semantics* **8**: 77–155.
Lewis D (1979) Scorekeeping in a language game. *Journal of Philosophical Logic* **8**: 339–359.
Reinhart T (1983) *Anaphora and Semantic Interpretation*. Chicago, IL: University of Chicago Press.
Roberts C (1989) Modal subordination and pronominal anaphora in discourse. *Linguistics and Philosophy* **12**: 683–721.
Van der Sandt R (1992) Presupposition projection as anaphora resolution. *Journal of Semantics* **9**: 333–377.

Anaphora, Processing of

Introductory article

Andrew Barss, University of Arizona, Tucson, Arizona, USA
Janet L Nicol, University of Arizona, Tucson, Arizona, USA

CONTENTS

INTRODUCTION: REFERENCE AND REFERENTS

Whenever a speaker utters a sentence, he or she has a particular message in mind, which the utterance conveys to the listener. An ongoing aspect of communication is the act of pointing out to the listener what objects in the world the speaker is trying to talk about. The speaker chooses certain words and phrases, and uses them to indicate to the listener what thing or things are being described. When a *phrase* of language is used to pick out a *thing* or *entity* in the world (more precisely, either the real world or the inner world of our thoughts), we say that the phrase *refers* to the thing or entity, and that the thing or entity is the *referent* of the phrase. Reference to objects in the world is one of the most fundamental, and common, things we do with language. Virtually every sentence, in any language, contains phrases that refer to things in the world or in our thoughts.

There is a fundamental distinction to be made between *directly referring* and *indirectly referring* phrases. The first type consists of a phrase whose lexical content – the meanings of the word(s) of the phrase – contains sufficient information to allow the listener to understand what entity is referred to. The phrase by itself carries enough information to refer to, or 'pick out', the entity. Typically, a directly referring phrase contains a noun (which describes the kind of entity being referred to) together with various modifiers the speaker might use to further specify the referent. (In this discussion, we use 'speaker' to mean whoever is producing the utterance, and 'listener' to mean whoever is trying to comprehend the utterance, independent of the medium of expression.) Because the phrase is built around the noun, we call it a Noun Phrase, or

NP. For example, each of the italicized expressions below is an NP, and in each case it refers to an entity or group of entities:

The fat dog is barking. (1)

The fat dog that lives next door is barking. (2)

Dogs are barking. (3)

These phrases refer to their referents directly, independently of any other words in the sentence (thus the term *independent reference*). If you know what 'the', 'fat', and 'dog' all mean (i.e. if you are a speaker of English), you will know what is being referred to in sentence (1).

DEPENDENT PHRASES AND INDIRECT REFERENCE

The second type of reference, *indirect reference*, works differently. It always involves a pair of NPs. One will belong to a small class of semantically incomplete NPs, called the *proforms*, which includes the personal *pronouns* (English pronouns are listed in (4)); *reflexives* (listed in (5), which always consist (in English) of a personal pronoun plus the suffix '-self' (in the singular) or '-selves' (in the plural)); and *reciprocals*, seen in (6).

English personal pronouns (4)

	First person	Second person	Third person
Singular	*I, me*	*you*	*he, she, it, him, her*
Plural	*we, us*	*you*	*they, them*

English reflexives (5)

	First person	Second person	Third person
Singular	*myself*	*yourself*	*himself, herself, itself*
Plural	*ourselves*	*yourselves*	*themselves*

English reciprocals: *each other; one another* (6)

The other NP in the pair is the *antecedent* of the proform, the phrase that the dependent element 'looks back to' to determine its reference. Antecedents will typically be independently referring NPs. Proforms are nouns that are scaled down in meaning and, in English, convey information about person and about number, gender, and humanness. In other languages, proforms may convey additional information, including the familiarity of the speaker and listener, whether the listener is included or excluded, and information about fundamental aspects of the antecedent, such as its orientation in a plane, or whether it is flat. Regardless of the internal information expressed by the proforms, they are typically used in a sentence to 'echo', or repeat, the reference established by using the antecedent. For example, in (7), the reflexive 'themselves' and reciprocal 'each other' refer again to the dogs down the street:

a. The dogs down the street are looking at *themselves*.

b. The dogs down the street are looking at *each other*. (7)

The term *anaphora* refers to this pairing-up of a proform and its antecedent. As an aside on terminology, we note that the ancient Greek grammarians distinguished between cases where the antecedent preceded the dependent element, as in (7), which they termed 'anaphora', from cases where the antecedent followed the proform, which they termed 'cataphora', as in, for example, 'His mother talked about John'. In most modern literature the term 'anaphora' is used to cover both cases, and we follow that convention here. (Cases of the latter type are statistically very infrequent in written and spoken English, a fact which might follow from the increased processing burden placed on the language comprehension system by receiving the proform in advance of its antecedent.)

For the discussion below, it is crucial to keep in mind that sentences such as (7) contain two separate references to objects: the antecedent (e.g. 'the dogs down the street') refers directly to the referent; and then the proform connects back to the antecedent, and refers to whatever the antecedent did. Thus, in (7a), there is no way to determine what 'themselves' refers to without first figuring out what the antecedent is; the reference by the word 'themselves' to the dogs is indirect, mediated by the connection of the reflexive to its antecedent phrase. Proforms are also called *dependent expressions*. The two NPs are said to co-refer when they have such a dependent connection. Anaphora, then, is the link between a proform and its antecedent, which establishes co-reference. Figure 1 illustrates this graphically.

CONSTRAINTS ON ANAPHORA

There is a major contrast between the pronouns in (4) and the reflexives and reciprocals in (5, 6). Typically, reflexives and reciprocals appear only in very limited positions within a sentence, and pronouns typically appear in cases of anaphora where a reflexive would be unacceptable.

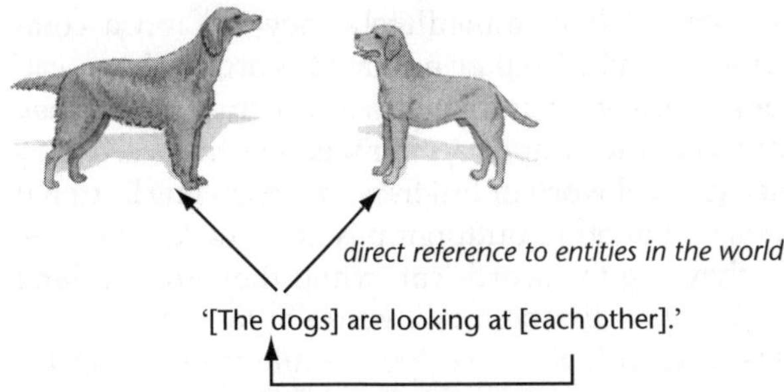

Figure 1. The antecedent–proform relation, creating indirect reference to the dogs.

The antecedent of a reflexive or a reciprocal will usually be structurally close to the proform, typically occurring as the subject of the sentence in which the reflexive occurs (square brackets are used here to mark the boundaries of propositions or clauses; '*' indicates that the sentence with the intended interpretation is not grammatically acceptable):

a. [John thinks that [Mary admires herself]].

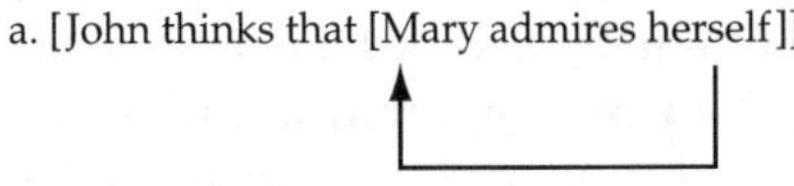

b. *[John thinks that [Mary admires himself]].

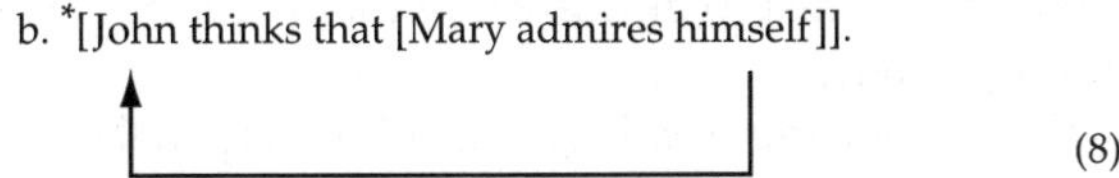

(8)

By contrast, the antecedent of a personal pronoun has a different distribution, and cannot be in the same position as the antecedent of a reflexive or reciprocal:

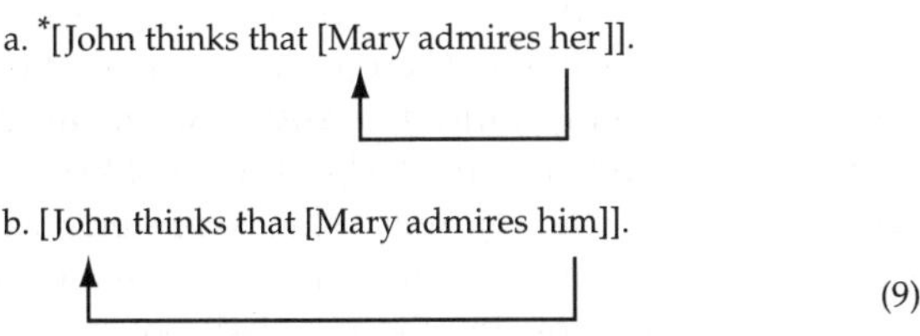

a. *[John thinks that [Mary admires her]].

b. [John thinks that [Mary admires him]].

(9)

Notice the complementariness of the distribution of proform–antecedent pairs. (8b) is ungrammatical because the antecedent, 'John', is structurally too far away from the reflexive. This then permits the use of a pronoun to co-refer with 'John', as in (9b). By contrast, in (8a), the antecedent is close enough to the proform to permit use of a reflexive. This consistently blocks the use of a pronoun: (9a) cannot be understood to mean co-reference between 'her' and 'Mary'.

Consequently, in the course of understanding language, the listener must keep track of (1) the type of proform being used (pronoun versus reflexive/reciprocal), (2) where it occurs in the sentence, (3) the position(s) in which other NPs, which are candidate antecedents, occur, and (4) the person, number, and gender features of each NP (including the proform), which further constrain the pairing up of a proform with an antecedent.

Above we noted that a proform and its antecedent must match in person, number, and gender. Notice that (10a) is ambiguous (since both NPs match the pronoun in features), while (10b) and (10c) are unambiguous, since only one NP matches:

a. The boy told the man that [he was lucky].
b. The girl told the man that [he was lucky].
c. The boy told the woman that [he was lucky]. (10)

There are some cases in which feature mismatch is permitted. One is the use of the genderless, normally plural 'they/them' as an indefinite singular pronoun, when the speaker does not wish to seem gender-biased in using the singular (English contains no singular, gender-unmarked pronoun), as in (11a):

a. Every student must make sure they check in with the nurse.
b. Every student must make sure he checks in with the nurse. (11)

A second case is the use of the normally second-person pronoun to mean 'one', 'anyone', in casual speech:

a. You shouldn't do that.
b. One shouldn't do that. (12)

(12a) is ambiguous. It may be understood as directed at the listener(s), in which case the second-person feature is required. But it may also be taken as a casual variant of (12b), in which case the reference is generic, and not restricted to the listener(s).

In general, these mismatches can be seen as a type of compromise: a way of avoiding gender commitment or formal register.

THE PROCESSING OF ANAPHORA

The term *processing* refers to how the human mind registers and makes sense of sensory information over time. Cognitive processes, such as language production and comprehension, are largely unconscious and automatic. When listeners hear a sentence, they must first recognize the words contained in the sentence. For instance, for the sentence 'The dogs are looking at the mail carrier', the listener first must determine that the words 'the' and 'dogs' occurred, then figure out what the words mean, and then combine the meanings. Word meanings seem to just pop into mind: when we hear the words 'the dogs', we automatically think of the concept DOG. This process – of identifying words, thinking about their meanings, and combining meanings together – occurs over and over in the course of communication. One might wonder what happens to all this information: if the DOG concept is in mind, what happens to it when 'mail carrier' is uttered? Clearly, a listener would have to keep both concepts (THE DOG and THE MAIL CARRIER) in mind simultaneously (along with the LOOKING concept) in order to understand the sentence. And, of course, the concepts must be kept separate, so that we do not think of the dogs as being mail carriers, or the carriers carrying both mail and dogs. There is some evidence that words and their meanings are kept in mind until the end of a proposition (usually the end of a sentence), and then they are converted into a less detailed memory representation.

The presence of anaphora complicates matters considerably. In a sentence such as 'The dogs are looking at each other', what meaning 'pops into mind' when 'each other' appears? Do listeners automatically think about 'the dogs' again? Or does it require time and effort to make the connection between 'each other' and 'the dogs'? Before we address this question, we will describe how research on the processing of anaphora has been conducted.

Methods

There have been three basic approaches to the study of how proforms are interpreted. (1) Ask people. The most direct approach is to have people read sentences and then ask them about their interpretations (e.g. presenting a sentence like 'John told Bill that Frank admired him' and asking 'Who does *him* refer to?'). The trouble with this approach is that it does not reveal anything about the process over time; it probes only the final interpretation of the sentence, and so does not answer any of the questions raised just above. There are two types of methods that do provide information about how a listener or reader understands a sentence as it unfolds.

(2) Probing word meanings. One variant of this method probes the activation of word meanings by having people listen to sentences and simultaneously make judgments about items they see on a computer screen. Here is how it works. Listeners hear the phrase 'The dog' at the beginning of the sentence. Then immediately they see (on a computer monitor) either a related word such as 'cat' or a completely unrelated word such as 'pen'. They are asked to simply decide whether the word they see is a real word of English, and press one button if it is and another button if it is not. The logic is this: if they see the word 'cat' while they are thinking DOG, then listeners will be quite fast to respond, faster than if they see 'pen' while thinking DOG, because the concepts CAT and DOG are connected in our word system, and PEN and DOG are not. By measuring the time it takes listeners to respond, experimenters can get some idea of what people are thinking about. Now suppose that listeners hear the sentence 'The dog down the street hurt itself yesterday'. If we probe what listeners are thinking about when they hear the word 'itself', we should be able to tell whether they are thinking about the concept DOG. If they are, this means that the connection between the reflexive 'itself' and the antecedent 'the dog' is established very quickly indeed. Another variant on this technique simply repeats the antecedent, and listeners or readers must decide if the word appeared in the sentence or not. The logic is similar: if people are thinking about DOG after the reflexive 'itself', they should be relatively fast to indicate that the word 'dog' was in the sentence.

(3) Measuring reading. Another approach has been to have people read sentences and to determine at what points in the sentences they slow down. In one variant of this approach, researchers construct sentences that have a 'surprise' ending IF the reader has adopted a particular interpretation. For example, given the sentence 'The actress liked the queen because she thought the queen did a good job', readers could initially assume at the point where 'she' occurs ('The actress liked the queen because she…') that 'she' co-refers with 'the queen'. Such an assumption is reasonable because an explanation for the proposition that 'the actress liked the queen' is likely to involve something that the queen did. If readers do make this assumption when they read the pronoun 'she', then they will be surprised when material after the pronoun reveals that 'she' is really 'the actress'. What happens when readers are surprised by how a sentence unfolds is a slowdown in reading (compared to the reading of a sentence that does not contain a surprise ending, for example 'The actress admired the queen because she thought the actress did a good job'). Reading time can be measured with the use of a device called an *eye tracker*, which, when it is linked to a computer, determines where on a computer screen the eye fixates and for how

long (and whether the reader backtracks in order to re-read a section of text). A simpler way to measure reading time is to (have a computer) present pieces of a written sentence in sequence, in such a way that only one section is visible at a time. The reader controls the presentation by pressing a designated key on a computer keyboard. The key press makes the current sentence fragment disappear and the next fragment appear. The time between key presses provides a measure of reading time.

The Experimental Findings

Experiments that have focused on the questions raised above suggest that listeners do automatically think about the antecedent of the proform – immediately, or very soon after the proform appears – if they are reasonably attentive to what they are hearing. Hence, it appears to be automatic that proforms serve to reinstantiate their antecedents. Such findings raise a number of questions: does the information about sentence position have an effect on what people think about when they encounter a proform? What happens when there is more than one possible antecedent for a proform? Must a single antecedent be selected right away, or does a whole set of possible antecedents get computed, awaiting final narrowing down? What sources of information enter into the process? We will consider each of these questions in what follows.

First, recall the facts about the complementarity of the distribution of antecedents of pronouns on one hand and reflexives/reciprocals on the other. Examples are shown in (13).

a. The actress thinks that the queen admires her.
b. The actress thinks that the queen admires herself. (13)

In (13a), 'the queen' cannot be the antecedent of the pronoun 'her', but 'the actress' can be. In (13b), 'the queen' must be the antecedent of the reflexive 'herself' and 'the actress' cannot be. Some research suggests that knowledge about these restrictions constrains the process: listeners and readers think about only those antecedents that are 'allowed'. Further, they think about only those antecedents that agree in number and gender (this point is discussed in greater detail below).

Now consider the case in which there is more than one possible antecedent for a proform, as in the following examples.

The actress told the queen that the director wanted to meet her after dinner. (14)

The actress told the queen that she would be sitting next to the director. (15)

In (14), both 'actress' and 'queen' are possible antecedents of the pronoun 'her'. It might be slightly more likely that the director would want to sit with a queen rather than an actress (this is information about how the world works, or information about *plausibility*), or people might tend to think that since 'her' is the object of a verb (the verb 'meet'), an object ('the queen' in the first clause) is a better antecedent (this is called the *parallel function* interpretation). These kinds of information do matter, but evidence from word-probe experiments suggests that they do not come into play immediately. That is, listeners think about both 'actress' and 'queen' after hearing 'her'. Then they might use a variety of cues to settle on one antecedent or the other. (Typically, listeners do attempt to link proforms with antecedents right away; it would be difficult to follow a conversation without doing so.)

In (15), again, both 'actress' and 'queen' are potential antecedents of the pronoun, and both would come to mind after the occurrence of 'she'. Then other cues would be used to narrow down the set: parallel function would dictate that since 'she' is a subject, another subject ('the actress') should be the antecedent. Another factor that appears to affect pronoun interpretation is the order of potential antecedents: the first NP in the sentence has a sort of privileged status, and, given a choice, people are more likely to fix on that NP as the antecedent. A third factor has to do with plausibility: the likelihood that an actress would inform the queen about where the queen would be seated at dinner; the likelihood of the queen being seated next to the director, and so on. Again, it appears that such information acts to eliminate antecedents from a candidate set: the proform makes a listener think about a number of antecedents, selected via inspecting just the sentence structure, and other information acts later on to eliminate potential antecedents from the set.

It should be emphasized here that such instances of antecedent ambiguity are far from uncommon. This is because speakers are unaware when they say a pronoun that it may not be clear who the antecedent is; after all, for the speaker, there is no confusion at all. (Sometimes they do catch themselves, and identify the antecedent: e.g. '… and then she… that is, Susan, disappeared…'. Other times a listener might ask 'she, who?'.)

Now consider example (16).

The actresses told the queen that they were going to be late. (16)

Here there are also multiple antecedents for the pronoun 'they': just the 'actresses' or both 'the actresses' and 'the queen'. Research shows that listeners consider just 'the actresses', which matches in number with 'they'. When a matching antecedent is available, as it is here, listeners tend not to consider a joint referent consisting of both the actresses and the queen. (Note that, in contrast, a sentence such as 'The actress told the queen that they were going to be late' requires the listener to infer that 'the actress' and 'the queen' together constitute the antecedent of 'they'. It is not known at present whether or not this inference takes additional time.)

The examples above contain nouns that are gender-specific: 'actress' and 'queen' are both feminine. What happens when a gender-neutral noun appears, as in the following?

The supervisor warned the actress that
she was going to be late. (17)

Only a handful of English nouns are specified for gender. These include kinship terms (e.g. 'aunt, grandfather, daughter'), royal titles (e.g. 'king, princess, duchess'), and miscellaneous others, many of which are disappearing from English (e.g. 'aviatrix, murderess'). Of the miscellaneous set of feminine-marked forms, the male counterparts are not clearly marked for gender: although 'aviatrix', 'murderess', and 'actress' must refer to women, 'aviator', 'murderer', 'actor' can be applied to both men and women. In general, most occupation nouns are gender-neutral: e.g. 'doctor, nurse, lawyer, director, president, manager, supervisor, teacher, professor, student, programmer, architect, gardener, coach'. Of this set, there is considerable variation with respect to the probability with which a term is likely to refer to a man or a woman. While 'student' seems to be equally likely to refer to a male or a female, some nouns are more likely to refer to men (e.g. 'astronaut') and some to refer to women (e.g. 'nurse'). How does this type of information affect the processing of anaphora? There is not much research in this area, but the research to date suggests that listeners initially consider as a potential antecedent for a proform any noun that does not actually clash in terms of gender. Given a sentence like (17), and no additional information about the gender of the referent of 'supervisor', both 'actress' and 'supervisor' would initially be considered potential antecedents. Then the gender-probability information, along with information about real-world plausibility, parallel function, and order of mention, are all likely to come into play to guide the selection of a single antecedent. In a sentence such as (17), there will be conflicting information: parallel function and first-mention weight 'the supervisor' more heavily. But gender marking favors 'the actress'. Pragmatics arguably favors both equally, since the plausibility of 'being late' applies equally to both. The processing of sentences such as this should be slower than for sentences in which all the information favors the same antecedent.

Overall, the research suggests that the processing of anaphora involves a number of stages. First, the occurrence of a proform evokes a candidate set of antecedents. This set includes only those antecedents that are grammatically allowed, and only those that are not incongruent with respect to gender and number. Next, other types of information (about order of mention, sentence function, gender-probability, and plausibility) act to eliminate nouns from the candidate set.

Actually, the picture is more complicated than this because sometimes potentially helpful information about the identity of the antecedent is not available until later in the sentence. Take a sentence such as 'The director told the actress that she couldn't recommend her for the role'. At the point in the sentence at which 'she' occurs, it really is not clear which is the correct antecedent. Competition of the other types of information could lead the listener to settle on 'actress' (possibly because there are many more male directors than female ones). But the appearance of 'her' throws this interpretation into question: if 'she' co-refers with 'actress', and if 'her' cannot co-refer with 'she', then 'her' co-refers with 'director', and here's where the sentence goes awry. It is unlikely (but not impossible) that someone would tell someone else whom he or she could recommend. So, 'she' ought to co-refer with 'director', and 'her' with 'actress'. And when it is revealed that the recommendation is for a role, then this confirms the revised interpretation.

Given that late-occurring material can inform the interpretation of proforms, it might make more sense for listeners to postpone co-reference until the end of the sentence. Listeners and readers could simply take the information conveyed by a proform (e.g. third person, female, plural, subject) and plug that into the ongoing meaning of the sentence. But it appears to be the case that, as in language comprehension in general, a 'wait-and-see' approach carries too high a cost. The cost is that we might forget some of the critical details of an utterance and lose track of what the sentence is conveying. After all, for the most part, human beings talk about people and things. During a discourse, a speaker and listener both construct a *mental model* of the participants and events being discussed. In order to follow a discourse listeners and readers must figure out who is doing what to whom, and it appears that they make these calculations quickly, lest they forget.

Further Reading

Cloitre C and Bever T (1988) Linguistic anaphors, levels of representation, and discourse. *Language and Cognitive Processes* **3**(4): 293–322.

Ehrlich K and Rayner K (1983) Pronoun assignment and semantic integration during reading: eye movements and immediacy of processing. *Journal of Verbal Learning and Verbal Behavior* **22**: 75–87.

Garnham A, Oakhill J and Cruttenden H (1992) The roles of implicit causality and gender cue in the interpretation of pronouns. *Language and Cognitive Processes* **7**: 231–255.

Gernsbacher M, Hargreaves D and Beeman M (1989) Building and accessing clausal representations: the advantage of first mention versus the advantage of clause recency. *Journal of Memory and Language* **28**: 735–755.

MacDonald MC and MacWhinney B (1990) Measuring inhibition and facilitation from pronouns. *Journal of Memory and Language* **29**: 469–492.

Matthews A and Chodrow M (1988) Pronoun resolution in two-clause sentences: effects of ambiguity, antecedent location, and depth of embedding. *Journal of Memory and Language* **27**: 245–260.

McDonald J and MacWhinney B (1995) The time course of anaphor resolution: effects of implicit verb causality and gender. *Journal of Memory and Language* **34**: 543–566.

McKoon G, Greene S and Ratcliff R (1993) Discourse models, pronoun resolution, and the implicit causality of verbs. *Journal of Experimental Psychology: Learning, Memory & Cognition* **19**(5): 1040–1052.

Nicol J and Swinney D (2002) The psycholinguistics of anaphora. In: Barss A (ed.) *Anaphora*. Oxford, UK: Blackwell.

Shillcock R (1982) The on-line resolution of pronominal anaphora. *Language and Speech* **25**: 385–401.

Stevenson RJ, Nelson AWR and Stenning K (1995) The role of parallelism in strategies of pronoun comprehension. *Language and Speech* **38**(4): 393–418.

Animal Cognition

Introductory article

Valerie A Kuhlmeier, Yale University, New Haven, Connecticut, USA
Sarah T Boysen, Ohio State University, Columbus, Ohio, USA

CONTENTS

The study of how animals learn, behave, and think, often from a comparative perspective, is the crux of animal cognition research. Recent topics explored in this field include understanding of abstract concepts, spatial learning and memory, imitation, representation of social relations, and examining the similarities between nonhuman primates and human children.

INTRODUCTION

The field of animal cognition has emerged since the 1980s as a rich, interdisciplinary area representing animal learning and behavior, comparative psychology, ethology, and behavioral ecology. It is different from the behaviorist tradition in that researchers do not focus primarily on conditioning processes. Interest in animal cognition cuts across numerous academic fields, including psychology, zoology, biology, anthropology, and primatology, among others. In psychology, several topic areas have been particularly important, although they represent only a portion of ongoing studies in animal cognition. In particular, studies of abstract concepts, spatial learning and memory, imitation, representation of social relationships, and developmental comparisons between human children and animals (typically primates) represent a growing literature. These areas encompass research issues and approaches that are currently being studied with different animal species, including (but not limited to) rats, pigeons, primate species such as rhesus monkeys, capuchins, and squirrel monkeys, as well as great apes such as chimpanzees, bonobos, and orangutans.

ABSTRACT CONCEPTS

One of the most basic questions being investigated with animals is whether any species is able to reason abstractly. That is, can an animal understand an abstract concept such as 'same versus different', recognize itself in a mirror (possess a 'concept of self'), or even count using numbers through a process similar to numerical skills shown by young children? Are they able to use their understanding of a concept when presented with new examples? Finally, and most importantly, how might such capacities for abstract thinking and concept formation serve a particular species in the wild when they are confronted with real problem-solving opportunities and/or life-and-death situations?

What is meant by an abstract concept, particularly for an animal, requires careful thought, and has been a contentious issue in animal learning and cognition. If concepts are defined in such a way that only humans can acquire them, as some investigators have proposed, and especially if conceptual understanding can be acquired only if mediated by language, then clearly concepts must be unique to *Homo sapiens*. However, investigators in animal cognition have suggested that an abstract concept should be operationally defined. For example, if an animal is presented with complex stimuli such as photographs of trees and other scenes, learns to respond only to pictures of trees, and subsequently chooses new pictures of trees from among a novel set of photographs, the animal may be said to have an understanding of the 'tree concept'. Similarly, if an animal with training on counting and other number skills is able to correctly assign a number to a novel collection of objects, despite learning to count using only gumdrops, it could be argued that the subject, whether it was a chimpanzee, rat, or pigeon, has a concept of number. In this case, the ability to generalize an understanding of numbers to brand-new counting opportunities would suggest that the animal has some representation of an abstract concept of number and numerical relationships.

According to the current literature in animal cognition, both types of capacities for abstract concepts have been demonstrated in animal subjects. Such studies include pigeons' abilities to recognize a range of items conceptually, demonstrated by identifying slides depicting trees, people, or even the letter 'a' presented in very different typefaces, when the animals were presented with novel sets of visual stimuli. In the case of number concepts, Sarah Boysen has demonstrated that chimpanzees can assign Arabic numerals to quantities of candies and other collections of objects. More impressively, individual chimpanzees have been shown to invoke spontaneous addition algorithms that enabled them to count different-sized arrays that were hidden in several locations around a test room. The animals were able to report the total number of objects found during their search of the sites when they were given the very first opportunity with this novel task, even though they had no prior training or testing with items separated in both time and space.

In this case, previous associative training with numerals and candy arrays most probably contributed to emergent conceptual skills with numbers and counting which went far beyond the animals' specific training. Thus, this study of rudimentary addition in chimpanzees supported the notion that the subjects had acquired a 'concept of number' and could use it spontaneously with completely novel problems and in a new setting. Such cognitive flexibility and generalizability of skills, above and beyond explicit training, provides evidence for animals' capacities for conceptual understanding.

SPATIAL LEARNING AND MEMORY

As early as the 1930s, an experimental psychologist, Edward Tolman, studied the ability of rats to learn the spatial organization of a maze. As the animals explored the maze, they learned the various turns, blind alleys, and eventually, the location of the goal box. Tolman proposed that, during the course of their exploration, the rats came to represent the spatial configuration of the maze internally as a kind of 'cognitive map'. The term is still used today to refer to a mental representation of territory and the landmarks within it, sites where food has been cached, and a host of other spatially mediated information which may contribute to survival. Since Tolman proposed the idea of a cognitive representation of the external environment within the rat, scientists from numerous disciplines, including comparative psychology, neuroscience, cognitive science, biology, and ethology, have designed experiments to test ideas about such mental representations in a wide range of species, including many types of passerine birds and rodents, nonhuman primates, as well as comparative studies with children and adult humans from differing cultures.

Studies of spatial learning and memory are providing a wealth of clues towards understanding basic mechanisms related to the processing of spatial information. Most animals must be able to learn and remember information about their location within their territory in order to forage for food, seek mates, migrate, select nest or den sites, care for offspring, store food (cache), and avoid predators. For example, hoarding food and relocating it later, sometimes months after the original caching, places exceptional demands on spatial learning and memory. Among the caching birds, several species of jays and nutcrackers store large numbers of nuts and seeds, and are dependent upon finding stored food in order to survive the winter months. During experimental laboratory studies, efforts to control possible odor and visual cues did little to decrease the birds' accuracy in locating cached food. However, displacement of the cached food to a site a short distance away resulted in the birds' inability to locate it. Furthermore, while there have been some suggested alternative explanations for the birds' success, including mere chance encountering of the sites, their ability to locate and retrieve hoarded food has repeatedly been far better than would have been predicted by chance alone. Such findings have been replicated in numerous laboratories, with several different species, with the same results.

Another factor influencing spatial learning and memory is a species' particular social structure and mating system. For example, males and females of a monogamous species (i.e. those in which males and females mate exclusively) live in the same territory together. Consequently, both sexes use similar spatial abilities for getting around their range. On the other hand, in species that are polygynous (i.e. those in which males have access to more than one female), males have additional spatial demands: they must keep track of females' locations within the territory and avoid rival males. These critical behaviors might translate to better spatial processing abilities in polygynous males relative to females of polygynous species and to males and females of monogamous species. This hypothesis has been explored in studies of sex differences in neural structure and spatial cognition in different rodents, specifically prairie voles, a monogamous species, and meadow voles, which are polygynous. Males and females of these two closely related species were compared on their performance on a series of mazes, in order to test their spatial abilities. As predicted, polygynous males outperformed females and monogamous males. Studies such as these demonstrate how particular spatial abilities may develop in a species and subsequently help to explain observed differences within and between species.

It is likely that the mechanisms and processes that subserve spatial learning and memory in animals will continue to be an active area of research, as a host of critical and intriguing questions remain about how animals acquire and utilize spatial information. For example, new research in animal spatial cognition is examining what types of cues in the environment (e.g. landmarks, configurations of landmarks) animals attend to, learn, and remember in order to effectively navigate and orient. Researchers are even exploring animals' understanding of symbols of their environment, such as maps and scale models. The research continues to be comparative in nature, looking for similarities and differences among animal species and between humans and other animals. Indeed, the study of animal spatial learning and memory promises to have an exciting future in the field of animal cognition.

REPRESENTATION OF SOCIAL RELATIONS

Animals that live in long-lasting social groups interact with the individuals of their group in a complex manner. Required for these interactions are cognitive abilities that may include recognizing others as individuals, remembering which individuals have aided one in the past, and monitoring interactions among the other members of the group. Indeed, some researchers in the 1960s and 1970s proposed that the evolution of general problem-solving abilities was driven by the need for complex cognitive mechanisms for social interactions. But what exactly do animals understand and represent of their social world? Do animals recognize and interact with others in certain ways only because of past experiences with these individuals (i.e. by making associations), or can animals also distinguish classes of relationships and understand concepts such as kinship and dominance (by having mental representations)?

Although the social lives of many animals with long-lasting social groups have been examined (e.g. wolves, lions, elephants, marine mammals, birds), most research in the representation of social relations has focused on primate species. For example, Verena Dasser tested whether Java monkeys can discriminate pairs of animals based on their familial relationship. One monkey subject was trained to choose a photograph of a mother and daughter from the subject's social group and ignore a photograph of an unrelated pair. After training trials using this same mother–daughter pair, the monkey subject was able to correctly choose new photographs of other mother–offspring pairs from the social group. How did the monkey do this? It is possible that the monkey relied on the perceptual similarity between the mother and her offspring to represent the relationship between the members of the pair and solved the task by choosing the pair of animals that looked alike. Results from a recent study by Lisa Parr and Frans deWaal support this possibility. They tested chimpanzees with a similar task; however, their subjects had to match photographs of unfamiliar chimpanzee mothers with their equally unfamiliar offspring, while ignoring a third photograph of an unrelated chimpanzee. The chimpanzees were successful; they correctly selected the photographs of the sons of the sample mothers. Since the subjects had no past experience with the pictured animals, successful matching implied a recognition of physical similarity between individuals.

Dorothy Cheney and Robert Seyfarth have used vocal playback experiments to study how monkeys represent their social world. In one experiment, when an infant vervet monkey's cries were played out of hidden speakers, the other monkeys in the group looked to the direction of the infants' mother. Their behavior indicated that they associate particular infants with the infants' mothers. But can monkeys go beyond simple associations like these and demonstrate the existence of mental representations of their social world?

A second vocal playback experiment with baboons suggested that this might indeed be the case. The experimenters took advantage of natural vocal exchanges between baboons at different levels of the dominance hierarchy. Normally, when a dominant female approaches a subordinate female, the dominant monkey will grunt and the subordinate monkey will fear bark. However, when a subordinate female approaches a dominant female, there is no vocal exchange. The experimental procedure relied on the fact that monkeys tend to look longer at sources of unfamiliar or strange sounds. The experimenters played sequences of consistent and inconsistent grunt/fear bark exchanges. In the inconsistent exchange, the sound of a subordinate female's grunt was played, followed by a dominant female's fear bark, an unlikely event. However, in the consistent exchange, a subordinate female's grunt was followed by a high-ranking female's fear bark and a higher-ranking female's grunt. This sequence is consistent with natural baboon vocal behavior due to the last part of the sequence; the high-ranking animal's fear bark could have been caused by the higher-ranking female's approach. The baboons looked longer in the direction of the hidden speaker that played the inconsistent sequence, indicating that they recognized the strangeness of the event and had some representation of their dominance hierarchy and the appropriate behavior of the members within it. This understanding of social causation suggests a mental representation of their social environment.

Thus, research has demonstrated that some social animals may interact differentially with the individuals of their group with the aid of associations such as physical similarity and past experience. Furthermore, they might also develop mental representations of the underlying social relations based on these associations. It is possible, then, that the learning of associations between individuals and other individuals, or even associations between individuals and certain behaviors, may help give rise to mental representations such as kinship and dominance status.

IMITATION

Since the time of Darwin, there has been an interest in whether animals can imitate, that is, perform an action after seeing the same action being performed by another animal. Darwin concluded that this ability was one we share with other animals; however, researchers today are not so quick to grant animals this ability. Indeed, many terms other than 'imitation' have been coined to describe behavior that may appear to be truly imitative, but that actually occurs through much simpler cognitive processes. But why the careful attribution of imitation? Many researchers believe true imitation to be a precursor to theory of mind (the ability to attribute mental states to others), and thus involves processes such as self-awareness and perspective-taking. Thus, to grant that a behavior is imitative is to grant the actor the precursors to highly complex cognitive capacities.

For most researchers, demonstrating true imitative behavior typically requires that an animal has a representation of another animal's action and uses the representation to behave in a manner that matches that of the other. Observations of animals doing human-like actions (e.g. a pet opening the back door) are not enough to demonstrate true imitation. Without knowing past training experience, it is difficult to determine how the behavior developed. Also, it may be that an animal is not attending to the actions of another, but attending to the object or location the other animal is acting on. For example, when animal A behaves similarly to nearby animal B (perhaps by digging for food in a certain area), it may be the result of A's attention being drawn to the location of the food, not a direct imitation of B's action. This type of behavior has been called 'stimulus enhancement' or 'local enhancement', and it is often difficult to separate this type of explanation from one implicating true imitation.

One promising paradigm has been the 'two-action task', in which one animal watches the actions (usually a trained, unusual behavior) of another animal and is then placed in a similar situation to examine whether it completes the same actions. For example, Thomas Zentall and his colleagues have trained pigeons and Japanese quail to either step or peck on a lever to receive food. Then, untrained birds were allowed to observe the trained demonstrator as it pecked or stepped on the lever. After the observation period, the untrained observer was placed in the cage with the lever and its behavior was measured. These birds displayed almost 90 percent imitative responses during testing. That is, if they had observed the demonstrator bird step on the lever, they too stepped, but if they observed pecking, they pecked. One criticism that has been raised is that the pecking and stepping behaviors themselves were not necessarily unusual for the birds in natural contexts, and thus, these experiments may not have demonstrated true imitation, but stimulus enhancement to the lever. However, despite this criticism, the results are very suggestive of imitative behavior and point to the value of the two-action experimental design for examining animal imitation.

The study of nonhuman primate imitative ability has consisted of many suggestive anecdotes, but recently, controlled experiments have also been conducted. Many of these have incorporated object manipulation and tool-using, taking advantage of the animals' dexterity and, in some cases, their natural tool-using abilities. For example, Andrew Whiten and his colleagues developed a two-action task for both chimpanzees and young children. The chimpanzee and human subjects observed an adult human either twist or poke out bolts that secured the opening of a box containing a desired fruit. When given access to the box, all subjects used the action demonstrated, suggesting imitation. However, the chimpanzees exhibited fewer instances of imitation than the oldest children tested (four years old). Indeed, the children seemed to imitate the fine motor details of the demonstrator more closely than the chimpanzees.

Similar observations have been made by Michael Tomasello and his colleagues. In their study, chimpanzees tended to achieve the same goal as a demonstrator (e.g. use a rake to attain food), but the techniques the chimpanzees used differed slightly from those of the demonstrators. Tetsuro Matsuzawa and Masako Myowa-Yamakoshi have interpreted these results and results from their own similar experiments to suggest that apes may have difficulty in transforming the demonstrator's motions into their own matching motor acts. That is, in these imitation experiments, the apes may be less sensitive to the demonstrator's body movements than to his or her underlying goals and the objects used to attain them.

In summary, new methodologies have brought us closer to understanding the extent of animals' imitative capacities. The results so far are highly suggestive of true imitative behavior, yet many skeptics are still not convinced. The two-action task has proven to be an effective method of testing, although experimenters will have to address concerns that performance is due to stimulus enhancement and not true imitation. Furthermore, the creative studies with nonhuman primates, especially those simultaneously testing young human children, are valuable and should be explored further to determine possible species differences and similarities.

COMPARISONS BETWEEN NONHUMAN PRIMATES AND HUMAN CHILDREN

Animal cognition research is often comparative, and very often the comparisons are made between humans, specifically infants and children, and nonhuman primates. These specific comparisons are often made to examine the cognitive capacities that can exist in the absence of language, to examine the possibility of innate modules for certain cognitive or perceptual processes, and to chart the evolution of our cognitive capacities and examine conditions that may have supported their development. Research in animal cognition and human cognitive development has often delved into the same questions, yet direct comparisons on a given topic have sometimes been hard to make owing to different experimental procedures. Of late, however, there has been an increase in experimental studies that directly compare children and nonhuman primates by using very similar methodologies. Some of these studies have been discussed above, including the imitation studies by Andrew Whiten and his colleagues. Two more examples, studies of numerical ability and scale model comprehension, will be mentioned here.

Karen Wynn has used the 'violation of expectation' procedure to examine the numerical abilities of infants. This procedure relies on the fact that infants tend to look longer at events that are unexpected, or involve some sort of violation. In one study, infants watched as an object was placed on a stage. A screen was then raised to hide the object. The infants then saw a second object being placed behind the screen. Now, the screen was lowered, revealing one of two outcomes: one object (the unexpected outcome) or two objects (the expected outcome). Infants looked longer at the unexpected outcome of one object. In fact, in subsequent studies, they looked longer at a $1 + 1 = 3$ event than at $1 + 1 = 2$. This effect was also seen for simple subtraction events, and together, the studies suggest that infants have some understanding of number estimation and quantity. Using similar experimental procedures, Marc Hauser and his colleagues have examined wild rhesus monkeys' numerical abilities. The experiment was unchanged from Wynn's except that large eggplants were used as the items to be counted. The monkeys responded to the events in a manner similar to the infants, looking longer at operations that yielded unexpected numbers of eggplants. Thus, the use of identical experimental procedures has demonstrated that infants and rhesus monkeys seem to have similar numerical estimation abilities.

How and when humans come to understand the correspondence between a physical representation of space, such as a map or a model, and its real-world referent has also been the focus of much research in developmental psychology. Judy DeLoache has approached this question in her studies exploring children's ability to understand the representational nature of a scale model. She and her colleagues have found that after witnessing a miniature item being hidden in a scale model of a room, three-year-old children can locate a full-size item hidden in the analogous location in the real room. However, slightly younger children, 2.5-year-olds, have difficulty with the task. Their difficulty with the task implies the lack of 'representational insight', or knowledge that the model and room are related as symbol and referent. DeLoache and her colleagues have suggested that many factors can contribute to the development of this understanding, including the perceptual similarity between the model and the room, experience with other symbol systems, instruction on the nature of the model–referent relationship, and the ability to form a 'dual representation' (i.e. understanding the model as an object unto itself as well as a symbol for something else).

Until recently, it was not known if a nonhuman species could understand a physical representation of space such as a scale model and use it as a source of information regarding the environment. Valerie Kuhlmeier and Sarah Boysen found that chimpanzees were able to solve a scale model task that was similar to DeLoache's procedure. After watching an experimenter hide a miniature bottle of juice within a scale model of an outdoor enclosure, three chimpanzees readily found the real juice bottle that was hidden in the analogous location in the actual enclosure. That is, they went to the correct site and retrieved the bottle immediately upon entering the enclosure. These chimpanzees performed similarly to the three-year-old children in DeLoache's task. However, the performance of the other four chimpanzees tested was poor, or at best, varied. These four chimpanzees often relied on a search strategy that consisted of searching the hiding site in the front left corner of the enclosure, and continuing to search each site successively as they circled the room clockwise. The search strategy that was observed with these chimpanzees has not been reported in studies by DeLoache and suggests that, although some chimpanzees solve the task in a manner similar to young children, there may be other factors that influence some chimpanzees' performance. These factors will no doubt be the focus of future study.

Thus, these studies of numerical ability and scale model comprehension illustrate how similar test procedures can be used to examine the similarities and differences between nonhuman primates and human children. They demonstrate that the dialogue between those who study cognitive development and those who study animal cognition is increasing, with many researchers flexibly moving back and forth from one to the other during their careers.

CONCLUSION

The study of animal cognition is a growing one, with increasing dialogue among researchers in different academic fields. The topics discussed above represent only a subset of research in animal cognition but provide evidence for flexible and complex cognitive abilities in nonhuman animals. Future research of these topics, and all areas of animal cognition, is necessary to further examine the manner in which animals learn, behave, and think.

Further Reading

Boysen ST and Berntson GG (1989) Numerical competence in a chimpanzee (*Pan troglodytes*). *Journal of Comparative Psychology* **103**: 23–31.

Cheney DL and Seyfarth RM (1990) *How Monkeys See the World*. Chicago, IL: University of Chicago Press.

Dasser V (1988) A social concept in Java monkeys. *Animal Behaviour* **36**: 225–230.

DeLoache JS (1987) Rapid change in the symbolic functioning of very young children. *Science* **238**: 1556–1557.

Hauser MD, Carey S and Hauser LB (2000) Spontaneous number representation in semi-free ranging rhesus monkeys. *Proceedings of the Royal Society, London*, **267**: 829–833.

Kuhlmeier VA, Boysen ST and Mukobi KL (1999) Scale model comprehension by chimpanzees (*Pan troglodytes*). *Journal of Comparative Psychology* **113**: 396–402.

Myowa-Yamakoshi M and Matsuzawa T (1999) Factors influencing imitation of manipulatory actions in chimpanzees (*Pan troglodytes*). *Journal of Comparative Psychology* **113**: 128–136.

Nagell K, Olguin R and Tomasello M (1993) Processes of social learning in the imitative learning of chimpanzees and human children. *Journal of Comparative Psychology* **107**: 174–186.

Parr LA and deWaal FBM (1999) Visual kin recognition in chimpanzees. *Nature* **399**: 647–648.

Shettleworth S (1998) *Cognition, Evolution, and Behavior*. New York, NY: Oxford University Press.

Whiten A, Custance DM, Gomez J-C, Teixidor P and Bard KA (1996) Imitative learning of artificial fruit processing in children (*Homo sapiens*) and chimpanzees (*Pan troglodytes*). *Journal of Comparative Psychology* **110**: 3–14.

Wynn K (1992) Addition and subtraction by human infants. *Nature* **58**: 749–750.

Zentall TR, Sutton JE and Sherburne LM (1996) True imitative learning in pigeons. *Psychological Science* **7**: 343–346.

Animal Consciousness

See **Consciousness, Animal**

Animal Language

Introductory article

Duane M Rumbaugh, Georgia State University, Atlanta, Georgia, USA
Michael J Beran, Georgia State University, Atlanta, Georgia, USA

CONTENTS

Animal language references the field in which animals' capabilities for various dimensions and functions of language are researched. Language is here defined as a neurobehavioral system that provides for the construction and use of symbols to enable the conveyance and receipt of information and novel ideas between individuals. The meanings of symbols in this system are basically defined and modulated through social interactions.

INTRODUCTION

Although fascinated for centuries by the prospect of 'talking with animals', humans have only recently begun studying language as a part of the behavioral repertoire of nonhuman animals. Although there are abundant data indicating that many animal species have evolved various forms of both simple and complex communication systems (such as the vocalizations of vervet monkeys), language, as it is defined and promoted by humans, has been a more difficult phenomenon to demonstrate when studying nonhuman animals. Songs of the whales and birds, dances of the honeybee, vocalizations of monkeys, and other such forms of communication all fall short of the requirements of a formal language system in that they lack some or all of the formal components of language as it is defined. For example, the meaning of the dance of the honeybee is disrupted with landmark orientation changes. Also, monkeys give alarm calls to individuals already aware of the danger, and bird songs are predominantly ritualized and do not frequently change to meet new circumstances. However, some experimental attempts to demonstrate that nonhuman animals can learn and use language systems have been successful.

SYNTAX USE AND COMPREHENSION

There are two primary structural components to a language system: its semantics and its syntax. Syntax refers to the rules and guidelines governing how linguistic units (typically words) can be combined and the order in which those units must be used to convey the meaning of the speaker. Syntax is therefore highly tied to productive use of a language. Semantics refers to the word meanings of a language, and is intricately tied to comprehension of a language. Animal language studies have focused on both of these aspects of language.

APES

Numerous experimental investigations of language have been conducted with great apes. In these projects, the focus has shifted across time. Initial attempts were aimed at teaching apes to speak. Later, attempts were made to teach apes the formal syntactic structure of a language. Then, the focus shifted to comprehension of symbol systems. Most recently, apes' acquisition of language has occurred most robustly when infant apes are provided with a linguistic environment in which to mature, and the apes can observe and integrate themselves into this language-rich surrounding. In these attempts, the focus is on the environment and its role in both comprehension and production of language by apes.

It was discovered very early on that apes had a limited ability for producing speech sounds. This handicap would have to be overcome through the use of 'artificial' symbol systems. These systems included sign language, plastic tokens used on magnetic boards, and embossed geometric symbols called lexigrams. The sign language studies were among the first conducted, and they were conducted in four separate research projects. The Gardners worked with the chimpanzee Washoe, Terrace worked with the chimpanzee Nim, Patterson worked with the gorilla Koko, and Miles worked with the orangutan Chantek. In these projects, the apes were taught to make signs, and they were exposed to human caretakers' use of sign as well. Although initial reports indicated that the apes quickly learned to use and respond to others' use of the signs, doubt was cast on the projects by Terrace himself. Terrace concluded that the apes were simply imitating the sign use of the humans around them, and that there was not an originality or appropriateness (nor any indication of syntax) in the sign language of the apes.

Premack worked with a chimpanzee, Sarah, using plastic tokens that the chimpanzee responded to, based on the 'questions' posed to her. Sarah showed a great affinity for properly answering the questions posed to her using this system, but she rarely used the tokens to communicate her intentions or desires. Her use of the tokens was in response to posed questions only. Thus, although Sarah demonstrated comprehension of the 'linguistic' problems, she did not develop language in the sense of operating in a two-way, give-and-take communicative context.

Rumbaugh worked with the chimpanzee Lana through use of a computerized system. The lexigram keyboard responded to the requests made by Lana, provided those requests were in the proper grammatical order. Lana learned to string together lexigrams into stock sentences that produced desired outcomes. On her own, she learned to finish sentences correctly when those sentences were started by others.

Despite these novel approaches to studying language acquisition in apes, concerns and criticisms remained that, although the apes were learning what was expected of them, they did not have a true language. Rather, claims were made that apes such as Lana learned only to chain sequences of lexigrams appropriately without understanding the meanings of the symbols that were used. To address these concerns, future projects centered on the issues of intentional communication, reference, and semantics, and syntax became of secondary concern. Receptive competence became the important capacity to demonstrate.

Savage-Rumbaugh made several important findings when she began work with two chimpanzees, Sherman and Austin. These chimpanzees were taught not only to use lexigrams (which were no longer linked solely to a computer system but were used in the everyday interactions between humans and apes) but also to comprehend each other's (and human caretakers') use of those symbols. Sherman and Austin displayed a sense of symbolic thought through the novelty of their lexigram use as well as the use of other symbols (such as labels to containers and food items). The two chimpanzees made statements about their future actions, they requested items from each other, and they responded to each other's requests appropriately. Both chimpanzees also learned not only to categorize real-world items into functional categories, but also to categorize the lexigrams for those items as well. Lexigrams were categorized as either foods or tools on the very first trial in the same manner in which the real-world objects represented by those lexigrams were categorized. In other words, the lexigrams had meanings for these apes, and the lexigrams functioned as symbols for these apes.

Fortuitously, the next finding in ape-language research occurred as a result of the failure of an adult female bonobo (*Pan paniscus*, a species closely related to the chimpanzee) to replicate the findings with Sherman and Austin. After repeatedly trying to teach this bonobo to use the lexigram symbols, the attempts were interrupted so that she could be bred with another bonobo. During this time, it was discovered that her son, who had never been instructed in the use of lexigrams but who had constantly been in the area while his mother was instructed, had acquired not only comprehension of some of the symbols, but also comprehension of spoken English. This bonobo, Kanzi, responded appropriately to novel spoken requests of humans as well as to their lexigram use. This finding, of spontaneous language acquisition that occurs when young apes are raised in a language-rich environment in which there is structure in the use of lexigrams and spoken English, has since been replicated with two additional bonobos and a chimpanzee. Additionally, research with bonobos has demonstrated a much greater complexity in both language comprehension and production than had previously been demonstrated in any other study of nonhuman animals. The bonobo Kanzi responded appropriately to sentences in which word order was of vital importance in correct interpretation of the speaker's meaning. Additionally, Kanzi and other bonobos raised in a similar language-enriched environment displayed comprehension of far more spoken English than lexigrams; this comprehension is the result of their living in a linguistically rich environment in which information is available, provided that the apes can glean it from both the context of an interaction and the words spoken by humans in that context.

These findings provided clear evidence that both comprehension (of spoken English and lexigrams) and productive use (of lexigrams) was a part of the cognitive competence of these animals when raised in a structured, language-rich environment in which language learning was latent (i.e. not taught through discrete trials training). This environment is much like that in which human children are raised, and it is almost certainly this environmental context which provides the necessary stimulation for language to develop in humans and apes.

In ape-language studies, those focused on sign language have been primarily concerned with the extent to which apes could learn the meanings of different signs (the semantic aspects of sign language). Washoe, Nim, and Chantek learned the names of numerous items as well as actions and locations. The work of Premack and Rumbaugh, however, was more focused on the grammatical requirements of language. Lana learned not only to construct grammatically correct sequences of lexigrams that followed the grammar of her language (called 'Yerkish'), but she also finished sentences started by experimenters by retaining the necessary grammar needed for the sentence to be correct. However, if the provided sentence fragment was already grammatically incorrect (as produced by the experimenter), Lana would erase it and start from scratch with a correct sequence. Thus, the emphasis was on Lana's productive language skills.

As the emphasis shifted to the productive and receptive skills demonstrated with lexigrams by the chimpanzees Sherman and Austin, elements of receptive language were more evident. However, a focus on the syntactic understanding of these chimpanzees was absent. In contrast, it became clear with later research that at least one bonobo, Kanzi, responded appropriately to spoken requests dependent not just on the words within that request, but also on the order in which the words were produced. Kanzi, therefore, could respond appropriately to sentences such as 'Pour the Coke in the lemonade' and 'Pour the lemonade in the Coke'. Additionally, Kanzi responded appropriately to sentences with embedded phrases. For example, when presented with the sentence 'Kanzi, get the ball that is outdoors', Kanzi would walk past another ball indoors to retrieve the ball that was outdoors. Kanzi demonstrated syntactic understanding of word-order rules for spoken English, and he responded appropriately based on the referential and relational aspects of spoken sentences.

DOLPHINS

Dolphins also made good candidates for language acquisition research because of their complex use of vocal communication (in the form of whistles) as well as their large brain size. Research with dolphins by Herman has focused on comprehension rather than production. In one experimental paradigm, the dolphin Akeakamai (Ake) was taught to respond to gestural signals produced by humans using their arms and hands (another paradigm used by Herman involved an acoustic language with acoustic signals). Each signal representing an object, an action or a modifier (such as 'left' or 'right') could be combined with other signals so that the dolphin's comprehension could be measured through its response. Initially, the question was whether the dolphins could respond appropriately to requests either asking for an action to be done to a single object or asking that the dolphin perform an action with two objects. Later, semantic categories could be combined according to syntactic rules in such a way that three-, four-, and five-word relational sequences could be directed to the dolphin. The dolphins organized their responses in such a way as to take into account both the syntactical aspect of the sentence and the meanings of all symbols within that sentence. For example, syntactic understanding was demonstrated through correct responding to sentences containing the same words but different word order (such as LEFT HOOP PIPE FETCH, which asked the dolphin to take the pipe to the hoop on her left, versus HOOP LEFT PIPE FETCH which asked the dolphin to take the pipe on her left to the hoop). Semantic understanding was demonstrated through selection of the correct items.

In addition to responding to the requests of humans who were present, the dolphins responded appropriately to video displays of this gestural language even when the clarity of the image was degraded, and to anomalous gestural signals in which the semantic rules and syntactic constraints of the language were violated. These abilities indicate that the dolphins have a referential understanding of the signs used in their language. As with apes, dolphins utilize mental representations when responding to language-mediated tasks. Comprehending degraded images without training suggests that the dolphins have a network of semantic and gestural representations in memory. The dolphins recognized degraded images by comparing those images to a set of representational exemplars in memory rather than through simple stimulus generalization. The degree of generalization in responding to even highly degraded gestures indicates that the dolphins represent the gestures in memory.

The research with dolphins, as well as research by Schusterman with sea lions, has focused heavily on syntactic aspects of language comprehension as well as semantic understanding of word meaning. As noted, the dolphins respond appropriately not only to the meaning of individual gestures and signals, but also to the relation of those gestures and signals to each other within a sentence. Syntactic understanding was shown through responses to semantic contrasts in reversible sentences, to syntactically anomalous sentences, to structurally novel sentences, to sentences with word modification within the sentence (where one word modified the meaning of another), to interrogative and imperative sentence form contrasts, and to sentences that contained variations in the placement of modifiers.

SUMMARY

We have provided a working definition appropriate in defining the aspects of communication that differentiate language from nonlanguage. This definition allows us to investigate language competence in nonhuman animals. Such a definition is needed to establish the comparative and evolutionary bases of fully elaborated human language. Although humans clearly do more with their language capacities (especially in the area of verbal speaking) than do any other animals, comparative language acquisition research is vital to understanding the basic mechanism of language. Animal competencies for certain dimensions of human language clearly reflect comprehension and use of meanings accrued by symbols.

Further Reading

Gardner RA, Gardner BT and van Cantfort TE (1989) *Teaching Sign Language to Chimpanzees*. Albany, NY: State University of New York Press.

Herman LM (1988) The language of animal language research: reply to Schusterman and Gisiner. *Psychological Record* **38**: 349–362.

Herman LH, Morrel-Samuels P and Pack AA (1990) Bottlenosed dolphin and human recognition of veridical and degraded video displays of an artificial gestural language. *Journal of Experimental Psychology: General* **119**: 215–230.

Herman LM, Richards DG and Wolz JP (1984) Comprehension of sentences by bottlenosed dolphins. *Cognition* **16**: 129–219.

Patterson FL and Linden E (1981) *The Education of Koko*. New York, NY: Holt, Rinehart & Winston.

Premack D and Premack AJ (1983) *The Mind of an Ape*. New York, NY: Norton.

Roitblat HL, Herman LM and Nachtigall PE (1993) *Language and Communication: Comparative Perspectives*. Hillsdale, NJ: Lawrence Erlbaum.

Rumbaugh DM (1977) *Language Learning by a Chimpanzee: The LANA Project*. New York, NY: Academic Press.

Rumbaugh DM and Savage-Rumbaugh ES (1994) Language in a comparative perspective. In: Mackintosh NJ (ed.) *Animal Learning and Cognition*, pp. 307–333. San Diego, CA: Academic Press.

Savage-Rumbaugh ES (1986) *Ape Language: From Conditioned Response to Symbol*. New York, NY: Columbia University Press.

Savage-Rumbaugh ES (1991) Language learning in the bonobo: how and why they learn. In: Krasnegor NA, Rumbaugh DM, Schiefelbusch RL and Studdert-Kennedy M (eds) *Biological and Behavioral Determinants of Language Development*, pp. 209–233. Hillsdale, NJ: Lawrence Erlbaum.

Savage-Rumbaugh ES and Lewin R (1994) *Kanzi: The Ape at the Brink of the Human Mind*. New York, NY: Wiley & Sons.

Savage-Rumbaugh ES, Murphy J, Sevcik RA *et al.* (1993) Language comprehension in ape and child. *Monographs of the Society for Research in Child Development* **1**: 1–221.

Schusterman RJ and Gisiner R (1988) Artificial language comprehension in dolphins and sea lions: the essential cognitive skills. *Psychological Record* **38**: 311–348.

Terrace HS (1979) *Nim*. New York, NY: Knopf.

Animal Learning

Introductory article

Armando Machado, University of Minho, Braga, Portugal
Francisco J Silva, University of Redlands, Redlands, California, USA

CONTENTS

The field of animal learning studies the behavioral mechanisms and processes that animals use to adapt to changes in their environment. Its emphasis is on environment–behavior interactions.

INTRODUCTION

Learning refers to a heterogeneous set of processes that evolved in animals to accommodate changes in their environments. These processes can produce relatively permanent changes in behavior and are brought into play by some form of interaction between the animal and its surroundings. For instance, a moving nematode (a tiny roundworm) that stops momentarily or reverses its motion when it experiences a vibration will cease doing so if the vibration occurs repeatedly. Foraging bees perform an intricate dance, the orientation and speed of which change with the direction and distance of the food source from the hive, such that other bees also can locate the food site. Hungry domestic cats mew in the presence of their owner, who will then give them food. In all of these examples, the organism's behavior is showing the effects of particular interactions between itself and its environment. Classifying distinct types of interactions, identifying their elements, quantifying their static and dynamic properties, and describing how their cumulative effects are expressed is the goal of people who study learning. Before classifying interactions between organisms and their environments, it is important to place the process of learning within an evolutionary context – that is, to understand why learning evolved.

EVOLUTION AND LEARNING

It is widely assumed that learning evolved in specific contexts, such as gathering food, eluding predators, capturing prey, attracting mates and avoiding poisons. Despite this variety in contexts, countless experiments indicate that the same principles of learning hold across many different species, tasks and behaviors. How can we reconcile the assumption that learning evolved in specific contexts with the fact that it occurs similarly in many contexts? The answer is, by conceiving of learning processes as mixtures in varying proportions of both context-specific and general mechanisms. For example, taste aversion learning occurs when animals avoid gustatory and olfactory cues associated with foods that made them ill, even when the flavor is separated from the illness by many hours. That animals can learn which cues predict biologically significant events is commonplace; that animals can learn to avoid cues separated by many hours from a biologically significant event typically happens only when flavor is the predictive cue and illness is the significant event.

To clarify further the relationship between context-specific and general features of learning, consider the following analogy. A house built for living near the Arctic Circle will differ considerably from one built for living in south Florida: the former requires thick insulation, double-paned windows and a furnace; the latter needs mildew-resistant insulation, shaded glass and air conditioning. Despite these differences, there are general features common to both houses, such as the presence of windows, doors, rooms, walls and a roof, and a general function for both houses, such as sheltering and protecting its inhabitants from weather and predators. Because the function of a house is similar in both regions, there is some commonality to the solutions.

The presence of general features in learning is illustrated by the fact that animals from widely separated taxa respond to environmental stimuli in similar ways. For example, they ignore repetitive harmless stimuli, a process called habituation; they detect correlations among biologically important events, a process called Pavlovian conditioning; and they learn causal relations between their behavior and its consequences, a process called operant conditioning.

Central to these processes is temporal integration, for only by tracking and integrating events across time can animals determine whether an event is repeating, whether it occurs before, during or after other events, or whether it is a reliable consequence of responding or not responding. It should come as no surprise, then, that temporal variables are often critical determinants of learning.

HABITUATION: ADAPTING TO REPETITIVE, HARMLESS EVENTS

When a harmless stimulus occurs repeatedly and there are no other events associated with it, there might be an advantage to ignoring the stimulus. For example, imagine a land snail on a small wooden platform that vibrates briefly while the animal moves. Typically, this stimulus (the vibration) elicits a protective response, contracting the antennas. If these vibrations are repeated, say every 30 s, then the contractions decline. Eventually, the vibrations are ignored in the sense that the antennae are not contracted and the snail keeps moving. This waning of a response to repeated presentations of a stimulus is termed habituation.

To interpret habituation we can appeal to the concept of response threshold, which is defined roughly as the minimum stimulus intensity required to elicit a reflexive response. As the stimulus is repeated, the threshold increases, which makes it more difficult to produce a response. Eventually the threshold is greater than the current stimulus intensity and the response fails to occur.

Properties of Habituation

Habituation is present in virtually every animal species, from single-celled animals such as the ciliate *Vorticella* to humans. It has even been observed in individual motor neurons. This widespread phenomenon deserves attention for two related reasons. First, it introduces some of the key variables and functional relations that psychologists have identified in most learning processes. Second, habituation reveals the amazing complexity of even the simplest of learning processes.

Recovery from habituation

In the example above, if the platform ceases to vibrate after habituation has occurred, then with the passage of time since the last vibration, the snail's antennae are increasingly likely to contract when the platform again vibrates. In other words, habituation seems to 'wear off' when the stimulus that produced it is not presented. This recovery of the response corresponds with a return of the threshold to its initial value.

Stimulus intensity

If the vibration of the platform is sufficiently intense, then the snail may withdraw into its shell. This response might also habituate if the strong stimulus is repeated. Typically, however, the courses of habituation and recovery from habituation are slower for strong than for weak stimuli. When the stimulus is more intense, the rise of the threshold takes longer to surpass the stimulus intensity, and the fall of the threshold in the absence of the stimulus takes longer to return to its initial, baseline value.

Time between stimulus presentations

All else being equal, stimuli closer in time produce faster habituation than stimuli farther apart. This finding is consistent with the threshold account of habituation: longer intervals give the threshold more time to decrease, which partly offsets the effect of the stimulus presentations. Interestingly, high rates of stimulation may also lead to faster recovery from habituation. This result, unlike the preceding ones, does not follow from the view that changes in threshold are responsible for habituation, unless the rate at which the threshold returns to its baseline value depends on the rate of stimulation.

Relearning effect

Imagine the following experiment. After we record the course of habituation on day 1, we stop the vibrations and let recovery occur. On day 2 we vibrate the platform again and record the new course of habituation. Typically, the rate of habituation is faster during the second day. The importance of this finding is that the difference in the rates of habituation shows that the animal's internal state on day 2 is different from its state on day 1 despite the similarity in its initial responses. Again, a simple change in the response threshold cannot accommodate this finding.

Stimulus generalization and stimulus specificity

The contraction of the antennas ceases not only in the presence of the original vibration, but also in the presence of similar stimuli (stimulus generalization). However, it is readily elicited by different stimuli such as a blast of air (stimulus specificity). These properties are two sides of the same coin; generalization focuses on the fact that habituation to one stimulus extends to some of the other stimuli that can also elicit the response; specificity focuses on the fact that habituation to one stimulus does not extend to all stimuli that can elicit the response. Careful empirical work is needed to identify the stimulus properties (e.g. intensity, duration, rate) along which generalization proceeds.

Functions of Habituation

As noted above, habituation has been observed in almost every animal species. Why is it so prevalent – and why do the properties of habituation hold true across many different species, responses and *a fortiori* physiological mechanisms? Probably, habituation occurs because it is sometimes safe and economical to ignore a repetitive stimulus. An animal that continued to respond to every stimulus impinging on its receptors would be overwhelmed by stimulation and incapable of acting appropriately. However, the animal would pay a high price if the effects of habituation were permanent, for what was once a harmless vibration caused by the running of a distant animal might now be caused by an approaching predator. In the same vein, assuming stronger stimuli are potentially more harmful than weaker stimuli, it makes sense that they should be ignored less quickly than weaker stimuli. Classifying incoming stimuli as 'The same!' 'The same!' 'The same!' also seems safer when the stimuli are close in time.

PAVLOVIAN CONDITIONING: LEARNING ABOUT CORRELATIONS

Food and water, predators and prey, mates and offspring, and escape routes and shelter, are some of the primary determiners of survival and reproduction. As such, it is reasonable to attribute great evolutionary advantage to animals capable of anticipating them. For all animals, specific sounds, sights or odor trails, places or times of occurrence, or more complex sequences and configurations of stimuli might be reliably correlated with biologically important events. If an animal could learn the correlational texture of its world (i.e. the relationships among events), then it would have the advantage of responding one way when a stimulus predicts an important event and in another way when a stimulus does not.

The correlations that an animal can learn depend on the animal and the types of stimuli and events in its environment. In terms of the animal, there might be reliable cues that it cannot detect simply because it has not evolved the required biological machinery (e.g. sensory receptors). In terms of the environment, a stimulus might be detectable but its frequency of occurrence or its reliability as a cue might be too low to support the evolution of an ability to fully exploit it; the cost would outweigh the benefit, as it were. Learning about the cueing function of a stimulus is therefore constrained both by the animal's physiology and the specific arrangements of events in the animal's world.

Constraints notwithstanding, how does an animal learn the correlation between a neutral and a biologically important stimulus? The pioneering work on this question was conducted by Ivan Petrovich Pavlov (1849–1936), the famous Russian physiologist and 1904 Nobel prizewinner. In good scientific fashion, Pavlov reduced the problem to its bare essentials: a tone reliably preceded a bit of meat powder delivered to the mouth of a hungry dog. Of interest was the animal's behavior during the tone. Initially, when the tone was presented the dog pricked up its ears and looked in the direction of the source of the tone, but, critically, it did not salivate. After a few pairings of the tone and food, the orienting response elicited by the tone ceased (habituation had set in) and a new response during the tone began to occur – salivation. Because, 'food in the mouth' elicited copious salivation without any previous training, Pavlov called it the unconditional stimulus (US) and 'salivation in the presence of food' the unconditional response (UR). As the quantity and quality of salivation to the tone depended on the prior predictive history of the tone, Pavlov called the salivation to the tone a conditional response (CR) and the tone a conditional stimulus (CS). The study of how behavior changes when two or more stimuli are paired, as in the preceding example, is known as Pavlovian or classical conditioning.

With this and similar laboratory preparations, Pavlov and many subsequent researchers have tried to understand how animals learn the cueing function of stimuli. Some of their experiments showed the following results, many of which resemble those obtained in studies of habituation.

Extinction

If, after the tone elicits salivation reliably, it is presented without the food, then the dog will eventually stop salivating during the tone. That is, when the CS no longer predicts the US, the CR weakens and may eventually disappear. Through acquisition and extinction processes, animals adjust to changes in the pattern of events in their environment.

Spontaneous Recovery

If the experimenter allows the dog to rest for, say, 24 h after the extinction training, and then presents the tone again, the animal that had stopped salivating to the tone may again salivate to it; that is, the CR spontaneously recovers. The passage of time undoes some of the effects of extinction. Why spontaneous recovery of the CR happens is still poorly understood.

Stimulus Generalization

Having learned to salivate to a specific tone, the dog also will salivate to similar tones. That is, a CR will be elicited by the original stimulus as well as similar stimuli; however, the more different these other stimuli are from the original CS, the weaker the CR they elicit. Because no stimulus ever recurs in precisely the same way (e.g. the rustling of the leaves announcing a lion is different in different situations), it is advantageous to extend newly acquired responses to similar stimuli.

Stimulus Discrimination

When Pavlov alternated two tones during training and paired one but not the other with food, his dogs eventually salivated only to the tone paired with food. That is, if one stimulus (CS+) is paired with a US, but another stimulus (CS−) is not, then the CR will occur only or mainly in the presence of the CS+. Stimulus discrimination helps ensure that a response occurs in particular environments, rather than indiscriminately across situations and time.

Contingency Effects

Assume that during the original training the food only follows the tone on 50 percent of the trials. On the remaining 50 percent the tone occurs alone. Under this circumstance, the amount of salivation to the tone during training is smaller than when the food always followed the tone. Similarly, if food also occurs occasionally in the absence of the tone, the CR is weaker than when food only follows the tone. In the extreme, if food occurs more often in the absence of the tone than in its presence, the tone will actively suppress salivation instead of eliciting it. In summary, the results of many experiments show that animals are sensitive to the direction (positive or negative) and the strength of the correlation, or contingency, between the CS and the US.

The effect of contingency shows that temporal contiguity between the tone and food is insufficient to ensure that the tone will become a CS. Much depends on what else the animal has been experiencing, both during the presence and the absence of the tone. That is, the animal seems to integrate events that are temporally extended, and to behave according to the actual correlation value between the tone and the food. Both temporal and probability relations between the CS and US, or contiguity and contingency, are important in Pavlovian conditioning.

In fact, the process is even more complex than stated above. Consider an experiment in which a tone is paired with food until it elicits salivation reliably. Next, the tone is presented along with a light, and this compound stimulus is followed with food. Will the light elicit salivation when it is presented alone and without the food? Because food always occurs after the light and never in its absence, the light is maximally (and positively) correlated with food. Moreover, because the food closely follows the light, the two stimuli also are temporally contiguous. Hence, one might predict a strong association between the light and food and, therefore, salivation to the light. However, routinely little or no salivation to the light is found. Control experiments indicate that because the tone already predicted the food at the end of the first part of the experiment, it somehow blocked the association 'light–food'. We could say that the light provided no new information about the food beyond that already provided by the tone and, hence, the light did not help the animal anticipate the US any better than the tone. The important point is that such blocking highlights the fact that an animal's prior experiences can modulate the effects of contiguity and contingency.

Relevance of Pavlovian Conditioning

Since Pavlov's pioneering work, the study of Pavlovian conditioning has revealed many other complex relations among the CR and temporal variables, the sequential arrangements of the various stimuli, the context in which conditioning occurs, and the animal species and the particular response system under consideration. Pavlovian conditioning is fundamental to understanding drug addictions, phobias and a variety of sexual responses in humans and other animals. Its domain of study also has become increasingly quantitative. Real-time, dynamic models of the learning process have started to replace verbal accounts. However, much remains unknown about the process through which stimuli that are insignificant when considered in isolation become significant when they signal biologically important events.

OPERANT CONDITIONING: LEARNING ABOUT CAUSATION AND CONTROL

The preceding discussion focused on how an animal's behavior is changed by repeated presentations of single stimuli (habituation) or by relationships among stimuli (Pavlovian conditioning). In both of these situations, behavior changes as a result of the stimuli that precede it. However, it is also the case that things happen after a response. For example, a young male cowbird sings one of its song variants and elicits a subtle wing flick from a female. An adult male cowbird sings a variant that stimulates a precopulatory display in a female and a vigorous attack from a dominant male. However, if the same adult cowbird sings a less stimulating song, then it avoids being chased by the dominant male. Operant or instrumental learning results when an animal's behavior causes a stimulus change, which in turn changes the animal's subsequent behavior: the young cowbird is more likely to sing the variant that caused the positive female reaction; the adult cowbird is less likely to sing the song that caused the attack and more likely to sing the one that avoided it. This capacity to change behavior because of its consequences enables animals to learn about control and to exploit the causal texture of their social and physical worlds.

In the examples above, the operant response produced different types of consequences. Psychologists classify these consequences by means of their effect on behavior. Consequences of an action that increase the likelihood of that action recurring are termed 'reinforcers' – positive reinforcers if the consequence is the occurrence of a stimulus (e.g. the wing flick display from the female), and negative reinforcers if the consequence is the cessation or avoidance of a stimulus (e.g. the threat and attack avoided by the adult cowbird when it sang the less stimulating song). Consequences of an action that decrease the probability of that action recurring are 'punishers' (e.g. the attack suffered by the low-ranking cowbird when it sang its most stimulating song). By modifying its behavior to produce reinforcers and eliminate or avoid punishers, an animal shapes its world while its behavior is shaped by its world. This closed feedback system is the hallmark of operant conditioning.

The laboratory study of operant conditioning began with the work of Edward L. Thorndike (1874–1949), who showed that cats placed in a puzzle box (a wooden cage with a door that could be operated by the animal) become quicker at escaping with repeated successes. Later, B. F. Skinner (1904–1990) studied how behavior is shaped by its consequences, and how new response forms emerge when variations in behavior have different consequences. To conduct his experiments, Skinner invented the operant chamber, a box with a lever that hungry rats could press to receive food dispensed into a tray. The operant chamber soon became the microscope of learning psychologists.

What sort of consequences function as reinforcers or punishers? An agreed-upon theory that predicts which stimuli reinforce or punish behavior, the circumstances in which they do so, and why they do so, has eluded experimental psychologists. However, researchers have identified several factors that influence the behavioral effects of reinforcers and punishers.

Contiguity and Contingency

As in Pavlovian conditioning, contiguity and contingency are important. Other things being equal, long intervals between a response and a consequence weaken the strengthening effect of the latter; the more immediately a consequence follows a response, the more likely it is that the response will be affected by the consequence. However, short intervals may have weak effects if the correlation between the response and the consequence is low. This can occur in two ways – when a response is followed by the consequence too infrequently, or when a reinforcer occurs independently of the response that normally produces it. An adult, subordinate cowbird might continue to sing its most stimulating song if that song rarely produces aggression from other males; a young cowbird might spend less time singing if the positive female display occurred in the absence of the song.

Extinction

The behavioral changes brought about with operant conditioning are reversible. When the environment changes and a response that used to be followed by a positive outcome is no longer followed by it, the probability of that response occurring declines. That is, the animal ceases emitting behavior that is no longer functional. However, extinction may be rapid or slow. On some occasions, the animal quickly changes its behavior, whereas on other occasions it perseveres for long periods of time. Whether extinction is rapid or slow depends largely on whether every instance of a particular response was reinforced (rapid) or only occasionally reinforced (slow).

Schedules of Reinforcement

In the natural environment, it is rare that every instance of a particular response is followed by a consequence. To understand the effects of intermittent reinforcement psychologists have studied different rules specifying when a response will produce a consequence. Two examples of these rules, collectively known as schedules of reinforcement, are ratio and interval schedules. In ratio schedules the reinforcer depends solely on the occurrence of behavior (e.g. a rat receives food each time it completes five lever presses); in interval schedules the reinforcer depends on the occurrence of behavior and the passage of time (e.g. a rat receives food following the first lever press after 15 s since the previous occurrence of food). Because there are no restrictions to when the rat can press the lever, the experimenter can study how the rate of a response changes across time as a function of how it produces a consequence.

Typically, ratio schedules support higher rates of responding than comparable interval schedules. Why? Because the two types of schedule induce different feedback functions, that is, different relations between response rate and reinforcement rate. As an illustration, consider a ratio schedule in which five responses produce one reinforcer. In this schedule, how often reinforcers occur depends exclusively on how rapidly the animal responds; reinforcement rate will always equal one-fifth of the response rate. In contrast, consider an interval schedule in which a response is reinforced if it occurs at least 15 s since the previous reinforcer. In this schedule, a response rate of four responses per minute matches the reinforcement rate. Slower response rates produce proportional changes in reinforcer rates, but faster response rates do not. That is, reinforcement rate ceases to vary with response rate. Differences in the feedback function explain why ratio schedules typically produce higher rates of responding than interval schedules.

Choice

Just as simple processes combine to produce complex phenomena such as weather, geological formations and evolution, so too do basic processes of learning combine to produce more complex behavior. Choice among options is an example that illustrates how reinforcement rates interact to affect behavior.

In the simplest situation, an animal faces two response keys, each of which delivers a reinforcer according to a schedule of reinforcement. For example, a pigeon might peck one key and receive a morsel of food with probability p, or peck another key and receive food with probability q. Another example might consist of a rat that presses one lever that delivers reinforcers with rate r, or another lever that delivers them with rate s. Studies such as these with pigeons, rats, rhesus monkeys and humans, among other species, have yielded a robust empirical finding known as the matching law. The law states that the proportion of choices on one alternative equals the proportion of reinforcers obtained from that alternative. In symbols:

$$x/(x+y) = R_x/(R_x+R_y) \tag{1}$$

where x and y are the total numbers of choices of each alternative, and R_x and R_y are the corresponding total numbers of obtained reinforcers.

Much less understood is how basic behavioral processes combine to yield the matching law. Some researchers propose that the equality is the outcome of the cumulative strengthening effect of individual reinforcers on the two response alternatives. Others suggest that the law results from the animal's sensitivity to global rates of reinforcement and its ability to maximize these rates under constraint. Still others suggest that matching derives from the tracking of the intervals between successive reinforcers on the two alternatives. In these three hypotheses we see, once again, the difficulty of determining the timescale of the learning process. Equally poorly understood is the acquisition of preference and how it relates to the various parameters of the choice situations (e.g. how the values of p and q, or r and s, in the examples above, determine how fast the animal comes to prefer the best alternative).

Stimulus Control

Because no response occurs in a vacuum, a response–consequence relation is always context-specific. In the laboratory, if a pigeon is reinforced for pecking a green key but not a red one, then the bird will restrict its pecking to the green key. This differential responding occurs because the two stimuli are correlated with different response–consequence relations: but, as with habituation and Pavlovian conditioning, the extent that stimuli different from the ones used in training control operant behavior depends on a variety of factors. For example, the amount of pecking in the presence of stimuli similar to the green key (e.g. a blue key) and stimuli similar to the red key (e.g. an orange key) depends on how much training the pigeon received with the original stimuli. More extensive stimulus discrimination training promotes discrimination, whereas less training promotes stimulus generalization. Also, reinforcing behavior differentially in the presence of different stimuli produces sharper discriminations (less generalization) than reinforcing a response in the presence of a single stimulus.

Moreover, when two or more stimulus elements signal that a response is likely to be reinforced, some form of stimulus competition may ensue. Consider the following experiment: a pigeon is trained to peck a green key with a black horizontal line, but not to peck a red key with a vertical line. The degree of stimulus discrimination and generalization is then tested by recording the amount of pecking at a white key during presentations of the line in various degrees of orientation. During the test, the pigeon pecked all line orientations similarly (i.e. stimulus generalization). However, when tested with keys of different colours but without the line, the pigeon pecked most at the green key and least at the red one (i.e. stimulus discrimination). This example shows that, for reasons that are poorly understood, some features of a stimulus may overshadow others. For the bird in this example, color overshadowed line orientation; but the reverse could have happened. A similar effect occurs in Pavlovian conditioning.

Timing

Temporal variables play a fundamental role in habituation and in Pavlovian and operant conditioning. Time may also be more directly involved in learning, as when animals learn to act according to the temporal attributes of a stimulus. For example, rats, pigeons, monkeys and other vertebrates can learn to behave in one manner after a 2 s stimulus and in another manner after an 8 s stimulus. When a reinforcer such as food is available periodically, say every 30 s (an example of an interval reinforcement schedule), animals learn to pause immediately after food and then, after about 15 s, respond at an increasingly faster rate until reinforcement. The study of the temporal regulation of behavior is one of the most developed areas in the study of animal learning.

A major empirical finding that has emerged from these studies is the scalar property of temporal discrimination, which states that all temporal judgments are relative. Hence, how a rat behaves at 10 s when reinforcement occurs every 30 s is similar to the way it behaves at 20 s when reinforcement occurs every 60 s. As another example, assume that a rat is trained to press a lever on the left after a 2 s signal and a lever on the right after an 8 s signal. Empirical tests show that the rat will be indifferent (i.e. it is just as likely to press the left as the right lever) when presented with a 4 s signal, because 2 is to 4 as 4 is to 8. However, if the two training stimuli were 4 s and 16 s long, then the rat would be indifferent when presented with an 8 s signal. The scalar property derives its name from the fact that when the intervals of a temporal discrimination change, the animal's performance is scaled (stretched or shrunk) by the same factor. Why the scalar property holds remains a matter of controversy.

Avoidance Learning

Although operant and Pavlovian conditioning seem clearly distinguishable, components of each are often present in the procedures of the other. In this sense, it is perhaps better to conceive of operant and Pavlovian conditioning as analogous to elemental hydrogen and oxygen. These two elements are rare in nature, but their combination in the form of water is common. Similarly, most learned behavior is a mixture in varying proportions of operant and Pavlovian conditioning.

The clearest examples of operant–Pavlovian interaction can be seen in situations where animals have to avoid aversive outcomes. A gopher that sees a hawk overhead will not wait to see what the bird will do; upon sensing the hawk, the gopher will retreat to its burrow. In the laboratory, a dog will jump over a hurdle during a tone if this response prevents the delivery of shock. In these and similar circumstances, the sight of the hawk or the sound of the tone predict an aversive outcome (being attacked or shocked) unless a certain response occurs (retreating to a burrow or jumping a hurdle). The relation between the signal and the aversive event is a Pavlovian CS–US relation; but because responding during the CS allows the animal to avoid the aversive event, this response is negatively reinforced (an operant response–consequence relation).

Relevance of Operant Conditioning

Since Thorndike's and Skinner's early work, the study of operant conditioning has been extended in many different directions. Neuroscientists, pharmacologists and clinical psychologists, for example, have used the techniques and conceptual tools of operant conditioning to understand the functioning of the nervous system, the behavioral effects of drugs, and the intricacies of behavioral disorders such as depression. Artificial intelligence researchers also have borrowed ideas from the domain of operant conditioning to design systems that learn through the consequences of their actions. The study of operant conditioning also has become more quantitative. As in Pavlovian conditioning, real-time, dynamic models of the operant process have started to replace purely verbal accounts. However, much remains unknown. For example, although it is reasonable to assume that a consequence with survival value (or which is closely associated with a stimulus that has survival value) will have a strong effect on the response that produced it, research has yet to yield a general theory of reinforcement and punishment.

CONCLUSION

The processes reviewed above represent only a fraction of the most basic categories of the taxonomy of learning. Much else has been done, but still more remains to be investigated. Despite a century of research, three central questions of learning theory remain largely unanswered. First, how does an animal's evolutionary history constrain the sorts of things that it can learn? Second, how are processes that occur on different timescales integrated? Third, why are seemingly simple processes so complexly organized? These questions are likely to set the research agenda of learning psychologists for the next decades.

Further Reading

Abramson CI (1994) *A Primer of Invertebrate Learning: The Behavioral Perspective*. Washington, DC: American Psychological Association.

Hearst E (1988) Fundamentals of learning and conditioning. In: Atkinson RC, Herrnstein RJ, Lindzey G and Luce RD (eds) *Steven's Handbook of Experimental Psychology*, 2nd edn, vol. 2, Learning and Cognition, pp. 3–109. New York, NY: John Wiley.

Mackintosh NJ (1974) *The Psychology of Animal Learning*. New York, NY: Academic Press.

Mackintosh NJ (1983) *Conditioning and Associative Learning*. Oxford, UK: Oxford University Press.

Mazur J (1998) *Learning and Behavior*, 4th edn. London: Prentice-Hall.

Pavlov IP (1927) *Conditioned Reflexes*, translated by GV Anrep. London, UK: Oxford University Press.

Rescorla RA (1988) Pavlovian conditioning: it's not what you think it is. *American Psychologist* **43**: 151–160.

Skinner BF (1961) Selection by consequences. *Science* **213**: 501–504.

Staddon JER (2001) *Adaptive Dynamics: The Theoretical Analysis of Behavior*. Cambridge, MA: MIT Press.

Thorndike EL (1911) *Animal Intelligence*. New York, NY: Macmillan.

Williams BA (1988) Reinforcement, choice, and response strength. In: Atkinson RC, Herrnstein RJ, Lindzey G and Luce RD (eds) *Steven's Handbook of Experimental Psychology*, 2nd edn, vol. 2, Learning and Cognition, pp. 167–244. New York, NY: John Wiley.

Animal Navigation and Cognitive Maps

Intermediate article

Bruno Poucet, National Centre for Scientific Research, Marseille, France

CONTENTS

Many species have impressive spatial navigation capabilities which seem to rely on the existence of representations, or cognitive maps, of the spatial environment.

INTRODUCTION

One of the most challenging questions for contemporary neuroscience is to understand the mechanisms by which the brain processes, encodes and stores information. In this respect, our understanding of memory has considerably improved in recent decades. Such progress is largely due to the use of animal models which allow us not only to tackle the problem of memory from a comparative perspective, but also to raise empirical issues that cannot be addressed in humans.

Spatial memory is a popular model of memory. One reason is that spatial memory is ubiquitous: almost every action takes place in space and requires some form of spatial memory. Another reason is that it can easily be studied in animals that often have outstanding spatial capabilities. Consider, for instance, the memory capabilities of rats that solve the radial arm maze task (Olton, 1979). In this task, the animal has to gather food at the end of the eight arms of a radial maze (Figure 1). As arms depleted of food are not rebaited, the rat learns to avoid locations that have already been visited. Once the animal is trained, a delay can be inserted between the fourth and fifth choices: the rat is blocked at the central choice point of the apparatus and required to wait for some time before completing the trial. Even with delays lasting up to 30 min, the rat is still able to perform the last four choices without returning to arms visited before the delay. This ability implies that it has stored the memory of the locations of depleted arms and uses this memory to visit the remaining baited arms. The current interpretation of such performance is that the rat's choice is based on a cognitive map; that is, a representation of the food locations relative to the configuration of visual cues within the testing environment.

COGNITIVE MAPS AND PLACE NAVIGATION

Historically, the concept of spatial cognitive maps is important because it suggests that animals do not

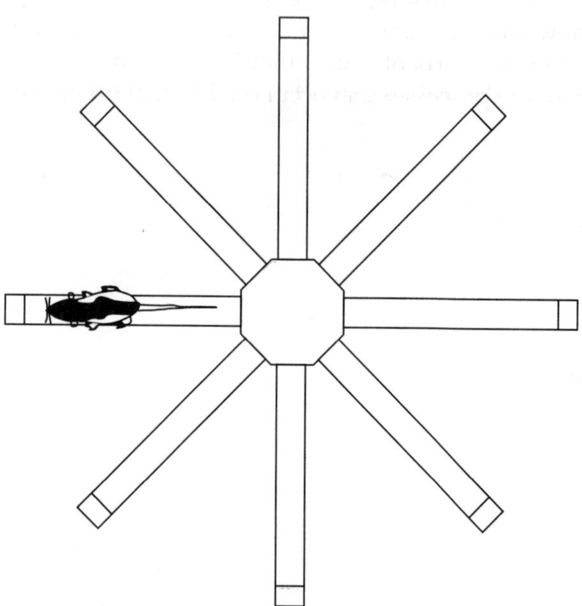

Figure 1. Overhead view of the radial arm maze. Food is located in small containers at the end of each arm and cannot be seen by the rat until it has reached the container. The task for the animal is to visit each arm without returning to an arm previously visited during the trial.

merely base their actions on specific stimulus–response associations (as strongly advocated by the early behaviorist theory in the twentieth century), but also internally reorganize acquired spatial information so as to form cognitive representations of the environment (Tolman, 1948). An important property of such representations is that they allow a reaction to stimuli that are not immediately present, since the relationship of such stimuli to those actually perceived is maintained in a cognitive representation. In other words, animals are aware of the properties of the environment beyond their field of perception. Thus, cognitive maps confer greater flexibility and efficacy to behavior. These properties depart considerably from the rigidity in behavior that results from the gradual acquisition of fixed relationships between stimuli and responses. (*See* **Tolman, Edward C.**)

A prototypical illustration of the behavioral flexibility afforded by cognitive maps is provided by the ability of rats to navigate efficiently in the water maze task (Morris, 1984). In this spatial task, the rat must find a safe platform in a pool filled with water (Figure 2). As the start position is changed from trial to trial and the platform is not visible, the animal cannot apply rigid solutions to the problem. Instead, it must rely on the visual cues located outside the swimming pool so as to infer the platform location. The rat quickly learns to swim directly towards the platform from all starting positions. More importantly, it shows immediate transfer when novel start points are used. Thus, the animal's knowledge of the platform location is independent of its current location.

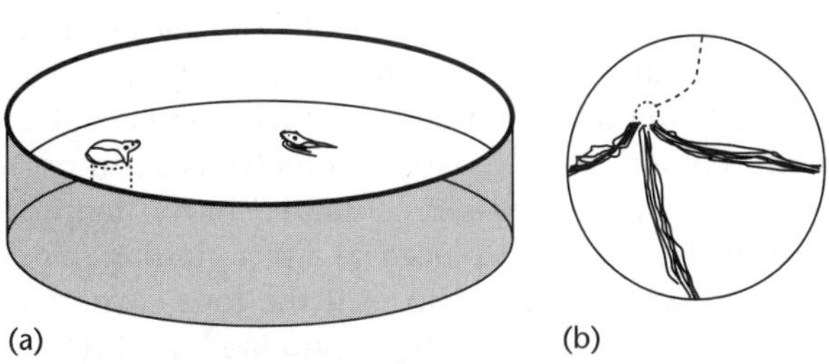

Figure 2. The water maze. (a) The pool is filled with water made opaque by the addition of powdered milk, preventing the rat from seeing the platform which lies under water level. Only distal visuospatial cues can be used to locate the goal platform. After 10 days of training, the rat performs direct trajectories. (b) Superimposed trajectories for the last six trials from the three start points used for training (black lines) and the almost direct path used by the rat when started from a start position never experienced before (dotted line).

ALLOCENTRIC CODING AND PLANNING

The rat's performance in the radial maze and water maze navigation tasks provide nice illustrations of two major functions of spatial cognitive maps. Namely, such maps are useful to memorize the positions of potential goals and to perform place navigation. These capabilities rely on an allocentric coding process, which allows the organism to memorize a location in relation to the spatial layout of surrounding landmarks, independently of its own current position.

The allocentric coding of spatial locations stands in contrast to an 'egocentric' (self-centred) coding process. This process allows an animal to memorize a location in relation to its own position. The goal location is stored in egocentric coordinates as a vector that specifies the distance and head-referred direction of the goal from the animal's current location. The egocentric coding process requires a path integration mechanism which updates the memory of the goal location as the animal moves. Although egocentric coding has the major advantage that the information to be stored is limited to two values (distance and direction of the goal relative to current position), its major drawback is that the computation of these parameters is subject to cumulative error. If the cumulative error is great enough, the animal may miss its target. Although the egocentric coding process is unreliable, its precision can be improved if the path integration information can be recalibrated from visual or other sensory information available from the environment. It is not clear, for rats or humans, if any reliance is placed on path integration except when external sensory information is missing or inadequate.

Place navigation in the water maze cannot be accounted for by an egocentric coding process, because the experimental design precludes the use of route information. So, how can we explain this performance? One possibility is that the rat takes a 'snapshot' of the cue configuration when it is on the goal platform. It could return to the platform by moving in a direction that reduces discrepancies between its current views of environmental landmarks and the views of the same landmarks from the goal. However, analyses of paths suggest that the rat does not behave in this way. The straightness of swimming paths when the rat is started from different locations in the water maze indicates that it is not just continually adjusting its current position in reference to a memorized snapshot taken at the goal; such a solution, which relies on step by step movements, would result in more erratic movements. Rather, right from the start of its movement, the rat seems to have some knowledge of the distance and direction of its final destination, and sets its course immediately to reach the goal as soon as possible.

This ability to generate a plan requires a stored representation of the spatial relationship between the goal and rat's current position relative to the environment. This representation makes behavior relatively independent from immediately available sensory information, thus allowing adaptive changes in trajectories when the circumstances so require. Rats, cats, dogs as well as many other species are able to take short cuts when a new, less circuitous, path is made available, or to make detours around obstacles in their way to a target. For example, if a previously available path is suddenly blocked, the animal quickly reorganizes its trajectory to select the next most appropriate path leading to the goal. This ability shows that place navigation directed at a specific goal location takes into account the overall connectivity of space. (*See* **Navigation**; **Spatial Cognition, Models of**; **Animal Learning**)

EXPLORATION AND COGNITIVE MAPS

Anyone who watches a rat in a novel environment immediately notices the exploratory responses displayed by the animal. Exploratory behavior is vigorous and is presumably necessary to acquaint the animal with its new environment. As familiarization goes on, exploratory activity decreases, to stabilize at a low asymptotic level. This habituation process suggests that the animal comes to know various aspects of its environment. If the animal is deprived of the opportunity to explore its environment, it is unable to successfully solve spatial problems. Thus, exploration is required for the emergence of place navigation, presumably because cognitive maps are built during this phase through active collection of information about the spatial environment. Eventually, the map comes to match the real environment as closely as possible, therefore providing the animal with spatial invariants which allow it to perform efficient place navigation. Knowing the layout of its current space, the animal is also in a position to detect any changes that might occur. To do so, it performs routine patrolling to check the stability of the environment and to update the contents of the map if a change has been detected.

Thus, exploration may lead either to a new representation or to the updating of a former spatial representation. An illustration of the updating process is provided by studies which show that subtle changes in an otherwise familiar environment induce strong re-exploration. If such changes only involve the spatial arrangement of environmental components, it becomes possible to study the way the animal encoded the initial arrangement.

INFORMATION CONTENT OF SPATIAL COGNITIVE MAPS

Recognizing that animals are able to use representations of their spatial environment does not say much about the nature of such representations. In fact, there is some evidence that these representations might not encompass all aspects of the environment.

Configural Cues and Geometry

Based on the re-exploration technique, it can be demonstrated that changing the spatial configuration of small objects in a previously explored arena induces a renewal of exploratory activity generally aimed at the displaced objects (Thinus-Blanc *et al.*, 1987). Further analyses of exploratory responses as a function of the object arrangement show that although the animal keeps a record of the spatial situation, this record is possibly specific to certain classes of spatial relationships. In general, changes that induce the strongest responses are those that affect either the overall geometrical arrangement of the object set, or the topological relationships among the objects. The configuration seems to be privileged over the absolute position of objects.

This conclusion is in agreement with the observation that performance in both the radial maze and the water maze tasks relies on configural cues rather than on individual landmarks. It also agrees with the notion that spatial geometry is an important piece of information contained in the map. In one study, rats were placed in a rectangular chamber and were required to visit the four corners of the chamber. Each corner contained a different amount of food and was associated with a distinctive visual insert. Two opposite corners had the most bait while the remaining two opposite corners had the least bait. After the animals mastered the task (i.e. visited each corner according to a decreasing order from the most baited to the least baited arm), a number of probe tests were conducted based on transformations of the initial distribution of the inserts. These tests revealed that the rats' patterns of visits to the food locations were remarkably insensitive to the modification brought to the spatial arrangement of the inserts. Rather, the animals were using the rectangular shape of the experimental chamber as a means to locate the various food sources (Gallistel, 1990). Thus, rats ignore obvious landmarks and attend to landscape features instead in reorienting themselves in symmetric environments. Although they are aware of the landmarks, they simply do not appear to use them for certain purposes. The rules governing when landmarks or landscapes are used in behavior remain to be fully explicated. Current thinking suggests that local landmarks are not used to determine directions in space, nor to define a spatial framework within which such directions can be nested. Rather, landscape features are seemingly used for these purposes.

Distal Landmarks

Landscape features cannot be considered to be the sole markers of the spatial arrangement. In fact, distal landmarks provide another major source of information on which spatial navigation can rely. In general, animals will preferentially use such distal information when it is available, to the detriment of local cues. Distal information also predominates when it conflicts with movement-related information such as that provided by path integration mechanisms. While path integration-based homing behavior in hamsters is barely altered by the manipulation of local cues, it can be easily altered by manipulating distal information. To understand why distal cues are so effective in controlling place navigation, it is necessary to realize that only distal cues maintain their reciprocal relationships with respect to the animal during its motion. Thus the perceived reciprocal relationships between distal cues are minimally affected by an animal's movements. In contrast, the perceived reciprocal relationships between local cues are subject to strong changes during movements. Since a cognitive map contains absolute topographical information rather than information about egocentric locations relative to the animal, it becomes evident that distal cues provide a more reliable source of information for localization since they retain relatively stable relationships as the animal moves about the environment.

Path Information

Since the primary function of spatial cognitive maps is to allow efficient navigation in space, a question that arises is whether animals are aware of the structural properties of space, so that their knowledge is based on a representation of possible paths (and therefore not limited to the start and the goal). Studies in cats which had to choose between several paths all leading to the same goal location provide an unambiguous answer to this question. The paths differed in several ways, such as their length or angular deviation (i.e. how much the start of a path deviated from the goal direction). Cats displayed preferences for particular paths primarily on the basis of length (with short paths being preferred over long paths), and secondarily on the basis of their angular deviation (with paths whose starting direction was closer to the goal direction being preferred over less direct paths). Thus, spatial knowledge is not limited to representation of the start and goal, but extends to more indirect relationships provided by the rest of the environment. There appears to be an encoding of properties of the structure of the environment. The validity of the hierarchy of properties depends, however, on the goal location being hidden. If the goal is visible, the animals' choices appear to be constrained by the perceived direction of the goal from the start point. The visible goal seems to act as an anchor, shifting control over behavior from spatially based information to sensory guidance. The animals no longer make optimal choices about path length, but instead tend to take the path that most closely approximates the direction from the start to the goal.

Metrics and Topology

Evidence indicates that rats possess a representation of the geometry of their environment. The representation, however, seems not to be complete, as certain spatial relationships are better handled than others. Also, the representation is not homogeneous: animals process certain locations in a more detailed way during exploration. This is not surprising, because space is not homogeneous. It is more surprising, however, that the topological relationships among such locations (e.g. whether they are in the same vicinity) often have stronger control over spatial behavior than do metric relationships (e.g. their absolute distance in Euclidean space). It is therefore possible that spatial representations can be both topological, therefore affording relatively unstructured information about the connectivity within space (for example, place A is directly connected to place B but not to place C), and metric (for example, place A is a certain distance and direction from place B), affording more detailed information about specific relationships among places (Poucet, 1993). The advantage of this dual format is that topological information is more rapidly acquired than metric information because metric encoding would largely rely on motion-related signals provided by repeated movements between places. However, direct empirical support for a dissociation between topological and metric encoding of spatial information is still awaited.

NEUROBIOLOGY OF SPATIAL COGNITION

Since rats remember various aspects of their spatial environment, it seems natural to look for the brain processes that underlie this ability. One locus that has received much attention in the recent years is the hippocampal formation, a structure lying below the cortex in mammals. Although other parts of the brain are known to participate in spatial behavior, the role of the hippocampus has been demonstrated on many occasions with lesion experiments. Thus, removal of the hippocampus induces dramatic and permanent deficits in a wide variety of spatial abilities. Rats with such lesions are impaired in the water maze navigation task and have an impaired spatial memory in the radial maze. Their spatial patterns of exploration are also strongly altered following hippocampal damage, and they fail to detect spatial changes in a familiar environment.

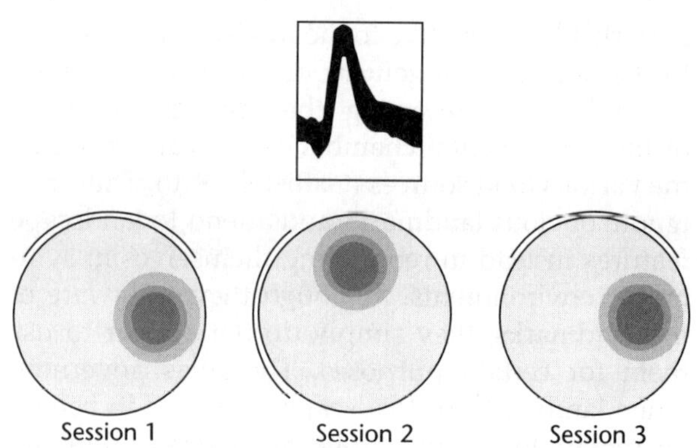

Figure 3. Firing rate maps of a hippocampal place cell 047
during a cue rotation experiment. The cell activity is characterized by the production of action potentials whose waveforms (shown in the inset) are its signature. The cell was recorded for three successive sessions during which the rat freely moved in a cylindrical apparatus one meter in diameter (the outer circle). The firing field of the cell is shown as a set of smaller concentric circles. Darker shading indicates increasing activity of the place cell. The only available visual cue was a white cue card attached to the inner wall of the cylinder (shown as an arc). When the cue card was rotated 90° counterclockwise from the first to the second recording session, the cell firing field also rotated 90°. When the cue card was returned to its original location, the firing field also returned to its initial position. This shows that the cue card has control over the position of the field.

Spatial Signals in the Rat Brain

The critical evidence for the spatial function of the hippocampal formation is the existence of cells that carry a spatial signal. Such cells can be classified as 'place cells' and 'head direction cells'. (*See* **Hippocampus**)

Place cells

Place cells were first identified in the hippocampus of rats. With appropriate methods, it is possible to record the firing of a single hippocampal neuron while tracking the position of a rat as it moves freely inside a circular arena. It is then possible to display the spatial firing of the neuron by plotting a map of the number of action potentials fired by the neuron per time unit and in each location in the environment. The activity of many pyramidal cells in the dorsal hippocampus is strongly correlated with the rat's position (O'Keefe, 1976). A given place cell is virtually silent except when the animal is in a region called the firing field (Figure 3). Each place cell has its own specific firing location. Place cell discharge is often independent of the direction faced by the rat and varies only with location.

Two properties of place cells are important. First, when rats are placed in new surroundings, place cells become progressively active while the rat is at a given location and, once established, the locations of their firing fields are stationary for weeks and months. Second, place cell firing fields are controlled by environmental visual cues. For example, a rotation of these cues induces a corresponding rotation of firing field. Additional experiments show, however, that this control is more complex than a mere sensory triggering. Thus, if the landmark is removed, place cells usually display firing fields remarkably similar to those observed when the cue is present. This property suggests that these cells encode information about locations in the environment rather than information about sensory views of the environment (Muller, 1996). (*See* **Navigation and Homing, Neural Basis of**; **Place Cells**)

Head direction cells

Head direction cells are primarily found in the postsubiculum. However, cells with similar properties have been found in other brain areas such as the anterior and lateral dorsal nuclei of the thalamus as well as in specific cortical areas. All these regions have extensive connections with the hippocampus. The firing pattern of head direction cells depends only on the heading of the animal, independently of its location. Each head direction cell has its own specific preferred firing direction. Both head direction cells and place cells share many properties, including being under the control of visual and idiothetic (motor-related) inputs. As for place cells, the activity of head direction cells is maintained when the environmental cues are removed. Their activity is therefore not simply visually triggered. Also, head direction cells do not function independently of each other. Changes in activity of a given cell are accompanied by corresponding changes in activity of other cells, which suggests that they are part of a tightly connected functional neural network. This network includes place cells since their firing fields react in the same manner to environmental manipulations as head direction cells. Thus, the two types of cells have access to the same information, and appear to form a tightly connected functional neural network whose function is to provide the animal with information about both its location and its heading. Their cooperative function is understood to allow the rat to navigate efficiently in the current environment.

Spatial Signals in the Primate Brain

Interestingly, cells that carry somewhat similar spatial signals have been found in the hippocampus of nonhuman primates, suggesting a comparable function in higher species. In addition, evidence based on functional neuroimaging of brain activity during navigation in familiar yet complex virtual-reality environments suggests that the human hippocampus also has a special role in spatial navigation. Activation of the right hippocampus is strongly associated with knowing accurately where places are located and navigating between them (Maguire *et al.*, 1998). In addition, specific regions of the hippocampus are found to be enlarged in London taxi drivers, who need to make extensive use of their navigation abilities. This finding bears a direct relation to animal studies and supports the hypothesis that the role of the hippocampal formation in spatial memory should be extended to other mammalian species including humans.

Contribution of the Cerebral Cortex

Clear-cut spatial deficits are found after specific cortical lesions. Rats with parietal cortical lesions are impaired in maze learning, place navigation, spatial working memory and response to spatial novelty, while having no gross visual impairment. However, the magnitude of these deficits is usually smaller than that produced by hippocampal damage. Thus, although the involvement of the parietal cortex in spatial processing is undoubted, its specific contribution is not yet understood. The situation is complicated because other cortical areas are important in spatial processing, as shown by lesion and electrophysiological studies. For example, damage to the posterior cingulate cortex produces a severe spatial deficit in place navigation, even though spatial memory does not seem to be affected. Lesions of the medial frontal cortex result in spatial navigation impairments, although animals are not impaired in their response to spatial novelty. Rather, their deficit is best explained as caused by impaired working memory, which would preclude them from appropriately planning their trajectories. Further work will be necessary, however, to understand how these different cortical areas interact with the hippocampal formation to provide the organism with a spatial representation useful for navigation.

CONCLUSION

Although the navigation capabilities of mammals are impressive, they are not unique since they are found in many other species as well. Consider, for instance, the food-storing birds. During autumn, they can store seeds in hundreds of locations scattered throughout their home range, and yet retrieve them several months after the storing episode. Similarly, the homing ability of pigeons is a well-described behavioral system that allows birds to orient efficiently within huge territories. Even phylogenetically lower species such as bees have outstanding spatial navigation abilities. Do all these organisms rely on a cognitive map to find their way in their environment ? Answering this question would first require that cognitive maps are unambiguously defined. While the literature is replete with the concept, there is still uncertainty about what it means exactly. In fact, while there is little doubt that many species are able to construct internal models of their environment, the spatial extent of such representations is still a questionable issue. For example, are animals able to represent remote portions of space and to use new routes when there is no overlap in the perception of the landmarks available at the origin and goal of the trajectory? So far, this ability seems to be specific to humans. However, uncovering the premises of such an ability in nonhuman animals is likely to help us understand how our brain creates complex representations of the world.

References

Gallistel CR (1990) *The Organization of Learning*. Cambridge, MA: MIT Press.

Maguire EA, Burgess N, Donnett JG *et al* (1998) Knowing where and getting there: a human navigation network. *Science* **280**: 921–924.

Morris RGM (1984) Developments of a water-maze procedure for studying spatial learning in the rat. *Journal of Neuroscience Methods* **11**: 47–60.

Muller RU (1996) A quarter of century of place cells. *Neuron* **17**: 813–822.

O'Keefe J (1976) Place units in the hippocampus of the freely moving rat. *Experimental Neurology* **51**: 78–109.

Olton DS (1979) Mazes, maps and memory. *American Psychologist* **34**: 583–596.

Poucet B (1993) Spatial cognitive maps in animals: new hypotheses on their structure and neural mechanisms. *Psychological Review* **100**: 163–182.

Thinus-Blanc C, Bouzouba L, Chaix C *et al.* (1987) A study of spatial parameters encoded during exploration in hamsters. *Journal of Experimental Psychology: Animal Behavior Processes* **13**: 418–427.

Tolman EC (1948) Cognitive maps in rats and men. *Psychological Review* **55**: 189–208.

Further Reading

Clayton NS (1998) Memory and the hippocampus in food-storing birds: a comparative approach. *Neuropharmacology* **37**: 441–452.

Hermer L and Spelke E (1994) A geometric process for spatial orientation in young children. *Nature* **370**: 57–59.

McNaughton BL, Barnes CA, Gerrard JL *et al* (1996) Deciphering the hippocampal polyglot: the hippocampus as a path integration system. *Journal of Experimental Biology* **119**: 173–185.

O'Keefe J and Nadel L (1978) *Hippocampus as a Cognitive Map*. Oxford, UK: Clarendon.

Poucet B, Save E and Lenck-Santini PP (2000) Sensory and memory properties of place cells firing. *Reviews in the Neurosciences* **11**: 95–111.

Roitblat HL (1987) *Introduction to Comparative Cognition*. New York, NY: Freeman.

Taube JS (1998) Head direction cells and the neurophysiological basis for a sense of direction. *Progress in Neurobiology* **55**: 225–256.

Thinus-Blanc C (1996) *Animal Spatial Cognition: Behavioral and Neural Approaches*. Singapore: World Scientific.

Trullier O, Wiener SI, Berthoz A and Meyer JA (1997) Biologically based artificial navigation systems: review and prospects. *Progress in Neurobiology* **51**: 483–544.

Wallraff HG (1996) Seven hypotheses on pigeon homing deduced from empirical findings. *Journal of Experimental Biology* **199**: 105–111.

Anomalous Monism

Intermediate article

Louise M Antony, Ohio State University, Columbus, Ohio, USA

CONTENTS

According to the theory of anomalous monism, every individual mental event is identical with some physical event, but there are no strict laws relating physical kinds to mental kinds.

INTRODUCTION

Anomalous monism (AM) is a theory of the mind–body relationship developed by Donald Davidson (1970, 1974, 1980). The theory states that while every individual mental event is identical with some physical event, there are no strict laws relating physical kinds to mental kinds.

HISTORY

AM grew out of Davidson's attempt to defend a causal theory of intentional action. In Davidson (1963) he considered the question of how appeal to an agent's reasons could serve as an explanation of the agent's actions. Earlier accounts of these 'rational explanations' (or, as Davidson referred to them, 'rationalizations') had it that the explanatory power of such appeals lay entirely in their revealing the action performed to be one that was reasonable for the agent. Advocates of such accounts argued that rational explanations could not be causal in character, for two reasons: (1) the normative element in rational explanation is not present in nonteleological causal explanation; and (2) rationalizations involve mere redescriptions of the action explained, instead of citing prior events contingently connected to the action explained.

Davidson's main objection to such views was that they could not adequately account for the force of the 'because' in rational explanation. A belief and desire of mine may make it reasonable for me to perform a certain action, but something more is needed for it to be true that it is because of the belief and desire that I do it. This 'something more', Davidson argued, could only be a causal connection.

It remained for Davidson to answer objections to the suggestion that reasons were causes. In addition to the objections noted above, there was the following: causation involves laws, and there are no laws relating beliefs and desires to the behavior they rationalize. Davidson's response contained the germs of the theory of anomalous monism: for a law to back a singular causal claim, it is not necessary that the law describe the subsumed events in the same vocabulary as is used in the causal claim. This set the stage for anomalous monism: Davidson would accept his opponents' contention that there was an irreducibly normative element in rational explanations, but would argue that this counted only against there being psychological laws, not against psychological explanation's being causal.

AM is one version of what is now known as 'nonreductive materialism', a view of mind that emerged in the early 1970s, partially in reaction to the 'strong' or 'type–type' reductionism of such philosophers as U. T. Place (1956) and J. J. C. Smart (1962), who argued that mental state types could be identified with, and hence reduced to, neurophysiological state types. Davidson is widely credited with one set of arguments (to be discussed below) against the possibility of such strong reductions for at least one important class of mental states, namely propositional attitude states. A different kind of argument against strong reductionism for mental states came from Hilary Putnam (1967), who argued that, because mental states were functional states, they could be 'multiply realized', that is, realized in a variety of physically different systems: pain might be a matter of c-fibres firing in humans, but of something altogether different in a Martian. (See also Fodor, 1974.) Putnam's arguments were designed to show that there could be no physical necessary conditions for the instantiation of a mental state type; Davidson's were meant to show that there could be no physical sufficient conditions. Both views are properly classed as 'token-reductionist' materialist theories, for according to both, it makes sense to identify individual mental events with physical events.

ARGUMENTS FOR ANOMALOUS MONISM

Davidson's (1970) argument for AM was that the theory provided an attractive solution to an important puzzle, namely, how the following three principles could all be true together:

Principle of causal interaction (PCI): At least some mental events interact causally with other events.
Principle of the nomological character of causality (NCC): Events related as cause and effect fall under strict deterministic laws.
Principle of the anomalism of the mental (AOM): There are no strict deterministic laws on the basis of which mental events can be predicted and explained.

According to anomalous monism, every mental event is identical to some physical event. In fact, all it is for an event to be a 'mental event' is for there to be a true description of the event in mentalistic terms: the same event could be just as accurately described using purely physicalistic terminology. Thus, every interaction between a mental event and a physical event is equally an interaction between two physical events. This secures PCI. To reconcile PCI with NCC and AOM, Davidson argues that there is a crucial difference between causation and subsumption under laws. The former is an extensional relation, holding of events however described, whereas the latter is an intensional relation, sensitive to the way in which events are picked out. AOM can then be understood as saying that there are no strict deterministic laws covering events described mentalistically, or that there are no such laws utilizing mentalistic predicates. Interpreted in this way, AOM can be seen to be consistent with NCC, which simply states that any pair of causally related events is subsumed under some law or other. If we assume, as Davidson seems to, that all physical predicates are apt for use in the statement of strict, deterministic laws, then the token-identity of every mental event with some physical event suffices to ensure that whenever a mental event interacts causally with a physical event, there is a way of describing the events so that they are covered by an appropriate law.

Perhaps the most controversial of the three principles is AOM (although NCC is also open to challenge – see below). Davidson offers brief and somewhat enigmatic remarks in its defence. Broadly, there are two arguments. The first is that psychological ascriptions answer to different kinds of evidence, and are constrained by different kinds of norms, from nonpsychological claims. As a result, Davidson (1970) says, 'mental and physical predicates are not made for each other'. Since laws can only 'bring together' predicates that are made for each other, there can be no psychophysical laws. The second argument is based on the 'holistic character of the cognitive field' (Davidson, 1974). According to Davidson, the content of a belief or desire depends on its role in the overall pattern of the agent's psychology and behavior. This makes it impossible for there to be local sufficient conditions for the correct ascription of a propositional attitude.

Possibly Davidson has something like the following arguments in mind. The norms that govern mental predicates are the norms of cogency and rationality: we are constrained to attribute beliefs and desires to agents in such a way as to make agents' behavior appear (as much as possible) rational, or else we will fail in ascribing psychological states at all – we will be 'changing the subject' (Davidson, 1970). Since physical predicates are not subject to this normative constraint, the possibility of laws connecting the mental to the physical entails the possibility of competition between the evidence relevant to the physical attributions and the considerations of 'overall cogency' appropriate to the psychological realm. If we could know, for example, that all *x*-neuron firings were, as a matter of law, beliefs that it is raining, then the physical evidence that someone's *x*-neurons were firing might warrant the attribution of this belief even if that attribution was not supported by the kinds of behavioral evidence we otherwise rely on, and even if the belief did not cohere properly with others of the agent's mental states. Moreover, if such a law were available, then we would have a localistic basis for the attribution of a belief with the content 'it is raining', in violation of the supposition that what constitutes a mental state's having a particular content is its playing a certain role in the agent's overall patterns of thought and action. Such a situation, Davidson thinks, would amount to the abandonment of our conception of ourselves as rational agents: 'we must conclude that nomological slack between the mental and the physical is essential as long as we conceive of man as a rational animal' (Davidson, 1970).

CRITICISMS OF ANOMALOUS MONISM

There are three major objections to Davidson's theory. The first concerns the conception of causation on which Davidson's original puzzle depends. NCC expresses a commitment to a regularity theory of causation. But there are competing accounts, according to which singular causal claims need not be 'backed' by general laws. If one adopts one of these alternative accounts of causation, then the original problem that was supposed to motivate anomalous monism is dissolved.

The second objection is that the theory does not, in fact, solve the problem of accounting for the explanatory force of rational explanations, because it leaves the rationalizing part of the explanations detached from the causal part. That is, because, according to AM, there is no possibility of relating, in a systematic way, the physical properties of a mental event with its mental properties, and because it is only the physical properties of the mental event that figure in causal laws, the mental properties appear to be causally inert, or 'epiphenomenal'. Davidson's response to these objections can be found in Davidson (1993). (That essay and many others on the issue of the causal relevance of mental properties can be found in Heil and Mele, 1993.) The issues raised in this connection led to a general debate as to whether the mental properties of mental events are causally relevant to the production of action.

The third objection challenges the adequacy of the arguments for AOM. Critics contend that if agents are in fact rational – and the degree to which they are is, arguably, an empirical matter – then the constraints of 'overall cogency' do in fact come into play in determining the adequacy of a proposed physical account of propositional attitude states. For suppose that we have some candidate reductive account of beliefs and desires. If we have independent justification for thinking that agents' behavior is usually reasonable given their beliefs and desires, then there will be a constraint on the adequacy of our candidate reduction theory that it have the consequence that the beliefs and desires we attribute on the basis of purely physical evidence coincide with the ones that we would attribute on the basis of considerations of overall cogency. There is thus no need to fear a competition between norms and evidence pertinent to the respective domains. (See Antony, 1989 and Van Gulick, 1980.)

A number of essays containing critical discussion of AM and the arguments for it can be found in McLaughlin and LePore (1985).

ANOMALOUS MONISM AND COGNITIVE SCIENCE

An important philosophical question in cognitive science is whether the concepts and generalizations of our ordinary mentalistic talk – 'folk psychology' – can be preserved and explained within a scientific psychological theory. If Davidson's theory is correct, then the prospects for such a scientific vindication of our pretheoretic notions seem poor.

Consider, for example, the computational model of mind. According to this view, psychological states are functional states of the brain involving physically realized mental representations. Cognitive processes like rational deliberation are computations defined over these representations. In this way, the causal transactions within the agent's brain can be shown to mirror rational relations among the agent's mental states. If this computational model is accurate, then there would be lawful connections between physical state-types of the brain and mental state-types, of the sort that would support inferences, at least in principle, from local physical evidence to conclusions about an agent's beliefs and desires. And this, it appears, would be incompatible with AOM.

If one rejects the computational model of the mind in favor of, say, a connectionist architecture, then the relevance of AM will depend on whether one thinks that the alternative architecture permits a type reconstruction of propositional attitude states.

However, Davidson is not altogether clear about whether AOM is supposed to rule out any kind of nomic connection involving mental types, or only strict (i.e., exceptionless) nomic connections. The former claim seems to be what the arguments in (Davidson, 1970) mean to establish; if the claim is weakened in the latter way, then the thesis seems considerably less interesting, and AM itself would appear to amount to little more than the claim that psychology is not a basic science. See Davidson (1993) for further discussion.

References

Antony L (1989) Anomalous monism and the problem of explanatory force. *Philosophical Review* **98**: 153–187.

Davidson D (1963) Actions, reasons, and causes. *Journal of Philosophy*, **60**: 685–700. [Reprinted in Davidson, 1980.]

Davidson D (1970) Mental events. In: Foster L and Swanson J (eds) *Experience and Theory*. Amherst, MA: University of Massachusetts Press/Duckworth Press. [Reprinted in Davidson, 1980.]

Davidson D (1974) Psychology as philosophy. In: Brown S (ed.) *Philosophy of Psychology*. New York, NY: Macmillan Press/Barnes, Noble. [Reprinted in Davidson, 1980.]

Davidson D (1980) *Essays on Actions and Events*. Oxford: Oxford University Press.

Davidson D (1993) *Thinking causes*. In: Heil J and Mele A (eds) *Mental Causation*. Oxford: Oxford University Press.

Fodor JA (1974) Special sciences. *Synthese* **28**: 97–115.

Heil J and Mele A (eds) (1993) *Mental Causation*. Oxford: Oxford University Press.

McLaughlin BP and LePore E (eds) (1985) *Action and Events*. Oxford: Blackwell.

Place UT (1956) Is consciousness a brain process? *British Journal of Psychology* **47**(1): 44–50.

Putnam H (1967) The nature of mental states. In: Capitan WH and Merrill DD (eds) *Art, Mind, and Religion*. Pittsburgh, PA: University of Pittsburgh Press.

Smart JJC (1962) Sensations and brain processes. In: Chappell VC (ed.) *The Philosophy of Mind*. Englewood Cliffs, NJ: Prentice Hall.

Van Gulick R (1980) Rationality and the anomalous nature of the mental. *Philosophy Research Archives* **7**: 1404.

Further Reading

Antony L (1994) The inadequacy of anomalous monism as a realist theory of mind. In: Preyer G *et al.* (eds) *Language, Mind and Epistemology*, pp. 223–253. Dordrecht: Kluwer.

Campbell N (1997) The standard objection to anomalous monism. *Australasian Journal of Philosophy* **75**: 373–382.

Davidson D (1973) The material mind. In: Suppes P, Henkin L, Mosil GC and Joja A (eds) *Logic, Methodology and the Philosophy of Science*. Amsterdam: North-Holland. [Reprinted in Davidson, 1980.]

Evnine S (1991) *Donald Davidson*. Stanford, CA: Stanford University Press.

Honderich T (1982) The argument for anomalous monism. *Analysis* **42**: 59–64.

Johnston M (1985) Why having a mind matters. In: McLaughlin and LePore, 1985.

Kim J (1985) Psychophysical laws. In: McLaughlin and LePore, 1985.

Latham N (1999) Davidson and Kim on psychophysical laws. *Synthese* **118**: 121–144.

LePore E and Loewer B (1987) Mind matters. *Journal of Philosophy* **84**: 630–642.

Lycan WG (1981) Psychological laws. *Philosophical Topics* **12**: 9–38.

McDowell J (1985) Functionalism and anomalous monism. In: McLaughlin and LePore, 1985.

McLaughlin BP (1985) Anomalous monism and the irreducibility of the mental. In: McLaughlin and LePore, 1985.

Patterson SA (1996) The anomalism of psychology. *Proceedings of the Aristotelian Society* **96**: 37–52.

Anosognosia

Introductory article

Alfred W Kaszniak, University of Arizona, Tucson, Arizona, USA

CONTENTS

Anosognosia is a clinical syndrome in which a patient with brain dysfunction appears unaware of a neurological or neuropsychological impairment, such as paralysis or amnesia.

INTRODUCTION

In 1914, Joseph Babinksi described several patients who were paralyzed on the left side of their body due to damage in the right hemisphere of the brain. Most startling was his observation that these patients appeared to have no awareness of their paralysis and never complained about their impairment. Although others had made similar observations in the late 1800s and early 1900s, it was Babinski who coined the term 'anosognosia' to describe this syndrome. Subsequently, the term has been extended to encompass similar unawareness of deficit phenomena in persons suffering from a variety of neurological impairments. Anosognosia has been documented for deficits as diverse as blindness, impaired recognition of familiar faces, hemiplegia (paralysis of one side of the body), and aphasia (acquired disorder of language comprehension and expression).

Historically, there has been debate about how best to conceptualize anosognosia. In their influential book entitled *Denial of Illness*, published in 1955, neurologist Edwin Weinstein and psychologist Robert Kahn argued that anosognosia is a motivated unawareness, reflecting the operation of psychodynamic defense mechanisms that block symptoms or deficits from awareness. Although this conceptualization is intuitively appealing, neurologist and neuroscientist V. S. Ramachandran has recently noted that the psychodynamic interpretation does not account for two important aspects of anosognosia. First, anosognosia typically occurs only when there is damage to particular brain structures. These structures include lower portions of the parietal cortex, frontal lobes, and the subcortical structures connecting to these structures, and anosognosia is seen predominantly following right, rather than in left hemisphere brain damage. Anosognosia is typically not observed when peripheral, spinal, lower brain stem, or cortical primary sensory or motor areas alone are affected. Second, the unawareness is often quite specific. For example, a patient may deny his/her hemiplegia but readily admit to other disabling or distressing symptoms. Occasionally, persons with hemiplegia of both their upper and lower extremities may admit that their leg is paralyzed but insist that their arm is not. In contrast to earlier psychodynamic conceptualizations, most current theories of anosognosia adopt a neuropsychological explanation, attributing the impaired awareness of deficit to dysfunction in brain circuits important for self-monitoring. In a frequently cited review paper published in 1989, neuropsychologists Susan McGlynn and Daniel Schacter emphasized the specificity and dissociations (i.e. intact awareness for some impairments and impaired awareness for other impairments) found in different forms of anosognosia in individuals with focal neurological damage. Based on their review of clinical-pathological correlation studies, they concluded that damage to right parietal and/or frontal brain regions is of particular importance in causing anosognosia.

CLINICAL STUDIES OF ANOSOGNOSIA FOR AMNESIA AND DEMENTIA

Some of the most informative recent research for understanding the neuropsychology of anosognosia comes from studies of amnesia and dementia. Amnesia refers to an acquired loss of memory. Although amnesia may be due to psychological factors, such as intensely frightening personal experiences, it is most often due to neurological causes. In 1889, the physician S. S. Korsakoff described the amnesic disorder that is now called Korsakoff's syndrome and observed that most of his patients had little apparent awareness of their memory deficits. Korsakoff's syndrome most typically occurs in persons with a history of chronic alcohol abuse and thiamine deficiency, with postmortem examinations of the brain revealing damage in the dorsomedial nuclei of the thalamus, the mammillary bodies, and other nearby brain structures. Computerized tomographic (CT) and magnetic resonance imaging (MRI) studies have confirmed the presence of thalamus and mamillary body damage in persons with Korsakoff's syndrome and have also revealed damage to the orbitofrontal and mediotemporal areas of the cerebral cortex. In addition to severely impaired memory (thought to be associated with the thalamic, mamillary body, and mediotemporal cortex damage), persons with Korsakoff's syndrome also show other cognitive deficits, such as impairments in shifting and dividing attention, in performing tasks requiring hypothesis generation, testing, and problem solving, and in preserving the temporal order of information. These deficits appear to be associated with the frontal cortical damage.

In contrast to persons with Korsakoff's syndrome, individuals with equally severe memory impairment, due to restricted medial temporal (particularly hippocampus) damage, do not show frontally related cognitive dysfunction and are generally well aware of their memory deficits. Persons with amnesia due to medial temporal damage may spontaneously comment about their memory difficulty and rely upon mnemonic aids such as reminder notes and schedule books. Thus, amnesic patients without frontal lobe damage appear to have intact awareness of their memory deficits, while those with frontal damage (i.e. those with Korsakoff's syndrome) are anosognosic for their amnesia.

The study of persons with missile wound brain injuries has supported the association of impaired awareness of memory deficit with frontal lobe damage. Those who lack awareness of their memory impairment have been found to have damage involving both frontal lobes (among other areas), whereas those with intact awareness of their memory deficits show no evidence of frontal lobe damage. Other persons with amnesic syndromes due to brain damage that includes the frontal lobes (e.g. anterior communicating artery aneurysm rupture, closed head injury) have also been found to be anosognosic for their memory deficits.

Dementia refers to progressive impairment of multiple cognitive functions (including memory) due to acquired brain disease. Clinical accounts of anosognosia in dementia syndromes began appearing in the early 1980s. The most prevalent cause of dementia is Alzheimer disease (AD), a neurodegenerative disorder with neuron loss and other microscopic brain changes that are most prominent in the medial temporal, posterior temporal, parietal, and frontal brain regions. Persons with AD show relatively severe memory impairment early in the course of their dementia, along with milder deficits in other cognitive functions (e.g. perception, language, judgment). These other cognitive functions become increasingly impaired with disease progression. The memory and other cognitive impairments correlate with the pattern of observed neuropathological changes, particularly involving medial temporal and frontal brain regions. Many persons with AD, particularly as their illness progresses, show impaired awareness of their memory and other cognitive deficits.

Impaired awareness of deficits has also been noted in clinical descriptions of other types of dementia besides AD. These include vascular dementia (typically associated with multiple and bilaterally distributed small cerebral strokes), frontotemporal dementia (presenting with impaired impulse control, poor social judgment, and cognitive deficit, and typically involving severe atrophy of frontal and temporal cortices), and Huntington's disease (an autosomal dominant genetic neurodegenerative disorder characterized by motor, cognitive, and emotional dysfunctions associated with caudate nucleus atrophy and frontal lobe changes).

AD and other dementia types involve impairments in a wide range of cognitive, behavioral, emotional, and functional areas. Thus, the study of deficit awareness in dementia provides an opportunity for examining the degree to which impaired awareness is general or limited to particular deficits. Clinical observations have suggested that awareness of deficit can be selective in dementia. For example among persons with AD some deny all cognitive deficits, while others claim that their memory is good but admit to difficulty in finding words, and still others deny memory impairment but admit to reading difficulties.

Some clinical observers have suggested that loss of awareness of deficits occurs earlier in the course of illness for frontotemporal dementia than for AD. Both of these progressive dementia types are typically associated with signs of frontal lobe pathology, but frontal damage is generally more severe in the early stages of frontotemporal dementia than AD. This is consistent with the association between frontal lobe pathology and impaired awareness of deficit that has been observed in amnesic syndromes.

Studies of the neuropsychological correlates of anosognosia for dementia in AD have also supported a relationship to frontal lobe damage. For example, an association has been demonstrated between decreased fluency of speech (as measured in tests requiring the generation of words within particular categories or beginning with particular letters) that is typically seen following left or bilateral frontal cortex damage, and clinical ratings of anosognosia for memory deficit in persons with AD. Similarly, anosognosia for dementia in AD has been shown to be associated with both lower general mental status test performance and specific impairment on measures of 'executive functions' (involving the capacity to plan and carry out complex, goal-oriented behavior) that are sensitive to frontal lobe dysfunction. More direct evidence of a relationship to frontal brain dysfunction is provided by research showing that clinical ratings of anosognosia for memory loss in persons with AD are associated with decreased blood flow (as measured by single photon emission computed tomography) in the right dorsolateral frontal brain region. (*See* **Alzheimer Disease**; **Amnesia**; **Huntington Disease**)

LABORATORY STUDIES OF ANOSOGNOSIA FOR AMNESIA AND DEMENTIA

Although important, clinical observations alone allow for only limited and tentative conclusions concerning both the neurobiological correlates and nature of anosognosia. There are several reasons for the inferential limitations of clinical observations. First, the methods employed in making clinical observations of impaired awareness (e.g. clinical rating scales, patients' responses to interview questions) are generally unsystematic and have varied across investigators. This creates problems for any attempt to compare findings across studies. Further, clinical observations alone do not allow for the development of any articulated theory of anosognisia. One goal of research on anosognosia is to make theoretical inferences concerning human metamemory functions. The term 'metamemory' refers to those processes involved in the conscious monitoring and control of, as well as beliefs about, one's own memory functioning. Theoretically important questions such as whether anosognosia represents inaccurate self-efficacy beliefs (e.g. a person's beliefs about how well their memory generally functions), poor self-monitoring, or some combination of these or other factors, cannot be adequately addressed by purely clinical observations. Recently, systematic laboratory studies have become available, allowing quantitative measurement of different aspects of awareness in amnesia and dementia. Most of these laboratory studies have focused on awareness of deficits in Korsakoff's and other amnesic syndromes or AD.

Laboratory approaches have typically used one of three different methods to study awareness of memory deficit: comparisons of patient self-report and others' ratings of patient disability; evaluation of the accuracy of patients' predictions for their performance on specific cognitive tasks; or feeling-of-knowing paradigms (e.g. confidence ratings or rankings regarding the likelihood that recently learned information or long-term knowledge which the individual failed to recall would later correctly be recognized from among multiple alternatives). These different methods provide information relevant to theoretical formulations concerning the nature of metamemory impairment. Psychologists Christopher Hertzog and Roger Dixon, reviewing metamemory research involving healthy adult participants, have distinguished three aspects of metamemory: (1) knowledge about how memory functions and the utility of different strategies in memory tasks; (2) memory self-monitoring, defined as awareness of the current state of one's own memory system; and (3) self-referent beliefs about memory (memory self-efficacy beliefs). In general, the first two methods described above (patient self-report versus others' ratings; performance prediction accuracy) have provided information relevant primarily to those aspects of metamemory concerning knowledge of how memory functions and memory self-efficacy beliefs. The third method (feeling-of-knowing paradigms) provides information that can be interpreted as more specifically relevant to questions about memory self-monitoring.

Results of studies using these different methods are consistent with the conclusion that anosognosia for amnesia occurs in memory-impaired persons who also have frontal lobe dysfunction. Further, evidence suggests that it is the memory self-monitoring aspect, rather than knowledge of how memory operates, that breaks down in disorders such as Korsakoff's syndrome and AD. However, the question of whether at least some of the relevant research findings may reflect a failure to update self-efficacy beliefs (due to the memory impairment itself), rather than impaired self-monitoring, awaits an answer from future research.

Prospective studies are needed to determine whether premorbid patient characteristics may be related to anosognosia. Further research is also needed to simultaneously compare different methods of assessing anosognosia, particularly contrasting those approaches relevant to theoretically distinct aspects of metamemory (e.g. memory self-monitoring versus memory self-efficacy beliefs). Finally, there is a need for additional studies concerning the practical implications of anosognosia for amnesia and dementia: for example relationships may exist between anosognosia for dementia and the tendency to engage in potentially risky behavior such as driving. It has been hypothesized that impairment in the ability to recognize cognitive and behavioral limitations may play a role in both driving and other risky behavior among persons with AD. Given the practical implications, research designed to systematically test this hypothesis is of high priority.

Further Reading

Duke LM and Kaszniak AW (2000) Executive control functions in degenerative dementias: a comparative review. *Neuropsychology Review* **10**: 75–99.

Hertzog C and Dixon RA (1994) Metacognitive development in adulthood and old age. In: Metcalfe J and Shimamura AP (eds) *Metacognition: Knowing about Knowing*. Cambridge, MA: MIT Press.

Kaszniak AW and Zak MG (1996) On the neuropsychology of metamemory: contributions from the study of amnesia and dementia. *Learning and Individual Differences* **8**: 355–381.

McGlynn SM and Schacter DL (1989) Unawareness of deficits in neuropsychological syndromes. *Journal of Clinical and Experimental Neuropsychology* **11**: 143–205.

Prigatano GP and Schacter DL (eds) (1991) *Awareness of Deficit after Brain Injury: Clinical and Theoretical Issues*. New York, NY: Oxford University Press.

Ramachandran VS (1995) Anosognosia in parietal lobe syndrome. *Consciousness and Cognition* **4**: 22–51.

Shimamura AP (1994) The neuropsychology of metacognition. In: Metcalfe J and Shimamura AP (eds) *Metacognition: Knowing about Knowing*. Cambridge, MA: MIT Press.

Weinstein EA and Kahn RL (1955) *Denial of Illness: Symbolic and Physiologic Aspects*. Springfield: Charles C. Thomas.

Antisocial Personality and Psychopathy

Introductory article

Peter R Finn, Indiana University, Bloomington, Indiana, USA

CONTENTS

Antisocial personality is a mental disorder that begins in childhood and continues into adulthood, and involves the persistent violation of the rights of others and social norms in general. Deficits in emotional, motivational, and cognitive processes contribute to the development of the disorder.

INTRODUCTION

Antisocial personality is a personality disorder that affects about 3% of men and 1% of women. Personality disorders are mental disorders that reflect fundamental problems in the early development of personality, resulting in enduring psychological and behavioral problems that begin in childhood or adolescence and continue into adulthood. Antisocial personality is characterized by a range of symptoms that involve difficulties controlling impulses, forming caring interpersonal attachments, learning from experience, and experiencing guilt or remorse. The specific symptoms of antisocial personality disorder include deceitfulness, stealing, destroying property, aggression, impulsivity, engaging in unlawful and other rule-breaking behavior, consistent irresponsibility, and a disregard for others. Conduct disorder is the childhood precursor to adult antisocial personality.

Although antisocial personality is defined as a specific disorder, there are different subtypes, such as primary and secondary psychopathy (or sociopathy). In primary psychopathy antisocial behavior is associated with a deficit in the capacity to experience negative emotions such as fear, anxiety, and guilt, and a superficial and manipulative manner of relating to others. Secondary psychopathy involves high levels of impulsivity and negative affect (guilt and anxiety) combined with a pattern of chronic antisocial behavior. Regardless of the variations in the pattern of symptoms, antisocial individuals demonstrate a failure to inhibit behavior that violates others, transgresses social norms, or increases the risk for generally negative consequences to self or others. Research indicates that emotional, motivational, and cognitive deficits all contribute to the inhibitory problems experienced by those with antisocial personality.

EMOTIONAL PROCESSES

The Role of Emotion in Self-control and Socialization

Because people generally want to avoid experiences that result in fear, anxiety, guilt, or shame, these emotional experiences play an important role in facilitating self-control and socialization. For instance, experiencing anticipatory anxiety when one considers engaging in a specific behavior that might lead to a negative outcome should bolster self-control and lead to an inhibition of that behavior. Also, a common method of socialization involves using the threat of punishment to manipulate fear and anxiety as a means of encouraging or enforcing conformity.

Emotion also has an important role in regulating interpersonal behavior through the experience of sympathy and empathy. The process of experiencing, or sharing, the negative emotion of another contributes to a sense of identification with that person. Since identification involves incorporating aspects of the other into one's own self construct, people are unlikely to do harm to someone with whom they identify because they would be symbolically doing harm to themselves in the process.

Deficits in the ability to experience fear, anticipatory anxiety, or negative emotions in general would compromise the processes that contribute to self-control, socialization, and the regulation of interpersonal behavior.

Emotional Deficits in Psychopathy

A key feature of primary psychopathy is a lack of fear and arousal. Research on the skin conductance response (SCR), a physiological measure of arousal, indicates that people with primary psychopathic disorder have a deficit in the ability to experience anxiety as a response to threatening stimuli. The SCR reflects changes in the level of electrical conductivity of the skin on the palm of the hand resulting from activity of the sweat glands, which are influenced solely by the sympathetic nervous system. In studies where SCRs are classically conditioned to stimuli that signal aversive events, such as electric shock, people with primary psychopathic disorder consistently show smaller SCRs when compared with those with secondary psychopathy and normal volunteers. Research also shows that they have smaller anticipatory SCRs when waiting for a noxious stimulus to be administered (Figure 1).

Another hallmark of primary psychopathy is a deficit in caring, loving, and showing empathy for others. People with this disorder show a peculiar pattern of emotional detachment, in which they can correctly label the emotions of another with words, but do not experience the emotion. Such people show blunted physiological responses to pictures of mutilated accident victims but can correctly describe the pictures as depicting unpleasant scenes. This same pattern of emotional detachment has been observed in patients who developed psychopathic traits after sustaining damage to the frontal lobes of the cortex of their brain.

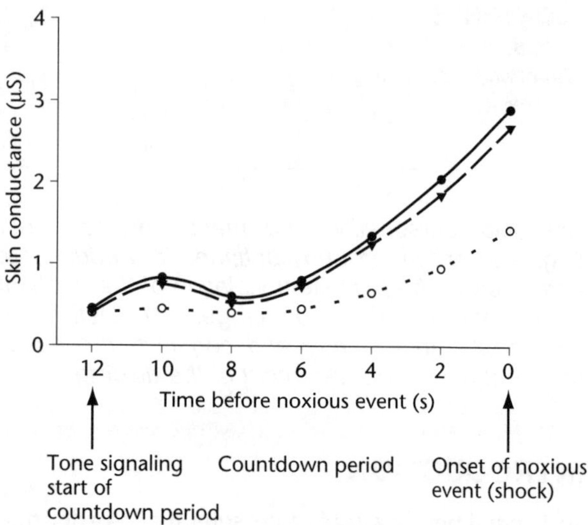

Figure 1. The typical pattern of anticipatory skin conductance while anticipating a noxious stimulus in people with primary psychopathy (open circles), secondary psychopathy (triangles), and normal controls (solid circles).

IMPULSIVITY

Impulsivity is associated with a wide range of problems, including antisocial personality, substance abuse, suicide, overeating, gambling, and aggressive behavior. Rather than indicating a single trait, impulsivity refers to a range of characteristics that include preferences for immediate versus delayed gratification, acting without thinking, acting quickly, paying more attention to the present than the future, having poor inhibitory control, and being overresponsive to reward. These characteristics promote behavior that often leads to unforeseen negative consequences to self and others, and behavior that disregards the rights, feelings, and safety of others in the service of immediate gratification for self. Both primary and secondary psychopathy are associated with impulsivity, but impulsive behavior is associated with different qualities and probable causes in the two subtypes of antisocial personality. In secondary psychopathy, impulsivity is often associated with negative emotions, cognitive deficits (lack of planning and future orientation), and poor inhibitory control. Impulsive behavior in primary psychopathy is probably due more to a lack of concern about negative outcomes and a greater interest in obtaining immediate gratification of impulses.

Laboratory studies indicate that when given the opportunity to choose between an immediate smaller monetary reward or a larger, but delayed, reward, individuals with antisocial or impulsive traits display a greater preference for the immediate reward and tend to discount the value of future rewards. Other studies indicate that impulsive personality traits and an inability to tolerate delay of gratification are associated with lower levels of brain serotonin, a neurochemical found in areas of the brain associated with emotion and basic motivational functions.

MOTIVATIONAL PROCESSES

Antisocial personality is associated with abnormalities in the motivational processes underlying approach and avoidance behavior. In this broad theoretical perspective, antisocial personality is the result of strong approach tendencies that override the inhibitory influences that affect avoidance behavior. For example, excessively strong desires for wealth, power, dominance, or sexual gratification (i.e. strong approach motivations) may override the normal inhibitory effects that social sanctions (e.g. arrest, humiliation, rejection) have on attempts to unlawfully gain these rewards. On the other hand, antisocial personality could also result from an insensitivity to punishment that results in weak inhibitory influences. In this scenario, individuals commit antisocial acts because they do not care about – or do not pay attention to – the possibility that their behavior may be punished. Research suggests that both perspectives are true: some with antisocial personality are less sensitive to punishment (weak avoidance), some are more sensitive to reward (strong approach), and still others have deficiencies in processing and attending to information about the potential for reward and punishment.

Response to Reward

The majority of studies of motivational deficits in antisocial personality use go/no go tasks where the person is required to learn when to make a response (go trial) and when to inhibit a response (no go trial). The studies manipulate the rewards and punishments for responses and nonresponses on each type of trial. Typically 'go' responses result in rewards (winning money) and 'no go' responses (failures to inhibit) result in punishment (losing money). In these types of laboratory studies antisocial individuals consistently show deficits in the ability to learn to inhibit their response on no go trials, termed a passive avoidance learning deficit. Figure 2 illustrates the typical pattern of passive avoidance deficits in antisocial personality.

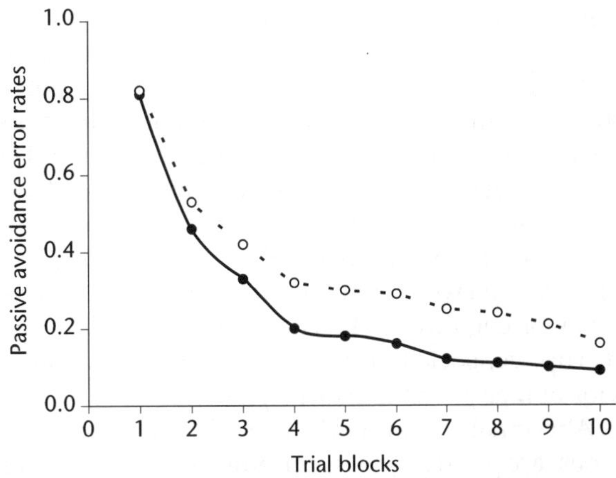

Figure 2. The typical pattern of passive avoidance learning for people with antisocial personality (open circles) and normal participants (solid circles) in a go/no go task where passive avoidance errors (responses on no go trials) result in losing money (e.g. 20 cents) and correct responses on go trials win the same amount of money. The learning curves reflect gradual reductions in the error rates within blocks of four no go trials.

Additional research indicates that these passive avoidance learning deficits are only apparent when the antisocial person is trying to obtain a reward, and when the reward stimuli are more salient (i.e. approach behavior is the dominant response). Other research indicates that antisocial individuals are more sensitive to rewards in general. Overall, these studies indicate that when actively seeking out rewards, those with antisocial personality have difficulty inhibiting their behavior to avoid negative consequences. In other words, they are less able to develop an optimum strategy for behavior that maximizes positive outcomes while minimizing negative consequences.

Response to Punishment

Studies that use aversive consequences, such as electric shock as punishments, indicate that people with primary psychopathic disorder are uniquely insensitive to punishment. Although few studies use noxious stimuli as punishments, these studies indicate that such people show substantial deficits in their ability to learn to avoid electric shocks. Rather than being associated with increased response to reward, insensitivity to noxious punishments is associated with a fearless, low harm-avoidant temperament.

COGNITIVE PROCESSES

Cognitive processes such as reflection, planning, and problem-solving have a central role in self-regulation and are important in the motivational deficits and impulsive traits of those with antisocial personality. When abilities to reflect, plan, or solve the problems of strategizing for behavior are deficient, the immediate situation tends to control a person's behavior. In such situations, people tend to act according to their strongest feelings, impulses, and concerns, and fail to consider the effect of their behavior on others or the future. These types of cognitive deficits are associated with the disruptive, socially deviant, and aggressive behavior of many with antisocial personality.

Abnormal, or underdeveloped, frontal cortical lobes are thought to contribute to the cognitive deficits observed in those with antisocial personality. Those with damage to the dorsolateral and orbital areas of the prefrontal portion of the frontal lobes (Figure 3), display both the antisocial behavior and cognitive deficits observed in antisocial personality. In fact, those with antisocial personality show evidence of cognitive deficits on the neuropsychological tests that were developed to measure the deficits resulting from frontal brain damage. Brain scanning studies also indicate that people with chronic antisocial behavior show evidence of abnormal frontal lobes.

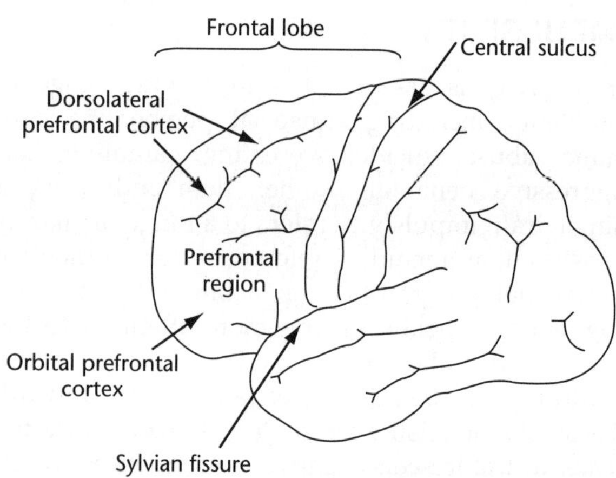

Figure 3. Lateral view of the human brain illustrating the location of specific regions in the frontal lobe of the left hemisphere. The frontal lobe is bounded by the central sulcus and the sylvian fissure.

Cognitive Deficits in Antisocial Personality

Deficits in verbal ability and executive cognitive functions are the most commonly reported cognitive abnormalities associated with conduct disorder and antisocial personality. Executive cognitive functions refer to the cognitive abilities involved in the planning and execution of effective goal-oriented behavior. Executive functions include abstract reasoning, planning, associative learning, problem-solving, attentional processes, self-monitoring, set-shifting, and working memory. Deficits in verbal ability and executive functions are associated with a greater severity in antisocial symptoms and poor outcomes in children with a history of antisocial behavior.

Verbal ability

Verbal and language deficits associated with antisocial traits include low levels of fluency (ease of finding words to label experience or concepts), limited vocabulary, poor language expressive skills, problems in understanding and following verbal instructions, reading problems, and poor concept formation.

Executive functions

Antisocial persons perform most poorly on laboratory tests of executive function that tap: (a) the ability to learn arbitrary stimulus–response associations and rules for behavior; (b) the ability to change behavior to adapt to changes in the rules for appropriate behavior (i.e. response set-shifting); and (c) the ability to plan behavior in complex tasks that require the development of strategies for effective performance. In addition, on complex decision-making tasks that tap these executive processes and manipulate rewards and punishments, both antisocial personality and damage to the orbital prefrontal cortex are associated with decisions that favor a larger immediate reward even when it leads to long-term losses. This disadvantageous decision-making style reflects a preference for immediate gratification and a lack of consideration for long-term consequences.

CONCLUSION

Antisocial personality is a disorder associated with a range of symptoms and apparent causes. The factors that contribute to the development of antisocial personality are deficits in the capacity to experience fear and anxiety, preferences for immediate versus long-term rewards, increased sensitivity to reward, deficits in the capacity to inhibit previously rewarded behavior that leads to long-term negative consequences, and cognitive deficits that compromise the ability to reflectively learn and develop effective strategies for goal-directed behavior.

Further Reading

Bechara A, Damasio AR, Damasio H and Anderson SW (1994) Insensitivity to future consequences following damage to human prefrontal cortex. *Cognition* **50**: 7–15.

Cleckley H (1982) *The Mask of Sanity*. St Louis, MO: Mosby.

Evenden JL (1999) Varieties of impulsivity. *Psychopharmacology* **146**: 348–361.

Fowles DC (1993) Electrodermal activity and antisocial behavior: empirical findings and theoretical issues. In: Roy JC, Boucsein W, Fowles DC and Gruzelier JH (eds) *Progress in Electrodermal Research*, pp. 223–237. New York, NY: Plenum Press.

Hare RD (1993) *Without Conscience: The Disturbing World of the Psychopaths Among Us*. New York, NY: Pocket Books.

Lykken DT (1995) *The Antisocial Personalities*. Hillsdale, NJ: Erlbaum.

Mazas CA, Finn PR and Steinmetz JE (2000) Decision making biases, antisocial personality, and early-onset alcoholism. *Alcoholism: Clinical and Experimental Research* **24:** 1036–1040.

Moffitt TE (1993) The neuropsychology of conduct disorder. *Development and Psychopathology* **5**: 135–151.

Newman JP and Schmitt WA (1998) Passive avoidance in psychopathic offenders: a replication and extension. *Journal of Abnormal Psychology* **107**: 527–532.

Patrick CJ, Bradley MM and Lang PJ (1993) Emotion in the criminal psychopath: startle reflex modulation. *Journal of Abnormal Psychology* **102**: 82–92.

Anxiety Disorders

Introductory article

Patrick A Palmieri, University of Illinois at Urbana-Champaign, Champaign, Illinois, USA
Wendy Heller, University of Illinois at Urbana-Champaign, Champaign, Illinois, USA

CONTENTS

Anxiety is an emotional state generally characterized by fear and worry. When anxiety levels become excessively high owing to a combination of biological, psychological and social influences, one or more anxiety disorders may be diagnosed.

INTRODUCTION

Anxiety is one of the most important topics in abnormal psychology. It is an emotional state that all of us experience at some point in our lives. For some individuals anxiety reaches excessive levels, resulting in significant impairment of functioning in several areas of life – occupational, academic and social. Depending on the specific symptoms shown, such psychopathology would probably be diagnosed as one or more of the anxiety disorders listed in the *Diagnostic and Statistical Manual of Mental Disorders*, 4th edition (DSM-IV). These psychological disorders are described primarily by maladaptive levels of anxiety and associated emotional responses such as fear, panic and worry. Several effective treatment options that alleviate the human suffering due to these conditions are available. These interventions are designed to target the biological, psychological or social aspects of the anxiety disorders. (*See* **Emotion**)

What is Anxiety?

Imagine you are walking across a stage to a podium where you will soon deliver a keynote address to a large audience of respected members of your community. How do you feel? What is going through your mind? In such a situation, most of us would experience one or more of the following: a sense of uneasiness, upset stomach, sweating, trembling, worry about the audience's reaction to your speech, rapid heartbeat, and other similar phenomena. These are all examples of symptoms of anxiety.

Anxiety *per se* is not a bad thing. It is an adaptive response: a low or even moderate level of anxiety would serve to motivate you to prepare for your speech. In evolutionary terms, an appropriate fear response to a threatening situation enhances the likelihood of taking an action that would promote survival. When anxiety levels become excessively high, however, they become maladaptive, often leading to persistent problems in various areas of life. For example, individuals who are deathly afraid of flying (a common phobia) may encounter difficulties in pursuing their career. (*See* **Evolutionary Psychology: Theoretical Foundations**)

Related Constructs

Any discussion of the nature of anxiety must address the related but distinct concepts of fear, panic, worry, obsessions and compulsions. Anxiety can be described in terms of categorical disorders, as defined by the DSM-IV. It can also be described in terms of symptom dimensions such as worry. Worry (anxious apprehension) is a future-oriented emotional response. It involves cognitive activity that is best described as uncontrollable negative thoughts about future potentially threatening events. Obsessions are repetitive, intrusive, anxiety-provoking thoughts. Whereas worry is typically brought on by everyday difficulties, obsessions come more 'out of the blue'. Compulsions are repetitive behaviors engaged in to reduce or avoid anxiety caused by obsessions. Fear (anxious arousal), which involves physiological hyperarousal, is an emotion experienced in the face of immediate danger that serves to prepare the individual to react adaptively to threatening situations. Panic, on the other hand, is essentially a fear response that can happen at an inappropriate time – that is, when no real danger is present. All of these constructs are combined in various ways in the anxiety disorders.

The phenomenon of anxiety is also closely linked to depression. In fact, there is some debate about whether they are distinct constructs, partly because of their high rate of co-occurrence. David Watson and Lee Anna Clark's tripartite model of anxiety and depression posits that negative affect – a general distress dimension – is a component of both anxiety and depression. This general distress may be responsible for the high comorbidity and also may signify a common etiology. Their model also proposes unique aspects of depression (low positive affect) and anxiety (physiological hyperarousal). Jack Nitschke and colleagues have found that negative affect is distinct from anxious arousal and anxious apprehension, both of which, according to their research, were robust factors in their own right. Thus, although depression and anxiety are commonly found together, it appears that each has separable components. (*See* **Depression**)

CLASSES OF ANXIETY DISORDER

History of Anxiety Classification

Sigmund Freud and his followers provided some of the earliest clinical descriptions of anxiety. Because they believed the underlying causes of pathological anxiety were similar, they did not place much value on differentiating types of anxiety. This conceptualization of anxiety persisted through the first two editions of the DSM, in which all forms of anxiety were lumped into one category called 'neuroses'. Over time, however, this method of classification was discontinued, because the psychoanalytic principles on which it was based fell from favor.

The DSM-III marked a major change in the classification of psychopathology. This classification system was based on clinically descriptive symptoms of abnormal psychology, rather than on unproved theories of their etiology. Thus, the category of neuroses was eliminated, and several new categories were formed based on clinical features. Several of these categories were contained under the newly created class of anxiety disorders. Further differentiation of the anxiety disorders is found in the fourth edition of the DSM. Its class of anxiety disorders contains several related but distinguishable psychological disorders.

Specific Phobia

Specific phobia involves a persistent and excessive fear elicited by the presence or anticipation of a particular object or situation. This is the most common anxiety disorder, occurring in approximately 9 percent of the adult population in any given year. Some common phobic objects or situations are spiders, snakes, flying and heights. This diagnosis requires that the individual actively avoid the stimulus, and if avoidance is not possible, that exposure to the object or situation evoke an immediate fear response. Importantly, the fear must be irrational. Fear of truly dangerous situations is adaptive, and therefore not abnormal. The fear and avoidance also must be of sufficient severity to disrupt performance in important areas of the person's life, such as at work or in personal relationships.

Social Phobia

Social phobia involves the fear and avoidance of social situations, such as speaking in public or attending a party. It is similar to specific phobia, but often involves performance (e.g. speaking in front of a group) and the fear of being evaluated negatively, causing embarrassment or humiliation.

Agoraphobia

Another form of phobic anxiety is agoraphobia, a fear of public spaces. Commonly feared situations of the agoraphobic individual are being in a crowded supermarket or theater, or traveling on public transport. The main source of fear is the feeling that one will be unable to escape the situation. People suffering from agoraphobia feel increasingly uncomfortable the farther they are from a safe place, such as home. In its most severe form, the individual might be unable to leave the house.

Panic Disorder

Panic disorder is defined primarily by the presence of recurrent and unexpected panic attacks. A panic attack is an overwhelming experience of fear that develops suddenly 'out of the blue', reaches peak intensity quickly, and is described mostly in terms of physical or somatic symptoms such as rapid heart rate, sweating, dizziness, nausea, trembling, chest pain or discomfort, or chills or hot flashes. Cognitive symptoms also may be present, such as thinking that one is dying or losing one's mind. A misinterpretation of somatic experiences may be largely responsible for panic attacks. For example, a rapid heart rate may be misinterpreted as an impending heart attack (even immediately following aerobic exercise, such as running up a flight of stairs), and this conclusion can trigger a panic attack. To meet the criteria for panic disorder, panic attacks must be followed by a significant period during which there is worry about the effects of a panic attack or the possibility of future panic attacks, or a change in behavior (e.g. avoidance) designed to prevent future attacks. Often, panic attacks are experienced in settings like those feared in agoraphobia. Because of this, there are two types of panic disorder: with agoraphobia and without agoraphobia.

Obsessive–Compulsive Disorder

Obsessive–compulsive disorder involves the presence of obsessions or compulsions (but usually both) that the individual recognizes as being excessive. The obsessions cause significant levels of distress and anxiety. Two common obsessional themes are contamination (e.g. disease, germs) and loved ones being involved in an automobile accident. Compulsions are repeated behaviors designed to reduce or avoid the anxiety caused by obsessions. One common compulsion is excessive hand-washing; in its most severe form, the individual washes so much that the hands start to bleed. Another common compulsion is excessive checking: for example, an individual might get out of bed dozens of times before falling asleep to check repeatedly that the front door is locked. An individual suffering from this disorder cannot stop engaging in these repetitive behaviors, even though attempts are made to resist them.

Generalized Anxiety Disorder

Generalized anxiety disorder is mainly characterized by excessive worry that is difficult to control and that precipitates impairment in one or more areas of life, such as personal, academic or occupational functioning. The worry must be related to numerous concerns and be present consistently for at least 6 months. In addition, the concerns should not be related solely to events associated with other anxiety disorders, such as having one's performance evaluated negatively, worrying about panic attacks, or fearing embarrassment in a social situation.

Posttraumatic Stress Disorder

Epidemiological studies have shown that most people experience or witness at least one traumatic event in their lifetime. A trauma is a stressful situation involving actual or threatened death or serious injury, or a threat to the physical integrity of self or others; it also must evoke in the individual a sense of intense fear, helplessness, or horror. Examples of such traumatic events are combat experiences, sexual assault and serious motor vehicle accidents. Sometimes such experiences result in a wide variety of physical and psychological symptoms. Posttraumatic stress disorder (PTSD) is diagnosed when those symptoms include, but are not necessarily limited to, reexperiencing or reliving the trauma (flashbacks, nightmares), persistent avoidance of stimuli associated with the trauma (avoiding people, places or activities that evoke memories of the trauma) and increased arousal (difficulty falling or staying asleep, exaggerated startle response). Once the symptoms appear, they must be present for at least 1 month. Interestingly, the symptoms do not always appear immediately after the traumatic experience; instead, onset may be delayed for months or even years. (*See* **Post-traumatic Stress Disorder**)

Acute Stress Disorder

Acute stress disorder is similar to PTSD but with two important differences. First, there is the additional symptom requirement of dissociation during or following the trauma. Dissociative symptoms include such experiences as being unable to recall important aspects of a trauma (dissociative amnesia), feeling detached from or an outside observer of one's own body (depersonalization), perceiving the external world as unreal or dreamlike (derealization), experiencing a reduction in awareness of surroundings, and having a sense of emotional detachment or numbing. Second, symptoms must start within 4 weeks of the traumatic event and must last no longer than 4 weeks. If they persist, then the diagnosis would be converted to PTSD.

ETIOLOGY OF ANXIETY

Anxiety is a complex biopsychosocial phenomenon and thus has no single type of causative factor. In some cases anxiety disorders originate in some stressful life event or series of events that is experienced by the individual as unpredictable or uncontrollable. These events can include social situations that involve fear or perceived physical or emotional threat. The presence of such events interacts with psychological and biological factors to lead to the development and maintenance of anxiety disorders.

Psychological Factors

An abundance of evidence indicates that phobias can develop through classical conditioning, a learning mechanism described by Ivan Pavlov. This is the process in which a previously neutral stimulus, when paired with an unconditioned stimulus (US) that evokes fear (unconditioned response, UR), can acquire the ability (conditioned stimulus, CS) to elicit the fear response (conditioned response, CR) even in the absence of the US. Martin Seligman proposed the idea of preparedness, which states that we are biologically prepared to learn certain CS–US associations very easily, sometimes with only one pairing, because there is evolutionary advantage in doing so. For example, relevant to specific phobias, humans seem hard-wired to develop a conditioned fear response to snakes. (*See* **Conditioning**)

Conditioning also appears to be critical for the development and maintenance of PTSD. Consider the example of a soldier in a fear-evoking combat situation. When stimuli such as loud sounds of machine-gun fire are repeatedly paired with the fear of being in combat, similar loud noises become able to elicit the fear response in noncombat contexts (classical conditioning). This fear is then maintained through operant conditioning, a learning theory introduced by B. F. Skinner, that states that behavior is a function of its consequences (i.e. the frequency of a behavior increases if it is reinforced and decreases if it is punished). Avoidance of fear-evoking situations reduces one's anxiety level. This decrease in anxiety reinforces, or makes more likely, the avoidance behavior, thereby maintaining the post-traumatic stress response. (*See* **Reward, Brain Mechanisms of**)

Information-processing models of anxiety identify cognitive biases as potential etiological and maintenance factors of anxiety disorders. A wealth of research using a variety of cognitive tasks shows that anxious individuals exhibit a selective attention towards threatening stimuli. This attentional bias may result in certain stimuli becoming more salient, thereby affecting the individual's assessment of the level of threat. An increase in perceived threat leads to heightened anxiety which, in turn, can further increase the salience of the stimuli and the anxiety level, in a self-perpetuating cycle. David Clark uses such a cognitive approach to model the etiology of panic disorder. An individual's heart might skip a beat, which might elicit an anxiety response (including the typical physiological symptom of increased heart rate), thereby making the heartbeat more salient, leading to the misinterpretation that a heart attack is imminent. This cycle may continue until it ultimately induces a panic attack. (*See* **Information Processing; Selective Attention**)

Genetic Factors

Several twin and family behavioral genetics studies have helped elucidate the role of genes in the development of anxiety disorders. In one such study Russell Noyes found that relatives of patients with panic disorder had higher levels of panic disorder than the general population, suggesting the possibility that a predisposition to this disorder is inherited. Relatives did not have elevated rates of generalized anxiety disorder, however, suggesting that these disorders have some unique etiological factors. A family study of obsessive–compulsive disorder conducted by Donald Black and colleagues suggested that what is inherited is a general predisposition toward developing any anxiety disorder, rather than a specific predisposition for developing OCD. (*See* **Behavior, Genetic Influences on**)

Because family behavioral genetic studies are confounded by shared environment, twin studies are needed to provide converging evidence of the role of genes in anxiety disorders. A large twin study of several anxiety disorders conducted by Kenneth Kendler and colleagues revealed that concordance rates were higher for monozygotic (MZ) or identical twins than for dizygotic (DZ) or fraternal twins but were not exceptionally high, suggesting both moderate genetic and environmental etiological influences. In a separate twin study, William True and colleagues found that concordance rates for experiencing traumatic events and for developing PTSD were higher for MZ than for DZ twins. This genetic effect may be acting through psychological characteristics known to increase risk for trauma and PTSD, such as antisocial personality. In addition to genetic factors, there are other biological factors that may contribute to the etiology of anxiety disorders.

TREATMENT OF ANXIETY

Given the magnitude of anxiety problems as evidenced by epidemiological studies, it is fortunate that several methods of effective treatment exist. These treatments are designed to intervene by affecting the psychological and biological mechanisms responsible for causing and maintaining anxiety. Depending on the nature of the anxiety disorder and possible comorbid conditions, the optimal treatment might be psychotherapy, pharmacotherapy, or a combination of the two.

Psychotherapy

Behavioral and cognitive-behavioral interventions involving exposure to feared and avoided objects or situations enjoy the strongest empirical support among psychotherapeutic treatments, despite their seemingly counterproductive nature. Joseph Wolpe developed a specific exposure technique called 'systematic desensitization'. In this intervention the first step is to teach relaxation techniques. Then the patient and therapist collaboratively construct a hierarchy of the patient's fears, arranged from least to most anxiety-provoking. Next, the patient imagines the least provoking situation while maintaining the relaxation response. Exposure is repeated or maintained until the imagined situation no longer evokes the fear response. This sequence is repeated with each step up the hierarchy. Empirical evidence indicates that this strategy is very effective in treating phobias, and the treatment gains are maintained after treatment is terminated. In the treatment of obsessive–compulsive disorder, the exposure component is combined with response prevention. Patients are exposed to anxiety-provoking thoughts (their obsessions) and prevented from engaging in the compulsive behavior typically used to reduce their anxiety. This prevention component is necessary for maintaining exposure to the anxiety-provoking situation. (*See* **Imagery**)

Cognitive therapy is another common approach to treating anxiety disorders and is often used in conjunction with the behavioral technique of exposure. The therapist helps the patient identify illogical thoughts (e.g. exaggerated estimates of the likelihood of various negative outcomes of a situation) that may be connected to the anxiety response. For instance, a patient with social phobia might believe the likelihood of a social interaction going poorly is 90 percent. It is possible that this is an overestimate of the actual probability. Having the patient record the outcomes of a number of interactions can be a useful way of gathering evidence to convince the patient that the estimate is exaggerated. Hopefully, the ensuing reduction in the probability estimate will help decrease the fear associated with the situation. Edna Foa and Barbara Olasav Rothbaum have developed an effective cognitive-behavioral treatment for patients with PTSD following rape. It includes, among other things, imaginal exposure to the traumatic event, cognitive restructuring and relaxation training. Another example of an effective cognitive-behavioral intervention is David Barlow's approach to the treatment of panic disorder. In addition to relaxation and exposure it includes a cognitive component to address faulty logic such as jumping to conclusions, overgeneralizing, all-or-none thinking and grossly exaggerating 'worst case' scenarios.

Antianxiety Medications

The benzodiazepines are a commonly prescribed class of medication for alleviating anxiety symptoms. Examples of these tranquilizers are diazepam and alprazolam. They work by binding to receptor sites in the brain for the neurotransmitter γ-aminobutyric acid (GABA) and inhibiting the activity of GABA neurons. They seem to reduce many somatic symptoms of anxiety but are less effective at reducing cognitive symptoms such as worry. There is some empirical evidence for the effectiveness of these drugs for treating social phobia, panic disorder and (ironically) generalized anxiety disorder. On the other hand, they do not seem to provide much relief for individuals suffering from other phobias and obsessive–compulsive disorder. (*See* **Psychoactive Drugs**)

Although there is moderate success for this drug treatment, there are several disadvantages. Common side effects include motor and cognitive impairments. In addition, it is common for gains to be lost once the medication is discontinued. Perhaps most importantly, though, there is a risk of addiction. This risk is highest for individuals who have a history of abusing alcohol and other substances. Unfortunately, such histories are common among people with anxiety disorders, who often self-medicate themselves with these substances to alleviate their anxiety symptoms.

Given the high comorbidity of anxiety and depression it is perhaps not surprising that antidepressant medications show some effectiveness in treating anxiety disorders. This improvement might be due to the alleviation of depressive symptoms, but there also seem to be more direct effects on some anxiety symptoms, which suggests there are some shared etiological factors for depression and anxiety. Traditional (tricyclic) antidepressants such as imipramine and clomipramine can be effective in treating panic disorders and obsessive–compulsive disorder, respectively. However, selective serotonin reuptake inhibitors (SSRIs), a newer class of antidepressants including such drugs as fluoxetine and paroxetine, have shown effectiveness in treating several anxiety disorders and have a more manageable side-effect profile.

ANIMAL MODELS OF ANXIETY

Animal research is an important subset of anxiety research. Such research enables us to investigate questions that are not researchable with humans for ethical reasons. In doing so, much of this work provides a valuable complement to our understanding of the nature of anxiety. (*See* **Neuropsychological Disorders, Animal Models of**)

Controllability and Unpredictability

Early research on experimental neuroses provides a good example of the benefits of animal research. When exposed to inescapable stressful situations or tasks, animals exhibit behavior that resembles anxious behavior in humans. Results of animal experiments show that uncontrollable and unpredictable stressful events reliably elicit animal behavior analogous to intense and persistent fear and physiological arousal in humans. Cognitive theories of anxiety used these findings in stressing the importance of the relation between anxiety and perception of control. Anxiety is less likely to be shown by individuals who feel in control of events than it is by individuals who feel helpless.

Observational Learning, Preparedness and Phobias

Michael Cook and Susan Mineka have conducted animal research on observational learning and preparedness in the development of phobias. Rhesus monkeys raised in their natural habitat are frightened of snakes. Those raised in captivity do not exhibit this fear response when initially exposed to a toy snake. After they observe a monkey, in real life or on videotape, act fearfully in the presence of a snake, however, they do react to the snake with fear. Interestingly, similar results were not found in monkeys observing a model reacting with fear to stimuli not evolutionarily associated with fear. Thus, this animal model of phobic response is consistent with Seligman's idea of preparedness in humans. (*See* **Social Learning in Animals**)

NEURAL BASES OF ANXIETY

Increasing attention is being paid to the neurobiological aspects of the anxiety disorders. Some especially useful research findings deal with particular brain structures, patterns of brain activity associated with anxiety, and the effects of psychological trauma on brain physiology. (*See* **Emotion, Neural Basis of**)

The Amygdala

An abundance of neurobiological research on emotion has focused on the amygdala, a subcortical brain structure involved in processing emotion-related information. Joseph LeDoux and Michael Davis have conducted extensive research on the role of the amygdala within the neural circuitry underlying fear conditioning. This brain structure is thought to be responsible for attaching emotional meaning to incoming sensory stimuli and triggering anxiety responses. Indeed, damage to the amygdala results in a loss of the fear response. (*See* **Amygdala**)

Patterns of Brain Activity Associated with Anxiety

Neuropsychological studies have provided discrepant results regarding the patterns of hemispheric brain activity associated with anxiety. Some suggest there is more right hemisphere activity, others that there is more left hemisphere activity. As it turns out, the studies implicating the left hemisphere tended to include participants characterized primarily by excessive worry, or anxious apprehension. On the other hand, studies implicating the right hemisphere tended to include individuals with more somatic symptoms, or anxious arousal. In a direct test of the patterns of brain activity associated with these proposed subtypes of anxiety, Wendy Heller and colleagues showed that they were distinguishable. These results highlight the need for studies to distinguish between different types of anxiety and depression, since they often occur together, if we hope to identify the precise neurobiological patterns associated with each anxiety disorder. (*See* **Electroencephalography (EEG)**; **Brain Asymmetry**)

Effects of Psychological Trauma on the Brain

The experience of psychological trauma can have profound biological consequences, some of which may be related to the etiology and maintenance of PTSD. Research in this area is growing rapidly with the more widespread availability of technologies for measuring the structure and function of specific brain regions. (*See* **Neuroimaging**; **Neuropsychological Development**; **Neurotransmitters**)

One response to trauma involves an increase in the production of noradrenaline (norepinephrine). It is believed that elevated levels of this neurotransmitter are partly responsible for the emergence of the hyperarousal symptoms associated with PTSD. In addition, the dysregulation of this neurotransmitter system may elicit a normal stress response that is more intense and more easily triggered. This may be one mechanism by which traumatic events increase the likelihood of future traumatic events leading to full-blown PTSD. There may be a cumulative effect of traumatic experiences.

A complementary response to trauma involves the release of cortisol, a stress hormone that serves as the parasympathetic counterpart to the sympathetic activity of noradrenaline. Whereas noradrenaline serves to mobilize the body's resources in stressful situations, cortisol is released to moderate that response. Cortisol levels are reliably elevated shortly after a stressful event, reflecting the shutting off of the stress response. However, if the trauma is repeated or long-term, such as extended combat or chronic child sexual abuse, cortisol levels are paradoxically low. This may be due to a sensitization of the stress response system, resulting in lower levels of cortisol being needed to shut off the stress response. (*See* **Stress and Cognitive Function, Neuroendocrinology of**)

The hippocampus is a brain structure important for memory. It contains many cortisol receptor sites, and it appears that it may be damaged by elevated levels of cortisol (prior to sensitization) in response to trauma. In fact, some trauma studies show significantly smaller hippocampal volumes in trauma victims with PTSD than in trauma victims without PTSD. This finding is interesting, because one of the symptoms of PTSD is a memory deficit for details of the traumatic event. Owing to the correlational and retrospective nature of these studies, though, it is not yet known whether a small hippocampus is a risk factor for developing PTSD in response to a traumatic experience, or whether the volume reduction is a direct result of experiencing (and possibly reexperiencing) traumatic events. (*See* **Hippocampus**; **Amnesia**; **Hormones, Learning and Memory**; **Memory Distortions and Forgetting**)

CONCLUSION

The anxiety disorders are among the most commonly diagnosed psychological disorders, affecting millions of individuals and their families. Research on humans and nonhumans has provided a great deal of information about the biological, psychological and social factors responsible for the escalation of anxiety from adaptive to maladaptive levels. Several effective psychological and pharmacological treatments have been developed to help ease the human suffering associated with anxiety and its disorders. Further elucidation of the etiology of anxiety will pave the way for even more effective treatments.

Further Reading

American Psychiatric Association (1994) *Diagnostic and Statistical Manual of Mental Disorders*, 4th edn. Washington, DC: American Psychiatric Association.

Barlow DH (1988) *Anxiety and Its Disorders: The Nature and Treatment of Anxiety and Panic*. New York: Guilford.

Black DW, Noyes R, Goldstein RB and Blum N (1992) A family study of obsessive-compulsive disorder. *Archives of General Psychiatry* **49**: 362–368.

Clark LA and Watson D (1991) Tripartite model of anxiety and depression: psychometric evidence and taxonomic implications. *Journal of Abnormal Psychology* **100**: 316–336.

Davis M (1992) The role of the amygdala in conditioned fear. In: Aggleton JP (ed.) *The Amygdala: Neurobiological Aspects of Emotion, Memory, and Mental Dysfunction*, pp. 255–306. New York, NY: Wiley-Liss.

Foa EB and Kozak MJ (1986) Emotional processing of fear: exposure to corrective information. *Psychological Bulletin* **99**: 20–35.

Foa EB and Rothbaum BO (1997) *Treating the Trauma of Rape: Cognitive-Behavioral Therapy for Posttraumatic Stress Disorder*. New York, NY: Guilford.

Heller W, Nitschke JB, Etienne MA and Miller GA (1997) Patterns of regional brain activity differentiate types of anxiety. *Journal of Abnormal Psychology* **106**: 376–385.

Kendler KS, Neale MC, Kessler RC, Heath AC and Eaves LJ (1992) The genetic epidemiology of phobias in women: the interrelationship of agoraphobia, social phobia, situational phobia, and simple phobia. *Archives of General Psychiatry* **49**: 273–281.

LeDoux JE (1996) *The Emotional Brain: The Mysterious Underpinnings of Emotional Life*. New York, NY: Simon & Schuster.

Mineka S (1985) Animal models of anxiety-based disorders: their usefulness and limitations. In: Tuma AH and Maser JD (eds) *Anxiety and The Anxiety Disorders*, pp. 199–244. Hillsdale, NJ: Lawrence Erlbaum.

Mineka S, Watson D and Clark LA (1998) Comorbidity of anxiety and unipolar mood disorders. *Annual Review of Psychology* **49**: 377–412.

Nitschke JB, Heller W and Miller GA (2000) Anxiety, stress, and cortical brain function. In: Borod JC (ed.) *The Neuropsychology of Emotion*, pp. 298–319. New York, NY: Oxford University Press.

Noyes R, Clarkson C, Crowe RR, Yates WR and McChesney CM (1987) A family study of generalized anxiety disorder. *American Journal of Psychiatry* **144**: 1019–1024.

Skinner BF (1953) *Science and Human Behavior*. New York, NY: Macmillan.

True WR, Rice J, Eisen SA *et al.* (1993) A twin study of genetic and environmental contributions to liability for

posttraumatic stress symptoms. *Archives of General Psychiatry* **50**: 257–264.
Wolpe J (1958) *Psychotherapy and Reciprocal Inhibition*. Stanford, CA: Stanford University Press.
Yehuda R and McFarlane A, eds (1997) *Psychobiology of Posttraumatic Stress Disorder*. New York, NY: New York Academy of Sciences.
Zinbarg RE, Barlow DH, Brown TA and Hertz RM (1992) Cognitive-behavioral approaches to the nature and treatment of anxiety disorders. *Annual Review of Psychology* **43**: 235–267.

Aphasia

Intermediate article

Argye Elizabeth Hillis, Johns Hopkins University School of Medicine, Baltimore, Maryland, USA
Alfonso Caramazza, Harvard University, Cambridge, Massachusetts, USA

CONTENTS

Aphasia is impairment of language caused by brain damage. The term may be restricted to acquired disorders of language caused by focal brain lesions such as stroke, or more broadly applied to developmental language disorders or impairments due to diffuse brain damage such as head trauma or dementia.

INTRODUCTION

The study of language disturbance resulting from brain lesions has fascinated researchers over the years. In 1836, Dax reported one of the first cases of aphasia and proposed that the disorder reflected damage to a language center in the left frontal lobe. Nearly thirty years later, Paul Broca made a similar proposal on the basis of several patients with minimal speech output who were found to have lesions in the left, posterior frontal lobe at autopsy. That report, published in 1861, set the stage in the following decades for the theories of Wernicke and Lichtheim, which postulated the existence of various 'brain centers' that were critical for separate aspects of language. However, over subsequent decades the Wernicke–Lichtheim proposal of aphasia classification based on damage to specific brain centers was severely criticized by Freud, Marie, Goldstein, Luria and others. At the same time, competing classification schemes were being proposed (Table 1). The resulting confusion concerning the relationship between brain damage and aphasia led to a virtual abandonment of aphasia research until 1965, when Norman Geschwind revived and expanded the Wernicke–Lichtheim theory. In this scheme, three distinct domains of language were said to be mediated by specific regions of the left cortex: fluency of speech production mediated by Broca's area (posterior-inferior frontal lobe); comprehension of language mediated by Wernicke's area (posterior superior temporal gyrus); and repetition mediated by the arcuate fasciculus – a white-matter tract between Broca's area and Wernicke's area. The relative sparing or impairment of these domains formed the basis of the 'classical aphasia' syndromes: Wernicke, Broca, conduction, transcortical sensory, transcortical motor, and mixed transcortical aphasias. (Note that a similar effort to revive this same classification scheme had been made by Nielsen in 1936, but with far less impact.) These aphasia syndromes, identified in chronic stroke patients, roughly coincide with distinct cortical lesions on radionuclide brain scans or computed tomographic head scans. Perhaps because of these early claims of a correspondence between the site of brain damage and the resulting clinical aphasia syndrome, Geschwind's

Table 1. Proposed aphasia classifications: syndrome clusters comparable to Geschwind's 1965 classification

Other classification systems (year)	*Geschwind classification*							
	Broca aphasia	*Wernicke aphasia*	*Conduction aphasia*	*Global aphasia*	*TCM*	*TCS*	*MTC*	*Anomic aphasia*
Broca (1865)	Aphemia	Verbal						Amnesia
Wernicke (1881)	Cortical	Cortical	Conduction	Total	TCM	TCS		
Lichtheim (1885)	Motor	Sensory						
Head (1926)	Verbal	Syntactic				Nominal		Semantic
Pick (1931)	Expressive	Impressive		Total				Amnestic
Weisenberg and McBride (1935)	Expressive	Receptive		Expressive/ receptive				Amnestic
Kleist (1934)	Word muteness	Word deafness	Repetition					Amnestic
Neilsen (1936)	Broca	Wernicke	Conduction	Global	TCM	TCS		Amnestic
Wepman (1964)	Syntactic	Jargon				Pragmatic		Semantic
Luria (1966)	Efferent motor	Sensory	Afferent motor		Dynamic	Acoustic		Semantic

MTC, mixed transcortical; TCM, transcortical motor; TCS, transcortical sensory.

proposal gained credibility. It has since formed the basis for much of the current conceptualizations of aphasia in neurology, speech–language pathology and neuropsychology. Although it is not evident that Geschwind's classification scheme is any more valid than other proposed classification schemes, it merits description because of its historical influence.

CLASSICAL APHASIA SYNDROMES

Wernicke Aphasia

Wernicke aphasia is characterized by prominent impairment in the understanding of spoken words and sentences, and effortless production of utterances with the basic structure and melody of sentences but mostly devoid of clear meaning. Speech in Wernicke aphasia is described as 'empty', consisting of strings of real words in apparently meaningless combinations ('yeah, that was the pumpkin furthest from my thanks') or phrases replete with word-like neologisms ('the scroolish prastimer ate my spanstakes'). Repetition resembles spoken output – mostly fluent jargon. Naming is generally poor. The syndrome has been ascribed to lesions of the posterior-superior temporal cortex (Wernicke's area), often along with the nearby parietal cortex.

Broca Aphasia

The hallmark of Broca aphasia is nonfluent, 'agrammatic' speech output with relatively preserved comprehension ability. Spoken output is characterized by the omission of grammatical morphemes, such as prepositions, conjunctions and verb inflections. To illustrate, a patient with Broca aphasia described her recent onset of neurological impairments in this way: 'Stroke…Sunday…arm, talking – bad.' Often there is concomitant effortful production of words with distorted articulation. Repetition of sentences mirrors spontaneous speech – it is effortful and telegraphic, but the gist of the sentence is mostly retained. Naming is generally impaired, with verbs named more poorly than nouns. Comprehension of single words is mostly spared, but comprehension of syntactically complex sentences is often impaired (Caramazza and Zurif, 1976). The full syndrome is typically ascribed to damage to Broca's area, along with adjacent frontal fields, and the underlying white matter and basal ganglia. Damage to Broca's area, or the underlying white matter, alone causes selective impairment in planning and executing the complex movements to articulate words, a disorder known as 'aphemia'. Often, cases of Broca aphasia evolve from an initial global aphasia, with severely impaired comprehension as well as speech production. In at least some cases, this 'evolution' may be due to initial poor blood flow to Wernicke's area posterior to the stroke, causing impaired comprehension. If blood flow is restored to that area, comprehension improves (Figure 1).

Global Aphasia

The combination of severely impaired comprehension of language and extremely limited speech output is known as global aphasia. Spontaneous speech and repetition are limited to a few perseverative words or nonword utterances (e.g. 'dee dee dee'). Cursing or the production of frequently

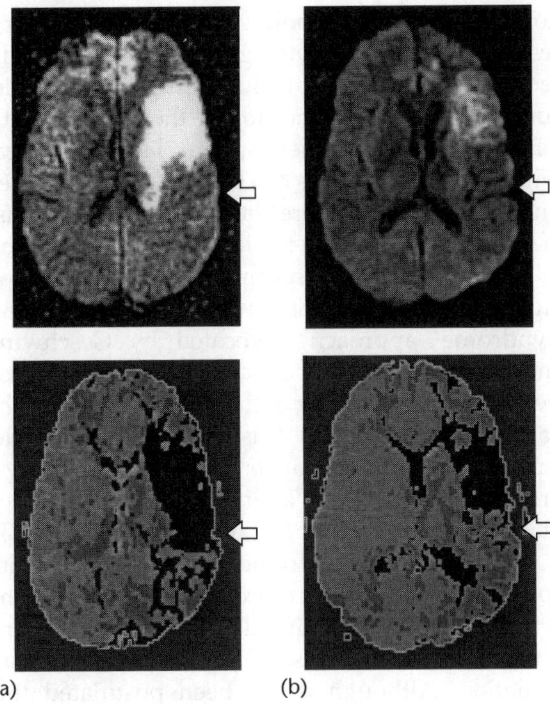

Figure 1. (a) Magnetic resonance diffusion-weighted (top) and perfusion-weighted imaging showing acute stroke involving Broca's area (and other frontal regions, and caudate), with hypoperfusion of Wernicke's area (arrow) in a patient with early global aphasia. Darker areas are regions of poor perfusion. (b) The same patient, after partial reperfusion of Wernicke's area and concomitant resolution of the comprehension deficit.

used phrases, such as 'I don't know' or 'How do you do?', may be spared in this and other aphasia types. Global aphasia is usually ascribed to large lesions involving both Broca's and Wernicke's areas, and the basal ganglia, but can also result from two lesions – one anterior and one posterior.

Transcortical Aphasia

In transcortical aphasias the fluency and content of sentence repetition far exceed those of spontaneous speech. In transcortical sensory aphasia, comprehension of language and content of speech are markedly impaired (as in Wernicke aphasia) but sentence repetition is relatively accurate. A lesion posterior to Wernicke's area, near the temporooccipital junction, is frequently implicated. In transcortical motor aphasia, comprehension and content of speech are relatively intact, while fluency and articulation are impaired in spontaneous speech but not in sentence repetition. Stroke or poor blood flow to the left dorsolateral frontal lobe, anterior and superior to Broca's area, has been associated with this form of aphasia. Finally, in mixed transcortical aphasia, also known as 'isolation syndrome', all language functions are impaired except repetition. Such patients can repeat sentences verbatim, with no evidence of comprehending them. A combination of lesions causing transcortical motor and transcortical sensory aphasia has been postulated as the cause of mixed transcortical aphasia.

Conduction Aphasia

In contrast to the transcortical aphasias, conduction aphasia is associated with disproportionately impaired sentence repetition. Speech is fluent and grammatical, although phonemic paraphasias are often present. The frequently observed phenomenon of progressive self-correction of speech errors, such as 'tormano, tornano, tornado', has been called '*conduit d'approche*'. In at least some cases the impaired repetition has been shown to be due to severely limited auditory short-term memory, such that the patient is unable to retain the entire sentence in short-term memory while articulating it (Warrington and Shallice, 1969). Conduction aphasia was initially claimed to result from lesions of the arcuate fasciculus (resulting in a putative disconnection between Wernicke's area and Broca's area), but there is little evidence for this particular lesion–deficit association. More recently, conduction aphasia has been ascribed to lesions of the left supramarginal gyrus and/or the insula or Wernicke's area.

Reading and Writing Disorders

In all of the above aphasia syndromes, reading comprehension is generally impaired at least to the degree of auditory comprehension, and written output is typically impaired at least as much as spoken output, often mirroring the content of speech. However, there are cases of pure (auditory) word deafness, in which comprehension of spoken language is severely impaired, but comprehension of written language is intact (Denes and Semenza, 1975). There have also been reported cases in which written naming accuracy far exceeds spoken naming accuracy. There are also a variety of patterns of pure reading impairment (alexia), writing impairment (agraphia) or both (alexia with agraphia), associated with different lesion sites.

Anomic Aphasia

In anomic aphasia the predominant problem is in naming, or word retrieval. Occasionally, naming is selectively impaired for certain categories, such as proper names, verbs or nouns (Goodglass *et al.*, 1966). More often in category-specific aphasias, both naming and comprehension are impaired or spared in selective semantic categories, such as that of living things (Warrington and Shallice, 1984). In pure anomic aphasia, words are understood but poorly retrieved. Anomic aphasia was initially considered to be nonlocalizing and to be the residual state of other aphasic syndromes, or broadly overlapping with Wernicke aphasia. However, one type of naming disorder that has a well-documented association with lesion site is that of optic aphasia, in which naming of pictures and other visual stimuli is severely disrupted, but naming in response to definitions and naming of stimuli presented in auditory or tactile form are spared. The patient appears to recognize the visual stimulus, and will often mime how to use the object (Lhermitte and Beauvois, 1973). This clinical syndrome was initially described in 1881 by Dejerine, who proposed that optic aphasia occurred when there were two lesions: one in the left occipital lobe causing right homonymous hemianopia, such that all visual processing takes place in the right occipital lobe; and a second lesion in the splenium of the corpus callosum causing a disconnection of the right occipital lobe from the left hemisphere language areas (Figure 2). Thus, objects can be seen and recognized (in the right hemisphere) but not named (because of the disconnection to the left perisylvian region).

This classification scheme is summarized in Figure 3. It is important to note, however, that this 'syndrome' approach advocated by Geschwind and his colleagues is a clinical nosology, based on frequently co-occurring deficits in individuals who have sustained strokes. It is not based on any defensible theory of language: that is, it has not been proposed that there is a single underlying deficit that gives rise to all of the symptoms observed in a single aphasia syndrome. For example, Broca aphasia refers to the co-occurrence of nonfluent, agrammatic speech, difficulty in comprehending syntactically complex sentences, and effortful articulation. Although it has been postulated that damage to a 'central syntactic processor' might

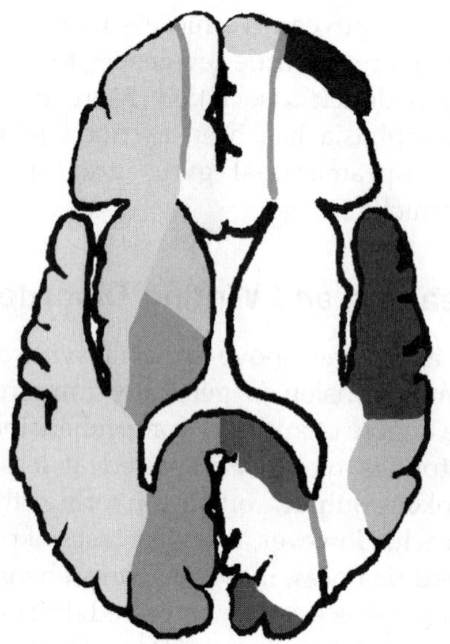

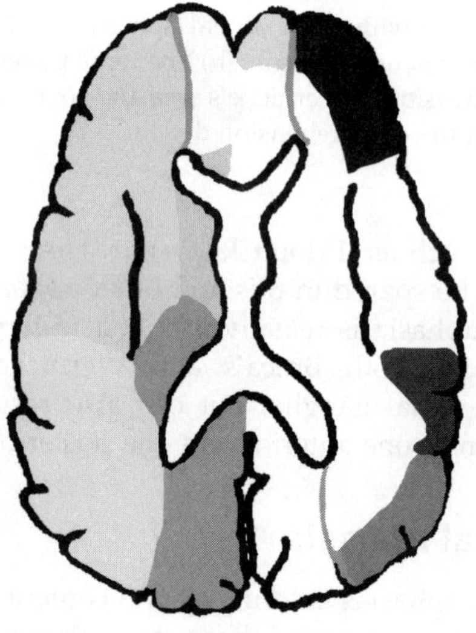

Figure 2. [*Figure is also reproduced in color section.*] Approximate vascular distributions of major arteries supplying language cortex are shown on the left side of each brain cut (the right hemisphere on computed tomographic and magnetic resonance imaging). Pale yellow, anterior cerebral artery (ACA); beige, anterior choroidal branch of internal carotid artery; light blue, superior division of the middle cerebral artery (MCA); pink, inferior division of the MCA; lavender, posterior cerebral artery (PCA); gray, border of ACA/MCA and MCA/PCA (potential 'watershed' areas). Regions of the cortex implicated in classical aphasia syndromes are shown on the right side of each brain cut (the left hemisphere). Bright yellow, Brodmann's area (BA) 10, associated with some cases of transcortical motor aphasia; dark blue, BA 44 and 45, Broca's area; red, BA 22, Wernicke's area; dark green, BA 39, the angular gyrus; light green, BA 37, the posterior middle temporal gyrus; purple, BA 31, the splenium of the corpus callosum and BA 18, visual cortex (areas associated with optic aphasia); gray, 'watershed' areas associated with the transcortical aphasias (the width of this territory depends on the degree of diminished flow in the ACA/MCA or MCA/PCA). Adapted from Damasio and Damasio (1989).

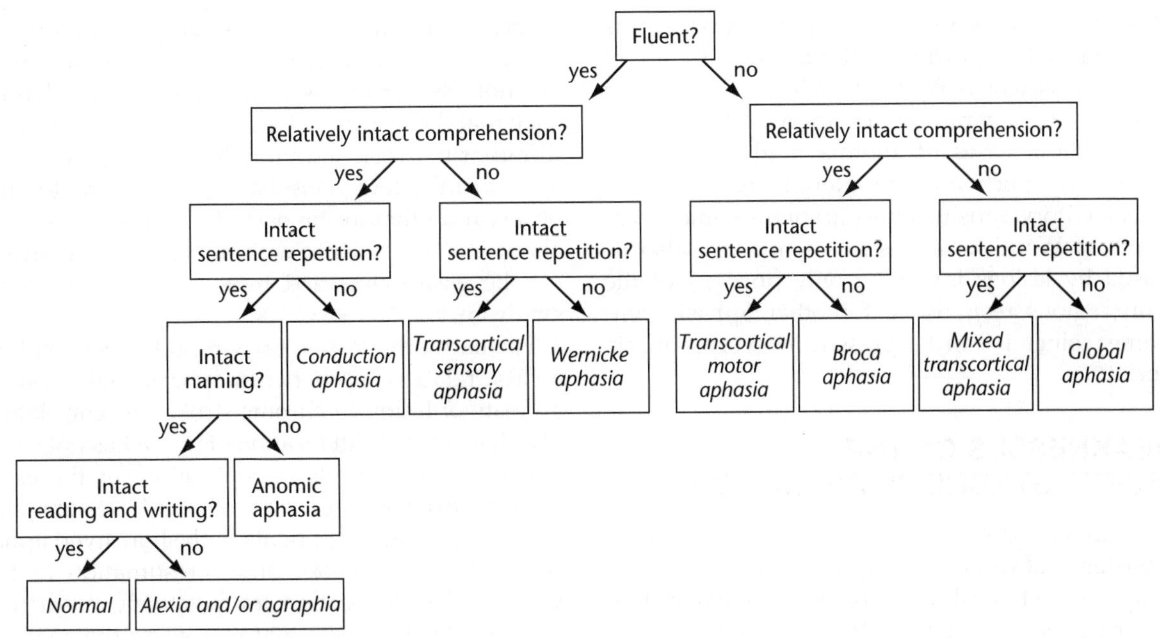

Figure 3. A decision tree for classifying aphasia according to Geschwind's 1965 scheme.

give rise to both the agrammatic speech and the impaired comprehension of syntactically complex sentences, such a proposal does not account for impaired speech articulation. Rather, it is likely that the frequent co-occurrence of these various symptoms reflects the fact that large, consistent regions of the brain are typically supplied by distinct cerebral arteries, the occlusion of which results in stroke. Suppose the large area supplied by a vessel such as the superior division of the left middle cerebral artery (MCA) consists of a number of smaller regions each responsible for a specific language function (e.g. grammatical sentence formulation, computation of syntactic relations, and articulation); occlusion of this vessel would typically result in impairment of all three functions. According to this hypothesis, occlusion of a different cerebral artery would spare these functions and damage other functions. Thus, the clinical syndromes might serve to localize the region of the brain affected by the lesion, and identify the artery involved.

ANATOMICAL CORRELATES OF APHASIA SYNDROMES: VASCULAR DISTRIBUTIONS

There is a wealth of data consistent with the hypothesis that the syndromes described above are manifestations of occlusion of specific arteries. There is evidence that Broca aphasia reflects a blockage of the superior division of the left MCA which supplies the left posterior, inferior frontal lobe and much of the basal ganglia (Figure 2). Because this branch also serves the cortex responsible for motor function of the face and arm on the contralateral side, most patients with Broca aphasia have right face and arm weakness. In contrast, Wernicke aphasia is thought to reflect blockage of the inferior division of the left MCA, resulting in stroke in the left posterior temporoparietal area. This branch also supplies the optic tract (visual pathway), so that patients with Wernicke aphasia often have a contralateral visual field cut. Global aphasia typically reflects occlusion of the left proximal MCA before it divides, resulting in a large stroke involving the left frontotemporoparietal cortex and subcortical structures. Transcortical motor aphasia has most often been attributed to left anterior cerebral artery strokes, involving the left medial frontal lobe, but in other cases has been attributed to strokes in the watershed distribution on the border between the left anterior cerebral artery (ACA) and left MCA (see Figure 2). In these cases, blood flow in both the ACA and MCA is so diminished that it does not reach the borders of the normal territories. Similarly, the collection of symptoms that characterize transcortical sensory aphasia are frequently caused by watershed strokes between the left MCA and left posterior cerebral artery (PCA). Not surprisingly, mixed transcortical aphasia results from strokes involving both watershed regions (ACA–MCA and MCA–PCA), which can occur in the presence of low blood pressure resulting in poor blood flow through all of these vessels. Of course, there can be strokes involving only a portion of the territory of any of these vessels, which would plausibly result in only a subset of the symptoms that form a clinical syndrome. Thus, cases of anomia might be due to damage to a portion of the inferior division of the MCA territory, causing only part of the clinical syndrome of Wernicke aphasia. Anomia can also be caused by lesions involving only portions of the brain regions responsible for other aphasic syndromes, since it can be present in every aphasia type.

WEAKNESSES OF THE LESION–SYNDROME APPROACH

If the above clinical syndromes reflect the relative consistency of vascular beds, such that strokes generally affect a typical collection of brain regions that may each be crucial for different language functions, it follows that brain damage caused by nonvascular lesions, such as tumor or trauma, would be likely to affect different brain areas and, consequently, different sets of language functions. In other words, profiles of language disturbance would not be expected to correspond with the classical aphasia syndromes. Indeed, tumor (primary or metastases), trauma, abscesses and other infections can all cause aphasia, but the patterns of impairment frequently do not fit the classification scheme outlined above. For example, herpes encephalitis tends to affect the mesial temporal and frontal lobes. When the encephalitis predominantly occurs in the left temporal lobe, a selective impairment in naming and comprehending animals and other living things is frequent. This type of category-specific language deficit has also been reported after closed head trauma, which preferentially affects the temporal lobes, and in the temporal variant of frontotemporal dementia (also called semantic dementia). Patients with frontotemporal dementia may have a primary progressive aphasia with deterioration in language functions before other cognitive abilities. In the temporal variant speech is often fluent, but progressively devoid of content or meaning. In contrast, the frontal variant of frontotemporal dementia is often heralded by a nonfluent, primary progressive aphasia, characterized by increasingly halting, telegraphic speech, with relatively spared comprehension. Tumors and abscesses cause widely disparate language disturbances, depending on both the location of the lesion and the rate of growth. Slow-growing lesions may cause no neurologic dysfunction until late in the disease, when associated brain swelling may acutely compress a ventricle or crucial structure. Even in stroke, there are notable variations in the human vasculature, such that the territory of a given vessel is somewhat different between individuals and may be markedly so in a few. This fact may account for the observation that in the best of circumstances, only about 50% of patients with aphasia due to stroke can be easily classified into one of the classical syndromes.

Furthermore, despite early reports documenting a close relationship between aphasia classification and site of lesion in chronic stroke, the correlation of aphasia type with location of lesion has not withstood recent attempts at replication. In the early studies, exclusion of patients with lesions but without aphasia, or patients with short-lived aphasia, may have resulted in overestimation of the value of lesion location for positively predicting aphasia. Another potential reason for the conflicting results is that early studies may have included only the 'best' or 'cleanest' cases of each syndrome type, whereas other studies may have included cases that were not as easily classifiable. Nevertheless, large-scale studies that have used similar measures to classify patients, and have included a category of 'others' for patients who do not fit well into any of the syndromes, have not provided evidence for a strong localization of these aphasia types (Kreisler *et al.*, 2000). Kreisler and colleagues found that the most common site of lesion in patients with Wernicke aphasia and those with Broca aphasia was the insula-external capsule region in both groups. Only 60% of patients with Broca aphasia had strokes involving the inferior frontal gyrus (Broca's area), and only 70% of patients with Wernicke aphasia had strokes involving the left posterior temporal gyrus (Wernicke's area). However, stronger associations were found between impairment of specific language tasks (naming, repetition) or speech characteristics (fluency) and the site of lesion. For instance, 92% of patients with poor auditory comprehension had lesions involving the posterior superior and middle temporal gyri. This study indicates that there may be stronger localization of impairment for specific language tasks than for aphasia syndromes.

However, even the separation of impairments to different tasks may be too gross a characterization of language disorders to identify the neuroanatomical substrates. Kreisler *et al.* (2000) found, for example, that damage to any one of several regions of the brain results in impaired naming. The presence of damage to any one of five areas – the

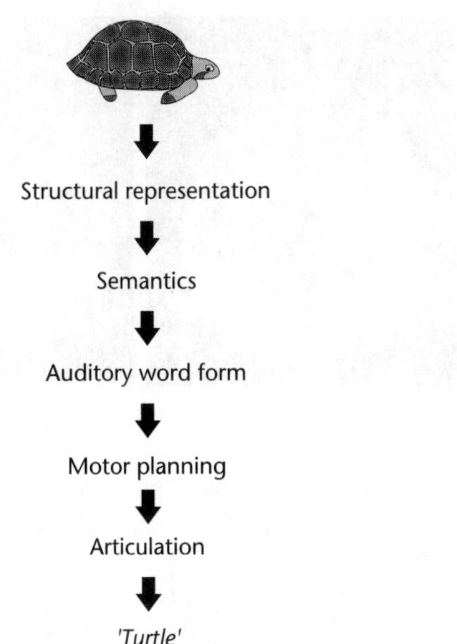

Figure 4. The cognitive processes underlying naming.

insula–external capsule plus the white matter underlying temporoparietal cortex, the medial temporal gyrus, superior frontal gyrus, middle frontal gyrus, or the thalamus – was associated with impaired picture naming. These observations probably reflect the fact that naming is a complex process, consisting of a number of cognitive processes that may take place in different brain regions. Naming a picture involves, at the very least, early visual processing resulting in recognition of the picture as something familiar; accessing the semantic representation of the object (the collection of features that define how that object is distinguished from other objects with different names); accessing the phonological form of the word (the lexical representation, or stored pronunciation of the word); programming the movements of the lips, tongue, jaw and palate to produce the name; and finally, articulating the word by implementing the planned movements (Figure 4). It is plausible that these components of the naming task are carried out by separate regions of the brain. In this case, damage to any one of the regions would disrupt the person's ability to name an object, although the mechanism would be different across cases.

LOCALIZATION OF SPECIFIC LEXICAL PROCESSES AND REPRESENTATIONS

Studies using advanced neuroimaging of function or dysfunction have indicated that there may be more reliable localizations of each of these component processes (e.g. access to semantic information) than of broader language tasks (e.g. naming).

Structural Representations for Object Recognition

Evidence from surgical ablation studies in primates and functional activation imaging in humans indicate that visual object recognition entails a neural network consisting of retinotopic visual representations in primary visual cortex; visual feature perception, such as perception of color and movement in visual association cortex; unimodal visual/structural descriptions of the shape, color and motion of familiar objects in the medial superior temporal, lateral intraparietal, and the posterior and anterior inferior temporal cortices; and a polymodal representation of familiar objects in the midtemporal cortex.

Semantic Representations

Some lesion studies indicate that impairments of semantic representations (word meaning) are associated with lesions in Wernicke's area and the left middle temporal gyrus. Likewise, electrical stimulation of Wernicke's area interferes with lexical-semantic processing. Positron emission tomography (PET) studies, which show areas of 'activation' or increased regional blood flow (rCBF) associated with increased metabolism, also show activation of Wernicke's area and the left midtemporal cortex in semantic processing. Similar results are found with another method of functional imaging, perfusion-weighted imaging (PWI), which shows regions of poor perfusion, or blood flow, resulting in tissue dysfunction in the setting of cerebrovascular disease. Studies confirm a strong correlation between impaired access to semantics and poor perfusion of Wernicke's area (Hillis *et al.*, 2001). Furthermore, when blood flow was restored to Wernicke's area in a subset of these patients, access to semantics was restored, indicating that this region is involved in semantic processing (see Figure 1).

Such findings do not allow the conclusion that Wernicke's area (or any other brain region) is alone responsible for a given cognitive process. For example, although Wernicke's area appears to be important for the understanding of words, it is unclear what – if any – semantic information is represented in this area. It is more likely that semantic representation is distributed across a variety of temporal and parietal brain regions and that the role of Wernicke's area is in linking phonological (spoken) word forms to meanings. Although it is often difficult to establish that patients with damage to Wernicke's area are impaired in mapping words to their meanings, rather than impaired at the level of meanings themselves, rare cases of dissociation between these two functions have been reported in patients with extraordinarily small lesions. For example, Hillis *et al.* (1999) described a patient, J. B. N., who spoke in fluent jargon and failed to comprehend spoken words or sentences, despite normal hearing and 'early' auditory processing. Nevertheless, J. B. N. had intact writing and comprehension of written language. Thus, this patient did not have impaired semantics, or word meanings, but was impaired in linking spoken words to their meanings and vice versa, owing to poor blood flow in the sylvian branch of the anterior temporal artery, supplying Wernicke's area. Interestingly, this proposal about the role of Wernicke's area in linking words to their meanings was first put forward by Carl Wernicke himself in the 1870s.

Auditory Word Forms

Most functional imaging studies fail to distinguish between access to auditory word forms and mapping auditory word forms to semantics, since recognizing or saying a word is likely to 'automatically' activate its meaning. Hence, not surprisingly, PET studies show increased blood flow in Wernicke's area during tasks of auditory word-form processing. As discussed above, this region may be where auditory word forms are linked to semantics, not where auditory word forms are represented. Evidence from PWI in acute stroke patients indicates that impaired naming without impaired comprehension occurs with hypoperfusion (poor blood flow) just posterior and inferior to Wernicke's area. To illustrate, the patient whose scans are shown in Figure 5 had selective impairment in accessing auditory word forms for oral naming and oral reading when the posterior middle temporal gyrus was hypoperfused (receiving poor blood flow), but this impairment resolved when this region was reperfused the following day. Such studies indicate that the posterior middle temporal gyrus was crucial for access to auditory word forms for output, but not for semantics, in this patient.

Motor Speech

Studies of stroke patients indicate that Broca's area and the medial third of periventricular white matter are critical for motor speech. Similarly, intraoperative cortical stimulation, causing temporary, focal dysfunction of regions within Broca's area, disrupts motor speech.

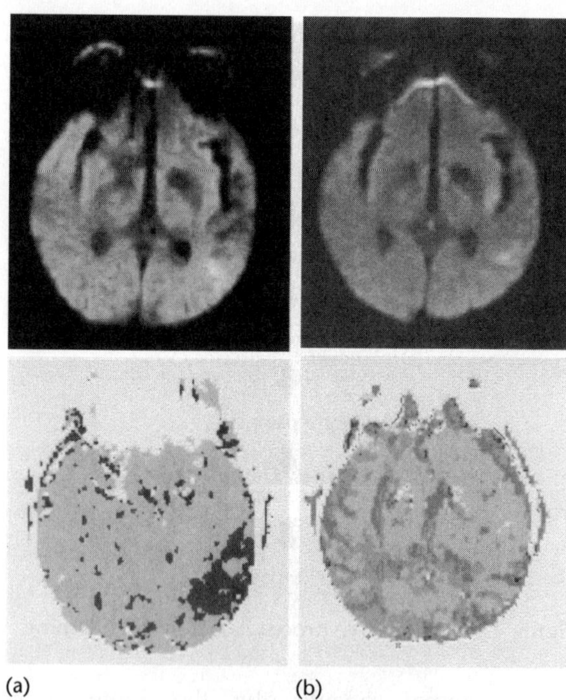

Figure 5. [*Figure is also reproduced in color section.*] (a) Magnetic resonance diffusion-weighted imaging (top) and perfusion-weighted imaging (bottom) showing hypoperfusion of the posterior middle temporal gyrus on day 1 in a patient with impaired access to auditory word forms for output, causing poor oral naming and oral reading, but intact semantics. (b) Day 2, showing reperfusion of the posterior middle temporal gyrus when naming and oral reading abilities had recovered.

WEAKNESSES OF THE LESION–DEFICIT APPROACH

We have discussed the role of various cortical regions that seem to be involved in specific components of the naming process. However, subcortical structures may also play a role in carrying out these processes, since subcortical stroke can cause aphasia. Studies of dysfunctional tissue, using PET, single photon emission computed tomography or PWI, have shown that at least some cases of aphasia associated with lesions restricted to subcortical structures show perfusion abnormalities in the temporoparietal cortex (Figure 6). These studies indicate that language deficits in such cases might be due to low blood flow to the cortex, rather than to the subcortical lesion itself.

Most lesion–deficit studies are complicated by the fact that the neural substrates of a given function may change after stroke. Reorganization of motor and sensory (and perhaps language) 'maps' shift over the course of days to months after brain

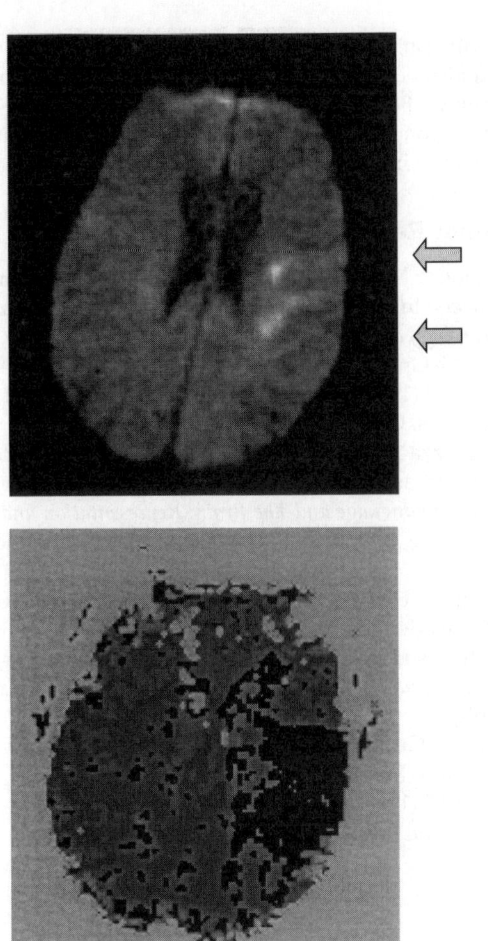

Figure 6. Diffusion-weighted imaging (top) showing tiny subcortical strokes (arrows), and perfusion-weighted imaging done at the same time, showing cortical hypoperfusion of the left temporal lobe, in a patient with global aphasia. Darker areas show regions of poor perfusion.

injury or peripheral lesions. Therefore, the neural regions responsible for any given language process are probably identified best by experimental temporary lesions (such as those produced by transcranial magnetic stimulation) or by imaging stroke and dysfunctional tissue in the first day of stroke symptoms before significant reorganization occurs. Investigations of the correlation between language deficits in the first day of stroke and concurrent regions of abnormality defined by diffusion-weighted imaging (DWI), which shows areas of acute stroke within minutes of onset, and PWI are likely to be less contaminated by varying degrees of reorganization or recovery than traditional studies of chronic deficits and lesions. Preliminary studies using these advanced imaging techniques have revealed some strong associations between areas of dysfunction (due to hypoperfusion) and impairments of specific lexical functions in acute stroke before reorganization. These data converge with results from activation studies (using functional magnetic resonance imaging and PET) in support of the hypothesis that language tasks such as naming involve a network of brain regions that are each responsible for relatively discrete types of language representations or processes.

CONCLUSION

A profusion of reports spanning the last century and a half of brain lesions associated with patterns of language disturbance has shown that occlusion of a given cerebral artery or branch often results in a particular aphasia syndrome – a set of frequently co-occurring language deficits. More recent studies, using advanced functional imaging to show areas of activation in normal people engaged in a particular task or to show regions of dysfunctional brain tissue at the time of onset of specific language deficits, have demonstrated that distinct processing components of any language task, such as naming, are subserved by separate brain areas. These findings are consistent with research in cognitive neuropsychology and other branches of cognitive science, showing that each language task entails numerous levels of representation and processing which can be differentially impaired by brain damage. However, we are still far from defining a theory of the neural substrates for language processing that goes much beyond a coarse taxonomy of the major components of language: semantics, auditory (phonological) word forms and the like. This is in contrast to the rich data that aphasia provides for inferring the structure of normal language processing. For example, there are many reports of exquisitely fine-grained dissociations involving selective damage or sparing of spoken but not written word retrieval, nouns relative to verbs, content words (nouns and verbs) relative to function words, inflectional (e.g. number and tense) relative to derivational morphology (such as the '-er' rule of noun formation from verbs, e.g. 'hunter'), consonants relative to vowels, and so on. Dissociations such as these impose biologically motivated constraints on the theories of lexical processing. However, we still do not have an equally fine-grained analysis of the neural mechanisms corresponding to the functional units of language processing revealed through the detailed analysis of language deficits in aphasia.

References

Caramazza A and Zurif E (1976) Dissociation of algorithmic and heuristic processes in language comprehension. *Brain and Language* **3**: 572–582.

Damasio H and Damasio A (1989) *Lesion Analysis in Neuropsychology*. New York, NY: Oxford University Press.

Denes F and Semenza C (1975) Auditory modality-specific anomia: evidence from a case of pure word deafness. *Cortex* **11**: 401–411.

Goodglass H, Klein B, Cary P and James KJ (1966) Specific semantic word categories in aphasia. *Cortex* **12**: 145–153.

Hillis AE, Boatman D, Hart J and Gordon B (1999) Making sense out of jargon: a neurolinguistic and computational account of jargon aphasia. *Neurology* **53**: 1813–1824.

Hillis AE, Kane A, Barker P *et al*. (2001) Neural substrates of the cognitive processes underlying reading: evidence from magnetic resonance perfusion imaging in hyperacute stroke. *Aphasiology* **15**: 919–932.

Kreisler A, Godefroy O, Delmaire C *et al*. (2000) *Neurology* **54**: 1117–1123.

Lhermitte E and Beauvois MF (1973) A visual-speech disconnexion syndrome: report of a case with optic aphasia, agnosic alexia and colour agnosia. *Brain* **96**: 695–714.

Warrington E and Shallice T (1969) The selective impairment of auditory verbal short-term memory. *Brain* **92**: 885–896.

Warrington E and Shallice T (1984) Category specific semantic impairments. *Brain* **107**: 829–853.

Further Reading

Alexander MP (1997) Aphasia: clinical and anatomical aspects. In: Feinberg TE and Farah MJ (eds) *Behavioral Neurology and Neuropsychology*, pp. 133–150. New York, NY: McGraw-Hill.

Caplan D (1992) *Language: Structure, Processing and Disorders*. Cambridge, MA: MIT Press.

Caramazza A (2000) Aspects of lexical access: evidence from aphasia. In: Grodzinsky Y, Shapiro L and Swinney D (eds) *Language and The Brain: Representation and Processing*, pp. 203–228. San Diego, CA: Academic Press.

Goodglass H (1993) *Understanding Aphasia*. San Diego, CA: Academic Press.

Goodglass H and Kaplan E (1972) *The Assessment of Aphasia and Related Disorders*. Philadelphia, PA: Lea & Febiger.

Ojemann GA (1994) *Cortical Stimulation and Recording in Language*. London, UK: Academic Press.

Sarno MT (1998) *Acquired Aphasia*. San Diego, CA: Academic Press.

Apraxia

Intermediate article

Michael Peters, University of Guelph, Guelph, Ontario, Canada

CONTENTS

Apraxias are movement disorders that cannot be attributed to primary motor problems such as paralysis or weakness, or to mental incompetence. Apraxias manifest in the failure to perform, accurately and smoothly, simple or complex movements. In studying apraxias we learn much about purposive motor behavior in general.

INTRODUCTION

In the absence of a generally agreed definition, apraxia is identified here as a higher-order movement disturbance that involves many facets of motor behavior (Freund, 1992). However, more restrictive definitions are often used in clinical applications. A widely used definition states that apraxia is 'a disorder of skilled movement that is not caused by weakness, akinesia, deafferentation, abnormal tone or posture, movement disorders – such as tremors or chorea – intellectual deterioration, poor comprehension, or uncooperativeness' (Heilman and Rothi, 1993). It manifests itself in the failure to perform, accurately and smoothly, simple or complex movements by imitation, or in response to verbal command.

In approaching the problem of apraxia, we can subdivide the production of skilled movement into several stages. First, there is the general idea of an action plan. For instance, we may decide to peel an apple. Second, there is the translation of the general 'I want to...' idea into a specific plan of action. In this case, we need to find a suitable knife, and then the apple. Once we have the knife and the apple, we can execute the action plan 'peel the apple'. There are two different aspects to execution. First, we need to have a clear idea as to how we peel an apple. One hand will have to hold the apple and this hand will have to position the apple so that the knife can act on it. This requires us to monitor the orientation of the apple in space, with constant adjustments of the position as the act of peeling progresses. Simple as the act of peeling is, it requires us to orient the apple so that the peeling strokes are not wasted. We do not wish to peel the same area twice, or rotate the apple so that bits of peel are missed. The second aspect of execution lies in the actual motor aspects of the peeling action. That is, we know what we want to do, but now we must do it as well as we can. The angle of the knife must be appropriate and the strokes must be such that they remove as little of the apple flesh and as much of the skin as possible. In addition, the actions of the two hands have to be coordinated so that the knife stroke does not begin before the apple is oriented properly. This example illustrates the very simple case where an action is implemented out of the person's own volition. When a person is asked to demonstrate an action, for example how to use a hammer, there is the important element of memory which is implicated both in the association of the object with hammering as a concept as well as the association of the object with a specific type of action. Here, access to motor planning areas through visual or auditory or even tactile avenues becomes an important component of performance.

Considering the number of variables involved, it is easy to see that problems in carrying out this task can be affected by disturbances at different levels, from the initial development of an action plan to the technical aspects of carrying out the movement. By consensus, the definition of apraxia excludes problems that arise in the actual implementation of movement, or what might be called the very last stage of producing skilled movement. In addition, the definition also excludes problems at the 'top end' of the process, so that difficulties in carrying out skilled movement due to general intellectual deterioration, an inability to understand, or a lack of willingness to cooperate, are also excluded from the definition of apraxia.

The definition places the problem of apraxia somewhere between actual motor implementation at one extreme and general mental competence at the other. The variety of factors that can produce apraxia of skilled movement, once these two end points are excluded, is still dauntingly large. Indeed, because the causes of the disturbance can be so varied, investigators have attempted to provide a classification of apraxia in terms of both the body parts affected, and the particular level at which the disturbance occurs. There is no ideal or compelling classification of apraxia, but in the following we shall discuss the major categories of apraxia that have been recognized. In particular, we shall see that it is not always possible to separate general mental state and the state of the motor system from intervening variables that are thought to cause apraxia.

VARIETIES OF APRAXIA

Ideomotor Apraxia

By far the best-researched and understood type of apraxia is ideomotor apraxia, especially that involving the upper limbs. For this reason, we will consider some of the general aspects of how to test for apraxia, and the principal categories of disturbances, within the framework of this form of apraxia.

Testing for apraxia

Recognition of the different kinds of apraxia depends on clinical examination and experimental measurement of motor behavior. Although there is no standardized methodology, certain approaches are common to many investigations. In examining motor performance, individuals can be asked verbally to perform a task, or they can be asked to imitate a movement that is demonstrated for them. Investigators distinguish between transitive movements, where the individual acts on an object or operates a tool, and intransitive movements where no object is acted on. Examples of intransitive movements would be saluting, taking up the stance of a boxer, or waving goodbye. A further distinction is occasionally made between movements that are representational and those that are not.

Part of testing for ideomotor apraxia involves an assessment of actual motor performance, to evaluate the possible role of 'lower level' causes such as tremor or weakness. In the past, data were mostly collected by recording observations of the performance of the patients, and subtle deficits in motor execution were often missed or ignored. More recently, precise recording of movement in space and time has become possible, and studies in which precise measurement of movement supplements clinical observation are becoming more widespread.

General disturbances in apraxia

Because of the complex neurological machinery that underlies skilled motor behavior, one can expect that lesions in different areas will produce different kinds of problems. Such disturbances can manifest themselves in whatever motor systems are involved (such as movements of the speech musculature, the body axis, or limbs). For the limbs, the following problems have been noted.

Problems in spatial function

Current consensus holds that much of motor planning – which by definition includes spatial factors because movement implies movement in space – relies on regions in the parietal lobe of the cortex. We can expect that lesions in this region will lead to problems both in the planning and execution of movement, and in spatial problems. A distinction is made between personal and extrapersonal space, the former referring to the spatial relations of parts of the body to each other, while the latter refers to external objects and their spatial relations to each other and the body of the observer. Common problems in personal space would involve the incorrect positioning of a tool: thus, if asked to pantomime cutting a loaf of bread, an individual might position the hand and imaginary knife in the horizontal rather than the vertical plane. An incorrect coordination of joints will also lead to spatial errors during movement. Spatial errors also result when there is a faulty integration of proximal and distal parts of the limb: for instance, an individual might position the arm correctly for a salute while assuming an unrelated posture of the hand.

Apraxic patients may show problems not only in the execution of movement in space but also in imagining movement. Interestingly, such difficulties can be quite specific to imagined movement, leaving spatial imagery involving objects intact (Ochipa *et al.*, 1997).

Problems in timing, sequencing and initiation of movement

Timing is often confounded with sequencing because in sequencing of movement the timing of the initiation and cessation of component movements determines successful 'running off' of a sequence. It may be helpful to distinguish between two kinds of sequencing. First, there is the rapid sequencing of progressive movements, such as in those involving several joints. Much of the timing is automatic, and may be considered 'process' timing. There is also the sequencing of separate functional components in the gain of a goal-oriented action. For example, in preparing an envelope for mailing, the separate components involve inserting the letter, sealing the envelope, attaching the stamp and writing the address. Problems in timing and sequencing in ideomotor apraxia can be observed following parietal or frontal damage, with different causation of timing errors for each location. Modern techniques allow a clear visual presentation of apraxic sequencing problems in three-dimensional space (Poizner *et al.*, 1990). Timing anomalies are also of great importance in the documentation of apraxia of speech. Clinical observations suggest that not only is the sequencing of movements across joints in the same limb affected by apraxia, but that sequencing of movement between hands (bimanual movements) is often the most conspicuously impaired activity in apraxic patients, to the point that bimanually coordinated movements may not just be poor but impossible to perform.

Simple movements

The study of simple movements is important for methodological reasons, because such movements can be measured with some precision. The many regions in the brain that are responsible for the production of skilled movement add to the controversy about the execution of 'simple' movements. For instance, rapid finger-tapping has been said to be affected in apraxia by some, while others fail to find an effect (Heilman, 1975; Haaland *et al.*, 1980). The initiation of movement as such may also be affected in apraxia, but here too the evidence is not clear. It is in the study of simple movements where the dividing line between apraxia and problems in motor execution becomes difficult to define. We may draw a parallel to agnosia, where individuals have difficulties with the recognition of objects. While it is felt that such difficulties cannot be attributed to problems in sensory processing as such, it is extremely rare to encounter an agnosia without some degree of sensory impairment. Similarly, it is rare indeed to encounter a person with apraxia who does not have some degree of motor impairment, however subtle. Nevertheless, because impairments in motor execution do not lead to apraxia in themselves, it is felt that impairment in movement execution may coexist with, but does not cause apraxia.

Ideational Apraxia

Ideational apraxia is rare but striking, and involves problems in sequencing the separate acts that lead to goal-directed behavior. For instance, when asked to prepare a cup of tea, the difficulties would not lie so much in the execution of the individual acts – such as reaching for and filling a kettle, placing tea in the teapot, letting it steep, and then pouring the tea – as in the correct sequencing of the component acts. We note here that the component acts are in themselves complex, and they may be performed quite well in isolation. For instance, the individual might place the teabag reasonably well – but in the kettle, not in the teapot. It is true that ideational apraxia is rarely seen without ideomotor apraxia, but they vary in severity independently from each other. Ideational apraxia affects normally occurring activities while ideomotor apraxia is more often observed in precisely the opposite setting, when movements are to be performed out of context. Some see ideational apraxia as part of a disconnection syndrome, where the motor planning substrate has impaired or no access to the motor execution machinery. The specificity of ideational apraxia to sequencing of motor acts has been illustrated by showing that patients will not be able to arrange in sequence a number of pictures that show progressive phases of actions with given objects, but are nevertheless able to correctly sequence pictures that show a chain of events that did not involve a sequence of movements (Lehmkuhl and Poeck, 1981).

Conceptual Apraxia

Conceptual apraxia and ideational apraxia are often considered to be synonymous. However, it is reasonable to distinguish between a sequencing problem for a succession of movements, and a problem that involves the proper selection of a specific movement. Heilman and Rothi (1993) emphasize that conceptual apraxia often involves the conceptual aspects of tool use, such as impairment in performing the appropriate motion with a given tool, not being able to pair a tool with the object that the tool would normally act on, choosing substitute tools that do not possess the essential features needed for a given application, or the ability to fashion a tool. As in the case of ideational apraxia, conceptual apraxia is often seen together with ideomotor apraxia, but not necessarily so. The converse, ideomotor apraxia without conceptual apraxia, is quite common.

Apraxia of Speech

Apraxia of speech, also referred to as 'anarthria', has been controversial because it is difficult to disentangle anarthria from Broca aphasia, where patients have difficulties in producing fluent speech, or dysarthria, where the final elements of speech articulation are affected. In speech apraxia, problems arise regardless of the context in which speech is produced. This contrasts with Broca aphasia, where it does matter under what conditions vocalizations are produced. For instance, in Broca aphasia singing may be relatively much less affected than spontaneous speech, while in speech apraxia the patient will show no such difference. If the person with limb apraxia has difficulties in accessing skilled movements and postures of the limbs, the person with speech apraxia has difficulties in accessing and organizing the 'postures' and movements necessary in speech production. The problem lies at a level one step removed from the actual production of speech sounds. This can be demonstrated in patients who have buccofacial apraxia (problems in assuming tongue and mouth postures) without speech apraxia. However, more often than not speech apraxia is accompanied by such problems (Martin, 1974). As might be expected, speech apraxia is sensitive to the length of utterances.

Apraxia in Dementia and Other Brain Disorders

Apraxias are part of the defining clinical symptoms of dementias and of Alzheimer disease in particular. Strictly speaking, because such apraxias are part of a disease that involves general intellectual deterioration, they do not fall under the general exclusionary definition. However, because apraxias in Alzheimer disease are an important part of the clinical picture, they are considered under this label. It is not surprising, considering the widespread neurological changes in Alzheimer disease, that numerous varieties of apraxia have been described. Conceptual apraxia, ideomotor apraxia, dressing apraxia, constructional apraxia and ideational apraxia have all been described. Systematic work in this area is only now beginning, and some claims have been made that patients with Alzheimer disease are more impaired in performing intransitive than transitive movements, compared with patients with specific left hemisphere brain damage. Contrary claims are also made. While apraxia in Alzheimer disease is common, it also occurs in other dementias and degenerative brain diseases (Leiguarda *et al.*, 1997).

Other Varieties of Apraxia

Limb-kinetic apraxia

In limb-kinetic apraxia, we find that the idea of the movement plan is spared but that the execution of even simple and practiced movements is coarse. Limb-kinetic apraxia is found contralateral to the lesion (in contrast to ideomotor limb apraxia, where damage may be ipsilateral or contralateral to the apraxic limb). Because limb-kinetic apraxia is close to the implementation aspect of movement, some have suggested the label 'apraxia of execution' (Freund, 1992). In prototypical patients, it has been claimed that there is no evidence of direct primary defects in motor function and the defect is one of 'the breakdown of fine skillfulness of fingers' in the absence of other types of apraxia (Denes *et al.*, 1998).

Apraxia of lid opening

Apraxia of lid opening shares with limb-kinetic apraxia the debate of whether it should be considered a proper apraxia. This apraxia presents itself as inability to close the eyelids voluntarily even though they close or open during reflex blinking (Chapanis and Gropper, 1968; Boghen, 1997; Defazio *et al.*, 1998).

Axial apraxia

In axial (or truncal) apraxia neck and trunk movements are affected. Most interesting are cases where aphasia (in this case language comprehension) affects limb movements, while truncal and gait movements can be performed without the patient having any clear comprehension of the verbal command asking for trunk and gait movements. This is probably due to some capacity of the right hemisphere to perform bilateral trunk and gait movements.

Gait apraxia

Whether or not gait apraxia should be considered an apraxia is not clear. Because gait movements have been observed in spite of intact sensation or limb weakness some clinicians feel that a true gait apraxia may exist. However, a predominant aspect of gait apraxia is a lack of initiative to move the legs and a lack of spontaneity in movement. To the extent that difficulties with gait can be caused by many factors it is likely that cases of true gait apraxia are very rare. The point has been made that the problem with gait apraxia is not so much one of incorrect or inappropriate movements but one of problems with initiation (Brown, 1972).

Dressing apraxia

Dressing apraxia may denote a true apraxia, such that the organization of the learned and skilled movements required for dressing are affected, but the failure to dress properly may also be due to more general problems, such as lateral neglect.

Constructional apraxia

In the clinical demonstration of constructional apraxia, individuals fail to arrange objects such as building blocks according to a visually presented scheme. Such cases, often associated with right hemisphere lesions, are more appropriately considered a secondary outcome of visuospatial processes.

NEUROANATOMY

The Role of the Left Hemisphere

There is little doubt that the left hemisphere has a predominant role in apraxia. Beginning with the earliest descriptions of apraxia, in the vast majority of cases the patients have left hemisphere damage. This supports the general understanding that in right-handed people especially, the conception and generation of movements relies heavily on left hemisphere function. Beginning with Liepmann in 1905, attempts have been made to separate different types of apraxia in terms of cortical lesion location. Thus, Liepmann suggested that lesions associated with limb-kinetic apraxia would be straddling the primary motor and sensory cortex, lesions causing ideomotor apraxia would be posterior to this in the higher-order sensory cortex, and lesions causing ideational apraxia even further posterior, close to the visual cortex.

Subsequent work is less confident of a precise allocation of cortical region for specific apraxias. Three general regions of damage are associated with apraxia. First, there is the parietal region in general and the angular and supermarginal gyri in particular, which are associated with the representation of movement schemes in time and space. Large lesions in this region would lead not only to apraxia, but also to an inability to judge whether a movement demonstrated to the patient is flawed or not. Lesions that leave this region intact but are slightly more anterior (without infringing on premotor regions in the frontal cortex) would lead to apraxia, but affected patients would have the ability to judge whether a movement demonstrated to them is flawed or not. Such lesions would be responsible for 'disconnection' apraxias, where the problem arises from a lack of access from the posterior areas involved in planning and movement image generation to the anterior areas that are more directly involved in implementation. Small lesions that sever the arcuate fasciculus, a tract that connects posterior to anterior cortex, can produce apraxia. This type of disconnection apraxia is probably the only kind where a circumscribed small lesion can cause apraxia (Tanabe *et al.*, 1987).

In contrast to posterior lesions, anterior lesions (anterior to the primary motor cortex) are implicated in disturbances of implementation and specific sequencing of consecutive motor acts. The supplementary motor area (SMA) has long been suspected of involvement in apraxia and a number of cases are known where the SMA has been specifically implicated. It is of interest that these cases involve bilateral apraxia; the SMA is known to be involved in bimanual coordination. In addition, the premotor cortex has been identified with limb-kinetic apraxia, also involving both arms.

The Role of the Right Hemisphere

The role of the right hemisphere in apraxia is contested. Soon after the emphasis on left hemisphere lesions in apraxia was pointed out, cases of apraxia after right hemisphere lesions were described (Brun, 1922). Isolated cases are known of right-handed people who are fully apraxic in the right hand after right brain lesions. However, there is a debate about whether in larger series of brain-damaged individuals, some degree of apraxic impairment is seen in those with right hemisphere lesions. It appears that in such series, apraxia after right brain damage is rare. However, a number of researchers show that subtle problems exist. For instance, patients with right-side lesions performed consistently much better than patients with left-side lesions on most tests for apraxia, but they performed worse than control subjects on verbal command when asked to make intransitive movements. In addition, of the 11 patients with right hemisphere lesions, five performed at a level below that of the worst control subject. Other researchers similarly found that while individuals with right-brain damage showed fewer apraxic problems, they performed less well than normal participants on specific tests (Goldenberg and Hagmann, 1998). It is clear that much of the negative evidence on right hemisphere contributions cannot be considered conclusive because it matters what movements were tested, and to what extent the right hemisphere would be involved in a given movement (Roy *et al.*, 1991). An important distinction is between the frequency of apraxic disturbances and the severity; the frequency of apraxia tends to be less after right hemisphere lesions while the severity of apraxia when it does occur is comparable after right and left hemisphere lesions.

The Corpus Callosum

Implicit in the idea that the left hemisphere is predominant in the formulation of movement plans is the assumption that such plans reach the right hemisphere regions responsible for the movements of the left arm via the corpus callosum, from the left hemisphere. 'Pure' lesions in the anterior regions of the corpus callosum that carry motor commands to be implemented by the right hemisphere might therefore lead to apraxia in the left arm – even though the right arm will not be affected. Some cases corresponding to this scheme have been described.

Subcortical Involvement

In general, lesions of subcortical structures that lead to apraxia also involve cortical damage, and it is difficult to disentangle the effects of such multiple lesions. Perhaps the strongest evidence for a distinct subcortical lesion associated with apraxia (usually ideomotor apraxia) comes from apraxic patients with restricted thalamic lesions (Pramstaller and Marsden, 1996). Speculative interpretations implicate part of the thalamus that is intimately connected with a cortical region known to be involved in apraxia. One possible involvement of the basal ganglia in particular may relate to reports of slight difficulties in the initiation of movement in some apraxic patients. Here, it may be suggested that in addition to the cortically derived signs of apraxia, subcortical contributions may be involved. The failure to see slowness in initiation of movement in some apraxic patients may therefore stem from a noncortical source.

APRAXIA AND APHASIA

The association between aphasia and apraxia is well known and documented. In the demonstration of ideomotor apraxia a language problem was implicit in the very first descriptions, because it was the inability to perform in response to verbal commands that defined the disturbance. In addition, even if patients are selected not on the basis of apraxia but on the basis of brain lesions in the left or right hemispheres, patients with left-brain damage will show a strong association of aphasia with ideomotor apraxia. Similarly, in the much rarer cases of ideational apraxia, lesions that tend to produce this apraxia will almost always also produce aphasia. Thus, there are strong links between apraxia and aphasia for different apraxias. However, it is here where individuals with anomalous cerebral lateralization of praxis and language functions provide important information. Despite their rarity, over time a considerable number of patients have been described who have dissociations between aphasia and apraxia. Such patients usually have what is called 'crossed apraxia', where apraxia is caused by right hemisphere lesions. In some of these aphasia is also produced by right hemisphere lesions, but in others there is no aphasia. Of greatest importance here are left-handers who will tend to have language presentation in the left hemisphere. While there is some debate as to whether left-handers show apraxia after right hemisphere lesions, specific and careful testing suggests that this is the case. Careful testing is indicated because left-handers may show less severe apraxia than right-handers and faster recovery after lesions in either hemisphere.

While it is tempting to speculate that the strong general association of aphasia with apraxia is due to higher-order mechanisms that are involved in generation action in either language or body movement, it is more likely that the coincidence of disturbances in both domains is due to lesions that are large enough to affect both substrates involved with language and body movement. Even allowing for the fact that some apraxias often are unrecognized, apraxias are less common after general brain damage than aphasias.

Aphasias therefore appear to be related to apraxias in two ways. First, directly, as in apraxias that emerge when movement to verbal command is tested and when the access from language areas to motor implementation is impaired. Second, there is a general correlation between presence and severity of aphasia and presence and severity of apraxia. Two possible explanations can be considered. First, the regions involved with language comprehension and motor planning in the posterior cortex are coextensive and lesions that affect one function are likely to affect the other. This is slightly less of a problem in the frontal lobe, but the same principle obtains. For instance, lesions producing Broca aphasia are likely to impinge on regions associated with buccofacial apraxia. Second, it is possible that especially at the top level of motor planning and language preparation, there are higher-order mechanisms that operate at a level above the specific modality and when these are damaged, both apraxia and aphasia ensue because mechanisms common to both are affected.

CONCLUSION

The systematic study of apraxia is in its infancy. If we adopt a broad definition of apraxia, we can see that damage in many parts of the central nervous system can lead to apraxia, and that the causes for the observed 'higher order motor disturbance' are manifold, ranging from relatively low-level problems in the timing and sequencing of unfolding movements to problems in the conceptual plan for movements. Future work will probably focus on an aspect of apraxia that has been relatively neglected by research: the representation of and access to motor memory.

In the course of trying to understand apraxia, we are also trying to understand purposeful motor behavior and its generation by analyzing the stages from conception of a movement plan to the final realization of that plan with a specific set of muscles. This will benefit neural network applications of movement control. Neural network approaches have been extremely successful in simulation of language recognition and the recognition of stimuli in the various sensory modalities. There is little doubt that some of the neural network techniques applied to recognition of incoming information can also be of use in the guidance of robotic movement. For instance, there is no reason why specified motor concepts such as 'reaching' and 'grasping' cannot be used in algorithms that control simple movements. Nevertheless, there is a large gap between neural networks that can implement a concept such as 'reaching' and neural networks that can generate such concepts.

References

Boghen D (1997) Apraxia of lid opening: a review. *Neurology* **48**(6): 1491–1494.

Brown JW (1972) *Aphasia, Apraxia and Agnosia*. Springfield, IL: Charles C Thomas.

Brun R (1922) Klinische und anatomische Studien über Apraxie. II Zur Lokalisation der Apraxie. *Schweizer Archiv für Psychiatrie und Neurologie* **10**: 186–209.

Chapanis A and Gropper BA (1968) The effects of operator's handedness on some directional stereotypes on control-display relationships. *Human Factors* **10**: 303–320.

Defazio G, Livrea P, Lamberti P *et al.* (1998) Isolated so-called apraxia of eyelid opening: report of 10 cases and a review of the literature. *European Neurology* **39**: 204–210.

Denes G, Mantovan MC, Gallana A and Cappelletti JY (1998) Limb-kinetic apraxia. *Movement Disorders* **13**: 468–476.

Freund HJ (1992) The apraxias. In: Ashbury AK, McKahann GM and McDonald WJ (eds) *Diseases of the Nervous System*, pp. 751–767. Philadelphia, PA: WB Saunders.

Goldenberg G and Hagmann S (1998) Tool use and mechanical problem solving in apraxia. *Neuropsychologia* **36**: 581–589.

Haaland KY, Porch BE and Delaney HD (1980) Limb apraxia and motor performance. *Brain and Language* **9**: 315–323.

Heilman KM (1975) A tapping test in apraxia. *Cortex* **11**: 259–263.

Heilman KM and Rothi LJG (1993) *Apraxia*. In: Heilman KM and Valenstein E (eds) *Clinical Neuropsychology*, pp. 141–163. New York, NY: Oxford University Press.

Lehmkuhl G and Poeck K (1981) A disturbance in the conceptual organization of actions in patients with ideational apraxia. *Cortex* **17**: 153–158.

Leiguarda RC, Pramstaller PP, Merello M *et al.* (1997) Apraxia in Parkinson's disease, progressive supranuclear palsy, multiple system atrophy and neuroleptic-induced parkinsonism. *Brain* **120**: 75–90.

Martin AD (1974) Some objections to the term apraxia of speech. *Journal of Speech and Hearing Research* **39**: 53–64.

Ochipa C, Rapcsak SZ, Maher LM *et al.* (1997) Selective deficit of praxis imagery in ideomotor apraxia. *Neurology* **49**: 474–480.

Poizner H, Mack L, Verfaellie M, Rothi LJG and Heilman KM (1990) Three-dimensional computergraphic analysis of apraxia. Neural representations of learned movement. *Brain* **113**: 85–101.

Pramstaller PP and Marsden CD (1996) The basal ganglia and apraxia. *Brain* **119**: 319–340.

Roy EA, Square-Storer P, Hogg S and Adams S (1991) Analysis of task demands in apraxia. *International Journal of Neuroscience* **56**: 177–186.

Tanabe H, Sawada T, Inoue N *et al.* (1987) Conduction aphasia and arcuate fasciculus. *Acta Neurologica Scandinavica* **76**: 422–427.

Further Reading

Brown JW (1972) *Aphasia, Apraxia and Agnosia*. Springfield, IL: Charles C Thomas.

Goldenberg G (1995) Imitating gestures and manipulating a mannikin – the representation of the human body in ideomotor apraxia. *Neuropsychologia* **33**: 63–72.

Goldenberg G, Hermsdorfer J and Spatt J (1996) Ideomotor apraxia and cerebral dominance for motor control. *Cognitive Brain Research* **3**: 95–100.

Heilman KM and Rothi LJG (1997) *Apraxia: The Neuropsychology of Action*. Hove, UK: Psychology Press.

Leiguarda RC and Marsden CD (2000) Limb apraxias: higher-order disorders of sensorimotor integration. *Brain* **123**: 860–879.

Liepmann H (1905) Die linke Hemisphäre und das Handeln. *Münchener Medizinische Wochenschrift* **52**: 2322–2326, 2375–2378.

Poeck K (1986) The clinical examination for motor apraxia. *Neuropsychologia* **24**: 129–134.

Rothi LJG, Ochipa C and Heilman KM (1997) A cognitive neuropsychological model of limb praxis and apraxia. In: Heilman KM and Rothi LJG (eds) *Apraxia: The Neuropsychology of Action*, pp. 29–49. Hove, UK: Psychology Press.

Roy EA, Black SE, Blair N and Dimeck PT (1998) Analysis of deficits in gestural pantomime. *Journal of Clinical and Experimental Neuropsychology* **20**: 628–643.

Roy EA, Heath M, Westwood D *et al.* (2000) Task demands and limb apraxia in stroke. *Brain and Cognition* **44**: 253–279.

Schnider A, Hanlon RE, Alexander DN and Benson DF (1997) Ideomotor apraxia: behavioral dimensions and neuroanatomical basis. *Brain and Language* **58**: 125–136.

Seddoh SA, Robin DA, Sim HS *et al.* (1996) Speech timing in apraxia of speech versus conduction aphasia. *Journal of Speech and Hearing Research* **39**: 590–603.

Arousal

See **Reticular Activating System**

Articulation: Dynamic Models of Motor Planning

Advanced article

Kevin G Munhall, Queen's University, Kingston, Ontario, Canada

CONTENTS

The complex muscular system of the human vocal tract changes states rapidly and flexibly during fluent speech. The planning and control system responsible for this remarkable behavior must coordinate the spatial and temporal aspects of movement for a large number of independent articulators. Current models suggest that detailed representation of vocal tract movements and the acoustic consequences of those movements is required.

INTRODUCTION

All human cultures communicate primarily by using spoken language, and the children of these cultures acquire a remarkable level of articulatory skill by their second birthday. Although talking may thus be the most common and natural of human activities, it is still very much a scientific mystery. The linguistic, cognitive, motoric and neural mechanisms responsible for talking have remained beyond the grasp of a multidisciplinary research effort. As Gallistel (1999) has stated, there appears to be an inverse relationship between the apparent ease of performing an activity and the computational apparatus required to account for the behavior.

It is now clear that a complete analytic understanding of the production of speech will require the solution of a number of general motor control problems, as well as some problems that are relatively unique to oral language production. These problems include accounting for the mapping between a symbolic linguistic representation and a biomechanical system, the learning and production of sound categories, the coordination of a large number of movement systems so that the temporal and spatial requirements of articulation are met, and the accomplishment of this complex movement in real time. This article briefly summarizes articulatory phonetics and then considers current views of the way in which the nervous system solves the speech motor control problem.

ARTICULATORY PHONETICS

Articulatory phonetics is traditionally defined as the study of the way in which the sounds of a language are physically produced. This includes a description of vocal tract acoustics and the key aspects of sound production for consonants and vowels. Briefly, speech production involves the generation of sound (e.g., by the vocal folds) and its filtering by the acoustic properties of the vocal tract (size, shape, wall characteristics, etc.). This separation of source and filter has been the standard model of speech research since the middle of the twentieth century (Chiba and Kajiyama, 1941/1958; Fant, 1960), but the precise details of the process are still being elucidated.

SOURCE CHARACTERISTICS

During speech there are a number of different sources of sound. The primary source of sound is the vibration of the vocal folds. Through the activity of a complex muscle framework, the frequency, amplitude and mode of vocal fold vibration can be controlled. The muscular control of voicing has been reviewed elsewhere (Honda, 1995), as has the laryngeal anatomy and physiology (Titze, 1994). These attributes of laryngeal behavior produce the perceived pitch, loudness, and quality of voicing, and are used to signal segmental, prosodic, and paralinguistic information.

Our understanding of this subsystem of speech has depended on converging evidence from different experimental techniques and approaches. Electromyographic studies of humans during voicing (Honda, 1995), animal and cadaver studies of vibration (Kakita *et al.*, 1981), photographic and imaging research during speech (Hirose, 1988), and sophisticated biomechanical models (Titze, 1973, 1974) have revealed a vocal articulation system the behavior of which results from tuning the patterns of muscular tension to control the tissue biomechanics of the vocal folds themselves.

In addition to vocal fold vibration, sound is produced at a number of locations in the vocal tract above the larynx. This mainly occurs during consonant production (e.g., stop consonants and fricatives), and it involves sound being created through the manipulation of the air flowing through the oral cavity. The sudden release of air during the production of stop consonants produces an acoustic burst that varies depending on the air pressure, position of constriction in the vocal tract, and release movements. Frication is generally produced by air rushing though a small opening.

The characteristics of fricative sounds are similarly determined by the air pressure and position of constriction, but also by the details of the small channel through which air rushes, and whether or not the air jet strikes an obstacle such as the teeth. The articulatory conditions that are required for producing these sounds have been studied using a range of different techniques, such as electropalatography (Hardcastle and Hewlett, 1999), physical and computational models (Shaddle, 1985), electromyography, and kinematic measurements of articulator movement (Löfqvist and Gracco, 1997).

FILTER CHARACTERISTICS

The sounds that are produced in the vocal tract are modified or filtered by the acoustic properties of that structure. These filtering characteristics are known as the transfer function of the vocal tract, and include the cross-sectional area, length, and wall characteristics. Although there has been a general understanding of both vowel and consonant production for more than 100 years, the details of the filtering aspect of articulation are still being revealed. This slow progress is partly due to the relative inaccessibility of the speech articulators, as most of the determinants of the vocal transfer function are internal and not directly accessible to measurement without invasive recording techniques.

According to the simple textbook view, vowels are distinguished by a small number of articulation parameters (e.g., tongue height, front–back location of the constriction, lip roundedness) and consonants by a different but similarly small number of articulation parameters (e.g., place of articulation, manner of articulation, voicing). However, this textbook view significantly under-represents the full complexity of articulation. At least two aspects of this simplification should be noted. First, vowel and consonant acoustics are determined by the shape of the entire vocal tract. The gross details of this have been known since the first X-ray studies of speech (for a summary of these studies see Moll, 1960). Figure 1 shows two-dimensional magnetic resonance imaging (MRI) views of the three English vowels /i/, /u/, and /a/. As can be seen, the surface of the tongue, lips, jaw, pharynx, velum, and larynx changes position for each vowel to produce a unique transfer function. New developments in imaging of the vocal tract have revealed intricate three-dimensional shape changes associated with different phonemes (for a review of progress in imaging the shape of the vocal tract, see Munhall, 2001). Secondly, speech sound production involves movements, not a sequence of vocal tract configurations. According to the textbook

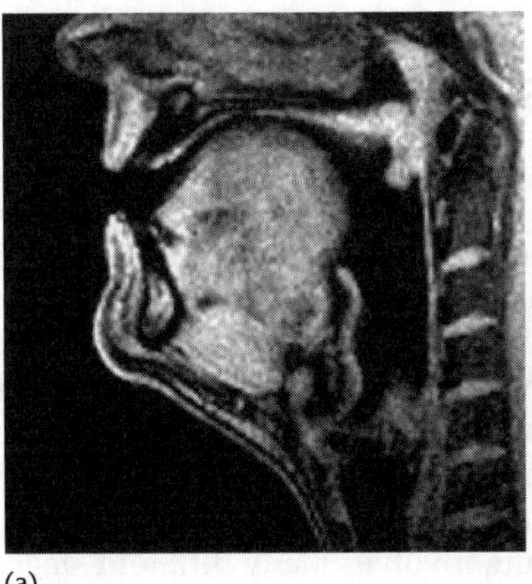
(a)

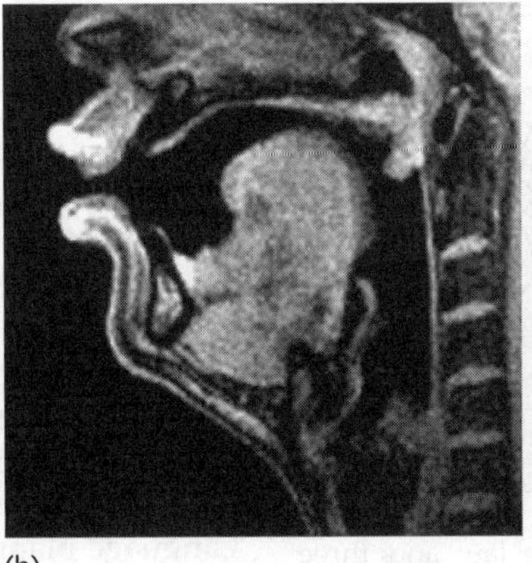
(b)

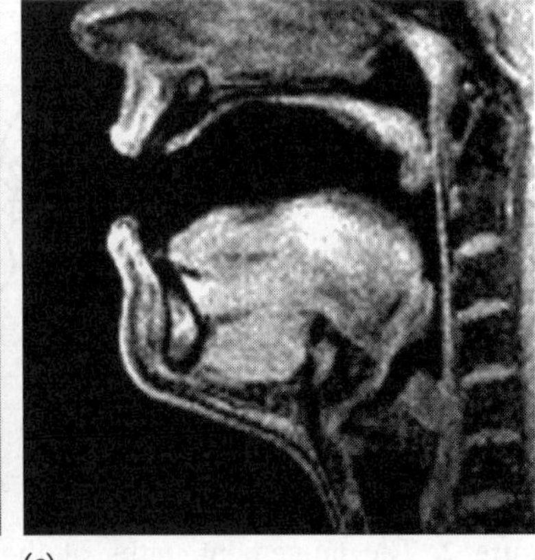
(c)

Figure 1. Mid-sagittal view of the vocal tract during vowel production. The vocal tract shapes for the vowels (a) /i/, (b) /u/, and (c) /a/. Figure provided by M. Tiede.

view, the position of a few key articulators at one point in time is used to describe the sound. In contrast, what really characterizes fluent speech is the seamless flow of gestures of the entire vocal tract.

Both of these factors contribute to the difficulty in mapping between traditional phonetic descriptions of speech and the reality of articulation. During speech the vocal tract is continuously changing in many dimensions, with the articulations for a given sound varying according to the context. However, the traditional phonetic description depicts speech as a series of static, invariant postures where only a few key dimensions are essential. An account of articulatory phonetics that recognizes its dynamic nature must depict speech as a motor skill.

Timing and Coordination

In order to control the vocal tract during speech, the nervous system must pattern the muscle activations of a large number of articulators so that the resultant movement paths shape the vocal tract properly. Because of the large number of degrees of freedom of the vocal tract, the spatial paths in articulation can be complicated. Figure 2 shows a mid-sagittal view of movement paths of the jaw and three points on the tongue. The speaker is uttering the following phrase: 'the oleander is a flower'. As can be inferred from the changes in articulator positions, the vocal tract shape is changing continuously.

The timing of muscle activations required to produce speech such as this also requires precise

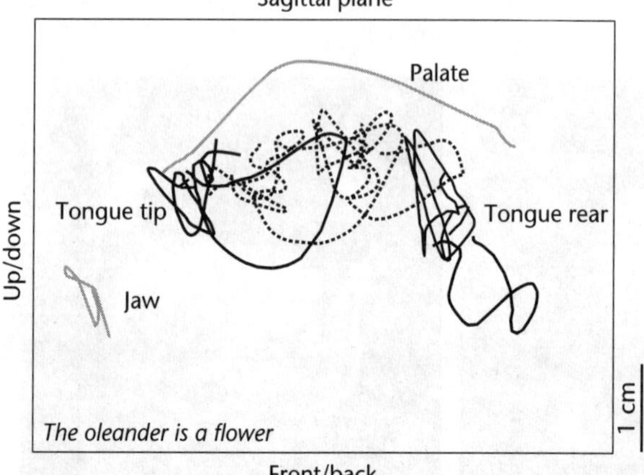

Figure 2. The movement paths of the jaw and three positions on the tongue during production of the utterance 'the oleander is a flower'. Figure provided by V. Gracco.

control, and the muscle activation generates patterns at many temporal scales. During the production of consonants and consonant clusters, coordinated movements between articulators can occur within tens of milliseconds, and even within a single articulator such as the tongue tip, successive movements can be coordinated to occur within 50 ms (Kent and Moll, 1975). At the syllable level, there is evidence of temporal organization that spans hundreds of milliseconds (for a recent example see Shaiman, 2001). At longer temporal spans there is a tradition of characterizing languages based on their rhythm class. For example, Japanese is classified as a mora-timed language in which strings of equal-duration mora are considered to be produced. However, Warren and Arai (2001) have recently published a critical review of mora timing.

Explanations for this complex behavior have been hindered by high levels of variance in the measured data. For both temporal and spatial descriptions of speech, the consistent impression is one of coordination but also of variability from one trial to another. The potential sources of such variance are manifold, including biomechanical, social, linguistic, and environmental contexts. Speaking rate, lexical and emphatic stress, and syllabification can vary from one repetition to another. In natural conversation the talker's mood, intent and social attention, as well as the conceptual and emotional meaning of a message, are all transmitted in parallel with phonetic information by subtle differences in the way in which words are spoken. These subtle differences are produced by systematic modifications to the movement timing and paths of the oral articulators. This variability that characterizes natural speech suggests that the control system must be programming articulation in real time and balancing many simultaneous requirements.

MODELING SPEECH PRODUCTION

In order to understand such complex behavioral organization as that of the speech motor control system, both experiments and sophisticated models are required. A common strategy is to identify the significant behavioral patterns and then decide how best to account for those patterns. Part of the difficulty that faces researchers is knowing what parts of behavior should be attributable to what parts of the planning and production process. Language planning involves many different stages and processes (e.g., semantic, syntactic, lexical, phonological; for one account of these stages, see Levelt, 1989), and speech motor control involves additional stages of planning (Munhall *et al.*, 2000; Perkell *et al.*, 2000). Any single observation about articulation could thus be attributed to many parts of this complex chain of events. For example, context sensitivity or coarticulation in speech has been attributed to serial planning processes that take into account upcoming phonemes in order to plan smooth trajectories (Henke, 1966). However, David Ostry and his colleagues (Ostry *et al.*, 1996; Perrier *et al.*, 1996) have demonstrated that kinematic patterns due to speaking rate changes and coarticulatory context could, in principle, fall out from the physiology of the articulators. In their simulations, apparent context effects can be reproduced by modeling realistic physiological structures such as the jaw. The input to their jaw model does not take context into account, but the output is a smooth trajectory in which the articulator and muscle dynamics significantly shape the movement form to reproduce coarticulation effects.

The success of the research by Ostry and his colleagues depended on their dynamic modeling of the jaw and its muscle system. The term 'dynamics' is used here in a number of senses. First, it is used to refer to the general study of movement. Secondly, the term is used to refer to a subcomponent of the general study of movement that deals with forces (as opposed to kinematics, which deals with spatial descriptions of movement). Finally, it is used to refer to the behavior of complex nonlinear systems (so-called dynamical systems). A dynamic model of the vocal tract in all of these senses is necessary to test linguistic models of timing, phonological organization, etc. This model would represent the 'plant' or the physiology and biomechanics that are actually being controlled. It would also represent the complex interactions of control signals, biomechanics, and reflex coupling that are a part of stable movement. Unfortunately, such a complete model is currently beyond the scope of the field. To model the plant alone accurately is a daunting task. Histological, kinematic, dynamic (force), and myoelectric parameters are required for all of the musculoskeletal systems in the vocal tract, as well as good estimates of the three-dimensional morphology of the articulators. Many of these parameters are not available at present, and some are difficult if not impossible to acquire using current technology.

Models of individual articulators such as the vocal folds (Titze, 1973, 1974, 1994), the tongue (Wilhelms-Tricarico, 1995), the jaw (Laboissière *et al.*, 1996), the tongue–jaw complex (Sanguineti *et al.*, 1998; Dang and Honda, in press) and the face (Lucero and Munhall, 1999; Pitermann and Munhall, 2001) have been constructed with simulations of the tissue biomechanics (e.g., mass, stress/strain characteristics), muscle physiology (e.g., length tension characteristics of force development) and morphology. Although these simulations all involve some simplification of the articulator being studied (e.g., simplified geometry or finite element approximation of tissue structure), the models have played an important role in advancing the study of speech motor control. They have revealed the extent to which articulatory phonetics can be influenced by the physiological and biomechanical substrate of the vocal tract (Shiller *et al.*, 1999). Furthermore, they have indicated that issues such as force, stability, and feedback processing are central problems for articulatory phonetics.

More global models of articulation have been proposed, which involve even more extensive simplifications of the oral motor mechanisms than the individual articulator models. However, these models address more complex phonetic output by controlling the vocal tract as a whole over time.

Saltzman's task dynamic model (Saltzman, 1986; Saltzman and Munhall, 1989) controls a set of stylized model articulators in a mid-sagittal vocal tract. The model describes the behavior of a set of 'tract variables' that refer to constrictions along the longitudinal axis of the vocal tract. The behavior of these tract variables is determined by a number of influences on the model articulators associated with the constrictions. For example, the behavior of the lip aperture tract variable is determined by the manner in which the lips and jaw are coordinated, as well as by the sequence of movements involving those articulators. The strengths of this model include its explanation of the coupling between articulators during articulation, and its focus on a task-level frame of reference in motor planning. The model can account for a number of important articulatory phenomena, such as some aspects of coarticulation and compensation following unexpected perturbations. When loads are applied to the lips or jaw, compensation from other articulators to achieve a goal at the task level is observed. For example, loads applied to the lips result in compensatory changes in the lips and jaw (Abbs and Gracco, 1984; Gracco and Abbs, 1985, 1988) as well as in the larynx (Munhall *et al.*, 1994). The task dynamics model reproduces these findings and others. However, the model's vocal tract and articulator degrees of freedom are greatly simplified, and the articulators themselves are massless, with no physiology or biomechanics represented.

Another global model has been proposed by Guenther (Guenther, 1994; Guenther *et al.*, 1998). The DIrections in auditory space to Velocities in Articulator space (DIVA) model represents a series of mappings between different frames of reference in speech motor control (e.g., auditory to articulatory mapping, articulatory to auditory mapping, phoneme to auditory mapping). These mappings are learned using neural networks trained in a 'babbling' phase in which random settings are applied to the articulator system and the acoustic consequences are used as feedback (for a similar approach, see Bailly *et al.*, 1997). Like the task dynamics model, a series of key speech motor control phenomena are accurately replicated by the model. For example, when subjects are asked to speak with a bite block between their teeth, they are quickly able to produce normal vowel acoustics, even though the configuration of individual articulators may be unique. The DIVA model can reproduce this finding.

The DIVA model has a number of important features. It includes mappings between perceptual and motor systems as well as kinesthetic and acoustic feedback representations of speech. It involves learning, and it has been extended into a developmental model (Callan *et al.*, 2000). Finally, it is consistent with current views on the detailed representations required for motor control (Wolpert and Kawato, 1998; Wolpert *et al.*, 1998). Perkell *et al.* (2000) have described these detailed representations in DIVA. However, like the task dynamics model, the DIVA model is not controlling a realistic vocal tract. The work to date has used Maeda's articulatory synthesizer (Maeda, 1990), which is a two-dimensional mid-sagittal representation with stylized articulators that have no biomechanics or physiology.

An ambitious statistical approach has been developed by researchers at Advance Telecommunications Research (ATR) laboratories (Kawato, 1989; Hirayama *et al.*, 1993, 1994; Vatikiotis-Bateson *et al.*, 1993, 1994; Wada *et al.*, 1995; Yehia *et al.*, 1998, 2000). This model differs in many ways from existing models of speech motor control (Saltzman, 1986; Saltzman and Munhall, 1989; Guenther, 1994) in that it attempts to take into account the dynamics of the actual articulatory system and the fact that all parameters are derived from observed data. Briefly, the current model attempts to simulate the complete process from linguistic primitives to the dynamics and kinematics of the articulatory system, and finally to the output acoustics. The computation of motor commands (and thus movement trajectories) is constrained by global settings of the speech apparatus that last for a number of segments (e.g., for speaking rate). Linguistic units are provided as input to the speech-planning system and converted to a sequence of targets in vocal tract, acoustic, and articulator planning spaces.

The actual computation of the trajectories requires a knowledge of the physiological and biomechanical structures of the vocal tract in order to deal with the complex and nonlinear mappings between muscle activity and force, muscle force and movement, movement and speech acoustics, etc. It has been postulated that the control system builds up this knowledge though experience during speech development. In other words, the nervous system learns dynamic models of its own motor system as an aid to controlling the complex dynamics of articulation. The ATR group was among the first to recognize the importance of these 'internal models' of the articulator system as a major component of the control system in speech. An internal model is a neural representation of the spatial (kinematic), force (dynamic), and feedback (e.g., acoustic, proprioceptive) characteristics of movements that could be used by the nervous system to predict movement outcome (Miall and Wolpert, 1996; Kawato, 1999).

Like Guenther's model, the structure of the ATR model is composed of a series of learned mappings, and the plausibility of this statistical mapping has been demonstrated for a number of different dimensions and articulator systems (Yehia *et al.*, 1998). This is a complex modeling program, but one that is addressing the full complexity of the articulation process (for an overview of the approach, see Munhall *et al.*, 2000).

CONCLUSION

The speech motor system is one of the most complex human movement systems. The study of its control requires detailed models of the articulators in action, including simulations of their physiology and biomechanics. Surprisingly, recent studies suggest that the nervous system itself requires models of its own articulator system to accomplish trajectory planning. Research progress in this area will depend on the integration of detailed individual articulator models into global models of the full articulation process.

References

Abbs JH and Gracco VL (1984) Control of complex motor gestures: orofacial muscle responses to load perturbations of the lip during speech. *Journal of Neurophysiology* **51**: 705–723.

Bailly G, Laboissière R and Galván A (1997) Learning to speak: speech production and sensori-motor representations. In: Morasso P and Sanguineti V (eds) *Self-Organization, Computational Maps and Motor Control*, pp. 593–615. Amsterdam, Netherlands: Elsevier.

Callan DE, Kent RD, Guenther FH and Vorperian HK (2000) An auditory-feedback-based neural network model of speech production that is robust to developmental changes in the size and shape of the articulatory system. *Journal of Speech, Language and Hearing Research* **43**: 721–736.

Chiba T and Kajiyama M (1941/1958) *The Vowel: its Nature and Structure*. Tokyo, Japan: Phonetic Society of Japan.

Dang J and Honda K (in press) A physiological model of a dynamic vocal tract for speech production. *Journal of the Acoustical Society of Japan*.

Fant G (1960) *Acoustic Theory of Speech Production*. The Hague, Netherlands: Mouton.

Gallistel CR (1999) Coordinate transformations in the genesis of directed action. In: Bly BM and Rumelhart DE (eds) *Cognitive Science*, pp. 1–42. New York, NY: Academic Press.

Gracco VL and Abbs JH (1985) Dynamic control of the perioral system during speech: kinematic analyses of autogenic and nonautogenic sensorimotor processes. *Journal of Neurophysiology* **54**: 418–432.

Gracco VL and Abbs JH (1988) Central patterning of speech movements. *Experimental Brain Research* **71**: 515–526.

Guenther F (1994) A neural network model of speech acquisition and motor equivalent production. *Biological Cybernetics* **72**: 43–53.

Guenther FH, Hampson M and Johnson D (1998) A theoretical investigation of reference frames for the planning of speech movements. *Psychological Review* **105**: 611–633.

Hardcastle WJ and Hewlett N (eds) (1999) *Coarticulation: Theory, Data and Techniques*. Cambridge, UK: Cambridge University Press.

Henke W (1966) *Dynamic Articulatory Models of Speech Production Using Computer Simulation*. Unpublished doctoral dissertation, Massachusetts Institute of Technology, Cambridge, MA.

Hirayama M, Vatikiotis-Bateson E and Kawato M (1993) Physiologically based speech synthesis using neural networks. *Institute of Electronics, Information and Communication Engineers (IEICE) Transactions* **E76-A**: 1898–1910.

Hirayama M, Vatikiotis-Bateson E and Kawato M (1994) Inverse dynamics of speech motor control. In: Hanson SJ, Cowan JD and Giles CL (eds) *Advances in Neural Information Processing Systems*, vol. 6, pp. 1043–1050. San Mateo, CA: Morgan Kaufmann.

Hirose H (1988) High-speed digital imaging of vocal fold vibration. *Acta Oto-Laryngologica* **458**: 151–153.

Honda K (1995) Laryngeal and extra-laryngeal mechanisms of F0 control. In: Bell-Berti F and Raphael LJ (eds) *Producing Speech: Contemporary Issues*, pp. 215–232. New York, NY: American Institute of Physics.

Kakita Y, Hirano M and Ohmaru K (1981) Physical properties of the vocal fold tissue: measurements on excised larynges. In: Stevens K and Hirano M (eds) *Vocal Fold Physiology*, pp. 377–397. Tokyo, Japan: University of Tokyo Press.

Kawato M (1989) Motor theory of speech perception revisited from the minimum torque-change neural network model. In: *Eighth Symposium on Future Electron Devices*, pp. 141–150. Tokyo.

Kawato M (1999) Internal models for motor control and trajectory planning. *Current Opinions in Neurobiology* **9**: 718–727.

Kent R and Moll K (1975) Articulatory timing in selected consonant sequences. *Brain and Language* **2**: 304–323.

Laboissière R, Ostry D and Feldman A (1996) Control of multi-muscle systems: human jaw and hyoid movements. *Biological Cybernetics* **74**: 373–384.

Levelt WJM (1989) *Speaking: from Intention to Articulation*. Cambridge, MA: MIT Press.

Löfqvist A and Gracco VL (1997) Lip and jaw kinematics in bilabial stop consonant production. *Journal of Speech, Language and Hearing Research* **40**: 877–893.

Lucero JC and Munhall KG (1999) A model of facial biomechanics for speech production. *Journal of the Acoustical Society of America* **106**: 2834–2842.

Maeda S (1990) Compensatory articulation during speech: evidence from the analysis and synthesis of vocal tract shapes using an articulatory model. In: Hardcastle WJ and Marchal A (eds) *Speech Production and Speech Modeling*, pp. 131–149. Boston, MA: Kluwer Academic.

Miall RC and Wolpert DM (1996) Forward models for physiological motor control. *Neural Networks* **9**: 1265–1279.

Moll K (1960) Cinefluorographic techniques in speech research. *Journal of Speech and Hearing Research* **3**: 227–241.

Munhall KG (2001) Functional imaging during speech production. *Acta Psychologica* **107**: 95–117.

Munhall K, Löfqvist A and Kelso JAS (1994) Lip–larynx coordination in speech: effects of mechanical perturbations to the lower lip. *Journal of the Acoustical Society of America* **96**: 3605–3616.

Munhall KG, Kawato M and Vatikiotis-Bateson E (2000) Coarticulation and physical models of speech production. In: Broe M and Pierrehumbert J (eds) *Papers in Laboratory Phonology. V. Acquisition and the Lexicon*, pp. 9–28. Cambridge, UK: Cambridge University Press.

Ostry D, Gribble P and Gracco V (1996) Coarticulation of jaw movements in speech production: is context sensitivity in speech kinematics centrally planned? *Journal of Neuroscience* **16**: 1570–1579.

Perkell JS, Guenther FH, Lane H *et al.* (2000) A theory of speech motor control and supporting data from speakers with normal hearing and with profound hearing loss. *Journal of Phonetics* **28**: 233–272.

Perrier P, Ostry DJ and Laboissière R (1996) The equilibrium point hypothesis and its application to speech motor control. *Journal of Speech and Hearing Research* **39**: 365–378.

Pitermann M and Munhall KG (2001) An inverse dynamics approach to face animation. *Journal of the Acoustical Society of America* **110**: 1570–1580.

Saltzman EL (1986) Task dynamic coordination of the speech articulators: a preliminary model. Generation and modulation of action patterns. In: Heuer H and Fromm C (eds) *Experimental Brain Research*, series 15, pp. 129–144. New York, NY: Springer-Verlag.

Saltzman EL and Munhall KG (1989) A dynamical approach to gestural patterning in speech production. *Ecological Psychology* **1**: 333–382.

Sanguineti V, Laboissière R and Ostry DJ (1998) A dynamic biomechanical model for neural control of speech production. *Journal of the Acoustical Society of America* **103**: 1615–1627.

Shadle C (1985) *The acoustics of fricative consonants*. Research Laboratory of Electronics, Technical Report 506. Cambridge, MA: Massachusetts Institute of Technology.

Shaiman S (2001) Kinematics of compensatory vowel shortening: the effect of speaking rate and coda composition on intra- and inter-articulatory timing. *Journal of Phonetics* **20**: 89–107.

Shiller DM, Ostry DJ and Gribble PL (1999) Effects of gravitational load on jaw movement in speech. *Journal of Neuroscience* **19**: 9073–9080.

Titze IR (1973) The human vocal chords: a mathematical model. Part I. *Phonetica* **28**: 129–170.

Titze IR (1974) The human vocal cords: a mathematical model. Part II. *Phonetica* **29**: 1–21.

Titze IR (1994) *Principles of Voice Production*. Englewood Cliffs, NJ: Prentice Hall.

Vatikiotis-Bateson E, Hirayama M, Wada Y and Kawato M (1993) Generating articulator motion from muscle activity using artificial neural networks. *Annual Bulletin of the Research Institute of Logopedics and Phoniatrics (RILP)* **27**: 67–77.

Vatikiotis-Bateson E, Tiede M, Wada Y, Gracco V and Kawato M (1994) Phoneme extraction using via point estimation of real speech. In: *Proceedings of 3rd International Conference on Spoken Language Processing, ICSLP-94*, pp. 531–534. Yokohama, Japan.

Wada Y, Koike Y, Vatikiotis-Bateson E and Kawato M (1995) A computational theory for movement pattern recognition based on optimal movement pattern generation. *Biological Cybernetics* **73**: 15–25.

Warren N and Arai T (2001) Japanese mora-timing: a review. *Phonetica* **58**: 1–25.

Wilhelms-Tricarico R (1995) Physiological modeling of speech production: methods for modeling of soft-tissue articulators. *Journal of the Acoustical Society of America* **97**: 3085–3098.

Wolpert DM and Kawato M (1998) Multiple paired forward and inverse models for motor control. *Neural Networks* **11**: 1317–1329.

Wolpert DM, Miall C and Kawato M (1998) Internal models in the cerebellum. *Trends in Cognitive Sciences* **2**: 338–347.

Yehia HC, Rubin PE and Vatikiotis-Bateson E (1998) Quantitative association of vocal-tract and facial behavior. *Speech Communication* **26**: 23–44.

Yehia H, Kuratate T and Vatikiotis-Bateson E (2000) Facial animation and head motion driven by speech acoustics. In: Hoole P (ed.) *Fifth Seminar on Speech Production: Models and Data*, pp. 265–268. Munich, Germany.

Artificial Intelligence, Gödelian Arguments against

Intermediate article

Peter Slezak, University of New South Wales, Sydney, Australia

CONTENTS

Gödel's incompleteness theorem says that there are arithmetical truths expressible in certain formal systems which cannot be proven within those systems. This result reveals certain inherent limitations on what is computable, which, according to some theorists, show that minds are not machines because they are not subject to the same limitations.

WHAT ARE GÖDELIAN ARGUMENTS?

Gödelian arguments attempt to exploit the famous incompleteness theorem of Kurt Gödel (1931) to show that the mind cannot be a machine. In slightly different forms, such arguments have been advanced by J. R. Lucas (1961) and R. Penrose (1989, 1995). Gödel himself expressed sympathy for such views, but there are significant differences among these accounts regarding the relevance of the mathematics to the nature of the mind.

In general terms, Gödelian arguments conclude that minds cannot be machines because there is something that minds can do but no machine can do. A similar kind of argument was made by René Descartes (1637), who claimed that no purely mechanical contrivance could show the infinite, creative novelty revealed in human language or action guided by knowledge. Descartes' argument concerned machines as understood at that time – namely, clockwork devices with a finite repertoire of behaviors. It was only with the work of Alan Turing (1936) that we arrived at a fundamentally different conception of machines that are not subject to the limitations recognized by Descartes.

Gödel's incompleteness result holds for formal, axiomatic systems such as that described in the *Principia Mathematica* of Russell and Whitehead (1910–1913), but it may be stated as a limitation on any computing machine that realizes a formal system. Gödel's theorem shows that there are true arithmetical propositions that are neither provable nor disprovable; that is, they are undecidable within their own system. Gödel's proof specifies such an arithmetical statement, but the claim that it is undecidable is not itself an arithmetical proposition, but a metamathematical one, since it is concerned with the notion of provability. That is, Gödel's theorem concerns statements about mathematical systems rather than statements of, or within, such systems. The undecidability of propositions within arithmetic is a syntactical claim of such metamathematical proof theory. However, since the undecidable proposition is, in fact, true, the formal axiomatic system is said to be 'incomplete', meaning that there are truths expressible in the system that are unprovable in it.

Gödel himself favored an anti-mechanist position. However, Hao Wang (1987, p. 146) says that Gödel recognized that 'his theorem does not settle the question of mind surpassing matter' and 'unlike certain ignorant philosophers, G[ödel] realizes that his incompleteness theorem does not by itself imply that the human mind surpasses all machines' (1987, p. 197). Wang cites the additional premises that Gödel required, including (1) that the mind is separate from matter, and (2) that the mind is not static, but constantly developing and therefore there is reason to believe the mind's states might converge to infinity (see Wang, 1974, p. 325).

Regarding the first of the above premises Gödel believes that materialism is a 'prejudice of our time' which will be disproved with the progress of science. His skepticism about the thesis of mechanism depends essentially on a belief in dualism, indeed a vitalism (Wang, 1974, p. 326; 1996, p. 193) and not directly on the implications of his incompleteness theorem.

The second of the above premises is the more significant. It is further explained by Wang (1974, p. 325; 1996, p. 194), who cites Gödel's conviction that 'Turing's argument for the adequacy of his definition [of computation] includes an erroneous proof of the stronger conclusion that minds and machines are equivalent'. Gödel takes the argument of Turing's (1936) landmark paper as supporting the mechanist thesis, but Gödel suggests that the case is inconclusive because it depends on the assumption that a finite mind is capable of only a finite number of distinguishable states. Nevertheless, Gödel recognizes that 'the incompleteness results do not rule out the possibility that there is a theorem-proving computer which is in fact equivalent to mathematical intuition' (Wang, 1996, p. 186). In Wang's opinion, 'clearly he himself realized that such a refutation of mental computabilism is not convincing, as we can infer from his continued efforts to find other ways to achieve the desired refutation' (1996, p. 185). While inclining towards the anti-mechanist position, Gödel says that his incompleteness results imply only that if we are machines then 'either we do not know the exact specification of the computer or we do not know that it works correctly' (Wang, 1996, p. 186). Related conclusions about the limitations of self-knowledge have been drawn from Gödelian arguments by Whiteley (1962), Benacerraf (1967) and Slezak (1982, 1984).

J. R. Lucas (1961) articulated the Gödelian argument against mechanism in a detailed form which has given rise to a considerable critical literature. Lucas states the fundamental difference between minds and machines by saying that it follows from Gödel's theorem 'that given any machine which is consistent and capable of doing arithmetic, there is a formula which it is incapable of producing as being true – i.e., the formula is unprovable-in-the-system – but which we can see to be true' (1968, p. 44). From this alleged difference between the abilities of minds and machines, Lucas concludes 'that no machine can be a complete or adequate model of the mind, that minds are essentially different from machines' (1968, p. 44).

Against critics, Lucas (1999) has continued to insist that his argument must be understood as having a special, essentially 'dialectical' form in which he does not claim the superiority of minds in all respects over all machines, but only that a mind can surpass any particular machine that might be proposed as a model in each case. Lucas claims that he can refute each particular instance of a machine offered by the 'mechanist' as a model of Lucas's mind. However complicated a machine we construct, it will always be vulnerable to the Gödel procedure for finding a formula unprovable-in-the-system that corresponds to the machine. 'This formula the machine will be unable to produce as being true, although a mind can see that it is true.' Lucas concludes that, since this procedure may be repeated endlessly with any machine proposed by the mechanist, 'the mind always has the last word' (1961, p. 48).

RESPONSES TO GÖDELIAN ARGUMENTS

Alan Turing (1950) considers the 'mathematical objection' to intelligent machines specifically arising from Gödel's theorem. In his brief treatment, Turing acknowledges that Gödel's theorem shows that there are certain questions which any machine must fail to answer correctly. However, Turing asks whether this 'proves a disability of machines to which the human intellect is not subject ... it has only been stated, without any sort of proof, that no such limitations apply to the human intellect' (1950, p. 16). Nevertheless, Turing concedes that 'a certain feeling of superiority' over machines is 'no doubt quite genuine' and seems to acknowledge the force of the anti-mechanist argument, while discounting its importance.

Lucas (1961, p. 49) notes the irrelevance of such an argument to the central question at issue, which is not whether minds are superior to machines, but whether they are the same. Conceding that machines may be superior to minds in many, perhaps even most, respects, Lucas notes that it is enough to show that they are unable to do one thing, however trivial, that a mind is able to do, to establish that minds and machines are essentially different.

Putnam (1960) has argued that Lucas's argument rests on a misapplication of Gödel's theorem, which in fact asserts the conditional 'if *T* is consistent, then *G* is true', where *T* is the formal system and *G* is the undecidable sentence. Putnam's point is that Lucas cannot assert the truth of the Gödel sentence *G* alone, but only the entire conditional of which it is the consequence. Lucas (1968, p. 158) has conceded that the question of consistency is a matter of 'faith' and not susceptible to formal proof. However, Lucas points to the normative character of consistency in regulating our deliberately asserted statements. In any case, Putnam's objection is not decisive, because he says that the consistency of *T* is only 'unlikely if *T* is very complicated'.

Pertinently, Benacerraf (1967, p. 19) asks what exactly it is that Gödel's theorem precludes the machine from doing. The impossibility is clearly to prove *G* from the machine's own axioms according to the machine's own rules. Benacerraf asks: 'but can Lucas do that?' Evidently not. Chihara (1972), too, has noted a certain unclarity in Lucas's claimed ability to 'produce as true' the

Gödel sentence. Lucas uses the expression 'produce as true' to capture both what a machine is precluded from doing and also what a mind can do. Slezak (1982) suggests that Lucas's phrase appears to conflate the notions of proof and truth, which Gödel showed must be distinguished. What minds can do in following Gödel's theorem is to establish the truth of *G* at the level of a metamathematical argument, whereas what the machine cannot do is to generate the sentence *G* from the system's axioms. Thus, the notion of 'produce as true' is being used in two different senses, to describe both what minds can do and what machines cannot do. In spite of the crucial difference between truth and provability, Lucas incorporates them both in his notion of 'produce as true', and relies on the resulting equivocation to establish his desired conclusion. Whatever may be the admitted difficulty in understanding how minds can determine the truth of sentences such as *G* (Lucas, 1968, p. 148), Gödel's theorem has no bearing on this, nor does Gödel's theorem suggest any reason why machines might be incapable of doing it too.

Lucas insists that his argument depends on the mechanist proferring a specific machine which purports to model Lucas's mind before he can refute it by 'producing as true' its Gödel sentence. Lucas (1968, p. 146) says: 'If the mechanist maintains any specific thesis, I show that a contradiction ensues. But only *if*. It depends on the mechanist making the first move and putting forward his claim for inspection.' However, the relevant facts for assessing the mechanist thesis are objective ones concerning the relative abilities of minds and machines. These could not be inherently dependent on whether the mechanist chooses to make the first move by proposing a specific model. Lucas's dialectical strategy relies on the fact that he is able to determine the truth of a machine's unprovable sentence only by virtue of being outside the system; that is, by operating in the metalanguage. However, this amounts to merely restating Gödel's result, and not to demonstrating any essential difference between minds and machines. As J. Webb (1980, p. 230) has pointed out, 'the real source of Lucas's feeling of superiority here is just the *effectiveness* with which he can find the Achilles' heel of any machine'. Gödel's diagonal argument is itself a formal procedure which can be carried out by a machine.

Benacerraf's (1967) reconstruction of Lucas's argument leads him to conclude that, if we are Turing machines, then it seems we may be precluded from obeying the injunction 'know thyself'. Benacerraf construes this as a limitation on empirical psychology. However, Slezak (1982, 1984) has suggested that, if the indexical, self-referential features of the relevant statements are properly taken into account, Benacerraf's own account suggests a limitation for self-knowledge only in the sense of first-person, introspective knowledge. As we saw earlier, this was precisely the alternative to an anti-mechanist conclusion that Gödel himself suggested. The appropriate question for Lucas is not what he can know about the machine, but rather what he can know about himself. Slezak (1982) suggests that the close affinities of Gödel's result with familiar paradoxes of self-knowledge (Popper, 1950; Gunderson 1970) are highly suggestive, so that, far from refuting mechanism, Gödel's theorem may even provide persuasive support for it. As Webb (1980, p. vii) suggests in his detailed study, quite the reverse of Lucas's claims, 'Gödel's work was perhaps the best thing that ever happened to both mechanism and formalism'.

A foundational tenet of cognitive science is the computational character of the mind. The 'functionalist' view of so-called 'strong AI' (Searle, 1980) asserts that minds may be not only simulated on machines, but embodied or realized in them. Accordingly, as Penrose (1995) asserts, if Gödelian arguments are sound, then the ambitions of cognitive science and artificial intelligence are unattainable in principle. In a slight variant of Lucas's argument, Penrose (1995, p.75) takes the computational insolubility of the 'halting problem' to imply that mathematical intuition cannot be formalized in any algorithm. In particular, Penrose sees Gödelian arguments as showing that certain features of minds, such as consciousness, free will, creativity, and insight, are beyond the realm of ordinary rule-governed – and therefore psychological or even physical – explanation. However, Dennett (1989) suggests that such arguments constitute a *reductio ad absurdum* which counts against the Gödelian arguments themselves. Moreover, Dennett points out that it is a non sequitur to conclude that no algorithm can characterize a human ability such as playing chess, or understanding mathematics, just because there is no rule for checkmate, or mathematical insight. These tasks may be performed at any level, including our own, by heuristic problem-solving methods, which are algorithms in the sense of being rule-governed computer programs, but are not recipes that guarantee success.

References

Benacerraf P (1967) God, the Devil, and Gödel. *The Monist* **51**: 9–32.

Chihara C (1972) On alleged refutations of mechanism using Gödel's incompleteness results. *Journal of Philosophy* **64**: 507–526.

Dennett D (1989) Murmurs in the Cathedral. *Times Literary Supplement* **4513**: 1055–1056.

Descartes R (1637/1985) *The Philosophical Writings of Descartes*, vol. I: *Discourse and Essays*, translated by J. Cottingham, R. Stoothoff and D. Murdoch. Cambridge, UK: Cambridge University Press.

Gödel K (1931) Über formal unentscheidbare Sätze der Principia Mathematica und verwandter Systeme I. *Monatshefte für Mathematik und Physik* **38**: 173–198. [Translated as 'On formally undecidable propositions of the *Principia Mathematica* and related systems I'. In: Davis M (ed.) (1965) *The Undecidable*. New York, NY: Raven Press.]

Gunderson K (1970) Asymmetries and mind–body perplexities. In: Radner M and Winokur S (eds) *Minnesota Studies in the Philosophy of Science*, vol. IV, pp. 273–309. Minneapolis, MN: University of Minnesota Press.

Lucas JR (1961) Minds, machines and Gödel. *Philosophy* **36**: 112–127. [Reprinted in: Anderson AR (ed.) (1964) *Minds and Machines*, pp. 43–59. Englewood Cliffs, NJ: Prentice-Hall.]

Lucas JR (1968) Satan stultified: a rejoinder to Paul Benacerraf. *The Monist* **52**: 145–158.

Lucas JR (1999) *The Gödelian Argument: Turn Over the Page*. http://users.ox.ac.uk/~jrlucas/turn.html.

Penrose R (1989) *The Emperor's New Mind: Concerning Computers, Minds and the Laws of Physics*. Oxford, UK: Oxford Unversity Press.

Penrose R (1995) *Shadows of the Mind: A Search for the Missing Science of Consciousness*. London, UK: Vintage.

Popper K (1950) Indeterminism in quantum physics and in classical physics. *British Journal for the Philosophy of Science* **1**: 117–133.

Putnam H (1960) Minds and machines. In: Hook S (ed.) *Dimensions of Mind*, pp. 138–164. New York, NY: New York University Press. [Reprinted in: Anderson AR (ed.) (1964) *Minds and Machines*, pp. 72–97. Englewood Cliffs, NJ: Prentice-Hall.]

Russell B and Whitehead AN (1910, 1912, 1913) *Principia Mathematica*, 3 vols. Cambridge, UK: Cambridge University Press.

Searle J (1980) Minds, brains and programs. *Behavioral and Brain Sciences* **3**: 417–424.

Slezak P (1982) Gödel's theorem and the mind. *British Journal for the Philosophy of Science* **33**: 41–52.

Slezak P (1984) Minds, machines and self-reference. *Dialectica* **38**(1): 17–34.

Turing AM (1936) On computable numbers with an application to the entscheidungsproblem. *Proceedings of the London Mathematical Society, Series 2* **42**: 230–265. [Reprinted in: Davis M (ed.) (1965) *The Undecidable*. New York, NY: Raven Press.]

Turing AM (1950) Computing machinery and intelligence. *Mind* **59**: 433–460. [Reprinted in: Anderson AR (ed.) (1964) *Minds and Machines*, pp. 4–30. Englewood Cliffs, NJ: Prentice-Hall.]

Wang H (1974) *From Mathematics to Philosophy*. New York, NY: Humanities Press.

Wang H (1987) *Reflections on Kurt Gödel*. Cambridge, MA: Bradford/MIT Press.

Wang H (1996) *A Logical Journey: From Gödel to Philosophy*. Cambridge, MA: MIT Press.

Webb JC (1980) *Mechanism, Mentalism and Metamathematics: An Essay on Finitism*. Dordrecht, Netherlands: Reidel.

Whiteley CH (1962) Minds, machines and Gödel: a reply to Mr Lucas. *Philosophy* **37**: 61–62.

Further Reading

Dennett D (1990) Betting your life on an algorithm. *Behavioral and Brain Sciences* **13**: 660–661.

Hofstadter DR (1979) *Gödel, Escher, Bach: An Eternal Golden Braid*. New York, NY: Basic Books.

Nagel N and Newman JR (1958) *Gödel's Proof*. New York, NY: New York University Press.

Penrose R (1997) *The Large, the Small and the Human Mind*. Cambridge, UK: Cambridge University Press.

Smullyan R (1987) *Forever Undecided: A Puzzle Guide to Gödel*. New York, NY: Alfred A. Knopf.

Artificial Intelligence, Philosophy of

Introductory article

Eric Dietrich, Binghamton University, Binghamton, New York, USA

CONTENTS

The philosophy of artificial intelligence examines the foundational assumptions, methodologies, and consequences of the field of artificial intelligence.

WHAT IS THE PHILOSOPHY OF ARTIFICIAL INTELLIGENCE?

The philosophy of artificial intelligence is a collection of issues primarily concerned with whether or not artificial intelligence (AI) is possible; that is, with whether it is possible to build an intelligent thinking machine. Of ancillary concern is the issue of whether humans and other animals are best thought of as machines themselves.

The most important of the 'whether-possible' problems lie at the intersection of theories of the semantic contents of thought and the nature of computation. One view of human thinking is that it is the manipulation of contentful thoughts. When we engage in such cognitive processes as making inferences, recognizing patterns, planning and executing activities, etc., the thoughts we have and manipulate pick out or refer to various things in our world. This view seems so plausible that it is difficult to imagine how it could be false; but when it is translated to computers, its truth is much easier to doubt, for it entails that machine cognition would be just algorithmic manipulation (computing) of certain kinds of contentful data structures. Here is where the philosophical problem enters: while thoughts obviously have content, it is far from clear that the computations and data structures do. When a computer adds 1 to 1, do its internal states actually denote, or refer to, the number 1? If a computer derives 'Fido will chase cats' from 'Dogs chase cats' and 'Fido is a dog,' do any of its internal states actually pick out Fido, dogs, cats, and the act of chasing? If computations are radically different from thoughts in that they cannot have semantic content, then it is unlikely that computers can think.

Such problems are usually referred to in AI and cognitive science as 'the problem of mental content' or 'the problem of representational content.' In philosophy, the problem of mental content is sometimes called 'the problem of intentionality.' Often this latter term is used when problems of content are combined with the problems of epistemology and consciousness. The Chinese Room problem fits in here (see later).

A second set of 'whether-possible' problems surrounds the nature of rationality. Humans constantly evaluate ways of achieving goals and rank them according to various measures such as the probability of success, efficiency, and consequences. Humans also evaluate the goals themselves, as well as the means needed to achieve the goals. In so doing, humans constantly gauge the relevance of one piece of information to another, the relevance of one goal to another, and the relevance of evidence to achieving a goal or holding a belief. Humans do this evaluation with varying degrees of success, but often are quite successful at it. However, some philosophers believe that computers cannot do such evaluation at all; or that if they can, their methods of evaluation are too brittle to capture anything as robust as human rationality. But if human-level rationality is not obtainable, then the project of building an intelligent machine will fail.

A third set of problems revolves around the seemingly transcendent reasoning powers of the human mind. All these problems begin with Kurt Gödel's famous 'incompleteness theorem'. This theorem states that logic together with number theory (and consistent extensions) contains true statements that are unprovable. Gödel's theorem basically states that any suitably robust, formal system is incomplete; but all computer systems precisely instantiate formal systems, hence all computer systems are incomplete. Moreover, from within the given formal system, the incompleteness cannot be proved; to do that, one has to step outside the formal system and prove a metamathematical result (this is exactly what Gödel did). Computer systems, being instantiated formal systems, cannot step outside of themselves. Hence humans can do something computers cannot, so humans aren't computers. And since humans' ability to step outside formal systems is a crucial part of their intelligence, it seems to follow that computers cannot be intelligent, at least not in the same way as humans.

A final collection of 'whether-possible' problems concerns the architecture of an intelligent machine. Currently there are four main positions in this debate, each claiming that its approach to building intelligent machines is the best (or the only) way to achieve success. Accordingly, the philosophical issues here are methodological: What are the real differences between the four main approaches? Is it likely that one could succeed where the others could fail? And if so, why?

There are many other important philosophical questions apart from the whether-possible type. Can a computer be conscious? Can a computer have a moral sense? Is it moral for humans to attempt to build an intelligent machine? If we did build such a machine, would turning it off be the equivalent of murder? If we had a race of such machines, would it be immoral to force them to work for us as slaves?

EARLY WORK ON AI

AI research is dedicated to the proposition that it is possible to build a machine, a computer, that is as intelligent as, if not more intelligent than, a human. Why would anyone believe that such an endeavor is possible? Computers are not even alive. A brief examination of the historical beginnings of AI will help to put into context the doubts about AI that philosophers raise.

The roots of AI are often said to lie in the work of Alan Turing and Alonzo Church whose development of the mathematical theory of computation was one of the most important developments in the entire history of mathematics. However, the relevance to AI of the theory of computation was established after AI had already emerged. The primary influence on the early AI researchers was the work of cyberneticists and neurophysiologists.

The first idea was that the computer was not just capable of numerical calculations; it was a general symbol manipulator capable of performing virtually any task having to do with information. The symbols manipulated could represent anything: words, propositions, concepts, dogs, cats, rocks, etc. The idea that computers could be more than 'number crunchers' was revolutionary. ENIAC, one of the first computers, was intended solely for numerical calculations. And Howard Aiken, who built the first electromechanical computer in the United States, once said: 'If it should ever turn out that the basic logics of a machine designed for the numerical calculation of differential equations coincide with the logics of a machine intended to make bills for a department store, I would regard this as the most amazing coincidence that I have ever encountered.' Aiken, of course, was amazed.

From the work of the neurophysiologists, around the middle of the twentieth century, came the idea that at least some neural activity was information processing, and hence that some kinds of thinking could be explained in terms of processing information.

Then, late in 1955, Alan Newell and Herbert Simon (and others) brought these two ideas together. Computers manipulate information; thinking is manipulating information; therefore, thinking might be computing. This insight resulted in a profound hypothesis that is the heart of AI. Newell and Simon also provided researchers with the fundamental ingredient of information manipulation, namely, the *symbol*. A symbol, or what is now called a *representation*, is considered to be a fundamental constituent of both thinking and computing.

THE PROBLEM OF REPRESENTATIONAL CONTENT

Philosophers have been intensely interested in symbols (words, primarily) and their meanings since before Socrates. Essentially, the puzzle is: how can we think about things in the world? It is clear that our thoughts about the things around us (such as a pet dog) are really quite different from the things themselves, if for no other reason than thoughts abstract and leave out information relative to what is being thought about (dog thoughts are not dogs). How then do these thoughts manage to represent what they do if they leave out information? How can one's thought, or mental representation, of one's pet dog be about *this* dog rather than some other dog if the representation leaves out information? Moreover, we can think about things that are not immediately present (the dark side of the moon), things that do not exist (unicorns), and things that cannot exist (round squares). How is this possible?

However it is that human thoughts have content, many doubt that the computations going on inside a computer are about anything at all. When a computer adds 1 and 1 to get 2, its data structures do not seem to refer to numbers and adding. It is noted sometimes that computers do not seem to *know* anything about the numbers 1 and 2 and the function 'plus'. When a computer adds, a cascade of causal processes occurs which implements an algorithm which in turn, if followed exactly, guarantees that two numbers will be added. Consider an analogy: When a mousetrap catches a mouse, the mousetrap does not represent the mouse; it merely catches it (the mousetrap clearly does not know anything about mice or even the mouse it is catching). Why is the situation any different for computers and their processes? If there is no difference, perhaps computations are contentless. But thoughts are not. Hence thinking cannot be computing. Hence a computer cannot think: it can merely compute. Yet we saw above that there are good reasons for hypothesizing that thinking is computing, and that a computer can think.

One way out of this dilemma is to attempt to develop a philosophical theory of mental content that clearly explains how thoughts get the content that they do. Then we could just check to see whether computations could get content in the same way. If they can, then AI is on firm ground; if they cannot, then AI is without hope.

Much work is going into this project. For many years, there were two general strands, depending on how one views the nature of semantics. One strand investigated world–mind relations. It saw semantics as essentially associated with truth, causation, and getting along in the world. The second strand investigated mind–mind relations. It saw semantics as essentially associated with being able to draw certain inferences, construct plans, and in general determine how one thought or representation relates to another. Philosophers are now tying these two strands together in an effort to develop a complete theory of representational content. Various camps emphasize the different strands, but most agree that both strands are needed.

Neither strand seems beyond the reach of computers. Computers can be causally connected to their environments – such machines are called *robots* or *agents*, depending on whether the relevant environments are virtual or not. And computers can easily implement vast networks of representations and determine how the representations relate to one another. This strand, in particular, receives much attention in AI. So, to many AI scientists and AI-friendly philosophers, the future looks bright: on the best proto-theories of semantics they have, nothing about content seems beyond the capacities of computers.

THE FRAME PROBLEM AND HUMAN RATIONALITY

The human mind does more than merely think about things. When it thinks, the mind is governed, at least in part, by the rules of logic and inductive reasoning, and by analyses of the strength of evidence for and against certain beliefs, desires, and actions. In short, the mind is rational, at least from time to time. It is plausible that computers would be quite good at calculating evidential strength, using various logics. So, on the face of it, computers could be rational, maybe even more rational than humans. But again, closer analysis reveals serious difficulties.

Suppose a human is given a new piece of information: it is snowing outside. Such new knowledge occasions some new inferences and updating of older information, such as: 'If I go outside, I'll need my coat,' 'I will have to shovel the sidewalk in a few hours,' 'The roads will be slick when I drive, so I should drive slowly and hence leave earlier for my meeting,' etc. However, given the information that it is snowing outside, it would not occur to a human to check to make sure her phone number is still the same, or to check whether her bank account is still active. Snowing is not the sort of thing of that affects phone numbers or bank accounts. This is so obvious that it seems silly even to wonder about it. But when it comes to programming an intelligent machine, this problem is not silly; indeed it is quite serious.

When a computer is given a new piece of information – it is snowing outside – it *does* have to check everything else in its store of knowledge to see what has to be updated and what does not. In order to know that a phone number does not have to be updated given that it is snowing outside, the computer has to check the phone number in order to determine that it is not the sort of thing that has to be updated because of the snow. But what is true for snow is true in general. Hence, any time a new piece of information is input, an intelligent computer has to check everything it knows in order to find out what has to be updated. Such checking takes a lot of time because, typically, the computer's knowledge base is large. So an intelligent computer will spend most of its time checking its knowledge base finding out that most things do not need updating.

This problem is known as the *frame problem*. It was first formulated by John McCarthy and Pat Hayes in 1969. The alleged implications of the frame problem for machine rationality were first discussed by Jerry Fodor in 1983; he revisited the issue in 2000. Since then, narrow aspects of the problem have been solved using nonmonotonic logics, but the general philosophical problem remains. We know why humans are not subject to the frame problem. Humans use the following rational heuristic to update their knowledge: given a piece of new information, update the knowledge that is *obviously relevant* to the new knowledge and leave everything else alone. This is only a heuristic because it does not guarantee that *all and only* the relevant knowledge will be updated, but as a heuristic it works quite well. The problem is that AI researchers do not have, but philosophers insist on, a universal, precise definition of 'obviously relevant.' The answer to the philosopher, however, is that AI does not need a universal, precise definition of 'obviously relevant.'

What gives the 'obviously relevant' heuristic its power in humans is that humans can alter what they consider relevant information; their definition of 'obviously relevant' can change. For example, we can learn that smoking is relevant to lung cancer, that driving is relevant to global weather changes, that computers are relevant to understanding the human mind, etc. We can see analogies such as the one between the structure of the solar system and the structure of the atom, which may suggest that the structure of the solar system is relevant to the structure of the atom. We can construct new categories 'on the fly', relevant to new situations: given that the house is on fire, we can construct the new category 'things to save from a burning house' immediately. Understanding how humans do these things requires understanding how humans do their science, how analogical thinking works, and how humans form new categories and concepts. All of these are difficult questions, but they do appear tractable. Computers, too, successfully use the 'update the information that is obviously relevant' heuristic. It is relatively easy to program simple versions of this heuristic; and, using machine learning techniques, what counts as 'obviously relevant' can change over time. AI researchers have not yet implemented a machine that is as flexible as a human in defining what counts as 'obviously relevant', but there are reasons to believe that they might.

Many philosophers want to know what 'obviously relevant' means across all cases and uses of the heuristic. They want, for example, *one metric* that tells them, for any question and any domain, which information in that domain is relevant to that question. Other philosophers insist that such a thing cannot be obtained. They point out that in-principle solutions are rare in science, and are almost completely unknown outside of basic theoretical physics. Many AI researchers now believe that 'obviously relevant' is going to have only local, pragmatic definitions that change constantly. The AI research task now is to implement the capability to pragmatically use and alter such definitions.

GÖDEL AND METAMATHEMATICS

The logical problems surrounding self-reference and the incompleteness of certain logical systems seem to some to present insurmountable problems to building an intelligent machine. In 1931, Kurt Gödel proved that every consistent, formal, logical system which includes some number theory contains true but undecidable propositions; that is, propositions that cannot be proved true nor proved false *within the given formal system*. Gödel constructed a proposition in logic, which rendered in English is: 'The proposition with the Gödel number G cannot be proven.' (This proposition, often referred to as a *Gödel sentence* for the given formal system, was actually written in first-order logic with identity, the natural numbers, and the arithmetic operations added in.) In Gödel's construction, the sentence with Gödel number G is the proposition 'The proposition with the Gödel number G cannot be proven.' When coupled with the assumption that the formal system, logic with number theory, is consistent, Gödel's theorem establishes that the formal system is necessarily incomplete: there are true propositions that cannot be proven, namely G itself.

This stunning result in logic is thought by some to present problems for AI and intelligent machines because it is assumed that computers are formal systems, no stronger than logic with number theory and hence susceptible to Gödel's results. Hence, it seems, humans know something the machine cannot know, namely that G is unprovable. Hence, humans can do something which computers cannot. To the extent that this kind of insight is crucial to intelligence, it seems that computers cannot be intelligent.

The most common reply to this argument is that it misses the role of the assumption of consistency. What we humans actually know, and what Gödel proved, is that *if* a suitably powerful formal system is consistent, *then* it contains an undecidable proposition like G, above. And for complex, robust formal systems like logic with number theory, we cannot know if they are consistent (and their consistency cannot be proved within the system, either), so we merely assume consistency. Gödel's famous theorem is not that formal systems are incomplete; it is that if the given system is consistent, then it is incomplete. An intelligent computer can prove Gödel's theorem, it seems. All that is required is the concept of consistency (given the rest of the formal apparatus). Let T be a computer and *G* its Gödel sentence (i.e., *G* is the Gödel sentence of the formal system describing T). It is not the case that we humans know that *G* is true, but T cannot know *G* is true. We know that if the system describing T is consistent, *G* is true. But T can know this, too, provided that it has the concept of consistency. And, it seems, there is nothing inherently noncomputational about this concept.

In spite of this reply, the Gödelian whether-possible argument continues to strike many as a serious problem for AI.

COGNITIVE ARCHITECTURE AND AI METHODOLOGY

A fourth major problem for philosophers is what is the best architecture for an intelligent machine. Deciding on an architecture goes hand in hand with which methodology to use to build an intelligent machine. Currently, there are four major architectures that AI researchers and cognitive scientists are using:

1. Symbolic architectures based on high-level computational representations and algorithms. High-level representations are those that are relatively easy to translate into sentences in a natural language, such as English.
2. Connectionist architectures (sometimes called artificial neural networks) based on some level of neural organization within the brain.
3. Dynamic system architectures based not on the 'hardware' of the brain, but on the dynamical properties of neurons.These architectures are based on the theory of nonlinear dynamics.
4. Embodied robotic architectures based on layering on more and more abstract kinds of robotic control until abstract thought is achieved.

The issues surrounding these four approaches to building an intelligent machine are fascinating. Briefly they are:

- To what extent can intelligence emerge out of non-intelligent processing?
- What is the relationship between perception and cognition?
- What is the role of high-level representations in intelligent thought?
- What is the role of representations, at any level, in intelligent thought?

We can sum up the fourth philosophical problem as follows. The last three architectural approaches discussed above are closely tied to what is known about human cognitive architecture. The idea is that since humans are intelligent, perhaps AI ought to model them. But the first architectural approach is committed to the view that the details of human cognitive architecture (e.g. that we have brains) is merely an implementational detail and not relevant (at least not in theory, though it may be practically). Is this view plausible?

THE CHINESE ROOM ARGUMENT

Probably the most famous 'whether-possible' argument is the anti-AI position called the 'Chinese Room argument'. This does not fall neatly into any of the four categories discussed so far. Instead, it is an amalgam of philosophical issues concerning representational content, a system (machine or human) knowing and understanding its environment, and consciousness.

Imagine someone locked in a room. Chinese texts written on sheets of paper are slipped through a slot in the door. The person's task is to take the sheets, write further Chinese characters in response to them, and pass the responses back out through the slot. Imagine also that the person knows no Chinese at all, either written or spoken. Fortunately, the room includes a large book written in English (or the language in which the person is fluent) which stipulates what to write in response to certain Chinese characters in certain sequences. The person goes about laboriously reading and following the rules in the book, completely ignorant of what the characters mean. Unbeknownst to the person in the room, outside are several Chinese scholars discussing, say, Chinese history. Because of the apparent insightfulness of the comments coming out through the slot, they believe that the person in the room is an expert on China's history.

It has been argued that the person in the room is precisely analogous to a computer, engaging in computations by following the rules. Since there is no difference in this respect between the person in the room and a computer, and since the person knows nothing about Chinese history, it must be the case that no computer can know anything about Chinese history, either. Since there is nothing special about Chinese history in this situation, it must be the case that computers in general can know nothing.

The replies to this argument would fill a library. One of the most interesting concedes that the person in the room knows no Chinese history, but claims that the *virtual machine* consisting of the person together with the book of rules (and perhaps the interlocutors outside the room) constitutes a genuine understander or knower. Other replies introduce robotic connections to the external environment, or try to pick apart knowing (or understanding), representational semantics, and consciousness. The current consensus is that the Chinese room argument is flawed, but beyond that there is little agreement as to why.

CONSCIOUSNESS AND MORAL ISSUES IN AI

Beyond the whether-possible problems, there are moral issues surrounding the project to build an intelligent machine. Is this something we should be doing at all? For example, computers are not alive, so currently there is nothing morally wrong with tossing one out of an airplane. But would this still be true of an intelligent computer? To answer this, we would have to know whether an intelligent computer would have emotions, dreams, hopes, and plans. Consciousness is crucial to any notion of morality. Would an intelligent computer be conscious? How could we tell? Would an intelligent computer know and care what happened to it? An embodied robotic architecture might well know and care, in a very real way. So would it not be immoral to push such a robot off a cliff on a whim?

Further Reading

Aiken H (1956) The future of automatic computing machinery. In: *Elektronische Rechenmaschinen und Informationsverarbeitung*, pp. 32–34. Braunschweig: Vieweg. [Proceedings of a symposium published in *Nachrichtentechnische Fachberichte* 4.]

Church A (1936) A note on the entscheidungsproblem. *Journal of Symbolic Logic* **1**: 40–41, 101–102.

Crevier D (1993) *AI: The Tumultuous History of the Search for Artificial Intelligence*. New York, NY: Basic Books.

Dietrich E (ed.) (1994) *Thinking Computers and Virtual Persons: Essays on the Intentionality of Machines*. San Diego, CA: Academic Press.

Fodor J (1983) *The Modularity of Mind*. Cambridge, MA: MIT Press.

Fodor J (1987) Modules, frames, fridgeons, sleeping dogs, and the music of the spheres. In: Pylyshyn Z (ed.) *The Robot's Dilemma*, pp. 139–149. Norwood, NJ: Ablex.

Fodor J (2000) *The Mind Doesn't Work That Way*. Cambridge, MA: MIT Press.

Gödel K (1931) On formally undecidable propositions of *Principia Mathematica* and related systems I. In: van Heijenoort J (ed.) (1967) *From Frege to Gödel: A Source Book in Mathematical Logic, 1879–1931*, pp. 596–616. Cambridge, MA: Harvard University Press.

McCarthy J and Hayes P (1969) Some philosophical problems considered from the standpoint of artificial intelligence. In: Metzer B and Michie D (eds) *Machine Intelligence*, vol. IV, pp. 463–502. New York, NY: Elsevier.

McCorduck P (1979) *Machines Who Think*. San Francisco, CA: WH Freeman.

Morgenstern L (1996) The problem with solutions to the frame problem. In: Ford K and Pylyshyn Z (eds) *The Robot's Dilemma Revisited*. Norwood, NJ: Ablex.

Partridge D and Wilks Y (eds) *The Foundations of Artificial Intelligence: A Source Book*. Cambridge, UK: Cambridge University Press.

Penrose R (1994) *Shadows of the Mind*. New York, NY: Oxford University Press.

Putnam H (1960) Minds and machines. In: Hook S (ed.) *Dimensions of Mind: A Symposium*, pp. 138–164. New York, NY: New York University Press.

Searle JR (1980) Minds, brains, and programs. *The Behavioral and Brain Sciences* **3**.

Turing A (1936) On computable numbers with an application to the entscheidungsproblem. *Proceedings of the London Mathematical Society, Series 2*, **42**: 230–265 and **43**: 544–546.

Artificial Life

Intermediate article

Norman H Packard, Prediction Company, Santa Fe, New Mexico, USA
Mark A Bedau, Reed College, Portland, Oregon, USA

CONTENTS

Artificial life is the study of life and life-like processes through simulation and synthesis.

INTRODUCTION

Artificial life literally means 'life made by human artifice rather than by nature'. It has come to refer to a broad, interdisciplinary endeavor that uses the simulation and synthesis of life-like processes to achieve any of several possible ends: to model life, to develop applications using intuitions and methods taken from life, or even to create life. The aim of creating life in a purely technological context is sometimes called 'strong artificial life'.

Artificial life is of interest to biologists because artificial life models can shed light on biological phenomena. It is relevant to engineers because it offers methods to generate and control complex behaviors that are difficult to generate or control using traditional approaches. But artificial life also has many other facets involving *inter alia* various aspects of cognitive science, economics, art, and even ethics.

There is not a consensus, even among workers in the field, on exactly what artificial life is, and many of its central concepts and working hypotheses are controversial. As a consequence, the field itself is evolving from year to year. This article provides a snapshot and highlights some controversies.

HISTORY

The roots of artificial life are quite varied, and many of its central concepts arose in earlier intellectual movements.

John von Neumann implemented the first artificial life model (without referring to it as such) with his famous creation of a self-reproducing, computation-universal entity using cellular automata. At the time, the construction was surprising, since many had argued its impossibility, for example on the grounds that such an entity would need to contain a description of itself, and that description would also need to contain a description, *ad infinitum*. Von Neumann was pursuing many of the very issues that drive artificial life today, such as understanding the spontaneous generation and evolution of complex adaptive structures; and he approached these issues with the extremely abstract methodology that typifies contemporary artificial life. Even in the absence of modern computational tools, von Neumann made striking progress.

Cybernetics developed at about the same time as von Neumann's work on cellular automata, and he attended some of its formative meetings. Norbert Wiener is usually considered to be the originator of the field (Wiener, 1948). It brought two separate foci to the study of life processes: the use of information theory and a deep study of the self-regulatory processes (homeostases) considered essential to life. Information theory typifies the abstractness and material-independence of the approach often taken within both cybernetics and artificial life. Both fields are associated with an extremely wide range of studies, from mathematics to art. As a discipline, cybernetics has evolved in divergent directions; in Europe, academic departments of cybernetics study rather specific biological phenomena, whereas in America cybernetics has tended to merge into systems theory, which generally aims toward formal mathematical studies. Scientists from both cybernetics and systems theory contribute substantially to contemporary artificial life.

Biology (i.e. the study of actual life) has provided many of the roots of artificial life. The subfields of biology that have contributed most are microbiology and genetics, evolution theory, ecology, and development. To date there are two main ways that artificial life has drawn on biology: crystalizing intuitions about life from the study of life, and using and developing models that were originally devised to study a specific biological phenomenon. A notable example of the latter is Kauffman's use of random Boolean networks (Kauffman, 1969, 1993). Biology has also influenced the problems studied in artificial life, since artificial life's models provide definite answers to problems that are intractable by the traditional methods of mathematical biology. Mainstream biologists are increasingly participating in artificial life, and the methods and approaches pioneered in artificial life are increasingly accepted within biology.

The most heavily represented discipline among contemporary researchers in artificial life is computer science. One set of artificial life's roots in computer science is embedded in artificial intelligence (AI), because living systems exhibit simple but striking forms of intelligence. Like AI, artificial life aims to understand a natural phenomenon through computational models. But in sharp contrast to AI, at least as it was originally formulated, artificial life tends to use bottom-up models in which desired behavior emerges in a number of computational stages, instead of top-down models that aim to yield the desired behavior directly (as with expert systems). In this respect, artificial life shares much with the connectionist movement that has recently swept through artificial intelligence and cognitive science. Artificial life has a related set of roots in machine learning, inspired by the robust and flexible processes by which living systems generate complex useful structures. In particular, some machine learning algorithms such as the genetic algorithm (Holland, 1975) are now seen as examples of artificial life applications, even though they existed before the field was named. New areas of computer science (e.g., evolutionary programming, autonomous agents) have increasingly strong links to artificial life. (*See* **Artificial Intelligence, Philosophy of**)

Physics and mathematics have also had a strong influence on artificial life. Statistical mechanics and thermodynamics have always claimed relevance to life, since life's formation of structure is a local reversal of the second law of thermodynamics, made possible by the energy flowing through a living system. Prigogine's thermodynamics of dissipative structures is the most modern description of this view. Statistical mechanics is also used to analyze some of the models used in artificial life that are sufficiently simple and abstract, such as random Boolean networks. Dynamical systems theory has also had various contributions, such as its formulation of the generic behavior in dynamical systems. And physics and dynamical systems have together spawned the development of synergetics and the study of complex systems (Wolfram, 1994), which are closely allied with artificial life. One of artificial life's main influences from physics and mathematics has been an emphasis on studying model systems that are simple enough to have broad generality and to facilitate quantitative analysis.

The first conference on artificial life (Langton, 1989), where the term 'artificial life' was coined, gave recognition to artificial life as a field in its own right, although it had been preceded by a similar conference entitled 'Evolution, Games, and Learning' (Farmer *et al.*, 1986). Since then there have been many conferences on artificial life, with strong contributions worldwide (e.g., Bedau *et al.*, 2000). In addition to the scientific influences described above, research in artificial life has also come to include elements of chemistry, psychology, linguistics, economics, sociology, anthropology, and philosophy.

CONCEPTS AND METHODOLOGY

Most entities that exhibit lifelike behavior are complex systems – systems made up of many elements simultaneously interacting with each other. One way to understand the global behavior of a complex system is to model that behavior with a simple system of equations that describe how global variables interact. By contrast, the characteristic approach followed in artificial life is to construct lower-level models that themselves are complex systems and then to iterate the models and observe the resulting global behavior. Such lower-level models are sometimes called agent- or individual-based models, because the whole system's behavior is represented only indirectly and arises merely out of the interactions of a collection of directly represented parts ('agents' or 'individuals').

As complex systems change over time, each element changes according to its state and the state of those 'neighbors' with which it interacts. Complex systems typically lack any central control, though they may have boundary conditions. The elements of a complex system are often simple compared to the whole system, and the rules by which the elements interact are also often simple. The behavior of a complex system is simply the aggregate of the changes over time of all of the system's elements. In rare cases the behavior of a complex system may actually be mathematically derived from the rules governing the elements' behavior, but typically a complex system's behavior cannot be discerned short of empirically observing the emergent behavior of its constituent parts. The elements of a complex system may be connected in a regular way, such as on a Euclidean lattice, or in an irregular way, such as on a random network. Interactions between elements may also be without a fixed pattern, as in molecular dynamics of a chemical soup or interaction of autonomous agents. When adaptation is part of a complex system's dynamics, it is sometimes described as a complex adaptive system. Examples of complex systems include cellular automata, Boolean networks, and neural networks. Examples of complex adaptive systems include neural networks undergoing a learning process and populations of entities evolving by natural selection.

One of the simplest examples of a complex system is the so-called 'game of life' devised by the mathematician John Conway (Berlekamp *et al.*, 1982). The game of life is a two-state two-dimensional cellular automaton with a trivial nearest-neighbor rule. You can think of this 'game' as taking place on a two-dimensional rectangular grid of cells, analogous to a huge checkerboard. Time advances in discrete steps, and a cell's state at a given time is determined by the states of its eight neighboring cells according to the following simple 'birth–death' rule: A 'dead' cell becomes 'alive' if and only if exactly three neighbors were just 'alive', and a 'living' cell 'dies' if and only if fewer than two or more than three neighbors were just 'alive'. From inspection of the birth–death rule, nothing particular can be discerned regarding how the whole system will behave. But when the system is simulated, a rich variety of complicated dynamics can be observed and a complex zoo of structures can be identified and classified (blinkers, gliders, glider guns, logic switching circuits, etc.). It is even possible to construct a universal Turing machine in the game of life and other cellular automata, by cunningly configuring the initial configuration of living cells. In such constructions gliders perform a role of passing signals, and analyzing the computational potential of cellular automata on the basis of glider interactions has become a major research thrust.

Those who model complex adaptive systems encounter a tension resulting from two seemingly conflicting aims. To make a model 'realistic' one is driven to include complicated realistic details about the elements, but to see and understand the emergent global behavior clearly one is driven to simplify the elements as much as possible. Even though complex adaptive systems include systems whose elements and dynamical rules are highly complicated, the spirit of most artificial life work is to look for the complexity in the emergent global behavior of the system, rather than to program the complexity directly into the elements.

Computation is used extensively in the field of artificial life, usually to simulate models to generate data for studying those models. Simulation is essential for the study of complex adaptive systems for it plays the role that observation and experiment play in more conventional science. Having no access to significant computational machinery, Conway and his students first studied the game of life by physically mapping out dynamics with go stones at teatime. Now thousands of evolutionary generations for millions of sites can be computed in short order with a conventional home computer. Computational ability to simulate large-scale complex systems is the single most crucial development that enabled the field of artificial life to flourish and distinguish itself from precursors (such as cybernetics or systems theory).

The dependence of artificial life on simulation has led to debate within the field over the ontological status of the simulations themselves. One version of strong artificial life holds that life may be created completely within a simulation, with its own virtual reality, yet with the same ontological status as the phenomenon of life in the real world. Some hold, however, that simulated, virtual reality cannot possibly have the same ontological status as the reality we experience. These point out that a simulated hurricane can never cause us to become wet. They also believe that if artificial life is to achieve the status of reality, it must include an element of embodiment, an extension into the real, non-simulated world enabling an interaction with that world. Believers in the reality of simulation point out that a simulation has its own embodiment within a computer, that a simulation is not an abstract formula specifying a program but the actual running of a program in a real physical medium using real physical resources. The belief that artificial life has its own bona fide reality is particularly strong among those who generate experimental data with simulations.

Both living systems and artificial life models are commonly said to exhibit emergent behavior – indeed, many consider emergent behavior to be a hallmark of life – but the notion of emergence remains ill-defined. There is general agreement that the term has a precise meaning in some contexts, most notably to refer to the resultant aggregate global behavior of complex systems. The higher-level structures produced in Conway's game of life provide a classic example of this kind of emergent behavior. In spite of clear examples like the game of life, there is no agreement regarding how one might most usefully define emergence. Some believe that emergence is merely a form of surprise. On this view, emergence exists only in the eye of the beholder and whether a phenomenon is emergent or not depends on the mindset of the observer. Others believe that there is an objective, observer-independent definition of emergence in terms of whether a phenomenon is derivable from the dynamical rules, even if it is often difficult to tell *a priori* what can be derived from the dynamical rules underlying complex systems. These difficulties lead some to argue that the term 'emergence' should simply be dropped from the vocabulary of artificial life. However, this advice is not widely heeded at present.

Complexity is another commonly recognized hallmark of life, and this notion has also so far eluded satisfactory definition. Apparently several different concepts are involved, such as structural complexity, interaction complexity, and temporal complexity. To some, it seems obvious that the biosphere is quite complex at present and that its complexity has increased on an evolutionary timescale. But the difficulties of defining complexity lead others to claim that life's present complexity and its increase over time are either illusory or a contingent artifact of our particular evolutionary history. Understanding complexity and its increase through the course of evolution are at the center of much research in artificial life. In fact, one of the field's main goals at present is to produce and then understand open-ended evolution, an ongoing evolutionary process with continually increasing complexity.

Darwin's view of evolution, with its emphasis on survival of the fittest, implied that the process of adaptation was the key to the creation of intelligent design through life's evolution. However, the role and significance of adaptation is controversial today. Some hold that adaptation is the main force driving the changes observed in evolution. Others maintain that most of evolution consists of non-adaptive changes that simply explore a complex space of morphological forms. Still others claim that much of the apparent intelligence of complex systems is a necessary result of certain complex system architectures. Artificial life may shed light on this debate by providing many diverse examples of evolutionary processes, with an attendant ability to analyze the details of those processes in a way that is impossible with the biosphere, because the analogous assaying of historical data is currently impractical and much of the historical data is simply unavailable.

Analysis of adaptation has led to the idea of a fitness landscape. Organisms (or agents in an artificial life model) are considered to be specified by a genome (or sometimes a set of model parameters). The interaction of the organism with other organisms as well as with its environment yields an overall fitness of the organism, which is often thought of as a real-valued function over the space of possible genomes (or model parameters). In various applications of evolutionary algorithms, such as the genetic algorithm, specifying a fitness function is an essential part of defining the problem. In such cases, adaptation is a form of optimization, 'hill climbing in the fitness landscape'. In artificial life models, however, fitness is often not specified explicitly, but is a property emerging from the interactions of an organism with its world.

The concept of a fitness landscape as an analytical device suffers various limitations. One is that a fitness landscape is generally an approximation; the fitness landscape itself can evolve when organisms in a population interact strongly with each other. Another reason is that on an evolutionary timescale, the space on which a fitness function is defined is changing with the advent of new elements to the genome or new model parameters for artificial organisms. Simulating agent-based artificial life models is a natural and feasible way to study these more general situations.

MODELS AND PHENOMENA

Generally, artificial life models choose a level of biological life to model. The lowest stratum may be thought of as analogous to the chemical level; higher stages include modeling of simple organisms such as bacteria, constituents of more complex organisms such as cells, complex organisms themselves, and varieties of complex organisms that can give rise to ecologies. One might consider a holy grail of artificial life to be the discovery of a single model that can span all these levels; so far the field has had difficulty producing a model that spans even one connected pair of levels.

The most primitive phenomenon explored by some artificial life models is self-organization. Such models study how structure may emerge from unstructured ensembles of initial conditions. Naturally, one aim is to discover the emergence of lifelike structure; some models explicitly aim to model the origin of life – such as chemical soups from which fundamental structures such as self-maintaining autocatalytic networks might be seen to emerge. Models for the immune system are another example of a lifelike process emerging from chemical interactions. Self-organization has also been studied in models for higher-level living structures, such as metabolisms and cell networks, with Boolean networks whose dynamics converge to different structures depending on model meta-parameters (Kauffman, 1969, 1993).

A host of models target the organismic level, sometimes with significant interactions between organisms. These models typically allow changes in the organisms as part of the system's dynamics (e.g., through a genetic mechanism), and the most common goal of research using these models is to identify and elucidate structure that emerges in the ensuing evolutionary process. Some models fit in between the chemical level and the organismic level, aiming to understand development by modeling interacting cells. Other models are inter-organismic, in the sense that they aim explicitly to model interactions between different types of organisms or agents. These models often contain elements of game theory.

Many of the models studied in artificial life should be viewed as 'purely digital' models. Purely digital models drop any pretense of modeling any pre-existing biological structures; their elements are digital constructs having no direct biological reference. Such models seek to produce novel, purely digital instances of biological phenomena in their emergent behavior. Conway's game of life is a purely digital model at the physical or chemical level, embodying an extremely simple and unique form of 'chemical' interactions (the birth–death rule). The self-organization exhibited in the game of life is not a representation of chemical self-organization in the real world but a wholly novel instance of this phenomenon. Another chemical-level model is AlChemy (Fontana, 1992), which consists of a mixture of 'reacting chemical molecules' that are actually simple programs that produce new programs as output when one program is given as input to another program.

One example of a purely digital model on the 'organismic' level is Tierra (Ray, 1992), which consists of 'organisms' that are actually simple self-replicating computer programs populating an environment consisting of computer memory. Tierra was a mature version of earlier efforts of a model called Core Wars (Dewdney, 1984) and has been followed by more developed versions such as Avida (Adami and Brown, 1994). In Tierra, the world is a one-dimensional ring of computer memory, which may be populated with instructions that are much like idealized microprocessor assembly language instructions (e.g., copy, jump, conditional branch, etc.). The instructions are the microscopic components of the model, and the model's central processing unit (CPU) implements the instructions in memory, creating a chemistry from which structure in the model can emerge. The model is generally seeded with a primordial organism consisting of a group of instructions that can copy itself to another place in memory. The copying is accompanied by errors (mutations) that can enhance the functionality of the organisms.

The accomplishments and shortcomings of most artificial life models are exemplified by those of Tierra. On the side of accomplishments, Tierra shows clear evidence of evolution, and the resulting emergence of structure and organization that were not 'programmed' into the model explicitly. Careful analysis of the evolutionary results reveals computational features such as evolution of subroutines and versions of parasitism. On the negative side, the model shows only one level of emergence (e.g., the model must be seeded by a primordial organism; evolution of an unstructured soup has not yet produced an emergent viable organism). Secondly, the evolution of the digital organisms appears to 'level off', reaching a stage where increasingly insignificant innovations are absorbed into the population, instead of displaying the open-ended evolution of natural systems. Reasons for this limitation include (1) simplicity of the model's evolutionary driving force (the evolutionary value of replication with minimal CPU time), (2) structural limitations on the space of innovations possible, which create limitations on organism functionality, and (3) structural limitations on organisms' ability to interact with each other and their environment. Different artificial life models have different detailed reasons for the two limitations we have discussed in Tierra, but the limitations are generally prevalent.

Another important area of artificial life is not so much a modeling activity as much as an implementation activity. This work aims to produce hardware implementations of lifelike processes. Some of these implementations are practical physical devices. But some of this activity is primarily theoretical, motivated by the belief that the only way to confront the hard questions about how life occurs in the physical world is to study real physical systems. Again, there is an analogy with biological levels. The 'chemical' level is represented by work on evolvable hardware, often using programmable logic arrays (e.g., Breyer *et al.*, 1998). The 'organismic' level is represented by recent work in evolutionary robotics (e.g., Cliff *et al.*, 1993). An 'ecological' level might be represented by the Internet along with its interactions with all its users on computers distributed around the world.

Artificial life, like its antecedent, cybernetics, has a peculiarly broad cultural scope extending beyond cut and dried scientific progress. This breadth is best exemplified by the work of Karl Sims (Sims, 1991), who has coupled rich image-producing computational environments with interactions between those environments and people watching the images at an exhibit. The result is an evolutionary system that is not constrained to live within the confines of a particular model's framework, but rather that is a coupling of two evolutionary subsystems, one of which is natural (the audience). Sims' interactive evolutionary art has produced several visually striking results, and human interaction seems to give the evolutionary system an open-ended quality characteristic of natural evolution.

FUTURE DIRECTIONS

One broad direction artificial life will continue to take in the future is that of synthesis: the synthesis of significant biological phenomena either within the context of model simulation or hardware implementation. A grave difficulty facing progress in this area is the lack of any quantitative basis of comparison for many of the biological phenomena artificial life aims to model. An example of this difficulty is modeling open-ended evolution. How could we know when this is achieved? In general, measurable characterization of phenomena is a prerequisite to quantitative comparison, and much progress is needed in order to achieve this for many target phenomena.

Probably the largest goal of the field is to understand the nature of life itself. This will be furthered to some extent with the quantitative comparisons just mentioned, but there is also a broader goal of discerning what the boundaries of life are, and how the idea of life might be extended to phenomena beyond biological life. Is there a sense in which financial markets or sociotechnical networks are alive, independent of the lives of their biological constituents? Many in the field of artificial life believe that, if the concept of life is properly framed and understood, such questions may well have a precise affirmative answer.

References

Adami C and Brown CT (1994) Evolutionary learning in the 2D artificial life system 'Avida'. In: Brooks R and Maes P (eds) *Artificial Life*, vol. IV, pp. 377–381. Cambridge, MA: Bradford/MIT Press.

Bedau MA, McCaskill JS, Packard NH and Rasmussen S (eds) (2000) *Artificial Life VII: Proceedings of the Seventh International Conference on Artificial Life*. Cambridge, MA: MIT Press.

Berlekamp ER, Conway JH and Guy RK (1982) *Winning Ways*, vol. II. New York, NY: Academic Press.

Breyer J, Ackermann J and McCaskill JS (1998) Evolving reaction–diffusion ecosystems with self-assembling structures in thin films. *Artificial Life* **4**: 25–40.

Cliff D, Harvey I and Husbands P (1993) Explorations in evolutionary robotics. *Adaptive Behavior* **2**: 73–110.

Dewdney A (1984) In a game called core wars hostile programs engage in a battle of bits. *Scientific American* **250**: 14–23.

Farmer JD, Lapedes A, Packard NH and Wendroff B (eds) (1986) *Evolution, Games, and Learning: Models for Adaptation for Machines and Nature*. Amsterdam, Netherlands: North-Holland.

Fontana W (1992) Algorithmic Chemistry. In: Langton CG, Taylor CE, Farmer JD and Rasmussen S (eds) *Artificial Life II: Proceedings of the Workshop on Artificial Life*, pp. 159–209. Reading, CA: Addison Wesley.

Holland JH (1975) *Adaptation in Natural and Artificial Systems*. Ann Arbor, MI: University of Michigan Press. [Revised and expanded edition (1992) Cambridge, MA: MIT Press.]

Kauffman SA (1969) Metabolic stability and epigenesis in randomly constructed genetic nets. *Journal of Theoretical Biology* **119**: 437–467.

Kauffman SA (1993) *The Origins of Order: Self-Organization and Selection in Evolution*. New York, NY: Oxford University Press.

Langton CG (ed.) (1989) *Artificial Life: The Proceedings of an Interdisciplinary Workshop on the Synthesis and Simulation of Living Systems*. Redwood City, CA: Addison Wesley.

Ray T (1992) An approach to the synthesis of life. In: Langton CG, Taylor CE, Farmer JD and Rasmussen S (eds) *Artificial Life II: Proceedings of the Workshop on Artificial Life*, pp. 371–408. Reading, CA: Addison Wesley.

Sims K (1991) Artificial evolution for computer graphics. *Computer Graphics* **25**: 319–328. (ACM SIGGRAPH '91 Conference Proceedings, Las Vegas, Nevada, July 1991.)

Wiener N (1948) *Cybernetics, or Control and Communication in the Animal and the Machine*. New York, NY: Wiley.

Wolfram S (1994) *Cellular Automata and Complexity: Collected Papers*. Reading, MA: Addison Wesley.

Further Reading

Adami C (1998) *Introduction to Artificial Life*. New York, NY: Springer.

Adami C, Belew RK, Kitano H and Taylor CE (1998) *Artificial Life VI: Proceedings of the Sixth International Conference on Artificial Life*. Cambridge, MA: MIT Press.

Bedau MA (1997) Weak emergence. *Philosophical Perspectives* **11**: 375–399.

Bedau M, McCaskill J, Packard N *et al.* (2000) Open problems in artificial life. *Artificial Life* **6**: 363–376.

Boden MA (1996) *The Philosophy of Artificial Life*. Oxford, UK: Oxford University Press.

Brooks RA and Maes P (1994) *Artificial Life IV: Proceedings of the Fourth International Workshop on the Synthesis and Simulation of Living Systems*. Cambridge, MA: MIT Press.

Floreano D, Nicoud J-D and Mondada F (1999) *Advances in Artificial Life: 5th European Conference, ECAL'99*. Berlin, Germany: Springer.

Husbands P and Harvey I (1997) *Fourth European Conference on Artificial Life*. Cambridge, MA: MIT Press.

Langton CG (ed.) (1995) *Artificial Life: An Overview*. Cambridge, MA: MIT Press.

Langton CG, Taylor CE, Farmer JD and Rasmussen S (eds) (1992) *Artificial Life II: Proceedings of the Workshop on Artificial Life*. Reading, CA: Addison Wesley.

Levy S (1992) *Artificial Life, The Quest for a New Creation*. New York, NY: Pantheon.

Koza JR (1992) *Genetic Programming: On the Programming of Computers by Means of Natural Selection*. Cambridge, MA: MIT Press.

Mitchell M (1996) *An Introduction to Genetic Algorithms*. Cambridge, MA: MIT Press.

Varela FJ and Bourgine P (1992) *Towards a Practice of Autonomous Systems: Proceedings of the First European Conference on Artificial Life*. Cambridge, MA: MIT Press.

Asset Market Experiments

Intermediate article

Daniel Friedman, University of California, Santa Cruz, California, USA
Hugh M Kelley, Indiana University, Bloomington, Indiana, USA

CONTENTS

Laboratory evidence indicates that prices of financial assets such as stocks and bonds respond to changes in the assets' fundamental value but are also sometimes distorted by investors' cognitive and other biases.

INTRODUCTION

Any society allocates some resources to current consumption and some to investment, to building a better future. Asset markets determine the extent and form of investment in modern economies. Non-market allocation procedures such as those once used in Communist countries clearly worked less well and became less prevalent in the late twentieth century. Asset markets now have global scope and significance.

By definition, asset markets are efficient when asset prices reflect all relevant information about investment opportunities. Theory shows that efficient asset markets lead society to choose only the most productive investment prospects, and to choose the best overall level of investment.

The efficient asset price is called *fundamental value*. Actual asset prices are set by fallible human investors in imperfect markets, and thus may contain other components, called *bubbles*, that can lead to inefficient resource allocation and impair future wellbeing.

Laboratory and field evidence sheds light on asset market efficiency. Asset markets sometimes compensate for investors' cognitive biases, but at other times they amplify them and produce large bubbles. Laboratory experiments help to test policies intended to improve asset market efficiency.

FUNDAMENTAL ASSET VALUE

An asset is anything that provides its owner with a stream of benefits over time. Its economic value is the monetary equivalent of the net benefits it provides. Valuation of a real asset (such as a house, a pizza delivery car or a microprocessor production facility) involves estimating prices for the services the asset generates and for the resources required to maintain its productivity. This article will focus on financial assets, such as stocks and bonds, which are easier to value since they directly promise a monetary income stream. A stock, for example, promises annual per share dividends chosen by the company's board of directors.

Valuing a known income stream is straightforward. One computes the discounted present value: the amount of cash that, if put on deposit now and withdrawn over time, could replicate the income stream. For example, if the asset is an annuity (a simple bond) that promises \$1000 every year for 30 years, and the annual interest rate is 7%, then the asset value is $(\$1000)((1.07)^{-1} + (1.07)^{-2} + \ldots + (1.07)^{-30})$ or, summing the geometric series and simplifying, $(\$1000)(1 - (1.07)^{-30})/(0.07) = \$12{,}409$.

Valuation is more interesting when the income stream is uncertain, either because the promise is vague (as for a stock) or because the promise might not be kept (as in the case of a 'junk' bond). The economic value is defined to be the mathematical expectation of the income stream, discounted at the interest rate appropriate for the associated risk. For example, if the \$1000 annual payment in the previous example had a 50% independent probability of being canceled every year, and this requires a 2% risk premium or 9% interest rate, then the expected annual income is \$500 and the asset value is $(\$500)(1 - (1.09)^{-30})/(0.09) = \$5137$.

Two questions arise. Firstly, what is the appropriate interest rate (or risk premium)? This question stimulated the development of modern finance theory in the 1950s and 1960s, but we will not pursue it here. Secondly, how can people with diverse information arrive at a common expectation of the income stream (or its discounted present value)? Some people may know a lot about the technical performance of a company's new product, others about customer demand, or competing products, or production costs, or the company's management and financing capacity. Usually nobody has all the relevant knowledge.

'Fundamental asset value' is the present value expectation incorporating all existing information regarding future income. The definition seems to assume that people immediately share all knowledge and combine it without bias or distortion. But people may not share information; they often are cognitively biased. Cognitive biases might cause the market price to deviate from fundamental value.

COGNITIVE BIASES

Which cognitive biases might distort people's estimates of asset value? Some of the main biases are summarized below (Camerer, 1993; Rabin, 1998; Thaler, 1992).

Firstly, judgment biases may distort the aggregation of diverse information (Massaro and Friedman, 1990) and income stream estimates. In particular, investors may:

- neglect some pieces of information and over-weight others, as in cue competition;
- overestimate the resemblance of the future to the immediate past, as in the well-known availability and representativeness heuristics or the anchoring and inertia biases;
- over-weight new information and neglect old information, as in overreaction to news and base rate neglect;
- regard ambiguous news as reinforcing current beliefs, as in the confirmatory bias;
- overrate the precision of their own information, relative to other traders' information, as in overconfidence;
- indulge in the gambler's fallacy or magical thinking, perceiving patterns in random data;
- react incorrectly to increasing information precision, or switch biases depending on state, for example, by overreacting to news when asset prices are volatile but underreacting otherwise.

Secondly, when moving from income estimates to asset value estimates, investors may apply hyperbolic discounting, using too high interest rates when comparing current income with near future income, and too low interest rates when comparing income received at two distant dates (Ainslie, 1992).

Thirdly, investors may make decision errors when they buy and sell assets, such as overvaluing assets they currently hold, as in the endowment effect, or making inconsistent risky choices, as in prospect theory, regret theory, or (more generally) decision field theory (Busemeyer and Townsend, 1993).

Fourthly, investors may learn by trial and error, some faster than others, creating additional asset price deviations from fundamental value (Kitzis *et al.*, 1998; Busemeyer *et al.*, 1993).

Note, however, that individual investors' biases do not translate directly into asset price biases. Market prices are set by subtle interactions between investors that may either attenuate (e.g., Friedman, 2001) or amplify (e.g., Akerlof and Yellin, 1985; DeLong *et al.*, 1990) the individual biases and learning heterogeneities. Thus, whether (or when) asset price follows fundamental value is an empirical question in its own right.

FIELD EVIDENCE

Evidence from existing asset markets is interesting but inconclusive. Historians point to dramatic episodes of asset price increase and collapse, from the South Sea bubble and 'tulipmania' in the sixteenth century to Japan's 'bubble' economy of the late 1980s, various 1990s financial crises (in Western Europe, Latin America, East Asia, and Russia), and the 'dot com' bubble of 2000.

Some economists argue that these episodes are merely unusual movements in fundamental value (Garber, 2000). Economists cannot observe the private information held by traders in the field, and therefore have no direct measure of fundamental value or bubbles. Some indirect field evidence, however, favors the bubbles interpretation.

Shiller (1981) and later writers surveyed in LeRoy (1989) show that stock market indices are much more volatile than would be justified by subsequent changes in dividends. Roll (1984) argues that changes in the fundamental value of US orange juice futures come predominantly from two observables (Florida weather hazard and Brazil supply) but account for only a small portion of the actual price variability. Additionally, nonfundamental effects consistently observed, such as the day of the week or month and January effects, also suggest that stock prices are more volatile than fundamental values. Finally, in a natural field experiment, the stock prices (but arguably not the fundamental values) are far less volatile on weekends and days when the exchange is closed for upgrades. On the other hand there is also field evidence suggesting efficient information aggregation. The Iowa Electronic Market has consistently outperformed major polls in predicting election outcomes (Forsythe *et al.*, 1999) and similarly with the Hollywood Stock Exchange for box office revenues and Oscar winners (Pennock *et al.*, 2001).

Field evidence is indirect and can seem either to support or to undermine claims of market efficiency.

LABORATORY EVIDENCE

Laboratory asset markets provide direct evidence, because the experimenter can control the information available to traders and can always compute the fundamental value. The first generation of asset market experiments in the early 1980s used oral double auction trading procedures similar to the traditional Chicago trading pits. To sharpen inferences, the laboratory markets are much simpler than field markets: for example, they often allow only 8 to 16 subjects to trade a single asset. Most of the results described below involve experienced traders who are fully adapted to the laboratory market. See Sunder (1993) for an excellent early survey, and Holt (1999) for an online bibliography.

Market Attenuation of Traders' Biases

There are several distinct forces that tend to attenuate biases. Firstly, people can learn to overcome their biases when the market outcomes make them aware of their mistakes. Secondly, to the extent that biased traders earn lower profits (or make losses), they will lose market share and will have less influence on asset price. Thirdly, institutions evolve to help people overcome cognitive limitations: for example, telephone books mitigate the brain's limited digital storage capacity. Trading procedures such as the oral double auction evolved over many centuries and seem to enhance market efficiency.

Oral double auctions allow all traders to observe other traders' attempts to buy and sell, and might enable them to infer other traders' information. Moreover, the closing price is not set by the most biased trader or even a random trader. The most optimistic traders buy (or already hold), and the most pessimistic traders sell (or never held), the asset, so the closing price reflects the moderate expectations of 'marginal' traders, the most reluctant sellers and buyers.

These attenuating forces may explain the surprisingly rapid convergence of asset price to fundamental value in the first generation of laboratory experiments. Forsythe *et al.* (1982) obtained such convergence for assets that paid different dividends to different traders over two periods. Likewise, Plott and Sunder (1982) obtained convergence to an efficient asset price for a single-period asset even when some traders had inside information. Friedman *et al.* (1984) found that simultaneous operation of spot and futures markets improved convergence to an efficient asset price and allocation when assets paid different dividends to different traders over three periods and traders knew only their own dividend schedule. Generally, convergence was first observed in the last dividend period, then in the middle period as traders correctly anticipated last-period prices, and finally in the first trading period as traders learned the asset's present value.

Market Amplification of Traders' Biases

Later experiments detected systematic discrepancies between price and fundamental value in more complex environments. Copeland and Friedman (1991) found that in a computerized double auction with three information events and eight states, a model of partial aggregation predicted price better than full aggregation or fundamental value.

Several experimental teams found that insider information was incorporated into asset price less reliably and less quickly when the asset paid the same dividend stream to each trader and the number (or presence) of insiders was not publicly known. Some data suggest the following scenario. An uninformed trader *A* observes trader *B* attempting to buy (due to some slight cognitive bias, say) and mistakenly concludes that *B* has favorable inside information. Then *A* tries to buy. Now trader *C* concludes that *A* (or *B*) is an insider and tries to mimic their trades. Other traders follow, creating a price bubble.

Such 'information mirages', or 'herding' bubbles, amplify the biases of individual traders, but they cannot be produced consistently, since incurred losses teach traders to be cautious when they suspect the presence of better-informed traders. (This lesson does not necessarily improve market efficiency, however, since excessive caution impedes the aggregation of information.)

Smith *et al.* (1988) found large positive bubbles and crashes for long-lived assets and inexperienced traders. Their interpretation invokes the 'greater fool' theory, another bias amplification process. Traders who themselves have no cognitive bias might be willing to buy at a price above fundamental value because they expect to sell later at even higher prices to other traders dazzled by rising prices. Subsequent studies confirmed that such dazzled traders do exist, and that bubbles are more prevalent when traders are less experienced (individually and as a group), have larger cash endowments, and have less conclusive information.

Other mechanisms that amplify biases in laboratory asset markets include firm managers' discretionary release of information and fund managers' rank-based incentives (James and Isaac, 2000).

Policy Studies

We have only a tentative and fragmentary understanding of when asset markets amplify or attenuate investors' cognitive biases. The impact of proposed policy changes often cannot be predicted reliably, and must be assessed empirically by regulators, asset market makers contemplating reform, and entrepreneurs creating new asset markets. Laboratory markets offer helpful guidance at low cost. Recent relevant research includes performance assessment of:

- alternative market formats, including oral (or face-to-face) double auctions, anonymous electronic double auctions, call or uniform price periodic auctions, and fragmented opaque (or bilateral search) markets;
- trader privileges, such as price posting and access to order flow information;
- transaction taxes, price change limits, and trading suspensions intended (usually ineffectively) to mitigate price bubbles and panics;
- new derivative securities such as call and put options and state-contingent claims.

Current Research

Research continues at a rapid rate along all the lines mentioned. One promising new line of research investigates learning, and judgment and decision biases, in environments similar to asset markets. These environments clearly do attenuate some sorts of biases (Kelley and Friedman, forthcoming) and amplify others (Ganguly *et al.*, 2000); eventually such work should clarify the patterns.

Other promising new lines of research integrate agent-based simulation models (e.g. Epstein and Axtell, 1996) into asset markets that may include human traders. The simulated agents, or 'bots', incorporate specified cognitive limitations and the simulations examine the market-level influence of these (e.g. Arthur *et al.*, 1997). Gode and Sunder (1993) showed that simple (non-asset) double auction markets are efficient even when populated by 'zero intelligence' agents, bots that are constrained not to take losses but are otherwise quite random. Research is beginning to show how these and more intelligent bots affect efficiency in various sorts of asset markets and how they interact with humans.

The laboratory asset market evidence is more direct than the field evidence, but is still mixed. The laboratory evidence clearly demonstrates that asset price bubbles exist and persist under some circumstances, but that under other circumstances asset prices closely track fundamental values. Future work promises to identify more clearly the circumstances that promote or impair market efficiency. This should lead to improved policy and a better-functioning economy.

References

Ainslie G (1992) *Picoeconomics: The Strategic Interaction of Successive Motivational States Within the Person*. New York, NY: Cambridge University Press.

Akerlof G and Yellen J (1985) Can small deviations from rationality make significant differences to economic equilibria? *American Economic Review* **75**: 708–720.

Arthur WB, Holland J, LeBaron B, Palmer R and Tayler P (1997) Asset pricing under endogenous expectations in an artificial stock market. In: *The Economy as an Evolving Complex System*, vol. II, pp. 15–44. Reading, MA: Addison-Wesley.

Busemeyer JR, Myung IJ and McDaniel MA (1993) Cue competition effects: theoretical implications for adaptive network learning models. *Psychological Science* **4**: 196–202.

Busemeyer J and Townsend J (1993) Decision field theory: a dynamic cognition approach to decision making. *Psychological Review* **100**: 432–459.

Camerer C (1993) Individual decision making. In: Kagel JH and Roth AE (eds) *The Handbook of Experimental Economics*, pp. 587–704. Princeton, NJ: Princeton University Press.

Copeland T and Friedman D (1991) Partial revelation of information in experimental asset markets. *Journal of Finance* **46**: 265–295.

DeLong JB, Shleifer A, Summers L and Waldmann RJ (1990) Noise traders and risk in financial markets. *Journal of Political Economy* **98**: 703–738.

Epstein JM and Axtell R (1996) *Growing Artificial Societies: Social Science From the Bottom Up*. Cambridge, MA: MIT Press.

Forsythe R, Palfrey T and Plott C (1982) Asset valuations in an experimental market. *Econometrica* **50**: 537–568.

Forsythe R, Rietz TA and Ross TW (1999) Wishes, expectations and actions: price formation in election stock markets. *Journal of Economic Behavior and Organization* **39**: 83–110.

Friedman D (2001) Towards evolutionary game models of financial markets. *Quantitative Finance* **1**: 177–185.

Friedman D, Harrison G and Salmon J (1984) The informational efficiency of experimental asset markets. *Journal of Political Economy* **92**: 349–408.

Ganguly AR, Kagel JH and Moser DV (2000) Do asset market prices reflect traders' judgement biases? *Journal of Risk and Uncertainty* **20**: 219–245.

Garber PJ (2000) *Famous First Bubbles: The Fundamentals of Early Manias*. Cambridge, MA: MIT Press.

Gode D and Sunder S (1993) Allocative efficiency in markets with zero intelligence (ZI) traders: market as a partial substitute for individual rationality. *Journal of Political Economy* **101**: 119–137.

Holt C (1999) *Y2K Bibliography of Experimental Economics and Social Science: Asset Market Experiments*. http://www.people.virginia.edu/~cah2k/assety2k.htm.

James D and Isaac M (2000) Asset markets: how are they affected by tournament incentives for individuals? *American Economic Review* **90**: 995–1004.

Kelley H and Friedman D (forthcoming), Learning to forecast price. *Economic Inquiry*.

Kitzis S, Kelley H, Berg E, Massaro D and Friedman D (1998) Broadening the tests of learning models. *Journal of Mathematical Psychology* **42**(2): 327–355.

LeRoy S (1989) Efficient capital markets and martingales. *Journal of Economic Literature* **27**: 1583–1621.

Massaro D and Friedman D (1990) Models of decision making given multiple sources of information. *Psychological Review* **97**: 225–252.

Pennock DM, Lawrence S, Giles CL and Nielsen FA (2001) The real power of artificial markets. *Science* **291**: 987–988.

Plott C and Sunder S (1982) Efficiency of experimental security markets with insider information: an application of rational expectations models. *Journal of Political Economy* **90**: 663–698.

Rabin M (1998) Psychology and economics. *Journal of Economic Literature* **36**: 11–46.

Roll R (1984) Orange juice and weather. *American Economic Review* **74**: 861–880.

Shiller RJ (1981) Do stock prices move too much to be justified by subsequent changes in dividends? *American Economic Review* **71**: 421–436.

Smith V, Suchanek G and Williams A (1988) Bubbles, crashes, and endogenous expectations in experimental spot asset markets. *Econometrica* **56**: 1119–1151.

Sunder S (1993) Experimental asset markets: a survey. In: Kagel JH and Roth AE (eds) *The Handbook of Experimental Economics*, pp. 445–500. Princeton, NJ: Princeton University Press.

Thaler RH (1992) *The Winner's Curse: Paradoxes and Anomalies of Economic Life*. Princeton, NJ: Princeton University Press.

Further Reading

Friedman D (1998) Monty Hall's three doors: construction and deconstruction of a choice anomaly. *American Economic Review* **88**: 933–946.

Hogarth R and Reder M (eds) (1987) *Rational Choice: The Contrast Between Economics and Psychology*. Chicago, IL: University of Chicago Press.

Mackay C (1841) *Memoirs of Extraordinary Popular Delusions and the Madness of Crowds*. London, UK: Bentley.

Shefrin H (2000) *Beyond Greed and Fear*. Boston, MA: Harvard Business School Press.

Shiller R (2000) *Irrational Exuberance*. Princeton, NJ: Princeton University Press.

Attention

Introductory article

Daniel Gopher, Technion Institute of Technology, Haifa, Israel
Cristina Iani, Technion Institute of Technology, Haifa, Israel

CONTENTS

Attention is the scientific term primarily used to describe all processes and mechanisms that govern the subjective constraints imposed by the human organism on the flow and interpretation of external and internal information, and on the organization and selection of responses, in the service of goal-directed behavior. In some cases, attention can also be automatically captured by sudden changes in the situation, or by well-trained stimulus–response tendencies.

WHAT IS ATTENTION?

At any single moment in life, humans perceive, attend to, respond to, and make use of only a very small fraction of the information from the outside world that stimulates their sensory systems (vision, audition, touch, motion, balance, etc.), or from the internal information and skills that are stored in their memory and acquired through past experiences. This fraction is described as the focus of their attention. Humans are not passive subjects of their environment. Rather, they actively seek, bias, and attribute significance to outside and inside events. Attention represents this active process. Like any other physical system, humans are limited in the amount and rate at which they can process information and perform tasks. The study of attention is the study of the processes and mechanisms that govern the 'top-down', subjective constraints, that are imposed by the human organism, on the flow and interpretation of information from the outside world, the use of internal information, and the organization and selection of responses in the service of goal-directed behavior. In some cases, it has been shown that attention can be automatically captured by sudden changes in the environment or by well-trained stimulus–response tendencies ('bottom-up' processes). (*See* **Attention, Neural Basis of**; **Visual Attention**)

To illustrate this process, think for example of a person reading a novel in a train, while travelling back home from work. He has to focus attention on the book page and ignore all other visual information, noises of the train, announcements, or passenger conversations. He also has to stabilize the book in his hands and combat interference from train motion, vibration, and acceleration changes. On the book page he needs to focus on one line, ignore all other lines, and progress continuously with reading. Note also that the task of reading calls upon his knowledge of the language, reading ability, and memory of the content of previously read chapters of the book. All are knowledge bases stored in memory. Reading is one of many alternative tasks in which the person can engage himself. He can hum a song, participate in a conversation, or operate a laptop computer. He can also switch from one task to another. Note the difference in the sources of information and responses that have to be attended to or ignored in each of these tasks. Attending to a task is engaging, and limits giving attention to other tasks. The person should be careful to monitor the train route from time to time, so as not to miss his stop.

The above example illustrates many of the phenomena signifying the act of attending. It is also accompanied by distinct subjective experience. Attended information appears vivid and intense, while unattended and ignored information may fade away completely, or appear attenuated and weak.

The study of attention has been a major topic in scientific psychology since its early days. Research has questioned the nature of the underlying processes; the span and limits of attention; the ease and cost of attending to different environmental features and internal events; the ability to control, mobilize, and divide attention; and the linkage between attention and consciousness. The knowledge and understanding of human attention have greatly benefited from behavioral and neurophysiological studies using animals – in particular, mammals – where direct invasive measurement of brain activity was possible.

ATTENTION AND CONSCIOUSNESS

The relationship between attention and consciousness is of special significance, because early models in psychology equated attention with the content of consciousness. William James, one of the forefathers of scientific psychology, wrote in 1890:

> Every one knows what attention is. It is taking possession by the mind, in a clear and vivid form, of one of what seems several simultaneously possible objects or trains of thoughts. Focalization, concentration of consciousness are of its essence. It implies withdrawing from some things in order to deal effectively with others ...

However, we now know that only a very limited portion of the ongoing processing and response activities is admitted to consciousness. Even if triggered by a voluntary, conscious intention, the influences, consequences, and products of this intention on perception, processing, and response, are mostly not admitted to consciousness. The nature of the relationship between conscious and nonconscious products of attention is still a topic of contemporary theoretical debate. It also has an important methodological implication. The study of attention and its influences cannot limit itself to verbal reports of people. Specific research paradigms, employing behavioral and physiological measures, were developed to complement verbal reports and assess the different aspects of attention. Behavioral measures focus on changes in performance speed and accuracy, resulting from a manipulation of attention demands. These measures are complemented by a variety of neurophysiological indices (e.g. heart rate, pupil dilation, brain-evoked potentials), which have been shown to accompany changes in attention requirements. (*See* **Consciousness and Attention**; **Consciousness, Cognitive Theories of**; **Consciousness, Disorders of**)

ASPECTS OF ATTENTION

The study of attention has followed several major task categories: selective attention, divided attention, intensive aspects of attention, and mobilization of attention. They are briefly reviewed below.

Selective Attention

The study of selective attention has addressed two major questions. One is concerned with factors that influence the ease and efficiency of selection. The second examined the consequences of selection for the processing of and response to the selected and the rejected or unattended stimuli. Most of the experiments in selective attention require subjects to perform some kind of a filtering task: that is, subjects are instructed to process, memorize, judge, listen to, or respond only to stimuli that satisfy a specified criterion, for example, 'listen to female and ignore male voices', 'read blue words and ignore red and yellow', 'vocalize and memorize only animal names in a sequence containing animal, plant, and artifact names'. (*See* **Attention, Models of**; **Selective Attention**)

Overall, humans seem to be able to utilize and selectively focus on any group of stimuli that are consistently distinguished by a signifying attribute. Selection is shown to be efficient and easy when the signifying properties are highly distinguishable physical attributes, such as visual location or auditory pitch. In this case there is no long-term memory and little trace of the ignored information. The ease of selection depends on the discriminability of stimuli, with simple physical dimensions easier than semantic dimensions. Indirect behavioral measures indicate that unattended stimuli are also analyzed to a certain level, but in the majority of cases do not reach the level of full semantic analysis.

An interesting aspect of selection is the influence of a preparatory set. Research has shown that when a single stimulus is presented, processing efficiency can derive considerable benefit from advance information on the nature or likelihood of this stimulus. Similarly, performance is impaired when invalid information is presented. These effects are stronger when the number of possible stimuli and their confusability increase. Selective attention is tuned and biased, in anticipation of the forthcoming information.

Divided Attention

In divided attention tasks individuals are asked to attend to or monitor multiple information sources arriving at the same time, or to perform more than one task concurrently. Under such conditions research demonstrated good, parallel, unlimited-capacity monitoring and search capabilities in two major cases. One case is when targets differ from non-targets on a simple, distinct physical feature, such as color or size. The other case is when targets belong to a very well learned and frequently used symbolic category, and are perceptually easy to distinguish from non-targets (e.g. letters and digits). In all other cases time-sharing and concurrent performance show considerable deficit, compared to single task performance. An interesting phenomenon has been labeled 'attention blink', in which detection of a difficult target impairs performance for a short period thereafter. (*See* **Selective Attention**)

Capacity limitations in dividing attention have been shown to be more severe when targets are defined by complex discriminations (e.g. conjunction of features), rely heavily on working memory (e.g. mental arithmetic), or require coordinated response. The great sensitivity to attention limitations has made divided attention conditions a popular paradigm in the study of mental workload and the cost of mental operations.

Intensive Aspects of Attention

Selection and division of attention do not manifest themselves only in the sources of information and type of events that are attended to or ignored. They also influence the intensity and invested effort in processing and response. This is the intensive dimension of attention. There is ample evidence to show that selection and division do not operate in an all-or-none fashion. Rather, invested effort, which operates like an amplifier or gain factor, influences the response rate and quality of performance on the attended task. Humans can allocate graded levels of effort to the performance of tasks. When fully concentrated on an interesting and demanding task, there is little attention for anything else. When bored by a task, little attention is invested in attending to it; attention is easily distracted, and captured by new events. In divided attention conditions, more than one task is attended to simultaneously, and effort is allocated according to priorities. Think, for example, of the multiple elements comprising the task of driving a car in heavy traffic and the change in the relative importance of each element in different conditions.

Intensity modulations of attention have an autonomous and a voluntary component. The autonomous component is closely linked with the physiological cycles of metabolism, hormones, body temperature, and rest–activity: They jointly influence the arousal, vigilance, and efficiency of the human processing and response system. Research has shown that time of day, and phase in the circadian rhythm, may lead to a change of up to 30 per cent in the efficiency of performance. We are also well aware of the effects of fatigue, exhaustion, jet lag, and sleep deprivation on attention capabilities. The second intensive component is voluntary effort: this represents intensity modulations attributed to the ability of humans to concentrate, focus, and invest graded effort at will. The influence of intensive aspects is of special significance in sustained attention tasks when performers have to attend to a task for extended periods, such as driving for long distances, control room operations, and studying for exams. Such tasks are extremely sensitive to variation in the autonomous component. In addition, it has been shown that the voluntary component can operate at full force for only a limited period, ranging from a few minutes to about half an hour, depending on the demands of the performed task. Tasks performed for extended periods demonstrate an increased number and longer durations of lapses of attention.

Mobilization of Attention versus Modulations of Processing Intensity

An interesting distinction has emerged in the study of attention, between the act of moving attention from one focal point to another, and the act of modulating the intensity of processing once attention has been locked and engaged in a task. Evidence from neurophysiological research on visual attention has shown that these acts are associated with separate brain mechanisms, localized at different brain areas. Corroborating evidence comes from the study of individual differences in attention. It shows that the ability to focus and divide attention, both representing different levels of attention allocation to a task, is distinct from the ability to switch attention between tasks, which represents disengagement, mobilization, and reconfiguration efforts.

THE LIMITS OF ATTENTION

An important topic in the study of attention has been the attempt to identify the limits of attention. How many things can a person do in parallel? How much information can be processed and responded to concurrently? What are the limits on the rate of processing? Theoretical models of attention have advocated two major views of the source of limitations. 'Bottleneck' models associate the limit with the existence of a central, limited-capacity processor that can deal with only one task at a time. This processor constitutes a bottleneck, because a new task cannot access it before processing of the previous task has been completed. Two variants of this approach have located the bottleneck either early, when only partial or limited analysis of a stimulus has occurred, or late, at the stage of response selection, after stimuli have been fully identified. (*See* **Selective Attention**)

An alternative view of the limitation is proposed by resource models. According to this approach parallel processing is possible, but there is a limited pool of processing and response resources. Resources are allocated to the performance of tasks, performed singly or in parallel, as long as their total demand does not exceed the limit of resources. One variant of this approach is a single general pool of resources which are utilized by all tasks. Another variant advocates a multiple resource view, according to which there are several types of resources which are distinguished by the types of information, processing activities, or response modes that they serve. Performance of tasks is limited only to the extent that jointly performed tasks compete for and exhaust the limits of the same resource.

Proponents of both models conducted extensive experimentation, and brought ample evidence to support their claims. Can both views coexist? A third, and more recent, concept of the limitation of attention proposes to view it as strategic. According to this view there is no mandatory architecture to the flow of task performance, nor a strict universal central limitation and scarcity of processing and response facilities. Rather, given the nature of the task involved, the characteristics of the environment, the specific abilities and skills of an individual, and his motivations and intentions, the best strategic solution to the task's performance is developed. This solution represents the most advantageous combination of all composites at that moment. Once a strategy is developed, it does have its costs and limitations; however, those may change as the conditions are changed, experience accumulates, or utilities vary. The basic idea of the strategic approach is that there are many redundancies in the performance of every task, many degrees of freedom (elasticity) of the human processing system, and a considerable flexibility in using them. Hence, there are many ways in which a single task may be performed, which in turn may change its demand profile. Repeated or very common experiences associated with the performance of some mundane tasks may create the impression of a more 'hard-wired' limitation. However, even this can be changed with extensive training. This view of limitation shifts the focus from the study of the sources of limitation and universal limits, to the study of the degrees of freedom of the human processing system, and the ability of humans to control and efficiently channel their attention capabilities.

ATTENTION AND TRAINING

There is robust evidence to show that experience and practice lead to a dramatic reduction in the attention requirements of tasks. Think, for example, of the differences in the attention requirements of driving, for expert drivers and novice trainees. The difference is attributed to the increased organization of behavior with practice, and the development of well-organized sequences of behavior (schemas) that can operate as one, by a single command. This mode of behavior is labelled 'automatic'. It is contrasted with a 'controlled' mode of operation, which characterizes early stages of responding to new tasks and unexpected events. Controlled processes impose high demands on attention, and in the majority of cases result in less efficient performance.

This relationship between attention demands and training highlights some of the aforementioned functional and phenomenological aspects of attention. First, it underlines the important role of attention during early stages of training, in focusing on relevant information, linking external information with internal knowledge bases, coordinating and synchronizing activities, and developing the overall architecture of a given task. Second, the development of well-organized sequences of processing and response activities that are activated by a single command hints at the possible break between attention and the content of consciousness. It is possible that only the beginning, but not the entirety, of an organized sequence is admitted to consciousness. A related observation is that automated components can capture attention involuntarily and compete with voluntary intents. This finding is easy to interpret, once it is realized that automatic sequences, which are not conscious and only partially controlled, bias the processing and response flow, and attribute significance to events. Finally, training illustrates the way in which attention and performance strategies are linked together, can be developed, and acquire potency. Such strategies may then compete with other tasks, and require considerable effort to change.

The Skill of Attention Control

Intentions and goals are the major drivers of attention. We have seen that humans can focus, divide, assign priorities, and invest differential efforts in the performance of tasks. However, research has also shown that there are many difficulties and failures in the application and maintenance of attention policies. Can humans be taught to improve the management of their attentional capabilities? The study of this question has shown that there are two major difficulties in the control of attention. One is that people do not have sufficient knowledge of the efficiency of their invested efforts (this is not surprising in view of the limits of consciousness). The second is an execution problem: some attention policies and priority settings are difficult to establish and maintain. Studies that have targeted the development of these components with training have shown that attention control is a skill that can be developed and improved with proper training protocols. Moreover, the skill can be generalized to other situations in which performers face difficult and demanding tasks. Training of attention control was shown to improve the flight performance of trainee pilots, and to help older people to cope with demanding time-sharing tasks.

INDIVIDUAL DIFFERENCES IN ATTENTION

Are there consistent individual differences in attention? Do some people have better attention capabilities than others? The study of these questions showed that there are indeed consistent individual differences in attention, which are generic over and above the specific properties of the tasks that are performed, the modality, and the mode of information and response. Some individuals are better able to deal with high load than others. These studies served for the development of attention ability tests, which were shown to predict performance in flight tasks and the accident-proneness of bus drivers and others. More recent research distinguished between the attention acts of focusing, dividing, and switching between tasks. Consistent individual differences were revealed on all three dimensions. However, focusing and dividing capabilities were shown to be highly correlated, while attention-switching appears to constitute a separate ability. The research has also shown the existence of a general factor of attention control ability, which is common to all acts of attention.

NEUROPHYSIOLOGY OF ATTENTION

Most of our knowledge of the neurophysiology and brain mechanisms of attention in normal-performing humans derives from two classes of measurement techniques. One is the recording from the scalp of the electrical response of large groups of neurons to specific events (event-related brain potentials or ERPs). The second is the measurement of regional cerebral blood flow (positron emission tomography, PET, and functional magnetic resonance imaging, *f*MRI).

Data from ERP and neuroimaging studies together with the behavioral observation of patients with brain damage suggest that several brain areas are involved in attentional processing (Figure 1). The brain systems that have been associated with attention processes can be described as an interconnected network of cortical and subcortical structures which include the prefrontal and posterior parietal lobes, the cingulate gyrus, the basal ganglia, the pulvinar and reticular nuclei of the thalamus, and the midbrain and superior colliculus. These areas have been shown to be sensitive to attention-related manipulations, and control attention through their inputs to perceptual processing areas. ERP studies have also been very useful in identifying the stages of information processing at which attention begins to exert control.

Recent studies have attempted to connect different brain areas to specific aspects of attention. The posterior attentional system, including the parietal cortex, the superior colliculus and the pulvinar, has been shown to be involved in aspects of attention mobilization (directing attention to what has been selected as the focus of attention, engaging and disengaging attention from locations or objects). The two brain hemispheres seem to be involved in different ways in attention control, with the right hemisphere controlling the mobilization of attention to both sides of space and the left hemisphere controlling only the orienting of attention to the right side of space. The anterior attentional system, including the prefrontal cortex and the cingulate gyrus, is responsible for executive control processes. In particular, the prefrontal cortex exerts top-down influence on early sensory processing,

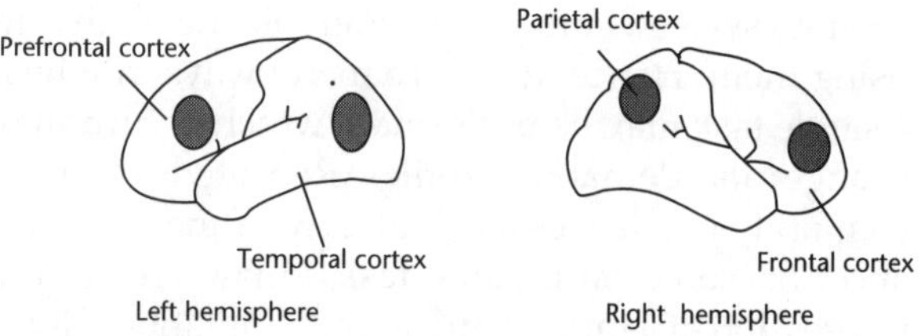

Figure 1. Lateral view of the right and left hemispheres of the human brain. The figure shows the main areas involved in attention control (gray patches).

determining which stimuli entering the perceptual pathway have to be facilitated or suppressed. This area is extensively connected to numerous cortical, limbic, and subcortical areas and seems to play a crucial role not only in attention to external events, but also to the internal mental representations that contribute to working memory, decision-making, and planning. The vigilance system, including the right frontal lobe and the brain stem, is responsible for maintaining readiness during sustained attention and vigilance tasks. (*See* **Attention, Neural Basis of**; **Spatial Attention, Neural Basis of**)

Attention Disorders

As already mentioned, attentional processing seems to involve several brain areas. Research with patients has related deficits in specific aspects of attention to impairment in one or other of these areas. Two major classes of disorder have been documented: impairment in spatial attention, and deficits of attention control. These are briefly described below.

Neglect

Visual neglect is a neurological syndrome that commonly results from lesions of the right cerebral hemisphere, especially from lesions involving the temporoparietal cortex. Although vision is intact, neglect patients are not aware of and fail to report or respond to objects or parts of objects presented controlateral to the lesion (that is, in the left visual field). This disability affects many aspects of their life. For example, patients typically fail to eat the food located on the left side of their plate. They shave or make-up only one half of their face. When asked to copy a picture, they draw only one half of that picture. Patients may no longer recognize the left side of their body as their own. Neglect has been observed in auditory, tactile, and proprioceptive modalities and can also affect mental imagery; for instance, patients may neglect the left side of an imagined mental map. Despite the fact that neglect is considered a disorder of spatial attention, the exact explanation of neglect is still a controversial matter. The combination and severity of symptoms shown by different patients seems to depend mostly on the extent and location of the lesion. (*See* **Consciousness and Attention**; **Attention, Neural Basis of**)

Extinction

This deficit is often associated with unilateral neglect to the extent that some accounts consider it a mild form of neglect. Like neglect, extinction follows unilateral brain damage. Patients suffering from extinction have no difficulty in identifying a single object presented on either side of the visual field, but when two stimuli are presented simultaneously, they do not seem to 'see' the object presented in the visual field controlateral to the lesion.

Balint syndrome

This neurological syndrome is associated with symmetrical lesions to the hemispheres (mostly involving posterior parietal areas or the parietal–occipital junction). It is less common than neglect. It involves three major deficits. First, subjects have difficulties in orienting to visual stimuli or in tracking a moving object. Second, subjects are unable to orient their arm and hand correctly when trying to reach or grasp objects. Third, subjects are able to see only one object at a time (simultagnosia), even when objects overlap. For example, if presented with an apple and a knife and asked to peel the apple, they are unable to follow the instructions because they can see only one of the items. (*See* **Consciousness and Attention**; **Attention, Neural Basis of**)

Frontal lobe syndrome

After lesions of the frontal lobe, patients show disorganized and incoherent behaviour. People suffering from this syndrome, also known as dysexecutive syndrome, are unable to plan actions and to solve problems. They show behavioural rigidity and perseveration, that is they are unable to change their behaviour when required to; increased distractibility accompanied by difficulties in focusing and maintaining concentration; difficulties in inhibiting unwanted behaviours; and difficulty in maintaining goal-directed behaviour. These symptoms seem to be caused by the impairment of areas involved in the top-down control of information processing, such as the prefrontal cortex.

Attention deficit hyperactivity disorder

This disorder appears in childhood; the major features are excessive and impairing levels of activity, inattention, and impulsiveness. The diagnosis is often reached because of the difficulties shown by these children at school. In general, the inattention symptoms include difficulties in remaining seated when required to (for example, at the dinner table or in the classroom); inability to follow instructions; inability to concentrate and focus attention on details; easy distractibility by external stimuli; inability to interrupt an action once initiated; and difficulty in organizing tasks and activity. Some current accounts of ADHD suggest the possibility that children with ADHD may be impaired in executive control processes. (*See* **Attention Deficit Hyperactivity Disorder**)

Further Reading

Gopher D and Koriat A (eds) (1998) *Attention and Performance XVII: Cognitive Regulation of Performance: Interaction of Theory and Application*. Cambridge, MA: MIT Press.

Inui T and McClelland L (eds) (1996) *Attention and Performance XVI: Information Integration in Perception and Communication*. Cambridge, MA: MIT Press.

Kahneman D (1973) *Attention and Effort*. Englewood Cliffs, NJ: Prentice-Hall.

Meyer DE and Kornblum S (eds) (1992) *Attention and Performance XIV: Synergies in Experimental Psychology, Artificial Intelligence, and Cognitive Neuroscience*. Cambridge, MA: MIT Press.

Monsell S and Driver J (eds) (2000) *Attention and Performance XVIII: Control of Cognitive Processes*. Cambridge, MA: MIT Press.

Parasuraman R (ed.) (1998) *The Attentive Brain*. Cambridge, MA: MIT Press.

Pashler H (1998) *The Psychology of Attention*. Cambridge, MA: MIT Press.

Posner M, Petersen SE, Fox PT and Raichle ME (1988) Localization of cognitive operations in the human brain. *Science* **240**: 1627–1631.

Styles AE (1997) *The Psychology of Attention*. Hove, UK: Psychology Press.

Attention Deficit Hyperactivity Disorder

Introductory article

James M Swanson, University of California, Irvine, California, USA
Nora D Volkow, Brookhaven National Laboratory, Upton, New York, USA
Jeffrey Newcorn, Mount Sinai School of Medicine, New York City, New York, USA
BJ Casey, Sackler Institute, Weill College of Medicine at Cornell University, New York City, New York, USA
Robert Moyzis, University of California, Irvine, California, USA
David Grandy, University of Oregon Health Sciences Center, Portland, Oregon, USA
Michael Posner, University of Oregon, Eugene, Oregon, USA

CONTENTS

Attention deficit hyperactivity disorder is a childhood syndrome characterized by developmentally inappropriate inattention, impulsivity, and hyperactivity which produces impairment at home and school. Long-term outcome is poor, but treatment with stimulant medication and behavior modification is effective. Investigations of the disorder and its treatments suggest a dopamine deficit exists that may be corrected by stimulant medication.

INTRODUCTION

The combination of inattentive, hyperactive, and impulsive behavior in children has been recognized as a syndrome since the start of the twentieth century, dating back to Still's description in 1902 of children with 'marked inability to concentrate and sustain attention' and impaired 'inhibitory volition'. The term now used as a label for this syndrome, attention deficit hyperactivity disorder (ADHD), is defined in the fourth edition of the *Diagnostic and Statistical Manual of Mental Disorders* (DSM-IV), published by the American Psychiatric Association. The DSM-IV definition lists 18 behaviors as grounds for diagnosis (Table 1), which fall into two domains: inattention, and hyperactivity/impulsivity. These are behaviors of normal

Table 1. Alignment of symptom domains, cognitive processes and neural networks

Symptom domain	*Cognitive process*	*Neural network*
Inattentive – Alerting	Sustained attention	Alerting
difficulty sustaining attention	vigilance level/decrement	cortical: right frontal
fails to finish	persistence	midbrain: locus ceruleus
avoids sustained effort	performance	thalamic:?
Inattentive – Orienting	Selective attention	Orienting
distracted by stimuli	visual cueing	cortical: parietal
does not seem to listen	auditory cueing	thalamic: pulvinar
fails to give close attention	visual search	other:?
Inattentive – Memory	Memory/planning	Executive control
has difficulty organizing tasks	planning	cortical: prefrontal
loses things	memory for objects	striatal: basal ganglia
is forgetful	memory for time	other:?
Impulsivity – Executive control	Cognitive regulation	Executive control
blurts out answers	conflict resolution	cortical: anterior cingulate
interrupts or intrudes	behavioral inhibition	striatal: nucleus accumbens
cannot wait	delay aversion	other:?
Hyperactivity – Fine motor	Motor/vocal control	Fine motor control
fidgets	fine motor control	cortical: left frontal
cannot play quietly	nonverbal control	striatal: cerebellar vermis
talks excessively	verbal	other:?
Hyperactivity – Gross motor	Activation level	Gross motor control
leaves seat	gross motor control	cortical: right frontal
runs about and climbs	novelty seeking	striatal: caudate
always on the go	arousal level	other:?

childhood when they occur infrequently or at a low intensity, so they represent symptoms of a psychiatric disorder only when they are developmentally inappropriate, severe, and produce significant impairment in multiple settings. For a diagnosis of ADHD–combined type, both symptom domains must be present and contribute to impairment. Partial syndromes (ADHD–inattentive type or ADHD–hyperactive/impulsive type) are diagnosed if symptoms in only one domain are present, and comorbid conditions (such as anxiety and depression) are diagnosed if they are also present. Based on these criteria, about 3–5% of the population of children in American elementary schools are diagnosed and treated for ADHD.

The identification of a specific cognitive deficit unique to ADHD has been elusive. Some researchers have suggested that deficiencies of children with ADHD are due to their inability to control their behavior, rather than a structural deficit of attention. Others have concluded that there is no attentional deficit in ADHD, but instead that the core deficit is in behavioral inhibition. However, ADHD children do clearly have abnormal performance on several tasks such as the Stroop color-word naming task, the Matching Familar Figures test of comparison of almost identical complex figures, the Tower of Hanoi test of planning and stacking colored rings to match a pattern, and the Trails B test of search for characters on a page in the face of distraction.

Advances in the field of cognitive neuroscience led to new concepts of attention linked to specific brain circuitry. For example, Posner and Raichle's neuroanatomical network theory of attention is based on the concepts of alerting (supressing background neural noise by inhibiting ongoing or irrevevant activity or mental effort to establish a state of vigilance), orienting (mobilizing specific neural resources toward a source of sensory stimulation), and executive control (coordinating multiple specialized neural processes by detecting targets, starting and stopping mental operations, and resolving conflict among responses), each with a well-defined neural circuitry (anterior cingulated, prefrontal cortex and basal ganglia). This cognitive neuroscience approach can be used to constrain the definition of attention, and it offers modern terminology for describing its components. The application of three levels of analysis (behavioral, cognitive, and neural) may provide some new insights about the cognitive component of this disorder. Each symptom can be classified based on its relationship to alerting, orienting, and executive control (see Table 1). The nine symptoms of inattention listed in Table 1 logically split into three groups when aligned with the three concepts of attention and the underlying neural networks. The three symptoms of impulsivity are behavioral manifestations of deficits in self-regulation, which align with the executive control network, and the six symptoms of hyperactivity fall into two groups based on deficits in fine motor and gross motor control.

ONSET AND COURSE

The DSM-IV diagnostic criteria specify the onset of symptoms by the age of 7 years, but in most cases the symptoms of ADHD are present much earlier. Impairment tends to increase during the elementary-school years in response to the cognitive and behavioral demands of the classroom setting, and this is when most diagnoses are made. At this age, more boys than girls are recognized and treated for ADHD (reported male to female ratios range from 3:1 to 9:1), but this may be due to referral biases related to disruptive behaviors (aggression, opposition, and defiance) that often coexist in boys. In most cases, ADHD symptoms decline with age, especially for the domain of hyperactivity and impulsivity. When symptoms no longer produce impairment, the diagnostic label changes to ADHD–residual type. In about one-third of the cases, the full criteria are still met in adulthood, and in another third symptoms are present but at a subthreshold level.

The subjective nature of the assessment process raises legitimate questions about the validity of the diagnosis of ADHD, but evidence for validity has accumulated from follow-up studies showing that a childhood diagnosis of ADHD is associated with extremely poor outcome in many areas, including juvenile delinquency. Children identified by the ADHD diagnosis have a serious disorder that demands recognition and deserves treatment.

TREATMENT

Since the 1930s ADHD has been treated with stimulant drugs. The first of these was amphetamine, but over the years this has been superseded by methylphenidate. Immediate-release formulations of these drugs are short-acting and must be given two or three times a day. Newer sustained-release formulations based on 'osmotic pump' and 'coated bead' delivery systems have been developed that have long duration of action and avoid the mid-day dose at school (which is often associated with embarrassment). The stimulant medications are effective in reducing the symptoms of ADHD (about 80% of children with this diagnosis show clinically meaningful benefits) and are safe (despite some common side effects such as decreased appetite and sleep). With the immediate-release formulations, the therapeutic effects emerge within 1–2 h after each oral dose, but then dissipate within 3–6 h. The sustained-release formulations have a duration of action of 8–12 h. These stimulant medications exert a profound cognitive effect, characterized by focused attention to tasks (even those with low intrinsic interest) and maintenance of attention over time (even in the face of repetitive or boring tasks). The behavioral effects are also profound: inappropriately high levels of activity and inattention in the classroom setting are reduced and compliance with typical requests and rules is increased dramatically. Stimulants do not produce a paradoxical response in ADHD children: normal children and adults respond in the same way on most measures (e.g. by decreasing normal levels of activity and increasing normal levels of attention). It is important to note that in individuals who do not have ADHD there is no impairment in these areas, so the response to stimulants does not alleviate impairment, which is the hallmark of clinical response.

In addition to pharmacotherapy, contingency management programs have been developed based on the general principles of behavior modification (reinforcement, punishment, extinction, and stimulus control). Typically, these interventions use token systems in the home and at school to prompt and shape appropriate target behavior (e.g. getting started, staying on task, interacting appropriately with others, completing work, and shifting activities on schedule) and to extinguish inappropriate behaviors (e.g. getting out of seat, talking without permission).

The most recent information on treatment comes from the Multimodality Treatment of ADHD (MTA) study, a large, six-site randomized clinical trial designed to evaluate the long-term effects of pharmacological and psychosocial interventions. Over 500 children with ADHD aged 7–9 years were recruited from a variety of sources and randomly assigned to a treatment group for a period of 14 months. In this study methylphenidate administered three times a day was more effective than psychosocial treatment (intensive behavioral intervention at home and school), and combinations of these two therapies were little better than medication alone. The success rates defined by a reduction of symptoms to a subthreshold level reflected this also: psychosocial 34%, pharmacological 56%, and combination 68%. This study provides empirical evidence of the long-term effectiveness of these two most common treatments for ADHD.

ETIOLOGY

The most prominent current theory about the cause of ADHD implicates dysfunction of brain dopamine (DA), a neurotransmitter involved in the regulation of motoric, attentional and motivational circuits. One variant of this theory suggests that ADHD is the result of a DA deficit at the neural level, which results in inattentiveness and distractibility at the cognitive level. This theory is supported by the mechanism of action of methylphenidate, which blocks DA transporters, the primary mechanism for removing DA from the synapse. Imaging studies in humans have demonstrated that therapeutic doses of stimulants block more than half of the DA transporters, markedly enhancing DA neurotransmission in the brain. Similar findings have been obtained in animal studies, which have shown that stimulants given at therapeutically relevant dosages increase extracellular DA and activate DA-regulated circuits. In animals, gene 'knockout' studies have suggested that the DA transporter gene (*DAT*) located on chromosome 15 and the DA type 4 receptor gene (*DRD4*) located on chromosome 11 are involved in basic underlying processes of activity and attention that may contribute to ADHD.

What might produce a DA deficit? Acquired and inherited factors have been proposed. One suggestion is that bouts of hypoxia and hypotension during fetal development might selectively damage striatal neurons, which are the main target for DA cells. Inherited factors have also been implicated by molecular genetic studies of ADHD. Initial investigations focused on two candidate genes involved with DA regulation: the *DAT* and *DRD4* genes. Several research groups have documented association of these two genes with ADHD.

NEURAL CORRELATES OF ADHD

In the early 1990s several teams of investigators used imaging techniques to investigate brain anatomy in groups of children with ADHD compared with children free from this disorder. Abnormalities in size of specific brain regions were observed across multiple studies. Even though groups of children with ADHD were recruited from very different clinical settings by independent research teams, research teams showed a moderate reduction in size (about a 10% decrease compared with a normal group) for measures of frontal lobes and basal ganglia (caudate nucleus and globus pallidus). Functional brain imaging studies have provided converging information implicating basal ganglia and frontal lobe abnormalities in ADHD. Imaging studies based on single photon and positron emission tomography and performed during baseline (resting) conditions documented a reduction in blood flow and metabolism in striatal and frontal brain regions. Studies using [^{18}F]-labeled dopa as a marker of DA synthesis in brain showed significant reductions in prefrontal cortex in people with ADHD when compared with normal controls. Functional magnetic resonance imaging has shown hypoactivity of frontal circuits during activation by cognitive tasks, including blunted activation in the anterior cingulate gyrus, a brain region with a central role in executive attention that has been linked to the behavioral symptoms of inattention and impulsivity in ADHD (see Table 1).

CONCLUSION

The phenomenology of ADHD has been refined over the years, and clinical manuals now agree on the specific symptoms of this disorder. At a cognitive level, abnormalities in neuropsychological performance suggest that children with ADHD are inefficient in information processing, resulting in slow and inaccurate performance. Advances in brain imaging and molecular biology have started to reveal functional and biochemical brain abnormalities associated with ADHD, localized predominantly in dopamine pathways and frontal and striatal circuits modulated by DA. Since these circuits regulate attention, executive function, motivation, response inhibition and motor activity, research on ADHD has focused on how their dysfunction could result in the cognitive deficits and behavioral symptoms of this disorder.

Further Reading

American Psychiatric Association (1994) *Diagnostic and Statistical Manual of Mental Disorders*, 4th edn. Washington, DC: APA.

Barkley RA (1997) Behavioral inhibition, sustained attention, and executive functions: constructing a unifying theory of ADHD. *Psychological Bulletin* **121**: 65–94.

Barkley RA, Fischer M, Edelbrock CS and Smallish L (1990) The adolescent outcome of hyperactive children diagnosed by research criteria 1. An 8-year prospective follow-up study. *Journal of the American Academy of Child and Adolescent Psychiatry* **29**: 546–557.

Bradley C (1937) The behavior of children receiving benzedrine. *American Journal of Psychiatry* **94**: 577–585.

Bush G, Frazier JA, Rauch SL *et al.* (1999) Anterior cingulate cortex dysfunction in attention-deficit/hyperactivity disorder revealed by fMRI and the Counting Stroop. *Biologicol Psychiatry* **45**: 1542–1552.

Castellanos FX (1997) Toward a pathophysiology of attention-deficit/hyperactivity disorder. *Clinical Pediatrics* **36**: 381–393.

Castellanos FX, Giedd JN, Marsh WL *et al.* (1996) Quantitative brain magnetic resonance imaging in attention-deficit hyperactivity disorder. *Archives of General Psychiatry* **53**(7): 607–616.

Collier D, Curran S and Asherson P (2000) Mission: not impossible? Candidate gene studies in child psychiatric disorders. *Molecular Psychiatry* **5**: 457–460.

Ernst M, Zametkin AJ, Matochik JA, Jons PH and Cohen RM (1998) DOPA decarboxylase activity in attention deficit hyperactivity disorder adults. A [fluorine-18] fluorodopa positron emission tomographic study. *Journal of Neuroscience* **18**: 5901–5907.

Faraone S, Doyle A, Mick E and Biederman J (2001) Meta-analysis of the association between the dopamine D4 gene 7-repeat allele and ADHD. *American Journal of Psychiatry* **158**: 1052–1057.

Greenhill LL, Halperin JM and Abikoff H (1999) Stimulant medications. *Journal of the American Academy of Child and Adolescent Psychiatry* **38**(5): 503–512.

Levy F and Swanson JM (2001) Timing, space and ADHD: the dopamine theory revisited. *Australian and New Zealand Journal of Psychiatry* **35**: 504–511.

Lou HC (1996) Etiology and pathogenesis of attention-deficit hyperactivity disorder (ADHD): significance of prematurity and perinatal hypoxic-haemodynamic encephalopathy. *Acta Paediatrica* **85**: 1266–1271.

Lou HC, Henriksen L and Bruhn P (1990) Focal cerebral dysfunction in developmental learning disabilities. *Lancet* **335**(8680): 8–11.

Mannuzza S, Klein RG, Bessler A, Malloy P and LaPadula M (1993) Adult outcome of hyperactive boys: educational achievement, occupational rank, and psychiatric status. *Archives of General Psychiatry* **50**: 565–576.

MTA Cooperative Group (1999) Multimodal Treatment Study of Children with ADHD. A 14-month randomized clinical trial of treatment strategies for attention-deficit/hyperactivity disorder. *Archives of General Psychiatry* **56**: 1073–1086.

NIH Consensus Conference (2000) National Institutes of Health Consensus Development Conference statement: diagnosis and treatment of attention-deficit/hyperactivity disorder (ADHD). *Journal of the American Academy of Child and Adolescent Psychiatry* **39**: 182–193.

Pelham WE and Fabiano G (2000) Behavior modification. *Child and Adolescent Psychiatric Clinics of North America* **9**: 671–688.

Pennington BF and Ozonoff S (1996) Executive functions and developmental psychopathology. *Journal of Child Psychology and Psychiatry and Allied Disciplines* **37**(1): 51–87.

Posner MI (2001) Developing brains: the work of the Sackler Institute. *Clinical Neuroscience Research* **1**: 258–266.

Posner MI and Raichle ME (1994) *Images of Mind*. New York, NY: Scientific American Library.

Posner MI, Rothbart MK, Farah M and Bruer J (2001) Human brain development: introduction to the report to the McDonnell Foundation. *Developmental Science* **4**/3 (special issue): 253–384.

Rubia K, Overmeyer S and Taylor E (1999) Hypofrontality in attention deficit hyperactivity disorder in higher order motor control: a study with functional MRI. *American Journal of Psychiatry* **156**: 891–896.

Satterfield J, Swanson JM, Schell A and Lee F (1994) Prediction of antisocial behavior in attention-deficit hyperactivity disorder boys from aggression/defiance scores. *Journal of the American Academy of Child and Adolescent Psychiatry* **33**: 185–190.

Sagvolden T and Sergeant JA (1998) Attention deficit/hyperactivity disorder – from brain dysfunctions to behaviour [editorial]. *Behavioural Brain Research* **94**(1): 1–10.

Still GF (1902) Some abnormal psychical conditions in children. *Lancet* **1**: 1008–1012; 1077–1082; 1163–1168.

Swanson JM, McBurnett K, Wigal T *et al.* (1993) Effect of stimulant medication on children with attention deficit disorder: a 'review of reviews'. *Exceptional Children* **60**: 154–162.

Swanson JM, Sergeant JA, Taylor E *et al.* (1998) Attention-deficit hyperactivity disorder and hyperkinetic disorder. *Lancet* **351**: 429–433.

Swanson J, Castellanos FX, Murias M, LaHoste G and Kennedy J (1998) Cognitive neuroscience of attention deficit hyperactivity disorder and hyperkinetic disorder. *Current Opinion in Neurobiology* **8**: 263–271.

Swanson J, Posner M, Cantwell D *et al.* (1998) Attention-deficit/hyperactivity disorder: symptom domains, cognitive processes and neural networks. In: Parasuraman R (ed.) *The Attentive Brain*, pp. 445–460. Boston, MA: MIT Press.

Swanson JM, Kraemer H, Hinshaw S *et al.* (2001) Clinical relevance of the primary findings of the MTA: success rates based on severity of symptoms at the end of treatment. *Journal of the American Academy of Child and Adolescent Psychiatry* **40**: 168–179.

Swanson JM, Deutsch C, Cantwell D *et al.* (2001) Genes and ADHD. *Clinical Neuroscience Research* **1**: 207–216.

Taylor E, Chadwick O, Heptinstall E and Danckaerts M (1996) Hyperactivity and conduct problems as risk factors for adolescent development. *Journal of the American Academy of Child and Adolescent Psychiatry* **35**: 1213–1216.

Volkow ND, Ding YS, Fowler JS *et al.* (1995) Is methylphenidate like cocaine? Studies on their

pharmacokinetics and distribution in human brain. *Archives of General Psychiatry* **52**: 456–463.
Volkow ND, Wang G, Fowler JS *et al.* (2001) Therapeutic doses of oral methylphenidate significantly increase extracellular dopamine in the human brain. *Journal of Neuroscience* **21**(2): RC121.
Volkow ND, Wang GJ, Fowler JS *et al.* (2002) Relationship between blockade of dopamine transporters by oral methylphenidate and the increases in extracellular dopamine: therapeutic implications. *Synapse* **43**: 181–187.

Attention, Models of

Introductory article

Rajesh PN Rao, University of Washington, Seattle, Washington, USA

CONTENTS

Attention is the ability of an organism to select and process only the relevant or 'interesting' parts of its sensory inputs, while discarding other potentially irrelevant parts. Models of attention seek to explain the mechanisms and neuronal substrates underlying this evolutionarily important sensory ability.

INTRODUCTION

Animals are confronted with a vast amount of sensory information in their day-to-day interactions with the natural world. Only a fraction of this information can be processed at any given moment in time owing to the brain's limited processing resources. Fortunately, only a fraction of the total information received is typically relevant to any particular task at hand and to the animal's continued survival. Thus, the fundamental problem faced by an animal's perceptual system is to select and process only the relevant or 'interesting' parts of its sensory inputs, discarding the other potentially irrelevant parts. Attention is nature's solution to this problem. (*See* **Attention**; **Selective Attention**)

Traditionally, attention is defined as the phenomenon by which animals preferentially process parts of their input, shifting their focus of processing from one part of their input to another in a serial fashion. The metaphor that is frequently used, especially in psychology, is that of a spotlight that can be moved to different parts of a scene independent of eye movements. Any information lying outside the spotlight is filtered out. This form of attention is generally referred to as 'covert' attention, to distinguish it from the 'overt' attentional shifts due to eye movements. (*See* **Eye Movements**)

Although the spotlight metaphor is not a computational model of attention, it has proved useful in characterizing various properties of attention. For example, events occurring within a pre-cued 'spotlight' region in the visual field can be detected faster and more accurately than events occurring outside the spotlight region. The spotlight metaphor has also been used to distinguish between two forms of visual search: parallel search, corresponding to a search for an object that differs from neighboring objects by a single feature (e.g. an X among O's), and serial search, where the target differs from neighboring objects by a conjunction of features (e.g. a green X among red X's and green O's). It has been suggested that serial search requires sequential analysis of the scene by an attentional spotlight that 'binds' different features of an object together, while parallel search is the result of 'pop-out' of the target item due to differences in a single feature. Although considerable psychophysical data exist, for instance, on the speed and size of the hypothetical attentional spotlight, neurobiological evidence for such a spotlight has remained inconclusive. The notion of a spotlight has, however, been influential in the formulation of several computational models of attention, some of which are discussed below.

GATING MODELS OF ATTENTION

One of the early quantitative models of attention inspired by the spotlight metaphor is the attention gating model, which characterizes how information is processed within the attentional spotlight. It was originally formulated to explain results from rapid serial visual presentation experiments. In these experiments, participants would fixate on a location on a screen and focus attention on a stream of letters displayed rapidly and sequentially to the left of the fixation point. Upon detection of a specified target letter, the participant shifted attention as quickly as possible to a stream of numerals being displayed to the right of the fixation point, and reported (for example) the four earliest occurring numerals after the target. This paradigm is based on the view that the spotlight of attention is first focused on the stream of letters, allowing the target to be detected, and then is moved to the stream of numerals, where the spotlight allows the next four numerals to be registered in short-term memory. Thus, attention acts as a 'gate' into short-term memory, regulating the flow of sensory information into conscious awareness. (*See* **Working Memory, Computational Models of; Neural Basis of Memory: Systems Level; Working Memory, Neural Basis of; Memory Models; Working Memory**)

The attention gating model is depicted in Figure 1. In this model, when the attentional cue (for example, the target letter) is detected, the attention 'gate' is opened and the items from the to-be-attended stream of inputs is admitted into memory. To avoid overflow of memory capacity, the gate is automatically closed a short time after opening. This prevents other stimulus items from entering memory. The strength of an item stored in memory depends on the gating function G and the time at which the item occurred with respect to gate opening. The interaction is multiplicative: a stimulus S passes through the attention gate with output strength $G \times S$. The items stored in memory are then reported in decreasing order of strength. The gating function G can be estimated experimentally based on the performance of the subjects and has been shown to be a function with exponential rise and decay (Figure 1).

The attention gating model has been quite successful in quantitatively modeling psychophysical results. However, it is a purely phenomenological model (compare with the description below of

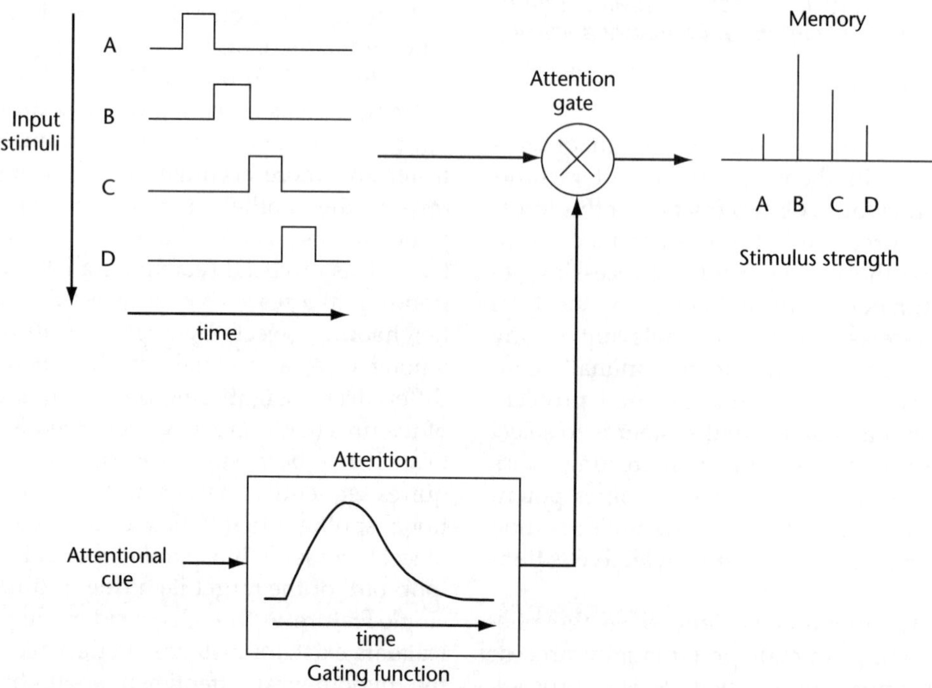

Figure 1. Attention gating model. When an attentional cue is detected, the attention gate is activated with a time course given by the attention gating function (shown within the box). The input stimuli A, B, C and D pass through the attention gate with a strength given by the product of the stimulus with the current value of the gating function. Each input stimulus is stored in memory according to its strength (indicated by the vertical bars on the right). Stimuli from memory are recalled and output as responses in decreasing order of strength.

Crick's model). For example, the gating model does not address how a spatial region is selected for attentional processing and how information within this region is processed before reaching short-term memory and awareness. These issues have been addressed using saliency maps and dynamic routing circuits respectively.

SALIENCY MAPS

A saliency map is a topographic representation of an input sensory field: each location in a saliency map contains a value that represents how different or salient that location is, compared with other locations in the sensory field. A 'winner takes all' mechanism is then used to select the current most salient location (peak) in the map. Attention is directed to the selected location by a gating mechanism that allows the stimuli near the selected location to be conveyed to higher processing centers and suppresses stimuli at other locations.

The saliency of a location may be computed based on 'bottom up' and/or 'top down' information. In 'bottom up' calculations the saliency of a location may be defined in terms of the difference between the stimulus at that location and the stimuli in neighboring locations. For example, the presence of an X in a field of O's would cause the location containing the X to attain a higher saliency value than locations containing O's, owing to the large difference in features between the X and the surrounding O's. Thus, 'bottom up' saliency maps provide a computational mechanism for implementing the 'pop-out' inherent in the types of visual search characterized as parallel search. Certain forms of serial search may also be modeled using 'bottom up' saliency maps where saliency is computed based on conjunctions of stimulus features (such as 'red/green' and 'X/O'). In this case, attention is directed sequentially to successively smaller peaks in the saliency map and previously visited locations are suppressed for a short duration, a process known as 'inhibition of return'.

'Top down' saliency maps are based on the differences between sensory stimuli at different locations and a stored prototype target object. The more similar the stimulus at a particular location is to the target object, the higher the saliency value of that location. Such saliency maps provide a mechanism for implementing object-based serial search, where attention is directed sequentially to different objects in a visual scene in the order of their similarity to the prototype target object. This type of search is frequently performed by our visual system, for example when we are looking for a

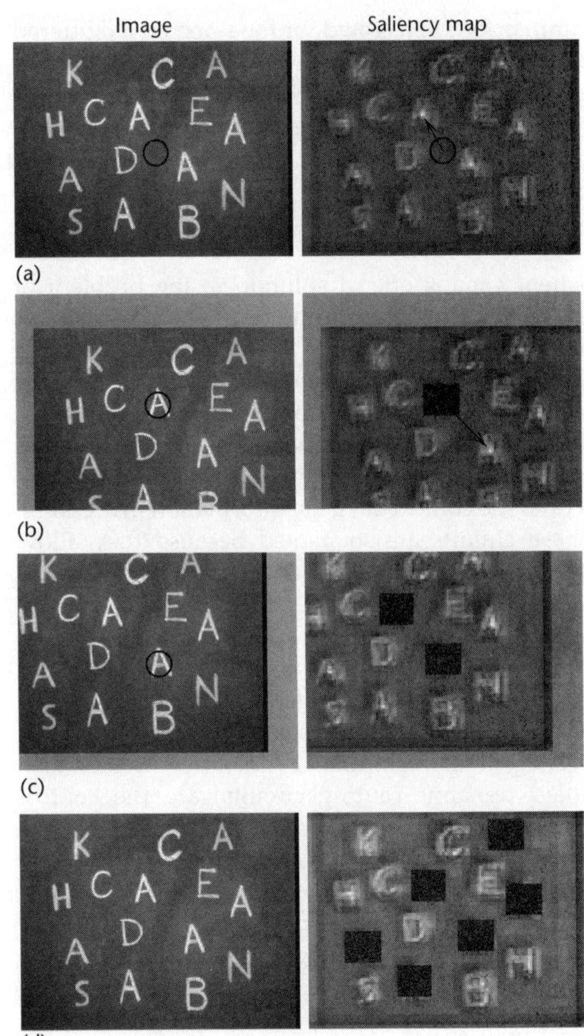

Figure 2. Saliency maps. The use of 'top down' saliency maps in a visual counting task. The goal is to count the number of occurrences of the letter A on a blackboard containing a collection of letters. The image of the board is shown on the left and the corresponding saliency map on the right. Each letter is represented by the vector of outputs of a set of oriented spatial filters modeled after the receptive fields of visual cortical neurons. The brightness at each location in the saliency map indicates the degree of closeness (saliency) of the letter at that location to the letter A, as given by the similarity in their filter-based representations. The brightest spot, indicated by the arrow, is chosen to be the next location to be attended (a). The circle indicates the original focus of attention. (b, c) The result of shifting attention to two different locations containing the letter A. In both cases, once the letter has been attended to and processed, the location containing the letter is inhibited (dark square) to prevent this location from competing again for the focus of attention during the course of the task. This implements the process of 'inhibition of return' during visual search. (d) The final result of counting all six occurrences of the letter A in the input image. Figure adapted from Rao and Ballard (1997).

familiar face in a crowd, or for a pen on a cluttered desk. Figure 2 demonstrates how 'top down' saliency maps could be used to perform an object-based serial search during a visual counting task.

SHIFTER CIRCUITS AND DYNAMIC ROUTING

Saliency maps offer a solution to the problem of selecting relevant portions of a sensory field for further processing. They do not, however, solve the problem of routing this information to higher centers in a position- and size-invariant way. Shifter circuits were proposed as a computational model for how the brain may accomplish this within the context of visual information processing. These circuits are so named because they allow information in one location of the visual field to be 'shifted' and fed to a different location in a higher processing module. Such a process leads to the formation of higher-level visual representations that remain stable regardless of actual object position and size at the lowest levels.

The primary components of shifter circuits (also called dynamic routing circuits) are the control neurons, which set the size and position of the attentional window. The control neurons modulate the strength of synaptic connections from the lower level to a higher level such that only the information from the attended region in the lower level is routed to the higher level. The control neurons are driven by a saliency map. Only neurons that correspond to highly salient locations are activated, causing information in only these locations to be routed to higher processing centers.

Position and size invariance is achieved by arranging multiple inputs from a lower level to converge onto a single higher-level unit (Figure 3a). This allows units at higher levels to have access to information from increasingly large areas of the input field. By having the control neurons select appropriate subsets of input connections at each level, one can 'focus attention' on any local region in the input field and route the information in this local region to the same set of processing units at the highest level (Figure 3b). This makes the circuit invariant to the actual position of objects in the input field. Similarly, size invariance can obtained by integrating several lower level inputs into a single higher level output, repeating this strategy at each level until the very highest level (Figure 3c). Once again, the control neurons dictate which inputs are integrated.

As mentioned above, saliency maps control attention in routing circuits. Once a saliency map has been computed, a competitive mechanism is used to activate the control units that correspond to the most salient parts of the input. These control units in turn select the appropriate input connections at each level for routing the selected information to the highest level. The exact equations governing the dynamics of the control neurons can be derived from well-known optimization principles. Information received at the highest level can be fed into an associative memory for recognition. In addition,

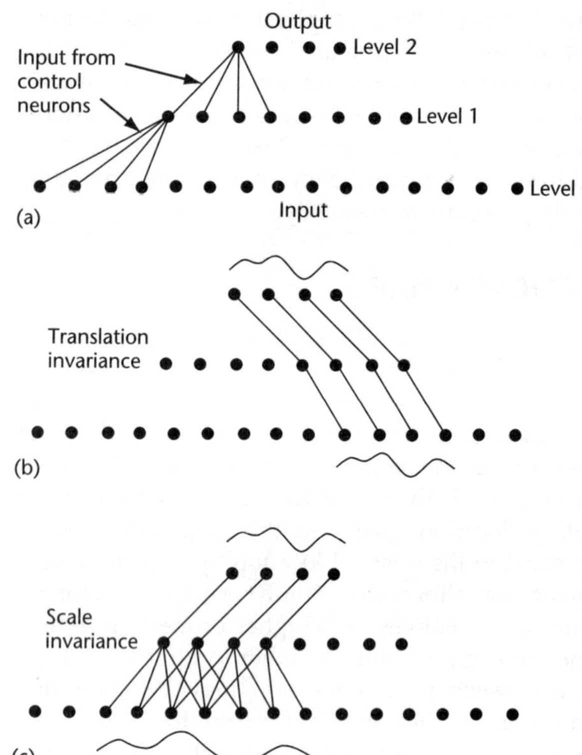

Figure 3. Routing circuits. (a) A dynamic routing circuit with three levels. Each neuron, represented by a black circle, receives inputs from several lower level neurons via dynamically modifiable connections. Only the connections for the leftmost neurons in levels 1 and 2 are shown. The connections for the other neurons are identical except for a rightwards shift. The strength of these connections is determined by the control neurons, which modulate the connections multiplicatively (indicated by the two arrows) and set the position and size of the window of attention. (b) Attention can be focused on a local input region. Information lying within the attended region (depicted as a wavy line) is routed to the output layer by the control neurons. The pattern obtained at the output level is invariant to translations of the pattern at the input level. (c) The window of attention can be enlarged to process large patterns. In this case, the control neurons set the connections for a net convergence from input to output. As a result, the pattern obtained at the output level is invariant to changes in input pattern size.

during the recognition process, the output of the associative memory can be used to readjust the position and scale of the attention window to enhance recognition accuracy. Small-scale dynamic routing circuits of up to three levels have been simulated on desktop computers. Promising results have been obtained for simple pattern recognition tasks such as recognizing letters in binary images, but the performance of routing circuits in more realistic tasks involving natural images has not yet been thoroughly investigated. (*See* **Vision: Object Recognition**)

SPATIAL VERSUS OBJECT-BASED ATTENTION

The models we have considered so far have emphasized how attention may be directed towards specific spatial locations, and how sensory information from these locations may be routed to higher centers for further processing. Behavioral and single-cell studies have suggested that attention may also select parts of a sensory input based on visual features (such as motion or color) or even whole objects, a phenomenon known as feature-based or object-based attention. (*See* **Visual Attention; Selective Attention**)

Evidence for feature- and object-based attention comes from studies showing that participants can reliably track a target object that is superimposed with a distracter object in the same spatial location. For example, if the image of a face is transparently superimposed on the image of a house and the face image is moved back and forth, subjects are able to track the face despite the presence of the image of the house. In addition, brain imaging studies reveal higher neural activity in the face-selective and motion-selective areas of the brain when tracking the face, compared with the case where the house is the target of attention. Other evidence for object-based attention comes from studies involving patients with damage to their right parietal lobe. Many of these patients are unable to attend to the left side of objects, where the definition of an object is dependent on the task at hand. (*See* **Parietal Cortex**)

Computational models of object-based attention are still in their infancy, partly because of the broad range of phenomena that fall into this category. Preliminary attempts at modeling object-based attention have focused on cooperative networks that comprise two complementary subnetworks, one that estimates objects and their features (such as color and shape) and another that estimates object transformations (such as translation, dilation and rotation). In such networks, attention can be focused on a particular object or a particular location by biasing 'top down' signals from higher processing centers or short-term memory. Object-based attention corresponds to keeping the 'top down' signal for the object-estimating network fixed: this causes the transformation network to converge to estimates of the attended object's transformations. Spatial attention corresponds to keeping the 'top down' signal for the transformation network fixed: this causes the object network to converge to the identity of the object in the attended location. The inspiration for such a model comes from neuroanatomical studies showing a rough dichotomy in the mammalian brain between networks specialized for object identification and networks geared towards spatial transformations and action.

NEUROANATOMICAL SUBSTRATE

The spotlight metaphor inspired some of the early attempts at identifying possible neuronal substrates of attention. In 1984, Francis Crick, one of the co-discoverers of the structure of DNA, proposed that the thalamic reticular nucleus (TRN) may have an important role in implementing a spotlight. The thalamus is a nucleus that receives input information from a majority of the senses, including visual, auditory and somatosensory information. This information is conveyed to the neocortex by relay cells in the thalamus. The TRN is a single-layer network of inhibitory neurons strategically located between the thalamus and the cortex. The network receives both ascending inputs from the thalamus as well as descending inputs from the cortex. Its neurons process these inputs and inhibit neighboring neurons as well as corresponding neurons in the thalamus. Thus, the TRN is well suited to regulating the flow of input sensory information from the thalamus to higher cortical centers. In other words, it could implement an attentional spotlight by allowing only selected parts of the sensory inputs to be conveyed to the cortex, inhibiting all other neurons in the thalamus that correspond to other sensory inputs. Despite considerable progress in our knowledge of the TRN and the thalamus, evidence for Crick's spotlight model remains inconclusive. The main stumbling block remains our poor understanding of the feedback loop between the cortex and the thalamus in alert behaving animals, and the role of the TRN in modulating this feedback loop.

Several possibilities exist for biophysically implementing the gating mechanisms that are integral parts of not only the attention gating model, but also other models such as the routing circuit model. The primary requirement here is a mechanism that can inhibit or filter out input channels that are not being attended, allowing only the attended inputs to proceed to the next stage of processing. Presynaptic inhibition (inhibition of an input fiber before it can affect a cell) allows precise gating of inputs to a single neuron, but there is little evidence for its existence in the neocortex. Postsynaptic inhibition (inhibition of a cell after an input has been delivered) can also be used to gate the inputs to a neuron but at coarser level. Various cellular mechanisms have been suggested for this form of gating, most of which are based on particular characteristics of the ionic channels and receptors that are embedded in the membrane of neurons.

The visual cortical areas V1, V2, V4 and inferotemporal cortex (IT) are assumed to be the major neural substrates of dynamic routing circuits in the visual cortex. These areas are organized in a roughly hierarchical manner, with neurons in each area receiving convergent inputs from up to a thousand neurons in the preceding area. This is consistent with the multiple-level architecture of a routing circuit. The control neurons that direct the flow of information from one level to the next are hypothesized to be located in the pulvinar, a subcortical nucleus of the thalamus. The pulvinar sends and receives connections from each of the areas from V1 to IT, making it a suitable candidate for controlling attentional routing. Neurophysiological and brain imaging studies support the hypothesis that the pulvinar is involved in filtering out unattended stimuli. The competitive interactions between the control neurons are assumed to be mediated by lateral inhibition within the pulvinar and/or through the TRN. Finally, the control neurons in the pulvinar are assumed to be driven by saliency maps, whose neural implementation is discussed below. (*See* **Occipital Cortex**; **Temporal Cortex**)

Several different cortical and subcortical areas have been suggested as possible substrates for implementing saliency maps. The posterior parietal cortex (PP) is one such area. Neurons in PP show either elevated or suppressed activity when attention is directed to a visual target. In addition, damage to PP impairs the ability to disengage attention from a currently attended location. These results are consistent with a saliency map-based interpretation of PP. Another possible neural substrate for a saliency map is the superior colliculus, a subcortical nucleus that plays a major role in targeting eye movements to different parts of a scene. The superior colliculus receives direct input from the retina and can influence activity in higher cortical areas via its connections to the pulvinar. Other possible areas that may encode saliency and/or behavioral relevance of targets include the frontal eye fields and the inferior/lateral divisions of the pulvinar itself. It is currently unclear whether these different areas encode different types of saliency or different types of sensorimotor modalities (such as attentional shifts, or eye or hand movements). (*See* **Parietal Cortex**)

Models that jointly address spatial and object-based attention usually rely on the approximate division of visual processing in the visual cortex into the tasks of 'what' (object identification) and 'where' (motion/spatial reasoning). Object-related processing has traditionally been associated with the so-called 'form' pathway, comprising the hierarchy of areas V1, V2, ..., IT in the ventral part of the visual cortex. Spatial reasoning and action-related processing is traditionally ascribed to areas in the dorsal part of the visual cortex, particularly the intraparietal and PP cortex. One model for spatial and object-based attention assumes that the cortical areas in the frontal lobe store task-relevant information and apply either spatial or object-related constraints to the dorsal or ventral networks respectively. A spatial constraint causes the object ('what') networks in the ventral pathway to respond strongly to the objects or features in the input image with that spatial property (such as position or motion). Similarly, an object-based constraint on ventral networks would cause the dorsal networks ('where') to focus upon the spatial properties of the object expressed in the constraint. Such a model, which is based on interactions between dorsal, ventral and frontal areas of the brain, integrates the neuronal and systems-level views of attention. It is, however, hard to validate, given the current difficulty in recording from multiple brain areas simultaneously. (*See* **Temporal Cortex**; **Parietal Cortex**; **Frontal Cortex**)

At the single neuron level, neurophysiological recordings in awake behaving monkeys have provided some important clues to understanding the mechanisms of attention. For example, the response of many neurons in the visual cortical areas V2 and V4 are modulated by attention. In these experiments, a visual stimulus that can activate a recorded neuron is first found and then a second stimulus is placed within the activating region (receptive field) of the neuron. This typically causes the neuron's response to shift significantly from its original response. However, when the monkey is made to focus attention on the original stimulus, the neuron's response becomes almost identical to its original response, as if the attended stimulus appeared all by itself. This provides direct evidence for attention filtering out irrelevant stimuli. A model based on 'biased competition' between two populations of neurons, one representing the attended stimulus and the other representing the distracter stimulus, has been suggested to explain these results. Attention is assumed to increase the strengths of the inputs from the population representing the attended stimulus. Such a model has been shown to account quantitatively for the neurophysiological results. However, the lack of details regarding its neurobiological implementation makes it hard to distinguish this mathematical model from the computationally motivated models discussed above, most of which use biased competition in one form or another to focus attention. (*See* **Attention, Neural Basis of; Spatial Attention, Neural Basis of**)

CONCLUSION

Computational models of attention provide useful insights into how the brain selectively processes portions of its inputs based on measures of saliency and task relevance. The process by which parts of a scene are selected and processed without movement of the eyes is often explained by a spotlight metaphor. Saliency maps provide a mechanism for implementing a spotlight by allowing the selection of a single stimulus based on either its 'bottom up' saliency compared with competing stimuli or its 'top down' relevance to the task at hand. Once selected, a stimulus can be routed to higher processing centers and eventually to memory using dynamic routing circuits. These circuits allow information from any portion of the sensory field to be selectively routed from one stage of processing to another in a manner independent of size and position. Once information is routed to memory, attention gating models provide a quantitative characterization of how attended items are transferred to short-term memory. Recent models of attention have stressed global interactions between networks specialized for object identification and spatial reasoning. Such models may provide a unified framework for modeling object-based and spatial attention.

The neural substrates of attention have not yet been completely identified, but the effects of attention on single neurons have been studied in several visual cortical areas, including V2, V4, the frontal eye fields and posterior parietal cortex. The pulvinar nucleus of the thalamus, the thalamic reticular nucleus and the superior colliculus are some of the subcortical nuclei implicated in attentional control. How these different areas interact to produce the emergent phenomenon of attention is currently the subject of both modeling as well as experimental studies.

Further Reading

Crick F (1984) Function of the thalamic reticular complex: the searchlight hypothesis. *Proceedings of the National Academy of Science of the USA* **81**: 4586–4590.

Desimone R and Duncan J (1995) Neural mechanisms of selective visual attention. *Annual Review of Neuroscience* **18**: 193–222.

Kastner S and Ungerleider LG (2000) Mechanisms of visual attention in the human cortex. *Annual Review of Neuroscience* **23**: 315–341.

Koch C and Ullman S (1985) Shifts in selective visual attention: towards the underlying neural circuitry. *Human Neurobiology* **4**: 219–227.

Olshausen BA, Anderson CH and Van Essen DC (1993) A neurobiological model of visual attention and invariant pattern recognition based on dynamic routing of information. *Journal of Neuroscience* **13**: 4700–4719.

Parasuraman R (ed.) (1998) *The Attentive Brain.* Cambridge, MA: MIT Press.

Pashler HE (1998) *The Psychology of Attention.* Hillsdale, NJ: Erlbaum.

Posner MI and Raichle ME (1994) *Images of Mind.* New York, NY: Scientific American Books.

Rao RPN and Ballard DH (1997) A computational model of spatial representations that explains object-centred neglect in parietal patients. In: Bower JM (ed.) *Computational Neuroscience: Trends in Research 1997,* pp. 779–785. New York, NY: Plenum Press.

Reeves A and Sperling G (1986) Attention gating in short-term visual memory. *Psychological Review* **93**: 180–206.

Treisman AM and Gelade G (1980) A feature-integration theory of attention. *Cognitive Psychology* **12**: 97–136.

Attention, Neural Basis of

Introductory article

Peter De Weerd, University of Arizona, Tucson, Arizona, USA

CONTENTS

Attention, the ability to selectively process sensory information, is generated by the modulation of sensory processes by frontoparietal neural networks.

INTRODUCTION

The neural basis of attention resides in the ability of structures in the brain representing the behavioral relevance of stimuli to alter sensory processing, so that relevant stimuli are processed effectively and irrelevant ones are ignored. Attention, however, is not a unitary cognitive ability. Attention can alter the sensory processing of stimuli in all sensory systems. In addition, there are different kinds of attention which are related to specific behavioral requirements in different cognitive tasks. Thus, depending on the cognitive task one faces, and the sensory systems involved, different neural networks in the brain will become active during attentional operations.

Definitions of Attention

At the end of the nineteenth century the psychologist William James described attention as a 'concentration of consciousness', with the purpose of 'withdrawing from some things in order to deal effectively with others'. An experimental demonstration of that idea was given by Hermann von Helmholtz in 1894. He filled a large screen with letters, which were spread too widely to be read at once without moving the eyes. During a brief illumination of the screen with a flash of light, which made it impossible to use eye movements to explore the screen, he measured the number of letters that could be read. Only a few letters (up to about five) could be read during each flash. Interestingly, the letters that could be read were not necessarily located at the center of gaze. Letters in any region of the screen away from the center of gaze could be read if the region of interest was selected in advance. This experiment demonstrates several aspects of attention which are still being investigated today: the dissociation of directed attention from eye gaze ('covert' attention), the selective nature of attention, and the limited capacity of attention.

The experiment of Helmholtz indicates that the ability to selectively attend to information is an answer to limitations in sensory information processing. The ability to select information also avoids our behavior being determined by the strength of random stimuli in our environment. Behavior otherwise would inevitably become uncoordinated and chaotic. Hence, the ability to select information is crucial, because it allows us to willfully give precedence to behaviorally relevant information, at the cost of irrelevant information. The 'grabbing' of attention by powerful stimuli is referred to as 'bottom up' (exogenous) attention, and the intentional selection of behaviorally relevant stimuli is referred to as 'top down' (endogenous) attention.

There are different kinds of top-down attention. Selective attention, the main topic discussed here, is the ability to detect or discriminate relevant stimuli (targets) in the presence of competing irrelevant stimuli (distracters). Selective attention is required to find a familiar face in a crowd at a party. A second attentive process is vigilance: the ability to orient attention and respond to randomly occurring, relevant events in the environment over an extended period. A guard who is keeping an eye on the entrance of a building, monitoring for suspicious activity throughout the night, is carrying out a vigilance task. In addition, it has been proposed that there are attentional functions that maintain 'executive control' over goal-directed behavior and thought processes. Executive control functions are especially important when stimuli contain conflicting information. The onset of reading of the word 'green' will be faster when the word is printed in green than when it is printed in red, a phenomenon known as the Stroop effect. Applying a level of analysis to a stimulus that contains conflicting information, appropriate for the required behavioral response, is a function of executive processes.

Circuits of Selective Attention in the Visual System

The ability to attend to a stimulus is best understood as an emergent property of the functional architecture of sensory systems. This concept will be explored within the context of the visual system (Figure 1).

The primary visual area (V1) receives retinal information through the lateral geniculate nucleus, a relay station in the thalamus. Following lesion and anatomical studies in the monkey, Ungerleider and Mishkin proposed in 1982 that forward projections from V1 give rise to two visual processing streams. One is directed ventrally to the temporal lobe, and is involved in the analysis of objects (ventral pathway, or 'what?' pathway). The other one is directed dorsally to the parietal lobe, and is involved in the analysis of locations of objects relative to each other and relative to the observer (dorsal pathway, or 'where?' pathway). A relative segregation between ventral and dorsal streams is maintained in their projections to the prefrontal cortex in the frontal lobe (Figure 1). Prefrontal cortex is involved in the maintenance and manipulation of information in working memory, response selection and a variety of executive processes.

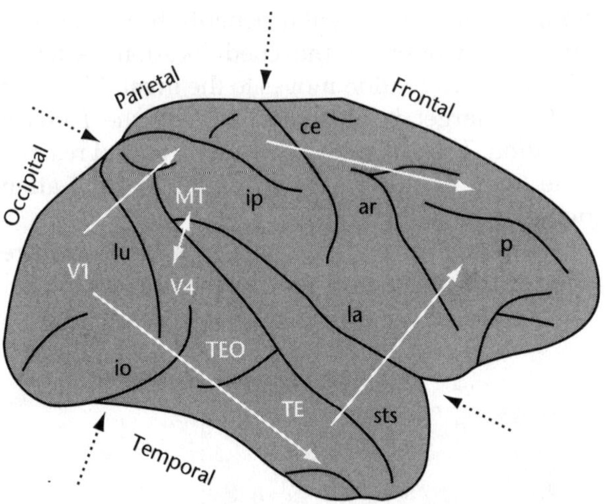

Figure 1. A neural network for attention in the primate brain. Shown is a lateral view of the brain of a rhesus macaque, with the back of the brain towards the left. Dotted arrows correspond to the boundaries between occipital, temporal, parietal, and frontal lobes. White labels indicate visual areas, white arrows indicate anatomical pathways. Visual area V1 gives rise to a ventrally directed pathway which includes V2 (buried in the lunate sulcus), V4, and areas in the temporal lobe (TEO and TE). It also gives rise to a dorsally directed pathway which includes V3 (buried in the lunate sulcus), MT (buried inside the superior temporal sulcus), and areas in the parietal lobe. There is cross-talk between areas belonging to these two visual processing pathways (e.g. double arrow between V4 and MT). Roughly 40 visual areas have been described, and only a few are indicated here. Projections from parietal and temporal lobes towards the prefrontal region of the frontal lobe retain a significant degree of segregation. Forward projections (white arrows) from one to the next area are always returned by backward projections (not shown). ar, arcuate sulcus; ce, central sulcus; io, inferior occipital sulcus; ip, intraparietal sulcus; la, lateral sulcus; lu, lunate sulcus; p, principal sulcus; sts, superior temporal sulcus.

Each of the two processing streams is composed of a number of areas, which are hierarchically organized. Lower-order areas (closer to V1) each contain an orderly retinotopic map of the environment, and are composed of neurons with small receptive fields (RFs), which process simple aspects of the stimulus, such as the orientation of edges. Higher-order areas do not show retinotopy, their neurons have large RFs, and they process complex aspects of the stimulus, such as its shape. Owing to this hierarchical organization, lesions at low levels result in severe sensory deficits, while lesions at intermediate and higher levels leave elementary sensory processes intact and interfere in more subtle ways with visual processing. Lesions at higher levels of the ventral stream induce a disorder in visual object recognition that is referred to as 'agnosia'. Parietal damage causes severe spatial deficits. Because these lesion effects reflect at least in part attention deficits, they are discussed in the following section.

When we look at an object it is automatically perceived as a whole, with a particular location, color, shape, and orientation. Because different aspects of a stimulus are processed in a distributed manner in different cortical regions, this raises the question of how the different features of a stimulus are combined in order to generate a holistic perception. This question is referred to as the 'binding' problem, and the quest for a solution to that problem is intimately related to the quest to understand mechanisms of selective attention. A fundamental insight that will help to resolve both issues is the fact that the separation of the different processing streams is only relative: at various levels in the system, there are anatomical connections between pathways that permit cross-talk. Furthermore, feedforward projections are returned by feedback projections. Hence, neural activity in a given area can be modulated by activity in other areas or structures to which it is connected. These modulatory influences are a functional property that reflects the structural layout of the system, which is crucial to understanding attention and other cognitive abilities.

PARADIGMS OF ATTENTION IN EXPERIMENTAL PSYCHOLOGY

Before explaining how modulation of activity creates the cognitive ability of attention, it is important to review the different paradigms psychologists have used to investigate attention. These paradigms have led to a precise characterization of attentional behavior, and have become the cornerstone of most experiments on attention in the broad field of cognitive neuroscience (Figure 2).

Spatial Cueing

The spatial cueing paradigm was developed by Posner and colleagues to measure the effects of directed attention (Figure 2(a)). Participants are instructed to keep their eyes on a central fixation cross presented on a computer monitor, and to covertly attend to a location in the left or right half of the screen, indicated by an arrow close to the fixation cross (the cue). The cue is followed in time by a target (e.g., a small square) briefly presented away from fixation. The target can be presented in the cued location (valid cue), or in a minority of cases in a location in the other half of the screen (invalid cue). Participants have to make a button response as soon as they detect the target. On trials in which the target is preceded by a valid cue, response times are shorter (about 250 ms) than on trials in which the target is preceded by an invalid cue (about 300 ms). These results support the idea of a 'focus of attention'. After a valid cue, participants move their focus of attention to the expected location in anticipation of target presentation, and target detection benefits from the presence of attention at the cued location. After an invalid cue, attention moves to the invalid location, and the target is presented outside the focus of attention. The associated cost, an increased reaction time, may be due to the reorienting of the attentional focus to the target.

The reorienting of attention to a location in response to a cue is controlled by the participant, and

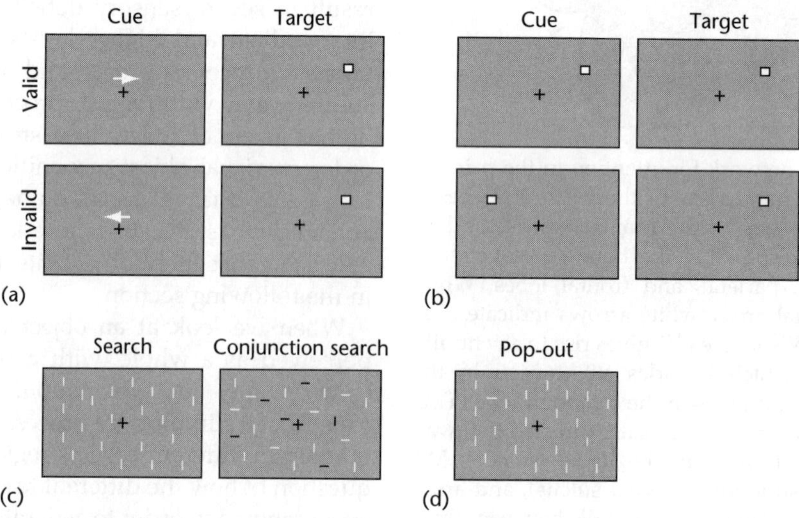

Figure 2. Paradigms in attention research. (a, b) Spatial cueing paradigms. (a) The top two panels show a symbolic cue (arrow) close to a fixation spot (cross) on a computer monitor (gray rectangle) which validly predicts the location of the target (square), presented briefly after offset of the cue. The bottom two panels show target presentation after an invalid cue. When cues are valid on most trials, attention is shifted to the cued location in anticipation of target presentation, and this speeds up responses to the target. (b) Valid (top row) and invalid (bottom row) cueing of attention by a physical stimulus. Even if half of the cues are invalid (rendering the cue unpredictive of target location), the cue has specific attention effects on targets presented in the cued location. (c, d) Search paradigms. (c) Searching for a target among distracters is slow and time-consuming when the target is similar to the distracters (left panel), or if the target is defined by a conjunction of features on more than a single stimulus dimension (right panel). (d) Search is fast and seemingly effortless (pop-out) when the target differs from distracters on a single dimension, and when the difference is large.

this type of cueing is referred to as endogenous cueing ('top down' attention). On the other hand, attention is often automatically oriented to new stimuli. When a student enters class late, it is almost inevitable that everybody turns attention to the latecomer. This exogenous cueing effect ('bottom up' attention) has been investigated in a variation of the spatial cueing paradigm (Figure 2(b)). In this variation, small cues are presented in random locations on the computer screen while the participant fixates a small cross in the middle of the screen. The cue location is unpredictive for the location of the subsequently presented target stimulus. Nevertheless, the cue affects processing of the target if the target happens to be presented in the location of the preceding cue. This effect depends upon the time interval between the cue and target presentation. A target that follows a cue within 250 ms will be responded to with a decreased reaction time, while a target lagging behind the cue by more than about 300 ms will be responded to with an increased reaction time, compared with target presentations not preceded by a cue.

Thus, the mere presentation of a stimulus in a given location can attract attention to that location for a limited time. After that time, attention is disengaged from the stimulus location, and a reorientation of attention to the same location is suppressed. This effect is referred to as 'inhibition of return'. It is important that attention can be attracted easily to unexpected but significant events in the environment, but it is equally important that attention then can be freed from that event to make it available for new, potentially important stimuli. If there were no easy disengagement of attention, attention could get stuck to a stimulus, and new incoming stimuli might go unnoticed. The brief period during which attention is engaged in the processing of a particular stimulus, and during which limited or no attention is available for the processing of new stimuli, is often referred to as the 'attentional blink'.

Search Paradigms

The metaphor of a 'focus of attention' has also been used to explain performance in visual search tasks, a second major paradigm to study attention. In a search task, participants are presented with displays filled with distracter stimuli, which may or may not contain a specific target. Participants are instructed to respond with a button press as soon as they find the target, or to press another button to indicate its absence. Performance is assessed by measuring the time required to find the target as a function of the number of distracters in the display. The resulting plot is referred to as a 'search curve'. A typical experiment suggesting serial search would show an increase in search time by about 30 ms per item.

Search performance becomes dependent on the number of distracters in the display when the target is difficult to distinguish from the distracters (Figure 2(c)). An example of such a search display is one in which the target is a line element slightly tilted away from vertical, surrounded by vertical distracters. Another example is one in which the target is defined by a conjunction of features: the target could be a vertical red line, surrounded by horizontal red, vertical green, and horizontal green lines. The latter task (conjunction search) is difficult because different properties of the objects have to be analyzed and combined before it can be determined whether the object matches the target being searched. The time-consuming nature of these types of search suggests that a focus of attention is continually relocated during search to scan items in the display serially, and in the case of conjunction search also suggests that focal attention plays a role in solving the binding problem.

While search displays have been used to demonstrate serial search, 'pop-out' displays have been used to demonstrate the existence of 'parallel search'. In pop-out displays, the target (e.g. a horizontal line) differs strongly from the surrounding distracters (e.g. vertical lines) and the time required to detect the target appears independent from the number of distracters (Figure 2(d)). The identification of a target in pop-out displays is often considered to occur 'preattentively', or to result from a parallel attentional operation.

A COGNITIVE NEUROSCIENCE APPROACH TO ATTENTION

While the experimental paradigms introduced above characterize important aspects of attentional behavior, they do not address the important question of how modulations of sensory processes produce attentive behavior. Recent neurophysiological findings have brought us closer to an answer.

Findings in the Temporal Lobe

In a series of ground-breaking recording studies, Desimone and colleagues flashed pairs of stimuli inside the RF of single neurons of temporal lobe areas V2, V4, and TE in monkeys trained to identify one of the stimuli and ignore the other.

The neurons' responses were determined by the attended stimulus (target), while the influence from unattended stimuli inside the RF upon the neurons' responses was greatly attenuated (Figure 3). In the period during which the monkeys were waiting for the presentation of each stimulus pair, increased baseline activity was observed compared with the spontaneous firing rate. Depending on the task, increased baseline activity might represent the instruction to the monkey to attend in a particular location, or to expect a stimulus of a given identity, and it might facilitate processing of the target stimulus.

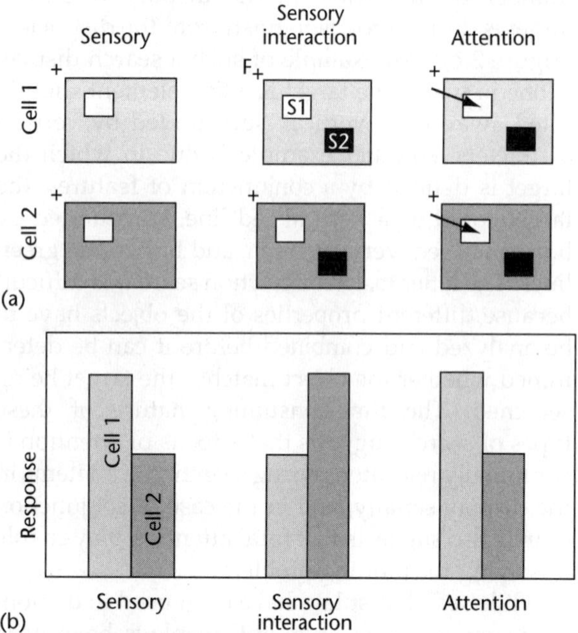

Figure 3. Elimination of sensory interaction (competition) by attention. (a) Different configurations of stimuli and covert attention inside the receptive fields (RF) of two cells (cell 1 and 2). The monkey fixates a cross (F) while covertly attending (arrow) or ignoring stimuli away from fixation placed in the RF of cells 1 or 2, whose activity is recorded. The RFs of cell 1 and 2 are shown as the light and dark gray squares, respectively. Stimuli are symbolized by small white (S1) or black (S2) squares. (b) Pattern of responses illustrating competition and effects of attention. For cell 1, stimulus S1 is more effective than S2 (response to S2 alone not shown). Because stimuli compete for control over the firing rate of the cell, the addition of the less effective stimulus (S2) to the more effective stimulus (S1) drives the response down (sensory interaction). That suppressive effect is eliminated when attention is covertly directed to S1 (arrow in A). Cell 1 now responds as if only S1 were present in its RF, despite the presence of S2 in the RF. Thus, attention eliminates competitive effects caused by behaviorally irrelevant stimuli. In cell 2, stimulus S2 is more effective than S1, and adding S2 to S1 in the RF increases the response compared with the response obtained with S1 alone. Attention to the less effective stimulus S1 will screen out the effect of the more effective stimulus S2, resulting in a decreased response. Thus, whether attention will increase or decrease the activity of a given neuron depends upon the selectivity of the neuron for the stimuli in its RF.

These data provide a neural foundation for the behavioral finding that target stimuli are processed more efficiently than nonattended distracters. In 1995, Duncan and Desimone proposed a 'biased competition' model of attention to account for these findings. In this model, objects presented simultaneously in the visual field compete with each other for control over the firing rate of cells within whose RF they are presented. The competition is decided in favor of stimuli that receive a biasing signal which can be generated 'bottom up' (by a salient stimulus) or 'top down' (by an instruction to identify a particular target among distracters, or to monitor objects in a particular location). The increased spontaneous activity reported during attention tasks may be a correlate of such a biasing signal. Thus, the process of attention is viewed as a biased competition between populations of cells representing different objects or locations (Figure 4).

In agreement with the idea that temporal lobe areas are important for attention, cortical lesions of monkey areas V4 and TEO in the temporal lobe cause an inability to discriminate properties of target objects surrounded by distracters, especially when the target is weak (e.g. smaller or dimmer than the distracters). Without distracters, deficits in discriminating the target are absent or limited. These lesion deficits in the monkey resemble agnosia, a neurological syndrome found after occipitotemporal damage in humans. Humans with agnosia have trouble recognizing visually presented objects embedded in a complex scene, or visually presented complex objects consisting of multiple parts; yet simple stimuli presented by themselves can be discriminated with great accuracy. These general characteristics of agnosia suggest that attention deficits contribute to the syndrome. In the terminology of Duncan and Desimone, agnosia may reflect, at least in part, a failure of biased competition, because of damage to the substrate in which the competition takes place.

Findings in the Parietal Lobe

The description of attention as a process of biased competition leads to the question: where does the

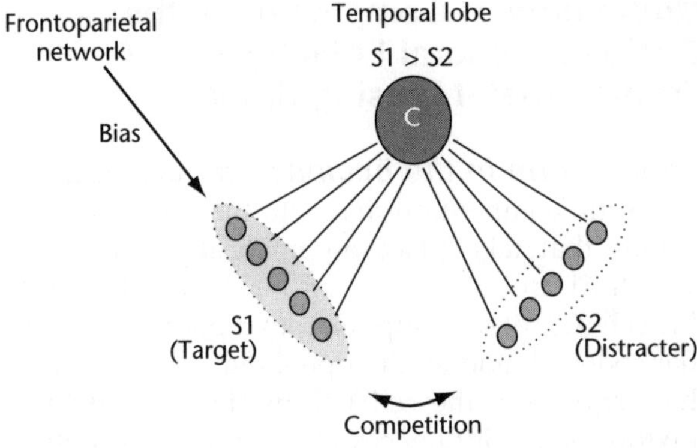

Figure 4. Competition in the temporal lobe is biased by signals generated in a frontoparietal network. Cell C receives input from a population of cells coding S1, and another population of cells coding S2. Each population provides a mixture of excitatory and inhibitory inputs to cell C. The balance of excitatory and inhibitory inputs to C provided by each population determines the size of C's response to S1 and S2 presented alone in C's RF. The response to S1 and S2 presented together in C's RF is determined by the total balance of all excitatory and inhibitory inputs from both populations taken together (competition). When one of these two stimuli (e.g. S1) is made behaviorally relevant (a target), and is therefore attended to, a biasing signal renders inputs from cells coding the target more efficient compared with the distracter. As a result, C's activity will reflect the balance of excitatory and inhibitory inputs provided by S1 alone, rather than the balance of excitatory and inhibitory inputs provided together by the two stimuli in the RF, and S1 wins a competition with S2 for control over the firing rate of cell C (S1 > S2). 'Bottom up' stimulus-driven bias can have effects similar to 'top down' bias generated in a frontoparietal network. This formalization of biased competition theory was first proposed by Reynolds and Desimone in 1999.

bias come from? To answer this question, we turn briefly to a parallel line of work on attention in the parietal cortex. Mountcastle in 1976 and Wurtz in 1982 showed that neurons in parietal cortex of the monkey were more active when the stimulus in their RFs was behaviorally relevant (attended) than when it was not. Several imaging studies in humans have now confirmed that covertly orienting one's attention to a stimulus or location leads to enhanced activity in parietal cortex. Further neurophysiological work in monkeys suggests that the effects of behavioral relevance on parietal activity can depend upon the type of action planned by the monkey towards the target (e.g. an arm versus an eye movement).

Lesions in the parietal cortex induce a complex set of deficits in the representation of space and action. Unilateral parietal damage leads to neglect of the contralateral side of the body, and of stimuli in contralateral extrapersonal space. Neurologists often use the line bisection test to assess neglect: patients with neglect will bisect a horizontal line towards one end of the line, as the other part of the line will become neglected when they look at it. When a strong stimulus is presented on the neglected side, patients can recognize the stimulus perfectly, but when two such stimuli are presented, one in the neglected hemifield and one in the other hemifield, then the stimulus in the neglected hemifield will remain unnoticed, a phenomenon referred to as *extinction*. Extinction and neglect do not always co-occur after parietal damage, indicating that both phenomena may constitute dissociable syndromes after different types of parietal damage.

Using a spatial cueing test, Posner and colleagues found that patients could shift attention to the neglected hemifield when cued correctly, and engage in attentional operations in that hemifield, but had trouble redirecting attention from the ipsilateral to the contralateral (neglected) hemifield. It was therefore postulated that parietal lesions induce a failure of a disengagement operation. Other experiments suggest that the pulvinar nucleus, a thalamic nucleus that provides weak but direct projections to the parietal lobe, plays a part in the shifting and engaging of attention.

Bilateral parietal damage leads to a constellation of symptoms, including optic ataxia (deficit in visually guided reaching), paralysis of gaze, inattention to peripheral stimuli, and sustained hyperattention to single objects or locations (Balint syndrome). Patients with this syndrome do not know where they are, and have no idea of spatial relations between objects. However, once attention is directed to an object, they can identify it perfectly. In pop-out displays, their report of the presence of the target is not influenced by the number of distracters, as in non-afflicted individuals, but they cannot tell where the target is once it has been reported. In conjunction search tasks, people with Balint syndrome require extraordinary amounts of time to find the target (about 1 s per item). In addition, they often make 'illusory conjunctions' (e.g. joining the orientation of a given item with the color of a neighboring item). These illusory conjunctions can also be demonstrated in non-afflicted individuals when the search display is presented very briefly (and followed by a mask). In both situations, there may be not enough attention available to focus accurately on one of the items.

A compelling account of the function of focused attention was given in 1980 by Treisman and Gelade in a model referred to as the 'feature integration' theory. According to this theory, different elementary features in the image (e.g. color, orientation, motion) are analyzed in separate feature maps – an idea at least in part supported by physiological evidence – and the read-out of activity within single maps occurs fast and in parallel. This would explain simple pop-out phenomena. However, when targets have to be identified that are defined by a combination of two or more features, the activity in different feature maps must be combined, which is a lengthy operation requiring focused attention. Hence, the time-consuming nature of conjunction search may reflect a mixture of factors, including the disengagement, orienting, and engagement of attention, as well as the time required to conjoin features. Imaging data support the idea that parietal cortex is involved in the directing of attention to targets, in the shifting of attention in response to both endogenous and exogenous cues, and in conjunction search, but not in feature search or pop-out. Since conjunction search implies shifts of attention and conjoining of features, parietal activity could indicate either or both.

Findings in the Frontal Lobe

An additional line of research that is related to the question where the biasing signal comes from was initiated by Fuster and Alexander in 1971. They found that neurons recorded around the principal sulcus in prefrontal cortex showed enhanced activity during delayed-response tasks, in which the monkey was required to remember a previously cued location that would become the target of a directed action after a delay of a few seconds. Later research has demonstrated that neurons in prefrontal cortex can be activated during tasks that require working memory for both location and object identity.

Many attentional operations require working memory. For example, while a participant looks for a target in a search array, incoming sensory information is matched against a template of the target that is kept on-line in working memory. Many physiological and imaging studies have confirmed that working memory and attention both rely heavily on prefrontal cortex. Limitations on attentional capacity, witnessed by the impossibility of identifying more than a few targets at once, are reminiscent of limitations to hold more than a few stimuli in working memory.

Modulation of Competition in the Occipitotemporal Lobe by a Frontoparietal Biasing Signal

Competition in the occipitotemporal lobe

The term 'competition' refers to the sensory interactions that take place automatically when two stimuli S1 and S2 are placed inside a cell's RF. If this cell (C) receives inputs from a pool of neurons that codes S1, and another pool that codes S2, then the response of the cell will be determined by a mixture of the influences of S1 and S2. Attention biases the efficacy of one set of inputs relative to the other set, such that the activity of the cell will reflect preferentially the attended stimulus (Figure 4). Cells encoding a target could enhance their influence on the activity of cell C by slightly depolarizing their cell membranes. This enhances the probability that these cells will generate action potentials whenever the target is presented in their RFs. Alternatively, the pool of neurons that represents the target may synchronize its spontaneous and stimulus-driven activity. This enhances the probability of coinciding spikes in the pool of neurons coding the target. Because of the summation properties of neurons, rate enhancements and synchronization could both increase the control of the pool of neurons coding the target over the activity of the postsynaptic cell C. The pool of neurons coding the target is thought to include neurons coding various aspects of the stimulus, distributed in temporal lobe areas and other parts of the brain. The synchronization of the activity of neurons coding different properties of an object could be a mechanism for the binding of those object properties, a proposal championed by Singer and colleagues.

A frontoparietal source for the biasing signal

Where does the biasing signal come from? Alternatively, what makes a stimulus behaviorally relevant? A stimulus is behaviorally relevant when there is a requirement to make a saccade to it, to reach and grab it, and to keep it in working memory. It is precisely under those conditions that specific regions in parietal and frontal cortex become active. Thus, to the extent that requirements in an attention task engage frontoparietal regions, activity in those regions could bias activity in ventral stream areas through connections between retinotopically matched regions in ventral and frontoparietal areas.

Evidence suggests that parietal activation during endogenous directing, shifting, and maintenance of spatial attention may be controlled by the prefrontal cortex, while parietal activity related to exogenous cueing may be controlled by subcortical thalamic input to the parietal cortex. From the perspective of a theory of binding, the bias would be considered part of the representation of the attended object. Indeed, a complete object representation would consist of a synchronized ensemble of neurons distributed in the frontal, parietal, and temporal lobes, representing the object's identity and location, and possible actions towards it. Similarly, the appropriate binding of features (e.g. color and orientation) in conjunction stimuli may be related to the synchronization of activity in neurons coding the color and orientation of those stimuli by a common biasing signal. Thus, selective attention, synchronization, and binding may be intimately related.

Implications for the Serial versus Parallel Search Debate

Attentional bias signals affect sensory processing at the earliest levels of sensory systems. This settles a long debate between 'early selection' proposals (pioneered by Broadbent in 1958) and other proposals advocating 'late selection'. Not unlike feature integration theory, early selection theory predicts that nonattended stimuli are incompletely processed, and that full processing of all aspects of a target stimulus requires focusing of attention, which implies that search must have a serial component. However, strong arguments have been made against the existence of serial search. It has been argued that target search can be mediated by biasing sensory processing towards the target everywhere in the visual field in parallel. Increases in search time as a function of increases in the number of distracters would merely reflect a thinner spread of a limited amount of attention over all items in the display.

Evidence is mounting, however, that performance in search tasks depends on a mixture of serial and parallel operations. For instance, it has been reported that detection of a 'pop-out' target is impaired when an attention-demanding task is carried out at fixation during stimulus presentation. This does not exclude a parallel component in the detection of pop-out, but it does argue against the idea that pop-out is entirely preattentive. Thus, some degree of directed attention may have a role, even in pop-out. Furthermore, performance in spatial cueing tasks, designed to reveal serial relocations of a focus of attention, is influenced by a segmentation of visual space into surfaces and objects, which takes place in parallel across the field. Specifically, experiments by Driver and colleagues show that reorienting attention to a target location, after an invalid positional cue, occurs faster if the invalid cue and target are both presented within the same shape, compared with when the target and cue are presented in different shapes. Although those experiments are set up to equate the distance over which attention has to be reoriented, reorientation takes significantly longer between than within objects. Hence, spatial attention does not operate within unsegmented space.

Findings in patients with neglect due to unilateral parietal damage support this hypothesis. Neglect is often described as a purely spatial imbalance in the distribution of attentional operations between hemifields. However, experiments in a patient with right parietal damage showed that details on the right side of an object placed in the left (neglected) hemifield are perceived more accurately than details on the left side of an object placed in the right (normal) hemifield. The idea that parietal cortex manipulates the distribution of attention in an object-driven way, rather than a purely spatial way, is compatible with the role of parietal cortex in representing potential actions towards target objects.

In sum, processing of a visual scene may begin with a fast quasiautomatic segmentation based on parallel processes that require little attention. Subsequently, attention may be distributed across this roughly segmented scene, in ways that reflect behavioral demands, and which may not be exclusively parallel or serial. When relevant objects are expected in a single location, attentional resources will be focused on that location. When switches of attention between locations are required, this serial redistribution of attention will be influenced by the preceding segmentation of space. Furthermore, when a specific target object is searched for without knowledge of its location, a 'top down' parallel bias may contribute to the identification of potential targets, but does not exclude the possibility that potential targets are serially inspected. Thus, while items in a search display may not be scanned individually, parallel processes (including automatic 'bottom up' segmentation and parallel 'top down' biases) may lead to the determination of likely targets, which may then be inspected serially. The exact distribution of attention in response to task demands is controlled by executive processes. These and similar ideas have led to various mixed parallel–serial models of attention.

CONCLUSION

The combination of behavioral paradigms developed by psychologists and new methodological approaches developed in the field of cognitive neuroscience has greatly enhanced our understanding of attention and of the neural processes that underlie it. Attention can be best understood as a modulation of sensory processes driven by regions in the brain that represent the behavioral relevance of stimuli. These modulations ensure that behaviorally relevant objects (or locations) in the environment are preferentially processed at the cost of irrelevant ones.

Further Reading

Andersen RA, Snyder LH, Bradley DC and Xing J (1997) Multimodal representation of space in the posterior parietal cortex and its use in planning movements. *Annual Review of Neuroscience* **20**: 303–330.

Broadbent DE (1958) *Perception and Communication*. London, UK: Pergamon Press.

Colby CL and Goldberg ME (1999) Space and attention in parietal cortex. *Annual Review of Neuroscience* **22**: 319–349.

Desimone R and Duncan J (1995) Neuronal mechanisms of selective attention. *Annual Review of Neuroscience* **18**: 193–222.

Desimone R and Ungerleider LG (1989) Neural mechanisms of visual processing in monkeys. In: Boller F and Grafman J (eds) *Handbook of Neuropsychology*, vol. 2, pp. 267–299. New York, NY: Elsevier.

Desimone R, Wessinger M, Thomas L and Schneider W (1990) Attentional control of visual perception: cortical and subcortical mechanisms. In: *Cold Spring Harbor Symposia on Quantitative Biology*, vol. 55, pp. 963–971. Cold Spring Harbor Press.

Farah MJ (1990) *Visual Agnosia: Disorders of Object Recognition and What They Tell Us About Normal Vision*. Cambridge, MA: MIT Press.

Gazzaniga MS (1995) *The Cognitive Neurosciences*. Cambridge, MA: MIT Press.

Gazzaniga MS, Ivry RB and Mangun RM (1998) *Cognitive Neuroscience: The Biology of the Mind*. New York, NY: WW Norton.

Hillyard SA and Picton TW (1987) Electrophysiology of cognition. In: Plum F (ed.) *Handbook of Physiology*, Section 1: The nervous system, vol. 5, Higher functions of the brain, part 2, pp. 519–584. Bethesda, MD: American Physiological Society.

Miller EK and Cohen JD (2001) An integrative theory of prefrontal cortex function. *Annual Review of Neuroscience* **24**: 167–202.

Parasuraman R (1998) *The Attentive Brain*. Cambridge, MA: MIT Press.

Pashler HE (1999) *The Psychology of Attention*. Cambridge, MA: MIT Press.

Singer W and Gray CM (1995) Visual feature integration and the temporal correlation hypothesis. *Annual Review of Neuroscience* **18**: 555–586.

Treisman A (1999) Solutions to the binding problem: progress through controversy and convergence. *Neuron* **24**: 105–110.

Wolfe JM (1994) Guided search 2.0: a revised model of visual search. *Psychonomic Bulletin and Review* **1**: 202–238.

Attitudes

Introductory article

George Y Bizer, Eastern Illinois University, Charleston, Illinois, USA
Jamie C Barden, Ohio State University, Columbus, Ohio, USA
Richard E Petty, Ohio State University, Columbus, Ohio, USA

CONTENTS

An attitude is a global and relatively enduring evaluation (e.g. good or bad) of a person, object, or issue. Attitudes can be based on affective, cognitive, or behavioral information and can vary in their strength (e.g. how enduring, how resistant to change, and how predictive of behavior they are).

INTRODUCTION

What drives our behavior? When we choose a candy bar at the grocery store or decide for whom to vote in an election, what determines the choices that we make? Attitudes, the mental representations of what we like and dislike in our world, help to explain these choices.

Attitudes are one of psychology's fundamental concepts because they help to explain people's decisions and actions. An attitude is a global and relatively enduring evaluation of a person, object, or issue – a representation of whether we think the target is generally good or bad, desirable or undesirable. We can hold attitudes towards tangible objects such as ice cream or trees, people such as the President or a sister, ideas such as democracy or wealth, and issues such as the death penalty or tax increases. Simply put, the more favorably we evaluate something, the more positive our attitude towards the object is; the more negatively we evaluate something, the more negative our attitude is.

Attitudes serve various functions. As noted by Daniel Katz and others, some attitudes serve a utilitarian function in that they help us to achieve rewards and avoid punishments (e.g. having the correct evaluation of one's mortgage company can save you money). Other attitudes serve an ego-defensive function in that they foster our own self-images (e.g. holding prejudiced attitudes might make some people feel superior). A number of additional functions have also been identified.

MEASUREMENT

Researchers have developed a wide array of tools to measure attitudes. These techniques can be categorized into two broad groups. *Explicit* measures directly ask people to report their attitudes; in contrast, *implicit* (or indirect) measures are assessments that allow inferences about a person's attitude without having to ask him or her directly. The latter method is commonly used in situations in which people either do not want to or are unable to provide their true evaluations of an object. No measure of attitudes is perfect, however, as assessments can be affected by the measurement context. Seemingly innocent influences like the weather or answers to previous questions can have a considerable impact on people's reported attitudes.

Explicit Measures

Two common explicit measures are the *Likert scale* and the *semantic differential*. The Likert procedure presents respondents with series of evaluative statements along with a series of response options for each. For example, an attitude scale on ice cream might contain the statement 'Ice cream tastes good' and choices of various degrees of agreement from which the respondent can choose ('strongly agree', 'agree', 'neither agree nor disagree', 'disagree', and 'strongly disagree'). Participants report the extent to which they agree or disagree with each statement. Likert scales include a wide variety of evaluative statements regarding the object, and scores from all the statements are combined to create a measure of the attitude.

With the semantic differential technique, developed by Charles Osgood and colleagues, respondents are presented with the name of the attitude object and some evaluative adjectives that might describe that object. Participants then rate how well the adjectives describe the object. For example, a series of items might prompt respondents to report the extent to which they think ice cream is beneficial versus harmful, good versus bad, and pleasant versus unpleasant. These scores are combined to form one global attitude measure.

Implicit Measures

Implicit measures come in various shapes and sizes. They range from monitoring simple behaviors from which evaluation can be inferred (e.g. how close one person chooses to sit next to another) to complex physiological techniques. A good example of an implicit measure is Russell Fazio's priming procedure. To assess racial attitudes with this technique, participants are presented with images of Caucasian or African-American faces to make the concept of one or the other race more accessible. Immediately after being shown a face, participants are asked to judge whether a particular concept (e.g. ice cream) is 'good' or 'bad'. Over many pairings of faces and concept words, the amount of time it takes the participant to report 'good' or 'bad' for each word following a face is measured. The pattern of reaction times is used to infer the person's implicit attitude. Since one negative attitude tends to activate or prime others, if a participant dislikes African Americans, showing an African-American face should make the evaluations of other negatively perceived objects (e.g. 'dirt') faster, but make the evaluations of positively perceived objects (e.g. 'ice cream') slower. Thus, if a person needs more time to report that a good word like 'ice cream' is 'good' after seeing an African-American face and less time to report that a bad word like 'dirt' is 'bad' (compared to seeing a Caucasian face), there is evidence that the person holds a more negative attitude towards African Americans than Caucasians.

TRIPARTITE MODEL OF ATTITUDE STRUCTURE

As global and enduring evaluations, attitudes can be based on up to three separate types of input: affective, cognitive, and behavioral. An attitude can be based on any one or a combination of these three information sources. Attitudes, once formed, also guide affective, cognitive, and behavioral reactions to the object.

The *affective* basis of an attitude is made up of feelings, moods, and emotions that have become associated with the attitude object through past or current experience. It is possible to have multiple affective responses to an object based on the same, or different, experiences with it. Each affective response has an evaluation made up of valence (positive to negative) and magnitude (strong to weak). Researchers often measure the affective basis of the attitude by asking to what extent individuals *feel* good or bad about the object, or the extent to which the attitude object makes them feel 'happy', or 'disgusted', or 'angry'.

The *cognitive* basis is made up of particular attributes that are ascribed to the object. An *attribute* is any characteristic, quality, trait, concept, value, or goal associated with the object. The impact of an attribute is determined by the evaluation of whether the attribute is good or bad, and the perceived likelihood that this attribute applies to the object. Thus, if the attitude object is 'ice cream', one attribute associated with this object might be 'fattening'. If the person thought this attribute was negative and highly associated with ice cream, the attribute would contribute to a generally unfavorable evaluation of ice cream. Of course, any one attitude object can be associated with many attributes that contribute to the overall evaluation.

In practice, researchers such as Martin Fishbein and Icek Ajzen suggest obtaining a listing of attributes about an object and then the evaluation and likelihood associated with each attribute. The evaluation and likelihood are multiplied together for each attribute and the products are added across attributes to estimate the cognitive component of the attitude. One implication of this approach is that two individuals endorsing the same attributes can have different attitudes if they evaluate the attributes differently, and individuals believing in different attributes can hold the same attitude.

The *behavioral* basis is made up of two kinds of information, past behaviors and intentions to commit future behaviors. Daryl Bem's *self-perception theory* holds that we sometimes infer our attitudes directly from our past behaviors towards an object. For example, if a person looks back on his or her life and realizes that he or she has never eaten at a Chinese restaurant even though he or she had many chances to do so, the person might infer that he or she does not like Chinese food. This inference occurs as long as there is no memory of external forces compelling the behavior – the behavior needs to be seen as voluntary.

CONSTRUCTED VERSUS STORED ATTITUDES

When an object is encountered for the first time, there is no information about it in memory. An attitude must therefore be *constructed* by making inferences from the behaviors, thoughts, and feelings that occur in the current social environment. Irrelevant features of the current context can bias the constructed attitude even though they have little to do with the attitude object itself. For example, one's attitude towards the economy might be more favorable on a sunny than a rainy day. Norbert Schwarz and colleagues have documented a wide variety of contextual influences on attitude reports.

After information is gathered about an object and an evaluation is formed, the attitude can be stored in memory and subsequently retrieved directly. A *stored attitude* is an evaluation that is linked to the object in the form of a thought (e.g. 'I like candy'). If the attitude object is brought to mind again, and the object–evaluation link is strong enough, the stored attitude is brought to mind as well.

Generally, the attitudes people report fall somewhere in between purely constructed and purely retrieved. That is, even if a person has already formed a global evaluation, the specific evaluation that is reported at any moment in time is dependent on a wide variety of factors. In general, the stored attitude acts as an anchor point and is adjusted based on affective, cognitive, and behavioral information that is currently salient in memory or in the immediate context. That is, when an attitude is retrieved, some information related to that attitude may also be retrieved and pull the attitude report in its direction. For example, on one occasion the 'taste' attribute of ice cream might be especially salient, but on another occasion, the 'fattening' attribute might be more salient. The immediate context can influence which attitude-relevant information is retrieved, providing a source of contextual bias. Because 'strong' attitudes are less likely to be influenced by context effects than are 'weak' attitudes, the study of attitude strength is also important in attitude research.

STRONG AND WEAK ATTITUDES

Attitudes fall along a continuum from weak to strong, such that stronger attitudes are more durable and impactful. A durable attitude is persistent over time, meaning that it does not decay in memory, and is resistant to change when faced with counter-attitudinal information. An attitude has impact when it influences information processing and guides behavior. Attitudes can possess these strength properties to varying degrees. Also, these strength properties can be independent. Thus, it is possible for an attitude to persist over time but not influence behavior, or to greatly influence thought at a given point in time, but not resist attempts to change it.

A number of variables contribute to making attitudes stronger or weaker. First, extreme attitudes (i.e. when people rate the object as intensely good or intensely bad) tend to be stronger than more moderate attitudes. This may be because extreme attitudes tend to have a number of structural properties that contribute to this strength. For example, extreme attitudes may be based on high amounts of consistent knowledge, and they may come to mind more rapidly (i.e. are more accessible) than more moderate attitudes.

Subjective beliefs about our attitudes are also related to strength. For example, strength can result from perceptions of how much knowledge we have on a topic (regardless of actual knowledge), how important the attitude object is to us personally, or how confident we are in the validity of our attitudes. Finally, the manner in which an attitude is formed can contribute to its strength. Most notably, if an attitude is created through extensive thinking and careful scrutiny of information, it tends to be stronger than if it was formed by means requiring less effort.

EXPLICIT AND IMPLICIT ATTITUDES

To this point, we have discussed attitudes as global evaluations we are aware of and can control. These conscious or *explicit attitudes* result from integrating information from one or more components into an evaluation. These attitudes can vary from strong to weak, and from mostly stored to mostly constructed. Retrieval or construction of these explicit attitudes can either be relatively automatic or require considerable cognitive effort.

In addition to these explicit attitudes, researchers such as Anthony Greenwald and Mazarin Banaji have argued that people can also hold *implicit attitudes* – attitudes of which they are generally not aware. Implicit and explicit attitudes can sometimes be in opposition to each other, such that the implicit attitude can lead people to think and behave in ways they do not consciously intend. For example, a person who holds a prejudiced implicit attitude based on negative stereotypes learned as a child, but now consciously rejected, may also hold an explicit unprejudiced (and conscious) attitude learned later in life. In such situations, conscious attitudes direct behaviors that are generally under constant conscious control (such as deciding the guilt or innocence of a black defendant). However, more automatic behaviors such as one's body language and eye contact can reflect a person's implicit attitudes.

ATTITUDE CHANGE

Although attitudes are generally considered to be relatively stable and enduring, they are subject to change over time. Being exposed to new information and new experiences can lead people to change their attitudes. Numerous studies have demonstrated the processes by which attitudes change.

Central and Peripheral Routes to Change

Much contemporary research is guided by the idea that attitudes are sometimes changed thoughtfully, but sometimes are changed with very little cognitive effort. The *elaboration likelihood model* (ELM) of persuasion developed by Richard Petty and John Cacioppo presents a framework that helps explain the various processes and outcomes of attitude change. Although the amount of thinking involved in attitude change forms a continuum from none to extensive, the model divides the specific processes of attitude change conveniently into two 'routes' to persuasion.

The first or *central route* to attitude change occurs when people are relatively careful in scrutinizing the issue-relevant information available. If, after careful consideration, a person finds the information to be compelling, attitudes change accordingly. If, however, the information is deemed specious, attitudes will not change, or can even change in a direction opposite to that advocated – a boomerang effect. For example, a person following the central route when processing a magazine advertisement for a car will carefully assess the perceived validity of the information presented in the ad. The person might examine the information presented about the horsepower, price, resale value, safety record, and so forth. The person will be influenced if his or her *cognitive responses* – the thoughts generated during message processing – are positive. The person is not likely to be influenced, however, by the beautiful sunset pictured in the background or the cute puppy sitting in the driver's seat because these are peripheral cues that are unrelated to the central merits of the car.

Sometimes, however, people follow the *peripheral route* when exposed to a persuasive communication. In such cases, people are not likely to pay attention to all of the issue-relevant information within the message. Rather, people seek a shortcut way to evaluate the ad. In this case, they might be influenced by the mere number of arguments in the ad (regardless of their quality). Or, the cute puppy in the driver's seat might lead to a positive feeling that becomes associated with the car.

An important consequence of the route to persuasion that a person takes is the strength of the attitudes that result. Specifically, when people change or form an attitude through the central route to persuasion, attitudes tend to be stronger than those created or changed through the peripheral route. Attitude changes that occur because a person has carefully considered issue-relevant information have a substantive backing which contributes to the durability and impact of the attitude. In contrast, attitudes formed under the peripheral route do not have this substantive support. Because they lack supporting cognitions, these attitudes are much less durable and impactful. This does not mean that peripheral route changes are completely unimportant. For example, advertisers can take advantage of the short-term effects of the peripheral route by continual pairing of peripheral cues with their products in repeated messages. Also, in some cases, what starts out as a peripheral cue can become an argument if people subsequently think about the cue in a way that gives it substantive meaning.

Amount of elaboration

What determines whether a person will follow the central or the peripheral route? This depends on whether the person has the *ability* and the *motivation* to think carefully about the message. Variables influencing ability (whether a person is able to think) include distraction and time pressure. If a person is distracted or under great time pressure while exposed to a persuasive communication, it is simply not possible to follow the central route, and thus the peripheral route to persuasion is more likely. Variables influencing motivation (whether a person wants to think) include personal relevance and accountability. For example, if people are told that a message is of low relevance to them (the product advertised is available only in a faraway country), or that they will not be accountable for the attitude they report on the topic (questionnaires will be completed anonymously), it is likely that they will feel little motivation to think carefully. They are likely to follow the less taxing peripheral route to persuasion instead. All else being equal, people prefer to save their cognitive energy for the most important tasks and decisions in life. People who are high in their *need for cognition* tend to enjoy thinking about a wide variety of topics and thus tend to follow the central route to persuasion. People who are low in this need tend to follow the peripheral route.

Objective and biased processing

In addition to the amount of information processing that takes place, it is also important to consider whether that processing is relatively objective or biased. Objective processing refers to the case in which thinking is guided by the qualities of the information at hand. If the information is cogent, people's thoughts are favorable, but if the information is specious, their thoughts are largely negative. However, people can process messages in a biased manner. For example, people may be forewarned that a message will attempt to change their attitudes. In such cases, people tend to think negatively about all of the arguments – regardless of their actual quality – in an attempt to assert their individual freedom not to be influenced. There are a number of motivations besides asserting freedom that can induce biased processing, such as the motive to be consistent, or to maintain one's self-esteem. Each of these motivations selectively directs people's information-processing activity to favor one attitudinal position over another.

Multiple roles of variables

The ELM highlights the fact that variables can influence attitudes by serving in different roles in different situations. For example, the physical attractiveness of the source of a persuasive message might influence attitudes in a number of ways. First, such a source might serve as a simple peripheral cue when the situation constrains people's motivation or ability to think about the message to be low. For example, when people are distracted, they might go along just because the source is attractive, regardless of the merits of what the source says (peripheral route). On the other hand, if people are highly motivated and able to think about the message, an attractive source might bias that thinking in a favorable way, or the source itself might be scrutinized to see whether it provides information central to the merits of the issue (central route). Finally, if thinking is not already constrained to be high or low, an attractive person might encourage recipients to pay more attention to the message – leading to more agreement if the message is sound, but to less agreement if the message is not. Under this scenario, then, attractiveness would serve as a determinant of elaboration.

Thus, although a variable can serve as a determinant of elaboration in one scenario (unconstrained elaboration), it can serve as a cue in others (low elaboration) or can bias processing or serve as an argument in still other situations (high elaboration). The ELM thus limits the fundamental roles a variable can play in persuasion situations and provides a guiding framework for assessing when variables take on each role.

Mood and Persuasion

One persuasion variable that has been studied extensively in its multiple roles is a person's mood state – whether he or she is feeling good or bad. As with other variables, the effect of mood on persuasion depends on the amount of elaboration taking place during the message presentation. Under low-elaboration conditions, a person's mood can serve as a peripheral cue. In such situations, people may associate their mood with the message's object. The pairing of a good mood with the object can produce positive attitudes towards that object, but bad mood can produce negative attitudes. Second, when elaboration is high, a positive mood can bias people's reactions to the message arguments. In particular, positive mood states make good consequences (e.g. living longer if you stop smoking) seem more likely than when in a neutral or negative mood state, but make negative consequences (e.g. getting cancer if you don't stop smoking) seem less likely. Negative mood states have the opposite consequences. Finally, under moderate elaboration conditions, positive moods affect the amount of thinking people do about the message. If the message appears to be negative or depressing, positive moods decrease information processing compared to negative moods. People in positive moods do not want to think about negative information. On the other hand, if the message appears to be positive or uplifting, positive moods increase thinking over negative moods.

Persuasion from Our Own Behavior

Leon Festinger's theory of *cognitive dissonance* suggests that sometimes our own behavior can lead to attitude change. Specifically, the theory holds that cognitive conflict occurs when people believe that they have behaved in a way that is inconsistent with their attitudes. This cognitive conflict produces tension that people are motivated to reduce in order to restore consistency. Since behavior is often difficult to undo, one way to restore attitude–behavior consistency is to change one's attitude to be in line with one's behavior. This is not the only way to reduce dissonance (people could reduce the importance of the conflicting attitude or behavior), but it is a common one. Dissonance theory explains such processes as why people come to favor products more after they purchase them, and why people come to like groups more if they have voluntarily exerted considerable effort to join them. Dissonance can also lead people to process information in a biased fashion. That is, they think about attitude objects in a way that restores consistency.

Resisting Persuasion

Although there are many different ways in which people can be persuaded, there are some techniques through which attitudes can be made more resistant to change. One way to create resistant attitudes is simply to make those attitudes stronger. This can be done by increasing issue-relevant thinking prior to the attacking message. For example, if a positive attitude towards obeying the speed limit is weak, a person could spend time thinking and learning about why he or she holds the attitude. This additional thinking and learning should serve to strengthen the once-weak attitude.

Perhaps surprisingly, an attitude can in some situations be made more resistant to persuasion through attempted counter-persuasion! Some attitudes are weak because they have very little substantive basis at all. These attitudes may have been created through peripheral processes or may simply be 'cultural truisms' – attitudes we hold just because we have always been taught to think that way (such as favorable attitudes towards freedom of speech). If cultural truisms are mildly attacked, they can actually become stronger. This process of *inoculation*, as outlined by William McGuire, occurs because although the weak attack may not be enough to change the attitude, it may be strong enough to make the recipient think (often for the first time) about why he or she holds that attitude in the first place. This additional thought can serve to create a basis for holding the attitude and motivate individuals to effectively counter-argue subsequent attacking messages.

ATTITUDE – BEHAVIOR CONSISTENCY

One reason why attitudes are a principal area of research in psychology is that, under the right circumstances, attitudes guide people's behavior, and thus are useful to know in order to predict voting, consumer purchases, jury decisions, and so forth. According to Russell Fazio's model of attitude–behavior consistency, exactly *how* attitudes guide behavior depends on the type of behavior in question – is the behavior one that is engaged in spontaneously or one that elicits reflection prior to action?

Some behaviors in which we engage are not well thought out. When it comes to impulse purchases, such as the candy people buy while waiting at the checkout line, people may not spend much time in making decisions. In such cases, people simply look to their attitudes to make a choice. In such situations, whether our attitudes drive behavior is determined by whether we can recall the attitude easily and quickly (i.e. if the attitude is accessible).

Other behaviors are not as spontaneous. When we have the motivation and opportunity to choose our behaviors more carefully, accessibility alone is less important. Instead, according to Fishbein and Ajzen's theory of reasoned action, behavior is determined by one's behavioral intention, which is in turn determined by several factors. First, intention is determined by one's attitude towards the particular behavior under consideration. Attitude towards the behavior in any given situation (e.g. baking a cake for your spouse's birthday on Wednesday) will depend on the beliefs that come to mind in assessing this action – if an attitude relevant to this behavior is not readily retrievable. Intentions are also determined by subjective norms – what other people we admire would want us to do in the situation. Perceptions of our own abilities to carry out some action also play an important role in determining deliberative behaviors.

CONCLUSION

Attitudes have a profound impact on virtually every aspect of our lives. From fundamental issues, such as how attitudes can be measured, to more complex ones, such as the nuances of how attitudes can be changed, a long and rich array of research has helped us to understand the nature of the attitude construct. Space limitations have allowed us to provide only a sampling of what is known about attitudes; the extensive literature on this topic requires further exploration.

Further Reading

Bem DJ (1972) Self-perception theory. In: Berkowitz L (ed.) *Advances in Experimental Social Psychology*, vol. 6, pp. 1–62. New York, NY: Academic Press.

Eagly AH and Chaiken S (1993) *The Psychology of Attitudes*. Fort Worth, TX: Harcourt Brace Jovanovich.

Eagly AH and Chaiken S (1998) Attitude structure and function. In: Gilbert DT, Fiske ST and Lindzey G (eds) *Handbook of Social Psychology*, pp. 269–322. Boston, MA: McGraw-Hill.

Fazio RH (1990) Multiple processes by which attitudes guide behavior: the MODE model as an integrative framework. In: Zanna MP (ed.) *Advances in Experimental Social Psychology*, vol. 23, pp. 75–109. San Diego, CA: Academic Press.

Festinger L (1954) *A Theory of Cognitive Dissonance*. Stanford, CA: Stanford University Press.

Fishbein M and Ajzen I (1975) *Belief, Attitude, Intention, and Behavior: An Introduction to Theory and Research*. Reading, MA: Addison-Wesley.

Greenwald AG and Banaji MR (1995) Implicit social cognition: attitudes, self-esteem, and stereotypes. *Psychological Review* **102**: 4–27.

Katz D (1960) The functional approach to the study of attitudes. *Public Opinion Quarterly* **24**: 163–204.

McGuire WJ (1964) Inducing resistance to persuasion: some contemporary approaches. In: Berkowitz L (ed.) *Advances in Experimental Social Psychology*, vol. 1, pp. 191–229. New York, NY: Academic Press.

Osgood CE, Suci GJ and Tannenbaum PH (1957) *The Measurement of Meaning*. Urbana, IL: University of Illinois Press.

Petty RE and Cacioppo JT (1986) The Elaboration Likelihood Model of persuasion. In: Berkowitz L (ed.) *Advances in Experimental Social Psychology*, vol. 19, pp. 123–205. New York, NY: Academic Press.

Petty RE and Krosnick JA (eds) (1995) *Attitude Strength: Antecedents and Consequences*. Hillsdale, NJ: Lawrence Erlbaum.

Petty RE and Wegner DT (1998) Attitude change: multiple roles for persuasion variables. In: Gilbert DT, Fiske ST and Lindzey G (eds) *Handbook of Social Psychology*, pp. 323–390. Boston, MA: McGraw-Hill.

Schwarz N (1999) Self-reports: how the questions shape the answers. *American Psychologist* **54**: 93–105.

Zimbardo PG and Leippe MR (1991) *The Psychology of Attitude Change and Social Influence*. New York, NY: McGraw-Hill.

Attractor Networks

Intermediate article

Garrison W Cottrell, University of California, San Diego, California, USA

CONTENTS

Attractor networks are types of neural network that are often used to represent human subjects' content-addressable memory. That is, given part of a memory, such as a name, we can complete the pattern and recall other information about the individual associated with that name.

INTRODUCTION: COGNITION AS ACTIVATION DYNAMICS

How is it we recall information? Many psychologists have suggested that our memory is *redintegrative* – that is, we 're-integrate' the information each time we recall something. The correlate in computer science is called *content-addressable memory* (CAM). Standard retrieval of information from a computer's memory requires an *address*, the location of the information. One can, however, buy an expensive piece of hardware that will retrieve the bits at any address in the memory that match a partial pattern of bits; i.e., based on the *content* of that memory. What is expensive to a computer scientist seems effortless for a human. For example, I can tell you I am thinking of someone who is a former actor, a former President of the United States, a Republican, quite elderly, and a Rhodes scholar. You probably retrieved 'Ronald Reagan', even though part of the description is incorrect. You retrieved this memory through *pattern completion*: given part of the memory, you retrieved the rest of the memory.

In neural networks, systems that provide for the storage and retrieval of such patterns are called *attractor networks,* and have been used to model such diverse phenomena as memory, lexical access, reading, aphasia, dyslexia, associating names and faces, and face recognition. This article very briefly reviews some of the mathematics of such networks and how they are trained.

HOPFIELD NETWORKS

Hopfield networks, named after the physicist John Hopfield who studied them and proved many of their properties, are a particular kind of neural network in which the units are symmetrically connected. (Such networks had been studied earlier; see Cowan and Sharp (1988) for a review. Hopfield (1982) became associated with these networks by proving their stability properties using an energy function and by promoting the idea of them as memory models.) Hopfield networks are important because there is a great deal of elegant theory surrounding them. Unfortunately, the theory shows that as memories, they have a very small capacity. However, the important role they played in the development of the theoretical properties of neural networks makes them worthy of consideration. The following discussion owes much to Hertz *et al.* (1991).

A binary Hopfield network is a collection of N *units,* connected by weighted links. A unit is represented by its activation value, y_i:

$$y_i = g(u_i) \tag{1}$$

$$u_i = \sum_{j=0}^{N} w_{ij} y_j \tag{2}$$

$$g(x) = 1 \text{ if } x \geq 0, \text{ else } -1 \tag{3}$$

where $g(x)$ is called the *activation function* of the units (here, it is also known as the sign function), u_i is often called the *net input* to the unit, and w_{ij} is the weight from unit j to unit i. Also, $y_0 = 1$ by definition. w_{i0} is sometimes called the *bias* of the unit, and is equivalent to the negative of a threshold (which it is explicitly when y_0 is defined as -1 instead of 1). In Hopfield networks, there is a constraint that $w_{ij} = w_{ji}$; that is, the network is symmetrically connected. This is explained later.

This system has a *dynamics.* That is, if we start the system in some state (a pattern of activation on the y_i), it updates its state over time based upon the above equations. For a 'standard' Hopfield network, this is done asynchronously. That is, on each time step of the system, we randomly choose an i between 1 and N, and update the activation of unit i using the above equations.

Given this formulation, the question is: how can we store a set of patterns of activation such that, when presented with a new pattern, the system activations evolve to the closest stored pattern? This idea is illustrated abstractly in Figure 1. The coordinates of the graph are the activations of two continuously-valued units (note this is easier to draw than if we used -1, 1 units, as we are here). The 'X's represent stored activation patterns. The arrows represent the idea that, starting from nearby patterns, the system should move towards the nearest one. The boundaries around any pattern are called the *attractor basin* for that pattern.

Following Hertz *et al.* (1991), we start with one pattern, call it $\vec{x} \in \{-1, 1\}^N$, a vector of 1s and -1s of length N. We would at least like this pattern to be *stable;* that is, if we impose this activation pattern on the units by setting all of the y_i, the network should not move from that pattern. The requirement for stability is simply:

$$g\left(\sum_j w_{ij} x_j\right) = x_i \quad \forall i \tag{4}$$

If the update rule is applied, then, nothing changes. The question, then, is how to set the weights w_{ij} to guarantee this. One common rule is:

$$w_{ij} = \frac{1}{N} x_i x_j \tag{5}$$

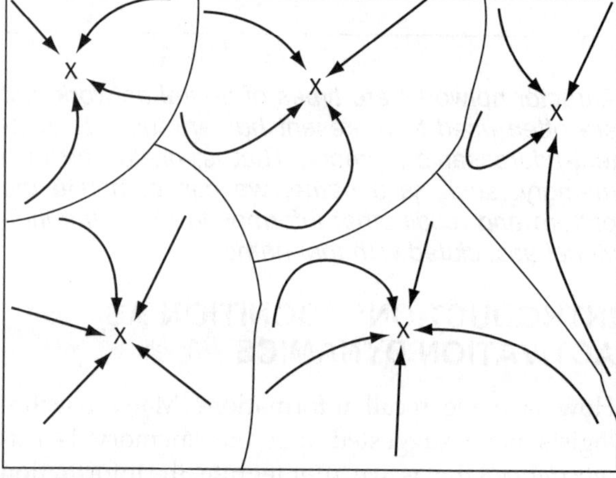

Figure 1. Attractor basins in an imaginary network. The x, y axes represent activations of two of the units, in this case, continuous valued.

Then (assuming no bias):

$$\begin{aligned} u_i &= \sum_j w_{ij} x_j \\ &= \frac{1}{N} \sum_j x_i x_j x_j \\ &= \frac{1}{N} \sum_j x_i = x_i \end{aligned} \tag{6}$$

as required (note: $g(x_i) = x_i$). In fact, if fewer than half of the elements are 'wrong', they will be overruled by the majority that are correct. Thus this system has an attracting state that is the desired pattern. Given a partially correct pattern, it will 'complete' the pattern, just as the reader completed the 'Reagan' pattern. It should be noted that it also possesses another attractor, $-\vec{x}$, called the *reversed state* (Hertz *et al.*, 1991). This is the other attracting state of the system.

For multiple patterns, we just overlay the weight prescription of each pattern:

$$w_{ij} = \frac{1}{N} \sum_p x_i^p x_j^p \tag{7}$$

where p ranges over all of the patterns. This is called the *outer product rule* or *Hebb rule* after Donald Hebb (1949), who proposed a similar idea in his classic book, *The Organization of Behavior*. If one imagines that these patterns are imposed (say, via perception) on the set of units, then this rule sets the weights according to the correlation between their firing. In neuroscience, the slogan is, 'neurons that fire together wire together'.

In a learning setting, we can imagine that upon presentation (or imposition) of a pattern (i.e., all y_i are set to the corresponding x_i^p), starting from some random initial weights, each weight is updated by:

$$w_{ij} := w_{ij} + \eta y_i y_j \tag{8}$$

where η is a (usually small) learning rate parameter. Over many presentations of the patterns, the weights will become proportional to the correlation between the elements of the patterns. They will also grow without bound, a problem that can be addressed by artificially limiting the size of the weights, or by adding a 'weight decay' (or 'forgetting') term that moves the weights slowly back towards 0.

Intuitively, we can think of the weights in the network as *constraints* between the units. Suppose, for the sake of exposition, some unit represents the feature 'happy' and another unit represents the feature 'has a big smile'. We would like both of these units to be 'on' ($y_{happy} = 1$, $y_{smile} = 1$), or both of these units to be 'off' ($y_{happy} = -1$, $y_{smile} = -1$) in a stable pattern. Then they should have a positive weight between them, since the features are correlated. On the other hand, these units should have a negative connection to a unit representing 'sad'. Then if $y_{sad} = 1$, it will tend to try to turn off y_{happy} and vice versa. Hence the weights represent the way that features may 'vote' for their friends (positive weights) and vote against their enemies (negative weights).

Returning to the rule for the weights given in equation 7, the requirement for stability of a pattern becomes:

$$g\left(\sum_j w_{ij} x_j^p\right) = x_i^p \quad \forall i, p \tag{9}$$

Note that we can decompose the sum into:

$$\begin{aligned} \sum_j w_{ij} x_j^p &= \frac{1}{N} \sum_j \sum_{p'} x_i^{p'} x_j^{p'} x_j^p \\ &= x_i^p + \frac{1}{N} \sum_j \sum_{p' \neq p} x_i^{p'} x_j^{p'} x_j^p \end{aligned} \tag{10}$$

The second term is called the *crosstalk* term. The pattern will be stable if the crosstalk term is the same sign as x_j^p for all j, or if its magnitude is less than 1, so that it does not overwhelm x_j^p.

The question now becomes: what is the capacity of such a network? That is, how many such patterns can reliably be stored? First, by 'stored' we simply mean that the pattern is stable, not that a partial pattern will complete to a full pattern. Also, there are several ways in which we might define 'reliably'. For example, we might require that the patterns remain exactly as stored, or that they change only a small percentage of their bits, or that they are stable with some probability and some amount of distortion. It also matters how correlated the patterns are with one another. The details are beyond the scope of this article, but the general result agrees with Hopfield's empirical finding in his original article that the capacity is about $0.15N$. The most quoted figure for the capacity is $0.138N$, which corresponds to about 1.6% of the bits in a pattern changing to an incorrect setting before it stabilizes (Hertz *et al.*, 1991). This means that to store, say, one pattern from every day of our lives until we reach 50 years, we would need of the order of 132,000 units. Looked at another way, given that we have 10^{11}–10^{12} brain cells, we could store of the order of 10^{10}–10^{11} memories – if the brain were a Hopfield network, and if it had to do nothing but store memories. However, even if we used only 1% of our brains for memory, we would still be able to store around 100 million memories by this calculation. So perhaps the view that they have small capacity is not such an worry.

ENERGY FUNCTIONS AND CONVERGENCE

One of the novel contributions of Hopfield's original paper was the proof of convergence of a Hopfield network. *Convergence* here means that when the network starts in some state of activation, using the update equations 1–3, the network will reach a point where no activations change. Note that this does not mean that it will converge to one of the stored memories! It simply means that the network will not oscillate, or become chaotic. The proof relies on the notion of an *energy function*, also known in physics as a *Lyapunov function*. The idea is that the energy function is a real-valued function of the state of the network, that it is bounded from below, and that the update equations of the network always make this number stay the same or go down. If one can come up with a Lyapunov function for a system, one can infer that the system reaches a point where nothing changes. (This is speaking rather loosely. There may be updates that move the state of the system to another state with the same energy value. However, in the proof below, we assume that only one unit changes its state, which avoids this problem.) The particular Lyapunov function Hopfield proposed is:

$$E = -\frac{1}{2}\sum_{ij} w_{ij} y_i y_j \tag{11}$$

This expression, without the negation sign, has also been called the 'goodness' function of the network by Rumelhart *et al.* (1986b), in which case it always goes up or stays the same.

Here we have excluded the contribution of any external input to the network, which can be easily incorporated into this expression. Given a set of weights, E clearly has a minimum value (it is bounded from below). Intuitively, we can think of this as a representation of how many constraints have been violated by the current activation state. That is, suppose two units are connected by a positive link. Then the constraint between them is that they should both be on or both be off, as in the 'happy/smile' example above. This 'constraintlet' is represented by two terms (redundantly) in the above sum, $w_{happy\ smile} y_{happy} y_{smile}$ and its dual. If both units were on, then that would be a positive component of the sum, which would mean a more negative overall energy (more constraints satisfied). If one of them was 1 and the other -1, then the energy would be higher, all other things being equal. Similarly, if w_{ij} is negative, then the

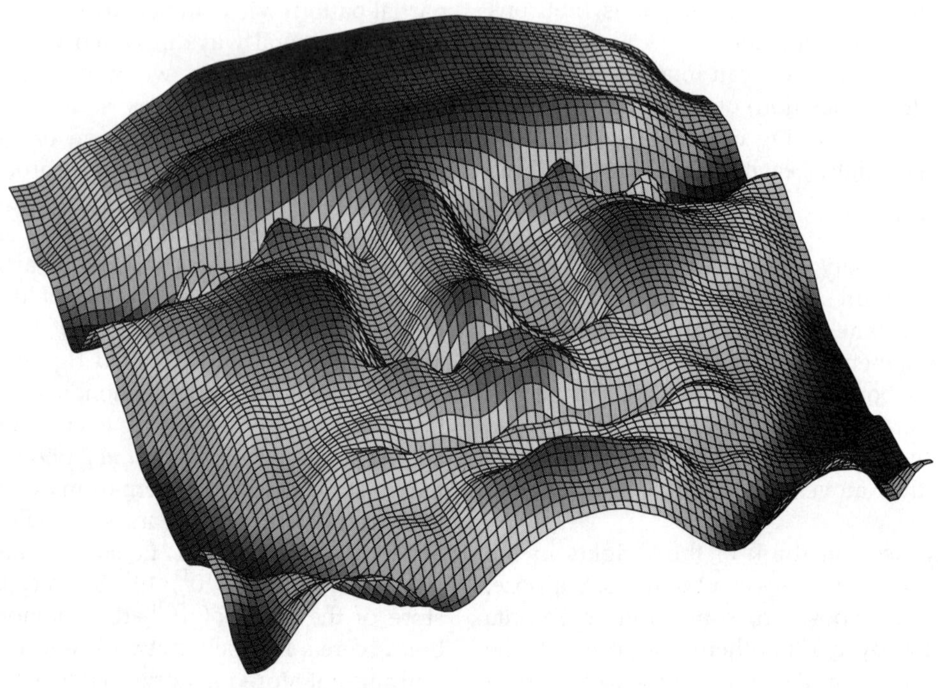

Figure 2. [*Figure is also reproduced in color section.*] An imaginary energy landscape.

energy is lower when y_i and y_j are of opposite sign, versus when they have the same sign.

Another way to conceive of this is to imagine the values of E for different activation values of the network. If there are just two units in the network, we can lay out the possible states of the system in a plane. For some values of the units, E will be high, and for others, E will be low. Doing some injustice to the fact that we have only four discrete states, we can imagine that E forms an *energy landscape* over this two-dimensional state space of the system, as shown in Figure 2. The surface shows the height of the energy. The state of the network can be thought of as a ball on this surface. Updating the network state will cause the ball to roll downhill. This is just a more detailed picture of the same phenomenon shown in Figure 1.

Going back to the equation for E, it is intuitively clear why the operation of the network dynamics (equations 1–3) lowers this number. For example, if a unit's current value is -1, and the input from the rest of the network is positive, the unit will update its state to be $+1$, which will violate fewer constraints. To see this formally, suppose that a unit changes state upon updating (if it does not change state, the energy remains the same). Let y'_i represent the new state of unit i. Then the difference in energy with y_i in its new state, E', and the previous E only involves the terms with y_i and y'_i in them:

$$\begin{aligned} E' - E &= -\sum_j w_{ij} y'_i y_j + \sum_j w_{ij} y_i y_j \\ &= -y'_i \sum_j w_{ij} y_j + y_i \sum_j w_{ij} y_j \\ &= (y_i - y'_i) \sum_j w_{ij} y_j \\ &= (y_i - y'_i) u'_i \end{aligned} \tag{12}$$

By assumption, u'_i is of opposite sign to y_i, and must be the same sign as y'_i. Hence the difference is negative, and E' must therefore be of lower energy than E, as required.

BOLTZMANN MACHINES

One possible problem with Hopfield networks is that, since the update equation is deterministic, the energy always decreases. This means that the network will always move to the lowest nearby minimum, even if there is a better minimum somewhere else. The situation where this matters is where some of the units in the network are *clamped* – that is, their activation values are held fixed – and the problem is to find the 'best completion' of the pattern, given the constraints encoded in the network weights. One possible fix to this problem is to have the network sometimes go *uphill* in energy with some small probability. *Boltzmann machines* are one embodiment of this notion (Hinton and Sejnowski, 1986). The formulation of Boltzmann machines is somewhat different from the standard Hopfield network. First, the units are *stochastic*, and have states that are either 0 or 1:

$$s_i = \begin{cases} 1 \text{ with probability } g(u_i) \\ 0 \text{ with probability } 1 - g(u_i) \end{cases} \tag{13}$$

The probability of a unit being 1 is usually taken to be a *sigmoidal* function of the input:

$$g(u_i) = \frac{1}{1 + exp(\frac{-u_i}{T})} \tag{14}$$

The parameter T is called the *temperature*, because of an analogy between the operation of the system and spin glass models in statistical physics. It controls the steepness of the function g. T can be thought of as a noise parameter. The bigger T is, the less each unit will respond to the input from other units (the function g will be very flat). If T were infinite, each unit would flip between 0 and 1 with 50% probability. Thus, the basic idea in the operation of a Boltzmann machine is to start with a high T, and slowly lower T until it is near 0 and the system is practically deterministic (the function g will be very close to a threshold unit as in equations 1–3). Hinton has described the idea as follows (paraphrasing). Suppose you have a black box with a surface like that in Figure 2 inside, and a ball is dropped into the box. Your job is to get the ball to the lowest spot on the surface, obviously without being able to see inside the box. One way to do this is to shake the box vigorously, then slowly shake it less and less. The energetic shaking should get the ball into the largest well in the box, and as the shaking subsides, it should get into the lowest spot. One can prove that if you shake the box for an infinitely long time, and slow your shaking appropriately, then the ball will end up in the lowest spot in the box.

Another way to think about what T does is that when it is big, it smooths out the bumps in the energy landscape, as if you are squinting at it. As it is lowered, the smaller bumps will emerge, and the ball's behavior will depend more on the fine details of the landscape. It should be clear that no one is proposing that your brain 'heats up' as you are recalling memories! However, the idea that noise may help in reaching better states is not so far-fetched. It has been shown that laughing in the middle of a test improves performance.

Another novelty in Boltzmann machines is the idea of *hidden units* in the network. These are units that are not part of the stored patterns (which are placed on the *visible units*), but can be used to differentiate between patterns that might otherwise be confused because they are too similar. The introduction of a set of units that are not part of the stored patterns makes the use of the Hebb rule problematic, as the states of the hidden units are not specified. However, learning in Boltzmann machines nevertheless turns out to be relatively simple conceptually, but tends to be very slow. (Recently, Hinton has developed a relatively fast learning algorithm for a special case of Boltzmann machines (Hinton, 2002), but it is beyond the scope of this article.) The basic idea is to have two phases. In one phase, the patterns that are to be stored are clamped on the visible units. The network is run for a long time, while statistics are collected on how often units are on and off together. Then the network is run again without the patterns clamped. The weights are updated according to:

$$w_{ij} = \eta(< y_i^+ y_j^+ > - < y_i^- y_j^- >) \qquad (15)$$

where the angle brackets mean averages over time, η is a learning rate parameter, y_i^+ refers to the clamped phase, and y_i^- refers to the unclamped phase. Notice that this rule will still result in symmetrically connected networks. Intuitively, one is subtracting off the statistics of the network running 'on its own' from the statistics of the network when the desired pattern is present. This has been compared to an 'awake' state (where the network is clamped by its perceptions of the environment) versus a 'sleep' state (the unclamped phase). Such evocative imagery has not yet been verified by neuroscientists, but one really wants it to be true.

One advantage of such networks is that, if they learn successfully, operating them without inputs will result in the network displaying on the visible units the entire *distribution* of the training environment. Unlike deterministic networks, then, these networks can be thought of as *generative models* of their environment. Also, if one thinks of the networks as simply learning by exposure to an environment, they can be thought of as *unsupervised* models that learn (on the hidden units) efficient encodings (features) of their environment. This has appeal as a model of how humans learn from 'mere exposure' to the environment. Of course, we start with a highly structured neural network, based upon millions of years of evolution. Integrating such pre-structuring of the network with appropriate learning rules is the subject of a great deal of current research.

SEQUENTIAL ATTRACTORS: BACKPROPAGATION THROUGH TIME

While the networks described so far are attractive as models of pattern completion memory, they also seem flawed as a model of how brains might work. First of all, they require the connections in the network to be symmetric. There is little evidence that the neurons in brains, human or otherwise, are so connected. Second, if one takes seriously the notion of finding a stable state of the network, the idea that our neurons settle to a stable state is not particularly palatable. As Walter Freeman has remarked, 'the only stable neuron is a dead neuron'. One would like a model that perhaps reaches stable states transiently, and then progresses to a new state. While there has been some work in this area for Hopfield networks (see Hertz *et al.* (1991) for examples), there has been more work on training networks to go through sequences of states using *supervised*, or *error-correction*, learning techniques. The standard approach is called *backpropagation through time* (BPTT). While the full details are beyond the scope of this article, we can summarize some of the main points.

First, note that if we eschew stable states, the networks must *not* be symmetrically connected. Otherwise, there would be a Lyapunov function to describe their dynamics. Second, if one wants the network to go through different states, it must be told, in some way or another, what those states are. Hence, the training must be supervised – states are specified for the trajectory of the network, and the network is required to pass through those states in the order specified. We may retain the notion of attractors if we generalize it to cyclical behaviors. This means that, for example, if the network is somehow pushed out of the trajectory it has been trained to produce, it will move back towards that trajectory. Finally, such systems usually use *continuous* units, that take values in the range [0,1]. A standard equation for the activity of such units is the logistic equation:

$$g(u_i) = \frac{1}{1 + exp(-u_i)} \qquad (16)$$

which bears a remarkable resemblance to the equation for the probability of a unit being 'on' in a Boltzmann machine. However, here this produces not a probability, but the actual activation of the unit. Otherwise, the u_i is calculated in exactly the same way as in equations 1–3.

BPTT uses the idea illustrated in Figure 3. On the left of the figure, there is a simple asymmetric recurrent network. This network is converted into a

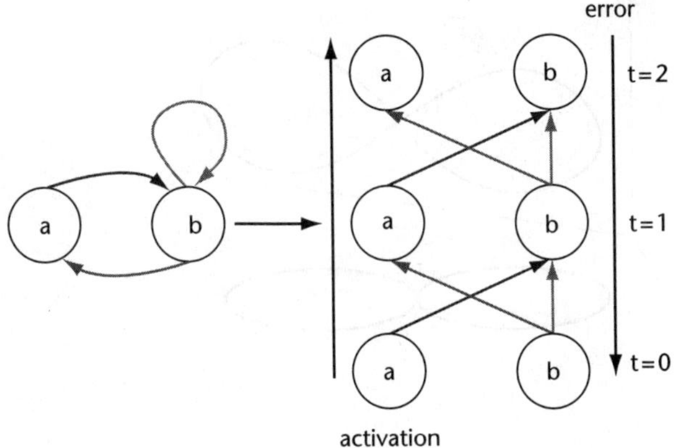

Figure 3. [*Figure is also reproduced in color section.*] Back-propagation in time.

so-called feedforward network by 'unrolling' it in time, as shown on the right. Backpropagation (Rumelhart *et al.*, 1986a) is a supervised learning technique that adjusts the weights in a network in order to reduce the error in the state of the network. Errors, in the form of target states, can be 'injected' into the network at any time step, and the weights are adjusted to make the state of the network closer to the target state. Errors are then propagated backwards through the network (hence the name 'back-propagation'), in this case through time (hence the name, BPTT).

The weighted links are color-coded to show how the links in the feedforward version relate to the links in the recurrent version. Also, since links of the same color in the feedforward version are, conceptually, the *same* link, this means that any weight adjustments to one link must be made to all of the links of the same color. Essentially, in BPTT, the weight changes to each link are added together for links of the same color. One of the most striking examples of BPTT in action was its use to train a system to back up a semi-tractor trailer, which required many steps and is a complex task that requires a very skilled human operator to perform (Nguyen and Widrow, 1989).

PHASE-SPACE LEARNING

BPTT in its basic form has problems in training systems with complicated attractors, especially if one wants multiple sequential attractors in the same system (see Tsung and Cottrell (1995) for a discussion). A mild variation on BPTT produces surprising results. Tsung and Cottrell (1995) introduced a variation that applies in the case where the desired system dynamics are deterministic, which is often the case (an example where this is not the case is in machine translation; for example, the Spanish *casa* can be translated into English as *house* or *home*). It uses two ideas. The first, borrowed from time-series prediction, is to perform a *delay-space embedding* of the desired trajectory. This is useful in cases where the observed behavior does not appear deterministic, but could be, if more dimensions were introduced. Take, for example, a figure 8 shape. If the two visible units of a system (represented by the two axes in Figure 4(a)) are supposed to describe a figure 8 with their behavior, what happens at the middle point of the 8 is not determined – the system could go one of two ways. However, if there was a third dimension that basically raised one of the curves above the other (think of a raised highway), then the system would be deterministic. The idea is shown in Figure 4(b). Note that the 'shadow' of the trajectory in 2-D is still a figure 8 shape, but the system has a third variable that allows the crossing point to be separated in space.

Delay-space embedding is a technique for adding more state variables to avoid such crossing points, basically by taking the desired trajectory and adding more variables that are simply the ones we already have, but delayed in time. The skill is in picking the amount of delay, and the number of extra variables formed this way (Kennel *et al.*, 1992). An important property of a proper embedding is that each point on the trajectory uniquely determines the next point on the trajectory. The space that the system is embedded into is called a *phase space*. In this space, time is represented by movement through the space. For example, the picture given in Figure 1 is a phase space picture: movement along the arrows is movement in time for the system. In this case, the memories given by the Xs in that picture are called *fixed point* attractors. We are interested here in attractors that may involve, for example, closed loops in phase space, which correspond to oscillations. An example of this has already been shown in Figure 4.

The second idea is, once the system has been made deterministic in this way, the mapping from one point to another is a *map*; i.e., there is some function, let's call it f, that produces the next point on the trajectory given the current point. Introducing t, a time variable, $f(\vec{y}(t)) \rightarrow \vec{y}(t+\delta t)$, for some increment δt of our choosing. We can use a feedforward neural network to learn f from examples. Essentially, we are doing BPTT only one δt step back in time. However, given the way we have arranged things via delay-space embedding, this is all that is needed. Now, given this map, we may *iterate* it. That is, once we start from some

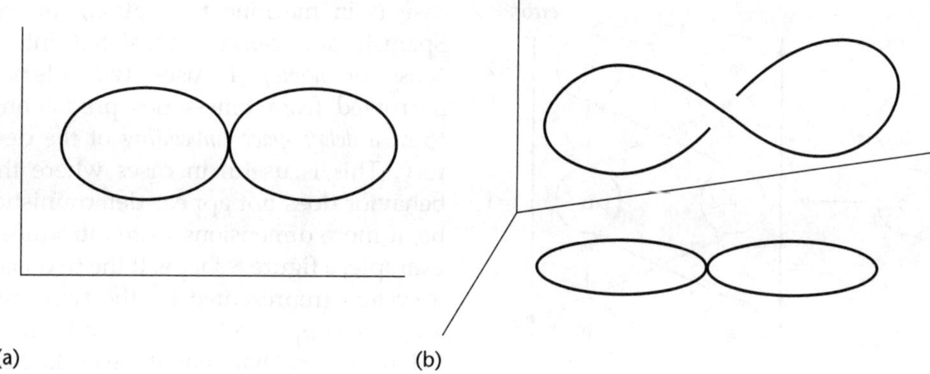

Figure 4. Embedding a '∞' from 2-D into 3-D.

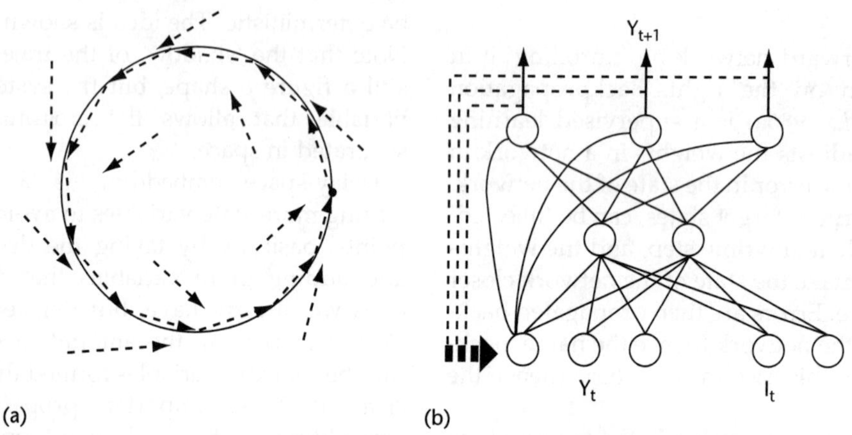

Figure 5. Phase-space learning. (a) The training set is a sample of the trajectory elements. (b) Phase-space learning network. Dashed connections are used after learning. From Tsung and Cottrell (1995), reprinted by permission of MIT Press.

point, $\vec{y}(0)$, and obtain $f(\vec{y}(0)) = \vec{y}(\delta t)$, we can apply f again, to get the next point, and so on. The final idea is that we can make any trajectory an attractor by training the network to start from points near the desired attractor and making the target closer to the attractor. Tsung and Cottrell (1995) noted that this function f may be arbitrarily complex, so that hidden units between the input and the output may be needed. The idea is shown in Figure 5.

Thus, *phase-space learning* consists of: (1) embedding the trajectory to avoid crossing points, (2) sampling trajectory elements near the desired trajectory, and (3) training a feedforward network on these trajectory elements. Since feedforward networks are universal approximators (Hornik *et al.*, 1989), we are assured that, at least locally, the trajectory can be represented. The trajectory is recovered from the iterated output of the pre-embedded portion of the visible units (the ones we started with – e.g., the 'shadow' of the embedded system in the original space, as in Figure 4(b)). Additionally, we may extend the phase-space learning framework to also include time-varying inputs that affect the output of the system, as shown in Figure 5, bottom right.

The phase-space learning approach has no difficulties storing multiple attractors. Learning multiple attractors can be done in the same way a single attractor is trained; one simply includes a sufficient number of trajectory segments near all of the desired attractors. Figure 6 shows the result of training four coexisting oscillating attractors, one in each quadrant of the two-dimensional phase space. The underlying feedforward network has two inputs, two layers of 20 hidden units each, and two outputs. The network will remain in one of the oscillating regimes until an external force pushes it into another attractor basin. Such oscillating attractors are called *limit cycles* in dynamical systems theory.

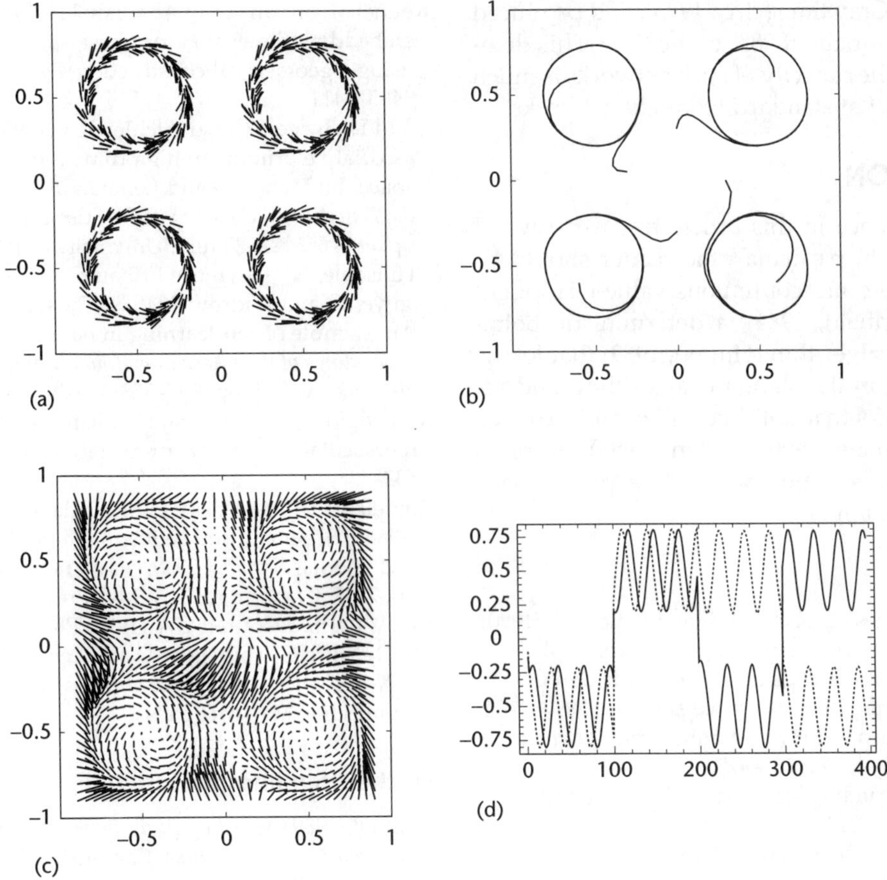

Figure 6. Learning four coexisting periodic attractors. The network had 2-20-20-2 units and was trained using back-propagation with conjugate gradient for 6000 passes through the training set: (a) the training set: 250 data pairs for each of the attractors; (b) eight trajectories of the trained network delineate the four attractors; (c) vector field of the network: this shows, for every little arrow, where the network would go next; (d) graph of the activations of the two visible units over time, as they are 'bumped' into different attractor basins. From Tsung and Cottrell (1995), reprinted by permission of MIT Press.

Similarly, such a system can be used to avoid the problems inherent in standard Hopfield networks. The author has used phase-space learning to create a standard fixed-point attractor network to store 'meaning' patterns derived from co-occurrence counts. Specifically, 233 word vectors were used that were processed versions of vectors obtained from Curt Burgess at UC Riverside (Lund *et al.*, 1995). The words were those used by Chiarello *et al.* (1990) in their priming experiments. They represented various words from 'ale' to 'wool'. The structure of the vectors was such that, for example, all of the food words were similar, all of the clothing words were similar, etc. The vectors were 36-dimensional (hence 36 visible units would be required to store them), and the elements were +/−1. Hence they were perfect for storing in a Hopfield network, except for one thing: there were over six times as many vectors as units, and a Hopfield network with 36 units should be able to store about five vectors. Instead, the author used phase-space learning, in the following way: a 36-70-36 feedforward network was used – that is, there were 36 inputs, 70 hidden units, and 36 outputs. On every training trial, one of the vectors was chosen to present on the input. 25% of the bits in the vector were probabilistically set to 0. The network was trained to produce the original vector from this nearby vector.

Once trained, it could be iterated by copying the outputs back to the inputs. One can think of this network as a recurrent network of 36 units, with 70 hidden units, by 'folding over' the output onto the input. The activation starts at the input, flows to the hidden units, and back to the input. There were also direct connections from the units to themselves (from the input to the output in the original network). This network was trained in about

10 minutes of Cray time (circa 1995), and produced the correct vector about 98% of the time. This demonstrates that the capacity of such networks is much higher than that of standard Hopfield networks.

CONCLUSION

There is not space in this article to cover several related topics. In particular, the reader should be aware that there are continuous-valued Hopfield networks (Hopfield, 1984); a deterministic Boltzmann learning algorithm (Hinton, 1989) that learns much faster than the standard algorithm; and recurrent networks that both recognize and produce sequences (Elman, 1990; Jordan, 1986) in cases where deterministic methods such as phase-space learning do not apply.

References

Chiarello C, Burgess C, Richards L and Pollock A (1990) Semantic and associative priming in the cerebral hemispheres: Some words do, some words don't... sometimes, some places. *Brain and Language* **38**: 75–104.

Cowan J and Sharp DH (1988) Neural nets. *Quarterly Reviews of Biophysics* **21**: 365–427.

Elman JL (1990) Finding structure in time. *Cognitive Science* **14**(2): 179–212.

Hebb D (1949) *The Organization of Behavior*. New York, NY: John Wiley.

Hertz J, Krogh A and Palmer RG (1991) *Introduction to the Theory of Neural Computation*, volume I of *Lecture Notes in the Santa Fe Institute Studies in the Sciences of Complexity*. Redwood City, CA: Addison-Wesley.

Hinton G (1989) Deterministic boltzmann learning performs steepest descent in weight space. *Neural Computation* **1**: 143–150.

Hinton G (2002) Training products of experts by minimizing contrastive divergence. *Neural Computation* **14**(8): 1771–1800.

Hinton GE and Sejnowski TJ (1986) Learning and relearning in Boltzmann machines. In: Rumelhart D and McClelland J (eds) *Parallel Distributed Processing*, vol. I, chap. VII, pp. 282–317. Cambridge, MA: MIT Press.

Hopfield JJ (1982) Neural networks and physical systems with emergent collective computational abilities. *Proceedings of the National Academy of Sciences of the USA* **79**: 2554–2558.

Hopfield J (1984) Neurons with graded responses have collective computational properties like those of two-state neurons. *Proceedings of the National Academy of Sciences of the USA* **81**: 3088–3092.

Hornik K, Stinchcombe M and White H (1989) Multilayer feedforward networks are universal approximators. *Neural Networks* **2**: 359–366.

Jordan MI (1986) *Serial Order: A Parallel Distributed Processing Approach*. Technical report, Institute for Cognitive Science, UCSD.

Kennel M, Brown R and Abarbanel H (1992) Determining embedding dimension for phase-space reconstruction using a geometrical construction. *Physical Review A* **45**: 3403–3411.

Lund K, Burgess C and Atchley RA (1995) Semantic and associative priming in high-dimensional semantic space. In: Moore JD and Lehman JF (eds) *Proceedings of the 17th Annual Conference of the Cognitive Science Society*, pp. 660–665. 22–25 July, University of Pittsburg. Hillsdale, NJ: Lawrence Erlbaum.

Nguyen D and Widrow B (1989) The truck backer-upper: An example of self-learning in neural networks. In: *Proceedings of the International Joint Conference on Neural Networks*, vol. II, pp. 357–363. Washington, DC.

Rumelhart D, Hinton G and Williams R (1986a) Learning representations by backpropagating errors. *Nature* **323**: 533–536.

Rumelhart D, Smolensky P, McClelland J and Hinton G (1986b) Schemata and sequential thought processes in PDP models. *Parallel Distributed Processing*, vol. II, pp. 7–57. Cambridge, MA: MIT Press.

Tsung F-S and Cottrell GW (1995) Phase-space learning. In: Tesauro G, Touretzky D and Leen T (eds) *Advances in Neural Information Processing Systems 7*, pp. 481–488. Cambridge, MA: MIT Press.

Further Reading

Amit DJ (1989) *Modeling Brain Function: The World of Attractor Neural Networks*. Cambridge, UK: Cambridge University Press.

Anderson JA (1993) The BSB Model: a simple nonlinear autoassociative neural network. In: Hassoun M (ed.) *Associative Neural Memories*. New York, NY: Oxford University Press.

Anderson JA and Rosenfeld E (eds) (1988) *Neurocomputing:Foundations of Research*. Cambridge, MA: MIT Press.

Cottrell GW and Plunkett K (1994) Acquiring the mapping from meanings to sounds. *Connection Science* **6**: 379–412.

Kawamoto AH, Farrar WT and Kello CT (1994) When two meanings are better than one: modeling the ambiguity advantage using a recurrent distributed network. *Journal of Experimental Psychology, Human Perception and Performance* **20**: 1233–1247.

McClelland JL and Rumelhart DE (1988) *Explorations in Parallel Distributed Processing*. Cambridge, MA: MIT Press.

Plaut DC (1995) Semantic and associative priming in a distributed attractor network. In: *Proceedings of the 17th Annual Conference of the Cognitive Science Society*, pp. 37–42. Hillsdale, NJ: Lawrence Erlbaum.

Plaut DC and Shallice T (1993) Deep dyslexia: a case study of connectionist neuropsychology. *Cognitive Neuropsychology* **10**: 377–500.

Rumelhart DE, McClelland JL and the PDP Group (1986) *Parallel Distributed Processing: Explorations in the Microstructure of Cognition*, vols 1 and 2. Cambridge, MA: MIT Press.

Audition, Neural Basis of

Introductory article

Shihab Shamma, Institute for Systems Research, University of Maryland College Park, College Park, Maryland, USA

CONTENTS

Sound is translated into neural signals by the organs of the auditory system. Key to this process is the cochlea, which converts sound pressure waves into spatially ordered patterns of membrane vibrations and then transforms these vibrations into neural patterns in the auditory nerve.

INTRODUCTION

Sounds in the environment are produced when a force excites a structure and causes it to vibrate. Sounds may be sudden and non-repetitive (e.g. a clap or the snapping of a twig), sustained and irregular (e.g. the burbling of water or the whisper of a friend) or sustained and regular (e.g. a singing voice or the rhythm of a drumbeat). In all cases the emitted sound carries information about the exciting force, the vibrating structure, and other physical features, such as the resonating chambers that modify the sound on its way to our ears. Our auditory systems, and those of most other animals, have evolved ingenious ways of extracting all of this information and they use it to detect, locate and identify sound sources of danger, food, courtship and companionship.

THE NATURE OF SOUND CUES

When the driving force causes a structure to vibrate (e.g. a plucked string), or is itself repetitive in nature, the sound produced consists of a succession of almost identical copies of the same emission. The fact that the emissions are so similar is an excellent clue to their common origin, so we might expect organisms to have evolved special mechanisms for recognizing repetition. Furthermore, the time intervals between repetitions are meaningful, as they characterize both the driving force and the vibrating structure of the sound source. When repetitions are slow (less than 50 repetitions per second – or Hz), they are heard as distinct sounds that reflect the dynamics of the driving force. For example, the sequence of notes in a melody or the rhythmic tapping of a drum result from finger movements, and the succession of speech sounds in an utterance reflect movements of the mouth. When the dynamics or frequency of the driving force become faster (50–4000 Hz), we begin to perceive a sustained sound with a distinct pitch that is proportional to the frequency of vibrations. This percept is critical to our appreciation of melody in music, and to the recognition of human voice. Finally, excited structures often exhibit complex patterns of vibrations consisting of numerous simultaneous frequencies that may extend to very high rates (even exceeding the audio range of humans at 20 000 Hz). These complex vibrations evoke distinct sound qualities (or timbres) that reflect the shape of the structure (e.g. a violin versus a cello), its material composition (e.g. wood versus brass) and dynamics (e.g. a muted versus a free string).

COCHLEA FUNCTION

When sound reaches the ears as pressure waves in air or water, it causes the eardrum to vibrate. It in turn transmits the vibrations to the cochlea of the inner ear via an attached chain of three tiny bones located in the middle ear. The cochlea is the key hearing organ. It consists of an elongated fluid-filled cavity with elaborate sensory cells that are embedded within exquisitely sensitive membranes that extend along its entire length. The cochlea has two main functions as illustrated in Figure 1.

First, it converts the sound pressure wave into a spatially ordered pattern of membrane vibrations. Specifically, the mechanics of the cochlear membranes gradually change their electromechanical properties along its length in a manner that causes different places to vibrate best (or be tuned) to

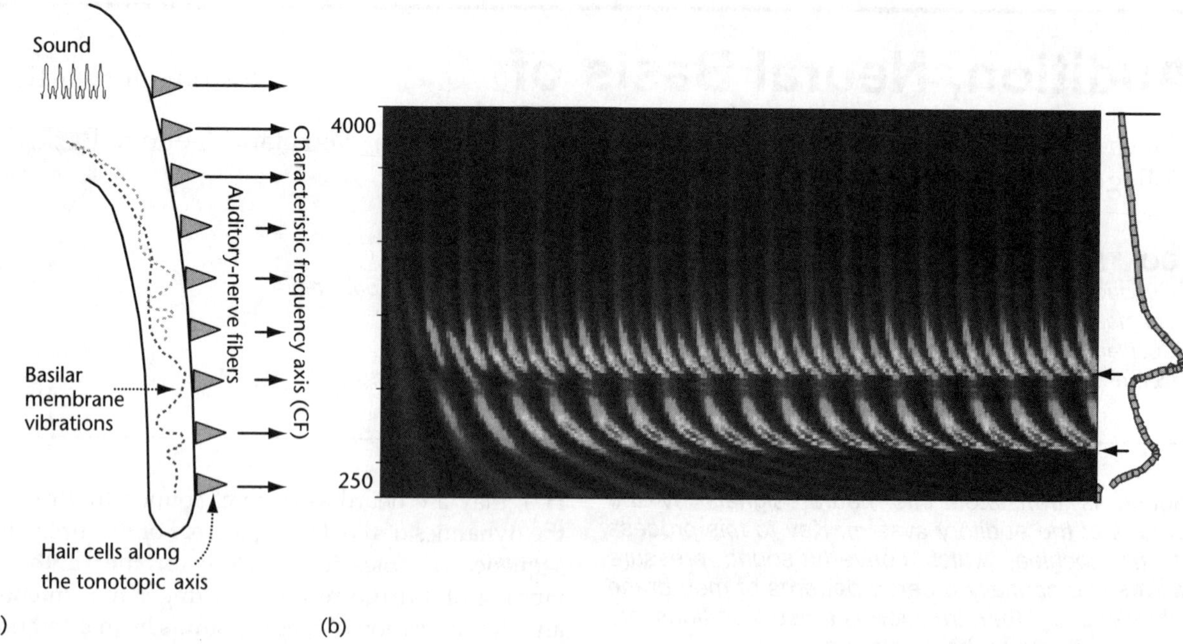

Figure 1. [*Figure is also reproduced in color section.*] Schematic model of the early stages in auditory processing. (a) Cochlear analysis. Sound enters the cochlea via the eardrum and the middle ear, initiating travelling wave displacement patterns on the basilar membrane. Vibrations due to low frequencies propagate and achieve their maximum amplitude further down the cochlea (broken red line) compared to high frequencies (broken green line). This mapping of sound frequency components onto different places along the cochlea creates the tonotopically ordered (spatial) axis of the auditory system. Basilar membrane vibrations are transduced into spatiotemporal responses on the auditory nerve by an array of hair cells distributed along the length of the cochlea. (b) Auditory nerve responses. The spatiotemporal response patterns in the auditory nerve due to a two-tone stimulus (300 and 600 Hz). The ordinate represents the tonotopic axis (labeled by the characteristic frequency axis (CF) at each location). The response at each CF represents the instantaneous probability of firing in the nerve fiber at that CF. Note that each component in the stimulus initiates a localized travelling wave pattern that abruptly ends creating a prominent discontinuity near the appropriate CF (marked by the arrow heads to the right of the panel). The stimulus responses depicted here are at a low sound level, such that it does not saturate the nerve responses. The response amplitudes are therefore strongest near the CFs of the two tones, resulting in clear peaks in the average response curve (red plot to the right).

different frequencies. Consequently, sounds with very high frequencies cause large membrane vibrations near the entrance to the cochlea, whereas those with low frequencies cause maximal vibrations near the end of the cochlea. In this way, the cochlea effectively separates the different components of the sound according to their frequency, sending them off to different places and creating a frequency-organized axis known as the tonotopic axis of the cochlea. Each complex sound therefore creates a unique spatial pattern of strong and weak vibrations along the tonotopic axis that reflects the amplitudes of its different frequency components – or its frequency spectrum.

The second main function of the cochlea is to transform these vibrations into neural patterns on the auditory nerve, to be interpreted by the brain subsequently. This is accomplished by more than 3000 specialized sensory cells (known as the hair cells) that are distributed along the cochlea. Hair cells possess channels that open and close rapidly, modulating the flow of electric current into them. The currents initiate a cascade of electrochemical events, culminating in neural signals on the auditory nerve that faithfully encode the phase (or are phase locked to the time course) of the vibrations at each point up to fairly high frequencies (4000 Hz in some mammals, and 9000 Hz in some birds). Since stronger vibrations also lead to a more vigorous neural response, the auditory nerve in effect encodes the spectrum of the sound both by the level and by the phase-locked structure of the responses along the tonotopic axis.

NEURAL PATHWAYS

The first neural structure beyond the auditory nerve is the cochlear nucleus, which consists of several anatomically elaborate subdivisions that receive parallel direct projections from the nerve. Multiple pathways emerge from the cochlear nucleus up through the midbrain and thalamus to the auditory cortex, each passing through different neural structures, repeatedly converging on to and diverging from other pathways along the way. This complexity reflects the rich and varied auditory percepts extracted from the sound, the integration of these percepts into a total auditory sensation, and its final fusion with vision and other sensory modalities, and with motor actions.

Although there is still much to be learned about the exact mechanism whereby all of these neural pathways and structures process sound, it is nevertheless clear which signal cues the nervous system must extract in order to give rise to a few important auditory percepts that include loudness, pitch, timbre and sound location.

LOUDNESS, PITCH, AND TIMBRE

Perhaps the most intuitive of these percepts is that of loudness. It is normally associated with increasing the volume (amplitude or intensity) of the sound. However, there is another physical dimension that correlates strongly with loudness, namely the range of frequencies that make up the sound, or its bandwidth. Loudness can therefore be generally viewed as being mediated by the total volume of neural activity in the auditory nerve, and thus increasing the activity (by raising either the sound intensity or the bandwidth) leads to a louder sound.

The sensation of pitch is also a readily understood attribute of sound that is normally associated with musical scales and melodies, or with the low and high voices of men and women, respectively. Pitch is strongly correlated with the repetition rates or frequencies in a sound. However, unlike loudness, the neural basis of pitch is a much more contentious topic, largely because 'pitch' is an imprecise term that is ascribed to multiple sensations which have distinct origins, and most probably different neural mechanisms. For example, the pitch of a pure tone is directly related to its frequency, and is felt over a very broad range of approximately 50 to 20 000 Hz. From a neural perspective, this percept is readily encoded by the location of best tone-evoked response along the tonotopic axis. A second example is so-called rattle pitch, namely a relatively weak sensation that is correlated with the modulation rate of the amplitude of a noise or a tone. This pitch is typically heard only up to a few hundred hertz (<400 Hz), and is likely mediated by neural responses (in the auditory nerve and beyond) that are explicitly entrained to the modulation rate. The final and most salient sensation of pitch is that of musical instruments and voices. This percept exists over a moderate range of frequencies (<4000 Hz), and is exclusively associated with harmonic sounds composed of frequencies that are integer multiples of a common fundamental frequency. An interesting fact about this pitch is that its value remains that of the fundamental frequency, even if the fundamental component in the sound is missing (hence the common description of this percept as the 'pitch of the missing fundamental'). That is, the pitch value is derived from the harmonic relationship between the components, and not simply from the fundamental frequency *per se* as illustrated in Figure 2a. The neural basis of this percept remains uncertain. One plausible theory proposes that the brain stores (or learns) harmonic templates of all pitch values, and the percept is derived according to which templates best match the spectrum of the incoming sound (Figure 2b). According to another theory, the pitch is computed directly from the incoming sound without resort to any templates, and as such it does not explicitly distinguish between this percept and the 'rattle' pitch described earlier (Figure 2b). Both of these theories (and many other variations) can account for most relevant psychoacoustical findings. The major missing piece in the pitch puzzle is the lack of firm biological understanding of the mechanisms underlying pitch processing in general.

Timbre is best regarded as the quality of a sound. It is the percept that allows us to distinguish between a violin, an oboe and a piano playing at the same pitch, or to perceive the difference between vowels (or other phonemes) in spoken language. Timbre is a multidimensional percept that is difficult to reduce to a simple scale (e.g. the low-to-high scale of pitch and loudness). Instead, it has been customary to propose several 'descriptive' scales to quantify it, using intuitive notions such as 'sharp to dull' and 'continuant to transient'. However, experimental evidence makes it abundantly clear that the shape and dynamics of a sound spectrum directly influence its timbre. Spectral shape is extracted early in the auditory system, perhaps as early as the cochlear nucleus, where the neural activity of a specific type of neuron has been found to represent faithfully and robustly the acoustic spectrum along its tonotopic axis. Further elaboration of this representation occurs in the midbrain and cortex, where more complex spectral features are selectively detected. The neural correlates of

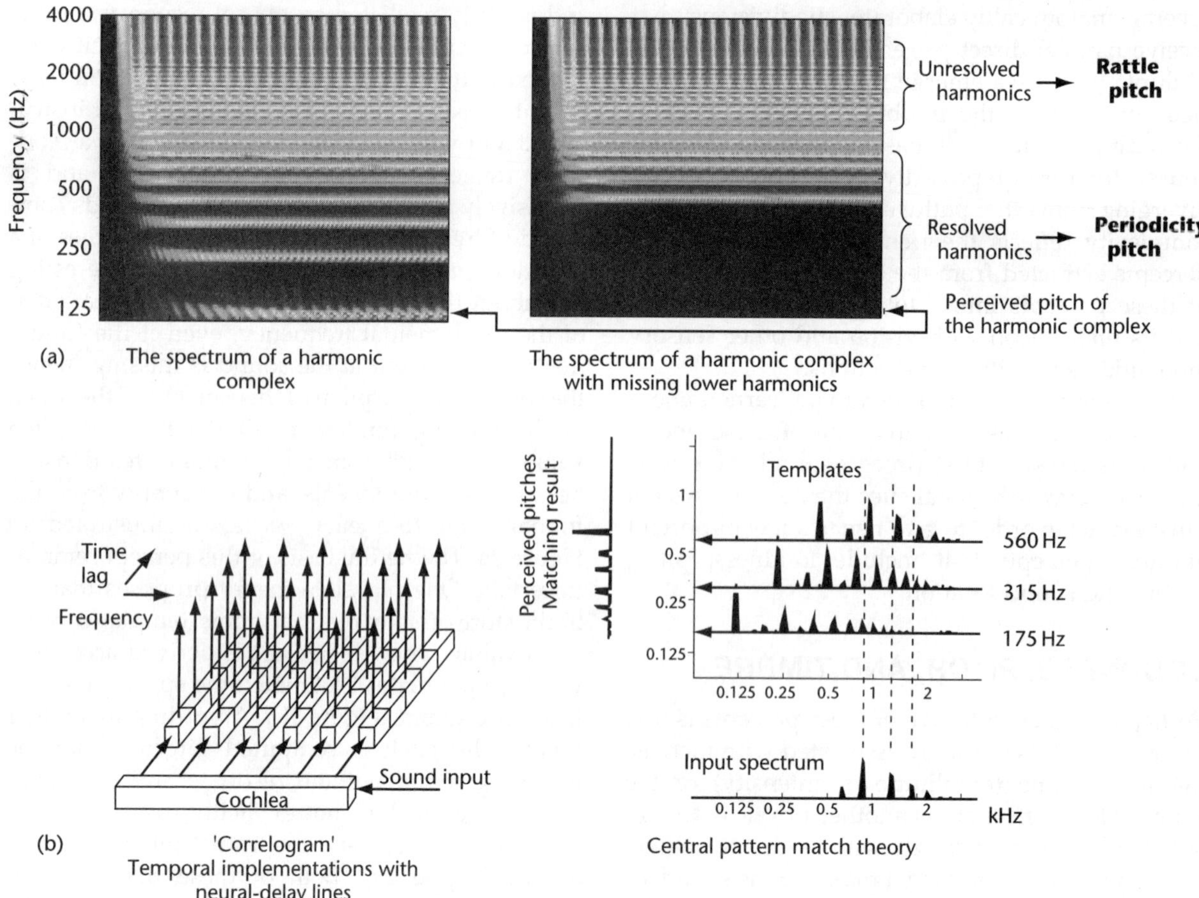

Figure 2. [*Figure is also reproduced in color section.*] Representation and extraction of stimulus periodicity of a harmonic complex. (a) The auditory spectra of two harmonic series of a 125 Hz fundamental. The low-order harmonics are well *resolved* along the tonotopic axis. These harmonics evoke a pitch sensation at the fundamental frequency of the series (i.e. at 125 Hz). The high-order harmonics (> 8th harmonic or 1 kHz) are largely *unresolved* and they instead evoke a pattern that 'beats' at the difference frequency of 125 Hz. These harmonics evoke a weaker pitch (called the 'rattle pitch') at the beating frequency (i.e. 125 Hz in this case). (Left) The harmonic complex here contains all 40 harmonics and evokes a strong pitch at 125 Hz. (Right) The harmonic complex here lacks the lowest three harmonics (125, 250, 375 Hz); nevertheless, it still evokes a strong pitch at the 'missing fundamental' frequency of 125 Hz. (b) Two algorithms for extracting this missing fundamental pitch. (Left) The schematic illustrates an autocorrelogram implementation. It presumes the existence of organized delay lines to compute the autocorrelation of the responses from each auditory nerve channel (or fiber) prior to computing the pitch. (Right) Schematic illustrates a *template-matching idea*. It presumes the existence of harmonic templates in the brain that are matched to incoming spectra so as to measure the pitch.

spectral dynamics also undergo significant transformations in different auditory stages, with faster temporal features (> 50 Hz) mainly being evident in the pre-cortical stages. In the auditory cortex (as in other sensory cortices) the dynamics of the responses explicitly represent the temporal evolution of the sound spectrum over relatively slow rates (< 50 Hz). These timescales are commensurate with the dynamics of the vocal tract in speech, the rate of change in pitch in musical melody, the transient dynamics that differentiate a string that is struck from one that is bowed (e.g. a piano versus a violin), and the rhythms of percussion instruments. Finally, there are other more complex representations of sound combining spectral and dynamic features which have been found to endow cortical cells with elaborate response selectivity. Examples of these include selectivity to downward or upward frequency-swept sounds in bats, to phrases of species-specific calls in primates, to phoneme or phonemic clusters in humans, or even to entire songs in birds. However, it is still unclear what neural mechanisms and architectures give rise to these representations and how, and the way in which these representations ultimately relate to our perception of timbre.

SPATIAL HEARING

Finally, we consider the perception of sound location in space. This auditory function is critical for survival – for example, in escaping predators, following prey and finding mating partners. Despite the enormous variability in the nature of the cues and mechanisms involved in this task, some principles are common to most species. The first of these is the use of differences in the sound impinging on the two ears, especially differences in time of arrival and in sound intensity. Specifically, when a sound source is centered in front of the head, it reaches the two ears simultaneously. If it moves to the right, the path to the right (relative to the left) ear shortens, and thus the sound reaches the right ear sooner – by a fraction of a thousandth of a second, a detectable difference for many animals (especially those with larger heads). An analogous difference in sound level occurs when a source is closer to one ear, especially for high-frequency sounds where the head shadow is more effective. Most animals have evolved neural mechanisms for detecting, extracting and utilizing these inter-aural differences in order to locate the sound source. For example, there are specialized coincidence cells in all mammals and birds that receive synchronized inputs from the two ears, and which are tuned to detect a particular time-delay between them. In barn owls, such cells are highly organized topographically so as to create an optimum time-delay axis. Another commonly found cell type is tuned to detect specific level differences between the two ears.

The second important localization principle concerns the use of special spectral cues (from one or both ears) to locate the elevation of a sound source or to characterize the acoustic environment. These spectral cues are usually introduced by auxiliary structures such as the pinna (the external ear), the shoulders, and nearby walls and floors. In the case of the pinna, its highly convoluted cavities function as mini-resonators that absorb or amplify certain sound frequencies depending on the direction of arrival of sound. This is extremely useful because, when a source is located on the midline, the two paths to the ears are equal regardless of the source elevation, and thus the only way to localize it is based on pinna-originated spectral cues. In mammals, there is some evidence that specialized neural pathways have evolved as early as the cochlear nucleus to detect these unique cues and process them in conjunction with the binaural cues. Another useful function of certain spectral cues is to convey information about the reverberant qualities of a room, and hence (indirectly) about its size and material structure. This issue is extremely important with regard to both the architectural design of music halls and auditoriums and the assessment of the quality of communication channels and equipment.

CONCLUSION

There are many other auditory attributes and tasks in the animal world that we have not touched upon, and that are as important to these animals as the perception of sound timbre, pitch and location is to us. Examples include the ultrasonic echolocation in bats and dolphins that enables them to locate prey and avoid obstacles in cluttered environments, the infrasonic (very-low-frequency) communication signals among many terrestrial animals (e.g. elephants), and the unique auditory adaptations of many animals that help them to deal with the limitations of their small size (e.g. insects) or their aquatic environments (e.g. fish). Finally, we have not considered a number of key questions. For example, how does the auditory system assemble all of these disparate percepts into an integrated whole that identifies the sound source as an entity amidst the clutter of other simultaneous sound sources in the environment? And how do auditory percepts become integrated with visual, motor and other neural processes of attention and memory so as to give rise to the typical active auditory behaviors that we normally associate with this amazing modality? The answers to these and countless other auditory mysteries lie in the findings of future exciting research.

Further Reading

Blauert J (1996) *Spatial Hearing: The Psychophysics of Human Sound Localization*, translated by J Allen. Cambridge, MA: MIT Press.

Moore BCJ (1997) *An Introduction to the Psychology of Hearing*. London, UK: Academic Press.

Pickles JO (1998) *An Introduction to the Physiology of Hearing*. London, UK: Academic Press.

Auditory Event-related Potentials

Intermediate article

Terence W Picton, Rotman Research Institute, Toronto, Ontario, Canada

CONTENTS

Auditory event-related potentials are electrical changes recorded from the brain in association with auditory stimuli.

INTRODUCTION

Human auditory event-related potentials (ERPs) are evoked by an acoustic stimulus and indicate the cerebral activity occurring as that stimulus is processed in the brain. When recorded from the human scalp, ERPs are intermixed with other electrical potentials. Several techniques may be used to distinguish ERPs from these other activities (Picton *et al.*, 1995): one of the most common is averaging – when the responses to multiple stimuli are averaged together, the auditory ERP (which is generally constant from one stimulus to the next) remains the same, whereas the other activity (which varies from one stimulus to the next) attenuates. (*See* **Electroencephalography (EEG)**)

Auditory ERPs can be classified in many ways. In relation to the stimulus, the ERPs may be transient, steady state, or sustained. Transient responses are evoked by rapid changes such as the onset or offset of a sound. Transient evoked potentials are often named according to their polarity and latency, e.g. the N100 is a negative wave recorded with a typical peak latency of 100 ms. Steady state responses are evoked by a regularly changing stimulus such as an amplitude-modulated tone. These can be described in terms of the frequencies at which they are elicited, e.g. the 40 Hz ERP is best recorded when stimuli are presented at rates near 40 Hz. Sustained potentials are recorded through the duration of a stimulus. Some ERPs are characterized by where they are generated in the brain, e.g. the auditory brainstem response, others by where they are maximally recorded, e.g. the vertex potential, and others by their purported function, e.g. the mismatch negativity. The plethora of nomenclatures reflects the paucity of our understanding.

In order for potentials to be recorded at a distance from where they are generated, many neurons must respond, their activation must be synchronous, and the neurons must be oriented in a similar direction so that their fields combine rather than cancel. Much of what goes on in the auditory nervous system does not fulfill these criteria and passes unnoticed in the scalp-recorded ERP. What is recorded, however, can tell us much about the processing of sounds in the brain.

SENSORY PROCESSING

A brief transient stimulus evokes a sequence of potentials that indicate the processing of stimulus information all the way from cochlea to cortex (Figure 1). The upper part of the figure shows an early sequence of vertex-positive waves (numbered with Roman numerals). These waves are usually lumped together as the auditory brainstem response, even though wave I is generated in the auditory nerve rather than the brainstem. Since it can be recorded down to intensities near threshold, wave V can demonstrate that the brain is receiving sounds without the person being tested having to make a subjective response to those sounds. This 'objective' audiometry is important when evaluating the hearing of infants and others who cannot provide reliable behavioral responses.

The middle part of Figure 1 shows a sequence of waves named according to their polarity and alphabetical sequence. The most prominent of these middle-latency waves is Pa, with a peak latency of about 25 ms. The Pa is most likely to be generated in or near the primary auditory cortex. The initial activation of the human primary auditory cortex probably occurs at about 15 ms, but this may not be reliably recorded from the scalp. If stimuli are presented at rapid rates, the middle-latency waves evoked by one stimulus superimpose on those evoked by preceding stimuli to give a periodic response at the frequency of stimulation.

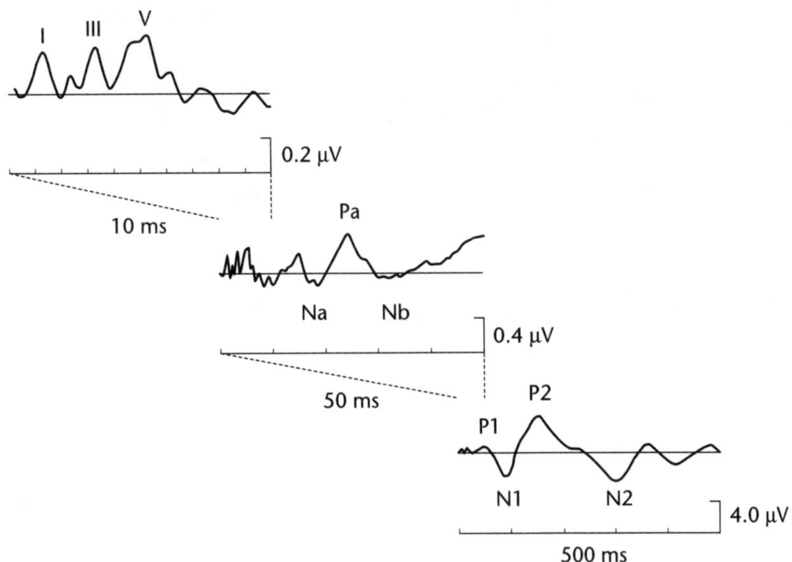

Figure 1. Auditory event-related potentials recorded from the vertex relative to the right mastoid in response to clicks at 60 dB above threshold presented at a rate of 1 s^{-1}. This recording was from a single person showing typical responses. The largest waves occur with latencies between 50 ms and 500 ms, as seen in the lower right waveform. The first 50 ms of this recording can be expanded in time and amplitude to show the middle-latency responses. The first 10 ms of this recording can then be expanded to show the auditory brainstem response.

The most prominent of these steady state responses occurs at 40 Hz (Galambos *et al.*, 1981). This response is attenuated by sleep and blocked by anesthesia. The 40 Hz ERP is probably related to bursts of cortical activity that occur at frequencies of 20–50 Hz (the 'gamma band'). These rhythmic responses may reflect the oscillatory transfer of information between sensory areas (Singer, 2000).

The lower part of Figure 1 shows a sequence of waves named according to polarity and numerical sequence. Since these waves are much larger than the earlier waves, they were historically the first to be recognized, and were called 'vertex potentials' because they were maximally recorded from the top of the head. However, since the earlier waves are often also largest at the vertex, the terminology is more nostalgic than definitive. The most prominent of the late waves is N1, which has a peak latency of about 100 ms. Several different intracerebral generators contribute to the scalp-recorded N1 (Picton *et al.*, 1999). As well as being evoked by the onset of a stimulus, an N1 also follows the offset of a stimulus or a change in any of its attributes. The N1 wave remains an enigma. Why does such a large response occur so late in the processing of auditory information? The N1 may represent a process whereby the nonspecific detection of a change in the auditory world, perhaps through the brainstem reticular system, initiates a read-out of specific information from the auditory cortex to other regions of the cortex.

SENSORY MEMORY

Memory for a stimulus can be demonstrated by a change in the response to that stimulus when it occurs at a later time. A common demonstration of memory involves the habituation of a response when a stimulus is repeated. This is most efficiently measured in ERPs by recording the responses at different interstimulus intervals. The N1 wave of the auditory ERP decreases quite strikingly with decreasing interstimulus intervals. This decrease is specific to the stimulus attributes; for example, if the tonal frequency of a test stimulus is different from the frequency of a repeating stimulus, the N1 is larger than when the stimuli are the same.

In addition to a change in the N1, a later mismatch negativity (MMN) occurs when a repeating auditory stimulus changes. In this context, the repeating stimulus is usually called the 'standard' and the changed stimulus the 'deviant'. The MMN is typically measured in a difference waveform obtained by subtracting the response to the standard stimulus from the response to the deviant. This subtraction removes elements of the response that are common to both stimuli, leaving the MMN (Figure 2). The MMN differs from the N1 in many ways. First, it has a longer latency, typically

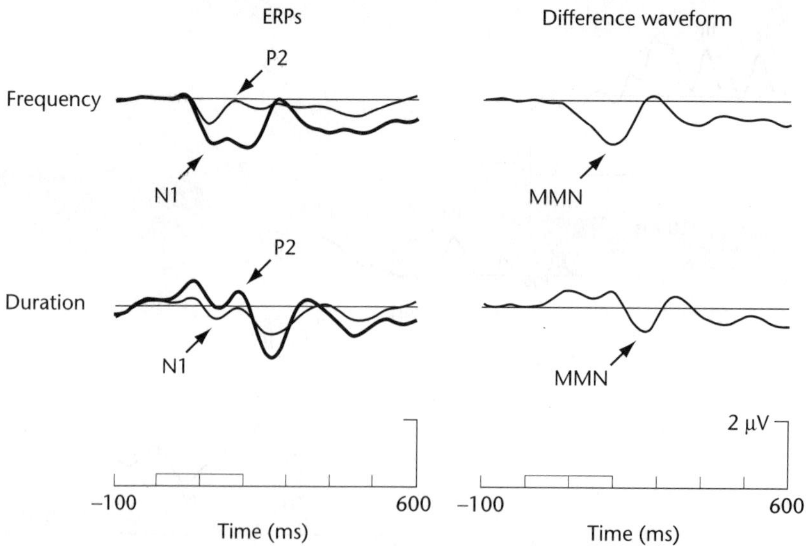

Figure 2. The mismatch negativity (MMN) in response to changes in the frequency or duration of a brief tone. The standard stimuli were 1000 Hz tones with a duration of 200 ms and an intensity 60 dB above threshold. Recordings were obtained from the midfrontal scalp using an average reference. Deviant stimuli occurred with a probability of 0.2. For the upper tracings the deviants had a frequency of 1100 Hz; for the lower tracings the deviants had a duration of 100 ms. Deviant–standard difference waveforms are shown on the right. These event-related potentials (ERPs) are averaged over ten test participants. Deviant ERP, thick line; standard, thin line.

peaking about 50 ms later than the N1. Second, this latency increases with decreasing difference between the deviant and the standard. Third, the latency is determined not by the onset of the stimulus but by the time when the deviant can be distinguished from previous stimuli. When the deviance is a change in duration, the MMN occurs approximately 150 ms after the time of the shorter stimulus, regardless of whether the deviant is the shorter or the longer stimulus. Fourth, the MMN is more related to the size of the deviance than to the parameters of the stimuli. A decrease in the intensity of a stimulus can thus elicit a MMN even though the N1 is smaller for the less intense stimulus. Fifth, the MMN has a different scalp topography, with a maximum amplitude anterior to that of the N1. The intracerebral sources for the MMN may vary with the nature of the deviance, for example being larger in the left hemisphere for changes in speech sounds (Näätänen, 2001).

The MMN is related to sensory (or echoic) memory through the time during which a sensory regularity must be detected. The system generating the MMN must recognize that the deviant does indeed break some perceived regularity in the stimuli. It can only do this when the information is maintained in sensory memory (Picton *et al.*, 2000). Since the MMN occurs whether or not the person is attending to the stimuli, it seems to represent an automatic processing of stimulus changes. However, attending to the stimuli can enhance the wave (Woldorff *et al.*, 1998), particularly when detecting the deviance is more complicated, such as noting a change in the temporal pattern of the stimuli (Alain and Woods, 1997).

The role of the MMN is unknown. It may indicate the increased processing that is evoked by a deviant stimulus, either to detect it as different from the standard or to initiate processing once it is detected. In the latter case, it might serve to alert other parts of the brain, particularly the right frontal regions, that a deviant has occurred.

AUDITORY ATTENTION

The effects of attention on the auditory ERP can be studied by presenting two or more trains of information, like hearing different conversations at a party, and asking the participant to attend to one and ignore the others. Attention can be monitored by measuring how well the participant detects occasional changes in the stimuli ('targets'). The difference between the ERPs evoked by the attended and ignored stimuli then indicates a specific effect of selective attention, independent of nonspecific changes in arousal. Despite some contrary reports, the bulk of the evidence indicates that the early auditory ERPs are unaffected by whether the participant pays attention to the stimuli or ignores them (Hackley *et al.*, 1990). This suggests that the brain automatically processes auditory information to the level of the auditory cortex. The earliest undisputed evidence of an attentional effect on the auditory ERPs is a change in the middle latency potentials between 20 ms and 50 ms after the stimulus (Woldorff and Hillyard, 1991). This change occurs in demanding selective attention tasks and probably reflects some facilitated processing of the attended information in the auditory cortex. (*See* **Attention**)

The most striking effect of auditory attention is an increased negativity beginning at about 50 ms after the onset of a sound and overlapping the N1 wave. This effect can also be demonstrated as a negative difference wave or Nd, obtained by subtracting the response to ignored stimuli from the response to the same stimuli when they are attended. The effect probably indicates both an enhancement of the processing normally represented by the N1 wave and extra processing independent of the N1. The size and duration of the Nd varies with the amount of processing needed for the attentional task and the time pressure under which it occurs. Two parts of the Nd wave have been distinguished: an early wave, which is probably related to processing attended information, and a later wave, which may be related to task monitoring. The early Nd wave is mainly generated in the auditory cortices of the supratemporal plane. The neurons contributing to this wave probably vary with the different attentional tasks, e.g. whether the person is attending to the loudness of the stimulus or its frequency. Differences in Nd between tasks are difficult to demonstrate because neurons specific to different tasks have similar locations on the supratemporal plane. Nevertheless, the Nd for attended stimuli clearly varies with the location of the stimulus in space (Teder-Sälejärvi *et al.*, 1999).

WORKING MEMORY

Once the information in an attended stimulus is processed, it is available to working memory for further evaluation. The ERP evoked by the detection of an improbable auditory target (or 'oddball') in an attended train of standard stimuli contains, in addition to the P1–N1–P2 waves, an N2–P3 complex and a later 'slow wave' (Figure 3). The P3, the largest of these attention-dependent waves, typically occurs with a peak latency of about 300 ms, and therefore also goes by the name of P300. It is maximally recorded from the vertex and midparietal regions of the scalp. The amplitude of the wave is

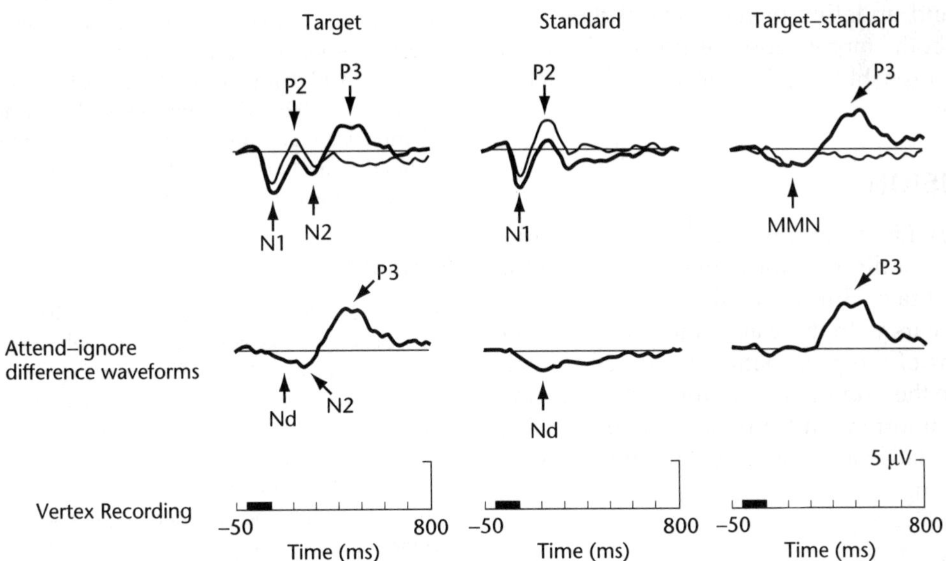

Figure 3. Effects of attention on the auditory event-related potentials (ERPs) to tones presented 65 dB above threshold at a rate of 2 s^{-1}. The participant's task was to detect occasional ($p = 0.2$) targets with a slightly different frequency or to ignore the stimuli and attend to a concurrent set of stimuli in the other ear. The ERPs for ten participants were averaged across the ears according to whether the stimuli were targets versus standards, and attended (thick line) versus ignored (thin line). The attend–ignore difference waveforms show a Nd wave for both targets and standards, and a P3 wave for the targets. The target–standard waveforms show a mismatch negativity (MMN) for the target regardless of attention and a P3 wave during attention.

inversely related to the probability of the stimulus. Its peak latency increases and its amplitude decreases with increasing difficulty in distinguishing the target from the standard. The peak of the P3 may occur before a motor response in difficult tasks, but often occurs at the same latency or later than the motor response in easy tasks. It therefore probably represents cerebral activity that is unrelated to the motor response. In general, the latency is more closely related to the duration of sensory processing and is relatively unaffected by manipulations that alter the subsequent selection of perceptual or motor responses. Several hypotheses have been proposed for the function of the P3 wave, including the updating of working memory, the access of information to conscious processing, and the resetting of perceptual analyzers once their processing has finished. (*See* **Working Memory; Event-related Potentials and Mental Chronometry**)

Intracerebral recordings and blood flow studies indicate that many different regions of the brain are active during the P3, most prominently regions of the hippocampus and the parietal lobe (Linden *et al.*, 1999). All these regions probably contribute to the scalp-recorded waveform during the P3 and the subsequent slow wave. A network of intracerebral events overlapping in time and in potential occur in relation to recognizing the target, associating it with the required response, initiating the response, and updating memory about the occurrence of both target and response. Working memory is probably manifest in these interactive connections.

CONCLUSION

The auditory ERPs can time the different activities that occur as auditory information is processed in the human brain. Studies of the increased blood flow that occurs with these activations can indicate the locations of the processing. Studies of animals can indicate the underlying neuronal mechanisms. Combining information from all these approaches should elucidate the 'when', 'where' and 'how' of human hearing.

References

Alain C and Woods DL (1997) Attention modulates auditory pattern memory as indexed by event-related brain potentials. *Psychophysiology* **34**: 534–546.

Galambos R, Makeig S and Talmachoff PJ (1981) A 40 Hz auditory potential recorded from the human scalp. *Proceedings of the National Academy of Sciences USA* **78**: 2643–2647.

Hackley SA, Woldorff M and Hillyard SA (1990) Cross-modal selective attention effects on retinal, myogenic, brainstem and cerebral evoked potentials. *Psychophysiology* **27**: 195–208.

Linden DEJ, Prvulovic D, Formisano E *et al.* (1999) The functional neuroanatomy of target detection: an fMRI study of visual and auditory oddball tasks. *Cerebral Cortex* **9**: 815–823.

Näätänen R (2001) The perception of speech sounds by the human brain as reflected by the mismatch negativity (MMN) and its magnetic equivalent MMNm. *Psychophysiology* **38**: 1–21.

Picton TW, Lins O and Scherg M (1995) The recording and analysis of event-related potentials. In: Boller F and Grafman J (eds) *Handbook of Neuropsychology*, vol. 10, sect. 14, Johnson R (ed.) Event-Related Brain Potentials and Cognition, pp. 3–73. Amsterdam: Elsevier.

Picton TW, Alain C, Woods DL *et al.* (1999) Intracerebral sources of human auditory evoked potentials. *Audiology and Neurootology* **4**: 64–79.

Picton TW, Alain C, Otten L, Ritter W and Achim A (2000) Mismatch negativity: different water in the same river. *Audiology and Neurootology* **5**: 111–139.

Singer W (2000) Response synchronization: a universal coding strategy for the definition of relations. In: Gazzaniga MS (ed.) *The New Cognitive Neurosciences*, 2nd edn, pp. 325–338. Cambridge, MA: MIT Press.

Teder-Sälejärvi WA, Hillyard SA, Röder B and Neville HJ (1999) Spatial attention to central and peripheral auditory stimuli as indexed by event-related potentials. *Cognitive Brain Research* **8**: 213–227.

Woldorff MG and Hillyard SA (1991) Modulation of early auditory processing during selective listening to rapidly presented tones. *Electroencephalography and Clinical Neurophysiology* **79**: 170–191.

Woldorff MG, Hillyard SA, Gallen CC, Hampson SR and Bloom FE (1998) Magnetoencephalographic recordings demonstrate attentional modulation of mismatch-related neural activity in human auditory cortex. *Psychophysiology* **35**: 283–292.

Further Reading

Hillyard SA, Mangun GR, Woldorff MG and Luck SJ (1995) Neural systems mediating selective attention. In: Gazzaniga MS (ed.) *The Cognitive Neurosciences*, pp. 665–681. Cambridge, MA: MIT Press.

Näätänen R (1992) *Attention and Brain Function*. Hillsdale, NJ: Lawrence Erlbaum.

Näätänen R and Picton TW (1987) The N1 wave of the human electric and magnetic response to sound: a review and an analysis of the component structure. *Psychophysiology* **24**: 375–425.

Picton TW (1990) Auditory evoked potentials. In: Daly DD and Pedley TA (eds) *Current Practice of Clinical Electroencephalography*, 2nd edn, pp. 625–678. New York, NY: Raven Press.

Picton TW (1992) The P300 wave of the human event-related potential. *Journal of Clinical Neurophysiology* **9**: 456–479.

Starr A and Don M (1988) Brain potentials evoked by acoustic stimuli. In: Picton TW (ed.) *Handbook of Electroencephalography and Clinical Neurophysiology*, vol. 3, Human Event-Related Potentials, pp. 97–157. Amsterdam: Elsevier.

Woods DL (1990) The physiological basis of selective attention: implications of event-related potential studies. In: Rohrbaugh JW, Parasuraman R and Johnson R (eds) *Event-related Brain Potentials: Basic Issues and Applications*, pp. 178–209. New York, NY: Oxford University Press.

Woods DL (1995) The component structure of the N1 wave of the human auditory evoked potential. *Electroencephalography and Clinical Neurophysiology* supplement **44**: 102–109.

Auditory Perception, Psychology of

Introductory article

Stephen Handel, University of Tennessee, Knoxville, Tennessee, USA
Mark Hedrick, University of Tennessee, Knoxville, Tennessee, USA

CONTENTS

The production of all sounds creates regularities in the air-pressure sound wave that reaches the listener. The physiological mechanisms and cognitive processes involved in locating and identifying the source of those sounds take advantage of these regularities.

INTRODUCTION

We can identify our friends, rainfall, doorbells, and screeching chalk on a blackboard by sound alone. The key to understanding the diversity of auditory perceptions is to comprehend the complementary nature of the production and subsequent perception of those sounds. The production of all sounds creates regularities in the physical air-pressure wave that reaches the listener. The physiological mechanisms and cognitive processes necessary to locate and identify the source of those sounds take advantage of these regularities.

The first step is to describe the regularities in the sound wave, namely 'what is out there to perceive'. The second step is to describe the physiological adaptations of the auditory system that transform the amplitude of the air-pressure wave into the neural signal which is transmitted to the cortex. The third step is to describe the cognitive heuristics (where a *heuristic* is a rule that usually works but which can lead to a wrong outcome in some instances) that are used to make sense of the neural signal in order to perceive what is happening in the external world.

THE PRODUCTION OF SOUND

Source-filter Model

The basic notion is that a 'source' (e.g., a violin string, or puffs of air coming through the vocal cords) is excited by energy. The source then imposes its vibration pattern on the filter (e.g., the wooden sound body of the violin, or the shape of the mouth), which modifies the vibration of the source into the pressure wave that is radiated to the environment.

The source can vibrate at several component frequencies which are determined by its physical construction, material, and method of excitation. The frequencies of the vibrations usually form a harmonic series (the frequencies of the components are integer multiples of the lowest frequency: 1:2:3:4:5…). The amplitudes of each component vary depending on the above characteristics and, within limits, component vibration starts and ends at the same time. The source vibrations then stimulate the filter. The latter also has specific vibration frequencies (resonances) and is thus capable of being excited when the source vibration frequencies match the filter's vibration frequencies. Owing to the match and mismatch of frequencies, certain source vibrations are transmitted if the source and filter frequencies match, and others are 'stilled' if the frequencies do not match.

Thus the vibrations in the sound pressure wave coming from one source are typically harmonically related (albeit at different amplitudes), start at the same time, evolve slowly across their duration due to frictional decay or changes such as vibrato, and finally end at approximately the same time. All of these regularities can be used to detect and identify the source of the sound.

Transparency of Sound Waves

If only one sound occurred at any particular time, then the problem of auditory perception would be easy. If each unique sound wave represented one source, it would merely be a problem of memorization. Unfortunately, the sound waves coming from two or more overlapping sources (e.g., a radio, street noise, and someone talking) are added together to give one wave in which the individual sound waves from each source are lost in the composite (Figure 1). The frequency components from each sound source are completely intermingled. The unique problem for auditory perception is to untangle this mixture in order to recover the individual waves created by each separate source. We hear each source easily without conscious calculation, but this effortlessness belies the difficulty involved. No computer can do this task. In contrast, nearly all visual objects are opaque rather than transparent, so that the light ray from each point in space invariably comes from light reflecting off just one object.

The goal of the auditory system must be to analyze the composite sound pressure wave into its frequency components, and to assign sets of components to different sound sources. The heuristics used to make this assignment are based on the regularities inherent in the production of sound, namely that the frequency components from one sound source tend to be harmonically related, start and end at the same time, change in quality slowly and continuously, and occur at one location in space. Thus we would expect that the physiological mechanisms would be designed to analyze the sound wave into the amplitude pattern of its component frequencies, to maintain the onset and offset timing of the components, and to maintain the location of the components. Essentially, this is what the auditory pathways accomplish.

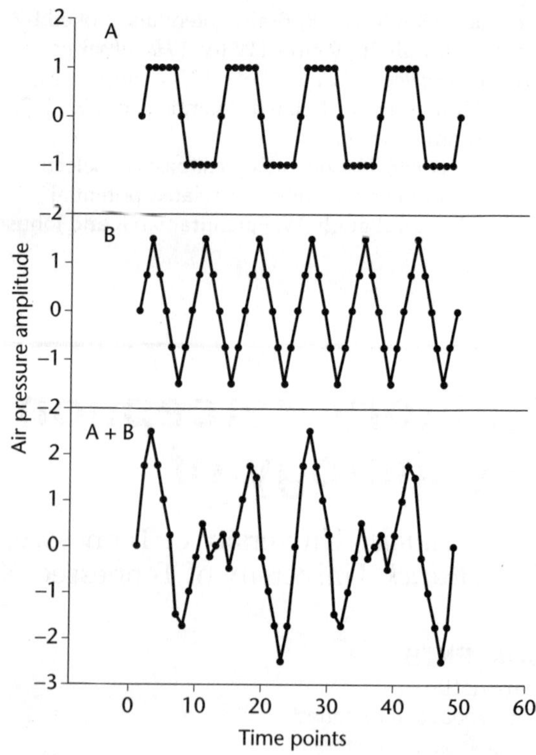

Figure 1. The addition of two waves gives rise to a combination wave that is the sum of the two individual waves. The 'square wave' in A takes 8 time periods to go through one cycle, repeating at 9, 17, 25, and so on. The 'triangle wave' in B takes 12 time periods to go through one cycle, repeating at 13, 25, and so on. The sum of A and B takes 24 time periods to go through one cycle, repeating at 25, 49, and so on.

PHYSIOLOGICAL STRUCTURES

The Ear

The fundamental problem is to convert the changes in air pressure into neural impulses that travel to the auditory cortex. The physical construction of the ear is designed to overcome the large mismatch in physical properties between the air medium of the sound and the fluid-filled medium in the vertebrate inner ear.

The outer ear consists of the visible pinna and the hollow ear canal terminating at the ear-drum that captures sound energy at frequencies mainly used for speech (Figure 2). The middle ear consists of three tiny bones that provide a bridge between the vibration in air at the ear-drum and the fluid vibration at the base of the inner ear. Together, the motion of the bones and the relative size of the ear-drum compared with the oval window at the base of the inner ear convert most of the air pressure vibration into fluid vibration. The inner ear consists of the cochlea, which is responsible for transforming the vibration pattern into neural firing (Figure 3). The cochlea is a coiled fluid-filled tube, about the size of a small bean, with a complex membrane that divides the tube into two halves. Vibration of the middle ear bones creates fluid pressure waves that travel down the cochlea,

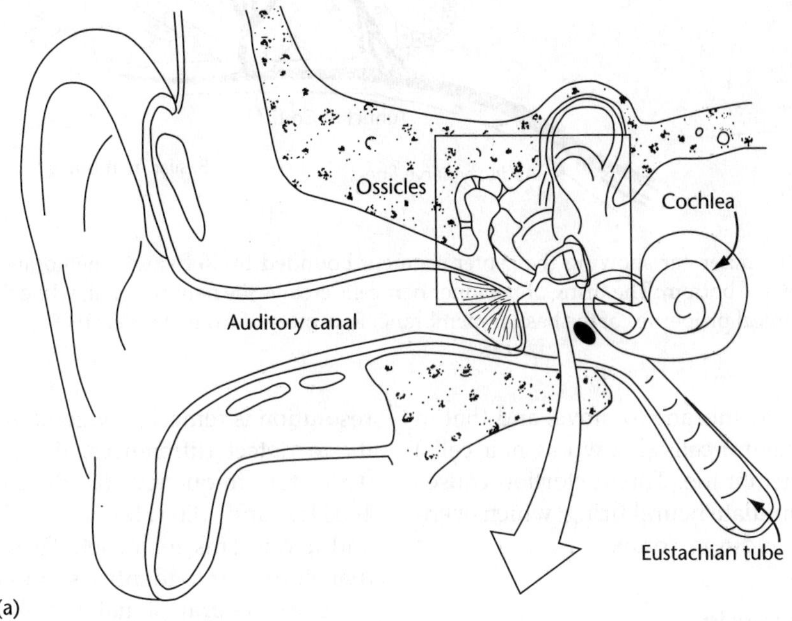

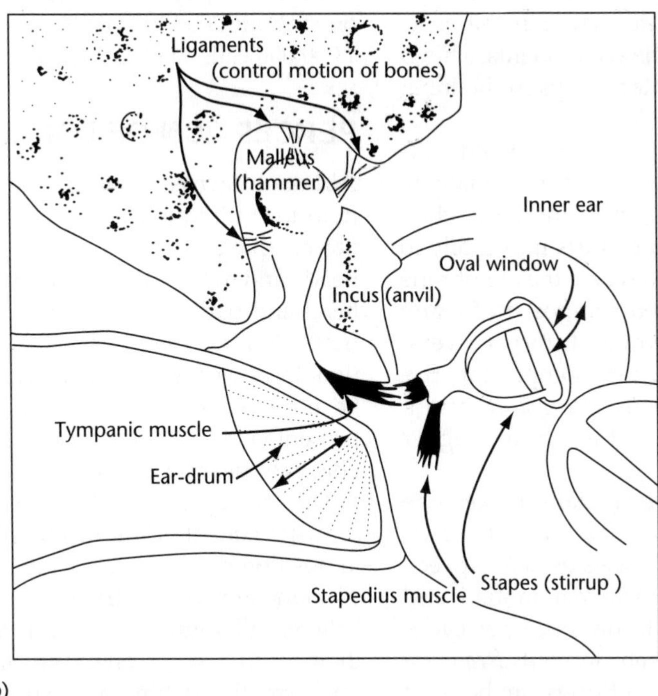

Figure 2. (a) A schematic view of the outer, middle, and inner ear. (b) A detailed view of the middle ear. The middle ear consists of three connected bones, namely the malleus, incus, and stapes. The stapedius and tympanic muscles control the acoustic reflex that protects the ear from loud prolonged sounds by loosening the connections between the middle ear bones. Reproduced from Handel (1989, p. 466) with permission.

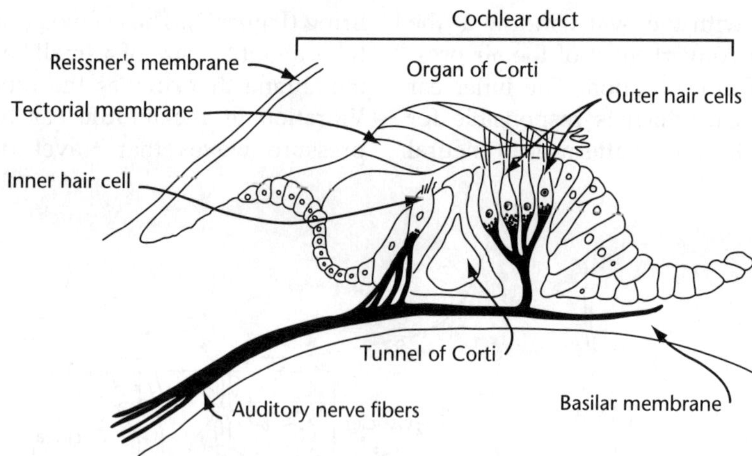

Figure 3. View of the inner ear showing the center segment bounded by Reissner's membrane on the top and the basilar membrane on the bottom. The firing of the inner hair cells creates the neural signal, while the outer hair cells act to change the mechanical properties of the basilar membrane. Reproduced from Handel (1989, p. 471) with permission.

causing the center membrane to move, and that in turn distorts the neural cells (known as hair cells) attached to the membrane. This distortion causes the hair cells to stimulate neural firing, which eventually reaches the auditory cortex.

Frequency Analysis

The goal of the auditory system is to isolate the frequency components and to maintain the onset and offset timing between the components. This is achieved in two complementary ways in the inner ear.

First, the membrane changes its shape and stiffness along its length so that it distorts maximally to higher frequencies of vibration at the base, close to the vibrating bones, and it distorts maximally to lower frequencies at the apex, where the membrane ends. For a complex vibration composed of many frequencies, the membrane would distort at several points, each representing one component frequency. This has been termed *place coding*, and a larger amount of distortion will result in a higher rate of neural firing.

Secondly, the membrane essentially makes one 'up-and-down' movement for one cycle of the vibration. For most of the frequencies used for speech and music, the hair cells track that motion and cause the attached neurons to fire once per cycle. This firing pattern has been termed *frequency coding*, and the timing between firings can be used to code frequency. If the hair cells fire at the same point in the movement, this is termed *phase-locking*.

Even though the inner ear seems to be physically rather crude, its frequency, intensity and temporal resolution is remarkably good. In terms of the ability to detect differences, the resolution is about 0.5% for frequency (a difference of between 1000 Hz and 1005 Hz), about 10% for intensity, and about 2 ms for onset. These values are more than sufficient to identify sound objects.

As the neural signal travels to the cortex, it passes through many brain nuclei, where there is much neural elaboration. Although there are only about 2000 hair cells per ear, there are about 1 000 000 cells in each side of the auditory cortex.

PERCEPTION OF LOCALIZATION

Although each ear analyzes the pressure wave only in terms of frequency, the auditory image is one of objects that appear to be fixed in space outside the head, and which do not appear to move despite head and body movements. It is the human body that generates the physical cues to object localization. If we were simply points in space, there would be no way of localizing sounds, and if sound is presented directly to the ears using headphones, objects appear inside the head and not out in space.

The perception of direction depends on two acoustic cues that are produced by the difference in position of the two ears. First, with the exception of sounds exactly in front of or exactly behind the listener, all sounds will reach the near ear before they reach the far one. The time differences are quite small, reaching a maximum of about 0.5 ms if the sound is directly opposite one ear. Secondly, the head casts a sound shadow so that the intensity of the sound reaching the far ear will be less than that of the sound reaching the near ear (Figure 4).

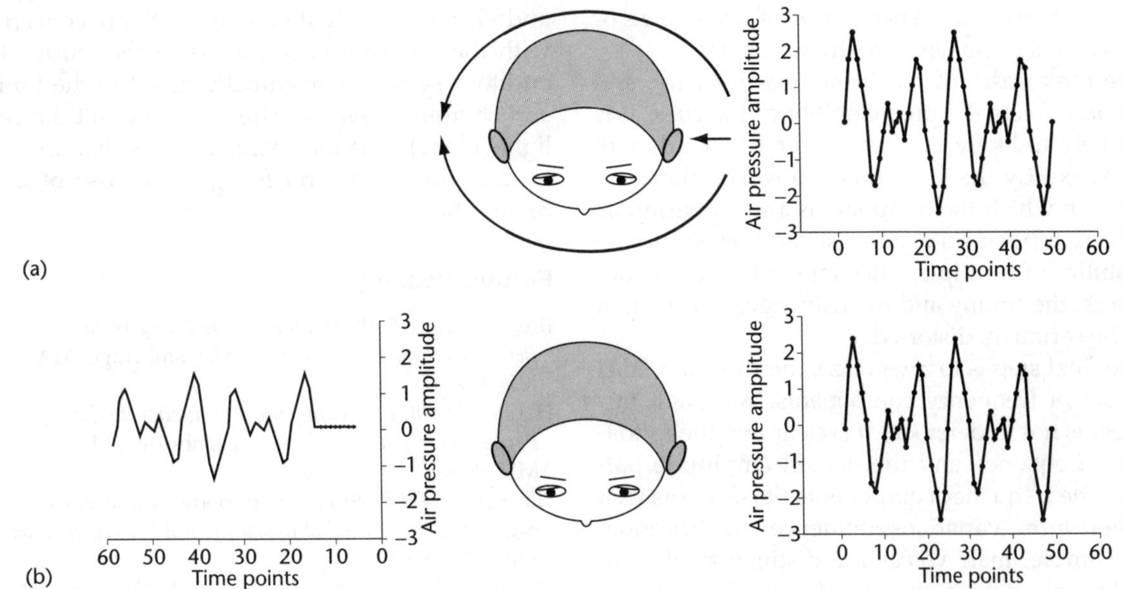

Figure 4. Sound localization. (a) A sound pressure wave is directly opposite the left ear. The wave travels directly to the left ear, and circles the head to travel to the right ear. (b) The wave to the left ear is unchanged, but the wave arriving at the right ear is delayed in time and reduced in amplitude. Note that the right ear wave is reversed in time.

Both time and intensity differences are used to assess direction. Time differences appear to be more important for frequencies lower than 1500 Hz, whereas intensity differences appear to be more important for higher frequencies. The head reflects all frequencies, but at the lower frequencies the sound wave can 'bend' around the head, reducing the intensity difference. At frequencies higher than 2500 Hz the sound waves still bend around the head, but they converge at a distance behind it. Therefore there is a strong intensity shadow that can be used to detect the direction of the sound. There are always ambiguities. For example, a source 45° in front of one ear creates the same time and intensity cues as a source 45° behind the same ear. These types of ambiguities can be resolved by simple head motions.

MAKING SENSE OF THE SOUND WAVE: WHAT ARE THE OBJECTS IN THE WORLD?

To hear the singer of a jazz combo, the listener must group together the frequency components produced by the singer separately from those produced by the instruments. Surprisingly, the most important acoustic cue leading to the grouping of the components is synchrony of the onsets. Remember that the auditory system has very fine temporal acuity – it can distinguish differences of less than 2 ms. Several experiments have placed the various cues in conflict and have found that listeners will group frequency components that begin at the same time, even if the components are not harmonically related or they are presented to different ears. If the components start at slightly different time points, listeners do not report that they hear two sounds starting at different time points. Instead, they simply report hearing two sounds, and they are unable to report the order of those sounds. Thus, curiously, short time asynchronies are converted into source information and not order information.

The second most important cue is that of harmonic relatedness. If all of the frequency components are harmonically related, the obvious decision would be that all of the components come from a single source, and the percept is that of one complex tone. If one of the components is mistuned so that its harmonic relationship is lost (e.g., 200, 415, 600, 800 Hz), the percept is divided – the mistuned component is heard as one emergent tone, and the remaining components are heard as a second tone that has changed in sound quality because the mistuned harmonic has been segregated out. If the frequency components are inharmonic (e.g., 100, 125, 200, 250, 300, 375 Hz…), the auditory system partitions the components so that each set is harmonically related – one complex tone based on a fundamental frequency of 100 Hz and a second one based on a fundamental of 125 Hz. For speech, fundamental frequency differences as small as 2% are sufficient to lead to the perception of two vowels or syllables, each spoken by a different voice.

The rank order of synchrony, harmonicity, and location reflects the predictability of the cues. It is extremely unlikely that two different sounds will start at exactly the same time. However, there are sounds in which the components are not harmonic (e.g., any type of static or noise) and, owing to the possibility of multiple reflections off hard smooth surfaces, the timing and intensity cues to direction may be seriously distorted.

The final step is to categorize, identify, and label each set of frequency components. Although this process is not understood, it is clear that the fundamental frequency and the overall amplitude pattern of the frequency components (its *shape*) as well as short-term variations influence identification. For example, male voices are distinguished from female ones on the basis of fundamental frequency, but female (and male) voices are often distinguished within gender on the basis of the pattern of frequency components.

CONCLUSION

Although there are striking differences in the stimulus energies and physiology of the auditory and visual perceptual systems, both are concerned with the description of objects in the world. The auditory system is maximally tuned to the timing and rhythm of events (after all, we don't dance to lights alone), and the cognitive rules that are used to interpret the neural firings make use of those regularities.

Further Reading

Bregman AS (1990) *Auditory Scene Analysis: the Perceptual Organization of Sound*. Cambridge, MA: MIT Press.

Handel S (1989) *Listening: an Introduction to the Perception of Auditory Events*. Cambridge, MA: MIT Press.

Hartmann WH (1996) Pitch, periodicity and auditory organization. *Journal of the Acoustical Society of America* **100**: 3471–3502.

McAdams S and Bigand E (eds) (1993) *Thinking in Sound*. Oxford, UK: Clarendon Press.

Moore BCJ (1997) *An Introduction to the Psychology of Hearing*, 3rd edn. London, UK: Academic Press.

Taylor C (1976) *Sounds of Music*. New York, NY: Scribner's Son.

Warren RM (1999) *Auditory Perception: a New Analysis and Synthesis*. Cambridge, UK: Cambridge University Press.

Yost WA (2000) *Fundamentals of Hearing*, 4th edn. San Diego, CA: Academic Press.

Autism

Introductory article

Karen Pierce, University of California, San Diego, California, USA
Eric Courchesne, University of California, San Diego, California, USA

CONTENTS

Autism is a developmental disorder, five times more common in males, with clinical onset during the first years of life. It is characterized by abnormalities in social behavior, language, and cognition, and is now known to be biological rather than psychogenic in origin.

INTRODUCTION

Autism is a pervasive developmental disorder, affecting males five times as often as females, with clinical onset during the first years of life. The prevalence is approximately 1 out of every 600 live births. This biological disorder is characterized by abnormalities in social behavior, language, cognition, and environmental interests that persist throughout the affected individual's lifetime. Although symptoms are probably mediated by diverse brain defects, no physical abnormalities are apparent in individuals with this disorder.

BEHAVIORAL SYMPTOMS OF AUTISM

Social Behavior

One of the first indicators of social abnormality in autism is a deficit in the ability to engage in joint social attention, which in normal infants is the merging of attention between two people and an object or activity of interest. Joint social attention is normally achieved by 14 months of age, and is an important precursor to both social and language development. The absence of such a skill in autism has thus been implicated as fundamental to the cascade of developmental problems that ensue. As the autistic toddler matures, other social abnormalities become apparent, such as low rates of eye contact and reciprocal social interaction, and difficulties in identifying and interpreting the emotions of others. Social abnormalities in autism have often been referred to as the hallmark of the disorder, probably because such abnormalities are not only severe, but also obvious in affected individuals.

Language

Approximately 50% of all individuals with autism fail to develop functional speech; in those who do, language is characterized by several abnormalities including pronoun reversals (saying 'he went to the market' instead of 'I went to the market'), use of neologisms (nonsensical or made-up words), stereotyped or rigid speech and abnormalities in intonation. The speech of autistic individuals is also characterized by echolalia, which is the repetition of words either immediately after someone has spoken them or after a delay of hours, days, or even months. For example, an autistic individual may repeat the phrase 'How old are you?' hundreds of times in a single day, after hearing the phrase only once.

Cognition

Seventy-five per cent of autistic individuals also suffer from mental retardation, ranging from mild to severe. Since social interactions and language ability are important for learning about the world and developing cognitive skills, it has been difficult to establish whether impaired mental development in the autistic child occurs independently of the main symptoms of autism (social and language impairment) or occurs in part because of the early failure of normal social and language skills. Although it is difficult to disentangle secondary effects of early social and language impairment from mental retardation *per se*, some scientists believe that deficits in higher-order memory abilities, conceptual reasoning (e.g., categorization skills), executive function (e.g., switching between two or more mental sets) and auditory information processing may exist as important features of this disorder.

One robust finding, however, has been in the domain of attention. Behavioral as well as functional neuroimaging studies have consistently demonstrated dysfunction in three primary attention abilities: disengaging attention from one source of information, orienting attention to a new source, and shifting attention back and forth between two separate sources of stimulus information. Autistic individuals are slow and inefficient in each of these abilities. Also, they are unable to properly adjust their 'spotlight of attention' so that they may have an excessively narrow focus of attention on visual details or an abnormally broad focus. Understanding attention deficits in autism is important because the ability to attend is required for an infant to follow the rapid and unpredictable ebb and flow of human social activity, such as words, gestures, and facial expressions. Dysfunctions in attention thus significantly interfere with the general ability to learn, as well as being likely to amplify other areas of difficulties for autistic individuals, such as language and social behavior.

Restricted and Repetitive Interests

Individuals with autism commonly display restricted, repetitive and stereotyped patterns of interests and activities. This general category of behavior manifests itself in many ways, such as an inflexible adherence to specific routines, stereotyped and repetitive motor mannerisms such as hand-flapping (Figure 1) or a preoccupation with an object or part of an object. In general, autistic individuals display a limited interest in their environment, instead focusing their attention on one specific aspect or obsessive idea (e.g., amassing facts about cars). Further, individuals with autism may insist on sameness and show distress over trivial changes in their surroundings, such as movement of a piece of furniture. Such restricted and repetitive environmental interests are likely to interfere with learning and may have significant developmental implications for autistic children, because they may miss many learning opportunities that fall outside their scope of interest. Combined with attention deficits described above, the autistic child is ill-prepared to learn from the environment.

Figure 1. A 7-year-old girl with autism. The left picture illustrates a commonly found repetitive motor behavior in children with this disorder, known as hand-flapping. Other odd hand mannerisms such as finger posturing (right) are also common.

ETIOLOGY

Autism is among the most heritable of neuropsychiatric disorders. Twin studies report pairwise concordance rates as high as 90% for monozygotic (identical) twins but only 5–10% for dizygotic (fraternal) twins, suggesting the disorder is strongly genetic. Studies of the location of chromosomal abnormalities and break points can be extremely useful in the identification and mapping of genes that predispose an individual to disease. Although chromosomal abnormalities have been reported on many chromosomes in autism, the most consistent site is on chromosome 15. For example, reports indicate a duplication in a specific region on this chromosome (15q11–13) in approximately 2–4% of autism cases. This finding has prompted scientists to look closely at this chromosome for a particular candidate gene or genes that might contribute to autism.

Another common approach in genetic studies of autism is the 'genomic screen', where the whole genome in multiplex families (families with more than one person with autism) is screened in order to identify autism susceptibility loci. As with studies of chromosomal abnormalities, genomic screening studies have reported linkage to specific locations on many chromosomes; however, the strongest evidence seems to point to linkage to chromosome 7 and also chromosome 2. Although such linkages are encouraging, the genes related to autism have not yet been identified.

In addition to genetic inquiries, the search for the causes of autism has led some researchers to suggest a viral, toxic, or teratogenic etiology. Although certain viral or teratogenic agents (e.g., environmental toxins, anticonvulsant medicines) are known to produce brain defects similar to some of those present in autism, research has not yet shown these agents to be significant in the etiology of most individuals with autism.

CLINICAL ONSET AND COURSE

There is almost complete scientific consensus that autism is a disorder with biological onset prenatally or shortly after birth, but clinical symptoms may not be recognized until late infancy or early childhood. Prompted by their child's failure to achieve normal developmental milestones in speech, language and social behavior, most parents seek professional help when their child is aged 2–3 years. Documentation of behavioral characteristics prior to this age is therefore sparse. There is, however, some evidence based on retrospective analyses (e.g., videotapes of children at their first birthday parties) of defects in attention patterns as well as in motor behaviors such as walking or crawling in infants and toddlers with autism. As the autistic child develops, the symptoms described above become pronounced, and many children enter a specialized treatment program in the home, clinic or school setting. Treatment may take the form of behavioral intervention (e.g., repeated practice of a target skill, such as pointing, followed by reward), occupational therapy (e.g., integrating fine and gross motor skills) or pharmacological intervention (e.g., drugs that facilitate serotonin transmission). Most children with the disorder live at home with their families and attend ordinary schools, although some children (usually those with severe symptoms, including mental retardation) may be placed in residential settings under the exclusive care of professionals. As adults, however, many with the disorder move into assisted living facilities.

NEURAL DEFECTS

Interestingly, when autism was first described by Leo Kanner in 1943, people formulated the idea that autism was a psychological disorder, caused by poor parenting or other environmental factors. By the 1970s, however, scientists began to investigate biological explanations for the bizarre behaviors noted in this disorder.

Among all types of biological abnormalities in autism, evidence for defects in the structure of the brain is the strongest. This evidence comes from two primary sources: magnetic resonance imaging

(MRI) and autopsy studies. Magnetic resonance imaging is a relatively new technology that allows scientists to image the living brain. The size of specific brain regions in autism can be easily obtained from MRI scans and compared with the brains of nonautistic people. Autopsy cases of autism are rare, but provide essential information not only about size, but also about more microscopic details, such as the type, density or number of individual neuronal cells, that are not possible to obtain with MRI.

Studies show that in autism most major brain structures may be affected (Figure 2); these include the frontal lobes, parietal lobes, cerebellum, brainstem, corpus callosum, basal ganglia, amygdala, and hippocampus. Within the cerebellum and cerebrum, abnormalities have been found in gray matter (where neuronal cell bodies are located) and also in white matter (which contains axons carrying signals from neurons in one location to those in another) in the youngest autistic children. These cerebellar and cerebral growth abnormalities are so commonly found in these patients and so extreme, that quantitative measures of them could potentially be used in clinical settings in conjunction with psychological information to assist in the diagnosis, prognosis and treatment recommendations at the youngest possible ages. Such widespread anatomic abnormality may explain why autism involves pervasive and persistent neurological and behavioral dysfunction. It is important to note, however, that not every autistic person has every neuroanatomic abnormality. For example, MRI studies suggest that approximately half of autistic adults have decreased volume of brain tissue in the parietal lobes, whereas the other half do not. One consistent neuroanatomic finding, however, has been that the cerebellum is abnormal in the majority of individuals with autism, both young and old. For example, in brain autopsy studies, 95% of all autism cases had reduced numbers of cerebellar Purkinje cells, a type of neuron crucial to cerebellar learning functions.

Certain defects in the autistic brain are signs of prenatal maldevelopment, which makes it likely that autism has a biological onset prior to birth. Such signs include, for example, incomplete formation of a group of neurons in the brainstem called the inferior olive, which is part of a learning circuit involving cerebellar Purkinje neurons.

Abnormal brain growth appears to continue after birth. For example, although whole brain volumes appear to be normal at birth, by the time an autistic child is 2–3 years old, the brain is far larger than normal. This pattern of inflated growth early on in the disorder is followed by a period of reduced growth so that eventually the normal brain outgrows the autistic brain. This illustrates the likelihood that neuroanatomic abnormalities in autism both interact and compound over time, and is consistent with the complex symptom profile seen in this disorder.

Two rules should guide interpretation of neuroanatomical findings in autism. First, apparently 'normal' measures of a particular brain structure (e.g., the parietal lobes in about 50% of autistic patients) do not necessarily imply normal function. The results from functional neuroimaging techniques such as functional magnetic resonance imaging (fMRI) suggest that several regions in the autistic brain may be functionally abnormal, in spite of appearing structurally normal. Second,

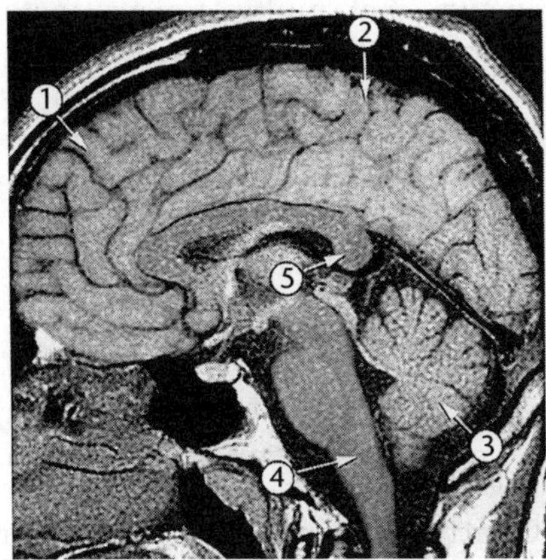

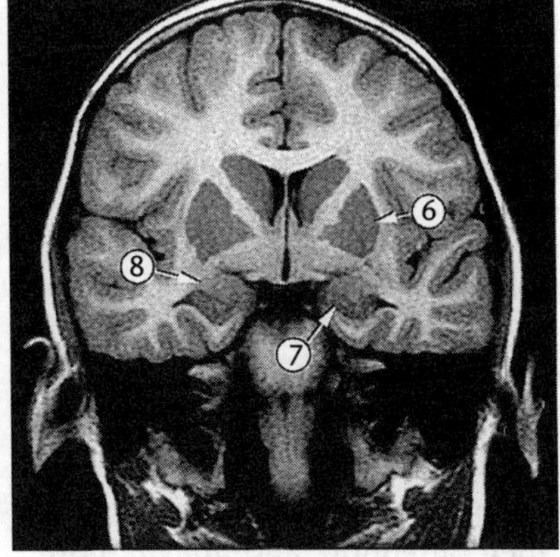

Figure 2. [*Figure is also reproduced in color section.*] Cortical and subcortical structures reported as abnormal in autism from autopsy or magnetic resonance imaging data from laboratories across the USA and Europe. 1, frontal lobes; 2, parietal lobes; 3, cerebellum; 4, brainstem; 5, corpus callosum; 6, basal ganglia; 7, hippocampus; 8, amygdala.

'normal' macroscopic structure does not necessarily mean normal microscopic structure. That is, given that much of what we know about the neurobiology of autism is obtained from MRI – an excellent technique for macroscopic analysis but less suitable for microscopic investigation – subtle defects in structure, such as dendritic or synaptic density, may be missed. As the number of histological cases examined increases, important and detailed questions about the neurobiology of autism will be answered.

Biochemical markers have also recently been reported in autism. For example, increased levels of several brain proteins, specifically neurotrophic factors, have been reported in the blood samples of newborn babies who were later diagnosed with autism. Neurotrophic factors are known to regulate cell growth and proliferation, and thus this finding is provocative in light of reports of increased brain growth early on in the disorder. The relationship between these two biological excesses (increased brain chemicals and increased brain growth) in autism may afford important insights into biologically based treatments and ultimately prevention of the disorder.

In conclusion, developmental brain defects, abnormal brain protein levels from birth, combined with the strong heritability component as shown by twin studies, together provide clear evidence that autism is a biological disorder, not a psychogenic one as was thought in the past.

Relationship Between Neural Defects and Behavioral Symptoms

Only a few studies have investigated the relationship between the brain and behavior in autism. Given the consistency of anatomical defects noted in the cerebellum, it should not be surprising that much of what is known about brain–behavior relationships in autism relates to this structure. For example, it is known that individuals with autism with one type of cerebellar defect, namely smaller vermis lobule VI–VII area measures, are less likely to explore their environment and more likely to engage in repetitive and stereotyped behavior than those with a more normal area measure. Similar relationships between the cerebellum and other behaviors such as attention have also been reported. For example, individuals with autism with more vermis VI–VII lobule abnormality take longer to orient their attention and make more errors when asked to respond to an attention-orienting stimulus than those with less damage. As another example, autistic patients with parietal defects are abnormally slow in disengaging their attention from a source of visual information in order to attend to an unexpected different source of information.

Interesting work has also been done investigating the neural basis of social abnormalities in autism using fMRI. One finding has been that when autistic individuals look at the faces of strangers, neural activity in the expected brain region (i.e., the 'fusiform face area') is drastically reduced. However, when autistic subjects look at the faces of the people closest to them (e.g., mother or classmate), many brain regions are active including those essential to emotion processing. This fMRI evidence suggests that autistic individuals may not be as socially detached as once thought, and is one example of how neuroimaging techniques can add insight into our understanding of the behaviors found in autism. Many functional abnormalities, however, have been found in the autistic brain when processing more complex social stimuli, and many believe that social dysfunctions are related to abnormalities in the amygdala.

CONCLUSION

The constellation of symptoms that constitute autism are severe and affect almost every domain of functioning including language, cognition, social behavior, and environmental interests. Such symptoms are likely to be mediated by equally complex and diverse systems of biological abnormalities. Although treatment efforts have been successful at ameliorating some symptoms for some children with the disorder, currently there is no cure. Significant strides have been made, however, in understanding the neurobiology as well as the etiology of this disorder. For example, it is now known that multiple neuroanatomical sites are affected both structurally and functionally in autism, brain chemical levels are abnormal at birth, and some individuals present defects at the chromosomal level. Such insights may lead to biologically based interventions, and ultimately prevention of this disorder.

Further Reading

Bailey A, Luthert P, Dean A *et al.* (1998) A clinicopathological study of autism. *Brain* **121**: 889–905.

Courchesne E, Yeung-Courchesne R, Press GA, Hesselink JR and Jernigan TL (1988) Hypoplasia of cerebellar vermal lobules VI and VII in autism. *New England Journal of Medicine* **318**: 1349–1354.

Courchesne E, Yeung-Courchesne R and Pierce K (1999) Biological and behavioral heterogeneity in autism: role of pleiotropy and epigenesis. In: Broman S and Fletcher J (eds) *The Changing Nervous System: Neurobehavioral Consequences of Early Brain Disorders*, pp. 292–338. New York, NY: Oxford University Press.

Kanner L (1943) Autistic disturbances of affective contact. *Nervous Child* **2**: 217–250.

Lamb JA, Moore J, Bailey A and Monaco AP (2000) Autism: recent molecular genetic advances. *Human Molecular Genetics* **9**: 861–868.

Lewy AL and Dawson G (1992) Social stimulation and joint attention in young autistic children. *Journal of Abnormal Child Psychology* **20**: 555–566.

Lovaas OI (1987) Behavioral treatment and normal educational and intellectual functioning in young autistic children. *Journal of Consulting and Clinical Psychology* **55**: 3–9.

Mundy P and Sigman M (1989) Theoretical implications of joint-attention deficits in autism. *Development and Psychopathology* **1**: 173–183.

Pierce K, Müller RA, Ambrose J, Allen G and Courchesne E (2001) People with autism process faces outside the 'fusiform face area': evidence from fMRI. *Brain* **124**: 2059–2073.

Townsend J and Courchesne E (1994) Parietal damage and narrow 'spotlight' spatial attention. *Journal of Cognitive Neuroscience* **6**: 220–232.

Autism, Psychology of

Introductory article

Christopher Jarrold, University of Bristol, Bristol, UK
Francesca Happé, Institute of Psychiatry, London, UK

CONTENTS

Autism is a disorder that has a biological cause but is diagnosed on the basis of problems in socialization, communication, and imagination. Psychological explanations of the condition aim to account for this pattern of behavioral difficulties, and for the strengths that individuals with autism show in other areas.

BACKGROUND

Autism is a disorder that is most often diagnosed in childhood. It affects five to 10 in every 10,000 individuals, and is about four times more common among males than females. It is clear that autism has a biological origin, rather than being the result of psychogenic factors, but as yet the precise cause or causes of autism are unknown. One thing that is clearer is that the condition has a genetic component. This does not mean that it only occurs among children of people who themselves have autism. However, if a mother has one child with autism then the chances of a subsequent child having autism are increased approximately 50-fold – although of course this is still relatively unlikely.

The absence of a clearly specified biological cause of autism means that the condition is currently diagnosed on the basis of individuals' behavior. The term 'autism', which comes from the Greek word *autos* for 'self', was first used to refer to the condition by an American clinician, Leo Kanner. Writing in 1943, Kanner identified the key feature of 'autistic aloneness' among his patients, describing their apparent reluctance to engage in social interaction with other people. At approximately the same time, and independently of Kanner, an Austrian physician, Hans Asperger, identified 'autistic psychopathy' among a sample of children showing similar patterns of social peculiarity.

Difficulties in socialization are now seen as one of the key behavioral features of autism, along with problems of communication and imagination, and this 'triad' of impairments forms the basis of most current diagnostic schemes. Problems in *socialization* are seen in individuals' reluctance or inability to engage in social interactions with others. Some individuals with autism do attempt to talk and interact with other children and adults, but often do so clumsily, without a real awareness of the social nuances that underpin our normal everyday contact with others. Problems of *communication* often take the form of severe language delay. Those individuals with autism who have good language skills still have difficulties in understanding subtle aspects of speech, such as the use of irony, metaphor, or jokes, and as a result can often appear rude or blunt when they speak to people. Finally, problems of *imagination* are reflected in the inflexibility and repetitive nature of thoughts and actions. Younger individuals with autism tend not to engage in pretend play, and may show repetitive behaviors such as hand-flapping or rocking. Older individuals tend to be inflexible and like to stick to precise routines. Individuals with autism often develop intense, narrow interests, usually based around predictable or repetitive themes, such as an interest in bus timetables.

Two other aspects of the condition are also worth noting. First, the majority of individuals with autism suffer from a degree of intellectual handicap, which in some cases can severely affect day-to-day functioning. However, it is certainly possible to have autism and to be of normal, or above average, intelligence. The children identified by Hans Asperger had a higher level of intelligence than those seen by Leo Kanner, and today the term 'Asperger syndrome' refers to a condition on the autism spectrum, but which is not associated with mental disability. However, it is not entirely clear whether these two conditions are really distinguishable, and the characteristics which might differentiate them (lack of language or cognitive delay) are a subject of current debate. A second point to emphasize is that autism is characterized by strengths as well as weaknesses. In addition to their problems, individuals with autism perform well in certain (nonsocial) areas, described below.

THEORY OF MIND

Arguably, the most influential psychological explanation of autism at present is the 'theory of mind deficit' hypothesis. The term 'theory of mind' refers to our normal ability to predict and explain what others are doing on the basis of their beliefs and desires. For example, if we see someone leave their house, walk down the road, stop, check the pockets of their coat, turn around, and re-enter their house, then we assume that they have realized that they have forgotten something which they believe to be still inside the house. On the basis of people's behavior we infer what they are thinking, and equally, if we know what people are thinking we can predict what they will do.

Psychologists often assess individuals' theory of mind by asking them to predict what someone will do on the basis of a 'false belief'. For example, if someone doesn't see that an object is moved from one location to another, they will incorrectly believe that it is still in the original location. Typically developing children younger than around four years of age fail to appreciate that someone in this situation will have a false belief, and instead predict that they will look for the object where the child themselves knows it to be. Around the age of four children come to realize that others can have beliefs which differ from their own – an important marker of theory of mind.

Many studies have shown that individuals with autism have severe difficulties on this kind of task, even when they are much older than four years of age or have intellectual abilities well above the four-year-old level. The majority of individuals fail to appreciate that others can have false beliefs, and those that do pass this kind of test show more subtle problems in 'reading other people's minds'. This has led to the claim that autism is associated with a failure to acquire a theory of mind, or at least, is linked to severely delayed acquisition of this ability.

A major strength of this hypothesis is its potential to explain the triad of symptoms seen in autism. Problems of socialization are the result of a failure to appreciate what motivates social interaction, as most of our dealings with others rest on an appreciation of what people know or don't know, and what they want or don't want. Similarly, communication requires an understanding that others intend to communicate ideas to us, that others' beliefs can be changed by what you tell them, and that others may or may not know what you know. Some of the problems in imagination may arise from an inability to make sense of the mental state of 'pretence'. It has also been suggested that repetitive behaviors may reflect an attempt to impose order on what is, to individuals with autism, a confusing and unpredictable social world. However, there is no direct evidence to support this suggestion, and while this may explain a preference for routines among individuals with autism it is not clear how a 'theory of mind' problem would lead to more basic repetitive behaviors such as hand-flapping.

EXECUTIVE FUNCTION

The theory of mind hypothesis can therefore account for many of the features of autism, particularly problems in socialization and communication, but the explanation it offers for problems of imaginative flexibility is less strong. Instead, lack of imagination and inflexibility appear more likely to be related to problems of 'executive functioning' which are also thought to be associated with the condition. The psychological notion of executive function concerns our ability to control our actions. In the same way that a chief executive in a company guides the 'behavior' of the company by instigating procedures, stopping other actions, monitoring progress, and planning for the future, so we need executive control to allow us to consciously control our actions, and to prevent us from making inappropriate automatic responses. This ability appears to be linked to the frontal regions of the brain, as individuals who have suffered damage to these regions become impulsive and show inappropriate repetitive behaviors. Individuals with autism often show similar problems, and make impulsive and inappropriate responses on the kinds of tasks used to assess executive control. There is also some evidence of frontal lobe abnormalities in individuals with autism (see below).

SAVANT ABILITIES AND CENTRAL COHERENCE

The theory of mind account and executive function theory struggle to explain why individuals with autism show strengths in certain areas, as noted above. One example of this is the 'savant abilities' seen in a small minority of individuals with autism. This term refers to skills that are far superior to what would be expected given an individual's general level of intellectual ability. For example, some individuals with autism can tell you the day of the week that corresponds to any given date – for example, the 3rd of October 1907 was a Thursday – even though they might find simple maths problems difficult. Other individuals with autism have an incredible memory for music and can play a piece note-perfect after hearing it only once. Others have remarkably precise drawing abilities, and can recreate complex scenes accurately from memory.

Savant abilities are seen in only a few individuals with autism, but most, if not all, individuals show strengths in visual and spatial areas, rote memory, or attention to detail. Many people with autism are good at doing jigsaw puzzles, and individuals with autism perform relatively well on psychological tests that require them to look for details in a complicated visual image. These strengths are thought to relate to a particular bias among individuals with autism. When ordinary people are presented with a visual image or a story, they typically remember the overall pattern of the picture, or the gist of the story, at the expense of the details. In other words they focus on the 'whole' and not the 'parts'. This tendency has been termed a 'drive for central coherence'. In contrast, individuals with autism focus on the parts of a stimulus and not the whole, and are said to have 'weak central coherence'. They therefore find it difficult to extract the overall meaning from complex information, but are very good at perceiving and remembering the details of this information.

This bias towards parts rather than wholes may explain some of the savant skills described above. In some of these cases it may be that perceiving information in terms of its constituent parts is the best way to remember it accurately. For example, if an individual perceives and remembers a visual image in terms of every individual line that makes up the picture, rather than by remembering the overall pattern of the image, then they may be able to recreate it in great detail.

NEUROLOGICAL UNDERPINNINGS

The exact brain basis of autism is as yet unclear, despite several decades of investigation. Many brain regions have been proposed as the site of abnormality, and studies in this area have produced somewhat contradictory findings, perhaps due to differences of participants, comparison groups, and techniques employed. As a result, there is as yet little agreement as to what differences might be specific and universal to the brains of people with autism. Current suggestions for the site of key abnormalities include the cerebellum and parietal cortex, which have been linked to attentional control; the amygdala and limbic system, important in processing of emotional information; and the prefrontal cortex, responsible for higher-order control functions such as planning and flexibility. There is also a suggestion of greater cell density and larger, heavier brains in autism, perhaps reflecting a failure of synaptic pruning, thought to be an important part of normal brain maturation. Future progress in this area may be made using functional brain imaging techniques.

Studies of the brain basis of normal theory of mind have begun to suggest specific regions for further investigation in autism. Specifically, medial and orbital regions of the frontal lobes, the amygdala, and temporo-parietal regions have all been shown to be more active during theory of mind tasks than during control tests. Recent research suggests that some of these regions (specifically the amygdala, and the medial-frontal region) are not activated when people with autism attempt theory of mind tasks. Future brain scanning studies with children with autism may clarify the role of these and other brain regions in the development of the psychological characteristics of the condition. At present, however, biological or genetic therapy appears a long way off, and the most effective current interventions are educational and behavioral.

Further Reading

Bailey AJ (1993) The biology of autism. *Psychological Medicine* **23**: 7–11.

Bailey A, Phillips W and Rutter M (1996) Autism: towards an integration of clinical, genetic, neuropsychological, and neurobiological perspectives. *Journal of Child Psychology and Psychiatry* **37**: 89–126.

Baron-Cohen S (1992) The theory of mind hypothesis of autism: history and prospects of the idea. *The Psychologist* **5**: 9–12.

Baron-Cohen S (1995) *Mindblindness*. Cambridge, MA: Bradford Books/MIT Press.

Frith U (1989) *Autism: Explaining the Enigma*. Oxford, UK: Basil Blackwell.

Happé F (1994) *Autism: An Introduction to Psychological Theory*. London: UCL Press.

Happé FGE (1994) Annotation: current psychological theories of autism: the 'theory of mind' account and rival theories. *Journal of Child Psychology and Psychiatry* **35**: 215–229.

Happé F (1999) Autism: cognitive deficit or cognitive style? *Trends in Cognitive Sciences* **3**: 216–222.

Jarrold C, Butler DW, Cottington EM and Jimenez F (2000) Linking theory of mind and central coherence bias in autism and in the general population. *Developmental Psychology* **36**: 126–138.

O'Connor N and Hermelin B (1988) Low intelligence and special abilities. *Journal of Child Psychology and Psychiatry* **29**: 391–396.

Ozonoff S (1995) *Executive functions in autism*. In: Schopler E and Mesibov GB (eds) *Learning and Cognition in Autism*, pp. 199–219. New York, NY: Plenum Press.

Russell J (1997) *Autism as an Executive Disorder*. Oxford, UK: Oxford University Press.

Autobiographical Memory

Intermediate article

David Rubin, Duke University, Durham, North Carolina, USA

CONTENTS

Autobiographical memory refers to the store of memories of events that have happened to an individual.

A TAXONOMY: RECOLLECTIVE MEMORY AND AUTOBIOGRAPHICAL FACTS

There is no universally agreed definition of autobiographical memory, but there is a taxonomy based on philosophical and behavioral considerations developed by Brewer (1986, 1996) that makes many issues clear. Two factors underlie the classification: (1) whether what was recalled was a single or a repeated occurrence and (2) whether the memory involves having an image. What is often called an *autobiographical memory*, a *personal memory* or a *recollective memory* occurs when a single event is recalled that involves an image. Thus you may have a memory of typing an email at a computer on one particular occasion in which you have an image, although it may be quite vague, of the email, the computer and your surroundings. A memory of the same event without the image would be an *autobiographical fact*.

In terms of the remember/know distinction commonly used in experiments of laboratory recognition, you know that you typed the letter but you would not remember, or more precisely you do not recollect, doing so. If the memory involves an image but is not for one instance of typing a particular email, but rather of sitting in your usual surroundings typing an email the way you usually do, you would have a *generic personal memory*. If all that you remember is the fact that you generally type emails on a particular computer, then you would have a *self-schema*. The four types of memories seem different to the lay person and the courts, and for a particular event that is one of a series of similar events these four types of memories can be lost independently either with the simple passage of time or as a result of neurological damage. Experimental psychologists who studied memory strived for a long time not to make introspection or phenomenology a part of the definitions of their terms or objects of study, but as I hope Brewer's analysis demonstrates, it appears to be needed.

RECOLLECTIVE MEMORY: MEMORY FOR THE PERSONAL PAST

From the previous description, autobiographical or recollective memory can be regarded as a small subset of memory – memory for an event that comes with an accompanying image. However, when one considers what is needed in order to have an autobiographical memory, the latter becomes a synthesis of many cognitive systems that are often put to other uses. To have a recollection one needs a memory for some details of an event, a sensory image (usually visual, but often in several modalities), a spatial sense of the location of actors and objects in the event, and usually emotional connotations. One does not need a sense of when the event occurred, as this is inferred or remembered independently of the other types of information just listed (Brewer, 1986, 1996; Larsen *et al.*, 1996). The sensory, spatial and emotional information must be strong enough to lead to the metacognitive judgment that you recollect, as opposed to just know, that the event occurred. Moreover, there is almost always a belief that the event really happened to you – a belief that can exist in the face of counter evidence (Brewer, 1986, 1996). For many autobiographical memories there is a coherent narrative that links the memory to one's self or life narrative and helps to organize it (Conway and Pleydell-Pearce, 2000; Habermas and Bluck, 2000). The memories may arise as a result of a conscious search or unbidden as involuntary memories through associations with ongoing thoughts or environmental cues without effort (Berntsen, 1998). Thus much of our cognitive and emotional ability comes into play in an autobiographical memory.

The visual image, sense of reliving or recollection, and belief that are part of autobiographical memories cause problems in the real world. For example, eyewitnesses and people who recover childhood memories of trauma have a strong belief in the accuracy of their memories, but at least on some occasions these beliefs have been shown to be unjustified. Although autobiographical memory is generally accurate, there are well-documented cases in which it is not. In many cognitive tasks, people use visual imagery as a basic form of mental models – objects and their locations in images can be transformed and manipulated at will. For instance, if you see yourself in an autobiographical memory, you have transformed the view seen out of your own eyes to that seen by an outside observer (Nigro and Neisser, 1983). However, in autobiographical memory people tend to view images as permanent, unchanging and accurate photographs. Thus you can manipulate an image yet later believe that it is unchanged. The sense of recollection and the belief that one's memories are generally true add to this problem.

VIVID OR 'FLASHBULB' MEMORIES

Brown and Kulik (1977) coined the phrase *flashbulb memory* to describe vivid memories of important public events, because it suggests surprise, relatively indiscriminate although not necessarily complete illumination, and brevity. According to Brown and Kulik, flashbulb memories are 'memories for the circumstances in which one first learned of a very surprising and consequential (or emotionally arousing) event' (Brown and Kulik, 1977, p. 73). Two issues raised by Brown and Kulik's paper, namely whether flashbulb memories are different in kind from other autobiographical memories, and whether they are more accurate, have fueled much research. There is no strong evidence that flashbulb memories are a different type of memory to other recollective memories; they might be nothing more than the most extreme case of vivid recollection outside the flashbacks discussed in the section below on memory for traumatic events. Nonetheless, the extreme sense of reliving that characterizes them makes them a special and interesting topic of study and a battleground for the issue of whether recollective memories are accurate. Here it appears that flashbulb memories are often accurate, but that major distortions can occur. A common error involves recalling two different times when one 'first learned' of an event; a relatively minor error in noting where one first heard of the event can result in large differences in the details and even major points of what is recalled. For example, in reporting when they first learned of the explosion of the *Challenger* space shuttle, people who initially reported hearing about it from a person later reported, often with great clarity and confidence, that they initially learned about it by watching television (Winograd and Neisser, 1992).

AUTOBIOGRAPHICAL MEMORY FOR EMOTIONAL AND TRAUMATIC EVENTS

The relationship between emotions and autobiographical memory is complex. The literature on this issue, although as rigorous as any on autobiographical memory, produces few generalizations (Christianson and Safer, 1996). For instance, the flashbulb memory literature shows good recall of the circumstances and context in which a shocking event occurred, while studies of similarly shocking events tend to show poor recall of the context and more focus on and better recall of the central events. Whereas in diary studies pleasant events tend to be much better recalled than unpleasant ones, most studies of emotion and autobiographical memory use unpleasant events, and these studies have found that increased intensity of emotion increases recall. At more extreme levels, such as traumatic memories, concepts from clinical psychology (e.g. dissociation and repression) are often invoked.

Traumatic events often violate one's expectations, making them difficult to integrate into a life narrative. They involve strong emotions, and when they return in memories they can bring with them an overpowering and unwanted sense of reliving in the form of intrusive memories. Thus autobiographical memories for traumatic events are at the extremes of the systems for and phenomenology of autobiographical memory noted earlier. In some case, traumas lead to post-traumatic stress disorder, a syndrome that is defined by reliving, avoidance and arousal symptoms (American Psychiatric Association, 1994). The reliving symptoms can occur in the form of intrusive memories, which makes autobiographical memory part of the diagnosis and, for the patient, part of the problem. The intrusive memories may occur as flashbacks – an extreme form of flashbulb memories which takes the patient back to the time and setting of the original trauma.

THE DISTRIBUTION OF AUTOBIOGRAPHICAL MEMORY OVER THE LIFESPAN

One of the most regular quantitative findings in the literature on autobiographical memory is the distribution of memories over the lifespan. Autobiographical memories for such analyses have been obtained in many different ways, such as having people give their lives in narrative form, just list events, or provide memories of specific types (e.g. important memories, or memories cued by odors). Although the results are similar, most work has been done with word cues, so this approach is the easiest to synthesize over many studies. This method was developed by Galton over a century ago and revived by Crovitz and Schiffman (1974). Individuals give a memory to each of a set of randomly chosen words, and after all autobiographical memories have been obtained, they date them. The distribution of memories over the lifespan of undergraduates can be described as a power function of the time since the event occurred (i.e. $y = at^{-b}$) (Crovitz and Schiffman, 1974). The fits to the curve are surprisingly good, with correlations usually being greater than 0.95. If we assume that undergraduates encode an equal number of events each day of their lives, then the curve is a retention function. Because the power function is a common choice for a retention function, it appears that as a first approximation, laboratory and autobiographical memory have similar patterns of forgetting.

The data for older adults are more complex, requiring three components. The first component is the retention function, which can be described mathematically as a power function, that covers the whole lifespan, but which has its main quantitative contribution over the two most recent decades of life; for periods longer than two decades ago, it is too small to produce a measurable number of memories. The second component is childhood amnesia. This component is very stable over numerous studies, having the same basic shape irrespective of the method used to produce the data, so long as the participants come from the USA (Rubin, 2000). Based on an average of the available data, the percentage of memories from before the age of 8 years for when a person was 0, 1, 2, 3, 4, 5, 6 and 7 years old, respectively, are 0.1, 0.4, 1.7, 5.5, 13.1, 22.1, 27.0 and 30.1% (or only 2.2% before the third birthday). Figure 1 shows this component.

The third component, known as the bump, is an increase in the number of memories from when older adults were between 10 and 30 years of age compared with what would be expected from the other two components or from any monotonically decreasing function. The initial description of the bump in 1986 was based on data from several laboratories. Since that time there have been consistent findings using the word cue technique with older adults (Rubin, 2002; Rubin *et al.*, 1998). Minor differences exist in the shape of the distribution with changes in procedure, but the bump appears

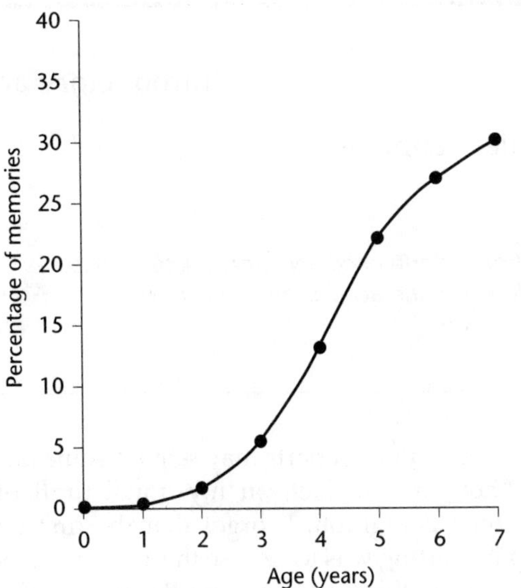

Figure 1. The distribution of 10 118 memories dated as occurring before the age of 8 years from published studies. From Rubin (2002).

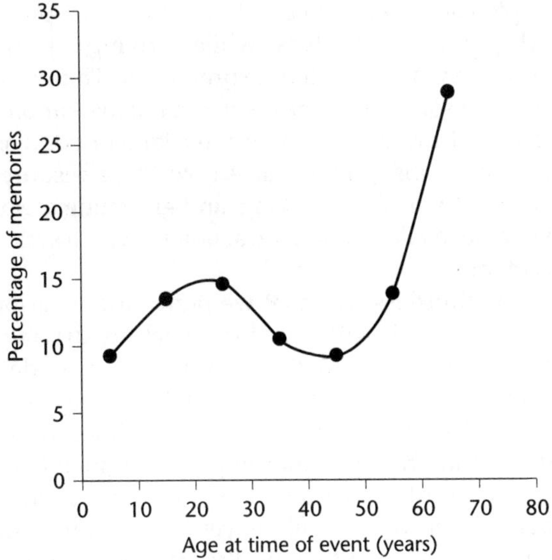

Figure 2. The distribution of autobiographical memories over the lifespan for older adults; events dated as occurring in the most recent year are excluded. From Rubin *et al.* (1986).

repeatedly, even for individuals. Figure 2 shows this component.

References

American Psychiatric Association (1994) *Diagnostic and Statistical Manual of Mental Disorders*, 4th edn. Washington, DC: American Psychiatric Association.

Berntsen D (1998) Voluntary and involuntary access to autobiographical memory. *Memory* **6**: 113–141.

Brewer WF (1986) What is autobiographical memory? In: Rubin DC (ed.) *Autobiographical Memory*, pp. 25–49. Cambridge, UK: Cambridge University Press.

Brewer WF (1996) What is recollective memory? In: Rubin DC (ed.) *Remembering our Past: Studies in Autobiographical Memory*, pp. 19–66. New York, NY: Cambridge University Press.

Brown R and Kulik J (1977) Flashbulb memories. *Cognition* **5**: 73–99.

Christianson S-A and Safer MA (1996) Emotional events and emotions in autobiographical memory. In: Rubin DC (ed.) *Remembering our Past: Studies in Autobiographical Remembering*, pp. 218–243. Cambridge: Cambridge University Press.

Conway MA and Pleydell-Pearce CW (2000) The construction of autobiographical memories in the self-memory system. *Psychological Review* **107**: 261–288.

Crovitz HF and Schiffman H (1974) Frequency of episodic memories as a function of their age. *Bulletin of the Psychonomic Society* **4**: 517–551.

Habermas T and Bluck S (2000) Getting a life: the emergence of the life story in adolescence. *Psychological Bulletin* **126**: 748–769.

Larsen SF, Thompson CP and Hansen T (1996) Time in autobiographical memory. In: Rubin DC (ed.) *Remembering our Past: Studies in Autobiographical Memory*, pp. 129–156. New York, NY: Cambridge University Press.

Nigro G and Neisser U (1983) Point of view in personal memories. *Cognitive Psychology* **15**: 467–482.

Rubin DC (2000) The distribution of early childhood memories. *Memory* **8**: 265–269.

Rubin DC (2002) Autobiographical memory across the lifespan. In: Graf P and Ohta N (eds) *Lifespan Development of Human Memory*, pp. 159–184. Cambridge, MA: MIT Press.

Rubin DC, Rahhal TA and Poon LW (1998) Things learned in early adulthood are remembered best. *Memory and Cognition* **26**: 3–19.

Winograd E and Neisser U (eds) (1992) *Affect and Accuracy in Recall: Studies of 'Flashbulb' Memories*. New York, NY: Cambridge University Press.

Further Reading

Conway MA (1990) *Autobiographical Memory: An Introduction*. Milton Keynes, UK: Open University Press.

Conway MA (1995) *Flashbulb Memories*. Hove: Erlbaum.

Conway MA, Rubin DC, Spinnler H and Wagenaar WA (eds) (1992) *Theoretical Perspectives on Autobiographical Memory*. Dordrecht, The Netherlands: Kluwer Academic Publishers.

Rubin DC (ed.) (1986) *Autobiographical Memory*. Cambridge, UK: Cambridge University Press.

Rubin DC (ed.) (1996) *Remembering our Past: Studies in Autobiographical Memory*. New York, NY: Cambridge University Press.

Winograd E and Neisser U (eds) (1992) *Affect and Accuracy in Recall: Studies of 'Flashbulb' Memories*. New York, NY: Cambridge University Press.

Automaticity

Introductory article

Thomas J Palmeri, Vanderbilt University, Nashville, Tennessee, USA

CONTENTS

Automaticity refers to the way we perform some mental tasks quickly and effortlessly, with little thought or conscious intention. Automatic processes are contrasted with deliberate, attention-demanding, conscious, controlled aspects of cognition.

INTRODUCTION

Try to think back to when you first learned how to drive a car. Your primary aim was to steer the car clear of other vehicles, pedestrians, and trees – a difficult task by itself. But you also had to control the pressure applied to the accelerator pedal to keep within posted speed limits. You needed occasionally to apply the brake to obey traffic signals and to avoid plowing into the car in front of you. Added to this, if you first learned to drive a car with a manual transmission, you had to decide when to change gear and then you needed to coordinate the complex movements involved in doing so – releasing the accelerator pedal, depressing the clutch, shifting to the appropriate gear, carefully releasing the clutch while applying some gas. And you had to do this while continuing to pay attention to the road ahead. On top of that, you probably had to linguistically process the commands, pleas, and screams of the poor soul who (perhaps regrettably) agreed to teach you how to drive. You had to direct all your mental energies to controlling and coordinating the complex sequence of movements involved in safely driving a car. Trying to simultaneously steer, accelerate, brake, shift, and listen was an exceedingly difficult task.

Contrast this scenario with how you may be able to drive after many years of practice. On long trips, you find yourself daydreaming and may not even remember what happened during the last several uneventful miles of highway driving. Shifting gears becomes one smooth continuous action. Indeed, breaking up this complex action into its component parts may require some deliberate thought – in fact, on my initial draft of the previous paragraph, I forgot that the first critical step in shifting was to release the accelerator pedal; this is something I have done thousands of times during my twenty years of driving cars with manual transmissions, but this basic action did not initially come to mind when I tried consciously to decompose the act of shifting gears. Experienced drivers use so few mental 'resources' that some people can drink coffee, talk on a cellular phone, and groom themselves while driving at high speeds on a congested expressway. Things are fine until something unexpected happens – another distracted driver veers into their lane or someone stops very abruptly ahead – now those resources diverted to drinking, talking, and grooming are not available to take immediate action to avert a serious accident.

That effortless way that we perform the various components of skilled actions, such as driving a car, is termed *automaticity*. Many routine daily events become so automatic that we may seem unconscious of them – how many times have I lathered my hair this morning, did I remember to put the freshly ground coffee in the coffee maker, have I checked my mailbox yet this morning? Literate adults read automatically – try not to read the billboards and signs that bombard you when driving through suburban commercial developments. When skilled at playing a musical instrument, reading musical notation, translating notes into finger and hand movements, controlling breathing and embouchure (mouth position), are all automatized procedures, allowing the musician to focus on higher levels of musicality such as style, phrasing, and coordination with the conductor and other musicians. Skilled professionals automatically execute complex tasks that demand years of training. Experienced radiologists may be able to tell automatically, at a glance, whether a patient has a benign growth or a malignant tumor.

Experienced pilots control complex aircraft automatically. Landing a commercial jetliner in good weather is performed with nearly the same fluency as driving to the neighborhood grocery store. This automaticity allows the pilot to monitor for unexpected events – an unauthorized aircraft on the runway, an approaching flock of geese, an engine fire, or wind sheer – and be able to take corrective action immediately to avert potential disaster.

This article describes the properties that distinguish automatic processes from those that require conscious mental control, describes factors necessary for achieving automaticity, illustrates the effects of automaticity with some classic experimental paradigms, and describes some psychological models of the acquisition of automaticity.

CHARACTERISTICS OF AUTOMATIC PROCESSES

A number of characteristics have been emphasized to distinguish *automatic processes* from those that require some kind of overt mental control, what have been referred to as *controlled processes*. Theorists disagree on what particular characteristics are most important for describing a process as being automatic, and disagree on whether some particular properties appropriately characterize automaticity at all. In addition, some theorists have argued that perhaps the concept of automaticity itself should be abandoned entirely, since no cognitive process is ever truly automatic given most lists of critical characteristics. Automaticity is a current topic of active research in the cognitive sciences, and ideas of how best to characterize automatic processing are still evolving.

Table 1. Some proposed characteristics of automatic and controlled processes

Automatic processes	*Controlled processes*
obligatory	intentional
stimulus-driven	executive-driven
stereotypic	reconfigurable
rigid	flexible
no monitoring	monitoring
no dual-task interference	dual-task interference
parallel	serial
well practiced	novel
expert	novice
fast	slow
effortless	effort
unconscious	conscious
no attention	attention

The aim of this section is to survey most of the various characteristics of automaticity that have been proposed. These characteristics, summarized in Table 1, will be elaborated upon below. These characteristics should certainly not be considered independent dimensions of automatic processes because many of them may overlap in some respects.

- Automatic processes are *obligatory*. Given the presence of particular stimuli within particular contexts, automatic processes can execute without the conscious intention of the individual. Automatic processes seem to occur reflexively. Controlled processes require conscious intention to become initiated.
- For this reason, automatic processes are said to be *stimulus-driven*. Given the appropriate triggering conditions, automatic processes execute without intention. Controlled processes are intentionally initiated by the individual, often with the guidance of central executive processes.
- Automatic processes are often *rigid* and *stereotypic*. Controlled processes can be reconfigured to deal with novel events, allowing for a far greater degree of flexibility.
- Once initiated, automatic processes require *no monitoring*. They run to completion without any need for overt executive control. Controlled processes require monitoring, and distractions can lead to breakdowns in performance.
- Automatic processes are *free from dual-task interference*; that is, automatic processes are not influenced by other tasks that are executed concurrently. Controlled processes suffer from dual-task interference. It is often extremely difficult to perform more than one controlled process at the same time.
- Because automatic processes can execute simultaneously, they are said to be processed in *parallel*. Not only can independent automatic processes be executed in parallel, but the various component processes of a complex skill may overlap one another in a parallel manner. Controlled processes execute serially. They are processed one step at a time and cannot be processed simultaneously.
- Many automatic cognitive processes are *well practiced*. Controlled processes may be novel or less practiced.
- Automatic processes often characterize *expert* performance. Controlled processes often characterize novice performance.
- Because automatic processes can be performed in parallel without conscious monitoring, automatic processes are often *fast* compared to controlled processes.
- Automatic processes seem *effortless*. Controlled processes require mental effort.
- Automaticity is often discussed in the context of consciousness. Automatic processes may be *unconscious*. Controlled processes are conscious.
- Automaticity is also often discussed in the context of attention. Automatic processes may require *no attention*. Controlled processes do require attention.

FACTORS NECESSARY FOR AUTOMATIC PROCESSES

Some processes may be automatic because the human brain is equipped with special-purpose neural mechanisms for carrying out certain critical aspects of perception and cognition. Such automatic processes are obligatory because a specialized neural 'module' operates autonomously, triggered by particular stimulus events in the environment. These are hard-wired mechanisms, making them rigid and stereotypic. Because these modules operate independently, they are not influenced by other concurrent processes operating within other parts of the brain, they do not require monitoring, they operate unconsciously, and they require no overt deployment of attentional resources.

Let us illustrate an example of an automatic process that may reflect the operation of one such hard-wired perceptual mechanism. Search for a yellow **X** in each panel of Figure 1. The target automatically 'pops out' from the distractors in the left panel but an active search is required to locate the target among the distractors in the right panel. In the left panel, the yellow **X** differs from the red **X**s by a single feature, but in the right panel, the yellow **X** differs from the yellow **O**s and red **X**s by the particular combination of color and form features. Salient singleton features are thought to pop out automatically because of the way early stages of the visual system process elementary visual information. Indeed, visual search tasks have often been used to distinguish automatic, pre-conscious processing of elementary visual features from the more high-level, attention-demanding processing of conjunctions of multiple visual features necessary for object recognition. Similarly, we may also automatically notice other kinds of perceptual events such as abrupt onsets of visual stimuli (a flash of lightning), auditory stimuli (a clap of thunder), or somatosensory stimuli (a crawling insect), because our perceptual systems may be hard-wired to process sudden unexpected changes in the environment automatically. So some aspects of perception and cognition may be automatic, and truly reflexive in nature, because there exist special-purpose neural mechanisms that operate autonomously, below the level of conscious awareness and control.

Clearly there do not exist innate hard-wired mechanisms for reading a book, driving an automobile, or flying an airplane. Yet people can become automatic at the elements of these tasks with sufficient practice. Therefore, a great deal of automaticity must be learned. How can a process go from being one that requires overt cognitive control to one that is automatic? And are there limitations on what kinds of tasks can become automatized?

For most aspects of human cognition that can become automatized, no one achieves automaticity without a great deal of practice. But some tasks may become automatized more quickly than others. Clearly, a simple task may be automatized more quickly than a complex task. Some real-life tasks may take only a few hours of practice to become automatized. Others require many years of training. But complexity is not the only factor, nor the most critical factor in determining how rapidly a task can become automatized.

To illustrate, let us consider another example of a search task that has been used to study the development of automaticity. The visual display can contain between one and four letters (call this variable the display size, D). You can be asked to search for between one and four possible target letters (call this variable the memory set size, M). The task is to decide whether a target is present or absent in each display as quickly as possible without making any errors. The time to make a correct response will be recorded. A display size of one (D = 1) and a memory set size of one (M = 1) is a relatively easy search. As shown in the top of Figure 2, if the target is an **X**, the 'search' is simply a matter of deciding whether the single presented letter on each display is an **X** or not. As the number of items in the display is increased, the task gets harder, and as the number of items in the memory set increases, the task gets harder. As shown in the bottom of Figure 2, suppose I tell you that the target memory set is now **T**, **L**, **Z**, and **V** (M = 4). Each display will contain four letters (D = 4) and you must decide if any of those four letters is one of the four target letters in the memory set. This search is quite hard. To accomplish this task, people generally search through each item in the display one at a time and compare it with each item in the memory set one at a time until a target is found. As such, search times increase systematically as a function of both the display size and the memory set size. This is a slow, deliberate, attention-demanding, serial search process.

Can this controlled search become automatized through training? Imagine that the set of targets and distractors changes throughout training such that a target on one trial may be a distractor on another trial. In such *varied mapping* conditions, it is very difficult, if not impossible, for the search task ever to become automatized, even with

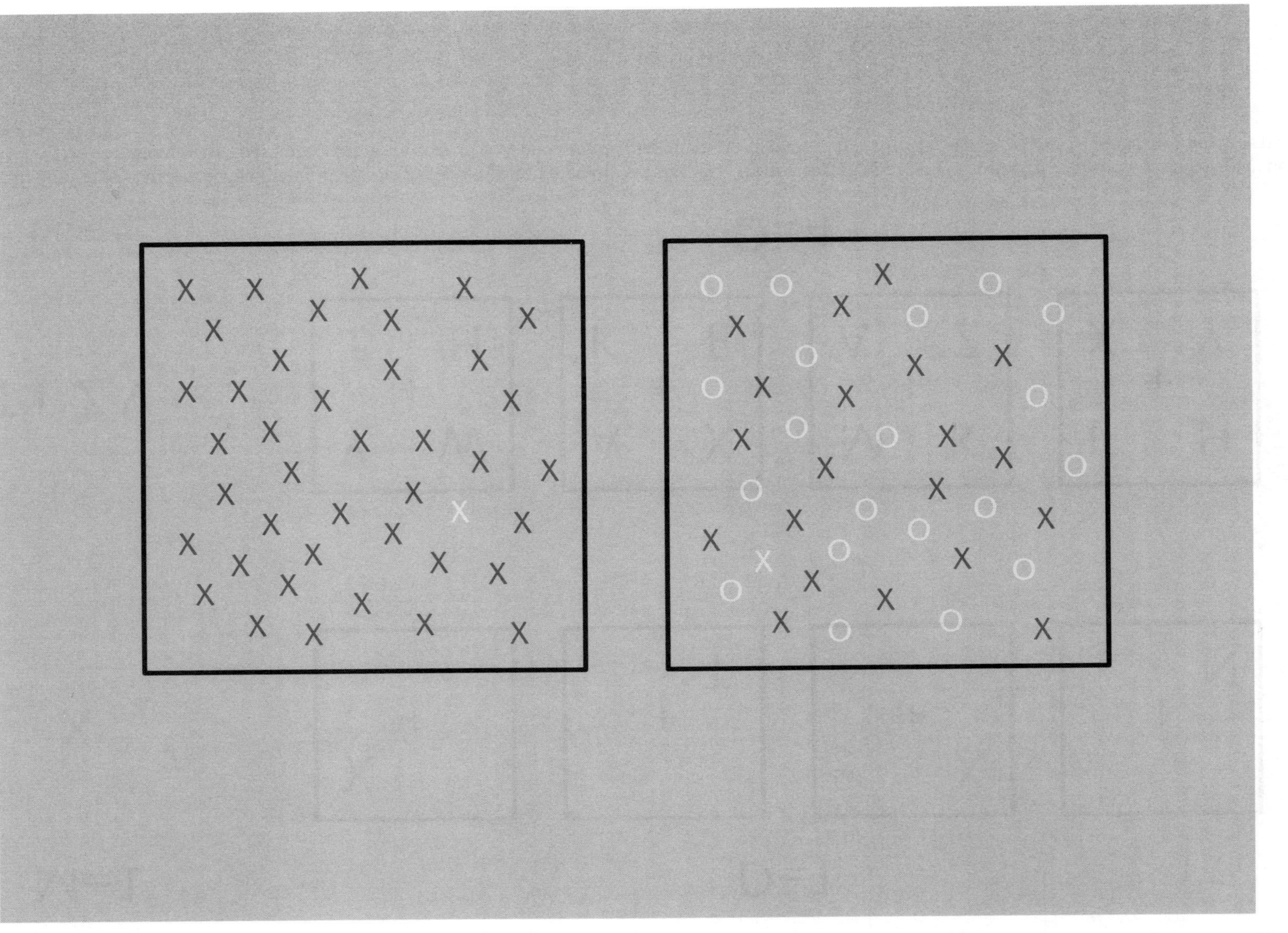

Figure 1. [*Figure is also reproduced in color section.*] Find the yellow **X**. The left panel illustrates a feature search task in which the target automatically 'pops out' from the field of distracters. The right panel illustrates a conjunction search task in which the target must be actively searched for with deliberate shifts of attention.

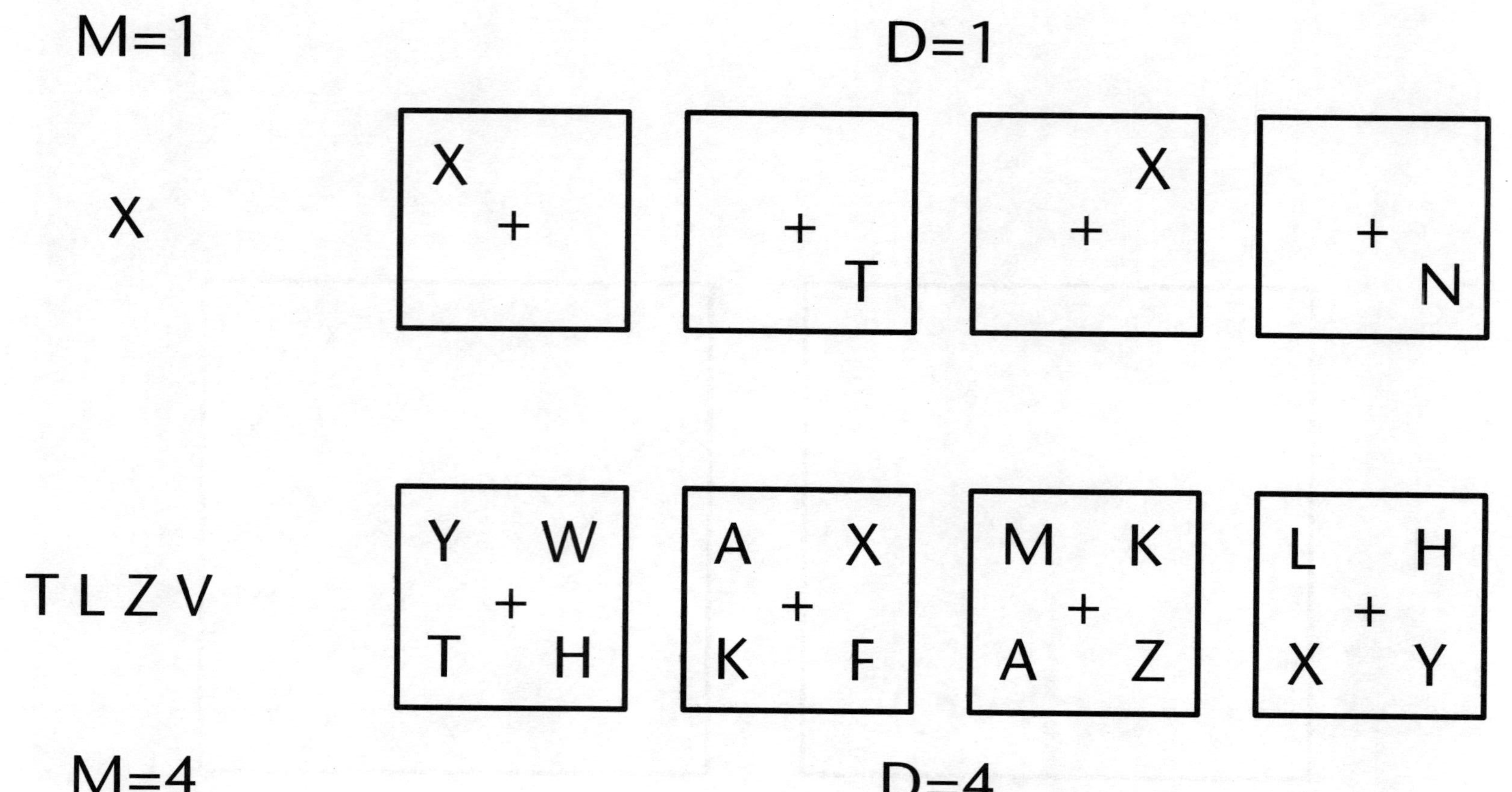

Figure 2. Illustration of search task that manipulates display size (D) and memory set size (M). The top search has a single target (M = 1) and a single display item on every trial (D = 1). The bottom search has four possible targets (M = 4) and four display items on every trial (D = 4). The task is to detect a target as quickly as possible without making errors.

extended practice over several weeks. So practice by itself is not guaranteed to produce an automatic process.

Instead imagine that the set of targets and distractors remains consistent, such that the targets must be drawn from one set of letters and the distractors must be drawn from a different set of letters, and this differentiation is maintained throughout the entire course of training. In such *consistent mapping* conditions, automaticity can be achieved with practice. Indeed, after extended practice, the time taken to search for targets does not vary with display size or memory set size. That is, a target pops out from the display much like the color pop-out shown in the left panel of Figure 1. But this is a learned automaticity, not a hard-wired one. This automaticity is immune to dual-task interference. This automaticity is rigid and inflexible in that switching to a varied mapping condition causes the search to revert back to a slow, deliberate, attention-demanding, serial process. Moreover, switching targets to distracters and distracters to targets causes performance to become even worse than it was before any training whatsoever, and it takes a long time to 'unlearn' the original automatization of target searches. So one important criterion for developing automaticity is that there is a consistent mapping between stimuli and responses. This may be one reason for the stimulus-driven nature of much automatic processing.

STROOP INTERFERENCE AND OTHER RELATED MEASURES

A different manifestation of automaticity can be seen in the classic Stroop effect. Named after John Ridley Stroop, the psychologist who developed it as part of his doctoral dissertation in the 1930s, the Stroop task has been used in thousands of experiments to study automaticity. First, find a stopwatch or a clock with a second hand. Now, time how long it takes you to *name the ink color* of the words in the first column of Figure 3 (i.e. BLUE, RED, PURPLE, etc.) – name the ink color, don't read the words. Next, time how long it takes to name the ink colors in the second column (i.e. RED, BLUE, ORANGE, etc.). And then do the same with the third column (i.e. PINK, RED, YELLOW, etc.). In all cases, try to respond as quickly as possible without making errors.

The classic Stroop interference effect is that the identity of the word can have a large effect on the speed of color naming. In the first column, the words themselves have no color association. In the second column, each word is congruent with its ink color, such as 'red' in RED ink or 'green' in GREEN ink. People are generally a bit faster to name the colors in the second column (congruent condition) than to name the colors in the first column (control condition). In the third column, each word is incongruent with its ink color, such as 'red' in GREEN ink or 'blue' in YELLOW ink. People are generally far slower to name the colors in the third column (incongruent condition) than to name the colors in the other columns. In the original paper by Stroop, subjects took nearly twice as long to name colors in the incongruent condition than in the control condition, a finding that has since been replicated thousands of times across numerous experimental variations. Even without a stopwatch, you surely found naming ink colors in the third column quite difficult and perhaps a bit frustrating. This is the fundamental Stroop interference effect.

Stroop interference is not simply caused by having an incongruency between words and their ink color. Time how long it takes you to *read the words* in the first column of Figure 3 (i.e. TRUCK, HOUSE, GRASS, etc.). Then time how long it takes to read the words in the second column (i.e. RED, BLUE, ORANGE, etc.). Then do the same with the third column (i.e. GREEN, PINK, BLUE, etc.). Ink color has little or no effect on the speed of reading words. Even for color words like 'red', the speed of word reading is not influenced by whether the word 'red' is written in RED ink or GREEN ink.

Stroop interference is asymmetric. In the incongruent condition, words interfere with color naming but colors do not interfere with word reading. One acknowledged explanation for this is that word reading is a more highly automatized process than color naming. Word reading happens rapidly and effortlessly, without conscious intention, and cannot generally be suppressed. Naming colors requires more attention, conscious intention, and effort. Even when the task is to name the colors, and to ignore the words, word reading happens anyway, automatically, and can interfere with color naming.

The Stroop effect is not limited to interference of word reading on color naming. Figure 4 shows incongruent conditions from three variants of the Stroop task. In the first column, the task is either to read the digits (i.e. 4, 3, 5, etc.) or to count the number of digits (i.e. THREE, FIVE, FOUR, etc.). Reading digits is more automatized than counting, so digit identity interferes with counting, but the number of digits does not interfere with digit naming. In the second column, the task is either to

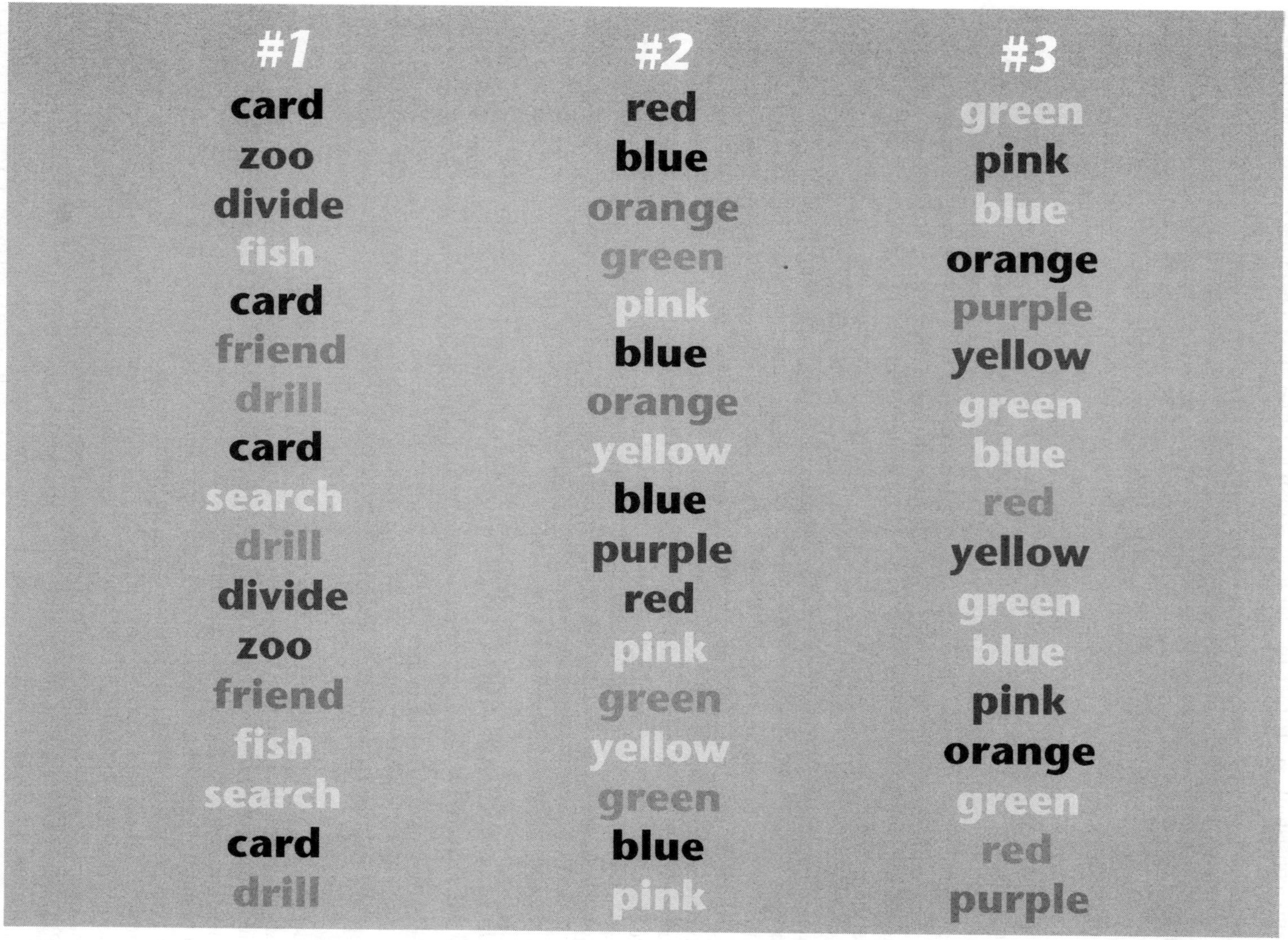

#1	#2	#3
card	red	green
zoo	blue	pink
divide	orange	blue
fish	green	orange
card	pink	purple
friend	blue	yellow
drill	orange	green
card	yellow	blue
search	blue	red
drill	purple	yellow
divide	red	green
zoo	pink	blue
friend	green	pink
fish	yellow	orange
search	green	green
card	blue	red
drill	pink	purple

Figure 3. [*Figure is also reproduced in color section.*] Demonstration of the Stroop task. Using a stopwatch, separately time how long it takes to *name the color* of each printed word in column 1, column 2, and column 3. Then separately time how long it takes to *read each word* in column 1, column 2, and column 3. Column 1 is a *control condition* in which the word and the color bear no relationship. Column 2 is a *congruent condition* in which the word and the color match. Column 3 is an *incongruent condition* in which the word and the color mismatch.

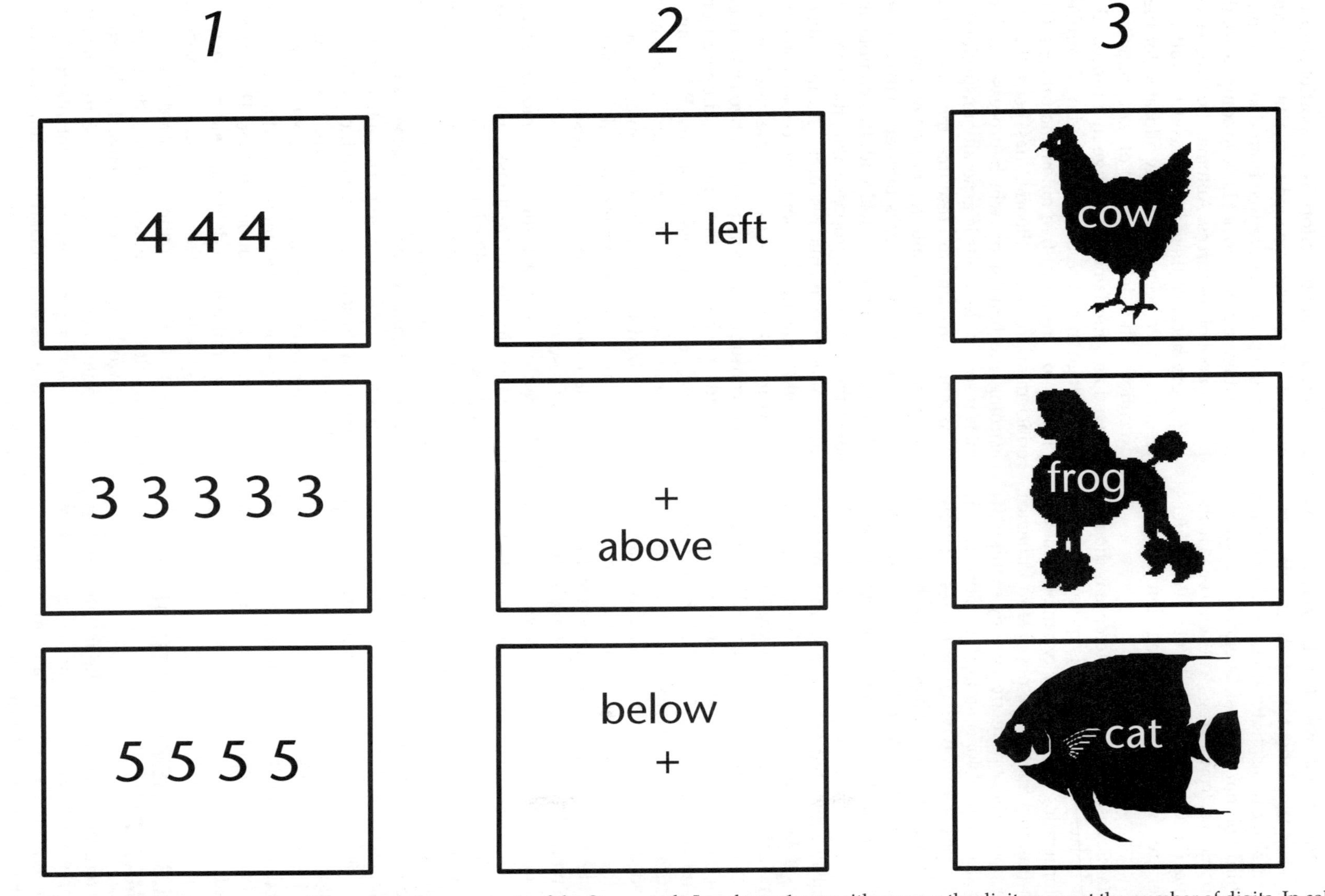

Figure 4. Illustration of incongruent conditions from three variants of the Stroop task. In column 1, you either name the digit or count the number of digits. In column 2, you either read the word or describe the spatial position of the word with respect to the central cross. In column 3, you either read the word or name the picture.

read the spatial terms (i.e. 'left', 'above', 'below', etc.) or to specify the location of the term with respect to the central cross (i.e. RIGHT, BELOW, ABOVE, etc.). Word identity interferes with specifying spatial locations, but not vice versa. Finally, in the third column, the task is either to read the animal name (i.e. 'cow', 'frog', 'cat', etc.) or to name the animal (i.e. BIRD, DOG, FISH, etc.). Word identity interferes with object naming, but not vice versa.

The classic case of Stroop interference is thought to occur because word reading is more automatic than color naming. If automaticity can be achieved through training, might it be possible to influence the direction of Stroop interference by manipulating practice with color naming? In principle, it should be possible to have color names interfere with word reading if color naming has been sufficiently practiced. But even with practice, it is extremely difficult to overcome the great prior advantage of word reading over color naming.

Instead imagine that you have just memorized that the symbols shown in Figure 5 are glyphs in some ancient language for the concepts blue, yellow, green, and red, respectively. The glyphs can be filled with various colors, creating congruent stimuli (e.g. 'blue' glyph in BLUE) or incongruent stimuli (e.g. 'red' glyph in YELLOW), as illustrated in the figure. When asked to name the color of the glyph, color naming is not influenced by the identity of the glyph, but when instead asked to name the glyph, glyph naming is strongly influenced by the color of the glyph. Because color naming is much more automatized than glyph naming, color interferes with glyph naming, but not vice versa. Now imagine that you are trained on glyph naming for several weeks, causing glyph naming to become more automatized than color naming. The direction of Stroop interference now reverses. Glyph identity interferes with color naming, but not vice versa. Results such as these suggest a continuum of automaticity, with the direction of Stroop interference a potential marker for which cognitive process is more automatized.

MODELS OF THE ACQUISITION OF AUTOMATICITY

Resource theories are based on the intuitive notion that people seem to have a limited amount of mental 'energy' that can be allocated to performing various tasks. Controlled processes require a certain amount of these limited mental resources whereas automatic processes do not. Automatic processes are fast because they are not limited by available resources. Automatic processes are effortless because mental 'effort' is proportional to the amount of resources needed to execute a process. Automatic processes are free from dual-task interference because they do not have to compete for the limited pool of resources. Automatic processes are obligatory because they do not need to wait until resources have been specifically allocated for their execution. The development of automaticity is viewed as a fundamental change in a process that makes it go from a resource-demanding controlled process to a resource-free automatic process. One criticism of resource theories has been that the learning mechanism by which processes reduce their resource demands is generally unspecified.

Another problem for resource theories is that complex patterns of interference have sometimes been observed. Some tasks interfere with one another, others do not. To deal with this complexity, some theorists have proposed that there may be multiple pools of mental resources. As an analogy, we could imagine that some processes consume electricity, other processes consume gasoline, and others consume coal. Any time that two tasks interfere with one another, they must be dipping from the same pool of limited resources. While intuitively appealing, *multiple resource theories* have been criticized as being inherently untestable assertions. Any complex pattern of task interference effects is explained *post hoc* by positing multiple pools of resources.

Instead of viewing resources as mental energy that is allocated to different tasks, resources may instead be conceptualized as specific processing components of the cognitive system that different tasks may need to share. As an analogy, we could imagine a mental toolbox, with some tasks requiring a screwdriver, others requiring a hammer, and others a saw. When two tasks need to use the same tool, they have to wait their turn. For example, working memory is limited. To the extent that two tasks both store information in working memory, they may interfere with one another. There is evidence for multiple modality-dependent working memory systems for verbal, spatial, and object information, so complex patterns of interference may be the result of different tasks placing demands on different working memory systems. To the extent that an automatic process is divorced from its reliance on working memory, it will not interfere with other processes that demand those limited processing resources.

In an extreme case, there may be some central process that must be shared by all aspects of

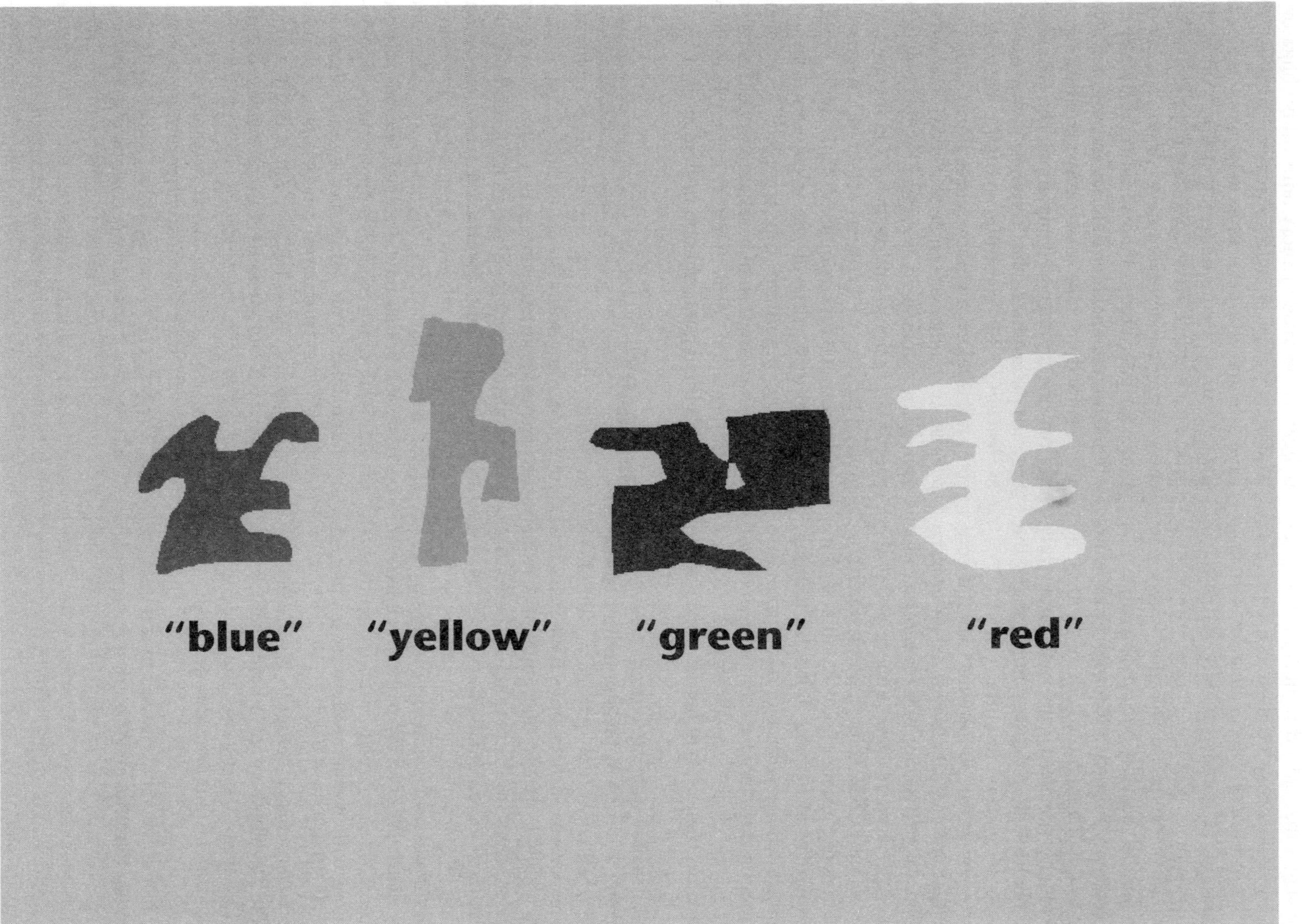

Figure 5. [*Figure is also reproduced in color section.*] Illustration of novel stimuli used to manipulate the direction of Stroop interference through training. Each shape (glyph) is associated with one of four color names that must be learned. Each shape can also be filled with one of four colors. Subjects either name the shape ('blue', 'yellow', 'green', or 'red') or name the color of the shape (RED, GREEN, BLUE, or YELLOW). Early in training, color interferes with shape naming. Later in training, shape interferes with color naming.

cognition that require selection among competing responses, what has been termed a *central bottleneck theory*. All tasks can be decomposed into a series of processing stages that extend from stimulus to response. Bottleneck theory posits that a particular one of these stages, that responsible for selecting among competing responses, can only be dedicated to one task at a time. All other stages prior to and subsequent to the response selection stage may proceed in parallel, but only one process can access response selection – other processes must wait. According to this theory, no cognitive process, no matter how highly practiced, can ever become truly automatic because all processes must share the limited response selection resource.

Our discussion of the Stroop effect should convince you that automaticity is not an all-or-none phenomenon. Word reading interferes with color naming because word reading is more automatic than color naming. But color naming interferes with shape naming because color naming is more automatic than shape naming. But with training, shape naming interferes with color naming because shape naming is now more automatic than color naming. It is not clear how a resource account could explain these asymmetric interference effects, nor how the direction of interference effects can be modulated by training. *Strength theories* represent learning in terms of the strength of association within pathways from particular stimuli to particular responses. Such theories have been implemented within a variety of frameworks from production systems to connectionist networks. The development of automaticity is seen as the strengthening of particular associations – be they production rules or connection weights – as a function of experience. Where these pathways intersect, interference can be observed. Stronger pathways interfere more with weaker pathways, leading to asymmetric interference effects.

Finally, *instance theories* propose a different account of the development of automaticity. Controlled processes are the result of the execution of some explicit algorithm whereas automatic processes are the result of memory retrieval. The development of automaticity is caused by a transition from algorithm to retrieval. When first engaged in some task, people may use an algorithm or rule to execute that behavior. For example, when first learning to add single digits, children typically adopt a strategy of starting with one of the digits and counting the requisite number of additional digits to generate the answer. Instance theory equates automaticity with memory retrieval. With experience, children (and adults) just remember that $2 + 2$ equals 4 without needing to explicitly count. Automatic processes are fast because memory retrieval is fast. Automatic processes are obligatory because memory retrieval is obligatory. Automaticity is effortless because memory retrieval seems effortless, especially compared with the execution of a multistep algorithm. Automatic processes are free from dual-task interference because a single memory retrieval offers less opportunity for interference than a multistep algorithm.

Execution of the algorithm and memory retrieval are assumed to take place concurrently, racing against one another to completion. The winner of the race determines what response is made. Early in learning, the algorithm is used because it completes before memory retrieval can finish (or because no memories can be retrieved). The development of automaticity is caused by the obligatory encoding of stimuli and responses in memory. As more memories of solutions are stored, memories can be retrieved more quickly. Thus, with experience, memory retrieval can eventually complete before the algorithm can complete. Consistent mappings are important because they yield consistent information from memory; varied mappings yield conflicting information from memory.

SUMMARY

Automatic processes are the autopilots of human cognition. They seem to execute outside our awareness and without our conscious control. They seem to execute quickly, and we may be entirely unaware of the steps involved in their execution. They can execute while we are doing other things at the same time. Some processes are automatic because our brains have evolved special-purpose mechanisms that respond without our conscious intention, and even sometimes against those intentions. Other processes can become automatic because of our experiences. Automaticity can be learned. But automaticity can be achieved only under certain circumstances: it may well be that some processes may just never become automatic, regardless of how much experience a person has had. The concept of automaticity is intimately tied with concepts of attention, consciousness, learning, and memory. Some research aims to relate attention and automaticity, with some attempts to relate both to the far more elusive concept of consciousness. Other research aims to relate the development of automaticity to what we know about learning and memory more generally, examining how they all manifest themselves across the full spectrum of human cognition.

Acknowledgements

This work was supported by NIMH Grant MH61370 and NSF Grant BCS-9910756.

Further Reading

Dulany DE and Logan GD (1992) Special Issue: Views and varieties of automaticity. *The American Journal of Psychology*. **105**(2) (Summer).

Groeger JA (2000) *Understanding Driving: Applying Cognitive Psychology to a Complex Everyday Task*. Philadelphia, PA: Psychology Press.

Kirsner K, Speelman C, Maybery M *et al.* (1998) *Implicit and Explicit Mental Processes*. Mahwah, NJ: Lawrence Erlbaum Associates.

Logan GD (1988) Toward an instance theory of automatization. *Psychological Review* **95**: 492–527.

MacLeod CM (1991) Half a century of research on the Stroop effect: an integrative review. *Psychological Bulletin* **109**: 521–524.

MacLeod CM and Dunbar K (1988) Training and Stroop-like interference: evidence for a continuum of automaticity. *Journal of Experimental Psychology: Learning, Memory and Cognition* **10**: 304–315.

Pashler H, Johnson JC and Ruthruff E (2000) Attention and performance. *Annual Review of Psychology* **52**: 629–651.

Schneider W and Shiffrin RM (1977) Controlled and automatic human information processing: I. Detection, search, and attention. *Psychological Review* **84**: 1–66.

Shiffrin RM and Schneider W (1977) Controlled and automatic human information processing: II. Perceptual learning, automatic attending, and a general theory. *Psychological Review* **84**: 127–190.

Wyer RS (1997) *The Automaticity of Everyday Life: Advances in Social Cognition*, vol. X. Mahwah, NJ: Lawrence Erlbaum Associates.

Autonomic Nervous System

Intermediate article

Gary G Berntson, Ohio State University, Columbus, Ohio, USA
Martin Sarter, Ohio State University, Columbus, Ohio, USA
John T Cacioppo, University of Chicago, Illinois, USA

CONTENTS

The autonomic nervous system has been viewed as a reflexive system for maintaining internal homeostasis. It is now clear, however, that the autonomic nervous system has reciprocal interactions with higher neurobehavioral substrates, influencing both autonomic control and cognitive/behavioral processes.

INTRODUCTION

The autonomic nervous system (ANS) is the designation applied by John Langley to a complex network of peripheral nerves and ganglia, together with associated regulatory systems of the brain and spinal cord, which serve to control smooth muscles and glands of the viscera (Langley, 1921). Implicit in the term 'autonomic' is the view that the ANS is rigidly regulated, and not subject to the vagaries of 'volitional' control like the somatic nervous system. This was even more strongly implied by an earlier name: the involuntary nervous system.

The influential physiologist Walter Cannon argued that the ANS was specialized for what he termed 'homeostasis', or the maintenance of stability of the internal fluid matrix, necessary to sustain life (Cannon, 1939). This function was suggested to be implemented by an array of feedback-regulated autonomic reflexes responding to perturbations in visceral states with compensatory adjustments to restore homeostatic balance. An example is the baroreceptor reflexes, whereby an increase in blood pressure, signaled by baroreceptor afferent activity, triggers reflex responses including relaxation of vascular smooth muscle, as well as decreases in heart rate and myocardial contractility that reduce cardiac output. Together, these responses serve to decrease blood pressure and compensate for the initiating perturbation (Figure 1).

Early research focused on the plethora of autonomic reflexes organized at lower brainstem and spinal cord levels. An important historical development, however, was an emerging recognition that central regulatory systems extended well above these lower levels of the neuraxis. Although many aspects of ANS regulation may be based on simple reflexes, autonomic adjustments are also closely linked to cognitive and behavioral processes that arise from higher neurobehavioral substrates, including the limbic system and cerebral cortex. Another important recognition is the fact that the ANS is as much a sensory system as a motor system. In addition to contributing to lower-level reflexive regulation, visceral afferent information now appears to bias processing in higher neural systems.

ANATOMY AND PHYSIOLOGY

Peripheral Components

In addition to the intrinsic enteric system that is sometimes considered to be part of the ANS, the ANS consists of two major peripheral divisions, the sympathetic and parasympathetic branches. These two branches have distinct central origins, and differ in their peripheral anatomy and physiology. The sympathetic nervous system has its central origins in the intermediolateral cell column of the thoracic and lumbar divisions of the spinal cord (Figure 1), and so has also been termed the thoracolumbar division. Spinal sympathetic motor neurons give rise to preganglionic efferents which exit the spinal cord in the ventral roots and enter an interconnected set of sympathetic chain ganglia which lie along each side of the cord. On entering the chain ganglia, preganglionic fibers may ascend or descend before terminating on sympathetic ganglion cells, which give rise to postganglionic axons that in turn project to visceral organs. Because of the extensive ganglionic interconnections within the sympathetic nervous system, this division was often considered to discharge as a whole (i.e., in sympathy). We now know, however, that the sympathetic system can exert much more precise and organ-specific actions. The primary neurotransmitter at the ganglionic synapse is acetylcholine (ACh), whereas the postganglionic synapse employs the catecholamine neurotransmitter noradrenaline (norepinephrine). There are some deviations from

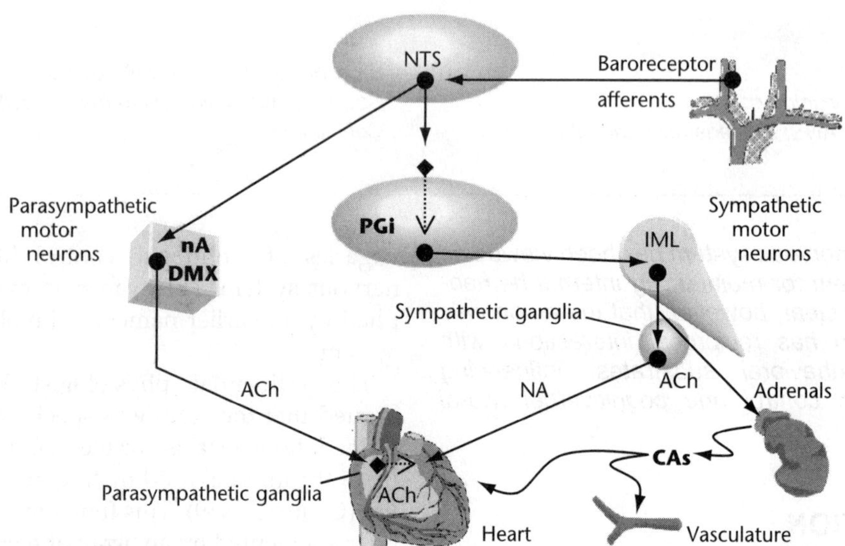

Figure 1. Some features of the baroreflex circuits and peripheral organization of the autonomic nervous system. Baroreceptor afferents project to the nucleus of the tractus solitarius (NTS), by which baroreceptor activity leads to activation of parasympathetic motor neurons in the nucleus ambiguus (nA) and dorsal motor nucleus of the vagus (DMX). The NTS also indirectly inhibits the nucleus paragigantocellularis (PGi) within the rostral ventrolateral medulla, leading to a withdrawal of excitatory drive on the sympathetic motor neurons in the intermediolateral cell column of the cord (IML). ACh, acetylcholine; CAs, catecholamines; NA, noradrenaline (norepinephrine). (Adapted from Cacioppo JT, Tassinary LG and Berntson GG (2000) *Handbook of Psychophysiology*, p. 465, with permission from Cambridge University Press.)

this general anatomical plan, as preganglionic fibers innervate the adrenal medulla directly, without synaptic interruption in the chain ganglia. Hence, sympathetic synapses onto the adrenal medulla release ACh, and the adrenal secretory cells release the catecholamines noradrenal and adrenaline (epinephrine). In contrast to postganglionic sympathetic innervation, however, these adrenomedullary catecholamines are released into the general circulation where they can act humorally on many organ systems, including some that do not receive direct innervation. Because many of the peripheral actions of the sympathetic system are activational and promote energetic metabolism, this division has been considered to be a mobilization system involved in responding to adaptive challenges.

The other branch of the ANS, the parasympathetic division, differs from the sympathetic in its central origin, peripheral anatomy, neuropharmacology, and functions. The lower central motor neurons of the parasympathetic division lie in the intermediolateral column of the sacral spinal cord, and in numerous nuclei within the brainstem (e.g. the nucleus ambiguus, dorsal motor nucleus of the vagus, and salivatory nuclei; see Figure 1). Because of this anatomy, the parasympathetic division has been termed the craniosacral branch, and is also sometimes referred to as the 'vagal' branch, after the vagus nerve (10th cranial nerve) that carries parasympathetic efferents. Strictly, however, the latter term applies only to the vagal component of the parasympathetic branch. Like the sympathetic division, the parasympathetic system includes peripheral ganglia, but these are not collected into coherent ganglionic chains, but rather are generally located in or near the visceral organs innervated. Because of this anatomical difference the parasympathetic system has been thought to be capable of more localized action, although considerable regional specificity can also be demonstrated for the sympathetic branch. As in the sympathetic system, the preganglionic axons of the parasympathetic branch employ ACh as a neurotransmitter (both divisions acting primarily via nicotinic cholinergic synapses on the ganglia); but in contrast to the sympathetic division, postganglionic axons of the parasympathetic system also employ ACh (generally acting via muscarinic cholinergic receptors). This simple distinction in neurochemical coding among the peripheral autonomic branches belies the tremendous complexity of neurotransmitter, neuromodulatory, and neurohormonal interactions within the peripheral ANS.

Opposing versus Synergistic Actions and Modes of Autonomic Control

Many visceral organs are dually innervated by both autonomic branches, and the two divisions are often opposing in their actions. For example, the sympathetic cardiac innervation increases heart rate via β adrenergic receptors that speed the depolarization of the sinoatrial pacemaker potential. In contrast, the parasympathetic innervation slows the beat of the heart via ACh, acting at muscarinic receptors on the sinoatrial node, which increases potassium conductance and slows the rate of pacemaker depolarization. More generally, in contrast to the mobilizing functions of the sympathetic branch, the parasympathetic system has been viewed as a conservation system that functions to promote energy intake, reduce energy expenditure, and preserve energy reserves. Historically, the autonomic branches sometimes have been considered to be subject to reciprocal central control, with increased activity of one division associated with decreased activity of the other (Berntson *et al.*, 1991). Indeed, to the extent to which branches are functionally opposed, the reciprocal mode of regulation would yield the widest dynamic range of autonomic control over target organs (Berntson *et al.*, 1991).

All organs are not dually innervated, however, and it is often the case that actions of the two branches on a given organ are not precisely opposite. Major arterioles, for example, receive only sympathetic innervation, and both sympathetic and parasympathetic activity can stimulate salivary secretion. Even when generally opposing in their actions, as in the control of heart rate, the two branches may operate by different cellular mechanisms with distinct features and temporal dynamics. These differences can lead to distinct functional states, with differing levels of activity of the two branches, which cannot be duplicated by simple variations along a reciprocal bipolar continuum extending from maximal sympathetic to maximal parasympathetic activity. Penile erection and ejaculation, for example, require the coactivation of both autonomic branches.

Higher Central Controls and Neurobehavioral Systems

Many basic autonomic homeostatic reflexes, such as baroreceptor reflexes, do display a reciprocal pattern of control over the autonomic branches. However, the ANS is far from being a simple homoeostatic mechanism controlled by reflex systems of the lower brainstem. Indeed, it is now apparent that autonomic outflow is regulated by neurobehavioral systems at the highest levels of the neuraxis, including the cerebral cortex. Rostral brain systems not only modulate lower reflex mechanisms, but issue descending projections that terminate directly on autonomic source nuclei in the brainstem and spinal cord. In accord with the general increase in flexibility and integrative capacity of higher neural systems, rostral neurobehavioral mechanisms appear to exert more variable and flexible control over autonomic outflows. Consequently, in addition to the reciprocal mode of control often seen in reflex regulation, a wider range of control modes can be observed in behavioral contexts, including independent changes in the autonomic branches as well as coactivation or coinhibition of both divisions.

Autonomic Afferents and Ascending Central Pathways

An important aspect of the autonomic nervous system is its sensory function (Dworkin, 2000). In fact, over 75% of the fibers in the largest autonomic nerve, the vagus, are afferents. Visceral afferents carry a range of information concerning the internal state of the body, from baroreceptors, chemoreceptors, and other interoceptors. Some visceral afferents enter the spinal cord via the dorsal root (along with somatic afferents) and terminate in the dorsal horn, where second- and higher-order neurons may participate in local autonomic reflexes, or relay visceral information to higher central structures. One such structure is the nucleus of the tractus solitarius (NTS), a major visceral relay station in the brainstem (Figure 1). Additional visceral afferents, such as those carried by the vagus and other cranial nerves, terminate directly in the NTS. The NTS is a key structure in brainstem autonomic reflexes and serves as an important relay in ascending pathways to higher levels of the neuraxis where they can modulate the processing of rostral neural systems. Although the functional contributions of this ascending visceral information have not been fully elucidated, it has been shown, for example, that baroreceptor activation can reduce cortical arousal, suppress spinal reflexes, and attenuate pain transmission (Dworkin, 2000). The impact of this ascending information on rostral neurobehavioral mechanisms, and its role in cognitive and behavioral processes, has become an active area of research.

FUNCTIONS OF THE ANS: HOMEOSTASIS AND ALLOSTASIS

A historically recognized role of the ANS is in the maintenance of internal homeostasis. Central and ganglionic autonomic reflex circuits react to perturbations in internal states and generate responses that compensate for these perturbations and restore internal conditions. On standing up from a sitting position, for example, gravitational forces result in a pooling of blood in the legs, which could lead to a dangerous drop in blood pressure and circulatory compromise. This orthostatic challenge becomes a serious problem in autonomic failure and other conditions of impaired autonomic function, where it may lead to syncope (fainting). In healthy individuals, however, this postural maneuver results in the unloading of the carotid baroreceptors and an associated decrease in baroreceptor afferent activity. Baroreceptor reflexes then trigger a compensatory increase in sympathetic outflow and a reciprocal decrease in parasympathetic activity. The resulting increase in heart rate and cardiac output, together with sympathetically mediated vasoconstriction, serves to restore normal blood pressure. This illustrates a classical feedback-regulated (servocontrolled) homeostatic reflex. However, the homeostatic model of autonomic function is overly restrictive, and does not adequately reflect the adaptability and flexibility of autonomic regulation. Feedback-regulated systems, for example, can respond only after a disturbance has taken place, and hence do not effectively prevent these disturbances. Fortunately, central autonomic control systems provide for a broader range of regulatory adjustments, including anticipatory responses.

The early work of Pavlov, demonstrating the conditioning of autonomic responses, foreshadowed current models of autonomic regulation that extend well beyond simple feedback-regulated control mechanisms. Through central associative processes, both exteroceptive and interoceptive stimuli can come to control ANS activity in an anticipatory fashion, and can effectively prevent or minimize perturbations prior to their occurrence (Dworkin, 2000). Conditioned anticipatory autonomic responses, for example, contribute to cardiovascular adjustments to orthostatic stress prior to severe blood pressure perturbations, and trigger anticipatory insulin release prior to the onset of a meal. Specific associations between an innocuous stimulus (conditioned stimulus, CS) that is paired with an aversive event (unconditioned stimulus,

US) can result in conditioned fear reactions to the CS in which preparatory somatic and autonomic responses can be emitted in anticipation of the aversive stimulus or event (Ohman *et al.*, 2000). Similarly, the exaggerated affective and autonomic reactions to a wider (and sometimes poorly defined) range of stimuli and contexts in anxiety states probably reflects a pattern of hyperattentional processing of threat-related stimuli based on a broader and less specific associative structure (Berntson *et al.*, 1998).

Moreover, although central autonomic regulatory systems do contribute to homeostasis, steady state conditions are not always optimal, and central regulatory systems may also promote explicit deviations from homeostasis. Psychological stress, for example, can lead to an inhibition of the baroreceptor heart rate reflex, allowing an increase in both heart rate and blood pressure that might contribute to an adaptive response (see Berntson and Cacioppo, 2000). This is indicative of what has been termed 'allostasis' (McEwen, 1999), rather than homeostasis, and represents a class of autonomic control by which central mechanisms can actively adjust internal states in accord with adaptive demands (see McEwen, 1999; Berntson and Cacioppo, 2000). This is important, as it is not always optimal to maintain homeostatic, steady state conditions. During physical exercise or in the face of a survival threat, for example, there would be considerable adaptive advantage to increasing cardiac output and blood pressure, to enhance blood perfusion of muscles. In such instances, integrative neural systems shift the regulatory set point to a new level (allostasis), rather than maintaining a single homeostatic state. In fact, regulation of autonomic activity is often more dynamic than static, especially in behavioral contexts, so central neurobehavioral systems could more appropriately be considered to exert a pattern of allodynamic regulation over autonomic states (Berntson and Cacioppo, 2000). The construct of allodynamic regulation more appropriately encompasses the complexity and flexibility of central autonomic control and its integration with neurobehavioral function.

EMOTION, ANXIETY, AND ANS ACTIVITY

Autonomic reactions have long been recognized as integral aspects of emotional states such as fear and anxiety, and it is now apparent that there is no clear distinction between higher neural systems underlying behavioral, autonomic, and neuroendocrine regulation. Increasingly, a close integration of these diverse response domains is apparent, so that behavioral responses, for example, are accompanied by the requisite autonomic, and neuroendocrine support. An illustration of this comes from the broad actions of corticotrophin releasing hormone systems at many levels of the neuraxis, which appear to coordinate and integrate the affective, behavioral, autonomic, and neuroendocrine states underlying stress reactions.

In addition, cerebral cortical areas, including the medial prefrontal cortex, have been implicated in the cognitive aspects of affective states, and have been shown to have relatively direct projections to lower autonomic mechanisms (Berntson *et al.*, 1998). These descending pathways (Figure 2) represent important routes by which rostral neural systems can modulate and regulate ANS activity.

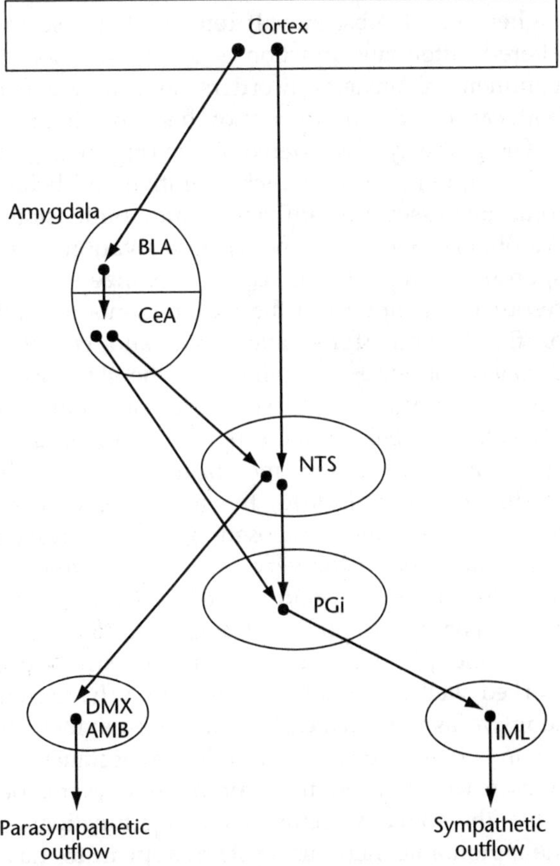

Figure 2. Descending neural systems that affect fear, anxiety, and autonomic control. AMB, nucleus ambiguus; BLA, basolateral amygdala; CeA, central nucleus of the amygdala; DMX, dorsal motor nucleus of the vagus; IML, intermediolateral cell column of the cord; NTS, nucleus of the tractus solitarius; PGi, nucleus paragigantocellularis.

Indeed, autonomic adjustments are ubiquitous in cognitive and behavioral contexts, and these adjustments frequently do not adhere to expectations based on simple metabolic demands. Rather, these autonomic adjustments are likely to reflect an integrated pattern of central control over cognitive, behavioral, and autonomic functions. Consequently, it is not surprising that psychological dysfunctions are often associated with altered autonomic control.

The fourth edition of the *Diagnostic and Statistical Manual of Mental Disorders* (DSM-IV) published by the American Psychiatric Association recognizes altered autonomic activity as a frequent feature of anxiety disorders. However, there is no single autonomic feature that uniformly characterizes anxiety states; indeed, there appear to be considerable differences between the anxiety disorders, and even between individuals within a given diagnostic category (Berntson *et al.*, 1998). This variability attests to the flexibility of autonomic control by higher neural substrates. It remains the case that altered autonomic function is a common accompaniment of anxiety disorders, and may reflect both cause and consequence of these conditions.

The pathways considered above (Figure 2) provide ample routes by which cognitive and behavioral processes can influence autonomic states. Autonomic and neuroendocrine activation in response to stressors serves to mobilize visceral resources in support of the requirements of 'fight or flight', and of the attentional and cognitive demands of adaptive challenges. The stressors of contemporary society, however, do not require – or even allow – a behavioral response of fight or flight, and hence this visceral mobilization may not be readily resolved. Indeed, the metabolic demand for somatic activation imposed by chronic stressors has diminished at the very time that the requirement for visceral support of attentional, cognitive, and behavioral adaptation strategies has increased. Thus, the physiological response to stress that worked well in human evolution may have maladaptive aspects, which become more evident as urban societies develop and life expectancy increases well beyond the reproductive years. Because the somatovisceral mobilizing functions of the autonomic nervous system appear to have evolved to deal with transient challenges (Dworkin, 2000), the prolonged somatovisceral activation associated with chronic psychological stressors in today's society may have long-term health costs (McEwen, 1999). Consequently, the cortical and cognitive processes that underlie this chronic activation are especially important to understand.

COGNITIVE INTERACTIONS WITH ANS ACTIVITY

The descending pathways of Figure 2 represent important substrates for 'top down' regulation of the ANS, whereby cognitive and behavioral states can affect autonomic function. Indeed, cognitive imagery alone is sufficient to trigger autonomic responses. Because cognitive and attentional biases lie at the core of anxiety disorders such as post-traumatic stress disorder and generalized anxiety disorders, these descending pathways probably contribute to the altered autonomic function in anxiety states. The interactions between cognitive and autonomic functions are not one-way, however, and the sensory functions of the ANS may allow autonomic states to bias cognitive processing.

The nineteenth-century psychologist William James proposed that sensory feedback from physiological responses might constitute emotional experience (see Cacioppo *et al.*, 1992). Although it now appears that somatovisceral feedback is not essential for emotional reactivity, this feedback may importantly bias cognitive and emotional processing (see Berntson *et al.*, 1998). Thus, visceral afference, carried by vagal afferents with a relay in the NTS, has been shown to enhance emotional memory in animals (McGaugh *et al.*, 2000), and feedback from autonomic responses may be sufficient to trigger panic attacks in people with panic disorder. More generally, somatovisceral and environmental feedback may have a crucial role in organizing and regulating broad aspects of behavior, affect, and cognition (see, for example, Bechara *et al.*, 2000).

One anatomical route by which this type of 'bottom up' priming may be effected is suggested by work on the nucleus paragigantocellularis (PGi) and the locus ceruleus (LC) in the brainstem (Aston-Jones *et al.*, 1996). These structures could be considered as parts of an ascending visceral afferent system. The PGi receives direct input from the NTS, and modulates sympathetic outflow via descending projections and LC activity via ascending projections (Figure 3). Thus, activity of LC neurons would be expected to be highly responsive to the state of activity of the sympathetic branch. Noradrenergic neurons of the LC in turn issue excitatory projections to the basal forebrain cortical cholinergic system, as well as to the cortex directly.

The ascending pathways of Figure 3 represent a potentially important route by which autonomic activity and associated visceral afference might modulate cognitive and emotional processing. In

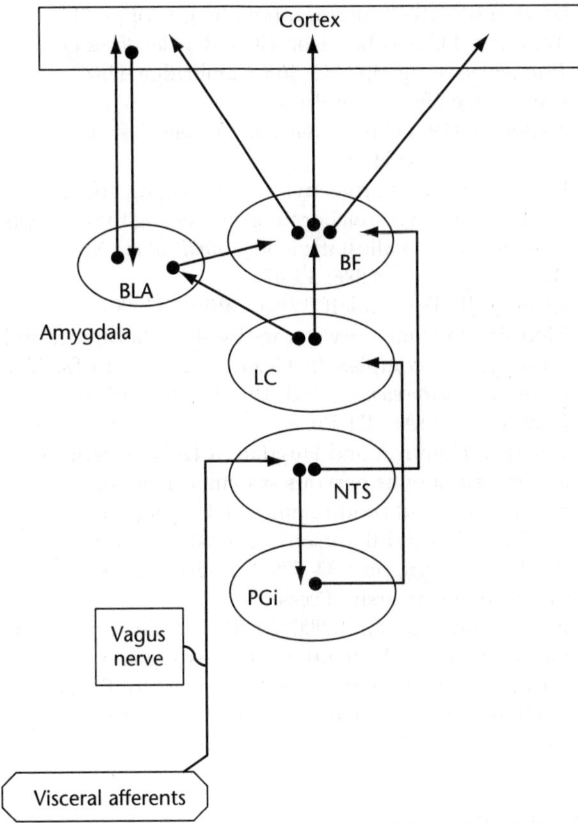

Figure 3. Ascending neural systems that may affect cognitive and emotional processing. BF, basal forebrain; BLA, basolateral amygdala; LC, locus ceruleus; NTS, nucleus of the tractus solitarius; PGi, nucleus paragigantocellularis.

addition to the known descending projections from rostral neurobehavioral systems, this ascending pathway suggests an additional class of interactions between autonomic activity and cortical/cognitive processes. Moreover, the potential for both 'top down' (Figure 2) and 'bottom up' (Figure 3) influences may represent the substrate for a vicious circle of reciprocal cognitive/autonomic priming that might account for the apparent irrationality and exaggerated hyperattentional processing in anxiety states.

Anxiety, Anxiolytics, and the Basal Forebrain Cortical Cholinergic System

Because cognitive processes appear to contribute substantially to the clinical features of anxiety, including the hyperattentional processing of threat-related stimuli, it is not surprising that cortical systems have been implicated in anxiety disorders. An illustration comes from studies on anxiety and the basal forebrain cholinergic system (Berntson *et al.*, 1998). The widespread corticopetal projections of the basal forebrain provide the primary source of cortical ACh, and ACh appears to enhance cortical processing generally (Sarter and Bruno, 2000). This is illustrated by the global dementia of Alzheimer disease, which is associated with degeneration of the basal forebrain cholinergic system. In view of these considerations, cortical cholinergic activity would be expected to exaggerate the cognitive processing underlying anxiety. This is consistent with the finding that selective immunotoxic lesions of the cholinergic neurons of the basal forebrain in the rat attenuate the exaggerated behavioral and autonomic reactions in anxiogenic contexts (Berntson *et al.*, 1998).

Additional evidence supports this view and explains the antianxiety actions of benzodiazepine receptor agonists such as chlordiazepoxide and diazepam (for a review, see Berntson *et al.*, 1998). The activity of basal forebrain cholinergic neurons is regulated in part by an inhibitory input mediated by γ-aminobutyric acid (GABA), which in turn is bidirectionally modulated through a benzodiazepine binding site on the GABA receptor complex. The anxiolytic benzodiazepine receptor agonists enhance GABA-mediated inhibition and reduce basal forebrain cholinergic activity. Conversely, benzodiazepine receptor inverse or partial inverse agonists, which decrease GABA-mediated inhibition and enhance basal forebrain cholinergic activity, have notable anxiogenic properties (see Berntson *et al.*, 1998). Especially relevant in the link between anxiety and autonomic function may be the medial prefrontal cortex, an area that has been repeatedly implicated in both affect and autonomic control. Selective immunotoxic lesions of the basal forebrain cholinergic projections to the medial prefrontal cortex, like lesions of the basal forebrain itself, eliminate the exaggerated autonomic reactions in anxiety-related contexts (see Berntson *et al.*, 1998).

The ascending pathways in Figure 3 are intended to be representative rather than exhaustive, and there are many important questions that remain in this area. It is increasingly apparent, however, that comprehensive models of cognitive processing or autonomic functions will require a deeper understanding of the reciprocal interactions between these functional domains. In contrast to the Jamesian view that somatovisceral feedback constitutes emotion, the ascending pathway of Figure 3 provides a route by which visceral afference may more subtly trigger or modulate cognitive/emotional processing, even in the absence of conscious perception of the visceral stimulus.

CONCLUSION

Classical views of the autonomic nervous system often focused on brainstem reflexes and the autonomic regulation of internal homeostasis. With the elucidation of higher neurobehavioral influences over autonomic outflow, and the important sensory functions of the ANS, these limited perspectives are no longer tenable. Beyond lower sensory and motor neurons, there is no clear distinction between central systems that regulate behavioral, autonomic, neuroendocrine, and immune processes. Moreover, it appears that the higher the level of neural organization, the greater is the integration among these response domains, culminating in the supreme integrative functions of cerebral cortical systems. Consequently, an understanding of cortical/cognitive processes and the systems and mechanisms that regulate this processing is a prerequisite for a comprehensive model of the autonomic nervous system and its central regulation.

References

Aston-Jones G, Rajkowski J, Kubiak P, Valentino RJ and Shipley MT (1996) Role of locus coeruleus in emotional activation. *Progress in Brain Research* **107**: 379–402.

Bechara A, Damasio H and Damasio AR (2000) Emotion, decision making and the orbitofrontal cortex. *Cerebral Cortex* **10**: 295–307.

Berntson GG and Cacioppo JT (2000) From homeostasis to allodynamic regulation. In: Cacioppo JT, Tassinary LG and Berntson GG (eds) *Handbook of Psychophysiology*, pp. 459–481. Cambridge, UK: Cambridge University Press.

Berntson GG, Cacioppo JT and Quigley KS (1991) Autonomic determinism: the modes of autonomic control, the doctrine of autonomic space, and the laws of autonomic constraint. *Psychological Review* **98**: 459–487.

Berntson GG, Sarter M and Cacioppo JT (1998) Anxiety and cardiovascular reactivity: the basal forebrain cholinergic link. *Behavioural Brain Research* **94**: 225–248.

Cacioppo JT, Berntson GG and Klein DJ (1992) What is an emotion? The role of somatovisceral afference, with special emphasis on somatovisceral 'illusions'. *Review of Personality and Social Psychology* **14**: 63–98.

Cannon WB (1939) *The Wisdom of the Body*. New York, NY: WW Norton.

Dworkin BR (2000) Interoception. In: Cacioppo JT, Tassinary LG and Berntson GG (eds) *Handbook of Psychophysiology*, pp. 482–405. Cambridge, UK: Cambridge University Press.

Langley JN (1921) *The Autonomic Nervous System*. Cambridge, UK: Heffer.

McEwen BS (1999) Protective and damaging effects of mediators of stress: elaborating and testing the concepts of allostasis and allostatic load. *Annals of the New York Academy of Sciences* **896**: 30–47.

McGaugh JL, Roozendall B and Cahill L (2000) Modulation of memory storage by stress hormones and the amygdala complex. In: Gazanniga MS (ed.) *The New Cognitive Neurosciences*, 2nd edn, pp. 1081–1098. Cambridge, MA: MIT Press.

Ohman A, Hamm A and Hugdahl K (2000) Cognition and the autonomic nervous system: orienting, anticipation, and conditioning. In: Cacioppo JT, Tassinary LG and Berntson GG (eds) *Handbook of Psychophysiology*, pp. 533–579. Cambridge, UK: Cambridge University Press.

Sarter M and Bruno JP (2000) Cortical cholinergic input mediating arousal, attentional processing and dreaming: differential afferent regulation of the basal forebrain and brainstem afferents. *Neuroscience* **95**: 933–952.

Further Reading

Appenzeller O (1999) *The Autonomic Nervous System. Part I, Normal Functions*. New York, NY: Elsevier.

Appenzeller O (2000) *The Autonomic Nervous System. Part II, Dysfunctions*. New York, NY: Elsevier.

Cacioppo JT, Tassinary LG and Berntson GG (2000) *Handbook of Psychophysiology*. Cambridge, UK: Cambridge University Press.

Goehler LE, Gaykema RP, Hansen MK *et al.* (2000) Vagal immune-to-brain communication: a visceral chemosensory pathway. *Autonomic Neuroscience* **85**: 49–59.

Schulkin J, Gold PW and McEwen BS (1998) Induction of corticotropin-releasing hormone gene expression by glucocorticoids: implication for understanding the states of fear and anxiety and allostatic load. *Psychoneuroendocrinology* **23**: 219–243.

Williams CL and Clayton EC (2001) Contribution of brainstem structures in modulating memory storage processes. In: Gold PE and Greenough WT (eds) *Memory Consolidation: Essays in Honor of James L. McGaugh*, pp. 141–162. Washington, DC: American Psychological Association.

Backpropagation

Intermediate article

Paul Munro, University of Pittsburgh, Pittsburgh, Pennsylvania, USA

CONTENTS

Backpropagation is a supervised training procedure for feedforward connectionist networks that is widely used by cognitive scientists to model learning phenomena.

LEARNING AND GENERALIZATION IN NEURAL NETWORKS

Initial approaches to developing general-purpose learning machines have generally been restricted to a form of learning known as supervised learning, i.e. abstracting the properties that underlie an input–output relation given a sample of input–output pairs. A standard test of successful learning is the ability of the system to generalize, which is measured by the average correctness of the system's responses to a set of novel stimuli.

Since the response properties of a neural network depend on the weights, or connections between pairs of units, models of learning are generally framed in terms of how the weights change as a function of experience. Donald Hebb (1949) suggested that changes in the biological correlates of these weights, the synapses between neurons, underlie human learning. In the late 1950s and early 1960s, learning rules were developed for networks. These rules were computationally limited, since both the units and the network architectures were simple (Rosenblatt, 1958; Widrow and Hoff, 1960). While the design of more powerful networks was well within the scope of scientific knowledge at that time, there was no known method for training them. (*See* **Hebb Synapses: Modeling of Neuronal Selectivity; Hebb, Donald Olding**)

A simple change to the processing function in the individual units led directly to the development of a learning rule for feedforward networks of arbitrarily complex connectivity. Notions from statistics can be applied to derive a learning procedure for networks with complex architectures. This was first done by Paul Werbos (1974), but his results went unnoticed by the scientific community until their almost simultaneous rediscovery by LeCun (1985), Parker (1982), and Rumelhart *et al.* (1986).

SINGLE-LAYER CONNECTIONIST NETWORKS WITH LINEAR THRESHOLD UNITS

The computational power of a connectionist network depends on the computational power of each unit and on the connectivity among the units. Early learning rules applied to the 'linear threshold unit' (LTU) model, which can respond to a stimulus pattern with one of just two possible values (usually these are 0 and 1). Each LTU performs a 'categorization' function on the space of possible stimuli: it divides the set of stimuli into those that generate a response of 1 and those that generate a response of 0. For an LTU, the boundary between these regions is linear. A 'layer' of LTUs performs independent categorization tasks. A category whose stimuli can be separated from stimuli outside the category by a linear boundary, and which is thus computable by an LTU, is called linearly separable (LS) in the stimulus space. Some LS categorization tasks are shown in Figure 1, along with some that are not LS. (*See* **Connectionism**)

LEARNING IN SINGLE-LAYER NETWORKS

Learning by an LTU is a process whereby the weights change in order to improve the placement of the category boundary. A 'supervised learning rule' operates on the parameters of a system (in this

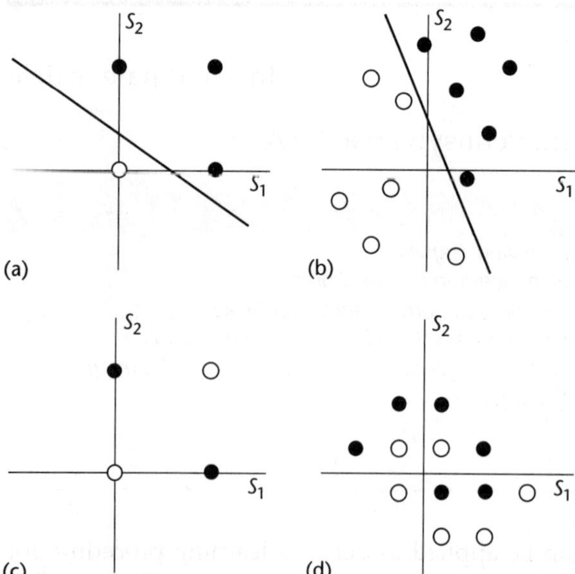

Figure 1. Linear separability. The graphs illustrate four categorization tasks in which patterns are plotted according to the stimulus coordinates (s_1, s_2). Filled dots represent stimuli that are in the category, and open dots represent stimuli not in the category. Categorization boundaries are drawn for the two linearly separable tasks: (a) the 'Boolean OR' task, and (b) a real-valued task. Two tasks that are not linearly separable are also shown: (c) the 'Boolean XOR' task, and (d) a 'double spiral' task.

case, the weights of the network) under the assumption that a set of labeled data (i.e. stimulus points for which the correct categorizations are known) is available. This 'training set' is used to tune, or train, the network weights in the hope that the system will generalize from the training so that it will classify novel stimuli appropriately. This approach resembles standard regression techniques from statistics.

Figure 2 illustrates weight changes of a linear threshold unit and the resulting categorizations on a two-dimensional stimulus space. Note that, in the initial state, some responses of the network are correct and some are incorrect. Eventually, the system finds a classification boundary that solves the given task as well as possible.

MULTI-LAYER NETWORKS

A network built of LTUs can compute a more general class of functions than a single LTU, which is restricted to computing functions that are LS. A typical network structure is the 'multi-layer perceptron' (MLP), originally proposed by Rosenblatt (1958). The MLP architecture first computes the responses of several units, each with different weights (the first layer) to a common stimulus. The pattern of responses is fed as a stimulus to a second layer, and so forth until the final (output) layer. The layers that precede the output layer are known as 'hidden layers'.

An MLP can compute functions that are not LS. Note that each unit in an MLP performs a linear separation on its direct input, but the category it computes on the network input might be more complex. By introducing one or more layers between the network stimulus and the ultimate response, the stimulus pattern is transformed to another 'representation'. A task that is not LS using the representations given at the stimulus level may become LS at a hidden layer. Thus, an MLP can compute a complex categorization task by changing the representation of the task. What is needed is to find a transformation under which the hidden-layer representations are LS. An example is shown in Figure 3.

BACKPROPAGATION OF ERROR

Until the publication of the 'backprop' procedure, there was no technique for training an MLP with more than a single layer. Like standard regression techniques, the derivation of backprop begins with the definition of an 'error measure' E which quantifies how closely the network approximates the given data as a function of its weight parameters.

The well-known 'gradient descent' technique is used to modify each weight, by an amount that is proportional to the derivative of E with respect to that weight. This approach eluded researchers in the 1960s, because of the abrupt shift in the value of the threshold function at the threshold, which renders the required derivatives undefined. The insight that enables the use of gradient descent is to replace the threshold function with a function that is differentiable but retains the important features of the threshold function (Figure 4).

The backpropagation learning rule is derived by applying the gradient descent technique to fit a feedforward network of sigmoid units to a set of data. The rule can be implemented as a process whereby an error value is first computed for each output unit. Subsequently, these errors are 'propagated backwards' through the weights of the network to determine an 'effective error' for all the hidden units in the network.

In its simplest form, the backprop procedure is implemented as follows (compare with the single-layer learning procedure described above):

	s_1	s_2	r	T
A	0	6	1	1
B	2	0	0	1
C	3	2	0	1
D	4	5	1	1
E	1	–4	0	0
F	–1	3	1	0
G	–3	4	1	0
H	–5	–1	0	0

	s_1	s_2	r	T
A	0	6	1	1
B	2	0	0	1
C	3	2	0	1
D	4	5	1	1
E	1	–4	0	0
F	–1	3	1	0
G	**–3**	**4**	**0**	**0**
H	–5	–1	0	0

	s_1	s_2	r	T
A	0	6	1	1
B	2	0	1	1
C	3	2	1	1
D	4	5	1	1
E	1	–4	0	0
F	–1	3	0	0
G	–3	4	0	0
H	–5	–1	0	0

(a) Initial (b) One iteration (c) Final

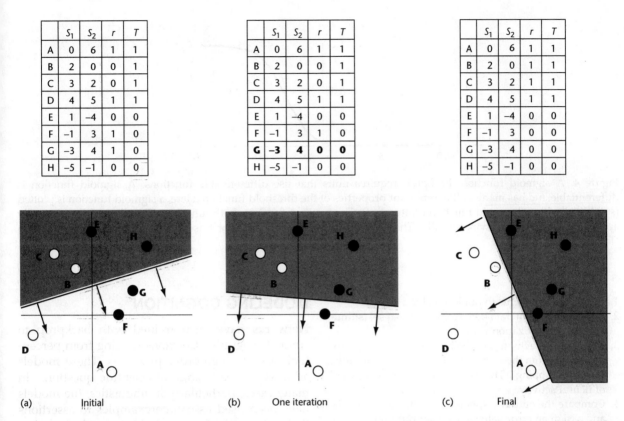

Figure 2. Learning by an LTU. Three stages during the learning process are shown. At each stage a table lists the stimulus values (s_1, s_2), the target (T), and the actual response (r), for the eight elements (A to H) of the training set. Each element is plotted according to its stimulus values, and is either open or filled depending on its target value. The line shows the discrimination boundary at each stage. (a) The initial (random) state. (b) The state after presentation of the first pattern, G. (c) The final state: all patterns are correctly classified.

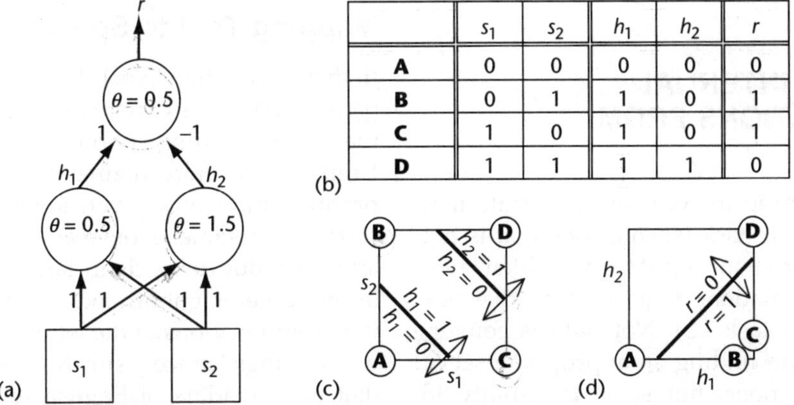

	s_1	s_2	h_1	h_2	r
A	0	0	0	0	0
B	0	1	1	0	1
C	1	0	1	0	1
D	1	1	1	1	0

Figure 3. A simple MLP. (a) The network shown computes the 'XOR' function of the stimulus, which is not a linearly separable function. The intermediate (hidden) layer has two LTUs, which compute responses using the weights (arrow labels) and thresholds (θ) shown in the diagram. (b) The table shows the responses (h_1, h_2) of the hidden units and the response r of the output unit for each stimulus (s_1, s_2). (c) The linear classification boundaries of the hidden units are shown in 'S-space', in which the stimulus patterns are plotted. (d) The linear classification boundary of the output unit is shown in 'H-space', in which the representations of the patterns at the hidden layer are plotted.

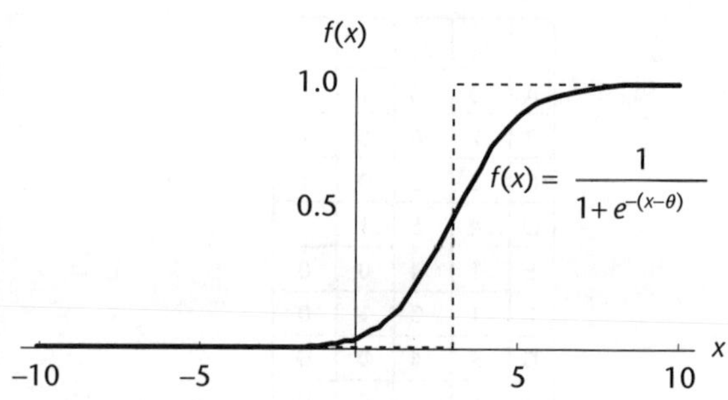

Figure 4. A sigmoid function. Backprop requires units that use differentiable functions. A sigmoid function is differentiable and has many of the important properties of the threshold function. Here, a sigmoid function is plotted (solid line) with a threshold function (dotted line). The value of θ is 3 for both functions. The sigmoid function only approaches its bounds asymptotically. The most common function used for this purpose is the 'logistic' function $f(x) = \frac{1}{1+e^{-x}}$.

1. Initialize the weights to random values.
2. Choose a random data item from the given set (stimulus and categorization value).
3. Compute the activities of the hidden units (first mapping), and from these the activities of the output units (second mapping). This is the 'forward propagation' of neural activity.
4. Compare the output responses with the target values and assign an error value to each output unit.
5. Compute the 'effective error' for each hidden unit as a function of the output unit errors and the hidden–output weights. This is the 'backward propagation' of error.
6. Modify the weights and biases.
7. Test the network on all items in the set of labeled data. If the number of incorrect classifications is acceptable, or if the number of iterations hits the maximum allowed, then stop – otherwise, go back to step 2.

Table 1 gives a more detailed description of the computations.

DEVELOPING INTERNAL REPRESENTATIONS FROM EXPERIENCE

In order for backprop to converge to a state that computes a 'difficult' task (such as one that is not LS), the mapping from the input to the hidden layer must give representations that simplify the task (e.g. rendering the task LS). Not only is convergence dependent on finding an appropriate set of internal representations, but so is the ability to generalize appropriately. Although backprop is not guaranteed to find an appropriate mapping of this kind, it does converge to a solution in many cases. In addition, it should be noted that an 'almost perfect' solution is acceptable for modeling many phenomena.

MODELING COGNITION

Networks have been trained with backprop to simulate cognitive functions ranging from perceptual tasks to high-order processes. These models address a broad range of scientific questions. In many cases, particularly in linguistics, the models have been used as counterexamples to assertions that certain cognitive capabilities must be innate or must require specific types of symbolic manipulation. Generally, backprop is used as an example of how a 'neural-like' system can extract statistical regularities from the environment, so that the performance of the system appears to follow 'rules' without any explicit encoding of those rules. The following examples from linguistics demonstrate the facility of backprop for developing models of cognitive tasks at several levels.

Mapping Text to Speech

In their simulation 'NetTalk', Sejnowski and Rosenberg (1987) trained a network to map English text to the corresponding phonological representation. Unlike some more regular languages, the correct pronunciation of a given letter in English is not always the same. However, it is not completely arbitrary, but depends in large part on the letters in the same neighborhood. Consider, for example, the pronunciations of the letter 'c' in the three nonword strings 'stince', 'stinch', and 'stinck'. The fact that most readers of English would agree on the pronunciation, even though they have never heard the words read aloud, indicates that there are 'rules' for pronunciation rather than arbitrary correspondences between letters and phonemes. NetTalk is trained on a corpus of text, and eventually is able to pronounce not only text from the

Table 1. The backprop algorithm

Algorithmic step	*Formulae*	*Pseudocode*

Hidden unit activities	$h_j = \frac{1}{1+e^{-x}}$
	where $x = b_j^{hid} + \sum_i w_{ij} s_i$
		`for j := 1 to nhid`
		`arg[j] := hidb[j] ;`
		`for i := 1 to ninput`
		`arg[j] := arg[j] + w[i,j] * s[i];`
		`h[j] := f(arg[j]);`
		`end`

Output unit responses	$r_k = \frac{1}{1+e^{-x}}$
	where $x = b_k^{out} + \sum_j v_{jk} h_j$
		`for k := 1 to noutput`
		`arg[k] := outb[k] ;`
		`for j := 1 to nhid`
		`arg[k] := arg[k] + v[j,k] * h[j] ;`
		`r[k] := f(arg[k]) ;`
		`end`

Output unit errors	$\delta_k^{out} = (T_k - r_k) r_k (1 - r_k)$	`for k := 1 to noutput`
		`dout[k] :=`
		`(T[k] - r[k]) * r[k] * (1 - r[k]) ;`

Hidden unit errors	$\delta_j^{hid} = \left[\sum_k v_{jk} \delta_k\right] h_j (1 - h_j)$	`for j := 1 to nhid`
		`dhid[j] := 0 ;`
		`for k := 1 to noutput`
		`dhid[j] := dhid[j] + v[j,k] * d[k] ;`
		`dhid[j] := dhid[j] * h[j] * (1 - h[j]) ;`
		`end`

Weight and bias adjustments	$\Delta w_{ij} = \eta \delta_j s_i$
	$\Delta v_{jk} = \eta \delta_k h_j$
	$\Delta b_j^{hid} = \eta \delta_j$
	$\Delta b_k^{out} = \eta \delta_k$
		`for j := 1 to nhid`
		`bhid[j] := bhid[j] + q * dhid[j] ;`
		`for i := 1 to ninput`
		`w[i,j] := w[i,j] + q * dhid[j] * s[i] ;`
		`for k := 1 to noutput`
		`bout[k] := bout[k] + q * dout[k] ;`
		`for j := 1 to nhid`
		`v[j,k] := v[j,k] + q * dout[k] * h[j] ;`

training corpus, but also text unseen during the training process. It extracts the mapping from text to speech without an explicit representation of the rules (see Figure 5).

Generating the Past Tense

The generation of past-tense verbs has been the subject of study by developmental psychologists because children almost universally go through similar stages on the path to adult competence in this task. A network simulation developed by Rumelhart and McClelland (1986), which did not use backprop, required a carefully designed representation in order to learn the task of mapping verbs from their present-tense forms to their past-tense forms. Their simulation successfully generalized from the training examples to novel verbs, both regular (e.g. *jump* and *jumped*) and irregular (e.g. *sing* and *sang*). In addition, it was able to mimic certain developmental stages of language learning in children over the course of its training. With the development of backprop, these results have been replicated in multi-layered networks that develop the requisite representations, rather than having them specified (Plunkett and Marchman, 1991).

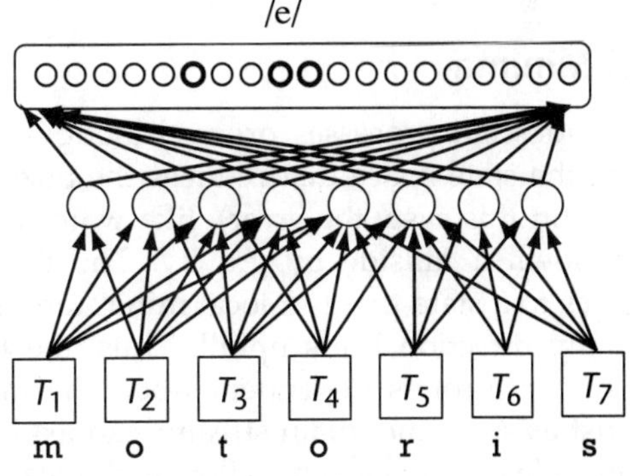

Figure 5. NetTalk. A sliding 7-character window of text (T_1, T_2, T_3, T_4, T_5, T_6, T_7) is presented as input to a network that is trained to generate a phonological representation of the central character (T_4). There are 29 input nodes for each of the 7 character positions (26 letters, space, period, and comma), a single hidden layer, and output units representing phonemic features.

Evolution of Language

Over the course of centuries, languages undergo subtle incremental changes that tend to reinforce regularity (Quirk and Wrenn, 1957) – that is, exceptions to rules are gradually lost, especially for verbs that occur with low frequency. Hare and Elman (1993) offer an explanation and support it by training a succession of networks on the past-tense task using backprop. The first network in their simulation is trained on present–past verb pairs from Old English. Before it achieves perfect performance, a second 'child' network is trained using the first network's computed past tenses. This process proceeds iteratively, each network learning imperfectly from its parent, and so the language evolves. While the process does not precisely follow the evolution of English, it exhibits similar properties with respect to increased regularization and the influence of word frequency.

ENHANCEMENTS TO BACKPROP

As a technique for training feedforward networks for classification tasks in many domains and for developing models of cognitive processes, backprop has been very successful. However, as a gradient-descent procedure, the pure form of the backprop procedure (commonly known as 'vanilla backprop') has some flaws, which can make it an inelegant, or even useless, approach. Some of these are discussed below.

Local Minima

Gradient-descent processes proceed along a path through the state space that reduces the objective function (in this case, the error), like water being driven down a hillside by gravity. The process converges to a state that is a local minimum, from which any direction leads uphill. Thus, the final value of the error is dependent upon the initial state, just as some mountain streams can lead to a high-altitude lake, while others flow to the sea. One imperfect, but simple, remedy is to add a 'momentum' term to the learning rule, which reinforces those components of weight changes that are common across learning trials.

Overfitting Training Data

The input–output items used for training are generally 'noisy'; that is, the output is not perfectly dependent on the input. A system with many adjustable parameters can be trained to fit noisy training data exactly, but only at the expense of its ability to generalize. By monitoring network performance on a set of 'test data' (consisting of items not used for training, but representative of the same sample set), training can be stopped before the system overfits the training data (Figure 6).

Slow Convergence

Gradient-descent processes tend to be slow. Learning of this kind is often criticized as a model of human learning for its inability to learn an item from a single exposure. The slow convergence is exacerbated by the complexity of the input–output relationship and the number of parameters. The speed of convergence is closely related to the value of the learning rate η.

Network Architecture

The determination of the best number of hidden units and their connectivity is a challenge, for which guesswork and trial-and-error are often relied upon. There are two primary approaches to optimizing network architecture. Networks can be 'grown' by starting with a minimal architecture and incrementally adding hidden units (Ash, 1989; Fahlman and Lebiere, 1990; Hanson, 1990) when the generalization error stops decreasing. At some point, the addition of more hidden units no longer benefits the network, or may even be a negative contribution.

Alternatively, one can begin with many more hidden units than are required and then 'prune'

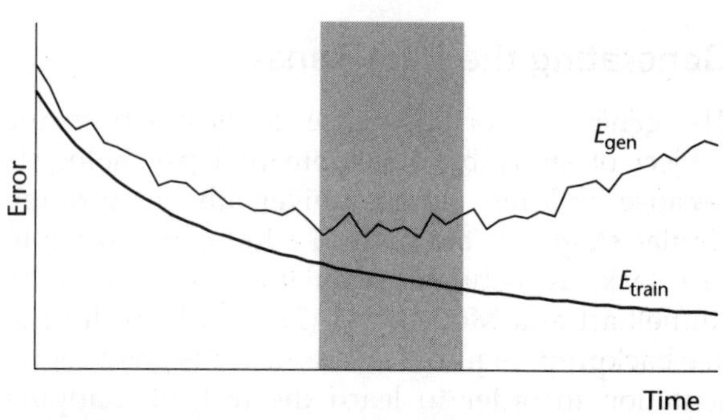

Figure 6. Two error measures. The error measured over the training set, E_{train}, steadily decreases over time. The error with respect to a set sampled from the same population, E_{gen}, also decreases during the first period of learning. Typically, for data sets with noise, E_{gen} eventually begins to steadily increase. Learning should be stopped when E_{gen} is at a minimum (at some point in the shaded region).

units that are deemed to have a low 'relevance' to the network task (Chauvin, 1989; Mozer and Smolensky, 1989; Le Cun *et al.*, 1990; Hassibi and Stork, 1991; Demers and Cottrell, 1993). With each removal, the network is retrained and the generalization error is measured. The cycle of pruning and retraining is continued until the generalization performance begins to deteriorate. The objective function (the function minimized by the gradient descent) is typically augmented by an additional term related to the number of active hidden units.

LEARNING TO PERFORM TEMPORAL TASKS

The first implementations of backprop were confined to feedforward networks with static inputs and static outputs. Temporal processes are more naturally accommodated by networks that have recurrence (i.e. they are not feedforward). The fact that many cognitive processes are temporal in nature has led to the development of strategies to enable the application of backprop to temporal tasks. Three main approaches are described below.

Time Delay Neural Networks

In a 'time delay neural network' (TDNN), the input nodes encode consecutive items from a discrete temporal sequence. The sequence is shifted one item at a time, presenting the network with a sliding window on the entire temporal pattern. The NetTalk architecture (Figure 5) is a TDNN that processes a sliding window of text as input and produces a stream of phonemes as output. While a TDNN combines signals from different points in time to interact at the input level, the temporal interaction is limited by the size of the window.

Simple Recurrent Networks

Jordan (1986) introduced the idea of cycling the output back to the input in order to learn sequences. The next step was Elman's (1990) *simple recurrent network* (SRN), which learns to recognize patterns within a sequence by storing 'temporal context' in the hidden layer. With each iteration of the learning procedure, the hidden unit activities from the previous iteration are treated as if they were part of the input to the network (Figure 7(a)). Thus, the hidden layer acts as a memory that can retain information over several time steps. Servan-Schreiber *et al.* (1989) explored grammar learning by an SRN. They showed that the SRN could not only detect errors (with 100% accuracy in a simple grammar), but could generate novel sequences as well.

Backpropagation Through Time

The functionality of a given recurrent network **R** with N nodes can be approximated for T timesteps by a feedforward network **F** that has T layers with N nodes per layer (see Figure 7(b)). In the 'backpropagation through time' procedure (Williams and Zipser, 1989), each layer of **F** represents one timestep of **R**. Thus there are T units in **F** corresponding to each unit in **R**. Each connection from a unit i to a unit j in **R** is replicated by a number of copies in **F**. The implementation of backprop on **F** is subject to the constraint that the replicants of a given weight in **R** have the same value.

UNSUPERVISED LEARNING WITH BACKPROPAGATION

Simply defined, the *autoencoder* is a network trained to compute an identity map; that is, the target pattern is the same as the input. The standard form is a strictly-layered network (i.e. there are only connections between adjacent layers) with a single hidden layer of K units, and an equal number (N) of input and output units. Typically, $K<N$, and since the output layer only has access to the hidden layer, it must reconstruct the input pattern from the reduced (encoded) representation. In order to learn this task successfully, the hidden unit representations must evolve such that each pattern is unique.

One of the most obvious applications of this (assuming $K<N$) is to reduce the dimensionality of data representation. Any data item that can be successfully generated by the network can be stored in a compressed form, simply by presenting it as input to the network and storing the hidden-unit representation. The well-known data compression technique of principal components analysis is mathematically similar, though not exactly the same (Baldi and Hornik, 1989). The original item can be reconstituted at the output by activating the hidden layer with the compressed representation (see Figure 8). With this technique, the data compression is 'lossy': that is, the reconstructed data are not guaranteed to be accurate.

Some applications of auto-encoders trained with back propagation are described below.

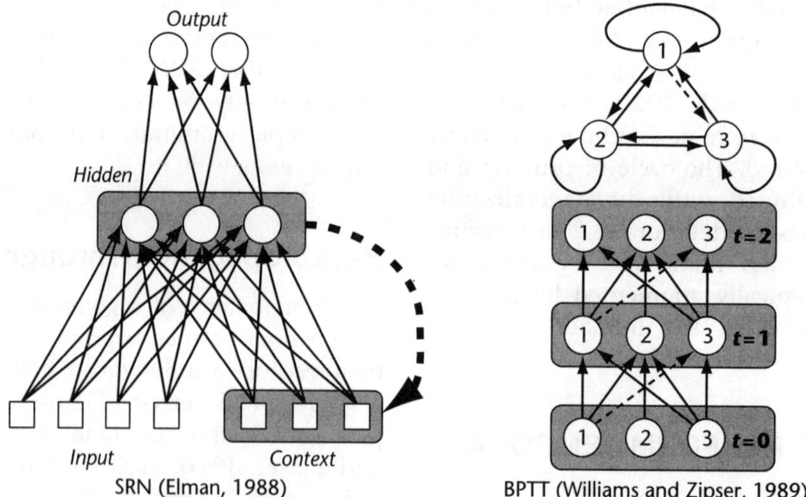

Figure 7. Architectures for temporal tasks. *Left* The simple recurrent network (SRN) maintains a history of the input sequence at the hidden layer by including the previous pattern of hidden unit activity as if it were part of a new input. *Right* A recurrent network (top) is approximated by a three-layer feed-forward architecture (bottom), where every node is replicated at every layer, and each layer corresponds to a different time step.

Image Compression

Cottrell *et al.* (1989) trained an auto-encoder on image data. After training, the network was able to represent the image using only 25% of the space required for the bitmap version. The reconstructed images had a very small deviation from the originals.

Novelty Detection

For some classification tasks, there are not enough data of one class to train a standard classifier (whether connectionist or not). After training an auto-encoder on items from the one class with sufficient data, any test item that shows low reconstruction error is classified as a member of the class used for training. If the reconstruction is poor, the item is classified as a member of the other class (we assume only two classes here). This approach has been used for predicting failure of electric motors using data from normally functioning motors only (Petsche *et al.*, 1996).

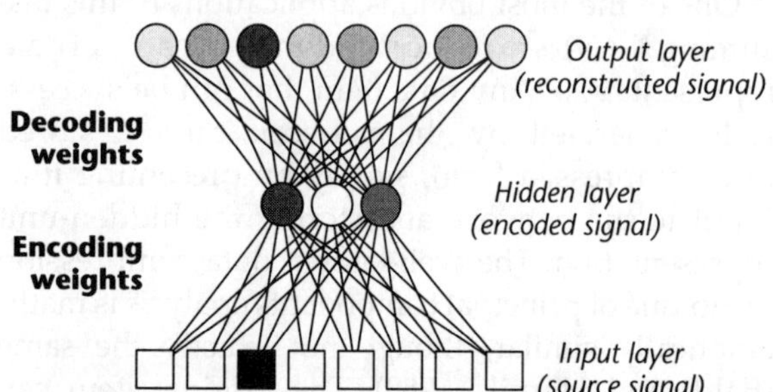

Figure 8. The autoencoder. This network is trained to reconstruct the input pattern at the output layer. Accurate reconstruction depends upon sufficient information in the hidden units, since they supply the only information available to the output units. Thus, the input is encoded by the input-hidden weights. The hidden unit representation is decoded by the hidden-output weights.

Random Access Auto-associative Memory

Pollack (1990) presented a technique by which an auto-encoder could develop a representation for a binary tree. In his scheme, each node in the tree is represented by an n-dimensional pattern of activity. An auto-encoding network with $2n$ input units, n hidden units, and $2n$ output units is used. Such 'random access auto-associative memories' have been applied as models of grammar learning and mental representations of music (Large *et al.*, 1995).

BACKPROPAGATION AS A MODEL OF HUMAN LEARNING

Generally, there are two aspects of backprop training that are of potential interest to cognitive scientists: the dynamic process of learning, and the properties of the network after learning. While it is recognized as an important technique for cognitive modeling, the application of backprop to account for cognitive phenomena has been criticized on several grounds. These include: biological implausibility, the nature of the teaching signal, and interference between learning old and new information.

Biological Plausibility

In part, the appeal of the neural-network approaches in artificial intelligence and cognitive science is their connection with biology. While there is ample neurobiological support for the notion that synaptic modification underlies learning, there is no specific mechanism known that corresponds to the error transmission implied by backprop. Furthermore, backprop has the inherent property of allowing weights to change sign, which would be analogous to an excitatory synapse becoming inhibitory or vice versa. A more biologically plausible procedure that is similar to backprop has been suggested by O'Reilly (1996). (*See* **Long-term Potentiation and Long-term Depression**)

Teaching Signals

In its pure form, backprop requires an explicit teaching signal to every output unit with every pattern presentation. However, a great deal of learning takes place with feedback that is much less specific, or entirely absent. A network with multiple output units can be trained with the minimal feedback of a scalar reward signal by introducing a second network that predicts the reward as a function of the first network's response (Munro, 1987).

Catastrophic Interference

If a network is trained with backprop on a set A of input–output pairs, and then trained on an independent set B, with no further training on the items from A, the performance on A deteriorates quickly. Eventually the items in A are forgotten (although they are on average more easily relearned than completely novel patterns). Partial remedies to the problem include: using two types of weights that change at different speeds (Hinton and Plaut, 1987); various forms of rehearsal (Ratcliff, 1990; Robins, 1995); and enforcing sparser hidden-layer representations (French, 1992; Krushke, 1993).

SUMMARY

Backprop has generated much attention, both inside and outside the cognitive science community. As a tool for cognitive modeling, it is still the best technique for abstracting the statistics of a task into a structure, and studying the internal representations that emerge and their influence on generalization performance. Thus, the acquisition of knowledge by the artificial system can be compared with human learning in several ways and over many modalities.

Aside from any relevance it has to cognition, backprop is also now a standard tool for machine learning, and performs well compared with other techniques. Variants of backprop are used in software in a broad range of application domains, from recognition of handwritten characters, to financial forecasting, to medical diagnosis. Of course, developers of commercial software are not concerned with cognitive and biological plausibility. They are generally more concerned with minimizing error rates than with emulating human patterns of error.

References

Ash T (1989) Dynamic node creation in backpropagation networks. *Connection Science* **1**: 365–375.

Baldi P and Hornik K (1988) Neural networks and principal component analysis: learning from examples without local minima. *Neural Networks* **2**: 53–58.

Chauvin Y (1989) A back-propagation algorithm with optimal use of hidden units. In: Touretzky DS (ed.) *Advances in Neural Information Processing Systems 1*. San Mateo, CA: Morgan Kaufmann.

Cottrell G, Munro P and Zipser D (1989) Image compression by back propagation: an example of extensional programming. In: Sharkey NE (ed.) *Models of Cognition: A Review of Cognitive Science*, vol. I, pp. 208–240. Norwood, NJ: Ablex.

Demers D and Cottrell G (1993) Non-linear dimensionality reduction. In: Hanson SJ, Cowan JD and Giles CL (eds) *Advances in Neural Information Processing Systems 5*. San Mateo, CA: Morgan Kaufmann.

Elman J (1990) Finding structure in time. *Cognitive Science* **14**: 179–211.

Fahlman S and Lebiere C (1990) The cascade-correlation learning architecture. In: Touretzky DS (ed.) *Advances in Neural Information Processing Systems 2*. San Mateo, CA: Morgan Kaufmann.

French R (1992) Semi-distributed representations and catastrophic forgetting in connectionist networks. *Connection Science* **4**: 365–377.

Hanson S (1990) Meiosis networks. In: Touretzky DS (ed.) *Advances in Neural Information Processing Systems 2*. San Mateo, CA: Morgan Kaufmann.

Hare M and Elman JL (1993) From weared to wore: a connectionist account of language change. In: *Proceedings of the 15th Meeting of the Cognitive Science Society*, pp. 265–270. Princeton, NJ: Erlbaum.

Hassibi B and Stork DG (1993) Second order derivatives for network pruning: optimal brain surgeon. In: Hanson SJ, Cowan JD and Giles CL (eds) *Advances in Neural Information Processing Systems 5*. San Mateo, CA: Morgan Kaufmann.

Hebb D (1949) *The Organization of Behavior*. New York, NY: Wiley.

Hinton G and Plaut D (1987) Using fast weights to deblur old memories. In: *Proceedings of the 9th Meeting of the Cognitive Science Society*, pp. 177–186. Princeton, NJ: Erlbaum.

Jordan M (1986) Attractor dynamics and parallelism in a connectionist sequential machine. In: *Proceedings of the 8th Meeting of the Cognitive Science Society*, pp. 531–546. Princeton, NJ: Erlbaum.

Krushke J (1993) Human category learning: implications for backpropagation models. *Connection Science* **5**: 3–36.

Large E, Palmer C and Pollack J (1995) Reduced memory representation for music. *Cognitive Science* **19**: 53–96.

LeCun Y (1985) *Modeles Connexionnistes de l'Apprentissage*. PhD thesis, Université Pierre et Marie Curie, Paris, France.

LeCun Y, Denker J and Solla S (1990) Optimal brain damage. In: Touretzky DS (ed.) *Advances in Neural Information Processing Systems 2*. San Mateo, CA: Morgan Kaufmann.

Mozer MC and Smolensky P (1989) Skeletonization: a technique for trimming the fat from a network via relevance assessment. In: Touretzky DS (ed.) *Advances in Neural Information Processing Systems 1*. San Mateo, CA: Morgan Kaufmann.

Munro P (1987) Dual backpropagation: a scheme for self-supervised learning. In: *Proceedings of the 9th Meeting of the Cognitive Science Society*. Princeton, NJ: Erlbaum.

O'Reilly R (1996) Biologically plausible error-driven learning using local activation differences: the generalized recirculation algorithm. *Neural Computation* **8**: 895–939.

Parker DB (1982) *Learning-Logic*. Invention Report S81-64, File 1. Stanford, CA: Office of Technology Licensing, Stanford University.

Petsche T, Marcantonio A, Darken C *et al.* (1996) A neural network autoassociator for induction motor failure prediction. In: Touretzky DS, Mozer MC and Hasselmo ME (eds) *Advances in Neural Information Processing Systems*, vol. VIII, pp. 924–930. Cambridge, MA: MIT Press.

Plunkett K and Marchman V (1991) U-shaped learning and frequency effects in a multi-layered perceptron: implications for child language acquisition. *Cognition* **38**: 43–102.

Pollack J (1990) Recursive distributed representations. *Artificial Intelligence* **46**: 77–105.

Quirk R and Wrenn CL (1957) *An Old English Grammar*. London: Methuen.

Ratcliff R (1990) Connectionist models of recognition memory: constraints imposed by learning and forgetting functions. *Psychological Review* **97**: 285–308.

Robins A (1995) Catastrophic forgetting, rehearsal, and pseudorehearsal. *Connection Science* **7**: 123–146.

Rosenblatt F (1958) The perceptron: a probabilistic model for information storage and organization in the brain. *Psychological Review* **65**: 386–408.

Rumelhart DE, Hinton GE and Williams RW (1986) Learning internal representations by error propagation. In: (Rumelhart and McClelland, 1986), pp. 318–364.

Rumelhart DE and McClelland JL (1986) *Parallel Distributed Processing: Explorations in the Microstructure of Cognition*, vol. I 'Foundations'. Cambridge, MA: MIT Press/Bradford.

Sejnowski TJ and Rosenberg CR (1987) Parallel networks that learn to pronounce English text. *Complex Systems* **1**: 145–168.

Servan-Schreiber D, Cleermans A and McClelland J (1989) Learning sequential structure in simple recurrent networks. In: Touretzky DS (ed.) *Advances in Neural Information Processing Systems 1*. San Mateo, CA: Morgan Kaufmann.

Werbos P (1974) *Beyond Regression: New Tools for Prediction and Analysis in the Behavioral Sciences*. PhD thesis, Harvard University.

Widrow B and Hoff M (1960) Adaptive switching circuits. In: *1960 IRE WESCON Convention Record*, pp. 96–104. New York, NY: IRE.

Williams R and Zipser D (1989) A learning algorithm for continually running fully recurrent neural networks. *Neural Computation* **1**: 270–280.

Further Reading

Allman W (1990) *Apprentices of Wonder: Inside the Neural Network Revolution*. New York, NY: Bantam.

Anderson J (1995) *Introduction to Neural Networks*. Cambridge, MA: MIT Press.

Anderson J (2000) *Talking Nets: An Oral History of Neural Networks*. Cambridge, MA: MIT Press.

Ballard D (1999) *An Introduction to Natural Computation*. Cambridge, MA: MIT Press.

Bishop C (1996) *Neural Networks for Pattern Recognition*. Oxford: Oxford University Press.

Hebb D (1949) *The Organization of Behavior*. New York, NY: Wiley.

Reed R and Marks R (1999) *Neural Smithing*. Cambridge, MA: MIT Press.

Rumelhart DE and McClelland JL (1986) *Parallel Distributed Processing: Explorations in the Microstructure of Cognition*, vol. I 'Foundations'. Cambridge, MA: MIT Press/Bradford.

Bartlett, Frederic Charles

Introductory article

Henry L Roediger, Washington University, St Louis, Missouri, USA

CONTENTS

Frederic C. Bartlett (1886–1969) was a distinguished British psychologist who spent most of his career at Cambridge University. He is chiefly known today for his book Remembering: A Study in Experimental and Social Psychology, *which laid the foundation for schema theory.*

INTRODUCTION

Frederic C. Bartlett (1886–1969) was a distinguished British psychologist who spent most of his career at the University of Cambridge. He was trained as an experimental psychologist and became the most prominent English psychologist of his generation through the influence of his writings, his work on applied problems, and the great students he trained who continued work in his tradition. He is chiefly remembered today for his 1932 book, *Remembering: A Study in Experimental and Social Psychology*, which laid the foundation for schema theory and pioneered the study of memory distortions. Bartlett was knighted in 1948 for his great accomplishments, which are described briefly below.

BIOGRAPHICAL DETAILS

Early Life and Education

Bartlett was born in Stow-on-the-Wold, a small country town in Gloucestershire. His father was a successful businessman who made shoes and boots, but the educational opportunities in town were slim. A severe illness when he was 14 years old made it impossible for him to attend a boarding school, so young Bartlett stayed in Stow and educated himself with the aid of his family (his father had a great library) and friends. He eventually took a distance course at the University of London and settled on psychology, logic, sociology, and ethics as topics of study. He received an MA from London in 1911 and continued to Cambridge, where he came under the influence of W. H. R. Rivers, Cyril Burt, and C. S. Myers. He obtained his doctorate with first-class honors in 1914, just as Burt decided to leave Cambridge. Myers then offered Bartlett Burt's vacated position, so Bartlett stayed in Cambridge.

The First World War and Bartlett's Development

The First World War broke out soon after Bartlett took up his position at Cambridge. Most of Bartlett's colleagues left to aid the war effort, but poor health prevented him from joining them. However, the absence of people senior to him thrust him into the role of leading the psychological laboratory. He threw himself into teaching and began writing a book based on his dissertation, although it would not appear for many years. Much of his research during this time focused on practical problems driven by the war, such as detecting weak auditory signals in noise (to help with the problem of detecting German submarines). His war work eventually culminated in a book, *The Psychology of the Soldier* (1927).

After the war, Rivers and Myers returned to Cambridge and became Bartlett's associates. However, in 1922 Rivers died suddenly and Myers retired, so Bartlett became director of the Cambridge Laboratories and built them into a research powerhouse over the years. In the 1920s Bartlett's research turned to social anthropology, an early interest, and he wrote *Psychology and Primitive Culture* (1923). His international reputation expanded and he came to know distinguished psychologists from around the world.

REMEMBERING

In 1932 Bartlett published his great book, which is still in print today. *Remembering* actually grew out of his dissertation experiments begun in 1913, so the gestation period was nearly 20 years. The book introduced a very different tradition for studying memory from the scientific methods of Ebbinghaus with their emphasis on careful control and measurement of memory in rather unnatural conditions. Bartlett's methods were casual, almost anecdotal, compared with those of Ebbinghaus, yet he uncovered powerful truths about remembering that reverberate through the field even today. Bartlett tested people under fairly relaxed conditions and his 'data' consisted largely of verbal reports with which he sprinkled his writing. (*See* **Ebbinghaus, Hermann**)

The early chapters of *Remembering* actually consist of studies of perceiving. The great middle part of the book is directly concerned with memory. The last section of the book deals with social and anthropological factors in cultural transmission. The general thrust of the book is to emphasize the constructive nature of cognition. Perceiving, remembering, and all of thinking involve the individual as part and parcel of the cognitive process. For example, in perceiving an ambiguous stimulus that is briefly presented, one's past background and experience determine what is perceived as much as (or even more than) the stimulus that is presented.

Bartlett devised two methods to study remembering: repeated reproduction and serial reproduction. In his most famous work he read a native American folk tale, *The War of the Ghosts*, to his British participants and then later tested their memories. This bizarre and supernatural story was usually read twice, aloud. In the repeated reproduction technique Bartlett would have his listeners recall the story after an interval of about 15 min. Next he would test their memory for the story at various later times, but with no further presentations of the story. Thus, repeated reproduction involves the same individual repeatedly reproducing the story, as the name implies. Bartlett's interest centered on how people remembered the story and how their memories would change over time and repeated retellings.

Not surprisingly, people remembered less about the story as time passed – their reports became increasingly short. Of more interest was the content of what they did remember and what these recollections indicated about the workings of memory. Besides becoming shorter, the stories became simpler, supernatural elements dropped out and other bizarre items would be reinterpreted. Bartlett called this process 'rationalization' because people added material to explain unnatural elements, or dropped them out altogether if they did not seem to fit the person's past experience. Rationalization over repeated retellings caused the story 'to be robbed of all its surprising, jerky and inconsequential form, and reduced it to an orderly narration' (p. 153 of the 1932 edition of *Remembering*). Bartlett also referred to the 'effort after meaning' that occurred in his perception and memory experiments, whereby people try to convert or recode elements that are difficult to perceive or understand into forms that can be comprehended. People try to impose structure and order to understand the world around them, even when their experience does not conform neatly to their prior categories.

Bartlett wrote that 'the most general characteristic of the whole of this group of experiments was the persistence, for any single subject, of the form of his first reproduction', and the use of 'a general form, order and arrangement of material seems to be dominant, both in initial reception and in subsequent remembering' (p. 83). He named this general form that people use to encode and to remember experiences a 'schema', a term now used throughout the cognitive sciences. A schema is a general organization of a story of a typical event. So, for example, many old films about the American wild west follow a schema involving 'good guys', 'bad guys', crisis, and resolution. The schema can aid encoding and retention of details that are consistent with it, but details that do not fit may be forgotten or distorted to fit the schema. In remembering *The War of the Ghosts* some English participants seemed to use the schema of a fairy tale, a genre to which they were more accustomed. Some even tacked on a moral at the end of the story.

The method of serial reproduction, the other major technique Bartlett introduced, is like the children's game of rumor or telephone. One person hears *The War of the Ghosts* (or is exposed to some other material) and recalls it after a set period. This person's recollections are then read to a second person, who recalls it in turn. This second recall is then read to a third person for later recall, and so on, through as many instantiations as desired. The changes in recall across repeated tests using the serial reproduction method are much greater than those in repeated reproduction, although Bartlett thought the same types of memory processes were at work (but in greater force). The serial reproduction technique involves a human chain, and if there were to be one weak link in the chain – someone who was wildly inaccurate in recall – then there would be no hope of a person later in the chain correcting the false memory of the material because that person would never have been exposed to the correct version. Reading through the lengthy samples that Bartlett provided in

Remembering (chapters 7 and 8) leads to agreement with his basic claims. The serial reproduction technique was later championed by psychologists studying the transmission of rumors.

The serial reproduction technique also served, Bartlett believed, as a useful analogy for the way information might be handed down from one generation to another within a society or even for the spread of ideas from culture to culture. He dealt with these issues in some detail, although with anecdotal evidence, in the last section of his book.

Bartlett's *Remembering* provides many interesting ideas and quotable passages. The book was well known at the time, but his research tradition did not really catch on. Part of the reason for this is that, in his hands, the research was more anecdotal than experimental (despite the subtitle of his book). He has been criticized for this lack of careful empirical research to document his points, and it was not until recently that a successful replication of his basic findings using the repeated reproduction technique appeared in print. Bartlett's book came to the forefront of the field when Neisser adopted Bartlett's theme of the constructive nature of cognition for his 1967 text, *Cognitive Psychology*, which helped to usher in the cognitive revolution in psychology. In the early 1970s psychologists such as Elizabeth Loftus, John Bransford and Marcia Johnson became interested in errors of memory and Bartlett's ideas were invoked and his book was once again read by a new generation.

Throughout *Remembering*, Bartlett's message ran counter to the idea that memory should be conceived of as static memory traces that are called to mind and read off in a more or less accurate fashion. Memory does not work like a video recorder, tape recorder, or computer. In his words, 'Remembering is not the re-excitation of innumerable fixed, lifeless and fragmentary traces' (1932, p. 213). Rather, 'remembering appears to be far more decisively an affair of construction rather than reproduction' (p. 205). 'It is an imaginative reconstruction, or construction, built out of the relation of our attitude towards a mass of organized past reactions or experiences' (p. 213). This credo still guides the field today in many ways.

LATER CONTRIBUTIONS

The Second World War confronted psychologists with many more practical problems to be solved. Bartlett and Kenneth Craik worked during the war on problems of skill acquisition, and Bartlett served on the Royal Air Force's Flying Personnel Research Committee, focusing his work on pilot training. They also studied related topics such as the effects of fatigue on performance. When Craik was tragically killed in an automobile accident two days before the war in Europe ended, Bartlett felt the loss keenly, because the men had become best friends as well as close collaborators.

After the war, Bartlett applied notions of skill learning to those of higher-order thinking, capitalizing on the insight that just as experts in a physical skill develop their exquisite expertise after many hours of practice, so do experts in thinking skills – problem-solving, reading X-ray graphs and so on. In 1958 he published *Thinking: An Experimental and Social Study*, which provided his insights on these topics. However, this book did not enjoy the earlier success of *Remembering*, although it too is an interesting treatise.

Bartlett retired from the chair of experimental psychology in Cambridge in 1952, but maintained his affiliation with the applied psychology unit which he had helped to found. His many students frequently called on him for advice and he continued to serve on national committees. Despite his early health difficulties, he remained generally robust in his later years, although he was bothered by hearing loss. He died after a brief illness on 30 September 1969.

CONCLUSION

Frederic Charles Bartlett wielded tremendous influence both nationally and internationally. Some commentators have remarked that this influence was out of proportion to his actual scholarly work. His contributions were good, but only one (*Remembering*) was of enduring importance. Rather, Bartlett's own charismatic character drew people to him and established his leadership, the power of his personality infecting those around him with his wit, his wisdom, his generosity, and his good nature.

Knighted, in 1948, Sir Frederic Bartlett received many other honors, including honorary doctorate degrees from seven universities in six countries. In Britain he was elected to the Royal Society in 1932 and received its Baly and Huxley medals in 1943. He was awarded the Royal Medal in 1952, the highest distinction a scientist in Britain can receive. In the USA Bartlett was elected to the American Philosophical Society, the National Academy of Sciences (as a foreign fellow) and the American Association of Arts and Sciences. Bartlett was a towering figure of twentieth-century psychology, and in recent years the study of human memory has come around to the approach he advocated so strongly in the 1930s.

Further Reading

Allport GW and Postman L (1947) *The Psychology of Rumor*. New York, NY: Holt.

Bartlett FC (1916) An experimental study of some problems of perceiving and imaging. *British Journal of Psychology* **8**: 222–266.

Bartlett FC (1923) *Psychology and Primitive Culture*. Cambridge, UK: Cambridge University Press.

Bartlett FC (1927) *Psychology and the Soldier*. Cambridge, UK: Cambridge University Press.

Bartlett FC (1932) *Remembering: A Study in Experimental and Social Psychology*. Cambridge, UK: Cambridge University Press.

Bartlett FC (1958) *Thinking: An Experimental and Social Study*. London, UK: Allen & Unwin.

Bergman E and Roediger HL (1999) Can Bartlett's repeated reproduction experiments be replicated? *Memory and Cognition* **27**: 937–947.

Kintsch W (1995) Foreword. In: Bartlett FC, *Remembering: A Study in Experimental and Social Psychology*. [reprint] Cambridge, UK: Cambridge University Press.

Neisser U (1967) *Cognitive Psychology*. New York, NY: Appleton-Century-Crofts.

Roediger HL (2000) Sir Frederic Charles Bartlett: experimental and applied psychologist. In: Kimble GA and Wertheimer M (eds) *Portraits of Pioneers in Psychology*, vol. 4, pp. 149–161. Mahwah, NJ: Lawrence Erlbaum.

Basal Forebrain

Intermediate article

Eve De Rosa, Stanford University School of Medicine, Stanford, California, USA
Mark G Baxter, Harvard University, Cambridge, Massachusetts, USA

CONTENTS

The basal forebrain is a complex of subcortical nuclei that project widely to cortical and limbic areas involved in cognitive function. Damage to the basal forebrain is associated with cognitive deficits. The contributions of particular neuroanatomical and neurochemical components of the basal forebrain to different aspects of cognitive function can be dissociated to some extent.

INTRODUCTION

The term 'basal forebrain' commonly refers to an extended continuum of subcortical neurons that projects to diverse limbic and neocortical areas implicated in various aspects of cognitive function. Damage to the basal forebrain region can result in global cognitive impairments. For instance, aneurysms of the anterior communicating artery that injure the basal forebrain are associated with amnesia and impairments in executive function. Cognitive deficits in both normal aging and age-related pathological conditions have also been associated with the basal forebrain. The severity of cognitive impairment observed in Alzheimer disease is correlated with the extent of deterioration of cholinergic neurons in the basal forebrain. A similar relationship between cognitive impairment and alterations in basal forebrain cholinergic neurons is seen in normal aging. For these reasons, cholinergic neurons have been central to most explanations of the cognitive effects of basal forebrain damage. Hypotheses regarding the involvement of the basal forebrain cholinergic system in global aspects of cognitive function have been gradually revised as more and more selective lesion methods have become available for experimental studies of this region (Wenk, 1997).

Attempts to identify the specific cognitive functions of basal forebrain cholinergic neurons have suggested that these neurons do not play a specific part in memory processing *per se*, or a general role in sustaining the function of their cortical targets. Instead, damage limited to basal forebrain cholinergic neurons produces highly restricted impairments in aspects of sensory information processing and attention.

ANATOMY AND CONNECTIONS OF THE BASAL FOREBRAIN

Cholinergic basal forebrain neurons are intermingled with a substantial population of noncholinergic neurons that share similar projection patterns (Gritti *et al.*, 1997). Estimates of the proportion of cortically projecting basal forebrain neurons that are cholinergic vary from study to study but are generally in the range 30–50 percent. Many noncholinergic neurons in the basal forebrain may be local circuit neurons, receiving cortical input and modulating activity of cortically projecting cholinergic and noncholinergic neurons (Zaborszky *et al.*, 1997).

The basal forebrain can be divided into four groups of cells: the medial septum, projecting primarily to the hippocampus; the vertical limb of the diagonal band of Broca, projecting primarily to the hippocampus and cingulate cortex; the horizontal limb of the diagonal band of Broca, projecting primarily to the olfactory bulb, piriform cortex and entorhinal cortex; and the nucleus basalis magnocellularis or substantia innominata, projecting primarily to the neocortex and amygdala (Figure 1). These cholinergic nuclei have also been designated Ch1 to Ch4 (Mesulam *et al.*, 1983); this nomenclature approximately corresponds to the above divisions. The organization of the basal forebrain is similar in the primate and in the rat, although subdivisions of the nucleus basalis can be identified reliably in the primate but not in the rat. Some areas of the basal forebrain are reciprocally connected with their targets; anatomical experiments in rats have shown that the medial septum and diagonal band project to the hippocampus and medial prefrontal cortex and receive projections back from these structures. Similarly, the nucleus basalis is reciprocally connected with the amygdala. The basal forebrain also receives inputs from the hypothalamus, as well as from the midbrain and upper pons (Wainer and Mesulam, 1990). Inhibitory projections from the nucleus accumbens to the basal forebrain may have a particular role in regulating cortical acetylcholine release (Sarter and Bruno, 2000).

INVOLVEMENT OF THE BASAL FOREBRAIN IN MEMORY

Considerable attention has been devoted to the magnocellular cholinergic neurons of the basal forebrain and their function in memory, because of the hypothesis that cholinergic dysfunction specifically is partially or wholly responsible for the memory deficits seen after damage to the basal forebrain (the 'cholinergic hypothesis'). Alzheimer disease involves loss of these neurons. Postmortem examinations of brains from people with Alzheimer disease have found a marked cell loss in the basal forebrain relative to the brains of healthy people. In Alzheimer disease there is a loss of choline acetyltransferase and acetylcholinesterase in the cortical targets of the basal forebrain. The enzyme choline acetyltransferase synthesizes acetylcholine; once released from the presynaptic terminal, acetylcholine is degraded by the enzyme acetylcholinesterase. Cognitive impairment in

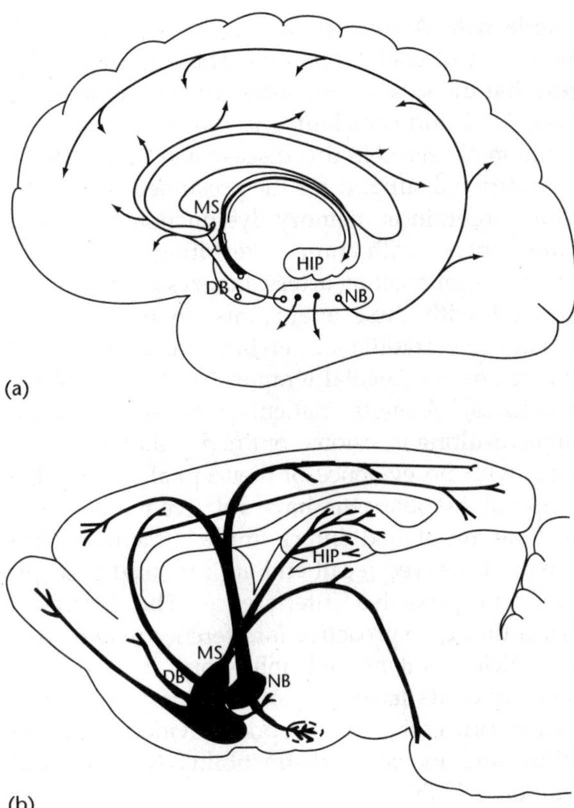

Figure 1. The basal forebrain cholinergic system, schematically represented in sagittal views of (a) human and (b) rat brain. The basal forebrain can be roughly divided into three major divisions (rostral to caudal): the medial septum (MS), projecting primarily to hippocampus (HIP); the diagonal band nuclei (DB), consisting of the vertical limb of the diagonal band of Broca, projecting to the hippocampus and cingulate cortex, and the horizontal limb of the diagonal band of Broca, projecting to the olfactory bulb and entorhinal cortex; and the nucleus basalis (NB), projecting to neocortex and amygdala. These cell groups share similar projection patterns in both species. The substantia innominata, which forms a less discrete nucleus in the rat than in the primate, sends cholinergic projections to the neocortex.

people with Alzheimer disease is correlated with the extent of cholinergic loss. These findings suggest that the loss of cholinergic transmission in the basal forebrain contributes to the mnemonic dysfunction observed in this disease (Collerton, 1986).

Additional support for the presumed role of the basal forebrain in memory dysfunction is derived from patients with amnesia resulting from an anterior communicating artery aneurysm. Infarcts associated with these aneurysms do not typically damage the traditional cerebral areas implicated in amnesia (i.e. medial temporal and diencephalic structures). Amnesic patients who sustained infarcts resulting in lesions confined to the basal forebrain with no evidence of diencephalic or medial temporal involvement have relatively intact immediate recall and intact implicit memory, but impaired delayed recall and an increased susceptibility to proactive interference. The increased susceptibility to proactive interference may be due to deficient attentional inhibitory mechanisms. These patients have also shown a reduced information processing speed, poor divided attention ability and increased distractibility (DeLuca and Diamond, 1995).

The findings in the above disorders suggest that the basal forebrain contributes to mnemonic dysfunction, motivating the development of animal models of basal forebrain damage to elucidate the exact role of the cholinergic basal forebrain system in mediating such dysfunction. In these models, the basal forebrain cholinergic system or its cholinergic targets are damaged and behavioral tests are performed to determine the consequent pattern of mnemonic function. The validity of these models depends on the similarity of the cognitive processes being tested in animals and in humans, as well as on the anatomical homology of the basal forebrain in animals and humans.

To confirm the cholinergic hypothesis in an animal model, two challenges must be met: any memory deficits observed must be due to a specific impairment in memory processes, as opposed to disruption of other cognitive or noncognitive processes; and the deficits must not be attributable to destruction of other populations of noncholinergic neurons in the basal forebrain (Sarter and Bruno, 1997).

Most initial studies used stereotaxic placement of electrolytic or excitotoxic lesions in different regions of the basal forebrain in rats. Interpretation of these studies has been made difficult by the fact that electrolytic lesions damage mixed populations of neurons as well as passing fibers, and that excitotoxic lesions – produced, for example, by ibotenic acid, kainic acid, quisqualic acid or α-amino-3-hydroxy-5-methylisoxazole-4-propionic acid (AMPA) – preserve the passing fibers but still damage mixed populations of neurons. Ibotenate and other excitotoxin-induced lesions of the nucleus basalis magnocellularis disrupt learning in working memory tasks (such as delayed alternation) and reference memory tasks (measuring spatial or conditional visual discrimination learning).

The cholinergic hypothesis was called into question, however, because lesions of the basal forebrain made with other excitotoxins (quisqualic acid and AMPA) resulted in greater destruction of cortically projecting cholinergic neurons of the nucleus basalis magnocellularis than that seen following ibotenate lesions, but induced only a few of the learning and memory deficits caused by ibotenate. Consequently, many of the deficits induced by ibotenate lesions of the basal forebrain cannot be attributed to a primary destruction of cholinergic neurons in the nucleus basalis magno cellularis (Everitt and Robbins, 1997).

Because these lesion techniques are not selective for cholinergic neurons and therefore the resulting behavioral deficits may be due in part to damage in some neighboring neuronal system, explicit tests of the cholinergic hypothesis require a neurotoxin specific for basal forebrain cholinergic neurons. The immunotoxin, 192 IgG-saporin, provides a route for directly targeting the toxin saporin to cholinergic neurons *in vivo* in the rat without damaging noncholinergic neurons at the lesion site. Immunotoxic lesions restricted to specific cholinergic nuclei of the basal forebrain have generally failed to produce mnemonic deficits. It is noteworthy that the more specific the cholinergic lesion, the less dramatic the effects on learning and memory (Everitt and Robbins, 1997; Baxter and Chiba, 1999). The studies with less specific excitotoxic lesions underscore the importance of careful validation that basal forebrain lesions intended to be specific to cholinergic neurons do actually spare noncholinergic neurons at the lesion site (for related discussion see Chappell *et al.*, 1998).

The foregoing experiments in rats suggesting a role for the basal forebrain in memory function but no specific role for cholinergic basal forebrain neurons are supported by experiments in nonhuman primates. These animals have a more spatially distinct nucleus basalis of Meynert, and comparisons of cognitive deficits between humans and monkeys are more plausible than between humans and rats. Voytko (1996) and her colleagues extensively and systematically investigated the effect of administration of ibotenic acid lesions placed stereotaxically into the basal forebrain cholinergic nuclei of Old World (macaque) monkeys. Their battery of tests (delayed nonmatching to sample, delayed response, simultaneous discrimination, and a visuospatial attention task) revealed no mnemonic deficits. There is some contradiction between the findings of ibotenic lesions of the basal forebrain nuclei in Old and New World monkeys. Ibotenate lesions to the nucleus basalis of Meynert led to impairments in retention of preoperatively learned visual discriminations and reversal learning of visual discriminations postoperatively in New World monkeys. This was accounted for by possible nonspecific damage in the lesions made in New World monkeys. Administration of the putative cholinergic toxin ME20.4 IgG-saporin to the basal forebrain nuclei of New World monkeys led to perceptual impairments on retention and acquisition of visual discriminations (Fine *et al.*, 1997), although sparing of noncholinergic neurons at the lesion site in these monkeys was not verified.

Aims to increase acetylcholine levels in the clefts of the cholinergic synapse with acetylcholinesterase inhibitors, or to directly replace acetylcholine with cholinomimetic drugs, have largely been unsuccessful in reversing cognitive deficits in people with Alzheimer disease. This may be because direct enhancement of acetylcholine levels may disrupt any remaining intact cholinergic transmission. An alternative approach, trans-synaptic modulation, describes the effects of changes in the activity of an afferent neuronal system on the excitability of its target neurons, e.g. the effects of afferents mediated by γ-aminobutyric acid (GABA), originating locally or distally, on the cholinergic neurons of the basal forebrain. Another complementary, pharmacological strategy to treatment of cognitive deficits consequent to cholinergic loss would be to modulate the cholinergic system by trans-synaptic mechanisms, for example by affecting cholinergic transmission with GABA. This approach may be more effective in alleviating the cognitive deficits since this acts to amplify the endogenous physiological cholinergic signal (Sarter and Bruno, 1997). However, given that loss of cholinergic neurons does not seem to be a primary cause of memory impairment in these conditions, even pharmacological treatments aimed at enhancing function of remaining cholinergic neurons by trans-synaptic modulation might be expected to have little beneficial effect on memory.

In summary, the most recent and stringent tests of the cholinergic hypothesis in animal models have not supported an essential role for cholinergic involvement in learning and memory functions *per se*. In contrast, basal forebrain cholinergic neurons do appear to have an essential role in regulating attentional processing capacity.

INVOLVEMENT OF THE BASAL FOREBRAIN IN ATTENTION

Mechanisms of attention are thought to enable the efficient allocation of processing resources in order to respond to important events in the environment. There are different types of attentional processes: selective or focused attention targets attentional resources on a restricted number of sensory stimuli and excludes other sensory stimuli; sustained attention or vigilance reflects a state of readiness to detect and respond to unpredictable or rare events; and divided attention refers to simultaneous monitoring of several different channels of sensory information. Attention filters what sensory information is introduced into memory. The basal forebrain appears to regulate many of these attentional functions; recent studies of lesions of the cholinergic forebrain in rats and monkeys have suggested deficits in several different paradigms for measuring attention, including selective spatial, divided and sustained attention.

The five-choice serial reaction time task, based on a human test of continuous performance, measures sustained spatial attention in rats. In this task, rats are required to monitor the occurrence of a light stimulus in one of five locations. With this paradigm, excitotoxic lesions (e.g. ibotenate acid, quisqualic acid and AMPA) of the nucleus basalis magnocellularis (NBM) produce impairments in performance, which are ameliorated by increasing the duration of the stimuli. In support of the involvement of the cholinergic basal forebrain, this deficit is alleviated by an acetylcholinesterase inhibitor, physostigmine (Everitt and Robbins, 1997). A role for the NBM in attention has also been demonstrated in a temporal divided attention task, where at times rats are required to divide their attention between two simultaneous stimuli: rats with excitotoxic lesions of the NBM are specifically impaired in their ability to time two stimuli simultaneously (Olton *et al.*, 1988). In a cross-modal divided attention task, infusion of 192 IgG-saporin into the basal forebrain selectively increases the response latencies in a bimodal condition, without affecting the accuracy or the retrieval and execution of the stimulus–response action in either the visual or auditory modality (Turchi and Sarter, 1997).

Basal forebrain neurons, including the cholinergic neurons, receive direct contacts from GABA-mediated afferents and are inhibited by GABA.

Like cholinergic lesions, infusions of a benzodiazepine receptor agonist into the basal forebrain impair the performance of rats tested in a sustained attention without altering the efficacy of perceptual processes. Infusion of the $GABA_A$ receptor agonist, muscimol, directly into the basal forebrain both decreases the activity of the cortically projecting NBM neurons and impairs accuracy on the five-choice task. Two-lever vigilance tasks present the rat with successive trials of changes in the intensity of continuously delivered auditory stimuli (e.g. white noise) or visual stimuli (e.g. illumination). One lever measures the 'hits' and the other lever measures the correct rejections. In two-lever vigilance tasks for rats, both 192 IgG-saporin lesions of the NBM and intra-NBM administration of $GABA_A$ benzodiazepine receptor agonists have impaired vigilance performance (Sarter and Bruno, 1997).

In a visuospatial attention task, Old World monkeys with basal forebrain lesions showed impairments in shifting attention (Voytko, 1996). A brief cue indicates the expected location of a target stimulus. The difference in response time to a target that appears at expected and unexpected locations is used as the measure of shifting of attention. A slower response time on invalid trials, in which the target appears at an unexpected location, compared with valid trials in which the target appears at the expected location, indicates that the subject has paid attention to the cue that is signaling the likely location of a target. Monkeys with basal forebrain lesions showed disproportionately longer response times on invalid trials, suggesting an impairment in disengaging attention from the invalidly cued location. This was not due to a general impairment in motor skills, since the response to the target following the initiation of a response was comparable in monkeys with and without basal forebrain lesions. A similar result has been described in rats with selective lesions of cholinergic neurons in the NBM (Chiba *et al.*, 1999). It is important to note that this attentional impairment was selective; the monkeys with basal forebrain lesions in the study by Voytko were not impaired on tests of visual learning and memory, and rats with selective cholinergic lesions of the NBM were not impaired on a variety of cognitive tasks including tests of spatial learning and memory (reviewed by Baxter and Chiba, 1999).

Experiments that examine the regulation of attentional processing in associative learning paradigms have identified dissociable roles for rostral (septal) and caudal (NBM) regions of basal forebrain cholinergic neurons in these aspects of attention (reviewed by Baxter and Chiba, 1999). Increases in conditioned stimulus processing, brought about by violations of learned contingencies, require the integrity of cholinergic neurons in the NBM, but not cholinergic neurons in the medial septum or vertical limb of the diagonal band of Broca. Decreases in conditioned stimulus processing, which occur when conditioned stimuli are preexposed in the absence of reinforcement, or are consistent predictors of another event, require the integrity of cholinergic neurons in the latter structures, but not in the NBM.

The basal forebrain has reciprocal connections with the central nucleus of the amygdala, which has potential for widespread influences on cortical processing directly through projections to both lower-order and higher-order sensory areas and indirectly through projections to cholinergic neurons in the basal forebrain. Stimulation of the amygdala central nucleus has been found to influence basal forebrain electroencephalographic patterns consistent with acetylcholine release in rats and humans (Kapp *et al.*, 1994). Cortical acetylcholine release stimulated by the amygdala is thought to induce a state of cortical readiness for processing sensory information. This suggests that the basal forebrain cholinergic system might provide an output pathway for the regulation of attention and cortical information processing by the amygdala. Indeed, evidence from crossed-lesion studies suggests that projections from the central nucleus to the NBM are critical for producing increases in conditioned stimulus processing: disconnection of the NBM from the central nucleus of the amygdala by crossed unilateral lesions also disrupts the ability to increase attentional processing in response to a violation of expectancy (Baxter and Chiba, 1999).

Thus far, selective cholinergic lesions in the basal forebrain system appear to have a limited effect on measures of learning and memory, but rather impair attentional processing. It is assumed that attentional processes are involved in the primary stages in information processing, so it is remarkable that no dramatic effects in various learning and memory tasks have been observed following damage to basal forebrain cholinergic neurons. If attentional processes are impaired, then a performance deficit would be expected in most learning and memory tasks, but it appears that this is not the case after selective cholinergic lesions. However, the statement that basal forebrain damage impairs attention is an oversimplification. Instead, the effects of these lesions might be better characterized as producing an impairment in the ability to regulate attentional processing appropriately in response to task or environmental demands (Baxter and Chiba, 1999), rather than a reduction in the ability to attend to stimuli *per se*.

CONCLUSION

The basal forebrain is anatomically situated to regulate information processing in cortical and limbic structures involved in a wide array of cognitive functions. Indeed, extensive damage to the basal forebrain (to both cholinergic and noncholinergic components) results in a broad array of cognitive impairments. In contrast, damage limited to basal forebrain cholinergic neurons produces more restricted impairments in regulation of attentional processing. Hence, basal forebrain cholinergic neurons appear to have a selective role in regulating cortical information processing, rather than a generalized role in supporting the functions of their target areas. This apparently restricted function of basal forebrain cholinergic neurons suggests that a loss of these cells is probably not the core factor in producing cognitive deficits in conditions such as Alzheimer disease, or in the amnesia consequent to aneurysms in the basal forebrain.

The characterization of the function of basal forebrain cholinergic neurons as 'attentional' may fail to capture the array of functions performed by these neurons. These neurons are probably involved in other aspects of sensory processing and representational plasticity in primary sensory cortical areas (Baskerville *et al.*, 1997; Kilgard and Merzenich, 1998). The involvement of basal forebrain cholinergic projections to other cortical areas in similar types of functions has not yet been investigated. Hence, a general operating principle for the role of these cholinergic neurons in cognition remains elusive, as does a role for the basal forebrain generally in cognitive function. The basal forebrain, particularly the cholinergic component, may be more broadly involved in gating cortical information processing and regulating aspects of conscious experience (Sarter and Bruno, 2000); hence abnormalities in basal forebrain function could contribute to a wide variety of cognitive impairments in both neurodegenerative disease and in psychopathology. These hypotheses represent a useful guiding principle for future studies of basal forebrain function, both in animal models and in human clinical populations.

References

Baskerville KA, Schweitzer JB and Herron P (1997) Effects of cholinergic depletion on experience-dependent plasticity in the cortex of the rat. *Neuroscience* **80**: 1159–1169.

Baxter MG and Chiba AA (1999) Cognitive functions of the basal forebrain. *Current Opinion in Neurobiology* **9**: 178–183.

Chappell J, McMahan R, Chiba A and Gallagher M (1998) A re-examination of the role of basal forebrain cholinergic neurons in spatial working memory. *Neuropharmacology* **37**: 481–487.

Chiba AA, Bushnell PJ, Oshiro WM and Gallagher M (1999) Selective removal of cholinergic neurons in the basal forebrain alters cued target detection in rats. *Neuroreport* **10**: 3119–3123.

Collerton D (1986) Cholinergic function and intellectual decline in Alzheimer's disease. *Neuroscience* **19**: 1–28.

DeLuca J and Diamond BJ (1995) Aneurysm of the anterior communicating artery: a review of neuroanatomical and neuropsychological sequelae. *Journal of Clinical and Experimental Neuropsychology* **17**: 100–121.

Everitt BJ and Robbins TW (1997) Central cholinergic systems and cognition. *Annual Review of Psychology* **48**: 649–684.

Fine A, Hoyle C, Maclean CJ *et al.* (1997) Learning impairments following injection of a selective cholinergic immunotoxin, ME20.4 IgG-saporin, into the basal nucleus of Meynert in monkeys. *Neuroscience* **81**: 331–343.

Gritti I, Mainville L, Mancia M and Jones BE (1997) GABAergic and other noncholinergic basal forebrain neurons, together with cholinergic neurons, project to the mesocortex and isocortex in the rat. *Journal of Comparative Neurology* **383**: 163–177.

Kapp BS, Supple WF and Whalen PJ (1994) Effects of electrical stimulation of the amygdaloid central nucleus on neocortical arousal in the rabbit. *Behavioral Neuroscience* **108**: 81–93.

Kilgard MP and Merzenich MM (1998) Cortical map reorganization enabled by nucleus basalis activity. *Science* **279**: 1714–1718.

Mesulam MM, Mufson EJ, Wainer BH and Levey AI (1983) Central cholinergic pathways in the rat: an overview based on an alternative nomenclature (Ch1–Ch6). *Neuroscience* **10**: 1185–1201.

Olton DS, Wenk GL, Church RM and Meck WH (1988) Attention and the frontal cortex as examined by simultaneous temporal processing. *Neuropsychologia* **26**: 307–318.

Sarter M and Bruno JP (1997) Trans-synaptic stimulation of cortical acetylcholine and enhancement of attentional functions: a rational approach for the development of cognition enhancers. *Behavioural Brain Research* **83**: 7–14.

Sarter M and Bruno JP (2000) Cortical cholinergic inputs mediating arousal, attentional processing and dreaming: differential afferent regulation of the basal forebrain by telencephalic and brainstem afferents. *Neuroscience* **95**: 933–952.

Turchi J and Sarter M (1997) Cortical acetylcholine and processing capacity: effects of cortical cholinergic

deafferentation on crossmodal divided attention in rats. *Cognitive Brain Research* **6**: 147–158.
Voytko ML (1996) Cognitive functions of the basal forebrain cholinergic system in monkeys: memory or attention? *Behavioural Brain Research* **75**: 13–25.
Wainer B and Mesulam MM (1990) Ascending cholinergic pathways in the rat brain. In: Steriade M and Biesold D (eds) *Brain Cholinergic Systems*, pp. 65–119. Oxford, UK: Oxford University Press.
Wenk GL (1997) The nucleus basalis magnocellularis cholinergic system: one hundred years of progress. *Neurobiology of Learning and Memory* **67**: 85–95.
Zaborszky L, Gaykema RP, Swanson DJ and Cullinan WE (1997) Cortical input to the basal forebrain. *Neuroscience* **79**: 1051–1078.

Further Reading

Baxter MG and Murg SL (2001) The basal forebrain cholinergic system and memory: beware of dogma. In: Squire LR and Schacter DL (eds) *Neuropsychology of Memory*, 3rd edn, pp. 425–436 (2000). New York, NY: Guilford Press.
Berger-Sweeney J, Stearns NA, Murg SL *et al.* (2001) Selective immunolesions of cholinergic neurons in mice: effects on neuroanatomy, neurochemistry, and behavior. *Journal of Neuroscience* **21**: 8164–8173.
Burk JA and Sarter M (2001) Dissociation between the attentional functions mediated via basal forebrain cholinergic and GABAergic neurons. *Neuroscience* **105**: 899–909.
De Rosa E, Hasselmo ME and Baxter MG (2001) Contribution of the cholinergic basal forebrain to proactive interference from stored odor memories during associative learning in rats. *Behavioral Neuroscience* **115**: 314–327.
Himmelheber AM, Sarter M and Bruno JP (2001) The effects of manipulations of attentional demand on cortical acetylcholine release. *Cognitive Brain Research* **12**: 353–370.
McGaughy J, Everitt BJ, Robbins TW and Sarter M (2000) The role of cortical cholinergic afferent projections in cognition: impact of new selective immunotoxins. *Behavioural Brain Research* **115**: 251–263.
Sarter M and Bruno JP (1999) Abnormal regulation of corticopetal cholinergic neurons and impaired information processing in neuropsychiatric disorders. *Trends in Neurosciences* **22**: 67–74.
Turchi J and Sarter M (2000) Cortical cholinergic inputs mediate processing capacity: effects of 192 IgG-saporin-induced lesions on olfactory span performance. *European Journal of Neuroscience* **12**: 4505–4514.
Waite JJ, Wardlow ML and Power AE (1999) Deficit in selective and divided attention associated with cholinergic basal forebrain immunotoxic lesion produced by 192-saporin; motoric/sensory deficit associated with Purkinje cell immunotoxic lesion produced by OX7-saporin. *Neurobiology of Learning and Memory* **71**: 325–352.

Basal Ganglia

Introductory article

Lucy L Brown, Albert Einstein College of Medicine, Bronx, New York, USA
Samuel M Feldman, New York University, New York, USA

CONTENTS

Nuclei near the base of the brain integrate information from the entire cerebral cortex and serve a regulatory function for movement and cognition.

INTRODUCTION

The basal ganglia are a group of brain structures that regulate movement in mammals. Damage to these subcortical structures is the primary cause of involuntary, abnormally sequenced movements such as tremor, rigidity, athetosis, tics, chorea and ballism. However, the role of the basal ganglia goes well beyond the regulation of the sequencing of movements. They also have a major role in the learning of motor and cognitive skills. They control the expression of voluntary behaviors that are essential for normal mammalian interaction with an ever-changing and challenging environment. Unlike the spinal cord, which is necessary for withdrawal reflexes and the final stages of the movements we execute, the basal ganglia influence the selection and sequence of the final movement, and participate in adaptive motor control.

A neuroscientist's gag is: 'Why are the basal ganglia like the Dean's office?' Answer: 'Because they take up a lot of space and nobody knows what they do.' This old joke turns out to be surprisingly on target. A serious answer might be that they facilitate the global operations of a large 'institution', planning and making decisions based on many varied inputs. Such an executive role in brain function is difficult to analyze, and our knowledge about the function of these nuclei is still evolving. Accordingly, the descriptions of sensorimotor and cognitive functions of the basal ganglia that follow must be seen as descriptions of work in progress.

Anatomically, the basal ganglia are a collection of nuclei in the mammalian forebrain and midbrain. The forebrain nuclei include the caudate nucleus and putamen, the nucleus accumbens, the globus pallidus (internal and external segments) and the subthalamic nucleus. Two subunits of the substantia nigra, the pars reticulata and pars compacta, make up the midbrain component. Similar structures exist in other vertebrates, including reptiles, amphibians, birds and fish. Although the amygdala and other basal forebrain nuclei were once included in discussions of the basal ganglia, current concepts of anatomy and function restrict the definition now used.

Alternative designations are used for some of the basal ganglia nuclei. For example, in primates the caudate and putamen are anatomically distinct, but lower mammals such as the rat lack this dichotomy and have a single undifferentiated nucleus called the striatum. In the primate brain the putamen and globus pallidus are physically close and often hard to distinguish, and are referred to collectively as the lenticular nucleus.

The striatum receives its major synaptic input from the entire extent of the cerebral cortex. This diverse cortical input reflects the many functions in which the basal ganglia participate, ranging from sensorimotor feedback and motor control to spatial working memory. Both the anatomy and physiology suggest that the basal ganglia act collectively as a 'suppress and release' mechanism for a wide range of behaviors. Dysfunction of these mechanisms is seen in basal ganglia disorders; for example, people with Parkinson disease have difficulty initiating movements, while people with Huntington chorea cannot stop. These disorders are involuntary: people with Huntington chorea have little or no control over their flailing movements, nor can people with Parkinson disease voluntarily overcome their 'frozen' state. Thus, the suppress and release mechanism operates at the unconscious level.

ANATOMY AND CHEMOARCHITECTURE

Location and Gross Structure

The caudate–putamen and globus pallidus are situated on either side of the thalamus, beneath the mantle of the cerebral cortex. The corpus callosum and external capsule form a border around the caudate–putamen. In primates, the caudate and putamen are separated by the internal capsule, which contains axon projections between the cortex, brainstem and spinal cord. In the primate, the caudate has a head, body and tail. The head borders the anterior lateral ventricle while the body and tail follow the contour of the lateral ventricle into the temporal lobe (Figure 1). The nucleus accumbens is continuous with the caudate and putamen, extending anteriorly and downward to the inferior surface of the brain. The subthalamic nucleus, at the junction of the midbrain and forebrain, and substantia nigra, in the midbrain, are smaller than the caudate–putamen. The right basal ganglia control the left side of the body and vice versa. Accordingly, parkinsonian tremor of the left hand is evidence of pathological changes in the right basal ganglia.

Neuroanatomists have described connections of the basal ganglia in the mammal as loops that originate in different regions of the cortex, and project back to subregions of the originating cortex (Figure 2). For example, a sensorimotor loop has its origins in sensorimotor cortex. Excitatory sensorimotor cortex projections to striatum activate cells whose output to the globus pallidus is inhibitory. The pallidal neurons also have an inhibitory effect on their target neurons in the ventral thalamus, which in turn has an excitatory projection to the supplementary motor cortex, a region known to be involved in the planning of complex movements. The loop circuits are examples of brain regulatory systems that use inhibition of a tonic inhibitory drive as a gate to release normally suppressed output. Other important basal ganglia circuits include an oculomotor loop, which has a major output pathway to the superior colliculus, via the reticulata region of the substantia nigra, in addition to the cortical frontal eye fields, via the thalamus; a dorsolateral prefrontal cortex loop; and a so-called limbic loop.

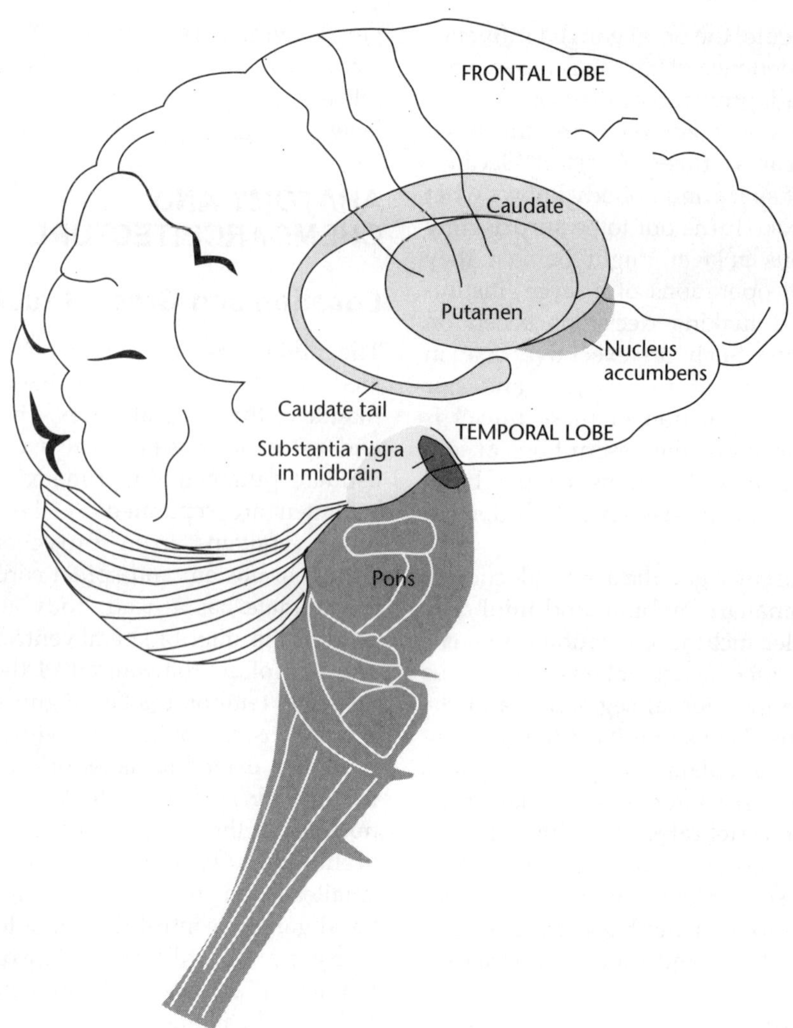

Figure 1. Sagittal (side) view of the human brain shows the central locations of the largest basal ganglia nuclei. The caudate, putamen, globus pallidus and nucleus accumbens are in the forebrain. The substantia nigra is in the midbrain while the subthalamic nucleus (not shown) is at the junction of the midbrain and forebrain.

In addition to the fundamental circuit described above, a critically important input to the striatum comes from the compacta region of the substantia nigra via the nigrostriatal tract (Figure 2). One of the major dopamine pathways in the brain, its integrity is essential for normal functioning of the basal ganglia. Parkinsonism is the direct result of interference with nigrostriatal transmission. Finally, complex and specialized interactions among basal ganglia structures involve the subthalamic nucleus through the 'indirect pathway' (Figure 2).

Chemoarchitecture

The basal ganglia are a collection of heterogeneous nuclei. The striatum contains small zones with a dense concentration of opioid receptors, called 'striosomes' or 'patches' (Figure 3), within a 'matrix' rich in cholinesterase. Many other neurally active substances (such as somatostatin, calbindin and substance P) and receptor proteins (such as the $GABA_A$ receptor) are unevenly distributed between the striosome and matrix compartments. Although the functional significance of this chemoarchitectonic heterogeneity is largely unknown, cells in the patch compartment are active following administration of some dopaminergic drugs, and may play a major part in regulation of dopamine in substantia nigra.

Other components of the basal ganglia have a distinctive complex of cell types, neurochemicals, receptor proteins and synaptic connections that interact in a highly structured fashion. For example,

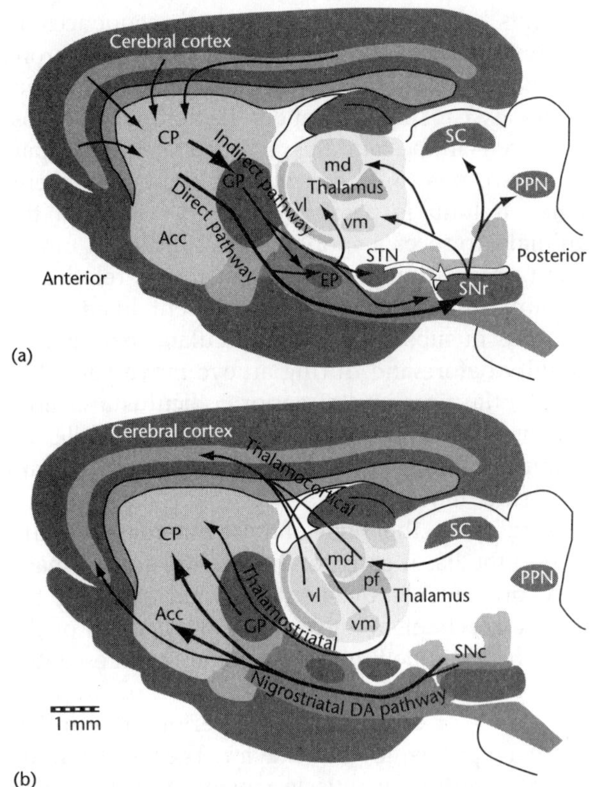

Figure 2. Nuclei and major pathways of the basal ganglia in the rat. The pathways exist also in primates. The view is sagittal, thus showing the extent of the basal ganglia from the forebrain to the midbrain. (a) Descending pathways. All of the cerebral cortex projects to the caudate–putamen (CP). The direct pathway projects to the internal segment of the globus pallidus, called the entopeduncular nucleus (EP) in rats, and to the substantia nigra reticulata before projecting out of the basal ganglia nuclei to the thalamus, superior colliculus and pedunculopontine nucleus. The indirect pathway projects to the subthalamic nucleus and globus pallidus externa (GP in rats) before going to the entopeduncular nucleus and substantia nigra and leaving the basal ganglia. (b) Feedback pathways. The substantia nigra compacta, which contains dopaminergic cells, sends an important projection called the nigrostriatal dopamine pathway to the caudate–putamen and nucleus accumbens. The thalamocortical projection completes the loop from cortex to basal ganglia to thalamus and back to cortex. Acc, nucleus accumbens; CP, caudate–putamen; DA, dopamine; EP, entopeduncular nucleus; GP, globus pallidus; md, mediodorsal nucleus; pf, parafascicular nucleus; PPN, pedunculopontine nucleus; SC, superior colliculus; SNc, substantia nigra compacta; SNr, substantia nigra reticulata; vl, ventrolateral nucleus; vm, ventromedial nucleus. Adapted from Gerfen and Wilson, 1999.

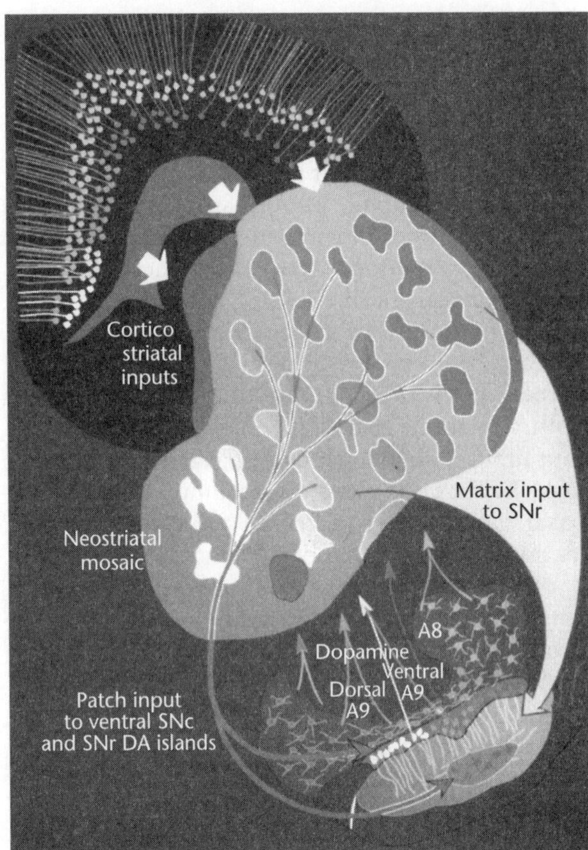

Figure 3. [*Figure is also reproduced in color section.*] The patch–matrix design of the striatum in the rat, seen in the right side of the rat cortex, striatum, nucleus accumbens and substantia nigra. The patches (red, orange and yellow zones), also called striosomes, are rich in an opioid receptor. The surrounding matrix is rich in acetylcholinesterase and calbindin. The anatomical connections of the patches and matrix are color-coded to show, for example, that the cells in the deep layers of lateral cortex project to the lateral patches in the striatum, which in turn project to the ventral substantia nigra compacta dopaminergic cells. SNr, substantia nigra reticulata; SNc, substantia nigra compacta; A9, nomenclature that refers to substantia nigra compacta cells; A10, refers to dopaminergic cells of the midbrain tegmentum that project to the nucleus accumbens. Diagram courtesy of C. Gerfen.

regions thought to have more cognitive than motor functions are rich in calbindin, a calcium binding protein, while those thought to regulate motor functions are calbindin-poor. In addition, the major target of the neostriatum, the globus pallidus, has two segments, internal and external, divided by different inputs; the external segment also has a much higher concentration of enkephalin. Finally, the substantia nigra has two components: reticulata and compacta (Figure 3). The reticulata component contains GABA-mediated neurons, while compacta neurons are dopaminergic, their axons being the nigrostriatal pathway.

FUNCTIONS OF THE BASAL GANGLIA

Sensorimotor Functions

Tennis professionals and pianists practice constantly. Eventually their skills become unconscious habits, and during performance they do not think about them in any detail. Indeed, we all have large repertoires of highly practised motor skills (buttoning a shirt, riding a bicycle, driving a car) that require sensory feedback during learning, but are eventually executed without thought or verbalization. Consider the task of driving a car with a manual gear shift, in which each of the four limbs is simultaneously engaged in a different task! Evidence is currently accumulating that the basal ganglia have a critical role in regulation, learning and execution of such motor habits.

We know from animal studies that electrical stimulation of small areas of the caudate–putamen produces movement in localized body regions such as the tongue or contralateral wrist. Stimulation of larger areas produces contralateral head movements or turning, or flexion of the limbs. This reflects the fact that the output of the basal ganglia affects the elements of which the skilled behaviors are constructed. They play a critical part in innate motor sequences (e.g. grooming in rodents), voluntary movements (e.g. limb trajectory) and the sensory feedback that guides directional movement, all of which are essential for developing complex trained sequences or motor habits.

Injection of dopamine directly into the rat neostriatum produces contralateral turning, while small neostriatal lesions prevent the normal sequencing of movements seen during grooming. In rodents and subhuman primates, an abnormal increase in dopamine at synapses in the basal ganglia causes repetitive, stereotyped behaviors: uncontrolled, purposeless, repetitive movements. Rodents will continually chew or sniff, or move back and forth repetitively. In humans, stereotypies may be observed in following administration of dopaminergic drugs, including cocaine.

Note, however, that all of these affected elements of motor behavior are organized at the spinal cord and brainstem level. It is their release, suppression and timing that seem to be regulated by the basal ganglia. Rather than programming these behaviors directly, the basal ganglia apparently implement sequences of behavioral elements that achieve complex behavior.

Additional insight comes from electrophysiological studies. Striatal cells fire in relation to the serial order of innate grooming sequences in rodents, or to learned movement sequences in primates. Inactivating the globus pallidus neurons by cooling disrupts smooth performance of movements by causing continuous flexion of the limbs. Also, cells in the caudate–putamen and globus pallidus fire before or during a learned movement, when a limb is passively moved, or when the animal produces a spontaneous, unlearned movement. Interestingly, cells do not fire in relation to the amplitude or speed of a movement. In addition, neurons in substantia nigra reticulata change their activity before and during an eye movement. The substantia nigra and superior colliculus are more involved in head and eye movements, while the globus pallidus appears to be involved in limb and trunk movements.

Note, again, that none of these functions is exclusive to the basal ganglia. Basal ganglionic damage may prevent a movement by compromising the release mechanism, even while other parts of the brain (e.g. cerebellum, cortex) retain both essential and redundant circuits for execution of the task. The result of such damage is serious disruption and even prevention of behavior. The basal ganglia can also influence muscle tension and extensor/flexor balance, which are largely controlled by the brainstem and spinal cord. Rather than acting as a motor control system, the basal ganglia and its related structures apparently function as a gating mechanism for innate and complex movement patterns that are organized at lower levels.

Cognitive Functions

Just as the basal ganglia have a role in the unconscious aspects of motor control, they also affect the unconscious aspects of cognition. Such processes are difficult to study and require subtle investigative approaches. Spatial working memory, one of the first cognitive functions of the basal ganglia to be recognized, is linked closely to similar functions of the cortex. Recall that the targets of basal ganglia projections are regions of the cortex involved in motor planning and higher-order cognitive functions, such as spatial working memory. Caudate–putamen lesions consistently produce deficits in spatial learning tasks, and imaging studies in humans show that learning a complex spatial task such as the Tower of London puzzle involves the caudate. In addition, cells of the basal ganglia are involved in detecting and evaluating the learned context of an environmental stimulus. They fire in relation to conditioned stimuli and also when a stimulus becomes relevant, which suggests that the basal ganglia may affect cognitive behavior on a global scale. For example, detecting a primary reward such as a sweet taste, or learning that money is rewarding, or expecting a reward, all profoundly affect our behavior and and its planning. In human brain-mapping studies, several categories of reward such as cash, the 'rush' and 'high' of cocaine, and verbal feedback, activate the caudate, putamen and nucleus accumbens. Activation in the nucleus accumbens tracks the value of a monetary reward during a gambling task, even when there is only a prospect of winning money. Thus, the anticipation of reward, a highly influential – and global – role in behavior is also represented in the basal ganglia.

Studies of people with basal ganglia disease add further insight. People with Parkinson disease experience attention deficits and difficulty in shifting mental set. They are poor at learning to predict probabilistic classifications, which are similar to unconscious, learned, motor habits. This involves unconscious memory and selection of a group of objects that are correct more often than not during a training series. People with globus pallidus lesions describe their mental life as empty; they have no spontaneous thoughts. In Tourette syndrome, people experience racing thoughts. People who suffer Huntington disease exhibit symptoms of obsessive–compulsive disorder prior to the onset of severe motor symptoms.

CONCLUSION

The basal ganglia are subcortical forebrain and midbrain nuclei that process information from the cortex to affect movement and cognition. Physiological and pathophysiological studies indicate that the basal ganglia influence stopping, starting and switching behaviors at the unconscious level; that they play a role in complex visuospatial tasks; and that they have access to learned motor and cognitive habits. An understanding of the normal functions of the basal ganglia nuclei remains unclear, perhaps because global, executive, decision-making processes are based on so many factors. 'Adaptive motor control' may be a good description of their functions, not only for motion, but also for strategic planning of movements and tasks.

Further Reading

Breiter HC, Aharon I, Kahneman D, Dale A and Shizgal P (2001) Functional imaging of neural responses to expectancy and experience of monetary gains and losses. *Neuron* **30**: 619–639.

Brown LL, Schneider JS and Lidsky TI (1997) Sensory and cognitive function of the basal ganglia. *Current Opinion in Neurobiology* **7**: 157–163.

Cromwell HC and Berridge KC (1996) Implementation of action sequences by a neostriatal site: a lesion mapping study of grooming syntax. *Journal of Neuroscience* **16**: 3444–3458.

Gerfen CR and Wilson CJ (1996) The basal ganglia. In: Swanson LW, Bjorklund A and Hokfelt T (eds) *The Handbook of Chemical Neuroanatomy* vol. 12, Integrated Systems of the CNS, Part III, pp. 371–468. New York, NY: Elsevier.

Jog MS, Yaso K, Connolly CI, Hillegaart V and Graybiel AM (1999) Building neural representations of habits. *Science* **286**: 1745–1749.

Kawagoe R, Takikawa Y and Hikosaka O (1998) Expectation of reward modulates cognitive signals in the basal ganglia. *Nature Neuroscience* **1**: 411–416.

Kermadi I and Joseph JP (1995) Activity in the caudate nucleus of monkey during spatial sequencing. *Journal of Neurophysiology* **74**: 911–933.

Knowlton BJ, Mangels JA and Squire LR (1996) A neostriatal learning system in humans. *Science* **273**: 1399–1402.

Knutson B, Adams CM, Fong GW and Hommer D (2001) Anticipation of increasing monetary reward selectively recruits nucleus accumbens. *Journal of Neuroscience* **21**: RC159.

McDonald RJ and White NM (1994) Parallel information processing in the water maze: evidence for independent memory systems involving dorsal striatum and hippocampus. *Behavioral Neural Biology* **61**: 260–270.

Bayesian and Computational Learning Theory

Intermediate article

David H Wolpert, NASA Ames Research Center, Moffett Field, California, USA

CONTENTS

THE MATHEMATICS OF INDUCTIVE LEARNING

Inductive learning is the process of coming to statistical conclusions based on past experiences. Unlike deduction, with induction one is never perfectly sure of one's conclusion, instead arriving at a (hopefully highly probable) guess. Inductive learning is performed by the human brain continually: almost all of a brain's conclusions, from the 'simplest' ones involved in sensor-motor decisions, to the most 'sophisticated' ones concerning how one should live one's life, are based at least in part on inductive learning. Even science is ultimately inductive in nature, with the 'past experiences' its conclusions are based on being previous experimental data, and its 'conclusions' being theories that are always open to revision.

A lot of work has been directed at implementing inductive learning algorithmically, in computers. 'Adaptive computation', involving neural networks, fuzzy logic, and computational statistics, can be viewed as a set of attempts to do this. The topic of algorithmic induction also looms large in other fields, like artificial intelligence and genetic algorithms. Recently this work has fostered new research on the mathematical underpinnings of inductive learning. A thorough understanding of those would not only result in improvements in our applied computational learning systems; it would also provide us with insight into the scientific method, as well as human cognition.

This article surveys 'Bayesian learning theory' and 'computational learning theory'. These are the two principal mathematical approaches that have been applied to supervised learning, a particularly important branch of inductive learning. The form of supervised learning considered in this article is simplified, the aim being to highlight the distinctions between these two learning theories rather than to present either in its full form.

A mathematical framework that can encapsulate both learning theories is the 'extended Bayesian framework' (EBF) (Wolpert, 1997). A simplified version of it, sufficient for current purposes, can be roughly described as follows. Say we have a finite *'input space'* X and a finite *'output space'* Y, and a set of m input–output pairs $d = \{d_X(i), d_Y(i)\}$. Call d a *'training set'*, and assume it was created by repeated noise-free sampling of an $X \to Y$ *'target function'* f. More formally, assume that the *'likelihood'* governing the generation of d from f is $P(d \mid f) = \prod_{i=1}^{m} \pi(d_X(i))\delta(d_Y(i), f(d_X(i)))$, where $\delta(.,.)$ is the Kronecker delta function, which takes the value 1 if its arguments are equal and equals 0 otherwise, and π is the *'sampling distribution'*. $P(f)$ is known as the *'prior distribution'* over targets, and $P(f \mid d)$ is known as the *'posterior distribution'*.

Let h be the $X \to Y$ function our learning algorithm produces in response to d. As far as learning accuracy is concerned, that learning algorithm is described by $P(h \mid d)$: the details of how the algorithm generates h from d are irrelevant. (One of the major reasons why formalisms other than EBF have limited scope is that they do not use $P(h \mid d)$ to describe the learning algorithm; there is no other quantity that can capture all possible learning algorithms.) When discussing multiple learning algorithms – i.e., multiple distributions $P(h \mid d)$ – we will sometimes distinguish the different algorithms with the notation $\gamma_1, \gamma_2, \ldots$. Note that learning algorithms only ever see d, never f (although they often make assumptions concerning f). Accordingly, $P(h \mid d, f) = P(h \mid d)$. Also note that $P(h) = \sum_{d,f} P(h \mid d)P(d \mid f)P(f)$, and in general need not equal the prior $P(f)$ evaluated for $f = h$.

Take s to be the fraction of $d_X(i)$ such that $d_Y(i) = h(d_X(i))$; i.e., s is the learning algorithm's average accuracy on the training set. We use c to indicate an error value, with its dependence on the other variables indicated by $c(h,f,d)$. In particular, we write c_O for the average (according to π) across all $x \in X$ lying outside the training set of whether h and f agree on x. We call c_O the *'off training set'* (OTS) error; it is a measure of how well our learning algorithm generalizes from the training set. An alternative error function, indicated by c_I, is the 'independent, identically distributed' (IID) error function. It is the same average, but not restricted to $x \notin d_X$, so that a learning algorithm gets some credit simply for memorizing what it's already seen.

Extensions of these definitions to allow for other kinds of error functions – noise in the target, uncertain sampling distributions, different likelihoods, infinite input and output spaces, etc. – are all straightforward, though laborious; see Wolpert (1997). The next section presents some theorems which will help us to compare the Bayesian and computational theories of supervised learning.

A FORMALIZATION OF INDUCTIVE BIAS

We start with the following theorem (Wolpert, 1995), which specifies the expected generalization error after training on some particular training set:

Theorem 1. The value of the conditional expectation $E(c \mid d)$ can be written as a (non-Euclidean) inner product between the distributions $P(h \mid d)$ and $P(f \mid d) : E(c \mid d) = \sum_{h,f} c(h,f,d)P(h \mid d)P(f \mid d)$.

(Similar results hold for $E(c \mid m)$, etc.)

Theorem 1 says that how well a learning algorithm $P(h \mid d)$ performs is determined by how 'aligned' it is with the actual posterior, $P(f \mid d)$, where 'alignment' is quantified by the error function. This theorem allows one to ask questions like 'for what set of posteriors is algorithm γ_1 better than algorithm γ_2?' It also means that, unless one can somehow prove from first principles that $P(f \mid d)$ has a certain form, one cannot prove that a particular $P(h \mid d)$ will be aligned with $P(f \mid d)$, and therefore one cannot prove that the learning algorithm generalizes well.

There are a number of ways to formalize this impossibility of establishing the superiority of some particular learning algorithm with a proof from first principles, i.e. with a proof that is not implicitly predicated on a particular posterior. One of them is in the following set of *'no free lunch'* theorems (Wolpert, 1996a):

Theorem 2. Let $E_{\gamma i}(.)$ indicate an expectation value evaluated using learning algorithm i. Then for any two learning algorithms γ_1 and γ_2, independent of the sampling distribution:

1. Uniformly averaged over all f,
   $E_{\gamma_1}(c_O \mid f,m) - E_{\gamma_2}(c_O \mid f,m) = 0$.
2. Uniformly averaged over all f,
   $E_{\gamma_1}(c_O \mid f,d) - E_{\gamma_2}(c_O \mid f,d) = 0$ for any training set d.
3. Uniformly averaged over all $P(f)$,
   $E_{\gamma_1}(c_O \mid m) - E_{\gamma_2}(c_O \mid m) = 0$.
4. Uniformly averaged over all $P(f)$,
   $E_{\gamma_1}(c_O \mid d) - E_{\gamma_2}(c_O \mid d) = 0$, for any training set d.

According to these results, by any of the measures $E(c_O \mid d)$, $E(c_O \mid m)$, $E(c_O \mid f,d)$, or $E(c_O \mid f,m)$, all algorithms are equivalent, on average. The uniform averaging that goes into these results should be viewed as a calculational tool for comparing algorithms, rather than as an assumption concerning the real world. In particular, the proper way to interpret statement 1 is that, appropriately weighted, there are 'just as many' targets for which algorithm 1 has better $E(c_O \mid f,m)$ as there are for which the reverse is true. Accordingly, unless one can establish *a priori*, before seeing any of the data d, that the f that generated d is one of the ones for which one's favorite algorithm performs better than other algorithms, one has no assurances that that algorithm performs any better than the algorithm of purely random guessing.

This does not mean that one's algorithm must perform no better than random guessing in the real world. Rather it means that, formally, one cannot establish superiority to random guessing without making some assumptions. Note in particular that you cannot use your prior experience – or even the billion years or so of 'prior experiences' of your genome, reflected in the design of your brain – to circumvent this problem, since all that prior experience is, formally, just an extension to the training set d.

As an important example of the foregoing, consider assessing the validity of a hypothesis by using experimental data that were not available when the hypothesis was created. In the form of 'falsifiability', this is one of the primary tools commonly employed in the scientific method. It can be viewed as a crude version of a procedure that is common in applied supervised learning: choose between the two hypothesis functions $h_{\gamma 1}$ and $h_{\gamma 2}$, made by running two generalizers γ_1 and γ_2 on a training set d_1, by examining their accuracies on a distinct 'held out' training set d_2 that was generated from the same target that generated d_1.

Such a procedure for choosing between hypotheses seems almost unimpeachable. Certainly its crude implementation in the scientific method has resulted in astonishing success. Yet it cannot be justified without making assumptions about the real world. To state this more formally, take any two learning algorithms γ_1 and γ_2, and consider two new algorithms based on them, S and T. S uses an extension of the choosing procedure outlined above, known as *'cross-validation'*: given a training set d, S breaks d into two disjoint portions, d_1 and d_2; trains γ_1 and γ_2 on d_1 alone; sees which resultant hypothesis is more accurate on d_2; and then trains the associated learning algorithm on all of d and uses the associated hypothesis. In contrast, T uses anti-cross-validation: It is identical to S except that it chooses the learning algorithm the accuracy of whose associated hypothesis on d_2 was worst. By the 'no free lunch' theorems, we know that T must outperform S as readily as vice versa, regardless of γ_1 and γ_2. It is only when a certain (subtle) relationship holds between $P(f)$ and the γ_1 and γ_2 one is considering that S can be preferable to T (see Theorem 1). When that relationship does not hold, T will outperform S.

This result means in particular that the scientific method must fail as readily as it succeeds, unless there is some *a priori* relation between the learning algorithms it uses (i.e. scientists) and the actual truth. Unfortunately, next to nothing is known formally about that required relation. In this sense, the whole of science – not to mention human cognition – is based on a procedure whose assumptions not only are formally unjustified, but have not even been formally stated.

BAYESIAN LEARNING THEORY

Intuitively, the Bayesian approach to supervised learning can be viewed as an attempt to circumvent the 'no free lunch' theorems by explicitly making an assumption for the posterior. Usually, to do this it first restricts attention to situations in which the likelihood is known (which in the context of this article means there is no 'noise'). It then makes an assumption about the prior distribution, $P(f)$. Next Bayes' theorem is invoked to combine the prior with the likelihood to give us our desired posterior: $P(f \mid d) \propto P(d \mid f)P(f)$, where the proportionality constant is independent of f. (Besides these kind of assumptions concerning the prior, there are other kinds of assumptions which, when combined with the likelihood, fix the posterior (Wolpert, 1993). However, such assumptions have not yet been investigated in any detail.)

Given such a posterior, the value of $E(C \mid d)$ that accompanies any particular learning algorithm $P(h \mid d)$ is determined uniquely (see Theorem 1). In particular, one can solve for the $P(h \mid d)$ that minimizes $E(C \mid d)$. This is known as the *'Bayes-optimal'* learning algorithm. This algorithm is given by the following theorem (which is rather more general than we need):

Theorem 3. Let $c(h,f,d) = \sum_{x \in X} \pi'(x)G(h(x),f(x))$ for some real-valued function $G(.,.)$ and some real-valued $\pi'(.)$ that is nowhere negative (and may or may not equal the distribution $\pi(.)$ arising in $P(d \mid f)$). Then the Bayes-optimal $P(h \mid d)$ always guesses the same function h^* for the same d:

$$h^* = \{x \in X \to \arg\min_{y \in Y} \Omega(x,y)\}, \text{where}$$
$$\Omega(x,y) \equiv \sum_f G(y,f(x))P(f \mid d).$$

(The function $\pi'(.)$ is allowed to vary with d, as it does in OTS error.)

$\Omega(x,y)$ is the contribution to the posterior expected error that arises if the learning algorithm outputs (an h having) the value y at point x. So intuitively, Theorem 3 says that for any x, one should choose the $y \in Y$ that minimizes the average 'distance' from y to $f(x)$, where the average is over all $f(.)$, according to the distribution $P(f \mid d)$, and 'distance' is measured by $G(.,.)$. Note that this result holds regardless of the form of $P(f)$, and regardless of what (if any) noise process is present: all such considerations are taken care of automatically, in the $P(f \mid d)$ term. Note also that h^* might be an f with zero-valued posterior: in the Bayesian framework, the output h of the learning algorithm does not really constitute a 'guess for the f which generated the data'.

This is all there is to the Bayesian framework, as far as foundational issues are concerned (Berger, 1985; Loredo, 1990; Buntine and Weigend, 1991; Wolpert, 1995). Everything else in the literature concerning the framework involves either philosophical or calculational issues. The philosophical issues usually concern what $P(f)$ 'means' (Wolpert, 1993). In particular, some Bayesians do not view the $P(f)$ they use to derive their learning algorithm as an assumption for the actual $P(f)$, which may or may not correspond to reality. Rather, in general they interpret the probability of an event as one's 'personal degree of belief' in that event, and therefore in particular they interpret $P(f)$ that way. According to this view, probability theory is simply a calculus for forcing consistency in one's use of probability to manipulate one's subjective beliefs. Accordingly, no matter how absurd a Bayesian's prior, under this interpretation practitioners of non-Bayesian approaches to supervised learning are by definition always going to perform worse than that Bayesian (since the Bayesian determines $P(f)$ and therefore $P(f \mid d)$, and accordingly guesses in an optimal manner – see Theorem 1).

Unfortunately, there are algorithms that cannot be cast as the Bayes-optimal algorithm for some implicit prior and likelihood (Wolpert, 1996b). Accordingly, even if one accepts the 'fundamentalist' Bayesian's view of what $P(f)$ 'means', the rigidity of the framework makes it ill-suited to broad analysis of algorithms. More generally, often our prior knowledge does not concern targets directly, but rather concerns the relative performances of various (possibly non-Bayesian) algorithms, or the efficacy of a scheme (like cross-validation) for choosing among those algorithms. The conventional Bayesian framework provides no way to exploit that prior knowledge. In general, we need to introduce the random variable h for such an analysis – which is what is done in EBF. (See, however, Wolpert (1993) for a discussion of how one can sometimes employ Bayes' theorem to exploit such knowledge directly.)

Some of the calculational issues in the Bayesian framework involve evaluating the Bayes-optimal algorithm, given knowledge of the posterior $P(f \mid d)$. The problem is that using the Bayes-optimal algorithm requires evaluating (and then minimizing) the sum giving $\Omega(x, y)$. Since this can be difficult, people often settle for approximations to finding $\arg\min_{y \in Y} \Omega(x, y)$. For example, the 'maximum *a posteriori*' (MAP) estimator is $h(x) = [\arg\max_f P(f \mid d)](x)$. Evaluating it involves finding a peak of a surface (namely $P(f \mid d)$) rather than performing a sum, and is often simpler.

Even if one is willing to use a MAP estimator, there might still be difficulties in evaluating the surface $P(f \mid d)$. For example, if we have a 'hierarchical' likelihood, then $P(d \mid f) = \sum_\lambda P(d \mid \lambda, f)\ P(\lambda)$, where λ is a 'hyperparameter'. (An example is where we know that the data are generated via a particular kind of noise process, but don't know the noise level in advance – that noise level is a hyperparameter.) Sums being difficult, often we can't even evaluate such a hierarchical likelihood (and therefore can't evaluate the associated posterior, $P(f \mid d)$). Instead, often one makes an ad hoc estimate of the hyperparameter, λ', and replaces $P(d \mid f)$ with $P(d \mid \lambda', f)$. (See the discussion of empirical Bayes and ML-II in Wolpert (1995).)

COMPUTATIONAL LEARNING THEORY

The computational learning framework takes a number of forms, the principal ones being the statistical physics, PAC and VC (uniform convergence) approaches (Baum and Haussler, 1989; Vapnik, 1982; Wolpert, 1995). All three can be cast as bounds concerning a probability distribution that involves IID error, and that is conditioned on f (in contrast to the Bayesian framework, in which f is not fixed). In their most common forms they all have m rather than d fixed in their distribution of interest (again, in contrast to the Bayesian framework). This means that they do not address the question of what the likely outcome is for the training set at hand. Rather, they address the question of what the likely outcome would be if one had different training sets from the actual d. Such varying of quantities that are in fact fixed and known has been criticized by Bayesian practitioners on formal grounds, as violating any possible self-consistent principles for induction. (See Wolpert (1993), and the discussion of the 'honesty principles' in Wolpert (1995), for an overview of the conflict between the two learning theories.)

For purposes of illustration, we will focus on (a pared-down version of) the VC framework. Start with the following simple result, which concerns the '*confidence interval*' relating c and s, for the case where $H^\sim$, the h-space support of a learning algorithm's $P(h)$, consists of a single h (Wolpert, 1995):

Theorem 4. Assume that there is an h' such that $P(h \mid d) = \delta(h - h')$ for all d. Then:

$$P(c_I > s + \varepsilon \mid f, m) < 2e^{-m\varepsilon^2}$$

(Recall that s is the empirical misclassification rate.) Note that this bound is independent of f, and therefore of the prior $P(f)$.

If $H^\sim$ instead consists of more than one h, the bound in Theorem 4 still applies if one multiplies the right-hand side by $\mid H^\sim \mid$, the number of functions in $H^\sim$. The major insight behind the '*uniform convergence*' framework was how to derive even tighter bounds by characterizing $P(h \mid d)$ in terms of its VC dimension (Baum and Haussler, 1989; Vapnik, 1982; Wolpert, 1995). (It is important to distinguish between this use of the VC dimension and its use in other contexts, as a characterization of $P(f)$.) For $Y = \{0, 1\}$ and our error function, the VC dimension is given by the smallest m such that, for any d_X of size m, all of whose elements are distinct, there is a d_Y for which no h in $H^\sim$ goes through d. (The VC dimension is this smallest number minus one.)

Common to all such extensions of Theorem 4 is a rough equivalence (as far as the likely values of c are concerned) between: (1) lowering s; (2) lowering the expressive power of $P(h \mid d)$ (i.e., shrinking its

VC dimension, or shrinking $|H^{\sim}|$); and (3) raising m. Important as these extensions of Theorem 4 are, to understand the foundational issues underpinning the uniform convergence framework it makes sense to restrict attention to the scenario in which there is a single h in $H^{\sim}$.

In general, since we can measure s and want to know c_{I} (rather than the other way around), a bound on something like $P(c_{\mathrm{I}} > k \mid s, m)$, perhaps with $k \equiv s + \varepsilon$, would provide some useful information concerning generalization error. With such a bound, we could say that if we observe the values of m and s to be M and S, then with high probability c_{I} is lower than $U(M, S)$ for some appropriate function $U(.,.)$. However, since both f and (for our learning algorithm) h are fixed in the probability distribution in Theorem 4, c_{I} is also fixed there, for IID error. (By contrast, in the Bayesian framework, c_{I} is only probabilistically determined.) In fact, in Theorem 4 what is varying is d_X (or more generally, when there is noise, d). So Theorem 4 does not directly give us the probability that c_{I} lies in a certain region, given the training set at hand. Rather, it gives the probability of a d_X (generated via experiments other than ours) such that the difference between the fixed c_{I} and (the function of d_X) s lies in a certain region.

It might seem that Theorem 4 could be modified to provide a bound of the type we seek. After all, since the value of c_{I} is fixed in Theorem 4, that theorem can be written as a bound on the 'inverse' of $P(c_{\mathrm{I}} > k \mid s, m)$, $P(s < \kappa \mid c_{\mathrm{I}}, m)$, where $\kappa \equiv c - \varepsilon$. How does $P(s \mid c_{\mathrm{I}}, m)$ relate to what we wish to know, $P(c_{\mathrm{I}} \mid s, m)$? The answer is given by Bayes' theorem: $P(c_{\mathrm{I}} \mid s, m) = P(s \mid c_{\mathrm{I}}, m) P(c_{\mathrm{I}} \mid m) / P(s \mid m)$.

Unfortunately, this result has the usual problem associated with Bayesian results: it is prior-dependent. Nor does it somehow turn out that that prior has little effect. Depending on $P(c_{\mathrm{I}})$, $P(c_{\mathrm{I}} > s + \varepsilon \mid s, m)$ can differ markedly from the bound on $P(s < c_{\mathrm{I}} - \varepsilon \mid m, c_{\mathrm{I}})$ given in Theorem 4. Even if, given a truth c_{I}, the probability of an s that differs substantially from the truth is small, it does not follow that given an s, the probability of a truth that differs substantially from that s is small.

To illustrate this point, suppose we have two random variables, A and B, which can both take on the values 'low' and 'high'. Suppose that the joint probability distribution is proportional to: $P(A = \text{high}, B = \text{high}) = 100$, $P(A = \text{high}, B = \text{low}) = 2$, $P(A = \text{low}, B = \text{high}) = 1$, $P(A = \text{low}, B = \text{low}) = 1$. Then the probability that A and B differ is quite small (3/104); we have a tight confidence interval relating them, just as in Theorem 4. Nonetheless, $P(A = \text{high} \mid B = \text{low})$ is 2/3: despite the tight confidence interval, if we observe $B = \text{low}$, we cannot infer that A is low as well. Replace 'A' with 'c_{I}', and 'B' with 's', and we see that results like Theorem 4 do not imply that having observed a low s, one can conclude that one has a low c_{I}.

A more concrete example of this effect in the context of supervised learning is the following result, established in Wolpert (1995):

Theorem 5. Let $\pi(x)$ be flat over all x and $P(f)$ flat over all f. For the noise-free IID likelihood, and the learning algorithm of Theorem 4:

$$P(c_{\mathrm{I}} \mid s, m) = \left[\binom{m}{sm} c_{\mathrm{I}}^{sm} (1 - c_{\mathrm{I}})^{m - sm}\right] \times \left[\binom{n}{nc_{\mathrm{I}}} (|Y| - 1)^{nc_{\mathrm{I}}}\right]$$

where $|Y|$ is the number of values in Y.

Theorem 5 can be viewed as a sort of compromise between the likelihood-driven 'something for nothing' results of the VC framework, and the 'no free lunch' theorems. The first term in the product has no c_{I}-dependence. The second and third terms together reach a peak when $c_{\mathrm{I}} = s$; they 'push' the true misclassification rate towards the empirical misclassification rate, and would disappear if we were using OTS error. These two terms are closely related to the likelihood-driven VC bounds. However, the last two terms, taken together, form a function of c_{I} whose mean is $1/|Y|$. They reflect the fact that any f is allowed with an equal prior probability, and are closely related to the 'no free lunch' theorems (despite the fact that IID error is being used). In this sense, our result for $P(c \mid s, m)$ is just a product of a 'no free lunch' term with a VC-type term.

In response to the formal admonitions of these theorems, one is tempted to make the following intuitive reply: 'Say we have a function f and a given hypothesis function h' that have no *a priori* relation with one another. A sample point is drawn from f, and it is found that h' correctly predicts that point. Then another sample point is drawn, with the same result. Based on such a sequence of points, you guess that h' will correctly predict the next sample point. And lo and behold, it does. You keep extending the original sequence this way, always getting the same result that h' makes the correct prediction (since s is small, and the full training set d consists of the extended sequence, not the original one). In other words, the generalizer given by the rule "always guess h'" has excellent cross-validation error. In this situation, wouldn't you believe that it is unlikely for h' and f to disagree on future sample points, regardless of the "no free lunch" theorems?'

To disentangle the implicit assumptions behind this argument, consider it again in the case where h' is some extremely complex function that was formed by a purely random process. The claim in the intuitive argument is that h' was fixed independently of any determination of f, d, or anything else, and is not biased in any way towards f. Then, so goes the claim, f was sampled to generate d (the sequence of points), and it just so happened that f and h' agree on d. According to the intuitive argument presented above, we should conclude in such a case that h' and f would agree on points not yet sampled. Yet in such a situation our first suspicion might instead be that the claims that were made are wrong, that cheating has taken place and that h' is actually based on prior knowledge concerning f. After all, how else could the 'purely random' h' agree with f? How else could there be agreement when h' was supposedly fixed without any information concerning d, and therefore without any coupling with f?

If, however, we are assured that no cheating is going on, then 'intuition' might very well lead one to say that the agreements between f and h' must be simple coincidence. They have to be, since, by hypothesis, there is nothing that could possibly connect h' and f. So intuition need not proclaim that the agreements on the data set mean that f and h' will agree on future samples. Moreover, if cheating did occur, then to formulate the problem correctly, we have to know about the *a priori* connection between f and h' in order to properly analyze the situation. This results in a (prior-dependent) distribution different from the one investigated in the uniform convergence framework. (In the real world, of the two alternatives of coincidence and cheating, the reason for low s is almost always 'cheating'. Almost always one uses prior knowledge of some sort to guide the learning, rather than generate hypotheses purely at random.)

CONCLUSION

In all forms of reasoning that do not proceed by strict logical deduction, some kind of statistical algorithm must be employed. One of the major types of such reasoning is 'supervised learning'. In this type of reasoning one is provided with a training set of input–output pairs, and must make a guess for the entire input–output function in such a way as to minimize the error between that guess and the actual function that generated the data.

Apart from conventional sampling theory statistics, there are two principal mathematical approaches to supervised learning: the Bayesian framework and the computational learning framework. We have examined the foundations of these two approaches, especially in light of the 'no free lunch' theorems which limit what *a priori* formal assurances one can have concerning a learning algorithm without making assumptions concerning the real world.

In the Bayesian framework the assumptions concerning the real world arise explicitly, as 'prior probabilities' to be used to calculate the optimal guess. Unfortunately, the fact that the formalism concentrates solely on this assumption-driven calculation prevents it from allowing easy and broad investigation of what happens when those underlying assumptions are incorrect. In particular, Bayesian analysis cannot be used to analyze most learning algorithms that do not make their assumptions explicit; its scope is limited by construction. In particular, this restricts the framework's ability to analyze perhaps the most common algorithm in science, cross-validation.

In contrast, the simplest version of the computational learning framework makes no explicit assumptions about the nature of one's learning algorithm, or about the priors. In this sense it is universally applicable. Whereas the Bayesian framework skirts the 'no free lunch' theorems by forcing the underlying assumptions concerning the problem domain to be explicit, the strategy of the computational learning framework is to instead focus on the counterfactual scenario in which one's data are not fixed (to whatever the data set currently in front of you happens to be), but are averaged over. Unfortunately, the resulting bounds on learning error cannot be modified to concern some particular scenario a learning practitioner is confronted with; they are by their nature concerned with an average over multiple scenarios, when only one actually exists.

References

Baum E and Haussler D (1989) What size net gives valid generalization? *Neural Computation* **1**: 151–160.

Berger J (1985) *Statistical Decision Theory and Bayesian Analysis*. New York, NY: Springer-Verlag.

Buntine W and Weigend A (1991) Bayesian back-propagation. *Complex Systems* **5**: 603–643.

Loredo T (1990) From Laplace to Supernova 1987a: Bayesian inference in astrophysics. In: Fougere P (ed.) *Maximum Entropy and Bayesian Methods*. Dordrecht: Kluwer.

Vapnik V (1982) *Estimation of Dependences Based on Empirical Data*. New York, NY: Springer-Verlag.

Wolpert DH (1993) Reconciling Bayesian and non-Bayesian analysis. In: Heidbreder G (ed.) *Maximum Entropy and Bayesian Methods 1993*. Dordrecht: Kluwer.

Wolpert DH (1995) The Relationship between PAC, the statistical physics framework, the Bayesian framework, and the VC framework. In: Wolpert DH (ed.) *The Mathematics of Generalization*, pp. 117–214. Reading, MA: Addison-Wesley.
Wolpert DH (1996a) The lack of a priori distinctions between learning algorithms. *Neural Computation* **8**: 1341–1390.
Wolpert DH (1996b) The bootstrap is inconsistent with probability theory. In: Hanson K and Silver R (eds) *Maximum Entropy and Bayesian Methods 1995*. Dordrecht: Kluwer.
Wolpert DH (1997) On bias plus variance. *Neural Computation* **9**: 1211–1243.

Bayesian Belief Networks

Intermediate article

Brendan J Frey, University of Toronto, Toronto, Canada

CONTENTS

A "Bayesian belief network" is a directed acyclic graph that specifies how stochastic variables in a complex system interact. The joint distribution is equal to the product of a set of conditional distributions. There is one conditional distribution for each node of the graph and each conditional distribution is conditioned on the parents of the corresponding node. Unlike Markov random fields, Bayesian networks do not require a partition function.

REASONING UNDER UNCERTAINTY

An intelligent agent that makes decisions in a realistic environment should take into account the uncertainties in the environment and the uncertainties introduced by incomplete knowledge of the environment. Also, a mathematical description of a physical system for inference should account for the uncertainties in physical systems.

Probability theory provides a way to account for uncertainty. A system is described by a set of random variables, and an instantiation of the variables is called a *configuration*. A numerical probability between 0 and 1 is associated with each configuration and this number corresponds to the relative frequency with which the configuration occurs, or possibly, in the Bayesian view, the chance that the configuration *will* occur. The sum over the probabilities of all configurations must be 1. If we are interested in only a subset of the variables, we can derive the probability of each sub-configuration by summing the probabilities of all configurations that have matching sub-configurations. In this way, a system can be viewed as being 'consistent' with a larger system. So, when building or inferring a system, we need not include all variables in the universe, but can instead include only a smaller, more tractable, subset.

For example, we may use $P(T=1)$ to represent the probability of the event that there is a tiger in the field of view of an intelligent agent, and $P(T=0)$ to represent the probability that there is not a tiger. In this case, $P(T=1)$ is equal to the sum over the probabilities corresponding to all configurations of the universe for which there is a tiger in the field of view of the intelligent agent.

It is often convenient to express probabilities in functional form. For example, the probabilities that $T=1$ and $T=0$ can be written $P(T)$. If we set $T=1$, then $P(T)$ is the probability that $T=1$. We refer to $P(T)$ as a *probability distribution*.

We use the conditional probability distribution

$$P(T|V=v) \tag{1}$$

to represent the probabilities that there is and is not a tiger, given that the random variable V representing the agent's visual input has the value v. If $P(T=1|V=v) >> P(T=0|V=v)$, there is very probably a tiger in the scene and the agent ought to seriously consider running away.

It is natural to represent the random variable for the agent's visual input by a continuous vector (e.g., a vector of real-valued pixel intensities). For brevity, we will assume that all variables are discrete. In the case of continuous variables, probabilities can be replaced with probability density functions, and summations can be replaced by integrals. Many computations involving seemingly continuous quantities are in fact discrete (e.g., floating point computations on a computer), so not too much is lost by assuming the variables are discrete.

Suppose v is a visual scene containing black and orange stripes, and the probability of a scene containing black and orange stripes is greater if there is a tiger than if there is no tiger:

$$P(V = v|T = 1) > P(V = v|T = 0) \tag{2}$$

After observing $V{=}v$, it is tempting to conclude that there is probably a tiger in the scene, but this may not be so.

The probability that there is a tiger in the scene is given by the posterior probability distribution $P(T|V = v)$. The posterior distribution can be computed from the values of $P(V = v|T = 1)$ and $P(V = v|T = 0)$ using Bayes' rule:

$$P(T|V = v) = \frac{P(V = v|T)P(T)}{\sum_{t=0,1} P(V = v|T = t)P(T = t)} \tag{3}$$

In this expression, $P(T|V = v)$ is called the *posterior distribution*, $P(T)$ is called the *prior distribution*, and $P(V = v|T)$ is called the *likelihood function*. Notice that all three expressions are functions of the unknown random variable T.

For example, on a jungle tour in India, we may have $P(T = 1) = 0.2$ and $P(T = 0) = 0.8$. However in Canada, we may have $P(T = 1) = 0.001$ and $P(T = 0) = 0.999$. So, for the scene with orange and black stripes, even though $P(V = v|T = 1) > P(V = v|T = 0)$, it may turn out that $P(T = 1|V = v) < P(T = 0|V = v)$, depending on the prior distribution over T.

MAKING DECISIONS UNDER UNCERTAINTY

Even though $P(T = 1|V = v) < P(T = 0|V = v)$, it may be a good idea to leap aside if orange and black stripes appear in the visual scene.

For every configuration of the unknown random variable ($T{=}1$ and $T{=}0$), we can specify a utility (benefit, negative cost) for every action, say 'Leap' and 'NoLeap'. Leaping aside when there is a tiger may mean survival, whereas not leaping aside when there is a tiger may mean death:

$$U^{\text{Leap}}(T = 1) >>> U^{\text{NoLeap}}(T = 1) \tag{4}$$

Leaping aside when there is no tiger will waste some energy, so

$$U^{\text{Leap}}(T = 0) < U^{\text{NoLeap}}(T = 0) \tag{5}$$

However, notice that there is much less difference in this inequality than in the previous inequality.

The agent can maximize its expected utility by choosing the decision that has highest expected utility. For Leap and NoLeap, we have

$$EU^{\text{Leap}} = \sum_{T=0,1} U^{\text{Leap}}(T)P(T|V = v) \tag{6}$$

$$EU^{\text{NoLeap}} = \sum_{T=0,1} U^{\text{NoLeap}}(T)P(T|V = v) \tag{7}$$

So, even in Canada, the second term in each formula may dominate, in which case the agent should leap aside if orange and black stripes appear in the visual scene.

BAYESIAN NETWORKS

Instead of a real tiger, orange and black stripes could be caused, say, by a child's stuffed toy tiger. Further, in a toy store, the cause is more likely to be a stuffed toy. In a zoo, the cause is more likely to be a real tiger. These random variables influence each other in a structured way, and Bayesian networks provide a graphical description of this structure.

A *Bayesian network* (Pearl, 1988) is a graphical description of the probability distribution on a system of random variables. It is a directed acyclic graph (DAG) on vertices corresponding to the random variables, along with a specification of the conditional distribution for each variable given its parents in the graph. In the context of directed graphs, 'acyclic' means there aren't any cycles *when edge directions are followed.* The probability distribution on the system of random variables is equal to the product of all of the conditional distributions. If the conditional distributions are not specified, the Bayesian network refers to the set of all probability distributions that can be derived by choosing numbers for the conditional distributions in the network. Further, all distributions described by the network satisfy certain conditional independence properties that can be determined directly from the graph.

Suppose $S = 1$ indicates there is a stuffed toy tiger in the scene, whereas $S{=}0$ indicates there is not a stuffed toy tiger in the scene. The presence of orange stripes in the scene can be caused by either a tiger or a stuffed toy tiger. Figure 1(a) shows one possible Bayesian network that can be used to describe the relationship between the random variables T, S, and V. (Pearl uses an example where a burglar alarm is tripped by either a burglar or an earthquake.) This network describes the set of probability distributions that satisfy

$$P(T,S,V) = P(T)P(S)P(V|T,S) \quad (8)$$

the product of the conditional distributions for each child given its parents. Note that since T does not have parents, its 'conditional distribution' is written $P(T)$.

As described in Pearl (1988), the network can be examined to determine that all distributions described by the network have the property that T and S are independent random variables. This example is simple enough that we can show this

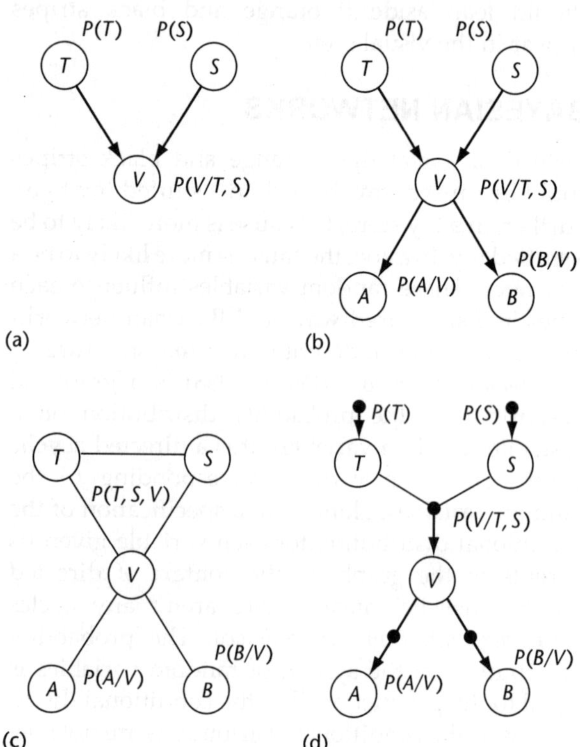

Figure 1. (a) A Bayesian network describing the probability distribution $P(T, S, V) = P(T)P(S)P(V|T, S)$ for the binary indicator variables tiger T, stuffed toy S, and a variable V that indicates the presence of orange stripes in the visual scene. (b) The network in (a) is extended to include variables that indicate whether Alice (A) and Bob (B) leap aside when orange stripes are present in the same scene. The joint distribution is $P(T, S, V, A, B) = P(T)P(S)P(V|T, S)\ P(A|V)P(B|V)$. (c) A Markov network, and (d) a factor graph that describe the same joint distribution.

is true by summing over V to obtain the distribution $P(T, S)$:

$$P(T,S) = \sum_V P(T,S,V) = \sum_V P(T)P(S)P(V|T,S)$$
$$= P(T)P(S) \quad (9)$$

It follows that $P(T|S) = P(T)$ and $P(S|T) = P(S)$, so T and S are independent. For example, the distribution over T is the same, whether or not S is known.

A particular distribution $P(T, S, V)$ is determined by specifying the numerical values of the conditional probability tables; e.g., $P(V = v|T = t, S = s)$ for all values of v, t, and s. Suppose $V = 1$ indicates the presence of orange stripes in the visual scene and $V = 0$ indicates that orange stripes are not present. In Canada, where toy tigers are more abundant than real tigers, we may have the following conditional probability tables:

P(T)	
T = 0	T = 1
0.999	0.001

P(S)	
S = 0	S = 1
0.999	0.001

		P(V|T,S)	
T	S	V = 0	V = 1
0	0	0.999	0.01
0	1	0.5	0.5
1	0	0.1	0.9
1	1	0.05	0.95

Now, suppose the agent is accompanied by two other agents, Alice and Bob, and that in response to the presence of orange stripes in the scene, Alice and Bob independently choose to leap aside. $A = 1$ indicates that Alice has leapt aside, whereas $A = 0$ indicates that Alice has not leapt aside. Similarly, B indicates Bob's behavior. Figure 1(b) shows a Bayesian network that includes the behavior of Alice and Bob. The conditional probability table for $P(A|V)$ gives the probability that Alice leaps aside given the presence or absence of orange stripes in the scene.

This Bayesian network indicates that the joint distribution factorizes as follows: $P(T,S,V,A,B) = P(T)P(S)P(V|T,S)P(A|V)P(B|V)$. Also, various conditional independencies can be determined by studying the graph, as described in Pearl (1988). For example, Alice's and Bob's behaviors, A and B, are generally not independent. Their behavior is a consequence of a common cause, V. However, Alice's and Bob's behaviors are independent given the visual scene, V.

MARKOV RANDOM FIELDS

Like Bayesian networks, *Markov networks* (Markov random fields) provide a graphical description of the structure of the joint distribution. Unlike Bayesian networks, they are undirected.

A Markov network (Kinderman and Snell, 1980) is an undirected graph on vertices corresponding to the variables, along with a specification of potential functions defined on the variables in maximal *cliques*. A clique is a set of variables that are completely connected. A maximal clique is a clique that cannot be expanded to include an additional variable, without violating the condition that the variables be completely connected. Each potential function is a nonnegative function of the appropriate set of variables. Assuming all potentials are strictly positive, the probability distribution on the system of random variables is equal to the product of all of the clique potentials, multiplied by a normalizing constant.

The conditional independencies indicated by a Markov network are quite different from those indicated by a Bayesian network. In a Markov network, given the neighbors of a variable (the Markov blanket), the variable is independent of all other variables in the network. For a given set of variables, it is possible that there is a Bayesian network that can represent conditional independencies that cannot be represented by a Markov random field. The converse is also true.

For example, the Markov network for the above example is shown in Figure 1(c). The factor $P(V|T,S)$ requires that at least one maximal clique contain V, T, and S. Consequently, by examining the network without reference to the potentials, it is not possible to determine that T and S are independent.

FACTOR GRAPHS

Factor graphs subsume Bayesian networks and Markov networks, in the sense that there is a unique Bayesian network (or Markov network) for every factor graph. However, for a given Bayesian network (or Markov network), there may be multiple factor graphs. Since the Bayesian networks or Markov networks corresponding to the multiple factor graphs are the same, the different factor graphs do not indicate different conditional independencies. However, they can indicate more detailed factorizations of the joint distribution. Also, factor graphs have an extra set of nodes that identify factorization sites, and, in the algorithm described in the next section, computation sites for message-passing algorithms.

A *factor graph* (Kschischang *et al.*, 2001) is a bipartite graph on *function vertices* and *variable vertices*. For each function vertex, a *local function* is specified, which is a function of the variables connected to the function vertex. 'Bipartite' means that each edge connects a variable vertex and a function vertex. The local functions may correspond to the conditional distributions in the Bayesian network, the clique potentials in the Markov network, or something else. The probability distribution on the system of random variables is equal to the product of all of the local functions, multiplied by a normalizing constant, if necessary.

If a local function is a conditional distribution over a variable, the edge connecting the local function to the variable may be directed toward the variable, to graphically indicate that the local function is a conditional distribution. This notation is useful for converting a factor graph to a Bayesian network.

Figure 1(d) shows a factor graph corresponding to the Bayesian network in Figure 1(b), where variable vertices are shown as white discs and function vertices are shown as black discs. This graph indicates that the joint distribution factors into the product $P(T)P(S)P(V|T,S)P(A|V)P(B|V)$.

A Bayesian network is converted to a factor graph by creating one variable vertex for each variable, creating one function vertex for each variable, connecting the function vertex for each variable to the variable and its parents, and setting the local function for each function vertex to P (variable|parparents) from the Bayesian network. A factor graph is converted to a Bayesian network by removing the function nodes and using the directed edges to identify the child–parent relationships. Compare Figure 1(b) and Figure 1(d).

A Markov network is converted to a factor graph by creating one variable vertex for each variable, creating one function vertex for each clique potential, setting the local function for each function vertex to the corresponding clique potential, and connecting the function vertex to the variables on which it depends (i.e., to the variables in the corresponding clique of the Markov network). A factor graph is converted to a Markov network by considering each local function in turn, and creating a maximal clique from all variables connected to the local function. Compare Figure 1(c) and Figure 1(d).

From the above descriptions, it is clear that a factor graph has a unique Markov network, and it has a unique Bayesian network as well, if edge directions are provided.

PROBABILISTIC INFERENCE

From a joint distribution, we can compute the conditional distribution of one subset of variables given another subset of variables. To do this, we use the *chain rule*

$$P(A,B) = P(A|B)P(B) \tag{10}$$

and the rule for marginalization

$$P(A) = \sum_B P(A,B) \tag{11}$$

For the example where $P(T,S,V) = P(T)P(S)P(V|T,S)$, we can compute $P(T|V=v)$ as follows:

$$\begin{aligned} P(T|V=v) &= \frac{P(T,V=v)}{P(V=v)} \\ &= \frac{\sum_s P(T,S=s,V=v)}{\sum_{t,s} P(T=t,S=s,V=v)} \\ &= \frac{\sum_s P(V=v|T,S=s)P(T)P(S=s)}{\sum_{t,s} P(V=v|T=t,S=s)P(T=t)P(S=s)} \end{aligned} \tag{12}$$

While this 'direct' approach works for small problems, the number of additions needed grows exponentially with the number of variables. So, the direct approach becomes intractable when there are more than a few dozen binary variables, since the summations will then involve many billions of terms.

PROBABILITY (BELIEF) PROPAGATION, OR THE SUM–PRODUCT ALGORITHM

Probability propagation provides a way of computing the distribution of one variable given a set of observed variables. This computation is usually not as straightforward as the direct application of Bayes' rule, since there may be many other variables that must be properly accounted for. For example, to compute $P(A|B)$ from a distribution $P(A,B,C)$, we first sum over C to get $P(A,B) = \sum_c P(A,B,C)$.

The algorithm is easily understood from an example. Suppose the distribution over variables $A, B, \ldots, F$ factorizes as follows:

$$\begin{aligned} &P(A,B,C,D,E,F) \\ &\quad = P(A|B)P(B|C)P(C|D)P(D|E)P(E|F)P(F) \end{aligned} \tag{13}$$

(Note that the Bayesian network, Markov network, and factor graph for this distribution all have the form of a chain.) Say we would like to compute the marginal distribution, $P(A)$:

$$\begin{aligned} P(A) = &\sum_B \sum_C \sum_D \sum_E \sum_F (P(A|B)P(B|C)P(C|D) \\ &P(D|E)P(E|F)P(F)) \end{aligned} \tag{14}$$

Computing this distribution directly takes roughly 2^5 multiplications and additions.

The distribution can be computed more efficiently by distributing the summations over the products:

$$\begin{aligned} P(A) = &\sum_B P(A|B)\Big(\sum_C P(B|C)\Big(\sum_D P(C|D) \\ &\Big(\sum_E P(D|E)\Big(\sum_F P(E|F)P(F)\Big)\Big)\Big)\Big) \end{aligned} \tag{15}$$

Computing the distribution $P(A)$ by successively computing 'partial distributions' takes roughly 2×5 summations, an exponential speed-up over the direct approach.

The computation of each 'partial distribution' can be thought of as a procedure that takes in messages, combines them with a conditional probability (or a potential or local function), performs a summation, and produces a new message. In the above example, the summation $\sum_F P(E|F)P(F)$ produces a message that is a real-valued function of E.

Probability propagation has a very simple form in factor graphs, so we describe the algorithm using factor graphs. Using the procedures described above, Bayesian networks and Markov networks can easily be converted to factor graphs. Also, the Bayesian network or Markov network corresponding to a given factor graph is quite obvious, so working with the factor graph does not obfuscate the original model.

Probability propagation consists of passing messages (implemented in a computer as short vectors of real numbers) on edges in the factor graph. Both function vertices and variables vertices combine incoming messages to produce outgoing messages on each of their edges. Each edge in the factor graph can pass a message in either direction, but *the number of values in a message is equal to the number of values its neighboring variable can take on*. For any edge, this number is unique, since in a factor graph, each edge is connected to only one variable vertex. So, we can think of each message as being a function of its neighboring variable.

For now, we'll assume that we are given a *message-passing schedule* that specifies which messages should be updated at each timestep. Think of each edge in the factor graph as having two message buffers (memory to store two messages) – one for each direction. Initially, we set all the messages (all the elements of all the vectors used to store the messages) to 1.

There are three types of computation that are performed in probability propagation:

1. *Propagating variable-to-function messages.* To produce an outgoing message on an edge, a variable computes the element-wise product of incoming messages on the other edges. For example, in Figure 2(a), $f(V)$ is the message sent from function 4 to variable V (note that it is a function of its neighboring variable, V). Let $g(V)$ be the message sent from function 5 to variable V. The message sent from variable V to function 3 is computed from:

$$h(V) = f(V)g(V) \tag{16}$$

That is, for each value of V, the corresponding elements $f(V)$ and $g(V)$ are multiplied together. Note that if a variable has just one edge (e.g., A in Figure 2(a)) its outgoing message is set to 1.

Propagating observations. If variable V is observed and has the value v, then an outgoing message is computed in the same fashion as described above, *except* that for all values $V \neq v$, we set the outgoing message to *zero*. So, in the above example, we compute $h(V)$ as follows:

$$h(V) = \begin{cases} f(V)g(V) & \text{if } V = v \\ 0 & \text{if } V \neq v \end{cases} \tag{17}$$

2. *Propagating function-to-variable messages.* To produce an outgoing message on an edge, a function takes the product of its associated local function (conditional probability function, in the case of Bayesian networks, potential, in the case of Markov networks) with the incoming messages and sums over all variables in the conditional probability function, *except* the variable to which the message is being sent. For example, in Figure 2(b), $f(T)$ is the message sent from variable T to function 3 (note that it is a function of its neighboring variable), and $g(S)$ is the message sent from variable S to function 3. The message sent from function 3 to variable V is computed from

$$h(V) = \sum_T \sum_S P(V|T,S)f(T)g(S) \tag{18}$$

3. *Fusion.* Suppose the *incoming* messages to variable V are $f(V)$, $g(V)$ (from Figure 2(a)) and $h(V)$ (from Figure 2(b)). Variable V can fuse its incoming messages to compute an estimate of the *joint* probability of V and the observed variables:

$$\hat{P}(V, \text{Observations}) = f(V)g(V)h(V) \tag{19}$$

If V is observed to have the value v, we use

$$\hat{P}(V, \text{Observations}) = \begin{cases} f(V)g(V)h(V) & \text{if } V = v \\ 0 & \text{if } V \neq v \end{cases} \tag{20}$$

If $\hat{P}(V, \text{Observations})$ is then normalized with respect to V, we obtain an estimate of the *conditional* probability of V *given* the observed variables:

$$\hat{P}(V|\text{Observations}) = \hat{P}(V, \text{Observations}) / \sum_V \hat{P}(V, \text{Observations}) \tag{21}$$

Exact Inference Using Probability Propagation

If the factor graph is a tree, and if the messages arriving at a variable, say V, are based on the input from every other vertex in the graph (variable vertices and function vertices), then

$$\hat{P}(V, \text{Observations}) = P(V, \text{Observations}) \tag{22}$$

and

$$\hat{P}(V|\text{Observations}) = P(V|\text{Observations}) \tag{23}$$

That is, the fused messages give *exact* probabilistic inferences.

The Generalized Forward–Backward Algorithm

Suppose we wish to infer the probability for each and every variable in the network, given the observations. Clearly, for each and every variable in the network to receive messages from each and every other variable, at least roughly $2E$ messages must be computed, where E is the number of edges in the

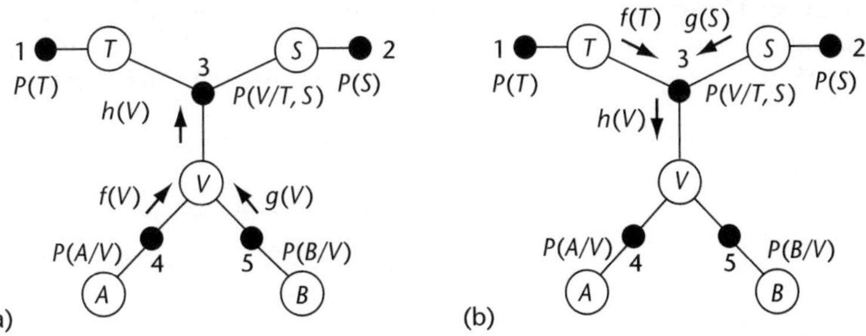

Figure 2. (a) Computing a variable-to-function message in the factor graph from Figure 1(d). The edge directions in the factor graph are dropped for visual clarity. (b) Computing a function-to-variable message.

factor graph. (Slightly less than $2E$ messages may be needed, since, for example in Figure 2(a), we needn't pass a message from variable T to function 1.) The following procedure, called the *generalized forward–backward algorithm*, shows how we can achieve this bound – infer the probability for each and every variable in the network by passing $2E$ messages.

First, arbitrarily pick a vertex in the factor graph and call it the 'root'. Form a tree by arranging the vertices in layers, with the root at the top. Now, pass messages layer by layer *up* from the bottom to the top and then pass messages layer by layer *down* from the top to the bottom. Clearly, this procedure computes $2E$ messages and the messages arriving at each variable contain the input from each and every other vertex in the factor graph. Note that in this case, we need not initialize the messages.

Approximate Inference Using Probability Propagation

If the messages arriving at a variable do not contain the input from each and every other vertex in the factor graph, then $\hat{P}(A, \text{Observations})$ may not equal $P(A, \text{Observations})$. However, it *may* be a good estimate and in some cases can even be exactly correct. A more interesting case is when the factor graph is not a tree, but contains lots of cycles. In this case, even if the messages arriving at a variable contain the input from each and every other vertex in the factor graph, $\hat{P}(A, \text{Observations})$ will usually not be equal to $P(A, \text{Observations})$, because of the cycles. However, there are some very impressive applications where the approximation is astonishingly good (Frey and MacKay, 1998; Freeman and Pasztor, 2000; Frey *et al.*, 2001). Also, new analysis is emerging that partly explains the approximation (Weiss and Freeman, 2001; Yedidia *et al.*, 2001; Wainwright *et al.*, 2002).

References

Freeman W and Pasztor E (1999) Learning low-level vision. *Proceedings of the International Conference on Computer Vision*, 1182–1189.

Frey BJ and MacKay DJC (1998) A revolution: {B}elief propagation in graphs with cycles. In: Jordan MI, Kearns MI and Solla SA (eds) *Advances in Neural Information Processing Systems, vol. 10*. Cambridge, MA: MIT Press.

Frey BJ, Koetter R and Petrovic N (2002) Very loopy belief propagation for unwrapping phase images. In: Dietterich TG, Becker S and Ghahraman Z (eds) *Advances in Neural Information Processing Systems 14*. Cambridge, MA: MIT Press.

Kinderman R and Snell JL (1980) *Markov Random Fields and Their Applications*. Providence state, USA: American Mathematical Society.

Kschischang FR, Frey BJ and Loeliger HA (2001) Factor graphs and the sum-product algorithm. *IEEE Transactions on Information Theory* **47**(2): 498–519.

Pearl J (1988) *Probabilistic Reasoning in Intelligent Systems*. San Mateo, CA: Morgan Kaufmann.

Wainwright MJ, Jaakkola T and Willsky AS (2002) Tree-based reparameterization for approximate estimation on loopy graphs. In: Dietterich TG, Becker S and Ghahraman Z (eds) *Advances in Neural Information Processing Systems 14*. Cambridge, MA: MIT Press.

Weiss Y and Freeman W (2001) On the optimality of solutions of the max-product belief propogation algorithm in arbitrary graphs. *IEEE Transactions on Information Theory* **47**(2): 736–744.

Yedidia J, Freeman WT and Weiss Y (2001) Generalized belief propogation. In: Dietterich TG, Becker S and Ghahraman Z (eds) *Advances in Neural Information Processing Systems 14*. Cambridge, MA: MIT Press.

Further Reading

Neapolitan RE (1990) *Probabilistic Reasoning in Expert Systems*. New York, NY: John Wiley and Sons.

Frey BJ (1998) *Graphical Models for Machine Learning and Digital Communication*. Cambridge, MA: MIT Press.

Hinton GE, Dayan P, Frey BJ and Neal RM (1995) The wake-sleep algorithm for unsupervised neural networks. *Science* **268**: 1158–1161.

Hinton GE and Sejnowski TJ (1986) Learning and relearning in Boltzmann machines. In: Rumelhart DE and McClelland JL (eds) *Parallel Distributed Processing: Explorations in the Microstructure of Cognition*, 282–327. Cambridge, MA: MIT press.

Jordan MI (2001) *Learning in Graphical Models*. Cambridge, MA: MIT Press.

Bayesian Learning in Games

Intermediate article

JS Jordan, Pennsylvania State University, University Park, Pennsylvania, USA

CONTENTS

Bayesian learning is a method by which players in a game attempt to infer each other's future strategies from the observation of past actions. Under certain assumptions, learning is successful in the sense that players' expectations converge to Nash equilibria of the game.

INTRODUCTION

A Nash equilibrium of a game consists of a strategy for each player that maximizes the player's expected pay-off against the strategies of the other players. It is natural to assume that players seek to maximize their expected pay-offs, but the assumption that each player correctly anticipates the strategies of the others is more problematic. If the players' pay-off functions are public knowledge, then presumably each player could compute the equilibrium strategies, provided that some selection criterion is adopted in the case of multiple equilibria. In most applications of game theory to economics, however, individual characteristics are private information not directly observable by others. This raises the question of whether players might learn from experience.

Suppose that the game is repeated over time while pay-off functions remain fixed. If the actions taken by each player at each repetition are publicly observable, then players might, through some form of inductive inference, learn to form the correct expectations. In game theory, the canonical model of expectation formation is Bayes' theorem, so Bayesian learning provides a natural model of inductive inference. In a Bayesian learning model, each player has probabilistic beliefs about the sequence of actions that the repeated game will generate, and updates those beliefs at each iteration in response to the actions observed up to that point. The question is whether, over time, each player's beliefs about the future actions of the others converge to the correct Nash equilibrium expectations.

If all of the players except one were machines that take the same, possibly randomized, action each time, the inference problem for the single real player would be a straightforward problem of consistent statistical inference. Instead, the fact that all players are learning from each other's actions means that Bayesian learning produces complex interactive dynamics. Despite this complication, the Bayesian learning model provides some encouraging results on the possibility of learning Nash equilibrium expectations.

It is useful to distinguish between two different versions of the Bayesian learning model, which in this article will be termed 'sophisticated' and 'naive'. In the sophisticated Bayesian learning model, players are assumed to be knowledgeable game theorists who derive their expectations of the other players' actions from prior beliefs about the other players' pay-off functions. After each repetition, the observed actions cause them to revise their beliefs and update their expectations. In the naive Bayesian learning model, the players' prior beliefs are arbitrary probability distributions over sequences of actions. Expectations of future actions are updated directly from observed actions via Bayes' theorem, with no knowledge of game theory required.

SOPHISTICATED BAYESIAN LEARNING

Myopic Behavior

Since the game is repeated, players receive a stream of pay-offs over time. If the players ignore future pay-offs in each period, and seek to maximize their one-period pay-off against the expected actions of the other players, then it is natural to ask whether expectations approach correct Nash equilibrium expectations for the one-shot game that is being played each period. In contrast, if players anticipate future repetitions and seek to maximize the expected discounted sum of future pay-offs, then it is more natural to ask whether expectations approach correct Nash equilibrium expectations for the full repeated game. The case of myopic behavior will be discussed first.

The actions available to player i at each iteration lie in a finite set S_i. There are n players, and, following interation t, each player observes the entire n-tuple $s_t = (s_1, \ldots, s_{nt})$ of chosen actions. A learning process for player i is a sequence of expectation functions e_{it} which associate with any observed finite history of play $h_t = (s_{1t}, \ldots, s_t)$ an expectation $e_{it}(h_t) \in \Delta(S_{-i})$, where $\Delta(S_{-i})$ is the set of probability distributions over $(n-1)$-tuples $s_{-i} = (s_1, \ldots, s_{i-1}, s_{i+1}, \ldots, s_n)$ of actions to be chosen by the other players in period $t+1$. After each iteration t, each player i receives the pay-off $u_i(s_t)$. Player i knows the pay-off function u_i, which remains fixed through time, but does not know the pay-off function of any other player. Player i chooses s_{it} in period t to maximize the one-period expected value of u_i against the expectation $e_{i(t-1)}$ (h_{t-1}) of the other players' actions. The question is whether the expectations sequence $(e_{it}(h_t))_{t=1}^{\infty}$ approaches the set of Nash equilibrium expectations.

A Nash equilibrium consists of a strategy $\sigma_i^* \in \Delta(S_i)$ for each player i that is expected-pay-off-maximizing against the strategies of the other players, that is,

$$\sum\nolimits_{s_i \in S_i} \sigma_i^*(s_i) \sum\nolimits_{s_{-i} \in S_{-i}} u_i(s_i, s_{-i}) \sigma_{-i}^*(s_{-i})$$
$$\geq \sum\nolimits_{s_i \in S_i} \sigma_i(s_i) \sum\nolimits_{s-i \in S-i} u_i(s_i, s_{-i}) \sigma_{-i}^*(s_{-i})$$

for every other strategy $\sigma_i \in \Delta(S_i)$. (We write $u_i(s_i, s_{-i})$ for the pay-off resulting from player i using strategy s_i and the other players using strategies s_{-i}.) Given a Nash equilibrium $(\sigma_1^*, \ldots, \sigma_n^*)$, player i's Nash equilibrium expectation is the probability distribution $\sigma_{-i}^* = \sigma_1^* \times \cdots \times \sigma_{i-1}^* \times \sigma_{i+1}^* \times \cdots \times \sigma_n^* \in \Delta(S_{-i})$. Learning has the desired convergence property if, as $t \to \infty$, the distance between $e_{it}(h_t)$ and the nearest Nash equilibrium expectation σ_{-i}^* goes to zero for every player i. Since many games have multiple Nash equilibria, the sequence of expectations may have multiple limit points because it has subsequences converging to different Nash equilibrium expectations. The desired convergence property is that each limit point be a Nash equilibrium.

The expectation functions that constitute Bayesian learning can be derived as follows. The fact that player i knows u_i but not u_j for all $j \neq i$ means that the players face a game of incomplete information, in which each player's private type is the player's pay-off function. The incomplete information is modeled by supposing that player i believes that, prior to the first iteration, Nature chooses each pay-off function u_j according to a probability distribution μ_j over pay-off functions. Since the domain of pay-off functions S is finite, pay-off functions lie in the finite-dimensional vector space $\mathbb{R}^S$. Since best-response strategies are invariant with respect to positive linear transformations, we can reduce the space of possible pay-off functions to the unit sphere $B \subset \mathbb{R}^S$, which is an innocuous but convenient normalization. The pay-off functions determine the game that is being learned, so let $G = B^n$ denote the space of all possible games. All players believe that a game $(u_1, \ldots, u_n) \in G$ is initially chosen by nature according to the prior distribution $\mu = \mu_1 \times \cdots \times \mu_n$ over G. In particular, it is assumed that players share the common belief μ, under which the pay-off functions of different players are independently distributed. These assumptions can be relaxed, as will be discussed below.

The initial expectations e_{i0} are derived as a Bayesian Nash equilibrium of the incomplete information game in which each player i knows u_i and believes that each u_j is distributed according to μ_j. Given an initial expectation $e_{i1} \in \Delta(S_{-i})$, there is a best response $b_i(u_i) \in \Delta(S_i)$ associated with each possible pay-off function u_i. Together with the initial distribution of pay-off functions μ_i, a best-response function $b_i(\cdot)$ determines a distribution of player-i actions $\sigma_i \in \Delta(S_i)$, as $\sigma_i(s_i) = \int_B b_i(u_i)(s_i)$ $\mu_i(du_i)$. The Bayesian Nash equilibrium condition is that, for each i, $e_{i0} = \sigma_{-i}$, that is, each player's expectation is simply the derived distribution of the best responses of the other players to their expectations.

The actions $(s_{11}, \ldots, s_{n1})$ are chosen at the first iteration as best responses to the expectations e_{i0} for the true pay-off functions $(u_1, \ldots, u_n)$. Each player i then observes each s_{j1} and revises the prior distribution μ_j according to Bayes' theorem. The revised distribution $(\mu_1 = \mu_{11} \times \cdots \times \mu_{n1})$ over G results in new Bayesian Nash equilibrium expectations $e_{i1}(h_1)$, where h_1 is the one-period history $h_1 = (s_{11}, \ldots, s_{n1})$, and so on. At each iteration t, the expectations $e_{it}(h_t)$ are derived as a Bayesian Nash equilibrium for the revised beliefs μ_t. Bayesian Nash equilibria need not be unique, so the expectation functions are not generally determined uniquely. The Bayesian learning model encompasses all expectation functions derived in this way from all initial beliefs $\mu = \mu_1 \times \cdots \times \mu_n$, although many of the convergence results mentioned below place restrictions on the initial beliefs.

Although the expectation functions are not unique, the players' best-response strategies have a useful deterministic property. A mixed strategy can only be a best response if there are two or more actions that each maximize the player's expected pay-off. In particular, there must be actions s_i and s_i' that yield the same expected pay-off, $\sum_{s_{-i}} u_i(s_i, s_{-i}) e_{it}(s_{-i}|h_t) = \sum_{s_{-i}} u_i(s_i', s_{-i}) e_{it}(s_{-i}|h_t)$. Given the expectations $e_{it}(h_t)$, this equation is satisfied only for a proper linear subspace of the space of pay-off functions $\mathbb{R}^S$, and therefore only for a set of pay-off functions having Lebesgue measure zero, i.e., probabilistically negligible. For each t, the set of possible t-period histories is finite; so the set of all histories is countable. Therefore, given any sequence of expectation functions, all pay-off functions except for a set of Lebesgue measure zero have unique best-response actions in every period for every history. If the prior distributions μ_i are absolutely continuous with respect to Lebesgue measure on the sphere $B \subset \mathbb{R}^S$, then the set of pay-off functions having unique best responses throughout the learning process has probability one. Given the expectation functions $e_{it}(\cdot)$, Bayesian-learning players never play mixed strategies, except for a set of games having Lebesgue measure zero.

2×2 games

In general, the derivation of expectations as Bayesian Nash equilibria makes their explicit computation problematic. However, in the case of 2×2 games with uniform prior beliefs, the derivation of expectations is straightforward and provides a useful illustration.

Let player 1 be the row player, player 2 the column player, $S_1 = \{U, D\}$, and $S_2 = \{L, R\}$. Then player i's pay-off function is a 2×2 matrix with the entries $u_i(U,L)$, $u_i(U,R)$, $u_i(D,L)$, and $u_i(D,R)$. However, it will be convenient to employ a normalization, which reduces each player's space of possible pay-off functions to the unit circle $C \subset \mathbb{R}^2$. To motivate this normalization, suppose that player 1 anticipates that player 2 will play L with probability $\sigma_2(L)$. Then player 1's optimal strategy is U, D, or both, according as the quantity

$$\sigma_2(L)[u_1(U,L) - u_1(D,L)] + (1 - \sigma_2(L))[u_1(U,R) - u_1(D,R)]$$

is positive, negative, or zero. Therefore we can subtract the second row of player 1's pay-off matrix from each row, so that the top row is $(u_1(U,L) - u_1(D,L), u_1(U,R) - u_1(D,R))$ and the bottom row is $(0,0)$. Applying the same normalization to the columns of player 2's pay-off matrix produces the pay-off bimatrix in Table 1. If we ignore the measure-zero possibility that $a = b = 0$, that is, $u_1(U,L) = u_1(D,L)$ and $u_1(U,R) = u_1(D,R)$, then we can further normalize (a,b) to the unit circle without affecting player 1's best response to any mixed strategy played by player 2. Thus we can assume that $a^2 + b^2 = 1$, and (for player 2) that $\alpha^2 + \beta^2 = 1$. Under this normalization, each player's pay-off function is a point on the unit circle, and each 2×2 game is a point on the torus $C \times C$ (we will continue to exclude the degenerate cases $a = b = 0$ and $\alpha = \beta = 0$).

An obvious choice for the priors is the uniform distribution on the unit circle C. For this prior distribution, we will derive the expectations along the two-period history $((U,R), (D,L))$. More precisely, we will compute the first-period expectations $e_0(\cdot)$ and second-period expectations $e_1(\cdot|U,L)$, which are uniquely determined, and show that there are three possible choices for the third-period expectations $e_2(\cdot|(U,L), (D,R))$. First, let σ_2 denote the initial probability distribution on $S_2 = \{L, R\}$ facing player 1. That is, $\sigma_2(L)$ is the probability that player 2 will play L in the first period. Then player 1 will play U in period 1 if player 1's pay-off function, represented by (a,b), satisfies $a\sigma_2(L) + b(1 - \sigma_2(L)) > 0$. That is, the set of player-1 'types' that play U in period 1 is a semicircle $\{(a,b) : a\sigma_2(L) + b(1 - \sigma_2(L)) > 0\}$. The set $\{(a,b) : a\sigma_2(L) + b(1 - \sigma_2(L)) = 0\}$ has prior probability zero and thus can be ignored. Hence, for any expectation $\sigma_2(L)$, the measure of the set of player-1 types that play U, which is simply the prior probability of a semicircle, equals $\frac{1}{2}$. Since this reasoning applies to both players symmetrically, the unique first-period Bayesian Nash equilibrium expectations are $e_{20}(U) = e_{20}(D) = \frac{1}{2}$ and $e_{10}(L) = e_{10}(R) = \frac{1}{2}$.

Now suppose that the actions (U,R) are played in period 1. This reveals that player 1's pay-off function lies in the semicircle $\mathcal{T}_1 = \{(a,b) : \frac{1}{2}a + \frac{1}{2}b > 0\}$, and that player 2's pay-off function lies in the semicircle $\mathcal{T}_2 = \{(\alpha,\beta) : \frac{1}{2}\alpha + \frac{1}{2}\beta < 0\}$. To solve for the Bayesian Nash equilibrium conditional expectations $e_1(\cdot|(U,R))$, let $x = e_{21}(U|(U,R))$, the conditional measure of the set of

Table 1. Normalized pay-off bimatrix for a 2×2 game

	L	R
U	a, α	$b, 0$
D	$0, \beta$	$0, 0$

player-1 types that play U in period 2, and let $y = e_{11}(L|(U,R))$. Given y, x is simply $\mu_1(\{(a,b) \in \tau_1 : ay + b(1-y) > 0\})/\mu_1(\mathcal{T}_1)$, where μ_1 denotes the uniform distribution on the unit circle. Thus x is simply the relative arc length given by the formula

$$x = (\pi - |\theta(y)|)/\pi$$

where $\theta(y)$ is the angle between the expectations vectors $(\frac{1}{2}, \frac{1}{2})$ and $(y, 1-y)$ (in radians). The analogous formula for y as a function of x is

$$1 - y = (\pi - |\theta(x)|)/\pi$$

These two equations have a unique solution, $x^* \approx 0.82$ and $y^* \approx 0.18$, so the unique Bayesian Nash equilibrium expectations are $e_{21}(U|(U,R)) = x^*$ and $e_{11}(L|(U,R)) = y^*$.

Now suppose that the strategies (D, L) are played in period 2. This reveals that (a, b) and (α, β) satisfy the following conditions:

$$\frac{1}{2}a + \frac{1}{2}b > 0$$

$$y^*a + (1-y^*)b < 0$$

$$\frac{1}{2}\alpha + \frac{1}{2}\beta < 0$$

$$x^*\alpha + (1-x^*)\beta > 0$$

Since $y^* > \frac{1}{2}$ and $x^* > \frac{1}{2}$, equations 5 to 8 imply that $a > 0$, $b < 0$, and $\alpha > 0$, $\beta < 0$. It follows that (U, L) and (D, R) are pure-strategy Nash equilibria for every game $((a, b), (\alpha, \beta))$ that generates the two-period history $((U, R), (D, L))$. It also follows that there is a mixed-strategy Nash equilibrium, but the equilibrium mixed strategies are not yet revealed.

Thus there are three possible Bayesian Nash equilibrium expectations:

- $e_{22}(U|(U,R),(D,L)) = e_{12}(L|(U,R),(D,L)) = 1$
- $e_{22}(U|(U,R),(D,L)) = e_{12}(L|(U,R),(U,L)) = 0$
- $e_{22}(U|(U,R),(D,L)) \approx 0.61,\ e_{12}(L|(U,R),(U,L)) \approx 0.39$

The first two are 'pure strategy' Bayesian Nash equilibria. They correspond to the pure-strategy Nash equilibria and reveal no further information about each player's pay-off function. The third is a 'quasi-mixed' Bayesian Nash equilibrium. It further partitions the pay-off types revealed by the history $((U, R), (D, L))$.

This multiplicity of Bayesian Nash equilibria continues for all future periods. If the 'quasi-mixed' equilibrium is selected infinitely often, the expectations in those periods will converge to the mixed-strategy Nash equilibrium determined by the true pay-off types (a, b) and (α, β).

Far-sighted Behavior

The learning model described above addresses the question of whether players can learn from repeated experience to form correct Nash equilibrium expectations for the true game determined by the pay-off functions $(u_1, \ldots, u_n)$. However, the repetitions introduce the possibility that players may anticipate future pay-offs and take account of the effect that their current actions may have on the future actions of the other players. This can be modeled by supposing that player i has an additional pay-off characteristic consisting of a discount factor $1 > \delta_i \geq 0$, and chooses an action s_{it} in period t to maximize the expected discounted sum of current and future pay-offs $(1 - \delta_i) \sum_{\tau=t}^{\infty} \delta_i^{\tau-t} u_i(s_\tau)$. Myopic behavior corresponds to the special case $\delta_i = 0$ (under the convention $0^0 = 1$). The prior beliefs μ_i can be extended to cover the discount factors δ_i as well as the stage-game pay-off functions u_i. The game that the players are learning about is now the full repeated game determined by the discount factors together with the stage-game pay-off functions. Like the stage game, the true repeated game is stationary over time, so the true repeated game is itself being repeated over time in the sense that each period is the first period of the game to be played from that period on.

A strategy for player i in a repeated game is a sequence of functions $f_{it} : S_1 \times \cdots \times S_t \to \Delta(S_i)$ that determine player i's (possibly randomized) action in period $t+1$ as a function of the previous history of play. The sequence of functions $(f_{it})_{t=0}^{\infty}$ is called a 'behavior strategy'. A Nash equilibrium for a repeated game consists of a behavior strategy for each player that maximizes the expected discounted sum of pay-offs against the behavior strategies of the other players. Any Nash equilibrium of the stage game, if repeated every period, is also a Nash equilibrium of the repeated game. More precisely, if $(\sigma_1^*, \ldots, \sigma_n^*) \in \Delta(S_1) \times \cdots \times \Delta(S_n)$ is a Nash equilibrium of the stage game, then the constant-behavior strategies in which player i plays σ_i^* in every period after every history is also a Nash equilibrium of the repeated game. However, repeated games typically have many more Nash equilibria, in which players can influence one another's actions over time, especially if the discount factors are near unity.

The derivation of expectations as Bayesian Nash equilibria extends directly to repeated games

(Jordan, 1995). In this setting, player i's initial expectation e_{i0} is a probability distribution over the behavior strategies of the other players. A theorem due to Kuhn simplifies the analysis by-establishing that any probability distribution over behavior strategies generates the same distribution over action sequences as a single appropriately chosen behavior strategy. Hence the expectation e_{i0} can be represented as an $(n-1)$-tuple of behavior strategies $((f^i_{jt})^{\infty}_{t=1})_{j \neq i}$, and the subsequent expectations $e_{it}(h_t)$ reduce to the behavior strategies $((f^i_{j\tau}(h_t, \cdot))^{\infty}_{\tau=t})_{j \neq i}$. The question is whether these expected behavior strategies approach Nash equilibrium behavior strategies for the true pay-off characteristics $((u_1, \delta_1), \ldots, (u_n, \delta_n))$.

NAIVE BAYESIAN LEARNING

The naive Bayesian learning model drops the assumption that expectations are derived as Bayesian Nash equilibria from underlying prior beliefs about private pay-off characteristics. Instead, players are assumed to have arbitrarily given prior beliefs about the strategies of the other players, and to form expectations by conditioning their prior beliefs on the histories of observed actions. This model of expectation formation accommodates both far-sighted and myopic behavior. As in the sophisticated Bayesian learning model, the players choose actions in each period as best responses to their expectations, which generate the histories that the players observe over time.

In the case of myopic behavior, naive Bayesian learning accommodates all possible expectation functions. Given an arbitrary sequence of expectation functions $e_{it}(\cdot)$, one can construct a prior distribution μ_i over S^{∞} using the expectations $e_{it}(h_t)$ as successive conditional distributions over $s_{-i(t+1)}$. The only restriction on the expectation functions is that player i expects players j and k to choose their next-period actions independently, that is, $s_{j(t+1)}$ and $s_{k(t+1)}$ are independently distributed under $e_{it}(h_t)$. Subject to this restriction, naive Bayesian learning formally includes all learning models in which players choose actions in each period as one-period best responses to expectations. The naive Bayesian learning model that is perhaps most in the spirit of Bayesian inference is fictitious play (e.g. Krishna and Sjöström, 1998). In fictitious play, player i expects player j's next-period action to be drawn randomly, and uses the observed frequency distribution of player j's past actions as the expected distribution.

A general model of naive Bayesian learning with far-sighted players is formulated by Kalai and Lehrer (1993a). Player i has an arbitrarily given prior belief μ_i over the behavior strategies of the other players. Given player i's pay-off characteristics (u_i, δ_i) and the expected behavior strategies $((f^i_{jt})^{\infty}_{t=1})_{j \neq i}$, player i chooses a best-response behavior strategy $(f^i_{it})^{\infty}_{t=1}$. Kalai and Lehrer do not assume that prior beliefs are common across players, so players i and j may have different expectations about the behavior strategies of player k. A behavior strategy specifies what a player would do in response to every possible history, and Nash equilibrium requires that each player have the correct expectations about the behavior strategies of all other players. In particular, this requires players i and j to have the same expectations about what player k would do in response to every possible history. Kalai and Lehrer (1993b) broaden the concept of Nash equilibrium to 'subjective equilibrium' by allowing players to have different expectations about the behavior strategies of third players for histories that occur with probability zero in equilibrium. Asymptotic expectations are more appropriately compared to subjective equilibria than Nash equilibria, since the observed history may not be sufficient to resolve disparities in expectations about behavior for all histories.

CONVERGENCE TO NASH EQUILIBRIUM

In each of the Bayesian learning models described above, players form expectations about the actions of the other players and choose their own actions in each period as best responses to their expectations. The best responses, which may be mixed, constitute the probability distribution of action sequences that the players will observe over time. In the case of sophisticated Bayesian learning, the expectations are derived as Bayesian Nash equilibria from a prior belief over the pay-off characteristics that constitute the true game. Each game generates a best-response sequence, so the prior distribution over games, together with the best-response process for each game, generates a joint distribution over games and action sequences. In this context, one can ask whether a random draw of a game and action sequence from this joint distribution will produce expectations that approach Nash equilibria of the drawn game.

In the case of naive Bayesian learning, the true game is taken as given and players' expectations are given initially rather than derived as Bayesian Nash equilibria. In this case, one can ask whether a random draw from the best-response distribution over action sequences will produce expectations that approach Nash equilibria of the true game. One can also ask this question in the case of sophisticated Bayesian learning. That is, one can take the view, common in Bayesian statistics, that the derivation of expectations from a prior distribution is merely a heuristic device for obtaining expectations, and regard the expectations and the true game as given in the same sense as in myopic Bayesian learning.

The answers that have been obtained thus far to these and other questions are described below. For games with multiple Nash equilibria, convergence to Nash equilibrium will be understood to mean that the sequence of expectations has one or more limit points, all of which are Nash equilibria.

Sophisticated Bayesian Learning

In the model of sophisticated Bayesian learning with far-sighted behavior, with probability one, a randomly drawn game and best-response path has expectations that converge to Nash equilibrium. This was proved by Jordan (1995) under the assumptions that all players have the same prior beliefs about other players' pay-off characteristics, and that the pay-off characteristics of different players are independently distributed under the common prior. The common-prior assumption implies that any two players have the same expectations about the behavior strategies of third players, ensuring that limit points are Nash equilibria as opposed to subjective equilibria. This result is inherited under myopic behavior as a special case of far-sighted behavior with zero discount factors, but in the case of myopic behavior, Nyarko has substantially generalized the assumptions on prior beliefs. Nyarko (1998) retains the independence assumption but generalizes the common prior assumption to a condition ensuring that each player's prior belief is absolutely continuous with respect to a common distribution. The concept of prior belief is also extended to a hierarchy of beliefs, beliefs about beliefs and so on. Nyarko (1994) drops the independence assumption. In this case, a player's knowledge of his or her own pay-off characteristics may provide information about the pay-off characteristics of other players. Nyarko shows that with probability one, expectations converge to correlated equilibria.

These results imply that the set of games for which expectations do not converge to equilibria has probability zero according to the prior beliefs from which the expectations are derived. For the case of myopic behavior, a stronger assumption on prior beliefs makes it possible to derive expectations that converge for every game. Suppose that the prior beliefs, in addition to being the same for all players and satisfying the independence of pay-off characteristics across players, are 'smooth', in the sense that the prior distribution has a density function with respect to Lebesgue measure that is bounded from above and bounded from below away from zero. The uniform prior, as in the 2×2 example discussed above, is the natural example. Jordan (1991) establishes that in this case, for every game, the distribution of best-response paths generates expectations that converge to Nash equilibria of the game with probability one. Except for a set of games having Lebesgue measure zero, best-response actions are unique in every period (even when there are multiple Nash equilibria), and convergence is guaranteed along the unique best-response path. Moreover, under the same assumptions, except for a set of games having Lebesgue measure zero, expectations converge to Nash equilibria at an exponential rate (Jordan, 1992).

These results can be illustrated in the 2×2 game represented by the pay-off bimatrix in Table 2. The unique Nash equilibrium of this game is the mixed equilibrium in which the row player chooses T with probability 0.4 and the column player chooses L with probability 0.6.

Under the expectations derived from the uniform prior distribution (with pay-offs normalized as above), Table 3 shows the expectations, rounded to four decimal places, and the best-response actions for the first 12 iterations.

The table illustrates the rapid convergence of expectations to the unique Nash equilibrium. However, it also illustrates a typical disparity between expectations and actions in the case of convergence to mixed equilibrium. Player 1 is rapidly learning to predict that player 2 will choose L with probability 0.4, but in every period, player 2 actually chooses L with either probability one or probability zero. Since $e_{2(t-1)}(U \mid h_{t-1})$ is never exactly 0.6, player 2 always has a unique best-response action. This raises the question whether player 1 could eventually recognize that s_{2t} is not a random draw from the expected distribution $e_{1(t-1)}(h_{t-1})$. There is a precise sense in which the answer to this question is 'no'.

Table 2. A pay-off bimatrix representing a 2×2 game

	L	R
U	−2, 3	3, 0
D	0, −2	0, 0

Table 3. Expectations and best-response actions for the first 12 iterations of the game in Table 2

t	s_t	$e_{1(t-1)}(L \mid h_{t-1})$	$e_{2(t-1)}(U \mid h_{t-1})$
1	*U, L*	0.5000	0.5000
2	*D, L*	0.8192	0.8192
3	*D, R*	0.7624	0.1494
4	*D, R*	0.6514	0.3917
5	*U, L*	0.5808	0.4552
6	*D, L*	0.6203	0.4314
7	*D, L*	0.6037	0.4158
8	*U, L*	0.5944	0.4061
9	*U, L*	0.6000	0.4004
10	*D, R*	0.6022	0.3970
11	*D, R*	0.6013	0.3990
12	*D, R*	0.6007	0.3998

The actual performance of any probabilistic forecasting procedure can be evaluated by means of 'calibration tests'. For example, one can ask whether it actually rained on more than half the days for which the forecast probability of rain was greater than 50%. In the case of myopic behavior, any given calibration test comparing sophisticated Bayesian expectations with the actual sequence of best-response actions will be passed for every game with the exception of a set of games having Lebesgue measure zero. This result, due to Turdaliev (2002), assumes that the prior distribution is common, independent, and smooth.

Naive Bayesian Learning

In the naive Bayesian learning model, the true game and players' expectations of each other's actions are both arbitrary, so no convergence results are possible without further assumptions. In the case of myopic behavior, fictitious play is known to converge in zero-sum games and 2×2 games, but otherwise can fail to converge (e.g. Krishna and Sjöström, 1998).

For the more general case of far-sighted behavior, Kalai and Lehrer (1993a) prove convergence under the assumption that the distribution of best-response paths determined by the true game is absolutely continuous with respect to each player's expectations. Under this assumption, they prove that each player's expectations approach the true best-response distribution, and therefore that the players' best responses are approximately Nash. In the case of myopic behavior, this is enough to ensure that each player's expectations converge to Nash equilibria. In the more general case of far-sighted behavior, two players may continue to differ in their expectations of a third player's future response to actions off the best-response path, so that expectations are only guaranteed to converge to subjective equilibria.

The absolute-continuity assumption ensures that players' best-response strategies converge to Nash equilibrium, unlike the situation demonstrated in Table 3. If, as in Table 3, expectations converge to a mixed equilibrium while the best-response actions in each period are unique, the absolute-continuity assumption is clearly violated. If the true game is the game in Table 2, then the absolute-continuity assumption requires that, after some finite number of periods, each player's expectations are equal to the unique Nash equilibrium. Otherwise the players' best-response strategies would fail to be approximately Nash infinitely often.

The convergence of best-response strategies as well as expectations means that any calibration test comparing expectations with a randomly drawn sequence of best-response actions will be passed with probability one (Kalai *et al.*, 1999). This result applies to the general case of far-sighted behavior, for every game, provided the expectations and best-response distributions satisfy the absolute-continuity assumption.

WHAT IS BEING LEARNED?

Bayesian learning enables players to learn Nash equilibrium expectations over time without knowing the pay-off functions of the other players. It is natural to ask what players must already know in order to be capable of Bayesian learning. In all Bayesian learning models, players choose their actions as best responses to their expectations, so players must know their own pay-off functions. Naive Bayesian learning takes the players' expectations about future play paths as given, although each player is aware of the separate identities of the other players, in the sense that different players are expected to choose their actions independently conditional on past play. No additional knowledge is explicitly assumed, but the absolute-continuity assumption imposes a strong implicit relation between the players' expectations and the paths of best-response actions determined by the true pay-off functions.

Sophisticated Bayesian learning can be viewed as assuming that the players are knowledgeable game theorists who derive their expectations as Bayesian Nash equilibria from a common prior distribution over the unknown pay-off functions. In addition to having a common (or at least mutually consistent) prior distribution, players share a common selection process for choosing among multiple Bayesian Nash equilibria. The resulting structure of expectations avoids the need for the absolute-continuity assumption, albeit at the expense of losing the convergence of best-response strategies in the case of mixed equilibria.

Alternatively, the sophisticated Bayesian expectations could be viewed as having been derived by a game theorist who does not know the players' payoff functions. If the players rely on the theorist's predictions, their faith will be vindicated by the convergence of the expectations to Nash equilibria of the true game, at least in the case of myopic behavior and the theorists' use of a prior distribution satisfying the smoothness and independence conditions. Thus the sophisticated Bayesian learning model can be viewed simply as providing a class of expectation functions that can be used by myopic players in any game to learn Nash equilibrium expectations.

References

Jordan J (1991) Bayesian learning in normal form games. *Games and Economic Behavior* **3**: 60–81.

Jordan J (1992) The exponential convergence of Bayesian learning in normal form games. *Games and Economic Behavior* **4**: 202–217.

Jordan J (1995) Bayesian learning in repeated games. *Games and Economic Behavior* **9**: 8–20.

Kalai E and Lehrer E (1993a) Rational learning leads to Nash equilibrium. *Econometrica* **61**: 1019–1045.

Kalai E and Lehrer E (1993b) Subjective equilibrium in repeated games. *Econometrica* **61**: 1231–1240.

Kalai E, Lehrer E and Smorodinsky R (1999) Calibrated forecasting and merging. *Games and Economic Behavior* **29**: 151–169.

Krishna V and Sjöström T (1998) On the convergence of fictitious play. *Mathematics of Operations Research* **23**: 479–511.

Nyarko Y (1994) Bayesian learning leads to correlated equilibria in normal form games. *Economic Theory* **4**: 821–841.

Nyarko Y (1998) Bayesian learning without common priors and convergence to Nash equilibria in normal form games. *Economic Theory* **4**: 821–841.

Turdaliev N (2002) Calibration and Bayesian learning. *Games and Economic Behavior*.

Further Reading

Fudenberg D and Levine D (1998) *The Theory of Learning in Games*. Cambridge, MA: MIT Press.

Jordan J (1997) Bayesian learning in games: a non-Bayesian perspective. In: Bicchieri C, Jeffrey R and Skyrms B (eds) *The Dynamics of Norms*, pp. 149–174. Cambridge, UK: Cambridge University Press.

Myerson R (1991) *Game Theory: Analysis of Conflict*. Cambridge, MA: Harvard University Press.

Behavior, Genetic Influences on

Intermediate article

M Frank Norman, University of Pennsylvania, Philadelphia, Pennsylvania, USA

CONTENTS

Studies of twins have established that individual differences in intelligence, personality, and psychopathology are strongly associated with genetic variations. Environmental variations are also important.

INTRODUCTION

The program of traditional (i.e. nonmolecular) behavioral genetics is to use correlations of relatives (especially twins) to make inferences about genetic and environmental contributions to behavioral variation. There are two major conclusions from such studies. First, a substantial percentage of behavioral variation seems to be due to genetic variation. Second, although environmental variation is also important, it appears that unique, idiosyncratic, experiences are more important for behavioral variation than experiences shared by twins.

Both findings represent challenges to traditional social science, the first because social science does not emphasize biological factors, and the second because the environmental factors that are the focus of social science research (parents, homes, socioeconomic status, neighborhoods, and schools) are shared by twins and are thus seen by behavioral genetics as being in a sense less potent than social science supposes. The 'in a sense' qualification is necessary since twins obviously have unique experiences with common environmental factors. A 'strict' or 'cultured' home environment may be experienced in different ways by different children (Maccoby, 2000).

TWIN STUDIES

The following text provides a moderately detailed example of a behavioral genetic analysis that supports the first and second conclusions outlined above; it exposes the weaknesses as well as the strengths of such analyses. The analysis applies to studies that include twins reared apart as well as twins reared together. Thus these studies involve twins separated early in life by adoption. Analysis of twin studies without adopted subjects is considered later.

Fundamental Equations

Behavioral genetic analyses typically lead to equations expressing correlations in terms of genetic and environmental parameters. Consideration of 'identical' (monozygotic, MZ) and 'fraternal' (dizygotic, DZ) twins reared apart (A) and together (T) leads to four equations, the character of which varies somewhat with the precise assumptions that are in force. The following set is representative:

$$r_{\mathrm{MZA}} = h^2 + d^2 + i^2 \tag{1}$$

$$r_{\mathrm{DZA}} = 0.5h^2 + 0.25d^2 \tag{2}$$

$$r_{\mathrm{MZT}} = h^2 + d^2 + i^2 + c^2_{\mathrm{MZ}} \tag{3}$$

$$r_{\mathrm{DZT}} = 0.5h^2 + 0.25d^2 + c^2_{\mathrm{DZ}} \tag{4}$$

where h^2 is the additive genetic variance, d^2 is the dominance genetic variance, i^2 is the epistatic genetic variance, c^2_{MZ} is the common or shared environmental variance for MZ twins raised together, and c^2_{DZ} is the common or shared environmental variance for DZ twins raised together. These equations instantiate the model that is the focus of this section. In the present interpretation there are four population correlations (r) on the left, identified by the subscript abbreviations defined above, and five population parameters on the right. Our objective is to use these equations to develop estimators of the parameters based on sample correlations. Before doing this, however, we must make a long digression to define the parameters and discuss the kinds of assumptions that lead to the equations.

All of these variances are relative to the total phenotypic variance of the trait under discussion. Additive genetic variance is associated, fundamentally, with the sum of effects of the gene derived from the mother and the gene derived from the father at a single genetic locus. *Dominance* is the interaction of these effects. However, it is clear that most if not all of the traits discussed here are influenced by many genetic loci, and our h^2 and d^2 correspond to the sums of additive and dominance effects over all contributory loci. Variance i^2 associated with multilocus interaction is epistatic. Dominance and epistasis are termed 'nonadditive effects'. Remarkably, it is possible to see the signature of such low-level effects in molar correlational data.

Additive genetic variance, h^2, is called 'heritability' or 'narrow heritability'. The sum of all three genetic variances is termed 'broad heritability' and denoted $h_b{}^2$. According to eqn [1], this quantity equals r_{MZA}.

Both genetic variation and environmental variation contribute to behavioral variation. Behavioral genetics usually treats environmental variation as the sum of two uncorrelated components. The first reflects parts of the environment such as parents, home, neighborhood, and schools that are shared by twins reared together. The second represents such things as unique friends and unique experiences that are not shared. The extent of the latter is indexed by yet other parameters, u^2_{MZ} and u^2_{DZ}. It is assumed that only shared experiences contribute to psychological similarity. That is why only shared environmental variances appear on the right in eqns [3] and [4].

Obviously, different twins may have unshared experiences with common parents, houses, neighborhoods, and schools, and such experiences would be recorded in the unshared column, if it were really necessary to tally them up. Fortunately, it is not. Although effects of particular environmental variables can be studied (e.g. Caspi *et al.*, 2000), behavioral genetics can and usually does make inferences about aggregate shared and unshared environmental effects without explicitly measuring their constituents.

Measurement error (m^2) is a source of differences between twins' test scores. It is sometimes implicitly included in unshared environmental variance, but it can be separated from other sources of uniqueness if the reliability of the test is known.

Newcomers to this subject may be puzzled by the 'squares' in these parameters. The unsquared genetic and common environmental parameters are correlations of a single individual's test score with an underlying genetic or environmental component. The squares arise in correlations between relatives because the 'path' between the relatives has two links: from one relative to a common genetic or environmental factor and then on to the other relative.

Now that we have defined the parameters that appear in eqns [1] to [4], we can begin to discuss the equations themselves. The 0.5 coefficient of h^2 in eqns [2] and [4] reflects the fact that DZ twins share half their genes, on average. The 0.25 coefficient of d^2 is explained by the observation that, at a single locus with only two alleles, 0.25 is the probability that two twins receive copies of exactly the same gene from both parents, and thus inherit precisely the same genotype at this locus.

Absence of covariances on the right-hand sides reflects the assumption that all sources of variation are uncorrelated. Also absent by assumption are genotype–environment interaction and effects of assortative mating. It is easy to think of concrete environmental factors (such as parental education, in the case of cognitive ability) that are definitely correlated with children's genotype. This brings out the point that the environmental factors in our equations should not be thought of as composites of concrete, directly measurable, environmental factors, but rather as akin to regression residuals, so that they are, by construction, uncorrelated with genetic 'main effects'. It is possible that the oversimplified model instantiated in our equations may allow interactions and correlations involving concrete environmental factors to be disproportionately absorbed into genetic terms.

Substantial assortative mating is known to occur for some of the traits to which we will apply the model (e.g. cognitive ability). The main defense that can be offered for omitting it, and for our other questionable assumptions, is that the model gives an adequate fit to most of the data to which we will apply it. This is not a strong justification, since MZA twin correlations are based on relatively small numbers of twin pairs, so the corresponding tests of goodness of fit are not very powerful. Thus, successful fits can be regarded only as showing that our assumptions are probably not flagrantly inappropriate.

Fitting the Model

One's first impulse is to replace population correlations in the equations by sample correlations, and solve the resulting linear equations for the quantities on the right. The solutions would then be estimators of the corresponding population parameters. However, this is not feasible because there are five unknowns but only four equations. Fortunately, it is possible to eliminate one unknown by reparameterization, in such a way that our ability to test interesting hypotheses is not seriously compromised. This involves appending half of d^2 to h^2 and the other half to i^2, yielding the new parameters $h_d^2 = h^2 + 0.5d^2$ and $i_d^2 = i^2 + 0.5d^2$. These parameters can respectively be called 'intermediate heritability' (since it lies between broad and narrow heritability) and 'nonadditivity'. In terms of these quantities, eqns [1] to [4] take the following attractive form:

$$r_{\mathrm{MZA}} = h_d^2 + i_d^2 \tag{5}$$

$$r_{\mathrm{DZA}} = 0.5h_d^2 \tag{6}$$

$$r_{\mathrm{MZT}} = h_d^2 + i_d^2 + c_{\mathrm{MZ}}^2 \tag{7}$$

$$r_{\mathrm{DZT}} = 0.5h_d^2 + c_{\mathrm{DZ}}^2 \tag{8}$$

These equations are easily solved for the four parameters on the right in terms of the correlations on the left, and it is somewhat amusing to apply the resulting formulas to sample correlations to obtain parameter estimates. However, this 'equation-solving' approach is a sterile exercise, since it does not yield a test of the framework within which the estimation takes place. Consequently, one does not know whether the estimates are of any interest. Thus we will proceed directly to a more modern model-fitting approach, in which we explicitly test various hypotheses about the parameters, and implicitly test the underlying framework. The specific hypotheses to be tested are:

- equal environments assumption (EEA): $c_{\mathrm{MZ}}^2 = c_{\mathrm{DZ}}^2$
- no common environmental effects: $c_{\mathrm{MZ}}^2 = c_{\mathrm{DZ}}^2 = 0$
- no nonadditive effects (NNE): $i_d^2 = 0$
- no additive or dominance effects: $h_d^2 = 0$
- no genetic effects: $h_b^2 = 0$

Note that $h_b^2 = h_d^2 + i_d^2$, so the last hypothesis is equivalent to the conjunction of the two preceding it. Note also that, if EEA fails, one expects $c_{\mathrm{MZ}}^2 > c_{\mathrm{DZ}}^2$.

Any of these hypotheses reduces the number of parameters to less than four, so eqns [5] to [8] cannot be solved exactly. Instead, one accepts approximate equality in place of exact equality, and seeks parameter values that yield the best approximate solution. Different estimation methods correspond to different overall measures of the approximation error. We will use the error function and associated tests described by Loehlin (1989) in a slightly different context. The computer programs MX and LISREL provide other approaches to model fitting (Neale and Cardon, 1992).

INTELLIGENCE AND PERSONALITY

The Swedish Adoption/Twin Study of Aging

The Swedish Adoption/Twin Study of Aging (SATSA) is a study of elderly twins. The basic design and many results are summarized by Pedersen *et al.* (1991), and twin correlations and analyses of different types of variables appear in separate papers (Pedersen *et al.*, 1992; Bergeman *et al.*, 1993). Table 1 summarizes the results of a reanalysis of a few variables using the approach described above.

Cognitive ability is the first principal component of a number of tests of special abilities, and is thus a variant of intelligence quotient (IQ). The next seven variables in the table are standard dimensions of personality. The type A variable is derived from the famous Framingham type A scale, which measures the degree to which an individual is hard-driving, ambitious, and feels as if he or she is under pressure. Variables 'F-Cohesion' and 'F-Control' relate to the twins' recollections of the warmth and strictness of the families in which they were raised. Bear in mind that, for twins reared apart, these were different families, so these variables explore the possibility that twins' perceptions of family warmth and strictness may, to some extent, derive from the twin instead of from the family. Variable BMI is the body mass index, a measure of fatness. This is not a personality variable, but relates to eating habits, which are of great psychological interest in connection with eating disorders.

The χ^2 and p-values in the last two columns correspond to tests of the equal environments assumption. This assumption is rejected only for the last variable, 'F-Control'. This rejection confirms that the tests of EEA are not hopelessly insensitive. The parameter estimates given in the table are optimal assuming EEA, and are thus meaningful for all variables except 'F-Control', which will not be considered further. The c^2 parameter in Table 1 is the common value of c^2_{MZ} and c^2_{DZ} under EEA. Asterisks refer to tests of hypotheses that the corresponding parameters are zero, with one, two, and three asterisks indicating $p < 0.05$, 0.01, and 0.001, respectively. These tests are based on increments in the χ^2 goodness of fit index, beyond its value assuming just EEA. The $u^2 + m^2$ values cannot be tested for significance within this framework, but these values are large for all variables except cognitive ability and BMI.

The table confirms the overall conclusions presented in the introduction, which are, in turn, consistent with those of the original SATSA papers. In 10 out of 12 cases, broad heritability is statistically significant whereas common environmentality is statistically insignificant. The reversal of this pattern for 'Agreeableness' indicates that this pattern is not forced by an artefact of the method. The very high estimate of broad heritability for cognitive

Table 1. Components of variation for variables in the Swedish Adoption/Twin Study of Aging

Variable	h^2_a	i^2_a	h^2_b	c^2	$u^2 + m^2$	χ^2	p
Cognitive ability	0.55***	0.24	0.79***	0.00	0.21	0.54	0.46
Extraversion	−0.02	0.40***	0.38***	0.12	0.50	2.27	0.13
Neuroticism	0.49***	−0.16	0.33***	0.04	0.63	1.81	0.18
Openness	0.38**	0.11	0.49***	−0.01	0.52	1.24	0.26
Agreeableness	−0.06	0.21	0.15	0.26**	0.59	0.00	1.00
Conscientiousness	0.08	0.21	0.30**	0.12	0.58	2.65	0.10
Impulsivity	0.25*	0.18	0.44***	−0.01	0.57	0.56	0.46
Monotony avoidance	0.27*	−0.05	0.22*	0.03	0.75	0.06	0.80
Type A	0.34**	−0.07	0.27**	0.08	0.65	0.40	0.53
BMI (male)	0.41*	0.26	0.67***	0.08	0.25	0.63	0.43
BMI (female)	0.51***	0.14	0.65***	0.01	0.34	0.02	0.90
F-Cohesion	0.61***	−0.16	0.44***	0.14*	0.41	0.31	0.58
F-Control						7.89	0.01**

BMI, body mass index. Probability: *, $p = 0.05$; **, $p = 0.01$; ***, $p = 0.001$. See text for details of variables and parameters.

ability is consonant with values obtained for IQ in other studies involving adult MZAs (Neisser *et al.*, 1996).

Negative estimates arise because the χ^2 minimization routine varied $a = h_d^2$, $b = i_d^2$, etc., without restricting these quantities to positive values. In no case would the fit have been significantly worse if the negative quantity had been assumed to be zero, so the negative values should simply be regarded as negligible.

Of the 11 variables with significant h_b^2, 9 had h_d^2 but not i_d^2 significant, suggesting that genetic variation is mainly additive. One, 'Conscientiousness', had neither h_d^2 nor i_d^2 significant, leaving us little basis for inference about the distribution of broad heritability among its three components. Finally, 'Extraversion' had i_d^2 but not h_d^2 significant, suggesting that genetic variation is mainly epistatic. (Dominance and epistasis do not contribute to parent–child correlation, so they contradict the common misconception that genetic effects are always revealed by parent–child resemblance.)

We close this section with the results of analysis of a variable from the Minnesota study of twins reared apart (Bouchard *et al.*, 1990), by the methods used in Table 1. There is a personality scale called 'well-being' that is, roughly speaking, a measure of happiness. For this scale, EEA was not rejected, $c^2 = -0.04$ is negligible, and $h_b^2 = 0.48$. Happiness in adults thus appears to be highly heritable (Lykken and Tellegen, 1996).

Studies Involving Only Twins Reared Together

Monozygotic twins reared apart are uniquely informative, but they are scarce. The analyses reported in Table 1 involved between 44 and 95 pairs of such twins. This leads to high variability of estimates and low power of tests. On the other hand, dizygotic and monozygotic twins reared together are plentiful, so it is not surprising that studies involving only twins reared together are the mainstay of traditional behavioral genetics. Unfortunately, the gain in precision from large samples is matched by a loss of generality due to the necessity of extra assumptions. For a feeling for the difficulties involved, consider

$$h_{\text{est}}^2 = 2(r_{\text{MZT}} - r_{\text{DZT}}) \tag{9}$$

the traditional rough-and-ready estimator of heritability using sample correlations of twins reared together. For population correlations,

$$2(r_{\text{MZT}} - r_{\text{DZT}}) = h_b^2 + i_d^2 + 2(c_{\text{MZ}}^2 - c_{\text{DZ}}^2) \tag{10}$$

as a consequence of eqns [7] and [8], so h_{est}^2 will tend to overestimate h_b^2 when EEA fails or nonadditive effects are present.

There is a companion formula, $2r_{\text{DZT}} - r_{\text{MZT}}$, that is traditionally used to estimate common environmental variance. According to eqns [7] and [8],

$$2r_{\text{DZT}} - r_{\text{MZT}} = c_{\text{DZ}}^2 - i_d^2 - (c_{\text{MZ}}^2 - c_{\text{DZ}}^2) \tag{11}$$

so the companion formula has a tendency to underestimate c_{DZ}^2 when EEA or NNE fails. Sizeable negative values of the companion formula strongly suggest that use of h_{est}^2 is inappropriate, but one sees many instances in the literature where this warning has not been heeded.

Modern behavioral genetics uses model-fitting techniques in place of h_{est}^2 (see, for example, Loehlin, 1992), but model-fitting analyses, like their less sophisticated predecessors, typically assume EEA and sometimes also NNE.

MENTAL ILLNESS

A characteristic feature of many studies of mental illness is a categorical 'sick versus well' classification of each patient's condition. In place of twin correlation for (say) IQ, we have concordance for (say) schizophrenia, estimating the likelihood that a second twin is affected given that the first twin is affected. Concordances carry some of the same intuitions as correlations, but they are not the kinds of correlations to which behavioral genetic theory can be directly applied.

The liability threshold model provides a simple bridge from concordances to behavioral genetics. According to this model, there is a normally distributed liability, L, to the condition under consideration, and the condition is manifested if and only if L exceeds a threshold parameter T. Assuming that the distribution of liability has a mean of zero and a standard deviation of 1, one can estimate T from the prevalence of the condition.

Assuming a bivariate normal distribution of twins' liabilities, computer programs like PRELIS can estimate the liability correlations corresponding to concordances. These are sometimes referred to as *tetrachoric* correlations, and their variances are different from those of ordinary, Pearson, correlations. Taking account of this, behavior genetic analyses of liability correlations can be done in a manner analogous to behavior genetic analyses of Pearson correlations. In particular, there are older studies that calculate h_{est}^2 from liability correlations.

Modern studies apply model-fitting techniques via LISREL or MX to twins reared together, usually assuming EEA and, often, NNE. Substantial differences between MZ and DZ liability correlations are invariably associated with substantial heritability in these analyses. Such differences have been found for autism, attention deficit hyperactivity disorder, depression, bipolar disorder, and schizophrenia (McGuffin and Martin, 1999).

There is great uncertainty concerning the extent of possible genetic contributions to bulimia and anorexia (Fairburn *et al.*, 1999). Though it is not a mental illness, it is interesting to note that McGue and Lykken (1992) have reported a heritability estimate of 0.525 for liability to divorce.

THE SEARCH FOR DEFINITE GENES

Traditional behavioral genetics operates at a tremendous level of abstraction (some might call it vagueness). It provides information only about aggregate quantities, though these quantities are of considerable interest. Some results are available showing behavioral effects of specific genes. In the future, one expects increasing emphasis on studies seeking to demonstrate such effects.

Specific gene effects on behavior vary greatly in size. One imagines that many genes affect each of the personality and ability dimensions, as well as most of the psychopathologies, considered above. If many genes contribute to a dimension, most of these contributions will be small and thus relatively hard to detect.

There are a number of cases where variation at a single genetic locus has drastic effects on the organism, including mental retardation. An especially interesting case of such a single major gene effect is phenylketonuria, in which neither of the alleles at a certain locus on chromosome 12 supports production of an enzyme that breaks down phenylalanine. The consequent build-up of the latter substance in the brain causes mental retardation. Such build-up and retardation can be prevented by a special diet. Since the genetic deficiency leads to a behavioral deficiency only in the presence of a certain environment (normal diet), this is an example of a genotype–environment interaction. At a more basic level, phenylketonuria illustrates that genetic involvement in a behavioral deficiency does not imply that the deficiency is immutable. One of the major thrusts of modern medical science is to discover environmental compensations (medicines) for genetic conditions.

Definite genes, or small chromosomal regions, have been implicated with various degrees of certainty in the following conditions: early-onset and late-onset Alzheimer disease, autism, bipolar disorder, dyslexia, and schizophrenia (McGuffin and Martin, 1999; Owen and Cardno, 1999). For example, the apolipoprotein E ε4 allele appears to be associated with late-onset Alzheimer disease. Information from the human genome project will doubtless make great contributions to this rapidly developing area.

EVOLUTION

Although we have considered genetic variability in a number of psychologically relevant dimensions, we have not considered genetic variability in fitness or reproductive success, the additive component of which directly controls evolution. In so far as certain traits are prevalent today, it is natural to think that they might somehow have conferred enhanced fitness long ago. Such evolutionary speculation is among the most powerful heuristics in biological science in general and evolutionary psychology in particular. Although the disciplines that incorporate it are flourishing, it is important to be aware that inference from present predominance to past superior fitness is fallible. This can be shown by examples involving variation controlled by a single genetic locus with only two alleles, A_1 and A_2. Assuming random mating, the heterozygotic genotype A_1A_2 has relative frequency $2p(1-p)$, where p is the relative frequency of the A_1 allele. Thus, regardless of its fitness, the relative frequency of A_1A_2 cannot exceed 1/2. In fact, it is not difficult to construct examples where the homozygote A_1A_1 predominates after many generations of evolution, even though the heterozygote is the fittest genotype.

Variants of such simple genetic examples can be constructed within the frameworks of W. D. Hamilton's kin selection theory and J. Maynard Smith's (1982) evolutionary game theory, two of the pillars of evolutionary psychology (Norman, 1981). A frequently cited part of Hamilton's theory suggests that one can predict the evolutionary fate of altruistic and selfish genotypes just by examining their inclusive fitnesses, but it turns out that superior inclusive fitness is not sufficient for asymptotic predominance of, say, an altruistic genotype, if that genotype is heterozygotic. Similarly, the standard version of Maynard Smith's criterion for an evolutionarily stable strategy (ESS) depends entirely on fitness 'payoffs', irrespective of genetic structure. However, a behavioral strategy exhibited by a heterozygote cannot be evolutionarily stable, regardless of associated payoffs, since a population composed entirely of heterozygotes will give rise to a mixed population in the next generation.

These examples represent relatively minor blemishes on these theories, but the examples do suggest that the theories should be applied with caution.

References

Bergeman CS, Chipuer HM, Plomin R *et al.* (1993) Genetic and environmental effects on openness to experience, agreeableness, and conscientiousness: an adoption/twin study. *Journal of Personality* **61**: 159–179.

Bouchard TJ, Lykken DT, McGue M, Segal NL and Tellegen A (1990) Sources of human psychological differences: the Minnesota study of twins reared apart. *Science* **250**: 223–228.

Caspi A, Taylor A, Moffitt TE and Plomin R (2000) Neighborhood deprivation affects children's mental health: environmental risks identified in a genetic design. *Psychological Science* **11**: 338–342.

Fairburn CG, Cowen PJ and Harrison PJ (1999) Twin studies and the etiology of eating disorders. *International Journal of Eating Disorders* **26**: 349–358.

Loehlin JC (1989) Partitioning environmental and genetic contributions to behavioral development. *American Psychologist* **44**: 1285–1292.

Loehlin JC (1992) *Genes and Environment in Personality Development.* Thousand Oaks, CA: Sage.

Lykken D and Tellegen A (1996) Happiness is a stochastic phenomenon. *Psychological Science* **7**: 186–189.

Maccoby EE (2000) Parenting and its effects on children: on reading and misreading behavior genetics. *Annual Review of Psychology* **51**: 1–27.

Maynard Smith J (1982) *Evolution and the Theory of Games.* Cambridge, UK: Cambridge University Press.

McGue M and Lykken DT (1992) Genetic influence on risk of divorce. *Psychological Science* **3**: 368–373.

McGuffin P and Martin N (1999) Science, medicine, and the future: behaviour and genes. *British Medical Journal* **319**: 37–40.

Neale MC and Cardon LR (1992) *Methodology for Genetic Studies of Twins and Families.* Dordrecht, Netherlands: Kluwer.

Neisser U, Boodoo G, Bouchard TJ *et al.* (1996) Intelligence: knowns and unknowns. *American Psychologist* **51**: 77–101.

Norman MF (1981) Sociobiological variations on a Mendelian theme. In: S Grossberg (ed.) *Mathematical Psychology and Psychophysiology*, pp. 187–196. Providence, RI: American Mathematical Society.

Owen MJ and Cardno AG (1999) Psychiatric genetics: progress, problems, and potential. *Lancet* **354**(suppl. 1): 11–14.

Pedersen NL, McClearn GE, Plomin R *et al.* (1991) The Swedish Adoption Twin Study of Aging: an update. *Acta Geneticae Medicae et Gemellologiae: Twin Research* **40**: 7–20.

Pedersen NL, Plomin R, Nesselroade JR and McClearn GE (1992) A quantitative genetic analysis of cognitive abilities during the second half of the life span. *Psychological Science* **3**: 346–353.

Further Reading

Bailey MJ (1998) Can behavior genetics contribute to evolutionary behavioral science? In: Crawford C and Krebs DL (eds) *Handbook of Evolutionary Psychology: Ideas, Issues, and Applications*, pp. 211–233. Mahwah, NJ: Lawrence Erlbaum.

Buss DM (1999) *Evolutionary Psychology: The New Science of the Mind.* Boston, MA: Allyn & Bacon.

Kendler KS (1993) Twin studies of psychiatric illness: current status and future directions. *Archives of General Psychiatry* **50**: 905–915.

Lykken DT, McGue M, Tellegen A and Bouchard TJ (1992) Emergenesis: genetic traits that may not run in families. *American Psychologist* **47**: 1565–1577.

Plomin R (1994) *Genetics and Experience: The Interplay Between Nature and Nurture.* Thousand Oaks, CA: Sage.

Plomin R, DeFries JC, McClearn GE and McGuffin P (2001) *Behavioral Genetics*, 4th edn. New York, NY: Worth.

Reiss D, Neiderhiser JM, Hetherington EM and Plomin R (2000) *The Relationship Code: Deciphering Genetic and Social Influences on Adolescent Development.* Cambridge, MA: Harvard University Press.

Rowe DC (1994) *The Limits of Family Influence: Genes, Experience, and Behavior.* New York, NY: Guilford Press.

Wahlsten D (1999) Single-gene influences on brain and behavior. *Annual Review of Psychology* **50**: 599–624.

Behavioral Neuropharmacological Methods

Introductory article

Roy A Wise, National Institute on Drug Abuse, Baltimore, Maryland, USA

CONTENTS

The neural systems responsible for cognitive function are segregated by anatomical and neurochemical specificity. Characterization of the effects of drugs on thought and action helps us to identify the basic functional units of neuronal and cognitive organization.

INTRODUCTION

Neuropharmacology deals with the 'chemical coding' of brain circuitry and the ability to influence and study this circuitry selectively, either through the use of drugs or in an effort to understand their action. Neurons are normally activated or inhibited by endogenous substances released from nerve cells and variously termed neurotransmitters, neuromodulators or neurohormones. These chemical messengers act at specialized receptor proteins embedded in the surface membranes of nerve cells. Because of peculiarities in their geometry, the receptors bind their appropriate messenger very selectively; transmitters and hormones fit their receptors in much the same way as a key fits a lock. Neuropharmacology deals with the activation or inhibition of neurons by chemicals, including exogenous substances – drugs – that can mimic, augment or block the functions of endogenous transmitters. Traditionally neuropharmacologists used electrophysiological recording techniques to measure the responses of single neurons to chemicals delivered systemically or locally. Behavioral neuropharmacology is the study of neurochemical control of brain function as it is reflected in behavior.

Behavioral neuropharmacological methods have key roles in the analysis of behavior. Much of behavior is dominated by neuropharmacological variables. The roles of hormones in stress and sexual behavior, the roles of nutrients and hormones in feeding and drinking behavior, the roles of endogenous opioids in control of pain, and the roles of drugs in addiction illustrate the importance of neuropharmacology in behavioral analysis. Behavioral neuropharmacology not only teaches us ways to control behavior; it contributes to our understanding of neuronal organization and function. The chemical selectivity of brain circuitry gives us major clues to the structure and function of the brain. As pointed out several decades ago by Donald Hebb, our best theories about the nature of cognition are always constrained by our current understanding of the structure and activity of the functional units of the brain.

MODIFYING NEUROTRANSMITTER SYSTEMS

The neurotransmitter systems of the brain can be selectively manipulated in several ways. The most direct ways involve drugs that bind to the receptors for endogenous neurotransmitters. Such drugs share the geometrical features that allow the neurotransmitter to bind to receptors on a target neuron. If the drug binds to the receptor for a given transmitter but does not trigger a transmitter-like action there, it is an antagonist of the transmitter, often termed a 'receptor blocker' because it physically blocks the access of the endogenous transmitter to its normal binding site. The effects of drugs on a neurotransmitter system are usually temporary.

If the drug's geometry is sufficiently similar to that of the transmitter it will not only bind to the receptor but also trigger the biological action of the transmitter; in such cases the drug is termed an 'agonist', because it mimics the transmitter action. Nicotine is an agonist at a subset of acetylcholine receptors; morphine is an agonist at receptors for endogenous 'opioid' peptide neurotransmitters. Phencyclidine is an antagonist at a subset of glutamate receptors.

Amphetamine and cocaine are termed 'indirect' monoamine agonists because although they do not bind to monoamine receptors, they act at the transporter molecules that dispatch or take back up the monoamine transmitters. By reversing or blocking the reuptake mechanism, they elevate the synaptic concentrations of the three monoamine transmitters – dopamine, noradrenaline (norepinephrine), and serotonin. These drugs have the same behavioral effects as the transmitters because they increase the transmitter concentration in the local extracellular fluid. Injections of drugs often have more dramatic effects than injections of the transmitters themselves because the drugs are generally much more resistant than the transmitters to the normal deactivation mechanisms.

Neurotransmitter agonists and antagonists have varying degrees of selectivity for different receptors and different receptor subtypes. Cholinergic receptors (defined by their common sensitivity to the neurotransmitter acetylcholine) are of two major subtypes: nicotinic (binding and responding to nicotine but not to muscarine) and muscarinic (binding and responding to muscarine but not to nicotine). There are five or more variations of muscarinic receptors: five slightly different receptor molecules that each bind and respond to muscarine and acetylcholine. There are five known subtypes of dopamine receptor and 13 known subtypes of serotonin receptors. The selectivity of a given agonist or antagonist depends on sometimes subtle geometric peculiarities of the neurotransmitter, the drug and the receptor. One goal of neuropharmacology is to identify drugs that are highly selective for a given receptor molecule.

It is possible to cause permanent damage to single neurotransmitter systems or portions of single neurotransmitter systems by the use of neurotoxins. The substance 6-hydroxydopamine is a general neurotoxin if it is given in sufficient concentration. However, if given in low concentration, this molecule (which resembles dopamine) is taken up selectively by neurons that express and take up the closely related monoamines dopamine and noradrenaline. These neurons will concentrate the toxin intracellularly. Thus the toxin can be used to cause selective degeneration of dopaminergic and noradrenergic neurons. If a selective blocker for the noradrenergic uptake mechanism is given, the drug will be taken up and damage only dopamine systems. If the toxin is microinjected into a local brain region, it will damage only those dopaminergic or noradrenergic neurons found in that region. There are analogous neurotoxins for serotonergic neurons. Other toxins selectively damage noradrenergic systems; one substance, MPTP (1-methyl-4-phenyl-1,2,3,6-tetrahydropyridine), is metabolized to a selective toxin for dopamine neurons by an enzyme found in monkeys but not in rats.

Another selective degeneration method involves coupling the general ribosome-inactivating protein saporin to an antibody that gives it selectivity for neurons of a given neurotransmitter type. Saporin-linked antibodies have been used successfully to destroy forebrain cholinergic neurons and, more recently, noradrenergic pathways. The linkage of saporin to other antibodies will allow it to be used to target other transmitter systems. Other neurotoxins, particularly excitatory amino acids, have little selectivity for particular neurotransmitter systems but, when injected locally, can be used to damage the cell bodies of a given region while sparing passing fiber systems.

In addition to neuropharmacological methods for selective modulation of neurotransmitter systems, new molecular biological approaches are rapidly being developed. The expression of receptors, enzymes or neurotransmitters can be blocked at the level of gene transcription by antisense oligonucleotides, and viral vectors can be used to insert genetic material that transiently increases expression of such gene products. There is thus an expanding range of methods for selective activation, blockade or destruction of particular neurotransmitter systems. Each of these methods can be used to study the behavioral role of a given neurotransmitter or the behavioral sensitivity to a given drug.

MEASURING NEUROTRANSMITTER ACTIVITY

The spontaneous activity of various neurotransmitter systems can be measured in several ways. Extracellular recordings of the electrophysiological activity of single neurons are useful when the cell type can be identified by anatomical location, firing patterns or sensitivity to different drugs. Unique firing patterns have been characterized from intracellular recordings in simplified preparations *in vitro* where a dye can be injected into the cell for subsequent identification of neurotransmitter type. Unfortunately, not all cell types have unique electrophysiological signatures, and the electrophysiological characteristics of some cell types differ considerably between conditions *in vitro* where identification is positive and conditions *in vivo* where it is not.

There are two approaches to measuring or estimating the concentrations of various neurochemicals in the extracellular fluid of a given brain region. One involves sampling extracellular fluid from some local region and subjecting that fluid to bench-top assay. The extraction of extracellular fluid was originally accomplished with 'push-pull' cannulas where artificial cerebrospinal fluid was injected through one line and withdrawn through the other. Neurochemicals from the region mixed with the injected fluid and were withdrawn with it. This method is now infrequently used because it collects not only the transmitters but also the enzymes that continue to degrade them. In a more recent method the tips of push-pull cannulas are encapsulated in microdialysis tubing which allows the small neurotransmitters to diffuse into the perfusate, but blocks entry of the larger enzymes. This technique for collecting brain chemicals is termed 'microdialysis' (Figure 1).

When microdialysis samples are assayed, unambiguous identification of neurochemicals is possible. Microdialysis is currently used widely in conjunction with high-performance liquid chromatography to assay acetylcholine, glutamate, γ-aminobutyric acid (GABA) and the monoamine transmitters dopamine, noradrenaline (norepinephrine) and serotonin. The many neuropeptide transmitters tend to be released in very low concentrations, and are more difficult to assay with current methods. Mass spectrometric methods with much greater sensitivity are currently under development for this purpose.

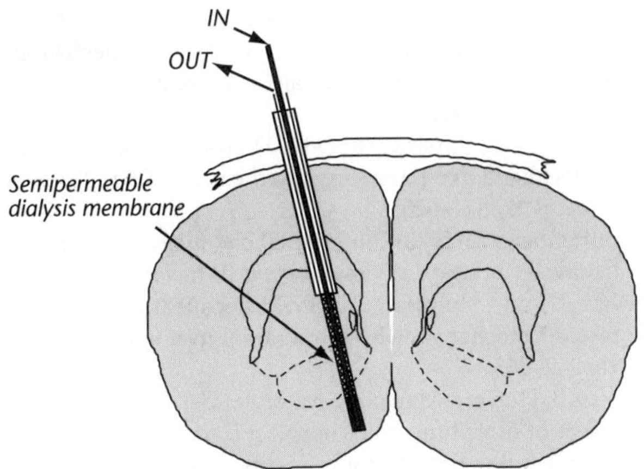

Figure 1. A microdialysis probe is inserted into a rat brain. The membrane portion of the probe is semipermeable, and (like a blood vessel) allows the exchange of brain chemicals between the artificial extracellular fluid which is perfused through it and the brain's extracellular fluid which surrounds it. The perfusing fluid is pumped slowly into the internal cannula, passes back up between the inner cannula and the membrane itself – where it absorbs neurotransmitters and other substances from the endogenous brain fluids – and is collected in a vial connected by flexible tubing to the outer cannula. The neutral perfusion fluid and the collected brain chemicals that it carries out of the brain are then analyzed to determine what neurotransmitters are being released under the conditions of testing.

The microdialysis approach allows confident identification of a given chemical species, but the temporal resolution is poor for behavioral studies. It often takes several minutes to collect detectable levels of a given transmitter, and it is not possible to identify peaks of neurotransmitter release within a given sampling period. Thus, even though it is clear which chemicals were released, it is difficult to determine precisely when they were released. A technique with greater temporal resolution is *in-vivo* voltammetry. Here a neurochemical assay is performed within the brain. Different chemicals oxidize at different voltages, and a voltage appropriate for the chemical of interest is applied locally. The voltage-specific oxidation of the chemical in question induces a measurable increment in the current flow between the electrodes. Fluctuations of these oxidation currents are used to estimate fluctuations in the concentration of the chemical of interest. Unfortunately, more than one chemical species may oxidize at a given voltage, and thus neurochemical resolution is not as good as in microdialysis.

Each of these methods is undergoing rapid refinement. The temporal resolution of microdialysis is being improved by development of more sensitive laboratory assays. The neurochemical resolution of voltammetry is being improved by development of different electrodes and by the use of pharmacological tools in conjunction with locally stimulated release of transmitters. The development of more precise methods for monitoring neurochemicals in freely moving animals is receiving active attention in several laboratories.

Another use for microdialysis is to infuse substances into local brain regions. Just as local neurochemicals diffuse from the brain across the dialysis membrane and into dialysis samples, neurochemicals in the dialysis fluid can diffuse into the brain. Drug infusion by dialysis is more localized and less stressful for local tissue than are bolus injections by the more traditional hydraulic pressure. If two dialysis probes are implanted in appropriate regions, it is possible to infuse a drug at one site and observe the consequences on behavior and on neurotransmitter release at the second site. Such studies begin to identify the connections between sites of drug action and help characterize the distal transmitter release caused by a drug.

THE STUDY OF COGNITION

The coding of information flow in the brain is both spatial and neurochemical. Our eventual understanding of the brain processes underlying various cognitive processes will require us not only to identify the portion of the brain that is sending and receiving relevant messages, but also which chemical messenger – among the dozens that can be found in most brain regions – is carrying the signals.

CONCLUSION

The realization that the circuits of the brain contain a wide range of chemical messengers, and that dozens of chemically coded messages can be transmitted simultaneously in the same brain region, has motivated the development of methods for distinguishing the neurochemical constituents of brain fluid, measuring their fluctuations during thought and action, and determining their consequences by microinfusing them into the brains of freely moving animals. Anatomists have come far in identifying the cells of different brain regions, the source of their inputs, and the targets of their outputs. They have also made great progress in identifying – through fluorescence and immunohistochemical techniques – the chemical messengers in various cell groups.

Neuropharmacologists build on such information to probe the functions of chemically distinct neuronal populations. Neuropharmacological methods help us to understand the effectiveness of various drugs in alleviating the symptoms of such conditions as schizophrenia, depression and Parkinson disease. Such methods are used increasingly to develop new drugs – with greater potency and fewer side effects – for the treatment of mood and thought disorders. Neuropharmacological methods also help us to understand basic brain function; they help us to parse and discover the syntax of the activity of the brain. Relative to the complexity of brain function, current neuropharmacological methods are still crude; none the less, the rate of technological advance is rapid and neuropharmacological methods have already given us major insights into the functional units of mood and movement.

Further Reading

Benveniste H (1989) Brain microdialysis. *Journal of Neurochemistry* **52**: 1667–1679.

Boulton AA, Baker GB and Adams RN, eds (1995) *Neuromethods. Voltammetric Measurements in Brain Systems*. Totowa, NJ: Humana Press.

Emmett MR and Caprioli RM (1994) Micro-electrospray mass spectrometry: ultra-high-sensitivity analysis of peptides and proteins. *Journal of the American Society for Mass Spectrometry* **5**: 605–613.

Hebb DO (1955) Drives and the CNS (conceptual nervous system). *Psychological Review* **62**: 243–254.

Kiyatkin EA and Rebec GV (1998) Heterogeneity of ventral tegmental area neurons: single-unit recording and iontophoresis in awake, unrestrained rats. *Neuroscience* **85**: 1285–1309.

Miller NE (1965) Chemical coding of behavior in the brain. *Science* **148**: 328–338.

Morari M, O'Connor WT, Darvelid M *et al.* (1994) Functional neuroanatomy of the nigrostriatal and striatonigral pathways as studied with dual probe microdialysis in the awake rat – I. Effects of perfusion with tetrodotoxin and low-calcium medium. *Neuroscience* **72**: 79–87.

Quan N and Blatteis CM (1989) Microdialysis: a system for localized drug delivery into the brain. *Brain Research Bulletin* **22**: 621–625.

Routtenberg A (1972) Intracranial chemical injection and behavior: a critical review. *Behavioral Biology* **7**: 601–641.

Wise RA and Hoffman DC (1992) Localization of drug reward mechanisms by intracranial injections. *Synapse* **10**: 247–263.

You ZB, Herrera-Marschitz M, Nylander I *et al* (1996) Effect of morphine on dynorphin B and GABA release in the basal ganglia of rats. *Brain Research* **710**: 241–248.

Behaviorism, Philosophical

Intermediate article

Max Hocutt, University of Alabama, Tuscaloosa, Alabama, USA

CONTENTS

Philosophical behaviorism, the belief that states and traits of mind are behavioral dispositions, was the dominant philosophy of mind for much of the twentieth century.

INTRODUCTION

Reduced to a slogan, behaviorism is the belief that states and traits of mind are behavioral dispositions. Behaviorism in psychology began with the recommendation that introspective analysis of consciousness be replaced by systematic observation of behavior. Behaviorism in philosophy was the attempt to formulate a theory of mind and personality that would justify this new methodology. According to philosophical behaviorism, thoughts and feelings are not private states of invisible but introspectible minds; they are observable dispositions of visible animals. Thus, fear is a disposition to flee; anger a disposition to destroy; desire a disposition to prefer; belief a disposition to assent; and so on.

HISTORY

Although it had roots in the seventeenth-century empiricism of Thomas Hobbes and the nineteenth-century pragmatism of Charles Peirce, behaviorism as an explicit doctrine began in the first quarter of the twentieth century at the University of Chicago with J. B. Watson, whose colleagues James Angell and John Dewey advocated a kindred idea which they called functionalism. Watson criticized the dependence of psychologists on introspection because its claims could not be independently confirmed (Carnap, 1931; Watson, 1913). As the philosopher of science Arthur Pap would later explain:

> Scientific propositions must be intersubjectively verifiable; the verdicts of introspection, however, are not intersubjectively testable; therefore a science based on introspection is not really a science. (Pap, 1962)

Watson concluded that psychology should cease being a first-person study of private consciousness and become a third-person study of public behavior.

An early formulation of behaviorist philosophy was by the logical positivist Rudolph Carnap, who argued that, if consciousness cannot be studied scientifically, the reason must be that minds and their contents are unreal: the reality is bodies and their behavior. Carnap concluded, to quote Pap again, that 'there is but behavior, dispositions to respond in specific ways to specific stimuli, and neurophysiological processes within the human and animal body'. 'Behaviorism' was the name of the first half of this conjunction; Carnap named the second half 'physicalism'.

To Carnap, physicalism meant the belief that everything is describable in the language of physics. An advocate of the unity of science, Carnap thought that psychology could be reduced to physics, in two stages. First, statements about the mind would be translated into statements about behavior; then these would be translated in their turn into statements about the motions of bodies in space. Thus, mentalistic statements such as 'Sam feels hungry' would give way to behaviorist remarks such as 'Sam is disposed to eat'; then these remarks would themselves be reformulated using the more austere terminology of the physicist.

This programme of analysis – which has been called logical, or analytic, behaviorism – encountered difficulties at the very first step. A glance at our behaviorist analysis of hunger will show why. 'Mary feels hungry' is clearly not synonymous with 'Mary is disposed to eat'. The two statements do not even have the same truth conditions. Mary might feel hungry without being disposed to eat, perhaps because she is fasting; or be disposed to eat without feeling hungry, perhaps because she wants to gain weight.

The usual behaviorist reply to this problem was that it demonstrates not the inadequacy of behavioral analysis as such but only the inadequacy of a particular analysis: in this case, the analysis of hunger, which could be improved by defining hunger as a disposition to eat that usually results from deprivation of food and will usually manifest itself in eating unless there is some stronger disposition to the contrary. In short, a hungry person is one who has not eaten recently and is, as a result, disposed to eat, other things being equal. Since *ceteris paribus* clauses such as this are common in the physical sciences, behaviorists saw no reason why they should not exist in psychology too; but critics protested that the resulting analyses were illegitimately ad hoc. Worse, they sounded tautological, like saying that people who are disposed to eat are disposed to eat unless they are not. Because the logical behaviorists were seeking analyses that are true by definition, they were not greatly perturbed by this complaint, but it seemed to their critics to be damning.

Perhaps seeking to deflect this criticism, Ludwig Wittgenstein formulated behaviorism more cautiously, by claiming that inner states require outward criteria (Wittgenstein, 1953). What Wittgenstein meant by this was that differences in states of mind must manifest themselves in some way, if not always in a particular way; the mind cannot be a fifth wheel, which spins idly, making no difference to the progress of the vehicle. To see how this Wittgensteinian observation applies to a particular case, consider pain, which a logical behaviorist might define as a disposition to moan and groan. Hilary Putnam's observation that there might be Spartans who grit their teeth and grimace but do not moan may refute Carnap's logical behaviorism, but it does not refute Wittgenstein's criterial behaviorism. To refute criterial behaviorism, one would have to show that there can be a state of mind that manifests itself in no observable response, overt or covert; and it is not obvious how this could be so.

After Wittgenstein had made this point in Cambridge, another blow for behaviorism was struck in Oxford by Gilbert Ryle. Influenced by Wittgenstein, Ryle mounted an assault on behaviorism's *bête noire*: Cartesian dualism, the belief that a human being is a mind, which thinks, within a body, which takes up room. Ryle (1949) argued that this Cartesian belief is the result of a 'category mistake'. What Ryle had in mind can be brought out by using an analogy. Suppose that Jones is a fast runner. Then she may be said to 'have speed'. It would, however, be an error to ask: 'Where does Jones keep her speed?' This would mistake an attribute for a thing; a predicate adjective for a noun. A runner's speed is not one of her possessions; it is one of her characteristics. It is not a thing that she owns; it is a way that she is – namely, able to run fast.

Ryle had the same view of statements about the mind. To say that visible people have minds does not mean that they possess invisible and intangible things that do their thinking and feeling for them. It means that they have visible capacities for visible feeling and thought. Thus, saying that Smith has great intelligence means not that Smith possesses a large quantity of some mysterious stuff, but that Smith can usually be observed to behave in ways that are well adapted to his ends. As Watson's colleague Dewey had put the point: mind is not substantival; it is adjectival. In other words, it is not our minds that do our thinking. It is not even we who do our thinking with our minds. Rather, it is simply we – i.e., we bodily beings – who do our thinking.

Most behaviorists agreed; but there was a problem. Could not someone think silently, without anybody else being able to discover the fact just by watching her? Indeed, would it not make sense to suppose that intelligence never manifested itself in visible behavior? Faced with these questions, behaviorists divided into two groups. Some, reverting to Carnap's physicalism, took the position that mental abilities are rooted not in an invisible mind but in visible brains, glands and muscles. Thus, Ullin Place and Jack Smart held that sensations are events in the brain (Place, 1956; Smart, 1959). As such, they are observable in principle, if not always in practice (because they take place under the skin, which is opaque).

Ryle avoided this conciliatory line because it seemed to him to entail too mechanistic a view of human beings. An Aristotelian, Ryle did not share the physicalist hope of reducing psychology, which concerns purposive action, to physics, which recognizes only mechanical motion. Unaware that B. F. Skinner had undertaken to explain purposive behavior mechanistically by invoking Thorndike's 'law of effect' – which says that the likelihood of future behavior is modified by the effects of past behavior of the same sort – Ryle protested that human beings are animals, not machines. Noting that the physiologist may have an explanation for the patellar tendon reflex, Ryle doubted that he could explain the act of kicking a football. Here Ryle disagreed with Carnap.

In defence of his scepticism, Ryle argued that statements about psychological states and traits are 'mongrel categoricals': conditional statements in categorical disguise. To see what he meant, consider again the statement 'Mary is hungry'. This resembles 'Mary is short', but if Ryle was right, the similarity of grammar is misleading. In Ryle's view, there is nothing 'iffy' about the statement that Mary is short; but, analysed behaviorally, the statement that she is hungry means that she will eat if food is available. Believing that conditional statements could not be reduced to categorical statements, Ryle did not believe that psychology could be reduced to physiology.

Few philosophers have been convinced by this argument. To see why not, consider the statement: 'The bar is magnetic.' It implies that the bar will attract iron filings if any are present, which is a conditional remark. But this is true only because the constituent molecules of magnets are arranged in dipole, a fact that is in no way 'iffy'. In this case, the conditional fact presupposes the categorical fact; it does not rule it out. Similarly, although '*x* is hungry' implies the conditional statement '*x* is likely to eat if offered food', it also implies the categorical statement '*x*'s blood sugar level is low'. In opposing the conditional to the categorical, Ryle had posed a false dichotomy.

Later, Daniel Dennett, a disciple of Ryle, proposed a variant on behaviorism that would correct Ryle's error. Dennett argued that states of mind are best regarded as states of the body having certain typical causes and effects (Dennett, 1978). Thus, hunger is that state of the body that usually results from food deprivation and usually gives rise to eating; pain is that state of the body that is usually caused by tissue damage and usually elicits moaning; and so on. This showed how categorical facts about the body might be not only connected with but even equivalent to conditional facts about behavior.

Hilary Putnam proposed a somewhat different view that he called functionalism. According to this view, functionally equivalent states of mind might have both different physical embodiments and different behavioral manifestations (Putnam, 1978). Thus, suppose that a Martian and I both feel pain. Because the Martian will have a different nervous system, his pain will take a different form in his brain; and because he is brave while I am a coward, he will suffer in silence while I moan and groan. This possibility – the multiple realizability of mental states – had apparently been suggested to Putnam by the fact that two computers might be constructed with different chips, use different operating systems, and display their result in different ways, yet be performing the very same operation – say, adding two and two.

By taking this line, Putnam changed behaviorism, which had grown out of one kind of functionalism, into a different kind of functionalism, thereby allowing cognitive science – the study of the mental workings of the physical brain – to mature.

ARGUMENTS FOR PHILOSOPHICAL BEHAVIORISM

As noted above, the distinctive feature of behaviorism was its resolve to make psychology scientific by confining it to observable responses and their observable causes. This resolve, which required the psychologist to abstain from discussing what could not be observed, worked well for several decades, turning the analysis of behavior into a flourishing science with useful applications in psychotherapy, advertising, education, and industry. The variety and utility of this technology is evidence of the great value of behavioral psychology and the legitimacy of the philosophy behind it.

Nevertheless, many people believe that behaviorism has been decisively refuted by, among others, the linguist Noam Chomsky in his review (1959) of B. F. Skinner's *Verbal Behavior*. Challenging Skinner's account of speech as reinforced operant, Chomsky argued that the rules of grammar, which are innate, give us the capacity to formulate an unlimited number and variety of sentences. Chomsky's criticism is valid if the rules of grammar are innately known and if Skinner's theory implies that our verbal facility consists of a limited and predetermined repertoire; but the first of these claims has not been proved and the second is not obviously true. Besides, even if Chomsky is right, it is one thing to refute Skinner, another to refute behaviorism, which is not committed to Skinner's analysis of language.

A second criticism of behaviorism, one emphasized most recently by John Searle but stated earlier by Roderick Chisholm, is that it gives no satisfactory account of the mind's intentionality: the directedness of desires towards ends and thoughts towards objects. This criticism is accurate, but in fact intentionality is an unsolved puzzle for every philosophy, behaviorist and non-behaviorist alike. Thus, Searle declares it a biological given. The case may be comparable to that which once existed in biology. Intentionality is closely related to purpose. Until Darwin showed how natural selection could mimic deliberate design, there was no explanation of the teleology, or purposiveness, that seems to be present in nature.

Recently, behaviorism has been the object of two further charges. Psychologists have charged that it discourages the scientific study of the brain; and philosophers have charged that it discourages the introspective study of consciousness.

To begin with the first charge: emphasis on studying overt behavior has indeed distracted behaviorists' attention from the brain, the workings of which are observable in principle but not, until recently, easily observed in practice. Thus, the psychologist B. F. Skinner actively discouraged speculation about the brain. However, not even Skinner doubted that understanding behavior would eventually require understanding its biological, including its neurophysiological, underpinnings. Watson emphasized biology from the beginning, and, as noted earlier in this article, Carnap endorsed Watson's approach. Among important behaviorist philosophers, only Ryle can be accused of believing that biology is irrelevant, and his error was soon corrected by Dennett. In short, the first charge is aimed at a straw man.

In fact, so is the second: the charge that, by disallowing introspection, behaviorism entails neglect of consciousness. The problem with this claim is that talk of introspection is ambiguous. No behaviorist has ever denied the uses of introspection if by this is meant the familiar practice of discerning one's feelings, sizing up one's motives, giving voice to one's opinions, noticing what one perceives, or observing one's conduct and appraising one's character. What behaviorists have emphatically denied is that introspection yields knowledge that is either privileged or infallible. The behaviorist acknowledges that you can know whether you are angry, but he insists that other people can know it too – by observing what you say and do. Furthermore, against the Cartesian he adds that you can also fail to know that you are angry – for example, by being so angry that you do not pause to recognize the fact.

Now consider consciousness. Again there is ambiguity. No behaviorist has ever denied that there is a difference between someone who is awake and someone who is asleep, or between someone who is alert and someone who is drugged. Nor, to come back to introspection again, has any behaviorist denied the reality of self-consciousness, if that means being aware of one's own mental condition. What behaviorists have denied are the claims made for consciousness by followers of Descartes, who held that consciousness has features that can be known only by means of introspection and cannot fail to be known by it.

Followers of Descartes are legion. The commonplace view that nobody can know what another person feels is supported both by Thomas Nagel (1974), who argues that no one can know 'what it is like to be a bat', and by Frank Jackson, who argues that there is something which a color-blind physicist could not know – namely, what colors look like.

Behaviorists believe that these Cartesians have been victimized by an ambiguity in the concept of knowledge. Grant that I cannot have, or share, your (or a bat's) feelings. It does not follow that I cannot recognize or identify them. On the contrary, if the behaviorist is right, your feelings (or the bat's) can be known in two ways: by studying your (or its) behavior and by examining your (or its) nervous system, glands, and musculature. As to the color-blind physicist Mary, it is true that she lacks knowledge that other people have, but if the functionalist David Lewis (1980) is right, her problem is not that she is ignorant of some facts; it is that she cannot make certain discriminations, which is a behavioral disability.

Cartesians often buttress their position by arguing that, since only I can know my thoughts and feelings from the inside, only I can know their phenomenal properties. To see what this metaphor and this piece of philosophical jargon mean, suppose that Smith and Jones both look at a red fire engine. David Chalmers and Galen Strawson maintain that each of them will know what sensations he is having but that neither can know what the other's sensations are like, because although every person can look into his or her own mind, no person can look into the mind of another. Therefore, even if Smith and Jones agree that they are both looking at a red fire engine, they may not be having the same sensation. Let us call this the argument from introspection.

U. T. Place said that this argument is based on the phenomenological fallacy: treating the appearances of things as if they were themselves things that appear, only to a limited audience of one person. The idea seems to be that each of us has a personal moving picture screen, which nobody else can view. Objects in the world project pictures onto that screen for the viewer's private edification. Each viewer knows what his or her own pictures look like, but can have no idea whether the pictures on other screens resemble them.

The fallacy lies in the belief that people see not things but their appearances. What we call the thing's appearance is not the thing we see; it is how we see it. To quote Wittgenstein, it is 'what we see the thing as'. Thus, Smith and Jones see a red fire engine; they do not see its appearances. Seeing something requires you to have your eyeballs aimed at it, and by hypothesis, Smith and Jones are aiming their eyeballs at the red fire engine, which is in front of their eyes, not at their sensations, which, being events in their brains, are behind their eyes.

What about the redness of the sensation of red and the painfulness of the sensation of pain? There are no such things. It is the fire engine, not the sensation of red, that is red; it is the hot poker, not the sensation of pain, that is painful.

To say so is not to deny that the fire engine might present different appearances to different people. For example, it might look blurry to Smith, who has astigmatism, and small to Jones, who is far away. To explain this, however, we do not need to suppose that Smith and Jones are seeing their private sensations rather than public objects. On the contrary, if sensations are events in their brains, their intrinsic nature and character will be better known to the neurophysiologist than to Smith and Jones. It is only if sensations are events occurring in invisible and intangible minds that they become inscrutable, and in that case we have good reason to regard them as unreal. What is real can be detected by more than one person, in more than one way.

PHILOSOPHICAL BEHAVIORISM AND COGNITIVE SCIENCE

Confusion about this may have been augmented in recent years by an idea that has become the basis of much cognitive science: the comparison of the human mind to an electronic computer. According to the usual accounts, coded information is input by peripherals such as a keyboard through a cable to the central processing unit (CPU). This message, a representation of something in the external world, exists in the CPU as a pattern of electrical impulses and, after interpretation by the computer's program, is converted into appropriate outputs to peripherals such as a printer, or a robot welding a plate to an automobile.

As Jerry Fodor has noted, this way of describing the workings of a computer closely resembles the Cartesian account of the mind. According to this account, our sense organs send physically coded messages to the brain, where our mind reads and interprets them as appearances of things external. Having done this, the mind then orders the brain to activate the appropriate muscles or glands in order to produce suitable behavior. The mind is thus a calculating machine.

The behaviorist objection to this account is not that it seeks to understand the brain by comparing it to a computer, but that it seeks to understand the computer by comparing it to a mind. We do this when, speaking metaphorically about neurophysiological processes that are still largely unknown to us, we describe the brain as an 'interpreter' of 'information'. Having set out to mechanize man, we end up anthropomorphizing the machine, thus coming full circle back to Cartesianism, the very outcome Skinner warned against. Dennett aptly calls this Cartesian materialism.

What is needed, if Dennett is right, is not to reject the computational theory of the mind but to purify it of Place's internal cinema screen. As Place himself liked to point out, this can be done by regarding the brain as a connectionist machine, or parallel distributed processor, of the sort described by Rummelhart and McClellan. Instead of having what might be thought of as internal representations, these machines convert inputs (stimuli) to outputs (responses) in accordance with connection weights that are gradually altered using an algorithm designed to respond to feedback by eliminating unsuccessful responses in favor of successful ones – providing a working model of the workings of the behavioral law of effect.

The behaviorist likes this approach, for three main reasons. First, it has been purified of those anthropomorphic notions that the behaviorist finds objectionable in taking literally the idea of the computer as a mind engaged in processing information. Second, it is in better accord with our present knowledge of the brain, which is almost certainly a parallel processor. Third, and most importantly, the connectionist understanding of cognition as adaptation to external stimuli retains the externalist emphasis that is missing from Cartesianism, which is essentially solipsistic. Let us close by saying something about this third point.

Behaviorism came into being shortly after the publication of Darwin's theory of biological evolution, and was inspired by it. According to this theory, nature weeds out organisms that are ill-adapted to cope with their environments. This implies that cognition is best regarded not just as something going on in our heads but also as something that enables us to cope with our surroundings. So what matters in cognition is that internal processing manifests itself in behavior that is appropriate to the external world. What Wallace Matson calls an 'outside-in' philosophy sits better with this than the 'inside-out' philosophy of Cartesianism, and behaviorism is the pre-eminent outside-in philosophy.

References

Carnap R (1931) Psychology in the language of physics. In: Lyons W (ed.) *Modern Philosophy of Mind*, pp. 43–79. London: J. M. Dent.

Chomsky N (1959) A review of B. F. Skinner's *Verbal Behavior*. In: Geirsson H and Losonsky M (eds) *Readings in Language and Mind*, pp. 413–441. Oxford: Blackwell.
Dennett D (1978) *Brainstorms: Philosophical Essays on Mind and Psychology*. Cambridge, MA: MIT Press.
Lewis D (1980) Mad pain and Martian pain, and knowing what it's like. In: Rosenthal D (ed.) *The Nature of Mind*, pp 229–235. New York, NY: Oxford University Press.
Nagel T (1974) What is it like to be a bat? In: Lyons W (ed.) *Modern Philosophy of Mind*, pp. 159–174. London: J. M. Dent.
Pap A (1962) Mind and behaviorism. In: *Introduction to the Philosophy of Science*, pp. 374–409. New York, NY: The Free Press of Glencoe.
Place UT (1956) Is consciousness a brain process? In: Lyons W (ed.) *Modern Philosophy of Mind*, pp. 106–116. London: J. M. Dent.
Putnam H (1973) Philosophy and our mental life. In: Lyons W (ed.) *Modern Philosophy of Mind*, pp. 133–147. London: J. M. Dent.
Ryle G (1949) *The Concept of Mind*. London: Hutchinson and Company.
Smart JJC (1995) Sensations and brain processes. In: Lyons W (ed.) *Modern Philosophy of Mind*, pp. 117–132. London: J. M. Dent.
Watson JB (1931) Psychology as the behaviorist views it. In: Lyons W (ed.) *Modern Philosophy of Mind*, pp. 24–42. London: J. M. Dent.
Wittgenstein L (1953) *Philosophical Investigations*. Oxford: Blackwell.

Further Reading

Armstrong DM (1963) Is introspective knowledge incorrigible? *Philosophical Review* **72**(4): 417–432.
Hocutt MO (1996) Behaviorism as opposition to Cartesianism. In: O'Dononue and Kitchner (eds) *Psychology and Philosophy: Interdisciplinary Problems and Responses*, pp. 81–95. London: Sage Press.
Quine WV (1985) States of mind. *Journal of Philosophy* **82**(1): 5–8.
Zuriff GE (1985) *Behaviorism: A Conceptual Reconstruction*. New York, NY: Columbia University Press.

Belief Networks

(*See* **Bayesian Belief Networks**)

Bidding Strategies in Single-unit Auctions

Intermediate article

Bart Wilson, George Mason University, Fairfax, Virginia, USA

CONTENTS

Economists have developed models of auctions to enhance our understanding of one of the most commonly observed exchange institutions. Auction models are used to examine the allocation of products from an auctioneer to a number of bidders. The optimal bidding strategies and efficiency of four different auction institutions (first-price sealed bid, second-price sealed bid, English, and Dutch) can be analysed.

INTRODUCTION

Economists have developed models of auctions to enhance our understanding of one of the most commonly observed exchange institutions. Auction models are used to examine the allocation of products from an auctioneer to a number of bidders. For example, a manufacturer may procure inputs by soliciting sealed bids that name each supplier's contract price. In addition to their use in industrial procurement, auctions have been used to allocate products as varied as art, cattle, produce, government securities, and offshore mineral rights. As the internet reduces the costs of conducting auctions, this type of exchange will become still more pervasive.

AUCTION INSTITUTIONS

In this introduction to auctions and bidding, the discussion is limited to a few simplifying, yet often realistic, assumptions that will facilitate our analysis. The first is that an auctioneer is selling a single unit of a good to only one of several potential buyers, who are the bidders in the auction. The second assumption is that each bidder i knows the value v_i from consuming the item. This assumption allows each bidder to ignore any information from rivals in establishing a value estimate. The buyer's value indicates the maximum amount that the bidder is willing to pay for the item. The third assumption is that the bids b_i are statistically independent and derived from a known distribution, presumably estimated from history. Lastly, we will assume that the bidders are risk-neutral; i.e., the bidders are indifferent between playing a lottery and receiving the expected pay-off from the lottery with certainty.

The winning bidder w wins the auction, pays the price p, and receives a pay-off of $v_w - p$. All other buyers receive no pay-off. The seller's profit is $p - c$, where c is the (opportunity) cost to the seller of supplying the item. Hence, the total surplus from the exchange is $v_w - c$. The 'efficiency' of the auction – the metric by which we evaluate auction mechanisms – is defined as $(v_w - c)/(v_h - c)$, where v_h is the highest realized v_i. If the bidder with the highest value wins the auction, then the auction is perfectly efficient, or equivalently, the auction maximizes the gains from exchange.

The auctioneer chooses the institutional rules by which the bids determine who is the winning bidder and what price the winning bidder pays. As an indication of the extent to which the details of the institution matter in an auction, consider the following four different auction institutions: first-price sealed bid, second-price sealed bid, English, and Dutch auctions.

In a first-price sealed bid auction, the buyers independently submit a private sealed bid to the seller. The seller then awards the item to the highest bidder at a price equal to the highest bid. Ties are broken randomly. Note how this institution forces the buyers to simultaneously condition their bids on the other buyers' expected bids. The bidders do not use any information from the actual auction process in forming their bids.

The 'outcry' version of an English auction differs significantly from the first-price sealed bid auction in that the bidders call out increasing bids until only the highest bidder remains. Again, the highest bidder wins the auction and pays a price equal to his or her bid, but in this case the bids are conditioned on buyers' actual bids.

The second-price sealed bid auction, as its name suggests, is similar to the first-price auction in that the seller awards the item to the bidder with the highest (simultaneously submitted) bid. However, the price that the highest bidder pays is equal to the second-highest submitted bid. (In the case of a tie, one of the bidders is randomly chosen and pays the price of the other tied bidder's identical bid.)

In a Dutch auction, the auctioneer begins with a very high price which none of the bidders is willing to pay, and then in real time lowers the bid until one bidder claims the good, paying a price exactly equal to the bid called by the auctioneer.

OPTIMAL BIDDING STRATEGIES

Let us first analyze the optimal bidding strategy for the first-price sealed bid auction. For ease of exposition suppose that there are two bidders ($i=1,2$) whose values are independently and uniformly distributed on the interval [0, 1]. We constrain the bids to be nonnegative. Bidder i's pay-off is

$$u_i = \begin{cases} v_i - b_i & \text{if } b_i > b_j \\ \frac{(v_i - b_i)}{2} & \text{if } b_i = b_j \\ 0 & \text{if } b_i < b_j \end{cases} \qquad (1)$$

A player's strategy is a function that maps values (v_i) to actions (b_i). To find the optimal bidding strategy we will assume that player 1's strategy $b_1(v_1)$ is a best response to player 2's strategy $b_2(v_2)$, and vice versa. Player 1 thus chooses the strategy b_1 to maximize the expected pay-off from the auction, or $(v_1 - b_1)\ P(b_1 > b_2(v_2))$. (Technically, we should include the pay-off $\frac{1}{2}(v_1 - b_1)\ P(b_1 = b_2(v_2))$, but since the values are continuously distributed, the probability of identical values is zero.) That is, in choosing a bid b_1, player 1 weighs the probability that b_1 will be greater than player 2's bid, which is a function of player 2's value. (*See* **Games: Coordination**)

Rasmusen (2001) provides an accessible derivation of the optimal bidding function in equilibrium, namely that a player's best-response strategy is $b_i(v_i) = \frac{v_i}{2}$. That is, each bidder submits a bid equal to one-half of the value. Notice how this optimal bidding function weighs the bidder's fundamental trade-off in the auction. The lower the bid, the more likely that the bidder will win the gain from an increased pay-off from winning the auction; but the higher the bid, the more likely that the bidder will actually win the auction. Note that the bidder with the highest value will always win the auction, so that the auction is perfectly efficient.

In contrast to a risk-neutral bidder, a risk-averse bidder will prefer to increase the probability of winning the auction at the cost of a lower pay-off by submitting a bid greater than $\frac{v_i}{2}$. In particular, it is possible that a more risk-averse bidder with a lower value will submit a bid greater than that of the bidder with the highest value who is less risk-averse. This results in an inefficient allocation if the highest bidder does not win the auction.

Now consider the English auction. If the auctioneer starts the bidding at the price of zero, every bidder is willing to purchase the item and is willing to submit a bid of some minimum increment, say 0.01. Each bidder continues to raise the bid until the standing bid exceeds his or her value. This process continues until the bidder with the second-highest value drops out of the auction at price equal to v_i (± 0.01). Hence, the bidder with the highest value wins the auction and pays a price equal to the second-highest value (± 0.01). This auction is also perfectly efficient. Furthermore, risk aversion does not change this result. Because the bidding occurs in real time, a bidder always knows whether or not he or she holds the highest bid and what is necessary to become the current highest bidder. Trying to determine the optimal bid as a function of this value is unnecessary: a bidder can always respond with a highest bid until the current highest bid exceeds his or her value. Hence, the cognitive costs of participating in this auction are much lower. (*See* **Markets, Institutions and Experiments**)

Even though it is a one-shot game, bidding incentives in a second-price auction lead to outcomes identical to those in an English auction. The optimal bidding strategy in a second-price sealed bid auction is a simple one: $b_i = v_i$. To understand why, first note that it is never profitable for a bidder i to submit a bid b_i greater than v_i, because if the second-highest bid happens to be less than b_i but greater than v_i then bidder i wins the auction but incurs a negative pay-off since the price (the second-highest bid) is greater than the value v_i. To eliminate such a possibility, the bidder should reduce his or her bid to v_i. A bidder should also never submit a bid less than the value, because the bidder gains nothing in terms of pay-off should he or she win: the price is determined by the second-highest bid, submitted by another bidder. Moreover, lowering a bid only reduces the likelihood that the bidder actually submits the highest bid and is declared winner. Hence, the optimal bid in a second-price sealed bid auction is $b_i = v_i$.

An interesting feature of this result is that bidding one's value is optimal, independently of the other bidders' actions. A strategy that is always a best response to any strategy employed by the other players is known as a dominant strategy (see Rasmusen (2001) for a discussion). Notice that when every bidder plays the dominant strategy, the bidder with the highest value wins the auction and pays the price equal to the second-highest bid, which is equivalently the second-highest value. This outcome is identical to that of the English auction. Notice also that because setting $b_i = v_i$ is optimal regardless of what the other players do, the outcomes in this auction institution are robust to risk aversion: a bidder can never lose and only gain by playing the dominant strategy. Hence, the second-price sealed bid auction is perfectly efficient with risk-averse or risk-neutral bidders.

While seemingly quite different from the first-price sealed bid auction, the Dutch auction is strategically equivalent to it. We will call the successively lower prices offered by the auctioneer the 'bid' in a Dutch auction. When the auction begins at a very high bid, a bidder will first consider claiming the item when the bid falls to v_i. If another bidder has not already claimed the item by the time the bid reaches v_i, then the bidder must consider allowing the auctioneer to continue lowering the bid to increase the pay-off to the bidder from winning the auction, but the further the bid drops, the more likely it becomes that another bidder will actually win the auction at a price less than the bidder's own value. A bidder evaluates precisely the same trade-off in the first-price auction. Because the winner in a Dutch auction pays a price equal to his or her bid, the optimal bidding strategy, and concomitant results for efficiency and prices, in a Dutch auction are isomorphic to those in a first-price sealed bid auction.

EXPERIMENTAL RESULTS

There have been many experimental studies of bidding behavior in auctions (for a more comprehensive discussion, see Kagel and Roth (1995)). Two early studies test the strategic equivalence of first-price and Dutch auctions. Coppinger *et al.* (1980) and Cox *et al.* (1982) both find that prices are higher in first-price auctions than in Dutch auctions, and that bidding is consistent with risk-averse behavior. Cox *et al.* (1982) suggest two alternative models for the lower bidding in Dutch auctions. The first conjecture is that the bidders derive utility from the 'suspense' associated with the anticipation of purchasing at a lower and lower price, which is added to the pay-off from buying at a price less than the value for it. The second conjecture is that the real-time nature of the Dutch auction leads bidders to mistakenly update the estimates of their rivals' values to be lower when no one has taken the item at successively lower prices. Cox *et al.* (1983) test these alternative explanations and find that the probability miscalculation hypothesis cannot be rejected in favor of the suspense hypothesis. (*See* **Asset Market Experiments**; **Choice under Uncertainty**)

The predicted isomorphism between English and second-price auctions also fails to be observed. Coppinger *et al.* (1980) and Kagel *et al.* (1987) find that bidding in the English outcry auction conforms to the theoretical predictions, while in the one-shot second-price auction bidders consistently bid higher than the dominant-strategy prediction, even with experience in the auction mechanism (Kagel and Levin, 1993). One explanation for the deviation in behavior is that the real-time nature of the English auctions produces strong, immediate, and overt feedback as to what a bidder should and should not bid. The English auction makes immediately transparent the potential of a negative pay-off from bidding above one's value. The one-shot nature of the second-price auction obscures this realization. If the highest-value buyer submits a bid slightly greater than his or her value and still wins the auction, but the price (equal to the second-highest bid) is still less than that value, then the winning bidder is no worse off for having bid above the value. Since the price equals the second-highest bid, Kagel *et al.* speculate that bidders are deceived by the illusion that bidding higher is a low-cost means of increasing the probability of winning. This raises an interesting question for further research. Is this illusion based on familiarity with the standard first-price sealed bid and English auctions, where the buyer's own bid simultaneously determines the price and improves his or her chances of winning? Or does bidding one's own value present a cognitive impediment?

References

Coppinger V, Smith V and Titus J (1980) Incentives and behavior in English, Dutch and sealed-bid auctions. *Economic Inquiry* **43**: 1–22.

Cox J, Roberson B and Smith V (1982) Theory and behavior of single object auctions. In: Smith V (ed.) *Research in Experimental Economics*, vol. II, pp. 1–43. Greenwich, CT: JAI Press.

Cox J, Smith V and Walker J (1983) A test that discriminates between two models of the Dutch–first auction non-isomorphism. *Journal of Economic Behavior and Organization* **14**: 205–219.

Kagel J, Harstad R and Levin D (1987) Information impact and allocation rules in auctions with affiliated private values: a laboratory study. *Econometrica* **55**: 1275–1304.

Kagel J and Levin D (1993) Independent private value auctions: bidder behavior in first-, second-, and third-price auctions with varying numbers of bidders. *Economic Journal* **103**: 868–879.

Kagel J and Roth A (1995) *The Handbook of Experimental Economics*. Princeton, NJ: Princeton University Press.

Rasmusen E (2001) *Games and Information: An Introduction to Game Theory*. Malden, MA: Blackwell.

Further Reading

Davis D and Holt C (1993) *Experimental Economics*. Princeton, NJ: Princeton University Press.

Gibbons R (1992) *Game Theory for Applied Economists*. Princeton, NJ: Princeton University Press.

Milgrom P (1989) Auctions and bidding: a primer. *Journal of Economics Perspectives* **3**(3): 3–22.

Smith V (1991) *Papers in Experimental Economics*. Cambridge, UK: Cambridge University Press.

Binding

See **Thalamocortical Interactions and Binding**

Binding Problem

Intermediate article

Jacques P Sougné, University of Liège, Liège, Belgium

CONTENTS

The binding problem is the problem of how a cognitive system (a brain or a computational model) groups a set of features together, associates a filler with a role, a value with a variable, an attribute with a concept, etc.

OVERVIEW OF THE BINDING PROBLEM

Humans effortlessly recognize objects, faces, sounds, tastes and so on, all of which are composed of many features. For example, a red ball has a particular shape and color. Since shape and color are not processed in the same cortical areas, there must be a mechanism able to bind the round shape with the red color and differentiate them from a blue cube nearby. A particular face is composed of a set of properties that have to be linked but must also be bound to a particular name and be distinguished from other faces and other names. Representing a predicate and its arguments in 'John loves Mary' requires correctly binding the filler 'John' to the role of 'lover', and the filler 'Mary' to the role of 'lovee', without confusing them. Representing a rule 'if *a* then *b*' requires correctly binding '*a*' to the antecedent and '*b*' to the consequent part of the rule. As these examples show, binding is an essential mechanism in a wide range of cognitive tasks.

How can an artificial neural network perform binding? Early critiques of connectionism raised this problem. How could a connectionist network represent the simple fact that 'the red rose is on a green table'? Since 'red', 'green', 'rose' and 'table' have distributed overlapping representations in the system, the problem is to correctly associate 'rose' with 'red' and 'table' with 'green' while avoiding 'crosstalk', i.e. avoiding the spurious associations between 'table' and 'red' and between 'rose' and 'green'.

The first idea that comes to mind for solving the problem in a connectionist setting is to increase the connection weights between 'rose' nodes and 'red' nodes and between 'table' nodes and 'green' nodes while decreasing other connection weights. But once these connection weights have been set, linked nodes cannot individually participate in other representations. Nodes must be reusable for representing another object like 'yellow rose'. Binding, therefore, must occur dynamically.

Fodor and Pylyshyn (1988) point out a difficulty that arises from the binding problem. They question the value of connectionism as a model of cognition since, according to them, connectionist models cannot display what they call 'systematicity'. The examples they give make reference to the binding problem. They point out that there are no people able to think that 'John loves Mary' but unable to think that 'Mary loves John'. Nobody is able to infer that 'John went to the store' from 'John, Mary, Susan and Sally went to the store' but unable to infer that 'John went to the store' from 'John, Mary and Susan went to the store'. One can define systematicity as the ability to apply a particular structure to any content.

Symbolic systems achieve systematicity very efficiently, but human systematicity does not prevail for complex logical forms. The ability to deduce $\sim A$ from the two statements $\sim B$ and $(A \supset B)$ is not linked to the ability to deduce A from the two statements B and $(\sim A \supset \sim B)$. These two inferences are both obtained by applying the same deduction rule (modus tollens), but they are not cognitively equivalent. Research in cognitive psychology has demonstrated that the first inference is easier (e.g. Evans, 1977; Wildman and Fletcher, 1977). Furthermore, children learning to talk do not display systematicity in their predicate use. (*See* **Symbolic versus Subsymbolic**)

BINDING IN SYMBOLIC SYSTEMS

Binding is not a problem in symbolic systems. A variable is defined and a value is associated with ('bound to') that variable. The lack of constraints on binding means that symbolic systems would not encounter the problems that humans sometimes have when doing binding.

For example, when the time of visual presentation is short, 'illusory conjunctions' are frequent. Illusory conjunction occurs when a particular feature of one object is incorrectly bound to another object. When people are rapidly shown a scene containing a blue ball and a yellow vase they may report having seen a yellow ball. Illusory conjunction is even more frequent if objects share common features.

While most connectionist systems of the 1980s were unable to do binding at all, symbolic systems were clearly too efficient in solving the binding problem compared with humans. Even if one could constrain the performance of a symbolic system with a parameter that would stochastically perturb binding, this would be far less psychologically persuasive than having binding be constrained as an emergent consequence of the architecture of the system.

Before examining the binding problem for connectionist systems, we will briefly review how the brain might achieve binding.

HOW THE BRAIN MIGHT ACHIEVE BINDING

How are neuron assemblies constituted? How are different assemblies bound and differentiated? How can we, as external observers, understand the messages involved in neural patterns of activity – and what code is used by the brain? There are two main hypotheses concerning this code: rate codes and pulse codes.

Rate Codes

There are three ways of considering rate coding, each of which uses a different averaging procedure.

Rate can be computed for a single cell firing over time. In this case, spikes are counted and their number divided by the time elapsed. The objection to this code is that behavioral reaction times are sometimes too small to allow the system to compute an average. If neurons fired at regular intervals, averages could theoretically be computed after two spikes. But noise is also a factor. To obtain a good estimate of the rate, it is necessary to compute the average over a longer period.

The second procedure for evaluating neural firing rate is to repeatedly average single cell spikes. The experimenter repeatedly records the spikes of a cell before and after a stimulation. The average is obtained by dividing the total number of spikes by the number of repetitions and the length of the recording intervals. However, this measure cannot be the code used by the organism to process information, since the reactions of most organisms are the consequence of single stimulus presentations.

The third procedure for computing rate involves recording several neurons before averaging. This rate represents the activity of a population of neurons. Some populations seem to react to particular classes of stimuli. If, after stimulation, an experimenter records the firing of each of these neurons and divides the sum by the number of neurons and by the length of the recording time window, a rate measure is obtained. The advantage of this procedure is that it allows the rate to be calculated for a short time window.

Rate coding has received empirical support. For example, Thomas *et al.* (2001) found that what was crucial for categorization was not category-specific neurons, but rather those neurons that respond more (at a higher rate) to one category than to another.

Pulse Codes

Pulse codes are based on precise timing of spikes. One possible pulse code is latency. The idea behind this is that the time separating a stimulus from the first spike of a neuron can carry information. Gawne *et al.* (1996) recorded activity of striate cortex cells and showed that spiking latency was a function of the visual stimulus contrast.

A second possible pulse code is phase coding. Oscillation of a population activity has been found in the hippocampus and cortical areas. This background oscillation can serve as a reference signal. A particular firing of a neuron can be compared to this background oscillation, and its location on the oscillation curve can serve as a code. This coding scheme has received empirical support. O'Keefe and Recce (1993), for example, showed that phase codes contained spatial information independently of spike rate in the rat hippocampus.

A third possible pulse code is synchrony. Neurons corresponding to microfeatures of a stimulus fire at the same time, and thereby allow the representation of the whole stimulus. This hypothesis has also received empirical support. Engel *et al.* (1991) showed that if several objects make up a scene, distinct clusters of synchrony are formed, each associated with a particular object in the scene. Synchrony is often associated with oscillation since it has been shown that gamma oscillations enable synchronization.

It is important to note that rate codes and pulse codes are not mutually exclusive. Synchrony within a population of neurons over a short period of time also means that the population firing rate is high. It is also possible that different kinds of information could be coded by different coding schemes.

A final problem is how the brain reads the code expressed by neurons. People are capable of describing and observing their own thoughts, but exactly how this is done is not known.

CONNECTIONIST SOLUTIONS TO THE BINDING PROBLEM

Grandmother Cells

The first solution to the binding problem is to use one node for each possible binding. According to this purely localist solution, a binding is represented by a single cell which responds whenever this particular binding is used. This unique cell is called a 'grandmother cell'. This kind of representation poses a number of problems. First, imagine a soccer player banging his head against the goalpost and losing his 'soccer ball' cell. What would this player do after getting up? Would he no longer know anything at all about soccer balls? Second, an enormous number of feature combinations are necessary for representing the multitude of objects, concepts, etc., that humans deal with. If every specific combination had a particular corresponding cell, such a representation would require an impossibly large number of neurons. A final problem is generalization. If every new object would need a new representing neuron, similar objects cannot help in representation building.

Coarse–Fine Coding

At the opposite end of the spectrum from grandmother cells representations are fully distributed representations in which every node may participate in every object representation. Such representations also pose problems. Suppose that a particular object (a ball) has a representation and a particular color (blue) has another representation. Since every node participates in every representation, the conjunction 'blue ball' cannot be represented, since activations coming from 'blue' will be added to activations coming from 'ball' and the subsequent activations will not necessarily represent 'blue ball', 'blue' or 'ball', but could represent nothing or a ghost representation which can correspond to 'pear' or anything else. A solution to this problem is to use a finer coding in which a subset of the population is used to code a particular value. A node becomes active if the input falls in its receptive field. This partially distributed coding needs fewer nodes than purely localist codes to represent the same amount of information, therefore avoiding the problem faced by grandmother cells representation of needing an impossibly large number of cells. However, this solution, at least in its simplest form, only allows one variable to be bound at a time. This solution can represent the conjunction 'blue ball' as the set of 'blue' – responding nodes and 'ball'-responding nodes, but adding a 'yellow vase' to the scene to be represented would make everything bind together. Consequently, we would not be able to detect which color is associated with which object.

Tensor Products

Smolensky (1990) proposed the use of a tensor product representation of binding. Tensor products are similar to outer products of vectors (Figure 1a)). The outer product of two n-dimensional vectors **u** and **v** is the product $\mathbf{uv}^{\mathrm{T}}$ of **u** and the transpose of **v** ($\mathbf{v}^{\mathrm{T}}$ is a matrix with one row and n columns). This is an $n \times n$ matrix. If one has to encode 'John loves Mary', the 'John' filler must be bound to the 'lover' role. John in the role of lover will be represented by the outer product of 'John' and 'lover' vectors. This solution can encode bindings, but it does not allow inferencing. It is also not clear what the neural correlate of outer products might be. One disadvantage of the tensor product is the increase of dimensionality for each additional tensor product (the tensor product of two vectors is a 2D array, the tensor product of a vector and a 2D array is a 3D array, etc.). To overcome this problem, other techniques have been proposed in which the binding of two vectors results in a vector (e.g. convolution – correlation: Plate, 1995). (*See* **Convolution-based Memory Models**)

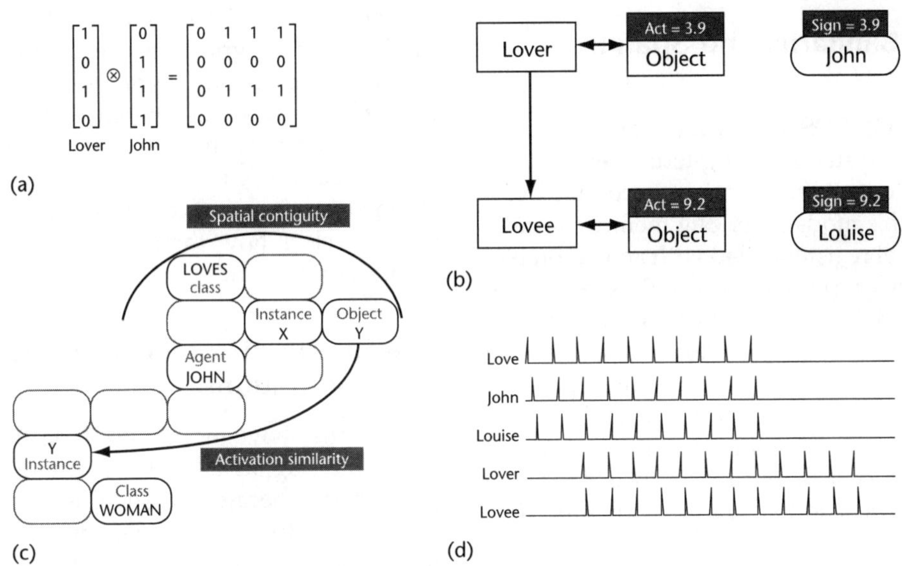

Figure 1. Connectionist solutions to the binding problem. (a) Tensor product representation (Smolensky, 1990) of 'John' bound to 'lover': 'lover' is represented by the (column) vector [1010] and 'John' by the (column) vector [0111]; their binding is represented by their outer product. (b) Values associated with role and filler, as in the model ROBIN (Lange and Dyer, 1989) in which binding is achieved by a match between activation of role object node and filler signature. For representing 'John loves Louise', the activation of the 'lover' object node has the same value as the signature of the 'John' node while the activation value of the 'lovee' object node has the same value as the signature of the 'Louise' node. (c) COMPOSIT binding by activation similarity and spatial contiguity, the representation of 'John loves some woman' in working memory. The register that contains the 'instance' flag with the X symbol denotes a particular instance of the adjacent 'loves' class. The agent of this loving situation is found in the adjacent register containing the 'agent' flag: 'JOHN'. The object of that loving instantiation is found in the adjacent register that contains the 'object' flag 'Y'. By activation similarity, 'Y' points to another register with the same 'Y' symbol and the 'instance' flag. This instance is adjacent to another register that contains the 'class' flag and the 'WOMAN' symbol, denoting that the object of the loving relation is some undefined woman. (d) Binding by synchrony. For representing 'John loves Louise', the firings of the nodes associated with the predicate 'Love' are followed by the 'John' nodes firing in synchrony with the 'lover' nodes while the 'Louise' nodes fire in synchrony with the 'lovee' nodes.

Values Associated with Roles and Fillers

In the model ROBIN (Lange and Dyer, 1989) each filler has an associated node that outputs a particular constant value (called its signature). Each role has an associated object node or binding node. When a role object node has the same activation as that of a symbol's signature, this symbol is bound to the role (Figure 1(b)). A similar solution was proposed by Sun (1992) with his hybrid model CONSYDERR, which consists of a localist network and a distributed network. The localist network is composed of nodes representing symbols and links between these symbols representing rules. Each symbol node is linked to several nodes in the distributed network. The nodes of the distributed network represent the features of the symbol. Variable binding is achieved by the use of a particular value which is passed from a role node to a filler node along a link. These solutions lack psychological plausibility since the number of separated bindings is not constrained.

Activation Similarity and Spatial Contiguity

The model COMPOSIT (Barnden and Srinivas, 1991) uses two systems, a long-term memory and a working memory, both of which are connectionist networks. In this model, working memory is composed of several registers filled with activation patterns from long-term memory. Fillers and their roles are stored in registers as two vectors. One vector (the symbol vector) represents the filler and the other vector (the highlighting vector) represents the role. Predicates, related roles, and fillers are stored contiguously and thus constitute a distinguishable set that can be linked to a particular role pertaining to another predicate (Figure 1(c)). A role can be linked to another role by the similarity of their highlighting vectors. This is a very efficient solution that permits recursive predication. However, it is not clear what neural mechanism might correspond to the loading of registers.

The above solutions have various advantages and limitations, but they all lack neural plausibility. We will now explore more neurally plausible solutions.

Binding by Temporal Frequency

Lange and Dyer (1989) proposed the use of temporal frequency (instead of signatures) for binding, each signature being an unique frequency of node spike. Nodes having the same firing frequency would be bound together. This model has not yet been explored by cognitive scientists, though there are neurobiological data that seem to be consistent with it. However, if the activation of a node is considered to be its firing rate, then binding by activation, as in CONSYDERR and ROBIN, can fall into this category.

Binding by Synchrony

For systems using temporal synchrony for variable binding, nodes can be in two different states: they can be firing ('on'), or they can be at rest ('off'). A node fires at a precise moment and transmits activation to other connected nodes. When a node's activation reaches threshold, it fires. Whenever two nodes (or two sets of nodes) representing two objects fire simultaneously, these objects are temporarily associated (Figure 1(d)). On the other hand, if two nodes (or two sets of nodes) fire in succession, they are distinguished. This is how the systems Shruti (Shastri and Ajjanagadde, 1993) and INFERNET (Sougné, 2001) solve the binding problem.

Synchrony has been used as a binding mechanism in various cognitive models, for perception, for attention, for spatial cognition, for memory, and for different types of inference. These models fit human data well, mainly because the number of distinguishable entities in these systems is constrained by the precision of synchrony. However, it is not clear how these models could represent recursive structures.

The Binding Problem and Distributed Representations

Most of the above models used localist representations. Binding in a fully distributed network is problematic, because as soon as two symbols are required at the same time, their representations may overlap, which could lead to crosstalk.

This problem is greatly reduced by using partially distributed representations, where each symbol is represented by an assembly. This solution avoids problems of 'grandmother' cells, and, since some assemblies share nodes, similarity effects related to binding can be achieved (Sougné, 2001).

THE PROBLEM OF MULTIPLE INSTANTIATION

Multiple instantiation involves the simultaneous use of the same parts of the knowledge base in different ways. Knowing that 'John is the father of Peter' and that 'Peter is the father of Paul', one can readily infer that John is the grandfather of Paul. To derive this conclusion, one must simultaneously instantiate the predicate 'father of' and the object 'Peter' twice. Precisely how this is done is the problem of multiple instantiation. This problem is also called the type–token problem. It is closely related to the binding problem. Solving the binding problem is not by itself sufficient to solve the problem of multiple instantiation; it is, however, necessary.

Symbolic models load copies of pieces of knowledge into a working area before processing them. For these models, there is no problem of multiple instantiation: they simply make several copies of the same content from the long-term knowledge base (LTKB) and store them in the working area. By contrast, for connectionist models that use the knowledge base itself as the place where symbols are associated and processed, multiple instantiation is a serious problem. How can the same object be associated with different roles at the same time without making several copies of this object? In general, multiple instantiation poses significant problems for distributed representations. Two closely related objects will, in principle, share nodes. Therefore, if both objects are needed simultaneously (for different roles), their common nodes must be associated with two different clusters of nodes.

People are able to cope with multiple instances of the same concept, unlike most connectionist models, but their performance when doing so is diminished. There is no naturally arising decrease in performance for symbolic models doing multiple instantiation.

Multiple Instantiation in Neural Networks

For localist networks (where each symbol is represented by a single node) that can represent predicates with more than one argument, the problem of multiple instantiation arises if two instantiations of a predicate have different bindings of their arguments. For example, 'John likes tennis and John likes football' does not require separate instances of the predicate 'likes' since this statement is equivalent to 'John likes tennis and football', in which 'tennis' and 'football' are bound to the same role. However, when two sets of two fillers must be associated with identical pairs of roles, the system must be able to handle two copies of the predicate and argument slots. For example, 'John likes football and Mary likes tennis' cannot be reduced to 'John and Mary like football and tennis', because if it is one can no longer distinguish who likes what.

In distributed networks the problem of multiple instantiation arises as soon as one node must be used by clusters that have to be differentiated. If a predicate with more than one argument has to be represented, and if either the predicate's arguments or any of the arguments' fillers need to use a common node, then this node will have to be bound to different roles.

Relevance for Cognitive Science

In some sense, the connectionist limitations regarding the problem of multiple instantiation could be a blessing in disguise, because some tasks involving multiple instantiation are precisely those tasks that cause problems for humans. Empirical evidence of these difficulties comes from relational reasoning, from repetition blindness, and from the effects of similarity on working memory and on perception. These data show that multiple instantiation can indeed cause problems for humans and animals. In short, when confronted with multiple instantiation, people tend to be slower or to make more mistakes. A good model should not only be able to deal with multiply instantiated symbols, but should also reflect the difficulties that humans have with them.

Connectionist Solutions

There are three types of solution to the problem of multiple instantiation. The first type of solution is to use two systems, an LTKB and a working area into which copies of pieces of the LTKB are loaded. The second type of solution uses several copies of the same symbol in the LTKB. The third relies on superposing frequencies of oscillation.

Multiple copies loaded in a working area

The first solution is borrowed from symbolic models. Each additional instance of a symbol that is required will be represented by an additional copy of the symbol inside short-term memory (STM). An example is the COMPOSIT model (Barnden and Srinivas, 1991), illustrated in Figure 2(a). This solution is probably the most powerful one, and allows a broad range of high-level inferencing capabilities.

Unfortunately, however, this solution inadequately reflects the difficulties people have when they perform multiple instantiation. For these models, even if the STM has a limited capacity, it is as easy to load one copy as to fill the STM with copies of the same content (unless, of course, this is explicitly prevented). Other solutions have been

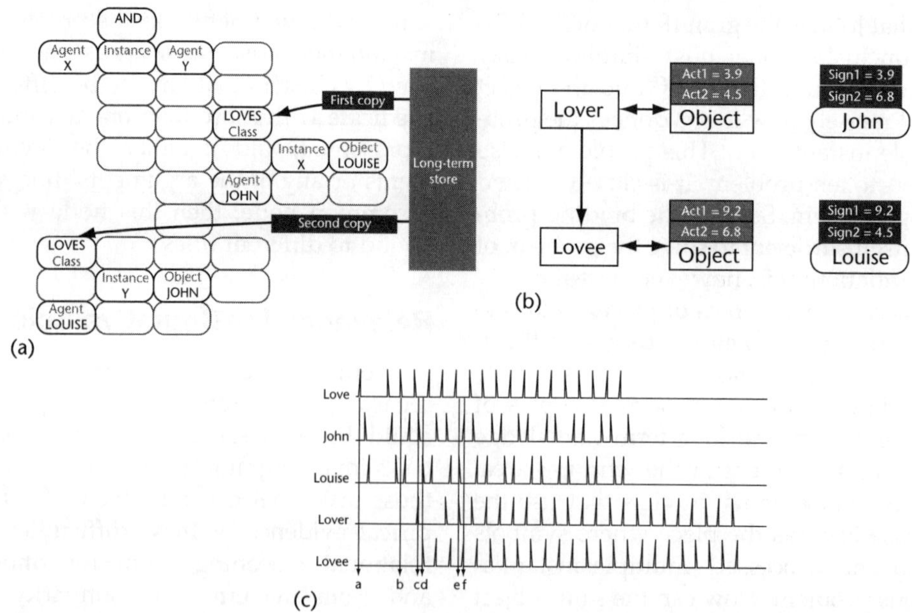

Figure 2. Connectionist solutions to the problem of multiple instantiation. Examples of how different solutions could encode 'John loves Louise and Louise loves John'. (a) The solution of the COMPOSIT model (Barnden and Srinivas, 1991), borrowed from symbolic artificial intelligence. Each additional instance occupies a new place in the 'working memory' area. The 'AND' symbol is instantiated and has two adjacent agents 'X' and 'Y' each pointing to their own instantiation of 'loves'. The register that contains the 'instance' flag with the 'X' symbol denotes a particular instance of the adjacent 'loves' class. The agent of this loving situation is found in the adjacent register containing the 'agent' flag: 'JOHN'. The object of that loving instantiation is found in the adjacent register that contains the 'object' flag: 'LOUISE'. The register that contains the 'instance' flag with the 'Y' symbol denotes a particular instance of the adjacent 'loves' class. The agent of this loving situation is found in the adjacent register containing the 'agent' flag: 'LOUISE'. The object of that loving instantiation is found in the adjacent register that contains the 'object' flag: 'JOHN'. (b) Multiple copies inside a long-term knowledge base. Handling multiple instantiation inside ROBIN requires that each symbol has more than one signature and each role more than one activation (Lange, 1992). For representing 'John loves Louise' and 'Louise loves John', one activation of the 'Lover' object node has the same value as one of the signatures of the 'John' node, and one activation of the 'Lovee' object node has the same value as one of the signatures of the 'Louise' node. Additionally, one activation of the 'Lover' object node has the same value as one of the signatures of the 'Louise' node, and one activation of the 'Lovee' object node has the same value as one of the signatures of the 'John' node. (c) Multiple instantiation by period doubling (Sougné, 2001). The fact 'John loves Louise' is introduced at time a. This pattern starts oscillating. At time c 'John' begins to fire in synchrony with 'Lover' and at time d 'Louise' starts firing in synchrony with 'Lovee'. The statement 'Louise loves John' is introduced at time b. This pattern starts oscillating. At time e 'Louise' begins to fire in synchrony with 'Lover' and at time f 'John' starts firing in synchrony with 'Lovee'. Therefore, the 'Love', 'John', 'Louise', 'Lover', and 'Lovee' nodes sustain two oscillations, enabling the 'John' and 'Louise' nodes to bind alternatively to 'Lover' and 'Lovee' nodes.

developed, however, in which the STM is the activated part of the LTKB.

Multiple copies inside the LTKB

Lange (1992) describes a potential solution designed to handle multiple instantiation inside ROBIN. This solution involves each symbol having more than one signature and each role more than one activation (Figure 2(b). Multiple instantiation in Shruti (Shastri and Ajjanagadde, 1993) is achieved by the use of a bounded set of copies or banks of predicates and their argument slots (usually at most three).

One difficulty with the solution involving multiple copies inside LTKB is that the number of allowable copies is arbitrary. Another difficulty is the abrupt loss of performance. As long as a copy is available, performance will be perfect, but when no more copies are available, performance will decrease abruptly. This behavior does not accurately reflect human performance.

Period doubling

Sougné (2001) describes another solution to the problem of multiple instantiation. In INFERNET, nodes pertaining to a doubly instantiated object will support two oscillations while those singly instantiated will support one oscillation (Figure 2(c)). This makes doubly instantiated object nodes fire twice while singly instantiated objects fire once. This means that each new instance will occupy a new place in STM, thus avoiding crosstalk.

This solution can be compared to the neural phenomenon of bifurcation by period doubling, whereby a stable oscillatory state can lose its stability, giving rise to a new stable state with doubled period. Similar period doubling was obtained experimentally, *in vivo*, by Ishizuka and Hayashi (1996). They recorded cell activity in the somatosensory cortex of rats while increasing the stimulation frequency on a medial lemniscus fiber. As stimulation frequency increased, successive period doublings occurred, finally leading to a chaotic firing pattern.

It is therefore reasonable to assume that multiply instantiated symbol nodes could receive a higher frequency of input than singly instantiated ones; so that increasing the number of instantiations would increase the chance of period doublings.

Another advantage of this solution is its psychological plausibility. As the number of instantiations increases, cognitive performance decreases. The model can also simulate various similarity effects that humans display. However, it has not yet been proved to be able to simulate recursive predication.

SUMMARY

When looking at a field of poppies on a sunny day, how can we correctly associate the color red with the poppies, green with the grass, and blue with the sky, and avoid associating the color red with the grass and the color blue with the poppies? If we see Louise picking a red poppy, how can we correctly associate Louise with the picker and the red poppy with the picked object, without making the opposite and incorrect association? As long as the number of distinct ensembles remains small, these associations are easy for us, but how does the brain perform them correctly? The question of how a connectionist system binds a set of features together, associates a filler with a role, a value with a variable, an attribute with a concept, etc., is what is called 'the binding problem'. Related to this problem is the problem of multiple instantiation.

References

Barnden JA and Srinivas K (1991) Encoding techniques for complex information structures in connectionist systems. *Connection Science* **3**: 269–315.

Engel AK, Kreiter AK, König P and Singer W (1991) Synchronisation of oscillatory neuronal responses between striate and extrastriate visual cortical areas of the cat. *Proceedings of the National Academy of Sciences* **88**: 6048–6052.

Evans JStB (1977) Linguistic factors in reasoning. *Quarterly Journal of Experimental Psychology* **29**: 297–306.

Fodor JA and Pylyshyn ZW (1988) Connectionism and cognitive architecture: a critical analysis. *Cognition* **28**: 3–71.

Gawne TJ, Kjaer TW and Richmond BJ (1996) Latency: another potential code for feature binding in striate cortex. *Journal of Neurophysiology* **76**: 1356–1360.

Ishizuka S and Hayashi H (1996) Chaotic and phase-locked responses of the somatosensory cortex to a periodic medial lemniscus stimulation in the anesthetized rat. *Brain Research* **723**: 46–60.

Lange TE (1992) Lexical and pragmatic disambiguation and re-interpretation in connectionist networks. *International Journal of Man–Machine Studies* **36**: 191–220.

Lange TE and Dyer MG (1989) High-level inferencing in a connectionist network. *Connection Science* **1**: 181–217.

O'Keefe J and Recce M (1993) Phase relationship between hippocampal place units and the hippocampal theta rhythm. *Hippocampus* **3**: 317–330.

Plate T (1995) Holographic reduced representations. *IEEE Transactions on Neural Networks* **6**: 623–641.

Shastri L and Ajjanagadde V (1993) From simple associations to systematic reasoning: a connectionist representation of rules, variables and dynamic bindings using temporal synchrony. *Behavioral and Brain Sciences* **16**: 417–494.

Smolensky P (1990) Tensor product variable binding and the representation of symbolic structures in connectionist systems. *Artificial Intelligence* **46**: 159–216.

Sougné J (2001) Binding and multiple instantiation in distributed networks of spiking nodes. *Connection Science* **13**: 99–126.

Sun R (1992) On variable binding in connectionist networks. *Connection Science* **4**: 93–124.

Thomas E, Van Hulle MM and Vogels R (2001) Encoding of categories by noncategory-specific neurons in the inferior temporal cortex. *Journal of Cognitive Neuroscience* **13**: 1–11.

Wildman TM and Fletcher HJ (1977) Developmental increases and decreases in solutions of conditional syllogism problems. *Developmental Psychology* **13**: 630–636.

Further Reading

Barnden JA and Pollack JB (1991) Introduction: problems for high-level connectionism. In: Barnden JA and Pollack JB (eds) *Advances in Connectionist and Neural Computation Theory*, vol. I 'High-Level Connectionist Models', pp. 1–16. Norwood, NJ: Ablex.

Dyer MG (1991) Symbolic neuroengineering for natural language processing: a multilevel research approach. In: Barnden JA and Pollack JB (eds) *Advances in Connectionist and Neural Computation Theory*, vol. I

'High-Level Connectionist Models', pp. 32–86. Norwood, NJ: Ablex.
Hadley RF (1996) Connectionism, systematicity and nomic necessity. In: Cottrell GW (ed.) *Proceedings of the Eighteenth Conference of the Cognitive Science Society*, pp. 80–85. Mahwah, NJ: Erbaum.
Maass W and Bishop CM (1999) *Pulsed Neural Networks*. Cambridge, MA: MIT Press.
Norman DA (1986) Reflections on cognition and parallel distributed processing. In: McClelland JL and Rumelhart DE (eds) *Parallel Distributed Processing*, vol. II, pp. 531–546. Cambridge, MA: MIT Press.
Singer W (1993) Synchronization of cortical activity and its putative role in information processing and learning. *Annual Review of Physiology* **55**: 349–374.
Sougné J (1998) Connectionism and the problem of multiple instantiation. *Trends in Cognitive Sciences* **2**: 183–189.
Treisman A (1996) The binding problem. *Current Opinion in Neurobiology* **6**: 171–178.
Von der Malsburg C (1995) Binding in models of perception and brain function. *Current Opinion in Neurobiology* **5**: 520–526.

Binding Theory

Intermediate article

Eric Reuland, Utrecht University, Utrecht, The Netherlands

CONTENTS

Languages contain elements such as pronouns and anaphors or reflexives, which lack lexical/semantic content and may or must depend on another element for their interpretation. These dependencies cannot always be established freely, but are subject to structural conditions.

INTRODUCTION

In any language, distinct expressions may be used to refer to the same object. For instance, English 'morning star' and 'evening star' both refer to the planet Venus. Such expressions are said to *co-refer*. Here, co-reference holds on the basis of an empirical fact discovered by astronomers. In other cases speakers' intentions suffice to establish co-reference. The pronominal 'he' can be used to refer to any object that is linguistically classified as masculine and singular, as in 'John's mother thought he was infallible'. Here, 'he' may refer to John but also to some other masculine individual. However, in 'no one believes he is infallible' there is no individual such that both 'no one' and 'he' refer to that individual. Yet under the most salient interpretation 'he' does depend on 'no one' for its interpretation: the interpretive dependency is linguistically encoded, and instantiates *binding*. Binding is subject to constraints which cannot be explained on the basis of logic alone. Rather, they provide us with a window into the computational principles underlying language.

THE BINDING CONDITIONS

The theory of A(rgument)-binding explains the interpretive dependencies between phrases in *argument positions*, or *A-positions*, briefly *arguments*. A-positions are positions for phrases with grammatical functions, such as subject, object, etc., to which a predicate assigns a semantic role (agent, patient, beneficiary, etc.), or of which a predicate governs the case (nominative, objective, etc.) In (1), for instance, all the nominal expressions are arguments in A-position:

> *The old baron* was crossing the bridge at dusk with a ramshackle carriage. *The driver* was visibly tired. Suddenly, the carriage tipped over and *the man* fell into the swamp. When *he* had pulled *him*/*himself* out there came no end to *his* tall tales. (1)

Arguments can be dislocated, ending up in a non-A-position (by topicalization, question formation, etc.), as in (2); *t* indicates their canonical position:

a. *Him*, I never believed the baron to have pulled out *t*
b. *Which man* did he think *t* fell from the bridge
c. *Himself*, the driver pulled *t* out immediately (2)

Binding theory treats the italicized phrases as if they are in the position of *t* (see also below).

Arguments are classified as R-expressions, pronominals, or anaphors. If the head of a phrase has lexical features it is an *R-expression*. Thus 'the old baron', 'no one', 'everyone', 'which man', etc., are all R-expressions. R-expressions are interpretively independent. *Pronominals* ('I', 'you', 'he', etc.) are only specified for *person*, *gender*, and *number* (the ϕ-features). They may, but need not, depend on another argument for their interpretation and they can be accompanied by a pointing gesture, that is, used deictically. *Anaphors* are referentially defective nominal elements: they cannot be used deictically. (A word of caution: some of the literature uses *anaphor* for a class including pronominals, and *reflexive* for anaphor in the present sense.) Anaphors are generally interpreted by binding. Under certain conditions anaphors can, nevertheless, remain unbound (see below). Anaphors come in two types: *simplex anaphors* and *complex anaphors*. Also reciprocals, such as *each other*, behave as anaphors, although their semantics is more complex (see Heim *et al*., 1991). Also, elements such as '(his/her) own', '(the) other', '(the) same' are in some sense anaphoric. Here, we discuss only simplex and complex anaphors.

Simplex anaphors are like pronominals that are underspecified for certain features. Quite generally, specifications for *number* and *gender* are lacking; a *person* specification may be lacking as well (as in Russian *s'eb'a*, (Mandarin) Chinese *ziji*, or Japanese *zibun*). English lacks simplex anaphors, but cross-linguistically they abound. Some well-studied examples are Dutch *zich*, Icelandic *sig*, Chinese *ziji*, and Japanese *zibun*. Their interpretation often corresponds to English 'himself'.

Complex anaphors generally consist of a pronominal or a simplex anaphor and some other element. These other elements may be of varied provenance. Some are historically intensifiers, and currently semantically virtually empty, such as English *self* in 'himself', Dutch *zelf* (*zichzelf*), and Icelandic *sjalfan* (*sjalfan sig*). Many languages use body-part reflexives. These are based on an element that occurs independently as a nominal head designating a part of the body such as 'head', 'bones'; but also, designations such as 'soul, spirit, reflection' are found (see Schladt, 2000). For instance, in Basque 'the father killed himself' is literally expressed as 'the father killed his head'. The form *bere burua* 'his head', which in this sentence means 'himself', is also used in 'he put the cap on his head'. Other languages double a pronominal form, as in Cachur (spoken in Daghestan) and Old Syriac (a Semitic language), or put a special marker on the verb as in Kannada, a Dravidian language (see Lust *et al*., 2000).

Anaphor binding and pronominal binding differ with respect to allowing 'split antecedents', as in 'John talked to the girls about themselves' and 'John talked to the girls about them'. The pronominal 'them' can refer to John (subject) and the girls (indirect object) together (a 'split' antecedent); the anaphor 'themselves' only allows 'the girls' as an antecedent. (Like many other linguistic tests, this one must be used with caution.)

If *a* binds *b*, *a* is the *antecedent* of *b*. Since binding relations cannot be directly read off linguistic expressions they are indicated by a system of *indexing*. Each argument is assigned an integer as its index: in practice, one uses subscripts such as *i*, *j*, *k*, etc. If *a* and *b* have identical subscripts they are *co-indexed*. Thus, in (a_i ... b_i), *a* and *b* are co-indexed. Since indices are just linguistic markers, two expressions can still be assigned the same object in some outside world if they are not co-indexed ('morning star' and 'evening star' are not necessarily co-indexed). Binding without co-indexing is not possible, though. In order for *a* and *b* to be co-indexed, (3) must hold:

a and *b* are nondistinct in features for person, number, and gender (3)

Nondistinctness, rather than identity of features, is required for co-indexing, since in many languages one anaphoric element can have masculine or feminine, singular or plural antecedents. Dutch *zich* and Icelandic *sig* are like that; but they are specified as third person, since they cannot have first- or second-person antecedents. In other languages (for instance, Slavic languages such as Russian) a person specification is also lacking, and we find one anaphoric form for all persons.

Binding relations can also be represented in a *logical syntax notation*. In such a notation, pronouns and anaphors are represented as variables; R-expressions, such as 'every old man', are analyzed containing a *determiner* ('every') and a *set expression* which consists of a *variable* (not overtly represented) and a *restriction* ('old man'). A sentence such as 'Every old man sleeps' can then be represented as 'for all individuals x, provided you pick your individuals from x's that are old and that are men, it is also the case that x sleeps', or, in formula, $\forall x\ ((old\ (x)\ \&\ man\ (x)) \rightarrow sleeps\ (x))$. Often an informal notation is used, in which determiner and restriction stay together, as in *Every old man*$_x$ *(x sleeps)*, or *John*$_y$ *(y sleeps)* (= John sleeps). Here, 'Every old man' and 'John' bind their respective variables. Using this notation, A-binding can also be represented by variable binding. 'Translating' 'himself' in 'John saw himself' as the variable x, yields *John*$_x$ *(x saw x)* as a logical syntax representation. Similarly, in 'Every boy expected Mary to see him', 'him' can be translated as a variable bound by 'Every boy', as in the informal structure *Every boy*$_x$ *(x expected Mary to see x)*.

For binding to obtain, the binder must *c-command* the bindee in the syntactic structure. The standard definition is (4):

> a c-commands b if and only if a does not contain b and the first branching (or maximal) projection dominating a also dominates b (4)

Binding by a non c-commanding antecedent is impossible, as illustrated by the ungrammaticality of **John$_i$'s mother loves himself$_i$*. Putting both conditions together yields (5) as the standard condition on binding ('iff' = 'if and only if'):

> a binds b iff a and b are co-indexed and a c-commands b (5)

Outline of the Canonical Binding Theory

Over the last few decades, the English system has served as a standard model of the binding theory. Its canonical form is presented in Chomsky (1981), and elaborated in Chomsky (1986).

Anaphors and pronominals impose specific conditions on their binders. An anaphor must be locally bound: its antecedent must be sufficiently 'nearby'. A pronominal may be bound, but not locally: its antecedent must be sufficiently 'far away'. An R-expression cannot be bound at all; that is, it must be *free*. The notion of a *governing category* provides a measure for the relevant distance. As a first approximation, pronominals and anaphors are in complementary distribution. This is captured by the binding conditions in (6):

> Binding Conditions
> (A) An anaphor is bound in its governing category
> (B) A pronominal is free in its governing category
> (C) An R-expression is free (6)

Governing category is defined in (7):

> γ is a governing category for α if and only if γ is the minimal category containing α, a governor of α, and a SUBJECT accessible to α (7)

The SUBJECT of a category is its most prominent nominal element (including the agreement features in finite clauses). This is illustrated in (8). Binding is indicated by italics; $[_{GC\text{-}\alpha}]$ stands for the *governing category* of α.

> a. *John* expected $[_{GC\text{-}himself/him}$ the queen to invite him/*himself for a drink]
> b. $[_{GC\text{-}himself/him}$ *John* expected $[_{Clause}$ *himself* / **him* to be able to invite the queen]] (8)

(8) exemplifies the *Specified Subject Condition* (SSC); the governing category is the domain of the nearest subject to α. For 'him/himself' this is 'the queen' in (8a) and *John* in (8b).

Finite clauses are governing categories for their major arguments, including their subjects. Noun phrases with possessive phrases are governing categories for any other object they contain. There is a class of exceptions to this generalization, such as (9):

> a. They expected that pictures of themselves would be on sale.
> b. Max expected the queen to invite Lucie and *himself* for a drink. (9)

Sentences of the type *John$_i$ saw a snake behind him$_i$/?? himself$_i$* are also difficult to fit in with the binding theory of (6) (Chomsky, 1981, 1986). (See below for further discussion and references.)

Binding and Co-reference

A pronominal can be bound by an antecedent, but also be *co-referential* with it. Hence, there is a potential ambiguity when the antecedent is referential (if the antecedent is not referential no ambiguity can arise). The ambiguity surfaces in the two interpretations of (10):

> a. Only Lucie loves her husband
> b. Binding: (Of all the women) Only Lucie has the property x loves [husband of (x)])
> c. Co-reference: (Of all the women) Only Lucie has the property (x loves a) & a = *the individual (for instance, Jack) who happens to be Lucie's husband* (10)

Readings as in (10b) are also called *sloppy readings*, readings as in (10c) *strict readings*. Consider next:

a. **John* saw *him*
b. We all know what's wrong with Oscar. Everyone hates him. [ok]Even *Oscar* hates *him* (11)

Binding Condition B rules out (11a), but what about (11b)? Suppose 'Oscar' and 'him' are *co-referential*, being assigned identical values directly, what prevents this in (11a)? Rule I (Reinhart, 1983) regulates binding and co-reference:

Rule I : Noun phrase (NP) A cannot co-refer with NP B if replacing A with C, C a variable A-bound by B, yields an indistinguishable interpretation. (12)

In (11b), under co-reference Oscar is ascribed the property of *Oscar*-hatred: (*x hates him &* him = Oscar). Under binding the property *(x hates x)* = *self*-hatred is ascribed. *Oscar*-hatred and *self*-hatred are different properties. The fragment is only felicitous for *Oscar*-hatred. So, Rule I allows the co-reference interpretation. In (11a) the two interpretations are indistinguishable, hence Rule I rules out co-reference, leaving no well-formed interpretation since binding is ruled out by Condition B (as it is in (11b)).

Applying (12) requires comparing two different derivations. The processing difficulties this entails have been proposed to explain the fact that children master Condition B at a substantially later age than Condition A (the 'delayed Condition B effect').

Binding and Reflexivity

Many languages have been investigated with binding systems outside the scope of the canonical binding theory. Results indicate that binding uses a modular system, in which the canonical binding conditions arise as special cases. Only a selection of results can be discussed here.

Many languages have a three-way distinction between pronominals, simplex anaphors, and complex anaphors, instead of the two-way distinction covered by the canonical binding theory; some (the Scandinavian languages) even have a four-way system, not counting reflexive possessives.

These systems show that binding interacts with reflexive properties of predicates. A predicate is *reflexive* iff two of its arguments (e.g. subject and object) are co-indexed (Reinhart and Reuland, 1993). A predicate can be *marked* as reflexive by its intrinsic lexical properties (exemplified in English by 'behave' or 'wash'). In English such predicates allow the direct object to be absent; Dutch has a simplex anaphor (*zich*) in such cases. If a predicate is not intrinsically reflexive, argument co-indexing is not sufficient for a licit reflexive interpretation. Reflexivity must be licensed by an extrinsic reflexive marker operating on the lexical entry of the predicate. This is the effect of SELF-anaphors, that is, elements with English *self,* or Dutch *zelf.* For instance, nonreflexive *bewonderen* 'admire' requires the complex anaphor *zichzelf* 'himself', as in *George$_i$ bewondert zichzelf$_{i}$/*zich$_i$* 'George admires himself'. When the anaphor and its antecedent are not co-arguments, a complex anaphor is not required, since no reflexive predicate is formed, as in *Jan$_i$ voelde* [*zich$_i$ wegglijden*] 'John felt [himself slide away]', where the anaphor is a small clause subject.

Cross-linguistically, licensing may also involve clitics, pronoun doubling, body parts, verbal affixes, etc., with varying further syntactic and semantic effects.

Whether a SELF-anaphor must be locally bound is in part determined by its content, in part by its relation to a predicate. In Dutch, *zichzelf* must always be locally bound. In English, a SELF-anaphor is exempted from this requirement if it does not exhaustively occupy the argument position of a predicate. Compare the ungrammatical (8a) where it does, to (9b) where it does not. A discussion of bound versus exempt anaphors based on grammatical functions can be found in Pollard and Sag (1992).

In other languages, for instance Malayalam, the element licensing reflexivity need not be locally bound at all (Jayaseelan, 1997).

Local Binding of Pronominals

In many languages (including all Romance and Germanic languages except English), first- and second-person pronominals can be locally bound (as in French *Je me lave* 'I wash myself' or German *Du sahst dich im Spiegel* 'You saw yourself in the mirror'). An important factor allowing local binding of pronominals is underspecification. Benvéniste (1966) has shown that first- and second-person pronouns are not grammatically, but lexically, marked for number ('we' is not a plurality of 'I's). So, these pronominals are grammatically underspecified. This allows them to be locally bound, despite being true pronominals in all other respects.

Frisian (spoken in a northern province of The Netherlands) also fits this generalization. It has a two-way distinction: (1) a complex anaphor *himsels*, when the predicate is nonreflexive, i.e. where

Dutch has *zichzelf*; (2) a pronominal (*him*/etc.) in all environments where Dutch has *zich*, violating Condition B for third person as well. However, Frisian *him*/etc. is also underspecified; not for number, but for Case. This is the reason local binding is allowed in all persons.

LONG-DISTANCE ANAPHORA

Many languages allow anaphoric elements with an antecedent beyond their governing category as defined in (7), or without any linguistic antecedent whatsoever. Icelandic has become a classical case (see Thráinsson, 1979), but there is also long-distance anaphora in English (see Reinhart and Reuland, 1993 for an overview and references). It is often claimed that long-distance anaphors are simplex (i.e. they consist of only one morpheme) and require a subject as their antecedent, but this is a rather rough characterization.

Icelandic *sig* requires an antecedent within an indicative clause, but if *sig* is in an infinitival clause its antecedent may be outside it. The same holds for the cognate forms of *sig* in other Scandinavian languages. Also, *sig* in a subjunctive clause may have a long-distance antecedent. Yet, there are differences between subjunctives and infinitives. If the antecedency relation crosses a subjunctive, binding is not required: the antecedent need not c-command the anaphor, and the existence of a discourse antecedent (not linguistically expressed) may suffice:

María var alltaf svo andstyggileg. ψegar Olafur$_j$ kaemi segδi hún sér$_{i/*j}$ áreidanlega aδ fara... (Thráinsson, 1991)
Mary was always so nasty. When Olaf would come, she would certainly tell himself [the person whose thoughts are being presented – not Olaf] to leave
[NB. (13) could not begin a story] (13)

Many languages admit anaphor binding which violates the SSC (see (8)). For instance, Russian allows binding across infinitival boundaries; Dutch allows binding into perception verb complements. Yet, in all these cases c-command must be respected.

There is evidence that *sig* in the subjunctive domain behaves like a pronominal instead of an anaphor. Such pronominal use of an anaphoric form is often called *logophoric*.

The term *logophor* was introduced by Hagège (1974) to characterize a paradigm of specialized pronominal elements in languages of the Niger-Congo group. Subsequently, this term has been generalized to all elements with the following characteristics (Clements, 1975, pp. 171–172):

1. logophoric pronouns are restricted to *reportive contexts* transmitting the words or thoughts of an individual or individuals other than the speaker/narrator;
2. the antecedent does not occur in the same reportive context as the logophoric pronoun;
3. the antecedent designates the individual or individuals whose words or thoughts are transmitted in the reported context in which the logophoric pronoun occurs.

Conditions (1) and (3) are not structural, but involve the discourse status of the antecedent. In the following Icelandic example these conditions are met and c-command is not necessary (in all other cases c-command and subject orientation are strictly enforced in Icelandic):

Skoδun Jóns$_i$ er [aδ ζδ hafir svikiδ sig$_i$]...
(Thráinsson (1991) opinion John's is that you have betrayed self (14)

Here, *Jón* holds the opinion expressed. (In (13) above, *ser* refers to the person whose thoughts are being presented.) If these conditions are not met, logophoric elements are infelicitous. The situation is more complex than the above quote indicates. In some languages logophoricity is restricted to verbs of saying, excluding thoughts; there may be special logophoric forms with respect to the adressee instead of the speaker, etc. (For more discussion of logophoricity the reader is referred to Sells (1987), and the extensive literature on Icelandic – see the Further Reading list for references.) Well-known further cases of elements that vary between a bound and a referential, logophoric, use are Japanese *zibun* and Chinese *ziji* (see Huang and Tang (1991) and Cole *et al.* (2000) for discussion and references).

Also, English allows a logophoric use of 'himself' (which is not mono-morphemic). Its sensitivity to the discourse status of the antecedent is illustrated by the contrast in (15). (15a) is presented from John's perspective, (15b) from Mary's:

a. John$_i$ was going to get even with Mary. That picture of himself$_i$ in the paper would really annoy her, as would the other stunts he had planned
b. *Mary was quite taken aback by the publicity John$_i$ was receiving. That picture of himself$_i$ in the paper had really annoyed her, and there was not much she could do about it (Pollard and Sag, 1992) (15)

It is an important result of these investigations that a systematic distinction exists between true long-distance binding and a logophoric use. A language may allow long-distance binding without admitting a logophoric use of anaphors; the converse holds as well.

CROSSOVER PHENOMENA AND PARASITIC GAPS

Crossover

Since binding is defined in terms of c-command and co-indexing, movement is expected to feed binding. (16) shows this with anaphor binding:

a. — seemed to himself [John to have been incompetent]
b. $John_i$ seemed to $himself_i$ [t_i to have been incompetent] (16)

The R-expression 'John', which cannot bind the anaphor 'himself' from its base position, can do so after *A-movement*. (16) involves an antecedent moving from one A-position to another.

Wh-movement, *topicalization*, *relativization*, etc., move a phrase to an *non-A- (= A′-) position*. The process of assigning scope to expressions such as 'everyone' has properties of covert A′-movement.

Pronominals can be bound by *wh*-phrases and by 'everyone':

a. Who_i t_i complained that Mary damaged his_i car?
b. $Everyone_i$ complained that Mary damaged his_i car (17)

When A′-movement crosses a pronominal, two cases are to be considered. In (18) the pronominal c-commands the trace of the moved element:

a. He saw who?
b. *Who_i did he_i see t_i? (18)

(18) instantiates *strong crossover*. The *wh*-trace is an R-expression. 'He' binds the trace; this is a Condition C violation, which is strongly ungrammatical. (See Chomsky, 1982, p. 35 for an alternative account.)

In (19b) the pronoun does not c-command the base position of the *wh*-phrase that crossed over it:

a. His_i mother loves who_i
b. ??who_i does his_i mother love t_i (19)

In (20) the object quantifier 'everyone' is assigned scope over the subject, resulting in a similar configuration:

a. His mother loves everyone
b. ??Every x_i [his_i mother loves x_i] (20)

Both (19b) and (20b) are not felicitous, but also not as ungrammatical as (18b). Hence this phenomenon is called *weak crossover*. Reinhart (1983) argues that weak crossover violates the requirement that at surface structure the antecedent c-command the bound element from an A-position. It is easily seen that this requirement is violated in both (19) and (20). An alternative is based on the *bijection principle* (Koopman and Sportiche, 1982). This principle requires that a quantifier bind precisely one variable and that a variable be bound by precisely one quantifier. In (19b) and (20b) the quantifier ('who' or 'every') has to bind both the pronominal and its trace, which violates the bi-uniqueness requirement of the bijection principle.

Parasitic Gaps

A configuration that is reminiscent of weak crossover is found in so-called *parasitic gap* constructions (Chomsky, 1982 and references cited there). The construction is illustrated in (21). There are gaps in the object positions of both 'file' and 'reading' in (21a), and, similarly, in the object position of 'cook' and 'eat' in (21b), yet only one phrase has been extracted:

a. ?Which $article_i$ did John file t_i without reading e_i
b. ?This is the kind of $food_i$ you must cook t_i before you eat e_i (21)

Furthermore, only the direct object gaps of 'file' and 'cook' are in a position from which an element could have been moved. No elements can be moved out of a *without*-clause or a *before*-clause (these are so-called *islands for extraction*). This is brought out by the contrast resulting from filling one or the other position by a pronominal:

a. *Which article did John file it without reading *e*
b. Which article did John file *t* without reading it (22)

Filling the object position of 'file' with a pronominal, as in (22a), makes the other gap the only possible source for the *wh*-phrase, which results in full ungrammaticality. Hence, only one of the gaps can be a trace; the other one has a different status.

The licensing of the gaps in (23a, b), which correspond to the rightmost gaps in (21a, b), is parasitic on the existence of an extraction site resulting from A′-movement. If no such extraction takes place the structure is degenerate:

a. *John filed those articles$_i$ without reading e_i
b. *You must cook this food$_i$ before you eat e_i (23)

The main factors in the distribution of parasitic gaps are given in (24):

In the construction (A), where order is irrelevant and α, t, e, are co-indexed, in order for the parasitic gap e to be licensed (B) must hold :

(A) ... α ... t ... e ...
(B) i. α c-commands t and e
ii. t does not c-command e and conversely
iii. t is a variable (24)

From (24Bii) it follows, for instance, that a trace in subject position will not be able to license a parasitic gap, as illustrated in (25):

*Which articles$_i$ t_i were filed without reading e_i (25)

The reason for the non c-command requirement in (24Bii) is that the 'parasitic gap' must be interpreted as a variable; i.e. it must be A′-bound. If it is bound by an operator such as the *wh*-phrase, the latter can assign it the content necessary for its interpretation. It must be licensed by an operator, since English does not generally allow pure null-pronominals. If the extraction site c-commands the parasitic gap the latter is bound by the trace, not by the operator. Hence, it will not be licensed. In (23) no phrase is able to bind the gap. In (25), on the other hand, the gap is bound by the trace. In neither case can the gap be interpreted. If the gap is assigned content by an operator, and interpreted as a variable, this violates the bijection principle, hence the relative marginality of (21).

CONNECTIVITY AND RECONSTRUCTION PHENOMENA

Can phrases satisfy or violate certain conditions via their traces, or not? The issue is called *connectivity*, and arises in binding, but also in the licensing of so-called *polarity* items. (26) illustrates polarity:

a. Nobody could see anything
b. *Anything$_i$, nobody could see t_i (26)

'Anything' must be in the domain of a negative (or, more generally, downward-entailing) quantifier such as 'nobody'. However, when 'anything' is fronted, this requirement cannot be satisfied via its trace.

For binding, however, (2) showed that pronominals, anaphors, and R-expressions behave as if they are in the position of their traces, when they are moved to an A′-position. That is, they *reconstruct*. Reconstruction is limited to overt A′-movement. A-movement does not show connectivity effects for binding, as illustrated in (27). Despite the fact that 'the claim …' originates from the trace position there is no Condition C violation in (27a), although there is in (27b):

a. The claim that *John* was asleep seems to *him* [t to be correct] (ok. *him* = *John*)
b. It seems to *him* that the claim that *John* was asleep is correct (*him* ≠ *John*) (27)

Where the moved phrase is complex and contains elements that are themselves in A-position more than just reconstruction may seem to be involved.

(28a–c) are often taken to show that movement enlarges the domain of anaphor binding:

a. John said [that Bill liked that picture of himself best]
b. John wondered [[which picture of himself] Bill liked t best]
c. [which picture of himself] did John say [t' that Bill liked t best] (28)

In (28a) 'Bill' is the antecedent of 'himself', in accordance with Condition A. In (28b) and (28c) 'John' is also possible as antecedent of 'himself'. Moving 'which picture of himself' causes 'John' to be added as a possible antecedent. This follows if the governing category of 'himself' can be calculated from the source position of the moved phrase (t), its derived position, and also the intermediate t' position. (It is assumed that in (28c) 'which picture of himself' moves to its derived position via the C-projection of the embedded clause.)

Yet, sentences modeled on (28a) do not consistently disallow the matrix subject as an antecedent. Moreover, the latitude found here does not extend to pronominals and R-expressions, which can only be interpreted in their base position. Alternatively, then, reconstruction applies in all cases equally. The additional interpretations for 'himself' would still follow since, being the object of a picture noun, it is in an exempt position and can be interpreted logophorically.

References

Benveniste E (1966) *Problèmes de linguistique générale.* Paris: Gallimard.

Chomsky N (1981) *Lectures on Government and Binding*. Dordrecht: Foris.
Chomsky N (1982) *Some Concepts and Consequences of the Theory of Government and Binding*. Cambridge, MA: MIT Press.
Chomsky N (1986) *Knowledge of Language: Its Nature, Origin and Use*. New York: Praeger.
Chomsky N (1995) *The Minimalist Program*. Cambridge, MA: MIT Press.
Clements GN (1975) The logophoric pronoun in Ewe: its role in discourse. *Journal of West African Languages* **10**: 141–177.
Cole P, Hermon G and Huang J (2000) *Long Distance Reflexives*. Syntax and Semantics, vol. 33. San Diego, CA: Academic Press.
Evans G (1980) Pronouns. *Linguistic Inquiry* **11**(2): 337–362.
Hagège C (1974) Les Pronoms logophoriques. *Bulletin de la Société de Linguistique de Paris* **69**: 287–310.
Heim I, Lasnik H and May R (1991) Reciprocity and plurality. *Linguistic Inquiry* **22**: 63–101.
Huang J and Tang J (1991) The local nature of long-distance reflexives in Chinese. In: Koster J and Reuland E (eds) *Long-distance Anaphora*, pp. 263–283. Cambridge, UK: Cambridge University Press.
Jayaseelan KA (1997) Anaphors as pronouns. *Studia Linguistica* **51**(2): 186–234.
Koopman H and Sportiche D (1982) Variables and the bijection principle. *Linguistic Review* **2**: 139–160.
Lust B, Wali K, Gair JW and Subbarao KV (2000) *Lexical Anaphors and Pronouns in Selected South Asian Languages*. Berlin: Mouton de Gruyter.
Pollard C and Sag I (1992) Anaphors in English and the scope of the binding theory. *Linguistic Inquiry* **23**: 261–305.
Reinhart T (1983) *Anaphora and Semantic Interpretation*. London: Croom Helm.
Reinhart T and Reuland E (1993) Reflexivity. *Linguistic Inquiry* **24**(4): 657–720.
Schladt M (2000) The typology and grammaticalization of reflexives. In: Frajzyngier Z and Curl T (eds) *Reflexives: Forms and Functions*, pp. 103–124. Amsterdam: Benjamins.
Sells P (1987) Aspects of logophoricity. *Linguistic Inquiry* **18**: 445–479.
Thráinsson H (1979) *On Complementation in Icelandic*. New York: Garland.
Thráinsson H (1991) Long-distance reflexives and the typology of NPs. In: Koster J and Reuland E (eds) *Long-distance Anaphora*, pp. 49–76. Cambridge, UK: Cambridge University Press.

Further Reading

Baltin M and Collins C (eds) (2000) *The Handbook of Contemporary Syntactic Theory*. Oxford, UK: Blackwell.
Barss A (1986) *Chains and Anaphoric Dependence*. Doctoral dissertation, MIT.
Bennis H, Pica P and Rooryck J (eds) (1997) *Atomism and Binding*. Dordrecht: Foris.
Burzio L (1991) The morphological basis of anaphora. *Journal of Linguistics* **27**: 81–105.
Chien Y-C and Wexler K (1991) Children's knowledge of locality conditions in binding as evidence for the modularity of syntax and pragmatics. *Language Acquisition* **1**: 225–295.
Cole P, Hermon G and Sung L-M (1990) Principles and parameters of long-distance reflexives. *Linguistic Inquiry* **21**: 1–22.
Everaert M (1986) *The Syntax of Reflexivization*. Dordrecht: Foris.
Faltz LM (1977) *Reflexivization: A study in Universal Syntax*. Doctoral dissertation, University of California at Berkeley. Distributed by University Microfilm International, Ann Arbor, MI and London.
Fiengo R and May R (1994) *Indices and Identity*. Cambridge, MA: MIT Press.
Gelderen E van (2000) Bound pronouns and non-local anaphors: the case of earlier English. In: Frajzyngier Z and Curl T (eds) *Reflexives: Forms and Functions*, pp. 187–225. Amsterdam: Benjamins.
Gelderen E van (2000) *A History of English Reflexive Pronouns*. Amsterdam: Benjamins.
Grodzinsky Y and Reinhart T (1993) The innateness of binding and coreference. *Linguistic Inquiry* **24**: 69–101.
Heim I (1998) Anaphora and semantic interpretation: a reinterpretation of Reinhart's approach. In: Sauerland U and Percus O (eds) *The Interpretive Tract*, pp. 205–246. MIT Working Papers in Linguistics, vol. 25. Cambridge, MA: MIT.
Hornstein N (2000) *Move! A Minimalist Theory of Construal*. Oxford, UK: Blackwell.
Keller RE (1961) *German Dialects: Phonology and Morphology*. Manchester, UK: Manchester University Press.
Koopman H and Sportiche D (1989) Pronouns, logical variables, and logophoricity in Abe. *Linguistic Inquiry* **20**: 555–589.
Lasnik H (1989) *Essays on Anaphora*. Dordrecht: Reidel.
Lebeaux D (1988) *Language Acquisition and the Form of Grammar*. Doctoral dissertation, University of Massachusetts at Amherst.
Lidz J (1995) Morphological reflexive marking: evidence from Kannada. *Linguistic Inquiry* **26**(4): 705–710.
Pollard CJ and Sag IA (1994) *Head-Driven Phrase Structure Grammar*. Chicago: University of Chicago Press.
Reinhart T (2000) Strategies of anaphora resolution. In: Bennis H, Everaert M and Reuland E (eds) *Interface Strategies*, pp. 295–324. Amsterdam: Royal Academy of Arts and Sciences.
Reuland E (2001) Primitives of binding. *Linguistic Inquiry* **32**: 439–492.
Reuland E and Koster J (1991) Long-distance anaphora: an overview. In: Koster J and Reuland E (eds) *Long-distance Anaphora*, pp. 1–27. Cambridge, UK: Cambridge University Press.
Reuland E and Reinhart T (1995) Pronouns, anaphors and case. In: Haider H, Olsen S and Vikner S (eds) *Studies in Comparative Germanic Syntax*, pp. 241–269. Dordrecht: Kluwer.

Reuland E and Sigurjónsdóttir S (1997) Long-distance 'binding' in Icelandic: syntax or discourse? In: Bennis H, Pica P and Rooryck J (eds) *Atomism and Binding*, pp. 323–340. Dordrecht: Foris.

Safir K (1996) Semantic atoms of anaphora. *Natural Language and Linguistic Theory* **14**: 545–589.

Wexler K and Chien Y-C (1985) The development of lexical anaphors and pronouns. *Papers and Reports on Child Language Development* **24**. Stanford, CA: Stanford University.

Blindsight

Introductory article

Robert W Kentridge, University of Durham, Tyneside, UK

CONTENTS

Patients with damage to primary visual cortex or its afferents report that they are blind in the area of the visual field corresponding to this damage. Blindsight refers to the ability demonstrated by some of these patients to perform a variety of visual tasks despite denying awareness of the stimuli to which they are responding – a dissociation between performance and consciousness.

WHAT IS BLINDSIGHT?

Blindsight is the term given to the remarkable abilities found in a small number of neurological patients who have damage affecting striate cortex, the first cortical area of the brain which normally processes visual information. Despite its rarity, the condition has profound implications for our understanding of consciousness. As a consequence of its rarity and the importance of its implications it is a condition surrounded by controversy. (*See* **Blindsight, Neural Basis of**)

As a result of their brain damage patients with blindsight deny being aware of visual stimuli in the area corresponding to their damage. For example, a patient with damage to the left side of striate cortex reports that he cannot see stimuli presented to the right of his direction of gaze. When tested using standard procedures these patients are classified as clinically blind in the area corresponding to their damage (that is, they have a scotoma). However, if the patients are tested in a way which forces them to make decisions about stimuli presented in their scotoma then, even though the patients deny seeing anything and maintain that their decisions are simply guesses, they usually make the correct response to the unseen stimuli on a variety of visual tasks.

Blindsight, then, is the dissociation between awareness of visual stimuli and the ability to respond appropriately to them found in patients with damage to striate cortex or the neural connections leading directly to it. It is clear that blindsight subjects can detect whether a spot of light within their scotoma accompanies an auditory signal, whereabouts it is, and, if it is moving, in which direction and how fast it is going. The evidence for more complex residual abilities is less strong.

HISTORY

Striate cortex gets its name from a fine white line identifiable near its surface in slices of the brain. This 'stripe of Gennari', discovered in 1782, was the first evidence that the anatomy of the cortex was not uniform and hence that different areas of cortex may be specialized to serve particular functions. Striate cortex lies at the occipital pole of the brain; in humans much of it is hidden on the adjoining lateral surfaces of the cerebral hemispheres (Figure 1).

In addition to being the first identified anatomically specialized cortical area, it was also the focus of the earliest work on functional specialization. Observations of stroke patients dating back to the 1850s suggested that damage to the brain's occipital pole had specific effects on vision. Towards the

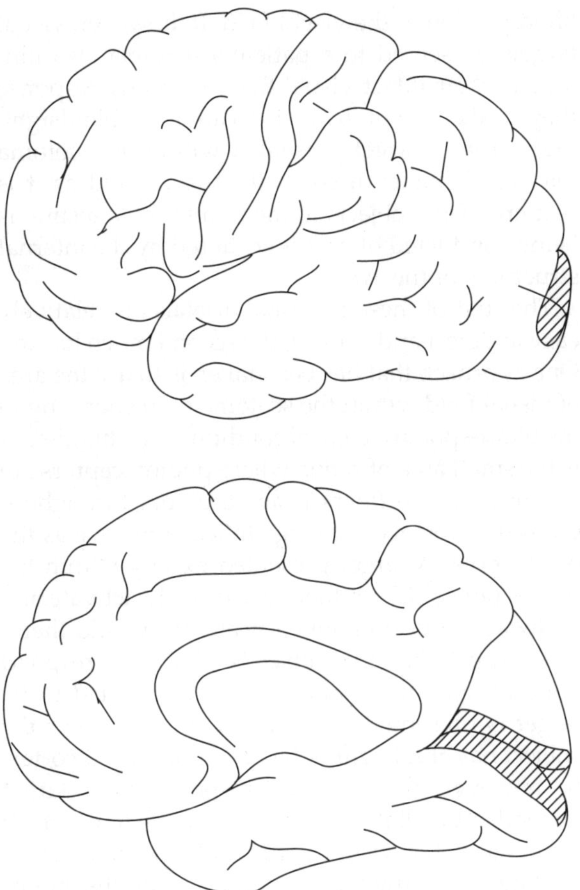

Figure 1. Lateral (upper panel) and medial (lower panel) views of the human cerebral cortex showing primary visual cortex (hatched). Note how little of primary visual cortex is exposed on the surface of the brain. Most of primary visual cortex lies on the medial surface of the brain and is therefore hidden between the two cerebral hemispheres.

end of the nineteenth century, experiments on monkeys showed that lesions of occipital pole large enough to include all of striate cortex rendered animals blind, and it was generally agreed that the occipital lobes were indispensable for vision. From the mid-1930s, however, it became apparent that animals with lesions restricted to striate cortex and not impinging upon other parts of the occipital lobes retained some visual abilities – they could be conditioned to respond to flashes of light and could follow moving spots of light with their eyes.

Starting in the mid-1960s, Nicholas Humphrey studied a single monkey, named Helen, who had bilateral striate cortex lesions. On the basis of many years of observation, Humphrey concluded that Helen retained many (but not all) visual abilities, despite her lesion. For example, she would routinely pick up very small objects with great precision; however, it was clear that she could not identify what these objects were until she explored them with her mouth. Helen apparently retained the ability to detect and locate visual stimuli despite her lesion, but she could no longer identify them.

Although the animal studies of Humphrey allowed the visual abilities remaining after striate cortex lesions to be identified, they could not provide any insight into the subjective nature of visual experience without striate cortex. To do so one must be able to ask a human patient lacking striate cortex to describe what they see. Such patients had been studied for many years and reported that they saw nothing in the region corresponding to their brain damage. One exception, to which we will return later, was the perception of movement. During the First World War, George Riddoch found that soldiers with injuries to the occipital cortex, although blind to stationary stimuli, reported that vigorously moving stimuli did elicit visual experience. Studies of wounded soldiers feature prominently in the history of visual neuropsychology. Careful collation of the locations of gunshot wounds and areas of lost vision in soldiers from both World Wars provided the evidence for maps of the representation of the visual field in striate cortex.

In 1973 the team of Ernst Pöppel, Richard Held, and Douglas Frost, working at Massachusetts Institute of Technology, decided to test whether soldiers (and one stroke patient) with visual scotomata as a result of damage to the visual cortex could, nevertheless, move their eyes so as to direct their gaze at spots of light presented in their regions of blindness. Pöppel, Held, and Frost were prompted to attempt this experiment by earlier work which, amongst other things, had shown intact responses of the pupil and intact optokinetic nystagmus (a slow drift of eye-gaze in one direction, interrupted by occasional flicks back in the opposite direction, induced by presentation of a continually moving pattern) in patients with occipital lesions. Since both of these responses are mediated by midbrain structures, it might be the case that neural pathways transmitting information directly from the retina to the midbrain without passing through striate cortex could support a range of simple visual abilities in these patients. As at least one circuit used in the control of eye movements is entirely subcortical, eye-movement control was a clear candidate for such a potentially spared function. Although the patients found the task puzzling, one remarking 'how can I look at something

I haven't seen', there was a consistent relationship between the location of visual targets and the eye movements the patients produced when asked to look at the locations where they 'guessed' these targets had been presented. The appropriate behavioral response of patients to visual targets shown in this task, coupled with their complete denial of awareness of those targets, is acknowledged as the first systematic experimental demonstration of blindsight.

The term was not, however, coined until a year later when Lawrence Weiskrantz described similar work he had carried out on a patient who, as a result of surgery to alleviate pain caused by abnormalities in the blood supply to the occipital pole of the brain, had lost most of the striate cortex on one side of his brain. Weiskrantz found that not only did this patient (known as DB) move his eyes appropriately towards unseen targets, but he could also point towards target locations accurately with his finger, detect the presence of a luminance grating (a smoothly varying pattern of light and dark stripes), discriminate the orientation of lines and discriminate between the shapes 'X' and 'O' in his blind field, all while denying any visual experience. Blindsight was clearly a complex phenomenon requiring considerable work, both to evaluate the range of visual functions spared after damage to striate cortex, to determine the extent of the dissociation between behavior and visual consciousness, and to test models of the anatomical basis of residual function.

EXPERIMENTAL WORK ON BLINDSIGHT

Four questions need to be addressed in the experimental study of blindsight. Apart from evaluating the anatomical basis of blindsight, the range of spared functions, and the dissociation between behavior and awareness, it is crucial to demonstrate that blindsight is a real phenomenon and that the results obtained cannot be explained by experimental artefacts which allow subjects to perform tasks using the intact portion of their visual field or in some other unintended manner.

Artefacts

Blindsight patients are quite rare. Moreover, in virtually all reported cases, visual field loss is not total. These patients therefore retain normal conscious vision in part of their visual field. The residual visual abilities of interest in blindsight are those used in response to stimuli presented in the blind portion of the visual field. If, however, visual targets presented to a patient's scotoma also illuminate their intact visual field, then any response they make is not truly indicative of blindsight. Light from a target presented within the scotoma may reach intact areas of the visual field as it is scattered from objects in the room where testing is being conducted or as it is scattered by the internal structures of the eye.

The first of these potential artefacts is relatively easy to detect and control, the second much harder. One approach that has been taken is to use the area of visual field within the scotoma corresponding to the blind-spot in a control condition. The blind-spot is the small area of retina where photoreceptors are absent as nerve fibers from receptors throughout the rest of the eye converge to leave the eye as the optic nerve. A target presented exactly within the blind-spot could not therefore directly activate any pathway, cortical or subcortical. One would therefore expect that the subject's ability to respond appropriately to a target will be eliminated if the target is presented in the blind-spot, whether the subject has blindsight or has an undamaged cortex. If, however, the subject's response to a target depends upon light scattered to remote (and intact) portions of the visual field, it should not matter whether the target is presented over the blind-spot or an adjoining area of retina – the presence of receptors at the target location is neither here nor there. The performance of blindsight subjects does indeed fall to chance when targets are presented to the blind-spot, suggesting that residual performance in blindsight does not depend upon a scattered light artefact. This does not, however, mean that scattered light can be ignored. It may still provide cues to a subject unless steps are taken to control it. The most common of these is to use dark targets against a bright background wherever possible, and to flood the subject's intact visual field with bright light.

Light-scatter is not the only means by which information from stimuli intended to reach the scotoma alone can travel to intact regions of the visual field. The most common method of presenting stimuli to patients is with a computer display screen. Stimuli presented in one part of a computer display can produce unintended but visible effects in other parts of the display. Presentation of a bright spot, for example, can cause a small brightening in a narrow horizontal band at the same height as the spot across the entire width of the screen. Care must, therefore, be taken to mask portions of the screen visible outside a patient's scotoma when using such stimuli.

Anatomical Bases of Blindsight

The processing of visual stimuli starts in the array of interconnected photoreceptors of the retina at the back of the eye (Figure 2). The most prominent output from the retina projects to a midbrain structure called the lateral geniculate nucleus (LGN) and from there to striate cortex. This is not, however, the only output from the retina which projects to many other structures. Initially it was supposed that blindsight was mediated by such structures which controlled basic responses to light without any cortical involvement. For example, the superior colliculus can control reflexive eye movements which direct gaze towards a visual target without involving cortex. Although subcortical circuits mediate very specialized responses, blindsight patients might learn to monitor these specialized responses in the course of performing more general tasks. It might, for example, be possible to monitor the location towards which one is about to move one's eyes and use this information to choose whether or not to press a button even if the eye movement itself is suppressed. According to this scenario, blindsight may be mediated by subcortical visual pathways.

Although the bulk of visual input to the cortex passes through the striate cortex, there are ways in which visual information can reach the cortex while bypassing the geniculo-striate route. The superior colliculus sends projections, via the pulvinar, to a number of cortical areas involved in vision (V2, V3, V4, and MT). These are parts of cortex involved in visual processing which normally receive their major input via the striate cortex. Since these areas can receive visual input in the absence of the geniculo-striate projection, it is possible that blindsight may be mediated by visual pathways outside the striate cortex.

In addition to mediation by subcortical or extrastriate cortical routes, there remains the possibility that damage to striate cortex in blindsight patients is not, in fact, complete. Rather than demonstrating that circuits other than the major geniculo-striate route support visual function but do not give rise to visual awareness, residual visual function in blindsight would then essentially be a demonstration that the magnitude of stimulation required to

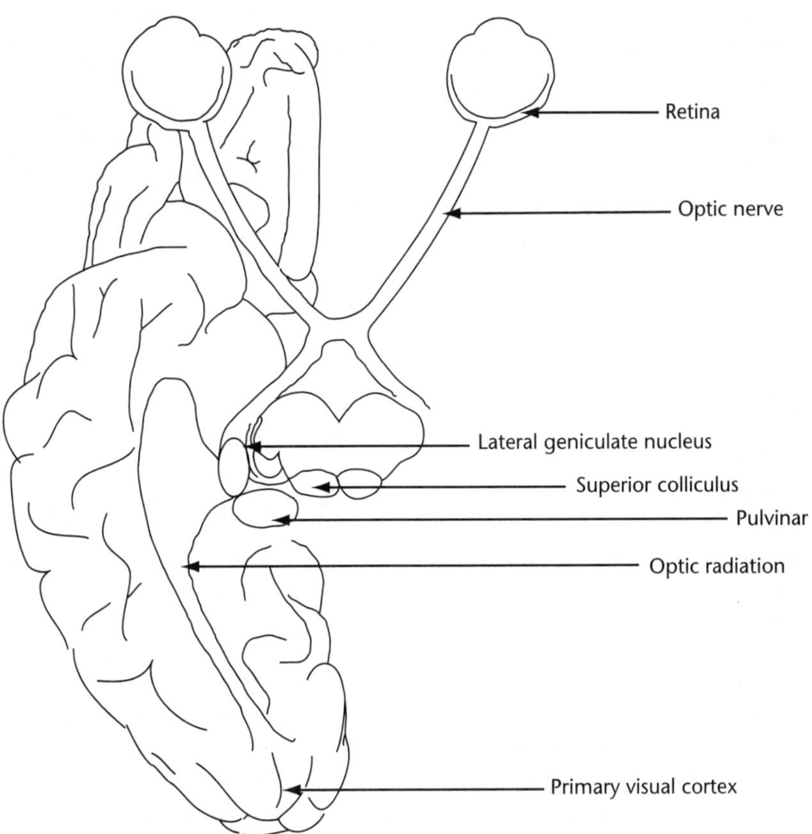

Figure 2. A basal view of the human brain showing major components of the visual system which may be involved in the mediation of blindsight.

evoke awareness from the geniculo-striate system is greater than that required to support simple behavioral responses. Blindsight would not, under these circumstances, be a particularly special phenomenon since it would differ little from the abilities of normal subjects when presented with stimuli near the limits of their visual abilities (e.g., very faint or very short duration stimuli).

A number of studies have been made of patients who have had small spared regions of striate cortex surrounded by damage. These patients did not experience stimuli falling in these spared regions, yet, as in blindsight, they could perform simple visual discriminations in these regions. Can such spared cortex explain the apparently extensive region of blindsight found in other blindsight subjects? Patches of residual vision surrounded by areas of complete blindness might not be revealed in most studies if random eye movements fortuitously brought stimuli into a region of the retina which activated a patch of spared cortex. If, however, one ensures that eye movements cannot affect the location in the cortex which a visual stimulus potentially activates, then any patchiness should become apparent. It is possible to do this by using eye-movement measurements to yoke stimulus position to the direction of gaze. A study using this technique in a patient with blindsight covering a large proportion of one visual field did not reveal patches of residual vision surrounded by blindness, suggesting that an explanation of all blindsight in terms of islands of spared cortex is untenable.

Diffuse, as opposed to patchy, subtotal damage is harder to detect behaviorally. The undamaged neurons in a diffusely damaged region of cortex should, however, still be metabolically active. Functional neuroimaging, which detects changes in blood flow or blood oxygen levels indicative of metabolic activity, has not revealed activity in the striate cortex of blindsight patients when a visual stimulus was presented, although changes did occur in extrastriate cortex.

If blindsight relies on a visual pathway used in normal vision, albeit seriously damaged, the implication is that blindsight should be like very poor normal vision. The apparent dissociation between the abilities of blindsight patients and their reports of awareness may be explained in terms of a change in their willingness to report that they have seen a stimulus – not a surprising change given their knowledge that they have a serious visual impairment. It is, however, possible to disentangle the effects of such biases from the underlying visual sensitivity. The results of such experiments indicate that, for normal subjects presented with stimuli near the limits of visual ability, there is no difference between mechanisms which serve conscious report and those which serve the 'forced-choice' discrimination tasks typically used in assessing blindsight. A similar comparison in a blindsight patient showed quite different properties for conscious report and forced-choice discrimination, indicating behaviorally that blindsight is not simply near-threshold normal vision.

Ingenious experiments have been devised which show that monkeys with unilateral visual cortex lesions treat stimuli in their 'blind' and normal visual fields quite differently, even though they are quite capable of making behavioral responses to those blind-field stimuli. The monkeys were first trained to point at visual targets presented in either their blind or normal hemifields and the minimum brightness contrast required was measured for each hemifield. The target contrasts were then adjusted so that they easily exceeded these thresholds for the rest of the experiment. The monkeys now learned a new task in which they had to make different responses depending on whether one or two stimuli were presented. They performed accurately when both stimuli were presented in the intact hemifield. However, when two targets were presented but one of them fell in the lesioned hemifield, the animals made the 'one target' response. They behaved as if they had seen only one target even though the target they ignored was easily bright enough for them to point at accurately.

There is no question that these monkeys had no spared cortex – striate cortex was surgically removed and the completeness of the damage verified at the end of the experiment. Unless one accepts that there are fundamental differences in the anatomy of vision and awareness between monkeys and man, these results suggest that blindsight cannot rely on spared striate cortex.

Blindsight and Awareness

Although blindsight is the dissociation between awareness of visual stimuli and the ability to respond appropriately to them, it is not the case that blindsight subjects are unaware of all visual stimuli presented in their scotoma. We have already seen that injuries to the occipital cortex leave patients able to report conscious experience of vigorously moving stimuli, as Riddoch discovered at the end of the First World War. Blindsight subjects also report some experience of rapidly moving stimuli or stimuli with sudden onsets or offsets. It is not clear whether these experiences are anything like *visual* sensations. Blindsight subjects differ in the descriptions they give of these experiences, ranging from a feeling that the response they are making is not quite a guess, to descriptions of movement being like a black hand moving across a black background.

Some authors have argued that the fact that blindsight subjects sometimes have an experience induced by visual stimuli, even if it is quite dissimilar to a normal visual experience, invalidates the contention that visual processing and visual consciousness are dissociated in blindsight. Weiskrantz has suggested that blindsight be divided into two subtypes:

- Type 1 blindsight conforms to the 'classical' definition and is residual visual function in the absence of any acknowledged awareness.
- Type 2 blindsight is defined as residual vision accompanied by an acknowledged experience of events in the blind field but in the absence of acknowledged 'seeing'.

It can be hard to draw broad conclusions about the nature of awareness from type 2 blindsight, as distinguishing between visual and nonvisual experience involves a difficult subjective decision about the nature of experience. Interesting results have, however, been obtained by comparing brain activation in blindsights patient when they do and do not report this type 2 nonvisual experience. These results suggest that frontal areas of the brain are activated during the experience of knowing but not seeing, whereas subcortical structures are primarily active during trials in which there is no report of experience whatsoever.

It is important to point out that the distinction between type 1 and type 2 blindsight is not based on performance. It is possible to show that the ability to perform a task and awareness of the stimuli involved are quite dissociated in type 1 blindsight. For example, as task difficulty is varied the performance of blindsight subjects can increase from chance to being near 100 percent correct without any change in their reported absence of awareness. With appropriate stimuli the dissociation in blindsight between awareness and performance remains unequivocal.

Residual Abilities in Blindsight

The early work of Weiskrantz with patient DB showed that a range of visual functions were spared in blindsight. Since then some controversial new claims have been made about the abilities of blindsight patients. First we shall look at some uncontroversial findings.

There is little doubt that blindsight patients can localize single bright or dark visual targets in their blind fields. Similarly, they can discriminate when such targets appear in a task where the subject is required to indicate in which of two time intervals a target is presented (a temporal two-alternate forced choice task). There is evidence from a number of sources that blindsight patients retain some ability to discriminate the color of stimuli presented in their blind fields, although they are impaired in comparison with normal subjects. Blindsight patients can also detect the presence of a pattern of alternating bright and dark stripes even if the average brightness of the pattern does not differ from the background. Their ability to detect these patterns is much poorer than normal in their blind field – they are unable to detect very fine or faint patterns of stripes. The ability to discriminate between stimuli composed of lines with different orientations is also preserved, albeit in a severely impaired guise and with some variations between patients. GY, for example, can discriminate the orientation of single lines but not patches of stripes. His performance becomes poorer than normal as the lines get shorter than 10 degrees of visual angle.

The ability to discriminate the orientation of lines is one of the basic building blocks of form perception. The extent to which blindsight patients can discriminate between complex forms is, however, a vexed question. Weiskrantz found that his patient DB could discriminate reliably between circles and crosses. As he showed, however, this discrimination may be based on discrimination of differences in the components of these shapes, such as the orientation of the line segments that make them up, rather than discrimination of the shapes *per se*. This is borne out by findings that blindsight subjects fail to discriminate between different shapes constructed from the same line segments, for example equilateral triangles with the point either at the top or at the bottom (Δ versus ∇) and are poor at discriminating between rectangles differing in the ratio of side lengths but not orientation.

Early results indicated that form discrimination is absent or severely impaired in blindsight. More recent studies which have tested form-processing abilities indirectly appear to tell a different story. Studies of the manual responses of blindsight subjects to objects placed wholly or partially in their blind fields indicate that shape, orientation, or size properties which could not elicit appropriate verbal or forced choice discriminations nevertheless influenced hand movement and grasp. Other studies have sought to identify whether shapes presented in the blind field influence subsequent responses to stimuli presented to the conscious good field. Although a number of groups have apparently failed to find any such effects, there have been at least two reports of positive results. In one case words presented to the blind field were reported to influence the interpretation of ambiguous words in the good field. For example, if the word 'money' was presented to the blind field then the subject was more likely to describe the word 'bank' in the good field as a financial institution than as the edge of a river. Unfortunately, relatively few short ambiguous words could be used in this study and so the result, which is of great interest given the weakness of simple form processing in blindsight, is based on relatively few observations. Further evidence derives from a study in which the similarity between the shape of stimuli (in this case single letters) presented in the blind and good fields influenced reaction time to the good field stimulus in a letter discrimination task. The most dramatic evidence supporting the existence of complex shape discrimination without awareness comes from a study on the perception of emotion in blindsight. The blindsight patient GY correctly attributed one of four emotions (happiness, sadness, fear, or anger) to video clips presented to his blind field of an actress expressing one of these emotions.

What are we to make of the apparent contradiction between the limited shape-processing abilities indicated by studies of simple geometric shapes and the abilities necessary to make the complex discriminations required in order to be influenced by letters, words, and facial emotions presented in the blind field? One possibility is that most of the latter tasks did not involve the subject in responding directly to the stimulus in the blind field. By assessing blind-field shape-processing through its effects on seen targets, subjects are relieved of the problem of making decisions about stimuli they do not believe they can see. Perhaps removing the conflict for the subjects between their conscious blindness and the demands of a task in which they must respond to stimuli they cannot see uncovers abilities hidden in direct tasks. It may also be the case that certain properties of stimuli and methods of response are mediated by specialized neural circuits. Perhaps the processing of emotion is of such basic evolutionary importance that facial cues to emotion are processed by systems independent of the brain's general shape identification system. These are open questions; at present there is insufficient evidence to come to a firm conclusion about why and whether blindsight subjects can discriminate complex shapes without awareness.

The basis of another residual ability in blindsight is also controversial. The ability to detect the direction or speed of moving stimuli has been studied in blindsight for many years, and there is good evidence that such discriminations can be made both with and without an accompanying experience (rapidly moving high-contrast stimuli are particularly likely to elicit reports of awareness). There are, however, two ways in which motion can be inferred from the stimuli typically used in these experiments. One of these is not strictly a matter of motion perception. One can infer the direction and speed of a moving dot or line by noting its position at one instant and comparing this with its position some time later. Unfortunately there are stimuli with which such a positional comparison method will not work. For example, one can construct a stimulus comprising many dots, in which each dot is displayed for only a short time before it disappears and another dot appears at a different place. If each of these dots moves in a different direction, but on average the dots move more in one direction than any other, then a normal observer will easily be able to report the average direction and speed of the pattern (this is an example of a random dot kinematogram). In some experiments (but not all) the blindsight subject GY failed to discriminate the direction of motion when stimuli which precluded the use of position comparison were used. He could, however, still distinguish moving from stationary stimuli. It is therefore not safe to assume that motion processing is fully preserved in blindsight, even though some forms of motion can be discriminated by blindsight patients.

Some recent studies indicate that residual abilities in the blind field can be modulated by processes of alerting and spatially selective attention. GY's ability to perform a spatial localization task is enhanced if the visual stimuli are immediately preceded by an auditory warning. He is also faster at responding to a visual target if it appears in the location indicated by a preceding cue. This effect can be found even when the cue itself is also presented in the blind field. Such results may have profound implications for our understanding of the relationship between consciousness and attention.

RELEVANCE OF BLINDSIGHT FOR CONSCIOUSNESS AND COGNITIVE SCIENCE

For years consciousness was a taboo word in psychology. If blindsight has done one thing for psychology and cognitive science it is to make the scientific study of consciousness respectable once again. As well as providing insights about the relationship between processing visual stimuli and visual awareness, blindsight offers some insight into the modularity of psychological processes and the extent to which apparently complex processes can, in fact, occur essentially automatically, without awareness.

Blindsight is often used in philosophical arguments about the nature of consciousness. In particular, the apparent dissociation between access to visual information and visual experience in blindsight has been used to explore the role, and even existence, of experiences as something distinct from the properties of stimuli in the outside world, our knowledge of them, and our responses to them. One of the attractions of blindsight to philosophers is that it appears to offer a real, albeit partial, example of a favorite of the philosophical thought experiment – the zombie.

The philosophical zombie is a being whose behavior is indistinguishable from that of real people, but who is supposed to have no inner experience at all of the world in which it is behaving. These inner experiences are often referred to as 'qualia'. Thought experiments about zombies sometimes hinge on a *reductio ad absurdum*, purporting to show that presupposing the existence of zombies leads to some paradoxical difference between our observation of the world of real people and that of zombies. It is argued that if zombies and beings with inner experience differ behaviorally, zombies who lack inner experience and yet are indistinguishable from us behaviorally must be an impossibility. Since the only difference between ourselves and zombies is the presence or absence of qualia and zombies cannot exist, then qualia have no explanatory power and hence no existence outside an individual's mind. On the other hand, one might argue that inner experiences are real (it makes sense to discuss them, as I am doing here, for example); perhaps they just do not have causal consequences for the physical world. (*See* **Zombies**)

Blindsight appears to make zombiehood concrete. Blindsight has been used to argue that inner mental states are real and correspond to physical, that is neural, states. In fact blindsight adds an extra twist to zombiehood – well-tested blindsight subjects come to know consciously that they respond appropriately to visual stimuli even though they do not know what they see and have no inner experience of seeing. Of course, it might be the case that blindsight people do have inner experiences, it is just that they do not know they have them. Unfortunately, many of these arguments are weakened when they either ignore some abilities of real blindsight people or use thought experiments which go far beyond the actual abilities of blindsight subjects. Philosophical consideration of the real properties of blindsight (as opposed to those of nonexistent super-blindsighters) does suggest that inner experiences are real, can be investigated scientifically, and make a difference in the real world, even if they do not solve the problem of telling us what such inner experiences are and why they feel the way they do.

Further Reading

Cowey A and Stoerig P (1991) The neurobiology of blindsight. *Trends in Neurosciences* **29**: 65–80.

Cowey A and Stoerig P (1995) Blindsight in monkeys. *Nature* **373**: 247–249.

Holt J (1999) Blindsight in debates about qualia. *Journal of Consciousness Studies* **6**: 54–71.

Kentridge RW, Heywood CA and Weiskrantz L (1997) Residual vision in multiple retinal locations within a scotoma: implications for blindsight. *Journal of Cognitive Neuroscience* **9**: 191–202.

Marcel AJ (1998) Blindsight and shape perception: deficit of visual consciousness or of visual function? *Brain* **121**: 1565–1588.

Morland AB, Jones SR, Finlay AL, Deyzac E, Le S and Kemp S (1999) Visual perception of motion, luminance and colour in a human hemianope. *Brain* **122**: 1183–1198.

Sahraie A, Weiskrantz L, Barbur JL, Simmons A, Williams JCR and Brammer ML (1997) Pattern of neuronal activity associated with conscious and unconscious processing of visual signals. *Proceedings of the National Academy of Sciences USA* **94**: 9406–9411.

Weiskrantz L (1986) *Blindsight: A Case Study and Implications*. Oxford, UK: Oxford University Press.

Weiskrantz L (1997) *Consciousness Lost and Found*. Oxford, UK: Oxford University Press.

Blindsight, Neural Basis of

Intermediate article

Carlo A Marzi, University of Verona, Verona, Italy

CONTENTS

Blindsight is the presence of unconscious visually guided behavior elicited by stimuli presented within the visual field loss of patients with damage to the primary visual cortex.

INTRODUCTION

The term 'blindsight' to describe the presence of unconscious visually guided behavior in people with a lesion of the primary visual cortex was coined by Weiskrantz *et al.* (1974) and should not be confused with the term 'residual vision', which defines conscious visually guided behavior following a visual cortical lesion. The essence of blindsight lies primarily in the lack of conscious awareness in the presence of an above-chance visual performance. The first demonstration of blindsight was the target localization by eye movements reported by Poeppel *et al.* (1973), followed by target localization by manual pointing and target detection (Weiskrantz *et al.*, 1974). Many other functions have been since then tested with different success in humans and in nonhuman primates (Stoerig and Cowey, 1997). The importance of blindsight in neuroscience research is threefold. First, it has opened the way to the scientific investigation of conscious awareness, a topic before relegated to the domain of philosophy. Second, it has reconciled the discrepancy between human and nonhuman primates as far as the effect of the lesion of the primary visual cortex is concerned; lesions of the primary visual cortex in monkeys (or in cats), although severely impairing visual acuity, leave some visually guided behavior intact, while this is typically not the case in humans. Research has shown, however, that visually guided behavior in humans may persist following a primary visual cortex lesion, but it remains unconscious. Third, blindsight might be an initial stage in the return of vision; unfortunately, the correlation between presence of blindsight and successful rehabilitation of conscious visual function is not good. However, blindsight can improve with training and there is evidence of an increase in unconscious visual sensitivity over the years (Stoerig and Cowey, 1997). Whether this may lead to recovery of conscious vision is still an open question.

An important distinction is between direct and indirect methods of testing blindsight. The former include a forced-choice procedure similar to that used in animal testing, in which the person is asked to guess despite lack of stimulus awareness. In contrast, in the indirect procedure, the person does not have to guess but the presence of blindsight is inferred from the influence of unseen stimuli on the response to stimuli presented to the normally sighted hemifield.

LESION SITE

Unilateral complete lesions of the primary visual cortex (also known as striate cortex, area 17 in the Brodmann nomenclature, or area V1 in the jargon of electrophysiology) result in contralateral homonymous hemianopia: that is, the entire hemifield on the side opposite to the lesion is blind as assessed by clinical perimetric examination. Partial lesions, result in a scotoma which may affect various portions of the contralesional hemifield. Finally, bilateral complete lesions of the visual cortex result in cortical blindness that affects the whole visual field.

Which Regions Must Be Affected to Produce the Deficits?

Damage to visual centres other than V1 may result in a hemianopia but such lesions may not be compatible with blindsight when crucial centres are visually deafferented. This is the case with lesions of the optic tract, which funnels visual information not only to the geniculostriate system but also to the superior colliculus and other visual midbrain areas, or to thalamic areas such as the pulvinar which relay visual information to cortical areas bypassing the primary visual cortex. When these areas are deafferented in addition to primary visual cortex, no blindsight is possible.

WHICH REGIONS MEDIATE SPARED FUNCTION?

There are two main candidates as centres mediating the spared unconscious functions characterizing blindsight: the superior colliculus and the extrastriate visual cortex. The superior colliculus (SC) projects indirectly through the thalamic pulvinar to extrastriate cortical areas such as V2, V3, V4 and V5 (also known as MT, see below) and to visually responsive temporal areas. Therefore, in the absence of V1, visual input can still activate these cortical areas. In addition, a small number of cells – mainly located in interlaminar zones of the dorsal lateral geniculate nucleus (dLGN) – project directly to extrastriate cortex, rather than to V1 like the majority of dLGN neurons. The difference between the relative contribution of these structures has been investigated by comparing the effects of lesions restricted to the primary visual cortex with the effects of lesions including the extrastriate cortex and those of loss of an entire hemisphere (Azzopardi *et al.*, 2001). An intriguing conclusion emerging from those studies is that following hemispherectomy – that is, in the absence of both V1 and extrastriate cortex – the remaining SC and pulvinar cannot subserve voluntary responses such as those employed in the direct methods to test blindsight described above. In contrast, there are indications that some hemispherectomy patients can still show blindsight when tested with indirect methods, as reported by Tomaiuolo *et al.* (1997).

More evidence on the areas mediating blindsight comes from brain imaging studies in which visual stimuli are presented within the perimetrically blind area of the visual field. It has been shown with functional magnetic resonance imaging (fMRI) that, despite absence of activation of their lesioned V1, hemianopic patients show preserved responsiveness in area V5 with moving stimuli, and in areas V4 and V8 within the fusiform gyrus with colored images of objects (Goebel *et al.*, 2001). The former is an area, or rather a complex of areas, which belong to the so-called dorsal system including a series of cortical regions mediating spatial perceptual and visuomotor functions. The area MT is named for its anatomical location in the monkey's middle temporal sulcus in the proximity of the junction of temporal, parietal and occipital lobes. It contains neurons selectively sensitive to the direction and velocity of motion stimuli. The human homolog of area MT is also known as area V5: its bilateral lesion results in a severe impairment in the detection of moving stimuli: see Zeki (1991) for a review.

In contrast to the dorsal stream, that is, a series of cortical regions, the ventral stream includes a host of cortical visually responsive areas whose neurons respond preferentially to forms, colors and natural objects. Areas within the fusiform gyrus (V4 and V8) belonging to this system have been found to be selectively activated by appropriate stimuli presented within the hemianopic field. It is important to consider that both the dorsal and the ventral system activations were not accompanied by conscious awareness of the stimuli. This shows that activation of extrastriate cortex *per se* is not sufficient to yield stimulus awareness.

WHICH FUNCTIONS ARE SPARED?

Blindsight functions can be divided into two categories: direct or voluntary responses usually elicited with a forced-choice procedure, and indirect responses usually tested by assessing the influence of unseen stimuli (presented to the hemianopic field) on stimuli presented to the normal hemifield.

Direct Response

Direct responses include basic functions such as simple detection of stationary or moving stimuli, spatial localization and discrimination of direction of motion, stimulus displacement, wavelength and orientation. The latter ability seems to be crucial for the apparent form discrimination exhibited by some people with blindsight. In fact, it has been shown that shape discrimination in the hemianopic field is impossible when orientation cues are eliminated. One interesting dissociation has been repeatedly described, namely that between action and perception. Some people who are unable to discriminate forms or objects because of cortical damage can none the less show reaching or grasping hand movements that are appropriate for the size and orientation of the stimuli presented. This means that information that is not available for conscious perception can be used for motor action on the same object.

Indirect Response

An example of this approach is the use of the redundant target effect (RTE), with bilateral stimuli presented across the vertical meridian of the visual field. With this paradigm, simple manual reaction time in response to bilateral brief visual stimuli is typically faster than for unilateral single stimuli. It has been shown that this is the case even when one stimulus in the pair has been presented to the hemianopic hemifield of patients with a V1 lesion. Despite their claim of having seen only one stimulus they show an RTE with bilateral stimuli. The speeding up of reaction time by an unseen stimulus can thus be taken as indirect evidence of blindsight (Marzi *et al.*, 1986). A similar implicit RTE has been found in patients with an hemianopia resulting from hemispherectomy performed as an extreme therapy for intractable epilepsy (Tomaiuolo *et al.*, 1997). Another example of indirect approach is the demonstration that distractor signals in the blind half of the visual field could inhibit saccades toward targets in the intact visual field (Rafal *et al.*, 1990).

RELEVANCE TO THEORIES OF CONSCIOUSNESS

One of the merits of research on blindsight is to have given impulse to a neuroscientific study of consciousness and to have aroused the interest of philosophers in the neural basis of conscious experience. A few years ago an experiment was attempted to answer this fundamental question (Sahraie *et al.*, 1997). In an fMRI experiment on a blindsight patient extensively investigated in other studies (GY), stimulation of the blind hemifield yielded conscious or unconscious above-chance visual performance depending upon the velocity of motion stimuli. The patient was required to discriminate the direction of a moving stimulus; with slower stimuli, discrimination was as good as with faster stimuli but stimulus awareness was lost. The main thrust of the study was to provide evidence for a shift in the pattern of neural activation from cortical to subcortical neural structures associated with the change from conscious to unconscious vision. Notably, the superior colliculus was activated in the unconscious mode alone. These results confirm earlier views that blindsight may be mediated by subcortical structures. However, as pointed out by Searle (2000), they cannot be extended to consciousness in general because the patients only exhibit blindsight if they are already conscious. For a clue to the neural mechanisms of consciousness in general it is necessary to demonstrate that there are neural structures that are crucial for shifting from unconsciousness to a conscious state. So far, this evidence has not been provided.

OTHER POSSIBLE INTERPRETATIONS

In principle, blindsight effects could be the result of various spurious factors (Campion *et al.*, 1983). One possibility is light-scattering: this possible source of artefact has been taken care of by minimizing the light intensity of the stimuli and, ingeniously, by introducing control trials in which stimuli are presented to the blind spot of the hemianopic hemifield, that is, to an area corresponding to a retinal zone without photoreceptors. Real blindsight cannot survive a blind spot presentation, and this is what has been found in the majority of cases; see for example Tomaiuolo *et al.* (1997).

Another possible source of spurious blindsight comes from a shift in the decision criteria adopted by the hemianopic patient in moving from clinical visual field assessment to experimental blindsight testing. In the former situation the patient may adopt a more conservative criterion and deny having seen something, despite some near-threshold residual vision. In contrast, under laboratory conditions, especially when using forced-choice procedures, the patient may adopt a more liberal criterion and performance might improve. A specific answer to this type of criticism has been provided by Azzopardi and Cowey (1997) who measured the sensitivity of a hemianopic patient independently of his response criterion. They found that, in contrast to normal control subjects, sensitivity was higher during the forced-choice task than during a procedure similar to that used in clinical visual field testing in which a conscious report is required ('Do you see the stimulus?'). This means that blindsight cannot be simply assimilated to normal near-threshold vision and that a mere shift of response criterion cannot explain it.

Finally, it has been proposed (Campion *et al.*, 1983; Fendrich *et al.*, 1992) that blindsight may be related to islands of spared visual cortex yielding correspondingly small areas of visual field preservation that can be documented only with special techniques (Fendrich *et al.*, 1992). In contrast to this possibility, however, it has been found that two blindsight patients studied with fMRI did not show any activation of V1 (Goebel *et al.*, 2001) and therefore it is unlikely that their blindsight might have been related to spared V1, although this might explain other cases. All in all, one can safely conclude that blindsight is a genuine phenomenon, but its investigation requires careful control of all possible sources of artefact.

References

Azzopardi P and Cowey A (1997) Is blindsight like normal, near-threshold vision? *Proceedings of the National Academy of Science USA* **94**: 14190–14194.

Azzopardi P, King SM and Cowey A (2001) Pattern electroretinograms after cerebral hemispherectomy. *Brain* **124**: 1228–1240.

Campion J, Latto R and Smith Y M (1983) Is blindsight an effect of scattered light, spared cortex, and near-threshold vision? *Behavioural Brain Sciences* **6**: 423–486.

Fendrich R, Wessinger CM and Gazzaniga MS (1992) Residual vision in a scotoma. Implications for blindsight. *Science* **258**: 1489–1491.

Goebel R, Muckli L, Zanella FE, Singer W and Stoerig P (2001) Sustained extrastriate cortical activation without visual awareness revealed by fMRI studies of hemianopic patients. *Vision Research* **41**: 1459–1474.

Marzi CA, Tassinari G, Aglioti S and Lutzemberger L (1986) Spatial summation across the vertical meridian in hemianopics: a test of blindsight. *Neuropsychologia* **24**: 749–758.

Poeppel E, Frost D and Held (1973) Residual visual function after brain wounds involving the central visual pathways in man. *Nature* **243**: 295–296.

Rafal R, Smith J, Krantz J, Cohen A and Brennan C (1990) Extrageniculate vision in hemianopic humans: saccade inhibition by signals in the blind field. *Science* **250**: 118–121.

Sahraie A, Weiskrantz L, Barbur JL *et al.* (1997) Pattern of neuronal activity associated with conscious and unconscious processing of visual signals. *Proceedings of the National Academy of Science USA* **94**: 9406–9411.

Searle JR (2000) Consciousness. *Annual Review of Neuroscience* **23**: 557–578.

Stoerig P and Cowey A (1997) Blindsight in man and monkey. *Brain* **120**: 535–559.

Tomaiuolo F, Ptito M, Marzi CA, Paus T and Ptito A (1997) Blindsight in hemispherectomized patients as revealed by spatial summation across the vertical meridian. *Brain* **120**: 795–803.

Weiskrantz L, Warrington EK, Sanders MD and Marshall J (1974) Visual capacity in the hemianopic field following a restricted occipital ablation. *Brain* **97**: 709–728.

Zeki P (1991) Cerebral akinetopsia (visual motion blindness). A review. *Brain* **114**: 811–824.

Further Reading

Holmes G (1945) Ferrier lecture. The organisation of visual cortex in man. *Proceedings of the Royal Society (London) Series B* **132**: 348–361.

Marzi CA (1999) Why is blindsight blind? *Journal of Consciousness Studies* **6**: 12–18.

Milner AD and Goodale MA (1995) *The Visual Brain in Action*. New York: Oxford University Press.

Ungerleider LG and Mishkin M (1982) Two cortical visual systems. In: Ingle DJ, Goodale MA and Mansfield RJW (eds) *Analysis of Visual Behavior*, pp. 549–586. Cambridge, MA: MIT Press.

Weiskrantz L (1986) *Blindsight: A Case Study and Implications*. Oxford: Clarendon Press.

Weiskrantz L (1997) *Consciousness Lost and Found: A Neuropsychological Exploration*. Oxford: Oxford University Press.

Zeki S (1993) *A Vision of The Brain*. Oxford: Blackwell.

Zihl J (2000) *Rehabilitation of Visual Disorders after Brain Injury*. Hove, UK: Psychology Press.

Bloomfield, Leonard

Introductory article

Stephen R Anderson, Yale University, New Haven, Connecticut, USA

CONTENTS

Leonard Bloomfield (1887–1949) was an American linguist whose contributions to general linguistics as well as to the study of a number of language families make him one of the central figures in the history of this field of study. His name is virtually synonymous with the American Structuralist approach to language through the 1950s.

INTRODUCTION

Few figures in the history of linguistics stand out as prominently as incarnations of their time and place as Leonard Bloomfield. Linguistics in America from the publication of his book *Language* in 1933 until the development of Generative Grammar in the 1960s is practically identifiable with his approach. This was in part because he represented the desire of linguists to be treated seriously as pursuing a scientific discipline with its own methods, goals, and results. Edward Sapir and Franz Boas, other major figures in the history of linguistics whose activity overlapped with Bloomfield's, studied languages within the theoretical framework of anthropology or psychology. Others studied particular languages and language families for their own sake. In contrast, Bloomfield thought of himself as a linguist, studying language for its own sake. In the process, he aligned himself with contemporary positions in philosophy (positivism) and psychology (behaviorism) that were seen as paving the road to a genuinely scientific view of language, in contrast to humanistic approaches.

As a result, linguistic theory as it developed during this time was largely formed either through Bloomfield's own work or by what his students and colleagues did in the name of his views. Although a good deal of 'post-Bloomfieldian' linguistics was not particularly close to Bloomfield's own positions, it was nonetheless felt that a scientific approach to language could be largely identified with the task of working out Bloomfield's views. American structural linguistics largely *was* Bloomfieldian linguistics.

BLOOMFIELD'S LIFE

Leonard Bloomfield was born in Chicago in 1887, and moved to Elkhart Lake, Wisconsin, in 1896 when his father bought a resort lodge there. During his childhood in Wisconsin he came into contact with the Menomini people and their language, a member of the Algonquian family, which would occupy much of his later attention. His father's brother, Maurice Bloomfield, was a noted Sanskritist and no doubt had an influence on Bloomfield's subsequent interest in this language and its grammatical tradition.

Bloomfield entered Harvard in 1903, received his AB degree in 1906, and went to the University of Wisconsin for graduate study. One of the first scholars Bloomfield met there was Edward Prokosch, one of the major names in Germanic studies, who interested him in historical work within the framework of Indo-European linguistics. In 1908 he moved to the University of Chicago, where he received his PhD in 1909 for a thoroughly traditional, philologically oriented thesis: *A Semasiological Differentiation in Germanic Secondary Ablaut.*

Most of Bloomfield's academic career was spent as a teacher of German: although he practiced general linguistics within the limits of such positions, it was not until he came to Yale in 1940 that he actually held a professorship of linguistics, as opposed to German. His first job was at the University of Cincinnati, from which he moved to the University of Illinois in 1910. He was told early on that while his department was enthusiastic about promoting him, a competing candidate had the edge by virtue of having studied in Germany, and that if Bloomfield wanted to get ahead, he would have to study in Germany too. Taking this advice to heart, he spent the year 1913–1914 in Leipzig and Göttingen, studying with such notable Indo-Europeanists as Leskien and Brugmann. In the process, he rubbed elbows (quite literally) with a number of other students who would later be important names in linguistics, such as Nikolai Trubetzkoy.

In 1914 he published his first book, *An Introduction to the Study of Language*. This general survey was based solidly in the introspectionist psychology of Wundt, influential at the time but virtually the antithesis of the approach he would later champion. This book is little read today, but interesting for understanding the later development of Bloomfield's views on, for instance, morphology (inflection and word formation).

The outbreak of the First World War led to an immediate and precipitous decline in German studies in the USA, and Bloomfield no doubt had a certain amount of time on his hands as a teacher of German. During the war, he worked with a student at Illinois who spoke Tagalog, the principal indigenous language of the Philippines. This work resulted in a book *Tagalog Texts with Grammatical Analysis*, which contains an extensive grammar of the language, though one that is difficult to use as a result of Bloomfield's explicit, conscious avoidance of traditional categories and terminology in describing a system far from the familiar structure of Indo-European languages.

In 1921, he moved to Ohio State University (not having been offered tenure at Illinois, despite having studied in Germany!) where he immediately became a full professor of German. Here one of his colleagues was Albert Weiss, a major figure in the early development of behaviorism, whose views on the mind largely determined Bloomfield's own for the rest of his career. In 1927, he was invited to the University of Chicago, again in the German department. Here one of his colleagues was Sapir, in the anthropology department: the two were professional collaborators (but uneasy friends) in the emerging discipline of general linguistics.

In addition to his work on German and general linguistics, Bloomfield was also occupied during this time with comparative Algonquian studies. This was not just an escape from the rigors of Germanic linguistics. Bloomfield brought the methods of Indo-European studies to work on American Indian (and by extension, other indigenous) languages. This was unusual: others had suggested that the methodology of comparative reconstruction, developed with respect to Indo-European, was substantially dependent on the fact that several languages (Vedic Sanskrit, Gothic, Homeric Greek, Hittite, Old Church Slavonic, etc.) of the family are attested at considerable time depth, and that the same techniques would not be effective in dealing with unwritten languages. Bloomfield showed that the methodology could be applied in establishing the comparative grammar of Algonquian (especially its central branch, based on data from Fox, Cree, Menomini and Ojibwa).

The clinching demonstration of this came when, in working out the system of consonant clusters in the system ancestral to these languages, he was left with one correspondence set that did not fit any known combination of segments. For this, he postulated an additional proto-Algonquian cluster which he wrote as *çk. Later, however, as data from other languages and dialects of the family became available, it became clear that exactly the words for which Bloomfield had posited this cluster showed consistent unique reflexes across the family; and indeed other words came to light that illustrated the same correspondence set. This was widely seen as providing a dramatic confirmation of the correctness and generality of the assumptions of comparative linguistics – as dramatic, in its way, as the confirmation provided by the analysis of Hittite for the prior assumption of 'laryngeal' segments in the phonology of proto-Indo-European.

Bloomfield's role in the professionalization of linguistics in the 1930s and 1940s was tremendously important. He worked hard for the establishment of the field's distinctive institutions, especially the Linguistic Society of America, its journal *Language*, and the annual summer institutes which it organized (at the time, virtually the only occasions when linguists gathered in significant numbers). In 1940 Bloomfield was invited to Yale, after the death of Sapir (who had preceded him there as Sterling Professor of Linguistics). He never really settled in New Haven: he and his wife were both attached to Chicago, and she suffered from severe depression when they left. To this was of course added the dislocation provoked by the Second World War, but he turned this to advantage, working actively in the army's Intensive Language Program during the war years and thereby providing useful work for a new generation of descriptive linguists. In 1946, Bloomfield suffered a severe stroke from which he never really recovered. He died in 1949.

BLOOMFIELD'S VIEW OF LANGUAGE AND THE MIND

Bloomfield's first writing dealing with general issues in the study of the mind was in his 1914 book *An Introduction to the Study of Language*, but he soon lost confidence in the explanatory power of the Wundtian psychology underlying that book. That point of view was soon supplanted by an ardent embrace of the behaviorist (or 'mechanist' as he preferred to call it) psychology of his Illinois colleague Weiss. This was already evident in his 1926 paper, 'A set of postulates for the science of language' (*Language* **2**: 153–164), intended as a fairly direct calque on a paper by Weiss laying out an axiomatization of psychology, although it also shows considerable influence of the study of the Sanskrit grammatical tradition. More important perhaps than Bloomfield's intent to emulate Weiss's point of view, the paper's terminology in referring to psychological factors is enthusiastically behaviorist in tone, as when he defines the meanings of utterances as their 'corresponding stimulus–reaction features'.

A product of his times, Bloomfield's notion of a scientific explanation was one based solely on propositions relating observable events by principles of logic and mathematics alone. Throughout his career, he repeatedly ridiculed 'mentalistic' explanations as they appeared in the supposedly scientific literature on language and linguistics. Subsequent commentators (as well as many of his contemporaries) took this to imply a rejection of the existence of a mental life, but this is not at all what he intended. Rather, he meant to reject the notion that linguistic (or any other) phenomena are causally affected by a mysterious and unobservable entity (the 'mind') whose principal property is its nonobedience to normal laws of physical causation.

Early behaviorists insisted that if the mind were to be taken seriously as an object of scientific inquiry, it must be reduced to special cases of the activity of some observable physical system. There are, of course, alternatives to this, as the 'cognitive revolution' has made clear, but for Bloomfield, an attempt to ground the study of language in the properties of mental and cognitive organization seemed like an effort to evade the constraints of rational inquiry. Considering the excesses of romanticist approaches to the nature of language, and indeed the introspectionist psychology Bloomfield himself followed in his early years, these concerns were not entirely illusory. For Bloomfield, the only sensible alternative to antirational speculation about the mysteries of the soul was a denial of the scientific relevance of anything but the material embodiment of mind.

This restriction of scientific discourse, including all talk about 'meanings' apart from the framework of observable stimuli and responses, was not, as it is commonly seen, intended to deny that minds and meanings exist, or even that they might play a central role in human life. His point, rather, was that in the present state of science we have no way of cashing these notions out in strictly observable terms, and thus that talk about them necessarily falls outside real science. He did believe that a satisfactory account of meaning would need to be based on an encyclopedic knowledge of the world and its laws, down to the last detail – something obviously well beyond the scope of linguistics or perhaps any science. This belief that meaning ultimately has a comprehensive explanation in terms of sufficiently minute details concerning (potentially observable) electrochemical events within the nervous system is just as much a matter of faith on his part as the 'mentalist' picture is for others.

A language can be seen as a system that relates sounds and meanings, and it would thus seem that some account of meaning is necessary to linguistics. For this reason, Bloomfield introduces a mechanistic picture that seems naive even by comparison with other behaviorist work, but he also denies that the difference between such a view and the mentalist one has any significance. For him, the structural properties of language can be investigated perfectly well even if meaning is simply reduced to the status of a postulate, not treated in its substance (whatever that might be). In a 1944 article, he compared his 'antimentalism' to 'a community where nearly everyone believed that the moon is made of green cheese, [in which] students who constructed nautical almanacs without reference to cheese would have to be designated by some special term, such as *non-cheesists*.'

With complete impartiality, Bloomfield maintained that the concrete properties of sound are also, strictly speaking, irrelevant to an understanding of language; and thus neither phonetics nor semantics played a role in the sort of structuralist accounts he advocated. His actual practice involved appeals to our understanding both of sound and of meaning that were not significantly different from those he opposed: he simply maintained that these matters were not essential to an understanding of linguistic systems.

Bloomfield's views on the nature of mind and cognitive organization were surely much too simplistic, as generations of commentators have maintained. Nonetheless, his repeated insistence that the methodology and results of linguistics are independent in principle of any particular theory of psychology (his or another) should be taken at face value. The radical behaviorist views he advocated had much less influence on his own practice with regard to central areas such as phonology and morphology than his pronouncements would have on his own students and their immediate successors. Bloomfield was a solid scholar in a number of areas, and one whose intuitions about linguistic structure took him far beyond the limitations of his stated basic principles.

Further Reading

Anderson SR (1985) *Phonology in the twentieth century.* Chicago: University of Chicago Press (pp. 250–276).
Bloomfield L (1914) *An Introduction to the Study of Language.* New York, NY: Henry Holt.
Bloomfield L (1933) *Language.* New York, NY: Henry Holt.
Hall RA (ed.) (1987) *Leonard Bloomfield: Essays on his life and work.* Amsterdam: Benjamins.
Hall RA Jr. (1990) *A life for Language.* Amsterdam: Benjamins.
Hockett CF (ed.) (1970) *A Leonard Bloomfield Anthology.* Bloomington, IN: University of Indiana Press.
Hymes DH and Fought J (1981) *American structuralism.* The Hague: Mouton.
Matthews PH (1993) *Grammatical theory in the United States from Bloomfield to Chomsky.* Cambridge, UK: Cambridge University Press.

Body Image

See **Disorders of Body Image**

Brain

See **Language and Brain**

Brain Asymmetry

Introductory article

Albert M Galaburda, Harvard Medical School, Boston, Massachusetts, USA
Glenn D Rosen, Harvard Medical School, Boston, Massachusetts, USA

CONTENTS

The two cerebral hemispheres are specialized for different functions. The discovery of anatomic asymmetries in the brain has given new light to our understanding of the cognitive differences between the left and right hemispheres.

INTRODUCTION

It is generally accepted that in humans the left hemisphere is specialized for the processing of some aspects of language while the right hemisphere dominates over many spatial, emotional and musical functions. The exact relationship between brain asymmetry and side differences in function, however, is not known. In fact, a century and a half after the initial discoveries of lateralization of language the neural basis for language is still not clearly understood, and as part of this incomplete knowledge the relationship between language lateralization and cerebral asymmetry is at best tentative. Moreover, the biological substrates underlying non-language-based functional lateralization is perhaps even less clear. That being said, there is a wealth of information concerning asymmetries in the brain, giving intriguing insights into cognitive function.

GROSS BRAIN ASYMMETRIES

Sylvian Fissure and Planum Temporale

Left–right asymmetries in the sylvian fissure (Figure 1) have been noted since the end of the nineteenth century. In general, the left sylvian fissure tends to be longer and have a flatter trajectory than its right hemisphere counterpart, which deviates dorsally at the posterior end. This obvious asymmetry led the neurologists Norman Geschwind and Walter Levitsky to examine the portion of the sylvian fissure subsumed by a structure known as the planum temporale in 100 autopsied human brains. The planum temporale, which contains several auditory association cortices, is thought on the left side to be an important portion of the language network of the left hemisphere (Figure 1(b)). Lesions affecting significant portions of the planum temporale, usually on the left side in right-handed people, often lead to Wernicke's aphasia. Geschwind and Levitsky found that 65% of the sample showed a longer left planum temporale, whereas the right planum was longer in only 11% and both plana were equal in the remaining 24%. The results were highly significant and were essentially confirmed by subsequent studies using both postmortem and *in vivo* imaging techniques. (*See* **Aphasia; Geschwind, Norman**)

Inferior Frontal Area

The inferior frontal area, composed of the pars opercularis, triangularis and orbitalis of the inferior frontal gyrus, the frontal operculum and the subcentral cortex, is complex, highly folded, and variable among individuals (Figure 1(a)). Asymmetries have been demonstrated in this part of the brain. The ascending limb of the sylvian fissure, which separates the pars opercularis posteriorly from the pars triangularis anteriorly, has been reported to be more often branched on the left side, giving off the diagonal sulcus. Others have found the pars opercularis to be more developed on the left, and still others have found greater surface area on the left side. Wada and colleagues, on the other hand, measured the surface portion of the pars opercularis together with part of the pars triangularis and found it to be larger on the right. (*See* **Frontal Cortex**)

Asymmetries in Fetal and Infant Brains

Gross anatomical asymmetries are present in the cerebral cortex before birth. Even though clearly cerebral dominance can be modified after birth

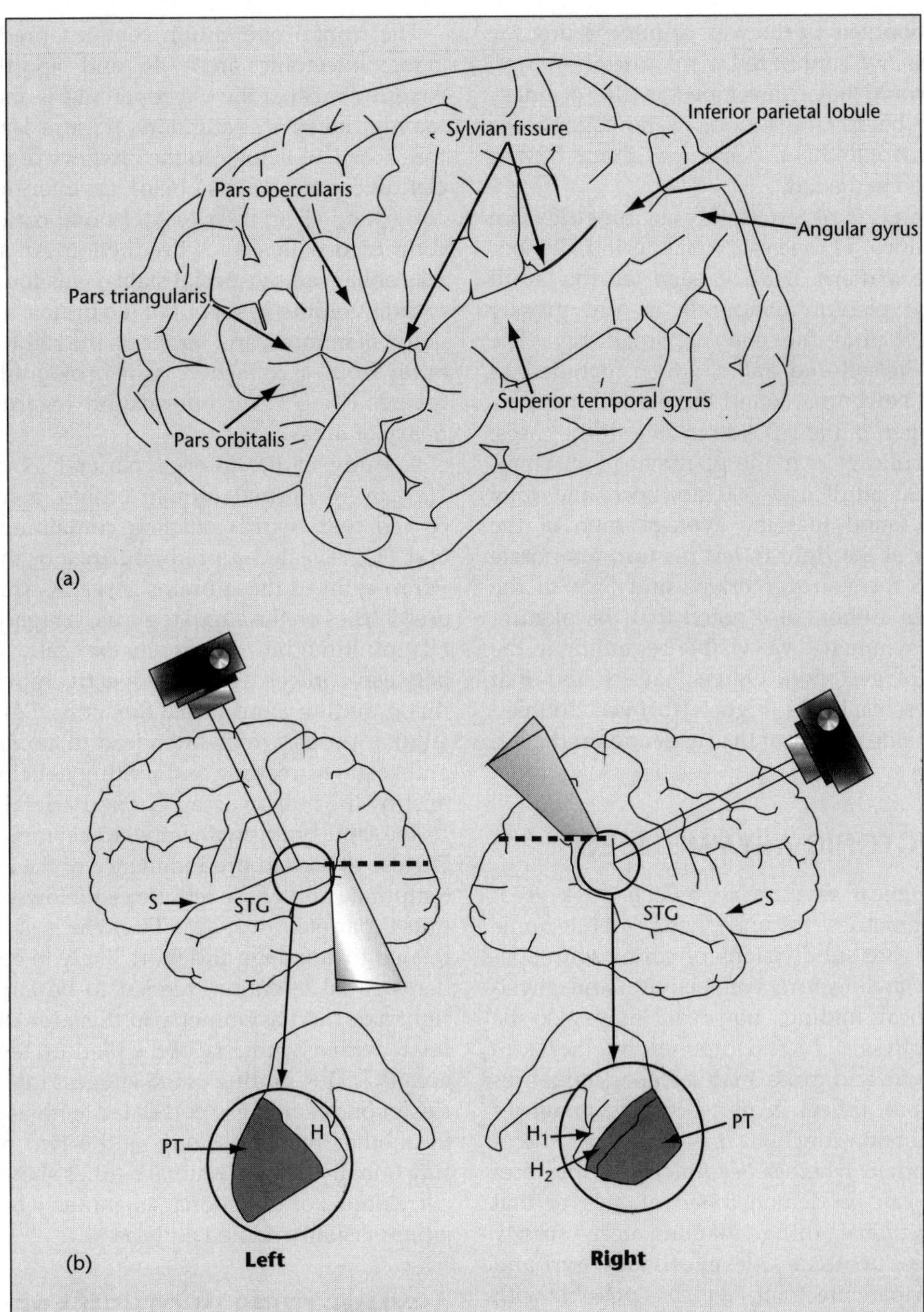

Figure 1. Locations of various structures in the human brain. (a) Lateral view of the left hemisphere identifying areas known to be asymmetric. (b) Lateral views of the right and left hemispheres: note the asymmetric patterns of the sylvian fissure (S), with the left fissure being longer and the right more angled posteriorly. To view the planum temporale (PT), a cut (dotted line) is made at the end of the sylvian fissure (encircled) exposing the superior surface of the superior temporal gyrus (STG). The resulting images are shown below. Here, the planum temporale is larger on the left than the right, and there are two Heschl's gyri (H) on the right side.

(e.g. recovery from early hemispheric injury and ability to switch handedness), the anatomical asymmetries that may underlie functional lateralization are fixed, at least in their gross design, before birth. This is not to say that recovery from early lesions is perfect at any age or that switching handedness leads to equivalent degrees of dexterity with the new hand. We do not know about this because controlled experiments cannot be performed in individual patients and variability in the population gets in the way of interpreting the results. Thus, we cannot tell in advance how well language would have developed in the ordinary hemisphere before the lesion, or how good the right hand would have become without having had to switch to the left.

The sylvian fissures are visibly asymmetric from about the middle of the gestational period, demonstrating the pattern usually seen in the adult human. The planum temporale is also grossly asymmetric before the end of pregnancy. The study by Witelson and Pallie, which included 14 brains from newborns, found the planum temporale to be larger on the left side in 79% of the cases. Wada and colleagues made planimetric measurements of 100 adult and 100 newborn and fetal brains and found that the average ratio of the surface area of the right to left planum temporale was 67% in the younger brains and 55% in the adults. These authors also noted that the planum temporale asymmetry was visible beginning in the 29th week of gestation. Others have found that the right Heschl's gyrus (Figure 1(b)) was doubled on the right side in 54% of the cases and on the left in 18%.

ARCHITECTONIC ASYMMETRIES

Gross anatomical asymmetry reflects to a great extent asymmetries in underlying architectonic areas, which are subdivisions of cortex with specific cellular architecture, connectivity, and physiology. Cortical folding, however, may also be related to physical forces imposed by the skull during growth, and gross anatomical asymmetries may, therefore, reflect in part bony asymmetries unrelated to brain function. It is important, therefore, to ascertain whether asymmetry in the cerebral cortex can be demonstrated at a level that heralds functional differentiation more strictly than the gross anatomical level of lobes, gyri and sulci – the cytoarchitectonic level. It is possible with experience to draw architectonic borders around many cytoarchitectonic areas and to measure their volume in each hemisphere. This has shown asymmetries in several cortical areas implicated in language function. For instance, area Tpt, which is located on the posterior third of the superior temporal gyrus extending onto the planum temporale, is larger on the left side in the majority of brains. Asymmetry of area Tpt correlates positively with asymmetry of the planum temporale, and in some cases the left can be several times the size of the right. Damage to Tpt may result in problems with language comprehension.

The frontal operculum contains predominantly cytoarchitectonic areas 44 and 45, the former covering most of the pars opercularis and the latter mainly the pars triangularis (Figure 1(a)). Both of these cortices belong to the category of motor association cortex, with 44 being an inferior premotor cortex and 45 an inferior prefrontal cortex, and are concerned with speech production. Area 44, which was enhanced by special stains, was found to be of greater volume in six out of ten brains, nearly symmetrical in three, and larger on the right side in one brain from a collection of neurologically normal brains. There is no information regarding asymmetry of area 45.

A study of the inferior parietal lobule (Figure 1(a)) in 10 normal human brains, a region concerned with word meaning containing areas PF and PG, revealed a predominance of the left area PG in eight of those brains. Area PG (Brodmann's area 39) lies on the angular gyrus, is highly functionally multimodal, and is anatomically interposed between cortices dealing respectively with somesthetic, auditory and visual functions. Lesions of the angular gyrus, which often lead to anomic aphasia and acquired reading and writing deficits probably destroy the bulk of area PG. (*See* **Parietal Cortex**)

The same brains with leftward asymmetry of area PG also exhibited predominance of the left planum temporale and of the left area 44. However, a more dorsal parietal area, area PG, which is less clearly related to language and more likely to be related to hemispatial attention, tended to be larger on the right side, and asymmetry in this area did not correlate with asymmetry of the planum temporale or area PG. This finding could suggest that asymmetries in one region are correlated with asymmetries in another region as long as the two regions are functionally linked. Animal studies showed similar correlations of directional asymmetry between adjacent, visually related cortices.

ASYMMETRIES IN OTHER SPECIES

The cerebral cortex emerges for the first time in its six-layered organization in mammals. Asymmetries in structures other than cortex are present, however, in birds, fish and amphibians. In the case of birds, the asymmetries are mainly in the functional domain and consist of differential effects of neural lesions on the ability of birds to sing, with the left predominating. Slight left anatomical superiority of one of the song-relevant nuclei has also been demonstrated. Fish and amphibians often show asymmetries in the habenular nuclei, the functional significance of which is not clear.

In nonhuman primates, the types if not the degree of brain asymmetry are similar to those found in the human brain, which raises questions about what lateralized functions, if any, these structural asymmetries might serve. For instance, sylvian fissure asymmetries have been found in other primate brains. If one is permitted to attach significance to the sylvian asymmetry in the human *vis à vis* linguistic capacity, what then is the meaning of asymmetry in the same structure in the baboon, the orangutan and the chimpanzee? Adding fuel to the fire, recent reports suggest that the pattern of human planum temporale asymmetry is present in chimpanzees. This again has potentially important implications for the role of cortical asymmetry and language function.

Asymmetries have also been noted in fossil skulls. The best-known of these is the asymmetry in the Neanderthal fossil from La Chapelle-aux-Saints, about 60 000 years old, which showed a sylvian fissure asymmetry similar to that seen in modern humans. There was a suggestion of a comparable asymmetry in the endocast of Peking man, which dates from 500 000–600 000 years ago. Asymmetries in perisylvian cortex also appear to exist in australopithecines, *Homo habilis* and *Homo erectus*, which date to up to 3.5 million years. Again, as is the case in nonhuman primates, asymmetries in the sylvian fissure are found in individuals whose language capacity is in question. Whether these asymmetries represent linguistic capacities in early humans or some preadaptive behavior is likely to remain unknown. It will help to find out what it is about asymmetry of the modern human brain, if anything, that can explain linguistic capacity.

MECHANISMS OF CEREBRAL ASYMMETRY

The phrenological approach to neuropsychology suggests that the basis for cognitive capacity is size. For example, our brains are uniquely able to carry out complex cognitive tasks because they are bigger than the brains of other less cognitively gifted primates. Similarly, the left hemisphere is able to support language because some of the perisylvian cortices involved in language are larger than corresponding areas on the right. An alternative to the phrenological concept suggests that highly specialized functions may depend more on specificity of organization. Thus, building a phonologically competent brain area may require increasing the specificity of the brain substrate by customized reduction rather than just enlargement of its components. In this sense, an asymmetric language area would result from the pruning of one side rather than from the enlargement of the other.

We have gained some insight as to the underlying substrates of brain asymmetry through the use of an animal model. Rats, as it turns out, also exhibit brain asymmetry. While individual rats may exhibit a large degree of asymmetry (magnitude), there is no leftward or rightward directional bias of the population as a whole. Exploiting this distinction, researchers have begun to ask questions concerning what distinguishes a symmetric from an asymmetric brain, rather than what factors bias a population to one side or the other. Interestingly, Collins has convincingly demonstrated in mice that while magnitude of lateralization, in this case paw preference, is under genetic control, direction is not. Using rats, in combination with examination of human asymmetries, we can begin to gain a hold on the phrenologic debate outlined above.

Symmetry of cortical areas is associated with changes in the volume of architectonic areas – symmetric brain regions are larger than asymmetric ones. Further, side differences in overall volume are the result of different numbers and densities of some neurons. Also, with increasing asymmetry, the proportion of the targeted area receiving callosal connections diminishes. In other words, the more asymmetrical areas have relatively fewer callosal connections than similar areas that are more symmetric, suggesting that the one key component of anatomic asymmetry may be greater intrahemispheric – as opposed to interhemispheric – connectivity. Indirect evidence suggests, furthermore, that asymmetry is determined during the earliest stages of brain development.

CONCLUSION

The histological characteristics of symmetric and asymmetric cortical areas do not support the notion that cerebral dominance represents simply a case of storing functional areas in one hemisphere or the other, but rather that there are storage factors as well as factors of network size and detailed connectivity. Thus, cerebral dominance reflects variation in functional properties of symmetric and asymmetric cortical areas, which provides the species with desirable and sometimes problematic individual variation.

For the first time, it is possible to assess the normally functioning human brain with sophisticated methods that can address issues about localization and lateralization of function. With improving anatomical imaging it will be possible to extract individual information rather than sample averages, which will be useful for studies on individual variation of localization and lateralization. This knowledge can then be applied to studies of variability in response to brain injury and of developmental variation in cognitive style, response to brain injury, and learning disorders. (*See* **Neuroimaging**)

Further Reading

Denenberg VH (1981) Hemispheric laterality in animals and the effects of early experience. *Behavioral and Brain Sciences* **4**: 1–49.

Gannon PJ, Holloway RL, Broadfield DC and Braun AR (1998) Asymmetry of chimpanzee planum temporale: humanlike pattern of Wernicke's brain language area homolog. *Science* **279**: 220–222.

Geschwind N and Galaburda AM (1987) *Cerebral Lateralization. Biological Mechanisms, Associations, and Pathology*. Cambridge, MA: MIT Press/Bradford Books.

Geschwind N and Levitsky W (1968) Human brain: left-right asymmetries in temporal speech region. *Science* **161**: 186–187.

Rosen GD (1996) Cellular, morphometric, ontogenetic and connectional substrates of anatomical asymmetry. *Neuroscience and Biobehavioral Reviews* **20**: 607–615.

Shapleske J, Rossell SL, Woodruff PW and David AS (1999) The planum temporale: a systematic, quantitative review of its structural, functional and clinical significance. *Brain Research Reviews* **29**: 26–49.

Brain Damage, Treatment and Recovery from

Intermediate article

Barbara A Wilson, MRC Cognition and Brain Sciences Unit, Cambridge, and the Oliver Zangwill Centre for Neuropsychological Rehabilitation, Ely, UK

CONTENTS

Natural recovery can and does occur in children and adults following brain injury. Rehabilitation can also result in improvements of functioning.

IS THERE RECOVERY FROM NONPROGRESSIVE DAMAGE?

The term 'recovery', when applied to nonprogressive brain damage, is interpreted in different ways by people according to their professional background, knowledge, expectations, and experience. Some will focus on biological repair of brain structures, others may regard survival rates as a measure of recovery, while others will be looking for signs of recovery of cognitive functions or motor skills. Some might argue that recovery can be thought of only in terms of the complete restoration of functions lost or impaired after brain injury. However, recovery in this sense is rarely achievable for the majority of brain-injured people. Others regard a good measure of recovery as resumption of normal life, even though there might be minor neurological and psychological deficits that do not disappear over time; this is certainly achievable for some people with nonprogressive brain damage. Yet another interpretation of recovery sees it as a diminution of impairments in behavioral or physiological functions over time; and experience suggests this is certainly true for the majority of people with brain injury. Observations by those involved in the treatment of brain injury suggest that recovery from nonprogressive brain damage typically involves partial recovery of function together with substitution of function, and can be operationally defined as complete or partial resolution of deficits incurred as a result of an insult to the brain.

People who sustain brain damage from a moderate or severe traumatic head injury (the most common cause of brain damage in people under the age of 25 years) usually undergo some – and often considerable – recovery. This is likely to be fairly rapid in the early weeks and months after the injury, followed by a slower recovery which can continue for many years. A typical pattern is an initial period of coma (the patient makes no verbal response, does not obey commands or open the eyes spontaneously or after stimulation), followed by posttraumatic amnesia (PTA), a period in which the patient is confused and disoriented, suffers from retrograde amnesia, and seems to lack the capacity to store and retrieve new information. The next stage is when the patient emerges from PTA, possibly with a number of motor, cognitive, and behavioral problems; these may resolve or partially resolve over time. Variations on this pattern may occur in other kinds of nonprogressive brain injury such as encephalitis (a viral infection of the brain), hypoxia (brain damage caused through shortage of oxygen), or cerebral vascular event.

A number of factors influence the extent of recovery, some of which we can do nothing about once the damage has occurred. These include the age of the person at the time of injury, the severity of damage, the location of damage, the status of undamaged areas of the brain, and the premorbid cognitive status of the brain. Other factors such as motivation, emotions, and the quality of rehabilitation programs available can be manipulated.

Age, often thought to be an important factor, has a less clear-cut influence than many believe. There appears to be a general belief that younger people recover better from injury to the brain than older people; this is known as the 'Kennard principle' after Kennard (1940), who showed that young primates with lesions in the motor and premotor cortex exhibited sparing and partial recovery of motor function. Even Kennard, however, recognized that such sparing did not always occur and some problems became worse over time for certain younger individuals. Once severity, etiology and other demographic factors are taken into account, age is not always predictive of good outcome and younger people sustaining severe head injury often do worse than older people in terms of behavior problems and social deficits. Age, then, must be regarded as just one factor in the recovery process, to be considered alongside other factors such as whether the lesion is focal or diffuse, the severity of the injury, and the time since acquisition of the function under consideration: for example, someone who has only just learned to read is more likely to show reading deficits than someone who learned to read many years earlier.

One factor thought to interact with brain injury is level of 'cognitive reserve', suggesting that people with more education and higher intelligence may show less impairment than those with poor education and low intelligence. Most clinicians are aware of the fact that any injury of the same severity can produce profound damage in one patient and minimal damage in another. The concept of cognitive reserve has been used in studies of HIV (human immunodeficiency virus) infection and Alzheimer disease, and may also prove useful in understanding recovery from nonprogressive brain injury. A neurologist once said, 'It is not only the kind of head injury that matters but the kind of head.' Support for the idea of cognitive reserve comes from the field of language therapy, where severity of aphasia and site of lesion are not unfailing predictors of improvement.

Individuals with high intelligence and possibly superior education may process tasks in a more efficient way. Consequently, in cases of Alzheimer disease, task impairment may occur later in people with superior cognitive reserve than in those who are less intelligent and less well educated when both groups are matched for severity.

TREATMENT, REHABILITATION, AND OTHER FACTORS THAT INFLUENCE RECOVERY

Before considering intervention, one needs to have some understanding of the natural course of recovery from nonprogressive brain injury. This has been reported in a number of studies (Wilson, 1998). For example, a young girl developed meningococcal meningitis at the age of 14 months and, as a result, became prosopagnosic. When last seen at the age of 11 years and 7 months, there was no change in her prosopagnosia. Another study reported the case of a child who developed viral encephalitis at the age of 9 years, and was left with a visual object agnosia. When the girl was last seen at the age of 16 years, there had been a limited degree of recovery that the authors put down to her intact spatial abilities, which enabled her to compensate for her recognition difficulties. Scans using imaging techniques such as computerized tomography (CT) and magnetic resonance imaging (MRI) showed little change over time, leading the authors to believe that neural plasticity for visual processes is limited.

In contrast, a boy with Sturge–Weber syndrome showed dramatic recovery of language. The boy was mute until he underwent a left hemispherectomy at the age of 9 years. Language functioning then developed, and he was reported to have achieved clearly articulated speech with well-structured and appropriate language. At the age of 15 years he had the language skills appropriate for a child of 8–10 years. Possibly language skills are more likely to recover than other cognitive functions: children who develop hippocampal damage and consequent memory impairments early in life appear to show little, if any, improvement in memory.

These findings are confirmed in adults. Some people with language deficits go on to make a reasonable recovery, unlike those who sustain memory impairments. The patient HM, first reported by Scoville and Milner (1957), is perhaps the most famous amnesic patient in the world and has shown no recovery since his operation to relieve epilepsy in the 1950s. An amnesic musician who survived encephalitis, CW, also showed no recovery of memory functioning over a 10-year period. One study found that about two-thirds of memory-impaired people showed no change in memory functioning 5–10 years following discharge from a rehabilitation center, although most were compensating better.

As well as the nature of the cognitive deficit, the cause of the brain damage probably influences recovery. Thus people with head injuries often do better than people with other diagnoses: for example, a cohort of Second World War servicemen who sustained missile wounds to the brain showed considerable preservation of ability as a group despite some selective impairments related to the specific loci of the lesions. Most people with head injuries do not sustain penetrating wounds. Age and duration of coma are often found to be significant predictors of recovery from closed head injury (although recall the earlier caveats about age). Different patterns of recovery are to be expected following severe head trauma, with some people doing well and others remaining in a minimally conscious state even after several years. In contrast to the results of head injury, people with bilateral surgical lesions, such as HM, and people with encephalitis such as CW, often show less recovery and little change over time. People with hypoxic brain damage also tend to show little improvement. On the other hand, recovery from a cerebrovascular event can be considerable for some people: Taub *et al.* (1993) showed it was possible to improve function in the hemiparetic upper limbs of stroke patients by preventing them using the corresponding unaffected limbs. After two weeks of training with the affected limbs (and immobilization of the unaffected ones), a significant improvement in motor functioning occurred.

According to some research, the recovery of stroke patients can be predicted by attentional skills. It is possible that restitution of functioning following stroke may be possible after small lesions while compensatory procedures are more likely to underlie recovery from larger lesions. We will return to this issue later. Meanwhile, note that in addition to natural recovery, treatment or rehabilitation of people with nonprogressive brain damage can also result in improvement of functioning.

Rehabilitation is a process whereby people disabled by injury or disease work together with professional staff, relatives and members of the wider community to achieve their optimum physical, psychological, social, and vocational wellbeing. Rehabilitation programs may work by teaching people to compensate for their problems, by helping them to learn more efficiently, or by achieving restoration (or partial restoration) of functioning through plasticity or regeneration. In clinical practice there appears to be a tendency to aim for restitution of function in the early days and weeks, with compensatory strategies coming into play when natural recovery stops or slows. Robertson (1999), however, believes that the choice of restitution or compensation as an aim depends on the extent (and possibly the location) of the lesion. He believes that restitution of function is achievable for some people with unilateral neglect, and states that 'there is growing evidence that restitution of basic function may be influenced by appropriate behavioural and cognitive inputs' (p. 691).

It is not clear whether such restitution is possible for other skills. The lack of change in memory function over time has already been mentioned. This may also be true for other cognitive functions, such as object recognition skills and spatial localization abilities. For such patients, compensation techniques taught during rehabilitation may offer the best hope of reducing everyday problems, and plenty of evidence exists to demonstrate that this is possible, as we shall see later.

EVIDENCE OF THE RECOVERY OF SENSORIMOTOR FUNCTIONS

Like other deficits, sensorimotor problems may show spontaneous recovery after nonprogressive brain damage. Zihl (2000) discusses spontaneous recovery of visual field disorders as well as recovery following training. He reports a finding that nearly a third of 41 patients with unilateral homonymous field loss showed recovery over an eight-month period. In several studies of people with cerebral blindness, only about 6% made a complete recovery but a further 67% showed some recovery of vision of varying degrees. Recovery tended to follow a particular pattern, with light perception recovering first, then perception of 'vague' contours with 'foggy vision', and finally perception of objects, faces, and surroundings. An interesting self-report of recovery of a visual field deficit following an occipital stroke is that of Bryan Kolb, a neuropsychologist (Kapur, 1997). Kolb kept a diary of his visual field deficit, his recovery, and his emotional reactions to this over a period of four years.

With regard to the recovery of motor functions, it has been found that the average patient receiving a program of focused stroke rehabilitation performed better than the majority of patients in control groups. The work of Taub *et al.* (1993) has already been mentioned, in which the unaffected upper limbs of hemiplegic stroke patients were immobilized for a period of two weeks while the patients were encouraged to use the affected limbs; considerable recovery was achieved. Furthermore, the recovery was maintained over the subsequent two-year period. Others have found that improvements in function could be obtained long after the natural recovery process was over and the hemiparetic limb had ceased to improve. Taub's team felt that temporary loss of use of a limb resulted in 'learned nonuse': the patients lost the habit of using the limb even though the underlying mechanisms enabling limb use had been repaired.

Some studies suggest that the incidence of visual, tactile, proprioceptive, and motor deficits is higher in patients whose stroke occurred in the right hemisphere rather than the left. This is thought to be due to a greater incidence of attentional deficits in patients with a right-sided injury, attention having an important role in influencing the function of primary sensory and motor circuits, and consequently in recovering function within these circuits after a brain lesion. The functional recovery of 47 brain-damaged patients after a right-hemisphere injury was monitored over a two-year period, and it was found that sustained attention (assessed two months after the stroke using tasks from the Test of Everyday Attention) predicted motor recovery two years after the stroke. In addition, in comparison with a group of patients with left brain damage, those with right brain damage showed less functional ability and poorer attention skills.

Positron emission tomography (PET) has been used to look at the recovery of stroke patients. There appears to be considerable plasticity in the recovery of motor functions and this recovery is mediated by (a) recruitment of cortical areas in the undamaged hemisphere, and (b) extension of specialized areas adjacent to the site of the lesion. Another kind of imaging study looked at a 17-year-old girl who had sustained a unilateral brain injury at birth. Functional MRI revealed that the healthy hemisphere was controlling motor movements via direct ipsilateral corticospinal projections together with the contralateral cerebellum.

Relatively little has been written on the recovery of motor functioning following traumatic brain injury (TBI), although physical impairments are common resulting from both the brain damage and concomitant orthopedic trauma. It is believed that nearly half of people with TBI have physical problems, with about one quarter of these having orthopadic and/or muscular skeletal injuries as well as problems caused by the brain damage. A year after injury, however, most people with TBI are ambulant, and a study of Vietnam war veterans with penetrating head injuries and associated movement disorders found that at follow-up all were independent despite continuing mobility problems. It would appear that motor recovery after TBI follows, to a large extent, a developmental sequence.

EVIDENCE OF THE RECOVERY OF COGNITIVE FUNCTIONING

There appear to be four major approaches to cognitive rehabilitation: cognitive retraining through stimulation or exercises; strategies derived from cognitive–neuropsychological theoretical models; techniques combining methodologies and theories from a number of different fields, particularly behavioral psychology, neuropsychology, and cognitive psychology; and holistic approaches that address social and emotional problems alongside the cognitive ones. The nature of these approaches and an examination of their strengths and weaknesses are addressed in further detail by Wilson (1997).

There is increasing evidence that rehabilitation can improve cognitive functioning. In the field of memory disorders, the method of choice for reducing everyday problems is probably to teach compensatory approaches. Several publications show that people with memory impairments can function independently if they are able to use strategies to bypass their difficulties. Over a 10-year period, JC, a young man who became densely amnesic following a ruptured aneurysm on the left posterior cerebral artery, developed a sophisticated system of memory aids enabling him to live alone, hold down a job, and be totally independent despite remaining severely amnesic. The natural history of the development of his compensatory system shows that he began by writing on scraps of paper a few weeks after his stroke before gradually developing his highly successful system.

Another encouraging study describes the rehabilitation of a young woman who became amnesic following status epilepticus. She was taught to use a personal organizer (datebook), to refer to it regularly, and to monitor a number of daily events. Over a period of several weeks the young woman was able to learn the different sections of the datebook and use this to manage her life. After leaving the rehabilitation center, she worked in a voluntary capacity (still using her system) and eventually was taken on as a paid employee.

Even for people who are unable to return to work, memory compensations can assist independent living. One device that has been helpful to a number of people with memory impairment is a specifically designed alphanumeric pager worn on a belt, which sends out daily reminders. People with memory and/or planning problems all showed statistically significant improvements in achieving everyday target behaviors, such as taking medication, feeding the cat, and preparing meals, when using this pager, in comparison with a baseline period.

In addition to compensations in the form of external memory aids, techniques for improving learning in people with impaired memory are a major part of rehabilitation. It has been demonstrated that people with amnesia learn better if they are prevented from making mistakes during the learning process. Most of us can benefit from trial-and-error learning if we can remember our previous mistakes. People with amnesia, of course, cannot do this. Once a mistake has been made, it may be strengthened or reinforced or be indistinguishable from the correct response. Consequently it is better to prevent the incorrect response being made in the first place. Several studies have applied the 'errorless learning' principle to teaching real-life tasks. Although a number of questions, both clinical and theoretical, remain to be answered about errorless learning, it appears to be more effective than trial-and-error learning when used to impart useful information to people with memory difficulties.

One group of people often considered difficult to treat are those with frontal lobe or executive deficits such as problems with planning, organization, and problem-solving. Even this group, however, have been helped by appropriate rehabilitation. Specific problem-solving training can help such people. Furthermore, the learning of patients with a dysexecutive syndrome and severe behavior problems can be enhanced through the provision of exaggerated feedback on their performance. A stroke patient with problems of attention, planning, distractibility, and perseverative behavior was able to overcome a number of her everyday problems through a combination of use of a pager and a checklist. More recently, a problem-solving strategy called Goal Management Training has been designed to help people with executive deficits and disorganized behavior manage their daily lives. This involves five stages, each corresponding to an important aspect of goal-directed behavior (e.g. 'define the main task', 'list the steps').

Several studies have demonstrated improvement of language functioning in both children and adults using training exercises presented through computer games. In one study children with language impairments had 8–16 hours of training over 20 days. This resulted in marked improvements of the children's ability to recognize sequences of non-speech and speech stimuli. In another study, also using computer-based exercises, language comprehension improved with acoustically modified speech. Computer training was also carried out with 55 adults each of whom had sustained a left-hemisphere stroke resulting in aphasia. A reading treatment (visual matching and reading comprehension tasks) led to better results than computer stimulation (involving nonverbal tasks and games) or absence of treatment.

Numerous studies discuss the treatment of unilateral visual neglect, and a number of these discuss the rehabilitation of patients with visual disorders after brain injury. As with the studies quoted above, substantial evidence accrues to suggest that improvement of cognitive functioning can be achieved through rehabilitation.

MECHANISMS OF RECOVERY

The process of recovery from brain injury is not well understood and probably involves different biological processes. Changes seen in the first few minutes (for example after a mild head injury) presumably reflect the resolution of temporary dysfunction without accompanying structural damage. Recovery after several days is more likely to be due to resolution of temporary structural abnormalities such as edema or vascular disruption, or to the depression of metabolic enzyme activity.

Recovery after months or years is even less well understood. There are several ways this might be achieved, including regeneration, diaschisis, and plasticity. Regeneration in the central nervous system can occur and is more likely to do so early in life, although it does occur in adults. Thus the view held for many years that cerebral plasticity is severely restricted in the adult human brain is no longer credible. Although the limits of neuro-rehabilitation have been significantly influenced by the basic premise that brain cells can never regenerate, this is now known to be false and our horizons may well be extended. What is less clear is the extent to which regeneration can lead to functional gains in coping with real-life problems.

Diaschisis assumes that damage to a specific area of the brain can result in neural shock or disruption elsewhere in the brain. The secondary neural shock can be adjacent to the site of the primary insult or much further away. In either case, the shock follows a particular neural route. Similar to this, but not identical, is the idea of inhibition. In inhibition, however, the shock is more diffuse and affects the brain as a whole. Robertson and Murre (1999) interpret diaschisis as 'a weakening of synaptic connections between the damaged and undamaged sites, contingent on the reduced level of activity in the lesioned area' (p. 547). Because cells in the two areas are no longer firing together, synaptic connectivity between them is weakened and this results in the depression of functioning in the undamaged but partly disconnected remote site.

Plasticity implies anatomical reorganization based on the idea that undamaged areas of the brain can take on the functions subserved by a damaged area. Until recently this idea was discredited as an explanation for recovery in adults, although views are now changing. Robertson and Murre (1999), in a thought-provoking paper, suggest that plastic reorganization may occur initially because of a rapidly occurring alteration in synaptic activity taking place over seconds or minutes, followed by structural changes taking place over days and weeks. The authors focus in particular on people who are likely to show recovery provided they have assistance and rehabilitation. According to Robertson and Murre, some individuals show autonomous recovery, others show little recovery even over a period of years, while others show reasonably good recovery provided they receive rehabilitation. They refer to this as a triage of spontaneous recovery, assisted recovery, and no recovery. Robertson and Murre believe that the strategy of choice for people in the 'no recovery' group is to teach compensatory approaches, and that the 'spontaneous recovery' group do not need rehabilitation as they will get better anyway. Consequently, they focus on the 'assisted recovery' group to address issues about brain plasticity. They also believe that the severity of the lesion relates to this triage, with mild lesions resulting in spontaneous recovery, moderate lesions benefiting from assisted recovery, and severe lesions necessitating the compensatory approach.

Although heuristically useful, this idea may be too simplistic. For example, people with mild lesions in the frontal lobes could be more disadvantaged in terms of recovery than people with severe lesions in the left anterior temporal lobe. The former group might have attention, planning, and organization problems precluding them from gaining the maximum benefit from the rehabilitation on offer, whereas the latter group with language problems could show considerable plasticity by transferring some of the language functions to the right hemisphere. Nevertheless, Robertson and Murre make some interesting arguments and present a model of self-repair in neural networks based on a connectionist model of recovery of function.

One technique that has potential to help us understand the nature of recovery is the employment of imaging techniques. Grady and Kapur (1999) suggest that imaging studies could enable us to measure specific changes occurring in the brain during recovery and allow us to determine whether recovery is the result of reorganization of functional interactions within an existing framework, recruitment of new areas into the network, or plasticity in regions surrounding the damaged area.

A few studies using imaging techniques to look at recovery following brain injury have been reported. Perhaps the first paper in this area reported on changes in regional cerebral blood flow (rCBF) following cognitive rehabilitation for people who had sustained encephalopathy after exposure to toxins. Later in 1997, PET was used to identify the neural correlates of stimulation procedures employed in language rehabilitation in people with dysphasia. Single photon emission computed tomography (SPECT) has been used to evaluate rCBF during recovery from brain injury. Specific changes in rCBF appeared to be related to (a) the location of the injury, and (b) strategies used in cognitive rehabilitation. Continued improvements in three patients were documented in rCBF, functional abilities, and cognitive skills up to 45 months following the injury.

The following year functional imaging was employed to monitor the effects of rehabilitation for unilateral neglect. The brain regions most active after recovery were almost identical to the areas active in control subjects engaged in the same tasks. This would appear to support the view that some rehabilitation methods repair the damaged network and do not simply benefit patients through compensation or behavioral change.

The use of imaging studies is certain to grow in research into recovery from brain injury. To what extent findings from such imaging studies will help us plan or improve rehabilitation remains an open question.

CONCLUSION

The human brain is capable of more plasticity than previously thought, since natural recovery can and does occur over time not only in children but also in adults. In addition to natural recovery there is increasing evidence that rehabilitation can result in improvements of both sensorimotor and cognitive functioning. Despite this, there is considerable variability in the nature and extent of such recovery and there are clearly limits to the amount of recovery and improvement possible in people with non-progressive brain injury. We are now faced with a number of questions. What factors limit the recovery process? Can neurogenesis, for example, lead to cells that can survive in sufficient numbers and integrate in ways to improve everyday functioning? When should rehabilitation programs begin, during the spontaneous recovery process or later? How should we determine whether to aim for plasticity, regeneration, or compensation? Should we be influenced solely by the severity of the lesion, the cognitive function affected, the time after injury, or by all of these?

References

Grady CL and Kapur S (1999) The use of imaging in neurorehabilitative research. In: Stuss DT, Winocur G and Robertson I (eds) *Cognitive Neurorehabilitation: A Comprehensive Approach*, pp. 47–58. New York, NY: Cambridge University Press.

Kapur N (1997) *Injured Brains of Medical Minds: Views from Within*. Oxford, UK: Oxford University Press.

Kennard MA (1940) Relation of age to motor impairment in man and in subhuman primates. *Archives of Neurology and Psychiatry (Chicago)* **44**: 377–397.

Robertson IH (1999) Theory-driven neuropsychological rehabilitation: the role of attention and competition in recovery of function after brain damage. In: Gopher D and Koriat A (eds) *Attention and Performance XVII: Cognitive Regulation of Performance: Interaction of Theory and Application*, pp. 677–696. Cambridge, MA: MIT Press.

Robertson IH and Murre JMJ (1999) Rehabilitation after brain damage: brain plasticity and principles of guided recovery. *Psychological Bulletin* **125**: 544–575.

Scoville WB and Milner B (1957) Loss of recent memory after bilateral hippocampal lesions. *Journal of Neurology, Neurosurgery and Psychiatry* **20**: 11–21.

Taub E, Miller NE, Novack TA *et al.* (1993) Technique to improve chronic motor deficit after stroke. *Archives of Physical Medicine and Rehabilitation* **74**: 347–354.

Wilson BA (1997) Cognitive rehabilitation: how it is and how it might be. *Journal of the International Neuropsychological Society* **3**: 487–496.

Wilson BA (1998) Recovery of cognitive functions following non-progressive brain injury. *Current Opinion in Neurobiology* **8**: 281–287.

Zihl J (2000) *Rehabilitation of Visual Disorders After Brain Injury*. Hove, UK: Psychology Press.

Further Reading

Broman M, Rose AL, Hotson G and Casey CM (1997) Severe anterograde amnesia with onset in childhood as a result of anoxic encephalopathy. *Brain* **120**: 417–433.

Fawcett JW, Rosser AE and Dunnett SB (2001) *Brain Damage, Brain Repair*. New York, NY: Oxford University Press.

Katz RC and Wetz F (1997) The efficacy of computer-improved reading treatment for chronic aphasic adults. *Journal of Speech and Language Hearing Research* **40**: 493–507.

Newcombe F (1996) Very late outcome after focal wartime brain wounds. *Journal of Clinical and Experimental Neuropsychology* **18**: 1–23.

Schiavetto A, Decarie J-C, Flessas J, Geoffroy G and Lassonde M (1997) Childhood visual agnosia: a seven year follow-up. *Neurocase* **3**: 1–17.

Tallal P, Miller SL, Bedi G *et al.* (1996) Language comprehension in language-learning impaired children improved with acoustically modified speech. *Science* **271**: 81–84.

Vargha-Khadem F, Carr LJ, Isaacs E *et al.* (1997) Onset of speech after left hemispherectomy in a nine-year-old boy. *Brain* **120**: 159–182.

Vargha-Khadem F, Gadian DG, Watkins KE *et al.* (1997) Differential effects of early hippocampal pathology on episodic and semantic memory. *Science* **227**: 376–380.

Wilson BA (1999) *Case studies in Neuropsychological Rehabilitation*. New York, NY: Oxford University Press.

Young AW and Ellis HD (1989) Childhood prosopagnosia. *Brain and Cognition* **9**: 16–47.

Brain Self-stimulation

Introductory article

Peter M Milner, McGill University, Montreal, Quebec, Canada

CONTENTS

Most vertebrates and some invertebrates learn to make responses that deliver electrical stimulation to parts of the nervous system. This provides a tool for studying the anatomy and neuropharmacology of reward mechanisms. Knowledge so gained has proved useful for understanding motivation and addiction.

INTRODUCTION

The observation that rats seek electrical stimulation of certain brain areas was made in 1953 by James Olds and Peter Milner, working in the laboratory of Donald Hebb at McGill University. The publication in 1949 of Hebb's influential book, *The Organization of Behavior*, brought a group of scientists to Hebb's laboratory, keen to explore his theories. James Olds was a postdoctoral fellow with a degree in social psychology from Harvard; I was a graduate student with a degree in communication engineering from Leeds University. A year before the discovery Seth Sharpless, a philosophy student from the University of Chicago, and I, seeking to improve on Hebb's account of motivation, performed a pilot study to test our notion that reward involved the recently discovered reticular activating system (RAS). Our experiments indicated, however, that RAS stimulation was punishing. As pain pathways pass through the RAS this outcome was not deemed sufficiently exciting to justify a change of thesis research.

When Olds arrived on the scene it fell to me to show him how to implant electrodes in the rat brain; Olds intended to implant his first practice electrode in the RAS. When the rat recovered from the operation he placed it on a large table and tested the effect of short bursts of stimulation. In the initial experiments the stimulation was a few volts of 60 Hz alternating current from a transformer connected to a power outlet. Most later experiments used brief (> 1 s) trains of short (0.1–1.0 ms) rectangular current pulses of negative polarity. It was soon obvious that the rat found any place where it had been stimulated attractive. If removed from the place it would quickly return, sniffing and searching. Later tests, in which the rat learned to press a lever to deliver electric current to its brain, gave further reason to believe that the stimulation served as a reward.

This was very exciting. I immediately operated on another rat, using the same stereotaxic settings but, as before, without success. Suspecting that the rewarding electrode of the successful rat was not at the intended site we X-rayed its head, confirming our suspicion. Probably the dental cement holding the electrode to the skull had not set before the rat was released from the stereotaxic instrument. Unfortunately, when the rat finally came to autopsy, its brain was too badly damaged for precise localization of the electrode. Based largely on the x-ray, we estimated that the electrode was in the vicinity of the septal area, but after the relation of self-stimulation behavior to electrode site had been established by subsequent mapping studies, it became evident that the behavior exhibited by the first rat had been more similar to that of rats with hypothalamic electrodes than to rats with septal area electrodes. Before he embarked on a systematic search for reward areas in the brain, Olds designed an electrode that was held to the skull by screws rather than by dental cement.

MAJOR CHARACTERISTICS OF SELF-STIMULATION

The characteristics of self-stimulation depend greatly on the site of the electrode, owing in part to concomitant stimulation of paths not related to reward. Rewarding sites have been reported in many areas of the brains of most, if not all, vertebrates that have been tested, and even in some invertebrates. In mammals, stimulation is very rewarding in the vicinity of the medial forebrain bundle (MFB) where it passes through the lateral hypothalamus (LH), a nucleus at the base of the brain involved in appetitive behaviors (Figure 1). Most rats with electrodes in that region quickly learn to press a lever for stimulation. Typically, after a few successful responses, they become excited and assail the lever vigorously, pausing momentarily at intervals to run in a tight circle. Some gnaw on the lever or other parts of the apparatus. If given the opportunity they may continue working at a high rate for as long as 24 h, without sleeping or eating.

Stimulation at some sites in the ventral midbrain (near the MFB) and the ventral pons elicits similar enthusiastic responding, but at most other reward sites, including parts of the frontal cortex, septal nuclei, basal ganglia, thalamus, and amygdala (too lateral to be shown in Figure 1), the results are less dramatic. At these sites, learning to bar-press may take several sessions, and responding is usually intermittent with long pauses. Sometimes the pressing is slow and deliberate and in some cases the rat recoils from the bar after pressing it, as if frightened or hurt, but nevertheless returns for more.

Stimulation at almost all reward sites appears to become aversive if the train lasts for more than a second or two, depending on its intensity. In most experiments the duration of each train of stimulation is determined by a timer, but if it is not, rats almost always release the lever within half a second. If stimulation is turned on automatically, rats will learn to press a lever to turn it off, so the effect is not due to stimulation-induced hyperactivity.

Rewarding brain stimulation can replace food or water reward in most animal learning experiments, but there are some differences. Rats with LH electrodes, for example, quickly learn to press a lever or run through a maze to receive stimulation, but they appear to forget equally quickly. They also show severe 'overnight decrement', initially ignoring, possibly even avoiding, the lever or goal-box for which they showed great enthusiasm the previous day. 'Priming' by a few bursts of stimulation delivered by the experimenter restores the motivation.

MECHANISMS OF SELF-STIMULATION

Quite soon after mapping the reward system, Olds discovered that self-stimulation was profoundly depressed by antipsychotic drugs. Subsequent discovery of dopamine fibers in the MFB led to the theory that stimulation fires these fibers, releasing dopamine in forebrain structures to reinforce synaptic connections. This theory was too simple to be the whole story. Dopamine fibers, which are all unmyelinated, are not easily fired by the short pulses of current commonly used for self-stimulation.

It was next postulated that the current fires myelinated fibers, which then synapse with dopamine neurons in the ventral tegmental area (VTA), thus indirectly increasing the delivery of dopamine to forebrain structures. However, such an increase has proved difficult to measure. Dopaminergic activity seems to increase mainly during unexpected

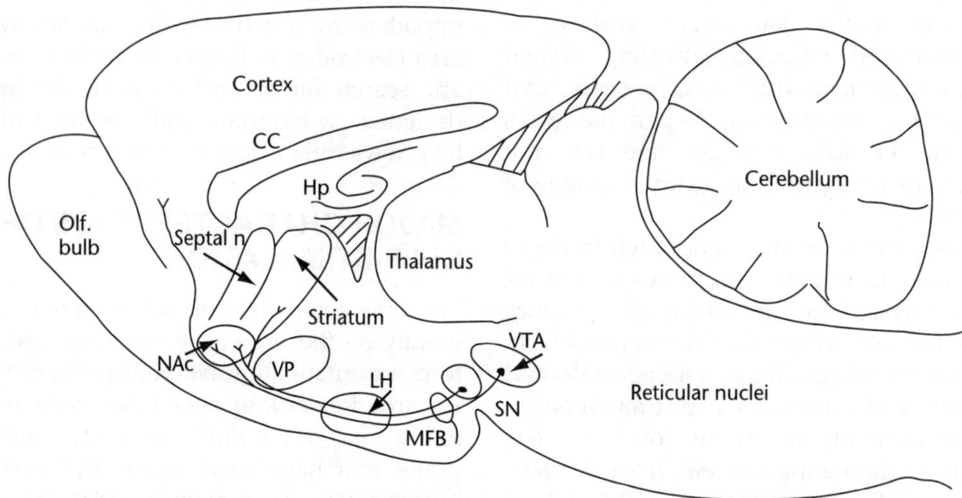

Figure 1. Longitudinal section of rat brain showing the locations of some brain-stimulation reward sites. CC, corpus callosum; Hp, hippocampus; LH, lateral hypothalamus; MFB, medial forebrain bundle; NAc, nucleus accumbens; Olf., olfactory; SN, substantia nigra; VP, ventral pallidum; VTA, ventral tegmental area.

events, which is of course when most learning takes place.

It certainly appears that dopamine is essential for brain stimulation reward from MFB sites; blocking its action in the accumbens nucleus by injecting a small quantity of a dopamine-blocking agent quickly diminishes self-stimulation. The accumbens is a nucleus that until the 1970s was considered part of the septal area, but its cell type and connections indicate that it is functionally part of the ventral striatum. It receives dopaminergic input via the MFB.

Using the fact that there is a rapid increase in the expression of the protein Fos in neurons when they are active, it has been found that self-stimulation of the LH or VTA activates neurons in many subcortical nuclei. Lesions indicate, however, that only a few of them play a vital role in self-stimulation. Some neurons are activated directly by the electrical current, others via synaptic connections from directly stimulated neurons. Some of the neurons that express Fos may be those involved in bar-pressing. An activated area that seems important for reward is the ventral pallidum, which lies just anterior to the LH. Some of its connections are to midbrain nuclei, including those that deliver dopamine to forebrain areas.

Signals from hunger, thirst, and other motivational states are delivered to the ventral part of the striatum, where they elicit appropriate response plans. It is there that incipient response activity can be either suppressed, or amplified to become an overt movement. The amplification circuit, which is modulated by dopamine, normally receives input from innately rewarding stimuli (such as water, or the smell of food) via the hypothalamus or the amygdala. The response suppressing system, which may also be modulated by dopamine, receives innately aversive input (pain or predator smell) mainly via the thalamus or the amygdala.

Innately neutral sensory input reaches the striatum via the cerebral cortex. In the striatum this input may acquire associations (by classical conditioning) with either the response amplifier or the response inhibitor, depending on which type of innate reinforcer influences the ventral striatum at the time. Thus an animal may at first be attracted only by the smell of food in a dish, but in time it is also attracted by the sight of the dish.

Rewarding brain stimulation is assumed to stimulate pathways that would normally be activated by innate rewards such as food, thereby triggering the response release mechanism in the striatum. As a consequence, sensory input and plans for motor activity that reach the striatum during rewarding stimulation acquire the ability to amplify and release a planned response.

It is generally assumed that brain self-stimulation has an effect similar to a conventional reward except that the modulatory effect of bodily need is bypassed. If an animal is not hungry, for example, the smell and taste of food do not amplify response plans, but self-stimulation, by acting directly on the reward system, is not greatly influenced by the intensity of bodily need. Self-stimulation has proved a useful tool for studying the basic mechanism of reward, providing direct access to the anatomical and physiological features of the system.

SELF-STIMULATION AND THE STUDY OF ADDICTION

Harmful addictions are the penalty we pay for an effective motivation system. Understanding the reward system in the brain is highly relevant to the way we approach the problems of drug abuse and other obsessive indulgences. At one time the most widely held view of drug dependency was that aversive after-effects of abused substances are alleviated by taking more of the substance, leading to a vicious spiral. Neither the sites of action nor the pharmacological processes involved were known. After the discovery of brain-stimulation reward, parallels were drawn between the behavior of the self-stimulating rat and that of addicts.

All creatures are in a sense addicted to certain essentials like air, water, heat, and food, all of which can be harmful in excess. Fortunately, protective regulators evolved along with the appetites. Automatic regulation of stimulants that act directly on the reward system is more difficult. Electrical stimulation has a similar direct effect on reward mechanisms, but with the advantage that it is easier to identify the brain structures involved, so that their connections and pharmacological properties may be studied.

When the importance of dopamine for self-stimulation was discovered, dopamine soon became a prime focus of investigations into drug abuse. It was pointed out that enhancers of dopaminergic action, such as amphetamine and cocaine, are addictive. Many dopaminergic neurons are sensitive to other abused substances such as opiates and nicotine.

The knowledge concerning the anatomy of reward systems was helpful in designing experiments to explore addictive processes; the techniques used for electrical stimulation were also adapted for use in these experiments. Rats learn to press a lever to deliver many of the substances that are addictive to humans directly to a point in the brain via implanted cannulae. This makes it possible to determine or confirm the sites of action of a drug.

The dopamine theory of reward changed the direction of addiction research, but like most new theories it proved to be an oversimplification. Opiate receptors, for example, are widespread in the brain and almost certainly affect learning and other systems directly, as well as via dopamine. The mechanism by which dopamine itself influences reward is still not fully understood. Today the incentive for a great deal of the research on self-stimulation is the desire to resolve the problem of drug dependency. This quest, having been reoriented by the discovery of electrical self-stimulation, is now the tail that wags the dog.

SELF-STIMULATION AND THE STUDY OF COGNITION

Although philosophers used to draw a sharp distinction between knowledge and the will, behavior cannot be so clearly compartmentalized. Attention provides a strong link between motivation and perception, for example. Research on brain stimulation reward contributes to cognition by providing a better understanding of motivational mechanisms.

It is probably true to say that most perceptual research and theory during the twentieth century ignored attention. Sensory paths were treated as one-way streets from receptors to some internal representation of a stimulus in memory, or to the response mechanism. The basic function of sensory systems to extend the capacity of the motor system went unnoticed, or was forgotten.

Decisions to make a response are made on the basis of the kind of information we call motivational. Once the decision to act has been established, motor adjustments required to implement it are guided by the sort of stimuli that are most usually studied in perceptual experiments: the shape, color, distance and so on of objects. Of the very large number of stimuli usually present in the environment, the perceptual systems select for processing only those that pertain to the task in hand. In order to study the mechanism by which this is achieved it is necessary to understand how motivating stimuli influence response plans, and how these plans then interact with the reception and transmission of sensory signals.

Incentive learning is another example of the interplay between perceptual input and motivation. Important motivational paths meet sensory input in the striatum, suggesting that it is a place where stimuli that are not innately motivating may acquire value. Patterns of sensory input acquire the ability to release behavior or inhibit it. By the reverse process, motivational signals become associated with sensory patterns that are correlated with innately reinforcing events. Hunger becomes associated with the image of a food dish, for example. It is this image that determines the pattern of stimulation to which the sensory system is sensitized. In other words, reverse pathways in the sensory systems carry attentional facilitation. They can lower the perceptual threshold for goal-related stimuli and the motor threshold for intention-related actions. Thresholds for distracting stimuli are raised.

Further Reading

Bozarth MA (1987) Ventral tegmental reward system. In: Engel J and Oreland L (eds) *Brain Reward Systems and Abuse*, pp. 1–17. New York, NY: Raven Press.

Gallistel CR, Shizgal P and Yeomans JS (1981) A portrait of the substrate for self-stimulation. *Psychological Review* **88**: 228–273.

Milner PM (1989) The discovery of self-stimulation and other stories. *Neuroscience and Biobehavioral Review* **13**: 61–67.

Milner PM (1991) Brain-stimulation reward: a review. *Canadian Journal of Psychology* **45**: 1–36.

Milner PM (1999) *The Autonomous Brain*. Mahwah, NJ: Lawrence Erlbaum.

Olds J (1973) Commentary: the discovery of reward systems in the brain. In: Valenstein ES (ed.) *Brain Stimulation and Motivation*, pp. 81–99. Glenview, IL: Scott, Foresman.

Olds J and Milner PM (1954) Positive reinforcement produced by electrical stimulation of septal area and other regions of rat brain. *Journal of Comparative and Physiological Psychology* **47**: 419–427.

Rolls ET (1975) *The Brain and Reward*. Oxford, UK: Pergamon Press.

Shizgal P and Murray B (1989) Neuronal basis of intracranial self-stimulation. In: Liebman JM and Cooper SJ (eds) *The Neuropharmacological Basis of Reward*, pp. 106–163. Oxford, UK: Clarendon Press.

Vaccarino FJ, Schiff BB and Glickman SE (1989) Biological view of reinforcement. In: Klein SB and Mowrer RR (eds) *Contemporary Learning Theories: Instrumental Conditioning Theory and the Impact of Biological Constraints on Learning*, pp. 111–142. Hillsdale, NJ: Lawrence Erlbaum.

White NM and Milner PM (1992) The psychobiology of reinforcers. *Annual Review of Psychology* **43**: 443–471.

Wise RA, Bauco P, Carlezon WA and Trojniar W (1992) Self-stimulation and drug reward mechanisms. *Annals of the New York Academy of Sciences* **654**: 192–197.

Yeomans J (1988) Mechanisms of brain-stimulation reward. *Progress in Psychobiology and Physiological Psychology* **13**: 227–266.

Brazier, Mary A. B.

Introductory article

Mary Brown Parlee, Massachusetts Institute of Technology, Cambridge, Massachusetts, USA

CONTENTS

Mary A. B. Brazier was an authority on electroencephalography who after World War II pioneered the use of correlational techniques and high-speed computers to study brain activity and behavior. She also authored influential books and articles on the history of neurophysiology which stimulated historical interest in the brain and behavioral sciences.

INTRODUCTION

Mary A. B. Brazier (1904–1995), an authority on electroencephalography, pioneered the use of computers to study brain physiology and brain–behavior relationships. She was also an influential editor, organizational leader and historian of neurophysiology (Figure 1). In collaboration with colleagues at Harvard University and the Massachusetts Institute of Technology, she was the first to use modern high-speed computers and correlational techniques for frequency analysis to study the electrical activity of the brain in relation to behavior. Author of a classic textbook, *The Electrical Activity of the Nervous System* (see Further Reading section), she edited several volumes of Macy Foundation-sponsored conferences on brain and behavior. These brought the latest research to the attention of a broad scientific readership at a time (the late 1950s and early 1960s) when the modern cognitive and neurosciences were emerging as new sciences from their interdisciplinary roots. Brazier's influence on these fields continued through her editorship (1975–84) of the journal *Electroencephalography and Clinical Neurophysiology* (of which she was a founder in 1949) and through her leadership in the International Brain Research Organization (IBRO) and other national and international scientific organizations concerned with brain and behavior. She also authored numerous articles and two books on the history of neurophysiology in the seventeenth, eighteenth and nineteenth centuries. Prior to her move in 1940 from England to the USA, where she made her reputation as a brain researcher and historian of science, Brazier conducted award-winning research in endocrinology. In the USA she was usually the first and only woman in the post-World War II scientific circles of neurophysiologists, neurologists, electrical engineers and mathematicians among whom she traveled and thrived.

LIFE AND WORK

Mary Agnes Burniston Brown, known as 'Mollie' to her friends, was born in Weston-super-Mare, near Bristol, England on 17 May 1904, the second of two children (she had an older brother) in a Quaker family. As a child she developed what would be lifelong loves of the sea and of science, and after attending school at Sidcot she entered the all-women Bedford College of the University of London, where she studied physiology, obtaining her BSc in 1926. In 1927 she took up an appointment as a Research Fellow in the Imperial College of Science and Technology, working with Frederick Golla at the Maudsley Hospital on separation of the products of protein hydrolysis. She continued this research for more than a decade, obtaining a PhD in biochemistry from the University of London in 1930. In 1928 she married electrical engineer Leslie J. Brazier, with whom she had a son, Oliver.

Figure 1. Mary A. B. Brazier (1904–1995). Photo supplied courtesy of the History and Special Collections Division of the Louise M. Darling Biomedical Library, UCLA.

Mary Brazier's research at the Maudsley Hospital focused on measurement of the electrical changes in the skin which occurred in patients with diseases of the thyroid gland. The technique for measuring these changes, which Brazier named the *impedance angle test*, proved useful for diagnosing thyrotoxicosis. It led to her receiving research awards in 1934 from the (British) Institute of Electrical Engineers and the American Association for the Study of Goiter. A few years later, also at the Maudsley Hospital, research by W. Grey Walter demonstrated the diagnostic usefulness of another electrophysiological measure, namely the electroencephalogram (recordings of electrical activity of the brain through the unopened scalp and skull of humans). EEGs could be used to locate tumors of the brain. This finding, together with a report that petit mal epilepsy was associated with changes in the EEG, generated considerable interest in EEGs and their potential clinical applications. References to these and other articles – more than 100 are covered – can be found in Brazier's 1958 account of the development of concepts relating to electrical activity of the brain (see Further Reading).

In 1940, with London under heavy attack, Mary Brazier and her son moved to Boston, while Leslie Brazier remained in England. With a Rockefeller fellowship she secured an appointment as a neurophysiologist at Massachusetts General Hospital (MGH), in Harvard Medical School's Department of Psychiatry, headed by Stanley Cobb (she later held appointments in the Departments of Anesthesia and Neurology). One of the earliest EEG laboratories in the USA had been established at MGH in 1937 by Robert Schwab, and Brazier soon found her way there. She remained at MGH, later heading her own Neurophysiological Laboratory, for the next 20 years. During the war she collaborated with her new colleagues in national defense research on a variety of topics, including peripheral nerve injuries, aircraft pilot selection, war neuroses, electromyograms and muscle dysfunction in poliomyelitis. Her first publication on the subject of EEGs appeared in 1942, and during the next 7 years such research became the focus of her work, including the characteristics of normal EEGs: comparison of the EEGs of psychoneurotic patients and normal adults; and the effects of blood sugar levels, anoxia and various anesthetic agents on EEGs.

The first International EEG Congress was held in London in 1947, and there Brazier was a founding member of the International Federation of Societies for Electroencephalography and Clinical Neurophysiology. In 1950, her comprehensive 178-page *Bibliography of Electroencephalography, 1875–1948* (see Further Reading) was published as the first supplement to the novel field's new (two-year-old) journal, *EEG and Clinical Neurophysiology*. It became, as three of her long-time friends and colleagues later put it, 'a guiding beacon to the then newcomers to the field'. The same would prove true of Brazier's 1951 textbook, *The Electrical Activity of the Nervous System* (see Further Reading), which went through four editions and was translated into seven languages. During this period (the late 1940s to early 1950s), in addition to her research, writing and organizational work, Brazier began to develop her thinking about the relationships between nervous activity, consciousness and behavior.

Until the end of the 1940s, most research on EEGs involved 'eyeballing' the data, and it became apparent to Brazier and others that more precise and reliable techniques were needed. In 1946, Schwab and Brazier persuaded Grey Walter to visit Boston with the automatic low-frequency analyzer he had developed for use in his EEG research. He demonstrated the Walter analyzer at MGH at a meeting of the Eastern Association of Encephalographers, and the analyzer remained in the MGH Clinical EEG Laboratory. It was used there by

Brazier and others to analyze human EEGs until the early 1950s, when it was superceded by new technologies that were developed in the post-World War II ferment of excitement about cybernetics, signal analysis and communication (information) theory.

This ferment was particularly intense in the Boston and Cambridge area, with MIT mathematician Norbert Wiener at its epicenter. After the war, Wiener published two influential books based largely on his wartime work on prediction theory, namely *Cybernetics or Control and Communication in the Animal and the Machine* (published in 1948) and *Extrapolation, Interpolation and Smoothing of Stationary Time Series* (published in 1949). Scientists and engineers in a wide range of fields regarded these texts as providing mathematically specifiable concepts (e.g. information, feedback, communication system, signal/noise) which could be used to analyze complex systems of all kinds (machine, biological, human–machine, social), including the central nervous system. Wiener's influence was particularly strong at MIT's interdepartmental Research Laboratory of Electronics (RLE), where in 1951 Walter Rosenblith established a Communications Biophysics Group to investigate sensory (primarily auditory) systems in animals and humans using the latest engineering technologies.

Beginning in 1948, when Brazier invited Wiener to speak to a group of MGH researchers in psychiatry and physiology, she became increasingly interested in using mathematical techniques developed by Wiener (autocorrelation and cross-correlation analyses, used to detect the presence of a weak signal embedded in noise) to analyze EEGs. These correlational techniques could be used both to identify naturally occurring rhythms in the brain's 'resting' state and to detect responses evoked by sensory stimulation, making them useful general tools for exploring questions about brain functioning and behavior. Stimulated by Wiener's work and his new-found interest in EEGs, Brazier and James Casby, an MIT undergraduate working in Brazier's MGH laboratory, published a paper on 'cross-correlation and autocorrelation studies of electroencephalographic potentials' in 1952. Brazier and Wiener discussed the application of these techniques to EEGs at the Third International Congress of Electroencephalography and Clinical Neurophysiology, which was held in Cambridge, Massachusetts in 1953.

'Application' of the mathematical techniques in EEG research required instrumentation, in particular devices for filtering the electrical activity recorded through EEG electrodes and 'correlators', namely machines for calculating the cross-correlations (between a stimulus presentation and brain activity, over repeated stimulus presentations) and autocorrelations (between a segment of the recorded activity and the same segment overlaid but displaced in time). (The latter technique enables the detection of cycles of unknown periodicities and small amplitudes relative to background activity.) Brazier's MGH group, which after 1951 included the then third-year Harvard Medical School student John Barlow, began a mutually fruitful collaboration with Walter Rosenblith's Communications Biophysics Group. An analog electronic correlator had been developed at the RLE in the late 1940s, and reliable, high-speed digital electronic computers were being developed by computer designers at MIT's Lincoln Laboratory, which had close ties to Rosenblith's laboratory. Both analog and digital machines could be used to perform the correlations required for the quantitative analyses of EEGs which Brazier sought, and she used both in research published in the early 1950s. When the first of a new generation of general-purpose digital computers was developed at Lincoln Laboratory in the mid-1950s (the TX-0), one of its earliest applications was in a device called the Average Response Computer (ARC-1). The latter was used in RLE's Communications Biophysics Laboratory by Rosenblith, Brazier, Barlow, Nelson Kiang and others to analyze electrophysiological recordings of nervous system activity (including EEGs). A technically detailed, first-hand account of the early history of EEG data processing by the MIT–MGH collaborators was published by Barlow in 1997 (see Further Reading). Brazier herself published a more extended overview with illustrations from her experiments on animals, normal human subjects, and neurological patients.

Interest in electroencephalography as a technique for investigating brain activity and consciousness continued to grow during the 1950s. Beginning in 1953, after Stalin's death, international exchanges between brain scientists in the West and in the former Soviet Union generated considerable excitement. Brazier participated in one of the earliest of these, namely the 1953 Laurentian Symposium on Brain Mechanisms and Consciousnesss (held in Ste Marguerite, Quebec), and in the subsequent Moscow Colloquium on Electroencephalography of Higher Nervous Activity, to which Brazier was invited as one of five US scientists. From the Moscow Colloquium, plans emerged for the formation of the International Brain Research Organization (IBRO), the first international organization to encompass all of the areas of what are now termed the neurosciences, including investigations of brain–behavior relationships. The first meeting was held in 1960, and Brazier served as Secretary General of IBRO from 1978 to 1982. Throughout her career Brazier actively promoted internationalism in brain research, not only through IBRO, but by serving successively as Treasurer, Secretary and President of the International Federation of Societies for Electroencephalography and Clinical Neurophysiology between 1953 and 1965. After her death, the International Federation established a scientific award in clinical neurophysiology in her honor, the M.A.B. Brazier Young Investigator International Award.

The objective of Brazier's EEG research, in the words of long-time collaborators and friends, 'was to try to understand the nature of the EEG, as reflected in its statistical properties, as a signal in a communication system, i.e. the brain'. Throughout her career, Brazier continued to consider the implications of cybernetics and communication theory for brain research, sometimes contrasting models drawn from information theory with classical approaches ('deterministic models') in ways that presaged neural network models by many years. Brazier's interactions with Soviet brain scientists also influenced her research, and in the late 1950s and early 1960s she used correlational techniques to investigate EEGs in relation to the orienting, conditioning and habituation responses (phenomena of considerable interest to brain scientists working in the Pavlovian tradition). Brazier's bibliography of EEG research, published in 1950, had shown an interest in the historical background of her science, and in the 1950s she began to publish articles about the history of neurophysiology. In the 1980s she published her landmark books, *A History of Neurophysiology in the Seventeenth and Eighteenth Centuries* and *A History of Neurophysiology in the Nineteenth Century* (see Further Reading).

In 1958, the Macy Foundation sponsored the first of two invited Conferences on the Central Nervous system and Behavior, organized by Horace Magoun of the University of California at Los Angeles (UCLA). Magoun had been instrumental in establishing a brain and nervous system research unit when UCLA's School of Medicine was organized in 1950, and the unit was established as the Brain Research Institute (BRI) in 1959. Brazier's skillful editing of these conference proceedings led to a continuation of the series – now called Brain and Behavior Conferences – under the auspices of the American Institute of Biological Sciences, again with Magoun as organizer and Brazier as editor. These interdisciplinary conferences had international participation (including distinguished Soviet and East European brain scientists), and their rapid publication brought the latest research in the brain sciences to broad scientific audiences. Concurrently, between 1961 and 1965 the US Air Force Office of Scientific Research sponsored conferences on brain function. These, too, were organized by Magoun, edited by Brazier, and influential in the new field that was becoming known as 'neuroscience'. In 1961, Brazier left her MGH laboratory and her long-time MIT collaborators to join Magoun, Donald Lindsley, Louise Marshall and others at the BRI. Always impossible to pigeon-hole in a neat grid of established disciplines, Brazier was appointed professor in the Departments of Anatomy, Biophysics and Nuclear Medicine, and Physiology of UCLA's Medical School. Prior to her arrival, computers were not widely used by BRI researchers, and Brazier, who served on a National Institutes of Health (NIH) advisory committee on computers in research, led the development of a data-processing facility. Her EEG research continued at the BRI, expanding to include new clinical applications. She was active in university-wide affairs at UCLA, and also served as a consultant or advisor on several NIH and National Science Foundation committees, as well as those of national non-governmental organizations.

In 1988, with her eyesight failing, Brazier moved back to the East Coast, to the sea and to the house she had built on Cape Cod while she was living in Boston. There she gardened, sailed, traveled (at least once every year to Paris and London) and visited and corresponded with friends and family – her son Oliver and his family were nearby. She died on 9 May 1995, nine days before what would have been her ninety-first birthday.

HONORS AND AWARDS

Mary A. B. Brazier was elected to the American Academy of Arts and Sciences in 1956. In 1962 she received a Career Research Award from the National Institutes of Health, one of four scientists so honored in the first year of these awards. The University of London honored her with a doctorate (DSc, on the basis of her published works) in 1960, and she received an MD (honoris causa) from the University of Utrecht in 1976. In 1985, the British EEG Society awarded her the Grey Walter Medal. The most complete and detailed account of Brazier's life and work to date is the 1996 memorial tribute written by John Barlow, Robert Naquet and Hans van Duijn (see Further Reading).

Brazier's papers are archived in the History and Special Collections Division of the Louise M. Darling Medical Library, UCLA. Contemporary researchers are fortunate that her published work (she was the author or editor of almost 250 articles and books) speaks clearly for itself. She wrote well, and her sense of history – even when she was working within the conventional genres of scientific articles and chapters – enabled her to place her own work in an unusually broad and detailed scientific context. Reading Brazier's writings, both scientific and historical, will give contemporary cognitive and brain scientists a richer and deeper understanding of their field and its conceptual and technical roots.

Acknowledgments

Dr John Barlow has very kindly allowed me to see some of the extensive correspondence he had with Mary Brazier's friends and colleagues while he was preparing the 1996 memorial tribute to her life and work. I thank him for doing so.

Further Reading

Barlow JS (1997) The early history of EEG data-processing at the Massachusetts Institute of Technology and the Massachusetts General Hospital. *International Journal of Psychophysiology* **26**: 443–454.

Barlow JS, Naquet R and van Duijn H (1996) In memoriam: Mary A.B. Brazier (1904–1995). *Electroencephalography and Clinical Neurophysiology* **98**: 1–4.

Brazier MAB (1950) Bibliography of electroencephalography, 1875–1948. *EEG and Clinical Neurophysiology* **Supplement 1**: 1–178.

Brazier MAB (1950) Neural nets and integration. In: Richter K (ed.) *Perspectives in Neuropsychiatry*, pp. 35–45. London, UK: HK Lewis.

Brazier MAB (1951) *The Electrical Activity of the Nervous System*. London, UK: Pitman.

Brazier MAB (1954) The Laurentian Symposium on the electrical activity of the cortex as affected by the brainstem reticular formation in relation to states of consciousness. *EEG and Clinical Neurophysiology* **6**: 355–359.

Brazier MAB (1958) The development of concepts relating to the electrical activity of the nervous system. *Journal of Nervous and Mental Diseases* **126**: 303–321.

Brazier MAB (1959) The historical development of neurophysiology. In: Field J, Magoun HW and Hall VE (eds) *Handbook of Physiology – Neurophysiology*, vol. 1, pp. 1–58. Washington DC: American Physiological Society.

Brazier MAB (1960) Some uses of computers in experimental neurology. *Experimental Neurology* **2**: 123–143.

Brazier MAB (1960) Long-persisting electrical traces in the brain of man and their possible relationship to higher nervous activity. *EEG and Clinical Neurophysiology* **Supplement 13**: 347–358.

Brazier MAB (1963) How can models from information theory be used in neurophysiology? In: Fields WS and Abbott W (eds) *Information Storage and Neural Control*, pp. 230–242. Springfield, IL: Thomas.

Brazier MAB (1984) *A History of Neurophysiology in the Seventeenth and Eighteenth Centuries: From Concept to Experiment*. New York, NY: Raven.

Brazier MAB (1988) *A History of Neurophysiology in the Nineteenth Century*. New York, NY: Raven.

French JD, Lindsley DB and Magoun HW (1984) *The Brain Research Institute, UCLA: An American Contribution to Neuroscience*. Los Angeles, CA: UCLA Brain Research Institute.

Marshall LH (1996) Early history of IBRO: the birth of organized neuroscience. *Neuroscience* **72**: 283–306.

Rosenblith WA (1966) From a biophysicist who came to supper. In: *R.L.E.: 1946 + 20*, pp. 42–50. Cambridge, MA: Research Laboratory of Electronics, Massachusetts Institute of Technology.

Wiesner JB (1966) The communication sciences – those early days. In: *R.L.E.: 1946 + 20*, pp. 12–16. Cambridge, MA: Research Laboratory of Electronics, Massachusetts Institute of Technology.

Broca, Paul

Intermediate article

Stanley Finger, Washington University, St Louis, Missouri, USA

CONTENTS

Paul Broca (1824–1880) was a French surgeon, pathologist, anatomist and anthropologist. He is remembered for localizing speech in the frontal lobes, recognizing cerebral dominance, and performing the first surgery based on localization, as well as his research on the brain and intellect, and for his insights and theories on ancient trepanned skulls.

INTRODUCTION

Paul Broca (Figure 1) was born in France on 29 June 1824 in Sainte-Foy-la-Grande, a small town near Bordeaux, to Protestant parents. Following in the footsteps of his father, he opted to study medicine. This decision led him to Paris, where he excelled in his studies and completed his medical degree in 1848. From the beginning he had a reputation for being extremely thoughtful, thorough, and for looking at medical and scientific problems from many perspectives. He was also considered a liberal – not by today's standards, but in the culture in which he lived. Broca first made a name for himself by showing that cancer cells can be spread through the blood, and with his studies on various diseases, including muscular dystrophy and rickets. His early research in pathology, coupled with his strong belief that laboratory and clinic must join forces to improve medicine, helped him to secure several desirable Paris hospital appointments, such as surgeon at the Bicêtre.

SPEECH AND THE FRONTAL LOBE

In 1859 Broca founded the world's first anthropological society, the Société d'Anthropologie. It was at the meetings of this fledgling society, where Broca served as secretary, that scientists discussed human groupings, intelligence, and the brain. It was also here that some French physicians interested in the effects of brain damage started to make the case for cortical localization of function. Two such individuals were Jean-Baptiste Bouillaud and his son-in-law Simon Alexandre Ernest Aubertin. They contended that damage towards the front of the cerebrum is more likely to disrupt speech than damage towards the back of the massive cerebral hemispheres. Others, however, including Pierre Gratiolet, disagreed, arguing that the cerebral hemispheres function as an indivisible unit.

Figure 1. Paul Broca (1824–1880).

On 12 April 1861 a 51-year-old man suffering from cellulitis and gangrene was transferred to Broca's surgical service at the Bicêtre. Described as mean and vindictive, the patient (named Leborgne) had suffered from epilepsy since youth and had been hospitalized at the age of 31, after losing his power to speak. He then developed a paralysis on the right side of his body with loss of sensitivity on the same side. Broca invited Aubertin to examine Leborgne with him, to assess his speech and to see if he would exhibit damage to his frontal cortex when he died. Indeed, the patient was found to have great difficulty speaking. After he died an autopsy showed a chronic softening in the third frontal convolution, near the rolandic fissure of the left hemisphere. Broca first presented this patient's brain to the Société d'Anthropologie. A more detailed report was given to the Société d'Anatomie later in 1861. Broca used the French word *aphemie* (or aphemia) for Leborgne's inability to speak. It was Armand Trousseau who coined the more popular word 'aphasia' in 1864. This form of aphasia became known as Broca aphasia or motor aphasia. As for Monsieur Leborgne, since 'tan' was once of the few sounds he was able to make before he died, he was often to be called Tan in the later literature.

Broca used this case to argue for a special frontal cortical area that is responsible for fluent speech. He went out of his way to explain that this is located behind and below the one proposed by phrenologists, such as Gall and Spurzheim, who associated character and abilities with bumps on the skull and whose theories were in disrepute. Today we refer to this specialized cortical region as 'Broca's area'. Hence, two eponyms that helped make Broca famous stemmed from this one case. With the landmark case of Leborgne, Broca became fully committed to the cortical localization revolution and was looked upon as its champion. As a careful investigator, however, he worried that he might have gone too far on the basis of only a single case study. Over the next few years he was gladdened to find additional cases that were supportive of his frontal lobe localization for speech, beginning with the case of an old man named Lelong later in 1861.

CEREBRAL DOMINANCE

On 2 April 1863 Broca lectured about eight cases of loss of fluent speech. He remarked that, to his surprise, all exhibited lesions of the left hemisphere. Still, he felt that more cases were needed before he could make a definitive statement about the left hemisphere being special. The idea seemed likely to generate even more of a storm than cortical localization of function.

Broca's clearest statements and most important thoughts about cerebral dominance appeared in 1865, in an article in the *Bulletin de la Société d'Anthropologie*. In this he theorized that the left hemisphere is, in fact, dominant for language. Because it matures faster than the right hemisphere, it is better suited to take the lead. As was true with the concept of cortical localization of function, Broca was not the first to argue for this new way of looking at the brain, although his role was significant. That honor goes to Marc Dax, a physician from the south of France who wrote a memoir on the left hemisphere and speech in 1836, but never published it (he died a year later). Whether his material had indeed been presented orally at a congress in 1836, as was claimed by his son Gustave, is not certain. Broca himself could find no evidence that it was. What we do know is that the Marc Dax paper was sent to the Académie de Médecine in Paris in 1863. It was a part of a larger report by Gustave Dax which contained his own collection of cases supportive of cerebral dominance. The paper arrived and was announced by title (but not made public) just before Broca made his first tentative remark about the eight cases with left hemispheric lesions. The Dax report was then sent to a 'secret' committee, where it languished before it was severely criticized by the committee chairman (Lelut) late in 1864. Upset by the Lelut report, Gustave Dax then saw to it that both his father's report and his own data were published elsewhere. They appeared as two short but separate articles in a medical periodical in 1865, just weeks before Broca's own celebrated paper appeared in a different journal.

Today, both the Daxes and Broca are recognized for their seminal contributions to the development of the concept of cerebral dominance. For a while, however, Broca was given most – if not all – of the credit for this important discovery.

AGE, BRAIN DAMAGE AND THERAPY

In his 1865 paper on cerebral dominance, Broca was forced to deal with exceptions to the idea that the center for articulate language resides in the third left frontal convolution. One such case was that of a woman with epilepsy who was a patient at the largest Paris hospital, the Salpêtrière. This patient was probably born without a left Broca's area, but was able to learn to read, speak fairly well, and express her ideas without difficulty. Broca suggested that the healthy right hemisphere had taken over the role of the compromised left hemisphere, something that is accomplished more readily when the brain damage occurs early in life. He also postulated that a small percentage of healthy people might be born 'right-brained'. He then considered the question of why we do not see more sparing and recovery following damage to this part of the brain, postulating that one limiting factor might be that most aphasic patients also suffer from intellectual deficiencies, limiting their ability to relearn (see below). This would be especially likely after strokes and injuries that affected more than just Broca's area in the frontal lobes.

Broca also pointed out that professionals did little to retrain their aphasic patients. He suggested teaching people with aphasia in the same way that a child learns to speak: therapy should begin with sounds of the alphabet, then words, then phrases, and eventually sentences. By working from the simple to the complex, Broca suggested, the right hemisphere might find it easier to take over from its injured counterpart on the left side. He tried speech therapy with one of his own adult patients, who was successful in relearning the alphabet and in working with syllables, but did not do well when it came to constructing longer words. Nevertheless, Broca was optimistic and expressed the hope that others would be able to devote more time to speech therapy than he had been able to do owing to his busy schedule.

SURGERY BASED ON LOCALIZATION

In 1865 Broca was elected president of the Paris Surgical Society and 3 years later he became professor of clinical surgery. In 1868 he introduced cranial cerebral topography, a technique that uses skull and scalp landmarks to localize underlying parts of the brain. Broca used his new method to open the skull in the right place and drain an abscess in a patient whose speech had become impaired after a closed head injury. Although the operation took place late in the 1860s, it was not reported until 1876. This was probably the first brain surgery to be based on the new theory of cortical localization of function.

INTELLECT, BRAIN AND RACE

Beginning in 1861, Broca also raised the possibility that the frontal lobes might serve executive functions other than speech, including judgment, reflection and abstraction. Indeed, when Leborgne's lesion was spreading throughout the frontal lobes, this patient showed signs of losing his intellect, not just his fluent speech. By arguing that the front of the brain is more 'intellectual' than the back, Broca was able to explain why there was not more relearning and recovery after large frontal lobe lesions that affect speech. He and the other localizationists who accepted this idea also had a good explanation for why some individuals with large skulls were not as intelligent as others with smaller skulls. For example, in 1873 he examined some recently unearthed Cro-Magnon specimens from central France. They had cranial capacities that far exceeded those of the modern French. To Broca and his colleagues in Paris, it was not that Cro-Magnon men and women were geniuses; they most certainly were not. Instead, the greater overall size of their crania only reflected the greater development of the more pedestrian back of the brain. Broca made precisely the same point when referring to some old exhumed Basque skulls that were sent north to Paris. They were also large relative to the skulls of modern Parisians, but this too was attributed to growth in the back of the brain, not the intellectual front.

Thus, although Broca initially believed that cranial capacity was a good physical correlate of intellect, he abandoned this view as he learned more about cortical localization of function. In addition, like many others at the time, Broca believed in multiple creations for the different human races. However, once caught up in the Darwinian revolution of 1859, he also embraced evolution, rejecting the older notion that the human groups are fixed entities, and with it the popular belief that only pure races could prosper. Moreover, he found slavery, even for people with small brains and low intelligence, inexcusable and repulsive.

TREPANATION

Broca's interest in trepanned skulls began in 1867 when he was asked to examine an Inca skull with cross-hatched cuts. This unusual cranium had recently been obtained in Peru by American diplomat-archeologist E. George Squier. Broca agreed with Squier that the cuts on the skull had been made on a living person prior to the European conquest, and that this individual had survived the operation by a few weeks. Thanks to Broca's help, this was the first case of trepanation from an ancient culture to be correctly and widely recognized as such.

Broca then became involved with the discovery of much older trepanned skulls in France. Many late Neolithic (New Stone Age) crania that had been trepanned were found, most of which are now estimated to be approximately 5000 years old (the Peruvian skull was judged to be only around 500 years old). Broca visited burial sites and unearthed some cranial specimens himself, but mostly studied the findings presented to him by others, especially one of his associates, Prunières. Broca postulated that that the openings in the Neolithic skulls had been made by scraping with a sharp stone, such as a piece of flint or obsidian, and that the fibrous dura mater covering the brain was left intact. As for the smoothed surfaces where bone had been removed, they were the result of an extended period of healing. He further posited (from one skull in particular) that the operation was probably performed early in life.

During the mid-1870s Broca gave many talks and published a large number of papers on trepanation. One of his goals was to convince people that the holes in many of the unearthed French skulls were not due to accidents, combat, nature, or gnawing animals. Another was to associate the surgery with some sort of therapy and with the primitive mind. Broca held that the openings were not the result of surgical treatment of head wounds, since there would have been more openings over the facial areas, which were carefully avoided. Moreover, he did not see signs of fractures. Instead, he thought it more likely that the operations were done on the living to treat 'internal maladies'. After much thought, and after considering newer anthropological evidence, he suggested that the surgery might have originated as a way to treat benign infantile convulsions, such as seizures caused by fever spikes or teething. These were disorders that primitive people might have attributed to demons. Moreover, the children would have recovered anyway: an illusion of success would have been achieved, and the practice would have spread and perhaps generalized.

Broca's theory about trepanation, demonology and seizure disorders is still widely cited in books and papers on trepanation. Most researchers agree that he was probably on the mark when he suggested that the practice had something to do with medicine, the brain, and abnormal behaviors; but he was wrong to think that the operations were confined to children.

LATER YEARS

Paul Broca made his last statements about speech and the brain in 1877. At this time, he was much more interested in the family of man than in cortical localization of function. In addition, he was intrigued by the limbic lobe, a collection of brain parts then thought to be associated with olfaction, a subject on which he wrote in 1877 and 1878.

Broca died in 1880, only months after he was elected to the French Senate as a representative of science and medicine. He had been a perfectionist and a 'workaholic' who published over 500 books and articles during his intense scientific career. When he succumbed to heart disease, his wife, three children, and scientists and physicians around the world mourned the passing of a man who had contributed monumentally to many fields. In the neural and cognitive sciences he is best remembered for his theory of the cortical localization of speech, his recognition of cerebral dominance, his thoughts about intelligence and the races, and for his discoveries and insights bearing on the ancient practice of cranial trepanation.

Further Reading

Broca P (1861) Remarques sur le siège de la faculté du langage articulé; suivies d'une observation d'aphémie (perte de la parole). *Bulletins de la Société Anatomique (Paris)* **6**: 330–357, 398–407. Translated as 'Remarks on the seat of the faculty of articulate language, followed by an observation of aphemia' in von Bonin G (1960) *Some Papers on the Cerebral Cortex*, pp. 49–72. Springfield, IL: Charles C. Thomas.

Broca P (1865) Sur le siège de la faculté du langage articulé. *Bulletins de la Société d'Anthropologie* **6**: 337–393. Translated as 'Localization of speech in the third left frontal convolution' by Berker EA, Berker AH and Smith A (1986) in *Archives of Neurology* **43**: 1065–1072.

Broca P (1878) Anatomie comparée des circonvolutions cérébrales. Le grand lobe limbique et la scissure limbique dans la série des mammifères. *Revue d'Anthropologie Sériè 2* **1**: 385–498.

Clower WT and Finger S (2001) Discovering trepanation: the contributions of Paul Broca. *Neurosurgery* **49**: 1417–1425.

Dax M (1865) Lésions de la moitié gauche de l'encéphale coïncidant avec l'oubli des signes de la pensée (lu au Congrès méridional tenu à Montpellier en 1836). *Gazette Hebdomadaire de Médecine et de Chirurgie* **2** (series 2): 259–260.

Finger S (1994) *Origins of Neuroscience*. New York, NY: Oxford University Press.

Finger S (2000) *Minds Behind the Brain*, chap. 10, Paul Broca, pp. 137–154. New York, NY: Oxford University Press.

Finger S and Roe D (1996) Gustave Dax and the early history of cerebral dominance. *Archives of Neurology* **53**: 806–813.

Joynt RJ and Benton AL (1964) The memoir of Marc Dax on aphasia. *Neurology* **14**: 851–854.

Schiller F (1992) *Paul Broca: Founder of French Anthropology, Explorer of the Brain*. New York, NY: Oxford University Press.

Stone JL (1991) Paul Broca and the first craniotomy based on cerebral localization. *Journal of Neurosurgery* **75**: 154–159.

Catastrophic Forgetting in Connectionist Networks

Intermediate article

Robert M French, University of Liège, Liège, Belgium

CONTENTS

Unlike human brains, connectionist networks can forget previously learned information suddenly and completely ('catastrophically') when learning new information. Various solutions to this problem have been proposed.

INTRODUCTION

The connectionist paradigm in artificial intelligence came to prominence in 1986 with the publication of Rumelhart and McClelland's two-volume collection of articles entitled *Parallel Distributed Processing: Explorations in the Microstructure of Cognition*. Some 20 years earlier, research on an elementary type of neural network, known as perceptrons, had come to a sudden halt in 1969 with the publication of *Perceptrons*, Minsky and Papert's careful mathematical analysis of the capacities of a particular class of single-layered perceptrons. Minsky and Papert's work demonstrated a number of fundamental theoretical limitations of elementary perceptrons. Multi-layered perceptrons and new learning algorithms, which overcome these limitations, were developed over the course of the next two decades. These new networks were able to do many tasks that presented serious problems for traditional symbolic artificial intelligence programs. For example, they were able to function appropriately with degraded inputs, they could generalize well, and they were fault-tolerant. In the late 1980s there were many attempts to apply these networks to tasks ranging from underwater mine detection to cognition, from stock market prediction to bank loan screening.

At the end of the 1980s, however, a problem with these multi-layered networks came to light. McCloskey and Cohen (1989) and Ratcliff (1990) showed that the very property – namely, using a single set of weights as the network's memory – that gave these networks such power caused an unsuspected problem: catastrophic interference. Grossberg (1982) had previously cast this problem in the more general context of 'stability–plasticity'. In short, the problem was to determine how to design network architectures that would be sensitive to new input without being overly disrupted by it.

Catastrophic interference occurs when a network has learned to recognize a particular set of patterns and is then called upon to learn a new set of patterns. The learning of the new patterns modifies the weights of the network in such a way that the previously learned set of patterns is forgotten. In other words, the newly learned patterns suddenly and completely – 'catastrophically' – erase the network's memory of the previously learned patterns.

CATASTROPHIC VERSUS NORMAL FORGETTING

Catastrophic forgetting is significantly different from normal human forgetting. In fact, catastrophic forgetting is almost unknown in humans. Barnes and Underwood (1959) conducted a well-known series of experiments in human forgetting. Subjects begin by learning a set of paired associates *A–B*, each consisting of a non-word and a word (e.g., *pruth–heavy*). Once this learning is complete, they learn to associate a new word with each of the original non-words (*A–C*). At various points during the learning of the *A–C* pairs, they are asked to recall the originally learned *A–B* associates. McCloskey and Cohen (1989) conducted a similar experiment using addition facts on a standard connectionist network. After five learning trials in the *A–C* condition, the network's knowledge of the *A–B* pairs had dropped to 1 per cent, and after 15 trials it had disappeared. The newly learned pairs had catastrophically interfered with the previously learned pairs.

The problem for connectionist models of human memory – in particular, for those models with a single set of shared multiplicative weights (e.g. feedforward back propagation networks) – is that catastrophic interference is hardly observed in humans. This raises a number of issues of significant practical and theoretical interest. Arguably, the most important issue for cognitive science is understanding how the brain manages to overcome the problem of catastrophic forgetting. The brain is, after all, a distributed (or partially distributed) neural network, yet it does not exhibit anything like the catastrophic interference seen in connectionist networks. What neural architecture allows the brain to overcome catastrophic interference, and what characteristics of neural networks in general will allow them to overcome it?

At present, in order to avoid catastrophic interference, most connectionist architectures rely on learning algorithms that require the network to cycle repeatedly through all the patterns to be learned, adjusting the weights by a small amount at a time. After many cycles (called 'epochs') through the entire set of patterns, the network will (usually) converge on an appropriate set of weights for the set of patterns that it is supposed to learn. Humans, however, do not learn in this way. In order to memorize ten piano pieces, we do not play each piece once and then cycle repeatedly through all the pieces until we have learned them all. We learn piano pieces – and just about everything – sequentially. We start by learning one or two pieces thoroughly, then learn a new piece, then another, and so on. If a standard connectionist network were to do this, each new piece learned by the network would probably erase from its memory all previously learned pieces. By the tenth piece, the network would have no recollection whatsoever of the first piece. In order for connectionist networks to exhibit anything like human sequential learning, they must overcome the problem of catastrophic interference.

MEASURES OF CATASTROPHIC INTERFERENCE

The two most common measures of catastrophic interference are known as 'exact recognition' and 'relearning'. In both cases, the network is first trained to a given level on a set of patterns. It is then given a second set of patterns to learn. Once it has learned this second set of patterns, we use one of the two measures of forgetting to determine the effect on the original learning of having learned the second set of patterns. The exact recognition measure depends on the percentage of the original patterns that can still be recognized by the network. (We give the input part of the pattern to the network; and if it produces the correct output, it is considered to have 'recognized' the pattern.) The relearning measure, first proposed by Ebbinghaus for human memory in the late nineteenth century, depends on how long it takes the network to relearn the originally learned patterns. Thus, even if the rate of exact recognition is very low, the knowledge might lie 'just below the surface' and the network might be able to relearn it very quickly. The studies by McCloskey and Cohen (1989) and Ratcliff (1990) used an exact recognition measure. A study by Hetherington and Seidenberg (1989) using the relearning measure showed that, at least in some cases, catastrophic interference might be less of a problem than was thought because the network, even if it could not recognize the originally learned patterns exactly, could relearn them very quickly.

SOLUTIONS TO THE PROBLEM

As soon as the problem came to light, there were many attempts to solve it. One of the first was the suggestion by Kortge (1990) that the problem was due to the back propagation learning algorithm. He proposed a modified learning algorithm using what he called 'novelty vectors' that did, in fact, decrease catastrophic interference. The basic idea of novelty vectors was 'blame assignment'. Kortge's learning rule was developed for autoassociative networks, i.e., networks that, starting from a random weight configuration, learn to produce on output the vectors that they received on input. Each pattern to be learned was fed through the network and the output was compared with the intended output (i.e., the input). Kortge called the resulting difference vector a 'novelty vector' because the bigger the activation differences at each node, the more novel the input – for vectors that the network had already learned there would be little difference between output and input. The novelty vector indicated by how much to change the weights: the greater the novelty activation, the more the weights were changed. This technique significantly reduced catastrophic interference, but it applied only to auto-associative networks.

French (1992) argued that catastrophic forgetting was in large measure due to excessive overlap of internal representations. He claimed that the problem lay with the fully distributed nature of the network's internal representations, and suggested that by developing algorithms that produced 'semi-distributed' internal representations (i.e., representations whose activation was spread only over a limited subset of hidden nodes) catastrophic interference could be reduced. To this end he suggested a learning algorithm, 'node sharpening', that developed much sparser internal representations than standard back propagation. The result was a significant reduction in catastrophic interference. However, the overly sparse representations developed by this technique resulted in a significant decrease in the network's ability to discriminate categories. What was needed was a means of making representations both highly distributed and well enough separated.

Brousse and Smolensky (1989) and McRae and Hetherington (1993) showed that the problem was closely related to the domain of learning. In domains with a high degree of internal structure, such as language, the problem is much less acute. McRae and Hetherington managed to eliminate the problem by pretraining the network on a random sample of patterns drawn from the domain. Because of the degree of structure in the domain, this sample was enough to capture overall regularities of the domain. Consequently, the new patterns to be learned were perceived by the network to be variants of patterns, already learned and did not interfere with previous learning.

The early attempts to solve the problem of catastrophic interference concentrated on reducing representational overlap on input or internally. Kortge (1990) and Lewandowsky (1991) modified the input vectors in an attempt to achieve greater mutual orthogonalization (this is equivalent to reducing the overlap among input vectors). French (1992), Murre (1992), Krushke (1992) and others developed algorithms that reduced internal representational overlap, thereby significantly reducing the amount of catastrophic interference.

Some researchers (e.g. Carpenter, 1994) have laid the blame for the problem of catastrophic interference on a particular architectural feature of the most widely used class of connectionist networks, namely their use of multiplicative connection weights. In the ART family of networks (Carpenter and Grossberg, 1987), new input does not interfere with previously learned patterns because the network is able to recognize new patterns as being new and assigns a new set of nodes for their internal representation.

Hopfield networks, and related architectures, have been shown to have critical saturation limits beyond which there is a steep fall in memory performance. For these networks, forgetting is gradual until the memory becomes saturated, at which point it becomes catastrophic.

REHEARSAL AND PSEUDOREHEARSAL

Most connectionist networks learn patterns concurrently. In terms of human cognition, this is a contrived type of learning. For a given set of n patterns $\{P_1,\dots P_n\}$, the network will successively adjust its weights by a very small amount for all of the patterns and then will repeat this process until it has found a set of weights that allow it to recognize all n patterns. This, in itself, is a way of learning foreign to humans. In addition, if a new set of patterns $\{P_{n+1},\dots P_m\}$ must then be learned by the network, the standard way of handling the situation is to mix the original set of patterns with the new set of patterns to be learned, creating a new set $\{P_1,\dots P_n, P_{n+1},\dots P_m\}$, and then train the network on this new expanded set. In this way, the new patterns will not interfere with the old patterns, but there is a major problem with this technique: namely, that in the real world, the originally learned patterns are often no longer available and cannot simply be added to the set of new patterns to be learned.

In 1995 Anthony Robins made a major contribution to research on catastrophic forgetting with a technique based on what he called 'pseudopatterns' (Robins, 1995). His idea was simple and elegant. Suppose that a connectionist network with n inputs and m outputs has learned a number of input–output patterns $\{P_1, P_2,\dots P_N\}$ generated by some underlying function f. Assume that these original input–output vectors are no longer available. How could one determine, even approximately, what function the network had originally learned? One way would be to create a number M of random input vectors of length n, $\{\hat{\imath}_1,\dots \hat{\imath}_M\}$. These pseudo-input vectors would be fed through the previously trained network, producing a corresponding set of outputs $\{\hat{o}_1,\dots \hat{o}_M\}$. This would result in a set of 'pseudopatterns' $S = \{\psi_1, \psi_2,\dots \psi_M\}$ where $\psi_k\colon \hat{\imath}_k \rightarrow \hat{o}_k$. This set of pseudopatterns would approximate the prior learning of the network. The accuracy of the pseudopatterns in describing the originally learned function would depend on the nature of the function. Thus, when the network had to learn a new set of patterns, it would mix in a number of pseudopatterns with the new patterns.

The pseudopattern technique was the basis of the dual memory models developed by French (1997) and Ans and Rousset (1997), which loosely simulate the hippocampal–neocortical separation, considered by some to be the brain's way of overcoming catastrophic interference (McClelland *et al.*, 1995). These models incorporate two separate, continually interacting, pattern processing areas, one for early processing and one for long-term storage, information being passed between the areas by means of pseudopatterns. This technique allows them to forget gradually and to perform sequential learning appropriately. Somewhat unexpectedly, these dual memory networks also exhibit over time a gradual representational 'compression' (i.e., fewer active nodes) of the long-term internal representations. If this can be shown to occur also in humans, it might help explain certain types of category-specific deficits commonly observed in amnesiacs (French and Mareschal, 1998).

OTHER TECHNIQUES FOR ALLEVIATING CATASTROPHIC FORGETTING IN NEURAL NETWORKS

A number of other techniques have been developed to address the problem of catastrophic interference. Notably, there have been attempts to combine auto-associative architectures with sparse representations. Some architectures use two different kinds of weights on the connections between nodes, one that decays rapidly to zero and another that decays much more slowly. Convolution-correlation models such as CHARM and TODAM, which are mathematically equivalent to certain types of connectionist networks (sigma–pi networks) seem to be relatively immune to catastrophic interference, at least up to a point. Cascade-correlation learning algorithms have also been tried as a means of alleviating catastrophic interference, with some success. For a more complete review of the various models that have been developed to handle the problem of catastrophic interference in connectionist networks, see (French, 1999).

SUMMARY

The problem of catastrophic interference in connectionist networks has been known and studied since the early 1990s. Sequential learning of the kind done by humans cannot be achieved unless a solution is found to this problem. In other words, network models of cognition must, as Grossberg has stressed, be sensitive to new input but not so sensitive that the new input destroys previously learned information. Certain types of patterns, such as those found in highly structured domains, are less susceptible to catastrophic interference than patterns from less well structured domains. Nature seems to have evolved a way of keeping new learning (hippocampal learning) at arm's length from previously learned information stored in the neocortex (neocortical consolidation), thus physically preventing new learning from interfering with previously learned information. Connectionist models have been developed that simulate this cerebral separation. This is certainly not the only way to tackle the problem of catastrophic interference, but its close relationship with the way in which the brain may have solved the problem makes further exploration of these dual memory models of particular interest.

References

Ans B and Rousset S (1997) Avoiding catastrophic forgetting by coupling two reverberating neural networks. *Academie des Sciences: Sciences de la Vie* **320**: 989–997.

Barnes J and Underwood B (1959) 'Fate' of first-learned associations in transfer theory. *Journal of Experimental Psychology* **58**: 97–105.

Brousse O and Smolensky P (1989) Virtual memories and massive generalization in connectionist combinatorial learning. In: *Proceedings of the Eleventh Annual Conference of the Cognitive Science Society*, pp. 380–387. Hillsdale, NJ: Erlbaum.

Carpenter G (1994) A distributed outstar network for spatial pattern learning. *Neural Networks* **7**: 159–168.

Carpenter G and Grossberg S (1987) A massively parallel architecture for a self-organizing neural pattern recognition machine. *Computer Vision, Graphics and Image Processing* **37**: 54–115.

French RM (1992) Semi-distributed representations and catastrophic forgetting in connectionist networks. *Connection Science* **4**: 365–377.

French RM (1997) Pseudo-recurrent connectionist networks: an approach to the 'sensitivity–stability' dilemma. *Connection Science* **9**: 353–379.

French RM (1999) Catastrophic forgetting in connectionist networks. *Trends in Cognitive Sciences* **3**(4): 128–135.

French RM and Mareschal D (1998) Could category-specific anomia reflect differences in the distributions of features within a unified semantic memory? In: Gernsbacher A and Derry SJ (eds) *Proceedings of the Twentieth Annual Conference of the Cognitive Science Society*, pp. 374–379. Hillsdale, NJ: Erlbaum.

Grossberg S (1982) *Studies of Mind and Brain: Neural Principles of Learning, Perception, Development, Cognition, and Motor Control*. Boston, MA: Reidel.

Hetherington P and Seidenberg M (1989) Is there 'catastrophic interference' in connectionist networks? In: *Proceedings of the Eleventh Annual Conference of the Cognitive Science Society*, pp. 26–33. Hillsdale, NJ: Erlbaum.

Kortge C (1990) Episodic memory in connectionist networks. In: *Proceedings of the Twelfth Annual Conference of the Cognitive Science Society*, pp. 764–771. Hillsdale, NJ: Erlbaum.

Krushke J (1992) ALCOVE: an exemplar-based model of category learning. *Psychological Review* **99**: 22–44.

Lewandowsky S (1991) Gradual unlearning and catastrophic interference: a comparison of distributed architectures. In: Hockley W and Lewandowsky S (eds) *Relating Theory and Data: Essays on Human Memory in Honor of Bennet B. Murdock*. pp. 445–476. Hillsdale, NJ: Erlbaum.

McClelland J, McNaughton B and O'Reilly R (1995) Why there are complementary learning systems in the hippocampus and neocortex: insights from the successes and failures of connectionist models of learning and memory. *Psychological Review* **102**: 419–457.

McCloskey M and Cohen N (1989) Catastrophic interference in connectionist networks: the sequential learning problem. In: Bower GH (ed.) *The Psychology of Learning and Motivation*, vol. XXIV, pp. 109–164. New York, NY: Academic Press.

McRae K and Hetherington P (1993) Catastrophic interference is eliminated in pretrained networks. In: *Proceedings of the Fifteenth Annual Conference of the Cognitive Science Society*, pp. 723–728. Hillsdale, NJ: Erlbaum.

Murre J (1992) *Learning and Categorization in Modular Neural Networks*. Hillsdale, NJ: Erlbaum.

Ratcliff (1990) Connectionist models of recognition memory: constraints imposed by learning and forgetting functions. *Psychological Review* **97**: 285–308.

Robins A (1995) Catastrophic forgetting, rehearsal, and pseudorehearsal. *Connection Science* **7**: 123–146.

Further Reading

French RM (1999) Catastrophic forgetting in connectionist networks. *Trends in Cognitive Sciences* **3**(4): 128–135.

Hetherington P (1991) *The Sequential Learning Problem*. Master's thesis, McGill University.

Categorial Grammar and Formal Semantics

Intermediate article

Michael Moortgat, Utrecht Institute of Linguistics, Utrecht University, Utrecht, Netherlands

CONTENTS

Categorial grammar is a lexicalized grammar formalism based on logical type theory. A categorial lexicon assigns one or more types to the atomic elements of a language; the assembly of form and meaning is accounted for in terms of the rules of inference for these types, seen as formulae of a grammar logic. Cross-linguistic variation results from extending the invariant core of the grammar logic with facilities for structural reasoning.

INTRODUCTION

Categorial grammar, a linguistic framework with firm roots in type theory and constructive logic, is well represented in the logical and mathematical literature. This article puts the emphasis more on the categorial modeling of the cognitive abilities underlying the acquisition, use, and understanding of natural language. The sections below address two central questions. First of all, what are the invariants of grammatical composition, and how do they capture the uniformities of the form–meaning correspondence across languages? Secondly, how can we reconcile the idea of grammatical invariants with structural variation in the realization of the form–meaning correspondence?

The slogan 'parsing as deduction' concisely expresses the categorial perspective on these questions. A grammar, essentially, is given by an assignment of types to the elementary units in the lexicon. The type-forming operations have the status of logical connectives: determining whether an expression is well formed amounts to presenting a derivation, or proof, in the logic for these connectives. Natural-language expressions are signs, with a form and a meaning dimension. The categorial type language, consequently, is model-theoretically interpreted with respect to these two dimensions, and a derivation encodes an effective procedure for building up the structural organization of an expression, and for associating this structure with a recipe for meaning assembly.

The article is organized as follows. First, we focus on the form dimension of expressions. We identify the logical constants of the computational system, and study how the base logic for these constants can be extended with facilities for structural reasoning. Then, we see how the logical rules of inference for the type-forming operations can be read as instructions for meaning assembly, and how the structural rules determine which components of an expression can enter into the assembly process.

FORM

The Base Logic

Natural-language expressions are structured objects with a linear order and a hierarchical grouping. In categorial grammar, the traditional parts of speech assume the form of type formulae. The structure of these types mirrors the composition of the expressions they categorize. The set *Type* of type formulae is obtained as the closure of a small set *Atom* of *basic* types under a number of type-forming operations. Individual categorial grammars will differ with respect to the type-forming operations they employ. For our present purposes, the following clauses will be representative:

1. *Atom* is a subset of *Type*.
2. If A is a formula in *Type*, then $\Diamond A$ and $\Box A$ are too.
3. If A and B are formulae in *Type*, then $A \bullet B$, A/B, and $A \backslash B$ are too.

Basic types play a role similar to that of major constituents in phrase-structure grammar: they categorize expressions one can think of as 'complete'. Examples could be the types np for proper names, s for sentences, and n for common noun phrases. Languages can differ as to which basic type distinctions they make. The unary and binary operations provide a vocabulary to categorize expressions in terms of their constituent parts. Informally, a formula $A \bullet B$ categorizes an expression that can be decomposed into a constituent of type A followed by a constituent of type B. An expression with a fraction type A/B (or $B \backslash A$) is incomplete: it combines with an expression of type B on its right (or left, respectively) into an expression of type A. The unary type-forming operations are more recent additions to the categorial vocabulary. They can be thought of as features: an expression of type $\Box A$ issues a request for a feature to be checked; such an expression can be used as a regular A as soon as the $\Box$ feature is eliminated. The operation $\Diamond$ provides the means to perform the required feature-checking.

Frame semantics

To make this informal description precise, Došen (1992) and Kurtonina (1995) make use of *frame-based* models familiar from possible-world semantics for modal logics. For the categorial type language, a *frame* is a tuple $(W, R_\Diamond, R_\bullet)$. W is a nonempty set, the set of expressions. $R_\Diamond$ and $R_\bullet$ are binary and ternary relations over W, interpreting the unary and binary type-forming operations, respectively. One can think of $R_\bullet$ as the 'merge' relation: $R_\bullet xyz$ holds in case x is the composition of the parts y and z. Similarly, $R_\Diamond xy$ holds if the feature-checking relation connects y to x. One obtains a *model* by adding a *valuation* V assigning subsets of W to the atomic formulae. For complex types, the valuation respects the conditions below:

$x \in V(\Diamond A)$ iff there exists a y such that $R_\Diamond xy$ and $y \in V(A)$.
$x \in V(\Box A)$ iff for all y, $R_\Diamond yx$ implies $y \in V(A)$.
$x \in V(A \bullet B)$ iff there are y and z such that $y \in V(A)$, $z \in V(B)$ and $R_\bullet xyz$.
$x \in V(C/B)$ iff for all y and z, if $y \in V(B)$ and $R_\bullet zxy$, then $z \in V(C)$.
$x \in V(A \backslash C)$ iff for all y and z, if $y \in V(A)$ and $R_\bullet zyx$, then $z \in V(C)$.

Type computations, soundness, and completeness

On the proof-theoretic level, we are interested in a deductive system to perform type computations $A \rightarrow B$ ('type B is derivable from type A'). We want this system to be faithful to the interpretation of the type-forming operations, in the sense that $A \rightarrow B$ is provable iff $V(A) \subseteq V(B)$, for every frame F and valuation V. Such a system is 'sound' and 'complete'.

An axiomatization satisfying the soundness and completeness requirements starts with an identity axiom $A \rightarrow A$, and an inference rule allowing one to conclude $A \rightarrow C$ from premises $A \rightarrow B$ and $B \rightarrow C$. Semantically, these express the reflexivity and transitivity of the derivability relation. In addition, one has the inference rules below, establishing the relationship between the interpretation of $\Diamond$ and $\Box$, and between $\bullet$ and left and right division $\backslash$ and $/$. These patterns turn ($\diamond,\Box$), ($\bullet,/$), and ($\bullet,\backslash$) into what are known as 'residuated pairs' in algebra, or 'adjoint functors' in category theory:

(R0) $\diamond A \to B$ if and only if $A \to \Box B$
(R1) $A \bullet B \to C$ if and only if $A \to C/B$
(R2) $A \bullet B \to C$ if and only if $B \to A\backslash C$

Elementary theorems

Let us look at some elementary theorems of the grammatical base logic. From the identity axioms of line 1 below, one obtains the Application schemata of line 2 in one step, using the residuation inferences in the 'if' direction; from Application, one derives the Lifting schemata of line 3, this time reasoning in the 'only if' direction:

1. $A\backslash B \to A\backslash B$ (Ax) $\quad B/A \to B/A$ (Ax)
2. $A \bullet (A\backslash B) \to B$ (R2 $\Leftarrow$) $\quad (B/A) \bullet A \to B$ (R1 $\Leftarrow$)
3. $A \to B/(A\backslash B)$ (R1 $\Rightarrow$) $\quad A \to (B/A)\backslash B$ (R2 $\Rightarrow$)

The Application schemata are no doubt the most familiar laws of categorial combinatorics. In fact, the original categorial grammars of Ajdukiewicz and Bar-Hillel were restricted to Application. Using the Application schemata, one can 'lexicalize' the rules of a context-free phrase-structure grammar. Take the productions '$S \to NP\ VP$' and '$VP \to TV\ NP$' for the derivation of a Subject–Transitive-Verb–Object (SVO) pattern. In categorial terms, one types the Transitive Verb as $(np\backslash s)/np$, thus projecting the SVO pattern in two Application steps: rightward application consumes the Object, leftward application the Subject. The auxiliary label *VP* disappears; the complex type $np\backslash s$ expresses its combinatory role.

Instances of Lifting would be type transitions from *np* (the type assigned to simple proper names) to $s/(np\backslash s)$ or $((np\backslash s)/np)\backslash(np\backslash s)$. These lifted types are appropriate for noun phrases with a distribution restricted to the subject position, in the case of $s/(np\backslash s)$, or the direct-object position, in the case of $((np\backslash s)/np)\backslash(np\backslash s)$. What the derivability arrow says here is that any expression that is assigned the type *np* will be able to occur in the subject or object position, but that there can be expressions with a restricted subject or object distribution, expressed through the higher-order types. One can think of case-marked pronouns, as Lambek (1958) pointed out. With $s/(np\backslash s)$ as the lexical type assignment for 'he'/'she', but $((np\backslash s)/np)/(np\backslash s)$ for 'him'/'her', we correctly rule out 'him irritates she' while allowing 'he irritates her'.

Elementary theorems for the unary type-forming operations are established below:

1. $\Box A \to \Box A$ (Ax) and $\diamond A \to \diamond A$ (Ax)
2. $\diamond\Box A \to A$ (R0 $\Leftarrow$) and $A \to \Box\diamond A$ (R0 $\Rightarrow$)

An illustration of the added expressivity of the unary operators can be found in Bernardi (2002), where they are used to control the distribution of polarity-sensitive items. Consider the contrast between 'nobody left yet', with the negative-polarity item 'yet', and '*somebody left yet'. In a type language with just the binary type-forming operations, both 'somebody' and 'nobody' would receive the subject type $s/(np\backslash s)$, and 'yet' the modifier type $(np\backslash s)\backslash(np\backslash s)$. Such type assignment is too crude to block the ungrammatical '*somebody left yet'. In the extended type language, the negative-polarity trigger 'nobody' can be assigned the type $s/\Box\diamond(np\backslash s)$, whereas 'somebody' keeps the undecorated type $s/(np\backslash s)$. By typing the negative-polarity item 'yet' as $(np\backslash s)\backslash\Box\diamond(np\backslash s)$ one expresses the fact that it requires a trigger such as 'nobody' to check the $\Box\diamond$ decoration in its numerator subtype. For the derivation of the simple sentence 'nobody left' (with no polarity item to be checked), we rely on the fact that in the base logic, we have $s/\Box\diamond(np\backslash s) \to s/(np\backslash s)$; i.e., the $\Box\diamond$ decoration on argument subtypes can be simplified away, allowing the combination (in terms of the Application schema) of 'nobody' with a simple verb phrase 'left' of type $np\backslash s$.

Monotonicity properties

Apart from these theorems, the base logic has several derived rules of inference. With respect to the derivability relation, the operations $\diamond$ and $\Box$ are order-preserving (isotone). The $\bullet$ operation is order-preserving in its two arguments; the division operations / and \ are order-preserving in their numerator, and order-reversing (antitone) in their denominator argument.

$A \to B$ implies:

$\diamond A \to \diamond B$ and $\Box A \to \Box B$
$A/C \to B/C$ and $C\backslash A \to C\backslash B$
$C/B \to C/A$ and $B\backslash C \to A\backslash C$
$A \bullet C \to B \bullet C$ and $C \bullet A \to C \bullet B$

From a combinatorial point of view, these rules produce an infinite number of type transformations from some small inventory of 'primitive' ones. Consider the Lifting schema. From it, one obtains the transformations known as Value Raising (for example, lifting a determiner type np/n to $(s/(np\backslash s))/n$) and Argument Lowering (for example, lowering a third-order verb phrase type $(s/(np\backslash s))\backslash s$ to first-order $np\backslash s$).

Alternative presentations, and natural deduction

The categorial base logic allows many alternative axiomatizations, each serving its own function. The essential point is that the different presentations must find their justification in the model-theoretic interpretation of the connectives; i.e., one has to prove that they are equivalent syntaxes for performing valid type computations. In the Gentzen sequent calculus, one replaces the arrows $A \to B$ by statements $\Gamma \Rightarrow B$ ('structure Γ is of type B'). The antecedent Γ is built out of formulae by means of the structure-building operations $\langle\cdot\rangle$ and $(\cdot\circ\cdot)$, counterparts of the logical connectives $\Diamond$ and $\bullet$. The purpose of this presentation is to show that the transitivity rule (the Cut rule) can be eliminated. Every logical rule of inference in the Gentzen calculus introduces a connective either in the antecedent or in the succedent, so that backward-chaining, cut-free proof search immediately yields a decision procedure for categorial derivability, as shown in Lambek (1958) for the binary and in Moortgat (1996) for the unary connectives.

The derivational format of 'combinatory categorial grammar' (CCG) (e.g., Steedman, 2000b) is a Hilbert-style presentation. Functional Application here is taken as the basic, primitive schema for type combination. To the Application schema are added extra schemata, such as Lifting, also known as combinator T. The CCG format of derivations is related to the Gentzen style as the combinator presentation of intuitionistic logic is to its Gentzen presentation. The recursive generalization of the primitive type transformations under monotonicity is important for such 'combinatory' presentations of categorial derivability: without this generalization, one loses completeness.

In a third format, 'natural deduction' (ND), every type-forming connective has an introduction and an elimination rule. As a result, ND doesn't have the pleasant proof search properties of the Gentzen calculus, but it provides a perspicuous presentation of a derivation once it has been found. For this reason, ND is often used in linguistic discussion of categorial analyses. Also, ND is the most transparant format to associate meaning assembly with a derivation, as we will see. Figure 1 shows the ND rules for the base logic, using the Gentzen sequent style, which is explicit about the structural configuration of the antecedent assumptions.

$$\frac{\Gamma \vdash \Box A}{\langle\Gamma\rangle \vdash A}\,(\Box E) \qquad \frac{\langle\Gamma\rangle \vdash A}{\Gamma \vdash \Box A}\,(\Box I)$$

$$\frac{\Gamma \vdash A}{\langle\Gamma\rangle \vdash \Diamond A}\,(\Diamond I) \qquad \frac{\Delta \vdash \Diamond A \quad \Gamma[\langle A\rangle] \vdash B}{\Gamma[\Delta] \vdash B}\,(\Diamond E)$$

$$[/I]\,\frac{\Gamma \circ B \vdash A}{\Gamma \vdash A/B} \qquad \frac{\Gamma \vdash A/B \quad \Delta \vdash B}{\Gamma \circ \Delta \vdash A}\,[/E]$$

$$[\backslash I]\,\frac{B \circ \Gamma \vdash A}{\Gamma \vdash B\backslash A} \qquad \frac{\Gamma \vdash B \quad \Delta \vdash B\backslash A}{\Gamma \circ \Delta \vdash A}\,[\backslash E]$$

$$[\bullet I]\,\frac{\Gamma \vdash A \quad \Delta \vdash B}{\Gamma \circ \Delta \vdash A \bullet B} \qquad \frac{\Delta \vdash A \bullet B \quad \Gamma[A \circ B] \vdash C}{\Gamma[\Delta] \vdash C}\,[\bullet E]$$

Figure 1. Natural deduction. $\Gamma \vdash A$ stands for the deduction of a conclusion A from a configuration of assumptions Γ. Axioms are of the form $A \vdash A$. Antecedent structures are built from formulae with the structure-building operations $\langle\cdot\rangle$ and $(\cdot\circ\cdot)$. These are the structural counterparts of $\Diamond$ and $\bullet$, respectively, as the $\Diamond$ and $\bullet$ Introduction rules show.

Multimodal generalization

One can straightforwardly generalize the base logic to a system where one has not just one single merge and feature-checking relation, but families of them. In terms of modal logic, this means moving from a unimodal to a multimodal system, with frames $(W, \{R_i^2\}_{i \in I}, \{R_j^3\}_{j \in J})$ where the different relations are kept apart by indexing them with a composition mode label. Similarly, in the formula language, we index the connectives for these composition modes. The concept of multiple composition modes is not unfamiliar. For the binary operations, one can think of a distinction between the structure of words (morphology) and the structure of phrases (syntax): one can give a categorial analysis of morphology and syntax in terms of $/$, $\bullet$, and $\backslash$, but still one will want to keep these grammatical levels distinct, say as $\bullet_w$ versus $\bullet_\phi$. For the unary connectives $\Diamond$ and $\Box$, multimodality makes it possible to distinguish a number of named features in the grammar, so that they can play different roles in controlling composition.

The multimodal perspective turns out to be particularly useful once we move beyond the base logic and consider its structural extensions, where one can then have interaction between different binary composition modes (between morphology and syntax, in the case of complement inheritance, for example), and between specific unary control features and binary composition operations. Such interaction principles are discussed below.

The Structural Module

The laws of the base logic do not depend on specific structural properties of the merge and feature-checking relations: the completeness condition does not impose any restrictions on the interpretation of $R_\bullet$ and $R_\Diamond$. In this sense, the base logic can be said to capture the invariants of grammatical composition. Although the base logic already has a rich deductive structure, the system also has its limitations. If an expression can occur in different structural configurations, one would like to relate these configurations. In the base logic, this cannot be done: type assignment is structurally rigid, in the sense that different structural environments will lead to different type assignments. To overcome the problem of structural rigidity, one extends the base logic with facilities for structural reasoning. Technically, such facilities have the status of non-logical axioms, or postulates. They can be introduced in a global or in a controlled fashion. We look at these in turn.

Global structural rules

The postulates below create a hierarchy of categorial systems: with the addition of structural options, the flexibility of type combination increases, but structural discrimination deteriorates.

$$(\mathrm{A_l})\ (A \bullet B) \bullet C \rightarrow A \bullet (B \bullet C)$$
$$(\mathrm{A_r})\ A \bullet (B \bullet C) \rightarrow (A \bullet B) \bullet C$$
$$(\mathrm{C})\ A \bullet B \rightarrow B \bullet A$$

The rebracketing postulates $\mathrm{A_l}$ and $\mathrm{A_r}$, added to the fragment of the base logic formed by $\bullet$, $/$, and $\backslash$, produce the system known as L, the associative calculus of Lambek (1958). The fragment of the base logic formed by $\bullet$, $/$, and $\backslash$ is known as NL: in Lambek (1961) this system was obtained by dropping the associativity postulates from L. Characteristic theorems of L are the type transitions below: the Geach laws $\mathrm{G_r}$ and $\mathrm{G_l}$, and the functional composition schemata (known as combinator **B** in CCG) of which $\mathrm{B_r}$ and $\mathrm{B_l}$ are the simplest forms.

$$(\mathrm{G_r})\ A/B \rightarrow (A/C)/(B/C)$$
$$(\mathrm{G_l})\ B\backslash A \rightarrow (C\backslash B)\backslash(C\backslash A)$$
$$(\mathrm{B_r})\ (A/B) \bullet (B/C) \rightarrow A/C$$
$$(\mathrm{B_l})\ (C\backslash B) \bullet (B\backslash A) \rightarrow C\backslash A$$

Adding the commutativity postulate to L produces LP (Lambek calculus with permutation), a system coinciding with the multiplicative fragment of linear logic, which has a commutative product operation matched by a single linear implication. The distinction between left-incompleteness and right-incompleteness collapses in the presence of the commutativity postulate C.

Extending the base logic with facilities for structural reasoning has consequences for the interpretation of the type-forming operations (Došen, 1992; Kurtonina, 1995). An interpretation with respect to arbitrary frames is obviously not available anymore. Instead, each postulate introduces a corresponding frame constraint restricting the interpretation of the merge relation $R_\bullet$, and completeness is stated with respect to frames respecting the relevant constraints. A Commutativity postulate, for example, would impose the semantic constraint that for all $x, y, z \in W$, $R_\bullet xyz$ implies $R_\bullet xzy$. Similarly for the other postulates discussed. In the presence of such semantic constraints, it will often be the case that one can specialize the abstract relational interpretation to more concrete models. A good example is the system L with its associative composition relation $R_\bullet$. In this case, one can read $R_\bullet xyz$ as concatenation; i.e., $x = y \cdot z$. Pentus (1994) proves that L is indeed complete with respect to this concatenation interpretation.

Controlled structural reasoning

There are many natural-language phenomena that seem to require some of the flexibility offered by the postulates $\mathrm{A_l}$, $\mathrm{A_r}$, and C. Cases of nonconstituent coordination can be naturally handled with the possibilities for type combination that follow from the rebracketing postulates. Displacement phenomena are ubiquitous in natural language, and seem to require some form of commutativity. At the same time, it is clear that in a global form, these structural options overgenerate. Commutativity would entail that well-formedness is preserved under arbitrary changes in word order; free rebracketing makes constituent structure irrelevant for determining grammaticality.

To obtain controlled structural extensions of the base logic, various strategies have been pursued. In the rule-based approach of combinatory categorial grammar, one augments the Application–Lifting basis with structural combinators which, in an unconstrained form, would be overgenerating. One then imposes type restrictions on these extra combinators. In addition, the set of rule schemata (combinators) is kept finite, so that one can avoid the consequences of the recursive generalization of rules under monotonicity. The alternative is to exploit the intrinsic logical instruments for structural resource management offered by richer type systems with unary control features and multimodal interaction principles. To compare these two strategies, consider the following cases of extraction: 'what Alice found' and 'what Alice found there' (see Figure 2).

In CCG, the peripheral case of extraction ('what Alice found') is derived from an assignment $wh/(s/np)$ to the *wh*-pronoun by lifting the type for 'Alice' to $s/(np\backslash s)$ which is then composed with the

Alice: np ⇒ (T) $s/(np\backslash s)$
found: $(np\backslash s)/np$; there: $(np\backslash s)\backslash(np\backslash s)$ ⇒ ($B_{l\times}$) $(np\backslash s)/np$
$s/(np\backslash s)$, $(np\backslash s)/np$ ⇒ (B_r) s/np
what: $wh/(s/np)$, s/np ⇒ wh

Figure 2. *wh*-extraction: combinator-style derivation. The clause body 'Alice found there' is assigned type s/np by means of the backwards crossed composition combinator $B_{l\times}$. The rule can apply because the cancelled $(np\backslash s)$ satisfies the type restriction on $B_{l\times}$.

transitive verb type $(np\backslash s)/np$ for 'found' by means of B_r. To obtain the nonperipheral case of extraction ('what Alice found there'), one needs the combinator $B_{l\times}$, a form of composition which depends on the commutativity postulate. To avoid collapse into LP, one imposes a side condition on the rule, restricting the middle term B to certain verbal categories, in this case $(np\backslash s)$:

($B_{l\times}$) $(B/C) \bullet (B\backslash A) \rightarrow A/C$ where B is a predicate category

The $\Diamond$ and $\Box$ connectives make it possible to avoid extralogical type restrictions. The postulates P1 and P2 below implement a controlled form of rebracketing and reordering for formulae carrying the $\Diamond$ control feature (Moortgat, 1999). With a lexical type assignment $wh/(s/\Diamond\Box np)$ to the *wh*-pronoun, one obtains peripheral and medial extraction from right branches. Under this analysis, one does not attribute any associativity or commutativity to the $\bullet$ operation itself; displacement effects arise through the interaction of the merge operation with a gap hypothesis carrying the licensing $\Diamond$ feature. A derivation is given in Figure 3.

(P1) $(A \bullet B) \bullet \Diamond C \rightarrow (A \bullet \Diamond C) \bullet B$
(P2) $(A \bullet B) \bullet \Diamond C \rightarrow A \bullet (B \bullet \Diamond C)$

Generative Capacity and Computational Complexity

The modular view on grammatical invariants and structural variation invites a comparison between the categorial landscape and the Chomsky hierarchy. For a recent survey, see Buszkowski (1997). The discovery in the 1980s of dependency patterns that cannot be adequately captured by context-free grammars has led to an interest in 'mildly context-sensitive' formalisms; i.e., systems with an expressivity beyond the context-free, but sufficiently restricted to have polynomial parsing algorithms. The Ajdukiewicz–Bar-Hillel grammars have long been known to be weakly equivalent to context-free grammars, hence to be too poor to serve as models of universal grammar. The same is true for the base logic described above. The correctness of Chomsky's conjecture that context-free equivalence extends to the Lambek calculus L was finally established in Pentus (1993). This result does not have a direct corollary for polynomial parsability, because the construction of a context-free grammar from an L grammar is of exponential complexity.

For the structural extensions of the base logic discussed above, the challenge is to identify appropriate constraints: it is clear that arbitrary combinator extensions, or structural rule packages, lead to excessive expressivity. But Vijay-Shanker and Weir (1994) show that an appropriately restricted version of CCG is weakly equivalent to the linear indexed grammars, hence polynomially parsable. In a similar spirit, Moot (2002) shows how, with appropriate restrictions on lexical assignments and structural postulates, one can carve out a class of multimodal categorial grammars equivalent to lexicalized tree adjoining grammars and inheriting the polynomial parsability of these systems. The

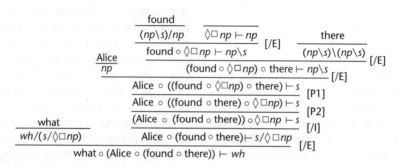

found: $(np\backslash s)/np$; $\Diamond\Box np \vdash np$
found $\circ$ $\Diamond\Box np \vdash np\backslash s$ [/E]
there: $(np\backslash s)\backslash(np\backslash s)$
(found $\circ$ $\Diamond\Box np$) $\circ$ there $\vdash np\backslash s$ [\E]
Alice: np
Alice $\circ$ ((found $\circ$ $\Diamond\Box np$) $\circ$ there) $\vdash s$ [\E]
Alice $\circ$ ((found $\circ$ there) $\circ$ $\Diamond\Box np$) $\vdash s$ [P1]
(Alice $\circ$ (found $\circ$ there)) $\circ$ $\Diamond\Box np \vdash s$ [P2]
Alice $\circ$ (found $\circ$ there) $\vdash s/\Diamond\Box np$ [/I]
what: $wh/(s/\Diamond\Box np)$
what $\circ$ (Alice $\circ$ (found $\circ$ there)) $\vdash wh$ [/E]

Figure 3. *wh*-extraction: $\Diamond$ control. The type assignment to the relativizer 'what' expresses the fact that the relative clause body is a sentence built with the help of a 'gap' hypothesis of type $\Diamond\Box\ np$. The feature-marked hypothesis has to be withdrawn at the right periphery, but it is not selected in that position. It is related to the nonperipheral direct-object position within the relative clause body by virtue of the postulates P1 and P2. Once it has found the direct-object position, the licensing feature $\Diamond$ has done its work and can be cleaned up by the law $\Diamond\Box\ np \vdash np$. The 'gap' hypothesis is then used as a regular direct object with respect to the selecting verb 'found'.

general theory of ◇ and □ as control operators has been investigated in Kurtonina and Moortgat (1997). These authors establish a number of embedding theorems showing that the full logical space between the base logic and LP can be navigated in terms of the control connectives, both in the 'licensing' direction illustrated above (allowing structural inferences that would be unavailable without the control features) and in the 'constraining' sense (blocking structural options that would be licit in the absence of the control features).

More important than weak generative capacity are issues of strong capacity, which in the categorial tradition would mean the proof structures (or their lambda terms, discussed below) that produce a certain string. In this area, Tiede (2001) has obtained interesting results, showing that while the Lambek systems L and NL are weakly context-free, their expressivity in terms of strong capacity goes beyond that of context-free grammars.

Language Learning

Kanazawa (1998) has studied formal learning theory for categorial grammar within Gold's paradigm of identification in the limit on the basis of positive data. The focus is on classical categorial grammars, using only the Application rules, and on combinatory extensions with extra rule schemata. On the input side, Kanazawa considers learning both from strings and from function-argument structures. On the output side, the class of 'rigid' grammars (where the grammar assigns a unique type to each word) is compared to the class of k-valued grammars (where at most k types are assigned to a lexical item). It is a matter of dispute whether Gold's very abstract formulation of the learning problem is directly relevant for first-language acquisition. An alternative, purely inductive, approach – learning a subclass of the shallow context-free languages – is presented in Adriaans (2001).

The discussion above suggests some directions for further research in this area. First of all, one would like to obtain learnability results for classes of Lambek-style categorial grammars, where the learner has access to both the Elimination rules and the Introduction rules for the type-forming operators. Secondly, one would like to go beyond systems with a 'hard-wired' structural component, in order to investigate the learnability effects of different choices of structural packages, in combination with an invariant base logic. The work of Foret (2001) is promising in this respect: she mixes unification and substitution with Lambek-style deduction, so suggesting modulation of learnability questions in terms of different structural postulates.

Finally, the role of semantic information in learning needs further investigation. The challenge here is to find a level of informativity that would be realistic in the setting of first-language acquisition.

MEANING ASSEMBLY

Categorial grammar adheres to the truth-conditional theory of semantics: the interpretation process establishes a systematic relationship between linguistic expressions and states of affairs in the world in such a way that specifying the meaning of a sentence comes down to giving its truth conditions. Model theory provides the tools to carry out this program. For semantic interpretation this involves the construction of a set-theoretic model of 'the world' in terms of objects and configurations of objects; these set-theoretic constructs then serve as the semantic values of natural-language expressions.

The integrated treatment of syntax and semantics, which is now seen as the most attractive aspect of categorial grammar, is of relatively recent origin. The original Lambek systems (Lambek, 1958, 1961) were presented as syntactic type calculi. At about the same time, Curry (1961) was advocating the use of purely semantic types in natural-language analysis. Curry in fact criticized Lambek for the admixture of syntactic considerations in his category concept, coining the famous distinction between 'tectogrammatic' and 'phenogrammatic' organization. The tectogrammatic level, in Curry's view, provides the appropriate information for meaning composition; the phenogrammatic pertains to the way this abstract grammatical structure is represented in terms of surface expressions. About the actual mapping between the two levels, Curry provides no specific information.

The design of the syntax–semantics interface becomes of central importance in Richard Montague's work. The cornerstone of his 'universal grammar' program is a precise implementation of Frege's *compositionality principle*. Informally, this fundamental principle of natural-language semantics requires that the meaning of a complex expression be given as a function of the meanings of its constituent parts, and the way they are put together. In Montague's algebraic system, compositionality takes the form of a homomorphism, that is, a structure-preserving mapping, between a syntactic and a semantic algebra. Ironically, when van Benthem (1987) reintroduced semantic interpretation in the discussion of Lambek's syntactic calculi, it was by establishing the connection between categorial derivations and Curry's own 'formulae-as-types' program which we describe below. (The discussion below is restricted to functional types; the full Curry–Howard interpretation involves extension to the other type-forming operations.)

Model-theoretic Semantics, Type Theory, and the Lambda Calculus

For semantic interpretation, we associate every type A with a semantic domain D_A. Expressions of type A find their denotations in D_A. Semantic domains can be set up in two ways: directly on the basis of the syntactic types discussed in the previous section, or indirectly, via a mapping from syntactic to semantic types. The indirect option is attractive for a number of reasons. On the level of atomic types, one may want to make different basic distinctions depending on whether one uses syntactic or semantic criteria. For complex types, a map from syntactic to semantic types makes it possible to forget information that is relevant only for the way expressions are to be configured in the form dimension. Finally, the semantic type system naturally fits the language of the typed lambda calculus, which we can then use, together with its standard interpretation, to specify the instructions for meaning assembly.

Semantic and syntactic types

For a simple extensional interpretation, the set of atomic semantic types *SemAtom* could consist of types e and t, with D_e the domain of discourse (a nonempty set of entities, objects), and $D_t = \{0, 1\}$, the set of truth values. The full set of semantic types *SemType* is then obtained by closing *SemAtom* under the rule that if A and B are in *SemType* then $A \rightarrow B$ is also. $D_{A \rightarrow B}$, the semantic domain for a functional type $A \rightarrow B$, is the set of functions from D_A to D_B. The mapping from syntactic to semantic types $(\cdot)^*$ could now stipulate for basic syntactic types that $np^* = e$, $s^* = t$, and $n^* = (e \rightarrow t)$. Sentences, in this way, denote truth values; (proper) noun phrases individuals; common nouns functions from individuals to truth values. For complex syntactic types, we set $(A/B)^* = (B \backslash A)^* = B^* \rightarrow A^*$. On the level of semantic types, the directionality of the 'slash' connective is no longer taken into account. The distinction between numerator and denominator – domain and range of the interpreting functions – is kept. Notice that both verb phrases with syntactic type $np \backslash s$ and common nouns are mapped to the semantic type $e \rightarrow t$.

The language of the simply-typed lambda calculus

Below, a procedure is presented to associate a derivation $A_1,\ldots,A_n \vdash B$, with a term t of type B representing a recipe for meaning assembly, with parameters $x_1,\ldots,x_n$ for the lexical assumptions $A_1,\ldots,A_n$. To prepare the ground, we build up the set of meaningful expressions (terms) of semantic type A, starting from a denumerably infinite set of variables for each type. For each expression t of type A, we specify its interpretation $[\![t]\!]^g$ relative to an assignment function g which assigns to each variable of type A a member of D_A.

1. *Variables.* Let x be a variable of type A. Then x is a term of type A. Interpretation: $[\![x]\!]^g = g(x)$.
2. *Application.* Let t and u be terms of type $A \rightarrow B$ and A respectively. Then $(t\ u)$ is a term of type B. Interpretation: $[\![(t\ u)]\!]^g = [\![t]\!]^g([\![u]\!]^g)$; i.e., the result of applying the function $[\![t]\!]^g$ to $[\![u]\!]^g$.
3. *Abstraction.* Let x be a variable of type A and t a term of type B. Then $\lambda x.t$ is a term of type $A \rightarrow B$. Interpretation: $[\![\lambda x.t]\!]^g$ is that function h from D_A into D_B such that for all objects $k \in D_A$, $h(k) = [\![t]\!]^{g'}$, where g' is the assignment that is exactly like g except for the possible difference that it assigns the object k to the variable x.

Given this interpretation, certain equalities hold between terms. One can see them as syntactic simplifications, replacing a more complex term (the *redex*) by a simpler one with the same interpretation (the *contractum*):

$(\lambda x.t)\ u \rightsquigarrow_\beta t[u/x]$ provided u is free for x in t.
$\lambda x.(t\ x) \rightsquigarrow_\eta t$ provided x is not free in t.

Formulae as Types, Proofs as Programs

Curry's basic insight was that one can see the functional types of type theory as logical implications, giving rise to a one-to-one correspondence between typed lambda terms and natural deduction proofs in positive intuitionistic logic. A natural deduction presentation for $\rightarrow$ starts from identity axioms $A \vdash A$ and has the introduction and elimination rules below, where Γ and Δ represent finite lists of formulae, and where $\Gamma - A$ results from dropping some or all occurrences of A from Γ.

$$\frac{\Gamma \vdash A \rightarrow B \quad \Delta \vdash B}{\Gamma, \Delta \vdash B} \rightarrow \text{Elim}$$

$$\frac{\Gamma \vdash B}{\Gamma - A \vdash A \rightarrow B} \rightarrow \text{Intro}$$

Let us write $\Gamma(t)$ for the string of types of free occurrences of variables in a term t. Each term t of type A now encodes a natural deduction proof of the sequent $\Gamma(t) \vdash A$. The Variable clause in the definition of well-formed terms corresponds to the axiom sequent, the Application clause to $\rightarrow$ Elimination, and the Abstraction clause to $\rightarrow$ Introduction, where the dropped A assumption corresponds to the variable bound by the lambda abstractor. In the opposite direction, every natural-deduction proof is encoded by a lambda term. The normalization of natural-deduction proofs corresponds to the β–η reductions of terms.

Translating Curry's 'formulae-as-types' idea to the categorial type logics we are discussing, we have to take the differences between intuitionistic logic and the grammatical resource logic into account. Below we repeat the natural-deduction presentation of the base logic, now taking term-decorated formulae as basic declarative units. Judgments take the form of sequents $\Gamma \vdash t{:}A$. The antecedent Γ is a structure with leaves $x_1{:}A_1,\ldots,x_n{:}A_n$. The x_i are unique variables of type A_i^*, where $(\cdot)^*$ is the mapping from syntactic to semantic types. The succedent is a term t of type A^* with exactly the free variables $x_1,\ldots,x_n$, representing a program which given inputs $k_1,\ldots,k_n$ produces $[\![t]\!]^g$ under the assignment that maps the variables x_i to the objects k_i. The x_i, in other words, are the parameters of the meaning-assembly procedure. A derivation starts from axioms $x{:}A \vdash x{:}A$. The Elimination and Introduction rules have versions for the right and the left implications. On the meaning-assembly level, this syntactic difference is ironed out, as we already know that $(A/B)^* = (B\backslash A)^*$. As a consequence, we don't have the isomorphic (one-to-one) correspondence between terms and proofs of Curry's original program. But we do read off meaning assembly from the categorial derivation. (See Figure 4.)

A second difference between the programs and computations that can be obtained in intuitionistic implicational logic and the recipes for meaning assembly associated with categorial derivations has to do with the resource management of assumptions in a derivation. The formulation of the $\rightarrow$ Introduction rule makes it clear that in intuitionistic logic the number of occurrences of assumptions (the 'multiplicity' of the logical resources) is not critical. One can make this style of resource management explicit in the form of

$$[/I]\ \frac{\Gamma \circ x : B \vdash t : A}{\Gamma \vdash \lambda x.t : A/B} \qquad \frac{\Gamma \vdash t : A/B \quad \Delta \vdash u : B}{\Gamma \circ \Delta \vdash (t\ u) : A}\ [/E]$$

$$[\backslash I]\ \frac{x : B \circ \Gamma \vdash t : A}{\Gamma \vdash \lambda x.t : B\backslash A} \qquad \frac{\Gamma \vdash u : B \quad \Delta \vdash t : B\backslash A}{\Gamma \circ \Delta \vdash (t\ u) : A}\ [\backslash E]$$

Figure 4. Natural-deduction rules: term labeling.

structural rules of Contraction and Weakening, allowing for the duplication and waste of resources:

$$\frac{\Gamma, A, A \vdash B}{\Gamma, A \vdash B}\ [C]$$

$$\frac{\Gamma \vdash B}{\Gamma, A \vdash B}\ [W]$$

In contrast, the categorial type logics are resource-sensitive systems where each assumption has to be used exactly once. At the level of LP, we have the following correspondence between resource constraints and restrictions on the lambda terms coding derivations:

No empty antecedents: each subterm contains a free variable.

No Weakening: each λ operator binds a variable free in its scope.

No Contraction: each λ operator binds at most one occurrence of a variable in its scope.

Moving from LP to the grammatical base logic imposes even tighter restrictions on binding: in the absence of Associativity and Commutativity, the 'slash' introduction rules responsible for the λ operator can only reach the immediate children of a structural domain.

The syntax–semantics interface

Applied to the composition of natural-language meaning, the 'proofs-as-programs' approach has some interesting consequences for the syntax–semantics interface.

A first point to notice is the strictly modular treatment of derivational versus lexical semantics. The proof term that is read off a derivation is a *uniform instruction* for meaning assembly that fully abstracts from the contribution of the particular lexical items on which it is built. As a result, no assumptions about lexical semantics can be built into the meaning assembly process as represented by a derivation. The interplay between lexical and derivational semantics is illustrated in Figures 5 and 6. Whereas the proof term in Figure 5 is a faithful encoding of the derivation (modulo directionality and structural operations), the term one obtains in Figure 6 after substitution of lexical meaning programs and β-simplification has lost transparency with respect to the derivation.

The second feature is the limited semantic expressivity of a structure-sensitive type logic: many forms of meaning assembly that can be straightforwardly expressed in the language of the lambda calculus cannot be obtained as Curry–Howard

$$\dfrac{\dfrac{\text{Noun}}{z_0 : n} \quad \dfrac{\dfrac{\text{that}}{x_1 : (n\backslash n)/(s/np)} \quad \dfrac{\dfrac{\dfrac{\dfrac{\text{Subj}}{x_2 : np} \quad \dfrac{\dfrac{\text{TV}}{y_2 : (np\backslash s)/np} \quad [np \vdash y_1 : np]^1}{\text{TV} \circ np \vdash (y_2\ y_1) : np\backslash s}\,[/E]}{\text{Subj} \circ (\text{TV} \circ np) \vdash ((y_2\ y_1)\ x_2) : s}\,[\backslash E]}{(\text{Subj} \circ \text{TV}) \circ np \vdash ((y_2\ y_1)\ x_2) : s}\,[P_2]}{\text{Subj} \circ \text{TV} \vdash \lambda y_1 \cdot ((y_2\ y_1)\ x_2) : s/np}\,[/I]^1}{\text{that} \circ (\text{Subj} \circ \text{TV}) \vdash (x_1\ \lambda y_1) \cdot ((y_2\ y_1)\ x_2)) : n\backslash n}\,[/E]}{\text{Noun} \circ (\text{that} \circ (\text{Subj} \circ \text{TV})) \vdash ((x_1\ \lambda y_1) \cdot ((y_2\ y_1)\ x_2))\ z_0) : n}\,[\backslash E]$$

Figure 5. Computation of the proof term for the pattern 'Noun that Subj Transitive-Verb'. Leaves are labeled with variables. The derivation produces a meaning recipe with parameters for the lexical meaning programs. The recipe can be applied to any choice of lexical items fitting the type requirements: 'biscuit that Alice ate', 'book that Carroll wrote', and so on.

1.	biscuit : n – **biscuit**	Lex
2.	that : $(n\backslash n)/(s/np)$ – $\lambda z_{15}.\lambda x_{16}.\lambda y_{16}.((z_{15}\ y_{16}) \wedge (x_{16}\ y_{16}))$	Lex
3.	alice : np – **a**	Lex
4.	ate : $(np\backslash s)/np$ – **eat**	Lex
5.	$np : np - y_1$	Hyp
6.	ate $\circ\ np : np\backslash s$ – (**eat** y_1)	/E (4, 5)
7.	alice $\circ$ (ate $\circ\ np$) : s – ((**eat** y_1) **a**)	\E (3, 6)
8.	(alice $\circ$ ate) $\circ\ np : s$ – ((**eat** y_1) **a**)	P2 (7)
9.	alice $\circ$ ate : s/np – λy_1.((**eat** y_1) **a**)	/I (5, 8)
10.	that $\circ$ (alice $\circ$ ate) : $n\backslash n$ – $\lambda x_{16}.\lambda y_{16}$.(((**eat** y_{16}) **a**) $\wedge$ ($x_{16}\ y_{16}$))	/E (2, 9)
11.	biscuit $\circ$ (that $\circ$ (alice $\circ$ ate)) : n – λy_{16}.(((**eat** y_{16}) **a**) $\wedge$ (**biscuit** y_{16}))	\E (1, 10)

Figure 6. Substitution of lexical semantics in the pattern 'Noun that Subj Transitive-Verb'. Bold-face is used for non-logical constants. In steps 10 and 11, β-conversion is applied on the fly to the application terms obtained from the 'slash' elimination rules. The derivation is presented in the linear or Fitch-style natural-deduction format.

images of the Introduction–Elimination inferences of the categorial base logic.

To resolve the tension between structure sensitivity and semantic expressivity, categorial grammars can exploit a combination of two strategies. Structural reasoning (in terms of combinators or structural postulates) makes it possible to explicitly determine which positions are accessible for semantic manipulation (binding). The example of controlled *wh*-extraction in Figure 3 is an illustration. Secondly, lexical-meaning programs do not have to obey the resource constraints of the derivational semantics. Specifically, we do not impose the single-bind condition on lexical meanings (although the ban on vacuous abstraction does make sense, also in the lexicon.) An example of multiple binding is the lexical lambda term for the relative pronoun 'that' in Figure 6, a program which computes property intersection. Another example would be a reflexive pronoun like 'himself'. With a type $((np\backslash s)/np)\backslash(np\backslash s)$, it consumes its transitive-verb argument in a resource-sensitive way. The identification of subject and object arguments of the verb is realized through its lexical lambda term $\lambda x \lambda y.((x\ y)\ y)$.

The interplay between these two strategies in current research is nicely illustrated by the construal of quantifier scope ambiguities and antecedent–anaphor dependencies. Generalized quantifier expressions, like 'everyone', 'someone', and 'nobody', require an interpretation as sets of properties; i.e., they find a denotation in $D_{(e \to t) \to t}$. A syntactic type compatible with such denotations would be $s/(np\backslash s)$. But there are two problems with such a type. First of all, it is restricted to subject position, and one wouldn't like to resort to multiple type assignments for non-subject occurrences. Secondly, it doesn't allow nonlocal scope readings, as in line 3 below, where the embedded quantifier takes scope at the main clause level.

1. Alice thinks someone left.
2. ((**think** ($\exists\ \lambda x$.(**leave** x))) **a**)
3. ($\exists\ \lambda x$.((**think** (**leave** x)) **a**))
4. *Alice* thinks *she* dreams.
5. ((**think** (**dream a**)) **a**)

The construal of antecedent–anaphora relations, like that of quantifier scope, involves nonlocal dependencies beyond the reach of the grammatical base logic, as in line 4 above, where the anaphor in the subordinate clause can pick up its antecedent in the main clause. In addition, meaning composition for anaphora resolution involves a duplication of resources, in the sense that one would like to make the pronoun 'she' in the example above responsible for the copying of the antecedent meaning.

Proposals for dealing with these problems rely either on combinator-style type-shifting rule schemata or on structural extensions of the Lambek calculus. For quantifier scope construal, these options are discussed in depth in Carpenter (1998). For anaphora resolution, Jäger (2001) offers a comparison of the CCG approach of Jacobson (1999) with a type-logical treatment based on identity semantics for anaphora, in combination with a restricted copying rule in syntax, in the form of a controlled structural rule of Contraction. An alternative perspective on scope and anaphora, more in the spirit of Curry's tectogrammatic program, simplifies the categorial type theory to a nondirectional LP system, and enforces structural control by introducing lambda term labeling also for the form dimension of grammatical signs. Oehrle (1994) provides an early formulation of this approach, which has recently found new advocates.

Processing Issues

The interpretation procedure discussed above is essentially dynamic: interpretations are assembled 'online' in the course of the derivation process, rather than being computed *post hoc* from a given static structure. This has led to a distinctly 'categorial' view on processing issues.

Incrementality and information structure

The flexible notion of derivational constituency engendered by type-changing principles makes left-to-right parsing directly compatible with incremental interpretation. The resulting categorial modeling of natural-language processing has been worked out in Steedman (2000a). This work shows that derivational constituency is guided by 'prosodic' articulation (intonation contour). To do justice to this dimension of grammatical organization, one needs a richer notion of semantic interpretation, accommodating notions of focus and information structure. Steedman's proposals are formulated in the CCG style; Hendriks (1999) analyzes information packaging and intonation contour in multimodal type-logical terms.

Proof nets

A novel computational view on natural-language processing derives from the 'proof net' approach. Proof nets were originally developed in the context of linear logic, where they elegantly capture the essence of resource-sensitive derivations in graph-theoretical terms. Moot and Puite (2002) refine the proof-net techniques for use with the grammatical type logics discussed in this article, where, apart from resource multiplicity, structural patterns also have to be taken into account.

Johnson (1998) and Morrill (2000) have pointed out that proof nets offer an attractive perspective on performance phenomena. A net can be built in a left-to-right incremental fashion by establishing

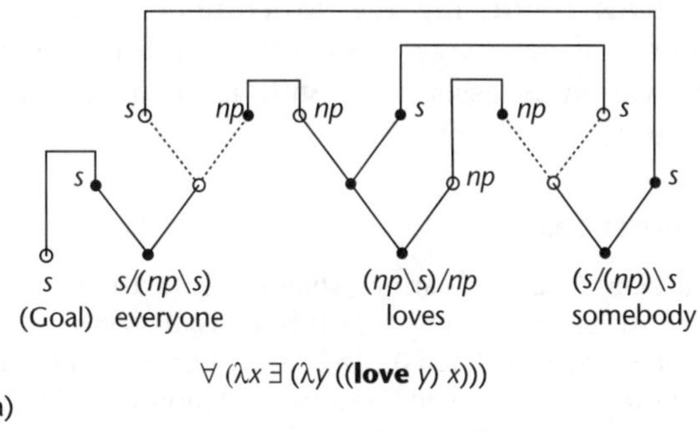

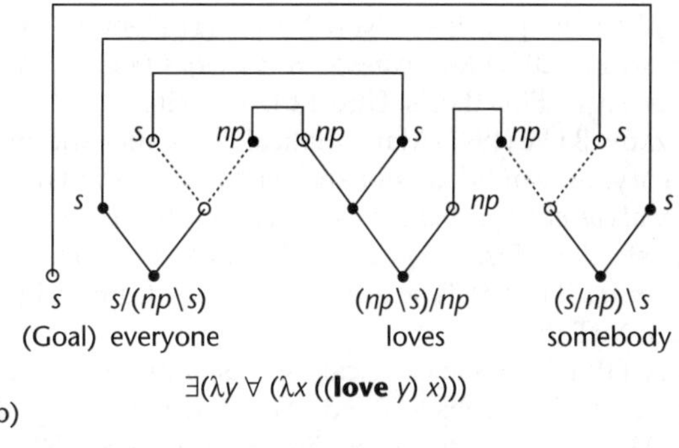

Figure 7. Incremental proof-net construction. The diagrams show a proof net for the sentence 'everyone loves somebody'. Formula decomposition trees have polarized vertices (black for input, white for output). Solid edges represent positive slashes; dotted edges represent negative slashes. A linking of leaves with opposite polarities is well-formed if it produces a graph that is connected, acyclic (for each removal of a dotted edge from a pair), and planar. The net is constructed in a left-to-right incremental fashion. Processing complexity is measured in terms of the number of unresolved dependencies. (a) The subject-wide-scope reading for 'everyone loves somebody' (maximum of 3 unresolved dependencies). This is preferred over (b), the object-wide-scope reading (maximum of 4 unresolved dependencies). Adapted from Johnson (1998) and Morrill (2000).

possible linkings between the input–output connectors of lexical items as they are presented in real time. This suggests a simple complexity measure on a traversal, given by the number of unresolved dependencies between literals. This complexity measure on incremental proof-net construction makes the right predictions about a number of well-known processing issues, such as the difficulty of center embedding, garden-path effects, attachment preferences, and preferred scope construals in ambiguous constructions. An illustration is presented in Figure 7.

CONCLUSION

In this article we have presented the core part of the categorial vocabulary: a set of type forming connectives controlling the structure-building operations of language, and the way in which a derivation is associated with an instruction for meaning assembly.

References

Adriaans P (2002) Learning shallow context-free languages under simple distributions. In: Vermeulen K and Copestake A (eds) *Algebras, Diagrams and Decisions in Language, Logic and Computation*, Stanford, CA: CSLI.

van Benthem J (1987) Categorial grammar and lambda calculus. In: Skordev D (ed.) *Mathematical Logic and Its Applications*, pp. 39–60. New York, NY: Plenum Press.

Bernardi R (2002) *Reasoning with Polarities in Categorial Type Logic*. PhD thesis, Utrecht University.

Buszkowski W (1997) Mathematical linguistics and proof theory. In: van Benthem J and ter Meulen A (eds) *Handbook of Logic and Language*, chap. XII, pp. 683–736. Amsterdam: Elsevier/Cambridge, MA: MIT Press.

Carpenter B (1998) *Type-Logical Semantics*. Cambridge, MA: MIT Press.

Curry HB (1961) Some logical aspects of grammatical structure. In: Jacobson R (ed.) *Structure of Language and Its Mathematical Aspects*, pp. 56–68. Providence, RI: American Mathematical Society.

Došen K (1992) A brief survey of frames for the Lambek calculus. *Zeitschrift für mathematischen Logik und Grundlagen der Mathematik* **38**: 179–187.

Foret A (2001) On mixing deduction and substitution in Lambek categorial grammars. In: de Groote P, Morrill G and Retoré C (eds) *Logical Aspects of Computational Linguistics*, pp. 158–174. Berlin, Germany: Springer.

Hendriks H (1999) The logic of tune: a proof-theoretic analysis of intonation. In: Lecomte A, Lamarche F and Perrier G (eds) *Logical Aspects of Computational Linguistics*, pp. 132–159. Berlin, Germany: Springer.

Jacobson P (1999) Towards a variable-free semantics. *Linguistics and Philosophy* **22**(2): 117–184.

Jäger G (2001) Anaphora and quantification in categorial grammar. In: Moortgat M (ed.) *Logical Aspects of Computational Linguistics*, pp. 70–90. Berlin, Germany: Springer.

Johnson M (1998) Proof nets and the complexity of processing center-embedded constructions. *Journal of Logic, Language and Information* **7**(4): 443–447.

Kanazawa M (1998) *Learnable Classes of Categorial Grammars*. Stanford, CA: CSLI.

Kurtonina N (1995) *Frames and Labels: A Modal Analysis of Categorial Inference*. PhD thesis, OTS Utrecht, ILLC Amsterdam.

Kurtonina N and Moortgat M (1997) Structural control. In: Blackburn P and de Rijke M (eds) *Specifying Syntactic Structures*, pp. 75–113. Stanford, CA: CSLI.

Lambek J (1958) The mathematics of sentence structure. *American Mathematical Monthly* **65**: 154–170.

Lambek J (1961) On the calculus of syntactic types. In: Jacobson R (ed.) *Structure of Language and its Mathematical Aspects*, pp. 166–178. Providence, RI: American Mathematical Society.

Moortgat M (1996) Multimodal linguistic inference. *Journal of Logic, Language and Information* **5**(3–4): 349–385.

Moortgat M (1999) Constants of grammatical reasoning. In: Bouma G, Hinrichs E, Kruijff GJ and Oehrle RT (eds) *Constraints and Resources in Natural Language Syntax and Semantics*, pp. 195–219. Stanford, CA: CSLI.

Moot R (2002) *Proof Nets for Linguistic Analysis*. PhD thesis, Utrecht University.

Moot R and Puite Q (2002) Proof nets for the multimodal Lambek calculus. *Studia Logica* **71**. [Special issue on the occasion of Lambek's eightieth birthday, edited by Wojciech Buszkowski and Michael Moortgat.]

Morrill G (2000) Incremental processing and acceptability. *Computational linguistics* **26**(3): 319–338.

Oehrle RT (1994) Term-labeled categorial type systems. *Linguistics and Philosophy* **17**(6): 633–678.

Pentus M (1993) Lambek grammars are context free. In: *Proceedings of the 8th Annual IEEE Symposium on Logic in Computer Science*, pp. 429–433. IEEE Computer Society Press.

Pentus M (1994) Language completeness of the Lambek calculus. In: *Proceedings of the 9th Annual IEEE Symposium on Logic in Computer Science*, pp. 487–496. IEEE Computer Society Press.

Steedman M (2000a) Information structure and the syntax–phonology interface. *Linguistic Inquiry* **31**(4): 649–689.

Steedman M (2000b) *The Syntactic Process*. Cambridge, MA: MIT Press.

Tiede HJ (2001) Lambek calculus proofs and tree automata. In: Moortgat M (ed.) *Logical Aspects of Computational Linguistics*, pp. 251–265. Berlin, Germany: Springer.

Vijay-Shanker K and Weir D (1994) The equivalence of four extensions of context free grammars. *Mathematical Systems Theory* **27**(6): 511–546.

Further Reading

Ajdukiewicz K (1935) Die syntaktische Konnexität. *Studia Philosophica* **1**: 1–27. [The seminal work on categorial grammar. English translation in: McCall S (ed.) (1996) *Polish Logic, 1920–1939* pp. 207–231. Oxford, UK: Oxford University Press.]

Bar-Hillel Y (1953) A quasi-arithmetical notation for syntactic description. *Language* **29**: 47–58. [Important continuation of Ajdukiewicz' work.]

Buszkowski W, Marciszewski W and van Benthem J (eds) (1998) *Categorial Grammar*. Amsterdam, Netherlands: Benjamins. [Includes several early papers, including Lambek's seminal paper of 1958.]

Dowty D, Wall R and Peters S (1981) *Introduction to Montague Semantics*. Dordrecht, Netherlands: Reidel.

Girard J-Y (1987) Linear logic. *Theoretical Computer Science* **50**: 1–102.

Girard J-Y, Lafont Y and Taylor P (1988) *Proofs and Types*. Cambridge, UK: Cambridge University Press. [A good source for the Curry–Howard interpretation.]

Kruijff G-J and Oehrle R (2002) *Resource Sensitivity in Binding and Anaphora*. Dordrecht, Netherlands: Reidel. [A reflection of current categorial views on anaphora and binding.]

Lambek J (1999) Type grammar revisited. In: Lecomte A, Lamarche F and Perrier G (eds) *Logical Aspects of Computational Linguistics*, pp. 1–27. Berlin, Germany: Springer.

Montague R (1974) *Formal Philosophy: Selected papers of Richard Montague*. Yale, CT: Yale University Press.

Moortgat M (1997) Categorial type logics. In: van Benthem J and ter Meulen A (eds) *Handbook of Logic and Language*, chap. II, pp. 93–177. Amsterdam: Elsevier / Cambridge, MA: MIT Press. [The primary source for this article.]

Moot R (2002) *Grail*. http://www.let.uu.nl/~Richard.Moot/personal/grail.html. [A grammar development environment that provides a versatile computational tool for categorial exploration. The user interacts with the kernel via a graphical user interface, which provides control over the lexicon and the structural module, and which gives access to a fully-fledged proof-net-based debugger. A number of sample fragments can be accessed online at http://www.grail.let.uu.nl/tour.pdf.]

Morrill G (1994) *Type Logical Grammar: Categorial Logic of Signs*. Dordrecht, Netherlands: Kluwer. [A rich fragment of syntactic and semantic phenomena in the grammar of English, using a variety of type-forming operations (Boolean and quantificational) in addition to the composition operators.]

Oehrle R, Bach E and Wheeler D (eds) (1988) *Categorial Grammars and Natural Language Structures*. Dordrecht, Netherlands: Reidel. [A good picture of categorial research in the 1980s, in both the rule-based and the logical traditions.]

Restall G (2000) *An Introduction to Substructural Logics*. New York, NY: Routledge. [An accessible textbook.]

Retoré C and Stabler E (eds) (2002) *Resource Logics and Minimalist Grammars. Proceedings ESSLLI'99 workshop*. [Special issue on language and computation. An exploration of the connections between linear logic, categorial grammar, and computational formulations of minimalist grammars.]

van Benthem J (1995) *Language in Action: Categories, Lambdas and Dynamic Logic*. Cambridge, MA: MIT Press. [A detailed study of the relations between categorial derivations, type theory and lambda calculus, and of the place of categorial grammars within the general landscape of resource-sensitive logics.]

van Benthem J and ter Meulen A (eds) (1997) *Handbook of Logic and Language*. Amsterdam: Elsevier/Cambridge, MA: MIT Press. [Includes chapters on the connections between categorial type systems and mathematical linguistics and proof theory, formal learning theory, type theory, and Montague grammar.]

Categorical Perception

Intermediate article

Stevan Harnad, University of Quebec, Montreal, Canada

CONTENTS

CATEGORIES: CATEGORICAL AND CONTINUOUS

A category, or kind, is a set of things. Membership in the category may be (1) all-or-none, as with 'bird': something either is a bird or it isn't a bird; a penguin is 100 percent bird, a platypus is 100 percent not-bird. In this case we would call the category 'categorical'. Or membership might be (2) a matter of degree, as with 'big': some things are 'more big' and some things are 'less big'. In this case the category is 'continuous' (or rather, degree of membership corresponds to some point along a continuum). There are range or context effects as well: elephants are relatively big in the context of animals, relatively small in the context of bodies in general, if we include planets.

Many categories, however, particularly concrete sensorimotor categories (things we can see and touch), are a mixture of the two: categorical at an everyday level of magnification but continuous at a more microscopic level. Color categories are good examples: central reds are clearly reds, and not shades of yellow. But in the orange region of the spectral continuum, red/yellow is a matter of degree; context and contrast effects can also move these regions around somewhat. Perhaps even with 'bird', an artist or genetic engineer could design intermediate cases in which their 'birdness' was only a matter of degree.

RESOLVING THE 'BLOOMING, BUZZING CONFUSION'

Categories are important because they determine how we see and act upon the world. As William James noted, we do not see a continuum of 'blooming, buzzing confusion' but an orderly world of discrete objects. Some of these categories are 'prepared' in advance by evolution: the frog's brain is born already able to detect 'flies'; it needs only normal exposure rather than any special learning in order to recognize and catch them. Humans have such innate category-detectors too: the human face itself is probably an example. So too are our basic color categories; although, according to the 'Whorf hypothesis' (Whorf 1956; also called the 'linguistic relativity' hypothesis) colors are determined by how our culture and language happens to subdivide the spectrum (we will return to this).

But if one opens up a dictionary at random and picks out a content word, it is probable that it names a category we have learned to detect, rather than one that our brains were innately prepared by evolution to detect. The generic human face may be an innate category for us, but surely all the specific people we know and can name are not. 'Red' and 'yellow' may be inborn, but what about 'scarlet' and 'crimson'?

THE MOTOR THEORY OF SPEECH PERCEPTION

And what about the very building blocks of the language we use to name categories. Are our speech sounds – ba, da, ga – innate or learned? The first question we must answer about them is whether they are categories at all, or merely arbitrary points along a continuum. It turns out that if one analyzes the sound spectrogram of ba and pa, for example, both turn out to lie along an acoustic continuum called 'voice-onset-time'. With a technique similar to the one used in morphing visual images continuously into one another, it is possible to 'morph' a ba gradually into a pa and beyond by gradually increasing the voicing parameter.

Liberman *et al.* (1957) reported that when people listen to sounds that vary along the voicing continuum, they hear only ba's and pa's, nothing in between. This effect – in which perception jumps abruptly from one category to another at a certain point along a continuum, instead of changing gradually – he called 'categorical perception' (CP). He suggested that CP was unique to speech, that it made speech special, and that its explanation lay in the anatomy of speech production. This came to be called 'the motor theory of speech perception'.

According to the (now abondoned) motor theory, the reason we perceive an abrupt change between ba and pa is that the way we hear speech sounds is mediated by the way we produce them when we speak. What is varying along this continuum is voice-onset-time: the 'b' in ba is voiced and the 'p' in pa is not. Unlike the synthetic morphing apparatus, our natural vocal apparatus is not capable of producing anything in between. So when I hear a sound from the voicing continuum, my brain perceives it by trying to match it with what it would have had to do to produce it. Since the only thing I can produce is ba or pa, I will perceive any of the synthetic stimuli along the continuum as either ba or pa, whichever it is closer to. A similar CP effect is found with ba/da; these too lie along a continuum acoustically, but vocally ba is formed with the two lips, da with the tip of the tongue and the hard palate, and our anatomy does not allow any intermediates.

The motor theory of speech perception explained how speech was special and why speech sounds are perceived categorically: sensory perception is mediated by motor production. Wherever production is categorical, perception will be categorical; where production is continuous, perception will be continuous. And indeed vowel categories like a/u were found to be much less categorical than ba/pa or ba/da (less categorical, but not altogether continuous either; we will return to this).

ACQUIRED DISTINCTIVENESS

If motor production mediates sensory perception, then one assumes that this CP effect is a result of learning to produce speech. Early research, however, found that infants show speech CP before they begin to speak. Perhaps, then, it is an innate effect, evolved to 'prepare' us to learn to speak. But Kuhl found that chinchillas also show 'speech CP' even though they never learn to speak, and presumably did not evolve to do so (Kuhl, 1987). Lane (1965) went on to show that CP effects can be induced by learning alone, with a purely sensory (visual) continuum in which there is no motor production discontinuity to mediate the perceptual discontinuity. He concluded that speech CP is not special after all but merely a special case of Lawrence's (1950) classic demonstration that stimuli to which you learn to make a different response become more distinctive, and stimuli to which you learn to make the same response become more similar.

It also became clear that CP was not quite the all-or-none effect Liberman had originally thought it was: it is not that all pa's are indistinguishable and all ba's are indistinguishable. We can hear the differences, just as we can see the differences between different shades of red. It is just that the within-category differences (pa1/pa2 or red1/red2) seem to be much smaller than the between-category differences (pa2/ba1 or red2/yellow1), even when the size of the underlying physical differences (voicing, wavelength) are the same.

WITHIN-CATEGORY COMPRESSION AND BETWEEN-CATEGORY SEPARATION

This evolved into the contemporary definition of CP that is no longer peculiar to speech or dependent on the motor theory: CP occurs whenever perceived within-category differences are compressed and/or between-category differences are separated, relative to some baseline. The baseline might be the actual size of the physical differences involved, or, in the case of learned CP, it might be the perceived similarity or discriminability within and between categories before the categories were learned.

A typical learned CP experiment would be the following. A set of stimuli is tested for pairwise similarity or discriminability. In the case of similarity, multidimensional scaling might be used to scale the rated pairwise similarity of the set of stimuli. In the case of discriminability, same/different judgments and signal detection analysis might be used to estimate the discriminability of a set of stimuli. Then the same subjects or a different set are trained, using trial and error and corrective feedback, to sort the stimuli into two or more categories. After the categorization has been learned, similarity or discriminability are tested and compared against the untrained data. If there is significant within-category compression and/or between-category separation, this is operationally defined as CP (Harnad, 1987).

THE WHORF HYPOTHESIS

We can now return both to the 'Whorf hypothesis' and the 'weaker' CP for vowels: according to the

Whorf hypothesis (of which Lawrence's acquired similarity/distinctiveness effects would simply be a special case), colors are perceived categorically only because they happen to be named categorically; our subdivisions of the spectrum are arbitrary, learned, and vary across cultures and languages (Whorf, 1964). But Berlin and Kay (1969) showed that this was not so: not only do most cultures and languages subdivide and name the color spectrum the same way, but even for those who don't, the regions of compression and separation are the same. We all see blues as more alike and greens as more alike, with a fuzzy boundary in between, whether or not we have named the difference. So there is no Whorfian learning effect with colors; or is there?

EVOLVED CP

First, let us go back to vowels. The signature of CP is within-category compression and/or between-category separation. The size of the CP effect is merely a scaling factor; it is this compression/separation 'accordion effect' that is CP's distinctive feature. In this respect, the 'weaker' CP effect for vowels, whose motor production is continuous rather than categorical but whose perception is by this criterion categorical, is every bit as much of a CP effect as the ba/pa and ba/da effects. But, as with colors, it looks as if the effect is an innate one. Our sensory category detectors for color and speech sounds are born already 'biased' by evolution: the perceived color and speech sound spectrum is born 'warped' with these compression/separations.

LEARNED CP

Is that all there is to it? Apparently not. There are still the Lane/Lawrence demonstrations, lately replicated and extended by Goldstone (1994, 1999), that CP can be induced by learning alone. And there are also the countless categories cataloged in our dictionaries that could not possibly be inborn (though nativist theorists such as Fodor (e.g. 1983) have sometimes seemed to suggest that all of our categories are inborn). There are even recent demonstrations that although the primary color and speech categories are probably inborn, their boundaries can be modified or even lost as a result of learning, and weaker secondary boundaries can be generated by learning alone (Roberson *et al.*, 2000).

Perhaps CP performs some useful function in categorization? In the case of innate CP, our categorically biased sensory detectors pick out their prepared color and speech sound categories far more readily and reliably than if our perception had been continuous. Could something similar be the case for our repertoire of learned categories too?

COMPUTATIONAL AND NEURAL MODELS OF CP

Computational modeling (Tijsseling and Harnad, 1997; Damper and Harnard 2000) has shown that many types of category-learning mechanisms (e.g. both backpropagation and competitive networks) display CP-like effects. In backpropagation nets, the hidden-unit activation patterns that 'represent' an input build up within-category compression and between-category separation as they learn; other kinds of net display similar effects. CP seems to be a means to an end: inputs that differ among themselves are 'compressed' onto similar internal representations if they must all generate the same output; and they become more separate if they must generate different outputs. The network's 'bias' is what filters inputs onto their correct output category. The nets accomplish this by selectively detecting (after much trial and error, guided by error-correcting feedback) the invariant features that are shared by the members of the same category and that reliably distinguish them from members of different categories; the nets learn to treat all other variation as being irrelevant to the categorization.

Very little is known yet about the brain mechanisms of category perception and learning. The computational models are really causal hypotheses about what the brain might be doing. Neural data provide correlates of CP and of learning (Sharma and Dorman, 1999). Differences between event-related potentials recorded from the brain have been found to be correlated with differences in the perceived category of the stimulus viewed by the subject. Neural imaging studies have shown that these effects are localized and even lateralized to certain brain regions in subjects who have successfully learned the category, and are absent in subjects who have not (Seger *et al.*, 2000).

LANGUAGE-INDUCED CP

Both innate and learned CP are sensorimotor effects. The compression/separation biases are sensorimotor biases, and presumably had sensorimotor origins, either during the sensorimotor life history of the organism, in the case of learned CP, or during the sensorimotor life history of the species, in the case of innate CP. The neural net I/O models are also compatible with this fact: their I/O biases derive from their I/O history. But when we look at our repertoire of categories in a dictionary, it is highly unlikely that many of them had a direct sensorimotor history during our lifetimes, and even less likely in our ancestors' lifetimes. How many of us have seen unicorns in real life? We have seen pictures of them, but what had those who first drew those pictures seen? And what about categories I cannot draw or see (or taste or touch)? What about the most abstract categories, such as goodness and truth?

Some of our categories must originate from a source other than direct sensorimotor experience, and here we return to language and the Whorf hypothesis. Can categories, and their accompanying CP, be acquired through language alone? Again, there are some neural net simulation results suggesting that, once a set of category names has been 'grounded' through direct sensorimotor experience, they can be combined into Boolean combinations (man = male and human) and into still higher-order combinations (bachelor = unmarried and man). These combinations not only pick out the more abstract, higher-order categories in much the way the direct sensorimotor detectors do, but also inherit their CP effects, as well as generating some of their own. Bachelor inherits the compression/separation of unmarried and man, and adds a layer of separation/compression of its own (Cangelosi *et al.*, 2000; Cangelosi and Harnad 2001).

These language-induced CP-effects remain to be directly demonstrated in human subjects; so far only learned and innate sensorimotor CP have been demonstrated (Pevtzow and Harnad 1997; Livingston *et al.*, 1998). The latter show the Whorfian power of naming and categorization, in warping our perception of the world. That is enough to rehabilitate the Whorf hypothesis from its apparent failure on color terms (and perhaps also from its apparent failure on eskimo snow terms – see Pullum, 1989); but to show that it is a full-blown language effect, and not merely a vocabulary effect, it will have to be shown that our perception of the world can also be warped, not just by how things are named but by what we are told about them.

References

Berlin B and Kay P (1969) *Basic Color Terms: Their Universality and Evolution.* Berkeley, CA: University of California Press.

Damper RI and Harnad S (2000) Neural network modeling of categorical perception. *Perception and Psychophysics* **62**(4): 843–867. [http://cogprints.soton.ac.uk/documents/disk0/00/00/16/20/index.html.]

Eimas PD, Siqueland ER, Jusczyk PW and Vigorito J (1971) Speech perception in infants. *Science* **171**: 303–6.

Fodor J (1983) *The Modularity of Mind.* Cambridge, MA: MIT Press.

Goldstone RL (1994) Influences of categorization on perceptual discrimination. *Journal of Experimental Psychology: General* **123**: 178–200.

Goldstone RL (1999) Similarity. In: Wilson RA and Keil FC (eds) *MIT Encylopedia of the Cognitive Sciences*, pp. 763–765. Cambridge, MA: MIT Press.

Kuhl PK (1987) The special-mechanisms debate in speech perception: nonhuman species and nonspeech signals. In: Harnad S (ed.) *Categorical Perception: the Groundwork of Cognition.* New York, NY: Cambridge University Press.

Lane H (1965) The motor theory of speech perception: a critical review. *Psychological Review* **72**: 275–309.

Lawrence DH (1950) Acquired distinctiveness of cues: II. Selective association in a constant stimulus situation. *Journal of Experimental Psychology* **40**: 175–188.

Liberman AM, Harris KS, Hoffman HS and Griffith BC (1957) The discrimination of speech sounds within and across phoneme boundaries. *Journal of Experimental Psychology* **54**: 358–368.

Pullum GK (1989) The great eskimo vocabulary hoax. *Natural Language and Linguistic Theory* **7**: 275–281.

Seger CA, Poldrack RA, Prabhakaran V, Zhao M, Glover GH and Gabrieli JDE (2000) Hemispheric asymmetries and individual differences in visual concept learning as measured by functional MRI. *Neuropsychologia* **38**(9): 1316–1324.

Sharma A and Dorman MF (1999) Cortical auditory evoked potential correlates of categorical perception of voice-onset time. *Journal of the Acoustical Society of America* **106**(2): 1078–1083.

Tijsseling A and Harnad S (1997) Warping similarity space in category learning by backprop nets. In: Ramscar M, Hahn U, Cambouropolos E and Pain H (eds) *Proceedings of SimCat 1997: Interdisciplinary Workshop on Similarity and Categorization*, pp. 263–269. Edinburgh, UK: Department of Artificial Intelligence, Edinburgh University. [http://cogprints.soton.ac.uk/documents/disk0/00/00/16/08/index.html.]

Whorf BL (1964) *Language, Thought and Reality.* Cambridge MA: MIT Press.

Further Reading

Andrews J, Livingston K and Harnad S (1998) Categorical perception effects induced by category learning. *Journal of Experimental Psychology: Learning, Memory, and Cognition* **24**(3): 732–753.

Belpaeme T (2002) *Factors Influencing the Origins of Colour Categories.* PhD thesis. Artificial Intelligence Lab, Free University of Brussels. [http://arti.vub.ac.be/~tony/phd/index.htm.]

Bimler D and Kirkland J (2001) Categorical perception of facial expressions of emotion: evidence from multidimensional scaling. *Cognition and Emotion* **15**: 633–658.

Calder AJ, Young AW, Perrett DI, Etcoff NL and Rowland D (1996) Categorical perception of morphed facial expressions. *Visual Cognition* **3**: 81–117.

Campanella S, Quinet O, Bruyer R, Crommelinck M and Guerit JM (2002) Categorical perception of happiness and fear facial expressions : an ERP study. *Journal of Cognitive Neuroscience* **14**(2): 210–227.

Cangelosi A, Greco A and Harnad S (2000) From robotic toil to symbolic theft: grounding transfer from entry-level to higher-level categories. *Connection Science* **12**(2): 143–162. [http://cogprints.soton.ac.uk/documents/disk0/00/00/16/47/index.html.]

Cangelosi A and Harnad S (2001) The adaptive advantage of symbolic theft over sensorimotor toil: grounding language in perceptual categories. *Evolution of Communication* **4**(1): 117–142. [http://cogprints.soton.ac.uk/documents/disk0/00/00/20/36/index.html.]

Goldstone RL, Lippa Y and Shiffrin RM (2001) Altering object representations through category learning. *Cognition* **78**: 27–43.

Guest S and Van Laar D (2000) The structure of colour naming space. *Vision Research* **40**: 723–734.

Harnad S (ed.) (1987) *Categorical Perception: the Groundwork of Cognition*. New York, NY: Cambridge University Press. [http://cogprints.soton.ac.uk/documents/disk0/00/00/15/71/index.html.]

Harnad S (1990) The symbol grounding problem. *Physica D* **42**: 335–346. [http://cogprints.soton.ac.uk/documents/disk0/00/00/06/15/index.html.]

Kotsoni E, de Haan M and Johnson MH (2001) Categorical perception of facial expressions by 7-month-old infants. *Perception* **30**: 1115–1125.

Pevtzow R and Harnad S (1997) Warping similarity space in category learning by human subjects: the role of task difficulty. In: Ramscar M, Hahn U, Cambouropolos E and Pain H (eds) *Proceedings of SimCat 1997: Interdisciplinary Workshop on Similarity and Categorization*, pp. 189–195. Edinburgh, UK: Department of Artificial Intelligence, Edinburgh University. [http://cogprints.soton.ac.uk/documents/disk0/00/00/16/07/index.html.]

Roberson D, Davies I and Davidoff J (2000) Color categories are not universal: replications and new evidence from a stone-age culture. *Journal of Experimental Psychology: General* **129**: 369–398.

Rossion B, Schiltz C, Robaye L, Pirenne D and Crommelinck M (2001) How does the brain discriminate familiar and unfamiliar faces? A PET study of face categorical perception. *Journal of Cognitive Neuroscience* **13**: 1019–1034.

Schyns PG, Goldstone RL and Thibaut J (1998) Development of features in object concepts. *Behavioral and Brain Sciences* **21**: 1–54.

Steels L (2001) Language games for autonomous robots. *IEEE Intelligent Systems* **16**(5): 16–22.

Steels L and Kaplan F (1999) Bootstrapping grounded word semantics. In: Briscoe T (ed.) *Linguistic Evolution through Language Acquisition: Formal and Computational Models*. Cambridge, UK: Cambridge University Press.

Categorization, Development of

Intermediate article

Jean M Mandler, University of California, San Diego, California, USA

CONTENTS

Categorization occurs when individual items that can be discriminated from one another are considered to be the same or belong together because they are alike in some way.

INTRODUCTION

Categorization is an automatic part of perceiving and conceptualizing. Whether perceiving something or thinking about it, the mind tends to subsume individual items into larger classes. The term 'categorization' is also applied to a task that individuals perform. Such tasks require deliberate choice as to the basis of similarity that is used to do the grouping: for example, in a set of red and green dogs and cats, one might choose to group them on the basis of similar color or similar kind. One can categorize in this way on many bases ranging from taxonomic categories to categories made up on the spot.

Behavior on deliberate categorization tasks changes dramatically in the early years, but the reasons for the changes are complex and varied. Children's understanding of the instructions, their construal of what is wanted, and the salience of different aspects of objects vary depending on school and home experiences. These variables influence the developmental change that has been reported on these tasks, from the preference of young children to categorize objects by their overall similarity to the preference of older children and adults to concentrate on dimensions of the objects such as size or shape (Smith and Kemler, 1978). Schooling is also implicated in the tendency of children and young adults to categorize taxonomically on these tasks, compared with preschool children and older adults, who are more likely to categorize objects by their functions or the events in which they take part (Smiley and Brown, 1979).

There is a good deal of research on categorization in infancy using instructionless tasks that rely on infants' spontaneous categorizing behavior, and that therefore provide information about the earliest bases of categorization. A typical experimental procedure familiarizes infants with one category and then measures preferential looking or examining to a member of another category. This article concentrates on the infancy period when the first perceptual and conceptual categories are formed.

MODELS OF CATEGORIZATION

One model of categorization posits that encountering similar stimuli leads to the formation of a prototype of the presented items. For example, upon seeing many dogs, one extracts the central tendency or principal components (as in factor analysis) of 'dogginess' in terms of physical features and presumably in terms of behavior as well. Even 3-month-old babies extract a perceptual prototype from a set of similar stimuli. For example, if shown a number of squares or triangles, each of which is somewhat distorted in shape, they will treat a regular square or triangle as more familiar than any of the distorted figures they have seen (Bomba and Siqueland, 1983).

Another model of categorization is that we accumulate many instances and compare new instances with this accumulated knowledge base. This procedure enables us to recognize a prototypical instance of a category because it is most similar to the largest number of examples. Although much published research differentiates these two theories of categorization, there is little to choose between them at a behavioral level and in terms of development almost no relevant data to decide between them.

PERCEPTUAL VERSUS CONCEPTUAL CATEGORIES

People categorize perceptually ('this dog looks like that dog') and conceptually ('courage and honesty are both virtues'). Both kinds of categorization can be characterized as involving similarity (e.g. in shape or value system), and so some researchers describe both kinds of categorization in terms of a single process of computing similarity (either in terms of similarity to a prototype or similarity of exemplars). However, it is not always easy to specify the similarities among exemplars of conceptual categories such as virtues (exactly how are courage and kindness similar?), so some researchers find it preferable to say that conceptual categories are based on a definition or rule. On this view, then, there are two different processes involved in categorization: assessing similarity, and applying a rule. (*See* **Concept Learning**; **Conceptual Change**)

It seems plausible that current debate on this topic reflects differences in the way that perceptual and conceptual categories are formed. The items being categorized differ greatly in the two cases, and there appear to be a number of differences in their formation, suggesting that a two-process account may be necessary. Perceptual categories are almost certainly based on similarity, but many conceptual categories may be more rule-like in nature. For example, there is evidence that the perceptual categorization of faces as male or female is an automatic part of the operation of the visual system and is implicit in nature. We have all been able to categorize male and female faces since infancy, but we cannot say accurately what the differences are. Conceptual categories, on the other hand, are based on notions that are open to our inspection and we can often give a rule for why two things belong to different classes. Even a category of objects such as animals can be described as rule-based. For example, for infants an animal might be anything that moves itself and acts on other objects, regardless of what it looks like.

Perceptual categorization happens automatically and does not require conscious attention. The process, which enables generalization from prior instances of a category to new instances, is a fundamental aspect of perception and is present very early in life. For example, using a familiarization/preferential-looking technique, 3-month-old babies have been found not only to extract perceptual prototypes but also to categorize realistic pictures of cats after only a few exposures and differentiate them from pictures of dogs (Quinn *et al.*, 1993). If the stimuli are more variable, the process takes longer; hence, it takes more exposure to categorize pictures of dogs than of cats.

Unfortunately, most of the research on perceptual categorization of realistic stimuli at this young age has been conducted with pictures of animal kinds, which may not be representative of category formation in general. It has also tended to contrast objects at what is known as the basic level (Rosch *et al.*, 1976). For example, dogs have been contrasted with cats, as opposed to a subordinate comparison of poodles contrasted with terriers, or a superordinate comparison of animals contrasted with vehicles. However, there are some data indicating that babies at 3 months and 6 months of age can also categorize pictures of tables, differentiating them from chairs. They also discriminate pictures of mammals from vehicles at this age, although more trials are required, presumably because of the greater perceptual variability of these global categories. (Superordinate categories are usually termed 'global categories' in the infancy literature, because infants typically make few conceptual subdivisions within such large classes.)

By about 6 months, as infants begin to manipulate objects, a familiarization/preferential-examining technique is used with models of objects to assess categorization. Seven-month-old infants categorize animals as different from vehicles. By 9 months, they categorize animals as different from furniture as well, and by 11 months categorize both these classes as different from plants and utensils. The range of perceptual variation in these global categories is much greater than for basic-level categories and even for four-legged mammals (compare dogs with horses versus dogs with birds). Nevertheless, 9-month-old infants have no difficulty in differentiating models of birds with outstretched wings from airplanes in spite of the high degree of similarity in appearance of the two classes. Because of the apparent lack of difficulty posed by within-category variability in the case of animals, and between-category similarity in the case of birds and airplanes, it has been suggested that perceptual similarity alone cannot account for this kind of categorization, and more conceptual information, as opposed to the physical appearance of the exemplars, is being used (Mandler, 2000).

GLOBAL VERSUS BASIC-LEVEL CATEGORIES

The infants who at age 7–11 months categorize animals, vehicles, and furniture as different kinds do not categorize dogs versus cats or tables versus chairs on object examination tests. Nevertheless, infants aged 3–6 months who categorize pictures of dogs versus cats and tables versus chairs have more difficulty categorizing furniture versus vehicles. Although the dissociations shown on these various categorization tests do not perfectly differentiate global from basic-level categories, there are enough differences in categorizing behavior to make it likely that different processes are being measured when infants categorize objects rather than pictures. Eventually, of course, pictures and objects produce similar data, but this is not the case in infancy, which makes the study of categorization at this period of life particularly informative.

Perceptual processes are capable of categorizing differently shaped pictures, even if the patterns are novel or meaningless. Hence, 3-month-old infants who have had no experience with dogs and cats nevertheless quickly learn to tell pictures of them apart. However, when perceptual variation is great, as is often the case for superordinate (global) categories, it may not be possible for the perceptual system to extract the principal components of the displayed patterns, and so this kind of perceptual categorization fails.

In contrast to the automatic perceptual processes involved in perceptual categorization, global categorization of animals, vehicles, furniture, and plants appears to be a response to a conceptual difference between these very different classes (Mandler, 2000). It is more selective in nature (and hence more 'rule-like'), acting on those aspects of events that have attracted infants' attention and interest. Young infants are particularly interested in differences in the way that animals and inanimate things move and the kinds of interactions in which they engage. Even relatively superficial analysis of the activities of animate and inanimate objects results in concepts that differentiate these classes. For example, such analysis leads to categorization of animals as things that start themselves and interact with other objects from a distance in a contingent fashion, and categorization of inanimate objects as things that either do not move at all or, if they do, do not start themselves and do not act on other objects from a distance. In the early months of life infants may not be able to conceptualize global categories more finely than this. It is relatively easy to differentiate the behavior of an animal from a piece of furniture, but it requires considerably more analysis to conceptualize how a dog differs from a cat or a car from a truck. (*See* **Infant Cognition**; **Object Perception, Development of**)

TIME COURSE OF CATEGORIZATION

Perceptual categorization begins at or near birth. To date, all the research indicating conceptual categorization has involved manipulating objects, which means that no data are available for infants younger than about 6–7 months. One piece of evidence for the conceptual nature of the global categories found in the second 6-month period of life is that it is these categories that are used for purposes of inductive inference. By 9–11 months of age, infants generalize the behavior of an observed animal, such as a dog drinking, to all animals including fish. They generalize using a key on a car to all vehicles, including airplanes (Mandler and McDonough, 1996). At the same time they do not differentiate between cups and pans as appropriate containers from which to drink, or between beds and bathtubs as appropriate places to sleep. The initially global categories become gradually differentiated during the second year, with current evidence suggesting that (at least in American culture) artefacts are differentiated earlier than natural kinds. The vocabulary that children are learning contributes to this development (Waxman, 1999). By the age of 2 years, children's conceptual categories become increasingly differentiated and so basic-level inductions result. By 4 years of age, the adult tendency to be more certain of inductions made on the basis of smaller classes is already established (Gelman and O'Reilley, 1988).

CONCLUSION

Although more research is needed, the preponderance of evidence indicates that beginning in infancy at least two kinds of categorization occur. The first is an automatic part of perception that computes the perceptual similarity of objects. This leads to implicit categorizing at the basic level. In addition, infants attempt to make sense of what they perceive – to construe the meaning or significance of the events they observe. This leads to explicit conceptual categorization, which tends to be global and overly general at the start.

It is not known for certain whether the two kinds of categorization involve the same or different processes. However, there are some distinctive differences that suggest different processes are at work. First, perceptual categorization occurs automatically and seems to use all the information encoded, whereas conceptual categorization depends on selective attention to only certain kinds of information. Second, perceptual categorization initially makes finer groupings than does conceptual categorization, which begins at a more global level. Third, the functions served by the two kinds of categories differ. Perceptual categorization is used to identify exemplars of conceptual categories. Infants may respond in a global conceptual way to animals as a class, but must learn to identify them by their features. For example, they learn that 'self-starting objects that do things to other objects' tend to have legs or wings, and so forth. Conceptual categories, on the other hand, are created to understand the world, to characterize the important differences among object kinds. Hence it is these categories that are used for purposes of inductive inferences.

References

Bomba PC and Siqueland ER (1983) The nature and structure of infant form categories. *Journal of Experimental Child Psychology* **35**: 294–328.

Gelman SA and O'Reilley AW (1988) Children's inductive inferences within superordinate categories: the role of language and category structure. *Child Development* **59**: 876–887.

Mandler JM (2000) Perceptual and conceptual processes in infancy. *Journal of Cognition and Development* **1**: 3–36.

Mandler JM and McDonough L (1996) Drinking and driving don't mix: inductive generalization in infancy. *Cognition* **59**: 307–335.

Quinn PC, Eimas PD and Rosenkrantz SL (1993) Evidence for representations of perceptually similar natural categories by 3-month-old and 4-month-old infants. *Perception* **22**: 463–475.

Rosch E, Mervis CB, Gray W, Johnson D and Boyes-Braem P (1976) Basic objects in natural categories. *Cognitive Psychology* **3**: 382–439.

Smiley SS and Brown AL (1979) Conceptual preference for thematic or taxonomic relations: a nonmontonic age trend from preschool to old age. *Journal of Experimental Child Psychology* **28**: 249–257.

Smith LB and Kemler DG (1978) Levels of experienced dimensionality in children and adults. *Cognitive Psychology* **10**: 502–542.

Waxman SR (1999) Specifying the scope of 13-month-olds' expectations for novel words. *Cognition* **70**: 35–50.

Further Reading

Carey S (1985) *Conceptual Change in Childhood.* Cambridge, MA: MIT Press.

Gelman SA and Wellman HM (1991) Insides and essences: early understandings of the nonobvious. *Cognition* **38**: 213–244.

Keil FC (1989) *Concepts, Kinds, and Cognitive Development*. Cambridge, MA: MIT Press.

Mandler JM (1998) Representation. In: Kuhn D and Siegler R (eds) *Cognition, Perception, and Language*, vol. 2 of *Handbook of Child Psychology*, pp. 255–308. New York, NY: John Wiley.

Mandler JM (2000) Perceptual and conceptual processes in infancy. *Journal of Cognition and Development* **1**: 3–36.

Markman EM (1989) *Categorization and Naming in Children: Problems of Induction*. Cambridge, MA: MIT Press.

Waxman SR (1999) The dubbing ceremony revisited: object naming and categorization in infancy and early childhood. In: Medin DL and Atran S (eds) *Folkbiology*, pp. 233–284. Cambridge, MA: MIT Press.

Causal Attribution

See **Cultural Differences in Causal Attribution**

Causal Perception, Development of

Intermediate article

Lisa M Oakes, University of Iowa, Iowa City, Iowa, USA

CONTENTS

The perception of causality develops during the first year of life. In general, this development progresses from infants perceiving the individual elements of events (e.g. particular objects, whether or not those objects touch) to their perceiving the relationship between those objects.

INTRODUCTION

Perceiving cause-and-effect relationships is important for understanding how things work, how to produce outcomes and how to link events that co-occur. Philosophers and psychologists have long debated the origins of causal understanding. Hume, for example, argued that real causal connections are unknowable. According to this view, humans determine causality through experience with regularities in the environment and causal perception should develop gradually. Infants may perceive the causality of some events from an early age, but a sophisticated appreciation of causality in a wide range of events develops as they gain experience of the world. Others have argued that the idea of cause and effect is an innate predisposition of the mind. According to this view, causality itself can be perceived without prior experience of events. Therefore young infants should be able to perceive causality in a wide range of events from an early age, and little development should be observed.

PERCEPTION OF LAUNCHING AND COLLISION EVENTS

Psychologists have primarily studied the development of infants' causal perception by assessing their perception of launching events or collisions (e.g. Leslie, 1984; Oakes and Cohen, 1990). At the start of a launching event, an object moves from one side of a display toward a second object sitting at rest in the middle of the display (Figure 1). The first object hits the second object, which begins to move immediately upon contact. Adults report that the first object appears to cause the second object to move (Michotte, 1963). The perception of causality can be interrupted by imposing a delay (i.e. the two objects remain stationary momentarily after they have made contact and before the second object begins to move) or a spatial gap between the two objects (i.e. the second object begins to move before the first object contacts it). Infants' sensitivity to causality is tested using habituation procedures. They are first shown one event on several trials until their looking time habituates, or decreases to some specified criterion (e.g. 50% of their original level of looking). Then they are shown one or more new events, and if they perceive those events as being different from the familiarization event, they will dishabituate or increase their looking. For example, if infants perceive the causality of events, then infants who are habituated to a delayed launching event, a noncausal event, will dishabituate to a novel causal event but not to a novel no-collision event. However, if infants do not attend to causality, but instead attend to the particular features of the event (e.g. whether or not the two objects touched), then they will dishabituate to novel events that differ in terms of those features regardless of changes in causality.

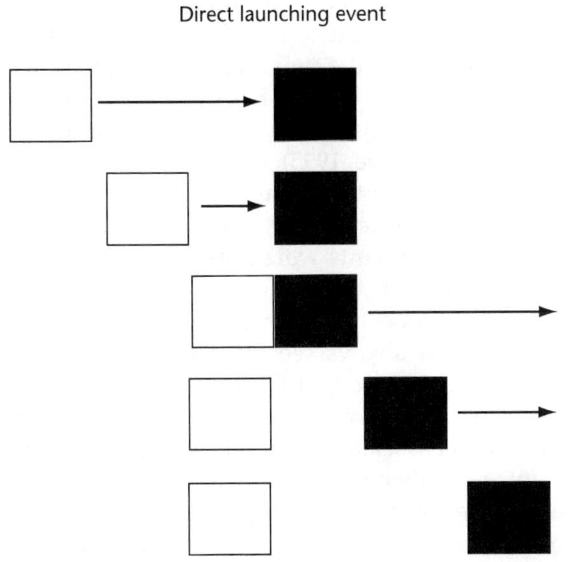

Figure 1. The sequence of actions in a typical launching event. The white object moves from the left of the screen towards the black object, which is initially stationary in the middle of the screen. When the white object makes contact with the black object, the black object immediately begins to move towards the right side of the screen. Reprinted from Oakes LM and Cohen LB (1995) with permission.

Using this type of procedure, studies have revealed that young infants are sensitive to causality. Infants aged 6 to 7 months treat the causal launching event as different from noncausal events, and they treat different noncausal events as if they are equivalent (Leslie, 1984; Oakes, 1994; Cohen and Amsel, 1998). Thus we might conclude that some aspects of causal perception are innate. However, causal perception develops considerably during infancy. Infants under 6 months of age do not perceive the causality of launching events. Instead, following habituation with one event, these younger infants dishabituate to changes in other types of features of the event (e.g. whether or not the two objects touched) (Cohen and Amsel, 1998). Even once infants begin to perceive causality, their perception is limited. Infants aged 6 to 7 months only perceive the causality of launching events if the objects are simple (e.g. colored squares). They do not perceive the causality of events if the objects in the event are complex (e.g. multicolored, multi-featured objects). By 10 months of age, infants perceive the causality of launching events involving both complex and simple objects. However, if the objects do not move along the same trajectory, 10-month-old infants fail to perceive causality (Oakes, 1994; Oakes and Cohen, 1995). In general, therefore, the perception of launching events develops over time, and salient perceptual features of the events (or the objects in the events) may overwhelm young infants' ability to perceive the causal relationship in those events.

CAUSALITY VERSUS INDEPENDENT FEATURES MODELS

Clearly, infants over 6 months of age, at least under some conditions, perceive causality (Oakes and Cohen, 1995). The causal event is treated as different from any noncausal event, and different noncausal events are treated as being equivalent. This pattern is consistent with the causality model (Figure 2b). This model is based on the idea that events can be organized in terms of causality, with

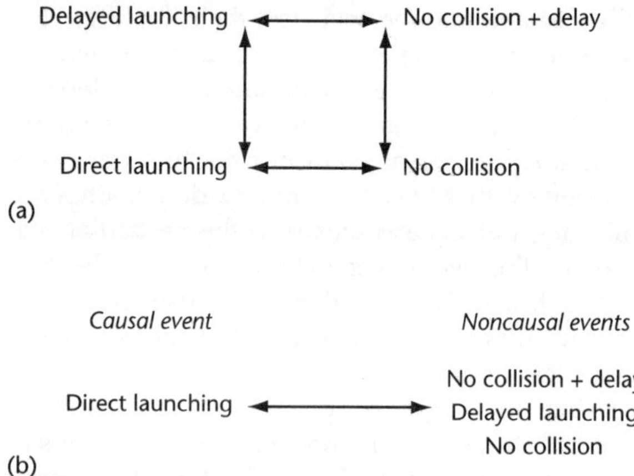

Figure 2. Abstract representations of (a) the independent features model and (b) the causality model of launching events. Reprinted from Oakes LM and Cohen LB (1995) with permission.

causal events being perceived as distinct from all noncausal events. According to this model, causal events or spatio-temporally contiguous launching events (events in which the objects touch and there is no delay imposed) are perceived as being different from noncausal events, or launching events with violations of spatio-temporal contiguity. Although this view is called the 'causality model', it is not clear whether causality *per se* is perceived. The same outcome is expected whether observers perceive the causality or are responding to the spatio-temporal contiguity of the events. That is, infants would respond in the same way if they treated a spatio-temporally contiguous event as different from any event that violates spatio-temporal contiguity. What is important is that differences between events are determined along a single dimension that corresponds to causality or spatio-temporal contiguity. As a result, two noncausal events that differ in two ways (e.g. in terms of both the temporal and spatial features) would be treated as the same, and a causal event and a noncausal event that differ in only one respect (e.g. only in terms of temporal features) would be treated as different.

However, infants younger than 6 months do not perceive the causality of events (Cohen and Amsel, 1998). Rather, these younger infants respond to differences in the independent features of the event (e.g. whether or not the two objects touch). The perception of launching events by infants younger than 6 months is consistent with the independent features model (Figure 2a). According to this model events are organized in terms of the presence or absence of specific features (e.g. whether or not the two objects touch). If infants process events as sets of independent features, then the perceptual difference between any two events can be represented by an additive combination of the lines shown in the rectangle. Infants would not treat launching events that violate spatio-temporal contiguity as equivalent. Rather, they would respond to differences in the events, such as the presence or absence of a gap or a delay.

In summary, therefore, infants first perceive events according to the independent features model, and in the middle of the first year of life they begin to perceive events according to the causality model. Importantly, this developmental transition is not 'all or none'. Infants can perceive the causality of collisions involving simple objects by approximately 6½ months, but it is not until 10 months of age that they perceive the causality of collisions involving more complex objects.

AGENT VERSUS PATIENT DISTINCTION

Perceiving causality does not simply mean recognizing the difference between causal and noncausal events. Rather, in causal events the two objects have meaningful roles, and a full appreciation of causality requires a recognition of the difference between those roles. Consider the events depicted in Figure 3. For adults, in the initial event the white object appears to cause the black object to move, and it is therefore the causal agent. What makes an object an agent? The agent has the force to cause an outcome, and may be thought of as acting in pursuit of goals (Leslie, 1995). Infants seem to be sensitive to the different roles that objects have in events (Cohen and Oakes, 1993). In this study, 10- to 12-month-old infants were habituated to a causal and a noncausal event in which either the first object was associated with the type of event (e.g. object A was always seen in the agent role of a causal event, and object B was always seen in the agent role of a noncausal event), or the second object was associated with the type of event. The infants were then tested with an event in which the roles were switched (e.g. object A was in the agent role of a noncausal event). Infants dishabituated to this role switch when the agent was associated with the type of event, but not when the patient was associated with the type of event. In other words, they linked the agent with whether or not the event was causal, but not the patient. This is an important step in distinguishing between the agent and the patient in the events. However, simply linking the first object with the type of event does not reflect a

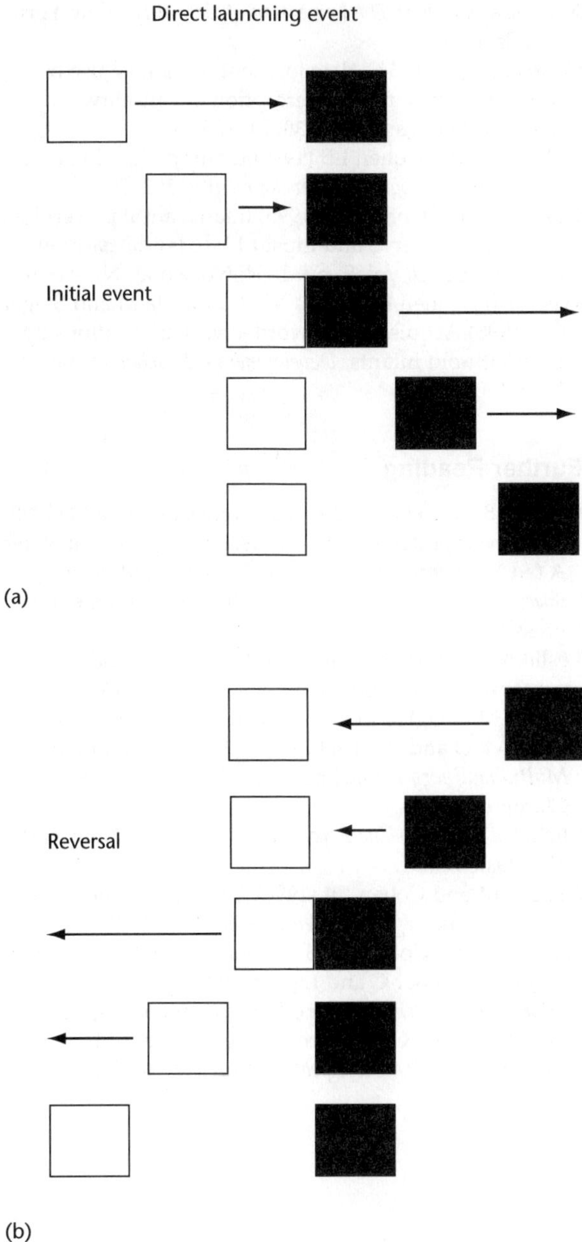

Figure 3. The sequence of actions in (a) a typical launching event and (b) a reversal of that event. Note that when the event is causal, a reversal not only involves a reversal of the direction of movement, but also a change in the roles of the objects in the events.

full understanding of the distinction between the agent and the patient.

The difference in the roles of the two objects is illustrated by the reversal of the event (Figure 3b). In this event, the action occurs in the opposite direction to that in the initial event and, importantly, the roles of the black and white objects have changed. In this reversal, the black object appears to cause the white object to move, and it is now the agent. A reversal of a causal event should therefore be more compelling than a reversal of a noncausal event. Reversals of causal events involve changes in the agent–patient roles, and reversals of noncausal events do not. In fact, infants do find reversals of causal events more interesting than reversals of noncausal ones, which suggests that they are sensitive to the agent–patient distinction (Lesile and Keeble, 1987; Cohen *et al.*, 1998). However, their recognition of this distinction develops during infancy. Six-and-a-half-month-old infants notice the agent–patient distinction when the objects involved are simple red and green bricks (Leslie and Keeble, 1987), but it is not until they reach 14 months of age that infants notice the agent–patient distinction when the objects are complex and multi-featured (Cohen *et al.*, 1998).

RELATION TO LANGUAGE

The development of general concepts such as causality is believed to be a prerequisite for learning language. Indeed, infants perceive causal relationships long before they learn the corresponding linguistic concepts. Thus we have evidence that the development of cognitive concepts precedes language development. For example, infants perceive the distinction between 'pushing' and 'pulling' at between 10 and 14 months of age. However, they do not associate labels with the type of action until 18 months of age (Cohen *et al.*, 1998). In other words, infants first perceive the type of action (pushing and pulling), and only later learn a label for that action. Thus they appear to be able to learn the general concept earlier than they can learn the linguistic concept. Interestingly, by 14 months of age infants can associate labels with objects (Werker *et al.*, 1998), which suggests that they can first associate labels with the individual objects in events, and can only later associate labels with the relationships between the objects. In general, therefore, infants are able first to learn words that refer to parts or features of the event, and only later are they able to learn words that refer to the relationships between those parts.

CONCLUSION

In summary, the perception of causality develops gradually during the first years of life. Infants' perception of causality itself develops from initially perceiving the independent features of events, such as the particular objects or some aspects of the relationships between those objects (e.g. whether or not they touch), to distinguishing causal events from noncausal ones. However, the perception of causality does not emerge fully developed. Rather, infants perceive causality earlier in some events than they do in others. They are sensitive to the agent–patient roles in events, but they recognize this distinction earlier in events that involve simple objects. Finally, it is only relatively late in infancy that they become able to associate labels with the actions in the events. Thus, in general, infants first learn about particular objects and only later learn about the relationships between those objects. Therefore an understanding of causality develops gradually with experience.

References

Cohen LB and Oakes LM (1993) How infants perceive a simple causal event. *Developmental Psychology* **29**: 421–433.

Cohen LB and Amsel G (1998) Precursors to infants' perception of the causality of a simple event. *Infant Behavior and Development* **21**: 713–732.

Cohen LB, Amsel G, Redford MA and Cassasola M (1998) The development of infant causal perception. In: Slater A (ed.) *Perceptual Development: Visual, Auditory and Speech Perception in Infancy*, pp. 167–209. Psychology Press Ltd.

Leslie AM (1984) Spatiotemporal contiguity and perception of causality in infants. *Perception* **13**: 287–305.

Leslie AM (1995) A theory of agency. In: Sperber D, Premack D and Premack AJ (eds) *Causal Cognition: A Multidisciplinary Debate*, pp. 121–149. Oxford, UK: Clarendon Press.

Leslie AM and Keeble S (1987) Do six-month-olds perceive causality? *Cognition* **25**: 265–288.

Michotte A (1963) *The Perception of Causality*. New York, NY: Basic Books.

Oakes LM (1994) The development of infants' use of continuity cues in their perception of causality. *Developmental Psychology* **30**: 748–756.

Oakes LM and Cohen LB (1990) Infant perception of a causal event. *Cognitive Development* **5**: 193–207.

Oakes LM and Cohen LB (1995) Infant causal perception. In: Rovee-Collier C and Lipsitt LP (eds) *Advances in Infancy Research* vol. 9, pp. 1–54. Norwood, NJ: Ablex.

Werker JF, Cohen LB, Lloyd VL, Cassasola M and Stager CL (1998) Acquisition of word–object associations by 14-month-old infants. *Developmental Psychology* **34**: 1289–1309.

Further Reading

Cohen LB, Amsel G, Redford MA and Cassasola M (1998) The development of infant causal perception. In: Slater A (ed.) *Perceptual Development: Visual, Auditory and Speech Perception in Infancy*, pp. 167–209. Psychology Press Ltd.

Leslie AM (1984) Spatiotemporal contiguity and perception of causality in infants. *Perception* **13**: 287–305.

Leslie AM (1995) A theory of agency. In: Sperber D, Premack D and Premack AJ (eds) *Causal Cognition: A Multidiciplinary Debate*, pp. 121–149. Oxford, UK: Clarendon Press.

Michotte A (1963) *The Perception of Causality*. New York, NY: Basic Books.

Oakes LM and Cohen LB (1990) Infant perception of a causal event. *Cognitive Development* **5**: 193–207.

Oakes LM and Cohen LB (1995) Infant causal perception. In: Rovee-Collier C and Lipsitt LP (eds) *Advances in Infancy Research*, vol. 9, pp. 1–54. Norwood, NJ: Ablex.

White PA (1988) Causal processing: origins and development. *Psychological Bulletin* **104**: 36–52.

Causal Reasoning, Psychology of

Introductory article

Barbara A Spellman, University of Virginia, Charlottesville, Virginia, USA
David R Mandel, University of Victoria, Victoria, British Columbia, Canada

CONTENTS

Causal reasoning is an important universal human capacity that is useful in explanation, learning, prediction, and control. Causal judgments may rely on the integration of covariation information, pre-existing knowledge about plausible causal mechanisms, and counterfactual reasoning.

INTRODUCTION

The subject of causal reasoning has engaged psychologists of many kinds (e.g. cognitive, social, animal, clinical, developmental), other cognitive scientists (e.g. philosophers, computer scientists, anthropologists), and others outside the cognitive science community (e.g. lawyers). The question that psychologists want to answer is: how do we go from the information that the world provides us in the form of events occurring (seemingly at random sometimes) to our beliefs about what causes what? Sometimes causal judgments are made in formal settings: in the laboratory, scientists try to find out what causes cancer or heart disease; in the legal system, before liability or punishment is imposed, jurors are required to determine who caused the accident or who caused someone's death. But more informally, we all reason about causality daily. Why did I fail this exam (or all of the exams in this course)? Why is my friend unhappy? Why is my computer likely to crash in the next five minutes? To answer these kinds of questions we may rely on repeated observations, pre-existing knowledge, thought experiments, or all of these cues to causality.

THE ROLE OF REPEATED OBSERVATIONS

One kind of information that we use to assess causality is information from repetitions of the same events: watching causes and effects as they repeatedly occur. Sometimes people use the word 'cause' deterministically, so that '*A* causes *B*' means that every time *A* occurs *B* must follow. But people also use the word 'cause' probabilistically, so that '*A* causes *B*' means that *A* increases the chances that *B* will occur. For example, someone might say that a baseball team has a winning record early in the season 'because' they have played most of their games at home. The argument is that playing at home increases the chances of winning; however, it's still possible for the team to lose some games at home and win some away games. Similarly, scientists claim that smoking causes lung cancer because it increases the chance of getting lung cancer. Yet there are people who smoke and don't get lung cancer, and people who get lung cancer without smoking.

In order to determine whether something might be causal when we have repeated observations, we divide our observations into the following categories (see Figure 1): cause present and effect present (cell *A*); cause present and effect absent (cell *B*); cause absent and effect present (cell *C*); cause absent and effect absent (cell *D*). We then use that information to decide whether the cause increases the probability of the effect. We do this by comparing the proportion of times the effect occurs when the cause is present, $A/(A+B)$, with the proportion of times the effect occurs when the cause is absent, $C/(C+D)$. The difference between these proportions is called the contingency (symbolized by Δp – read 'delta *p*'), which is a measure of the strength or effectiveness of a cause. If the proportion of times the effect occurs is greater when the cause is present than when the cause is absent then the difference will be positive, and we speak of a 'generative' or 'facilitative' cause – it makes the effect more likely to happen. If the proportion of times the

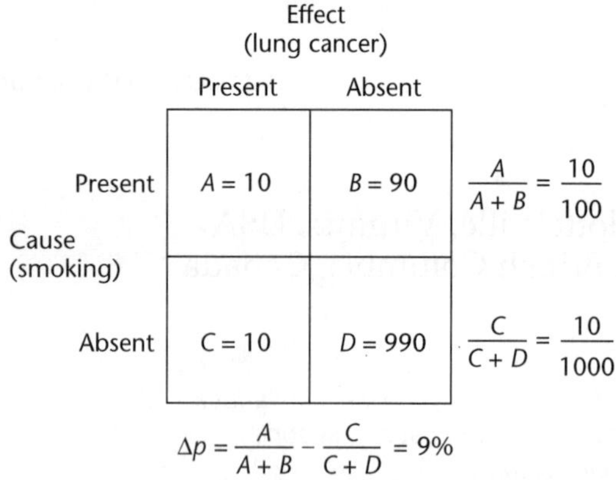

Figure 1. We need four kinds of information to understand the relation between smoking and lung cancer. Note that we do not directly compare the number of people who smoke and get lung cancer with the number of people who don't smoke and get lung cancer. Rather, we look at whether smoking increases the chances of getting lung cancer. For people who smoke, the proportion who get it is 10/100 or 10%, whereas for people who don't smoke, the proportion who get it is 10/1000 or 1%.

effect occurs is smaller when the cause is present than when the cause is absent then the difference will be negative, and we speak of a 'preventive' or 'inhibitory' cause – it makes the effect less likely to happen. (Contingencies can range from −1 to +1.) Usually people are concerned with generative causes.

Using all the Information

The above computation of Δp suggests that all of the information should be equally important in evaluating the effectiveness of a cause. However, studies have shown that people tend to overweight the information in cell *A* – the 'present–present' cell. This overweighting may be one reason why people believe in superstitions or horoscopes. For example, there are people who believe that if they walk under a ladder they will have bad luck. And, in fact, once or twice when they did walk under a ladder they did have bad luck. However, they may fail to remember or consider the times they walked under a ladder and didn't have bad luck (cell *B*), the (possibly many) times they had bad luck without walking under a ladder (cell *C*), and the very many times they didn't walk under a ladder and didn't have bad luck (cell *D*). All of that information is needed before one can correctly evaluate whether walking under a ladder causes bad luck.

Considering the Base Rate of the Effect

Although the Δp computation captures the idea that a cause is something that changes the probability of an effect, the number that results from the computation may be deceptive when one is trying to evaluate the effectiveness of a particular cause. Besides looking at the contingency, people also consider how often an effect occurs in general (called the 'base rate' of the effect).

For example, suppose you have 100 plants in your garden but only 80 of them have flowers. You buy a special plant food, and soon all 100 have flowers. How effective is this product? It may only work 20 percent of the time (and have happened to work on the flowers that hadn't already bloomed); or it may work 100 percent of the time (but you couldn't tell because some flowers had already bloomed anyway).

It appears that people are sensitive to this problem when judging the effectiveness of a cause. For example, compare two plant foods. One was given to 100 plants where there were initially no flowers, and then 20 bloomed ($20/100 - 0/100 = 0.20$). The other, as above, was given to 100 plants where there were initially 80 flowers, and then all 100 bloomed ($100/100 - 80/100 = 0.20$). Both plant foods have a contingency of 0.20, yet the first worked only on 20 percent of the plants that didn't already have a flower whereas the second worked on 100 percent of the plants that didn't already have a flower.

People judge the second plant food as more effective than the first even though the contingencies are equal. Thus, it seems that people take into account the base rate of the effect and adjust their estimates according to how much influence a cause had above and beyond the influence of other causes.

THE ROLE OF PRE-EXISTING KNOWLEDGE

We have seen how we can use statistical covariation information to assess the relation between a potential cause and effect. However, we don't only use statistical information when making causal judgments; our pre-existing knowledge of the world will influence which statistical information we will pick up and use, and how we will limit and interpret the statistical relations we discover.

The most important limitation to note is that even though finding a contingency means that there is a correlation between the two events, that does not mean that the first event causes the second event. As scientists often say: 'correlation does not prove causation'. For example, every morning, just before dawn, the rooster crows. Then the sun rises. Yet we do not believe that the rooster crowing causes the sun to rise. In many cities, when ice cream sales go up, the murder rate goes up; when ice cream sales go down, the murder rate goes down. Yet we do not believe that eating ice cream causes people to commit murder. Why not? Because we have other knowledge.

Causal Mechanisms

Why don't we believe that the rooster causes the sun to rise even though there is a perfect correlation? One idea is that we don't believe it because we can't conceive of a causal mechanism. How could the noise of a tiny animal affect a powerful celestial object? In fact, belief (or non-belief) in a mechanism can direct, or misdirect, searches for potential causes. For example, in the mid-nineteenth century a physician named Ignaz Semmelweis had a very difficult time convincing physicians to wash their hands after examining a cadaver before turning to deliver a baby (to lessen mortality of the mothers) – a practice that seems obvious and obligatory now – because no one then could imagine a causal mechanism.

So where do our beliefs about causal mechanisms come from? One possibility is that they come from our knowledge of similarities, categories, and other statistical relations. We don't believe that the rooster causes the sun to rise because we know that lions roaring don't cause rain and dogs barking don't cause full moons. There are no statistical relations between one 'kind' of event (animal noise) and the other 'kind' of event (weather), so we never developed the idea that there could be a causal mechanism. On the other hand, these days we are willing to accept that many new ailments can be caused by unseen microorganisms which can be transmitted in many ways (e.g. breathing, touching) because we have noted other similar relations in the past.

Labeling a Cause as a Cause

Given the same combination of events, which gets labeled as a 'cause' of the outcome may differ between situations, individuals, and cultures. For example, suppose a fire breaks out in a nearby warehouse and you are explaining the cause of the fire to a friend. You are likely to mention that there was an arsonist or a stroke of lightning. You are unlikely to mention the presence of combustible material or oxygen, even though both of those are necessary for the fire. Under 'normal' circumstances we just assume that they are present, and so their presence or absence does not covary with the outcome and we do not consider them causes. Now imagine a special furniture factory in which an area is kept free of oxygen so that high-temperature welding can take place. One day there is an oxygen leak, and when the usual welding begins a fire ensues. Under these special circumstances, you would call the oxygen a cause of the fire. Thus, what we point to as being causal is not just something that increases the probability of an effect, but rather something that increases it relative to some background assumption of what is stable or normal.

Which of the many relevant factors a person chooses to pick out from the background and label as a cause may also depend on that individual's beliefs. For example, suppose a young man robs a shop. What caused this behavior? Some people would argue that it was because he was brought up in a bad neighborhood, citing the fact that children brought up in his neighborhood are more likely to go on to commit such a crime than children brought up in better neighborhoods. Other people would argue that it was because he was a 'bad apple', citing the fact that there are many other children brought up in his neighborhood who do not commit such crimes. In such a case, what you label as a 'cause' may influence what you believe is the best treatment for the problem.

Different cultures also tend to pick out different factors as causal. For example, in individualist cultures (such as the United States and Australia) people are more likely to attribute causality for an action to the actor's personality or 'disposition', whereas in collectivist cultures (such as China and India) people are more likely to attribute causality to the situation or circumstances. (*See* **Cultural Differences in Causal Attribution**)

Alternative Causes

Although it seems easy enough to figure out the statistical relationship between one cause and one effect once you know where to look, the world is complicated and it is not always possible to examine one potential causal relation at a time. Sometimes potential causes are independent, so you can evaluate each separately. But often potential causes covary with each other, making it difficult to distinguish the causal contribution of each.

Controlling for alternative causes

When two or more potential causes of an effect act at once, and not independently, we have to control for one cause while evaluating the other. As a simple example, suppose you rush into your favorite coffee shop and assert loudly that drinking coffee must cause lung cancer because people who drink lots of coffee get lung cancer more often than those who do not (a positive Δp). Probably none of the coffee drinkers there would be alarmed: they would point out to you that perhaps people who drink more coffee also smoke more, so although it may look as if coffee drinking causes lung cancer, it is really smoking that is doing the causal work. Here people see an alternative causal mechanism to explain the lung cancer: smoking. To evaluate whether coffee drinking causes lung cancer while controlling for smoking you need to (1) consider all people who don't smoke and ask whether coffee drinking increases their probability of getting lung cancer, and (2) consider all people who do smoke and ask whether coffee drinking increases their probability of getting lung cancer. If the answer is negative in both cases, then coffee drinking is not a cause of lung cancer: it is only because it covaries with smoking that it seems to raise the probability of lung cancer.

When evaluating whether something is a cause of an effect, it is important to control for alternative causes. Obviously, it is never possible to know for certain that one has considered all potential alternative causes, but controlling for known alternative causes is a technique intentionally used by psychologists and other scientists to improve scientific reasoning.

However, controlling for alternative causes is difficult without a theory of what those alternative causes might be. In the coffee example, people don't believe there is a way in which coffee drinking could cause lung cancer, so they seek an alternative causal mechanism. But in the case of ice cream sales and murder rates, it might be plausible to think that increased sugar consumption causes violence, and leave it at that. A mechanism is not necessarily correct just because it is plausible. When temperature is controlled for, there is no correlation between ice cream sales and murders; they only seem related because hot weather leads to more of both. Many experiments have shown that people do control for known alternative causes when judging causal effectiveness. If experimenters tell people about an alternative cause, or if they have a reason to believe that some alternative factor (e.g. smoking) might be causal, then they will think of controlling for that cause. However, in experiments where the alternative factor (like temperature) is not so obvious, people are less good at controlling for it.

Thus, pre-existing knowledge of a causal mechanism may affect what information people acquire from the environment, what they control for, and, ultimately, what they will judge as the causes of other events.

Discounting

Although people do control for known alternative causes, the presence of an alternative cause may lead to a misjudgment of causal strength – known as 'discounting'. Discounting occurs when someone learns about two causes at the same time, and the judgment of the strength of the first cause is affected by the strength of the second cause. For example, suppose you have allergies and your doctor prescribes two medications *A* and *B*. Sometimes you take one, sometimes the other, and sometimes both. Medication *A* works ($\Delta p = 0.33$), but medication *B* doesn't work ($\Delta p = 0$). You tell the doctor that *A* is fairly effective in relieving your allergies. Now consider what would have happened had the doctor prescribed medications *A* and *C* instead. *A* still works ($\Delta p = 0.33$), but *C* works even better ($\Delta p = 0.67$). In your report to the doctor, you are likely to judge *A* as being less effective when the alternative cause is strong (*C*) than when the alternative cause is weak (*B*) – even though *A*'s effectiveness is the same.

Experiments have shown that discounting occurs even in simple cases when the two causes are independent. Discounting might result from a strategic decision or a belief that once we have found a good cause of an effect we need not invest in reliably assessing other potential causes.

JUDGING CAUSALITY IN SINGLE INSTANCES

We have considered how people make causal judgments when they have information about many instances of the cause and effect occurring. But how do we make causal judgments for events that occur only once (e.g. an accident, a crime, or a big promotion)? One theory is that we use counterfactual reasoning to make these judgments.

Counterfactual Reasoning

In many situations, people look back on a past episode and wonder what might have happened if some change had occurred leading up to its conclusion. For instance, suppose you decided to drive home from work by a scenic route one day because it was particularly beautiful outside, and along the way your car was hit by a reckless driver. Many, if not most, people in this situation would replay the episode in their mind in such a way that the accident is somehow 'undone'. For instance, you might imagine that if only you had taken your usual route home, or if only you had left a few seconds later, the accident would have been avoided. These imaginings of how the past might have been different involve counterfactual (or contrary-to-fact) thinking because the mentally replayed episode differs in some respects from the real episode.

Some philosophers and psychologists have proposed that counterfactual thinking plays an important role in causal reasoning. To explore the possible causes of a particular outcome, a person may mentally change one of the events preceding the outcome (an 'antecedent') and observe whether the outcome still occurs in the mental replay. If it is easy to imagine that the outcome would also be undone, then the antecedent is likely to be seen as one of its necessary causes. If, however, the outcome still seems inevitable, then the antecedent is unlikely to be seen as causally relevant.

Political scientists, historians, and legal scholars sometimes run counterfactual thought experiments to examine the causal implications of a particular change to a complex system. For example, some historians have considered what might have happened if Archduke Franz Ferdinand had not been assassinated in Sarejevo; and some have concluded that if that event had not happened, then the First World War would not have happened either. Although counterfactual thought experiments can be informative, it is usually impossible to verify whether the causal inferences drawn from them are justified – precisely because history cannot literally be replayed.

Similarities and Differences Between Causal and Counterfactual Reasoning

Counterfactual and causal reasoning sometimes focus on different events. For instance, in the car accident scenario, if people are asked directly about the cause of the accident they tend to identify the reckless driver, whereas if they are asked to generate counterfactuals that would undo the accident they tend to mention the route taken home. The counterfactual thoughts that come to mind may represent one's after-the-fact understanding of how the bad outcome could have easily been prevented. Thus, counterfactual thoughts often focus on behaviors that individuals can control. On the other hand, causal explanations often focus on antecedents that our knowledge of the world indicates would covary with the outcome over a set of similar cases. Thus, people say that the reckless driver was the cause of the accident because they realize that reckless driving is predictive of car accidents in general.

Even though the two forms of reasoning sometimes diverge, there is nevertheless a strong interplay between counterfactual and causal reasoning. If one cannot imagine that an antecedent might have been different, then it is unlikely that the antecedent will be identified as a cause. This is why, under normal conditions, oxygen makes a poor causal explanation for fire, even though it is a necessary condition. Similarly, if one cannot imagine an outcome having been different, then it is questionable whether it could be causally explained other than by recourse to concepts such as fate or destiny, which by definition, emphasize the immutability and inevitability of past episodes. (*See* **Counterfactual Thinking**)

DEVELOPMENT OF CAUSAL REASONING

Causal reasoning is necessary for human survival and, not surprisingly, the ability to perform such reasoning develops early. However, it is difficult to study causal reasoning in infants, because researchers cannot ask them direct questions about their judgments. Instead, researchers often use a technique called the 'habituation paradigm'. This technique takes advantage of the fact that when infants see the same events repeatedly (e.g. pictures on a video monitor, animations, objects moving in real life), they gradually get bored and will look at the events for shorter durations. If, however, they see a new or different event, they will 'dishabituate' and look for longer. Researchers can infer what infants count as 'the same thing' or 'a different thing' using this technique.

When using the habituation paradigm to study causal reasoning, researchers may show infants videotapes of collision events. In the 'causal launching' event, object *A* moves across the screen and hits stationary object *B*. When *A* strikes *B*, *A* stops, and *B* immediately begins to move with the same speed and direction as *A* had previously. This event looks natural to an adult, as if *A*'s collision with *B* caused *B* to move. Researchers can then modify these events. For example, a delay may be introduced, so that after *A* hits *B*, *B* remains in place for a second or two before it begins moving. Or, *B* may begin moving before being hit by *A*. With these modifications, adults will claim that it doesn't look as if *A* caused *B* to move. Do infants treat these modified events as the same as, or different from, the natural-looking causal launching event? Research has shown that by the age of about seven months, infants do perceive a difference between causal launching events and noncausal events. However, they perceive the difference only if the objects involved in the events are simple; for more complicated objects infants may have to be 10 months old to make the distinction.

If it takes infants longer to understand causality regarding more complicated objects, what about more complicated events? Often in life, we don't just say that an outcome was caused by the action that immediately preceded it; rather, we look back in time to see what caused that particular action. For example, when Mum sees milk spilled on the floor and yells at Little Sister to wipe it up because it fell from her tilted glass, if Little Sister says 'Big Brother pushed me' then he, rather than she, is seen as the cause and gets Mum's wrath. It is considered a sign of sophisticated reasoning in adults to be able to look back into the past for causes; this ability also develops over time in infants: at 10 months old they don't look back at earlier causes, whereas at 15 months old they do.

Thus, we see that causal reasoning starts to develop early, but not all at once. Rather, both the complexity of the objects involved in the events and the complexity of the relations between events affect causal understanding. (*See* **Causal Perception, Development of**)

CONCLUSION

Causal reasoning is a pervasive and important form of thinking that begins at an early age. People engage in causal reasoning in order to explain past outcomes, to achieve control over their natural and social environment in the present, and to forecast, plan and prepare for the future. However, reasoning about causality is complicated. It is complicated, in part, because causal judgments depend on multiple cues, such as covariation, spatial and temporal contiguity, and our beliefs about what is normal. It is also complicated because information about such cues may be obtained in a variety of ways, such as by observing new cause–effect sequences, recalling knowledge about the world, and mentally imagining counterfactuals. Moreover, the answer to the question 'What is the cause?' may depend on one's choice of causal background – which may be affected by motivation, knowledge, and culture. Despite the complexity of the concept, the power of such knowledge, when it is accurate, is formidable. Without the ability to understand causality and use causal knowledge, both our internal mental world and the external physical world in which we live would be radically different.

Further Reading

Cheng PW and Wu M (1999) Why causation need not follow from statistical association: boundary conditions for the evaluation of generative and preventive causal powers. *Psychological Science* **10**: 92–97.

Cohen LB, Rundell LJ, Spellman BA and Cashon CH (1999) Infants' perception of causal chains. *Psychological Science* **10**: 412–418.

Hart HL and Honoré AM (1985) *Causation in the Law*, 2nd edn. Oxford, UK: Oxford University Press. [First edition published in 1959.]

Hilton DJ (ed.) (1988) *Contemporary Science and Natural Explanation: Commonsense Conceptions of Causality*. New York, NY: New York University Press.

Mandel DR and Lehman DR (1996) Counterfactual thinking and ascriptions of cause and preventability. *Journal of Personality and Social Psychology* **71**: 450–463.

Mandel DR and Lehman DR (1998) Integration of contingency information in judgments of cause, covariation, and probability. *Journal of Experimental Psychology: General* **127**: 269–285.

McGill AL (1989) Context effects in judgments of causation. *Journal of Personality and Social Psychology* **57**: 189–200.

Spellman BA (1997) Crediting causality. *Journal of Experimental Psychology: General* **126**: 323–348.

Spellman BA and Mandel DR (1999) When possibility informs reality: counterfactual thinking as a cue to causality. *Current Directions in Psychological Science* **8**: 120–123.

Spellman BA, Price CM and Logan J (2001) How two causes are different from one: the use of (un)conditional information in Simpson's paradox. *Memory and Cognition* **29**: 193–208.

Sperber D, Premack D and Premack AJ (eds) (1995) *Causal Cognition: A Multidisciplinary Debate*. New York, NY: Oxford University Press. [Symposia of the Fyssen Foundation.]

White P (1990) Ideas about causation in philosophy and psychology. *Psychological Bulletin* **108**: 3–18.

Cell Assemblies

See **Hebbian Cell Assemblies**

Cerebellum

Intermediate article

Frank A Middleton, State University of New York Upstate Medical University, Syracuse, New York, USA
Stephen I Helms Tillery, Arizona State University, Tempe, Arizona, USA

CONTENTS

The cerebellum ('little brain') contains more neurons than the rest of the entire nervous system, and its complex organization has been the subject of much research. Although it is not essential for cognitive thought processes, the cerebellum is clearly involved in the performance of some cognitive tasks and in the improvement of motor control.

INTRODUCTION

The human cerebellum is a fist-sized structure that sits in the back of the skull, located just behind the midbrain and below the cerebral hemispheres (Figure 1). It has been the subject of considerable scientific attention for much of the past two hundred years from researchers in such diverse fields as phrenology, psychiatry, evolutionary and comparative biology, neurology, genetics, and developmental biology. Cerebellum is translated literally as 'little brain' and indeed, despite its relatively small size, this single structure contains more neurons than the rest of the nervous system combined. While this feature alone might seem daunting at first glance, there are several aspects of cerebellar organization that have greatly simplified its study and led to its widespread use as a model system for addressing a large number of structural and functional issues in brain research.

ANATOMY

The cerebellum is often depicted as a composition of three basic elements: the cerebellar cortex (which consists of a granular layer, Purkinje layer and molecular layer), the deep cerebellar nuclei, and the large white-matter tracts that run between these structures and connect the cerebellum with other brain structures. In cerebellar cortex, five main types of neurons have been identified – the Purkinje, Golgi, granule, basket and stellate cells (Table 1, Figure 2) – as well as three minor classes of neurons – Lugaro, unipolar brush and pale cells (not shown). In the deep cerebellar nuclei, there are only two main types of neurons – projection neurons and local interneurons – although several minor classes have also been identified. In both the cortex and deep nuclei, the connections and neurochemical markers of these neurons have been well characterized (Table 1, Figure 2). In addition to these neuronal classifications, there are also three different types of nonneuronal cells in the cerebellum: radial (or Bergmann) glial cells in the Purkinje and molecular layers of cerebellar cortex; bushy astroglia in the granular layer; and oligodendrocytes in the white-matter layer between the cerebellar cortex and deep nuclei. Perhaps the most salient feature of cerebellar organization is the orderly

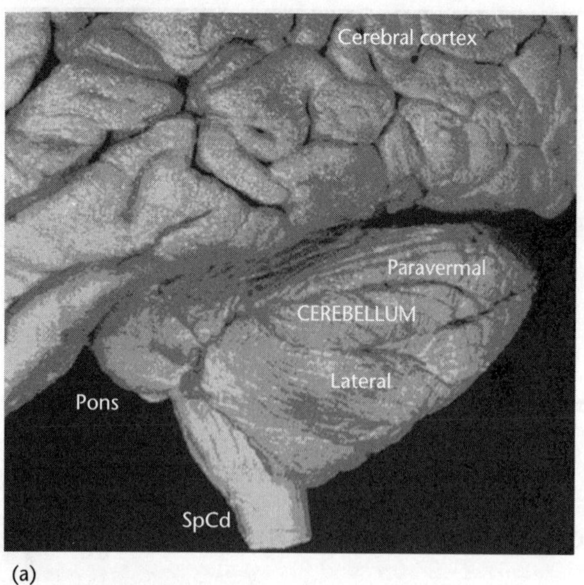

(a)

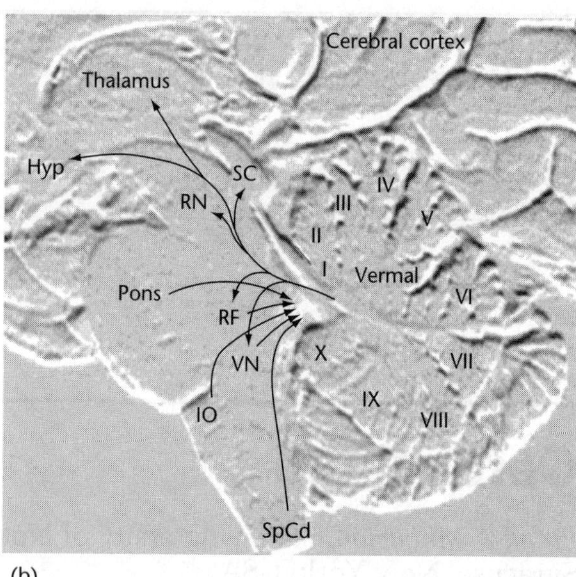

(b)

Figure 1. The cerebellum. (a) External view. (b) Midline view showing major inputs and outputs. Roman numerals refer to Larsell's nomenclature of the vermal lobules as listed in Table 2. Hyp, hypothalamus; IO, inferior olive; RF, reticular formation; RN, red nucleus; SC, superior colliculus; SpCd, spinal cord; VN, vestibular nuclei.

Table 1. Main cell types in the cerebellum

Location	*Name*	*Type*	*Markers*	*Afferents*	*Efferents*
Cortex, Purkinje layer	Purkinje	Inhibitory neuron	GABA, GAD, GABA-T, zebrin, motilin	Inferior olive, via climbing fibers Granule cells, via parallel fibers Other cortical neurons (basket, stellate, Lugaro)	Deep nuclei
Cortex, Purkinje layer	Basket	Inhibitory neuron	GABA, GAD	Granule cells, via parallel fibers	Purkinje cell somas, initial axon segments
Cortex, Purkinje layer	Stellate	Inhibitory neuron	GABA?	Granule cells, via parallel fibers	Purkinje cell dendrites
Cortex, granular layer	Golgi	Inhibitory neuron	GABA	Granule cells, via parallel fibers	Granular layer cells
Cortex, granular layer	Granule	Excitatory neuron	GABA, glutamate	Pontine and brainstem nuclei via mossy and mossy-like fibers	Dendrites of other cerebellar cortical cells
Deep nuclei	Projection	Excitatory neuron	Glutamate	Purkinje cells, collaterals of other afferent systems	Thalamus, magnocellular red nucleus, brainstem, reticular nuclei, superior colliculus, cerebellar cortex
Deep nuclei	Projection	Inhibitory neuron	GABA	Purkinje cells, other afferent systems	Inferior olive
Deep nuclei	Interneuron	Inhibitory neuron	GABA, glutamate	Other deep nuclei cells	Other deep nuclei cells

GABA, gamma-aminobutyric acid; GABA-T, gamma-aminobutyric acid transaminase; GAD, glutamate acid decarboxylase.

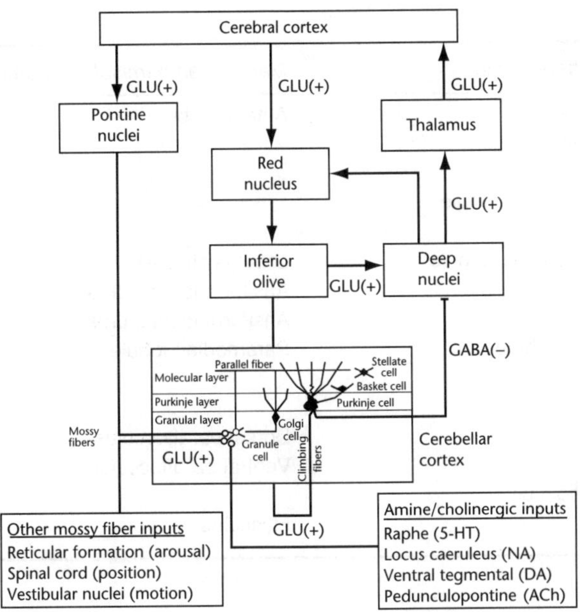

Figure 2. Connections and microcircuitry of cerebellar structures. 5-HT, 5-hydroxytryptamine (serotonin); ACh, acetylcholine; DA, dopamine; GABA, gamma-aminobutyric acid; GLU, glutamate; NA, noradrenaline (norepinephrine). Plus and minus signs refer to excitatory and inhibitory projections, respectively.

arrangement of connections between all of these different cell types in the cerebellum (Figure 2). This circuitry was worked out by careful and time-consuming anatomical and electrophysiological mapping studies (see Ito, 1984). Because this basic circuitry exists throughout the entire cerebellum, researchers have been able to understand how the whole cerebellum works by studying only small parts of it.

Superimposed on the repetitive microcircuitry of the cerebellum are larger structural and functional subdivisions which have proved extremely useful for understanding cerebellar organization. At the gross anatomical level, the cerebellum is divided into multiple lobes or lobules, often smooth in appearance in smaller species but composed of numerous folds and fissures in more complex animals (Table 2, Figure 1). Each of these subdivisions contains essentially the same microcircuitry shown in Figure 2. The somewhat complicated nomenclature listed in Table 2 (and defined and described in detail by Larsell and Jansen, 1972) was actually developed as a simplifying scheme of the transverse structural organization of the cerebellum that holds across all species of animals. However, the functional organization of the cerebellum as it is currently understood is simpler, and consists of different transverse 'zones' (see Jansen, 1972; Voogd, 1975). In the sagittal plane, the cerebellum can be divided into midline (vermal), intermediate (paravermal) and lateral (hemispheric) zones (Table 3, Figure 1). Each of these zones receives a certain type of functional input at the level of the cerebellar cortex and sends its output largely through a single deep nucleus. For example, most of the vermal zone is concerned with maintaining posture and balance and coordinating reflexes (including reflexive eye movements and some autonomic functions) and directs its output through the medial deep cerebellar nucleus. In a similar manner, most of the intermediate cerebellum is concerned with sensory and motor processing that facilitates and coordinates voluntary movement, and directs its output through the intermediate deep nuclei. Lastly, most of the lateral cerebellum (also called 'neocerebellum' by some investigators) appears to be concerned with the planning, generation and control of complex behavior and directs its output through the lateral deep nucleus (Table 3).

The cerebellum is known to be connected with a large variety of different brain regions, including the spinal cord, brainstem, vestibular and reticular nuclei, as well as several major nuclei that are known to be the site of synthesis of specific neurotransmitters, such as the raphe, ventral tegmental area, locus caeruleus, and pedunculopontine nuclei (Figures 1 and 2). The sole output structures of the cerebellum are the four deep cerebellar nuclei and the vestibular nuclei (Table 3). These nuclei send signals back to many of the same brain regions that provide input to the cerebellum. Not surprisingly, the area of the cerebellum that is reciprocally connected with these extracerebellar regions has often been shown to be involved in functions that are related to that structure. For example, vestibular signals from the inner ear regarding movement of the head are sent primarily to the vermis. The activity of neurons in certain parts of the vermis has been shown to be related to both these sensory signals and to the commands for movement that are sent from the cerebellum to the vestibuloocular system to help maintain orientation of the head and eyes while the body is in motion. A similar type of arrangement has been shown to exist for the connections between the cerebral cortex and cerebellum. Most of the cerebral cortex sends fibers to the pontine nuclei that are a source of input to the cerebellar cortex. In turn, the output nuclei of the cerebellum project back, via the thalamus, to many of the cortical areas from which they receive input. Defining the full scope of areas that participate in these types of reciprocal cerebellar loops has

Table 2. Main structural divisions of the cerebellum

Lobule	*Vermal name*	*Human hemisphere name*	*Mammalian hemisphere name*
I	Lingula	Vinculum lingulae	Anterior lobe
II	Central	Ala lobuli centralis	
III			
IV	Culmen	Anterior quandrangular lobule	
V			
VI	Declive	Posterior quandrangular lobule	Lobulus simplex
VIIA	Folium	Superior semilunar lobule	Ansiform lobule, crus I
			Ansiform lobule, crus II
VIIB	Tuber	Inferior semilunar lobule	Paramedian lobule
		Gracile lobule	
VIII	Pyramis	Biventral lobule	
IX	Uvula	Biventral lobule	Dorsal paraflocculus
		Tonsilla	Ventral paraflocculus
		Accessory paraflocculus	
X	Nodulus	Flocculus	Flocculus

Lobule numbers according to Larsell and Jansen (1972).

Table 3. Functional zones of the cerebellum

Cerebellar zone	*Cortex region*	*Deep nuclei nomenclature*		*Putative motor function*	*Putative cognitive function*
		Human	*Mammalian*		
Midline	Vermis	Fastigial Vestibular	Fastigius or medialis Vestibularis	Balance, eye movement, reflexes	Autonomic arousal, limbic regulation
Intermediate	Paravermal hemisphere	Globose Emboliform	Interpositus anterior Interpositus posterior	Sensorimotor integration, movement execution	Simple verbal responses to commands
Lateral	Lateral hemisphere	Dentate	Dentatus or lateralis	Preparation and planning of movements, fine motor dexterity, eye movements, imagined movements	Verbal association, rule-based learning, working memory, problem-solving, monitoring performance, temporal perception

important implications for understanding the functions with which the cerebellum is most likely to be involved.

The precise nature of the input projection to the cerebellum depends on the structure from which it originated. Signals from the pontine nuclei, vestibular nuclei, trigeminal nuclei and spinal cord are conveyed to the cerebellum by mossy fiber projections to granule cells. Signals from the red nucleus are conveyed by a projection to the inferior olive, which sends climbing fibers to Purkinje cells (Figure 2). Within the cerebellum, a great deal of processing of these signals takes place using cerebellar microcircuitry. For many functions, this processing has been shown to improve the accuracy of the signals, so that the behavior or function that modulates the activity in the circuit is seen to improve. For this reason, the cerebellum has often been thought of as a type of external teaching device that improves the performance of any brain region participating in cerebellar feedback loops. In addition, the cerebellum has also been depicted as a type of feedforward processor that integrates much of the low-level sensory information it receives into signals that help direct the activity of higher-level brain areas.

Three types of evidence are used to determine the functions that the cerebellum is involved in. First, anatomical data help define what areas of the brain participate in cerebellar loops, and thus what types of signals the cerebellum is likely to process. Second, physiological data from imaging studies and recordings of cerebellar cells have indicated which cerebellar regions and neurons are active during specific tasks. Finally, data from clinical, pathological and behavioral studies have revealed what the functional consequences of cerebellar damage are.

ROLE OF THE CEREBELLUM IN MOVEMENT

Theories about the role of the cerebellum as a component of the motor control system have a venerable history, dating back at least to the eighteenth century. In the nineteenth and early twentieth centuries, Flourens, Luciani and Holmes separately described the results of cerebellar damage (for a thorough review of these and later studies see Dow and Moruzzi, 1958). During the First World War, Holmes described several discrete movement disorders that resulted from gunshot wounds of the cerebellum, and attempted to characterize these disorders based on the site of damage. In so doing, Holmes essentially defined the classic 'cerebellar syndrome' that has guided much of the research on the cerebellum ever since.

Cerebellar Lesions and Motor Function

Following unilateral lesions limited to the lateral lobes of the cerebellum, Holmes and others noted the loss of muscle strength and tone, most often involving the arms (but occasionally the legs), along with ataxia, rebound disorder, oculomotor disorders, postural disorders and scanning speech. The most conspicuous of these symptoms is ataxia, a generalized disorder of coordinating and executing voluntary movements. In their attempts to coordinate movement, cerebellar patients often exhibit asynergia, or difficulty coordinating muscular actions, and decomposition of movement, in which normal complex movements become broken down into single movements involving single joints. In executing movements, patients with cerebellar damage suffer from dysmetria, poorly directed movements which often miss their targets, and deviations from the line of movement, where a movement path does not follow the shortest line between two points. Possibly related to dysmetria is intention tremor, or involuntary shaking of the reaching hand or limb as it performs an action. Examples of these types of problems are easily seen when people with cerebellar damage are asked to touch their nose with one of their fingers. At first, the movement becomes decomposed. To reach the finger to the nose requires coordinated movements of the shoulder, elbow and hand, but is nonetheless readily performed in one smooth motion by normal individuals. People with cerebellar damage, however, might perform this movement by first lifting the shoulder until the arm is horizontal, and then flexing the elbow separately to bring the hand close to the nose. Finally, when the finger approaches the nose, involuntary alternating contractions of the muscles in the hand and wrist cause the finger to successively undershoot and overshoot the nose. Interestingly, high levels of ethanol consumption produce a similar effect, probably due to its action on the cerebellum, forming the basis of the modern sobriety test.

Another specific problem of coordination often seen in people with cerebellar damage is dysdiadochokinesis, or difficulty in performing rapidly alternating movements, such as slapping the surface of a desk repeatedly first with the palm of the hand and then the back of the hand. In dysdiadochokinesis the movements of the affected limb slow down and decrease in amplitude.

Midline cerebellar lesions often also lead to oculomotor and postural disorders. Some of these may be analogous to the weakness and dysmetria of the limbs seen with lateral cerebellar lesions. For example, people with cerebellar damage often exhibit difficulty maintaining their gaze on moving objects, and may display nystagmus – continuous oscillations of the eyes as if they were watching a train pass. Speech is often also affected in cerebellar lesions, causing scanning speech in which articulation is difficult and often staccato. Importantly, however, none of the effects of cerebellar lesions on motor performance that have been described implies that the cerebellum is essential for the direct control of movement. Indeed, lesions of the cerebellum do not prevent the initiation of movement, but rather cause problems in the optimal regulation and performance of ongoing movement.

Another aspect of cerebellar organization revealed by Holmes's early work, and reinforced by many anatomical and physiological mapping studies, has been the realization that the functions of the cerebellum are somatotopically organized: that is, the regions of the cerebellum concerned with the sensation and movement of the arms, legs, eyes and face are largely separate from each other. This is true even though these body parts are each represented multiple times throughout the cerebellar cortex and deep nuclei.

Cerebellar Physiology

Knowledge of the effects of cerebellar damage formed a backdrop to the earliest physiological studies of the cerebellum. Research by Eccles and his colleagues (Eccles *et al.*, 1967) and more recently by Rodolfo Llinas (Llinas and Sotelo, 1992) has dissected the basic physiology of the cerebellum and particularly the cerebellar cortex. Indeed, the basic firing properties of the major cell types in the cortex have been worked out in detail. The main output cell of the cerebellar cortex is the Purkinje cell; these distinctive cells are notable in that they fire two distinct types of action potentials, simple spikes and complex spikes (Thach, 1968). (*See* **Single Neuron Recording**)

Simple spikes are typical neuronal action potentials, in which the neuron's membrane voltage changes rapidly from about −50 mV to +50 mV, and then returns again to its resting membrane potential of about −50 mV. This entire cycle occurs in the space of less than 3 ms when the sum of a Purkinje cell's inputs from other sources, notably parallel fibers, exceeds a threshold voltage. These simple spikes can occur at rates of up to $100\,s^{-1}$. Simple spikes are known to occur in a predictable fashion with a variety of volitional movements. The relations between behavior and simple spike discharge have been especially well characterized during arm movements, head and neck movements, smooth pursuit and vestibuloocular reflex eye movements, and during walking. Generally, the firing rates of Purkinje cells during movement can be related to specific aspects of the movement, such as velocity of joint rotation or eye movement, or direction of arm movement.

Complex spikes, on the other hand, are not typical at all. During a complex spike, the neuron's membrane voltage increases rapidly, but then stays at a high value for an extended period, lasting perhaps 20 ms or longer. While the membrane potential is depolarized, the neuron typically fires many action potentials. The complex spikes occur in a one-to-one relationship with the arrival of action potentials on the climbing fibers. Thus, a single action potential from the inferior olive produces a profound and long-lasting depolarization in the membrane potential of a Purkinje cell. Typically, complex spikes occur only once or twice per second. Complex spiking in Purkinje cells can often be reliably produced by application of stimuli to distinct patches of skin. Interestingly, during behavior, complex spiking can often be most readily related to the occurrence of unexpected stimuli. During walking, for example, if an unexpected obstacle prevents the swing of the leg, complex spikes are often seen in the lateral lobes.

Explaining the relations between simple spikes, complex spikes and behavior in terms of the operation that the cerebellum performs, however, has proved remarkably difficult. Two general types of models, not necessarily exclusive, have been proposed to describe the operations of the cerebellum in motor control: cerebellar learning hypotheses and cerebellar timing hypotheses. According to learning hypotheses, the cerebellum learns maps between input patterns and output patterns. The core idea, proposed independently by David Marr and James Albus around 1970, is that the strength of synapses between parallel fibers and Purkinje cells can be modified, and that the changes in the strength of those synapses are regulated by the action of climbing fibers. While evidence for this idea has proved both elusive and controversial (Bell *et al.*, 1996), a great deal of cerebellar research has been inspired by this theory. Work by Masao Ito and colleagues (Ito, 1984) suggested that complex spikes induce a long-term change in the responsiveness of Purkinje cells to parallel fiber inputs, termed 'long-term depression' (LTD). While LTD is reliably produced in experimental situations, it is not clear whether it also occurs in more physiological conditions – see work by Bloedel and Kelley, in Llinas and Sotelo (1992). It also seems likely that the cerebellum plays a part in classical conditioning such as the nictitating membrane reflex, but the original idea that the engram for this conditioning lies in the cerebellum itself has not been borne out by all researchers (see Bell *et al.*, 1996, and special issues of *Trends in Neurosciences* and *Trends in Cognitive Science* listed under 'Further Reading'). In contrast, there is accumulating evidence that the cerebellum is a critical storage site for the storage of the engram established during the learning of some operantly conditioned behaviors (see Milak *et al.*, 1997). (*See* **Long-term Potentiation and Long-term Depression**)

Another idea consistent with the observations of asynergia and decomposition of movement is that the cerebellum participates in coordination by handling problems of timing. The core idea was initially proposed by Valentino Braitenberg, who noted that the parallel fibers stretching across a row of Purkinje cells looked something like a delay line, and proposed that this feature might enable the cerebellum to select the relative timing of output events. Research by Ivry and Keele has indeed shown that cerebellar patients often have difficulty judging and reproducing temporal intervals. More recently, evidence from several laboratories, notably the experimental work of Fred Miles and Steve Lisberger in eye movements and Jim Bloedel in limb movements, as well as the theoretical work of Chris Miall, has suggested that complex spikes might serve as a generalized error signal, reporting that in some way behavioral output is not meeting expected outcomes. Thus, the inferior olive would communicate an error signal to the cerebellar cortex that it needed to modify its output. The specific mechanism by which this signal leads to a change in the processing of information by the cerebellum is not yet clear, but it does seem to be consistent with a number of lines of evidence, including the error signal idea. (*See* **Time Perception and Timing, Neural Basis of**)

DOES THE CEREBELLUM PLAY A PART IN COGNITION?

The cerebellum has received a great deal of attention regarding its potential involvement in cognitive functions. In discussing this possibility, it is helpful to define precisely what is meant by the term 'cognition'. For many researchers, the term is often operationally defined as an 'awareness' of behavior and the ability to voluntarily modify it. Using this definition, the evidence for the cerebellum's involvement in cognitive behavior is compelling. However, this involvement does not imply that the cerebellum is essential for that behavior. Animals and humans lacking a functioning cerebellum can interact meaningfully and demonstrate awareness of their behavior. Nonetheless, studies have shown that when the cerebellum is badly damaged, there are alterations in certain types of cognitive behavior that can be assessed with well-designed formal tests. When evaluating such studies, however, it is important to keep in mind the nature of the damage and the nature of the behavior being measured before reaching any conclusions. For example, it has been reported that individuals with localized cerebellar damage have deficits on many standardized intelligence and psychological tests. However, a close inspection of the results in some of these studies reveals that most of the tests that the cerebellar patients performed badly placed considerable demands on hand–eye coordination (visuomotor abilities) and the speed of responding, parameters that have little to do with formal cognitive processing. Thus, from these types of studies it is often not clear what type of cognitive deficits are present that are independent of motor deficits. In order for the cerebellum to be accepted as a true nodal point for normal cognitive processing, it is necessary to show that cerebellar damage can produce cognitive deficits without affecting motor performance. To date, very few studies have reported this type of finding for cerebellar patients.

Fiez and colleagues performed psychological testing on a patient with a localized lesion resulting from a small stroke in the lateral portion of one cerebellar hemisphere (Fiez *et al.*, 1992). This patient had excellent scores on all standard intelligence and language measures. However, when he was given tasks to perform that involved rule-based learning, verbal association and planning, he showed deficits in the ability to improve his performance with practice, and mild deficits in the actual tasks themselves. Similar types of practice-related and performance deficits were also reported in studies of cerebellar patients by Jordan Grafman and colleagues (1992) at the National Institutes of Health. In addition, Akshoomoff and Courchesne (1992) studied a group of people with widespread cerebellar damage and concluded that one of the main consequences of cerebellar damage in humans was reduced verbal fluency and verbal association capacity, and impaired visuospatial skills. Finally, work by Ivry and Keele (1989) has shown that patients with lesions of the lateral cerebellum, but not the medial cerebellum, display impaired perception of temporal intervals. Thus, it appears that certain cognitive tasks do involve the cerebellum, particularly the lateral portions of it, and that improving the performance of these tasks, or detecting performance errors, requires the activity of an intact cerebellum. Moreover, in some cases the cognitive deficits appear to be present without considerable motor deficits. Therefore, it is possible that the deficits in verbal association, visuospatial skills, rule-based learning, planning and error detection form the core cognitive symptoms of cerebellar damage.

In addition to studies of cerebellar damage, perhaps the best evidence in support of the cerebellar involvement in cognitive function is derived from studies of cerebellar activation during cognitive tasks using brain imaging techniques such as positron emission tomography (PET) or functional magnetic resonance imaging (fMRI); for a review of this topic see Desmond and Fiez (1998). Importantly, with these approaches, it is possible to compare the activation seen during cognitive tasks with that seen during motor tasks, and thus reach more accurate conclusions regarding the patterns of activity one sees. In one of the early PET studies of cerebellar activation, Fiez and colleagues (1992) examined the potential cerebellar involvement in cognitive and language tasks. Participants in the study were asked to vocalize a verb after being presented with a noun; for example, given the noun 'dog', a participant might generate the verb 'bark'. These nouns were presented by auditory or visual means, and the levels of cerebellar activity during this task were compared with the activity seen during two types of control task. In the sensory control task, participants only looked at or listened to the noun, and in the motor control task for speech output, participants immediately repeated the noun they were presented with. In the motor control task, activation was confined to the midline cerebellar hemisphere and to motor and premotor areas of the cerebral cortex normally seen activated during speech. In the verb generation task, however, in addition to the areas activated during the motor control task, there was activation of the lateral cerebellar cortex and prefrontal regions of the cerebral cortex. Whether this additional activation was related to purely language functions or a working knowledge of the strategy involved in performing the task is not clear, and both are likely possibilities. This same task, however, was the one that the cerebellar patient studied by Fiez and colleagues was found to be profoundly deficient on. In another imaging study, Decety *et al.* (1990) reported activation of the cerebellum in participants who were asked to imagine swinging a tennis racket, and to silently count the number of times they hit the ball. In comparison with people at rest, the people performing the mental imagery had a 10% increase in lateral cerebellar activity during silent counting and a 20% increase during the mental imagery condition. Finally, using fMRI, Kim *et al.* (1994) examined the activation of the lateral output nucleus of the cerebellum (the dentate) while participants sat in a scanner and either thought about ways to solve a difficult pegboard puzzle, or made visually guided movements of pegs in a pegboard. The participants had only mild increases in activity during the motor task, but displayed extensive activation of the dentate while mentally working out possible solutions to the puzzle. Taken together, these imaging studies show that whether or not observable movement takes place, there are regions of the cerebellum that display prominent activation during cognitive tasks, and that these areas appear to be different from those activated during simple motor tasks. This finding supports the idea that the cerebellum can process both cognitive and motor information using parallel circuits. (*See* **Planning: Neural and Psychological; Neuroimaging**)

ROLE OF THE CEREBELLUM IN COGNITIVE FUNCTION

The cerebellum is not essential for cognitive thought processes. However, the cerebellum is clearly involved in the performance of some cognitive tasks. The demonstration that cerebellar damage can alter affect behaviors that are known to rely on specific cognitive areas of the cerebral cortex suggests that the cerebellum must be communicating with these areas to exert its influence. In monkeys and cats, reciprocal cerebellar connections have been described with regions of the parietal cortex concerned with spatial orientation and attention and also areas of prefrontal cortex concerned with planning, rule-based learning, short-term working memory and verbal association. There are also reciprocal cerebellar connections with neuroendocrine areas of the brain, including parts of the hypothalamus. Together, these connections allow the possibility of direct cerebellar involvement in modulating limbic and autonomic functions as well as cognitive and motor functions. Such connections with primitive and highly advanced brain areas indicate that the cerebellar involvement in modulation of nonmotor function has a long-standing precedent and is not unique to humans. (*See* **Spatial Attention, Neural Basis of; Frontal Cortex; Hypothalamus**)

Although we know a great deal about the functions of the cerebellar microcircuitry and how it operates within the domain of vestibular, motor and sensory processing, we do not yet have any real insight into what the cerebellum might be doing during cognitive processing, or any idea at all how it utilizes the signals it receives from limbic, neuroendocrine and neurotransmitter synthesizing nuclei. We assume that the cerebellum does the same thing for cognitive tasks that it does for noncognitive tasks – namely, that it improves overall task performance by using its circuitry to fine-tune the speed, efficiency and accuracy of the response or planned response. However, one of the challenges of future research will be to show how this is achieved at the cellular, molecular and physiological levels for all types of behavior that fall under the influence of the 'little brain'.

References

Akshoomoff NA and Courchesne E (1992) A new role for the cerebellum in cognitive function. *Behavioral Neuroscience* **106**: 731–738.

Bell C, Cordo P and Harnad S (eds) (1996) *Controversies in Neuroscience IV: Motor Learning and Synaptic Plasticity in the Cerebellum*. Special issue of *Behavioral and Brain Sciences* **19**(3).

Decety J, Sjoholm H, Ryding E, Stenberg G and Ingvar DH (1990) The cerebellum participates in mental activity: tomographic measurements of regional cerebral blood flow. *Brain Research* **535**: 313–317.

Desmond JE and Fiez JA (1998) Neuroimaging studies of the cerebellum: language, learning and memory. *Trends in Cognitive Sciences* **2**: 355–362.

Dow RS and Moruzzi G (1958) *The Physiology and Pathology of the Cerebellum*. University of Minnesota Press, Minneapolis.

Eccles JC, Ito M and Szentagothai J (1967) *The Cerebellum as a Neuronal Machine*. New York, NY: Springer.

Fiez JA, Petersen SE, Cheney MK and Raichle ME (1992) Impaired non-motor learning and error detection associated with cerebellar damage. *Brain* **115**: 155–178.

Grafman J, Litvan A, Manaquoi S, Stewart M, Sirigu A and Hallett M (1992) Cognitive planning deficit in patients with cerebellar atrophy. *Neurology* **42**: 1493–1496.

Ito M (1984) *The Cerebellum and Neural Control*. New York: Raven Press.

Ivry RB and Keele SW (1989) Timing functions of the cerebellum. *Journal of Cognitive Neuroscience* **1**: 136–152.

Jansen J (1972) Features of cerebellar morphology and organization. *Acta Neurologica Scandinavica* **51**: (supplement) 197–217.

Kim SG, Ugurbil K and Strick PL (1994) Activation of a cerebellar output nucleus during cognitive processing. *Science* **265**: 949–951.

Larsell O and Jansen J (1972) *The Human Cerebellum, Cerebellar Connections, and Cerebellar Cortex*. University of Minnesota Press, Minneapolis.

Llinas RL and Sotelo C (eds) (1992) *The Cerebellum Revisited*. New York, NY: Springer.

Milak MS, Shimansky Y, Bracha V and Bloedel JR (1997) Effects of inactivating individual cerebellar nuclei on the performance and retention of an operantly conditioned forelimb movement. *Journal of Neurophysiology* **78**: 939–959.

Thach WT (1968) Discharge of Purkinje and cerebellar nuclear neurons during rapidly alternating arm movements in the monkey. *Journal of Neurophysiology* **31**: 785–797.

Voogd J (1975) Bolk's subdivision of the mammalian cerebellum. Growth centres and functional zones. *Acta Morphologica Neerlando-Scandinavica* **13**: 35–54.

Further Reading

Altman J and Bayer SA (1997) *Development of the Cerebellar System: In Relation to Its Evolution, Structure, and Functions*. CRC Press.

De Zeeuw CI, Strata P and Voogd J (eds) (1997) *The Cerebellum: From Structure to Control*. Special issue of *Progress in Brain Research* **114**.

Middleton FA and Strick PL (eds) (1998) *The Cerebellum*. Special issues of *Trends in Neurosciences* (**21**) and *Trends in Cognitive Science* (**2**).

Schmahmann JD (ed.) (1997) *The Cerebellum and Cognition*. Special issue of *International Review of Neurobiology* (**41**).

Cerebral Commissures

Intermediate article

Sandra F Witelson, McMaster University, Hamilton, Ontario, Canada
Debra L Kigar, McMaster University, Hamilton, Ontario, Canada
Alison Walter, McMaster University, Hamilton, Ontario, Canada

CONTENTS

The cerebral commissures house the fibers that interconnect the two hemispheres of the brain. The main fiber tract, namely the corpus callosum, varies greatly in size in the human brain. Factors such as sex of the individual and chronological age affect its size. Relationships exist between callosal size, degree of functional asymmetry and cognitive ability. The callosum may play a crucial role in the experience of conscious unity.

INTRODUCTION

With the increasing prominence of the neocortex in the mammalian brain, a system of neocortical commissures developed which provide a direct route for interhemispheric exchange of sensory and stored information. The importance of these commissures for cognitive function is now well established, but this issue has had a chequered history. As early as the eighteenth century, when neuroanatomists still did not know of the existence of neurons and their axons, the corpus callosum was believed by some to be the seat of the soul or the center of rationality or of higher thought. However, during the first half of the twentieth century, most authors considered the corpus callosum to have no function at all, except perhaps as a mechanical support for the cerebral cortex. There were exceptions, like Ramón y Cajal, who postulated that since memory centers for language are unilateral, but perceptual centers are present on both sides, the corpus callosum must be necessary to allow information to flow between the two hemispheres. By the 1960s, experimental research in animals and then studies of patients in whom the corpus callosum was surgically divided demonstrated the functional necessity of the callosum. Most recently of all, the role that the corpus callosum may play in the experience of human consciousness and personal identity is being reconsidered.

STRUCTURE

Gross Anatomy and Functional Topography

The corpus callosum (CC) is the main fiber tract connecting the cortical neurons between the two cerebral hemispheres. It is a major feature of the mammalian brain. The human CC is a large structure, about 7 cm in length from front to back, and it contains about 300 million fibers or axons. Axons from callosal neurons pass through the CC in a systematic pattern. Figure 1 is a photograph of a postmortem brain showing the CC and some anatomical regions. In general, the most anterior curved region of the CC, namely the genu, interconnects the prefrontal regions. The body of the CC interconnects the precentral and postcentral regions. The posteriormost part of the CC body, referred to as the isthmus due to its narrowing in height relative to the other regions, connects the posterior temporal and anterior parietal cortical regions between the hemispheres. The size of the isthmus has been particularly linked to the asymmetry in function between the two hemispheres, and to sex differences in CC anatomy (issues which will be discussed in more detail later). The posterior bulbous region is the splenium, which connects mainly occipital lobe cortex. The splenium has also been the subject of numerous studies concerning differences in CC size between the sexes.

Experimental research using methods of retrograde degeneration following cortical lesions and autoradiographic tracing techniques in monkeys have shown the specificity of the systematic pattern of the location where axons course through the CC to the opposite hemisphere (Figure 2). This topography is consistent with results obtained from neuropsychological studies of commissurotomized (split-brain) patients who underwent sectioning of

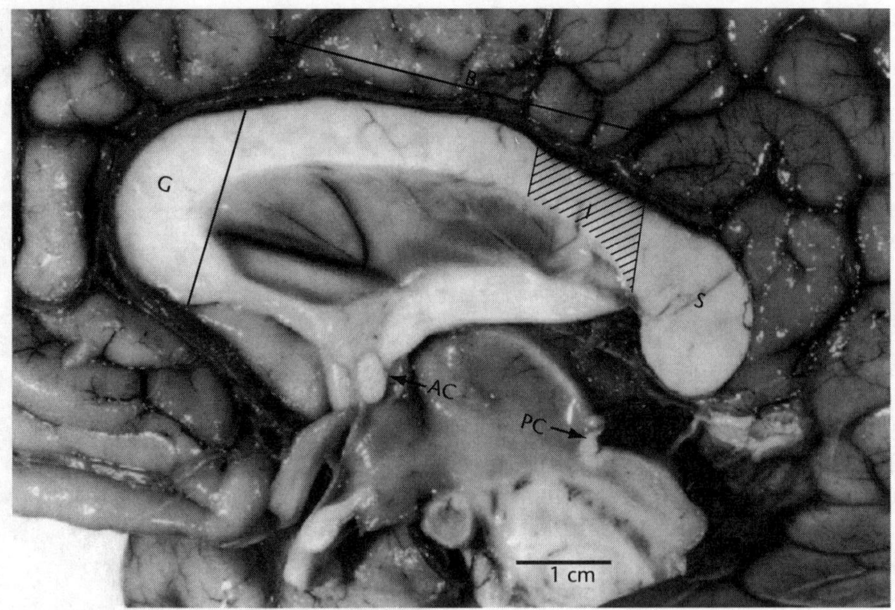

Figure 1. Photograph of the midsagittal view of an adult brain highlighting the three commissures, namely the corpus callosum (CC) and its main subregions [G, genu (most anterior region); B, body; I, isthmus (shaded); S, splenium] the anterior commissure (AC) and the posterior commissure (PC). Scale bar represents 1 cm.

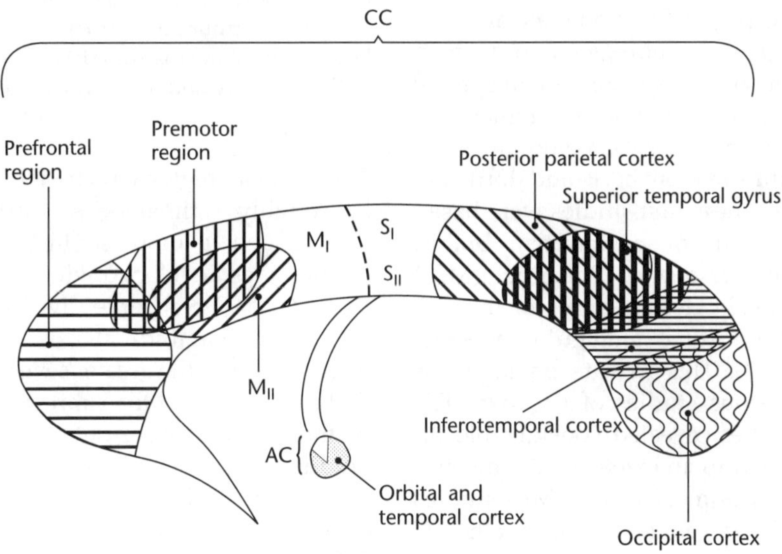

Figure 2. Diagrammatic representation of the midsagittal section of the corpus callosum (CC) and anterior commissure (AC) of the rhesus monkey, showing the topography of commissural fibers from different parts of the cerebral cortex. M_I, precentral motor region; M_{II}, supplementary motor region; S_I and S_{II}, postcentral somesthetic gyri. (Adapted from Pandya DN and Seltzer B (1986) The topography of commissural fibers. In: Leporé F, Ptito M and Jasper HH (eds) *Two Hemispheres – One Brain: Functions of the Corpus Callosum*, pp. 47–73. New York: Alan R. Liss.)

the CC for the relief of epilepsy that was uncontrollable by drug treatment. In some cases, the complete CC is not cut. Sectioning the posterior CC regions results in loss of transfer of visual or tactual information between the hemispheres. Sectioning only the anterior regions does not interfere with modality-specific cognition, but does interfere with more general cognitive functions such as semantic processing and aspects of memory (Sperry, 1974). The precise topography of the CC, as well as the fractionation of cognition that study of the CC makes possible, are revealed by other clinical cases, such as those with callosal lesions. The latter indicated that the anterior to middle splenium is involved in the transfer of pictorial information from the language-nondominant hemisphere to the language-dominant side, and that the ventroposterior part is involved in the transfer of letter information.

Sex Differences

There is considerable variation in the size and shape of the CC between individuals. Similar to the larger volume and weight of the male brain compared with that of the female, anatomical studies at the turn of the last century revealed that the CC was larger in area in the midsagittal plane in men than in women. However, during the last few decades, several studies reported that some subregions, mainly the splenium (de Lacoste-Utamsing and Holloway, 1982) and the isthmus (Witelson, 1989), were larger in women than in men relative to total callosal size. These sex differences have been interpreted as a possible anatomical correlate of the sex differences in the pattern and degree of hemispheric functional asymmetry. The relatively larger CC subregions in women would be consistent with the greater bihemispheric representation of some aspects of language in women than in men. A larger CC could allow for more transfer of information back and forth between hemispheres. These first studies were based on direct measurement of postmortem brains. However, magnetic resonance imaging (MRI) scans provide relatively clear pictures of the CC, although not without some ambiguity of exact boundaries (Figure 3). This new technology has been used in numerous studies of CC size. The results of these studies revealed considerable inconsistency, which led to an extensive review and meta-analysis of 49 postmortem and MRI studies conducted since 1980. It was concluded that the CC is larger in absolute size in men than in women, and that subregions (when CC size is taken into account) are not significantly different in size between the sexes, with the exception of the isthmus, for which there is some consistency of evidence for a larger isthmus in women, albeit with a difference of small magnitude (Bishop and Wahlsten, 1997).

Most studies have used one of two methods to measure the subregions of the CC. The earlier method employs an arithmetic parcellation of the CC into seven subregions that are determined mainly by dividing it into halves and thirds along the maximal longitudinal CC length (Figure 4a).

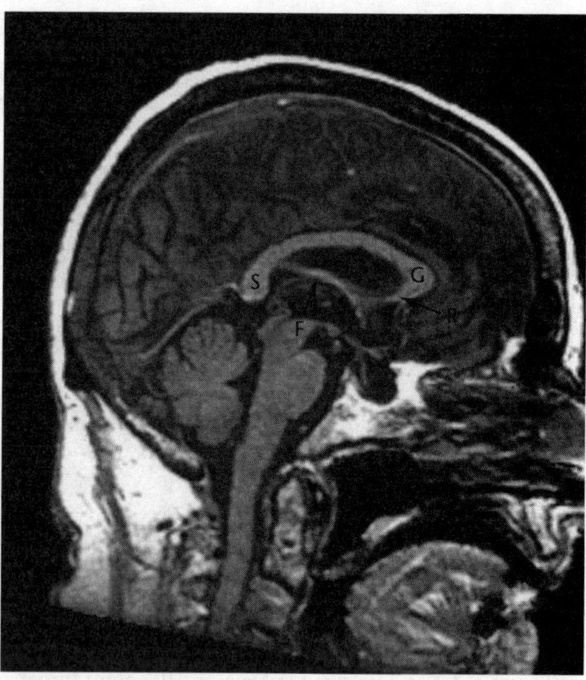

Figure 3. MR image showing the midsagittal view of a normal adult brain, illustrating the clear depiction of the corpus callosum (CC) that is obtained on scans. However, some regions of the CC are difficult to distinguish from the surrounding structures in scans, such as the boundaries of the rostrum (R) of the CC, and the junction of the CC body and fornix (F). G, genu; S, splenium.

The cortical regions related to each region may be roughly estimated. As further study of CC size ensued, a more sophisticated computerized method was developed which obtained the widths (height) of 99 percentile divisions of the CC along its curved longitudinal axis (Figure 4b). Studies then measured the area of several groups of widths that approximated the subregions of the arithmetic method, or that were obtained by factor analysis. Studies of CC size raise the important issue for anatomical studies in general of how to correct or control for variation in brain size as a baseline. Different normalization techniques using ratio scores, analyses of covariance, or stereotaxic methodology which scales MRI scans into standardized space were used on the same data sets and revealed different results with regard to the differences in CC size (Bermudez and Zatorre, 2001). (*See* **Sex Differences in the Brain**)

Microscopic Anatomy

Variation in size of the CC could reflect various histological differences, such as a different total number of axons, a different proportion of thick

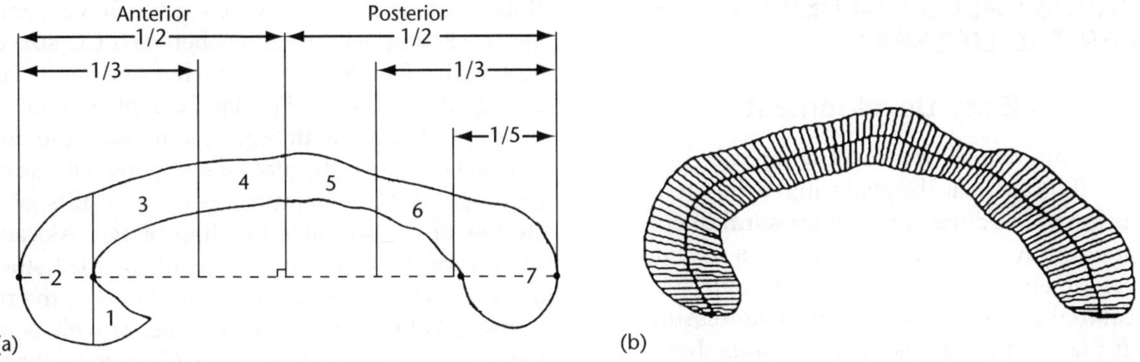

Figure 4. Sketches showing the two different methods for dividing the CC into subdivisions for measurement. (a) Arithmetic method: diagram of the midsagittal view of the CC of the human adult, showing seven subregions. The broken line is the linear axis used to subdivide the CC arithmetically into anterior and posterior halves, as well as anterior, middle and posterior thirds, and the posterior one-fifth region (region 7) which is roughly congruent with the splenium. The boundary line for the genu (region 2) is drawn perpendicular to the linear axis. Regions 3, 4, 5 and 6 constitute the body of the CC. (Adapted from Witelson SF (1989) Hand and sex differences in the isthmus and genu of the human corpus callosum: a postmortem morphological study. *Brain* **112**: 799–835.) (b) Computerized method: a tracing of a human CC showing the division into 99 minimum widths along the curved longitudinal axis. The widths are determined by dividing the dorsal and ventral CC perimeters into 99 percentiles and connecting the corresponding points. (Adapted from Denenberg *et al.* (1991) A factor analysis of the human's corpus callosum. *Brain Research* **548**: 126–132, with permission from Elsevier Science.)

(large-diameter) and thin fibers, or variation in the amount of myelination. A larger CC could have the same number of fibers as a smaller CC, but with a smaller packing density, or it could have a greater number of fibers with the same packing density. Different microscopic anatomy could have functional consequences. Using light microscopy, Aboitiz and colleagues found that the overall density of callosal fibers was not significantly correlated with callosal area, indicating that an increased CC area housed an increased total number of fibers (Aboitiz *et al.*, 1992). However, there was regional differentiation in the density of fibers of different size. Thin fibers were most dense in the anterior third of the CC, and their density decreased to a minimum in the posterior body. Thicker fibers had a complementary pattern, having a peak density in the posterior body. Thicker and more myelinated fibers have faster rates of conductivity. Relatively more thin fibers interconnect higher-order association cortex. It may be that the longer interhemispheric transfer time over thin fibers may be irrelevant for higher-order cognition compared with more direct sensory information.

Small Commissures

Fibers also cross the midline in two other commissures, namely the anterior and posterior commissures (Figure 1). Each is small, on average less than $10\,mm^2$ in cross-section. The anterior commissure houses fibers from the orbitofrontal regions and from the anterior and midregions of the temporal lobe. The posterior commissure connects the diencephalic and midbrain structures. These two commissures are important features in brain imaging, as they are readily viewed and serve as stereotactic landmarks to define the horizontal axis of the brain.

The anatomy of the anterior commissure has undergone an interesting evolutionary change which has functional relevance. Phylogenetically it is older than the neocortex. In lower mammals (e.g. the lemur), it primarily connects regions related to olfaction. Marsupials (e.g. kangaroos) are the one group of mammals that do not have a CC, and in this case the anterior commissure plays a key role in transmitting information between the hemispheres. In humans and other higher primates, the anterior commissure has anterior and posterior limbs, the anterior limb being part of the olfactory system. The posterior limb has neocortical fibers from auditory and visual association regions of the temporal lobe. In cases of agenesis or lack of development of the CC, the anterior commissure appears to play a role in functional plasticity and to mediate some information that is usually transmitted via the CC. The anterior commissure is highly variable in size, which appears to be related to gender and the pattern of hemispheric functional asymmetry. (*See* **Reorganization of the Brain**)

CORPUS CALLOSUM DEVELOPMENT OVER THE LIFESPAN

Fetal and Early Development

Early in brain development, a temporary bridge of glial cells (known as the glial sling) forms across the hemispheric midline, and commissural fibers cross supported by the sling. In some cases of abnormal development this process does not occur, resulting in callosal agenesis. The anterior commissure may then play a greater role in interhemispheric connectivity, but the pattern of hemispheric functional asymmetry is atypical. Initially in development there is an exuberance of callosal fibers, and subsequently some axons are eliminated. For example, this was demonstrated dramatically by electron microscopic analysis in the monkey (LaMantia and Rakic, 1990). In addition, axons are eliminated in a systematic pattern (e.g. in the monkey, up until 1 month before birth there are callosal axons connecting the primary and secondary somesthetic cortex, which are then eliminated, resulting in the adult pattern). Research on fetal and infant brains has been conducted via postmortem study. A review of such studies has indicated that the size of the CC doubles in fetal life and then doubles again within the next 2 years. CC size was found to correlate with gestational age, a feature that is useful for identifying normal versus abnormal brain development *in utero*. The CC is thin during the first few months of postnatal life, but the genu and splenium grow rapidly over the next few months. During this period myelination begins, and the adult shape and orientation of the CC are achieved by about 8 months of age.

There has been relatively little research on CC anatomy during childhood and adolescence. In a recent large longitudinal MRI study, it was found that the area of the CC increases substantially from 5 to about 18 years of age (Giedd *et al.*, 1999), when it reaches adult size. This increase is probably due to myelination. Now that high-resolution structural MRI scans are possible, growth of the CC and its subregions can be studied in relation to cognitive development over periods of marked cognitive changes.

Adulthood

Even after the mature CC has developed, there is great variation among individuals, some of which is related to the sex of the person (see above), while some may be associated with variation in cognitive ability (see below). Age is another major factor. Both postmortem and MRI studies have demonstrated negative correlations between CC size and age, with men showing an earlier and more rapid change than women. Frontal CC regions show the greatest change with age, paralleling the more marked change in frontal lobe regions with advancing age. Table 1 summarizes the results of the studies of CC size as a function of age. As can be seen from the table, the results of the MRI studies are somewhat inconsistent with regard to the relationship between CC area and age, as well as with regard to the larger overall CC in men than in women. Although the CC is one of the most clearly visualized structures in MRI scans, the exact boundaries may be ambiguous in some regions, depending on the contrast of voxels in the surrounding structures (Figure 3). (*See* **Neuroimaging**)

Changes in the overall size of the CC must reflect some microscopic change. Only electron microscopy, whose demanding requirements for tissue preparation are problematic in non-experimental situations, can reveal all unmyelinated axons reliably. Consequently, little research has been done in this area. However, one such study reported no relationship between age and the total number of CC fibers, suggesting that the decreasing size of the CC with age may be due to loss of myelin or to a reduction in axon diameter, rather than being due to loss of axons.

Hormonal and Environmental Effects on Corpus Callosum Anatomy

Experimental research has demonstrated that sex hormones contribute to the larger overall CC in the male rat brain compared with that of the female. The rat brain is sensitive to the effects of hormones after birth. During this period of sexual differentiation of the brain, testosterone results in an increase in CC size and, in contrast, estrogen actively inhibits CC growth (Mack *et al.*, 1996).

Several recent studies have documented the strong genetic component in CC size. Studies of monozygotic versus dizygotic twin pairs have shown that heritability may account for as much as 70–90% of the variation. A related issue is whether environmental/experiential factors affect callosal anatomy. Later in this article it will be noted that although left-handers or non consistent-right-handers (people who use their left hand for even a few tasks) have a larger CC than consistent-right-handers, the experience of an early forced shift from left- to right-hand writing does not appear to affect CC size (Witelson, 1989). However, there is some evidence that rats living in enriched

Table 1. The corpus callosum and age

Study	*Subjects*	*Method*[a]	*Findings*[b]
Witelson, *Brain*, 1989	50 brains from cognitively normal subjects 25–70 years 15M; 35W	Post mortem Total CC	M: $r = -0.6$ W: ns CC area: M > W
Witelson, *New England Journal of Medicine*, 1991	62 brains from cognitively normal subjects 25–70 years 23M; 39W	Post mortem Total CC Three CC subregions	M: total CC, $r = -0.7$ Genu: $r = -0.7$ Body: $r = -0.5$ Splenium: $r = -0.5$ W: ns
Doraiswamy *et al.*, *Journal of Neuropsychiatry*, 1991	36 normal subjects 26–79 years 16M; 20W	MRI Total CC	M: $r = -0.6$ W: ns No sex difference in CC area
Cowell *et al.*, *Developmental Brain Research*, 1992	73 pairs of age-matched M and W 2–79 years	MRI Total CC 99 widths (see Figure 4b)	M: CC area peaks at about 20 years W: CC area peaks at about 50 years Peaks followed by decline, especially in genu and posterior splenium Little decline in body, isthmus and anterior splenium
Parashos *et al.*, *Journal of Neuropsychiatry*, 1995	80 normal subjects 30–91 years 28M; 52W	MRI Total CC Five CC subregions	CC areas decrease with age No sex difference in CC areas
Salat *et al.*, *Neurobiology of Aging*, 1997	76 subjects 65–95 years 31M; 45W	MRI Three CC subregions	M: ns W: anterior region, $r = -0.4$ mid region, $r = -0.4$ Posterior CC region: W > M
Sullivan *et al.*, *Neurobiology of Aging*, 2001	92 subjects 22–71 years 51M; 41W	MRI Total CC Three CC subregions	No decrease with age CC areas: M > W

M, men; W, women.
[a]Method: midsagittal areas.
[b]Findings: r-values are CC area as a function of age.

as opposed to deprived environmental conditions have enlarged callosa. In humans, literacy (as opposed to being illiterate) and early musical training have been associated with larger CC size. Thus environment as well as heredity may affect CC size.

RELATIONSHIP OF THE CORPUS CALLOSUM TO FUNCTIONAL ASYMMETRY AND COGNITION

The Corpus Callosum and Hemispheric Functional Asymmetry

The pattern and degree of hemispheric specialization varies between individuals. It is possible that variation in CC anatomy is related to brain lateralization. Several studies have examined the relationship between callosal anatomy and neuroanatomical asymmetries, with varying results. Studies of asymmetries of the Sylvian fissure or the posterior surface of the superior temporal gyrus (planum temporale) in relation to CC size have found that the anatomical asymmetries were correlated with posterior CC regions. Since the posterior CC houses fibers which connect right and left posterior temporo-parietal regions which are asymmetrical in function between the two hemispheres, these results support the hypothesis that the extent of callosal connectivity is related to the direction and degree of anatomical asymmetry

Table 2. The corpus callosum and hemispheric functional asymmetry

Study	*Subjects*	*Method*[a]	*Lateralization measure*	*Findings*
Witelson, *Science*, 1985	42 brains from cognitively normal subjects 25–65 years 12M; 30W 27 CRH; 15 nonCRH	Post mortem Total CC Seven CC subregions (see Figure 4a)	Hand preference (tested, 12 items)	Total CC, anterior half, posterior half and posterior fifth areas Total group: NonCRH > CRH
O'Kusky *et al.*, *Annals of Neurology*, 1988	50 normal controls (for study of 50 epileptic subjects) 17–60 years M and W	MRI Total CC Five CC subregions	Hand, foot and eye preference Verbal dichotic listening 50 epileptic subjects, 44 had Wada testing for speech lateralization	No association of CC area with hand preference No sex difference in CC In patients, those with right-hemisphere speech lateralization had larger CC
Witelson and Goldsmith, *Brain Research*, 1991	22 brains from cognitively normal subjects Mean age 54 years All M 13 CRH; 9 nonCRH	Post mortem Total CC Seven CC subregions (see Figure 4a)	Hand preference (tested, 12 items)	Total CC: nonCRH > CRH Isthmus and handedness score: $r = -0.5$
Habib *et al.*, *Brain and Cognition*, 1991	53 normal subjects 18–51 years 35M; 18W 26 CRH; 27 nonCRH	MRI Total CC Six CC subregions	Hand preference (10-item questionnaire)	Total CC, subregions: M: NonCRH > CRH W: ns No sex difference in CC
Hines *et al.*, *Behavioral Neuroscience*, 1992	28 subjects All W 20–45 years	MRI Total CC Three CC subregions	Hand preference (tested, 18 items) Language lateralization	Posterior CC area and language lateralization: $r = -0.3$
Cowell *et al.*, *Neurology*, 1993	104 normal subjects 18–49 years 51M; 53W 52 RH; 52 LH	MRI Total CC CC divided into 99 widths, and widths grouped into 7 regions (see Figure 4b)	Hand preference (defined by writing hand only)	Rostral body (widths 22–39) and posterior midbody (widths 49–62) W: RH > LH M: LH > RH
Yazgan *et al.*, *Neuropsychologia*, 1995	11 normal subjects Mean age 34 years 11M; 2W RH	MRI Total CC Five CC subregions	Verbal dichotic listening Verbal–manual interference (VMI) test	CC and REA, $r = -0.7$ CC and VMI, $r = -0.6$
Jancke *et al.*, *Cerebral Cortex*, 1997	120 normal subjects 18–45 years 71M; 49W 54 CRH; 28 CLH; 38 MH	MRI Total CC Four CC subregions	Hand preference (tested, 12 items)	No sex difference for absolute CC areas Total CC and subregions: W > M for relative size Middle third CC area: CRH > CLH, MH

Moffat *et al.*, *Brain*, 1998	16 subjects Mean age 23 years All M, LH 10 REA; 6 LEA RH male archival data	MRI Total CC Six CC subregions	Verbal dichotic listening Hand preference (tested, 8 items)	Total CC and posterior regions: LH > RH LH-REA subjects > LH-LEA or RH subjects Smaller isthmus associated with LEA

M, men; W, women; CRH, consistent-right-handed; nonCRH, non consistent-right-handed; RH, right-handed; LH, left-handed; MH, mixed-handed; REA, right-ear advantage; LEA, left-ear advantage.
[a]Method: midsagittal areas.

and possibly functional asymmetry. (*See* **Brain Asymmetry**)

One of the first behavioral measures that was found to be related to CC size was hand preference (Witelson, 1989). This result indicated that structure and function may be related in very direct ways (i.e. that the size of parts of the human brain may be correlated with behavioral measures). Since left-handers have a greater degree of bihemispheric representation of cognitive functions, it was suggested that the larger CC may provide a substrate for greater interhemispheric communication in left-handers. Subsequent studies using MRI measures of CC generally replicated and supported the finding of a larger CC and particularly some subregions in left-handers or in individuals who did not have the typical pattern of left-sided speech representation. As research progressed to address further issues, it appeared that the CC varies with both hand preference and speech lateralization independently to some degree. The CC was smaller in individuals who had what may be called 'ipsilateral' (same-sided) representation of hand preference and speech than in people who had 'contralateral' representation (e.g. left-handers with right-sided hand representation and left-sided speech representation), but it was not smaller than in left-handers with right-sided speech (also 'ipsilateral' cases) (Table 2).

The Corpus Callosum and Cognitive Correlates

Corpus callosum anatomy may provide a window on the anatomy and function of the cerebral cortex, and consequently it may reflect variation in cognitive processing. Several MRI studies have focused on CC size and its relationship to performance on verbal and spatial tasks in predominantly right-handed individuals (Table 3). In general, small but statistically significant positive correlations were observed. If larger callosa have a greater number of fibers and provide greater interhemispheric communication, these results suggest that, at least in strongly lateralized individuals, such anatomy may confer some functional advantage.

Although it seems remarkable to find relationships between ability and brain size – in this case CC area and, by inference, cortical anatomy – much further research is needed to elucidate the structure–function relationships and possible mechanisms. Such research also raises major ethical issues. MRI scans have the potential to be used as tests or indicators of ability by educational and professional bodies. Clearly caution and proactive consideration are needed with regard to such applications of basic research findings, which will likely only increase with time.

Split-brain Studies and Theoretical Implications

Commissurotomy is an operative procedure which permanently severs the CC or all tracts between the right and left hemispheres. Each hemisphere then only has access to the stimulation it receives and the neural processes that are established on that side. The neuropsychological study of split-brain patients carried out by neuroscientist Roger Sperry and colleagues yielded unambiguous information about functional lateralization of the brain. In such patients, the two sides of the brain are isolated from each other, so that neuropsychological testing can determine not only which side is dominant for specific cognitive functions, but also the limitations of each hemisphere, by determining which functions it cannot support. This has been done in a series of ingenious studies (Sperry, 1974). (*See* **Split-brain Research**)

The research with split-brain patients also raised the issue of the nature of conscious unity. For the development of this issue, in addition to

Table 3. The corpus callosum and cognition

Study	*Subjects*	*Method*[a]	*Cognitive measure*	*Findings*
Hines *et al.*, *Behavioral Neuroscience*, 1992	28 W 20–45 years	MRI Total CC Three CC subregions	Verbal fluency tests Visuospatial ability	Posterior CC and verbal fluency: $r = 0.6$ Anterior CC and visuospatial ability: $r = 0.4$
Strauss *et al.*, *Journal of Clinical and Experimental Neuropsychology*, 1994	47 epileptic subjects 12–57 years	MRI Total CC Five CC subregions	Wechsler Adult Intelligence Test (WAIS) (Full-scale IQ – FSIQ)	Splenial area and FSIQ: $r = 0.4$
Yazgan *et al.*, *Neuropsychologia*, 1995	11 normal subjects Mean age 34 years 11 M; 2W RH	MRI Total CC Five CC subregions	Line bisection WAIS-R	CC and line bisection: $r = 0.6$ CC and WAIS: ns
Atkinson *et al.*, *Journal of Neuroimaging*, 1996	20 age- and sex-matched controls (for epilepsy study) 12–47 years 8M; 12W	MRI Total CC area	WAIS-R Wechsler Memory-R	CC and FSIQ: $r = 0.3$ CC and Performance IQ: $r = 0.4$
Salat *et al.*, *Neurobiology of Aging*, 1997	76 subjects 65–95 years 31M; 45W RH	MRI Total CC Three CC subregions	Weschler Memory-R (logical memory and visual reproduction subscales) WAIS-R (block design subscale)	W: CC and visual memory: $r = 0.3$ M: ns in all tests
Davatzikos and Resnick, *Cerebral Cortex*, 1998	114 highly educated subjects 56–85 years 68M; 46W RH	MRI Total CC and subregions	Card rotations Boston naming test Letter fluency Verbal recognition memory test Figural recognition memory test	W: CC areas and all five neuropsychological tests: positive *r*-values M: ns
Peterson *et al.*, *Human Brain Mapping*, 2001	138 control subjects (for a study of multiple syndromes) 6–88 years M and W CRH = 86%	MRI Total CC area, widths, bending angle, splenial bulbosity, etc.	WAIS-R Performance IQ and Verbal IQ	A factor representing a thinner, more concave anterior body of the CC predicted higher IQ scores

M, men; W, women; CRH, consistent-right-handed; nonCRH, non consistent-right-handed; RH, right-handed; LH, left-handed; MH, mixed-handed; REA, right-ear advantage; LEA, left-ear advantage.
[a]Method: midsagittal areas.

experimental neurobiological work, Dr Sperry was awarded the Nobel Prize for Medicine in 1981. More recently, it has been suggested that this tract of white matter enables the human condition. The CC may have eliminated the necessity for having two redundant systems, by linking the two hemispheres to such a degree that information can be easily shared and transmitted to appropriate sensory or motor systems. It allows for specialization of the two hemispheres, and it provides the anatomical substrate for the development of completely new functions. Gazzaniga (2000) suggested that it is processing by the left hemisphere which interprets events and allows one to attempt to find reason or logic in life events. It may be left-hemisphere processing which is crucial for generating hypotheses to explain inexplicable events. This capacity of the left hemisphere may be what makes a person feel as if they control their own actions. Although the right hemisphere has access to what is going on in the outside world, its role may be to keep track of this information so that the left hemisphere can process and interpret it. Thus the two hemispheres may work synergistically to provide an adaptive system of experience and to interpret a person's environment and internal thoughts. (*See* **Neural Correlates of Consciousness as State and Trait)**

References

Aboitiz F, Scheibel A, Fisher RS and Zaidel E (1992) Fiber composition of the human corpus callosum. *Brain Research* **598**: 143–153.

Bermudez P and Zatorre RJ (2001) Sexual dimorphism in the corpus callosum: methodological considerations in MRI morphometry. *NeuroImage* **13**: 1121–1130.

Bishop KM and Wahlsten D (1997) Sex differences in the human corpus callosum: myth or reality? *Neuroscience and Biobehavioral Reviews* **21**: 581–601.

de Lacoste-Utamsing C and Holloway RL (1982) Sexual dimorphism in the human corpus callosum. *Science* **4553**: 1431–1432.

Gazzaniga MS (2000) Cerebral specialization and interhemispheric communication. Does the corpus callosum enable the human condition? *Brain* **123**: 1293–1326.

Giedd JN, Blumenthal J, Jeffries NO *et al.* (1999) Development of the human corpus callosum during childhood and adolescence: a longitudinal MRI study. *Progress in Neuropsychopharmacology and Biological Psychiatry* **23**: 571–588.

LaMantia A-S and Rakic P (1990) Axon overproduction and elimination in the corpus callosum of the developing rhesus monkey. *Journal of Neuroscience* **10**: 2156–2175.

Mack CM, McGivern RF, Hyde LA and Denenberg VH (1996) Absence of postnatal testosterone fails to demasculinize the male rat's corpus callosum. *Developmental Brain Research* **95**: 252–254.

Sperry RW (1974) Lateral specialization in the surgically separated hemispheres. In: Schmitt FO and Worden FG (eds) *The Neurosciences: Third Study Program*, pp. 5–19. Cambridge, MA: MIT Press.

Witelson SF (1989) Hand and sex differences in the isthmus and genu of the human corpus callosum: a postmortem morphological study. *Brain* **112**: 799–835.

Further Reading

Gazzaniga MS and LeDoux JE (1978) *The Integrated Mind*. New York, NY: Plenum Press.

Leporé F, Ptito M and Jasper HH (eds) (1986) *Two Hemispheres – One Brain: Functions of the Corpus Callosum*. New York, NY: Alan R. Liss.

Rakic P and Yakovlev PI (1968) Development of the corpus callosum and cavum septi in man. *Journal of Comparative Neurology* **132**: 45–72.

Witelson SF and Kigar DL (1988) Anatomical development of the human corpus callosum: implications for individual differences and cognition. In: Molfese DL and Segalowitz SJ (eds) *Developmental Implications of Brain Lateralization*, pp. 35–57. New York, NY: Guilford Press.

Zaidel E and Iacoboni M (eds) *The Parallel Brain: Cognitive Neuroscience of the Corpus Callosum*. Cambridge, MA: MIT Press.

Change Blindness

Intermediate article

J Kevin O'Regan, Centre National de Recherche Scientifique, Université Paris 5, France

CONTENTS

Change blindness is a phenomenon in visual perception in which very large changes occurring in full view in a visual scene are not noticed.

WHAT IS CHANGE BLINDNESS?

A number of studies have shown that under certain circumstances, very large changes can be made in a picture without observers noticing them. What characterizes the experiments showing such 'change blindness' in visual scenes is the fact that the changes are arranged to occur simultaneously with some kind of extraneous, brief disruption in visual continuity, such as the large retinal disturbance produced by an eye movement, a shift of the picture, a brief flicker, five or six small, localized disturbances flashed briefly on the picture, an eye blink, or a film cut in a motion picture sequence. These phenomena are attracting an increasing amount of attention from experimental psychologists and from philosophers, because they suggest that humans' internal representation of the visual world is much sparser than usually thought.

EXPERIMENTAL WORK ON CHANGE BLINDNESS

In the first experiments that triggered interest in change blindness (McConkie and Currie, 1996), observers viewed high-resolution, full-color everyday visual scenes presented on a computer monitor, while their eye movements were being measured. The computer could make changes in the scene as a function of where the observer looked. For example, when the observer looked from the door of a house to the window, the window (or some other element of the scene such as the sky, or the car parked in front of the house) changed in some way: it could disappear, be replaced by a different element, change color, change position, etc. It was found that when the change occurred *during* an eye movement, surprisingly large changes could be made without observers noticing them. Elements of the picture that occupied as much as a fifth of the picture area would not be seen. At first, the explanation of the phenomenon was assumed to have something to do with the mechanisms the brain uses to combine information from successive eye fixations to form a unified view of the visual world. In particular, every time the eye moves, the retinal image shifts. Some mechanism in the brain may correct for such shifts in order to create a stable view of the world. However, the mechanism could be imperfect and not take into account certain differences in the visual content across the shift, thereby explaining why changes made during saccades might sometimes go unnoticed.

But a subsequent set of experiments showed that, in fact, the change blindness phenomenon was not specifically related to eye movements. Rensink *et al.* (1997) used what they called the 'flicker' technique, in which, instead of an eye movement, a brief flicker was introduced between successive images. A first picture (picture A) would be shown for, say, 250 ms, followed by a modified picture (picture B). In between A and B, a brief blank screen (bl) would be shown. This would cause a flicker, lasting about 80 ms, that is, a duration similar to that of an eye movement. The cycle A-bl-B-bl-... was then repeated. Observers were told that something was changing in the picture every time the flicker occurred, and they were asked to search for it.

Under conditions where no flicker was inserted in between the pictures (A-B-A-B-...) the change was immediately visible and totally obvious (animated gif (158 kb): http://nivea.psycho.univ-paris5.fr/ECS/bagchangeNoflick.gif). However, with the flicker, it was often extremely difficult to locate the change. This was particularly true for changes which concerned aspects of the scene which were not of 'central interest'. For example, the reflection of houses in a lake scene, though occupying a very large part of the picture, would not be considered to be what the picture was about. Observers sometimes were unable to see such changes at all, even after searching actively for as long as one minute. On the other hand, the changes were perfectly visible once they were pointed out to observers (animated gif (158 kb): http://nivea.psycho.univ-paris5.fr/ECS/kayakflick.gif).

The flicker technique is very easy to implement using widely available computer software (for example, software to make video presentations), and so lends itself to easy experimentation. Pictures as well as symbolic or text material can be used. The timing of the flicker between original and modified images is not critical. What is important about the flicker technique is that it shows that change blindness can be obtained without the change being synchronized with eye movements. This shows that in the earlier experiments where changes *were* synchronized with eye movements, the inability to detect the change was probably not specifically related to the eye movement and to the mechanisms that the brain uses to combine images of the world during eye explorations.

Following the discovery that change blindness was not specifically related to eye movements, but to the brief disruption that is inserted between the two versions of the picture, considerable interest in the phenomenon developed, and a large number of further experiments have been performed. These can be classified according to the nature of the disruption that is used between successive images: global disruptions, local disruptions, and progressive changes. (A review of different change blindness experiments can be found in Simons and Levin (1997).)

Global disruptions are ones in which the picture change is accompanied by a disruption which covers the whole area of the picture. The experiments where the changes occurred during eye movements were global disruption experiments, since the whole retinal image is completely smeared during the time of approximately 20 to 80 ms that it takes for the eye to move from one fixation point to the next. The flicker experiments are also global disruption experiments, since the blank displayed briefly between the original and modified images covers the whole picture. Other examples of experiments with global disruptions are experiments involving eye blinks, picture shifts, and film cuts. In the blink experiments, observers' eye blinks, registered by online computer monitoring, are used to trigger the picture change. The blink produces a global disruption similar, though somewhat longer in duration, to the disruption caused by an eye movement. In the picture shift experiments, a picture is suddenly shifted in position, and a change made at the same time. Here a global disruption is caused by the retinal smearing that accompanies the eye movement that observers make to refixate the shifted picture. In film cut experiments, observers view motion picture extracts, and at the moment when the camera 'cuts' from one view to another, a large change is made – for example, an actor is replaced by a different actor. The camera cut produces a global disruption similar to the blank in the flicker experiments.

An additional, amusing, variant of the experiments with global disruptions are experiments in which the change occurs in real life. In a typical scenario described by Simons and Levin (1998), the experimenter stops a person in the street and asks for directions. While the person is speaking to the experimenter, workers carrying a door pass between the experimenter and the person, and an accomplice takes the place of the experimenter. The person usually goes on giving directions after the interruption, and very often does not notice that the experimenter has been replaced by the accomplice.

Change blindness experiments with *local disruptions* are experiments in which, at the moment of the change, five or six small, localized disturbances are superimposed on the picture, like mudsplashes on a car windscreen (O'Regan *et al.*, 1999). The disturbances can be small in comparison to the size of the change and they need not coincide with the location of the change: the change takes place in full view. As for change blindness with global interruptions, changes are very often not noticed (animated gif (378 kb): http://nivea.psycho.univ-paris5.fr/ECS/dottedline.gif).

Experiments with *slow changes* are experiments in which there is no local or global disruption at all. Instead, the change is made so slowly that the attention-grabbing processes that would normally cause attention to be attracted to the change location can no longer operate. Again, it is found that in many cases, changes are hard to detect (Quicktime video (1.4 Mb): http://nivea.psycho.univ-paris5.fr/ECS/sol_Mil_cinepack.avi).

RELATED PARADIGMS

The change blindness phenomenon is strongly related to a well-established line of research in experimental psychology that started in the 1970s and concerns visual short-term memory (for a review, see Haber, 1983).

In this literature, experiments analogous to the change blindness experiments had been performed using briefly displayed arrays of simple elements such as letters. These experiments showed that although observers have the impression of seeing all the letters in, say, a 12-element array, in fact they notice changes to or report the identity of only about four or five letters. It appears that there is a kind of attentional 'bottleneck' which limits information transfer into memory: only a fraction of the information available in a scene is transferred into visual storage for later report or comparison. Further work additionally showed that the code in which the information is stored in visual short-term memory is not a visual code, but a code in which only the category or identity of the elements is available. This work was also coherent with another line of research showing that information from successive eye fixations is combined only in categorical form, and not as a picture-like composite image (for a review, see Irwin and Andrews, 1996).

We shall see in the next section that the conclusion from these experiments, showing that visual storage is sparse and categorical, is also applicable to the change blindness results. Because in the case of change blindness, natural, highly detailed visual scenes are used as stimuli, the conclusion is more striking than it was in the older literature using simple stimuli.

Change blindness, in addition to links with research on visual short-term memory, also has relations with several more recent lines of research showing that attentional capacity in short-term visual processing is severely limited, both in spatial extent, and in the way it extends over time. Thus in 'inattentional blindness' (Mack and Rock, 1998), observers do an attention-demanding visual task. At a given moment, a large, unexpected visual event takes place. Even though such an event would be totally obvious under normal circumstances, and even though the event takes place in full view, it is often not noticed. For example, Simons and Chabris (1999) used a task in which observers look at a film of two groups of players, a black-clad and a white-clad group, each playing with their own ball in the same small room. The observer's task is to try to track the number of times one group exchanges the ball. While the observer is doing this task a woman with an umbrella walks through the room, in full view. Observers often fail to notice this totally obvious event.

An example of a temporal restriction on the deployment of attention is the 'attentional blink' (for a review, see Shapiro and Terry, 1998): in this, an observer is required to identify a target letter in a stream of rapidly presented letters. It is found that the observer often fails to report the occurrence of a second target letter if the second target follows the first by less than about 450 ms: it is as though attention had to recover for a brief period after having been solicited. In 'repetition blindness' (Kanwisher, 1987), a visual stimulus such as a letter, symbol, picture, or word tends not to be noticed if it is the second of two identical occurrences of the item in a rapidly presented series.

Another field of research that has connections to change blindness is the extensive literature on memory and cognitive descriptions (for a review, see Pani, 2000). Part of the explanation for change blindness may reside in memory limitations rather than in perceptual limitations. If this is so, then we expect that change blindness will be affected by factors similar to those that affect memory. This is compatible with the finding that changes made to elements in a scene which are of 'central interest' will in general be easier to detect than 'marginal interest' changes. Other work has shown that variables such as semantic coherence, observer familiarity, and task to be achieved, affect change blindness in a way similar to how they affect normal memory.

THEORIES OF CHANGE BLINDNESS

The currently accepted explanation of change blindness owes much to the work done in the 1960s and 1970s showing how visual information is transferred via an attentional 'bottleneck' to a very low capacity short-term visual storage (e.g. Gegenfurtner and Sperling, 1993). Within this context, the explanation of change blindness involves two components: a component related to what is called 'visual transients', and a component related to the way a scene is encoded in memory.

Visual transients are fast changes in luminance or color in the retinal image, such as would be produced by a sudden appearance or disappearance, or through motion of an element of the scene. It is known that such transients are detected in the first levels of the visual system, and that attention is automatically attracted to the location where they occur. Under normal viewing conditions, therefore, when a change occurs, it produces a visual transient which attracts attention to the change location. The transient thus provides information *that* a change has occurred, and it says *where* it occurred, but it does not provide information about *what* the change was.

In order for an observer to be able to determine *what* the change was, he or she will have had to have encoded into visual memory what was at the change location before the change occurred, and compare it to what is there after the change. There are thus two things that can go wrong in change detection: either the transient that attracts attention to the change location may be interfered with (thereby causing a deficit in detection *that* or *where* the change has occurred), or the encoding and comparison process may be interfered with (causing a deficit in determining *what* the change was).

Both these mechanisms may be at work in the change blindness experiments. In the paradigms using global disruptions, such as the flicker, blink, and film cut experiments, the global disruption presumably creates a large number of transients all over the picture, which mask or compete with the local transient corresponding to the sought-for change, and which prevent attention being automatically drawn to it. The change will be immediately noticed only if an observer happens to have been attending to the changing element at the moment it changes. Failing this, in order to find a change, the observer must search through the scene looking for an element which is different from what was previously encoded about the scene. However, because of the limitations in short-term visual memory, very little of the scene is likely to have been previously encoded, and the chances of success are very limited.

In change blindness paradigms using local disruptions like 'mudsplashes', the situation is very similar, with the difference that the local transient corresponding to the change location is missed by observers, not because it is swamped by a global transient, but because the mudsplashes act as 'decoys', attracting attention to locations other than the true change location.

In change blindness paradigms with slow changes, the change occurs so slowly that no local transient is generated. Attention is thus not attracted to the change location, and again, the observer must rely on the very sparse information that he or she has encoded about the scene in order to locate the change.

Whereas researchers working in change blindness will broadly agree on the explanation just outlined of the phenomenon, further work is necessary to ascertain the relative roles of the different component mechanisms involved. To what extent does the flicker in the flicker paradigm act to mask or 'wipe out' the internal representation? Or does it act essentially like the mudsplashes in the mudsplash paradigm to create local transients that act as decoys? Exactly how much information is encoded concerning the initial and final views of the scene? Is the overall 'gist' of a scene coded in some way? Does what is encoded depend on the observer's attentional state, on the task, on viewing strategies, on the semantic relation between the gist of the scene and the element that is changing? Are certain aspects of elements (their layout? their color?) automatically encoded and easier to detect when they change? Even if little information is available to make conscious judgments about display changes, could it be that some information is retained unconsciously? A number of recent lines of research are investigating these issues.

RELEVANCE OF CHANGE BLINDNESS FOR CONSCIOUSNESS AND COGNITIVE SCIENCE

Change blindness raises an important question: if the information that is encoded about a visual scene is so sparse, how is it that we have the subjective impression of visual richness, that is, of seeing everything there is to see in our field of view, so to speak in 'glorious technicolor and cinemascope'?

Perhaps the most natural view to take is to suppose that what we have the subjective impression of seeing is not the very sparse, more semantically coded, content of visual memory, but the content of a shorter-lived but higher quality, image-like replica or 'icon' of the visual scene. The impression of richness that we have from the world would derive from this high-quality icon. On the other hand, only a small portion of the icon's contents, namely the parts that have been attended to, would at any moment be transferred into memory and be available for doing such tasks as change detection – the rest would be forgotten. This view of visual processing has been called 'inattentional amnesia' (Wolfe, 1999): the idea is that we see everything, but forget most of it immediately.

The notion that what underlies the richness of vision is a high-quality internal replica of the outside world is the basis of some of the current work in neurophysiology and neuroanatomy, where cortical sites are being sought which provide the 'neural correlate of consciousness'. Indeed, area V1 of the visual cortex contains a distorted map of the visual field which might be a plausible locus for visual consciousness, possibly in relation to other brain structures.

A more radical answer to the question of why we have the impression of continuously seeing everything in our visual field has also been suggested

(O'Regan and Noë, 2001). The idea is that in fact the experience of seeing does not derive from the activation, inside the brain, of an 'icon' of the outside world. Rather, the experience of seeing is somewhat like the temporally extended, multifaceted experience of driving a car, involving a kind of 'give and take' between the observer and the environment, a kind of attunement to the laws that link the observer's actions to the changes in sensory input.

Under this view, the outside world serves as a form of 'external memory'. Only those aspects of the environment that are currently being 'visually manipulated' are actually available for conscious processing at a given moment. We have the impression of seeing everything because we know we have access to everything, even though without actually accessing something, no detailed information is available about it. This explains the apparent paradox between the feeling of richness we have of our visual environments, and our striking inability, in change blindness experiments, of knowing what has changed.

References

Gegenfurtner KR and Sperling G (1993) Information transfer in iconic memory experiments. *Journal of Experimental Psychology: Human Perception & Performance* **19**(4): 845–866.

Haber RN (1983) The impending demise of the icon: a critique of the concept of iconic storage in visual information processing. *Behavioral and Brain Sciences* **6**: 1–54.

Irwin DE and Andrews RV (1996) Integration and accumulation of information across saccadic eye movements. In: Inui T and McClelland JL (eds) *Attention and Performance XVI: Information Integration in Perception and Communication*, pp. 125–155. Cambridge, MA: MIT Press.

Kanwisher NG (1987) Repetition blindness: type recognition without token individuation. *Cognition* **27**(2): 117–143.

Mack A and Rock I (1998) *Inattentional Blindness*. Cambridge, MA: MIT Press.

McConkie GW and Currie CB (1996) Visual stability across saccades while viewing complex pictures. *Journal of Experimental Psychology: Human Perception & Performance* **22**(3): 563–581.

O'Regan JK and Noë A (2001) A sensorimotor account of vision and visual consciousness. *Behavioral and Brain Sciences* **24**(5): 883–917.

O'Regan JK, Rensink RA and Clark JJ (1999) Change-blindness as a result of 'mudsplashes'. *Nature* **398**: 34.

Pani JR (2000) Cognitive description and change blindness. *Visual Cognition* **7**(1/2/3): 107–126.

Rensink RA, O'Regan JK and Clark J (1997) To see or not to see: the need for attention to perceive changes in scenes. *Psychological Science* **8**(5): 368–373.

Shapiro K and Terry K (1998) The attentional blink: the eyes have it (but so does the brain). In: Wright RD (ed.) *Visual Attention*, pp. 306–329. New York, NY: Oxford University Press.

Simons DJ and Chabris CF (1999) Gorillas in our midst: sustained inattentional blindness for dynamic events. *Perception* **28**(9): 1059–1074.

Simons DJ and Levin DT (1997) Change blindness. *Trends in Cognitive Sciences* **1**(7): 261–267.

Simons DJ and Levin DT (1998) Failure to detect changes to people in a real-world interaction. *Psychonomic Bulletin and Review* **5**(4): 644–649.

Wolfe JM (1999) Inattentional amnesia. In: Coltheart V (ed.) *Fleeting Memories*. Cambridge, MA: MIT Press.

Further Reading

Coltheart V (ed.) (1999) *Fleeting Memories: Cognition of Brief Visual Stimuli*. Cambridge, MA: MIT Press.

O'Regan JK (2001) Thoughts on change blindness. In: Harris LR and Jenkin M (eds) *Vision and Attention*, pp. 281–302. New York, NY: Springer.

Pashler HE (1998) *The Psychology of Attention*. Cambridge, MA: MIT Press.

Special issue on Change Blindness in *Visual Cognition* (2000) **7**:(1/2/3). DJ Simons' change detection database: http://viscog.beckman.uiuc.edu/change

Change Blindness, Psychology of

Intermediate article

George McConkie, University of Illinois, Champaign, Illinois, USA
Lester Loschky, University of Illinois, Champaign, Illinois, USA

CONTENTS

Change blindness is the tendency to fail to detect changes in a stimulus array while actively exploring it. This happens when the perception of the motion that typically accompanies stimulus change is prevented or disrupted.

INTRODUCTION

As we look around our visual world, we have a sense that we are continuously perceiving most or all of the available information. Thus, it came as some surprise in 1992 when scientists from the University of Illinois reported research indicating that if parts of a rich photographic image are changed while an observer makes a saccadic eye movement, these changes frequently go unnoticed (Grimes, 1996). Such intrasaccadic changes of the presence, location, size, orientation, color, category, or identity of a prominent object often go undetected as a person looks around a picture. Normally a person notices an image change because of the stimulus motion accompanying it; however, when the change occurs during a saccade, that motion is hidden by saccadic suppression. Thus, people cannot notice the change unless information they acquire following the change conflicts with that obtained prior to the change. Such detection failures have been taken as evidence that making a saccade (lasting only 20–80 ms) destroys the iconic image from the prior eye fixation, thus ruling out integration of iconic retinal images across fixations (Irwin, 1992), and further, that only a small amount of information (e.g., an object's identity or category, or perhaps the scene's gist) is actually retained from each eye fixation. The mental representation of a scene that is built up, incrementally, across eye fixations is apparently quite sparse; although visual memory cannot store the rich retinal image that is present during a fixation, it may not need to, since (as Kevin O'Regan has argued) the brain can easily access that information again by simply making an eye movement, so long as the stimulus array remains unchanged. Failure to detect a change in the visual stimulus was later labeled 'change blindness' by Daniel Simons. A lively research area has arisen, investigating what is, and is not, retained from a complex display during its viewing, with numerous theories about the significance this has for understanding perception, attention and cognition.

REQUIREMENTS FOR CHANGE BLINDNESS

Change blindness occurs when an initial stimulus is presented, followed by a change to a different stimulus, but with the transition being perceptually hidden in some way. Normally a local or global change in a picture is easy to detect, since the change induces the perception of motion, which automatically captures attention. In the original paradigm, making the transition during a saccade hid the change. Studies were conducted in which an image was changed for a single eye fixation, allowing a precise examination of whether the changed information was acquired during that particular fixation (McConkie and Currie, 1996). However, a later study (Rensink *et al.*, 1997) showed that a saccade was not necessary to elicit change blindness. Rather, the same effect can be produced by alternating between two versions of a picture, blanking the screen for roughly 100 ms or more between them. Here, the visual disruption produced by the blank screen hides the transition. By continuously 'flickering' between images, the experimenter can measure the time taken by the viewer to discover what is changing. Many such changes are surprisingly difficult to find, as illustrated in Figure 1. Soon other methods for hiding the transition were being employed: changing the image during a blink, briefly introducing irregular blotches ('mud splashes') that compete for attention simultaneously with the image change, morphing between the two images so slowly that the motion is not detected, using film editing techniques to 'cut' from one camera shot to another in motion pictures, or even physically blocking the person's view in a real-world setting.

(a)

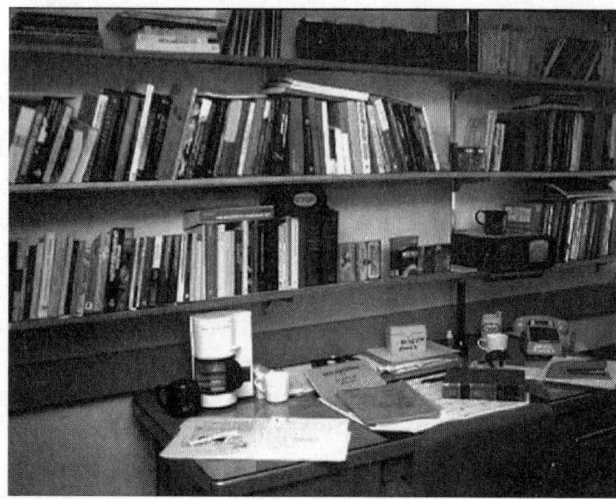

(b)

Figure 1. Two versions of a photograph used to study change blindness. These two versions (a) and (b), are repeatedly presented, one after the other, on a computer screen for about 0.25 s each, with a 0.10 s blank period between. After a period of searching, people typically detect a change in one small area, and only later realize the full extent of the change taking place. This demonstrates that people are able to retain only a small amount of information from each presentation of the picture. (used with permission of Gregory J. Zelinsky).

In a dramatic demonstration of the last method, while an experimenter asked a person on the street for directions, their view of each other was briefly occluded by two men who rudely walked between them carrying a door, and an experimenter of different dress and build quickly replaced the questioner (Figure 2). Roughly half of the people tested failed to notice the change in their interlocutor (Simons and Levin, 1998).

PREVIOUS OBSERVATIONS OF CHANGE BLINDNESS

On reflection, it is obvious that the change blindness phenomenon has long been utilized: magicians developed ways of diverting attention so a change is not noticed, and the familiar 'spot the difference' cartoons require people to make saccades (with the attendant disruptions) between two pictures to find what has been changed. In psychology, from the 1950s on, various studies tested people's memory by asking them to look first at one picture, then at another, and indicate any difference. However, only recently has the change blindness phenomenon elicited particular theoretical interest among those studying perception and cognition.

EXPLANATIONS OF CHANGE BLINDNESS

There are several types of explanations for the failure to detect image changes. First, the critical prechange information might never have been attended to, or it might have been forgotten afterwards. Alternatively, after the change, the critical information might not be attended to. Finally, the information both before and after the change might have been attended to but the discrepancy between them might not be noticed, indicating a lack of integration of information acquired at different times. Thus, researchers try to determine the extent to which change blindness in a given situation results from attentional selectivity (information from before or after the change is not attended and stored), forgetting, or failure to integrate stored information. Explanations of instances of change blindness are given below.

Spatial or Object-based Inattention

Selective visual attention is critical in change blindness. Factors that draw attention to an object increase the likelihood of detecting changes to it,

Figure 2. A real-world example of change blindness. (a) An experimenter stops someone (an unwitting 'subject') on the street and asks for directions. (b) In the middle of their interaction, two men rudely pass between them carrying a door; during this brief interval, the original experimenter is replaced by a second experimenter. After the door passes, the second experimenter continues the conversation with the subject as if nothing had happened (c). About half the time, subjects fail to notice that they are now talking to a different person, even though the two experimenters differ in height, build, hair style, and clothing (d). From Simons and Levin (1998), with permission.

while reducing the detection of changes elsewhere. If a verbal cue indicates the identity of a change before its occurrence, or a visual cue draws attention to the changed object, detection is greatly facilitated (Rensink *et al.*, 1997; Scholl, 2000). Having more objects in the display delays detection of a change in one object due to increased attentional scanning (Zelinsky, 2001), and when several objects change location, people usually detect the movement of only one (McConkie and Loschky, 2000). People who have a wider attentional breadth, measured by their ability to detect briefly presented targets in peripheral vision, are better at detecting changes under flicker conditions (Pringle, 2000). Changes occurring in central vision, the region most likely to be attended to during a fixation, are better detected than changes in the visual periphery. Finally, changes to objects of greater interest in a picture are detected faster than changes to objects of lesser interest (Rensink *et al.*, 1997). All these phenomena argue for the importance of selective attention, though attending to an object does not necessarily result in detection of a change to it.

Temporal Inattention

During reading, there is evidence of attention to words at only selected times during fixations (Blanchard *et al.*, 1984). In one study, at the beginning of each fixation the text being read contained one letter at a specified location. After a short period the text was briefly masked, and then reappeared but with the critical letter being changed, for example changing the word 'leaks' to 'leans'. Thus, one word was present at that location at the beginning of each fixation, and a different word during the latter part of each fixation. Readers were usually unaware of this change, reporting having read only one of the words: the perceived word was sometimes the earlier word and sometimes the later. This suggests temporal inattention. Whether this inattention to a word only occurs because attention is being given to a different word is not known.

Memory

The initial information that will be changed must be retained if the change is to be noticed. Pringle

(2000) found a strong relationship between people's composite scores on a battery of visuo-spatial working memory (VSWM) tests and their ability to detect changes, providing evidence for the role of memory in change detection. In fact, the VSWM measure predicted people's detection ability better than a test of attentional breadth did.

Expectations

Viewer's expectations and world knowledge also play a part. Changes of task-related aspects of a display (color of a traffic light for a driver) are detected more quickly than changes of unrelated aspects (location of a light pole) (Pringle, 2000). Changes to objects inappropriate in their context are detected faster than when they are appropriate (Hollingworth and Henderson, 2000).

Stimulus Characteristics

The more objects or parts of an object that change, the greater is the likelihood that a change will be detected (McConkie and Loschky, 2000; Williams and Simons, 2000). On the other hand, some features of the stimuli to which the gaze is directed can be changed during a saccade without detection. For example, size changes are often not detected (Grimes, 1996; McConkie and Currie, 1996). When reading text printed in AlTeRnAtInG cAsE (every other letter in upper case), the case of every letter can be changed during a saccade, thus changing the shape of every letter and word, without the readers' awareness or any effect on their eye movements. This indicates that letter and word shapes are not preserved across saccades during reading. Just directing the gaze toward an object does not guarantee that changing it will be detected: there are cases in which the viewer makes two successive eye fixations on an object in a picture, with the object changing substantially between those fixations, with no detection of the change (Grimes, 1996). These examples are all consistent with the proposal that the visual system retains only a limited set of the features of an attended object.

Coordinating Perception Across Saccades

A different mechanism, related to basic processes involved in coordinating perception across saccades, has been proposed for detecting intrasaccadic shifts, or displacements, of the stimulus (McConkie and Currie, 1996). It is proposed that, prior to making a saccade, the visual system chooses a target to send the eyes to (the 'saccade target object') and stores some identifying features (the 'locating features'). On the next fixation, the first task is to search for the locating features within the region of central vision (the 'search region'), to find the retinal location of the saccade target object. Locating that object establishes a mapping function between current retinal information and information obtained during prior fixations, enabling further perceptual activities to occur. If an intrasaccadic displacement of the stimulus moves the saccade target object out of the search region, this disrupts further processing and produces a conscious awareness of the image displacement, or change detection. Similarly, if the locating features are changed, this will interfere with the finding of the saccade target object.

The importance of the saccade target object in perception is confirmed by the fact that spatially displacing only that object is far more detectable than is displacing everything else in a picture except the saccade target (Currie *et al.*, 2000). This work suggests that intrasaccadic display changes can be detected on different bases. Changes may be detected by early visual processes that are involved in locating the saccade target at the beginning of a fixation, or later as information acquired in the fixation is recognized as conflicting with information retained from earlier fixations.

CONSCIOUSNESS AND CHANGE BLINDNESS

An important limitation of current research is that it has depended almost exclusively on conscious report: did the viewer detect the change? However, it is possible that image changes produce effects on processing even if not consciously perceived. Studies are needed to search for effects of undetected display change on implicit measures of processing, such as eye movements or brain waves. Results from such studies could challenge current conclusions about the sparsity of information retained from brief stimulus exposures.

CONCLUSION

Although change blindness is an intriguing phenomenon, in some sense it is only incidentally a topic of study. Rather, it is a research paradigm that can be used to study issues regarding the selection, acquisition, retention and integration of information. Its primary contribution has been to call attention to how little information is retained from eye fixations and other brief exposures to complex stimulus patterns such as photographs or real-world scenes, and how sparse our mental representations actually are.

References

Blanchard HE, McConkie GW, Zola D and Wolverton GS (1984) Time course of visual information utilization during fixations in reading. *Journal of Experimental Psychology: Human Perception and Performance* **10**(1): 75–89.

Currie CB, McConkie GW, Carlson-Radvansky LA and Irwin DE (2000) The role of the saccade target object in the perception of a visually stable world. *Perception and Psychophysics* **62**(4): 673–683.

Grimes J (1996) On the failure to detect changes in scenes across saccades. In: Atkins KA (ed.) *Perception*, vol. 5, pp. 89–110. New York, NY: Oxford University Press.

Hollingworth A and Henderson JM (2000) Semantic informativeness mediates the detection of changes in natural scenes. *Visual Cognition* **7**(1–3): 213–235.

Irwin DE (1992) Memory for position and identity across eye movements. *Journal of Experimental Psychology: Learning, Memory, and Cognition* **18**(2): 307–317.

McConkie GW and Currie CB (1996) Visual stability across saccades while viewing complex pictures. *Journal of Experimental Psychology: Human Perception and Performance* **22**(3): 563–581.

McConkie GW and Loschky LC (2000) Attending to objects in a complex display. In: Benedict ME (ed.) *Advanced Displays and Interactive Displays Consortium ARL Federated Laboratory Fourth Annual Symposium Proceedings*, pp. 21–25. Adelphi, MD: Army Research Laboratory.

Pringle HL (2000) *The Roles of Scene Characteristics, Memory and Attentional Breadth on the Representation of Complex Real-world Scenes* [unpublished doctoral dissertation]. Urbana, IL: University of Illinois.

Rensink RA, O'Regan JK and Clark JJ (1997) To see or not to see: the need for attention to perceive changes in scenes. *Psychological Science* **8**(5): 368–373.

Scholl BJ (2000) Attenuated change blindness for exogenously attended items in a flicker paradigm. *Visual Cognition* **7**(1–3): 377–396.

Simons DJ and Levin DT (1998) Failure to detect changes to people during a real-world interaction. *Psychonomic Bulletin and Review* **5**(4): 644–649.

Williams P and Simons DJ (2000) Detecting changes in novel, complex three-dimensional objects. *Visual Cognition* **7**(1–3): 297–322.

Zelinsky GJ (2001) Eye movements during change detection: implications for search constraints, memory limitations, and scanning strategies. *Perception and Psychophysics* **63**(2): 209–225.

Further Reading

Hayhoe MM, Bensinger DG and Ballard DH (1998) Task constraints in visual working memory. *Vision Research* **38**(1): 125–137.

O'Regan JK, Rensink RA and Clark JJ (1999) Change-blindness as a result of 'mudsplashes'. *Nature* **398**(6722): 34.

O'Regan K (1992) Solving the 'real' mysteries of visual perception: the world as an outside memory. *Canadian Journal of Psychology* **46**(3): 461–488.

Ryan JD, Althoff RR, Whitlow S and Cohen NJ (2000) Amnesia is a deficit in relational memory. *Psychological Science* **11**(6): 454–461.

Simons DJ (ed.) (2000) Change blindness. *Visual Cognition* **7**(1–3) [special triple issue].

Simons D and Levin D (1997) Change blindness. *Trends in Cognitive Sciences* **1**: 261–267.

Chinese Room Argument, The

Intermediate article

John Searle, University of California, Berkeley, California, USA

CONTENTS

The Chinese room argument is a refutation of 'strong artificial intelligence' (strong AI), the view that an appropriately programmed digital computer capable of passing the Turing test would thereby have mental states and a mind in the same sense in which human beings have mental states and a mind. Strong AI is distinguished from weak AI, which is the view that the computer is a useful tool in studying the mind, just as it is a useful tool in other disciplines ranging from molecular biology to weather prediction.

SUMMARY OF THE ARGUMENT

Strong AI is often expressed in the formula: 'Mind is to brain as program is to hardware.' On this view, the human mind is a program running in the hardware, or 'wetware', of the brain. The Chinese room argument against strong AI proceeds by a thought experiment. If strong AI were true, then one could acquire any cognitive capacity that one does not have by simply implementing the program for that cognitive capacity in a way that would enable one to pass the Turing test.

Imagine that I, who am a native English speaker, unable to speak any Chinese at all, am locked in a room containing several boxes of Chinese symbols (the database). Imagine that I have in the room a set of instructions for manipulating Chinese symbols (the program). I receive, through a window in the room, Chinese symbols which, unknown to me, are in the form of questions. I follow the instructions in the program, and give back through the window Chinese symbols which, unknown to me, are answers to the questions. For the purposes of the thought experiment we may suppose that the programmers get so good at writing the programs, and I get so good at shuffling the symbols, that after a time my answers are indistinguishable from those of the native Chinese speaker. I pass the Turing test for understanding Chinese, and I do so by implementing the program. But I do not understand a word of Chinese. This is the point of the thought experiment: if I do not understand Chinese by virtue of implementing the Chinese-understanding program, then neither does any other digital computer by virtue of doing so.

Why is it that I do not understand Chinese? The answer seems obvious. Though I manipulate the symbols, I have no knowledge of what any of the symbols mean. One can see this by contrasting my manipulation of Chinese symbols with my answering questions in English. Suppose that the people on the outside of the room also submit written questions in English and I submit written answers to the questions. My answers to the questions in Chinese are as good as those of a native Chinese speaker because I have been appropriately programmed. My answers to the questions in English are as good as a native English speaker because I am a native English speaker. From the outside, from the third-person behavioral point of view, my behavior is equally good in Chinese and in English. From the inside it is obviously quite different: in English I understand perfectly both the questions and my answers; while in Chinese I understand nothing – I am just a digital computer.

Construed as a deductive argument, the Chinese room argument has three steps and a conclusion. We may formulate these as follows.

> Computer programs are defined entirely in terms of symbolic or syntactic operations. (1)

The implemented program consists entirely of symbol manipulations. To put this somewhat more precisely: the notion 'same implemented program' defines an equivalence class that is specified entirely in symbolic or syntactic terms, and independently of the physics of the underlying medium. There is nothing more to the implemented program, qua implemented program, than symbol manipulations.

> Minds – actual human minds such as yours and mine – have mental contents or semantics. (2)

For example, when I understand a sentence in English I have more than just symbols going through my head: I know what the symbols mean.

> By themselves, the implemented syntactic steps of the program are neither constitutive of mental content nor sufficient to guarantee the presence of mental content. (3)

This is what was shown by the Chinese room thought experiment. I went through the appropriate syntactic steps, but I had no Chinese thought content, no Chinese semantics, associated with them.

> Conclusion: the implemented computer program is insufficient by itself to constitute or to guarantee the presence of the appropriate mental states. (4)

I went through the right steps of the program, I had the right behavior, but I did not have the appropriate mental states. Therefore, strong AI is false.

The argument rests on two fundamental logical principles: firstly, syntax is not semantics, and secondly, simulation is not duplication. Any problem-solving process that can be described as an effective procedure, that is, a procedure going through a finite number of exactly specifiable discrete steps, can be programmed on a computer. That is why the computer is so powerful: we can represent any domain that we can describe precisely. Thus we can represent the stages of the weather, the flow of money in the economy, or the understanding of Chinese sentences. The syntax of the program states can be used to represent anything. They can be used to represent weather changes, economic developments, and even semantics. But the simulation of the process, whether it be atmospheric, economic, or semantic, is not a duplication of the process. You do not produce a rainstorm by doing a computer simulation of a rainstorm. You do not produce wealth by doing a computer simulation of the production of wealth. And you do not produce understanding and thought processes by doing a computer simulation of understanding and thought processes.

RESPONSES TO THE ARGUMENT

A number of responses have been presented against the Chinese room argument. We will consider four of these.

The Systems Reply

Perhaps the most commonly presented answer is this: the person in the room does not understand, but the person is only an element in a larger system. The system consists of the room, the program, the database, etc. So the understanding should be found in the entire system, not in the person, because the person is only the central processing unit.

Just as we would not say of a single neuron in the brain that it understood English, so we should not say of a single element, the person, in the whole system that that person understands Chinese.

Answer to the systems reply

The answer to this reply is that the reason the person does not understand is that he has no way to attach any meaning to the symbols. But if he has no way to attach meaning to the symbols, neither does the whole room. The whole room has no way to get from the syntax to the semantics any more than the person does. To see this, simply imagine that the person internalizes the entire system. Imagine that the person memorizes the database, memorizes the program, does all of the calculations in his head, and works outdoors in the middle of an open field. In this variation there is nothing in the room that is not in the person, and still there is no understanding in the person.

The Robot Reply

The robot reply is based on a variation of strong AI whereby the unit of understanding is not the computer, but the computer within a motorized system that will be able to process sensory inputs and produce motor outputs computationally. The robot would move about, with video cameras attached to its head; it would take in information from the video cameras, and adjust its movements accordingly.

According to the robot reply, the computer by itself does not have semantic content, but the causal relations between a robot and the external world would be sufficient to give semantic content to the symbols processed by the robot.

Answer to the robot reply

The robot reply tacitly abandons the thesis of strong AI, which is that the implemented computer program by itself is sufficient to guarantee or constitute understanding. The idea behind the robot reply is that the addition of causal relations between the system and the external world would be sufficient to produce semantic content or understanding.

The answer to the robot reply is that even this amendment to the strong AI thesis will not be sufficient to produce understanding. Imagine that the robot has a very large cranium, and inside the cranium is a room, and I am inside the room. I receive inputs in the form of Chinese symbols. I process them according to the program, and I produce outputs in the form of Chinese symbols. Unknown to me, the input symbols are the product of video cameras and other sensors attached to the outside of the robot. The input stimuli are converted by transduction into Chinese symbols, and the output that I provide is converted into motor output of the entire robot system. But I have no way of understanding what is going on because I have no way of attaching any meaning to any of the symbols, or to anything else that is going on in the robot's brain. I am the robot's homunculus, but unlike the usual homunculi of philosophical literature, I understand nothing, because I have no way of attaching any meaning to any of the symbols that I process.

The robot reply tries to defeat the Chinese room argument by adding causal relations. But the causal relations will produce semantic content only if there is some conscious agent who can become aware of the causal relations. I, as a human being, can become aware of the causal relations between the Chinese symbol for chicken chow mein and the actual food type of chicken chow mein if I can *see* chicken chow mein associated with this symbol. But in the robot, I am just a computer and, like any other computer, I function by processing meaningless symbols. The symbols in the computer brain are meaningless in a way that is quite different from the symbols passing through my mind when I think in English. When I think in English, symbols do indeed go through my mind, but I know what they mean.

The Brain Simulator Reply

Suppose we simulated the actual operations of a Chinese person's brain when that person understands sentences in Chinese. Suppose we produced a perfect computer simulation of all of the synaptic transmissions in the Chinese brain. Then we would have to say that the system understood, otherwise we would have to deny that the Chinese person understood. Since the brain operates, like a computer, with a series of state transitions, there is no reason why we could not produce a perfect replica of these state transitions on a digital computer.

Answer to the brain simulator reply

The computer simulation of the brain is not duplicating the relevant features of the brain. It is merely duplicating the formal pattern. The actual human brain, like any other organ, is a causal mechanism, and it causes consciousness and intentionality by quite specific neurobiological processes. The computer merely produces a model or representation of these processes, but the model or representation lacks the causal features of the original.

To see this, compare the brain to any other organ. We can do a perfect simulation of the digestive processes in the stomach on a digital computer. But even if we have a perfect simulation on the computer, to any degree of accuracy, we do not produce actual digestion. When we run the digestion program, we cannot put a pizza into the computer and expect the computer to digest it. The computational simulation is merely a matter of zeros and ones, not a matter of the enzymes and other chemicals that actually carry out digestion. The situation in the brain is similar. Specific biochemical processes cause consciousness and intentionality. We cannot reproduce those by doing a simulation with zeros and ones, any more than we can reproduce digestion with zeros and ones.

The Parallel Distributed Processing Reply

The Chinese room argument works against the von Neumann symbolic digital computer, but recent developments in computer technology have created new types of computational systems which are immune to the Chinese room argument. These new types of systems are known variously as 'parallel distributed processing' (PDP), 'neural net modeling, 'connectionism' or 'new connectionism'.

PDP systems function in a way that is quite different from the traditional von Neumann system. They have a series of computational processes going on in parallel, distributed over a network. Whereas the traditional von Neumann machine works in a series of discrete steps, PDP systems do massively parallel distributed processing.

Answer to the parallel distributed processing reply

There is an ambiguity in the PDP reply. It is not clear which of the differences between the connectionist machine and the von Neumann machine are being appealed to in order to claim that the connectionist machine is not subject to the Chinese room argument. Either what is claimed is that there is some computational power of the connectionist machine lacking in the von Neumann machine, or it is claimed that there is some hardware feature of the connectionist architecture which is superior to the von Neumann architecture. But neither of these approaches is successful in answering the Chinese room argument.

According to Church's thesis, there is no computation that can be performed on a connectionist machine that cannot be performed on a von Neumann machine. According to this thesis, any computable function whatever, any problem that can be solved algorithmically, can be computed on a Turing machine. All effective computability is Turing computability. Church's thesis is one of the foundational principles of the modern theory of computation and is universally accepted by the parties to this dispute. It has the consequence that there cannot be any computational power possessed by a PDP system that is not possessed by a von Neumann machine.

The other possibility is that something is being claimed for the connectionist architecture: for the actual structure of the wiring and the hardware. But if this is so, it is no longer strong AI. Strong AI is a thesis about the powers of computation. If it is claimed that the particular hardware of the connectionist machines can duplicate the powers of the brain to cause mental content, then the thesis is no longer strong AI, but is rather a form of speculative neurobiology. The Chinese room argument is not intended to answer any claims in speculative neurobiology, but is intended as a logical thesis about the distinction between the syntax of the implemented program and the semantics of actual human minds.

Either we are to think of the essential feature of the system as being its computational power, or we are to think of it as some causal property of the specific hardware in which the computation is implemented. If it is a matter of computational power, no new power is added by the connectionist architecture. If we are to think of it as an architectural feature, then it is no longer the thesis of strong AI. Actual human brains cause consciousness and other mental phenomena by way of specific neurobiological processes operating in a 'bottom-up' fashion. That is, processes at the level of neurons and synapses cause consciousness and other mental phenomena that are features of much larger elements of the brain system.

The thesis of the PDP reply, if followed to its logical conclusion, would have to be that the connectionist architecture is capable of duplicating and not merely simulating the causal powers of the brain to cause higher-level consciousness, etc., by way of bottom-up causation. Nothing in the neurobiological literature would tend to support this thesis. In any case, it is not strong AI, and in consequence, is irrelevant to answering the Chinese room argument.

COMMON MISUNDERSTANDINGS OF THE CHINESE ROOM ARGUMENT

The Chinese room argument is sometimes misinterpreted, and several of these misinterpretations are common in the literature. Firstly, it is sometimes supposed that the argument is intended to show that 'machines cannot think'. But that is not the point of the argument. The argument assumes that the brain is a machine. The problem with computation is that in the relevant sense it does not name a machine process. It names an abstract mathematical process that we can implement on machines, but computation is not defined in terms of machine processes such as energy transfer. Thus, on the view implicit in the Chinese room argument, the brain is a machine, and brain processes are machine processes. 'Computers' of the ordinary kind are indeed machines, but computation is not essentially a machine process.

Another misinterpretation of the Chinese room argument is that it is attempting to show that only human brains have the power of thinking. But that is not the point of the argument. Whether or not we can build an artifact out of some other type of material capable of producing consciousness is an empirical scientific question. In principle there is no more serious logical obstacle to building an artificial brain than there is to building an artificial heart. The point of the argument is that we do not produce the same causal powers by simply duplicating the formal pattern. The computer gives us a picture, or a model, of thought processes, but it does not actually produce thought processes.

CONCLUSION

In the early days of cognitive science, the computationalist model of cognition was the dominant paradigm. At present there is a gradual paradigm shift away from computational cognitive science towards cognitive neuroscience. As we learn more about the brain we see that cognition is essentially a matter of a certain sort of brain processing. We may be able to simulate this on a digital computer, and we may eventually be able to duplicate it in some other medium. But the Chinese room argument shows that simulation by itself does not guarantee duplication. To guarantee duplication, the artificial creation of a real mind, we would have to duplicate, and not merely simulate, the powers that actual brains have to cause consciousness and cognition.

Further Reading

Dietrich E (ed.) (1994) *Thinking Computers and Virtual Persons*. San Diego, CA: Academic Press.

Preston J and Bishop M (eds) (2002) *Views Into the Chinese Room: New Essays on John Searle's Arguments Against 'Strong AI'*. Oxford, UK: Oxford University Press.

Searle JR (1980) 'Minds, brains, and programs'. *Behavior and Brain Sciences* **3**: 417–457.

Searle JR (1982) The Chinese room revisited. *Behavior and Brain Sciences* **5**: 345–348.

Choice Selection

Intermediate article

John W Payne, Duke University, Durham, North Carolina, USA
James R Bettman, Duke University, Durham, North Carolina, USA

CONTENTS

Choice selection involves a set of alternatives, each described by some attributes or consequences for the decision-maker's goals. Decision-makers choose among such options using a variety of different psychological strategies.

INTRODUCTION

Multiattribute Choice Problems

Decision-making is a fundamental cognitive behavior which depends on a person's values and beliefs and can range from mundane problems (such as selection of a menu item for lunch) to more substantial and infrequent decisions (such as which automobile to purchase), and life-or-death decisions (such as a choice between alternative medical treatments). At the heart of the decision-making process is choice or selection among alternative courses of action. A simplified automobile choice task is illustrated in Table 1. Each choice option i (alternative) is described by a vector of attribute values ($x_{i1}, x_{i2}, \ldots, x_{in}$), reflecting the extent to which each option meets the objectives (goals) of the decision-maker.

A key feature of almost all choice problems is the presence of conflict, since no single alternative is best (most preferred) on all attributes. Conflict is a major source of decision difficulty; negative emotions can arise from the need to accept less of one valued attribute (e.g., safety) in order to achieve more of another valued attribute (e.g., cost savings). The presence of conflict and the fact that a rule for resolving the conflict often cannot be drawn from memory are reasons that even simple choice tasks can lead to tentativeness and the use of novice problem-solving methods rather than the kind of pattern recognition methods that are typically associated with expertise in various domains.

Choice Among Risky Options

Another key feature of choice problems is the degree of certainty of the consequences associated with an attribute value. For example, one might (or might not) be certain about the level of reliability for a particular car in Table 1. Another example of uncertainty and choice is selecting between two gambles, one of which offers a higher probability of winning a lower amount of money while the other offers a lower probability of winning a larger amount of money. A real-world example of a risky choice problem with such trade-offs is selecting among investment options (gambles) such as bonds versus computer stocks for the next year when one is uncertain whether the state of the economy will improve or worsen.

Table 1. An example of a choice task

Car	Reliability	Price	Safety	Horsepower
A	Worst	Best	Good	Very poor
B	Best	Worst	Worst	Good
C	Poor	Very good	Average	Average
D	Average	Poor	Best	Worst
E	Worst	Very poor	Good	Best

Attributes are scored on seven-point scales ranging from 'best' (the most desirable value for the attribute) to 'worst' (the least desirable value).

CHOICE MODELS

The study of decision processes has long been of interest to psychologists, economists, and researchers in many other fields. How do people make preferential choice decisions of the types described above? The rational choice view is that people solve all (most) decision problems by obtaining all the relevant information about the decision problem, incorporating uncertainties into their reasoning, making trade-offs where necessary, and eventually selecting the alternative that maximizes their values. That is, people are presumed to be exquisitely rational beings who have, if not perfect, at least clear and voluminous information, who are able and willing to make trade-offs, and who always select the best course of action (Simon, 1955). The view that people are generally rational decision-makers means that one can start by trying to work out the best way a decision problem should be solved, assume that people try to do the best and are capable of calculating the best option, and therefore that the same model that provides a normative definition of rational choice also provides a reasonable model of actual decision behavior.

Mathematical (Rational) Decision Models

Two classic rational choice models of decision behavior are the weighted additive value model (WADD) and the subjective expected utility model. The WADD model is often used to represent in a mathematical form the trading-off process in decision-making. A measure of the relative importance (weight) of an attribute is multiplied by the attribute's value for a particular alternative and the products are summed over all attributes to obtain an overall value for that alternative, WADD(X):

$$\text{WADD}(X) = \Sigma\, W_i V(X_i) \text{ for } i = 1 \text{ to } n \quad (1)$$

where W_i is the weight given to attribute i, $V(X_i)$ is the value of option X on attribute i, and n is the total number of relevant attributes. The WADD model represents a normative procedure for dealing with multiattribute decision problems (see Keeney and Raiffa, 1976) in that it uses all the relevant information, explicitly resolves conflicting values, and selects the option with the highest overall evaluation. Note that it also assumes that the effects of the attributes are independent (i.e., there are no interactions), which may not always be appropriate. The WADD model underlies many of the techniques used by economists, market researchers, and others to assess preferences.

A model for making choices under risk and uncertainty is the subjective expected utility (SEU) model. The subjective expected utility of a risky option (gamble) is given by

$$\text{SEU}(X) = \Sigma\, S(P_i)U(X_i) \text{ for } i = 1 \text{ to } n \quad (2)$$

where $S(P_i)$ is the subjective probability of occurrence for outcome X_i, and $U(X_i)$ is the utility to the individual of receiving amount X_i, e.g., an amount of money. Note that the SEU model is similar in structure to the WADD model, weighting each possible outcome by its probability of occurrence, summing those products over all possible outcomes, and then selecting the gamble with the highest SEU.

Cognitive Limitations and Heuristic Decision Processes

Although people sometimes make decisions in ways consistent with the WADD and SEU models, it has become obvious over years of decision research that people often make decisions using simpler decision processes (heuristics), more consistent with the idea that people are, at best, only boundedly rational. The bounded rationality view of decision-making emphasizes a decision-maker's limited information processing capabilities and the interaction of those computational limits with the complexity of task environments: 'human rational behavior is shaped by a scissors whose two blades are the structure of task environments and the computational capabilities of the actor' (Simon, 1990: p. 7).

In part because of limitations in information processing capacity, preferences for objects are often constructed at the time people are asked to make choices. That is, people may construct preferences on the spot when needed rather than simply retrieving well-defined values for objects when they are asked to make a choice (March, 1978). In addition, because of limited processing capacity and the need to be adaptive to task demands, decisions are often made not by using some invariant rule such as SEU, but rather by using a variety of decision strategies contingent upon the demands of the task environment (Payne *et al.*, 1993). Such contingency upon a variety of task and context factors is consistent with and implied by the constructive nature of preferences. Finally, although the concept of bounded rationality does not mean that people are poor decision-makers, the combination of limited cognitive capabilities and difficult decision problems means that people sometimes make systematic errors when facing choice problems. For example, people will sometimes say that they prefer option A to option B, B to option C, and C to A (Tversky, 1969), or that they prefer A to B but would be willing to pay more for B than for A (Tversky *et al.*, 1988).

Several commonly used decision-making heuristics have been identified. If a lexicographic strategy (LEX) is used, the alternative with the best value on the most important attribute is simply selected (assuming that there are no ties on this attribute); for example, if a decision-maker faced with the choice problem in Table 1 thought that reliability was the most important attribute for cars, he or she could use a lexicographic strategy, examine reliability (and no other information) for all five cars, and choose car B. If two alternatives have tied values, the second most important attribute is considered, and so on until the tie is broken. The LEX strategy is a choice heuristic, consistent with the bounded rationality notion that limited capacity for processing information implies that people generally cannot process all of the available information in a particular situation and must therefore be selective in what information is used. Even when the amount of available information is within the bounds of processing capacity, the processing of that information imposes cognitive costs. Thus, selective processing of information is generally necessary, and the information that is selected for processing will have a major impact on choice. Under some task conditions, for instance, the LEX choice heuristic can be almost as accurate a decision rule as more information-intensive strategies like WADD (Payne *et al.*, 1993). Finally, the LEX strategy is a good example of a conflict-avoiding decision strategy that may minimize the emotional aspects of making a decision, because the focus is on only a single attribute at a time. The LEX strategy is a form of 'one reason' decision-making emphasized by Gigerenzer *et al.* (2001).

Satisficing (SAT) is another classic strategy for coping with bounded rationality. With a satisficing strategy, alternatives are considered sequentially, in the order in which they occur in the choice set. The value of each attribute for the option currently under consideration is compared to a predetermined cutoff level for that attribute. If any attribute fails to meet the cutoff level, the option is rejected and the next option is considered. For example, car A might be eliminated rapidly because it has the worst level of reliability. The first option passing the cutoffs for all attributes is selected. If no option passes all the cutoffs, the levels can be relaxed and the process repeated. Like the LEX strategy, satisficing does not involve explicitly considering trade-offs and is therefore a noncompensatory model of decision-making – that is, a good value on one attribute cannot compensate for a below cutoff (poor) value on another attribute. One of the key differences across choice processes is the extent to which a compensatory (e.g., WADD) or noncompensatory decision process is utilized. Another property of the SAT strategy is that the option chosen will be a function of the order in which the options are processed. Thus, one can potentially influence choice by structuring the order in which options are considered.

Elimination by aspects (EBA) is a commonly used choice heuristic combining elements of both the LEX and SAT strategies. It eliminates options that do not meet a minimum cutoff value for the most important attribute. This elimination process is repeated for the second most important attribute, with processing continuing until a single option remains (Tversky, 1972). In our car example, suppose that the decision-maker's two most important attributes were reliability and safety, in that order, and that the cutoff for each was an average value. This individual would first process reliability, eliminating any car with a below-average value (cars A, C, and E). Then the person would consider safety for cars B and D, eliminating car B. Hence, car D would be selected. Elimination by aspects focuses on the attributes as the basis for processing information, is noncompensatory, reflects rationality in the ordered use of the attributes, and does not use all potentially relevant information. The extensiveness and selectivity of processing will vary when using EBA depending upon the exact pattern of elimination of options. As a general rule, the preferences expressed when using choice heuristics like LEX, satisficing, and EBA are subject to potentially 'irrelevant' task variables such as the order in which options and/or attributes are considered.

Decision-makers also use combinations of choice strategies. A typical combined strategy has an initial phase in which some alternatives are eliminated, and a second phase where the remaining options are analyzed in more detail. One frequently observed combination is an initial use of EBA to reduce the choice set to two or three options, followed by a compensatory strategy such as weighted adding to select among those. An important implication of the use of combined strategies is that the 'properties' of the choice task itself may change as the result of using a particular strategy first. For example, the use of a process for eliminating dominated alternatives from a choice set, an often advocated procedure, will make the conflict among attribute values more extreme, perhaps then triggering the application of a new strategy on the reduced set of options.

TASK AND CONTEXT EFFECTS

Constructed Preferences

Research supports the idea that people use choice heuristics and the more general constructive view of preferences. For example, people use a variety of different strategies to solve choice problems contingent upon the nature of the task, e.g. how many options are available, and the context of the choice problem, e.g. is the choice among a set of generally good or generally poor options. Task factors are general characteristics of a decision problem, such as number of alternatives available, response mode (e.g. choice or judgment), time pressure, or information format, that do not depend on the particular values of the alternatives. One well-replicated and important task effect is that people use compensatory (trade-off based) types of decision strategies (e.g. WADD) when faced with a decision problem involving only two or three alternatives. However, when faced with a more complex (multialternative) decision task, people tend to use noncompensatory strategies such as EBA. People also tend to use more noncompensatory strategies when asked to make choices rather than judgments, and when choosing under time pressure.

Context factors, such as the similarity of alternatives, are associated with the particular values of the alternatives in a choice set and the relationships among those values. Context variables affect both how decisions are made and which option is chosen from a set of alternatives. For example, the more negative the correlations among the attribute values (i.e., the more one has to give up something on one attribute to obtain more of another attribute), the more likely it is that people will use a WADD (expected value) strategy when choosing among gambles. Interestingly, when the attributes are more emotional, such as safety, greater conflict among attribute values tends to lead to greater use of strategies like the LEX rule (Luce *et al.*, 2001).

A context effect that shows how the nature of the choice set can influence which option is chosen is illustrated by the problem in Table 2. Comparing options A and B, one is faced with a trade-off between ride quality (better with A) and fuel consumption (better with B). A basic principle of most choice models, regularity, states that adding a new option to the choice set containing A and B cannot increase the probability of choosing one of the original options. However, imagine that you are faced with the choice of A, B, or C rather than just A or B. Note that C is 'dominated' by option B: B is better than C on one attribute (ride quality) and at least equal to C on the other attribute (fuel economy). Option C is not dominated by A, so that in comparing options A and C there would still be a trade-off to be considered. Thus, there is an asymmetric dominance relationship among the three options A, B, and C. Given that C is dominated by B, it is highly unlikely that a person would select C. However, does the presence of C influence the choice between A and B? The answer is yes. Adding option C to the original choice set of A and B tends to increase the probability of selecting B (Huber *et al.*, 1982). This increase in the probability of choosing the dominating option B with the addition of C violates the principle of regularity. Such a context effect indicates that 'people do not maximize a precomputed preference order, but construct their choices in the light of available options' (Tversky, 1996: p. 17).

Reason-based Choice

A number of explanations have been offered for the asymmetric dominance effect. One that has received support is that people use the relations among options as reasons for justifying their choices; that is, one can easily see that B is better

Table 2. An example of an asymmetric dominance task

Car	*Ride quality*	*Fuel consumption (miles per gallon)*
A	83	24
B	73	33
C	70	33

than C, and this relationship provides a good reason for choosing B and not A. The size of the asymmetric dominance effect increases when people are asked to explicitly justify their choices (Simonson, 1989). Bettman *et al.* (1998) speculated that the fact that option B dominates option C is a good reason for choosing option B at an outcome level of explanation (it is clearly a better outcome than C). There is no need to refer to a process-level explanation (e.g. I chose B over A because of the trade-offs I prefer between ride quality and fuel economy). Outcomes are likely to be more salient than process in the wake of a decision, so arguments based on outcomes may provide better reasons.

Using easily detected relationships among the options in a choice set as a reason for choice allows a person to avoid cognitively and emotionally difficult trade-offs. For example, selecting the option in a set of three that is between the more extreme options can be justified as a 'compromise' choice without having to explicitly consider trade-offs. For more on a reason-based view of choice processes, see Shafir *et al.* (1993).

CONCLUSION

The study of choice processes has been a subject of long-standing interest for psychologists, economists, and researchers in many other fields. How people make choices frequently departs from a purely rational decision process, reflecting the interplay of cognitive processing limits and task demands. People use a variety of simplifying choice heuristics and search for easy-to-justify reasons for preferring one option over another. As a consequence, the option people choose can be influenced by a variety of predictable task and context factors. More generally, in many situations the preferences people exhibit are constructed at the time of choice rather than reflecting precomputed values.

While much has been learned about choice behavior, there is still much to be learned about how decisions are made and how decision-making can be improved.

References

Bettman JR, Luce MF and Payne JW (1998) Constructive consumer choice processes. *Journal of Consumer Research* **25**: 187–217.

Gigerenzer G, Todd PM and the ABC Research Group (2001) *Simple Heuristics That Make Us Smart*. New York, NY: Oxford University Press.

Huber J, Payne JW and Puto CP (1982) Adding asymmetrically dominated alternatives: violations of regularity and the similarity hypothesis. *Journal of Consumer Research* **9**: 90–98.

Keeney RL and Raiffa H (1976) *Decisions with Multiple Objectives: Preferences and Value Tradeoffs*. New York, NY: John Wiley.

Luce MF, Bettman JR and Payne JW (2001) Emotional decisions: tradeoff difficulty and coping in consumer choice. *Monographs of the Journal of Consumer Research*.

March JG (1978) Bounded rationality, ambiguity, and the engineering of choice. *Bell Journal of Economics* **9**: 587–608.

Payne JW, Bettman JR and Johnson EJ (1993) *The Adaptive Decision Maker*. Cambridge, UK: Cambridge University Press.

Shafir E, Simonson I and Tversky A (1993) Reason-based choice. *Cognition* **49**: 11–36.

Simon HA (1955) A behavioral model of rational choice. *Quarterly Journal of Economics* **69**: 99–118.

Simon HA (1990) Invariants of human behavior. *Annual Review of Psychology* **41**: 1–19.

Simonson I (1989) Choice based on reasons: the case of attraction and compromise effects. *Journal of Consumer Research* **16**: 158–174.

Tversky A (1969) Intransitivity of preferences. *Psychological Review* **76**: 31–48.

Tversky A (1972) Elimination by aspects: a theory of choice. *Psychological Review* **79**: 281–299.

Tversky A (1996) Contrasting rational and psychological principles in choice. In: Zeckhauser RJ, Keeney RL and Sebenius JK (eds) *Wise Choices: Decisions, Games, and Negotiations*, pp. 5–21. Boston, MA: Harvard Business School Press.

Tversky A, Sattath S and Slovic P (1988) Contingent weighting in judgment and choice. *Psychological Review* **95**: 371–384.

Further Reading

Baron J (2001) *Thinking and Deciding*. Cambridge, UK: Cambridge University Press.

Goldstein WM and Hogarth RM (1997) Judgment and decision research: some historical context. In: Goldstein WM and Hogarth RM (eds) *Research on Judgment and Decision Making: Currents, Connections, and Controversies*, pp. 3–68. Cambridge, UK: Cambridge University Press.

Hastie R (2001) Problems for judgment and decision making. *Annual Review of Psychology* **52**: 653–683.

Hastie R and Dawes RM (2001) *Rational Choice in an Uncertain World: The Psychology of Judgment and Decision Making*. Thousand Oaks, CA: Sage.

Kahneman D and Tversky A (2000) *Choices, Values, and Frames*. New York, NY: Cambridge University Press.

Plous S (2001) *The Psychology of Judgment and Decision Making*. New York, NY: McGraw-Hill.

Choice under Uncertainty

Advanced article

Mark J Machina, University of California, San Diego, California, USA

CONTENTS

The standard theory of individual choice under uncertainty consists of the joint hypothesis of expected utility risk preferences and probabilistic beliefs. Experimental work by both psychologists and economists has uncovered systematic departures from both hypotheses, and has led to the development of alternative, usually more general, models.

INTRODUCTION

Decisions under uncertainty take place in two types of settings. In settings of 'objective uncertainty', the probabilities attached to the various outcomes are specified in advance, and the objects of choice consist of 'lotteries' of the form $P = (x_1, p_1; \ldots; x_n, p_n)$, which yield outcomes or monetary pay-offs x_i with probability p_i, where $p_1 + \ldots + p_n = 1$. Examples include games of chance involving dice and roulette wheels, as well as ordinary lotteries.

In settings of 'subjective uncertainty', probabilities are not given, and the objects of choice consist of 'bets' or 'acts' $f(\cdot) = [x_1 \text{ on } E_1; \ldots; x_n \text{ on } E_n]$, which yield outcomes or pay-offs x_i in event E_i, for some mutually exclusive and exhaustive collection of events $\{E_1, \ldots, E_n\}$ which can be thought of as a partition of the set S of all possible 'states of nature'. Examples include bets on horse races or the weather, as well as standard insurance contracts.

Under objective uncertainty, choices are determined by an individual's attitudes towards risk. Under subjective uncertainty, they are additionally determined by the individual's subjective beliefs about the likelihoods of the various states and events.

EXPECTED UTILITY THEORY AND EXPERIMENTAL EVIDENCE

Axiomatic and Normative Foundations of Expected Utility Theory

The earliest formal hypothesis of individual attitudes towards risk, proposed by Pascal, Fermat and others in the seventeenth century, was that individuals evaluate monetary lotteries $P = (x_1, p_1; \ldots; x_n, p_n)$ simply on the basis of their mathematical expectation $E[P] = \sum_{i=1}^{n} x_i \cdot p_i$. This hypothesis was dramatically refuted by Daniel Bernoulli's 'St Petersburg paradox'. In this game, a fair coin is repeatedly flipped until it lands heads. If it lands heads on the first flip, the player wins \$1; if it does not land heads until the second flip, the player wins \$2; and in general, if it does not land heads until the ith flip, the player wins $\$2^{i-1}$. Most people would prefer to receive a sure payment of, say, \$50 than a single play of the St Petersburg game, even though the expected pay-off of the game is $\frac{1}{2}\cdot\$1 + \frac{1}{4}\cdot\$2 + \frac{1}{8}\cdot\$4 + \ldots = \$\frac{1}{2} + \$\frac{1}{2} + \$\frac{1}{2} + \ldots = \$\infty$. In the first of what has turned out to be a long series of such developments, Bernoulli weakened the prevailing expected-value hypothesis by positing that individuals instead evaluate lotteries on the basis of their 'expected utility' $\sum_{i=1}^{n} U(x_i)\cdot p_i$, where the utility $U(x)$ of receiving a monetary amount x is probably subproportional to x. Bernoulli himself proposed the form $U(x) = \ln(x)$, which leads to an evaluation of the St Petersburg game consistent with typical actual play.

The expected utility hypothesis came to dominate decision theory on the twin bases of its elegant and highly normative axiomatic development (von Neumann and Morgenstern, 1944; Marschak, 1950)

and its analytical power (Arrow, 1965; Pratt, 1964). In the modern approach, risk preferences are denoted by the individual's 'weak preference' relation $\succeq$ over lotteries, where $P^* \succeq P$ reads 'P^* is weakly preferred to P', and its implied 'strict preference' relation $\succ$ (where $P^* \succ P$ iff $P^* \succeq P$ but not $P \succeq P^*$) and 'indifference' relation $\sim$ (where $P^* \sim P$ iff $P^* \succeq P$ and $P \succeq P^*$). The preference relation $\succeq$ is said to be 'represented' by an expected utility preference function $V_{EU}(P) = \sum_{i=1}^{n} U(x_i)\cdot p_i$ if $P^* \succeq P \Leftrightarrow V_{EU}(P^*) \geq V_{EU}(P)$. $U(\cdot)$ is called the 'von Neumann–Morgenstern utility function'.

The axiomatic and normative underpinnings of expected utility theory are based on the notion of a 'probability mixture' $\alpha\cdot P + (1-\alpha)\cdot P^*$ of two lotteries $P = (x_1, p_1; \ldots; x_n, p_n)$ and $P^* = (x_1^*, p_1^*; \ldots; x_{n^*}^*, p_{n^*}^*)$, which is the lottery that would be generated by a coin flip yielding the lotteries P and P^* as prizes with respective probabilities α and $1-\alpha$, and where both stages of uncertainty (the coin flip and the resulting lottery) are realized simultaneously, so that we can write $\alpha\cdot P + (1-\alpha)\cdot P^* = (x_1, \alpha\cdot p_1; \ldots; x_n, \alpha\cdot p_n; x_1^*, (1-\alpha)\cdot p_1^*; \ldots; x_{n^*}^*, (1-\alpha)\cdot p_{n^*}^*)$. A preference relation $\succeq$ will then be represented by an expected utility preference function $V_{EU}(\cdot)$ for some utility function $U(\cdot)$ if and only it satisfies the following axioms:

- *Completeness.* For all lotteries P and P^*, either $P \succeq P^*$ or $P^* \succeq P$, or both.
- *Transitivity.* For all lotteries P, P^* and P^{**}, if $P \succeq P^*$ and $P^* \succeq P^{**}$ then $P \succeq P^{**}$.
- *Mixture Continuity.* For all lotteries P, P^* and P^{**}, if $P \succ P^*$ and $P^* \succ P^{**}$ then $P^* \sim \alpha\cdot P + (1-\alpha)\cdot P^{**}$ for some $\alpha \in (0,1)$.
- *Independence Axiom.* For all lotteries P, P^* and P^{**} and all $\alpha \in (0,1)$, if $P \succeq P^*$ then $\alpha\cdot P + (1-\alpha)\cdot P^{**} \succeq \alpha\cdot P^* + (1-\alpha)\cdot P^{**}$.

Completeness and Transitivity are standard axioms in preference theory, and Mixture Continuity serves as the standard Archimedean property in the context of choice over lotteries. The key normative and behavioral axiom of the theory is the Independence axiom. Behaviorally, it corresponds to the property of separability across mutually exclusive events. Normatively, it corresponds to the following argument: 'Say you weakly prefer P to P^*, and have to choose between an α:$(1-\alpha)$ coin flip yielding P if heads and P^{**} if tails, or an α:$(1-\alpha)$ coin flip yielding P^* if heads and P^{**} if tails. Now, either the coin will land tails, in which case your choice won't have mattered, or it will land heads, in which case you are back to a choice between P and P^*, so you should weakly prefer the first coin flip to the second.'

The tension between the compelling nature of the Independence axiom and its systematic violations by experimental subjects has led to a sustained debate over the validity of the expected utility model, with some researchers continuing to posit expected utility maximization, and others developing and testing alternative models of risk preferences.

Analytics of Expected Utility Theory

Analytically, the expected utility hypothesis is characterized by the simplicity of its representation (involving the standard concepts of utility and mathematical expectation) as well as by the elegance of the correspondence between standard features of risk preferences and mathematical properties of $U(\cdot)$. The most basic of these properties is 'first-order stochastic dominance preference', which states that raising the level of some pay-off x_i in a lottery $P = (x_1, p_1; \ldots; x_n, p_n)$ – or alternatively, increasing its probability p_i at the expense of a reduction in the probability p_j of some smaller pay-off x_j – will lead to a preferred lottery. An expected utility maximizer's preferences will exhibit first-order stochastic dominance preference if and only if $U(\cdot)$ is an increasing function of x.

A second property is 'risk aversion'. Originally, this was defined as the property whereby the individual would always prefer receiving the expected value of a given lottery with certainty, rather than bearing the risk of the lottery itself. This is equivalent to the condition that the individual's 'certainty equivalent' CE(P) of a nondegenerate lottery $P = (x_1, p_1; \ldots; x_n, p_n)$ – that is, the value that satisfies $U(\text{CE}(P)) = \sum_{i=1}^{n} U(x_i)\cdot p_i$ – is always less than the expected value of P. In modern treatments, risk aversion is defined as an aversion to all 'mean-preserving spreads' from any (degenerate or nondegenerate) lottery, where a mean-preserving spread consists of a decrease in the probability of a pay-off x_i by some amount Δp, and increases in the probabilities of some higher and lower pay-offs $x_i + \alpha$ and $x_i - \beta$ by the respective amounts $\Delta p\cdot\beta/(\alpha+\beta)$ and $\Delta p\cdot\alpha/(\alpha+\beta)$. This 'spreads' the probability mass of the lottery in a manner that does not change its expected value, so it can be thought of as a 'pure increase in risk'. An expected utility maximizer will be risk-averse in both the original and the modern senses if and only if $U(\cdot)$ is a strictly concave function of x. If $U(\cdot)$ is twice continuously differentiable, strict concavity is equivalent to a negative second derivative $U''(\cdot)$. Although the widespread purchase of actuarially unfair state lottery tickets is evidence of the opposite property of 'risk preference', the even more widespread purchase of insurance and the prevalence of other risk-reducing instruments has led researchers to hypothesize that individuals are for the most part risk-averse.

After these basic characterizations, the most important analytical result in expected utility theory is the Arrow–Pratt characterization of 'comparative risk aversion', which states that the following four conditions on a pair of risk-averse von Neumann–Morgenstern utility functions $U_A(\cdot)$ and $U_B(\cdot)$ are equivalent:

- *Comparative Concavity.* $U_A(\cdot)$ is an increasing concave transformation of $U_B(\cdot)$, that is, $U_A(x) \equiv \rho(U_B(x))$ for some increasing concave function $\rho(\cdot)$.
- *Comparative Arrow–Pratt Measures.* $-U_A''(x)/U_A'(x) \geq -U_B''(x)/U_B'(x)$ for all x.
- *Comparative Certainty Equivalents.* For any lottery $P = (x_1, p_1; \ldots; x_n, p_n)$, if $\text{CE}_A(P)$ and $\text{CE}_B(P)$ satisfy $U_A(\text{CE}_A(P)) = \sum_{i=1}^n U_A(x_i)\cdot p_i$ and $U_B(\text{CE}_B(P)) = \sum_{i=1}^n U_B(x_i)\cdot p_i$, then $\text{CE}_A(P) \leq \text{CE}_B(P)$.
- *Comparative Demand for Risky Assets.* For any initial wealth W, constant $r > 0$, and random variable $\tilde{x}$ such that $E[\tilde{x}] > r$ but $P(\tilde{x} < r) > 0$, if γ_A^* and γ_B^* respectively maximize $E[U_A(\gamma\cdot\tilde{x} + (W - \gamma)\cdot r)]$ and $E[U_B(\gamma\cdot\tilde{x} + (W - \gamma)\cdot r)]$, then $\gamma_A^* \leq \gamma_B^*$.

(Note: here and elsewhere we write $P(\cdot)$ for the probability of an event. This should not be confused with the use of P to stand for a lottery.)

Each of these conditions can be interpreted as saying that $U_A(\cdot)$ is at least as risk-averse as $U_B(\cdot)$. The first condition extends the above characterization of risk aversion by the concavity of $U(\cdot)$ to its comparative version across individuals, and the second shows that this can be expressed in terms of a numerical index $-U''(x)/U'(x)$, known as the 'Arrow–Pratt index of absolute risk aversion'. The third condition extends the original notion of risk aversion as low certainty equivalents (lower than the mean) to its comparative form.

The fourth condition involves comparative optimization behavior. Consider an individual with initial wealth W to be divided between a riskless asset yielding gross return r, and a risky asset whose gross return $\tilde{x}$ has a higher expected value, but offers some risk of doing worse than the riskless asset. This condition states that the less risk-averse utility function $U_B(\cdot)$ will always choose to invest at least as much in the risky asset as will the more risk-averse $U_A(\cdot)$.

The equivalence of the above four conditions, the first two mathematical and the second two behavioral, and their numerous additional behavioral equivalencies and implications, has made the Arrow–Pratt characterization one of the central theorems in the analytics of expected utility theory, with applications in insurance, financial markets, auctions, the demand for information, bargaining, and game theory.

Experimental Evidence on the Independence Axiom

Experimental testing of the expected utility hypothesis has centered on the Independence axiom, either directly or via its implication that the expected utility preference function $V_{\text{EU}}(P) = \sum_{i=1}^n U(x_i)\cdot p_i$ is linear in the probabilities p_i. One of the best-known tests is the 'Allais paradox' (Allais, 1953). An individual is asked to rank each of the following two pairs of lotteries (where \$1M = \$1,000,000):

- $a_1 = \{1.00$ chance of \$1M versus

  $a_2 = \begin{cases} .10 \text{ chance of \$5M} \\ .89 \text{ chance of \$1M} \\ .01 \text{ chance of \$0} \end{cases}$

- $a_3 = \begin{cases} .10 \text{ chance of \$5M} \\ .90 \text{ chance of \$0} \end{cases}$ versus

  $a_4 = \begin{cases} .11 \text{ chance of \$1M} \\ .89 \text{ chance of \$0} \end{cases}$

The expected utility hypothesis implies that the individual's choices from these two pairs must either be a_1 and a_4 (whenever $.11\cdot U(\$1\text{M}) > .10\cdot U(\$5\text{M}) + .01\cdot U(\$0)$), or else a_2 and a_3 (whenever $.11\cdot U(\$1\text{M}) < .10\cdot U(\$5\text{M}) + .01\cdot U(\$0)$). However, when presented with these choices, most subjects choose a_1 from the first pair and a_3 from the second, which violates the hypothesis. Only a small number violate the hypothesis in the opposite direction, by choosing a_2 and a_4.

Although the Allais paradox was originally dismissed as an 'isolated example', subsequent experimental work by psychologists, economists and others has uncovered a similar pattern of violation over a range of probability and pay-off values, and the Allais paradox is now seen to be just one example of a type of systematic violation of the Independence axiom known as the 'common-consequence effect'. It is observed that for lotteries P, P^* and P^{**}, pay-off c, and mixture probability $\alpha \in (0,1)$, such that P^{**} first-order-stochastically dominates P^* and c lies between the highest and lowest pay-offs in P, preferences depart from the Independence axiom in the direction of exhibiting $\alpha\cdot P + (1-\alpha)\cdot P^* \succ \alpha\cdot c + (1-\alpha)\cdot P^*$ yet $\alpha\cdot P + (1-\alpha)\cdot P^{**} \prec \alpha\cdot c + (1-\alpha)\cdot P^{**}$. (In the Allais paradox, these constructs are $P = (\$5\text{M}, 10/11; \$0, 1/11)$, $P^* = (\$0, 1), P^{**} = (\$1\text{M}, 1), c = \$1\text{M}$ and $\alpha = .11$.)

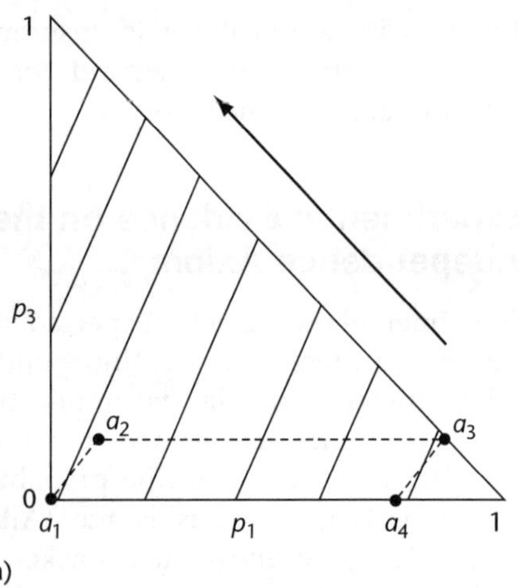

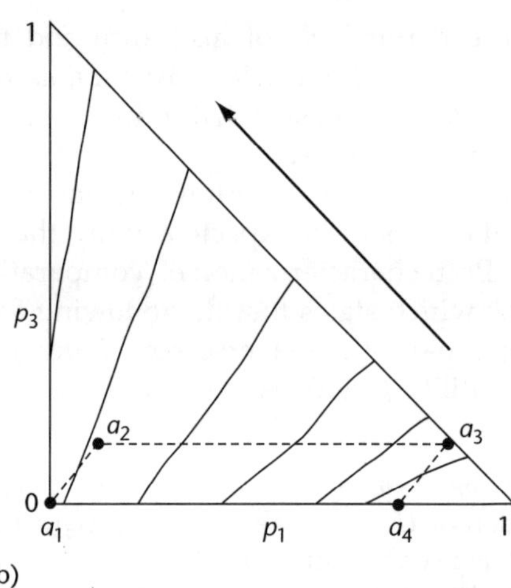

Figure 1. Indifference curves in the probability triangle. (a) Expected utility indifference curves, which are parallel straight lines. (b) Non-expected utility indifference curves, which 'fan out', illustrating the common-consequence effect.

Both the implications of the Independence axiom and the nature of this violation can be illustrated in the special case of all lotteries $P = (\bar{x}_1, p_1; \bar{x}_2, p_2; \bar{x}_3, p_3)$ over a triple of fixed pay-off values $\bar{x}_1 < \bar{x}_2 < \bar{x}_3$. Since we can write $P = (\bar{x}_1, p_1; \bar{x}_2, 1 - p_1 - p_3; \bar{x}_3, p_3)$, each such lottery is uniquely associated with a point in the (p_1, p_3) triangles of Figures 1(a) and 1(b). Since we can write $V_{\mathrm{EU}}(P) = U(\bar{x}_1)\cdot p_1 + U(\bar{x}_2)\cdot(1 - p_1 - p_3) + U(\bar{x}_3)\cdot p_3$, the loci of constant expected utility ('expected utility indifference curves') consist of parallel straight lines as in Figure 1(a). Since upward shifts in the triangle represent increases in p_3 at the expense of p_2, and leftward shifts represent reductions in p_1 to the benefit of p_2, first-order stochastic dominance preference implies that these indifference curves will be upward-sloping, with increasing levels of preference in the direction indicated by the arrows.

Fixing the pay-offs at $\bar{x}_1 = \$0$, $\bar{x}_2 = \$1\text{M}$ and $\bar{x}_3 = \$5\text{M}$, the Allais paradox lotteries a_1, a_2, a_3 and a_4 are seen to form a parallelogram when plotted in the probability triangle, which explains why parallel straight-line expected utility indifference curves must either prefer a_1 and a_4 (as illustrated for the relatively steep indifference curves of Figure 1(a)) or else prefer a_2 and a_3 (for relatively flat expected utility indifference curves). Figure 1(b) illustrates 'non-expected utility indifference curves' that 'fan out', and are seen to exhibit the typical Allais paradox rankings of $a_1 \succ a_2$ and $a_3 \succ a_4$.

Another type of systematic experimental violation of the Independence axiom that has been uncovered is known as the 'common-ratio effect'. For pay-offs $x^* > x > 0$, probabilities $p^* < p$ and $r \in (0, 1)$, preferences depart from the Independence axiom in the direction of exhibiting $(x^*, p^*; 0, 1 - p^*) \prec (x, p; 0, 1 - p)$ yet $(x^*, r\cdot p^*; 0, 1 - r\cdot p^*) \succ (x, r\cdot p; 0, 1 - r\cdot p)$. For losses $0 > -x > -x^*$, with $p^* < p$ and $r \in (0, 1)$, preferences depart in the reflected direction of $(-x^*, p^*; 0, 1 - p^*) \succ (-x, p; 0, 1 - p)$ yet $(-x^*, r\cdot p^*; 0, 1 - r\cdot p^*) \prec (-x, r\cdot p; 0, 1 - r\cdot p)$.

With the pay-offs $\bar{x}_1 = 0$, $\bar{x}_2 = x$ and $\bar{x}_3 = x^*$, the line segment between the lotteries $b_1 = (x^*, p^*; 0, 1 - p^*)$ and $b_2 = (x, p; 0, 1 - p)$ in the probability triangle of Figure 2(a) is seen to be parallel to that between $b_3 = (x^*, r\cdot p^*; 0, 1 - r\cdot p^*)$ and $b_4 = (x, r\cdot p; 0, 1 - r\cdot p)$, and the common-ratio-effect rankings of $b_1 \prec b_2$ and $b_3 \succ b_4$ again suggests that indifference curves depart from expected utility by fanning out. For losses, with $\bar{x}_1 = -x^*$, $\bar{x}_2 = -x$ and $\bar{x}_3 = 0$ (to maintain the ordering $\bar{x}_1 < \bar{x}_2 < \bar{x}_3$), the reflected rankings of $-b_1 \succ -b_2$ and $-b_3 \prec -b_4$ again suggest fanning out, as in Figure 2(b). Fanning out is consistent with other observed forms of departure from the Independence axiom, although it is not universal across subjects, and seems to be more pronounced near the edges of the triangle than in its central region.

GENERALIZATIONS OF EXPECTED UTILITY THEORY

Non-Expected Utility Functional Forms

The above phenomena, as well as other systematic departures from linearity in the probabilities, have prompted researchers to develop more general models of preferences over lotteries, primarily by generalizing the functional form of the lottery preference function $V(P) = V(x_1, p_1; \ldots; x_n, p_n)$. The earliest of these attempts, which used the form $V(P) = \sum_{i=1}^{n} U(x_i)\cdot\pi(p_i)$, was largely abandoned

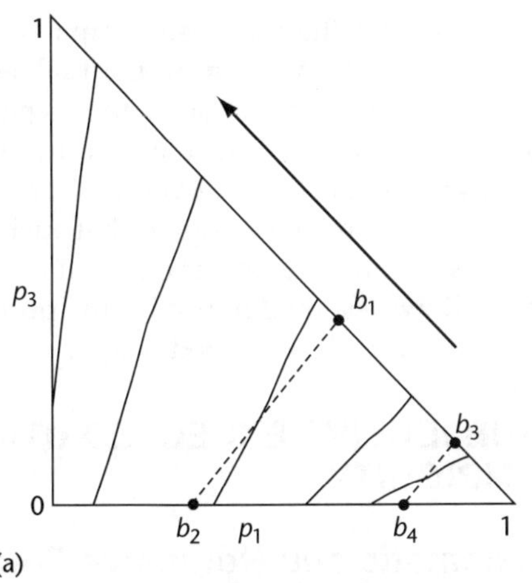

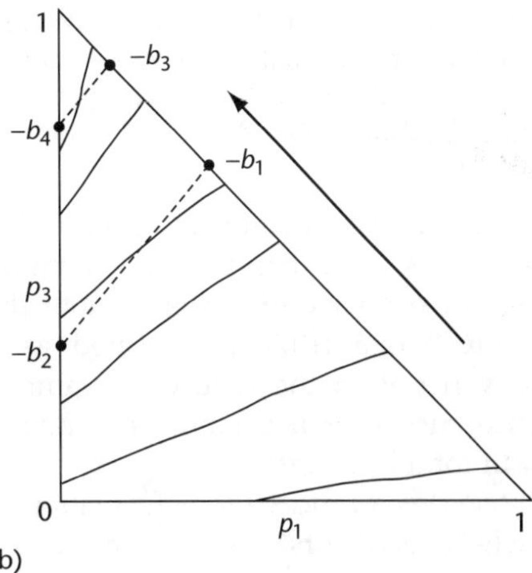

Figure 2. Probability triangles illustrating the common-ratio effect. (a) Positive pay-offs. (b) Negative pay-offs (losses).

when it was realized that, except for the case $\pi(p) \equiv p$ when it reduced to expected utility, it was inconsistent with the property of first-order stochastic dominance preference. Current models include the following:

- *Weighted Utility.* $V(P) = \sum_{i=1}^{n} U(x_i)\cdot\pi(p_i)/\sum_{i=1}^{n} S(x_i)\cdot\pi(p_i)$
- *Moments of Utility.* $V(P) = F(\sum_{i=1}^{n} U(x_i)\cdot p_i, \sum_{i=1}^{n} U(x_i)^2\cdot p_i, \sum_{i=1}^{n} U(x_i)^3\cdot p_i)$
- *Rank-Dependent Expected Utility.* $V(P) = \sum_{i=1}^{n} U(x_i)\cdot(G(\sum_{j=1}^{i} p_j) - G(\sum_{j=1}^{i-1} p_j))$for $x_1 \preceq \ldots \preceq x_n$
- *Quadratic in the Probabilities.* $V(P) = \sum_{i=1}^{n}\sum_{j=1}^{n} T(x_i, x_j)\cdot p_i\cdot p_j$

Under the appropriate monotonicity or curvature assumptions on their constituent functions $U(\cdot)$, $\pi(\cdot)$, $G(\cdot)$, etc., each of these forms is capable of exhibiting first-order stochastic dominance preference, risk aversion and comparative risk aversion, as well as many of the types of observed systematic violations of the Independence axiom. Researchers have also used these forms to revisit many of the applications previously modeled by expected utility theory (e.g. insurance, financial markets, auctions), to determine which of the earlier expected utility-based results are crucially dependent on preferences exhibiting the expected utility functional form, and which are robust to departures from expected utility.

Generalized Expected Utility Analysis

An alternative branch of research on non-expected utility preferences does not rely on any specific functional form, but links properties of attitudes towards risk directly with the probability derivatives of a general (i.e. not necessarily expected utility) preference function $V(P) = V(x_1, p_1; \ldots; x_n, p_n)$ over lotteries. Such analysis reveals that the basic analytics of the expected utility model as outlined above are in fact quite robust to general smooth departures from linearity in the probabilities. It proceeds from the correspondence between the properties of a linear function as determined by its coefficients and the properties of a nonlinear function as determined by its partial derivatives – in this case, between the 'probability coefficients' $U(x_1), \ldots, U(x_n)$ of the expected utility form $\sum_{i=1}^{n} U(x_i)\cdot p_i$ and the 'probability derivatives' $\partial V(x_1, p_1; \ldots; x_n, p_n)/\partial p_1, \ldots, \partial V(x_1, p_1; \ldots; x_n, p_n)/\partial p_n$ of a general smooth preference function $V(x_1, p_1; \ldots; x_n, p_n)$. Under such a correspondence, most of the fundamental analytical results of expected utility theory pass through directly (Machina, 1982). For example:

- *First-Order Stochastic Dominance Preference.* Under expected utility, this is equivalent to $U(x)$ (the coefficient of $P(x)$) being an increasing function of x. For a general smooth $V(\cdot)$, if $\partial V(P)/\partial P(x)$ is an increasing function of x at every lottery P, then for any pay-offs $x_i > x_j$ we will have $\partial V(P)/\partial p_i > \partial V(P)/\partial p_j$ at each P, so any (small or large) rise in p_i and matching fall in p_j will lead to an increase in $V(P)$ and hence will be preferred.
- *Risk Aversion.* Under expected utility, this is equivalent to $U(x)$ being a strictly concave function of x. For a general smooth $V(\cdot)$, if $\partial V(P)/\partial P(x)$ is a strictly concave function of x at each P, then for any pay-offs $x_i - \beta < x_i < x_i + \alpha$ we will have $[\partial V(P)/\partial P(x_i) - \partial V(P)/\partial P(x_i - \beta)]/\beta > [\partial V(P)/\partial P(x_i + \alpha) - \partial V(P)/\partial P(x_i)]/\alpha$ at each P, which implies that each mean-preserving spread over the pay-offs $x_i - \beta < x_i < x_i + \alpha$ will lead to a reduction in $V(P)$ and hence will be dispreferred.
- *Comparative Risk Aversion.* Under expected utility, this is equivalent to $U_A(\cdot)$ being an increasing concave transformation of $U_B(\cdot)$. For general smooth $V_A(\cdot)$ and

$V_B(\cdot)$, if at each P the function $\partial V_A(P)/\partial P(x)$ is some increasing concave transformation of $\partial V_B(P)/\partial P(x)$, then $V_A(\cdot)$ and $V_B(\cdot)$ will exhibit the above conditions for comparative certainty equivalence and comparative demand for risky assets.

In addition to the above theoretical results, this approach also allows for a direct characterization of the fanning-out property in terms of how the probability derivative $\partial V(P)/\partial P(x)$, treated as a function of x, varies with the lottery P. Namely, the indifference curves of a preference function $V(\cdot)$ will fan out for all pay-offs $\bar{x}_1 < \bar{x}_2 < \bar{x}_3$ if and only if $\partial V(P^*)/\partial P(x)$ is a concave transformation of $\partial V(P)/\partial P(x)$ whenever P^* first-order-stochastically dominates P.

Regret Theory

Another type of non-expected utility model dispenses with the assumption of an underlying preference order $\succeq$ over lotteries, and instead derives choice behavior from the underlying psychological notion of 'regret' – that is, the reaction to receiving an outcome x when an alternative decision would have led to a preferred outcome x^* (Loomes and Sugden, 1982). The opposite experience, namely of receiving an outcome that is preferred to what the alternative decision would have yielded, is termed 'rejoice'. The primitive for this model is a 'rejoice function' $R(x, x^*)$ which is positive if x is preferred to x^*, negative if x^* is preferred to x, and zero if they are indifferent, and satisfies the skew-symmetry condition $R(x, x^*) \equiv -R(x^*, x)$.

In the simplest case of pairwise choice over two lotteries $P = (x_1, p_1; \ldots; x_n, p_n)$ and $P^* = (x_1^*, p_1^*; \ldots; x_{n^*}^*, p_{n^*}^*)$ that are realized independently, the individual's expected rejoice from choosing the lottery P over the alternative lottery P^* is given by the formula $\sum_{i=1}^{n}\sum_{j=1}^{n^*} R(x_i, x_j^*)\cdot p_i \cdot p_j^*$, and the individual is predicted to choose P if this value is positive, to choose P^* if it is negative, and to be indifferent if it is zero. Various proposals for extending this approach beyond pairwise choice have been made, including a formal result that shows that for any finite collection of lotteries, one of these lotteries or some randomization over them will exhibit nonnegative expected rejoice with respect to every other lottery or randomization.

As with the non-expected utility functional forms listed above, various monotonicity and curvature assumptions on the rejoice function $R(\cdot,\cdot)$ can be shown to correspond to various properties of risk preferences, such as risk aversion and comparative risk aversion, as well as to the general fanning-out property. Since this model derives from the pairwise comparison of lotteries rather than from their individual evaluation by some preference function, it allows pairwise choice to be intransitive, so that an individual could choose P over P^*, P^* over P^{**}, and P^{**} over P. Although some have argued that such cyclic choice allows for the phenomenon of 'money pumps', it also allows the model to solve the problem of 'preference reversal' described below.

SUBJECTIVE EXPECTED UTILITY AND AMBIGUITY

Axiomatic and Normative Foundations of Subjective Expected Utility

The expected utility model of choice under subjective uncertainty – sometimes called the 'subjective expected utility' model – hypothesizes that the individual's preference relation $\succeq$ over subjective acts $f(\cdot) = [x_1 \text{ on } E_1; \ldots; x_n \text{ on } E_n]$ can be represented by a preference function of the form $W_{\text{SEU}}(f(\cdot)) = \sum_{i=1}^{n} U(x_i)\cdot \mu(E_i)$ for some von Neumann–Morgenstern utility function $U(\cdot)$ and 'subjective probability measure' $\mu(\cdot)$ over events. Thus, both attitudes towards risk and subjective beliefs are specific to the individual, and the values $\mu(E_1), \ldots, \mu(E_n)$ are sometimes called 'personal probabilities'. By virtue of its independent representation of risk attitudes by the utility function $U(\cdot)$, and beliefs by the subjective probability measure $\mu(\cdot)$, the subjective expected utility model is sometimes described as achieving a 'separation of preferences and beliefs'.

By analogy with the probability mixture of two objective lotteries, the axiomatic and normative underpinnings of the subjective expected utility model are based on the notion of a 'subjective mixture' $[f(\cdot) \text{ on } E; f^*(\cdot) \text{ on} \sim E]$ of two acts $f(\cdot) = [x_i \text{ on } E_1; \ldots; x_n \text{ on } E_n]$ and $f^*(\cdot) = [x_1^* \text{ on } E_1^*; \ldots; x_{n^*}^* \text{ on } E_{n^*}^*]$, which is the act that would yield the same outcome as $f(\cdot)$ should the event E occur, and the same outcome as $f^*(\cdot)$ should the event $\sim E$ occur, so that we can write $[f(\cdot) \text{ on } E; f^*(\cdot) \text{ on} \sim E] = [x_1 \text{ on } E \cap E_1; \ldots; x_n \text{ on } E \cap E_n; x_1^* \text{ on} \sim E \cap E_1^*; \ldots; x_{n^*}^* \text{ on} \sim E \cap E_{n^*}^*]$. An event E is said to be 'null' for the individual if $[x^* \text{ on } E; f(\cdot) \text{ on} \sim E] \sim [x \text{ on } E; f(\cdot) \text{ on} \sim E]$ for all outcomes x and x^* and all acts $f(\cdot)$, so that the individual effectively treats E as if it had zero likelihood. Since we can identify each outcome x with the 'constant act' $[x \text{ on S}]$, we can write $x^* \succeq x$ if and only if $[x^* \text{ on S}] \succeq [x \text{ on S}]$. The individual's preferences over subjective acts can then be represented by a subjective expected utility preference function

$W_{SEU}(f(\cdot)) = \sum_{i=1}^{n} U(x_i) \cdot \mu(E_i)$ for some $U(\cdot)$ and $\mu(\cdot)$ if and only they satisfy the following axioms (Savage, 1954):

- *Completeness*. For all acts $f(\cdot)$ and $f^*(\cdot)$, either $f(\cdot) \succeq f^*(\cdot)$ or $f^*(\cdot) \succeq f(\cdot)$, or both.
- *Transitivity*. For all acts $f(\cdot), f^*(\cdot)$ and $f^{**}(\cdot)$, if $f(\cdot) \succeq f^*(\cdot)$ and $f^*(\cdot) \succeq f^{**}(\cdot)$ then $f(\cdot) \succeq f^{**}(\cdot)$.
- *Eventwise Monotonicity*. For all outcomes x^* and x, non-null events E and acts $f(\cdot)$, $[x^*$ on $E; f(\cdot)$ on $\sim E] \succeq [x$ on E; $f(\cdot)$ on $\sim E]$ if and only if $x^* \succeq x$.
- *Weak Comparative Probability*. For all events A and B and outcomes $x^* \succ x$ and $y^* \succ y$, $[x^*$ on A; x on $\sim A] \succeq [x^*$ on B; x on $\sim B]$ implies $[y^*$ on A; y on $\sim A] \succeq [y^*$ on B; y on $\sim B]$.
- *Small-Event Continuity*. For all acts $f(\cdot) \succ g(\cdot)$ and outcomes x, there exists a partition $\{E_1,\ldots,E_n\}$ such that $f(\cdot) \succ [x$ on E_i; $g(\cdot)$ on $\sim E_i]$ and $[x$ on E_i; $f(\cdot)$ on $\sim E_i] \succ g(\cdot)$ for each $i = 1,\ldots,n$.
- *Sure-Thing Principle*. For all events E and acts $f(\cdot), f^*(\cdot)$, $g(\cdot)$ and $h(\cdot)$, $[f^*(\cdot)$ on E; $g(\cdot)$ on $\sim E] \succeq [f(\cdot)$ on E; $g(\cdot)$ on $\sim E]$ implies $[f^*(\cdot)$ on E; $h(\cdot)$ on $\sim E] \succeq [f(\cdot)$ on E; $h(\cdot)$ on $\sim E]$.

Completeness and Transitivity are as before, and Eventwise Monotonicity is the subjective analogue of first-order stochastic dominance preference. Weak Comparative Probability essentially ensures that the individual's 'revealed likelihood ranking' of a pair of events A and B, as given by their preference for staking the more preferred of a pair of prizes on A versus staking it on B, is stable in the sense that it does not depend on the particular prizes involved. Small-Event Continuity serves as the standard Archimedean property in the context of choice over subjective acts. The key normative and behavioral axiom of subjective expected utility theory is the Sure-Thing Principle. Behaviorally, it once again corresponds to the property of separability across mutually exclusive events. Normatively, it corresponds to the same argument as for the Independence axiom, with the objective randomization by the α:$(1-\alpha)$ coin replaced by the 'subjective randomization' via the events E and $\sim E$.

State-dependent Utility

In some subjective settings, the individual's valuation of outcomes may depend on the source of uncertainty itself. Thus, for the mutually exclusive and exhaustive events ('rain','shine') and prizes ('umbrella', 'sun lotion'), each of which is preferred to \$0, the individual may well exhibit the preferences [umbrella on rain; \$0 on shine] $\succ$ [umbrella on shine; \$0 on rain] and [sun lotion on rain; \$0 on shine] $\prec$ [sun lotion on shine; \$0 on rain], which violates the Weak Comparative Probability axiom for $x^* =$ umbrella, $y^* =$ lotion, $x = y = \$0$, $A =$ rain and $B =$ shine, and hence is inconsistent with the subjective expected utility preference function $W_{SEU}(\cdot)$. This phenomenon, known as 'state dependence', can occur even when the outcomes are monetary pay-offs: if the state of nature is the individual's health, the utility of a \$50,000 prize may be very high in states where the individual requires a \$50,000 operation to survive, much lower in states where the individual requires much more than that for the operation, and somewhere in between in states of good health.

The subjective expected utility model can be easily adapted to accommodate the phenomenon of state dependence, by allowing the utility function $U(\cdot|E)$ to depend upon the event or state of nature, so that the preference function over acts takes the 'state-dependent expected utility' form $W_{SDEU}(x_1 \text{ on } E_1; \ldots x_n \text{ on } E_n) = \sum_{i=1}^{n} U(x_i|E_i) \cdot \mu(E_i)$. Most of the analytics of the standard (i.e. 'state-independent') form $W_{SEU}(\cdot)$ extend to the state-dependent case (Karni, 1985). However, under state dependence, subjective probabilities cannot be uniquely inferred from preferences over acts: for any state-dependent preference function $W_{SDEU}(f(\cdot)) = \sum_{i=1}^{n} U(x_i|E_i) \cdot \mu(E_i)$, and any distinct subjective probability measure $\mu^*(\cdot)$ that satisfies $\mu^*(E) > 0 \Leftrightarrow \mu(E) > 0$, $W_{SDEU}(\cdot)$ is indistinguishable from the preference function $W^*_{SDEU}(f(\cdot)) = \sum_{i=1}^{n} U^*(x_i|E_i) \cdot \mu^*(E_i)$ with $U^*(\cdot|\cdot)$ defined by $U^*(x|E) = U(x|E) \cdot [\mu(E)/\mu^*(E)]$.

Ambiguity and Nonprobabilistic Beliefs

A more serious departure from the notion of well-defined probabilistic beliefs arises from the phenomenon of 'ambiguity', which is distinct from the phenomenon of state dependence and much more difficult to model. The best-known example is the 'Ellsberg paradox' (Ellsberg, 1961). An individual must draw a ball from an opaque urn that contains 30 red balls and 60 black or yellow balls in an unknown proportion, and is offered four possible bets on the color of the drawn ball, as shown in Figure 3.

Most individuals exhibit the preference rankings $f_1(\cdot) \succ f_2(\cdot)$ and $f_4(\cdot) \succ f_3(\cdot)$. When asked why, they explain that the probability of winning under $f_2(\cdot)$ could be anywhere from 0 to $\frac{2}{3}$ whereas the probability of winning under $f_1(\cdot)$ is known to be exactly $\frac{1}{3}$, and they prefer the act that offers the known probability. Similarly, the probability of winning under $f_3(\cdot)$ could be anywhere from $\frac{1}{3}$ to 1 whereas the probability of winning under $f_4(\cdot)$ is known to be exactly $\frac{2}{3}$, so it is preferred. However, these preferences are inconsistent with any assignment of

	30 balls	60 balls	
	Red	Black	Yellow
$f_1(\cdot)$	\$100	\$0	\$0
$f_2(\cdot)$	\$0	\$100	\$0
$f_3(\cdot)$	\$100	\$0	\$100
$f_4(\cdot)$	\$0	\$100	\$100

Figure 3. Four possible bets on the color of the drawn ball in the 'Ellsberg paradox'. The proportion of black to yellow balls is unknown. Most individuals prefer $f_1(\cdot)$ to $f_2(\cdot)$ and $f_4(\cdot)$ to $f_3(\cdot)$.

numerical subjective probabilities μ(red), μ(black), μ(yellow) to the three events: if the individual were choosing on the basis of such probabilistic beliefs, the ranking $f_1(\cdot) \succ f_2(\cdot)$ would reveal that $\mu(\text{red}) > \mu(\text{black})$, but the ranking $f_4(\cdot) \succ f_3(\cdot)$ would reveal that $\mu(\text{red}) < \mu(\text{black})$.

This phenomenon be cannot be accommodated by simply allowing the event to enter the utility function and working with the state-dependent form $\sum_{i=1}^{n} U(x_i|E_i) \cdot \mu(E_i)$, since this form still satisfies the Sure-Thing Principle, whereas the preferences $f_1(\cdot) \succ f_2(\cdot)$ and $f_4(\cdot) \succ f_3(\cdot)$ violate this axiom (for $E = \text{red} \cup \text{black}$ and $\sim E = \text{yellow}$). The Ellsberg paradox and related examples are attributed to the phenomenon of 'ambiguity aversion', whereby individuals exhibit a general preference for bets based on probabilistic partitions such as {red, black $\cup$ yellow} rather than on ambiguous partitions such as {black, red $\cup$ yellow}.

Just as the Allais paradox and related violations of the Independence axiom led to the development of non-expected utility models of risk preferences, the Ellsberg paradox and related examples have led to the development of nonprobabilistic models of beliefs. The most notable of these involves replacing the additive subjective probability measure $\mu(\cdot)$ over events by a 'capacity' $C(\cdot)$, which is similar to $\mu(\cdot)$ in that it satisfies the properties $C(\emptyset) = 0$, $C(S) = 1$, and $E \subseteq E^* \Rightarrow C(E) \leq C(E^*)$, but differs from $\mu(\cdot)$ in that it is not necessarily additive. By labeling the outcomes in any act so that $x_1 \preceq \ldots \preceq x_n$ and writing the subjective expected utility formula as $W_{\text{SEU}}(f(\cdot)) = \sum_{i=1}^{n} U(x_i) \cdot \mu(E_i) = \sum_{i=1}^{n} U(x_i) \cdot (\mu(\cup_{j=1}^{i} E_j) - \mu(\cup_{j=1}^{i-1} E_j))$, we can generalize from an additive $\mu(\cdot)$ to a non-additive $C(\cdot)$ to obtain the 'Choquet expected utility' preference function $W_{\text{Choquet}}(f(\cdot)) = \sum_{i=1}^{n} U(x_i) \cdot (C(\cup_{j=1}^{i} E_j) - C(\cup_{j=1}^{i-1} E_j))$ over subjective acts (Schmeidler, 1989). Selecting $U(\$100) = 1$, $U(\$0) = 0$, $C(\text{red}) = \frac{1}{3}$, $C(\text{black} \cup \text{yellow}) = \frac{2}{3}$, $C(\text{black}) = \frac{1}{2}$, $C(\text{red} \cup \text{yellow}) = \frac{3}{4}$ yields the values $W_{\text{Choquet}}(f_1(\cdot)) = \frac{1}{3}$, $W_{\text{Choquet}}(f_2(\cdot)) = \frac{1}{4}$, $W_{\text{Choquet}}(f_3(\cdot)) = \frac{1}{2}$, $W_{\text{Choquet}}(f_4(\cdot)) = \frac{2}{3}$, which correspond to the typical Ellsberg rankings.

Another alternative to the subjective expected utility model of act preferences, also capable of exhibiting ambiguity aversion, is the 'maxmin expected utility' form, which involves a family $\{\mu_\tau(\cdot) \,|\, \tau \in T\}$ of additive probability measures over the events, and the preference function $W_{\max\min}(f(\cdot)) = \min_{\tau \in T} \sum_{i=1}^{n} U(x_i) \cdot \mu_\tau(E_i)$.

DESCRIPTION AND PROCEDURE INVARIANCE

Although the alternative models described above drop or weaken many of the axioms of standard objective and subjective expected utility theory, they typically retain the primary implicit assumptions of the standard theory, namely that: the objects of choice (objective lotteries or subjective acts) can be unambiguously described; net changes in wealth are combined with any initial endowment and evaluated in terms of the final wealth levels they imply; and situations that imply the same set of final opportunities (the same set of objective lotteries or same set of subjective acts over final wealth levels) will lead to the same choice. They also assume that the individual is able to perform the mathematical operations necessary to determine this opportunity set, e.g. to calculate the probabilities of compound or conditional events and add net changes to initial endowments. However, psychologists have uncovered several systematic violations of these assumptions.

Framing Effects

Effects whereby alternative descriptions of the same decision problem lead to systematically different responses are called 'framing' effects. Some framing effects in choice under uncertainty involve alternative representations of the same likelihood. For example, contingency of a gain or loss on the joint occurrence of four independent events, each with probability p, is found to elicit a different response from contingency on the occurrence of a single event with probability p^4. In comparison with the single-event case, making a gain contingent on the joint occurrence of events seems to make it more attractive, and making a loss contingent on the joint occurrence of events seems to make it more unattractive (Slovic, 1969).

Other framing effects in choice under uncertainty involve alternative representations of the same final wealth levels. Consider the following two proposals.

- 'In addition to whatever you own, you have been given 1,000 (Israeli pounds). You are now asked to choose between a ½:½ chance of a gain of 1,000 or 0 or a sure chance of a gain of 500.'
- 'In addition to whatever you own, you have been given 2,000. You are now asked to choose between a ½:½ chance of a loss of 1,000 or 0 or a sure loss of 500.'

These two problems involve identical distributions over final wealth. However, when put to two different groups of subjects, 84% chose the sure gain in the first problem but 69% chose the ½:½ gamble in the second (Kahneman and Tversky, 1979).

Response-Mode Effects and Preference Reversal

Effects whereby alternative response formats lead to systematically different inferences about underlying preferences are called 'response-mode' effects. For example, under the expected utility hypothesis, an individual's von Neumann–Morgenstern utility function can be assessed or elicited in a number of different ways, which typically involve a sequence of prespecified lotteries $P_1, P_2, P_3, \ldots$, and ask for the individual's certainty equivalent $\mathrm{CE}(P_i)$ for each lottery P_i, or else the 'gain equivalent' G_i that would make the lottery $(G_i, \frac{1}{2}; \$0, \frac{1}{2})$ of equal preference to P_i, or else the 'probability equivalent' $\wp_i$ that would make the lottery $(\$1000, \wp_i; \$0, 1-\wp_i)$ of equal preference to P_i. Although such procedures should be expected to generate equivalent assessed utility functions, they have been found to yield systematically different ones (Hershey and Schoemaker, 1985).

In an experiment that demonstrates what is now known as the 'preference reversal phenomenon', subjects were first presented with a number of pairs of bets and asked to choose one bet out of each pair. Each pair of bets took the form of a 'p-bet', which offered a p chance of \$$X^*$ and a $1-p$ chance of \$X, versus a '\$-bet', which offered a q chance of \$$Y^*$ and a $1-q$ chance of \$Y, where $X^* > X, Y^* > Y, p > q$ and $X^* < Y^*$. The names 'p-bet' and '\$-bet' derive from the greater probability of winning in the first bet, and greater possible gain in the second bet (in some cases, X and Y took on small negative values). Subjects were next asked for their certainty equivalents of each of these bets, via a number of standard elicitation techniques.

The expected utility model, and most of the aforementioned alternative models, predict that for each such pair, the bet that was selected in the direct-choice problem would also be the one assigned the higher certainty equivalent. However, subjects exhibit a systematic departure from this prediction in the direction of choosing the p-bet in a direct choice but assigning a higher certainty equivalent to the \$-bet (Lichtenstein and Slovic, 1971). Although this finding initially generated widespread scepticism, especially among economists, it has been widely replicated by both psychologists and economists in a variety of different settings involving real-money gambles, patrons of a Las Vegas casino, group decisions, and experimental market trading. By viewing it as an instance of intransitivity (\$-bet $\sim$ CE(\$-bet) $\succ$ CE(p-bet) $\sim$ p-bet $\succ$ \$-bet), some economists have explained the phenomenon in terms of the regret theory model. However, most psychologists and a growing number of economists regard it as a response-mode effect, specifically, that the psychological processes of valuation (which generates certainty equivalents) and choice are differentially influenced by the probabilities and pay-offs involved in a lottery, and that under certain conditions this can lead to choice and valuation that reveal opposite 'underlying' preference rankings over a pair of gambles.

SUMMARY

Since the work of Bernoulli, the theory of choice under uncertainty has seen both a tension and a scientific interplay between theoretical models of decision making and experimentally observed violations of these models. Current research in the field continues to reflect this tension, while the degree of interplay has increased, with theorists now more willing to model experimentally generated phenomena, and experimenters more willing to provide constructive feedback on these attempts.

References

Allais M (1953) Le comportement de l'homme rationnel devant le risque: critique des postulats et axiomes de l'école américaine. *Econometrica* **21**: 503–546.

Arrow K (1965) *Aspects of the Theory of Risk Bearing*. Helsinki: Yrjö Jahnsson Säätiö.

Ellsberg D (1961) Risk, ambiguity, and the Savage axioms. *Quarterly Journal of Economics* **75**: 643–669.

Hershey J and Schoemaker P (1985) Probability versus certainty equivalence methods in utility measurement: are they equivalent? *Management Science* **31**: 1213–1231.

Kahneman D and Tversky A (1979) Prospect theory: an analysis of decision under risk. *Econometrica* **47**: 263–291.
Karni E (1985) *Decision Making Under Uncertainty: The Case of State Dependent Preferences*. Cambridge, MA: Harvard University Press.
Lichtenstein S and Slovic P (1971) Reversals of preferences between bids and choices in gambling decisions. *Journal of Experimental Psychology* **89**: 46–55.
Loomes G and Sugden R (1982) Regret theory: an alternative theory of rational choice under uncertainty. *Economic Journal* **92**: 805–824.
Machina M (1982) 'Expected utility' analysis without the Independence Axiom. *Econometrica* **50**: 277–323.
Marschak J (1950) Rational behavior, uncertain prospects, and measurable utility. *Econometrica* **18**: 111–141. [Errata: *Econometrica* **18**: 312.]
von Neumann J and Morgenstern O (1944) *Theory of Games and Economic Behavior*. Princeton, NJ: Princeton University Press. [Second edition 1947; third edition 1953.]
Pratt J (1964) Risk aversion in the small and in the large. *Econometrica* **32**: 122–136.
Savage L (1954) *The Foundations of Statistics*. New York, NY: Wiley. [Revised and enlarged edition, 1972. New York, NY: Dover.]
Schmeidler D (1989) Subjective probability and expected utility without additivity. *Econometrica* **57**: 571–587.
Slovic P (1969) Manipulating the attractiveness of a gamble without changing its expected value. *Journal of Experimental Psychology* **79**: 139–145.

Further Reading

Camerer C and Weber M (1992) Recent developments in modeling preferences: uncertainty and ambiguity. *Journal of Risk and Uncertainty* **5**: 325–370.
Einhorn H and Hogarth R (1985) Ambiguity and uncertainty in probabilistic inference. *Psychological Review* **92**: 433–461.
Epstein L (1999) A definition of uncertainty aversion. *Review of Economic Studies* **66**: 579–608.
Fishburn P (1982) *The Foundations of Expected Utility*. Dordrecht: Reidel.
Heath C and Tversky A (1991) Preferences and belief: ambiguity and competence in choice under uncertainty. *Journal of Risk and Uncertainty* **4**: 5–28.
Kahneman D, Slovic P and Tversky A (eds) (1982) *Judgment Under Uncertainty: Heuristics and Biases*. Cambridge, UK: Cambridge University Press.
Kelsey D and Quiggin J (1992) Theories of choice under ignorance and uncertainty. *Journal of Economic Surveys* **6**: 133–153.
Starmer C (2000) Developments in non-expected utility theory: the hunt for a descriptive theory of choice under risk. *Journal of Economic Literature* **38**: 332–382.
Tversky A and Fox C (1995) Weighing risk and uncertainty. *Psychological Review* **102**: 269–283.
Weber M and Camerer C (1987) Recent developments in modeling preferences under risk. *OR Spektrum* **9**: 129–151.

Circadian Rhythms

Introductory article

Ralph E Mistlberger, Simon Fraser University, Burnaby, British Columbia, Canada

CONTENTS

Circadian rhythms are daily (about 24 h) rhythms of behavior, physiology and biochemistry that are controlled by internal clocks. These rhythms are entrained by environmental cues, and modulate cognitive performance.

INTRODUCTION

The rotation of the earth about its axis creates daily cycles of light, temperature, humidity and other geophysical variables that have had a profound impact on the evolution of life. Most living organisms, from single-celled bacteria to fungi, plants and animals, exhibit daily rhythms in their biochemistry, physiology and behavior that mirror the dramatic environmental changes that define the solar day. Some daily rhythms may represent a direct response to environmental stimuli, but most are controlled by one or more internal, 'circadian' clocks (from the Latin *circa*, 'about' and *dies*, 'day'). The primary function of these circadian clocks is to organize and synchronize the organism's cellular and behavioral activities with the outside world. By internalizing the mechanism for rhythmicity, the organism can anticipate and prepare in advance for predictable changes in its environment. Circadian clocks have been further exploited in some species to regulate seasonal reproductive cycles by measuring day length, and to enable certain cognitive operations that require internal representations of time of day.

While daily rhythms in the activity of plants and animals have undoubtedly always been recognized, the existence of endogenous circadian clocks was a matter of contention until the latter half of the twentieth century. The most compelling evidence for these clocks is the observation that daily rhythms persist (free run) when organisms are maintained in environments lacking 24 h time cues, such as in a controlled laboratory, at the poles, or in orbit aboard the space shuttle. The average duration ('period') of one complete cycle under these conditions usually deviates slightly from 24 h, hence the designation 'circadian'. The period varies both within and between species; in temporal isolation, the circadian clock of the typical mouse completes a cycle in less than a day (about 23.5 h), whereas in most humans the clock cycles at a rate slower than one solar day (the most current estimate is about 24.1 h).

Research on circadian rhythms attempts to answer the following fundamental questions. How are circadian rhythms synchronized with the environment? What are the neural and genetic mechanisms of the circadian clock? What are the consequences of circadian rhythmicity for human performance, health and welfare?

ZEITGEBERS AND ENTRAINMENT

According to the terminology used in the study of biological clocks, a rhythm is any process that repeats itself at regular intervals. A device that produces a rhythm can be said to 'oscillate'. An oscillator can be used like an hourglass to measure the passage of time. If the oscillator is synchronized to the solar day, it can also be used as a 24 h 'clock' to recognize local time, analogous to a sundial or digital wristwatch. A circadian clock is therefore an oscillator that has a mechanism by which it is synchronized to the solar day. Without such a mechanism, the circadian clock would generate daily rhythms that would drift in and out of optimal alignment with the environment. The mechanism for synchronization is therefore vital to the adaptive function of circadian clocks.

Circadian rhythms are strongly synchronized to 24 h light–dark (LD) cycles by a process of 'entrainment', defined as phase and period control of one oscillation by another. 'Phase' refers to any discrete point in the cycle, such as the wake-up time within the daily sleep–wake cycle. Thus, when the circadian sleep–wake rhythm is entrained to a 24 h LD cycle, its period is exactly 24 h, and wake-up time occurs at about the same time within each LD cycle (e.g. when the lights come on in diurnal species, and when the lights go off in nocturnal species). Any stimulus that can entrain another periodic process is a 'zeitgeber' (from the German, 'time-giver').

The mechanism of LD entrainment has been investigated by exposing animals to brief pulses of light (from seconds to hours in duration) during prolonged recordings in constant dark. Light exposure early in the 'subjective night' (that portion of the circadian cycle when the animal acts as if it were night) resets the circadian cycle back to an earlier phase (a 'delay' phase shift; Figure 1). Light exposure late in the subjective night resets the clock forward to a later phase (an 'advance' phase shift). The brighter or longer the light pulse, the larger the shift. Light exposure during most of the 'subjective day' (when the sun would normally be up) typically has little or no effect. Entrainment is accomplished by small, light-induced phase shifts at the dawn and dusk transitions that precisely compensate for the difference between the periodicity of the LD cycle (normally exactly 24 h) and that of the circadian cycle (normally about 24 h). In humans, light during the first half of the night (before the body temperature reaches its minimum, about 2–4 h before the habitual wake-up time) induces delay shifts, while light later in the night induces advance shifts.

Although the LD cycle is the dominant zeitgeber for most species, several nonphotic zeitgebers have also been identified. Daily cycles of temperature can entrain rhythms in some species, although even in plants and poikilothermic animals this does not involve a direct effect of temperature on the rate at which the clock cycles. The biochemical machinery of the circadian clock includes a mechanism for temperature compensation, such that the circadian cycle is relatively constant across a range of tissue temperatures. If this were not the case, the 'clock' of poikilotherms would serve better as a thermometer than as a timepiece.

A number of vertebrate species can be entrained by daily schedules of food intake, and some can

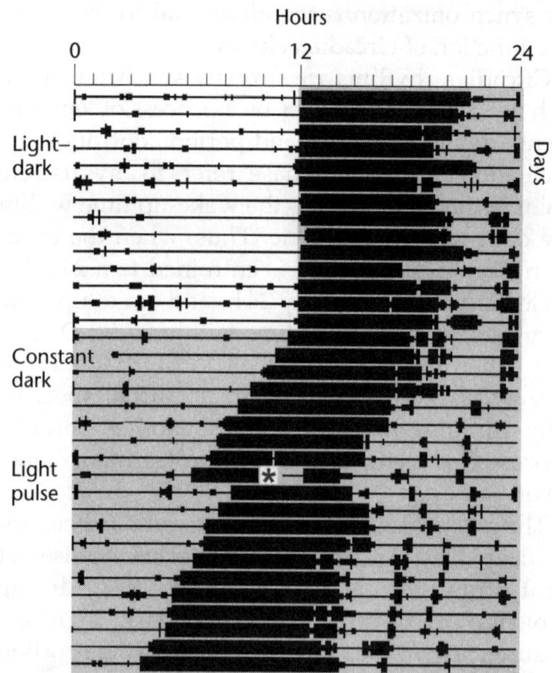

Figure 1. Wheel-running activity of a mouse in a light–dark cycle (12 h dark period indicated by shading), and in constant darkness. Each line represents one day, plotted in 10 min time bins from left to right. Vertical deflections (heavy bars) indicate time bins when wheel running occurred. In the light–dark cycle mouse activity is nocturnal and has a periodicity of exactly 24 h. In constant dark, the activity rhythm 'free runs' with a period of less than 24 h. A 30 min light pulse (asterisk) on day 10 of constant darkness induced a 'delay' phase shift of approximately 2 h.

also be entrained or phase shifted by scheduled bouts of physical exercise or nonspecific arousal. These behavioral zeitgebers appear to be most effective at times of day when the animal would normally be asleep. The circadian timekeeping system thus appears to have the flexibility to adjust daily rhythms to important nonphotic stimuli that might be of more immediate consequence to survival than is the LD cycle.

NEURAL MECHANISMS

The biological system generating circadian rhythmicity has at its core three integrated components: a self-sustaining circadian clock, an entrainment pathway by which zeitgebers can influence the clock, and one or more output pathways by which the clock confers rhythmicity to other systems (Figure 2). In mammals, the 'master' circadian clock is located in the suprachiasmatic nucleus (SCN) in the hypothalamus. The SCN receives LD information directly from the retina, via a pathway that does not participate in form vision. Damage to the SCN disrupts or eliminates circadian rhythms in all mammals so far studied. Electrical or chemical stimulation of SCN neurons can phase shift circadian rhythms. Individual SCN neurons, grown from embryonic cells in culture, oscillate with a circadian periodicity. Remarkably, transplants of embryonic SCN tissue to adult animals can rapidly restore circadian rhythmicity lost by SCN damage. These observations converge in support of a hypothesis that the SCN is the master circadian clock in mammals. This is an exceptional example of the use of complementary techniques to establish the function of a discrete brain structure.

Because of its importance for normal circadian organization of behavior and physiology, the SCN is accorded the status of 'master' oscillator, or 'pacemaker'. However, other circadian oscillators may also exist. One example is the retina, which contains a circadian oscillator that drives daily rhythms of its own local processes, such as photoreceptor disc renewal. In birds and reptiles, the pineal gland, retina and hypothalamus may all contain circadian oscillators, although the role of these as pacemakers or secondary oscillators varies across species.

The SCN receives major inputs from several other brain structures, some of which are important for nonphotic entrainment (e.g. inputs from the thalamic intergeniculate leaflet and pontine raphe nuclei). Neurons of the SCN in turn send axons to a limited number of structures, primarily within the hypothalamus, which presumably disperse circadian timing information more widely. The SCN may also send rhythmic signals by a diffusible factor, conveyed in the extracellular fluid, cerebrospinal fluid or blood.

GENETIC MECHANISMS

Circadian rhythms are genetically programmed and, although sensitive to environmental cycles, do not require these cycles to develop normally. Circadian phenotypes can be selected by breeding, and can be altered by gene mutations. Genetic differences are thought to underlie variability in human rhythms, such as the 'night owl' versus 'early bird' phenotypes.

In all species studied so far, circadian rhythms appear to be generated by an intracellular feedback loop involving a set of 'clock' genes and their protein products. Activation of these clock genes induces the synthesis of proteins, which then feed back upon and temporarily inhibit further clock

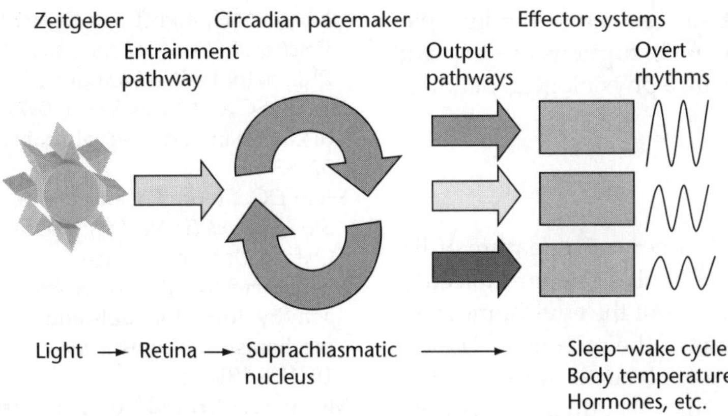

Figure 2. Conceptual model of the core elements of the system generating circadian rhythms in mammals. Overt rhythms may vary in amplitude and phase, but when entrained to light, all exhibit the same 24 h periodicity and a relative stable phase relation with other rhythms. This defines internal temporal order.

gene expression. As clock proteins gradually degrade, clock genes are released from inhibition and protein production is renewed. Positive and negative regulators of clock gene expression are articulated in such a way as to produce an approximately 24 h cycle of cellular activity. Zeitgebers shift and entrain the clock by altering clock protein levels. Putative clock genes have been identified and cloned in fungi, fruit flies and mammals, and the feedback principle and some of the specific genes appear highly conserved across a range of phylogenetic groups. A complete molecular description of the circadian 'clockworks' can be expected soon for several species.

CONSEQUENCES OF CIRCADIAN RHYTHMS FOR COGNITION

Consistent with its ubiquitous role in physiological regulation, the circadian timing system also influences performance on cognitive tasks. This is in part secondary to circadian regulation of arousal or alertness, which in LD-entrained humans is low early in the morning and peaks in the early evening, in parallel with the circadian rhythm of body temperature. Performance speed on cognitive tasks that stress simple, repetitive throughput of information, as in tests of vigilance (e.g. detection of an infrequent signal), serial search, card dealing, additions, and reaction times, follows the same circadian function, with the best scores achieved in the evening. Performance accuracy on some of these tasks varies inversely with body temperature. Tasks stressing short-term memory also generate better scores earlier in the day, before the body temperature reaches its maximum. In addition, performance on some tests of long-term memory is best when participants are tested at the same time of day at which they were trained, even when retest intervals are weeks apart. This implies that circadian time may be incorporated within the neural representations that mediate learning and recall.

The circadian clock has been adapted by many species to provide the sense of time necessary for foraging, navigating and migrating. These species can use timing information provided by the circadian clock to compensate for the movement of the sun and stars, and thereby use the position of these celestial landmarks to guide travel over great distances. Some species, including bees, fish, birds and at least some mammals, can learn and remember the time of day when food is available at specific places in the environment, without the aid of external time cues. According to one theory, circadian time may be encoded within all memories, for use in cognitive computations that guide the temporal aspects of many behaviors.

In humans, the ability to estimate the passage of time is also regulated in part by the circadian system. In temporal isolation, subjective estimates of hourly intervals are proportional to the duration of the circadian sleep–wake cycle. Estimation of intervals in the seconds to minutes range, however, are independent of circadian time.

Disruption of circadian rhythms, induced by transmeridian jet travel, shift-work rotation or a change in laboratory LD cycles, is associated with impaired performance on many cognitive tasks in humans and animals. Some of these effects can be attributed to the partial sleep deprivation that typically accompanies travel and shift rotation in humans, but impairments are also evident in laboratory studies that minimize sleep disruptions.

Cognitive deficits that emerge across the life span may also be related to the disruptions of sleep and circadian rhythms that are a physiological hallmark of old age.

CONCLUSION

Circadian rhythmicity is a pervasive feature of life on earth. At all levels of analysis, from molecules to behavior, the daily cycles of the environment are reflected in the functions of the organism. The circadian clocks that have evolved to meet the challenges of the solar day modulate cognitive processes and mediate forms of animal behavior that require precise recognition of time of day or day length.

Further Reading

Aschoff J (ed.) (1981) *Handbook of Behavioral Neurobiology*, vol. 4, Biological Rhythms. New York: Plenum Press.

Aschoff J (1989) Temporal orientation: circadian clocks in animals and humans. *Animal Behaviour* **37**: 881–896.

Brown FM and Graeber RC (1982) *Rhythmic Aspects of Behavior*. London: Lawrence Erlbaum.

Dunlap J (1999) Molecular basis for circadian clocks. *Cell* **96**: 271–290.

Harrington ME and Mistlberger RE (2000) Anatomy and physiology of the mammalian circadian system. In: Kryger MH, Roth T and Dement WC (eds) *Principles and Practise of Sleep Medicine*, 3rd edn. pp. 334–345. Philadelphia: WB Saunders.

Hinton SC and Meck WH (1997) The 'internal clocks' of circadian and interval timing. *Endeavour* **21**: 82–87.

Klein DC, Moore RY and Reppert SM (1991) *Suprachiasmatic Nucleus: The Mind's Clock*. New York: Oxford University Press.

Mistlberger RE (1994) Circadian food anticipatory activity: formal models and physiological mechanisms. *Neuroscience and Biobehavioral Reviews* **18**: 171–195.

Mistlberger RE and Rusak B (2000) Circadian rhythms in mammals: formal properties and environmental influences. In: Kryger MH, Roth T and Dement WC (eds) *Principles and Practise of Sleep Medicine*, 3rd edn. pp. 321–333. Philadelphia: WB Saunders.

Monk T (1994) Circadian rhythms in subjective activation, mood and performance efficiency. In: Kryger MH, Roth T and Dement WC (eds) *Principles and Practise of Sleep Medicine*, 2nd edn. pp. 321–333. Philadelphia: WB Saunders.

Refinetti R (2000) *Circadian Physiology*. Boca Raton: CRC Press.

Takahashi JS, Turek FW and Moore RY (eds) (2001) *Handbook of Behavioral Neurobiology*, vol.12, *Circadian clocks*. New York: Kluwer Academic.

Winfree AT (1987) *The Timing of Biological Clocks*. New York: Scientific American Books.

Classifier Systems

Intermediate article

Lashon B Booker, MITRE Corporation, McLean, Virginia, USA

CONTENTS

A classifier system is a parallel, message-passing, rule-based system designed to learn and use internal models of complex environments. Classifier systems are useful to cognitive modelers because they build representations that have both connectionist and symbolic qualities.

OVERVIEW

Real-world environments seldom provide salient, timely, complete, and unambiguous information. Therefore the correspondence between the unfolding complexity of the world and the representations used by a cognitive system cannot be taken for granted. Environmental properties, particularly complexity and uncertainty, are an important constraint on cognitive behavior. A complex environment may overwhelm a system with large amounts of information, not all of which is directly relevant to the system's appointed task. In order to function at all, a cognitive system must be discriminating and selective about what information it stores and uses. An uncertain environment is one in which it is unlikely that input configurations can be discerned or predictions made with accuracy. Under such circumstances, the only viable information-processing strategies are those capable of making pragmatic 'good guesses' about the true state of the world. In order to cope with both complexity and uncertainty, one must resist the temptation to try to know the environment in explicit detail. A system must focus on learning the basic concepts and regularities in the environment, their relationships, and their relevance to system goals. It is precisely this kind of economical, orderly arrangement of knowledge that constitutes an internal model of the environment. The notion that organisms can benefit from the use of internal models has long been recognized by psychologists as a powerful idea (Craik, 1943; Tolman, 1948).

Inductive processes generate and revise the constituent elements of internal models. Learning that leads to the development of an internal model can therefore be viewed as a pragmatic cognitive strategy for successful functioning in the real world. For this reason, induction is a central topic in cognitive science. In a comprehensive discussion of induction from this perspective, Holland *et al.* (1986), viewing induction broadly as any inferential process that expands knowledge in the face of uncertainty, describe a rule-based framework that seeks to answer the question: 'How can a cognitive system process environmental input and stored knowledge so as to benefit from experience?' Classifier systems are general-purpose rule-based systems designed to learn and use internal models in a manner consistent with this framework.

CLASSIFIER SYSTEMS AND COGNITIVE MODELING

Several properties of this framework for induction, and classifier systems in particular, are well suited to support cognitive modeling. In order to see why, it is helpful to begin with a broad characterization of internal representations and the mental operations that use them. The discussion by William James (1892) remains one of the most useful such characterizations available. James distinguishes two important aspects of cognitive representations and processes that are relevant here. Firstly, internal representations are dynamic, composite descriptions of the current situation and the expectations associated with it. The constituent elements of representations are associated with the many different aspects of the current stimuli and the overall context. The simultaneous, context-dependent activation of some combination of these elements constitutes the system's interpretation of the current situation. Secondly, the various elements that are candidates to participate in a representation interact with each other to determine which elements become active. The nature of these interactions is determined by the network of associations connecting the elements together, and by mechanisms that direct the flow of activity from one element to another.

This notion of distributed representations emerging from dynamic patterns of interactions among large numbers of primitive elements is common to many connectionist approaches to cognitive science. The framework for induction proposed by Holland *et al.* takes a similar notion of representation, but uses simple condition–action rules as the basic elements. Rules interact by passing messages, and many rules can be active simultaneously. Inductive mechanisms organize rules into clusters that provide a multifaceted representation of the current situation and the expectations that flow from it. The structure of a concept is modeled by the organization, variability, and strength of the rules in a cluster. Simple rules thereby become building blocks for representing complex concepts, constraints, and problem-solving behaviors. Knowledge can be represented at many levels of organization, using rules and rule clusters as building blocks of different sizes and complexities. Because constituent elements compete to become active, aspects of a representation are selected only when they are relevant in a given context. Moreover, since rules are activated in parallel, new combinations of existing rules and rule clusters can be used to dynamically represent novel situations. (*See* **Connectionism**; **Distributed Representations**)

Because the framework for induction uses symbolic rules as representational primitives, there are important differences between this approach and the typical connectionist approach to cognitive modeling. The most obvious difference is that it is easier to construct representations having symbolic expressiveness. Another significant difference is in the way inductions are achieved. Modification of connection strengths is the only inductive mechanism available in most connectionist systems. Moreover, the procedures for updating strength are part of the initial system design, and cannot be changed, except perhaps by adjusting a few parameters. The framework for induction, on the other hand, permits a range of inductive mechanisms, from strength adjustments to analogies. In principle, many of these mechanisms can be controlled by or easily expressed in terms of condition–action rules. These rules can be evaluated, modified and used to build higher-level concepts. Overall, this paradigm offers a set of potential cognitive modeling capabilities that occupy an important middle ground between subsymbolic connectionist systems and the traditional symbolic paradigms of artificial intelligence.

Classifier systems were originally proposed as a rule-based implementation of this approach to cognitive modeling. The earliest proposals (Holland, 1976) and experiments (Holland and Reitman, 1978) described simple cognitive systems designed to adapt to their environment under sensory guidance. Most subsequent work has focused less on developing cognitive architectures and more on the algorithms and computational techniques needed to implement architectures of this kind. The remainder of this article discusses these algorithms and their properties from the standpoint of cognitive modeling.

ALGORITHMIC DESCRIPTION OF CLASSIFIER SYSTEMS

A classifier system is a parallel, message-passing, rule-based system designed to permit nontrivial modifications and reorganizations of its knowledge as it performs a task. The typical operating principles for a classifier system can be briefly summarized as follows. In the simplest version, all information processing is accomplished using messages that are encoded as binary strings of fixed length k. The set of messages to be processed at any given moment is stored on a 'message list'. Classifier systems process these messages using a population of rules, called 'classifiers'. Every classifier is a fixed-length string having the form

$$s_1, s_2, \ldots, s_n \Rightarrow m \quad (1)$$

Each s_i is an input condition, represented as a string of length k in the ternary alphabet {0, 1, #}. The action part m is also a string of length k in this ternary alphabet. A condition is satisfied by a message whenever there is a 'match' – that is, whenever the 0 and 1 symbols in the condition are identical to the bit values at corresponding positions of the message. The # symbol is a 'don't care' place holder that matches any message bit value at the designated position. A classifier is eligible to become active only if each of its conditions matches a message on the message list. Once a classifier is activated, the action part of the rule specifies a message m which the rule will 'post' on the message list on the next timestep. When the # symbol appears in the action part it is interpreted as a 'pass through' place holder. This means that the designated position in the posted message is assigned the value of the corresponding bit from a message that matched one of the classifier's conditions.

In more detail, the typical classifier system can be described in terms of three interacting subsystems: a 'performance' system, a 'credit assignment' system, and a 'rule discovery' system (see Figure 1). The performance system is responsible for interacting with the external environment and generating behavior. It is assumed that an input interface is available to translate the state of the environment into messages (e.g. by using a set of detectors) and that an output interface interprets relevant messages as action specifications (e.g. by using a set of effectors). It is also assumed that the environment occasionally gives the system explicit performance feedback in the form of pay-off or reinforcement. Given these assumptions, the basic execution cycle for the performance system proceeds as follows:

1. Place messages from the input interface on the current message list.
2. Compare all messages to all conditions and conduct a competition among matching classifiers to determine which ones will become active.
3. For each active classifier, generate one message for the new message list.
4. Replace the current message list with the new message list.
5. Process the current message list through the output interface to produce system output.
6. Return to step 1.

Two important aspects of the performance system are worth noting here. Firstly, the system is highly parallel. Many classifiers can be active simultaneously. Since different classifiers match different subsets of messages, the classifiers can be thought of as building blocks which can be activated in many combinations to represent a variety of situations. Secondly, matching classifiers must compete to become active. This competition makes it possible to avoid imposing any consistency requirements on posted messages, and allows the system to insert new rules smoothly without disrupting existing capabilities. Classifiers are treated as tentative hypotheses about the effects of posting a message given the current conditions. Each of these 'hypotheses' is repeatedly tested according to the system's experience with the environment.

The competition mechanism assumes that a reliable assessment of the plausibility of each hypothesis is available. The credit assignment system is responsible for computing those assessments. It is

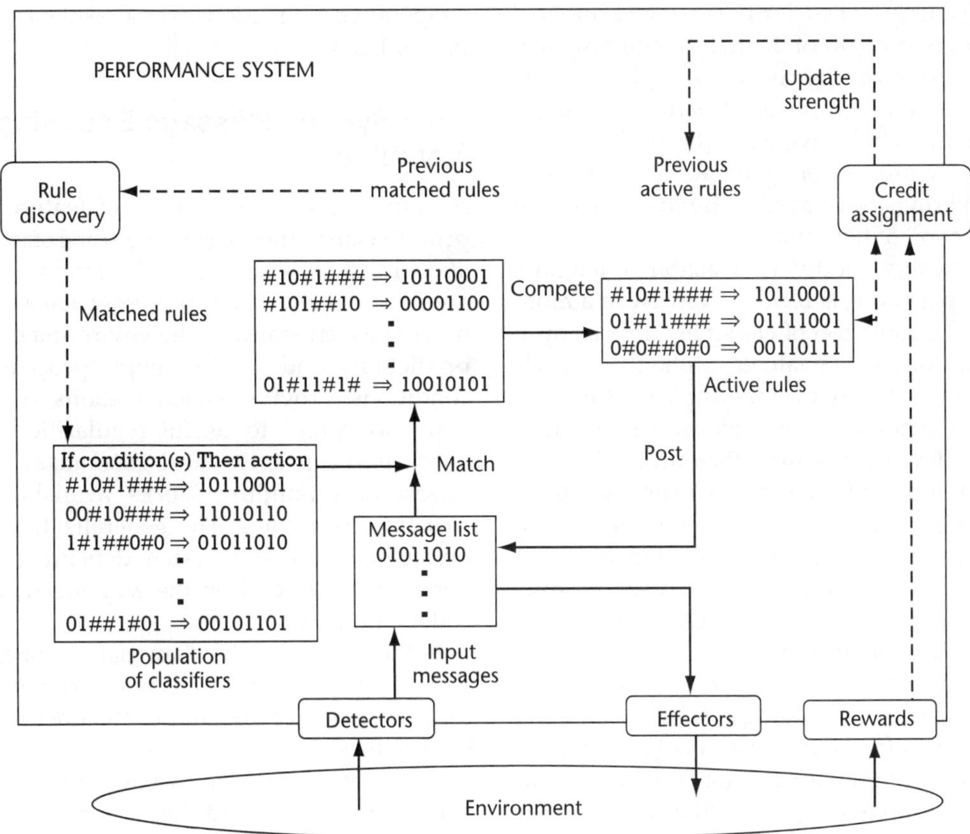

Figure 1. The organization of a typical classifier system. See the text for a detailed explanation.

difficult to assign credit for successful problem-solving behavior in a system that uses many rules over several timesteps to generate behavior. The difficulty is even more pronounced when overt feedback from the environment is intermittent or rare. The only realistic strategy for classifier systems is to evaluate performance for behavioral sequences in local terms for each classifier involved. The most straightforward approach is to rely on a simple reinforcement principle: strengthen a hypothesis whenever the associated message leads to a favorable or rewarding situation. A considerable amount of theoretical work in psychology, adaptive control, and machine learning has used this principle as a starting point for understanding how to solve difficult credit-assignment problems. One computational approach designed for classifier systems is the 'bucket brigade' algorithm (Holland *et al.*, 1986), one of many algorithms from the reinforcement learning literature that are suitable for this purpose. (*See* **Reinforcement Learning: A Computational Perspective**)

Credit assignment is only one of several learning mechanisms that are needed for a successful inductive process. It is also necessary to have a mechanism that can generate plausible new hypotheses. For rule-based systems this means generating new rules as candidates to replace existing rules that have not been particularly useful. In the classifier system framework, the rule discovery system is responsible for generating plausible new classifiers. 'Plausibility' in this context is tied to the notion of 'building blocks' mentioned in the discussion of parallelism above. The simple syntax used in classifier conditions makes it straightforward to view classifiers as strings composed of readily identifiable parts. One easily-specified and useful set of parts can be defined as follows. In each classifier condition and action, there are many 'components' which can be identified by specific combinations of symbols at designated positions. For example, the condition 0#011### ... 1# begins with a 0 in position 1 and a # in position 2. This combination defines a template, or 'substring schema', which can be denoted by the string 0#****** ... ** (where the symbol * indicates positions not involved in the definition). Any condition beginning with this two-symbol prefix contains the corresponding substring schema as one of its parts. The utility of such a part can be estimated by the average strength of the rules containing that part. A new rule can be considered 'plausible' to the extent that it is composed of parts, or building blocks, having above-average utility. Thus, the rule-discovery process can generate plausible new rules by favoring above-average building blocks in the construction of new rules.

Note that every condition or action of length k uses 2^k distinct building blocks. Even given a moderately sized population of classifiers, it is computationally infeasible to evaluate explicitly and use information about all of the building blocks in all of the classifier conditions and actions. It is possible, however, to devise procedures that implicitly make use of this information. The mechanism used most often for this purpose in classifier systems is a 'genetic algorithm'. A genetic algorithm is a general-purpose search procedure that uses sample-based induction (Holland *et al.*, 1986) to conduct the search. The algorithm repeatedly selects a sample of rules (the 'parents') from the population and recombines their building blocks to construct new rules (the 'offspring'). The new elements are constructed using genetic operators such as recombination and mutation. The selection criterion for parents is biased to favor high-strength classifiers, and new classifiers replace low-strength classifiers in the current population. (*See* **Evolutionary Algorithms**)

REPRESENTATIONS IN CLASSIFIER SYSTEMS

The starting point for representing knowledge in the classifier system framework is the use of simple message-passing rules as primitive elements. Each rule condition specifies a basic equivalence class or category of messages that match the condition. Because of parallelism and recombination, higher-level representations built using these primitives have an inherently composite nature. Rather than construct a syntactically complex representation of a symbolic concept that might be difficult to use or modify, a classifier system is designed to use clusters of rules as representations. The implementation of this approach to representing knowledge in classifier systems relies on principles and mechanisms operating at two levels of organization: rules and rule clusters. Each rule offers a range of representational power that is determined by the way messages are matched, encoded and processed. The representational power of rule clusters is determined by the way rules are organized in relation to each other and by their competitive and cooperative interactions. These issues are discussed briefly below.

Rule Syntax, Message Encoding and Matching

An input message is a string of feature values or primitive attributes describing some state configuration in the environment. Condition–action rules are a convenient way to represent states and transitions between states in the environment, generate predictions, and specify simple procedures. Rule conditions provide generalizations of messages that correspond to useful regularities and attribute-based concepts. These generalizations are the lowest-level building blocks available for constructing new rules. The generalizations that are possible in a classifier system depend on the classifier rule syntax and on the way messages are encoded and matched.

In the basic classifier 'language', conditions are fixed-length ternary strings that correspond to generalizations given by simple conjunctions of attribute values. Because the syntax is so simple, no extra mechanisms are needed to specialize or generalize these conditions. Specialization only requires changing a # to a 0 or 1, while generalization is accomplished by changing a 0 or 1 to a #. Changes of this kind are routinely generated by the genetic algorithm in the rule discovery system. Note, however, that simple classifier conditions of this kind cannot be used to express arbitrary, general relationships among structured attributes. There are two ways to increase the expressive power of these classifier conditions.

First, by using multiple conditions and multiple rules, it is possible to represent concepts involving the logical conjunction and disjunction of simple clauses. A classifier with multiple conditions can be activated only when each condition matches a message on the message list. The posting of that classifier's message therefore indicates an input configuration in which the conjunction of those conditions is true. When two or more classifiers have different conditions but identical actions, the posting of that particular message indicates an input configuration in which at least one of the conditions is true (disjunction) Additional expressive power is obtained by allowing some conditions to be satisfied when no matching message is on the list. This provides a way to represent logical negation, making it possible to represent arbitrary Boolean expressions.

The second way to increase the expressive power of individual classifiers is to use more powerful encoding schemes for messages. In the simplest encoding, each message bit corresponds to a single attribute–value pair or predicate. This encoding is adequate for categories defined by a set of critical features whose presence or absence is mandatory for category membership. More sophisticated encodings support generalizations about the range of ordinal values and disjunctions of nominal values. These encodings provide capabilities that compare favorably to the primitive language constructs used in many symbolic learning paradigms. Moreover, the basic building blocks supporting these generalizations can be easily recombined and otherwise manipulated by the genetic algorithm and other local, syntactic rule modification algorithms.

Even with more expressive encodings, however, there are limits to what can be represented with conditions using the standard syntax. For example, although continuous input values can be represented as bit strings, classifier conditions can only represent a subset of the generalizations that may be useful or required. Moreover, the standard matching algorithm does not allow for generalizations that require variable binding or parametrization. This means that it is not possible to express simple relations between subfields in a single condition or across multiple conditions. In order to overcome these limitations, some classifier system implementations have utilized more sophisticated symbolic expressions in rule conditions. This can be problematic, though, because it is not always clear what the buildings blocks are in these complex representations. One of the important open research questions is how to increase the expressive power of classifier conditions without sacrificing the efficiency and well-chosen building blocks characteristic of the basic classifier language syntax.

Multiple Rules and Default Hierarchies

As noted above, the expressive power of rule conditions can be enhanced by using a set of rules to represent disjunctive clauses in the description of an attribute-based concept. This is a form of implicit cooperation among rules, whereby different subsets of rules are responsible for different aspects of the overall representation. Many forms of cooperative interactions among rules are possible.

For example, another useful form of implicit cooperative interaction is tied to the competition mechanism. The competition for the right to post messages is usually implemented with a bias towards selecting the 'specific' rules whose conditions contain the most detail about the current situation. This makes it possible for rules to become organized into 'default hierarchies'. The simplest example of a rule-based default hierarchy consists of two rules. The first ('default') rule has a relatively general condition and provides an action that is sometimes incorrect. The second (exception) rule is satisfied only by a subset of the messages satisfying the default rule, and its action generally corrects errors committed by the default. That is, the exception rule uses additional information (its more specific condition) to distinguish situations that lead the more general default rule astray. When the exception wins the competition it prevents the default from making a mistake and losing strength under the credit assignment mechanism. There is consequently a kind of symbiotic relationship between the two rules. Note that the default may have other exceptions, and each exception may, in turn, have exceptions, resulting in a hierarchy of interactions. Default hierarchies are an important feature of classifier systems from the standpoint of cognitive modeling. Classifier systems with default hierarchies have been used to model a wide variety of conditioning phenomena (Holyoak *et al.*, 1990) and human performance on simple discrimination tasks (Riolo, 1991).

Another useful implicit interaction among rules is based on patterns of conditions and actions. If the action posted by rule R_1 matches the condition of another rule R_2, then the activation of R_1 will tend to result in the activation of R_2. In this case the two rules are said to be 'coupled'. Since rule coupling implies sequential activation, classifier systems can use representations in which the constituent rules are activated over more than one match–compete–post execution cycle. This is the first step needed to go beyond simple reactive behavior to a mode in which internal information processing can be used to generate responses. Only a few classifier systems have exploited implicit coupling for cognitive modeling purposes. Booker (1988) describes a system that learns a simple internal model in which the goal-relevance of environmental states can be retrieved from memory using internal messages.

Complex Knowledge Structures

The coupling mechanism becomes even more useful when the rule interactions are organized explicitly. Particular bits incorporated into a rule's condition – such as a suffix or prefix – can be used as a kind of identifier or address, called a 'tag'. Messages can be directed to a rule explicitly by incorporating the appropriate tag into those messages. If there are several rules sensitive to the same tag, then they will be activated together as a cluster. Consequently, any rule that contains that tag in its action part will be explicitly coupled to the entire cluster. Tags are the building blocks for representing sequences and associations. Because tags are simple parts of messages and rules, associations among rules can be constructed and modified with the standard repertoire of rule discovery mechanisms.

Moreover, a tag can be given semantics as a label indicating the origin (e.g. the input interface) or destination (e.g. the output interface) of a message. By identifying tags with semantic content, we can impose a hierarchical organization on classifier representations. Tags become identifiers for relationships within and between levels in a hierarchy, and for relationships between hierarchies. In principle, the use of semantically useful tags to couple rules makes classifier systems a powerful framework for representing complex knowledge structures. For instance, it has been shown (Forrest, 1991) that the knowledge contained in a standard semantic network description (e.g. KL-ONE or NETL) can be mapped into a set of classifiers that support the same information-retrieval operations. It remains to be seen whether these kinds of complex knowledge structures can be learned by a classifier system as a result of its experiences in an environment. However, experiments have demonstrated that classifier systems can learn simple associative knowledge structures (Riolo, 1990; Stolzmann and Butz, 2000). (*See* **Semantic Networks**)

PROBLEM SOLVING USING CLASSIFIER SYSTEMS

Problem solving in the classifier systems framework differs from more conventional approaches to problem solving in that a strong emphasis is placed on flexibly modeling the problem-solving context. The rationale for this emphasis on flexible problem representations is that cognitive systems are often confronted with problems that are poorly defined – that is, various aspects of the initial problem specification may be unknown or partially known. Instead of relying on stand-alone procedures for reformulating such problems, classifier systems can use building blocks, together with the simultaneous activation and combination of multiple representations, to recategorize the problem components. In this way, the system conducts a search for solutions to a problem both in the problem space and in the space of alternative problem representations.

The starting point for this approach is a good model that allows for prediction-based evaluation of the knowledge base, and the assignment of credit to the model's building blocks. This makes it possible to modify, replace, or add to existing rules via inductive mechanisms such as the recombination of highly rated building blocks. After repeated experiences solving instances of a problem, the inductive mechanisms generate rules which are specialized or generalized as needed to adequately represent the typical elements of the problem space. Maintaining a varied repertoire of useful specific and general rules is essential to achieving problem-solving flexibility. When a novel situation is encountered, there is some chance that it will be matched by general rules that provide some useful, though perhaps imperfect, guidance about how to proceed. The simultaneous activation of specific and general rules defines an implicit default hierarchy in which default expectations can be overridden whenever a specific exception occurs. Novel situations can also be handled by activating associations that suggest recategorizations of structured concepts and relations. By coordinating multiple sources of knowledge, hypotheses and constraints in this way, problem solving can proceed opportunistically, guided by the integration of converging evidence and building on weak or partial results to arrive at confident conclusions.

In simple cases, problem solving can be achieved by learning a direct mapping between inputs and outputs. A large body of research within the reinforcement learning paradigm has been devoted to solving Markovian decision tasks using this strategy, and classifier systems can achieve comparable results. However, classifier systems used in this way are simple reactive systems that do not require any cognitive processing beyond basic categorization. One particularly sophisticated problem-solving mechanism that has been implemented in a classifier system is based on Baum's (Baum and Durdanovic, 2001) approach, which views rule clusters as 'post-production systems'. The system learns algorithms that solve instances of Rubik's cube and arbitrary block-stacking problems. (*See* **Markov Decision Processes, Learning of**)

SUMMARY

Classifier systems can be thought of as connectionist systems that use rules as the basic epistemic unit. Using simple rules in this way makes it possible to enjoy the advantages of distributed representations, and at the same time to represent nontrivial symbolic concepts and employ flexible problem-solving mechanisms. Consequently, classifier systems have the potential to occupy an important middle ground between the symbolic and connectionist paradigms. In order to realize this potential, more work must be done to understand how classifier systems might dynamically construct and modify the kinds of multifaceted representations described here. Research on small systems and simple problems has been a promising first step towards that goal.

References

Baum EB and Durdanovic I (2001) An artificial economy of post production systems. In: Lanzi PL, Stolzmann W and Wilson SW (eds) *Advances in Learning Classifier Systems*, pp. 3–20. Berlin, Germany: Springer.

Booker LB (1988) Classifier systems that learn internal world models. *Machine Learning* **3**: 161–192.

Craik KJW (1943) *The Nature of Explanation*. Cambridge, UK: Cambridge University Press.

Forrest S (1991) *Parallelism and Programming in Classifier Systems*. London, UK: Pitman.

Holland JH (1976) Adaptation. In: Rosen R and Snell FM (eds) *Progress in Theoretical Biology*, vol. IV, pp. 263–293. New York, NY: Academic Press.

Holland JH, Holyoak KJ, Nisbett RE and Thagard PR (1986) *Induction: Processes of Inference, Learning, and Discovery*. Cambridge, MA: MIT Press.

Holland JH and Reitman JS (1978) Cognitive systems based on adaptive algorithms. In: Waterman DA and Hayes-Roth F (eds) *Pattern-Directed Inference Systems*, pp. 313–329. New York, NY: Academic Press. [Reprinted in: Fogel B (ed.) (1998) *Evolutionary Computation: The Fossil Record*. IEEE Press.]

Holyoak KJ, Koh K and Nisbett RE (1990) A theory of conditioning: inductive learning within rule-based default hierarchies. *Psychological Review* **96**: 315–340.

James W (1892) *Psychology: The Briefer Course*. New York, NY: Henry Holt. [Reprinted edition edited by G. Allport, published by Harper & Row, New York, 1961.]

Riolo RL (1990) Lookahead planning and latent learning in a classifier system. In: Meyer JA and Wilson SW (eds) *From Animals to Animats 1. Proceedings of the First International Conference on Simulation of Adaptive Behavior*, pp. 316–326. Cambridge, MA: MIT Press/ Bradford Books.

Riolo RL (1991) Modelling simple human category learning with a classifier system. In: Booker LB and Belew RK (eds) *Proceedings of the 4th International Conference on Genetic Algorithms (ICGA91)*, pp. 324–333. San Mateo, CA: Morgan Kaufmann.

Stolzmann W and Butz M (2000) Latent learning and action-planning in robots with anticipatory classifier systems. In: Lanzi PL, Stolzmann W and Wilson SW (eds) *Learning Classifier Systems: From Foundations to Applications*, pp. 301–317. Berlin, Germany: Springer.

Tolman EC (1948) Cognitive maps in rats and men. *Psychological Review* **55**: 189–203.

Further Reading

Booker LB, Goldberg DE and Holland JH (1989) Classifier systems and genetic algorithms. *Artificial Intelligence* **40**: 235–282.

Booker LB, Riolo RL and Holland JH (1994) Learning and representation in classifier systems. In: Honavar V and Uhr L (eds) *Artificial Intelligence and Neural Networks*, pp. 581–613. San Diego, CA: Academic Press.

Donnart J-Y and Meyer JA (1996) Learning reactive and planning rules in a motivationally autonomous animat. *IEEE Transactions on Systems, Man and Cybernetics – Part B: Cybernetics* **26**(3): 381–395.

Holland JH (1986) A mathematical framework for studying learning in a classifier system. In: Farmer D, Lapedes A, Packard N and Wendroff B (eds) *Evolution, Games and Learning: Models for Adaptation in Machines and Nature*, pp. 307–317. Amsterdam: North-Holland.

Holland JH (1990) Concerning the emergence of tag-mediated lookahead in classifier systems. *Physica D* **42**: 188–201. [Republished in *Emergent Computation*, edited by S. Forrest, MIT Press/Bradford Books.]

Lanzi PL, Stolzmann W and Wilson SW (eds) (2000) *Learning Classifier Systems: From Foundations to Applications*. Berlin: Springer.

Lanzi PL, Stolzmann W and Wilson SW (eds) (2001) *Advances in Learning Classifier Systems*. Berlin, Germany: Springer.

Stolzmann W, Butz M, Hoffmann J and Goldberg DE (2000) First cognitive capabilities in the anticipatory classifier system. In: Meyer JA *et al.* (eds) *From Animals to Animats 6: Proceedings of the Sixth International Conference on Simulation of Adaptive Behavior*, pp. 287–296. Cambridge, MA: MIT Press.

Wilson SW (1995) Classifier fitness based on accuracy. *Evolutionary Computation* **3**(2): 149–175.

Wilson SW and Goldberg DE (1989) A critical review of classifier systems. In: Schaffer JD (ed.) *Proceedings of the 3rd International Conference on Genetic Algorithms (ICGA89)*, pp. 244–255. San Mateo, CA: Morgan Kaufmann.

Cognitive Assessment

Intermediate article

Jonna M Kulikowich, University of Connecticut, Storrs, Connecticut, USA
Patricia A Alexander, University of Maryland, College Park, Maryland, USA

CONTENTS

Cognitive assessment encompasses a wide variety of tests, tasks, and methods used to monitor and evaluate knowledge acquisition, strategic processing, and development of complex thinking.

INTRODUCTION

Measuring how we think and learn is not easy. Unlike behavior that can be observed directly, human cognitive processes are internal. Despite the difficulty in assessing cognitive processes, psychologists throughout history have designed many creative ways to represent them. In this article, we will introduce some of the assessments used by cognitive psychologists. We will consider assessment of domain-specific knowledge, strategic processing, and conceptual change. We will also discuss how researchers evaluate the effects of instruction, that is, academic achievement.

COGNITIVE PROCESSES

Anyone with an interest in science may wonder how experts like Albert Einstein or Marie Curie were able to make so many important contributions in physics and chemistry. Can we learn to think and solve problems like them?

It takes many years of hard work to become an expert. Experts possess vast amounts of knowledge related to their field of study, and this knowledge base has been acquired by working on complex problems, communicating with other experts, and making many mistakes. Fields of study, like physics and chemistry, are called 'domains'. Most researchers agree that domain-specific knowledge is of two primary kinds: 'declarative' and 'procedural' knowledge (Anderson, 1983). Declarative knowledge includes our knowledge of concepts, facts, and details. Experts possess an abundance of declarative knowledge, and this knowledge is structured in memory in representations, called 'schemata' in such a way that experts can quickly access the important principles of their domain. Procedural knowledge is knowing how to apply declarative knowledge. When we adjust the knob on a microscope to focus our view on a certain part of a cell, we use our procedural knowledge.

As well as possessing domain-specific knowledge, experts are also very strategic. 'Strategic learning' is effortful processing that helps experts link schemata and monitor and evaluate their progress while they are solving problems. Not all information can be immediately recalled or accessed. So, experts need a system of strategies that helps them put their existing knowledge to use. Some strategies are considered to be domain-specific because they are commonly used in one field of study while they may not be used to such extent in another field of study (mnemonics usually fall into this category). Other strategies can be used in any domain (for example, strategies for summarizing the main ideas in a passage of text).

We discuss how researchers have measured several important variables in the field of cognitive psychology. We begin with domain-specific knowledge and strategic learning and describe how these have been assessed. Then, we discuss how the study of expertise has influenced assessment of school students' performance. It is important to understand how students' cognitive processes develop over time. The topics of dynamic assessment and conceptual change pertain to developmental changes in students. Finally, we present some new developments in the analysis of student achievement. 'Achievement' relates specifically to what students know, and how they are able to solve problems, as a result of instruction. Performance-based assessment and portfolio assessment are two testing techniques useful in the measurement of complex student achievement.

DOMAIN-SPECIFIC KNOWLEDGE

In studying how experts organize their knowledge and use it while solving problems, one of the simplest ways to assess domain-specific knowledge is by asking. Interview techniques are commonly used when studying experts. Psychologists develop lists of interview questions to ask experts, regarding what kinds of problems they like to solve, how long it takes them to solve problems, the challenges they face, and their evolving interests.

Two different types of interviews are popular: 'open-ended' and 'structured' interviews. In open-ended interviews, respondents are asked a general question that allows for a large variety of responses. Thus, we might ask Marie Curie: 'When did you realize that you were becoming an expert in chemistry?' Cognitive psychologists would hope to hear about how Curie realized that she knew a lot of information about chemistry that was more extensive than many of her peers. She might well mention that she could solve difficult problems quickly. She might mention that she recognized that her research was actually redefining the content of the field. In effect, Curie's discoveries, such as new chemical elements, became the declarative knowledge that others would have to know.

In structured interviews, cognitive psychologists seek very specific answers to detailed questions. These can be used in conjunction with observational records of the experts while they work. Thus, someone might observe an expert in chemistry like Curie while she is performing an experiment. After completion, a structured interview might be conducted, based on specific procedures that the expert used to solve a problem. Thus, we might ask her why she used one piece of equipment rather than another. Further, we might ask why and how she adjusted the Bunsen burner while heating a particular substance. We would compose these questions in an effort to assess the declarative and procedural knowledge of the expert based on our observations of her performance.

Expert–Novice Differences

Interview techniques are very useful when one has already identified an expert. But cognitive psychologists are also interested in assessing the differences between those who are experts and those who are not. Such work is usually referred to as the study of 'expert–novice differences' (Ericsson and Smith, 1991). Several methods of assessment have been very useful in studying these differences.

Concept maps

Concept maps allow cognitive psychologists to study how individuals organize their conceptual (primarily declarative) knowledge. Given a list of words that represent concepts in a domain, experts and novices are asked to create diagrams that show how these concepts are interrelated and how they lead to understanding of other concepts. Directional links show that concepts lead to one another, while nondirectional links simply show that concepts are related.

Expert statisticians, for example, know a lot about the normal distribution. Concepts related to the study of the normal distribution include central tendency, variability, and *z*-scores.

Figure 1 shows two concept maps. Figure 1(a) represents an expert's schematic understanding of the relations among a given list of statistical concepts, while Figure 1(b) represents a novice's understanding. The expert was also able to show how concepts lead to other concepts (for example, how means and standard deviations lead to computation of *z*-scores). The novice was not able to make these links. In general, cognitive maps demonstrate that experts organize their knowledge around important principles while novices rely on a rather fragmented organization of knowledge.

Think-aloud procedures

Think-aloud procedures can help in the assessment of procedural knowledge. Cognitive psychologists seeking to assess the processing that differentiates expert and novice performance might ask them to 'think aloud' while they are generating their solutions. Unlike interviews, which take place before or after the task, the think-aloud procedure takes place during the task. It can inform psychologists about such things as the information to which individuals attend as they solve problems (Ericsson and Simon, 1993). Further, it can provide hints as to the sequence of steps that problem solvers follow.

For example, if asked to think aloud while solving three essentially similar statistical problems all to do with examination results but expressed in different terms, experts in statistics are apt to study the problem structure in each task and compare those structures across tasks. Thus, an expert might say: 'All problems involve univariate distributions and require estimates of central tendency'. Novices, who have a more fragmented knowledge

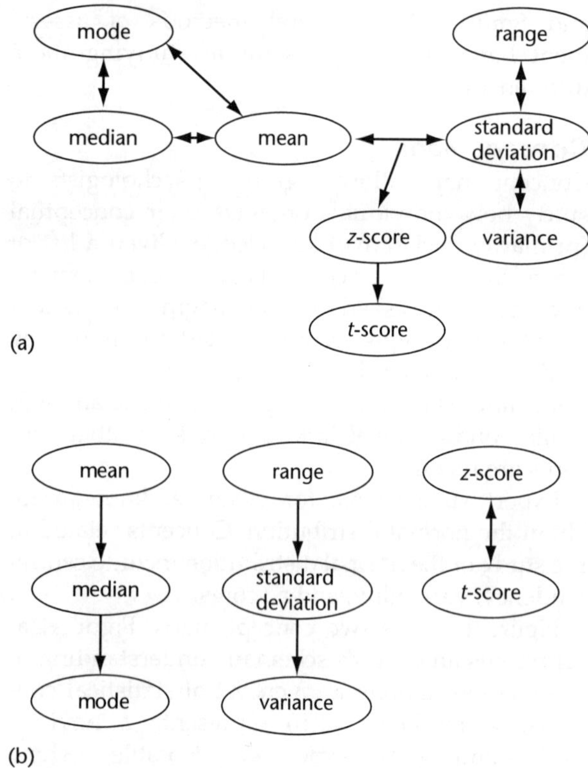

Figure 1. Concept maps in the domain of statistics. The same list of eight concepts was given to an expert and to a novice. Directional links show that concepts lead to one another, while nondirectional links simply show that concepts are related. (a) The expert's concept map. (b) The novice's concept map.

base, do not tend to mention the principles that are common to all the problems as they think aloud. Instead, they tend to focus on surface features which are usually specific to a given problem. If we were to ask novices to think aloud while solving three statistics problems, they might mention that they all relate to experiences in school, particularly examinations; but they would probably not say that all involve the principle of central tendency.

Rating tasks

Analyzing interview and think-aloud protocols can take a long time. Studying them is like grading very long essay tests. A more simple way to assess cognitive processes is to ask individuals to rate the similarity between problems based on a specific characteristic. For example, experts and novices might be asked to rate the similarity between pairs of statistics problems as regards the principle of central tendency. Such ratings reveal that the experts see stronger association between pairs of problems that are essentially similar, compared with novices.

STRATEGIC KNOWLEDGE

Linking Domain-Specific Knowledge to Strategic Knowledge

Analogical reasoning relates the familiar to the unfamiliar. It is therefore relevant to the relationship between domain-specific knowledge and basic forms of strategic knowledge. Robert Sternberg (1977) has researched analogical reasoning and proposed that four component processes are used to solve problems: 'encoding', 'inferring', 'mapping', and 'applying'.

For example, consider a verbal analogy problem, such as: '*boy* is to *man* as *girl* is to…?'. First, one has to define what the three terms *boy*, *man* and *girl* (in general, *A*, *B* and *C*) mean. This is the encoding step. Then, one has to infer the relationship between the *A* and *B* terms. The relationship between *boy* and *man* is that a boy is a male child while a man is a male adult. Then, one maps the relationship between the *A* and *C* terms. The common principle between boys and girls is that they are both children. The final component of analogical reasoning is to apply the rules learned in inferring and mapping to generate a solution, or construct a *D* term.

Essentially, the experts' ability to say that one problem is 'like' another problem because of the underlying theme of central tendency is a form of analogical reasoning.

Analogies and General Strategic Processing

Analogies are also relevant to general strategic processing, particularly when they are constructed in nonverbal, figural form. The same component processes (encoding, inferring, mapping and applying) seem to operate for these nonverbal analogies as for verbal analogies. Figure 2 is an example characteristic of those included on the 'Ravens Standard Progressive Matrices' tests (Raven, 1958). Items like this contain no content information, so they make for good measures of general strategic processing. Respondents have to track how the figures change in the matrix. Sometimes the strategic rules simply involve adding or deleting elements; sometimes the rules are very complex, involving rotations and transformations as well. Tracking complex rules requires considerable effort and concentration.

Psychologists have recently discovered that nonverbal analogy tests like the Ravens Standard Progressive Matrices assess not only general strategic

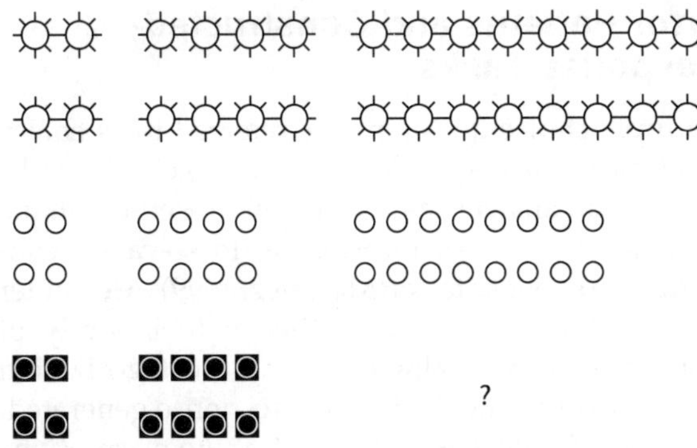

Figure 2. A nonverbal analogy matrix.

processing, but also metacognitive ability. Put simply, metacognition is thinking about our thinking. We use metacognition as we evaluate our progress, make plans, and consider possible new approaches to problem solving. As one works from matrix to matrix in solving the nonverbal analogies, one monitors how the rules change, from simple addition and deletion to more complex rotation and transformation problems.

The Link Between Metacognition and Complex Strategic Processing

For some, the analogy problems discussed above might seem easy to solve. Certainly, these types of problems are highly structured, and many items can be presented on one assessment tool. However, there are other types of tasks used to assess metacognition that are complex and less structured. These tasks reveal the degree to which project completion or problem solution require complex strategic processing. Consider the common situation where experts in many disciplines compose research papers to share their knowledge with others. Locating research material, taking good notes when reading research material, organizing notes into themes or summary statements, writing a draft of the paper so that main ideas are highlighted, and revising and editing the draft to produce a final version, are some of the many strategies that expert authors use to compose their research papers. Metacognition helps experts to select, monitor, and evaluate these strategies in an effort to produce papers of high quality that contribute new knowledge in their fields of study.

Cognitive psychologists have employed many methods to assess domain-specific and strategic knowledge. These methods have shed light on how expert processing differs from novice processing within and across domains. We will now turn our attention to assessment practices as they pertain to academic learning.

DYNAMIC ASSESSMENT

The goal of 'dynamic assessment' is to study cognitive development together with effectiveness of instruction. Dynamic assessment is therefore long-term: it is the study of one's potential to know more information in time. Dynamic assessment involves intermittent interventions of instruction. It is essential not only to provide feedback to a student, but also to instruct the student about concepts and strategies that contribute to proficient understanding. In their extensive review of dynamic testing, Grigorenko and Sternberg (1998) traced the roots of this assessment paradigm to the work of Lev Vygotsky (1962) in the 1930s. Vygotsky introduced the concept of the 'zone of proximal development', which reflects one's potential for growth in cognitive and social functioning facilitated through social interaction, such as learning from a teacher. The zone of proximal development is not to do with how one has developed, but with how one can become what one is not yet. Thus, rather than reflecting how an expert became proficient, it indicates how a novice might become proficient as a result of a synchronous interplay of evaluation and instruction.

CONCEPTUAL CHANGE

Like dynamic assessment, 'conceptual change' focuses on development. But while dynamic assessment attends to one's potential to learn, conceptual change attends to how knowledge is restructured as a result of learning and instruction (Pintrich *et al.*, 1993). The study of human error patterns provides important insights as to how knowledge is restructured over time. From their study of the errors that participants produced given various types of tasks for various domains, Alexander *et al.* (1998) showed that the mistakes or errors we make are typically nonrandom; they occur in systematic patterns (e.g. we tend to make similar mistakes repeatedly); and they can provide the means for instructional intervention.

Multiple-Choice Questions and Error Patterns

Multiple-choice questions, for example, can be used to study how individuals choose options that vary in their relatedness to the target domain. Consider the following example from the domain of statistics:

The standard deviation of a univariate distribution is:

1. The average distance from the mean. [correct statistics option]
2. The square root of the range. [incorrect statistics option]
3. A difference between a negative and a positive value. [incorrect mathematics option]
4. A common mistake made when spelling words. [incorrect non-mathematics option]

The four options have varying degree of correctness. All items on a test will employ the same hierarchy: there will always be a correct response in the domain of statistics; an incorrect response in the domain of statistics; an incorrect response in mathematics (not statistics) and an incorrect response outside the domain of mathematics.

Analysis of students' responses to such tests indicates that learners gravitate toward one error category rather than another, according to their familiarity with the domain. Such error patterns allow psychologists to profile learners according to their schematic knowledge. Thus, some students know a lot about the domain of statistics, and even when incorrect, they choose options that are representative of the correct domain. Other students do not have an initial sense of the domain: their errors indicate this. These error patterns are valuable to instructors, for they help them to modify the curriculum to meet the needs of learners with a wide range of abilities.

Error Patterns and Constructed-Response Tasks

Multiple-choice questions are referred to as choice tasks since the respondent is asked to check, circle, or mark an answer that is already presented on the test. Errors are also informative in so-called constructed-response tasks (Martinez, 1999). Alexander *et al.* (1998) demonstrated this in their study of domain-specific analogies. Using a categorization system to evaluate the types of response generated, they noted that even incorrect analogy solutions hinted at whether students lacked domain knowledge, strategic knowledge, or both. Table 1 presents this categorization scheme, using one analogy problem (in statistics) to illustrate the errors that inform researchers of the degree to which a student lacks knowledge. As with multiple-choice questions, error patterns help teachers to design intervention strategies. When such tests are applied at different times in studies of development, the transitions between error categories, leading toward correct responses, assist cognitive psychologists in their assessment of conceptual change: they can determine exactly when students' knowledge has been restructured to allow for principled use.

ACHIEVEMENT

'Achievement' refers to the assessment of knowledge acquisition, reading comprehension, or problem solving resulting from instruction. Traditionally, standardized assessments using multiple-choice formats have been used to evaluate student

Table 1. Categorization scheme for analysis of analogy problem responses

Category	*Example*	*Description*
No response	sample: statistic : : population:	No evidence of domain-specific or strategic knowledge.
Repetition	sample : statistic : : population : *statistic*	No strategic understanding of analogical relationships.
Non-domain response	sample : statistic : : population : *people*	Response is not related to target domain of statistics.
Structural dependency	sample : statistic : : population : *large statistic*	Response indicates use of component processes and some idea of the relationships among terms in statistics.
Domain response	sample : statistic : : population : *sigma*	Response indicates use of component processes and a response that is a term related to the correct response in statistics.
Target variant	sample : statistic : : population : *parametric*	Strong evidence of both domain-specific knowledge and strategic processing. The answer is a high-level error because it is merely a variation of the correct response. (The correct response should be a noun, not an adjective.)
Correct response	sample : statistic : : population : *parameter*	

achievement. However, poorly-constructed multiple-choice assessments encourage students to value rote memorization over more complex forms of thinking (Martinez, 1999). Constructed-response tasks in the forms of 'performance-based' and 'portfolio' assessments, are generally recommended by cognitive psychologists.

Performance-Based Assessment

Performance-based assessments require students to build objects, design plans, conduct experiments, deliver speeches and presentations, or stage plays and debates that are indicative of the kinds of tasks and procedures that experts perform. Psychologists develop 'rubrics' or scoring schemes, and train raters, or judges, to assign scores to student performance. Rubrics can take many forms, from simple checklists to complex rating scales. Imagine an instructor in physics who wants to assist students in their use of the internet to locate resources to conduct science experiments. The instructor finds a website that will be interesting to students. There are links at the site that allow students to study original scientific documents, visit virtual science museums, and communicate online with experts. Using the links on the site, the instructor hopes that students will learn to appreciate that finding information in the form of documents or dialogs with experts, and comprehending this information, is as important as working with equations or performing experiments in the laboratory. Knowing how to share this information with others in the form of a report is also important. Thus, the teacher creates the rubric presented in Figure 3 to evaluate student performance. This rubric is a simple rating scale that the teacher can use, the student can use for self-evaluation, or peers can use to provide feedback to one another.

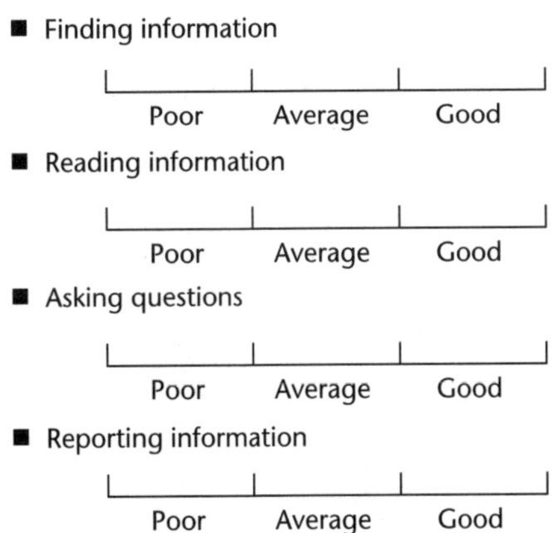

Figure 3. A possible rubric for a performance-based assessment of a research task.

Portfolio Assessment

Sometimes psychologists assess achievement over time using combinations of test formats that allow students to 'showcase' their best work. These 'portfolio assessments' not only provide information on the cognitive development of students, but also present examples of students' creative products and their interests. Interests and motivation are important to cognitive psychologists: according to models of academic development such as that proposed by Patricia Alexander (1997), one does not activate knowledge use and strategic processing without being motivated by one's domain of study. Indeed, being interested in the subject matter contributes as much to becoming an expert as does domain-specific knowledge, strategic processing, and metacognition. Portfolio assessments are a good way for students to express what is of interest to them, and they can show how these interests have changed over time in relation to changes in their knowledge.

FUTURE PROSPECTS

There are many more types of assessment than we have discussed in this article. With emergent technologies in areas like robotics and computer-aided and simulation design, and opportunities afforded by the internet, new ways to assess cognitive processes are emerging rapidly. Perhaps the only limitation of cognitive psychologists' ability to assess is their own creativity.

References

Alexander PA (1997) Mapping the multidimensional nature of domain learning: the interplay of cognitive, motivational, and strategic forces. *Advances in Motivation and Achievement* **10**: 213–250.

Alexander PA, Murphy PK and Kulikowich JM (1998) What responses to domain-specific analogy problems reveal about emerging competence. *Journal of Educational Psychology* **90**(3): 397–406.

Anderson JR (1983) *The Architecture of Cognition.* Cambridge, MA: Harvard University Press.

Ericsson KA and Simon HA (1993) *Protocol Analysis: Verbal Reports as Data.* Cambridge, MA: MIT Press. [Revised edition.]

Ericsson KA and Smith J (1991) *Toward a General Theory of Expertise: Prospects and Limits.* New York, NY: Cambridge University Press.

Grigorenko EL and Sternberg RJ (1998) Dynamic testing. *Psychological Bulletin* **124**(1): 75–111.

Martinez ME (1999) Cognition and the question of test item format. *Educational Psychologist* **34**(4): 207–218.
Pintrich PR, Marx RW and Boyle RA (1993) Beyond cold conceptual change: the role of motivational beliefs and classroom contextual factors in the process of conceptual change. *Review of Educational Research* **63**: 167–199.
Raven JC (1958) *Standard Progressive Matrices: Sets A, B, C, D, and E*. London: H. K. Lewis.
Sternberg RJ (1977) *Intelligence, Information Processing and Analogical Reasoning: The Componential Analysis of Human Abilities*. Hillsdale, NJ: Erlbaum.
Vygotsky LS (1962) *Thought and Language*. Cambridge, MA: MIT Press. [First published 1934.]

Further Reading

Alexander PA (1992) Domain knowledge: evolving themes and emerging concerns. *Educational Psychologist* **27**: 33–51.
Chi MTH, Feltovich PJ and Glaser R (1981) Categorization and representation of physics problems by experts and novices. *Cognitive Science* **5**: 121–152.
Dole JA and Sinatra GM (1998) Reconceptualizing change in the cognitive construction of knowledge. *Educational Psychologist* **33**(2–3): 109–128.
Embretson SE and Prenovost LK (2000) Dynamic cognitive testing: what kind of information is gained by measuring response time and modifiability? *Educational and Psychological Measurement* **60**(6): 837–863.
Ericsson KA, Patel V and Kintsch W (2000) How experts' adaptations to representative task demands account for the expertise effect in memory recall: comment on Vicente and Wang (1998) *Psychological Review* **107**(3): 578–592.
Feuerstein R, Rand Y, Jensen MR, Kaniel S and Tzuriel D (1987) Prerequisites for testing of learning potential: the LPAD model. In: Lidz CZ (ed.) *Dynamic Testing*, pp. 35–51. New York, NY: Guilford Press.
Hall BW and Hewitt-Gervais CM (2000) The application of student portfolios in primary-intermediate and self-contained-multiage team classroom environments: implications for instruction, learning, and assessment. *Applied Measurement in Education* **13**(2): 209–228.
Hunt E (1978) The mechanisms of verbal ability. *Psychological Review* **85**: 109–130.
Kelderman H (1996) Multidimensional Rasch models for partial-credit scoring. *Applied Psychological Measurement* **20**: 155–168.
Kulikowich JM and Alexander PA (1994) Evaluating students' errors on cognitive tasks: applications of polytomous item response theory and log-linear modeling. In: Reynolds CR (ed.) *Cognitive Assessment: A Multidisciplinary Perspective*, pp. 137–154. New York, NY: Plenum.
Shavelson RJ, Ruiz-Primo MA and Wiley EW (1999) Notes on sources of sampling variability in science performance assessments. *Journal of Educational Measurement* **36**(1): 61–71.

Cognitive Development

See **Culture and Cognitive Development**

Cognitive Development, Computational Models of

Intermediate article

Denis Mareschal, Birkbeck College, London, UK

CONTENTS

Computational models of cognitive development are formal models of the information processing, and changes in information processing, that take place during cognitive development. They are generally implemented as running computer simulations. They are tools for exploring mechanisms of transition (development) from one level of competence to the next during the course of cognitive development.

COGNITIVE DEVELOPMENT AND THE COGNITIVE SCIENCES

Jean Piaget, the father of cognitive development research, saw himself as an empirical philosopher. His goal was to answer the fundamental questions of epistemology through rigorous experimentation. He asked how knowledge (especially abstract conceptual knowledge and logic-based reasoning) could emerge from a child's interactions with the world. Piaget produced a vast body of work exploring the development of cognitive components such as the concepts of Space, Time, Number, and Causality. He is generally recognized as having identified the key questions that have set the agenda for cognitive development research since the 1950s.

Piaget was greatly influenced by the philosophies of Kant and Bergson, and by the Cybernetics movement of the early twentieth century. He believed that children constructed an understanding of the world through active engagement with the world and that feedback played a crucial role in learning and development. Unfortunately, he failed to ground many of his proposals because he lacked an appropriate vocabulary with which to express his dynamic and mechanistic ideas. The advent of cognitive science has provided powerful conceptual tools for addressing many of Piaget's original queries.

Contemporary theories of cognitive development lie along two distinct – albeit related – dimensions. One of these is the Nativist vs. Empiricist dimension (i.e. the nature vs. nurture debate). Radical Nativists believe that all knowledge is available to the infant prior to any learning experiences. Radical Empiricists believe that the infant is born with no prior knowledge, and that all knowledge is acquired through some form of experience with the world. Although no developmentalist will admit to holding either of these extreme views, the field is nevertheless polarized into two camps with strong, sometimes extreme, biases towards one or the other of these poles.

A second, more recent, dimension is the symbolic vs. subsymbolic dimension. Those in the symbolic camp believe that cognition is best characterized as a rule-governed physical symbol system. In this view, cognitive development consists in the modification rules. Those in the subsymbolic camp see cognition as a highly interactive dynamic system (e.g. an artificial neural network) that does not operate as a symbol processing system. In this view development consists in the continuous tuning of the underlying parameters of the cognitive system.

The rest of this article presents a number of models that illustrate the different developmental domains in which modeling has been undertaken. It also illustrates fundamental differences in the developmental mechanisms proposed. As will become apparent in this review, most symbolic models have emphasized the tractability of the knowledge representations involved in cognitive development at the expense of implementing explicit transition mechanisms. In contrast most subsymbolic models have emphasized the specification of a developmental mechanism at the expense of the tractability of the knowledge representations.

WHY BUILD COMPUTATIONAL MODELS OF COGNITIVE DEVELOPMENT

The Computer Modeling Methodology

Computer models complement experimental data gathering by placing constraints on the direction of future empirical investigations. First, developing a computer model forces the user to specify precisely what is meant by the terms in his or her underlying theory. Terms such as representations, symbols, and variables must have an exact definition to be implemented within a computer model. The degree of precision required to construct a working computer model avoids the possibility of arguments arising from the misunderstanding of imprecise verbal theories.

Secondly, building a model that implements a theory provides a means of testing the internal self-consistency of the theory. A theory that is in any way inconsistent or incomplete will become immediately obvious when trying to implement it as a computer program. The inconsistencies will lead to conflict situations in which the computer program will not be able to function. Such failures point to a need to reassess the situation and re-evaluate the theory.

A positive corollary of these two points is that the model can be used to work out unexpected implications of a complex theory. Because the world is highly complex with a multitude of information sources constantly interacting, even a simple process theory can lead to uninterpretable behaviors. Here again, the model provides a tool for teasing apart the nature of these interactions and corroborating or falsifying the theory.

Perhaps the main contribution made by computational models of cognitive development is to provide an account of the representations that underlie performance on a task that also incorporates a mechanism for representational change. One of the greatest unanswered questions of cognitive development is the nature of the transition mechanisms that can account for how one level of performance is transformed into the next level of performance at a later age. This is a difficult question because it involves observing how representations evolve over time and tracking the complex interactions between the developing components of a complex cognitive system. Building a model and observing how it evolves over time provides a tangible means of doing this.

Formulating development in computational terms forces the theoretician to be explicit about the mechanisms that underlie information processing. Piaget's own mechanistic theory provides an excellent example of why this is necessary. He described cognitive development in terms of assimilation, accommodation, and equilibration. Assimilation consisted in adapting or filtering incoming information to make it more compatible with existing knowledge representations. In contrast, accommodation consisted in adapting one's knowledge representations to make them more consistent with novel information. Equilibration was the process by which assimilation and accommodation interacted to cause cognitive development. While assimilation and accommodation capture intuitive notions of what might be involved in cognitive development, they are too loosely defined to be of any explanatory value. Some connectionist computational models of cognitive development have tried to solve this problem by providing computational implementations of assimilation and accommodation in terms of activation flow and weight updates respectively.

MODELS OF DEVELOPMENT IN INFANCY

Infancy is an ideal age range to begin modeling because infant behaviors are not complicated by the presence of language and sophisticated metacognitive strategies. Infant abilities are closely tied to their developing sensorimotor skills.

Object-Directed Behaviors

Kant identified objects as a fundamental category of cognition. The ability to represent hidden objects liberates infants from the tyranny of direct perception. It is the first step towards representational thought. Piaget suggested that infants progressed through six stages on the way towards an adult level of understanding of object permanence at the age of two. While many of Piaget's original findings have been replicated, changes in methodology (e.g. relying on visual attention measures rather than manual retrieval measures) have suggested that infants have a far more precocious understanding of hidden objects. These studies have all focused on infant competence at different ages but not on the mechanisms of development from one level of competence to the next.

There are relatively few computational models of infant object-directed behaviors (Mareschal, 2000). Early models took a strong cognitivist stance on behavior and were thus implemented in rule-based production systems (e.g. Luger *et al.*, 1983).

Unfortunately, they were basically competence models that described infant behaviors but did not provide a mechanistic account of development. They proposed different sets of rules to describe behavior at different ages but did not explain how new rules could be acquired or how one set of rules was transformed into another set of rules. More recent (cognitivist) models have turned to attention-based accounts of object processing in an attempt to explain infant behaviors (Simon, 1998). Unfortunately, these models still fail (by and large) to implement any account of *how* development might occur.

One mechanistic learning model has implemented a parallel processing version of Piaget's sensorimotor theory of infant development. Drescher (1991) tried to show how the coordination of intra- and intermodal perceptual motor schemas could lead to a single unified representation of an object. Perceptual motor schemas were encoded as 'context-action-result' rules and implemented in a parallel processing machine. Learning consisted in using marginal probabilities to fill in context and results slots in appropriate perceptual motor schemas. Although this system developed an intricate network of intra- and intermodal schemas that mimic the infant's sensorimotor integration, it did not develop according to the pattern described by Piaget.

A number of connectionist models have also been proposed. In one family of models, a partially recurrent autoencoder network learns to predict the reappearance of a stationary object from behind a moving screen that temporarily occludes the object (Munakata *et al.*, 1997). Network performance is measured by taking the difference in response of the nodes coding the location of the hidden object when an object should be revealed and subtracting it from the response of the nodes when an object should not be revealed. An increase in this difference is interpreted as increased knowledge of hidden objects. What this model shows is that object representations that guide action can be graded and arise though interactions within an environment.

Mareschal *et al.* (1999) describe an alternative connectionist model that is more closely tied to the neuropsychological finding that visual object information is processed down two separate routes. This model uses a combination of modules to implement dual route processing. One route learns to process spatial-temporal information while the other route learns to process feature information. Finally, a response module recruits and coordinates the representations developed by the other modules as and when required by a response task. The route specializations emerge as a result of the different associative mechanism in each module.

Perceptual Categorization

Because categorization lies at the heart of cognition it is not surprising to find that great effort has been exerted in trying to understand the early roots of category formation. Many infant categorization tasks rely on preferential looking or habituation techniques, based on the finding that infants direct more attention to unfamiliar or unexpected stimuli (Mareschal and Quinn, 2001). Connectionist autoencoder networks have been used to model the relation between sustained attention and representation construction (Mareschal *et al.*, 2000). The successive cycles of training in the autoencoder are an iterative process by which a reliable internal representation of the input is developed. This approach assumes that infant looking times are positively correlated with the network error. The greater the error, the longer the looking time, because it takes more training cycles to reduce the error.

The perceptual categories formed by infants are not always the same as the corresponding adult categories. For example, when shown a series of cat photographs three- to four-month-olds will form a category of CAT that includes novel cats and excludes dogs (as will adults). However, when shown a series of dog photographs, the same infants will form a category of DOG that includes novel dogs but also includes cats (in contrast to adults). Many aspects of early infant perceptual categorizations (including the asymmetric exclusivity of CAT and DOG categories) are captured by the connectionist autoencoder model. In contrast to adults who apply top-down schemas when recognizing photographs of cats and dogs, three- to four-month-olds, like the autoencoder networks, simply process the bottom-up information in these images. Hence, their internal category representations are yoked to the distributional properties of features in the images. The model demonstrates that categorical representations can self-organize in a neural system as a result of exposure to the familiarization exemplars encountered within the test session itself.

Early Word Learning

Categorization is equally important in early language acquisition, both in terms of learning which sounds in the environment are relevant to speech and in terms of learning the domain of applicability of a new word.

Infants are better at discriminating phonemes that do not belong to their linguistic environment at seven months than at 14 months. This has been interpreted as evidence for a shift in lexical processing between these age groups. The suggestion is that seven-month-olds are processing the sounds of the stimuli whereas the 14-month-olds are processing the stimuli as words. The process of word learning itself changes the way the stimuli are processed. In contrast, Schafer and Mareschal (2001) present a connectionist autoencoder model suggesting that these behaviors reflect two stages of the same processing mechanism. The model learns to associate sound representations with image representations (i.e. simple word learning). Early in training, the model is not committed to any meaningful internal representations for sound and is therefore better able to learn novel speech sounds in testing than later in training when it has firmly committed to representations tailored to its native linguistic environment. This model illustrates how discontinuities in behavior can emerge by the slow tuning of continuous parameters.

Once infants have segmented the sound stream into word units, there is the further problem of identifying the extension of the category referred to by the word (the Symbol Grounding Problem). One early model (Schyns, 1991) used a combination of self-organizing (kohonen feature maps) and supervised (backpropagation) connectionist networks to model the interactive process of identifying the category underlying the word, and then associating it with an appropriate label. This model showed many of the same developmental effects as young children learning novel words. In particular, the categories showed prototype effects and mutual exclusivity constraints. It demonstrated how the acquisition of a new word arose as an interaction of bottom-up and top-down effects. A more recent (autoencoder) model (Plunkett *et al.*, 1992) shows how over extension and under extension errors, typical of children's early vocabulary, can arise through simple associative mechanisms in a system with shared sound and image internal representations.

MODELS OF DEVELOPMENT IN CHILDHOOD

Language acquisition marks the end of infancy and the beginning of childhood. Reasoning and conceptual development are the hallmark of cognitive development in childhood. The models in this section all focus on some aspect of reasoning development. We begin by reviewing models that have explicitly tried to implement Piagetian ideas. This is followed by a review of work that breaks away from the Piagetian tradition.

Modeling Piagetian Stage Development

There have been several attempts to explain the apparent stage-like growth of competence in children in terms of self-organization in dynamic systems, competition between cognitive growers, and bifurcation theory (e.g. van Geert, 1998). However such accounts have tended to rely only on mathematical descriptions that are either not implemented in running computer models or grounded in measurable information processing components.

As early as the 1960s an effort was made to use neural networks to operationalize the Piagetian notions of assimilation and accommodation. Several interpretations of connectionist learning in terms of assimilation and accommodation have been proposed. One such interpretation suggests that weight changes constitute a form of accommodation whereas the transformation (by the network weights) of input patterns into internal patterns of activation constitutes assimilation (McClelland, 1995). Another interpretation suggests that the adaptation of a network architecture constitutes a form of accommodation whereas the adaptation of weights is a form of assimilative learning (Shultz *et al.*, 1995).

Models that try to implement Piagetian notions of development have attempted to model children's performance on key tasks (e.g. conservation: Klahr and Wallace, 1976; Shultz, 1998). One such task is the seriation (or sorting) task. Piaget found that children's ability to order a set of sticks developed through a number of stages. In a first stage, children were unable to sort the sticks. In a second stage, they were able to apply local ordering relations but could not extend the order to the set as a whole. In the third stage, they were able to sort the set of sticks, but only by applying a costly trial and error strategy. Finally, in the fourth stage children were able to sort the set quickly and efficiently by applying a systematic selection strategy.

Young (1976) approached this task from an information processing perspective. He carried out detailed analyses of the actions children carried out at different ages when sorting blocks. Based on the results of protocol analyses, he developed a rule-based production system that captured children's performance at each stage of development.

Progress from one stage to the next was modeled by the (hypothesized) modification of the rules. Although this model provided a good fit to children's behavior at individual stages, it fails to provide a working account of how those rules are modified. A recent connectionist model of the seriation task suggests how development could occur (Mareschal and Shultz, 1999). This model argues that development consists in the gradual tuning of connection weights, and the gradual extension of knowledge about small sets to larger sets. The model not only captures the stage progression described by Piaget, but it also captures the variability in sorting behaviors observed both within and between subjects.

Beyond Piaget: The Balance Scale Task

A recent benchmark of cognitive development, first developed by Inhelder and Piaget and later significantly extended by Robert Siegler, is the balance scale task. Siegler explored children's developing abilities to reason about balance scales. In these problems, children were presented with a symmetric balance scale with five equally spaced pegs on either side of the fulcrum. Weights were then placed on pegs to the left and the right of the fulcrum and children were asked to predict whether the balance scale would tip to the left, to the right, or remain balanced.

Seigler used a rule assessment methodology to infer the rules that govern children's performance. He found that rule one children relied only on a dominant dimension (weight) to predict which side the balance scale would tip. They predicted that the side with the most weights would be the side that the balance scale would tip. Rule two children applied the same rule as the preceding children, but had an additional rule stating that if the number of weights was equal on both sides, the side with the greatest distance would predict the side to which the balance scale would tip. Rule three children behaved like the rule two children with the exception that if the weight and distance cues provided conflicting answers they would guess which side went down. Finally, rule four children had a set of rules that effectively computed the torque on both sides of the balance scale and chose the side with the greatest torque. Seigler suggested that this knowledge was represented in the form of a growing decision tree. Klahr (1992) pointed out the equivalence of this representation to that of an increasing rule; a representation consistent with production system models of cognitive development.

Both connectionist (McClelland, 1995; Shultz *et al.*, 1994) and decision tree models (Schmidt and Ling, 1996) of development on the balance scale task have been proposed. The connectionist models construe the problem as one of integrating information from two sources. They capture stage development in terms of microgenetic weight changes and hidden unit recruitment. An assumption of these models is that children have greater experience with weight comparisons than distance comparisons. The decision tree model uses the C4.5 tree-inducing algorithm. Children were hypothesized to have an increasing memory capacity and to care increasingly about the detail of correctness of their answers. While this latter model captures the main features of children's performance, the decision trees developed did not map onto those proposed by Siegler to reflect children's knowledge at different ages.

Modeling of the Development of Reasoning

Many of the successful developmental models described above are connectionist models. Such models process information based on the surface similarity between different exemplars. However, there are cases when children's (and adults') reasoning does not follow surface similarity. Analogical reasoning requires the child to distance his or herself from the surface similarity between the target and vehicle domains. Indeed between six and nine years of age children move from basing their analogies on surface similarity (such as color) between the two domains to structural similarity (such as the function of an object) between the two domains. This reliance on structural similarity is very difficult for connectionist systems to capture.

Gentner has suggested that adults and children solve analogical problems by comparing mental representations via a structure-mapping process of alignment of conceptual representations. A structurally consistent match conforms to a one-to-one mapping constraint between the domains. The process is implemented in the Structural Mapping Engine (SME) model. The SME is used to model the relational shift in children's analogical reasoning in terms of increased domain knowledge. As their knowledge of domain relations increases, children's relation representations within a domain become richer and deeper, increasing the likelihood that their comparisons will focus on matching relations. Thus, what develops between six and nine years is only knowledge and not processing.

Siegler and Shipley present a model of strategy choice. The intention is to provide some account of the range and variability of strategies observed in young children's problem solving. Their model is based on the strategies used by children when adding integers. The strategies are explicitly represented in terms of rules. Strategy choice is probabilistic. The probability of retrieving and executing a strategy depends on the previous association of that strategy with an outcome in conjunction with considerations of cost and efficiency. The strategy pool evolves according to a Darwinian procedure in which infrequently used strategies die off and new strategies enter the pool via random perturbations of existing strategies.

CHALLENGES TO CURRENT MODELS OF COGNITIVE DEVELOPMENT

The 'poverty of the stimulus' argument was most effectively put forward by Chomsky in response to Skinner's account of language acquisition. It is not possible for an unconstrained inductive learner to acquire a particular target grammar within a reasonable time. This argument has been wielded against connectionist models of language and cognitive development in general. However, it is important to understand that the 'poverty of the stimulus' argument holds for all inductive learning systems and in all domains (its application is not unique to connectionism). This is why most contemporary scholars of learning (whether studying learning in children or machines) believe that the key to understanding cognitive development is to identify the nature of the constraints on the learner that will allow knowledge to emerge.

Fodor's paradox claims that an inductive learner can never acquire any truly novel concept (Fodor, 1980). Indeed, in order to test the domain of applicability of a concept, the inductive learner must be able to represent that concept prior to having identified it. Hence, any learning simply involves the recombination of existing representational tokens in a system. While this may be true of inductive learning systems, it is not true of systems that increase their representational power in response to environmental pressures. Such systems include neural networks that construct their own architecture as part of learning and development (Mareschal and Shultz, 1996).

Finally, while computer models allow us to formulate questions about what can possibly cause cognitive development (e.g. processing capacity, processing speed, knowledge, and strategy choice) more constraints are required to identify the actual mechanisms involved in children's cognitive development. Recent advances in neuroimaging techniques have allowed us to place greater constraints on how information is processed in the brain. Many of the models above make little use of these constraints and, in the future, such constraints should be incorporated in any functional models of development. Furthermore, cognitive development does not occur in a social vacuum. Vygotsky has emphasized the role of social interactions in cognitive development. Society provides a kind of cognitive scaffolding that nurtures and aids the child's cognitive development by actively selecting and filtering the type of problems the child is faced with at any age. Future models will need to consider these constraints to reflect the child's learning environment more accurately.

References

Drescher GL (1991) *Made-up Minds. A Constructivist Approach to Artificial Intelligence*. Cambridge, MA: MIT Press.

Fodor J (1980) Fixation of belief and concept acquisition. In: Piatelli-Palmarini M (ed.) *Language and Learning: The Debate between Chomsky and Piaget*, pp. 143–149. Cambridge, MA: Harvard University Press.

van Geert P (1998) A dynamic systems model of basic developmental mechanisms: Piaget, Vygotsky, and beyond. *Psychological Review* **105**: 634–677.

Klahr D (1992) Information processing approaches to cognitive development. In: Bornstein MH and Lamb ME (eds) *Developmental Psychology: An Advanced Textbook*, 3rd edn, pp. 273–335. Hillsdale, NJ: Lawrence Erlbaum Associates.

Klahr D and Wallace JG (1976) *Cognitive Development: An Information Processing View*. Hillsdale, NJ: Erlbaum.

Luger GF, Bower TGR and Wishart JG (1983) A model of the development of the early object concept. *Perception* **12**: 21–34.

Mareschal D (2000) Infant object knowledge: current trends and controversies. *Trends in Cognitive Science* **4**: 408–416.

Mareschal D and Quinn PC (2001) Categorisation in Infancy. *Trends in Cognitive Science* **5**: 443–450.

Mareschal D and Shultz TR (1996) Generative connectionist networks and constructivist cognitive development. *Cognitive Development* **11**: 571–604.

Mareschal D and Shultz TR (1999) Children's seriation: a connectionist approach. *Connection Science* **11**: 153–188.

Mareschal D, French RM and Quinn P (2000) A connectionist account of asymmetric category learning in infancy. *Developmental Psychology* **36**: 635–645.

Mareschal D, Plunkett K and Harris P (1999) A computational and neuropsychological account of object-oriented behaviors in infancy. *Developmental Science* **2**: 306–317.

McClelland JL (1995) A connectionist perspective on knowledge and development. In: Simon TJ and Halford GS (eds) *Developing Cognitive Competence: New Approaches to Process Modeling*, pp. 278–304. Hillsdale, NJ: Erlbaum.

Munakata Y, McClelland JL, Johnson MH and Siegler RS (1997) Rethinking infant knowledge: towards an adaptive process account of successes and failures in object permanence tasks. *Psychological Review* **104**: 686–713.

Plunkett K, Sinha C, Møller MF and Strandsby O (1992) Symbol grounding or the emergence of symbols? Vocabulary growth in children and a connectionist net. *Connection Science* **4**: 293–312.

Schafer G and Mareschal D (2001) Modeling infant speech sound discrimination using simple associative networks. *Infancy* **2**: 7–28.

Schmidt WC and Ling CX (1996) A decision-tree model of balance scale development. *Machine Learning* **18**: 1–30.

Schyns P (1991) A modular neural network model of concept acquisition. *Cognitive Science* **15**: 461–508.

Shultz TR (1998) A computational analysis of conservation. *Developmental Science* **1**: 103–126.

Shultz TR, Mareschal D and Schmidt WC (1994) Modeling cognitive development on balance scale phenomena. *Machine Learning* **16**: 59–88.

Shultz TR, Schmidt WC, Buckingham D and Mareschal D (1995) Modeling cognitive development with a generative connectionist algorithm. In: Simon TJ and Halford GS (eds) *Developing Cognitive Competence: New Approaches to Process Modeling*, pp. 205–261. Hillsdale, NJ: Erlbaum.

Simon TJ (1998) Computational evidence for the foundations of numerical competence. *Developmental Science* **1**: 71–78.

Young R (1976) *Seriation by Children: An Artificial Intelligence Analysis of a Piagetian Task*. Basel: Birkhauser.

Further Reading

Boden MA (1995) *Piaget*, 2nd edn. Modern Masters. London, UK: Fontana Press.

Elman JL, Bates EA, Johnson MH *et al.* (1996) *Rethinking Innateness: A Connectionist Perspective on Development*. Cambridge, MA: MIT Press.

Karmiloff-Smith A (1992) *Beyond Modularity*. Cambridge, MA: MIT Press.

Lewandowsky S (1993) The rewards and hazards of computer simulations. *Psychological Science* **4**: 236–243.

Rogoff B (1990) *Apprenticeship in Thinking*. Oxford, UK: Oxford University Press.

Siegler RS (1996) *Emerging Minds*. Cambridge, MA: MIT Press.

Siegler RS (1997) *Children's Thinking*, 3rd edn. Prentice Hall.

Simon TJ and Halford GS (eds) (1995) *Developing Cognitive Competence: New Approaches to Process Modeling*. Hillsdale, NJ: Erlbaum.

Cognitive Linguistics

Intermediate article

Gilles Fauconnier, University of California, San Diego, California, USA

CONTENTS

Cognitive linguistics is a theoretical and empirical programme that goes beyond the visible structure of language to investigate the complex background operations of cognition that create grammar, conceptualization, discourse, and thought itself.

INTRODUCTION

Cognitive linguistics is a powerful approach to the study of language, conceptual systems, and human cognition. It addresses the structuring of basic conceptual categories such as space and time, scenes and events, entities and processes, motion and location, force and causation, and also the structuring of ideational and affective categories, such as attention and perspective, volition and intention (Talmy, 2000). In doing so, it develops a rich conception of grammar that reflects fundamental cognitive abilities: the ability to form structured conceptualizations with multiple levels of organization, to conceive of a situation at varying levels of abstraction, to establish correspondences between facets of different structures, and to construe a situation in different ways (Langacker, 1987, 1991).

Much of traditional linguistics (including structural and generative linguistics) focuses on the ways in which words are combined into sentences. For example, in looking at the four English sentences *the plane flies over the city*, *the post office is over the hill*, *the war is over the oil wells*, and *this topic is over my head*, a grammarian might see a single structure, and leave it at that. For cognitive linguistics, these four sentences would present a far greater challenge: to explain how the same structure, and the same word *over*, gives rise to such different kinds of conceptualizations: a plane moving 'above' a city, a post office located 'at the end of' a path on the hill, a war 'caused by' something to do with the oil wells, and a topic 'high' on some scale of difficulty. Hidden behind simple words and everyday language are vast conceptual networks manipulated unconsciously through the activation of powerful neural circuits.

Cognitive linguistics recognizes that the study of language is the study of language use, and that when we engage in any language activity, we draw unconsciously on vast cognitive and cultural resources, call up models and frames, set up multiple connections, coordinate large arrays of information, and engage in creative mappings, transfers, and elaborations. Language does not 'represent' meaning: it prompts for its imaginative construction. Very sparse grammar guides us along rich mental paths, by prompting us to perform complex cognitive operations. A large part of cognitive linguistics centers on the creative 'online' construction of meaning as discourse unfolds in context (Fauconnier and Sweetser, 1996; Sweetser, 2000).

Aspects of language and expression that have often been consigned in formal work to the rhetorical periphery of language, such as metaphor (Lakoff and Johnson, 1980, 1999; Sweetser, 1990) and metonymy (Panther and Radden, 1999; Barcelona, 2000) are central within cognitive linguistics. They are understood to be powerful conceptual mappings essential to human thought, important for the understanding not just of poetry, but also of science, mathematics, religion, philosophy, and everyday speaking and thinking.

Thought and language are embodied. Conceptual structure arises from our sensorimotor experience and the neural structures that give rise to it. Reason is embodied and imaginative. A grammar is ultimately a neural system, and general cognitive capacities drive language (Fauconnier and Turner, 1998, 2002).

The stage was set for cognitive linguistics in the 1970s and early 1980s with Len Talmy's work on figure and ground, Ronald Langacker's cognitive grammar framework, George Lakoff's research on metaphor, gestalts, categories and prototypes (Lakoff, 1987), Fillmore's frame semantics (Fillmore, 1982) and Fauconnier's mental spaces (Fauconnier, 1994). A wealth of discoveries, empirical studies, and applications have since emerged (Janssen and Redeker, 2000; Tomasello, 1998; Cuyckens and Geeraerts, forthcoming).

GRAMMAR AND COGNITION

The relation of grammar to cognition is studied in fine detail in the foundational work of Talmy (2000) and Langacker (1987, 1991). Talmy shows that there are great restrictions on the conceptual categories that grammatical systems actually specifiy. For example, languages often require that number be marked with forms like singular, plural, or dual, but no language has a system for marking the color of nouns. Topological configurations are marked grammatically, but not Euclidean metrics. (Thus, prepositions like *across* indicate the same configuration regardless of the size of the landmarks (*across the sky*, *across the table*), but absolute size or range of size is not marked or differentiated grammatically.) Talmy singles out multiplexing, and states of boundedness and dividedness, as strongly markable in grammatical systems. He also singles out perspective and sequentializing. Thus, *the door slowly opened and two men walked in* and *two men slowly opened the door and walked in* describe the same event from two different perspectives; *there are some houses in the valley* and *there is a house now and then in the valley* describe the same objective state of affairs, but the second sentence takes us along a sequential path where we 'see' the houses one at a time.

Langacker shows how grammar imposes 'trajector–landmark' organization on scenes and events. In *the table is below the lamp*, the table (the 'trajector') is located with respect to the lamp (the 'landmark'). In *the lamp is above the table* this relationship is reversed: the table serves as landmark. But both sentences reflect the same spatial configuration. 'Profiling' is another important construct of Langacker's cognitive grammar. The word *hypotenuse* evokes a right-angled triangle and 'profiles' a particular part of it: the same line segment without the rest of the triangle is no longer a hypotenuse. In *I melted it*, the word *melted* profiles an entire action chain with causation and change leading to a liquid state. In *it melted easily*, only the change is profiled, although the causation is still evoked. In *it is finally melted*, only the resultant state is profiled, but the unprofiled change is evoked. Langacker analyzes in considerable detail the ways in which component structures are integrated through correspondences and elaboration to form composite structures: for example, how the phonological integration *jar lid* symbolizes the semantic integration of 'jar' and 'lid' (Langacker, 1987, 2000; Van Hoek, 1997).

Other basic aspects of conceptual structuring, as reflected by grammar and found in many languages (Talmy, 2000), include 'fictive motion' (*the blackboard goes all the way to the wall*), 'event integration' (*the ball rolled in, the candle blew out, I kicked the door shut*), and 'force dynamics' (*the ball kept rolling, he refrained from closing the door*). (Force dynamics also applies to social organization and abstract reasoning (Sweetser, 1990).) Fictive motion allows stationary scenes (the position of the blackboard) to be construed in terms of the motion of a fictive trajector (*goes all the way to*). Event integration compresses complex chains of events. Force dynamics is based on our experience of physical forces that cause or prevent motion: *kept rolling* indicates an absence of force to impede the ball's movement; *refrained* indicates a clash of counteracting forces. *Max must be home by ten* indicates a social force (e.g. parental rules, curfew) directed towards an outcome (Max being home by ten). *Max must be home by now* uses the same modal *must* to indicate a logical force directed towards the conclusion that Max is now home.

The way in which language structures space is interesting both linguistically and psychologically. No two languages are quite alike in this respect, although there are some general principles. Deceptively simple-seeming prepositions like *in*, *out* or *over* define elaborate networks of spatial meaning with hundreds of linked schemata, some of which are prototypical and central. Compare the very different senses of *over* in *the plane flew over the field, the post office is over the hill, the log rolled over, the party is over, he had to do it over again, he overlooked it, he looked it over, he oversaw it*. A native speaker of English unconsciously masters a vast network of related schemata linked to the single, simple word *over*. Remarkable work on this topic has been done by cognitive linguists (Lindner, 1982; Brugman, 1981; Herskovits, 1986; Vandeloise, 1991; Talmy, 2000). Explicit computational models (Regier, 1996) reflect the cognitive complexity of the human capacity to structure space linguistically.

METAPHOR THEORY

A second strand of work in cognitive linguistics since the 1980s has been the development of metaphor theory. Launched by George Lakoff and Mark Johnson, this line of research begins with the insight that metaphor is basic and constitutive for all the thinking that we do. Metaphor theory is based on source domains of human experience and on neural connections to our embodied sensations, actions, and emotions. Metaphors create the possibility of 'abstract' reasoning, scientific and mathematical thought, and language and culture generally.

Source domains seem to be used systematically to structure target domains by means of metaphorical mappings. For example, our general way to talk and think about event structure is in terms of motion. In this metaphorical mapping, states are locations, change of state is change of location, causes are forces, purposes are destinations, means are paths to destination, guided action is guided motion, etc. This metaphor is reflected in the language we use to express event structure: *he went crazy, she entered a state of euphoria, the clothes are somewhere between wet and dry, the home run threw the crowd into a frenzy, she walked him through the problem, I've hit a brick wall, we're moving ahead, we're at a standstill* (Lakoff and Johnson, 1999). The structure and inferences of the source domain of motion are projected to the target domain of events and action to create a rich emergent conceptualization.

Time itself is conceptualized in terms of space and motion. In English, times can be represented as objects moving towards and then past a stationary observer, or as objects that are stationary with respect to a moving observer: *the time will come; Christmas is approaching; the summer just zoomed by; we're getting close to Christmas; we've reached the end of May already*.

Conventional metaphors such as these can be extended to enrich conceptual understanding. Time can *fly* and *crawl* and *disappear*. In a line by Shakespeare (*Troilus and Cressida*, IV. v. 202–203; cited in Gibbs (1994)), where Hector greets Nestor, Time becomes a moving person, who holds the hand of the venerable Nestor:

Let me embrace thee, good old chronicle,
That hast so long walk'd hand in hand with time.

MENTAL SPACES AND CONCEPTUAL INTEGRATION

Mental spaces are small conceptual packets, constructed as we think and talk, for purposes of local understanding and action. They are partial assemblies of elements, structured by frames and cognitive models. They are interconnected and can be modified as thought and discourse unfold.

Mental spaces proliferate in the unfolding of discourse, map onto each other, and provide abstract mental structure for shifting viewpoint and focus, allowing us to direct our attention at any time onto partial and simple structures while maintaining an elaborate web of connections in working and long-term memory.

For example, if we say *in reality, Richard Burton loves Elizabeth Taylor, but in the movie, he kills her*, we set up two mental spaces, one for reality and one for the movie. Richard Burton in reality has a counterpart (say Mark Antony) in the movie, and Elizabeth Taylor in reality has a counterpart (say Cleopatra) in the movie. Connections between mental spaces allow access to elements in one mental space via counterparts in other mental spaces (e.g. Mark Antony via Burton). Mental spaces offer a general and elegant means of dealing with opacity, presupposition, counterfactuals, and tense and mood in language.

Take for example the sentence *in 1957, the president was a baby*, appearing in a discourse where a base mental space with G. W. Bush as current US president has been set up. The phrase *in 1957* sets up a new '1957' space. If we take *the president* to describe Bush in the base, its counterpart 'Bush in 1957' will be accessed, and the sentence will mean that Bush was a baby back in 1957. If, on the other hand, we take *the president* to describe someone in the new mental space of '1957', then that person will be both a baby and a president in 1957, and the sentence will mean that a baby was president in 1957. Multiple access possibilities of this kind allow the same sentence to prompt for different connection paths in different situations. A wide range of puzzling reference phenomena can be explained in terms of this general underspecification of connecting paths.

Behind the idiosyncrasies of language, cognitive linguistics has uncovered much evidence for the operation of more general cognitive processes. 'Conceptual integration' is an example of a general cognitive operation on mental spaces that is reflected universally in the way we think.

Conceptual integration consists in setting up networks of mental spaces that map onto each other and blend into new mental spaces in various ways. Some of the integrations are novel, others are more entrenched, and we rarely pay conscious attention to the process, because it is so pervasive. In a conceptual integration network, partial structure from input mental spaces is projected to a new blended mental space which develops dynamic (imaginative) structure of its own.

For example, the counterfactual *in France, Watergate would not have done Nixon any harm* is intended to prompt inferences on the difference between the American and French political systems. It requires the listener to construct input spaces for American politics and for French politics and to project selectively into a blended space in which Nixon and Watergate are embedded into French politics. The imaginative emergent structure of that mental space (why Nixon is not harmed, etc.) provides insight into the political realities of the two countries.

Most aspects of human life evoke conceptual integration networks. This remarkable cognitive capacity has been studied in a variety of domains, including mathematics, music, action and design, distributed cognition, magic and religion, anthropology and political science (Zbikowski, 2002; Hutchins, in preparation; Sorensen, 2000; Lakoff and Núñez, 2000; Liddell, 1998; Turner, 2001). It has been suggested that the capacity of conceptual integration evolved biologically to reach a threshhold, 'double-scope creativity', that constitutes a necessary condition for the cognitively modern human singularities of art, creative tool-making, religious thought, and grammar (Fauconnier and Turner, 2002).

SUMMARY

Cognitive linguistics goes beyond the visible structure of language and investigates the complex background operations of cognition that create grammar, conceptualization, discourse, and thought itself. The theoretical insights of cognitive linguistics are based on extensive empirical observation in multiple contexts, and on experimental work in psychology and neuroscience (Gibbs, 1994; McNeill, 2000; Coulson, 2001; Mandler, 1992; Gentner, 2001). Results of cognitive linguistics, especially from metaphor theory and conceptual integration theory, have been applied to wide ranges of nonlinguistic phenomena.

References

Barcelona A (ed.) (2000) *Metaphor and Metonymy at the Crossroads*. Berlin: Mouton de Gruyter.

Brugman C (1981) *Story of Over: Polysemy, Semantics, and the Structure of the Lexicon*. New York, NY: Garland.

Coulson S (2001) *Semantic Leaps*. New York, NY and Cambridge, UK: Cambridge University Press.

Cuyckens H and Geeraerts D (forthcoming) *Handbook of Cognitive Linguistics*. Oxford: Oxford University Press.
Fauconnier G (1994) *Mental Spaces*. New York, NY: Cambridge University Press. [First published 1985. Cambridge, MA: MIT Press.]
Fauconnier G and Sweetser E (eds) (1996) *Spaces, Worlds, and Grammar*. Chicago, IL: University of Chicago Press.
Fauconnier G and Turner M (1998) Conceptual integration networks. *Cognitive Science* **22**(2): 133–187.
Fauconnier G and Turner M (2002) *The Way We Think*. New York, NY: Basic Books.
Fillmore C (1982) Frame semantics. In: Linguistic Society of Korea (eds) *Linguistics in the Morning Calm*, pp. 111–137. Seoul: Hanshin.
Gentner D (2001) Spatial metaphors in temporal reasoning. In: Gattis M (ed.) *Spatial Schemas in Abstract Thought*, pp. 203–222. Cambridge, MA: MIT Press.
Gibbs R (1994) *The Poetics of Mind*. Cambridge, UK: Cambridge University Press.
Herskovits A (1986) *Language and Spatial Cognition: An Interdisciplinary Study of Prepositions in English*. Cambridge, UK: Cambridge University Press.
Hutchins E (in preparation) Material anchors for conceptual blends.
Janssen T and Redeker G (eds) (2000) *Scope and Foundations of Cognitive Linguistics*. The Hague: Mouton de Gruyter.
Lakoff G (1987) *Women, Fire, and Dangerous Things*. Chicago, IL: University of Chicago Press.
Lakoff G and Johnson M (1980) *Metaphors We Live By*. Chicago, IL: University of Chicago Press.
Lakoff G and Johnson M (1999) *Philosophy in the Flesh*. New York, NY: Basic Books.
Lakoff G and Núñez R (2000) *Where Mathematics Comes From*. New York, NY: Basic Books.
Langacker R (1987) *Foundations of Cognitive Grammar*, vol. I. Stanford, CA: Stanford University Press.
Langacker R (1991) *Foundations of Cognitive Grammar*, vol. II. Stanford, CA: Stanford University Press.
Langacker R (2000) Assessing the cognitive linguistic enterprise. In: Janssen and Redeker (2000), pp. 13–60.
Liddell S (1998) Grounded blends, gestures, and conceptual shifts. *Cognitive Linguistics* **9**(3): 283–314.
Lindner S (1982) What goes up doesn't necessarily come down: the Ins and Outs of opposities. *Chicago Linguistic Society* **18**: 305–323.
Mandler JM (1992) How to build a baby II: Conceptual primitives. *Psychological Review* **99**: 587–604.
McNeill D (2000) *Language and Gesture*. Cambridge, UK: Cambridge University Press.
Panther KU and Radden G (eds) (1999) *Metonymy in Language and Thought*. Amsterdam: John Benjamins.
Regier T (1996) *The Human Semantic Potential: Spatial Language and Constrained Connectionism*. Cambridge, MA: MIT Press.
Sorensen J (2000) *Essence, Schema, and Ritual Action: Towards a Cognitive Theory of Magic*. PhD thesis, University of Aarhus.
Sweetser E (1990) *From Etymology to Pragmatics: Metaphorical and Cultural Aspects of Semantic Structure*. Cambridge, UK: Cambridge University Press.
Sweetser E (2000) Compositionality and blending: working towards a fuller understanding of semantic composition in a cognitively realistic framework. In: Janssen and Redeker (2000), pp. 129–162.
Talmy L (2000) *Toward a Cognitive Semantics*. Cambridge, MA: MIT Press.
Tomasello M (ed.) (1998) *The New Psychology of Language: Cognitive and Functional Approaches to Language Structure*. Mahwah, NJ: Erlbaum.
Turner M (2001) *Cognitive Dimensions of Social Science*. New York, NY: Oxford University Press.
Vandeloise C (1991) *Spatial Prepositions: A Case Study from French*. Chicago, IL: University of Chicago Press.
Van Hoek K (1997) *Anaphora and Conceptual Structure*. Chicago, IL: University of Chicago Press.
Zbikowski L (2002) *The Conceptualizing Music: Cognitive structure, theory and analysis*. New York, NY: Oxford University Press.

Further Reading

Fauconnier G (1997) *Mappings in Thought and Language*. Cambridge, UK: Cambridge University Press.
Israel M (forthcoming) *The Rhetoric of Grammar*. Cambridge, UK: Cambridge University Press.
Jackendoff R (1983) *Semantics and Cognition*. Cambridge, MA: MIT Press.
Kemmer S and Verhagen A (1994) The grammar of causatives and the conceptual structure of events. *Cognitive Linguistics* **5**: 115–156.
Mandelblit N (1997) *Grammatical Blending: Creative and Schematic Aspects in Sentence Processing and Translation*. PhD thesis, University of California, San Diego.
Moore T and Carling C (1982) *Language Understanding: Towards a Post-Chomskyan Linguistics*. New York, NY: St Martin's Press.
Robert A (1998) Blending in the interpretation of mathematical proofs. In: Koenig J-P (ed.) *Discourse and Cognition*. Stanford, CA: CSLI.
Turner M (1996) *The Literary Mind*. New York, NY: Oxford University Press.

Cognitive Maps

See **Animal Navigation and Cognitive Maps**

Cognitive Organisation

See **Early Experience and Cognitive Organization**

Cognitive Processing Through the Interaction of Many Agents

Intermediate article

Chris Jones, University of Southern California, Los Angeles, California, USA
Maja Matarić, University of Southern California, Los Angeles, California, USA
Barry Werger, Jet Propulsion Laboratory, Pasadena, California, USA

CONTENTS

A collection of interacting autonomous entities, called 'agents', may be capable of creating complex cognitive processes and behaviors, which could not be achieved by a single agent, without the need for outside centralized coordination or control.

INTRODUCTION

Several theories of cognition, most notably Minsky's 'society of mind', posit that intelligent behavior can be seen as the result of the interaction of simple processes. Minsky states: 'Very few of our actions and decisions come to depend on any single mechanism. Instead, they emerge from conflicts and negotiations among societies of processes that constantly challenge one another' (Minsky, 1986). The central tenet of such theories of cognition and behavior is that complex system-level behavior can emerge from the interaction of multiple, possibly numerous, components.

A canonical example of such emergence is the function of the human brain. The brain itself is made up of billions of simple neurons organized into a massively connected network (Nicholls *et al.*, 2001). In general, an individual neuron acts as a comparatively simple processing unit that receives signals from a set of neighboring input neurons, and under appropriate conditions transmits signals to a set of its neighboring output neurons. From this network of interacting neurons emerges the complexity of human cognition and behavior. No single neuron or subset of neurons is responsible for this complexity; rather, it is the result of their interactions.

Several disparate research fields are actively involved in investigating the principles of interaction among a collection of components. These include cognitive science, computer networks, distributed systems, artificial life, collective robotics, multi-agent systems, as well as others. The rest of this article aims to explain conceptually, and show through examples, how complex cognition and behavior can emerge from the interaction of individual components and how those emergent behaviors can be used in a variety of ways.

AGENTS

The term 'agent' has become a popular choice for a nontrivial component of a system with many interacting components that result in emergent behavior. Precisely defining an agent remains difficult, as agents come in many guises. An agent could be a piece of software, a specific computer on the Internet, a mobile robot, or even a person. In general, an agent is an autonomous entity, situated in an environment, and equipped with some degree of intelligence.

Being 'situated' places a number of constraints on how the agent can operate (Brooks, 1991; Maes, 1990). It implies that the agent has some means of sensing its environment, but the sensing may be limited and inaccurate. For example, a mobile robot may be situated in an office environment and have a means of sensing the distance to nearby objects; thus its sensing capabilities are both limited and prone to error and noise. Situatedness also implies an interaction with the environment. Thus, the same robot may be able to drive around, affecting the environment with its placement, and perhaps move objects, intentionally or otherwise. Conversely, the actions of a situated agent are influenced by the environment. The objects the robot encounters affect what it senses and how it behaves. Another implication of being situated is that an agent is constrained by environmental characteristics. For example, a mobile robot cannot drive through walls nor can it avoid falling if it drives down a set of stairs. Finally, the agent's characteristics – its computational, sensory, and actuation capabilities – influence how it interacts with its environment, which includes other agents.

SOCIETIES OF AGENTS

A collection of interacting agents is referred to as a 'society of agents'. Using this metaphor, a brain is a society of agents, as is a team of mobile robots cooperating on a task. Such agent societies are interesting for a number of reasons. First, for certain tasks and/or environments, a society of agents is the only viable or efficient solution. Second, even for tasks that can be handled by an individual agent, there may be more efficient, adaptive, and robust solutions performed by a society of agents.

A society of agents may consist of a homogeneous or heterogeneous collection of agents. In a homogeneous society, all agents are identical, while in a heterogeneous society, agents may have different characteristics. The variations in capabilities may result in hierarchies, specializations, or various other forms of social organization. Consequently, heterogeneous societies are generally more complex to control but are typically capable of a larger set of tasks.

The human immune system (Segel and Cohen, 2001) is an excellent example of a society of heterogeneous yet simple agents. It is capable of protecting the body against infection and invasion by foreign substances of all kinds, whether bacteria, virus, parasite, etc., which can be viewed as a very large set of different defensive 'tasks' to be accomplished. Each agent in this society is very specific, but the large number of agents of each kind and in total, combined with the ability to generate additional agents when needed, produces an unprecedented defensive functionality.

AGENT INTERACTION

Agents situated in a shared environment have ample opportunity to interact, by directly sensing each other, communicating, coordinating actions, and even competing. The shared environment in which the interactions take place may be an abstract data space, the physical world, or anything in between. A society of software agents may interact through a personal computer or even the entire Internet. Likewise, a society of mobile robots may interact in an office environment or on the surface of Mars. Furthermore, multiple environment types may be spanned in a single multi-agent system. For example, in multi-robot systems using behavior-based control (Matarić, 1994), individual robots are controlled by a collection of internal agents that interact through a computational environment, while the society of physical robots that contains them interacts in and through the physical world. The nature of the interactions in a society of agents depends upon such factors as agent capabilities, environmental constraints, and desired local and global behavior.

The spectrum of agent interaction is broad. At one end are methods that employ large numbers of simple, identical agents, connected together in patterns which lead to useful computation as a result of data flow through the system. At the other end are systems of complex, specialized agents which explicitly negotiate for task assignments and resources. Mechanisms for agent interaction can be broadly classified as fitting into the following, often overlapping categories: interaction through the environment, interaction through sensing, and interaction through communication. Each is described in turn.

Interaction Through the Environment

The first mechanism for interaction among agents is through their shared environment. This form of interaction is indirect in that it consists of no explicit communication or physical interaction between agents. Instead, the environment itself is used as a medium of indirect communication. This is a powerful method of interaction that can be utilized by very simple agents with no capability for complex reasoning or for direct communication.

Stigmergy is an example of interaction through the environment employed in a variety of natural insect societies. Originally introduced in the biological sciences to explain some aspects of social insect nest-building behavior, stigmergy is defined as 'the process by which the coordination of tasks and the regulation of constructions does not depend directly on the workers, but on the constructions themselves' (McFarland, 1985; Holland and Melhuish, 1999). The notion was originally used to describe the nest-building behavior of termites and ants (Franks and Deneubourg, 1997). It was shown that the coordination of building activity in a termite colony was not inherent in the termites themselves. Instead, the coordination mechanisms were found to be regulated by the task environment, in this case the growing nest structure. A location on the growing nest stimulates a termite's building behavior, thereby transforming the local nest structure, which in turn stimulates additional building behavior of the same or another termite.

Examples of artificial systems in which agents interact through the environment include distributed construction (Bonabeau *et al.*, 1994), sorting (Deneubourg *et al.*, 1990), clustering (Beckers *et al.*, 1994), optimization problems (Dorigo *et al.*, 1999), object manipulation (Donald *et al.*, 1993), analysis of network congestion (Huberman and Lukose, 1997), and phenomena such as the spread of computer viruses (Minar *et al.*, 1998).

Interaction Through Sensing

The second mechanism for interaction among agents is through sensing. As described by Cao *et al.* (1997), interaction through sensing 'refers to local interactions that occur between agents as a result of agents sensing one another, but without explicit communication'. This form of interaction is also indirect, as there is no explicit communication between agents; however, it requires each agent to be able to distinguish other agents from miscellaneous objects in the environment.

Interaction through sensing can be used by an agent to model the behavior of another agent or to determine what another agent is doing in order to make decisions and respond appropriately. For example, flocking birds use sensing to monitor the actions of other birds in their vicinity in order to make local corrections to their own motion. It has been shown that effective flocking results from quite simple local rules followed by each of the birds in the society (flock), responding to the direction and velocity of the local neighbors (Reynolds, 1987). Such methods of interaction through sensing can be found in use in mobile robot flocking, following, and foraging (Matarić, 1995), robot soccer (Werger, 1999), robot formations (Fredslund and Matarić, 2002), and simulations of behaviorally realistic animations of fish schooling (Tu and Terzopoulos, 1994). Other applications of interaction through sensing include human-like physical or visual interaction between physical agents (Murciano and del R. Millan, 1996; Michaud and Vu, 1999; Nicolescu and Matarić, 2000), including the ability to understand and influence the motives of other physical agents (Breazeal and Scassellati, 1999; Ogata *et al.*, 2000).

Interaction Through Communication

The third mechanism for interaction among agents is through direct communication. Unlike the first two forms of interaction described above, which were indirect, in interaction through communication agents may address other agents directly, either in a system-specific manner or through a standard agent communication protocol such as KQML (Finin *et al.*, 1996) or CORBA (Vinoski, 1997). Such agent-directed communication can be used to request information or action from others or to respond to requests received from others. Communication may be task-related rather than agent-directed, in which case it is made available to all (or a subset) of the agents in the environment. Two common task-related communication schemes are blackboard architectures (Schwartz, 1995; Gelernter, 1991) and publish/subscribe messaging (Arvola, 1998). In blackboard architectures, agents examine and modify a central data repository; in publish/subscribe messaging, subscribing agents request to receive certain categories of messages, and publishing agents supply messages to all appropriate subscribers. In some domains, such as the Internet, communication is reliable and of unlimited range, while in others, such as physical robot interaction, communication range and reliability are important factors in system design (Arkin, 1998; Gerkey and Matarić, 2001).

FROM AGENT INTERACTION TO COGNITION AND BEHAVIOR

Given a society of interacting agents, how is complex system-level behavior achieved? Interaction among the society members is not sufficient in itself to produce an interesting or useful global result. In order for the interacting agents to produce coherent global behavior, there must be some overarching coordination mechanism that appropriately organizes the interactions in both space and time.

There are many coordination mechanisms by which to organize the various interactions among agents in order to produce coherent system-level behavior. Self-organization techniques are based on a 'set of dynamical mechanisms whereby structures appear at the global level of a system from interactions among its lower-level components. The rules specifying the interactions among the system's constituent units are executed on the basis of purely local information, without reference to the global pattern, which is an emergent property of the system rather than a property imposed upon the system by an external ordering influence' (Bonabeau *et al.*, 1997). Methods such as genetic algorithms (Holland, 1975), machine learning techniques such as reinforcement learning (Sutton and Barto, 1998), and distributed constraint satisfaction (Clearwater *et al.*, 1991) can all be used to design agents and their interactions such that the resulting behavior meets desired system-level goals. Agents may also explicitly negotiate with each other for resources and task assignments in order to coordinate their behavior.

One such approach, employed in human as well as synthetic agent societies, is 'market-based' coordination, where individual agents competitively bid for tasks, which they must either complete or report as broken contracts. Auctions are a common coordination method in market-based techniques. In auctions, the most appropriate agents are continuously selected and (re)-assigned to various non-terminating roles (Tambe and Jung, 1999; Werger and Matarić, 2000). In contrast, more symbolic negotiation protocols based on distributed planning involve multiple stages, in which agents first share their plans, then criticize them, and finally update them accordingly (Bussmann and Muller, 1992; Kreifelt and von Martial, 1991; Lesser and Corkill, 1981). Game-theoretic approaches to negotiation have proven effective in situations where agents may be deceptive in their communication (Rosenschein and Zlotkin, 1994). In most complex models of negotiation-based coordination, agents reason about the beliefs, desires, and intentions of other agents, and influence those using specialized techniques (Brandt *et al.*, 2000).

CONCLUSION

As was stated earlier, in systems where complex global behavior emerges from the interactions of a society of simple agents, as the complex function of the brain emerges from the interactions of a large society of neurons, the resulting complexity cannot be attributed to any single agent but instead to the interaction of all agents. Agents and their, often local, interactions with each other and with the environment generate the resulting global behavior of the system – no additional external coordination mechanism is needed. Interaction in the agent society can take place through several mechanisms, including interaction through the environment, through sensing, and through communication. In lieu of a central coordinator, a society of agents coordinates its interactions to produce desired system-level behavior through such mechanisms as self-organization, machine learning techniques, or more complex negotiation mechanisms.

The notion that complex behavior can arise from the interaction of simple agents has powerful and far-reaching implications. Many fields, ranging from biology, to artificial intelligence, computer networking, and business management, all find inspiration and motivation from these principles.

References

Arkin R (1998) *Behavior-Based Robotics*. Cambridge, MA: MIT Press.

Arvola C (1998) *Transactional Publish / Subscribe: The Proactive Multicast of Database Changes*. SIGMOD Conference.

Beckers R, Holland OE and Deneubourg JL (1994) *Proceedings, Artificial Life IV*. In: R Brooks and P Maes (eds), pp. 181–189. Cambridge, MA: MIT Press.

Bonabeau E, Theraulaz G, Arpin E and Sardet E (1994) The building behavior of lattice swarms. In: Brooks R and Maes P (eds) *Artificial Life* **4**: 307–312.

Bonabeau E, Theraulaz G, Deneubourg J-L and Camazine S (1997) Self-organisation in social insects. *Trends in Ecology and Evolution* **12**(5): 188–193.

Brandt F, Brauer W and Weiss G (2000) *Task Assignment in Multiagent Systems based on Vickrey-type Auctioning and Leveled Commitment Contracting*. Proceedings of the Fourth International Workshop on Cooperative Information Agents.

Breazeal C and Scassellati B (1999) *How to Build Robots that Make Friends and Influence People*. Proceedings of the IEEE/RSJ International Conference on Intelligent Robots and Systems (IROS-99), Kyongju, Korea.

Brooks RA (1991) *Intelligence without Reason*. Proceedings of the International Joint Conference on Artificial Intelligence, Sydney, Australia.

Bussmann S and Muller JA (1992) *Negotiation Framework for Cooperating Agents*. In: Deen SM (ed.) Proceedings CKBS-SIG, Dake Centre, University of Keele.

Cao Y, Fukunaga A and Kahng A (1997) Cooperative mobile robotics: Antecedents and directions. *Autonomous Robots* **4**: 7–27.

Deneubourg JL, Goss S, Franks, Sendova-Franks A, Detrain C and Chretien L (1990) *Proceedings, Simulation of Adaptive Behavior*, pp. 365–363, Cambridge, MA: MIT Press.

Donald BR, Jennins J and Rus D (1993) Proceedings, International Symposium on Robotics Research, Hidden Vallen, PA.

Dorigo M, Di Caro G and Gambardella LM (1999) Ant algorithms for discrete optimization. *Artificial Life* **5**(3): 137–172.

Finin T, Labrou Y and Mayfield J (1996) KQML as an agent communication language. In: Bradshaw JM (ed.) *Software Agents*. Cambridge, MA: AAAI/MIT Press.

Franks NR and Deneubourg J-L (1997) Self-organising nest construction in ants: individual worker behaviour and the nest's dynamics. *Animal Behaviour* **54**: 779–796.

Fredslund J and Matarić M (2002) *A General, Local Algorithm for Robot Formations*. IEEE Transactions on Robotics and Automation.

Gelernter D (1991) *Mirror Worlds*. New York, NY: Oxford University Press.

Gerkey BP and Matarić MJ (2001) Principled communication for dynamic multi-robot task allocation. In: Rus D and Singh S (eds) *Experimental Robotics VII*, pp. 253–362. Springer Verlag: Berlin.

Gerkey BP and Matarić MJ (2002) Sold! Auction methods for multi-robot coordination. *IEEE Transactions on Robotics and Automation*, special issue on multirobot systems.

Holland JH (1975) *Adaptation in Natural and Artificial Systems*. Ann Arbor, MI: University of Michigan Press.

Holland OE and Melhuish C (1999) Stigmergy, self-organisation, and sorting in collective robotics. *Artificial Life* **5**(2): 173–202.

Huberman A and Lukose RM (1997) Social dilemmas and Internet congestion. *Science* **277**: 535–537.

Kandel ER, Schwartz JH and Jessell TM (1995) *Essentials of Neural Science and Behavior*. Norwalk, CT: Appleton and Lange.

Kreifelt T and von Martial FA (1991) A negotiation framework for autonomous agents. In: Demazeau Y and Muller JP (eds) *Decentralized A. I. 2*. Oxford, UK: Elsevier Science.

Lesser V and Corkill D (1981) Functionally accurate, cooperative distributed systems. *IEEE Transactions on Systems, Man and Cybernetics* **11**(1): 81–96.

Maes P (1990) Situated agents can have goals. *Robotics and Autonomous Systems* **6**: 49–70.

Matarić MJ (1994) *Interaction and Intelligent Behavior*. MIT EECS PhD Thesis, MIT AI Lab Tech Report AITR-1495.

Matarić MJ (1995) Designing and understanding adaptive group behavior. *Adaptive Behavior* **4**(1): 51–80.

McFarland D (1985) *Animal Behavior*. Menlo Park, CA: Benjamin Cummings.

Michaud F and Vu MT (1999) Managing robot autonomy and interactivity using motives and visual communication. In Proceedings Conference Autonomous Agents. Seattle, Washington.

Minar N, Kwindla HK and Pattie M (1999) *Cooperating Mobile Agents for Mapping Networks*, Proceedings of the First Hungarian National Conference on Agent Based Computation.

Minsky M (1986) *The Society of Mind*. New York, NY: Simon & Schuster.

Murciano A and del R Millán J (1996) Learning signaling behaviors and specialization in cooperative agents. *Adaptive Behavior* **5**(1).

Nicolescu M and Matarić M (2000) *Learning Cooperation From Human-Robot Interaction*. Proceedings, 5th International Symposium on Distributed Autonomous Robotic Systems (DARS), 4–6 Oct, Knoxville, TN.

Ogata T, Matsuyama Y, Komiya T *et al.* (2000) Development of emotional communication robot: WAMOEBA-2R -Experimental evaluation of the emotional communication between robots and humans. Proceedings of IEEE/RSJ International Conference on Intelligent Robots and Systems (IROS' 2000), pp. 175–180, November. Takamatsu, Japan.

Reynolds C (1987) Flocks, herds, and schools: a distributed behavioral model. *Computer Graphics* **21**(4): 25–34.

Rosenschein JS and Zlotkin G (1994) *Rules of Encounter: Designing Conventions for Automated Negotiation among Computers*. Cambridge, MA: MIT Press.

Schwartz DG (1995) *Cooperating Heterogeneous Systems*. Dordrecht: Kluwer Academic.

Segel LA and Cohen IR (2001) *Design Principles for the Immune System and Other Distributed Autonomous Systems*. Oxford, UK: Oxford University Press.

Sutton and Barto (1998) *Reinforcement Learning: An Introduction*. Cambridge, MA: MIT Press.

Tambe M and Jung H (1999) The benefits of arguing in a team. *AI Magazine* **20**(4).

Tu X and Terzopoulos D (1994) *Artificial Fishes: Physics, Locomotion, Perception, Behavior*. Computer Graphics, SIGGRAPH 94 Conference Proceedings, pp. 43–50, July.

Vinoski S (1997) CORBA: Integrating diverse applications within distributed heterogeneous environments. *IEEE Communications Magazine*, February.

Werger B (1999) Cooperation without deliberation: a minimal behavior-based approach to multi-robot teams. *Artificial Intelligence* **110**: 293–320.

Werger B and Matarić MJ (2001) From insect to internet: Situated control for networked robot teams. *Annals of Mathematics and Artificial Intelligence* **31**(4): 173–198.

Further Reading

Bonabeau E, Dorigo M and Theraulaz G (1999) *Swarm Intelligence: From Natural to Artificial Systems*. Oxford, UK: Oxford University Press.

Brooks R (1999) *Cambrian Intelligence*. Cambridge, MA: MIT Press.

Clearwater HS, Huberman BA and Hogg T (xxxx) Cooperative solution of constraint satisfaction problems. *Science* **254**: 1181–1183.

Cognitive Science

See **History of Cognitive Science and Computational Modeling**

Cognitive Science: Experimental methods

Introductory article

Raymond S Nickerson, Tufts University, Medford, Massachusetts, USA

CONTENTS

Cognitive activity can be studied in a variety of ways, including observation, simulation by computer modeling, and controlled experimentation.

INTRODUCTION

Cognitive science is a broad topic. It encompasses cognitive – or cognitive-like – activity wherever it is found, in humans, animals or machines. It is studied in a variety of ways, including observation, simulation (notably by efforts to give computers the ability to do things that when done by people are considered cognitive), and controlled experimentation.

PURPOSES OF EXPERIMENTATION

Scientific experimentation involves the investigation of how the controlled manipulation of one or more (independent) variables affects one or more other (dependent) variables.

Experiments are performed for several purposes, including testing hypotheses, establishing the values of parameters of process models, comparing the predictive power of competing theoretical accounts of specific phenomena, and evaluating the effectiveness of operational devices or procedures in realizing the intents of their designers. Sometimes experiments are done for such purposes as investigating hunches that are not sufficiently precise to be treated as testable hypotheses, looking for relationships or regularities that are worthy of more focused study, checking the adequacy of experimental designs, or fine-tuning setups of experimental equipment in anticipation of their use for more formal purposes. Experiments of these types are sometimes referred to as exploratory, pilot or calibration studies, and they are less likely than more formal studies to be reported in scientific journals – but they are experiments nonetheless. Experiments are also sometimes done to demonstrate already known relationships among variables; although these may replicate the conditions of experiments that have already been conducted for purposes of discovery, their purpose is strictly educational.

All these distinctions pertain to experimentation on cognition as well as to experimentation in other fields.

COMPARISON OF METHODS

Laboratory Versus Field Methods

Most studies of cognition take place in university laboratories or classrooms, or in the research facilities of industrial or government organizations. Some, however, are done in the field. An example of laboratory research is an experiment designed to investigate the effects of the spacing of rehearsal on the memorization and recall of verbal material. An example of field research is a study of the effects of a secondary task, such as carrying on a telephone conversation, on how well an automobile driver performs the driving task.

The choice of laboratory or field research in any particular case is likely to involve consideration of a trade-off between control and realism: between precision and applicability to real-world contexts. Generally speaking, it is possible to exercise much greater control over variables – both those whose effects the investigator is interested in studying, and those that are better thought of as nuisance factors – in a laboratory setting than in the field. However, this greater degree of control is usually bought at the price of making the situation so artificial that generalization of the findings from the laboratory to the real world may be difficult or impossible.

For these reasons it is desirable when possible to verify the generalizability of findings obtained in the laboratory to the real-world situations to which they are believed to apply before drawing firm conclusions about their applicability. The laboratory findings can serve as tentative conclusions that need to be confirmed in the real-world situation of interest. This approach is illustrated by efforts to determine whether what students have learned in a laboratory context designed to teach specific aspects of flying an airplane transfer to performance in an actual flight situation.

Cross-sectional Versus Longitudinal Studies

Most experiments in psychology involve comparisons between measurements that are made at approximately the same time. To determine the immediate effects on hearing of short-term exposure to noise, for example, one might simply determine the ability of people to detect weak auditory stimuli after exposure to noise of different intensities. Sometimes, however, the interest is in how people's abilities (attitudes, values, beliefs) change over long periods.

One experimental approach to the study of such changes is to investigate people of different ages; another is to study the same group of people over many years. The advantage of the first (cross-sectional) approach is that the study can be done quickly; a disadvantage is that people in the different groups are likely to differ in ways other than age (they were born and grew up in different times), which can complicate the interpretation of results. The advantage of the second (longitudinal) approach is the opportunity to see how people change with respect to abilities and other characteristics of interest with the passing of time; a disadvantage is that such studies take many years to perform, during which participants may leave the study, funding may be lost, and so on.

Both cross-sectional and longitudinal studies have proved useful in the study of aging, the findings of the one type complementing those of the other.

Single-case Studies Versus Averaging Over Multicase Samples

The vast majority of experiments on cognition involve the averaging of experimental data over a group of participants. Case studies, however, can sometimes provide extensive and in-depth information about individuals that is generally not available in data from multicase samples. Often case studies are opportunistic in the sense that they capitalize on the occurrence of a unique event or the chance discovery of an individual with an unusual ability. The highly publicized case of

Phineas Gage, the railroad worker who in 1848 suffered a horrendous but nonfatal injury when an iron bar used for tamping explosive powder was propelled by an accidental explosion completely through his head, illustrates the first situation. Gage's experience was unusual, if not unique, and documentation of the long-term cognitive and affective effects of his injury contributed to a better understanding of how certain functions depend on specific parts of the brain.

There are many published examples of case studies of individuals with unusual memories. John Dean's recorded recollection of events as given in testimony before the Watergate committee of the US Senate in 1973 provided the basis for one such study; discovery of a man (Hideaki Tomoyori) who was able to recite from memory the first 40 000 digits of pi provided another. Experiments with Tomoyori showed his ability to recall the details of narrative stories to be quite ordinary.

Comparing and Evaluating Models with Empirical Evidence

As in all experimental sciences, the acid test of the tenability of a theory in cognitive science is its ability to predict the results that would be obtained in carefully controlled experiments. In practice, progress occurs in a cyclic fashion. Predictions are derived from a theory. The predictions are tested by experimentation. Depending on the outcomes of experiments, theories may be strengthened and made more precise, or they may be shown to be false or in need of modification. Sometimes experimental results match theoretically derived predictions closely, in which case the theories from which the predictions were derived are considered to have been corroborated – not to have been proved, but to have gained greater credibility. Sometimes the results match the predictions only marginally or not at all. In such a case, if the experiment has been carefully done, the theory from which the prediction was derived might have to be modified or replaced.

A type of theorizing that has been used to great advantage in science involves the building of models of processes of interest. Attempting to build a working model of a process – an artifact that behaves in the same way as the process of interest – is an especially fruitful way to develop an understanding of that process. It has often been pointed out that centuries of observing birds in flight were not nearly as effective a means of learning about aerodynamics as attempts to build machines that could fly. Efforts to give computers the ability to do things that human beings do with apparent ease has revealed the hidden complexity underlying many human capabilities. The point is illustrated by the history of efforts to give computers the ability to understand natural language. Despite decades of intensive work on this problem, it remains unsolved in any general sense. It is now clear that the complexity of the problem was grossly underestimated when the quest was first engaged, but the effort has revealed much about human language comprehension, answering many questions while raising others that no one knew enough to ask before the effort was made.

The case for computational modeling has been articulated by several psychologists and cognitive scientists. Mathematical and computer models of human performance have been applied to great advantage to the design and study of complex systems in which humans function as hands-on operators or as supervisors of largely automated processes. Modeling typically involves an iterated series of steps: a model is developed that will accommodate experimental data that have already been collected, that model is then used as a basis for making predictions about the outcomes of additional experiments, and the actual outcomes of those experiments are used to modify the model or adjust its parameters so as to increase its predictive power or accuracy. This process of testing and refining can be continued indefinitely, or at least until the model is capable of predicting outcomes with a desired degree of accuracy or it becomes clear that it should be discarded for a qualitatively different one.

CONVERGING METHODOLOGIES

Many factors are relevant to the selection of an experimental methodology: the nature of the phenomenon of interest; the availability of resources (equipment, experimental participants, time); the practicalities of controlling the variables that must be controlled; the types of measurements that are feasible (brain potentials, skin conductance, pupil size, eye fixation, motor response times, verbal responses); the options one has for analyzing experimental data; the kinds of inferences one wishes to draw; and the nature of the population to which one wants to generalize the results. Often it is necessary to make trade-offs, gaining control, for example, at the expense of limited generalizability. Similarly, it typically is much easier for academic researchers to conduct experiments with students as participants than to conduct them with samples that are more representative of the general population, or of nonstudent populations to which they may wish to generalize results.

In general, experimenters tend to think of variance in data, other than that produced by intentional manipulation of the independent variables involved, as 'noise' that must be treated statistically in order to determine the effects of the experimental manipulations. However, one of the ways of limiting this noise (of controlling within-condition variability) is to use relatively homogeneous samples of participants; although this increases the chances of obtaining statistically significant results, it can also preclude the generalizability of the findings to populations that are less homogeneous than the sample studied.

For these and other reasons, it is good that experimentation can be done in a variety of ways. Generally, advances are made when insights into relationships are confirmed by experimental evidence of more than one type. Seldom, if ever, is any question of more than trivial importance, theoretical or practical, settled decisively with a single experiment. The general rule is one of gradual increase in understanding of phenomena of interest resulting from the convergence of evidence from a variety of sources.

Further Reading

Damasio H, Grabowski T, Frank R, Galaburda AM and Damasio AR (1994) The return of Phineas Gage: clues about the brain from the skull of a famous patient. *Science* **264**: 1102–1105.

Gopher D, Weil M and Bareket T (1994) Transfer of a skill from a computer game trainer to flight. *Human Factors* **36**: 1–19.

Meyer DE and Kieras DE (1999) Précis to a practical unified theory of cognition and action: some lessons from EPIC computational models of human multiple-task performance. In: Gopher D and Koriat A (eds) *Attention and Performance*, vol. 17, pp. 17–88. Cambridge, MA: MIT Press.

Neisser U (1981) John Dean's memory. *Cognition* **9**: 1–22.

Recarte MA and Nunes LM (2000) Effects of verbal and spatial-imagery tasks on eye fixations while driving. *Journal of Experimental Psychology: Applied* **6**: 31–43.

Cognitive Science: Philosophical Issues

Introductory article

Barbara Von Eckardt, University of Nebraska-Lincoln, Lincoln, Nebraska, USA

CONTENTS

Numerous philosophical questions can be raised about cognitive science, including what cognitive science is, what counts as representation, computation, and explanation in cognitive science, and what the limits of cognitive science are.

INTRODUCTION

Cognitive science is a multidisciplinary approach to the study of cognition and intelligence. It emerged in the late 1950s and early 1960s when researchers in psychology, linguistics and computer science worked together to develop an alternative to behaviorist approaches to mind and language. At its core, it holds that the human mind is a kind of computer which processes information in the form of mental representations (in this article, references to 'mind' are in the sense of a functional description of the brain). The major disciplinary participants in the cognitive science enterprise are psychology, linguistics, neuroscience, computer science, and philosophy. Other fields that are sometimes included are anthropology, education, mathematics, biology, and sociology.

Broadly speaking, there are two kinds of philosophical issue associated with cognitive science:

theoretical and conceptual issues raised within or about cognitive science to which philosophers have made a contribution; and traditional philosophical issues to which cognitive science research is relevant. Examples of the first would be: By virtue of what do mental representations have semantic properties? Are 'symbolic' or connectionist models more likely to account for our cognitive capacities? Examples of the second would be: Are there innate ideas? Is color an objective or a subjective property?

This distinction is somewhat blurred because theoretical and conceptual issues in cognitive psychology are sometimes very similar to those that have historically been discussed by philosophers. This article deals only with issues of the first type.

WHAT IS COGNITIVE SCIENCE?

The Core Research Framework of Cognitive Science

Because cognitive science is immature, complex, and continually changing, the question 'What is cognitive science?' is not easy to answer. One way to start is to identify a relatively stable 'core' and then describe various ways in which some cognitive science research deviates from that core.

The aim of core cognitive science is to explain the human cognitive capacities. These include our capacity to use language (perceive it, comprehend it, produce it, translate it, communicate with it, and so on), to perceive visually, to apprehend music, to learn, to solve problems (reason, draw inferences), to plan actions, to act intentionally, to remember, and to imagine. Each of these capacities can be explored in several ways. Consider, for example, the capacity for language. Cognitive scientists study language in normal adults; language development; language variation between cultures, groups (e.g. men and women, first and second language learners), and individuals; pathologies of language (speech disorders, aphasia); and how language is realized in the brain.

In studying the cognitive capacities, cognitive scientists make three basic assumptions: that the human mind is a kind of computer, that it has mental representations, and that it exhibits both conscious and unconscious processing. A popular way of describing cognitive science research is in terms of the three levels proposed by the vision scientist David Marr: a computational level, an algorithmic level, and an implementation level. At the computational level, we ask what precisely any given capacity is as a (mathematical) function from inputs to outputs; at the algorithmic level, we attempt to explain how a person executes the capacity characterized at the computational level; and at the implementation level, we ask how the capacity is implemented in the human brain.

Core cognitive science is also characterized by methodological assumptions. The most important are: that the research methods of cognitive science are scientific; that a complete theory of human cognition will not be possible without a substantial contribution from each of the contributing disciplines of cognitive science; that human cognition can be successfully studied by focusing on the individual cognizer and his or her place in the natural environment; and that answers to the basic questions of cognitive science in terms of information processing are constrained by the findings of human neuroscience.

The research framework of core cognitive science is, in principle, a framework to which each of the contributing disciplines of cognitive science conforms. However, each of these disciplines contributes to the research program built on this framework in its own distinctive way (see list below). (Note that not all subdisciplines of the contributing disciplines are part of cognitive science. For example, philosophy encompasses ethics and political philosophy, which are not considered part of cognitive science.)

Psychology

Enhances our understanding of the nature and limits of our cognitive capacities.

Proposes hypotheses concerning normal adult cognition, including hypotheses about the mental representations and computational processes involved in any given cognitive capacity, and devises experiments for testing these hypotheses.

Studies the acquisition of cognition in children and the development of cognition throughout our lifetime.

Studies individual and group differences in cognition.

Studies the psychopathology of cognition.

Computer science

Proposes computationally detailed models of cognition and tests these models against the data provided by psychology and neuroscience.

Develops hypotheses regarding the computational nature of the representation-bearers for the human system of mental representation.

Neuroscience

Studies the realization of our capacities in the brain.

Studies what happens when our cognitive capacities are exercised abnormally due to some neural abnormality or lesion.

Develops computational and representational hypotheses concerning cognition in normal adults, children, and abnormal individuals on the basis of low-level information about the brain.

Linguistics

Provides a theory of ideal capacity (a competence model) for language comprehension, production, and acquisition.

Proposes hypotheses regarding the representations involved in language comprehension and production.

Anthropology

Studies cognition across cultures.

Philosophy

Articulates the foundations of the field.

Explores the viability of the cognitive science research program.

Contributes to the development of a theory of content determination for the system of representation posited by cognitive science theories.

Contributes to the discussion of controversial theoretical and empirical issues in cognitive science, often involving adjudication between competing claims.

Helps to develop theories of ideal capacity (competence models) for reasoning and language use.

Areas of Disagreement and Evolution Within Cognitive Science

Within core cognitive science, the primary area of disagreement and change has been the conception of a computer. Initially, computers were thought to be 'physical symbol systems', a view proposed by Allen Newell and Herbert Simon in 1976. Research based on this assumption has been given various labels: 'classical', 'conventional', 'symbolic', and 'rules- and representation-based'. In the mid-1980s a new class of 'connectionist' or 'parallel distributed processing' computers began to attract the attention of cognitive scientists, resulting in heated disagreements about whether symbolic or connectionist computers were the best model of the human mind. Recently, two further views have emerged. The 'dynamic' approach models cognition in terms of a set of quantitative variables that change over time in accordance with dynamical mathematical laws (generally expressed by differential or difference equations). And many neuroscientists hold that the brain is not a physical symbol system nor a connectionist device nor a dynamic system; rather, it is a fourth, specifically biological kind of computational device, still poorly understood.

Deviations from core cognitive science can be found with respect to its domain, its approach, and the role it assigns to neuroscience. Since the late 1970s and early 1980s, when cognitive scientists started making statements about what cognitive science is, there has been a split (largely along disciplinary lines) over whether the domain of cognitive science is human cognition, or intelligence in general, including human, machine, and possibly animal intelligence. A further area of disagreement concerns whether cognitive science is concerned only with cognition or whether it includes all aspects of the mind, including touch, taste and smell, emotion, mood, motivation, and personality.

Although there is fairly widespread agreement that the core working assumption of the cognitive science 'approach' is that the mind is both a computational and a representational device, there are researchers, who consider themselves to be under the cognitive science umbrella, who have rejected either one of these assumptions: There are those who hold that the mind performs computations without representations; and there are also those who posit representations without computations.

A final area of disagreement concerns the role of neuroscience. Because cognitive science originally emerged from cognitive psychology, artificial intelligence research, and generative linguistics, in the early years neuroscience was often relegated to a secondary role. Citing the fact that descriptions of a system in purely functional terms can always be 'multiply realized', some cognitive scientists even declared that neuroscience, as the science of the physical realization of the functional mind, was irrelevant to cognitive science. Most researchers adopted a more moderate position: that the particular neural realization of the human mind imposes important constraints on what functions the mind can compute. But even such moderates sometimes held that research on the mind could (and should) proceed in a 'top-down' fashion. Most cognitive scientists, however, now believe that research on the mind should proceed in an interactive way: simultaneously top-down and bottom-up. The prevailing view in the field of cognitive neuroscience seems to be that neuroscience will be at the centre of efforts to develop a computational or information-processing theory of the mind driven primarily by bottom-up considerations.

Cognitive Science and Folk Psychology

The relationship between the conceptual frameworks of cognitive science and folk psychology is complex. As one might expect of an immature science, cognitive science has strong roots in common sense, but is also striving to develop its own empirically-based theoretical categories. Thus, cognitive science has clearly taken on board the ordinary notion of a cognitive capacity and the ordinary taxonomy of the specific cognitive capacities we take ourselves to have, including many of their associated states (such as states of perceiving, understanding and intending). However, since folk psychology has little to say about the unconscious processing that goes on when we exercise those capacities, it has been necessary for cognitive science to introduce a variety of subpersonal unconscious information-processing states to describe that processing. Furthermore, to complete the description at the subpersonal level, subpersonal information-processing states are also posited as 'underlying' the conscious states described in folk psychology. Thus, for example, recognizing that a given string of letters is a word of English becomes, in the theoretical language of cognitive science, successfully accessing the appropriate entry in the mental lexicon on the basis of a graphemic representation of the word.

REPRESENTATION IN COGNITIVE SCIENCE

Mental Representations

A mental representation is a structure or state in the mind that has semantic properties, such as referring to phenomena in the world, expressing a proposition, or predicating some property of something. In addition to having semantic properties (content), cognitive scientists generally assume that mental representations are carried by some representation bearer, that their content is 'grounded' in naturalistic properties and relations, and that they have significance for the individual that has them.

Given the assumption that the mind is a kind of computer, the bearers of mental representations are hypothesized to be computational structures or states. The specific nature of these structures or states depends on the kind of computer one takes the mind to be. The representation bearers of classical ('symbolic') computers are typically data structures; in contrast, the representation bearers of connectionist computers are activation states of nodes or sets of nodes (corresponding to occurrent mental states), or states consisting of links having certain weights (corresponding to dispositional mental states).

In attempting to explain cognition and intelligence, cognitive scientists have posited many kinds of mental representation. There is no neat taxonomy of these kinds. Sometimes representations are grouped together based on what they represent (phonological, lexical, syntactic and semantic representations in psycholinguistics), sometimes on their computational characteristics (local and distributed representations in connectionist systems), and sometimes on both (sentences, frames, schemas and scripts).

Theories of Content Determination

Philosophers interested in mental representation have focused primarily on the problem of what determines the semantic content of mental representations. Such theories are sometimes misleadingly referred to as theories of 'semantics' (e.g. 'informational semantics', 'functional role semantics'). It is more accurate to call them theories of 'content determination'. It is generally assumed that mental content is not a basic fact about the world, hence, that it comes about because the bearers of our mental representations have other naturalistic (non-intentional, non-semantic) properties. These content-determining properties can be considered the 'ground' of representational content.

There are currently a variety of proposals about this ground. They appeal either exclusively or in some combination to: the structure of the representation bearer (Palmer); actual historical or counterfactual causal relations between the representation bearer and phenomena in the world (Fodor, Devitt, Dretske); actual and counterfactual (causal, computational, inferential) relations between the state of the representation bearer and other states in the mind (Block, Cummins); the information-carrying or other functions of the state of the representation bearer and associated components (based on the qualities for which they were selected in evolution or learning) (Millikan); and the phenomenal properties of the representation. All current theories of content determination have problems. A major problem is how to account for the fact that we can falsely represent. For example, we can have a perceptual experience of a state of affairs that does not actually exist, or entertain a thought that is false.

COMPUTATION IN COGNITIVE SCIENCE

Philosophers have discussed two issues concerning the computational assumption of cognitive science. Cognitive science assumes that the mind is a computer. But what exactly is a computer? This question becomes particularly important given that alternative 'machine' models have been proposed (e.g. physical symbol system, connectionist device). Are these all different kinds of computer, as suggested above, or did cognitive science move away from the computational assumption when it became interested in connectionism? To answer these questions we need to know what a computer is.

The second issue concerns the dispute between the 'symbolic' and connectionist approaches. Which is the most promising approach? And can we decide this question on the basis of currently available evidence about cognition?

Computation and Computers

The term 'computer' is used in many different ways. Arguably, the ideal conception of a computer for cognitive science would be one that: (1) builds on the mathematical theory of computation; (2) encompasses classical 'symbolic' models, connectionist devices, and talk of a 'computational' brain; (3) isn't vacuous, and (4) is sufficiently informative to provide some guidance for construction of a theory. Point (3) is important: both Putnam and Searle have argued that the current conception of a computer is so vague as to be vacuous, in the sense that we can interpret any physical system as instantiating any computational characterization.

The theory of computation in mathematics defines a variety of different kinds of abstract machines capable of 'computing' functions in the mathematical sense. Although, in principle, functions can be defined over any domain (e.g. 'father of' maps people to people), mathematicians generally study functions defined over the natural numbers; theorems concerning such functions can then be generalized to functions over other domains. An important discovery has been that different kinds of automata (abstract machines) are associated with different sets of functions distinguished by various interesting mathematical properties.

The kind of machine of greatest interest to cognitive science is the Turing machine, named after the British mathematician Alan Turing. Such machines are important because it is believed that any function that can be computed by an *effective* method, that is, a method specified by a finite number of exact instructions, in a finite number of steps, to be carried out by a human unaided by any machinery except paper and pencil, and demanding no insight or ingenuity, can be computed by a Turing machine. (This claim is called the 'Church–Turing thesis'.) The so-called 'universal Turing machine' can compute all such Turing-computable functions.

Many cognitive scientists think that the notion of a computer relevant to cognitive science can simply be borrowed from the mathematical theory of computation. Unfortunately, the matter is not so straightforward. To see why, we must first distinguish between a physical system *P* being *equivalent* to an abstract automaton *A* and its being an *implementation* of *A*. Assuming we have some way of systematically mapping the inputs and outputs of *A* onto the inputs and outputs of *P*, we take *P* to be equivalent to *A* if and only if it can compute the same functions as *A*. To determine what counts as a legitimate implementation is much more difficult. Basically, the idea is that *P* counts as an implementation of *A* if and only if *P* is equivalent to *A* and there is a mapping from the formal states of *A* onto the physical states of *P* such that the formal state transitions of *A* are mirrored, both actually and counterfactually, by causal sequences in *P*.

There are many problems with defining a computer in terms of equivalence to some kind of abstract automaton: principally that such a definition fails to satisfy criterion (4) above, since it tells us nothing about the internal structure of the system. In contrast, while a definition in terms of implementation can be theoretically informative, saying that the mind is an implementation of a Turing machine seems false: at least, on any 'natural' way of doing the mapping. A Turing machine, even a universal Turing Machine, is in only one state at a time and uses a very small number of basic operations. In contrast, the causal structure of the brain relevant to cognition is extremely complex, and brains can change their structure in response to the information being processed. Some researchers, such as Copeland, have even suggested that the brain may have primitive operations that can compute functions that are not Turing-machine computable.

An alternative to defining a computer in terms of one or another class of abstract automaton is to provide a characterization that generalizes over different architectures but also, possibly, excludes some. One such characterization (due to Von Eckardt) is that a computer is a device capable of automatically receiving, storing, manipulating, and outputting information, by virtue of receiving, storing, manipulating and outputting representations of that information in accordance with a finite, effective set of rules that are, in some sense, in the machine itself. Another characterization (due to Copeland), specifically designed to encompass non-classical computation, is that a computer is a device, capable of solving a class of problems, that can represent both the problems and their solutions, contains some primitive operations (which may include operations that are not Turing-computable), can sequence these operations in some predetermined way, and has provision for feedback.

Computational Models and Human Cognition

A major dispute within cognitive science, to which philosophers have contributed in important ways, has been whether the mind is a classical or 'symbolic' computer ('classicism') or a connectionist computer ('connectionism'). Much of the discussion has centered around a challenge to connectionism put forward in 1988 by two proponents of classicism, Jerry Fodor and Zenon Pylyshyn. Fodor and Pylyshyn argued as follows:

1. An adequate theory of cognition must explain the fact that our cognitive capacities exhibit systematicity, that is, roughly, that some capacities are intrinsically connected to others. (If a native speaker of English knows how to say 'John loves the girl' she will know how to say 'The girl loves John'.)
2. Classical models can explain this property of cognition.
3. But it is not clear that connectionist models can, unless they are implementations of a classical model – in which case connectionism wouldn't constitute an alternative to classicism.
4. Therefore, classicism is to be preferred to connectionism.

Their reason for advocating premises 2 and 3 is that it is 'characteristic' of classical systems but not of connectionist systems to be 'symbol processers', that is, systems that posit mental representations with constituent structure (parts ordered in a certain way) and process these representations in a way that is sensitive to that structure.

Since 1988 there has been a flood of responses to this argument; in addition, the argument itself has evolved in two significant ways. First, it has been emphasized that to be a counterexample to premise 2, a connectionist model must not only exhibit systematicity; it must also explain it. Second, it has been claimed that what needs explaining is not just that human cognitive capacities are systematic; it is that they are *necessarily* (on the basis of scientific law) systematic.

Every aspect of the Fodor–Pylyshyn argument has been subjected to scrutiny. Numerous authors have noted the need for, and proposed, a more precise conception of systematicity. There has also been discussion of what it means to model cognition at the cognitive level, under what conditions one model counts as an implementation of another, and what is required for a model to explain (rather than merely model) either the fact or the necessity of systematicity.

In addition, each of the premises has been questioned. Several computer scientists and philosophers have sought to demonstrate that connectionist systems can model systematicity without being classical implementations, thus refuting premise 3. Some have accepted the Fodor–Pylyshyn conceptualization of the challenge whereby any connectionist model that processes symbols is, *ipso facto*, a classical implementation, and have sought to demonstrate that systematicity can be modelled by a system that is not a symbol processor. Others have resisted the Fodor–Pylyshyn dichotomy, arguing that a system can be connectionist, and have structured representations and structure-sensitive processes, without also being classical. Although proponents of classicism have not explicitly conceded defeat on this point, recent defences of the Fodor–Pylyshyn argument have concentrated on the claim that connectionists will never have the resources to explain either the fact or the necessity of systematicity, rather than that they will not be able to simply model systematicity.

Premise 1 has been questioned by researchers who have variously argued: that our cognitive capacities aren't, in fact, systematic; that that systematicity is a matter of conceptual necessity and, hence, not something that an empirical theory needs to explain; or that the explanation of their systematicity need not come from a theory of cognition. Another point is that the empirical facts about systematicity may be more complicated than Fodor and Pylyshyn assume.

The questions regarding premise 2 are especially important. The Fodor–Pylyshyn argument rests on classical systems being able to do something that connectionist systems (at a cognitive, non-implementational level) cannot. However, it has been pointed out that the fact that a system is a symbol processer (and, hence, classical) does not by itself explain systematicity, much less the lawfulness of systematicity: additional assumptions must be made about the system's computational resources. Thus, if it is true that certain kinds of connectionist system can also explain systematicity, then the explanatory asymmetry between classicism and connectionism will no longer hold (that is, both classical and connectionist systems may or may not explain systematicity when appropriately supplemented) and the Fodor–Pylyshyn argument will be unsound.

EXPLANATION IN COGNITIVE SCIENCE

From the late 1960s onwards, several philosophers have noted that explanation in cognitive science has certain distinctive features. Scientific explanation had commonly been taken to be an answer to a 'why' question ('Why did this event occur?' 'Why do these regularities hold?'), and to involve subsumption of the phenomenon to be explained under laws. In contrast, it was noted, cognitive scientists seek to explain capacities, rather than events or regularities – either by virtue of what we have them (Haugeland's 'systematic explanation') or how we exercise them (Cummins's 'functional analysis') – by appeal to the organized interaction of subcapacities.

While these early discussions succeeded in drawing philosophical attention to some important distinctive features of explanation in cognitive science, the full story turns out to be more complicated. If one adopts the view that explanations are answers to certain kinds of questions and that kinds of explanation are distinguished ('individuated') both by what is being explained and by the kind of explaining being done, then, as Barbara von Eckardt has emphasized, there are many different kinds of explanation to be found in cognitive science. Each of the various questions associated with each type of cognitive capacity has its own kind of explanation. For any given capacity *C* (say, the capacity to perceive visually), cognitive scientists attempt to explain how a normal adult generally exercises *C*, how a child generally acquires *C*, in what ways *C* breaks down under various conditions of psychopathy, what kinds of cultural variation there are in the exercise of *C*, and so on. Further, insofar as cognitive science entertains alternative ways of answering questions – for example, by means of a classical AI program, connectionist model, or dynamical system – even more forms of explanation can be distinguished. Classically explaining how a child acquires the capacity to speak a natural language involves a different kind of explanation than explaining the same phenomenon in a connectionist way.

Do the accounts offered by cognitive science succeed in being explanatory? Do they at least provide possible explanations for what they are intended to explain? Philosophers have raised two sorts of concern. First, the human cognitive capacities that cognitive science seeks to explain are ultimately rooted in our commonsense conception of ourselves. Thus, among the properties of these capacities that require an explanation is their intentionality, that is, the fact that they have content (we perceive or understand that something is the case). The explanatory strategy adopted by cognitive science is to explain the intentionality of mental states, as ordinarily construed, by appealing to the existence of mental representations with content at the subpersonal level. The concern is that this strategy will not work because none of the possible ways of construing subpersonal representational content gives this content the required explanatory role.

A second concern arises from the fact that although most forms of explanation in cognitive science involve 'how' or 'what' questions, cognitive scientists do sometimes seek to explain why people behave in certain ways. (The ability to explain a subject's behavior under experimental conditions constitutes the evidence for proposed answers to the 'how' and 'what' questions.) Typically, of course, these explanations purport both to be causal and to appeal to individual mental representations. But, it has been argued, there are reasons for thinking that it is not a representational state's having a certain content that is doing the causal work; rather, it is the computational or neural structure which 'has' that content (just as what causes me to perceive a word on the printed page is not the word's meaning but its physical embodiment in ink). So there really isn't any genuinely mental causal explanation.

There is one final worry. Granted that explanations that appeal to mental representations could provide possible explanations, are such explanations really needed? For example, couldn't cognitive science simply posit processing that is purely formal, or purely low-level computational (number crunching), or purely neurophysiological?

LIMITS OF COGNITIVE SCIENCE

For more than twenty years, critics of cognitive science, such as John Haugeland, have put together lists of mental phenomena that, they claim, cognitive science will never be able to explain. These include: that people 'make sense' in their actions and speech; that they have sensations, emotions, and moods; that they have a self and a sense of self; that they have consciousness; that they are capable of insight and creativity; that they can possess highly developed intellectual and artistic skills; and that they interact closely and directly with the world in which they live. When cognitive science consisted simply of a top-down, classical ('symbolic') approach, it was quite plausible to view some (or even all) of these phenomena as representing limits of the field. However, given the increasing integration of the non-neural cognitive sciences with neuroscience and the apparent flexibility of the notions of representation and computation, it is now far less clear whether these phenomena are really 'limiting'. Perhaps the most difficult aspect of mentality for cognitive science to explain is what philosophers have called 'phenomenal consciousness', that is, the 'feeling' or experience we have in connection with various kinds of mental state. In particular, as Joseph Levine has emphasized, there seems to be an 'explanatory gap' here: even if we can determine what the neural basis of a certain kind of experience is, cognitive science doesn't seem able to explain why that neural state gives rise to this kind of experience. However, even if this problem of phenomenal consciousness does represent an absolute limit to what cognitive science can explain, it is not obviously a limit to the *scientific* research program of cognitive science, that is, to the programme of developing a scientific theory of the mind.

Further Reading

Albright TD and Neville HJ (1999) Neurosciences. In: Wilson RA and Keil FC (eds) *The MIT Encyclopedia of the Cognitive Sciences*. Cambridge, MA: MIT Press.

Bechtel W and Abrahamsen A (1991) *Connectionism and the Mind*. Cambridge, MA: Blackwell.

Block N (1986) Advertisement for a semantics for psychology. In: French PA, Uehling TE and Wettstein HK (eds) *Midwest Studies in Philosophy, Studies in the Philosophy of Mind*, vol. X. Minneapolis, MN: University of Minnesota Press.

Churchland PS and Sejnowski TJ (1994) *The Computational Brain*. Cambridge, MA: MIT Press.

Copeland BJ (1997) The broad conception of computation. *American Behavioral Scientist* **40**: 690–716.

Cummins R (1983) *The Nature of Psychological Explanation*. Cambridge, MA: MIT Press.

Cummins R (1989) *Meaning and Mental Representation*. Cambridge, MA: MIT Press.

Cummins R and Cummins DD (eds) (2000) *Minds, Brains, and Computers: The Foundations of Cognitive Science*. Malden, MA: Blackwell.

Dawson MRW (1998) *Understanding Cognitive Science*. Malden, MA: Blackwell.

Devitt M (1981) *Designation*. New York, NY: Columbia University Press.

Dretske F (1981) *Knowledge and the Flow of Information*. Cambridge, MA: MIT Press.

Dretske F (1986) Misrepresentation. In: Bogdan R (ed.) *Belief*. Oxford, UK: Oxford University Press.

Fodor JA (1975) *The Language of Thought*. Cambridge, MA: MIT Press.

Fodor JA (1987) *Psychosemantics*. Cambridge, MA: MIT Press.

Gazzaniga MS (ed) (2000) *Cognitive Neuroscience*. Malden, MA: Blackwell.

van Gelder T (1995) What might cognition be, if not computation? *Journal of Philosophy* **92**: 345–381.

Haugeland J (1978) The nature and plausibility of cognitivism. *Behavioral and Brain Sciences* **2**: 215–260.

Haugeland J (1985) *Artificial Intelligence: The Very Idea*. Cambridge, MA: MIT Press.

Horst SW (1996) *Symbols, Computation, and Intentionality*. Berkeley, CA: University of California Press.

Levine J (1983) Materialism and qualia: the explanatory gap. *Pacific Philosophical Quarterly* **64**: 354–361.

Macdonald C and Macdonald G (1995) *Connectionism: Debates on Psychological Explanation*. Oxford: Blackwell.

Mathews RJ (1994) Three-concept monte: explanation, implementation and systematicity. *Synthese* **101**: 347–363.

Millikan R (1989) Biosemantics. *Journal of Philosophy* **86**: 281–297.

Newell A (1980) Physical symbol systems. *Cognitive Science* **4**: 135–183.

Newell A and Simon HA (1976) Computer science as empirical inquiry: symbols and search. *Communications of the Association for Computing Machinery* **19**: 113–126.

Osherson DN (1990) *Invitation to Cognitive Science*, 3 vols. Cambridge, MA: MIT Press.

Palmer SE (1999) *Vision Science: Photons to Phenomenology*. Cambridge, MA: MIT Press.

Putnam H (1988) *Representation and Reality*. Cambridge, MA: MIT Press.

Rumelhart DE, McClelland JL and the PDP Research Group (eds) (1986) *Parallel Distributed Processing: Explorations in the Microstructures of Cognition*, vol. I. Cambridge, MA: MIT Press.

Searle J (1992) The critique of cognitive reason. In: *The Rediscovery of the Mind*. Cambridge, MA: MIT Press.

Von Eckardt B (1993) *What Is Cognitive Science?* Cambridge, MA: MIT Press.

Color

See **Inverted Spectrum**

Color Perception, Psychology of

Intermediate article

Michael D'Zmura, University of California, Irvine, California, USA

CONTENTS

The psychology of color perception concerns color appearance and the visual processing of light spectral information. Researchers in this area seek to determine the relationships among visual stimuli, activity in the human nervous system and the conscious representation of color.

INTRODUCTION

Color is the psychological representation of light spectral properties. Normal human color vision is served by three classes of retinal photoreceptor, which differ in their abilities to respond to photons of varying wavelength. Comparing the responses of the three kinds of photoreceptor to a light provides an estimate of the light's spectral properties: how its energy varies with wavelength. Color depends also on the perceived cause of light reaching the eye, such as emission by a light source or reflection by a surface. Color thus helps to identify objects and their material properties, to the extent that object color remains constant under varying conditions of viewing and that color is handled appropriately by memory and other cognitive systems.

PHOTORECEPTORAL RESPONSES TO LIGHT

The transduction of light energy into neural responses is accomplished by the approximately 5 million cones and 100 million rods arrayed across the retina. Found primarily within the central 10° of the visual field, cones serve photopic vision under viewing conditions when colors are visible. Color vision depends on the responses of more than one spectral class of cone. The spectral sensitivity of a class of cones is determined by the photopigment molecule responsible for photon absorption. Knowledge of cone spectral sensitivities lets one predict the initial chromatic response of the visual system to a light.

Trichromatic Color Matching

That normal human color vision is served by a system with three spectral classes of sensors may be inferred from the results of color-matching experiments. A prototypical experiment uses quasi-monochromatic lights, each possessing energy in just a narrow band of visible wavelengths. The visible spectrum ranges in wavelength from about 400 nm (blue, with ultraviolet lights at shorter wavelengths) to about 700 nm (red, with infrared lights at longer wavelengths). Results show that a single monochromatic test light can be matched in appearance by combining additively three primary lights R, G and B, of appropriate intensity. It is sometimes necessary for one of the primaries to be added to the test light rather than to the other primaries. By varying the wavelength of the test light, one generates three color-matching functions

$\bar{r}(\lambda)$, $\bar{g}(\lambda)$ and $\bar{b}(\lambda)$, which describe the intensity of each primary required to match the test of a particular wavelength (Figure 1a). The dependence of such color-matching functions on the arbitrary choice of primary wavelengths led to the development in 1931 by the Commission Internationale de l'Eclairage (CIE) of the XYZ colorimetric system. The $\bar{x}(\lambda)$, $\bar{y}(\lambda)$ and $\bar{z}(\lambda)$ color-matching functions in this system have no negative values, and the $\bar{y}(\lambda)$ function is identical to the photopic luminosity function $V(\lambda)$, which describes the relative brightness of monochromatic lights presented at photopic intensity levels (Figure 1b).

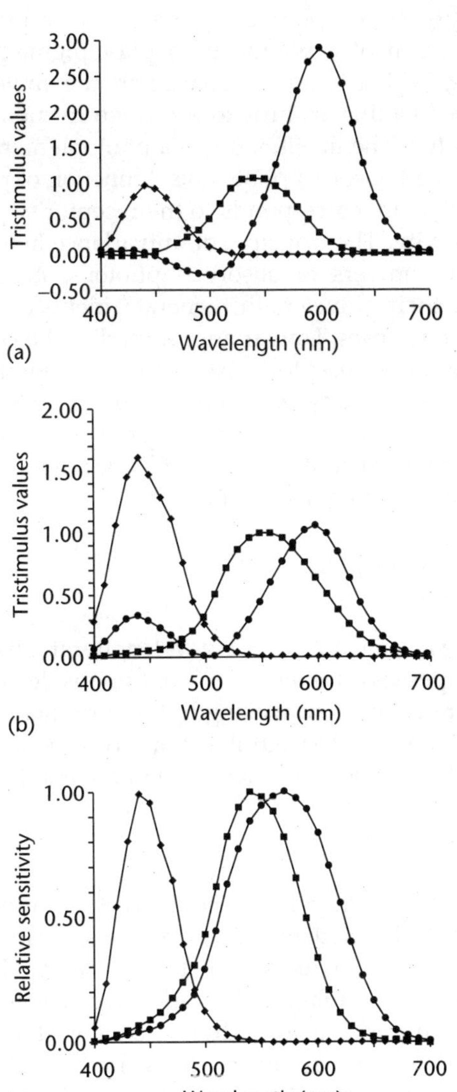

Figure 1. Trichromatic spectral sensitivity. (a) Values of the Stiles 2° color-matching functions $\bar{r}(\lambda)$ (●), $\bar{g}(\lambda)$ (■) and $\bar{b}(\lambda)$ (♦) with monochromatic primaries at 645.2 nm, 526.3 nm and 444.4 nm, plotted at 10 nm intervals. (b) The color-matching functions $\bar{x}(\lambda)$ (●), $\bar{y}(\lambda)$ (■) and $\bar{z}(\lambda)$ (♦) of the Judd-modified CIE 1931 standard 2° observer *XYZ* system. (c) The Stockman and Sharpe (2000) 2° cone fundamentals $L(\lambda)$ (●), $M(\lambda)$ (■) and $S(\lambda)$ (♦). The L and M cones have similar spectral sensitivities; S cones are almost completely insensitive to lights at wavelengths longer than 570 nm.

Cone Spectral Sensitivities

The three physiological response systems underlying normal color matching are the long (L), medium (M) and short (S) wavelength-sensitive cones, each class with a different photopigment. Their spectral sensitivities, taken in combination with knowledge of the spectral absorption characteristics of the eye's lens and macular pigment, provide color-matching functions that determine the initial response of the visual system to spectral variation in lights. Rods possess a fourth photopigment, rhodopsin, which differs from those of the three classes of cones, and so potentially support a tetrachromatic response to lights of mesopic intensity.

The first accurate estimates of the cone spectral sensitivities were provided by an analysis of color matching by dichromatic observers (Smith and Pokorny, 1975). A dichromat can match in appearance any light by combining additively just two primaries. A dichromat confuses lights that are distinguishable to trichromatic observers in one of three ways, depending on whether the dichromat lacks L cones (a protanope), M cones (deuteranope) or S cones (tritanope). The color-matching confusions made by the three classes of dichromat let one derive a linear transformation relating cone spectral sensitivities to the CIE system. The psychophysical estimates were soon confirmed by the results of physiological experiments using microspectrophotometric and electrophysiological techniques. Refinements of these estimates based on molecular genetic considerations (Figure 1c) have now been introduced (Stockman and Sharpe, 2000).

Variation Among Individuals

Differences among individuals in chromatic sensitivity can be traced to five primary factors. The first is the pigmentation of the eye's lens, which tends to become more yellow (absorb more light at shorter wavelengths) as one ages and with exposure to ultraviolet light. The second is the amount of macular pigment in an eye. This yellow pigment covers the region of the retina serving central (foveal) vision and varies in density from one person to the next. The third is the numerosity of each class of cone. Anatomical measurements show that S cones form about 7% of the cone population; there are none in the very center of the fovea. There is some controversy about the average ratio of L cones to M cones, which may be about 1.5–2; however, this ratio varies considerably from person to person, and is thought to influence both photopic luminosity, to which S-cone signals do not contribute, and estimates of unique yellow. The fourth factor is photopigment optical density, which influences absolute photoreceptor responsivity. The fifth factor influencing individual chromatic sensitivity is photoreceptor relative spectral sensitivity. There are polymorphisms in the genes lying on the X chromosome which code for the L and M cone opsins (Sharpe *et al.*, 1999). There are also hybrid genes which underlie anomalous trichromacy, evident when one of the L or M cone pigments has an abnormal spectral sensitivity. Dichromacy is exhibited when there is only one X-chromosome-linked cone photopigment (that of either the normal L or M cone, or of a hybrid), or when there are polymorphic versions of a single gene. Estimates of the peak spectral sensitivity of the M cone pigment lie in the range 528–532 nm, and those for the two primary L cone polymorphisms in the ranges 553–558 nm and 557–563 nm. Hybrid genes lead to anomalous M cone peak sensitivities, estimated to lie in the range 529–538 nm and, for anomalous L cones, in the range 545–559 nm. In males of European descent, the behavioral expression of red–green color deficiencies occurs at about a rate of 7.4%; for males of Asian and African descent, the rates are 4.2% and 2.6%, respectively. Owing to the X chromosome placement of the genes for the L and M cones, behavioral expression of red–green color deficiency is much less common among females (about 0.5%).

Models of color perception by dichromats as reduced forms of normal trichromatic perception have been validated by reports of observers who are born with one normal and one color-deficient eye. Rather than perceiving the hues of the visible spectrum in the sequence red, orange, yellow, green, blue, indigo and violet, in passing from long to short wavelengths, both protanopes (1% incidence in males of European descent) and deuteranopes (1.3%) see the sequence yellow, blue, with the transition occurring at a wavelength just shorter than 500 nm. They are able to distinguish lights normally perceived as red and green only in terms of saturation and brightness variations. Tritanopes perceive the sequence red, blue–green, with the transition at about 560 nm; they are unable to discriminate lights at short and middle wavelengths. Inherited tritan defects are rare, because the gene sequence for the opsin component of the S cone photopigment molecule resides on chromosome 7. Acquired defects are more common, because S cone function deteriorates in a variety of medical conditions.

Chromatic Response

The first step in photoreceptoral transduction is the absorption of a photon by a photopigment molecule, which causes a change in the molecule's shape, leading in turn to a change in membrane potential. The likelihood that a photopigment molecule will absorb a photon, as a function of photon wavelength, corresponds to the receptor's spectral sensitivity. The principle of univariance holds that equal numbers of absorbed photons, no matter what their wavelength, generate identical receptoral responses. Univariance underlies the computation of a receptor's response to a light with photons at many wavelengths. In discrete form, one sums over wavelength the product of a photoreceptor's spectral sensitivity $P[\lambda]$ and light energy $E[\lambda]$ to compute the response p:

$$p = \sum_{i=1}^{n} P[\lambda_i]E[\lambda_i] \qquad (1)$$

where n is the number of wavelengths at which the light possesses energy. In continuous form, one integrates over wavelength the product of the photoreceptoral spectral sensitivity function $P(\lambda)$ and the function $E(\lambda)$ describing light energy:

$$p = \int P(\lambda)E(\lambda)\mathrm{d}\lambda \qquad (2)$$

One often works with spectral sensitivity functions that are tabulated at discrete intervals, for instance, from 400 nm to 700 nm in steps of 10 nm. The 31 numbers describing such a function can be thought of as comprising a 31-dimensional vector $\mathbf{p}$. By describing the light energy in a similar fashion to provide a vector $\mathbf{e}$, one finds that the response p is given by the dot product of the two vectors:

$$p = \mathbf{p} \cdot \mathbf{e} \qquad (3)$$

The response of a trichromatic system to a light is represented by three numbers. For instance, one can use the L, M and S cone spectral sensitivities in place of the spectral sensitivity P above to determine the responses l, m and s, respectively. One consequence of the three-dimensionality of trichromatic visual response is that lights that differ physically may produce the same response. Such different but indistinguishable lights are known as metamers, and exist in consequence of the fact that a three-dimensional representation of lights cannot possibly represent accurately variation in the high-dimensional space of lights.

A light's chromaticity describes its chromatic properties independently of its intensive properties. For instance, if X, Y and Z are the responses to a light determined using the CIE $\bar{x}(\lambda)$, $\bar{y}(\lambda)$ and $\bar{z}(\lambda)$ functions, then the light's chromaticity (x, y) and its luminance L are given by the following formulae:

$$x = X/(X + Y + Z) \quad (4)$$

$$y = Y/(X + Y + Z) \quad (5)$$

$$L = Y \quad (6)$$

Perceived hue and saturation vary in the two-dimensional color space described by chromaticity; equiluminant lights vary in chromaticity alone.

COLOR OPPONENCY

The Young–Helmholtz theory that the responses of the L, M and S cone systems are tied directly to the perception of red, green and blue, respectively, is of historical interest alone. Subsequent work has substantiated Hering's idea that color appearance is due to the activity of opponent mechanisms that compare photoreceptoral system responses. Color-opponent processing commences in the retina, as is evident in the responses of bipolar cells and ganglion cells, and continues in the lateral geniculate nucleus (LGN) and in visual cortex. The mismatch between psychologically determined color-opponent functions and physiological opponent-mechanism sensitivities is of current research interest. Higher-order color-opponent mechanisms, although implicated in the results of chromatic detection experiments and organized in a manner consistent with physiological studies of visual cortex, have yet to be linked to appearance.

Standard Mechanisms of Color Appearance

Color-opponent theory is based on the analysis of hue. Under simple viewing conditions, one cannot perceive a light that appears both reddish and greenish, or one that appears both bluish and yellowish. This suggests that the responses of two opponent mechanisms underlie color appearance. The first is a red–green mechanism, the response of which can generate red or green, but not both simultaneously; the second is a similarly organized blue–yellow mechanism. A second critical observation concerns the existence of lights with unique hues: unique red and unique green appear neither yellowish nor bluish, while unique yellow and unique blue appear neither reddish nor greenish. All other hues appear to combine two of the primary hues. For instance, orange appears both reddish and yellowish. One can integrate the observation of unique hues into the opponent model as follows: first, unique red and unique green are seen when the red–green mechanism responds red or green, respectively, and when the response of the blue–yellow mechanism is zero (signals neither blue nor yellow); and second, unique blue and unique yellow are seen when the blue–yellow mechanism responds blue or yellow, respectively, and when the response of the red–green mechanism is zero. Graded responses by the mechanisms can account for perceived color saturation (chroma). For instance, a highly saturated red will be seen if the red–green mechanism produces a strong red response; less-saturated reds correspond to weaker responses.

Hurvich and Jameson used a hue-cancellation technique to derive spectral sensitivities for the two opponent mechanisms (Hurvich and Jameson, 1957). For instance, the relative amounts of red in monochromatic lights of equal energy can be measured by adding a green light of fixed wavelength and varying its intensity to produce a unique yellow light, which appears neither reddish nor greenish. The intensity of the green light needed to cancel the redness found in lights of long and short wavelength provides a measure of the red half of the red–green mechanism's spectral sensitivity function. Similar procedures provide the green half of the red–green mechanism's sensitivity, and the blue and yellow halves of the blue–yellow mechanism's sensitivity (Figure 2a).

Opponent-mechanism sensitivities can be described in terms of cone responses by combining cone spectral sensitivity functions to provide the best fit to the opponent functions. Such fits suggest, first, that the responses of L and S cones are opposed to those of the M cones by the red–green mechanism, and second, the responses of S cones are opposed to those of L and M cones by the blue–yellow mechanism. The contribution by L and S cones to red causes not only lights at long wavelengths to appear red but also lights at very short wavelengths, a feature required to account for the redness in the violet lights at short wavelengths. The L and M cones combine in the blue–yellow

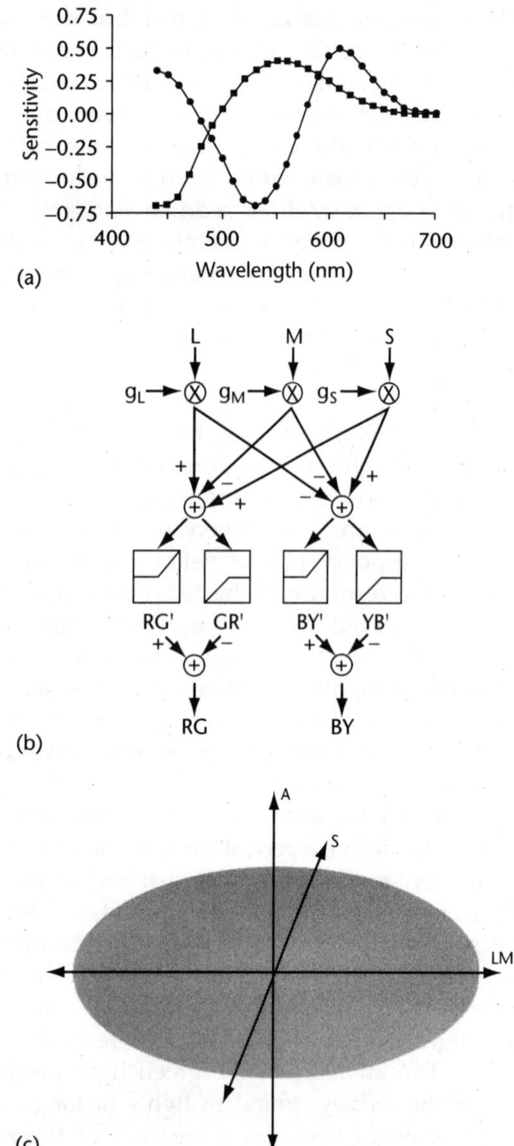

Figure 2. [*Figure is also reproduced in color section.*] Color-opponent transformation of photoreceptoral signals. (a) Hurvich and Jameson hue-cancellation functions. The red–green function (•) codes redness through positive values and greenness through negative values, while the yellow–blue function (▪) codes yellowness through positive values and blueness through negative values. (b) Wiring diagram for color-opponent mechanisms. Signals from L, M and S cone photoreceptors with multiplicative gain controls are combined by half-wave-rectified opponent mechanisms to produce mechanisms that signal red (RG′), green (GR′), blue (BY′) and yellow (YB′) which are then combined to form RG and BY opponent channels. (c) Equiluminant colors in a plane defined by LM and S axes. Modulation among lights along LM and S axes changes only L and M cone signals and only S cone signals, respectively. These axes correspond to the peak chromatic sensitivities of the retinogeniculate pathways.

mechanism to produce a yellow response that is opposed to the S cone's blue.

The opposition of cone responses results in color-opponent sensitivities that take on both positive and negative values. For instance, if the L and S cones are taken to excite and the M cones to inhibit the red–green mechanism, then the 'red' portion of the red–green spectral sensitivity will take on positive values while the 'green' portion will take on negative values. A positive response by such a mechanism would cause red to be perceived, and a negative response green.

A common emendation to the standard model, consistent with physiological evidence, uses rectification to produce separate red, green, yellow and blue mechanisms (as well as black and white) (Figure 2b). Nonlinear combination of cone inputs by such mechanisms can account for the spectral nonlinearity in blue–yellow processing; stated simply, too much of the spectrum appears blue and too little yellow to be consistent with linear combination of cone responses.

Chromatic Sensitivity

Chromatic sensitivity depends on adaptation by the visual system to viewing conditions. Absolute sensitivity to changes in light level about some reference light declines as the intensity of the reference light increases. This change occurs, in part, through the action of multiplicative gain controls within individual cone photoreceptors, although such adaptation of cones is significantly less than that predicted by psychophysical measurements. Chromatic adaptation to a steady background light also occurs at color-opponent sites in the visual pathways and is thought to involve both multiplicative and additive components.

The sensitivity of color-opponent mechanisms to modulations about a steady background light is high. The red–green mechanism supports the detection of red signals with L cone contrasts as small as 0.1%, a sensitivity more than five times greater than that found with stimuli of varying intensity (e.g. black–white). Equiluminant stimuli (Figure 2c) are generally used to probe the sensitivity of chromatic processing, because chromatic mechanisms are more likely to detect equiluminant stimuli than are mechanisms sensitive specifically to variation in light intensity. Equiluminant stimuli provide poor inputs to visual subsystems that handle motion, stereopsis, Vernier acuity and the interpretation of shadows. This is thought to be due, in part, to the strong coloration possible with very small contrast signals.

Chromatic sensitivity depends also on the spatiotemporal pattern of light modulation about a steady background light. For instance, prolonged viewing of a red–green temporal modulation reduces red–green sensitivity but leaves blue–yellow sensitivity relatively unscathed, and vice versa. Such long-term habituation to chromatic modulation is complemented by the activity of rapidly acting gain controls which operate on the spatial pattern of color-opponent contrast signals. Both habituation and color contrast gain control are thought to have a cortical locus.

Higher-order Mechanisms

Two spectral classes of color-opponent neuron are found in macaque retina and LGN. The first opposes L and M cone inputs linearly. Such a neuron does not have the sensitivity of the standard red–green opponent mechanism, which requires S cone input to produce red at short wavelengths. The second opposes S cone inputs to those of L and M cones. This class of neuron is largely insensitive to lights that correspond to unique red, so bearing one of the characteristics of the standard blue–yellow mechanism, but typically combines cone responses linearly in a steady state of adaptation.

The mismatch between retinogeniculate processing and the standard color-opponent model takes a different form in areas V1, V2 and V3 of the visual cortex. Color-sensitive neurons in cortex have spectral sensitivities which are scattered more uniformly in the color plane (Lennie *et al.*, 1990). Many such neurons are most sensitive to hues lying along axes in the color plane that lie between the LM and S axes that characterize retinogeniculate processing (Figure 2). Psychophysical work with habituation, visual search and noise-masking paradigms suggests that these higher-order mechanisms operate in everyday visual detection tasks. For example, observers can deploy a violet-sensitive mechanism when looking for a violet signal rather than looking for simultaneous red and blue signals. How these mechanisms contribute to color appearance, if at all, is an open question.

COLOR CONSTANCY

The utility of color in object recognition depends on the stability of an object's color appearance under change in its viewing conditions. An important aspect of this is the stability of surface color under change in the spectral properties of illumination. Observers can discount illumination spectral change nearly completely under appropriate circumstances (Kraft and Brainard, 1999). Physics-based analyses have identified conditions under which this is possible, and have explored a variety of cues that a trichromatic visual system can use to find descriptors of an object's surface reflectance function from reflected lights (Maloney, 1999). Transparency provides a further avenue into the study of surface color appearance; observers assign colors to surfaces behind a filter in a manner consistent with filter-specific changes in color-opponent signals (D'Zmura *et al.*, 2000). The parallel representation of filter and surface chromatic properties in perceived transparency suggests that low-level chromatic signals are shunted into parallel, layered representations of scene color.

Change in Illumination Spectral Properties

A white piece of paper reflects very different lights towards the eye when viewed outdoors under direct sunlight or under a yellow tungsten lightbulb (Figure 3a). Perfect color constancy entails identical colors for the white paper under the two viewing conditions. This is possible if the visual system represents directly the chromatic properties of the paper's surface. These are specified by a reflectance function $R(\lambda)$, which describes the fraction of light reflected by a surface as a function of wavelength (Figure 3b).

Principal components analysis of collections of surface reflectance functions shows that much of their variation (greater than 99% in sets such as the Munsell chips) can be captured by just three components; to a good approximation, surface reflectance functions lie in a three-dimensional space (Figure 3d). A particular surface reflectance is identified by three descriptors which tell how much each principal component contributes to the reflectance. A trichromatic visual system can thus represent surface reflectance properties directly through a linear transformation of cone responses, similar to that by the standard black–white and color-opponent mechanisms, which produces surface reflectance descriptors.

Nevertheless, reflected light confounds illumination and surface spectral properties. The reflected light $E(\lambda)$ is the product of the illumination spectral power distribution $D(\lambda)$ and the surface reflectance function $R(\lambda)$. The linear transformation of cone responses which provides surface reflectance descriptors must change in response to illumination change if color constancy is to be possible.

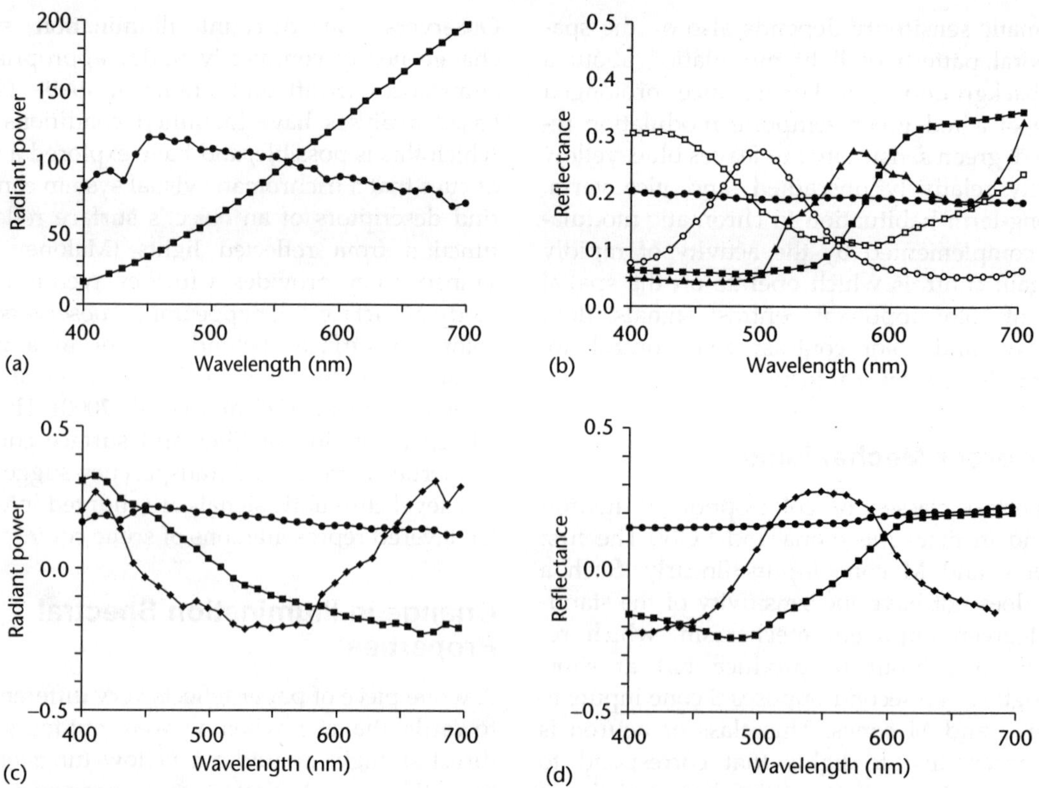

Figure 3. Finite-dimensional linear models for illuminant spectral power distributions (SPDs) and surface reflectance functions. (a) The SPDs of CIE illuminant D65 (●) and illuminant A (■), which match those of average daylight and a yellow tungsten light bulb, respectively. (b) Reflectance functions of five Munsell chips: N 5 (●), 10R 4/6 (■), Y 5/6 (▲), BG 4/6 (○) and 10PB 4/6 (□). (c) First three functions in the CIE daylight illumination model (Gram–Schmidt orthogonalized). The function $D_1(\lambda)$ (●) captures variation in the mean level of illumination; the function $D_2(\lambda)$ (■) captures 'blue–yellow' variation like that found between blue sky and the yellow solar disk, while the function $D_3(\lambda)$ (♦) captures minor 'red–green' variation. Daylight with SPD $D(\lambda)$ can be approximated accurately as a linear combination of these three functions: $D(\lambda) \approx d_1\ D_1(\lambda) + d_2\ D_2(\lambda) + d_3\ D_3(\lambda)$. (d) First three functions in the Cohen principal components analysis of Munsell chip reflectance functions (Gram–Schmidt orthogonalized). The function $R_1(\lambda)$ (●) captures variation in the mean level of reflectance; the function $R_2(\lambda)$ (■) captures 'red–green' variation, while the function $R_3(\lambda)$ (♦) captures 'blue–yellow' variation. A reflectance R can be approximated by a linear combination of these three functions: $R(\lambda) \approx r_1\ R_1(\lambda) + r_2\ R_2(\lambda) + r_3\ R_3(\lambda)$. The coefficients $\{r_1, r_2, r_3\}$ describe the reflectance function and do not vary with illumination, so that a visual system able to recover these descriptors can exhibit color constancy.

Principal components analysis of collections of daylight spectral power distributions (SPDs) shows that almost all of their variation can be captured by just three components (Figure 3c). For daylight, then, knowledge of an illuminant's chromaticity lets one reconstruct its SPD, up to a single scale factor.

Potential cues to the chromatic properties of scene illumination are numerous. If one assumes that the brightest surface in a scene is white, with a spectrally flat reflectance function, then the photoreceptoral responses to the white surface correspond to those of the illuminant itself. This knowledge can be used to adjust the linear transformation of photoreceptor responses to provide color-constant reflectance descriptors. This idea works more generally for reference surfaces with known reflectance properties. The gray-world assumption is similar; one assumes that the space-averaged surface reflectance in a scene corresponds to a spectrally neutral gray, so that illuminant chromatic properties can be inferred from the space-averaged photoreceptor responses. Specularity provides a further such cue; the lights from two or more surfaces with highlights provide sufficient information to infer illuminant chromatic properties. Of these three cues, only the gray-world assumption is thought to have perceptual relevance; eye movements let mechanisms with sufficiently long integration times adjust their sensitivity to space-averaged inputs. The scaling of photoreceptoral responses by their average inputs, a form of

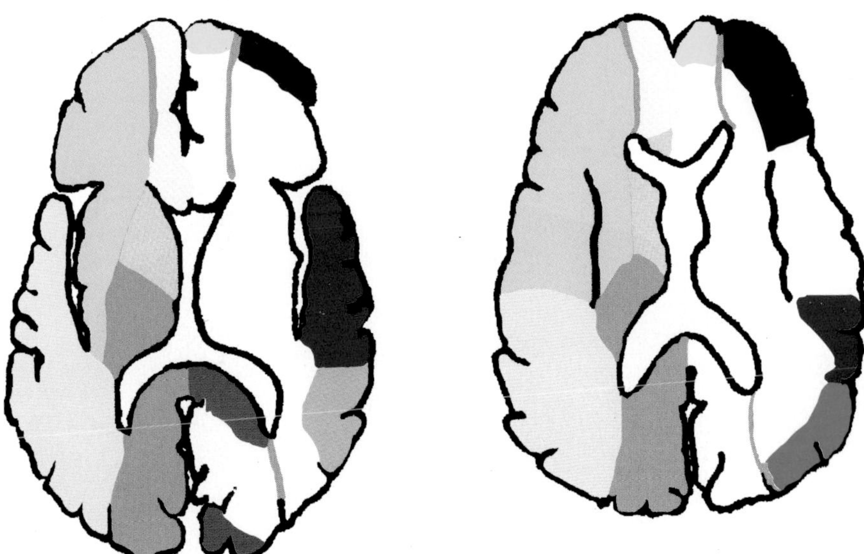

Plate 1 [Aphasia] Approximate vascular distributions of major arteries supplying language cortex are shown on the left side of each brain cut (the right hemisphere on computed tomographic and magnetic resonance imaging). Pale yellow, anterior cerebral artery (ACA); beige, anterior choroidal branch of internal carotid artery; light blue, superior division of the middle cerebral artery (MCA); pink, inferior division of the MCA; lavender, posterior cerebral artery (PCA); gray, border of ACA/MCA and MCA/PCA (potential 'watershed' areas). Regions of the cortex implicated in classical aphasia syndromes are shown on the right side of each brain cut (the left hemisphere). Bright yellow, Brodmann's area (BA) 10, associated with some cases of transcortical motor aphasia; dark blue, BA 44 and 45, Broca's area; red, BA 22, Wernicke's area; dark green, BA 39, the angular gyrus; light green, BA 37, the posterior middle temporal gyrus; purple, BA 31, the splenium of the corpus callosum and BA 18, visual cortex (areas associated with optic aphasia); gray, 'watershed' areas associated with the transcortical aphasias (the width of this territory depends on the degree of diminished flow in the ACA/MCA or MCA/PCA). Adapted from Damasio and Damasio (1989).

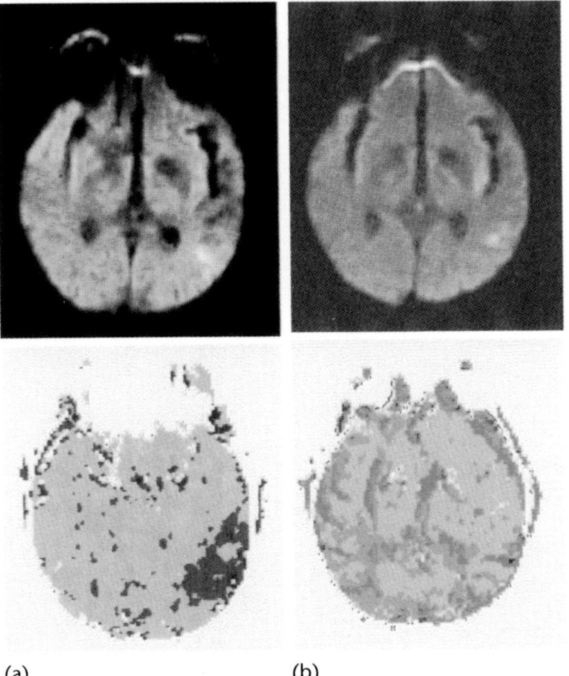

Plate 2 [Aphasia] (a) Magnetic resonance diffusion-weighted imaging (top) and perfusion-weighted imaging (bottom) showing hypoperfusion of the posterior middle temporal gyrus on day 1 in a patient with impaired access to auditory word forms for output, causing poor oral naming and oral reading, but intact semantics. (b) Day 2, showing reperfusion of the posterior middle temporal gyrus when naming and oral reading abilities had recovered.

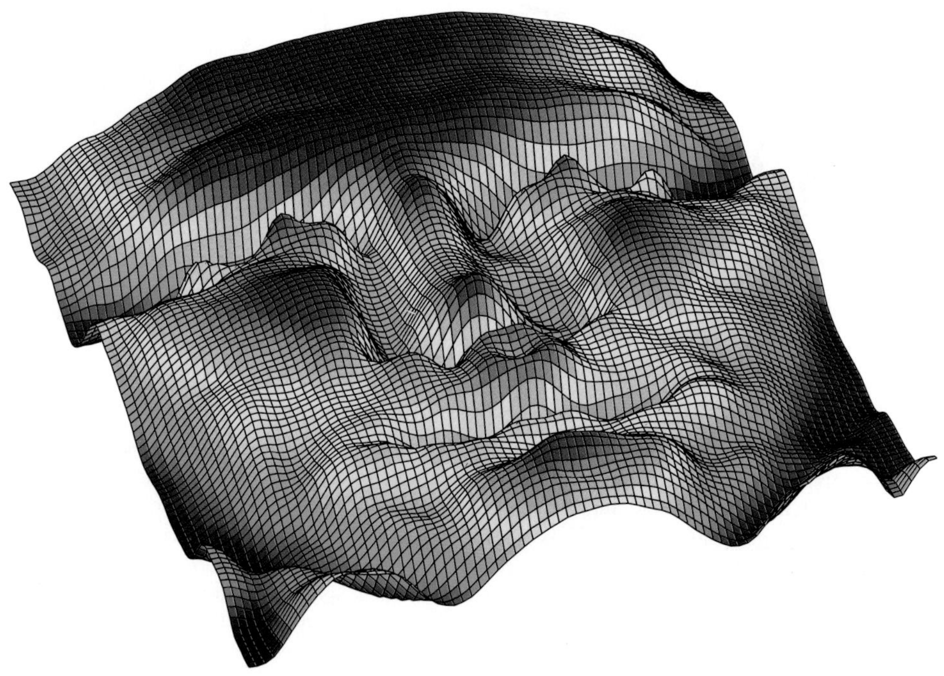

Plate 3 [Attractor Networks] An imaginary energy landscape.

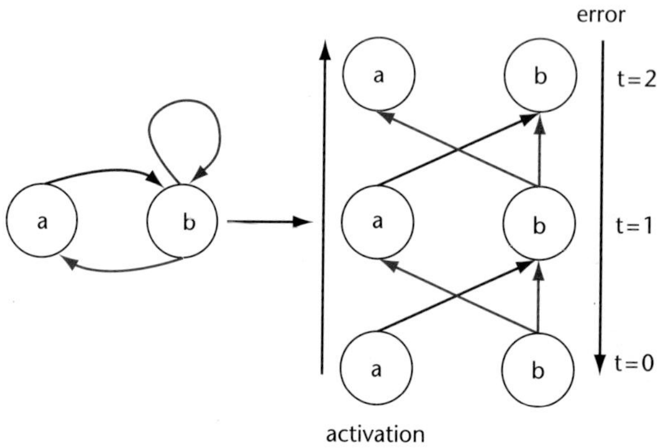

Plate 4 [Attractor Networks] Back-propagation in time.

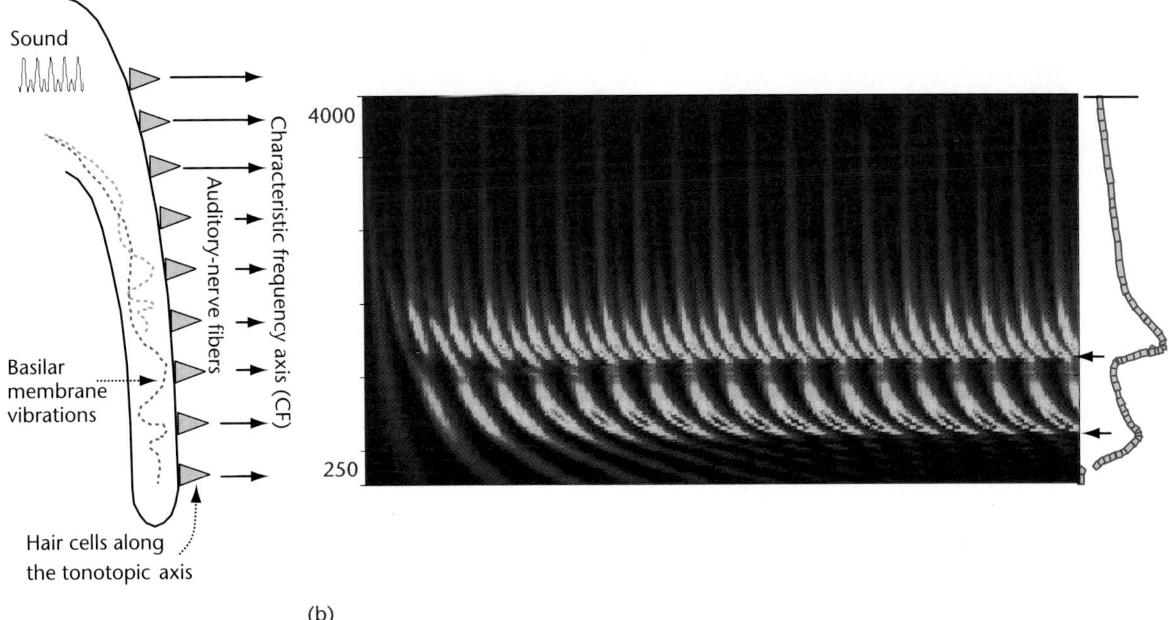

Plate 5 [Audition, Neural Basis of] Schematic model of the early stages in auditory processing. (a) Cochlear analysis. Sound enters the cochlea via the eardrum and the middle ear, initiating travelling wave displacement patterns on the basilar membrane. Vibrations due to low frequencies propagate and achieve their maximum amplitude further down the cochlea (broken red line) compared to high frequencies (broken green line). This mapping of sound frequency components onto different places along the cochlea creates the tonotopically ordered (spatial) axis of the auditory system. Basilar membrane vibrations are transduced into spatiotemporal responses on the auditory nerve by an array of hair cells distributed along the length of the cochlea. (b) Auditory nerve responses. The spatiotemporal response patterns in the auditory nerve due to a two-tone stimulus (300 and 600 Hz). The ordinate represents the tonotopic axis (labeled by the characteristic frequency axis (CF) at each location). The response at each CF represents the instantaneous probability of firing in the nerve fiber at that CF. Note that each component in the stimulus initiates a localized travelling wave pattern that abruptly ends creating a prominent discontinuity near the appropriate CF (marked by the arrow heads to the right of the panel). The stimulus responses depicted here are at a low sound level, such that it does not saturate the nerve responses. The response amplitudes are therefore strongest near the CFs of the two tones, resulting in clear peaks in the average response curve (red plot to the right).

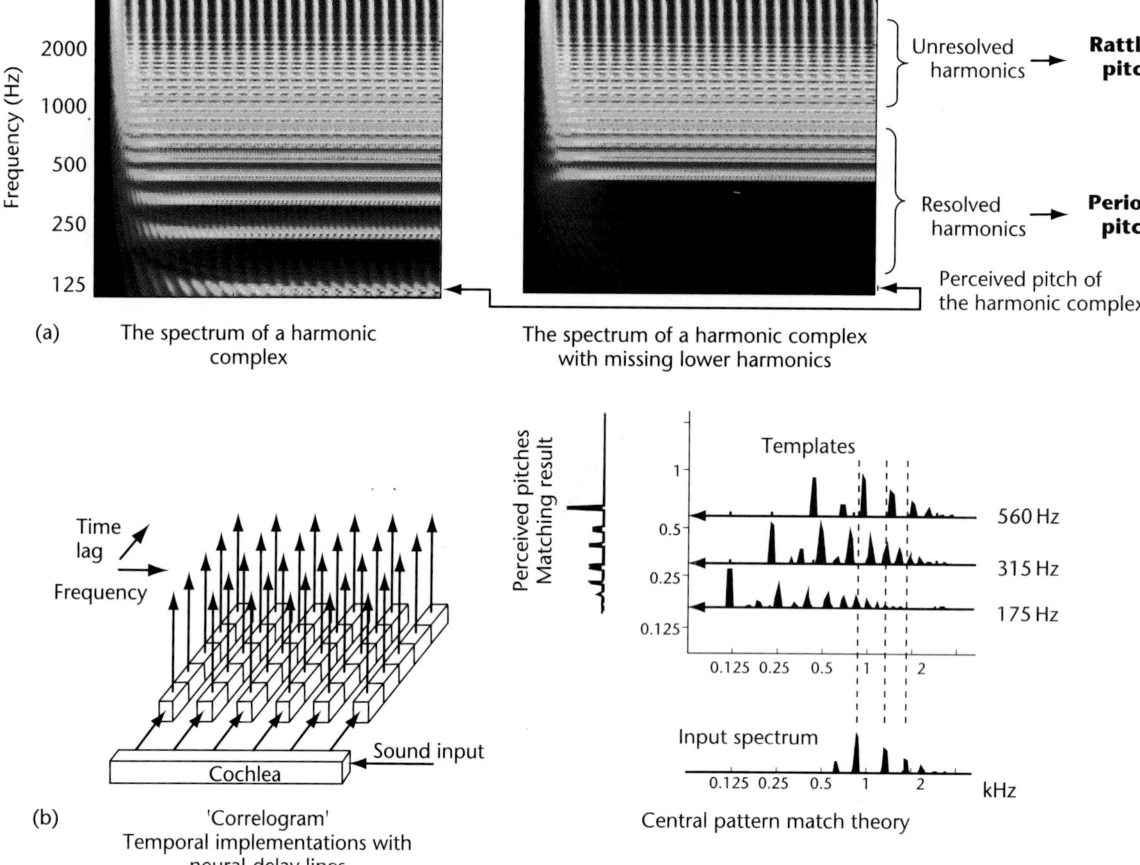

Plate 6 [Audition, Neural Basis of] Representation and extraction of stimulus periodicity of a harmonic complex. (a) The auditory spectra of two harmonic series of a 125 Hz fundamental. The low-order harmonics are well *resolved* along the tonotopic axis. These harmonics evoke a pitch sensation at the fundamental frequency of the series (i.e. at 125 Hz). The high-order harmonics (> 8th harmonic or 1 kHz) are largely *unresolved* and they instead evoke a pattern that 'beats' at the difference frequency of 125 Hz. These harmonics evoke a weaker pitch (called the 'rattle pitch') at the beating frequency (i.e. 125 Hz in this case). (Left) The harmonic complex here contains all 40 harmonics and evokes a strong pitch at 125 Hz. (Right) The harmonic complex here lacks the lowest three harmonics (125, 250, 375 Hz); nevertheless, it still evokes a strong pitch at the 'missing fundamental' frequency of 125 Hz. (b) Two algorithms for extracting this missing fundamental pitch. (Left) The schematic illustrates an autocorrelogram implementation. It presumes the existence of organized delay lines to compute the autocorrelation of the responses from each auditory nerve channel (or fiber) prior to computing the pitch. (Right) Schematic illustrates a *template-matching idea.* It presumes the existence of harmonic templates in the brain that are matched to incoming spectra so as to measure the pitch.

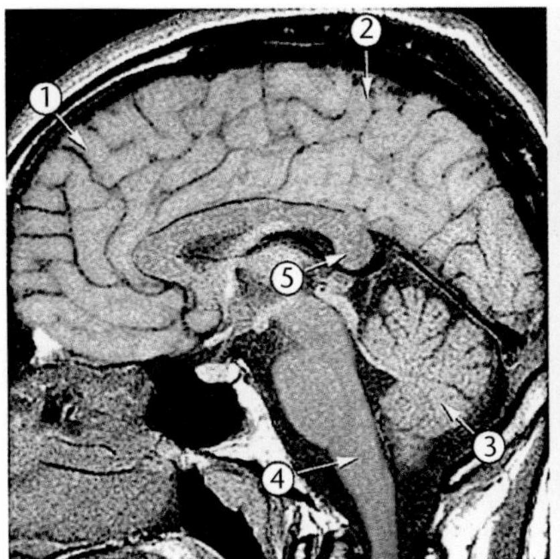

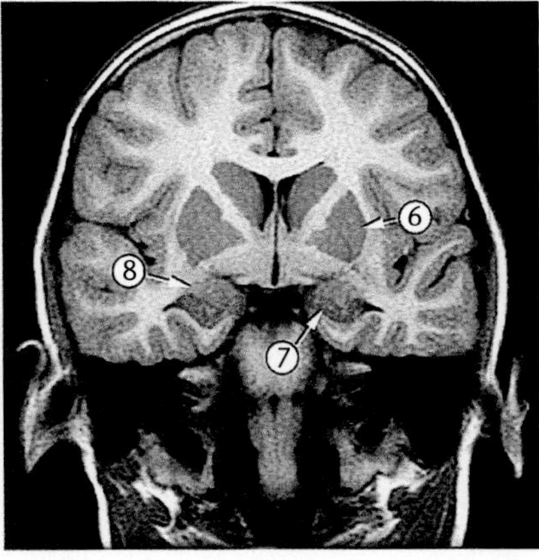

Plate 7 [Autism] Cortical and subcortical structures reported as abnormal in autism from autopsy or magnetic resonance imaging data from laboratories across the USA and Europe. 1, frontal lobes; 2, parietal lobes; 3, cerebellum; 4, brainstem; 5, corpus callosum; 6, basal ganglia; 7, hippocampus; 8, amygdala.

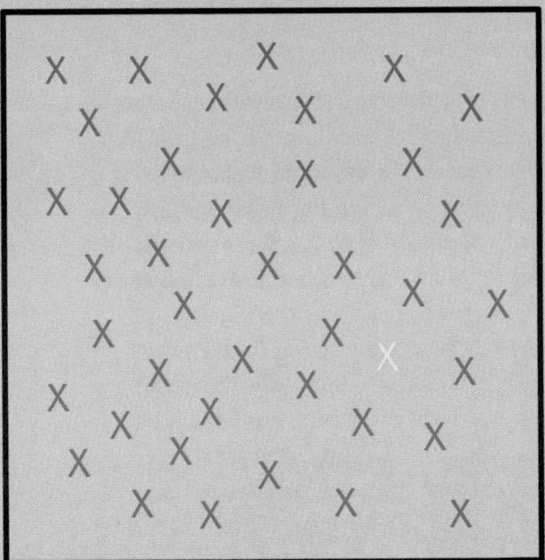
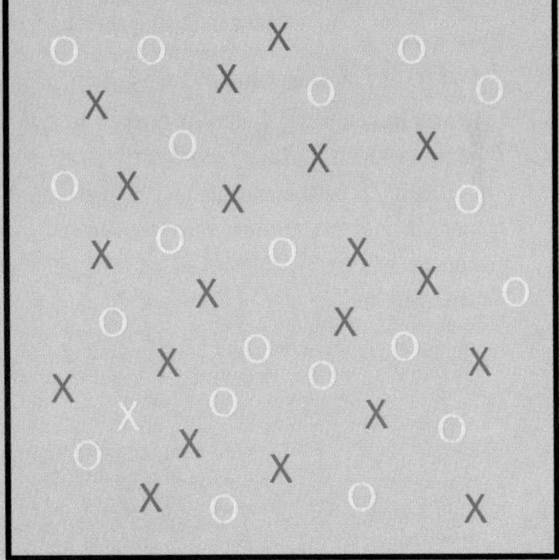

Plate 8 [Automaticity] Find the yellow **X**. The left panel illustrates a feature search task in which the target automatically 'pops out' from the field of distracters. The right panel illustrates a conjunction search task in which the target must be actively searched for with deliberate shifts of attention.

#1	#2	#3
card	red	green
zoo	blue	pink
divide	orange	blue
fish	green	orange
card	pink	purple
friend	blue	yellow
drill	orange	green
card	yellow	blue
search	blue	red
drill	purple	yellow
divide	red	green
zoo	pink	blue
friend	green	pink
fish	yellow	orange
search	green	green
card	blue	red
drill	pink	purple

Plate 9 [Automaticity] Demonstration of the Stroop task. Using a stopwatch, separately time how long it takes to *name the color* of each printed word in column 1, column 2, and column 3. Then separately time how long it takes to *read each word* in column 1, column 2, and column 3. Column 1 is a *control condition* in which the word and the color bear no relationship. Column 2 is a *congruent condition* in which the word and the color match. Column 3 is an *incongruent condition* in which the word and the color mismatch.

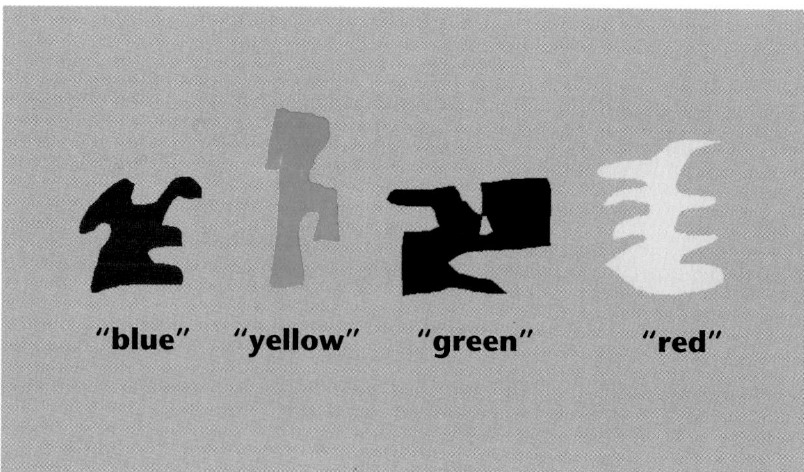

Plate 10 [Automaticity] Illustration of novel stimuli used to manipulate the direction of Stroop interference through training. Each shape (glyph) is associated with one of four color names that must be learned. Each shape can also be filled with one of four colors. Subjects either name the shape ('blue', 'yellow', 'green', or 'red') or name the color of the shape (RED, GREEN, BLUE, or YELLOW). Early in training, color interferes with shape naming. Later in training, shape interferes with color naming.

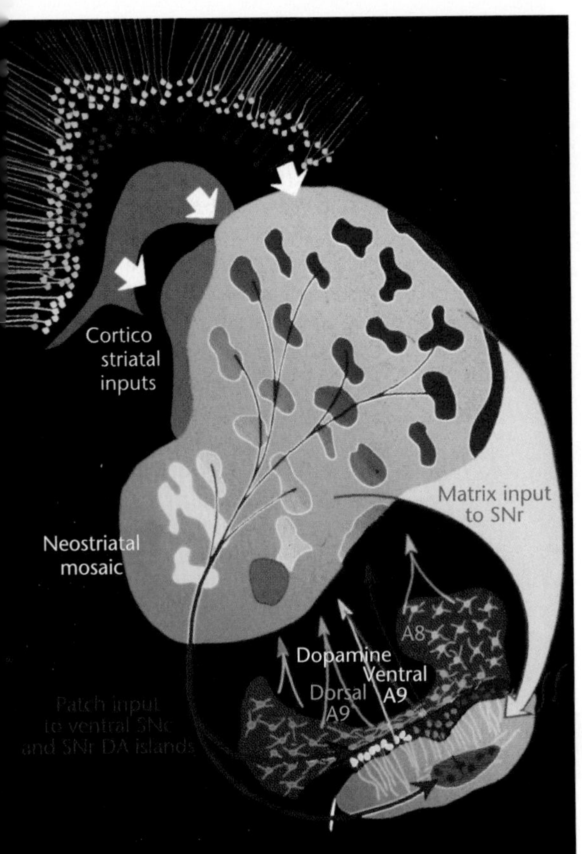

Plate 11 [Basal Ganglia] The patch–matrix design of the striatum in the rat, seen in the right side of the rat cortex, striatum, nucleus accumbens and substantia nigra. The patches (red, orange and yellow zones), also called striosomes, are rich in an opioid receptor. The surrounding matrix is rich in acetylcholinesterase and calbindin. The anatomical connections of the patches and matrix are color-coded to show, for example, that the cells in the deep layers of lateral cortex project to the lateral patches in the striatum, which in turn project to the ventral substantia nigra compacta dopaminergic cells. SNr, substantia nigra reticulata; SNc, substantia nigra compacta; A9, nomenclature that refers to substantia nigra compacta cells; A10, refers to dopaminergic cells of the midbrain tegmentum that project to the nucleus accumbens. Diagram courtesy of C. Gerfen.

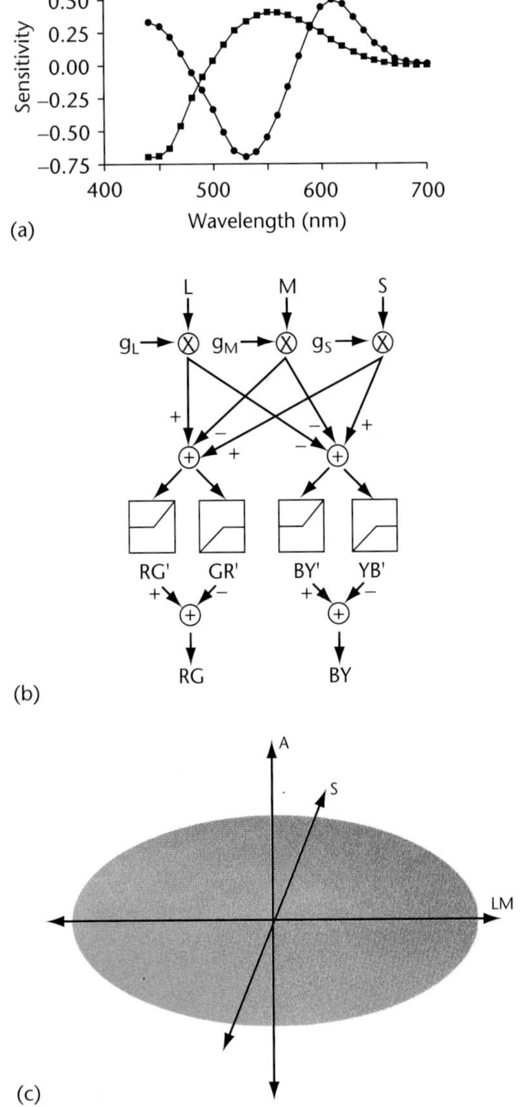

Plate 12 [Color Perception, Psychology of] Color-opponent transformation of photoreceptoral signals. (a) Hurvich and Jameson hue-cancellation functions. The red–green function (•) codes redness through positive values and greenness through negative values, while the yellow–blue function (■) codes yellowness through positive values and blueness through negative values.
(b) Wiring diagram for color-opponent mechanisms. Signals from L, M and S cone photoreceptors with multiplicative gain controls are combined by half-wave-rectified opponent mechanisms to produce mechanisms that signal red (RG'), green (GR'), blue (BY') and yellow (YB') which are then combined to form RG and BY opponent channels.
(c) Equiluminant colors in a plane defined by LM and S axes. Modulation among lights along LM and S axes changes only L and M cone signals and only S cone signals, respectively. These axes correspond to the peak chromatic sensitivities of the retinogeniculate pathways.

(a)

(b)

Plate 13 [Color Vision, Neural Basis o (a) Henri Matisse, *Femme au Chapeau* (Paris, autumn 1905). Oil on canvas, 80 cm × 60 cm; San Francisco Museum of Modern Art (bequest of Elise S. Haas). The gray-scale reproduction (b) shows that the surprising color transitions do n interfere with an accurate representation of the woman's face. Yet the color reproduction is obviously more appealin – why? The dissociation of color and form, clear in this picture, shows that color is processed by the visual system separately from other stimulus attributes like form.

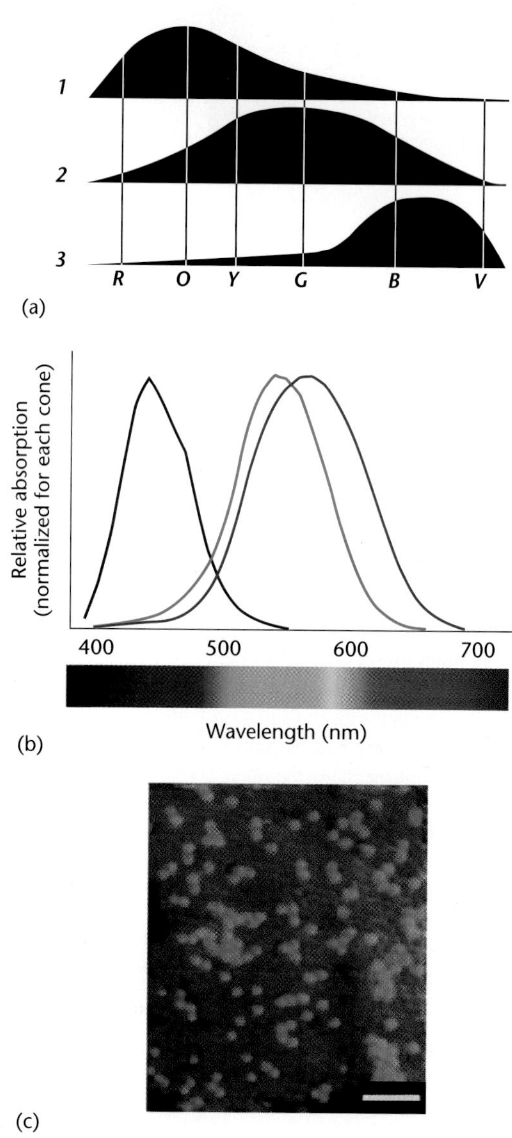

Plate 14 [Color Vision, Neural Basis of] Color perception begins in the retina of the eye with three classes of photoreceptors called cones. (a) The absorption spectra of the three detectors proposed by Helmholtz in 1866: shorter wavelengths (V, or violet) were represented on the right. (b) The actual cone absorption spectra of the three cone classes, L, M and S, based on the cone fundamentals of Smith and Pokorny (1972). Convention today puts shorter wavelengths on the left. Below the plot is the visible spectrum. (c) The cone mosaic of a patch of living human retina made visible with adaptive optics, from Roorda and Williams (1999). The S cones are represented by blue, M by green and L by red. Scale bar, 5 arc minutes of visual angle.

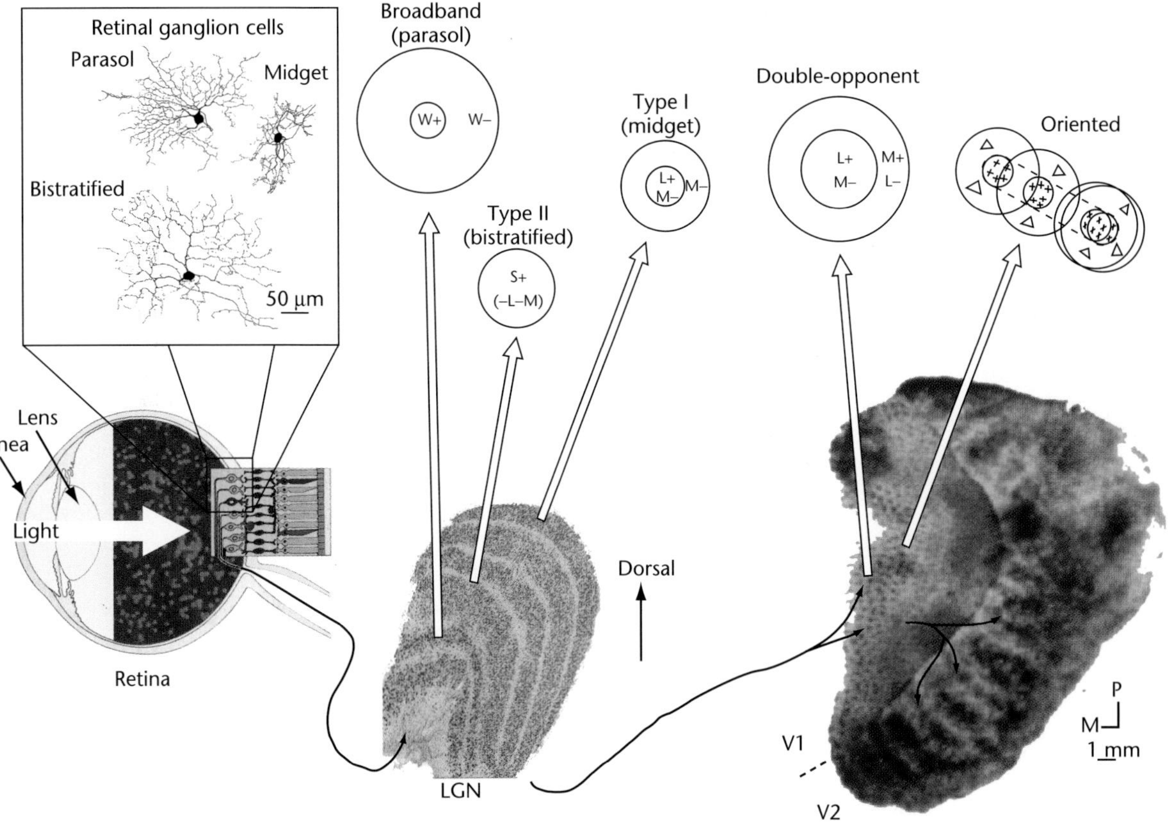

Plate 15 [Color Vision, Neural Basis of] A summary of color processing in the visual system. Light enters the eye and is focused on the retina by the cornea and lens. The three classes of cones respond to the light. Different retinal ganglion cells (inset; adapted from Dacey and Lee, 1994) sample the cone mosaic and provide input to the lateral geniculate nucleus (LGN). The retinal ganglion cell names 'midget' and 'parasol' reflect the relative sizes of their dendritic fields, which in turn reflect the relative sizes of their receptive fields. The cells of the LGN, here stained with Nissl substance, comprise six well-defined layers: four dorsal (or parvocellular) layers and two more darkly staining ventral (or magnocellular) layers. Each purple dot is a single cell, about 10 µm in diameter. The parvocellular layers contain type I cells; the receptive field of an L-ON center/M-OFF surround type I cell is given. The magnocellular layers contain the broadband cells. Between the darkly staining parvocellular and magnocellular layers are the koniocellular layers. Type II cells reside in these layers. Broadband cells, type II cells and type I cells are the LGN targets of parasol, bistratified and midget ganglion cells, respectively.

Neurons in the LGN send their axons to the primary visual cortex (V1). In this figure, V1 is represented by a tangential section of one hemisphere of unfolded and flattened squirrel monkey cortex that has been stained with the metabolic enzyme cytochrome oxidase (M, midline; P, posterior). Cytochrome oxidase (CO) staining clearly demarcates the border between V1 and the second visual area, V2, and reveals CO blobs in V1 and the CO stripes in V2. Color information is processed by the double-opponent cells, which reside in the V1 blobs and send their axons to the thin CO stripes of V2 (arrows). Between the blobs are cells that are sensitive to the orientation of a visual stimulus.

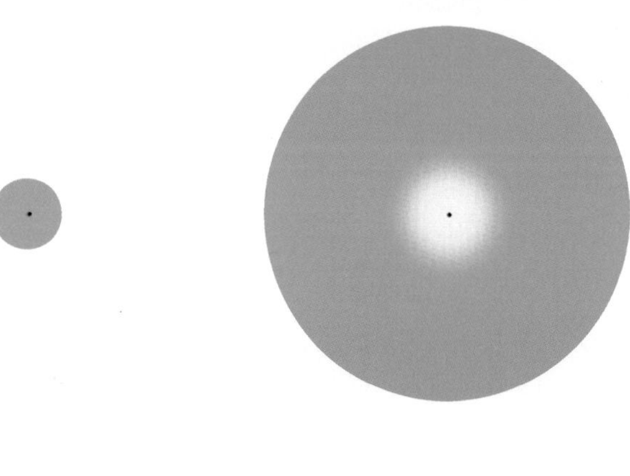

(a) (b)

Plate 16 [Color Vision, Neural Basis of] There is more to color than meets the eye! Stare at the fixation dot in the middle of the green disk (a), being careful to hold your gaze steady. After 20 s or so, transfer your gaze to the fixation dot below; you should see a reddish afterimage. Now consider the small, fuzzy gray disk in the center of the colored annulus (b). After prolonged viewing the gray seems to adopt a weak reddish tinge. Such induced colors are much more striking when the colored annulus occupies the entire visual field surrounding the central gray spot. Try generating an afterimage to the gray spot. The afterimage to the gray spot is surprisingly green! This shows that the spatial configuration of a scene affects both the color of perceived images and the color of afterimages.

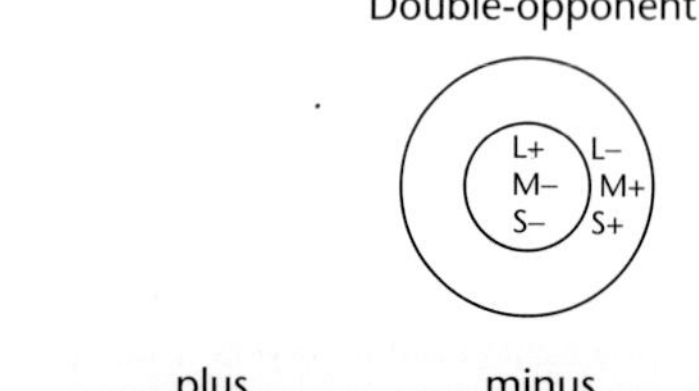

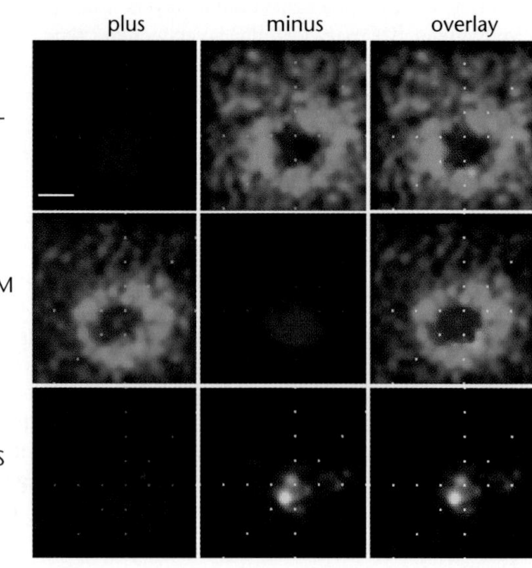

Plate 17 [Color Vision, Neural Basis of] The receptive field of a double-opponent cell in monkey V1. The left-hand column shows the spatial extent of the cell's response to increasing activity of the three cone classes (L, top; M, middle; S, bottom); the middle column shows the same cell's response to decreasing the activity of the three cone classes. Comparing the maps (right-hand column) shows that this cell's receptive field is both spatially and chromatically opponent. This double-opponent structure is critical to color constancy – our ability to determine an object's color despite changing illumination conditions. From Conway (2001).

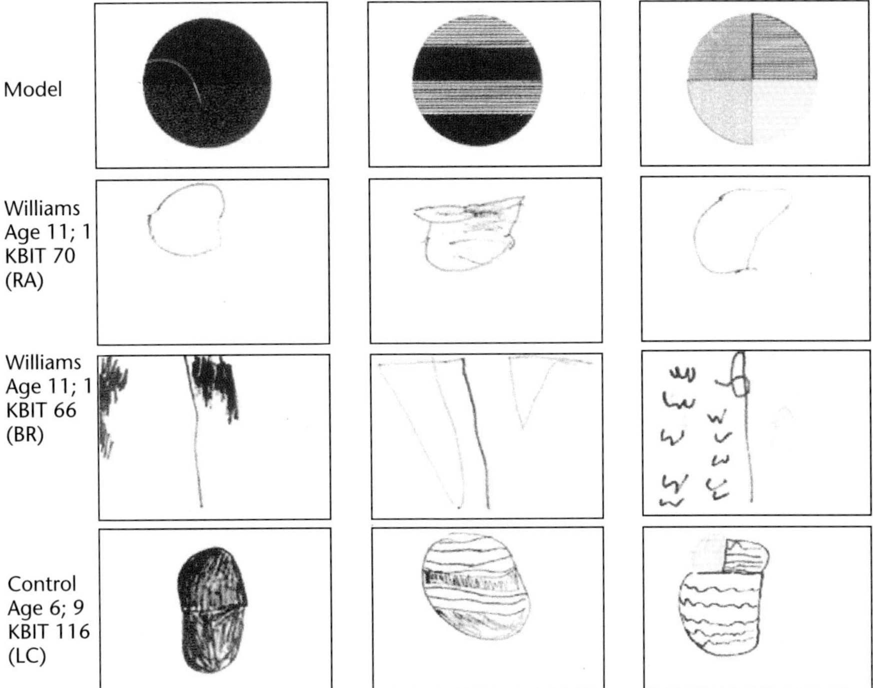

Plate 18 [Early Experience and Cognitive Organization] Copies made by two 11-year-old individuals with Williams syndrome and a normally developing 6-year-old who was matched for mental age. Each model contains between two and four colors, and these are correctly copied by the children with Williams syndrome as well as by controls. However, there is considerable impairment of the spatial organization of Williams syndrome copies, resulting in unrecognizable configurations. Ages are stated in years and months. KBIT scores are from the Kaufman Brief Intelligence Test, and represent approximate IQ equivalents.

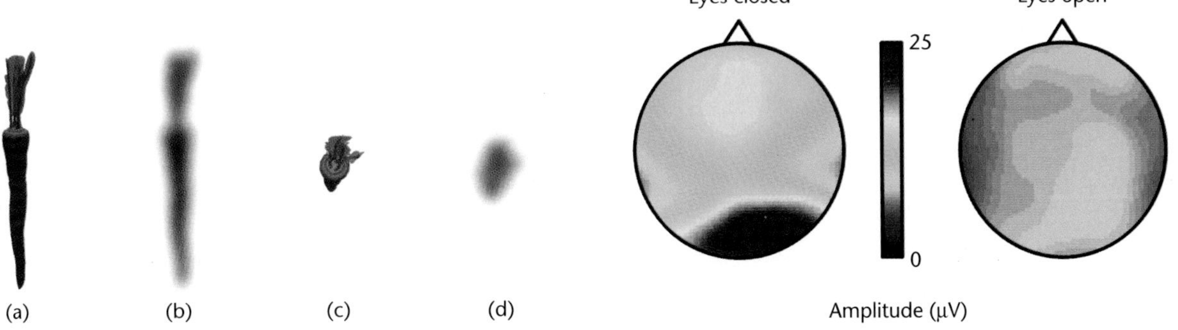

late 19 [Early Experience and Cognitive Organization] hese objects are easily recognized and named by children with Villiams syndrome, despite the fact that they have profound patial impairments. The objects are all the same carrot, but vary n whether they are common viewpoints (a and b) or unusual nes (c and d), and also in whether they are clear images (a and) or blurred ones (b and d).

Plate 20 [Electroencephalography (EEG)] Topography of the alpha rhythm. The scalp is viewed from above using an azimuthal equidistant projection centered at the vertex. The outside of the circle reaches the level of the ear canal. The α activity when the eyes are closed is maximally recorded from the posterior regions of the scalp, and is slightly greater on the right than on the left. These maps were based on activity from 65 scalp electrodes (10 of which are shown in Figure1).

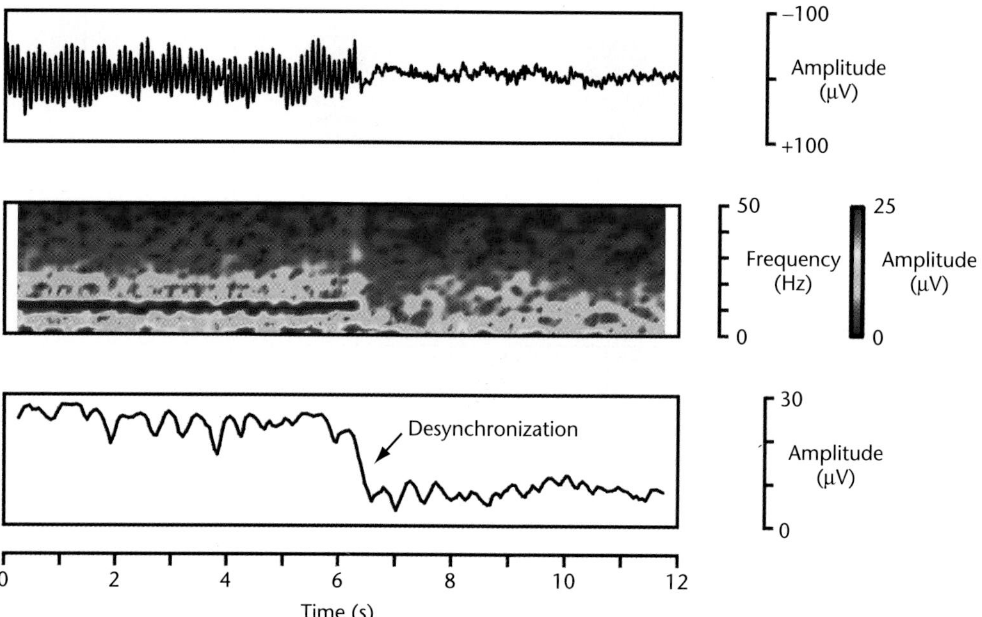

Plate 21 [Electroencephalography (EEG)] Electroencephalographic changes over time. The upper tracing shows the EEG signal recorded from the O_2 electrode in Figure 1. The middle panel shows the spectrogram. This plots the changes in the spectrum over time: the frequency (on the x axis in Figure 2) is plotted vertically (y axis) and the amplitude of the activity is plotted on the z axis as color. A dark red line at 10 Hz represents t α rhythm. This line is clearly recognizable when the eyes are closed and disappears when the eyes are open. The lower tracing represents the amount of activity present in the 8–13 Hz frequency band as it changes over time. When the eyes open, this activity decreases (event-related desynchronization).

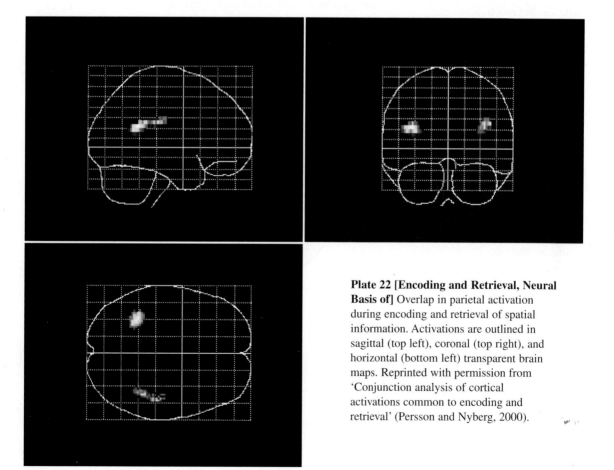

Plate 22 [Encoding and Retrieval, Neural Basis of] Overlap in parietal activation during encoding and retrieval of spatial information. Activations are outlined in sagittal (top left), coronal (top right), and horizontal (bottom left) transparent brain maps. Reprinted with permission from 'Conjunction analysis of cortical activations common to encoding and retrieval' (Persson and Nyberg, 2000).

von Kries adaptation, is considered an important influence on perceived surface color. Yet numerous examples show that von Kries adaptation is insufficient to account for surface color appearance.

The gray-world assumption is a statistical one that may be generalized to assumptions concerning the prior distributions of possible surface reflectances and illuminant SPDs in visual scenes. Bayesian algorithms for color constancy take as data the photoreceptoral responses from a set of surfaces and determine the illuminant and surface chromatic properties most likely to have given rise to the data.

Models of photoreceptor response that depend linearly on both surface reflectance and illumination spectral properties are bilinear models. Analysis of such bilinear models shows that viewing several surfaces under an unknown daylight illuminant does not provide enough information to find three reflectance descriptors per surface. Yet if the same surfaces are seen under first one illuminant and then another, so that two views of the surfaces are provided, sufficient information exists to determine both surface and illuminant chromatic properties. A key assumption in the use of multiple views is that the visual system must know the correspondence between surfaces viewed under the first and second illuminants.

Transparency

Surface correspondence is readily determined in viewing situations involving transparent filters. Surfaces that lie along the edge of a filter are viewed both directly and through the filter. At the intersections of surface and filter edges are X junctions, where color changes are cues to surface and filter chromatic properties. Psychophysical studies show that a filter of uniform color properties is perceived best when surface chromatic changes are characterized by a single shift (translation) in color space and/or a single change in contrast. The shift in surface responses corresponds to filter color, while the change in contrast corresponds to filter cloudiness, which ranges from clear to opaque.

The perceived separation of a spatiochromatic pattern into surface and filter layers is an example of scission, the layered representation of the visual field. Standard color theory allows for a trichromatic representation of scene spectral properties at every point, and this is seemingly insufficient to account for our ability to perceive two (or more) colors simultaneously at the same point. One possibility is that mechanisms involved in scene segmentation feed back on standard representations to shunt trichromatic signals into multiple, concurrent color representations.

Color Coding

While color provides information about an object's material properties, it also serves as a flexible visual code well suited to investigations in visual attention, category learning, verbal production (e.g. the Stroop effect) and other cognitive tasks. Performance based on color coding depends on memory and language, which depend, in turn, on categorization. All major modern languages have equivalents for the achromatic color names white, gray and black; the primary color names red, yellow, green and blue; and the secondary color names pink, brown, orange and purple. These names correspond to volumes in three-dimensional color space that bound the chromatic stimuli that elicit the names. Stimuli in the center of such a volume typify the color category better than those close to a boundary. Studies of color memory suggest that remembered colors often tend to shift towards their prototypes, revealing perhaps the action of verbal encoding on color memory. Color coding has many important roles in society through semantic binding (e.g. the red, orange and green of traffic lights), perceived emotional content, and through esthetic considerations expressed in art and design.

CONCLUSION

The psychophysical study of color relies on careful distinctions among physical light stimuli, chromatic properties related to photoreceptoral and opponent mechanism responses, and perceived color. Relating perceived color to physical light stimuli is difficult, not only because of the intermediate chromatic mechanisms, but also because of philosophical issues in understanding color and comparing color among individuals. These difficulties notwithstanding, considerable progress has been made in characterizing chromatic processing and in identifying sources of individual differences in chromatic sensitivity.

References

D'Zmura M, Rinner O and Gegenfurter K (2000) The colors seen behind transparent filters. *Perception* **29**: 911–926.

Hurvich LM and Jameson D (1957) An opponent-process theory of color vision. *Psychological Review* **64**: 384–404.

Kraft JM and Brainard DH (1999) Mechanisms of color constancy under nearly natural viewing. *Proceedings of the National Academy of Sciences USA* **96**: 307–312.

Lennie P, Krauskopf J and Sclar G (1990) Chromatic mechanisms in lateral geniculate nucleus of macaque. *Journal of Physiology (London)* **357**: 649–669.

Maloney LT (1999) Physics-based approaches to modeling surface color perception. In: Gegenfurtner KR and Sharpe LT (eds) *Color Vision: From Genes To Perception*, pp. 387–416. New York: Cambridge University Press.

Sharpe LT, Stockman A, Jaegle H and Nathans J (1999) Opsin genes, cone photopigments, color vision, and color blindness. In: Gegenfurtner KR and Sharpe LT (eds) *Color Vision: From Genes To Perception*, pp. 3–51. New York: Cambridge University Press.

Smith VC and Pokorny J (1975) Spectral sensitivity of the foveal cone photopigments between 400 and 500 nm. *Vision Research* **15**: 161–171.

Stockman A and Sharpe LT (2000) Spectral sensitivities of the middle- and long-wavelength sensitive cones derived from measurements in observers of known genotype. *Vision Research* **40**: 1711–1737.

Further Reading

Backhaus WGK, Kliegl R and Werner JS (eds) (1998) *Color Vision: Perspectives from Different Disciplines*. New York: Walter de Gruyter.

Byrne A and Hilbert DR (eds) (1997) *Readings on Color*, vol. 1. *The Philosophy of Color*. Cambridge, MA: MIT Press.

Byrne A and Hilbert DR (eds) (1997) *Readings on Color*, vol. 2. *The Science of Color*. Cambridge, MA: MIT Press.

Gegenfurtner KR and Sharpe LT (1999) *Color Vision: From Genes to Perception*. New York: Cambridge University Press.

Kaiser PK and Boynton RM (1996) *Human Color Vision*, 2nd edn. Washington, DC: Optical Society of America.

Katz D (1935) *The World of Colour*, translated by RB MacLeod. London: Kegan Paul.

Nassau K (1983) *The Physics and Chemistry of Color: The Fifteen Causes of Color*. New York: John Wiley.

Stiles WS (1978) *Mechanisms of Colour Vision*. London: Academic Press.

Wandell BA (1995) *Foundations of Vision*. Sunderland, MA: Sinauer.

Wyszecki G and Stiles WS (1982) *Color Science. Concepts and Methods, Quantitative Data and Formulae*. New York: John Wiley.

Color Vision, Neural Basis of

Intermediate article

Bevil R Conway, Harvard Medical School, Boston, Massachusetts, USA
Margaret S Livingstone, Harvard Medical School, Boston, Massachusetts, USA

CONTENTS

The brain determines color from different wavelengths. Specialized cells in the retina, thalamus, primary visual cortex and higher brain areas achieve color perceptions by taking into account chromatic context, which explains various color illusions and color constancy.

INTRODUCTION

A world without color is bleak. Despite this, the benefits of color vision are difficult to quantify. Picasso said, 'When I run out of blue I use red', by which he meant that it is the brightness of a pigment and not its color that describes the three-dimensional shape of objects. Matisse demonstrated this point beautifully in his painting *Femme au Chapeau* (Figure 1). A gray-scale reproduction shows that the values of the pigments do not interfere with an accurate representation of the play of light across his subject's face. That the painting reads well as a face despite the radical color transitions shows that color is not an important cue to shape. In fact, object shapes are easily recognizable even in dim light when color vision is

Figure 1. [*Figure is also reproduced in color section.*] (a) Henri Matisse, *Femme au Chapeau* (Paris, autumn 1905). Oil on canvas, 80.5 cm × 60 cm; San Francisco Museum of Modern Art (bequest of Elise S. Haas). The gray-scale reproduction (b) shows that the surprising color transitions do not interfere with an accurate representation of the woman's face. Yet the color reproduction is obviously more appealing – why? The dissociation of color and form, clear in this picture, shows that color is processed by the visual system separately from other stimulus attributes, like form.

absent. Moreover, many people function perfectly well with impaired color perception: about 1 in 12 men are red–green color-blind and many of them are unaware of it. However, color cues are useful. In monkeys, they assist the discrimination of ripe fruit and of suitable procreative partners, and in humans, color is more than a cue for discriminating objects, for unlike shape and texture, color has emotional significance. One is 'green' with envy, 'red' with anger, 'blue' with sadness. Indeed, Matisse used this to push his portrait past mere representation. Moreover, his picture is much more appealing in color than in black and white. It is probably the emotional quality of color that has fueled color vision research, and it also helps explain the passionate controversies that fill this field's history.

COLOR VISION: WHAT IS IT?

Color vision is the ability to discriminate surfaces based on the spectral content of the light reflected from them, taking into account the light reflected from surrounding objects. Color vision, which has evolved in many animal groups such as insects and birds, is crude in most mammals except some primates, such as humans. Research has focused on color vision in Old World monkeys because their color vision is virtually identical to that in humans (De Valois *et al.*, 1974).

THE ROLE OF CONES

In 1802 the English physician Thomas Young proposed that color was subserved by three classes of sensors, each maximally sensitive to a different part of the visible spectrum (Figure 2a). This trichromatic theory, extended by Helmholz in 1866, resolved a profound problem. Though we can see millions of different colors, our retinas simply do not have enough space to accommodate a separate detector for every color at every retinal location, as proposed by Newton. Given that almost all hues can be matched by the combination of three primary colors, and that the number of receptors for color at every retinal location must be small, Young's proposal of three sensors was reasonable. Moreover, it changed the way we think about color: the bottleneck imposed by the small number of sensors implied that color was a neural construction reflecting both physical properties of light and biological properties of photopigments, neurons and networks of neurons. Color is a perception, and not a property of the world.

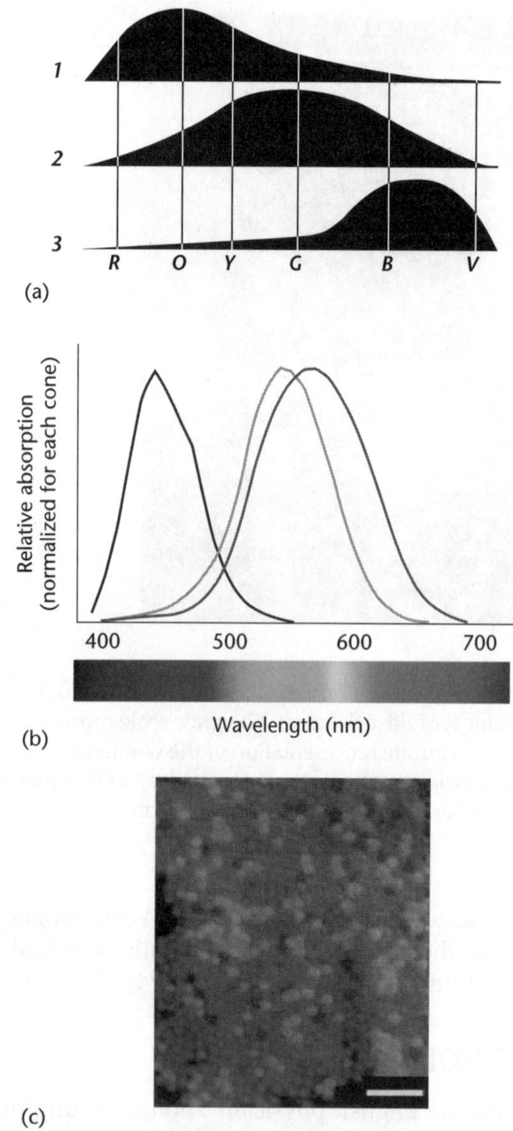

Figure 2. [*Figure is also reproduced in color section.*] Color perception begins in the retina of the eye with three classes of photoreceptors called cones. (a) The absorption spectra of the three detectors proposed by Helmholtz in 1866: shorter wavelengths (V, or violet) were represented on the right. (b) The actual cone absorption spectra of the three cone classes, L, M and S, based on the cone fundamentals of Smith and Pokorny (1972). Convention today puts shorter wavelengths on the left. Below the plot is the visible spectrum. (c) The cone mosaic of a patch of living human retina made visible with adaptive optics, from Roorda and Williams (1999). The S cones are represented by blue, M by green and L by red. Scale bar, 5 arc minutes of visual angle.

We now know that color perception begins in the retina with photoreceptors called cones (Figure 2b, c). Cones are more densely packed in the portion of the retina corresponding to the center of gaze (the fovea), and become less dense in the periphery. A second type of photoreceptor, rods, are absent from the fovea. Rods function best in dim light, when the cones do not function well. Because we only have one class of rods, and a comparison between at least two classes of photoreceptor is required for color vision, rods for the most part are not involved in color perception, and we do not see color in very dim light.

Cones are divided into three classes according to their peak absorptions: the S cones absorb shorter wavelengths optimally (peak 440 nm); the M cones absorb middle wavelengths (peak 535 nm) and the L cones absorb long wavelengths (peak 565 nm). All cone classes are somewhat sensitive to wavelengths throughout most of the spectrum (Figure 2b). Thus a single class of cones is color-blind because it cannot distinguish between a dim light of optimal wavelength and an intense light of less optimal wavelength. Moreover, at any given point in the retina there is only one cone, so the retina is color-blind on a spatial scale of single cones. Despite these facts, the cones are often loosely called 'blue', 'green' and 'red', because these names are somehow more intuitive, but we must be cautious because these are not even the color names that we assign to the region of the spectrum to which each class is maximally sensitive. That single cone classes do not code the perception of single colors is proof that the simple trichromatic theory cannot fully explain color perception.

It was once assumed that cones in the primate retina would be regularly distributed (to facilitate uniform sampling of wavelength), as is the case in the goldfish. Primate S cones are distributed fairly regularly (Curcio *et al.*, 1991), but L and M cones are surprisingly patchy (Roorda and Williams, 1999) (Figure 2c). The resulting clumpiness may help us detect fine-grained luminance variations, but only at the cost of color resolution. Indeed, variations in color are harder to resolve than variations in luminance (for a review see Livingstone and Hubel, 1987). Two objects that differ only in color are described as equiluminant, and you can find examples of them in the Matisse painting (equiluminant colors will come out roughly the same gray in a gray-scale copy).

The very center of the fovea (0.1° of visual angle) is devoid of S cones. Our eyes are focused for about 550 nm light, where the L and M cones have their peak sensitivities. Consequently shorter-wavelength light will be blurred. Evolution may have selected against having S cones in the center of the fovea where high spatial acuity is the goal because the short-wavelength light to which the

S cones are most responsive will be an unreliable source of spatial information. The absence of S cones in the center of the fovea has surprisingly little impact on color vision, probably because the spatial extent of the S cone hole is finer than the coarse resolution of color vision.

The wavelength sensitivity of a given cone cell is attributed to the specific photopigment protein that it expresses (Nathans, 1999). The M and L photopigment genes, which are encoded on the X chromosome, are fairly similar in sequence, suggesting that the M and L photopigments arose from a common ancestral gene that duplicated not so long ago – around 30–40 million years ago, shortly after the continents of Africa and South America separated. The similarity of the M and L gene sequences predisposes them to recombination during meiosis. This has led to a polymorphism of the L and M photopigments, which is more commonly manifest in males because they only have one X chromosome on which to rely for their M and L photopigments. The polymorphism can be a complete loss of L cones (protanopia), loss of M cones (deuteranopia) or, more frequently, the expression of a mutant M/L hybrid. It is these polymorphisms that underlie the range of so-called red–green color blindness, the most famous case of which is that of Sir John Dalton (a deuteranope), who in 1794 was the first to describe the condition. The deletion of the S cone gene, on the seventh chromosome, is possible (tritanopia), but rare.

COLOR IS AN OPPONENT PROCESS

In 1880 the German psychologist Ewald Hering proposed that color was mediated not by trichromacy but by opponency. Hering observed that we cannot perceive a continuous mixture of colors as predicted by the trichromatic theory – we cannot perceive (or even conceive of) reddish-greens or bluish-yellows. Some colors are mutually exclusive of others. So, Hering argued, color must be determined by the activity of opponent mechanisms, and he proposed three: a red–green mechanism, a blue–yellow mechanism and a black–white mechanism. Today we can appreciate another reason why color must be an opponent process: the L and M cone fundamentals are so similar that to perceive long-wavelength light as distinct from middle-wavelength light (i.e. red and not yellow) the responses of the M cone must be subtracted from those of the L cone.

Each opponent mechanism can be thought of as one axis in a three-dimensional space that encompasses all colors. So any given color could be uniquely defined by three variables: the activity along the red–green axis, the activity along the blue–yellow axis and the activity along the black–white axis. The scientific dispute between Hering and Helmholtz and their supporters is the source of much animosity in the field of color, even today, although studies have now reconciled them.

The retinal ganglion cells, which receive input from the cones, project to neurons in the lateral geniculate nucleus (LGN). These in turn project to neurons in the primary visual cortex (Figure 3). The most common retinal ganglion cells are the midget cells and the parasol cells. Midget cells project to type I cells in the four parvocellular layers of the LGN. Parasol cells project to broadband cells in the two magnocellular layers. Most type I cells receive antagonistic inputs from M and L cones and therefore respond in opposite ways to different colors (DeValois *et al.*, 1958; Wiesel and Hubel, 1966; Reid and Shapley, 1992). A type I cell may be excited by long-wavelength light and suppressed by middle-wavelength light. These cells represent the sort of building block for Hering's red–green opponent process; the three cone types represent Young and Helmholtz's trichromacy.

A third type of retinal ganglion cell is the bistratified cell (Dacey and Lee, 1994). Bistratified cells are excited by S cones and suppressed by a mixture of L and M cones, making them likely candidates for Hering's blue–yellow mechanism. Bistratified cells project to the (S versus M + L) type II cells in the koniocellular layers of the LGN (Figure 3).

Black–white cells must contribute to our perception of color. For example, the addition of black changes the color of orange to brown. Many parvocellular cells do not show cone opponency, and could therefore represent black–white, but it is unclear whether these or the broadband cells of the magnocellular layers (or an unidentified class of cells) underlie the black–white color axis (Wiesel and Hubel, 1966).

COLOR CONSTANCY

Perhaps the greatest misperception about color is that it is equated to wavelength. This misperception is cultivated early in our education when we are taught (incorrectly!) that long-wavelength light is 'red' and short-wavelength light is 'blue'. Though wavelength is the critical determinant of color, it is not the only determinant. For example, our perception of white depends on responses from all three cone classes, which is normally achieved when we see broadband light. However, after viewing a colored surface, one class of cones

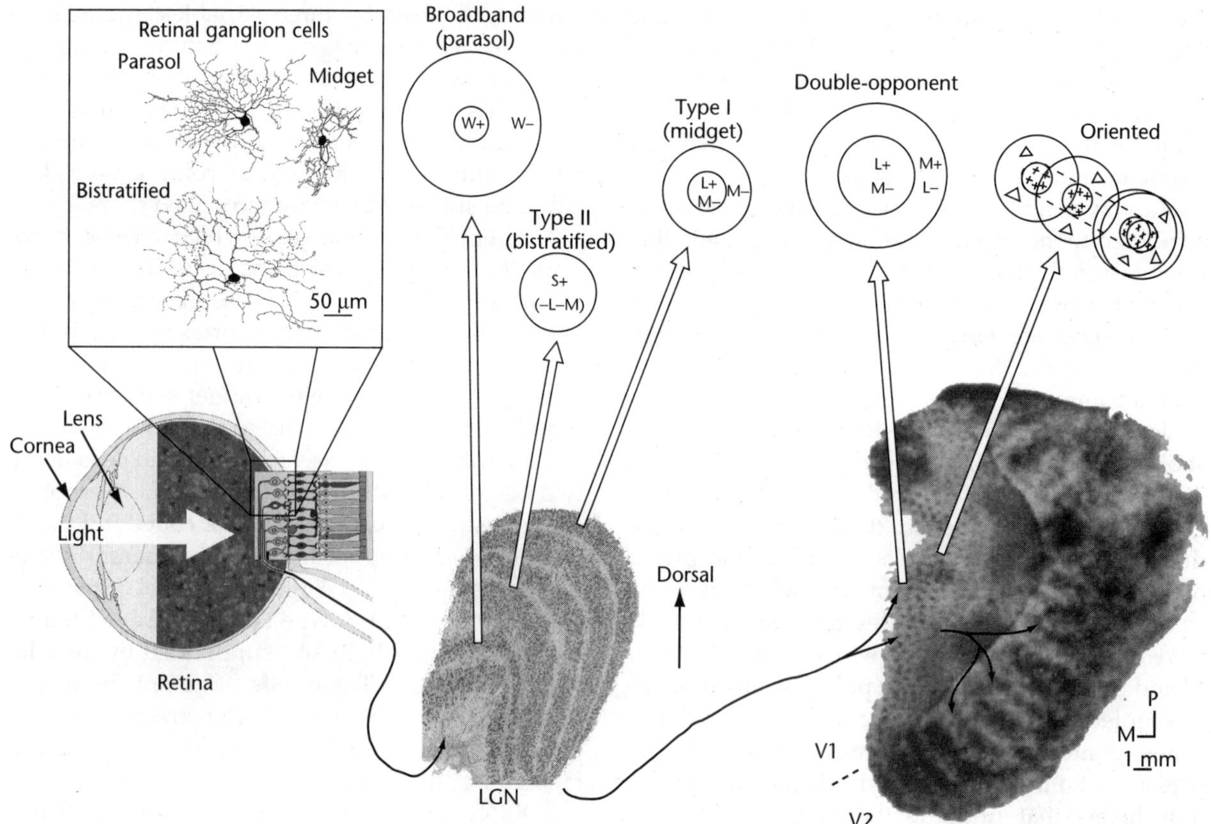

Figure 3. [*Figure is also reproduced in color section.*] A summary of color processing in the visual system. Light enters the eye and is focused on the retina by the cornea and lens. The three classes of cones respond to the light. Different retinal ganglion cells (inset; adapted from Dacey and Lee, 1994) sample the cone mosaic and provide input to the lateral geniculate nucleus (LGN). The retinal ganglion cell names 'midget' and 'parasol' reflect the relative sizes of their dendritic fields, which in turn reflect the relative sizes of their receptive fields. The cells of the LGN, here stained with Nissl substance, comprise six well-defined layers: four dorsal (or parvocellular) layers and two more darkly staining ventral (or magnocellular) layers. Each purple dot is a single cell, about 10 µm in diameter. The parvocellular layers contain type I cells; the receptive field of an L-ON center/M-OFF surround type I cell is given. The magnocellular layers contain the broadband cells. Between the darkly staining parvocellular and magnocellular layers are the koniocellular layers. Type II cells reside in these layers. Broadband cells, type II cells and type I cells are the LGN targets of parasol, bistratified and midget ganglion cells, respectively.

Neurons in the LGN send their axons to the primary visual cortex (V1). In this figure, V1 is represented by a tangential section of one hemisphere of unfolded and flattened squirrel monkey cortex that has been stained with the metabolic enzyme cytochrome oxidase (M, midline; P, posterior). Cytochrome oxidase (CO) staining clearly demarcates the border between V1 and the second visual area, V2, and reveals CO blobs in V1 and the CO stripes in V2. Color information is processed by the double-opponent cells, which reside in the V1 blobs and send their axons to the thin CO stripes of V2 (arrows). Between the blobs are cells that are sensitive to the orientation of a visual stimulus.

may become fatigued, perhaps by bleaching of the photopigment, and a previously 'white' surface will appear as a colored afterimage (Figure 4a).

Chromatic context also affects our perception of color. This is well known by artists, who place red against a green background to make it redder. Another example where wavelengths do not correlate directly with color is the phenomenon of induced colors: a gray spot can be made to appear colored if it is surrounded by a large colored annulus (the larger the annulus and the less sharp the boundaries, the stronger the effect). In fact spatial context even colors afterimages (Figure 4b). The discrepancies between physical cues and our perceptions can leave us confused. However, as Edwin Land pointed out, one should avoid asking the question, 'What color is it really?' as if to imply that our visual systems are deceiving us. Color is a product of our physiology and its interaction with the physical world; visual illusions simply point out what

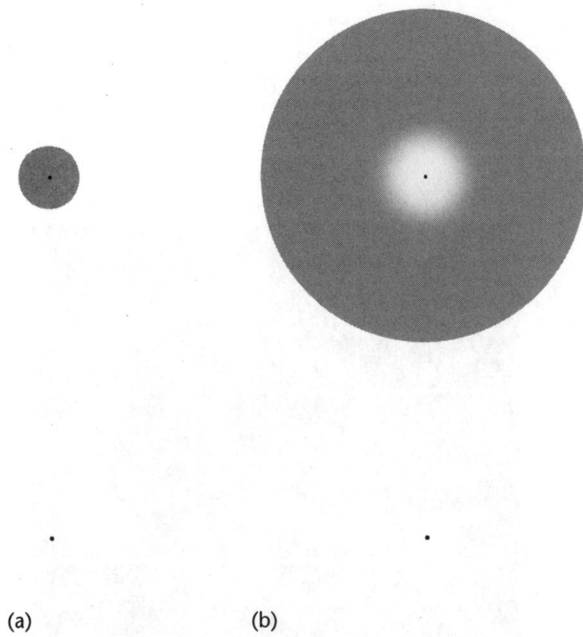

Figure 4. [*Figure is also reproduced in color section.*] There is more to color than meets the eye! Stare at the fixation dot in the middle of the green disk (a), being careful to hold your gaze steady. After 20 s or so, transfer your gaze to the fixation dot below; you should see a reddish afterimage. Now consider the small, fuzzy gray disk in the center of the colored annulus (b). After prolonged viewing the gray seems to adopt a weak reddish tinge. Such induced colors are much more striking when the colored annulus occupies the entire visual field surrounding the central gray spot. Try generating an afterimage to the gray spot. The afterimage to the gray spot is surprisingly green! This shows that the spatial configuration of a scene affects both the color of perceived images and the color of afterimages.

our visual systems are constantly (and usually effectively) doing.

An object will appear colored if it selectively absorbs some wavelengths and reflects others. The spectral distribution of reflected light is a product of the absorptive properties of the object's surface and the spectral properties of the light source (the illuminant). So if the illuminant changes, the reflected light will change too. Illuminants are constantly changing. A bright sunny day, under a blue sky, will contain a large proportion of shorter wavelengths, while a tungsten light bulb will produce longer wavelengths. The paradox of color vision is that despite these different illumination conditions, and the resulting difference in reflected light, the color of objects is fairly constant. A red apple is red, for example. It is not that a red apple is red only when viewed under a certain illumination condition such as a blue sky. This color constancy is mostly a property of our visual systems and not a function of memory.

It is easy to see why our visual systems have evolved in this way. If we were to assign a color to an object based solely on the light reflected from it then we would assign different colors to the same object depending on the conditions under which the object was viewed. Color constancy means that colors are properties of objects (which are constant) and not viewing conditions (which are continually changing).

Edwin Land, the inventor of instant photography, went to great lengths to reiterate that the color we assign an object is largely independent of the spectral content of the illuminant but is correlated with the absorption properties of the object or surface (Land, 1977). He was prompted by the familiar problem faced by color photographers: a scene photographed under tungsten light comes out with a reddish cast, and one photographed outside on a sunny day with a bluish cast. This is in contrast to our perceptions of the scene under the two illumination conditions – we see neither a reddish cast nor a bluish cast. Land concluded that our visual systems do not simply equate color and reflected wavelengths. In his experiments, Land used different light sources to illuminate different patches of a colored 'Mondrian' display, and varied the spectral content of his illuminants. He was able to show that two differently pigmented patches could be made to reflect the same spectral distribution and yet, remarkably, the patches still appeared as different colors. He also showed that an identical surface could be made to reflect a different spectral distribution and yet appear the same color.

The puzzle remained. How could the visual system achieve different color judgments for two areas if the light from the two areas were the same? Land devised the retinex algorithm, which is capable of determining illuminant-independent colors (Land, 1977). According to this algorithm, the critical determinant of the color of a surface is the chromatic context in which the surface appears. This might sound like a reiteration of what artists already knew empirically, but it went further. It provided a testable hypothesis about the visual system. It claimed that color is determined by abrupt changes in the relative cone activities across a scene. For example, retinex would identify a region as 'red' only when the long wavelength light reflected from it is surrounded by regions reflecting shorter wavelengths. Thus we would not expect to 'see' the reddish cast of a tungsten light because the cast is diffuse.

DOUBLE-OPPONENT COLOR CELLS

The cone-opponent retinal ganglion and LGN cells could subserve wavelength discrimination. However, color is not simply wavelength discrimination (see above). Rather, color is achieved through a spatial comparison of wavelengths across an image. A cell having a receptive field fed by a single cone class in a spatially opponent fashion might be the building block for such a comparison: for example, a cell excited by L cones in one part of visual space but suppressed by L cones in an adjacent part of visual space (an L-ON center/L-OFF surround cell); but despite intensive searches, no retinal ganglion cells or LGN cells like this have been found. The cone-opponent type I cells (see Figure 3) have spatially opponent receptive fields, but the centers and surrounds are fed by different cone classes, and the opponency is in the wrong direction to subserve color constancy. So where in the primate visual system is such a comparison made?

In 1968, Nigel Daw showed that some cells in goldfish retina have receptive fields that are both chromatically and spatially opponent, and therefore capable of computing simultaneous color contrast (Daw, 1968). Computational studies have shown that such 'double-opponent' cells could subserve color constancy in primates: a single double-opponent cell exceeds the requirement of a spatial comparison for one cone class: it is a spatial comparison for two cone classes. A common type of double-opponent cell, for example, is L-ON center/L-OFF surround and M-OFF center/M-ON surround.

The existence of double-opponent cells in the primary visual cortex (V1) of primates has been controversial, but there is now a consensus that they do exist (Conway, 2001; Johnson *et al.*, 2001) (Figure 5). Cortical cells that show simple chromatic opponency (such as LGN type II cells) also exist, although they probably do not represent a distinct cell class but rather the end of a continuum of cone-opponent cells that show very weak surrounds.

In addition to mediating spatial color contrast, double-opponent cells may also play an important part in temporal color contrast (a red spot is redder if preceded by a green spot) because they respond to both the onset and cessation of a stimulus (Cottaris and DeValois, 1998) and show stronger responses to sequences of oppositely colored stimuli, e.g. green and then red (Conway *et al.*, 2002).

Cortical cone-opponent cells represent about 10% of the total population of cells in V1. Both red–green double-opponent cells – i.e. L versus M

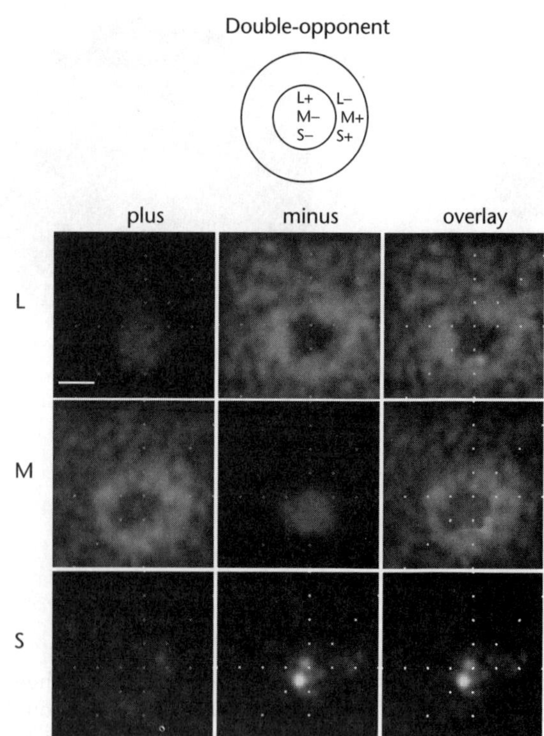

Figure 5. [*Figure is also reproduced in color section.*] The receptive field of a double-opponent cell in monkey V1. The left-hand column shows the spatial extent of the cell's response to increasing activity of the three cone classes (L, top; M, middle; S, bottom); the middle column shows the same cell's response to decreasing the activity of the three cone classes. Comparing the maps (right-hand column) shows that this cell's receptive field is both spatially and chromatically opponent. This double-opponent structure is critical to color constancy – our ability to determine an object's color despite changing illumination conditions. From Conway (2001).

– and blue–yellow double-opponent cells – i.e. S versus (L + M) – are found, and these, with a class of opponent achromatic cells, could be the sole basis for color perception despite their relative scarcity, because color perception is coarse and would require many fewer cells than (say) form perception. Perhaps not by coincidence, color in color televisions requires only about 10% of the bandwidth.

Curiously, some red–green cells appear to receive S cone input. This needs to be studied further. It also remains to be shown how double-opponent cells are constructed from the LGN inputs and why there are more red–green double-opponent cells than blue–yellow ones.

Double-opponent cells reside in the cortex in clusters; moreover, these clusters are coarsely localized in metabolically distinct regions of cortex called 'blobs' (Figure 3, bottom right) (Livingstone and Hubel, 1984). Blobs are easily identified in primate visual cortex by staining with the metabolic enzyme cytochrome oxidase. Why double-opponent cells are localized to the cytochrome oxidase blobs remains a mystery – perhaps these cells require more energy and therefore express higher levels of this metabolic enzyme. Surprisingly, other animals with poor color vision have cytochrome oxidase blobs (although the blobs in some of these mammals, such as cats, are not as prominent). Thus, it may be that the blobs represent regions of cortex dedicated to a more generic function, like parsing surfaces, of which color is only one component.

Nevertheless, the segregation of color cells in the LGN (in the parvocellular and koniocellular layers) and in the primary visual cortex (in the cytochrome oxidase blobs) shows that color is largely processed separately from other visual attributes. This segregation of color processing is evident perceptually (see below) and was used to advantage by Matisse (see Figure 1).

COLOR IS PROCESSED SEPARATELY FROM FORM AND MOTION

Color signals carried by the blobs of V1 are relayed to V2 and then to higher visual areas. Like V1, V2 displays an interesting pattern of staining for the enzyme cytochrome oxidase; unlike V1, the staining consists of alternating thick and thin stripes separated by interstripes (see Figure 3). Cells residing in the blobs of V1 send their axons to the thin stripes of V2 (Livingstone and Hubel, 1984). Not surprisingly, cells in the thin stripes are more likely to be color selective than cells in the thick stripes. The color cells in the V2 thin stripes respond best to colored spots, but they do so over a larger area of visual space. They are not responsive to a large field of color that encompasses the entire region over which small spots are effective. Such 'complex' color cells may be useful in identifying color boundaries present anywhere within a large area.

Color signals carried by cells in V1 and V2 are relayed to subsequent areas where, presumably, color percepts are elaborated. The V2 thin stripes project to V4; the V2 thick stripes, on the other hand, project to the middle temporal area, an area specialized for analyzing motion. Many V4 neurons respond better to some wavelengths than to others (Zeki, 1983), suggesting they are involved in color vision. Their receptive fields are much larger than the receptive fields of V1 cells, suggesting they are involved in elaborating color constancy.

Extrastriate visual areas are better described in the macaque monkey than in the human, but it becomes difficult to compare the areas of humans and monkeys the further the areas are from V1. In humans, for example, an area (V8) situated in the inferotemporal cortex is specialized for computing color (Hadjikhani *et al.*, 1998); it is debated if monkeys have a homolog of this area. Perhaps V4 is the monkey equivalent of human V8. Neurons in V4 are color biased (Zeki, 1980) but they are also selective for other attributes of a stimulus, such as the stimulus orientation (Schein and Desimone, 1990), suggesting that V4 is not simply a color area. To complicate the matter further, unlike lesions of V8 in humans, partial lesions of V4 in monkeys have little effect on tasks requiring color vision. Perhaps V4 is involved in piecing together form and color information; conversely, V4 may actually be a complex of areas, one devoted to form processing and another to color processing. The increasing resolution of functional magnetic resolution imaging may soon make it possible to address these issues.

Certain people who have damage to V8 following a stroke show a profound loss of color perception. Remarkably, this acquired achromatopsia does not interfere with their perception of form and motion – further suggesting that color is processed separately from other visual attributes. Oliver Sacks described one such stroke patient who was an artist (Sacks, 1995). After the patient had lost his color vision, he made peculiar color choices in his paintings; but he was still able to represent luminance and shape, because the areas of his brain dedicated to processing those aspects of the visual world were unaffected by the stroke. Moreover, he had no loss of motion perception.

CONCLUSION

An observer would say that the color and the shape of an object are inextricably linked. If color and form are processed separately by the cortex, how do they then become bound? How would this binding be manifest in the brain? The binding might simply be found in the correlated activity of the two pathways: an orange ball would produce separate sensations of 'round' in the form pathway and 'orange' in the color pathway, but the sensations would be elicited simultaneously. The reliability of the simultaneous activation of the two pathways, possibly reinforced by neural connections joining separate areas, could be enough to bind the 'round' ball with its 'orange' color. But how to test this hypothesis, or any hypothesis for binding for that matter, is challenging.

Parallel processing is an efficient means of computing information because it is fast: multiple aspects of a scene can be processed simultaneously. Although many lines of evidence support a parallel processing model for visual perception (Livingstone and Hubel, 1988), some anatomical and physiological studies show that the visual system does not operate according to such a simple model. There are cells in the superficial layers of V1 (not confined to blobs) that are both selective for stimulus orientation and also more responsive to some wavelengths than others. What contribution do such 'oriented color' cells make to color perception? What are the LGN inputs to red–green double-opponent cells? Do the same LGN cells provide input to sharply orientation-tuned cells? How is the activity of the three chromatic axes in V1 integrated to bring about the perception of specific hues, and how is hue then bound with form? Perhaps most compelling, how do colors bring about emotional responses? These and many more questions will keep the field of color vision research alive for many years to come.

References

Conway BR (2001) Spatial structure of cone inputs to color cells in alert macaque primary visual cortex (V-1). *Journal of Neuroscience* **21**: 2768–2783.

Conway BR, Hubel DH and Livingstone MS (2002) Color contrast in macaque V-1. *Cerebral Cortex* **12**: 915–925.

Cottaris NP and De Valois RL (1998) Temporal dynamics of chromatic tuning in macaque primary visual cortex. *Nature* **395**: 896–900.

Curcio CA, Allen KA, Sloan KR *et al.* (1991) Distribution and morphology of human cone photoreceptors stained with anti-blue opsin. *Journal of Comparative Neurology* **312**: 610–624.

Dacey DM and Lee BB (1994) The 'blue-on' opponent pathway in primate retina originates from a distinct bistratified ganglion cell type. *Nature* **367**: 731–735.

Daw N (1968) Goldfish retina: organization for simultaneous color contrast. *Science* **158**: 942–944.

De Valois RL, Smith CJ, Kitai ST and Karoly AJ (1958) Response of single cells in monkey lateral geniculate nucleus to monochromatic light. *Science* **127**: 238–239.

De Valois RL, Morgan HC, Polson MC, Mead WR and Hull EM (1974) Psychophysical studies of monkey vision. I. Macaque luminosity and color vision tests. *Vision Research* **14**: 53–67.

Hadjikhani N, Liu AK, Dale AM, Cavanagh P and Tootell RB (1998) Retinotopy and color sensitivity in human visual cortical area V8. *Nature Neuroscience* **1**: 235–241.

Johnson EN, Hawken MJ and Shapley RM (2001) The spatial transformation of color in the primary visual cortex of the macaque monkey. *Nature Neuroscience* **4**: 409–416.

Land EH (1977) The retinex theory of color vision. *Scientific American* **237**: 108–128.

Livingstone MS and Hubel DH (1984) Anatomy and physiology of a color system in the primate visual cortex. *Journal of Neuroscience* **4**: 309–356.

Livingstone MS and Hubel DH (1987) Psychophysical evidence for separate channels for the perception of form, color, movement, and depth. *Journal of Neuroscience* **7**: 3416–3468.

Livingstone M and Hubel D (1988) Segregation of form, color, movement, and depth: anatomy, physiology, and perception. *Science* **240**: 740–749.

Nathans J (1999) The evolution and physiology of human color vision: insights from molecular genetic studies of visual pigments. *Neuron* **24**: 299–312.

Reid RC and Shapley RM (1992) Spatial structure of cone inputs to receptive fields in primate lateral geniculate nucleus. *Nature* **356**: 716–718.

Roorda A and Williams DR (1999) The arrangement of the three cone classes in the living human eye. *Nature* **397**: 520–522.

Sacks O (1995) *An Anthropologist on Mars*. New York, NY: Knopf.

Schein SJ and Desimone R (1990) Spectral properties of V4 neurons in the macaque. *Journal of Neuroscience* **10**: 3369–3389.

Smith and Pokorny (1972) Spectral sensitivity of color-blind observers and the cone photopigments. *Vision Research* **12**: 2059–2071.

Wiesel TN and Hubel DH (1966) Spatial and chromatic interactions in the lateral geniculate body of the rhesus monkey. *Journal of Neurophysiology* **29**: 1115–1156.

Zeki S (1980) The representation of colours in the cerebral cortex. *Nature* **284**: 412–418.

Zeki S (1983) The relationship between wavelength and color studied in single cells of monkey striate cortex. *Progress in Brain Research* **58**: 219–227.

Further Reading

Conway BR (2002) *Neural Mechanisms of Color Vision*. Boston: Kluwer.

Gegenfurtner KR and Sharpe LT (1999) *Color Vision: From Genes to Perception*. Cambridge, UK: Cambridge University Press.

Hubel DH (1995) *Eye, Brain and Vision*. New York, NY: Scientific American Library.

Hurvich LM (1981) *Color Vision*. Sunderland, MA: Sinauer.

Livingstone MS (2002) *Vision and Art: The Biology of Seeing*. New York, NY: Abrams.

Zeki S (1993) *A Vision of the Brain*. Cambridge, MA: Blackwell.

Color Vision, Philosophical Issues about

Intermediate article

Alex Byrne, Massachusetts Institute of Technology, Cambridge, Massachusetts, USA
David R Hilbert, University of Illinois, Chicago, Illinois, USA

CONTENTS

The main philosophical issues about color concern whether objects have colors, and what sorts of properties colors are. Some philosophers hold that nothing is colored, others that colors are powers to affect perceivers, and others that colors are physical properties.

INTRODUCTION

According to our everyday experience, many things are colored. Roses are red and violets are blue. On the other hand, according to physical science, roses and violets are composed of colorless particles (or if not particles, something equally colorless). These two pictures of the world are not obviously compatible, and some thinkers have considered them to be plainly incompatible. Galileo, for example, thought that physical science had shown that objects are not really colored, and that colors are 'in the mind'. Philosophical theories of color in modern times have attempted either to reconcile the two pictures, or else to explain why one of them should be rejected.

Until recently, philosophers drew most of their data about color and color perception from their own experience of color. Although personal experience is a valuable source, a good deal of information relevant to abstract philosophical questions about color and the world as revealed by science is to be found in the work of color scientists. Many contemporary philosophers take the view that the physical, biological, and behavioral sciences place significant constraints on philosophical theories of color.

CENTRAL PHILOSOPHICAL ISSUES ABOUT COLOR

The Problem of Color Realism

If someone with normal color vision looks at a lemon in good light, the lemon will appear to have a distinctive property – a property that bananas and grapefruit also appear to have, and which we call 'yellow' in English. However, it does not follow from the fact that an object visually appears to have a certain property that the object has that property. To use an example known to the ancient Greeks, a straight oar half-immersed in water appears bent, but of course it does not have the property of being bent. Ordinarily we take for granted that lemons and other objects are as they appear, but in a philosophical mood we naturally wonder whether we are right to do so.

Such reflections give rise to the central philosophical problem concerning color, the problem of color realism. It is posed by the following two related questions. First, do objects like lemons, bananas and grapefruit really have the distinctive property that they appear to have? That is, are any objects yellow? Second, what is this property? Is it a physical property of some sort (for example, a certain way of reflecting light)? Or is it some kind of property that is specified partly psychologically (for example, a power to produce certain sorts of sensations)? One can ask similar questions for other color properties that objects appear to have.

These questions seem especially important when we consider the physically diverse nature of objects that look colored, and the way the visual system processes color information. As noted above, color is not a property attributed to objects in fundamental physics. In addition, the perception of color is a complex process and the relation between the physical properties of an object and the color it appears to have is far from straightforward. For example, it is not true that yellow-looking objects are those that reflect 'yellow light': there is no simple relationship between the light reaching the eye from an object and the color that object is perceived to have.

The problem of color realism is fundamentally a problem about perception – vision, in particular. The questions concern the nature of certain properties that objects visually appear to have, and whether reality is in these respects as it appears to be. The problem is not primarily about how we talk and think, although of course facts about color language and color concepts may be relevant. It is plausible to assume that the way we use color language is closely connected to the character of human color vision, but the problem of color realism itself concerns perception and not language. These points have to be stressed because it is always tempting to think that philosophical questions are basically questions about how words should be defined, and so are of little relevance to science. The problem of color realism is no more about the definitions of words than is the problem of why the dinosaurs disappeared.

The Representational Content of Visual Experience

It is helpful to put the problem of color realism in terms of what visual experience represents, its representational content. When someone has a visual experience, the scene before his or her eyes visually appears a certain way: for example, it might visually appear to someone that there is a yellow ovoid object in the bowl, or that two lines are the same length. In other words, the experience represents that there is a yellow ovoid object in the bowl, or that two lines are the same length. The experience represents the world correctly if there is a yellow ovoid object in the bowl, or if the two lines are the same length; otherwise, the experience represents the world incorrectly. In the former case the subject's experience will be veridical, and in the latter case illusory. For example, if the subject is looking at lines of the same length with appropriately oriented arrowheads on the ends (the Müller–Lyer illusion), the subject's experience will represent the lines to be of different lengths (even if the subject believes that they are of the same length). The experience represents the world incorrectly and is therefore an illusion.

In this terminology, the problem of color realism concerns certain properties (the colors) represented by visual experience, and whether such experiences represent the world correctly. There is a large philosophical (and, of course, psychological) literature on mental representation, and some philosophical discussions of color draw heavily on it (e.g. Boghossian and Velleman, 1991).

Why Color Matters to Philosophy

Although color is of interest in its own right, in philosophy it mainly serves as a tractable example that can be used to investigate problems of more general scope. One reason why color is particularly suitable for these purposes is that a great deal is known about the relevant physical properties of objects, and the way in which color information is extracted and processed.

One of these more general problems concerns the relation between appearance and reality – whether, or to what extent, the world is as it appears. This problem may be investigated in a reasonably manageable way by restricting attention to a specific instance of it, namely the problem of color realism.

There are a number of other philosophical problems that can be usefully addressed by focusing (not necessarily exclusively) on color. Examples include many central issues in the philosophy of perception: how to distinguish the various sensory modalities; the relationship between perception, thought, and action; and whether we see objects like lemons 'directly', or via our awareness of mental intermediaries. And the famous 'inverted spectrum' thought experiment, which (with some qualifications) supposes that objects that look green to me look red to you, and vice versa, has been used to illuminate a variety of philosophical topics, from the nature of consciousness (Block, 1990) to our knowledge of others' minds (Palmer, 1999).

PHILOSOPHICAL VIEWS AND THEORIES OF COLOR

Eliminativism

Eliminativism is the view that nothing is colored – at least, not ordinary physical objects like lemons. The eliminativist therefore regards much ordinary experience as erroneous. Historically, eliminativism has been the dominant philosophical view; it has its roots in the ancient Greek atomists.

Some eliminativists are 'projectivists': they hold that we 'project' colors that are 'in us' onto objects in our environment. According to the projectivist, some things are colored (for example, neural events, or mental entities like sensations or visual experiences), which we then mistakenly take for properties of objects like lemons. Projectivism is often found in psychology textbooks: Stephen Palmer (1999, p. 95), for instance, writes that 'color is a *psychological* property of our visual experiences when we look at objects and lights, not a *physical* property of those objects and lights'.

An obvious problem with projectivism is that the 'inner' things the projectivist says are colored do not have the right colors, if indeed they have any color at all. Nothing inside the brain becomes yellow when one is looking at a lemon, and it is hard to imagine how a visual experience could be colored.

But an eliminativist need not be a projectivist. One may simply take the view that nothing is colored, not even sensations or visual experiences.

The main line of argument for eliminativism proceeds by claiming that science has shown that objects like lemons do not in fact have colors. The surface of a lemon has a reflectance, various microphysical properties, and is disposed to affect perceivers in certain ways. No other properties are required to explain causally our perceptions of color. But the color properties, whatever they are, do causally explain our perceptions of color. So there is no reason to suppose that the lemon is yellow.

This argument represents a challenge to those who think that lemons are yellow but that this property is not to be identified with a reflectance (the percentage of light reflected by a surface), a microphysical property, or a disposition to affect perceivers (see the discussion of primitivism below). But it begs the question if one simply identifies the property yellow with (say) a reflectance.

The case for eliminativism therefore depends on showing that colors cannot be identified with properties that do causally explain our perceptions of color.

Dispositionalism

Dispositionalism is the view that colors are dispositions (powers, tendencies) to cause certain visual experiences in certain perceivers in certain conditions; that is, colors are psychological dispositions. (Strictly speaking we should add that, according to dispositionalism, at least sometimes our perceptions of color are veridical. This qualification should also be added to the three other views discussed below.) A simple version of dispositionalism is this: the property yellow is just the disposition to look yellow to typical human beings in daylight (for other versions, see Byrne and Hilbert, 1997).

Dispositionalism is often associated with the seventeenth-century English philosopher John Locke who, incidentally, also invented the 'inverted spectrum' thought experiment mentioned above. Locke, like other seventeenth-century philosophers, drew a distinction between 'primary' and 'secondary' qualities. Primary qualities have been characterized in a number of different (and often incompatible) ways, but the essential idea is that they comprise a set of fundamental properties in terms of which all material phenomena can be explained. For Locke, the primary qualities included shape, size, motion, and solidity. Because objects have certain primary qualities, they will be disposed to affect perceivers in certain ways: these dispositional properties are the secondary qualities. In this terminology, dispositionalism is the view that colors are secondary qualities. It is not clear whether Locke was himself a dispositionalist: his view sometimes interpreted as projectivism.

Poisonousness is a straightforward example of a dispositional property. To be poisonous is to be disposed to cause bodily harm if ingested or otherwise taken into the body. According to dispositionalism, yellowness is like poisonousness: to be yellow is to be disposed to cause certain visual experiences if placed in certain viewing conditions. The comparison with poisonousness reveals the relational nature of dispositionalism. Just as a substance may be poisonous for certain organisms and harmless to others, many dispositionalists hold that lemons are only yellow 'for us', and might even be blue relative to some other class of perceivers.

One objection to dispositionalism is that 'certain perceivers' and 'certain conditions' cannot be specified in a principled way. Perhaps a more fundamental problem is that it is not obviously well motivated. It is certainly plausible that yellow objects are disposed to look yellow (at least, once various qualifications are made). However, it is equally plausible that square objects are disposed to look square. It is not very tempting to conclude from this that squareness is a disposition to look square – as noted above, shape properties were supposed by Locke to be paradigmatic examples of properties that are not secondary qualities. It is not clear why color should be treated differently. The dispositionalist needs to explain why perceivers should be mentioned in the account of color, but not in the account of shape. Here the arguments are varied and complex. Dispositionalists often draw their arguments from Locke's own discussion of primary and secondary qualities.

The Ecological View

Thompson *et al.* (1992) have recently developed an 'ecological' view of color, inspired by J. J. Gibson. The ecological view rejects the orthodox account of vision as 'inverse optics' – the process of extracting information about the scene before the eyes from retinal stimulation together with built-in assumptions about the environment. On the other hand, the ecological view stresses the connection between perception and action, and insists that the animal and environment should not be treated as 'fundamentally separate systems'; environmental properties are supposed to be partly 'constituted' by visual perception. Colors, in particular, 'are not already labelled properties in the world which the perceiving animal must simply recover. ... Rather, colours are properties of the world that result from animal–environment codetermination' (Thompson *et al.*, 1992, p. 21).

The ecological view is perhaps best seen as a version of dispositionalism, identifying the colors with 'ecological-level dispositions' to affect perceivers (Thompson, 1995, p. 751). However, it must be emphasized that the ecological view's proponents see the comparison to traditional dispositionalism as somewhat superficial.

One obvious criticism of the ecological view is that it relies on controversial claims about perception. But there is a more basic difficulty, namely that crucial components of the theory are hard to understand. In particular, the meaning of the claim that colors are 'codetermined' by the perceiver and its environment is unclear.

Physicalism

Physicalism is the view that colors are physical properties of some kind, for example, microphysical properties, or reflectances.

Physicalism has been disputed on a number of grounds. First, it is argued that physicalism cannot account for the apparent similarities and differences between colors. In other words, the physicalist cannot explain the structure of phenomenal color space (Boghossian and Velleman, 1991).

Second, and relatedly, it is argued that physicalism cannot account for important observations about the way colors appear to us. For example, it is argued that physicalism cannot explain why orange is a 'binary' hue (every shade of orange is seen as reddish and yellowish), while yellow is a 'unique' hue (there is a shade of yellow that is neither reddish nor greenish) (Hardin, 1993).

Thirdly, it is argued that studies of color vision in animals show that there is no single kind of physical property (e.g. a reflectance) detected by color vision. So, since colors are whatever is detected by color vision, it is concluded that colors are not physical properties (Thompson, 1995).

Primitivism

According to primitivism, objects are colored, but the colors are neither dispositions to affect perceivers, nor physical properties (Yablo, 1995). In fact, the primitivist claims that there is no specially informative account of the nature of the colors. If primitivism is correct, the colors are analogous to fundamental physical properties, like the property of being electrically charged. Given the reductive cast of mind in cognitive science, it is not surprising that few cognitive scientists subscribe to primitivism.

Like eliminativism, primitivism is unmotivated if there are already good candidates to be the colors, for instance, physical properties of some sort, or psychological dispositions. The basic argument for primitivism, then, is similar to the argument for eliminativism: alternative explanations must first be eliminated.

There is little consensus on the best approach to the problem of color realism. The arguments for and against particular theories are rarely conclusive. But it would be wrong to think that no progress has been made. Philosophy is often advanced by showing that apparently unrelated theses are after all closely related, thereby forcing a proponent of, say, dispositionalism to take on a particular burden of commitments. Much recent work in the philosophy of color is of this kind.

INFLUENCE OF COGNITIVE SCIENCE ON ISSUES ABOUT COLOR

One notable feature of recent philosophical work on color is the attempt to integrate philosophical concerns with what is known empirically about color and color vision. Below we discuss a few areas in which this interaction either has been or has the potential to be especially significant.

Color Spaces and Opponent Processes

The colors stand in a complex web of similarity relations to each other. For instance, purple is more similar to blue than to green; and shades of red can be more or less similar to each other. Relations of color similarity also have an opponent structure. Red is opposed to green in the sense that no reddish shade is greenish, and vice versa; similarly for yellow and blue. Further, there is a shade of red ('unique red') that is neither yellowish nor bluish, and similarly for the three other 'unique' hues yellow, green and blue. Thus, in experiments summarized by Hurvich (1981), a normal observer looking at a stimulus produced by two monochromators (light sources that emit in a narrow band of wavelengths) is able to adjust one of them until he reports seeing a yellow stimulus that is not at all reddish or greenish. In contrast, every shade of purple is both reddish and bluish, and similarly for the other three 'binary' hues orange, olive and turquoise. The binary hues are sometimes said to be 'perceptual mixtures' of the unique hues. These sorts of observations, supplemented with physiological data, form the basis of the opponent process theory of color vision. The main idea of this theory is that color perceptions are the result of two opponent processes (red–green and yellow–blue) and one non-opponent process (light–dark).

As mentioned above, two objections to physicalism start from these facts. We can now elaborate somewhat on the second of these. If physicalism is true, the objection runs, then the difference between the unique and binary hues must be explained in terms of 'unique' and 'binary' physical properties of objects like lemons and oranges. However, the correct explanation is in terms of neural opponent processes, and does not involve the physical properties of objects at all. This objection is controversial, but at least the opponent process theory has led philosophers to a renewed interest in understanding relations of similarity and difference among the colors.

Animal Color Vision

One active area of empirical research is comparative color vision: the study of color vision in non-human animals. Color vision is very widely distributed among animals: some degree of it appears to exist in all the major groups of vertebrates, and it is also common among invertebrates. But there is great variation in the precise type of color vision and the purposes for which it is used. Traditionally, philosophers have restricted their attention to human color vision, although this seems now to be changing. As mentioned above, the kinds of variation in color vision found in animals have been used to argue against physicalism, and also to support the ecological view.

Variation in Perceived Color

The perceived color of an object depends in complicated ways on both the illumination and the other objects in the scene before the eyes. As many people have discovered, even lightness relationships can be reversed by sodium lighting of the sort often found in parking lots. And interior decorators know that the perceived color of an object can depend on the color of its surroundings.

The perceived color also depends on the perceiver: color vision in human beings is surprisingly variable for a basic perceptual ability. Approximately 10 percent of men suffer from a substantial deficit of red–green color vision, and complete red–green color blindness is not rare. There is also substantial variation, among people classified as having 'normal' color vision, in, for example, the spectral location of unique green – ranging over 30 nanometers, nearly 10 percent of the visible spectrum. In addition to variation between subjects, there is variation within subjects. For example, the optical media of the eye, in particular the macula and lens, tend to yellow with age, producing shifts in perceived hues.

These facts have implications for philosophical theories of color. For example, dispositionalism appeals to a notion of 'certain perceivers'. Given the degree of normal variation, slightly different specifications of the privileged class of perceivers will lead to big differences in the resulting dispositionalist theories. The dispositionalist needs to give a principled reason for selecting one group rather than another to be the 'certain perceivers'. Similar remarks apply to the dispositionalist's 'certain viewing conditions'.

Color Constancy

As mentioned above, perceived color depends on the illumination. It is less well known that in many circumstances perceived color is relatively insensitive to changes in illumination. This feature of human (and animal) color vision is known as approximate color constancy. It implies that in many circumstances the perceived color of an object is more closely dependent on its (illumination-independent) surface properties than on the spectral character of the light reaching the eye. Some cognitive scientists have attempted to explain color constancy by treating color vision as a system that extracts information about the surface properties of objects, in particular their reflectances, from the light reaching the eye. This provides one of the inspirations for physicalist theories of color (Hilbert, 1987).

The Inverted Spectrum

It is often supposed that there is no reason to believe that people are 'spectrally inverted' with respect to each other. Indeed, under the influence of the logical positivists of the early twentieth century, it used to be a common philosophical opinion that spectrum inversion makes no sense at all, because it is not 'verifiable'. However, with improved understanding of the genetic and physiological basis of color blindness, since the 1970s some color scientists have speculated that there might be actual cases of spectrum inversion in the human population. Red–green color blindness is caused by genetic defects that result in two of the three types of cones containing pigments that are very similar in their spectral sensitivity (their readiness to absorb light of different wavelengths). There are two types of red–green color blindness, depending on whether the spectral sensitivities of the two relevant cone types match that of the normal middle-wavelength-sensitive pigment or that of the long-wavelength-sensitive pigment. A person who has inherited the genes for both forms of red–green color blindness would have the two pigments switched from the normal condition. If the development of the neural circuitry for color vision depends only on the cell type and not on the pigment it contains, then this scenario should result in spectrum inversion. In fact, though, there does not seem to be the required independence between circuitry and pigment. Still, the fact that spectrum inversion is treated as a testable empirical hypothesis shows that it cannot be dismissed as a philosopher's fiction.

Metamers

Lights with quite different spectra, and surfaces with quite different reflecting characteristics, can appear to be identical in color. This phenomenon, known as metamerism, is a consequence of the fact that all the information available for perception of color derives from just three cone types with broad spectral sensitivity. If the light reaching the eye from two objects produces the same response in each of these three cone types then they will appear to have exactly the same color, no matter how their spectra or reflectances differ. This fact about (human) color vision has been significant in philosophical discussions of color since the 1970s. The central question is whether metamerism is incompatible with taking colors to be physical properties, because the phenomenon seems to show that there is no single spectrum (or reflectance) that all objects with a particular color share. For a variant of this argument see (Hardin, 1993, pp. 63–64); for a response see (Jackson, 1996).

References

Block N (1990) Inverted earth. *Philosophical Perspectives* **4**: 53–79.

Boghossian PA and Velleman JD (1991) Physicalist theories of color. *Philosophical Review* **100**: 67–106.

Byrne A and Hilbert DR (1997) Editors' introduction. In: Byrne A and Hilbert DR (eds) *Readings on Color*, vol. I 'The Philosophy of Color', pp. xi–xxviii. Cambridge, MA: MIT Press.

Hardin CL (1993) *Color for Philosophers: Unweaving the Rainbow (Expanded Edition)*. Indianapolis, IN: Hackett.

Hilbert DR (1987) *Color and Color Perception: A Study in Anthropocentric Realism*. Stanford, CA: CSLI.

Hurvich LM (1981) *Color Vision*. Sunderland, MA: Sinauer Associates.

Jackson F (1996) The primary quality view of color. In: Tomberlin J (ed.) *Philosophical Perspectives* vol. x, pp. 199–219. Cambridge, MA: Blackwell.

Palmer SE (1999) Color, consciousness, and the isomorphism constraint. *Behavioral and Brain Sciences* **22**: 923–943.

Thompson E (1995) *Colour Vision*. New York, NY: Routledge.

Thompson E, Palacios A and Varela FJ (1992) Ways of coloring: comparative color vision as a case study for cognitive science. *Behavioral and Brain Sciences* **15**: 1–74.

Yablo S (1995) Singling out properties. In: Tomberlin J (ed.) *Philosophical Perspectives* vol. IX, pp. 477–502. Atascadero, CA: Ridgeview.

Further Reading

Byrne A and Hilbert DR (1997a) *Readings on Color*, vol. I 'The Philosophy of Color'. Cambridge, MA: MIT Press.

Byrne A and Hilbert DR (1997b) *Readings on Color*, vol. II 'The Science of Color'. Cambridge, MA: MIT Press.

Hilbert DR (1992) What is color vision? *Philosophical Studies* **68**: 351–370.

Jackson F (1977) *Perception: A Representative Theory*. Cambridge, UK: Cambridge University Press.

Lewis DK (1997) Naming the colors. *Australasian Journal of Philosophy* **75**: 325–342.

Maund JB (1995) *Colours: Their Nature and Representation*. Cambridge, UK: Cambridge University Press.

McGinn C (1983) *The Subjective View: Secondary Qualities and Indexical Thoughts*. Oxford: Oxford University Press.

McGinn C (1996) Another look at color. *Journal of Philosophy* **93**: 537–553.

Stroud B (2000) *The Quest for Reality: Subjectivism and the Metaphysics of Colour*. New York, NY: Oxford University Press.

Tye M (2000) *Consciousness, Color, and Content*. Cambridge, MA: MIT Press.

Westphal J (1991) *Colour: A Philosophical Introduction*. Oxford: Blackwell.

Commissures

See **Cerebral Commissures**

Comparative Psychology

Introductory article

Edward A Wasserman, University of Iowa, Iowa City, Iowa, USA

CONTENTS

Comparative psychology explores the behaviors of different species of animals with a special interest in any similarities and differences that may reveal the evolutionary origins of those behaviors.

INTRODUCTION

Are animals intelligent? How can we learn about animal intelligence? Do different species differ in intelligence? How can we measure species differences in intelligence? Does animal intelligence resemble human intelligence? What would it mean if animals were indeed intelligent and if their intelligence approached, or even eclipsed, our own? (*See* **Intelligence**; **Animal Cognition**; **Animal Learning**)

Most people, young and old, have asked these intriguing questions. Answering them is the business of comparative psychology, a field that explores similarities and differences in the behavior of human and nonhuman animals. Comparative psychologists are interested in a wide range of behaviors: from mating to migrating, from feeding to fighting, from sleeping to scratching. These scientists have been especially interested in intelligent behaviors: those actions whose acquisition advances an animal's chances of surviving and reproducing in an environment that is fraught with danger and uncertainty.

Historical Foundations

The comparative psychology of intelligence is only a century or so old if we date the origin of the field as an exact experimental science with the 1898 publication of Edward L. Thorndike's pioneering monograph, *Animal Intelligence: An Experimental Study of the Associative Processes in Animals*.

But, weren't people interested in and didn't they know about animal intelligence for much more than the past century? Certainly. Ancient texts and drawings suggest that humans have trained animals for work and amusement for centuries; doing so meant that humans understood and exploited the modifiability of animal behavior. Nevertheless, a practical understanding of animal intelligence is decidedly different from a scientific understanding. Formulating precise laws of learning and developing effective technologies of teaching are crucial to determining the nature of intelligence and to exploring its generality throughout the animal kingdom. These critical additions to our knowledge of animal intelligence came only at the beginning of the twentieth century with the work of Thorndike and other trailblazers in the new science of comparative psychology. (*See* **Learning, Psychology of**)

Even before this modern experimental era, famous scholars had deemed animal intelligence to be central to comprehending the nature and origin of humankind. Two key points in the history of human thought place the study of animal behavior and learning at the very center of philosophical and scientific inquiry: (1) Descartes' distinction between humans and brutes, and (2) Darwin's hypothesis of mental continuity between human beings and nonhuman animals.

The seventeenth-century French philosopher René Descartes believed that human beings were profoundly different from brutes. Animals were mere machines; they had intricate bodily systems that controlled their physiology and behavior, but they lacked what humans alone possessed – a rational soul. The rational soul was divinely created, it was not made of matter, nor did it reside in the human body. (*See* **Descartes, René**)

The operation of the rational soul had two unique behavioral consequences: (1) it allowed us to communicate our private thoughts and feelings to other human beings, and (2) it permitted us to suitably tailor our behaviors to a vast variety of complex and ever-changing environmental situations. Descartes believed that animals had no thoughts to communicate; they were thus forced to respond as their sensory and motor systems demanded, without the involvement of intelligence.

Against this backdrop of Cartesian thinking, the nineteenth-century English biologist Charles Darwin proposed that the nature and descent of human beings was not a matter for theology or philosophy, but biology. Scores of naturalistic and anecdotal observations convinced Darwin that humans and animals were not fundamentally different from one another, nor did they have different origins; all beings were the products of organic evolution. In contrast to Descartes, Darwin viewed both communication and intelligence from a natural scientific perspective; primitive or even highly advanced forms of each of these abilities were to be found throughout the animal kingdom, thus disclosing what Darwin called 'mental continuity' between human and nonhuman animals. (*See* **Evolutionary Psychology: Theoretical Foundations**)

Darwin's bold evolutionary ideas made the study of animal behavior crucial to understanding human behavior: if humans arose from lower forms of animals, then the study of animal intelligence is essential to elucidating the biological precursors of the human mind.

Thorndike's Contributions

When he began his research, Thorndike accepted Darwin's evolutionary perspective, but he was skeptical of Darwin's evidence of animal intelligence. Indeed, Thorndike devised his famous 'puzzle box' method in order to provide an objective and experimental antidote to the subjective and anecdotal accounts of animal intelligence that were in vogue during the final decades of the nineteenth century, and that Darwin used to make his case for mental continuity between human and nonhuman animals. These anecdotes were tall tales of animal genius that were spun by pet owners, zookeepers, and amateur naturalists. Most of these astounding and amusing anecdotes proved to be of dubious accuracy and reliability; but, they did provoke great interest and debate in popular and scientific circles.

In his innovative research, Thorndike placed animals into small boxes from which they could escape and receive food by solving a simple behavioral puzzle. For instance, a hungry cat might have to pull a loop of string; doing so opened a door through which the feline could exit the box and nibble a tasty titbit of fish. By measuring the time that it took the cat to claw the string after each placement into the box, Thorndike found that this time progressively fell with successive trials.

Thorndike had discovered a basic law of learning – the law of reinforcement. When a response is followed by a reward, an organism is more likely to make that response again in that situation. Further study by Thorndike and later comparative psychologists showed that this law of reinforcement is not limited to cats, to pulling loops of string, or to fish snacks. Pigeons more quickly peck a button when grains of seed ensue. Rats more rapidly rotate a wheel when draughts of water follow. And, human infants kick with added alacrity when a motorized mobile turns afterwards. Such learned behaviors need not continue indefinitely; if the reward is revoked following the response, then the behavior returns to its initial level – the law of extinction.

By 1911, Thorndike's own research and that conducted by several other investigators led him to conclude that most vertebrate animals learn in the same general way: stimulus–response associations are automatically strengthened by reward and weakened by extinction. (*See* **Reinforcement Learning: A Biological Perspective**)

Learning by consequence was positively Darwinian. Natural selection leads to the retention of fit organisms and to the elimination of unfit ones; the laws of reinforcement and extinction lead, respectively, to the retention of effective behaviors and to the elimination of ineffective ones.

Most relevant to the evolution of intelligence, Thorndike believed that species differences in learning are matters of degree, not kind: stimulus–response associations may increase in number, may be formed more quickly, may last longer, and may become more complex. Growth in the number, speed of formation, permanence, and complexity of associations reaches its high point in human beings. Thorndike also suggested that a parallel exists between individual development (ontogeny) and species evolution (phylogeny): the development of the infant's intelligence to the adult condition may be viewed as progressing from animal to human competence.

COMPARATIVE PSYCHOLOGY OF LEARNING

Thorndike's groundbreaking research and theorizing set the stage for the work of succeeding generations of comparative psychologists.

Critical to this later work was the crafting of advanced methods which could sensitively and reliably measure animals' learning of new behaviors. Most famous among these new methods was I. P. Pavlov's conditioned reflex procedure and B. F. Skinner's 'Skinner Box' procedure, a refinement of Thorndike's puzzle box procedure. These new methods have been creatively adapted to the unique behavioral repertoires of different species in the quest to uncover similarities and differences in their intelligence. (*See* **Pavlov, Ivan Petrovich**; **Skinner, Burrhus Frederic**; **Conditioning**)

Quantitative Differences in Learning

Despite the large amount of experimentation into quantitative differences among animal species that has been conducted since Thorndike's investigations, most reviewers have concluded that this line of research has not been productive. Quite simply, animal behavior is affected by so many other factors – differences in sensory and motor capabilities, variations in motivation and reward, differences in daily activity levels – that valid quantitative cross-species comparisons of intelligence have proven to be virtually impossible to document. (*See* **Motivation**)

Consider an illustrative example: is a cat smarter than a dog? To begin, we have to select some common behavioral technique that can properly measure intelligence in these two very different animals. But, which technique? The senses of cats and dogs are obviously not equally keen. Cats have excellent audition (hearing); dogs have excellent olfaction (sense of smell). By requiring the animals to use one sense or the other in the experimental task, we could stack the deck in favor of one species over the other. The two species' motor abilities also differ, with cats being outstanding jumpers and dogs being outstanding runners. Different response requirements could favor one species over the other. What about the reward? How much of which kinds of foods will equate the value of the rewards for the two species? What will be the effect of different levels of food deprivation in the two species? And, at what time of day should the animals be trained? Should we train in the day for the diurnal dog or in the night for the nocturnal cat?

These and many additional considerations make it abundantly clear that, however interesting and well-intentioned the original question may have been, it is practically impossible to say whether the cat or the dog is the more intelligent species of animal.

Does it therefore follow that there are no differences in intelligence among different species to be documented? Not necessarily. We may simply have to adopt a different experimental tactic in order to discover them.

Qualitative Differences in Learning

The comparative psychologist M. E. Bitterman has suggested that we investigate qualitative differences in learning among different species of animals. Bitterman's proposed plan of investigation has been to compare species according to their orderly responses to changes in the conditions of reinforcement.

For example, some species of animals may come to learn discrimination reversals faster than they learned the original problem, whereas others may not show such improvement. To illustrate: the choice of a white circle over a black one might initially lead to reward. Then, the discrimination is reversed: the choice of the black circle over the white one now leads to reward. Later, the discrimination is reversed over and over again from problem to problem. In fact, some species of animals improve in the speed of reversal learning, whereas others do not. Such a dramatic difference suggests that qualitatively different learning processes may be producing these effects.

As another example, some species may exclusively select the better response alternative ('maximizing'), whereas other species may match their choices to the reward probabilities ('matching'). So, if responses to one button produce food with a probability of 0.75, whereas responses to a second button do so with a probability of 0.25, then the receipt of reward will be maximized by choosing the first button 100 per cent of the time (100(0.75) + 0(0.25) = 75.00), but not by choosing the first button 75 per cent of the time and the second button 25 per cent of the time (75(0.75) + 25(0.25) = 62.50). These two patterns of choice responding may represent qualitatively different decision strategies: a potentially important species difference in learning and behavior. (*See* **Choice Selection**)

COMPARATIVE PSYCHOLOGY OF MEMORY

Beyond the comparative study of learning in animals, other advanced intellectual processes have attracted experimental attention, among them memory. Memory is obviously necessary for learning to occur. If an organism in Thorndike's puzzle box were to forget what response it last performed in that situation, then there would be no way for food to strengthen that particular stimulus–response bond. (*See* **Memory**)

Given how central memory is to the acquisition of intelligent action, it should not be surprising that comparative psychologists have long been interested in the experimental investigation of memory. Furthermore, given Darwin's hypothesis of mental continuity and Thorndike's suggestion that species differences in learning are primarily quantitative and not qualitative in character, it should not be surprising that researchers have endeavored to determine if some species can remember prior acts and events for different lengths of time.

Delayed Response

The pioneer in the study of animal memory was W. S. Hunter, who in 1913 devised the delayed response paradigm. In this paradigm, an animal might see one of three potential sites baited with food or otherwise marked by a brief stimulus such as a light. Then, after a delay of a few or several seconds, the animal would be required to select the earlier baited or marked site. If the animal succeeded in choosing the correct site and in receiving the food that was to be found there, then this result might constitute evidence for memory of the baited location. The longer the delay over which the animal is able to respond correctly, the better is its memory.

But, a rival interpretation presents itself. Successful performance in the delayed response task might not be the result of some enduring cognitive or neural process; it might instead be due to the animal merely maintaining its bodily orientation to the baited site during the delay period. Postural mediation was an especially obvious possibility for one of Hunter's contingent of different experimental subjects – the dog, an animal that is well known for its pointing behavior.

Hunter and other memory researchers cleverly strove to eliminate this postural possibility by rotating the animal on a turntable or by removing the animal from the apparatus during the delay period. These measures did sometimes succeed in disrupting the animal's delayed discrimination performance, as they should if pointing were all that there was to accurate performance after a delay. But, these measures did not always disrupt the animal's performance. Clearly, then, there are at least some bona fide cases of memory in the delayed response paradigm that cannot be due to mere postural mediation.

Delayed Matching-to-Sample

More recent research has exploited an alternative testing method for measuring animal memory – delayed matching-to-sample. This method is not plagued by the problem of positional mediation; also, delayed matching-to-sample is much more versatile in its application to a wide variety of issues in animal memory and intelligence than is the delayed response paradigm. Further contributing to the popularity of delayed matching-to-sample has been its extremely successful application to the behavior of the pigeon, which – because of its long life, excellent vision, and ready adaptation to laboratory captivity – has become a favorite of experimental psychologists interested in animal cognition.

In this procedure for the pigeon, two simultaneously presented testing stimuli (for example, red and green lights) follow the presentation of the sample stimulus (randomly, a red or a green light on alternate trials). Only one testing stimulus (whose color matches the sample stimulus) is correct and leads to food if it is chosen; the other testing stimulus (whose color does not match the sample stimulus) is incorrect and does not lead to food if it is chosen. With a short delay between the sample and the testing stimuli, pigeons show a strong tendency to choose the correct (matching) testing stimulus. However, as the delay is lengthened, choice accuracy declines towards the indiscriminate selection of the correct (matching) and the incorrect (nonmatching) testing stimuli, thereby documenting the forgetting of the sample stimulus. Not only is there a decline in memory as the sample–test interval is lengthened; choice accuracy is also positively related to the duration of the sample stimulus and negatively related to the time between trials.

Because these simple temporal parameters of delayed matching-to-sample exert such a strong influence on memory performance within a species, any claims about memory differences between species are extremely difficult to prove. The parallel to the case of quantitative species differences in learning is obvious.

An important variant of the delayed matching-to-sample procedure involves sample and testing stimuli that are drawn from different pools of stimuli. For instance, pigeons might be shown different colors as sample stimuli and different forms as testing stimuli. No true matches are possible; only arbitrary or symbolic matches can hold. So, the selection of circle after red or the selection of square after green might be the correct responses, whereas the selection of square after red or the selection of circle after green might be the incorrect responses. This so-called 'delayed symbolic matching-to-sample' procedure has afforded researchers special opportunities to expand the investigation of animal memory.

With true matching-to-sample procedures, it has been shown that pigeons can remember the color of a sample stimulus, its shape, its orientation, and its spatial location. With the symbolic matching-to-sample procedure, it has also been possible to show that the duration of a stimulus can be remembered. Thus, pigeons remember different durations of a red sample stimulus and report that memory during testing stimuli of differing line orientations.

Beyond the attributes of single sample stimuli, pigeons that have been given embellished versions of the symbolic matching-to-sample paradigm have also been shown to remember the temporal order (e.g. red–green) of two differently colored sample stimuli, the spatial order (e.g. left–right) of two identically colored sample stimuli, and the relative duration (e.g. short–long) of two differently colored sample stimuli. As well, pigeons and rats have successfully been trained to make one of two different responses depending on the number of prior visual or auditory stimuli (for example, two versus four).

Another way in which the memory of complex information has been studied has involved sample stimuli that comprise two or more elements. Pigeons were thus shown two-element sample stimuli that were composed of color (red or green) and line orientation (horizontal or vertical) elements. Tests with just color comparisons or just line comparisons each yielded highly accurate testing performance, indicating that the pigeons discriminated and remembered both the color and the line orientation of the compound sample stimuli. Significantly, however, accuracy on these compound sample (color and line) trials was lower than on other trials involving only single-element samples (color or line); this result suggests that the two sample elements on compound sample trials competed with one another for what in humans is commonly called *attention*. (*See* **Attention; Selective Attention**)

All of this evidence shows that animals are amazingly sensitive to the richness and subtlety of their environment. That sensitivity is also enduring, as witnessed by the fact that stimuli can be retained for substantial periods of time, sometimes for durations as long as half a minute.

Rehearsal

Many theorists of human memory have proposed the operation of control processes: means by which memories are changed in accord with the needs of the individual or the demands of the task. One key control process is *rehearsal*: a covert activity that helps to sustain the memory of a prior event. Engaging in rehearsal should aid in retaining earlier information, whereas terminating rehearsal should impair retention.

In order to investigate the role of rehearsal in human memory, researchers have devised the directed-forgetting paradigm. In one version of the directed-forgetting paradigm, people are given one of two cues (e.g. red or green colors) either to remember or to forget a previously presented stimulus. When the remember cue is given the memory test is presented afterwards; when the forget cue is given the memory test is not presented afterwards.

Finally, a trick is played on the individual; a forget cue is given and is followed by a retention test. When these unexpected retention tests are given, researchers generally find that memory performance is worse on forget-cued trials than on remember-cued trials; this result implies that the post-stimulus cues were affecting the rehearsal process and thereby modulating memory.

Researchers have further found that postponing the forget cue in a delay interval of fixed duration leads to a loss in its effectiveness; memory performance for earlier information improves the later into the delay interval the forget cue is given. This result suggests that the spontaneous or uncued rehearsal that occurs prior to the forget cue effectively protects memory from the decremental effect of the forget cue.

Several workers in the area of animal memory have attempted to see whether directed forgetting is uniquely human. Research with both pigeons and monkeys has adapted the delayed matching-to-sample paradigm to this objective by adding brief post-sample cues during the delay interval in order to signal that a test for sample memory either would or would not be given. As in the case of human memory, animal memory proved to be much lower on forget-cued trials than on remember-cued trials. In addition, memory was more markedly reduced if the post-sample forget cue was presented early than if it was presented late in the delay interval.

These results suggest that animals may indeed have active control over memory processing. Rehearsal may not be uniquely human nor necessarily verbal.

COMPARATIVE PSYCHOLOGY OF CONCEPTUALIZATION

Human beings and other animals are incessantly bombarded by an extraordinarily complex array of external stimuli; yet, they somehow make sense of these varied and varying stimuli. One way to reduce the demands on an organism's sensory and information-processing systems is for it to treat similar stimuli as members of a single class; by doing so, considerable behavioral economy can be achieved, thus freeing its adaptive machinery to deal with other competing demands of survival. In addition, categorical processing permits an organism to identify novel stimuli as members of a particular class and to generalize knowledge about that category to these new members. So, an organism need not be bound to respond only to those stimuli with which it has had prior experience, further enhancing its ability to cope with a continually changing world. (*See* **Generalization**)

Theorists often trumpet these adaptive virtues of categorization and conceptualization; yet, we remain far from understanding precisely how organisms partition the world into classes of related objects and events. Indeed, theorists have historically doubted whether nonhuman animals are even capable of conceptual behavior. More than a century ago, the famous English comparative psychologist C. Lloyd Morgan denied animals the ability to behave conceptually. Morgan believed that only adult humans, and not even children, are capable of conceptualization. Recent research is changing that opinion. (*See* **Categorical Perception**)

Object Concepts

One familiar case of conceptual behavior involves the kinds of open-ended categorization responses that we make when we label different natural (e.g. cat) and artificial (e.g. car) objects with different nouns. Such verbal behaviors are occasioned by specific instances of wide variability and individuality. Indeed, accurate classification even extends to categorical exemplars that we have never seen before. Is it at all possible for nonhuman animals lacking language to engage in this form of conceptual or classificatory behavior?

In order to answer this question, a new technique was devised to train pigeons concurrently to discriminate stimuli from several human language categories. The specific method parallels the technique that parents often use in order to teach their children to label objects in a picture book. When the page is turned, the child is first asked to look at the object and then she is requested to name it. If she is correct, then she is praised. If she is incorrect, then she is told 'no' and is encouraged to try again. If self-correction fails, then she is provided with the correct name. (*See* **Concept Learning**)

In order to implement this method with pigeons, a color snapshot was displayed on a small screen and the pigeon was required to peck a clear plastic key covering the screen in order to guarantee that it was looking at the snapshot. Then, four differently colored keys were illuminated just beyond the corners of the screen. A single choice response was permitted. If the response was to the correct key for reporting the stimulus on the viewing screen, then the pigeon was fed grain; if the response was to any of the three incorrect report keys, then no grain was given and the pigeon had to repeat that trial until it made the correct response. The slides that were shown in each daily session depicted several different examples of cats, flowers, cars, and chairs. The pictures contained one or more instances of the critical stimulus object; the objects were indoors or outdoors, near or far away, centered or off-center, and in different colors, orientations, and backgrounds.

In a representative experiment, pigeons attained a level of discriminative performance that averaged about 75 percent correct at the end of a month of training, after beginning the investigation near the chance level of 25 percent correct. Most important were the results of two later days of testing performance with the original training slides and with brand-new slides of cats, flowers, cars, and chairs. Accuracy to the old slides averaged about 80 percent and accuracy to the new slides averaged about 65 percent.

These results suggest that the pigeons had learned general object concepts that permitted them to categorize both old and new stimuli from four human language classes. Although the pigeons' testing performance was highly discriminative to both old and new stimuli, accuracy was reliably higher to the old pictures than to the new ones, perhaps because the birds memorized some or all of the old slides. (*See* **Concept Learning and Categorization: Models**)

In another project, three groups of different pigeons were also trained to categorize photographic slides. The three groups were given 48 daily training trials comprising: 12 copies of one example from the categories cat, flower, car, and chair (group 1); three copies of four examples from the same categories (group 4); or one copy of 12 examples from the same categories (group 12). The rate of learning was a negative function of the number of examples per category. Of additional importance were the results of a generalization test with 32 novel stimuli: eight from each category. Now, accuracy was a positive function of the number of training examples.

Although increasing the difficulty of original learning, greater numbers of training examples per category enhanced the accuracy of generalization performance, perhaps because of the increased likelihood that any given test stimulus resembled one or more of the remembered training stimuli. These data are not only orderly, but they neatly correspond with a large body of research on categorization in human adults and children. Empirical parallels like these are unlikely to be coincidental; rather, they suggest a basic behavioral similarity. Whether there are other correspondences in conceptualization by humans and animals is a topic of current research.

Abstract Concepts

One of the empirical hallmarks of conceptualization is that discriminative responding is independent of the specific details of the prevailing stimuli. Note that it was imperative in research on object concepts to show that the discriminative responding that was established to a familiar set of training stimuli also extended to a novel set of testing stimuli. To have conceptualized 'chairs' requires that new chairs occasion the same response as old ones. (*See* **Representations, Abstract and Concrete**)

An even more advanced level of conceptualization may be achieved when organisms respond 'same' or 'different' to several simultaneously or successively presented stimuli. Again, the critical test comes when novel stimuli are given, in order to see whether the organism appropriately responds to them.

Here, too, great doubt has historically been expressed that animals can learn abstract concepts. The seventeenth-century English philosopher John Locke, in particular, believed that abstract conceptualization represented the key intellectual divide between humans and animals.

Despite his firm conviction that nonhuman animals were incapable of abstraction and conceptualization, Locke may have been premature in his assessment. Mounting behavioral evidence suggests that nonhuman animals – even pigeons – can form abstract concepts.

In this research, pigeons received food reward for pecking one button (for example, red) when they were shown any displays that pictured 16 copies of the same computer icon, and for pecking a second button (for example, green) when they were shown any other displays that pictured one copy of 16 different computer icons. Incorrect responses led to nonreward.

After the pigeons had reached a high level of discrimination accuracy (exceeding 80 percent correct choices when the chance score was 50 percent correct), in testing sessions the birds were shown brand-new Same displays and brand-new Different displays that pictured icons that they had never seen before. In various experiments, discrimination accuracy averaged from 83 to 93 percent correct to the Same displays and to the Different displays from the training set; accuracy averaged from 71 to 79 percent correct to the Same displays and to the Different displays from the testing set. These high levels of discrimination accuracy to both familiar and novel displays are consistent with the pigeons' having learned an abstract same–different concept.

Not only has same–different conceptualization been demonstrated for pigeons that were given simultaneous displays of visual items, but also for pigeons that were given successive lists of visual items. In the latter case, the 16-list icons were presented one at a time, thereby requiring the memory of prior items in order to decide whether the just-presented list comprised identical or nonidentical items.

Despite this important procedural change, discrimination accuracy averaged 94 percent correct to the Same lists and to the Different lists from the set of training icons; further, accuracy averaged

72 percent correct to the Same lists and to the Different lists from the set of testing icons. Abstract conceptualization may thus be within the ken of even the pigeon.

DIFFERENT APPROACHES TO COMPARATIVE INTELLIGENCE

The foregoing has described a sampling of investigations whose results suggest that advanced intellectual processes – including learning, memory, and conceptualization – can be comparatively studied. The results of these investigations disclose many salient similarities in the behavior of human and nonhuman animals.

Should we thus agree with Thorndike that intelligence is pretty much the same throughout the animal kingdom? Perhaps not. But the many similarities that have been uncovered so far are simply too provocative not to merit further investigation. However, although they are surprisingly few in number, species differences in intelligence have been found and many more may ultimately surface if we assiduously seek them.

The Process Approach

Despite the prominence of the above line of research, some authors have questioned the value of the comparative study of such distantly related species as pigeons, rats, cats, dogs, monkeys, and humans. These critics have insisted that any comparisons among species that are not close evolutionary relatives of one another are capricious and of doubtful biological significance.

These critics have also argued that the seemingly expedient selection of distantly related species is an anthropocentric or human-centered strategy that springs directly from Darwin's hypothesis of mental continuity. Indeed, to such anthropocentric theorists, the true vision of comparative psychology is as a distinctly human psychology whose prime purpose is to understand and to specify the rules of human psychological functioning, as well as their degree of zoological generality.

A different and perhaps more accurate picture of this line of research in the comparative psychology of intelligence focuses on the matter of process. Thus, we are interested in establishing whether some intellectual process is involved in human and nonhuman behavior and the degree to which that process reflects common behavioral and biological mechanisms.

For example, the comparative study of the laws of reinforcement and extinction could speak to the nature and generality of these adaptive processes. If these behavioral processes are indeed general, then animal models and subsequent neuroanatomical and neurophysiological research might shed great light on their underlying biology.

Furthermore, if most extant species of animals similarly obey the laws of reinforcement and extinction, then there is good reason to believe that a common ancestor of all of these species must also have been subject to the same behavioral laws, perhaps because of common selection pressures. It would not make much sense for each species to have independently devised the same evolutionary solution to a common survival problem.

The Phylogenetic Approach

It is also possible more directly to develop the comparative psychology of intelligence from a phylogenetic perspective. In this approach, researchers study the behavior of more strategically selected species in order to elucidate the natural and evolutionary histories of the chosen animals. Adopting this more ethological and ecological approach means that one may compare either closely related species that face divergent survival demands or unrelated species that face similar survival demands. Such phylogenetic study may not only help to identify the selection pressures for complex behavior and intelligence, but it might also help to identify any specialized behavioral or intellectual adaptations. (*See* **Learning and Memory, the Ecology of**)

Take, for example, the spatial memory of animals that do not always eat the food that they find while they are foraging; rather, these animals may store the food in small reserves to which they can return for later meals. Such a food-storing strategy makes especially good sense when food is abundant at some times of the year, but scarce at other times.

One famous food-storing animal is Clark's nutcracker, a bird that lives in high mountainous regions of the Southwestern United States. These nutcrackers collect pine seeds in a small pouch under their tongue; the birds later drive those seeds into the soil with their beak. Field observations indicate that the nutcrackers return to small feeding caches many months later in order to retrieve those seeds, even under snow cover; in autumn, 33,000 seeds may be stored in 2500 caches for later recovery in winter and spring.

This is truly a remarkable feat of spatial memory. But what does it imply about the general memory ability of this particular species? In order to find out, laboratory experiments were conducted with this species and with three other bird species that are not proficient at storing and recovering food. Two general types of memory tasks were devised that were variants of delayed matching-to-sample: one task required memory for the location of a prior stimulus and the other task required memory for the color of a prior stimulus.

Clark's nutcrackers easily won the contest for spatial memory; but, they were in the middle of the pack in the contest for color memory. These data suggest that Clark's nutcrackers do not have generally exceptional memory ability; rather, they possess a more highly advanced spatial memory that may be a very special adaptation to their particular evolutionary niche. Other special intellectual adaptations may include the pigeon's homing ability, the rat's rapid learning of taste–illness associations, and the human's propensity to learn language.

Reconciliation

These two different approaches to the comparative study of intelligence may seem to be at odds with one another. But, there really is no inherent conflict because the two programs ask different questions about animal intelligence. The process approach concentrates on a few taxonomically distant 'focus' animals in order to understand the general processes that are shared among species, whereas the phylogenetic approach concentrates on more closely related species in order to elucidate the intellectual mechanisms of specific adaptive behaviors and their ecological determinants. Together, the parallel use of both strategies should make for a powerful and complementary alliance for the future study of animal intelligence.

CONCLUSION

This brief review of research in the comparative psychology of intelligence suggests that we have merely scratched the surface in our quest to understand the nature of animal intelligence. Are animals capable of even greater cognitive feats? Can intelligence be demonstrated in animals without backbones? Are the same behavioral results the product of common brain mechanisms? (*See* **Learning in Simple Organisms**)

These questions are among the most important for the science of comparative psychology as it embarks upon its second century. Answering them will help us to tackle the most challenging question of all: how did intelligence evolve? This question was famously posed by Darwin, who hypothesized that humans and animals not only share common natural origins, but that they share common emotions and intelligence. Answering these questions with hard scientific evidence, not with beguiling anecdotes, is the business of our field.

Of course, the more we learn about the intelligence of animals, the more similar to us they seem to be. Such similarity often gives rise to empathy, leading some individuals to question whether animals should ever be the objects of scientific investigation. But, continued scientific inquiry is not only critical to answering key theoretical questions, it is also crucial to solving vexing practical problems.

Basic research into behavioral principles is vital to remedying such maladies as obesity, drug abuse, and AIDS – all the result of human behavior. That research may as well help us to contend with other serious societal woes such as crime, pollution, and overpopulation – also the products of human action.

Animals too may profit by such study, as we try to preserve and to expand their natural habitats. Knowledge of animal behavior and intelligence may also enable us to protect natural animal populations or even to increase them through repopulation efforts coordinated through zoos and animal sanctuaries.

Further Reading

Bitterman ME (1975) The comparative analysis of learning. *Science* **188**: 699–709.

Budiansky S (1998) *If a Lion Could Talk*. New York, NY: Free Press.

Domjan M (1987) Comparative psychology and the study of animal learning. *Journal of Comparative Psychology* **101**: 237–241.

Gallistel CR (1989) Animal cognition: the representation of space, time, and number. *Annual Review of Psychology* **40**: 155–189.

Macphail EM (1985) Vertebrate intelligence: the null hypothesis. *Philosophical Transactions of the Royal Society of London* **B308**: 37–51.

Roberts WA (1998) *Principles of Animal Cognition*. New York, NY: McGraw-Hill.

Roitblat HL, Bever TG and Terrace HS (eds) (1984) *Animal Cognition*. Hillsdale, NJ: Lawrence Erlbaum.

Shettleworth SJ (1993) Varieties of learning and memory in animals. *Journal of Experimental Psychology: Animal Behavior Processes* **19**: 5–14.

Spear NE, Miller JS and Jagielo JA (1990) Animal memory and learning. *Annual Review of Psychology* **41**: 169–211.

Wasserman EA (1993) Comparative cognition: beginning the second century of the study of animal intelligence. *Psychological Bulletin* **113**: 211–228.

Wasserman EA (1995) The conceptual abilities of pigeons. *American Scientist* **83**: 246–255.

Wasserman EA (1997) Animal cognition: past, present, and future. *Journal of Experimental Psychology: Animal Behavior Processes* **23**: 123–135.

Weiskrantz L (ed.) (1985) *Animal Intelligence*. New York, NY: Oxford University Press.

Computability and Computational Complexity

Introductory article

Patrick Doyle, Stanford University, Stanford, California, USA

CONTENTS

Processes that can be specified precisely can be formalized as algorithms. The theory of algorithm analysis determines important properties of these algorithms, such as the resources they consume, and the theory of computational complexity categorizes the problems these processes solve according to those properties, revealing the fundamental limitations of computation.

MEASURING THE DIFFICULTY OF INFORMATION PROCESSING PROBLEMS

The major tasks of a computational system are the retrieval, processing, and presentation of information. The processing of information involves some set of operations designed to transform it from one form into another. A spreadsheet program may add up a column of numbers to determine a month's expenses. A mathematics package might compute the area under a curve. A buyer of a new car might analyze several alternative models to decide which has the best combination of features, safety, speed, and cost.

Some of these tasks are simpler than others. Most people would find it easier to add up a column of numbers than to compute the area under a curve. There are also usually many different ways to solve a particular problem, but some are preferable to others because they consume fewer valuable resources.

In order to solve a problem, there must be a procedure that will generate the answer to the problem, and that procedure must be executed. The branch of computer science known as the theory of algorithms deals with the analysis of such procedures, and attempts to find meaningful measures for comparing one procedure with another. The study of computational complexity extends this theory by organizing problems into categories according to the difficulty of the procedures needed to solve them.

WHAT IS AN ALGORITHM?

In order to solve a problem, some organized sequence of operations must be performed – some sort of procedure, process, routine, or recipe. These terms all capture an intuitive notion of algorithm, but in order to apply formal analysis, it is necessary to be more precise about just what such an activity entails.

An algorithm is a finite sequence of effective and exact instructions for transforming any set of appropriately expressed pieces of information (the input) into another set of pieces of information (the output) in a finite amount of time.

Consider the problem of sorting a hand of playing cards. The goal is to sort the cards from lowest rank to highest. One algorithm to solve this problem works as follows: given a hand of cards, go through the hand and find the card with the lowest rank (if there is more than one, take the first one). Place that card on top of a separate pile. Repeat this process until no cards are left in the hand. Now the pile contains the sorted cards.

This procedure, while inefficient, meets the criteria of an algorithm. The inputs and outputs are well defined as sequences of playing cards. The instructions are effective for a human being to execute, since they involve moving cards in a hand. They are also exact, since they precisely and unambiguously explain what to do next in each case. The algorithm is guaranteed to terminate since there are only finitely many cards in a hand, and at each repetition of the loop there is one fewer card to examine.

The particular language in which an algorithm is written does not matter much. Generally, in computer science one finds algorithms described in formal programming languages that are designed for this purpose. However, so long as it meets the above requirements, it does not matter whether the algorithm is written as a flowchart, a Java program, English prose, or a Japanese haiku. It is only important that the environment in which the algorithm is to be executed can interpret and follow its instructions exactly and unambiguously. The above description of the card sorting algorithm, for example, is sufficient for an English-speaking human being, but not for a computer.

There are several natural questions about algorithms that have important and surprisingly complex answers. First: does every problem have an algorithm that will solve it? Second: are some algorithms 'better' than others for solving a certain problem? Last, and most practically: is there a way to measure the 'goodness' of an algorithm? The modern theory of computation is a result of the attempt to answer these questions.

COMPUTABILITY AND DECIDABLE PROBLEMS

In the early twentieth century, before the first electronic computers were built, several different formal models of computation were proposed. One of the simplest, and the one most used in computer science today, is the Turing machine, introduced by the mathematician Alan Turing in 1936. (*See* **Turing, Alan**)

A Turing machine is an imaginary mechanical device with three components: an infinitely long tape of paper, divided into cells that may each hold one of a finite set of symbols; a read–write head that can read and modify the contents of a cell; and a controller that may be in any of a finite set of states. The machine's behavior is directed by a fixed set of rules (the machine's program) that depend only on the current state and the symbol under the read–write head. The machine operates by repeatedly cycling through four steps:

1. Examine the symbol in the current cell.
2. Change the internal state (possibly to the same state it already occupies).
3. Write a symbol in the cell (possibly the same symbol already there).
4. According to the rules, move the head one cell to the left or to the right, leave it where it is, or halt the computation.

Although a Turing machine is a simple device, it has all the power and generality of a modern computer. That is, any problem that a modern computer can solve, a Turing machine could also solve. The Church–Turing thesis asserts that the Turing machine (or any equivalent model) exactly captures the notion of 'effective computability'.

A computable function is one for which there is an algorithm that will, for any input it is designed to understand, produce the correct output in a finite amount of time. Thus, according to the Church–Turing thesis, a Turing machine can solve any problem for which an algorithm can be provided. This thesis is only a hypothesis, however. It cannot be proven, since the notion of algorithm is not mathematically rigorous. The Church–Turing thesis says that our intuition about computability is captured by these formalisms.

Because their operations are so simple, Turing machines are frequently used in proving properties of computation. One important area of research into these properties deals with a special class of problems called decision problems. These are problems that have only 'yes' or 'no' answers. Any problem can be cast as a decision problem, and it is often simpler to prove properties of the decision version of a problem than to prove properties of the original problem directly. A decision problem may be 'decidable', 'semi-decidable', or 'undecidable'.

A problem is decidable if there exists some algorithm that is guaranteed either to accept (with a 'yes') or reject (with a 'no') any instance of the problem. Many obvious problems have decidable algorithms. Even the game of chess is a decidable problem: since there are only finitely many games (assuming no boundless repetition of useless moves), it is possible to find a strategy that will play perfectly just by exhaustively examining every possible game. The amount of computation is entirely impractical, but in principle, questions such as 'is white guaranteed to win in this position?' can be answered.

Semi-decidable problems have algorithms that are guaranteed to accept all 'yes' instances. If the instance is a 'no', the algorithm may reject the instance, or it may run forever without being able to determine that the instance should be rejected. An example of a semi-decidable problem is the problem of determining whether a given number is the difference of two primes. If a problem has no algorithm that can be guaranteed to accept all 'yes' instances, it is said to be fully undecidable. Many important questions, even some about algorithms themselves, are undecidable.

GÖDEL'S INCOMPLETENESS RESULTS AND THE ALGORITHMIC NATURE OF COGNITION

At the beginning of the twentieth century, the mathematician David Hilbert posed the problem of finding an algorithm that could determine whether any mathematical proposition, expressed in the language of logic, was true or false. This question remained open until 1931, when the mathematician Kurt Gödel published his famous 'Incompleteness Theorem'. This theorem states that, in any sufficiently powerful logical system such as mathematics, there are propositions that cannot be proved true or false within the system itself. Gödel showed that mathematics is fundamentally incomplete because there are true propositions in mathematics that cannot be proved using the axioms of mathematics. Thus Hilbert's algorithm could not exist.

When Alan Turing introduced the Turing machine and its definition of computability in 1936, he used a version of Gödel's argument to show that there are certain problems that are not computable, in the sense that there exist no algorithms that can compute their solutions. Specifically, Turing proved that no Turing machine could solve the so-called halting problem, a decision problem that asks whether a given Turing machine will eventually halt on a given input.

The proof can be described informally. Suppose there does exist a Turing machine $D(M, x)$ that decides whether Turing machine M would halt when run on input x. Now construct a Turing machine $N(M)$ that takes a machine M as input. N operates as follows. First, $N(M)$ runs $D(M, M)$; that is, N uses D to decide whether machine M would halt when fed its own design as its input. If D replies that M would halt on input M, N enters an infinite loop. If D replies that M would never halt on input M, N halts.

What happens if N is run on itself? If D says that N would halt on input N, N actually enters an infinite loop. If D says that N is would not halt, N actually halts. This is a contradiction of the assumption that D can correctly decide the halting problem, so that assumption must be false. No such machine D can exist.

These remarkable results have led to philosophical debates about algorithms and their relationship to human information processing. Do human beings think in ways that can be expressed in algorithms, or is there some other kind of computation taking place? If humans do use algorithms, are they therefore bound by the same limitations as computers, and logically incapable of performing certain kinds of tasks? (*See* **Artificial Intelligence, Gödelian Arguments against**)

Several prominent philosophers have argued against this. Their arguments range from John Searle's 'Chinese room' argument that computation cannot produce understanding, to Roger Penrose's hypothesis that the brain has certain quantum-mechanical properties that allow it to function in ways no computer could mimic. These arguments have not convinced many computer scientists, and the branch of computer science known as artificial intelligence is devoted to building computers with abilities that equal or exceed those of human beings. (*See* **Artificial Intelligence, Philosophy of**; **Chinese Room Argument, The**; **Computation, Philosophical Issues about**)

ORDER OF MAGNITUDE AND COMPLEXITY MEASURES IN SPACE AND TIME

Given a problem together with an algorithm for solving it, the next task is to analyze the resources that the algorithm consumes. With each instance of the problem, we associate a 'size', which is the number of symbols required to describe the instance. The resources the algorithm consumes when operating on an instance of the problem can then be expressed in terms of the size of the instance. Generally the resource of interest is either the time it takes to run or the space the algorithm needs to perform its computations.

Recall the simple algorithm for sorting a hand of playing cards. It looks for the highest-ranked card and moves it to a separate pile. It repeats this process with each succeeding card until all the cards are sorted. If there are n cards in the hand, the first card must be compared against all $(n-1)$ remaining cards, the second card against $(n-2)$ cards, and so on down to the last two cards, which require only one comparison. This approach requires $(n-1)+(n-2)+\ldots+1=(n^2-n)/2$ comparisons altogether.

To determine how much time this algorithm would actually take would require detailed information about the computer on which it is run, including how much time it takes to read the input, make a comparison, and so on. Ordinarily algorithm analysis is not interested in this level of detail; it is enough to determine the rate of growth, or order of growth, of the algorithm, a more abstract measure that will be identical on all machines that share the same fundamental principles of operation.

Several simplifying assumptions are made to find the algorithm's order of growth. First, constant-time overhead costs are ignored. Only operations that grow in number as the size of the input grows are considered. Second, only the dominant term ($\frac{1}{2}n^2$ in the above example) in the number of operations is used. As the size of the input grows, this term overwhelms the others: when $n=10$, $\frac{1}{2}n=5$ and $\frac{1}{2}n^2=50$; when $n=1000$, $\frac{1}{2}n=500$ but $\frac{1}{2}n^2=500\,000$. Finally, any coefficients on the dominant term are ignored, since such constants are less important than the overall rate of growth. This leaves n^2 as the term of interest. This order-of-growth bound is written as $O(n^2)$. This O notation is commonly used to describe the worst-case complexity of an algorithm, which is the amount of effort it will require on the most difficult possible input.

This formulation makes it straightforward to compare algorithms without having to consider the details of a particular computer or minor factors in the algorithm that do not have a significant effect on its overall efficiency. For the card-sorting problem, for example, there are many known algorithms, and the most efficient comparison algorithms have an order of growth of $O(n \ln n)$, a considerable improvement over $O(n^2)$ for large hands.

PROBLEM REDUCTION AND COMPLEXITY EQUIVALENCE CLASSES

The kind of analysis used in the previous section is helpful when examining a single algorithm, or comparing several algorithms, but the theory of computational complexity is concerned with understanding fundamental distinctions between classes of algorithms. There are two especially important classes, known as P and NP, which intuitively divide problems into 'easy' and 'hard'.

P is the class of all decision problems that take an amount of time that is polynomial or better in the size of the input, such as $O(n \ln n)$, $O(n^2)$ or $O(n^{357})$. These problems are regarded as 'easy' or 'tractable' because the amount of time grows relatively slowly with the size of the input.

The other important basic complexity class is called NP, for nondeterministic polynomial time. NP consists of decision problems whose answers can be verified in polynomial time. That is, given some instance of a problem and a guess for an answer, there is an algorithm that can check in polynomial time whether that guess is correct.

Clearly P is a subset of NP, since any problem that can be solved in polynomial time can have its solution checked in polynomial time. However, there are many NP decision problems for which no polynomial-time algorithm has been found: it is known how to check their solutions in polynomial time, but the best-known algorithms for solving them take exponential time, such as $O(2^n)$, or worse. These problems are in NP but seemingly not in P.

Many other complexity classes have been studied. Some deal with different time bounds: for example, EXP consists of problems solvable by exponential-time algorithms, and is a strict superset of NP. Others measure different resources, such as space. PSPACE is a class of problems whose algorithms are polynomially bounded (in the size of the problem instance) in the amount of space they use rather than in the number of operations they perform.

Proving that a decision problem is in a certain complexity class can be difficult. An important concept in complexity analysis is that of reducing one problem to another. In order to determine the complexity of some problem A, one can often show that it can be transformed into another problem B

whose complexity is already known. Then the complexity of A is no worse than the complexity of B plus the complexity of turning instances of A into instances of B. Computer scientists have built up a large library of problems with known complexities, making this approach a common way to determine the complexity of a new problem.

NP-COMPLETE PROBLEMS AND THE 'P = NP' QUESTION

In 1971, Stephen Cook proved that there is a certain problem in NP, called the satisfiability problem, such that any other problem in NP can be reduced, in polynomial time, to an instance of it. Hence, if there is a fast algorithm for satisfiability, then there would be a fast algorithm for every problem in NP; conversely, if any problem in NP is intractable, then satisfiability is intractable.

The satisfiability problem can be stated informally as follows. Given a set of variables that can be either 'true' or 'false', and a formula combining those variables with the logical connectors 'and', 'or' and 'not', is there an assignment to each variable that makes the entire formula true? For example, '((v_1 and v_2) or (v_2 and not v_3))' is true when either both v_1 and v_2 are true or when v_2 is true and v_3 is false. Any instance of any problem in NP can be rewritten, in polynomial time, as an instance of satisfiability.

Since Cook's proof, many other problems in NP have also been shown to have this property. Such problems are called NP-complete. They are, in a sense, the 'hardest' problems in NP, since finding a fast algorithm for any one of them would mean finding a fast algorithm for all problems in NP.

As mentioned above, the class P is contained within NP. Surprisingly, it is still not known whether P is actually equal to NP: that is, whether every problem in NP actually has a polynomial-time algorithm, and we have just not yet found one for any NP-complete problem.

If P = NP, then the huge range of important problems in NP for which no tractable algorithms have been found must all have tractable algorithms. After decades of research, however, it is widely believed that P $\neq$ NP. Many modern cryptographic systems base their security on the assumption that certain problems in NP are too difficult to solve in any reasonable span of time. Whether these classes are equivalent or not, a proof either way will have widespread implications for complexity analysis and algorithm design. This is one of the great unsolved questions in computer science.

COMBINATORIC EXPLOSIONS IN COMMON COGNITIVE TASKS AND THE USE OF APPROXIMATION ALGORITHMS

There are many important problems that are in NP. Many common cognitive tasks fall into this category, and yet human beings are able in their daily lives to perform tasks that are theoretically too difficult to solve in any reasonable time. One explanation is that humans perform a fundamentally different kind of computation than computers do. But there are some other possibile explanations.

A problem that is in NP may still be tractable in practice. There exist problems in NP for which algorithms with very low exponents are known; for example, $O(2^{0.00001n})$ might be an acceptable order of growth in most cases.

Another possibility is using a fast algorithm that provides an approximate solution to the problem. Often it is possible to find an algorithm in P that is guaranteed to provide solutions that are within some bounded range of the best solution. One well-known example is the travelling salesman problem, which provides a map of cities and the distances between them, and asks for the most efficient route for visiting all the cities. This problem is in NP, but there is a known polynomial-time algorithm that is guaranteed to find a route no worse than twice the length of the most efficient route. In many cases these approximation algorithms are good enough for practical needs, and they are useful when the exact algorithms are impractical.

STRONG COMPLEXITY CONSTRAINTS ON COGNITIVE MODELS

Complexity measures ordinarily assume that computations are being performed on a serial machine – that is, steps in the algorithm are executed one after another and one at a time. However, there are strong arguments that this is not a reasonable model of human cognition. One prominent argument, given by Jerome Feldman and Dana Ballard, is contained in what is called the '100-step rule'.

Simple cognitive tasks, such as identifying and naming an object, take human beings something on the order of half a second (500 ms) to perform. Since neurons, which are the basic computational components of the brain, take around 5 ms to act, the 100-step rule declares that these cognitive tasks cannot take more than about 100 sequential neuronal operations. (*See* **Computational Models of Cognition: Constraining**)

However, although a single neuron may take several ms to act, our brains contain many billions of neurons. Inspired by the structure of the brain, connectionist models of computation consist of many independent processors that are connected together, and that perform their computations in parallel, rather than serially. As long as the serial algorithm can be redesigned so that each one of the individual parallel processors has something to do in each time step, its speed can be improved by as many times as there are processors – in the case of the human brain, a parallel algorithm might run billions of times faster than a serial one. (See **Connectionist Architectures: Optimization**; **Connectionism**)

Although parallel algorithms can be more complex to design, the same tools that are used for ordinary serial algorithms can be applied. The complexity analyses on time and space are the same for both parallel and serial systems, since a parallel machine only improves over a serial one by a constant factor, and such constant terms are ignored in order of magnitude analysis. However, in practice a linear speed-up of billions of times significantly increases the range of problems that can be solved.

SUMMARY

The design of algorithms is fundamental to computer science. The theory of algorithm analysis makes it possible to determine the amount of a resource that an algorithm requires. This allows the comparison of different algorithms to determine which are best in what situations. The theory of computational complexity has led to categorizing problems according to their difficulty, with the polynomial-time P problems being intuitively 'easy' and the nondeterministic polynomial-time NP-complete problems 'hard'. Many important problems have been shown to be NP-complete, but often it is possible to develop approximation algorithms that will give good, if not necessarily the best, answers to these problems. These concepts can be applied to algorithms that attempt to explain human brain processes. However, since the brain contains many billions of slow neurons operating in parallel, rather than a few fast processors, some argue that they may not be appropriate measures for human cognition.

Further Reading

Cook S (1971) The complexity of theorem-proving procedures. In: *Proceedings of the Third Annual ACM Symposium on Theory of Computing*, pp. 151–158. New York, NY: ACM Press.

Cormen T, Leiserson C and Rivest R (1996) *Introduction to Algorithms*. Cambridge, MA: MIT Press.

Hopcroft J and Ullman J (1979) *Introduction to Automata Theory, Languages, and Computation*. Reading, MA: Addison-Wesley.

Feldman J and Ballard D (1982) Connectionist models and their properties. *Cognitive Science* **6**: 205–54.

Garey M and Johnson D (1979) *Computers and Intractability: A Guide to the Theory of NP-Completeness*. New York, NY: W. H. Freeman.

Johnson-Laird P (1988) *The Computer and the Mind: An Introduction to Cognitive Science*. Cambridge, MA: Harvard University Press.

Lewis H and Papadimitriou C (1981) *Elements of the Theory of Computation*. Englewood Cliffs, NJ: Prentice-Hall.

Papadimitriou C (1994) *Computational Complexity*. Reading, MA: Addison-Wesley.

Pinker S (1997) *How the Mind Works*. New York, NY: W. W. Norton.

Computation, Formal Models of

Introductory article

Arun Jagota, University of California, Santa Cruz, California, USA

CONTENTS

This is the study of abstract machines that compute, and of what they compute. This is also a study that characterizes what problems can be computed in principle, and which of the computable problems can be computed in practice.

THE MIND AS MACHINE?

For centuries, philosophers have wondered whether the mind is a 'mere' machine. We are no closer to answering this question now, at the beginning of the twenty-first century, than we were then. On the other hand, we have made tremendous progress in building and understanding *machines*. Now we are even able to put together ones that work in ways that might be considered somewhat intelligent. (*See* **Computational Models: Why Build Them?**)

THE BENEFITS OF A FORMAL MATHEMATICS OF COMPUTATION

The advances in building and understanding mind-like machines would not have occurred without the theoretical and algorithmic foundations provided by formal models of computation. Formal models of computation establish the limits of what can and cannot be computed. For a problem that is computable, complexity theory tells us whether it can be computed *efficiently* or not. Automata theory tells us what types of computations different types of machines can perform. Formal language theory presents to us a hierarchy of languages of increasing sophistication. A remarkable correspondence is found between the automata of automata theory and the languages of formal language theory. One of the most powerful abstract models of computation is the Turing machine. The study of its properties has led to major results, including the theory of NP-completeness.

Upto this point we have discussed only sequential models of computation. Parallel computation on the other hand can be much faster. Brain-inspired models of massively parallel computation have been developed and are now widely studied and used. A fundamentally different type of model of parallel computation – the quantum computer – is being studied in the abstract. Should it be realizable, it would radically transform the nature of computation as we know it today.

AUTOMATA

The word *automaton* really only means 'machine'. However, it is used to convey a sense that our interest lies in the computational properties of the machine, especially those that may be characterized formally. This section considers automata that recognize languages. Such an automaton reads a string of input symbols and returns 'yes' if the string is in the language associated with the automaton and 'no' if not.

There are three broad classes of automata: *finite-state automata (FSA)*, *pushdown automata (PDA)*, and *Turing machines (TMs)*. Finite-state automata are the simplest and, not surprisingly, the least powerful. Pushdown automata have more features; this makes them more powerful than finite-state automata. Turing machines have even more features; this makes them the most powerful. The power of a class of automata is measured by the richness of the class of languages associated with it (see the last paragraph of this section for more on this).

To appreciate the issues we need to understand what we mean by *features* in a machine, and what we mean by *power* of a machine. The features of a machine may be roughly classified as type of *memory*, and the *instruction set*. For instance, the only memory in a finite-state machine resides in its *states*. The only instruction that a finite-state automaton has is one that reads the next input symbol in the current state, and makes a transition to the next state. Both issues are illustrated in Figure 1.

In this diagram, let us imagine that we are presently in state A. This is the memory we are talking about. That is, the FSA *knows* that we are in state A. However, the FSA does *not* know how we got there; i.e. it does not remember states, if any, that were visited prior to reaching state A. Next, the machine reads the next letter in the input. If the letter is a 0, it moves to state B. If the letter is a 1, it moves to state C. These are the only types of actions this machine can take.

A pushdown automaton on the other hand has a more sophisticated memory mechanism – its memory resides not only in the present state it is in, but also in a *stack* that it is allowed to use. To make effective use of the stack, the PDA also has a richer instruction set, specifically, instructions to push items onto, and pop items off, the top of the stack as illustrated in Figure 2.

In this diagram, let us imagine that we are presently in state A. If the next letter in the input is a 0 and the top of the stack contains a Y, then we replace the Y on the top of the stack by Z and move to state B. If, on the other hand, the next letter in the input is a 1 and the top of the stack contains a W, then we replace the W on the top of the stack by X and move to state C. From this example, we see that the instructions of a PDA are more sophisticated than those of an FSA. It may appear that, like an FSA, the PDA also does not remember which states it visited before reaching A. This is not true. We can record these states in the stack. In fact, creative use of the stack is what endows a PDA with its added power over an FSA.

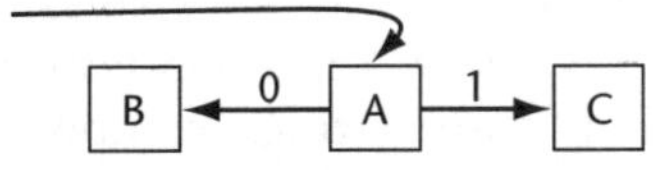

Figure 1. Finite-state automaton.

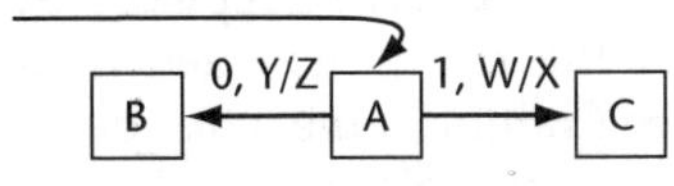

Figure 2. Pushdown automaton.

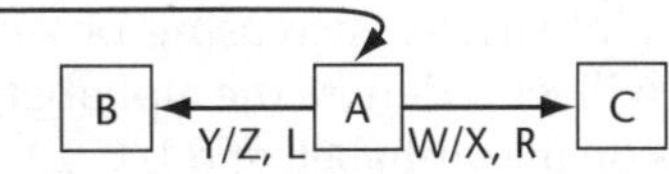

Figure 3. Turing machine.

One wonders whether a stack is enough to compute anything that is computable. Interestingly, it turns out that it is not. The Turing machine, which *is* enough, has an even more flexible memory, and instructions to go with it. A TM has a *tape*, which can be read in *both* directions. (By contrast, one can only push items onto, and pop items off, the top of a stack.) Figure 3 illustrates the use of this tape, especially the instructions that read, write, and use it. (*See* **Turing, Alan**)

In this diagram, let us imagine that we are presently in state A. Also, the tape head is on top of a certain cell on the tape (the tape is divided into cells). If the current cell – that the tape head is scanning – contains a Y, we replace this cell's content (Y) by Z, and move the tape head one cell to the left (this is what L means). If the current cell contains a W, we replace this cell's content by X, and move the tape head one cell to the right. From this example we see that, unlike an FSA or a PDA, a TM does not explicitly read the external input – it is assumed to be already on the tape before the TM execution is begun. We see that the TM has a richer feature set than a PDA, both in its memory and in its instructions.

Now one would hope that the automata with the richer feature sets are strictly more powerful than the ones with the poorer feature sets. Indeed this is the case. We explain this in the following setting. We measure computing power of an automaton in its use as a *language recognizer*. In this use, TMs recognize all languages that are decidable (the recursive languages), PDAs recognize only a *proper* subset of these languages (the context-free languages), and FSAs recognize only a *proper* subset of the languages that PDAs recognize (the regular languages). Various classes of languages mentioned above – recursive, context-free and regular – are discussed in the next section.

FORMAL LANGUAGES AND THE CHOMSKY HIERARCHY

As everyone knows, language is essential to communication. What one might not realize is that language is also essential to computation.

The Chomsky hierarchy of formal languages was a landmark event in linguistics and, as it turns out, also in computation. This hierarchy defines classes of languages of strictly increasing power. In what follows we will let Σ denote the alphabet, Σ^* the set of all strings on the alphabet, and $L \subseteq \Sigma^*$ a language on Σ. (Notice that a language is a *particular subset* of all possible strings on an alphabet.)

At the lowest level in the hierarchy is the class of *regular languages*. This class, R, may be defined as follows. The language $L = \emptyset$ that contains no strings is in R. The languages $L = \{a\}, a \in \Sigma$ are in R. (Each of these languages contains a one-letter string, comprised of an alphabet symbol.) The language $L = \{\wedge\}$, which contains just one element – the empty string – is in R. If languages L_1, L_2 are in R then so are the languages $L_1 \cup L_2$, $L_1 \circ L_2$, and $L_1{}^*$. Here $L_1 \circ L_2 = \{x \circ y | x \in L_1, y \in L_2\}$ where the small circle symbol ($\circ$) denotes the concatenation of two strings, and $L_1{}^* = \bigcup_{i=0,\ldots,\infty} L_1^i$ where $L_1^i = \underbrace{L_1 \circ L_1 \circ \ldots \circ L_1}_{i \text{ times}}$. In other words, a regular language is built from simpler languages by the union of two regular languages, the concatenation of two regular languages, or the repeated concatenation – zero or more times – of the same language.

This simple structure suggests that regular languages should be easy to parse. Indeed this is the case. Regular languages are parsed by *regular expressions*. Regular expressions mirror the structure of the definition of the class of the regular languages; specifically they are composed of operators that correspond to the union and the concatenation of two languages and an operator for repeated concatenation of the same language.

Despite being at the lowest rung of the Chomsky hierarchy, a regular language can have a rich structure. On the other hand, precisely because the language is regular, this structure can be efficiently parsed by computer methods. These two facts collectively explain why regular languages are so popular, having many applications. These include free-text searches, especially in Unix and on the web, and computational biology. (*See* **Finite State Processing**)

At the next level in the hierarchy is the class of *context-free languages*. Every regular language is a context-free language. On the other hand, there are context-free languages that are not regular. The language formed by the set of all palindromes (say on the alphabet $\Sigma = \{0, 1\}$) is one such example. (A palindrome is a string that coincides with its reverse.)

Before we give a formal definition for this class of languages, it will help us to examine the closely related notion of a grammar. A *grammar* for a language is a set of rules which collectively allow us to determine which strings are in the language and which are not. A grammar is itself written in a *meta-language*. By imposing different restrictions on what forms the rules can take in this meta-language, we get different classes of grammars, hence different classes of languages.

In a *context-free grammar* we can have rules only of the type $A \Rightarrow x$. Here A is a single non-terminal symbol and x a string, possibly empty, composed of non-terminal and/or terminal symbols. The terminal symbols in the grammar are the alphabet symbols in the associated language.

Let us use an example to explain what a non-terminal symbol is, and how the set of rules comprising a grammar is applied to test whether a given string is in the language associated with the grammar or not. Here is one context-free grammar for the language of palindromes mentioned earlier.

$$S \Rightarrow 0S0 \quad S \Rightarrow 1S1 \quad S \Rightarrow 0 \quad S \Rightarrow 1 \quad S \Rightarrow \wedge$$

In this grammar, S is the only non-terminal symbol, 0 and 1 are the terminal symbols, and $\wedge$ denotes the empty string. This grammar has five rules. Let us now apply this grammar to verify that 0110 is a palindrome. We do this by starting with the non-terminal symbol S and applying a certain sequence of rules which transforms this S to the string 0110. This process is called a *derivation*. The derivation in this example is:

$$S \rightarrow 0S0 \rightarrow 01S10 \rightarrow 0110$$

Which rules were applied when is not made explicit in this derivation. In our example though, we can easily infer this from our derivation.

A context-free language may now be defined as one that is recognized by a context-free grammar.

At the next level in the Chomsky hierarchy is the class of *context-sensitive languages*. Every context-free language is a context-sensitive language. On the other hand, there are context-sensitive languages that are not context-free. The language $L = \{xx | x \in \Sigma^*\}$ is one such example.

A context-sensitive language is associated with a grammar in which we can have rules only of the type $xAy \Rightarrow xzy$. Here A is a non-terminal symbol and x, y, and z are strings composed of non-terminal and/or terminal symbols. (x and/or y can be empty, but z cannot be.) The meaning of this rule is that 'when A occurs with x on its left and y on its right, it may be replaced by the string z'. Notice that this rule is more general than one for a context-free grammar; the latter does not permit context-dependent replacement.

Finally, a *recursive language* is one whose characteristic function is decidable. That is, given a string x in Σ^*, there is an effective procedure for determining whether x is in the language or not.

THE CORRESPONDENCE BETWEEN AUTOMATA AND FORMAL LANGUAGES

What solidifies the connection between languages and computation is the precise correspondence between formal languages of the Chomsky hierarchy and the automata that we saw in a previous section. Specifically, it turns out that FSAs recognize regular languages, PDAs recognize context-free languages, and TMs recognize recursive languages. In fact, even context-sensitive languages have their automaton counterpart – the linear-bounded automaton, which is realized by imposing certain restrictions on a TM.

THE TURING MACHINE

To this point, we have described the Turing machine as a language recognizer. Unlike FSAs and PDAs, however, a Turing machine may also be used as a more conventional type of computer, to read input, do some task, and produce output. This is done in much the same way as a language would be recognized, the main difference being that after reading the input and doing its computation, the TM places the output on the tape at a specific location.

THE UNIVERSAL TURING MACHINE

A TM runs a particular program. This is cumbersome; to run different programs, one needs to construct different TMs, one for each program. The universal Turing machine (UTM) is a fix to this. A UTM accepts, as its input, both the program P and the input x to P. The UTM then runs the program P on input x, and writes out, to the tape, the output that P would have produced, on input x.

THE VON NEUMANN ARCHITECTURE

The von Neumann architecture may be thought of as a practically-inclined realization of the idea of a universal Turing machine. This architecture uses the key notion of a 'stored program concept'. A program, like data, is stored in memory. A program is indistinguishable from data, as it resides in memory. (*See* **von Neumann, John**)

In the von Neumann architecture, there is a 'central processing unit' that reads and executes the instructions of a program stored in memory, one by one. The data needed by an instruction is read from memory; the result produced by the instruction is written back to memory. For example, the instruction $z = x + y$ would be executed as follows. First, the instruction is read from memory. Next, the data items needed by the instruction, x and y, are read from memory. Then, the values of x and y are added together. Finally, the result is stored back into memory at location z.

INSTRUCTION SETS, COMPUTER LANGUAGES, AND THE IDEA OF THE 'VIRTUAL MACHINE'

The von Neumann architecture is the basis for every sequential computer built. The huge significance of the stored program concept – that it facilitates the execution of countless programs on the same computer – was recognized early on and profitably exploited since. Computers were developed with carefully designed instruction sets. The instruction sets had to be powerful enough so that any programs could be written in them. On the other hand, they had to be simple enough to keep the central processing unit from becoming overly complex.

These conflicting objectives were resolved by adopting 'minimalist' designs. Specifically, the instruction set was kept small, and composed of simple and 'orthogonal' instructions. The instructions were designed to be combinable. Thus, a user wanting a more elaborate instruction set could in effect create one by combining various instructions from the smaller set in various ways.

While this strategy solved the problem it was designed to solve, a different one remained. The instructions in a computer's instruction set were at a very low level; just right for the computer to understand them, but very tedious for humans to program in. A breakthrough came in the development of computer languages that humans could write in more easily. These languages had very high-level instructions. On the other hand, computers could not understand them. This dilemma was resolved by building translators, programs that take a program written in a high-level language and translate it to a program in the computer's (low- level) language.

Once we had freed ourselves from having to write programs at the level of the computer, we had also freed ourselves from its constraints to a considerable extent. Specifically, we no longer needed to write programs that would fit in its memory, or use only its instructions. The translator in effect created a 'virtual machine' around the actual computer, one that was vastly richer. This notion of a 'virtual machine' revolutionized software development, freeing it to a great extent from the confines of a particular computer.

WHAT IS AN ALGORITHM?

An algorithm is a step-by-step procedure to solve a particular problem. An algorithm is required to terminate in any one run. How does this notion differ from that of a program? Well, for one thing, a program may not always terminate – loops in its control flow may cause it to run forever at times. Second, a program typically means a sequence of instructions (to do something) written in a particular language, while an algorithm is a higher-level, language-independent recipe for solving a problem in a particular way.

THE NOTIONS OF COMPUTABILITY AND COMPUTATIONAL COMPLEXITY

Is it possible to develop a program that can check whether or not any given program will terminate? This would be nice; we could use it to find 'infinite loops' in our programs. Unfortunately, it is impossible to do so. (On a practical computer with bounded resources such a program can be developed, though it will be impractical.) That is, the so-called halting problem is noncomputable.

More precisely, let us define a partial function f:N → N as a function that takes a natural number as input and returns a natural number $f(n)$ if f is defined on n, and returns 'undefined' if f is undefined on n. We say that a Turing machine T computes f if T reads the input n as 1^n and produces output $1^{f(n)}$ if $f(n)$ is defined and does not halt if $f(n)$ is undefined. We say that a partial function is computable if there exists a Turing machine that computes it. Since any given TM can compute at most one partial function, the number of TMs is countable, and the number of partial functions is uncountable, we see that there are many partial functions that are not computable.

The notion of computability establishes the limits of what can be computed in principle. It turns out that the limit of what can be computed in practice is far lower. The characterization of this limit is the subject of computational complexity.

Computational complexity is the study of the time needed to solve a particular problem as a function of the size of an instance of the problem. For example, consider the problem of factoring a composite number. An instance of this problem is specified by a particular positive integer. The size of an instance – the number n – is $\lceil \log_2 n \rceil$, the number of bits needed to describe n. The fastest known factoring algorithm runs in time exponential in n. The apparent difficulty of factoring has a very significant application: the development of public key cryptosystems such as those used at banks (and now for e-commerce over the internet) that seem unbreakable.

Now consider a different problem – of testing whether a given number occurs in a list of numbers. This problem can be easily solved in time proportional to n. If we assume that the instance size is also proportional to n – a valid assumption when the numbers are required to fit in a fixed word size – then this problem's computational complexity is linear.

Thus, we would conclude that factoring is computationally hard, while search is computationally easy.

How do we know that factoring is indeed computationally hard? Perhaps there is an efficient algorithm – we just have not found it. In the early 1970s this very sort of questioning led to the remarkable theory of NP-completeness. Informally speaking, the class NP-complete is the set of computable problems, none of which is known to be efficiently solvable, yet all are efficiently interconvertible (i.e. any NP-complete problem can be efficiently transformed into any other). This still does not mean that an NP-complete problem is definitely hard. However, the efficient interconvertibility yields an intriguing and significant property: If *any* NP-complete problem is easy, then *every* NP-complete problem is easy.

The practical import of this theory is as follows. Thousands of problems are known to be NP-complete. If one can show that a new problem of interest is NP-complete, then this suggests that it really is hard. If this problem were easy, then so would the thousands of other problems be. But since many able researchers have tried to devise efficient algorithms for many of these problems, over the past thirty years, and all have failed, this is highly unlikely.

PARALLEL COMPUTATION, ASSOCIATIVE NETWORKS, AND CELLULAR AUTOMATA

The von Neumann architecture, while pivotal to the evolution of computers, has one major drawback – it has a sequential bottleneck. Instructions must be executed one at a time. On the other hand, everyone realizes that many algorithms are intrinsically parallel. For example, to test if a number x occurs in a list L of numbers, we could compare x against each number in L in parallel.

The observation that many algorithms are inherently parallel, coupled with the incessant need of humans to 'have things done faster', has spurred research on parallel computation. Many models of parallel computation have been proposed. Some have even been implemented. These days every computer – even a sequential one – is parallel. On your computer you can compose an e-mail message in one window at the same time that the computer is downloading a video off the internet. This is an example of parallel computation.

The brain has been one source of inspiration for many models of parallel computation. Many computations in the brain are massively parallel (many others are sequential). Groups of neurons fire in synchrony, or in asynchrony. Brain-inspired models such as artificial neural networks and cellular automata are composed of primitive computing elements interconnected together in various ways. (In the case of artificial neural networks, these are the neurons.) The power of these models arises from the precise way in which these primitive elements are interconnected. If we 'add up' the computations performed by the computing elements, they don't amount to much. If we now interconnect these same elements in particular ways, suddenly we have gained a lot of computing power. In these cases, the whole is indeed greater than the sum of its parts.

In artificial neural networks, the net input to neuron i is $n_i = \sum_j w_{ij} S_j$ where j indexes the neurons connected to neuron i, S_j is the output of neuron j, and w_{ij} is the weight of the connection from neuron j to neuron i. The output of neuron i is $S_i = g(n_i)$ where g is the neuron's transfer function, typically a sigmoid.

In a cellular automaton, the primitive computing element is a cell. By contrast to neural networks, the cells are usually interconnected in a lattice, typically with one or two dimensions (neural networks are typically more flexibly connected). A cell updates its state by examining the states of its neighbors on the lattice. Although a cellular automaton typically has more restricted connectivity than a neural network, the cells in a cellular automaton are more flexible in what they can compute. Specifically, their computations are not restricted to taking the weighted combinations of the outputs of their neighbors and following these by a transfer function.

QUANTUM COMPUTATION

Complexity theory tells us that there are many problems that seem to be inherently intractable. These results hold only on models of classical computation, however. Quantum computers, on the other hand, are capable of performing exponentially many computations in parallel, exponential in the size of the input. They are therefore able to break the intractability barrier of classical computation. However, not all NP-complete problems are known to be tractable on quantum computers. The class of tractable problems on quantum computers is still not well understood in relation to classical models.

Let us revisit the factoring problem discussed earlier. Recall that we mentioned that the best classical algorithm takes exponential time in the size $\lceil \log_2 n \rceil$ of the binary representation of the composite number n that is to be factored. By contrast, there exists an algorithm for factoring on a quantum computer that takes only roughly quadratic time in the size $\lceil \log_2 n \rceil$ of the input. Should quantum computation become a practical reality at some point in time, this would suggest that public key cryptosystems might be breakable. (The first famous paper demonstrating tractability of prime factorization on quantum computers is due to P. Shor.)

At this time, quantum computers are already a little more than an abstract model in that they have been realized in toy devices. It remains to be seen whether practical quantum computers can be built.

Further Reading

Davis MD and Weyuker EJ (1983) *Computability, Complexity, and Languages: Fundamentals of Theoretical Computer Science*. New York, NY: Academic Press.

Garey MR and Johnson DS (1979) *Computers and Intractability. A Guide to the Theory of NP-Completeness*. New York, NY: WH Freeman.

Hopcroft J, Ullman JD and Motwani R (2000) *Introduction to Theory of Neural Computation*. New York, NY: Addison-Wesley.

Hertz J, Krogh A and Palmer R (1991) *Introduction to the Theory of Neural Computation*. New York, NY: Addison-Wesley.

Martin JC (1991) *Introduction to Languages and the Theory of Computation*. New York, NY: McGraw-Hill.

Nielsen MA and Chuang IL (2000) *Quantum Computation and Quantum Information*. New York, NY: Cambridge University Press.

Toffoli T and Margolus N (1987) *Cellular Automata Machines: A New Environment for Modeling*. Cambridge, MA: MIT Press.

Computation, Philosophical Issues about

Intermediate article

Matthias Scheutz, University of Notre Dame, Notre Dame, Indiana, USA

CONTENTS

'Computation' is a cluster concept and has been characterized in many different ways (e.g. 'the execution of algorithms'). It underwrites philosophical analyses of what can be done in principle by a mechanism, and is intrinsically connected to the idea of manipulating symbols or representations by formal rules.

INTRODUCTION

The notion of computation is undoubtedly one of the very central, increasingly influential notions of our time. It has captured the attention of researchers from many disciplines for different reasons. In cognitive science it was the capacity of computers to process information that inspired cognitive psychologists to think of cognitive functions in terms of programs and of the brain as a computer running these programs. To be able to appreciate this view of cognition and the central role of computation within it, one needs a clear understanding of what 'computation' means and what computations are.

WHAT IS COMPUTATION?

An Intuitive Perspective

Like many widely used notions 'computation' does not have a single, clear-cut meaning, but rather, *qua* cluster concept, takes on different meanings depending on the context in which it is used. A glance in Webster's dictionary reveals the ordinary language conception of 'to compute': derived from the Latin 'com + putare' – to consider, it means something like 'to determine or to calculate especially by mathematical means'. However, this definition is rather vague and furthermore too restrictive to do justice to the variety of uses to which the notion of computation is put in computer science alone.

More to the point is defining computation as 'the execution of algorithms', which, in turn, puts the burden on the notion of *algorithm* and what 'executing an algorithm' means. Roughly speaking, an algorithm consists of a finite set of instructions, which operate on certain entities (symbols, representations of numbers, etc.) and can be *implemented* in some mechanism. To execute an algorithm then intuitively means to have the mechanism carry out the instructions for any given input in a deterministic, discrete, stepwise fashion (without resorting to random or analog methods and devices). The mechanism goes through a sequence of atomic steps in such a way that (one or more of) these steps correspond to some instruction, for all the instructions specified by the algorithm. Note that nothing is said about the nature of the mechanism yet: it could be concrete or abstract, natural or artificial. Depending on the kind of mechanism, the algorithmic specification will take different forms: in the case of computers, it is expressed in a *programming language*; in the case of humans, instructions may be given in ordinary language (as long as the individual steps are clearly distinguishable and described at a sufficient level of precision) – just think of cooking recipes or the instructions on public phones for making phone calls.

Computation defined as the execution of algorithms does not commit one as to what the computation is about or what it is supposed to achieve. Rather, it ties algorithmic descriptions to *mechanically realizable processes*. This leaves two issues to be addressed: first, it needs to be made clear what a mechanism is, and second, a precise specification of the notion of algorithm is required. The following brief historical overview reveals the origin of the idea of mechanism as well as that of using representations for calculations.

A Historical Perspective

The history of computation traces back to Leibniz and before, when daring philosophers pondered mechanical systems that could aid humans in performing calculations, and possibly even calculate by themselves without any assistance. The first functioning mechanical calculators were built in the seventeenth century and were composed of various mechanical parts (such as gears, cogs, etc.). Leibniz, having constructed calculators himself, was one of the first to envision an application quite different from their typical commercial and military use, that of 'mechanical reasoners'. His view that calculations, in particular, and logical reasoning (i.e. thinking), in general, could be *mechanized* lies at the heart of the notion of computation as used in cognitive science today.

Another crucial contribution to the modern notion of computation is also a product of that time (due to Descartes, Hobbes, Locke, and others), namely the idea that reasoning or, more generally, thinking involves *representations*. The mathematical practice of using marks and signs as representations in calculations became a paradigm for thought itself, as expressed by Hobbes' famous dictum that everything done by our mind is a computation (Pratt, 1987).

Computation was, therefore, already very much tied to the idea of mechanically manipulating representations, and prototypical manipulators were found in the mechanical calculators of those days. While many attempts were made at building mechanical calculators up to the end of the nineteenth century (with varying success: e.g. see Williams, 1997), the computing capabilities of these machines remained very modest. It was only the twentieth century that witnessed major progress in the construction of computers and the conception of computing. This was largely due to two quite independent developments: (1) the thorough logical analysis of the notions 'formal system' and 'formal proof' (leading to further studies of notions such as 'effectively computable function' and 'algorithm'), and (2) the rapid progression in the engineering of electronic components (from vacuum tubes, to transistors, to integrated circuits, and beyond).

A Logico-philosophical Perspective

In the 1930s, logicians laid the main philosophical groundwork for a well-defined formal notion of computation in their attempt to make the intuitive notion of computation, then called 'effective calculability', formally precise. Being logicians, they were solely concerned with the class of functions (over the positive integers) that can be effectively calculated *in principle* – besides, digital computers did not exist yet. Church (1936) was the first to give this class of effective calculable functions a formal characterization through a definition postulate, which later came to be known as 'Church's Thesis' (CT): 'We now define the notion… of an effectively calculable function of positive integers by identifying it with the notion of a recursive function on positive integers (or of a λ-definable function of positive integers)' (Church, 1936, p. 356). Note that by virtue of relating an intuitive notion and a formal notion, CT cannot be proved in principle, as mentioned by Church himself: 'This definition is thought to be justified by the considerations which follow, so far as positive justification can ever be obtained for the selection of a formal definition to correspond to an intuitive notion' (Church, 1936, p. 356).

While this was a first step to capture the meaning of 'computable', it was not quite satisfactory, as CT is silent about what 'effectiveness' of a calculation means. As it stands, the notion of 'effectively calculable function' implies that two ingredients are needed to understand computation: a notion of 'effective procedure or algorithm' and a notion of 'function computed by an algorithm'. The latter can be straightforwardly explicated: it is the mapping obtained by pairing all possible inputs with the corresponding outputs resulting from applying the algorithm to them. The former, however, received a satisfactory account only after Turing (1936) had introduced his machine model of a 'computer', which resulted from his analysis of the possible processes a human – what he then called 'the computer' – can go through while performing a calculation using paper and pencil applying rules from a given finite set. It was crucial to Turing's conception of computation that the human computer follow the rules 'blindly', that is, without using insight or ingenuity. In his analysis of the limitations of the human sensory and mental apparatus five major constraints for doing 'automatic computations' crystallize: (1) only a finite number of symbols can be written down and used in any computation; (2) there is a fixed bound on the amount of scratch paper (and the symbols on it) that a human can 'take in' at a time in order to decide what to do next; (3) at any time a symbol can be written down or erased (in a certain area on the scratch paper called 'cell'); (4) there is an upper limit to the distance between cells that can be considered in two consecutive computational steps;

(5) there is an upper bound to the number of 'states of mind' a human can be in, and the current state of mind together with the last symbol written or erased determine what to do next.

Turing then defined a mathematical model of an 'imagined mechanical device' that satisfies all of the above, later referred to as a 'Turing machine' (TM) by Church. A TM consists of an unbounded tape divided into squares, each of which can hold exactly one symbol, a tape head for reading and writing symbols from a given alphabet on the squares, and a controller, which is in exactly one of finitely many states at any given time. Each computational step of the machine first involves reading the symbol under the tape head and then, depending on the current state of the controller, writing a new symbol on the square, possibly switching to another state and possibly moving the tape head one square to the left or to the right. 'The computation proceeds by discrete steps and produces a record consisting of a finite (but unbounded) number of cells, each of which is blank or contains a symbol from a finite alphabet. At each step the action is local and is locally determined, according to a finite table of instructions' (Gandy, 1988, p. 81). This way the TM became a model of human computing, an *idealized* model to be precise, since it could process and store *arbitrarily long, finite sequences of symbols*. The TM model is also a very abstract model, for it only captures high-level processes that take place in humans when they compute (as opposed to low-level neuronal processes, for example).

Turing intended his analysis to show that *any function computable by a human being following fixed rules can be computed by a TM*. And, furthermore, he also believed the converse, that every function computed by a Turing machine could (in principle) be computed by a human computer. Note that this equivalence *per se* does not preclude humans from being able to find answers to problems (expressed in terms of functions) which no TM can compute (e.g. using intuition).

PHILOSOPHICAL VIEWS OF COMPUTATION

Turing Computability and Beyond

The logico-philosophical analyses of the intuitive notion of computation led to the crucial insight that different attempts to characterize it can all be proven extensionally equivalent: recursive functions, λ-definable functions, and TM-computable functions all define the same class of functions. These equivalence results are possible, because what 'computing' means with respect to any of the suggested formalisms is expressed in terms of functions from inputs to outputs, which are used as mediators in the comparison of the various classes of functions defined by the different formalisms. Later, other formalisms such as Markov algorithms, Post systems, universal grammars, PASCAL programs, as well as various kinds of automata, were also shown to give rise to the same class of functions. Hence, by CT, any of the above mentioned formalisms captures our intuitive notion of computation, that is, *what it means to compute*. (Some disagree with this conclusion, arguing that the equivalence results capture only a restricted notion of computation as shared by certain philosophers of mathematics and logicians, e.g. Sloman (1996).)

Common to all the above computational formalisms (besides their attempts to specify formally the intuitive notion of 'computation') is their property of being independent from the physical: computations in any of these formalisms are defined *without* recourse to the nature of physical systems that (potentially) realize them. Even the TM model, which is often considered the prototype of a 'mechanical device', does not incorporate physical descriptions of its inner workings, but abstracts from the mechanical details of a *physical realization*. The first to incorporate physically motivated mathematical constraints into a formal model of computation was Gandy (1980) in his attempt to define a notion of computation for any discrete, deterministic, physical machine. He formulated five conditions to determine whether any system qualifies as a 'mechanical machine' and proved that any function computable by a discrete deterministic device (in his sense) is effectively computable and vice versa. Hence, TM-computability (i.e. effective computability) and computability by mechanical devices are equivalent notions. Some even extend the claim by suggesting that the behavior of any finitely realizable physical system can be 'computed' (in the sense of 'perfectly simulated') by a TM (e.g. see Deutsch, 1985).

It is not clear, however, whether computation should be equated with 'effective computability', since there are, at least in principle, imaginable computing devices that give rise to 'Super Turing computability' (i.e. compute functions that no TM can compute). An example of such a device is Turing's 'oracle machine' (O-TM), which is a TM with additional atomic operations to query an 'oracle'. The oracle itself is a device that somehow produces values of a particular (possibly TM-uncomputable) function – how the results are obtained is left unspecified. It is easy to see that any O-TM with an oracle for any uncomputable function can compute more functions than any TM. Whether such a machine could be physically realized is an open question (maybe there are physical quantities that happen to encode some TM-uncomputable function). The interesting point is simply that an O-TM would be perfectly mechanistic in the classical sense without being effective as it uses some noneffective device, namely the oracle. O-TMs, hence, drive a wedge between the notions of 'effectiveness' and 'mechanism' (e.g. see Copeland, 2000). A similar point can be made with respect to the notions of 'effectiveness' and 'algorithm'.

There are other suggestions along the same lines coming from neural network research: it can be shown, for example, that certain neural networks (consisting of about 1000 neurones) with rational-valued connection weights between neurones can compute any TM-computable function (Siegelmann and Sontag, 1995). And if real-valued weights are allowed, they can compute any function whatsoever.

Other Construals of Computation

Although TMs have become the canonical models of computation and permeate various academic disciplines in that role, there are other construals targeted more towards possible philosophical merits and potential practical applications of computation. Following Smith (forthcoming), for example, the following views should all be distinguished as they emphasize and capture different aspects of computation:

1. *formal symbol manipulation*: the manipulation of symbols by virtue of their formal properties (without regard to possible interpretations or semantic content);
2. *effective computability*: what can be done effectively by a mechanism;
3. *rule-following* or *execution of an algorithm*: what is involved in following rules or instructions;
4. *finite (digital) state machines*: automata with a finite set of internal states;
5. *information processing*: what is involved in storing, manipulating, and displaying information;
6. *interactive systems*: computation as interaction and communication embedded in an environment;
7. *dynamical systems*: computation expressed in the language of dynamic systems (using concepts like state space, trajectory, attractors, etc.).

To some extent all of the above notions play a role in various disciplines (especially in computer science), but some of them are more dominant in specific intellectual areas: (1) figures mainly in philosophical debates and meta-mathematics, where (2) and (3) are tied to logical investigations; (4) is largely an engineering concept, while (5)–(7) have become increasingly important in cognitive science, the theory of complex systems, and, of course, computer science.

While the above list is far from exhaustive, it is intended to give a flavor of the wealth of different aspects the notion of computation has acquired, especially in the course of the last century. For that very reason, it is argued, we are still lacking the 'grand unified theory' (similar to physics) that can accommodate all these multiple facets, if such a theory is possible in the first place.

Real-life Computation

Despite the theoretical success of TM-computability, computer science *qua* practice is concerned not so much with the limits of what can be computed in theory, but rather with the more modest, mundane question of what can be computed within reasonable limits (using given resources). A whole new discipline within computer science called 'complexity theory' – an offspring of the classical investigations of effective computability – is dedicated to the study of what is computationally feasible. Still other issues arise from computational practice with which the TM model, for example, can hardly cope, in particular, the need for computational systems (embedded in various kinds of devices) to continually interact with their environments: what function does an operating system compute (or the world wide web, for that matter)? According to the classical view, such questions cannot be answered easily as the underlying functions are simply not defined for inputs on which computational processes run forever. (A simple example of a program that loops forever on all of its inputs is the following control code of a 'router' for the internet, which simply copies messages from its input to its output port.) Yet, there is a strong intuition that computational processes as they occur in operating systems or web browsers do have a purpose, can accomplish certain tasks or fail at achieving them. As a consequence the notion of 'computation of a function', and with it the classical notion of algorithm, had to make room for the notion of interaction:

> Interaction is shown to be more powerful than rule-based algorithms for computer problem-solving, overturning the prevalent view that all computing is expressible as algorithms. The radical notion that

> interactive systems are more powerful problem-solving engines than algorithms is the basis for a new paradigm for computing technology built around the unifying concept of interaction. (Wegner, 1997)

This paradigm shift from programs to processes renders many of the old reservations to the notion of computation obsolete, which were a consequence of taking computation to be defined solely in abstract syntactic terms thereby abstracting over physical realization, real-world interaction, and semantics. The new approach reveals computation, contrary to standard orthodoxy, as interactive and embodied, hence very much concerned with the constraints imposed on computational processes by the real world.

ROLE OF COMPUTATION IN COGNITIVE SCIENCE

The Midwife: Computation and the Birth of Cognitive Science

The independence of computations (in the sense of TM-computations) from their physical realizers was one major source of attraction for cognitive psychologists in the late 1950s. The information-processing capabilities of computers, an ability thought to underlie human cognition, and the potential of computer programs to specify exactly *how* information is processed was another. Together they led to the thought that cognition, viewed as 'the processing of information', could be completely understood and explained in terms of computations: if cognitive functions *are* computations, then explanations of mental processes in terms of programs are scientifically justifiable without having to take the 'implementing' neurological mechanisms into account, similar to computers, where it is the programs implemented on the computer hardware, not the hardware itself, that explain (if not entirely, then at least for the most part) what the computer does. The *computer metaphor* implicit in this view has been summarized as the claim that 'the mind is to the brain as the program is to the hardware' (Johnson-Laird, 1988) (note that this should really read 'the mind is to the brain as *computational processes* are to the hardware' to avoid conflating the program–process distinction). Its guiding ideas eventually became so prominent (originally in psychology, later in artificial intelligence) as to assist in the birth of cognitive science and establish *the computational claim about mind,* also called *computationalism,* as a genuine research paradigm.

The Paradigm: Computation and the Computational Claim about Mind

As with the notion of computation, computationalism is not a unified view, but construed differently by philosophers, psychologists, or neuroscientists. Various condensed slogan-like phrases such as 'the brain is a computer', 'the mind is the program of the brain', or 'cognition is computation' can be found in the literature, to name just a few. Yet, they cannot be taken at face value, for if they were read together, they would equivocate essentially distinct notions (such as program and process, mind and cognition, etc.). Furthermore, depending on their subdiscipline within cognitive science, researchers stress different aspects of computations: their information-processing capabilities, their formal nature, their control functions, their potential to have semantics, and so on.

Common to different views of computationalism are the assumptions that (1) mental processes are computational processes and (2) the same kind of relation that obtains between programs and computer hardware (i.e. the *implementation* relation) obtains between mental descriptions and brains too. It follows that cognitive functions can be described by and explained in terms of programs, and that the right level of abstraction at which to understand cognition is the computational level and not the level of the implementing mechanism (i.e. the brain), even though it might be helpful to know the functional organization and role of certain brain areas in determining what they implement.

Computationalism has many appealing facets, especially when it comes to high-level cognition: many features related to logic and language (such as systematicity, productivity, compositionality, and interpretability of syntax or the compositionality of meaning, e.g. see Fodor and Pylyshyn, 1988) are supported by computations 'almost for free', and many mental operations on various kinds of representations such as rotating three-dimensional images, searching for a good move in a chess game, reasoning about other people's behavior, planning a route through a city avoiding construction sites, etc. can be described computationally and implemented on computers. After all, this is what computers do: they manipulate symbol tokens (e.g. strings of bits), some of which are representations of the subject matter the computation is about. These representations, in turn, have both formal and semantic properties, of which the former are causally efficacious. Computational processes then manipulate symbols by virtue of their formal and not their semantic properties (e.g. Fodor, 1981).

While computationalists take this to be a virtue of their approach, it is a major shortcoming for others and various arguments have been advanced to establish that formal symbol manipulation is not sufficient for human intentionality and semantics (e.g. Searle's Chinese Room, 1980) or that minds are not TM-computable (e.g. Lucas's Gödelian argument, 1961). More recently, connectionists and dynamicists have tried to replace the notion of computation with alternatives, arguing that the representational level of description of a cognitive system so crucial to computationalism cannot be taken for granted. In fact, most dynamicists find the symbolic/representational level of description superfluous altogether and argue instead for an explanation of cognition in terms of *dynamic systems* (e.g. Port and van Gelder, 1995).

The Method: Computation and the Simulation of Cognition

While there are undoubtedly tendencies in cognitive science to replace the classical notions of computation, either by dynamic systems or by more adequate notions of computation (e.g. notions based on interaction, real-time constraints, etc.), even those opposed to computationalism agree at least that computation is still a valuable tool in the study of cognition (regardless of its explanatory success). In particular, computer simulations and computational models (of aspects) of cognition have become increasingly important in cognitive science. While computational models, at least to some extent, presuppose that whatever is modeled is computational, simulation models do not have to make such an assumption. Rather, they implement a computational approximation of the mathematical description of the phenomenon under scrutiny, and as long as any resultant error is within predetermined bounds the simulations are considered 'models'. In particular, they might elucidate complex dynamical relations between various parts of the simulated model, which are difficult to see (and often not 'visible' at all) from the formal, mathematical description. From trajectories through complex state spaces of dynamic systems to evolutionary processes in artificial environments, computer simulations provide a testbed for cognitive scientists to evaluate their hypotheses without always having to study them in 'real systems'. Furthermore, simulations can focus on different aspects of cognitive systems at different levels of description, they can be reproduced, slowed down, sped up, and modified in various other ways (e.g. simulating damage, disease, and various other disorders), in which no real cognitive system could be manipulated while preserving its normal functionality (obviously, ethical considerations would enter the picture here as well).

A crucial difference between simulation and computational models is, however, that the former usually does not share all the relevant causal properties with the modeled system, whereas the latter, being a computational model of computational processes, can in principle have the right causal structure (depending on various constraints on inputs, outputs, real-time, etc.).

SUMMARY

'Computation' is a multifarious notion, which defies a single, simple characterization. Yet, it is often explicated as 'executing an algorithm', presupposing some sort of mechanism able to 'execute' instructions as specified by the algorithm. For many logicians and philosophers it was the notion of Turing machine computability that for the first time gave precise meaning to the intuitive notion of computation understood as 'blindly following rules or instructions', thereby answering the question of what 'effective calculability' is supposed to mean. The connection of 'effectiveness' and computation goes back at least to the seventeenth century, when 'calculation' was very much tied to mechanical devices. Only in the twentieth century did effectiveness, mechanism, and computation become separated, when alternative models of computations such as interactive systems were considered. Various construals of the notion of computation (such as 'formal symbol manipulation' or 'information processing') emphasize different aspects of computation, although none of them seems to capture what computation may signify in its entirety. In cognitive science, computation played a crucial role right from the start. It figured prominently in the emergence of the discipline and became the basis of computationalism, the paradigmatic view that mental processes are computational, leading to the development of computational models of cognitive functions. Even for researchers objecting to computationalism, computations can be of great utility when used to simulate cognitive processes.

References

Church A (1936) An unsolvable problem of elementary number theory. *American Journal of Mathematics* **58**: 345–363.

Copeland BJ (2000) Wide vs. narrow mechanism. *Journal of Philosophy* **97**: 5–32.

Deutsch D (1985) Quantum theory, the Church–Turing principle and the universal quantum computer. *Proceedings of the Royal Society*, Series A, **400**: 97–117.

Fodor JA (1981) *Representations*. Cambridge, MA: MIT Press.

Fodor JA and Pylyshyn ZW (1988) Connectionism and cognitive architecture: a critical analysis. *Cognition* **28**: 3–71.

Gandy R (1980) Church's thesis and principles for mechanism. In: Barwise J, Keisler HJ and Kunen K (eds) *Proceedings of the Kleene Symposium*, pp. 123–148. New York, NY: North-Holland Publishing Company.

Gandy R (1988) The confluence of ideas in 1936. In: Herken R (ed.) *The Universal Turing Machine: A Half-Century Survey*, pp. 55–111. Berlin: Kammerer & Unverzagt.

Johnson-Laird PN (1988) *The Computer and the Mind*. Cambridge, MA: Harvard University Press.

Lucas JR (1961) Minds, machines, and Gödel. *Philosophy* **36**: 122–127.

Port R and Van Gelder T (1995) *Mind as Motion: Explorations in the Dynamics of Cognition*. Cambridge, MA: MIT Press.

Pratt V (1987) *Thinking Machines – The Evolution of Artificial Intelligence*. Oxford, UK: Basil Blackwell.

Searle J (1980) Minds, brains and programs. *The Behavioral and Brain Sciences* **3**: 417–424.

Sloman A (1996) Beyond Turing equivalence. In: Millican PJR and Clark A (eds) *Machines and Thought: The Legacy of Alan Turing*, vol. I, pp. 179–219. Oxford, UK: Clarendon Press.

Siegelmann HT and Sontag ED (1995) On the computational powers of neural nets. *Journal of Computer System Sciences* **50**: 132–150.

Smith BC (forthcoming) *The Age of Significance. An Essay on the Foundations of Computation and Intentionality*, vols I–VII. Cambridge, MA: MIT Press.

Turing AM (1936) On computable numbers, with an application to the Entscheidungsproblem. *Proceedings of the London Mathematical Society*, Series 2, **42**: 230–265.

Wegner P (1997) The paradigm shift from algorithms to interaction. *Communications of the Association for Computing Machinery* 1997 **40**(5).

Williams MR (1997) *A History of Computing Technology*, 2nd edn. Los Alamitos, CA: IEEE Computer Society Press.

Further Reading

Cleland CE (1993) Is the Church–Turing thesis true? *Minds and Machines* **3**: 283–312.

Copeland BJ (1996) What is computation? *Synthese* **8**(3): 335–359.

Davis M (1958) *Computability and Unsolvability*. New York: McGraw-Hill Book Company.

Dietrich E (1990) Computationalism. *Social Epistemology* **4**(2): 135–154.

Gardner H (1985) *The Mind's New Science: A History of the Cognitive Revolution*. New York: Basic Books.

Haugeland J (1985) *Mind Design I*. Cambridge, MA: MIT Press.

Haugeland J (1996) *Mind Design II*. Cambridge, MA: MIT Press.

Herken R (ed.) (1988) *The Universal Turing Machine: A Half-Century Survey*. Berlin: Kammerer & Unverzagt.

Hopcroft JE and Ullman JD (1979) *Introduction to Automata Theory, Languages, and Computation*. Reading, MA: Addison-Wesley Publishing Company.

Searle J (1992) *The Rediscovery of Mind*. Cambridge, MA: MIT Press.

Smith BC (1996) *The Origin of Objects*. Cambridge, MA: MIT Press.

Sterelny K (1990) *The Representational Theory of Mind*. Oxford, UK: Blackwell.

Van Gelder TJ (1998) The dynamical hypothesis in cognitive science. *The Behavioral and Brain Sciences* **21**: 615–665.

Webb J (1980) *Mechanism, Mentalism, and Mathematics: An Essay on Finitism*. Boston, MA: Reidel.

Computational Modelling

See **History of Cognitive Science and Computational Modeling**

Computational Models of Cognition: Constraining

Intermediate article

Terry Regier, University of Chicago, Chicago, Illinois, USA

CONTENTS

Computational models of cognition may be constrained in various ways. The three primary benefits of constraining a computational model are: (1) parsimony; (2) avoiding over-flexibility; and (3) potentially stronger motivation for the elements of the model.

OVERVIEW

Computational models of cognition are an essential tool in the study of the mind. One advantage of such models is their explicitness. The computational researcher must specify a mental mechanism in sufficient detail to allow the resulting model to be instantiated on a computer, and run as a cognitive simulation. This often requires that theoretically important elements of the model be described very clearly: the theory must be explicit enough to be described as a machine.

But what sort of machine? Computational machines range from the inflexible to the very flexible. An inflexible machine performs only a single predetermined task, such as addition, or multiplication, or accepting 65 cents in coins and returning a can of soda. A very flexible or general machine, in contrast, may be used for a wide variety of functions. Significantly, there exists an abstract machine, the Turing machine, that is so general that it has been taken as the formal equivalent of the very broad informal notion of an algorithm – thus, a Turing machine is taken to be capable of implementing any algorithm at all (Lewis and Papadimitriou, 1981). Modern programmable computers are approximations to Turing machines. Thus, the flexibility of modern computers gives a sense of the generality of Turing machines, and of the potential generality of computational machines as a class.

What is the significance of these issues of computational generality and flexibility for models of cognition? Often, generality is taken as a strength in a scientific theory. One would like to be able to account for a broad array of data, from a variety of sources, rather than provide a micro-theory of a very limited domain of cognition. There is, however, a danger of over-generality in model construction. If data are accounted for by a model that is very general and flexible, it is not always clear what to make of the model's success. The problem is that such a general model could also perhaps have fit other data, of a sort never empirically observed. That is, it may have succeeded in accounting for the empirical data because of its extreme flexibility, rather than because of a systematic match between the structures of the model and the mental processes at play. In this sense, a model's computational power may undermine its explanatory power. A more constrained, less computationally powerful model may in some instances provide more insight into the mental processes under study.

THE BENEFITS OF CONSTRAINTS

There are three primary benefits of a constrained model. These are: a more parsimonious characterization, an avoidance of over-flexibility, and potentially stronger motivation for the elements of the model. To illustrate these benefits, let us turn to a simple example.

Consider the task of silently counting from zero up to a given positive integer, and then pressing a button when done. This task involves the mental process of counting, with no overt behavioral clues as to the nature of the process other than the total time required. This time is shown as a function of the target number in Figure 1. These are informally collected data, meant for illustrative purposes only. The data represent averages over five testing sessions, from one subject. The figure also shows the fits of two models, one more constrained than the other.

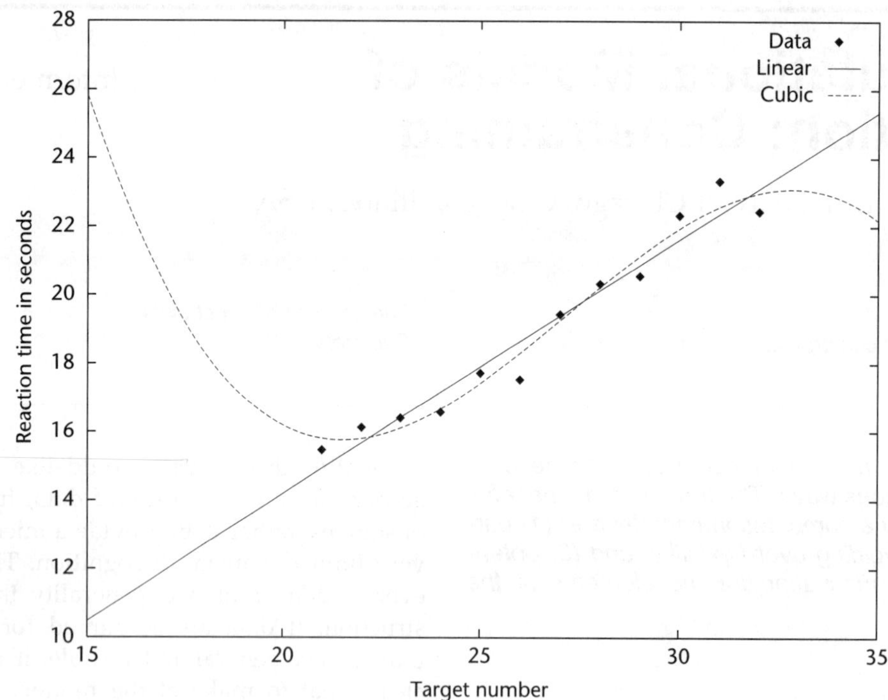

Figure 1. Mental counting. The data represent average time taken to count silently up to a given number. The best linear fit and best cubic fit are also shown.

The obvious computational characterization of mental counting is a simple iterative loop: begin with zero, then add 1 repeatedly until the target number is reached. We may reasonably assume that the counting will assume a rhythmic character, such that each iteration requires the same amount of time. Thus, overall reaction time T should be a linear function of the target number X:

$$T = aX + b \tag{1}$$

Here a and b are free parameters. Note that the parameter a has a straightforward psychological interpretation: the time taken for a single counting iteration. The parameter b denotes the amount of time needed to count from zero to zero – ideally this would be zero. Fitting this linear model to these data, we obtain a good fit, as shown in the figure ($R^2 = 0.9504$, $p < 0.0001$).

Let us compare this with the performance of a less constrained model, one based on a cubic function of the target number:

$$T = a_1X + a_2X^2 + a_3X^3 + b \tag{2}$$

This more general model provides a somewhat closer fit to the data ($R^2 = 0.9713$, $p < 0.0001$). The two new terms in the model approach significance. However, a comparison between the two models also highlights the three benefits of more constrained models.

Perhaps the most obvious benefit is that of parsimony. The linear model is simpler than the cubic model, which contains two terms in addition to those of the linear model. Model simplicity is aesthetically pleasing, and also makes the model easier to analyze.

A related issue is that the constrained model is less likely to be overly flexible. In this example, the less constrained cubic model overfits the data. With the obtained parameter settings, the cubic model predicts the observed data very well. But in doing so it also fits some of the noise, so that it generalizes poorly outside the observed range. As can be seen in the figure, this model predicts that it will take longer to count to 15 than to 30: an obviously false prediction. This problem is avoided by the more constrained linear model.

Finally, as we have seen, in this case the elements of the constrained linear model are motivated: they readily admit clear psychological interpretations, in the context of an iterative counting process. This is not true of the additional terms in the cubic model. The addition of such elements to a model, in the hope of increasing flexibility and thereby improving the fit to the data, comes at the price of arbitrariness.

These various benefits of constrained models need not all correlate perfectly. One could imagine, for instance, an ill-motivated model with few free parameters, or a well-motivated one with many. But the appeal of a constrained model lies in the prospect of accurately explaining a psychological phenomenon using a small number of well-motivated components.

These issues are relevant to a range of different styles of cognitive modeling, not just to simple models like that used in the example. Concerns about unconstrained models have been voiced in the context of symbolic models, and also in the context of connectionist or neural network models. For example, Miller *et al.* (1960) proposed a symbolic mental structure called a plan, which might be recursively created by other plans, yielding an overall system that was potentially quite complex and flexible. They anticipated that some might find their proposal too general and open-ended, and imagined their critics arguing as follows (quoted in Seidenberg, 1993):

> A good scientist can draw an elephant with three parameters, and with four he can tie a knot in its tail. There must be hundreds of parameters floating around in this kind of theory and nobody will ever be able to untangle them.

While Miller *et al.* defended their proposal against this critique, on the grounds that a complex system may in some instances be required, their sensitivity to the issue demonstrates an early concern with underconstrainedness in symbolic models. Another manifestation of the same concern may be found in Newell's (1990, p. 220) discussion of the computational universality of his production system framework SOAR, and the need for constraints in such a framework.

Some connectionist models have raised similar concerns. Massaro (1988) claimed that the multi-layer perceptron – a commonly-used connectionist architecture – was too computationally powerful to be psychologically meaningful. His argument was based on the finding that a single connectionist model could simulate results generated by three mutually exclusive process models. Thus, the connectionist model appeared to Massaro to be overly flexible, and potentially unfalsifiable. And Cybenko (1989) showed formally that multi-layer perceptrons are in principle flexible enough to approximate arbitrarily well any continuous function over inputs that range from 0 to 1. Further discussion of constraints in connectionist models may be found in (McCloskey, 1991; Regier, 1996; Seidenberg, 1993; Siegelmann, 1995). We shall return to some of these issues below.

THE SOURCES OF CONSTRAINTS

There are many possible sources for constraints in a cognitive model. Four particularly important ones are treated here: constraints derived from known psychological structure, those derived from known biological structure, those derived from task structure, and those based on the nature of the input to the model.

Psychological Structure

It is natural for a computational model of one psychological phenomenon to be informed by – and constrained by – independent observations concerning another, related phenomenon. This would afford an explanation of the one mental process in terms of the other. For example, Gluck and Bower (1988) accounted for aspects of human categorization using a simple connectionist learning rule, the delta rule, which is ultimately motivated by studies of Pavlovian conditioning in animals (Rescorla and Wagner, 1972). This rule has known constraints: for example, it predicts blocking and overshadowing in learning. The strength of Gluck and Bower's presentation is that they find empirical evidence of these constraints in human category learning – thus mechanistically linking human categorization and animal conditioning. Another example of the use of psychological constraints may be found in Nosofsky's (1986) model of categorization. This model builds on an existing model of identification, or the discrimination of stimuli from one another (Shepard, 1957). Thus, the constraints of the original model – such as a monotonic decrease in generalization with increasing psychological distance – are incorporated into its successor. This implicity grounds an account of categorization in an existing account of identification.

Biological Structure

Biological constraints may also be brought to bear on cognitive models. This idea holds the potential of reduction, of explaining cognitive processes in terms of the neural structure that underlies them. Wilson *et al.* (2000) provide a concrete example. They present a model of how humans perceive the orientation of another person's head: whether the other person is facing one directly, or in partial profile. They are able to account for their empirical findings in this domain using a model based on a population code of neurons found in area V4 of cortex, neurons sensitive to concentric and radial visual structure. Thus, the constraints of the underlying neural structure are used to explain a psychological phenomenon. Another example can be found in Regier and Carlson's (2001) study of spatial language. Participants were asked to rate the acceptability of linguistic spatial descriptions such as 'the dot is above the triangle', when shown pairs of objects in various spatial configurations. Their linguistic responses were well described by a model based in part on a neurobiological finding: the representation of overall direction as a vector sum. Thus, the linguistic data are partially grounded in a neurobiological constraint. More generally, one of the major appeals of connectionism as a modeling framework has been the prospect of bringing neural constraints to bear on psychological models (Feldman and Ballard, 1982; Seidenberg, 1993).

An important source of biological constraints is timing (Feldman and Ballard, 1982). Complex mentally-guided behavior can occur at a timescale of seconds. Assume, for example, that someone were to ask: 'Do you know where the registrar's office is?' It would take only a few seconds to hear the question, extract the intended meaning, recall where you believe the office is, prepare a linguistic response to the question, and deliver that response. Thus, a good amount of cognitive computational work is accomplished in a short timespan. But neurons, which are widely assumed to be the ultimate implementation of this computational work, operate on a timescale of a few milliseconds – relatively slowly by the standards of modern computers. Thus, the entire process of hearing and responding to the question must take place within only a few thousand neural computational timesteps. This constrains the number of iterations a model may realistically take when accomplishing such a behavioral or cognitive task. Since very little can be computed in a few thousand time steps using serial computation, this constraint strongly motivates parallel computation in cognitive models.

Task Structure

The nature of the psychological or behavioral task under study may also provide constraints on potential models. An example is the simple motor task of moving one's hand (or a pointer) to a target area of a specific size. Empirically, it has been found that the time required for this task varies as log (D/S), where D is the distance from starting point to end point of the motion and S is the size of the target region. This regularity is known as Fitts' law. An elegant model explaining this law has been given, based on the nature of the behavioral task. On this model, the most natural solution to the task is to launch the hand in the right general direction, and then iteratively correct during movement so as to bring the hand closer and closer to the target. This defines a recursion which, when solved analytically, yields the formula for Fitts' law (Keele, 1968; Meyer *et al.*, 1988; Newell, 1990). Another example is the simple mental counting task described earlier. Here, it is the intuitively obvious iterative nature of the counting task that motivates the linear model, the more constrained of the two models considered.

Input

An important source of constraint lying outside the model itself is the nature of the input supplied to it. Researchers investigating very flexible or general models often emphasize the constraints that reside in the model's input, at least as much as those residing in the structure of the model itself. Much connectionist modeling takes this approach. Specifically, as we have seen, multi-layer perceptrons are computationally quite general, and this generality has been a point of criticism. However, flexible mechanisms of this sort can be very useful scientifically, when considered together with the nature of the model input. A concrete example may be found in language acquisition. For many years, the study of language acquisition was dominated by the 'argument from poverty of the stimulus' (Chomsky, 1986). This view holds that the linguistic input heard by the language-learning child is too sparse, too impoverished to eventually give the child full knowledge of the syntactic structure of the language. Therefore, on this account, some elements of this knowledge must be innate. This account predicts that general-purpose learning mechanisms would not be able to learn the syntactic structure of natural language – as such mechanisms lack the requisite language-specific innate structure. This stance has been challenged recently. Very flexible, general-purpose connectionist networks have succeeded in learning artificial languages that resemble natural language in some important respects (Elman, 1993; Rohde and Plaut, 1999). This suggests that the input may not be as impoverished as had earlier been asserted, and that there may be no need to posit innate language-specific structure. Thus, the importance of these demonstrations lies precisely in the unconstrainedness of the mechanism itself, and the substantial constraints present in the input.

Given this variety of possible sources of model constraints, is there a most appropriate source? Is there any reason to prefer a model constrained in one manner over a model constrained in another? The answer to this question ultimately lies in the nature of the scientific question being asked. A general mechanism with constrained input is useful in addressing the alleged necessity of innate structure. For other sorts of questions, however, it may be more informative to demonstrate that elements of already acknowledged mechanistic structure can explain a novel phenomenon, one to which they were not originally tied. In both cases, the data at hand are explained in terms of known and independently motivated constraints, whether these constraints reside in the mechanism itself or in environmental input.

SUMMARY

Over-generality is a potential danger in computational models of cognition. It is conceivable that a model may account well for a set of empirical data because of its extreme flexibility, rather than because of a clear match between its structures and those of the cognitive process under study. This possibility has been a concern in both symbolic and connectionist models of cognition. However, the problem can be avoided by constraining models in principled ways. This can yield a more parsimonious, less over-flexible, and more convincingly motivated model. There are several possible sources for constraints on cognitive models: existing knowledge concerning psychological or biological structure, the nature of the cognitive task itself, or the input to the model. The scientific question being posed will indicate the most appropriate source of constraint in a given modeling enterprise.

References

Chomsky N (1986) *Knowledge of Language: Its Nature, Origin, and Use*. New York: Praeger.

Cybenko G (1989) Approximations by superpositions of a sigmoidal function. *Mathematics of Control, Signals, and Systems* **2**: 303–314. [Also available as report number 856, Center for Supercomputing Research and Development, University of Illinois at Urbana-Champaign, IL, USA.]

Elman JL (1993) Learning and development in neural networks: the importance of starting small. *Cognition* **48**: 71–99.

Feldman J and Ballard D (1982) Connectionist models and their properties. *Cognitive Science* **6**: 205–254.

Gluck M and Bower G (1988) From conditioning to category learning: an adaptive network model. *Journal of Experimental Psychology: General* **117**(3): 227–247.

Keele SW (1968) Movement control in skilled motor performance. *Psychological Bulletin* **70**: 387–403.

Lewis HR and Papadimitriou CH (1981) *Elements of the Theory of Computation*. Englewood Cliffs, NJ: Prentice-Hall.

Massaro D (1988) Some criticisms of connectionist models of human performance. *Journal of Memory and Language* **27**: 213–234.

McCloskey M (1991) Networks and theories: the place of connectionism in cognitive science. *Psychological Science* **2**(6): 387–395.

Meyer D, Abrams R, Kornblum S and Wright C (1988) Optimality in human motor performance: ideal control of rapid aimed movements. *Psychological Review* **95**(3): 340–370.

Miller G, Galanter E and Pribram K (1960) *Plans and the Structure of Behavior*. New York, NY: Holt, Rinehart, and Winston.

Newell A (1990) *Unified Theories of Cognition*. Cambridge, MA: Harvard University Press.

Nosofsky R (1986) Attention, similarity, and the identification–categorization relationship. *Journal of Experimental Psychology: General* **115**(1): 39–57.

Regier T (1996) *The Human Semantic Potential: Spatial Language and Constrained Connectionism*. Cambridge, MA: MIT Press.

Regier T and Carlson L (2001) Grounding spatial language in perception: an empirical and computational investigation. *Journal of Experimental Psychology: General* **130**: 273–298.

Rescorla RA and Wagner AR (1972) A theory of Pavlovian conditioning: variations in the effectiveness of reinforcement and non-reinforcement. In: Black AH and Prokasy WF (eds) *Classical Conditioning II: Current Research and Theory*. New York, NY: Appleton-Century-Crofts.

Rohde D and Plaut D (1999) Language acquisition in the absence of explicit negative evidence: how important is starting small? *Cognition* **72**(1): 67–109.

Seidenberg M (1993) Connectionist models and cognitive theory. *Psychological Science* **4**(4): 228–235.

Shepard R (1957) Stimulus and response generalization: a stochastic model relating generalization to distance in psychological space. *Psychometrika* **22**: 325–345.

Siegelmann HT (1995) Computation beyond the Turing limit. *Science* **268**: 545–548.

Wilson HR, Wilkinson F, Lin L-M and Castillo M (2000) Perception of head orientation. *Vision Research* **40**: 459–472.

Computational Models: Why Build Them?

Introductory article

Herbert A Simon, Carnegie Mellon University, Pittsburgh, Pennsylvania, USA

CONTENTS

The term 'computational model' is used for a theory that is stated definitely enough that precise reasoning (not necessarily numerical) can be performed on it to infer the course of system behavior. Nowadays, such a theory is often expressed as a computer program, the computer tracing the system's path.

INTRODUCTION

A computational model traces a system's processes over time, determining its path from its present state and the influences impinging on it at each moment. By this definition, a clock, whose hour hand moves one degree of arc for every half-degree of rotation of the earth, is a computational model in astronomy. In physics, computational models commonly take the form of systems of differential equations: for example, Maxwell's equations for electromagnetism.

HOW COMPUTERS MODEL THE HUMAN MIND

Computational models are not new to the social sciences. Economics has been using them for more than a century and a half. Because much of economics is concerned with quantities and prices of commodities, many economic situations and can be translated directly into numerical form, and standard analytic methods, together with computer simulations, can be used to study them.

In cognitive science, simple numerical computational models have long been used in psychophysics and psychometrics: for example, in scaling stimulus–response relations and in factor analysis. A powerful new class of computational models, *physical symbol systems* (PSS), emerged when it was recognized (around 1950) that the patterns stored in computers could be used to represent not only numbers, but also symbols of any kind, including words in natural language or pictures and diagrams. A PSS, implemented by a computer program, may represent the successive states of the disks and pegs of a Tower of Hanoi puzzle while it is being solved, or successive states of the thought processes of the person solving it. This article focuses upon physical symbol systems, both serial and parallel (connectionist), which are the predominant forms of computational model in contemporary cognitive science; but it will comment on their relation to mathematical models as well.

To behave as a PSS, a computer carries out certain fundamental processes: receiving (*sensing*) patterns, encoding (*perceiving*) them, storing (*remembering*) them, evoking (*recognizing*) them on appropriate occasions, modifying symbol structures in memory (*reasoning*), and outputting symbol structures that initiate actions (*behaving*). Finally, the system's actions may depend upon what particular symbols it finds in memory (*branching* or *choosing*). In each case the symbolic process can be matched with the corresponding class of psychological processes.

Symbols, in this context, are any kinds of distinguishable patterns, which may be linguistic, diagrammatic, pictorial, or purely abstract-relational in character. Although 'symbol' is popularly used specifically for linguistic patterns, these have no privileged position in cognitive modeling. The essential characteristic of a pattern is that it be capable of denoting (pointing to) other patterns, as a line drawing may denote a cat, or a DNA sequence, or a protein.

The *physical symbol system hypothesis* postulates that any system possessing these capabilities, and only such a system, can be programmed, or can program itself, to act intelligently. This is an empirical hypothesis to be confirmed or rejected by comparing the behaviors of such systems with human behavior, exactly as differential equations in physics are matched with physical systems.

A computer program, whether it processes numbers or non-numerical symbols, is formally a system of *difference equations*. Difference equations differ from differential equations only in that they move ahead by discrete increments of time instead of continuously (the increment being the time required to execute an instruction). Thus, the PSS hypothesis asserts that cognitive processes can be modeled by difference equations, numerical or symbolic or both in combination.

Computational models of cognition often not only simulate but actually carry out the solution process, and arrive at the solution (or fail to); they are both models of problem solving (or, more generally, thinking) and problem solvers. They describe and predict the course of thought, explaining its mechanisms while seeking to solve the problem. In this respect, they are different from the equations of physics, for equations representing the path of a planet do not move in such a path, but merely represent it, symbolically or graphically.

CONSTRUCTING AND TESTING MODELS

Some examples will show how models are constructed and what they do. Beginning with a very specific problem in algebra, the model is then generalized to a wide range of problems. This raises the question of how to accommodate the different ways in which different people may respond to the same problem: what invariants in cognitive processes constitute scientific laws of thinking?

An Algebraic Example

A program can model a student solving a linear equation in algebra, say, $7x-15=3x+9$. The solution takes the form $x=N$, where substitution of the number N for x satisfies the original equation. To discover the solution process, data are obtained from a student who writes down successive steps while talking aloud. In this case, she first adds 15 to both sides of the equation, obtaining $7x=3x+24$; then subtracts $3x$ from both sides, obtaining $4x=24$. Finally, she divides both sides by 4, obtaining $x=6$. At each step, she also collects terms (e.g. $9+15=24$). Each equation is transformed into another that has a closer resemblance to the final result.

A computer program can be constructed to follow this same procedure. The program applies a succession of *operators* that remove the differences between the current expression and the goal expression, leaving unchanged the number that satisfies the equation. The first step removes the unwanted number (-15) from the left side of the equation; the second step removes the unwanted term in x ($3x$) from the right side; the third step divides out the non-unitary coefficient of x (4). At each step the program performs the same operation on both sides of the equation and collects terms to simplify the result.

Deriving a More General Theory

Means–ends analysis and heuristic search

The method just described, *means–ends analysis*, is applicable to all sorts of problems, numerical or not. It is a powerful form of an even more general method called *heuristic search*. In heuristic search a problem is defined by a set of possible situations, or *problem space*. One of these is the *starting situation*, and any situation that satisfies a specified set of criteria is a *solution*. By applying a sequence of operators, the system can move through the space in search of a solution.

In the Tower of Hanoi we begin with three pegs *A*, *B* and *C* and a pyramid of disks, all stacked, say, on peg *A*. The disks must be moved, one by one, to other pegs, never placing a larger disk on a smaller one, until all have been moved to peg *C*. Each time a disk, starting with the largest, is moved successfully to peg *C*, a difference is removed between the initial and the final situation: fewer disks now remain to be relocated. The theory that people solve problems by using heuristic search, and means–ends analysis in particular, is captured in the computational model called the general problem solver (GPS). The theory has been shown to explain much human problem solving, but it clearly does not tell the whole story. More recently, GPS has been broadened into even more comprehensive models of cognition, including SOAR and the ACT-R family of programs.

Production systems

At a still more general level, the computational model proposes that the problem solving search is implemented by a mechanism called a *production system*. The choice of successive search steps results from the execution of 'if–then' rules (called *productions*), successors to the stimulus–response (S–R) pairs of behaviorist psychology. Whenever the conditions of a rule ('ifs') are satisfied, its actions ('thens') are executed. The principal advance over S–R is that if–then rules, unlike S–R connections, can contain variables that are instantiated in each problem situation, and the stimuli and responses can be symbol structures within the brain as well as external sensory stimuli and motor responses. Thus, short-term memory at a given moment can supply one or more of the ifs of an if–then rule, and the action, the 'then', or part of it, may invoke knowledge in long-term memory. The algebraic problem described above might be solved mentally, holding the intermediate symbol structures internally and only writing down the answer.

Many problem spaces that humans search are very large. In chess, about 10^{20} branches emanate from each move. In the time available to make a move, a master can explore perhaps 1000 of these, a minute fraction. Similarly, in the algebraic problem, an infinite number of equations have the same value of x as the given one, and x has an infinite number of possible values. How does the student, after generating only three equations, find the right value?

The computer program of the example shows that knowledge of the operations of adding or subtracting the same quantity from both sides of an equation, and of multiplying or dividing both sides by the same quantity, combined with knowledge of means–ends analysis, is sufficient to solve the equation rapidly. The student's written work and verbal statements will show whether this is the process she used to solve it. Thus the computational model is a theory that can be tested in detail, and at several levels of generality, against the verbal and other behavior of human subjects, with a time resolution of about five seconds per action.

The model both tests the sufficiency of the theory's mechanisms to produce the observed behavior, and compares its information processes, point by point, with the observed processes of the human subject. This is precisely how we test any theory against data; a computational model is not special in this respect. The comparison tests its sufficiency to perform the task and the degree of its correspondence with human processes.

Model invariants

The model described above can only solve algebraic equations; but the architecture of the model, ignoring its knowledge of algebra, constitutes a broad theory of problem solving. The theory postulates that a problem solver defines a problem space for the given problem; then, operations appropriate to the problem domain, implemented as productions, search selectively through the space, guided by means–ends analysis. The validity of this architecture as a general theory of problem solving can be tested by constructing the appropriate problem spaces and if–then productions for various problem domains, then simulating the system's behavior for specific problems in these domains and comparing the behavior with that of human subjects.

The problem solving models, which make much use of information stored in semantic memory, can be joined to several programs, for example, the EPAM program, which serves as a perceptual 'front end', UNDERSTAND and ISAAC, which are capable of translating verbal problem statements into inputs to GPS-like problem solvers, and CaMeRa, which reasons from pictorial or diagrammatic information. In this way the theory can undergo successive stages of generalization.

Individual differences

Whereas laws of human behavior presumably describe invariants over our species, different people perform the same task differently, and a single person behaves differently on different occasions. A major source of variation is the difference between people's memory stores, which are augmented by new learning, and different subsets of which are invoked in each problem situation. There is also genetic and other variability in the biological structure of brains. Theories in cognitive science, including computational models, must somehow extract invariants from this variability.

Invariant relations may be obtained in several ways. Laws may describe nonidentical but similar behaviors: 'When people are hungry, they search for food.' This is a (weak) law of behavior. The food varies greatly between cultures, with availability, and from one time to another, as do the methods of search (hunting, fishing, visiting the supermarket, etc.). The invariant is that people take actions that are relevant to the goal of food gathering. A theory of problem solving (like those already discussed) must deal with goals, and with processes for attaining them, taking as boundary conditions the goal-arousing circumstances and priorities among competing goals, and as initial conditions the actor's knowledge of ways of attaining goals and skills in pursuing the paths that are seen.

Because much variability derives from differences in what subjects know and attend to, an important approach to variability, and to the independent estimation of the parameters needed for testing models, is to design experiments that explicitly manipulate the knowledge available to subjects or modify the focuses of their attention or their goals, inducing them to adopt particular strategies.

These considerations are relevant whether behavior is studied at the level of symbolic processes or of neural activity. All human behavior is ultimately implemented by neurons, just as computer behavior is by chips. The neuronal activity is just as susceptible to individual variation deriving from inheritance, experience, and focus of attention as is the more aggregated symbolic activity it produces.

The variations in a system's structure, knowledge, and attention serve as initial and boundary conditions for a computational model, and (like parameter values in numerical models) must be supplied before the model can simulate the behavior of a particular subject in a particular situation. Of course, in many tasks, members of a group (say, college sophomores) will all behave in much the same way. To the extent that this is so, we can construct a model of a 'typical' subject by averaging over subjects in each experimental condition.

Unified theories of cognition

A model may contain so many parameters that it can fit almost any data, hence predict nothing. One solution to this problem is to identify a range of tasks throughout which a certain set of parameters play a significant role, then fit a single set of parameter values to data from all the tasks. This strategy can be pursued within unified theories of cognition, which aim to embody the whole range of cognitive processes, hence are testable in almost all task environments. For such theories, the ratio of number of data points to number of parameters will increase with the range of tasks.

A more conservative strategy is to take advantage of the cognitive system's subsystems, and to construct computational models of the subsystems, with the aim of ultimately joining them. This strategy reaps the usual dividends of 'divide and conquer', but needs to be carried out with due attention to the compatibility of the component models, so that they can later be joined, and to defining components that are capable of performing a variety of tasks, the more the better. As an example of design for compatibility, the UNDERSTAND system was constructed some years after GPS with the condition that the output of the former should be a suitable input to the latter. With respect to range of tasks, the EPAM system, using the same set of parameter values throughout, has been applied to perceptual, learning and expert memory tasks, and tasks of categorization and concept attainment.

The two most significant efforts to move towards a unified system are the Soar system of Newell, Rosenbloom and Laird, and J. R. Anderson's ACT systems. The 'divide and conquer' strategy has motivated the development of the set of programs that includes GPS, EPAM, UNDERSTAND, CaMeRa, and a model of scientific discovery.

All the problems that human variability, the complexity of the cognitive system and the multiplication of parameters pose for cognitive modeling are present to the same degree in other modes of theorizing about cognition. Vagueness may create the illusion of lawfulness; it cannot create the reality. Without the discipline of modeling, the sources of variability are not likely to be recognized as clearly or dealt with as carefully. The rigor and unambiguity of cognitive models brings these problems forcibly to mind, and allows them to be addressed with precision and clarity.

MODELS OF LEARNING

Because human (and animal) responses are highly modifiable, learning processes play a major role in cognitive psychology, and many computational models are theories of learning. In fact, learning processes are generally closer than task behaviors to being invariants of behavior, although learning processes are also modifiable as people learn to learn. Among the models of learning mechanisms are *adaptive production systems* (APS), *parallel distributed processors* (PDP), the 'elementary perceiver and memorizer' (EPAM), the *chunking* mechanism of Soar, and a number of similar learning processes in the ACT-R system. Several or all of these different mechanisms (and perhaps others) may be required to account for the full range of human learning capabilities.

Adaptive Production Systems

The algebraic example discussed earlier will illustrate how an adaptive production system can learn. An APS constructs new productions, and inserts them in its memory along with those already there. These new productions will henceforth be executed along with those previously stored. In each new task environment, the APS mechanism calls for a distinct set of differences and operators for building the new if–then rules.

Suppose an APS is presented with a set of worked-out examples of problems like the algebraic problem examined above, having learned previously what operations can legitimately be performed on an equation. It examines the first two lines of a worked-out example to discover what change was made in the equation, what operator was used to accomplish it, and what difference between initial expression and solution was removed. It then constructs a new production whose 'if' clause is the difference that was removed and whose 'then' clause is the operator used to remove it. The clauses are then generalized to permit any numbers to replace the specific numbers in the example. Thus, the APS initially constructs the production: 'If -15 appears on the left side of the equation, then add 15 to both sides and simplify.' It then generalizes it to: 'If a numerical term, N, appears on the left side of the equation, then add N to both sides and simplify.' The first APS with these capabilities was constructed by D. Neves.

Parallel Distributed Processors

Parallel distributed processors, which derive their original inspiration from Hebb's 'cell assemblies', use elements that somewhat resemble simplified neurons, or collections of neurons. Each such element consists of a link connecting two nodes. Each node has a modifiable level of stimulation, and each link an activation level. The links are typically organized into an input layer, one or more 'hidden' layers, and an output layer, the layers being arranged sequentially. A stimulus (a set of features) activates one or more nodes (each corresponding to a feature) in the input layer to varying levels of stimulation; each node then contributes to the activation of the links it is connected to; these links contribute to the stimulation of the input nodes of the hidden layer; and so on. Each final node in the output layer corresponds to a possible output, and the PDP responds with the output of the node that has the strongest total stimulation. A PDP model can learn to recognize distinct stimuli, and to conceptualize, by grouping them together, stimuli that share certain properties.

To learn, for example, to associate each letter of the Roman alphabet with a distinct set of input features, the system is given feedback on the correctness of its response to each set of features presented. A correct response increases the activation of the links that contributed to it, and decreases the activation of the other links; after an incorrect response, the reverse occurs. This simple scheme has been used to construct systems that can discriminate and conceptualize.

Discrimination Nets

Other kinds of learning which are closely related but organized in serial instead of parallel fashion, are modeled by the 'elementary perceiver and memorizer'. EPAM, in learning, grows a network of nodes and links. Each node contains a test for the value of some feature of the stimulus; the system tests this feature, then moves to a new node along the link that represents the stimulus's value on the test. If, for example, a test discriminated among various sizes of birds (wren-size, robin-size, chicken-size, etc.), then EPAM would move to a new node, following the link corresponding to the size of the bird being recognized. At the new node, EPAM tests another feature (the bird's color, say), and repeats the process until a terminal node was reached.

If given feedback when it is right or wrong, EPAM can learn to recognize large numbers of different patterns (an EPAM net has learned to distinguish 300 000 different patterns of chess pieces on chessboards). It can also store relevant information at each node in the net; hence it can serve, for example, as a medical diagnosis system which, at each node, has information about the disease whose symptoms would cause it to be sorted there. EPAM is a computational model of substantial breadth and power which has modeled a wide range of the phenomena that have been observed in memorization, discrimination, and categorization experiments.

Learning by Compilation ('Chunking')

Yet another learning mechanism is employed by Soar, an adaptive production system that can improve the functionality of new productions it has acquired by 'chunking' them. When two or more processes are frequently performed in sequence, their performance can be greatly speeded up by 'automating' them: that is, by removing time-consuming perceptual tests that would otherwise be used to determine what to do next, but which are unnecessary if the sequential process becomes automatic.

Comparing Learning Models

The existing evidence suggests that human learning probably employs all of the methods that have been examined here, and perhaps others as well. Nevertheless, close comparison of the behavior of different learning systems in approaching the same task can advance our understanding of the contribution of each mechanism to their performance, and can detect commonalities of mechanism among them.

For example, the EPAM and PDP models perform almost identically in a well-rehearsed experiment that found that letters embedded in four-letter words are recognized more rapidly than letters embedded in nonsense words. An investigation into how two such different architectures (serial versus parallel, symbolic versus numerical) could arrive at the same predictions showed that both models recognize words by recognizing the letters of which they are composed: EPAM by first recognizing letters, then words, by a recursive process; PDP systems using two serially arranged hidden layers with feedback from the 'word' layer to the 'letter' layer. Comparisons like this, of different models in performing the same task, shed valuable light on the functional equivalence of apparently quite different processes, and separate those features of theories that are important for accounting for the phenomena, from those that are merely artefacts of implementation. Such comparisons are important for choosing among proposed models and mechanisms.

SYMBOLIC AND NUMERICAL MODELS

This article has emphasized symbolic rather than numerical models. Removing the necessity of expressing the qualitative differences among architectures in numbers greatly extends the range of systems and realistic detail of mechanisms that can be modeled. For example, traditional mathematical learning models applied to conceptualization tasks generally compare feature vectors of two stimuli, using some kind of correlation between them as a single measure of similarity. As we have seen, EPAM and PDP systems employ much more detailed mechanisms of comparison, hence can answer much more specific questions about the process. No one would propose that computation of correlations is the actual psychological mechanism of comparison, whereas EPAM and PDP systems embody plausible candidate mechanisms. Observations of behavior may allow such claims to be proven.

LEVELS IN COMPLEX SYSTEMS

Not all computational models describe cognition at the same level of detail or in terms of processes on the same timescale. A similar diversity of theory is familiar in the other sciences. In physics, we have classical thermodynamics, which theorizes about heat and temperature averaged over bodies of substantial size. On the other hand, kinematic theories describe interactions of individual particles like atoms or molecules; while statistical mechanics treats of similar detail, but only on a probabilistic basis.

Levels in Scientific Theory

In any science there exist theories (and usually important theories) at each of several levels. This is possible because most complex systems observed in the world are constructed in a hierarchy of levels, and the elements are joined in such a way that the interaction between elements at any single level can be described without specifying any but very general properties of the elements at the next level below, and without considering dynamics at the next level above.

For example, a car can be described in terms of ignition, engine, transmission, and wheels, without describing anything about these components except their functions and their mutual connections. The ignition, at the proper time, fires the fuel from which the engine obtains energy to turn the drive shaft. The latter, connected to the wheels by the transmission, turns them, moving the car down the road. However, we can also fill a book describing each of these components in terms of their subcomponents, but still without mentioning details about their atomic structure.

Levels in Cognitive Theory

The human mind and brain are also organized as a hierarchy of components, each of these components made up of subcomponents, and so on. Thus we speak of the brain, then of the cerebrum, the cerebellum, the hippocampus, and so on; or of long-term memory, short-term memory, visual senses, auditory senses, and so on. Components can be recognized because there are stronger and more frequent interactions among the elements of a component than between distinct components on the same level.

The elementary actions and events among components at any given level usually take place on the same general timescale, shorter times as we go lower in the hierarchy, longer times as we go higher. Thus, a simple reaction to a sensory stimulus may take half a second, a step in solving an algebraic problem a couple of seconds, solution of a homework problem ten minutes, performance of a sequence of scientific experiments three weeks.

As in the biological and physical sciences, a theory of cognition can be built at each broad level to describe and explain the phenomena and processes at that level without attention to the detailed structure of the levels below, treating the latter as already in steady state and treating the levels above as constant constraints for the time interval of interest. This near-independence between levels follows mathematically from the comparative independence of components from each other as compared with their high level of internal interaction. Typically, descriptions refer to events at a particular level, whereas explanations account for events at one level in terms of processes at the next level below.

Three major levels of theory have been identified in cognitive science: the knowledge (or representational) level, of events that are minutes or more in duration, the symbolic level, whose events are seconds or tenths of seconds in duration, and the neural level, whose events may endure for only milliseconds or tens of milliseconds. Both the symbolic and the neural levels are possibly divisible into sublevels. Computational models of various aspects of thinking have been built at all of these levels.

The knowledge level

The behavior of a goal-oriented organism can often be predicted, in first approximation, by asking what actions would be effective, in its given environment, to attain its current goals. This is like predicting the shape of jelly from the shape of the mould in which it set. It requires information only about the goal and the environment, and none about the processes used by the organism to select the goal-attaining behavior. In ignoring the mechanisms, it is predictive and descriptive, rather than explanatory.

For example, the dominant theory in twentieth-century economics was that economic decisions maximize achievement of a goal called *utility*. A real system can behave in this way only if it has capabilities for determining in any situation which of all the available choices would yield this maximum. To predict its behavior, the goals embodied in the utility function and their relative weights have to be specified. The term 'rational' is often reserved for behavior that satisfies these criteria.

In some situations in real life, where we know what people value (what has utility), and they know what alternative actions are available and how to implement them, rationality defined in this way can be a powerful tool for prediction. This level of theory is called the 'knowledge level', calling attention to the centrality of an actor's knowledge of the environment as both generating and limiting the possible actions for reaching goals and deriving utility. If the environment is simple enough and the actors are rational, then their behavior (like the behavior of the jelly) can be predicted from the shape of the environment (the mold), specifying no other characteristics of the actors than their goals. The fulfillment of knowledge-level predictions provides no information about the mechanisms of human thought, except that there exist mechanisms of making simple goal-oriented decisions in these situations.

Thus, if we bring a college student into the laboratory with instructions to point to the taller of a five-foot and a six-foot person standing in the room, we can generally predict which person will be selected. We are assuming that the student has the goal of performing the task and sensory and mental capabilities for picking the taller person. However banal these conditions, they are at the core of descriptions of behavior at the knowledge level.

Two examples of knowledge-level theory in psychology are J. J. Gibson's theory of affordances, and the 'rational' (R) component of J. R. Anderson's ACT-R model. Gibson, for example, characterized a pilot landing a plane in terms of the characteristics of the landing field's surface (e.g. the size variation of ground patterns with distance), implicitly assuming that these characteristics were visible to the pilot, who understood their implications for the position of the plane. Similarly, Anderson's ACT-R model has built-in assumptions that certain learning processes will use information optimally. No strong claim is made that the mechanisms in the model for these processes correspond to the real mechanisms (those that would be discovered by exploration at the symbolic or neural levels).

An important argument for analysis at the knowledge level is that evolutionary processes tend to produce a fit between environmental properties and the behaviors of adaptive organisms in response to them. If sitting is sometimes an important goal, then the organism will become able to recognize objects that afford 'sittability', perhaps logs and chairs among them.

The symbolic level

At the next level below the knowledge level – the information processing, or symbolic, level – the theory takes account not only of the goals and external environment, but also of the actor's knowledge, and ability to use that knowledge to recognize, search and reason. For example, most means–ends analysis is at this level. PDP and other 'neural net' models are sometimes said to belong to the neural level, or sometimes to combine both levels, denying any separation in this range.

Information processing theories have been especially successful in modeling complex, often ill-structured, professional-level problem solving and design tasks, including chess playing, medical diagnosis, and scientific discovery. Such tasks require intuition, insight, and even creativity. Similar programs compose music, draw, or paint. All of these theories share a common core: they employ recognition and heuristic search as the basic mechanisms in problem solving. Hence, they are instances of a more general theory.

Bounded rationality

Knowledge-level theories, particularly those that assume optimization, have been criticized as ignoring the boundedness of human rationality: limited knowledge, and limited ability to compute consequences and compare incommensurate goals. Defenders of optimization reply that the bounded rationality viewpoint merely calls attention to the occurrence of human irrationality without proposing an alternative; and bounded rationality has sometimes been interpreted incorrectly as declaring that human behavior is largely irrational.

The principal contact between the proponents of knowledge-level optimization and the proponents of bounded rationality has been through research, often published in the fields of economics or decision theory, that attacks specific assumptions of economic theory, especially the consistency of the utility function.

By introducing the variability of human behavior, symbolic-level theories become less general and require the specification of numerous parameters before they can be applied to specific situations. This issue has already been discussed, and ways described of maintaining a high ratio of data points to parameters. It should also be noted that, in actual applications, most optimization models that ignore the limits on human rationality are (and must be) applied to complex real-world situations only after much abstraction and simplification of details, and that these approximations are generally carried out in a very unsystematic way without direct measurement of their effects.

The neural level

It is an important goal of science to connect theories at adjoining levels: for example, symbolic-level psychology with neuroscience, or neuroscience with biochemistry. The major explanatory theories in science use the mechanisms at one level to account for the mechanisms at the next level above without challenging the value of the theory at either level. The second half of the twentieth century saw a rapid growth in knowledge about human cognition at both symbolic and neural levels, as well as at the linkage between neural and biochemical levels. Understanding of the connections between symbolic and neural levels has developed more slowly, but in recent years, more attention has been paid to these connections, and the instruments for discovering them have been greatly improved. The modeling of such connections between levels is likely to receive much attention in the years immediately ahead.

SUMMARY

Cognitive science aims to understand systems as complex as physical and other biological systems. To reach that understanding, scientific representations are needed that provide the same kind of precision and reasoning power that physics, chemistry and biology get from mathematics and from the formal notations that describe chemical structures and processes. Psychology was severely handicapped by the looseness of the language and associated inference processes traditionally employed in its theories. In the past, a major deterrent to the employment of clear representations was that they could only be used if the phenomena were translated into mathematical language.

The modern computer created, to the surprise even of its inventors, a new set of formalisms appropriate for describing and analyzing information processing systems, without needing to cast these descriptions in numerical form. Computer programs can simulate cognition at the level of its information processes, even to the point of actually carrying out the cognitive tasks. Today, we have theories, in the form of operative programs, that give clear and precise accounts of most of the human cognitive processes at the symbolic level that enable people to perform complex, often ill-structured, tasks at professional, and sometimes creative, levels. These accounts may be incorrect in places. But we know that expert behavior derives from knowledge-based recognition and search processes, and we know how this is accomplished in an important collection of task domains.

In cognitive science today, our capabilities for stating theories at the symbolic level far exceed the power of our tools for actually observing the cognitive phenomena in the detail required to test the theories rigorously. As observational means improve, for example, by advances in the temporal resolution of functional magnetic resonance imaging (FMRI) studies, one can hope that the balance can be restored between our theory building and our observational capabilities.

Further Reading

Anderson JR (1990) *The Adaptive Character of Thought.* Hillsdale, NJ: Erlbaum.

Anderson JR and Lebiere C (1998) *The Atomic Components of Thought*. Mahwah, NJ: Erlbaum.

Feigenbaum E (1961) The simulation of verbal learning behavior. In: *Proceedings of the Western Joint Computer Conference* **19**: 121–132.

Friedman M (1953) *The Methodology of Positive Economics.* Chicago, IL: University of Chicago Press.

Gibson JJ (1979) *The Ecological Approach to Visual Perception*. Boston, MA: Houghton-Mifflin.

Gigerenzer G and Todd PM (1999) *Simple Heuristics That Make Us Smart*. Oxford: Oxford University Press.

Gobet F, Richman H, Staszewski J and Simon HA (1997) Goals, representations, and strategies in a concept attainment task: the EPAM model. In: Medin DL (ed.) *The Psychology of Learning and Motivation*, vol. xxxvii. San Diego, CA: Academic Press.

Hayes JR and Simon HA (1974) Understanding written problem instructions. In: Gregg LW (ed.) *Knowledge and Cognition*. Potomac, MD: Erlbaum.

Hebb DO (1949) *The Organization of Behavior: A Neuropsychological Theory*. New York, NY: John Wiley.

Hiller LA and Isaacson LM (1959) *Experimental Music: Composition With an Electronic Computer*. New York, NY: McGraw-Hill.

Iwasaki Y and Simon HA (1994) Causality and model abstraction. *Artificial Intelligence* **67**: 143–194.

Langley P, Simon HA, Bradshaw GL and Zytkow JM (1987) *Scientific Discovery: Computational Explorations of the Creative Processes*. Cambridge, MA: MIT Press.

McCorduck P (1991) *Aaron's Code: Meta-Art, Artificial Intelligence, and the Work of Harold Cohen*. New York, NY: Freeman.

Medin DL and Smith EE (1981) Strategies and classification learning. *Journal of Experimental Psychology: Human Learning and Memory* **7**(4): 241–253.

Neves DM (1978) A computer program that learns algebraic procedures by examining examples and working problems in a textbook. In: *Proceedings of the Second Conference of Computational Studies of Intelligence*, pp. 191–195. Toronto: Canadian Society for Computational Studies of Intelligence.

Newell A (1973) You can't play 20 questions with nature and win. In: Chase WG (ed.) *Visual Information Processsing*, pp. 283–308. New York, NY: Academic Press.

Newell A (1990) *Unified Theories of Cognition*. Cambridge, MA: Harvard University Press.

Newell A and Simon HA (1961) GPS: a program that simulates human thought. In: Billings H (ed.) *Lernende Automaten*, pp. 109–124. Munich: Oldenbourg.

Newell A and Simon HA (1976) Computer science as empirical inquiry: symbols and search. *Communications of the Association for Computing Machinery* **9**(3): 113–126.

Novak GS (1977) Representations of knowledge in a program for solving physics problems. In: *Proceedings of the Fifth International Joint Conference on Artificial Intelligence*. Cambridge, MA.

Richman H and Simon HA (1989) Context effects in letter perception: comparison of two theories. *Psychological Review* **96**: 417–432.

Rumelhart DE and McClelland JL (1986) *Parallel Distributed Processing: Explorations in the Microstructure of Cognition*, 2 vols. Cambridge, MA: MIT Press.

Shortliffe EH (1974) *MYCIN: A Rule-Based Computer Program for Advising Physicians Regarding Antimicrobial Therapy Selection*. Stanford, CA: Stanford University Press.

Simon HA (1983) *Reason in Human Affairs*. Stanford, CA: Stanford University Press.

Simon HA and Ando A (1961) Aggregation of variables in dynamic systems. *Econometrica* **29**: 111–138.

Tabachneck-Schijf HJM, Leonardo AM and Simon HA (1997) CaMeRa: a computational model of multiple representations. *Cognitive Science* **21**(3): 305–350.

Tversky A and Kahnemann D (1981) The framing of decisions and the psychology of choice. *Science* **211**: 353–358.

Computational Neuroscience: From Biology to Cognition

Intermediate article

Randall C O'Reilly, University of Colorado, Boulder, Colorado, USA
Yuko Munakata, University of Denver, Colorado, USA

CONTENTS

Computational neuroscience involves the construction of explicit computational models that implement neural mechanisms to simulate cognitive functions such as perception, learning and memory, motor function, and language.

INTRODUCTION

This article describes computer models that simulate the neural networks of the brain, with the goal of understanding how cognitive functions (perception, memory, thinking, language, etc.) arise from their neural basis. Many neural network models have been developed over the years, focused at many different levels of analysis, from engineering, to low-level biology, to cognition. Here, we consider models that try to bridge the gap between biology and cognition. Such models deal with real cognitive data, using mechanisms that are related to the underlying biology.

THE RELATIONSHIP BETWEEN COGNITIVE AND NEURAL THEORIES

Computational models provide an important tool for linking data across multiple levels of analysis. The cognitive implications of cellular and network properties of neurons are often not immediately apparent: there are too many factors at many different levels interacting in complex ways. Trying to develop behavioral predictions that capture the complexity of the neural level can be like trying to predict the weather from a number of satellite measurements. A computational model, of the weather or of the brain, can help by formalizing information and relating it through complex, emergent dynamics. Cognitive properties can thus be understood as the product of a number of lower-level interactions, and neural properties can be understood in terms of their functional role in cognitive processes. Further, the effects of manipulations of lower-level interactions (e.g. through genetic knockouts or lesions) can be simulated and reconciled with the observed behavioral effects. Importantly, these simulations can make sense of much more subtle behavioral effects than the generic impairment of behavior on a cognitive task.

Although models thus have the potential to clarify brain–behavior relations, they do not always do so. Models can be underconstrained by neural and behavioral data, and thus be of questionable value in showing how the brain actually subserves behavior. Moreover, models may be devised merely as demonstrations that a behavior can be simulated, but this is insufficient for understanding why the models behave as they do. Thus, models must be evaluated in a balanced way for whether they advance understanding of specific phenomena, provide general principles, and make useful links between brain and behavior.

This article reviews a number of neuroscience-based computational models of various cognitive phenomena, with an emphasis on the general principles embodied by these models and their implications for understanding the general nature of cognition. Specifically, we examine models of: vision, including topography and receptive fields in primary visual cortex and spatial attention emerging from interactions between parietal and temporal streams of processing; episodic memory subserved by the hippocampus; conditioning and skill learning subserved by the basal ganglia and cerebellum; working memory and cognitive control subserved by the prefrontal cortex; and language processing guided by neuropsychological cases. For a more comprehensive treatment of many of these models and the ideas behind them, see O'Reilly and Munakata (2000).

COMPUTATIONAL MODELS OF VISION GUIDED BY NEUROSCIENCE

Vision is one of the best-studied domains in cognitive neuroscience, having a long tradition of integrating biological and psychophysical levels of analysis. Computational models of vision have been influential in both the vision and computational research communities. We review two areas of visual modeling here: topography and receptive fields in primary visual cortex; and spatial attention and the effects of parietal lobe damage. Other major areas of visual processing that have been modeled include object recognition, motion processing, and figure–ground segmentation.

Topography and Receptive Fields in Primary Visual Cortex

The primary visual cortex (V1) provides an interesting target for computational models, because it has a complex but relatively well-understood organization of visual feature detectors (a 'representational structure') subject to considerable experience-based developmental plasticity (Hubel and Wiesel, 1962; Gilbert, 1996). Thus, the major question behind many of the V1 models has been: can we reproduce the complex representational structure of V1 through principled learning mechanisms exposed to realistic visual inputs?

First, we summarize the complex representational structure of V1. V1 neurons are generally described as edge detectors, where an edge is simply a roughly linear separation between regions of relative light and dark. These detectors differ in their orientation, size, position, and *polarity* (i.e. whether they detect transitions from light to dark or dark to light, or dark–light–dark or light–dark–light). The different types of edge detectors (together with other neurons that appear to encode visual surface properties) are packed into the two-dimensional sheet of the visual cortex according to a particular topographic organization. The large-scale organization is a 'retinotopic map', which preserves the topography of the retinal image in the cortical sheet. At the smaller scale are 'hypercolumns' (see Figure 1), containing smoothly varying progressions of oriented edge detectors, among other things (Livingstone and Hubel, 1988). The hypercolumn also contains 'ocular dominance columns', in which V1 neurons respond preferentially to input from one eye or the other.

Many computational models have emphasized only one or a few aspects of the many detailed properties of V1 representations; for reviews, see Swindale (1996) and Erwin *et al.* (1995). For example, models have demonstrated how ocular dominance columns can develop based on a Hebbian learning mechanism, with greater local correlations in the neural firing coming from within one eye than from across eyes (Miller *et al.*, 1989). Hebbian learning encodes correlational structure by strengthening the weights between neurons that fire together, and decreasing the weights between those that do not. (See Oja (1982) and Linsker (1988) for mathematical analyses of Hebbian correlational learning.)

Several models have demonstrated how a realistic set of oriented edge-detector representations can develop in networks presented with natural visual scenes, preprocessed in a manner consistent with the contrast-enhancement properties of the retina (e.g. Olshausen and Field, 1996; Bell and Sejnowski, 1997; van Hateren and van der Schaaff, 1997; O'Reilly and Munakata, 2000). The Olshausen and Field (1996) model demonstrated that sparse

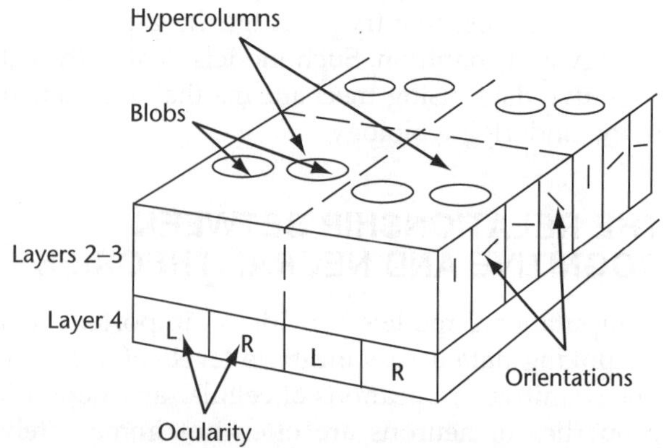

Figure 1. Structure of a cortical hypercolumn, representing a full range of orientations (in layers 2–3), ocular dominance columns (in layer 4, one for each eye), and surface features (in the blobs). Each such hypercolumn is focused within one region of retinal space, and neighboring hypercolumns represent neighboring regions.

representations (with relatively few active neurons) provide a useful basis for encoding real-world (visual) environments, but this model was not based on known biological principles. Subsequent work has shown how biologically-based models can develop oriented receptive fields, through a Hebbian learning mechanism with sparseness constraints in the form of inhibitory competition between neurons (a known property of cortex) (O'Reilly and Munakata, 2000). Furthermore, lateral excitatory connections within this network (another known property of cortex) produced a topographic organization consistent with several aspects of the hypercolumn structure (e.g. gradients of orientation, size, polarity, and phase tuning and pinwheel discontinuities: see Figure 2).

To summarize, these V1 models demonstrate how Hebbian learning mechanisms exposed to naturalistic stimuli, with certain kinds of biological prestructuring (e.g. connectivity patterns and inhibition), can produce aspects of the observed representational structure of V1. However, many complex aspects of early visual processing remain to be addressed, including motion, texture, and color sensitivity of different populations of V1 neurons.

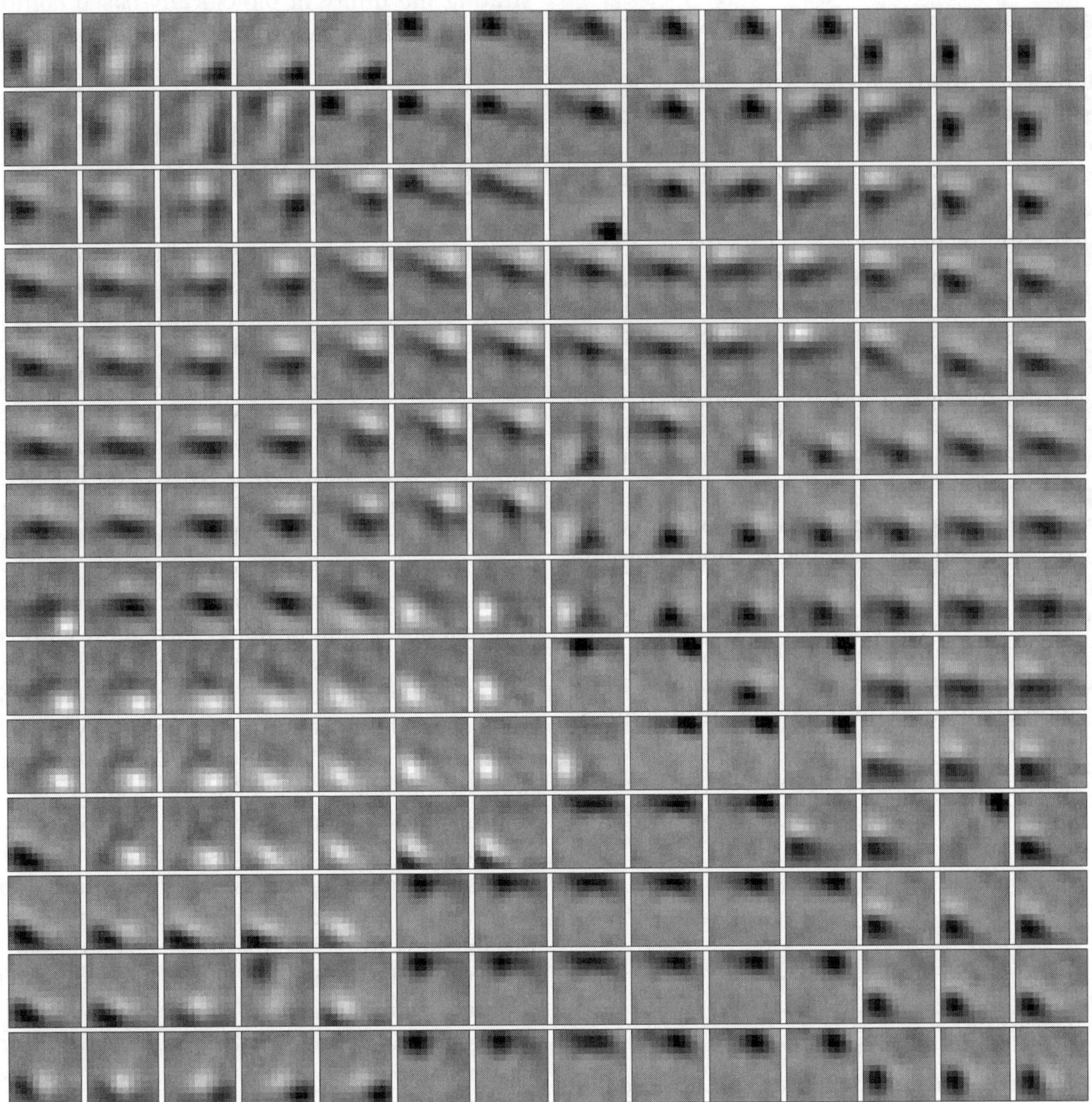

Figure 2. The receptive fields of model V1 neurons (O'Reilly and Munakata, 2000). Lighter shades indicate areas of on-center response, and darker shades indicate areas of off-center response. Individual units are shown by smaller grids (showing weights into those units from different locations in the retinally-organized input). These are organized into a larger grid representing the location of each unit within the simulated V1 hypercolumn.

Spatial Attention and the Effects of Parietal Lobe Damage

Many computational models of higher-level vision have explored object recognition (e.g. Mozer, 1991; Fukushima, 1988; LeCun *et al.*, 1989) and spatial processing (e.g. Pouget and Sejnowski, 1997; Mozer and Sitton, 1998; Vecera and O'Reilly, 1998). Here we describe a model of spatial attention (Cohen *et al.*, 1994) that demonstrates how biologically-based computational models can provide alternative interpretations of cognitive phenomena. Spatial attention has traditionally been operationalized according to the Posner spatial cuing task (Posner *et al.*, 1984: see Figure 3). When attention is drawn, or cued, to one region of space, participants are faster to detect a target in that region (a validly cued trial) than a target elsewhere (an invalidly cued trial). Patients with damage to the parietal lobe have particular difficulty with invalidly cued trials.

According to the standard account of these data, spatial attention involves a 'disengage' module associated with the parietal lobe (Posner *et al.*, 1984). This module typically allows one to disengage from an attended location to attend elsewhere. This process of disengaging takes time; hence the slower detection of targets in unattended locations. Further, the disengage module is impaired with parietal damage, so that patients have difficulty disengaging from attention drawn to one side of the space.

Biologically-based computational models, based on recurrent excitatory connections and competitive inhibitory connections, provide an alternative explanation for these phenomena (Cohen *et al.*, 1994; O'Reilly and Munakata, 2000). In this framework, the facilitative effects of drawing attention to one region of space result from excitatory connections between spatial and other representations of that region: this excitatory support makes it easier to process information in that region. The slowing observed in the invalid trials results from inhibitory competition between different spatial regions. Under this model, damage to the parietal lobe simply impairs the ability of the corresponding region in space to have sufficient excitatory support to compete effectively with other regions.

The two models make distinct predictions (Cohen *et al.*, 1994; O'Reilly and Munakata, 2000). For example, following bilateral parietal damage, the disengage model predicts disengage deficits on both sides of space (Posner *et al.*, 1984), but the competitive inhibition model predicts reduced attentional effects (smaller valid and invalid trial effects). Data support the latter model (Coslett and Saffran, 1991; Verfaellie *et al.*, 1990), demonstrating the utility of biologically-based computational models for alternative theories of cognitive phenomena.

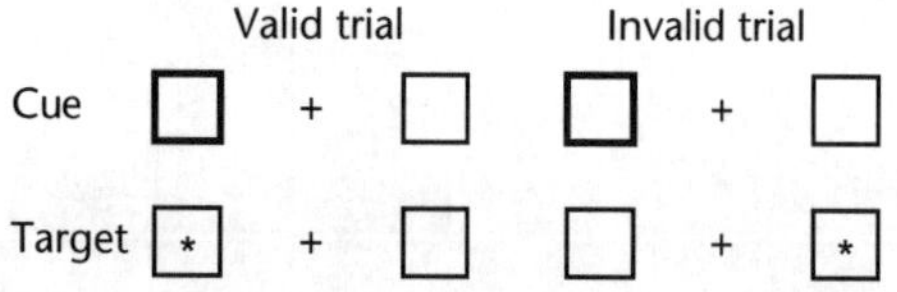

Figure 3. The Posner spatial attention task. The cue is a brightening or highlighting of one of the boxes, which focuses attention on that region of space. Reaction times to detect the target are faster when this cue is valid (the target appears in that region) than when it is invalid (the target appears elsewhere).

COMPUTATIONAL MODELS OF EPISODIC MEMORY AND THE HIPPOCAMPUS

Damage to the hippocampus, in the medial temporal lobe, can produce severe memory deficits, while leaving unimpaired certain kinds of learning and memory (Scoville and Milner, 1957; Squire, 1992). Many computational models have been developed to explore the exact contribution of the hippocampus, and these models have had a major influence (e.g., Marr, 1971; Treves and Rolls, 1994; Hasselmo and Wyble, 1997; Moll and Miikkulainen, 1997; Alvarez and Squire, 1994; Levy, 1989; Burgess *et al.*, 1994; Samsonovich and McNaughton, 1997).

One framework has combined known biological features of the hippocampal formation with computationally motivated principles about learning and memory to further clarify the unique contributions of the hippocampus in memory (McClelland *et al.*, 1995; O'Reilly and Rudy, 2000, 2001; O'Reilly and McClelland, 1994; O'Reilly *et al.*, 1998). The central idea is that there are two basic types of learning that an organism must engage in – learning about specifics and learning about generalities – and that because the computational mechanisms for achieving these types of learning are in direct conflict, the brain has evolved two separate brain structures to achieve them. The hippocampus appears to be specialized for learning about specifics, while the neocortex is good at extracting generalities.

Learning about specifics requires keeping representations separated (to avoid interference), whereas learning about generalities requires overlapping representations that encode shared structure across many different experiences. Furthermore, learning about generalities requires a slow learning rate to gradually integrate new information with existing knowledge, while learning about specifics can occur rapidly. This rapid learning is particularly important for episodic memory, where the goal is to encode the details of specific events as they unfold.

These computational principles provide a satisfying and precise characterization of the division of labor between the hippocampus and the neocortex. The models that implement these principles have been shown to account for a wide range of specific learning and memory findings, including nonlinear discrimination, incidental conjunctive encoding, fear conditioning, and transitive inference in rats (O'Reilly and Rudy, 2001) and human recognition memory (O'Reilly *et al.*, 1998). However, these models fail to incorporate important aspects of the hippocampal formation (e.g. the subiculum and the mossy cells in the hilus), and many more complex behaviors that depend on the hippocampus (and its interactions with other brain areas) remain to be addressed.

COMPUTATIONAL MODELS OF CONDITIONING AND SKILL LEARNING IN THE BASAL GANGLIA AND CEREBELLUM

A convergence between biological, behavioral and computational approaches has been achieved in the domain of conditioning (learning to associate stimuli and actions with rewards). In the computational domain, reinforcement learning mechanisms can change the behavior of a simulated animal according to reward contingencies in the environment (Sutton and Barto, 1998). Such learning mechanisms, including the 'temporal differences' algorithm (Sutton, 1988), not only work well mathematically (e.g. Dayan, 1992), but also correspond with aspects of neural recordings made in the reward-processing area of the brain (Montague *et al.*, 1996; Schultz *et al.*, 1997).

Specifically, a straightforward neural implementation of the temporal differences algorithm involves a systematic transition of reward-related neural firing similar to that observed in dopamine neurons in the midbrain. During a simple conditioning task where a sensory stimulus (e.g. a tone) reliably predicts a subsequent reward (e.g. orange juice), these neurons initially fire in response to the reward, but then after some trials of learning they respond to the sensory stimulus that predicts the reward and no longer to the reward itself (Schultz *et al.*, 1993; Schultz *et al.*, 1995). This transfer of reward-related firing from the actual reward to predictors of the reward is an essential property of the temporal differences mechanism as implemented by Montague *et al.* (1996), which thus provides a principled, provably effective explanation for why the brain appears to learn in this manner.

Models of motor performance and skill learning have been developed based on the biological properties of the relevant underlying brain areas, including the basal ganglia (which includes the striatum, globus pallidus, substantia nigra, subthalamic nucleus, and nucleus accumbens) and the cerebellum (e.g. Beiser *et al.*, 1997; Wickens, 1997; Houk *et al.*, 1995; Berns and Sejnowski, 1996; Schweighofer *et al.*, 1998a, b; Contreras-Vidal *et al.*, 1997). These models accord well with detailed neural properties of these areas, but tend to focus on simpler aspects of motor performance: complex motor skills remain to be addressed.

COMPUTATIONAL MODELS OF WORKING MEMORY, COGNITIVE CONTROL, AND PREFRONTAL CORTEX

The prefrontal cortex is important for a range of cognitive functions, which can be described generally as higher level cognition, in that they go beyond basic perceptual, motor, and memory functions. For example, frontal cortex has been implicated in problem-solving tasks like the Tower of Hanoi (e.g. Shallice, 1982; Baker *et al.*, 1996; Goel and Grafman, 1995), which requires executing a sequence of moves to achieve a subsequent goal. Many theoretical perspectives summarize the function of frontal cortex in terms of 'executive control', 'controlled processing', or a 'central executive' (e.g. Baddeley, 1986; Shallice, 1982; Gathercole, 1994; Shiffrin and Schneider, 1977), without explaining at a mechanistic level how such functionality could be achieved. Computational models provide an important tool for exploring specific mechanisms that might achieve 'executive-like' functionality.

Working Memory and Active Maintenance

One proposal is that the fundamental mechanism underlying frontal function is 'active maintenance', which then enables all the other 'executive' functionality ascribed to the frontal cortex (Cohen *et al.*, 1996; Goldman-Rakic, 1987; Munakata, 1998;

O'Reilly *et al.*, 1999; O'Reilly and Munakata, 2000; Roberts and Pennington, 1996). For example, a flexible, adaptive, active maintenance system can meet information processing challenges by using the strategic activation and deactivation of representations (activation-based processing) instead of weight changes (weight-based processing) (O'Reilly and Munakata, 2000). There are tradeoffs between these types of processing (e.g. activations can be more rapidly switched than weights, but they are also transient); so using both kinds of processing is better than using either alone.

There is considerable direct biological evidence that the frontal cortex subserves the active maintenance of information over time, as encoded in the persistent firing of frontal neurons (e.g. Fuster, 1989; Goldman-Rakic, 1987; Miller *et al.*, 1996). Many computational models of this basic active maintenance function have been developed (Braver *et al.*, 1995; Dehaene and Changeux, 1989; Zipser *et al.*, 1993; Seung, 1998; Durstewitz *et al.*, 2000; Camperi and Wang, 1997). Some models have further demonstrated that active maintenance can account for frontal involvement in a range of different tasks that might otherwise appear to have nothing to do with maintaining information over time.

Inhibition, Flexibility, and Perseveration

For example, several models have demonstrated that frontal contributions to 'inhibitory' tasks can be explained in terms of active maintenance instead of an explicit inhibitory function. Actively maintained representations can support (via bidirectional excitatory connectivity) correct choices, which will therefore indirectly inhibit incorrect ones via standard lateral inhibition mechanisms within the cortex. A model of the Stroop task provided an early demonstration of this point (Cohen *et al.*, 1990). In this task, color words (e.g. 'red') are presented in different colors, and people are instructed to either read the word or name the color in which the word is written. In the conflict condition, the ink color and the word are different. Because we have so much experience of reading, we naturally tend to read the word, even if instructed to name the color, so that responses are slower and more error-prone in the color-naming conflict condition than in the word-reading one. These color-naming problems are selectively magnified with frontal damage. This frontal deficit has usually been interpreted in terms of the frontal cortex helping to inhibit the dominant word-reading pathway. However, Cohen *et al.* (1990) showed that they could account for both normal and frontal-damage data by assuming that the frontal cortex instead supports the color-naming pathway, which then collaterally inhibits the word-reading pathway. Similar models have demonstrated that, in infants, the ability to inhibit perseverative reaching (searching for a hidden toy at a previous hiding location rather than at its current location) can develop simply through increasing ability to actively maintain a representation of the correct hiding location (Dehaene and Changeux, 1989; Munakata, 1998). Again, such findings challenge the standard interpretation that inhibitory abilities *per se* must develop for improved performance on this task (Diamond, 1991).

The activation-based processing model of frontal function can also explain why frontal cortex facilitates rapid switching between different categorization rules in the Wisconsin card sorting task and related tasks. In these tasks, subjects learn to categorize stimuli according to one rule via feedback from the experimenter, and then the rule is changed. With frontal damage, patients tend to perseverate in using the previous rule. A computational model of a related intradimensional/extradimensional (ID/ED) categorization task demonstrated that the ability to rapidly update active memories in frontal cortex can account for detailed patterns of data in monkeys with frontal damage (O'Reilly *et al.*, 2002; O'Reilly and Munakata, 2000).

Computational models of frontal function can provide mechanistic explanations that unify the various roles of the frontal cortex, from working memory to cognitive control and planning and problem-solving. However, it remains to be shown whether complex 'intelligent' behavior can be captured using these basic mechanisms.

COMPUTATIONAL MODELS OF LANGUAGE USE GUIDED BY NEUROPSYCHOLOGICAL CASES

Damage to language-related brain areas causes a wide variety of impairments. One class of such impairments, the dyslexias (also known as alexias), have been the subject of a series of influential computational models of the normal and impaired reading process (Seidenberg and McClelland, 1989; Plaut and Shallice, 1993; Plaut *et al.*, 1996). These models simulate the pathways between visual word inputs (orthography), word semantics, and verbal word outputs (phonology), and can account for different kinds of dyslexias in terms of differential patterns of damage to these pathways.

These models have been influential in part because they suggest an alternative, somewhat counterintuitive, interpretation of how words are represented and how language processing works. Traditional models have assumed that the brain contains a 'lexicon', with distinct representations for different words. Furthermore, these models assume that reading a word aloud (i.e., mapping from orthography to phonology) can occur via two different routes: pronunciation rules (for 'regular' words like 'make'), or a mechanism like a lookup table (for 'exception' words like 'yacht') (Pinker, 1991; Coltheart *et al.*, 1993; Coltheart and Rastle, 1994). In contrast to these dual-route models, the neural network models posit a single pathway to process both regular and exception words, and they employ a distributed lexicon without centralized, discrete lexical representations. Lexical processing occurs in pathways that map between different aspects of word representations (see Figure 4).

In general, neural networks can learn all kinds of different mappings: fully regular ones, like the spelling-to-sound mapping of the 'a' in words like 'make' and 'bake', as well as irregular mappings that occur in words like 'yacht'. Nevertheless, networks are sensitive to both the degree of regularity and the frequencies of different mappings. Specifically, neural network models predict frequency-by-regularity interactions that would not be expected in dual-route models, and which are observed in behavioral tests (Plaut *et al.*, 1996). Furthermore, these network models can account for patterns of deficit with brain damage that would seem improbable under dual-route models. For example, people with surface dyslexia can read

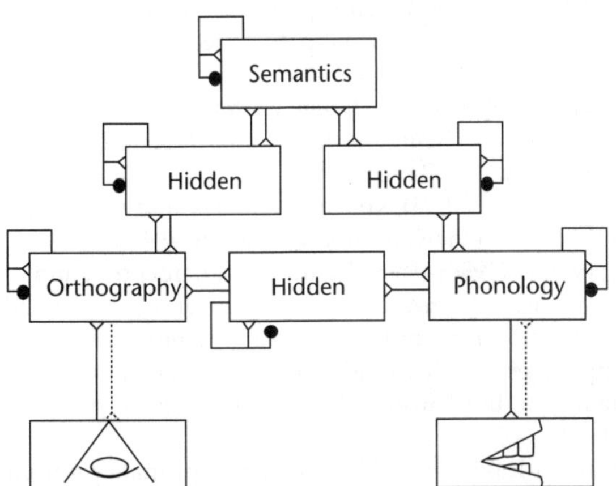

Figure 4. A neural network model of reading aloud. Words are represented in a distributed fashion across orthographic (visual word recognition), phonological (speech output), and semantic areas.

non-words (e.g. 'nust'), but they are impaired at retrieving semantic information from written words, and have difficulty in reading exception words. It would be natural in neural network models to interpret this as damage in the pathway between orthography and semantics. Interestingly, surface dyslexics' difficulty with exception words is generally limited to low-frequency exceptions (e.g. 'yacht'); they do not have difficulty reading high-frequency exceptions (e.g. 'are'). This pattern suggests that the remaining 'direct' pathway between orthography and phonology can handle both regular words and high-frequency exceptions, as in the network models. This pattern of data is not easily explained by the dual-route models: with two pathways, either regular words or exceptions should be affected, but not both, and independently of frequency.

Neural network models of language can provide alternative, counterintuitive ways of explaining some of the complex patterns of deficits that occur with brain damage. Nevertheless, such models remain controversial: neural network accounts are challenged by revised versions of dual-route models, and by the complexity of different neuropsychological profiles associated with damage to different language areas.

CONCLUSION

Computational models based on the neural networks of the brain can provide important insights. Many models have applied a set of basic principles to a range of phenomena, and arrived at explanations completely different from those based on purely verbal cognitive theories. Hence, these models have played an important role in guiding empirical research and theorizing in a number of domains.

Despite these successes, many researchers remain skeptical of models. A common concern is that different models may employ different sets of mechanisms to explain the same data, so that it may not be very significant that a given model can simulate a set of data. Several points have been made in response to this concern.

Firstly, it applies not only to computational models, but to scientific theorizing in general (several theories can account for the same data). Competing theories and models can be evaluated by many other criteria than simply accounting for a set of data, such as the accuracy of their predictions, the coherence of their theoretical framework, and the ease of accounting for new data (Munakata and Stedron, 2002).

Secondly, mechanisms developed independently can turn out to be equivalent (e.g. O'Reilly, 1996), providing converging evidence for their utility, and indicating more coherence to principles than might otherwise be evident.

Thirdly, a common set of mechanisms appears to be emerging as the field continues to mature. For example, over 40 different phenomena (including most of what has been described above) have been modeled using a common set of mechanisms (O'Reilly and Munakata, 2000). This set of mechanisms was developed over many years by many different researchers, and has now been consolidated and integrated into one coherent framework (O'Reilly, 1998).

Therefore, there is a largely consistent set of ideas underlying many neural network models, and this framework provides an important way of understanding the connections between cognition and underlying neural systems.

References

Alvarez P and Squire LR (1994) Memory consolidation and the medial temporal lobe: a simple network model. *Proceedings of the National Academy of Sciences* **91**: 7041–7045.

Baddeley AD (1986) *Working Memory*. New York, NY: Oxford University Press.

Baker SC, Rogers RD, Owen AM *et al.* (1996) Neural systems engaged by planning: a PET study of the Tower of London task. *Neuropsychologia* **34**: 515–526.

Beiser DG, Hua SE and Houk JC (1997) Network models of the basal ganglia. *Current Opinion in Neurobiology* **7**: 185–190.

Bell AJ and Sejnowski TJ (1997) The independent components of natural images are edge filters. *Vision Research* **37**: 3327–3338.

Berns GS and Sejnowski TJ (1996) How the basal ganglia make decisions. In: Damasio A, Damasio H and Christen Y (eds) *Neurobiology of Decision-Making*. Berlin, Germany: Springer-Verlag.

Braver TS, Cohen JD and Servan-Schreiber D (1995) A computational model of prefrontal cortex function. In: Touretzky DS, Tesauro G and Leen TK (eds) *Advances in Neural Information Processing Systems*, pp. 141–148. Cambridge, MA: MIT Press.

Burgess N, Recce M and O'Keefe J (1994) A model of hippocampal function. *Neural Networks* **7**: 1065–1083.

Camperi M and Wang XJ (1997) Modeling delay-period activity in the prefrontal cortex during working memory tasks. In: Bower J (ed.) *Computational Neuroscience*, chap. XLIV, pp. 273–279. New York, NY: Plenum Press.

Cohen JD, Dunbar K and McClelland JL (1990) On the control of automatic processes: a parallel distributed processing model of the Stroop effect. *Psychological Review* **97**: 332–361.

Cohen JD, Romero RD, Farah MJ and Servan-Schreiber D (1994) Mechanisms of spatial attention: the relation of macrostructure to microstructure in parietal neglect. *Journal of Cognitive Neuroscience* **6**: 377–387.

Cohen JD, Braver TS and O'Reilly RC (1996) A computational approach to prefrontal cortex, cognitive control, and schizophrenia: recent developments and current challenges. *Philosophical Transactions of the Royal Society, Series B* **351**: 1515–1527.

Coltheart M and Rastle K (1994) Serial processing in reading aloud: evidence for dual-route models of reading. *Journal of Experimental Psychology: Human Perception and Performance* **20**: 1197–1211.

Coltheart M, Curtis B, Atkins P and Haller M (1993) Models of reading aloud: dual route and parallel-distributed-processing approaches. *Psychological Review* **100**: 589–608.

Contreras-Vidal JL, Grossberg S and Bullock D (1997) A neural model of cerebellar learning for arm movement control: cortico-spino-cerebellar dynamics. *Learning and Memory* **3**: 475–502.

Coslett HB and Saffran E (1991) Simultanagnosia. To see but not two see. *Brain* **114**: 1523–1545.

Dayan P (1992) The convergenece of TD(λ) for general λ. *Machine Learning* **8**: 341–362.

Dehaene S and Changeux JP (1989) A simple model of prefrontal cortex function in delayed-response tasks. *Journal of Cognitive Neuroscience* **1**: 244–261.

Diamond A (1991) Neuropsychological insights into the meaning of object concept development. In: Carey S Gelman R (eds) *The Epigenesis of Mind*, chap. III, pp. 67–110. Mahwah, NJ: Lawrence Erlbaum.

Durstewitz D, Seamans JK and Sejnowski TJ (2000) Dopamine-mediated stabilization of delay-period activity in a network model of prefrontal cortex. *Journal of Neurophysiology* **83**: 1733–1750.

Erwin E, Obermayer K and Schulten K (1995) Models of orientation and ocular dominance columns in the visual cortex: a critical comparison. *Neural Computation* **7**: 425–468.

Fukushima K (1988) Neocognitron: a hierarchical neural network capable of visual pattern recognition. *Neural Networks*: **1**: 119–130.

Fuster JM (1989) *The Prefrontal Cortex: Anatomy, Physiology and Neuropsychology of the Frontal Lobe.* New York, NY: Raven Press.

Gathercole SE (1994) Neuropsychology and working memory: a review. *Neuropsychology* **8**: 494–505.

Gilbert CD (1996) Plasticity in visual perception and physiology. *Current Opinion in Neurobiology* **6**: 269–274.

Goel V and Grafman J (1995) Are the frontal lobes implicated in 'planning' functions? Interpreting data from the Tower of Hanoi. *Neuropsychologia* **33**: 623–642.

Goldman-Rakic PS (1987) Circuitry of primate prefrontal cortex and regulation of behavior by representational memory. In: Brookhart JM and Mountcastle VB (eds) *Handbook of Physiology. The Nervous System*, vol. V, pp. 373–417. Baltimore, MD: American Physiological Society.

Hasselmo ME and Wyble B (1997) Free recall and recognition in a network model of the hippocampus: simulating effects of scopolamine on human memory function. *Behavioural Brain Research* **67**: 1–27.

van Hateren JH and van der Schaaff A (1997) Independent component filters of natural images compared with simple cells in primary visual cortex. *Proceedings of the Royal Society, Series B* **265**: 359–366.

Houk JC, Davis JL and Beiser DG (eds) (1995) *Models of Information Processing in the Basal Ganglia*. Cambridge, MA: MIT Press.

Hubel D and Wiesel TN (1962) Receptive fields, binocular interaction, and functional architecture in the cat's visual cortex. *Journal of Physiology* **160**: 106–154.

LeCun Y, Boser B, Denker JS *et al.* (1989) Backpropagation applied to handwritten zip code recognition. *Neural Computation* **1**: 541–551.

Levy WB (1989) A computational approach to hippocampal function. In: Hawkins RD and Bower GH (eds) *Computational Models of Learning in Simple Neural Systems*, pp. 243–304. San Diego, CA: Academic Press.

Linsker R (1988) Self-organization in a perceptual network. *Computer* **21**(3): 105–117.

Livingstone M and Hubel D (1988) Segregation of form, color, movement, and depth: anatomy, physiology, and perception. *Science* **240**: 740–749.

Marr D (1971) Simple memory: a theory for archicortex. *Philosophical Transactions of the Royal Society, Series B* **262**: 23–81.

McClelland JL, McNaughton BL and O'Reilly RC (1995) Why there are complementary learning systems in the hippocampus and neocortex: insights from the successes and failures of connectionst models of learning and memory. *Psychological Review* **102**: 419–457.

Miller EK, Erickson CA and Desimone R (1996) Neural mechanisms of visual working memory in prefontal cortex of the macaque. *Journal of Neuroscience* **16**: 5154–5167.

Miller KD, Keller JB and Stryker MP (1989) Ocular dominance column development: analysis and simulation. *Science* **245**: 605–615.

Moll M and Miikkulainen R (1997) Convergence-zone episodic memory: analysis and simulations. *Neural Networks* **10**: 1017–1036.

Montague PR, Dayan P and Sejnowski TJ (1996) A framework for mesencephalic dopamine systems based on predictive Hebbian learning. *Journal of Neuroscience* **16**: 1936–1947.

Mozer MC (1991) *The Perception of Multiple Objects: A Connectionist Approach*. Cambridge, MA: MIT Press.

Mozer MC and Sitton M (1998) Computational modeling of spatial attention. In: Pashler H (ed.) *Attention*, pp. 341–393. London, UK: UCL Press.

Munakata Y (1998) Infant perseveration and implications for object permanence theories: a PDP model of the A-not-B task. *Developmental Science* **1**: 161–184.

Munakata Y and Stedron JM (forthcoming). Memory for hidden objects in early infancy. In: Fagen J and Hayne H (eds) *Advances in Infancy Research*, vol. XIV. Norwood, NJ: Ablex.

Oja E (1982) A simplified neuron model as a principal component analyzer. *Journal of Mathematical Biology* **15**: 267–273.

Olshausen BA and Field DJ (1996) Emergence of simple-cell receptive field properties by learning a sparse code for natural images. *Nature* **381**: 607–609.

O'Reilly RC (1996) Biologically plausible error-driven learning using local activation differences: the generalized recirculation algorithm. *Neural Computation* **8**: 895–938.

O'Reilly RC (1998) Six principles for biologically-based computational models of cortical cognition. *Trends in Cognitive Sciences* **2**: 455–462.

O'Reilly RC and McClelland JL (1994) Hippocampal conjunctive encoding, storage, and recall: avoiding a tradeoff. *Hippocampus* **4**: 661–682.

O'Reilly RC and Munakata Y (2000) *Computational Explorations in Cognitive Neuroscience: Understanding the Mind by Simulating the Brain*. Cambridge, MA: MIT Press.

O'Reilly RC and Rudy JW (2000) Computational principles of learning in the neocortex and hippocampus. *Hippocampus* **10**: 389–397.

O'Reilly RC and Rudy JW (2001) Conjunctive representations in learning and memory: principles of cortical and hippocampal function. *Psychological Review* **108**: 311–345.

O'Reilly RC, Norman KA and McClelland JL (1998) A hippocampal model of recognition memory. In: Jordan MI, Kearns MJ and Solla SA (eds) *Advances in Neural Information Processing Systems*, vol. X, pp. 73–79. Cambridge, MA: MIT Press.

O'Reilly RC, Braver TS and Cohen JD (1999) A biologically based computational model of working memory. In: Miyake A and Shah P (eds) *Models of Working Memory: Mechanisms of Active Maintenance and Executive Control*, pp. 375–411. New York, NY: Cambridge University Press.

O'Reilly RC, Noelle D, Braver TS and Cohen JD (2002) Prefrontal cortex and dynamic categorization tasks: representational organization and neuromodulatory control. *Cerebral Cortex* **12**: 246–257.

Pinker S (1991) Rules of language. *Science* **253**: 530–535.

Plaut DC and Shallice T (1993) Deep dyslexia: a case study of connectionist neuropsychology. *Cognitive Neuropsychology* **10**: 377–500.

Plaut DC, McClelland JL, Seidenberg MS and Patterson KE (1996) Understanding normal and impaired word reading: computational principles in quasi-regular domains. *Psychological Review* **103**: 56–115.

Posner MI, Walker JA, Friedrich FJ and Rafal RD (1984) Effects of parietal lobe injury on covert orienting of visual attention. *Journal of Neuroscience* **4**: 1863–1874.

Pouget A and Sejnowski TJ (1997) Spatial transformations in the parietal cortex using basis functions. *Journal of Cognitive Neuroscience* **9**: 222–237.

Roberts RJ and Pennington BF (1996) An interactive framework for examining prefrontal cognitive processes. *Developmental Neuropsychology* **12**(1): 105–126.

Samsonovich A and McNaughton BL (1997) Path integration and cognitive mapping in a continuous attractor neural network model. *Journal of Neuroscience* **17**: 5900–5920.

Schultz W, Apicella P and Ljungberg T (1993) Responses of monkey dopamine neurons to reward and conditioned stimuli during successive steps of learning a delayed response task. *Journal of Neuroscience* **13**: 900–913.

Schultz W, Apicella P, Romo R and Scarnati E (1995) Context-dependent activity in primate striatum reflecting past and future behavioral events. In: Houk JC, Davis JL and Beiser DG (eds) *Models of Information Processing in the Basal Ganglia*, pp. 11–28. Cambridge, MA: MIT Press.

Schultz W, Dayan P and Montague PR (1997) A neural substrate of prediction and reward. *Science* **275**: 1593–1599.

Schweighofer N, Arbib M and Kawato M (1998a) Role of the cerebellum in reaching quickly and accurately. I: A functional anatomical model of dynamics control. *European Journal of Neuroscience* **10**: 86–94.

Schweighofer N, Arbib M and Kawato M (1998b) Role of the cerebellum in reaching quickly and accurately. II: A detailed model of the intermediate cerebellum. *European Journal of Neuroscience* **10**: 95–105.

Scoville WB and Milner B (1957) Loss of recent memory after bilateral hippocampal lesions. *Journal of Neurology, Neurosurgery, and Psychiatry* **20**: 11–21.

Seidenberg MS and McClelland JL (1989) A distributed, developmental model of word recognition and naming. *Psychological Review* **96**: 523–568.

Seung HS (1998) Continuous attractors and oculomotor control. *Neural Networks* **11**: 1253–1258.

Shallice T (1982) Specific impairments of planning. *Philosophical Transactions of the Royal Society, Series B* **298**: 199–209.

Shiffrin RM and Schneider W (1977) Controlled and automatic human information processing. II: Perceptual learning, automatic attending, and a general theory. *Psychological Review* **84**: 127–190.

Squire LR (1992) Memory and the hippocampus: a synthesis from findings with rats, monkeys, and humans. *Psychological Review* **99**: 195–231.

Sutton RS (1988) Learning to predict by the method of temporal diferences. *Machine Learning* **3**: 9–44.

Sutton RS and Barto AG (1998) *Reinforcement Learning: An Introduction*. Cambridge, MA: MIT Press.

Swindale NV (1996) The development of topography in the visual cortex: a review of models. *Network: Computation in Neural Systems* **7**: 161–247.

Treves A and Rolls ET (1994) A computational analysis of the role of the hippocampus in memory. *Hippocampus* **4**: 374–392.

Vecera SP and O'Reilly RC (1998) Figure–ground organization and object recognition processes: an interactive account. *Journal of Experimental Psychology: Human Perception and Performance* **24**: 441–462.

Verfaellie M, Rapcsak SZ and Heilman KM (1990) Impaired shifting of attention in Balint's syndrome. *Brain and Cognition* **12**: 195–204.

Wickens J (1997) Basal ganglia: structure and computations. *Network: Computation in Neural Systems* **8**: 77–109.

Zipser D, Kehoe B, Littlewort G and Fuster J (1993) A spiking network model of short-term active memory. *Journal of Neuroscience* **13**: 3406–3420.

Computer Modeling of Cognition: Levels of Analysis

Introductory article

Michael RW Dawson, University of Alberta, Edmonton, Alberta, Canada

CONTENTS

According to Marr's three-level hypothesis, information processors must be described at the computational level, the algorithmic level, and the implementational level in order to be fully understood.

MODELING AT SEVERAL LEVELS: DESCRIBING COGNITION FROM MULTIPLE PERSPECTIVES

On 11 May 1997 a computer system called Deep Blue defeated world chess champion Garry Kasparov in a six-game match by a score of 3.5 to 2.5. For researchers interested in modeling high-level cognitive activities like thinking and problem solving, it is instructive to compare Deep Blue and Kasparov, and to consider in what ways these two chess players are different and in what ways they are similar.

At one level of analysis – that of their physical components – they are clearly quite different. Deep Blue is constructed from silicon circuitry, while Kasparov's chess playing ability is based in a biological brain. However, if we consider them from a more abstract perspective, we find that they are similar. At some level of description Deep Blue and Kasparov both fall into the same class of systems, because both can be described as 'chess players'. Furthermore, if one were to only observe a record of the moves made by these two systems, then it would appear that they were very similar because both played chess at a very high level of ability, and in many situations they would choose to make similar moves.

ANALYSIS OF TASK, ALGORITHM AND IMPLEMENTATION

From the example above, it is clear that when researchers wish to compare a computer model with a person or system being modeled, they must consider this comparison from a variety of perspectives. Similarly, when examining some phenomenon of interest in the hope of developing a computer simulation, the phenomenon must be investigated at a number of different levels of analysis.

The vision scientist David Marr argued convincingly that when information processors were being studied, three general levels of analysis were relevant. Each of these levels corresponds to asking a particular kind of question.

The first analysis proposed by Marr is at the *computational* level. At this level of analysis, a researcher is primarily concerned with the question: 'What information processing problem is being solved by the system?' In the chess example, this would involve specifying the abstract 'rules of chess', and classifying a system (e.g. Deep Blue or Kasparov) as being a chess player provided it obeyed those rules. Marr provided many examples in vision in which a computational analysis involved proving that the visual system could be described as solving a particular type of mathematical problem (such as a constrained optimization problem) when it was detecting some visual property (such as a pattern of motion). In general, because computational-level analyses often involve specifying abstract laws, they require researchers to use formal techniques such as mathematical or logical proof. In other words, the answer to the computational-level question is usually a mathematical statement that an information processing system is solving a particular problem; this statement can characterize the abstract laws that define the problem, or that impose constraints on how solutions to the problem are to be achieved. Such a statement can be applied either to a human subject or to a computer simulation.

The second analysis proposed by Marr is at the *algorithmic* level. At this level of analysis, a researcher is primarily concerned with the question: 'What procedures, program, or algorithm is being used by the system to solve the information processing problem?' There are many different ways of solving the same information processing problem. So, while Deep Blue and Kasparov may be equivalent when compared at the computational level, because they are both playing chess, they might at the same time be very different at the algorithmic level because they are using very different procedures to choose their next move.

Answering the algorithmic-level question is the primary goal of empirical cognitive scientists, such as experimental psychologists and psycholinguists. They use the research methodologies of the behavioral sciences to try and determine the procedures used by humans to solve information processing problems. This attempt usually separates into two different investigations. The first attempts to determine the general information processing steps (the program) being used to solve the problem. The second attempts to determine the primitive properties of symbols and the processes that act upon these symbols (the programming language). Many fundamental debates in cognitive science, such as the debate between proponents of symbolic connectionist models, are debates about the nature of the 'programming language' for human cognition. Zenon Pylyshyn has called this 'programming language' the functional architecture of cognition.

The third analysis proposed by Marr is at the *implementational* level. At this level of analysis, a researcher is primarily concerned with the question: 'What physical properties are responsible for carrying out the program or algorithm?' At the algorithmic level, it is sufficient to state what the components of the functional architecture are. At the implementational level, one has to explain how these components are 'brought to life' by an actual physical device. For human cognition, in which we assume that the brain is responsible for carrying out information processing, this requires researchers to determine how brain states and neural circuits are responsible for instantiating the functional architecture. Not surprisingly, the research methods of neuroscience provide the primary tools for answering the implementational-level question.

THE EXAMPLE OF COLOR PERCEPTION

To illustrate the three-level approach, let us consider color perception. In the seventeenth century, Sir Isaac Newton performed experiments with prisms and light and published the results in his book *Opticks*. As far as color vision was concerned, his great discovery was that some colors could be 'constructed' from combinations of others. Newton proposed that human color vision was based upon the perception of seven primary colors. Many alternative theories have been proposed. Goethe proposed a two-color theory in 1810; Young proposed a three-color theory in 1810; four-color theories were proposed by both Hering and Ladd-Franklin in the late nineteenth century.

The question of what is the minimum number of primary colors required for a complete theory of human color perception is a computational-level question. The physicist James Clerk Maxwell answered this question in 1856. Using a 'color solid' proposed by Newton, and an elegant geometric proof, Maxwell was able to show that any perceived color could be expressed in terms of three primary colors.

Maxwell's proof provides important information about color perception in principle, but leaves many unanswered questions about human color vision in practice. For instance, does the human visual system base itself upon the minimal number of primary colors? If so, what three primary colors does it use? These questions arise at the algorithmic level, and must be answered by experimental studies of human vision. In 1873, Helmholtz presented a series of lectures which popularized Young's three-color theory of color perception. This led to decades of psychophysical experimentation, such as having subjects differentiate pure colors from colors created by mixing combinations of light. The results of these experiments supported the Young–Helmholtz theory, indicated that Maxwell's proof applied to human color vision, and demonstrated that our perceptions of color arise from combining the sensations of the three primary colors red, green, and blue.

The answers to the algorithmic-level questions provide a great deal of information about the basic information processing steps used in color vision. However, they give no information about the physical mechanisms that are involved. Helmholtz himself was painfully aware of this. In 1873 he said: 'It must be confessed that both in man and in quadrupeds we have at present no anatomical basis for this theory of colors.' It wasn't until the second half of the twentieth century that a physical account of the Young–Helmholtz theory became possible. Researchers discovered three different kinds of cone cells in the retina, each containing a different light-sensitive pigment. A technique called microspectrophotometry, in which a small beam of light is passed through individual receptors that have been removed from the retina, revealed that these pigments are sensitive to different wavelengths of light: one is most sensitive to red light, one to green light, and one to blue light.

This example illustrates two general points. First, it shows that a psychological phenomenon can be examined at a number of different levels. Indeed, complete understanding of a phenomenon requires computational, algorithmic, and implementational accounts. Second, this degree of understanding is not easy to achieve. Our detailed modern theories of color perception are based upon several centuries of research.

INVESTIGATING MIND AT A VARIETY OF SCALES

The three levels of analysis that have been described above are not the only approach to understanding and modeling cognitive phenomena. For example, neuroscientists have argued that cognitive science requires a complete understanding of the brain, and that this understanding requires that the brain be examined at a number of different levels. This means analyzing the brain at a number of different scales. The neuroscientist Gordon Shepherd calls these levels of organization. At the largest scale, one studies the behavior of the whole brain. Reducing the scale to the next level of organization, one studies large systems and pathways in the brain (e.g. sensory pathways). At the next level, one studies the properties of specific centers and local circuits. Reducing the scale once more, one studies the properties of neurons (for example, through single cell recording). At even more microscopic scales, one studies structures within neurons (for example, systems of connectivity involving dendrites or axons). Reducing the scale further, one studies individual synapses. Finally, one can study the molecular properties of membranes and ion channels.

Patricia Churchland and Terry Sejnowski have cited these levels of organization as an argument against the three-level approach proposed by Marr. They suggest that because there are so many levels of organization, one cannot talk about a single implementational level. Furthermore, because different implementational levels might have different task descriptions, there will be a multiplicity of algorithms, so that it is inappropriate to talk about a single algorithmic level.

However, Marr's three levels were proposed as properties of an investigation, and not of the system being investigated (that is, they are an epistemology, not an ontology). Marr did not claim that information processors 'have' an algorithmic level. Rather, he claimed that to understand an information processor, one must answer the algorithmic-level questions. These questions need not be applied monolithically to an entire system. Thus, one can ask Marr's three questions of any level of organization in the brain, provided the level in question can be usefully considered as being involved in information processing. The 'levels of analysis' and 'levels of organization' approaches are not mutually exclusive, but are complementary.

SMOLENSKY'S SYMBOLIC AND SUBSYMBOLIC LEVELS

An alternative view of levels has arisen from attempts to relate symbolic models of cognition to connectionist simulations. Paul Smolensky has argued that in the context of connectionist simulations one can distinguish between a symbolic account and a subsymbolic account. To say that a connectionist network is subsymbolic is to say that the activation values of its individual hidden units do not represent interpretable features that could be represented as individual symbols. Instead, each hidden unit is viewed as indicating the presence of a microfeature. Individually, a microfeature is unintelligible, because its 'interpretation' depends crucially upon its context (i.e., the set of other microfeatures that are simultaneously present). However, a collection of microfeatures represented by a number of different hidden units can represent a concept that could be represented by a symbol in a symbolic model. From this perspective, a symbolic account of a network is only an approximate account. A more accurate and complete account of how a network solves a problem can only be found by analyzing it at the subsymbolic level.

The distinction between symbolic and subsymbolic levels is both interesting and important. However, this distinction too is compatible with Marr's approach. One could argue that a subsymbolic account of a network (e.g. a description of its microfeatures) would represent a description of its functional architecture while a symbolic account of the network would represent a description of an algorithm (e.g. a description of symbolic representations), as well as how this algorithm is composed from the functional architecture (the microfeatures).

This observation shows that the three-level approach can be applied to connectionist models as well as to more traditional symbolic computer simulations. Michael Dawson has argued that the three-level hypothesis applies equally to these two streams within cognitive science because connectionist and symbolic modelers both agree with the general assumption that cognition is information processing. This is an important point because many researchers are concerned that connectionist networks are not fully-fledged cognitive theories, but only provide an implementational account of cognitive phenomena.

SUMMARY

Information processing systems are very complex. In order to provide a complete account of an information processor, or to build a computer simulation of a cognitive phenomenon, three different levels of analysis are required. The computational level addresses the question: 'What information processing problem is being solved?' The algorithmic level addresses the question: 'What information processing steps are being applied to solve the problem?'. Thus the algorithm is distinguished from the functional architecture. The implementational level addresses the question: 'What physical properties are responsible for bringing these information processing steps into existence?' One reason why cognitive science is such a multidisciplinary undertaking is that the answers to these questions require vocabularies and methodologies from such diverse disciplines as mathematics, psychology and biology.

Further Reading

Bechtel W and Abrahamsen A (1991) *Connectionism and the Mind*. Cambridge, MA: Blackwell.

Dawson MRW (1998) *Understanding Cognitive Science*. Oxford: Blackwell.

Dennett DC (1987) *The Intentional Stance*. Cambridge, MA: MIT Press.

Fodor JA (1968) *Psychological Explanation: An Introduction to the Philosophy of Psychology*. New York, NY: Random House.

Jackendoff R (1992) *Languages of the Mind*. Cambridge, MA: MIT Press.

Marr D (1982) *Vision*. San Francisco, CA: Freeman.

Medler DA (1998) A brief history of connectionism. *Neural Computing Surveys* **1**(2): 18–72.

Newell A (1990) *Unified Theories of Cognition*. Cambridge, MA: Harvard University Press.

Pylyshyn ZW (1984) *Computation and Cognition*. Cambridge, MA: MIT Press.

Smolensky P (1988) On the proper treatment of connectionism. *Behavioural and Brain Sciences* **11**: 1–74.

Computer Vision

Introductory article

John K Tsotsos, York University, Toronto, Canada

CONTENTS

Computer vision is a discipline whose major concern is the development of computational methods, in terms of both hardware and software, to acquire, process, and interpret visual images.

INTRODUCTION

Computer vision systems attempt to recover shapes of objects, illumination direction, motion, object structure and identity, composition, and so on, from a digital image of a scene. The purpose of such systems is either to help us understand the contents of the image or to elicit an appropriate response from an autonomous system such as a robot. The interpretative process often occurs in the context of a particular domain. The goal of research in computer vision is to devise methods that approach or even exceed the performance of human vision. In addition, many researchers use the language of computation to formalize theories of biological vision.

Computer vision is difficult. At least half of human (or primate) cerebral cortex is involved in visual processing: this represents a huge amount of neural processing power. Some specific reasons for the difficulty of computer vision are described below.

Problems Inherent in the Analysis of Images

A camera is used to acquire a two-dimensional digital image of a three-dimensional scene. Imagine viewing an unknown scene with one eye and trying to estimate distances in that scene: you cannot do so reliably. Explicit information about depth is not present in a single image. The depth of objects in the scene must be inferred either indirectly or by using a binocular camera system. Indirect inference may use cues in the image such as occlusion (one object being in front of another).

Nor does an image contain explicit information about where the source of illumination is located. The computer system must infer this by considering positions and shapes of shadows, highlights, and shading on surfaces.

Often knowledge of where the scene is located and what the participants in the scene are doing assists greatly with solving ambiguities. For example, if you see a picture of a soccer game that includes a number of players and a ball in the air, from what direction did the ball come and who kicked it? Knowledge of the objects outside the scene and of the timing of the actions that led up to the scene can help with such a question.

The geometry of the camera system presents a different sort of difficulty, because it distorts the image. Extreme perspective effects are commonly seen if a wide-angle lens is used. The image is distorted on a finer scale by sources of noise in the imaging system. Such noise adds small variations to the true values of image intensity and color. For example, image sensors quantize light levels; therefore they do not represent those levels precisely.

Finally, it must be remembered that digital images represent large amounts of information. These bits of information must be compared with one another and combined with one another, and decisions must be made on all possible combinations in order to arrive at an interpretation of the image contents.

Although vision seems effortless for normally-sighted humans, it requires a large amount of processing power. Because of these difficulties, many have questioned whether a computer vision system could ever perform at the level of the human visual system.

Can Vision be Modeled Computationally?

An often-debated question is whether or not human or primate vision – and, more broadly, perception – can be modeled computationally. If not, then research on models of biological vision based on computational principles is sure to fail.

A proof of *decidability* is sufficient to guarantee that a problem can be modeled computationally. A decision problem has the form: given an element of a countably infinite set A, does that element belong to the set B (a subset of A)? Such a problem is decidable if there exists a Turing machine that computes 'yes' or 'no' for each element of A.

Perception, in general, has not yet been formulated as a decision problem, but many subproblems have been so formulated. Visual search, an important subproblem, has been formulated as a decision problem, and is decidable: it is an instance of the 'comparing' Turing machine.

Even if some other aspect of perception were determined to be undecidable, it would not follow that all of perception is undecidable, or that other aspects of perception cannot be modeled computationally. For example, one of the most famous undecidable problems is whether or not an arbitrary Diophantine equation has integral solutions (Hilbert's 10th problem). But this does not mean that mathematics cannot be modeled computationally. Similarly, another famous undecidable problem is the halting problem for Turing machines: it is undecidable whether a given Turing machine will halt for a given initial specification of its tape. This has important theoretical implications, but it certainly does not mean that computation cannot exist.

THE COMPONENTS OF A COMPUTER VISION SYSTEM

The processing stages depend strongly on the problem to be solved. For example, methods for processing color images would not be needed if only grayscale images are received. Similarly, if images from a microscope are analyzed, there is no depth inherent in the domain, so depth-specific processes can be omitted. The following discussion will assume no such domain-specific restrictions, and will describe the stages and components one might consider in the design of a general-purpose vision system. Note, however, that although the phrase 'general-purpose vision system' is often used, and may represent the ultimate goal of research in the area, general-purpose functionality has so far been elusive.

Levels of Analysis

It has been claimed that any information processing system must be understood at three different levels: the computational level, the representational and algorithmic level, and the hardware implementation level. The computational level is concerned with the goal of the computation and with the logic of the solution strategy. The representational and algorithmic level deals with the representations for the input, the output, and the intermediate computations, and with the algorithm that implements the solution strategy. The hardware implementation level concerns the physical realizations of the representations and algorithms – whether in a computer or in a brain. These levels of analysis have become part of the basic computational framework for the study of perception.

Acquiring Images

Cameras with arrays of sensing elements capture digital images. Each sensing element captures the color and brightness over a small region of the scene and translates it to a single value. (This value does not correspond directly to human perception of color and light.) For example, the quantity measured as brightness is irradiance, or power of light per unit area.

The result is an array of picture element values, or pixels, representing the scene. Note that the image is not an exact representation of the scene. The imaging geometry determines the size of the cone of light that is digitized by each pixel sensor; if that cone is large then detail from the scene is lost. There are also noise sources from within the electronic sensors.

The geometry of the imaging system relative to the scene is important if one wishes to recover the elements of the scene from a single image. A computer vision system must assume a projection model. Usually, this is the perspective model, but sometimes, for convenience, one assumes an orthographic projection model whereby the image is formed by rays from the object that are parallel to the optical axis of the camera. Perspective and orthographic projections do not differ much if the distance to the scene is much larger than the variation in distance among objects in the scene. Whatever model is chosen, if computations are made with the origin of a coordinate system on the camera, the frame of reference for the vision system is known as the 'camera frame'.

Often a single image is insufficient to recover the characteristics of a scene. Even an object as simple as a cube can present difficulties: if, for example, you could see only one face of it, you could not know if it was a cube, a pyramid or some other kind of solid. Edges of the object align with one another, and this alignment obscures the object structure. Only if you rotated the cube, or moved your head, could you confirm the shapes and sizes of the other faces. Using computer-controlled camera systems, researchers have studied this seemingly natural and obvious tactic for deciding between competing recognition hypotheses. In 'active perception', a passive sensor is used in an active fashion, its state parameters being changed according to sensing strategies. In such systems, intelligent control is applied to the data acquisition process in a way that depends on the current state of data interpretation, including recognition.

Active vision is useful in order to see a portion of the visual field that is otherwise hidden, to increase spatial resolution (by moving in closer, or zooming), and to disambiguate aspects of the visual world (for example by analyzing induced motion or lighting changes). In all these uses, some hypothesize-and-test mechanism must be at work. Only if hypotheses are available can a particular action actually yield benefits; otherwise the number of possible interpretations and actions is too large. Furthermore, in practical implementations of active perception, the additional cost imposed on a perceptual system must be considered.

Preparing for Analysis

The image acquisition process leads to a scene representation that contains errors and noise. Usually one attempts to model noise in such a way that algorithms can account for it and compensate for its effects. If it can be assumed that the noise is additive and random, there are algorithms that suppress, smooth or filter it. Such algorithms are usually either applied before any further processing or integrated into other processing stages.

A second major issue is camera calibration. Without calibration, it is impossible to tell, for example, the length of a line in the scene from the length of its representation in the image. If the image acquisition system is calibrated, knowledge of the camera system parameters is related to known coordinates in the three-dimensional scene. As a result, measurements in images can be related to measurements in the scene. Usually, in order to calibrate a camera system, a known target in a known position and pose is captured and the projection equations are solved for the resulting image.

Extracting Features from an Image

Image features are meaningful structures of an image, that is, they are strongly related to elements of the scene being imaged. There are two types of features: global properties, such as the average gray level across the image, and local properties, such as lines, circles, or textures. Each feature corresponds to a particular geometric arrangement of image values. Features that are preserved under some transformation are known as invariants with respect to that transformation. The length of an object, for example, is invariant with respect to rigid movement. Invariants are especially valuable as indices for databases of object models. (*See* **Vision, Early**)

We will now look at some of the most important image features.

Edges

Edges are sets of pixels around which image values exhibit a sharp variation, i.e., discontinuity. Edge detectors are designed to detect such pixels. The first generation of edge detectors simply calculated a discrete approximation to the first derivative of image intensity with respect to position; peaks in this derivative are then candidates for edge points. However, noise can both mask true peaks in the derivative and give rise to false ones.

The second derivative of image intensity, for an ideal step edge, is zero at the point of the step. On one side of the step the value is negative while on the other it is positive. If the step is corrupted by noise, the positive and negative regions of the second derivative may be similarly corrupted, but the zero is much more stable. Edge detectors using this idea have somewhat better properties than the first generation of detectors. They usually smooth the image with a mathematical operation designed to remove noise and then compute an approximation to the second derivative. This is sometimes known as the 'difference of Gaussians' method because the overall process can be modeled by subtracting two Gaussian functions with appropriate parameters. The scale of intensity structure in the image must be carefully considered: for good detection, the scale of the edge detector must be matched to the scale of the image structure.

Regions

'Image segmentation' is the partitioning of an image into regions, or subsets of pixels, that satisfy some homogeneity criterion. The partitioning is constrained so that all pixels are accounted for, there is no overlap between regions, and no two adjacent regions can be merged into a larger one that still satisfies the criterion. The criterion may relate, for example, to intensity, color, texture, or combinations of these.

How can such a segmentation be computed? Start with an arbitrary segmentation – say, into individual pixels. At each iteration, each region may or may not satisfy the criterion. If it does, then look at those of its neighbors that also satisfy the criterion and try to merge those regions into one. If the new region still satisfies the criterion, keep it; otherwise do not.

If a region does not satisfy the criterion, then split it into smaller regions. Continue the merging and splitting process until no further merges or splits are possible. (Note that regions consisting of single pixels must satisfy the homogeneity criterion, so the process terminates.)

This is known as the split-and-merge algorithm. Such regions are also known as 'connected components'. Segmentation remains one of the more stubborn problems in computer vision owing to the difficulty in defining comprehensive homogeneity criteria that correspond to objects.

Texture

The texture of a surface can be described as a statistical property. Texture elements ('texels') are visual primitives with certain invariant properties that occur repeatedly in different positions, deformations, and orientations inside a given area. The simplest approach to texture is to consider the first-order gray-level statistics of the image ('first-order' statistics are statistics of single pixels, as opposed to groups of pixels). These can be captured by the gray-level histogram (which specifies the frequency of occurrence in the image of each gray-level value). However, such histograms are insensitive to permutations of pixel positions, and thus have limited utility.

Second-order statistics describe pixel pairs in fixed geometric relationships. For example, how many pixels in an image have a given gray level while another pixel, a given distance away along a given direction, has another given value? Such statistics can be computed and used to classify textures such as those corresponding to images of wood, corn, grass, and water. Edge pairs – perhaps requiring certain angular relationships – can also be used to represent textures. (A cloth with a herringbone weave illustrates this.) Texture characteristics can be used as part of a segmentation algorithm to represent properties of regions.

Color

Color can be represented by associating red, green, and blue values with each pixel. Intensity is then the average of these three values. Other color spaces, such as the CIE system or the hue–saturation–intensity system, are also in common use. Each has advantages for different applications.

A major challenge in color processing is to model color constancy, the fact that humans see object colors consistently under widely varying illumination. 'Physics-based' methods attempt to calculate the spectral power distribution of the illumination and the objects. The physics of how light interacts with objects, and how colors are reflected from surfaces, is important in the design of color processing algorithms. The appearance of a surface depends strongly on the type of illumination, the direction of the illuminant, and other ambient light reflected off nearby objects.

Color is often also used as an aid in image segmentation. One technique uses knowledge of object color to classify pixels. For example, in a scene containing a basket of fruit, yellow may indicate banana, red may indicate apple, and so on. The yellow pixels can be grouped together, and if they are in the proper geometric configuration, the region may represent a banana. (*See* **Color Vision, Philosophical Issues about**; **Color Vision, Neural Basis of**; **Color Perception, Psychology of**)

Shape from shading

A smooth opaque object produces an image with spatially varying brightness even if the object is illuminated evenly and is made of material with uniform optical properties. Shading thus provides essential information about object shape. A smooth yellow billiard ball, illuminated with a single light source, has a surface that appears increasingly shaded as the angle of incidence of the light from the perpendicular increases. This effect can be mathematically modeled, and the resulting equation, describing the amount of light emitted from the surface, is called the image irradiance equation. Surface orientation has two degrees of freedom, while brightness – what is measured by image irradiance – has only one. Therefore, one needs to add a constraint in order to solve the image irradiance equation. This constraint may come from an overall smoothness metric on the desired solution, or from the use of multiple images (a technique known as 'photometric stereo').

Shape from texture

If an object's surface is textured, the variation in texture can be used to determine surface shape. Imagine a large white cloth, patterned with blue polka dots. Drape the cloth over a surface so that it is not flat. You can easily deduce the shape of the cloth from the shapes of the polka dots and their variation from true circles in the image seen by your eye. Shape-from-texture algorithms use this variation. Uniformly textured surfaces undergo two types of distortions due to imaging: as a surface recedes from the observer, increasingly large areas of the surface are isotropically compressed onto a given area of the image; and as a surface tilts away from the frontal plane, foreshortening causes anisotropic compression of texture elements. The rate of change of the projected size of texture elements – called the texture gradient – constrains the orientation of the plane.

A surface orientation can be represented by the slant – the angle between the normal to the surface and the projection direction – and the tilt – the angle between the normal's orthographic projection onto the image plane and the x-axis. If we know that a surface has blue dots as texture, then the variation in appearance of those dots in an image can be expressed as a slant and tilt and used to represent changes of local orientation of the surface.

Stereoscopic disparity

Normally sighted humans have two eyes. If you hold a finger in front of you and look at it first with the right and then with the left eye, it appears to shift position. The amount of displacement is the horizontal 'disparity'. Your visual system compares the two images and relates the disparity to the depth of the object (in the simplest case, by solving for the triangle formed by your eyes and the object). If you observe a scene with both eyes (a convergent imaging system), there is a single point on which both eyes fixate. At that point, there is zero horizontal and vertical disparity. For convergent systems, vertical disparity is zero only on the $x = 0$ and $y = 0$ planes. Horizontal disparity is zero only on the circle which contains the fixation point and the two eyes. The major problem is how to match a feature in one eye's image with the same feature in the other eye's image (the correspondence problem). For example, if you view a white picket fence from a close-up position, so that the fence covers the entire field of view, and if you look with each eye separately, assuming there is a uniform scene behind the fence (say, blue sky, or a green lawn), then you do not have enough information to know where a given picket in either image lies.

The geometry of the binocular imaging system can help. Given a point in the left image, the corresponding point in the right image is constrained to lie within a line in the right image, called the epipolar line. (This can be derived using the perspective projection transformation from one image to the other image.) In general, three important constraints help with this problem: similarity of corresponding features in the two views; the viewing geometry which constrains corresponding features to lie on epipolar lines; and piecewise continuity of surfaces in scenes. In the picket fence example, if you are close enough to the fence, matching the edges of the pickets will solve the problem if you can make assumptions about the width of the pickets and your viewing distance. (*See* **Depth Perception**)

Motion

Movies are natural representations of changes in object position, shape, and so on, and humans have no difficulty interpreting the apparent motion in an image sequence as equivalent to real motion, if the images are taken close enough in time. The simplest strategy for computing motion from a sequence of images is to consider the differences between successive images. The problem is to determine which pairs of points correspond to each other from one image to the next. This is another correspondence problem. For example, if the image sequence depicts a car moving along a road, it is easy to see that the points corresponding to the door handle in one image belong to those of the same door handle in the second. What about points on a smooth surface? What about the points of light in a fireworks display? Which ones are in correspondence, and what is the algorithm that determines the correspondence? In general, there are six translation and rotation parameters to recover for an object in motion. From two perspective images in time sequence, a computer system can recover only relative translation and not absolute translation of objects in motion at the scene. If the depth of objects in the scene is known, then absolute translation can be recovered.

Velocity can be measured only perpendicular to the contour of the moving object (that is, only one component of the velocity vector can be measured). This is known as the aperture problem. If one can assume that all edges in an image belong to the same rigid surface patch and that motion has constant velocity on the patch, then one can determine the full velocity vector by solving for the correct velocity in the equations that minimize the error of the measured velocity relative to the correct velocity. The field of velocity vectors thus obtained is known as the optical flow associated with the apparent motion in an image sequence.

If an image sequence contains several moving objects as well as a background, and perhaps a moving camera system, the problem is more complicated. Suppose motion vectors could be defined along the contours and internal structures of each of the objects and of the background. This set of vectors will then contain 'groupings' for each motion: each object produces its own distribution of motion vectors. So in order to determine which groups of vectors correspond to single objects, a mixture of distributions made up of several separate motions can be proposed. The groups of vectors that best fit the distributions would describe single objects in motion. (*See* **Motion Perception, Neural Basis of**)

Representations

How a computer system represents images, extracted features, domain knowledge, and processing strategies is critical to its performance. A number of representational schemes have been proposed whose value has been proven in practice.

Image pyramids

In an image pyramid, an image is represented as a number of layers, each layer being an image. Successive images have smaller numbers of pixels representing the whole image. Each layer computes image properties (at successively coarser resolutions), and each computation communicates only with computations occurring in layers immediately above or below or with computations within the layer. This representation helps to reduce the amount of computation required: the coarser layers constrain the processing in the more detailed layers.

Intrinsic images

Intrinsic images form an intermediate level of representation, between images and object models. They consist of a number of separate feature maps that interact so that they can be computed unambiguously. The features they describe may include surface discontinuities, range, surface orientation, velocity, and color. These representations seem to be related to the processing stages in the primate cortex, where different kinds of computations seem to be grouped together in separate, but interacting, brain areas.

Image sketches and visual routines

Progressions of successively more abstract representations are often useful for coding complex visual information. One commonly-used sequence consists of the 'primal' sketch, the 'two-and-a-half-dimensional' sketch, and the 'three-dimensional' sketch. The primal sketch represents intensity changes and their organization in a two-dimensional image. The two-and-a-half-dimensional sketch represents the orientation and depth of surfaces, and discontinuity contours. Finally, the three-dimensional sketch represents shapes and their spatial organization.

Given the large number of feature types that might be computed on an image, and the representations that arise from them, how can a vision system integrate this information in a meaningful manner? The idea of visual routines addresses this question. These routines extract abstract shape properties and spatial relations from the early representations. Shape properties are characteristics of single items (e.g. length, orientation, area), and spatial relations are characteristics of two or more items (e.g. 'above', 'inside', 'longer than'). Shape properties and spatial relations are important for object recognition, visually guided manipulation, and abstract visualizations. Using a fixed set of basic operations, the visual system might assemble different visual routines to extract a variety of shape properties and spatial relations. Several specific operations have been proposed, including shifts of attentional focus, indexing to an 'odd man out' location, bounded activation, boundary tracing, and marking.

Object representations

Vision systems require the explicit representation of points, curves, surfaces, and volumes. There are a number of schemes that are employed, but there is no consensus yet on what constitutes an adequate set of primitives for spatial representations. Several popular methods exist for representing an object, including 'generalized cylinders', 'deformable models', 'geons', and 'aspect graphs'.

Strategies for Recognition

Several methods have proved useful for recognition of objects in images. The three most important are model-based, invariant-based, and appearance-based methods. The problem of recognition remains unsolved in general. At an abstract level, recognition of objects in an image given a database of object types seems solvable in principle. One needs only to locate groups of features that might be objects, and then match each group to a model in the database. This is a hypothesize-and-test strategy: you first make a hypothesis, then test it; if it is false, you refine it and test it again, and so on until you are successful. If the image object is an instance of one of the models, then this procedure will find it.

This is a search task. However, a problem common to many search tasks arises: if it is not known in advance which parts of an image match with which parts of a model in a database, then all possible combinations of pixels must be checked, which is computationally infeasible.

Selective attention is an important mechanism for dealing with this problem. A visual attention mechanism can help with the selection of a region of interest in the visual field, the selection of feature dimensions and values of interest, the shifting from one selected region to the next, the integration of successive attentional fixations, interactions with memory, indexing into model bases, etc. (*See* **Vision, High-level**; **Object Perception, Neural Basis of**; **Vision: Object Recognition**; **Audition, Neural Basis of**; **Spatial Attention, Neural Basis of**; **Selective Attention**; **Vision: Top-down Effects**)

Model-based recognition

It is common for systems to employ databases of models to help with recognition. Important subtasks in recognition include deciding which models in the database have instances in the image, where those instances are, and their orientations.

Recognition can be viewed as a process that proceeds from the general to the specific and that overlaps with, guides, and constrains the derivation of a description from the image. For example, a database of models can be constructed using volumetric primitives and organized using a specialization hierarchy as well as a decomposition hierarchy. Models can be selected according to the characteristics of the extracted volumetric primitives.

Interpretation trees represent methods that use features of objects for recognition. Each model in the database is described by its component features, and the description must include the constraints between features. The interpretation tree enumerates all possible ways in which a given set of features found in an image can be matched with features of particular objects in the model base. Methods for searching this tree focus on limiting search so that not all of the tree need be examined (the trees can grow exponentially). An assignment can be verified by back-projecting the resulting model into the image (using the positions of features, and modeling the image formation process to see whether the object actually looks like the one in the image).

Much work has been done on understanding scenes with polyhedral objects. Edges are found and connected, and the resulting sequences labeled as concave or convex with respect to the polyhedron; shadows, occluding boundaries, and so on are also hypothesized. Labeled edges can then be grouped into hypothesized whole objects, and these hypotheses can be confirmed within an interpretation tree.

Invariant-based recognition

Invariants are properties of geometric configurations that do not change under certain transformations. For example, the length of an edge of a solid, rigid object is an invariant property of that object. Suppose several such invariants for a set of objects are defined. The database of objects can then have a feature vector associated with each model, where the features of the vector are all invariants. These features can be used as indices into the database. For example, if objects have planar sides and lines are straight or simple conics, then sets of contours may participate as invariant features under projective transformation.

Appearance-based recognition

Images themselves can also be considered as features. In order to make such features usable, several images of each object, taken from different viewpoints (and illumination conditions, if relevant), are needed. Object models in a database are defined by such a set of images. Face recognition is a particularly good application of this method.

Suppose you have a set of pictures, say, passport photographs. The size of the face in each is about the same; there is only a single face in each; and the lighting is approximately the same in each. Suppose now that you compute the average image: add up the pixel values at each location from all the pictures and divide by the number of pictures. Each picture can then be redefined as the sum of this average picture and a representation of how the image differs from the average in terms of a relatively small set of 'principal components' of the image set. This provides an economical way of representing the images. For any new picture, you compute this representation. Representations can be compared: if the new image is sufficiently close to one of the stored images, then they are taken to represent the same person.

Choosing a recognition strategy

In each of the above methods, the computer system must include representations that formalize the knowledge of a domain. Performance depends critically on the representation, the control strategy that uses this knowledge, and the quality of the knowledge. Only if the knowledge of how to solve the task can be extracted, codified in a suitable formalism, and used in drawing conclusions during processing, will the method work. The choice of recognition strategy thus depends on the kinds and qualities of domain knowledge available. If only very general knowledge of the domain is possible then an invaraint strategy might be the best choice. If explicit models are known, then model-based methods are preferable. If control over imaging conditions and scene positioning is possible, then appearance-based methods may be appropriate. The methods may also be combined.

The role of knowledge and its application to guide processing is critical. It can be shown that basic problems in vision such as matching are potentially intractable if no knowledge is used to guide processing: that is, there is no guarantee that those problems can be solved using any existing or future computational resources. The intractability is due solely to the difficulty of selecting which parts of the input image are to be processed: without knowledge, the number of such image subsets increases exponentially.

Task guidance can be implicit (as in positioning a person so that the face is imaged in a particular way) or explicit (as in annotating interesting portions of a scene by hand before computer processing). Attentional selection, using knowledge to optimize processing, may determine which image parts to attempt to process first; if the first few selections are good ones, a great deal of searching can be avoided.

CONCLUSION

Although computer vision has been an active discipline since the early 1960s, progress has been slow. The 1990s saw some promising developments, in part due to the advent of sufficiently powerful computers and the resulting ability to search through more possibilities and to test more complex theories. Computer vision is now widely used in commercial applications. Still, progress on finding theories that address the capabilities of human vision remains slow. This represents the greatest intellectual challenge in the field.

Further Reading

Faugeras O (1993) *Three-Dimensional Computer Vision*. Cambridge, MA: MIT Press.

Fischler MA and Firschein O (1987) *Readings in Computer Vision: Issues, Problems, Principles and Paradigms*. Los Altos, CA: Morgan Kaufmann.

Marr D (1982) *Vision*. San Francisco, CA: WH Freeman.

Tsotsos JK (1990) Analyzing vision at the complexity level. *Behavioral and Brain Sciences* **13**: 423–445.

Tsotsos JK (1992) Image understanding. In: Shapiro S (ed.) *Encyclopedia of Artificial Intelligence*, 2nd edn, pp. 641–663. New York, NY: John Wiley.

Zucker SW (1992) Early vision. In: Shapiro S (ed.) *Encyclopedia of Artificial Intelligence*, 2nd edn, pp. 394–420. New York, NY: John Wiley.

Computers

See **Human–Computer Interaction**

Concept Learning

Introductory article

Bradley C Love, University of Texas, Austin, Texas, USA

CONTENTS

Concept learning is the process of acquiring knowledge structures that enable an agent to make predictive inferences.

INTRODUCTION

The human species evolves to meet challenges in the environment. Unfortunately, evolution is a slow 'learning' process. Evolution can only help us address aspects of our environment that are not very variable and that are stable over a long period of time. Of course, many aspects of our environment are constantly undergoing change. Accordingly, many concepts have to be learned *de novo* by each individual. For example, a radiologist is not born knowing how to interpret x-ray images. It is hard to imagine how that particular skill could evolve.

Concept learning is integral to the survival of any agent (e.g. a human, an animal, a robot, etc.) operating in a complex and changing environment. A concept is a mental representation that is often derived from experiences with specific instances. We often develop concepts of categories (i.e. collections of objects) in the world. Without acquired concepts, we would be unable to make sense of the world around us. Every new object encountered would appear completely novel and we would not know how to interact with it. For example, the first time a child encounters a hot stove he may get burned. When the child visits a friend's house and encounters another stove, it is unlikely the child will touch it, even though the new stove may differ in a number of ways from the original stove (e.g. size, color, design, etc.). If the child did not generalize from his experiences and form a concept of stoves, he would go through life with burned hands. (*See* **Categorization, Development of**; **Generalization**)

One basic question is how do we learn new concepts? Philosophers, psychologists, and computer scientists have all pondered this question. In the following sections, three basic views (i.e. models) of concept learning and concept representation (i.e. what is stored as a consequence of learning) will be examined. The first account posits that concepts consist of rules. A more recent account holds that concepts are represented as prototypes. A prototype can be thought of as the average example of a concept. A third account of concepts is the exemplar view. The exemplar view holds that concepts are nothing more than a collection of stored exemplars (i.e. examples of the concept). We will evaluate the relative merits of each of these accounts of human concept learning. All three accounts correctly characterize some aspects of human concept learning. After evaluating these three accounts, we will discuss more modern neural network models of concept learning. Neural network models embody some of the characteristics of rule, prototype, and exemplar approaches. (*See* **Concept Learning and Categorization: Models**; **Classifier Systems**; **Concepts, Philosophical Issues about**; **Conceptual Representations in Psychology**)

RULES

The classical view of concepts holds that categories are defined by logical rules. In Figure 1, any item that is a square is a member of category A. This simple rule determines category membership. According to the rule view, our concept of category A can be represented by this simple rule. Discovering this rule would involve a rational hypothesis-testing procedure. This procedure attempts to discover a rule that is satisfied by all of the positive examples of a concept, but none of the negative examples of the concept (i.e. items that are members of other categories). In trying to come up with such a rule for category A, one might first try the rule 'if dark, then in category A'. After rejecting this rule (because there are counterexamples), other rules would be tested (starting with simple rules and progressing towards more complex rules) until the correct rule is eventually discovered. For example, in learning about birds, one might first try the rule 'if it flies, then it is a bird.' This rule works pretty well, but not perfectly (penguins do not fly and bats do). Another simple rule like 'if it has feathers, then it is a bird' would not work either because a pillow filled with feathers is not a bird. Eventually, a more complex rule might be discovered like 'if it has feathers and lays eggs, then it is a bird'.

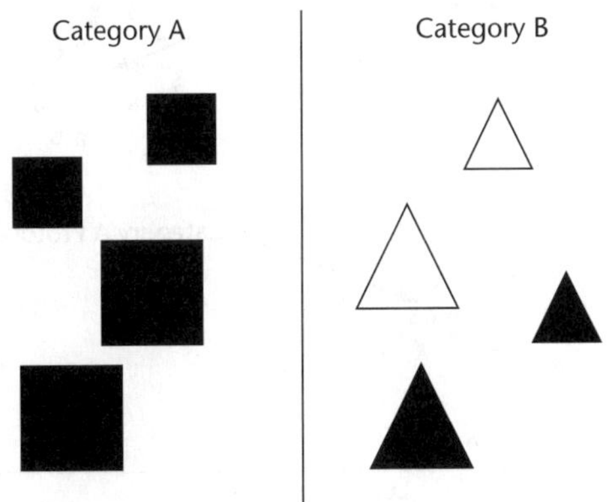

Figure 1. Examples of category A and category B.

Although rules can in principle provide a concise representation of a concept, often more elaborate representations would serve us better. Concept representation needs to be richer than a simple rule because we use concepts for much more than simply classifying objects we encounter. For instance, we often use concepts to support inference (e.g. a child infers members of the category stove can be dangerously hot). Using categories to make inferences is a very important use of concepts. Knowing something is an example of a concept tells us a great deal about the item. For example, if you can classify a politician from the USA as a Republican, you can readily infer the politician's position on a number of issues. The point is that our representations of concepts need to include information beyond what is needed to classify items as examples of the concept. For example, the rule 'if square, then in category A' correctly classifies all members of category A in Figure 1, but it does not capture the knowledge that all category A members are dark. One problem with rule representations of concepts is that potentially useful information is discarded.

The biggest problem with the rule approach to concepts is that most of our everyday categories do not seem to be describable by a tractable rule. To demonstrate this point, Wittgenstein noted that the concept game lacks a defining property. Most games are fun, but Russian roulette is not fun. Most games are competitive, but ring around the roses is not competitive. While most games have characteristics in common, there is not a rule that unifies them all. Rather, we can think of the members of the category game as being organized around a *family resemblance* structure (analogous to how members of your family resemble one another).

A related weakness of the rule account of concepts is that examples of a concept differ in their *typicality*. If all a concept consisted of was a rule that determined membership, then all examples should have equal status. According to the rule account, all that should matter is whether an item satisfies the rule. Our concepts do not seem to have this definitive flavor. For example, some games are better examples of the category game than others. Basketball is a very typical example of the category game. Children play basketball in a playground, it is competitive, there are two teams, each team consists of multiple players, you score points, etc.

Basketball is a typical example of the category of games because it has many characteristics in common with other games. On the other hand, Russian roulette is not a very typical game – it requires a gun and one of the two players dies. Russian roulette does not have many properties in common with other games. In terms of family resemblance structure, we can think of basketball as having a central position and Russian roulette being a distant cousin to the other family members. These findings extend to categories in which a simple classification rule exists. For example, people judge the number three to be a more typical odd number than the number forty-seven even though membership in the category 'odd number' can be defined by a simple rule.

The fact that category membership follows a gradient as opposed to being all or none affords us flexibility in how we apply our concepts. Of course, this flexibility can lead to ambiguity. Consider the concept mother. It is a concept that we are all familiar with that seems straightforward – a mother is a woman who becomes pregnant and gives birth to a child. But what about a woman who adopts a neglected infant and raises it in a nurturing environment? Is the birth mother who neglected the infant a mother? What if a woman is implanted with an embryo from another woman? Court cases over maternity arise because the concept of motherhood is ambiguous. The concept exhibits greater flexibility and productivity than is even indicated above. For example, is it proper to refer to an architect as the mother of a building? All the above examples of the concept mother share a family resemblance structure (i.e. they are organized around some commonalities), but the concept is not rule based. Some examples of the concept mother are better than others.

PROTOTYPES

The prototype approach to concept learning and representation was developed by Rosch and colleagues to address some of the shortcomings of the rule approach. Prototype models represent information about all the possible properties (i.e. *stimulus dimensions*), instead of focusing on only a few properties like rule models do. The prototype of a category is a summary of all of its members. Mathematically, the prototype is the average or central tendency of all category members. Figure 2 displays the prototypes for two categories, simply named categories A and B. Notice that all the items differ in size and luminance (i.e. there are two stimulus dimensions) and that the prototype is located amidst all of its category members. The prototype for each category has the average value of both the stimulus dimensions of size and luminance for the members of its category. (*See* **Prototype Representations**)

The prototype of a category is used to represent the category. According to the prototype model, a novel item is classified as a member of the category whose prototype it is most similar to. For example, a large bright item would be classified as a member of category B because category B's prototype is large and bright (see Figure 2). The position of the prototype is updated when new examples of the category are encountered. For example, if one encountered a very small and dark item that is a member of category A, then category A's prototype would move slightly towards the bottom left corner in Figure 2. As an outcome of learning, the position of the prototype shifts towards the newest category member in order to take it into account. A prototype can be very useful for determining category membership in domains where there are

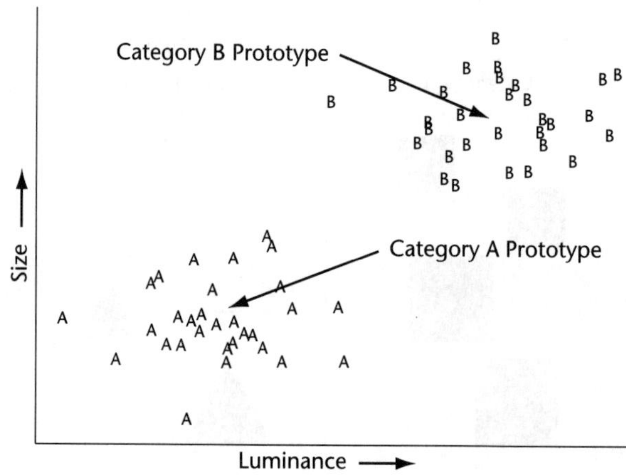

Figure 2. Two categories and their prototypes.

many stimulus dimensions that each provide information useful for determining category membership, but no dimension is definitive. For example, members of a family may tend to be tall, have large noses, a medium complexion, brown eyes, and good muscle tone, but no family member possesses all of these traits. Matching on some subset of these traits would provide evidence for being a family member. (*See* **Multidimensional Scaling; Similarity**)

Notice the economy of the prototype approach. Each cloud of examples in Figure 2 can be represented by just the prototype. The prototype is intended to capture the critical structure in the environment without having to encode every detail or example. It is also fairly simple to determine which category a novel item belongs to by determining which category prototype is most similar to the item.

Unlike the rule approach, the prototype model can account for typicality effects. According to the prototype model, the more typical category members should be those members that are most similar to the prototype. In Figure 2, similarity can be viewed in geometric terms – the closer items are together in the plot, the more similar they are. Thus, the most typical items for categories A and B are those that are closest to the appropriate prototype. Accordingly, the prototype approach can explain why robins are more typical birds than penguins. The bird prototype represents the average bird: has wings, has feathers, can fly, can sing, lives in trees, lays eggs, etc. Robins share all of these properties with the prototype, whereas penguins differ in a number of ways (e.g. penguins cannot fly, but do swim). Extending this line of reasoning, the best example of a category should be the prototype, even if the actual prototype has never been viewed (or does not even exist). Indeed, numerous learning studies support this conjecture. After viewing a series of examples of a category, human participants are more likely to categorize the prototype as a category member (even though they never actually viewed the prototype) than they are to categorize an item they have seen before as a category member.

Because the prototype approach does not represent concepts in terms of a logical rule that is either satisfied or not, it can explain how category membership has a graded structure that is not all or none. Some examples of a category are simply better examples than other examples. Also, categories do not need to be defined in terms of logical rules, but are rather defined in terms of family resemblance to the prototype. In other words, members of a category need not share a common defining thread, but rather can have many characteristic threads in common with one another.

The prototype approach, while preferable to the rule approach for the reasons just discussed, does fail to account for important aspects of human concept learning. The main problem with the prototype model is that it does not retain enough information about examples encountered in learning. For instance, prototypes do not store any information about the frequency of each category, yet people are sensitive to frequency information. If an item was about equally similar to the prototype of two different categories and one category was one hundred times larger than the other, people would be more likely to assign the item to the more common category (under most circumstances).

People are also sensitive to the variability along stimulus dimensions. To use Rips' example, a circular object with a 10 cm diameter may be more similar to a US quarter (which is about 2.5 cm in diameter) than to a pizza (which is much larger). Nevertheless, the novel object is more likely to be classified as a pizza than a quarter because quarters display very little variability in their diameters whereas pizzas can vary in size.

Finally, prototypes are not sensitive to the correlations and substructure within a category. For example, a prototype model would not be able to represent that spoons tend to be large and made of wood or small and made of steel. These two subgroups would simply be averaged together into one prototype. This averaging makes some categories unlearnable with a prototype model. One example of such a category structure is shown in Figure 3. Each category consists of two subgroups. Members of category A are either small and dark

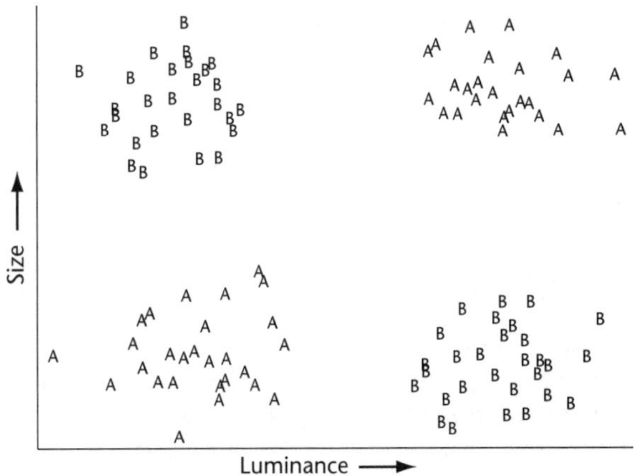

Figure 3. Two categories, each containing two subgroups.

or they are large and light, whereas members of category B are either large and dark or they are small and light. The prototypes for the two categories are both in the centre of the stimulus space (i.e. medium size and medium luminance). Items cannot be classified correctly by which prototype they are most similar to because the prototypes provide little guidance.

In general, prototype models can only be used to learn category structures that are *linearly separable*. A learning problem involving two categories is linearly separable when a line or plane can be drawn that separates all the members of the two categories. The category structure shown in Figure 2 is linearly separable because a diagonal line can be drawn that separates the category A and B members (i.e. the category A members fall on one side of the line and the category B members fall on the other side of the line). Thus, this category structure can be learned with a prototype model. The category structure illustrated in Figure 3 is nonlinear – no single line can be drawn to segregate the category A and B members. Mathematically, a category structure is linearly separable when there exists a weighting of the feature dimensions that yields an additive rule that correctly indicates one category when the sum is below a chosen threshold and the other category when the sum is above the threshold.

The inability of the prototype model to learn nonlinear category structures detracts from its worth as a model of human concept learning because people are not biased against learning nonlinear category structures. Some nonlinear category structures are actually easier to acquire than linear category structures. For example, it seems quite natural that small birds sing, whereas large birds do not sing. Many categories have subtypes within them that we naturally pick out. One way for the prototype model to address this learnability problem is to include complex features that represent the presence of multiple simple features (e.g. large and blue). Unfortunately, this approach quickly becomes unwieldy as the number of stimulus dimensions increases.

EXEMPLARS

Exemplar models address many of the shortcomings of the prototype model. Exemplar models store every training example in memory instead of just the prototype (i.e. the summary) of each category. By retaining all of the information from training, exemplar models are sensitive to the frequency, the variability, and the correlations among items. For the learning problem illustrated in Figure 2, an exemplar model would store every training example. New items are classified by how similar they are to all items in memory (not just the prototype). For the category structure illustrated in Figure 2, the pairwise similarity of a novel item and every stored item would be calculated. If the novel item tended to be more similar to the category A members (i.e. the item was small and dark) than the category B members, then the novel item would be classified as a member of category A.

One aspect of exemplar models that seems counterintuitive is their lack of any abstraction in category representation. It seems that humans do learn something more abstract about categories than a list of examples. Surprisingly, exemplar models are capable of displaying abstraction. For instance, exemplar models can correctly predict that humans more strongly endorse the underlying prototype (even if it has not been seen) than an actual item that has been studied (a piece of evidence previously cited in favor of the prototype model). How could this be possible without the prototype actually being stored? It would be impossible if exemplar models simply functioned by retrieving the exemplar in memory that was most similar to the current item and classified the current item in the same category as the retrieved exemplar (this is essentially how processing works in a prototype model, except that a prototype is stored in memory instead of a bunch of exemplars).

Instead, exemplar models engage in more sophisticated processing and calculate the similarity between the current item (the item that is to be classified) and every item in memory. Some exemplars in memory will be very similar to the current item, whereas others will not be very similar. The current item is classified in the category in which the sum of its similarities to all the exemplars is greatest. When a previously unseen prototype is presented to an exemplar model it can be endorsed as a category member more strongly than a previously seen item. The prototype (which is the central tendency of the category) will tend to be somewhat similar to every item in the category, whereas any given non-prototype item will tend to be very similar to some items (especially itself!) in memory, but not so similar to other items. Overall, the prototypical item can display an advantage over an item that has actually been studied. Abstraction in an exemplar model is indirect and results from processing (i.e. calculating and summing pairwise similarities), whereas abstraction in a prototype model is rather direct (i.e. prototypes are stored).

The exemplar model does seem to make some questionable assumptions. For example, exemplar models store every training example which seems excessive. Also, every exemplar is retrieved from memory every time an item is classified. In addition to these assumptions, one worries that the exemplar model does not make strong enough theoretical commitments because it retains all information about training and contains a great deal of flexibility in how it processes information. These issues are currently being resolved by researchers. On the whole, exemplar models seem to be a more viable approach to understanding human concept learning than existing prototype or rule-based approaches, but there is still room for further work. (*See* **Computational Models of Cognition: Constraining**)

NEURAL NETWORK MODELS

Neural network models are intended to learn in a manner analogous to how the brain learns. A neural network consists of layers of neuron-like units that connect to units in other layers. Units can excite and inhibit one another across these connections. An item is represented at the input layer (the first layer) and passes activity to more advanced layers in the network until it reaches the output layer which determines the category the item is a member of (e.g. if the unit in the output layer representing category B is the most activated, then the item is classified as a member of category B). Each unit integrates all the activity originating from the layer below via its connections and passes this summed activity through a transfer function to generate its own output which is passed on to the next layer. Figure 4 illustrates a feedforward neural network with an input, hidden, and output layer. (*See* **Connectionism**)

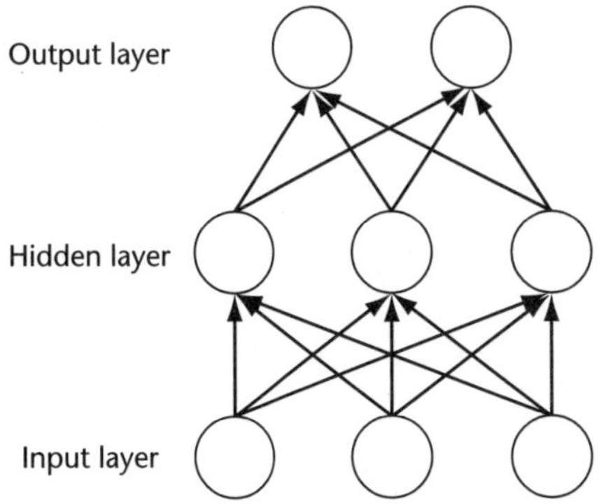

Figure 4. A typical feedforward neural network.

The connections between units are altered as a result of learning in order to minimize the prediction error (i.e. the weights are altered in order to correctly classify items). Sophisticated learning algorithms dictate how the weights should be altered as a result of learning. Neural networks with only an input and output layer share many of the limitations of the prototype model – they can only learn linearly separable functions (i.e. simple category structures). More complicated neural networks with a hidden layer (and nonlinear transfer functions) can learn just about any category structure. However, neural networks of this variety are not very good models of human concept learning because they tend to learn problems quickly that people learn slowly and vice versa.

Neural network models that are conceptually related to rule, prototype, and exemplar models have been successful as models of human concept learning. For example, the ALCOVE model replaces the hidden layer in Figure 4 with encoded exemplars. In other words, units in the hidden layer are added as exemplars are encountered. This exemplar neural network model, which combines an exemplar representation of concepts with the powerful learning algorithms of neural networks, does a good job of accounting for aspects of human concept learning. The SUSTAIN model is a neural network model that combines aspects of both exemplar and prototype models. SUSTAIN initially begins like a prototype model, but it can store exemplars (which themselves can later evolve into prototypes) when prediction errors occur. For the problem illustrated in Figure 3, SUSTAIN would form four prototypes that correspond to the four clusters of items. The ability to store multiple prototypes per category allows SUSTAIN to avoid the problems that plague prototype models. Both ALCOVE and SUSTAIN also incorporate rule-like dynamics. These models learn to attend to the most relevant stimulus dimensions and neglect the less meaningful dimensions, much like how rule models tend to focus on a limited number of stimulus dimensions (e.g. if it is *large*, then it is in category A).

CONCLUSIONS AND FUTURE DIRECTIONS

From this brief review of concept learning models we saw that the progression from rule models to prototype models to exemplar models was marked by a shift towards more concrete representations (i.e. more information about the training examples is retained), greater fluidity (i.e. category boundaries are not seen as rigid), and more sophisticated processing at decision time (exemplar models are the quintessential case – all abstraction is done after the training examples are encoded). Although all three approaches have their shortcomings, they all reflect some aspects of human concept learning. The successful neural network models of concept learning retain characteristics of all three approaches. Like the rule approach, these neural network models acknowledge the utility of strategically focusing on a subset of stimulus dimensions. If a stimulus dimension is irrelevant to a learning problem, the models will ignore the dimension and not be distracted by it. Like prototype models, some of these neural network models form abstractions which can assist generalization and reduce storage requirements. Like exemplar models, these neural network models are quite fluid, can encode individual exemplars, and engage in sophisticated processing at decision time.

One important aspect of concept learning that these models do not address is the influence of prior knowledge. Our prior knowledge exhibits strong influences on what we learn from a series of examples. For example, even if all the blue cars on a mechanic's lot have transmission problems and none of the red cars do, the mechanic would never predict that blue cars in general have transmission problems. Certainly, the mechanic would not paint a car red in the hope of repairing it. The mechanic's prior knowledge and theories of how cars function preclude this association. Instead, the mechanic is oriented towards more fruitful solutions. One important challenge for concept learning models is to illuminate how prior knowledge affects our interpretation of examples. Conversely, more work is needed in understanding how examples we encounter affect our theories of the world.

Further Reading

Lakoff G (1987) *Women, Fire, and Dangerous Things: What Categories Tell Us About the Nature of Thought*. Chicago, IL: University of Chicago Press.

Medin DL (1998) Concepts and conceptual structure. In: Thagard P (ed.) *Mind Readings*, pp. 93–126. Cambridge, MA: MIT Press.

Mervis CB and Rosch E (1981) Categorization of natural objects. In: Rosenzweig MR and Porter LW (eds) *Annual Review of Psychology* **32**: 89–115.

Rumelhart D (1989) The architecture of mind: a connectionist approach. In: Posner MI (ed.) *Foundations of Cognitive Science*, pp. 133–159. Cambridge, MA: MIT Press.

Wisniewski EJ (in press) Concepts and categorization. In: Medin DL (ed.) *The Steven's Handbook of Experimental Psychology*. New York, NY: John Wiley and Sons.

Concept Learning and Categorization: Models

Intermediate article

John K Kruschke, Indiana University, Bloomington, Indiana, USA

CONTENTS

Category learning involves generalizing from one learned case to another in appropriate ways. Models of category learning have been based on various representations, including exemplars, prototypes, rules, and hybrids thereof.

CATEGORIZATION IN COGNITION

Categories pervade our cognition. We classify variously shaped printed squiggles into different letter categories. We classify a spectrum of acoustic signals into phonemic categories. We categorize people, animals, plants, and artefacts, and we base our actions on how we categorize.

A central function of learned categories is generalizing from a particular learned instance to novel situations. If learned knowledge consisted merely of isolated facts with no generalization, then the knowledge would be inapplicable except for the unlikely exact recurrence of the learned situation. For example, learning that a four-legged, striped, 1.0-meter-tall animal is a tiger would not generalize to inferring that a four-legged, striped, 1.1-meter-tall animal is also a tiger. The consequence of this failure to generalize a category could be an eaten learner. At the opposite extreme, if learned knowledge were to generalize too broadly, then complementary errors could be committed: learning that a four-legged, striped, 1.0-meter-tall animal is a tiger would lead to inferring that a zebra is also a tiger. The consequence of this over-generalization could be a starved learner.

An equally crucial goal of learning categories is retaining previously learned knowledge while quickly acquiring new knowledge. For example, after having learned about zebras, it could prove disastrous if learning about tigers required dozens of exposures. It could also be disastrous if the learning about tigers erased valid knowledge about zebras.

Category learning is critically important because it underlies essentially all cognitive activities; yet it is very difficult because: (1) learned categories must generalize appropriately; (2) learning must occur quickly; and (3) new learning must not overwrite previous knowledge. Understanding how learners accomplish these feats is the topic of this article.

Any theory of category learning must specify: (1) what information from the world is actually retained in the mind; (2) how that information is used and learned; and (3) why that particular learning algorithm is useful. These three issues are addressed in turn, for different theories. Each type of theory is initially described informally, to convey the basic motivating principles of the theory. It is then described in formal, mathematical terms. By being expressed mathematically, the theory gains: quantitative precision rather than vague verbal description; publicly derivable predictions rather than theorist-dependent intuitively derived predictions; stronger support when predictions are confirmed in quantitative detail; greater explanatory power when the formal mechanisms in the model have clear psychological interpretations; and greater applicability because of precise specification of relevant factors.

EXEMPLAR THEORIES

Perhaps the simplest way to learn is just to memorize the experienced instances. For example, a learner's knowledge of the category *dog* might consist of knowing that the particular cases named Lassie, Rin-Tin-Tin, Old Yeller and Pongo are exemplars of dogs. There is no derived representation of a prototypical dog, nor is there any abstracted set of necessary and sufficient features that define what a dog is. As new cases of dogs are experienced, these cases are also stored in memory. Notice, however, that just because these exemplars of dogs are in memory, the learner need not be able to distinctly recall every dog ever encountered. Retrieving a memory might be quite different from using it for categorization.

According to these exemplar theories of categorization, a new stimulus is classified according to how similar it is to all the known instances of the various candidate categories. For example, a newly encountered animal is classified as a dog if it is more similar to known exemplars of dogs than it is to known exemplars of cats or horses, etc. The notion of similarity, therefore, plays a critical role in exemplar theories.

Selective attention also plays an important role in exemplar theories. Not all features are equally relevant for all category distinctions. For example, in deciding whether a novel animal is a dog or a cat, it might be more important to pay attention to size than to number of legs, because dogs and cats tend be of different sizes, but have the same number of legs.

In principle, exemplar encoding can accurately learn any possible category structure, no matter how complicated, because the exemplars in memory directly correspond with the instances in the world. This computational power of exemplar models is one rationale for their use. On the other hand, the uniform application of exemplar encoding can make learning slow, if highly similar instances belong to different categories. One way around this problem is to associate exemplars with categories only to the extent that doing so will improve accuracy of categorization. Analogously, features or stimulus dimensions may be attended to only to the extent that doing so will reduce error. Error reduction is one rationale for theories of learning.

Formal Models of Exemplar Theories

In a prominent exemplar-based model (Kruschke, 1992; Medin and Schaffer, 1978; Nosofsky, 1986), a stimulus is represented by its values on various psychological dimensions. For example, a tiger might be represented by a large numerical value on the dimension of size, and by another large numerical value on the dimension of ferocity, along with other values on other dimensions. The psychological value on the dth dimension is denoted $\psi_d{}^{\text{stim}}$. For the mth exemplar in memory, the psychological value on the dth dimension is denoted $\psi_{md}{}^{\text{ex}}$. These psychological scale values can be determined by methods of multidimensional scaling (e.g., Kruskal and Wish, 1978).

The similarity of the stimulus to a memory exemplar gets larger as the distance between the stimulus and the exemplar in psychological space gets smaller. For psychological dimensions that can be selectively attended to, the usual measure of distance between the stimulus, s, and the mth memory exemplar is given by $\text{dist}(s, m) = \sum_i \alpha_i |\psi_i{}^{\text{stim}} - \psi_{mi}{}^{\text{ex}}|$, where the sum is taken over the dimensions indexed by i, and $\alpha_i \geq 0$ is the attention allocated to the ith dimension. When attention on a dimension is large, then differences on that dimension have a large effect on the distance, but when attention on a dimension is zero, then differences on that dimension have no effect on the distance. The distance is then converted to similarity by an exponentially decaying function: $\text{sim}(s, m) = \exp(-\text{dist}(s, m))$. Therefore, when the stimulus exactly matches the memory exemplar, the similarity is 1.0, and as the distance between the stimulus and the memory exemplar increases, the similarity decreases towards zero. (Shepard (1987) provides a review of the properties of the exponential similarity function.)

Each exemplar then 'votes' for the categories. The strength of an exemplar's vote is its similarity to the stimulus, and the exemplar's selection of categories is a continuous weighting given by its associative strengths to the categories. The associative strength from exemplar m to category k is denoted w_{km}, and the total 'voting' for category k is $v_k = \sum_m w_{km}\text{sim}(s, m)$. The overall probability of classifying the stimulus into category k is the total vote for category k relative to the total of votes cast. Formally, the probability of classifying stimulus s into category k is given by $p_k = v_k / \sum_c v_c$.

In laboratory experiments on category learning, after the learner makes his or her guess as to the correct categorization of a given stimulus, he or she is given corrective feedback, and then tries to learn this correct answer. The same procedure applies to learning in the model. The model adjusts its associative weights and attention strengths to reduce the error between its vote and the correct answer. Error is defined as $E = \sum_k (t_k - v_k)^2$, where t_k is the 'teacher' value: $t_k = 1$ if k is the correct category, and $t_k = 0$ otherwise. There are many possible methods by which the associative weights and attention strengths could be adjusted to reduce this error, but one sensible method is 'gradient descent' on error. According to this procedure, the changes that make the error decrease most rapidly are computed according to the derivative of the error with respect to the associative weights and attention strengths. The resulting formula for weight changes is $\Delta w_{km} = \lambda(t_k - v_k)\ \mathrm{sim}(s, m)$, where λ is a constant of proportionality called the learning rate. This formula states that the associative weight between exemplar m and category k increases to the extent that the exemplar is similar to the current input and the category teacher is under-predicted. Notice that after the weight changes according to this formula, the predicted category will be closer to the correct category; i.e., the error will have been reduced. The formula for attentional changes is slightly more complicated, but essentially it combines information from all the exemplars to decide whether attention on a dimension should be increased or decreased (Kruschke, 1992; Kruschke and Johansen, 1999).

Variants of this exemplar model have been shown to fit a wide range of phenomena in category learning and generalization (e.g., Choi *et al.*, 1993; Estes, 1994; Kruschke and Johansen, 1999; Lamberts, 1998; Nosofsky and Kruschke, 1992; Nosofsky, Gluck *et al.*, 1994; Nosofsky and Palmeri, 1997; Palmeri, 1999).

PROTOTYPE THEORIES

Instead of remembering every exemplar of a category, the learner might construct a representation of what is typical of the category. For example, the mental representation of *dog* might be an average of all the experienced instances. The dog prototype need not necessarily correspond to any actually experienced individual dog. Alternatively, the representative summary could be an idealized caricature or extreme case that is maximally distinct from other categories, rather than the central tendency of the category.

According to prototype theories of categorization, a new stimulus is classified according to how similar it is to the prototypes of the various candidate categories. A newly encountered animal is classified as a dog if it is more similar to the dog prototype than it is to other category prototypes.

One rationale for this approach to categorization is that it is efficient: the entire set of members in a category is represented by just the small amount of information in the prototype.

Formal Models of Prototype Theories

Prototypes can be formally described in a similar way to exemplars. The prototype for category k has psychological value on dimension i denoted by ψ_{ki}^{proto}, and this value represents the central tendency of the category instances on that dimension. The model classifies a stimulus as category k in a manner directly analogous to the exemplar model, so that the probability of classifying stimulus s as category k is given by $p_k = \mathrm{sim}(s,k)/\sum_m \mathrm{sim}(s,m)$. The sum in the denominator is over all category prototypes, instead of over all exemplars.

In one kind of prototype model, each prototype must be tuned to represent the central tendency of the instances in its category. For the first experienced instance of a category, the prototype is created and set to match that instance. For subsequently experienced instances of the category, the prototype changes from its current values slightly towards those of the new case. By moving towards the instances of the category as they are experienced, the prototype gradually progresses towards the central tendency of the instances.

One way of formalizing the learning of central tendencies is the following algorithm, closely related to so-called 'competitive learning' or 'clustering' methods. The idea is that a prototype should be adjusted so that it is as similar as possible to as many instances as possible; in this way the prototype is maximally representative of the stimuli in its category. Define the total similarity of the prototypes to the instances as $S = \sum_{k,s} \mathrm{sim}(s,k)$, where $\mathrm{sim}(s,k) = \exp(-\sum_i \alpha_i[\psi_i^{\text{stim}} - \psi_{ki}^{\text{proto}}]^2)$. (This summation across all instances does not require that all the instances be stored in memory, nor that the instances be simultaneously available for learning.) The question then is how best to adjust ψ_{ki}^{proto} so that the total similarity increases. One way to do this is gradient ascent: the prototype values are adjusted to increase the total similarity as quickly as possible. The resulting formula, determined as the derivative of the total similarity with respect to the coordinates, yields $\Delta\psi_{ki}^{\text{proto}} = \lambda\ \mathrm{sim}(s,k)\alpha_i(\psi_i^{\text{stim}} - \psi_{ki}^{\text{proto}})$. This formula causes each prototype's values to move towards the currently experienced stimulus, but only to the extent that the prototype is already similar to the stimulus, and only to the extent that the dimension is being attended to. In this way, prototypes that do not represent the stimulus very well are not much influenced by the stimulus.

Some models allow multiple prototypes per category, to capture multimodal distributions (Anderson, 1991), and use other learning methods derived from Bayesian statistics. In the extreme case, there can be one prototype per exemplar, and such models become equivalent to exemplar models (Nosofsky, 1991). In exemplar models, however, the coordinates of the exemplars typically do not get adjusted from one trial to the next.

In several studies that compare prototype and exemplar models, it has been found that prototype models do not fit data better than exemplar models (e.g. Ashby and Maddox, 1993; Busemeyer *et al.*, 1984; Busemeyer and Myung, 1988; Nosofsky, 1992; but cf. Reed, 1972), but some have found evidence for prototypes either early or late in learning (Homa *et al.*, 1993; Smith and Minda, 1998). Prototype theory is, however, intuitively appealing, and a challenge for cognitive scientists is to discover phenomena that are naturally addressed by prototype models but that cannot be adequately accounted for by exemplar-based models, or by rule-based models, which are described next.

RULE THEORIES

Yet another way of representing categories is with rules that specify strict necessary and sufficient conditions for category membership. For example, something is a member of the category 'bachelor' if it human, male, unmarried and eligible. Many natural categories are very difficult to specify in terms of rules, however (e.g. Rosch and Mervis, 1975). For example, the category 'game' has no necessary and sufficient features (Wittgenstein, 1953). Nevertheless, people are prone to look for features that define category distinctions, and people tend to believe that such defining features exist even if in fact they do not (Brooks, 1978; Brooks *et al.*, 1998).

Rules for category definition are typically a single threshold on a single dimension: for example, a building is a skyscraper if and only if it is taller than 10 floors. Rules can also be logical combinations of such thresholds: for example, a building is a skyscraper if and only if it is taller than 10 floors and its facade is at least 60% glass. In some rule-based theories, rules can be more complicated boundaries: for example, a building is a skyscraper if and only if the number of floors multiplied by the percentage of glass in the facade exceeds the value 6.0. (By this multiplicative rule, a building only seven floors tall would be classified as a skyscraper if it had at least 86% glass in its facade, because $7 \times 86\% > 6.0$.)

In principle, categorization rules are absolute, and there is no 'gray area' around the boundary of the category. In practice, however, most rule-based models do incorporate some mechanism for blurring the category boundary, to accommodate real performance data.

Formal Models of Rule Theories

Traditionally, rule models have been referred to as 'hypothesis testing' or 'concept learning' models (for a review, see Levine, 1975). In these sorts of models, individual features are tested, one at a time, for their ability to account for the correct classifications of the stimuli. For example, the model might test the rule 'if it's red then it's in category K'. As long as the rule works, it is retained, but when an error is encountered, another rule is tested. As simple rules are excluded, more complicated rules are tried. A recent incarnation of this type of model is also able to learn exceptions to rules, by testing additional features of instances that violate an otherwise successful rule (Nosofsky, Palmeri and Mckinley, 1994). This model is also able to account for differences in behavior between people, because there can be different sets of rules and exceptions that equally well account for the classifications of the stimuli.

For stimuli that vary on continuous dimensions, there is a well-studied class of models for which the decision boundary is assumed to have a shape that can be described by a quadratic function, because a quadratic describes the optimal boundary between two multivariate normal distributions, and natural categories are sometimes assumed to be distributed normally (e.g. Ashby, 1992). In this approach, there are three basic postulates: (1) the stimulus is represented as a point in multidimensional space, but the exact location of this point is variable because of perceptual noise; (2) a stimulus is classified according to which side of a quadratic decision boundary it falls on; (3) the decision boundary is also subject to variability because of noise in the decision process. Thus, although the classification rule is strict and there is no explicit role in the model for similarity gradients, the model as a whole produces a gradation of classification performance across the boundary because of noise in perception and decision. There are many variations on this scheme of models, involving different shapes of boundaries, deterministic or probabilistic decision rules, and so on. (Ashby and Alfonso-Reese, 1995; Ashby and Maddox, 1993).

HYBRID REPRESENTATION THEORIES

It is unlikely that any one of these types of representation can completely explain the complexity of human category learning. A variety of work has shown that neither rule-based nor prototype models can fully account for human categorization (e.g. Ashby and Waldron, 1999; Kalish and

Kruschke, 1997). In particular, exemplar representation must be supplemented with rules to account for human learning and generalization (Erickson and Kruschke, 1998). Therefore, some recent theories combine different representations. A model constructed by Vandierendonck (1995) combines rectangular decision boundaries with exponentially decaying similarity gradients. A model constructed by Ashby *et al.* (1998) combines linear decision boundaries that involve single dimensions (corresponding to verbalizable rules) with linear decision boundaries that combine two or more dimensions (corresponding to implicitly learned rules). A model constructed by Erickson and Kruschke (1998) combines exemplars with single-dimension rules.

The challenge for hybrid representation theories is specifying the interaction of the various types of representation. If there are several representational types available, under which conditions is each type used? In the model of Erickson and Kruschke (1998), for example, which combines rules with examples, the representations compete for attention, so that the type of representation that reduces categorization error most quickly is the type that is used for that instance. Hybrid models will probably proliferate in the future.

ROLE OF SIMILARITY

Similarity is critical in exemplar and prototype theories, and also appears in hybrid rule-based theories as distance from the boundary (e.g. Vandierendonck, 1995). Some researchers have criticized the notion of similarity as being internally incoherent, or have argued that similarity does not always correlate with categorization.

Similarity can be empirically investigated in several different ways. One method is simply to ask people to rate the similarity of two items; another is to measure discriminability between items. Usually these different assessments agree, but sometimes they do not (e.g. Tversky, 1977). Similarity can be context-specific: in the context of hair, gray is more similar to white than to black, but in the context of clouds, gray is more similar to black than to white (Medin and Shoben, 1988). In general, models of similarity presume which features or dimensions are used for comparing the objects, without any explanation of why those features or dimensions are selected. Models of similarity do have parameters for specifying the attention allocated to different features, but the models do not describe how these attentional values arise (Goodman, 1972; Murphy and Medin, 1985).

Similarity is not always a clear predictor of categorization. Consider the category *things to remove from a burning house*. The items 'heirloom jewellery' and 'children' are both central members of this category, yet they have little surface similarity (Barsalou, 1983). On the other hand, if attention is directed only to the features *irreplaceable* and *portable*, then children and heirloom jewellery bear a strong similarity. Once again the question of what to attend to is crucial, but not addressed by current theories of similarity.

Despite these complexities, there are strong regularities in similarity and categorization data that should yield to formal treatment. Excellent reviews of these topics have been written by Goldstone (1994) and by Medin *et al.* (1993).

SUMMARY

Categorization is central to cognition. Different theories of category learning posit different representations for the information underlying categorization. Research has shown that no single type of representation can account for the full range of categorization observed in humans. Instead, recent models combine different types of representations in hybrid systems. Challenges for future research include determining how different representations interact, and how attention influences and is influenced by category learning.

References

Anderson JR (1991) The adaptive nature of human categorization. *Psychological Review* **98**: 409–429.

Ashby FG (1992) Multidimensional models of categorization. In: Ashby FG (ed.) *Multidimensional Models of Perception and Cognition*, pp. 449–483. Hillsdale, NJ: Erlbaum.

Ashby FG and Alfonso-Reese L (1995) Categorization as probability density estimation. *Journal of Mathematical Psychology* **39**: 216–233.

Ashby FG, Alfonso-Reese LA, Turken AU and Waldron EM (1998) A neuropsychological theory of multiple systems in category learning. *Psychological Review* **105**: 442–481.

Ashby FG and Maddox WT (1993) Relations between prototype, exemplar and decision bound models of categorization. *Journal of Mathematical Psychology* **37**: 372–400.

Ashby FG and Waldron EM (1999) On the nature of implicit categorization. *Psychonomic Bulletin and Review* **6**: 363–378.

Barsalou L (1983) Ad hoc categories. *Memory and Cognition* **11**: 211–227.

Brooks LR (1978) Nonanalytic concept formation and memory for instances. In: Rosch E and Lloyd BB (eds)

Cognition and Categorization, pp. 169–211. Hillsdale, NJ: Erlbaum.

Brooks LR, Squire-Graydon R and Wood TJ (1998) *The role of inattention in everyday concept learning: identification in the service of use.* [Available from L. R. Brooks, Department of Psychology, McMaster University, Hamilton, Ontario, Canada L8S 4K1.]

Busemeyer JR, Dewey GI and Medin DL (1984) Evaluation of exemplar-based generalization and the abstraction of categorical information. *Journal of Experimental Psychology: Learning, Memory and Cognition* **10**: 638–648.

Busemeyer JR and Myung IJ (1988) A new method for investigating prototype learning. *Journal of Experimental Psychology: Learning, Memory and Cognition* **14**: 3–11.

Choi S, McDaniel MA and Busemeyer JR (1993) Incorporating prior biases in network models of conceptual rule learning. *Memory and Cognition* **21**: 413–423.

Erickson MA and Kruschke JK (1998) Rules and exemplars in category learning. *Journal of Experimental Psychology: General* **127**: 107–140.

Estes WK (1994) *Classification and Cognition*. New York, NY: Oxford University Press.

Goldstone RL (1994) The role of similarity in categorization: providing a groundwork. *Cognition* **52**: 125–157.

Goodman N (1972) Seven strictures on similarity. In: Goodman N (ed.) *Problems and Projects*, pp. 437–447. New York, NY: Bobbs-Merrill.

Homa D, Goldhardt B, Burruel-Homa L and Smith JC (1993) Influence of manipulated category knowledge on prototype classification and recognition. *Memory and Cognition* **21**: 529–538.

Kalish ML and Kruschke JK (1997) Decision boundaries in one dimensional categorization. *Journal of Experimental Psychology: Learning, Memory and Cognition* **23**: 1362–1377.

Kruschke JK (1992) ALCOVE: an exemplar-based connectionist model of category learning. *Psychological Review* **99**: 22–44.

Kruschke JK and Johansen MK (1999) A model of probabilistic category learning. *Journal of Experimental Psychology: Learning, Memory and Cognition* **25**: 1083–1119.

Kruskal JB and Wish M (1978) *Multidimensional Scaling*. Beverly Hills, CA: Sage Publications.

Lamberts K (1998) The time course of categorization. *Journal of Experimental Psychology: Learning, Memory and Cognition* **24**: 695–711.

Levine M (1975) *A Cognitive Theory of Learning: Research on Hypothesis Testing*. Hillsdale, NJ: Erlbaum.

Medin DL, Goldstone RL and Gentner D (1993) Respects for similarity. *Psychological Review* **100**: 254–278.

Medin DL and Schaffer MM (1978) Context theory of classification learning. *Psychological Review* **85**: 207–238.

Medin DL and Shoben EJ (1988) Context and structure in conceptual combination. *Cognitive Psychology* **20**: 158–190.

Murphy GL and Medin DL (1985) The role of theories in conceptual coherence. *Psychological Review* **92**: 289–316.

Nosofsky RM (1986) Attention, similarity and the identification–categorization relationship. *Journal of Experimental Psychology: General* **115**: 39–57.

Nosofsky RM (1991) Relation between the rational model and the context model of categorization. *Psychological Science* **2**: 416–421.

Nosofsky RM (1992) Exemplars, prototypes, and similarity rules. In: Healy AF, Kosslyn SM and Shiffrin RM (eds) *Essays in Honor of William K. Estes*, vol. II 'From Learning Processes to Cognitive Processes', pp. 149–167. Hillsdale, NJ: Erlbaum.

Nosofsky RM, Gluck MA, Palmeri TJ, McKinley SC and Glauthier P (1994) Comparing models of rule-based classification learning: a replication of Shepard, Hovland, and Jenkins (1961). *Memory and Cognition* **22**: 352–369.

Nosofsky RM and Kruschke JK (1992) Investigations of an exemplar-based connectionist model of category learning. In: Medin DL (ed.) *The Psychology of Learning and Motivation*, vol. XXVIII, pp. 207–250. San Diego, CA: Academic Press.

Nosofsky RM and Palmeri TJ (1997) An exemplar-based random walk model of speeded classification. *Psychological Review* **104**: 266–300.

Nosofsky RM, Palmeri TJ and McKinley SC (1994) Rule-plus-exception model of classification learning. *Psychological Review* **101**: 53–79.

Palmeri TJ (1999) Learning categories at different hierarchical levels: a comparison of category learning models. *Psychonomic Bulletin and Review* **6**: 495–503.

Reed SK (1972) Pattern recognition and categorization. *Cognitive Psychology* **3**: 382–407.

Rosch EH and Mervis CB (1975) Family resemblances: studies in the internal structure of categories. *Cognitive Psychology* **7**: 573–605.

Shepard RN (1987) Toward a universal law of generalization for psychological science. *Science* **237**: 1317–1323.

Smith JD and Minda JP (1998) Prototypes in the mist: the early epochs of category learning. *Journal of Experimental Psychology: Learning, Memory and Cognition* **24**: 1411–1436.

Tversky A (1977) Features of similarity. *Psychological Review* **84**: 327–352.

Vandierendonck A (1995) A parallel rule activation and rule synthesis model for generalization in category learning. *Psychonomic Bulletin and Review* **2**: 442–459.

Wittgenstein L (1953) *Philosophical Investigations*. New York, NY: Macmillan.

Further Reading

Estes WK (1994) *Classification and Cognition*. New York, NY: Oxford University Press. [A mathematically oriented survey.]

Lamberts K and Shanks D (eds) (1977) *Knowledge, Concepts and Categories*. Cambridge, MA: MIT Press. [An accessible collection of tutorials.]

Rosch E and Lloyd BB (eds) (1978) *Cognition and Categorization*. Hillsdale, NJ: Erlbaum. [A collection of statements of fundamental results and theoretical perspectives.]

Shanks DR (1995) *The Psychology of Associative Learning*. Cambridge, UK: Cambridge University Press. [A lucid review of issues in category learning.]

Smith EE and Medin DL (1981) *Categories and Concepts*. Cambridge, MA: Harvard University Press. [A very readable introduction to the field of categorization.]

Concepts, Philosophical Issues about

Intermediate article

Jesse Prinz, Washington University, St Louis, Missouri, USA

CONTENTS

Concepts are generally defined as representations that allow us to think about properties or categories. Philosophers have debated the nature of these representations and the cognitive functions that they serve.

INTRODUCTION

Concepts (sometimes called 'ideas') are the tools by which we think about the world. They have been an object of philosophical scrutiny since Plato's time. New theories of concepts have been developed in recent decades, arising from interactions between philosophers and researchers in other disciplines. There is no consensus about which theory of concepts is correct, but there is wide agreement on the challenges that an adequate theory of concepts must meet.

FUNCTIONS OF CONCEPTS

Concepts are theoretical posits, introduced to play a variety of explanatory roles. This article will present the functions that concepts are most often alleged to serve. Some theorists doubt whether a single kind of entity can serve all of these functions (Rey, 1983), but others are more optimistic (Prinz, 2002). If all these different functions are served by different kinds of entities, there may be no nonarbitrary way to determine which of those entities deserve to be called concepts and the utility of the construct may be called into question.

One function that is often emphasized in philosophy involves reference: concepts are said to represent categories or properties. A category is a class of things consisting of zero or more members (e.g. the class of all actual and possible elephants); and a property can be characterized as that by virtue of which things form a cohesive class (e.g. elephanthood). Some philosophers think that concepts must represent properties rather than categories, because we can have distinct concepts corresponding to distinct properties of the same class (e.g. the class of triangles is the same as the class of trilaterals, but we can distinguish between these conceptually).

Concepts are also widely (though not universally) alleged to serve an epistemic function: they embody the information by which we understand categories. That information is often believed to take the form of 'features'. Features are usually construed as concepts themselves, designating properties possessed by category members. For example, an *elephant* concept may encompass the features *has a trunk* and *is an animal*. If concepts are associated with features, they can serve a categorization function. I determine that Jumbo is an elephant because he is an animal, has a trunk, and so on. Psychologists emphasize the categorization function above all others.

There are two main views about how concepts embody knowledge or information. According to 'decomposition' views, concepts are literally parts of other concepts (e.g. Smith, 1989). For example, an *elephant* concept might be construed as a data structure containing *trunk*. According to 'conceptual role' views, concepts are related to each other by inputs and outputs (e.g. Block, 1986). For example, an *elephant* concept might be construed as a mental predicate ('*X* is an elephant') that licenses certain inferences ('*X* has a trunk').

Decomposition and conceptual role views may be behaviorally indistinguishable. Defenders of decomposition views generally assume that constituent features can be used to draw inferences. If my *elephant* concept contains the feature *trunk*, and I believe that Jumbo is an elephant, I may infer that Jumbo has a trunk. On either approach, then, concepts can be said to serve an inference function.

Concepts also have a combination function: they combine to form compound concepts and thoughts. We often form thoughts that have never been entertained before, even when we have not acquired any new concepts. This suggests that concept combination is a compositional process. Concepts are said to combine compositionally when the content of a compound concept is a function of the concepts that comprise it together with the rules of combination. For example, one can form a concept of a clumsy elephant if one has an *elephant* concept and a concept of clumsiness, even if one has never considered or encountered clumsy elephants before. A compositional system can generate boundless novel compounds from a finite set of more basic concepts.

Concepts are also said to serve functions related to language. Firstly, they contribute to communication. Two people successfully communicate when they associate the same (or similar) concepts with their words. The communication function is related to a meaning function. The simplest version of this thesis says that concepts constitute the meanings of words. Philosophers have developed different theories of meaning, which lead to different interpretations of this hypothesis. 'Meaning-as-use' theories, for example, associate meanings with abilities. According to one version, the meaning of a word is determined by how that word is used in various conversational contexts (Wittgenstein, 1953). If concepts are meanings, it would follow that concepts are verbal abilities; and if concepts are verbal abilities, then one cannot possess a concept without language. On this view, creatures lacking language lack concepts.

Many philosophers reject that conclusion. It makes it more difficult to talk about apparent cognitive similarities between humans and non-human animals. It also makes it difficult to explain how language is acquired in the first place. How can one learn what our first words mean if understanding a meaning requires mastery of verbal abilities (Fodor, 1975)? Not all meaning-as-use theories face these difficulties. According to another meaning-as-use theory, the meaning of a word is determined by how a concept underlying that word is used in thought. This preserves the idea that knowing meanings involves mastering abilities, while denying that those abilities are necessarily verbal.

Proponents of a different class of theories equate the meaning of a term with a 'mode of presentation' (Frege, 1893). A mode of presentation can be a representation of features possessed by whatever the term represents. For example, the meaning of the word 'zebra' might be comprised of a representation of striped, horse-like animals roaming the African savannah. If concepts decompose into features, they are ideally suited to serve as modes of presentation and, thus, meanings. Even conceptual roles can be regarded as modes of presentation if the latter term is very broadly construed.

According to another theory, the meaning of a word is exhausted by its referent. The word 'zebra' simply means the class of all actual and possible zebras or the property of being a zebra. If this theory is correct, it makes little sense to identify concepts with meanings. It would be better to say that concepts are the bearers of meanings. More specifically, one might say that the meaning of a word is exhausted by the referent of the concept with which it associated. If meanings are exhausted by referents, then the features associated with a given concept are not part of linguistic meaning, as mode of presentation theories and meaning-as-use theories imply.

THEORIES OF CONCEPTS

Philosophers and psychologists have proposed a number of theories of concepts. One point of disagreement concerns the question of which concepts are 'primitive' (Fodor, 1981). As noted above, some theorists assume that many concepts are comprised of other concepts. On pain of regress, decomposition must end somewhere. A primitive concept is one whose identity conditions do not depend on any other concept. The British empiricists claimed that primitive concepts are sensory. This view has fallen out of favor, because it is very difficult to analyze abstract concepts into sensory features (but see Prinz, 2002). The issue of primitive concepts has received relatively little attention in recent times.

Another point of disagreement concerns the ontological status of concepts (Peacocke, 1992). According to some philosophers, concepts are abstract entities; while according to others they are mental tokens that reside inside our heads. There may be room for reconciliation here, however. If concepts are mental tokens, and if concept sharing is possible, it must still be possible to talk about them belonging to common types: two tokens of an *elephant* concept must be capable of belonging to a common type. Conversely, those who favor the view that concepts are abstract entities must admit that individuals grasp concepts, and, to the extent that grasping is a psychological process, token mental states must be involved.

A more divisive issue concerns the kind of information that concepts embody. One terminological caveat must be made before we survey these disputes: the word 'concepts' is sometimes used as shorthand for 'the majority of lexical concepts'. A lexical concept is a concept expressed by a single word in a natural language, such as English. A theory of concepts would thus be better described as a theory of the kind of information embodied in most of our lexical concepts.

The Definition Theory

Overview

The dominant theory in the history of philosophy claims that most lexical concepts are definitions (e.g. Katz and Fodor, 1963). A definition is a collection of features that are jointly sufficient and individually necessary for membership in a category. The concept *bachelor*, for example, is said to entail or decompose into the features *unmarried* and *male*.

Adherents of the definition theory (also called the 'classical' theory) draw a sharp distinction between those features that define a concept and those that are merely known to apply to the category designated by the concept. This generates two kinds of true sentences, those that are true by definition (e.g. 'bachelors are unmarried') and those that are true by virtue of how the world is (e.g. 'bachelors tend to like martinis'). The former are called 'analytic' and the latter 'synthetic'.

There are several different versions of the definition theory. According to the tradition deriving from Plato, each of us implicitly knows how to define the concepts we possess, but that information is not easily accessed. One must engage in arduous philosophical reflection (facilitated by dialogue) to reveal definitions. This process of discovering definitions is said to be *a priori*, because it relies on conceptual intuition rather than observation of experience. *A priori* conceptual analysis is one of the predominant methods used in philosophy.

According to a second version, definitions are discovered *a posteriori*, not by intuition and reflection (Rey, 1983). This account is thought to be most applicable to concepts that designate natural kinds (such as concepts of animals or substances found in nature). The definition of the concept *gnu* might be a description of the gnu genome, or some other scientifically determined conditions that are necessary and sufficient for being a gnu. One might hold a mixed view, according to which some definitions are discovered *a posteriori* and others are discovered *a priori*.

A third version of the definition theory has it that definitions are not discovered at all, but rather invented (Carnap, 1934). Defenders of this version recognize that many of our concepts are unclear, or variable between individuals. Communication is greatly facilitated, they argue, by stipulating definitions. These may roughly coincide with prior intuitions, but they are more precise. The primary motivation for this program of 'precisification' is to facilitate science. If different scientists agree on how their terms are defined, any remaining debates between them must be substantive (e.g., pertaining to data) rather than verbal.

Assessment

Despite its long philosophical pedigree, faith in definitions began to wane in the middle part of the twentieth century. One attack derives from Quine's (1951) famous critique of the distinction between the analytic and the synthetic. Believers in that distinction traditionally hold that analytic truths are known *a priori* and are invulnerable to empirical refutation.

Quine presents an alternative account of how sentences are confirmed, which contradicts this assumption. According to Quine, no sentence is empirically verified in isolation. When we encounter evidence that conflicts with a sentence held to be true, we have the option of either revising that sentence or revising various background assumptions, including those that may appear independent of experience. What we revise will depend on what we observe, what we take to be true, and a variety of pragmatic principles by which we strive to minimize disruption as we revise our theories. Because sentences face the 'tribunal of experience' collectively, any sentence is vulnerable to empirical refutation.

If Quine's 'confirmation holism' is correct, the traditional notion of analyticity is undermined. This has implications for the definition theory of concepts. Proponents of that theory say that some of the features associated with a concept define it and some do not. But which are the definitional features? They cannot be the features that are analytically related to the concept, because that notion lacks foundation. Without a principled way to distinguish defining features from collateral information, the definition theory is in jeopardy of being usurped by 'concept holism' (see below). This is damaging to Carnap's 'precisification' view. Science can stipulate definitions, but these will remain sensitive to the pressures of observation and discovery. No claim is purely verbal.

Another challenge to the definition theory was articulated by Wittgenstein (1953). Wittgenstein notices that certain terms cannot be captured by a single set of necessary and sufficient conditions. His famous example was the word 'game'. Every plausible defining condition on games has obvious counterexamples. For instance, games do not always have winners (consider catch), and they do not always have two or more sides (consider solitaire). This point may generalize. Every time one philosopher publishes an analysis of a concept, another publishes a convincing example of a case that the analysis fails to subsume. Even concepts that seem to have incontrovertible definitions are vulnerable. A *bachelor* is widely defined as an unmarried male, but there is a strong intuition that Catholic priests are not bachelors even though they satisfy the definition. Conditions thought to be defining often capture a salient range of cases, but they rarely capture all cases.

Psychologists have also criticised the definition theory (see Hampton, 1993, for a review). Experimental evidence has overwhelmingly shown that definitions do not figure prominently in conceptual tasks. People categorize on the basis of family resemblance, associate non-defining features with concepts, have difficulty learning definitions, and fail to rely on definitions once they have been learned. If concepts are mental representations, then it seems unlikely that concepts are definitions.

Defenders of the *a posteriori* version of the definition theory have a response to such arguments. On their view, definitions are scientifically discoverable facts unknown to most concept users. The fact that people have difficulty formulating adequate definitions only shows that most of us do not understand completed science. This might rescue the view that concepts are definitions, but it would not satisfy many cognitive scientists. If definitions are largely unknown, they will play little role in behavior. Even if concept users believe that definitions will someday be discovered, that article of faith cannot distinguish my *elephant* concept from my *walrus* concept. If a theory of concepts is to explain how different concepts affect behavior, then we would be better off identifying concepts with the actual knowledge that ordinary people use to categorize elephants and walruses.

Clusters, Stereotypes, and Prototypes

Overview

The arguments against the definition theory have spawned several alternative theories. Wittgenstein came to see many concepts as revolving around large clusters of features. Rather than having a universal essence, concepts like *game* are applied by determining whether something has a preponderance of the features in a cluster. All games exhibit a subset of features from the same cluster, but few if any cluster features are exhibited by all games. Wittgenstein calls *game* a 'family resemblance' concept, because each of its instances shares features with each other, but no features are universally shared.

Putnam (1975) explores a related idea. He argues that people think about categories by means of stereotypes. A stereotype is a collection of features that are thought to be highly characteristic of category instances. Such features are typically widespread, salient, and diagnostic. For example a *dog* stereotype might include such features as *barks, has a tail,* and *is a pet*. Stereotype features are also presumed to be contingent. There are wild dogs that lack tails and never utter a sound. Stereotypes can even contain features that are erroneously associated with categories (e.g., gorillas are stereotyped as ferocious). Putnam observes that when people explain the meaning of a word, they often list stereotypical features, rather than provide a definition. Putnam thinks that stereotypes are generally concise, in contrast to Wittgenstein's clusters. As with clusters, however, members of a category are often presumed to exhibit some sufficient proportion of the stereotype rather than the whole of it.

Stereotypes can be compared to what psychologists call prototypes. According to prototype theory, most lexical concepts are identified with representations of prototypical category instances (Hampton, 1993). Sometimes, a prototype is thought to be a representation of the actual instance that best captures the category's central tendency. More commonly, prototypes are regarded as weighted lists of features corresponding to the properties most frequently recognized in category instances. Features that are more frequently recognized or more diagnostic may be given higher weights. Categorization depends on passing a critical threshold of similarity, computed by comparing a target instance to the features contained in the weighted list. Prototype theorists emphasize the graded nature of category judgments. People regard certain category instances as more typical than others. These are said to reflect increased similarity to the prototype.

Assessment

There is ample evidence that people often categorize on the basis of judged similarity to collections of non-defining features. The psychological reality of something like clusters, stereotypes, or prototypes is rarely challenged. However, some researchers are reluctant to identify such cognitive structures with concepts. To see why, consider some of the objections to stereotypes (all of which apply to prototypes as well).

Putnam points out that stereotypes are insufficient to determine reference. To demonstrate, he has us imagine two very similar individuals living on different planets. One lives on Earth, where the thirst-quenching, clear liquid in rivers and streams is H_2O. His doppelgänger lives on Twin Earth, where the liquid that has those properties is a different chemical compound, XYZ. These individuals (who are ignorant of chemistry) have the same stereotypes associated with the word 'water', but apparently their concepts refer to different substances. Some researchers believe that concepts with different referents cannot be identical. One is a *water* concept, and the other is a *twin water* concept. If this is the case, then stereotypes can, at best, be one component of our concepts. In addition, we must individuate concepts by their referents. (Putnam draws this conclusion about meanings rather than concepts.)

The insufficiency of stereotypes can also be demonstrated by considering categorization judgments. While we often categorize on the basis of superficial similarity, we recognize that appearances can deceive. A wolf in sheep's clothing is still a wolf even if its disguise is good enough to resemble a stereotypical sheep. This is a psychological analogue of Putnam's observation; concept users know that similarity to their stereotypes is insufficient for reference. Stereotypes are good for rough and ready categorization, but we must be capable of transcending them.

This last observation is connected with a general concern. Stereotypes comprise only a small portion of the knowledge we have about categories. They generally capture superficial appearances. We also know a great deal about features hidden from view, the relations between superficial features, how category instances function, how they relate to instances of other categories, and so on. All of this information can potentially influence our judgments concerning categories. There may be no reason to privilege superficial features from this rich reservoir of information.

One of the most serious concerns about stereotypes is that they do not combine compositionally. For example, the stereotype for *pet bird* includes the feature *being caged*, which is not included in the stereotype for *pet* or the stereotype for *bird* (see Fodor, 1998, for a review). If concepts combine compositionally (as argued above), they cannot be identified with stereotypes.

The Theory Theory and Concept Holism

Overview

Psychologists have developed an alternative to the prototype theory, called the theory theory. Theory theorists maintain that most lexical concepts are similar to scientific theories in a number of ways.

Firstly, theory theorists say that concepts encode knowledge about causal and explanatory principles just as scientific theories encode knowledge about laws and mechanisms (Murphy and Medin, 1985). In addition to knowing that birds fly and that birds have wings (two stereotypical features), people know that wings enable flight. Features that enter into such explanatory relations are more likely to be included in concepts and highly weighted.

Secondly, theory theorists claim that concepts encode the knowledge that the features essential to category membership may be unobservable, like the postulates of many scientific theories. Something is identified as being a horse by virtue of its appearance, but it really counts as being a horse only by virtue of having a particular microstructure. Echoing the *a posteriori* definition theory view, theory theorists recognize that concept users have faith in such essences without knowing what they are in detail. Keil (1989) demonstrates this experimentally. He asked subjects to consider an animal that began looking like a horse but was painted to look just like a zebra, and acted like one. Young children are deceived by the transformation, but the rest of us recognize that such a creature would still count as a horse.

Keil's experiments also illustrate a third claim of theory theorists. In contrast to the horse case, mature concept users believe that artefacts can change their identity when they are superficially transformed. When a coffee pot is modified to look and function like a bird feeder, subjects say it has become a bird feeder. This suggests that we treat different kinds of concepts differently. Theory theorists speculate that concepts are parceled into distinct cognitive domains, each dedicated to knowledge about a distinct class of entities and driven by distinct principles. For example, we may have naive theories of artefacts, biological kinds, intentional agency, and physical mechanics.

The fourth claim of the theory theory involves conceptual change. Children often use words in very different ways from adults. For example, a young child might extend the word 'alive' to include cars and other inanimate objects (Carey, 1985). Such observations lead Carey to conclude that children's concepts may differ significantly from adults'. She argues that these differences lead to a certain degree of incommensurability, an idea she borrows from Kuhn's (1962) philosophy of science. Sometimes, a child's terms can be translated into our own without loss of meaning. This echoes the conclusion that Kuhn draws about terms used within different scientific theories.

The theory theory has been most explicitly defended by psychologists, but it is related to a class of theories that some philosophers defend. Quine emphasizes the continuity between theoretical knowledge and the totality of ordinary beliefs about a category. There is no obvious boundary between the knowledge that comprises our informal theory of a category and all other information associated with that category (recall Quine's confirmation holism). If concepts are identified via intuitive theories, then any given concept will be identified by a vast collection of beliefs comprised of other concepts, which will be identified by vast belief sets of their own. Following this logic, the identity conditions of any particular concept may depend on just about every other concept known to its possessor. This may be called 'concept holism'.

Without a clear boundary between theoretical knowledge and collateral information, theory theorists may be forced to embrace concept holism. Concept holism is often associated with meaning holism. If concepts are meanings, and concepts are identified by their place in a vast network of beliefs, then it is natural to conclude that the meaning of a term depends on a vast network of beliefs as well. As we saw earlier, concept holism is an unwelcome idea for defenders of the definition theory, who would wish to restrict concepts to discrete sets of necessary features. But it enjoys considerable support in its own right. Concept holists do not always emphasize the theoretical nature of concepts, and theory theorists do not always emphasize concept holism, but these two approaches may have a tendency to coincide.

Assessment

The theory theory and concept holism are open to objections. Like the stereotype theory, they face difficulties with concept combination: how do vast bodies of information combine compositionally? But there is a more pressing problem. If concepts are comprised of everything we know about a category, then no two people would share concepts, because different people's knowledge varies to some degree (Fodor and LePore, 1992). One might try to avoid this problem by saying that people have similar concepts rather than identical concepts. But Fodor and LePore reply that holism offers no way of quantifying similarity between people's concepts. We cannot say that concepts are similar by virtue of having some of the same constituent features, because each feature is itself a concept that, for the theory theorist, must be identified with an entire body of knowledge. If concepts cannot be shared, it is difficult to explain how communication ever occurs. To avoid such difficulties, theory theorists can try to distance themselves from concept holism by identifying a boundary between theoretical knowledge and collateral information.

Informational Atomism

Overview

All of the accounts considered thus far assume that most lexical concepts embody knowledge of the categories they represent. Fodor (1998) argues that any proposal of this kind is doomed to failure. Concepts cannot be definitions, because definitions are too scarce; they cannot be stereotypes, because stereotypes are not compositional; and they cannot be theories, because theories are not shared. The only way to avoid all of these problems is to deny that concepts embody knowledge.

Fodor calls his theory 'informational atomism'. It is called atomism, because it asserts that most of our lexical concepts are primitive. They neither entail nor decompose into any features. Instead, concepts are individuated by the properties to which they refer. They come to refer by means of 'informational semantics'. Informational semantics says that a concept refers to a property by virtue of being lawfully caused by that property. A *dog* concept refers to doghood because it becomes active when one encounters dogs. Fodor does not deny that we associate a considerable amount of knowledge with our *dog* concepts (including dog prototypes and dog theories), but he denies that this knowledge is conceptually constitutive. The *dog* concept can be regarded as an inner label in a language of thought. In principle, one could have the very same label even if one's dog prototypes and theories changed or disappeared entirely.

Inner labels can be compositionally combined, because their content is exhausted by their referents. There is no problem of compounds having features not found in their components. Inner labels can also be readily shared. Two people have the same concept by virtue of having labels that are lawfully caused by the same objects, even if their beliefs about those objects differ radically.

Assessment

Fodor's concepts are well suited to serve reference, combination, and communication functions, but these benefits come at a price. If concepts do not decompose into features, they cannot satisfy the knowledge, inference, or categorization functions: the very functions that motivate most psychological work on concepts. Fodor also faces the challenge of explaining concept acquisition. Traditionally, primitive concepts are presumed to be innate. If most lexical concepts are primitive, then most lexical concepts are innate (Fodor, 1975; but cf. Fodor, 1998). Many find this consequence unsettling.

CONCLUSION

Concepts are postulated to serve a variety of functions. These include roles in reference, knowledge, inference, categorization, combination, communication, and meaning. A number of theories have been proposed, but each seems to have serious limitations. Some researchers argue that we must reduce the list of functions that concepts ought to serve, while others hope for a more encompassing theory.

References

Block N (1986) Advertisement for a semantics for psychology. In: French PA, Uehling TE and Wettstein HK (eds) *Midwest Studies in Philosophy*, vol. X, *Philosophy of Mind*. Minneapolis, MN: University of Minnesota Press.

Carey S (1985) *Conceptual Change in Childhood*. Cambridge, MA: MIT Press.

Carnap R (1934/1959) *The Logical Syntax of Language*. London, UK: Routledge & Kegan Paul.

Fodor JA (1975) *The Language of Thought*. Cambridge, MA: Harvard University Press.

Fodor JA (1981) The current status of the innateness controversy. In: *Representations*. Cambridge, MA: MIT Press.

Fodor JA (1998) *Concepts: Where Cognitive Science Went Wrong*. Oxford, UK: Oxford University Press.

Fodor JA and LePore E (1992) *Holism: A Shopper's Guide*. Oxford, UK: Blackwell.

Frege G (1893) On Sinn and Bedeutung. In: Beaney M (ed.) (1997) *The Frege Reader*, pp. 151–171. Oxford, UK: Blackwell.

Hampton A (1993) Prototype models of concept representation. In: van Mechelen I, Hampton J, Michalski RS and Theuns P (eds) *Categories and Concepts: Theoretical Views and Inductive Data Analysis*, pp. 67–95. New York, NY: Academic Press.

Katz J and Fodor JA (1963) The structure of a semantic theory. *Language* **39**: 170–210.

Keil FC (1989) *Concepts, Kinds, and Cognitive Development*. Cambridge, MA: MIT Press.

Kuhn T (1962) *The Structure of Scientific Revolutions*. Chicago, IL: University of Chicago Press.

Murphy GL and Medin DL (1985) The role of theories in conceptual coherence. *Psychological Review* **92**: 289–316.

Peacocke C (1992) *A Study of Concepts*. Cambridge, MA: MIT Press.

Prinz JJ (2002) *Furnishing the Mind: Concepts and Their Perceptual Basis*. Cambridge, MA: MIT Press.

Putnam H (1975) The meaning of 'meaning'. In: Gunderson K (ed.) *Language, Mind and Knowledge*, pp. 131–193. Minneapolis, MN: University of Minnesota Press.

Quine WVO (1951) Two dogmas of empiricism. *Philosophical Review* **60**: 20–43.

Rey G (1983) Concepts and stereotypes. *Cognition* **15**: 237–262.

Smith EE (1989) Concepts and induction. In: Posner K (ed.) *Foundations of Cognitive Science*. Cambridge, MA: MIT Press.

Wittgenstein L (1953) *Philosophical Investigations*. New York, NY: Macmillan.

Further Reading

Armstrong SL, Gleitman LR and Gleitman H (1983) What some concepts might not be. *Cognition* **13**: 263–308.

Barsalou LW (1987) The instability of graded structure: implications for the nature of concepts. In: Neisser U (ed.) *Concepts and Conceptual Development: Ecological and Intellectual Factors in Categorization*, pp. 101–140. Cambridge, UK: Cambridge University Press.

Clark A (1993) *Associative Engines*. Cambridge, MA: MIT Press.
Gopnik A and Melzoff A (1997) *Words, Thoughts, and Theories*. Cambridge, MA: MIT Press.
Locke J (1690/1989) *An Essay Concerning Human Understanding*. Oxford, UK: Clarendon Press.
Margolis S and Laurence S (eds) (1999) *Concepts: Core Readings*. Cambridge, MA: MIT Press.
Millikan R (2000) *On Clear and Confused Ideas: An Essay About Substance Concepts*. Cambridge, UK: Cambridge University Press.
Smith EE and Medin D (1981) *Concepts and Categories*. Cambridge, MA: Harvard University Press.
Rosch E and Mervis C (1975) Family resemblances: studies in the internal structure of categories. *Cognitive Psychology* 7: 573–605.

Conceptual Change

Intermediate article

Paul Thagard, University of Waterloo, Waterloo, Ontario, Canada

CONTENTS

Conceptual change is the creation and alteration of mental representations that correspond to words. It is an important part of learning in science and everyday life.

INTRODUCTION

Concepts are mental representations corresponding to words. For example, the concept 'dog' is a mental structure that corresponds to the word 'dog' and refers to dogs in the world. Conceptual change is produced by mental processes that create and alter such mental representations. Explaining how conceptual change works is important for understanding the growth of scientific knowledge, the development of children's thinking, and the education of students in fields such as science and mathematics. In each of these kinds of learning, a theory of conceptual change is needed that can answer such questions as the following. What is the nature of the concepts that are learned? What kinds of changes do concepts undergo? What are the mental processes that produce different kinds of conceptual change? It is also interesting to inquire whether the processes of conceptual change in scientists, young children, and students are similar or different.

TYPES OF CONCEPTUAL CHANGE

The simplest type of conceptual change is when people learn a new concept. A more challenging type occurs when existing concepts must be adjusted and reorganized to accommodate new information: in such cases, the meaning of concepts changes in relation to other concepts and the world. In radical conceptual change, the development of knowledge involves a shift in which a collection of important concepts undergo alterations in meaning. In such cases, learning is not simply a matter of accumulating new concepts and beliefs; it also requires substantial revision and restructuring of mental representations.

CONCEPTUAL CHANGE IN SCIENTISTS

The problem of conceptual change in science was first highlighted in Thomas Kuhn's famous book, *The Structure of Scientific Revolutions* (Kuhn, 1962). He challenged the prevailing view that scientific knowledge grows cumulatively by progressively adding to the stock of available theories and concepts. Instead, Kuhn proposed that the development of science often involves revolutionary changes in which one theory or paradigm is replaced by a radically different one. For example, the acceptance of the Copernican theory that the earth revolves around the sun required the rejection of the Ptolemaic theory that the sun revolved around the earth. Replacement was not merely a matter of one theory being substituted for another, but also involved shifts in meaning of the concepts used in the theories. In the Copernican revolution, for example, the concept 'planet' shifted to include the earth and exclude the sun and moon. According to Kuhn, radical differences between theories make it difficult to establish rationally that one is better than another.

Kuhn distinguished between normal science, in which a dominant paradigm is taken for granted, and revolutionary science, in which the dominant paradigm is replaced by a radically new one. The main activity in normal science is puzzle solving, which deals with problems within the scope and constraints of the dominant way of thinking. Scientists pursue normal science until there is an accumulation of anomalies, which are problems that the paradigm fails to solve. For example, in the eighteenth century the prevailing theory of combustion based on phlogiston, a substance supposed to be given off by burning objects, encountered the anomaly that objects gain rather than lose weight during combustion. Scientists attempt to deal with individual anomalies as puzzles to be solved with the tools provided by the paradigm they accept, but the accumulation of anomalies produces a state of crisis in which scientists begin to consider the need for new theories. When a new paradigm is conceived that can solve the problems that were anomalous for the old one, a scientific revolution occurs and a new theory becomes accepted. Kuhn's favorite examples of scientific revolutions include the Copernican revolution, the chemical revolution in which Lavoisier's oxygen theory of combustion replaced the phlogiston theory, and the revolution in physics in which relativity theory was adopted.

Before Kuhn, science was generally viewed as a cumulative process in which new theories built on the successes of previous ones. Kuhn insisted that scientific revolutions are noncumulative episodes in which an older paradigm is replaced by an incompatible new one. He even suggested that the new and old theories are incommensurable with each other, that is, there may be no logical means for objectively choosing between them. A major source of incommensurability is the use by the different paradigms of very different concepts. For example, it might seem that the Newtonian physics and relativity theory both use the concept of mass, but Einsteinian mass can be converted into energy whereas Newtonian mass is conserved. Thus for Kuhn a major aspect of scientific revolutions was radical conceptual change.

In *Conceptual Revolutions*, Thagard (1992) offered a comprehensive account of the kinds of conceptual changes that have occurred in the major revolutions in the history of science. Most scientific revolutions involve the introduction of new concepts, such as Newton's gravitational force, Lavoisier's oxygen, Darwin's natural selection, and Wegener's continental drift. In addition, revolutions usually involve reclassification in which a concept changes its place in the hierarchy of kinds, just as Copernicus reclassified earth as a planet, Darwin reclassified humans as a kind of animal, and the cognitive revolution in psychology reclassified thinking as a kind of computation. Even more radically, the principle of classification sometimes changes, as when Darwin argued that species should be organized into kinds on the basis of evolutionary history rather than similarity. Like many other philosophers of science, Thagard argued that Kuhn had overestimated the conceptual differences between theories, so that conceptual change did not prevent one theory from being rationally preferred to another on the basis of its explanatory power. Nevertheless, he accepted Kuhn's basic contention that new theories often have very different conceptual systems from the ones they replace.

Philosophers and psychologists have discussed the cognitive mechanisms by which new conceptual systems in science are constructed. These include conceptual combination, in which a concept such as 'sound wave' is constructed out of the previously existing concepts 'sound' and 'wave'. New concepts are rarely derived directly from experience, but instead are built up from previously existing concepts. A concept produced by conceptual combination need not be a simple sum of the original concepts, but instead can involve emergent properties. For example, the concept 'blind lawyer' has characteristics not found in either 'blind' or 'lawyer': people use causal reasoning to conclude that a blind lawyer must be courageous.

Another creative mechanism is analogy, in which new scientific concepts are formed by adapting and transforming previous concepts. For example, Darwin's concept of natural selection was based in part on his familiarity with artificial selection practiced by breeders who produced new varieties of plants and animals. Maxwell developed concepts of electromagnetism using mechanical analogies (Nersessian, 1992), and Kepler extensively used analogies to develop new concepts concerning light and motion (Gentner *et al.*, 1997).

Once a new conceptual system has been constructed by mechanisms such as combination and analogy, it becomes a contender to replace an existing conceptual system. The major cognitive mechanism for such large-scale conceptual change is explanatory coherence: scientists adopt a new theory along with its conceptual system because it provides a better explanation of the evidence and is more coherent with other beliefs (Thagard, 1992). Of course, most conceptual change in science does not involve such large-scale shifts in which conceptual systems are substantially altered, but rather the introduction of new concepts that fit in with existing conceptual schemes and theories.

CONCEPTUAL CHANGE IN YOUNG CHILDREN

Young children acquire a wealth of new concepts as their knowledge of language and the world increases. The average high-school graduate in the USA knows around 60 000 root words, which must have been acquired at a rate better than 10 per day. Presumably, children have concepts that are mental representations corresponding to all these words, so how can we account for their acquisition in such large numbers? Much conceptual change is straightforwardly cumulative, as children simply add new concepts such as 'dog' and 'ice cream' to their mental systems. However, some developmental psychologists have argued that conceptual development in children is like conceptual change in science, in that it sometimes requires substantial revisions of existing conceptual schemes.

Susan Carey argued that children's acquisition of biological knowledge between the ages of 4–10 years involves considerable conceptual reorganization (Carey, 1985). In particular, the concepts 'alive' and 'animal' undergo substantial change during those years. Many 4-year-olds have difficulty naming any objects that are not alive, and take objects such as tables and clocks as being alive because they have activities or motions associated with them. By the age of 10 years, however, most children have acquired the adult concept of 'living thing'. Similarly, children under 7 years old often do not count people and insects as animals. According to Carey, children undergo a complete reorganization of knowledge of functions such as eating and sleeping and of organs such as the stomach and heart as the domain of biological knowledge becomes differentiated from the domain of knowledge of human activities. It is not just that the concepts of a 10-year-old have different relations among them than those of a 4-year-old, but more that the concepts themselves have changed as the result of additional biological knowledge. The concepts 'animal' and 'plant' coalesce into the concept 'living thing' by virtue of recognition that they are fundamentally alike. At the same time, children learn to differentiate 'dead' from 'inanimate' as two different senses of 'not alive'. Just as scientists had to learn to differentiate between heat and temperature, so children have to learn to differentiate weight from size and density. Like scientists, children have theory-like conceptual structures, and learning consists in radical alteration of such structures, not just additions to them.

Frank Keil reached similar conclusions from his studies of the development of children's concepts of biological kinds (Keil, 1989). As children gain an increasing appreciation of the biological principles that organize adults' intuitive theories of biology, they increasingly appeal to origins and internal parts in their biological classifications, reducing the impact of visible features. For example, older children are more likely to judge that a pear covered with apple skin is still a pear. In contrast, there was no similar shift for artefacts such as cup and nail, indicating that conceptual change was specific to biological kinds. Keil argues that concepts are part of coherent belief systems, so that conceptual change is closely tied in with theory change in children.

Gopnik and Meltzoff (1997) are even more emphatic in tying conceptual change to theory change. They advocate the 'theory theory', according to which the process of cognitive development in children is similar to and perhaps even identical to the process of theory development in scientists. They describe changes in understanding of objects in infants, who are born assuming a world of three-dimensional objects that have visual, auditory, and tactile features. By 6 months, infants have gained systematic, coherent knowledge about the movements of objects, but they still lack understanding of hidden objects, which develops around 9 months. Later, at around 18 months, infants acquire the ability to represent invisible movements. Gopnik and Meltzoff contend that these shifts are like theory change in science, and that there is a certain incommensurability between the concepts of the old and new theories held by the infants.

These and other studies of learning in children strongly suggest that conceptual development is not simply a matter of accumulating new concepts but also involves important changes in concepts and conceptual systems. However, the evidence is still limited for claims that children's conceptual systems are like those of scientists and that the cognitive mechanisms of change in children are like those that take place in the minds of scientists. It is possible that children's knowledge is much more fragmented than the conceptual systems that make up scientific theories such as relativity and evolution by natural selection. Scientific theories consist of hypotheses that provide unifying explanations of diverse empirical phenomena, but no one knows whether children's beliefs involve the same kind of explanatory hypotheses. Moreover, the process by which scientists come to realize that one theory is better than and should replace a previous one involves a systematic comparison of the explanatory coherence of the two theories. Belief change in children may be much more piecemeal, as isolated fragments of a new theory of objects and kinds are acquired from experience and teaching. It is possible that new ways of looking at things supplant previous ones by a process of gradual build-up of new concepts and progressive disuse of old ones, rather than by a dramatic replacement of the old theories by new ones. The view that conceptual change in children is similar to theory change in scientists has been heuristically useful in stimulating research on children's learning, but much more empirical research is needed before the analogy between children and scientists can be accepted as showing a common set of cognitive processes.

CONCEPTUAL CHANGE IN STUDENTS

Suppose it is true that learning in children and scientists involves radical conceptual change rather than mere accumulation of new concepts and beliefs; then teaching students cannot be understood as merely providing new material to mesh with what students already know. Rather, education in science and other subjects may require a much more challenging process of dealing with the prior concepts and hypotheses that guide students' thinking. If teachers are not aware that students come to science classes with misconceptions about living things and physical processes, the teachers will not understand many of the difficulties that the students have in learning. From the perspective of conceptual change, teaching requires an active approach in which children must be engaged in building explanations that challenge concepts and beliefs that they previously held. Effective teaching may require the use of the kinds of analogical models and thought experiments that have often facilitated conceptual change in the history of science.

Chi (1992) argues that physics education is often difficult because it requires conceptual change across fundamental ontological categories such as matter, events, and abstractions. For example, naive students start with concepts of force, light, heat and current that class them as kinds of material substances, but physics students must learn to reconceptualize them as fields, which are a complex kind of event. Vosniadou and Brewer (1992) studied the development of children's knowledge of astronomy and found that children have difficulty reconciling the teaching that the earth is round with their other beliefs and observations. Children develop models that reconcile their observation-based belief that the earth is flat with what they are taught about the earth being round. For example, first-graders often believe that there are two earths – a flat one on which we live and a round one up in the sky. Other children think that the earth is a sphere, but we live inside it rather than on top of it. Thus, teaching children that the earth is round is not just a matter of telling them an additional fact, but requires them to revise their basic beliefs about the nature of the earth and other planets.

Science education is thus in part a cognitive process involving conceptual change, but it is also being increasingly recognized as a social, contextual, and emotional process (Guzzetti and Hynd, 1998). Conceptual change is a kind of mental change, but this may come about because of social interactions that students have with teachers and each other, as well as with the physical world. Motivation and emotion can greatly influence conceptual change when students acquire the intention and enthusiasm to adopt new concepts and hypotheses rather than to remain entrenched in their previous frames of mind. Future research on conceptual change will have to find ways to integrate cognitive processes with social and emotional processes that interact with them continuously.

The last section raised the question of whether conceptual change in children is like that found in scientists undergoing major theoretical changes. It is also an open question whether students need to undergo conceptual revolutions, or whether instead they can learn by a more gentle process in which new conceptual systems come to predominate over previous ones without the explanatory conflicts that occur in science. More research is needed to determine whether the cognitive mechanisms of conceptual change and theory evaluation that operate in scientists are also responsible for educational progress in science students.

References

Carey S (1985) *Conceptual Change in Childhood.* Cambridge, MA: MIT Press/Bradford Books.

Chi M (1992) Conceptual change within and across ontological categories: examples from learning and discovery in science. In: Giere R (ed.) *Cognitive Models of Science*, Minnesota Studies in the Philosophy of Science, vol. 15, pp. 129–186. Minneapolis, MN: University of Minnesota Press.

Gentner D, Brem S, Ferguson R *et al.* (1997) Analogy and creativity in the works of Johannes Kepler. In: Ward TB, Smith SM and Vaid J (eds) *Creative Thought: An Investigation of Conceptual Structures and Processes*, pp. 403–459. Washington, DC: American Psychological Association.

Gopnik A and Meltzoff AN (1997) *Words, Thoughts, and Theories*. Cambridge, MA: MIT Press.

Guzzetti B and Hynd C (eds) (1998) *Perspectives on Conceptual Change*. Mahwah, NJ: Lawrence Erlbaum.

Keil F (1989) *Concepts, Kinds, and Cognitive Development*. Cambridge, MA: MIT Press/Bradford Books.

Kuhn T (1962) *The Structure of Scientific Revolutions*. Chicago: University of Chicago Press.

Nersessian N (1992) How do scientists think? Capturing the dynamics of conceptual change in science. In: Giere R (ed.) *Cognitive Models of Science*, vol. 15, pp. 3–44. Minneapolis, MN: University of Minnesota Press.

Thagard P (1992) *Conceptual Revolutions*. Princeton, NJ: Princeton University Press.

Vosniadou S and Brewer WF (1992) Mental models of the earth: a study of conceptual change in childhood. *Cognitive Psychology* **24**: 535–585.

Further Reading

Ball T, Farr J and Hanson RH (eds) (1989) *Political Innovation and Conceptual Change*. Cambridge, UK: Cambridge University Press.

Carey S (2001) *Science education as conceptual change*. [http://www.house.gov/science/carey_03-04.htm]

Dietrich E and Markman AB (eds) (1999) *Cognitive Dynamics: Conceptual and Representational Change in Humans and Machines*. Mahwah, NJ: Lawrence Erlbaum.

Feyerabend PK (1981) *Realism, Rationalism and Scientific Method*. Philosophical Papers, vol. 1. Cambridge, UK: Cambridge University Press.

Kunda Z, Miller D and Claire T (1990) Combining social concepts: the role of causal reasoning. *Cognitive Science* **14**: 551–577.

Nersessian N (1989) Conceptual change in science and in science education. *Synthese* **80**: 163–183.

Pearce G and Maynard P (eds) (1973) *Conceptual Change*. Dordrecht: Reidel.

Thagard P (1999) *How Scientists Explain Disease*. Princeton, NJ: Princeton University Press.

Conceptual Representations in Psychology

Introductory article

Arthur B Markman, University of Texas, Austin, Texas, USA

CONTENTS

Conceptual representation refers to the way that information about categories is stored and organized.

INTRODUCTION

Concepts are mental representations that are used to divide the world into groups that will be treated as equivalent for some purpose. Concepts may refer to objects, events, or ideas. Concepts may be used for reasoning, prediction, and communication. Some researchers have distinguished between concepts, which are the mental representations of information, and categories, which are sets of objects in the world that are grouped together. Often, however, these terms are used interchangeably.

Psychologists have explored concept representations in detail. This work has examined both within-category representation and between-category structure. Within-category representation refers to the information that describes a particular category such as 'dog'. Between-category structure refers to the relationships among different categories such as that between the categories 'dog', 'cat', and 'animal'.

WITHIN-CATEGORY REPRESENTATION

The central question about within-category representation involves the way people store information about particular concepts that enables them to classify new items (exemplars) as members of a category. Some work has looked at other uses of categories such as making predictive inferences, causal reasoning, and communication, but this discussion will focus on classification. Three broad types of within-category representation are rule-based models, similarity-based models, and theory-based models.

Rule-based Models

The classical approach to concept representation has been to seek a rule that specifies the necessary and sufficient conditions for something to be a member of a category. A property is a necessary condition for being in a category if all members of that category possess the property. A set of necessary conditions is sufficient to specify a category if all exemplars that have that set of properties are members of the category, and no exemplars that have that set of properties are members of any other category. For example, an object is a triangle if it is a three-sided closed figure. This set of features is necessary and sufficient, because all triangles are three-sided and closed. No object that has these properties can be anything but a triangle.

Unfortunately, outside formal domains like geometry, it is difficult (or perhaps impossible) to find a set of necessary and sufficient conditions that specify the members of a category. For example, it might seem at first glance that a bachelor is an unmarried adult man. While this rule correctly classifies most bachelors, there are many dubious cases. For example, Catholic priests and widowers are both unmarried adult men, but one might be hesitant to classify them as bachelors. While it is possible to continue to refine this definition, it is likely that exceptions could be found to any rule that was generated.

One approach that has been tried to save rule-based approaches has been to assume that people generate fairly simple rules that are good for classifying most exemplars, and then store exceptions to the rules separately. In the example above, the rule 'unmarried adult man' would be used to classify most bachelors, but exceptions such as priests and widowers would be considered separately.

Similarity-based Models

Intuitively, it seems that a new exemplar is classified based on its similarity to the category. For example, an object might be classified as a bird, because it looks like a bird. This intuition has been captured by similarity-based models, which assume that people classify a new exemplar based on its similarity to some stored category representation. Similarity-based models differ from each other primarily in their assumptions about the nature of the stored category representation.

Prototype models assume that people store some average representation of an object. The average need not be identical to any actual exemplar, but rather contains the features most frequently associated with that category. For example, the typical bird might be a small animal that flies, sings, and has feathers. Not all birds have this entire set of properties (e.g. penguins do not fly), but the more of these features an exemplar possesses, the more likely it is to be a bird.

One result often taken as evidence for prototype models is that categories have a graded typicality structure: that is, people have strong intuitions about which members of a category are typical members of that category and which members are atypical. For example, robins and sparrows are generally thought to be typical birds, while chickens and emus are thought to be atypical birds. Generally, the typicality of an exemplar is related to its similarity to the prototype of the category.

A second prominent similarity-based model is the exemplar model, which posits that people store representations of each category member rather than creating a prototype. New exemplars are then classified by comparing them with all of the known exemplars. The more similar a new exemplar is to the known exemplars of a particular category, the more likely it is that the exemplar will be classified as a member of that category. Exemplar models are also able to account for graded typicality structure, because typical exemplars are similar to many members of a category, but atypical exemplars are similar to only a few members of a category.

Theory-based Models

Despite the success of similarity-based models in predicting how people classify new items, there are situations in which people classify items in a manner that violates similarity. For example, at a party, a person might be classified as drunk if he or she dives headfirst into a cake. This person is not being classified on the basis of any similarity to known exemplars of drunk people; rather, common beliefs about drunken behavior are sufficient to classify the person as drunk.

There seems to be a developmental change in people's ability to use theory-based information. When children first learn categories, they often classify on the basis of surface characteristics. For example, if told about a black cat that has a white stripe painted on its back, and a bag of smelly stuff surgically placed inside it, young children will classify it as a skunk. Older children and adults, however, will classify it as a cat, suggesting that they are able to use a theory about biological categories to classify this (rather strange) exemplar.

Which Type of Model is Right?

Of the three types of models, the rule-based models are least often used in conceptual processing. There are some situations in which people must make repeated classifications that involve a rule and exception process. However, empirical studies suggest that even when people are asked to form rules, their ability to apply the rule is influenced by the similarity of a new exemplar to those seen before.

Both similarity-based and theory-based processes are often used in categorization. There are times when people must be able to identify a new item on the basis of the similarity of its properties to those of items seen in the past. In addition, there are cases in which people's theories about a domain influence categorization. Current research is focusing on how to integrate similarity-based and theory-based approaches.

BETWEEN-CATEGORY STRUCTURE

A second important aspect of category representation involves the relationships among categories. In this section, two aspects of between-category structure are examined. First, much research has examined the hierarchical organization of categories. This work is concerned with understanding how people may categorize objects at different levels of abstraction. A second area that has received attention is people's ability to generate categories based on goals. This study of goal-derived categories also provides a window into conceptual processing.

Hierarchical Organization of Categories

If you see a small, curly-haired, four-legged living creature being walked on a leash down the street, you can classify this thing as a poodle, a dog, or an animal. That is, for any given object, there are a variety of categories to which it belongs. Many of these categories differ from each other in their degree of abstraction: 'dog' is a more abstract category than 'poodle', because all poodles are dogs, but not all dogs are poodles. Similarly, 'animal' is a more abstract category than either 'dog' or 'poodle'.

A striking aspect of this category structure is that if you show people a picture of some object, they are most likely to identify it using a category at a middle level of abstraction. For example, shown a picture of the item described in the previous paragraph, people are likely to identify it first as a dog rather than as a poodle or as an animal. This tendency has led psychologists to refer to this middle level of abstraction as the basic level. Categories more abstract than those at the basic level (e.g. animal) are called superordinate categories, and categories more specific than those at the basic level are called subordinate categories (e.g. poodle).

Basic level categories have been shown to have a number of characteristics. First, they tend to be the most abstract categories whose members have a common shape, and whose shape differs from other contrasting basic level categories. For example, dogs tend to be shaped similarly to each other and differently from other animals; in contrast, animals come in many different shapes. Second, basic level categories tend to have shorter labels than either subordinate or superordinate categories; for example, 'car' is the label for a basic level category, and the labels 'vehicle' (for the superordinate) and 'sports car' (for a typical subordinate category) are both longer than the basic level label. Basic level categories are also the most abstract level for which the category members share the same set of parts: cars all have wheels, brakes, and engines, whereas there are many other vehicles (e.g. helicopters and boats) that do not share these parts. Finally, children tend to learn basic level labels for objects before learning the labels for categories at other levels of abstraction. The factors that characterize basic level categories can be summarized as follows:

- Objects are identified first at the basic level.
- Objects are classified fastest at the basic level.
- The basic level is the most abstract level at which the members tend to have the same shape.
- The basic level is the most abstract level at which the members tend to share parts.
- The basic level is the most abstract level at which people interact with the members using similar motor movements.
- The basic level is the most abstract level at which the category members tend to be similar.
- The basic level is the most abstract level at which category members tend to be dissimilar from members of contrasting categories.
- Children often learn basic level labels before labels at other levels of abstraction.
- Basic level labels are shorter than labels for categories at other levels of abstraction.

The hierarchical organization of categories seems to be strongest for object categories. Some research has been done on the between-category structure of abstract concepts such as events and ideas. People also have categories at different levels of abstraction for these concepts, but the basic level does not have as much of an advantage relative to subordinate and superordinate categories.

Goal-derived Categories

The previous section suggested that an object might belong to many different categories, and that these categories generally differ in their level of abstraction. There are some categories, however, that are organized around people's goals rather than around the overall shape and parts of objects. Some of these categories are ones that we use all the time, and their labels have become words. For example, a 'pet' is a domesticated animal that is kept as a companion. Thus, membership in this category is determined by whether an object serves a particular goal.

An important observation is that people can also generate goal-derived categories as they are needed. For example, you may never have considered the category 'Things to take out of a house in the event of a fire'. Now that this category has been suggested, however, it is easy to generate members of the category (e.g. children, jewellery, photographs).

Novel goal-derived categories are called *ad hoc* categories. A striking finding is that, although they are being generated on the fly, they share many characteristics with categories that were previously learned. For example, like regular categories, *ad hoc* categories exhibit a graded typicality structure: that is, people find it easy to determine which members of an *ad hoc* category are typical or atypical. For example, people might agree that old photographs of family members are good examples of things to take out of the house in the event of a fire, but that an old sofa is a poor example.

One key difference between regular categories and goal-derived categories is in the way that typicality is assessed. For regular categories, an object is typical to the extent that it is similar to the average member (or prototype) of the category. In contrast, for goal-derived categories there is often an ideal member, and items are more typical to the extent they are similar to the ideal. For example, someone might create the goal-derived category 'diet foods'. The ideal member of this category tastes great and has no energy content. A new object will be typical of a category to the extent that it is similar to its ideal.

CONCLUSION

Psychologists have explored the internal (within-category) representation and external (between-category) structure of category representations. Research on within-category representation has focused on the role of rules, similarity, and theory in determining category representation. Research on between-category structure has focused both on relationships among categories at different levels of abstraction and on goal-derived categories.

Further Reading

Barsalou LW (1983) Ad hoc categories. *Memory and Cognition* **11**: 211–227.

Keil FC (1989) *Concepts, Kinds, and Cognitive Development.* Cambridge, MA: MIT Press.

Medin DL, Lynch EB and Solomon KO (2000) Are there kinds of concepts? *Annual Review of Psychology* **51**: 121–147.

Morris MW and Murphy GL (1990) Converging operations on a basic level in event taxonomies. *Memory and Cognition* **18**(4): 407–418.

Murphy GL and Medin DL (1985) The role of theories in conceptual coherence. *Psychological Review* **92**(3): 289–315.

Nosofsky RM (1986) Attention, similarity, and the identification-categorization relationship. *Journal of Experimental Psychology: General* **115**(1): 39–57.

Nosofsky RM, Palmeri TJ and McKinley SC (1994) Rule-plus-exception model of classification learning. *Psychological Review* **101**(1): 53–97.

Rosch E and Mervis CB (1975) Family resemblances: studies in the internal structure of categories. *Cognitive Psychology* **7**: 573–605.

Rosch E, Mervis CB, Gray WD, Johnson DM and Boyes-Braem P (1976) Basic objects in natural categories. *Cognitive Psychology* **8**: 382–439.

Smith EE and Medin DL (1981) *Categories and Concepts.* Cambridge, MA: Harvard University Press.

Conditioning

Introductory article

Frances K McSweeney, Washington State University, Pullman, Washington, USA

CONTENTS

Classical and operant conditioning are procedures for changing behavior. Behavior changes when a previously neutral stimulus predicts an important stimulus (classical conditioning). Behavior also changes when a response produces a particular consequence (operant conditioning).

INTRODUCTION

Classical and operant conditioning provide powerful techniques for understanding and controlling behavior. In classical conditioning, the behavior towards a previously neutral stimulus changes when that stimulus predicts the occurrence of a stimulus that already evokes a response. In operant conditioning, the frequency of, or the time spent making, a response is changed by consequences (reinforcers or punishers) that follow that response.

CLASSICAL CONDITIONING

The discovery of classical conditioning is usually attributed to Ivan Pavlov, a Russian physiologist. Pavlov briefly turned on a metronome and then presented food to a dog. After several exposures to the metronome followed by food, the dog salivated when the metronome was presented alone. That is, the dog's behavior toward the metronome changed when the dog learned that the sound of the metronome was followed by food.

In Pavlov's procedure, the metronome or tone is usually referred to as the 'conditioned stimulus' (CS). The CS is originally neutral to the animal in the sense that it does not automatically evoke the response of interest (salivation). The food is referred to as the 'unconditioned stimulus' or US. It is a stimulus that already evokes a response. The response that is evoked by the US is called the 'unconditioned response' (UR). The response that occurs to the CS, as a result of its pairing with the US, is called the 'conditioned response' (CR).

The classical conditioning procedure presented above might be described by saying that when a CS is followed by a US several times, the CS comes to evoke a CR that resembles the UR. Even stated this way, classical conditioning is of some practical interest. For example, it may explain the acquisition of a fear or phobia (CR) when a stimulus (CS, e.g. a snake) precedes a frightening event (US: e.g. someone screams). It may play a part in the development of preferences for or aversions to food, and it may facilitate digestion because stimuli that precede food intake may help to prepare the body. Since the time of Pavlov, however, our understanding of classical conditioning has changed in ways that have greatly increased its practical usefulness.

The Form of the CR

Voluntary behaviors as CRs

Pavlov studied the reflexive response of salivation while his animals were immobilized by suspending them in a hammock. Reflexive responses are behaviors that are automatically evoked by a stimulus (e.g. you blink your eye in response to a threat). Reflexive responses are often distinguished from voluntary behaviors that are emitted rather than evoked by a particular external stimulus (e.g. you read this article). Pavlov's study of reflexive behaviors in restrained animals may have limited peoples interest in classical conditioning. Most people are more interested in the voluntary behavior of freely moving animals, including people.

The principle of 'sign tracking' suggests that classical conditioning does apply to the voluntary behavior of freely moving animals. Sign tracking states that animals approach and contact the stimulus that is the best predictor of a US, and withdraw from stimuli that signal the absence of a US. According to this idea, when a CS predicts a US, the behavior that is learned is movement (approach or withdrawal), not just reflexive behavior (e.g. salivation). For example, I once visited a wildlife park in Australia where a vending machine sold kangaroo chow. The machine made a loud noise when it operated and that sound (CS) predicted the availability of food (US). As expected from sign tracking, a stampede of kangaroos approached (CR) the food machine as soon as it operated (CS), an undesirable event for those standing by the machine. (Technically, without further tests, this behavior cannot be conclusively identified as classically conditioned. It might also be a discriminated operant, as described later.)

CRs may differ from URs

In Pavlov's experiment, similar responses served as the CR and UR. That is, dogs salivated when food was presented (UR) and they learned to salivate to the sound of a metronome that preceded food (CR). If the CR must be identical to the UR, then classical conditioning cannot be used to train a response unless a US can be found that automatically evokes that response. This is difficult in many cases (e.g. teaching someone to play the piano).

We now know that the CR may differ from the UR. For example, when some types of drugs are used as a US, the CR may be the opposite of the UR. To give one example, morphine is a painkiller (UR), but animals become hypersensitive to pain (CR) in the presence of an arbitrary stimulus (CS), e.g. the sight of a needle, that predicts a morphine injection (US). This means that classical conditioning may contribute to the build-up of tolerance for drugs and to the withdrawal symptoms that are observed when drugs are not available. Think of the UR to morphine as a 'high' (a pleasant state) and the CR to morphine as a 'low' (an unpleasant state). Classically conditioned responses gradually become stronger with each successive pairing of the CS and US. If a conditioned 'low' becomes stronger with each morphine injection, then more and more of the drug will be needed to overcome it – that is, tolerance will develop. If the conditioned stimuli that accompany a drug injection (e.g. time of day, sight of the needle) occur without the drug, then the animal will experience only the low (CR) without the high (UR) produced by the US. This low may contribute to withdrawal symptoms. Although classical conditioning may play a role in drug tolerance and withdrawal, it is undoubtedly only one of many contributors.

Information versus Temporal Contiguity

In the earlier description, classical conditioning occurred when a CS was followed by a US. More recently, it has been argued that a CS must provide information about, or predict, the occurrence of the US for conditioned responding to occur. Consider the following experiment. One group of animals receives a CS followed by a US 10 times. A second group receives the same 10 CS–US pairings, but also receives 10 extra unconditioned stimuli that are interspersed among the CS–US pairings. If conditioned responding develops when a CS is followed by a US, then conditioned responding should be strong in both groups. They both experience the CS followed by the US 10 times. If conditioned responding occurs when the CS provides information about the US, then the first group should respond more than the second group. In the first group, the CS is a perfect predictor of the US; in the second group, it is not. The evidence supports the information view.

Cue Competition Effects

The strength of conditioned responding to a CS depends on how well that CS predicts the US. It also depends on the strength of conditioning to other conditioned stimuli that also predict the US. Because the presence of other predictive conditioned stimuli usually weakens conditioned responding to any one of them, cues are said to compete for conditioning. Overshadowing and blocking are examples of cue competition.

Overshadowing

In most cases, more conditioned responding occurs to CS1 (e.g. a light) when CS1 alone predicts the US (CS1 → US) than when CS1 and CS2 (e.g. a light plus a tone) together predict the US (CS1 + CS2 → US). The stimulus CS2 is said to 'overshadow', and therefore to reduce conditioned responding to, CS1. Although overshadowing is the usual finding, the opposite, potentiation, is occasionally reported, especially when the stimuli are tastes and odors. Potentiation refers to greater conditioned responding to CS1 when it predicts the US in compound with CS2 than when CS1 predicts the US alone.

Blocking

If an animal learns that CS1 predicts a US (CS1 → US), then the animal will not later perform a CR to a second CS (CS2) that also predicts the US when presented in compound with CS1 (CS1 + CS2 → US). The prior conditioning to CS1 is said to block conditioned responding to CS2. The occurrence of blocking provides further support for the idea that conditioned responding develops only when a CS provides information about a US. Presumably, a CR does not occur to CS2 because it provides no information about the US beyond that already provided by CS1.

Again, blocking is the usual finding, but the opposite result of augmentation is sometimes reported. Augmentation refers to a situation in which prior conditioning to CS1 strengthens, rather than weakens, conditioned responding to a redundant CS2. Augmentation, as potentiation, is particularly likely when tastes and odors serve as CSs.

Cue to Consequence Effect

Some combinations of CSs and USs yield stronger CRs than others. For example, a strong CR of aversion will develop when a taste (CS) predicts illness (US), as well as when a light (CS) predicts an electric shock (US). Weaker or no aversion will result when a light (CS) predicts illness (US) and when a taste (CS) predicts an electric shock (US). Although it is not known why this occurs, some have argued that evolution favors animals that quickly learn to avoid foods whose tastes predict illness. Such animals are less likely to die of poisoning.

Significance of Classical Conditioning

Classical conditioning was once described in ways that made it appear to be a simple mechanical transfer of behavior from one stimulus to another. This view was challenged by many findings, including that the CR need not resemble the UR and that blocking may occur. More recently, classical conditioning has been seen as a mechanism through which the predictive relations among stimuli in the environment alter behavior. In many cases, receiving an early warning from an arbitrary stimulus may help an animal deal with a potentially harmful stimulus (e.g. a drug) to come.

OPERANT CONDITIONING

Operant conditioning is a change in behavior that occurs as a result of the consequences of that behavior. Its most prominent student was B. F. Skinner. Because of the power of operant techniques, they form the basis for behavior therapies that are effective in correcting many human behavioral problems. They are used to train nonhuman animals for performances in films or circuses. They are also often used in scientific studies to establish a baseline for assessing the effect of other manipulations (e.g. drug injections, physiological interventions). Operant techniques are useful as baselines because they provide stable and replicable control over behavior.

Positive Reinforcement

The most frequently studied form of operant conditioning is positive reinforcement. According to the principle of positive reinforcement, a response that is followed by a reinforcer will increase in frequency. For example, if lever pressing (response) yields food (reinforcer) for a hungry rat, the rat will press the lever more frequently in the future. Some behaviors are more easily measured in terms of the time devoted to them than in terms of their frequency (e.g. reading). In that case, following the response by a reinforcer will increase the amount of time spent making the response. For example, if practicing the piano (response) yields the opportunity to watch a favorite television program (reinforcer), then the time spent practicing will increase.

Notice that a response cannot be strengthened by reinforcement unless a reinforcer can be found. Over the years, many definitions for the term 'reinforcer' have been rejected. For example, reinforcers were once thought of as substances that are physiologically needed (e.g. food, water), but there are many reinforcers that are not physiologically needed (e.g. watching television, going to the cinema). Reinforcers were once defined as stimuli that reduce tension (e.g. sexual behavior), but many stimuli that increase tension also serve as reinforcers (e.g. watching a scary film, riding a roller coaster).

Because of these failures, a reinforcer is technically defined as any stimulus that increases the frequency of a response that it follows. This is an undesirable definition because it makes the principle of positive reinforcement circular. That is, the principle now states that a response followed by any stimulus that increases the frequency of the response that it follows will increase in frequency. This definition has been accepted because scientists can identify a stimulus as a reinforcer in one situation (e.g. they can show that the stimulus will increase the frequency of one response that it follows); they can then test the principle of positive reinforcement in another situation (e.g. they can ask whether that reinforcer also increases the frequency of other responses that it follows).

Some stimuli will serve as reinforcers for both human and nonhuman animals (e.g. food, water, access to conspecifics for herd animals). Other reinforcers will be more effective with people than with other animals (e.g. praise, money, the opportunity to watch television). Different items will serve as reinforcers for different people, and the

Premack principle provides a way to identify effective reinforcers. The Premack principle states that the opportunity to perform any higher probability response can serve as a reinforcer for any lower probability response. The probability of a response is measured by examining what the animal does when it has free time. According to the Premack principle, if a child often plays electronic games, then the opportunity to play such a game will serve as a reinforcer for less probable behaviors, such as doing homework.

Eliciting the first response

A response cannot be reinforced until that response occurs. When working with people, verbal instructions may be used to elicit the first response (e.g. 'please practice the piano'). For complex responses, such as a golf swing, physical guidance known as 'prompting' may also be needed. Shaping by successive approximations may be useful when dealing with nonhuman animals or with a nonverbal person such as a baby. During shaping, closer and closer approximations to the desired response are reinforced. For example, if you want to teach your dog to sit up, you could begin by following any movement by a reinforcer. Then you might reinforce only movements that involve some transfer of the dog's weight to its back paws. Then you might reinforce only movements that involve weight transfer to the back paws plus lifting the forepaws off the ground. By carefully choosing which behaviors to reinforce and when to alter the reinforced response, you should quickly have your dog sitting up.

The Four Basic Conditioning Procedures

Operant conditioning can be used to either increase (reinforcement) or decrease (punishment) the frequency of a response. The frequency of a response may change when the response produces something (positive) or when it escapes or avoids something (negative). It is called 'positive reinforcement' when a response increases in frequency because it produced something (e.g. you work because you have been paid for working). Negative reinforcement occurs when a response increases because it escaped or avoided something (e.g. you drive safely because safe driving has prevented accidents). Positive punishment occurs when a response decreases in frequency because it produced something (e.g. your cat no longer scratches the furniture because you squirted it with water whenever it scratched). Negative punishment occurs when a response decreases in frequency because it prevented or removed something (e.g. your child stops hitting his little brother because you took away his allowance or confined him to his room when he did so).

Technically, a stimulus is classified as a reinforcer or punisher depending on its effect on behavior. Reinforcers increase behaviors that they follow; punishers decrease behaviors. Terms related to pleasure and pain do not appear in these definitions. Nevertheless, stimuli that serve as positive reinforcers when they are delivered and as negative punishers when they are removed are usually described as pleasant. Stimuli that serve as negative reinforcers when they are removed and positive punishers when they are presented are usually described as aversive or painful.

Schedules of Reinforcement

Reinforcers are delivered according to schedules of reinforcement, which are rules specifying which response will be followed by a reinforcer. During a continuous reinforcement (CRF) procedure, every occurrence of a particular response is followed by a reinforcer. For example, in a perfect world, depositing an appropriate amount of money into a soft-drink machine (response) would yield a soft drink (reinforcer) each time that you did it. Continuous reinforcement procedures have drawbacks that reduce their usefulness. For example, the person administering a CRF procedure must be alert enough to identify and to reinforce every instance of the behavior when it occurs. Therefore, CRF is often used to initially teach a response but it is usually replaced by partial reinforcement as the response becomes stronger.

During partial reinforcement (PRF), some instances of a response do not yield the reinforcer. There are four basic schedules of partial reinforcement. In a fixed ratio schedule (FR x), a reinforcer is delivered after every x occurrences of a response. For example, in a piecework factory, you might be paid (reinforcer) every time you completed 10 widgets (10 responses). This would be an FR 10 schedule. In a variable ratio schedule (VR x), a reinforcer is delivered after every xth occurrence of the response on average. For example, a foraging pigeon probably does not find food (reinforcer) each time it pecks the ground (response), but it does find food after some variable number of pecks.

In a fixed interval (FI x min) schedule, a reinforcer follows the first response emitted after a fixed period of time since the last reinforcer. For example, if the post is delivered at approximately the same time every day, then the response of checking your mailbox will be followed by the reinforcer of finding mail approximately 24 h after the last time that you received mail. In a variable interval (VI *x* min) schedule, a reinforcer follows the first response emitted after a variable period of time since the last reinforcer. For example, the response of getting through when dialing a telephone number that is busy is probably reinforced on a VI schedule. Notice that that response must be emitted or the reinforcer will not be delivered during FI and VI schedules. Notice also that only one response is required to produce the reinforcer. In spite of this, animals emit many more than one response per reinforcer when responding on these schedules.

Psychologists distinguish between these four basic schedules because the schedules control behavior in different ways. Animals respond at a relatively high and steady rate when responding on VR and VI schedules. In contrast, animals pause after reinforcement when responding on FR and FI schedules. When responding begins, it either continues at a steady rate after the pause (FR schedules) or gradually accelerates to a peak rate just before the next reinforcer is delivered (FI schedule). The length of the pause following reinforcement is directly proportional to the size of the schedule requirement. For example, animals may pause for approximately 30 s when responding on an FI 1 min schedule that makes reinforcers available once per min. They may pause for approximately 5 min when responding on an FI 10 min schedule that makes reinforcers available only once every 10 min. If the requirement becomes too large, the animal may stop responding entirely. This is called 'ratio strain'. Ratio strain can be reversed by returning to a less difficult requirement for reinforcement.

CHARACTERISTICS OF CONDITIONED BEHAVIOR

Acquisition

Conditioned responses gradually gain strength as the reinforcer repeatedly follows the response or the CS repeatedly predicts the US. For example, the amount of saliva that Pavlov collected when he presented the CS (strength of the CR) would be greater after ten pairings of the CS with the US than it was after only two pairings. However, conditioned responses do not gain strength indefinitely. Eventually, a point is reached beyond which further pairings of the CS with the US or the response with the reinforcer do not further increase the strength of the conditioned behavior.

Extinction

The extinction of a classically conditioned response refers to the return of a CR to its baseline strength after the relation between the CS and US is broken. Baseline strength is the strength of the CR before conditioning. Extinction may be accomplished in either of two ways. The US may be removed entirely, or the CS and US may be presented randomly with respect to each other. In the earlier example, kangaroos approached (CR) a feeder because the sound of the feeder (CS) predicted that food (US) was available. Kangaroos would stop approaching the feeder (extinction) if either the feeder was empty so that the sound occurred with no food (US removal) or if the feeder was broken so that the sound occurred randomly with respect to the availability of food (random CS and US presentation).

The extinction of an operantly conditioned response refers to the return of that response to its baseline strength when the relation between the response and the reinforcer is broken. Again, this relation may be broken by removing the reinforcer or by presenting the reinforcer randomly with respect to the response. For example, if you work (response) because you have been paid for working (reinforcer), you would work less (extinction) if you were no longer paid (the reinforcer was removed) or if you won the lottery (received money regardless of whether you worked or not).

Generalization

Generalization of classical conditioning refers to the fact that a CR to one particular CS also occurs in response to other stimuli that resemble that CS. The greater the resemblance between the new stimulus and the CS that was actually paired with the US, the stronger the CR to the new stimulus. For example, if you are stung (US) by a bee (CS), you may learn to fear (CR) other insects, and your fear will be stronger the more closely the insect resembles a bee.

In operant conditioning, a response that has been reinforced in the presence of one stimulus will also occur in the presence of other stimuli that resemble the original stimulus. Again, the response to a new stimulus will be stronger, the more that stimulus resembles the stimulus in the presence of which the response was reinforced. For example, teachers do not have to teach appropriate classroom behavior at the beginning of each school term. The new classroom resembles the students' classrooms of the previous year. Therefore, behaviors (e.g. sitting quietly) that were reinforced in the old classroom occur in the new one without training.

Discrimination

During a classical conditioning discrimination procedure, one stimulus (CS+) predicts the US and another stimulus (CS−) does not. The CR occurs to CS+ but not to CS−. In our kangaroo example, the kangaroos approach (CR) the sound of the food magazine (CS) because it predicts food (US). They do not approach the sound of the gate opening (CS−) because it does not predict food (US).

During an operant discrimination procedure, a response is reinforced in the presence of one stimulus (S+) and not in the presence of another (S−). The response occurs in the presence of S+, but not in the presence of S−. For example, if a child learns that whining (response) gets him anything that he wants (reinforcer) when his father is around (S+), but not when his mother is around (S−), the child will whine only when his father is present.

Discrimination procedures provide useful techniques for asking questions of nonhuman animals or nonverbal people (e.g. infants). You may have heard that dogs do not see colors and wondered how we know. Part of the answer comes from discrimination training. Suppose you reinforce sitting up by giving the dog a treat in the presence of anything red, but not in the presence of anything green. If the dog can see colors, then you will quickly have a dog that sits up when a red, but not a green, stimulus is presented. This experiment must be done carefully. For example, if red and green stimuli reflect different amounts of light, then the dog will solve the discrimination on the basis of the brightness of the stimuli rather than on the basis of their color. However, when this experiment is done properly, dogs have difficulty forming a discrimination between red and green. As a result of many discrimination experiments, a great deal is known about the sensory and conceptual worlds of infants and many species of nonhuman animals.

Higher-order Conditioning

Some stimuli innately serve as unconditioned stimuli or reinforcers without additional training. These stimuli are called 'primary reinforcers' (or primary US). They include biologically important stimuli, such as food and water. Other stimuli acquire their ability to act as reinforcers through experience. These stimuli are called 'secondary', 'conditioned' or 'higher-order' reinforcers (or secondary US). Money provides the most obvious example of a secondary reinforcer.

Stimuli acquire the ability to act as secondary reinforcers in many ways. Two examples illustrate this. First, secondary reinforcers called 'tokens' are stimuli that can be exchanged for primary reinforcers. For example, money acquires the ability to act as a reinforcer because it can be exchanged for food, drink and other primary reinforcers. Second, classical conditioning pairing of a stimulus with a primary US or reinforcer will produce a secondary reinforcer or US. Therefore, a bell that is used to summon animals for feeding will gain the ability to act as a reinforcer itself.

The ability of stimuli to act as secondary reinforcers will be extinguished if their relation to the primary reinforcer or US is broken. Therefore, money would gradually lose its ability to reinforce if it was no longer exchangeable for goods, and the bell would lose its ability to reinforce if it was presented often without food.

Further Reading

Domjan M (1998) *The Principles of Learning and Behavior*, 4th edn. Pacific Grove, CA: Brooks/Cole Publishing.

Hearst E and Jenkins HM (1974) *Sign-tracking: The Stimulus-reinforcer Relation and Directed Action*. Austin, TX: Psychonomic Society.

Honig WK and Staddon JER (1977) *Handbook of Operant Behavior*. Englewood Cliffs, NJ: Prentice-Hall.

Mackintosh NJ (1974) *The Psychology of Animal Learning*. New York, NY: Academic Press.

Pavlov IP (1927) *Conditioned Reflexes*, translated by GV Anrep. London, UK: Oxford University Press.

Pear JJ (2001) *The Science of Learning*. Philadelphia, PA: Psychology Press.

Pearce JM (1997) *Animal Learning and Cognition*. Hove, UK: Psychology Press.

Pierce WD and Epling WF (1999) *Behavior Analysis and Learning*. Upper Saddle River, NJ: Prentice-Hall.

Rescorla RA (1988) Pavlovian conditioning: it's not what you think it is. *American Psychologist* **43**: 151–160.

Skinner BF (1938) *The Behavior of Organisms*. New York, NY: Appleton-Century-Crofts.

Conflict

See **Cooperation and Conflict, Large-scale Human**

Connectionism

Intermediate article

Jerome A Feldman, University of California, Berkeley, California, USA
Lokendra Shastri, University of California, Berkeley, California, USA

CONTENTS

Science must eventually explain how the properties of the brain determine cognitive behavior. Connectionist approaches to cognitive science are based on the belief that our present understanding of neural and brain physiology should guide cognitive research.

INTRODUCTION

Since the detailed anatomy of higher cognition is not known, most connectionist research involves mathematical and computational modeling of behavior. The same modeling techniques also have a wide range of biological, engineering and management applications, and the discipline called 'neural networks' or 'connectionism' now encompasses a wide range of efforts that have only a vague link to the brain or cognition. This article will describe the kinds of connectionist model used in cognitive science and their implications for how we think about thinking.

Connectionist computational models are almost always computer programs, but programs of a different kind from those used in, for example, word processing, or symbolic artificial intelligence (AI). Connectionist models are specified as networks of simple computing units, which are abstract models of neurons. Typically, a model unit calculates the weighted sum of its inputs from upstream units and sends to its downstream neighbors an output signal that is a nonlinear function of its inputs. Learning in such systems is modeled by experience-based changes in the weights of the connections between units.

EARLY DEVELOPMENTS

There is a distinguished prehistory to current connectionist modeling techniques. Many of the original papers are gathered in (Anderson and Rosenfeld, 1988).

Sigmund Freud and William James, arguably the greatest psychologists of the nineteenth century, both discussed explicit neural models of cognition. For much of the first half of the twentieth century, the Anglo-American study of cognition was dominated by the behaviorist paradigm, which rejected any investigation of internal mechanisms of mind. In the 1940s, there arose two new connectionist paradigms, which still shape the field. McCulloch and Pitts (McCulloch, 1988) emphasized the study of particular computational structures comprised of abstract neural elements, while Hebb (1949) focused on the properties of assemblies of cells. The study of bulk properties of abstract neural systems has continued through Hopfield networks (Hopfield, 1982) and the Boltzman machine

(Ackley *et al.*, 1985), but has had little impact in cognitive science. One lasting result of Hebb's work was the Hebbian model of learning, which will be discussed in the next section.

The notion of a computational model is now commonplace in all scientific fields and in many other areas of contemporary life. One builds a detailed software model of some phenomenon and studies the behavior of the model, hoping to gain understanding of the original system. While all fields use computational models, connectionist researchers also invent and study computational techniques for constructing models, presenting the results of simulations, and understanding the limitations of the simulation. From the time of the first electronic computers, people dreamed of making them 'intelligent' by two quite distinct means. The first is to build standard computer programs as models of intelligence. This remains the dominant paradigm in AI and has had considerable success. The second approach was to try to build hardware that was as brain-like as possible and have it learn the required behavior, this was the origin of connectionist modeling.

LEARNING CONNECTION WEIGHTS

Connectionist models consist of simple computing units (or nodes) connected via weighted links. A node communicates with the rest of the network by transmitting its output to all the nodes immediately downstream of it. This output is a monotonic (and often nonlinear) function of the node's total input. The contribution of each input link to the receiving node's total input is the output of the node at its source multiplied by the weight of the link. The most commonly used input–output transfer functions compute the total input by summing the inputs contributed by all the weighted links and then passing this total input through a nonlinear function. An example of such a nonlinear function is the step function, whereby the node produces an output of 1 (i.e. fires) if and only if the total input exceeds a threshold. Another commonly used function is the sigmoid function, whereby the output of a node varies smoothly between −1 and +1 as a function of the total input (the shape of this response curve resembles a stretched S).

The behavior of a connectionist network is completely determined by the input–output transfer function, the pattern of interconnection among the nodes, and the link weights. Typically, the input–output transfer function is assumed to remain fixed for a given model. It is possible to 'sculpt' the interconnection pattern by starting with a completely connected network and then pruning it by setting the weights of certain links to zero. Thus, changes in the functionality of a connectionist network (learning) can be effected by changing only the link weights; and most mechanisms and theories of learning in connectionist networks focus on changes in link weights.

One of the most influential proposals about learning in neural networks was made by Hebb (1949) and is known as Hebb's rule. In its simplest form, this rule states that the weight of the link from node i to node j is strengthened if i repeatedly participates in the firing of j. This rule and its variants lie at the core of many learning rules investigated by neural network modelers and theorists. In its simplest form, the rule suffers from several technical difficulties. These include lack of specificity, and saturation (all link weights would increase until they eventually reached their maximum possible values). To tackle this problem, Hebb's rule has been supplemented with an anti-Hebbian learning rule which says that the weight from node i to node j decreases if node j fires but node i does not; or alternatively, if node i fires but node j does not. The 'BCM' rule (Bienenstock, Cooper and Munro, 1982) states that the weight vector of a node is tilted in the direction of the input vector if the output exceeds a threshold, or in the direction opposite to that of the input vector if the output is below that threshold. The threshold can be variable, and may depend on the time-averaged output of the node. Note that the weight vector of a node with n input links is simply the n-tuple consisting of the n link weights.

Another framework for modifying link weights is competitive learning (von der Malsburg, 1973; Grossberg, 1976; Rumelhart and Zipser, 1985). Informally, the competitive learning algorithm may be described as follows. Let each node compute its output to be the weighted sum of its inputs, and assume that weight and input vectors are normalized to one. Then the output of a node is given by $\cos \alpha$, where α is the angle between the weight vector and the input vector. Weight modification in response to external inputs proceeds as follows: (1) An input pattern is presented to the network; (2) the node with the highest response is identified (this can be achieved by a 'winner takes all' configuration (Feldman and Ballard, 1982)); (3) the weight vector of this winning node is rotated towards the input vector; (4) the weight vector is renormalized. These steps are repeated for each input pattern. As a result of this weight modification regime, the nodes in the network modify their link weights so as to respond maximally to frequently occurring input patterns. In effect, each cluster (or category) in the input space 'recruits' one or more nodes in the network that produce a strong response to input patterns from the associated cluster.

The competitive learning algorithm can be augmented by requiring that (1) nodes in the network are arranged in a low-dimensional lattice (for example, in a linear sequence, or in a two-dimensional grid) and (2) changes in the weight vector of a node are accompanied by similar changes in the weight vectors of neighboring nodes (for example, via lateral connections between neighboring nodes). Such an augmented neural network learning algorithm is called a self-organizing map (Kohonen, 1982). Self-organizing maps have been applied to a variety of problems in pattern recognition. They have the important capability of developing topology-preserving maps from a high-dimensional space to a low-dimensional space. Examples of such maps abound in the brain: for example, several visual features (e.g., location in the visual field, orientation, direction of motion) are mapped to a layered two-dimensional cortex.

The 'adaptive resonance theory' architecture (Grossberg, 1980) also extends the competitive learning framework by incorporating several new features, such as separate layers for representing features and categories, top-down and bottom-up interactions between these layers, lateral inhibition, and mechanisms for regulating attention and detecting novelty. These mechanisms allow such models to control the granularity of classification, while achieving a balance between plasticity (the ability to learn new categories) and stability (the ability to retain categories that have already been learned).

All of the learning algorithms discussed above belong to the category of unsupervised, or self-organizing, algorithms, since their weight modification depends solely on the inputs. A different class of algorithms arises if the inputs are accompanied by some form of feedback indicative of the network's performance. This feedback can be used by the network to to guide the modification of link weights. Such feedback can be non-specific, providing a single measure of the network's performance (such as a simple positive or negative reinforcement signal); or highly specific, providing a detailed description of the desired network response (for example, a specification of the output pattern for each input pattern). Algorithms with non-specific feedback are called reinforcement learning algorithms (Sutton and Barto, 1998); while algorithms with specific feedback are called supervised learning algorithms (Rosenblatt, 1958; Block, 1962; Minsky and Papert, 1988; Rumelhart *et al.*, 1986; Werbos, 1994).

In the most general terms, reinforcement learning means learning to act in a manner that maximizes the expected future reinforcement (reward). A simplified version of this problem can be formulated as follows. In a trial T, the system observes an input pattern (stimulus) $x(T)$ and responds by performing action $a(x(T))$. After a number of such moves, the system receives a positive or negative reinforcement. The objective of the learning algorithm is to discover a policy whereby at each trial, the system chooses an action that maximizes the expected total reward. This is difficult for several reasons. First, there is uncertainty associated with actions: given the complexity and variability of the environment, the result of action a_i in response to input $x(T)$ is not fixed. Second, there can be a significant delay between choosing an action and receiving a reinforcement. Third, examples of optimal actions are not provided and must be discovered by the system through trial and error. Several reinforcement learning algorithms have been developed. These include the 'temporal difference method' and 'Q-learning' (Sutton and Barto, 1998). The central idea in these learning algorithms is, for each reward signal, to assign appropriate credit or blame to the decisions that led to that signal. After enough training, the system acquires tables that list the best action (that which maximizes the expected reward) for each combination of state and input. These tables are taken to stand for weighted connections in a neural network.

Reinforcement learning is more plausible biologically than supervised learning, which requires a teacher to provide detailed answers rather than just a numerical score. It has been applied to models of motor control and related phenomena. But it is much slower than supervised learning in complex tasks, and has not been widely used in cognitive models.

THE PERCEPTRON AND ITS SUCCESSORS

The most striking early result about supervised learning was the perceptron learning theorem (Rosenblatt, 1958; Block, 1962; Minsky and Papert, 1988). The perceptron model uses a single layer of linear threshold units for categorizing simple visual patterns. Each unit calculates the weighted sum of its inputs and fires (outputs 1) if this sum is greater than the unit's threshold. The associated learning rule involves changing each weight if the unit's prediction about an input pattern was wrong. The theorem showed that any classification that could be computed by a perceptron could be learned by this simple rule.

This theorem gave rise to the belief that ever more complex behavior could be learned directly from feedback. This turned out to be overly optimistic as other scientists, notably Minsky and Papert (1988), showed that many simple distinctions could not be captured by such simple networks. This led to a relatively quiet period for connectionist modeling.

Around 1980, a variety of ideas from biology, physics, psychology, computer science and engineering coalesced to yield a 'new connectionist' approach to modeling intelligence, which has become a core field of cognitive science and also the basis for a wide range of practical applications (McClelland and Rumelhart, 1986). Among the important advances was a mathematical technique (back propagation) that extended the early work on perceptron learning to a much richer set of network structures. Two ideas allowed error feedback to be used to train networks with multiple layers. The first idea was to replace the linear unit function of the perceptron with a smooth and bounded function, typically the sigmoid described above. The second idea exploits the chain rule for partial derivatives to assign appropriate amounts of credit and blame to model units that do not connect directly to the output nodes, where comparisons with training data are made. While it is theoretically possible to apply back propagation to arbitrary networks (McClelland and Rumelhart, 1986), this does not work well in practice, and essentially all of the back propagation work in cognitive science has been based on two architectural styles.

Most of the early connectionist learning models used strictly layered networks with no feedback at all (McClelland and Rumelhart, 1986). These have limited computational ability, but are still used as components of larger models, some of which can be quite elaborate (Plaut and Kello, 1999). A simple modification of the layer architecture (Elman *et al.*, 1996) involves adding fixed-weight connections from a hidden layer at one timestep to the same layer at the next timestep. Because the feedback weights remain fixed, the back propagation learning algorithm remains the same, but these networks can learn some serial tasks beyond the capabilities of layered networks without feedback. Most of the current work on learning in unstructured tasks uses this architecture.

THE CONNECTIONIST APPROACH TO COGNITIVE SCIENCE

The basic connectionist style of modeling is now being used in different ways in neurobiology, in applications, and in cognitive science. Neurobiologists who study networks of neurons employ a wide range of computational models, from very detailed descriptions of the internal chemistry of the neuron to the abstract units described above.

In cognitive science, connectionist techniques have been used for modeling all aspects of language, perception, motor control, memory and reasoning. This universal coverage represents a potential breakthrough: previous computational models of, for example, early vision and of problem solving used entirely different mathematical and computational techniques. Since the brain is known to use the same neural computation throughout, it is not too surprising that models based on this paradigm can be applied to all behavior. The existing models are neither broad nor deep enough to ensure that the current set of mechanisms will suffice to bridge the gap between structure and behavior, but the work remains productive.

Connectionist models in cognitive science belong to two general categories, often called structured (or localist) and layered (or parallel distributed processor (PDP)) networks. Most connectionists are primarily interested in learning, which is modelled as experience-driven change in connection weights. There is a great deal of research studying different models of learning with and without supervision, different rules for changing weights, and so on. Since the focus of such an experiment is on what the network can learn, any imposed structure will weaken the results of the experiment. The standard approach is to use networks with unidirectional connections arranged in completely connected layers, sometimes with a very restricted additional set of feedback links. This kind of network contains a minimal amount of imposed structure and is also amenable to efficient learning techniques such as the back propagation method described above. Most researchers using totally connected layered models do not believe that the brain shares this architecture, but there is a controversy, which we will discuss later, about the implications of PDP learning models for theories of mind.

Structured connectionist models are usually focused less on learning than on the representation and processing of information. Essentially all the modeling done by neurobiologists involves specific architectures, which are known from experiment. For structured connectionist models of cognitive phenomena, the underlying brain architecture is rarely known in detail and sometimes not at all at the level of neurons and connections. The methodology employed is to experiment with computational models of the behavior under study that are consistent with the known biological and psychological data and are plausible in terms of the resources (neurons, computing time, etc.) they require. This methodology is very similar to what are called spreading activation models and are widely used in psycholinguistics. Some studies combine structured and layered networks (Regier, 1996) or investigate learning in networks with initial structure dependent on the problem area or the known neural architecture.

Another important difference between the structured and the layered approaches is that the layered approach assumes that each 'item' (or mental object) is represented as a pattern of activity distributed over a common pool of nodes (van Gelder, 1992). This notion of 'holographic' representation suffers from some fundamental limitations. Consider the representation of 'John and Mary'. If 'John' and 'Mary' are represented as patterns of activity over the entire network such that each node in the network has two specific patterns for 'John' and 'Mary' respectively, then how can the network represent both 'John' and 'Mary' at the same time? The situation gets even more complex if the system has to represent relations such as 'John loves Mary', or 'John loves Mary but Tom loves Susan'. In contrast to the layered approach, the structured approach uses small clusters of nodes with distinct representational status.

An early example of a structured connectionist model was the interactive activation model for letter perception developed by McClelland and Rumelhart (1981). This model consisted of three layers of nodes, corresponding to visual features of letters, letters, and words. Nodes representing mutually exclusive hypotheses within the letter and word layers inhibited each other. For example, nodes representing letters in the same position inhibited each other, since only one letter can exist in a given position. A node in the feature layer was connected via *excitatory* connections (connections with positive weights) to nodes in the letter layer representing letters that contained that feature. Similarly, a node in the letter layer was connected via *excitatory* connections to nodes in the word layer representing words that contained that letter in the appropriate position. Additionally, there were reciprocal connections from the word layer to the letter layer. The interconnection pattern allowed bottom-up perceptual processing to be guided by top-down expectations. The model could explain a number of psychological findings about the preference of words and pronounceable non-words over other non-words and isolated letters. Other examples of early structured connectionist models included word sense disambiguation models (Cottrell and Small, 1983; Waltz and Pollack, 1985), and the semantic network model CSN (Shastri and Feldmanx, 1986). CSN encoded 'is a' relations using links, and property values using binder nodes that connected property, value and concept nodes. CSN could infer the most likely value of a specified property for a given concept, and find the concept that best matched a partial description.

The Binding Problem

A critical limitation of the early connectionist models was that they could not encode dynamic bindings. Consider the representation of the event 'John gave Mary a book'. This cannot be represented by simply activating the conceptual roles 'giver', 'recipient', and 'gift', and the entities 'John', 'Mary', and 'a book'. Such a representation would be identical to that of 'Mary gave John a book'. Unambiguous representation of an event requires the representation of bindings between the roles of an event (e.g., giver) and the entities that fill these roles in the event (e.g., John). In conventional computing, binding is carried out by variables and pointers, but these techniques have no direct neural counterparts.

Structured connectionist modelers have made significant progress towards neurally plausible solutions to the binding problem. Some of these models use the relative positions of active nodes and the similarity of firing patterns to encode bindings (Barnden and Srinivas, 1991). Some assign distinct activation patterns (signatures) and propagate these signatures to establish bindings (e.g., Lange and Dyer, 1989). Some models use synchronous firing of nodes to represent and propagate dynamic bindings (Shastri and Ajjanagadde, 1993; Hummel and Holyoak, 1997). The possible role of synchronous activity for feature binding had been suggested earlier by others (e.g., von der Malsburg, 1981), but a model called Shruti (Shastri and Ajjanagadde, 1993) offered the first detailed computational account of how such activity can be used to solve problems in the representation and processing of high-level conceptual knowledge and to carry out inference.

Shruti is a structured connectionist model that can encode semantic, causal, and episodic knowledge and perform inferences to establish referential and causal coherence. Shruti encodes relational knowledge using neural circuits composed of focal cell clusters. A systematic mapping between relations (and other rule-like knowledge) is encoded by highly efficacious links between focal clusters. Inference in Shruti results from the propagation of rhythmic activity across interconnected focal clusters. There is no interpreter or inference engine that manipulates symbols or applies rules of inference. In general, Shruti combines predictive inferences with explanatory (or abductive) inferences, instantiates new entities during inference, and unifies multiple entities by merging their phases of firing.

Recruitment Learning

In addition to incremental learning driven by repeated exposure to a large body of training data, structured models have also made use of one-trial learning using 'recruitment learning' (e.g., Feldman, 1982). Recruitment learning can be described as follows. Learning occurs within a structured network containing an unassigned group of randomly connected nodes. Recruited nodes in such a network acquire distinct 'meanings' (or functionality) by virtue of their strong connections to previously structured nodes. For example, a novel concept that is a conjunct of existing concepts x_1 and x_2 can be learned by (1) identifying free nodes that receive links from nodes representing both x_1 and x_2 and (2) 'recruiting' one or more such free nodes by strengthening the weights of links incident on such nodes from x_1 and x_2 nodes. In general, the recruitment process can transform a quasi-random network into a collection of nodes and circuits with specific functionalities.

It has been shown (Shastri, 1999) that recruitment learning can be firmly grounded in the biological phenomena of long-term potentiation and long-term depression, which involve rapid, long-lasting, and highly specific changes in synaptic strength.

Rules versus Connections

Perhaps the most visible contribution of connectionist computational models in cognitive science has been to provide a new conceptual framework for some long-standing debates on the nature of intelligence. Much of the current debate is being published in scientific journals. The question of nature versus nurture concerns how much of some trait, usually intelligence, can be accounted for by genetic factors and how much depends on postnatal environment and training. Some PDP connectionists have taken very strong positions suggesting that learning can account for everything interesting (Elman *et al.*, 1996). In the particular case of grammar, an important group of linguists and other cognitive scientists take an equally extreme nativist position, suggesting that humans only need to choose a few parameters to learn grammar. A related question is the whether human grammatical knowledge is represented as general rules or just appears as the rule-like consequences of PDP learning in the neural network of the brain. There is ample evidence against both extreme positions, but the debate continues to motivate a great deal of thought and experimentation.

CURRENT AND FUTURE TRENDS

Quantitative neural models are playing a major role in cognitive science. The mathematical and computational ideas underlying learning in neural networks have also found application in a wide range of practical problems, from speech recognition to financial prediction. Given current computing power, back propagation and similar techniques allow large systems of nonlinear units to learn reasonably complex probabilistic relationships using labelled data. This general methodology overlaps not only with AI but also with mathematical statistics, and is part of a unifying theory called computational learning theory. There is also a large community of scientists and engineers who are working on neural networks and related statistical techniques for various tasks, and there are conferences and journals to support this effort.

The explosion of activity on the internet is affecting the whole field of computing. Two application areas that seem particularly important to cognitive science are intelligent web agents and spoken language interaction. As the range of users and activities on the internet continues to expand, there is increasing demand for systems that are both more powerful and easier to use. This is leading to increased efforts on the design of human–computer interfaces, including the modeling of users' plans and intentions – clearly overlapping with traditional concerns of cognitive science. One particularly active area is interaction with systems using ordinary language. While machine recognition of individual spoken words is relatively successful, dealing with the full richness of language is one of the most exciting challenges facing computing and cognitive science, and a problem of great commercial and social importance.

Looking ahead, technical work on connectionist modeling is likely to remain closely linked to statistics and learning theory. From the scientific perspective, it is likely that most interdisciplinary connectionist research in cognitive science will remain focused on specialized domains such as language, speech and vision. With the rapid advances in neurobiology, the field will increasingly connect with the life sciences, yielding great mutual benefits (e.g. Carter *et al.*, 1998).

References

Ackley DH, Hinton GF and Sejnowski TJ (1985) A learning algorithm for Boltzman machines. *Cognitive Science* **9**: 147–169.

Anderson JA and Rosenfeld E (eds) (1988) *Neurocomputing: Foundations of Research.* Cambridge, MA: MIT Press.

Barnden J and Srinivas K (1991) Encoding techniques for complex information structures in connectionist systems. *Connection Science* **3**(3): 269–315.

Bienenstock EL, Cooper LN and Munro PW (1982) Theory for the development of neuron selectivity: orientation specificity and binocular interaction in visual cortex. *Journal of Neuroscience* **2**: 32–48. [Reprinted in Anderson and Rosenfeld, 1988.]

Block HD (1962) The perceptron: a model for brain functioning. *I Reviews of Modern Physics* **34**: 123–135. [Reprinted in Anderson and Rosenfeld, 1988.]

Carter CS, Braver TS, Barch DM *et al.* (1998) Anterior cingulate cortex, error detection and the on-line monitoring of performance. *Science* **280**: 747–749.

Cottrell GW and Small SL (1983) A connectionist scheme for modeling word sense disambiguation. *Cognition and Brain Theory* **6**: 89–120.

Elman J, Bates E and Johnson M (1996) *Rethinking Innateness: A Connectionist Perspective on Development (Neural Network Modeling and Connectionism).* Cambridge, MA: MIT Press.

Feldman JA (1982) Dynamic connections in neural networks. *Biological Cybernetics* **46**: 27–39.

Feldman JA and Ballard DB (1982) Connectionist models and their properties. *Cognitive Science* **6**: 205–254. [Reprinted in Anderson and Rosenfeld, 1988.]

van Gelder T (1992) Defining 'distributed representation'. *Connection Science* **4**(3,4): 175–191.

Grossberg S (1976) Adaptive pattern classification and universal recoding. *Biological Cybernetics* **23**: 121–134. [Reprinted in (Anderson and Rosenfeld, 1988).]

Grossberg S (1980) How does a brain build a cognitive code? *Psychological Review* **87**: 1–51. [Reprinted in Anderson and Rosenfeld, 1988.]

Hebb DO (1949) *The Organization of Behavior.* New York, NY: Wiley.

Hopfield JJ (1982) Neural networks and physical systems with emergent collective computational abilities. *Proceedings of the National Academy of Sciences* **79**: 2554–2558. [Reprinted in Anderson and Rosenfeld, 1988.]

Hummel JE and Holyoak KJ (1997) Distributed representations of structure: a theory of analogical access and mapping. *Psychological Review* **104**: 427–466.

Kohonen T (1982) Self-organized formation of topologically correct feature maps. *Biological Cybernetics* **43**: 59–69. [Reprinted in Anderson and Rosenfeld, 1988.]

Lange TE and Dyer MG (1989) High-level inferencing in a connectionist network. *Connection Science* **1**(2): 181–217.

von der Malsburg C (1973) Self-organization of orientation sensitive cells in the striate cortex. *Kybernetik* **14**: 85–100. [Reprinted in Anderson and Rosenfeld, 1988.]

von der Malsburg C (1981) The correlation theory of brain function. Internal Report 81–2, Department of Neurobiology, Max Planck Institute for Biophysical Chemistry, Göttingen, Germany.

McClelland JL and Rumelhart DE (1981) An interactive activation model of context effects in letter perception: part 1: an account of basic findings. *Psychological Reviews* **88**: 375–407. [Reprinted in Anderson and Rosenfeld, 1988.]

McClelland J and Rumelhart D (1986) *Parallel Distributed Processing.* Cambridge, MA: MIT Press.

McCulloch WS (1988) *Embodiments of Mind.* Cambridge, MA: MIT Press. [Reprint edition.]

Minsky ML and Papert SA (1988) *Perceptrons: Introduction to Computational Geometry.* Cambridge, MA: MIT Press. [Expanded edition; first published 1969.]

Plaut DC and Kello CT (1999) The emergence of phonology from the interplay of speech comprehension and production: a distributed connectionist approach. In: MacWhinney B (ed) *The Emergence of Language,* pp. 381–415. Mahwah, NJ: Erlbaum.

Regier T (1996) *The Human Semantic Potential: Spatial Language and Constrained Connectionism.* Cambridge, MA: MIT Press.

Rosenblatt F (1958) The perceptron: a probabilistic model for information storage and organization in the brain. *Psychological Review* **65**: 386–408. [Reprinted in Anderson and Rosenfeld, 1988.]

Rumelhart D, Hinton GE and Williams RJ (1986) *Learning internal representations by error propagation.* In: McClelland and Rumelhart, 1986.

Rumelhart D and Zipser D (1985) *Feature discovery by competitive learning.* In: McClelland and Rumelhart, 1986.

Shastri L (1999) Biological grounding of recruitment learning and vicinal algorithms in long-term potentiation and depression. Technical Report TR-99-009, International Computer Science Institute, Berkeley, CA.

Shastri L and Ajjanagadde A (1993) From simple associations to systematic reasoning: a connectionist encoding of rules, variables and dynamic bindings using temporal synchrony. *Behavioral and Brain Sciences* **16**(3): 417–494.

Shastri L and Feldman JA (1986) Neural nets, routines and semantic networks. In: Sharkey N (ed.) *Advances in Cognitive Science*, pp. 158–203. Chichester, UK: Ellis Horwood.

Sutton RS and Barto AG (1998) *Reinforcement Learning: An Introduction*. Cambridge, MA: MIT Press.

Waltz D and Pollack J (1985) Massively parallel parsing: a strongly interactive model of natural language interpretation. *Cognitive Science* **9**: 51–74.

Werbos P (1994) *The Roots of Back-Propagation: From Ordered Derivatives to Neural Networks and Political Forecasting*. New York, NY: Wiley.

Connectionism and Systematicity

Intermediate article

Robert J Matthews, Rutgers University, New Brunswick, New Jersey, USA

CONTENTS

Connectionists claim that human cognitive computational architecture is connectionist. Proponents of classical computational architectures have challenged this claim, arguing that a pervasive feature of human cognition, its 'systematicity', cannot be explained in connectionist terms.

INTRODUCTION

Connectionists claim that human cognitive computational architecture is connectionist, specifically that connectionist computational architectures provide an empirically more plausible foundation for computational theories of cognition than do more familiar classical computational architectures. Classicists, who conceive of cognition as implemented by a classical computational architecture, specifically as the structure-sensitive processing of syntactically structured mental representations, argue that the prospects for a connectionist theory of cognition are remote. Connectionist models, they contend, cannot explain, or at least have not shown that they can explain, any of the fundamental aspects of human cognition, most notably its seemingly pervasive 'systematicity', namely: the fact that many cognitive capacities are systematically related, so that as a matter of psychological law one possesses one capacity if and only if one possesses the other.

THE FODOR–PYLYSHYN CHALLENGE TO CONNECTIONISM

As first issued by Fodor and Pylyshyn (1988), the challenge to explain systematicity was one that connectionists purportedly could not meet because connectionist architectures lacked an essential feature found only in classical computational architectures, namely, representational states with a compositional constituent structure. Classical architectures, Fodor and Pylyshyn argued, postulate certain syntactically structured mental representations that, like the sentences of a natural language, exhibit a compositional syntax and semantics, such that the semantic content of a complex representation is a function of the semantic content of its constituents and of its syntax structure. (The sentence-like character of the postulated representations explains why classicists sometimes call these 'language of thought' architectures.) The computational processes that operate over these compositionally structured representations apply by virtue of their syntactic properties.

Connectionist representations, by contrast, lack such compositional constituent structure; hence, the computational processes postulated in connectionist architectures cannot, Fodor and Pylyshyn argued, be causally sensitive to constituent structure. But such sensitivity, they claimed, is essential to explaining systematicity. Fodor and Pylyshyn concluded not only that connectionist architectures could not explain systematicity, but that they were unable even to exhibit systematicity: 'the architecture of the mind is not a connectionist network'.

Fodor and McLaughlin (1990) put the challenge to connectionists this way: how can you 'explain the existence of systematic relations among cognitive capacities without assuming that cognitive processes are causally sensitive to the constituent structure of mental representations'? Fodor and McLaughlin argue that connectionists are faced with a dilemma: 'if connectionism can't account for systematicity, it thereby fails to provide an adequate basis for a theory of cognition; but if its account of systematicity requires mental processes that are sensitive to the constituent structure of mental representations, then the theory of cognition that it offers will be, at best, an implementation architecture for a "classical" (language of thought) model.' Given that a classical architecture could be implemented on a connectionist architecture (just as connectionist architectures can be, and often are, implemented on a classical architecture), the challenge to connectionists, Fodor and McLaughlin argue, is to demonstrate that connectionism can explain systematicity without implementing a classical architecture.

Fodor and Pylyshyn (1988) offer little by way of a characterization of the property – systematicity – that connectionists are challenged to explain. Instead they offer numerous examples of systematically related cognitive capacities. Many of their examples are drawn from language. They note, for example, that you do not find subjects who know how to say in English that John loves the girl, but who do not know how to say in English that the girl loves John; or who can understand the English sentence 'the monkey bit the lab technician', but who cannot understand the systematically related English sentence 'the lab technician bit the monkey'. In Fodor and McLaughlin (1990), and especially in McLaughlin (1993), systematicity is described more broadly, as a capacity for *intentional*, more specifically 'propositional attitude', states whose contents are systematically related, such that one has the capacity for one such intentional state (e.g. believing that the circle is above the square) just in case one has the capacity for systematically related states (e.g. believing that the square is above the circle).

PHILOSOPHICAL RESPONSES TO THE CHALLENGE

Some connectionists have been willing to accept the classicists' challenge, undertaking to demonstrate that their networks can, at least in principle, explain systematicity (and furthermore do so without implementing a classical architecture). Others have rejected the challenge, arguing that as posed the challenge is impossible to meet. They argue that the classicists' notion of what it would be to explain systematicity, indeed their notion of systematicity itself, at least as regards thought, is such as to insure that there can be no non-classical explanation of systematicity. Some also question the classicists claim to have an explanation themselves, thus challenging the presumption that connectionists are worse off in this respect than classicists.

The classicists' notion of systematicity in language processing capacity is relatively uncontentious: a language processor can be said to exhibit systematicity if when it can process a sentence *s* it can also process systematic variants of *s*, where systematic variation is understood in terms of permuting syntactic constituents or substituting constituents of the same syntactic category. Their notion of systematicity of thought is considerably more contentious. Fodor defines it in terms of being able to think thoughts of one form (e.g. *aRb*) just in case one can think other thoughts whose forms are systematically related to the first (e.g. *bRa*). But this characterization, as Cummins (1996) points out, begs the question, since it assumes that thoughts have the form of their classical linguistic representations. This objection can be avoided by reformulating the systematicity claim as follows:

> Anyone who can think a thought with the content c can think a thought with the content c^*, where c^* is a systematic variant of c. (1)

But this reformulation, Cummins notes, relativizes the systematicity of thought to the choice of some representational scheme for thought contents. Yet surely the systematicity of thought, which connectionists are challenged to explain, should not depend on the representational scheme that we, as theorists, choose to employ. For without some way of picking out systematically related thought contents (and hence systematically related thoughts), a way that is neutral regarding the representation scheme, it is not clear that any real property of thought has been identified for connectionists to explain. Nor is there any non-question-begging way of arguing from the systematicity of thought to the conclusion that human cognitive architecture is exclusively classical.

The challenge to connectionists is to *explain* systematicity, not simply to construct a connectionist network that exhibits it. Classicists have a narrow notion of what would count as an explanation of systematicity, one that connectionists might have difficulty satisfying even if they were successful in constructing a network that exhibited systematicity. McLaughlin (1993), for example, requires that explanations of cognitive capacities, including systematically related cognitive capacities, explain what the capacity 'consists in'. More precisely, he requires that the explanation take the form of a 'functional analysis'; i.e. an explanation of a complex cognitive capacity such as the capacity for systematic thought in terms of the cooperative interaction of certain more primitive cognitive capacities that are constitutive of the complex capacity. Systematically related cognitive capacities, McLaughlin assumes, are so related by virtue of certain shared constitutive capacities, so that to explain the former is to specify and describe these shared constitutive capacities and their interaction.

Functional analysis offers a plausible form of explanation if classicists are right in their assumptions about the nature of systematically related cognitive capacities. But these are not assumptions that connectionists accept; moreover, it is not clear, as Matthews (1997) argues, that any cognitive explanations that connectionists might provide could take this form. Connectionist networks do not have constitutive capacities of the sort that functional analysis envisions. Of course, the individual units that compose connectionist networks are constituents of those networks, but arguably they are not cognitive constituents of the networks in the classicists' sense of that term; i.e., they are not amenable to cognitive or intentional interpretation. And even if they were constituents in the requisite sense, there would still be a problem of explanatory 'grain': it would in most cases be very difficult, if not impossible, to grasp how the molar capacity of the network comprises ('consists in') the capacities of the individual units that constitute it. The specific contributions of individual units would typically be so diffuse as to preclude any claims about which units are responsible for which aspects of the networks' molar capacity; moreover, the number of units would typically be so large, and their interaction so complex, that it would be beyond our cognitive capacity to grasp how the molar capacity of the network could 'consist in' the capacities of the individual constitutive units.

To concede that connectionist explanations of cognitive capacities are unlikely to take the form of a functional analysis is not to concede that connectionists are unable to meet the classicists' challenge to explain systematicity (assuming that we can define in a non-question-begging way just what it is for thought to be systematic). Rather, it is to point out that *a priori* assumptions about the appropriate form of cognitive explanation are vulnerable to empirical discoveries about the nature of human cognitive architecture. Empirical discoveries about the bases of cognition may entail corresponding changes in what we take to be the appropriate form that explanations of cognition should take.

EMPIRICAL RESPONSES TO THE CHALLENGE

The empirical response to the classicist challenge has thus far been directed primarily towards demonstrating that connectionist networks can, in principle at least, *exhibit* systematicity in domains such as language processing. Thus, for example, Chalmers (1990) describes a connectionist network that is, he says, a 'direct counterexample' to the argument of Fodor and McLaughlin (1990) that to support structure-sensitive processing, representations of constituent structure must contain explicit tokens of the constituents. In part, this response has been prompted by suggestions to the effect that connectionist architectures are unable to explain systematicity because they cannot perform certain computational tasks that classical architectures can. No one challenges the well-known formal proof that connectionist architectures can compute (or at least approximate to any arbitrary degree) any computable function (Hornik *et al.*, 1989), but some have suggested that connectionist architectures may be limited computationally in ways that classical architectures are not. McLaughlin (1993), for example, speaks of connectionist architectures being able only to 'respect' certain syntactic operations that classical architectures can actually execute. Connectionists would naturally wish to dispel any suggestion that connectionist architectures are computationally limited in ways that would prevent them from exhibiting, and hence from explaining, systematicity.

Smolensky has argued that connectionists can explain systematicity, and furthermore can do so without implementing a classical architecture (Smolensky, 1991, 1995; Smolensky *et al.*, 1992). He argues that: (1) connectionist networks are naturally viewed as computing functions defined over vector product representations (i.e. over representations of a vector algebraic form); (2) such representations are adequate to express all constituency relations expressible by means of classical representations; (3) all computational operations definable over classical representations can be effected by connectionist operations defined over vector product representations; and hence (4) connectionists can explain systematicity in terms of such representations. Smolensky presents what he calls an 'integrated connectionist/symbolic architecture' that provides an algorithmic encoding scheme for moving between these two forms of representation. Fodor and McLaughlin (1990) criticize Smolensky's explanatory claims, largely on the grounds that vector product representations only encode but do not actually possess the constituent structure that a classicist explanation of systematicity presumes.

Whatever the merits of these criticisms, Smolensky's argument is clearly a non sequitur. It takes non-epistemological premises regarding the computational capacity of connectionist devices and the availability of a vector product representation of their operations, and reaches an epistemological conclusion regarding the availability of an explanation of the systematicity that connectionist networks so described can exhibit. But the argument nonetheless draws attention to both the known capacity of connectionist networks to compute arbitrary computable functions and the availability of a principled non-classical, specifically vector product, description of the computations of such networks.

CONCLUSION

Much of the empirical response of connectionists to the classicists' challenge has focused on establishing that connectionist networks can do the sorts of things that classicists have claimed that they cannot do. Connectionists have made a good case for the claim that connectionist networks can in principle exhibit systematicity, at least in domains such as language processing where it is tolerably clear what systematicity amounts to. But arguably connectionists have yet to demonstrate that what is in principle possible can in fact be accomplished. It also remains to be seen whether connectionists can actually explain systematicity, especially the systematicity of thought, assuming that classicists and connectionists can agree on what is to be explained and what is to count as an explanation. Ultimately, these are matters for empirical investigation, and cannot be decided *a priori*.

References

Chalmers D (1990) Syntactic transformations on distributed representations. *Connection Science* **2**: 53–62.

Cummins R (1996) Systematicity. *Journal of Philosophy* **93**: 591–614.

Fodor J and McLaughlin B (1990) Connectionism and the problem of systematicity: why Smolensky's solution won't work. *Cognition* **35**: 183–204.

Fodor J and Pylyshyn Z (1988) Connectionism and cognitive architecture: a critical analysis. *Cognition* **28**: 3–71.

Hornik K, Stinchcombe M and White H (1989) Multilayer feedforward networks are universal approximators. *Neural Networks* **2**: 359–366.

Matthews R (1997) Can connectionists explain systematicity? *Mind and Language* **12**: 154–177.

McLaughlin B (1993) The connectionism/classicism battle to win souls. *Philosophical Studies* **71**: 163–190.

Smolensky P (1991) Connectionism, constituency, and the language of thought. In: Loewer B and Rey G (eds) *Meaning in Mind: Fodor and His Critics*, pp. 201–227. London, UK: Blackwell.

Smolensky P (1995) Constituent structure and explanation in an integrated connectionist/symbolic cognitive architecture. In: MacDonald C and MacDonald G (eds) *Connectionism: Debates on Psychological Explanation*, pp. 223–290. London, UK: Blackwell.

Smolensky P, Legendre G and Miyata Y (1992) Principles for an integrated connectionist/symbolic theory of higher cognition. Technical Report 92–08, Institute of Cognitive Science, University of Colorado, Boulder, CO, USA.

Further Reading

Hadley R (1994) Systematicity in connectionist language learning. *Mind and Language* **9**: 247–272.

Hadley R (1997) Cognition, systematicity, and nomic necessity. *Mind and Language* **12**: 137–153.

Matthews R (1994) Three-concept monte: explanation, implementation, and systematicity. *Synthese* **101**: 347–363.

Niklasson L and van Gelder T (1994) On being systematically connectionist. *Mind and Language* **9**: 288–302.

Connectionist Architectures: Optimization

Advanced article

Marcus Frean, Victoria University of Wellington, Wellington, New Zealand

CONTENTS

A key issue in using connectionist methods is the choice of which network architecture to use. There are a number of ways this choice can be made automatically, driven by the problem at hand.

INTRODUCTION

If one takes a training set in the form of input–output pairs and trains a large connectionist network on it, the result is generally 'overfitting'. There are many functions which exactly fit the existing data and the act of learning arrives at just one of them, somewhat arbitrarily. The problem is compounded where the data is noisy, in which case the network uses its extra degrees of freedom to fit the noise rather than the underlying function generating the data. Conversely, if the network is too small it 'underfits', which is equally unsatisfactory. The real aim is usually not to get the training set correct, but to generalize successfully to new data. The model selection problem is to arrive at the network that gives the best possible predictions on new inputs, using only the available training data and prior knowledge about the task. (*See* **Machine Learning**)

There are several ways of controlling the complexity of mappings learned by neural networks. These include varying the number of weights or hidden units by building up or paring down an existing network, and direct penalties (otherwise known as regularization) on model complexity, such as weight decay. Other ideas include partitioning the input space into regions which are locally linear as in 'mixtures of experts', or using genetic algorithms to choose between different architectures.

CRITERIA FOR NETWORK OPTIMALITY

The optimality or otherwise of a network is, in many cases, determined by its ability to generalize. Almost by definition this ability is not directly observable, so in practice we have to make an educated guess at it and use that to choose between networks.

The simplest method takes part of the available data and sets it aside. Once the network has been trained on the remaining data it can be 'validated' by seeing how well it performs on the withheld data, thus giving an estimate of how well it will generalize. This estimate won't be very good unless the hold-out set is large, which wastes a lot of the data that could otherwise be used for training. To minimize this effect, 'cross-validation' applies the same idea repeatedly with different subsets of the data, retraining the network each time. In k-fold cross-validation, for example, the data is divided into k subsets. One at a time, these serve as hold-out sets, and the validation performance is then averaged across them to give an estimate of how well the network generalizes. 'Leave-one-out' cross-validation uses $k = N$, the number of samples, but $k = 10$ is typically used.

Another general approach, from conventional statistics, is to attempt to quantify the generalization performance of trained networks without a validation set at all. One prefers networks with a low 'prediction error'

$$C = C_{data} + C_{net} \tag{1}$$

Here C_{data} is the usual training error, such as the sum of squared errors, and C_{net} is taken to be a measure of the complexity of the network, proportional to the effective number of free parameters it has. Assuming a nonlinear network is locally linear in the region of the minimum, an approximation to C_{net} can be calculated (Moody, 1992; Murata *et al.*, 1994) given the Hessian matrix of second derivatives $H_{ij} = \partial^2 C_{data}/\partial w_i \partial w_i$, which can be found using a number of methods (Buntine and Weigend, 1994). For large training sets, leave-one-out cross-validation and the above are essentially equivalent, with the latter giving the same effect for much less computational effort. These approaches assume a single minimum however, so the estimate can be strongly affected by local minima. On the other hand, leave-one-out cross-validation also gets trapped in local minima, in which case 10-fold cross-validation is preferable.

A third approach is to use Bayesian model comparison to choose between networks, as well as to set other parameters such as the amount of weight decay. Bayesians represent uncertainty of any kind by an initial or 'prior' probability distribution, and use the laws of probability to update this to a 'posterior' distribution in the light of the training set. In this view we should choose between models based on their posterior probabilities given the available data – again this does not require that any data be set aside for validation (MacKay, 1995). In the fully Bayesian approach, ideally we should use not one set of weights and one structure but many sets and many architectures, weighting the prediction of each by their posterior probability. To the extent this averaging can be done, deciding on an 'optimal' model (and indeed all learning as it is usually thought of) becomes unnecessary. (*See* **Reasoning under Uncertainty**; **Pattern Recognition, Statistical**)

Generalization performance is not the only measure of usefulness of a given network architecture. Other potentially important measures are its fault tolerance, training time on the problem at hand, robustness to 'catastrophic forgetting', ease of silicon implementation, speed of processing once trained, and the extent to which hidden representations can be interpreted. (*See* **Catastrophic Forgetting in Connectionist Networks**)

PRUNING OF UNIMPORTANT CONNECTIONS OR UNITS

Pruning algorithms start by training an overly complex network before trimming it back to size. In other words, we knowingly overfit the data and then reduce the number of free parameters, attempting to stop at just the right point. Clearly the general model selection schemes described above (cross-validation, estimated prediction error, and Bayesian model comparison) can be applied to prune overly large networks; however a number of ideas have been formulated that are specific to pruning. Pruning algorithms can remove weights or whole units, and one can think of the choice of which element to remove as being driven by a measure of 'saliency' for that element. Each algorithm uses a different form for this saliency.

A simple measure of saliency to use for weights is their absolute value. However, while it may be true that removing the smallest weight affects the network the least, it doesn't follow that this is the best weight to remove to improve generalization. Indeed this seems completely opposite to weight decay (see below), which in effect 'removes' large weights by decaying them the most, and it performs poorly in practice.

A more principled idea, known as 'optimal brain damage' (Le Cun *et al.*, 1990), approximates the change to the error function that would be caused by removal of a given connection or unit, and uses this measure to decide which to remove. To make this approximation, one trains the network until it is at a minimum of the usual error function, and then calculates the Hessian H. Ignoring the off-diagonal elements of this matrix, the saliency of the weight is given by $H_{ii}w_i^2$.

This idea has been further developed in 'optimal brain surgeon' (Hassibi and Stork, 1993), which avoids the assumption that the Hessian is diagonal. Interestingly this gives a rule for changing all the weights, with the constraint that one of these involves the setting of a weight to zero. One must first calculate the full inverse Hessian matrix however, which can make the algorithm slow and memory intensive for large problems.

Statistical tests can also be applied to detect non-contributing units that could be made redundant. For example if two units are in the same layer and are perfectly correlated (or anti-correlated) in their activity, we know the network can perform the same mapping with one of them removed. A particularly simple case is when a unit has the same output all the time, making it functionally no different from the bias unit.

WEIGHT DECAY

In networks whose output varies smoothly with their input, small weights give rise to outputs which change slowly with the input to the net, while large weights can give rise to more abrupt changes of the kind seen in overfitting. For this reason one response to overfitting is to penalize the network for having large weights. The most obvious way to do this is by adding a new term

$$C_{net} = \frac{\beta}{2}\sum_i w_i^2 \quad 0 < \beta < 1 \qquad (2)$$

to the objective function being minimized during learning. The total cost $C = C_{data} + C_{net}$ can then be minimized by gradient descent. For a particular weight we have

$$\Delta w \propto -\frac{\partial C}{\partial w} = -\frac{\partial C_{data}}{\partial w} - \beta w \qquad (3)$$

The first term leads to a learning rule such as back propagation (depending on the form of C_{data}), while the second removes a fixed proportion of the weight's current value. Hence each weight has a tendency to decay toward zero during training, unless pulled away from zero by the training data. The 'decay rate' β determines how strong this tendency is. Clearly a major question is how to set β, for which the general methods described earlier are applicable. (*See* **Backpropagation**)

Weight decay helps learning in other ways as well as its effect on generalization – it reduces the number of local minima, and makes the objective function more nearly quadratic so quasi-Newton and conjugate gradient methods work better.

From a Bayesian perspective, weight decay amounts to finding a *maximum a posteriori* (MAP) estimate given a Gaussian prior over the weights, reflecting our belief that the weights should not be too large. Weight decay is not usually applied to bias weights, reflecting the intuition that we have no *a priori* reason to suppose the bias offset should be small. Depending on the nature of the problem, this may not be a particularly sensible prior – for instance we may actually believe that most weights should be zero but that some should be substantially nonzero. One expression of this to use a different weight cost such as

$$C_{net} = \beta \sum_i \frac{w_i^2}{c^2 + w_i^2} \qquad (4)$$

This has been called weight elimination, because it is more likely to drive weights towards zero than simple weight decay. Very small weights can then be eliminated. c is a second parameter which needs to be set by hand. An interesting alternative is 'soft weight sharing' (Nowlan and Hinton, 1992) which implements MAP with a prior that is a mixture of Gaussians. The means (which need not be zero) and variances of these Gaussians can be adapted by the learning algorithm as training proceeds.

GENERATIVE ARCHITECTURES

Generative architectures, also called constructive algorithms, build networks from scratch to suit the problem at hand. Once each unit is trained, its weights are 'frozen' before building the next unit. An important advantage of this is that only single layers of weights are being trained at any one time. Accordingly the learning rules involved need only be local to the unit in question (unlike back propagation), which tends to make learning particularly fast and straightforward.

For simplicity each algorithm is described here as it applies to a single output unit – multiple outputs are trivial extensions to this, as described in the cited papers.

Upstarts

The upstart algorithm (Frean, 1990) is a method for constructing a network of threshold units. Imagine a single linear threshold unit (perceptron) that is trained to minimize the number of errors it makes on the training set, and then frozen. This unit, which we will call u, makes two kinds of error: it is either wrongly on, or wrongly off. In the upstart algorithm these errors are dealt with separately by recruiting two new units, which we could call u_- and u_+, one for each type of error. These new units receive the same inputs, but their outputs go directly to u (see Figure 1(a)). The role of u_- is to correct the 'wrongly on' errors by the parent unit u so it has a large negative output weight, while the output weight of u_+ is large and positive since its function is to correct the wrongly off errors. (*See* **Perceptron**)

It is easy to derive appropriate targets for u_-, given the original targets and u's responses: u_- is to be given the target 1 whenever u is wrongly on, while in all other cases its target should be 0. Similarly the target for u_+ is 1 whenever u is wrongly off, and 0 otherwise. Notice that the output of u_- (u_+) does not matter if u was already correctly off (on), so these can be omitted from the child node's training set. Should the child units be free of errors on their respective training sets, u will itself be error-free. If, however, either u_- or u_+ still make mistakes of their own, these errors are likewise of two types and we can apply the same idea, recursively. The result is a binary tree of units, grown 'backwards' from the original output unit. Child nodes spend their time loudly correcting their parent's mistakes, hence the algorithm's name.

Suppose u_- has the output 1 for just one of the wrongly on patterns of u, and is zero in all other

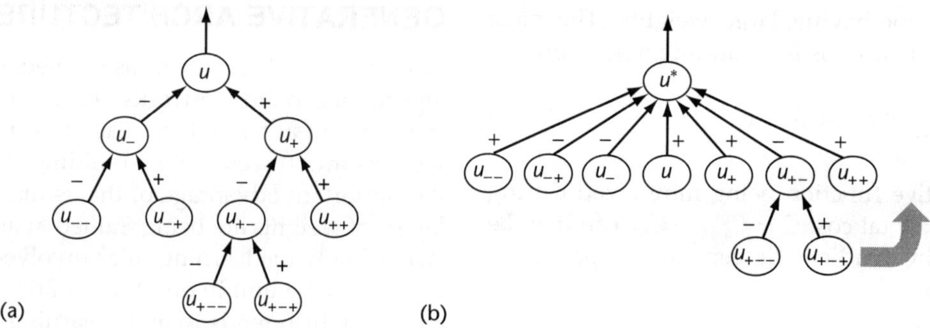

Figure 1. (a) A binary tree constructed by the upstart algorithm. All units have direct inputs, omitted here for clarity. The 'leaves' of the tree make no errors, so neither does the root node. (b) The same network being rearranged into a single hidden layer.

cases. For convex training sets (e.g. binary patterns) it is always possible to 'slice off' one pattern from the others with a hyperplane, so in this case it is easy for u_- to improve u by at least one pattern. Of course we hope that u_- and u_+ will confer much more advantage than this in the course of training. In practice a quick check is made that the number of errors by u_- is in fact lower than the number of wrongly on errors by u, to ensure convergence to zero errors.

Networks constructed using this method can be reorganized into a single hidden layer, if desired. That is, a new output unit can get zero errors by being connected to this layer with weights which are easily found, as shown in Figure 1(b).

For noise-free data this procedure usually produces networks that are close to the smallest that can fit the data, with attendant gains in generalization ability compared to larger networks. Notice however that the training set is learned without errors, so this is just as prone as any other algorithm to overfitting of noisy data (the idea has not been generalized to handle such noise, although there seems no reason why this couldn't be done). One can also use the same procedure to add hidden units to a binary attractor (Hopfield) network, thereby increasing its memory capacity from $\sim N$ to 2^N patterns.

The Pyramid Algorithm

Another algorithm for binary outputs is the pyramid algorithm (Gallant, 1993). One begins as before with a single binary unit, connected to the inputs and trained to minimize the number of errors. This unit is then 'frozen' and (assuming errors are still being made) a new unit is designated the output: this new unit sees both the regular inputs and any (frozen) predecessors as its input, as shown in Figure 2. (*See* **Perceptron**)

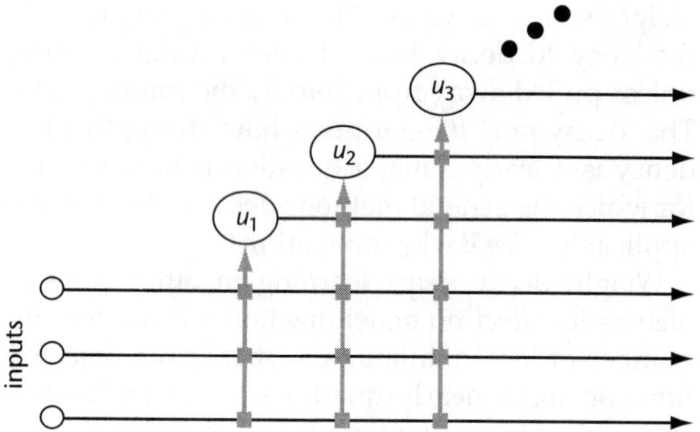

Figure 2. The architecture constructed by the pyramid algorithm. Vertical lines represent multiple connections (shown as squares) to the unit above. Each new unit assumes the role of output.

It is not hard to show that this new unit can achieve fewer errors than its predecessor, provided the input patterns are convex. If it sets its weights from the network inputs to zero and has a positive weight from the previous frozen unit, these two obviously make the same number of errors. As with upstarts, given convex inputs it could then easily reset its input weights so that this behavior was altered for just one input pattern where it was previously in error. This is the 'worst case' behavior, and appropriate weights can easily be predefined, to be improved by training (any method for arriving at good weights for a single unit is applicable). Despite its apparently 'greedy' approach to optimization and its extreme simplicity, the method can build concise networks. For example, given the N-bit parity problem (where the task is to output the parity of a binary input) the upstart algorithm generates a network with N hidden units, while the pyramid algorithm builds the apparently minimal network having only $(N+1)/2$

hidden units. Like the upstart algorithm, the method as it stands is prone to overfitting noisy data.

Cascade Correlation

Cascade correlation (Fahlman and Lebiere, 1990) can be applied to networks with real-valued outputs, and uses sigmoidal hidden units. We begin with a network having only direct connections between inputs and outputs, with no hidden units. These weights are trained using gradient ascent (the delta rule), or whatever learning procedure you like. We then introduce a hidden unit, with connections to the input layer. This unit sends its output via new weighted connections to the original output layer. In upstarts, the hidden unit is binary and is preassigned one of two roles, which determines how it is trained and the sign of its output weight – this is because being binary it can only correct errors of one type by the output unit, given its output weight. In this case, however, the hidden unit is real-valued and this means it can play a role in correcting errors of either sign by the output unit. Fahlman and Lebiere's idea is to train the hidden unit to maximize the covariance between its output and the existing errors by the output units. We can then use this gradient to learn input weights for the hidden unit in the usual way. When this phase of learning is deemed to have finished, all the output unit's connections are retrained, including the new ones. The process can now be repeated with a new hidden unit, with each such unit receiving inputs from the original inputs as well as all previous hidden units. Figure 3 shows the resulting cascade architecture.

A potential problem is that the output can make a lot of errors yet, after averaging, the correlation with a hidden unit can be very small. Despite this the method seems to work well in practice, and can be extended to recurrent networks.

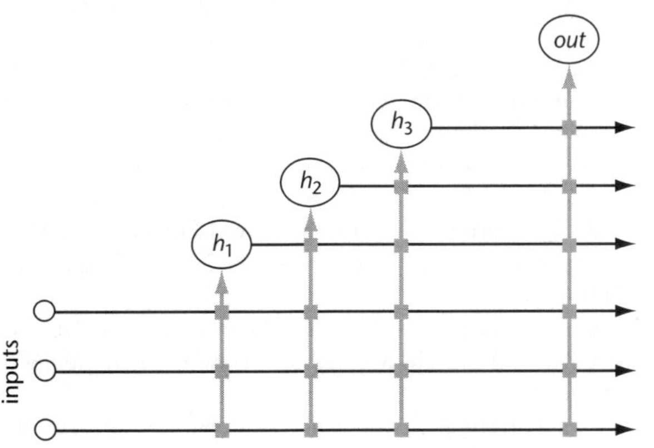

Figure 3. The architecture constructed by cascade correlation.

ADAPTIVE MIXTURES OF EXPERTS

In conventional back propagation networks, each sigmoid unit potentially plays a part in the network's output over its entire range of inputs. One way of restricting the power of the network is to partition the input space into distinct regions, and restrict the influence of a given unit to a particular region. Ideally we would like to learn this partition rather than assume it from the beginning.

A particularly appealing way to do this is known as the 'mixture of experts' architecture (Jacobs *et al.*, 1991). Each 'expert' consists of a standard feedforward neural network. A separate 'gating' network, with as many outputs as there are experts, is used to choose between them. The output of this network is chosen stochastically using the softmax activation function at its output layer,

$$Pr(i = 1) = \frac{e^{\phi_i}}{\sum_j e^{\phi_j}} \tag{5}$$

where ϕ_i is the weighted sum into the ith output unit. This reflects the fact that only one of the outputs is active at any given time. All of the nets (including the switch) are connected to the same inputs as shown in Figure 4, but only the expert that happens to be chosen by the switch is allowed to produce the output. A learning procedure can be found by maximizing the log likelihood of the network generating the training outputs given the inputs, in the same way as the cost function for back propagation is derived. Indeed the learning rule for the 'experts' turns out to be simply a weighted version of back propagation. During learning each expert network gets better at generating correct outputs for the input patterns that are

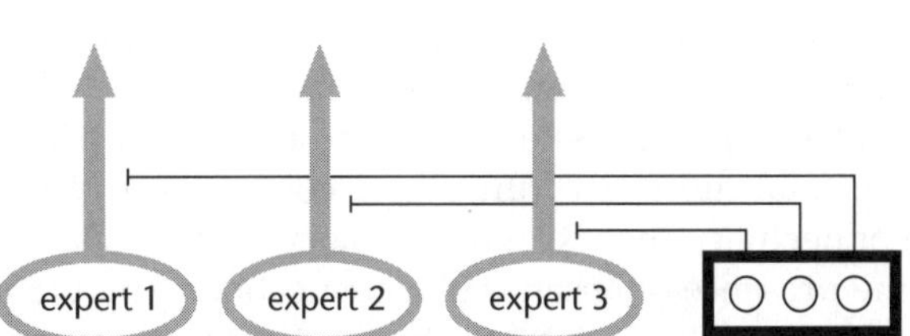

Figure 4. The mixtures of experts architecture. Each expert consists of a separate network, and may have multiple outputs. A separate gating network acts as a switch, allowing just one of the experts to generate the output. All of the experts, together with the gating network, have access to the input pattern.

assigned to it by the switch. At the same time, the switch itself learns to apportion inputs to the best experts. One can think of the switch as performing a 'soft' partition of the input space into sections which are learnable by individual experts. (*See* **Backpropagation**)

A further possibility is to treat each expert as a mixture of experts system itself (Jordan and Jacobs, 1994). A form of hierarchical decomposition of the task can thus be repeated for as many levels as desired. If simple linear units are used for the leaf nodes of the resulting tree-structured network, training can be achieved using a version of the expectation-maximization (EM) algorithm rather than gradient descent.

USING GENETIC ALGORITHMS TO EVOLVE CONNECTIONIST ARCHITECTURES

One criticism of both pruning and constructive algorithms is that they alter networks in only very limited ways, and as such they are prone to getting stuck in local optima in the space of possible architectures. Genetic algorithms offer a richer variety of change operators in the form of mutation and crossover between encodings (called 'chromosomes') of parent individuals in a population. The hope is that networks which are more nearly optimal may be found by evolving such a population of candidate structures, compared to making limited incremental changes to a single architecture. (*See* **Evolutionary Algorithms**)

In generating new candidate architectures, genetic algorithms choose parents based on their performance ('fitness'), which may be evaluated using the techniques described previously for determining network optimality. The main contribution of genetic algorithms then is their more general change operators, principally that of crossover, which operate on the chromosome rather than the network directly. Accordingly, the way in which architectures are mapped to chromosomes and vice versa is of central importance (Yao, 1999).

One approach is to assume an upper limit N to the total number of units and consider an $N \times N$ connectivity matrix, whose binary entries specify the presence or absence of a connection. Any units without outputs are effectively discarded, as are those lacking inputs. A population of such matrices can then be evolved, by training each such network using a learning algorithm initialized with random weights. To apply genetic operators, each matrix is simply converted to a vector by concatenating its rows. Restriction to feedforward networks is straightforward: matrix elements on and below the diagonal are set to zero, and are left out of the concatenation.

A drawback of this approach (though by no means unique to it) is that the evaluation of a given network is very noisy, essentially because the architecture is not evaluated on its own but in conjunction with its random initial weights. Averaging over many such initializations is computationally expensive, and one solution is to evolve both the connections and their values together. In this case an individual consists of a fully specified architecture together with the values of weights. On the other hand cross-over makes little sense for combining such specifications (unless the neural network uses localist units such as radial basis functions) because it destroys distributed representations.

Less direct encodings can be used, such as rules for generating networks, rather than the networks themselves. Evolutionary algorithms have also been used to change the transfer functions used by units (such as choosing between sigmoid and Gaussian for each unit), and even to adapt the learning rules used to set the weights.

References

Buntine WL and Weigend AS (1994) Computing second derivatives in feedforward networks: a review. *IEEE Transactions on Neural Networks* **5**(3): 480–488.

Fahlman SE and Lebiere C (1990) The cascade correlation learning architecture. In: Touretzky DS (ed.) *Advances in Neural Information Processing Systems*, vol. II, pp. 524–532. San Mateo, CA: Morgan Kaufmann.

Frean M (1990) The upstart algorithm: a method for constructing and training feedforward neural networks. *Neural Computation* **2**(2): 198–209.

Gallant SI (1993) Neural network learning and expert systems. Cambridge, MA: MIT Press.

Hassibi B and Stork DG (1993) Second-order derivatives for network pruning: optimal brain surgeon. In: Hanson SJ, Cowan JD and Giles CL (eds) *Advances in Neural Information Processing Systems*, vol. V, pp. 164–171. San Mateo, CA: Morgan Kaufmann.

Jacobs RA, Jordan MI, Nowlan SJ and Hinton GE (1991) Adaptive mixtures of local experts. *Neural Computation* **3**(1): 79–87.

Jordan MI and Jacobs RA (1994) Hierarchical mixtures of experts and the EM algorithm. *Neural Computation* **6**(2): 181–214.

Le Cun Y, Denker JS and Solla SA (1990) Optimal brain damage. In: Touretzky DS (ed.) *Advances in Neural Information Processing Systems*, vol. II, pp. 598–605. San Mateo, CA: Morgan Kaufmann.

Moody JE (1992) The effective number of parameters: an analysis of generalization and regularization in nonlinear learning systems. In: Moody JE, Hanson SJ and Lippmann RP (eds) *Advances in Neural Information Processing Systems*, vol. IV, pp. 847–854. San Mateo, CA: Morgan Kaufmann.

Murata N, Yoshizawa S and Amari S (1994) Network information criterion – determining the number of hidden units for artificial neural network models. *IEEE Transactions on Neural Networks* **5**: 865–872.

Nowlan SJ and Hinton GE (1992) Simplifying neural networks by soft weight sharing. *Neural Computation* **4**(4): 473–493.

Yao X (1999) Evolving artificial neural networks. *Proceedings of the IEEE* **87**(9): 1423–1447.

Further Reading

Bishop C (1995) *Neural Networks for Pattern Recognition*. Oxford: Clarendon Press.

Neal R (1996) *Bayesian Learning for Neural Networks*. New York: Springer-Verlag.

Read RD and Marks RJ (1999) *Neural Smithing – Supervised Learning in Feedforward Artificial Neural Networks*. Cambridge, MA: MIT Press.

Connectionist Implementationalism and Hybrid Systems

Intermediate article

Ron Sun, University of Missouri, Columbia, Missouri, USA

CONTENTS

We may incorporate symbolic processing capabilities in connectionist models, including implementing such capabilities in conventional connectionist models and/or adding additional mechanisms to connectionist models.

INTRODUCTION

Many cognitive models have incorporated both symbolic and connectionist processing in one architecture, apparently going against the conventional wisdom of seeking uniformity and parsimony of mechanisms. It has been argued by many that hybrid connectionist–symbolic systems constitute a promising approach to developing more robust and powerful systems for modeling cognitive processes and for building practical intelligent systems. Interest in hybrid models has been slowly but steadily growing. Some important techniques have been proposed and developed. Several important events have brought to light ideas, issues, trends, controversies, and syntheses in this area. In this article, we will undertake a brief examination of this area, including rationales for such models and different ways of constructing them.

MODELING DIFFERENT COGNITIVE PROCESSES WITH DIFFERENT FORMALISMS

The basic rationale for research on hybrid systems can be succinctly summarized as 'using the right tool for the right job'. More specifically, we observe that cognitive processes are not homogeneous: a wide variety of representations and processes seem to be employed, playing different roles and serving different purposes. Some cognitive processes and representations are best captured by symbolic models, others by connectionist models (Dreyfus and Dreyfus, 1987; Smolensky, 1988; Sun, 1995). Therefore, in cognitive science, there is a need for 'pluralism' in modeling human cognitive processes. Such a need leads naturally to the development of hybrid systems, in order to provide the necessary computational tools and conceptual frameworks. For instance, to capture the full range of skill-learning capabilities, a cognitive architecture needs to incorporate both declarative and procedural knowledge. Such an architecture can be implemented computationally by a combination of symbolic models (which capture declarative knowledge) and connectionist models (which capture procedural knowledge). The development of intelligent systems for industrial applications can also benefit greatly from a proper combination of different techniques, because currently no one technique can do everything successfully. This is the case in many application domains.

The relative advantages of connectionist and symbolic models have been argued at length. (See, for example, Dreyfus and Dreyfus, 1987; Smolensky, 1988 and Sun, 1995 for various views.) The advantages of connectionist models include: massive parallelism; graded representation; learning capabilities; and fault tolerance. The advantages of symbolic models include: crisp representation and processing; ease of specifying detailed processing steps; and the resulting precision in processing. With these relative advantages in mind, the combination of connectionist and symbolic models is easy to justify: hybrid systems seek to take advantage of the synergy of the two types of model when they are combined or integrated.

Psychologists have proposed many cognitive dichotomies on the basis of experimental evidence, such as: implicit versus explicit learning; implicit versus explicit memory; automatic versus controlled processing; incidental versus intentional learning. Above all, there is the well-known dichotomy between procedural and declarative knowledge. The evidence for these dichotomies lies in experimental data that elucidate various dissociations and differences in performance under different conditions. Although there is no consensus regarding the details of the dichotomies, there is a consensus on the qualitative difference between two types of cognition. Moreover, most researchers believe in the necessity of incorporating both sides of the dichotomies, because each side serves a unique function and is thus indispensable. Some cognitive architectures have been structured around some of these dichotomies.

Smolensky (1988) proposed a more abstract distinction of conceptual versus subconceptual processing; and he related the distinction to that between connectionist and symbolic models. Conceptual processing involves knowledge that possesses the following three characteristics: public access; reliability; and formality. This is what symbolic models capture. There are other kinds of cognitive capacities, such as skill and intuition, that are not expressible in linguistic forms and do not share the above characteristics. It has proved futile to try to model such capacities with symbolic models. These capacities should belong to a different level of cognition: the subconceptual level. The subconceptual level is better modeled by connectionist subsymbolic systems, which can overcome some of the problems faced by symbolic systems modeling subconceptual processing. Therefore, the combination of the two types of models can capture a significantly wider range of cognitive capacities. These ideas provide the justification for building complex hybrid cognitive architectures. For detailed accounts of a variety of examples of the synergistic combination of connectionist and symbolic processes, see Dreyfus and Dreyfus, 1987; Sun, 1995; Waltz and Feldman, 1986; and Wermter and Sun, 2000.

INTEGRATING CONNECTIONIST AND SYMBOLIC ARCHITECTURES

Hybrid models are likely to involve a variety of types of processes and representations, in both learning and performance. Therefore, they will involve multiple heterogeneous mechanisms interacting in complex ways. We need to consider how to structure these different components; in other words, we need to consider architectures. Questions concerning hybrid architectures include:

- Should hybrid architectures be modular or monolithic?
- For modular architectures, should we use different representations in different modules, or the same representations throughout?
- How do we decide whether the representation of a particular part of an architecture should be symbolic, localist, or distributed?
- What are the appropriate representational techniques for bridging the heterogeneity likely in hybrid systems?
- How are representations learned in hybrid systems?
- How do we structure different parts to achieve appropriate results?

Although many interesting models have been proposed, including some that correspond to the cognitive dichotomies outlined above, our understanding of hybrid architectures is still limited. We need to look at the proposed models and analyze their strengths and weaknesses, to provide a basis for a synthesis of the existing divergent approaches and to provide insight for further advances. Below we will provide a broad categorization of the existing architectures.

Architectures of hybrid models can be divided into 'single-module' and 'multi-module' architectures. Single-module systems can be further divided according to their representation types: symbolic (as in symbolic models); localist (i.e. using one distinct node for representing each concept – see, for example, Lange and Dyer, 1989; Sun, 1992 and Shastri and Ajjanagadde, 1993); and distributed (i.e. using a set of overlapping nodes for representing each concept – see, for example, Pollack, 1990 and Touretzky and Hinton, 1988). Usually, it is easier to incorporate prior knowledge into localist models, since their structures can be made to directly correspond to that of symbolic knowledge. On the other hand, connectionist learning usually leads to distributed representation (as in the case of back-propagation learning). Distributed representation has some useful properties.

Multi-module systems can be divided into 'homogeneous' and 'heterogeneous' systems. Homogeneous systems are similar to the single-module systems discussed above, except that they can contain several replicated copies of the same structure, each of which can be used for processing the same set of inputs, to provide redundancy for various reasons; alternatively, each module (of the same structure) can be specialized for processing inputs of a particular type (of content).

For heterogeneous multi-module systems, several distinctions can be made. First, a distinction can be made in terms of the representations of the constituent modules. There can be different combinations of types of constituent modules: for example, a system may be a combination of localist and distributed modules (as in CONSYDERR, described in (Sun, 1995), or it may be a combination of symbolic and connectionist modules, either localist or distributed (as in CLARION, described in (Sun and Peterson, 1998)).

Second, a distinction can be made in terms of the coupling of modules: a set of modules may be loosely or tightly coupled. In loosely coupled architectures modules communicate with each other, primarily through message passing, shared memory locations, or shared files. This allows for some loose forms of cooperation among modules. One form of cooperation is in terms of the type of processing: while one or more modules take care of preprocessing (e.g. transforming input data) or postprocessing (e.g. rectifying output data), another module focuses on the main part of the task. Preprocessing and postprocessing are commonly done using a neural network, while the main task is accomplished by symbolic methods. Another form of cooperation is through a master–slave relationship: while one module maintains control of the task at hand, it can command other modules to handle some specific aspects of the task. For example, a symbolic expert system, as part of a rule, may invoke a neural network to make a specific classification decision. Yet another form of cooperation is an equal partnership of multiple modules. In this form, the modules (the equal partners) may represent complementary processes; functionally equivalent but structurally and representationally different processes; or differentially specialized and heterogeneously represented 'experts'.

In tightly coupled architectures on the other hand, the constituent modules interact through multiple channels (for example, various possible function calls); or they may even have node-to-node connections between modules (as in CONSYDERR (Sun, 1995) and ACT-R (Anderson and Lebiere, 1988)). As in the case of loosely coupled systems, there are several possible forms of cooperation among modules.

TIGHTLY COUPLED ARCHITECTURES

Let us examine briefly a tightly coupled, heterogeneous, multi-module architecture: CONSYDERR (Sun, 1995). It consists of a concept level and a microfeature level. The representation is localist at the concept level, with one node for each concept, and distributed at the microfeature level, with an (overlapping) set of nodes for representing each concept. Rules are implemented, at the concept level, using links between nodes representing conditions and nodes representing conclusions, and weighted sums are used for evaluating evidence. Rules are diffusely duplicated at the microfeature level in a way consistent with the meanings of the rules. Rules implemented at the concept level capture explicit and conceptual knowledge that is available to a cognitive agent, and diffused representations of rules at the microfeature level capture (to some extent) associative and embodied knowledge. Figure 1 shows a sketch of the model. There are two-way (gated) connections between corresponding representations at the two different levels; that is, each concept is connected to all the related microfeature nodes, and vice versa. The operation of the model is divided into three phases: the top-down phase, the settling phase, and the bottom-up

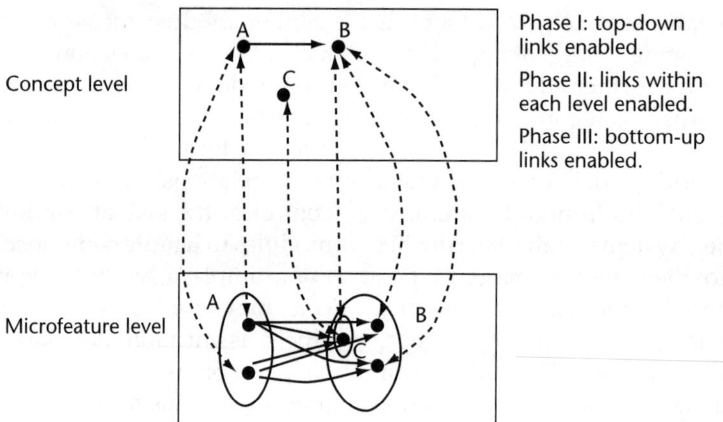

Figure 1. The CONSYDERR architecture.

phase. In the top-down phase, microfeatures corresponding to activated concepts are themselves activated, enabling similarity-based reasoning at the microfeature level. In the settling phase, rule-based reasoning takes place at each level separately. Finally, in the bottom-up phase, the results of rule-based and similarity-based reasoning at the two levels are combined.

Because of the interaction between the two levels, the architecture is successful in producing, in a massively parallel manner, a number of important patterns of common-sense human reasoning: for example, evidential rule application, similarity matching, mixed rule application and similarity matching, and both top-down and bottom-up inheritance (Sun, 1995).

COMPLETELY INTEGRATED ARCHITECTURES

An even tighter coupling between symbolic and connectionist processes exists in ACT-R (Anderson and Lebiere, 1998). ACT-R consists of a number of symbolic components, including declarative memory (a set of structured chunks), procedural memory (a set of production rules), and goal stacks. Retrieval in declarative memory is controlled by activations of chunks, which spread in a connectionist fashion and are affected by the past history of activations, similarity-based generalization, and stochasticity. Learning of associations among chunks and selection of procedural knowledge also happen in a connectionist fashion. Thus, the learning and the use of symbolic knowledge are partially controlled by connectionist processes. Through this tight integration of the two types of process, ACT-R has been successful in modeling human learning in areas such as arithmetic, analogy, scientific discovery, and human–computer interaction.

LOOSELY COUPLED ARCHITECTURES

Loosely coupled multi-module architectures, unlike the tightly coupled models discussed above, involve only loose and occasional interaction among components. For example, CLARION (Sun and Peterson, 1998), a model for capturing human skill learning, consists of two levels: a symbolic rule level and a connectionist network level. The two levels work rather independently, but their outcomes are combined in decision-making. The network level consists of back-propagation networks, which work through spreading activation and learn by reinforcement. The rule level works according to symbolic rules, which are learned by extracting information from the network level. Through the loose, outcome-based interaction of the two types of processes, the system is able to model a variety of types of human skill learning.

LOCALIST IMPLEMENTATIONS OF RULE-BASED REASONING

Among single-module or homogeneous multi-module models, localist implementations of symbolic processes, especially rule-based reasoning, stand out as an interesting compromise between connectionist networks and purely symbolic models. The representational techniques described below are shared by a number of localist models of rule-based reasoning (see, e.g. Lange and Dyer, 1989; Sun, 1992 and Shastri and Ajjanagadde, 1993).

The simplest way of mapping the structure of a rule set into that of a connectionist network is by associating each concept in the rule set with an individual node in the network, and implementing a rule by connecting each node representing a concept in the condition of the rule to each node representing a concept in the conclusion of the rule. The weights and activation functions can be set to carry out binary logic or fuzzy evidential reasoning.

To express relations, especially relations between large numbers of variables, we need to introduce variables into rules in connectionist implementations. We can represent each variable in a rule as a separate node. We assign values to these variable nodes dynamically and pass values from one variable node to another, based on links that represent variable binding constraints. Such values can be simple numerical signs (Lange and Dyer, 1989; Sun, 1992) or activation phases (Shastri and Ajjanagadde, 1993).

For example, in first-order predicate logic, each argument of a predicate is allocated a node as its representation; a value is assigned to represent an object (i.e., a constant in first-order logic) and thus is a sign of the object. This sign can be propagated from one node to other nodes, when the object that the sign represents is being bound to other variables from an application of a rule.

For each predicate in the rule set, an assembly of nodes is constructed. The assembly contains $k+1$ nodes if the corresponding predicate contains k arguments. We link up assemblies in accordance with rules. With this network, we can perform forward-chaining inference. We first activate the assemblies that represent known facts; then activations from these assemblies will propagate to other assemblies to which they are connected. This propagation can continue to further assemblies. For backward chaining, we first try to match the hypothesis with conclusions of existing rules; if a match is found, then we use the conditions of the matching rule as our new hypotheses to be proved: if these new hypotheses can be proved, the original hypothesis is also proved. To implement backward chaining in assemblies, we need, in addition to a predicate node, another node for indicating whether the predicate node is being considered as a hypothesis. Backward flow of activation through hypothesis nodes leads to backward-chaining inference.

Why should we use connectionist models (especially localist ones) for symbolic processing, instead of symbolic models? There are two reasons in particular why researchers explore such models. First, connectionist models are believed to be a more apt framework for capturing many (or even all) cognitive processes (Waltz and Feldman, 1986). The inherent processing characteristics of connectionist models often make them more suitable for cognitive modelling. Second, learning may be more easily incorporated into connectionist models than symbolic models: using, for example, gradient descent and its various approximations, expectation maximization, or the Baun–Welch algorithm. This is especially true of distributed models, but is also true of localist ones to some extent.

DISTRIBUTED IMPLEMENTATIONS OF RULE-BASED REASONING

A stronger notion of integration emphasizes developing symbolic processing capabilities in truly connectionist models, rather than juxtaposing symbolic codes with neural networks, or adopting a compromise as in localist implementations. This approach is more parsimonious explanatorily and thus potentially a more interesting form of cognitive modelling if it can be properly developed. Hence there is considerable interest in symbolic processing capabilities of distributed (or 'true') connectionist models.

An early example is Touretzky and Hinton's (1988) DCPS, which implements a production system in connectionist models. There is a working memory, which stores initially known facts and derived facts; there are two clause components, each of which is used to match one of the two conditions of a rule (each rule is restricted to have two conditions); there is a rule component, which is used to execute the action of a matching rule in the working memory; and a bind component is used to enforce constraints that may exist in a rule regarding variables. Each component is a connectionist network. See Figure 2.

The working memory consists of a large number of nodes, each of which has a randomly assigned 'receptive field'. A *triple* (a fact) is stored in the working memory by activating all the nodes that include the triple in their receptive fields. Many such triples can be stored in the working memory. The two clause components are used to 'pull out' two triples that can match two conditions of a rule. That is, they are used to match triples (in the working memory) with rules (in the rule component). Each node in working memory is connected to a corresponding node in each clause component. A clause component is a kind of 'winner takes all' network.

The rule component is made up of mutually inhibiting clusters. It is also a kind of 'winner

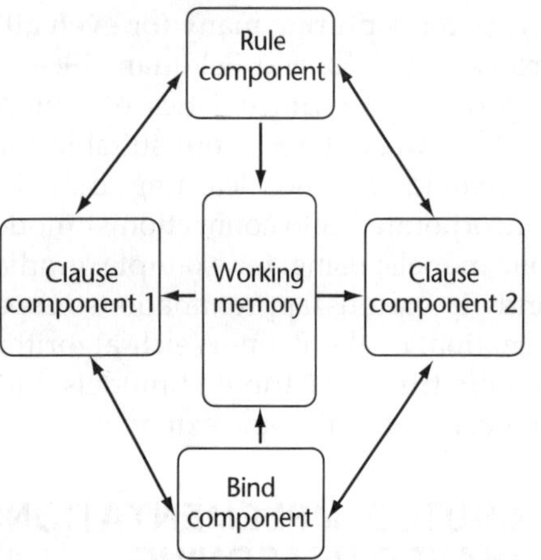

Figure 2. The overall structure of a connectionist production system.

takes all' network. Each rule is represented in the rule component by a cluster of identical nodes. The connections from the rule component to the clause components are used to help to pull out the triples that match a rule. In turn, these pulled-out triples also help a particular rule to win in the rule component. After successfully matching a rule with two triples in working memory, actions of the rule are carried out by the gated connections from rule nodes (in the rule component) to nodes in working memory. If the action of the rule includes adding a triple, then the gated connections will excite those nodes in the working memory that represent the triple; if the action includes deleting a triple, then the gated connections will inhibit those nodes in the working memory that represent the triple to be deleted. Overall, it is a complex system designed specifically to implement a limited production system.

EXTRACTION OF SYMBOLIC KNOWLEDGE FROM CONNECTIONIST MODELS

Many hybrid models involve extracting symbolic knowledge, especially rules, from trained connectionist networks. For example, some researchers proposed a search-based algorithm to extract conjunctive rules from networks trained with backpropagation (see Fu, 1989 and Wermter and Sun, 2000). To find rules, the algorithm first searches for all the combinations of positive conditions that can lead to a conclusion; then, with a given combination of such positive conditions, the algorithm searches for negative conditions that should be added to guarantee the conclusion. In the case of three-layered networks, the algorithm can extract two separate sets of rules, one for each layer, and then integrate them by substitution. Other researchers (e.g. Towell and Shavlik, 1993) tried rules of an alternative form, the '*N* of *M*' form: 'If *N* of the *M* conditions $a_1, a_2, \ldots a_M$ are true, then the conclusion *b* is true.' (It is believed that some rules can be better expressed in such a form, which more closely resembles the weighted-sum computation in connectionist networks, in order to avoid the combinatorial explosion and to discern structures.) A four-step procedure is used to extract such rules, by first grouping similarly weighted links and eliminating insignificant groups, and then forming rules with the remaining groups.

These early rule extraction algorithms are meant to be applied at the end of the training of a network. Once extracted, the rules are fixed; there is no modification 'on the fly', unless the rules are completely extracted again after further training of the network. In some more recent systems, rules can be extracted and modified dynamically. Connectionist learning and rule learning can work together, simultaneously. Thus the synergy of the two processes may be utilized to improve learning (Sun and Peterson, 1998). Dynamic modification is also suitable for dealing with changing environments, allowing the addition and removal of rules at any time.

SUMMARY

Overall, we can discern two approaches toward incorporating symbolic processing capabilities in connectionist models: combining symbolic and connectionist models; and using connectionist models for symbolic processing. In the first approach, the representation and learning techniques from both symbolic processing and neural network models are used to tackle complex problems, including modeling cognition, which involves modeling a variety of cognitive capacities. The second approach is based on the belief that one can perform complex symbolic processing using neural networks alone, with, for example, tensor products, RAAM, or holographic models (see Wermter and Sun, 2000). We may call the first approach 'hybrid connectionism' and the second 'connectionist implementationalism'.

Despite the differences between them, both approaches strive to develop architectures that bring together symbolic and connectionist processes, to achieve a synthesis and synergy of the two paradigms. Many researchers in this area share the belief that connectionist and symbolic methods can be usefully combined and integrated, and that such integration may lead to significant advances in our understanding of cognition.

References

Anderson J and Lebiere C (1998) *The Atomic Components of Thought*. Mahwah, NJ: Erlbaum.

Dreyfus H and Dreyfus S (1987) *Mind Over Machine*. New York, NY: The Free Press.

Fu L (1989) Integration of neural heuristics into knowledge-based inferences. *Connection Science* **1**(3): 240–325.

Lange T and Dyer M (1989) High-level inferencing in a connectionist network. *Connection Science* **1**: 181–217.

Pollack J (1990) Recursive distributed representation. *Artificial Intelligence* **46**(1,2): 77–106.

Shastri L and Ajjanagadde V (1993) From simple associations to systematic reasoning: a connectionist representation of rules, variables and dynamic bindings. *Behavioral and Brain Sciences* **16**(3): 417–494.

Smolensky P (1988) On the proper treatment of connectionism. *Behavioral and Brain Sciences* **11**(1): 1–74.

Sun R (1992) On variable binding in connectionist networks. *Connection Science* **4**(2): 93–124.

Sun R (1995) Robust reasoning: integrating rule-based and similarity-based reasoning. *Artificial Intelligence* **75**(2): 241–295.

Sun R and Peterson T (1998) Autonomous learning of sequential tasks: experiments and analyses. *IEEE Transactions on Neural Networks* **9**(6): 1217–1234.

Touretzky D and Hinton G (1988) A distributed connectionist production system. *Cognitive Science* **12**: 423–466.

Towell G and Shavlik J (1993) Extracting rules from knowledge-based neural networks. *Machine Learning* **13**(1): 71–101.

Waltz D and Feldman J (eds) (1986) *Connectionist Models and Their Implications*. Norwood, NJ: Ablex.

Wermter S and Sun R (eds) (2000) *Hybrid Neural Systems*. Heidelberg: Springer.

Further Reading

Barnden JA and Pollack JB (eds) (1991) *Advances in Connectionist and Neural Computation Theory*. Hillsdale, NJ: Erlbaum.

Giles L and Gori M (1998) *Adaptive Processing of Sequences and Data Structures*. New York, NY: Springer.

Medsker L (1994) *Hybrid Neural Networks and Expert Systems*. Boston, MA: Kluwer.

Sun R (1994) *Integrating Rules and Connectionism for Robust Commonsense Reasoning*. New York, NY: Wiley.

Sun R and Alexandre F (eds) (1997) *Connectionist Symbolic Integration*. Hillsdale, NJ: Erlbaum.

Sun R and Bookman L (eds) (1994) *Architectures Incorporating Neural and Symbolic Processes*. Boston, MA: Kluwer.

Wermter S, Riloff E and Scheler E (eds) (1996) *Connectionist, Statistical, and Symbolic Approaches to Learning for Natural Language Processing*. Berlin: Springer.

Consciousness

Introductory article

Adam Zeman, University of Edinburgh, Edinburgh, UK

CONTENTS

Consciousness refers both to wakefulness and to the contents of our experience. The subjective aspect of consciousness poses a philosophical problem for objective science.

INTRODUCTION

Since the early 1980s there has been a major effort to make better sense of consciousness. The current fascination with the subject flows from several sources: work by neuroscientists is steadily revealing details of the brain processes which make consciousness possible; psychologists have underlined the existence of a wide range of unconscious brain processes which can be contrasted informatively to conscious ones; computer scientists and engineers are designing sophisticated brain-like systems which can rival human intellectual performance, raising the question of whether such systems are conscious. It is clearly time to work out where consciousness belongs in the scientific scheme of things – and many philosophers are trying hard to do just that.

WHAT DO WE MEAN BY 'CONSCIOUS', 'AWARE', AND 'SELF-CONSCIOUS'?

Consciousness

Defining consciousness is tricky, but it is clearly important to clarify what we have in mind before we try to study it. Its linguistic origins deserve a moment's attention. The Latin source of 'consciousness', as of 'conscience', is the combination of 'cum', meaning 'with', and 'scio', meaning 'to know': in Latin 'conscire' meant to share knowledge, often guilty knowledge, with another person. This use was extended, metaphorically, to the sharing of knowledge with oneself. 'Conscientia' was the knowledge shared. In contemporary use, two senses of consciousness are particularly important.

Consciousness as the waking state

In everyday life, and particularly in medicine, if we ask whether someone is conscious we are generally asking whether he or she is awake – as opposed to asleep, anesthetized, very drunk, or in a coma. We are asking, in other words, about his or her 'state of consciousness'. (*See* **Consciousness, Disorders of**)

We tend to assume that if people are awake, they will also be capable of perceiving their surroundings and their bodies, and of interacting and communicating with others and with the environment. We are usually accurate observers of others' states of consciousness, and of the (normally) linked capacities to perceive, interact, and communicate: these are all 'objective' matters. Indeed, doctors use standardized scales, like the Glasgow Coma Scale, to assess patients' states of consciousness (Figure 1). These scales apply objective criteria to the assessment of consciousness, such as whether a a patient's eyes are open, whether he can speak, and his ability to move his limbs on request.

To be conscious in this sense is to be awake or vigilant. While we can usually come to a firm decision about whether someone is awake, asleep, or comatose, each of these states also admits of degrees: we can be wide awake or drowsy, half-asleep or stuporose.

Consciousness as experience

If someone is conscious in the sense of being awake, the person is usually conscious of something. In its second sense consciousness is the content of experience from moment to moment: what it feels like to be a certain person now, in a sense in which we suppose there is nothing it feels like to be a stone or in a coma. We can usually be much less sure of the contents of another person's consciousness than we can be that he or she is conscious. This second sense of consciousness is more inward, more subjective, than the first.

We can make several generalizations about the contents of consciousness in this second sense. They tend to be stable over short periods, from a few hundred milliseconds to a few seconds but changing over time; they have a foreground (at present the words in front of you), a background (the pressure of your clothes or the rumble of traffic), and a limited capacity (you can't simultaneously concentrate on Bach's first *Prelude* or

Name

Ward

Unit No:

COMA SCALE	Eyes Open	Spontaneously	4
		To speech	3
		To pain	2
		None	1
	Best Verbal Response	Orientated	5
		Confused	4
		Inappropriate words	3
		Incomprehensible sounds	2
		None	1
	Best Motor Response	Obey commands	6
		Localise pain	5
		Flexion to pain	4
		Abnormal flexion	3
		Extension to pain	2
		None	1
		GCS Total	

Figure 1. Glasgow Coma Scale. This scale is widely used in the clinical assessment of consciousness (it is of course fallible: how would someone who is fully conscious but completely paralyzed score?).

Britney Spears' last album and my article on consciousness); they are usually continuous over time, in the sense that memory allows us to connect the consciousness of the present with the consciousness of the past; they have an immense potential range, including information from our senses, and from all our major psychological processes including thought, emotion, memory, imagination, language, and action planning; and, above all, they are personal, conditioned by the perspective which our particular viewpoint supplies.

We all tend to consider ourselves to be experts on our experience: after all, who could know more than you do about the contents of your own consciousness? But perhaps we can be mistaken about what normally passes through our minds. This is an active area of consciousness research. For example, studies which require subjects to give instantaneous reports of their current experience, when a pocket-held buzzer sounds, produce some surprises: it can be difficult to spell out the contents of consciousness; 'inner thought', rather than sensation, tends to dominate awareness, and these thoughts are often clothed in neither images nor words. Other research, exploring our sensitivity to changes in our visual surroundings, suggests that our visual attention is much more narrowly focused than we normally imagine: we completely miss surprisingly large changes in the scene before our eyes unless our attention is on them at the moment they appear. This kind of study is important: it helps to clarify the data which the science of consciousness needs to explain.

Awareness

'Conscious' and 'aware' are used almost synonymously in ordinary speech, with the difference that 'awareness' tends to imply the occurrence of experience. The two words *can* be used to mark any of several subtly different distinctions: for example, wakefulness versus experience, the contents of experience versus the capacity for it, the objective versus the subjective aspects of experience. But these are all rather technical distinctions which should be explained when they are drawn. The two terms are used interchangeably here.

Self-consciousness

'Self-consciousness' is an even more slippery term than 'consciousness' itself. We can mean at least five different things when we say that someone is 'self-conscious'. (*See* **Self-consciousness**)

Self-consciousness as proneness to embarassment

In colloquial speech, if we say that someone is self-conscious we mean that the person is awkward in the company of others because he or she imagines that they are scrutinizing. In other words, we are self-conscious in this sense when we are excessively aware of others' awareness of ourselves.

Self-consciousness as self-detection

We sometimes say that a man or animal is 'self-conscious' when he detects a stimulus which impinges directly upon him (like an ant crawling over his hand), or when he behaves in a way that suggests an awareness of his own actions (like a dog with its tail between its legs after eating your supper). But this amounts to little more than perceptual awareness, directed towards events brought about by, or impinging directly upon, the creature in question.

Self-consciousness as self-recognition

This sense of self-consciousness was highlighted by the work of Gordon Gallup, in the 1970s, showing that chimpanzees and orangutans can recognize themselves in mirrors, but monkeys cannot. Human children become able to do so at around the age of 18 months. This ability suggests that apes, like small children, have an 'idea of me', a concept of 'self', although probably a very simple one. It is significant that over the few months after they come to recognize themselves in mirrors, children show a growing interest in self-adornment and master the use of the first-person pronoun, 'I'.

Self-consciousness as awareness of awareness

Between the ages of two and five, most children come to realize that they, like others, gain knowledge of the world from limited points of view, forming beliefs which can be mistaken. Their growing understanding of the nature of belief and the possibility of deception has been described as a developing 'theory of mind'. Once you possess this theory, your 'idea of me' has expanded to take in the notion that *you* are not merely a body, capable of reflection in a mirror, but also a mind, a subject of experience.

Self-consciousness as self-knowledge

In its broadest sense, our self-consciousness includes our knowledge of ourselves as members of particular families, schools, professions, social classes, language groups, and nations. We explore our peculiarly human fascination with ourselves in many forms of art: for example in Rembrant's astonishing life-long series of self-portraits.

THE SCIENCE OF CONSCIOUSNESS

Wakefulness

The electricity of the brain

We have learnt a great deal about the neurology of sleep and wakefulness in the past hundred years. In 1929 Hans Berger, a German psychiatrist, reported the first recordings of the electrical activity of the human brain made from the scalp: the 'electroencephalogram' or EEG. Berger and his followers went on to describe a series of brain rhythms (Figure 2) which correlate with states of consciousness: thus beta rhythm predominates while you read this article; alpha rhythm will become prominent if you relax and close your eyes, with increasing amounts of theta and delta if you let yourself drop off to sleep. (*See* **Consciousness, Sleep, and Dreaming**)

Research in the 1950s revealed that sleep itself has a complex structure: on falling asleep, our brain activity descends through a series of stages of deepening sleep, with progressive slowing of the EEG, but after half an hour or so of deep 'slow wave sleep' (SWS) it reascends through these stages. This ascent culminates in a period of 'rapid eye movement sleep' (REM), characterized by EEG appearances similar to those of wakefulness, rapid eye movements, deep relaxation of our muscles, and the experience of dreaming. The cycle is repeated three or four times each night with decreasing amounts of slow wave sleep and increasing amounts of REM in successive cycles (helping to explain why we so often wake in the morning with a dream in the mind's eye).

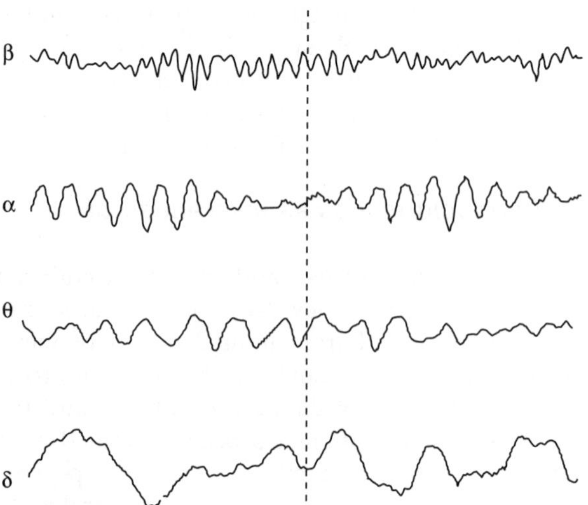

Figure 2. The rhythms of the EEG. This shows the four most commonly recognized EEG rhythms, obtained from four different clinical recordings: beta rhythm (> 14 Hz or cycles/second) characterizes active wakefulness: alpha (8–13 Hz) relaxed wakefulness with the eyes closed; theta (4–7 Hz) and delta (4–7 and < 4 Hz, respectively) occur in sleep and pathological states of depressed consciousness. A two-second period is shown. Reproduced with permission fom Zeman (2001) Consciousness. *Brain* **124**: 1263–1289.

Since his pioneering work, Berger's technique has been greatly refined. It is now possible to isolate the electrical activity associated with 'mental acts', as he had hoped: electrical correlates of sensation, attention, thought, and intention can all be identified at the scalp. There has been much interest recently in the idea that rapid synchronized activity in the gamma range (25–100 Hz) may be a hallmark of conscious processes.

The control of conscious states

In parallel with the exploration of the electrical correlates of consciousness, a series of discoveries has clarified the brain structures which control our conscious states. Observations of the effects of human brain disease, and experiments with animals, converged on the idea that regions of the brain stem and thalamus contain an 'activating system' that regulates the activity of the cerebral hemispheres (Figure 3).

Early models of this 'ascending reticular activating system' supposed that it was a nonspecific mechanism for maintaining wakefulness and alerting the hemispheres to the occurrence of significant events requiring their attention. This picture has been replaced by a much more complex one. It takes account of the existence of several chemical subsystems, employing different neurotransmitters – such as noradrenaline (norepinephrine), acetylcholine, serotonin, dopamine, and histamine – and of regions within the brainstem which serve specific functions, for example inducing REM sleep. But, the broad principle that regions of the brain stem and thalamus orchestrate our conscious states survives within this more sophisticated scheme.

Many details of the signals which switch the brain between wakefulness, SWS, and REM need to be clarified, but neuroscience can now give a plausible account of the neuronal basis of the distinction between wakefulness and sleep. Activating signals from the brainstem to the thalamus fall away as sleep begins, so that the neurons of the

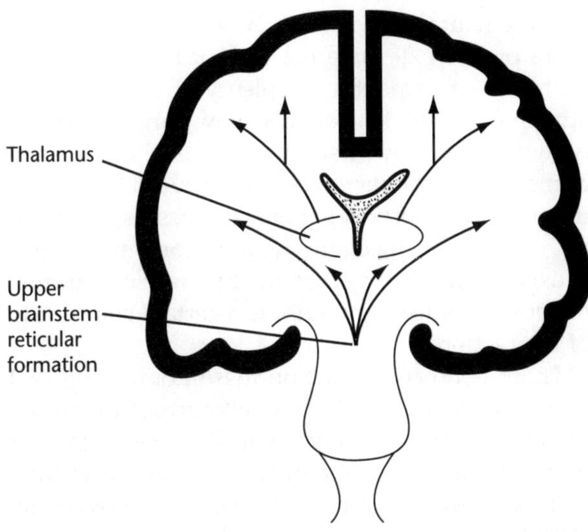

Figure 3. The reticular activating system. This simplified representation makes the points that the upper brainstem and the thalamus play a crucial role in activating the cerebral hemispheres and enabling wakefulness. Reproduced with permission fom Zeman (2001) Consciousness. *Brain* **124**: 1263–1289.

thalamus cease to transmit sensory signals faithfully to the cortex (their wakeful 'spike' mode of response). Instead, they enter into a series of rhythmic oscillations, detected at the scalp as the deepening stages of sleep (their 'burst' mode of response). The onset of REM corresponds to a partial reactivation of the thalamus and cortex, giving rise to a 'waking' EEG but with cerebral processing focused on internally generated events. (*See* **Consciousness, Sleep, and Dreaming**)

Pathologies of wakefulness

All the above points to the existence of three principal states of consciousness in health: wakefulness, SWS, and REM. Disease generates a number of further states: these include coma, a state of unresponsiveness resembling SWS but in which the subject is unrousable and the normal cycle of sleep and waking is lost; the vegetative state, a state of 'wakefulness without awareness', in which brainstem mechanisms continue to produce a sleep–wake cycle in the absence of the hemispheric function required to produce experience; and brain death, in which the brainstem is irrevocably destroyed. (*See* **Consciousness, Disorders of**)

Experience

Although brain scientists have sometimes fought shy of consciousness, a great deal of brain research is relevant to the neurology of experience: work exploring brain regions concerned with perception, attention, memory, language, emotion, and action often reveals correlations between brain events and features of awareness. Some scientists have been working to refine these correlations; others have taken a different, more roundabout, approach to understanding consciousness, by studying unconscious processes. (*See* **Neural Correlates of Consciousness as State and Trait; Perception, Unconscious; Unconscious Processes**)

Exquisite correlations

Vision is the most intensively studied human brain function and has provided the basis for much of the discussion of the neural basis of consciousness. A series of discoveries this century have revolutionized our picture of the brain events which underly conscious vision. Key findings include the discoveries that: the occipital cortex contains a detailed map of the visual world, in 'area V1'; within this map, cells inspecting each portion of visual space search for the presence of oriented edges; a further 30–40 visual 'maps' surround the primary visual area in the occipital cortex; parallel, though interconnected, streams of visual information flow through these maps, conveying information which defines visual form, color, depth, and motion; two broadly defined pathways fan out from the occipital cortex, an occipito-temporal pathway concerned with identifying objects and an occipito-parietal pathway concerned with visually guided action.

Detailed findings within this program of research have furnished remarkably close correlations between regional brain function and aspects of our experience. For example, human functional brain imaging studies indicate that perception of a colored scene selectively activates a particular region of visual cortex (often called V4); damage in this region can abolish the conscious perception of color. A distinct region plays a comparable role in the conscious perception of movement (V5).

Recently, several scientists have tried to home in on the neural correlate of consciousness (NCC) using a novel strategy. This stems from the thought that correlation need not imply cause: the fact that a brain area becomes active during visual perception does not imply that it causes our conscious experience – it might play some other role in the brain. One partial solution to this problem is to examine changes in brain activity which occur when experience changes without any change in the world. This happens, for example, when we summon up a visual image, or have an hallucination, or switch our attention without moving our eyes, or undergo a switch in the perception of an ambiguous figure. Examination of the neural correlates of these internally driven experiences is at the forefront of the quest for the NCC. (*See* **Neural Correlates of Consciousness as State and Trait**)

Unconscious processes

There is good evidence that our brains can register stimuli which we never consciously perceive, and that these in turn can influence our behavior. Understanding the events in the brain which enable these unconscious processes should help to sharpen the definition of the neurology of awareness. (*See* **Perception, Unconscious**; **Unconscious Processes**)

Conditions under which unconscious perception can be shown to occur include presentation of very brief, faint, or 'masked' stimuli to normal observers; presentation of stimuli to subjects during anesthesia or hypnosis, and neurological syndromes which impair conscious perception, such as blindsight and neglect ('blindsight' is the term given to a range of visually based abilities possessed by subjects who have no conscious vision after damage to area V1; 'neglect' is a disorder in which, most commonly, there is failure to pay attention to the left side of space following damage to the right side of the brain).

The investigation of these unconscious processes is an active area of research. It is too soon to reach firm conclusions about the key differences between conscious and unconscious brain activity. Two main types of explanation for the distinction have been proposed: that unconscious processes result when brain systems which sometimes give rise to consciousness are active at very low levels, and that unconscious processes occur in distinct brain systems, for example subcortical ones, where activity never gives rise to awareness. Recent experiments provide some support for both proposals.

THEORIES OF CONSCIOUSNESS

The data from the developing science of consciousness have spawned a number of overarching theories. They fall into three main types.

Neurobiological Theories

Theories of this type generally assume two broad principles which have emerged from the past century of research: that structures in the upper brainstem and thalamus play a key role in arousal, and that cortical activity supplies much or all of the contents of awareness. They tend, also, to assume that the neural correlate of consciousness will be a more or less extensive network of neurons. These points of agreement leave plenty of scope for disagreement over key details: how large must the network be to give rise to awareness? Need it incorporate particular types of neuron? Need it involve given cortical regions, or possess a particular range of connections with regions elsewhere? Must it engage in any particular pattern or duration of activity? Must it give rise to a certain complexity of interaction?

There is no consensus on these points. Figure 4 sketches a handful of the models currently on offer, underlining their variety. Semir Zeki has suggested that individual visual cortical areas generate their own 'microconsciousness' of color or of movement. David Milner proposes that only the occipitotemporal stream of visual processing is conscious, while the occipitoparietal is involved in action guidance. Gerald Edelman has emphasized the importance of reciprocal interactions between sensory cortical regions, limbic areas concerned with memory and 'value', and the thalamus. Francis Crick and Christof Koch have argued for the role of interactions between sensory areas and motor regions with which they directly interact. Larry Weiskrantz has developed the idea that visual consciousness arises from a secondary 'commentary' on visuomotor processes.

These theories focus mainly on the anatomy of consciousness: it is likely that a certain kind of neural activity is also required for consciousness. Current interest is focused on the role of rapid synchronized gamma-band activity, as there is evidence that this activity is abundant both in states of awareness generally and, specifically, in brain areas which are currently giving rise to experience.

Cognitive and Information-processing Theories

While neurobiological theories consider the nuts and bolts of consciousness, cognitive theories address its functions. Much of what we do, from brushing our teeth to riding a bike, can be achieved with little or no conscious attention. By contrast, novel challenges and unpredictable events force us to mobilize our psychological resources, engaging consciousness.

Cognitive theories take their lead from this everyday observation. Baars, for example, proposes that consciousness allows us to harness the resources of otherwise independent, unconscious, 'expert systems' in the brain to solve knotty problems as they arise: this sacrifices the high speed and high

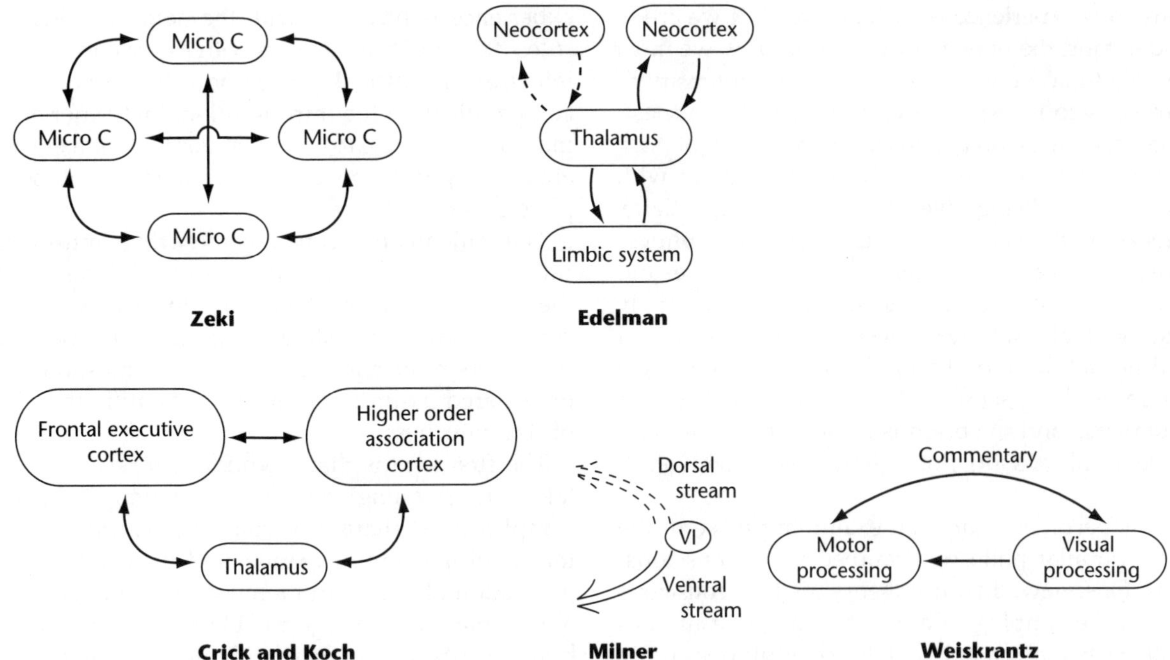

Figure 4. Neurobiological theories of consciousness. This highly schematic figure sketches the outlines of several current theories of consciousness: Zeki's model of interacting 'microconsciousnesses' within the visual system; Edelman's emphasis on interaction between shifting regions of neocortex concerned with perception and the limbic system, via the thalamus; Crick and Koch's proposal that only regions of cortex which 'directly' influence action can participate in conscious processing; Milner's distinction between a conscious 'ventral' and unconscious 'dorsal' stream of visual processing; Weiskrantz's suggestion that consciousness arises from a neural 'commentary' upon otherwise unconscious sensorimotor interactions.

capacity of automatic 'parallel' processing in the interests of flexible, deliberate behavior. This approach meshes with the widely held idea that consciousness arose in the course of evolution as flexible patterns of learned behavior emerged from the more rigid instinctive patterns of response seen in animals with simpler nervous systems. (*See* **Consciousness, Cognitive Theories of**)

If these theories are correct, they imply that the essence of consciousness lies not in its physical base but in the role it plays in processing information in the brain. If so, it follows that a machine which could reproduce the information flux in the human brain would necessarily be conscious. (*See* **Consciousness, Machine**)

Social Theories of Consciousness

The inspiration for social theories of consciousness is the thought that our awareness of our world and of ourselves is deeply influenced by other human beings: through social interactions from infancy on, through the acquisition of language, the greatest of all our social creations, and through our education in a common culture. In the course of our social development we acquire a 'theory of mind' which, as we have seen, supplies a distinctively human form of self-awareness. Yet while it is clearly necessary to take account of these social facts in giving a full description of human consciousness, it is doubtful whether social theories supply the right level of explanation for the simpler forms of consciousness which we share with animals, for example many of our sensory experiences, desires, emotions, intentions, pleasures, and pains.

THE PHILOSOPHY OF CONSCIOUSNESS

When philosophers discuss the nature of consciousness today, they are continuing an extremely ancient conversation about the relationship between mind and body, subject and object, the realm of experience and the realm of matter. This topic lies at the heart of philosophy: the views philosophers take on it cannot easily be prised apart from their views on a series of other thorny issues, such as the nature of meaning and of knowledge.

Three intuitions lie in the background of much of the philosophical discussion. The first is that conscious experience is 'rich and real': if we are to understand the universe we inhabit fully, we must be able to account for the variety and intensity of our conscious experience, from joy to sadness, from the hues of a sunset to the taste of salt. But, second, experience is clearly bound up with our physical being: everyone knows that fatigue, knocks on the head, too much beer, and countless other physical events can modify the state and contents of our consciousness. Third, we normally assume that awareness makes a difference: if I had not felt hungry, I would not have gone to the fridge. Making sense of the relationship between experience and the brain is difficult because these three intuitions do not square easily with each other.

This becomes clearer on examining three of the more popular philosophical theories of consciousness. One view, 'identity theory', is that conscious events are simply brain events. This idea meshes well with the second and third intuitions: brain events are physical, and well placed to cause our behavior. But does it do justice to the first? Some philosophers think not, arguing that one could know everything there is to know about a brain process and yet lack a full understanding of 'what it is like' to be the creature in whom it takes place. Take the famous example of 'what it is like to be a bat': knowing everything about the bat brain and bat behavior would not tell you what it is like to be a bat engaged in echolocation – or so the argument goes.

A second school of thought, 'functionalism', suggests that the essence of conscious states lies in the functions they serve in our behavior: the essence of vision is that it enables us to control our behavior using our eyes. Understand this control function and you understand visual experience: reproduce this function and you will have created visual consciousness. Once again this approach does justice to the second and third intutions but arguably fails to live up to the first: it is not immediately clear that a 'seeing machine' need be conscious; if it were conscious, it is not clear that its experience would necessarily resemble ours.

The third school of thought, 'dualism', is probably still dominant in our culture. It holds that mental and physical events are of radically different kinds: although the mental and physical realms must be linked in some way, neither can be reduced to the other. This approach appeals to those whose first intuition about consciousness is that it somehow 'goes beyond' the physical: that mental facts are 'further facts' about the world. It can be made compatible with the second intuition, that experience is physical, with the help of 'bridging rules' that link brain events with experience. But it is extremely difficult to see how it can be made compatible with the third intution, that experience makes a difference to behavior. For if mental events are nonphysical, how can they change the course of physical events?

The difficulty of reconciling these three intuitions calls for radically different ways of thinking about the philosophical problem of consciousness. So, consider two contrasting suggestions to indicate the diversity of current views. Each questions an undeclared assumption of most scientific theories of consciousness.

The first idea is that scientific theories are mistaken in assuming that consciousness *arises* from complexity. Perhaps consciousness is inherent in matter of the simplest kinds, and the complex organization of the brain merely allows this potentiality of matter to emerge and blossom. On this view brain events, in common with all physical events, have inherent mental and physical aspects: any attempt to reduce either aspect to the other would be mistaken. This view, panpsychism, is quite alien to our scientific culture, but it is an understandable response to the problem of explaining how consciousness can be conjured from brain events: on this view there is no need for any magic, as consciousness is present from the start.

The second idea is that we are mistaken in assuming that consciousness arises from the brain. At first sight this idea seems quite bizarre, a denial of the obvious. But it has some powerful backing. The argument runs: when we try to explain consciousness in terms of brain events we are doomed to failure, because we have denied ourselves precisely the resources the explanation requires. We need to expand the limits of our explanation to take in the world we inhabit and the means by which we explore it. For consciousness is a complex kind of interaction with the environment: 'seeing' for example is 'a way of acting'. This approach suggests – and gives reasons for believing – that much of what we take to be 'in our heads' is in fact out in the world, and that our ordinary picture of consciousness, as an internal representation of reality, is mistaken. On this view the brain is not the *source* of consciousness but a device which enables awareness, and awareness is not an invisible process but the use of a set of elaborate skills.

CONCLUSION

There has been huge progress in the scientific study of consciousness over the past century, and there is a growing clarity about the sources of conceptual difficulty in pinning down this elusive prey. The determination of contemporary scientists and philosophers to do justice both to the rich texture of experience and to its intimate relationship to brain events holds out great promise for the years to come.

Further Reading

Chalmers D (1996) *The Conscious Mind*. New York, NY: Oxford University Press.

Churchland PM (1984) *Matter and Consciousness*. Cambridge, MA: MIT Press.

Crick F (1994) *The Astonishing Hypothesis*. London, UK: Simon & Schuster.

Dennett D (1991) *Consciousness Explained*. London, UK: Penguin Press.

Edelman G (1992) *Bright Air, Brilliant Fire*. London, UK: Penguin Books.

Frith CD and Frith U (1999) Interacting minds – a biological basis. *Science* **286**: 1692–1695.

Frith C, Perry R and Lumer E (1999) The neural correlates of conscious experience: an experimental framework. *Trends in Cognitive Sciences* **3**: 105–114.

Thomas Nagel (1979) What is it like to be a bat? In: *Mortal Questions*. Cambridge, UK: Cambridge University Press.

O'Regan K and Noe A (in press) A sensorimotor account of vision and visual consciousness. *Behavioral and Brain Sciences*.

Rees G, Kreiman G and Koch C (2002) Neural correlates of consciousness in humans. *Nature Reviews* **3**: 261–270.

Schiff ND and Plum F (2000) The neurology of impaired consciousness: global disorders and implied models. http://www.phil.vt.edu/assc/niko.html.

Searle J (1992) *The Rediscovery of the Mind*. Cambridge, MA: MIT Press.

Weiskrantz L (1997) *Consciousness Lost and Found*. Oxford, UK: Oxford University Press.

Zeki S (1993) *A Vision of the Brain*. Oxford, UK: Blackwell Scientific.

Zeman A (2001) Consciousness. *Brain* **124**: 1263–1289.

Zeman A (2002) *Consciousness: A User's Guide*. London, UK: Yale University Press.

Consciousness and Attention

Intermediate article

Gregory J DiGirolamo, University of Cambridge, Cambridge, UK
Harry J Griffin, University of Cambridge, Cambridge, UK

CONTENTS

The concepts of consciousness and attention have been used in many ways and the processes that constitute them are not well agreed upon. Attention may be considered as an agency for bringing a stimulus into conscious awareness.

INTRODUCTION

It is hard to define consciousness or attention because these concepts have been used in many ways and the processes that fall under these umbrella terms are not well agreed upon. It is even more difficult to specify the functions and mechanisms of consciousness that lead to coherent behavior. Our strongest indication of consciousness remains the subjective experience. Nevertheless, advances in the scientific study of consciousness such as neuroimaging studies of perceptual awareness have given an anatomical and functional reality to both attention and consciousness in the human brain. This article looks at models of attention in order to explore the cognitive processes and neural substrates that may be shared between attention and consciousness. Evidence from cognitive neuroscience (the study of how the human brain carries out psychological processes) is applied to elucidate the complex relationship between consciousness and attention, and to enhance our understanding of both concepts.

THE RELATIONSHIP BETWEEN CONSCIOUSNESS AND ATTENTION

As early as the nineteenth century, psychologists and philosophers were suggesting a close relationship between attention and consciousness. William James (1890) succinctly captured this intuitive connection: 'Everyone knows what attention is ... Focalization, concentration of consciousness are of its essence ... My experience is what I agree to attend to.'

Attention and consciousness share many features, perhaps the most striking commonality being that at any given moment, one object or thought seems to predominate in the focus of attention and hence in our conscious awareness. In most situations in everyday life we are constantly bombarded by a variety of external stimuli from all sensory modalities (e.g. tactile, visual, and auditory) as well as by our own internal thoughts and memories. It would be difficult for an organism to achieve coherent, goal-directed behavior if all stimuli in the environment were processed and responded to in turn without prioritization. One can bring attention to bear on any of these objects (such as the background noises while you are reading this page), and suddenly this stimulus becomes the primary sensation or attribute in conscious awareness. It is, of course, possible to switch quickly between disparate thoughts or different objects in the environment; yet, the subjective experience is that only a single object is in attention or consciousness. Note that the perception of other objects in the environment remains, but when not the focus of your concentration the unattended objects have a vague, indistinct quality. (*See* **Consciousness, Unity of**)

With advances in neuroimaging it is now possible to investigate noninvasively the workings of the human brain as the subject attends to and becomes consciously aware of a stimulus. In addition, disorders of conscious awareness and attention following injury or psychosis have furthered our understanding of how these processes are carried out by the human brain as well as the relationship between attention and consciousness. (*See* **Consciousness, Disorders of**)

EXPERIMENTAL WORK ON ATTENTION AND CONSCIOUS AWARENESS

An important issue in the field of attention is how attention influences stimulus processing. Does attention manifest itself as alterations in the beta parameter (response biases), of a signal detection analysis or as changes in the d′ parameter (perceptual sensitivity)? In a typical experiment, participants were asked to keep their gaze fixed at a central point of a screen with four peripheral locations marked by small boxes. One location was cued (either by a brightening of that location's box, or by an arrow pointing to that location) so that participants could move their attention (but not their eyes!) to that location ahead of the actual target. A near-threshold target then appeared either at the cued location (valid) on 75% of trials, or at one of the uncued locations (invalid) on 25% of the trials. Reaction times to targets at the attended location were significantly faster and more accurate than targets at the uncued location. In addition, perceptual sensitivity changed at the location that was validly cued. Such studies demonstrate that attention to a location changes the sensitivity to incoming stimuli and makes these stimuli more perceivable. Targets at uncued locations were sometimes completely missed; that is, near-threshold targets at unattended locations did not receive sufficient attentional processing to enter conscious awareness. This simple behavioral paradigm suggests that items that are attended are processed more efficiently and enter consciousness; without attention, the individuals may not be aware of the item at all. (*See* **Attention, Neural Basis of**)

CONSCIOUSNESS AND ATTENTION IN NEUROPSYCHOLOGY

Neuropsychological evidence also elucidates the relationship between attention and awareness. One of the most striking findings is that lesions of the parietal lobe (particularly the right parietal lobe) produce specific deficits in attention. In the immediate aftermath of damage to the right parietal lobe, these patients will 'neglect' (not attend to or be aware of) information coming from the contralesional side of space (the side opposite to the hemisphere with damage). The side of space is important, as information from the left side of space is processed first by the right hemisphere, and vice versa; in addition, each hemisphere controls the opposite side of the body. In its severest form, patients with neglect fail to comb one side of their hair (the contralesional side) and eat off only one side of their plate (the ipsilesional side). If approached by a person on each side, these patients will look at the person on their ipsilesional side, even if the person on the contralesional side is the one who is speaking. They seem to fail to be consciously aware of information coming from the side of space processed by the damaged hemisphere. In fact, this deficit is so severe that, following a stroke producing neglect in the German painter Anton Räderscheidt, a self-portrait shows that, in the painting, he fails to represent the information from the contralateral side of space, including one side of his face. Fortunately, this deficit often resolves over time. (*See* **Attention, Neural Basis of; Attention**)

This deficit is not perceptual but rather attentional. For example, if patients with neglect are asked to create a visual mental image (e.g., the central square of their home town), they fail to report items on the left side of their mental image. However, if they are then asked to imagine walking to the opposite end of the square and turning around (so that the left and right sides are reversed), and to report what they see in their image, they will then report all of the items they failed to report from the previous view, and fail to report all the items they have just described. In this case, the patients are neglecting not incoming sensory information, but the represented visual image. Since the information is clearly present (as they report the entire representation between the two perspective shifts), these results suggest that the impairment is nonsensory in nature.

This deficit is not a general impairment, but specific to aspects of spatial attention. In people with neglect, their ability to shift attention to the contralesional side is severely impaired. Stimuli from the contralesional side appeared to have greater difficulty in summoning attention under conditions when the person is already processing something on the ipsilesional side. Thus, while normal people showed only a small deficit in reaction time when the target appeared in the opposite visual field from an attentional cue (invalid trials), people with neglect were often simply unaware of targets presented in the neglected field. Since these people were unable to report the mere presence of a target if their attention had been previously summoned to the good visual field, this finding suggests that attention is necessary for objects to come into conscious awareness. (*See* **Attention**)

Additional studies indicate that stimuli are processed in the neglected visual field even if unconsciously, and awareness is still possible under the appropriate conditions. For example, while experimental subjects might miss a contralesional target (a circle) in isolation, they might report the stimulus if it integrated with material on the ipsilesional side to make a single form, such as a dumbbell. Moreover, information in the neglected field can sometimes affect processing of items in both fields. Some patients will fail to be aware of an item in the neglected visual field if the item is also present in the good visual field (extinction). This deficit is ameliorated if the two items are different objects (e.g. a spoon and fork) rather than identical. Interestingly, these patients can tell if two objects (one in each visual field) are different, without being able to identify the object in the neglected field; this suggests that patients can process some information without attention, to the extent of telling differences between objects, while identity remains unavailable to conscious report. In addition, semantic information in the neglected visual field will affect processing in the good visual field – that is, the patient will respond faster to the word 'cat' in the good visual field if it has been preceded by the word 'dog' in the neglected visual field, than if it had been preceded by a neutral word in the neglected field. Hence, the word in the neglected visual field is processed to the level of meaning in the absence of attention and without conscious awareness.

These neuropsychological studies have outlined the depth of processing applied to an attended stimulus and implicated a strong relationship between spatial attention to and awareness of a stimulus. While most of these studies have dealt with deficits in spatial attention particular to disengaging and shifting attention in external space, we turn to one final syndrome following brain damage that suggests that attention also works on the object level to help bring stimuli into conscious awareness.

Damage to both parietal lobes (and the occipital lobes) can produce a condition known as Balint's syndrome in which patients can perceive only one object at a time. The visual system of these patients can 'see' the objects; however, attention can only be directed to one object at a time. Hence, these patients are only consciously aware of a single object. If presented with a comb, pen and fork, they will see only the comb, and then only the pen or only the fork. These results suggest that competition for conscious recognition is resolved through attentional processes, and without these processes objects cannot come into consciousness. Balint's syndrome demonstrates that objects excluded from awareness in favor of others need not lie within a specific region of external space. Rather, the simultanagnosia (inability to perceive two objects at the same time) indicates that a lack of awareness can be object-based. A single object may remain outside awareness despite being moved into a previously attended region of space. The return of this object to awareness may occur spontaneously; that is, the patient goes from seeing the fork to the pen. However, if the experimenter moves the object that is not currently in consciousness (the fork), the change in the attentional salience causes the patient to become aware of the moving object. This external change in the stimulus causes a shift in attentional bias from one internal representation (the pen) to the other (the fork). Likewise, a spontaneous shift in the object perceived is likely to be caused by a similar change in attentional bias between representations, but now in the absence of an external cue. (*See* **Attention; Attention, Neural Basis of**)

The lack of awareness in both neglect and simultanagnosia illustrates how the human brain chooses, from the representations offered to it, which objects enter into conscious awareness. When the neural areas underlying attention are damaged the number and breadth of these conscious interpretations becomes limited; these patients do not have even a vague, indistinct awareness of other objects in the environment (as normal people do even for objects outside their attentional focus). We now turn to studies in normal individuals that help to clarify the relationship between multiple representations, neural mechanisms of attention, and the entrance into conscious awareness.

DIFFERENT THEORIES OF CONSCIOUSNESS AND ATTENTION

Studies following brain damage have certain disadvantages as the lesions have ill-defined boundaries and may spread over areas that perform different functions. In addition, one must always be aware of the possibility of plasticity of function and the importance of patient strategies. However, a phenomenon exists that allows the study in normal people of the neural mechanisms of attention and its link to visual awareness: binocular rivalry.

Binocular rivalry occurs when dissimilar images (e.g. a face and a house) are presented to each eye. Instead of seeing a permanent mixture of the two monocular images, a multi-stable percept occurs in which the percept switches rapidly and involuntarily between the objects presented to each eye. Short periods of transition may also occur during which the percept is a fusion of the two images. The more different the two stimuli are in orientation, color, contrast or movement, the less prevalent are periods of piecemeal perception. Also, increasing the contrast of one of the stimuli increases its superiority, leading to it being perceived for longer periods. Despite being out of conscious awareness when suppressed, the subordinate stimulus can still influence cognitive processing (e.g. adaptation or priming).

Usually, functional magnetic resonance imaging (fMRI) studies compare brain response under equivalent rivalry and nonrivalry conditions. The rivalry condition is achieved by presentation of two images (one to each eye). The nonrivalry condition consists either of alternating binocular presentations of each of the two stimuli, or of alternating dichoptic presentation of an image to the relevant eye and a uniform gray field to the other. The rate of alternations is matched to the participant's pattern of awareness in the rivalry condition. In the nonrivalry condition, changes in awareness are caused by an alteration in the visual stimulation, whereas in the rivalry condition, changes in awareness are caused by involuntary shifts from the internal representation of one stimulus (the house) to the representation of the other (the face) in the absence of an external cue. By comparing these two conditions, we are able to study the neural mechanisms of shifts of visual awareness.

Binocular rivalry provides an opportunity to study shifts in conscious perception in the absence of any changes to the external stimuli. Using functional imaging with binocular rivalry, we can measure neural activity in visual awareness associated with both stable perception and perceptual transitions. It is also possible to distinguish neural areas of the visual system in which activation corresponds to the changing percept from areas in which activation corresponds to the fixed retinal stimulation. If the activation of a neural area corresponds to the reported percept, it suggests that the visual scene has been resolved at that stage of processing in the visual system (or, that feedback from attention is modulating the neuronal response in order to coherently resolve and stabilize the visual percept). (*See* **Attention, Neural Basis of**)

Activation of the right frontal and parietal cortices has been noted in perceptual transitions during the rivalry condition. The association of right frontal and parietal areas with tasks requiring shifts of spatial attention (both voluntary and involuntary) is well established. As we have discussed above, damage to the parietal cortex leads to disorders of visual awareness, and indeed, the similarity between binocular rivalry and extinction is worth noting. In both cases, two objects are presented in the environment and impinge upon the retina, but one of the objects is suppressed from visual awareness. Although a perceptual shift in binocular rivalry does not involve a shift in external space, it does involve a shift in object-based attention. The reason for perceptual shifts may be differential activity in the neurons representing each object caused by selective habituation to the perceived stimulus. When sufficient habituation has occurred, attention may shift to the object that now has greater neuronal activity. Once this shift has taken place and attention is now directed to the other object's representation, attention acts to stabilize the percept by increasing the neuronal activity associated with the previously suppressed object in lower visual levels and to push the visual system into a different but equally stable perceptual state. The threshold for the attentional shift is likely to be smaller than the threshold for a change in awareness; hence, the activation of neural areas involved in attention (e.g. the parietal cortex) amplifies the representation prior to conscious awareness. As suggested in the section on extinction, one gateway into conscious awareness is the relative amplification of the representation of an object through attention.

This cyclic habituation and attention capture could produce a multi-stable state, which would be beneficial as it allows accurate perception of at least one part of the visual scene rather than a confused interpretation of the entirety of the visual input. The constant shifts may be indicative of the tendency of the visual system to shift attention automatically towards a salient or novel stimulus in order to orient towards possibly important events (either external objects or internal representations). For example, movement of an object in the environment is salient and produces a capture of attention. As discussed above, patients with Balint's syndrome will become aware of an object currently out of awareness if it is moved about (as attention is drawn to the movement, and then the object enters consciousness). In binocular rivalry, the salience of a stimulus is not associated with changes in the external stimulus, but with its neural representation reaching a threshold level of activation that is sufficiently greater than the representation of the other monocular input. The function of parietal structures could be to disengage attention from the previously perceived object representation and shift it to the now more active representation.

Traditionally it was thought that the resolution of the ambiguous visual scene in binocular rivalry was carried out at a relatively late stage in the visual stream after the processing of each monocular image. Indeed single-cell studies have demonstrated that the activity of the majority of neurons in the striate cortex is unaffected across perceptual changes during rivalry; that is, activity in primary visual cortex correlates with the retinal stimulation whereas activity in later visual areas corresponds significantly to the changing percept. In the inferotemporal cortex, the activity of the majority of neurons follows the percept rather than the retinal stimulation. Functional MRI studies have shown that areas linked to perception of specific object categories increase their activity when an exemplar is perceived. For example, activation in part of the fusiform gyrus correlates with the presentation of a face and is relatively specific for faces or face-like stimuli. During a binocular rivalry condition in which a face and a house are presented to each eye, activation in this fusiform face area is seen only when the percept is that of a face. The changes in activity of these later visual areas, in contrast to that of V1, are as large in the rivalry condition as they are in the nonrivalry condition where the stimulus is presented binocularly (e.g. a face is presented to each eye). These results suggest that the awareness is resolved in a somewhat gradual fashion throughout the visual system. However, at least in later visual areas, neural activity correlates with conscious awareness, not the retinal stimulation. (*See* **Attention, Neural Basis of**)

CAN THERE BE ATTENTION WITHOUT CONSCIOUS AWARENESS?

Finally, we turn to one other neuropsychological disorder to help constrain the relationship between attention and consciousness. As previously explained, people who have sustained brain damage often acquire disorders of thought, perception, or even consciousness. People with damage to the primary visual cortex may deny any conscious sensation of visual stimulation presented in their blind visual field; yet, their 'guesses' of whether an object is in this part of the visual field yield accuracy rates that are often over 90%. This astonishing effect, 'blindsight', has been observed in both monkeys and humans with either unilateral or bilateral damage to the primary visual cortex. (*See* **Blindsight**)

The phenomenon of blindsight suggests that the disruption of primary visual cortex results in a disturbance of visual consciousness; hence, intact primary visual cortex seems necessary for visual consciousness. Although lacking conscious representations of visual stimuli, blindsight patients seem to be 'aware without being aware', which bears upon how attention and consciousness are related. For detailed information into the experiments on patients with blindsight, we refer the reader to the excellent summary by Weiskrantz (1997). (*See* **Consciousness, Function of**)

Although the incidence of damage confined to primary visual cortex is rare, the research that has been performed on blindsight patients is consistent with experimental work on nonhuman primates. Two types of tasks have been used with human blindsight patients. One relies on the patient's conscious report. In this method, one of two possible stimuli is presented on each trial: on half the trials the stimuli are presented to the blind visual field, and the person is instructed to decide which of the two stimuli were presented. The blindsight participants always report seeing nothing in the blind visual field; however, they are able to report some information about or respond to the stimulus in the blind visual field (although they report that there is nothing there, they will eventually guess). These patients can orient their eyes or hands to the approximate position of a stimulus; they can also discriminate stimulus orientation or stationary versus moving objects, as well as the direction of a moving stimulus.

The other method of testing blindsight patients involves measuring whether information presented to the patient's blind visual field influences (i.e. primes) the subsequent material presented to their intact hemifield. Results suggest that blindsight patients often experience implicit processing of stimuli presented in their blind hemifield. For example, information presented to a participant's blind field (e.g. the word 'river') significantly changes the way processing occurs for stimuli presented in the intact hemifield (e.g. the word 'bank'). In this case, when an ambiguous word ('bank') was presented in the good visual field, its conscious meaning was significantly prejudiced by a previous presentation of a semantic relative in the bad visual field. This finding suggests that the primary visual cortex may be unnecessary for some types of unconscious processing (such as priming) that influences our conscious experience.

These results suggest that much of the information that is sensed in the world can be processed without conscious awareness. Does the processing of this information rely on attentional processes even without conscious awareness? Alternatively, can attention work in the absence of conscious awareness? One telling experiment on a blindsight patient addressed this very issue (Kentridge *et al.*, 1999). The researchers asked whether orienting of attention would occur in the blind visual field. Using cuing experiments, blindsight patients were given an arrow cue that either correctly or incorrectly predicted the location of a subsequent target. The cue and the target could occur in either the good or blind visual field. As with normal participants, there was a benefit in performance for targets that occurred in the validly cued location, even when the target occurred in the blind visual field and the patient reported no target but 'guessed' that a target was present. Even more revealing, there was a benefit for the validly cued target location even when the cue itself occurred in the blind visual field and was not 'seen'. These results suggest that attention and consciousness are not absolutely linked. Even in the presence of attentional benefits (speeded and more accurate response to a validly cued target), conscious awareness of the stimulus is not guaranteed. Nor is conscious awareness of the cue necessary for attentional benefits. Hence, attention is not a sufficient conduit to ensure conscious awareness, and awareness is clearly not necessary for attentional processing. (*See* **Attention, Neural Basis of**)

In our view, attention is one agency for conscious awareness of each stimulus. Attention works to increase the neural response of salient, task-relevant stimuli. A shift of attention (and the increased neural activity) brings a stimulus into conscious awareness; however, attention can act on a stimulus without bringing that stimulus into consciousness. As suggested in the section on binocular rivalry, the level of neural activity to attract attention is likely to be below that necessary for conscious awareness. While it is clear that attention can be sufficient to bring a stimulus into consciousness, further research is required to determine what the necessary conditions are for conscious awareness of a stimulus.

References

James W (1890) *Principles of Psychology*. New York, NY: Holt.

Kentridge RW, Heywood CA and Weiskrantz L (1999) Attention without awareness in blindsight. *Proceedings of the Royal Society of London Series B* **266**: 1805–1811.

Weiskrantz L (1997) *Consciousness Lost and Found*. Oxford, UK: Oxford University Press.

Further Reading

Allport A (1988) What concept of consciousness? In: Marcel AJ and Bisiach E (eds) *Consciousness in Contemporary Science*, pp. 159–182. New York, NY: Oxford University Press.

Dennett D (1991) *Consciousness Explained*. Boston, MA: Little, Brown.

Farber IB and Churchland PS (1995) Consciousness and the neurosciences: philosophical and theoretical issues.

In: Gazzaniga MS (ed.) *The Cognitive Neurosciences*, pp. 1295–1306. Cambridge, MA: MIT Press.

Lumer ED (2000) Binocular rivalry and human visual awareness. In: Metzinger T (ed.) *Neural Correlates of Consciousness: Empirical and Conceptual Questions*, pp. 231–240. Cambridge, MA: MIT Press.

Marcel AJ (1983) Conscious and unconscious perception: experiments on visual masking and word recognition. *Cognitive Psychology* **15**: 197–237.

Norman DA and Shallice T (1986) Attention to action: willed and automatic control of behavior. In: Davidson RJ, Schwartz GE and Shapiro D (eds) *Consciousness and Self-regulation*, vol. 4, pp. 1–18. New York, NY: Plenum Press.

Posner MI (1994) Attention: the mechanisms of consciousness. *Proceedings of the National Academy of Sciences of the USA* **91**: 7398–7403.

Posner MI and Rothbart MK (1991) Attentional mechanisms and conscious experience. In: Milner AD and Rugg MD (eds) *The Neuropsychology of Consciousness*, pp. 91–111. London, UK: Academic Press.

Consciousness and Higher-order Thought

Intermediate article

David M Rosenthal, City University of New York Graduate School, New York, New York, USA

CONTENTS

The higher-order-thought hypothesis is a proposed explanation of what it is for a mental state to be a conscious state and hence of how conscious mental states differ from mental states that are not conscious.

INTRODUCTORY

Any satisfactory theoretical treatment of consciousness must begin by distinguishing several phenomena to which the term 'consciousness' applies. We describe people and other animals as being conscious when they are awake and responsive to sensory stimulation. What it is for a creature to be conscious in this sense is primarily a biological matter and peripheral to cognitive science and related theory.

We also describe creatures as being *conscious of* various things, for example, when they sense those things or think about them as being present. Sensing and thinking are central to cognitive functioning, but their nature is not what theorists typically have in mind in discussing consciousness. Rather, theorists have in mind primarily a third application of the term 'consciousness', by which we describe thinking and sensing itself as being conscious or not. It is this third use which dominates theoretical discussion about consciousness. The central issue is what it is for a mental state, such as thinking, sensing, and feeling, to be conscious, and more specifically what distinguishes the conscious cases from those which are not.

THEORIES OF CONSCIOUSNESS

It is fundamental to a mental state's being conscious that the individual in the state is aware of being in it. This is clear from consideration of mental states an individual is unaware of. If somebody is altogether unaware of thinking, feeling, or sensing something, that thinking, feeling or sensing does not count as conscious. Part of what it is for a state to be conscious is that one is conscious of being in that state.

Some theorists deny this, arguing that we are never conscious of our conscious states (Searle,

1992), or at least that conscious states occur without one's being conscious of them (Dretske, 1995). Thus Dretske, for example, urges that a state's being conscious consists not in one's being conscious of it, but in one's being conscious of something by virtue of being in that state.

This view has a disadvantage. Since sensing and thinking about things typically make one conscious of them, such states could not, on this view, occur without being conscious. Such theorists accordingly argue that the usual examples given of mental states that are not conscious are unconvincing. One especially common type of example does seem vulnerable to this charge. Armstrong (1978/1980) and others have appealed to the case of the long-distance driver who seems for a time not to notice the road consciously. But it may be that the driver notices the road consciously but simply does not at all remember doing so.

There are, however, other examples of mental states that more indisputably occur without being conscious. People often act in ways that betray some feeling, or belief, or desire of which they are wholly unaware until it is pointed out to them; this even happens with pains that are revealed by gestures or bodily movements. And people sometimes respond in a very fine-grained way to things that occur so far in the periphery of their visual field that they have no conscious perception of them.

Many experimental results confirm these commonsense observations (Merikle *et al.*, 2001). In masked-priming experiments, subjects presented very briefly with two successive stimuli report not being aware of the first at all, even though that stimulus has a demonstrable effect on mental processing (e.g. Marcel, 1983a, 1983b). And blindsight subjects, in whom part of the primary visual cortex has been destroyed, deny seeing visual stimuli in the relevant area of the visual field, though they can be prompted to guess the visible characteristics of such stimuli with startlingly high accuracy (Weiskrantz, 1997). Though conscious sensing is absent in these cases, subjects' behavior indicate the occurrence of sensing that is not conscious.

Some theorists have argued that it is circular to explain a mental state's being conscious in terms of an individual's being conscious of that state, since that would be to explain consciousness by appeal to consciousness (e.g. Goldman, 1993). But that explanation is not circular. Being conscious of something is sensing it or thinking about it as present. And, since we understand what it is to sense something or think about it even when that sensing or thinking is not conscious, we understand what it is to be conscious *of* something independently of knowing what it is for mental states to be conscious.

Even if a state's being conscious consists in one's being conscious of that state, theories divide about just how one is conscious of one's conscious states. The traditional and most widespread view is that one senses or perceives one's conscious states. But thinking about something can also make one conscious of that thing, and an alternative theory has been developed on which we do not sense our conscious states, but instead are conscious of them by having thoughts about them. It is useful to refer to the thoughts or sensations in virtue of which one is conscious of one's mental states as 'higher-order thoughts' or 'higher-order sensations'.

THE INNER-SENSE MODEL

The idea that we sense our conscious states has a long history. Locke (1700/1975) speaks of an 'internal sense' by which we are conscious of our mental states, and Kant (1787/1998) speaks of 'inner sense.' More recently, the idea has been defended by Armstrong (1978/1980) and by Lycan (1996).

Several factors suggest an account in terms of such higher-order sensing. For one thing, nothing seems to mediate between the the things we sense and our sensing them. And this intuitively unmediated character of sensing might explain why the way we are aware of our conscious states seems to be direct and immediate.

Another factor has to do with the qualitative character of conscious sensory experience. That qualitative character enters our mental lives through sensing; thinking has no qualitative character. So it may seem that the only way we could be conscious of this qualitative aspect of experience is by sensing it. A third source of the idea that we sense our conscious states is the sense we have that we are regularly and reliably conscious of many of our own mental states. And the best explanation for this may be that we monitor our mental states in the way that our exteroceptive senses monitor the environment (Armstrong, 1978/1980; Lycan, 1996).

But these considerations are far from decisive. Although nothing seems to mediate between our mental states and our consciousness of them, we need not appeal to higher-order sensing to explain that appearance of immediacy. Having thoughts about our mental states would also make us conscious of those states, and if we aware of nothing mediating between those thoughts and the states they are about, our consciousness of those states would also seem to be unmediated.

Perhaps some monitoring mechanism in the brain does subserve our being conscious of many of our mental states, but monitoring need not be sensory. The brain monitors many bodily functions in ways that do not at all resemble sensing. What differentiates sensing from other processes are the distinctive qualitative properties that occur when we sense. When sensing is conscious, we are conscious of these distinguishing qualities, qualities that vary with what is sensed, though these qualities also occur without our being at all aware of them.

The third consideration that seemed to support the inner-sense model, namely, the qualitative character of sensing, actually provides a compelling reason to reject the model (Rosenthal, 1997). Although sensations and perceptual states exhibit distinguishing qualitative properties, the way we are conscious of our own mental states does not. This is evident when the states we are conscious of are thoughts, beliefs, desires, and other so-called intentional states; these states have no qualitative properties, and there is no qualitative character to the way we are conscious of them. But even when the states we are conscious of are qualitative, as with our sensations and emotions, the qualities belong to the states we are conscious *of*, not to the way we are conscious of them.

Some theorists describe the inner-sense model in terms of higher-order perceiving of mental states (Güzeldere, 1995). Since perceiving, like sensing, has qualitative character, a higher-order perception view faces the difficulty that no higher-order qualities occur. But perceiving not only has qualitative character, but also resembles thinking in having conceptual content. So, if we had higher-order perceptions of our mental states, we would still need to determine whether the qualities or conceptual content of the perceptions were responsible for our being conscious of our mental states. Compare Güzeldere (1995), who argues that the higher-order-perception model collapses into a model that invokes higher-order thoughts.

THE HIGHER-ORDER-THOUGHT MODEL

The two ways of being conscious of things are sensing them and having thoughts about them as being present. Since we are not conscious of our mental states by sensing them, the best explanation of how we are conscious of some of our mental states is that we have higher-order thoughts (HOTs) about them (Rosenthal, 1986, 1993; in press a, b). It would be explanatorily empty to insist that we are conscious of them in some third way unless we have an independent grasp of what that third way consists in.

Difficulties with the inner-sense view actually suggest the HOT model. The higher-order states in virtue of which we are conscious of our mental states lack qualitative properties, and a thought that something is present makes one conscious of that thing in a way that involves no higher-order qualities. If there is a brain mechanism that monitors mental states, it might well make one conscious of those states by producing HOTs about them. And, if those HOTs seemed to arise independently of any inference, it would seem subjectively as though one is conscious of one's mental states in a way that is direct and unmediated.

It is important to distinguish between the ordinary way in which mental states are conscious and the focused, reflective way in which states can become conscious when we introspect them. The HOT model affords a natural explanation of this difference. The HOTs in virtue of which we are conscious of our mental states in ordinary, nonintrospective cases are not, themselves, conscious thoughts; HOTs make one aware of various mental states, but without one's being conscious also of the HOTs themselves. When one introspects a state, one deliberately focuses attention on it. One thereby becomes aware not only of the introspected state, but also of one's being conscious of it. So in these cases the relevant HOTs are themselves conscious thoughts (Rosenthal, 2000a).

Because HOTs need not be conscious, and indeed usually are not, people will normally be unaware of their presence. The occurrence of HOTs is not established by our being aware of them, since we are conscious of them only in the special case of introspection. HOTs are theoretical posits whose occurrence is established by theoretical considerations of the sort sketched above.

These considerations help dispel a certain misunderstanding. It is sometimes held (e.g. Block, 1995a; Chalmers, 1996) that the HOT model explains only introspective consciousness. That would be so if the model appealed only to conscious HOTs, establishing their occurrence by way of subjects' reports. But the HOTs the model invokes typically are not conscious, and they are intended to explain ordinary, nonintrospective consciousness.

A HOT is a thought to the effect that one is in a particular state, and so makes reference to oneself. Such reference to the self does not require any sophisticated concept of the self, but only a concept strong enough to distinguish oneself from everything else and to form thoughts about the self thus distinguished (Rosenthal, in press c). Nor do HOTs need to describe their target states in terms of some concept of the mind; they can describe those states simply in terms of their role in perceiving, or thinking or in information-processing terms.

VARIANT HIGHER-ORDER-THOUGHT THEORIES

A number of variants HOT theories have been put forth. Some differ in only minimal ways from the hypothesis just described. For example, Mellor (1977–78) appeals to second-order beliefs to explain only what it is for beliefs to be conscious, and denies that such an explanation works for any other types of mental state. But the foregoing arguments apply equally to all types of mental state. And Rolls (1998) has argued for a model on which the HOTs must be linguistic in character. Since Rolls construes being linguistic to cover any syntactically composite mode of representation, this again is at most a slight modification of the model.

Brentano (1874/1973) argued that the higher-order state in virtue of which one is conscious of a mental state is internal to the target state in question. Brentano's examples are largely perceptual, which leads Brentano to see the higher-order state as being perceptual as well; so his view may be best construed as a variant of the higher-order sensing model. Others have advanced views, however, that posit HOTs that are internal to their targets (Kobes, 1996; Gennaro, 1996).

But no view on which HOTs are internal to their targets is likely to succeed. Intentional states are individuated not only by their content but also by mental attitudes, such as mental affirmation, doubt, wonder, hope, and the like. Just as no single state can have two distinct contents, so no single state can exhibit two distinct mental attitudes. The mental attitude of HOTs is always assertoric; HOTs affirm that one is in a particular state. But, if HOTs were internal to their targets, then a conscious case of wondering something would exhibit the mental attitudes both of wondering and of affirming. So HOTs cannot in general be part of their targets. Similar considerations apply to Brentano's perceptual variant, since every perceptual state belongs to some sensory modality, and none of those standard modalities is suitable for making one conscious of one's mental states.

On another variant of the model developed by Carruthers (1996, 2000), a state need not be the object of an actual HOT to be conscious; it is enough that the state simply be *disposed* to cause a HOT about it. One main motive for this variant is that it avoids the high cost, both in computational capacity and cognitive space, of having actual HOTs for each of one's conscious states.

But that consideration is not all that compelling. HOTs very likely take less to implement cortically than their targets, since a HOT simply represents one as being in a particular state; so their causal connections will likely be far less complex than those of the perceptual and cognitive states the HOTs are about. And, since cortical capacity is known to be far from fully utilized, the cost of implementing actual HOTs is unlikely to be significant. Introspection makes this objection seem more pressing than it is. Because we are never conscious of many thoughts at once, it seems that we could not have many HOTs. But HOTs are seldom conscious, and we could have many at once that are not conscious. And introspection cannot be a reliable guide to the mind's nonconscious operations.

The principal reason for a higher-order account is to explain how we are conscious of all our conscious states. But being disposed to have a thought about something does not make one conscious of that thing. So the question arises about how a state's being disposed to cause a HOT could result in one's being conscious of that state.

Carruthers's answer appeals to a particular theory of mental content. On that theory, the content a state has is a matter of what it is disposed to cause. So Carruthers argues that a state's simply being disposed to cause a HOT can confer suitable higher-order content on the state itself. Both teleological (e.g. Millikan, 1984) and inferential-role (e.g. Block, 1986; Peacocke, 1992) theories of content might allow for this result. A state's having such higher-order content directed upon itself would then explain why one is conscious of being in that state.

This reply faces several difficulties. For one thing, these theories of content are far from uncontroversial, and it is preferable to have one's theory of consciousness committed to as little as possible that is not widely accepted. Moreover, since the higher-order content in virtue of which one is conscious of the state is internal to the state itself, the dispositional theory would face the difficulty about mental attitudes that faces any model on which the higher-order state is internal to its target.

Most important, any state with suitable first-order content would, on this model, have dispositional properties that result in its having higher-order content. So one would be conscious of any state that had that first-order content. Since a state's being conscious or not would depend on its first-order content, the model seems unable to explain how states of a given type can sometimes be conscious and sometimes not.

HIGHER-ORDER THOUGHTS AND SPEECH

It is widely accepted that, given a creature with suitable linguistic capacities, a mental state's being conscious coincides with that creature's ability to report noninferentially that it is in that state. Indeed, it is likely that this ability to report mental states noninferentially is what underlies the traditional intuition that we have special access to our mental states (Sellars, 1963).

This fits well with the HOT hypothesis. A noninferential report that one is in some mental state expresses one's thought that one is in that state, a thought that seems subjectively to rely on no inference. And it is arguable that the best explanation of this ability to report one's mental states noninferentially is that one actually has the HOTs that those reports would express (Rosenthal, 1993).

Some theorists hold that we cannot introspectively seem to be in a state that we are not in (e.g. Nelkin, 1996). Moreover, the seemingly noninferential character of our HOTs may suggest that they reflect some special access we have to our mental states. But thoughts need not be accurate, and thoughts about one's own mental states are no exception. Our consciousness even of what states we are in can be erroneous.

This is evident from compelling experimental findings in which subjects report thoughts and desires that they do not actually have. As with reports of thoughts and desires that do occur, these confabulations tend to make *ex post facto* sense of subjects' behavior, by rationalizing that behavior or by conforming to expectations or preconceived ideas. But in these cases evidence exists that subjects do not actually have the thoughts and desires they report (Nisbett and Wilson, 1977). Such confabulation appears to happen even with qualitative states, such as bodily or perceptual sensations (Staats *et al.*, 1998; Holmes and Frost, 1976).

These findings again fit well with the HOT model. When one confabulates being in some mental state, one is conscious *of oneself as being in that state*. And consciousness is a matter of how one appears to oneself. So, if one has a HOT that represents one as being in some state, there is nothing subjectively, from the point of view of consciousness, that could enable one to tell whether any such state actually occurs.

When one thinks that something has a certain property, one in effect interprets that thing as having the property. So having a noninferential HOT that one is in some mental state amounts to spontaneously interpreting oneself as being in that state. This echoes Dennett's (1991) interpretivist account of consciousness. But Dennett (1987, 1991) holds that one's being in a mental state at all, independent of whether that state is conscious, is a matter of one's being subject to some appropriate interpretation. The HOT model does not endorse that more general view.

Because conscious states are sometimes confabulated, the states one is conscious of oneself as being in do not always exist. So we cannot describe a conscious state as a state that bears some actual relation to a HOT. Rather, a state's being conscious must consist in its being the *intentional object* of a HOT, the object that the thought seems to be about. And, because the state may not actually occur, we also cannot require that it cause the HOT.

OBJECTIONS

The HOT hypothesis is sometimes seen as a claim about the *concept* of a mental state's being conscious (Goldman, 2000). Construed as a hypothesis about conceptual analysis the hypothesis is implausible, since it seems *conceivable* that a state accompanied by a HOT could fail to be conscious (Balog, 2000; Rey, 2000). But the HOT model is best taken not as a conceptual claim, but as an empirical hypothesis about the nature of consciousness. On that construal, though we can conceive of a state's being accompanied by a HOT without being conscious, it turns out empirically that this never happens. One might also object that any specification of the nature of consciousness purports to state a metaphysical necessity, and the HOT hypothesis is not metaphysically necessary. But, even apart from the difficulty of determining what is metaphysically necessary in a way that is not question begging, it is arguable that the HOT hypothesis is necessary if at all only in the way in which truths of natural science are.

It is sometimes argued that the stipulation that HOTs be noninferential is arbitrary, since it should not matter to a state's being conscious whether the accompanying HOT is caused by an inference (Byrne, 1997; Seager, 1999). But the aetiology of the HOT does not matter, only the appearance of aetiology. A state is conscious only if we are conscious of it in a way that *seems* spontaneous and noninferential. As long as it seems that way, it does not matter how it is caused. Nor is there any problem about establishing a causal or other connection between HOTs and their targets, as Natsoulas (1993) argues, since the targets are simply whatever states the HOTs are about.

Having a thought about something normally has no effect on it, and in particular does not make that thing conscious. So it may be objected that having a thought about a mental state could not result in that state's changing from not being conscious to being conscious (Block, 1995b; Byrne, 1997; Rey, 2000). But when a state becomes conscious that is not a change in the state itself, but only in whether one is noninferentially conscious of it; being conscious is not an intrinsic property of mental states.

Still, an objector might persist, since having a noninferential thought about a physical object does not result in that object or state's being conscious, why should having a HOT about a mental state result in that state's being conscious? But the only way objects might be conscious is the way a creature can be, by being awake and responsive to sensory input; objects cannot be conscious in the way mental states are. Still, having noninferential thoughts about states of one's liver presumably would not make those states conscious (Block, 1995b). But not every state can count as conscious. A state can be conscious only if being in it, even when the state is not conscious, results in one's being conscious *of* something, and states of the liver do not qualify (Rosenthal, 2000b).

Dretske has argued that a mental state's being conscious cannot consist in one's having a HOT about it, since there are cases in which a state is conscious without one's being conscious of it. Dretske offers the example of consciously seeing two scenes that differ in some single way without one's consciously noticing that they differ at all. Since one does not notice that the scenes differ, one also does not notice the difference between one's conscious visual experiences of the scenes. But every part of the two experiences is presumably conscious. So, if one is not conscious of that part of the experiences in respect of which they differ, that part is a conscious experience of which one is not conscious (Dretske, 1995). But all that Dretske's case shows is that one need not be conscious of that part *as* the part that makes a difference between the two experiences, not that one is not conscious of that part in some other way (Seager, 1999; Byrne, 1997; Rosenthal, 1999). It may well be that we are conscious of all our conscious experiences.

Conscious states presumably occur not just in humans, but in other animals as well. So perhaps conscious states occur even in animals whose mental functioning is too primitive to accommodate HOTs (Block, 1995a; Dretske, 1995; Byrne, 1997). Indeed, Carruthers (2000) actually argues that few if any nonhuman animals have HOTs. And he concludes that they lack conscious states, though many will resist that conclusion.

In any case, the reasons for thinking that few nonhuman animals have HOTs are not fully convincing. Carruthers (2000) argues that animals with HOTs would also have thoughts about the mental states of others. And he holds that having thoughts about the mental states of others would express itself in deceptive behavior, which he urges nonhuman animals do not engage in (cf. Povinelli, 1996). But it is arguable that many nonhuman animals do engage in deceptive behavior (Whiten, 1996; Whiten and Byrne, 1997). Nor, in any case, is it obvious that creatures would not have HOTs unless they had thoughts about the mental states of others (Ridge, 2001).

It might seem that nonhuman animals lack the conceptual resources needed to have HOTs. But HOTs do not require the elaborate conceptual apparatus characteristic of humans; they are simply thoughts that one is in states of particular types, states which we humans classify as mental.

At the same time, it is not obvious which nonhuman species do have mental states that are conscious. Though many such species plainly do sense and think, that does not show that their thinking and sensing are conscious; states can exhibit the characteristic causal roles of mental states without a creature's being conscious of being in those states. Some way independent of human subjective impressions is needed to establish which species do have mental states that are conscious.

As noted above, it is one thing for a creature to be conscious and another for its mental states to be conscious. Still, it might seem that a creature cannot be conscious unless at least some of its mental states are conscious; whenever humans are awake, after all, they are in some conscious states. But this may not hold generally. For a creature to be conscious it must function in characteristic mental ways, but it can do that without its mental states being conscious.

It is natural to think that a mental state's being conscious serves some useful function, such as enhancing the rationality of thinking and planning (Nelkin, 1996). But the function a mental state serves is a matter of its causal role, and the causal role a state has may well be largely unaffected by being accompanied by a HOT (Dretske, 1995).

Accompanying HOTs might, however, actually alter a state's causal role, and a state together with a

HOT will in any case have a different combined role from the state without any HOT. Indeed, it has been argued that HOTs enable the correction of plans that result from first-order processing (Rolls, 1998).

There are also experimental findings that subjects sometimes perform tasks better when stimuli are consciously perceived than when perceived nonconsciously (Merikle and Daneman, 1998). Since the relevant tasks require conscious thought, the difference may be due to operation of HOTs in the conscious cases.

There is a compelling intuition that our conscious states constitute some kind of unity, and one might object that a theory on which mental states are conscious in virtue of many distinct HOTs cannot do justice to that intuition (Shoemaker, in press). But since the content of each HOT is that one is, oneself, in some state, such reference to oneself will give rise to a conscious sense of unity (Rosenthal, in press c).

QUALITATIVE CONSCIOUSNESS

Perhaps the most important objection has to do with qualitative consciousness. It has been argued that HOTs cannot capture the enormous detail characteristic of conscious qualitative states (Byrne, 1997). And some have argued also that HOTs, which are nonqualitative, could not result in there being something it's like for one to be in various qualitative states (Byrne, 1997; Siewert, 1998; Balog, 2000).

It is easy to exaggerate the qualitative detail we are conscious of at any moment. It is well known that Parafoveal vision yields scant detail. More dramatically, recent work on change blindness shows that we often fail consciously to notice significant changes in a scene we are attentively looking at (Grimes, 1996; Rensink, 2000; Simons, 2000), which suggests that our impression of great conscious qualitative detail is erroneous (Dennett, 1991). And HOTs could presumably capture the detail present in any relatively small area of a sensory field on which one consciously focused.

We do not have concepts for all the individual qualities we are conscious of, but we have concepts for the ways those qualities vary. So HOTs can represent individual qualities comparatively. This may explain why we can judge whether qualities are the same far better when they are all present than when we must rely on memory (Raffman 1995). And, though concepts may be ill suited to capture the way qualitative states represent things, we are typically conscious of the relevant qualities in a way that lends itself to conceptualized description.

It is important not to place excessive demands on an explanation of qualitative consciousness. Very likely no explanation will reveal a conceptual or rational connection between nonconscious resources and conscious qualities (Levine, 2001), but scientific explanation seldom does that. Nor should we expect to discover an introspectible connection between conscious qualities and nonconscious resources, since nothing that is not conscious is available to introspection.

In any event, there is reason to think that HOTs do figure in there being something it's like to be in conscious qualitative states. We often come to be conscious of qualitative differences only when we come to have concepts fine-grained enough to draw those qualitative distinctions: for example as between similar musical instruments or tastes of wine. Such concepts would matter to how those experiences are conscious only if our thoughts about the experiences made a difference to how we are conscious of them (Rosenthal, in press a).

THE SCIENCE OF CONSCIOUSNESS

Although the foregoing arguments in support of the HOT hypothesis do not rely on empirical investigation, the hypothesis meshes fruitfully with scientific findings. Two examples already noted are change blindness and confabulated mental states. But there are others as well. As Weiskrantz (1997) has urged, a HOT model helps explain the phenomena of blindsight. Rolls (1998) argues for his linguistic version of the HOT hypothesis by appeal to different neural pathways that seem to subserve conscious and nonconscious stimulation. Dienes and Perner (2001) have appealed to the HOT model in distinguishing implicit from explicit knowledge and representation, and Dienes (in press) has applied the model in connection with implicit learning and subliminal perception.

The HOT model is particularly useful in explaining the finding by Libet (Libet, 1985) that the neural readiness potentials identified with subjects' decisions occur measurably in advance of subjects' awareness of these decisions, findings recently replicated and extended (Haggard and Eimer, 1999). It is natural to explain this result by supposing that the HOTs in virtue of which subjects become aware of their decisions occur measurably later than those decisions (Gomes, 1999; Rosenthal, in press d).

Frith and Frith (1999) report a number of studies in which functional brain imaging reveals neural activation in subjects who were asked to report their mental states. Strikingly, conscious monitoring of states results in activation of a single brain area, medial frontal cortex, even when the states monitored are as disparate as pain, tickles, emotions aroused by pictures, and spontaneous thoughts. This activation does not occur cortically where the monitored states occur; so a single, independent brain mechanism seems to subserve the monitoring that makes possible the reporting of mental states. Since reports of one's mental states express one's thoughts about those states, it is inviting to construe that activation as indicating the occurrence of HOTs.

References

Armstrong DM (1978/1980) 'What is consciousness?'. *Proceedings of the Russellian Society* **3**(1978): 65–76; reprinted in expanded form in Armstrong, *The Nature of Mind*, St. Lucia, Queensland, Australia: University of Queensland Press, pp. 55–67, 1980.

Balog K (2000) Comments on David Rosenthal's 'consciousness, content, and metacognitive judgments'. *Consciousness and Cognition* **9**(2) Part 1: 215–219.

Block N (1986) Advertisement for a semantics for psychology. *Midwest Studies in Philosophy* **X**: 615–678.

Block N (1995a) 'On a confusion about a function of consciousness'. *The Behavioral and Brain Sciences* **18**(2): 227–247.

Block N (1995b) How many concepts of consciousness? *The Behavioral and Brain Sciences* **18**(2): 272–287.

Brentano F (1874/1973) *Psychology from an Empirical Standpoint* edited by Kraus O, English edn edited by McAlister LL, translated by Rancurello AC, Terrell DB and McAlister LL. London, UK: Routledge & Kegan Paul, 1973.

Byrne A (1997) Some like it HOT: consciousness and higher-order thoughts. *Philosophical Studies* **86**(2): 103–129.

Carruthers P (1996) *Language, Thought, and Consciousness: An Essay in Philosophical Psychology*. Cambridge, UK: Cambridge University Press.

Carruthers P (2000) *Phenomenal Consciousness: A Naturalistic Theory*. Cambridge, UK: Cambridge University Press.

Chalmers DJ (1996) *The Conscious Mind: In Search of a Fundamental Theory*. New York, NY: Oxford University Press.

Dennett DC (1987) *The Intentional Stance*. Cambridge, MA: MIT Press/Bradford Books.

Dennett DC (1991) *Consciousness Explained*. Boston: Little, Brown and Company.

Dienes Z and Perner J (2001) When knowledge is unconscious because of conscious knowledge and vice versa. In: Moore JD and Stenning K (eds) *Proceedings of the Twenty-third Annual Conference of the Cognitive Science Society*, pp. 255–260. Mahwah, NJ: Lawrence Erlbaum Associates.

Dretske F (1995) *Naturalizing the Mind*. Cambridge, MA: MIT Press/Bradford Books.

Frith CD and Frith U (1999) 'Interacting minds – a biological basis'. *Science* **286**(i5445): 1692ff.

Gennaro RJ (1996) *Consciousness and Self-Consciousness: A Defense of the Higher-Order-Thought Theory of Consciousness*. Amsterdam and Philadelphia: John Benjamins.

Goldman AI (1993) Consciousness, folk psychology, and cognitive science. *Consciousness and Cognition* **2**(4): 364–382.

Goldman AI (2000) Can science know when you're conscious? epistemological foundations of consciousness research. *Journal of Consciousness Studies* **7**(5): 3–22.

Gomes G (1999) Volition and the readiness potential. *Journal of Consciousness Studies* **6**(8–9): 59–76.

Grimes J (1996) On the failure to detect changes in scenes across Saccades. In: Akins K (ed.) *Perception*, pp. 89–110. New York, NY: Oxford University Press.

Güzeldere G (1995) Is consciousness the perception of what passes in one's own mind?. In: Metzinger T (ed.) *Conscious Experience*, pp. 335–357. Exeter: Imprint Academic. Reprinted in: Block N, Flanagan O and Güzeldere G (eds) *The Nature of Consciousness: Philosophical Debates*, pp. 789–805. Cambridge, MA: MIT Press/Bradford Books, 1997.

Haggard P and Eimer M (1999) On the relation between brain potentials and awareness of voluntary movements. *Experimental Brain Research* **126**(1): 128–133.

Holmes DS and Frost RO (1976) Effect of false autonomic feedback on self-reported anxiety, pain perception, and pulse rate. *Behavior Therapy* **7**(3): 330–334.

Kant I (1787/1998) *Critique of Pure Reason*, translated and edited by P Guyer and AW Wood. Cambridge, UK: Cambridge University Press, 1998.

Kobes BW (1996) Mental content and hot self-knowledge. *Philosophical Topics* **24**(1): 71–99.

Levine J (2001) *Purple Haze: The Puzzle of Consciousness*. New York, NY: Oxford University Press.

Libet B (1985) 'Unconscious cerebral initiative and the role of conscious will in voluntary action'. *The Behavioral and Brain Sciences* **8**(4): 529–539.

Locke J (1700/1975) *An Essay Concerning Human Understanding*, edited from the 4th edn. by PH Nidditch. Oxford, UK: Clarendon Press.

Lycan W (1996) *Consciousness and Experience*. Cambridge, MA: MIT Press/Bradford Books.

Marcel AJ (1983a) Conscious and unconscious perception: experiments on visual masking and word recognition. *Cognitive Psychology* **15**: 197–237.

Marcel AJ (1983b) Conscious and unconscious perception: an approach to the relations between phenomenal experience and perceptual processes. *Cognitive Psychology* **15**: 238–300.

Mellor DH (1977–78) Conscious belief. *Proceedings of the Aristotelian Society*, New Series, LXXXVIII: 87–101.

Merikle PM, Smilek D and Eastwood JD (2001) Perception without awareness: perspectives from cognitive psychology. *Cognition* **79**(1–2): 115–134.

Merikle PM and Daneman M (1998) Psychological investigations of unconscious perception. *Journal of Consciousness Studies* **5**(1): 5–18.

Millikan RG (1984) *Language, Thought, and Other Biological Categories*. Cambridge, MA: MIT Press/Bradford Books.

Natsoulas T (1993) What is wrong with the appendage theory of consciousness? *Philosophical Psychology* **6**(2): 137–154.

Nelkin N (1996) *Consciousness and the Origins of Thought*. Cambridge, UK: Cambridge University Press.

Nisbett RE and Wilson TD (1977) Telling more than we can know: verbal reports on mental processes. *Psychological Review* LXXXIV(3): 231–259.

Peacocke C (1992) *A Study of Concepts*. Cambridge, MA: MIT Press/Bradford Books.

Povinelli DJ (1996) Chimpanzee theory of mind?: the long road to strong inference. In: Carruthers P and Smith PK (eds) *Theories of Theories of Mind*, pp. 293–329. Cambridge, UK: Cambridge University Press, 1996.

Raffman D (1995) On the persistence of phenomenology. In: Metzinger T (ed.) *Conscious Experience*, pp. 293–308. Exeter: Imprint Academic.

Rensink RA (2000) The dynamic representation of scenes. *Visual Cognition* **7**(1/2/3): 17–42.

Rey G (2000) Role, not content: comments on David Rosenthal's 'Consciousness, content, and metacognitive judgments.' *Consciousness and Cognition* **9**(2): 224–230.

Ridge M (2001) Taking solipsism seriously: nonhuman animals and meta-cognitive theories of consciousness. *Philosophical Studies* **103**(3): 315–340.

Rolls ET (1998) *The Brain and Emotion*. Oxford, UK: Clarendon Press.

Rosenthal DM (1986) Two concepts of consciousness. *Philosophical Studies* XLIX(3): 329–359.

Rosenthal DM (1993) Thinking that one thinks. In: Davies M and Humphreys GW (eds) *Consciousness: Psychological and Philosophical Essays*, pp. 197–223. Oxford, UK: Basil Blackwell.

Rosenthal DM (1997) Perceptual and cognitive models of consciousness. *Journal of the American Psychoanalytic Association* **45**(3): 740–746.

Rosenthal DM (1999) Sensory quality and the relocation story. *Philosophical Topics* **26**(1 and 2): 321–350.

Rosenthal DM (2000a) Introspection and self-interpretation. *Philosophical Topics* **28**(2): 201–233.

Rosenthal DM (2000b) Metacognition and higher-order thoughts. *Consciousness and Cognition* **9**(2): 231–242.

Rosenthal DM (in press a) *Consciousness and Mind*. Oxford, UK: Clarendon Press, 2003.

Rosenthal DM (in press b) Explaining consciousness. In: Chalmers DJ (ed.) *Philosophy of Mind: Contemporary and Classical Readings*. New York, NY: Oxford University Press, 2002.

Rosenthal DM (in press c) Unity of consciousness and the self. *Proceedings of the Aristotelian Society* **103**(3) (2003).

Rosenthal DM (in press d) The timing of conscious states. *Consciousness and Cognition* **11**(2) (2002).

Seager W (1999) *Theories of Consciousness: An Introduction and Assessment*. London and New York: Routledge.

Searle JR (1992) *The Rediscovery of the Mind*. Cambridge, MA: MIT Press/Bradford Books.

Sellars W (1963) Empiricism and the philosophy of mind. In: *Science, Perception and Reality*, pp. 127–196. London, UK: Routledge & Kegan Paul.

Shoemaker S (forthcoming) Consciousness and co-consciousness. In: Cleeremans A (ed.) *The Unity of Consciousness: Binding, Integration, and Dissociation*. Oxford: Clarendon Press.

Siewert CP (1998) *The Significance of Consciousness*. Princeton: Princeton University Press.

Simons DJ (2000) Current approaches to change blindness. *Visual Cognition* **7**: 1–16.

Staats PS, Hekmat H and Staats AW (1998) Suggestion/placebo effects on pain: negative as well as positive. *Journal of Pain and Symptom Management* **15**(4): 235–243.

Weiskrantz L (1997) *Consciousness Lost and Found: A Neuropsychological Exploration*. Oxford, UK: Clarendon Press.

Whiten A (1996) When does smart behaviour-reading become mind-reading? In: Carruthers P and Smith PK (eds) *Theories of Theories of Mind*, pp. 277–292. Cambridge, UK: Cambridge University Press.

Whiten A and Byrne RW (1997) *Machiavellian Intelligence, II: Extensions and Evaluations*. Cambridge, UK: Cambridge University Press.

Further Reading

Armstrong DM (1968/1993) *A Materialist Theory of the Mind*. New York: Humanities Press; 2nd revised edn. London, UK: Routledge & Kegan Paul, 1993.

Carruthers P (1989) Brute experience. *The Journal of Philosophy* LXXXVI(5): 258–269.

Dienes Z and Perner J (2001) The metacognitive implications of the implicit–explicit distinction. In: Chambres P, Izaute M and Marescaux P-J (eds) *Metacognition: Process, Function, and Use*, pp. 241–268. Dordrecht, Germany: Kluwer.

Dretske F (1993) Conscious experience. *Mind* **102**(406): 263–283; reprinted in Dretske, *Perception, Knowledge, and Belief*, pp. 113–137. Cambridge, UK: Cambridge University Press, 2000.

Haggard P (1999) Perceived timing of self-initiated actions. In: Aschersleben G, Bachmann T and Musseler J (eds) *Cognitive Contributions to the Perception of Spatial and Temporal Events*, pp. 215–231. Amsterdam, Netherlands: Elsevier.

Kobes BW (1995) Telic higher-order thoughts and Moore's paradox. *Philosophical Perspectives* **9**: 291–312.

Levine J (1993) On leaving our what it's like. In: Davies M and Humphreys GW (eds) *Consciousness: Psychological and Philosophical Essays*, pp. 121–136. Oxford, UK: Basil Blackwell.

Libet B, Gleason CA, Wright EW and Pearl DK (1983) Time of conscious intention to act in relation to onset of cerebral activity (readiness potential). *Brain* **106**(Part III): 623–642.

Lurz RW (in press) Advancing the debate between HOT and FO theories of consciousness. *Journal of Philosophical Research* **28** (2003).

Mellor DH (1980) Consciousness and degrees of belief. In: Mellor DH (ed.) *Prospects for Pragmatism*, pp. 139–173 Cambridge, UK: Cambridge University Press.

Perner J and Dienes Z (in press) Developmental aspects of consciousness: How much theory of mind do you need to be consciously aware? *Consciousness and Cognition*.

Rensink RA (2000) Seeing, sensing, and scrutinizing. *Vision Research* **40**(10–12): 1469–1487.

Rosenthal DM (2000) Consciousness and metacognition. In: Sperber D (ed.) *Metarepresentation: A Multidisciplinary Perspective*, pp. 265–295. New York, NY: Oxford University Press.

Rosenthal DM (2000) Content, interpretation, and consciousness. In: Ross D, Brook A and Thompson DL (eds) *Dennett's Philosophy: A Comprehensive Assessment*, pp. 287–308. Cambridge, MA: MIT Press/Bradford Books.

Rosenthal DM (in press e) Why are verbally expressed thoughts conscious?

Seager W (1994) Dretske on HOT theories of consciousness. *Analysis* **54**(1): 270–276.

Weiskrantz L (1986) *Blindsight: A Case Study and Implications*. Oxford, UK: Clarendon Press.

White PA (1988) Knowing more than we can tell: 'Introspective access' and causal report accuracy 10 years later. *British Journal of Psychology* **79**(1): 13–45.

Consciousness and Representationalism

Intermediate article

Benj Hellie, Sage School of Philosophy, Cornell University, Ithaca, New York, USA

CONTENTS

The representationalist theory of consciousness is the view that consciousness reduces to mental representation. This view comes in several variants which must explain introspective awareness of conscious mental states.

INTRODUCTION

Some mental states and processes are like something to their subjects; others are not. For instance, the states of seeing a stop sign, of hearing a screech, and of smelling gasoline are like something; as are the states of feeling fear, elation, or pain; as is the process of talking oneself through a problem. In contrast, states and processes that are not like anything to their subjects are accepted by both scientific and common sense psychology. Chomskian linguistic theories and Marrian theories of vision posit complex subpersonal operations, which make a difference to what one's mind is like to one only by their effects; common sense recognizes states of believing and intending that persist through dreamless sleep. States and processes that are like something to their subjects are conscious; otherwise not.

Among conscious states, what they are like to their subjects can differ: what seeing a red thing is like is standardly different from what seeing a green thing is like; what both are like differs from what smelling gasoline is like. A state has a 'phenomenal character' just in case it is conscious, or like something to its subject; two states have the same phenomenal character just in case what one is like to its subject is the same as what the other is like to *its* subject.

Phenomenal characters pose special problems for a fully naturalistic theory of the mind, for it may seem baffling how these properties can arise ultimately from interactions of particles and fields, or from processes in the brain. Wittgenstein famously wondered how *this* – his then current headache – could be a brain state; such bafflement is a proper reaction to the great difference in the ways in which phenomenal characters present themselves when thought of as phenomenal characters from the ways in which brain properties present themselves when though of as brain properties.

Representationalism is a view that attempts to naturalize phenomenal character without generating such bafflement by adopting a two-stage naturalistic reduction. The representationalist hopes that an intermediate reduction to certain representational properties will not generate bafflement; and that these representational properties may be reduced in turn, through one of the many ambitious projects for naturalizing mental representation.

WHAT IS REPRESENTATIONALISM?

Representationalism is the view that phenomenal characters somehow reduce to representational properties. The notion of a representational property deserves some expansion.

A state has a representational property when, to put it intuitively, it has a meaning or somehow stands in in some process for something else, such as an object, or a 'proposition' – a putative fact. Paradigmatic mental representational states are beliefs: one who believes that snow is green is in a state which means that snow is green, and which stands in for the putative fact that snow is green in a subject's reasoning. Another example is the state of thinking of Vienna: such a state means, or is about, Vienna; and stands in for Vienna itself in the subject's reasoning. Belief and thought-about are known to common sense psychology; scientific psychology also posits representational states: in some linguistic theories, for instance, in a many stage process of linguistic comprehension, a language-processing module goes into states which represent phonological, syntactic, and semantic properties of heard sentences.

Hoping that snow is green differs from believing that snow is green, although both are representational states and concern the proposition that snow is green. Standard philosophical theories consequently take representational states to involve a relation between a subject and a 'content' – what is meant – via an attitude or the relation borne to that meaning. When one believes that snow is green, the attitude is belief; when one hopes that snow is green, the attitude is hope. A representational property of a representational state may thus be characterized as a pair composed of an attitude and a content.

Representational states have correctness, or satisfaction, conditions partly determined by the correctness conditions for their contents. A proposition is correct just in case it is true; correctness conditions for other sorts of contents, such as Vienna, are less well understood by philosophy. So, for instance, a belief is correct just in case its content is; a desire or hope is satisfied just in case its content is correct.

VARIETIES OF REPRESENTATIONALISM

This crude formulation allows for a good deal of variation, along at least three dimensions.

What Sort of Reduction?

Any attempt at reduction may be more or less ambitious. This ambition influences the relation taken to hold between the reduced entity and the reducing entity.

The weakest interesting relation for reductive purposes is 'supervenience': the reduced entity cannot vary without variation in the reducing entity. Supervenience seems to be a necessary condition for reduction of any sort; whether it is sufficient is hotly debated.

A stronger thesis brings about an ontological reduction by identifying particular phenomenal characters with particular representational properties: for each phenomenal character ϕ there is a representational property ρ such that for a state to be ϕ is for it to be ρ; moreover, the property's status as representational is somehow fundamental, whereas its status as phenomenal is somehow more superficial.

A still stronger thesis brings about an epistemological or explanatory reduction by claiming the relevant identities to be *a priori* (under canonical ways of conceptualizing the properties in question).

Which Phenomenal Characters are Reduced?

The many different sorts of conscious states canvassed in the introduction have a wide variety of phenomenal characters: perceiving is unlike imagining; feeling sad is unlike feeling a physical pain. A representationalist may attempt to reduce all phenomenal characters, or only some privileged set of them, such as experiences of visually perceiving color.

There may be some purposes for which a limited theory would be interesting, such as that of avoiding a perceptual epistemology of sense-data (Russell, 1912; Harman, 1990). However, if the main purpose of representationalism is bringing consciousness under a unified naturalistic umbrella, less ambitious theories with narrower scope are less interesting; moreover, less ambitious theorists are forced to explain what it is about those special phenomenal characters which makes them susceptible to representationalist treatment when others are not.

First-Order and Higher-Order Representationalism

Not all representational states give rise to consciousness: e.g. sound sleepers continue to store memories. Which do?

Some mental states represent other mental states: I can think about my thinking about Vienna. Here the thought about Vienna is 'first order'; the thought about the thought is 'higher order'.

According to first-order representationalism (Harman, 1990; Tye, 1995) (sometimes called 'intentionalism'), some representational states that do not concern other mental states, such as seeing a green tree, are by their nature sufficient to give rise to phenomenal character. First-order representationalists identify phenomenal characters with a pair of a content and an attitude. First-order representation historically developed with the partial intent of avoiding a sense-datum epistemology (Harman, 1990), so that advocates of the view are often concerned to show that any conscious content must concern nonmental reality; but there is no obvious reason why a naturalist must hold this nonmentalist position: represented mental states might be themselves natural, and themselves represented as instancing natural properties.

By contrast, according to higher-order representationalism self-representation is necessary for consciousness, so that first-order states cannot by themselves give rise to consciousness (unless the first-order state is essentially such as to be self-representing; more below). There are several dimensions of variation in higher-order theory.

Perhaps the higher-order attitude is belief (Rosenthal, 1997); perhaps it is perception (Lycan, 1997; Lormand, 1994); perhaps it is a form of Russellian 'acquaintance' (Russell, 1912), which could be thought of as a relation which grounds the meanings of demonstrative concepts such as 'this' and 'thus'; further options are certainly available.

Moreover, there is room for variation in the causal relation the representing state bears to the represented state: perhaps no constraint is required, or perhaps a tight constraint, along the lines of that necessary for veridical perceiving is required. Or, alternatively, perhaps the representing and the represented states are in a tighter metaphysical relation of partial constitution.

Finally, though it would seem natural for the higher-order representationalist to take the phenomenal character of a state to be determined by how it is represented by the higher-order state, the fact that some conscious states are themselves representational gives rise to a choice here. What if the content of the lower-order state is misrepresented by the higher-order state – so that, for instance, an experience of seeing a red thing is misrepresented as an experience of seeing a green thing? Neither option is happy: if the higher-order content determines phenomenal character, although the subject would say 'that's red', he would seem irrational to himself in doing so; if the lower-order content determines phenomenal character, the higher-order content seems otiose. This dilemma can be dissolved if either the higher-order attitude is infallible, or if only noncontent properties of the lower-order state are represented.

A Mixed View

Finally, first- and higher-order views can be combined, if the 'special' first-order attitude is one which essentially involves self-representation: one bears this special attitude A to a content c only if one bears some further attitude A' to one's bearing A to c. If $A = A'$, an infinite hierarchy of bearings of A result (more on this point follows in the subsection: 'A Russellian view').

ARGUMENTS FOR REPRESENTATIONALISM

Higher-Order Representationalism

Higher-order representationalism can seem truistic. Intuitively, the phenomenal character of a mental state makes some impact on the subject's awareness: the idea of a state which has phenomenal character, but of which the subject is not in any way aware, is bizarre in the extreme. The impact need not consist in the presence of an occurrent opinion about which phenomenal character one is currently enjoying: a daydreamer might fail to notice subtle shifts in visual experience resulting from the gradual descent of the sun. However, in subjects with the conceptual capacity for such thoughts, the ground of such thoughts must be present (more on subjects without such capacity follows in the subsection: 'A Russellian view').

More must be done, of course, to specify what such grounding amounts to, and what it is to have an opinion about which phenomenal character one is currently enjoying. However, an analogy to perception may prove a fruitful source of investigation: just as perception provides the ground for occurrent thoughts about which colors and shapes are before one by serving as a stock of representations distinct from occurrent thought, so may awareness of phenomenal character do so by serving as a stock of representations distinct from occurrent thought.

Higher-order representationalism seems to be a commitment of the common idioms of consiciousness. We say that a conscious state is 'like something to its subject'; under analysis, this predicate is revealed to apply to a state just in case the subject represents the subject is acquainted with certain features of the state. On a slightly less common, but still natural, way of speaking of phenomenal character, we say that a state 'feels a certain way to its subject': here analysis is not needed to reveal that language draws an analogy between consciousness and perceptual representation. With a suitable theory of the link between truth-conditions and metaphysics, these observations could be extended to a proper argument for higher-order representationalism.

Phenomenology and First-Order Representationalism

The first source of support for first-order representationalism is an effect observed in phenomenological investigation, commonly known as 'transparency' (Harman, 1990; Tye, 1995): allegedly, when one sees a blue bead, one cannot detect any 'intrinsic' or nonrepresentational property (aside from the bead's apparent property of blueness) making a difference to the phenomenal character of this experience of seeing. If phenomenal characters are as they seem, no nonrepresentational property does make a difference; and what applies for this experience is held to apply for all experiences.

This argument does not show anything deep about consciousness, however. Even if transparency holds for states of seeing, the phenomenal character of one's total experience is complex, and there are contributions made by further mental states one is in: transparency may fail for these. There is nothing incoherent about the idea of a nonrepresentational property of a mental state or process: a mental process might procede at a certain rate, or be subject to voluntary control with a certain number of degrees of freedom. Nor is there anything incoherent about the idea of such a property being introspectively detectable, and thereby influencing phenomenal character. Consequently, if transparency holds, it does so at best contingently. Moreover, it does not even seem to hold generally for actual human phenomenal character (see subsection: 'Straightforward counter-examples to the first-order view').

Epistemology and First-Order Representationalism

The second source of support for the first-order view appeals to a 'recycling' theory of concepts of phenomenal character. Allegedly, when one forms an introspective judgment about which phenomenal character one is enjoying, one singles out that phenomenal character only by reusing discriminative capacities already conferred by undergoing an experience with a certain first-order content, together perhaps with a highly general concept of mental states of a certain sort (Evans, 1982). So for instance, when one sees a blue bead, one singles out the phenomenal character of the experience of seeing the blue bead roughly by taking it to be that phenomenal character had by experiences which represent things as thus, where one's grasp of 'thus' is grounded in the experience itself: here, the material before the 'thus' is responsible for the concept's application to one among the phenomenal characters; 'thus' is responsible for distinguishing the phenomenal character from all others. The end result is to distinguish phenomenal characters in general as representational properties. Hence, if our introspective concepts of phenomenal characters are true to and exhaustive of the natures of phenomenal characters, phenomenal characters just are properties involving representing nonmental reality as a certain way. Some terminology: the perceptual state responsible for grasp of the concept 'thus' contributes 'novel' content; whichever state is reponsible for the concept of phenomenal character contributes 'recycled' content.

The recycling argument has the same flaw as the transparency argument. Even if some discriminations of phenomenal characters recycle novel content concerning nonmental reality, this is compatible with there being introspective concepts with novel content concerning mental states and processes.

Mentalism and Nonmentalism

These objections only concern nonmentalist formulations of first-order representationalism. The recycling and transparency arguments do seem to go through once they have been weakened to allow for the sorts of phenomena to be discussed in the next section.

ARGUMENTS AGAINST REPRESENTATIONALISM

Spectral Inversion and the First-Order View

The first-order view must explain the data that suport the higher-order view: perhaps this can be done by adopting the mixed view described earlier. Once this challenge has been met, two other objections arise.

Consider first a sample inversion argument (Block, 1990). Perhaps there are three possible subjects s_1, s_2, and s_3, such that s_1 and s_2 are alike phenomenally and differ from s_3, and s_2 and s_3 are alike representationally and differ from s_1. For the first condition, suppose that s_1 and s_2 are alike intrinsically in those respects which matter for phenomenal character, while s_3 differs intrinsically enough to make phenomenal character differ.

In support of the possibility of such a trio, many have felt a powerful intuition that phenomenal character supervenes on a subject's intrinsic nature. For the second condition, suppose that internal constitution does not much matter for representational properties (the sign is arbitrary), so that the difference between s_2 and s_3 does not prevent them from being the same representationally – perhaps as a result of compensating divergences in their environments. A number of ways of establishing that there could be such compensating divergences in the presence of intrinsic likeness have been described in the literature; a typical attempt appeals to s_2 and s_3 being spectrally inverted with respect to one another but nonetheless deferring to the same experts in forming opinions about the colors of things. If thought content is deferential, and people standardly believe things are as they perceptually represent them to be, then s_2 and s_3 standardly perceptually represent things to be of the same color, violating supervenience.

The difficulty with this argument is that the premises both that thought content is deferential and that people standardly believe things are the way they perceptually represent them to be yield together the odd conclusion that perceptual representation is deferential: if perceptual content is nonconceptual, as many take it to be, it is unclear how deference could influence it. Moreover, however plausible deference may be for the concept 'red', it is less so for indexical predicate concepts such as 'thus'. Conversely, while 'thus' seems essentially tied to perceptual content, 'red', if subject to deference, is potentially less so.

The difficulties with this particular argument hold more generally: opinions about the narrowness of phenomenal character and the breadth of content may individually seem initially plausible, but the methodology for establishing such results has come under heavy attack in recent years. Moreover, even if color content is universally wide, the first-order representationalist may still retreat to the view that the phenomenal character-fixing first-order contents concern mental qualities, such as the degree to which certain retinal circuits are stimulated.

Straightforward Counterexamples to the First-Order View

Consider then a putative counterexample (for others see Peacocke, 1983): perhaps the most striking is double vision of the sort that results when one pokes one's eye with a finger. If one were to attempt to describe this experience, the most natural description would apply nonrepresentational predicates to one's own mental processes: one would say that one's visual field fragmented into two portions, which came to transparently overlay one another and move with respect to one another. If this description is correct, we seem to be introspectively aware of nonrepresentational properties of experience.

Mentalist first-order theorists may happily accept such examples. Anti-mentalist first-order theorists standardly reply to such counterexamples by redescribing the experience as one involving only ascription of properties to nonmental entities (Tye, 1995). For example: everything before me was suddenly replaced by a pair of transparent ghostly replicas of the scene before my eyes which then proceeded to move with respect to one another. Unfortunately, this description is implausible. The first-order representationalist should be concerned about this, for the transparency and recycling arguments both rely on a fairly high degree of privileged access to the nature of conscious mental states. Moreover, it is unclear what constrains such a strategy of redescription. Once initial plausibility has been set aside as a constraint, the position quickly threatens to become vacuous.

The Higher-Order View

Now consider the higher-order view. First, note that inversion-style objections can be readily modified to attack higher-order theory with more or less success.

Objections in the literature peculiar to the higher-order view have tended to focus less on the general higher-order approach than on particular hypotheses concerning the higher-order attitude. Of these, those receiving the most attention have been that the attitude is thought (the 'higher-order thought' view) and that it is perception (the 'inner-sense' view). A gamut of objections have been raised against each, too many to cover in detail. I will focus on the predictions these positions make concerning the nature of introspective knowledge of phenomenal character.

Epistemology and Higher-Order Thoughts

According to the higher-order thought view, a state is conscious just in case one has a belief about it; presumably the content of the belief concerns which phenomenal character the state has. Since one can open one's eyes without being flooded with an infinite hierarchy of conscious thoughts about one's perceptual states, the view must thus permit the higher-order beliefs to be themselves not conscious. The motivating idea behind higher-order representationalism was that a state has phenomenal character only if its subject is in some sense aware of it, in a way that grounds conscious introspective thought about which phenomenal character it has. The higher-order thought theorist should thus regard the possession of the nonconscious higher-order thought as the sort of awareness which provides such a ground. Forming such a conscious thought should then be a matter of bringing the unconscious thought to consciousness.

Compare standard cases in which one brings a nonconscious thought to consciousness: one example is simply bringing a belief to consciousness; another is straining to recall someone's name. Both processes seem distinct from the processes by which one forms conscious thoughts about the phenomenal characters of one's conscious states. As discussed above, one can do so either by recycling content, or by a perceptionlike process. Either case is, intuitively, a matter of making a discovery – perhaps a relatively banal discovery but a discovery still – whereas according to higher-order thought theory, it is a matter of mere rethinking something one already knew.

Recycling and Inner Sense

According to the inner sense view, conscious states are themselves perceived. As noted above, the attractiveness of the idea that conscious states are perceived is encoded in idioms for discussing consciousness ('states that feel a certain way'). Since the concept of perception is not wholly clear, however, nor is the adequacy of the view. Suffice it to say that while the idea that we perceive double vision and other distortions has some plausibility, the analogy becomes rather strained when applied to experiences whose phenomenal characters are known by recycling.

A Russellian View

Thought and perception do not exhaust the space of possible attitudes one might bear to one's conscious states. A view according to which distortions are perceived, and perceptual states are known by recycling, would evade the concerns raised against higher-order thought and inner-sense views. Such a view could avoid being *ad hoc*, if perceiving, and that grasp of concepts of phenomenal character which underlies the capacity for recycling perceptual contents, can both be plausibly taken as determinates of a more inclusive notion of Russellian 'knowledge by acquaintance' (Russell, 1912). There are in fact substantial cognitive similarities between perceptual knowledge and knowledge by recycling that justify so treating them: both are sorted by content and modality, both seem to ground demonstrative and recognitional indexical concepts, and so forth. Then Russell's observation that when one stands in a relation of acquaintance one is in addition acquainted with this relation generates an infinite hierarchy of relations of acquaintance.

The existence of such a hierarchy is plausible: we can introspect, and introspect our introspection, and introspect our introspection of our introspection, and so forth. Nor does this proposal face the concern that scotched the higher-order conscious thought proposal; relations of acquaintance serve as grounds of potential conscious thought, and need not give rise to actual or occurrent conscious thought.

An objection to this approach is that if concepts of phenomenal character are required to get consciousness off the ground, the experiences of dogs and children are not conscious. The correct reply is to bite the bullet: after all, do we have any way of knowing that they are?

Zombies and Nonreductive Representationalism

A final objection to representationalism appeals to zombies: one could perhaps conceive of creatures functionally like us but which lack phenomenal characters; together with suitable principles passing from conceivability to possibility, supervenience would fail. Whatever one might think of conceivability–possibility principles, it is important to note that this argument at best threatens reductive versions of representationalism: if one regards it as inconceivable that there could be consciousness without awareness of phenomenal character, one might do better to accept a nonnaturalistic conception of representation than to give up the consciousness–representation link.

REPRESENTATION IN THE COGNITIVE SCIENCES

It would not be far-fetched to say that the cognitive sciences are only the study of computational operations on mental representation. The Chomskian revolution in linguistics began with the recognition that the (or at least a central) goal of linguistics is the study of the means by which the mind determines the semantic, syntactic, and phonological properties of sentences by running computational operations on mental representations of the properties of those sentences. Marr's influential work on vision regards as the goal of the study of vision determining which operations must be performed on representations stemming ultimately from retinal stimulation, in order to generate representations of the properties of seen objects which have enough features to enable vision to do what it seems to do.

References

Block N (1990) Inverted earth. In: Tomberlin J (ed.) *Action Theory and the Philosophy of Mind*, vol. 4, *Philosophical Perspectives*, pp. 53–79. Atascadero: Ridgeview.

Byrne A (2001) Intentionalism defended. *The Philosophical Review* **110**: 49–90.

Evans G (1982) *The Varieties of Reference*. Oxford: Oxford University Press.

Harman G (1990) The intrinsic quality of experience. In: Tomberlin J (ed.) *Action Theory and the Philosophy of Mind*, vol. 4, *Philosophical Perspectives*, pp. 31–52. Atascadero: Ridgeview.

Lormand E (1994) Qualia! Now showing at a theater near you. In: Hill C (ed.) *The Philosophy of Daniel Dennett*, vol. 22, *Philosophical Topics*, pp. 127–156. Fayetteville, AR: University of Arkansas Press.

Lycan WG (1997) Consciousness as internal monitoring. In: Block N, Flanagan O and Güzeldere G (eds) *The Nature of Consciousness: Philosophical Debates*, pp. 755–771. Cambridge, MA: The MIT Press.

Peacocke C (1983) *Sense and Content: Experience, Thought, and Their Relations*. Oxford: Clarendon Press.

Rosenthal DM (1997) A theory of consciousness. In: Block N, Flanagan O and Güzeldere G (eds) *The Nature of Consciousness: Philosophical Debates*, pp. 729–753. Cambridge, MA: The MIT Press.

Russell B (1912) *The Problems of Philosophy*. Philadelphia: Hackett.

Tye M (1995) *Ten Problems of Consciousness*. Cambridge, MA: The MIT Press.

Further Reading

Armstrong DM (1968) *A Materialist Theory of the Mind*. London: Routledge and Kegan Paul.

Block N (1978) Troubles with functionalism. In: Block N (ed.) *Readings in the Philosophy of Psychology*, vol. i. Minneapolis: University of Minnesota Press.

Byrne A (1997) Some like it HOT: consciousness and higher-order thoughts. *Philosophical Studies* **86**: 103–129.

Carruthers P (2000) *Phenomenal Consciousness*. Oxford: Oxford University Press.

Dretske FI (1995) *Naturalizing the Mind*. Cambridge, MA: The MIT Press.

Hilbert DR and Kalderon MK (2000) Color and the inverted spectrum. In: Davis S (ed.) *Color Perception: Philosophical, Psychological, Artistic, and Computational Perspectives*. Oxford: Oxford University Press.

Shoemaker S (1982) The inverted specturum. *Journal of Philosophy* **79**: 357–381.

Shoemaker S (1994*a*) Phenomenal character. *NOÛS* **28**: 21–38.

Shoemaker S (1994*b*) Self-knowledge and 'inner-sense': the Royce lectures. *Philosophy and Phenomenological Research* **54**: 249–314.

Thau MA (2002) *Cognition and Consciousness*. Oxford: Oxford University Press.

Consciousness, Animal

Intermediate article

Colin Allen, Texas A & M University, College Station, Texas, USA

CONTENTS

Observable similarities between humans and other animals with respect to behavior and neurology, as well as considerations related to evolutionary continuity between species, underlie most opinions that animals have conscious experiences. The attempt to understand other forms of consciousness – other species of mind – may help us to understand the evolutionary roots of our own.

INTRODUCTION

In the last 30 years many innovative experiments by comparative psychologists and ethologists, both in the laboratory and in the field, have improved our understanding of the cognitive capacities of animals (see the edited collection by Bekoff *et al.*, 2002, for contributions by more than 50 leading researchers). As a result of this work we have acquired much knowledge about memory, learning, spatial navigation, social communication, and other cognitive capacities in a wide variety of species. Much of the work has concentrated on primates, and particularly on the great apes, reflecting a certain anthropocentric bias. But there has also been notable progress in understanding the cognitive capacities of other mammals, some birds, and even a few invertebrate species. Nonetheless there are large gaps in our understanding of the distribution of cognitive capacities across all taxonomic groups which limits our ability to understand the evolution of cognitive and mental abilities.

Opinions vary on the relevance of the work in cognition for the topic of consciousness *per se*. Some researchers regard consciousness as in internal, subjective state that is entirely beyond the range of scientific methodology. Others believe that existing scientific investigations shed light on conscious reasoning, self-awareness, and qualitative experience. By raising the issue of animal consciousness in a series of books, Donald Griffin (1976, 1984, 1992) deserves credit for having inspired the field that he named 'cognitive ethology'. Griffin promoted the idea, also to be found in Charles Darwin's work (1881; see Crist, 2002), that careful naturalistic observation and experiments under natural conditions can reveal the operation of mental processes in nonhuman animals, including invertebrates. Griffin's agenda has been strongly criticized, and his methodological suggestions often dismissed as anthropomorphic (see Bekoff and Allen, 1997, for a survey). But such criticisms may have overestimated the dangers of anthropomorphism (Fisher, 1990), and Griffin's suggestions have certainly acted as a catalyst for sophisticated work in cognitive ethology (Cheney and Seyfarth, 1990; Allen and Bekoff, 1997; Bekoff *et al.*, 2002). Although many scientists remain skeptical of Griffin's approach, and the question of whether other animals are conscious remains controversial among them, there are indications that the topic is no longer entirely taboo (see the edited collection by Bekoff, 2000).

There are two senses of consciousness that cause particular controversy when applied to animals: phenomenal consciousness and self-consciousness. Phenomenal consciousness refers to the qualitative, subjective, experiential, or phenomenological aspects of conscious experience, sometimes identified with qualia. To contemplate animal consciousness in this sense is to consider the possibility that, in Nagel's (1974) phrase, there might be 'something it is like' to be a member of another species. Self-consciousness refers to an organism's capacity to represent its own mental states, and is often related to the question of whether the organism possesses a theory of mind, can reason about its situation, and plan accordingly.

PHENOMENAL CONSCIOUSNESS

Observable similarities between humans and other animals with respect to behavior and neurology, as well as considerations related to evolutionary continuity between species, underlie most opinions that animals have conscious experiences. For instance, the behavior, neurology, and evolution of responses to noxious stimuli all appear to point to similar experiences of pain in humans and other animals. Noxious stimuli that humans would report as painful produce responses in animals that are easily mapped onto those of humans in similar circumstances. High-pitched vocalizations, fear responses, nursing of injuries, and learned avoidance are among the responses to noxious stimuli that are all part of the common mammalian heritage. The neural systems underlying these responses are virtually identical to those in humans, and animals respond to the same pain-relieving drugs, such as opiates. These responses are visible to some degree or other in organisms from several taxonomic groups.

Also in the realm of behavioral evidence, but less accessible to casual observation, are scientific demonstrations that members of other species, even of other phyla, are susceptible to the same visual illusions as we are (e.g. Fujita *et al.*, 1991) suggesting that their visual experiences are similar. Correspondingly, much of the basic research that is of direct relevance to understanding human visual consciousness has been conducted on the very similar visual systems of monkeys. Monkeys whose primary visual cortex is damaged even show impairments analogous to those of human blindsight patients (Stoerig and Cowey, 1997) suggesting that the visual consciousness of intact monkeys is similar to that of intact humans (see Carruthers, 2000 for dissent).

Such similarity arguments are somewhat weak for it is always open to critics to exploit disanalogies between animals and humans to argue that the similarities do not entail the conclusion that both are conscious (Allen, 1998). Even when bolstered by evolutionary considerations of continuity between the species, the arguments are vulnerable, for the mere fact that humans have a trait does not entail that our closest relatives must have that trait too. Thus, for instance, Povinelli and Giambrone (2000) argue that even quite similar behaviors in humans and chimpanzees are due to different underlying mechanisms, a point that Povinelli believes is demonstrated by his research into how chimpanzees use cues to track visual attention (Povinelli, 1996; but see Hare *et al.*, 2000, 2001 for a different view).

Despite the relevance of empirical research to similarity arguments, direct research into the phenomenological aspects of animal consciousness is hampered by at least two factors. One is the general problem that we lack a good theory of phenomenal consciousness, even in the human case. There is great uncertainty about the ontological status of phenomenal consciousness, whether in humans or in other animals. Accounts of consciousness in terms of basic neurophysiological properties, the quantum-mechanical properties of neurons, or *sui generis* properties of the universe are incomplete and controversial. More 'functionalist' accounts which attempt to explain the nature of experience in terms of the complex interactions between various cognitive subsystems fare no better. Consequently, no current theory of consciousness is secure enough to hang a decisive endorsement or denial of animal consciousness upon it.

A second factor hampering progress is that we lack a substantial account of what biological functions might be served by phenomenal consciousness. This point can be put another way: given any putative function of conscious experience, it seems we can imagine the same function being carried out without being accompanied by conscious experience. Thus, for example, while it is commonly asserted that the function of conscious pain is to 'tell' the organism when damage is occurring and thus promote behaviors which protect against further injury, it is also known that spinal reflexes alone, which are presumably quite unconscious, are sufficient to promote quite sophisticated kinds of withdrawal behavior and even associative learning (Grau, 2002). This raises the specter that phenomenal consciousness is epiphenomenal – completely devoid of physical effects – as some philosophers have asserted. If this were correct, then a search for the functions of consciousness would be doomed to futility. In fact, if consciousness is completely epiphenomenal then it cannot have evolved by natural selection, for selection can only operate on the effects of a trait on organismic fitness. On the assumption that phenomenal consciousness is an evolved characteristic of human minds, and therefore that epiphenomenalism is false, an attempt to understand the biological functions of consciousness would be useful for identifying its occurrence in different species by allowing us to search for organisms with the relevant functional capacities.

Many scientists remain convinced that no amount of empirical research can provide access to the subjective states of nonhuman animals. This remains true even among many scientists who are willing to invoke cognitive explanations of animal behavior that advert to internal representations. Opposition to dealing with consciousness can be understood as a legacy of behavioristic psychology: first, because of the behaviorists' rejection of terms for unobservables unless they could be formally defined; and second, because of the strong association in many behaviorists' minds between the use of mentalistic terms and the twin bugaboos of Cartesian dualism and introspectionist psychology (Bekoff and Allen, 1997). In some cases these scientists are even dualists themselves, but they are strongly committed to denying the possibility of scientifically investigating consciousness, and remain skeptical of all attempts to bring it into the scientific mainstream.

Because consciousness is assumed to be private or subjective, it is often taken to be beyond the reach of objective scientific methods (see Nagel, 1974). This claim might be taken in either of two ways. It might be taken to bear on the possibility of answering the question of whether members of another taxonomic group (e.g. bats) have conscious states. Or it might be taken to bear on the possibility of answering the question of what it's like to be a member of another species. The difference between believing with justification that a bat is conscious and knowing what it is like to be a bat is important because, at best, the privacy of conscious experience supports a negative conclusion only about the latter (Bekoff and Allen, 1997). To support a negative conclusion about the former one must also assume that consciousness has absolutely no measurable effects on behavior, i.e. one must accept epiphenomenalism. But such an assumption leads to the implausible conclusion that phenomenal consciousness did not evolve by natural selection.

REASONING AND SELF-CONSCIOUSNESS

René Descartes argued against animal minds on the grounds that animals do not use language conversationally or reason generally. While he was aware of the capacity of parrots to pronounce human words, he dismissed this as unintelligent, meaningless repetition. This judgment may have been appropriate for the few parrots he encountered, but it was not based on a systematic, scientific investigation of the capacities of parrots. Some would argue that Irene Pepperberg's multiple studies of the African Grey parrot 'Alex' (Pepperberg, 1999) should lay the Cartesian viewpoint to rest. This work, along with research into the acquisition of linguistic competence by chimpanzees (e.g. Gardner *et al.*, 1989; Savage-Rumbaugh, 1996; Fouts *et al.*, 2002), might be interpreted as undermining Descartes' assertions about lack of intelligent language use and general reasoning abilities in animals. There are now some serious questions about the interpretation of this work, however. Cartesians have pointed out the limitations shown by animals in such studies, and they are often joined by linguists who protest that the subjects of animal-language studies have not fully mastered the recursive syntax of natural human languages (e.g. Pinker, 1994).

Teaching human languages to animals tests their capacities on a task that is well outside the natural repertoire for the species. The same is true of testing animals on the capacity for mirror self-recognition (Gallup, 1970; Gallup *et al.*, 2002). It was long known that chimpanzees would use mirrors to inspect their images, but Gallup developed a protocol that appears to allow a scientific determination of whether it is merely the mirror image *per se* that is the object of interest to the animal inspecting it, or whether it is the image qua proxy for the animal itself that is the object of interest. Using chimpanzees with extensive prior familiarity with mirrors, Gallup anesthetized his subjects and marked their foreheads with a distinctive dye. Upon waking, marked animals who were allowed to see themselves in a mirror touched their own foreheads in the region of the mark significantly more frequently than controls who were either unmarked or not allowed to look into a mirror. Gallup's protocol has been repeated with other great apes and some monkey species, but besides chimpanzees only orangutans consistently 'pass' the test. (Shumaker and Swartz, 2002, cite preliminary evidence that the failure of gorillas may be one of motivation rather than basic intellectual capacity.) Gallup interprets this procedure as showing whether animals are self-aware and can infer the states of mind of another individual. (See Heyes, 1998, for a different opinion.)

The capacity to attribute mental states to others is often described as possession of a theory of mind. (But not always – see simulation theory.) The theory of mind debate has origins in the hypothesis that primate intelligence in general, and human intelligence in particular, is specially adapted for social cognition (see Jolly, 1966; Humphrey, 1976; Byrne and Whiten, 1988). Evidence for the theory of mind in great apes beside humans is mixed. Povinelli (1996) argues that in interactions with human food providers, chimpanzees apparently fail to understand the role of eyes in providing visual information to the humans, despite their outwardly similar behavior to humans in attending to cues such as facial orientation. The interpretation of Povinelli's work remains controversial. Hare *et al.* (2000) conducted experiments in which dominant and subordinate animals competed with each other for food, and concluded that 'at least in some situations chimpanzees know what conspecifics do and do not see and, furthermore, that they use this knowledge to formulate their behavioral strategies in food competition situations.' They suggest that Povinelli may have obtained negative results because his experiments involved cooperative chimp–human interactions. Hare and Wrangham (2002) argue that situations where chimpanzees are competing with each other for resources may be more ecologically relevant than cooperative tasks.

It is also likely that the mirror test is not an appropriate test for theory of mind in most species because of its specific dependence on the ability to match motor to visual information, a skill for which there may have been little selectional pressure. Consequently, it has been argued that evidence for the ability to attribute mental states in a wide range of species might be better sought in natural activities such as social play, rather than in laboratory-designed experiments which place the animals in artificial situations (Allen and Bekoff, 1997; see especially chapter 6; see also Hare *et al.*, 2000, 2001; Hare and Wrangham, 2001). Early attempts to find strong evidence of theory of mind in nonhuman animals under natural conditions generally failed to produce such evidence (see, e.g. Cheney and Seyfarth 1990). But anecdotal evidence (Byrne and Whiten 1988) and the more recent experimental results mentioned here tantalizingly suggest that researchers still have not managed to devise the right experiments.

CONCLUSION

Where does this leave questions about animal consciousness? While it may seem natural to think that we must know more about human consciousness before we try to determine whether other animals have it, such an approach may not be the most effective. The attempt to understand other forms of consciousness – other species of mind – may help us to understand the evolutionary roots of our own. Although research on primates is attractive because of their close evolutionary relationship to humans, much more needs to be learned about the cognitive abilities and forms of consciousness in less closely related species. It is important not to stifle research on consciousness because of worries about how to define the relevant notions. In the early stages of the scientific investigation of any phenomenon, putative samples are identified by rough rules of thumb (or working definitions) rather than complete theories. Early scientists identified gold by contingent characteristics rather than its atomic essence, knowledge of which had to await thorough investigation of many putative examples – some of which turned out to be gold and some not. Likewise, at this stage of the game, the study of animal consciousness may boldly investigate interesting cognitive capacities with no firm commitment to the idea that all these examples will involve conscious experience but absent of prejudice that none of them will.

References

Allen C (1998) The discovery of animal consciousness: an optimistic assessment. *Journal of Agricultural and Environmental Ethics* **10**: 217–225.

Allen C (2000) Animal consciousness. In: Zalta EN (ed.) *The Stanford Encyclopedia of Philosophy* (Winter 2000 Edition), [<http://plato.stanford.edu/archives/winter2000/entries/consciousness-animal/>]

Allen C and Bekoff M (1997) *Species of Mind*. Cambridge, MA: MIT Press. [See especially chapter 8.]

Bekoff M (2000) *The Smile of a Dolphin*. New York: Discovery Books.

Bekoff M and Allen C (1997) Cognitive ethology: slayers, skeptics, and proponents. In: Mitchell R *et al.* (eds) *Anthropomorphism, Anecdote, and Animals*, New York: SUNY Press.

Bekoff M, Allen C and Burghardt GM (eds) (2002) *The Cognitive Animal*. Cambridge, MA: MIT Press.

Byrne RW and Whiten A (eds) (1988) *Machiavellian Intelligence: Social Expertise and the Evolution of Intellect in Monkeys, Apes and Humans*. Oxford: Oxford University Press.

Carruthers P (2000) *Phenomenal Consciousness: A Naturalistic Theory*. Cambridge, UK: Cambridge University Press.

Cheney DL and Seyfarth RM (1990) *How Monkeys See the World: Inside the Mind of Another Species*. Chicago, IL: University of Chicago Press.

Crist E (2002) The inner life of earthworms: Darwin's argument and its implications. In: Bekoff M *et al.* (eds) *The Cognitive Animal*. Cambridge, MA: MIT Press.

Darwin C (1881/1985) *The Formation of Vegetable Mould, through the Action of Worms with Observations on Their Habits*. Chicago: Chicago University Press.

Fisher JA (1990) The myth of anthropomorphism. In: Bekoff M and Jamieson D (eds) *Interpretation and Explanation in the Study of Animal Behavior:* vol. 1, *Interpretation, Intentionality, and Communication*. Boulder, CO: Westview Press. [Reprinted in Bekoff M and Jamieson D (eds.) (1996) *Readings in Animal Cognition*. Cambridge, MA: MIT Press.]

Fouts R, Jensvold ML and Fouts D (2002) Chimpanzee signing: Darwinian realities and Cartesian delusions. In: Bekoff M *et al.* (eds) *The Cognitive Animal*. Cambridge, MA: MIT Press.

Fujita K, Blough DS and Blough PM (1991) Pigeons see the Ponzo illusion. *Animal Learning & Behavior* **19**: 283–293.

Gallup GG Jr (1970) Chimpanzees: self-recognition. *Science* **167**: 86–87.
Gallup GG Jr, Anderson JR and Shillito DJ (2002) The Mirror Test. In: Bekoff M *et al.* (eds) *The Cognitive Animal*. Cambridge, MA: MIT Press.
Gardner RA, Gardner BT and Van Cantfort TE (1989) *Teaching Sign Language to Chimpanzees*. Albany, NY: SUNY Press.
Grau J (2002) Learning and memory without a brain. In: Bekoff M *et al.* (eds) *The Cognitive Animal*. Cambridge, MA: MIT Press.
Griffin DR (1976) *The Question of Animal Awareness: Evolutionary Continuity of Mental Experience*. New York: Rockefeller University Press. [2nd edn, 1981.]
Griffin DR (1984) *Animal Thinking*. Cambridge, MA: Harvard University Press.
Griffin DR (1992) *Animal Minds*. Chicago: University of Chicago Press.
Hare B, Call J, Agnetta B and Tomasello M (2000) Chimpanzees know what conspecifics do and do not see. *Animal Behavior* **59**: 771–785.
Hare B, Call J and Tomasello M (2001) Do chimpanzees know what conspecifics know? *Animal Behavior* **61**: 139–151.
Hare B and Wrangham R (2002) The evolution of social cognition: comparative tests of the adapted cognition hypothesis. In: Bekoff M *et al.* (eds) *The Cognitive Animal*. Cambridge, MA: MIT Press.
Heyes C (1998) Theory of mind in nonhuman primates. *Behavioral and Brain Sciences* **21**: 101–148.
Humphrey N (1976) The social function of intellect. In: Bateson P and Hinde R (eds) *Growing Points in Ethology*. Cambridge, UK: Cambridge University Press. [Reprinted in Byrne and Whiten, 1988.]
Jolly A (1966) Lemur social behavior and primate intelligence. *Science* **153**: 501–506. [Reprinted in Byrne and Whiten, 1988.]
Nagel T (1974) What is it like to be a bat? *Philosophical Review* **83**: 435–450.
Pepperberg IM (1999) *The Alex Studies: Cognitive and Communicative Abilities of Grey Parrots*. Cambridge, MA: Harvard University Press.
Pinker S (1994) *The Language Instinct*. New York: William Morrow and Company.
Povinelli DJ (1996) Chimpanzee theory of mind? In: Carruthers P and Smith P (eds) *Theories of Theories of Mind*. Cambridge, UK: Cambridge University Press.
Povinelli DJ and Giambrone SJ (2000) Inferring other minds: failure of the argument by analogy. *Philosophical Topics* **27**: 161–201.
Savage-Rumbaugh S (1996) *Kanzi: The Ape at the Brink of the Human Mind*. New York: John Wiley and Sons.
Shumaker RW and Swartz KB (2002) When traditional methodologies fail: cognitive studies of great apes. In: Bekoff M *et al.* (eds) *The Cognitive Animal*. Cambridge, MA: MIT Press.
Stoerig P and Cowey A (1997) Blindsight in man and monkey. *Brain* **120**: 535–559.

Consciousness, Cognitive Theories of

Intermediate article

Bernard J Baars, Neurosciences Institute, San Diego, California, USA

CONTENTS

'Consciousness' operationally consists of all the things human beings report experiencing, from perception to mental images, inner speech, recalled memories, semantics, dreams, hallucinations, emotional feelings, and aspects of cognitive and motor control. In cognitive theory, consciousness appears to be a global access function, presenting an endless variety of focal contents to executive control and decision-making.

THE SCIENTIFIC REDISCOVERY OF CONSCIOUSNESS

Until recently consciousness was considered a scientifically intractable problem. That deep skepticism has faded with remarkable speed since the 1990s due to a flow of findings about the brain basis of conscious experience in perception, imagery, alertness, selective attention, working memory, episodic memory, and executive control (see Baars *et al.*, in press, for 70 scientific articles on the topic). Today a scientific race to consciousness is well under way. While the traditional philosophical paradoxes of mind and body are not resolved, many scientists now believe that significant progress can be made in the empirical study of conscious experience.

EVIDENCE TO BE EXPLAINED

To study anything in science we need to treat it as a variable. The concept of gravitational force would have been useless had Newton been unable to imagine zero gravity. Likewise, to understand consciousness we need to compare at least its presence to its absence; consciousness without unconsciousness is meaningless. In cases such as brain damage we can observe several degrees of consciousness, from full alertness to massive coma. But we need to compare at least two levels of consciousness to be able to ask the question 'what is the difference between them?' Only then can we deal with the issue of consciousness *as such*.

G. A. Mandler (1984) has made the penetrating observation that science is obliged to treat consciousness not as an observable datum but as an inferred concept based on public evidence. In science we can observe only the public *reports* people make about their conscious experience. Often we can make very reliable inferences about human conscious experience based on such reports. Almost 200 years of perceptual studies show that such reports correspond with exquisite sensitivity to the sensory stimulus array. Entire domains of research depend upon this well-established methodology.

Unconscious representations can also be inferred from public observations, though people cannot report them accurately. The simplest example is the vast multitude of memories that are currently unconscious. You may recall this morning's breakfast – but what happened to that memory before it was brought to mind? Apparently it was still somehow represented in the brain, though not consciously. Yet it has been known since Ebbinghaus that unconscious memories can shape mental processes without ever coming to mind; for example, it is easier to memorize material that was previously learned, even if you do not explicitly remember having learned it before. A case can be made for unconscious knowledge of many things: habituated stimuli, memories before and after recall, automatic skills, implicit learning and memory, the rules of syntax and semantics, unattended stimulation, presupposed knowledge, pre-conscious input processing, visual recognition of facial affect, and more.

To study consciousness as a variable, we can for example compare the reader's currently conscious contents, such as *these printed words*, to previous words in this sentence which are no longer conscious at the moment you are reading this. Notice that those previously conscious words must still be actively represented in memory to allow currently conscious words to be interpreted correctly. Thus the study of immediate memory presents many opportunities for treating consciousness as a variable. As in the case of Newtonian gravity, we can compare an event with its absence by keeping everything constant while varying the dimension of interest. This kind of contrastive study has now become routine, having been applied to waking consciousness (contrasted with sleep, coma, and general anesthesia), visual consciousness (compared to cortical blindness), consciousness of attended input (compared to unattended input), and much else (Baars *et al.*, in press). Taken together this sizable body of evidence serves to constrain theory in a very exacting way.

Even a simple set of contrasts brings out some important facts. The most prominent of these involve the remarkable limitations of conscious contents at any given moment, compared to the vastness and complexity of unconscious processes taking place at the same time. Consciousness is associated with limited capacity, seriality, and integration of multiple sources of information, while comparable unconscious processes involve much greater capacity, parallelism, and distributed autonomy. While conscious contents are limited at any given moment, they appear to facilitate access to multiple unconscious knowledge sources in the brain. Several current theories propose that a conscious event is made widely available to multiple brain mechanisms of memory, skill control, decision-making, semantics, anomaly detection, and the like. Thus consciousness may have a broad, architectural role as a gateway to multiple knowledge sources in the brain (e.g. Crick, 1984; Baars, 1988, 1997, 1998; Schacter, 1990; Chalmers, 1996; Damasio, 1989; John *et al.*, 1997; Edelman, 1989; Tononi and Edelman, 1999).

Fringe Conscious Events

William James (1890/1983) thought that vague or 'fringe conscious' events were at least as important as focally conscious ones. Fringe conscious phenomena include feelings of rightness, familiarity, beauty, coherence, anomaly, tip-of-the-tongue, attraction, repulsion, and the like. These phenomena can be operationally defined by a combination of high subjective certainty and high accuracy, but low experienced detail. Mangan (1993) has developed James's ideas about the fringe in modern terms, suggesting that fringe contents may not be subject to the classical capacity limitations of conscious experiences. Since focal conscious capacity is limited to one internally consistent experience at a time, Mangan sees fringe experiences as a very useful way of circumventing that limitation when needed.

COGNITIVE ARCHITECTURE THEORIES

A small cluster of theories has emerged to account for some of these aspects of conscious experience. Early theories were primarily cognitive, but quite sophisticated brain-based 'neuronal resonance' theories are now available (see below). Not surprisingly, the cognitive theories generally do a better job with psychological evidence, while neuronal resonance theories are better with the rapidly emerging evidence on the brain basis of conscious processes. A major theoretical aim for the future is to combine these approaches in a single, unified framework.

Cognitive theories can be divided into those that try to account for the evidence for consciousness as described above, and those that focus mainly on the various roles of consciousness in memory or executive functions. Baars' Global Workspace theory is the best-known example of the first class, while Schacter (1990) has developed a closely compatible approach to memory systems; Hilgard (1977), Johnson-Laird (1988), and Shallice (1988) focus on the role of consciousness in executive control.

Baars' Global Workspace (GW) theory extends the concepts of cognitive architectures to the problem of consciousness, in the tradition of A. Newell, H. A. Simon, and J. R. Anderson (see Baars, 1988, 1997, 1998). Specific mechanisms are proposed to deal with a sizeable array of psychological processes, from perception to imagery, spontaneous problem-solving, memory retrieval, goals and action control, and self. GW theory appears to be the most thoroughly worked out cognitive theory of conscious processes today, at a high level of description. Consciousness is associated with a global 'broadcasting system' that disseminates information widely throughout the brain. If this is true, then conscious capacity limits may be the biological price to be paid for the ability to make single momentary messages available to the entire system for purposes of coordination and control. Since at any moment there is only one 'whole system', a global dissemination facility must be limited to one momentary content.

GW theory relies on three theoretical constructs: unconscious specialized processors, a global workspace, and contexts. The first construct, the specialized unconscious processor, is an 'expert network' of which there are assumed to be many, working in parallel. There is direct evidence for many types of specialized systems in the brain. There are single cells, such as cortical feature neurons for color, line orientation, or faces, but also entire networks and arrays of neurons, such as cortical columns, cortical areas like Broca's or Wernicke's, large nuclei such as the thalamic relay nuclei, and so on. Like human experts, unconscious specialized processors may be quite limited in scope. They are extremely efficient in specific task domains, able to act autonomously or in coalitions with each other. They can receive global messages, and by mobilizing other experts by way of the global workspace they may be able to control mental or muscular activities. In routine tasks they may work autonomously, without conscious involvement, or they may display their output in the conscious global workspace. Answering a question such as 'What is your mother's maiden name?' requires a mission-specific coalition of unconscious experts, which return their answer to consciousness.

The second construct is the global workspace (GW) itself. A GW is an architectural capability for system-wide integration and dissemination of information. A global workspace is much like the bright spot of light on a theatre stage, cast by a spotlight in a darkened auditorium. Specialized unconscious processors are comparable to members of the audience sitting in the darkened theater. Groups of experts may interact locally, but in order to effect system-wide change they must compete for access to the bright spot on the theatre stage, perhaps supported by a coalition of other audience members. Once an expert reaches the bright spot on stage, it can broadcast a global message to the system as a whole. New links between unconscious experts are made possible by global interaction via the bright spot, and can then spin off to become new autonomous processors. The stage allows novel expert coalitions to form, to work on new or difficult problems which cannot be solved by existing experts and coalitions. Tentative solutions to new problems can then be globally disseminated, scrutinized, and modified. Since conscious experience seems to have a great perceptual bias, it is convenient to imagine that perceptual regions of cortex – visual, auditory, or multimodal – can compete for access to a brain version of a GW.

Theater models of consciousness have been criticized by Daniel Dennett and Marcel Kinsbourne on the ground that they are 'Cartesian' and conceptually flawed (Dennett and Kinsbourne, 1992). However, global workspace architectures have been implemented in artificial intelligence simulations for decades, and are not vulnerable to the Dennett–Kinsbourne critique. They do not assume a Cartesian point centre, for example, and both Dennett and Kinsbourne have refrained from applying their critique to global workspace theory.

'Context', the third construct in GW theory, refers to the powers behind the scenes of the theater of mind. Contexts are coalitions of expert processors that may function like a theatrical director, playwright, or stagehand who can influence the actors that appear in the spotlight. They can be defined empirically as *unconscious factors that shape conscious contents,* just as a director behind the scenes can influence the words and actions of actors on stage without being visible from the audience. Conceptually, contexts are defined as pre-established expert coalitions that can evoke, shape, and guide global messages without themselves entering the global workspace. Indeed, Dennett himself has recently endorsed a 'neuronal global workspace' approach to consciousnes (Dennett, 2001).

Contexts may be momentary or long-lasting. Momentary contexts may shape the reader's conscious interpretation of a word such as 'set', which has many different meanings. The word 'tennis' before 'set' shapes the interpretation of 'set', even when 'tennis' has already slipped from consciousness. But the word 'tennis' needed to be conscious initially to create the unconscious context that interprets 'set'. Contexts can also be long-lasting. The reader's ideas about consciousness from years ago may influence his or her current experience of this paragraph, even if those memories do not become conscious again. In general, major life experiences appear to set lifelong attitudes or character traits. Such major events typically influence current conscious experiences unconsciously, rather than being brought to mind. It is believed that a shocking or traumatic experience can also set up largely unconscious expectations that can shape subsequent conscious experiences.

Several proposals aim to cast the global workspace framework in a more neurally realistic form, giving special regard to thalamocortical mechanisms. Newman and Baars (1993) and Newman *et al.* (1997) described ways to integrate global workspace theory with the neuronal resonance theories discussed below. Newman *et al.* (1997, p. 1195) propose that 'One would expect the neural mechanism for global attention to be complex, and widely distributed. ... But the basic circuitry can be described, to a first approximation, in terms of repeating, parallel loops of thalamo-cortico-thalamic axons, passing through a thin sheet of neurons known as the nucleus reticularis thalami.' The overall framework suggests a neurocognitive model in which consciousness is viewed as a global integration and dissemination system operating in a large-scale, distributed array of specialized bio-processors, which controls the allocation of processing resources in the central nervous system.

Almost all current theories agree that consciousness involves system-wide functions, rather than local ones that are mostly unconscious, and that specialized 'expert' systems tend to be unconscious and relatively isolated. Some detailed implications have been worked out by way of computer simulations (Franklin and Graesser, in press).

Consciousness, Memory, and Self Functions

Other cognitive theories focus on functions that are associated with consciousness, such as working memory, episodic memory, and executive or 'self' functions. Hilgard's framework (e.g. 1977) is based on several decades of research on hypnotic dissociation, in which people under conditions of suggestion appear to lack conscious access to such things as sensory input, memories, normal voluntary control, or aspects of their own identity. Under experimental conditions it is often possible to show that these responses are not merely faked or simulated. Suggestible states are not merely oddities: a fifth of the normal population is highly suggestible, and all humans are suggestible under some circumstances. Hilgard points out a number of implications for normal executive control and access to self.

Johnson-Laird's (1988) operating system model of consciousness emphasized control functions such as directing attention, planning and triggering action and thought, and purposeful self-reflection. Johnson-Laird's cognitive architecture consists of a parallel processing system dominated by a control hierarchy. The system is a collection of largely independent processors (finite state automata), which cannot modify each other but which can receive messages from each other; each starts to compute when it receives appropriate input from any source. Each passes messages up through a hierarchy to the operating system, which sets goals for the subsystems. The operating system does not have access to the detailed operations of the subsystems – it receives only their output. Likewise, the operating system does not need to specify the details of the goals it transmits to the processors – these take the goal, abstractly specified, and elaborate it in terms of their own capabilities.

In this model conscious contents reside in the operating system or its working memory. Johnson-Laird believes his model can account for aspects of self-reflection, intentional decision-making, and action control.

Daniel Schacter has also proposed a compatible approach, to integrate evidence on neuropsychological disconnections from consciousness, particularly implicit memory and anosognosia, called the Dissociable Interactions and Conscious Experience (DICE) model. 'The basic idea motivating the DICE model ... is that the processes that mediate conscious identification and recognition – that is, phenomenal awareness in different domains – should be sharply distinguished from modular systems that operate on linguistic, perceptual, and other kinds of information' (1990, pp. 160–1).

Like Johnson-Laird, Schacter's DICE model assumes independent memory modules and a lack of conscious access to details of skilled, procedural knowledge. It is primarily designed to account for memory dissociations in normal and damaged brains. Schacter makes two main observations. First, with the exception of coma and stupor, failures of awareness in brain damage are usually restricted to the domain of the impairment; patients do not have difficulty generally in gaining conscious access to other knowledge. For example, amnesic patients do not necessarily have trouble reading words, while alexic individuals do not necessarily have memory problems.

However, implicit (unconscious) memory for lost conscious functions has been demonstrated in many conditions. For example, name recognition is facilitated in prosopagnosia (face-blind) patients when the name is accompanied by a matching face – even though the patient does not consciously recognize the face. Numerous examples are known of implicit knowledge in patients who do not have deliberate, conscious access to the information. These findings suggest an architecture in which various sources of knowledge function somewhat separately, since they can be selectively lost. These separable knowledge sources are not accessible to consciousness, even though they continue to shape voluntary action.

In DICE Schacter gives additional support to the idea of a system of separable knowledge sources, specifically to explain spared explicit knowledge in patients with brain damage. DICE does not try to explain the limited capacity of consciousness or the problem of selecting among potential inputs. In agreement with Shallice (below) the DICE model suggests that the primary role of consciousness is to mediate voluntary action under the control of an executive.

Tim Shallice's 1978 theory focused on conscious selection of a *dominant action system*, a set of current goals that work together to control thought and action. More recently Shallice (1988) modified and refined the theory to accommodate a broader range of conscious functions. Shallice's information-processing system also consists of a very large set of specialized processors, like Johnson-Laird's subsystems and Baars' (1988) specialized unconscious processors. A large set of action and thought schemata can 'run' on these modules. The schemata are well-learned, highly specific programs for routine activities, such as eating with a spoon, driving to work, etc. Competition and interference between currently activated schemata is resolved by *contention scheduling*, which selects among the schemata based on activation and lateral inhibition. Contention scheduling acts only during routine operations. A *supervisory system* modulates the operation of contention scheduling. It has access to representations of operations, of the individual's goals, and of the environment. It comes into play when operation of routinely selected schemata does not meet the system's goals, that is, when a novel or unpredicted situation is encountered or when an error has occurred. Finally, a *language system* can function either to activate schemata or to represent the operations of the supervisory system or specialist systems. More recently an *episodic memory* component with event-specific memory has been added to the set of control processes.

Shallice claims that consciousness cannot reside in any of these control systems taken individually. No single system is either necessary or sufficient to account for conscious events. Consciousness remains even when one of these control systems is damaged or disabled. And the individual control systems can all operate autonomously and unconsciously. Shallice suggests that consciousness may arise when there is concurrent and coherent operation of several control systems on representations of a single activity.

NEURONAL RESONANCE THEORIES

Several brain theories have now been developed. They have so much overlap that we will consider only a few in detail. All brain theories of conscious functions can be broadly characterized as 'neuronal resonance theories'. That is, they propose that in waking consciousness the thalamocortical core of the human brain is continuously cycling neuronal activity between the major thalamic relay nuclei and corresponding regions of cortex, supplemented by corticocortical and cortical-subcortical cycles of activity. When consciousness is at a very low level, as in deep sleep, some comas, deep general anaesthesia, and epileptic 'states of absence', the rapid and irregular electrical field activity characteristic of waking consciousness is replaced by slow, correlated, and high-amplitude waves. At the level of single neurons waking consciousness involves temporally uncorrelated firing, while deep sleep shows repetitive, highly correlated burst-pause firing in large cell populations. These basic facts and their underlying neurophysiology, neuroanatomy, and neurochemistry are so well established that they arouse little fundamental disagreement (e.g. Singer and Gray, 1995; John *et al.*, 1997; Edelman, 1989; Tononi and Edelman, 1999; Damasio, 1989). (*See* **Neural Correlates of Visual Consciousness**; **Thalamocortical Interactions and Binding**)

The development of parallel distributed processor (PDP) models is consistent with such large-scale brain models. Grossberg (in press) and Taylor (1992) have applied PDP concepts to aspects of consciousness. Grossberg's Adaptive Resonance Theory (ART) has been applied to specific brain functions including perception, recognition, attention, reinforcement, recall, and memory search. Grossberg writes that 'it is suggested that all conscious states are resonant states', especially those that occur 'between bottom-up and top-down processes as they reach an attentive consensus between what is expected and what is there in the outside world'.

Taylor (1992) also believes that conscious contents are determined by the intermingling of past and present. The reticular nucleus of the thalamus (nRt) appears to control the major sensory highways to cortex, and Taylor has provided a model for intersensory competition based on this evidence. Rather than a simple winner-take-all competition, he suggests that consciousness corresponds to a wave of activity of the coupled thalamic-nRt-cortical system, a multidimensional 'bubble' of neuronal firing patterns. Such a wave will show many regions over cortex that have nonzero activity. Recently Taylor has advocated the possibility that parietal cortex may act as a global workspace in the sense advocated by Baars (above).

Antonio Damasio (1989) suggests a role for consciousness in recognition and recall. Like other authors, Damasio's theory involves looping feedforward and feedback circuits of neurons, a massive resonant assembly of cells. To this mechanism he adds a specific role for sensory projection areas of the cortex (local convergence zones) and their neighboring higher-order association areas (nonlocal convergence zones). Temporal synchrony is proposed to account for retrieval of information from memory when neuron ensembles are activated in a time-locked fashion in the local and nonlocal convergence zones of cortex. Memories are stored as distributed sets of fragmentary features in large populations of neurons, and are retrieved by means of synchronous activation of related firing patterns in a subset of the same cell population.

CONCLUSION

Although consciousness has only recently returned as a central focus of the brain and cognitive sciences, current proposals capture a good deal of the evidence in a broad way. A critical mass of scientists is now collecting relevant evidence and developing theory. Conscious experience seems to create access to multiple, independent knowledge sources. While there are distinct pros and cons to each theoretical perspective, the general impression is of a surprising degree of convergence. A major goal for the future is to show how neuronal resonance could support the global cognitive functions that require consciousness.

Acknowledgements

This work was supported by the Neurosciences Research Foundation.

References

Baars BJ (1988) *A Cognitive Theory of Consciousness*. New York, NY: Cambridge University Press.

Baars BJ (1997) *In the Theater of Consciousness: The Workspace of the Mind*. New York, NY: Oxford University Press.

Baars BJ (1998) Metaphors of consciousness and attention in the brain. *Trends in Neurosciences* **21**(2): 58–62.

Baars BJ (2002) The conscious access hypothesis: origins and recent evidence. *Trends in Cognitive Sciences* **6**(1): 47–52.

Baars BJ (in press) Working memory requires conscious processes, not vice versa: a global workspace account. In: Osaka N (ed.) *The Neural Basis of Consciousness*. Amsterdam: Benjamins.

Baars BJ, Banks WP and Newman J (in press) *Essential Sources in the Scientific Study of Consciousness*. Cambridge, MA: MIT Press/Bradford Books.

Chalmers D (1996) *The Conscious Mind*. New York, NY: Oxford University Press.

Crick F (1984) Function of the thalamic reticular complex: the searchlight hypothesis. *Proceedings of the National Academy of Science of the USA* **81**: 4586–4590.

Damasio AR (1989) Time-locked multiregional retroactivation: a systems-level proposal for the neural substrates of recall and recognition. *Cognition* **33**: 25–62.

Dennett D and Kinsbourne M (1992) Time and the observer: the where and when of consciousness in the brain. *Brain and Behavioral Sciences* **15**(2): 183–247.

Dennett D (2001) Are we explaining consciousness yet? *Cognition* **79**: 221–237.

Edelman GM (1989) *The Remembered Present: A Biological Theory of Consciousness*. New York, NY: Basic Books.

Franklin S and Graesser A (in press) A software agent model of consciousness. In: Baars BJ, Banks WP and Newman J (eds) *Essential Sources in the Scientific Study of Consciousness*. Cambridge, MA: MIT Press/Bradford Books.

Grossberg S (in press) Brain learning, attention, and consciousness. In: Baars BJ, Banks WP and Newman J (eds) *Essential Sources in the Scientific Study of Consciousness*. Cambridge, MA: MIT Press/Bradford Books.

Hilgard ER (1977) *Divided Consciousness: Multiple Controls in Human Thought and Attention*. New York: Wiley.

James W (1890/1983) *The Principles of Psychology*. Cambridge, MA: Harvard University Press.

John ER, Easton P and Isenhart R (1997) Consciousness and cognition may be mediated by multiple independent coherent ensembles. *Consciousness & Cognition* **6**: 3–39.

Johnson-Laird PN (1988) A computational analysis of consciousness. In: Marcel AJ and Bisiach E (eds) *Consciousness in Contemporary Science*, pp. 357–368. Oxford, UK: Clarendon Press.

Mandler GA (1984) *Mind and Body*. New York: Basic Books.

Mangan B (1993) Taking phenomenology seriously: the 'fringe' and its implications for cognitive research. *Consciousness & Cognition* **2**(2): 89–108.

Newman J and Baars BJ (1993) A neural attentional model for access to consciousness: a Global Workspace perspective. *Concepts in Neuroscience* **2**(3): 255–290.

Newman JB, Baars BJ and Cho S-B (1997) A neural Global Workspace model for conscious attention. *Neural Networks* (Special Issue) **10**(2): 1195–1206.

Schacter DL (1990) Toward a cognitive neuropsychology of awareness: implicit knowledge and anosognosia. *Journal of Clinical and Experimental Neuropsychology* **12**(1): 155–178.

Shallice T (1988) Information-processing models of consciousness: possibilities and problems. In: Marcel AJ and Bisiach E (eds) *Consciousness in Contemporary Science*, pp. 305–333. Oxford, UK: Clarendon Press.

Singer W and Gray CM (1995) Visual feature integration and the temporal correlation hypothesis. *Annual Review of Neuroscience* **18**: 555–586.

Taylor J (1992) Towards a neural network model of the mind. *Neural Network World* **2**: 797–812.
Tononi G and Edelman GM (1999) Consciousness and complexity. *Science* **282**: 1846–1851.

Further Reading

Crick F (1995) *The Astonishing Hypothesis*. New York, NY: Touchstone Books.
Crick F and Koch C (1990) Towards a neurobiological theory of consciousness. *Seminars in Neuroscience* **2**: 263–275.
Edelman GM and Tononi G (2000) *A Universe of Consciousness*. New York, NY: Basic Books.
Flanagan O (1992) *Consciousness Reconsidered*. Cambridge, MA: MIT Press.
Hilgard ER (1986) *Divided Consciousness: Multiple Controls in Human Thought and Action*. New York, NY: Wiley-Interscience.
Llinás R and Paré D (1991) Commentary: of dreaming and wakefulness. *Neuroscience* **44**(3): 512–535.
Milner AD and Rugg MD (eds) (1992) *The Neurophysiology of Consciousness*. London: Academic Press.

Consciousness, Disorders of

Introductory article

Fred Plum, Weill Medical College of Cornell University, New York, USA
Nicholas D Schiff, Weill Medical College of Cornell University, New York, USA

CONTENTS

Disorders of consciousness include coma, stupor, confusion and other abnormal states of acute brief or moderately sustained unconsciousness.

INTRODUCTION

The brain generates the mind, and the healthy, wakeful mind formulates consciousness. The American psychologist William James in 1890 stated, 'Consciousness is the [indispensable] fundamental awareness of the self's internal ego.' He then expanded that self-centered focus to identify the self's greater qualities of memory, attention, intention, chronological time, emotion, learned behavior, and several other less general psychological qualities. At that early time, only philosophical thinking interpreted gross anatomic knowledge in trying to understand how the awake brain might lose conscious functions.

Modern neurological medicine has defined several distinct behavioral pathological states that arise from inherited and acquired brain injuries and lead to disorders of consciousness. Brain injuries that reflect global disorders of consciousness include stupor and coma, the vegetative state, akinetic mutism, absence and partial complex seizures, delirium, and severe dementia. These global disorders, described below, totally disable the capacity of the individual for intentional behaviors. Though different in pattern, 'focal' disorders of consciousness can exist in several serious illnesses. A patient suffering a focal disorder of consciousness can be awake and interact with the environment, and yet exhibit severe alterations in awareness. These disorders uniquely illustrate the constructed nature of conscious experience.

THE FORMULATION OF CONSCIOUSNESS

All people with a healthy brain and body can recognize themselves, their thoughts and their intentional conscious activity. Descriptions may vary in detail, but ask people what they think about the quality called their consciousness and the first reply is likely to be, 'I'm awake and I'm here'. Proof often follows with (for example) 'I'm John Smith!'

An educated person recognizes conscious awareness as a continuously unfolding, automatic sense of being awake, alive, and logically thoughtful. Actually, one's mind is being continuously filled with flowing thoughts, normal language, recent memories, learned motor behavior, or novel discoveries. Even the most educated person, however, sometimes wonders about how the brain automatically experiences normal emotions, how it generates logical thinking, and how it induces the smooth flow of relevant or original thoughts and coordinated deeds.

'Now, how did I come to think about that?' is an often-expressed question, but usually not one that is part of everyday conversation. Nor do we wonder what preceding activities our brain generates when we automatically take our daily walk down the same lane. Even when we 'instinctively' jump out of the way of an unseen, oncoming vehicle, we often fail to realize that our awake, preconscious frontal lobe thought and acted first. Only after we have jumped away from the danger do we become aware of our act and experience an emotional feeling of fright. This example illustrates how we often act or even speak before we consciously think.

The normal brain's cognitive processing systems organize intentional behaviors drawing on a rapidly accessed, vast store of relevant memories. It preconsciously formulates either incoming or spontaneous information in less than a quarter of a second. It is astonishing to realize that the functions of memory, intention, and perception may largely occur before any act or expressed thoughts enter immediate conscious awareness. We may think it strange that when we hold a conversation, our mind has preconsciously formulated what we are going to say a half-second or more before we actually say it. The Nobel laureate in medicine, Gerald Edelman, recognized this normal, instinctive preconscious formulation of thoughts, words and athletic acts in the ingenious title of his book, *The Remembered Present, A Biological Theory of Consciousness.*

Neuropsychological Dimensions of Consciousness

Consciousness is a time-ordered, egocentric process that interweaves inner and outer perceptions, stored memories, and immediately innovative thoughts. Emotional feelings imbue conscious awareness and sharpen intentional actions. Memory provides not only the ultimate storehouse of explicit conscious knowledge; it also develops the preconscious, implicit brain learning of motor skills and physical practice. Memory qualities and quantities depend on the combination of our innate cognitive talents, our subsequent schooling, our continuously thoughtful appraisal of new objects, and our interpersonal learning from and about people. The goal can be athletic, intellectual, or both. All evidence indicates that the earlier the young begin to learn and continue lifelong studies, the greater will be their future mental and behavioral capacities. Indeed, the longer a person's education and thought-requiring occupation last, the greater the brain's and body's functional longevity.

How this serially time-ordered, organized process of consciousness incorporates outer information with inner attention and immediate evaluation is the subject of intense neuroscientific investigation. Several distinct neuropsychological qualities can be ascribed to distributed networks of brain regions that selectively contribute to organized, wakeful human consciousness. These networks include the brainstem and allied arousal systems which control the sleep–wake cycles of the entire forebrain; prefrontal cortical regions (e.g. anterior cingulate cortex, frontal eye fields) which support continuous attention to self and environment and immediate intention; and posterior cortical regions of the temporal lobes (superior temporal gyrus) and parietal lobes (inferior parietal cortex) which support self-sensory perceptions and instinctive, and automatic awareness of inner and outer spatial relationships. Memory systems of the brain are widely distributed and depend on the integrity of the medial temporal lobe (hippocampus, entorhinal cortex) for initial storage, and on multiple cortical association areas (frontal, parietal, temporal, and occipital) and parts of the thalamus for functions of both storage and retrieval.

Additional neuropsychological qualities include the mind's chronological ordering of events (of unknown localization but disordered by injuries to the thalamus), moods and emotions (contributed by distributed regions of the 'limbic' brain). Learnt symbolic abstractions of verbal (left hemisphere), musical (right hemisphere), and geometric languages (left posterior parietal regions) contribute to humans' singular qualities of normal awareness.

SPECIFIC DISORDERS OF CONSCIOUSNESS

Coma

Coma is a totally unconscious and unarousable brain state resembling sleep, in which the eyes are closed and which lasts 24 h or more, due to any of several major causes. One is the use of sustained therapeutic anesthesia. More frequent causes are direct brain injury or diseases affecting the brain's cerebral hemispheres and arousal systems. Table 1 compares the loss of neuropsychologic components incurred in coma with those of other disorders of consciousness.

Table 1. Global disorders of consciousness

	Coma	PVS	ASZ	AKM	HKM	CPS	DEL
Arousal	–	+	+	+	+	+	+
Attention	–	–	–	+	+/–	+/–	+/–
Intention	–	–	–	–	+	+/–	+/–
Memory	–	–	–	–	–	–	+/–
Awareness	–	–	–	–/?	+/–	+/–	+/–

AKM, akinetic mutism; ASZ, absence seizures; CPS, complex partial seizures; DEL, delirium; HKM, hyperkinetic mutism; PVS, persistent vegetative state; –, absent; + present (in crude form for attention, AKM, and intention, HKM), +/– incompletely expressed; –/? apparently absent.

Stupor

Stupor is a condition of deep sleep or behaviorally similar unresponsiveness from which the person cannot be aroused except by vigorous and repeated exogenous stimulation. As soon as such stimulation ceases, the person relapses into the unresponsive state. Light stupor is typical in cases of overdoses of soporific drugs or alcohol. Deep stupor more frequently reflects severe pharmacological, metabolic, or traumatic injury to the brain. The term 'semi-coma' is occasionally used in nonmedical writing to describe patients in stupor or persistent vegetative state but is not considered a diagnostic category.

Persistent Vegetative State

The vegetative state is a condition in which physiologically active, systemic organs continue to sustain the life of a body that has become at least temporarily devoid of a conscious brain. In most cases of coma, wakefulness will return spontaneously in a matter of days or weeks; but in some people, despite a wakeful appearance, the mind may be absent for many weeks, months or forever. This tragedy has been named the *persistent vegetative state* (PVS). Such patients express irregularly timed sleep–wake patterns, but all feeding and bodily care must be provided by external sources. The term 'arbitrarily' identifies patients who remain psychologically unconscious for at least a month. They are alive, but totally unaware of self or their environment. The vegetative state presents the fundamental clinical dissociation of arousal from all other components of consciousness (Table 1).

The clinical judgment of unconsciousness in PVS has been supported by the results of positron emission tomography (PET) scan studies that reveal overall cerebral metabolism to be reduced by 50% or more below the normal rate. The observed metabolic levels are equivalent to those found in persons undergoing deep surgical anesthesia. In a study of behavioral and physiological variations in a few patients in the vegetative state, one woman randomly expressed occasional single, understandable words. Her PET studies revealed isolated islands of left frontotemporal cerebral structures that operated at an abnormally low metabolic rate but at nearly twice the rates of the remaining brain. Similar isolated expressions have been encountered in several other vegetative patients. Typically, the patients express easily identifiable, stereotypical, emotional-limbic responses. These emotional expressions probably reflect distinct and isolated limbic mechanisms; their preservation is likely to depend on integrative brainstem structures that lie outside the corticothalamic systems that typically undergo overwhelming injury in PVS patients.

Syncope

Syncope (fainting) consists of brief unconsciousness caused by reduction of systolic arterial blood flow through the brain. Most syncope is benign and occurs in persons younger than 50 years. Termed 'vasogenic', it reflects sudden dilation of the body's cholinergic and sympathetic neurovascular systems, reduces systemic blood pressure and deprives the erect brain's critical oxygen supply. A second type is less frequent and affects older people suffering from postural orthostatic hypotension. A third type affects elderly people with severe cardiac, cardiopulmonary, or systemic atheromatous illness. Such patients rarely regain normal brain function if they fail to gain accurate awareness in more than 4–5 minutes.

Concussion

Concussion is an unconscious state that immediately follows a severe traumatic head injury. Since its ultimate duration cannot be predicted accurately, some surgeons call post-traumatic lack of consciousness 'concussion' for 24 h; after that, the term is changed to 'coma'.

At its least, concussion interrupts the brain's organized thoughts and impairs or blocks its recent memory. Acute severe concussions may suddenly and briefly suppress vegetative brainstem functions, thereby invoking transient breathlessness, slowed heart rate, low blood pressure, and widening of the pupils. Boxing knockouts for 10 s or more vividly exemplify moderate to severe concussion, as the bewildered athlete staggers from the ring and sometimes falls. A few knocked-out boxers will remain unconscious after the count, and a very few may die from acute brain hemorrhage. A measurable group may gradually develop dementia during their early sixth decade. Many drivers or passengers in serious road traffic accidents can suffer brief knockouts of a few seconds, followed by several hours of confused memory and, frequently, light coma or unsteady behavior. Lack of arousal during this time is sometimes regarded as short-term concussion, but brief coma is a more accurate label to apply until the person awakens.

Confusion

Confusion can be either temporary or permanent. Temporary confusion refers to disturbed memory and an inexact orientation of time, place, or person. Awakening from deep sleep after moderate sedation, suffering the effects of using excessive alcohol or street drugs, or awakening in a strange room, are typical examples. Chronic, waking confusion relates to sustained difficulties in identifying time, date, the environment and the failure to recognize long-known persons. It is also a gentle term for dementia.

Absence and Complex Partial Seizures

Seizures reflect severe impairments of self-aware consciousness, accompanied by unique forms of behavior. Absence seizures typically occur in children and are often noted as 'staring spells'. During the event the eyes typically fix in a forward stare, motion ceases, and movements of the lips or eyelids may be noted. People who undergo frequent absence attacks (once called *petit mal*) may lose extended self-awareness for a matter of hours and sometimes longer. During these states they remain awake and usually continue vaguely purposeful behavior.

Absence seizures originate from the cerebral cortex but involve brainstem and thalamic neuronal networks. People suffering severe complex partial seizures, a different neurological disorder often emanating from the temporal lobes (see below), lose their cognitive memories, but may also express a variety of learned behaviors. Both types of event exhibit attentional and intentional failure, loss of working memory, and perceptual dissociation. In their classic form, absence seizures may be interpreted to represent momentary vegetative states (see Table 1).

Akinetic Mutism

The term 'akinetic mutism' covers different behavioral expressions that relate to damage of several cerebral and subcortical structures. While sometimes confused with the vegetative state, akinetic mutism may resemble a state of constant hypervigilance. Such patients appear attentive and vigilant but remain motionless. The preservation of visual tracking in the form of following persons or moving objects with smooth, roving eye movements can differentiate this condition from the vegetative state. Classic akinetic mutism as listed in Table 1 reflects the recovery of a crude wakeful attentiveness without the apparent recovery of any other neuropsychologic function.

A similar picture, but excluding absence of eye movements, can rarely be a feature of untreated, rigid Parkinson disease. A strong clinical resemblance to this syndrome has been identified in some forms of variant Creutzfeldt–Jakob disease. The hyperattentive form of this disorder is typically seen in patients with large bilateral injuries to the medial and ventral frontal lobes (see below).

Hyperkinetic Mutism

Hyperkinetic mutism is a wakeful, continuous movement disorder accompanied by at least partial loss of global self-awareness. Patients with hyperkinetic mutism exhibit totally unrestrained but coordinated motor activity in the absence of external evidence of awareness of the environment. The patients also demonstrate an inability to develop conditioned responses, and produce no apparent memory of self.

Hyperkinetic mutism is the converse of akinetic mutism, with preserved unconscious expression of frontal intentional mechanisms, loss of sustained directed attention presumably requiring posterior attentional components of the inferior parietal lobe or posterior temporal lobe (see below), and a state of behavioral unawareness despite a whirlwind of activity. In contrast to the akinetic mute state, these people demonstrate minimally expressed intention and attention. The fragment of intention expressed in the meaningless motor activity of the hyperkinetic mute person is a reciprocal of the crude form of attention seen in akinetic mutism. Both examples reveal the fundamentally unconscious nature of such fragmentary neuronal activity.

Similar examples of such unconscious motor activity include the repetitive, uncontrollable production of words in the neurological disorder known as Tourette syndrome.

Delirium

Delirium is generally perceived as an acute or semi-acute temporary deficit of attention and working memory. A salient component is temporal disorientation. Delirium may follow acute, moderately severe head injuries, encephalitis, bacterial meningitis, exceptionally high fever, heat stroke, or withdrawal from chronic alcoholism or drug misuse. Delirium in patients less than 45 years old usually subsides without serious reduction in intelligence, but in alcoholism the person must abandon alcohol completely after the first or second delirious bout or begin to lose intellectual capacities permanently. Elderly people suffering mild dementia often become delirious during acute systemic illness or frequently changed surroundings. Visual hallucinations or impaired perceptions often occur in systemic delirium, whereas auditory hallucinations appear more often, but not solely, in people with schizophrenia.

Dementia

Dementia is characterized by two different conditions. One is a permanent, sometimes fluctuating loss of short-term or long-term memory. It can follow severe brain trauma, a sudden, sustained loss of oxygen to the brain, or surgical removal of the anterior areas of both temporal lobes. The other consists of an insidious, gradual loss of (first) short-term and (later) long-term memory. This process results from degeneration and death of nerve cells in the cerebral cortex.

Focal Unconsciousness: Agnosia, Anosognosia, and Neglect

Agnosia is a term specifically applied to different types of focal losses of awareness. Examples include an inability to see or feel objects as a whole greater than the sum of several parts, and a loss of specific capacities to hear aspects of sounds. A rare bilateral injury to the ventral temporal occipital lobe may produce the loss of perception of motion, leading to an experience of life as if seen constantly through a stroboscope, never in continuous motion.

Anosognosia is a term specifically applied to a loss of awareness and an inability to consciously perceive. Examples of anosognosia include denying that one's hand is one's own and unable to move intentionally. This form of focal unconsciousness is also labeled 'neglect' and is typically applied to a syndrome arising from damage to the right parietal lobe. This normally provides automatic knowledge of the contralateral body as well as the immediate outer space that surrounds it. Neglect increasingly appears to be a disorder of entry of primary sensory information into the appropriate internal context to be integrated into the construction of consciousness. Neglect can be seen following damage to either frontal or posterior (inferior parietal or superior temporal) cortices (see below).

RELEVANCE TO UNDERSTANDING HUMAN CONSCIOUSNESS

Anatomic Relationships

Disorders of consciousness are often generated by selective brain injuries. Specific neuropsychological deficits accompanying these disorders reflect the relatively segregated cerebral neuronal networks that generate human consciousness and complex behavior. Autopsies over almost two centuries and the increasing knowledge of functional anatomy provided by modern brain imaging have greatly added to neuropsychological understanding of conscious or unconscious behavior. Several brain regions are implicated in these disorders, including the two cerebral hemispheres, each of which possesses approximately half of the cerebral cortex, the thalamus, and the basal ganglia. Near the mid-brainstem, they connect with the large cerebellum and the arousal systems that lie within the brainstem. To discover just how this network generates consciousness has become a major scientific effort. (*See* **Cerebellum**; **Basal Ganglia**)

Nonspecific arousal is generated largely in the brainstem and is indispensable to supporting sleep–wake cycling and the wakeful states of consciousness. Cholinergic (pedunculopontine, later dorsal tegmental nuclei), noradrenergic (locus ceruleus), and other neuronal populations located within the upper brainstem, hypothalamus, and basal forebrain have a key role in organizing this large-scale human behavior. By itself, however, arousal is independent of expressed neuropsychological qualities, as is evident in the vegetative state or in 'absence seizures' (see Table 1). Brain mechanisms that govern sleep and its various dreams and perambulations only partially overlap the circuitry of normal wakeful consciousness. The integrity of both distributed cortical and other subcortical structures as is necessary for the expression of integrated cerebral activity to generate consciousness.

Cognitive capacities expressed in the conscious state depend on the moment-to-moment continuity of short-term memory (disordered in delirium) with other neuropsychological components. Short-term or working memory appears to depend strongly on the integrity of the prefrontal and parietal cortices along with subcortical structures. The richness of mental life contained in the storage of long-term memories is a distributed capacity of the association regions of the cerebral cortex and is severely degraded in dementia.

The cortical regions indispensable for conscious behavior are the frontal lobes: these largely govern and express behavior, both immediate and learnt. Their functions provide the executive generator and dictator of consciousness, organizing mood, behavior, and mind. Within the frontal lobes the basal forebrain area has evolved from ancient mammalian brains and occupies most of the undersurface. It participates in generating emotional feelings and social behavior as well as stimulating the person's intentional purposes. The lateral and medial prefrontal areas (including the dorsolateral prefrontal regions, supplementary motor zones, and anterior cingulate cortices) largely influence physical coordination and participate in volitional and cognitive aspects of attention and working memory.

The most posterior regions of the lateral and medial frontal lobe generate and regulate coordinated expressions of logical manipulations, language, intended eye movements and, ultimately, all coordinated, intentional behavior. Examples include skilled athletics, the expression of well-learnt and practiced instrumental music, and other rapidly expressed activity.

Functional generation of self-directed attention and intention are mapped strongly in the ventral-medial frontal lobes and less frequently the posterior thalamus and rostral mesencephalon.

Akinetic mutism reflects the disabling of the ventral-medial and medial frontal and prefrontal networks (including a large contribution from the deep gray-matter structures of the basal ganglia, which interact with the cortex via long-loop connections with the thalamus and underpin much routine learnt behavior), providing volitional drive and self-directed (executive) attention. The crude aspect of attention remaining in this state of impaired consciousness may originate from automatic orienting systems driven by posterior parietal and subcortical structures (thalamus).

The posterior parts of the cerebrum, including the parietal, occipital and temporal lobes, in conjunction with the thalamus, generate the perceived contents of thoughtful consciousness. They receive their direct signals of attention and intention from the frontal lobes and express their immediate demands. The occipital lobes receive inputs of retinal vision, which are processed further within the adjacent temporal lobe. Auditory stimuli are also processed in the temporal lobe. Abstract cognitive icons represent the verbal, musical, mathematical, geometric, and pictorial languages that make up our intellectually conscious knowledge. Most of these particular cognitive qualities and contents are dominantly expressed by the left cerebral hemisphere:

The right inferior parietal lobe and adjacent superiorlateral temporal lobes, however, normally provide a person's dominant preconscious attentive perception and awareness of both the left side of the body and its surrounding environment. Severe acute damage to the right parietal-temporal areas as described in the paragraph on focal unconsciousness may cause total unawareness of the entire left side of the individual's personal universe. Lost is the memory of being able to see, or to remember normal vision; lost is the accurate perception of any existing left-spatial noises; lost is total awareness or memory of the absent hemiworld to the left and, remarkably, the person's ability to recognize his or her own left arm, leg or ear. This remarkable clinical syndrome demonstrates that our conscious experience is instinctual and can be lost in parts.

Evidence from neurological disorders of consciousness demonstrates that subcortical structures are also essential for normal integrative brain function associated with consciousness. Most causes of the global disorders of consciousness reviewed above appear to arise from either large bilateral injury to frontal (e.g. bilateral medial-basal frontal injuries and akinetic mutism) or posterior association cortices (bilateral temporal-parietal association areas and hyperkinetic mutism). In addition, it is known that selective subcortical injuries (generally damage to medial aspects of the thalamus or upper brainstem) may produce identical or very similar disorders. The subcortical injuries that may produce transient coma, vegetative state, akinetic mutism, or conditions resembling hyperkinetic mutism also implicate brainstem and thalamic structures. These include the brainstem arousal systems important for sleep and wake cycling and related brainstem and thalamic substructures that play a part in the complex, large-scale integration of many cerebral networks. The contribution of these deep brain structures may lie in the selective facilitation of activity patterns that allow widely separated brain regions to briefly communicate

Further Reading

Crick F (1964) *The Astonishing Hyphothesis. The Scientific Search for the Soul*. New York, NY: Charles Scribner's Sons.

Edelman GM (1987) *Neural Darwinism, the Theory of Neuronal Group Selection*. New York, NY: Basic Books.

Edelman GM (1989) *The Remembered Present A Biological Theory of Consciousness*. New York, NY: Basic Books.

Edelman GM and Tononi G (2000) *A Universe of Consciousness. How Matter Becomes Imagination*. New York, NY: Basic Books.

Plum F (1991) Coma and related global disturbances of the human conscious state. *Cerebral Cortex* **9**: 359–425. New York, NY: Plenum Press.

Plum F and Posner JB (1982) *Stupor and Coma*, 3rd edn. New York, NY: Oxford University Press.

Wilkinson IMS (1999) *Essential Neurology*, Chap. 11 Unconsciousness, pp. 171–186. London, UK: Blackwell Science.

Consciousness, Function of

Intermediate article

Thomas W Polger, University of Cincinnati, Ohio, USA

CONTENTS

To inquire about the function of consciousness is to ask what consciousness does, what it enables us to do that we might not be capable of otherwise, or why some creatures came to be conscious.

INTRODUCTION

Consciousness is perhaps the most salient feature of human mental life. The experience of tasting a red wine differs from the experience of tasting coffee, and from that of reading the label on a wine bottle. Whatever else can be said about these differences, they are manifested in us by different conscious experiences. And this seems to be good for us: different experiences are evidently important in our abilities to discriminate among foods, to avoid injury, to identify potential mates, and so on.

But upon reflection it is less obvious what, if anything, consciousness does, what it allows us to do that we might not be capable of otherwise, or why some creatures – like human beings – have come to be conscious. For it seems that conscious experience is not necessary for the ability to distinguish objects in the world, or avoid injury, or seek a mate. Even if the experience of pain, for example, is important to the way that humans detect and avoid noxious elements in the external environment, it seems that we or other creatures could avoid hazards without the experience of pain, or any conscious experience at all. There is little doubt that mindless mechanical devices can be constructed to detect heat, classify wavelengths of reflected light, or distinguish chemical substances. We do not feel compelled to say that such devices feel pain, see colors, or taste wines. And we need not think only of mechanical devices and the automata of science fiction, for there is ample evidence that biological creatures can evolve fairly sophisticated sensory and motor capacities without having conscious experiences. The natural world is rife with creatures (microorganisms, molluscs, insects, and so on) that interact with their environments effectively, at least some of which may lack conscious experiences altogether.

QUESTIONS ABOUT THE FUNCTION OF CONSCIOUSNESS

The question of the function of consciousness does not have an obvious answer. We are faced not with a single question but with a handful of more or less related inquiries about what consciousness (as a matter of fact) does for human beings, about what if anything consciousness enables us to do that we could not (possibly) do otherwise, and about what capacities consciousness allows that would explain why it should be favored by natural selection: why it would evolve. The question of the function of consciousness is ambiguous, and the ambiguity owes as much to the idea of function as it does to any special considerations having to do with consciousness. Questions about the functions of mechanical artifacts and biological organs meet many of the same problems. (*See* **Consciousness, Philosophical Issues about**)

What Does Conscious Experience Do?

We might ask what abilities consciousness in fact mediates in human beings. In this case we are treating consciousness as a mechanism with certain effects, and we are inquiring about those effects in the same way that we might ask about the function of a carburetor or a heart. Carburetors regulate and mix air and fuel in some combustion engines. Hearts pump blood. Conscious experience allows us to discriminate and identify objects in the environment, to avoid hazards, and so forth. This was the answer that made it at first seem obvious what the function of consciousness is. Such explanations tell us what the 'causal role functions' of carburetors, hearts, and sensations are. A causal role function of a trait or mechanism is an effect of that trait or mechanism that figures in an explanation of the overall capacities of the system of which it is a part. To explain the capacities of a system in terms of the causal role functions of its parts is to provide a 'functional analysis' of the system (Cummins, 1975). A special subset of causal role functions are those that can be characterized in terms of a computational device, such as a Turing machine.

Is Conscious Experience Necessary?

Even if we have a good explanation for what consciousness happens to do for us, we may still ask what it is that conscious experiences (likewise, carburetors or hearts) allow that could not be accomplished without them. This is not a question only about how human beings and cars are put together and what they are now capable of. It is also a question about how they might have been put together differently and what capacities they would have had under those circumstances. Could there be a car that does not have a carburetor? Certainly. Most automobiles these days use fuel injectors to mix air and fuel, rather than carburetors. Could there be a creature that does not have a heart? Mechanical devices are regularly used to circulate the blood of patients in the operating room. There is no reason to deny that some creature, however unlikely, could circulate its own blood without engaging a heart. So it may be with consciousness. The thesis that consciousness, though it may be crucial to our distinctively human way of interacting with the world, is not necessary for any of our capacities is called 'conscious inessentialism' (Flanagan, 1992).

There are interesting empirical phenomena that have seemed, at least to some, to support conscious inessentialism. Consider, for example, Bejamin Libet's (1985) experiments on the timing of conscious intentions to produce behavior. Libet's results purport to show that the muscular action potential that initiates movement occurs temporally prior to conscious awareness of an intention to move. These results have been interpreted as showing that consciousness does not play a role – or at least not the role traditionally envisioned – in the initiation of behavior. Daniel Dennett (e.g. 1991) has made much of the Libet experiments.

Blindsight is another phenomenon that has seemed, to some, to support some version of conscious inessentialism (e.g. Block, 1995). Lawrence Weiskrantz (1986) aroused the interest of many philosophers with his studies of patients with neural injuries who report no conscious visual experience in parts of their visual fields. Nevertheless, some of these patients perform much better than chance when they are forced to 'guess' about stimuli presented to the blind field. It seems that blindsight patients have residual information processing capacities despite lacking visual consciousness in the area of the scotoma, apparently supporting conscious inessentialism. This sort of phenomenon has led theorists to emphasize the importance of nonconscious visual processing (e.g. Milner and Goodale, 1995). But Weiskranz's results can also weigh against conscious inessentialism. After all, the tasks that blindsight patients perform better than chance – however remarkable that may be – are performed unerringly by normal subjects; and blindsight patients never initiate action based on the stimuli presented to the blind field (Van Gulick, 1985). These considerations suggest that consciousness does play an important role. (*See* **Blindsight**)

If conscious inessentialism is true, then there could be creatures that negotiate the world just as human beings do, but that nevertheless lack consciousness. Despite lacking conscious experience, such creatures (called 'zombies') make the same bodily movements as we do: they avoid fire, behave discriminately towards various wavelengths of light and chemical substances, etc. Sometimes it is claimed that conscious inessentialism entails that consciousness is epiphenomenal – that consciousness does not have any causal powers at all. (*See* **Zombies**; **Epiphenomenalism**)

But this is a mistake: from the fact that a carburetor is inessential to the operation of a car (because it could be replaced by a fuel injector) it does not follow that carburetors have no effects in those cars where they are found. Carburetors mix air and fuel in some cars; fuel injectors mix air and fuel in other cars. Neither a carburetor nor a fuel injector is necessary to the operation of automobiles in general; but carburetors and fuel injectors are not thereby epiphenomenal. (And just as there are reasons for generally preferring fuel injectors over carburetors, visual or painful experience may be better or worse ways of engaging with the world.)

Why Did Conscious Experience Arise?

In asking about the function of consciousness we might want to know not what conscious experiences enable us to do currently, but rather why we have come to be conscious at all. That is, we might be asking about the teleology of consciousness, about the purpose that it serves. If we understand teleology in terms of evolutionary history, then we are asking what the etiological function of consciousness is. The etiological function of a trait is the effect that the trait had in the ancestors of a creature that provided an adaptive advantage to creatures of that kind, so that evolutionary pressures favored creatures with the trait. The etiological or 'selected effect' function of a trait explains why the trait came to be present or maintained in creatures of a kind, why it was naturally selected (Millikan, 1989; Neander, 1991). (*See* **Evolutionary Psychology: Theoretical Foundations**)

Consider again the possibility of conscious inessentialism. If conscious inessentialism is false – if consciousness is necessary for some human capacities (e.g. detecting wavelengths, avoiding injury) – then it is easy to answer the question of why we are conscious: we are conscious because it is adaptively advantageous (e.g. to detect wavelengths or avoid injury). That is to say, consciousness evolved; the evolutionary history of conscious experience is just the same as that of the capacities that conscious experiences mediate. That history need not be obvious or simple; it may not even be knowable by us: we do not assume that we will know the evolutionary history of every (or perhaps any) biological trait. Further, we should not assume that every trait has adaptive value. Some traits are the result of chance alone – though assuming that particularly complex traits are products of evolution by natural selection may be a reasonable methodological stance (Brandon, 1990; Grantham and Nichols, 1999). Of course the contingencies of organism and environment are such that discrimination of wavelengths and chemicals, avoidance of flames, and so on are not themselves compulsory. But insofar as we could explain why a creature should avoid injury, we would be able to explain why it experiences pain. The research program known as evolutionary psychology proceeds on the assumption that most or all psychological traits are adaptations by natural selection that are required for capacities that would have conferred an advantage on hominids living in the late Pleistocene era (Barkow *et al.*, 1992).

On the other hand, if conscious inessentialism is true then the question of why consciousness has come to be is somewhat different. In that case we would need to explain not only why the capacity to avoid injury came to be enabled but also why, in some creatures, conscious experience mediates avoidance of injury. The answer might be that the presence of consciousness is a result of mere chance. That would perhaps be disappointing, but it would not undermine our belief that consciousness in fact has important effects in our lives. In particular, it would not force us to adopt epiphenomenalism. Just as an automobile might be built with a carburetor or a fuel injector, so creatures might evolve conscious experiences or some other mechanisms. Perhaps some forms of conscious experience have interesting evolutionary histories while others arose only by chance. We need not take consciousness to be a single phenomenon in order to meaningfully ask about its function.

CONSCIOUSNESS AND FUNCTIONAL KINDS

One might believe that whatever mixes air and fuel is a carburetor. That is to say, one might adopt a sort of metaphysical functionalism regarding carburetors. On this view, carburetors are functional kinds that are constituted by their capacity to mix air and fuel; thus fuel injectors are carburetors. Likewise one could think of hearts as blood pumps, and one could think of whatever mediates injury avoidance as pain experience. One must then regard the thesis of conscious inessentialism (likewise, carburetor inessentialism) as incoherent. According to a functionalist it would not even make sense to talk about something that did all the things (causal role functions) that pain does in human beings but that does not *ipso facto* have pain experiences. Two popular variations of functionalism about consciousness take conscious mental states to be a subset of representational states (e.g. Dretske, 1995) or to be meta-representations of first-order mental states – the higher-order thought theory championed by David Rosenthal (e.g. 1986). (*See* **Functionalism**)

The theory that consciousness is a functional kind is closely aligned with the view that all the facts about consciousness can be explained by reference to its functional role or roles (e.g. its causal role functions). One reason for holding such a view is general commitment to functionalist explanations, at least with respect to psychology; a widely-held theory is that all properties are causal role functional properties, and thus all explanations are functional explanations (Shoemaker, 1984). But in that case, if we cannot secure functional explanation of consciousness then we will be obliged to abandon the belief that consciousness is a physical phenomenon at all (Chalmers, 1996).

References

Barkow J, Cosmides L and Tooby J (eds) (1992) *The Adapted Mind: Evolutionary Psychology and the Generation of Culture*. New York, NY: Oxford University Press.

Block N (1995) On a confusion about the function of consciousness. *Behavioral and Brain Sciences* **18**: 227–247.

Brandon R (1990) *Adaptation and Environment*. Princeton, NJ: Princeton University Press.

Chalmers D (1996) *The Conscious Mind: In Search of a Fundamental Theory*. New York, NY: Oxford University Press.

Cummins R (1975) Functional analysis. *The Journal of Philosophy* **72**: 741–765.

Dennett D (1991) *Consciousness Explained*. Boston, MA: Little, Brown.

Dretske F (1995) *Naturalizing the Mind*. Cambridge, MA: MIT Press.

Flanagan O (1992) *Consciousness Reconsidered*. Cambridge, MA: MIT Press.

Grantham T and Nichols S (1999) Evolutionary psychology: Ultimate explanations and Panglossian predictions. In: Hardcastle V (ed.) (1999) *Where Biology Meets Psychology: Philosophical Essays*, pp. 47–66. Cambridge, MA: MIT Press.

Libet B (1985) Unconscious cerebral initiative and the role of conscious will in voluntary action. *Behavioral and Brain Sciences* **8**: 529–566.

Millikan R (1989) In defense of proper functions. *Philosophy of Science* **56**: 288–302. [Reprinted in: Allen C, Bekoff M and Lauder G (eds) (1998) *Nature's Purposes: Analyses of Function and Design in Biology*. Cambridge, MA: MIT Press.]

Milner B and Goodale MA (1995) *The Visual Brain in Action*. New York, NY: Oxford University Press.

Neander K (1991) Functions as selected effects: the Conceptual analyst's defense. *Philosophy of Science* **58**: 168–184. [Reprinted in: Allen C, Bekoff M, and Lauder G (eds) (1998) Nature's Purposes: Analyses of Function and Design in Biology. Cambridge, MA: MIT Press.]

Rosenthal D (1986) Two concepts of consciousness. *Philosophical Studies* **49**: 329–359.

Shoemaker S (1984) *Identity, Cause, and Mind*. New York, NY: Cambridge University Press.

Van Gulick R (1985) What difference does consciousness make? *Philosophical Topics* **17**: 211–230.

Weiskrantz L (1986) *Blindsight: A Case Study and Implications*. New York, NY: Oxford University Press.

Further Reading

Allen C, Bekoff M and Lauder G (eds) (1998) *Nature's Purposes: Analyses of Function and Design in Biology*. Cambridge, MA: MIT Press.

Amundson R and Lauder G (1994) Function without purpose: the uses of causal role function in evolutionary biology. *Biology and Philosophy* **9**: 443–469. [Reprinted in Allen *et al.* (1998).]

Cowey A and Stoerig P (1991) The neurobiology of blindsight. *Trends in Neuroscience* **29**: 65–80.

Cummins R (1983) *The Nature of Psychological Explanation*. Cambridge, MA: MIT Press. [Reprinted in Allen *et al.* (1998).]

Davies M and Humphreys GW (1993) *Consciousness: Psychological and Philosophical Essays*. Oxford, UK: Blackwell.

Fetzer J (ed.) (2002) *Evolving Consciousness*. Amsterdam, Netherlands: John Benjamins.

Flanagan O (2000) *Dreaming Souls*. New York, NY: Oxford University Press.

Flanagan O and Polger T (1995) Zombies and the function of consciousness. *Journal of Consciousness Studies* **2**(4): 313–321.

Güzeldere G, Flanagan O and Hardcastle V (1999) The nature and function of consciousness: Lessons from blindsight. In: Gazzaniga M (ed.) *The New Cognitive Neurosciences*, 2nd edn, pp. 1277–1284. Cambridge, MA: MIT Press.

Hardcastle V (ed.) (1999) *Where Biology Meets Psychology: Philosophical Essays*. Cambridge, MA: MIT Press.

Ito M, Miyashita Y and Rolls ET (eds) (1997) *Cognition, Computation, and Consciousness*. New York, NY: Oxford University Press.

Lycan W (1987) *Consciousness*. Cambridge, MA: MIT Press.

Lycan W (1996) *Consciousness and Experience*. Cambridge, MA: MIT Press.

Mack A and Rock I (1998) *Inattentional Blindness*. Cambridge, MA: MIT Press.

Marcel A and Bisiach E (eds) (1988) *Consciousness in Contemporary Science*. New York, NY: Oxford University Press.

Polger T (2000) Zombies explained. In: Ross D, Brook A and Thompson D (eds) *Dennett's Philosophy: A Comprehensive Assessment*. Cambridge, MA: MIT Press.

Polger T and Flanagan O (1999) Natural answers to natural questions. In: Hardcastle (1999, pp. 221–247).

Polger T and Flanagan O (2002) Consciousness, adaptation, and epiphenomenalism. In: Fetzer (2002).

Sober E (1985) Panglossian functionalism and the philosophy of mind. *Synthese* **64**: 165–193.

Tye M (1996) The function of consciousness. *Noûs* **30**(3): 287–305.

Weiskrantz L (1997) *Consciousness Lost and Found: A Neuropsychological Exploration*. New York, NY: Oxford University Press.

Consciousness, Machine

Introductory article

Keith Gunderson, University of Minnesota, Minneapolis, Minnesota, USA

CONTENTS

The widespread use of computers and robots within research programs in cognitive psychology has stimulated interest in the possibility of machine consciousness. As a result, many unsettled issues in the philosophy of mind and theory of knowledge have been reformulated in terms of, and in turn raised questions about, machine-oriented modeling.

INTRODUCTION

Consciousness is one of the most perplexing topics in the study of mind. There are many controversies surrounding its exact nature and relationship with the physical. For centuries, philosophers and others have argued about whether a machine could be conscious, partly so as to reach a better understanding of human and animal minds.

During the second half of the twentieth century, with the development of the digital computer, the topic of machine consciousness became intertwined with questions about artificial intelligence (AI). Many tasks that once seemed to require human conscious intelligence are now performed by computers. Machines with various kinds of programs and related robotic capabilities have been used to help explain some of the more baffling aspects of human mentality. (*See* **Artificial Intelligence, Philosophy of**)

Such approaches face a variety of problems. What counts as an example of consciousness? Some argue that consciousness is primarily a private (subjective) phenomenon. Nevertheless, attempts to construct machine models that would objectify it continue. Even when such models seem flawed or limited, an understanding of their shortcomings can provide fresh perspectives and stimulate lively debate on thinking, perception, awareness, purposive behavior, and the relationship between the mind and the body. (*See* **Mind–Body Problem**)

HISTORY

Could a machine be conscious? This question was being asked at least as long ago as the seventeenth century. Machines then (as now) were generally assumed to consist wholly of matter. To ask whether we might someday be able to build a conscious machine was one way of asking whether consciousness (animal or human) was made out of matter. The view that it was was called materialism.

The view that it was not, and that mind was essentially different from matter, was called dualism. (*See* **Dualism**)

Descartes was dualism's most influential proponent. He regarded the human mind (or soul) as connected to a physical body. This connection, in the case of human beings, made the use of language, and many other activities, possible. A great many of these activities, Descartes felt, could not be explained in a purely mechanical way as matter in motion. On the other hand, he believed that the activities of all nonhuman animals could be explained in a purely mechanical manner. So dualism was not true of animals. Their behaviors, upon close inspection, could be shown to be made up of reflexes, and were instinctual, not guided by reason. Animals were organic machines. They had no thoughts or sensations. This view was called 'animal automatism'. (*See* **Descartes, René**)

Most philosophers found the doctrine of animal automatism too extreme. But some of the arguments in support of it were challenging. Perhaps living organisms were more machine-like than had been previously thought. In the eighteenth century, the simile of the mind being like a machine became popular. The French philosopher La Mettrie tried to turn the tables on Descartes. In his book *The Man Machine*, he first applauded Descartes for being a pure materialist about animals, but then tried to show that Descartes' arguments for animal automatism could be used to show that human beings were just machines too.

Leibniz held the unusual view that both humans and animals were machines, but immaterial ones. Our bodies were said to be 'divine machines' or 'natural automata'. Human automata possessed the powers of perception and self-conscious memory. Animals had only a limited version of these abilities. Leibniz' doctrine, though odd and ambiguous, illustrates how flexible and abstract the idea of a machine can be. Almost anything, it seemed, might be a machine if it could be broken down into parts that functioned together in organized and predictable ways. (Somewhat surprisingly, Leibniz also claimed that we could never explain our perception in mechanical terms.) Leibniz agreed with the Aristotelean view that living created organisms, unlike windmills or clocks, could initiate purposive movement from within themselves, by virtue of a 'vital force' or 'entelechy'. These movements or activities were sometimes teleological, i.e., directed at ends or goals.

Long after Leibniz, philosophers, psychologists, biologists, and physiologists debated whether such behavior could be explained in purely materialistic mechanistic ways. Those who claimed that it could were called 'mechanists'. Those that claimed that it could not were often called 'vitalists'.

With the development of genetic theory, as well as discoveries about instinct and behavioristic psychology, proponents of vitalism nearly vanished from the intellectual scene. By the 1940s and 1950s, the prevailing view was wholly naturalistic.

The goal of the new field of 'cybernetics' (from the Greek word for 'steersman') was to understand in detail how control and communication worked in human and animal organisms as well as in self-regulating machines. The field was explicitly described as both behavioristic and functional. This meant that the focus of study was on inputs or stimuli from the environment to the animal or machine, together with states in between, leading to behavior. (*See* **Functionalism**; **Behaviorism, Philosophical**)

The centuries-old dream of being able to build a conscious machine seemed closer than ever. And it was at this time that the digital computer began to attract attention. Calculating machines had existed in both design and physical reality since the seventeenth century; but machines that could repeatably make use of their own computations in the production of further ones did not appear until the middle of the twentieth. This 'recursive' power had for some time been of intense logical and mathematical interest. It hinted at the possibility of self-reflection, and represented a new way of thinking about how a machine might be designed to imitate human thought processes.

PHILOSOPHICAL ISSUES ABOUT MACHINE CONSCIOUSNESS

Varieties of Consciousness

There is no tidy or uncontroversial concept of consciousness. But when trying to imagine a conscious machine, at least three (sometimes overlapping) aspects of consciousness should be considered. Firstly, there is the awareness involved in responding with purpose to stimuli from an external environment. Secondly (and often subsumed in the first aspect), there are simple perceptual experiences, such as seeing and touching, and inner thoughts, beliefs, intentions, feelings, emotions, moods, desires, and so on. Some of these mental states are 'about' other things (for example, a thought or belief about a cat). This 'aboutness' is sometimes called 'intentionality'. Other mental states are not about anything: they just are (for example, a visual experience of redness, or a bitter taste, or a sharp pain). Thirdly, there is self-consciousness, or the capacity of minds to reflect on their own conscious states, which might be any of the aforementioned ones. This 'second-order' consciousness is sometimes called 'introspective awareness'. (*See* **Consciousness, Philosophical Issues about**; **Introspection**; **Self-consciousness**)

Many of the aforementioned conscious states can be described as having a character or quality or 'feel' to them, such as 'what it's like' to see something red. These qualities are called 'qualia'. (*See* **Qualia**)

Many philosophers believe that, to some degree, some of these aspects of consciousness already exist in various self-regulating cybernetic mechanisms including computers. But it was the digital computer, with its many different forms of programming, that for many seemed to be the most promising candidate for machine mentality.

The Turing Test

What sorts of beings might be capable of thinking? It is often assumed that only beings with some degree of consciousness can think. It is also often assumed that the best evidence for conscious thinking existing in beings other than ourselves lies in various complex behaviors. The computer has shown impressive potential for performing complex tasks of which only thinking intelligent agents such as ourselves had previously seemed capable. Thus the question of whether machines can think has attracted much interest since the mid-twentieth century, and with it the question of whether machines could be conscious.

In 1950, the mathematician A. M. Turing, in an interesting 'thought experiment', proposed a test for whether a machine can think. He imagined a game being played (called the 'imitation game'), in which an interrogator asks two concealed participants questions and uses their answers as a means for discovering a hidden aspect of their identities (for example, which one is the man and which the woman). The aim of one participant (A) is to fool the interrogator (C). The aim of the other (B) is to help the interrogator guess the right answer. Turing asked whether a computer might do as well as a human participant A in fooling C. He argued that it could, in the foreseeable future, and by virtue of this ability he felt that such a machine should be credited with thinking. This test later became known as the Turing test. (*See* **Turing Test**; **Turing, Alan**)

Some philosophers, as well as scientists involved in devising programs for modeling psychological processes, gave at least qualified support to Turing's position. Others disagreed. One argument was that a computer might perform a task previously only accomplished by intelligent people, without necessarily being intelligent. Perhaps the computer's performance would simply illustrate that there can be a variety of processes – some conscious and intelligent, others not – that result in the same achievement. Electric eyes, like doormen, may open doors for people, but they don't rely on seeing anyone coming, nor are they polite or concerned in their doing it.

In spite of these objections, Turing's test remained popular. Debates concerning the test's merits and demerits are still continuing.

Early Experiments in AI and some Philosophical Reactions to Them

Some of the most influential work in AI during the 1960s and 1970s concerned computer simulation of cognitive processes (CS). Allen Newell and Herbert Simon and others adopted the following strategy. A simulation of intelligent human mental processes, such as proving a theorem or playing chess, was constructed on the basis of observations of a human problem-solver's behavior. These observations might include verbal reports, jottings on paper, and so on. Attempts would then be made to represent these in the programming vocabulary used for the simulation. When the program was run, a trace of its 'thinking' activity would be printed. This trace would then be compared to some of the general features assumed to be present in the mind or behavior of the human problem-solver. (*See* **Newell, Allen**; **Simon, Herbert A.**)

One would not know with precision beforehand what would result from running the computer program. Nor can one know beforehand exactly what will go on in a human problem-solver's mind. But certain parallels between program and person were pursued. First, the tasks performed by a person were compared to the computer's results (deciding whether it passed the Turing test). Other levels of comparison involved much more guesswork. Although the structure and processes of the computer simulation could be described, it was often far from obvious how much of that description could apply to the human subject. Furthermore, it seemed that whatever objections might be raised to the Turing test in its general form would also be relevant to the specific use of that test in CS contexts.

Nevertheless, there was one important principle that seemed to underly AI research in general. This was what Newell and Simon called the 'physical symbol system hypothesis'. It stated simply that a physical symbol system has all the necessary means for general intelligent action. Whether this hypothesis was plausible or not, it had a bearing not only on CS but on many other related AI projects.

SHRDLU was an ingenious AI project which also incorporated robotic features. SHRDLU was a simulation of a robot that could respond to human commands concerning the construction and manipulation of (imagined) blocks of varying sizes and shapes (cubes, pyramids) and colors in a limited environment, say on a table top. When asked to do something with a block with respect to some present arrangement, not only could SHRDLU comply with the explicitly requested move, but it could calculate and carry out on its own whatever moves had to be made in order to carry out the command. (*See* **SHRDLU**)

SHRDLU could also produce replies to questions about what it was doing. And if a new word was contained in a command to SHRDLU, the program might indicate its unfamiliarity, and acquire the new word.

The environment within which SHRDLU functioned consisted entirely of data structures. These structures were symbolically represented within the computer. This program contained much of psychological and philosophical interest. Here was a buildable mechanical being operating in three-dimensional space. (Another example of such a system, SHAKEY, was a mobile robot which negotiated its way through 'rooms' towards a goal.) In some sense, SHRDLU seemed to be sensitive to a variety of objects and able to interact with them. Its linguistic competence included an expandable vocabulary. But to what extent did SHRDLU, in its limited little world, reflect and illuminate our real world of perception, problem-solving, and discourse? Hubert Dreyfus has discussed the kinds of question–answer and command 'conversations' that SHRDLU had concerning its own blocky universe. He points out that SHRDLU's test for something being 'owned' is simply whether it is tagged as 'owned'. Dreyfus reminds us that SHRDLU couldn't own anything since it doesn't belong to a community in which ownership makes sense. This is just one example among many of how institutional and cultural facts could restrict how a subject like SHRDLU might act or what it could be 'conscious' of. Some form of embodiment seems to be required. (*See* **Embodiment**)

But questions remain about what SHRDLU can tell us about intelligent understanding, and perhaps about consciousness. At the heart of SHRDLU, as well as of other CS projects, is a program. But how plausible is Newell and Simon's physical symbol hypothesis?

Searle's Chinese Room Argument, and Other Problems for AI

Various AI projects durings the 1960s and 1970s seemed to exhibit some degree of understanding or comprehension (of, for example, language). In 1980, the philosopher John Searle published his famous 'Chinese room' argument, which tried to refute this. Searle claims he can imagine himself functioning just like an AI program. His mimicry consists of using a set of rules (written in his native language, English) whereby he manipulates a set of symbols in a language he doesn't understand at all (Chinese). The rules lead him to produce meaningful sentences in Chinese. Questions in Chinese are passed to him through a hole in the wall and he uses the rules to generate answers, which he passes back. To those outside the room where he is confined it will look as if something in the room is understanding questions in Chinese (about stories in Chinese) and answering them in Chinese. But no such understanding is taking place in the Chinese room. Nor, therefore, is it taking place in any AI program. (*See* **Chinese Room Argument, The**)

Symbol-based AI projects faced other serious problems. Conscious abilities such as recognizing patterns (like a face, or a house) or grouping things into kinds (like cats, dogs, or trees), or quickly sizing up situations, or performing complicated tasks, are often quickly done by people, and may involve many processes going on at once. Symbolic AI systems generally compute things step by step, in a serial way. Perhaps some model of consciousness based on the brain, which seems to carry out many different processes at the same time, would provide a better point of departure. This is called parallel processing. A number of 'connectionist' models have tried to capture this kind of processing. There has been considerable debate in recent years between researchers who favor symbolic AI models and those who favor connectionist ones. Serious philosophical questions arise for either approach to machine consciousness. (*See* **Connectionism**; **Computation, Philosophical Issues about**)

Qualia, Functionalism, Eliminative Materialism, and the Mind–Body Problem

Leibniz imagined a vast machine capable of perceptions, which we could go inside and walk around. No matter how hard we looked, he argued, we would never find anything in its mechanical make-up that would explain or disclose to us its perceptions. Leibniz saw this inevitable lack of disclosure as an important fact about the mind. The very same point could be made about consciousness and the brain. We could never 'see' consciousness displayed in a brain, or in any physical thing.

The philosopher Thomas Nagel has made a similar point by asking: What is it like to be a bat? Bats are very different from us. They are presumably conscious. But we don't know what it is like to be one. Nor would we know the answer to this question even if we knew everything that we possibly could about a bat's physical make-up.

Similarly, Frank Jackson has argued that a person brought up in a black-and-white environment, and thus deprived of any color experiences, might know everything there is to know about the material and functional facts underlying color vision. Nevertheless, when entering a multicolored environment for the first time, that person would be treated to an altogether new fact. The new fact would be what the experience of seeing red was like.

This character, or quality, or 'qualia' of our experiences, or 'what it's like' to have them, seems quite different from anything physical having to do with our brains or behavior. There seems to be a gap between brains and qualia, between machines and minds. The difficulty of trying to bridge this gap so that we can fit consciousness into a scientific picture of the world is sometimes called 'the hard problem'. In essence it is the mind–body problem itself. (*See* **Explanatory Gap**)

Some say we will never be able to bridge this gap. Others have suggested that it may be possible by some as-yet-unknown means. Or perhaps qualia have a special kind of physical nature, irreducible to other physical things such as brain processes. And some philosophers are not troubled by qualia or subjectivity at all.

All these questions arise in connection with the influential doctrine in philosophy and cognitive psychology known as 'functionalism'. Functionalism claims that what makes a mental state the kind of mental state it is is its function or role in a pattern of causal relations. A pain, for example, is viewed as something caused by an input from the environment, say a brick falling on one's toe. The pain in turn causes other mental states, such as wanting to ease the pain, which then cause certain behavior (the output), such as taking an aspirin, grimacing, moaning, and so on. Anything with this role will be a pain. In conscious human organisms, certain neural states of the brain are thought by many to be what pains are. In other kinds of creatures, physical events other than neural processes may have that same role. For example, some future robot may have a 'brain' made of silicon chips that can produce pain. It is not any specific kind of matter that defines pain: multiple realizations are possible. It is the causal role that determines which mental states are present or absent in the creature under consideration. (*See* **Multiple Realizability**)

If so, then qualia don't seem to matter much. Imagine two different people who have two different color experiences when they look at a fire engine. One sees red; the other sees green. But they talk in exactly the same way about it. They both call it 'red'. They have each learned to use the words 'red' and 'green' in the same way. Their behaviors with respect to fire engines have the same causal relations. (*See* **Inverted Spectrum**)

One would think that the experience of seeing red was a different mental state from the experience of seeing green. But for the functionalist it seems that this is not so. Similarly, if there were a robot that behaved in exactly the same way as a human in the presence of fire engines, it would be, according to the functionalist, in the same mental state as the person experiencing green and the person experiencing red – even if it had no color 'experiences' at all. (*See* **Zombies**)

How do qualia fit into the functionalist's theory of mind? A 'quick fix' to these problems is provided by the doctrine of 'eliminative materialism'. The eliminative materialist suggests that many or most of our ordinary terms for talking about minds are misleading. Terms such as 'thought', 'belief', 'desire', and 'mental image' have been passed on to us as part of our prescientific 'folk psychology'. Perhaps these terms will become obsolete, in the same way that the term 'vital force' became obsolete and was essentially eliminated from the biological vocabulary used to explain goal-directed behaviors. In its most extreme, and least popular, version, eliminative materialism may suggest that the question of whether there is or could be machine consciousness is settled by default, since there is no such thing as consciousness. (*See* **Folk Psychology**)

INFLUENCE OF COGNITIVE SCIENCE ON ISSUES ABOUT MACHINE CONSCIOUSNESS

Since the time of Descartes and La Mettrie, machines have 'evolved'. And in the wake of this robotic 'evolution' have come various bold claims, many made by cognitive scientists, about the potential of physical mechanisms to reflect or even share human mentality and that of other animals. Some philosophers have supported these claims; others have criticized them. The lively dialogue that has ensued has been good for both cognitive science and philosophy. We now have new contexts for asking detailed questions about both mental functions and qualia (the 'hard problem').

Functional problems in modeling constitute what David Chalmers has called the 'easy' problems of consciousness. They include aspects of self-reporting, a system's access to its own internal states, reactions to an environment, categorization, and so on. All these, he and others believe, can be directly approached by the computational and connectionist strategies of cognitive science. Some, however, would claim that the so-called 'easy' problems are not so easy. Certain functional, computational activities that seem at first to be free of qualia may not be so, at least when carried out by human beings. For example, on hearing an ambiguous remark, we may have an awareness, or sense, or 'feel' for what the right meaning for a word in a sentence is. We hear the word 'bank' and we know (or 'compute') that the speaker meant a river bank, not a financial organization. Consciousness (with qualia) may be present in virtually every intentional human activity. So the basic nature – the metaphysics – of qualia no doubt awaits further clarification.

Even if the 'easy' problems really are easy, we are left with the 'hard' questions – unless we adopt the extreme position of eliminative materialism. Cognitive scientists and philosophers alike want to know how 'outside' things (such as mechanisms, or observed machine behaviors) could ever depict subjective, 'inside' ('what it's like'), nonfunctional, mental features. If we imagine trying to build a machine with perspectives and perceptions step by step from scratch, at what point, if any, might we detect conscious experience entering the picture? Any interesting or useful machine model of consciousness should at least aim to incorporate whatever those features are that make the human mind, or any mind, so puzzling in the first place. Thus the model must be complex enough in what it contains that we are assured that something like the mind–body problem could arise for it. And such a model also needs to be instructive (or transparent) enough to give us some way of seeing through that problem. For suppose the machine model were complex enough to give rise to an analog of the mind–body problem, but it did not make consciousness more transparent than it now is. Then all we would learn from the construction (or imagined construction) of a conscious machine or robot would be that we could create things we didn't understand. We would not know if the machine model was itself a purely physicalistic model, or a dualistic one.

Ongoing attempts by cognitive scientists to model machine consciousness will, hopefully, contribute both to recreating the mind's complexity and to understanding its attendant perplexities. Of course, we cannot say what or 'who' the enhanced or combined 'offspring' of today's programs, robots, and neural networks are likely to be. But they will surely 'embody', as it were, various new research goals and presuppositions in cognitive science concerning mental function, qualia, and consciousness, which philosphers will want to assess.

Further Reading

Anderson A (ed.) (1964) *Minds and Machines*. Englewood Cliffs, NJ: Prentice Hall.

Block N, Flanagan O and Güzeldere G (eds) (1997) *The Nature of Consciousness*. Cambridge, MA: MIT Press.

Chalmers D (1996) *The Conscious Mind*. New York, NY: Oxford University Press.

Chomsky N (1966) *Cartesian Linguistics*. New York, NY: Harper & Row.

Dennett D (1991) *Consciousness Explained*. Boston, Toronto and London: Little, Brown.

Descartes R (1960) *Discourse on Method and Meditations*, translated by L. Lafleur. New York, NY: Bobbs-Merrill.

Leibniz G (1898) The monadology. In: *The Monadology and Other Philosophical Writings*, translated by R. Latta, pp. 215–277. London, UK: Oxford University Press.

Searle J (1992) *The Rediscovery of the Mind*. Cambridge, MA: MIT Press.

Shear J (1998) *Explaining Consciousness: The Hard Problem*. Cambridge, MA: MIT Press.

Shieber S (forthcoming) *The Turing Test*. Cambridge, MA: MIT Press.

Consciousness, Philosophical Issues about

Advanced article

Ned Block, New York University, New York, USA

CONTENTS

There are many problems of consciousness, but the most significant of them is: how is it possible to explain consciousness in terms of its neural basis?

THE 'HARD PROBLEM'

There are a number of different matters that come under the heading of 'consciousness'. One of them is phenomenality, the feeling of, say, a sensation of red, or a pain, that is, what it is like to have such a sensation or other experience. Another is reflection on phenomenality. Imagine two infants, both of which have pain, but only one of which has a thought about that pain. Both would have phenomenal states, but only one would have a state of reflexive consciousness. This article will first discuss phenomenality, reflexivity, and one other kind of consciousness, global availability.

The 'hard problem' of consciousness is how to explain a state of consciousness in terms of its neurological basis. If a neural state N is the neural basis of the sensation of red, why is N the basis of that experience rather than of some other experience, or of none at all? Chalmers (1996) distinguishes between the hard problem and 'easy' problems that concern the function of consciousness. The hard problem (though not by that name) was identified by Nagel (1974) and further analyzed in Levine (1983).

There are two reasons for thinking that the hard problem has no solution. The first is *'actual failure'*: the fact that no one has been able to think of even a speculative answer. The second is *'principled failure'*: the materials we have available seem ill suited to providing an answer – as Nagel says, an answer to this question would seem to require an objective account that necessarily leaves out the subjectivity of what it is trying to explain; and we do not even know what would count as such an explanation.

PERSPECTIVES ON THE HARD PROBLEM

Of the many perspectives on the hard problem, we will discuss four, all of which comport with a naturalistic framework.

Eliminativism

Eliminativism is the view that consciousness, as understood above, simply does not exist (Dennett, 1979; Rey, 1997). So there is nothing for the hard problem to be about.

Deflationism

Deflationists, also called philosophical reductionists (e.g. Dennett, 1991), move closer to 'common sense' by allowing that consciousness exists, but they 'deflate' this commitment – again on philosophical grounds – taking it to amount to less than 'meets the eye'. One prominent form of deflationism makes a conceptual reductionist claim: that consciousness can be conceptually analyzed in nonphenomenal terms. The main varieties of analysis are behavioral, functional, representational and cognitive.

Pitcher (1971) and Armstrong (1968) can be interpreted as analyzing consciousness in terms of beliefs. One type of prototypical conscious experience, for example of blue, is a matter of an inclination (perhaps suppressed) to believe that there is a blue object in plain view. (See Jackson (1977) for a convincing refutation.) A different kind of analysis appeals to higher-order thought or higher-order perception. Such analyses take the concept of a conscious pain to be the concept of a pain that is accompanied by another state that is about that pain. A pain that is not so accompanied is not a conscious state (Armstrong, 1968; Carruthers, 1992; Lycan, 1990). (Rosenthal (1997) advocates a higher-order thought view as an empirical identity rather than as a conceptual analysis.) Another deflationist view, compatible with analyses in terms of beliefs, concerns not the states themselves but their contents. Representationism holds that it can be established philosophically that the phenomenal character of experience is its representational content. Many representationists reject conceptual analysis, but still their accounts do not depend on any details of psychology or neuroscience (Harman, 1990, 1996; Dretske, 1995; Lycan, 1996; McDowell, 1994, Tye, 1995). (Shoemaker (1994) mixes phenomenal realism with representationism in an interesting way.) Conceptual functionalists say that the concept of consciousness is analyzable functionally (Lewis, 1994).

According to deflationism, there is such a thing as consciousness, but there is no hard problem, that is, there are no mysteries concerning the physical basis of consciousness that differ in kind from scientific problems about, for example, the physical or functional basis of liquidity, inheritance, or computation.

Inflationism

Inflationism, also called phenomenal realism, is the view that consciousness is a substantial property that cannot be conceptually reduced or otherwise reduced on armchair grounds in nonphenomenal terms. Logical behaviorists think that we can analyze the concept of pain in terms of certain kinds of behavior, but inflationists reject all such analyses of phenomenal concepts in nonphenomenal terms. According to most contemporary inflationists, consciousness plays a causal role and its nature may be found empirically as the sciences of consciousness advance. Inflationism is compatible with the empirical scientific reduction of consciousness to neurological or computational properties of the brain – just as heat was scientifically, but not philosophically, reduced to molecular kinetic energy. (It is not a conceptual truth that heat is molecular kinetic energy.) Inflationism accepts the hard problem but aims for an empirical solution to it (Block, 1995; Flanagan, 1992; Loar, 1997; Nagel, 1974). McGinn (1991) argues that an empirical reduction is possible but that we can't find or understand it. Searle (1992) endorses a roughly naturalistic point of view and rejects armchair reduction of phenomenality, but he also rejects any empirical reduction of phenomenal properties.

The inflationist regards all the deflationist accounts described above as leaving out the phenomenon. Phenomenality has a function and represents the world, but it is something over and above that function or representation. Something might function like a phenomenal state but be an ersatz phenomenal state with no real phenomenal character. The phenomenal character that represents red in you might represent green in me. Phenomenal character does represent but it also goes beyond what it represents. Pain may represent damage but that is not what makes pain painful.

Dualistic Naturalism

Dualistic naturalism is a broad category which includes Chalmers' (1996) view that standard materialism is false but that there are naturalistic alternatives to Cartesian dualism, such as panpsychism. Nagel (2000) proposes that there is a deeper level of reality that is the naturalistic basis both of consciousness and of neuroscience.

AN ATTEMPT AT A DISSOLUTION OF THE HARD PROBLEM

Suppose following a theory of visual experience proposed by Crick and Koch (1990) that corticothalamic oscillation (of a certain frequency) is the neural basis of an experience with phenomenal quality *Q*. There is a plausible argument that seems to solve, or rather dissolve, the hard problem from a physicalist point of view. According to this argument, the hard problem is illusory. One might as well ask why H_2O is the chemical basis of water rather than gasoline or nothing at all. Just as water just is its chemical basis, so *Q* just is its neural basis, and so the original question is vacuous.

This point is correct as far as it goes, but it does not go far enough. It begs another question: how could one property be both phenomenal property *Q* and corticothalamic oscillation? How is it possible for something subjective to be something objective, or for something first-personal to be something third-personal? The problem is not that we cannot find an explanation for this identity; indeed, in a sense there are no explanations for any identities (Block, 1978a; Block and Stalnaker, 1999). The problem is that the claim that a phenomenal property is a neural property seems just as mysterious as – maybe even more mysterious than – the claim that a phenomenal property has a certain neural basis. Explaining an identity is different from explaining how an identity can be true, where the latter involves removing a sense of puzzlement.

We can even see the same two obstacles stated above reappearing: 'actual failure' and 'principled failure'. In the first place, no one has even a speculative answer to the question of how something objective can be something subjective. In the second place, actual failure does not seem accidental. The objective seems to necessarily exclude the subjective. The third-personal seems to necessarily exclude the first-personal. Further, as McGinn (1991) notes, neural phenomena are spatial, but the phenomenal is prima facie nonspatial.

Thus, the reasons mentioned above for thinking that the explanatory gap resists closing seem to surface in a slightly different form. However, as we shall see, in this form they are more tractable.

AN APPROACH TO THE HARD PROBLEM

We now discuss a possible approach to the hard problem, whose main element is a distinction which is widely appealed to in discussions of Jackson's (1982) famous 'Mary' example. Jackson imagined a neuroscientist of the distant future who is raised in a black and white room but who knows everything scientific there is to know about color and the experience of it. When she steps outside the room for the first time, she learns what it is like to see red. Jackson argued that since the scientific facts do not encompass the new fact that Mary learns, dualism is true.

The most convincing response to Jackson's, argument (which derives from Loar (1990/1997)) involves an appeal to the distinction between a property and a concept of that property (Churchland, 1989; Loar, 1990/1997; Lycan, 1990; Van Gulick, 1993; Sturgeon, 1994; Tye, 1999; Perry, 2001).

A concept is a thought element in a way of thinking, a kind of representation. For our purposes, concepts can be interpreted as symbols in a 'language of thought'. This usage is different from another common philosophical usage in which a concept is something like a meaning. Concepts in our sense are individuated in part by meanings: x and y are instances of the same concept if and only if they are instances of the same representation and have the same meaning. 'Water' and 'H_2O' are instances of different representations, so the concept of water is distinct from the concept of H_2O.

Someone could believe that 'this color' is useful for painting pots but that red is not, even if 'this color' is red. Our experiential concept of red differs from our linguistic concept of red. An experiential concept involves a phenomenal element, a phenomenal way of thinking, for example a mental image that is in some sense of red, or an ability to produce such a mental image, or at least an ability to recognize red – which arguably could not be done without some phenomenal mental element.

Importantly, we can have an experiential concept of an experience (which we can call a phenomenal concept, phenomenal concepts being a subclass of experiential concepts), as well as an experiential concept of a color. And the very same mental image may be involved in both concepts. The difference between the phenomenal concept of the experience and the experiential concept of the color lies in the rest of the concept – in particular, in the way the phenomenal element functions in the concept. This can be a matter of further concepts explicitly invoked in the concept – the concept of a color in one case and the concept of an experience in the other. One type of experiential concept (of a color or of an experience) involves a demonstrative together with a mental image and a language-like representation, e.g. 'that [attention to a mental image] color' or 'that [attention to a mental image] experience' where the brackets indicate the use of attention to a nondescriptive element, a mental image, in fixing the demonstrative reference. Loar (1990/1997) gives an example involving two concepts in which something like a mental image of a cramp feeling is used to pick out a muscular knot in one concept, and in the other concept, the cramp experience itself. In our notation, the two concepts would be 'that [attention to a mental image] cramp' and 'that [attention to a mental image] cramp experience'.

An experiential concept uses a phenomenal property to pick out a related property. For example, an experiential concept of a color can use an aspect of the experience of that color to pick out the color. A phenomenal concept uses a phenomenal property to pick out a phenomenal property. The phenomenal property used in the concept need not be the same as the one picked out. For example, one could have a phenomenal concept of the experience of the missing shade of blue whose phenomenal elements are the flanking color experiences. Or, a phenomenal element involved in one's perception of a color could be used to pick out the experience of the complementary color. Importantly, the phenomenal element in a phenomenal concept need not be, and in general cannot be, conceptualized, at least if the conceptualization is meant to be itself phenomenal. For if a phenomenal concept had to make use of a phenomenal element via a distinct phenomenal concept of that element, there could be no phenomenal concepts. Thus we can define a phenomenal concept as a concept of a phenomenal property that uses a phenomenal property to pick out a phenomenal property, but not necessarily under a concept of the phenomenal property used.

In these terms, then, Mary acquired a new concept of a property she was already acquainted with via a different concept. In the room, Mary knew about the subjective experience of red via physical concepts. After she left the room, she acquired a phenomenal concept of the same property. So Mary did not learn a new fact: she learned a new concept of an old fact. She already had a third-person understanding of the fact of what it is like to see red. What she gained was a first-person understanding of the very same fact. She knew already that corticothalamic oscillation of a certain frequency is what it is like to see red. What she learned is that 'this [attention to a mental image] experience is what it is like to see red'. So the case does not demonstrate that there are facts that go beyond physical facts.

Recall that there is a principled reason why mind–body identity seemed impossible: that a first-person subjective property could not be identical to a third-person objective property. But the distinction between concepts and properties allows us to see that the above distinction between subjective and objective, and the distinction between first person and third person, are distinctions between kinds of concepts, not kinds of properties. There is no reason why a subjective concept and an objective concept cannot pick out the same property. Thus we can substitute a dualism of concepts for a dualism of properties.

There is another way in which the concept–property distinction helps with the hard problem. We can blame the explanatory gap and the hard problem on our inadequate concepts rather than on dualism. To take a variant on Nagel's (1974) example, we are like pre-Socratics who have no way of understanding how it is possible that heat equals mean molecular kinetic energy, lacking the concepts required to frame both sides of the equation. (Heat was not clearly distinguished from temperature until the seventeenth century.) What is needed is a concept of heat and a concept of kinetic energy that make it conceivable that there is a causal chain of the referential sort leading from the one magnitude to each concept. Or rather, since the phenomenal concept includes a sample of the relevant phenomenal property (on the Humean simplification we are using), there is no mystery about the mental side of the equation. The mystery is how the physical concept picks out that phenomenal property. This is the remaining part of the explanatory gap, which will be closed, if at all, by science. Is there a principled reason to think it cannot be? The hard problem itself does not contain such a reason. Perhaps our conceptual inadequacy is temporary, as Nagel sometimes appears to suppose, or perhaps it is permanent, as McGinn (1991) supposes.

EMPIRICAL FINDINGS ABOUT CONSCIOUSNESS

We will now turn to a discussion of recent empirical findings on consciousness. The most interesting line of experimental investigation of consciousness in recent years uses phenomena in which perception changes independently of the stimulus. One such paradigm uses binocular rivalry. If two stimuli – for example, horizontal and vertical stripes – are presented, a different stimulus to each eye, one does not see a blend, but rather, first horizontal stripes that fill the whole visual field, and then vertical stripes that fill the whole field. Logothetis (1998) trained monkeys to pull different levers for different patterns. They then presented different patterns to the monkeys' two eyes, and observed that, as with people, the monkeys switched back and forth between the two levers even though the sensory input remained the same. Logothetis recorded the firings of various neurons in the monkeys' visual systems. In the lower visual areas (e.g. V1), 80 percent of the neurons did not shift with the percept. But further along the occipital–temporal pathway, 90 percent shifted with the percept. So it seems that later areas in the occipital–temporal pathway (let us call it the 'ventral stream') are more dominantly part of the neural basis of (visual) consciousness than earlier areas. Recent work using imaging has extended and refined these findings. Kanwisher (2001) notes that 'neural correlates of perceptual experience, an exotic and elusive quarry just a few years ago, have suddenly become almost commonplace findings'. And she backs this up with impressive correlations between neural activation on the one hand and indications of perceptual experiences of faces, houses, motion, letters, objects, words, and speech on the other. Although the neural correlates of, say, faces and houses, are somewhat different, both are in the ventral stream, mainly in the higher areas. These results represent a major success: identification of the neural basis of visual consciousness in the ventral stream.

Paradoxically, what has also become commonplace is activation of the very same ventral stream pathways without 'awareness'. Damage to the inferior parietal and frontal lobes has long been known to cause visual extinction – in which subjects appear to lose subjective experience of certain stimuli on one side when there are stimuli on both sides. Extinction is associated with visual neglect, in which subjects do not notice stimuli on one side. For example, neglect patients often don't eat the food on the left side of the plate. Although subjects say they do not see the extinguished stimulus, the nature of the stimulus has various visual effects. For example, if the subject's task is to decide whether a letter string (e.g. 'saucer' or 'spiger') is a word, the subject is faster for 'saucer' if there is a picture of a cup or the word 'cup' in the neglected field, even though they perform no better than chance in guessing what the picture depicts (McGlinchey-Berroth *et al.*, 1996). So the stimulus, of which the subject is in some sense unaware, is processed semantically.

Driver and Vuilleumier (2001) point out that the ventral stream is activated for extinguished stimuli (i.e. stimuli that the subject claims not to see). Rees *et al.* (2000) report studies of a left-sided neglect and extinction patient on face and house stimuli. Stimuli presented only on the left side are clearly seen by the patient, but when there are stimuli on both sides, the subject says he sees only the stimulus on the right. However, the 'unseen' stimuli produce a pattern of activation of the ventral pathway that is the same in location and temporal course as the seen stimuli (though lower in magnitude). Furthermore, studies in monkeys have shown that a well-known 'blindness' syndrome is caused by massive cortical ablation which spares most of the ventral stream but not the inferior parietal and frontal lobes (Lumer and Rees, 1999). Kanwisher (2001) notes that dynamic visual gratings alternating with a gray field showed activation for V1, V2, V3A, V4v, MT and MST despite the subjects saying they saw only a uniform gray field.

Zeki and Ffytch (1998) hypothesize that the difference between conscious and unconscious activation of the ventral pathway is just a matter of the degree of activation. But Kanwisher (2001) mentions that evidence from event-related potential (ERP) studies using the attentional blink paradigm show that neural activation of meaning is no less when the word is blinked than when it isn't, suggesting that it is not lower neural activation strength that accounts for lack of awareness. Dehaene and Naccache (2001) note that in a study of neglect patients, it was shown that there is the same amount of semantic priming from both hemifields, despite the lack of awareness of stimuli in the left field, again suggesting that it is not activation strength that makes the difference.

The paradox, then, is that our success in identifying the neural correlate of visual experience in normal vision has led to the conclusion that in masking and neglect, that very neural correlate occurs without, apparently, subjective experience.

What is the missing ingredient *X* which, added to ventral activation, constitutes conscious experience? Kanwisher (2001) and Driver and Vuilleumier (2001) offer similar proposals as to the nature of *X*, namely, that the missing ingredient is binding perceptual attributes with a time and a place, a token event. Rees *et al.* (2000) make two suggestions as to what *X* is. One is that the difference between conscious and unconscious activation is a matter of neural synchrony at fine timescales. This idea is supported by the finding that ERP components P1 and N1 revealed differences between left-sided 'unseen' stimuli and left-sided seen stimuli. Their second suggestion is that the difference between seen and 'unseen' stimuli might be a matter of interaction between the 'visual stream' as we know it and the areas of parietal and frontal cortex that control attention.

Whether or not any of these proposals is right, the search for *X* seems to be the most interesting current direction for consciousness research. For the purposes of the discussion below, we will assume that *X* equals neural synchrony.

PHYSICALISM AND FUNCTIONALISM

There is another, very different, approach to the nature of consciousness. Dennett (1994) postulates that consciousness is 'cerebral celebrity'. What it is for a representation to be conscious is for it to be widely available in the brain. Dehaene and Naccache (2001) say consciousness is being broadcast in a global neuronal workspace.

The theory that consciousness is ventral stream activation combined with neural synchrony and the theory that consciousness is broadcasting in the global neuronal workspace are instances of the two major approaches to consciousness in the philosophical literature: physicalism and functionalism. The difference is that the functionalist says that consciousness is a role, whereas the physicalist says that consciousness is a physical or biological state that implements that role. For example, red may play the role of warning of danger – but green might also have played that role. The picture of consciousness as role could be characterized as computational in contrast to the biological picture of consciousness as implementer of the role.

Although functionalists are free to add restrictions, functionalism in its pure form is implementation-independent. Consciousness is defined as global accessibility, and although its human implementation depends on human biochemistry, silicon-based creatures without our biochemistry could implement the same computational relations. Functionalism and physicalism are incompatible doctrines, since a nonbiological implementation of the functional organization of consciousness would be regarded as uncontroversially conscious by the functionalist but not by the physicalist. The big question for functionalists is: 'how do we know that it is broadcasting in the global workspace that makes a representation conscious, as opposed to something about the human biological realization of that broadcasting?

The problem for functionalists could be put like this: the specifically human biochemical realization of global availability may be necessary to consciousness – other realizations of global availability being 'ersatz' realizations. The typical response to this 'ersatz realization problem' (Lycan, 1981) is that we can preserve functionalism by simply bringing lower-level causes and effects into our functional characterizations: for example, causes and effects at the level of biochemistry. But this response is inadequate because one can descend the hierarchy of sciences to the lowest level of all, that of basic physics, and find oneself vulnerable to the same point. Putting the point for simplicity in terms of the physics of the 1960s, the causal role of electrons is the same as that of anti-electrons. If you formulate a functional role for an electron, an anti-electron will realize it. Thus an anti-electron is an ersatz realizer of the functional definition of 'electron'. Physics is characterized by symmetries that allow ersatz realizations (Block, 1978b).

We have been talking about the two approaches of functionalism and physicalism as rivals, but we can instead see them as answers to different questions. The question that motivates the physicalist proposal of 'ventral activation plus *X*' is: what is the neural basis of experience? The question that motivates the 'global broadcasting' proposal is: what makes neuronal representations available for thought, decision, reporting and action? The former is a theory of phenomenal consciousness ('phenomenality'), and the latter of 'access consciousness'. (Theorists will differ on whether access consciousness is really a type of consciousness (Burge, 1997).) We can try to force a unity by postulating that it is a condition on *X* that it promote access, but that is merely a verbal maneuver, and only disguises the difference between the concepts and questions involved. Alternatively, we could hypothesize, rather than postulate, that *X* as a matter of fact is the neural basis of global neuronal broadcasting. Note, however, that the neural basis of global neuronal broadcasting might obtain but the normal channels of broadcasting none the less be blocked or cut, and this would again reveal the distinction between phenomenality and access, showing that we cannot think of the two as one. (As an analogy, rest mass and relativistic mass are importantly different from a theoretical point of view despite coinciding for all practical purposes at terrestrial velocities. Failure of coincidence, even if rare, is theoretically critical to the scientific nature of consciousness.)

Many of us have had the experience of suddenly noticing a sound (say, a drilling noise during an intense conversation), at the same time realizing that the sound has been going on for some time even though we were not attending to it. If there was a phenomenal representation of the sound before it was noticed, that is, if the sound was experienced first and noticed second, that state was not broadcast in the global neuronal workspace until it was noticed: there was a period of phenomenality without broadcasting. Of course, this is anecdotal evidence. But the starting point for work on consciousness is introspection, and we would be foolish to ignore it.

If we take seriously the idea of phenomenality without global accessibility, one theoretical option that we should consider is that ventral stream activation is visual phenomenality and the search for X is the search for the neural basis of what makes visual phenomenality accessible. The idea would be that the claims of extinction patients not to see extinguished stimuli are in a sense wrong: they really do have phenomenal experience of these stimuli without knowing it. A similar issue will arise in our discussion of the relation between phenomenality and a special case of global accessibility, reflexive or introspective consciousness, in which the subject not only has a phenomenal state but also has another state that is about the phenomenal state, say, a thought to the effect that he has the phenomenal state.

We have drawn a distinction between two concepts of consciousness, phenomenality and global accessibility. We will now add a third, reflexivity.

PHENOMENALITY AND REFLEXIVITY

Consider the 'false recognition' paradigm of Jacoby and Whitehouse (1989). Subjects are given a study list of 126 words presented for half a second each. They are then presented with a masked word, w_1 and an unmasked word w_2. Their task is to report whether w_2 was old (i.e. on the study list) or new (not on the study list). The variable was whether w_1 was lightly or heavily masked, the former presentations being thought of as 'conscious' and the latter as 'unconscious'. Confining our attention just to cases in which $w_1 = w_2$, subjects were much more likely to mistakenly report w_2 as old when w_1 was unconsciously presented than when w_1 was consciously presented. The explanation would appear to be that when w_1 was consciously presented, the subjects were able to use an internal monologue of the following sort (though perhaps not quite as explicit): 'the reason w_2 looks familiar is that I just saw it (as w_1)'. Thus they 'explained away' the familiarity of w_2. But when w_1 was unconsciously presented, the subjects were not able to indulge in this monologue and consequently attributed the familiarity of w_2 to its appearance in the study list.

Any monologue that can reasonably be attributed to the subject in this paradigm concerns why a word (w_2) 'looks familiar' to the subject. For it is only by 'explaining away' the familiarity of w_2 that the subject is able to decide that w_2 was not on the study list. Thus, in the 'conscious' case, the subject must have a state that is about the subject's own perceptual experience (looking familiar), and thus conscious in what might be termed a 'reflexive' sense. An experience is conscious in this sense just in case it is the object of another of the subject's states: for example, one has a thought to the effect that one has that experience. The reflexive sense of 'consciousness' contrasts with phenomenality, which may apply to some states that are not the objects of other mental states. Reflexive consciousness might better be called 'awareness' than 'consciousness'. Reflexivity is phenomenality together with something else (reflection), and there is the possibility in principle of phenomenality without reflection.

What is the relation between reflexivity and the notion of global accessibility discussed in the last section? Global accessibility does not logically require reflexivity, since global accessibility only requires access to all response modes that the organism actually has. (Perhaps a dog or a cat does not have the capacity for reflection.) Reflexivity is a special kind of access, one that requires intellectual resources that may not be available to every being that can have conscious experience.

There is another aspect of the experimental paradigm just discussed which motivates taking seriously the hypothesis that the reflexively unconscious case might possibly be phenomenally conscious. Consider a variant of the exclusion paradigm reported by Debner and Jacoby (1994). Subjects were briefly presented with pairs of words flanked by digits (e.g. '4reason5'), and then given stems consisting of the first three letters of the word ('rea___') to complete. Subjects were instructed to complete the stem, but not with the word that was briefly presented and flanked by digits. There were two conditions. In the 'conscious' condition, they were told to ignore the digits. In the 'unconscious' condition, they were told to report the sum of the digits before completing the stem. The results were that in the 'conscious' condition, the subjects were more likely than baseline to follow the instructions and complete the stem with a word other than 'reason', whereas in the 'unconscious' condition, subjects were more likely than baseline to violate the exclusion instructions, completing the stem with 'reason'. Merikle and Joordens (1997) report corresponding results for the false recognition paradigm with divided attention substituted for heavy masking.

Consider the hypothesis that there was a fleeting phenomenal consciousness of 'reason' as the subject's eyes moved from the '4' to the '5' in '4reason5'. There are two theoretical options that deserve serious consideration: either (1) the 'unconscious perceptions' are both phenomenally and reflexively unconscious (in this case, the exclusion and false recognition paradigms are about consciousness in both senses); or (2) they are fleetingly phenomenally conscious but reflexively unconscious. A third option, that (3) they are phenomenally unconscious but 'reflexively conscious' seems less likely because the reflexive consciousness would be 'false': subjects would have a state 'about' a phenomenal state without the phenomenal state itself. That hypothesis would require some extra causal factor that produced the false recognition, and would thus be less simple. One argument in favor of (2) is that subjects in experiments with near-threshold stimuli often report a 'mess' of partial perceptions that they 'can't hang on to'. Some critics (e.g. Dennett, 1991) have disparaged the idea of fleeting phenomenal consciousness. But they still do not provide a positive reason to think that (1) is the correct view.

It might seem that there is a principled argument that we could never find out about phenomenality in the absence of reflexive consciousness, for we require the subject's testimony about phenomenality, which requires the subject to have a state that is about the phenomenal state. We can see what is wrong with this reasoning by attention to some potential lines of evidence for phenomenality in the absence of reflexive consciousness.

Liss (1968) contrasted subjects' responses to brief unmasked stimuli (one to four letters) with their responses to longer lightly masked stimuli. He asked for judgments of brightness, sharpness and contrast as well as what letters they saw. He found that lightly masked 40 ms stimuli were judged as brighter and sharper than unmasked 9 ms stimuli, even though the subjects could report three out of four letters in the unmasked stimuli and only one out of four in the masked cases. Liss writes: 'The subjects commented spontaneously that, despite the high contrast of the letters presented under backward masking, they seemed to appear for such brief duration that there was very little time to identify them before the mask appeared. Although letters presented for only 7 ms with no masking appeared weak and fuzzy, their duration seemed longer than letters presented for 70 ms followed by a mask.'

Perhaps the subjects were phenomenally conscious of all the masked letter shapes, but could not apply the letter concepts required for reflexive consciousness of them. (The subjects could apply the concepts of sharpness, brightness and contrast to the letters, so they did have reflexive consciousness of those features, even if they did not have reflexive consciousness of the shapes themselves. There are two kinds of shape concepts that could have provided – but apparently did not provide – reflexive consciousness of the letters: the letter concepts that we all learn in school, and shape concepts of the kind we have for unfamiliar shapes.) In other words, perhaps phenomenal experience of shapes does not require shape concepts but reflexive consciousness, being an intentional state, does require shape concepts, concepts that the subjects seem unable to access in these difficult attentional circumstances. Alternatively, perhaps the phenomenal experience of shapes does involve shape concepts of some sort but the use of those shape representations in reflexive consciousness requires more attentional resources than were available to these subjects.

There is another hypothesis: that the contents of both the subjects' phenomenal states and their reflective states are the same, and include the features 'sharp', 'high contrast', 'bright' and 'letter-like' without any specific shape representation. On this hypothesis, there is no gap between phenomenal and reflexive consciousness. Both hypotheses have to be taken seriously, but the first is superior in at least one respect: anyone who has been a subject in this or in Sperling's (1960) similar experiment will feel that the last hypothesis does not really capture the experience, which is an experience of being able to see more letters than one can categorize.

Rosenthal (1997) defines reflexive consciousness as follows: *S* is a reflexively conscious state of mine if and only if *S* is accompanied by a thought – arrived at non-inferentially and non-observationally – to the effect that I am in *S*. He offers this 'higher-order thought' (HOT) theory as a theory of phenomenal consciousness. Both phenomenal consciousness without HOT and HOT without phenomenal consciousness are conceptually possible. For example, perhaps dogs and infants have phenomenally conscious pains without higher-order thoughts about them. Conversely, imagine that by biofeedback and imaging techniques of the distant future I learn to detect the state in myself of having the Freudian unconscious thought that it would be nice to kill my father and marry my mother. I could come to know – non-inferentially and non-observationally – that I have this Freudian thought even though the thought is not phenomenally conscious. Since there are conceptually possible counterexamples in both directions, the question again is whether reflexivity and phenomenality come to the same thing empirically.

If there are no actual counterexamples, the question arises of why. Is it supposed to be a basic law of nature that phenomenality and reflexivity cooccur? That would be a very adventurous claim that no one is in a position to make. Is it a contingent fact about us but not other phenomenally conscious creatures? Then reflexivity would not provide a basic account of phenomenality. Well then, is it supposed to be a contingent fact that phenomenality and reflexivity are coextensive in all creatures? What would be the evidence for such a far-reaching claim? Further, if phenomenality and reflexivity are correlated, then there must be a mechanism that explains the correlation, as the fact that both heat and electricity are carried by free electrons explains the correlation of electrical and thermal conductivity. But any mechanism breaks down under extreme conditions – as does the correlation of electrical and thermal conductivity at extremely high temperatures. So the correlation between phenomenality and reflexivity would break down too, showing that reflexivity does not yield the fundamental scientific nature of phenomenality.

Alternatively, it might be said that 'consciousness' is a 'natural kind' term, like 'water' or 'heat' or 'light'. We know that water is H_2O, as a matter of empirical fact without having to ask whether there might be a substance in another solar system that has a very different chemical constitution but nonetheless behaves exactly like water. The reason is that as a matter of the semantics of the word 'water', the question of whether there could be a substance that behaved exactly like water but that had a completely different chemical constitution does not matter since it would be wrong to call it 'water'. But this is a poor analogy for two reasons. First, if there are beings in another solar system who have something that feels like phenomenality but without reflexivity, then reflexivity does not capture the fundamental nature of phenomenality, whatever we call what the aliens have. Second, we call anything that feels like phenomenality 'phenomenality', so phenomenality is not a natural kind concept in the sense that 'water' is.

Rosenthal's definition of reflexivity has a number of *ad hoc* features. Non-observationality is required to rule out, for example, a case in which I know about a thought I have repressed by observing my own behavior. Non-inferentiality is required to avoid a somewhat different case in which I appreciate (non-observationally) my own psychic pain and infer a repressed thought from it. But why should the consciousness of a state depend on its causal history? Furthermore, Rosenthal's definition involves a stipulation that the possessor of the reflexively conscious state is the same as the thinker of the thought – otherwise my thinking about your pain would make it a conscious pain. All these *ad hoc* features can be eliminated by moving to the following definition of reflexivity: *S* is a reflexively conscious state if and only if *S* is phenomenally presented in a thought about *S*. This definition uses the notion of phenomenality; but this is no disadvantage unless one holds that there is no such thing apart from reflexivity itself. The new definition of reflexivity, requiring phenomenality as it does, has the additional advantage of making it clear that reflexivity is a kind of consciousness. (See Burge's (1997) critique of our notion of 'access consciousness' as constituting a kind of consciousness.)

We have encountered three concepts of consciousness: phenomenality, reflexivity, and global accessibility. The hard problem arises only for phenomenality. Imaging experiments on consciousness engage phenomenality and accessibility. But many psychological experimental paradigms mainly engage reflexivity. So empirical investigators of 'consciousness' may sometimes be talking at cross purposes.

References

Armstrong DM (1968) *A Materialist Theory of the Mind*. London: Routledge and Kegan Paul.

Block N (1978a) Reductionism. In: *Encyclopedia of Bioethics*, pp. 1419–1424. New York, NY: Macmillan.

Block N (1978b) Troubles with functionalism. In: Savage CW (ed.) *Minnesota Studies in the Philosophy of Science*, vol. IX, pp. 261–325. [Reprinted abridged in: Rosenthal DM (ed.) (1991) *The Nature of Mind*, pp. 211–229. Oxford: Oxford University Press.]

Block N (1995) On a confusion about a function of consciousness. *Behavioral and Brain Sciences* **18**(2): 651–726.

Block N, Flanagan O and Güzeldere G (eds) (1997) *The Nature of Consciousness: Philosophical Debates*. Cambridge, MA: MIT Press.

Block N and Stalnaker R (1999) Conceptual analysis, dualism and the explanatory gap. *Philosophical Review* **108**: 1–46.

Burge T (1997) Two kinds of consciousness. In: Block N, Flanagar O and Güzeldere G (eds) *The Nature of Consciousness: Philosophical Debates*, pp. 427–434. Cambridge, MA: MIT Press.

Carruthers P (1992) Consciousness and concepts. *Proceedings of the Aristotelian Society* **66**: 41–59.

Chalmers D (1996) *The Conscious Mind*. New York, NY: Oxford University Press.

Churchland P (1989) Knowing qualia: a reply to Jackson. In: *A Neurocomputational Perspective*, pp. 67–76. Cambridge, MA: MIT Press.

Crick F and Koch C (1990) Towards a neurobiological theory of consciousness. *Seminars in the Neurosciences* **2**: 263–275.

Debner JA and Jacoby LL (1994) Unconscious perception: attention, awareness and control. *Journal of Experimental Psychology: Learning, Memory, and Cognition* **20**: 304–317.

Dehaene S and Naccache L (2001) Towards a cognitive neuroscience of consciousness: basic evidence and a workspace framework. *Cognition* **79**(1–2): 1–37.

Dennett D (1979) On the absence of phenomenology. In: Gustafson D and Tapscott B (eds) *Body, Mind and Method: Essays in Honor of Virgil Aldrich*, pp. 93–113. Dordrecht: Reidel.

Dennett D (1991) *Consciousness Explained*. Boston, MA: Little, Brown.

Dennett D (1994) Get Real. In: *Philosophical Topics* **22** (1–2): 505–568.

Dretske F (1995) *Naturalizing the Mind*. Cambridge, MA: MIT Press.

Driver J and Vuilleumier P (2000) Perceptual awareness and its loss is unilateral neglect and extinction. *Cognition* **79**(1–2): 39–89.

Flanagan O (1992) *Consciousness Reconsidered*. Cambridge, MA: MIT Press.

Harman G (1990) The intrinsic quality of experience. In: Tomberlin J (ed.) *Philosophical Perspectives*, vol. IV

'Action Theory and Philosophy of Mind', pp. 31–52. Atascadero, CA: Ridgeview.

Harman G (1996) Explaining objective color in terms of subjective reactions. In: Villanueva E (ed.) *Philosophical Issues 7: Perception*. Atascadero, CA: Ridgeview.

Jackson F (1977) *Perception*. Cambridge, UK: Cambridge University Press.

Jackson F (1982) Epiphenomenal qualia. *Philosophical Studies* **32**: 127–136.

Jacoby LL and Whitehouse K (1989) An illusion of memory: false recognition influenced by unconscious perception. *Journal of Experimental Psychology: General* **118**: 126–135.

Kanwisher N (2001) Neural events and perceptual awareness. *Cognition* **79**(1–2): 89–113.

Levine J (1983) Materialism and qualia: the explanatory gap. *Pacific Philosophical Quarterly* **64**: 354–361.

Lewis D (1994) Reduction of mind. In: Guttenplan S (ed.) *Blackwell's Companion to Philosophy of Mind*, pp. 412–431. Oxford: Blackwell.

Liss P (1968) Does backward masking by visual noise stop stimulus processing? *Perception and Psychophysics* **4**: 328–330.

Loar B (1990/1997) Phenomenal states. In: Tomberlin J (ed) *Philosophical Perspectives*, vol. IV *'Action Theory and Philosophy of Mind'*, pp. 81–108. Atascadero, CA: Ridgeview. [A much-revised version of this paper is to be found in (Block *et al.*, 1997, pp. 597–616).]

Logothetis NK (1998) Single units and conscious vision. *Proceedings of the Royal Society of London, Series B* **353**: 1801–1818.

Lumer E and Rees G (1999) Covariation of activity in visual and prefrontal cortex associated with subjective visual perception. *Proceedings of the National Academy of Sciences* **96**: 1669–1673.

Lycan WG (1981) Form, function and feel. *Journal of Philosophy* **78**: 24–50.

Lycan WG (1990) Consciousness as internal monitoring, In: Tomberlin J (ed) *Philosophical Perspectives*, vol. IX, pp. 1–14. Atascadero, CA: Ridgeview. [Reprinted in (Block *et al.*, 1997).]

Lycan WG (1996) *Consciousness and Experience*. Cambridge, MA: MIT Press.

McDowell J (1994) The content of perceptual experience. *Philosophical Quarterly* **44**: 190–205.

McGinn C (1991) *The Problem of Consciousness*. Oxford: Blackwell.

McGlinchey-Berroth R, Milberg WP, Verfaellie M *et al.* (1996) Semantic processing and orthographic specificity in hemispatial neglect. *Journal of Cognitive Neuroscience* **8**: 291–304.

Merikle P and Joordens S (1997) Parallels between perception without attention and perception without awareness. *Consciousness and Cognition* **6**: 219–236.

Nagel T (1974) What is it like to be a bat? *Philosophical Review* **83**: 435–450.

Nagel T (2000) The psychophysical nexus. In: Boghossian P and Peacocke C (eds) *New Essays on the A Priori*, pp. 434–471. Oxford: Oxford University Press.

Perry J (2001) *Knowledge, Possibility and Consciousness*. Cambridge, MA: MIT Press.

Pitcher G (1971) *A Theory of Perception*. Princeton, NJ: Princeton University Press.

Rees G, Wojciulik E, Clarke K *et al.* (2000) Unconscious activation of visual cortex in the damaged right hemisphere of a parietal patient with extinction. *Brain* **123**: 1624–1633.

Rey G (1997) *Contemporary Philosophy of Mind*. Oxford: Blackwell.

Rosenthal DM (1997) A theory of consciousness. In: Block N, Flanagar O and Güzeldere G (eds) (1997) *The Nature of Consciousness: Philosophical Debates*, pp. 729–754. Cambridge, MA: MIT Press.

Searle J (1992) *The Rediscovery of the Mind*. Cambridge, MA: MIT Press.

Shoemaker S (1994) Self-knowledge and 'inner sense'. Lecture III: The phenomenal character of experience. *Philosophy and Phenomenological Research* **54**(2): 291–314.

Sperling G (1960) The information available in brief visual presentations. *Psychological Monographs* **74**(11): 1–29.

Sturgeon S (1994) The epistemic view of subjectivity. *Journal of Philosophy* **91**: 221–235.

Tye M (1995) *Ten Problems of Consciousness*. Cambridge, MA: MIT Press.

Tye M (1999) Phenomenal consciousness: the explanatory gap as a cognitive illusion. *Mind* **108**: 706–725.

Van Gulick R (1993) Understanding the phenomenal mind: are we all just armadillos? In: Davies M and Humphreys G (eds) *Consciousness: Psychological and Philosophical Essays*, pp. 137–149. Oxford: Blackwell. [Reprinted in (Block *et al.*, 1997).]

Zeki S and Ffytch DH (1998) The Riddoch syndrome: insights into the neurobiology of conscious vision. *Brain* **121**: 25–45.

Further Reading

Block N, Flanagan O and Güzeldere G (eds) (1997) *The Nature of Consciousness: Philosophical Debates*. Cambridge, MA: MIT Press. [A collection of papers on consciousness.]

Dehaene S (ed.) (2001) *Cognition* **79**(1–2). [A special issue on the cognitive neuroscience of consciousness.]

Güzeldere G (1997) The many faces of consciousness: *a field guide*. In: Block N, Flanagar O and Güzeldere G (eds) (1997) *The Nature of Consciousness: Philosophical Debates*, pp. 1–67. Cambridge, MA: MIT Press.

Huxley TH (1866) *Lessons in Elementary Physiology*. London: Macmillan.

Jackson F (1986) What Mary didn't know. *Journal of Philosophy* **83**: 291–295.

Jackson F (1993) Armchair metaphysics. In: O'Leary-Hawthorne J and Michael M (eds) *Philosophy in Mind*, pp. 23–42. Dordrecht: Kluwer.

Putnam H (1967) *Psychological predicates*. In: Capitan WH and Merrill DD (eds) *Art, Mind and Religion*. Pittsburgh, PA: Pittsburgh University Press. [Later entitled 'The nature of mental states.' Reprinted in: Putnam M (1975) *Mind, Language, and Reality*, pp. 429–440. Cambridge, UK: Cambridge University Press.]

Smart JJC (1959) Sensations and brain processes. *Philosophical Review* **68**: 141–156. [This paper has been reprinted many times, starting in 1962 in a somewhat revised form. See, for example: Rosenthal DM (ed.) (1971) *Materialism and the Mind–Body Problem*. Englewood Cliffs, NJ: Prentice-Hall.]

Strawson G (1994) *Mental Reality*. Cambridge, MA: MIT Press.

Consciousness, Sleep, and Dreaming

Intermediate article

J Allan Hobson, Harvard Medical School, Boston, Massachusetts, USA

CONTENTS

Advances in, and links between, neurobiology and psychology as applied in the study of sleep and dreaming compared to the waking state are increasingly revealing significant evidence for the brain basis of consciousness.

CONSCIOUS STATES AND BRAIN–MIND ISOMORPHISM

How can a material structure, like the brain, possibly possess awareness? Philosophers refer to this conundrum as 'the hard problem' and many are deeply pessimistic about its resolution. Countering such pessimism are recent developments in the cognitive neuroscience of waking, sleeping, and dreaming which have made it possible to understand – in considerable detail – how the brain changes its conscious state every day – and every night – of our lives. Even if we cannot yet detail how the brain becomes conscious in the first place, specification of the brain mechanisms responsible for the dramatic changes in consciousness that contrast waking and dreaming constitutes a partial solution of the brain–mind question and points the way to its more complete resolution in the foreseeable future (Hobson, 1999).

That dreaming might provide a privileged access to understanding the relationship of brain and mind has long been recognized by such diverse thinkers as the Enlightenment philosopher David Hartley and the Romantic poet Samuel Taylor Coleridge. Wilhelm Wundt, the father of experimental psychology, astutely speculated that to account for the difference between dreaming and waking consciousness, those brain functions supporting critical thought and self-reflective awareness must be in abeyance (in dreaming) while those supporting internal perceptions and emotions must be enhanced.

In declaring that dreaming was the royal road to the unconscious mind, Sigmund Freud was strongly influenced by his earlier attempt to create a scientific psychology based upon the structures and functions of the 100 billion nerve cells that constitute the brain. Until recently none of these theories could advance because so little was known about the activity of the brain during sleep.

Since the late 1920s brain science has seen dramatic advances that now make a solid neurology of dreaming possible. The first key step in this direction was the discovery of the electroencephalogram (EEG) by Adolf Berger in 1928 and the recognition by him that the pattern of brain waves changed when subjects became inattentive or drowsy. Figure 1 illustrates the normal human sleep cycle as it is understood today.

The idea that the brain waves of the cerebral cortex had to be kept electrically activated to support waking consciousness was crystallized in the discovery of the importance to arousal of the brain

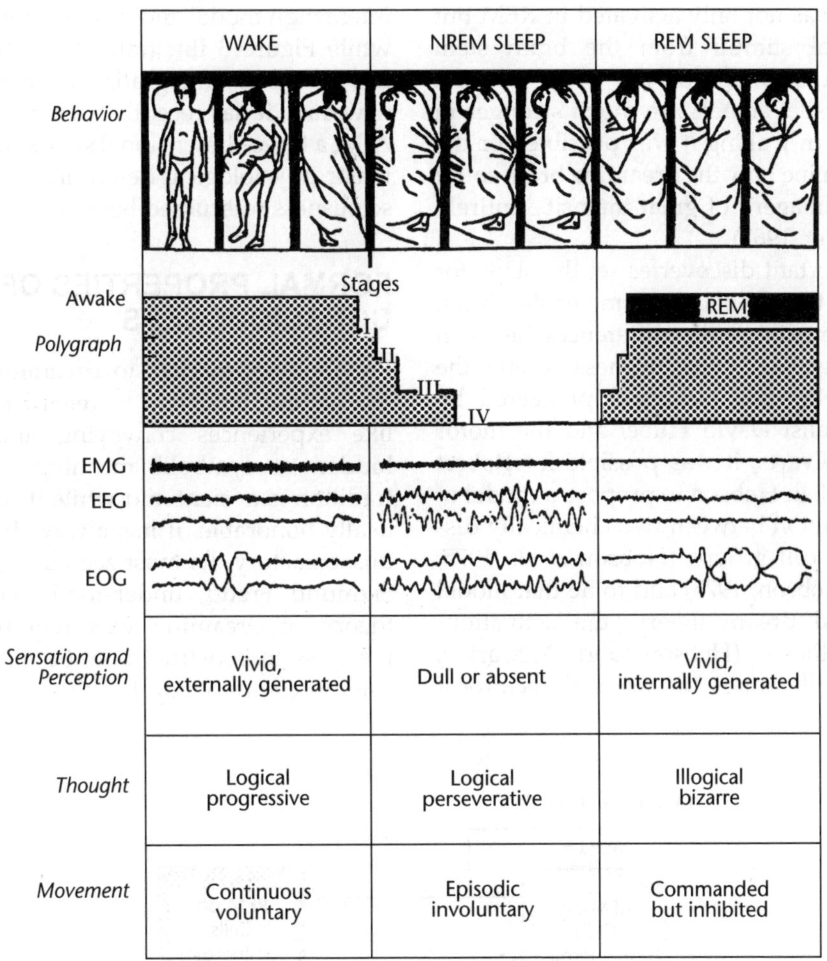

Figure 1. Behavioral states in humans. States of waking, rapid eye movement (REM) sleep, and non-REM (NREM) sleep have behavioral, polygraphic, and psychological manifestations. In the behavior channel, posture shifts (detectable by time-lapse photography or video) can occur during waking and in concert with phase changes of the sleep cycle. Two different mechanisms account for sleep immobility: disfacilitation (during stages I–IV of NREM sleep) and inhibition (during REM sleep). In dreams we imagine that we move but we do not. Sequence of these stages is represented in the polygraph channel. Sample tracings of three variables used to distinguish state are also shown: electromyogram (EMG), which is highest in waking, intermediate in NREM sleep, and lowest in REM sleep; and electroencephalogram (EEG) and electrooculogram (EOG), which are both activated in waking and REM sleep and inactivated in NREM sleep. Each sample record is 20 s. Three lower channels describe other subjective and objective state variables. (From Hobson and Steriade, 1986.)

stem reticular formation (Moruzzi and Magoun, 1949). Shortly thereafter, it was discovered that the brain was periodically reactivated during sleep and that this activation was associated with rapid eye movements (or REMs) and with intense and sustained dream consciousness (Aserinsky and Kleitman, 1953; Dement and Kleitman, 1957).

It remained to give an account of the mechanisms of electrical activation of the cortex by the brain stem and to distinguish between the brain processes associated with the electrical activation that occurred in waking and in sleep in such a way as to distinguish between dreaming and waking consciousness. These details emerged from studying the animal model of REM sleep provided by the discovery that REM sleep also occurred periodically in the sleep of cats (Dement, 1958).

Working with cats, Michel Jouvet had, by 1959, demonstrated that the brain activation of REM sleep (like that of waking) depended upon the brain stem. That the mechanisms of brain activation were quite different in REM sleep and waking became apparent from Jouvet's discovery that the postural tone of the muscles necessary to support the motor activity of waking was actively abolished in REM. Jouvet also showed that the cerebral cortex was not only activated in REM but received internal stimuli from the brain stem regarding the movement of the eyes. Because these internal visual signals are much stronger in REM sleep than in waking it was possible, for the first time, to imagine how the dreaming brain could produce visual imagery of great intensity entirely on its own (Jouvet, 1962).

All these important discoveries set the stage for the analysis of the cellular and molecular brain processes that mediated the differences between dreaming and waking consciousness. Using the movable microelectrode technique pioneered by the vision specialist David Hubel and the motor expert Edward Evarts, it was possible for Robert McCarley and Allan Hobson to propose a model of reciprocal interaction between two chemically distinct brain stem cell groups (Hobson *et al.*, 1975; McCarley and Hobson, 1975) and to tie that model to a brain-based dream theory, the activation-synthesis hypothesis (Hobson and McCarley, 1977). Figure 2 illustrates the original reciprocal interaction model and its behavioral consequences, while Figure 3 illustrates the activation-synthesis hypothesis. By integrating these two models with new data, it has recently been possible to elaborate AIM, a three-dimensional state space specifying the major physiological determinants of states of consciousness (discussed below).

FORMAL PROPERTIES OF DREAM CONSCIOUSNESS

Previous approaches to dreaming have reflected the natural tendency to regard dreams as story-like experiences conveying and/or concealing hidden or symbolic meaning. This interpretive tradition is ancient and while it remains humanistically honorable, it has always been scientifically unsatisfactory. Its most zealous modern exponent, Sigmund Freud, understood that a brain-based theory of dreaming was required but, lacking the tools to construct one, he instead created the pseudoscience of psychoanalysis.

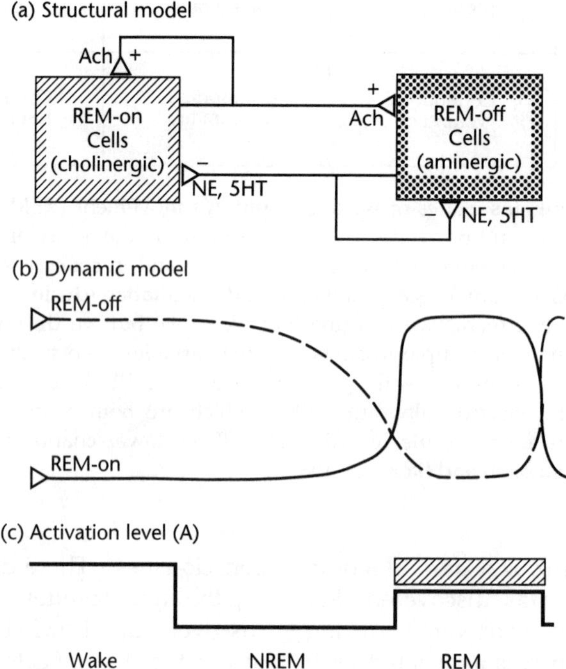

Figure 2. The original reciprocal interaction model of physiological mechanisms determining alterations in activation level. (a) Structural model of reciprocal interaction: REM-on cells of the pontine reticular formation are cholinoreceptively excited and/or cholinergically excitatory (ACH+) at their synaptic endings. Pontine REM-off cells are noradrenergically (NE) or serotonergically (5HT) inhibitory (−) at their synapses. (b) Dynamic model: during waking the pontine aminergic system is tonically activated and inhibits the pontine cholinergic system. During NREM sleep aminergic inhibition gradually wanes and cholinergic excitation reciprocally waxes. At REM sleep onset aminergic inhibition is shut off and cholinergic excitation reaches its high point. (c) Activation level: as a consequence of the interplay of the neuronal systems shown in (a) and (b), the net activation level of the brain is at equally high levels in waking and REM sleep and at about half this peak level in NREM sleep. (Taken from Hobson, 1992.)

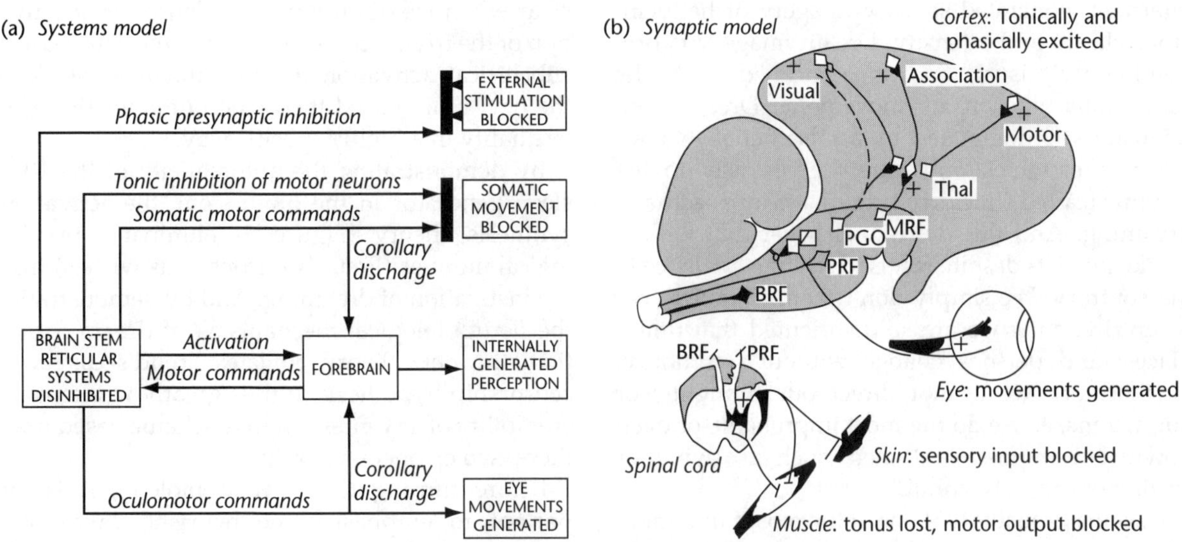

Figure 3. The activation-synthesis model. (a) Systems model: as a result of disinhibition caused by cessation of aminergic neuronal firing, brain stem reticular systems auto-activate. Their outputs have effects including depolarization of afferent terminals causing phasic presynaptic inhibition and blockade of external stimuli, especially during the bursts of REM, and postsynaptic hyperpolarization causing tonic inhibition of motor neurons that effectively counteract concomitant motor commands so that somatic movement is blocked. Only the oculomotor commands are read out as eye movements because these motor neurons are not inhibited. The forebrain, activated by the reticular formation and also aminergically disinhibited, receives efferent copy or corollary discharge information about somatic motor and oculomotor commands from which it may synthesize such internally generated perceptions as visual imagery and the sensation of movement, both of which typify dream mentation. The forebrain may, in turn, generate its own motor commands that help to perpetuate the process via positive feedback to the reticular formation. (b) Synaptic model: the midbrain reticular neurons (MRF) projecting to the thalamus convey tonic and phasic signals rostrally. PGO burst cells in the peribrachial region convey phasic activation and specific eye movement information to geniculate body and cortex (dotted line indicates uncertainty of direct projection). The pontine reticular-formation neurons (PRF) transmit phasic activation signals to oculomotor neurons (V1) and spinal cord which generate eye movements, twitches of extremities, and presynaptic inhibition. The bulbar reticular-formation neurons (BRF) send tonic hyperpolarizing signals to motor neurons in the spinal cord. As a consequence of these descending influences, sensory input and motor output are blocked at level of the spinal cord. At the level of the forebrain, visual association and motor cortex neurons all receive time and phasic activation signals for nonspecific and specific thalamic relays.

Whether dreams are stories or scenarios, and whatever their meaning might or might not be, dreaming is a state of consciousness with distinctive properties that can be defined and measured when the emphasis is placed upon the form of all dreams rather than the particular content of each one. These formal properties of dream consciousness can then be compared with those of waking. Any differences that emerge can then be mapped onto differences between REM sleep and waking state neurobiology (Hobson, 1988).

The emphasis upon the analysis of dream form and the definition and measurement of formal dream properties is every bit as important as the recording of individual nerve cells because it brings subjectivity itself into the scientific arena. In so doing it allows us to describe and quantify universal aspects of conscious experience, leaving aside for the moment individual differences which are more difficult to explain today by a brain-based theory.

The formal psychological feature which has been most extensively analyzed is the bizarreness that makes dreaming so strange, so puzzling, and by turns so amusing and so frightening. To our great surprise, we found that dream bizarreness was reducible to discontinuity and incongruity in the domains of time, place, and person, all three of which are quantitatively more unstable in dreaming than in waking consciousness. And one of them, the tendency of characters to have unstable or hybrid identities, seems to be absolutely unique to dreaming. Only in psychosis do people experience such disturbing phenomena when awake.

Dream consciousness is also characterized by internally generated sensory imagery of hallucinatory clarity and intensity. Dream imagery is predominantly visual, and is associated with the continuous illusion of movement. Dream consciousness is dominated by a false belief that we are awake, which only rarely gives way to the insight (called 'lucidity') that we are actually dreaming. And this delusional belief that we are awake persists despite robust cognitive evidence to the contrary. We simply don't seem to notice that when dreaming we are so disoriented that times, places, and persons change without our notice. Furthermore we cannot direct our thoughts, or our actions, as we do the most improbable or even downright impossible things, such as flying or cycling effortlessly uphill.

It seems probable that one of the most fundamental cognitive features of dreaming is a defect in episodic memory. When we dream we cannot remember prior events or even those that have just occurred in the dream. Furthermore, we cannot remember most of our dreams after we wake up. These perceptual and cognitive peculiarities are most often accompanied by strong negative emotions, such as anxiety, fear, and anger, but pleasantly intense elation may also be prominent. Table 1 summarizes these formal properties of dreaming.

Following Wilhelm Wundt, we agree that any dream theory thus needs to explain two classes of change in the state of consciousness from that of waking: (1) the enhancement of the perceptual and emotional components of consciousness, and (2) the impairment of its orientational, insightful, and memory components. Freud dealt with this dual aspect of dream consciousness by positing a defect in superego and ego functions and a reciprocal increase in instinctual id functions. As will be clear when we discuss new evidence for deactivation of the frontal cortex and a reciprocal increase in subcortical activation during human REM sleep, the new brain-based theory accounts for this kind of duality in a highly specific way.

By demonstrating the automaticity of the REM sleep generator in the brain stem, the activation-synthesis theory (Figure 3) eliminates psychological motives (Freud's unconscious wishes) from the instigation of dreaming. And by demonstrating the neurobiological mechanisms of dream formation, the new theory obviates Freud's disguised censorship hypothesis. It thus greatly weakens the credibility of any interpretative scheme based upon these two erroneous notions.

Before turning to the neurobiology, it is important to emphasize the heuristic value of a brain-based theory of dreams for the scientific understanding of abnormal as well as normal states of consciousness. This is because dream consciousness has so many features of major mental illness. And because it is both hallucinatory and delusional, normal dream consciousness is a psychotic state by definition. But to what abnormal psychosis is dreaming consciousness most similar? Not to schizophrenia where the hallucinations are primarily auditory and not visuomotor. And not to psychotic depression where the most prominent dysphoric emotions are sadness, guilt, and shame – not anxiety, anger, or elation. Because the hallucinations are visual, the thinking undirected, the orientation unstable, and episodic memory badly impaired, dream consciousness is most akin to those delirious states that are associated with toxic conditions which impair the brain organically. Table 2 summarizes these distinct differences between waking and dreaming consciousness.

Table 1. The formal features of REM sleep dreaming

Hallucinations – especially visual and motoric, but occasionally in any and all sensory modalities.

Bizarreness – Incongruity (imagery is strange, unusual, or impossible); discontinuity (imagery and plot can change, appear, or disappear rapidly); uncertainty (persons, places, and events often bizarrely uncertain by waking standards).

Delusion – we are consistently duped into believing that we are awake (unless we cultivate lucidity).

Self-reflection absent or greatly reduced relative to waking.

Lack of orientational stability – persons, times, and places are fused, plastic, incongruous, and discontinuous.

Narrative story lines explain and integrate all the dream elements in a confabulatory manner.

Emotions increased, intensified, and predominated by fear anxiety.

Instinctual programs (especially fight–flight) often incorporated.

Volitional control greatly attenuated.

Memory deficits across dream–wake, wake–dream and dream–dream transitions.

Table 2. Cognitive differences between wake and dream states

Function	*Dream compared to wake*
Perception (external)	Diminished
Perception (internal)	Enhanced
Attention	Lost
Memory (recent)	Diminished
Memory (remote)	Enhanced
Emotion	Episodically strong
Orientation	Unstable
Thought	Reasoning *ad hoc*, logical rigor weak, processing hyperassociative
Insight	Self-reflection greatly diminished
Narrative construction	Confabulatory
Volition	Weak

REM SLEEP NEUROBIOLOGY

We now know that the brain is in an organically altered state in REM sleep. Although it is electrically activated to a level equal to or even exceeding that of waking, the regional pattern of activation is quite different. And because the input–output gates are closed there is no discourse with the outside world. As a result there is no external space–time constancy against which to check the internally generated percepts and emotions. Finally, the percepts and emotions are engendered because of a shift in the chemical balance of the brain which also impairs our memory, our internal orientational stability, and our ability to think insightfully and critically. Three neurobiological factors affecting consciousness can thus be identified and defined as follows:

1. *Activation (A)*: activating the brain so that consciousness can be at least as intense in REM sleep dreaming as it is in waking is the responsibility of the brain stem reticular formation. This system turns off at sleep onset, allowing the thalamus and cortex to indulge in their own intrinsic rhythmic activity which are seen as the spindles and slow waves of NREM sleep when consciousness abates.
2. *Input–Output Gating (I)*: when the reticular formation turns on again in REM, the brain stem simultaneously inhibits sensory input and blocks motor output so that we do not wake up even though the brain is activated. This input–output blockade is an active process which naturally puts dream consciousness offline, as it were. The eyes move in REM because internal signals are spared the active inhibition that affects the other motor systems of the brain. They arise in the pons whence they are also directed to the upper brain where they stimulate visual and emotional centers in concert with the REMs and thus cause the visuomotor hallucinosis.
3. *Mode (M)*: the internally generated signals emanate from the lateral part of the pontine brain stem where neurons that manufacture acetylcholine are found. Acetylcholine is one of the chemicals involved in learning and memory and it is released in both waking and REM. In waking, the acetylcholine neurons are activated by external stimuli, whereas in REM they become spontaneously active. This is because they are then released from the inhibitory restraint of norepinephrine and serotonin neurons which are active and quell them in waking (see Figure 2). Norepinephrine and serotonin are two other brain chemicals that contribute to attention and memory, and their unavailability in REM contributes to the cognitive deficits in dream consciousness.

The net result of this change in neuronal traffic flow in REM sleep is an electrically activated, offline brain, whose chemical balance has been tipped in the acetylcholine direction. Because the brain is activated it is conscious of its own spontaneous information processing. But because it is offline it is cut off from external reality. Because of the shift in chemical balance, it generates its own signals and interprets them as if they came from the outside world. But the brain is not only deprived of external cues that help to structure waking consciousness. It is also deprived of two of the chemicals that it uses to organize its information in a coherent, logical, and directed manner. This is why dream consciousness is characterized by so much incongruity and discontinuity.

THE AIM MODEL

The perspicacious reader will now realize that the three neurobiological factors we have been discussing collaborate to determine our state of consciousness in the following way:

1. *The Activation level* (A) determines the *intensity* of consciousness. Factor A operates very much in the manner of a power supply. Its value which can be assessed from the EEG, is high in wake and REM but low in non-REM (NREM) sleep.
2. *The Input–Output gating level* (I) determines the source of the information that determines the *content* of consciousness. Its value can be determined by the degree of presynaptic inhibition on the sensory side and by the degree of inhibition of spinal reflex activity on the output side. When the input–output gates are open, as in waking, external information plays a major role in sleeping consciousness. When the input–output gates are closed, as they are in REM, only internal data can be processed.
3. *The Mode* (M) of the brain determines how the internally generated information is processed primarily via

its impact upon memory systems. Its value can be determined (so far only in animals) by the ratio of activity of norepinephrine- or serotonin-containing neurons to that of the acetylcholine cells. If the value of M is high, memories will be recorded, but if M is low – as it is in REM – memory will be deficient.

The range of values of A, I, and M can be laid out along the three dimensions of a cube. The resulting cubic space contains and bounds a virtual infinity of possible state points, each of which represents the instantaneous values of A, I, and M. As illustrated in Figure 4(a), the canonical states of waking, NREM sleep, and REM occupy specific subregions of the consciousness state space. With time as a fourth dimension, the normal diurnal trajectory through this space can be visualized as

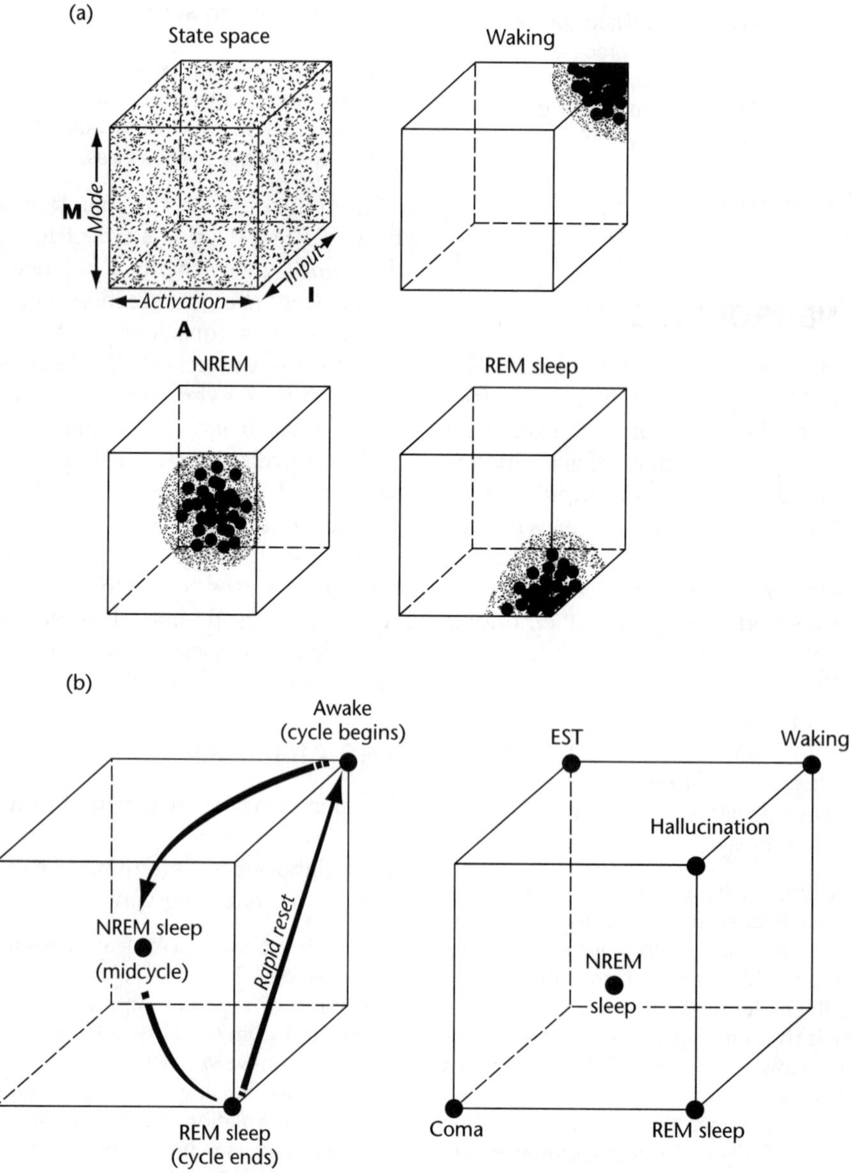

Figure 4. (a) Three-dimensional state space defined by the values for brain activation (A), input source and strength (I), and mode of processing (M). It is theoretically possible for the system to be at any point in the state space and an infinite number of state conditions is conceivable. In practice the system is normally constrained to a boomerang-like path from the back upper right in Waking (high A, I, and M), through the centre in NREM (intermediate A, I, and M) to the front lower right in REM sleep (high A, low I and M). (b) The canonical trajectory of the brain–mind is shown for each cycle of adult human sleep (left box). That this trajectory is an irregular ellipse is in keeping with the cyclical behavior of the brain stem neuronal control system for REM sleep. Besides the canonical states of wake, NREM, and REM, are "forbidden zones" which are only entered under pathological or exceptional normal conditions (right box).

the sequential points follow a large-diameter elliptical path (representing the circadian rhythm) with smaller dramatic ellipses within it (representing the NREM–REM sleep cycle).

At this early stage, the AIM model is only a heuristic, illustrative construct. But, for the student of consciousness, the three-dimensional model has several advantages over traditional two-dimensional graphic displays of brain–mind state.

The first is its ability to distinguish wake and REM which, despite their identity on the activation axis, are strongly contrasted along both the I and M axes. This contrast is congruent with the very different formal properties of consciousness in the two states.

The second advantage of AIM is its capacity to accommodate the myriad substates that constitute the conscious vicissitudes of everyday life. These include sleepiness, hyperalertness, fantasy, insomnia, and even creative imagination.

The third advantage of AIM is its recognition of those off-limit sectors of the state space into which consciousness may be pulled – or pushed – when humans suffer from spontaneous dysfunctional states such as epilepsy, schizophrenia, sleep disorders, or the comas that follow head injury. Conceptualization of such abnormal states using the three-dimensional AIM model is illustrated in Figure 4(b).

Finally, the AIM model is at home to the many intentional self-experiments on consciousness performed by humans using and abusing drugs like alcohol, amphetamine, LSD, and cocaine. Its explanatory generosity thus welcomes the vast family of recreational and therapeutic chemicals that play so powerfully – for good and evil – on the modulatory neurons of the brain.

HUMAN NEUROPSYCHOLOGY

Until recently, the experimental study of human REM sleep dreaming has been limited on the physiological side by the poor resolving power of the EEG. Even expensive and cumbersome evoked potential and computer averaging approaches have not helped to analyze and compare REM sleep physiology with that of waking in an effective way. This limitation has probably reinforced the erroneous idea that the brain activation mechanisms of REM sleep and waking are identical, or at least very similar. Fortunately, technological advances in the field of human brain imaging have now made it possible to describe a highly selective regional activation pattern of the brain in REM

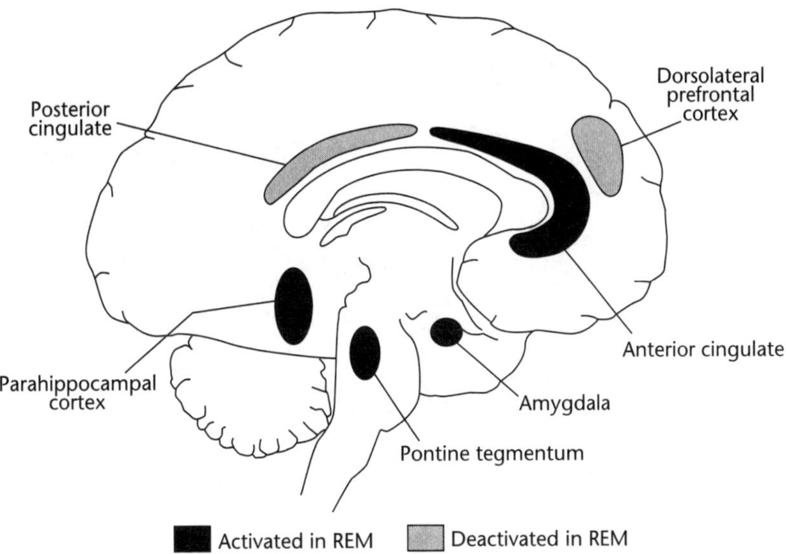

Figure 5. Convergent findings on relative regional brain activation and deactivation in REM compared to waking. A schematic sagittal view of the human brain showing those areas of relative activation and deactivation in REM sleep compared to waking and/or NREM sleep which were reported in *two or more* of the three PET studies published to date (Braun *et al.*, 1997; Maquet *et al.*, 1996; Nofzinger *et al.*, 1997). Only those areas which could be easily matched between two or more studies are schematically illustrated here, and a realistic morphology of the depicted areas is not implied. The depicted areas in this figure are thus most realistically viewed as representative portions of larger central nervous system areas subserving similar functions (e.g. limbic-related cortex, ascending activation pathways, and multimodal association cortex). (Source: Hobson *et al.*, 1998.)

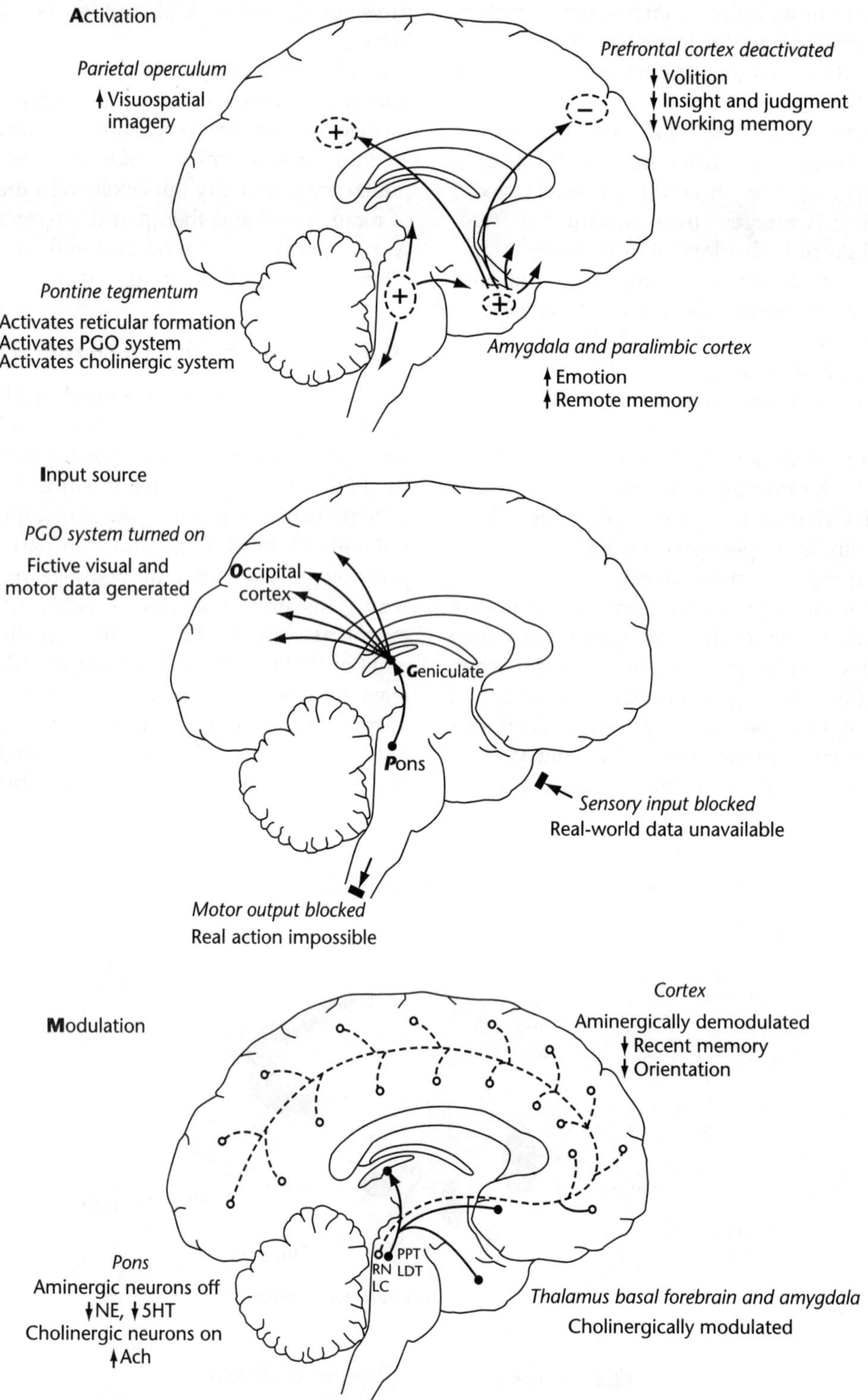

Figure 6. Physiological signs and regional brain mechanisms of REM sleep dreaming separated into the activation (A), input source (I), and modulation (M) functional components of the AIM model. Dynamic changes in A, I, and M during REM sleep dreaming are noted adjacent to each figure. Note that these are highly schematized depictions which illustrate global processes and do not attempt to comprehensively detail all the brain structures and their interactions which may be involved in REM sleep dreaming.

sleep. At the same time, experiments of nature – in the form of strokes – have allowed the locale of brain lesions to be correlated with the diminution or intensification of various aspects of dream experience in patients (Hobson *et al.*, 1998).

POSITRON EMISSION TOMOGRAPHY IMAGING STUDIES OF REM SLEEP DREAMING

Pierre Maquet and his co-workers at the University of Louvain in Belgium used an $H_2{}^{15}O$ positron source to study REM sleep activation in their subjects, who were subsequently awakened for the solicitation of dream reports (Maquet *et al.*, 1996). They observed a preferential activation of limbic and paralimbic regions of the forebrain in REM sleep (compared to either waking or to NREM sleep). One important implication of this finding is that dream emotion may be a primary shaper of dream plots rather than playing the secondary role in dream plot instigation that was previously hypothesized.

An equally interesting finding, relevant to the cognitive deficits in self-reflective awareness, orientation, and memory during dreaming, is the significant deactivation, in REM, of a vast area of dorsolateral prefrontal cortex. The fact that considerable portions of executive and association cortex are far less active in REM than in waking leads to the idea that in REM sleep there is a specific impairment of executive systems which normally participate in the highest order analysis and integration of neural information. See Figure 5 for results of this study integrated with those of several other recent neuroimaging studies.

LOSS OF DREAMING AFTER CEREBRAL LESIONS

An entirely complementary set of findings and conclusions has been reached following a neuropsychological survey of 332 clinical cases with cerebral lesions by Mark Solms at University College in London, England. The 112 patients who reported a 'global cessation of dreaming' either had damage in the limbic system or the parietal convexity, or suffered disconnections of the mediobasal frontal cortex from the brain stem and diencephalic limbic regions. With respect to the visual imagery aspect, a decrease in the 'vivacity' of dreaming was reported by two patients with damage to the visual associative area in the mediaoccipital-temporal cortex (Solms, 1997).

Taken together, these new neuroimaging and brain lesion studies strongly suggest that the forebrain activation and synthesis processes underlying dreaming are actually very different from those of waking. Not only is REM sleep chemically biased, but the enhanced cholinergic neuromodulation and diminished aminergic modulation are associated with selective activation of the subcortical and cortical limbic structures (which mediate emotion) and with relative inactivation of the frontal cortex (which mediates directed thought). These differences in regional activation obviously force the AIM model to consider still another dimension, that of brain space itself. Figure 6 is an initial mapping of AIM onto patterns of regional brain activation. The regional activation changes could be causally linked to the neuromodulatory dynamics in the following way: those areas which are inactivated in REM are those undergoing aminergic demodulation, while the activated areas are those heavily targeted by cholinergic modulatory neurons.

Whatever the link between the neuromodulatory and regional blood flow data, these findings greatly enrich and inform the integrated picture of REM sleep dreaming as emotion-driven consciousness with deficient memory, disorientation, diminished volition, and impaired analytic thinking. Now that we know that there is a close fit between the animal and human data regarding the mechanism and pattern of brain activation in REM sleep, we are in a much stronger position to strengthen the brain-based theory of dreaming that was first proposed in the early 1980s. And building upon this surprisingly strong base, we can begin to build a general theory of the brain basis of consciousness.

References

Aserinsky E and Kleitman N (1953) Regularly occurring periods of ocular motility and concomitant phenomena during sleep. *Science* **118**: 361–375.

Braun AR, Balkin TJ, Wesensten NJ *et al.* (1997) Regional cerebral blood flow throughout the sleep–wake cycle. *Brain* **120**: 1173–1197.

Dement WC (1958) The occurrence of low voltage fast electroencephalogram patterns during behavioral sleep in the cat. *Electroencephalography and Clinical Neurophysiology* **10**: 291–296.

Dement WC and Kleitman N (1957) Cyclic variations in EEG during sleep and their relation to eye movements, body motility, and dreaming. *Electroencephalography and Clinical Neurophysiology* **9**: 673.

Hobson JA (1988) *The Dreaming Brain*. New York: Basic Books.

Hobson JA (1992) A new model of brain–mind state: activation level, input source, and mode of processing

(AIM). In: Antrobus J and Bertini M (eds) *The Neuropsychology of Sleep and Dreaming*, pp. 227–247. Mahwah, NJ: Lawrence Erlbaum.
Hobson JA (1999) *Consciousness*. New York, NY: Scientific American Library. W. H. Freeman Co.
Hobson JA and McCarley RW (1977) The brain as a dream state-state generator: an activation-synthesis hypothesis of the dream process. *American Journal of Psychiatry* **134**: 1335–1348.
Hobson JA, McCarley RW and Wyzinski PW (1975) Sleep cycle oscillation: reciprocal discharge by two brainstem neuronal groups. *Science* **189**: 55–58.
Hobson JA and Steriade M (1986) Neuronal basis of behavioral state control. In: Mountcastle V and Bloom FE (eds) *Handbook of Physiology: The Nervous System*, vol. 4, pp. 701–823. Washington, DC; American Physiological Society.
Hobson JA, Stickgold R and Pace-Schott EF (1998) The neuropsychology of REM sleep dreaming. *Neuroreport* **9**: R1–R14.
Jouvet M (1962) Récherches sur les structures nerveuses et les mécanismes résponsables des différentes phases du sommeil physiologique. *Archives Italiennes de Biologie* **100**: 125–206.
Maquet P, Peters JM, Aerts J *et al.* (1996) Functional neuroanatomy of human rapid-eye-movement sleep and dreaming. *Nature* **383**: 163.
McCarley RW and Hobson JA (1975) Neuronal excitability modulation over the sleep cycle: a structural and mathematical model. *Science* **189**: 58–60.
Moruzzi G and Mogoun HW (1949) Brainstem reticular formation and activation of the EEG. *Electroencephalography and Clinical Neurophysiology* **1**: 455–473.
Nofzinger EA, Mintun MA, Wiseman MB, Kupfer DJ and Moore RY (1997) Forebrain activation in REM sleep: an FDG PET study. *Brain Research* **770**: 192–201.
Solms M (1997) *The Neuropsychology of Dreams: A Clinico-Anatomical Study*. Mahwah, NJ: Lawrence Erlbaum Associates.

Further Reading

Crick F (1995) *The Astonishing Hypothesis*. New York, NY: Touchstone Books.
Damasio A (1995) *Descartes' Error*. New York, NY: Avon Books.
Damasio A (1999) *The Feeling of What Happens*. New York, NY: Harcourt Brace.
Dennett D (1991) *Consciousness Explained*. Boston, MA: Little, Brown.
Hobson JA (1999) *Dreaming as Delirium*. Cambridge, MA: MIT Press.
Hobson JA, Pace-Schott E and Stickgold R (2000) Dreaming and the brain: toward a cognitive neuroscience of conscious states. *Behavioral and Brain Sciences* **23**.

Consciousness, Stream of

Intermediate article

Thomas Natsoulas, University of California, Davis, California, USA

CONTENTS

Except for time-gaps, when the stream comes to a stop and then starts again, a stream of consciousness is constituted, from the Jamesian perspective, of tightly adjacent states of consciousness occurring very briefly one at a time. Unless a second such stream flows in the individual simultaneously, his or her conscious mental life at any instant consists entirely of one integral state of consciousness that typically possesses many intrinsic features.

INTRODUCTION

We commonly speak of someone's being conscious. Also, we often say he or she is conscious *of* something, and we usually specify to some degree the object (or objects) of that consciousness. This object can be an environmental state of affairs or occurrence, the individual himself or herself, or some part of the latter, such as a temporal section of the individual's mental life. Moreover, what one is conscious of can be *merely apparent*: an item, occurrence, or state of affairs having no actual existence, whether present, past, or potentially in the future.

Even if one disbelieves in fire-breathing dragons, it is appropriate to assert, depending on the facts, that someone is undergoing a hallucination and is therein conscious of a fire-breathing dragon. The hallucinatory experiences themselves are certainly real; they are no less real than anything else in the universe. In the example, it is the *object* of the hallucinatory experience that happens to be something unreal, that possesses no kind of existence. Also, hallucinatory experiences may have real objects, such as a long-lost relative. But, of course, the hallucinated presence of real objects in the immediate environment is, with possible exceptions, illusory.

In addition to describing someone as conscious or conscious of something, we may describe certain, or even all, of the mental happenings in an individual as being of the *conscious* kind. Indeed, one sometimes encounters among cognitive scientists the thesis that *all* experiences are conscious in this latter sense. According to this thesis, controversial within cognitive science, to have or undergo any experience is, *ipso facto*, for one to be conscious of it, that is, for it to be an object of 'inner awareness' (Brentano, 1973).

Many cognitive scientists maintain that, in every healthy, intact human being who is functioning normally, there take place both conscious and unconscious mental occurrences. The unconscious mental occurrences are those that cannot be objects of inner awareness; that is, we cannot have first-hand, noninferential apprehension of any instance that transpires of any unconscious mental occurrence. In the case of the conscious mental occurrences, in contrast, all or some of the instances of any one of them are directly apprehended when they occur.

There is scientific disagreement concerning how inner awareness is accomplished. Some cognitive scientists propose that an 'appendage' is required to any mental-occurrence instance that is the object of inner awareness. Thus, inner awareness takes the form of a mental-occurrence instance distinct from the mental-occurrence instance it renders conscious (Natsoulas, 1993b). Other cognitive scientists hold that inner awareness is an *intrinsic* feature of any mental-occurrence instance that is apprehended first-hand. Accordingly, inner awareness is always a dimension of the phenomenological structure of the mental-occurrence instance that is its object (Natsoulas, 1996).

Cognitive scientists are rarely skeptical concerning the reality of conscious mental occurrences as existents in the natural world. They recognize that the pursuit of science itself, among much else in human life and society, *requires* consciousness and necessarily involves the scientist's undergoing mental-occurrence instances with inner awareness. Anyone who may be inclined to cast doubt upon conscious mental occurrences should recall the impossibility of 'mind-blind' science.

Imagine a physical scientist fully capable of perceptual awareness of those molar events in the environment of interest to his or her science. Now add what is less imaginable: all of the scientist's mental occurrences comprising this perceptual awareness are unconscious. That is, whatever the cause of this condition (of mind-blindness), these mental occurrences cannot be objects of the scientist's inner awareness. A mind-blind scientist could not function as such. It would be, for him or her, as though the environmental events of scientific interest that are objects of his or her perceptual awareness *do not occur*.

Some cognitive scientists attempt to treat of how conscious mental occurrences seem first-hand, to inner awareness, as *illusory* in some respects, although very rarely in respect to their existence. Briefly expressed, here are examples of two such attempts, the second less cogent than the first. (a) Whereas conscious mental occurrences may seem to the person to whom they occur to be mental in the sense of spiritual or nonphysical, they are, in reality, *only occurrences in the brain that,* like the many other kinds of brain occurrences, *possess only physical properties* (Sperry, 1976). (b) Whereas in the large majority of cases, mental-occurrence instances seem to inner awareness to be other than actions or behaviors, they actually are – this includes all of our perceptual experiences and feelings – *forms of self- or other-directed commentary*, either overt, covert, or incipient in any instance (Weiskrantz, 1997). However, cognitive scientists who diverge from the ordinary concept of a conscious mental occurrence do not generally intend thereby to commit a *referential displacement*: that is, to speak of something else, in place of mental occurrences, when they use terms commonly referring to the latter.

Cognitive scientists seldom deny the reality of consciousness, but some argue at length that *unconscious* mental occurrences do not transpire within us at all (e.g. O'Brien and Opie, 1997). In their view, we have brain occurrences that are mental, although not in the spiritual sense; but none of these mental brain occurrences are unconscious in the sense specified above, namely, incapable of being themselves objects of inner awareness. All mental occurrences are *open* to inner awareness, although this does not mean they are actually apprehended on every occasion of their occurrence (Natsoulas, 1995).

The above thesis entails a rejection of Freud's unconscious. According to Freud, an unconscious mental occurrence may 'become' conscious, but conscious only in the sense of evoking a counterpart of itself within a distinct subsystem (called 'perception–consciousness') of the mental apparatus (Natsoulas, 1993a). This counterpart must be of suitable type and instantiate much the same cognitive content as the unconscious mental occurrence. All mental occurrences that take place in the perception–consciousness system are of the conscious variety and, in all instances, are objects of inner awareness. Those mental occurrences that are not conscious transpire in a different subsystem of the mental apparatus in the brain.

The thesis of the nonexistence of unconscious mental occurrences entails, as well, rejection of a large body of contemporary scientific thought. Many hypotheses and theories circulating now inside cognitive science would explain particular instances of behaviors or of conscious mental occurrences by reference to mental happenings to which the person to whom they belong cannot have anything more than inferential access. This is the only kind of access cognitive scientists have to the person's mental life – with the probable exception that, today, some cognitive scientists are applying certain equipment directly to the brain and may be thereby observing mental occurrences by instrument.

The thesis that one's stream of consciousness is the entirety of one's conscious mental life will require modification if people, either normally or under certain conditions, are found to possess more than a single such stream (Puccetti, 1981). Note that the dual-consciousness hypothesis is not equivalent to countenancing the reality of unconscious mental occurrences. One's second conscious stream too would consist of conscious mental occurrences. Admittedly, however, neither of the two streams could have inner awareness of what is transpiring in the other stream and would depend for any information about the latter on inference. For example, the mental occurrences belonging to either stream might ascribe behavior issuing from the one body as causally connected in some instances to a separate consciousness also proceeding in the same body.

HISTORY

Cognitive scientists consider William James's classic contributions to be the most important to date in the history of the concept of the stream of consciousness. If one asks a cognitive scientist about the prevailing concept of the stream, the reply will assuredly contain some mention of James's detailed characterizations. These have a prominent place in *The Principles of Psychology*, in James's famous two-volume masterwork of basic psychology. In many psychologists' opinion, the ninth chapter, 'The Stream of Thought', is an achievement of great intellectual value and compares favourably to the best analyses of any topic in the scientific psychological literature.

The Principles was published over a century ago and was intended as an introductory textbook for undergraduates. Yet James's phenomenological account of the stream is anything but dated. Indeed, this account would not be out of place if it were included today under 'current work' in a compendium of materials required by anyone seriously interested in consciousness. That James's account of the stream of consciousness is still a central part of our developing scientific thought may be surprising. Its persisting relevance is owed not to cognitive scientists' extending James's empirical observations to any substantial degree. The account itself in its original form still serves to expand the understanding of much about consciousness that lies beyond the merely historical.

This will seem less surprising upon noticing that, soon after James, those who came to control the field of psychology, which James helped found, *refused to acknowledge* the importance of consciousness in psychological functioning. Their refusal to face facts was anti-empirical; their scientific behavior was inconsistent with a true vocation. These academics were motivated by a strong political desire to achieve a broad acceptance for the new psychology as a science among other natural sciences.

To include consciousness as a part of psychology's subject matter would have been, for them, to admit the causal function of something spiritual and to enable thereby religious belief to contaminate their field. The exclusion of consciousness was a joint, systematic, long-term project among people who saw themselves as 'running' the science. And, even with their eventual retirement, their past efforts continued to produce detrimental effects on the science for the remainder of the twentieth century. Beginning in the 1960s, however, it gradually became less difficult to publish scientific articles and books on consciousness and less objectionable to teach the forbidden topic.

The militancy of some psychologists was such that they even denied publicly the existence of consciousness. Others among them, although not expressing such doubts, demanded that all technical concepts be defined in terms only of what could be perceptually observed of the behavior of their experimental or research subjects and the causes of that behavior. Thus, all mental occurrences were to be excluded from playing any role in scientific explanation because they were not in themselves publicly observable. A number of the many psychologists who shared this discredited philosophy of science occupied powerful academic positions and managed to hold back for decades the kinds of research they did not approve of. An indication of their project's success is that, as a whole, James's account of the stream of consciousness has yet to be superseded by a more enlightening account.

THEORETICAL ANALYSES

In describing the stream of consciousness as James conceived of it, one begins traditionally with attention to his metaphor of flowing water. However, a close reading of *The Principles* shows that James's analogy pertains only to how it may well *seem* that one's conscious mental life proceeds. In James's phrase, this life 'does not appear to itself chopped up in bits'. Contrary to some of James's own statements, he did not consistently conceive of the stream of consciousness as a continuous, undivided, internally undemarcated process, albeit subject to stopping and starting. James's more consistent position was, rather, that the stream consists objectively of a succession of *discrete* instances, pulses, or states of consciousness (Natsoulas, 1992–1993).

States of consciousness take place one at a time, and each of them immediately succeeds the one just before it except if there is a 'time-gap' during which the stream goes out of existence sometimes only very briefly. According to James, time-gaps may actually be more numerous than commonly supposed, for they are only inferable, not directly noticeable in his view. However, a temporal section or segment of the conscious mental life that internally involves no time-gaps consists of a succession of pulses of mentality 'with absolutely nothing between'. A state of consciousness lasts very briefly and, thereupon, one's consciousness consists of another such state, and so on. At any moment, one of these states is the entirety of one's consciousness – setting aside the possibility of more than a single stream per person at the same time.

How do the states of consciousness, or basic durational components of the stream, come into existence one after another? Although, according to James, these mental pulses are nonphysical, they are nevertheless the immediate, automatic products of *the total, ongoing brain process*. This said, a major qualification is immediately in order. At different times, the occurrent parts of this completely physical process, which is proceeding in many brain structures, differ in their intensities and, consequently, are variably determinative of intrinsic features of the pulses that the total process produces.

The stream of consciousness is an accretion of basic durational components, rather than its being an ongoing process of its own whose course is merely influenced by the brain process. In contrast, the latter description is indeed applicable to how pulses of mentality are supposed to affect the total brain process. James's mind–body position expressed in *The Principles* was a dualist interactionism, but the existence of consciousness, of each one of those mental pulses constituting it, was held to depend on the brain. Mind issues from the body rather than the body's having effects on mind. Merely by its occurrence, the total brain process produces as a by-product something new at every point, whereas the pulses of mentality have their direct effects on the brain process as it is already taking place. They can only hinder or further this ongoing process.

Calling the stream's basic durational components 'pulses' of mentality, as James does, may convey the false impression that each of the pulses is proposed to be simple. This is belied by the fact that, soon enough after *The Principles*, James (1899) spoke of the stream of consciousness as consisting of a succession of 'fields' and insisted that these fields are always complex. He stated specifically that most of our concrete states of consciousness contain individually, in different proportions but in some positive degree, all of the following ingredients: sensations from the body, sensations owed to the impact of energies upon sense receptors that are outwardly directed, remembrances of past experiences, thoughts about faraway things, feelings of satisfaction and dissatisfaction, acts of will, desires and aversions, and other emotional conditions. Some of these ingredients are instantiated by a particular state more than by others and seem first-hand to have greater prominence in the state. Thus, although possessing many actual ingredients, one state of consciousness will seem to consist largely of sensation, another largely of remembrance, and so on. James pointed out that, for practical reasons, we tend to classify certain states of consciousness together, calling them states of emotion, sensation, abstract thought, volition, and the like, but a state of consciousness is not equivalent to any of the ingredients mentioned. Rather, these are to be understood as among the intrinsic features of a state, not as parts of it.

Speaking of an image and its 'surrounding' or 'penumbral' content, which is the source of the image's meaning, as though they were two distinct parts belonging to a particular state of consciousness, James in *The Principles* qualified his spatial metaphor in midstream: he described the surround as 'bone of the image's bone and flesh of its flesh'. If the penumbral contents of a state of consciousness had been, in any instance, different, any object of the state would have been taken and understood otherwise. This is not because the penumbral contents are among the causal determinants of a state, but because they are ingredients of the whole unitary, integral state and dimensions of how, specifically, the state apprehends its object or objects.

The ingredients of a single state, pulse, or field of consciousness listed above are not mutually distinct instances of being conscious of something. Such ingredients of a state as sensations, memories, thoughts, feelings, desires, aversions, emotions, and conations are not as they are traditionally considered: separate mental acts. Each state of consciousness is a unitary instance of consciousness. Any object of such a state that may be considered focal is apprehended in relation to all the other objects of the state. The above ingredients are abstractions from the concrete state that contains them. They are features of how the state apprehends its multiple objects together. This makes the ingredients of states no less real, but it does imply that they have no existence outside the states of which they are intrinsic features. For example, no auditory experience exists that is not a feature of one or more states of consciousness.

James found support for his conception of the states of consciousness in an unlikely source: Wilhelm Wundt, the founder of experimental psychology. Wundt assessed his three decades of introspective laboratory work as follows: 'From my inquiry into time-relations, etc. ... I attained an insight into the close union of all those psychic functions usually separated by artificial abstractions and names, such as ideation, feeling, will; and I saw the indivisibility and inner homogeneity, in all its phases, of the mental life' (translated and quoted by James, 1899, p. 21). James interpreted the passage in which this sentence appears as a total renunciation of the prevailing conception of the mental life as being made up of 'distinct processes and compartments'.

But James did not also characterize in such terms the neurophysiological source of the stream of consciousness. Although states of consciousness are integral and not compounded of smaller units or acts, the total brain process, which is responsible for the existence of states of consciousness, clearly consists of numerous distinct processes. Indeed, different sets of brain processes were held by James to be the proximate causes of different states of consciousness. The part-processes that constitute the total brain process of the moment somehow combine together to produce a unitary mental state.

EXPERIMENTAL APPROACHES

Introspection was the empirical ground for James's fundamental conception of consciousness. This conception included: (a) that the whole of one's mental life consists of a stream of consciousness, except insofar as a second stream of consciousness also flows within one; (b) that any stream of consciousness is constituted entirely of a sequence of tightly adjacent states, or pulses, of consciousness, except that the stream does stop and start again, and perhaps very frequently; (c) that each state of consciousness is of an integral character, a unitary awareness albeit often with many objects; and (d) that each such state typically instantiates a complexity of intrinsic features, some of which are traditionally conceived of, wrongly, as distinct mental acts.

The activity of introspecting can be carried out more or less adequately. James sometimes argued against views of consciousness with which he disagreed that these views have their basis in careless or biased introspection. But, in a chapter of *The Principles* devoted to methodology, James stressed that psychologists *always have to rely* on introspection, and that the most basic postulate of psychological science is that people do have inner awareness wherein they distinguish a state of consciousness from what it is about.

When James considered the matter more deeply, he became skeptical concerning our having the ability to apprehend our states of consciousness in a first-hand way. However, this skepticism did not deter him from proceeding on the assumption that the ability to introspect is among our most valuable powers. Indeed, much of what James is known for in psychology involves phenomenological descriptions made possible by inner awareness of his states of consciousness. James came to his skeptical position – which he quickly set aside for evidently practical reasons – as a consequence of introspecting and discovering thereby that inner awareness fails to reveal any spiritual activity at all going on within him.

Before moving on, James contemplated that, in point of fact, we may not have a stream of consciousness as he had been describing it. Better to say, perhaps, that the mental life that we do possess is a stream of 'sciousness' since its components are never objects of inner awareness (cf. Hebb, 1982). Our states of sciousness always have something else as their objects, that is, parts of the environment and body, except insofar as we infer the presence within us of those states from other, observable matters. Thus, states of sciousness may be among the non-immediate objects of some of our states of sciousness.

Although there are inconsistencies in James's reasoning, the only objection to his skeptical view requiring mention here pertains to his implicit notion that one can draw inferences – from perceptual observations to the occurrence of states of sciousness within one – in the complete absence of inner awareness. On James's skeptical view, perceptual awareness transpires not in a stream of consciousness but in a stream of sciousness every one of whose basic durational components are unconscious. The entirety of our mental life is proposed to take place in the 'dark' as we, at most, guess or infer about it. But how can one do any inferring based on the occurrence of a perceptual awareness of which one cannot be aware first-hand? If it is replied that we can know of the perceptual awareness by inference from something we objectively observe, the question is, again, how we can so infer if we cannot have inner awareness of the latter observation either.

More consistently with the skeptical position, which James seems to have preferred, he might have adopted a different methodology than the introspective one that he continued to practice and on whose results he relied for the rest of *The Principles*. This nonintrospective methodology would have him observing his behaviors and their objective context and making inferences about his mental life from what he observed. This is the kind of procedure on which present-day cognitive scientists often rely in studying the mental life of their research subjects, along with putting the latest instruments to use to detect properties of those of their subjects' brain processes that the scientists believe are the mental occurrences of current interest or closely associated with those mental occurrences. Of course, introspection helps them to formulate hypotheses to test experimentally, but they are very restrained in justifying their claims by reference to what they know first-hand by inner awareness.

However, even relying on perceptual observations exclusively – observations of behavior, of its objective context, and (by instrument) of the brain processes that are transpiring at the time – would not mean the cognitive scientist has succeeded in bypassing inner awareness so that inner awareness forms no part of his or her methodology. The mental processes of a scientist are an essential dimension of any objective methodology; surely, this is an unobjectionable statement. The cognitive scientist cannot proceed to study the states of consciousness of his or her research subjects (or to study anything else for that matter) unless the scientist's mental processes continue to include more than just unconscious mental occurrences. At least some of the scientist's mental states during the experimental researches must be objects of inner awareness; otherwise, the objective observations would be for naught. Although these observations may have unconscious effects upon the researcher, in the absence of all inner awareness of the observations, they cannot be put to use.

In their investigation of states of consciousness, cognitive scientists cannot limit their empirical database to their own introspections. They must seek to determine the properties of states of consciousness more generally, how these states are in other people. For this purpose, cognitive scientists carry out not only programs of objective observation, with some reference to how they find their own states of consciousness to be first-hand. Also, they secure self-reports from experimental subjects concerning their inner awareness of their own respective conscious states, notwithstanding the fact that the causation responsible for self-reports concerning consciousness is not well understood at the present time. Many cognitive scientists remain unwilling to put much evidential weight on such reports, not until cognitive science advances to the point where we know the conditions under which subjects' reports are to be trusted. However, it is important to realize that inference from objective observations to states of consciousness is not any less problematic. The validity of such inference depends on the understanding that we have of the causal relations between these observations and the conscious states that are among the causes of what we are observing about our research subjects.

References

Brentano F (1973) *Psychology from an Empirical Standpoint*. London: Routledge and Kegan Paul. [Original German corresponding edition published in 1911.]

Hebb DO (1982) Elaborations of Hebb's cell assembly theory. In: Orbach J (ed.) *Neuropsychology after Lashley*, pp. 483–496. Hillsdale, NJ: Lawrence Erlbaum.

James W (1899) *Talks to Teachers on Psychology: And to Students on Some of Life's Ideals*. New York, NY: Holt.

Natsoulas T (1992–1993) The stream of consciousness: I. William James's pulses. *Imagination, Cognition and Personality* **12**: 3–21.

Natsoulas T (1993a) Freud and consciousness: VII. Dimensions of an alternative interpretation. *Psychoanalysis and Contemporary Thought* **16**: 67–101.

Natsoulas T (1993b) What is wrong with appendage theory of consciousness. *Philosophical Psychology* **6**: 137–154.

Natsoulas T (1995) A rediscovery of consciousness. *Consciousness and Cognition* **4**: 223–245.

Natsoulas T (1996) The case for intrinsic theory: I. Introduction. *Journal of Mind and Behavior* **17**: 267–286.

O'Brien G and Opie J (1997) Cognitive science and phenomenal consciousness: a dilemma and how to avoid it. *Philosophical Psychology* **10**: 269–286.

Puccetti R (1981) The case for mental duality: evidence from split-brain data and other considerations. *Behavioral and Brain Sciences* **4**: 93–123.

Sperry RW (1976) Mental phenomena as causal determinants of brain function. In: Globus GG, Maxwell G and Savodnik I (eds) *Consciousness and the Brain*, pp. 163–177. New York, NY: Plenum.

Weiskrantz L (1997) *Consciousness Lost and Found: A Neurophysiological Exploration*. Oxford, UK: Oxford University Press.

Further Reading

Armstrong DM and Malcolm N (1984) *Consciousness and Causality: A Debate on the Nature of Mind*. Oxford, UK: Blackwell.

Dulany DE (1997) Consciousness in the explicit (deliberative) and implicit (evocative). In: Cohen JD and Schooler JW (eds) *Scientific Approaches to Consciousness*, pp. 179–212. Mahwah, NJ: Lawrence Erlbaum.

Freud S (1961) The Ego and the Id. In: *Standard Edition*, vol. 19, pp. 12–66. London: Hogarth. [Original work published 1923.]

James W (1950) *The Principles of Psychology*, 2 vols. New York, NY: Dover. [Originally published in 1890.]

Natsoulas T (1977) Consciousness: consideration of an inferential hypothesis. *Journal for the Theory of Social Behaviour* **7**: 29–39.

Natsoulas T (1981) Basic problems of consciousness. *Journal of Personality and Social Psychology* **41**: 132–178.

Natsoulas T (1984) Gustav Bergmann's psychophysiological parallelism. *Behaviorism* **12**: 41–69.

Natsoulas T (1997) Blindsight and consciousness. *American Journal of Psychology* **110**: 1–34.

Natsoulas T (1998) On the intrinsic nature of states of consciousness: James's ubiquitous feeling aspect. *Review of General Psychology* **2**: 123–152.

Searle JR (1992) *The Rediscovery of the Mind*. Cambridge, MA: MIT Press.

Woodruff Smith D (1989) *The Circle of Acquaintance: Perception, Consciousness, and Empathy*. Dordrecht, Netherlands: Kluwer.

Consciousness, Unity of

Intermediate article

Tim Bayne, Macquarie University, Sydney, Australia

CONTENTS

Consciousness is unified in various ways, but the nature and basis of its unity remains a matter of intense debate. This debate promises to illuminate the nature of experience, the self, and various pathologies of consciousness.

INTRODUCTION

At any particular point in time you might be enjoying a number of experiences. For instance, you might have the visual experience of seeing these words, the cognitive experience of understanding what these words mean, and bodily sensations of various kinds, such as the feeling of the ground beneath one's feet. These experiences are unified in various ways. In exactly what ways? How can these unity relations be explained and how are they related to each other? In what sense must experiences be unified? What implications might the unity of consciousness have for our understanding of the nature of consciousness and the nature of the self? These are some of the central questions raised by the unity of consciousness.

VARIETIES OF UNITY WITHIN CONSCIOUSNESS

There is no commonly accepted taxonomy for discussions of the unity of consciousness. The following taxonomy is just one path through what is a rather unstructured debate, and the reader is advised that some of the terms introduced below are used in very different ways by other authors.

Subject Unity

Experiences are *subject-unified* when they are had by the same subject of experience. Your experiences are subject-unified in so far as they are yours, while my experiences are subject-unified in so far as they are mine. It is important to distinguish subject unity from the unity of self-consciousness. Subject-unified experiences are had by a single subject of experience, but the subject in question need not be conscious of these experiences as his or her own; many creatures appear to be subjects of experience without being aware of themselves as subjects of experience. Although most mature human beings can – and often do – ascribe their experiences to themselves, this ability seems to be lost, or at least compromised, in certain pathologies of consciousness, such as depersonalization.

Object Unity

A second unity relation in consciousness is *object unity*. Experiences are object-unified when they are about the same object. When I see my dog barking at my neighbor's cat I unify my visual experience of my dog with my auditory experience of him, thus forming a multimodal percept of a single object. Object unity not only extends across perceptual modalities; it also binds perception and action together, as when one when reaches for something that one can see.

Spatial Unity

The intentional content of experience gives rise to a third unity relation, *spatial unity*. When I see my dog chasing the neighbor's cat I experience the dog and the cat as distinct but spatially related objects. Objects of perception are located in egocentric space – 'egocentric' because the structure of this space is given by the structure of one's own body. The experience of one's own body as a unitary physical object enables one to experience one's perceptual environment as a spatially unified domain.

Epistemic Unity

Sometimes a stream of consciousness is said to be unified to the degree that its contents are coherent, integrated, or comprehensive (Shoemaker, 1996). We might call this form of unity *epistemic unity*. Shoemaker has epistemic unity in mind when he writes that 'perfect unity of consciousness ... would consist of a unified representation of the world accompanied by a unified representation of that representation, the latter including not only information about what the former represents, but also information about the grounds on which the beliefs that make up the former are based, and about what the evidential relations between the parts of that representation are' (1996, p. 186). In a related use of the term, the 'unity of consciousness' is also used to refer to the consistency of consciousness. Although one can see the famous duck–rabbit as either a duck or a rabbit, one cannot simultaneously experience it as both – although priming experiments reveal that both interpretations can be simultaneously active (Baars, 1988).

Phenomenal Unity

There is a unity of consciousness that is arguably more primitive than any of the unity relations considered thus far. Consider again the set of experiences that you are currently enjoying: perceptual experiences, cognitive experiences, emotional experiences, and so on. All of these experiences seem to be contained within a single phenomenal field or stream of consciousness. They seem to be conscious together; they seem to be *phenomenally unified*. It has become common to talk about phenomenal unity in terms of the relation of *co-consciousness*. Experiences are co-conscious when they are experienced together: when they have a conjoint phenomenology. My current visual experiences are co-conscious with my current auditory experiences, but they are not co-conscious with your auditory experiences. (Note that psychologists often describe distinct streams of consciousness that concurrently belong to a single organism as 'co-conscious': a very different use of the term.)

Access Unity

A final type of unity of consciousness concerns the relation between consciousness and reportability or accessability. We can describe experiences that are reportable in exactly the same ways as being *access-unified*. My current auditory and visual experiences seem to be access-unified in that I am able to report my auditory experiences in any way in which I am able to report my visual experiences. Note that experiences can be access-unified without being co-reportable; that is, I may be able to verbally report either my auditory experience or my visual experience without being able to report both experiences together.

How are these various forms of unity related? In particular, how are subject unity, phenomenal unity, and access unity related? It is plausible to suppose that experiences can be access-unified or phenomenally unified only if they are had by the same subject of experience. The controversial questions are whether (simultaneous) co-subjective experiences must be phenomenally unified, and whether phenomenally unified experiences must be access-unified. The study of various pathologies of consciousness might help to answer these questions.

UNITY OF CONSCIOUSNESS IN PHILOSOPHY

History

Philosophical reflection on the unity of consciousness dates back at least as far as Aristotle, who argued that there must be a common sense in addition to the specific senses such as sight and hearing. This common sense was thought to be responsible for perceiving the common sensibles – properties such as motion, rest and number – that are perceivable by multiple sense modalities. It was also assigned the role of integrating the contents of the other senses.

In the seventeenth century, philosophical interest in the unity of consciousness shifted focus from what the unity of consciousness might tell us about the structure of the mind to what it might tell us about the ultimate nature of the mind. G. W. Leibniz argued that the unity of consciousness provides support for substance dualism, the view that the subject of experience is an immaterial substance. He argued that if the self is a material entity it must be spatially extended, and hence the different parts of an experience would be had by different subjects. Consequently, there would be no single subject that had the entire experience, as there obviously seems to be. (A similar argument for substance dualism appears in René Descartes' *Meditations*.) Contemporary functionalists respond to this argument by insisting that the subject of an experience is the entire system within which the experience occurs: although a functional system has parts, none of these parts is a subject of experience in its own right.

Another influential seventeenth-century discussion of the unity of consciousness can be found in John Locke's analysis of personal identity. Locke claimed that the identity of a person extends only so far as their consciousness extends. Although Locke draws an intimate connection between personal identity and the unity (and continuity) of consciousness, he says very little about what the unity of consciousness involves. Most work on Locke's accounts of personal identity has focused on the connection between personal identity through time and the continuity of consciousness; it is only recently that philosophers have begun to explore the relationship between the identity of a person at a time and the unity of consciousness.

Although the notion played little role in his treatment of the unity of consciousness, Locke held that experiences are had by a substantial self. David Hume, in contrast, was positively hostile to this idea. He claimed that since introspection reveals no sign of the self there is no good reason to posit such an entity. Hume claimed that the self is not something distinct from its experiences, something that has experiences, but is, in fact, nothing but a bundle of experiences. In response to Hume, Kant argued that there must be a transcendental ego that is responsible for synthesizing various representations into a unified consciousness. Kant's discussion of the unity of consciousness is notoriously obscure, and there is disagreement both over what Kant meant by the 'unity of consciousness' and over what he explained the unity (or unities) of consciousness in terms of (Brook, 1994; Hurley, 1998). Kant is variously said to have identified the unity of consciousness with, or explained it in terms of: the unity of concepts, the unity of agency, the unity of self-consciousness, and the unity of an objective world. (*See* **Split Brains, Philosophical Issues about**)

Current Issues

Current philosophical interest in the unity of consciousness is particularly concerned with two major questions.

Is consciousness unified?

The claim that consciousness is unified can be taken in a number of different ways. Some take the claim to mean that there is a single anatomical module or site for consciousness. Others take it to mean that co-conscious experiences are always intentionally integrated; that is, epistemically unified. It is doubtful that consciousness must be unified in either of these ways.

A more plausible sense in which consciousness might be unified is that co-conscious experiences must also be access-unified. We can call this the *accessibility thesis*. The accessibility thesis is attractive; indeed, it seems to be presupposed by certain accounts of consciousness (Baars, 1988). But there is reason to think that it might be false. Certain experimental results and pathologies of consciousness (discussed later in this article) suggest that experiences can be phenomenally unified without being accessible to the same report modalities.

Another conception of the unity of consciousness concerns the relationship between subject unity and phenomenal unity. According to the *unity thesis*, (simultaneous) co-subjective states must be co-conscious; that is, any experiences that a subject has at a time must be contained within a single fully unified stream of consciousness. The plausibility of the unity thesis depends in part on one's account of the subject of experience. The unity thesis seems implausible if, as many philosophers argue, subjects of experience should be individuated in biological terms, for it seems quite possible that a single organism might have two experiences at the same time without those experiences being co-conscious. On the other hand, one could also take the plausibility of the unity thesis as an argument against equating the self with a biological organism.

Finally, we can take the claim that consciousness is unified as a claim about the structure of co-consciousness. It is natural to suppose that the relation 'is co-conscious and simultaneous with' is reflexive, symmetrical, and transitive. (A relation R is transitive if, whenever aRb and bRc, then aRc.) Developing an idea that was inchoate in Nagel (1971), Lockwood (1989) suggested that synchronic co-consciousness might not be transitive, and that as a result some streams of consciousness may be only partially unified. If synchronic co-consciousness is not transitive then it would be possible to have three simultaneous experiences, e_1, e_2, and e_3, such that e_1 and e_2 are co-conscious, and e_2 and e_3 are co-conscious, but e_1 and e_3 are not co-conscious. We can call the claim that synchronic co-consciousness is transitive the *transitivity thesis*.

Philosophical debate over the transitivity thesis has focused on the question of whether partial unity might be possible. Most philosophers admit that it is difficult to conceive of what it would be like to have a partially unified consciousness, but they differ over what this might show. Dainton (2000) defends the transitivity thesis on the basis of the inconceivability of partial unity, while Hurley (2002) argues that this line of argument is mistaken. She claims that the 'what it's like' approach is relevant only when it comes to determining the content-based relations between experiences, and since the transitivity thesis concerns the token identity of states neither it nor its denial could be supported by appeal to phenomenological considerations.

Philosophers have also been interested in whether various pathologies of consciousness, such as those exhibited by split-brain patients, might yield counterexamples to the unity or transitivity theses (see below).

How can phenomenal unity be explained?

There are many accounts of phenomenal unity (or co-consciousness), but no standard way of categorizing them. One useful distinction is between theories that account for phenomenal unity in terms of factors internal to phenomenology (subjective theories) and theories that appeal to factors outside phenomenology (objective theories).

Some subjective theories appeal to introspection in order to explain phenomenal unity. Such appeals can take at least two forms, depending on the account of introspection offered. Introspection can be conceived of as a nonrepresentational act of awareness that unifies various experiences, or it can be thought of as an experience in its own right, a higher-order experience of first-order experiences.

An alternative subjectivist approach is to posit a primitive unity relation for consciousness. Dainton (2000) takes this approach, holding that co-consciousness itself is a primitive unity relation. Bayne and Chalmers (2002) adopt a similar approach, although they take subsumption as their primitive unity relation. (One experience subsumes another when, roughly, the subsuming experience entails the subsumed experience). Dainton's approach is 'bottom-up' – fully unified streams of consciousness are constructed out of particular experiences and relations of mutual co-consciousness – while Bayne and Chalmers take a 'top-down' approach, according to which particular experiences are phenomenally unified by virtue of being subsumed by a single total experience.

Current versions of objectivism tend to be functional in nature. Shoemaker's (1996, 2002) functionalism is a standard personal-level functionalism involving causal relations between content-bearing states, and is in part motivated by his functionalist account of consciousness. Hurley (1998) defends a sub-personal functionalism according to which the unity of consciousness involves a dynamic singularity centered on an active organism.

A number of approaches to the unity of consciousness can be developed along both objectivist and subjectivist lines. Consider, for instance, subject-based accounts of co-consciousness. It is sometimes suggested that experiences are co-conscious simply by virtue of being possessed by the same subject of experience at the same time. This account of phenomenal unity qualifies as a version of objectivism. However, an account which construes co-consciousness in terms of the phenomenology of self-consciousness – that is, in terms of the sense that certain experiences belong to oneself – is a version of subjectivism, for it locates the binding agent of consciousness within phenomenology itself.

It is important to note that objectivism and subjectivism are not mutually exclusive. Indeed, it seems plausible to suppose that a complete account of phenomenal unity will have both subjective and objective components.

UNITY OF CONSCIOUSNESS IN COGNITIVE SCIENCE

History

Cognitive science has had an interest in the unity of consciousness since its beginnings in the nineteenth century. A central strand of this concern has been to examine the conditions under which the human brain achieves perceptual consistency. In this regard, psychologists have devoted much attention to effects such as the phi phenomenon. Spots separated by a small visual angle (up to 4 degrees) that are briefly lit in rapid succession will produce the phenomenal effect of a single spot in motion. The brain assumes that it must be seeing a single source of light in rapid motion rather than two light sources lit in rapid succession. Closely related to the phenomenon of perceptual constancy are intermodal effects, such as the McGurk effect (Stein and Meredith, 1983). When the sound 'ba-ba' is dubbed onto the video of someone who is actually saying 'ga-ga' participants report hearing 'da-da'. Such intermodal effects are common, and reveal that the operations of the various modalities constrain each other in what are often surprising ways.

Object and spatial unity reveal that the brain binds disparate forms of information together. How does it do this? Many early models of integration attributed binding to specialized neurons – so-called 'grandmother cells' – and multimodal association areas. Although some theorists continue to accord such cells a role in binding, it is generally agreed that 'convergence' models are at best a partial solution to the binding problem. At a theoretical level it is hard to see how there could be a particular cell for each possible object of perception, while at a practical level the search for omnimodal cells and association areas has proved fruitless. Although there are many multimodal areas in the brain, there seems to be no one anatomical site of consciousness on which all information must converge in order to become conscious. Contemporary approaches to integration generally focus on functional integration rather than anatomical convergence.

While neurophysiologists have explored the unity of consciousness via the mechanisms of neuronal interaction, clinical psychologists have examined the unity of consciousness by studying the effects of brain damage and psychopathology. One class of such disorders can be loosely grouped under the heading of 'dissociative disorders': these include the 'hidden observer' effect in hypnosis, fugue states, and dissociative identity disorder (formerly 'multiple personality disorder') (Hilgard, 1986). It is often claimed that dissociative disorders reveal that consciousness is not unified, although the precise meaning of this claim is often unclear. Dissociation clearly involves a lack of integration in the contents of consciousness, but it is an open question whether this disintegration is accompanied by, or results in, multiple (or partially unified) streams of consciousness within the dissociated individual.

Split-brain research raises many of the same issues as do the dissociative disorders, but it has had a far greater impact on discussion of the unity of consciousness within cognitive science. In the mid-1960s surgeons sectioned the corpus callosum, the bundle of fibers linking the two cerebral hemispheres, of a number of patients in an attempt to alleviate their epilepsy. Although the everyday behavior of these patients was largely unaffected by the operation, under certain laboratory conditions these 'split-brain' patients behaved in ways that suggested to many, notably Sperry (1984), that they had two streams of consciousness. If, for instance, 'key-ring' is briefly projected onto the patient's visual field, the patient claims to see only 'ring', but when asked to pick out the object seen from a range of items, the left hand settles on a key (and rejects a ring). The standard explanation for this behavior proceeds as follows. Information concerning the left side of the visual field is conveyed to the right hemisphere of the brain, and vice versa. Further, in most patients linguistic ability is localized in the left hemisphere, and control over each hand is subserved by the contralateral hemisphere. Thus, it is often suggested that the right hemisphere locates a key with the left hand because it is aware only of 'key', while the patient claims only to have seen the word 'ring' because the left hemisphere was conscious only of the information contained in the right visual field. In other words, it has often been claimed that the best explanation of the patient's behavior is to regard each of the hemispheres as independently conscious. (*See* **Split Brains, Philosophical Issues about**)

Other interpretations of the split-brain data are possible. For instance, it might be suggested that at least some split-brain patients have a partially unified consciousness. For one thing, not all split-brain patients are equally split (Moor, 1982). Patients with complete commissurotomies are both tactually and visually split, while patients with central commissurotomies usually are tactually split but not visually split. Furthermore, even patients with complete commissurotomies are not completely split: they retain the ability to integrate certain types of information (e.g., olfactory, proprioceptive, and emotional) between hemispheres.

It is also noteworthy that the degree of disunity that split-brain patients manifest depends on how they are tested. Consider, for instance, the block design task. In the standard task the patient is asked to manually arrange four patterned cubes to match a sample design. The performance of each hand is timed, and the left hand consistently constructs the design much faster than the right hand. Gazzaniga and LeDoux (1978) found that replacing the standard free manipulospatial response with a response in which the patient was visually presented with three possible answers and asked to point to the correct one removed the asymmetry between left-hand and right-hand responses.

Current Issues

Much current research on the unity of consciousness in cognitive science is concerned with two major questions.

What are the mechanisms of integration and binding?

Although there is intensive discussion in cognitive science about the 'binding problem', there is little agreement about how best to approach it, or even what it is (Revonsuo, 1999). A number of binding problems can be distinguished, of which we will consider two. Both problems begin with the generally accepted claim that the mechanisms of consciousness are widely distributed throughout the brain.

The *object binding* problem is this: given that various visual features – colour, shape, location, etc. – are processed in various parts of the brain, how does the brain bind this information together to form a unified visual percept of a single object? When we look at a Swiss flag, why is that we see a white cross on a red background, rather than a red cross on a white background? The *global binding* problem is concerned with the unity of the entire field (or stream) of consciousness, rather than the unity of individual phenomenal objects. How are various experiences brought together into a single phenomenal field; by what means are they made mutually co-conscious?

Discussions of binding tend to restrict themselves to object binding (and usually to object binding in visual consciousness). As yet there is no commonly agreed approach to object binding. Some (e.g. Barlow, 1995) argue that single neurons play a significant role in object binding; others (e.g., Triesman, 1999) suggest that object binding might be subserved by selective spatial attention; still others argue that synchronized neural activity plays a role in integrating various features into unitary percepts (Crick and Koch, 1990; Engel *et al.*, 1999). It is unclear what implications these accounts of object binding might have for the global binding problem. Proponents of the neurophysiological approach sometimes suggest that neural synchrony might hold the key to all forms of phenomenal unity – indeed, to phenomenology itself – but such suggestions are little more than speculation at present.

It is often suggested that there is an intimate connection between the unity of consciousness and the computational architecture of consciousness. Some claim that because consciousness is unified it must be implemented in a serial or 'von Neumannesque' manner, while others claim that since consciousness is implemented in a parallel connectionist network it cannot be unified. Neither inference is unproblematic: there is no obvious reason why unity at a phenomenal level entails seriality in computational structure.

When and how is the unity of consciousness disrupted?

In addition to dissociative disorders and commissurotomy, a number of other 'pathologies' of consciousness seem to involve some form of disunity in consciousness, although it is often difficult to know exactly how consciousness is disunified in these cases.

Patients with anosagnosia fail to fully appreciate that they have an impairment of some kind. For instance, a patient with a paralysed limb may verbally deny that there is anything wrong. Nevertheless, anosagnosics may behave in ways that indicate that they may have 'dim knowledge' of their condition. They may, for example, agree that if the physician had the same complaint he or she would be unable to get out of bed (Bisiach and Berti, 1995).

Marcel (1993) elicited similar dissociations in response from normal subjects. He asked subjects to respond to the onset of a light in three ways: by blinking, by pushing a button, and by saying 'yes'. Marcel discovered that when subjects were asked to give all three responses simultaneously they were often inconsistent. For instance, a subject might give an affirmative response by blinking and saying 'yes' but a negative response by not pressing the button. Even more surprising, subjects were unaware that they had responded inconsistently. Marcel's subjects clearly suffered from a certain disunity of consciousness, but it is an open question how best to characterize this disunity.

References

Baars BJ (1988) *A Cognitive Theory of Consciousness*. Cambridge, UK: Cambridge University Press.

Barlow H (1995) The neuron doctrine in perception. In: Gazzaniga M (ed.) *The Cognitive Neurosciences*. Cambridge, MA: MIT Press.

Bayne T and Chalmers D (2002) What is the unity of consciousness? In: Cleeremans A (ed.) *The Unity of Consciousness: Binding, Integration, Dissociation*. Oxford, UK: Oxford University Press.

Bisiach E and Berti A (1995) Consciousness in dyschiria. In: Gazzaniga M (ed.) *The Cognitive Neurosciences*, pp. 1331–1340. Cambridge, MA: MIT Press.

Brook A (1994) *Kant and the Mind*. Cambridge, UK: Cambridge University Press.

Crick F and Koch C (1990) Towards a neurobiological theory of consciousness. *Seminars in the Neurosciences* **2**: 263–275.

Dainton B (2000) *Stream of Consciousness: Unity and Continuity in Experience*. London, UK: Routledge.

Engel AK, Fried P, König P, Brecht M and Singer W (1999) Temporal binding, binocular rivalry, and consciousness. *Consciousness and Cognition* **8**: 128–151.

Gazzaniga M and LeDoux J (1978) *The Integrated Mind*. New York, NY: Plenum Press.

Hilgard E (1986) *Divided Consciousness*. New York, NY: John Wiley. [Expanded edition.]

Hurley S (1998) *Consciousness in Action*. Cambridge, MA: Harvard University Press.

Hurley S (2002) Action, the unity of consciousness, and vehicle externalism. In: Cleeremans A (ed.) *The Unity of Consciousness: Binding, Integration, Dissociation*. Oxford, UK: Oxford University Press.

Lockwood M (1989) *Mind, Brain and the Quantum*. Oxford, UK: Blackwell.

Marcel A (1993) Slippage in the unity of consciousness. In: Bock GR and Marsh J (eds) *Experimental and Theoretical Studies of Consciousness*, pp. 168–179. Chichester, UK: John Wiley.

Moor J (1982) Split brains and atomic persons. *Philosophy of Science* **49**: 91–106.

Nagel T (1971) Brain bisection and the unity of consciousness. *Synthese* **22**: 396–413.

Revonsuo A (1999) Binding and the phenomenal unity of consciousness. *Consciousness and Cognition* **8**: 173–185.

Rosenthal DM (2002) Persons, minds, and consciousness. In: Hahn LE (ed.) *The Philosophy of Marjorie Grene*, in the Library of Living Philosophers. La Salle, IL: Open Court.

Shoemaker S (1996) Unity of consciousness and consciousness of unity. In: Shoemaker S *The First-Person Perspective and Other Essays*. Cambridge, UK: Cambridge University Press.

Shoemaker S (2002) Consciousness and co-consciousness. In: Cleeremans A (ed.) *The Unity of Consciousness: Binding, Integration, Dissociation*. Oxford, UK: Oxford University Press.

Sperry R (1984) Consciousness, personal identity and the divided brain. *Neuropsychologia* **22**: 661–673.

Stein BE and Meredith MA (1993) *The Merging of the Senses*. Cambridge, MA: MIT Press.

Treisman A (1999) Feature binding, attention and object perception. In: Humphreys GW, Duncan J and Treisman A (eds) *Attention, Space and Action*, pp. 91–111. New York, NY: Oxford University Press.

Further Reading

Bertelson P (1998) Starting from the ventriloquist: the perception of multimodal events. In: Michel S and Craik F (eds) *Advances in Psychological Science*, vol. II, *Biological and Cognitive Aspects*, pp. 419–439. Hove, UK: Psychology Press.

Braude S (1995) *First-Person Plural*. Lanham, MD: Rowman and Littlefield.

Hardcastle VG (1994) Psychology's binding problem and possible neurobiological solutions. *Journal of Consciousness Studies* **1**: 66–90.

Hill CS (1991) Unity of consciousness, other minds, and phenomenal space. In: *Sensations: A Defense of Type Materialism*. Cambridge, UK: Cambridge University Press.

James W (1981) The stream of thought. In: *The Principles of Psychology*, vol. I. Cambridge, MA: Harvard University Press.

Marks CE (1981) *Commissurotomy, Consciousness and Unity of Mind*. Cambridge, MA: MIT Press.

O'Brien G and Opie J (1998) The disunity of consciousness. *Australasian Journal of Philosophy* **76**: 378–395.

Radden J (1998) Pathologically divided minds, synchronic unity and models of the self. *Journal of Consciousness Studies* **5**: 658–672.

Rosenthal D (2001) Persons, minds, and consciousness. In: Hahn LE (ed.) *The Philosophy of Margorie Grene.*

Triesman A (1996) The binding problem. *Current Opinion in Neurobiology* **6**: 171–178.

Zaidel E (1995) Interhemispheric transfer in the split brain: long-term status following complete cerebral commissurotomy. In: Davidson RJ and Hugdahl K (eds) *Brain Asymmetry*. Cambridge, MA: MIT Press.

Constraint Satisfaction

Intermediate article

Rina Dechter, University of California, Irvine, California, USA
Francesca Rossi, Università di Padova, Padova, Italy

CONTENTS

Constraints are a formalism for the representation of declarative knowledge that allows for a compact and expressive modeling of many real-life problems. Constraint satisfaction and propagation tools, as well as constraint programming languages, are successfully used to model, solve, and reason about many classes of problems, such as design, diagnosis, scheduling, spatio-temporal reasoning, resource allocation, configuration, network optimization, and graphical interfaces.

INTRODUCTION

A constraint satisfaction problem (CSP) consists of a finite set of variables, each associated with a domain of values, and a set of constraints. Each constraint is a relation, defined on some subset of the variables, called its scope, specifying their legal combinations of values. Constraints may be described by mathematical expressions, or by computable procedures.

A solution is an assignment of a value to each variable from its domain such that all the constraints are satisfied. Typical constraint satisfaction problems are to determine whether a solution exists, to find one or all solutions, and to find an optimal solution relative to a given cost function.

An example of a constraint satisfaction problem is the well-known k-colorability problem. The task is to color, if possible, a given graph with k colors only, in such a way that any two adjacent nodes have different colors. A constraint satisfaction formulation of this problem associates the nodes of the graph with variables; the sets of possible colors are their domains; and the 'not equal' constraints between adjacent nodes are the constraints of the problem.

Another well-known constraint satisfaction problem in logic concerns 'satisfiability', which is the task of finding a truth assignment to propositional variables such that a given set of clauses are satisfied. For example, given the two clauses $(A \vee B \vee \neg C)$, $(\neg A \vee D)$, the assignment of 'false' to A, 'true' to B, 'false' to C and 'false' to D is a satisfying truth value assignment.

The structure of a constraint problem is usually depicted by a 'constraint graph', whose nodes represent the variables. Two nodes are connected if the corresponding variables participate in the same constraint scope. In the k-colorability problem, the graph to be colored is the constraint graph. In the satisfiability problem above, the constraint graph has A connected with D, and A, B and C connected with each other.

Constraint problems have proved successful in modeling many practical tasks, such as scheduling, design, diagnosis, and temporal and spatial reasoning. The reason is that constraints allow for a natural, expressive and declarative formulation of what has to be satisfied, without the need to say how it should be satisfied.

In addition, many cognitive tasks, such as language comprehension, default reasoning and abduction, can be naturally represented as constraint satisfaction problems. Historically, constraints and constraint satisfaction have been used in many cognitive tasks related to vision (Waltz, 1975).

When an observer looks at a two-dimensional representation of a three-dimensional geometrical scene, each line and line intersection can be interpreted in many ways, but the physical world together with the laws of geometry put restrictions. It is natural to model lines as variables, with three possible values, indicating that a line can represent a convex, concave, or boundary line, and to have constraints among the lines which are incident to the same junction, forcing the usual laws of three-dimensional geometry.

This representation provides a natural modeling, and is useful for finding a geometrically plausible three-dimensional interpretation of the two-dimensional scene. In fact, by looking at just one constraint, one can automatically eliminate values from the domains of its variables, if such values do not agree with the constraint. This can trigger a chain reaction in which other constraints are considered (one at a time) and other values are deleted in the domains of other variables, until nothing more can be deleted. At this point, we are often left with just one value for each variable, that is, the only plausible interpretation of the scene.

This (generally incomplete) method is usually called 'arc consistency' and belongs to the class of constraint propagation techniques that we will describe in greater detail below.

In linguistics, there are constraint-based views of a language (Chomsky and Lasnik, 1992), constraints over the logic of typed feature structures and relations which are used to describe principles of phrase grammars (Pollard and Sag, 1994), and constraint-based linguistic theories, which perform constraint-assisted searches for deductive proofs, in the spirit of constraint logic programming languages (see below).

It should be noted, however, that constraints and constraint satisfaction are sometimes understood in different ways by computer scientists on one hand and linguists and cognitive scientists on the other. In fact, in most cognitive science applications, a constraint is usually interpreted as being universally quantified over all its entities (e.g., linguistic signs), while this is not the way constraints are defined and used in computer science.

CONSTRAINT SATISFACTION AS SEARCH

Complexity of Constraint-related Tasks

In general, constraint satisfaction tasks (like finding one or all solutions, or the best solution) are computationally intractable (NP-hard). Roughly, this means that there are likely to be problem instances requiring all the possible variable instantiations to be considered before a solution (or best solution) can be found, and this can take a time exponential in the size of the problem. However, there are some tractable classes of problems that allow for efficient solution algorithms of all the problems in the class. Moreover, even for intractable classes, many techniques exhibit a good performance in practice in the average case.

Techniques for Solving CSPs

The techniques for processing constraint problems can be roughly classified into two categories: search (also called conditioning), and consistency inference (or propagation). However, techniques can be combined, and in practice, constraint processing techniques usually contain aspects of both categories.

These two methods have a cognitive analogy in human problem solving. Conditioning search uses the basic operation of guessing a value of a variable and trying to solve a subproblem with the guess. Inference corresponds instead to thinking and deduction, with a view to simplifying the problem.

Search algorithms traverse the space of partial instantiations, building up a complete instantiation that satisfies all the constraints, or else they determine that the problem is inconsistent. By contrast, consistency inference algorithms reason through equivalent problems: at each step they modify the current problem to make it more explicit without losing any information (that is, maintaining the same set of solutions). Search is either systematic and complete, or stochastic and incomplete. Likewise, consistency inference algorithms may achieve complete solutions (e.g., by variable elimination), or incomplete solutions. The latter are usually called local consistency algorithms because they operate on local portions of the constraint problem.

Backtracking search

The most common algorithm for performing systematic search for a solution of a constraint problem is the so-called backtracking search algorithm. This algorithm traverses the space of partial solutions in a 'depth first' manner, and at each step it extends a partial solution (that is, a variable instantiation of a subset of variables which satisfies all the relevant constraints) by assigning a value to one more variable. When a variable is encountered none of whose possible values are consistent with the current partial solution (a situation referred to as a dead end), backtracking takes place, and the algorithm reconsiders one of the previous assignments. The best case occurs when the algorithm is able to successfully assign a value to each variable without encountering any dead ends. In this case, the time complexity is linear in the size of the problem (often identified with the number of its variables). In the worst case, the time complexity of this algorithm is exponential in the size of the problem. However, even in this case the algorithm requires only linear space.

Look-ahead schemes

Several improvements to backtracking have focused on one or both of the two phases of the algorithm: moving forward to a new variable ('look-ahead' schemes) and backtracking to a previous assignment ('look-back' schemes) (Dechter and Pearl, 1987). When moving forward to extend a partial solution, some computation (e.g., arc consistency) may be carried out to decide which variable, and which of the variable's values, to choose next in order to either fail quickly if there is no solution, or not fail at all if there is one. Variables that maximally constrain the rest of the search space are usually preferred, and therefore, the most constrained variable is selected. This method follows the so-called 'first fail' heuristics, which aim to force failures as early as possible, in order to cut down on the amount of backtracking. By contrast, for value selection, the least constraining value is preferred, in order to maximize future options for instantiations (Haralick and Elliot, 1980). A well-known look-ahead method is 'forward checking', which performs a limited form of consistency inference at each step, ruling out some values that would lead to a dead end. A currently popular form of look-ahead scheme, called MAC (for 'maintaining arc consistency'), performs arc consistency at each step and uses the revealed information for variable and value selection (Gaschnig, 1979).

Look-back schemes

Look-back schemes are invoked when the algorithm encounters a dead end. These schemes perform two functions. The first is to decide how far to backtrack, by analyzing the reasons for the current dead end, a process often referred to as 'backjumping' (Gaschnig, 1979). The second is to record the reasons for the dead end in the form of new constraints so that the same conflict will not arise again, a process known as 'constraint learning' and 'no-good recording' (Stallman and Sussman, 1977).

Local search

Stochastic local search strategies were introduced in the 1990s and are popular especially for solving propositional satisfiability problems. These methods move in a hill-climbing manner in the space of all variables' instantiations, and at each step they improve the current instantiation by changing the value of a variable so as to maximize the number of constraints satisfied. Such search algorithms are incomplete, since they may get stuck at a local maximum and they are not able to discover that a constraint problem is inconsistent. Nevertheless, when equipped with some heuristics for randomizing the search, or for revising the guiding criterion function (e.g., constraint reweighting), they have been shown to be reasonably successful in solving many large problems that are too hard to be handled by a backtracking search (Selman *et al.*, 1992). A well-known local search algorithm for optimization tasks is called 'simulated annealing'.

Evaluation of Algorithms

The theoretical evaluation of constraint satisfaction algorithms is accomplished primarily by worst-case analysis, that is, determining a function of the problem's size that represents an upper bound of the algorithm's performance over all problems of that size. However, the trade-off between constraint inference and search is hardly captured by such analysis. This is because worst-case analysis, by its nature, is very pessimistic, and often does not reflect the actual performance. Thus in most cases an empirical evaluation is also necessary. Normally, an algorithm is evaluated empirically on a set of randomly generated problems, chosen in such a way that they are reasonably hard to solve (this is done by selecting them from the phase transition region (Selman *et al.*, 1992)). Several benchmarks, based on real-life applications such as scheduling, are also used.

CONSTRAINT PROPAGATION

Constraint propagation (or local consistency) algorithms (Montanari, 1974; Mackworth, 1977;

Freuder, 1982) transform a given constraint problem into an equivalent one which is more explicit, by inferring new constraints which are added to the problem. Therefore, they can make explicit inconsistencies that were implicitly contained in the problem specification. Intuitively, given a constraint problem, a constraint propagation algorithm will make any solution of a small subproblem extensible to some surrounding variables and constraints. These algorithms are interesting because their worst-case time complexity is polynomial in the size of the problem, and they are often very effective in discovering local inconsistencies.

Arc and Path Consistency

The most basic and most popular propagation algorithm, called arc consistency, ensures that any value in the domain of a variable has a legal match in the domain of any other variable. This means that any solution of a one-variable subproblem is extensible in a consistent manner to any other variable. The time complexity of this algorithm is linear in the size of the problem. Another well-known constraint propagation algorithm is path consistency. This algorithm ensures that any solution of a two-variable subproblem is extensible to any third variable. As would be expected, it is more powerful than arc consistency in discovering and removing inconsistencies. It also requires more time: its time complexity is cubic in the number of variables.

i-Consistency

Arc and path consistency can be generalized to *i*-consistency. In general, *i*-consistency algorithms guarantee that any locally consistent instantiation of $i - 1$ variables is extensible to any i^{th} variable. Thus, arc consistency is just 2-consistency, and path consistency is just 3-consistency. Enforcing *i*-consistency can be accomplished in time and space exponential in i: if the constraint problem has n variables, the complexity of achieving *i*-consistency is $O(n^i)$.

Global Consistency

A constraint problem is said to be globally consistent if it is *i*-consistent for every i. In this case, a solution can be assembled by assigning values to variables (in any order) without encountering a dead end, that is, in a backtrack-free manner.

Adaptive Consistency as Complete Inference

In practice, global consistency is not necessary to have backtrack-free assignment of values: it is enough to have directional global consistency relative to a given variable ordering. For example, an 'adaptive consistency' algorithm, which is a variable elimination algorithm, enforces global consistency in a given order only, so that every solution can be extracted with no dead ends along this ordering. Another related algorithm, called tree clustering, compiles the given constraint problem into an equivalent tree of subproblems whose respective solutions can be efficiently combined into a complete solution. Adaptive consistency and tree clustering are complete inference algorithms that can take time and space exponential in a parameter of the constraint graph called the 'induced width' (or 'tree width') (Dechter and Pearl, 1987).

Bucket Elimination

Bucket elimination (Dechter, 1999) is a recently proposed framework for variable elimination algorithms which generalizes adaptive consistency to include dynamic programming for optimization tasks, directional resolution for propositional satisfiability, Fourier elimination for linear inequalities, and algorithms for probabilistic inference in Bayesian networks.

Constraint Propagation and Search

Some problems that are computationally too hard for adaptive consistency can be solved by bounding the amount of consistency enforcing (e.g., applying only arc or path consistency) and embedding these constraint propagation algorithms within a search component, as described above. This yields a trade-off between the effort spent in constraint propagation and that spent on the search, which can be exploited and which is the focus of empirical studies.

TRACTABLE CLASSES

In between search and constraint propagation algorithms, there are the so-called structure-driven algorithms. These techniques emerged from an attempt to topologically characterize classes of constraint problems that are tractable (that is, polynomially solvable). Tractable classes are generally recognized by realizing that enforcing low-level consistency (in polynomial time) guarantees global consistency for some problems.

Graph-based Tractability

The basic constraint graph structure that supports tractability is a tree. This has been observed repeatedly in constraint networks, complexity theory and database theory. In particular, enforcing arc consistency on a tree-structured constraint problem ensures global consistency along some orderings. Most other graph-based techniques can be viewed as transforming a given network into a meta-tree. Among these, we find methods such as tree clustering and adaptive consistency, the cycle cutset scheme, and the biconnected component decomposition. These lead to a general characterization of tractability that uses the notion of induced width (Dechter and Pearl, 1987).

Constraint-based Tractability

Some tractable classes have also been characterized by special properties of the constraints, without any regard to the topology of the constraint graph. For example, tractable classes of temporal contraints include subsets of the qualitative interval algebra, expressing relationships such as 'time interval A overlaps or precedes time interval B'', as well as quantitative binary linear inequalities over the real numbers of the form $X - Y \leq a$ (Meiri *et al.*, 1990). In general, we exploit notions such as tight domains and tight constraints, row-convex constraints (van Beek and Dechter, 1995), implicational and max-ordered constraints, and causal networks. A connection between tractability and algebraic closure has been discovered and intensively investigated in recent years, yielding a nice theory of tractability (Cohen *et al.*, 2000).

SOFT CONSTRAINT SATISFACTION AND CONSTRAINT OPTIMIZATION

Constraint processing tasks include not only problems of constraint satisfaction, but also problems of constraint optimization. Such problems arise when the solutions are not equally preferred. The preferences among solutions can be expressed via a cost function (also called an objective function), and the task is to find the best-cost solution or a reasonable approximation to it.

For example, we may have the constraints $X \leq Y$ and $Y \leq 10$, with the objective function $f = X + Y$, to be maximized. The best solution (unique in this example) is: $X = 10$, $Y = 10$. All other solutions (e.g. $X = 5$, $Y = 6$), although satisfying all the constraints, are less preferred. Cost functions are often specified as a sum of cost components, each defined on a subset of the variables.

Branch and Bound

Adapting the backtracking search algorithm for the task of selecting the most preferred (best cost) solution yields the well-known branch and bound algorithm. Like backtracking, branch and bound traverses the search tree in a 'depth first' manner, pruning not only partial instantiations that are inconsistent, but also those that are estimated to be inferior to the current best solution. At each node, the value of the current partial solution is estimated (by an evaluation function) and compared with the current best solution; if it is inferior, search along the path is terminated. When the evaluation function is accurate, branch and bound prunes substantial portions of the search tree.

Soft Constraints

One way to specify preferences between solutions is to attach a level of importance to each constraint or to each of its tuples. This technique was introduced because constraints in real problems often cannot be described by a set of 'true or false' statements only. Often constraints are associated with features such as preferences, probabilities, costs, and uncertainties. Moreover, many real problems, even when modeled correctly, are overconstrained. Constraints that have varied levels of importance are called soft constraints. There are several frameworks for soft constraints, such as the semi-ring formalism (Bistarelli *et al.*, 1997), whereby each tuple in each constraint has an associated element taken from a partially ordered set (a semi-ring); and the valued constraint formalism, whereby each constraint is associated with an element from a totally ordered set. These formalisms are general enough to model classical constraints, weighted constraints, 'fuzzy' constraints, and overconstrained problems. Current research effort is focused on extending propagation and search techniques to these more general frameworks.

CONSTRAINT PROGRAMMING

The constraint satisfaction model is useful because of its mathematical simplicity on the one hand, and its ability to capture many real-life problems on the other. Yet, to make this framework useful for many real-life applications, advanced tools for modeling and for implementation are necessary. For this reason, constraint systems (providing some built-in propagation and solution algorithms) are usually embedded within a high-level programming environment which assists in the modeling phase and which allows for some control over the solution method.

Logic Programming

Although many programming paradigms have recently been augmented with constraints, the concept of constraint programming is mainly associated with the logic programming (LP) framework (Lloyd, 1993). Logic programming is a declarative programming paradigm whereby a program is seen as a logical theory and has the form of a set of rules (called clauses) which relate the truth value of an atom (the 'head' of the clause) to that of a set of other atoms (the 'body' of the clause). The clause

```
p(X,Y) :- q(X), r(X,Y,Z)
```

says that if atoms `q(X)` and `r(X, Y, Z)` are true, then also atom `p (X, Y)` is true. For example, the clauses

```
reach(X,Y) :- flight(X,Y)

reach(X,Y) :- flight(X,Z), reach
(Z,Y)
```

describe the 'reachability' between two cities (X and Y) via a sequence of direct flights.

Logic programming and search

Executing a logic program means asking for the truth value of a certain predicate, called the goal. For example, the goal `:- p (X, Y)` asks whether there are values for the variables X and Y such that `p (X, Y)` is true in the given logic program. The answer is found by recursively 'unifying' the current goal with the head of a clause (by finding values for the variables that make the two atoms equal). As with constraint solving, the algorithm that searches for such an answer in LP involves a backtracking search.

Constraint Logic Programming

To use constraints within LP, one just has to treat some of the predicates in a clause as constraints and to replace unification with constraint solving. The resulting programming paradigm is called 'constraint logic programming' (CLP) (Jaffar and Maher, 1994; Marriott and Stuckey, 1998). A typical example of a clause in CLP is

```
p(X,Y) :- X<Y+1, q(X), r(X,Y,Z)
```

which states that `p(X, Y)` is true if `q(X)` and `r(X, Y, Z)` are true and the value of X is smaller than that of Y + 1. While the regular predicates are treated as in LP, constraints are manipulated using specialized constraint processing tools. The shift from LP to CLP permits the choice among several constraint domains, yielding an effective scheme that can solve many more classes of real-life problems. Some examples of CLP languages are ECLiPSe (IC-PARC, 1999), CHIP (Dincbas *et al.*, 1988) SICStus Prolog (Carlsson and Widen, 1999), CHRs (Fruhwirth, 1995), and GNU Prolog (Codognet and Diaz, 1996; Diaz, 2000).

Specialized algorithms for CLP

CLP languages are reasonably efficient, due to their use of a collection of specialized solving and propagation algorithms for frequently used constraints and for special variable domain shapes. Global constraints and bounds consistency are two aspects of such techniques that are incorporated into most current CLP languages.

Global constraints

Global constraints are constraints, usually non-binary, for which there exist specialized and efficient solution methods. An example is the constraint `alldifferent`, which requires all the involved variables to assume different values, and which can be efficiently solved using a bipartite matching algorithm. Global constraints are used to replace a set of other constraints, usually involving fewer variables and belonging to some special class. For example, a set of binary inequality constraints seldom gives rise to useful constraint propagation and thus may require a complete search. This expensive search is avoided by replacing these constraints with a single `alldifferent` constraint that is accompanied by an efficient propogation algorithm. Most current CLP languages are equipped to handle several kinds of global constraints.

Bounds consistency

Bounds consistency is one of the major contributions of CLP to the field of constraint propagation. When the variable domains are sets of integers, they are usually represented, to save space, by intervals. In this way, one can store only their minimum and the maximum elements. However, constraint propagation techniques like arc consistency could destroy this representation by removing elements internal to the intervals. Therefore, CLP languages usually use an approximation of arc consistency, called bounds consistency, which removes an element from a domain only if the resulting domain is still an interval. This technique is used in most CLP languages, since it is both efficient (in time and space) and powerful in terms of propagation.

CONCLUSION

The study of CSPs is interdisciplinary, since it involves ideas and results from several fields, including artificial intelligence (where it began), databases, programming languages, and operations research. While it is not possible to cover all the lines of work related to CSPs, this article has covered the main ideas.

Current investigations include: identification of new tractable classes; studying the relationship between search and propagation; extending propagation techniques to soft constraints; and developing more flexible and efficient constraint languages.

References

van Beek P and Dechter R (1995) On the minimality and decomposability of row-convex constraint networks. *Journal of the ACM* **42**: 543–561.

Bistarelli S, Montanari U and Rossi F (1997) Semiring-based constraint solving and optimization. *Journal of the ACM* **44**(2): 201–236.

Carlsson M and Widen J (1999) *SICStus Prolog Homepage*. http://www.sics.se/sicstus/.

Chomsky N and Lasnik H (1992) Principles and parameters theory. In: Jacobs J, von Stechow A and Sternefeld W (eds) *Syntax: An International Handbook of Contemporary Research*. Berlin: Walter de Gruyter.

Codognet P and Diaz D (1996) Compiling constraints in clp(FD). In: *Journal of Logic Programming* **27**(3): 185–226.

Cohen D, Jeavons P, Jonsson P and Koubarakis M (2000) Building tractable disjunctive constraints. *Journal of the ACM* **47**(5): 826–853.

Dechter R (1999) Bucket elimination: a unifying framework for reasoning. *Artificial Intelligence* **13**(1–2): 41–85.

Dechter R and Pearl J (1987) Network-based heuristics for constraint satisfaction problems. *Artificial Intelligence* **34**: 1–38.

Diaz D (2000) *The GNU Prolog web site*. http://gnu-prolog.inria.fr/

Dincbas M, van Hentenryck P and Simonis M *et al.* (1998) The constraint logic programming language CHIP. In: *Proc. International Conference on Fifth Generation Computer Systems*. Tokyo: Ohmsha Ltd.

Freuder EC (1982) A sufficient condition for backtrack-free search. *Journal of the ACM* **29**(1): 24–32.

Fruhwirth T (1995) Constraint simplification rules. In: *Constraint Programming: Basics and Trends*. New York, NY: Springer-Verlag.

Gaschnig J (1979) *Performance Measurement and Analysis of Search Algorithms*. PhD thesis, Pittsburgh, PA: Carnegie Mellon University.

Haralick M and Elliot GL (1980) Increasing tree-search efficiency for constraint satisfaction problems. *Artificial Intelligence* **14**: 263–313.

IC-PARC (1999) *The ECLiPSe Constraint Logic Programming System*. http://www.icparc.ic.ac.uk/eclipse/

Jaffar J and Maher MJ (1994) Constraint logic programming: a survey. *Journal of Logic Programming* **19/20**: 503–581.

Mackworth AK (1977) Consistency in networks of relations. *Artificial Intelligence* **8**(1): 99–118.

Marriott K and Stuckey PJ (1998) *Programming with Constraints: An Introduction*. Cambridge, MA: MIT Press.

Meiri I, Dechter R and Pearl J (1990) Temporal constraint networks. *Artificial Intelligence* **49**: 61–95.

Montanari U (1974) Networks of constraints: fundamental properties and applications to picture processing. *Information Science* **7**(66): 95–132.

Pollard C and Sag IA (1994) *Head-Driven Phrase Structure Grammar*. Chicago, IL: University of Chicago Press.

Selman B, Levesque H and Mitchell D (1992) A new method for solving hard satisfiability problems. In: *Proceedings of the Tenth National Conference on Artificial Intelligence*, pp. 440–446. Menlo Park, CA: AAAI Press.

Stallman M and Sussman GJ (1977) Forward reasoning and dependency-directed backtracking in a system for computer-aided circuit analysis. *Artificial Intelligence* **9**(2): 135–196.

Waltz DL (1975) Understanding line drawings of scenes with shadows. In: Winston P (ed.) *The Psychology of Computer Vision*. New York, NY: McGraw-Hill.

Further Reading

Arnborg S and Proskourowski A (1989) Linear time algorithms for np-hard problems restricted to partial *k*-trees. *Discrete and Applied Mathematics* **23**: 11–24.

Beldicenau N and Contejean E (1994) Introducing global constraints in CHIP. *Journal of Mathematical and Computer Modeling* **12**: 97–123.

Dechter R (1990) Enhancement schemes for constraint processing: backjumping, learning and cutset decomposition. *Artificial Intelligence* **41**: 273–312.

Dechter R (1992) *Constraint networks*. In: *Encyclopedia of Artificial Intelligence*, 2nd edn, pp. 276–285. New York, NY: Wiley.

Golumbic MC and Shamir R (1993) Complexity and algorithms for reasoning about time: a graph-theoretic approach. *Journal of the ACM* **40**: 1108–1133.

Lloyd JW (1993) *Foundations of Logic Programming*. New York, NY: Springer-Verlag.

Mackworth AK (1992) Constraint satisfaction. In: *Encyclopedia of Artificial Intelligence*, 2nd edn, pp. 285–293. New York, NY: Wiley.

Maier D (1983) *The theory of relational databases*. Rockville, MD: Computer Science Press.

Tsang E (1993) *Foundation of Constraint Satisfaction*. London: Academic Press.

Wallace M (1996) Practical applications of constraint programming. *Constraints: An International Journal* **1**: 139–164.

Constraint-based Processing

Intermediate article

Bob Carpenter, SpeechWorks International, New York, NY, USA

CONTENTS

In a constraint-based grammar formalism, a class of constraints is used to reduce a class of potential representations to the representations that are well formed, or grammatical.

INTRODUCTION

Linguistic theories have primarily been concerned with the structures of languages and utterances within those languages. Extensionally, all of the mainstream linguistic theories concentrate on picking out the set of grammatical, or well-formed, structures (sounds, words or phrases, for instance) in a language. Intensionally, linguists debate the right way to pick out such structures, often arguing from a cognitive perspective based on evidence of human language development, comprehension or production, or typological variation. Computer scientists are typically concerned with the efficiency of processing in terms of time and space, often sidestepping cognitive issues in the interest of building effective software.

CONSTRAINT-BASED GRAMMAR FORMALISMS

Two basic approaches to characterizing well-formed structures can be usefully compared and contrasted. The first, more traditional, approach is based on the notion of inductive definitions. To use a logical example that motivated linguists, consider an inductive definition of the well-formed formulae (WFF) of propositional logic. First, there is the base case of propositional symbols, which are assumed to be WFFs. Then, if P and Q are WFFs then so is the conjunction (P and Q) and the negation (not P). Iterating these constructions generates the full set of WFFs. Any formula that can be built up from propositional symbols by applying conjunction or negation is taken to be well formed.

The second approach to picking out the well-formed structures is based on constraints. Rather than starting from a small set of structures known to be well formed, a large set of possibilities is considered and the ill-formed ones are removed. For instance, the prime numbers are naturally characterized in this way. Start with all of the natural numbers from 2 onwards, and eliminate the compound numbers, that is numbers that are the product of two smaller numbers greater than one. The remaining numbers are the prime numbers.

Bar-Hillel (1950), working from the tradition of logical languages in mathematics and philosophy, presented one of the first mathematically rigorous definitions of a system for generating the expressions of a natural language. He worked bottom-up from a lexicon associating words and categories, such as *dog* and 'singular common noun', and introduced rules to build larger phrases. For instance, by concatenating a determiner such as *the* with a common noun such as *dog*, the noun phrase *the dog* is generated. This was essentially an early instance of a phrase structure grammar. (*See* **Phrase Structure Grammar, Head-driven**)

Although there is no universally agreed definition, a generative theory is typically taken as one in which a mechanism is proposed for generating (usually by some formal mathematical means) the set of grammatical sentences of a language. Chomsky (1957) proposed perhaps the best-known example of an early generative theory of language. Chomsky introduced the notion of transformation, which he used to relate constructions in a language, such as interrogative and declarative forms of sentences, through derivation. He employed phrase-structure techniques similar to those of Bar-Hillel to generate 'base forms', and then applied various structure-permuting transformations to generate grammatical 'surface forms'. For instance, Chomsky produced the interrogative *will John run* from the delcarative *John will run* by means of a transformation that moves the matrix finite auxiliary to the front of the sentence.

Over the next 20 years, generative systems acquired all the accoutrements of a maturing science. Constraints were introduced to limit the applicability of some operations. For instance, noun compounding combines two nouns into a new noun, such as *coal oven*, but a constraint would be imposed that the first noun not be plural. By the late 1970s, in the 'government-binding theory' (Chomsky, 1980), the role of constraints on transformations had grown to the point where the transformations themselves were of the general form 'move anything anywhere', requiring constraints to control specifically what moved where. This resulting framework is thus explicitly constraint-based. (*See* **Government–Binding Theory**)

MATHEMATICAL FOUNDATIONS

The mathematics of constraint processing in general involves the distinction between inductive and co-inductive definitions (Barwise and Etchemendy, 1987). An inductive definition provides a base set, and then operations that expand the set. The set defined is then taken to be the smallest set containing the base cases and closed under the operations. A co-inductive definition, on the other hand, begins with a set of structures and provides constraints on the forms of elements. The set defined is the largest subset of the original set satisfying all of the constraints.

The standard general formulation of constraint problems is in terms of logical satisfiability. Consider the simple language of propositional formulae given above, under the standard interpretation. It is natural to ask, for a given formula, whether there is an assignment of truth values to its propositional symbols under which it is true. The formula is thus a kind of constraint on assignments of truth values. Unfortunately, even this simple constraint problem for this simple logical language is very complex computationally. The problem of determining propositional satisfiability (known as 'SAT') is NP-complete (Cormen *et al.*, 1990): every known algorithm to solve such problems can encounter cases whose processing time grows faster than any polynomial. (*See* **Computability and Computational Complexity**)

Robinson (1965) introduced the general technique of 'resolution' for proving theorems in first-order logic. Using the notion of resolution and the simple facts of linguistic agreement, Colmerauer (1993) introduced the notion of logic programming, motivated primarily by the search and constraint problems introduced by linguistic grammars. Colmerauer used logical rules such as the following to code the fact that the noun phrase and verb phrase must agree in number and that the sentence carries the verb form of its matrix verb:

```
s(Words1+Words2,VerbForm)
  if np(Words1, Number) and
     vp(Words2,VerbForm,
        Number)                (1)
```

The symbol `s` is taken to be a two-place predicate, whose first argument is a sequence of words and whose second argument is a verb form such as `finite` or `infinitive`. The symbols for noun phrases and verb phrases, `np` and `vp`, are to be read similarly, with an additional argument for number. Such rules are implicitly universally quantified, and thus the above rule can be read as saying that if `Words1` is a sequence of words that forms a noun phrase of number `Number` and `Words2` is a sequence of words that forms a verb phrase of the same number and a verb form `VerbForm`, then the concatenation of `Words1` and `Words2`, indicated as `Words1+Words2`, forms a sentence with the given verb form `VerbForm`. (*See* **Resolution Theorem Proving**)

The notion of constrained phrase-structure rules was introduced, along with a very general notion of feature, into mainstream linguistics by Harman (1963). Since then, every branch of the field, from phonetics to pragmatics, has adopted some kind of 'feature-based' analysis. In these theories, rather than relying on a positional encoding, as in first-order terms, the values are named.

In theories such as lexical–functional grammar and head-driven phrase structure grammar, feature structures are used to represent partial information. This more general data structure is particularly adept at representing the kind of sparse information found in linguistic constraints, which often indicate the behavior of a handful of linguistic features among hundreds. Feature structures of the form employed in head-driven phrase structure grammar also allow a natural form of knowledge representation through inheritance. Processing with such structures has essentially followed that of first-order terms, with systems being based on constraint resolution and, in particular, unification. (Unification is an operation that takes two terms or feature structures and produces a new term that contains all of the information in both, or returns failure if they are inconsistent.) There are also strong similarities to production systems operating over frames. (*See* **Production Systems and Rule-based Inference**; **Knowledge Representation**; **Lexical-Functional Grammar**)

CONSTRAINT-BASED PARSING

Natural language parsing, as a stage in more general natural language processing, involves generating some or all of the structures compatible with a given sequence of words. Given a context-free grammar and a sequence of words, a representation of all parses can be generated in polynomial time (slightly subcubic). The 'universal recognition problem' involves taking a grammar and a sequence of words and determining whether there is some parse for a sequence of words. Barton *et al.* (1987) showed that the introduction of agreement constraints like those of equation 1 yields a universal recognition problem that is NP-complete; the problem for any fixed grammar, though, remains polynomial. Similarly, word-order constraints added to a general context-free grammar result in NP-complete parsing problems. (*See* **Natural Language Processing**; **Parsing: Overview**)

Most constraint-based parsing systems are hybrids that involve a standard search-based parsing algorithm with side constraints. At any stage where constituents are combined or rules are expanded, a check is made to ensure that the set of constraints remains consistent. If not, the search path is abandoned. At each stage, the information known about variables and their values is represented by means of an assignment to variables. For instance, in parsing *the papers falls* from left to right and bottom up, the determiner *the* is encountered first. This is unspecified for number. When it is combined with the plural noun *papers*, the value of the noun phrase's agreement feature becomes 'plural'. Finally, the attempt to combine the plural noun phrase with the singular verb phrase fails due to a violation of an agreement constraint. Exponential growth in simple parsing arises because the number of possible assignments to variables can grow exponentially. (*See* **Agreement**)

In the field of artificial intelligence, the desire for richer, more structured systems of knowledge representation and reasoning has led to a generalized notion of features and values known as a *frame*. Frames generalize simple feature systems with a finite set of features and a finite set of values by allowing some features to take frames themselves as values. On top of this frame-based representational system, many forms of constraints can be introduced. These lead to a variety of reasoning systems for resolving the constraints. Of particular interest for natural language is the pure constraint-based representation of natural language grammars, which was first introduced in the system KL-ONE (Brachman and Schmolze, 1985).

Of particular interest is the notion of constraint resolution, which is often known as 'classification'. Early work in classification was applied to parsing, most notably in the system KL-ONE. In this purely constraint-based representation of parsing, features are used to represent phrase-structure trees in the same way in which trees are usually coded in computational data structures. For instance, the simple sentence *the kids laughed*, represented as [S [NP [Det the] [N kids]] [VP laughed]], would be represented in a frame as:

```
CAT: S
DTR1: CAT: NP
  DTR1: CAT: Det
    WORD: the
  DTR2: CAT: N
    WORD: kids
DTR2: CAT: VP
  WORD: laughed                    (2)
```

Note that features like CAT (for syntactic category) take simple atomic values, whereas features like DTR1 and DTR2 (for daughter constituents) take values that are themselves frames. A set of grammatical 'rules' can then be naturally modeled as a disjunctive constraint: every local frame must satisfy one of the phrase-structure rules, including lexical rules. In a logical language, this could be expressed as follows:

```
(CAT:S & DTR1:CAT:NP & DTR2:
  CAT:VP) |
(CAT:NP & DTR1:CAT:Det & DTR2:
  CAT:N) |
(CAT:Det & WORD:(the|a|every|
  some|...)) |
(CAT:N & WORD:(kid|boy|dog|...))   (3)
```

Representation of agreement is straightforward:

```
CAT: MAJOR: NP
     AGR: singular
DTR1: CAT: MAJOR: Det
           AGR: singular
      WORD: every
DTR2: CAT: MAJOR: N
           AGR: singular
      WORD: dog
```
(4)

The constraint on agreement can be represented using equations between paths of features:

```
CAT:NP & DTR1:CAT:Det & DTR2:
  CAT:N & (CAT:AGR = DTR1:CAT:
  AGR) & (CAT:AGR = DTR2:CAT:AGR)
```
(5)

Pollard and Sag's (1994) head-driven phrase structure grammar (HPSG) has taken this representational strategy to the limit, representing all grammatical structure by means of constraints. But rather than being a naive encoding of phrase structure grammars, HPSG followed the linguistic trend of factoring the rule-specific equations such as equation 5 into general theories of agreement, semantics, unbounded dependencies, etc. A typical constraint in HPSG would enforce agreement between a mother and a head daughter, and would apply to every phrase structure configuration. Another constraint might say that the unresolved dependencies in a phrase must be the union of all unresolved dependencies in the daughters. Thus rather than a disjunction of rules like

$$\text{Rule}_1|\text{Rule}_2|\ldots|\text{Rule}_n \qquad (6)$$

HPSG would have

$$\text{Principle}_1 \;\&\; \text{Principle}_2 \;\&\ldots\&\; \text{Principle}_n \qquad (7)$$

Of course, the principles themselves might involve disjunctions of cases. (*See* **Constraint Satisfaction**)

In general, constraint systems such as the one employed in HPSG are Turing-equivalent computational devices, and as such do not even have decidable parsing problems, much less tractable ones. However, the properties of actual models expressed in feature formalisms are much more tractable – the theory actually picked out by HPSG, for instance, is decidable, as are the other major feature-based linguistic theories. Even so, most parsers for HPSG and related constraint-based grammar formalisms are built on top of efficient parsers for phrase structure, with constraints being maintained on the side to filter search.

More recently, the trend in constraint-based theories of language has been to relax the all-or-nothing nature of constraints. In theoretical linguistics, optimality theory, for instance, relies on defeasible constraints that are ordered by strength (Archangeli and Langendoen, 1997). Within computational linguistics, statistical models have largely supplanted logical ones, with the advantage that constraints are not 'hard', but rather 'soft'. That is, like optimality theory, statistical algorithms search for the 'best' structure. In statistical models, that is assumed to be the one with the highest likelihood. In optimality theory, it is the one violating the fewest high-ranking constraints.

References

Archangeli D and Langendoen DT (eds) (1997) *Optimality Theory: An Overview*. Oxford, UK: Blackwell.

Bar-Hillel Y (1950) On syntactical categories. *Journal of Symbolic Logic* **15**: 1–16.

Barton GE, Berwick RC and Ristad ES (1987) *Computational Complexity and Natural Language*. Cambridge, MA: MIT Press.

Barwise J and Etchemendy J (1987) *The Liar: An Essay in Truth and Circularity*. Oxford, UK: Oxford University Press.

Brachman RJ and Schmolze JG (1985) An overview of the KL-ONE knowledge representation system. *Cognitive Science* **9**: 171–216.

Chomsky N (1957) *Syntactic Structures*. The Hague, Netherlands: Mouton.

Chomsky N (1980) *Rules and Representations*. New York, NY: Columbia University Press.

Colmerauer A (1993) Les systèmes-q ou un formalisme pour analyser et synthétier des phrases sur ordinateur. *Traitement Automatique des Langues* **33**: 105–148. [Written in 1970 as a technical report in the University of Montréal.]

Cormen TH, Leiserson CE and Rivest RL (1990) *Introduction to Algorithms*. Cambridge, MA: MIT Press.

Harman G (1963) Generative grammars without transformation rules: a defense of phrase-structure. *Language* **39**: 597–616.

Pollard C and Sag I (1994) *Head-Driven Phrase Structure Grammar*. Stanford, CA: CSLI.

Robinson JA (1965) A machine-oriented logic based on the resolution principle. *Journal of the ACM* **12**: 23–41.

Further Reading

Bird S (1995) *Computational Phonology: A Constraint-Based Approach*. Cambridge, UK: Cambridge University Press. [A purely constraint-based theory of morphophonology.]

Carpenter B (1992) *The Logic of Typed Feature Structures*. Cambridge, UK: Cambridge University Press. [A comprehensive analysis of logical constraints and constraint resolution.]

Manning CD and Schuetze H (1999) *Foundations of Statistical Natural Language Processing*. Cambridge, MA: MIT Press. [A thorough overview of statistical approaches to language.]

Saraswat VA (1993) *Concurrent Constraint Programming*. Cambridge, MA: MIT Press. [Discusses a general form of constraint logic programming.]

Sowa J (ed.) (1991) *Principles of Semantic Networks*. San Mateo, CA: Morgan Kaufmann. [An excellent collection of papers on frame-based knowledge representation systems.]

Constraints on Movement

Advanced article

Gert Webelhuth, University of North Carolina, Chapel Hill, North Carolina, USA

CONTENTS

Within a sentence, constituents may appear in the local context of the items they stand in a grammatical relation with or they may move to other positions. Such movement phenomena are subject to a number of grammatical constraints.

INTRODUCTION

Theories of syntactic movement try to capture the systematic relationship between groups of sentences such as

[$_S$ *Sandy* wants to show *those pictures* to Jill] (1)

[$_S$ Those pictures Sandy wants to show to Jill] (2)

[$_S$ *Those pictures* I am told [$_S$ Sandy wants to show to Jill]] (3)

In all three sentences the noun phrase *those pictures* is understood as the thing that Sandy wants to show to Jill, but only in sentence 1 does the noun phrase appear in the position immediately following the verb *show*, which is the canonical position for a noun phrase being interpreted as the thing whose state changes as a result of the event denoted by a verb like *show*. In sentences 2 and 3, the noun phrase appears intuitively too far to the left, at the left edge of a clause, a position frequently occupied by a characteristic set of elements in many languages of the world, namely question phrases (sentence 4), relative pronouns (sentence 5), and expressions receiving particular emphasis (sentence 6):

Which pictures does Sandy want to show to Jill? (4)

The pictures *which* Sandy wants to show to Jill are lying on the floor. (5)

The pictures on the left I like but the ones on the right I find ugly. (6)

In transformational grammar, it is assumed that sentences containing dislocated phrases are generated in two steps by a formal grammar. The first step occurs within the phrase structure component and puts each semantic dependent of a verb (or other part of speech) into a local relationship with that verb. This is often referred to as a 'base-generated' structure. For example, the base-generated structure of sentence 2 would be sentence 7, which is structurally identical to sentence 1:

[$_S$ Sandy wants to show *those pictures* to Jill] (7)

The second step, a 'movement' transformation, moves the NP *those pictures* to the beginning of the clause and thereby creates the visible surface order of sentence 2.

Sentence 1 would also be derived from the basic structure in sentence 7, the difference between sentences 1 and 2 being that the NP *those pictures* stays in its base-generated position in sentence 1 but moves to the left sentence periphery in the process of deriving sentence 2.

Not every potential movement of an expression like the movement of *those pictures* in sentence 2 leads to a grammatical sentence. This is illustrated by examples 8 and 9, and 10 and 11:

Sandy met the photographer who showed *those pictures to Jill.* (8)

**Those pictures* Sandy met the photographer who showed to Jill. (9)

Sandy questioned the assumption that the photographer showed *those pictures* to Jill. (10)

**Those pictures* Sandy questioned the assumption that the photographer showed to Jill. (11)

There must thus exist constraints on movement that allow sentence 2 to be derived from sentence 7 by movement of *those pictures* but which prevent the NP from moving to the left in sentences 8 and 10. This article will discuss the nature of these constraints.

THEMATIC RELATIONS AND ARGUMENT STRUCTURE

In order to address the question of what constraints on movement exist, one needs a way to tell which expressions have moved in a given sentence and where the movement originated. Rather than postulating sentence 7 as the common base structure of sentences 1 and 2 and deriving sentence 2 from sentence 7 by moving *those pictures* to the left, one might alternatively take the left-peripheral occurrence of the NP in sentence 2 as its base position and derive its position to the right of the verb in sentence 1 by a rightward movement operation. One consideration that favors the original derivation over this hypothetical alternative is that sentence 7 base-generates the moved NP next to the verb that it is semantically related to, in the following sense.

Verbs typically refer to events that contain a certain number of participants that stand in a relation specified by the verb. In the case at hand, the verb *show* refers to an event of showing that has three participants: somebody shows something to somebody. Other verbs may refer to events that contain fewer participants: thus, the verb *sneeze* refers to an event with just a single participant, and the verb *peel* refers to a two-place event where somebody peels something. However, across events, participants may share similar properties. For instance, the participant who peels potatoes is more actively involved in the peeling event (in the sense that he or she brings the event into existence) and as a result of this action changes the state of the potatoes. Likewise, whoever shows pictures to somebody else is typically more actively involved in the action and changes the state of the other participants of the event. There has been considerable effort in linguistics to develop a classification of participant roles.

Table 1 contains a number of participant roles that are widely used in the description of word meanings. (The 'theme' role assumed a prominent position when the concept of participant roles came into wide use, and therefore participant roles are often referred to as 'thematic' or 'theta' roles.)

Verbs (and other parts of speech) can be classified according to how many participants their events involve and which participant roles must be present. This information is often referred to as the 'argument structure' of a verb. Table 2 lists some verbs and the argument structures in which they are typically used.

Most verbs have more than one argument structure, i.e. they are compatible with more than one combination of arguments. For instance, *open* can

Table 1. Some common participant roles

Name	*Abbreviation*	*Approximate definition*
Agent (Actor)	Ag	A participant most actively involved in the event or a participant who causes a change of state in other participants.
Patient (Theme)	Pt	A participant who undergoes a change of state in the event.
Experiencer	Exp	A participant who undergoes a mental or perceptual change of state.
Source	So	A participant who is the source of a transfer or movement.
Goal	Go	A participant who is the goal of a transfer or movement.

Table 2. Typical argument structures of verbs

Verb	*Argument structure*	*Example sentence*
(a) *bark*	(Ag)	The dog *barked.*
(b) *die*	(Pt)	The dog *died.*
(c) *hallucinate*	(Exp)	The witness *hallucinated.*
(d) *kick*	(Ag, Pt)	The horse *kicked* the dog.
(e) *see*	(Exp, So)	The horse *saw* the dog.
(f) *show*	(Ag, Pt, Exp)	The owner *showed* the dog *to* her neighbor.
(g) *give*	(Ag, Pt, Go)	The owner *gave* the key *to* her neighbor.

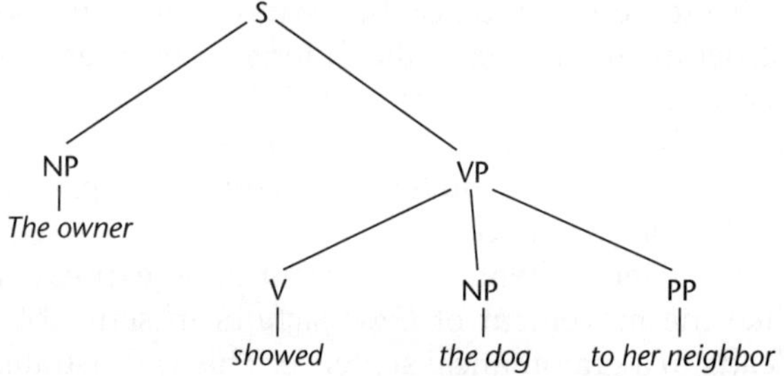

Figure 1. A syntactic structure determined by the argument structure (Ag, Pt, Exp).

be used with a single patient argument as in *the door opened* or with both an agent and a patient as in *Sue opened the door*.

LINKING BETWEEN LEXICAL AND SYNTACTIC STRUCTURE

The argument structure of a verb strongly influences the syntactic configuration in which its arguments are realized, as can be seen by examining the example sentences in Table 2. In each case, the subject precedes the verb. The intransitive verbs in (a), (b) and (c), i.e. those with a single participant role in their argument structure, make their single argument into a subject. In verbs with multiple participant roles the choice of subject is determined by the following two principles. (Argument-linking principles only apply to active verbs. The linking of arguments in passive verbs is determined by the linking in the active verbs from which the passive verbs are derived. See Dowty (1991), Levin (1993), and Davis (2001) for further discussions of argument linking.)

1. The participant that is most agent-like is realized as the subject in transitive verbs.
2. The participant that is most patient-like is realized as the direct object of transitive verbs.

In 'X-bar theory', subjects are analyzed as specifiers and direct objects as the NP immediately following the verb within the verb phrase. The argument structure (Ag, Pt, Exp) for the verb *show* thus determines the syntactic structure shown in Figure 1 for sentence (f) in Table 2.

The canonical position of the agent is before the verb, whereas the patient canonically follows the verb. This makes the order in sentence 1 a more natural candidate for the base order than the order in sentence 2. Indeed, sentence 1 is felt as more neutral by native speakers of English. The word order in sentence 2 obtains only under certain constructional conditions, as illustrated in sentences 4, 5 and 6.

ARGUMENTS FOR POSITING MOVEMENT

The motivation for analyzing some sentences in terms of base-generating an expression in one place and subsequently moving it to another place is that it elegantly captures certain grammatical generalizations which would otherwise be hard to account for. One class of such generalizations relates to the issue of argument structure which we have already discussed. For example, sentences based on the verb *show* are ungrammatical if the theme participant remains unexpressed:

*The owner showed to her neighbor. (12)

The ungrammaticality of example 12 can be captured elegantly by requiring that in the base structure of a sentence, the argument structure of every word must be fully realized by respecting linking constraints like the two principles given above. The verb *show* has a patient argument, which according to the second of those principles should be expressed as the first postverbal NP; this condition is not met in example 12, which is therefore ungrammatical. By deriving sentence 14 from 13 and sentence 16 from 15, we have an immediate explanation of why they are grammatical, in contrast to example 12:

Which dog did the owner show to her neighbor? (13)

The owner showed which dog to her neighbor. (14)

Which dog did Sandy say the owner showed to her neighbor? (15)

Sandy said that the owner showed *which dog* to her neighbor. (16)

In sentences 14 and 16, the structures from which sentences 13 and 15 are derived, the patient argument of *show* appears immediately behind the verb in base structure. In example 12, however, no patient NP appears in the sentence at all, either in postverbal position or in a possible moved position, and hence this syntactic structure does not fully realize the argument structure of the verb *show*.

Another powerful argument for movement operations comes from the locality of certain grammatical relations, such as agreement. For instance, present-tense verbs in English systematically agree with their subjects in person and number:

The *owner shows*/**show* the dog to her neighbor. (17)

But they do not systematically agree with any expression outside their clause, because subject–verb agreement is a local relationship:

*The *onlookers* think that the owner *show* the dog to the neighbors. (18)

Yet, in sentences like 19 where the subject of the subordinate clause headed by *show* appears at the beginning of the main clause, the verb *shows* still agrees with the dislocated constituent in person and number, just as it does in sentence 20 where *the owner* stays within the subordinate clause:

Which owner do the onlookers think [$_S$ *shows*/**show* the dog to the neighbors]? (19)

The onlookers think [$_S$ *the owner shows*/**show* the dog to the neighbors]. (20)

If subject agreement relations are determined before *wh*-extraction, then the verb *show* should agree with its subject in both sentences 19 and 20, no matter whether that subject is subsequently moved to a position in the main clause from which agreement with the embedded verb is otherwise impossible.

The co-occurrence restrictions on anaphoric pronouns provide similar motivation for movement. These expressions must find an antecedent within the immediate clause containing them:

The journalists said [$_S$ *the youngest participants* only paid attention to *themselves*]. (21)

In this sentence, *themselves* can only be understood as referring to the subject of the subordinate clause *the youngest participants*. If the reading were intended whereby the youngest participants only paid attention to the journalists, a personal pronoun would have to be substituted for *themselves*:

The journalists said [$_S$ the youngest participants only paid attention to *them*]. (22)

Compare sentence 23 to sentences 21 and 22:

Who did the journalists say [$_S$ only paid attention to *themselves*]? (23)

In this sentence, the NP *who*, moved from the subordinate clause to the initial position of the main clause, must still be interpreted as the antecedent of the anaphoric pronoun *themselves*, even though it no longer occurs within the immediate clause containing the anaphor. In fact, *who* must be chosen as the antecedent even though it is linearly further away from *themselves* than the subject of the main clause *the journalists*. All of these facts fall into place when we examine the base structure that underlies sentence 23:

The journalists said [$_S$ *who* only paid attention to *themselves*].

Of the two potentially available antecedents of *themselves* in sentence 23, only *who* is contained within the immediate clause containing the anaphor in base structure. If the co-occurrence restriction on anaphoric pronouns applies before the kinds of movement operations discussed here, then the empirical facts about sentence 23 immediately follow, because *who* has been moved out of the subordinate clause whereas *the journalists* was never contained in the same clause as *themselves* and therefore is not capable of anteceding it. The co-occurrence options of moved expressions in their base position should carry over to the new positions they occupy following movement operations.

ROSS'S CONSTRAINTS

Ross (1967), reacting to earlier proposals by Chomsky (1964), presented the first systematic investigation of movement constraints on a large scale. One movement rule that is systematically constrained in many languages is *wh*-movement, the rule that moves a *wh*-expression to the left periphery of a sentence in the formation of an interrogative clause. As Ross showed, this rule may operate in an unbounded fashion in English, i.e. the *wh*-expression can move an arbitrary number of clauses to the left:

What did Bill buy _? (25)

What did you force Bill to buy _? (26)

What did Harry say you had forced Bill to buy _? (27)

What was it obvious that Harry said you had forced Bill to buy_? (28)

And so on. Yet, we find that the following examples are all ungrammatical:

*What did Bill buy potatoes and _? (29)

*What did that Bill wore _ surprise everyone? (30)

*What did Cindy believe the claim that Otto was wearing _? (31)

*Whose did you find _ book? (32)

*What$_1$ did Jill wonder where$_2$ Sandy put $_{-1}$ $_{-2}$? (33)

The underscores in these examples mark the base positions of the moved *wh*-expressions and point to the intended interpretations. For instance, the intended interpretation of example 30 is similar to that of the echo-question in example 34:

That Bill wore WHAT surprised everyone? (34)

Clearly, the paradigm in examples 25 to 33 cannot be captured by putting an upper limit on the number of words or sentences that a *wh*-expression can move across on its way to the left periphery of the sentence. The movement path of the *wh*-word in the grammatical sentence 28 is much longer than the one we find in the ungrammatical sentence 32.

Ross proposed that what matters is the structural organization of the string that *wh*-movement crosses. For instance, to distinguish between examples 25 and 29, Ross formulates the 'coordinate structure constraint' namely, that in a coordinate structure, no conjunct may be moved, nor may any element contained in a conjunct be moved out of that conjunct.

In example 29, *what* is contained in the coordinate noun phrase *potatoes and what*:

Bill bought [$_{NP-1}$ [$_{NP-2}$ potatoes] and [$_{NP-3}$ what]] (35)

The coordinate structure constraint prevents the *wh*-movement transformation from pulling NP-3 out of the coordinate structure NP-1, and so example 29 is ungrammatical.

Example 30 may be contrasted with sentences like:

What did Cindy say that Bill wore _? (36)

These examples are structurally different in that *what* is extracted from a subject clause in example 30 but from an object clause in example 36. Ross formulates the 'sentential subject constraint' to prohibit any movement operation from extracting an element from a sentential subject. Example 31 involves a violation of yet another constraint, the 'complex NP constraint' which blocks extraction from complement clauses to nouns. Compare example 31 with the similar sentence:

What did Cindy believe that Otto was wearing _? (37)

What makes sentence 37 grammatical in contrast to 31 is that the clause *that Otto was wearing what* from which the interrogative element is extracted is a complement to a verb in the grammatical context in sentence 37 but a complement to the noun *claim* in 31, violating the complex NP constraint.

Ross formulates a substantial number of constraints on transformations, as well as empirical insights which have been elaborated on in later work. Ross's doctoral thesis is generally considered one of the most insightful and influential works in the history of generative grammar.

THE CONDITIONS FRAMEWORK

Chomsky (1973) sought to systematize a number of constraints on transformations, including constraints on movement and constraints on semantic interpretation. This effort to postulate general structural constraints that can replace groups of language-particular or construction-specific constraints marked the beginning of a research program that was highly influential and led to the theories of 'government' and 'binding' and the 'minimalist program' which dominated syntactic theorizing in the 1980s and 1990s. The three most far-reaching constraints that Chomsky proposes are the 'tensed sentence condition', the 'specified subject condition', and the 'subjacency condition'.

The tensed sentence condition captures the differences between examples 38, 39 and 40, under the assumption that these are derived from 41, 42 and 43 respectively by movement of the quantifier *each* to the right:

The candidates hated *each* other. (38)

The candidates expected [$_S$ *each* other to win] (39)

*The candidates expected [$_S$ that *each* other would win] (40)

The candidates *each* hated the other(s). (41)

The candidates *each* expected [$_S$ the other(s) to win] (42)

The candidates *each* expected [$_S$ that the other(s) would win] (43)

In example 40, unlike 38 and 39, the quantifier must move into a tensed sentence. In 39, the complement of *expected* is analyzed as a sentence in order to give this verb a uniform subcategorization frame. Yet, in 39 the quantifier moves into a nonfinite sentence and in 38 it moves into the noun phrase *the other*.

The tensed sentence condition prohibits a rule from applying to two constituents if one of the constituents is contained within a tensed sentence and the other one occurs outside that sentence.

The tensed sentence condition needs to be supplemented by the specified subject condition, since subjects of complement clauses behave differently from non-subjects, as a comparison between examples 38, 39 and 40 and the examples below illustrates:

The candidates expected [$_S$ PRO to defeat *each* other] (44)

*The candidates expected [$_S$ the soldier to shoot *each* other] (45)

The candidates$_1$ *each* expected [$_S$ PRO$_1$ to defeat the other] (46)

The candidates *each* expected[$_S$ the soldier to shoot the other] (47)

Both 44 and 45 are nonfinite, so the tensed sentence condition is unable to account for the grammaticality contrast. What differentiates them is that, in its rightward movement, *each* must cross the lexicalized subject *the soldier* in 45, whereas in 44 it must only cross the abstract anaphoric subject PRO. The specified subject condition prohibits a transformation from relating two constituents if they are separated by a specified subject, where lexical subjects always count as specified.

Finally, the subjacency condition generalizes a number of earlier conditions on movement out of sentences and noun phrases. This will be discussed below.

BARRIERS

Chomsky (1986) attempts to unify all movement constraints in terms of a phrase-structurally defined notion of 'barrier', which is also meant to play a role in the theory of government. Chomsky starts from the assumption that complements generally play a more active role in extraction constructions than either subjects or adjuncts. Consider the following examples:

Who did [$_{IP}$ Mary find [$_{NP}$ a picture of _]]? (48)

*Who did [$_{IP}$ [$_{NP}$ a picture of _] scare the children]? (49)

*Who did [$_{IP}$ Mary laugh [$_{PP}$ when she found a picture of _]]? (50)

??Which car did Mary sleep [$_{PP}$ while Jill fixed _]? (51)

*Which mechanic did Mary sleep [$_{PP}$ while _ fixed the car]? (52)

*How quickly did Mary sleep [$_{PP}$ while Jill fixed the car _]? (53)

(Note the intended interpretation in example 53: *how quickly* modifes *fixed*.)

Examples 48, 49 and 50 show that it is easier to extract a constituent from a complement than from a subject or an adjunct. In 49 *who* has been extracted from the subject of the sentence, and in 50 from an adjunct, and both sentences are worse than 48. Examples 51, 52 and 53 show that under certain conditions it is also easier to extract complements than to extract subjects or adjuncts. Each of these examples involves extraction from an adjunct, but the violation incurred by the subject in 52 and by the adjunct in 53 is felt to be stronger than the violation incurred by the complement in 51.

Complements are thus both easier to extract and easier to extract from, whereas subjects and adjuncts are harder to extract and harder to extract from. Chomsky develops a complex web of constraints and definitions in an attempt to capture these generalizations. Crucial to his approach are certain assumptions about the X-bar theory of phrase structure, including the tree structure shown in Figure 2.

One problem with these assumptions is that Chomsky requires this full abstract structure to be present even in sentences where the structure is not supported by observable morphological or lexical material. Thus, even simple sentences like *Mary worked* are given abstract C and I nodes, even though there is no empirical evidence for the

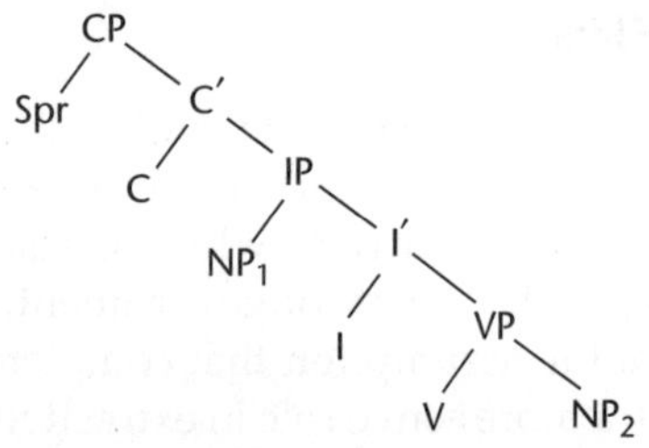

Figure 2. Chomsky's proposed uniform structure for sentences. 'C' stands for complementizers like *that* and *whether*; 'Spr' is a 'landing site' for *wh*-operators of any part of speech; 'I' stands for inflectional elements and auxiliaries; 'NP_1' is the clausal subject position; and 'NP_2' is the clausal direct object position.

existence of a complementizer or a self-standing inflectional element separate from the verb in these kinds of sentences.

In order to account for the contrasts in examples 48 to 53, which show that it is possible to extract from a complement but impossible to extract from subjects or adjuncts, Chomsky formulates several assumptions. Firstly, he defines 'blocking categories' as maximal projections that are not theta-governed by a lexical head. Here, lexical heads include the major parts of speech (verb, adjective, noun, etc.) but specifically exclude the minor parts of speech (complementizer and inflection).

IP is governed by C, while VPs and subjects are governed by the I. Adjuncts may in principle be governed by lexical heads, but they are, by definition, not theta-marked. Thus, IP, VP, subjects and adjuncts are blocking categories. Complements are the only class of expressions that are always theta-marked (this is a requirement of Chomsky's (1986) 'projection principle') and they are governed by a lexical head. Thus, complements are not blocking categories.

According to this definition, then, VPs should be barriers to movement, but this is empirically incorrect. Chomsky therefore proposes a mechanism that allows the 'barrierhood' of VP to become voided. He defines barriers to movement as follows: IP is a barrier for everything within its subject or a sentence-level adjunct; subjects and adjuncts are barriers for everything contained in them; and CP is a barrier for everything contained within IP.

The ungrammaticality of examples 49 and 50 now follows, given the subjacency condition, which stipulates that no movement step is allowed to cross more than one barrier. In 49, the subject NP is a barrier and so is the IP. The extraction of *who* from [$_{IP}$ [$_{NP}$ *a picture of* _] ...] thus crosses two barriers, implying that the sentence is ungrammatical. Example 50 is similar, except that in this case the second barrier is the adjunct PP. In contrast, example 48 is grammatical, since the complement NP is not a barrier, given that it is theta-marked by a lexical head.

Chomsky also addresses the data in examples 51, 52 and 53. While adjuncts are generally hard to extract from, the judgment of any given extraction depends on the kind of element that is extracted. In English, extraction of complements from movement islands often yields a slightly more favorable judgment than extraction of subjects or adjuncts from the same kind of island. Chomsky uses the 'empty category principle' (Chomsky, 1981) in an attempt to solve this problem. According to this principle, traces must be properly governed, i.e. they must be either theta-governed by an X^0 head or antecedent-governed. To be antecedent-governed, no barrier is allowed to intervene between the trace and its antecedent (i.e. the moved expression or another one of its traces).

From this definition of proper government it follows that the trace of a complement is always properly governed and hence need not be antecedent-governed. Subject and adjunct traces, in contrast, are not governed by X^0 heads that theta-mark them and thus need to be antecedent-governed in order to be properly governed. Extraction of a complement from an adjunct thus yields a violation only of the subjacency condition, as shown earlier, whereas extraction of a subject or an adjunct under the same conditions yields a violation of the empty category principle in addition to the subjacency violation incurred by complements. Examples 52 and 53 should thus be stronger violations than example 51.

The barriers framework has a number of conceptual and empirical weaknesses. Ultimately, Chomsky abandoned it to pursue even more abstract phrase-structural theories. Other researchers turned their back on the principles-and-parameters framework, believing that the syntactocentric, derivation-driven design of the theory as a whole, rather than individual definitions or principles, were to blame for its chronic empirical problems. The proliferation of abstract phrase-structure categories has already been mentioned. Furthermore, there are large amounts of data that are incompatible with the main ideas of the Barriers framework. There are languages in which the subject–object and object–adjunct asymmetries that Chomsky's theory is specifically designed to derive do not obtain. For instance, in Swedish subjects can be extracted much more easily than Chomsky's theory predicts. Moreover, with respect to adjuncts, von Stechow and Sternefeld (1988, p. 371) present constructions in German that show the precise opposite of what Chomsky's theory predicts: adjuncts can be extracted from certain phrases that do not allow complements to escape. In fact, even English is much less uniform than Chomsky's theory would lead one to expect. Santorini (2001) has collected a list of attested examples that violate a number of prominent constraints on movement that have been postulated, including the following example which contains an extraction from an adjunct:

a scenario that government agencies are spending billions of dollars preparing for (54)

Other data confirm that principles-and-parameters theories are an oversimplification of the data. Culicover (1999) presents the following contrast, illustrating that whereas prepositions generally allow their complement to move away in English, the preposition *since* is an idiosyncratic lexical exception to this generalization which cannot be captured by setting a category-wide parameter:

*Which party hasn't John called since _? (55)

As successive reforms of Chomsky's parametric framework have been unable to address these sorts of detailed empirical problems, constraint-based lexicalist theories of grammar have gained increasing acceptance in linguistics, psycholinguistics, language acquisition, and computational linguistics. We will now discuss such theories.

NON-TRANSFORMATIONAL APPROACHES

Non-transformational approaches to grammar are generally motivated by a belief that, as Chomsky has sought more elegant answers to a restricted set of problems, his theories have become less empirically realistic and the theoretical constructs he invokes have become less open to theory-neutral verification by other researchers. In the 1980s and 1990s, a number of Chomskyan linguists moved on to other theories, in particular constraint-based lexicalist theories.

One problem that systematically arises with Chomsky's theory is that the broad generalizations he attempts to derive are counterexemplified by small classes of items or individual words. In order to circumvent this problem, researchers have developed tools that are capable of deriving grammatical generalizations of variable scope, including the fully productive generalizations that Chomsky's theory can handle, but also semi-productive patterns and completely idiosyncratic phenomena.

Currently the most credible alternatives to Chomsky's syntactocentric principles-and-parameters approach invoke a sophisticated set of lexical and constructional tools in the analysis of grammatical phenomena, including movement phenomena. Most prominent among these theories are 'Head Driven Phrase Structure Grammar' (Pollard and Sag, 1994), 'Lexical Functional Grammar' (Bresnan, 2000), and 'Construction Grammar'. The remainder of this article will focus on the approach to movement proposed in the first of these theories.

Head Driven Phrase Structure Grammar developed from 'Generalized Phrase Structure Grammar' (Gazdar *et al.*, 1985). The desire for a linguistic theory that is simultaneously formally precise, empirically accurate and broad, and computationally implementable in an efficient manner, led Gazdar to avoid all transformations in his theory, including those that move constituents from a base-generated position to another position in a tree-altering fashion. Instead, he proposed that 'movement' is a metaphor for a featural dependency in a tree, encoding the information that a constituent that in principle can be expressed in one part of a tree can also be expressed elsewhere in that tree.

Sag and Fodor (1994) and Bouma *et al.* (2001) have combined Gazdar's theory of extraction with a theory of argument realization that gives words fine control over the syntactic realization options of their arguments. Together with constructional constraints on the flow of argument information in a syntactic tree, these assumptions yield an empirically powerful theory of extraction phenomena which its adherents believe to be capable of deriving the same broad generalizations that Chomsky's syntactocentric theory is able to capture, without sacrificing coverage of semi-productive and idiosyncratic phenomena.

Bouma *et al.* make use of the idea that a word can be represented by a data structure having an argument structure and a valence. The argument structure of a word is involved in interpretative properties, including the binding theory, while the valence determines the surface-syntactic context that the word must appear in (for principled reasons, the theory does not countenance any unobservable syntactic levels, such as Chomsky's D-structure or LF, nor any unobservable signs such as traces and phonologically empty heads). Part of the individual grammar of a language is a set of principles that determine how each of a word's arguments may be realized on the surface. Thus, if the language permits arguments to be realized morphologically, then they will not be projected into the syntax, which makes the dependence on phonologically empty subjects superfluous and keeps syntactic structures concrete and open to theory-independent verification. Extraction phenomena are handled in terms of permissible mappings between argument structure and valence. Just as words have the ability to determine whether or not their arguments can be spelled out morphologically, they have the ability to constrain the syntactic realization options of their arguments: arguments that appear in a word's valence are spelled out in the word's local syntactic domain, whereas the descriptions of arguments that appear in the word's GAPS specification are percolated upward in the syntactic tree until this information appears in a syntactic configuration where a filler sign is found which is featurally compatible with the argument description. For the verb *show* in examples 1 and 2, an argument structure like the following would be postulated:

$$\text{ARG-ST} < \text{NP}_{\text{agent}}, \text{NP}_{\text{patient}}, \text{PP}_{\text{goal}} > \quad (56)$$

Since *show* permits each of its arguments to be realized either locally or nonlocally, its valence and gaps specification may be any of the following, among others:

$$\begin{array}{l} \text{SUBJ} < \text{NP}_{\text{agent}} > \\ \text{COMPS} < \text{NP}_{\text{patient}}, \text{PP}_{\text{goal}} > \\ \text{GAPS} \{\} \end{array} \quad (57)$$

$$\begin{array}{l} \text{SUBJ} < \text{NP}_{\text{agent}} > \\ \text{COMPS} < \text{PP}_{\text{goal}} > \\ \text{GAPS} \{\text{NP}_{\text{patient}}\} \end{array} \quad (58)$$

$$\begin{array}{l} \text{SUBJ} < \text{NP}_{\text{agent}} > \\ \text{COMPS} < \text{NP}_{\text{patient}} > \\ \text{GAPS} \{\text{PP}_{\text{goal}}\} \end{array} \quad (59)$$

Whereas subjects and complements must be realized within the valence domain of the verb, the information in GAPS is percolated up the tree in a fashion that allows fillers to be found for these gaps outside the verb's local valence domain. This is how long-distance dependencies arise. Specification 57 is used to generate surface forms like sentence 1, where each of the three arguments of *show* is generated within its local valence domain. Specification 58 would underlie a sentence like sentence 2, where the subject and the PP object remain within the valence domain, but the direct object NP is realized as a filler at the beginning of the clause. Specification 59 would underlie such sentences as *To Jill I want to show these pictures* or *to whom do you want to show these pictures*, where the PP argument is realized at a distance.

This theory gives words full control over whether an argument may or must be realized within the local valence domain of the word. It is thus predicted that, as in other grammatical domains (e.g. inflection), some languages will be very homogenous in their constraints, whereas others will be less so. It is even predicted that different head types within a language may behave differently. There is impressive evidence to support this prediction. For instance, whereas verbs in English permit their specifier (their subject) to be extracted, nouns never do, as example 32 demonstrates. This generalization can be captured by imposing featural well-formedness conditions on the permissible relationship between members of the parts of speech noun and verb and their respective specifier arguments. The kind of lexical idiosyncrasy that is represented by example 55 in the extraction domain (i.e. that *since* does not allow its NP argument to be realized in GAPS) is parallel to idiosyncratic lexical requirements in other domains (e.g. that a verb exceptionally requires a different case on its direct object than most other transitive verbs in the language).

Other differences between languages and constructions can be captured elegantly by constraining the values of the GAPS attribute in different parts of the syntactic tree or in different constructions. For instance, in languages where extraction from subjects is impossible, it suffices for the grammar to contain a constraint that the GAPS value of any subject be the empty set. An analogous constraint will prevent extraction from adjuncts or *wh*-islands where these constraints are empirically called for. Finer distinctions can be drawn by imposing constraints on the content of GAPS in specific constructions and languages. Thus, it would be easy to derive the generalization that arguments are easier to extract from some constructions whereas adjuncts are easier to extract from others. Universal tendencies that are not the result of parsing or memory preferences can be encoded in terms of the grammatical archetypes proposed in Ackerman and Webelhuth (1998), which are conceived of as the set of grammatical concepts that guide the language learner's acquisition process.

References

Ackerman F and Webelhuth G (1998) *A Theory of Predicates*. Stanford, CA: CSLI Publications.

Bouma G, Malouf R and Sag IA (2001) Satisfying constraints on extraction and adjunction. *Natural Language and Linguistic Theory* **19**: 1–65.

Bresnan J (2000) *Lexical-Functional Syntax*. Oxford: Blackwell.

Chomsky N (1964) *Current Issues in Linguistic Theory*. The Hague: Mouton.

Chomsky N (1973) Conditions on transformations. In: Andersons S and Kiparsky P (eds) *Festschrift for Morris Halle*, pp. 232–286. New York, NY: Holt, Rinehart and Winston.

Chomsky N (1981) *Lectures on Government and Binding*. Dordrecht: Foris.

Chomsky N (1986) *Barriers*. Cambridge, MA: MIT Press.

Culicover P (1999) *Syntactic Nuts: Hard Cases, Syntactic Theory, and Language Acquisition*. Oxford: Oxford University Press.

Davis T (2001) *Linking by Types in the Hierarchical Lexicon*. Stanford, CA: CSLI Publications.

Dowty D (1991) Thematic proto-roles and argument selection. *Language* **67**: 547–619.

Gazdar G, Klein E, Pullum GK and Sag IA (1985) *Generalized Phrase Structure Grammar*. Cambridge, MA: Harvard University Press/Oxford: Blackwell.

Levin B (1993) *English Verb Classes and Alternations: A Preliminary Investigation*. Chicago, IL: University of Chicago Press.

Pollard C and Sag IA (1994) *Head Driven Phrase Structure Grammar*. Chicago, IL: University of Chicago Press.

Ross JR (1967) *Constraints on Variables in Syntax*. PhD thesis, MIT. [Reproduced by the Linguistics Club of Indiana University.]

Sag IA and Fodor JD (1994) Extraction without traces. In: *Proceedings of the Thirteenth Annual Meeting of the West Coast Conference on Formal Linguistics*, pp. 365–384. Stanford, CA: CSLI Publications.

Santorini B (2001) *(Un)expected Movement*. http://www.ling.upenn.edu/~beatrice/examples/movement.html

von Stechow A and Sternefeld W (1988) *Bausteine Syntaktischen Wissens*. Opladen, Germany: Westdeutscher, Verlag.

Construction Grammar

Introductory article

Adele E Goldberg, University of Illinois, Urbana, Illinois, USA

CONTENTS

Construction Grammar is a linguistic theory concerned with the nature of speakers' knowledge of language. Like traditional grammars, Construction Grammar takes the basic units of language to be form–meaning pairings, or constructions.

CONSTRUCTIONS

A *construction* is defined as a pairing of form with meaning/use, such that some aspect of the form or some aspect of the meaning/use is not strictly predictable from the component parts or from other constructions already established as existing in the language. On this view, phrasal patterns, including the constructions of traditional grammarians, such as relative clauses, questions, locative inversion, and so on, are given theoretical status. Words (or really, *morphemes*) are also constructions, according to this definition, since their form is not predictable from their meaning or use. Given this, it follows that the mental dictionary or *lexicon* is not neatly delimited from the rest of grammar, although phrasal constructions differ from lexical items in their internal complexity.

Both phrasal patterns and lexical items are stored in an extended 'constructicon'. Elements within the constructicon vary in degrees of idiomaticity. At one end of the idiomaticity continuum, we find very general, abstract constructions such as the subject–predicate construction; at the other end, we find simple lexical items and constructions with all of their lexical fillers specified but with noncompositional meanings (e.g. 'kick the bucket'). In between, we find the full range of possibilities: for example, idioms which have freely fillable positions (e.g. 'keep/lose X's cool'), compositional collocations with fixed word order (e.g. 'up and down'), phrasal patterns that are only partially productive (e.g. the English ditransitive), and phrasal patterns which are partially morphologically specified (e.g. 'The Xer, the Yer', as in 'The less it rains, the better the potatoes').

Construction Grammar shares with several other current theories, including Head-Driven Phrase Structure Grammar, Cognitive Grammar, and Montague Grammar, the basic and fundamental idea that the construction (or *sign*) is central to an account of language. This view of grammar can be contrasted with the claim made by Principles and Parameters theories that constructions are entirely epiphenomenal, a mere by-product of the interaction of the principles of Universal Grammar, once the values of the parameters are fixed. Although most aspects of language are highly motivated, in the sense that they are related to other aspects of the grammar and are non-arbitrary, Construction Grammar holds the view that much of language is idiosyncratic to varying degrees and must therefore be learned.

Declarative, Monostratal Representation

A given sentence is licensed by the grammar if and only if there exists in the language a set of constructions which can be combined (or superimposed) to produce an accurate representation of the surface structure and semantics of that sentence. An ambiguous sentence is a sentence for which there exists more than one set of constructions that can be assembled to produce a possible representation. Constructions are represented declaratively, and any constructions which do not conflict may be combined to give rise to grammatical expressions. Thus Construction Grammar is monostratal: no derivations are posited.

Typically, particular sentences (or *constructs*) instantiate several constructions simultaneously. For example, sentence (1) below instantiates the subject–predicate construction, the ditransitive construction, the determiner construction ('the letter'), the past tense morphological construction ('fax-*ed*'), and five simple morphological constructions, corresponding to each word in the sentence:

Elena faxed Ken the letter. (1)

Integrated Information

Conventionalized aspects of both meaning and use are directly related to particular syntactic patterns within individual constructions. Thus, Construction Grammar does not assume that syntax is generally isolated or isolatable from semantics or conditions of use. Construction Grammar also eschews a strict division between the pragmatic and the semantic. 'Frame-semantic' (encyclopedic) meaning is considered fundamental to an adequate understanding of linguistic entities, and as such is integrated with more traditional definitional characterizations. Generalizations about particular arguments being topical, focused, inferable, and so on, are also stated as part of the constructional representation. Facts about the use of entire constructions, including register, dialect variation, etc., are stated as part of the construction as well. Thus a construction may be posited because of something not strictly predictable about its frame-semantics, its packaging of information structure, or its context of use.

Relations among Constructions within a Language

Constructions do not form an unstructured set, but rather a highly integrated system, based on general principles of categorization. Constructions are typically closely related to other constructions, and are, in that sense, not arbitrary. Generalizations across constructions are captured within the theory via an inheritance hierarchy, which allows shared structure to be represented.

For example, an abstract *Left Isolate* construction is inherited by several different constructions, exemplified by the following:

a. the woman who she met yesterday (restrictive relative clause)
b. Abby, who she met yesterday (nonrestrictive relative clause)
c. Bagels, I like. (topicalization)
d. What do you think she did? (main clause nonsubject *wh* – question) (2)

Each of these patterns – restrictive and nonrestrictive relative clauses, topicalization, and *wh*-questions – requires a distinct construction of its own, owing to its particular formal and pragmatic properties. But each inherits from the more general Left Isolate construction, which specifies the properties that are shared. In particular, this construction has two 'sisters', with the specification that the left sister satisfies the valence requirement of some predicator at an undefined depth in the right sister; the right sister is a maximal verb phrase, with or without a subject. Thus the Left Isolate construction serves to capture the generalizations across these various patterns.

Formalization

Many practitioners of this theory have adopted the use of a unification-based formalism in order to rigorously detail the specifications of particular constructions. Thus each construction is represented by an Attribute–Value Matrix (AVM). Each attribute can have at most one value. Attributes may be *n*-ary, or may be feature structures themselves.

Any pair of AVMs can be combined to license a particular expression, as long as there is no value conflict on any attribute. When two AVMs unify, they map onto a new AVM, which has the union of attributes and values of the two original AVMs.

RESEARCH FOCUS

Data

Research in Construction Grammar has emphasized the importance of attested data, gathered from discourse or corpora. At the same time, Construction Grammarians routinely supplement corpus data with data gained from introspection, one obvious reason being that corpora do not contain sentences marked as unacceptable. Another source of data comes from psycholinguistic experimentation.

Full Coverage: Lexical Semantics and Marked Constructions

There has been a focus on the semantics and distribution of particular lexical items within the framework, owing to the belief that the rich semantic/pragmatic constraints on individual words or idiomatic phrases reveals much about our knowledge of language. There has been a great deal of attention paid to marked constructions within the theory. For example, consider the Covariational Conditional construction, exemplified by 'the more you think about it, the less you understand'. Independent knowledge of 'the' and grammatical comparison will not directly predict that this relevant class of expressions will exist or have exactly the form and meaning they have; therefore a distinct construction is posited. Other examples of marked constructions include the 'What's *X* doing *Y*?' construction, exemplified by sentences such as 'What's that fly doing in my soup?', and the Nominal Extraposition construction, e.g. 'It's amazing the difference!'.

As these examples indicate, Construction Grammar aims to account for the full range of facts of any language, without assuming that a particular subset of the data is part of a privileged 'core'. Researchers argue that marked constructions shed light on more general issues, and serve to illuminate what is required for a complete account of the grammar of a language. Construction Grammarians takes the point of view that the ordinary patterns of grammar do not differ qualitatively from these sorts of quantitatively more complex constructions.

Argument Structure Constructions

In many current linguistic theories, the form and general interpretation of basic sentence patterns of a language are taken to be determined by semantic and/or syntactic information specified by the main verb in the sentence. The sentence patterns given in (3) and (4) indeed appear to be determined by the specifications of 'give' and 'put' respectively:

Chris gave Pat a ball. (3)

Pat put the ball on the table. (4)

'Give' is a three-argument verb and is expected to appear with three complements corresponding to agent, recipient, and theme. 'Put', another three-argument verb, requires an agent, a theme, and a location, and appears with the corresponding three complements in (4). However, while (3) and (4) represent perhaps the prototypical case, the interpretation and form of sentence patterns of a language are not reliably determined by independent specifications of the main verb. For example, it is implausible to claim that 'sneeze' has a three-argument sense, and yet it can appear as in (5):

She sneezed her tooth across the yard. (5)

The following attested examples similarly involve sentential patterns that do not seem to be determined by independent specifications of the main verbs:

'She smiled herself an upgrade.' (Douglas Adams, *Hitchhiker's Guide to the Galaxy*; Harmony Books) (6)

'We laughed our conversation to an end.' (J. Hart, *Sin*; 1992, Ivy Books) (7)

Moreover, verbs typically appear with a wide array of complement configurations. Consider the verb 'sew' and the various constructions in which it can appear (labeled in parentheses):

a. Pat sewed all afternoon. (intransitive)
b. Chris sewed a shirt. (transitive)
c. Pat sewed Chris a shirt. (ditransitive)
d. Pat sewed the sleeve shut. (resultative)
e. Pat sewed a button onto the jacket. (caused – motion)
f. Chris sewed her way to fame and fortune. (way – construction) (8)

In Construction Grammar, instead of predicting the surface form and interpretation solely on the basis of the verb's independent specifications, the lexical verb is understood to combine with an argument structure construction (e.g. the ditransitive, resultative, the caused-motion construction, etc.). Verbs constrain the type of argument structure constructions with which they can combine by their frame-specific semantics and particular obligatory roles, but they typically can combine with constructions in several ways.

It is the argument structure constructions that provide the direct link between surface form and general aspects of the interpretation such as something causing something else to move, someone causing someone to receive something, something moving somewhere, someone causing something to change state, and so on. The argument structure constructions, which provide the basic sentence patterns of a language, directly reflect these types of basic frames of experience. That is, the skeletal patterns, independently of the main verb, designate such patterns of experience. Thus constructions are invoked both for marked or especially complex pairings of form and meaning and for many of the basic, unmarked patterns of language.

FUTURE PROSPECTS

Cross-linguistic Work

Constructions that are sometimes labeled as the 'same' in two languages typically differ subtly in their form, their meaning, and/or their use. Thus Construction Grammarians have generally been cautious about trying to explain generalizations that may not be exceptionless. There is, however, a growing body of work on constructions in various languages, and a growing focus on accounting for cross-linguistic tendencies, similarities, and implicational hierarchies.

Psycholinguistics: Processing and Acquisition

A central claim made by Construction Grammar is that words and phrases are the same basic type of entity: learned pairings of form and meaning/use. A good deal of interest in the theory has been generated within psycholinguistics, by researchers in both processing and acquisition, because they see in Construction Grammar the possibility of a psychologically plausible and testable linguistic theory.

Further Reading

Bates E and Goodman JC (1997) On the inseparability of grammar and the lexicon: evidence from acquisition, aphasia and real-time processing. *Language and Cognitive Processes* **12**(5–6): 507–584.

Bencini G and Goldberg A (2000) The contribution of argument structure constructions to sentence meaning. *Journal of Memory and Language* **43**: 640–651.

Fillmore CJ, Kay P and O'Connor MC (1988) Regularity and idiomaticity in grammatical constructions: the case of LET ALONE. *Language* **64**: 501–538.

Goldberg AE (1995) *Constructions: A Construction Grammar Approach to Argument Structure Constructions.* Chicago, IL: University of Chicago Press.

Jackendoff R (1997) Twistin' the night away. *Language* **73**(3): 534–559.

Jurafsky D (1996) A probabilistic model of lexical and syntactic access and disambiguation. *Cognitive Science* **20**: 137–194.

Kay P and Fillmore CJ (1999) Grammatical constructions and linguistic generalizations: the *What's X doing Y?* construction. *Language* **75**(1): 1–33.

Koenig J-P and Jurafsky D (1994) Type underspecification and on-line construction in the lexicon. *Proceedings of the Thirteenth West Coast Conference on Formal Linguistics.*

Lakoff G (1987) *There*-constructions. *Women, Fire and Dangerous Things: What Categories Reveal about the Mind.* Chicago, IL: University of Chicago Press.

Michaelis L and Lambrecht K (1996) Toward a construction-based theory of language function: the case of nominal extraposition. *Language* **72**(2): 215–247.

Tomasello M (1998) The return of constructions. *Journal of Child Language* **25**(2): 431–443.

Zwicky A (1994) Dealing out meaning: fundamentals of syntactic constructions. *Berkeley Linguistics Society* **20**: 611–625.

Constructivism

Intermediate article

Jacqueline Grennon Brooks, State University of New York, Stony Brook, New York, USA

CONTENTS

Constructivism is a learning theory based on the notion that learners generate meaning through iterative mental formulation and reformulation of theories that satisfy the search for understanding.

INTRODUCTION

Constructivism has roots in various research traditions. It asks the psychological question: how is learning self-regulated? It considers the epistemological query: what is knowledge? It poses the pedagogical problem: how can educators facilitate knowledge construction? It examines the philosophical dilemma: is there an objective truth that we struggle to know, or are their different truths dependent on perception? Amid an array of widely varied responses to these questions is the cohesive focus on the learner's active role in generating meaning.

ACTIVE PROCESSES OF MEANING MAKING

The cognitive processes involved in making meaning are active ones that require the learner to continually evaluate new information and experiences against the learner's current theories, rules or notions. This viewpoint stands in stark contrast to other assertions that the learner's mind is a clean slate ready for inscription through direct teaching. Constructivism states that the learner approaches new experiences with a set of pre-established beliefs and naive theories, and that the learner changes those beliefs and theories only when unable to reconcile new data with previously held conceptions. Often learners will dismiss data that do not fit their present thinking as irrelevant or unrelated, and will assimilate the new information into their previously established theories without any disequilibrium or cognitive conflict. When learners find new data compelling enough to reconsider old theories, however, they experience cognitive conflict. They conquer their disequilibrium by accommodating current theories to include adequate explanations of past and present data (Piaget, 1953, 1970a). Educational programs based on constructivist principles are premised on the idea that it is the learner's responsibility to construct meaning through reflection on experiences with objects, phenomena or people, and that it is the teacher's responsibility to scaffold learner reflection in a manner that may generate learner analysis, synthesis and insight. (*See* **Naive Theories, Development of**)

The Nature of Knowledge

When a learner can construct interrelationships among sets of factual information and apply those understandings of interrelationships in novel contexts, the learner has constructed knowledge. This knowledge is dependent on mental structures. If a learner does not have the precursor mental structures in place, for example, to understand proportions, no amount of repeating the statement 'density is a ratio of mass to volume' will help the learner establish the concept. The learner's search for relationships among variables is the mental activity critical for an understanding of density as a ratio and the comparison of densities as a proportion. The depth to which a learner can construct understandings is predicated not only on the precursor mental structures mentioned previously, but on the information to which the learner has access. For instance, with information about atomic structure, density can be further understood in terms of the degree and nature of the packing of atoms in crystalline form. Within the iterative process of cognitive growth, the learner can create new knowledge from an ever-expanding repertoire of information that gives rise to yet new mental structures and then possibly new mental stages. (*See* **Piagetian Theory, Development of Conceptual Structure**)

Mental Structures

One way of describing constructivism is through an analysis of mental structures.

> We may say that a structure is a system of transformations. Inasmuch as it is a system and not a mere collection of elements and their properties, these transformations involve laws. ... In short, the notion of structure is comprised of three key ideas: the idea of wholeness, the idea of transformation, and the idea of self-regulation. (Piaget, 1970b: p. 5)

Wholeness refers to the learner's quest to map the relationships among parts of a set. For instance, knowing the meaning of each word in a sentence is different from knowing the meaning of the entire sentence. It is the learner's ability to derive meaning from the relationships among the words in a sentence that forms the basis of the learner's understanding of the sentence. Transformation refers to the notion that as the learner constructs deeper understandings of the world, a mathematical necessity for grouping those understandings emerges. Thus, fundamental transformations of undergirding logical structures gives rise to new, more inclusive structures. Piaget's own career history provides an example. As a biologist studying the appearance of new structures in plants, Piaget transformed his understanding of how the hereditary aspects of the plant are affected by conditions in the environment into a more inclusive relationship between previous structures and environmental conditions that he could apply in a variety of settings. In the immediate case, he applied this relationship in the realm of psychology. Self-regulation refers to the learner's ability to engage in assimilation and accommodation in order to either maintain cognitive equilibrium or resolve cognitive conflict and reestablish cognitive equilibrium.

Personal knowledge construction is a key element of the constructivist paradigm. The eighteenth-century philosopher Vico was one of the earliest writers to put forth the notion that human beings can only know what their cognitive structures allow them to know. This notion surfaces for teachers in terms of readiness for learning. How far a learner can progress is a function of what the learner currently understands, and what the learner currently understands is a function of the learner's existing mental structures. The teacher plays a part in facilitating the learner's development of new knowledge and new structures through the creation of settings in which the learner may detect discrepancies and in which the teacher fosters discrepancy resolution. This iterative process gives rise to ensuingly more rigorous theory building.

Cognitive Conflict

Logical structures and processes of transformation, wholeness and self-regulation are terms from the philosophical and psychological literature. Constructivism, when described by educators, is associated with another set of correlated terms. Ausubel (1963) explains learning through a process he calls 'subsumption', in which new, more specific knowledge is linked with previous, more inclusive knowledge. His term 'obliterative subsumption', describes a process highly related to the law of transformation set out by Piaget in which concepts are modified so dramatically over time that learners can lose access to some of the specifics of their own earlier thinking. These more robust concepts characterize the superordinate learning that is typically called 'meaningful' learning.

Prior Knowledge

What a learner already knows and how the teacher scaffolds the current learning environment are important determiners of the future understandings the learner will be able to construct. When a learner is investigating a new phenomenon, whether it be how shadows are cast, how an odd number of supplies can be shared among an even number of students, or how one locates an entry in an encyclopedia, the constructivist teacher negotiates the phenomenon or concept with the learner using a subtle yet observable set of practices. The teacher seeks to understand the learner's readiness to generate certain types of knowledge by determining the learner's present hunches, conceptions, beliefs, etc. The teacher then provides opportunities for the learner to confirm or refute those initial thoughts. Unless the confirmations or refutations come from the learner, the likelihood that the learner will generate understanding is compromised. Unveiling the prior knowledge of the learner is an important aspect of constructivist teaching because constructivist theory stresses the importance of the learner's transforming current mental structures. It is the constant replacement of understandings with richer and deeper ones that characterizes the processes of cognitive growth and learning.

ROLE OF SOCIAL INTERACTION IN MEANING MAKING

The theory of constructivism postulates that learners come to know their world by interacting with it in ways that allow them to build the mental structures that 'explain' what is perceived. Ultimately, this process of meaning making is individually constructed, but it is an outgrowth of social interaction. The social interaction may be immediate, as in the case of discourse with others in a community, or may be contextual, as in the case of learners formulating ideas within historical and cultural moments. Some theorists view knowledge as the mechanism by which learners make sense of their experiences in their environment. Others view it as the outcome of learners making sense of their experiences in their environment; and yet others see knowledge as both a means and an end. Within these divergent views of knowledge, there is much debate as to the nature of the relative components of knowledge, and also over the merit of classifying particular knowledge as true or false (Philips, 1998). However, there is general agreement among constructivists that knowledge is socially constructed and a function of the culturally derived, community-sanctioned perspective of the knower. The goal of education is to foster the development of shared knowledge among community members. Peers play a powerful role in this shared knowledge construction.

Community of Learners

The theory of constructivism gives rise to a 'community of learners' model within educational settings, a model in which learners in search of understanding communicate current thinking with others by formulating and reformulating their thoughts based on peer and expert feedback and by reflecting on that feedback. This model presupposes a definition of knowledge as dynamic and socially constructed and rejects a definition of knowledge as static and 'passed on'. This model also requires a good deal of scaffolding on the part of the teacher to maximize the likelihood that meaningful learning will occur. Some level of dissonance must be established either through peers' sharing diverse perspectives or through teacher prompts to highlight sources of cognitive dissatisfaction. The task must be within the reach of the learners, more advanced than any individual within the group would be likely to complete independently, but not so far advanced that learning shuts down.

Zone of Proximal Development

What is the teacher's role in the community of learners model? As the 'expert', the teacher can guide the learner in intellectual arenas in which the learner could not independently navigate. Vygotsky (1962) referred to this arena as the zone of proximal development. The teacher provides an intellectual framework at the leading edge of the learner's current thinking on a topic. That framework can include questioning designed to help the learner see relationships, can include contradictions designed to help the learner examine subtleties, or can include hypothetical comments to help the learner extend an argument. Although the nature of the scaffolding may be diverse, the scaffolding has a unified purpose: to aid learners in restructuring their current theories. For Vygotsky, language is the basis of cognition. For Piaget, language is a mechanism to express cognition.

Teachers offer learners this guided participation to maximize the likelihood that restructuring will occur. To every learner, his or her present thinking holds a great deal of merit. Therefore, cognitive restructuring is resistant to casual interference. Furthermore, it is even resistant to direct instruction. Much research discusses the inability of direct instruction to dispel the misconceptions widespread over many topics to which learners cling (Clement, 1982). Fostering a learner's restructuring of present conceptions requires an analysis of the learner's current perspective with specific regard to the topic, concept or issue at hand. While a casual observer in such a classroom may not readily see the underlying pedagogy, the pedagogy none the less exists and is powerful. The teacher's pedagogy drives decisions concerning which responses to pursue, which student groupings to establish, which supplies to gather, and which follow-up questions to generate.

CONCEPTUAL CHANGE

The term 'constructivism' holds different meanings in many circles. The radical social constructivists discuss the illusoriness of objective truth (von Glasersfeld, 1998), the cognitive constructivists engage in structural analyses of knowledge generation (Piaget, 1953), and the human constructivists seek a synthesis of epistemological and psychological phenomena (Novak, 1993). Where there is significant intragroup and intergroup variation, a binding construct for all groups is the focus on conceptual change. The goals to which the learner aspires may differ, but the constructivist teacher and researcher are focused on better understanding the learner's conceptual changes over time, the nature of the changes and the contributing variables. (*See* **Conceptual Change**)

References

Ausubel DB (1963) *The Psychology of Meaningful Verbal Learning*. New York, NY: Grune & Stratton.

Clement J (1982) Algebra word problem solutions: thought processes underlying a common misconception. *Journal for Research in Mathematics Education* **13**(1): 16–30.

Novak J (1993) Human constructivism: a unification of the psychological and epistemological phenomena in meaning making. *International Journal of Personal Construct Psychology* **6**: 167–193.

Philips DC (1998) Coming to terms with radical social constructivism. In: Matthews MR (ed.) *Constructivism in Science Education*, pp. 139–158. London, UK: Kluwer.

Piaget J (1953) *Logic and Psychology*. Manchester, UK: Manchester University Press.

Piaget J (1970a) *Genetic Epistemology*. New York, NY: Columbia University Press.

Piaget J (1970b) *Structuralism*. New York, NY: Basic Books.

Von Glasersfeld E (1998) Cognition, construction of knowledge and teaching. In: Matthews MR (ed.) *Constructivism in Science Education*, pp. 11–30. London, UK: Kluwer.

Vygotsky L (1962) *Thought and Language*. Cambridge, MA: MIT Press.

Further Reading

Brooks JG and Brooks MG (1993) *In Search of Understanding: The Case for Constructivist Classrooms*. Alexandria, VA: Association for Supervision and Curriculum Development.

Copple C, Sigel L and Saunders R (1984) *Educating the Young Thinker*. New York, NY: Van Nostrand.

Davis RB, Maher CA and Nodding N (1990) Constructivist views on the teaching and learning of mathematics. *Journal for Research in Mathematics Education*, Monograph No. 4. Reston, VA: National Council of Teachers of Mathematics.

Driver R, Guesne E and Tiberghien A (eds) (1985) *Children's Ideas in Science*. Philadelphia, PA: Open University Press.

Duckworth E (1987) *'The Having of Wonderful Ideas' and Other Essays on Teaching and Learning*. New York, NY: Teachers College Press.

Fosnot CT (ed.) (1996) *Constructivism: Theory, Perspectives, and Practice*. New York, NY: Teachers College Press.

Piaget J and Inhelder B (1971) *Psychology of the Child*. New York, NY: Basic Books.

Sigel IE, Brodzinsky DM and Golinkoff RM (eds) (1981) *New Directions in Piagetian Theory and Practice*. Hillsdale, NJ: Lawrence Erlbaum.

Von Glasersfeld E (1995) A constructivist approach to teaching. In: Steffe L and Gale J (ed.) *Constructivism in Education*, pp. 3–16. Hillsdale, NJ: Lawrence Erlbaum.

Vygotsky LS (1962) *Thought and Language*. Cambridge, MA: MIT Press.

Conversation, Structure of

Introductory article

Herbert H Clark, Stanford University, Stanford, California, USA

CONTENTS

Conversations emerge as people use dialogue to coordinate on joint activities they engage in. People proceed turn by turn as they reach local agreements on the course of each section and subsection, including the opening and closing of the conversation itself.

INTRODUCTION

Conversations are the product of people engaged in joint activities. A joint activity is one in which two or more people have to coordinate with each other to succeed. When two people waltz, play a duet, or wrestle, they coordinate their individual actions largely by gesture, touch, and other techniques. When two people gossip, plan a vacation, or negotiate a contract, they coordinate largely through dialogue. The structure of these conversations emerges as the participants jointly manage their way through the gossip, the planning, or the negotiation.

Conversations, therefore, reflect the joint activities they coordinate. Every joint activity has participants who are distinct from bystanders, onlookers, or overhearers. In most joint activities, each participant has a role, such as clerk or customer, teacher or student, friend calling or friend called, and the roles help determine what the participants do and say. Most joint activities have mutually recognized goals such as exchanging gossip, planning a vacation, or negotiating a contract, and these have subgoals. Some goals are set from the start, but others get established in the course of the conversation. The participants also have private agendas – such as being polite, or finishing quickly – and these, too, constrain what they do and say. Often, people alternate between two or more joint activities – such as gossiping and eating dinner – and the structure of their conversation reflects the alternations.

ACTIONS OF DIALOGUE

It takes coordination to carry out a joint activity. Joint activities have boundaries – distinct beginnings and ends, and transitions from one part to the next – but these boundaries don't exist until the participants agree to them. To enter a planning session, for example, two people must agree on (1) what the joint activity is to be, (2) who is to take part, and (3) in what roles. They must also maintain or change these agreements at each transition point. People accomplish all this with dialogue, locally, turn by turn.

One basic method for reaching these agreements is the *adjacency pair*, as in this spontaneous example from Svartvik and Quirk (see Further Reading):

Ann where is your office,
Burton in the Strand,
Ann oh well, yes,

Adjacency pairs consist of two parts, by different speakers, where part 2 is conditionally relevant given part 1. Part 1 is a *proposal*, and part 2 is expected to be the *uptake* of that proposal. Here, in turn 1, Ann proposes that Burton tell her where his office is, and in turn 2, he takes up the proposal by saying that it is in the Strand. Ann and Burton use the two turns to agree on the content, participants, and roles of Ann's projected joint action. They would have failed to reach that agreement if, for example, Burton had replied 'What do you mean?' (failing to coordinate content) or 'You mean me?' (failing to coordinate participants). Turns 2 and 3 constitute a second adjacency pair, an assertion plus its uptake.

People in conversation use adjacency pairs for many types of joint actions. They use them for exchanges of information (as in Ann and Burton's question plus answer), greetings ('Hi,' 'Hi'), farewells ('Bye,' 'Bye'), offers ('Have a beer,' 'Thanks'), orders ('Sit down,' 'Yes, sir'), and apologies ('Sorry,' 'Oh, that's okay'), among others. They use them for even the simplest exchanges of information ('In the Strand,' 'Oh well, yes').

Adjacency pairs can also be used to *project* larger sections, as in this spontaneous example:

B I like tuh ask you something.
A Shoot.
B Y'know *I*'ad my license suspended fur six munts.
A Uh huh.
B Y'know for a reason which, I rathuh not, mention tuh you, in othuh words, – a *ser*ious reason, en I like tuh know if I w'd talk tuh my senator, or – somebuddy, could *they* help me get it back.

B's first turn is a *pre-question*. With it he proposes to ask A a question, and A agrees. B now has the freedom to take up preliminaries to his question, and it takes the two of them several turns to do that. Only then does he ask his question proper, 'Could they help me get it back?' Pre-questions project not only the eventual question but preliminaries to that question.

Pre-questions and their responses belong to a large family of so-called *pre-sequences*. Here are a few more examples:

Pre-request	Customer	Do you have hot chocolate?
	Waitress	Yes, we do.
Pre-invitation	Man	What are you doin?
	Woman	Nothin. What's up?
Pre-narrative	June	Did I tell you I was going to Scotland?
	Kenneth	No.
Pre-conversation	Caller	(rings telephone)
	Recipient	Miss Pink's office.

Each pre-sequence prepares the way for another joint action. The pre-request sets up a request ('I'll have one'); the pre-invitation sets up an invitation ('Would you like…'); the pre-narrative sets up a narrative; and the pre-conversation sets up an entire telephone conversation.

SECTIONS OF CONVERSATIONS

Conversations tend to emerge as a sequence of topics, or sections. Each section reflects a different phase in the overall joint activity – the next bit of gossip, the next segment of the vacation being planned, the next issue of the contract being negotiated. The participants must agree on the opening and closing of each section, and that is where pre-sequences are useful.

Sections that consist of narratives (jokes, anecdotes, recountings of events), for example, are often introduced by a pre-narrative and its response. The following is an instance from Svartvik and Quirk (see Further Reading):

Nancy: I acquired an absolutely magnificent sewing-machine, by foul means, did I tell you about that?
Kate: no,
Nancy: well when I was. doing freelance advertising – (proceeds to give a five minute narrative)

Nancy proposes to tell Kate a story ('Did I tell you about that'), and Kate accepts ('No'). That allows Kate to embark on her narrative – an extended section of the conversation. It takes both parties to agree, because the recipient can always decline, as in this example, also from Svartvik and Quirk:

Connie: did I tell you, when we were in this African village, and (- they were all out in the fields, - the)
Irene: (yes you did, yes, - yes)
Connie: babies left alone, -
Irene: yes.

Irene interrupts Connie (the speech in brackets is overlapping) to say that she *has* heard the story, and the two of them then go down a different path. So conversations are opportunistic: the paths people take depend on the opportunities that become available with each agreement. Nancy and Connie use their pre-narratives to find the best way to proceed and, receiving different replies, go in different directions.

People help signal which opportunities they are taking by using *discourse markers*. For example, Nancy used 'well' to signal that she was introducing a change in perspective as she began her story. Other discourse markers indicate such boundaries as the start of a new topic (e.g., 'so', 'then', 'speaking of that'), the start of a digression ('incidentally', 'by the way'), or the return from a digression ('anyway', 'so'). All help in coordinating what happens next.

Opening a conversation takes special coordination as two or more people move from not being in a conversation to being in one. The following is the opening of a conversation between acquaintances, again from Svartvik and Quirk:

Karen: (rings Charlie's telephone)
Charlie: Wintermere speaking? -
Karen: hello?
Charlie: h*el*lo
Karen: Charlie
Charlie: Yes
Karen: actually it's
Charlie: hello Karen
Karen: it's me
Charlie: M
Karen: I (- laughs) I couldn't get back last night, (continues)

First, Karen and Charlie coordinate contact through a proposal to have a conversation (the telephone ring) and its uptake ('Wintermere speaking?'). Next, they mutually establish their identities. Karen tells Charlie that she recognizes him in turn 5, but Karen has to say 'hello?' 'Charlie', and 'actually it's' before he identifies her in turn 8. Only then does Karen introduce the first topic. It took 10 turns for them to coordinate on the participants, roles, and content of the conversation.

Conversations are no easier to close, as illustrated in this ending to a telephone conversation:

June and I'll. I'll ring again, as soon as I can on the tenth, uhh to definite confirm it,
Kay right,
Kay okay,
June right,
June thanks a lot,
Kay r. right,
June bye bye,
Kay bye

Although June and Kay finish a topic in turns 1 and 2, they cannot hang up without agreeing to hang up. So in turn 3, Kay proposes to close the conversation ('Okay'), and although June could introduce a new topic, she agrees to Kay's proposal ('Right'). That opens up the closing in which the two exchange thanks ('Thanks a lot' 'Right') and then good-byes. The two must *agree* to close the conversation before they actually close it.

GROUNDING WHAT IS SAID

People carry out joint activities against their *common ground* – their mutual knowledge, mutual beliefs, and mutual assumptions. They infer their common ground from past conversation, joint perceptual experiences, and joint membership in cultural communities. When Ann asks Burton 'Where is your office?' she *presupposes* certain common ground – for example, that Burton works on computers and has an office in London, but that she doesn't know where. And with the question itself, she *adds to* their common ground that she wants to know. Conversations proceed by orderly increments to common ground – especially to the common ground relevant to the current joint activities.

So if conversations are to succeed, the participants must *ground* what they say. To ground what is said is to establish the mutual belief that the addressees have understood the speakers well enough for current purposes. One technique for grounding is the adjacency pair itself. When Burton said 'In the Strand', he displayed to Ann how he had interpreted her question. If Ann hadn't been satisfied with that interpretation, she could have corrected it, for example by replying 'No, I meant…'. By following up Burton's reply with 'Oh well, yes,' she displayed her acceptance of his interpretation. Another technique is the *back-channel response, acknowledgment*, or *continuer*. In two-party conversations, addressees are expected to add 'uh huh' or 'mhm' or 'yeah' at or near the ends of certain phrases. With these, they signal that they understand well enough for the speaker to continue.

Grounding is sometimes achieved through *side sequences*, as in this spontaneous example, once more from Svartvik and Quirk:

Roger	well there's no general agreement on it I should think,
Sam	on what?
Roger	on uhm - - on the uhm – the mixed up bits in the play, the
Sam	yes

When Sam didn't understand Roger's 'it', he initiated an embedded adjacency pair in turns 2 and 3, a side sequence, to clear up the problem. Only when he had cleared it up did he acknowledge or agree with 'Yes'. Side sequences are initiated to clear up not only mishearings and misunderstandings but other preconditions to taking up the first part ('Why do you want to know?'). Grounding is sometimes accomplished by overlapping speech. When Irene interrupted Connie's offer 'Did I tell you …' to say, 'Yes you did, yes, – yes', she was signaling to Connie that she already understood and Connie didn't need to go on.

CONCLUSION

The structure of conversations emerges step by step as people coordinate on each new move in their joint activities. People need to coordinate on the content, participants, and roles of each joint action, and they do that in a sequence of local, opportunistic agreements. It is these techniques that give conversations their structure.

Further Reading

Atkinson JM and Heritage J (eds) (1984) *Structures of Social Action: Studies in Conversation Analysis*. Cambridge, UK: Cambridge University Press.

Clark HH (1996) *Using Language*. Cambridge, UK: Cambridge University Press.

Drew P and Heritage J (eds) (1992) *Talk at Work : Interaction in Institutional Settings*. Cambridge, UK: Cambridge University Press.

Duncan S and Fiske DW (1977) *Face-to-Face Interaction*. Hillsdale, NJ: Lawrence Erlbaum.

Goffman E (1981) *Forms of Talk*. Philadelphia, PA: University of Pennsylvania Press.

Goodwin C (1981) *Conversational Organization: Interaction Between Speakers and Hearers*. New York, NY: Academic Press.

Kendon A (1990) *Conducting Interaction: Patterns of Behavior in Focused Encounters*. Cambridge, UK: Cambridge University Press.

Levinson SC (1983) *Pragmatics*. Cambridge, UK: Cambridge University Press.

Ochs E, Schegloff EA and Thompson SA (1996) *Interaction and Grammar*. New York, NY: Cambridge University Press.

Sacks H, Schegloff EA and Jefferson G (1974) A simplest systematics for the organization of turn-taking in conversation. *Language* **50**: 696–735.

Schegloff EA, Jefferson G and Sacks H (1977) The preference for self-correction in the organization of repair in conversation. *Language* **53**: 361–382.

Searle JR, Parret H and Verschueren J (eds) (1992) *(On) Searle on Conversation*. Amsterdam, Netherlands: J. Benjamins Pub. Co.

Stenström A-B (1994) *An Introduction to Spoken Interaction*. London, UK: Longman.

Svartvik J and Quirk R (eds) (1980) *A Corpus of English Conversation*. Lund, Sweden: Gleerup.

Convolution-based Memory Models

Intermediate article

Tony A Plate, Black Mesa Capital, Santa Fe, New Mexico, USA

CONTENTS

Convolution-based memory models are mathematical models of neural storage of complex data structures using distributed representations. Data structures stored range from lists of pairs, through to sequences, trees, and networks.

HOLOGRAPHIC MEMORY: THE BASIC IDEA

Convolution-based memory models (CBMMs) are mathematical models of storage for lists of paired items, and more complex data structures such as those needed to support language and reasoning capabilities. CBMMs use distributed representations in which items are represented as vectors of binary or real numbers (a *pattern*). (*See* **Distributed Representations**)

CBMMs can store information about items arranged in a great variety of relationships, such as lists of paired items, and sequences, networks, and tree structures. All CBMM storage schemes use a convolution operation to associate or *bind* two (or more) patterns together in a memory *trace*, which is also a pattern.

CBMMs are sometimes called *holographic* memory models because the properties and underlying mathematical principles of CBMMs and light holography are very similar. One of the most striking similarities is that both can reconstruct an entire pattern (in a noisy form) in response to a noisy or partial cue. This ability is a consequence of the *distributed* and *equipotential* nature of storage in both holograms and CBMMs: information about each element of an item or region of an image is distributed across the entire storage medium.

RAPIDLY BINDING TOGETHER COMPONENTS OF A MEMORY

CBMMs use two operations for composing patterns: superposition and binding. For patterns of real numbers, superposition is ordinary element-wise addition; for patterns of binary numbers, superposition is element-wise binary-OR. Superposition is useful for forming unstructured collections of items. However, associations or *bindings* between items cannot be represented using superposition alone because of the *binding problem*. (*See* **Binding Problem**; **Distributed Representations**)

CBMMs use *convolution* as a binding operation: convolution binds two patterns together into one. If $\mathbf{x}$ and $\mathbf{y}$ are n-dimensional pattern vectors (subscripted 0 to $n-1$), then the circular convolution of $\mathbf{x}$ and $\mathbf{y}$, written $\mathbf{z} = \mathbf{x} \otimes \mathbf{y}$, is also an n-dimensional pattern vector and has elements

$$z_i = \sum_{k=0}^{n-1} x_k y_{(i-k)\mathrm{mod}n} \quad (1)$$

Circular convolution can be viewed as a compression of the outer (or tensor) product of the two vectors, where compression is achieved by summing particular elements, as shown in Figure 1. (Other variants of convolution can be viewed as slightly different ways of compressing the outer product.)

A list of paired items can be represented as the superposition of pairs of items bound together by a convolution. For example, a simple way of representing the list of two pairs 'red-square and blue-circle' is as the pattern $(\mathbf{red} \otimes \mathbf{circle}) + (\mathbf{blue} \otimes \mathbf{square})$. This pattern is quite different from the one

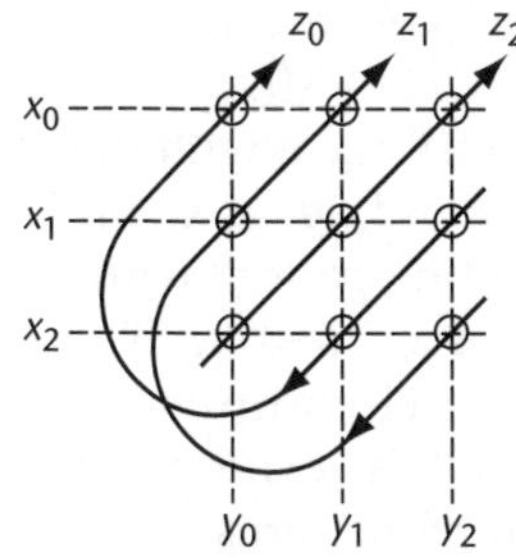

Figure 1. The *circular convolution* **z** of vectors **x** and **y** can be expressed as the sum of elements of their outer product.

that results from a different pairing of the same items such as (**blue** ⊗ **circle**) + (**red** ⊗ **square**).

In CBMMs, as in many other memory models that use vector or distributed representations, *similarity* is computed by either the *dot product* $\mathbf{x}\cdot\mathbf{y}$, or *cosine* (a scaled version of the dot product) of two pattern vectors:

$$\mathbf{x}\cdot\mathbf{y} = \sum_{i=0}^{n-1} x_i y_i \tag{2}$$

$$\text{cosine}(\mathbf{x},\mathbf{y}) = \frac{\sum_{i=0}^{n-1} x_i y_i}{|\mathbf{x}||\mathbf{y}|} = \frac{\sum_{i=0}^{n-1} x_i y_i}{\sqrt{\sum_{i=0}^{n-1} x_i^2}\sqrt{\sum_{i=0}^{n-1} y_i^2}} \tag{3}$$

One the most important properties of convolution is *similarity preservation*: if patterns **red** and **pink** are similar, then the bindings **red** ⊗ **square** and **pink** ⊗ **square** will also be similar, to approximately the same degree.

RAPID RETRIEVAL AND INTERFERENCE EFFECTS

Convolution bindings can be easily decoded using inverse convolution operations. For example, using exact inverses, $\mathbf{red}^{-1} \otimes \mathbf{red} \otimes \mathbf{circle} = \mathbf{circle}$. However, the exact inverse can be numerically unstable and is not always the best choice for decoding. For many vectors, such as those whose elements have independent Gaussian statistics with mean zero and variance $1/n$, an approximate inverse can be used. The approximate inverse of **x** is denoted by $\mathbf{x}^{\mathrm{T}}$ (this notation is chosen because the approximate inverse is closely related to the matrix transpose). It is a simple rearrangement of the elements of **x**: $x_i^{\mathrm{T}} = x_{(-i)\mathrm{mod}n}$. Reconstruction using the approximate inverse is noisy ($\mathbf{red}^{\mathrm{T}} \otimes \mathbf{red} \otimes \mathbf{circle}$ is only approximately equal to **circle**), but is usually more stable in the presence of noise than reconstruction using the exact inverse. If necessary, exact reconstructions can be provided by passing the noisy result through a clean-up memory, which returns the closest matching pattern among the patterns it contains.

Decoding still works when multiple associations are superimposed. For example:

$$\begin{aligned}\mathbf{blue}^{\mathrm{T}} &\otimes ((\mathbf{red} \otimes \mathbf{circle}) + (\mathbf{blue} \otimes \mathbf{square}))\\ &= (\mathbf{blue}^{\mathrm{T}} \otimes \mathbf{red} \otimes \mathbf{circle})\\ &\quad + (\mathbf{blue}^{\mathrm{T}} \otimes \mathbf{blue} \otimes \mathbf{square})\\ &\approx \mathbf{square}\end{aligned} \tag{4}$$

Because of the randomizing properties of convolution, the first term on the right in the expansion ($\mathbf{blue}^{\mathrm{T}} \otimes \mathbf{red} \otimes \mathbf{circle}$) is not similar to any of **blue**, **red**, **circle**, or **square** and can be regarded as noise. The second term on the right ($\mathbf{blue}^{\mathrm{T}} \otimes \mathbf{blue} \otimes \mathbf{square}$) is a noisy version of **square**. The sum of these two terms is an even noisier, but still recognizable, version of **square**. When larger numbers of bindings are superimposed together the interference effects can become significant, though increasing the vector dimension can reduce interference effects. For further discussion and quantitative analysis, see Murdock (1982), Metcalf-Eich (1982), or Plate (1995).

TODAM

Murdock's (1982) 'theory of distributed associative memory' model (TODAM) is intended to model patterns of human performance on memorization tasks, focusing on tasks involving lists of paired associates. For example, a subject might be asked to memorize the list 'cow-horse, car-truck, dog-cat, and pen-pencil' and then answer such questions as 'Did *car* appear in the list?' (recognition), or 'What was *cat* associated with?' (cued recall). Subjects' relative abilities to perform these and other tasks under different conditions, and the types of errors they produce, give insight into the properties of human memory. Some of the conditions commonly varied are the number of pairs, the familiarity of items, the similarity of items, and the position of recall or recognition targets within the list.

The TODAM formula for sequentially constructing a memory trace for a list of pairs ($\mathbf{x}_i$, $\mathbf{y}_i$) of item patterns is as follows:

$$\mathbf{T}_j = \alpha\mathbf{T}_{j-1} + \gamma_1\mathbf{x}_j + \gamma_2\mathbf{y}_j + \gamma_3\mathbf{x}_j \otimes \mathbf{y}_j \tag{5}$$

where $\mathbf{T}_j$ is the memory trace pattern (a vector) representing pairs 1 through j (with $\mathbf{T}_0 = \mathbf{0}$). The scalars α, γ_1, γ_2, and γ_3 are adjustable parameters of the model, taking values between 0 and 1.

TODAM uses an 'unwrapped' version of convolution which expands the size of vectors each time it is applied, but TODAM could use any convolution operation.

For example, the memory trace for the list of three pairs (**a, b**), (**c, d**), and (**e, f**) is built as follows:

$$\mathbf{T}_1 = \gamma_1\mathbf{a} + \gamma_2\mathbf{b} + \gamma_3\mathbf{a} \otimes \mathbf{b} \tag{6}$$

$$\mathbf{T}_2 = \gamma_1\mathbf{c} + \gamma_2\mathbf{d} + \gamma_3\mathbf{c} \otimes \mathbf{d} + \alpha(\gamma_1\mathbf{a} + \gamma_2\mathbf{b} + \gamma_3\mathbf{a} \otimes \mathbf{b}) \tag{7}$$

$$\mathbf{T}_3 = \gamma_1\mathbf{e} + \gamma_2\mathbf{f} + \gamma_3\mathbf{e} \otimes \mathbf{f} + \alpha(\gamma_1\mathbf{c} + \gamma_2\mathbf{d} + \gamma_3\mathbf{c} \otimes \mathbf{d}) + \alpha^2(\gamma_1\mathbf{a} + \gamma_2\mathbf{b} + \gamma_3\mathbf{a} \otimes \mathbf{b}) \tag{8}$$

Item recognition is done by comparing an item with the trace: item **x** was stored in trace **T** if $\mathbf{x} \cdot \mathbf{T} > t$ (if the dot product of **x** and **T** is greater than some threshold t).

Cued recall is accomplished by decoding the trace with the cue: if item **x** was stored in trace **T**, then $\mathbf{x}\#\mathbf{T}$ is a noisy reconstruction of the partner of **x** (where $\mathbf{x}\#\mathbf{T}$ is another way of writing $\mathbf{x}^{\mathrm{T}} \otimes \mathbf{T}$).

Some of the predictions of TODAM that are supported by evidence in the psychological literature are as follows:

- Performance decreases with increasing list length.
- Cued recall is symmetric: the recall of **x** given **y** from a trace containing the pair $\mathbf{x} \otimes \mathbf{y}$ is as accurate as the recall of **y** given **x** from the same trace.
- There is no primacy effect, only a recency effect, because forgetting is geometric in α.
- Cued recall for a particular item can be superior to recognition for that same item – it can be possible to recall an item that cannot be recognized. This is because weights can be defined so that associative information is stronger than item information.

CHARM

The 'composite holographic associative recall model' (CHARM) (Metcalfe-Eich, 1982) was specifically intended to address the effects of similarity among items in cued recall from lists of paired associates. CHARM uses an even simpler storage method than TODAM – it stores only associative information and no item information. The memory trace for a list of pairs ($\mathbf{x}_i$, $\mathbf{y}_i$) of item patterns is constructed as follows:

$$\mathbf{T} = \sum_{i=1}^{k} \mathbf{x}_i \otimes \mathbf{y}_i \tag{9}$$

CHARM uses a truncated version of the non-wrapped convolution used in TODAM so that the patterns for memory traces are the same size as for items.

As with TODAM, the process for performing cued recall in CHARM begins by correlating a composite memory trace with the cue; e.g., to find the item corresponding to $\mathbf{x}_1$ in **T**, $\mathbf{x}_1\#\mathbf{T}$ is computed. The resulting pattern will be a noisy version of the pattern associated with $\mathbf{x}_1$ in **T**, which is passed through a clean-up memory. For the purposes of Metcalfe's experiments, the clean-up memory contained patterns for items stored in the memory trace, and patterns for some other items not stored in the memory trace.

One type of retrieval phenomenon modeled with CHARM is the reduced ability to accurately recall items from a list whose members are similar, versus from a list whose members are dissimilar. For example, performance on a pair such as Napoleon–Aristotle is worse when the pair is embedded in a list of pairs of names of other famous people (a homogenous list) than when it is embedded in a list containing items conceptually unrelated to it, such as red–blue. Furthermore, with homogenous lists, incorrect recall of an item that is similar to the correct response and that was also in the list with an associated item similar to the cue is a frequent type of error in both CHARM and with human subjects.

BINDING VIA FULL TENSOR PRODUCTS

A list of paired items is a very simple set of relationships. Many cognitive tasks demand the ability to store more complicated relationships. For example, understanding language requires the ability to work with recursive structures: a phrase can have a verb, a subject and an object, but the object could be a phrase itself, which could even contain further subphrases. For example, the sentence 'I believe that politicians will say whatever will help them to get elected' contains at least three levels of recursion.

One of the first concrete descriptions of such a scheme was given by Smolensky (1990). Smolensky used tensor products to bind roles and fillers together in a recursive manner. For example, the sentence 'Politicians tell stories' could be represented as the rank-2 tensor $\mathbf{T} = \mathbf{politicians} \otimes \mathbf{tell}_{\mathrm{agent}} + \mathbf{stories} \otimes \mathbf{tell}_{\mathrm{object}}$, where **politicians** is a pattern representing politicians, $\mathbf{tell}_{\mathrm{agent}}$ is a pattern for the agent role of 'tell', etc., and $\otimes$ is the tensor product (a generalization of the outer product). Tensors can be superimposed and decoded in a manner similar to convolution traces; the role pattern $\mathbf{tell}_{\mathrm{agent}}$ can be used to decode the tensor **T** to retrieve the pattern **politicians**. What makes the use of tensors interesting is that the rank-2 tensor **T** can be used as the filler in some higher-level role-filler binding, such as representing the meaning of the sentence 'I know politicians tell stories'. This higher-level binding is a rank-3 tensor.

HOLOGRAPHIC REDUCED REPRESENTATIONS

Holographic reduced representations (HRRs) (Plate, 1995, 2000b) use convolution-based role-filler bindings to construct patterns representing a recursive structure.

The HRR for the proposition 'Politicians tell stories' is constructed as follows:

$$\mathbf{P}_{\text{tell}} = \mathbf{tell} + \mathbf{politicians} + \mathbf{stories} + \mathbf{tell}_{\text{agt}} \otimes \mathbf{politicians} + \mathbf{tell}_{\text{obj}} \otimes \mathbf{stories} \quad (10)$$

If we have the pattern $\mathbf{P}_{\text{tell}}$ and know the role patterns, then we can reconstruct a filler pattern by convolving $\mathbf{P}_{\text{tell}}$ with the approximate inverse of a role pattern. For example, $\mathbf{tell}^{\mathrm{T}}_{\text{agt}} \otimes \mathbf{P}_{\text{tell}}$ gives a noisy version of **politicians** which can be put through a clean-up memory to provide an accurate reconstruction.

The HRR pattern $\mathbf{P}_{\text{tell}}$ is a *reduced representation* for the proposition 'Politicians tell stories' and can be used as a filler in a higher-order proposition. For example, the HRR $\mathbf{P}_{\text{know}}$, representing 'Bill knows politicians tell stories', is constructed as follows:

$$\mathbf{P}_{\text{know}} = \mathbf{know} + \mathbf{bill} + \mathbf{P}_{\text{tell}} + \mathbf{know}_{\text{agt}} \otimes \mathbf{bill} + \mathbf{know}_{\text{obj}} \otimes \mathbf{P}_{\text{tell}} \quad (11)$$

Such higher-level HRRs can be decoded in the same way as first-order HRRs. For example, the filler of the know-object role is decoded as follows:

$$\mathbf{P}_{\text{know}} \otimes \mathbf{know}^{\mathrm{T}}_{\text{obj}} \approx \mathbf{P}_{\text{tell}} \quad (12)$$

This reconstructed filler is a proposition. To discover its fillers it could be cleaned up and then decoded again. (*See* **Distributed Representations**)

HRRs are similar if they merely involve similar entities or predicates. Because of the similarity-preserving properties of convolution, they will be even more similar if the entities are involved in similar roles. Thus it turns out that the similarity of HRRs can reflect both superficial and structural similarity in a way that neatly corresponds to the data on human analog retrieval (Plate, 2000).

IMPLEMENTING CONVOLUTION-BASED MEMORIES IN CONNECTIONIST NETWORKS

The various operations used in convolution-based memory models – convolution, correlation,

$$z_0 = x_0 y_0 + x_1 y_2 + x_2 y_1$$
$$z_1 = x_0 y_1 + x_1 y_0 + x_2 y_2$$
$$z_2 = x_0 y_2 + x_1 y_1 + x_2 y_0$$

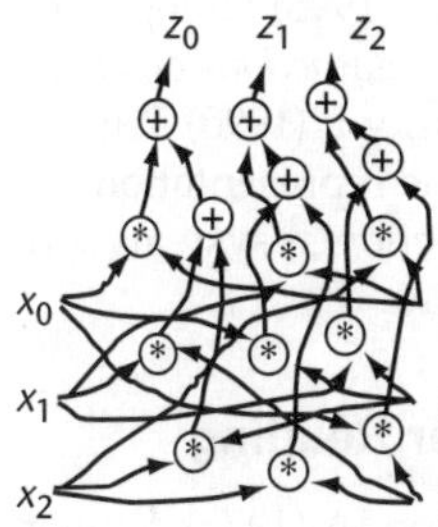

Figure 2. The *circular convolution* **z** of vectors **x** and **y** drawn as a network of three sigma–pi neurons. Each sigma–pi neuron computes the sum of three products as shown on the left.

approximate inverse, dot-product, and clean-up memory – are easily implemented in connectionist networks. Convolution encoding and decoding can be implemented by suitably connected networks of 'sigma–pi' neurons. Figure 2 shows a network that computes the circular convolution of two three-element vectors.

The pattern of connections in the sigma–pi network that computes circular convolution may seem unrealistically intricate and precise for a biological circuit. However, Plate (2000a) shows that sigma–pi networks that sum random products of pairs of elements from **x** and **y** can also function as encoding and decoding networks with similar properties to convolution.

For computation of similarity, a dot product can be computed by a single sigma–pi neuron. Clean-up memory can be implemented in several ways, such as with Kanerva's (1988) sparse distributed memory, or Baum *et al.*'s (1988) various associative content-addressable memory schemes.

References

Baum EB, Moody J and Wilczek F (1988) Internal representations for associative memory. *Biological Cybernetics* **59**: 217–228.

Kanerva P (1988) *Sparse Distributed Representations*. Cambridge, MA: MIT Press

Metcalfe-Eich J (1982) A composite holographic associative recall model. *Psychological Review* **89**: 627–661.

Murdock BB (1982) A theory for the storage and retrieval of item and associative information. *Psychological Review* **89**(6): 316–338.

Plate TA (2000a) Randomly connected sigma–pi neurons can form associator networks. *Network: Computation in Neural Systems* **11**(4): 321–332.

Plate TA (2000b) Structured operations with vector representations. *Expert Systems: The International Journal of Knowledge Engineering and Neural Networks* **17**(1): 29–40.

Plate TA (1995) Holographic reduced representations. *IEEE Transactions on Neural Networks* **6**(3): 623–641.
Smolensky P (1990) Tensor product variable binding and the representation of symbolic structures in connectionist systems. *Artificial Intelligence* **46**(1–2): 159–216.

Further Reading

Anderson JA (1973) A theory for the recognition of items from short memorized lists. *Psychological Review* **80**(6): 417–438.
Borsellino A and Poggio T (1973) Convolution and correlation algebras. *Kybernetik* **13**: 113–122.
Halford G, Wilson WH and Phillips S (1998) Processing capacity defined by relational complexity: implications for comparative, developmental, and cognitive psychology. *Behavioral and Brain Sciences* **21**(6): 803–831.
Kanerva P (1996) Binary spatter-coding of ordered k-tuples. In: von der Malsburg C, von Seelen W, Vorbruggen J and Sendhoff B (eds) *Artificial Neural Networks–ICANN Proceedings*, vol. 1112, pp. 869–873. Berlin, Germany: Springer.
Murdock B (1993) TODAM2: a model for the storage and retrieval of item, associative, and serial-order information. *Psychological Review* **100**(2): 183–203.
Rachkovskij DA and Kussul EM (2001) Binding and normalization of binary sparse distributed representations by context-dependent thinning. *Neural Computation* **13**(2): 411–452.
Van Gelder TJ (1999) Distributed versus local representation. In: Wilson R and Keil F (eds) *The MIT Encyclopedia of Cognitive Sciences*, pp. 236–238. Cambridge, MA: MIT Press.
Willshaw D (1981) Holography, associative memory, and inductive generalization. In: Hinton GE and Anderson JA (eds) *Parallel Models of Associative Memory*. Hillsdale, NJ: Lawrence Erlbaum.

Cooperation and Conflict, Large-scale Human

Advanced article

Francisco J Gil-White, University of Pennsylvania, Philadelphia, Pennsylvania, USA
Peter J Richerson, University of California, Davis, California, USA

CONTENTS

Along with a few other animals such as bees, termites, ants and wasps, humans live in societies that cooperate for a common goal. Serving the greater good seems to go against basic Darwinian theory, so the question is, how and why have complex, cooperative societies developed in some species? In spite of so much cooperation, why do human societies remain so conflict ridden?

INTRODUCTION

Suppose you stroll to the corner restaurant for breakfast: eggs, bacon, and a glass of orange juice. A simple activity? No. Mind-numbing complexity is more like it. A farmer in Virginia produced your egg, another in Florida your orange juice, and yet another in the Midwest your bacon. Different truckers brought each of these to a supermarket. The restaurateur then bought them there and had them prepared for you. Seven people are involved in your 'simple' activity? Well, no. This is a caricature. Just for starters, the egg farmer/capitalist hires several workers to operate considerable equipment, all of which was purchased from other companies, made up of capitalists and workers, who in turn bought their parts from yet other companies, which…(the mind reels). Your day has barely begun, and a few dollars' worth of breakfast has already brought an army of considerable size to your service.

Only a select few animal species have societies with extensive cooperation, fine coordination, and massive division of labor: the social insects (bees, termites, ants, and wasps), possibly naked mole rats (but their level of complexity is far below that of the advanced social insects, as is the scale of their societies) and us. This form of social organization is clearly evolutionarily successful; social insects are diverse and abundant, especially in the tropics, and human populations grow so fast that our rapid and energy-expensive expansion into every conceivable niche is a considerable threat to other species and the climate. Given this, one might be naturally inclined to ask: why is this adaptation not more common among the species of the world? After all, beneficial adaptations should proliferate, shouldn't they?

Actually, for modern students of evolutionary theory – trained as they are in the framework of what has been called the 'modern synthesis' and 'neo-Darwinism' – the puzzle is rather different. Modern biologists are trained to be surprised not by the rarity of this dramatic adaptation but by the fact that it is possible at all. How could something this strange evolve? Darwin himself worried that the self-sacrificial altruism of the social insects might be fatal to his theory. To see the problem his way, we must briefly develop the theoretical instincts of a modern evolutionary theorist.

DIFFICULTIES ACCOUNTING FOR HUMAN ULTRA-SOCIALITY IN DARWINIAN TERMS

Individuals attempt to reproduce before they die, and some do better than others. If the features of an individual are passed on to the offspring, then good reproducers will beget good reproducers, who in turn will beget good reproducers once again. And so on. Each time, good reproducers leave more descendants than other types, so after a number of generations the entire population will become of the type that reproduces best (with the exception of frequency-dependent effects, when selection will maintain several types at equilibrium). This is the basic Darwinian insight of 'natural selection'.

The mechanism responsible for stable similarities between parent and offspring is genetic inheritance. Mere individuals live and die, but genes can potentially keep going forever. Modern Darwinism focuses on changing genes in order to understand the processes responsible for historical change in organic populations.

The analytical focus of a modern Darwinian is the 'gene's eye view' heuristic, which relegates individual organisms to the status of temporary 'vehicles' conveying the potentially immortal genes from one generation to the next. Genes that have a better chance of proliferating are those that increase the reproductive success of their vehicles in competition with other vehicles. Finding 'unselfish' genes that cause their vehicles to suffer sacrifices to benefit another vehicle's reproduction is thus a major puzzle. But nothing in these arguments really depends upon there being a single gene for altruism; this is just a convenient way to strip the problem to its bare essentials. Darwin was right to worry.

This brings us to the social insects. Massive division of labor is impressive, but the reason it is possible in the first place is the truly big puzzle. Although many ants in a colony will famously give up their lives protecting it, for example, this is only because they have already given up their reproduction. The latter is, to a Darwinian, the really dramatic fact. How could they give up their reproduction? In human ultra-sociality, on the other hand, defense is the most dramatic puzzle because those who risk and often give their lives to defend their society are indeed capable of reproducing and by fighting give up some or all of this capacity.

For non-human altruism, twentieth-century evolutionary theory has provided two elegant and very successful explanations: kin selection and reciprocity. Before examining them, notice that the fundamental issue for any explanation in this domain is the problem of assortment. The gene's eye view allows us to state the obvious: since the gene is trapped inside its vehicle, the vehicle must reproduce if the gene is to proliferate. So how can a gene proliferate more than competing genes if it makes its vehicle transfer reproductively useful resources to other vehicles? At first glance this would seem impossible, and for most kinds of resource transfers it will be. But if the vehicle is making resource transfers to other vehicles also containing copies of that same gene, then the gene promotes its own proliferation at one remove.

The question therefore is: what could cause vehicles with altruistic genes to assort with one another?

KIN SELECTION

Green Beards

Imagine a gene – 'G' – producing two effects: (1) it gives you a green beard, and (2) it makes you help those with green beards (Dawkins, 1989, pp. 88–89). G's twin effects solve the problem of assortment: if you help those with green beards, then, because those individuals also have G (hence their beard), G is making you help other copies of itself. Copies of G can 'find each other' thanks to the beards, and therefore when G causes its vehicle to transfer resources to other vehicles it is nevertheless promoting the spread of G.

It is virtually impossible that the same gene will cause a discriminatorily altruistic behavior and also the cue used to discriminate, unless altruism itself is the cue. Theoretical considerations suggest that it is also highly improbably that "green beard" genes can arise as a result of two tightly linked loci where the gene at one locus would code for the green beard, and the gene at the other for the altruistic behavior. But the thought experiment brings the problem of assortment into sharp focus: if an 'altruistic' gene is to prosper, its vehicle must confer benefits disproportionately on other vehicles containing copies of the same gene. Something like a green beard must facilitate this nonrandom assortment for altruistic genes to evolve.

In one proposal (Hirshleifer, 1987; Frank, 1988), if altruism is mediated by emotions, and if emotions result inevitably in facial expressions and other bodily manifestations, and if such manifestations are hard to fake, then altruists can assort with each other by examining each other's expressions of emotion. In other words, those who 'look' altruistic probably are, so altruists can find and prefer each other for mutual benefit. Genes coding for altruistic emotion/displays will be favored.

But the problem with this kind of 'green beard' argument is that, once the signal is common, selection will favor selfish individuals who pretend to be altruists but don't help. Actors and confidence artists can fake emotions well enough to fool us. Darwinians indeed expect that the evolution of clever, green-beard-exploiting sociopaths will undermine the evolution of emotional signals. This theoretical embarrassment to the green beard argument is accompanied by an empirical one: emotions that appear very similar to ours occur in other mammals (as Darwin himself wrote in his book *The Expression of Emotions in Animals and Man*). Thus, if nonhumans can produce emotions and signal them, why can't they use this to assort for altruism and build ultra-social communities? Human emotions are no doubt involved in motivating and signaling cooperation in humans, but this is likely to be a secondary effect of other evolutionary processes, not something that can be shaped directly by natural selection to favor the original emergence of altruism. If it could, many nonhuman mammals should have it.

Kinship as a 'Green Beard' Substitute

If not emotional green beards, then what? Suppose that if you have the altruistic gene, then you can use an observable cue X to guess with some probability p that somebody else also has the gene. If so, then the altruistic gene – call it gene 'K' – will be helping itself so long as it specifies 'to individuals with cue X give a benefit size b, where b satisfies the following:

$$bp > c,$$

where c is the cost to the altruist of transferring the benefit. In other words, out of a large population of individuals bearing cue X and therefore receiving my help, only a proportion p will actually carry gene K. Thus – on average – the benefit that K's vehicle (me) confers on other copies of K is not b but the scaled down benefit bp. If this weighted payoff is greater than what it costs me to help, then K is giving itself a net benefit.

In 1964 William Hamilton argued persuasively that kinship can play the role of cue X. Consider two siblings, Higley and Bob. Bob carries a gene K that makes him an altruist. What is the probability that Higley also has gene K? Well, Bob's father passed down half of his genes to each sibling, who get the other half from their mother. These samples are subject to independent random assortment, so that Bob and Higley share a quarter of their father's genes and a quarter of their mother's. Thus, the probability that Bob and Higley share gene K is at least $p = \frac{1}{2}$. The probability may be higher if the gene is common in the population, but the critical value is the chance that siblings share the identical gene by common descent. This is the same as the probability of sharing the gene when it is rare. So suppose that K specifies a behavior that makes Bob give 5 units of benefit to siblings like Higley, for a cost to the actor of 2 units. Will K spread? Yes.

$$\frac{5}{2} = 2.5 > 2$$

On the other hand, if at the cost of 2 units K confers only 3 units of benefit to these recipients, then K will not spread.

What have we shown? That if a gene makes its vehicle assist its close kin, then it has found a way for its vehicle to assort (a fair amount of the time) with other vehicles carrying copies of the same gene. This assortment is what makes it possible for altruistic genes – within benefit/cost limitations – to evolve. This is, of course, far short of the perfect assortment that green beards would make possible, but it is what nature uses because green beards or their equivalents are usually impossible. This 'kin selection' argument explains the widespread observation of nepotistic altruism in humans and many other species. In particular, it explains the ultra-sociality of the social insects, for in, say, an ant colony, everybody is a close relative due to the fact that everybody is a child of the queen.

Washburn's Fallacy

The above insight is usually expressed as Hamilton's famous rule: $br > c$. Here r replaces p, and stands for 'coefficient of relatedness': the probability that two individuals have identical copies of the same gene, descended from the same, recent ancestor gene. Thus, recall that for Higley we calculated the probability that he has an identical gene to Bob's that is in fact descended from their father's or mother's copy.

The r in Hamilton's rule is often misinterpreted as 'the probability or proportion of genes shared in common between two individuals'. This is commonly referred to as 'Washburn's fallacy' because the anthropologist Sherwood Washburn used to argue – in critical fashion – that Hamilton's rule would imply altruism towards everybody and only slightly more altruism towards kin. Why? Because any of us shares about 80 percent of our genetic alleles with any other randomly chosen member of the human species, and 80 percent is a lot. If true, this argument would appear to solve the puzzle of human ultra-sociality, but it would create an even bigger puzzle: why aren't many more species ultra-social?

But Washburn's argument follows only if the r is interpreted as the proportion of genes shared in common, rather than as the probability of sharing identical copies of a gene descended from the same, recent ancestor.

Why is Washburn wrong? Even if 80 percent of the people in the population have the altruistic gene (and the others have a selfish alternative), since Washburn's altruistic gene says 'help anybody', having an altruistic gene will not make a vehicle disproportionately likely to get help. The 20 percent of people not sharing the altruistic gene will get the same benefit as the 80 percent that do, and since they don't pay the costs of helping others, they have higher fitness. Selfish genes will increase in frequency and drive out the altruistic genes. An altruistic gene that said 'help close kin', on the other hand, would make altruistic genes disproportionately likely to get help. Individuals with the altruistic gene are more likely than randomly chosen members of the population to have close relatives with copies of this gene, and are therefore more likely to get helped, than individuals with the selfish gene.

Washburn could have avoided his fallacy simply by imagining how his 'help anybody' gene could have become common in the first place. Here things become crystal clear: unless a gene codes for a behavior promoting its spread when it is a new and therefore rare mutation, the gene will wink out of existence as quickly as it appeared. When the 'help anybody' gene first appears virtually no other vehicles have copies of it, so 'helping anybody' confers no benefits on the gene's spread and the gene quickly goes extinct. A new mutant gene is, by definition, rare, and thus only close kin of its vehicle are likely to carry copies. An altruistic gene therefore has a chance of spreading from low frequency only if it discriminates in favor of close kin. Why not distant relatives? When the gene is rare, distant relatives are about as unlikely to have copies of the gene as a randomly chosen member of the population – in fact, at the limit, these are the same, because all members of a population are (very) distant relatives.

Kin selection can explain nepotism in many species, most spectacularly in the case of eusocial ants, bees, and termites, where huge numbers of close relatives cooperate. But in human social systems, even at their most simple, average r is so low that we may well say members are not, in fact, related. As Campbell (1983) rightly observed, human societies, unlike the social insects, exhibit cooperation among reproductive competitors. If kin selection can cause ultra-sociality with human levels of average r, then many more animal species should have such complex societies. Humans are probably a special case requiring a special explanation.

RECIPROCITY

In the logic of reciprocity (first explored by Robert Trivers (1971)) an 'actor' suffers a cost to benefit a 'recipient', expecting a return benefit at some other time (I'll scratch your back if you scratch mine). The time delay distinguishes this from 'trade' as commonly understood, for trade lacks the risk of no payback. Perhaps this explains the unfortunate popularity of Trivers's coinage 'reciprocal altruism', which has caused much confusion. If we stick to the gene's eye view, however, the terminological tangles quickly evaporate. When will a gene specify a transfer of reproductively relevant resources from its own to other vehicles? Kin selection can lead to this, as we have seen. Reciprocity can too, but it differs from kin selection in that, so long as the recipient pays back the favor, it matters little whether the recipient's motivation arises from a gene identical by recent descent (or indeed from some entirely different gene). Reciprocity may even occur between species, as in mutualisms. What matters is that there be some reasonable probability that the favor will be returned and a method for assessing this probability. If favors are made when they are relatively cheap for the actor but beneficial for the recipient, and if they are returned, then a gene making its vehicle do such favors will prosper.

How well will a rare reciprocity gene do? When it is rare, a vehicle carrying the gene is very unlikely to meet another such vehicle that will reciprocate its good turns. Thus, the evolution of reciprocity requires some initial assistance from kin selection. For example, since the individuals carrying a new and rare mutation will be close relatives, vehicles carrying the reciprocity gene will be likely to meet other such vehicles – even when rare – if individuals are organized in local kin groups. Once the gene for reciprocity becomes a little more common, such kin-biased population structure is unnecessary for the success of the reciprocity gene.

Even when reciprocators are common, it is important to ensure assortment to prevent 'cooperators' from being exploited by 'defectors', and this brings us to the question of the cognitive mechanisms involved. Theoretical considerations suggest that nice-but-not-gullible strategies like 'tit for tat' (if you cooperated with me last time, I will cooperate this time; if you didn't, I won't) are at the heart of our reciprocating psychology (Axelrod and Hamilton, 1981; Axelrod, 1984), but the actual mechanisms are complex and subtle.

The logic of reciprocity can easily explain cooperation in very small groups, especially dyads. However, reciprocity cannot so easily explain cooperation in larger groups (Boyd and Richerson, 1988). In a dyad, my help is a private benefit directed to one individual; if the partner does not reciprocate, I can ignore this individual in the future and direct my help towards another who *will* pay back my assistance. But when my benefit is consumed not by one partner but by two or more simultaneously (say, for example, that I build a wall which protects everybody who lives inside of it), the structure of the problem changes. (Notice, by definition, if the benefit is being consumed by a group, this means I cannot selectively withdraw the benefit from nonreciprocators, and am therefore producing a 'public good'. If I can discriminate, then we don't really have a 'group', but are back to dyadic interactions.) When everybody in the group returns my favor we all benefit, but if some don't return my favor, they create a dilemma for me: either (1) I can cooperate, and reward the defector (who gets the benefit without paying the cost of returning my favor); or (2) I can defect, giving up the benefits of reciprocity with those in the group who *are* reciprocators. The larger a group is, the less likely that just by chance it will have disproportionately large numbers of cooperators, so genes supporting (2) will do better with increasing group sizes. In particular, when the gene for reciprocity is new and therefore rare, the chances of having many reciprocators in a large group are vanishingly small. As groups get larger, then, kin selection is less and less effective at helping group-based reciprocity get started. For groups as small as 10, the potential to get group-based reciprocity off the ground becomes very small.

Some have considered the indirect benefits of reciprocity as a possible explanation for human ultra-sociality that sidesteps the public goods problem. Trivers (1971) speculated that given widespread dyadic reciprocity, selection would favor a strategy that used altruism towards third parties as a gauge of trustworthiness. Richard Alexander (1987) argued that the resulting structured webs could solve the problem of reciprocity in large groups. The argument is that humans are smart enough that each individual can keep track of who reciprocates with third parties; a strategy that prefers such reciprocators as partners will do well because it is better at picking low-risk partners. The resulting large webs of 'indirect reciprocity' can build much more complex societies of nonrelatives than in other species.

More recent models (Nowak and Sigmund, 1998a, b) challenge Boyd and Richerson's (1988) conclusion that large-group reciprocity cannot evolve. However, as Leimar and Hammerstein (2001) argue, the Nowak and Sigmund model makes a very unrealistic assumption: interactants never make mistakes. (see also Panchanathan, 2001). They show that when mistakes are allowed to occur, indirect reciprocity does not easily evolve because one needs information about people's intentions, not just their behavior (e.g., did Bob not reciprocate because he was punishing a nonreciprocator or because he himself is a nonreciprocator?). Indeed this is true even of dyadic reciprocity: if people make mistakes, we need to distinguish between honest mistakes and defections, and for that we need a gauge of people's intentions (Sugden, 1986; Boyd, 1989; Boerlijst *et al.*, 1997). Panchanathan (2001) concludes that language (in the form of gossip) can furnish people with very good information about the reputations of others, where reputation (based on the person's known record of interactions) works as a gauge of someone's probability of defection. Indirect reciprocity may thus help explain why a language-endowed social mammal was capable of organization on the scale of hunter–gatherer bands, which are larger and considerably more complex than other mammalian societies but small enough that people can keep track of reputation through gossip. Whether indirect reciprocity is a sufficient explanation for organization on the level of tribes, chiefdoms, and states is unclear.

Undoubtedly, dyadic and indirect reciprocity are importantly involved in the evolution of cognitive mechanisms such as guilt and shame, and their associated signals. For example, Fessler (1999) provides a detailed analysis of the situations that elicit shame. The purpose of the emotion/display is apparently to signal one's recognition of having made a 'mistake', with the implication that one is not really challenging the social norms. The importance of signaling contrition is evidence that people care about intentions, not merely behaviors.

Signaling

If large-scale organization does depend on generating public goods altruism, perhaps such behaviors can emerge through signaling. If I benefit from advertising my qualities to others, I will want a signal that cannot be faked by lower-quality competitors. This may explain the provision of expensive public goods as a form of signaling the quality of one's genes (Smith and Bliege-Bird, 2000). Male hunters, for example, may share difficult to catch prey items with everybody because they index the hunter's skill. Attention-getting sharing thus might ensure a strong broadcast of the 'hunting quality' signal. The benefits to such hunters would be things such as being preferred in the market for mates and greater political leverage.

The first benefit is obvious, as those who make themselves known as good hunters will be perceived, on average, as better providers, and their popularity in the marriage market will allow them to choose the most desirable (e.g., rich, healthy, hardworking, fertile) partners. This translates into healthier and more abundant progeny. The second benefit requires that there be a reason for other people to defer to the political interests of the hunter (and thus entails a form of trade or reciprocity, even if not a straightforward one). Since the prey is being shared collectively, one will not get more meat by deferring to the hunter, so why do it? Hawkes (1990) argues: in order to keep the hunter in the group (although she refers to the benefit that the hunter gets as 'social attention'). But this explanation does not solve the problem of selfishness, it merely places it elsewhere, as Smith and Bliege-Bird (2000) argue. Henrich and Gil-White (2001) suggest a reciprocal altruism hypothesis to explain deference to good hunters: sycophants who defer to the political interests of a hunter are buying access in order better to acquire the very skills the hunter has advertised.

The signaling hypothesis probably explains some altruism. However, it suffers from the same general problems as 'green beard' explanations. Why can't the selfish use the signals of altruists as a cue for whom to exploit selfishly? Why doesn't the signaling of qualities support complex societies in other species? Costly displays of good genes occur in many species, yet in no other species is aid to the group used to signal value as a mate. Emotional commitments to an altruistic moral order no doubt are a proximate explanation for such behaviors, but such emotions in turn have to be explained. The real puzzle is explaining how we came to be equipped with such emotional attachments to norms, and for that we probably need an explanation in terms of group selection generating the emergence of punishment for deviance, as argued below.

CULTURAL GROUP SELECTION

The Problem of Genetic Group Selection

Suppose we have two groups of the same species, one full of individuals with generalized altruism genes, and the other full of individuals with selfish genes. Which gene will do better evolutionarily? The fitness of a gene is equal to the average fitness of the vehicles carrying it, so here an average altruistic vehicle has higher fitness because it is surrounded by other such vehicles (which results in profitable mutual assistance). A selfish vehicle, on the other hand, has relatively lower fitness because it is surrounded by other selfish vehicles.

So the altruistic gene will win? The problem is maintaining sufficient variation between groups for group selection to be a potent force. Two forces erode variation in altruism between groups: the relative success of selfish individuals within groups and the migration of selfish individuals from group to group. Group selection can favor altruistic genes so long as (1) migration is sufficiently low; and (2) the fitness benefits of being in a group of mostly altruists is so large that new groups of altruists which competitively displace selfish groups are generated at a pace fast enough to more than compensate for the dilution of altruists by within-group processes and the arrival of selfish migrants.

Some students of altruism (Sober and Wilson, 1998) like to think of kin selection as a form of group selection in which relatedness creates sufficient variation between groups for group selection to operate. Terminological disputes aside, the kin selection view of groups illustrates the problem with large-scale group selection; if kin groups are reasonably outbred, relatedness falls dramatically with genealogical distance and the evolution of altruism is restricted to close kin. Outbreeding is equivalent to migration into the kin group. Observed rates of migration are generally too large to allow relatedness to build up in large groups, hence making group selection in them implausible. Ever since Williams's (1966) criticism of early attempts to explain adaptations as group selected, many evolutionists reject group selection as a plausible explanation almost as a matter of principle.

A Cultural Solution

Despite the problems with large-scale group selection explanations in outbred organisms, many, starting with Darwin, have speculated that some form of group selection is important in the special case of humans (Sober and Wilson, 1998). Humans certainly do compete as groups, and organized warfare is a spectacular example. But our groups are so porous (e.g., successful groups often induce a flow of mates from less successful ones) that one is brought back to the problem of migration. If some process could minimize the effects of migration – something quintessentially human – this would give us an elegant explanation simultaneously accounting for human ultra-sociality and also for the fact that other animal societies are restricted to forms of altruism derived from kin selection. That something might be culture, defined here as the intergenerationally stable, high fidelity, social transmission of information (socially transmissible packets of information are often referred to as 'memes', after Dawkins (1989, chap. 11)).

Theoretical models show that, given a capacity for acquiring information directly from others (which appears to be uniquely hypertrophied in humans), a bias for conformity will evolve. Conformity is adaptive because it helps individuals pick up useful memes that others have already converged upon (Boyd and Richerson, 1985; Henrich and Boyd, 1998). It is also advantageous to the degree that human societies often involve games of coordination in which direct advantages stem from doing what others do, such as driving on the agreed-upon side of the road (Gil-White, 2001). When in Rome, do as the Romans do. Many psychological studies have documented this cognitive bias (Miller and McFarland, 1991; Kuran, 1995; Asch, 1956, 1963). Conformity reduces the problem of migration (Boyd and Richerson, 1985; Henrich and Boyd, 1998) because when migrants absorb the memes in their host community they tend not to affect the local equilibrium. Rather the local equilibrium tends to convert *them*. Thus, selfish migrants arriving in an altruistic group will – if they are conformists – absorb the local altruistic norms even as their own are discriminated against, thus preserving rather than diluting the altruistic character of the group. This allows cultural group selection to generate new altruistic groups fast enough to overcome the rate at which spontaneous (cultural) mutations of individuals from altruistic to selfish erode altruism within groups (cf. Soltis, Boyd, Richerson, 1995). If cultural group selection operated over sufficiently long periods of time in the late Pleistocene, gene-culture coevolution might have resulted in the evolution of innate predispositions and skills adapted to participation in group selected social units (Richerson and Boyd, 2001).

WITHIN-GROUP COOPERATION AND BETWEEN-GROUP CONFLICTS

A complementary explanation maintains that if a norm for punishing deviations is adhered to by most members of a group, it can stabilize anything, including a norm for altruism (Boyd and Richerson, 1992). If much group competition is active rather than passive (e.g., violent combat for land), then within-group altruistic norms maintained by punishment will confer dramatic advantages. This could make the production of new altruistic groups faster than the processes which dilute altruism within the group (Boyd *et al.*, unpublished). The result would be a panhuman selection pressure for cognitive adaptations reducing the likelihood of 'mistakes' in order to avoid costly punishment (prosocial emotions such as duty, patriotism, moral outrage, etc. that commit us to predominant social norms even in the absence of coercion). These could easily form the basis for large-scale ultra-social organization, including dramatic cultural adaptations for collective defense. Such emotions could help explain why humans often engage in altruistic acts even in the absence of monitoring or reputational benefits and why they die anonymously in battlefields.

Clearly, the other side of the coin of group cooperation is group conflict. Groups that develop norms that channel their within-group cooperation towards outward bellicosity will force other groups to develop the same (or better) or become extinct. This process selects for ever stronger forms of within-group cooperation and outward aggression and is likely to be an important force responsible for the creation of ever larger and more complex social human groups.

IDEOLOGY, SYMBOLS, AND INGROUP MARKING

No society can exist without the acquiescence of its members to the roles they must play in the maintenance and reproduction of the social whole. Historically, anthropology and sociology were both centrally interested in the question of the functional organization of individuals into such roles (both disciplines owe much to the pioneering sociology of Emile Durkheim and pioneering anthropology of Bronislaw Malinowski), but these days the topic itself has fallen out of favor with the rise of 'methodological individualism' and 'rational choice' perspectives that insist on a picture of human nature as driven by selfish, individualistic considerations. Rational choice theorists, however, can account for high-cost altruism, such as soldiers being willing to die in battle, only by including in the concept of self-interest rewards and punishments that are in turn hard to explain on individual selection grounds. A soldier may not fight out of altruistic feelings (though at least a few undoubtedly do). But whatever the personal motives (glory, duty, shame, need for recognition from others, blind respect for authority), his behavior is more likely the result of adhering to a particular ideology, and the emotions which are inculcated as part of it, than a narrow calculation of the relative material costs and benefits to himself in the evolutionist's reproductive fitness sense. (The reader should note that group selected altruism is not saintly self-sacrifice. When the final tally is completed, altruists must do better at reproducing their genes or their culture than those adopting the selfish alternatives. One target of group selection may be systems of reward and punishment, especially culturally transmitted social institutions in the human case, that indeed motivate even the highly self-interested individual to cooperate. A relatively low frequency of altruistic moralistic punishers may be all that is necessary to keep reluctant cooperators cooperating.) If so, this means we must understand the cognitive processes by means of which ideas are acquired through social learning and emotions are attached to them. We must also understand why and how rendering ideas in the forms of reified symbols makes these ideas so attractive. Such work has barely begun.

In the domain of ethnic-group cognition, some first steps are being taken. It appears that the human brain is predisposed to essentialize ethnic and racial groups. One approach argues that essentialized 'human kinds' can be created out of any social category (Hirschfeld, 1996), depending on local cultural and historical circumstances.

Another approach argues that only those categories – such as, say, ethnic groups and castes – that superficially resemble biological species will tend to be essentialized (Gil-White, 2001). The salient resemblances to species categories are (1) normative endogamy; (2) descent-based membership; (3) characteristic marking (in ethnic groups this is outward marking in the form of dress, scarification, etc.); (4) a distinctive local social adaptation (in ethnic groups this is a local norm equilibrium). These surface resemblances fool the brain into thinking that it is looking at a species category, and the essentialism normally applied to biological kinds is activated. Features 1 and 2 are caused by 4 because interaction – especially in marriage – with outsiders who have different coordination norms is costly. A recent model shows that feature 3 also follows from 4 (McElreath *et al.*, in press). The model shows that everybody benefits from broadcasting the community of origin – if such communities differ in their norms – because in this way costly interactions between partners who will likely fail to coordinate properly will be avoided.

It is important to note that the above is insufficient for an explanation of, say, racial conflict. For most of history the antagonistic political units have often not been maximal ethnic units but smaller (e.g., subethnic tribes, clans) or larger (e.g., multiethnic chiefdoms, empires) units. Only with the recent advent of ethnonationalism – an ideology maintaining that political and ethnic boundaries should coincide – do we get a proliferation of conflicts where the antagonistic units are maximal ethnic groups. These conflicts appear especially difficult to contain and negotiate precisely because the groups in conflict perceive themselves as unalterably 'natural' groups. Smaller groups often recognize their 'inherent' similarities with coethnics, and larger ones usually find it impractical to motivate emotional adherence based on belonging to the same imperial system. We are still very far from understanding how and why ideologies such as ethnonationalism spread and remain stable and why they are so easily exportable into vastly different cultures. An understanding of the cognitive processes that make certain ideologies attractive in particular circumstances (i.e., become cultural selection pressures), and which commit us emotionally to such ideologies, is sorely needed.

CONCLUSION

Explanations that don't go beyond the mechanisms responsible for cooperation in nonhuman species fail to account in a satisfactory manner for the vast aggregations of cooperating nonrelatives that constitute human societies. Kin selection and reciprocity arguably need to be complemented by cultural group selection as the main driving force. While some work has been done to elucidate the formal properties of cultural group selection, the task of understanding the cognitive mechanisms that such processes have shaped, and their interactions, have only begun. As a result, we don't yet have a good theoretical handle on how the social brain creates selection pressures that affect the distribution and maintenance of ideologies central to large-scale human cooperation and conflict. However, we can now at least begin to ask the questions in a Darwinian framework, applied to culture as a system of inheritance in its own right.

References

Asch SE (1956) Studies of independence and conformity: I. Minority of one against a unanimous majority. *Psychological Monographs* **70**: (Whole No. 416).

Asch SE (1963 [1951]) Effects of group pressure upon the modification and distortion of judgments. In: Guetzkow H (ed.) *Groups, Leadership, and Men*. New York, NY: Russell & Russell.

Axelrod R (1984) *The Evolution of Cooperation*. London, UK: Basic Books/HarperCollins.

Axelrod R and Hamilton WD (1981) The evolution of cooperation. *Science* **211**: 1390–1396.

Boerlijst MC, Nowak MA and Sigmund K (1997) The logic of contrition. *Journal of Theoretical Biology* **185**(3): 281–293.

Boyd R (1989) Mistakes allow evolutionary stability in the repeated prisoner's dilemma game. *Journal of Theoretical Biology* **136**: 47–56.

Boyd R and Richerson PJ (1985) *Culture and the Evolutionary Process*. Chicago, IL: University of Chicago Press.

Boyd R and Richerson PJ (1988) The evolution of reciprocity in sizeable groups. *Journal of Theoretical Biology* **132**: 337–356.

Boyd R and Richerson PJ (1992) Punishment allows the evolution of cooperation (or anything else) in sizable groups. *Ethology and Sociobiology* **13**: 171–195.

Campbell DT (1983) Two distinct routes beyond kin selection to ultra-sociality: implications for the humanities and social sciences. In: Bridgeman D (ed.) *The Nature of Prosocial Development: Theories and Strategies*, pp. 11–39. New York, NY: Academic Press.

Dawkins R (1989 [1976]) *The Selfish Gene*, 2nd edn. Oxford and New York, NY: Oxford University Press.

Fessler DMT (1999) Toward an understanding of the universality of second order emotions. In: Hinton AL (ed.) *Biocultural Approaches to the Emotions*. New York, NY: Cambridge University Press.

Frank RH (1988) *Passions Within Reason: the Strategic Role of the Emotions*. New York, NY: WW Norton.

Gil-White FJ (2001) Are ethnic groups biological 'species' to the human brain?: essentialism in our cognition of some social categories. *Current Anthropology* **42**(4): 515–554.

Hawkes K (1990) Why do men hunt?: benefits for risky choices. In: Cashdan E (ed.) *Risk and Uncertainty in Tribal and Peasant Economies*, pp. 145–166. Boulder, CO: Westview Press.

Henrich J and Boyd R (1998) The evolution of conformist transmission and the emergence of between-group differences. *Evolution and Human Behavior* **19**(4): 215–241.

Henrich J and Gil-White FJ (2001) The evolution of prestige: freely conferred status as a mechanism for enhancing the benefits of cultural transmission. *Evolution and Human Behavior* **22**: 165–196.

Hirschfeld L (1996) *Race in the Making: Cognition, Culture, and the Child's Construction of Human Kinds*. Cambridge, MA: MIT Press.

Hirshleifer J (1987) On the emotions as guarantors of threats. In: Dupré J (ed.) *The Latest on the Best: Essays in Evolution and Optimality*. Cambridge, MA: MIT Press.

Kuran T (1995) *Private Truths, Public Lies: The Social Consequences of Preference Falsification*. Cambridge, MA: Harvard University Press.

McElreath R, Boyd R and Richerson P (In press) Shared norms can lead to the evolution of ethnic markers. *Current Anthropology*.

Leimar O and Hammerstein P (2001) Evolution of cooperation through indirect reciprocity. *Proceedings of the Royal Society of London B* **268**: 745–753.

Miller DT and McFarland C (1991) Why social comparison goes awry: the case of pluralistic ignorance. In: Suls J and Ashby T (eds) *Social Comparison: Contemporary Theory and Research*. Hillsdale, NJ: Lawrence Erlbaum.

Nowak MA and Sigmund K (1998a) The dynamics of indirect reciprocity. *Journal of Theoretical Biology* **194**: 561–574.

Nowak MA and Sigmund K (1998b) Evolution of indirect reciprocity by image scoring. *Nature* **393**: 573–577.

Panchanathan K (2001) *The Role of Reputation in the Evolution of Indirect Reciprocity*. Unpublished Master's Thesis, University of California, Los Angeles.

Richerson PJ and Boyd R (2001) The evolution of subjective commitment to groups: a tribal social instincts hypothesis. In: Nesse R (ed.) *Evolution and the Capacity for Commitment*, pp. 186–220. New York, NY: Russell Sage Foundation.

Smith EA and Bliege Bird RL (2000) Turtle hunting and tombstone opening: public generosity as costly signaling. *Evolution and Human Behavior* **21**(4): 245–261.

Sober E and Wilson DS (1998) *Unto Others: the Evolution and Psychology of Unselfish Behavior*. Cambridge, MA: Harvard University Press.

Sugden R (1986) *The Economics of Rights, Co-operation, and Welfare*. Oxford and New York: Basil Blackwell.

Trivers R (1971) The evolution of reciprocal altruism. *Quarterly Review of Biology* **46**: 35–57.

Williams GC (1966) *Adaptation and Natural Selection*. Princeton, NJ: Princeton University Press.

Further Reading

Richerson PJ (2001) Built for speed, not for comfort: Darwinian theory and human culture. *History and Philosophy of the Life Sciences* **23**: 423–463.

Nesse RM (2001) *The Evolution of the Capacity for Commitment*. New York, NY: Russell Sage.

Weingart P, Mitchell SD, Richerson PJ and Maasen S (1997) *Human by Nature: Between Biology and the Social Sciences*. Mahwah, NJ: Lawrence Erlbaum.

Wilson DS (2002) *Darwin's Cathedral: Evolution, Religion, and the Nature of Society*. Chicago, IL: University of Chicago Press.

Cooperative and Collaborative Learning

Intermediate article

Angela M O'Donnell, Rutgers, The State University of New Jersey, New Jersey, USA

CONTENTS

Forms of cooperative or collaborative learning have been used for centuries and a variety of peer learning techniques have emerged. The underlying premise of these techniques is that learning is enhanced by peer interaction.

THEORETICAL APPROACHES TO COOPERATION AND COLLABORATION

Cooperative/collaborative learning refers to a variety of instructional arrangements that have the common characteristic of students working together to help one another learn. The term 'cooperative learning' is often used to describe particular techniques such as Slavin's *Student Teams Achievement Divisions* or the Johnsons' *Learning Together*. However, cooperative learning is also a process involving collaboration.

'Collaboration' is a term generally used to describe the process of shared learning and understanding. Because of the close relationship of the two terms, they will be used interchangeably in this article. Cooperative and collaborative learning techniques have been used for instructional purposes for centuries and constitute some of the oldest forms of instruction. The most recent meta-analysis of cooperative learning studies included five hundred studies (Johnson and Johnson, 1989) and provides strong support for the positive benefits of working with peers, including positive effects on achievement, self-confidence, peer relationships, and the inclusion of children with special needs. Although a great deal of work has been conducted since 1989, this meta-analysis remains the most comprehensive integration of the literature.

A full discussion of the various forms of instructional uses of cooperation and collaboration is beyond the scope of this chapter. These approaches to understanding the benefits or other effects of cooperative and collaborative learning differ in whether they emphasize the individual (as in peer tutoring), the group (as in distributed cognition), or the reciprocal influence of the individual and the group (as in problem-based learning). The predominant approach to understanding the effects of collaborative or cooperative learning has been in terms of a focus on individual achievement. Other approaches that emphasize learning communities and apprenticeship models emphasize the group culture and the emergence of group cognition. A number of different approaches to understanding collaborative/cooperative learning are illustrated here. Specific cooperative and collaborative learning techniques may involve elements from more than one perspective.

SOCIAL PSYCHOLOGICAL APPROACHES

A variety of theories can be used to account for the benefits associated with cooperative/collaborative learning. Most of the published research on cooperative learning is influenced by social-motivational theory originating with the work of Morton Deutsch.

According to Deutsch, cooperation is one form of interdependence in which individuals' outcomes are linked. One person cannot succeed in a cooperative group unless all participants succeed. From this perspective, group learning is expected to be more productive than individual learning as a result of interdependence among group members that increases motivation within the group. The key mechanism by which cooperation can lead to successful outcomes is through motivation.

Interdependence can be created by providing group rewards (e.g., Slavin's *Student Teams' Achievement Divisions* (STAD)) or by developing group norms of caring and mutual helping (e.g., the Johnsons' *Learning Together*). In STAD, students work together in heterogeneous groups to help one another learn material presented by the teacher. Students' improvement scores on individual tests result in points earned for the team. Team averages are computed and the teams with the highest points are rewarded with some form of recomputed group average. The teams with the highest group points receive tangible recognition for their work in the form of certificates or other rewards valued by the group members. In contrast, cooperative learning techniques that depend on mutual care and concern as the basis for creating interdependence rarely use overt rewards. In these techniques, instructors spent a lot of time teaching students social skills and effective strategies for communicating and supporting one another.

DEVELOPMENTAL PSYCHOLOGICAL APPROACHES

Theories related to the mechanisms underlying effective collaboration can also be found in the developmental psychological literature.

Piaget described cognitive development as occurring through a process of adaptation in which a child's existing knowledge and concepts interacted with the world. Experience with the world provides the opportunity to create conflict with existing conceptual structures and the child will attempt to reduce the experience of conflict. From a Piagetian perspective, collaborative groups provide the possibility for cognitive development because group members may prompt cognitive conflict that can result in disequilibration and subsequent conceptual growth. Disequilibration occurs when the learner recognizes a conflict between new information or experience and existing conceptual structures. The effort to restore harmony to one's cognitive structures results in adaptation to the new information or experience. Conceptual growth may occur through this process, although other processes such as denial can achieve the restoration of equilibrium.

Many educators rely on the possibility that learners will bring different perspectives to a task. Difficulties can arise depending on the composition of a group. In groups in which there is a status hierarchy, students may defer to those with higher status, agreeing readily but experiencing little in the way of cognitive conflict. Without the experience of disequilibrium, conceptual growth is unlikely to occur. Piaget proposed that learning was most likely to occur when peers were mutually influential: that is a student was as likely to influence others or be influenced by them. In creating collaborative groups from a Piagetian standpoint, groups consisting of members who are relatively homogeneous are more desirable than very heterogeneous groups. Piaget did not write much about collaboration among peers and his focus was on the individual child's growth and development.

Collaboration among peers provided a context for development.

Another developmental psychologist provides a contrasting perspective. Vygotsky emphasized the crucial role of society and culture in shaping the cognitive skills of children. The community is the venue in which cognitive skills are first observed and subsequently internalized by the developing learner. Learning occurs from a Vygotskian perspective when a more skilled individual supports the performance of less skilled or younger learners. The weaker partner can perform tasks with the support of the more skilled partner that he or she could not do alone. The difference between what the child can accomplish alone and with a skilled partner is called the 'zone of proximal development'. This process is one way in which learners become participants in a community of practice. In other words, this is a process in which learners come to acquire the competencies already available in the community. Unlike Piaget, who viewed mutuality in power and influence as crucial to effective collaboration, Vygotsky seems to require an imbalance in power and expertise. The more expert student or adult scaffolds the learning of the more novice individual.

Peer Tutoring as an Example of Collaborative Learning

Peer tutoring is one of the most enduring forms of instruction and typically involves a more skilled individual teaching a less skilled individual. The relationship between learners is one of inequality with respect to knowledge of the target subject. Vygotskian theory might be drawn upon to explain the benefits of this strategy. Interactions between tutor and tutee have the goal of improving the performance of the tutee and are characterized by efforts to prompt the tutee to higher levels of achievement. Tutoring works – it is a unique form of individualized instruction that improves student achievement (Cohen *et al.*, 1982). However, understanding how tutoring works is not a simple task. Empirical studies of tutoring have not been conducted under the rubric of a common theory that details the processes underlying particular this particular form of instruction.

Effects on tutees and tutors

Cohen *et al.* conducted the only available comprehensive review of the effects of tutoring in 1982. Their meta-analysis provided convincing evidence that peer tutoring is effective. Strong effects on student learning were found in studies of short duration that involved structured tutoring related to lower-level skills, often in mathematics. Very few studies examined affective outcomes from tutoring, although in those studies that did, tutored students generally had positive attitudes. Tutees are not the only beneficiaries of tutoring. In 38 studies included in the 1982 meta-analysis related to tutor achievement, the average effect size for tutor achievement was a moderate one of 0.33.

Few studies examine the joint effects on tutors and tutees. One exception to this is the work of Connie Juel (1996). In this study, underachieving student athletes spent about four hours a week preparing materials for tutorial sessions with at-risk first grade students. Both tutors and tutees made substantial progress during the course of a year. The literacy levels of both groups of students increased.

Research on tutoring in the period since 1982 has not focused on achievement *per se* but has concentrated instead on the processes by which tutors engage in their task of scaffolding the learning of their tutees. Even when tutors were naive or provided wrong answers, tutees still benefited. There was general acceptance of the utility of tutoring as an instructional strategy and a shift towards studies designed to understand the processes involved in tutoring rather than documenting outcomes from such tutors. An important strand of this research concerned comparisons of human and computer tutors.

Human and computer tutors

Human tutors allow students to do much of the work and maintain a sufficient feeling of control, provide sufficient guidance to keep the tutee from being frustrated, monitor students' reasoning and intervene to keep problem solving on track, are flexible in how they interact, and are motivating (Merrill *et al.*, 1992). Naive tutors often provide indiscriminate feedback, use unsophisticated strategies, and employ politeness strategies that may interfere with tutoring effectiveness. Despite these inadequacies, even naive human tutors produce achievement gains for tutees. Many possible reasons might be offered to explain such findings. It may be that the one-to-one interaction is motivating and the tutee makes more of an effort to engage the material. Perhaps normal classroom instruction may provide such little benefit to those who end up as tutees that even a poor tutor is better than existing instruction.

Computer tutors are often designed to compare the student's problem-solving steps to those of a domain expert. In other words, the student's performance is compared to an expert model and the steps he or she uses are traced. Computer tutors attend to more of the components of recovering from error than humans and provide more diagnostic information, more accurate feedback, and provide interventions as the learner deviates from an expert model. They have limited flexibility in response to learner errors, although recent innovations in the design of computer tutors have improved this functionality. Recent innovations in the development of computer tutors use what is known about the human tutoring process and include strategies and prompts to student learning that are more typical of human tutors.

Future research

The published research on tutoring provides little analysis of how the subject matter influences the nature of the tutoring. Much of the work on tutoring is done in curricular areas that are highly constrained, such as mathematics or physics. Little is known about the effects of tutoring on more ambiguous domains such as social studies or writing. The kinds of outcome that were included in the 1982 meta-analysis were mostly low-level outcomes. The effect of tutoring on more high-level skills has not received the same kind of attention.

An additional area of potentially fruitful work is the comparison of the effectiveness of expert tutors and tutors who are subject matter experts. Tutors develop models of tutees and direct their efforts based on those models. Expert tutors are likely to have very refined models of tutee differences and have a range of tutorial strategies from which to select. Subject matter experts who are less experienced as tutors will likely bring a very different set of skills to the tutorial process.

INFORMATION-PROCESSING APPROACHES

Learning, from an information-processing perspective, occurs as a result of active processing of meaningful materials and experiences. Cooperative/collaborative learning approaches influenced by information-processing theory focus on using the group context for learning as a way for amplifying the active processing of information. Such approaches are often considered cognitive/elaborative approaches and suggest that the group interaction provides an opportunity for deeper processing of material and restructuring of ideas and understanding. Cognitive/elaborative approaches attempt to maximize the participation of each student in cognitive strategies and tasks.

Scripted Cooperation

Scripted cooperation (see O'Donnell and King, 1999) is an example of a cooperative technique that relies on elaborative processing of information by a cooperating dyad. In this strategy, pairs of students read a portion of a text. One partner then summarizes the information to his or her partner who, in turn, provides feedback on the accuracy and completion of the information summarized. Both partners work together to elaborate on the material read, generating connections to other knowledge they have and developing techniques for making the information more memorable. The students then proceed to the next section of the text in which they switch roles with the summarizer from the first section now acting as the person who detects errors and omissions. The use of this technique has been shown to be very successful. The size of the cooperating unit (pairs of students) promotes active participation by the partners, and the requirement to switch roles provides each student with practice of requisite cognitive skills.

Reciprocal Teaching

Another example of a cooperative technique that is influenced by information-processing theory, and also by Vygotskian theory, is reciprocal teaching.

Reciprocal teaching was designed for use with students who showed a significant disparity between their ability to decode and comprehend text (Palinscar and Herrenkohl, 1999). It involves guided instruction in the use of four strategies to assist text comprehension. The teacher and students take turns in leading discussions about a shared text. Before reading the text, students make predictions about what the text will be about. Second, everyone reads a segment of the text and participants then discuss the text content, asking and answering questions. The third strategy used is summarization. The fourth strategy involves clarifying difficult concepts and, finally, new predictions are made for the next segment of the text. The focus in this technique is on teaching students explicit strategies to aid comprehension that they can use with any new text. Instruction in strategy use proceeds through a dialogue between the teacher and the students. Students participate by elaborating on other students' responses or by commenting on one another's questions. Initially, the teacher is very much in charge, modeling the strategies and monitoring students' use of them. With practice, students assume a more central role in leading the discussions, with the teacher participating as a member of the group. In this role, the teacher can remind students about how to use the strategies. He or she provides a scaffold or support to the students' efforts. Finally, student leaders can conduct discussions about a segment of text without assistance from the teacher. The teacher's support has thus gradually been withdrawn from the group interaction.

The model of collaboration exemplified in reciprocal teaching is strongly influenced by Vygotskian theory. Skills are available in the social world before they can be internalized. The entire transition from the teacher-led discussions and modeling of strategies to the student-led discussions and execution of these strategies typifies the internalization of skill. Learners first observing these skills as performed and modeled by an expert practice the skills with guidance from the expert, and finally internalize them as part of their own cognitive repertoires.

Effects on student achievement

Rosenshine and Meister (1994) included 16 studies (only four were published) in their meta-analysis of studies using reciprocal teaching. Studies were selected if they included experimental and control groups and were considered high-, medium-, or low-quality studies, depending on the quality of implementation of the treatments, student assessment, and the quality of the reciprocal teaching dialogue. Across the 16 studies, the median effect size associated with reciprocal teaching was 0.32 when using standardized tests as the outcome measure, and 0.88 when using experimenter developed assessments. Reciprocal teaching was significantly better than a control comparison in only two of nine studies when standardized tests were used. When experimenter-developed assessments were used, reciprocal teaching was significantly better in four out of five studies. Based on an analysis of the reading materials provided in a standardized test and those provided in experimenter tests, Rosenshine and Meister concluded that the experimenter-developed texts were easier because they were longer, were almost always organized in a topic-sentence-and-supporting-detail form, and answering the questions required less background and searching of text.

No conclusions could be drawn from this meta-analysis about which strategies involved in reciprocal teaching were most effective in assisting students. Many of the studies included provided explicit instruction in each of the component strategies prior to engaging in reciprocal teaching, while others embedded the instruction in cognitive strategies in the use of reciprocal teaching. Both strategies seem to work.

Although the dialogue that occurs in a reciprocal teaching group is expected to play a key role in the benefits that may be derived from reciprocal teaching, few studies analyze this dialogue or use it to detect what kinds of cognitive strategy students are learning and using. Palinscar and Herrenkohl (1999) suggest that the explicit goal of jointly making sense of text encourages an intersubjective attitude in which learners feel they are part of a community. This notion of community is further emphasized by the need for each participant to take a turn in leading the discussion and contributing to the discussion.

DISTRIBUTED-COGNITION APPROACHES

Another view of collaboration/cooperation can be found in the perspectives drawn from distributed cognition that describes a process by which cooperation or collaboration can be accomplished. The types of collaborative/cooperative learning described previously are primarily concerned with individual cognition and the effect of collaborative interactions on individual achievement. The techniques described can be viewed as examples of distributed cognition in that cognitive activity is distributed within an interactive unit. A simple version of distributed cognition and collaboration is the use of scripted cooperation described previously. In this technique, the cognitive processing of information is divided among members of a dyad. One partner engages in rehearsal (summarizing information), while the other partner provides an elaboration of the information. In this way, cognitive processing is distributed.

Some views of distributed cognition consider the distribution to be an extension of the individual's competency, a person-plus approach. Others view the distribution of cognition within a group as inseparable from the activity and functioning of the group, a social-only perspective. The focus of the 'in-the-head' or person-plus approach is on the cognitive residue that remains from group interaction. The group can serve to distribute cognitive processing (e.g., metacognition) and reduce cognitive load. The role of scribe or recorder that is specifically built into some cooperative techniques may reduce working memory demands on other participants. Other cooperative techniques in which groups of students become expert in subtopics of a major topic and are then responsible to teach this content to other members of their group involve a distribution of expertise. This person-plus approach to distributed cognition is modeled on what is known about individual cognition, and functioning within the group is described in terms of individual cognitive functions.

The social-only perspective on distributed cognition views the group members and their work as indivisible. This perspective has strong influences from cultural-historical psychology that emphasizes the role of culture and history on cognition. Thus, distributed cognition approaches that involve the social-only perspective have much in common with a Vygotskian approach to cooperation/collaboration. Interpreted from this perspective, the activity of the individual is mediated by the tools available (to include both symbols and actual artefacts), the rules of the community, and the accepted divisions of labor. Tools represent and embody the expertise of others and interaction with these tools involves interaction with the wisdom and experience of others. For example, when a doctor conducts a history and physical examination of a patient and uses a checklist of steps to guide this process, the doctor is not simply performing an act of individual cognition. The steps included in the checklist represent the embodied expertise of other doctors whose expertise and wisdom in practice has been distilled to a checklist that can be used by others. Thus, the doctor in using this tool participates in the practices of a community that extend beyond the individual interaction with a particular patient.

The notion of distributed cognition has enormous appeal. We do not wish to fly in an aircraft where the work is considered divisible. There are products where the individual contributions to the final product should be seamless and invisible. Educational systems, however, focus on individual accomplishment, and viewing collaborative groups without concern for individual achievement is not realistic.

An Educational Example of Distributed Cognition

The examples of peer tutoring and scripted cooperation represent versions of distributed cognition and are used in many schools. There are fewer examples of the social-only perspective. One such example is the computer support for intentional learning environments (CSILE: Scardamalia *et al.*, 1994). The goal of this project was to support students' intentional and autonomous learning in a community of learners. An environment was created thought the use of a networked set of computers in which students contributed to a shared database. Students could add graphical or text notes to a shared database. They could add comments to notes in the database. Only the author of a note could edit the note or delete it. Authors were notified when a comment was added to a note they contributed. Students could use CSILE for a range of tasks from very traditional schoolwork in which they rehearsed information or for more open-ended student-initiated tasks. Knowledge was public and could be added to, critiqued, revised, and reformulated by anyone in the class. The task for students was the social construction of knowledge in which individual contributions were less important than the adequacy of the emerging knowledge base. Unlike most cooperative or collaborative learning techniques, the focus was not on students improving their own skills.

In one example of students' use of CSILE, students generated notes and comments on the inheritance of characteristics or why a child might look more like one parent than another. Students developed a complicated database of information about inheritance that connected to both current uses of genetics and previous history of research on genetics. Despite the fact that the focus in CSILE classrooms is on community knowledge, students in such classes do in fact improve their individual skills and perform better on standardized tests than do students in control classrooms (Scardamalia *et al.*, 1994).

Problem-based Learning

Both person-plus and social-only aspects of distributed cognition are present in problem-based learning (PBL). PBL must be distinguished from individual problem-solving even when the problems are authentic. Problem-based learning has three key features: a rich and authentic problem that supports student inquiry, student-centered learning, and learning that is done collaboratively with group members, scaffolded by the availability of a tutor or facilitator.

In medical schools the goals of PBL are to develop clinical reasoning, integrate clinical reasoning and basic biomedical knowledge, and promote self-directed learning. The use of PBL in medical schools involves a group of five to seven students who meet with a facilitator to discuss a problem. The facilitator provides students with a small amount of information about a patient's case. One student assumes the role of scribe, recording the decisions and hypotheses of the group. This record is typically visible to all. The group evaluates and defines various aspects of the case and develops an understanding of the causes of the problem. The students do so by identifying key information, generating and evaluating hypotheses, and formulating learning issues that they deem relevant and in need of further explanation. They conduct research to find information relevant to the identified learning issues and eventually draw conclusions about the patient's problem. The process concludes with reflections upon what was learned. The role of the facilitator is crucial to this process.

Problem-based learning exemplifies a number of aspects of distributed cognition. The knowledge needed for problem solution is distributed among the members of the group, library resources, other experts in the school, and prior records of patient cases. Learning requires connecting to a community of practice and prior experience in pursuit of a solution to a particular problem. In this sense, the work of the group is an example of social-only distributed cognition in that understanding of the patient case requires joint activity. The technique also provides an example of person-plus approaches to distributed cognition in that cognitive load is reduced by the use of a single scribe. Metacognition is partially offloaded to the facilitator who adopts some of the monitoring function typical of individual cognition.

Effects on student achievement

Early evaluations of problem-based learning examined traditional learning outcomes such as performance on the medical board examinations. Students from PBL programs score below those from more traditional curricula, but there is some evidence to suggest they do better on the clinical portions of the examinations. However, comparisons are confounded by selection problems associated with different kinds of students selecting to participate in different kinds of curricula. Hmelo (1998) compared students from a PBL program with those from a traditional curriculum on measures of problem solving. She concluded that the PBL students were more accurate, constructed better explanations, and were more likely to use hypothesis-driven reasoning strategies than students from the traditional curriculum.

Applications outside medicine

Problem-based learning is being used in contexts other than in medical training. Science departments in many universities in the United States have adopted PBL methods in basic courses at the undergraduate level as part of ongoing efforts to make science meaningful and relevant to large groups of students. Many inquiry-based programs in elementary and secondary schools include elements of PBL. In particular, anchoring knowledge acquisition in authentic problems is common and reflects a shift in theories of learning towards a view of knowledge acquisition that has social interaction as a key feature.

COMMON GROUND AMONG APPROACHES

All of the collaborative/cooperative learning techniques described here (reciprocal teaching, problem-based learning, scripted cooperation, and peer tutoring) involve some type of scaffolded instruction. Despite differences in approaches, most collaborative or cooperative techniques promote student engagement with one another or with tasks such that effective cognitive processing and productive discourse occur. Efforts to scaffold instruction are attempts to promote such processes.

Instruction can be scaffolded by setting up the initial context for learning, specifying the cognitive or other roles that students adopt during collaboration, delineating interaction rules such as taking turns, and monitoring or regulating the interactions. Task structures drive the nature and quality of interaction. Authentic tasks such as those found in problem-based learning scenarios engage student interest and provide meaningful challenges.

Instruction can also be scaffolded by the assignment of cognitive roles within a group. In problem-based learning groups, the roles of facilitator and scribe are important in assisting the group members to focus on the learning issues at hand. The scribe records the decisions of the group and this visible record of group hypotheses and decisions about needed information provides important metacognitive processing opportunities for the group. The external record of discussions allows group members to recheck their choices and monitor progress towards their goals. In many cooperative contexts (e.g., the Johnsons' *Learning Together*), specific cognitive roles are assigned (checker, recorder, question generator, etc.), and the performance of these roles provides a scaffold that moves the group towards goal completion. In some instances such as tutoring, the tutor quite deliberatively scaffolds the tutee's processing as he or she directs attention to features of the task and strategies for accomplishing the task. Depending on the content and pedagogical expertise of the tutor, the scaffolding that is provided will be more or less effective. Reciprocal teaching provides scaffolded instruction through dialogues between the teacher who models the desired cognitive strategies and the students who practice the strategies. Support is gradually removed as students become more expert in using the targeted cognitive strategies. Scaffolded instruction may include strategies for maintaining all group members' involvement in group dialogue. Examples of such strategies include requiring turn taking with specific roles or tasks such that the same individual is not always responsible for the same task. Thus, opportunity to participate is deliberately distributed.

Key features of scaffolded instruction include making the use of cognitive strategies visible and providing opportunity for modeling, practice, and feedback related to the performance of those strategies. Collaborative techniques vary depending on whose strategies are made visible (other students, experts) and how the modeling, practice, and feedback cycles occur. They share common goals of student engagement and enhancement of learning.

References

Cohen PA, Kulik JA and Kulik CC (1982) Educational outcomes of tutoring: a meta-analysis of findings. *American Educational Research Journal* **19**: 237–248.

Hmelo CE (1998) Problem-based learning: effects on the early acquisition of cognitive skill in medicine. *Journal of the Learning Sciences* **7**: 173–208.

Johnson DW and Johnson RT (1989) *Cooperation and competition: theory and research.* Edina, MN: Interaction Book Co.

Juel C (1996) What makes literacy tutoring effective? *Reading Research Quarterly* **31**: 268–289.

Merrill DC, Reiser BJ, Ranney M and Trafton JG (1992) Effective tutoring techniques: a comparison of human tutors and intelligent tutoring systems. *Journal of the Learning Sciences* **2**: 277–305.

O'Donnell AM and King A (1999) (eds) *Cognitive Perspectives on Peer Learning*. Mahwah, NJ: Lawrence Erlbaum.

Palinscar AS and Herrenkohl LR (1999) Designing collaborative contexts: lessons from three research programs. In: O'Donnell AM and King A (eds) *Cognitive Perspectives on Peer Learning*, pp. 151–177. Mahwah, NJ: Lawrence Erlbaum.

Rosenshine B and Meister C (1994) Reciprocal teaching: a review of the research. *Review of Educational Research* **64**: 479–530.

Scardamalia M, Bereiter C and Lamon M (1994) The CSILE project: trying to bring the classroom into the world. In: McGilly K (ed.) *Classroom Lessons: Integrating Cognitive Theory*, pp. 201–228. Cambridge, MA: MIT Press.

Further Reading

Dillenbourg P (1999) (ed.) *Collaborative Learning: Cognitive and Computational Approaches.* Oxford, UK: Elsevier.

Evensen DH and Hmelo CE (2000) (eds) *Problem-Based Learning: A Research Perspective on Learning Interactions.* Mahwah, NJ: Lawrence Erlbaum.

Graesser AC, Person NK and Magliano JP (1995) Collaborative dialogue patterns in naturalistic one-to-one tutoring sessions. *Applied Cognitive Psychology* **9**: 1–28.

Salomon G (1993) (ed.) *Distributed Cognitions: Psychological and Educational Considerations*, New York, NY: Cambridge University Press.

Webb NM and Palinscar AS (1996) Group processes in the classroom. In: Berliner DC and Calfee RC (eds) *Handbook of Educational Psychology*, pp. 841–873. New York, NY: Macmillan.

Coordinate Transformations

See **Modeling Coordinate Transformations**

Cortical Columns

Intermediate article

Geoffrey J Goodhill, Georgetown University Medical Center, Washington, DC, USA
Miguel A Carreira-Perpiñán, Georgetown University Medical Center, Washington, DC, USA

CONTENTS

In many regions of the cortex, neuronal response properties remain relatively constant in a direction perpendicular to the surface of the cortex, while they vary in a direction parallel to the cortex. Such columnar organization is particularly evident in the visual system, in the form of ocular dominance and orientation columns.

INTRODUCTION

The most prominent feature of the architecture of the cortex is its horizontal organization into layers. Each layer contains different cell types, and forms different types of connections with other neurons. However, a strong vertical organization is often also apparent: neurons stacked on top of each other through the depth of the cortex tend to be connected and have similar response properties despite residing in different layers. This type of vertical structure is called a *cortical column*, and has been hypothesized to represent a basic functional unit for sensory processing or motor output. Columnar organization has been most extensively studied in the somatosensory and visual systems.

DISCOVERY OF COLUMNAR ORGANIZATION

Cortical columns were first discovered electrophysiologically by Mountcastle (1957). When he moved an electrode obliquely to the surface of somatosensory cortex, he encountered neurons that responded to different sensory submodalities (e.g. deep versus light touch). However, when the electrode was moved perpendicularly to the cortical surface, all neurons had similar response properties. He summarized his findings as follows:

> These data ... support an hypothesis of the functional organization of this cortical area. This is that the neurons which lie in narrow vertical columns, or cylinders, extending from layer II through layer VI make up an elementary unit of organization, for they are activated by stimulation of the same single class of peripheral receptors, from almost identical peripheral receptive fields, at latencies which are not significantly different for the cells of the various layers.

Shortly following this, vertical uniformity was also found in the visual system by Hubel and Wiesel (1977). Here, response properties that vary across the cortical surface but not through the depth of the cortex include the location of the neuron's receptive field in visual space, and the degree to which neurons are dominated by one eye. Columnar organization has also since been found in the auditory cortices of cat and monkey, where alternating bands related to monaural or binaural responses occur. A number of techniques have been employed for the experimental determination of cortical columns since the original use of electrode penetrations. These include methods based on axonal transport of substances such as horseradish peroxidase; on the differential consumption of radioactive 2-deoxyglucose by neurons; on optical imaging techniques, where cortical activity is converted to a visual signal by changes in reflectance or by voltage-sensitive dyes; and most recently on functional magnetic resonance imaging (fMRI).

There are some difficulties with defining exactly what is meant by a column. In some cases it is relatively clear: for instance, 'barrels' in somatosensory cortex and ocular dominance columns in visual cortex have fairly discrete boundaries with neighboring columns. In other cases, for instance orientation columns, there is a smooth variation in response properties moving parallel to the cortical surface, rather than a series of discrete jumps. Another problem is that the term 'column' has been used to refer to structures at several different scales. At one extreme, from an anatomical point of view, are narrow vertical chains of neurons seen in Nissl-stained sections, barely more than one cell diameter wide, sometimes called *minicolumns*. At the other extreme, largely from a theoretical point of view, are complete functional units up to 1 mm in size, sometimes called *hypercolumns*. In between, Szentágothai (1978) specifies a generic column to be 200–300 μm wide. This article avoids such definitional issues by focusing on well-characterized examples of columnar organization. (*See* **Neuroimaging**)

COLUMNS IN THE VISUAL SYSTEM

Ocular Dominance Columns

Moving parallel to the surface of the primary visual cortex (V1) of several mammalian species, notably ferrets, cats, monkeys, and humans, there is a regular alternation between groups of neurons that respond best to input in the left eye and neurons that respond best to input in the right eye. The anatomical basis of this physiological pattern is the segregation of the thalamic input fibers – lateral geniculate nucleus (LGN) afferents – representing the left and right eyes to the visual cortex (Figure 1(a)). Although these fibers terminate primarily in layer 4 of the cortex, and this is where ocular preference is most sharply defined, a similar bias is also seen in higher and lower layers. This vertical structure of monocular preference is called an 'ocular dominance column' (reviewed by Hubel and Wiesel, 1977). When the entire pattern of eye preference is visualized in V1, for instance by injection of a radioactive tracer into one retina and its subsequent transport to the cortex, an alternating pattern of stripes is observed (Figure 1(b)). The periodicity of this pattern varies depending on the species and location in the cortex, and also varies substantially between individuals (Horton and Hocking, 1996):

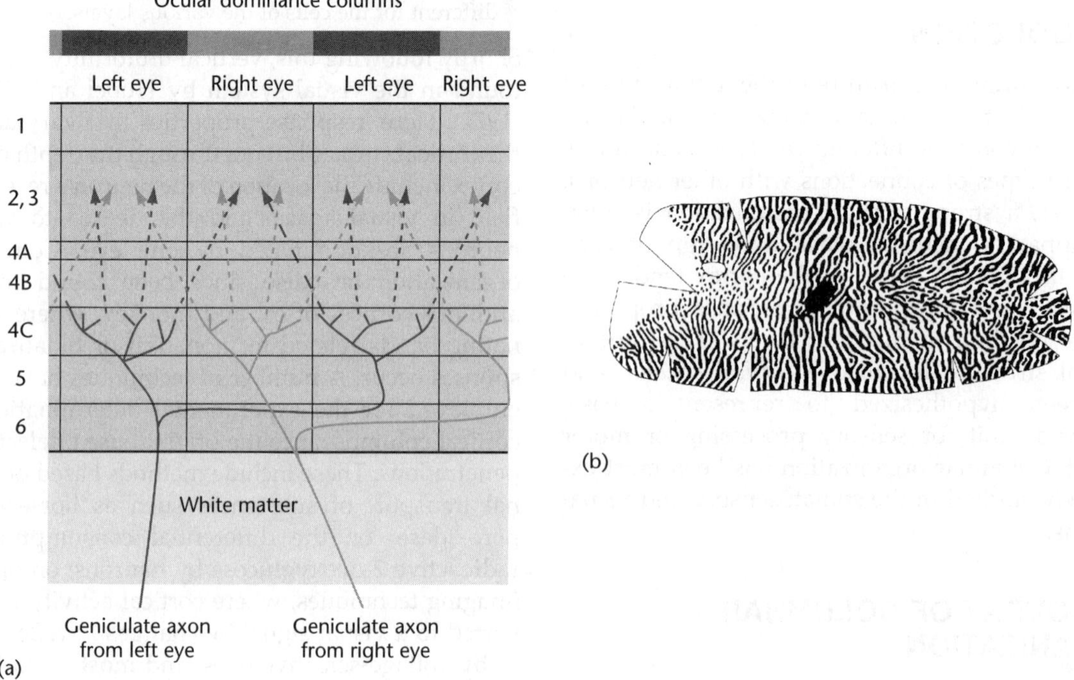

Figure 1. Ocular dominance columns in a monkey. (a) Anatomical basis. Each afferent axon from the lateral geniculate nucleus ascends through the deep layers of V1 (layers 5, 6) subdividing repeatedly and terminating in layer 4C in a couple of 0.5 mm-wide clusters separated by 0.5 mm gaps (approximately). Axons from the two eyes alternate, giving ocular dominance columns in 4C. The presence of horizontal connections and the arborization between different layers brings about overlapping and blurring of ocular dominance columns beyond layer 4: the ocular dominance of a given cell varies then between pure monocularity and pure binocularity. Adapted from Hubel (1995). (b) The pattern of ocular dominance columns from the primary visual cortex of a macaque monkey. White represents regions of cortex dominated by input from one eye, black the other eye. The width of individual columns is 0.5–1 mm. Source: LeVay *et al.* (1985) *Journal of Neuroscience* **5**: 486–501, © 1985 by the Society for Neuroscience.

in fact, each ocular dominance pattern is apparently as unique as a fingerprint. It can be seen from Figure 1 that these columns are in fact more like slabs, being long and relatively narrow rather than short and round.

Orientation Columns

Another type of columnar organization observed in the visual cortex is the orientation column. Many neurons in V1 respond best to an edge or bar of light at a specific orientation. This preferred orientation remains roughly constant through the depth of the cortex but varies mostly smoothly across the surface of the cortex. The overall pattern of orientation columns can be visualized by optical imaging methods. Cortical tissue changes its reflectance properties very slightly when neurons are active, and so by examining changes in reflected light from the cortical surface as visual stimuli of varying orientations are presented one can build up a picture of the complete map. An example is shown in Figure 2. A notable feature is the presence of pinwheels, point singularities around which all orientations are represented in a radial pattern. Superimposing the ocular dominance and orientation maps from the same animal, one observes regular geometric relationships between the two columnar systems. For instance, ocular dominance and orientation columns tend to meet at right angles, and orientation pinwheels tend to lie at the center rather than at the borders of ocular dominance columns.

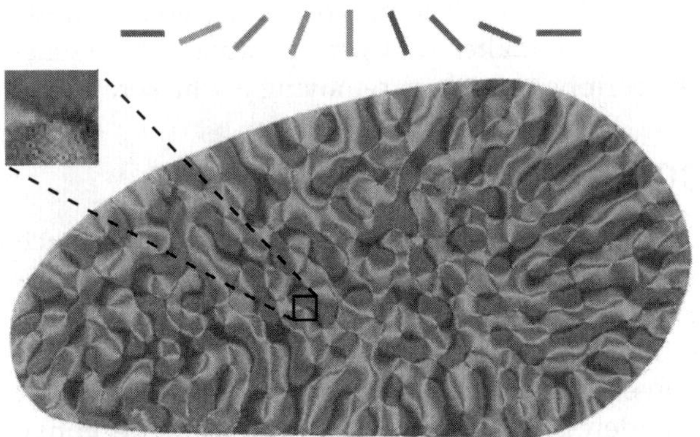

Figure 2. The orientation map in primary visual cortex of a tree shrew. The different degrees of shading represent patches that have different orientation preferences. The detail shows a pinwheel, where the orientation preference changes by 180° along a closed path around the center. Adapted from Bosking *et al.* (1997) *Journal of Neuroscience* **17**: 2112–2127, © 1997 by the Society for Neuroscience.

Other Types of Columns

Besides ocular dominance and orientation columns, several other types of columns are also present in the visual cortex. The most fundamental of these are what might be called position columns. Neurons in V1 have small receptive fields localized at specific positions in visual space. Moving vertically through the cortex, neurons have receptive fields at similar positions, while moving horizontally there is a smooth progression of visual field position versus cortical position, forming a topographic map of visual space in the cortex. This locality of information processing in visual cortex can also be seen from the fact that a small injury (e.g. a tumor or stroke) in part of V1 can cause blindness in a localized area in the visual field (a scotoma) with normal vision elsewhere, rather than an overall worsening of vision. Other receptive field properties that are organized into columns include preference for the spatial frequency of a stimulus across the receptive field, preference for the direction of movement of a stimulus, and disparity of inputs from the two eyes. All these columnar systems occupy the same cortical territory as the ocular dominance and orientation columns, and show complex geometric relationships that have yet to be fully characterized. Color-sensitive cells in layers 2–3 of monkey visual cortex (although not in other layers) are grouped in blobs, in which neurons respond to the color of a stimulus, but are mostly insensitive to orientation – unlike cells outside the blobs (the interblobs) which show marked orientation selectivity. (*See* **Color Vision, Neural Basis of**; **Depth Perception**)

Factors Driving the Formation of Columns

A number of different experiments on visual deprivation, where the visual experience that an animal receives is distorted, have shown that it is possible to produce physiological and structural changes in the columnar organization of visual cortex. For example, if one eye is sutured closed or strabismus is induced then most cells become monocular; if animals are presented with only bands at a specific orientation angle, then the proportion of cells that respond to that angle increases; if movement in a particular direction is excluded, the cells that would have responded to that movement direction no longer do so. Recovery to normal structure is also possible. However, both deprivation and recovery are effective only in an early period of the life when the connections are developing. These experiments suggest that the development of ocular dominance columns is the result of two competing processes: segregation is promoted when neural activity is equal in each eye but not correlated between both eyes; and binocular innervation of neurons and merging of the two sets of columns is promoted by the correlation in activity between corresponding retinal areas of the two eyes that results from normal binocular vision.

However, the relative importance of intrinsic, or genetically programmed, factors versus extrinsic, or activity-driven, factors is still not clear. On the one hand, Crowley and Katz (1999) found that total removal of retinal influence early in visual development did not prevent segregation of geniculocortical axons into ocular dominance columns of normal periodicity. They thus propose that ocular dominance column formation relies on molecular cues present on thalamic axons, cortical cells, or both. On the other hand, Sur and colleagues (e.g. Sharma *et al.*, 2000) have surgically rewired the optic nerve of newborn ferrets to feed into auditory thalamus (itself deprived of auditory inputs), which in turn projects to primary auditory cortex (A1) – rather than the normal pathway, optic nerve to LGN to V1. Such rewired ferrets develop in A1 a pattern of orientation columns with some similarities to that normally present in V1, though with a less regular periodicity. Such new cortical structure perceptually acts as visual; that is, the animals use the rewired A1 to see, rather than hear – although the resulting visual acuity is lower than normal. This suggests that retinal inputs can drive the formation of columns.

COLUMNS IN THE SOMATOSENSORY SYSTEM

The first physiological indication of cortical columns came from experiments by Mountcastle (1957) in the somatosensory cortex of cats. He found three types of neurons: those activated by light pressure on the skin, those activated by movement of hairs, and those activated by deformation of deep tissues (as occurs during for instance joint movement). As summarized by Mountcastle:

> Cells belonging to each subgroup were found in all the cellular layers. In 84 per cent of penetrations across the cellular layers which were directed perpendicularly, all the neurons encountered belonged to either cutaneous or deep subgroups. These modality-specific vertical columns of cells are intermingled for any given topographical region.

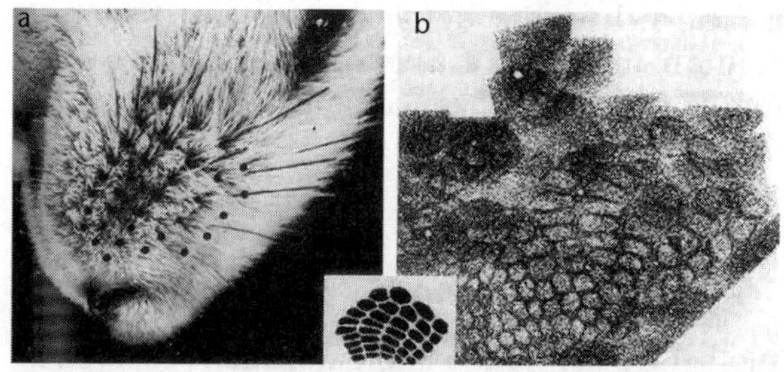

Figure 3. Posteromedial barrel subfield from a mouse's muzzle. Reprinted from Woolsey TA and van der Loos H (1970) The structural organization of layer IV in the somatosensory region (SI) of mouse cerebral cortex. *Brain Research* **17**: 205–242, © 1970, with permission from Elsevier Science.

More recent results have amplified this. For instance, Favorov and Diamond (1990) found discrete jumps in receptive field location between neighboring columns in cat primary somatosensory cortex, with no significant receptive field shifts within a column. However, the most striking example of columns in somatosensory cortex are the 'barrels' discovered by Woolsey and van der Loos (1970). In animals such as mice and rats, the long whiskers of the face are present in a stereotyped spatial pattern of rows. This is reflected in the posterior-medial barrel subfield of primary somatosensory cortex by a similar spatial pattern of columns, one column for each whisker (Figure 3). These are best defined in layer 4 where the thalamic afferents terminate. However, specialization to a single whisker is also apparent in higher and lower layers, and owing to their three-dimensional shape these columns were dubbed 'barrels'. The number and layout of these barrels can be altered by manipulations of the sensory periphery, such as removing a whisker.

COLUMNS IN OTHER SYSTEMS

The primary auditory cortex (A1) of animals such as cats and bats shows a systematic, spatially distributed representation of several independent auditory stimuli (reviewed in Schreiner, 1995). However, these auditory maps appear somewhat disordered because the local scatter of receptive field properties can vary over a wide range. The most regular map is that of preferred frequency, organized along a tonotopic axis without gradient reversals that mimics the tonotopic organization of the cochlea. Orthogonal to this axis, no systematic change of the preferred frequency is observed, with neurons being arranged along isofrequency contours. Other response parameters vary along the isofrequency contours in a systematic way, such as the bandwidth and shape of tuning curves. Further maps also appear to be represented in a columnar fashion, such as the coding of intensity and sound localization, but the details of their organization are still unclear. (*See* **Audition, Neural Basis of**)

Inferotemporal (IT) cortex is a visual area essential for object perception and recognition. Using intrinsic signal imaging and extracellular recording in macaque monkeys, Tsunoda *et al.* (2001) showed that the neural activity evoked in IT by complex objects is laid out spatially as distributed patches. This result suggests that an object is represented by a combination of cortical columns, each of which represents a visual feature. However, not all the columns related to a particular feature were necessarily activated by the original objects. Thus, objects would be represented by using a variety of combinations of active and inactive columns for individual features, rather than simply by the addition of feature columns. It is unclear, though, whether an object is represented by a combination of modules, each specific to a visual feature or a part of the object (feature-based or part-based representation), or whether modules are specific to the object (object-based representation). (*See* **Temporal Cortex; Object Perception, Neural Basis of; Vision: Object Recognition**)

Columnar maps also exist outside the cortical areas, such as in the brainstem (maps of interaural delay, of interaural intensity difference, and of auditory space) and the superior colliculus (map of motor space, or gaze direction). (*See* **Motor Areas of the Cerebral Cortex**)

INTRACOLUMNAR AND INTERCOLUMNAR CIRCUITRY

Cortical columns are also distinguished from each other by their patterns of circuitry. The majority of intracortical circuits are local, connecting neurons within the same columns, with only a minority of connections being between columns. Again, this organization has been most extensively studied in the visual system.

Callaway (1998) proposed a generic model of vertical connectivity linking layers within a column in primary visual cortex of cats and monkeys (Figure 4). The model is based on three simplifying assumptions: only excitatory synapses are considered; each cortical layer provides its primary output to only one layer; and only two types of connections are considered (feedforward and feedback). A direct path from inputs (coming from the LGN) to outputs (going mainly to other areas in the cortex) passes through cortical layers 4C and 2–4B, with layers 6 and 5 providing feedback (modulatory) connections, respectively. Since these dense connections are mostly confined within a column, this provides a great deal of purely intracolumnar – and therefore local – information processing. Long-range connections (generally up to a few millimeters long) between columns mostly project within layer 2/3. They are generally sparse and patchy, and tend to connect spatially separated columns with the same feature preference, such as the same orientation or ocular dominance preference (although some recent experiments do not fully agree with this cluster-like connection pattern). It is easy to imagine how such specific patterns could arise as a result of Hebbian learning, since columns with similar feature preferences would be expected to have highly correlated activity. Likewise, it has been suggested that color blobs are preferentially linked to color blobs of the same ocular dominance, and interblobs to interblobs.

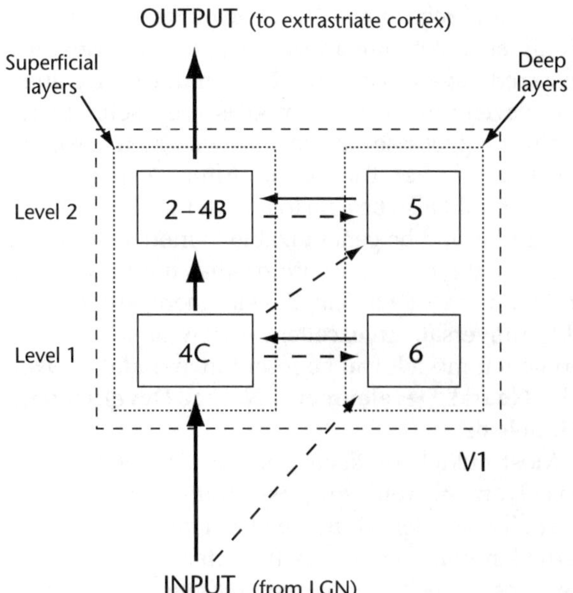

Figure 4. Two-level model of local cortical circuitry in V1. A direct path from input (from the lateral geniculate nucleus, LGN) to output (to extrastriate cortex) is provided by the two feedforward superficial layers 2–4B and 4C; feedback deep layers 5 and 6 modulate the activity of each level. Adapted from Callaway (1998).

COLUMNAR DEVELOPMENT AND COMPUTATIONAL MODELS

The high degree of order displayed by columnar structures and the large amount of data acquired, especially regarding the development of ocular dominance columns in primary visual cortex, has inspired several computational models of columnar development. Such models are useful to explain the processes at work as well as to produce predictions that can guide future experiments. They should also be able to account for interspecies variations and be generalizable to models for other areas of the cortex, assuming that the underlying mechanisms of cortical development are reasonably universal. An excellent review of such developmental models can be found in Swindale (1996). (*See* **Neural Development**; **Neural Development, Models of**)

Most models of visual cortical development are based on the following assumptions (which are partially supported by experimental data): patterned retinal activity in the afferents to cortical neurons; Hebb synapses; radially symmetric, short-range excitatory and long-range inhibitory lateral cortical connections; and normalization of synapse strength. Thus, most of these models largely disregard genetic factors and assume that the columns in the primary visual cortex appear during development from an apparently uniform cortical sheet by a process of activity-dependent self-organization that modifies synaptic strengths in response to patterns of visual stimulation. These patterns can be produced both externally by the world, and generated internally by spontaneous activity in the retina (Meister *et al.*, 1991). The rule by which synaptic strengths appear to change is roughly the one proposed by Hebb (1949): 'neurons that fire together wire together'. The models often represent the cortex as a two-dimensional array of neural units (each representing a collection of real neurons) and thus directly embody the definition of column. The visual stimulus is represented either in an abstract, low-dimensional way, as a vector of independent components representing ocular dominance, orientation preference, retinotopic position or direction preference; or in a concrete, high-dimensional way, as a vector containing the connection strengths between a cortical cell and a set of receptor cells in the retina. (*See* **Hebb Synapses: Modeling of Neuronal Selectivity**; **Hebb, Donald Olding**)

A common characteristic of these models is that they try to maximize coverage as well as continuity, as originally suggested by Hubel and Wiesel. Coverage refers to the fact that all combinations of eye and orientation preference occur at least once within any region (of a certain, small size) in stimulus space – otherwise, the animal might be blind to the unrepresented stimulus (although it has been suggested that higher cortical areas could interpolate between incomplete representations in lower cortical areas). Continuity refers to the fact that the preferences of neighboring neurons in cortex tend to be similar. Representing a high-dimensional stimulus space in a two-dimensional cortex results in coverage and continuity competing at the expense of each other, with the striped organization observed being perhaps a locally optimal solution to their trade-off. (*See* **Vision: Occlusion, Illusory Contours and 'Filling-in'**)

Two particularly important types of models are correlational (e.g. Miller *et al.*, 1989) and competitive (e.g. Goodhill, 1993). In correlational models the input–output function of neurons is linear, and receptive field development is driven by the eigenvectors of an operator dependent on the correlation of the input patterns, the intracortical connections, and the LGN arborization. In competitive models the input–output function of neurons is highly nonlinear, and such models implement something more akin to cluster analysis. Generally speaking, these models account for much of the observed phenomenology of cortical maps, including the striped structure of ocular dominance and orientation columns with the appropriate periodicity and interrelations, and the location of pinwheels and fractures. However, no model so far can account for all observed features for both ocular dominance and orientation maps at the same time, or for some of the more elusive data. (*See* **Pattern Recognition, Statistical**; **Receptive Fields**)

WHY A COLUMNAR ORGANIZATION?

The presence of a columnar organization in various regions of the cortex of many mammalian species has suggested that columns form the basic information processing elements of the cortex, with each column being responsible for analyzing a small range of stimuli, and the same modular unit being repeated multiple times to span the entire range of stimuli (e.g. Szentágothai, 1978). As such, columns have been considered to be a fundamental functional feature important for perception, cognition, memory, and even consciousness (Szentágothai, 1978; Eccles, 1981). However, there is no general agreement about the reason for the existence of such groupings. Such columnar structure has not been found in some mammalian species closely related to other species that do have columns (Purves *et al.*, 1992). Thus, it has been argued that the columnar organization of the cortex may not imply a functionally modular organization (Swindale, 1990; Purves *et al.*, 1992). In particular,

Purves *et al.* suggested that the production of iterated patterns of circuitry might be an incidental consequence of the activity-dependent elaboration of synaptic connections and be of little significance to cortical function. In other words, a given cortical system might work just as well if columns did not form. Purves *et al.* suggest several factors that could drive such origin. (*See* **Modularity in Neural Systems and Localization of Function**; **Synaptic Plasticity, Mechanisms of**)

CONCLUSION

Many areas of the cortex, particularly in the visual and somatosensory system, can be divided into repeating modules characterized by discrete patterns in both function and anatomy. The best-studied examples are 'barrels' and touch-modality columns in primary somatosensory cortex, and orientation and ocular dominance columns in primary visual cortex. There are many vertical connections linking neurons within a column, and a few horizontal connections linking different columns. Columnar development may be driven by activity-dependent self-organization, and can often be modeled using Hebbian learning rules – although the relative importance of genetic factors and patterned activity is not clear. As yet no compelling justification has emerged for why columnar structure exists. (*See* **Vision, Early**; **Pattern Vision, Neural Basis of**; **Cortical Map Formation**)

References

Callaway EM (1998) Local circuits in primary visual cortex of the macaque monkey. *Annual Review of Neuroscience* **21**: 47–74.

Crowley JC and Katz LC (1999) Development of ocular dominance columns in the absence of retinal input. *Nature Neuroscience* **2**: 1125–1130.

Eccles JC (1981) The modular operation of the cerebral neocortex considered as the material basis of mental events. *Neuroscience* **6**: 1839–1856.

Favorov OV and Diamond ME (1990) Demonstration of discrete place-defined columns – segregates – in the cat SI. *Journal of Comparative Neurology* **298**: 97–112.

Goodhill GJ (1993) Topography and ocular dominance: a model exploring positive correlations. *Biological Cybernetics* **69**(2): 109–118.

Hebb DO (1949) *The Organization of Behaviour*. New York, NY: John Wiley.

Horton JC and Hocking DR (1996) Intrinsic variability of ocular dominance column periodicity in normal macaque monkeys. *Journal of Neuroscience* **16**: 7228–7339.

Hubel DH and Wiesel TN (1977) Functional architecture of the macaque monkey visual cortex. *Proceedings of the Royal Society of London, Series B* **198**: 1–59.

Meister M, Wong ROL, Baylor DA and Shatz CJ (1991) Synchronous bursts of action potentials in ganglion cells of the developing mammalian retina. *Science* **252**: 939–943.

Miller KD, Keller JB and Stryker MP (1989) Ocular dominance column development: analysis and simulation. *Science* **245**: 605–615.

Mountcastle V (1957) Modality and topographic properties of single neurons of cat's somatic sensory cortex. *Journal of Neurophysiology* **20**: 408–434.

Purves D, Riddle DR and LaMantia AS (1992) Iterated patterns of brain circuitry (or how the brain gets its spots). *Trends in Neurosciences* **15**(10): 362–368.

Schreiner CE (1995) Order and disorder in auditory cortical maps. *Current Opinion in Neurobiology* **5**: 489–496.

Sharma J, Angelucci A and Sur M (2000) Induction of visual orientation modules in auditory cortex. *Nature* **404**: 841–847.

Swindale NV (1990) Is the cerebral cortex modular? *Trends in Neurosciences* **13**(12): 487–492.

Swindale NV (1996) The development of topography in the visual cortex: a review of models. *Network: Computation in Neural Systems* **7**(2): 161–247.

Szentágothai J (1978) The neuron network of the cerebral cortex: a functional interpretation. *Proceedings of the Royal Society of London, Series B* **201**: 219–248.

Tsunoda K, Yamane Y, Nishizaki M and Tanifuji M (2001) Complex objects are represented in macaque inferotemporal cortex by the combination of feature columns. *Nature Neuroscience* **4**: 832–838.

Woolsey TA and van der Loos H (1970) The structural organization of layer IV in the somatosensory region (SI) of mouse cerebral cortex. *Brain Research* **17**(2): 205–242.

Further Reading

Erwin E, Obermayer K and Schulten K (1995) Models of orientation and ocular dominance columns in the visual cortex: a critical comparison. *Neural Computation* **7**: 425–468.

Hubel DH (1995) *Eye, Brain, and Vision*. New York, NY: WH Freeman.

Jones EG and Diamond IT (eds) (1995) *The Barrel Cortex of Rodents*, vol. 11 of *Cerebral Cortex*. London, UK: Plenum Press.

Nicholls JG, Martin AR and Wallace BG (2000) *From Neuron to Brain: A Cellular and Molecular Approach to the Function of the Nervous System*, 4th edn. Sunderland, MA: Sinauer.

Peters A and Rockland KS (eds) (1994) *Primary Visual Cortex in Primates*, vol. 10 of *Cerebral Cortex*. London, UK: Plenum Press.

Cortical Map Formation

Intermediate article

Jon H Kaas, Vanderbilt University, Nashville, Tennessee, USA

CONTENTS

Cortical maps are orderly representations of sensory receptors or body movements that form during the development of the brain. Proper development depends on genetic information, molecular signals based on relative position in the brain, and information in spontaneous and evoked neural activity patterns.

INTRODUCTION

The functional machinery of the brain includes cortical areas and subcortical nuclei that are interconnected to form systems. Many of these nuclei and cortical areas systematically represent a sensory surface or the motor control of muscles. In the visual system, the inputs from the retina terminate in topographic patterns in thalamus and midbrain nuclei to form maps of the retina or visual space. The map in the laminated lateral geniculate nucleus in the thalamus projects to primary visual cortex, V1, to form a map of the hemiretinas of the two eyes that see the contralateral half of the visual world. Area V1 projects in turn to several additional visual areas to activate further maps of the contralateral visual hemifield. Monkeys and humans have over 25 visual areas, most of which can be described as containing maps of the contralateral visual hemifield (Kaas, 1989). The interconnected maps form a processing hierarchy for visual information in which the early maps, especially the one in V1, most precisely map the visual hemifield, while higher-order representations progressively become less retinotopic. However, these maps may represent other aspects of visual information in systematic ways, as they start to reflect more of the visual outcomes of cortical computations. The collections of interconnected cortical and subcortical maps constitute the visual system, and the auditory and somatosensory systems are constructed similarly.

The motor system is defined somewhat differently. Motor cortex includes a primary area and other areas where neural activity at any specific location in the map elicits a specific movement or set of movements. The movements can be elicited by natural neural activity or by neural activity evoked by focal electrical stimulation within these areas. By electrically stimulating a large number of sites across a motor area, the representation of movements can be experimentally revealed. For example, primary motor cortex, a mediolateral strip of cortex in the caudal portion of the frontal lobe, represents foot movements medially near the brain midline, trunk movements more laterally, forearm and hand movements next, and face movements most laterally (Penfield and Boldry, 1937). As motor areas also receive rather indirect sensory inputs, they can be said to represent somatosensory or other sensory inputs, but these sensory maps are seldom discussed. Likewise, movements can be evoked from somatosensory areas of cortex, at higher levels of stimulating current, and these somatosensory areas can also be said to contain motor maps. Monkeys and humans have as many as ten areas in each cerebral hemisphere that map movements of the muscles of the contralateral half of the body, as well as two main areas for moving the eyes and directing gaze into the contralateral visual hemifield. Because the internal representational organizations of the sensory and motor areas can be revealed by successively recording from neurons or stimulating neurons in many locations in each representation with microelectrodes, the recording or stimulating procedures are sometimes referred to as 'mapping' the brain.

In addition to recording or stimulating with microelectrodes, there are other ways of recording or imaging brain activity in response to sensory stimuli or repeated movements that also allow the organizations of brain maps to be revealed. The use of 'noninvasive' imaging procedures has revealed the locations and internal organizations of visual, auditory, somatosensory, and motor maps in humans. The existence of orderly maps in the cortex has been implied for some time by the nature of perceptual and movement deficits that follow focal lesions of regions of cortex in human patients. Motor maps were directly revealed in the 1870s when investigators exposed the surface of the brains of dogs and monkeys and described regions where body and eye movements could be evoked by electrical stimulation. Comparable evidence for sensory maps came later in the 1930s and 1940s when the invention of the oscilloscope and amplifiers made it possible to record the weak electrical activities of cortical neurons as they responded to sensory stimuli. These recording and stimulation methods have long been applied to humans when their brains were exposed during surgery. With the existence of such maps now well established, the interesting and challenging question of how they emerge in the developing brain can be addressed.

THE DEVELOPMENT OF CORTICAL MAPS

The primary sensory and motor maps occupy the same relative positions in the neocortex of most mammalian species. Thus, the visual cortex is at the back, auditory cortex is lateral, somatosensory cortex is toward the front, and motor cortex is just ahead of somatosensory cortex. There are several theories as to how this happens.

One theory is that the subdivision of the cortex into areas begins early in brain development, well before the neocortex has even been generated by the migration of cells from a generative zone along the cerebral ventricles. According to this theory, cellular interactions occur in the pool of dividing cells along the ventricle so that the cells become committed to certain fates and collectively form an early plan of the overall organization of neocortex. When the cells migrate in a point-to-point fashion to cortex, they carry with them the basic organization of where emerging cortical maps are relative to each other and even the internal organizations of maps. Thus, the basic organization of cortex is determined by interactions between progenitor cells before they even generate the cells that form cortex. This theory has been called the *protomap hypothesis* (Rakic, 1988), and it suggests that there is something inherent within each emerging cortical region that directs the region to become a specific cortical area.

A second view is that neocortex is generated as a *uniform sheet* and that patterns of inputs from the thalamus and later the emerging cortical connections subdivide the cortex into areas. Thus, each region of cortex starts out with no particular identity, but it acquires an identity when axons from another part of the brain arrive and tell it how to differentiate. The problem with the view that regions of cortex have no particular identity is that this does not explain by itself why visual cortex is always located toward the back and motor cortex toward the front of the neocortex. One possibility would be to utilize well-known differences in the front to back and lateral to medial neurogenic gradients in the emerging neocortex and thalamus to form patterns or connections. Axons could grow out in maturational order to arrive at cortical targets in maturation order, providing an overall organization in cortex. If axons maintained neighborhood relations with each other as they grew to cortex, as they largely seem to do, considerable order in cortex could result. Thalamic axons, according to this possibility, only need to be attracted to cortex, and the sorting mechanism would be a consequence of positional differences in developmental order, which in turn would be dependent on position effects on gene expression and the consequent molecules. Refinement of the initial order and the addition of local circuits in cortex could then follow.

Alternatively, and most likely, some sort of *positional signaling* is used. An early hypothesis of how growing axons reach and recognize an appropriate target was the chemoaffinity theory proposed by Roger Sperry in the 1960s (Sperry, 1963). According to this proposal, both the guidance of axons to the target region and the recognition of the appropriate target neurons are achieved by the operation of highly specific chemical affinities between substances in the growing axons and in the neurons of the target structure. Such a proposal nicely accounts for the formation of highly specific patterns of connections, but it fails to explain all observations. Most notably, lesions of neocortex very early in development (before the thalamic axons arrive) do not completely abolish some divisions of cortex and leave others intact; instead, smaller than normal areas form on the smaller sheet of neocortex. This type of observation is more compatible with a modified form of Sperry's chemoaffinity theory.

One proposal is that molecular gradients across developing neocortex indicate a general front to back and medial to lateral direction for patterns of thalamic connections, but not specific locations for connections (Fukuchi-Shimogori and Grove, 2001). Thus, axons from the visual thalamus would seek a high (or low) region of expression in the gradient and always would grow to the back of the neocortex (Huffman *et al.*, 1999). If the back part of neocortex had been removed, these axons would still grow to the back of the portion of neocortex that remained. Local competition between axons guided to positions along a molecular gradient would then lead to winners and losers in the reduced cortical sheet, but not all of the visual axons would be losers. In contrast, if appropriate target neurons were completely prespecified either in a map in the generative cells along the ventricle or in a similar protomap formed in the early cortical sheet, then removal of the target cells would completely abolish a cortical area. In principle, chemical gradients in cortex could provide the signals, given adjustable thresholding mechanisms and competition for synaptic space, for both cortical areas emerging in the correct places with the correct thalamic inputs, and at least approximately correct topographic patterns within cortical areas. Thus, molecular patterns intrinsic to cortex guide the development of cortex while allowing several outcomes. A pluripotential cortical sheet is differentiated by molecular gradients.

All cortical areas may not be specified in the same way. Possibly a few cortical areas, especially the primary areas, are specified very early in development and can only become those areas, although this does not seem to be the case for the visual cortex of opossums or somatosensory cortex of rats, where the outcomes have been experimentally altered. However, an argument has been made for primary visual cortex of monkeys and humans, as this cortex has more neurons across the thickness of cortex than other cortical areas, and the posterior ventricular zone that generates neurons for primary visual cortex generates more neurons. These observations suggest that the protomap of primary visual cortex already exists in the ventricular generative zone. Yet a major role in the specification of visual cortex by thalamic inputs is clearly indicated. In humans in whom the eyes fail to develop and in monkeys with eye removal early in fetal life, many of the neurons in the developing lateral geniculate nucleus of the thalamus degenerate without visual input, and the number of thalamic neurons projecting to visual cortex is greatly reduced. As a result the primary visual cortex is only a fraction of its usual size (Rakic, 1988). Some of the cortex that would normally become primary visual cortex develops features of nonprimary visual cortex. Thus, it appears that the region of primary visual cortex becomes primary visual cortex only if it receives an adequate input of visual axons from the thalamus.

The internal organization of cortical maps probably depends on a balance of intrinsic factors leading to regional differences in gene expression and gene products, and extrinsic factors including regional and local differences in neural activity patterns evoked by sensory stimuli, self-movement, or by correlated spontaneous activity. A combination of such molecular and activity-dependent mechanisms would best account for the basic features of cortical map development listed below.

- At least the crude topographic features of cortical maps appear to develop independently of any information from sensory receptors and afferents. In other words, the basic features of cortical maps develop without instruction from the receptor sheet. Activity patterns relayed from the periphery are not necessary. The most compelling evidence for this comes from studies of thalamocortical connections in genetically eyeless (anophthalmic) mice where the connections between the lateral geniculate nucleus of the visual thalamus and primary visual cortex are at least approximately normal in topography. Similar results have been reported in other mammals reared after the removal of retinal afferents early in development. Also, the basic pattern of connections between the ventroposterior nucleus of the somatosensory thalamus and primary somatosensory cortex of rats appears to form before the thalamus is activated by somatosensory inputs. The topographic order of the thalamocortical pattern is said to develop independently of a 'template' of the periphery.
- Detailed cortical maps of the receptor sheets often develop prenatally, and thus they develop without postnatal experience. Newborn mammals of many species have orderly normal maps of the body surface in primary somatosensory cortex at birth, and in newborn lambs and monkeys the retinotopic organization of primary visual cortex appears to be normal at or soon after birth.
- The internal organizations of sensory maps closely reflect the features of the sensory sheet. The maps reflect the densities of receptors across the skin or retina so that the maps are proportional to receptor densities rather than skin or retinal territories, although a behaviorally important receptor surface may achieve extra space in cortical maps. Even errors in the arrangement of receptors or the projections of afferents are reflected in the order of cortical maps. For example, the number and arrangement of the vibrissae on the face of rats and mice is highly consistent across individuals, and the maps in primary somatosensory cortex (S1) separately and precisely represent each whisker. When rare individuals with an extra whisker or two are examined, an extra anatomical module in S1 is found for each of the extra whiskers. Likewise, S1 of star-nosed moles with an array of 11 ray-like sensory appendages protruding from each side of the nose has 11 modules or bands in the face portion of S1, one for each ray. However, rare individuals who develop with 10 or 12 rays have 10 or 12 bands in S1 (Catania and

Kaas, 1997). Similarly, Siamese cats have a mutant gene that alters skin and eye pigment and also changes the projection pattern from the retina somehow so that the segment of the retina that projects to the contralateral lateral geniculate nucleus in the thalamus is extended by some 20° of visual angle. This extra input is often incorporated into the order of cortex maps so that primary visual cortex (V1) represents an extra 20° of visual space in a single retinotopic pattern (Kaas and Guillery, 1973).

- There is evidence from the study of rats that the segregation of afferents from the two eyes, or from different whiskers on the face, in sensory cortex depends on having intact inputs from the eyes or face in early development. In rats reared after section of sensory nerves at birth, the cortical arbors of thalamocortical neurons are unusually large and overlapping, with less focused distributions of synapses. In visual cortex, inputs relayed from the two eyes normally divide the space in visual cortex into alternating ocular dominance bands. If one eye is removed early in development, thalamocortical axons related to the remaining eye occupy nearly all of cortex. Similarly, if the activity of retinal neurons in one eye is simply reduced by rearing a cat or monkey with one eye sutured shut, the thalamocortical axons related to the normal eye acquire most of the cortical space (Hubel *et al.*, 1977).
- The development of cortical maps is altered if they are abnormally innervated. Several investigators have induced the sensory projections from one sensory system to innervate a thalamic nucleus for another sensory system early in development, thereby causing the sensory map in cortex to receive the wrong sensory input. When visual inputs from the retina were induced to innervate the medial geniculate nucleus of the auditory thalamus, the medial geniculate neurons thereby relayed visual rather than auditory information to primary auditory cortex (A1). This cortex then developed a map of the retina that had some, but not all, of the features of visual cortex. The researchers concluded that the nature of the sensory input, visual or auditory, influenced, but did not totally determine, cortical map development (Roe *et al.*, 1990).

MOLECULAR MECHANISMS OF CORTICAL MAP FORMATION

Regional differences in the expression of molecules in the developing cortex appear to have major consequences for the course of cortical development. Many molecular correlates of developmental stages and features have been described, including the expression of nerve growth factors and receptors for those factors, growth-associated proteins, structural proteins associated with synapses, membrane-associated glycoproteins, and axon guidance molecules. The underlying question concerns the factors that relate to differences in gene expression allowing various signaling and neuron construction molecules to be expressed at the proper time and place in the developing neocortex.

According to the early views of chemospecificity postulated by Roger Sperry, molecular signals – either regionally expressed or expressed as gradients across developing cortex – would at least set up the basic organization of cortex by guiding incoming connections to appropriate locations and maintaining synaptic contacts at those locations. Other local features of cortical organization might then emerge as a result of gene expression related to the neural activity patterns induced by the initial connections. The key assumption here is that the early regional differences in the expression of molecular signals depend on factors intrinsic to developing cortex.

Important evidence on intrinsic differences in the expression of molecular signals comes from studies of mutant mice in which thalamocortical afferents failed to develop. Despite a total lack of thalamic inputs, the emerging neocortex demonstrated both graded and sharp patterns of gene expression, including genes thought to be important in axon adhesion and in neuronal differentiation. Thus, developing neocortex has intrinsic signals that seem to be capable of directing regional differentiation. Further evidence comes from studies of abnormal thalamic connections in mutant mice lacking certain regulating genes. In these mice, the early expression of specific molecular markers are shifted toward the back in the developing cortex. As a result, thalamic connections are also shifted toward the back. For example, the somatosensory inputs that normally go to the middle of neocortex terminate instead in the back where visual inputs normally terminate. The visual inputs in turn are displaced to the very margin of the posterior neocortex. Thus, these experiments provide compelling evidence for the existence of intrinsic molecular positional cues for incoming thalamic axons.

In summary, it appears that differences in gene expression in cortical neurons based on their position in the cortex or their previous history as precursor cells in the ventricular generative zone provide the molecular guidance necessary for at least the gross pattern of thalamocortical connections to develop. This pattern includes a basic set of cortical sensory and motor areas, and at least the crude topographic order of the sensory and motor maps within those areas. However, refinements of crude patterns within maps appear to depend on information carried from the receptor sheet to the thalamus and then to cortex.

ROLE OF EXPERIENCE AND NEURAL ACTIVITY

The prevailing view is that cortical maps emerge in development as a result of a combination of mechanisms that require neural activity and those that do not. The activity-independent mechanisms are thought to be important in forming cortical areas and the crude topography of cortical areas, while activity-dependent mechanisms are thought to be essential in achieving precision in the connection patterns that make up cortical maps.

The early evidence of a role for activity in the formation of cortical maps came from the landmark experiments of David Hubel and Torsten Wiesel. In the early 1960s these investigators experimentally sutured together the lids of one eye in kittens whose eyes had not yet opened, thus depriving that eye of pattern vision. When the visual system of these cats was examined after they matured, few neurons in primary visual cortex were responsive to the previously closed eye and the anatomical relay of connections from the deprived eye to visual cortex were sparse. The interpretation of these results was that axons from neurons in the lateral geniculate nucleus of the thalamus that were activated by the deprived eye were less active than those related to the normal eye, and the two types of axons were in competition with each other for synaptic space on cortical neurons. The more active axons for the normal eye grew and took over most of the cortical neurons, while the axons for the deprived eye lost contacts with cortical neurons and failed to grow with the expanding brain and innervated little of visual cortex.

Subsequently there have been many deprivation studies that demonstrate that it is possible to alter the development of the visual system. Such results provide further evidence for activity-dependent mechanisms. Nevertheless, the roles of activity and especially experience continue to be questioned. For example, the separate alternating territories in primary visual cortex with inputs from one or the other eye, the ocular dominance columns, develop before birth in many mammals. Thus, they do not depend on pattern vision. However, the development of these columns has been attributed to spontaneous or nonvisual activity in the neurons of the two eyes before they are exposed to light. Neurons close together in any structure, including the eye, tend to be interconnected and spontaneously active at the same time. Thus the patterns of activity relayed to the thalamus from each developing eye would differ, and this difference would be relayed to cortex as a basis for segregating axons related to each eye in ocular dominance columns. Evidence suggests that such segregations of thalamocortical terminations develop even without eyes. However, this observation, if valid, does not eliminate the possibility of spontaneous activity being essential, as differences in spontaneous activity patterns could occur in the layers of the lateral geniculate nucleus normally related to each eye.

CONCLUSION

Over years of active investigation, a general theory of cortical map development has emerged. In brief, a few major factors appear to be highly important for the formation of map topology, modular organization within maps, and neuron response properties.

First, molecular mechanisms that are independent of external influences seem to be responsible for guiding axons to their approximate locations in cortex. Which molecules govern this guidance and establishment of a crude topographic order for connections remains largely uncertain, but a number of promising candidate substances are under investigation. Another uncertainty is how these molecules come to be expressed in regionally specific patterns. When and how are position cues recognized? Spatiotemporal gradients in neuron generation and maturation, and substrate cues for gene expression, are likely to be important. Several guidance and chemoaffinity factors have been demonstrated.

Neural activity patterns provide another important source of information, and this information is used to select some synaptic contacts over others. Neurons are induced by activity patterns to locally grow and form more connections, or to retract and lose connections. Much research supports the premise that, in both the mature and developing nervous system, synapses on a neuron that are active while the neuron itself is active are strengthened, while those that are inactive during such discharges are weakened. The physiological consequences of such changes in synaptic effectiveness, which are rapidly induced, have been called *long-term potentiation* (LTP) and *long-term depression* (LTD). The changes are often referred to as 'Hebbian' plasticity after the early proposal of cellular mechanisms for learning by Donald Hebb.

The key variable is whether overlapping inputs on cortical neurons are active at the same time. Activities that are correlated in time across receptors in the receptor sheet are based on proximity and receptor transducing factors. The same stimuli are likely to activate nearby receptors, but classes of receptors in the same location are differentially responsive to the same stimulus. In the central nervous system, proximity is additionally important because of local interconnections, and local computations add to response diversity by producing neurons that have new selectivities for features of sensory stimuli. Selections of synapses based on correlated and discorrelated activities fine-tune the internal representational order of sensory maps so they more closely reflect the proximities and disjunctions on the receptor sheets. Such selection also distributes and segregates inputs differing in correlated spontaneous activity based on proximity levels. Neurons projecting from one structure to the next would be more densely interconnected if adjacent than widely separate, and the local interconnections would induce correlations in activity. Thus, activity patterns could promote the formation of topographically matched connections between cortical areas. Neurons driven by different receptor classes would be separately activated and their overlapping projections would segregate. Similarly, neurons that became selective for different stimulus attributes, as a result of the computations of central neurons, would develop segregated projections. Thus, activity patterns would induce sharply defined modules such as ocular dominance columns, stimulus orientation modules, and light-on or light-off layers in cortical maps.

Early-maturing maps, largely the primary cortical areas, would be most influenced by prenatal activity patterns, especially correlated spontaneous activity but also sensory responses evoked by self-generated movements, while later-maturing maps would have the opportunity to use information from a more complex external environment and the processing outcomes of early maturing cortical areas and thalamic nuclei.

Finally, the susceptibility of neurons to both local substrate cues in the molecular environment and to the effects of neural activity patterns changes over the course of maturation, leading to the concept of critical or susceptible periods for developmental change.

References

Catania KC and Kaas JH (1997) The mole nose instructs the brain. *Somatosensory and Motor Research* **14**: 56–58.

Fukuchi-Shimogori T and Grove EA (2001) Neocortex patterning by the secreted signaling molecule FGF8. *Science* **294**: 1071–1074.

Hubel DH, Wiesel TN and LeVay S (1977) Plasticity of ocular dominance columns in monkey striate cortex. *Philosophical Transactions of the Royal Society of London Series B* **278**: 377–409.

Huffman KJ, Molnai Z, VanDellen A *et al.* (1999) Formation of cortical fields on a reduced cortical sheet. *Journal of Neuroscience* **19**: 9939–9952.

Kaas JH (1989) Why does the brain have so many visual areas? *Journal of Cognitive Neuroscience* **1**: 121–135.

Kaas JH and Guillery RW (1973) The transfer of abnormal visual field representations from the dorsal lateral geniculate nucleus to the visual cortex in Siamese cats. *Brain Research* **59**: 61–95.

Penfield W and Boldry E (1937) Somatic motor and sensory representation in the cerebral cortex of man as studied by electrical stimulation. *Brain* **60**: 389–443.

Rakic P (1988) Specification of cerebral cortical areas. *Science* **241**: 170–176.

Roe AW, Pallas SL, Hahm JO and Sur M (1990) A map of visual space induced in primary auditory cortex. *Science* **250**: 818–820.

Sperry R (1963) Chemoaffinity in the orderly growth of nerve fiber patterns and connections. *Proceedings of the National Academy of Science of the USA* **50**: 703–710.

Further Reading

Kaas JH (1988) Development of cortical sensory maps. In: Rakic P and Singer W (eds) *Neurobiology of Neocortex*, pp. 115–136. New York, NY: John Wiley.

Kaas JH (1997) Topographic maps are fundamental to sensory processing. *Brain Research Bulletin* **44**: 107–112.

Kaas JH (2000) Organizing principles of sensory representations. In: Bock G and Cardew G (eds) *Evolutionary Developmental Biology of the Cerebral Cortex*, Novartis Foundation Symposium 228, pp. 188–205. New York, NY: John Wiley.

Katz LC and Shatz CJ (1996) Synaptic activity and the construction of cortical circuits. *Science* **274**: 1133–1138.

Kennedy H and Dehay C (1993) Cortical specification of mice and men. *Cerebral Cortex* **3**: 171–186.

Krubitzer L and Huffman KJ (2000) Arealization of the neocortex in mammals: genetic and epigenetic contributions to the phenotype. *Brain, Behavior and Evolution* **55**: 322–335.

Levitt P (2000) Molecular determinants of regionalization of the forebrain and cerebral cortex. In: Gazzaniga MS (ed.) *The New Cognitive Neuroscience*, pp. 23–32. Cambridge, MA: MIT Press.

O'Leary DDM (1989) Do cortical areas emerge from a protocortex? *Trends in Neurosciences* **12**: 401–406.

O'Leary DDM, Schlaggar BL and Tuttle R (1994) Specification of neocortical areas and thalamocortical connections. *Annual Review of Neuroscience* **17**: 419–439.

Shatz CJ (1990) Impulse activity and the patterning of connections during CNS development. *Neuron* **5**: 745–756.

Wiesel TN (1982) Postnatal development of the visual cortex and the influence of environment. *Nature* **299**: 583–591.

Wong ROL (1999) Retinal waves and visual system development. *Annual Review of Neuroscience* **22**: 29–47.

Counterfactual Thinking

Intermediate article

Neal J Roese, University of Illinois, Champaign, Illinois, USA
James M Olson, University of Western Ontario, Ontario, Canada

CONTENTS

Mental constructions of alternatives to facts or events. These thoughts of 'what might have been' are linked to a variety of emotional and judgmental consequences.

INTRODUCTION

The term 'counterfactual' means contrary to established facts or actual events. Counterfactual thinking typically involves imaginative speculation about alternatives to past outcomes: that is, about what might have been. Counterfactuals often (though not always) take the form of conditional propositions, containing the dual components of antecedent and consequent. In everyday cognition, counterfactual thinking usually targets personal goals and desires, such that individuals focus on actions that might have brought about particular desired ends (e.g. 'If I had studied harder, I would have earned a higher grade'). Counterfactuals can also be deployed in everyday speech as arguments ('If not for Gorbachev, the Soviet Union and the Cold War would have persisted into the twenty-first century') or invitations to further speculation and elaboration, e.g., 'What if President Kennedy hadn't been assassinated?' Counterfactuals have intrigued philosophers throughout the twentieth century because of their implications for logic and epistemology, but more recently counterfactual thinking has inspired psychological research because such thought processes influence a wide range of emotional, judgmental, and behavioral outcomes.

The form and content of counterfactuals is limitless, and although they may conjure the bizarre and the fantastic, everyday counterfactual thinking is mundane. Indeed, an essential feature seems to be that counterfactuals preserve the integrity of the world as we know it, altering but one or two specific features, then unfurling immediate consequences against a backdrop that is essentially the same as actuality. Thus, one might wonder how the Second World War might have unfolded had Hitler attacked and defeated the British at Dunkirk rather than allowing them to escape, but background features, such as the previous history of Europe, the power of the respective nations' armaments, and for that matter the laws of physics, remain unchanged. Given this rule of restricted alteration, a key theoretical focus has been to specify which finite features of reality are perceived to be more changeable, or mutable, as opposed to the infinite background features that remain constant within one's mind. The sections below on determinants of counterfactual thought are descriptions of these patterns.

Counterfactual thinking is a rule-bound creative act, and as such has been construed as a principal ingredient of consciousness and language. Hofstadter (1985), for example, argued that a comprehensive attempt to create artificial intelligence must include some facility for production of counterfactuals that operates in a manner similar to that of human cognition. A further elaboration of this theme is that counterfactuals are constrained by reality because they are functional; that is, they often provide useful prescriptions for how a goal might have been achieved in the past, and hence how it might yet be achieved in the future (Roese, 1994).

TYPOLOGY

Counterfactuals have been classified in two main ways: direction and structure. Direction refers to whether the counterfactual specifies a state that is better than actuality (an upward counterfactual) or worse than actuality (a downward counterfactual). Counterfactuals are also described in terms of structure of their phrasing. The counterfactual antecedent may be an addition of some feature not in fact present (an additive counterfactual), or it may remove a feature that was present (a subtractive counterfactual). These two typologies have proved effective in delineating a variety of theoretical relations, described below.

PSYCHOLOGICAL CONSEQUENCES

Causation

Counterfactuals are intimately related to causal inferences. Causation may be defined as a relation between two variables (objects, states, etc.) in which one produces or generates changes in the other. A counterfactual conditional nearly always implies causation. Counterfactual conditionals denote an antecedent-consequent pair that diverges from a related, factual antecedent-consequent pair, thereby satisfying the logic of J. S. Mill's method of difference for inferring causation. For example, the observation that a match held motionless remains bereft of flame might be followed by the counterfactual supposition that 'if the match had struck a hard surface, it would have ignited.' The mental alteration of but one feature of actuality (striking as opposed to not striking the match), when accompanied by the imagined consequential variation in ignition, provides the basis for inferring that the antecedent of match strike causally influences ignition. The logic of the method of difference is the same as the covariation criterion for causation that forms the theoretical platform for many theories of causal attribution, in that counterfactuals present one datum, albeit imagined, that may be added to a set of divergent background observations. Although absence of covariation may be used to rule out causation, presence of covariation is not in itself sufficient to infer causation. Therefore, the same problems of induction that bedevil formal analyses of causation apply similarly to counterfactual reasoning (Spellman and Mandel, 1999). Nearly all psychological consequences of counterfactual thinking appear to be rooted either in this causal inference mechanism or in a contrast effect mechanism.

Contrast Effects

In comparative judgment, the juxtaposition of one object with a second can render judgments of the features of the latter more extreme. Thus, as demonstrated in classical psychophysics experiments, an object may be judged to be heavier after holding a lighter object, a color may be deemed darker if set against a lighter background, and so on. Counterfactual comparisons may similarly influence emotional appraisals of specific outcomes by making them, in contrast, seem better or worse. Thus, upward counterfactuals make an actual event seem less favorable, whereas downward counterfactuals make an actual event seem more favorable. This contrast effect underlies a variety of effects of counterfactual thinking on social judgment.

Social Judgment

A variety of social judgmental consequences of counterfactual thinking have been mapped; five are detailed here.

First, counterfactuals influence emotion, typically making emotional reactions more extreme (by way of a contrast effect) than would otherwise have been the case. Regret is an affective state predicated on upward counterfactual thinking and is the subject of much research in its own right. Counterfactual-induced affective changes can then influence judgment further. For example, in responses to victimization, inferring that a victim's misfortune could easily have been averted might create greater sympathy for the victim, but also greater recommendations for monetary compensation to the victim (Miller and McFarland, 1986).

Second, counterfactuals influence likelihood estimates in at least two ways, both rooted in the causal inference mechanism. The mental simulation of an alternative antecedent event can make future, similar events seem more likely. This would occur to the extent that the prior action is controllable and presumed to be sufficient to have brought about a favorable outcome. That is, the individual might intend to perform the action in the future to bring about a desired goal, in part because the individual infers that performing it in the past would have brought about that desired goal in the past (Roese, 1994). Counterfactuals can also make past events seem more predictable (the hindsight bias) to the extent that they clarify causal linkages, i.e. specify how an event was brought about and thus how it might have been improved or negated (Roese and Olson, 1996). For example, a student who reacts to a poor grade with the counterfactual, 'If only I had studied harder, I would have performed better', has used the counterfactual to articulate the causal power of studying to influence performance. This causal inference may then form the basis of a behavioral intention to study more thoroughly for the next examination, which then yields beliefs in the heightened probability of future success.

Third, and drawing directly on the previous description of heightened likelihood estimates, counterfactuals can heighten perceived control, again by way of causal inferences. To the extent that a desired event is seen to be attainable had one only acted in a certain way, it confers a belief in personal control (Nasco and Marsh, 1999). In other words, one may generalize from the specific instance of having been able to effect positive outcomes ('If I had studied harder, I would have performed better') to the beliefs regarding global personal efficacy ('I can accomplish many things with a little extra effort').

Fourth, counterfactuals can influence decision-making. If a decision is made but an alternative decision might have brought about clearly better rewards, the resulting emotion of regret may compel changes in decision-making strategy, altering the course of subsequent behavior. Research on cognitive dissonance theory specifies conditions under which individuals alter appraisals as a function of postdecisional regret, but theory linking dissonance to counterfactuals is underdeveloped.

Fifth, counterfactuals can make observers suspicious. If an event occurs but is surprising because it is easy to imagine it occurring differently, an observer might be more suspicious regarding ulterior goals of the actor than in cases in which it is easy to imagine the event occurring in many similar ways, even if the probability of event occurrence remains constant. Take the example of a child who loves chocolate-chip cookies: the child is permitted to have just one cookie before dinner, but is required to select the cookie with eyes closed from a jar containing one chocolate-chip cookie and nine oatmeal cookies. If the child happens to select the coveted chocolate-chip cookie, an observer might suspect that the child had peeked. If, however, the cookie jar contained ten chocolate-chip cookies and ninety oatmeal cookies, suspicion might be reduced as there are ten similar ways for the coveted cookie to be selected without intent. Even though the probability of selecting the chocolate chip cookie is identical in both cases, ease of generation of alternatives differs and results in variation in suspicion (Miller *et al.*, 1989).

DETERMINANTS OF COUNTERFACTUAL THINKING

Activation

When does counterfactual thinking occur? A principal trigger is negative affect resulting from an undesirable outcome (Sanna and Turley, 1996). When things go wrong, people often ruminate about how the outcome could have been avoided. Thus, thoughts about 'what might have been' are more common following defeats than victories, failures than successes, and penalties than rewards. A second activator of counterfactual thinking is surprise resulting from an unexpected event. Unexpected occurrences violate implicit predictions and thereby attract attention, which induces consideration of why the outcome occurred. A third trigger of counterfactual thinking is a near miss, or an event that almost occurred. When something nearly happens, it seizes the perceiver's imagination. An athlete who finishes second by a hair's breadth in a 100 m race is likely to experience vivid thoughts about the counterfactual outcome of winning, whereas finishing a distant second evokes fewer thoughts of hypothetical victory.

These triggers of counterfactual thinking correspond to situations where this activity is most useful. As noted earlier, counterfactual thinking provides causal information about an outcome. What kinds of outcomes are most important to understand? Negative outcomes demand comprehension for survival reasons (prevention). Unexpected outcomes, by definition, indicate failures of prediction. Outcomes that almost occurred might occur in the future. Thus, these triggers reflect adaptive coping and support a functional view of counterfactual thinking (Roese, 1994).

Content

Of the infinite number of possible alternatives to reality, which does the mind select for consideration? That is, what are the typical contents of mental reconstructions? Researchers have identified several qualities that render events or antecedents more mutable. As Hofstadter (1985: p. 239) argued, there are natural 'fault lines' of the mind along which reality is cognitively cleaved.

One variable influencing the content of counterfactual thoughts is the normality of the antecedents to an event (Kahneman and Miller, 1986). When considering alternative possibilities, perceivers often focus on unusual things preceding an outcome, rather than routine aspects of the situation, with the mental reconstruction transforming the unusual antecedent into a more normal form. For example, a student who spends less time than usual studying for an examination and performs poorly is likely to think, 'If only I had studied more, I would have done better', even though many other mutations are also theoretically possible (e.g. 'If only the test had been easier').

A second feature of antecedents that increases the probability that they will be selected for counterfactual mutation is controllability. Perceivers are more likely to mutate controllable than uncontrollable aspects of a situation. For example, following a car accident at high speed on a slippery winter road, the driver is more likely to think 'If only I had driven more slowly' than 'If only it hadn't been snowing'. Serial position also influences counterfactual content. Typically, the most recent antecedents are mutated. A missed shot at the buzzer of a one-point loss in basketball is more likely to be altered than preceding misses, even though all misses were equally responsible for the outcome. If, however, several antecedents constitute a causal chain, then early events are likely to be mutated. For example, if a truck blows a tire and hits a car, which then runs into a school bus injuring some children, perceivers will think 'If only the truck hadn't blown a tire', rather than 'If only the car hadn't hit the bus'.

CONCLUSION

Counterfactual thinking, or thoughts of alternatives to past outcomes, is a common feature of everyday mental life. It exerts a variety of effects on emotion and judgment, and is thought to do so primarily through underlying mechanisms rooted in causal inference effects or contrast effects.

References

Hofstadter DR (1985) *Metamagical Themas: Questing for the Essence of Mind and Pattern*. New York, NY: Basic Books.

Kahneman D and Miller DT (1986) Norm theory: comparing reality to its alternatives. *Psychological Review* **93**: 136–153.

Miller DT and McFarland C (1986) Counterfactual thinking and victim compensation: a test of norm theory. *Personality and Social Psychology Bulletin* **12**: 513–519.

Miller DT, Turnbull W and McFarland C (1989) When a coincidence is suspicious: the role of mental simulation. *Journal of Personality and Social Psychology* **57**: 581–589.

Nasco SA and Marsh KL (1999) Gaining control through counterfactual thinking. *Personality and Social Psychology Bulletin* **25**: 556–568.

Roese NJ (1994) The functional basis of counterfactual thinking. *Journal of Personality and Social Psychology* **66**: 805–818.

Roese NJ and Olson JM (1996) Counterfactuals, causal attributions, and the hindsight bias: a conceptual integration. *Journal of Experimental Social Psychology* **32**: 197–227.

Sanna LJ and Turley KJ (1996) Antecedents to spontaneous counterfactual thinking: effects of expectancy violation and outcome valence. *Personality and Social Psychology Bulletin* **22**: 906–919.

Spellman BA and Mandel DR (1999) When possibility informs reality: counterfactual thinking as a cue to causality. *Current Directions in Psychological Science* **8**: 120–123.

Further Reading

Ferguson N (ed.) (1997) *Virtual History: Alternatives and Counterfactuals*. London: Picador.

Gilovich T and Medvec VH (1995) The experience of regret: what, when, and why. *Psychological Review* **102**: 379–395.

Harris PL, German T and Mills P (1996) Children's use of counterfactual thinking in causal reasoning. *Cognition* **61**: 233–259.

Kahneman D and Tversky A (1982) The simulation heuristic. In: Kahneman D, Slovic P and Tversky A (eds) *Judgment Under Uncertainty: Heuristics and Biases*, pp. 201–208. New York, NY: Cambridge University Press.

Lewis D (1973) *Counterfactuals*. Cambridge, MA: Harvard University Press.

Roese NJ and Olson JM (eds) (1995) *What Might Have Been: The Social Psychology of Counterfactual Thinking*. Mahwah, NJ: Erlbaum.

Sanna LJ, Turley-Ames KJ and Meier S (1999) Mood, self-esteem, and simulated alternatives: thought-provoking affective influences on counterfactual direction. *Journal of Personality and Social Psychology* **76**: 543–558.

Tetlock PE and Belkin A (eds) (1996) *Counterfactual Thought Experiments in World Politics*. Princeton, NJ: Princeton University Press.

Creativity

Intermediate article

Thomas B Ward, Texas A&M University, College Station, Texas, USA
Katherine N Saunders, Texas A&M University, College Station, Texas, USA

CONTENTS

Creativity is the result of the convergence of basic cognitive processes, core domain knowledge, and environmental, personal, and motivational factors which allow an individual to produce an object or behavior that is considered both novel and appropriate in a particular context.

INTRODUCTION

One of the most salient features of the human mind is its capacity to generate novel ideas that are useful and appropriate for a given task or problem, that is, to exhibit creativity. Creativity is a complex phenomenon, determined by a wide range of factors, and requiring a multifaceted approach to arrive at even a partially complete understanding of the topic. This article addresses some of the issues that are important to that understanding, including a consideration of whether or not there are different types of creativity, what cognitive processes are most associated with creative outcomes, the extent to which machines can be said to be creative, the factors that limit creativity, the techniques that have been purported to enhance creativity, and the sources of individual differences in creative performance.

TYPES OF CREATIVITY

Although it is possible to describe creativity as the production of novel and useful outcomes from a convergence of skills, processes, knowledge, personal traits, environmental factors, and motivation, this general statement belies potentially important distinctions among types of creativity. These distinctions include contrasts between extraordinary and more mundane instances of creativity, and between general and specific manifestations of creativity.

Examples of the attempt to differentiate between extraordinary and commonplace forms of creativity include Boden's (1992) distinction between psychological (P) and historical (H) creativity, Gardner's (1993) contrast between 'little C' and 'big C' types of creativity, and Czikszentmihalyi's (1988) separation of personally creative and unqualifiedly creative individuals. Ideas that are P-creative are said to be novel in the mind of the individual currently having the idea, although the same ideas may have occurred to many other people before; in contrast, H-creative ideas are novel with respect to all of human history. Similarly, 'little C' creativity is manifested in everyday, small variations on themes, whereas 'big C' creativity occurs rarely and can represent a striking departure from what has come before. Personally creative people can adopt original perspectives, but unqualifiedly creative people radically alter whole domains of endeavor.

Sternberg's (1999) propulsion model introduces still more distinctions among various types of creative contributions. The model views creative work as propelling a field in different ways. These include replications that keep the field where it is, redefinitions that provide a new perspective, incrementations that move the field further in the direction it is already going, and redirections that take it in a new direction.

A question of some debate is how best to account for extraordinary versus everyday manifestations of creative behavior. One approach is to suggest that the minds of those who make notable creative contributions operate according to fundamentally different sets of rules than the minds of those whose generative accomplishments are more mundane. Alternatively, the cognitive processes may be similar, but major breakthroughs may occur only with very special convergences of personal, social, historical, and societal factors. The thought processes presumed to be involved in generating novel ideas (e.g. combining of concepts, analogical reasoning, imagery) are ones available to most humans, albeit on perhaps a lesser scale for most, but it remains for future research to delineate the ways in which these processes are invoked in everyday and extraordinary creative accomplishments.

Another long-standing issue in the field is whether creativity can best be characterized as domain-specific or domain-general. Do creative individuals, in general, possess some common, core set of traits and abilities that would allow them to function creatively in any of a variety of domains, or do the traits and abilities needed for creative accomplishment differ considerably from one domain to the next? One approach to providing evidence on this question has been to assess the personality traits of creative artists versus those of creative scientists. Data from these types of studies support a position between the extremes of pure domain-specificity and complete domain-generality. Creative artists and creative scientists appear to share some traits and differ on others (Barron and Harrington, 1981; Feist, 1999). For example, creative artists have been reported to be more open to experience, fantasy-oriented, imaginative, driven, and ambitious, and to demonstrate higher levels of anxiety, emotional sensitivity, and independence than non-artists. Creative scientists, on the other hand, have been described as possessing traits of arrogance, drive, introversion, flexibility of thought, ambition, and independence. Thus, in either domain, a basic level of unorthodox thought and behavior is characteristic, but achieving eminence in a scientific frontier may require a greater degree of conscientiousness, responsibility, and emotional stability than that which is found in the creative artist.

Another approach to the question of how different or similar artistic and scientific creativity are is to examine the cognitive processes associated with the production of novel ideas in each of the domains. Although specialized skills would be expected to contribute differentially to success in particular domains (e.g. visuo-spatial ability for art or pitch discrimination for music), many of the most basic generative processes, such as combining previously separate concepts and using analogies, are relevant in virtually all creative domains (e.g. Finke *et al.*, 1992).

INCUBATION, INSIGHT, AND OTHER CREATIVE PROCESSES

A widely noted creative phenomenon is incubation, defined as a temporary withdrawal from the problem at hand, which may culminate in an illumination or insight; that is, a sudden realization of a problem solution. Interestingly, there is much less experimental evidence regarding incubation than would be expected from the broad dissemination of the term. Historical anecdotes abound, including Archimedes' purported recognition of the principle of displacement while bathing, and Kekulé's realization regarding the circular structure of benzene while dozing by the fire. What such anecdotes have in common is a solution sequence in which the thinker devotes considerable deliberate effort towards solving a problem, reaches an impasse, withdraws temporarily, and is then struck with a sudden realization for a problem solution.

Although the phenomena of incubation and insight are broadly noted, the mechanisms by which incubation may facilitate insights are not well established. Theoretical mechanisms that have been proposed include conscious work, unconscious work, forgetting of interfering material, recovery from fatigue, and assimilation of cues encountered by chance during the incubation period.

According to the conscious work hypothesis, deliberate effort can continue on the problem while the thinker is engaged in routine tasks, such as bathing, which require only limited cognitive resources. Because the conscious thoughts that led to the solution may be quickly forgotten, the insight may appear to come 'from out of the blue'.

The unconscious work hypothesis also holds that work continues on the problem during the incubation phase, but the work occurs below the level of conscious awareness. That is, the effort is not consciously noted and then forgotten, but rather it is not available to consciousness at all.

The forgetting hypothesis states that inappropriate strategies adopted and ideas considered during initial work on a problem may be forgotten during incubation, which can facilitate the retrieval or generation of more appropriate ideas.

Recovery from fatigue holds that incubation serves as a kind of rest period during which the problem-solver can recover from the debilitating effects of an extended period of deliberate mental effort on the problem.

Finally, according to the opportunistic assimilation view, the problem-solver remains sensitive to cues in the environment that may relate to unsolved problems, even while not engaged in deliberate effort on the task.

There is little experimental research that clearly favors one view of incubation over the others, but at least some laboratory studies by S. Smith and his collaborators (e.g. Smith, 1995) are consistent with the forgetting hypothesis.

Some models of insight attempt to specify component processes that work in concert to produce the phenomenon. For example, Sternberg and Davidson's (1995) model includes subprocesses of selective encoding of problem-relevant information, selective comparison of new and old information, and selective combination of different pieces of information.

Experimental findings do reveal some differences between insight problem-solving and analytic or logical problem-solving. For example, Metcalfe (1986) has shown that feelings of 'warmth' or progress towards a solution increase gradually as subjects near solutions to analytic problems, whereas they jump dramatically for insight problems. In addition, J. Schooler has shown that verbalization can interfere with insight problem-solving, but not analytic problem-solving (Schooler and Melcher, 1995). Such results suggest that insight may be the result of special processes unlike those involved in noncreative problem-solving. However, Weisberg (1995) has attempted to show that insights, even those described in historical anecdotes, are the result of ordinary cognitive processes applied to existing knowledge. By this view, what appears as a dramatic change in awareness of a solution may well reflect a more incremental building of solution-relevant knowledge.

Although incubation occupies a special historical role in attempts to understand creative functioning, several fundamental cognitive processes have been either theorized or demonstrated to be central to the production of novel and useful ideas (see, e.g. contributions to Ward *et al.*, 1997). These include conceptual combination, analogical reasoning, and mental imagery.

In conceptual combination, the thinker merges two concepts that had previously been separate. Anecdotal accounts from creative individuals often include reference to a combining of concepts underlying some important creative advance. In addition, some theorists (e.g. Rothenberg, 1979) suggest that a simultaneous consideration of opposing concepts, termed 'Janusian' thinking, is a particularly important source of emergently creative ideas, and laboratory research on how people interpret novel combinations of concepts is beginning to provide support for this idea.

Analogical reasoning, in which a thinker uses information from a familiar domain to aid in understanding a less familiar domain, is also a central process underlying creative accomplishment. Historical cases abound in science, music, art, and literature. Recent analyses, such as Gentner's (1997) examination of Johannes Kepler's use of analogy in reasoning about the nature of the solar system, have related historical accounts directly to principles from contemporary process theories. That work has helped to establish the validity of claims that analogies between distant knowledge domains can underlie great creative advances. Studies of reasoning among contemporary scientists, such as Dunbar's (1997) look at the ongoing activities of molecular biology laboratories, also reveals that analogies to closely related domains (as opposed to distant domains) often dominate the day-to-day reasoning involved in creative breakthroughs.

MAKING MACHINES CREATIVE

A number of attempts have been made to get computers to function creatively, and Boden (1992) has provided a thorough account of these efforts. An important goal of such computational approaches is to develop a better understanding of human creativity by attempting to simulate it. In that sense, to the extent that creative outcomes spring from fundamental cognitive processes, even computational models of basic processes such as analogy (e.g. Structure Mapping Engine, or SME) are relevant to the issue of making machines creative.

In addition to computational attempts to understand broad processes such as analogy, there are also more direct attempts at simulating specific instances of creativity (e.g. scientific discovery). One of the best-known examples of such an attempt is BACON, which used heuristics to simulate the discovery of scientific laws (Bradshaw *et al.*, 1983). BACON was shown to be able to rediscover Kepler's laws of planetary motion from a set of heuristics and data on observations of planetary motion, although it has come under criticism for underrepresenting the complexities involved in real-world instances of discovery. Such programs, along with others concerned with creativity in drawing, literature, and music, still leave much to be desired, but they do represent important first steps.

LIMITS TO CREATIVITY

Both individual and environmental factors can provide limits to creativity. It is clear that below some minimum level of intellectual ability (e.g. an IQ of 85), a person would have a limited capacity to generate and express creative ideas, although studies tend not to examine creative functioning in those individuals. Studies on individuals with somewhat higher scores have shown that creative performance is linked with intellect in individuals with an IQ below 120, but this link all but disappears in individuals with IQs above 120 (Barron and Harrington, 1981).

Environmental factors also play a role in the expression of creativity. Extensive research by Amabile has found that the use of external rewards or evaluations decreases task motivation for creativity and overall creative performance in both adults and children (Hennessey and Amabile, 1988). This negative effect of reward on creative performance is so strong, in fact, that Amabile found that merely the expectation of some sort of external reward or evaluation diminished creative task motivation and performance in the same way that actually using such external constraints had.

TECHNIQUES FOR IMPROVING CREATIVITY

A wide variety of techniques have been developed with the goal of trying to improve creative performance. Some have emphasized the dynamics of group interaction, others the learning of specific idea-generation techniques, and still others the enhancement of intrinsic motivation.

One of the earliest and best-known techniques is brainstorming. Developed by Osborn (1953), this technique is designed to enhance creativity by encouraging groups (and individuals) to generate as many ideas as possible about a problem without expressing criticism towards those ideas. By eliminating criticism and allowing individuals to 'piggy-back' on ideas suggested by other group members, brainstorming is supposed to result in more ideas being generated, some of which may be extremely creative and provide excellent solutions to the problem being considered. Although there is some support for the usefulness of the procedure, a number of studies have actually shown a productivity loss in groups. That is, groups sometimes produce fewer ideas than the same number of individuals working independently. Thus, the question of where, when, and how brainstorming improves creative performance is yet to be resolved.

Another well-known attempt to enhance creativity is Edward deBono's (1970) lateral thinking approach, which encourages people to engage in thinking that moves off in different directions and to adopt many different perspectives on a problem, rather than thinking along a single narrow path. A major aspect of the approach is to teach people specific techniques designed to facilitate this type of broad attack on a problem, including the 'six hats' approach, in which people 'wear different types of hats', that correspond to different modes of thought (e.g. critical versus generative) (de Bono, 1985). An approach that makes use of idea-generation techniques and principles to facilitate group interaction is Gordon's (1961) synectics procedure, in which group members are coached to generate ideas using analogies and metaphor while also being instructed to suspend criticism of ideas generated by others.

Brainstorming, lateral thinking, synectics, and a host of other procedures have enjoyed a great deal of popular success, but do they make people more creative? To some extent the answer depends on how one defines and measures creativity. Although any given technique may be shown to facilitate performance on a particular task, the extent to which such changes in generative performance last or generalize beyond the immediate situation is less clear. Thus, it may be more appropriate to state that various training procedures can alter patterns of performance on a range of generative tasks, rather than to claim that they make people more creative.

Another avenue of training for improving creativity has been to increase creative motivation. Developed by Amabile and colleagues, this training paradigm, called inoculation training, seeks to increase creativity by training individuals to focus on the intrinsic joy that creative activities bring (Hennessey and Amabile, 1988). Developed as a way to counteract the negative effects of external reward on creative performance, inoculation training involves talking to groups about the internal or intrinsic rewards of behaving creatively. This is done in conjunction with watching videos demonstrating others behaving creatively in the face of external reward and finding pleasure in just engaging in the creative act alone. The use of this type of training has been shown to increase creative performance of both schoolchildren and adults on tasks which have an element of reward or evaluation associated with them.

MODELS OF CREATIVITY

Models of creativity differ in their scope, in the factors they emphasize, and in the tendency to view creativity as stable or malleable. Although historical models of creativity sought to explain creative behaviors as a reflection of differences in individual personalities, beginning with Guilford (e.g. 1967, 1968) creativity began to be viewed as a set of traits which, though stable, were influenced by motivation and temperament (Brown, 1989). Creativity was seen as the result of a set of traits such as problem sensitivity, fluency, flexibility, complexity, evaluation, the use of novel ideas, the ability to break down existing symbolic structure, and the general tendency to organize ideas into larger patterns. When conceptualized this way, variations in creativity could be measured using a variety of open-ended tests.

Torrance modified Guilford's definition slightly, viewing creativity as the combination of ability, skills, and motivation (Ford and Harris, 1992). By including skills in the account of individual differences, creativity became a teachable entity to the degree that a person's creative skills could be improved.

In a further departure from the focus on creativity as a personality trait, Amabile developed a model in which creativity cannot be simply the result of a single isolated personality trait or process, but rather must be accounted for by a constellation of personal characteristics, cognitive abilities and processes, and social environment factors. By this approach, creativity emerges from the confluence of domain-relevant skills, creativity-relevant skills, and task motivation (Amabile, 1990).

Gruber (1988), in what is known as the evolving systems approach, has also regarded creativity as the merging of personal knowledge, affect, and purpose. According to this developmental approach, creativity is the result of developmental changes in knowledge systems that result from the increasingly different situations that a person encounters over time. In this theory, creativity is an extended process, with a person having more than one insight or metaphor over time, and with multiple changes in thoughts and knowledge systems along the way.

A focus on a concert of factors as the root of creativity can also be seen in the burgeoning of research attempting to explain creativity as a multifaceted concept. Called *componential theories of creativity*, such theories hold that creativity occurs when a variety of biological, cognitive, and social factors merge or interact. As an example of one of the modern componential theories, Czikszentmihalyi (1988) regards creativity as an interaction of components both within and outside the individual. According to this model, creativity results from the interaction of the individual with any given domain of knowledge and those controlling the field of that domain. An individual is creative only to the extent that he or she can use cognitive processes, personality traits, and motivation to alter a particular domain in a way that is acceptable to the field at large. More recently, Sternberg and Lubart (1999) have also pursued this idea of creativity as the convergence of multiple components. In their *investment theory of creativity*, creative people are those who can 'buy low and sell high'; that is, generate or adopt ideas before they become popular, then popularize them, thus becoming associated with novel, impact-producing ideas. Such behavior is thought to require the merging of six resources: intellectual abilities, knowledge, thinking styles, personality traits, motivation, and environment.

In contrast to componential approaches, which provide a global account of the factors that interact to determine the creative impact of novel ideas, cognitive models focus more narrowly on the way in which basic cognitive processes operate on existing knowledge structures to produce those novel ideas. The models acknowledge that social and motivational factors can influence the likelihood or intensity of engaging in particular processes. Similarly, they acknowledge that factors outside the individual's thought processes will determine the extent to which an idea is judged acceptable or has an impact. However, they view cognitive processes as the crucial source of the ideas to be judged, and to some extent of the judgments as well.

Often called *process approaches*, these cognitive models focus on the acts of problem-identification and solution-generation as the keys to creative production. An example of this type of model is the Geneplore model of Finke *et al.* (1992), which characterizes the development of novel and useful ideas as resulting from an interplay between *generative* processes, that produce candidate ideas of varying degrees of creative potential, and *exploratory* processes that expand on that potential. Generative processes such as retrieval, conceptual combination, and analogical reminding are assumed to result in candidate ideas, which vary in their apparent novelty, surprisingness, aesthetic appeal, or other factors that would influence the creative person's perception that they hold promise for solving the current problem. People can use such properties to determine which ideas to develop by way of exploratory processes that modify, elaborate, consider the implications, assess the limitations, or otherwise transform the candidate ideas. The model also assumes that real-world constraints, such as the social acceptability of particular ideas, can influence the form of initially generated ideas, the person's judgment about which ideas to explore, or the way in which a candidate idea is modified through exploratory processes.

Other models focus on the production and retention of novel ideas, and make use of a Darwinian perspective: many variations on ideas may be developed but only the fittest will be selected and survive. Simonton (1999b) has extended the evolutionary view and claimed that the production of creative ideas should be viewed as akin to blind variation, in which the creator does not have any notion of whether a given generated idea will be successful or not. While others adopt a somewhat similar generation/selection view, they do not necessarily endorse the blind variation notion (e.g. Johnson-Laird, 1988; Perkins, 1998; Sternberg, 1998).

UNDERSTANDING INDIVIDUAL DIFFERENCES IN CREATIVITY

Traditional research concerned with individual differences in creative performance attributes those differences to one of two sources: differences in the ability or tendency to use particular creative thought processes, and differences in personality attributes thought to be related to creative behaviors. The work has made use of psychometric procedures as well as assessments of the historical record of the achievements of eminent creators.

Research concerned with thought processes has focused on individual differences in the ability to identify or recognize problems or solutions that have creative potential, to tap into broad thought networks, and to apply this expanded base of knowledge to the task at hand. Defining differences in creativity as the result of individual differences in the ability to associate or bring together different elements of thought to form new and useful creations is the hallmark of the associative approach to creativity (Brown, 1989). This associative approach is not a new one in psychology, as can be seen in the many introspective studies and historical anecdotes concerning the creative process (Barron and Harrington, 1981; Brown, 1989). Mednick (1962) extended this approach by defining creativity as the forming of associative elements into new combinations, which either meet specific requirements or are in some way useful. By this view, individual differences would be attributable to the ability to access remote associations, which in turn gives rise to the use of the Remote Associates Test as a measurement technique (Mednick and Mednick, 1967).

Individual differences in creative thought have also been explained by variations in divergent thinking ability, including fluency (the tendency to produce many ideas), flexibility (the tendency to produce differing ideas), and originality (the tendency to produce ideas that are normatively uncommon). A classic example of a divergent thinking task used to measure such differences is the Torrance Test of Creative Thinking (Torrance, 1974) in which people generate questions, unusual uses, and/or drawings in response to particular stimuli.

One question that can be raised about paper-and-pencil measures of divergent thinking ability is whether or not performance on those measures is indicative of real-world creative skill. Although various researchers have found a relationship between test performance and real-world indicators, such findings have not been consistent. Measures of divergent thinking, while related to some indices of creative achievement, are often unable to significantly predict creative achievement and behaviors in a real-world setting (Barron and Harrington, 1981; Brown, 1989). In addition, concerns have been raised about the domain-specificity of divergent thinking.

Another type of explanation for individual differences in creativity focuses on personality traits, and assumes that creativity differences are based on variations in personality attributes that are thought to contribute to creative production. The characteristics that have been identified as important to creativity are tolerance for ambiguity, openness, independence, positive sense of self, high energy, general curiosity, wide interests, as well as introversion, attraction to complexity, need for recognition, and a variety of others (Barron and Harrington, 1981). As indicated previously, however, the importance of each of the characteristics may vary according to the domain of creativity being pursued. In fact, the search for the single set of 'creative personality' traits that map onto real-world creative performance has, so far, been unsuccessful.

Variations in intrinsic motivation are another possible source of individual differences in creative performance. For instance, Amabile, in her seminal research on the relationship between intrinsic motivation and creativity, found that creative performance in such areas as writing and art can be both enhanced and hindered by changes in intrinsic interest in a task. Hennessey and Amabile (1988), in a continuation of this line of research, have proposed the intrinsic motivation principle of creativity which says that people will be the most creative when they feel motivated to perform primarily by the interest and enjoyment of the task, and not by external factors such as reward or punishment. Thus, all individual differences in creative performance are due to those differences in motivation towards the task at hand, and hence, all creativity is, at heart, domain-specific to the interests of the individual.

Although much contemporary work on individual differences relies on tests or laboratory observations of a broad sampling of participants, another approach involves detailed, narrative case studies of a small set of highly creative individuals in history. Somewhere between these extremes is Simonton's (1999a) historiometric approach in which historical data (e.g. number of publications, citations, performances, and so on) is sampled for a large number of contributors to a field, and statistical tests are performed to relate those measures to other indices. The approach can be used to examine a broad range of factors, including intellectual precocity, family background, and propensity towards mental illness. It goes beyond paper-and-pencil measures of the attributes of the many to a detailed look at individual differences among the eminent.

Acknowledgements

This work was supported by the National Science Foundation under Grant No. BCS-9983424.

References

Amabile TM (1990) Within you, without you: the social psychology of creativity and beyond. In: Runco MA and Albert RS (eds) *Theories of Creativity*, pp. 61–91. Newbury Park, CA: Sage.

Barron FX and Harrington DM (1981) Creativity, intelligence, and personality. *Annual Review of Psychology* **32**: 439–476.

Boden M (1992) *The Creative Mind: Myths and Mechanisms*. New York, NY: Basic Books.

Bradshaw GF, Langley PW and Simon HA (1983) Studying scientific discovery by computer simulation. *Science* **222** (4627): 971–975.

Brown RT (1989) Creativity: what are we to measure? In: Glover JA, Ronning RR and Reynolds CR (eds) *Handbook of Creativity*, pp. 3–32. New York, NY: Plenum.

Czikszentmihalyi M (1988) Society, culture, and person: a systems view of creativity. In: Sternberg RJ (ed.) *The Nature of Creativity*, pp. 325–339. New York, NY: Cambridge University Press.

De Bono E (1970) *Lateral Thinking*. New York, NY: Harper.

De Bono E (1985) *Six Thinking Hats*. Boston, MA: Little, Brown and Co.

Dunbar K (1997) How scientists think: on-line creativity and conceptual change in science. In: Ward TB, Smith SM and Vaid J (eds) *Creative Thought: An Investigation of Conceptual Structures and Processes*, pp. 461–493. Washington, DC: American Psychological Association.

Feist GJ (1999) The influence of personality on artistic and scientific creativity. In: Sternberg RJ (ed.) *Handbook of Creativity*, pp. 273–296. New York, NY: Cambridge University Press.

Finke RA, Ward TB and Smith SM (1992) *Creative Cognition: Theory, Research, and Applications*. Cambridge, MA: MIT Press.

Ford D and Harris JJ (1992) The elusive definition of creativity. *Journal of Creative Behavior* **26**: 186–198.

Gardner H (1993) *Creating Minds*. New York, NY: Basic Books.

Gentner D, Brem S, Ferguson RW and Wolff P (1997) Analogy and creativity in the works of Johannes Kepler. In: Ward TB, Smith SM and Vaid J (eds) *Creative Thought: An Investigation of Conceptual Structures and Processes*, pp. 403–459. Washington, DC: American Psychological Association.

Gordon WJ (1961) *Synectics: The Development of Creative Capacity*. New York, NY: Harper and Row.

Gruber HE (1988) The evolving systems approach to creative work. *Creativity Research Journal* **1**: 27–51.

Guilford JP (1967) Creativity: yesterday, today and tomorrow. *Journal of Creative Behavior* **1**(1): 3–14.

Guilford JP (1968) *Intelligence, Creativity, and their Educational Implications*. San Diego, CA: Robert R Knapp.

Hennessey BA and Amabile TA (1988) The conditions of creativity. In: Sternberg RJ (ed.) *The Nature of Creativity*, pp. 11–38. New York, NY: Cambridge University Press.

Johnson-Laird PN (1988) Freedom and constraint in creativity. In: Sternberg RJ (ed.) *The Nature of Creativity*, pp. 202–219. New York, NY: Cambridge University Press.

Mednick SA (1962) The associative basis for the creative process. *Psychological Review* **69**: 220–232.

Mednick SA and Mednick MT (1967) *Remote Associates Test, College and Adult, Forms 1 and 2 and Examiner's Manual*. Boston, MA: Houghton Mifflin.

Metcalfe J (1986) Feeling of Knowing in memory and problem solving. *Journal of Experimental Psychology: Learning, Memory, and Cognition* **12**(2): 288–294.

Osborn AF (1953) *Applied Imagination*. New York, NY: Scribner's.

Perkins DN (1998) Is the country of the blind an appreciation of Donald Campbell's vision of creative thought. *Journal of Creative Behavior* **32**(3): 177–191.

Schooler JW and Melcher J (1995) The ineffability of insight. In: Smith SM, Ward TB and Finke RA (eds) *The Creative Cognition Approach*, pp. 97–134. Cambridge, MA: MIT Press.

Simonton DK (1999a) Creativity from a historiometric perspective. In: Sternberg RJ (ed.) *Handbook of Creativity*, pp. 116–133. Cambridge, UK: Cambridge University Press.

Simonton DK (1999b) *Origins of Genius: Darwinian Perspectives on Creativity*. New York, NY: Oxford University Press.

Smith SM (1995) Getting into and out of mental ruts: A theory of fixation, incubation, and insight. In:

Sternberg RJ and Davidson JE (eds) *The Nature of Insight*, pp. 229–251. Cambridge, MA: MIT Press.
Sternberg RJ (1998) Cognitive mechanisms in human creativity: is variation blind or sighted? *Journal of Creative Behavior* **32**(3): 159–176.
Sternberg RJ (1999) A propulsion model of types of creative contributions. *Review of General Psychology* **3**: 83–100.
Sternberg RJ and Davidson JE (eds) (1995) *The Nature of Insight*. Cambridge, MA: MIT Press.
Sternberg RJ and Lubart TI (1999) The concepts of creativity: prospect and paradigms. In: Sternberg RJ (ed.) *Handbook of Creativity*, pp. 3–15. New York, NY: Cambridge University Press.
Torrance EP (1974) *Torrance Tests of Creative Thinking: Norms-technical Manual*. Bensenville, IL: Scholastic Testing Service.
Torrance EP (1988) The nature of creativity as manifest in its testing. In: Sternberg RJ (ed.) *The Nature of Creativity*, pp. 43–75. New York, NY: Cambridge University Press.
Ward TB, Smith SM and Vaid J (eds) (1997) *Creative Thought: An Investigation of Conceptual Structures and Processes*. Washington, DC: American Psychological Association.
Weisberg RW (1995) Prolegomena to theories of insight in problem solving: A taxonomy of problems. In: Sternberg RJ and Davidson JE (eds) *The Nature of Insight*, pp. 157–196. Cambridge, MA: MIT Press.

Further Reading

Amabile TM (1983) *The Social Psychology of Creativity*. New York, NY: Springer.
Czikszentmihalyi M (1996) *Creativity: Flow and the Psychology of Discovery and Invention*. New York, NY: HarperCollins.
Guilford JP (1950) Creativity. *American Psychologist* **5**: 444–454.
Perkins D (1988) The possibility of invention. In: Sternberg RJ (ed.) *The Nature of Creativity*, pp. 362–385. New York, NY: Cambridge University Press.
Rothenberg A (1979) *The Emerging Goddess*. Chicago, IL: University of Chicago Press.
Runco MA and Chand I (1995) Cognition and creativity. *Educational Psychology Review* **7**: 243–267.
Sternberg RJ and Lubart TI (1995) *Defying the Crowd: Cultivating Creativity in a Culture of Conformity*. New York, NY: Free Press.
Torrance EP (1988) The nature of creativity as manifest in its testing. In: Sternberg RJ (ed.) *The Nature of Creativity*, pp. 43–75. New York, NY: Cambridge University Press.
Ward TB, Finke RA and Smith SM (1995) *Creativity and the Mind: Discovering the Genius Within*. New York, NY: Plenum.

Cultural Differences in Abstract Thinking

Introductory article

Fons J R van de Vijver, Tilburg University, The Netherlands

CONTENTS

Abstract thinking is a central part of reasoning and the highest cognitive attainment in Piagetian theory. Studies of cross-cultural differences and similarities in abstract thinking show its relationship with culture.

INTRODUCTION

There are two research traditions in cognitive psychology for examining the relationship between culture and abstract thinking: the formal and the informal approach (Table 1). In formal research the scientific approach is the normative model of good problem-solving; there is an emphasis on the application of inductive and deductive reasoning, the solution of formalized problems that are unlikely to be met in everyday life, and the correctness of solutions (e.g., 'continue the following series: 1, 2, 4, 8, 16, …'). In the informal tradition the 'bricoleur' (jack-of-all-trades) is the implicit model of problem-solving. There is an emphasis on problem-solving in everyday life: an example would be,

Table 1. Differences between formal and informal research traditions

Formal tradition	*Informal tradition*
Closed problem spaces	Open problem spaces
Deterministic problems with one correct answer	Probabilistic problems with several correct answers
Contrived problems	Problems derived from everyday life
Academic intelligence	Practical intelligence
Focus on correctness of solution (is the solution correct?)	Focus on practical value of the solution (does the solution solve the problem?)
Problems and solutions are context-independent	Problems and solutions are context-dependent
Scientist as normative model of good problem-solver	Bricoleur as normative model of problem-solver
Algorithmic solutions	Heuristic solutions
Product-oriented (psychological tests, Piagetian tasks)	Process-oriented
Solution requires conceptual, theoretical knowledge	Solution requires practical intelligence
Cross-cultural comparison of test performance	Studies within a single culture

'What would you do when you are due for promotion at your work and a reliable source tells you that your direct colleague may be promoted instead of you?' These two more or less independent traditions have their own models of cross-cultural differences and similarities of abstract thinking.

FORMAL STUDIES OF ABSTRACT THINKING

Two types of study predominate in the formal tradition, differing in theoretical orientation and assessment procedures: psychometric research (based on Western models of intelligence and using psychological 'paper and pencil' tests) and the Piagetian approach (applying Piagetian theory and tasks).

Psychometric studies of abstract thinking have used a variety of tests (such as Raven's Progressive Matrices, Wechsler's Intelligence Scales and Cattell's Culture Fair Intelligence Test) and investigated a number of cultures. Many cultural comparisons involved participants from different countries; especially in the USA, research often compared the performance of different ethnic groups, mainly African Americans and Anglo-Americans. These studies have shown a remarkable consistency in results. Using advanced statistical techniques (mainly exploratory and confirmatory factor analysis), it has been shown that the structure of intelligence is identical across cultural groups; tests of abstract thinking tend to be related to reasoning and general intelligence. Broad cognitive abilities, such as reasoning, memory and visualization, are universal. Moreover, analyses of test performances have consistently shown that the difficulty order of items tends to be invariant across cultures; within the homogeneous domains used in the tests items that are easy in one culture are likely to be easy in another culture (though not necessarily solved by the same proportion of the population). The psychometric tradition does not support the alleged qualitative difference in abstract thinking between Western and non-Western individuals, historically often associated with Lévy-Bruhl. To the best of our knowledge abstract thinking is a universal attainment that can be found in all cultural groups.

Nevertheless, studies have reported consistent differences in scores on tests of abstract thinking between Western and non-Western groups, with the former groups usually obtaining higher scores. Similarly, comparisons of scores obtained by different ethnic groups in the USA have shown consistent differences in scores, ranking as follows (from high to low): East Asians (e.g. Chinese and Japanese), European Americans, Hispanics and African Americans. Interpretation of these differences has been controversial. Some researchers, such as Jensen and Eysenck, argue that the differences in performance are to be interpreted as cultural differences in cognitive ability. These authors attributed these performance differences to genetic differences between cultures. The current immature status of knowledge in behavior and molecular genetics does not yet allow for precise estimates of the role of genetics in cross-cultural differences in abstract thinking. Critics of Jensen and Eysenck often point to the influence of schooling and potential problems in the tests used to assess abstract thinking as a source of cross-cultural performance differences. Many psychological tests in the formal tradition have a format and test contents that are school-related. Schooling has been found to enhance the performance on tests of abstract thinking but has no formative influence on abstract thinking; however, the occurrence of abstract thinking among illiterate individuals and groups demonstrates that schooling is not a precondition for developing abstract thinking.

In Piagetian theory abstract thinking is part of formal-operational reasoning, which, according to Piaget, is acquired by Western subjects at age 12–15 years. Various measures of formal-operational reasoning are available; these are often based on elementary laws from physics. The person tested has to evaluate the impact of presumably relevant variables by experimental manipulation. One such task deals with the oscillation time of a pendulum. The variables that are presumably relevant are the length of the pendulum string, the weight of the object fastened to the string, the angle of the swing of the weight, and the momentum given to the weight at the start of the swing. The participant is asked to determine experimentally which of the four variables determines the oscillation time.

Many cross-cultural studies have addressed earlier Piagetian stages (e.g. the transition from preoperational to concrete-operational thinking), but formal-operational thinking has not been extensively studied. Studies among unschooled non-Western individuals invariably showed a poor performance, which was sometimes interpreted as evidence that non-Western individuals did not show abstract thinking. Some theoreticians even conjectured that abstract thinking was an achievement of Western, industrialized nation states. Later research redressed the picture. First, cognitive anthropological evidence argues strongly against the cultural specificity of abstract thinking; for example, Kalahari Bushmen show on average low scores on intelligence tests, yet their tracking and navigation skills in the desert far exceed those of Westerners and show great cognitive complexity. Moreover, even in Western societies many participants gave a poor performance on Piagetian tests. It was increasingly appreciated that the method of assessment may affect the test outcome; that individuals are unable to solve physics problems which are remote from their daily reality does not imply that their abstract thinking is undeveloped. A distinction was introduced between competence and performance: whereas the latter refers to actual test performance, the former refers to the performance in optimal conditions dealing with common tasks. The distinction implicitly points to the potential problems of Piagetian tasks in cross-cultural research, but subsequent research has not identified new procedures to assess the competence. It is widely believed now that although the performance on Piagetian tests of formal-operational thinking often points to the absence of abstract thinking, the competence is universal. Abstract thinking is a universal attainment, but the domains of application may vary across individuals and cultures. A car mechanic may be able to use abstract thinking in dealing with cars, but perform badly when applying laws of logic to other domains. It is paradoxical that abstract thinking, theoretically based on context-independent rules such as the laws of logic, turned out to be domain-specific.

INFORMAL STUDIES OF ABSTRACT THINKING

The label 'informal' is used here as a summary label for various traditions in the psychological literature, such as 'everyday cognition', 'indigenous cognition' and 'practical intelligence'. Common to these studies is their emphasis on the observation of cognitive processes in everyday problem-solving.

Mathematical thinking has been often studied. For example, shoppers were interviewed while they bought groceries in order to determine how they determined 'best buys'. It was consistently found that they seldom relied on the arithmetic competence they acquired in school. Rather, arithmetic in school and in everyday life seem to constitute different competencies. In one experiment a group of women were asked which of two cans of peanuts they would buy on the basis of a comparison of prices: can A weighing 10 oz for 90 cents, or can B weighing 4 oz for 45 cents. In another test the same women were asked to compare the ratios 90/45 and 10/4. From a mathematical perspective the two problems are identical, but from a psychological perspective they are very different. The former problem was correctly solved more often.

Some studies have examined the relationship between skills applied in everyday life and skills to solve tests in the formal tradition. It has been consistently found that, possibly contrary to expectation, there is no relationship between the scores on 'school' tests and 'everyday' tests in the same domain. Individuals who are capable of displaying highly skilled behavior in the context of their professional specialization are not necessarily the individuals with the highest scores on intelligence tests.

Another line of research has examined to what extent complex cognitive skills acquired in the course of learning a profession generalize beyond this professional context. For example, Zinancanteco women in Mexico can weave highly complex patterns. These women showed superior planning skills in a weaving task when they had to reproduce known patterns, but did not outperform nonweavers when the planning involved an unfamiliar task. Planning skills acquired in the context of professional training did not generalize broadly across the cognitive spectrum. This result exemplifies the findings in studies of specialized cognitive skills: the cognitive effects of mastery of a craft often do not go beyond the specific domain in which the craft is applied.

CONCLUSION

Abstract thinking has been studied from two different perspectives in cross-cultural psychology. The first, the formal tradition, focuses on the application of general principles such as the laws of logic and inductive schemes. Studies in this tradition have shown that abstract thinking is a universal attainment, but cultures may well differ in their areas of specialization. Cross-cultural differences in scores on tests of abstract thinking are typically open to multiple interpretations (e.g. valid cultural differences in abstract thinking, familiarity with testing procedures, differential cultural appropriateness of tests, confounding of cross-cultural differences in schooling). The second, the informal tradition, studies abstract thinking in action. This tradition has shed further light on the domain specificity as studied in the formal tradition. Within an area of expertise such as a profession, individuals can display remarkably high levels of performance. From a cognitive perspective these areas are often sharply delineated. The training of professional expertise often does not have a broad impact on cognitive functioning. Virtuoso performance is often restricted to one domain.

Abstract thinking shows both important similarities and differences across cultures. There is ample evidence for the universality of the basic structures of abstract thinking (general reasoning, Piagetian formal-operational thinking). There is no evidence for the existence of qualitatively different types of abstract thinking across cultures. On the other hand, there are massive cross-cultural differences on tests of abstract thinking. Moreover, domains in which individuals are able to use abstract thinking show some variation across cultures.

Further Reading

Dasen PR (ed.) (1977) *Piagetian Psychology. Cross-Cultural Contributions*. New York: Gardner.

Jensen AR (1998) *The G Factor. The Science of Mental Ability*. Westport, CT: Praeger.

Schliemann A, Carraher D and Ceci SJ (1997) Everyday cognition. In: Berry JW, Dasen PR and Saraswathi TS (eds) *Handbook of Cross-Cultural Psychology*, 2nd edn, vol. 2, pp. 177–216. Boston, MA: Allyn & Bacon.

van de Vijver FJR and Willemsen ME (1993) Abstract thinking. In: Altarriba J (ed.) *Culture and Cognition*, pp. 317–342. Amsterdam: North Holland.

Cultural Differences in Causal Attribution

Intermediate article

Douglas S Krull, Northern Kentucky University, Highland Heights, Kentucky, USA
Michael W Morris, Stanford University, Stanford, California, USA

CONTENTS

An attribution is a judgment about why an event (typically another person's behavior) occurred. Research suggests that cultures differ in the types of attributions that their members prefer.

ATTRIBUTION THEORY

An attribution is an explanation, a judgment about the cause of an event. Psychological research on attribution has primarily studied judgments about the cause of another person's behavior. Attributions for behavior are ubiquitous in everyday life (e.g. 'I think Luis achieved the highest score on the calculus exam because he has a talent for mathematics', 'Mariko is sad because her best friend moved away'). Moreover, attributions have important implications for social interaction in that they shape perceivers' expectations about others' future behavior (e.g. 'I helped Robert because I thought he was trying his best to succeed', 'I'm angry at Torsten because he took my car keys on purpose'). The ubiquity and importance of attributions has made them a central topic of social psychological research. (*See* **Judgment**)

A foundational premise of attribution theory is Heider's (1958) contention that perceivers seek to attribute fleeting behavior to stable dispositions in order to learn about the social environment. He wrote: 'It is an important principle of commonsense psychology ... that man grasps reality, and can predict and control it, by referring transient and variable behavior and events to relatively unchanging underlying conditions, the so-called dispositional properties of his world' (1958, p. 79). For example, upon noticing the anxious behavior of her new co-worker, Susan would make a judgment about something stable in the environment, either that the co-worker has an anxious personality or that the co-worker's job is stressful. Although most behaviors can reflect either situational or personal influences, Heider suggested that perceivers tend to trace action to dispositions of the actor. The existence and consequences of this tendency were documented by social psychology experiments (see Ross and Nisbett, 1991 for a review). Because of its apparent ubiquity and potentially important consequences, this tendency was designated the *fundamental attribution error* (Ross, 1977).

INDIVIDUALIST AND COLLECTIVIST CULTURES

Until recently, psychologists paid little attention to the possibility that attributions might differ across cultures, but lately the study of cultural differences has captured the attention of attribution researchers. Although the study of culture in causal attribution is in its infancy, research suggests that there are substantial and potentially important differences in how people from different cultures think about behavior, as well as potentially important similarities across cultures. (*See* **Cultural Psychology**; **Cultural Differences in Abstract Thinking**)

However, the attribution researcher who desires to learn something about culture is faced with a dilemma. Given that there are many cultures, how can one hope to reach general conclusions about the role that culture plays in attribution? Without denying the fact that there may be important differences between many different cultures, psychologists have found it useful to divide cultures into those that tend towards individualism and those that tend towards collectivism. In individualist cultures (e.g. Australia, Britain, the United States), personal autonomy is emphasized, and so, for the most part, people are seen as free agents who behave as they choose. In contrast, collectivist cultures (e.g. China, Guatemala, India) tend to emphasize supporting the goals of groups and behaving in a collectively appropriate manner, and so people are seen as constrained by social forces. As described in the sections that follow, these differences between individualist cultures and collectivist cultures have important implications for explanations of behavior.

CULTURAL DIFFERENCES IN ATTRIBUTION TENDENCIES

Judging Causes of Actions by Persons

Although the tendency to favor dispositional attributions for others' actions may well be fundamental in the sense of having many consequences, it does not seem to be fundamental in the sense of being universal. Ethnographers have long reported that lay people in some collectivist cultures refer to personality dispositions rarely and instead attribute behavior frequently to social roles (for a review, see Shore, 1996). More controlled, quantitative evidence for this claim first came in a study by Miller (1984) which asked participants of various ages from the USA and India to explain everyday actions that they had observed. Young children in both cultures were alike in the proportional frequency of their references to personality traits and to situational factors. However, as age increased, Americans showed an increasing reliance on attributions to personality, and Indians an increasing reliance on situational attributions. Thus, it seems that North Americans learn that behavior is primarily caused by personality, whereas Indians learn that behavior is primarily caused by the circumstances.

Although this initial evidence for cultural differences from everyday explanations of behavior had compelling external validity, it is open to multiple interpretations. It might reflect cultural differences in attributions, but it might reflect merely that everyday behaviors, and the actual causes thereof, differ across cultures. Indeed, there is evidence that personality traits do account for more variance in behavior in individualist cultures, whereas social roles and situations account for more variance in collectivist cultures (e.g. Argyle *et al.*, 1978; Triandis, 1995). To clarify the role of culture, Morris and Peng (1994) conducted several studies that examined attributions for the same event, such as a prominent crime covered by American and Chinese newspapers. Results showed that attributions for these events differed across cultures, suggesting that cultural differences indicate a difference in the interpretive tendencies of perceivers, not just in the events they typically explain. Insight about how perceivers differ is accumulating from studies of different kinds of attributions. (*See* **Causal Reasoning, Psychology of; Causal Perception, Development of**)

Judging Personal Traits from Behaviors

One of the most important research paradigms within attribution theory focuses on how perceivers judge traits from situationally constrained behavior. Strictly speaking, the task in these studies is not to explain the cause of the actor's behavior but to judge the degree to which an actor's personality corresponds to his or her behavior. A wealth of research with individualist participants suggests that people often infer that personality matches behavior, even when situational forces that are sufficient to explain the behavior are present. For example, although we know that television actors are only playing roles, we may assume that their personalities correspond to the characters that they play. This tendency has been called the *correspondence bias* (see e.g. Gilbert and Malone, 1995 for a review).

The correspondence bias seems to be a multiply determined phenomenon. Gilbert and Malone (1995) have suggested that the bias can arise through at least four distinct mechanisms. People may display correspondence bias because the situation is not immediately apparent (e.g. Bernard infers that Tia's anxiety reflects an anxious personality, because Bernard does not know that Tia is about to give an important speech). People may display correspondence bias because they fail to appreciate the power of the situation (e.g. Anne may infer that Tia has an anxious personality because Anne does not realize that making a speech can be anxiety-provoking). People may display correspondence bias because they are too busy to consider the influence of the situation (e.g. Dolf infers that Tia is an anxious sort of person because he is distracted and so does not fully consider the anxiety-provoking situation). Finally, people may display correspondence bias because their knowledge of the situation inflates their perceptions of the behavior (e.g. Roland knows how anxiety-provoking giving a speech can be, so he perceives greater anxiety in Tia's behavior, and infers that only a dispositionally anxious person would be so anxious).

Given these different mechanisms, one might expect that whether or not the correspondence bias is found across cultures depends on a variety of factors. Research indeed bears out that sometimes the correspondence bias is found across cultures (e.g. Choi and Nisbett, 1998, experiment 1;

Krull *et al.*, 1999). However, cultural differences arise under some conditions as a function of these mechanisms. First, although perceivers across cultures are capable of ignoring situational forces, when such forces are made salient, collectivists become less likely than individualists to show the bias (Choi and Nisbett, 1998, experiment 2). Second, collectivists recognize the strength of some situational forces, such as authority pressure, that individualists underestimate, and this is another source of cultural differences (Morris *et al.*, 2000). Third, collectivists, perhaps because they are more practiced in thinking about situational constraint, are not hindered by cognitive busyness as much as are individualists (Knowles *et al.*, 2001).

Judging Causes of Actions by Groups

The findings described heretofore might lead one to conclude that collectivists generally attribute to the context rather than to dispositions of actors. However, such a conclusion would be incomplete. Recent research suggests that collectivists readily attribute actions by groups to dispositions of the group, and make more reference to group dispositions when attributing events involving a combination of actions by individuals and groups (Menon *et al.*, 1999, Studies 1 and 2). In another study, perceivers in separate conditions read about an act of wrongdoing by an individual or a group (Study 3); results showed that Americans showed a stronger tendency towards dispositions in response to the individual actor, and Chinese, in response to the group actor.

Additional research by Chiu and colleagues (Chiu *et al.*, 2000) investigated how the need for cognitive closure affects perceivers. Results showed that time pressure increased Chinese perceivers' attributions to dispositions of groups, and increased American perceivers' attributions to dispositions of individuals. Another study that examined individuals who were chronically high or low on the need for cognitive closure found the same pattern. Thus, the desire to identify stable, dispositional properties of the social environment is by no means limited to perceivers in individualist cultures. However, perceivers in collectivist cultures are more likely to look for such properties in groups than in individuals.

TOWARDS A MODEL OF HOW CULTURE INFLUENCES ATTRIBUTION

Because research on culture and attribution is relatively new, there is no consensus on an explanatory model of how culture affects the attribution process. Broadly speaking, one model of cultural differences suggests that culture shapes general thinking principles or cognitive styles (Witkin and Berry, 1975). According to this view, collectivists are more holistic in their thinking and individualists are more analytical. In support of this view, research indicates that collectivists often seem to be more sensitive to context in a variety of perceptual and cognitive tasks (see, e.g. Nisbett *et al.*, 2001, for a review). Thus, collectivists may be more aware of the situational forces that influence behavior as part of a general contextual focus of attention.

A second model suggests that differences between individualists and collectivists stem from differences in implicit causal theories. Recent research on individual differences in attribution biases (Dweck *et al.*, 1995) has revealed that implicit theories can be proximal determinants of attributional biases. Accordingly, differences between cultures may or may not emerge, depending upon whether the cultures possess similar or dissimilar theories with regard to the specific domain and whether the circumstances foster theory-based processing. In support of this view, some clear boundary conditions on cultural differences have been identified, corresponding to the applicability of implicit theories. For example, Morris and Peng (1994) found that although Chinese individuals were more situational than North Americans in their judgments of behavior induced by social causality (e.g. an individual moving away from the pressure of a group), cultural differences were not obtained in judgments of behavior induced by mechanical causality (e.g. an object moving after being struck by another object). Further evidence that implicit theories are a mechanism through which cultural differences or similarities are produced comes from studies that demonstrate that cultural differences in attribution can be evoked by priming cultural knowledge (Hong *et al.*, 2000). (*See* **Implicit Learning; Memory: Implicit versus Explicit**)

In sum, attributions may only differ across cultures when different implicit theories are activated, and activation depends on the applicability of theories to the domain and the cognitive dynamics of the perceiver (Morris *et al.*, 2001). However, the two models – cognitive style and implicit theories – need not be considered as rivals, in that the former serves as a heuristic framework whereas the latter serves as a middle-range hypothesis-testing model. As described here, holism can be manifested in an implicit theory about the causal role of context in the behavior of individuals, or in an implicit theory of groups as actors. In sum, cultural differences depend on a variety of factors, but this complexity seems amenable to social cognition models.

References

Argyle M, Shimoda K and Little B (1978) Variance due to persons and situations in England and Japan. *British Journal of Social and Clinical Psychology* **17**: 335–337.

Chiu CY, Morris MW, Hong YY and Menon T (2000) Motivated cultural cognition: the impact of implicit theories on dispositional attribution varies as a function of need for closure. *Journal of Personality and Social Psychology* **78**: 247–259.

Choi I and Nisbett RE (1998) Situational salience and cultural differences in the correspondence bias and actor-observer bias. *Personality and Social Psychology Bulletin* **24**: 949–960.

Dweck CS, Chiu C and Hong Y (1995) Implicit theories and their role in judgments and reactions: a world from two perspectives. *Psychological Inquiry* **6**: 267–285.

Gilbert DT and Malone PS (1995) The correspondence bias. *Psychological Bulletin* **117**: 21–38.

Heider F (1958) *The Psychology of Interpersonal Relations*. New York, NY: Wiley.

Hong YY, Morris MW, Chiu CY and Benet-Martinez V (2000) Multicultural minds: a dynamic constructivist approach to culture and cognition. *American Psychologist* **55**(7): 709–720.

Knowles E, Morris MW, Chiu CY and Hong YY (2001) Culture and cognitive-process models of attribution: evidence for automatic situational correction among East Asians. *Personality and Social Psychology Bulletin* **27**: 1344–1356.

Krull DS, Loy MH–M, Lin J *et al.* (1999) The fundamental fundamental attribution error: correspondence bias in individualist and collectivist cultures. *Personality and Social Psychology Bulletin* **25**: 1208–1219.

Menon T, Morris MW, Chiu CY and Hong YY (1999) Culture and the construal of agency: attribution to individual versus group dispositions. *Journal of Personality and Social Psychology* **76**: 701–717.

Miller JG (1984) Culture and the development of everyday social explanation. *Journal of Personality and Social Psychology* **46**: 961–978.

Morris MW, Knowles E, Chiu CY and Hong YY (2000) *Culture and Judgment of Obedient and Disobedient Acts: Perceived Situational Force and Cultural Role Expectations.* Unpublished manuscript, Graduate School of Business, Stanford University.

Morris MW, Menon T and Ames DR (2001) Culturally conferred conceptions of agency: a key to social perception of persons, groups, and other actors. *Personality and Social Psychology Review* **5**: 169–182.

Morris MW and Peng K (1994) Culture and cause: American and Chinese attributions for social and physical events. *Journal of Personality and Social Psychology* **67**(6): 949–971.

Nisbett RE, Peng K, Choi I and Norenzayan A (2001) Culture and systems of thought: holistic vs. analytic cognition. *Psychological Review* **108**: 291–310.

Ross L (1977) The intuitive psychologist and his shortcomings: distortions in the attribution process. In: Berkowitz L (ed.) *Advances in Experimental Social Psychology*, vol. 10, pp. 174–221. New York, NY: Academic Press.

Ross L and Nisbett RE (1991) *The Person and the Situation. Perspectives of Social Psychology*. New York, NY: McGraw-Hill.

Shore B (1996) *Culture in Mind: Cognition, Culture, and the Problem of Meaning*. New York, NY: Oxford University Press.

Triandis HC (1995) *Individualism and Collectivism*. Bolder, CO: Westview Press.

Witkin HA and Berry JW (1975) Psychological differentiation in cross-cultural perspective. *Journal of Cross Cultural Psychology* **6**: 4–87.

Further Reading

Ames DR, Knowles ED, Rosati AD *et al.* (2001) The social folk theorist: insights from social and cultural psychology on the contents and contexts of folk theorizing. In: Malle BF, Moses LJ and Baldwin DA (eds) *Intentions and Intentionality: Foundations of Social Cognition*, pp. 307–329. Cambridge, MA: MIT Press.

Choi I, Nisbett RE and Norenzayan A (1999) Causal attribution across cultures: variation and universality. *Psychological Bulletin* **125**: 47–63.

Gilbert DT (1998) Ordinary personology. In: Gilbert DT, Fiske ST and Lindzey G (eds) *The Handbook of Social Psychology*, 4th edn, vol. 2, pp. 89–150. New York, NY: McGraw-Hill.

Jones EE (1990) *Interpersonal Perception*. New York, NY: Freeman.

Krull DS (2001) On partitioning the fundamental attribution error: dispositionalism and the correspondence bias. In: Moskowitz G (ed.) *Cognitive Social Psychology*, pp. 211–227. Mahwah, NJ: Lawrence Erlbaum Associates.

Morris MW, Ames DR and Knowles E (2001) What we theorize when we theorize that we theorize: examining the 'Implicit Theory' construct from a cross-disciplinary perspective. In: Moskowitz G (ed.) *Cognitive Social Psychology*, pp. 143–161. Mahwah, NJ: Lawrence Erlbaum Associates.

Rosati AD, Knowles ED, Gopnik A *et al.* (2001) The rocky road from acts to dispositions: insights for attribution theory from developmental research on theories of mind. In: Malle BF, Moses LJ and Baldwin DA (eds) *Intentions and Intentionality: Foundations of Social Cognition*, pp. 287–303. Cambridge, MA: MIT Press.

Cultural Processes: The Latest Major Transition in Evolution

Introductory article

Eörs Szathmáry, Collegium Budapest (Institute for Advanced Study), Budapest, Hungary

CONTENTS

The origin of humans and human society is linked to the appearance of a unique mechanism for cultural inheritance, namely language. Earlier major evolutionary transitions can shed light on the latest transition.

A REVIEW OF THE MAJOR TRANSITIONS IN EVOLUTION

What Are the Major Transitions?

Table 1 lists what have recently been termed the 'major transitions in evolution'. A few remarkable features can be seen. Some major transitions in evolution (e.g., the origin of multicellular organisms or that of social animals) occurred a number of times, whereas others (e.g., the origin of the genetic code, or language) appear to have been unique events. However, one must be cautious about using the word 'unique'. Owing to the lack of 'true' phylogeny of all extinct and extant organisms, one can give it only an operational definition. If all of the extant and fossil species which possess traits arising from a particular transition share a last common ancestor after that transition, then the transition is said to be unique. Obviously it is quite possible that there had been independent 'trials', as it were, but we do not have comparative or fossil evidence for them.

Important Common Features

There are a number of sufficiently common features of the major transitions for them to require special attention, namely the emergence of an evolutionary unit at a higher level from lower-level ones, an increase in complexity, the appearance of a novel inheritance system, and the 'freezing in' of the transition (often there is no way back). The means whereby these features are achieved include local interaction, synergy, contingent irreversibility, and central control. We will look at each of these in turn, but will first briefly consider evolutionary units.

Units of Evolution

A unit of evolution must be capable of *multiplication, heredity*, and *variation* (Figure 1). If some hereditary traits influence the likelihood of survival and/or reproduction of the unit, then in *a population* of such units, evolution by natural selection can take place. Note that this definition does not refer to living systems. Many consider that viruses are not alive (e.g., they lack metabolism), but they do evolve. Some computer programs evolve in the electronic environment, and they are not regarded as alive either. In addition, there are items of culture, which are passed on from individual to individual, that also behave as evolutionary units (referred to as 'memes', by analogy with genes; see below). It is important to bear the generality of this definition in mind, as it enables us to apply Darwinian reasoning to nontrivial cases as well. The reference to population is also crucial, as individuals metabolize, reproduce, run, behave, etc., but they do not evolve. Evolution takes place in populations through the generations of evolutionary units.

Emergence of higher-level evolutionary units

The origin of the eukaryotic cell is a good example. Eukaryotic cells are much more complex than bacteria. Humans are also built of such cells. Mitochondria are the cell organelles that serve as the power plant of the cell. They are very simple, tiny structures that look like bacteria. It turns out that this resemblance is not casual, as they are indeed descended from once free-living bacteria. They became captured by an ancestor of our cells

Table 1. The major transitions in evolution

Before	*After*
Replicating molecules	Populations of molecules in protocells
Independently replicating genes	Chromosomes
RNA as gene and enzyme	DNA genes, protein enzymes
Bacterial cells (prokaryotes)	Cells with nuclei and organelles (eukaryotes)
Asexual clones	Sexual populations
Single-celled organisms	Animals, plants, and fungi
Solitary individuals	Colonies with non-reproductive castes
Prelinguistic societies	Human societies with language

Reproduced from Maynard Smith J and Szathmáry E (1999) *The Origins of Life. From the Birth of Life to the Origin of Language.* Oxford: Oxford University Press.

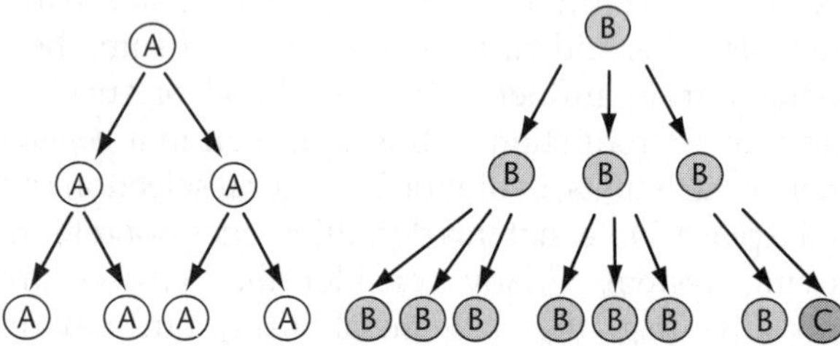

Figure 1. Units of evolution. A and B are reproducing entities of different types. Their average fecundity is shown to be different. Heredity is not exact (there is variability), as one B gives rise to a novel type C. Such units are not necessarily alive.

(around 2 billion years ago) and became enslaved for the production of ATP (the energy-storage molecule of all cells). Obviously, before this transition, the proto-eukaryote and the proto-mitochondrion were two types of unrelated, independently reproducing cell, but now they are integrated into one functional unit.

If such a transition is successful, then *adaptations*, which are discernible at the higher-level unit, evolve that suppress the competitive tendencies of the integrated lower-level units. Essentially, viewed from the lower level, a 'super-organism' is created. However, viewed from the higher level, it is just an organism.

Increase in complexity

Although natural selection does not guarantee that organisms will increase in complexity as they evolve, it is apparent that complexity along certain lineages, such as the one leading to humans, has increased during evolution. Is the number of genes in an organism's genome an appropriate measure of biological complexity? The recent flurry of completed genome sequences, including our own, suggests that this is not necessarily the case. There must be more sensible genomic measures of complexity than the mere number of genes. It is the *regulatory gene interactions* that seem to play a crucial role. In fact, *Drosophila* has more regulatory interactions than *Caenorhabditis elegans*, although mere gene numbers give the reverse order.

Figure 2 illustrates the ways in which genetic complexity can increase in evolution.

Novel inheritance systems

DNA is commonly referred to as the genetic material, with ample justification. However, there are other hereditary mechanisms. A good example is epigenetic inheritance in the cells of multicellular organisms. It is easy to see that something like that must operate in our bodies. Most animals start their lives as a fertilized egg. Cells divide and undergo differentiation in embryogenesis, so muscle, liver, nerve cells, etc. arise that look different and function differently. Some of them remain capable of proliferation. When a healthy liver cell divides, it gives rise to two liver cells – 'liver-cell-ness', as it were, is passed on. Note that the state of being a liver cell was not present in the fertilized egg, but rather it is generated in development. Thus it seems that a characteristic has been acquired and can be inherited at the level of the cell. Just so – this is a Lamarckian dimension of multicellular organisms. However, it extends very rarely from organism to organism during reproduction. To conclude, if a dual inheritance system were not active in us, we simply would not exist.

Obviously language, which is so central to our concern, is also a radically new method of information storage and retrieval. As we shall see, it has a Lamarckian component.

Local interactions

Whenever one develops a theory for a certain transition, one finds that some type of local interaction in the dynamics of the population is important. This can take several forms, and all of them are known to be important for the evolution of *cooperation*. *Reciprocal altruism* can lead to cooperation between unrelated individuals. Because of limited dispersal, cooperating individuals may remain close to each other, or they may remember past interactions with particular individuals. *Kin selection* is a mechanism in which genetic relatedness

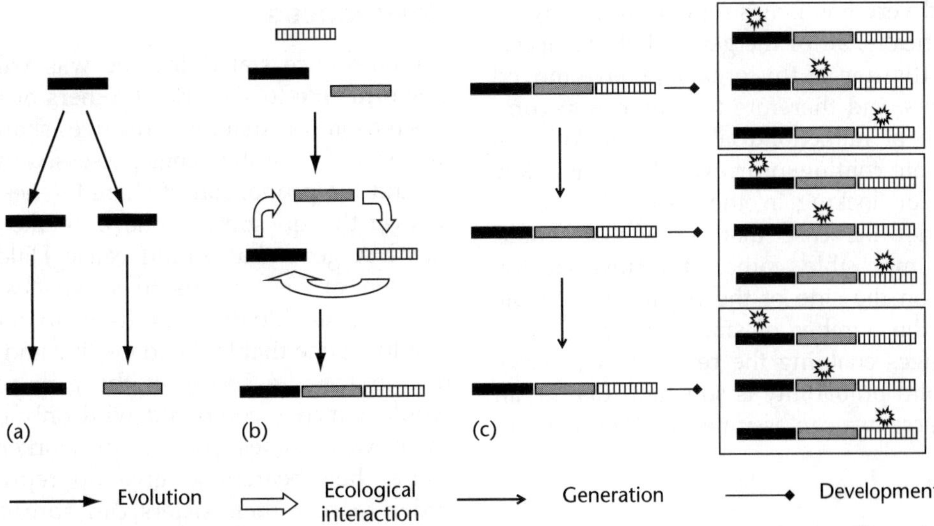

Figure 2. Means of increasing genomic complexity. (a) Duplication and divergence. (b) Symbiosis. First, independently reproducing units engage in an ecological interaction. Finally, the units cannot reproduce alone, and a new evolutionary unit has been formed. (c) Epigenesis. The genes remain the same, but they are differentially activated in the different cells. Modified from Maynard Smith J and Szathmáry E (1995) *The Major Transitions in Evolution*. Oxford: Freeman.

plays a crucial role. Hamilton's rule states that it pays to be an altruist if $br > c$, where b is the benefit received by the helped relative, c is the cost paid for by the helper, and r is the degree of genetic relatedness between them. Finally, *group selection* is a mechanism that applies when not only the individuals, but also groups formed of them, multiply and have heredity and variability. This criterion is readily satisfied when (1) the number of groups is much higher than the number of individuals per group, (2) each group is formed from one parental group only, and (3) there is no migration between groups.

The relative importance of the above mechanisms in accounting for the transitions varies from one case to another, but there is little doubt that all of them have been influential.

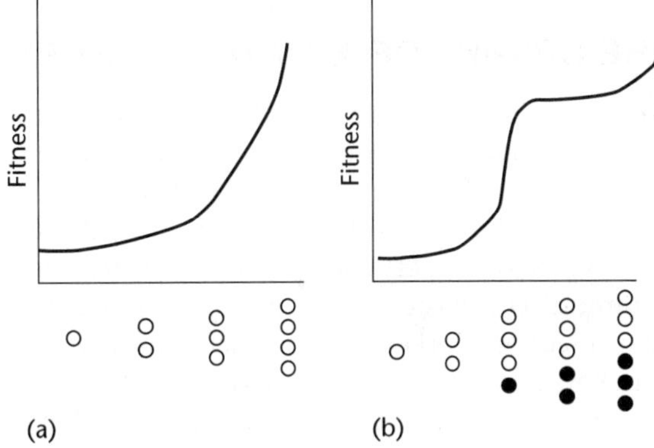

Figure 3. Types of synergy. (a) Quantitative synergy. (b) Qualitative synergy. In the latter case, combination of functions causes a very steep rise in performance, and hence fitness.

Synergy

Synergy can be both quantitative and qualitative (Figure 3). In both cases the performance (efficiency) of the interacting units increases non-linearly. In evolution, this translates into non-additive fitness interactions. In the case of qualitative synergy there are at least two types of interacting unit – they cannot substitute for each other. Cooperative guarding of the young is a good example of quantitative synergy. The interaction between different cell organelles, such as the mitochondrion and the plastid in a plant cell, is an example of qualitative synergy. *Economy of scale* and *combination of functions* are other terms used to refer to quantitative and qualitative synergy, respectively.

Contingent irreversibility

In many cases, once a transition has occurred there seems to be no way back. However, there are exceptions. For example, there are insects whose solitary state is secondary – all of their living relatives are highly social. Yet, in contrast, there is no mitochondrial cancer. This can be understood, given the fact that most mitochondrial genes had been lost in evolution, a fraction had been moved to the cell nucleus, and very few genes remain in the organelle. Emphatically, all of the genes that are necessary for the division of this organelle have moved to the nucleus, and therefore the latter is in complete control of mitochondrial division. We can thus appreciate contingent irreversibility as a key mechanism for 'locking in' the result of a transition. It is not the case that a reversal would be *logically* impossible, rather it is just far too demanding on the side of the requisite heritable variation – the number of simultaneous, chance genetic changes enabling the reversal is so large (and their joint probability is so small) that *for all practical purposes* we can assume that they will not occur.

Comparative 'Transitionology'

It is a striking feature that some transitions involved related individuals, whereas others involved unrelated ones (Table 2). This gives rise to the important distinction between 'fraternal' and 'egalitarian' types of transition. Kin selection does not work for the latter, but local interactions are crucial.

THE ORIGINS OF ANIMAL SOCIETIES

Animal societies with a complex division of labor between their members have evolved by different routes. Apart from the division of labor, and the economic advantages that result from it, the various types of society have one other feature in common. The existence of non-reproductive castes (the so-called workers) in the social insects and in some other social animals poses a formidable problem to the theory of evolution, as Darwin had already recognized. Why should worker bees give up reproduction? In what sense would this increase their fitness?

Relatedness

Haldane once stated that he was willing to lay down his life to save two brothers or ten cousins. His reason was that these relatives shared, on average, ½ and ⅛ of the genes possessed by him. Why should the proportion of shared genes matter? To answer this question, we have to take a *'gene's eye view'*. A gene that would cause Haldane to die but ten of his cousins to survive would cause more genes identical to itself to survive than would a gene that let Haldane live and his cousins die (in fact, $^{10}/_{8}$ copies of the gene, on average, would survive, compared with only one). In the same way, genes present in worker bees that cause their bearers to give up reproduction in order to rear their sisters can spread, provided that the advantage of cooperative breeding over individual reproduction is great enough. This consideration is expressed elegantly in Hamilton's inequality.

Eusociality

The degree of sociality in different species can be placed on a gradient. Most biologists are interested in what is called eusociality – 'real sociality'. By definition, eusocial animals must satisfy three criteria:

1. reproductive division of labor – that is, only some individuals reproduce;
2. an overlap of generations within the colony;
3. cooperative care of the young produced by the breeding individuals.

Eusociality is well known in ants, bees, wasps, and termites. It is less well known that a similar degree of eusociality can be observed in naked mole rats, spotted hyenas, African wild dogs, and some social spiders.

Table 2. Egalitarian and fraternal major transitions

	Egalitarian	*Fraternal*
Examples	Different molecules in compartments, chromosomes, nucleus and organelles, sex	Same molecules in compartments, organelles in the same cell, cells in individuals, individuals in colonies
Units	Unlike, nonfungible	Like, fungible
Reproductive division of labor	No	Yes
Control of conflicts	Fairness in reproduction, mutual dependence	Kinship
Initial advantage	Combination of functions	Economies of scale
Means of increase in complexity	Symbiosis	Epigenesis
Greatest hurdle	Control of conflicts	Initial advantage

Reproduced from Queller DC (1997) Cooperators since life began. *Quarterly Review of Biology* **72**: 184–188.

Super-organisms

Colonies of social animals can with some justification be regarded as 'super-organisms', in the sense that they display adaptations (traits that increase fitness) at the colony level. For example, the mound built by termites has a system of air channels that function as an air-conditioning system. By this analogy, the queen and the reproductive males are analogous to the germ line of multicellular organisms, and the non-reproductive individuals would be analogous to the soma of the super-organism.

THE ORIGINS OF HUMANS AND HUMAN SOCIETIES

Characteristics of Human Societies

Relatedness and individual recognition

Despite the obvious similarities between a termite mound and a human city, there are profound differences between the mechanisms that lead to cooperation in the two cases. One important feature of human societies, namely the recognition of individuals, already exists in some social mammals and birds. Although insects may recognize group membership, they do not recognize individuals. In contrast, a monkey recognizes other members of its troop as individuals, and behaves differently towards them. As the phrase 'pecking order' implies, the members of a flock of chickens sort themselves into a linear dominance hierarchy, and this probably requires individual recognition. Those who have studied baboons and other monkeys have observed the formation of alliances in which two or more individuals support one another in conflicts with other members of the group. Such alliances may be based on genetic relatedness, but this is not always so. The essential points are (a) that, in higher animals, social interactions within a group depend on individual recognition, and (b) that one individual's behavior towards another depends both on genetic relatedness and on a memory of previous interactions with that individual.

Cultural inheritance

It is often said that the defining characteristic of human societies is cultural inheritance – that is, individuals in a society acquire their beliefs and behavior, and their knowledge and skills, by learning from previous generations, and not by genetic inheritance. There is obviously much truth in this idea, particularly with regard to the differences between one individual and another, or between one society and another. At the level of the individual, differences in political opinions are not caused by differences between genes. Having said this, however, there are some reservations that need to be stated. First, there is some cultural inheritance in animals, a fact that is important when considering the origins of human culture (Table 3). Secondly, the ability of humans to learn, and to build societies that are dependent on cultural transmission, is genetic – human societies differ from chimpanzee societies because humans and chimps differ genetically. Thirdly, humans learn some things more readily than others – the human mind is not a blank slate upon which experience can write what it will.

Social learning

Different types of social learning are at the heart of cultural transmission (Table 4). Young rats can acquire a preference for a new food by smelling it on the coat of other rats. This is a type of cultural inheritance – two groups of rats feed on different foods, and the difference is transmitted by learning. This mechanism has been called *stimulus enhancement*, whereby the adults create an environment in which it is easier for the young to learn. This contrasts with *observational learning*, in which one animal watches what another is doing, and then copies it. It is difficult to believe that all culturally inherited traits in animals depend only on local enhancement. For example, in some areas of

Table 3. Criteria for culture

Invention of a new pattern, or modification of an existing pattern
Transmission from innovator to another
Consistent copying of pattern (often stylized)
Long-term persistence of pattern in the acquirer
Spread of pattern across social units
Pattern enduring across generations

Table 4. Examples of social learning

Type of learning	*Description*
Stimulus enhancement	B learns from A where to orient behavior
Observational learning	B learns to what circumstances a behavior should be a response
Imitation	B learns from A some part of the form of a behavior
Goal emulation	B learns from A the goal of an action

Greece, golden eagles feed mainly on tortoises. The bird is unable to break open the shell with its beak, so it picks up a tortoise, flies up to a considerable height, and then drops the tortoise on to the rocks below, thus breaking the shell. It would be absurd to suggest that in Greece, but nowhere else, this behavior in eagles is genetically programmed. A young bird could learn by local enhancement that tortoise shells contain meat that is good to eat, but how – other than by copying – could they learn to fly upward carrying a tortoise, and then drop it? A second example, involving chimpanzees, is given below.

The distinction between stimulus enhancement and observational learning is important, because only observational learning can lead to cumulative cultural change, which is the characteristic feature of human history. By observational learning, young individuals can learn from adults, but also, if one individual stumbles upon an improved way of doing something, that improvement can be copied. The result is that change can be continuing rather than occasional, and that an individual can learn, by copying, a skill that it could never have learned on its own.

Chimpanzee culture

It is clear that humans depend on observational learning, reinforced by teaching, including verbal instruction. As the above example of golden eagles shows, observational learning is not unknown in animals. There are examples in chimpanzees. Some (but not all) populations of chimps dip sticks into the nests of driver ants, and feed on the ants that crawl up the sticks. The chimps in Gombe use a different technique to those in Tai, and catch about four times as many ants per minute. Local enhancement could explain why some individuals in one population dip for ants, whereas others do not. However, it cannot explain why Tai chimps continue to use an inefficient technique, when there is no reason why they should not adopt a more efficient one. Yet continued use of an inefficient technique is what we would expect if young chimps copy their elders.

Language and culture

If higher animals are at least sometimes able to copy their elders, why is it that continuous cultural change does not occur among them, as it does in humans? Thus one population of chimps may differ from another for cultural reasons, but a given population is not continuously acquiring new habits. The likely explanation is that, in humans, the main mechanism whereby culture is transmitted is language. The nature and origin of language are discussed elsewhere. At the risk of repetition, two conclusions will be reiterated here, namely the close analogy between genetic and linguistic methods of transmitting information, and the implications of linguistics for the modular nature of the human mind.

From Ape to Human

Social groups

All Old World monkeys and apes live in social groups, with the single exception of the orangutan. Figure 4 shows a reconstructed phylogeny, or ancestral tree, of these animals. In Old World monkeys, females remain in the social group in which they were born, whereas males leave their natal group before sexual maturity, and must enter another group in order to breed. They are said to be 'female kin-bonded'. In chimpanzees, the situation is reversed – that is, males remain in their natal groups and females move. Other hominoids vary in their social systems, but none is female kin-bonded. The most parsimonious assumption is that male kin-bonding originated in the common

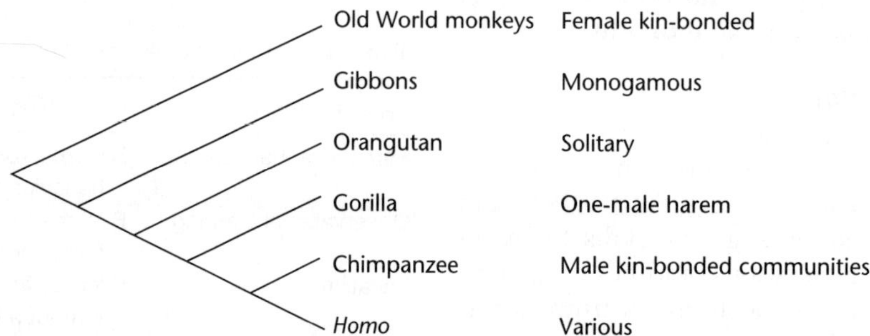

Figure 4. Primate phylogeny and social structure. Reproduced from Foley RA (1996) An evolutionary and chronological framework for human social behaviour. *Proceedings of the British Academy* **88**: 95–117.

ancestor of humans and chimps, since they are more closely related than either is to the gorilla. If this was so, then male kin-bonding is the ancestral condition for hominids. The social systems of modern humans are so varied that it is difficult to be sure whether this conclusion is correct, but it is the best we can do on the basis of the comparative evidence available.

Fossil records

The fossil record provides a second source of information about human origins (Figure 5). It is illuminating to compare this record with what is known of human technical achievements, if only because of the puzzles that the comparison raises. The australopithecines were bipedal and lived in open country. Their relative brain size was only slightly larger than that of the apes, and their tool kit was limited and uninventive. In the lineage from *Australopithecus* through *Homo habilis* and *Homo erectus* there was a gradual increase in brain size, but relatively little technical innovation. The most advanced tool used by *H. erectus* was the handaxe, made from a single block of stone worked on both surfaces, and symmetrical in shape. Such handaxes first appeared around 1.4 million years ago, and persisted almost unchanged for over a million years – hardly an example of cumulative cultural change.

Brain size

Although it seems that there was substantial brain evolution in *H. erectus*, the most rapid increase in relative brain size has occurred during the last

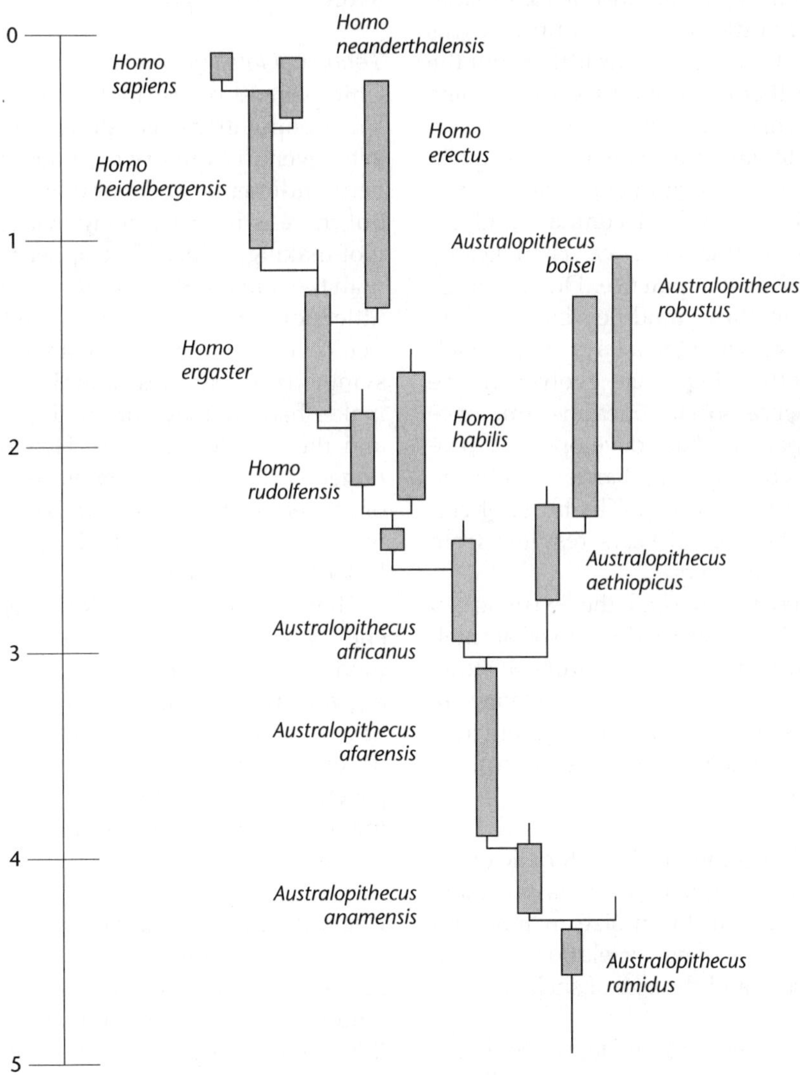

Figure 5. Fossil record of human origins.

300 000 years, culminating in the appearance of effectively modern humans about 100 000 years ago. Yet the acceleration in human technical inventiveness, with the appearance of a varied range of tools made of stone, bone and antler, dates back only 40 000–50 000 years. Burial of the dead, art in the form of cave paintings and musical instruments, personal adornment, and trade originated at much the same time. From around 40 000 years ago we are faced with evidence of continuing cultural innovation. This raises several questions. Why was there a delay of 50 000 years between the appearance of the first anatomically modern humans and the technological revolution? What selective force was responsible for the accelerated increase in brain size 300 000 years ago? When and why did language as we know it originate?

Modules

The problems are difficult, because a fossil skull can tell us relatively little about the brain that was once inside it, and stone tools tell us little about the society that made them. We may try to combine palaeontology, archaeology and psychology in order to obtain a tentative answer to these questions. The essence of this argument is as follows. The human mind does indeed contain modules that are adapted to particular tasks, as suggested by studies of linguistic competence. During much of human evolution these modules increased in efficiency, but they remained to a large degree isolated from one another. Language evolved in the first instance to serve social functions, but once grammatical competence had developed, it provided a means whereby the barriers between modules could be broken down. The burst of creativity during the last 50 000 years resulted from the breaking of these barriers.

Some authors have suggested the existence of three mental modules, concerned with social intelligence, technical intelligence, and natural history, respectively – that is, with the knowledge of animals and plants that is necessary for efficient foraging. We now look at each of these in turn.

Social intelligence

Social intelligence is a common characteristic of the primates. It has been argued that it is the main reason for the increase in brain size in monkeys and apes, as there is a striking correlation between brain size in a species, and the size of social groups in that species.

A crucial question concerns the degree to which apes and monkeys have what has been called a 'theory of mind'. To have a theory of mind is to be able to ascribe to others the possession of a mind like one's own, with similar desires and powers of reasoning. There is no convincing evidence that monkeys have such an ability. For example, a vervet monkey gives a different alarm call if it sees an eagle, a snake or a leopard, but it seems that the monkey does not have in mind the knowledge that another monkey may hear its call and respond appropriately. For example, a monkey may continue to call after all of the others have responded. However, many who have studied the social behavior of chimpanzees, and their skill in manipulation and deceit, are convinced that they do indeed have a theory of mind.

We can conclude that selection for social intelligence was a major cause of the increase in brain size in monkeys, apes, and humans, and that a theory of mind was likely to be present in the common ancestor of chimps and humans, around 5 million years ago.

Technical intelligence

Chimpanzees do use tools in the wild. For example, some populations use stones to crack nuts. However, even in captivity their ability to make tools is very rudimentary. Australopithecines used tools, but there is no convincing evidence for deliberate tool-making, which first appears to be associated with the remains of *H. habilis*, although the tools are little more than irregular chipped stones. *H. erectus* marks a clear advance, with the manufacture of symmetrical handaxes, indicating that the toolmaker had an image of the desired result in mind, and the skill to realize it. However, as was mentioned earlier, there remains an astonishing degree of conservatism. Thus there is evidence of a limited increase in technical intelligence, combined with a lack of inventiveness.

There is also evidence for a degree of independence of social and technical intelligence, even in modern humans. For example, researchers studying autism have suggested that autistic children have impaired understanding of the behavior of other humans (what has been called 'folk psychology') but better than average understanding of the behavior of inanimate objects ('folk physics').

Understanding of natural history

There was obviously selection for improved foraging skills, and hence for knowledge of the distribution and behavior of animals and plants. But was this achieved by an increase in general-purpose intelligence, or by the evolution of a specialized natural history module?

In favor of the latter, it has been argued that all human societies share certain ideas about the living world. First, all living things belong to one – and only one – 'natural kind'. An animal is a dog, or a cat, or a badger, and so on – it must belong to a particular 'species', it cannot belong to two, or to none, and it cannot change its species. Secondly, all human societies share the idea that natural kinds can be classified hierarchically into higher taxa. For example, a dog is a flesh-eater, a mammal (as opposed to a fish, reptile, etc.) and an animal (i.e., not a plant). These universal human attitudes to living things may reflect an innate predisposition.

The alternative is that they could be universally believed because they are true, or almost so, and would be learned by any human society to which knowledge of the living world was important. A second argument in favor of a special natural history module is the speed with which children acquire these beliefs. However, as yet the case for a special module is not decisive.

Coupling

The argument, then, is that the increase in human brain size prior to the emergence of modern humans around 100 000 years ago was associated with an increase in social, technical, and natural history skills, but that these abilities were to a large degree independent. Perhaps the competence in language, including grammar, also evolved during this period, although precise dating is obviously difficult. We may ascribe the cultural explosion that began around 50 000 years ago, and which has led to continuous and cumulative cultural change, to a breakdown of the isolation between mental modules as a result of the emergence of language. The essential point is that once words exist for social, technical, and living things, the same grammar can be used to say things about them.

THE ORIGINS OF LANGUAGE

Writing is much more recent than language. This is really unfortunate, because it is only by writing that language could have become 'fossilized'. We must therefore resort to comparative analyses in biology and linguistics, as well as to building theoretical models. Comparative analysis is also limited, because language is a uniquely human phenomenon. It is undoubtedly an adaptation, and a very complex one at that. Thus it is very unlikely to have arisen as a mere by-product of anything else, without considerable evolutionary fine-tuning by natural selection.

Generative Mechanisms

It is not the aim here to give a thorough description of language with all of its known components. Rather, this section will highlight some important features. First, the number of sounds, words, and grammatical rules that we use in any human language is finite. With these finite means we can potentially cover an indefinitely large domain of grammatically correct, possible sentences.

Symbols

Our vocabulary is finite but open-ended. Without the latter feature, cultural evolution would be impossible. Words are usually highly symbolic signs – they stand at the abstract end of the object–concept–sign triple. Because there is no *immediate* link between object and word, we can have words for purely imaginative concepts, such as *unicorn*.

The capacity for symbolic communication in overt or covert forms seems to be present in some other species as well as humans. For example, bottle-nosed dolphins, chimpanzees, and gray parrots are able to master protolanguage, defined bluntly as word use without grammar. Children under 2 years of age are also roughly at this level when they speak. This already suggests that the hardest nut to crack when contemplating the evolution of the language faculty is syntax.

Grammar

The syntax of every known human language can be characterized by a finite set of grammatical rules. By the application of these rules, all possible grammatically correct sentences can be generated – hence the term *generative grammar*. We are usually unaware of these rules, but we learn our language surprisingly fast, despite the fact that randomly assembled words hardly ever result in grammatical utterances. Our brain seems to be specially tuned, and genetically predisposed, to fast language learning – in short, we have an 'instinct' for language.

Biological Foundations

Genetics

Chimpanzees do not share the language instinct. This difference must ultimately be traced back to genetics. However, genetics of humans is notoriously difficult, because you cannot 'breed' humans as you can, say, breed fruitflies for genetic experiments. Instead, one must identify familial linguistic problems in the existing medical record. Such a syndrome, or rather a collection of them, has been described and is called *specific language impairment* (SLI). It is specific because, at least for certain individuals, other cognitive deficits are not apparent. Some authors claim that a subgroup of affected individuals has something even more special, known as *grammatical SLI*. It is indeed true that in some such people only aspects of grammar are affected.

In some cases we have the genetic description of this syndrome. In an English-speaking family the problems are associated with a single autosomal dominant allele. It is very unlikely that we shall find many such genes. In contrast, most of the genes that affect the language faculty are likely to be *liability genes* – that is, other things being equal, these genes increase the probability that we will have a normal language competence.

Our remarkable ability to learn a very complex system so fast, and the hints at a specific genetic predisposition, appear to be consistent with the idea of a *language organ* – that is, a genetically determined module in the brain that processes linguistic information. Obviously this organ must be absent from chimpanzees, and must have evolved somehow during the last 5 million years.

Quest for a 'language organ'

However, there is a snag. From what we know of the brain, the language organ cannot be a macroscopically distinct anatomical unit. The fact that neural localization of language can be plastic is now widely known. Studies of brain injury have revealed that damage to the left hemisphere which occurs before a critical period is not lifelong, as the right hemisphere can take over the necessary functions. This does not contradict the finding that in normal individuals Broca's area does appear to be specialized for syntax. It thus seems that the common left-hemisphere localization of language is just the *most likely* outcome when there is no genetic or epigenetic disturbance. Non-invasive brain studies have revealed a truly shocking feature of language development, namely the localization of linguistic processing shifts during normal ontogenesis. The outcome in 'normal' individuals is also highly variable.

The following conclusions can be drawn from brain studies.

- Localization of language is not entirely genetically determined, as even large injuries can be tolerated before a critical period.
- Language localization to certain brain areas is a highly plastic process, both in its development and in its end result.
- It does seem that a surprisingly large part of the brain can sustain language. There are (traditionally recognized) areas that seem to be most commonly associated with language, but they are by no means exclusive, either at the individual level or at the population level, during either normal or impaired ontogenesis.
- Whereas a large part of the human brain can sustain language, no such region exists in apes.

Evolution

It simply cannot be the case that the language faculty originated by a single crucial mutation. This also suggests that linguistic performance itself has evolved. Unfortunately, we can be only hypothetical at the moment. Figure 6 illustrates a possible scenario. Note that, according to this flowchart, components of language have co-evolved and – by inference – the underlying neural networks must also have co-evolved. A much deeper understanding of the brain during processing of linguistic information will be needed before we can establish whether this scenario – or any alternative – is nearly correct or not.

THE USES OF LANGUAGE

Language and Society

It is impossible to imagine our society without language. The society in which we live, day and night, depends on it. Even as we sleep, information about us is being stored and maybe processed. Imagine that we apply for a job on the other side of the world. We are confident, or at least we hope, that our application will be fairly treated, and that the country to which we would like to move is running properly – that is, that social contracts are observed. Our lives depend on the social division of labor and on detailed social contracts which could not exist without language. No ape or dolphin could comprehend, even in spoken form, a contract for a job.

Both the genetic and linguistic systems are able to transmit an infinitely large number of messages by the linear sequence of a small number of distinct units. In genetics, the sequence of four bases enables the specification of a large number of proteins, which in turn, by their interactions, can specify an indefinitely large number of morphologies. In language, the sequence of some 30 to 40 distinct unit sounds, or phonemes, specifies a large number of words, and the arrangement of these words in grammatical sentences can convey an infinitely large number of meanings.

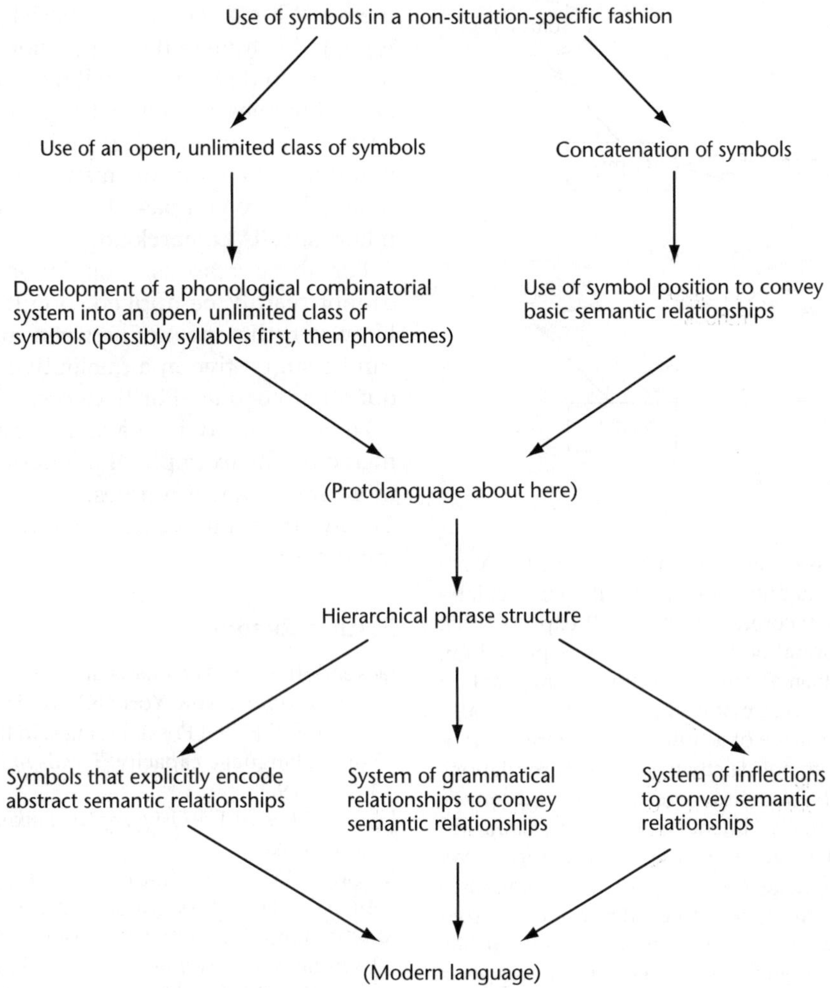

Figure 6. Scenario for the evolution of language. Reproduced from Jackendoff R (1999) Possible stages in the evolution of the human language capacity. *Trends in Cognitive Sciences* **3**: 272–279.

Language for Internal Representation

Language is not only for communication – it is also a means for powerful internal representation. If a person says to him- or herself 'I saw two leopards climbing up that tree; only one has come down, so it's better for me to stay away from that tree until the other one comes down, too', then this reasoning requires syntax that is functionally equivalent to syntax of natural language, even if it is not explicitly 'spoken' in one's head (although frequently it is). Some authors call such an internal language, that serves internal representation, *mentalese*. This emphasizes the idea that language, but not necessarily the spoken one, is a tool for thought as well.

Memes and Cultural Evolution

Richard Dawkins has emphasized the analogy between genetic and linguistic transmission, introducing the concept of a 'meme' – that is, the unit of cultural inheritance analogous to a gene. A meme, he argues, is a replicator. If we invent and tell you a limerick, you may tell it to your friends, and they may tell it to theirs. In this way a single original entity – the representation of the limerick in my brain – has replicated, as a gene might replicate. Clearly, there is room for selection. For example, if we invent a funny limerick, it is more likely to replicate than if we invent a boring one. Of course, whether a meme will replicate or fail depends on the nature of the human mind, and on the cultural milieu (i.e., on the other memes present in the population). However, the same is true of a gene – its increase depends on the environment and on what other genes are present.

There are of course differences. Genes are transmitted from parent to offspring, whereas memes can be transmitted horizontally, or even from offspring to parent. Yet there is a deeper difference

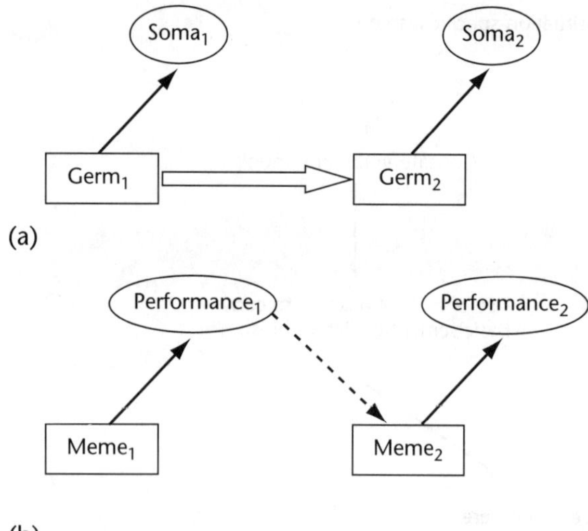

Figure 7. Transmission of genes and memes. (a) A so-called Weismann diagram, named after the famous late-nineteenth-century theoretical biologist. It expresses the idea that only information in the gametes is passed on, whereas the body (soma) cannot transmit its information by genetic means. Weismann thought that this also precludes the inheritance of acquired characteristics (Lamarckism). Although not strictly true, other formulations analogous to this diagram are more accurate and convey the message originally intended by Weismann. (b) Memes are passed on from the performance level ('phenotype'). Language itself is built up analogously from generation to generation. Reproduced from Szathmáry E (2002) Units of evolution and units of life. In: Pályi G, Zucchi L and Caglioti (eds) *Fundamentals of Life*, pp. 181–195. Paris, France: Elsevier.

between genes and memes. Genes specify structures or behaviors (i.e. phenotypes) during development. In inheritance, the phenotype dies and only the genotype is transmitted. The transmission of memes is quite different. A meme is in effect a phenotype – the analogue of the genotype is the neural structure in the brain that specifies that meme. When I tell you a limerick, it is the phenotype that is transmitted – I do not pass you a piece of my brain (Figure 7). It follows that, in the inheritance of memes but not that of genes, acquired characters can be inherited. If I tell you a limerick and you think of an improvement to it, you can incorporate it before you pass it on. In this sense, cultural inheritance is Lamarckian.

For these reasons, population genetic theory cannot readily be applied to cultural inheritance. However, the analogy between memes and genes can be suggestive in a qualitative sense if not in a quantitative one. Furthermore, although for the sake of simplicity we have illustrated the idea of a meme by the example of a limerick, it can refer to more important examples, such as a belief in the Trinity, or a knowledge of how to manufacture gunpowder.

Further Reading

Jackendoff R (1994) *Patterns in the Mind: Language and Human Nature*. New York, NY: Basic Books.

Jackendoff R (1999) Possible stages in the evolution of the human language capacity. *Trends in Cognitive Sciences* **3**: 272–279.

Leakey R (1995) *The Origin of Humankind*. New York, NY: Basic Books.

McGrew WC (1998) Culture in nonhuman primates? *Annual Review of Anthropology* **27**: 301–328.

Maynard Smith J and Szathmáry E (1995) *The Major Transitions in Evolution*. Oxford, UK: Freeman.

Maynard Smith J and Szathmáry E (1999) *The Origins of Life. From the Birth of Life to the Origin of Language*. Oxford, UK: Oxford University Press.

Mithen S (1996) *The Prehistory of the Mind: A Search for the Origins of Art, Religion and Science*. London, UK: Thames & Hudson.

Pinker S (1994) *The Language Instinct*. New York, UK: William Morrow.

Runciman WG, Maynard Smith J and Dunbar RIM (eds) (1996) *Evolution of Social Behaviour Patterns in Primates and Man*. Oxford, UK: Oxford University Press.

Cultural Psychology

Introductory article

Janxin Leu, University of Michigan, Ann Arbor, Michigan, USA
Nicole S Berry, University of Michigan, Ann Arbor, Michigan, USA
Lawrence A Hirschfeld, University of Michigan, Ann Arbor, Michigan, USA

CONTENTS

Cultural psychology is an interdisciplinary program of research that explores the relationship between individual minds and the complex environments in which they are deployed. The approach focuses on the contribution that content-rich, complex environments – ranging from workplaces to cultural traditions to nation states – make in shaping basic cognitive processes.

INTRODUCTION

Do cognitive theories predict the mental and behavioral processes of humans in all societies and communities? Is a scientific study of cultural variations in meaning, perceiving, thinking, and feeling across groups of people part of cognitive science? If the answer to the first question is 'no', and the second 'yes', how should cognitive science approach cultural differences? Cultural psychology is a loose amalgam of scholarship in psychology, anthropology, and linguistics that explores these questions. In broad terms, cultural psychology makes the empirical claims that (1) there exist substantial group differences in cognition, emotion, and motivation that cannot be understood without the study of cultural life; (2) aspects of cognition are extra-individual and need to be studied at the population level; and (3) differences in the content of thought result in differences in psychological processes in some, but not all, areas of cognition.

Cognitive scientists typically approach cognition by examining the processes that take place in individual minds or brains. Research in this vein has ordinarily assumed that the processes observed are primary. Higher-level phenomena such as categorization, perception, language use, and the capacity for culture are all supposed to be built upon universal, individual, cognitive processes. For these reasons, the content chosen to test cognitive processes is frequently assumed to be transparent. Stated differently, content is not considered to affect the process under examination.

In contrast, a central argument of much cultural psychological research is that content and process mutually constrain cognition. Myriad scholars from Darwin onwards have proposed that the human mind evolved to solve social problems. Therefore, it is not only plausible but probable that the brain is extremely sensitive to social context. Stated in these terms, much of cultural psychological research can be viewed as an attempt to isolate the effects of social context on cognition. Nevertheless, exactly what constitutes 'social context' is still debatable and evident in the three main methods used in cultural psychology: the comparative, situated/distributed, and epidemiological approaches.

The *comparative approach*, or cross-cultural approach, uses experimental methods to compare and contrast groups in their performance on a range of psychological tasks with the objective of demonstrating cultural difference and commonality on some psychological aspect in the lab and field. For example, Richard Nisbett and colleagues have shown differences among North Americans and members of various East Asian societies in their performance on tasks ranging from perception to categorization. Americans tend to perceive and recognize an object in isolation from its surrounding field and to categorize an object using its traits. In contrast, subjects from several East Asian societies tend to perceive and recognize an object in relation to its surrounding field and to categorize it according to its relationship with other objects. Nisbett and colleagues argue that these differences arise out of socialization in the home, school, and community that emphasize individuality and agency on the one hand, or harmony and compromise on the other.

The *situated/distributed approach* focuses primarily on cognition in specific, typically coordinated, contexts characterized by common sets of artifacts and habitual activities. This approach uses the distribution and locations of activities and their relation to the material environment as the units of analysis and has the advantage of examining how thought emerges as part of a system of local social practices in a continually evolving process of socialization. For example, Jean Lave studied arithmetic activity in grocery shopping using the supermarket as the arena for cognitive activity. She demonstrated that shoppers who made frequent arithmetic errors in formal testing situations did not make similar errors in best-buy problems in the supermarket. In a situated method, context is conceived as a relation between actors and the setting in which they act, as opposed to a singular 'cultural' unit determining behavior. The situated/distributed approach therefore de-emphasizes the independent study of the innate characteristics of individuals and the particular properties of the context in which action occurs. Instead, it focuses on everyday practices, the locus of interaction between a person and the environment as paramount to understanding cognition.

The *epidemiological approach* explores the processes that underlie the distribution of both mental and public representations, especially those representations that are widespread and enduring. A central question in the epidemiological approach is why some representations, such as myths and folk tales, are stable in a population, while others, such as gossip and rumors, are not. Dan Sperber, who developed the approach, argues that cultural beliefs, including apparently irrational ones, that resonate with common sense are more easily learned and remembered and hence are more likely to become widespread and enduring. In an empirical application of this proposal, Lawrence Hirschfeld has shown that underlying, and substantially constraining, historically recent and socially constructed concepts like 'race' and 'ethnicity' is a special-purpose knowledge structure treating input relevant to collectivities in the socio-political environment.

While the comparative, situated/distributed, and epidemiological approaches make transparent how group life and cultural content (i.e., beliefs, attitudes, domains of knowledge) shape psychological process (i.e., thinking, feeling, perceiving), they differ in how they highlight context. For example, the situated/distributed approach demonstrates not only context-specificity in thinking across everyday events, such as problem solving in the classroom versus the supermarket, but flexibility across everyday actions. The locus of motivating cognitive performance is in the dynamic rituals of daily habits. On the other hand, a comparative or cross-cultural approach centers the discussion of cognition-in-context on cognition in groups that have a set of shared, identifiable, and fairly stable practices. Comparative psychologists complain that the situated/distributed approach has lost sight of psychology and individual minds. Situated/distributed cognitive psychologists argue that comparative approaches mistake groups for daily contexts and fail to account for dynamics in cultural systems. Advocates of the epidemiological approach, in turn, caution that psychologists working in the comparative tradition often overestimate the coherence, stability, and boundedness of culture, hence misattributing causal properties to a fairly fluid environment. They similarly observe that the situated/distributed approach, while intriguing, has no principled way to delimit the context in which a task is situated or distributed.

COMPARATIVE STUDIES OF COGNITION

Language

Cultural psychologists are not the first to suggest that language, culture, and cognition are inextricably linked. Benjamin Lee Whorf and Edward Sapir each proposed, in what became known as the Whorf–Sapir hypothesis, that the cognition of a member of a particular culture is constrained by what can be said in the language the person speaks. For example, Whorf famously proposed that the Hopi have 'no general notion or intuition of TIME as a smooth flowing continuum … [because the Hopi language contains] no words, grammatical forms, constructions or expressions that refer directly to what we call "time," or to past, present, or future'. While this strong version of linguistic relativity is no longer seriously entertained, a number of scholars have explored more modest versions.

Counting

J. A. Lucy, in studies involving English and Yucatec Mayan speakers, has demonstrated that variations in grammatical form governing counting can influence categorization. Yucatec requires the speaker to use both numeric modifiers and classifiers when counting: whereas an English speaker counts, 'one candle, two candles', a Yucatec speaker counts, 'one, long-thin wax, two, long-thin wax'. Lucy contends that this grammatical difference accounts for differences in English and Yucatec Mayan-speaking participants' similarity judgments when asked to put together objects which could be sorted either by shape or substance. Yucatec speakers grouped substances together based on their construction from like substances, while English speakers grouped objects together based on shape.

Noun extension

Linguistic influences on the nature of categories were similarly demonstrated among English and Japanese speakers: children aged two, two-and-a-half, and four, and adults. As in Yucatec, nouns are not pluralized in Japanese. In an experimental procedure similar to the one described above, American children as early as two years of age extended a noun label to other objects based on shape, whereas Japanese children extended the label based on substance.

Early differences in language socialization and pragmatics may underlie linguistic variation in categorization. For example, Fernald and Morikawa compared Japanese and American mother–infant interactions at home in a study of lexical development. Infants were preverbal: aged six, 12, and 19 months. They found that American mothers labeled objects more often than Japanese mothers. While American mothers reported trying to direct their children's attention to objects and teaching them the proper noun labels, Japanese mothers reported emphasizing polite exchanges of the objects with the infant in fostering social exchange. Comparisons of American mothers with Korean and Chinese provide further evidence of socialization practices that differentially emphasize target objects and noun labels on the one hand, and social interactions and substance on the other.

Categorization

There has been a long tradition in both psychology and anthropology of trying to understand the origins of the categories humans use. Anthropologists long assumed that the systematic variation in lexicalized and covert categories reflected systematic variations in cognition. Cognitive psychologists long assumed that the representations of and computations over contrived unfamiliar categories reflected universal cognitive processes. Pioneering studies by Brent Berlin and Paul Kay, in anthropology, and Elinor Rosch, in psychology, refuted both assumptions.

Previous to Berlin and Kay's research, anthropologists assumed that each culture arbitrarily divided and named different colors on the color spectrum. After surveying color terms from over 80 different languages, both universal and culture-specific patterns were observed. Not all languages identified the same number of basic color terms. The number of color terms in any given language, however, was found to be a function of a closely organized system (in a comprehensive survey of languages Berlin and Kay found that only 11 of over 2000 combinations of basic color terms are actually used). Importantly, although Berlin and Kay found that there was great variation in the boundaries of color terms across (and within) languages, speakers of all languages converged on the same focal colors which best represented the color category.

Eleanor Rosch demonstrated that natural language categories are organized differently from contrived categories. As Berlin and Kay revealed for color terms, natural language categories are internally structured such that (a) some category members are better examples of the category than others, and (b) these prototypical members are linked to other category memberships by a relation of family resemblance, not common necessary and sufficient features.

Recently, it has been demonstrated that exemplar use can also vary across cultures, as can the principles around which objects are organized. For example, in one study, Chinese and American college students were asked to judge similarity among objects in which grouping by category (e.g., notebook and magazine) or relationship (e.g., pencil and notebook) was possible. Chinese were more likely to group objects by relationship and to justify their decision by invoking relationships, whereas Americans were more likely to group objects by shared categories and to refer to category membership as the explanation.

Reasoning

Cultural psychologists have demonstrated group differences in inductive and deductive reasoning. These differences have been explained as differences in expertise, as well as differences in intellectual traditions and resulting styles of reasoning.

Inductive reasoning

Scott Atran and Douglas Medin have carried out experiments comparing conceptualization of and reasoning about living beings across different cultures to demonstrate both universals and cultural specificity in inductive reasoning. In support of universal principles of folk biological categorization, Atran and Medin found that undergraduates from two Midwestern American universities and Itzaj Maya of Guatemala grouped different animals in ways that reliably correlated with scientific taxonomies. These groupings, organized around taxonomic rank, strongly influenced reasoning: across both groups, one taxonomic level (the folk-generic) was found to be the most inferentially rich despite differences in which level is most 'psychologically' basic in the sense identified by Rosch and her colleagues.

Having shown commonalties in the conceptual domain between the undergraduates and Itzaj Maya, Atran and Medin also examined specificity in category-based induction about living beings. For example, Atran and Medin asked American students and Itzaj Maya to judge which of two arguments is stronger: (1) all mammals will share a trait if it is known that two diverse mammals (e.g., a mouse and a jaguar) share the trait, or (2) all mammals will share a trait is if is known that two similar mammals (e.g., a jaguar and a leopard) share the trait. American students reasoned that (1) is the stronger argument, whereas Itzaj Maya reasoned that (2) is the stronger argument. Atran and Medin speculated that the Itzaj based their judgments on what they knew about the behavior and ecology of the species in the argument rather than reasoning from taxonomic positions. Choi, Nisbett, and Smith also found a cultural effect in inductions based on diversity. Koreans are less likely than their American counterparts to feel that increasing diversity of two given examples strengthens inductions about the higher category of which both examples are members. Nevertheless, when the higher category is made more salient to the Korean subjects through a manipulation, they perform the same as their American counterparts.

Deductive reasoning

Research also demonstrates cultural differences in how people use formal logic versus experiential knowledge in reasoning based on typicality. Sloman demonstrated that the typicality of an exemplar (e.g., eagles versus penguins as birds) influences the persuasiveness of deductive arguments for all subjects (a claim about all birds). Norenzaya found that the effect is stronger among Koreans than European Americans, with Asian American scores falling between the two populations. Further, Koreans, unlike their American counterparts, were less likely to dismiss conclusions as implausible based solely on logic. Rather, they more frequently used empirically derived beliefs about the world than logic to judge the plausibility of the conclusions.

Kaiping Peng and Richard Nisbett suggest that unlike Americans, subjects from several East Asian societies tolerate contradiction and are not as concerned with demands of consistency. Chinese and American students were compared in their preference for argument form and judgments about contradictory propositions. Whereas the Americans preferred logical arguments (i.e., God exists because there had to be a first cause), the Chinese preferred 'dialectical' arguments (i.e., God exists because there must be a truth that rises above all individual perspectives). Likewise, whereas the Americans rejected one proposition in face of a contradicting proposition, the Chinese embraced both propositions, finding merit in the middle ground.

Rather than concluding that individuals from these East Asian societies apply faulty logic, this work suggests that distinct, fully integrated systems of reasoning may exist among different groups. Nisbett and colleagues argue that American participants tend to prefer abstract logic over complex experience in highlighting how the world works. They stress the parts of a system that are coherent and noncontradictory, marginalizing and trying to resolve inconsistencies. It is argued that East Asian participants, on the other hand, tend to rely on empirical experience over logical models to understand their environment, which they view as constantly changing and inextricably interconnected. Within this framework, contradictions and inconsistencies are prevalent, natural, and even necessary.

Moral reasoning

Although a topic that has not traditionally attracted great attention in the cognitive sciences, the interest in evolutionary psychology has brought issues of morality to the forefront. This interest flows from morality's role in explanations of adaptations to social life – including the necessity to attend to and conceptually represent increasingly larger, more disperse, and more internally complex social groupings and the need to track exchange and acts of altruism. Morality is particularly salient in setting out, stabilizing, and justifying sanctions that underlie systems of complex social control.

Developmental psychologists have identified correlates of this salience in children's early reasoning, specifically about the differences between transgressions of social conventions and moral transgressions. Turiel and his colleagues have proposed that children spontaneously employ a universal moral appraisal involving expectations about harm and welfare. They provide experimental evidence that children everywhere discriminate among social conventions – which they believe vary readily from culture to culture and situation to situation, from moral transgressions – which they identify as harming others and believe to be wrong under any system of thought or in any situation.

For example, Smetana *et al.* presented American preschool children with accounts of both hypothetical and real transgressions. The children were asked to rate how 'bad' each was, whether it would be as equally bad if it occurred elsewhere, and whether it deserved punishment. Some were transgressions of social convention ('this child is not saying "please" when asking for something'); others were moral transgressions ('this child is hitting this child'). Even three-year-olds judged moral transgressions as more serious, more generalizable, and more deserving of punishment than social conventional transgressions. This pattern of reasoning appears to be robust across cultures and across individuals who personally suffered harm and violations of welfare (e.g., among abused, neglected, and maltreated children).

Other cognitive scientists have challenged this view, arguing that moral appraisals are culturally specific. Against Turiel's universalist position, Shweder and his colleagues contend that the distinction between morality and social convention is not universal and that moral appraisal is contingent on a particularly socio-historical context. For instance, when they asked Brahman and American children to rate how harmful various acts were, they found significant disagreement across the two populations. American children, but not Brahman children, accepted that it is wrong to cane a naughty child or eat with one's hands; whereas Brahman children, but not American children, deemed that eating beef or chicken or cutting hair after one's father's death is wrong. In short, in this view content is a function of context. Unlike other areas of reasoning, whether or not content affects the cognitive process is not addressed.

Spatial Cognition

Beginning with the last two decades of the twentieth century, a body of literature exploring the effects of language and culture on spatial reasoning has emerged. Several ways of identifying an object's relation in space to other objects have been identified: reference can be made with respect to the cardinal directions (the bus is to the south of the car), with respect to the position of the speaker (the bus is to the right of the car), and with respect to fixed geographical features (the bus is uphill from the car). Not all languages provide speakers with as ready means to express these relationships; nor do speakers of different languages employ all strategies with the same frequency. For example Steven Levinson and his colleagues found that Guugu Yimmithirr aboriginals are more likely to anchor an array of objects by reference to cardinality than speakers of Dutch, who are more likely to describe spatial arrangement with reference to their own body.

Attention and Visual Perception

All humans appear to be susceptible to most visual illusions. A well-studied exception is the Müller–Lyer illusion, which Herkovits *et al.* found did not affect the perception of members of some African societies. The researchers explained this by citing the absence of a susceptibility to the 'carpentered-world hypothesis'. On this hypothesis, susceptibility to the illusion is a function of experiencing a material world in which 90-degree angles are common (i.e., 'the carpentered world'), but in the African societies studied round shape architecture prevails.

In a study of change blindness, Masuda and Nisbett showed animations of scenes, which included stationary and moving objects, to Japanese and Americans; some animations were altered in later displays. The Japanese were more aware of changes in the background of a picture and in the changing location of objects in it, whereas the Americans were more aware of changes in the characteristics of objects.

Developmental evidence of group differences in attention suggests that differences can be learned. Chavajay and Rogoff demonstrated differences between Guatemalan Mayan and American 14–20 month-olds and their caregivers in attention management. Mayan caregivers and toddlers were more likely to attend simultaneously to competing events, unlike American caregivers and toddlers who were more likely to alternate their attention between competing events.

SITUATED/DISTRIBUTED STUDIES OF COGNITION

Cognitive scientists have typically examined how individual humans or their machine counterparts process information, traditionally through controlled investigations and laboratory studies. A number of cultural psychologists caution, however, that cognition cannot be fully understood as (1) taking place in individual minds, or (2) apparent place under controlled conditions. On the contrary, researchers who use the situated/distributed approach propose that the interactions between individual minds in their natural environments are mediated by extra-individual cognitive phenomena. Reflecting the influence of Lev Vygotsky's notion of 'the zone of proximal development' (and its variant, scaffolding), considerable work in this approach is developmental. According to Vygotsky, novices gain expertise by participating in contexts that extend their own skills sufficiently to boost performance and to stabilize enhanced competencies rather than acquiring skills through formal teaching.

Edwin Hutchins and others have proposed that cognition may be distributed among individuals in ways that argue for the existence of extra-individual cognitive systems. These systems are comprised of the habitual interactions between individuals and artifacts and, because of their emergent quality, cannot be understood from the perspective of the cognitions of the individuals participating. For example, Hutchins proposes that the speed of an airplane is remembered by the cockpit, arguing that the task is distributed over each of the crew and each of the instruments and manuals that they use, so that the cockpit itself is the level at which the cognitive task is being accomplished.

In a study of situated cognition, Geoffrey Saxe compared the mathematical competencies of three different groups of Brazilian children: urban street vendors, urban nonvendors, and rural nonvendors. None had received any formal schooling, and all three groups did poorly on abstract, orthographic tasks involving large number representation. When essentially the same tasks were repeated using large bills of money, the performance of urban children, who regularly deal with currency exchange, significantly improved, but the performances of both the rural children or nonvender urban children did not.

THE EPIDEMIOLOGICAL APPROACH

Both the comparative and situated/distributed approaches seek to demonstrate that culturally varying content affects underlying cognitive processes. The epidemiological approach, pioneered by Dan Sperber, in contrast understands cultural variation as a function of underlying and invariant cognitive processes. In particular, the approach focuses on the conditions by which beliefs and practices become distributed in populations, much as medical epidemiology focuses on how disease and pathogens are distributed.

Richard Dawkins proposed a similar notion, which he calls the 'meme', the cultural evolutionary analogue to a gene. Dawkins proposes that memes, like genes, replicate. He further suggests that the psychological processes underlying the reproduction of memes, specifically imitation, is fairly low level. In contrast, Sperber argues that cultural reproduction involves both ecological and higher-order psychological processes that in interaction determine the distribution of cultural phenomena in a given population. For Sperber to account for culture 'is to explain why some representations become widely distributed ... to explain why some representations are more successful – more contagious – than others'.

The epidemiological approach, accordingly, is closely linked with specific claims about cognitive architecture. The predominant view holds that humans are endowed with a general set of reasoning abilities that are brought to bear on a myriad of cognitive tasks. In the latter decades of the twentieth century researchers in psychology, linguistics, anthropology, and philosophy concluded that many cognitive abilities are specialized to handle specific types of information: the mind is composed of domain-specific or modular devices that, among other things, underlie much of common sense. For example, all normally endowed humans have a folk or naive theory of mind that interprets human behavior to be the result of mental states like belief and desire. Similarly, all humans have a naive physics that interprets the movement of inanimate objects. Those representations whose processing are subsumed by these various modularized dimensions of common sense are, according to Sperber, more likely to become widespread and relatively stable. Culture, then, is the totality of representations with these particular properties of distribution.

Building on the model of symbolism developed by Sperber, Pascal Boyer argues that supernatural entities are easily accepted and remembered not because they can be subsumed under common sense, but because they achieve a balance between commonsense expectations and violations of commonsense expectations. For example, Barrett and Keil have shown that folk notions of the Judeo-Christian concept of God tend to make God more humanlike (more consistent with commonsense expectations about humans), de-emphasizing aspects of formal theological beliefs that attribute omnipotence *and* omnipresence to God.

CONCLUDING REMARKS

Cultural psychologists and their counterparts in anthropology, cognitive anthropologists, are committed to developing an interdisciplinary program of research that explores the relationship between individuals' minds and the complex environments in which they exist. Much cultural psychological research is now focusing on topics that have tended to receive less attention in the cognitive sciences (e.g., the self and emotions). Cultural psychology is contributing significant insights into higher-order cognitive processes of central concern to cognitive science. Cultural psychology challenges the widely accepted assumption in cognitive science that the relationship between content and process is transparent, that the environment 'merely' provides content for cognitive processes. Instead, cultural psychology proposes that cognition must be understood in relation to population-level phenomena, ranging from local task-specific environments such as work groups to global, comprehensive domains such as cultures.

Further Reading

Barrett J and Keil FC (1996) Conceptualizing a nonnatural entity: anthropomorphism in God concepts. *Cognitive Psychology* **31**: 219–247.

Berlin B and Kay P (1969) *Basic Color Terms; Their Universality and Evolution*. Berkeley: University of California Press.

Boyer P (1994) *The Naturalness of Religious Ideas: A Cognitive Theory of Religion*. Berkeley: University of California Press.

Cole M (1996) *Cultural Psychology: A Once and Future Discipline*. Cambridge, MA: Harvard University Press.

Coley JD, Medin DL, Proffitt JB *et al.* (1999) Inductive reasoning in folkbiological thought. In: Medin DL (ed.) *Folkbiology*, pp. 205–232.Cambridge, MA: MIT Press.

Fernald A and Morikawa H (1993) Common themes and cultural variations in Japanese and American mothers' speech to infants. *Child Development* **64**: 637–656.

Herskovits M, Campbell D and Segall M (1969) *A Cross-cultural Study of Perception*. Indianapolis: Bobbs-Merrill.

Hutchins E (1995) *Cognition in the Wild*. Cambridge, MA: MIT Press.

Lave J (1988) *Cognition in Practice: Mind, Mathematics and Culture in Everyday Life*. Cambridge, MA: Harvard University Press.

Levinson SC (1996) Language and space. *Annual Review of Anthropology* **25**: 353–382.

Lucy JA (1992) *Language Diversity and Thought: A Reformulation of the Linguistic Relativity Hypothesis*. Cambridge, New York: Cambridge University Press.

Markus HR and Kitayama S (1991) Culture and the self: implications for cognition, emotion, and motivation. *Psychological Review* **98**(2): 224–253.

Nisbett RE, Peng K, Choi I and Norenzayan A (2001) Culture and systems of thought: holistic vs. analytic cognition. *Psychological Review* **108**: 291–310.

Rogoff B and Lave J (1984) *Everyday Cognition: Its Development in Social Context*. Cambridge, MA: Harvard University Press.

Saxe G (1988) The mathematics of child street vendors. *Child Development* **59**: 1415–1425.

Shweder R, Mahapatra M and Miller J (1987) Culture and moral development. In: Kagan J and Lamb S. *The Emergence of Morality in Young Children*, pp. 1–83. Chicago, IL: University of Chicago Press.

Sperber D (1996) *Explaining Culture: A Naturalistic Approach*. Oxford, UK: Blackwell.

Turiel E (1983) *The Development of Social Knowledge: Morality and Convention*. Cambridge, UK: Cambridge University Press.

Vygotsky LS (1978) *Mind in Society: The Development of Higher Psychological Processes*. Cambridge, MA: Harvard University Press.

Cultural Transmission and Diffusion

Advanced article

Robert Aunger, University of Cambridge, Cambridge, UK

CONTENTS

Cultural diffusion is the process by which information is disseminated through a population, typically by information exchanges among population members (cultural transmission), presumably with the involvement of social learning mechanisms.

INTRODUCTION

People rely heavily on shared beliefs and values to coordinate their social activities. Indeed, phenomena from ritualized greetings to the elaborate ceremonies that mark social events such as marriage can partly be explained by referring to the need to fulfill shared behavioral expectations. These expectations derive from norms and standards specific to each social group. The set of behavioral practices specific to a group constitute a pool of information which is typically what people mean by 'culture'. In anthropology textbooks, for example, culture is typically defined as a 'system of shared beliefs, values, customs, behaviours, and artifacts that the members of society use to cope with their world and with one another, and that are transmitted from generation to generation through learning' (Bates and Plog, 1990, p. 7). How these kinds of information and practices come to be shared by different people in the first place remains to be explained, however, even though it is arguably a central question in the social sciences.

CULTURAL TRANSMISSION PROCESSES

One explanation for shared culture is the exchange of information through social learning. However, considerable controversy attends this apparently common-sense notion. In particular, the recently burgeoning field of evolutionary psychology has claimed that much of what appears to be learned from others is in fact information already in place in people's brains – put there by a long history of natural selection for the retention of that information, which remains ready to be elicited by circumstances. In effect, some ecological trigger sets off an appropriate innate response, making so-called 'culture' a store of mental records – just like the way a jukebox stores musical records, any one of which can be chosen by punching the appropriate button (Tooby and Cosmides, 1992). The fact that people in different places do different things therefore cannot be taken as evidence that culture is transmitted, since they may be simply responding as individuals to subtle environmental differences. The crucial issue, then, is to distinguish the social transmission of information from individual phenotypic plasticity.

Advocates of the position that human beings and other social animals in fact depend significantly on social learning typically suppose that the need to acquire up-to-date information derives from the need to deal with quickly changing ecological conditions – including, in particular, the sometimes ephemeral nature of relationships with other organisms. Further, large brains had to evolve to support the ability to engage in such learning. Evolutionary psychologists counter, however, that big brains are necessary simply to store the many 'cultural' rules that might be elicited by varying circumstances. How can this deadlock be broken?

It seems likely that brains are not big enough to contain all the information required by the jukebox analogy. Many of the problems that contemporary urban life throws up, for example, could not yet have been incorporated into a genetic response: such fundamental experiences as living in large groups of unrelated people are simply too novel. Evolutionary psychologists acknowledge that modern conditions are likely to spawn maladaptive responses from 'Stone Age' minds. But the fact is that modern culture on the whole seems to be highly adaptive because it has massively extended the niche in which humans can live, and tremendously increased the total population of our species. It seems unlikely that brain mechanisms selected when humans lived under different circumstances would lead to the adoption of behavioral traits that currently contribute so greatly to genetic fitness.

Further, there are reasons to expect that evolution would naturally settle on social learning as an optimizing strategy for the acquisition of information relevant to changing environmental circumstances (Boyd and Richerson, 1985). So it seems reasonable to suppose that at least some of our behaviors are informed by rules acquired from other agents. Social learning should particularly be favored when the lessons from individual trial and error are expensive (either cognitively or in fitness terms), such as determining what is edible when a species can live in a variety of habitats.

Biology therefore determines our general capacity for cultural learning and is responsible for universal abilities like language. However, cultural variations among peoples are attributable to learnt traditions and not to innate or genetic propensities. At the same time, specific psychological adaptations have probably evolved to foster the selective but accurate acquisition of rules through social learning. Language itself can be seen as such an adaptation; it generates signals that allow the reliable transmission of complex, highly contextualized rules for behavior between people.

However, questions remain as to the nature of cultural transmission. In particular, is it like a copying process, or is it a reconstructive one? Some argue that even though cultural information must make its way from person to person in the coded form of messages, this process nevertheless results in a high-fidelity duplicate being produced in the receiver's brain – much as if it had been faxed or photocopied – thanks to evolved mechanisms for social communication. On the other hand, studies of social interactions suggest that the receiver – given the relative paucity of information that actually passes between people – must reconstruct a considerable proportion of a message's content. The nature of the process that underlies human communication has implications for the nature of human psychology. However, if reconstruction and fax-like transmission are equally reliable, then they may exhibit the same population-level dynamics.

CULTURAL SELECTION PROCESSES

Not all social messages are equally attended to or adopted by their receivers. In effect, selection among messages occurs. A selection process requires a population of entities whose frequencies increase or decrease according to their relative fitnesses. Cultural selection can be distinguished from natural selection by the kinds of units on which it operates (Cavalli-Sforza and Feldman, 1981). Just as natural selection is supposed to influence the evolutionary fate of genes, cultural selection works on the prevalence of cultural traits over time.

Cultural selection can occur at each point in the process of communicating information from one individual to another. At the source, there can be psychological selection among potential messages. Once a message has been sent into its channel, physical selection pressures can also affect the chances of that signal reaching its destination. For example, sometimes the message's code does not match well with the modality in which it is sent. Then, after the signal has been detected by the receiver, further psychological biases can exist for attending to and adopting the idea expressed by the signal. Some kinds of information acquired through social learning might not be consistent with other beliefs that an individual holds, for example, and would be rejected for that reason. Later, for further transmission to occur, performances of the related behaviors must also be motivated. Thus, certain traits are favored in effect by the social or physical circumstances in which they find themselves.

Analysts of cultural evolution must be careful to distinguish cultural selection from natural selection, since both can affect the frequency of cultural traits. Some beliefs, for example, cause people to engage in behavior that is detrimental to their health, survival, or likelihood of reproducing. Belonging to a religious group that forbids engaging in sex is only the most obvious case of such an effect. In this form of natural selection, the frequency of a culturally acquired belief is reduced not by changes in belief but by the culling of hosts with such beliefs from the population. This makes cultural evolution a process of 'dual inheritance', in which cultural selection and natural selection operate in parallel or in opposition on cultural traits (Boyd and Richerson, 1985).

Where do the mechanisms of cultural selection come from? How do they in turn evolve? The values used to discriminate between incoming messages can themselves be the product of earlier social learning, or constitute biases produced by a history of natural selection for discriminating between useful and harmful stimuli. Thus, some cultural traits can 'feather the nest' for later-arriving ones, suggesting that cultural evolution can engage in positive or negative feedback processes.

CULTURAL TRAITS

What is the unit of analysis in studies of culture? Traits – segregating particles of culture – or cultures themselves, taken as a whole?

Throughout its long history, anthropology has attempted to deal with the problem of identifying cultural traits. More recently, however, this attempt has largely been abandoned as impossible – and in any event unnecessary. According to this view, cultures are to be considered as unified wholes, or at least complexes, which are not necessarily divisible. Many ethnographers would argue that there is no 'atomic level' to culture, no way to uncover mutually exclusive entities with stable properties from which cultural compounds are formed. The tendency is now to describe culture as an ideal type, an artificial conglomeration of knowledge compiled from different people occupying varying social roles. However, this metaphysical Platonism – seeing culture as an integrated whole that transcends the minds of individuals – is analytically barren, since there is no contesting a representation built up by the imagination of the ethnographer.

Many would say that we have no legitimate basis for postulating that cultural transmission is intrinsically particulate. There may be no discrete variant that is reliably learned from others through observation or any other form of social interaction. Categories are imposed on a blended reality. You cannot count up cultural values in people's heads like votes for political parties. Traits are 'clumps' of culture content, not well-bounded entities (Gatewood, 2000). However, if culture is learned, and by individuals one at a time, then no one learns everything; culture winds up being distributed. Since learning a culture takes place through social interactions, only parts of culture can be acquired at any given moment: in effect, there must be units of transmission. Admittedly, after acquisition, these units may become amalgamated into complex mental representations which become difficult to tease apart. Nevertheless, these units of acquired information – the equivalent of the atom in physics, the molecule in chemistry, or the phoneme in linguistics – are the smallest possible meaningful unit of cultural information.

So culture must be analyzed as a set of traits, but these need not correlate with natural categories of things in the minds of those living in that culture. At the empirical level, an important question concerns the way in which one is supposed to pick out cultural traits. Where do the division points for defining categories come from – from 'outside' (researchers) or 'inside' (from the group members themselves)? Either is possible. Traits can be just some item of cultural content that the analyst finds it convenient to label – in effect, they become ethnographic conventions. Alternatively, much hard work can be devoted to ascertaining how expert informants themselves classify things.

Once the categories have been established, the question of how these traits are related to one another then arises. Computer scientists, who have thought hard about this problem, have come up with a variety of frameworks for representing knowledge bases, from nested hierarchies to network structures of nodes for concepts with links between them identifying kinds of relationships. What sort of representation is best? This may again be an empirical question, the answer to which depends on the analytical questions being addressed by a particular study.

What, then, is the locus of culture? What is the culture-bearing unit? The traditional solution of simply using the classification people apply to themselves retains considerable appeal, since it has the subjective authority of the participants in the study. However, the actual group influenced by some cultural process or knowledgeable about some domain of belief or practice may vary from domain to domain. This suggests that there is a fluid boundary to the social group to be identified as sharing cultural traits. At best, one might be able to identify a network of people who tend to be linked together for a wide range of cultural traits; but in the end, no easy solution to this problem has appeared.

ARTIFACTS

Human beings, along with many other species, engage in behaviors that significantly alter the environments in which they find themselves. This activity has been called 'niche construction' (Laland *et al.*, 2000). Its importance from an evolutionary viewpoint is that such alterations can subsequently influence the kinds of selection pressures that any species interacting with that modified feature of the environment will experience. As a result, population dynamics and the course of evolution can change. Niche construction introduces a feedback loop between behavior and its physical products, artifacts.

The production of artifacts requires 'technique', or knowledge and skills specific to that production process, typically acquired through social learning as well as individual practice. Technology is then the combination of artifacts and technique in a manufacturing context. A technology plus its supporting procedures and institutions, as well as environmental and sociopolitical conditions, can be considered a technological system.

Artifacts themselves can be divided into two general classes: tools and machines. Tools such as hammers or rulers can originate as ideas which are then turned into physical objects by people (or machines) as the expression of that representation. Alternatively, the first exemplar of a tool type might be produced by accident while manipulating some object, or through modification of an existing tool. In any case, it is an object that skilled people can use to extend their physical or mental capacities. Machines, on the other hand, can be considered artifacts with multiple component parts that together perform a novel function which any of the parts, taken individually, would not be able to accomplish. Only humans produce machines, probably because the planning involved in the execution of the multistep process of machine construction is beyond the cognitive abilities of other species.

How do such complex objects evolve? By playing an important part in cultural evolution. The existence of artifacts in the environment can have a significant influence on the course of cultural evolution, because they can constitute a store of information or a channel for information transmission between people (e.g. telephone wires). Artifacts, as transformed objects, should be distinguished from signals and tokens. A signal can be considered to be a patterned particle stream flowing through a channel. A token consists of a physical substrate with information inscribed as pattern in or on it; it is a template for the generation of signals. A signal is 'natural' if it is directly produced by the body (e.g. speech), or 'artificial' if produced by machines (e.g. laser beams or internet 'packets'). It is worth distinguishing between signals and tokens because they have different evolutionary roles: signals are short-lived, dynamic and energetic, designed for the transport of information. Tokens, on the other hand, are static and inert, because they are meant to be secure stores of information. Communicative artifacts contain a token, and hence are able to facilitate the communication of ideas. This means that artifacts can act as signal templates because signals can, after contact with some token, exhibit a new pattern (e.g. amplitude or frequency), to reflect the fact that it now carries information about that artifact.

This kind of communication – mediated by artifacts – has many benefits over traditional face-to-face communication: people separated by distance and time can exchange information (e.g. through email); the effective population reached by a given message can be increased (e.g. through mass media); new communication channels can have novel effects on message receivers (e.g. 'spamming' in email); and greater control over the distribution of messages can be achieved (e.g. through 'gatekeepers' such as the media corporations). The net effect is that this progress in the design of more and more complex artifacts allows the information available to human groups to accumulate beyond that available to any other species (Tomasello, 1999). It is the highly constructed nature of the human niche that arguably makes our cultural adaptation unique.

MEMES, MEMETICS, AND ASSOCIATED CONTROVERSIES

'Meme' is a word coined by Richard Dawkins (1976) to identify cultural traits with the ability to replicate themselves. The word has gained sufficient currency to be included in recent editions of the *Oxford English Dictionary*, where it is defined as 'an element of a culture that may be considered to be passed on by non-genetic means, especially imitation.' This reliance on imitation as a special form of social learning has proved crucial to the definition of memes, since many assert that only the relatively quick copying of instructions (not just behavior, but the directives for instigating that behavior) can support the chains of replication events necessary to sustain an evolutionary lineage, and particularly the curiously cumulative quality of human culture.

What distinguishes memetics from other evolutionary approaches to the understanding of cultural evolution (such as evolutionary psychology or sociobiology) is the insistence that what is transmitted during social learning is a replicator. Although the concept of replication is itself in need of some theoretical work, it generally is a process in which a copy is made of some source. This process must involve causation – that is, the source must be causally involved in the production of the copy (a causation condition). Second, the copy must be like its source in relevant respects (a similarity condition). Third, the process that generates the copy must obtain the information that makes the copy similar to its source from that same source (an information transfer condition). Finally, during the process, one entity must give rise to two or more (a duplication condition).

Thus, for a memeticist, not only are cultural traits identifiable as individualized units, but they have the power to cause their own duplication. Further, memes are cultural traits that behave as if they were interested in preserving themselves, using individual minds as hosts. Memes, in this view, are responsible for the persistence of certain cultural traits, including those that do not directly favor the biological fitness of the group in which those traits spread themselves.

It was originally thought that memes could encompass behaviors, artifacts or mental contents as varied as 'tunes, ideas, catch-phrases, clothes fashions, ways of making pots or building arches' (Dawkins, 1976). It seems unlikely, however, that replication could occur in a similar fashion in all of these contexts. In fact, behaviors generally do not replicate in the sense described above. It is true that one person can mimic the speech and even the accent of another, reproducing spoken phrases perfectly. Why is this not an example of cultural replication? The answer is that the spoken signal either dies 'in the air' prior to a second signal being constructed in response by the imitator; or it is converted into another form which circulates through the receiving brain before emerging again as the same spoken phrase. In either case, the signal is not duplicated; it either lives through a complex cycle of exchange between people, or a second example is produced after some lag. Thus, despite the facile appearance of a signal being replicated, behavioral mimicry does not hold up to our criteria for replication.

Similarly, the processes through which most artifacts are produced fail the information transfer condition. For example, on the factory floor, it is seldom the case that features of one car on the assembly line are determined by reference to those same characteristics in the previous exemplar. Instead, the assembler (a person or robot) relies on instructions from a centralized database such as a computer-derived datasheet to tell it what to do next. If one agrees with the majority of memeticists that behaviors and artifacts should be eliminated from consideration, then memes are restricted to minds, where the conditions for replication may hold.

As noted above, this means that imitation is the replication mechanism identified by memeticists. It has even been suggested that the human ability to imitate makes the existence of memes self-evident. The argument is that if imitation is defined as the ability to copy behavior or ideas by observing them, then surely the product of that observation must be replicated information. Although transmission is a process in which information is duplicated, it does not follow that the information itself is responsible. Other factors – such as common mental mechanisms for inferring mental or cultural content from sensory stimuli – could lead to the same result. The existence of memes therefore remains unproven (Aunger, 2002).

Neither is there any theoretical reason to suppose that just because culture evolves it must be founded on an independent replicator. Even though biological evolution is grounded in the replication of a biological entity (genes), this does not mean that every evolutionary process must be. In fact, evolution is a more general process than that. Why this is so requires some explanation.

The central concept in evolutionary theory, arguably, is that of a population. It is populations that evolve, and this evolution consists of shifts in the relative frequencies of various types of traits within the population. This population can consist of any collection of things. Simply divide these things up according to type. If these types have tendencies to increase or decrease in the local environment owing to the presence of instances of the same type, this is their relative fitness. If these types reproduce, that makes the process more 'biological', but it is not necessary. We only need to be able to establish objective criteria for determining what type of object something is, and ascertain it has fitness, to declare it subject to an evolutionary process. The replicator approach, however, makes additional demands – in particular, that changes in frequency result from the essential differentiating features of the type being copied from old tokens into the new ones.

Fitness, from this perspective, is just the tendency of a type to increase or decrease, given a specifiable suite of conditions in the local environment. Selection is change in the frequency of types owing to their relative fitness, and variation is change in a type's frequency independent of the type's fitness. Conspicuously missing from these definitions are the notions of replication, heredity (or copying fidelity) and the ancestor–descendant relation. This implies that we can understand evolution without invoking the concepts of either an evolutionary lineage, or replicators. Neither concept is a prerequisite for evolutionary theory to be fruitfully applied to some phenomena (Harms, 1996).

The natural conclusion, therefore, is that culture can evolve without memes. Support for memetics must come in the form of empirical studies that identify replicators at work in cultural reproduction. Until that happens, we can remain convinced that culture evolves, despite having no firm idea about the mechanisms underlying the reappearance of cultural traits in each generation.

CONCLUSION

How people come to share similar mental contents remains an essential problem for social science to solve. Does culture provide big-brained creatures with a system of informational inheritance which operates independently of, but in parallel with, genes? Alternatively, are supposedly cultural responses induced by mental algorithms that simply reflect a cumulative history of natural selection for genetically directed behaviors? Unfortunately, it will require additional research to find a conclusive answer to this question. However, the fact that cultural change exhibits the basic characteristics of an evolutionary process – inheritance, variation, and selection – seems less open to debate. Whether there will be overarching rules for describing cultural transmission processes remains to be determined, and it seems likely that who learns what from whom will turn out to be specific not only to particular groups, but also to particular periods within the life history of each group – perhaps even to each kind of trait being considered. The goal of establishing general rules of cultural transmission and selection applicable to everyone everywhere remains elusive, making the study of cultural evolution intrinsically complex.

References

Aunger R (2002) *The Electric Meme: A New Theory of How we Think and Communicate*. New York, NY: Free Press.

Bates DG and Plog F (1990) *Cultural Anthropology*. New York, NY: McGraw-Hill.

Boyd R and Richerson PJ (1985) *Culture and the Evolutionary Process*. Chicago, IL: University of Chicago Press.

Cavalli-Sforza LL and Feldman MW (1981) *Cultural Transmission and Evolution*. Princeton, NJ: Princeton University Press.

Dawkins R (1976) *The Selfish Gene*. Oxford, UK: Oxford University Press.

Gatewood JB (2000) Reflections on the nature of cultural distributions and the units of culture problem. *Ethnology* **35**: 293–303.

Harms W (1996) Cultural evolution and the variable phenotype. *Biology and Philosophy* **11**: 357–375.

Laland KN, Odling-Smee J and Feldman MW (2000) Niche construction, biological evolution and cultural change. *Behavioural and Brain Sciences* **23**: 131–75.

Tomasello M (1999) *The Cultural Origins of Human Cognition*. Cambridge, MA: Harvard University Press.

Tooby J and Cosmides L (1992) The psychological foundations of culture. In: Barkow JH, Cosmides L and Tooby J (eds) *The Adapted Mind: Evolutionary Psychology and the Generation of Culture*, pp. 19–136. Oxford, UK: Oxford University Press.

Further Reading

Aunger R (ed.) (2001) *Darwinizing Culture: The Status of Memetics as a Science*. Oxford, UK: Oxford University Press.

Blackmore S (1999) *The Meme Machine*. Oxford, UK: Oxford University Press.

Durham W (1991) *Coevolution: Genes, Culture and Human Diversity*. Stanford, CA: Stanford University Press.

Lumsden CJ and Wilson EO (1981) *Genes, Mind and Culture*. Cambridge, MA: Harvard University Press.

Plotkin HC (1993) *Darwin Machines and the Nature of Knowledge*. London, UK: Penguin.

Culture and Cognitive Development

Intermediate article

Michael Cole, University of California at San Diego, La Jolla, California, USA

CONTENTS

Although it is widely believed that culture is fundamental to cognitive development, theoretical and empirical advances are impeded by severe methodological problems which render firm conclusions elusive.

INTRODUCTION

Controversies over the relationship between culture and cognition began well before the formation of contemporary behavioral sciences, and continue to this day (Cole, 1996). Discussion of the topic was, and remains, complicated by confusion over each of the basic terms under discussion as well as by the severe limitations placed on the experimental method when it is used to deal with naturally occurring differences among human beings involving unknown mixtures of phylogenetic, cultural-historical and ontogenetic variations – all of which are involved in seeking to understand the relation of culture to cognitive development.

It is useful to begin by tracing the concept of culture as it has evolved since the term entered the English language from Latin many centuries ago. Modern conceptions of culture originate in terms that refer to the process of helping things to grow. From earliest times, this notion of culture included a general theory of how to promote growth: the creation of an artificial environment in which young organisms would have optimal conditions to develop. Such tending requires tools: both material (e.g. hoes) and mental (e.g. the knowledge of how and when to use a hoe). These tools were perfected over generations and designed for the special tasks to which they were put, and constitute culture in the present.

In contemporary social science writing, the term 'culture' is generally used to refer to the entire body of socially inherited past human accomplishments that serves as the resource for the current life of a social group ordinarily thought of as the inhabitants of a country or region. Although there is evidence of the rudiments of culture in nonhuman species, human beings are unique in their dependence upon the medium of culture and the forms of organism–environment interactions that culture supports to sustain and reproduce themselves.

Combining the historical notion of culture as a process of growing things with the modern conception of culture as social inheritance of prior generations' accomplishments, the study of culture in development can be seen to focus on the way in which biologically immature human beings incorporate and are incorporated into the cultural 'designs for living' that are their social heritage.

The interpenetration of cultural and phylogenetic contributions to human development was driven home several decades ago by Clifford Geertz, who noted the mounting evidence that the human body, and most especially the human brain, underwent a long (perhaps 3 million year) coevolution with the basic ability to create and use culture, and was led to conclude that

> man's nervous system does not merely enable him to acquire culture, it positively demands that he do so if it is going to function at all. Rather than culture acting only to supplement, develop, and extend organically based capacities logically and genetically prior to it, it would seem to be ingredient to those capacities themselves. A cultureless human being would probably turn out to be not an intrinsically talented, though unfulfilled ape, but a wholly mindless and consequently unworkable monstrosity (Geertz, 1973: p. 68).

As a consequence, those who wish to understand the role of culture in cognitive development are faced with the difficult interdisciplinary task of studying development in terms of antecedents that are tightly interwoven.

METHODOLOGICAL PROBLEMS

Culture-free versus Culture-based Measures of Cognitive Development

For most of the history of scholarly interest in the role of culture in development, research been based upon 'cross-cultural' comparisons. This phrase is placed in quotation marks because often the comparisons are cross-national in nature, involving social groups thought to differ in some theoretically interesting way (for example, with respect to language, natural ecology, or social institutions) such that cultural, biological, social and ecological factors all differ simultaneously.

The hazards of restricting such comparisons to only two naturally occurring groups has long been recognized. As a consequence, leading researchers into culture and cognitive development have routinely included a range of societies in their studies to reduce the risk that some factor other than the one under investigation is covertly influencing the results. However, such multisociety research is expensive, and also carries with it the difficulty that although it might reduce the probability of undetected covarying factors it does not eliminate it completely, often leading to such an apparently endless set of possibilities that definitive conclusions elude further research.

The problem of culturally valid cognitive assessment has been most extensively discussed with respect to the possibility of creating culture-free measures of cognition, seemingly a prerequisite for valid cross-cultural comparisons (the same issue applies, but is rarely addressed, when age comparisons within a single homogeneous group are of interest). A variety of cognitive measures used in cross-cultural comparisons have been the object of such analyses. Research using intelligence quotient (IQ) tests and Piagetian tests of conservation can serve as accessible examples.

Long ago, Florence Goodenough identified the crucial shortcoming of the cross-cultural use of IQ tests in a way that has broad – if rarely recognized – implications for all cross-cultural cognitive research, when she wrote that 'the fact can hardly be too strongly emphasized that neither intelligence tests nor the so-called tests of personality and character are measuring devices. They are sampling devices'.

Goodenough argued that when applied in American society, IQ tests may represent a reasonable sampling device because they are 'representative samples of the kind of intellectual tasks that American city dwellers are likely to be called upon to perform'. However, such tests are not representative of life in other cultural circumstances and hence their use as measuring devices for purposes of comparison is inappropriate. This injunction applies as much to variations among subgroups living in the USA as it does to people living in a wide variety of other societies.

It might be thought, on the basis of Goodenough's critique, that attempts to discover the influence of culture on the development of intelligence using IQ tests would have been abandoned long ago. However, the usefulness and theoretical implications of cross-cultural IQ testing continue to be heatedly debated. Logically, the only way to obtain a culture-free test is to construct items and procedures that are equally a part of the experience of all cultures. Following Goodenough's approach, this would require us to sample the valued adult activities in all cultures (or at least two!) and identify activities equivalent in their structure, their valuation and their frequency of occurrence. No one has carried out such a research program.

The same problem can be seen in the widespread use of Piagetian conservation tasks as measures of cognitive development during the late decades of the twentieth century. It was Piaget's initial hypothesis that the development of conservation would be a universal achievement, occurring in an invariant sequence at roughly equal ages across cultures because it represents a universally applicable logical principle (Piaget, 1974). However, application of standardized Piagetian procedures in different societies produced widely varying results, leading some to speculate not only that some cultures promote more cognitive development than others, but that without particular kinds of cultural experience, such as formal schooling, development might cease at the level of concrete operations.

However, this same literature contained within it anomalies that implicated cultural differences in interpretation of the task – not differences in logical development – as the source of cultural differences. The standard procedure, following methods developed in Geneva, was to present the participant with two beakers of equal circumference and height, filled with equal amounts of water. The water from one beaker was then poured into another, taller beaker with a smaller circumference, and the person was asked which of the two beakers containing water had the most water in it. Children and young adults from nonliterate societies of a given age were significantly more likely to assert that the taller, thinner beaker, contained more water; see Cole (1996) for a summary. However, an experiment on the effects of modifying testing procedures to match local cultural knowledge revealed a different pattern of results. Early research had shown that conservation was much more likely to be achieved if children were allowed to pour the liquid themselves instead of observing the experimenter. It was speculated that this change in procedure reduced the children's tendency to interpret the experiment as something of a magic show, allowing their real competence concerning the conservation of volume to be revealed.

In the revised procedure, participants were asked to solve the conservation task and then to act as informants whose job it was to clarify, for the experimenter, the local terms for resemblance and equivalence with respect to the task. When confronted with the critical test in which one beaker of water was poured into a narrower, taller beaker, the participants asked to play the role of informant gave the wrong response – they said that the beaker with water higher up its sides contained more liquid. However, in the role of linguistic informants these same people went on to explain that while the level of water was 'more', the quantity was the same. Such results provide nice examples of the kinds of performance factors that can block the actualization of competence and mislead researchers about the nature of cultural differences.

THEORIES OF CULTURAL CONTRIBUTIONS TO COGNITIVE DEVELOPMENT

Two major lines of theory have dominated discussions of culture and cognitive development, those of Jean Piaget and Lev Vygotsky. (*See* **Piaget, Jean**; **Vygotsky, Lev**)

Jean Piaget

According to Piaget, cognitive development occurs through a process of equilibration where accommodation to existing environmental circumstances constantly vies with assimilation of environmental structures to existing mental structures. In a widely cited article, Piaget (1974) divided potential influences on cognitive development into four main factors, each of which could be expected to influence the timing of developmental milestones resulting from the interplay of assimilation and accommodation.

- Biological factors: here Piaget mentions nutrition and general health, factors that influence what might be called the rate of physical maturation.
- Coordination of individual actions: this factor refers to equilibration, the active process of self-regulation resulting from the back-and-forth tug and pull of accommodation and assimilation. Equilibration is the proximal mechanism of development. All other factors operate through their influence on equilibration.
- The social factor of interpersonal coordination: the process by which children 'ask questions, share information, work together, argue, object, etc.' (Piaget, 1974: p. 302).
- Educational and cultural transmission. Piaget reasoned that children acquire specific skills and knowledge through interaction in culturally specific social institutions. In so far as some societies provide more overall experience relevant to discovering the nature of the world, true developmental differences would be created in either the rate or the final level of development.

Piaget recognized that the four contributing factors he identified are all tightly interconnected with each other in any given society. However, barring conditions of extreme malnutrition, he assumed that rates and levels of cognitive development would be universal across societies with respect to the first three factors, and would differ only with respect to the fourth. In early writings, he assumed that modern industrial societies would provide the requisite additional experiences to speed cognitive development, but it is interesting that Piaget was skeptical of schooling's development-enhancing properties, since this social institution varies greatly across cultures and in popular thinking ought to have a major influence on cognitive development. He argued that the asymmetrical power relations of teacher and student created an imbalance of equilibration, because the pressure to accommodate to teachers' views far outweighed the pressure for assimilation of instruction to the child's already existing schemas. The result was learning of a superficial kind that was unlikely to create fundamental cognitive change. He believed that fundamental change was more likely to occur in informal actions where the asymmetry of power relations was reduced, allowing for a more equal balance between assimilation and accommodation. Evidence on this matter was too inconclusive in the 1960s to support more than speculation.

Confronted by subsequent research with evidence of cultural variations in performance on his tasks, and particularly evidence that schooling enhanced performance within cultures, Piaget (1972) concluded that all individuals reach a universal stage of formal operational thinking, but that formal operations are attained first (and perhaps only) in fields of adult specialization or as a consequence of school-based training. This view offers an obvious line of reconciliation of Piagetian theory with the facts about cultural variability. The overall conclusion, however, accords a restricted role to culture in cognitive development, which is more a matter of individual invention through action on the environment than of environmental, and particularly cultural, influences, on the child.

Lev Vygotsky

The Russian psychologist Lev Vygotsky, unlike Piaget, accorded culture and the influence of children's social environments a central role in cognitive development (Vygotsky, 1978). The central thesis upon which his 'cultural-historical' school of psychology was founded is that the structure and development of human psychological processes emerge in the process of humanity's culturally mediated, historically developing, practical activity. Each term in this formulation is tightly interconnected with and in some sense implies the others, making it difficult at times to discern how each contributes to description and analysis of the dynamics of psychological experience.

1. Mediation through artifacts. The initial premise of the cultural-historical school is that the psychological processes of humans emerged simultaneously with a new form of behavior in which material objects were modified by human beings as a means of regulating their interactions with the world and each other.
2. Historical development. In addition to using and making cultural artifacts, human beings arrange for the rediscovery of the already created artifacts in each succeeding generation, which in turn adds its modifications to the culture pool of artifacts. The accumulated artifacts of a group – culture – is then seen as the species-specific medium of human development. Cognitive development, as a consequence, represents the capacity to develop within that medium and to arrange for its reproduction in succeeding generations.
3. Practical activity. The third premise of the cultural-historical approach, adopted from Hegel by way of Marx, is that the analysis of human psychological functions must be grounded in their everyday activities. It was only through such an approach, Marx claimed, that the duality of materialism versus idealism could be superseded, because it is in activity that people experience the ideal/material residue of the activity of prior generations.

Vygotsky's emphasis on the historical accumulation of artifacts and their infusion in activity led him to emphasize the social origins of human thought processes. Vygotsky argued it is only through the mediation of socialized others that the child can come to experience, and hence acquire, the cultural heritage which is the foundation of cognitive development. This view of social origins requires that special attention be paid to the power of adults to arrange the environments of children in a way that optimizes their development according to existing norms. It generates the idea of a 'zone of proximal development' which affords the proximal, relevant environment of experience for development.

CONTEMPORARY RESEARCH ON CULTURE AND COGNITIVE DEVELOPMENT

Because of the methodological problems inherent in cross-cultural approaches to the study of development, the pace of such work has now slackened. At the same time, there has been an increase in research within cultures that take advantage of naturally occurring contrasts associated with differential participation in particular cultural practices in order to show how culture enters into the process of development. This does not mean that cross-cultural research has come to a standstill. For example, faced with the ambiguities inherent in research carried out in different cultures, where differences in language, nutrition and social relations can so easily enter unbidden into comparisons that seek to highlight specific cultural factors, some researchers interested in studying the relation of culture to development have turned to studies within societies that greatly reduce the chances of undetected contamination by uncontrolled factors. One such method is referred to as the 'school cut-off strategy' (Christian *et al.*, 2001).

In many countries school boards require that in order to begin attending school, a child be a certain age by a particular date. For example, to enter grade 1 in September of a given year, children in Edmonton in Canada must have passed their sixth birthday by March 1 of that year. Six-year-olds born after that date must attend kindergarten instead, so their formal education is delayed for a year. Such policies allow researchers to assess the impact of early schooling on different domains of cognitive development while holding age virtually constant: they simply compare the intellectual performances of children who turn six in January or February with those who turn six in March or April, testing both groups at the beginning and at the end of the school year.

Researchers who have used this strategy find that the first year of schooling brings about a marked increase in the sophistication of some cognitive processes but not others. Frederick Morrison and his colleagues, for example, compared the ability of children to perform on tasks ranging from picture recall to word and number manipulation, and Piagetian number conservation tasks (Christian *et al.*, 2001). The first-grade children were, on average, only a month older than those in kindergarten, and at the start of the school year the performances of the two groups were virtually identical. At the end of the school year, however, the first-graders could remember twice as many pictures as they did at the beginning, whereas the kindergarten group showed no improvement at all. Significantly, the first-grade children engaged in active rehearsal during the testing, but the kindergarten children did not. Clearly, a year of schooling had brought about marked changes in memory strategies and performance. The same pattern of results was obtained for standardized reading and mathematics tests. However, similar differences were not obtained for a standard Piagetian test of number conservation. This evidence shows clear influences of a cultural factor, schooling, on cognitive development, while supporting Piaget's skepticism about the cognitive benefits of schooling.

Meanwhile, cross-cultural research with respect to a variety of cognitive functions including categorization, logical reasoning and memory continues to show significant effects. In some cases schooling appears to be critical to the results; in others it does not. For example, Alexander Luria (described in Cole, 1996) compared the performance of Central Asian peasants who had been organized into collective farms and exposed to a modicum of Western-style schooling with those who had not. He reported that categorization processes grounded in 'graphic, object-oriented experience' gave way to 'more complex processes which combine what is perceived into a system of abstract, linguistic, categories.' Thought processes grounded in 'practical, situational' thinking gave way to more abstract theory-driven modes of thinking, which Luria referred to as the 'the transition from the sensory to the rational'. While Luria attributed such findings to involvement in modern industrialized practices including schooling, more recent research into cultural influence on the development of categorization has shown that even among people without schooling, both density of experience and cultural variations in world views substantially influence the development of biological thinking in addition to physical similarity (Medin and Atran, 1999). This latter work fits well the idea that cultural variations build upon pan-human, phylogenetically derived perceptual processes to construct local theories appropriate to local circumstances.

A considerable body of work has been devoted to the study of cultural variations in syllogistic reasoning, one of the hallmarks of Piaget's stage of formal operations. According to Luria's data, as well as research conducted in other nonliterate societies in different parts of the world, syllogistic reasoning appears to be closely linked to the sociocultural institutions of Western-style schooling. However, cases where nonliterate peasant workers succeed in responding to formal syllogisms, as well as cases where college-educated adults do not, have been reported.

The tradition of studying the role of participation in culturally organized activities within a single culture has emphasized the many ways in which cultural practices beyond schooling enter into the process of cognitive development. Here we can mention research on the development of mathematical thinking among street-vendor children in Brazil as well as work among Nepalese youth who become engaged in new forms of economic activity (Saxe, 1994; Beach, 1995; Ueno, 1995).

CONCLUSION

Taken as a whole, research on culture and cognitive development persuasively demonstrates the centrality of culture and cultural variation to cognitive development. The current challenge facing researchers is to solve the methodological difficulties (which themselves involve cultural factors) in order to identify more clearly the necessary and sufficient conditions involved.

References

Beach K (1995) Activity as a mediator of sociocultural change and individual development. *Mind, Culture and Activity* **2**(4): 285–302.

Christian K, Bachnan HJ and Morrison FJ (2001) Schooling and cognitive development. In: Sternberg RJ and Grigorenko EL (eds) *Environmental Effects on Cognitive Abilities*. Mahway, NJ: Erlbaum.

Cole M (1996) *Cultural Psychology: A Once and Future Discipline*. Cambridge, MA: Harvard University Press.

Geertz C (1973) *The Interpretation of Cultures: Selected Essays*. New York, NY: Basic Books.

Medin DL and Atran S (eds) (1999) *Folkbiology*. Cambridge, MA: MIT Press.

Piaget J (1974) Need and significance of cross-cultural studies in genetic psychology. In: Berry JW and Dasen PR (eds) *Culture and Cognition: Readings in Cross-cultural Psychology*. London: Methuen.

Piaget J (1972) Intellectual evolution from adolescence to adulthood. *Human Development* **15**: 1–12.

Saxe G (1994) Studying cognitive developments in sociocultural context: the development of a practice-based approach. *Mind, Culture and Activity* **1**: 135–157.

Ueno N (1995) The social construction of reality in the artifacts of numeracy for distribution and exchange in a Nepalese bazaar. *Mind, Culture and Activity* **2**(4): 240–257.

Vygotsky LS (1978) *Mind in Society*. Cambridge, MA: Harvard University Press.

Further Reading

Berry J, Poortinga Y, Segall M and Dasen P (1992) *Cross-cultural Psychology: Research and Applications*, 2nd edn. Cambridge, UK: Cambridge University Press.

Greenfield PM (1976) Cross-cultural Piagetian research: paradox and progress. In: Riegel KF and Meacham JA (eds) *The Developing Individual in a Changing World: Historical and Cultural Issues* vol. 1. Chicago: Aldine.

Hallpike CP (1979) *The Foundations of Primitive Thought*. Oxford: Clarendon Press.

Irvine J (1978) Wolof 'Magical thinking: culture and conservation revisited.' *Journal of Cross-cultural Psychology* **9**: 300–310.

Jahoda G (1993) *Crossroads Between Culture and Mind*. Cambridge, MA: Harvard University Press.

Luria AR (1976) *Culture and Cognitive Development*. Cambridge, MA: Harvard University Press.

Decision-making

Introductory article

Barbara A Mellers, University of California, Berkeley, California, USA

CONTENTS

The field of decision-making focuses on normative questions – how should we make decisions if we want to do it right? – and description questions – how do we actually make decisions when dealing with uncertainties? Numerous demonstrations of how people violate the axioms of normative theory have given rise to a deeper understanding of the processes underlying actual decision-making.

INTRODUCTION

Expected utility theory is widely accepted as the standard for normative decision-making. It tells us how we 'should' make decisions if we want our choices to be internally consistent and coherent. However, we do not always follow those principles. Researchers in decision-making have constructed puzzles and paradoxes that show how people violate the axioms of expected utility theory. People focus on many other factors, including personal, social and cultural rules, justifiability, and emotional rewards. Despite these numerous other factors, actual choice behavior is often systematic and orderly.

EXPECTED UTILITY THEORY AND ITS VIOLATIONS

Expected utility theory has been, and continues to be, the dominant theoretical framework in the social sciences. The classical theory dates back to 1738 when Daniel Bernoulli suggested that people should make decisions that maximize their expected utilities (subjective values). Utilities reflect the psychological satisfaction of wealth, rather than wealth *per se*. Bernoulli suggested that the utility of wealth increased rapidly at first, then gradually slowed, consistent with a logarithmic function. A function of this shape describes the fact that people are often risk-averse in their preferences, although less risk-averse as wealth increases.

Modern utility theory began in 1947 with a book called *Theory of Games and Economic Behavior*. In this book, von Neumann and Morgenstern provided a mathematical foundation for expected utility theory. Suppose decisions can be represented as choices between gambles. If people can rank order their preferences for gambles, and their preferences are consistent with a small set of axioms, choices can be represented 'as if' they were based on the maximization of expected utilities.

Allais' Paradox

Soon after von Neumann and Morgenstern completed their work, researchers started to devise examples showing how people violate the axioms of expected utility theory. One famous example is Allais' paradox. Consider a choice between options A and B. Option A is \$1 million for sure. Option B is a 10% chance of winning \$2 million, an 89% chance of winning \$1 million, and a 1% chance of winning nothing. When presented with this hypothetical choice, most people prefer A to B. If the utility of nothing is zero, this preference can be expressed in the expected utility framework as:

$$1.0u(\$1\text{m}) > [0.10u(\$2\text{m}) + 0.89u(\$1\text{m})] \quad (1)$$

where $u(\$1\text{m})$ and $u(\$2\text{m})$ are the utilities of \$1 million and \$2 million respectively. The expression can also be written:

$$0.11u(\$1\text{m}) > 0.10u(\$2\text{m}) \quad (2)$$

Now consider a choice between options C and D: option C is an 11% chance of \$1 million, and option D is a 10% chance of \$2 million. Most people prefer to D to C, which implies:

$$0.10u(\$2m) > 0.11u(\$1m)$$

which directly contradicts the previous expression. A preference for A implies a preference for C, or a preference for B implies a preference for D; but one cannot prefer both A and D (or B and C). Examples such as this led researchers to question whether expected utility theory was both a normative and a descriptive theory of choice.

PROSPECT THEORY

In 1979 Kahneman and Tversky proposed a descriptive theory of risk choice called prospect theory. Prospect theory differs from expected utility theory in several respects. In expected utility theory, decision-makers evaluate the utility of total wealth. In prospect theory, utilities (which are referred to as values) are associated with changes in wealth relative to the status quo. Furthermore, losses have greater impact than gains of equal magnitude, an assumption known as loss aversion.

Endowment Effects

Loss aversion provides an explanation for a well-known finding called endowment effects. Consider an experiment that takes place in a college classroom. Half of the students are randomly assigned a gift, such as a university coffee mug. These students are sellers. Those without mugs are buyers. Sellers are asked to report the minimum amount of money they would accept to sell their mug. Buyers report the maximum amount of money they would be willing pay to buy a mug. An experimental market is conducted; if there are transactions, mugs and money are exchanged.

According to expected utility theory, the experimental market will ensure that mug owners are those who value mugs the most. Since mugs were randomly assigned to students, there is no reason to think that students designated as sellers would value the mugs more than those assigned the role of buyers. In order for those students who value mugs most to become mug owners, approximately half of the mugs would, on average, be exchanged. However, that is not what happens. Few mugs are ever traded. Selling prices are typically larger than buying prices by a factor of two or more. The explanation for the effect is loss aversion; the pain of losing the mug is greater in magnitude than the pleasure of gaining the mug.

The way in which an object is endowed also influences its value. People who are rewarded with an object for an exemplary performance tend to value that object more highly than people who obtain the same object through chance or poor performance. Furthermore, windfall gains, such as lottery winnings or inheritances, are spent more readily than other assets, presumably because they are valued less.

Framing Effects

Endowment effects demonstrate how shifts in the status quo can influence the value of objects. Framing effects demonstrate how shifts in the perception of the status quo can lead to preference reversals. Framing effects were initially demonstrated by Tversky and Kahneman in a story called the Asian disease problem. Participants were told, 'Imagine that the USA is preparing for the outbreak of an unusual Asian disease which is expected to kill 600 people. Two alternative programs to combat the disease have been proposed. Assume that the exact scientific estimates of the consequences of the programs are as follows: with program A, 200 people will be saved; with program B, there is a 1/3 chance that 600 people will be saved, otherwise no one will be saved.' Participants were asked to select a program, and the majority chose program A. Another group of participants were told the same story, except descriptions of the programs were presented in terms of lives lost. They were told, 'With program A, 400 people will die. With program B, there is a 1/3 chance that no one will die, otherwise 600 people will die.' The majority selected program B.

Although programs A and B are identical, preferences reverse with framing. The majority of participants preferred the safer option when alternatives were described as gains (lives saved), and the majority preferred the riskier option when alternatives were described as losses (lives lost). Prospect theory predicts these reversals. Both gains and losses have diminishing sensitivity. Saving 200 lives with certainty has greater value that a 1 in 3 chance of saving 600 lives. Furthermore, the certain death of 400 people is more painful than a 2 in 3 chance that 600 will die.

Framing effects go far beyond laboratory demonstrations. Researchers in the USA examined preferences for automobile insurance among drivers in New Jersey and Pennsylvania. Both states offered insurance at similar costs, and both states allowed drivers to cut costs by giving up their right to sue if an accident occurred. In New Jersey, the default coverage offered by insurance companies contained no right to sue, although a driver could buy that right at additional cost. In Pennsylvania, the default coverage contained the right to sue, although drivers could decline the right and reduce costs. The different reference points led to big differences in coverage. Only 20% of New Jersey drivers purchased the right to sue at additional cost, but as many as 75% of the Pennsylvania drivers purchased the right as part of the package. Drivers tended to accept the default coverage. If the default coverage in Pennsylvania had resembled the default coverage in New Jersey, Pennsylvania drivers would have saved approximately $200 million in annual insurance costs.

Contextual Effects

Expected utility theory implies that the relative preference for one option over another should not change when additional options are added in the choice set. This principle is called 'regularity'. Researchers have shown that relative preferences vary systematically with the context. In one study, decision-makers are asked to choose between two options, A and B, each described in terms of two attributes. At a later point, the same decision-makers choose among A and B, and a new option, C. Option C is worse than A on one attribute, and worse than B on both attributes. Suddenly in the context of C, option B starts to look better, and the relative preference for B over A increases.

Decisions can also be influenced by the global context or the implicit comparisons people make across many choices. A given change in an attribute has a greater effect on choice when the attribute range is narrow than when the attribute range is wide. In one experiment, students made choices between apartments that varied in monthly rent and distance to campus. When monthly rents ranged from $200 to $400, a $50 increase in monthly rent was much more aversive than when monthly rents ranged from $100 to $1000. Such contextual effects are also inconsistent with expected utility theory.

ALTERNATIVE FRAMEWORKS FOR CHOICE

It is clear that the decision strategies people use vary with the individual, the task, the context and the frame. Prospect theory describes some of these phenomena, such as endowment effects and framing effects, but not all of them. Contextual effects, for example, require other explanations. What other psychological processes describe choice processes? A number of researchers have explored alternative frameworks.

Rule Following

Researchers point out that some choices are based on the application of rules or norms to situations. People ask themselves, 'What kind of situation is this?', 'What kind of person am I?', and 'What does a person like me do in a situation like this?'. Rules do not involve the calculus of cost–benefit analysis. They are adopted simply because they seem right and reflect our social or professional identities. Generally, doctors make decisions that coincide with medical guidelines, teachers make decisions that follow academic codes, and lawyers make decisions that build on legal precedents.

Rules may also reflect our personal identities. Prudential rules of thumb are often used to cope with issues of self-control. Some courses of action have minimal effects when done once, but large effects when done repeatedly. Smoking a cigarette or eating a slice of chocolate cake has a small effect in the short run, but done habitually has large effects in the long run. In these cases, the short-term benefits loom large, and the long-term costs are remote. Following a rule minimizes effort and guards the decision-maker against both temptation and potentially painful trade-offs.

Some have argued that interactions among people can be categorized in terms of four fundamental social rules. These rules are communal sharing, authority ranking, equality matching, and market pricing. Communal sharing rules stress the common bonds among members of a group, such as families, lovers or nations. Rules based on authority ranking highlight asymmetries in rank, privilege or prestige, as found in the military or the workplace. Equality matching stresses reciprocity. Decisions to join babysitting cooperatives or car pools imply that one agrees to give back whatever one takes. Finally, market pricing rules are governed by supply and demand, expected utilities, or trade-offs between costs and benefits. People often find these rules intrinsically satisfying for their own sake. They also insist that others adhere to the rules and punish those who do not conform.

When decision-makers apply the wrong rule to a social situation, they may make 'taboo' trade-offs. To attach a monetary value to one's friendship, one's children or one's academic integrity is to demonstrate that one is not really a friend, a parent or a scholar. In fact, the mere mention of selling priceless things can degrade the reputation of the person suggesting it.

Reason-based Choice

When rules conflict or seem irrelevant, people look for reasons to justify their decisions to themselves or others. This approach to decision-making involves a balance of arguments for and against a course of action. Reasons might be lists of the pros and cons, or they might be stories. Jurors often construct stories to explain the facts. Courtroom evidence presented in the form of stories often leads to stronger decisions and more confident jurors than evidence presented in the form of issues.

Reason-based choices can be peculiar. In one study, the researchers asked students to imagine they were serving on a jury deciding the award of sole custody of an only child following a messy divorce. Parent A was described as having an average income, average health, average working hours, reasonable rapport with the child, and a relatively stable social life. Parent B was described as having an above-average income, a very close relationship with the child, an extremely active social life, lots of work-related travel, and minor health problems.

One group of students decided which parent would be awarded sole custody of the child. The majority selected parent B because of the positive features, such as the close relationship with the child and the good income. Another group decided which parent would be denied custody of the child. These students also selected parent B because of the negative features, such as the extensive travel schedule and health problems. Positive reasons can be more compelling when we select, and negative reasons can be more compelling when we reject. In these cases, we might accept and reject the same option.

Pattern Matching

Although many researchers have adopted the gamble as a template for a decision, other researchers reject this metaphor. They argue that decisions made by experienced people under time pressure in real-world settings are better represented by recognizing patterns. In these cases, decision-makers do not necessarily compare two or more options. Their experiences allow them to see situations as examples of prototypes, and they often know a course of action that is likely to succeed without making comparisons. This form of decision-making involves pattern matching.

Emotion-based Choice

Sometimes we make choices to feel good. Which film to watch, which book to read or which perfume to buy might depend on our view of what seems pleasurable. 'Feeling good' might mean avoiding regret. Researchers have shown that women who anticipated regret about their child dying as the result of a vaccination were less likely to have the child vaccinated, even when the mortality risks of the disease were greater than those of the vaccination. Consumers who imagined purchasing an unfamiliar product that later malfunctioned were more likely to buy a familiar product. Finally, students who were given a lottery ticket and asked if they would trade their ticket for a new one with objectively better odds tended not to trade because of the regret they anticipated if their original ticket were to win.

In other cases, 'feeling good' might mean anticipating the pleasure of positive outcomes and the pain of negative outcomes, and selecting the option that feels better on average. This rule is similar, though not identical, to expected utility theory. Expected utility theory predicts that people will choose the option with greater average utility, not greater average pleasure. Differences between the theories arise when anticipated pleasure deviates from utilities. This is often the case. Unlike utility, anticipated pleasure varies with multiple reference points. Comparisons with one's past performance, with other peoples' performance, with outcomes under different states of the world, and with outcomes under different choices might influence the anticipated pleasure of consequences, but they are typically not included in utilities.

Such comparisons have systematic effects on anticipated pleasure. Exceeding one's expectations adds to the pleasure of an outcome, and falling short of one's expectations detracts from the outcome. Furthermore, the impact of these comparisons is asymmetric. The incremental pain of doing worse than expected is greater in magnitude than incremental pleasure of doing better. In a gambling context, a small win can become more pleasurable if one avoids a big loss; but the same small win can be extremely disappointing if one fails to win an even larger amount. Another way in which anticipated pleasure and utilities differ involves our beliefs about outcomes. Pleasure depends on beliefs, while utilities are independent of beliefs. Unexpected outcomes are associated with more intense pleasure and pain. Utilities should be independent of likelihoods. Surprise effects are strong enough to make a smaller but surprising win more pleasurable than a larger win that was almost certain. In contrast, the utility of the smaller win could never exceed that of the larger win.

If decision-makers base their choices on the anticipated pleasure of future outcomes, the accuracy of their forecasts is essential. Inaccurate predictions can surely lead to suboptimal choice. Researchers have identified some systematic errors in forecasting. One source of error is our immediate emotions: feelings of joy, anger, and sadness influence our attention, perception, memory and information processing strategies. When happy, we are better at retrieving happy memories, and when sad, we are better at recalling sad events. When happy, we use more flexible and creative problem-solving strategies. Sadness can lead to greater analytical thinking and longer response times, while anger has been linked to faster and less discriminate use of information.

Another source of error is the tendency to focus on whatever is salient at the moment, even when it has little effect later. In one study, students in the American Midwest and in California were asked to assess how happy they were, and how happy students like them in the other region of the country would be. The comparison highlighted the cultural opportunities, better climate and greater natural beauty of California. Both groups thought that students in California were happier; but in fact, the students were equally happy.

Researchers also asked college professors who were coming up for tenure how they expected to feel if they did or did not receive tenure. The professors expected to be happy if given tenure and extremely unhappy otherwise. Later the researchers contacted the professors to ask them what had happened and how they felt about it. Those denied tenure were actually much happier than they expected to be.

People also overestimate the pleasure of future experiences. In a classic study, researchers examined the happiness of lottery winners, matched control subjects, and people who were paraplegic. Although the lottery winners might have been thrilled about their wins in the days or weeks immediately after the event, they were no happier than the control subjects approximately a year later. Furthermore, the control subjects were only mildly happier than the paraplegic subjects. The tendency to focus on a single event can lead to overestimates of both pain and pleasure.

Deciding How to Decide

What determines the strategy for making decisions? Researchers argue that people are limited information processors who have access to many possible choice rules. Holding all else constant, they want to minimize the effort involved in a decision, and make accurate decisions. The strategies they use depend on trade-offs between effort and accuracy. More recently, this framework has been extended to include trade-offs between accuracy and negative emotions. People want to make accurate choices, and simultaneously avoid painful trade-offs. These goals can also determine choice strategies.

THE PSYCHOLOGY OF BELIEFS

Most decisions are associated with uncertainty, and beliefs play an important role. Researchers investigated the extent to which people update their beliefs according to normative principles, such as Bayes' theorem. Some argued that people shift their beliefs in the appropriate direction, but not to the right extent. Others argued that people are not Bayesians at all; instead, they use at least three heuristics. Representativeness is judgment based on the similarity of the event to the category. Availability is judgment based on the ease with which instances come to mind. Anchoring and adjustment are judgment based on insufficient movement from a reference point. These heuristics are rules that influence judgments of probability.

Base Rate Neglect

Researchers have argued that people often neglect base rate information. Evidence comes from a famous story, the 'cab problem'. Participants are told, 'A cab was involved in a hit-and-run accident at night. Two cab companies, the Green and the Blue, operate in the city: 85% of the cabs in the city are green and 15% are blue. A witness identified the cab as blue. The court tested the reliability of the witness under the same circumstances that existed on the night of the accident and concluded that the witness correctly identified each one of the two colors 80% of the time and failed 20% of the time. What is the probability that the cab involved in the accident was blue rather than green?' The Bayesian solution is 41%, but the majority of participants say 80%. This judgment is sensitive to the accuracy of the witness, but not the base rate. People estimate probabilities based on the information that seems most salient and compelling. In the cab problem, that information is the accuracy of the witness.

Conjunctive Probabilities

Researchers have also argued that people violate the conjunctive rule, according to which the judged probability of the intersection of two events cannot exceed the judged probability of either single event. In these studies, participants are told a story about a woman named Linda who is described as 31 years old, single, outspoken, and very bright. She majored in philosophy and cared deeply about issues of discrimination and social justice. She also participated in antinuclear demonstrations. Participants are asked to rank the likelihood of various statements, including 'Linda is a bank teller' and 'Linda is a bank teller and a feminist'. Participants report that the statement, 'Linda is a bank teller and a feminist' is more probable than 'Linda is a bank teller'. Some researchers claim that representativeness, or the similarity of the target description to the category, governs these probabilities, and this heuristic can lead to mistakes.

Is it possible to 'fix' these mistakes? Some researchers have presented information using frequency formats. Participants are told to imagine 100 women who fit the description of Linda. Of those, how many are bank tellers? Feminists? Bank tellers and feminists? With this format, probability judgments are more likely to resemble the correct solutions with both base rate problems and conjunction problems.

Fast and Frugal Heuristics

Not all heuristics lead to errors. Some heuristics work well because they exploit the structure in the environment. Fast and frugal heuristics require a minimum of time, knowledge and computation, and tend to focus on one-reason decision-making. Such heuristics are based on easily computable search and stopping rules. The recognition heuristic involves the use of recognition to draw inferences. In some conditions, the lack of knowledge is beneficial. In one study, students at the University of Chicago were asked to decide which of two US cities had the larger population. University of Chicago students tend to be quite knowledgeable about US cities, yet despite their knowledge, they were less likely than German students to judge correctly whether San Diego or San Antonio had the larger population. Most of the German students had never heard of San Antonio, and using the recognition heuristic, they answered the question correctly.

Another heuristic called 'take the best' applies to predictions, inferences and decisions based on multiple imperfect cues. Suppose decision-makers are asked which of two cities is larger, and both cities are recognized. They compare cities on the basis of the most valid cue, and if that does not discriminate, they take the next best cue, and so on. With this rule they 'take the best, ignore the rest'. This heuristic can do almost as well as statistical rules in many cue environments.

CONCLUSION

Research on decision-making has demonstrated important strengths and limitations in human judgment and choice. These insights have led to descriptive theories that incorporate psychological concerns, such as social pressure, conformity, justifiability, and emotional satisfaction. Behavioral assumptions like these can greatly improve the prediction of human decision-making. Although people violate rational principles, their decisions are still regular and orderly. Uncovering this lawfulness will provide fertile ground for future research.

Further Reading

Dawes RM (1988) *Rational Choice in an Uncertain World*. Fort Worth, TX: Harcourt Brace.

Gigerenzer G, Todd P and the ABC Research Group (1999) *Simple Heuristics That Make Us Smart*. New York, NY: Oxford University Press.

Kagel JH and Roth AE (eds) (1995) *The Handbook of Experimental Economics*. Princeton, NJ: Princeton University Press.

Kahneman D and Tversky A (eds) (2000) *Choices, Values, and Frames*. New York, NY: Cambridge University Press.

Kahneman D, Slovic P and Tversky A (eds) (1982) *Judgment Under Uncertainty: Heuristics and Biases*. New York, NY: Cambridge University Press.

Klein G (1998) *Sources of Power*. Cambridge, MA: MIT Press.

March JG (1994) *A Primer on Decision Making*. New York, NY: Free Press.

Plous S (993) *The Psychology of Judgment and Decision Making*. New York, NY: McGraw-Hill.

Von Winterfeldt D and Edwards W (1986) *Decision Analysis and Behavioral Research*. New York, NY: Cambridge University Press.

Decision-making, Intertemporal

Intermediate article

David Laibson, Harvard University, Cambridge, Massachusetts, USA

CONTENTS

Intertemporal decisions imply trade-offs between current and future rewards. Intertemporal discounting models formalize these trade-offs by quantifying the values of delayed pay-offs.

INTRODUCTION

Decision-makers confront a wide range of critical choices that involve trade-offs between current and future rewards. For example, young workers save part of their paycheck to raise their quality of life in retirement. Habitual heroin users also make decisions with intertemporal consequences when they choose a short-term drug-induced pleasure that jeopardizes their long-term well-being.

To evaluate such trade-offs, decision-makers must compare the costs and benefits of activities that occur at different dates in time. The theory of discounted utility provides one framework for evaluating such delayed pay-offs. This theory has normative and positive content. It has been proposed as both a description of what people should do to maximize their well-being, and to describe what people actually do when faced with intertemporal decisions. Both applications of the model are controversial.

Discounted utility models typically assume that delayed rewards are not as desirable as current rewards or, similarly, that delayed costs are not as undesirable as current costs. This delay effect may reflect many possible factors. For example, delayed rewards are risky because the decision-maker may die before the rewards are experienced. Alternatively delayed rewards are more abstract than current rewards, and hence a decision-maker may not be able to appreciate or evaluate their full impact in advance. Some contributors have argued that delayed rewards should be valued no less than current rewards, with the sole exception of discounting effects that arise from mortality.

DISCOUNTED UTILITY

Formal discounting models assume that a consumer's welfare can be represented as a discounted sum of current and future utility. Specifically, the model assumes that at each point in time, t, the decision-maker consumes goods $c(t)$. These goods might be summarized by a single consumption index (say a consumption budget for period t), or these goods might be represented by a vector (say apples and oranges). The subjective value to the consumer is given by a utility function $u(c(t))$, which translates the consumption measure, $c(t)$, into a single summary measure of utility at period t.

To evaluate future consumption, the consumer discounts utility with a discount function $F(\tau)$, where τ is the delay between the current period and the future consumption. For example, if the current period is date t and a consumer evaluates consumption half a year from now, the consumer calculates the discounted utility value $F(\frac{1}{2})u(c(t+\frac{1}{2}))$.

Since future consumption is usually assumed to be worth less than current consumption, the discounted utility model posits that $F(\tau)$ is decreasing in τ. The more utility is delayed, the less it is worth. Since utility is not undesirable, $F(\tau) \geq 0$ for all values of τ. The model is normalized by assuming $F(0) = 1$. Combining these properties we have

$$1 = F(0) \geq F(\tau) \geq F(\tau') \geq 0, \qquad (1)$$

for $0 < \tau < \tau'$. For example if flows of utility a year from now are worth only $\frac{2}{3}$ of what they would be worth if they occurred immediately, then $F(1) = \frac{2}{3}F(0)$.

Continuous-time Discount Functions

Intertemporal choice models have been developed in both continuous-time and discrete-time settings.

Both approaches are summarized here. Readers without a calculus background may wish to skip directly to the discrete-time analysis.

In continuous-time, the welfare of the consumer at time t – sometimes called the objective function or utility function – is given by

$$\int_{\tau=0}^{\infty} F(\tau)u(c(t+\tau))d\tau, \tag{2}$$

where $F(\tau)$ is the discount function, $u(\cdot)$ is the utility function, and $c(\cdot)$ is consumption.

In both continuous-time and discrete-time models, discount functions are described by two characteristics: discount rates and discount factors. Discount rates and discount factors are normalized with respect to the unit of time, which is usually assumed to be a year.

A discount rate at horizon τ is the rate of decline in the discount function at horizon τ:

$$r(\tau) \equiv \frac{-F'(\tau)}{F(\tau)}. \tag{3}$$

Note that $F'(\tau)$ is the derivative of F with respect to time. Hence, $F'(\tau)$ is the change in F per unit time, so $r(\tau)$ is the rate of decline in F. The higher the discount rate, the more quickly value declines with the delay horizon.

A discount factor at horizon τ is the value of a util discounted with the continuously compounded discount rate at horizon τ:

$$f(\tau) \equiv \lim_{\Delta\to 0}\left(\frac{1}{1+r(\tau)\Delta}\right)^{1/\Delta} = \exp(-r(\tau)). \tag{4}$$

The lower the discount factor, the more quickly value declines with the delay horizon.

Discrete-time Discount Functions

Analogous definitions apply to discrete-time models. For this class of models the discount function, $F(\tau)$, need only be defined on a discrete grid of delay values: $\tau \in \{0, \Delta, 2\Delta, 3\Delta, \ldots\}$. For example, if the model were designed to reflect weekly observations, then $\Delta = \frac{1}{52}$ years.

Once the discrete-time grid is fixed, the discount function can then be written

$$\{F(0), F(\Delta), F(2\Delta), F(3\Delta), \ldots\}. \tag{5}$$

The welfare of the consumer at time t is given by

$$\sum_{\tau=0}^{\infty} F(\tau\Delta)u(c(t+\tau\Delta)). \tag{6}$$

At horizon τ, the discount function declines at rate

$$r(\tau) = -\frac{(F(\tau)-F(\tau-\Delta))/\Delta}{F(\tau)}. \tag{7}$$

The numerator of this expression represents the change per unit time.

The discount factor at horizon τ is the value of a util discounted with the discount rate at horizon τ compounded at frequency Δ.

$$f(\tau) = \left(\frac{1}{1+r(\tau)\Delta}\right)^{1/\Delta} = \left(\frac{F(\tau)}{F(\tau-\Delta)}\right)^{1/\Delta}. \tag{8}$$

As the time intervals in the discrete-time formulation become arbitrarily short (i.e. $\Delta \to 0$), the discrete-time discount rate and discount factor definitions converge to the continuous-time definitions.

Exponential Discounting

Almost all discounting applications use the exponential discount function: $F(\tau) = \exp(-\rho\tau)$. This discount function is often written $F(\tau) = \delta^{\tau}$, where $\delta \equiv \exp(-\rho)$. For the exponential discount function the discount rate is constant and does not depend on the horizon:

$$r(\tau) = \frac{-F'(\tau)}{F(\tau)} = \rho = -\ln\delta. \tag{9}$$

Likewise, the discount factor is also constant:

$$f(\tau) = \exp(-r(\tau)) = \exp(-\rho) = \delta. \tag{10}$$

Figure 1 plots three discount functions, including an exponential discount function. Note that the exponential discount function displays a constant rate of decline regardless of the length of the delay. Typical calibrations adopt an annual exponential discount rate of 5 percent.

Non-exponential Discounting

A growing body of experimental evidence suggests that decision-makers' valuations of delayed rewards are inconsistent with the constant discount rate implied by the exponential discount function. Instead, measured discount rates tend to be higher when the delay horizon is short than when the delay horizon is long. One class of functions that satisfies this property is generalized hyperbolas (Chung and Herrnstein, 1961; Ainslie, 1992; Loewenstein and Prelec, 1992). For example,

$$F(\tau) = (1+\alpha\tau)^{-\gamma/\alpha}. \tag{11}$$

For these functions, the rate of decline in the discount function decreases as τ increases:

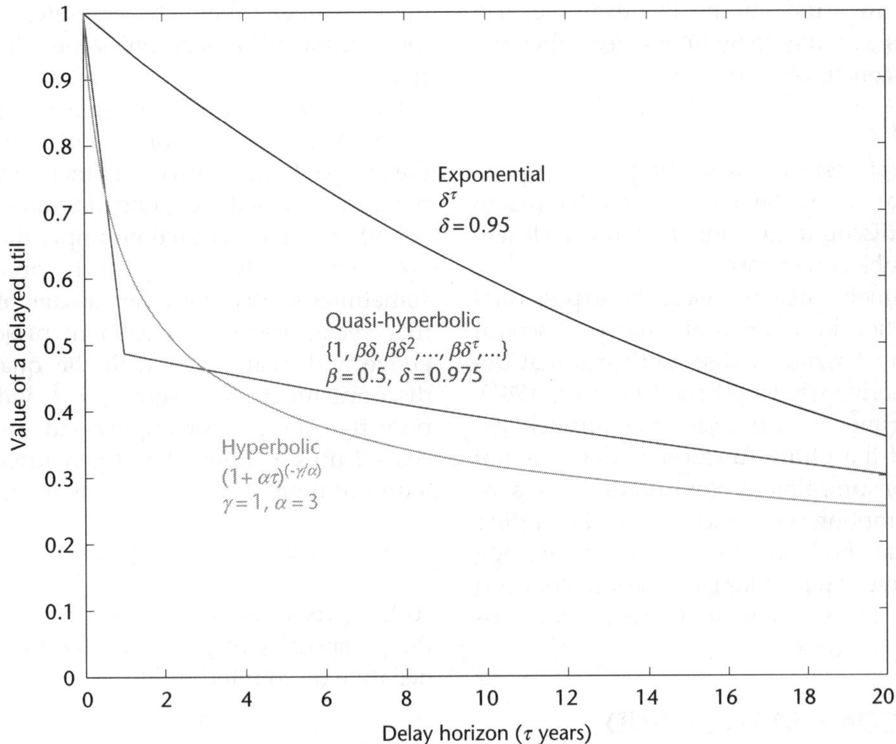

Figure 1. Three discount functions.

$$r(\tau) = \frac{-F'(\tau)}{F(\tau)} = \frac{\gamma}{1 + \alpha\tau}. \tag{12}$$

When $\tau = 0$ the discount rate is γ. As τ increases, the discount rate converges to zero. Figure 1 also plots this generalized hyperbolic discount function. Note that the generalized hyperbolic discount function declines at a faster rate in the short run than in the long run, matching a key feature of the experimental data.

To capture this qualitative property, Laibson (1997) adopts a discrete-time discount function, $\{1, \beta\delta, \beta\delta^2, \beta\delta^3, \ldots.\}$, which Phelps and Pollak (1968) had previously used to model intergenerational time preferences. This 'quasi-hyperbolic function' reflects the sharp short-run drop in valuation measured in the experimental time preference data and has been adopted as a research tool because of its analytical tractability. The quasi-hyperbolic discount function is only hyperbolic in the sense that it captures the key qualitative property of the hyperbolic functions: a faster rate of decline in the short run than in the long run. This discrete-time discount function has also been called 'present-biased' and 'quasi-geometric'.

The quasi-hyperbolic discount function is typically calibrated with $\beta \simeq \frac{1}{2}$ and $\delta \simeq 1$. Under this calibration the short-run discount rate exceeds the long-run discount rate:

$$r(1) = 1 > 0 = r(\tau > 1). \tag{13}$$

More generally, any calibration with $0 < \beta < 1$, implies that the short-run discount rate exceeds the long-run discount rate.

Measuring Discount Functions

Measuring discount rates has proved to be controversial. Most discount rate studies give subjects choices among a wide range of delayed rewards. The researchers then impute the time preferences of the subjects based on the subjects' laboratory choices.

In a typical discount rate study, the researchers make three very strong assumptions. First, the researchers assume that rewards are consumed when they are received by the subjects. Second, the researchers assume that the utility function is linear in consumption. Third, the researchers assume that the subjects fully trust the researchers' promises to pay delayed rewards.

If these assumptions are satisfied, it is then possible to impute the discount function by offering subjects reward alternatives and asking the subjects to pick their preferred option. For example, if a subject prefers \$x today to \$y in one year, then the experimenter concludes that

$$F(0) \cdot x > F(1) \cdot y. \tag{14}$$

Once the subject answers a wide range of questions of this general form, the researcher attempts to estimate the discount function that most closely matches the subject responses.

Such experiments usually reject the exponential discount function in favor of alternative discount functions characterized by discount rates that decline as the horizon is lengthened (Thaler, 1981). Related experiments also reject the implicit assumption that the utility function at date t is not affected by consumption at other dates. This 'separability' assumption is contradicted by the finding that subjects care both about the level *and* the slope of their consumption profiles (see Loewenstein and Thaler (1989) for findings that violate the discounted utility model).

DYNAMIC CONSISTENCY AND SELF-CONTROL

Exponential discount functions have the convenient property that preferences held at date t do not change with the passage of time. Consider the following illustration of this 'dynamic consistency' property. Suppose that at date 0 a consumer with an exponential discount function with discount factor δ, is asked to evaluate a project that requires investments that cost c utils at time 10 with resulting benefits of b utils at time 11. From the perspective of date zero, this project has utility value $\delta^{10}(-c) + \delta^{11}b$. Assume that this utility value is positive and that at date zero the consumer plans to execute the project.

Now imagine that 10 periods pass, and the consumer is asked whether she wishes to reconsider her decision to make the planned investment. From the perspective of period 10, the value of the project to the consumer is $-c + \delta b$. The costs are no longer discounted, since they need to be made in the current period. Likewise, the benefits are only discounted with factor $F(1) = \delta^1$ since they will now be available at only one period in the future.

Note that the consumer's original preference to pursue the project is unchanged by the passage of time, since $\delta^{10}(-c) + \delta^{11}b > 0$ implies that $-c + \delta b > 0$ (divide both sides of the original inequality by δ^{10}). This property of intertemporally consistent preferences is called 'dynamic consistency' and the property will always arise when the discount function is exponential. The passage of time will *never* cause the consumer to switch her preference regarding the investment project (unless new information arrives).

However, dynamic consistency is not a general property of intertemporal choice models. In fact, the only stationary discount function that generates this property is the exponential discount function. All other discount functions imply that preferences are dynamically inconsistent: preferences will sometimes switch with the passage of time. To see this, reconsider the investment project described above and evaluate it with the quasi-hyperbolic discount function (assume $\beta = \frac{1}{2}$ and $\delta = 1$). Suppose that the project requires an investment that costs 2 utils at time 10 and generates a pay-off of 3 utils at time 11. From the time 0 perspective,

$$\beta\delta^{10}(-c) + \beta\delta^{11}b = \frac{1}{2}(-2) + \frac{1}{2}3 = \frac{1}{2}, \tag{15}$$

so the project is worth pursuing. However, from the perspective of period 10, the project generates negative discounted utility

$$-c + \beta\delta b = -2 + \frac{1}{2}3 = -\frac{1}{2}. \tag{16}$$

Hence, the project that the consumer wished to pursue from the perspective of time 0 ceases to be appealing once the moment for investment actually arises in period 10.

This example captures a tension that many decision-makers experience. From a distance a project seems worth doing, but as the moment for sacrifice approaches the project becomes increasingly unappealing. For this reason, quasi-hyperbolic discount functions have been used to model a wide range of self-regulation problems, including procrastination, credit card spending, and drug addiction.

Sophistication, Commitment, and Naivité

The analysis above does not take a stand on whether consumers foresee these preference reversals. Strotz (1956) identifies two paradigms that can be used to analyze the question of consumer foresight: sophistication and naivité.

Sophisticated consumers will anticipate their own propensity to experience preference reversals. Such consumers will recognize the conflict between their early preference – i.e. the preference to undertake the investment project – and their later contradictory preferences. Such sophisticated consumers may look for ways to lock themselves into the investment activity. For example, consider a person who forces himself to exercise by making an appointment with an expensive trainer.

At the other extreme, Strotz also considered consumers who exhibit naivité about their future preference reversals. Such consumers fail to foresee these reservals and expect to engage in investments that they will not actually carry out (e.g. quitting smoking or completing a project with no deadline). Akerlof (1991) discusses such procrastination problems and O'Donoghue and Rabin (2001) propose a framework that continuously indexes the degree of naivité.

SUMMARY

The discounted utility model provides a way of formally evaluating intertemporal trade-offs. The principal component of the model is a discount function that is used to calculate the discounted value of future utility flows. The key characteristics of the discount function are the discount rate and the discount factor. The discount rate measures the rate of decline of the discount function. The discount factor measures the value of a discounted util. Exponential discount functions are commonly used in most applications of the discounted utility model. Exponential discount functions have a constant discount rate. Exponential discount functions also have the convenient property that they do not generate preference reversals. However, the experimental evidence contradicts the constant discount rate property. Most experimental evidence suggests that the discount rate declines with the length of the delay horizon. Such discounting patterns may play a role in generating self-control problems.

References

Ainslie G (1992) *Picoeconomics*. Cambridge, UK: Cambridge University Press.

Akerlof GA (1991) Procrastination and obedience. *American Economic Review Papers and Proceedings* **81**: 1–19.

Chung SH and Herrnstein RJ (1961) Relative and absolute strengths of response as a function of frequency of reinforcement. *Journal of the Experimental Analysis of Animal Behavior* **4**: 267–272.

Laibson DI (1997) Golden eggs and hyperbolic discounting. *Quarterly Journal of Economics* **62**: 443–478.

Loewenstein G and Thaler RH (1989) Anomalies: intertemporal choice. *Journal of Economic Perspectives* **3**: 181–193.

Loewenstein G and Prelec D (1992) Anomalies in intertemporal choice. Evidence and an interpretation. *Quarterly Journal of Economics* **57**: 573–598.

O'Donoghue T and Rabin M (2001) Choice and procrastination. *Quarterly Journal of Economics* **66**: 121–160.

Phelps ES and Pollak RA (1968) On second-best national saving and game-equilibrium growth. *Review of Economic Studies* **35**: 185–199.

Strotz RH (1956) Myopia and inconsistency in dynamic utility maximization. *Review of Economic Studies* **23**: 165–180.

Thaler RH (1981) Some empirical evidence on dynamic inconsistency. *Economic Letters* **8**: 201–207.

Further Reading

Angeletos G, Laibson DI, Repetto A, Tobacman J and Weinberg S (2001) The hyperbolic consumption model: calibration, simulation, and empirical evaluation. *Journal of Economic Perspectives* **15**(3): 47–68.

Frederick S, Loewenstein G and O'Donoghue T (in press) Time discounting and time preference: a critical review. *Journal of Economic Literature*.

Kirby KN (1997) Bidding on the future: evidence against normative discounting of delayed rewards. *Journal of Experimental Psychology* **126**: 54–70.

Laibson DI (2001) A cue-theory of consumption. *Quarterly Journal of Economics* **66**: 81–120.

Loewenstein G (1996) Out of control: visceral influences on behavior. *Organizational Behavior and Human Decision Processes* **65**: 272–292.

Loewenstein G and Elster J (eds) (1992) *Choice Over Time*. New York, NY: Russel Sage Foundation Press.

Loewenstein G and Prelec D (1991) Negative time preference. *American Economic Review Papers and Proceedings* **82**: 347–352.

Loewenstein G, Read D and Kalyanaraman S (1999) Mixing virtue and vice: the combined effects of hyperbolic discounting and diversification. *Journal of Behavioral Decision Making* **12**: 257–273.

Loewenstein G, Read D and Baumeister R (eds) (in press) *Intertemporal Choice*. New York, NY: Russell Sage Foundation Press.

O'Donoghue T and Rabin M (1999) Doing it now or later. *American Economic Review* **89**: 103–124.

Thaler RH and Shefrin HM (1981) An economic theory of self-control. *Journal of Political Economy* **89**: 392–410.

Decoding Neural Population Activity

Introductory article

Peter Foldiak, University of St Andrews, St Andrews, UK

CONTENTS

The brain uses the activity patterns of a large number of neurons to represent information about the world. Decoding methods allow us to read out this neural code and interpret the neural activity pattern.

INTRODUCTION

Brain activity can be observed using a variety of technical methods. Functional imaging techniques such as positron emission tomography and functional magnetic resonance imaging allow us to observe which brain areas show increased activity during the performance of various tasks. However, these methods have relatively poor resolution in space and time, which means that we can only observe average brain activity over many thousands of neurons and over many consecutive brain states. This prevents us from seeing functionally important details in brain activity. Yet we need to understand more than just which areas are involved in certain tasks; we are interested in the computations within these areas. As much of the interesting difference between brain states is visible on the level of much smaller groups of neurons and on a faster time scale, we need more detailed indicators of brain activity. In fact, the detailed response properties of individual neurons even within a small area of the brain seem to be substantially different from each other, which means that if we want to understand information processing in the brain, we need to observe the activity of single cells. Single cell recording is the only technique that gives us the precise times of each action potential (a 'spike train') from an isolated neuron.

In the sensory system, we try to make sense of these spike trains in terms of the sensory input stimuli, while in the motor system we try to understand the relationship between neural spikes and the resulting movements. One view of this relationship is captured by the concept of neural 'tuning curves'. It assumes that most of the information in the spike train can be summarized by a single number, the neural response. This number is often thought to be the firing rate, which is the number of action potentials generated by a neuron during a time interval, divided by the length of that interval. Other aspects of the spike train, such as the more precise timing of the spikes in relation to each other or to the onset of a stimulus, may also be carriers of information. A tuning curve is the (average) response of the neuron as a function of certain properties or parameters of the stimulus input or motor output. An example would be the orientation tuning curve of neurons in primary visual cortex, where the stimulus is a grating of equally spaced parallel dark and bright bars (or two-dimensional sine wave grating), and the neuron responds differently to gratings of different orientations. The orientation tuning curve plots the response as a function of the grating orientation, showing a peak at the 'preferred' orientation of the neuron and lower values elsewhere. A more complete description of the neural code also includes a measure of the response variability, as the neuron's response can be different on each stimulus repetition depending on unpredictable external and internal factors (e.g. stimulus and transduction noise, physiological state, attention, neural interactions).

The other, complementary way of asking questions about the relationship between the spike train and the stimulus (or movement) is by trying to identify the stimulus or movement based on the observed response from a neuron or responses from several neurons. This is the problem of decoding.

POPULATION VECTOR METHOD

A simple method of decoding the response of a neural population is based on the peaks of tuning curves. It was originally developed for predicting arm movements based on the responses of a population of neurons in motor cortex but it can also be applied to populations of sensory neurons to decode a sensory stimulus. Essentially it is a response-weighted sum of the preferred values of the encoding neurons. Each neuron has a vector associated with it that points in the direction that corresponds to the peak of its tuning curve. In the case of arm movements, the direction of this vector is the direction of the arm movement in physical space that maximized the neuron's response. The direction of this vector stays fixed but the length of this vector for each neuron is scaled at each moment to the size of the neuron's response. The population vector at any moment is the sum of the vector contributions of the individual neurons in the population (Figure 1). Under some assumptions about the shapes of the tuning curves and the distributions of their peaks, the population vector will be a reasonable prediction of the stimulus or movement parameters (e.g. the direction of arm movement). To understand this procedure, imagine that each neuron represents a single 'preferred' direction and the strength of each neuron's response indicates how close the actual direction is to the neuron's preferred direction. As the most active neurons will contribute the longest vectors to the vectorial sum, the neurons with the preferred directions closest to the actual direction will have the greatest weight in determining the resulting population vector. This weighted sum is then an estimate of the actual direction, and the estimate

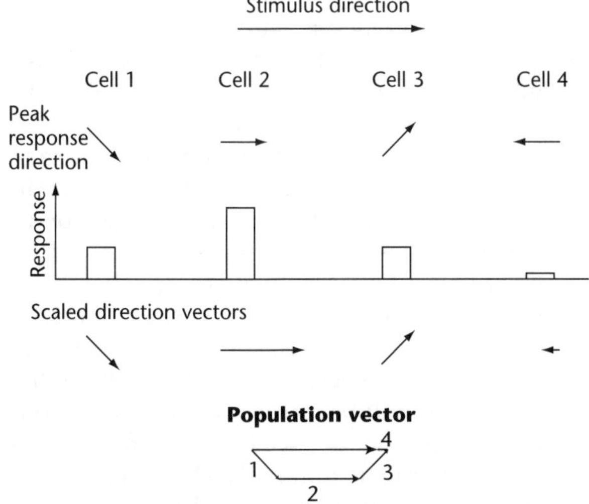

Figure 1. Calculation of the population vector for a horizontal stimulus (or movement) direction from the responses of four neurons.

should get better as the number of neurons in the population increases. While the population vector method is relatively simple to apply, there are several problems with it:

- Neurons in a population do not generally fulfill the condition that the peaks of the tuning curves of the neurons in the population have to be uniformly distributed in all directions (i.e. roughly the same number of neurons have to have peaks in each direction). If there is an excess of peaks in a certain direction, the population vector will be biased in that direction.
- It assumes a specific form of the tuning curve (falling off as a cosine function from the peak). In many cases, these assumptions are not good approximations of the actual tuning curves and their distribution in the population. The only information it uses about the actual tuning curve is the location of its peak, and we have no way of incorporating any additional information (e.g. breadth of tuning, or multiple peaks) we might have about the true tuning curves.
- Even if all conditions necessary for the population vector method are met, there are other methods that give better estimates.
- It has no way to take neural variability into account.
- It generates a single estimate, with no indication of confidence in the estimate or distribution of alternatives.
- In some cases the stimulus parameters do not naturally fall in a space where vectors make sense (e.g. discrete or nominal variables), and the population vectors cannot be calculated.
- Implementing the population vector method in neural machinery is not particularly easy or plausible. Alternative methods that take weighted sums of the neural responses are not equivalent to it, and are biologically much simpler and fit better with what is known about neural response properties.

BAYESIAN METHODS

Most of the above problems (except for the last) can be solved by thinking of both tuning curves and estimates as probability distributions. A description of a neuron's response properties is then given by a conditional probability distribution: $P(r|s)$. This distribution describes the probabilities of observing any response value (r) given that a particular stimulus (s) is presented. The conditional probability is also related to the joint probabilities $P(r,s)$ of the response and the stimulus:

$$P(r|s) = P(r,s)/P(s) \quad (1)$$

where $P(s)$ is the prior probability of seeing stimulus s, without any knowledge of the response. If these conditional distributions can be estimated for all stimuli, we have a full description of the neuron's tuning and variability. Decoding then becomes easy by applying Bayes' theorem, which allows us to calculate the probabilities of each of the possible stimuli given the observed responses: $P(s|r)$. Bayes' theorem follows from the definition of conditional probabilities

$$P(s,r) = P(r|s)\ P(s) = P(s|r)\ P(r) \tag{2}$$

and it states that

$$P(s|r) = P(r|s)\ P(s)/P(r) \tag{3}$$

where $P(r)$ is the probability of a given response, which is summed across all stimuli as

$$P(r) = \Sigma P(r|s)\ P(s) \tag{4}$$

and it is used to normalize the $P(s|r)$ distribution to sum to 1. An example of two discrete stimuli and two discrete responses from a single neuron is given in Figure 2. The formulas given above are applicable to discrete sets of stimuli and responses, but can also be expressed for continuous variables in terms of continuous probability distributions. When the responses of several neurons are available, the response can be considered to be a vector variable, and the distributions need to be estimated over such a vector space. The problem of estimating the density of a multidimensional probability distribution is usually a difficult one for several reasons. One is that the volume of response space increases exponentially with the number of neurons, and therefore the number of data points needed for estimation increases quickly. The other problem is experimental and is related to whether the responses can be recorded simultaneously from the neurons (e.g. using multiple electrodes). If simultaneous recording is not available then the trial-by-trial relationship between the responses is not observable. Estimates of this dependence and of the prior distribution can influence the results substantially. These apparent disadvantages of the Bayesian method are not avoided, only ignored by the alternative methods. The Bayesian method has a far greater flexibility and power than the population vector method. Not only is it free of assumptions about the shape and distribution of the tuning curves but it is also applicable in cases where vectors could not be defined at all (e.g. in some unknown shape space).

Observation: $P(r|s)$

Stimulus	Response $r = 0$	Response $r = 1$
↑	0.8	0.2
↗	0.6	0.4

Joint probability: $P(r,s)$

Stimulus	Response $r = 0$	Response $r = 1$	$P(s)$
↑	0.4	0.1	0.5
↗	0.3	0.2	0.5
$P(r)$	0.7	0.3	

Inference: $P(s|r)$

Stimulus	Response $r = 0$	Response $r = 1$
↑	0.57	0.33
↗	0.43	0.67

Figure 2. Bayesian decoding of a binary neural response (0 or 1) of a single neuron to a set of two possible stimuli (vertical or diagonal). Such binary responses could be observed if measured in a short time window. The conditional probability $P(r|s)$ is measured separately for the two stimuli, and these are the relative frequencies of the responses. The two stimuli are assumed to occur with equal (0.5) probability. Bayes' theorem (Eqn 3) gives the decoding separately for the two possible responses.

STIMULUS DISCRIMINATION USING POPULATION RESPONSES

The result of applying Bayes' theorem is $P(s|r)$. You can think of this distribution as the answer to the question of what the population response means. The Bayesian method is optimal in the sense that if you have the correct $P(r|s)$ values it gives you the true $P(s|r)$ values, and optimal decisions or discriminations can be made based on this. If the goal, for instance, is to make the smallest number of errors in guessing the stimulus, the most probable stimulus – the one corresponding to the peak of the $P(s|r)$ – should be chosen. Other values of this distribution provide possible alternatives with lower probabilities. A highly peaked probability distribution across the stimuli indicates a high degree of certainty, while a broad, flat distribution is a sign of uncertainty in the decoding due to the variability in the response and the overlap between the conditional distributions. Other optimality criteria can be expressed as a loss function $L(s,s')$, where s is the actual and s' is the chosen stimulus. It can be expected that confusing some pairs of stimuli (e.g. similar stimuli, or where both are members of the same category) will be much less costly than confusing other pairs (e.g. members of different categories, or dissimilar stimuli). The loss function can capture the relevant structure, and Bayesian inference gives the decision that minimizes expected cost.

APPLICATIONS

Decoding methods are not usually considered to be models of physiological processes, as it is not likely that explicit stimulus (or movement) decoding takes place in the brain. Information in inputs, outputs, as well as internal representations are distributed to some extent. There is no clear evidence that any component of such a representation could be considered a decoding of a stimulus. Decoding is still interesting from a theoretical perspective, as it tells us the amount and nature of information available for processing in a certain population, and sets upper limits on performance for a mechanism operating on inputs from this population, just as an 'ideal observer' is a useful theoretical concept in studying human psychophysical performance. It allows the calculation of the efficiency of an actual neural mechanism by providing the best possible level of performance for comparison and the calculation of the amount of information available. Such an ideal observer of the neural activity allows an investigation and interpretation of neural representations in real experimental situations. It does not, however, in itself tell us how the brain actually uses the information in a neural population.

Further Reading

Dayan P and Abbott LF (2001) *Theoretical Neuroscience: Computational and Mathematical Modeling of Neural Systems*, chap. 3, Neural Decoding, pp. 87–122. Cambridge, MA: MIT Press.

Oram MW, Foldiak P, Perrett DI and Sengpiel F (1998) The 'Ideal Homunculus': decoding neural population signals. *Trends in Neurosciences* **21**: 259–265.

Rieke F (1997) *Spikes*. Cambridge, MA: MIT Press.

Decoding Single Neuron Activity

Introductory article

James J Knierim, University of Texas–Houston Medical School, Houston, Texas, USA

CONTENTS

Neurons communicate with each other by firing trains of action potentials (spikes). Understanding what information is represented by this activity, and how to decipher the code in which the neural messages are encrypted, are the central issues of single neuron physiology.

INTRODUCTION

One of the most powerful techniques for deciphering the brain mechanisms that underlie cognition is the recording *in vivo* of single neuron activity with the use of microelectrodes. From low-level brain areas (such as primary sensory and motor cortex) to high-level areas (such as prefrontal cortex and hippocampus), single neuron recordings have provided a wealth of data about how information is processed and stored in the brain. One of the great challenges of neuroscience, however, is how to interpret the firing patterns of single neurons. Is information encoded in the average firing rate of neurons, in the precise temporal dynamics of a train of action potentials ('spikes'), or in some other parameter (or set of parameters)? Does the code differ depending on the brain area under study, the experimental task or stimulus at hand, or the amount of experience that the organism has? These questions are at the forefront of neurophysiological research.

REPRESENTING SPIKE TRAINS

Spike trains are usually represented in a few standard formats. Figure 1a shows an oscilloscope trace from an extracellular neuronal recording, with time on the x axis and spike amplitude (voltage) on the y axis. The line underneath the trace denotes the presentation time of the stimulus used to evoke a response from the cell. There are a number of

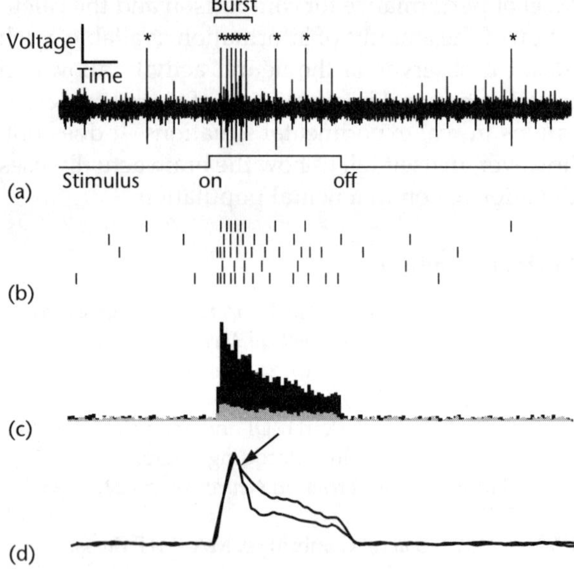

Figure 1. Representations of spike trains. (a) Oscilloscope trace of spikes. (b) Spike raster diagram. (c) Peristimulus time histogram. (d) Peristimulus time histogram averaged over a population of neurons.

discrete spikes of different amplitudes that rise above the baseline level of activity; each amplitude level presumably corresponds to the firing of a different neuron. Let us concentrate on the cell that fires the largest spikes in the example (denoted by asterisks). The neuron fired few spikes in the periods before and after the stimulus presentation; the rate of firing in the absence of stimulation is referred to as the background level or spontaneous activity level. When the stimulus was presented, the neuron fired a rapid series of spikes. There was a burst of activity shortly after the onset of the stimulus, and the activity decreased over time (although in some neurons the response might stay strong throughout the duration of the stimulus). These raw recordings are usually converted to a number of formats to represent the spike trains. Figure 1b is a spike raster plot from five repetitions of the same stimulus. The first line of the raster is a representation of the oscilloscope trace in Figure 1a. Each mark corresponds to the firing of a spike at that particular time. The neural responses to further repetitions of the stimulus are shown in the subsequent lines. As all responses are aligned to the onset of the stimulus, notice that the first spike of each response can occur after a variable delay period (see below). (*See* **Single Neuron Recording**)

Such spike rasters can be converted into a peristimulus time histogram (PSTH), representing the cumulative firing of the neuron over many trials (Figure 1c). The trial is divided into a series of time bins aligned to the onset of the stimulus, and all of the spikes that occurred within a time bin are summed to create the PSTH. Figure 1c illustrates two superimposed histograms (one black and one shaded), each corresponding to a different stimulus, making it clear that the neural response to one stimulus was greater than the response to the other. Finally, the PSTHs of many neurons can be averaged together, as in Figure 1d. Although the single neuron in Figure 1c had very different responses to the two stimuli, Figure 1d demonstrates that the average response of a population of many neurons did not distinguish the two stimuli as strongly; moreover, the average response was equal during the initial part of the response, and the difference in response became apparent only after a delay period (arrow). Such population averages provide a 'snapshot' of the overall neural population response to a stimulus or to a behavioral event, which sometimes may prove more informative than the response of any single neuron. (*See* **Decoding Neural Population Activity**)

As understanding of neural firing properties increases and theoretical advances occur, more sophisticated mathematical techniques are being employed to represent spike trains and to decipher the information they encode. Such approaches include the use of information theory to quantify rigorously the amount of information encoded in a spike train, and computational modeling to simulate how neurons or networks of neurons represent and store information. A growing legion of physicists, mathematicians, computer scientists and engineers are joining traditional neuroscientists to develop the analytical tools that will be necessary to understand fully the firing patterns of neurons. (*See* **Information Theory**)

SOURCES OF NEURONAL VARIABILITY

It is obvious from Figure 1b that there is a large degree of variability in the spike trains elicited on each of the five trials, both in the number of spikes fired on each trial and in the temporal pattern of spikes. Such variability is ubiquitous in the brain and derives from two broad sources. The first source is the set of uncontrolled variables that are present in a particular experiment. These uncontrolled variables may be behavioral variables in an alert animal preparation (e.g. changes in the animal's attentional state, changes in gaze direction, and changes in arousal or motivation); stimulus parameters that may change slightly from trial to trial; or other factors, such as changes in anesthesia level in an anesthetized preparation. The second source derives from properties that are intrinsic to the neuron or neural network. For example, the stochastic nature of the opening and closing of individual ion channels may contribute to whether a neuron that is near firing threshold actually generates a spike at a particular time. As a second example, repeated presentations of a stimulus can cause an adaptation of the response, in which the response strength decreases over time as the result of changes in biochemical cascades within a neuron or in the strengths of synaptic connections in the network. Of these intrinsic sources, it is useful to distinguish random sources (e.g. the stochastic properties of single ion channels) from deterministic sources (e.g. response habituation).

MOST EFFECTIVE STIMULI

The spike trains illustrated in Figure 1 are often reduced to a single number, the firing rate, by dividing the number of spikes in a certain period by the length of that period (e.g. the amount of time that the stimulus was presented). Neurons are often characterized by the type of stimulation that produces their greatest rate of firing. For example, the responses of a neuron in the visual cortex to different colors of light can be plotted as a tuning curve that illustrates the overall response selectivity of the cell. Figure 2 illustrates a hypothetical cell that fired strongly to a green light, with little or no response to light of other wavelengths. Such a cell is said to be tuned to, or selective for, a particular stimulus. In contrast, a second cell responded very similarly to wavelengths of all colors. This cell is

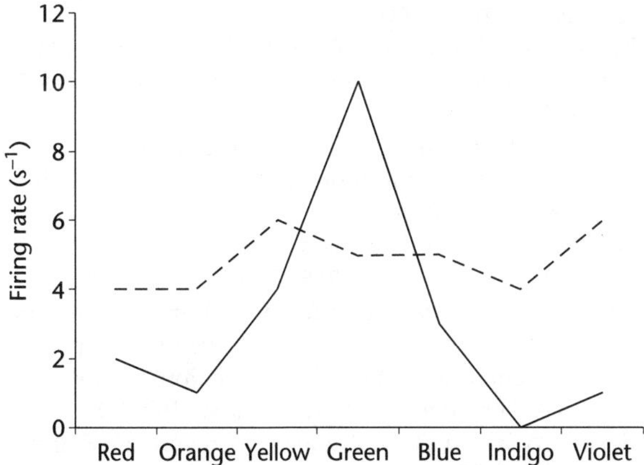

Figure 2. Hypothetical tuning curves for visual neurons. The solid line represents a neuron that is selective for the color green; the dashed line represents a neuron that is not selective for color.

said to be nonselective for color, although it might be selective for some other parameter (for example, the size of a stimulus). The characterization of the most effective stimulus for a particular neuron is one of the most prevalent and important analyses of single-unit data. Neurons have been identified in the visual cortex that are selective for color, size, depth, speed of motion, direction of motion, spatial frequency, and other, more complex, properties of visual stimuli. Neurons in the auditory system can be selective for sound frequency, loudness, or the spatial location of a sound source. Neurons in the thalamus and limbic cortex can be selective for the direction in which the animal is facing. These are a few examples of the many different types of selective responses displayed by neurons in different parts of the brain.

Although this type of research is important, a few caveats are necessary. First, finding a cell that is selective for color, for example, does not necessarily mean that the function of the cell is color perception *per se*. The neuron may be part of a network that is actually involved in extracting information about boundaries between objects based on differences in color between the two objects. Second, finding the most effective stimulus that drives a neuron is limited to the set of stimuli that are tested in a given experiment. Thus, a cell may respond best to a red spot of light when tested against other spots of light, but its true function may be related more closely to the perception of complex three-dimensional shapes. If the neuron's response to such shapes is never tested, however, the investigator may incorrectly attribute color selectivity as the primary correlate of that neuron. This point leads to the third caveat, which is that neurons often are tuned along multiple stimulus dimensions. Thus, a cell may be sharply tuned to red on the color dimension, horizontal on the orientation dimension, and far away on the depth dimension. Is such a cell specifically tuned for a horizontally oriented, red bar that is located far away from the observer? Or does this cell act as a filter for many different stimulus parameters in a multiplexed, population code? Most current thought favors the latter interpretation, but how the cell performs this task and how the brain interprets the code is not well understood.

REVERSE CORRELATION METHODS

The analyses described above are based on the methods of presenting a set of discrete stimuli to the animal and measuring the response to each member of the set. Another powerful method of finding the most effective stimulus for a cell is to record the activity of a neuron continuously, during behavior or during the presentation of a continuous stream of stimuli, and to use the method of reverse correlation to determine what aspect of the behavior or stimulus drove the cell to fire. For example, a neuron in visual cortex might be recorded while an animal views a succession of scenes on film. Whenever a spike occurs, the frame displayed at that moment is tagged. After the presentation of the whole film sequence, each tagged frame is analyzed to find the common element (or elements). One way to do this would be to create a prototypical tagged frame by averaging each pixel from all of the tagged frames. If the averaged frame showed a vertically oriented edge, one might infer that the cell was acting as a filter for vertical orientation. As another example, cells in the hippocampus of rats can be recorded as the animal freely explores an environment. For each spike, the spatial location of the animal is recorded, and by reverse correlation of the spikes of the cell with the position of the rat, it can be shown that the cell fires whenever the rat enters a particular part of that environment. Each pyramidal cell in the rat hippocampus fires in different selected regions of different environments, and this finding was one of the principal lines of evidence for the cognitive map hypothesis of hippocampal function. (*See* **Place Cells**)

ENCODING VERSUS DECODING

It is important to distinguish the encoding of neural activity from its decoding. This discussion has so far been concerned mainly with the former operation; that is, how does the brain take information (either from the external world via sensory receptors or from other brain areas) and transform it into a representation made of spiking neurons. The other side of the problem is understanding how the brain later on decodes this representation. To understand encoding, we want to know how a neuron will respond to a particular stimulus:

stimulus → neural response

To understand decoding, we want to know how a downstream neuron interprets this neural response in order to determine what stimulus produced the response:

neural response → stimulus

These two processes can be thought of in terms of probabilities, and they can be related by an equation known as Bayes' rule:

$$P(\text{stimulus}|\text{response}) = P(\text{response}|\text{stimulus}) \times P(\text{stimulus})/P(\text{response}) \quad (1)$$

where P(stimulus|response) is the probability that a particular stimulus was presented given that the neuron produced a particular response (decoding); P(response|stimulus) is the probability that the neuron produced that response given that the particular stimulus was presented (encoding); P(stimulus) is the general probability of the animal being presented with that particular stimulus out of all possible stimuli; and P(response) is the general probability of the neuron producing that response regardless of what stimulus is presented. (*See* **Neurons, Representation in**)

Understanding encoding is a prerequisite for understanding decoding, and much research focuses on exactly what coding schemes are used by neurons. For example, one of the great debates of neurophysiology is the degree to which the firing of neurons constitutes a rate code or a temporal code: is the information encoded by neurons contained in the firing rate of the neuron, averaged over a certain time interval, or is the information contained in the precise temporal pattern of spikes emitted by the neuron? How neurons perform the decoding operation is also a subject of much research. Although the complexity of the brain and the complexity of cognitive behavior combine to make these tasks daunting, they are necessary for a complete understanding of the brain mechanisms that underlie cognition. (*See* **Rate versus Temporal Coding Models**)

Further Reading

Abbott LF (1994) Decoding neuronal firing and modelling neural networks. *Quarterly Reviews of Biophysics* **27**: 291–331.

DeCharms RC and Zador A (2000) Neural representation and the cortical code. *Annual Review of Neuroscience* **23**: 613–647.

DeYoe EA and Van Essen DC (1988) Concurrent processing streams in monkey visual cortex. *Trends in Neuroscience* **11**: 219–226.

Laurent G (1999) A systems perspective on early olfactory coding. *Science* **286**: 723–728.

Optican LM and Richmond BJ (1987) Temporal encoding of two-dimensional patterns by single units in primate inferior temporal cortex. III. Information theoretic analysis. *Journal of Neurophysiology* **57**: 162–178.

Rieke F, Warland D, de Ruyter van Steveninck R and Bialek W (1997) *Spikes: Exploring the Neural Code.* Cambridge, MA: MIT Press.

Shadlen MN and Newsome WT (1994) Noise, neural codes and cortical organization. *Current Opinion in Neurobiology* **4**: 569–579.

Deductive Reasoning

Introductory article

PN Johnson-Laird, Princeton University, Princeton, New Jersey, USA

CONTENTS

Deductive reasoning yields conclusions that must be true given that the premises are true. The challenge to cognitive scientists is to discover the valid conclusions that human reasoners are competent to draw, and the mental processes underlying deductive performance.

INTRODUCTION

Deductive reasoning is a process yielding valid conclusions from premises, where a conclusion is valid if it must be true given that the premises are true. The premises may be assertions, perceptions, memories or thoughts. The conclusion is in the first instance a thought that may lead on to an assertion or to an action. A typical example of a valid deduction is:

Either government invests or a recession occurs.
Government doesn't invest.
Therefore, a recession occurs.

If the premises are true, then the conclusion must be true. If the premises are false, however, a valid deduction may yield a true or a false conclusion. Invalid inferences, such as inductions from experience, can be useful, but they lack a guarantee of truth even if their premises are true.

The business of life depends on deductive reasoning. Individuals differ in this ability, and those who are better at it also perform better on tests of intelligence. They also appear to be more successful in life. A person who is poor at deductive reasoning is liable to blunder. Conversely, without deductive reasoning, there would be no logic, mathematics, science, law or society. The challenge to cognitive scientists is to discover the valid conclusions that human reasoners are competent to draw, and the mental processes that underlie their performance.

LOGIC

Logic is the science of valid deduction, and logicians have systematized many different logical calculi. They approach the task in two distinct ways. The first way concerns the formal pattern or syntax of symbols, and it is known as 'proof' theory. Consider as an example the sentential calculus, which concerns negation and idealized versions of such connectives as 'if', 'or' and 'and'. The calculus can be formalized by specifying rules of inference that govern it. The following rule, for example, concerns disjunctions:

A or B.
Not-A.
Therefore, B.

This rule states that given a disjunction of the form *A or B*, such as:

Either government invests or a recession occurs.

and an assertion of the form *not-A*:

Government does not invest.

then one can derive a conclusion of the form *B*:

A recession occurs.

When logicians formalize a calculus, they bear in mind the intended meaning of the connectives, but the formal rules themselves make no use of this meaning. They are rules for writing new patterns of symbols, which are sensitive only to the syntax or form of sentences. Hence, formal rules operate like a computer program. When a computer program predicts the economy, for example, the computer has no idea of what an economy is or of what it is doing. It merely slavishly follows its instructions, and displays symbols that economists can interpret as predictions. Indeed, an intimate relation exists between computer programs and formal proofs: logic can be used to prove theorems about programs, and programs can be used to construct logical proofs.

Formalization can be carried out in different, though equivalent, ways. One way, which has been influential in cognitive science, is known as the method of 'natural deduction'. It uses formal rules of inference for each of the main sentential connectives. Some rules introduce connectives, such as the following three rules:

A		
B	A	A|– B
Therefore,	Therefore,	Therefore,
A and B	A or B, or both	If A then B

where A|– B signifies that the assumption of A for the sake of argument yields a proof of B. Other rules eliminate connectives, such as:

	A or B, or both	If A then B
A and B	Not-A	A
Therefore, B	Therefore, B	Therefore, B

Natural deduction had a vogue in textbooks, but still more intuitive methods exist for teaching logic.

The second way in which logicians systematize a calculus is semantic; i.e. it is concerned with meaning and truth. It is known as 'model' theory. For the sentential calculus, this method states the truth or falsity of a proposition containing a connective in terms of the truth or falsity of its constituent propositions. The ultimate constituents are atomic propositions, propositions that contain neither negation nor connectives. Consider a disjunction of the form *A or B, or both*, such as:

Either government invests or a recession occurs, or both.

Logicians assume that it is true if at least one of its two atomic propositions (government invests, a recession occurs) is true, and it is false if both of them are false. These conditions apply to many other disjunctions. Natural language, however, is not always so tidy. Just as there is poetic license, so there is logical license – a need for logicians to make simplifying assumptions about the meanings of logical terms. Logicians lay out the truth conditions for connectives in a truth table, such as that shown in Table 1. Each row in the table shows a possible combination of truth values of A and B, and the resulting truth value of the disjunction. The first row in the table, for instance, represents the case where A is true and B is true, and so the disjunction as a whole is true.

Table 1. Truth table for the disjunction *A or B, or both*

A	*B*	*A or B, or both*
True	True	True
True	False	True
False	True	True
False	False	False

To complete the semantic characterization of the sentential calculus, definitions of each connective need to be made. One tricky connective is 'if'. A conditional assertion such as:

If government invests then inflation occurs.

is true given that government invests and inflation occurs, and false given that government invests and inflation does not occur. What is its status when its antecedent is false – that is, when government does not invest? This puzzle has generated a vast literature. The simplest semantics, however, may govern usage in daily life. It treats the conditional as true when its antecedent is false. In other words, granted that the conditional is true but that the government does not invest, then inflation may or may not occur.

As logicians have proved, any conclusion that can be derived using the formal rules of the sentential calculus is also valid using truth tables, and vice versa. They have also proved that there is a decision procedure for the calculus; i.e. a method that takes a finite number of steps to decide whether an inference is valid or invalid. However, this happy state of affairs does not apply to all logical calculi. An important extension of the sentential calculus is known as the predicate calculus. It deals with proofs concerning quantifiers, such as 'some' and 'all', as in the following syllogism:

Some of those economies are inflationary.
All inflationary economies are unstable.
Therefore, some of those economies are unstable.

The predicate calculus is only semi-decidable. That is, any valid inference can be derived in a finite number of steps; but if an inference is invalid, no guarantee exists that its invalidity can be established in a finite number of steps. The most important logical discovery, as Kurt Gödel showed, is that logics exist in which not all valid inferences can be proved using formal rules. As he also famously proved, there are truths in arithmetic that cannot be derived in any consistent formal calculus. Some theorists have taken an analogous proof about computer programs to show that human creativity is not computable. Alan Turing, the intellectual father of the programmable digital computer, anticipated and rebutted this argument when he pointed out that human beings are unlikely to be consistent.

DEDUCTIVE COMPETENCE

Once, the task for cognitive scientists seemed to be to isolate the logic (or logics) that people have in their minds. The challenge was the variety of candidates. For example, there are infinitely many distinct modal logics, which deal with possibility and necessity. Nevertheless, theorists argued throughout the twentieth century that logic is *the* theory of deductive competence. Jean Piaget, the famous investigator of children's mental development, argued that 'reasoning is nothing more than the propositional [i.e. sentential] calculus itself'. Yet there is a flaw in this claim. The sentential calculus yields infinitely many different valid conclusions from any set of premises. For instance, the following premises:

If government invests then inflation occurs.
Government invests.

yield the valid conclusions:

Government invests and government invests.
Government invests and government invests and government invests.

... and so on *ad infinitem*.

Of course such conclusions are preposterous. No sane individual – other than a logician – is likely to draw them. Yet they are all valid. Given, say, the following two premises and asked what follows from them:

A villanelle is a poem.
Ruislip is in London.

most people respond, 'Nothing'. Yet, to repeat, logic allows infinitely many valid conclusions from any premises. So, the response is wrong. However, the conclusions that follow from these premises are of no interest or use:

A villanelle is a poem *and* Ruislip is in London.

People are sensible. They do not draw just any valid conclusion, and sometimes they respond that nothing follows. Logic is therefore not a theory of deductive competence (*pace* Piaget). It captures which conclusions are valid, but it has nothing to say about which particular valid conclusions, if any, individuals draw.

The conclusions that people do draw tend to conform to three general principles. First, the conclusion maintains the semantic information given in the premises; that is, individuals do not throw away semantic information by adding disjunctive alternatives. Thus, the inference:

Government invests.
Therefore, government invests *or* inflation occurs.

is valid. Yet logically untrained individuals baulk at it. They do so for the excellent reason that the premise conveys more information than the conclusion, i.e. the premise is compatible with fewer possibilities than the conclusion. Second, reasoners are parsimonious in their conclusions. They do not spontaneously make deductions such as:

Government invests.
Inflation occurs.
Therefore, government invests *and* inflation occurs.

This conclusion uses more words to say no more than the premises do. Third, reasoners try to draw conclusions that make explicit something that was not stated as such in the premises. For example, given the premises:

If government invests then inflation occurs.
Government invests.

they do not draw the conclusion:

Government invests.

The conclusion is parsimonious, but it does not say anything new. Valid deductions cannot add semantic information to the premises, and so sceptics sometimes suggest that valid deductions serve no useful purpose. In fact, they are wrong. Valid deductions can convey propositions that are not obvious consequences of the premises. They can make explicit a proposition that was not asserted as such by any of the premises. It is these inferences that naive individuals strive to make. Hence, given the preceding premises, they almost always infer:

Inflation occurs.

If there is no conclusion that satisfies these three principles, then people are sensible. They say that nothing follows.

In short, to deduce is to maintain semantic information, to simplify, and to reach a new conclusion. None of these principles can be derived from logic.

THEORIES OF DEDUCTIVE PERFORMANCE

How people make deductions is a matter of controversy. Do their mental processes depend on a single system, or, as evolutionary psychologists suppose, on a set of separate systems shaped by natural selection? Do the processes rely on formal rules like those of a logical calculus, or on rules with a specific content, or, as some researchers in artificial intelligence suggest, on calling to mind past cases of valid reasoning? Psychologists have been struggling with deduction for nearly a century; cognitive scientists have recently homed in on it, and defended each of the preceding positions. The controversy is hot because deduction is an excellent test case: if cognitive scientists cannot understand it, they are unlikely to understand much about any sort of thinking.

You can make valid deductions about matters of which you have no substantial knowledge. Even if you know nothing about poetry, for instance, you can still grasp the validity of the following inference:

If a poem is a villanelle then it has five tercets followed by a quatrain.
Elizabeth Bishop's *One Art* is a villanelle.
Therefore, it has five tercets followed by a quatrain.

Hence, theories based on knowledge of a domain can at best tell only part of the story of reasoning. Among the theories above, however, are two that can cope with inferences in general. These two theories run in parallel with the distinction in logic between proof theory and model theory. The first sort of theory postulates that your mind is equipped with formal rules of inference akin to those of a 'natural deduction' system. Lance Rips, Daniel Osherson, Martin Braine and others have proposed theories of this sort. You are not aware of these rules when you reason, and so if they exist, then they must be unconscious. One such rule is of the following form (see the earlier section on logic):

If A then B.
A.
Therefore, B.

You can make the preceding inference by matching its premises to the premises of this rule, and then by drawing the conclusion permitted by the rule.

The Selection Task

The British psychologist Peter Wason and his students pioneered the modern experimental study of deductive reasoning, and established that two main factors affect performance: the logical structure of inferences, and their semantic content. These studies showed, for example, that responses in Wason's well-known selection task were qualitatively different depending on the content of problems. In the abstract version of the task, the experimenter laid out four cards bearing letters or numbers as follows in front of the participants:

A B 2 3

The participants knew that each card had a letter on one side and a number on the other side. They were given a conditional assertion:

If a card has the letter 'A' on one side then it has the number '2' on the other side.

Their task was to select those cards that needed to be turned over to discover whether the rule was true or false about the four cards. Most people selected the A card alone, or the A and 2 cards. What was puzzling was their failure to select the 3 card: if it has an A on its other side, then the conditional assertion is false. Indeed, nearly everyone judges it to be false in this case. When the content of the assertion in the selection task was changed to a sensible everyday generalization, such as:

Every time I go to Manchester I travel by train.

many people made the correct selections. Each of four cards had the name of a destination on one side and the mode of transport for getting there on the other side, and the four cards shown to the participants were:

Manchester Sheffield train car

The participants tended to select the 'Manchester' and 'car' cards.

The selection task has been the most popular paradigm in the experimental study of deductive reasoning. Its large literature yields one overwhelming message: certain contents improve performance. Evolutionary psychologists argue that the mind has developed a set of specialized systems (or modules) as adaptations to the lives of our forebears. Thus, they have proposed an innate module for reasoning about cheating, because social exchanges were important to the hunter-gatherers of the Pleistocene epoch. Experiments have shown that selection tasks about potential cheaters do indeed improve performance; but, as the example above shows, such contents are not necessary to elicit insight into the task. Formal rules of inference cannot account for these phenomena either, because manipulations of content that have no bearing on logic can have a large effect on accuracy. These phenomena, however, did not deter subsequent theorists from postulating that reasoning relies on formal rules of inference.

Mental Models

As a reaction to the phenomena of the selection task, a second theory of deduction makes no use of formal rules of inference. This theory postulates that reasoning is not a formal matter (unless you have learned logic), but depends instead on your understanding of the meaning of the premises, on your perception of the situation, and on your general knowledge. You use this information to construct mental models of the situations described by the premises. You formulate a conclusion that holds in the models but that was not explicitly asserted by any premise. The strength of your inference depends on the proportion of the models in which the conclusion holds. A conclusion that holds in all the models is *necessary* given the premises, and a conclusion that purports to be necessary can be refuted by a counter-example; that is, by a model that satisfies the premises but not the conclusion. A conclusion that holds in most of the models is *probable* given the premises, and a conclusion that holds in at least one model is *possible* given the premises. Models accordingly go beyond formal rule theories to provide a unified account of logical, modal, and probabilistic reasoning.

The theory makes three principal assumptions. The first assumption is that possibilities, not truth values, are fundamental, and that each mental model represents a possibility. Thus, the disjunction:

> Either government invests or a recession occurs, or both.

calls for three mental models, shown here on separate lines, which represent the three possibilities:

Government invests	
	recession
Government invests	recession

where 'Government invests' denotes a model of the government investing, 'recession' denotes a model of the occurrence of a recession, and the third line denotes their conjunction in a single model. The diagram uses words, but that does not imply that mental models are made up of words. There is a critical difference between the diagram and what it denotes, namely a set of mental models. You understand what it means for government to invest. Your understanding is represented in your mind in a way that is independent of your language. Indeed, if such a thought comes to you, you may have difficulty in putting it into words. Moreover, if you also speak Spanish, then you are likely to build the same mental models from the corresponding assertion in Spanish.

The second assumption of the theory is known as the 'principle of truth'. It postulates that mental models represent only what is true, and not what is false. Hence, the preceding set of models does not represent the case in which the disjunction as a whole is false. In addition, however, each model represents the clauses in the premises only when they are true in a possibility. For instance, the first model in the preceding set represents the possibility that government invests, but it does not make explicit that in this possibility it is false that a recession occurs. People represent what is false in 'mental footnotes', but these footnotes are ephemeral and normally soon forgotten. The failure to consider what might falsify the conditional assertion accounts for the difficulty of the abstract selection task. The failure is mitigated, however, when the content of the task yields more familiar counter-examples.

The third assumption is that the parts of a mental model correspond to the parts of what it represents, and so the structure of the model corresponds to the structure of what it represents. A mental model is therefore like an architect's model of a building. Mental models underlie visual images, which are projections from them, but not all mental models can be visualized. Complex mental models of systems can be built up in long-term memory.

To illustrate reasoning by model, consider again the inference:

> If government invests then inflation occurs.
> Government invests.
> Therefore, inflation occurs.

The conditional calls for a model of the case in which its antecedent and consequent are both true. People realize that there are other possibilities consistent with the conditional – in particular, those in which it is false that government invests. However, they do not represent them explicitly. The conditional therefore elicits two models:

Government invests	inflation
...	

where the ellipsis denotes a model that has no explicit content. This implicit model is a 'place holder' for the cases in which it is false that government invests. The premise that government invests eliminates this second model, and the first model yields the conclusion that inflation occurs.

Illusions and Departures from the Norms of Logical Validity

Do individuals reason using formal rules or mental models? A recently discovered phenomenon may answer the question. Consider the following problem:

> Only one of the following assertions is true about a hand of cards:
> There is a king in the hand or an ace, or both.
> There is a queen in the hand or an ace, or both.
> There is a jack in the hand or a ten, or both.
> Is it possible that there is an ace in the hand?

Nearly everyone responds, 'Yes,' and the conclusion seems obvious. Yet the inference is illusory. It is impossible for an ace to be in the hand. If there were an ace then the first two assertions would both be true, contrary to the claim that only one of the three assertions is true. People succumb to the illusion because they think only about what is true, and the truth of the first assertion suggests that an ace is possible. But, when the first assertion is true, the other two assertions must be false. And the falsity of the second assertion means that there is neither a queen nor an ace in the hand.

Individuals are vulnerable to a variety of illusions in modal and probabilistic reasoning. The rubric 'Only one of the assertions is true' is equivalent to an exclusive disjunction, and a compelling illusion occurs in the following inference about a particular hand of cards:

> If there is a king in the hand then there is an ace in the hand, or else if there isn't a king in the hand then there is an ace in the hand.
> There is a king in the hand.

What, if anything, follows? Nearly everyone infers that there is an ace in the hand. It follows from the mental models of the premises. Yet it is a fallacy granted a disjunction between the two conditionals. One or other of the conditionals could therefore be false. And if, say, the first conditional is false, then there is no guarantee that there is an ace in the hand even though there is a king. Of course, skeptics can argue that 'or else' is treated as meaning 'and' in the problem, but that argument fails to explain the illusions based on the rubric 'only one of the assertions is true', and it also fails to explain which sentences yield an interpretation of 'or else' as 'and'. They seem to be precisely those for which the model theory predicts that illusions will occur because individuals neglect what is false.

The illusions are so compelling that they go unnoticed in daily life. For example, a professor cautioned students on his Web page:

> Either a grade of zero will be recorded if your absence [from class] is not excused, or else if your absence is excused other work you do in the course will count.

The mental models of this assertion represent the two possibilities that presumably he had in mind, assuming that a zero grade is incompatible with other work counting:

not-excused	zero-grade
excused	other-work-counts

The students probably made the same interpretation. Yet it is wrong. What really conveys these two possibilities is a conjunction:

> A grade of zero will be recorded if and only if your absence is not excused, *but* if and only if your absence is excused then other work you do in the course will count.

An inclusive disjunction of the two conditionals is a much weaker assertion compatible with more possibilities than those above. It allows that students with an excuse are no different from those without an excuse: both may get a zero grade or have other work on the course count, or neither. An exclusive disjunction of the two conditionals is even stranger: students with an excuse either get a grade of zero or not, but other work never counts; and students without an excuse never get a grade of zero, and other work may or may not count.

The illusions are a litmus test for mental models, because its principle of truth predicts them, but they jeopardize theories of reasoning based on formal rules. These theories currently postulate only valid rules, which cannot explain the systematic invalidity of illusory inferences. If these theories introduced invalid rules to account for the illusions, then it would follow that human beings are intrinsically irrational. However, the illusions have no such implication: people do understand the explanation of their errors. Given the limitations of human working memory, reasoners cannot cope with truth and falsity, that is, with complete truth tables. The principle of taking into account what is true and forgoing what is false is a sensible compromise. Truth is more useful than falsity. Just occasionally, however, truth alone leads human reasoners into the illusion that they grasp a set of possibilities that is in fact beyond them.

Further Reading

Baron J (1994) *Thinking and Deciding*, 2nd edn. New York, NY: Cambridge University Press.

Braine MDS and O'Brien DP (eds) (1998) *Mental Logic*. Mahwah, NJ: Lawrence Erlbaum.

Brooks M (2000) Fooled again. *New Scientist* **2268**: 24–28.

Davis M (2000) *The Universal Computer: The Road from Leibniz to Turing*. New York, NY: Norton.

Evans JSBT, Newstead SE and Byrne RMJ (1993) *Human Reasoning: The Psychology of Deduction*. Mahwah, NJ: Lawrence Erlbaum.

Garnham A and Oakhill J (1994) *Thinking and Reasoning*. Cambridge, MA: Blackwell.

Jeffrey R (1981) *Formal Logic: Its Scope and Limits*, 2nd edn. New York, NY: McGraw-Hill.

Johnson-Laird PN (1999) Deductive reasoning. *Annual Review of Psychology* **50**: 109–135.

Johnson-Laird PN and Byrne RMJ (1991) *Deduction*. Hillsdale, NJ: Lawrence Erlbaum.

Rips LJ (1994) *The Psychology of Proof*. Cambridge, MA: MIT Press.
Schaeken W, De Vooght G, Vandierendonck A and d'Ydewalle G (2000) *Deductive Reasoning and Strategies*. Mahwah, NJ: Lawrence Erlbaum.
Stanovich KE (1999) *Who is Rational? Studies of Individual Differences in Reasoning*. Mahwah, NJ: Lawrence Erlbaum.

Delusions

Introductory article

William O'Donohue, University of Nevada, Reno, Nevada, USA
Andrew Lloyd, University of Nevada, Reno, Nevada, USA

CONTENTS

Delusions are strongly held false beliefs about reality. Although there is only moderate agreement as to what counts as a delusion, delusional beliefs and associated behaviors have long been regarded as evidence of insanity.

INTRODUCTION

Delusions are strongly held false beliefs about reality. The belief that one's thoughts are being stolen by a clandestine government agency or that one's internal organs have been replaced with mechanical replicas are good examples of delusional beliefs. Delusions are associated with over 70 psychological conditions and disorders. Schizophrenia, Alzheimer disease, Parkinson disease (along with other dementing diseases), and many substance abuse-related disorders are often marked by delusions. Delusional beliefs, and the behaviors consistent with them, are frequently the most relevant factors leading to the conclusion that a person is mentally ill. Indeed, delusions have been considered to be the defining, or paradigmatic, feature of madness for centuries. However, despite the clear relevance of delusions to psychological, psychiatric, and general cognitive theories and models, only moderate agreement exists as to what counts as a delusion.

SYMPTOMATOLOGY

Standard definitions of delusions involve at least four components. First, a delusional belief is a false belief or error. Such false beliefs can run from the relatively mundane, such as the belief that a close friend is secretly in love with you, to the very bizarre, for example, that one's head is full of bees. Second, a delusional belief is maintained with great conviction or certainty. Such beliefs have the character of being fundamentally true and incapable of being questioned owing to their perceived importance. Third, a delusional belief is maintained despite, or in the face of, incontrovertible evidence to the contrary. Not only do such beliefs have no empirical support (because they are false), they also persist in the face of what most people would consider incontrovertible evidence of their falsity. The belief that the Earth is flat or that the Holocaust never took place are examples of beliefs maintained in the face of strong counter-evidence. Fourth, a delusional belief is the kind of belief that is not typically supported by the individual's culture. This condition, though it is of critical importance to the definition of delusions, admits of a great deal of complexity and is, by definition, relative to many context-dependent variables. Though certainly false, the belief that consuming the brain of a recently deceased family member can protect one from the cause(s) of that person's demise would not properly be considered delusional in the context of certain South American tribes.

These four components may be construed as the necessary conditions for labeling a belief delusional. That is, each of these criteria must be met before a belief can be considered a delusional belief. Because, however, each of these criteria poses potentially difficult diagnostic problems, it is possible that, even taken together, they do not constitute sufficient criteria for the classification of a belief as delusional. That is, even if it is judged that each of these four conditions has been met, it does not guarantee that the belief in question is delusional.

The number, type, and content of delusional beliefs are unlimited. Because of the unlimited complexity of language it is not possible to identify antecedently what particular beliefs might count as delusional. This further complicates the identification of delusional beliefs. Some beliefs, however, are straightforwardly identifiable as delusional. Some of these beliefs are so common in relation to other delusional beliefs that they have been named and studied in some detail. The Capgras delusion and the Cotard delusion are prominent examples. The Capgras delusion is marked by the individual's belief that his or her family (close acquaintances, co-workers, etc.) have been replaced by imposters. In the Cotard delusion the individual believes that he or she is dead or possesses no 'real' existence. Both of these exemplar delusions satisfy each of the four criteria outlined above. However, though both the Capgras and Cotard delusions are distinguished by their bizarreness, delusional beliefs need not be bizarre in their content.

One of the most prominent difficulties associated with correctly identifying a belief as delusional is encountered when the truth of the belief is assessed. People may express a number of beliefs that, at face value, seem clearly false. Believing one is dead falls squarely in this category. Little difficulty is associated with classifying such beliefs as false. However, unless care is taken to assess the truth of a belief, it would be hasty to classify the belief as delusional. In one classic example of a mistaken judgment of falsity, in the early 1970s Martha Mitchell, wife to the Attorney General of the United States, expressed seemingly false beliefs regarding illegal activities going on in the White House. Mrs Mitchell was believed to be suffering from some sort of psychopathology based on the bizarreness of her beliefs. As it turned out, however, Mrs Mitchell's accusations were accurate. Mrs Mitchell's claims pertained to the Watergate scandal of 1972. There are countless examples of situations similar to this one. A provisional name has been applied to situations in which the truth-value of a belief has been mistaken, the 'Martha Mitchell Effect'.

Delusional beliefs are distinguished from other forms of erroneous beliefs, such as overvalued ideas, obsessions, and confabulations. Overvalued ideas are recognized by their possessors as being potentially false and, as such, are not maintained with the degree of certainly associated with delusions. Obsessions are defined as persistent ideas, thoughts, impulses, or images. Obsessions are often experienced as distressing, and though the obsessional person attempts to ignore or suppress them, they are clearly recognized as his or her own thoughts. Because people attempt to suppress obsessional thoughts, they fail to count as delusional beliefs that are held with certainty. Confabulation is defined as filling in the gaps of memory and knowledge with false information. This information, however, need not involve beliefs that are generally considered questionable or bizarre by the members of the person's community.

COGNITIVE MECHANISMS

There are a number of cognitive theories that attempt to explain the presence of delusional beliefs. The most prominent features of these theories will be outlined. According to some thinkers, delusions may reflect errors in discriminating the origin of information. Mistakenly attributing one's belief to experiences, rather than to the operation of imagination, can lead to false beliefs. Internal imaginary events can generate local effects in the same pathways used by perception itself, thus increasing the difficulty associated with discriminating the original source of the belief. This phenomenon is similar to the one we experience when we are unable to discriminate accurately between reality and fiction while dreaming. The more elaborate imaginary scenarios are in terms of sensory detail, the more likely they are to be conflated with actual experiences. When we imagine a scene or event in elaborate detail (e.g. the color of the sunset and the odor of the ocean) the imagined event takes on more of the properties of our veridical experiences. This issue has been at the core of the debate regarding repressed memories.

Occasional confusion regarding the source of one's beliefs is, in itself, relatively innocuous. Most people have experienced this kind of confusion during their lives: 'Did somebody tell me that, or do I simply think it?' 'Do I really remember doing that when I was three years old, or am I relying on Mom's stories?' As the definition of delusional beliefs suggests, more than simple confusion or error with respect to the source of one's beliefs is required to classify them as delusional. Two other cognitive mechanisms are required to induce the individual to hold such false beliefs with certainty in the face of evidence otherwise: hypervigilance and an unfounded sense of profoundness.

Hypervigilance has been identified by many theories as one of the critical cognitive dispositions associated with delusional beliefs. Hypervigilance is marked by an increase in a person's awareness of the details of his or her environment. Under normal conditions of vigilance the environmental information to be processed is well within the individual's cognitive capacity to organize and contextualize into a manifold whole. States of hypervigilance, however, may induce a cognitive overload resulting in a sense of ambiguity with respect to environmental stimuli. Human cognitive mechanisms simply cannot contextualize and organize all of the environmental stimuli that are present at any given moment. The ambiguity that results from such experiences may lead to an unusually intense sense of mystery and profoundness. For example, an individual may notice that there are four pennies in his pocket, four steps in his staircase, four clients he must meet during the day, and four children playing tag on the lawn as he leaves for work. These four events might strike him as being in need of some sort of explanation. How often, after all, does one note four sets of four events on the way to work? The search for an explanation of this series of events may result in the belief that these four events are connected in some important way, despite the fact that they are simply chance observations. The meaning of these seemingly connected events may take on a significance that far outweighs any reasonably objective interpretation of the events. Does it mean that the man has only four years to live? Perhaps this means that the man will meet his soulmate at four o'clock that afternoon.

This profoundness of experience is best described as being similar to that which one has at the birth of a child, the death of a close friend, or the intense spiritual experiences common to many religions. In the normal course of events such experiences are rare and often life-altering. For the delusional individual, mundane daily experiences become increasingly entangled in a sense of mystery and profoundness that can seemingly only be explained in a profound manner. Such profundity may be associated with a feeling of certainty. Further complicating the issue is the increased likelihood that the individual will note that others do not react with a sense of awe at what seem to be awesome events. This, in turn, may incline the person to believe that only he or she is meant to understand, or see, these events as they are. This can lead either to isolation or to a sense of righteousness. When isolated, a person is unlikely to come into contact with enough evidence to dissuade him or herself of the meanings that have been mistakenly associated with random events. Similarly, self-righteousness may interfere with a willingness to accept the input of others when such input contradicts what he or she believes.

Thus, a person who takes close notice of his or her relatives may come to see things that were never noticed before. Behavioral irregularities, a small scar, and a slightly different vocal intonation may be attended to and seem to require explanation. For the individual suffering from the Capgras delusion, the explanation may be that his or her close relatives have been replaced by imposters. Such a belief is unlikely to be openly expressed to those who could counter it. As such, the belief may become more and more ingrained. By the time the individual expresses this belief to others it will be so thoroughly tangled in a web of spurious confirmatory evidence (e.g. Mom never used to like cabbage) that it will be difficult to dispel. The foregoing explanation of the predispositions and factors leading to the development of delusional beliefs are consistent with the view of delusional beliefs as unnecessary and errorful explanatory theories of events that required no explanation.

PERCEPTUAL MECHANISMS

Perceptual mechanisms are implicated in the development of delusional beliefs insofar as they are involved in the individual's experience of the world. The most relevant role that perceptions play with respect to delusions is in the area of hallucinations. Both visual (e.g. seeing snakes when no snakes are present) and auditory (e.g. hearing one's name called when no one is present) hallucinations may result from an elaboration of elementary sensations. For example, *muscae volitantes*, commonly known as 'eye floaters' (which are typically overlooked or ignored), have been experienced as rats, snakes, armies struggling for the person's soul, and so on. In such cases, the experience of something in the visual field is accurate, only the interpretation of what it is is inaccurate. Similarly, individuals suffering from alcohol withdrawal may incorrectly identify the sounds associated with intrinsic tinnitus with the buzzing of bees inside his or her head. Auditory hallucinations are more common than visual hallucinations because of the relative difficulty in identifying the source, either internal or external, of sounds as opposed to images.

NEURAL SUBSTRATES

Though little is known in this area, research continues in the identification of the neural substrates of delusional predispositions. Promising research has suggested that schizophrenic delusions and hallucinations may be associated with overactivity of the left hippocampus and ventral striatum.

Further Reading

American Psychiatric Association (1994) *Diagnostic and Statistical Manual of Mental Disorders*, 4th edn. Washington, DC: American Psychiatric Association.

Berrios GE (1991) Delusions as 'wrong beliefs': a conceptual history. *British Journal of Psychiatry* **159**: 6–13.

Jaspers K (1963) *General Psychopathology*, translated by J Hoenig and MW Hamilton. Chicago, IL: University of Chicago Press.

Oltmanns TF and Maher BA (1988) *Delusional Beliefs*. New York, NY: John Wiley.

Reed G (1988) *The Psychology of Anomalous Experience: A Cognitive Approach*. Buffalo, NY: Prometheus.

Roberts G (1992) The origins of delusion. *British Journal of Psychiatry* **161**: 298–308.

Dementia

See **Non-Alzheimer Dementias**

Depression

Introductory article

Lyn Y Abramson, University of Wisconsin-Madison, Madison, Wisconsin, USA
Lauren B Alloy, Temple University, Philadelphia, Pennsylvania, USA
Catherine Panzarella, The Philadelphia Behavioral Health System, Philadelphia, Pennsylvania, USA

CONTENTS

Depressive disorders are characterized by persistent depressed mood or loss of interest (normally for at least two weeks) and at least four other symptoms such as change in eating patterns or appetite, sleep disturbance, psychomotor agitation or retardation, fatigue or loss of energy, feelings of worthlessness or guilt, difficulty in concentration, and suicidal thoughts, plans, or attempts.

IMPORTANT FACTS ABOUT DEPRESSION

Descriptive research has produced some important facts about depression. First, depression is a common disorder. Recent estimates indicate that in many western countries, about 20 percent of the population will experience a clinically significant episode of depression at some point in their lives. Moreover, the rate of depression appears to be on the rise, especially among young people. The increased rate of depression in modern society may be due, in part, to increased focus on the self and its accomplishments. Although focus on the self may motivate achievement, it also may lead to depression when people fail to achieve their goals.

Second, depression is recurrent. Over 80 percent of depressed patients have more than one depressive episode in their lifetime. In fact, over 50 percent of depressed patients experience a relapse within two years of recovery. Such recurrence of depression suggests that some individuals are depression-prone because they exhibit a relatively stable vulnerability factor or diathesis for this disorder.

Third, the rates of depression surge during middle to late adolescence. Although young children can experience depression, rates of this disorder are relatively low during childhood. The surge in depression during middle to late adolescence may be due to the consolidation of vulnerability factors for the disorder such as negative cognitive styles as well as increased rates of negative life events within this age group.

Fourth, gender differences in depression exist among adults with twice as many women experiencing depression as men. This gender difference in depression emerges during adolescence.

Fifth, depression is associated with significant physical, vocational, and interpersonal impairment. Epidemiological studies have shown that depression is associated with poor physical health including cardiac problems and elevated rates of smoking. Moreover, depression lowers productivity in the workplace. In one year, it is estimated that in the United States the costs of depression-related lost productivity can exceed $33 billion. Finally, depression has high interpersonal costs. For example, depressed people have a higher divorce rate than nondepressed people.

Sixth, depression can be lethal as it clearly increases the risk for suicide. Research has suggested that hopelessness is the key factor contributing to suicide among depressed individuals.

Seventh, life events play a role in the development of depression. The occurrence of undesirable, major life events is associated with the onset of depression. However, not all people become clinically depressed when confronted with severe life stressors. Vulnerability–stress models posit that individuals exhibiting vulnerability factors (cognitive, genetic, biological, etc.) for depression are more likely to become depressed when confronted with severe stressors than individuals not exhibiting such vulnerability.

Finally, depression long has been viewed as heterogeneous with multiple causes. Thus, there are likely to be different causal pathways that culminate in depression. (*See* **Affective Disorders: Depression and Mania**)

COGNITION AND DEPRESSION

Although depression long has been recognized as an important form of psychopathology, experimental psychopathologists neglected this disorder until the 1970s. At that time, research on depression burgeoned within clinical psychology, and many investigators began to emphasize cognitive processes in the etiology, maintenance, and treatment of depression. A core idea within this cognitive perspective is that different people can perceive the same event differently. Psychologists focus on the characteristic ways that individuals perceive situations and how their perceptions, in turn, relate to their behaviors and emotions. From the cognitive perspective, emotional reactions, such as depression, to events are determined by a combination of characteristics of the event itself and the cognitive construal processes of the perceiver.

COGNITIVE THEORIES OF DEPRESSION

Why are some people vulnerable to depression whereas others never seem to become depressed at all? According to the two major cognitive theories of depression, the hopelessness theory (Abramson *et al*, 1999, 2002) and Beck's theory (Beck, 1987; Clark *et al*, 1999) the meaning or interpretation that people give to their experiences importantly influences whether they will become depressed and whether they will suffer repeated, severe, or long-duration episodes of depression. Indeed, the demonstrated efficacy of cognitive therapy for depression underscores the powerful clinical implications of a cognitive approach to depression.

According to the hopelessness theory, the expectation that highly desired outcomes will not occur or that highly aversive outcomes will occur and that one cannot change this situation – hopelessness – is a precipitant of depressive symptoms. How does a person become hopeless and, in turn, develop the symptoms of depression? As shown in Figure 1, negative life events (or the nonoccurrence of desired positive life events) are 'occasion setters' for people to become hopeless. However, the relation between negative life events and depression is imperfect; not all people become depressed when confronted with negative life events. According to the hopelessness theory, three kinds of inferences or conclusions that people may make when confronted with negative life events contribute to the development of hopelessness and, in turn, depressive symptoms: causal attributions, inferred consequences, and inferred characteristics about the self. In brief, hopelessness and, in turn, depressive symptoms are likely to occur when negative life

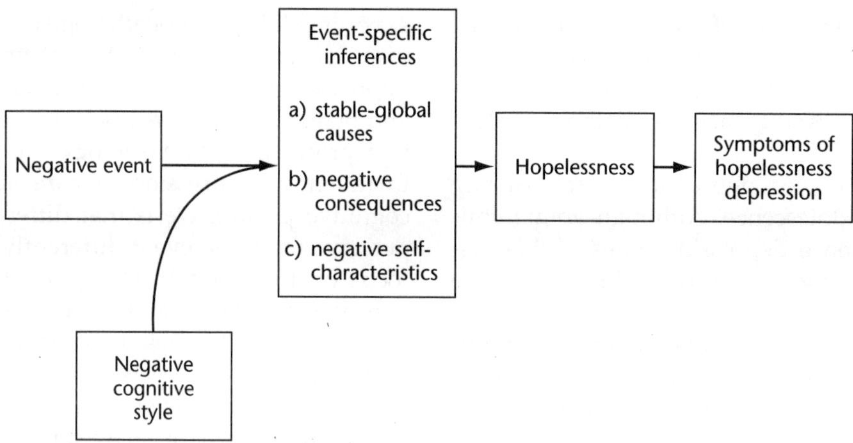

Figure 1. Causal chain in the hopelessness theory.

events are (1) attributed to stable causes (likely to persist over time) and global causes (likely to affect many areas of life) and viewed as important; (2) viewed as likely to lead to other negative consequences; and (3) construed as implying that the person is unworthy or deficient.

For example, suppose a student fails a test. According to the theory, the student is likely to become depressed if he/she believes that the failure: (1) was due to low intelligence; (2) will prevent him/her from pursuing a particular career; and (3) means that he/she is worthless. In contrast, another student who fails the same test will be protected from becoming depressed if he/she believes that the failure: (1) was due to not studying hard enough; (2) will motivate him/her to do especially well on the next test; and (3) has no implications for his/her self-worth.

In the hopelessness theory, informational cues in the situation as well as individual differences in cognitive style influence the content of people's inferences about cause, consequence, and self when negative life events occur. Individuals who exhibit a general tendency to attribute negative events to stable and global causes, to infer that current negative events will lead to further negative consequences, and to infer that the occurrence of negative events means that they are deficient or unworthy, are more likely to make these depressogenic inferences about a given negative event than individuals who do not exhibit this depressogenic cognitive style. However, in the absence of negative left events, those exhibiting a depressogenic cognitive tendency should be no more likely to develop hopelessness and, in turn, depressive symptoms than people not exhibiting this tendency. This aspect of the theory is a vulnerability–stress component: negative cognitive patterns are the cognitive vulnerability, and negative life events are the stress.

The etiological hypotheses of Beck's cognitive theory of depression are similar to those of the hopelessness theory. To understand Beck's theory, it is useful to describe cognitive schemata. Schemata are generalized core beliefs about the self and world that assist people in processing complex environmental information by: (1) selecting only a fraction of incoming stimuli for processing; (2) abstracting meaning from incoming information and favoring storage of the meaning rather than a veridical representation of the original stimulus; (3) using prior knowledge to assist in processing and interpreting information; and (4) integrating information to favor internal consistency over external accuracy. Schemata pertaining to the self are featured in Beck's theory.

As Figure 2 shows, in Beck's theory, maladaptive self-schemata containing dysfunctional attitudes involving themes of loss, inadequacy, failure, and worthlessness constitute the cognitive vulnerability for depression. Such dysfunctional attitudes often involve the theme that one's happiness and worth depend on being perfect or on other people's approval. Examples of dysfunctional attitudes include, 'If I fail partly, it is as bad as being a complete failure', or 'I am nothing if a person I love doesn't love me'. When these hypothesized depressogenic self-schemata are activated by the occurrence of negative life events (the stress), they generate specific negative cognitions (automatic thoughts) that take the form of overly pessimistic views of oneself, one's world, and one's future (the negative cognitive triad) that, in turn, lead to sadness and other symptoms of depression. In the

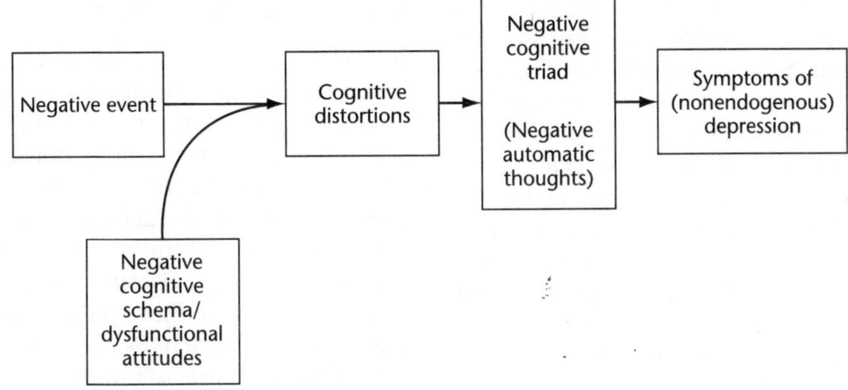

Figure 2. Causal chain in Beck's theory.

absence of activation by negative events, however, the depressogenic self-schemata remain latent, less accessible to awareness, and do not directly lead to negative automatic thoughts or depressive mood and symptoms.

Although differing in some specifics, hopelessness and Beck's theories share many important features. At the most basic level, both theories emphasize the role of cognition in the origins and maintenance of depression. In addition, both theories contain a cognitive vulnerability hypothesis in which negative cognitive patterns increase people's vulnerability to depression when they experience negative life events. Moreover, both theories also propose a mediating sequence of negative inferences that influence whether or not negative events will lead to depressive symptoms. Finally, both theories recognize the heterogeneity of depression and acknowledge that other causes of depression may exist.

Despite their similarities, there is one striking difference between hopelessness and Beck's theories. To understand this difference, it is useful to distinguish between cognitive 'processes' and cognitive 'products'. Cognitive processes involve the operations of the cognitive system such as information encoding, retrieval, and attentional allocation. Cognitive products are the end result of the cognitive system's information processing operations and consist of the cognitions and thoughts that the individual experiences. Inferences about cause, consequences, and self, as featured in the hopelessness theory, are examples of cognitive products. According to the hopelessness theory, depressive and nondepressive cognition differ in content (e.g., stable, global versus unstable, specific causal attributions for negative events) but not in process. In contrast, Beck's original theory emphasized that depressive and nondepressive cognition differs not only in content but also in process. Beck suggested that the inference process is 'schema driven' among depressed people and 'data driven' among nondepressed people. Thus, although both theories emphasize that depressed people's inferences are negative, Beck further proposed that depressed people's negative inferences are unwarranted given current information. Specifically, Beck suggested that depressed individuals ignore positive situational information and are unduly influenced by current negative situational information in making their negative inferences. In contrast, Beck hypothesized that nondepressed individuals appropriately utilize current information in making inferences. In short, Beck's original theory emphasized that depressive cognition is distorted whereas hopelessness theory is silent on the distortion issue.

EMPIRICAL FINDINGS ON COGNITION AND DEPRESSION

Supporting the cognitive theories of depression, many cross-sectional studies have established that negative cognitive patterns are associated with depression in adults, children, and adolescents. However, such cross-sectional studies do not provide strong tests of the cognitive vulnerability hypothesis, because they cannot distinguish between the possibility that the cognitive vulnerability came first and contributed to the occurrence of depression, as hypothesized in the cognitive theories of depression, and the alternative possibility that cognitive vulnerability does not contribute to depression and, instead, is a correlate or consequence of depression. Prospective behavioral high-risk studies are needed to establish that cognitive vulnerability actually precedes the occurrence of depression and contributes to its onset.

In the behavioral high-risk design, currently nondepressed people are selected who are at high versus low risk for depression based on the presence versus absence of the hypothesized depressogenic cognitive patterns. These cognitive high and low risk groups would then be followed and compared on their likelihood of developing depression in the future. Although there are some exceptions, in general recent studies with children, adolescents, and adults using the behavioral high-risk design and approximations to it have found that individuals who exhibit the hypothesized cognitive vulnerabilities are more likely to develop depressive moods, symptoms, and clinically significant episodes over time than individuals who do not exhibit cognitive vulnerability. Moreover, individuals exhibiting cognitive vulnerability also show elevated rates of suicidal thoughts when they are followed over time. Thus, work supports the hypothesis that ingrained negative patterns of thinking provide risk for depression.

In addition, some research suggests that attributional style for interpreting positive, rather than negative, events may be important in predicting recovery from depression as well as lower rates of relapse. For example, depressed people who show a tendency to attribute positive events to stable, global causes are more likely to recover from depressive symptoms following the occurrence of positive events than depressed people who attribute positive events to unstable, specific causes. Similarly, depressed patients with an internal, stable, global attributional style for positive events are less likely to relapse in the year following hospital discharge than depressed patients who do not show this attributional style. In other words, depressed people who did *not* show the tendency to minimize or discount positive information were more likely to recover and were less vulnerable to relapse.

In addition to work on cognitive styles as vulnerability and invulnerability factors for depression, much research has focused on documenting the nature of cognitive processes exhibited by depressed persons (Alloy *et al*, 1997). For example, this research shows that processing negative self-referent information is less effortful and more automatic for depressed than non-depressed people. Depressed people also exhibit more negative memory biases than nondepressed people. Taken together, a vast amount of research unequivocally shows that depressive cognition is more negative than nondepressive cognition. But are depressed people's negative cognitions also more unrealistic than nondepressed people's positive cognitions?

DEPRESSIVE REALISM?

In contrast to Beck's original characterization of 'depressive cognitive distortion' and 'nondepressive accuracy', research has demonstrated pervasive optimistic biases among nondepressed people and a 'depressive realism effect' in which depressed individuals actually are more accurate than nondepressed individuals. For example, nondepressed individuals often exhibit an 'illusion of control' in which they believe that they control outcomes over which they objectively have no control, whereas depressed individuals seem less susceptible to this illusion. Research on optimistic biases among 'normal' people suggests that in formulating theories of depressive cognition, clinical researchers may have been wrong to assume that accuracy is the baseline of normal cognitive functioning. Instead, laboratory work has demonstrated that both depressed and nondepressed people show cognitive biases and illusions that are consistent with their preconceived beliefs or schemas (Dykman *et al.*, 1989).

Although work on depressive realism and nondepressive optimistic illusions has not established that depressed people are *always* more accurate than nondepressed people, these studies nevertheless have posed an important challenge to the portrayal of depressed people as either impervious to information in their environments or hopelessly biased by pervasive negative schemata and of nondepressed people as completely data-driven and free from the influence of biasing schemata. Depressive and nondepressive cognition may differ more in content than in process. Thus, the negative cognitive biases that predict risk for depression may be no more distorted than the positive cognitive biases that predict protection from depression. Lively debate continues about the question of who are more accurate in perceiving reality over the long run – depressed people or nondepressed people?

DEVELOPMENTAL ORIGINS OF COGNITIVE VULNERABILITY

Given the growing body of work suggesting that negative cognitive styles may confer vulnerability for depression and suicidality, it is important to understand the developmental origins of these negative patterns of thinking. It is likely that a multitude of factors contributes to negative cognitive patterns such as genetic factors, biologically based temperamental factors, parental 'modeling' of negative cognitive patterns, and parental feedback, to name a few. Recent research has shown that adults who exhibit marked cognitive vulnerability to depression report growing up in environments characterized by maltreatment and neglect. Emotional maltreatment may be an especially potent contributor to cognitive vulnerability. Telling a child that he or she is incompetent, unlovable, or unattractive may 'program' the child to make very depressogenic interpretations of negative events later in life that, in turn, lead to depression. Further research is necessary to establish definitively that emotional maltreatment contributes to cognitive vulnerability to depression.

COGNITIVE-BEHAVIOR THERAPY FOR DEPRESSION

Consistent with the cognitive perspective on depression, cognitive–behavioral therapy (CBT) for depression has been developed that targets the negative cognitions such as hopelessness that are hypothesized to precipitate the onset of depressive symptoms and episodes as well as the negative cognitive patterns such as the tendency to make stable, global, causal attributions for negative events that are hypothesized to provide vulnerability for depression. CBT has been shown to be successful in remediating current depressive episodes and compares favorably with antidepressant medications for all but the most severely depressed patients. Given that depression is a recurrent disorder, it is especially noteworthy that preliminary work suggests that CBT may have an enduring effect that decreases the risk of relapse and recurrence among formerly depressed people. Finally, initial studies suggest that administration of CBT to currently nondepressed individuals at risk for depression can prevent the onset of first episodes of depression.

INTEGRATION OF COGNITIVE AND NEUROSCIENCE PERSPECTIVES ON DEPRESSION

Given the empirical support for the cognitive theories of depression and the success of cognitive therapy for depression, it is critical to integrate the cognitive approach with other successful approaches to depression. Much important work has been conducted demonstrating the fruitfulness of integrating cognitive and interpersonal approaches to depression. However, little has been done to integrate the cognitive perspective with biological approaches to depression.

Although much important work has been conducted on cognitive and biological vulnerability to depression, these two lines of research have proceeded in relative isolation from each other. Paving the way for an integration of cognitive and biological approaches to depression, recent research has begun to elucidate the neural circuitry involved in implementing the behavioral approach system (BAS) and the behavioral inhibition system (BIS) (Davidson *et al.*, 2002). Specifically, activation of the left frontal cortex is a key component of the neural circuitry implementing BAS function, and activation of the right frontal cortex is involved in BIS function. Consistent with this perspective, unipolar depression, which reflects low approach motivation, is associated with relative low left frontal cortical activity.

What is the nature of the relationship between cognitive vulnerability to depression and biological vulnerability to this disorder as indexed by patterns of cerebral asymmetry? To the extent that hopelessness, the expectation to which cognitively vulnerable individuals are predisposed, may be particularly powerful in signaling a shutdown of approach motivation (inactive BAS), cognitively vulnerable individuals may be characterized by relative low left frontal cortical activity. Consistent with this view, there are initial indications that attributionally vulnerable individuals exhibit relative low left frontal hemispheric activation. More generally, pessimism may be associated with low left frontal cortical activity, whereas optimism may show the reverse pattern. An exciting line of work has just begun to document a relationship between cortical activity and pessimism/optimism. For example, the autobiographical memories of patients with left hemisphere lesions, particularly to frontal regions, are more negative than those of patients with right hemisphere lesions. Moreover, activation of the left hemisphere by behavioral methods results in more self-serving attributions and more optimistic expectancies for the future.

Integration of the cognitive and neural approaches to depression promises to be an exciting and important direction for further theory and research.

EVOLUTIONARY APPROACHES TO UNDERSTANDING DEPRESSION

From an evolutionary perspective, depression may appear paradoxical. Depressed people show deficits in many basic human 'instincts' such as the pursuit of pleasure, sexual drive, appetite, sleep, parenting behavior, goal striving, and desire to be with other people. Perhaps most paradoxical of all, in defiance of the 'law of survival', depressed people sometimes actively try to terminate their own lives by suicide.

Despite these paradoxes, depression may be understood from an evolutionary perspective. Drawing on the fact that depression is common, some investigators have suggested that in our evolutionary past, depression may have had adaptational significance (Neese, 2000). For example, depression may facilitate disengagement from the pursuit of an unattainable goal. Alternatively, depression may involve a 'conservation of energy' principle in which a person slows down to reduce depletion of energy in pursuit of an unattainable goal or ceases activity that is likely to be futile or expose the self to some harm, disease, attack, or situation that is even worse than the current one. A complete account of depression is likely to involve an evolutionary perspective.

Further Reading

Abramson LY, Alloy LB, Hankin BL, Haeffel GJ, MacCoon DG and Gibb BE (2002) Cognitive vulnerability–stress models of depression in a self-regulatory and psychobiological context. In: Gotlib IH and Hammen CL (eds) *Handbook of Depression*, pp. 268–294. New York, NY: Guilford Press.

Abramson LY, Alloy LB, Hogan ME, Whitehouse WG, Donovan P, Rose D, Panzarella C and Raniere D (1999) Cognitive vulnerability to depression: theory and evidence. *Journal of Cognitive Psychotherapy: An International Quarterly* **13**: 5–20.

Alloy LB, Abramson LY, Murray LA, Whitehouse WG and Hogan ME (1997) Self-referent information-processing in individuals at high and low cognitive risk for depression. *Cognition and Emotion* **11**: 539–568.

Beck AT (1987) Cognitive models of depression. *Journal of Cognitive Psychotherapy: An International Quarterly* **1**: 5–37.

Clark DA, Beck AT and Alford BA (1999) *Scientific Foundations of Cognitive Theory and Therapy of Depression*. Philadelphia, PA: John Wiley.

Davidson RJ, Pizzagalli D and Nitschke JB (2002) The representation and regulation of emotion in depression: perspectives from affective neuroscience. In: Gotlib IH and Hammen CL (eds) *Handbook of Depression*, pp. 219–244. New York, NY: Guilford Press.

Dykman BM, Abramson LY, Alloy LB and Hartlage S (1989) Processing of ambiguous feedback among depressed and nondepressed college students: schematic biases and their implications for depressive realism. *Journal of Personality and Social Psychology* **56**: 431–445.

Hollon SD, Haman KL and Brown LL (2002) Cognitive-behavioral treatment of depression. In: Gotlib IH and Hammen CL (eds) *Handbook of Depression*, pp. 383–403. New York, NY: Guilford Press.

Nesse RM (2000) Is depression an adaptation? *Archives of General Psychiatry* **57**: 14–20.

Depth Perception

Introductory article

Myron L Braunstein, University of California, Irvine, California, USA

CONTENTS

The three-dimensional world that is immediately and effortlessly perceived is a product of inferential mechanisms that rely on many different types of information. Among these are binocular disparity, motion parallax, texture gradients, linear perspective, shading, interposition, accommodation and convergence.

INTRODUCTION

Depth perception has been described as a paradox. We experience a three-dimensional (3D) world immediately and effortlessly, yet our visual information about the external world comes from light imaged on the two-dimensional retinal surfaces at the backs of our eyes. Our experience of the 3D world is unambiguous, yet the information in these images is inherently ambiguous (Figure 1). Before reviewing specific sources of information (cues) for depth perception, it will be useful to state some general principles underlying our ability to perceive a 3D world:

1. Information is available in the retinal images, and in the mechanisms controlling the muscles that change the convergence angle of the two eyes and the shapes of the lenses, that can be used to recover the 3D structure of the world.
2. This information is inherently ambiguous and must be interpreted using constraints – in a sense, biases based on knowledge about how the 3D environment is structured and about the structure of the visual system itself. In interpreting binocular disparities, for example, the assumption of a constant interocular distance would be a constraint. In interpreting image motion, an assumption of rigid 3D motion is a possible constraint.
3. This process of interpretation can be formally regarded as inductive inference. It is important to understand that the use of inductive inference based on prior knowledge does not mean that the perceiver engages in a conscious or even in an unconscious thought-like process. Instead, the inferential processes in depth perception can be implemented by 'hard-wired' biological mechanisms with 'knowledge' of the structure of the 3D environment built into these mechanisms through evolution. These inferential processes may not match the optimum processes that would be theoretically available to an 'ideal' observer. Instead, the human observer appears to use heuristic processes – efficient shortcuts that usually lead to a correct interpretation of the 3D environment, but which can fail when presented with unusual stimuli, leading to illusions.

Different types of information about the 3D world have been labeled as 'cues', based on the concept that there are a small number of ways in which picture-like images on the retinas are interpreted by the brain. These cues are sometimes thought of as involving separate modules that function independently until their outputs are combined to determine what is perceived. Most of what have been regarded as unitary cues, however, have turned out to involve more than one type of information or more than one type of processing, and important interactions among the cues suggest that they may not function as separate modules. The cue concept is a convenient way to summarize what is known about the information used in depth perception, but it should be recognized that this

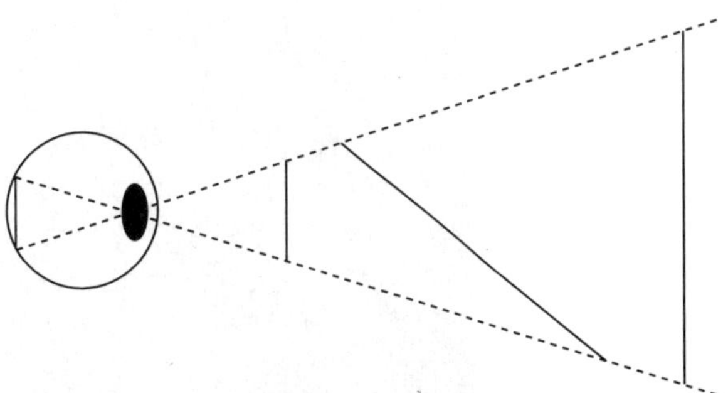

Figure 1. The ambiguity of the retinal image. An edge projected onto the retina is consistent with an infinite number of edges in three dimensions, varying in size, distance and orientation.

concept is a simplification that provides only a rough categorization of types of information used to infer a 3D world.

BINOCULAR DEPTH CUES

The two eyes are separated horizontally by about 6 cm. As a result of this separation the images projected onto the two retinas are different. The primary difference is that the relative positions of points in the images are shifted horizontally; this horizontal shift is a function of the interocular separation and the distances of the points from the eyes (Figure 2). It has been demonstrated using random dot stereograms (Figure 3) that horizontal disparities can lead to perceived variations in depth even in the absence of recognizable features in the individual images. A gradient of disparities, such as that produced by a slanted surface, is not, however, an effective depth stimulus. A discontinuity in disparities is needed, as when a scene contains an edge at which the depth or slant changes. Horizontal disparities appear to provide information about relative depth rather than absolute distance. Research suggests that vertical disparities in the two retinal images, when sufficiently large, can provide absolute distance information. Vertical disparities occur because one eye may be closer to an object, producing a different perspective projection. Another binocular cue is the convergence of the optic axes that occurs when the two eyes fixate a near object (also shown in Figure 2). Convergence is one of the first recognized depth cues, but is effective only at short distances.

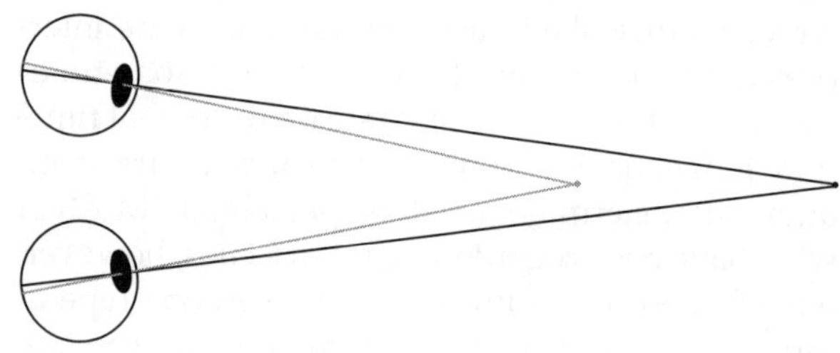

Figure 2. Binocular disparity and convergence. Note that the relative positions of the near and far points are reversed in the two retinal projections. The eyes are fixated on the near point and this determines the convergence angle.

MONOCULAR DEPTH CUES

Accommodation

Along with convergence, accommodation was one of the first recognized depth cues. When we fixate a nearby point the rays of light from this point diverge as they reach the lens of the eye and must be converged by the lens, so that a sharp point is imaged on the retina. This is accomplished by a change in the shape of the lens, which bulges to focus near objects and flattens to focus more distant objects (Figure 4). Information about the changes in the shape of the lens can therefore provide information about the distance of the fixated object. Accommodation is effective at short distances (within 2 m).

Monocular Perspective Cues

Perspective effects occur when the observation point (the eye or a camera) is relatively close to an object. The overall effect of perspective is that 3D distances closer to the observer project to larger

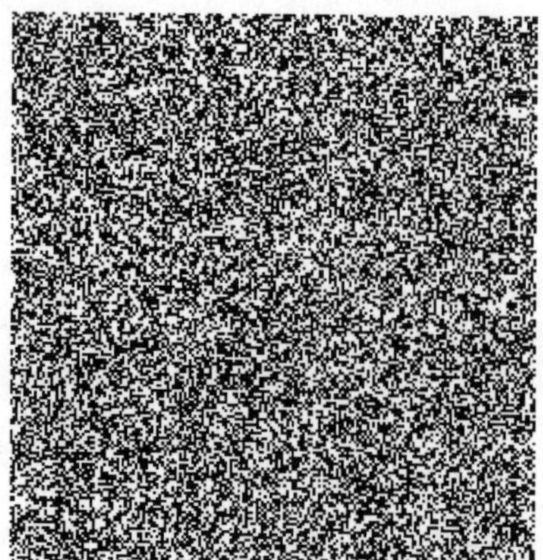
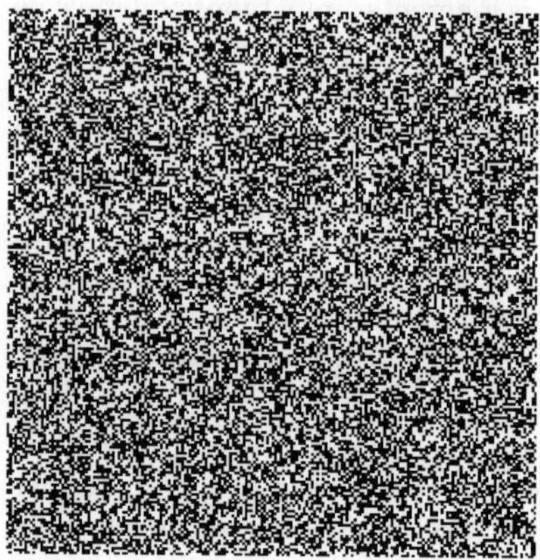

Figure 3. A random dot stereogram of the type introduced by Bela Julesz. If the images are fused by crossing the eyes, a small square should appear in front of the large square.

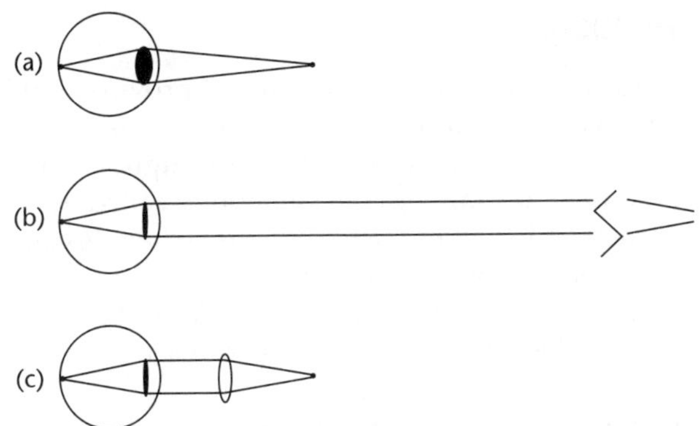

Figure 4. Accommodation to a near point (a) and far point (b). (c) Collimation: a lens is used to place a near point at optical infinity, making it appear to be a distant point.

distances in the retinal image than equivalent 3D distances further from the observer. This is reflected in several ways in a monocular image. Linear perspective, for example, refers to the convergence in the image of parallel lines extending into the distance in the 3D scene. It is a highly effective cue – merely drawing two converging lines can create an impression of a surface extending in depth. Other perspective cues are best described as gradients, as suggested by J. J. Gibson. The projection of a uniform 3D texture extending in depth results in a texture gradient. By itself, a texture gradient is not an especially effective cue unless the texture is regular (for example, a grid pattern), but texture gradients may combine with other information such as shading to provide an effective source of information about 3D layout. The projected sizes of similar objects describes another gradient that may result from perspective projection. This is closely related to the texture gradient. Usually a texture gradient describes spacing between texture elements (like blades of grass), whereas the size of similar objects describes the gradient in the projections of the elements themselves. Figure 5 shows several examples of monocular perspective cues.

Motion

Motion is not a single cue to depth but includes several ways in which changing patterns in the retinal projections over time provide information about 3D relationships. Motion parallax describes the changes that occur in the retinal image when there is relative motion between the eye and an object or a scene that is being observed. This usually occurs when the head moves relative to a stationary scene, but can also occur when an object or surface is moved relative to the head. It is primarily a perspective effect in that motions projected on the retina are more rapid for nearer objects than for more distant objects, assuming that the objects are moving in the same direction and at the same speed in 3D space. Another motion cue, structure from

Figure 5. Examples of perspective, shading and interposition. (a) The road illustrates linear perspective. (b) The rocks form a relative size gradient. (c) The arches form a regular texture gradient; specular highlights are found on the door at the back of the tunnel. Examples of interposition are found in all three photographs. (Images adapted from the Photo Library of the National Oceanic and Atmospheric Administration, US Department of Commerce.)

motion (also called the kinetic depth effect), occurs when an object is rotated relative to the direction of observation. Consider, for example, a cylinder rotating about a vertical axis. As the object rotates, the projections of imaginary lines connecting pairs of points on the cylinder will change in length (Figure 6). If we assume that the 3D distances between the points do not change as the object rotates (a rigidity assumption), the changing distances between points in the image can be used to compute the relative depths of the points on the 3D object. Unlike motion parallax, structure from motion does not depend on viewing distance and can provide information about the shape of a rotating object even when the object is at a great distance – for example, when an object is viewed through a telescope.

Shading

Consider a 3D shape and a light source illuminating that shape. At any point on this shape we can draw a tangent to the surface. The angle between this tangent and a line drawn from that point to the light source will vary over the surface and this will result in variation in the light reflected by points on the surface. The pattern of light reflected by the surface thus provides information about the shape of the surface and this source of depth information is referred to as 'shape from shading'. Note that shape from shading provides relative depth information within a surface, not information about distances from the eye. It is thus a source of object-centered depth information. Cast shadows and specular highlights also provide important information about object depth. Examples of shading cues are also seen in Figure 5. In dynamic scenes, important additional information about depth relationships is provided by changes over time in shading, in the positions of shadows, and in the locations of specular highlights.

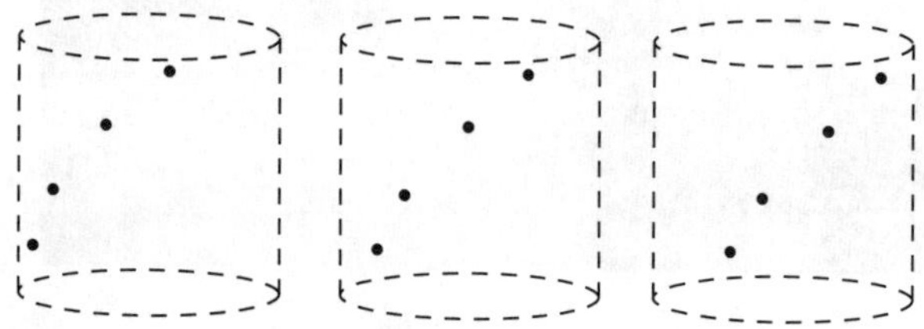

Figure 6. Structure from motion. Shimon Ullman showed that three-dimensional structure can be recovered mathematically from just three views of four non-coplanar points.

Occlusion

Occlusion is an unusual cue in that it provides only ordinal depth information. Occlusion of a far object by a near object, indicated by an interruption in the contours of the far object, is sometimes called interposition. Kinetic occlusion occurs when texture elements on one object disappear as they reach an implicit contour of another object, indicating that the first object is behind the second object.

Atmospheric Attenuation

Atmospheric attenuation refers to the reduced intensity and change in color of light reaching the eye that results from the scattering of light and absorption of light by particles in the atmosphere. Atmospheric effects can produce a gradient of intensity and color in the retinal projections that is informative about relative depth over large distances.

FLATNESS CUES

The retinal images should not be regarded as flat pictures that we can perceive as such in the absence of depth cues. To see a flat image we need information indicating that all points on the image are equally distant; this information is provided by flatness cues. Some flatness cues use the same sources of information as depth cues. For example, absence of differential accommodation to points on a near object indicates that these points are equally distant from the eye. In special applications, such as flight simulators, accommodation is removed as a flatness cue by collimating the image. Collimation, typically with a lens or a parabolic mirror, converts the diverging rays from a near object into parallel rays so that the visual system interprets the near object as a distant object (see Figure 4). Similarly, lack of binocular disparity within a near region suggests that the points within this region are equally distant. An especially interesting cue to flatness is a surrounding frame. Looking at a high-quality motion picture through a tube, which eliminates any explicit or implicit frame surrounding the image (and eliminates disparity if the tube is monocular) makes the picture appear less flat than if the frame were visible. Surrounding a real 3D scene with a frame can make that scene appear more flat, indicating that the frame is indeed a cue to flatness. What happens if there are neither cues to depth nor cues to flatness? This situation can be produced by creating a ganzfeld – a room with uniform illumination and no visible edges. The resulting perception is ambiguous. Usually a 3D fog is perceived.

CONCLUSION

The perceived structure of the 3D world is inferred from information available in the retinal projections and in the oculomotor system. There are many sources of information for depth which have been roughly categorized into 'cues', including binocular disparity, motion parallax, texture gradients, linear perspective, shading, interposition, accommodation and convergence.

Further Reading

Braunstein ML (1994) Decoding principles, heuristics and inference in visual perception. In: Jansson G, Bergström SS and Epstein W (eds) *Perceiving Events and Objects*. Hillsdale, NJ: Erlbaum.

Braunstein ML (1994) Structure from motion. In: Smith AT and Snowden RJ (eds) *Visual Detection of Motion*. New York: Academic Press.

Epstein W and Rogers S (eds) (1995) *Perception of Space and Motion*. San Diego, CA: Academic Press.

Gibson JJ (1950) *The Perception of the Visual World*. Boston, MA: Houghton Mifflin.

Pastore N (1971) *Selective History of Theories of Visual Perception: 1650–1950*. New York: Oxford University Press.

Descartes, René

Introductory article

Stephen Gaukroger, University of Sydney, Sydney, New South Wales, Australia

CONTENTS

Descartes pursued two different approaches to the question of the nature of the mind: one via psychophysiology and one via a theory of the different properties of mind and matter, construed as different substances.

DESCARTES' LIFE AND PHILOSOPHICAL DEVELOPMENT

René Descartes was born in 1596, and entered the Jesuit college of La Flèche as a boarder at the age of 10. He left it in 1614, and, after spending a year in Paris, completed his formal education by taking a degree in civil and canon law at the University of Poitiers in 1616. From 1619 onwards, he pursued a career as a gentleman soldier, first in the army of Maurice of Nassau and then in that of Maximillian I of Bavaria, before settling down to a life of science and scholarship in the early 1620s. He worked primarily in mathematics and natural philosophy in the 1620s and early 1630s, in Paris and elsewhere, moving to the Netherlands in 1628. In the mid-1630s he developed a skeptical form of epistemology, set out in the *Discourse on Method* (1637), in the *Meditations* (1641), and finally in the *Principles of Philosophy* (1644). It is this form of pure epistemological speculation for which Descartes is now principally remembered. In 1649 he moved to Sweden, where he died early in 1650.

We can trace three different strands of interest in Descartes' development. From 1619 to the late 1620s he pursued mathematics above all else. A precocious and original mathematician, his greatest contribution was to the discipline of analytic geometry, in which lines and curves are represented by equations through the use of coordinates. What he provided was a powerful unification of arithmetic and geometry, and it was from his treatise on the techniques that he had developed in this area, the *Geometry*, that Newton and others learned their advanced mathematics later in the seventeenth century.

Descartes also pursued an active research program in natural philosophy from 1619 onwards, moving from kinematics, hydrostatics, and optics to a general Copernican cosmology in the 1630s. Some time in the middle to late 1620s, he discovered a central law in geometrical optics, the law of refraction, which was crucial in the development of better telescope lenses. In the early 1630s, in *The World,* he developed a comprehensive physical cosmology – the most important seventeenth-century cosmological system before Newton – in which the problem of how the planets can revolve in stable orbits around a central sun was solved by proposing a model in which a revolving celestial fluid carries the planets along, their distance from the sun being a function of their size.

In the 1630s, Descartes began to develop a distinctive epistemology driven by skepticism. He focused on a number of problems, such as radical skepticism, the provision of foundations for knowledge, and the exact nature of the relation between mind and body, which were either new or treated in a new way. However, Descartes explicitly warned against an insulation of philosophy from empirical questions.

THE PSYCHOPHYSIOLOGY OF COGNITION

At various stages in his career, Descartes tried to describe the nature of various kinds of intellectual or psychological phenomena in psychophysiological terms. There are three that are of particular importance: mathematical cognition, perceptual cognition, and affective states.

Mathematical Cognition

In his *Rules for the Direction of the Mind,* the relevant parts of which were completed between about 1626 and 1628, Descartes was concerned with the question how a quantitative grasp of the world was possible. The question is how to connect the contents of the world, which consists of material objects, with the contents of the intellect, which, in the case of mathematical cognition, consists of abstract mathematical structures, which may have arithmetical or geometrical interpretations, but which are neither arithmetical nor geometrical in themselves. If a quantitative grasp of nature is to be possible, mathematics must somehow be mapped onto the material world.

Descartes' solution is to suggest that such a mapping cannot be direct: a determinate representation of the abstract mathematical structures is mapped onto a *representation* of the world. As regards the representation of the world, Descartes sets out to show how qualitative differences, such as differences in color, can be represented purely in terms of different arrangements of lines: red as vertical lines, blue as horizontal lines, green as a combination of these, yellow as diagonal lines, etc. We might think of this as a form of encoding: qualitative differences can be encoded in a very economical form, namely in terms of lines. As regards abstract mathematical structures, Descartes argued that these can be represented in terms of line lengths, or combinations of line lengths – the basic arithmetical operations of addition, subtraction, multiplication, division, and root extraction can all be performed using line lengths, for example, and geometrical operations present no problem in this respect. The contents of the intellect are represented in the 'imagination' as line lengths, and the contents of the material world are represented there as configurations of lines, and the former are mapped directly onto the latter, thus allowing a quantitative grasp of nature.

There are a number of interesting features of this account. Firstly, there is the idea that sensory information must be encoded in some way if we are to be able to engage with it cognitively. Secondly, note that the 'imagination' is a material organ (Descartes will later identify it with the pineal gland, this being the unique central, unduplicated organ in the brain, and hence ideally suited as a site for central cognitive processing), and that this, rather than the intellect, is where the cognition actually takes place. In other words, we seem to have a material site for cognition.

Perceptual Cognition

A distinctive feature of Descartes' natural philosophy is his commitment to mechanical explanation. This is evident in his physics and astronomy, but it goes further. Without appealing to vital forces of any kind, Descartes reasoned that, except in the case of the exercise of judgment and free will, which require consciousness, physiological processes – including such psychophysiological cognitive functions as visual perception, memory, and habitual and instinctual responses – can be accounted for mechanically. In the *Treatise on Man,* Descartes set out one of his most daring projects: the complete mechanization of physiology, from nutrition, excretion and respiration up to memory and perceptual cognition. His treatment of the last two is particularly ingenious.

In the case of visual perception, he argues – against a long and deeply entrenched tradition stretching back as least as far as Aristotle – that a visual image need not resemble the object perceived. Not only is there nothing in the optics or physiology of vision that requires resemblance, but the fact that the retinal image is inverted, that the retina where the visual image must be represented is a two-dimensional concave surface, that it must be transmitted through the nerves, and so on, all indicate a form of encoding of information. Descartes also shows awareness of fundamental problems of information recognition: he shows how we must employ an innate or unconscious geometry in order to be able to gauge the distance of objects, since our visual stimulation results from a light ray which cannot carry information about how far it has traveled from the object to the eye.

Descartes' account of animal cognition is very sophisticated. His aim is to show that the structure and behavior of animal bodies are to be explained in the same way as we explain the structure and behavior of machines. In doing this, he wants to show how a form of genuine cognition occurs in animals, and that this can be captured in mechanistic terms. He does not want to show that cognition does not occur, that instead of a cognitive process we have a merely mechanical one. In more modern terms, his project is a reductionist, not an eliminativist, one.

In the case of memory, Descartes offers an account in which the memory images do not have to resemble what caused them, and they do not have to be stored faithfully and separately but only in a way that enables the idea to be represented in a recognizable form. Unlike his contemporaries, who were largely preoccupied with identifying the physical location of memory storage (a favored location was in the folds of the surface of the brain, because of the large surface area such folds created), Descartes' concern is with just what is needed for recall, and he provides a rudimentary account of how memory works by means of association.

Affective States

In the *Passions of the Soul* (1649), Descartes provided an extensive account of affective states, or passions, in which he examined how the mind and the body interact to produce such states as fear, anger, and joy. One of his primary concerns here was to argue against the idea that there are higher and lower functions of the mind, that there is a hierarchy of appetites, passions, and virtues, with the will occupying a precarious position. What Descartes opposes is the idea of a fragmentation of the soul, whereby one loses a sense of how the agent can collect himself or herself together, and exercise true moral responsibility. This is particularly important for Descartes, because he sees the crucial part of ethics to be the difficult process of forming oneself into a fully responsible moral agent – this is what the control of one's passions is ultimately aimed at – rather than the question of how such an agent should behave.

THE METAPHYSICS OF MIND

The doctrine for which Descartes is most famous is 'Cartesian dualism'. He advocates a view of the mind whereby (1) the mind is a different substance from the body, by virtue of having different essential or defining properties, and (2) the mind can exist in its own right, independently of the body, and have an identity that distinguishes it from other minds. Modern versions of dualism usually restrict their claims to the first of the above claims, substance dualism, but Descartes' advocacy of substance dualism seems to be in large part motivated by the second claim, which, because mind is not subject to the physical processes that lead to death and corruption, is tantamount to the doctrine of personal immortality.

Substance dualism requires no commitment to the capacity for independent existence of the mind. The fact that mind and matter are separate substances does not in itself require us to imagine that mind might be able to exist independently of matter: we might conceive of the mind as the 'software' that runs the cognitive parts of the body, for example, thinking of it as something quite distinct from its material realization in a particular brain, or central nervous system, while at the same time arguing that it makes no sense to talk about such software independently of its being a program.

However, Descartes' concerns seem different. As he indicates in the dedicatory letter which prefaces the *Meditations*, he is concerned to defend the doctrine of personal immortality of the soul. This doctrine had been undermined by two different kinds of philosophical conception of the mind. The first was Alexandrism, which was in effect substance dualism without personal immortality. Alexander of Aphrodisias and his followers had argued that the mind or soul is the 'organizing principle' of the body, something essentially materially realized, so that with the death and corruption of the body, it goes out of existence. The second kind of conception was Averroism, whereby the mind can be separated from the body at death, but in undergoing such separation it loses any identifying features and becomes identical with 'mind' as such. The idea here is that what distinguishes my own mind from another is a set of features it has by virtue of being instantiated – my sensations, memories, and passions are easily sufficient to mark me out from everyone else, for example, but these are dependent on my having a body – and once it becomes separated from my body, it loses anything that might differentiate me from anything else, and so becomes one with a universal mind (God).

Descartes' challenge is to steer a middle path between Alexandrism, which denies immortality altogether to the soul, and Averroism, which denies it personal immortality. It is not clear that he is able to do this. His account of such processes as mathematical cognition, perceptual cognition, and affective states presuppose that the mind is instantiated in the body, and so are compatible with Alexandrism (conceived as a minimal substance dualism). He does not describe what a disembodied soul is like except to tell us that it contemplates universals, but that is what God does, and what Averroes' single mind does: there is nothing to distinguish disembodied souls from one another in this respect.

RELEVANCE OF DESCARTES' WORK TO COGNITIVE SCIENCE

Descartes was the first person to provide a comprehensive account of a mechanized psychophysiology. Although his ideas had some followers in the succeeding two centuries, the resources available – most importantly knowledge of brain physiology – were far from adequate, and Descartes' project looked like a dead end. As these resources were acquired, from the late nineteenth century onwards, the situation changed. There remain a number of deep philosophical problems about the nature of cognition and the mind – perceptual cognition in unintelligent animals, what is involved in memory retrieval, and so on – which Descartes, because of his limited empirical resources, was forced to focus on in a way that draws attention to some of the conceptual problems that need to be addressed if one is to orientate one's empirical investigations in a fruitful direction.

Further Reading

Baker G and Morris K (1996) *Descartes' Dualism*. London, UK: Routledge.

Descartes R (1984–1991) *The Philosophical Writings of Descartes*, edited and translated by J Cottingham *et al.*, 3 vols. Cambridge, UK: Cambridge University Press.

Descartes R (1998) *The World and Other Writings*, edited and translated by S Gaukroger. Cambridge, UK: Cambridge University Press.

Gaukroger S (1995) *Descartes: An Intellectual Biography*. Oxford, UK: Oxford University Press.

Gaukroger S (2002) *Descartes' System of Natural Philosophy*. Cambridge, UK: Cambridge University Press

Gaukroger S, Schuster J and Sutton J (eds) (2000) *Descartes' Natural Philosophy*. London, UK: Routledge.

Sepper D (1996) *Descartes's Imagination*. Berkeley, CA: University of California Press.

Sutton J (1998) *Philosophy and Memory Traces*. Cambridge, UK: Cambridge University Press.

Wolf-Devine C (1993) *Descartes on Seeing*. Carbondale, IL: Southern Illinois University Press.

Yolton JW (1984) *Perceptual Acquaintance from Descartes to Reid*. Oxford, UK: Blackwell.

Descending Motor Tracts

Introductory article

Michael Peters, University of Guelph, Guelph, Ontario, Canada

CONTENTS

The human ability to 'compose' new kinds of movement allows us to speak and to generate technology. Evolutionary changes in the descending motor tracts and especially the corticospinal tracts lie at the core of this ability.

INTRODUCTION

Action implies movement, and movement is implemented by nerve tracts that descend from the brain to the spinal cord, where they activate motor neurons. The activation can take place by direct contact of the terminals of the tracts on motor neurons, or indirectly, by termination on 'interneurons' in the spinal cord which then activate the motor neurons. Motor neurons are cells that directly activate muscle contraction. Movement ranges from simple reflexive responses to highly complex voluntary movement, and to serve this range of movement, a number of descending tracts are needed.

THE CORTICOSPINAL TRACTS

Of all descending motor pathways, the corticospinal pathways are by far the most massive. In humans there are about a million corticospinal fibers on each side, although there is considerable variability, with values as low as 750 000 and as high as 1 400 000. The vast majority of these fibers cross over from one side to the other, in the pyramidal decussation. The fibers in the corticospinal tract that cross over to the contralateral side form the dorsolateral corticospinal tracts (or funiculi) in the spinal cord (Figure 1). These tracts are thought to be primarily involved in the control of muscles in the distal parts of the limbs (especially the muscles that operate the hand). A minority of fibers originating from the left hemispheres of the brain descend in the left half of the spinal cord. These fibers are largely found in the ventral and medial portion of the spinal cord (Figure 1) and form the ventromedial or anterior corticospinal tracts. The proportion of fibers that remain uncrossed is variable from individual to individual, ranging from individuals in whom the normally crossed portion descends ipsilaterally, to individuals who lack an ipsilateral portion of that tract. The ventromedial corticospinal tracts are largely concerned with the central parts of the body, such as the torso.

Source of Corticospinal Tract Fibers

Where do the corticospinal tract fibers come from and where do they go to? The majority of the fibers come from the cortex that lies anterior to the central fissure. Immediately anterior to the central fissure is the primary motor cortex (usually referred to as M1). Primary motor cortex means the area of the anterior cortex that was first recognized as being specialized for motor function because stimulation in this area would produce movement in the limbs. This is the area from which the densest fiber projections are sent towards the spinal cord. Anterior to the M1 lie several other motor areas. Current work with primates suggests that there are at least seven of these areas. One of these, the supplementary motor area (SMA) also contains a rough topographical map of the body musculature, like the M1 area, and the SMA also sends fibers directly to the spinal cord. Other motor areas are collectively called premotor cortex. Some of these send fibers directly to the spinal cord and others connect to each other and the M1, without any direct output to the spinal cord. In humans, the corticospinal tract fibers that come from anterior of the central fissure amount to some 80% of the tract, with some 20% originating posterior to the central fissure. In monkeys the ratio is closer to 60:40, indicating a relatively greater weight on motor cortex origin in humans.

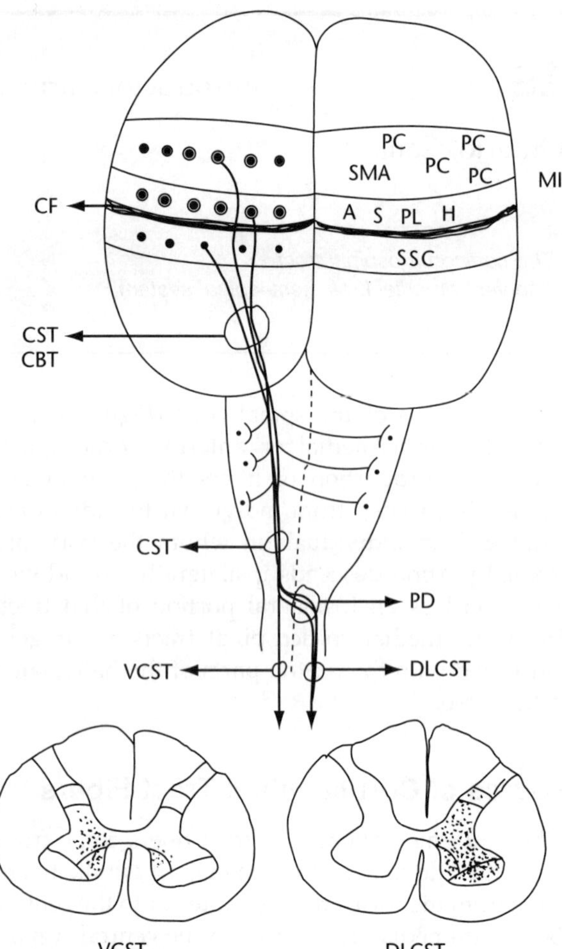

Figure 1. Origin and path of corticospinal and corticobulbar fibers. The bulk of fibers (80%) run from the cortex anterior to the central fissure and 20% from the somatosensory cortex. The bulk of corticobulbar fibers contact the motor nuclei of the nerves that control movements of the head and neck, eyes, facial musculature, jaws and tongue. Some corticospinal fibers do not cross the midline and descend on the ipsilateral (same) side, cross over to the contralateral side to descend the spinal cord as the ventral (anterior) corticospinal tract (VCST). The bulk of corticospinal fibers cross to the opposite side in the pyramidal decussation and descend as the dorsolateral corticospinal tract (DLCST). In the primary motor cortex (M1) there is a topographical relation to body parts, where 'A' denotes axial – the midline parts of the body, 'S' denotes the shoulders, 'PL' denotes the proximal limbs and 'H' denotes the hand. The two spinal cord sections (see Figure 2) show the terminations of the two corticospinal tracts. VCST ends mostly in the ipsilateral intermediate region, on premotor neurons concerned with the axial and proximal parts of the body. DLCST terminates throughout the grey matter, on sensory interneurons, premotor neurons in the intermediate region and on the motor neurons for proximal and distal body parts. Dots in the spinal cord section show predominant terminations of corticospinal fibers. The size of dots in the originating cortical areas indicates the relative contribution of those areas to the descending tracts. Abbreviations: CBT, corticobulbar tract: CF, central fissure; CST, corticospinal tract; MI, primary motor cortex; PC, premotor cortex; PD, pyramidal decussation; SMA, supplementary motor area; SSC, somatosensory cortex.

Destination of Corticospinal Tract Fibers

The termination of corticospinal fibers in the spinal cord reflects their region of origin (Figure 2). Fibers that terminate in the dorsal horn, where the sensory relay cells are located, come mostly from the sensory cortex, posterior to the central fissure. Fibers that terminate on the premotor neurons in the intermediate zone come mostly from the premotor cortex anterior to the primary motor cortex. Finally, fibers that terminate on the motor neurons in the spinal cord come mostly from the primary motor cortex. That fewer components of the corticospinal tract in humans originate from the sensory cortex is supported by the observation that fewer corticospinal tract fibers terminate in the dorsal horn of the spinal cord than is the case for monkeys. While humans share with other primates direct terminations of the corticospinal tract fibers on motor neurons that innervate finger muscles, humans are unique in also having direct terminations on the motor neurons that operate the thoracic musculature. This is needed because of the precise use of the thorax as a 'bellows' in speech.

Finally, descending corticospinal tract fibers end in the gray matter of the spinal cord in an orderly fashion, such that the termination on motor neurons that activate distal muscles, such as hand muscles, is located laterally (that is, near the left or right outer edges of the ventral part) in the gray matter. In contrast, fibers that activate muscles that operate the center of the body, such as the torso, terminate medially (that is, in the gray matter close to the center of the spinal cord).

Loss of the Corticospinal Tracts

Thorough observations about what happens after the corticospinal tracts are severed are only available for monkeys, because in humans accidental damage to the corticospinal tracts is rarely restricted to these tracts. In monkeys with destruction of the pyramids – and therefore of the cortical spinal tracts – there are extremely severe deficits immediately after the damage. However, recovery does take place. Researchers observed a striking

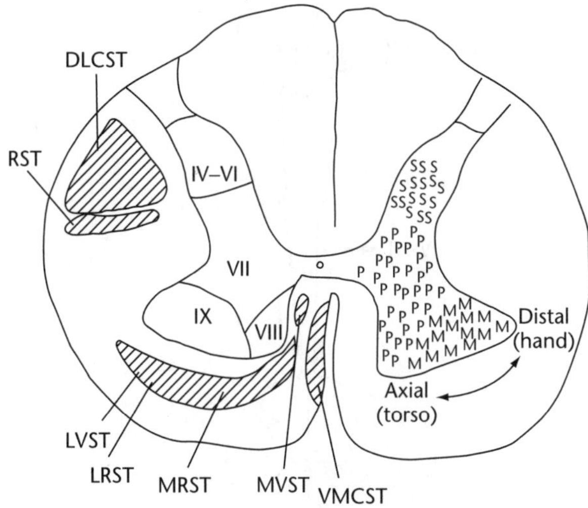

Figure 2. Cross-section of the spinal cord. The central butterfly-shaped region contains mostly cell bodies, and is know as the 'gray matter'. Around it lie the regions through which fiber tracts descend and ascend in the spinal cord, known as 'white matter'. The left half of the section shows the regions or Lamina of Rexed, and the right half shows the major contents of these regions. Cross-sections vary in appearance depending on the level at which the cord is cut. This level is roughly in the region of the shoulders. Motor neurons are arranged in a rough topographical order. The terms 'distal' and 'axial' denote the location of motor neurons that innervate distal portions of the body, such as the hand, and axial portions, such as the torso. Abbreviations: S, sensory neurons in the dorsal horn; P, intermediate region of the gray matter that contains mostly premotor neurons; M, motor neurons that connect directly to muscles. DLCST, dorsolateral corticospinal tract; LRST, lateral reticulospinal tract; LVST, lateral vestibular tract; MRST, medial reticulospinal tract; MVST, ventral vestibulospinal tract; RST, rubrospinal tract (very small in humans); VMCST, ventral or anterior corticospinal tract.

difference in recovery between movements that involved walking and climbing, and movements of the individual fingers. Ultimately the animals were able to walk and climb relatively well, even though the movements remained slow and there were signs of fatigue. In addition, independent movements of the arms recovered so that the animals could reach for food quite quickly and accurately. What did not recover, however, was the ability to use manipulative individual finger movements in, for instance, winkling a morsel of food out of a hole with a single finger. Interestingly, there were also difficulties with releasing an item of food once it was grasped with the whole hand even though the animals had no trouble grasping and releasing a bar of the cage during climbing. All of this suggests a prominent role of the corticospinal tracts in the voluntary execution of precise and independent finger movements.

A final observation concerns the capacity for recovery in animals with sparing of some corticospinal tract fibers. Individual finger movement seems to recover regardless of the precise location of the spared fibers in the tract. This suggests a remarkable plasticity in the role of descending corticospinal fibers, allowing considerable room for reorganization in the relation between corticospinal tract fibers and their target neurons.

THE CORTICORUBROSPINAL TRACT

While it is true that the corticorubrospinal connections show much less prominence in nonhuman primates and humans than in subprimate species, it is also true that the red nucleus itself remains undiminished in relative size. The answer to this puzzle lies in the nature of the red nucleus. It has a magnocellular (large-celled) part that is associated with the contralaterally descending corticorubrospinal tract, and appears to be involved in the direct control of limbs. It also has a parvocellular (small-celled) part that does not seem to be involved in projections to the spinal cord and also seems to play no part in the direct control of the muscles. Instead, this part is richly interconnected with the cerebellum. It is this part that has grown at the expense of the magnocellular part in primates, especially in humans (Figure 3).

Loss of the Corticorubrospinal Tract

In monkeys, selective lesions of the rubrospinal tract lead to slowing of movement as well as limpness and weakness in the hands. However, these problems show recovery to the point where no deficits are noted. If the rubrospinal tract is damaged on one side after loss of the corticospinal tracts, there are permanent problems with the ability to perform those hand and arm movements that showed recovery after corticospinal tract loss. Thus, the animals were no longer able to close the affected hand around a morsel of food, and reaching movements were poor. For instance, if animals were held close to the bars of the cage, they would try to move the affected arm into position by moving the shoulder rather than making use of the arm itself. In contrast, the animals could use the hand for climbing and clinging onto a bar even though there was weakness in the affected hand. In addition, animals showed the ability to right themselves and body posture was relatively normal.

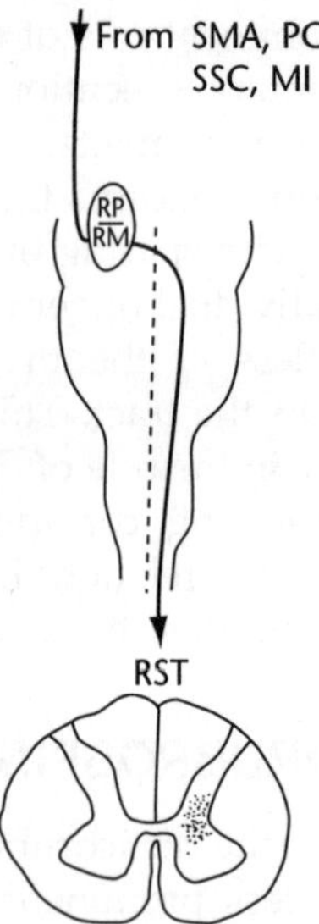

Figure 3. The cortical input from the supplementary motor area. The fibers from the RM cross the midline and descend as the rubrospinal tract (RST) in the spinal cord. They are shown to terminate mostly on premotor neurons in the intermediate region of the gray matter. PC, premotor cortex; SSC, somato-sensory cortex; MI, primary motor cortex to the large-celled portion of the red nucleus; RM, magnocellular portion of red nucleus; RP, parvocellular portion of red nucleus.

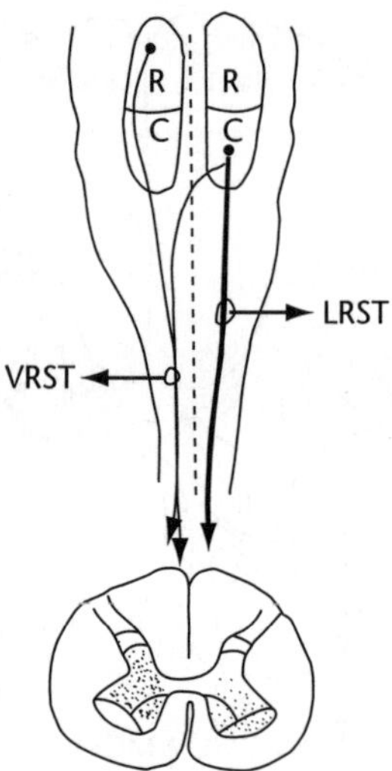

Figure 4. Two of the reticulospinal tracts are shown, in a much simplified form. VRST, ventral reticulospinal tract; LRST, lateral reticulospinal tract; R, rostral (toward the top of the brain) portion of the reticular formation; C, caudal (toward the spinal cord portion of the reticular formation).

However, the arm on the side affected by the lesion was not held in a normal position but hung loosely from the shoulder.

It is clear that postural control of the body is managed by a system other than the corticorubral or the corticospinal system, because such control survives loss of both of these systems.

THE VENTROMEDIAL BRAINSTEM–SPINAL SYSTEM

Several tracts originate in the brainstem, notably from the reticular formation in the region of the medulla oblongata and the region of the pons (Figure 4). Another important component of this system comes from the vestibular nuclei. There are two separate paths within the reticulospinal system; one originates more dorsally and rostrally in the reticular formation and descends ipsilaterally in the medial reticulospinal tract, while a more caudal and ventral portion descends bilaterally in the lateral reticulospinal paths. One hint as to the possible roles of these tracts comes from the observation that axons from neurons in the reticulospinal tract may make contact with target neurons in the region of the spinal cord where information flows out to the forelimbs (cervical level) and then continue to regions concerned with the hind limbs (lumbar levels). In addition, axons may also send collateral branches across the midline. It is likely that these pathways not only manage basic locomotion and postural adjustments but are also involved in the integration of these more basic motor behaviors with voluntary movement.

Loss of Function in the Reticulospinal System

Damage to the descending paths from the reticulospinal tracts and parts of the vestibulospinal tracts in monkeys leads to lasting impairment in body posture and movements. The animals walk unsteadily, and limbs and the trunk are flexed. The shoulders are raised; the head is not held upright in the normal position and falls forward. When the animals jump toward the bars of their cage, they often miss and when they reach for a morsel of food, they cannot move the hand smoothly toward the target. This is due to an ataxia (problem with muscular coordination) in the portions of the arm close to the body. In contrast, once the hands are near a target, the animals can perform delicate individual movements of the fingers; this would be expected because the corticospinal tracts are intact.

The description of the behavior of these animals provides the clearest illustration of the principle that in order to function properly, the corticospinal tracts need the support of the reticulospinal system.

Thus, the increased specialization of the corticospinal system in the guidance of voluntary and skilled movements in human as opposed to nonhuman primates does not render this 'older' system obsolete in humans. If anything, because the requirements of 'composing' and learning new movements demand strong support by the postural and integrative contributions of the brainstem–spinal systems, these systems are as important as ever.

Further Reading

Armand J (1982) The origin, course and terminations of corticospinal fibers in various mammals. In: Kuypers HGJM and Martin GF (eds) *Anatomy of Descending Pathways to the Spinal Cord*, vol. 57, pp. 229–360. Amsterdam: Elsevier Biomedical Press.

Canedo A (1997) Primary motor cortex influences on the descending and ascending systems. *Progress in Neurobiology* **51**: 287–335.

Donoghue JP and Sanes JN (1994) Motor areas of the cerebral cortex. *Journal of Clinical Neurophysiology* **11**: 382–396.

Galea MP and Darian-Smith I (1995) Postnatal maturation of the direct corticospinal projections in the macaque monkey. *Cerebral Cortex* **5**: 518–540.

Ghez C (1991) The control of movement. In: Kandel ER, Schwartz JH and Jessell TM (eds) *Principles of Neural Science*, 3rd edn, pp. 533–547. New York: Elsevier.

Kennedy PR (1990) Corticospinal, rubrospinal and rubro-olivary projections: a unifying hypothesis. *Trends in Neurosciences* **13**: 474–479.

Kuypers HGJM (1981) The descending pathways to the spinal cord, their anatomy and function. In: Brooks VB (ed.) *Handbook of Physiology*, sect. 1: The nervous system, vol. II, Motor control, part 1, pp. 597–666. Bethesda, MD: American Physiological Society.

Kuypers HGJM (1982) A new look at the organization of the motor system. In: Kuypers HGJM and Martin GF (eds) *Anatomy of Descending Pathways to the Spinal Cord*, vol. 57, pp. 381–403. Amsterdam: Elsevier Biomedical Press.

Lawrence DG and Kuypers HGJM (1968) The functional organization of the motor system in the monkey. I. The effects of bilateral pyramidal lesions. *Brain* **91**: 1–14.

Lawrence DG and Kuypers HGJM (1968) The functional organization of the motor system in the monkey. II. The effects of lesions of the descending brain-stem pathways. *Brain* **91**: 15–36.

Rizzolatti G, Luppino G and Matelli M (1998) The organization of the cortical motor system: new concepts. *Electroencephalography and Clinical Neurophysiology* **106**: 283–296.

Schwartzman RJ (1978) A behavioral analysis of complete unilateral section of the pyramidal tract at the medullary level in Macaca mulatta. *Annals of Neurology* **4**: 234–244.

Development

Introductory article

Nora S Newcombe, Temple University, Philadelphia, Pennsylvania, USA

CONTENTS

Development refers to the acquisition of mature knowledge states and cognitive capabilities. Cognitive scientists seek to characterize initial knowledge states and cognitive capabilities, and the processes involved in their transformation into adult competence.

BACKGROUND

Babies do not appear to think or act like adults. Thus, there seems to be a self-evident problem of development. How does the helpless infant change into the competent adult? Answers to this question have both practical and theoretical implications. On the practical side, knowledge about when and how children change can help parents to be aware of their children's capacities and sensitive to their needs. Understanding cognitive development can also help educators to work with children to learn optimally, and can help clinicians asked to deal with a variety of challenges to learning. On the theoretical side, the study of development is one of the crucial aspects of cognitive science. No theory of knowledge or skill would be complete without an approach to how the knowledge or skill is acquired. Indeed, thinking about acquisition can place important constraints on the theoretical enterprise and strengthen or rule out specific approaches.

Consideration of the nature of cognitive development predates psychology. Philosophers pondering the matter generally argued for one of two answers to the question of how the baby becomes the adult. The first kind of answer was to suggest that infants are molded into adults by the influence of the world in which they live. There are various versions of this kind of theory, generally known as empiricism and most famously associated with the name of the English philosopher John Locke. Empiricists postulate a baby born with very little knowledge or capability – Locke spoke of a 'tabula rasa', or blank slate – for whom responses are shaped by the associations of certain sensations in time and space with other sensations, and by reward and punishment.

The second kind of answer to the problem of development is to argue that infants know more than they may seem to know at first sight. Although they appear helpless, they may be endowed with capabilities and categories of thought that allow them to apprehend a world governed by immutable laws of space, time, and causality. Such a theory, emphasizing the importance of the human mind in creating the reality of the world around us, is most famously associated with the German philosopher Immanuel Kant. Kant's theory of knowledge has not one but two modern descendants. It can, in fact, be seen as leading to two quite different approaches to development within cognitive science: constructivism, as originally proposed by Jean Piaget, and nativism, as originally proposed for the case of language development by Noam Chomsky.

Philosophers discussed the nature of infants', children's, and adults' knowledge without actually observing any infants, children, or adults in the process of thinking. When the science of psychology began in the second half of the nineteenth century, it gave rise almost immediately to the empirical study of babies and young children and to theorizing about the nature of development based on observation. Some investigators, working in the empiricist tradition, emphasized the role of the environment in development. For example, John S. Watson made studies of conditioning that he argued showed the origins of certain thoughts and fears. B. F. Skinner analyzed language development as the product of contingencies in the world. Other investigators, working in a biological tradition that can be seen as an early version of nativism, portrayed development as the maturational unfolding of a preprogrammed biological being. For instance, Arnold Gesell and G. Stanley Hall created tables and charts showing just what children at typical ages could be expected to do and think, implying that functioning would be in a constant relation to chronological age. Some thinkers emphasized the Kantian theme of an active mind that structured understanding of the world. James Mark Baldwin and Heinz Werner are early examples of this type of constructivist thinking.

During this early period, it is interesting to note, writing about development already had a property that is still true today: it was seen to have relevance for thinking about applied problems of child rearing and education. Hence, for example, Edward Thorndike was simultaneously a learning theorist and a founder of the field of educational psychology.

CLASSIC APPROACHES

Prior to the 'cognitive revolution' of the 1960s, comparatively little was known about cognitive development, in large part because comparatively little was known about the nature of mature cognitive functioning. Psychological science, especially in the United States, was dominated by the view that there were universal laws of learning that applied equally to all animal species including humans. Therefore the problem of development tended to be conceptualized as the simple accumulation of learning achieved by means of the operation of the universal laws of learning (B. F. Skinner's cumulative record), or, slightly more developmentally, as the problem of learning to learn in an adult way. An example of research in the latter tradition would be experiments designed to examine exactly what children learned when they learned that a response to a particular stimulus was rewarded. For instance, children might learn that a reward would follow pressing a square of a particular absolute size that was also the middle-sized square of three. There was interest in determining whether there was a developmental change in children seeing the absolute size as controlling the reward as opposed to seeing the 'middle' relation as controlling the reward. Work by Jean Piaget and Lev Vygotsky was published in Europe in the first half of the twentieth century, but their writing was not widely appreciated when it first appeared.

The cognitive revolution led to a radical change in this situation. There were several events central to this revolution, many of them with developmental aspects. Chomsky's writings on language in the 1950s and 1960s led to a rethinking of the psychology of language by figures such as George Miller, and also to work on children's acquisition of language led by the pioneering investigations of Roger Brown. This research emphasized children as active constructors of a child grammar, in the Kantian tradition. Simultaneously, building on his earlier work on the constructive nature of perception and on the nature of adult concepts and thinking, Jerome Bruner started a research program on children's cognition that emphasized children as theorizers. Bruner's work, along with John Flavell's introductory writing on Jean Piaget, led to widespread interest in the United States in Piagetian research. By the mid-1970s, research on cognitive development was dominated by a focus on Piaget's theory and hypotheses. (*See* **Piagetian Theory, Development of Conceptual Structure**)

The cognitive revolution put an end to simple empiricism as an approach to development. Almost from the start, however, there were controversies between developmental theorists who emphasized the Kantian idea of the developing mind as an active constructor of knowledge and those who focused on the equally Kantian idea of innate capabilities. Piaget was the central example of the constructivist approach. But his theory quickly came under fire for several reasons. One criticism was that his thinking seriously underestimated the initial capabilities of infants. Piaget postulated that infants are born with only simple sensorimotor capabilities – they can see, hear, feel, taste, and smell the world, and they are also equipped with stimulus-response reflexes and other, less patterned motor exploration patterns. A large amount of research during the 1980s and 1990s showed, however, that infants and young children had far more substantial abilities than Piaget had seen. A second criticism was that various capabilities did not cohere in the way that Piaget's stage theory had envisioned. Piaget postulated that development consists of qualitative transformation through four distinct stages of thinking ability: sensorimotor, preoperational, concrete operational, and formal operational. Research on the abilities said to characterize these four stages showed, however, that children frequently achieved competence in some of the relevant tasks while lagging behind on others said to be characteristic of the stage, often for protracted periods of time. A third problem was that Piaget talked in general terms about the process of developmental change as being the operation of equilibration, a process of achieving cognitive equilibrium either by seeing the environment in terms of existing cognitive structures (assimilation), or, if needs be, changing cognitive structures in order to make sense of observations of the environment (accommodation). Many investigators argued that a more detailed specification of the mechanisms of change was needed.

The waning of support for Piagetian theory led to an increased interest in nativism as a solution to the problem of development. Nativism had long been espoused as a theory of language development, because Noam Chomsky had argued strongly that there must be an innate language acquisition device. Otherwise, he said, it would be impossible for children to learn language from the impoverished and error-laden speech sample that they would obtain from listening to adults. Beginning in the late 1970s, nativism also became popular when considering domains of cognitive development in addition to language, including understanding of the Kantian aspects of the world, such as causality, space, time, and number. Nativism was also applied to other aspects of cognitive development, such as the acquisition of categories, or the possibility of imitation of the actions of adults. In addition, many investigators who were not strict nativists became very intrigued by the demonstration of the surprising early competence of infants, such as their memory abilities.

A different response than nativism to the waning of support for Piagetian theory was to champion approaches that more strongly emphasized the role of the environment in development. Vygotsky's writings underwent a revival because their focus on cultural transmission through language and teaching seemed to emphasize the role of the environment in development, without representing a retreat to pure empiricism. Information processing approaches to development also stressed the role of the environment in development but conceptualized it somewhat differently than Vygotskyan theory, as a provider of informational feedback. Information processing theory melded a focus on environmental feedback with a developing part of the emerging field of cognitive science, namely computational modeling. Many information processing theorists also incorporated biological thinking in their models by using parameters in the models that could be conceptualized as changing maturationally, notably short-term memory and processing capacity. (*See* **Categorization, Development of; Memory, Development of; Culture and Cognitive Development**)

RECENT INFLUENCES

As the field of cognitive science developed during the 1980s and 1990s, in conjunction with the development of cognitive neuroscience during roughly the same period, the study of cognitive development changed concomitantly. The use of computational models expanded. In addition, the style of modeling changed from the predominant use of production systems, written line by line by modelers with certain assumptions about relations between input and processing, to an increasing use of parallel distributed processing (PDP) models, set up with fewer content-related assumptions. PDP modelers postulated a cognitive architecture and made assumptions about the nature of environmental input and feedback, but they then needed to run their systems as experiments to see what they actually did. Modeling in this tradition has led, in some hands, to a view of development as the self-modification of systems that can be identified as constructivist. However, in other models, there are *tabula rasa* assumptions and decisions about the nature of environmental feedback that bring the models close to old-fashioned empiricism. (*See* **Cognitive Development, Computational Models of**)

A second change in thinking about cognitive development in recent times has been the increasing attention given to the role of the neural substrate in thinking. The advent of developmental cognitive neuroscience has added new techniques and dependent variables to the armamentarium of investigators, as well as renewing interest in developmental disorders. In addition, neuroscience has been the source of certain broad insights about the probable nature of development. One important example is the attention received by Peter Huttenlocher's research on synaptogenesis in different areas of human cortex. Huttenlocher has shown that increases in synaptic connections occur during the early part of life, probably in a maturational way. When connections reach a critical level, one often sees the advent of new forms of functioning. From a peak far above the levels typical for adults, synaptic connections seem to be eliminated, or pruned, in a way that is dependent on environmental input. (*See* **Neural Development**; **Neuropsychological Development**; **Developmental Disorders of Language**)

Newer approaches to cognitive development are not only the products of modeling and the use of neuroscience – they have also been driven in traditional ways by the rethinking of existing data and the acquisition of new data. There are three theories that attracted much attention in the 1990s. Dynamic systems theory was developed in other areas of investigation, such as characterization of motor behavior and it has been productively applied in thinking about motor development. However, Esther Thelen and Linda Smith have argued that it can also be very useful in thinking about development in many other domains as well. In brief, this approach suggests that behavior is complexly determined by contextual variables, so that thinking about development as the acquisition of specific static competences is liable to be misleading and unproductive. Variation-and-selection theory is an updated version of information processing theory proposed by Robert Siegler. Siegler argues that various strategies for dealing with cognitive problems co-exist, with development consisting of the relative predominance of these theories at different points in time. Domain-specific interactionism is a constructivist approach to development that recognizes the fact that the initial starting points for cognitive development are often quite strong and specific, and that development proceeds in a way that is not strongly stagelike. Many key domains have been investigated from this point of view. (*See* **Motor Development**; **Lexical Development**; **Causal Perception, Development of**; **Intermodal Perception, Development of**; **Naive Theories, Development of**; **Object Concept, Development of**; **Space Perception, Development of**; **Speech Perception, Development of**; **Object Perception, Development of**)

SUMMARY

The central problem addressed by research on cognitive development is the issue posed philosophically as the opposition between nativism and empiricism. A wide variety of theories and research has addressed the question, and we have moved beyond the most extreme versions of the debate to consider more integrated and balanced solutions. In the course of this research, we have also learned a great deal about the nature of development in various domains and at various ages. Much of this knowledge is now mature enough to be useful in a variety of applied areas, such as designing intervention and remediation for children with developmental disabilities, or improving the effectiveness of education for normally developing children. In addition, from a theoretical point of view, the field of cognitive development is a vital part of cognitive science. Because no cognitive model can be considered complete if it cannot be acquired by human children, given sensible assumptions about the nature of their learning, developmental thinking offers both constraints on theories and opportunities for theory testing. (*See* **Early Experience and Cognitive Organization; Language Development, Critical Periods in**)

Further Reading

Chapman M (1988) *Constructive Evolution: Origins and Development of Piaget's Thought*. New York: Cambridge University Press.

Elman J, Bates E, Johnson M, Karmiloff-Smith A, Parisi D and Plunkett K (1996) *Rethinking Innateness: A Connectionist Perspective on Development*. Cambridge, MA: The MIT Press.

Hirsh-Pasek K and Golinkoff RM (1996) *The Origins of Grammar: Evidence from Early Language Comprehension*. Cambridge, MA: The MIT Press.

Karmiloff-Smith A (1992) *Beyond Modularity: A Developmental Perspective on Cognitive Science*. Cambridge, MA: The MIT Press.

Mix KS, Huttenlocher J and Levine SC (2002) *Quantitative Development in Infancy and Early Childhood*. New York: Oxford University Press.

Newcombe NS and Huttenlocher J (2000) *Making Space: The Development of Spatial Representation and Reasoning*. Cambridge, MA: The MIT Press.

Rakison DH and Poulin-Dubois D (2001) Developmental origin of the animate-inanimate distinction. *Psychological Bulletin* **127**: 209–228.

Rogoff B (1990) *Apprenticeship in Thinking: Cognitive Development in Social Context*. New York: Oxford University Press.

Siegler RS (1996) *Emerging Minds: The Process of Change in Children's Thinking*. New York: Oxford University Press.

Thelen E and Smith L (1994) *A Dynamic Systems Approach to the Development of Cognition and Action*. Cambridge, MA: The MIT Press.

Developmental Disorders of Language

Intermediate article

April A Benasich, Rutgers University, Newark, New Jersey, USA
Jennifer J Thomas, Rutgers University, Newark, New Jersey, USA

CONTENTS

A developmental disorder of language is a significant delay or impairment in the expression and/or comprehension of language in the absence of a known cause.

INTRODUCTION

A developmental disorder of language is a significant delay or impairment in language acquisition (for expression and/or comprehension) in the absence of a known cause. Here we will use the term 'specific language impairment' (SLI) to refer to such language disorders. Specific language impairment (also called developmental dysphasia) is diagnosed on the basis of exclusion, meaning that the child's language difficulties cannot be accounted for by readily identifiable factors such as hearing impairment, neurological disease, psychiatric disability (disorders such as autism or childhood schizophrenia), low intelligence, physical malformation of the vocal apparatus, or severe environmental deprivation. Individuals with SLI exhibit normal nonverbal intelligence and, with the exception of language, have otherwise uncompromised development.

There is no universally accepted criterion for the classification of SLI. The World Health Organization *International Classification of Diseases* (ICD-10) classifies children with language abilities (assessed using a standardized language test) in the lowest

3% of the population as language-impaired (WHO, 1993). The American Psychiatric Association's *Diagnostic and Statistical Manual of Mental Disorders* (DSM-IV) has similar diagnostic criteria to the ICD-10, but it also requires that a child's language difficulties interfere with normal, everyday activities – academic, occupational or social (APA, 1994). In addition, researchers sometimes create their own classification or selection criteria for studies of developmental language problems, resulting in a heterogeneous population with the diagnosis of SLI. There is also no general agreement regarding the classification of children with subtypes of SLI, categorizing different patterns of deficits in language skills. Based on psychometric evaluations of language ability, three broad subtypes have been defined: receptive only, expressive only and expressive-receptive mixed (Tomblin, 1996). These categories, reflective of the DSM-IV and ICD-10 classification systems, have been used in many research studies. However, some researchers use a subtype classification based on the work of Rapin and Allen (1987), in which clinical evaluation of a child's spontaneous language during play (including phonologic, syntactic, semantic and pragmatic skills) leads to assignment to one of six subgroups: verbal auditory agnosia, verbal dyspraxia, phonological programming deficit syndrome, phonological–syntactic deficit syndrome, lexical–syntactic deficit syndrome or semantic–pragmatic deficit syndrome. In sum, there are a number of subtypes of SLI that attempt to define precisely an individual's areas of difficulty.

Although there are advantages to using diagnostic criteria specific to the stated research or clinical objectives, consensus about the defining characteristics of SLI, and the classification of its different subtypes, is essential if there is to be meaningful comparison among the many research efforts investigating the etiological conundrum of whether SLI stems from a single common underlying mechanism, or has multiple genetic and or developmental origins.

The traditional view has been that the symptoms exhibited in SLI can be attributed to delays in the learning of linguistic-specific semantic and syntactic rules, which are critical to the development of language. In addition, there is substantial support for the role of poor phonological processing (perceiving and discriminating phonemes, which are the smallest units of sound in language that alone can differentiate meaning) in children with both language and reading deficits. However, there is also strong evidence that differences in basic auditory processing abilities of children with SLI may be related to their language deficits: see Leonard (1998) and Tallal (2000) for reviews. Leonard (1998) comments that findings of disordered processing of brief or rapidly presented stimuli in children with SLI are both ubiquitous and consistent across laboratories, tasks, and stimulus variations. Thus, it has been posited that basic difficulties in processing the brief, rapid, successive auditory cues that constitute speech could impair or delay the formation of distinct (categorical) representations of the sounds of language (phonemes). This would have an impact on emerging language through a cascade in which each disordered step (beginning with the most basic sound processing) affects subsequent levels of language processing. According to this hypothesis, auditory perceptual deficits lead to weak representations of phonemes, and these weak phonological representations may lead to oral language disabilities, and subsequent reading, writing and/or spelling deficits due to poor phonographic (spoken) to orthographic (visual/written) transfer (Tallal, 2000). However, there is still much debate about the etiology of SLI – in particular, which of the deficits simply co-occur and which are causative (Mody *et al.*, 1997; see also Denenberg, 1999, 2001).

Dyslexia is also considered to be a developmental disorder of language, although dyslexia is classically defined as the failure to develop age-appropriate reading ability in the presence of otherwise normal skills. Nevertheless, research has shown that approximately 80% of children classified as SLI go on to develop dyslexia, providing support for the view that developmental language and reading disorders may have a common etiology (APA, 1994). Individuals with SLI and dyslexia often exhibit similar behavioral deficits in rapid sensory processing, and neuroimaging studies have revealed alterations in neuroanatomical substrates common to these two disorders. Further, measurements of brain activation employing event-related potentials (ERPs) (electrical signals recorded from the scalp in response to external stimuli) of SLI and dyslexic individuals are comparable. Thus, it is likely that children with SLI and dyslexia form an overlapping, though not identical, population. For this reason, behavioral, neuroimaging and neuroanatomical studies of both SLI and dyslexia are discussed here.

NEURAL SUBSTRATES

The classic studies of the language areas in the brain described adult patients suffering from various types of aphasia (disturbances in language expression or comprehension) due to brain lesions sustained in adulthood. Disruption of language abilities in these patients was consistently linked to specific areas of damage in the brain, thus allowing a coarse topographic neural map of language function to be constructed. These landmark studies were the basis for the Wernicke–Geschwind model of language function. In this model, auditory information is first processed by the primary auditory cortex (A1; Brodmann's areas 41 and 42, located on Heschl's gyrus on the dorsal surface of the left temporal lobe), and is then sent to secondary auditory cortical areas: Wernicke's area (the left posterior perisylvian area, including Brodmann's area 22), shown to support language comprehension, and Broca's area (the inferior left temporal gyrus containing Brodmann's areas 44 and 45), which mediates speech production (Figure 1). These areas are linked by a number of subcortical fiber systems. The arcuate fasciculus, a fiber bundle that carries information from Wernicke's area to Broca's area, has been of particular interest as it has been implicated in subtypes of aphasia characterized by impairments of language expression.

These early studies were critical in laying the framework for subsequent research into the neural bases of language processing. This research has shown that there are many additional language centers in the brain and that these interact via complex neural networks. However, the precise mechanisms governing language function continue to be the subject of intense study. Many brain areas, including left hemisphere regions of temporal, parietal and frontal cortex, are the focus of contemporary research into the neural substrates of developmental language disorders.

In contrast to acquired language deficits such as aphasia, for which a specific brain lesion can often be identified, SLI and dyslexia (in which normal language skills fail to emerge in the context of otherwise normal development) present a far less clear picture of which brain areas might be disrupted. The search for the underlying neural substrates of these disorders has involved behavioral, neurophysiologic and neuroanatomical approaches and has been accelerated by methodological advances in functional neuroimaging and event-related potential (ERP) paradigms.

There is now a substantial literature reporting neurophysiological and neuroanatomic abnormalities in language-impaired children. Although there are generally concordant findings regarding differences in hemispheric asymmetry (see below), a number of studies report inconsistent or conflicting findings concerning other structural brain anomalies. These include alterations in corpus

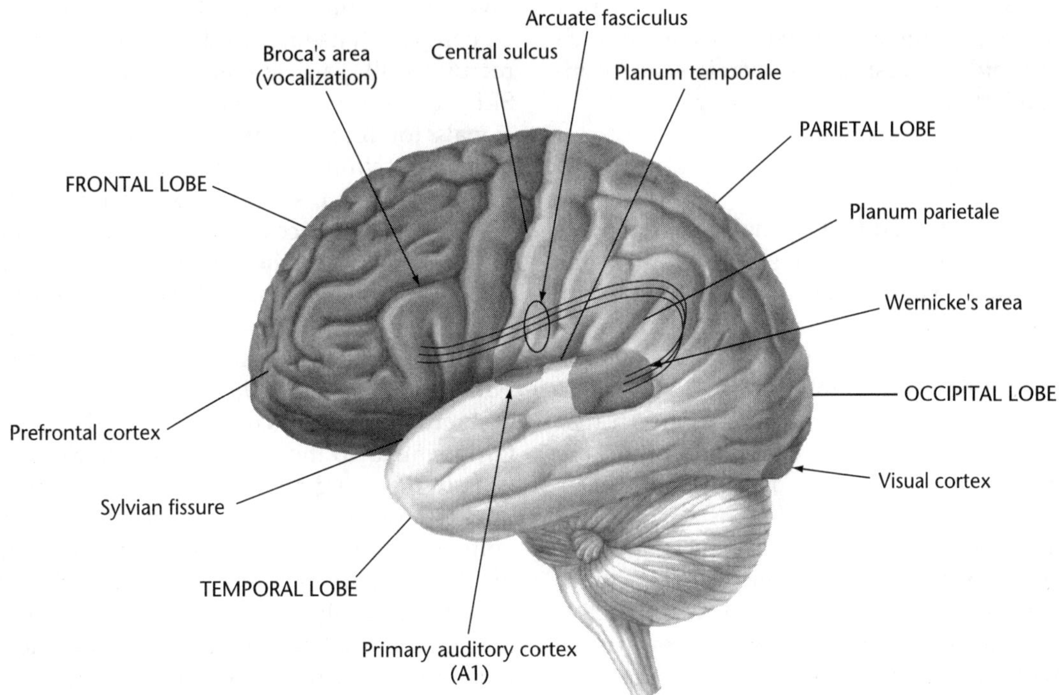

Figure 1. Areas of the human brain that are important for language, including the primary auditory cortex (A1), Wernicke's and Broca's areas, and the arcuate fasciculus.

callosum size, extra sulci, white-matter abnormalities, and cortical atrophies. Such findings suggest that a single, focal anatomical abnormality will not be sufficient to explain SLI, and may also account for the variation in the severity, nature and perhaps type of language deficits that individuals with SLI exhibit.

A number of hypotheses have been proposed that attempt to link the structural abnormalities and behavioral deficits found in individuals with SLI and dyslexia. One is that neuroanatomical disruptions in SLI may be bilateral (affecting both hemispheres) rather than unilateral (affecting only one hemisphere), thus reducing the possibility for recovery of language function. Subcortical structures such as the caudate and thalamus might also be involved, and lesions in these areas have been shown to cause lasting and profound language impairments. Another possibility is that a specific disruption of critical language processing systems might be due to delay in brain maturation, induced genetically or by some prenatal or perinatal brain insult. Such a disruption would particularly affect the left hemisphere (given the normal time course of early brain development) and could cause structural brain abnormalities. The resulting deviant brain organization and related alterations in connectivity might disturb critical periods of language development, resulting in delays as well as lasting impairments. Finally, a basic sensory integration deficit in SLI has been posited. This basic deficit would induce dysfunction of neural systems which mediate rapid processing of dynamic sensory information important to rate processing.

LATERALITY

It is generally accepted that, in the normal population, left hemisphere structures are dominant for language function. Evidence for this first came from lesion and psychophysical studies, and with the advent of neuroimaging techniques such as positron emission tomography (PET) and magnetic resonance imaging (MRI) it has been shown that the left hemisphere of the brain is larger than the right in the vast majority of normal people. Likewise, functional imaging studies (using PET and functional MRI) have revealed that language areas of the left hemisphere show more neural activity (reflected in cerebral blood flow or metabolic activity) during tasks that tap language processes, compared with the right hemisphere.

Interestingly, in individuals with dyslexia and SLI this asymmetry appears to be reversed or altered: right areas are larger or more active than left areas. For instance, differences have been found in the planum temporale, a structure located on the superior surface of the temporal lobe in the posterior perisylvian region (which includes a portion of Wernicke's area) that is important for language comprehension and phonological processing. The planum temporale is larger on the left than on the right in most people; however, in individuals with SLI and dyslexia this is not the case. In studies of dyslexia reversed asymmetry (right larger than left) of the planum temporale has been consistently reported, in both postmortem and neuroimaging studies. Evidence of structural abnormalities in the left hemisphere of individuals with dyslexia is also consistent with ERP studies reporting abnormal brain activation in language regions of the left hemisphere in this population. Other small, focal cellular abnormalities have also been reported in people with dyslexia, including ectopias (the presence of neurons in inappropriate cortical layers) and microgyria (abnormally small gyri). Such anomalies are thought to result from errors in neuronal migration caused by intrinsic genetic factors, or perhaps by prenatal or perinatal brain insult. Examination of the thalamus in dyslexics has also revealed cellular changes, suggesting that the processing of sensory information at this critical subcortical level may be compromised.

There have been few analogous findings in children with SLI owing to the fact that experiments involving imaging techniques are not often feasible in children (imaging studies in dyslexia usually recruit adult participants), and individuals with SLI are not routinely referred to medical professionals for brain scans. Additionally, there have been no postmortem studies of the brains of individuals with SLI, so it is impossible to confirm or rule out the presence of focal cellular abnormalities like those found in the sample of dyslexic individuals examined by Galaburda and colleagues (Galaburda *et al.*, 1994). The similar behavioral deficits in SLI and dyslexia suggest a similar neuropathology, but this remains speculative. Despite these limitations, the studies of children with SLI to date are fairly similar to those of adults with SLI and dyslexia in reporting abnormal asymmetry, especially in the perisylvian region. In one MRI study, the volume of the posterior perisylvian region, including the planum temporale, was found to be reduced bilaterally, though more dramatically in the left hemisphere, in language- and learning-impaired participants compared with controls (Jernigan *et al.*, 1991). Further, bilateral volumetric reductions were found in subcortical structures, including the caudate and putamen. A number of more recent MRI studies also report atypical asymmetry of perisylvian structures, with areas on the right larger than usual in individuals with SLI. In sum, neuroanatomical abnormalities were most often localized in the perisylvian region, which includes areas important for language processing, and atypical asymmetry in SLI compared with control subjects is consistently reported.

Functional imaging studies of individuals with SLI and dyslexia, using PET and fMRI, also report abnormal patterns of activation. Studies of children with language-related disorders report reduced activation (indexed by cerebral blood flow or glucose metabolism) in the perisylvian region in the left as compared with the right hemisphere during tasks requiring utilization of language areas. Such studies support the idea that the anomalies in anatomic asymmetry described above might be related to the inability of individuals with SLI to use these same areas effectively to process sounds. Brain activation studies of ERPs during visual and auditory sensory processing tasks add to these findings. Across studies, the ERP results support the premise that abnormal patterns of hemispheric activation can be documented in SLI populations.

Examination of results from studies of the brain's functional activity in response to rapidly presented and or brief acoustic cues (both speech and nonspeech) support the relevance of such cues to speech perception. This body of research further suggests that deficits in the underlying ability to process rapid acoustic change are often associated with abnormal patterns of brain activity during processing. It is important to note, however, that such relations are not always found. Some children with SLI do not exhibit abnormal asymmetry, and some individuals with verifiable atypical asymmetry exhibit no language difficulties. Moreover, individuals with SLI and dyslexia have been found to show abnormalities in brain areas outside of language centers. These deviations from the reported pattern of findings reinforce the view that SLI and dyslexia are disorders characterized by heterogeneous behavioral profiles, and thus may ultimately show variability in the underlying neurobiological substrates.

RATE OF PROCESSING

Individuals with SLI or dyslexia exhibit deficits in the ability to process successive, brief, sensory stimuli which are rapidly presented to the nervous system. Such stimuli may be tactile, visual, somatosensory or acoustic, including linguistic stimuli such as consonant–vowel (CV) syllables, and nonlinguistic stimuli such as tone pairs. The perception of human speech requires the decoding of brief, rapidly changing auditory stimuli that constitute language. Words are made up of phonemes, and phonemes are characterized by formants – frequency patterns created by sound resonating in the vocal tract. Formants provide acoustic cues in the perception of speech and represent sound waveforms across time. Formants of vowels are more or less constant over time, whereas stop consonants (/p, b, t, d, k, g/) are characterized by formant transitions. During a formant transition, frequency position changes rapidly as the stop consonant occlusion (vocal tract is in a closed position specific to a stop consonant) is released and the vocal tract shape changes to form the subsequent vowel (a stop consonant cannot be uttered alone, but must be combined with a vowel). So, in order to discriminate accurately and perceive stop consonants, the rapid formant transitions must be correctly processed by the auditory system. Frequency spectrographs (sound waveforms) for the CV syllables /ba/ and /da/ are shown in Figure 2. Note that the only differences between these CV syllables occur within the first 40 ms of formants 1 and 2. After that point, the spectrographs for /ba/ and /da/ are almost identical. Thus, in order for a listener to distinguish these CV syllables, the rapidly changing frequency information contained in the initial 40 ms of the speech sound must be properly processed by the auditory system. Such brief auditory cues, critical to discrimination of language, enter the central nervous

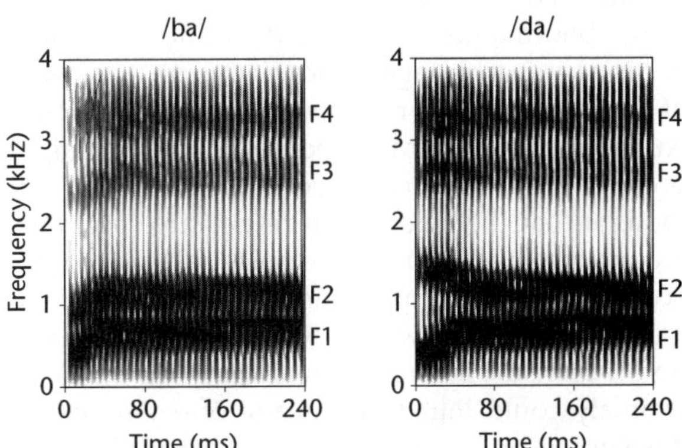

Figure 2. Spectrograph showing the formant transitions for the consonant–vowel (CV) syllables /ba/ and /da/. Notice the difference between /ba/ and /da/ in the 0–1 kHz range during the first 40 ms of formants 1 and 2. This brief sound difference must be detected in order to discriminate /ba/ from /da/. Formants 1 through 4 are noted on the right side of each spectrogram (F1–F4).

system in rapid succession, and an inability to process such transient, rapid, successive stimuli might ultimately result in deficits symptomatic of SLI and dyslexia. Indeed, in studies of adults with dyslexia, one of the most persistent deficits appears to be in phonological processing, suggesting that underlying auditory processing deficits are extremely long-lasting.

Studies of children with SLI provide further support for an underlying rapid auditory processing (RAP) deficit. Schoolage children with SLI were found to be impaired in processing both verbal and nonverbal stimuli that had brief and rapidly changing auditory components (Tallal *et al.*, 1985). The replication and extension of these findings, across many laboratories, supported the hypothesis that an impairment of basic-level RAP hinders the development of normal language and reading abilities. Although children and adults with deficits in RAP and associated language problems hear normally and can sequence sounds, such individuals are selectively impaired in their ability to both perceive and produce speech sounds characterized by brief or rapidly changing temporal cues. These auditory processing limitations may directly interfere with adequate perception of those speech sounds that include rapid acoustic changes – such as stop consonants – and disrupt processing of the speech stream.

Another well-established finding is that SLI occurs within families and that infants born to families with affected close relatives are at greater risk of the disorder. Children from families with a history of SLI are approximately four times more likely to develop SLI than children from control families. These facts, and the insight that the cortical abnormalities implicated in SLI probably occur at an early stage of development as a result of errors in neuronal migration, suggest that different neurophysiological responses in infancy might predict SLI in later childhood. For example, it is thought that the mechanism responsible for causing malformations such as ectopias and microgyria occurs during prenatal development, approximately between 18 weeks and 24 weeks of gestation. An insult to the prenatal brain or a genetic anomaly could initiate a cascade of events in the developing brain, ultimately resulting in the sensory processing profile characteristic of SLI associated with the underlying abnormal neurobiology described. If this hypothesis is valid, then RAP deficits should be observable in preverbal infants, and would predict subsequent language delay or impairments. Studies suggest that this may indeed be the case.

The links between infants' ability to process brief, rapidly presented stimuli and later language development has been examined in a series of studies by Benasich and colleagues, who found that infants with a family history of SLI performed more poorly on measures of RAP compared with a control group. In a longitudinal follow-up, it was found that the RAP thresholds measured in infancy were strongly related to later language comprehension and production through 36 months of age. Moreover, there are also indications that these basic acoustic processing skills in infancy (e.g. auditory gap detection and RAP) are predictive of later language development even in infants who are not at higher risk for language disorders. Such findings provide support for the notion that poor processing of rapidly presented and/or brief acoustic cues in infancy might be diagnostic of later language delays and impairments (for a review, see Benasich, 1998).

TEMPORAL INTEGRATION

It has been suggested that functional impairments in children with SLI might arise from basic sensory integration deficits, reflecting in turn dysfunctional neural processing of rapidly changing sensory information. So how might temporal integration deficits be propagated within the auditory processing pathways? Within the central auditory system, each neural stage of processing has a unique set of acoustic filtering and temporal integration properties. Thus, the critical linguistic cues of speech (phonemes) are first extracted from the raw acoustic waveforms produced by the vocal apparatus, and then processed by multiple neural stations (Figure 3). At the level of the auditory nerve, which carries sound information from the inner ear to the central nervous system, temporal information is probably encoded millisecond by millisecond. The acoustic information is then processed by the medial geniculate nucleus (MGN) of the thalamus, where the high resolution of the signal carried by the auditory nerve is largely preserved. The first cortical processing area is the primary auditory cortex (A1). At the level of the thalamus and A1, neural responses seem to occur primarily to the onset of temporal change. After A1, the acoustic information is further processed by secondary and association auditory areas, which appear to respond to increasingly 'segmented' components of the original temporal information. Thus, each area where acoustic information is processed may be specialized for extracting different temporal features from a given sound. This may

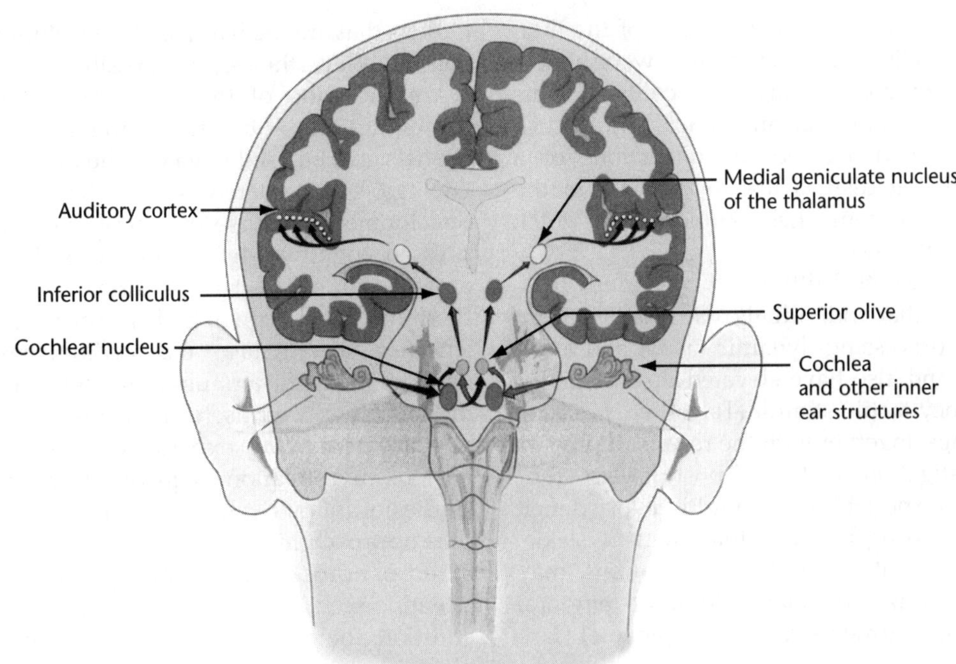

Figure 3. The ascending auditory processing pathways, as seen on a midcoronal section at about a 70° plane.

explain how a deficit at the most basic level of auditory processing could perturb the entire system. The subcortical structures implicated in SLI and dyslexia, such as the thalamus, are likely to be critical for encoding rapid temporal transitions (5–40 ms) embedded within longer acoustic sequences such as the speech stream. Therefore, if the rapid temporal integration capabilities of the thalamus are compromised (information is poorly encoded or incomplete), the cortical representations of the temporal features in a sound would degrade radically. Thus, the key elements of a neurobiological model in which auditory temporal processing defects lead, in a cascading fashion, to deficits in speech and language processing would be in place.

MULTIMODAL VERSUS AMODAL

Although SLI and dyslexia are defined as specific disorders of language in the presence of otherwise normal development (i.e. normal intelligence), there are reasons to question these diagnostic criteria as only applying to the auditory system. Research has shown that individuals with developmental language disorders exhibit processing deficits in other modalities, including motor coordination, and tactile and visual perception; see Tallal *et al.* (1985) for a review. It should be noted, however, that it has been proposed that some of these deficits seen in children with SLI and dyslexia might be the result of poor attention span (not explicitly assessed by traditional intelligence tests), and indeed quite a high proportion of these children meet the criteria for attention deficit disorder. To address this issue, future research must control for or eliminate variations in sustained attention in sample populations. Despite this caveat, there is accumulating evidence in support of the idea of a multimodal rapid sensory processing impairment in individuals with developmental language disorders.

Of particular interest are studies of visual processing that have revealed that the ability to process fast, low-contrast visual stimuli is impaired in people with dyslexia. The system responsible for processing this type of information is called the magnocellular pathway of the visual system. In contrast, no difference was found between participants with dyslexia and control participants in parvocellular visual processing. The parvocellular pathway of the visual system is responsible for processing high-contrast, fine detail, and color information. The magnocellular deficit is evident in behavioral assessment, and is also reflected in neurophysiological measurements: ERPs to rapidly presented visual stimuli have longer latencies in dyslexia, reflecting a slower neural response. Furthermore, postmortem analyses of the brains of individuals with dyslexia have revealed changes in the magnocells (larger cells which have faster conduction velocities than parvocells) in the lateral geniculate nucleus, the visual nucleus of the thalamus. Magnocells in dyslexic brains were abnormally small in comparison with controls. This convergence of behavioral, anatomical and electrophysiological evidence led to suggestions of a visual processing disturbance in dyslexia in the magnocellular system (Lehmkuhle *et al.*, 1993; Galaburda *et al.*, 1994).

Studies by Talcott, Witton and colleagues have demonstrated that individuals with dyslexia are impaired in processing dynamic visual and auditory events, and also have elevated thresholds for high-frequency tactile stimuli (Talcott *et al.*, 2000). These findings, together with the results of physiological investigations (ERP and postmortem studies) support the possibility of a multisensory deficit in rapid processing in individuals with dyslexia. Future experiments with SLI populations may help to characterize potential multimodal temporal processing impairments in developmental language disorders.

CONCLUSION

Much is still unknown about the developmental disorders of language, despite years of research. Debate continues as to whether disorders such as SLI and dyslexia result from deficits in basic sensory processing that affect developing language, or whether the difficulty is primarily one of acquisition of higher-level semantic, syntactic and phonological skills. However, with the focus on atypical or inefficient processing of basic acoustic input, progress has been made in delineating potential causative mechanisms. One scenario that could lead to such effects involves a confluence of genetic and/or environmental factors, which may occur in the perinatal period during key periods of neural development, disrupting the normal cascade of events. Atypical asymmetry, as compared with control subjects, is consistently reported in SLI and dyslexic groups, as well as a higher incidence of neuroanatomical abnormalities in areas important to language processing. A substantial number of contemporary functional imaging studies report reduced or abnormal patterns of activation in language areas in these populations. Moreover, close examination of the results of such studies supports the idea that processing rapidly presented and/or brief acoustic cues (both speech and nonspeech) is impaired in individuals with developmental language disorders.

Many questions remain unanswered regarding the developmental events and underlying pathological changes that lead to the specific behavioral deficits that are the hallmark of developmental disabilities. Nevertheless, it is possible that the neural characterization of these specific developmental disabilities, in combination with known behavioral profiles, can be used to gain some insight into their etiology. Unfortunately, studies that might link the developmental course of neural anomalies to expressed cognitive deficits in a causal fashion are exceedingly difficult to perform in humans. Furthermore, the nature of higher-order cognitive deficits (e.g. in language and reading) has not lent itself to study in nonhuman models, though such models have been useful in anatomical and basic sensory processing experiments. One promising area of investigation is prospective longitudinal studies of infants at high risk of SLI and dyslexia. This approach may allow the concomitant deficits in sensory processing, phonological processing and attention seen in children with SLI to be parsed. In addition, the emergence of more sensitive and sophisticated neuroimaging technology (both functional and anatomical) promises to provide additional evidence about the deficits implicated in developmental disorders of language. A better understanding of the developmental disorders of language will make the ultimate goal of remediation more attainable.

References

[APA] American Psychiatric Association (1994) *Diagnostic and Statistical Manual of Mental Disorders*, 4th edn. Washington, DC: American Psychiatric Association.

Benasich AA (1998) Temporal integration as an early predictor of speech and language development. In: von Euler C, Lundberg I and Llinás R (eds) *Basic Mechanisms In Cognition And Language – With Special Reference To Phonological Problems In Dyslexia* (Wenner-Gren International Series, vol. 70), pp. 123–142. Oxford, UK: Elsevier Science.

Denenberg VH (1999) A critique of Mody, Studdert-Kennedy, and Brady's 'Speech perception deficits in poor readers: Auditory processing or phonological coding?' *Journal of Learning Disabilities* **32**: 379–383.

Denenberg VH (2001) More power to them – statistically that is: a commentary on Studdert-Kennedy, Mody, and Brady's criticism of a critique. *Journal of Learning Disabilities* **34**: 299–301.

Galaburda AM, Menard MT and Rosen GD (1994) Evidence for aberrant auditory anatomy in developmental dyslexia. *Proceedings of the National Academy of Sciences USA* **91**: 8010–8013.

Jernigan TL, Hesselink JR, Sowell E and Tallal PA (1991) Cerebral structure on magnetic resonance imaging in language- and learning-impaired children. *Archives of Neurology* **48**: 539–545.

Lehmkuhle S, Garzia RP, Turner L, Hash T and Baro JA (1993) A defective visual pathway in children with reading disability. *New England Journal of Medicine* **328**(14): 989–996.

Leonard LB (1998) *Children with Specific Language Impairment*. Cambridge, MA: MIT Press.

Mody M, Studdert-Kennedy M and Brady S (1997) Speech perception deficits in poor readers: auditory processing or phonological coding? *Journal of Experimental Psychology* **64**(2): 199–231.

Rapin I and Allen D (1987) Developmental dysphasia and autism in preschool children: characteristics and subtypes. In: Martin J, Martin P, Fletcher P, Grunwell P and Hall D (eds) *Proceedings of the First International Symposium on Specific Speech and Language Disorders in Children*, pp. 20–35. London, UK: AFASIC.

Talcott JB, Witton C, McClean M *et al.* (2000) Dynamic sensory sensitivity and children's word decoding skills. *Proceedings of the National Academy of Sciences USA* **97**: 2952–2957.

Tallal P (2000) Experimental studies of language learning impairments: from research to remediation. In: Bishop DVM and Leonard LB (eds) *Speech and Language Impairments in Children: Causes, Characteristics, Intervention and Outcome*, pp. 131–155. Philadelphia, PA: Psychology Press.

Tallal P, Stark R and Mellits D (1985) Relationship between auditory temporal analysis and receptive language development: evidence from studies of developmental language disorders. *Neuropsychologia* **23**: 527–536.

Tomblin JB (1996) Genetic and environmental contributions to the risk for specific language impairment. In: Rice ML (ed.) *Toward a Genetics of Language*, pp. 191–211. Mahwah, NJ: Lawrence Erlbaum.

WHO (1993) *The ICD-10 Classification of Mental and Behavioural Disorders: Diagnostic Criteria for Research*. Geneva: World Health Organization.

Further Reading

Fitch RH, Read H and Benasich AA (2001) Neurophysiology of speech perception in normal and impaired systems. In: Jahn A and Santos-Sacchi J (eds) *Physiology of The Ear*, 2nd edn, pp. 651–672. San Diego, CA: Singular Publishing.

Benasich AA and Spitz RV (1998) Insights from infants: temporal processing abilities and genetics contribute to language development. In: Willems G and Whitmore K (eds) *A Neurodevelopmental Approach to Specific Learning Disorders*, pp. 191–210. London, UK: MacKeith Press.

Bishop DVM, Bishop SJ, Bright P *et al.* (1999) Different origin of auditory and phonological processing problems in children with language impairment: evidence from a twin study. *Journal of Speech, Language, and Hearing Research* **42**: 155–168.

Eden GF, VanMeter JW, Rumsey J *et al.* (1996) Abnormal processing of visual motion in dyslexia revealed by functional brain imaging. *Nature* **382**: 66–69.

Fitch RH, Miller S and Tallal P (1997) Neurobiology of speech perception. *Annual Review of Neuroscience* **20**: 331–353.

Kraus N, McGee TJ, Carrell TD *et al.* (1996) Auditory neurophysiologic responses and discrimination deficits in children with learning problems. *Science* **273**: 971–973.

Livingstone MS, Rosen GD, Drislane FW and Galaburda AM (1991) Physiological and anatomical evidence for a magnocellular defect in developmental dyslexia. *Proceedings of the National Academy of Sciences USA* **88**: 7943–7947.

Tallal P, Merzenich MM, Miller S and Jenkins W (1998) Language learning impairments: integrating basic science, technology, and remediation. *Experimental Brain Research* **123**: 210–219.

Tomblin JB (1996) Genetic and environmental contributions to the risk for specific language impairment. In: Rice ML (ed.) *Toward a Genetics of Language*. Mahwah, NJ: Lawrence Erlbaum.

Diffusion Models and Neural Activity

Intermediate article

Luigi M Ricciardi, Federico II University, Naples, Italy
Petr Lánský, Academy of Sciences of the Czech Republic, Prague, Czech Republic

CONTENTS

Neuronal interspike intervals can be characterized in terms of the first-passage time probability density of stochastic diffusion processes under steady state and periodic stimulation. The Wiener and Ornstein–Uhlenbeck models, and models with multiplicative noise, can be used to elucidate neuronal activity.

INTRODUCTION

One of the basic modes of signaling in the nervous system is by the frequency of action potentials. This is true even if the input signal is time-varying, in which case the firing rate is expected to be modulated in time to reflect the time course of the input. The rate coding that is introduced via the stochastic description of single neurons is the focus of this article. A common way to introduce stochasticity is by the assumption that the incoming signal includes a random component, generally denoted as 'noise'. Another source of stochasticity can originate in the neuron itself, where a random component is added to the signal. Taking these circumstances into account, one is led to the conclusion that the deterministic differential equations usually describing the membrane response should be completed by insertion of a noise term, thus becoming stochastic differential equations. The solutions of such equations, under certain regularity conditions on their coefficients, can be viewed as 'sample paths' or 'realizations' of stochastic processes belonging to the class of 'diffusion' processes. This is the rationale for the rise of diffusion models for the description of neuronal activity. Here, the center of interest is single-point models of interspike intervals generation. Such a one-point representation implies severe restrictions, discussed below. This type of simplification neglects the spike's duration and its detailed shape, so that the entire neuronal activity is schematized in the form of identical point-size signals occurring in time according to a stochastic point process. On the other hand, it permits us to quantify not only the mean of the spiking activity, but also its variability – even its probability distribution.

DETERMINISTIC LEAKY INTEGRATE-AND-FIRE NEURONAL MODEL

The electrical circuit model of a neuronal membrane is composed of a resistor and a capacitor in parallel charged by a battery (RC circuit). It can be described by the first-order differential equation

$$\frac{\mathrm{d}v(t)}{\mathrm{d}t} = -\frac{v(t)}{\tau} + i(t), \quad v(t_0) = v_0 \tag{1}$$

where $v(t)$ denotes the difference between the membrane potential at time t and the membrane potential in resting conditions (i.e. the membrane depolarization), τ is the membrane time constant, $i(t)$ is the input to the neuron and v_0 is the initial voltage after spike generation that for simplicity – and without loss of generality – is sometimes set to zero ($v_0 = 0$). This is also known as the Lapicque model, or the RC circuit model. In this model, under a constant input, $i(t) = i$, as time grows to infinity the depolarization tends to the asymptotic level $i\tau$, and if the input is removed at some instant, say $i(t) = 0$, the depolarization tends exponentially to zero (i.e. to the resting level), the speed of approach to zero depending on time constant τ. The extreme simplicity of eqn (1) witnesses that the action potential generation mechanism is not an inherent part of the model, so that a firing threshold S, with $S > v_0$, has to be postulated. Within this model the neuron is assumed to fire whenever the threshold voltage is reached, and the membrane depolarization $v(t)$ is assumed to be instantly reset to v_0 after each firing. The interspike intervals (ISIs) are identified with the first-passage time of $v(t)$ through S, namely with the variable

$$T = \inf\{t \geq t_0 : v(t) > S\}, \quad v(t_0) = v_0 < S \qquad (2)$$

In the case of a constant input, the condition for evoking a spike is $i\tau > S$, which defines suprathreshold stimulation. Otherwise the input is unable to produce a spike and the neuron remains silent. For a standard treatment of model described by eqn (1) see Keener *et al.* (1981) and Knight (1972).

STOCHASTIC LEAKY INTEGRATE-AND-FIRE NEURONAL MODEL

One of the stochastic versions of eqn (1), of interest in the present context, is formally obtained by adding to the right-hand side a term to account for the random components that are present as a part of the global signal acting on the neuron. This random component is usually identified, as a useful approximation, with the 'white noise' $\xi(t)$:

$$\frac{dv(t)}{dt} = -\frac{v(t)}{\tau} + i(t) + \sigma\xi(t), \quad v(t_0) = v_0 \qquad (3)$$

By definition, the white noise is a stationary Gaussian process characterized by zero mean and delta-type correlation function. In eqn (3) σ is a positive parameter representing the amplitude of the random fluctuations of the noise, a sort of measure of the degree of unpredictability of the deviations of the noise sample paths from its mean trajectory. Although the mathematical definition of the process $\xi(t)$ is a rather complicated and subtle issue, by analogy with white light as the superposition of all colors, the white noise may be intuitively envisaged as a random process in which spectral components of all frequencies are present and equally represented. For constant $i(t)$ the process defined in eqn (3) is called the Ornstein–Uhlenbeck model of the membrane potential. Its properties in the absence of the firing threshold are as follows: (1) at each time the probability density function of the membrane depolarization is Gaussian; (2) an equilibrium regime exists since in the limit, as $t \to +\infty$, the probability density function that describes the membrane depolarization attains mean $i\tau$, and variance $\tau\sigma^2/2$. Note that the asymptotic variance is independent of the input signal.

Generation of an action potential in the model defined in eqn (3) is again described by the first-passage time of the trajectories of the process across the firing threshold S, i.e. by the random variable (eqn (2)). However, since now the trajectories are different realizations of the stochastic diffusion process modeled by eqn (3), the ISIs are of different length even for constant input. Thus T is a nondegenerate random variable. For this model under constant input $i(t) = i$, as well as for others, the neuronal output is a renewal process, namely intervals between successive firings are independent and identically distributed random variables. Note that now, in contrast with model (1), spikes can be generated due to the presence of noise even for subthreshold stimulations.

Figure 1 depicts the spike generation process. It indicates the effects of a constant stimulation of magnitude i. In the absence of noise (Lapicque model), the depolarization grows exponentially towards the asymptotic value $i\tau$. However, as soon as the neuron's threshold S ($S < i\tau$) is reached, a spike is generated and the depolarization is instantly reset at its initial value. The process then starts afresh. Within such a model, a rigorously periodic sequence of spikes (arrows, Figure 1) at time T, $2T$,... is generated. In the case of the leaky integrate-and-fire model (cf. eqn (3)) the presence of noise alters dramatically the picture, in that the trajectories of the neuron's depolarization now exhibit an erratic time course. In Figure 1 three such trajectories are plotted and the instants t_1, t_2, t_3 of attainment of threshold S are indicated together with the corresponding generated spikes. The first-passage time of the process modeling the neuron's depolarization is now a random variable whose probability density function mimics the probability density of the ISIs. Figure 2 shows an ISI histogram for a finite number of trajectories, and the ISI probability density obtained when all possible trajectories are taken into account.

The properties of this model, including the moments of the random variable T, are complicated (Ricciardi, 1977; Ricciardi and Sacerdote, 1979). Inoue *et al.* (1995) have attempted to compare systematically this model with experimental data, and Lánský (1983) describes a method to estimate parameters based on intracellular recording of the trajectories of $v(t)$.

Until recently the models encountered in applications of stochastic diffusion processes to neuroscience have been predominantly time-homogeneous, reflected by the circumstance that i and σ in eqn (3) do not depend on time. An interest in stochastic resonance (a cooperative effect that arises out of the coupling between deterministic and random dynamics in nonlinear systems) evoked studies on diffusion neuronal models with time-dependent parameters. Analogous to the model in eqn (1), the Ornstein–Uhlenbeck model operates in two distinct regimes – the deterministic

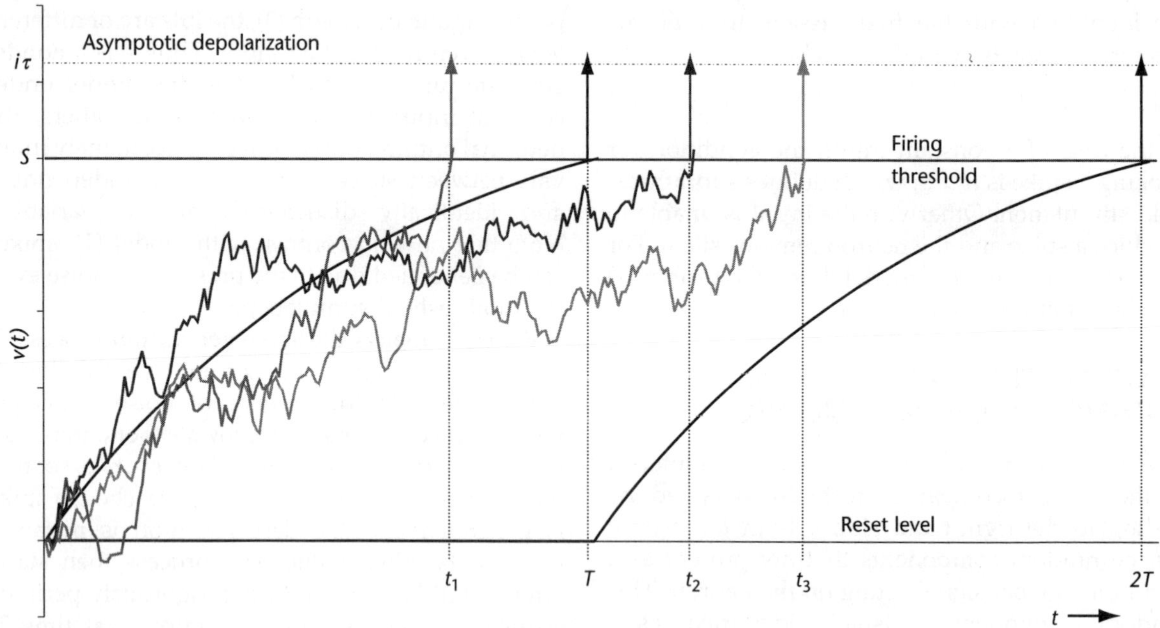

Figure 1. Spike generation mechanism for Lapicque and stochastic leaky integrate-and-fire models.

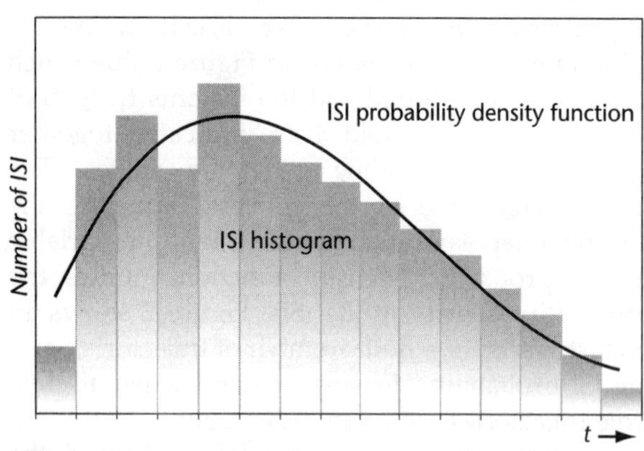

Figure 2. An interspike interval (ISI) histogram and the corresponding first-passage time probability density.

one (suprathreshold stimulation) and the stochastic one (subthreshold stimulation). In the first regime the signal $i(t)$ is large enough, and thus the firing events occur even in the absence of noise, the effect of which is merely to distort the output. In the second (noise-activated) regime the signal alone is insufficient to cause a firing, but the noise itself contributes to activating the neuron. The methods of stochastic resonance extend this view mainly for subthreshold periodic signals. Two sources of periodicity can be expected in the signal: either an endogenous periodicity, or a periodicity of external input, the exogenous periodicity (Lánský, 1997). In both cases, an optimum level σ of noise exists, for which the input frequency is best reflected in the output signal (Bulsara *et al.*, 1996; Shimokawa *et al.*, 1999). Another type of nonhomogeneity in model (3) is achieved if the natural assumption is made that the amplitude of the noise depends on the signal (Lánský and Sacerdote, 2001). All generalizations mentioned in this section can be applied to any stochastic neuronal diffusion model.

WIENER PROCESS AS A NEURONAL MODEL

If the time constant τ is very large ($\tau \to +\infty$), then eqn (3) takes the form

$$\frac{dv(t)}{dt} = i(t) + \sigma\xi(t), \quad v(t_0) = v_0 \tag{4}$$

This is called the Wiener neuronal model. Although eqn (4) can be taken as a definition of this model, it can also be obtained from first principles using the formalism of diffusion equations. To this purpose, let us initially assume that the neuron is subjected to a sequence of excitatory and inhibitory postsynaptic potentials of constant magnitudes $a > 0$ and $b < 0$ occurring in time in accordance with two independent Poisson processes. The membrane potential is thus viewed as a stochastic process $X(t)$ in continuous time with a discrete space consisting of the lattice $x_0 + ka + hb(h, k = \ldots, -1, 0, 1, \ldots)$ with the points of discontinuity randomly occurring in time. Ricciardi (1977) shows that if the input rates are taken larger and larger while simultaneously taking smaller and smaller postsynaptic potentials with suitable constraints, the membrane potential 'converges' to the diffusion process generated by eqn (4). This was the method initially used in the well-known paper by Gerstein and Mandelbrot (1964), in which the diffusion approach to neuronal modeling was first considered. By including the spontaneous decay of the membrane potential, which is reflected by a finite value of the time constant τ, in this model with discrete jumps, model (3) can be derived as well.

The basic properties of the Wiener model with constant input, $i(t) = i$, are the following: the distribution of the membrane potential at any instant is normal, with mean $it + v_0$ and variance $\sigma^2 t$. Hence, there is no steady state distribution of the membrane potential, in contrast to the Ornstein–Uhlenbeck model. By means of the methods outlined for instance in Ricciardi (1977), one can prove that the first-passage time density function of the Wiener model is given by

$$g(t|v_0, S) \equiv \frac{\partial}{\partial t} Prob\{T < t\} = \frac{S - v_0}{\sigma\sqrt{2\pi t^3}} \exp\left\{ -\frac{(S - v_0 - it)^2}{2\sigma^2 t} \right\} \qquad (5)$$

which is known in statistical literature as the inverse Gaussian distribution. In this model, for $i \geq 0$, neuronal firing is a sure event. If one takes $i < 0$, density (5) can be interpreted as the firing density conditional upon the event 'firing occurs'. The form of the density (5) permits evaluation of the moments of the ISI and the firing rate, as well as estimation of the parameters of the model. The assumptions of model (4) are oversimplified and many important electrophysiological properties of neuronal membrane are not taken into account. Therefore, the Wiener process is more suitable as a statistical descriptor of the data than as a realistic biological model.

GENERAL DIFFUSION MODELS

Consider a general deterministic model characterized by the dynamic equation

$$\frac{dv(t)}{dt} = h(v, t) + e(v, t)i(t), \quad v(t_0) = v_0 \qquad (6)$$

where $i(t)$ is an input, $e(v, t)$ describes the effect of this input on the depolarization and $h(v, t)$ is the function describing the rate of change of the depolarization in the absence of input. Both e and h are assumed to be sufficiently smooth functions. Then if $i(t)$ can be written as the sum of signal and white noise, more general models than those previously considered arise (Hanson and Tuckwell, 1983). It is indeed a well-known fact, also reflected in the Hodgkin–Huxley model, that the change of the membrane depolarization by a synaptic input depends on its actual value. Basically, the depolarization of the membrane caused by an excitatory postsynaptic potential decreases linearly with decreasing distance of the membrane potential from the excitatory 'reversal potential', say V_E. In the same manner, the hyperpolarization caused by inhibitory postsynaptic potential is smaller if the membrane potential is closer to the inhibitory reversal potential, V_I. In this way, unlike previous models, depolarizations are confined to finite interval (V_I, V_E). Natural conditions relating the reversal potentials, the initial depolarization and the firing threshold is $V_I < 0 < S < V_E$. The diffusion models, which take into account the existence of the reversal potentials result always in the multiplicative noise effect as in eqn (6). This is in agreement with the general notion that an additive noise is generated by external events that transmit messages, whereas the multiplicative noise is generated inside the processing unit, namely inside the neuron.

FEEDBACK, SPATIAL PROPERTIES AND REFRACTORINESS

The analogy of the process describing the time course of the neuron's membrane potential with the laws describing the diffusion of a substance in a liquid provides an intuitive justification for the use of the term 'diffusion model'. It must be stressed that the neuronal behavior described by diffusion models ultimately assumes that for time-constant input, the output is a renewal process. However, one can conceive models aimed, for instance, at simulating clustering effects in spike generation. Serial dependence among ISIs can be modeled in various ways, for instance by adjusting the reset value after each spike (afterhyperpolarization). Another possibility consists of inclusion into the model of some kind of feedback, usually self-inhibition, often experimentally observed. A further generalization is achieved by taking into account the spatial properties of a neuron. In the simplest way, it can be done by assuming that the model neuron is composed of two compartments: the dendritic compartment, described by a standard diffusion model, and the trigger zone compartment, including the spontaneous decay of depolarization and the firing mechanism. Also, the phenomenon of refractoriness can be included in stochastic diffusion models, usually by postulating the existence of time-varying thresholds, instead of constant thresholds as assumed in the foregoing.

References

Bulsara AR, Elston TC, Doering CR, Lowen SB and Lindberg K (1996) Cooperative behavior in periodically driven noisy integrate-and-fire models of neuronal dynamics. *Physical Review E* **53**: 3958–3969.

Gerstein GL and Mandelbrot B (1964) Random walk models for the spike activity of a single neuron. *Biophysical Journal* **4**: 41–68.

Hanson FB and Tuckwell HC (1983) Diffusion approximations for neuronal activity including synaptic reversal potentials. *Journal of Theoretical Neurobiology* **2**: 127–153.

Inoue J, Sato S and Ricciardi LM (1995) On the parameter estimation for diffusion models of single neurons' activities. *Biological Cybernetics* **73**: 209–221.

Keener JP, Hoppensteadt FC and Rinzel J (1981) Integrate-and-fire models of nerve membrane response to oscillatory input. *SIAM Journal of Applied Mathematics* **41**: 503–517.

Knight BW (1972) Dynamics of encoding in a population of neurons. *Journal of General Physiology* **59**: 734–766.

Lánský P (1983) Inference for diffusion models of neuronal activity. *Mathematical Biosciences* **67**: 247–260.

Lánský P (1997) Sources of periodical force in noisy integrate-and-fire models of neuronal dynamics. *Physical Review E* **55**: 2040–2044.

Lánský P and Sacerdote L (2001) The Ornstein-Uhlenbeck neuronal model with the signal-dependent noise. *Physics Letters A* **285**: 132–140.

Ricciardi LM (1977) *Diffusion Processes and Related Topics in Biology*. Berlin, Germany: Springer.

Ricciardi LM and Sacerdote L (1979) The Ornstein-Uhlenbeck process as a model for neuronal activity. *Biological Cybernetics* **35**: 1–9.

Shimokawa T, Pakdaman K and Sato S (1999) Time-scale matching in the response of a leaky integrate-and-fire neuron model to periodic stimulus with additive noise. *Physical Review E* **59**: 3427–3443.

Further Reading

Chhikara RS and Folks JL (1989) *The Inverse Gaussian Distribution: Theory, Methodology, and Applications*. New York, NY: Marcel Dekker.

Gardiner CW (1983) *Handbook of Stochastic Methods for Physics, Chemistry and the Natural Sciences*. Berlin, Germany: Springer.

Karlin S and Taylor HM (1981) *A Second Course in Stochastic Processes*. New York: Academic Press.

Lánský P and Sato S (1999) The stochastic diffusion models of nerve membrane depolarization and interspike interval generation. *Journal of Peripheral Nervous Systems* **4**: 27–42.

Ricciardi LM and Sato S (1994) Diffusion processes and first-passage-time problems. In: Ricciardi LM (ed.) *Lectures in Applied Mathematics and Informatics*, pp. 206–285. Manchester, UK: Manchester University Press.

Tuckwell HC (1988) *Introduction to Theoretical Neurobiology*. Cambridge, UK: Cambridge University Press.

Direct Reference

Intermediate article

David Braun, University of Rochester, Rochester, New York, USA

CONTENTS

The theory of direct reference says that the meaning of a proper name (such as 'Mark Twain') or indexical (such as 'he') is just its referent.

WHAT IS DIRECT REFERENCE?

The theory of direct reference is a theory about the meanings of two sorts of natural language expressions: proper names (such as 'Mark Twain' and 'Paris') and indexicals (such as 'I', 'today', 'you', 'he', and 'that'). It says (roughly) that the meaning of any such expression is just its referent. The traditional rival to this theory is the view that these expressions have descriptive meanings, in addition to their referents.

The theory of direct reference can be divided into negative and positive parts. The negative part says that the meanings of proper names and indexicals are fundamentally different from those of definite descriptions. Definite descriptions are expressions of the form 'the *F*', such as 'the first Postmaster General of the USA' and 'the even prime number'. A definite description expresses a property (for example, being-an-even-prime-number) which is part of the meaning of the definite description; the definite description refers to the object that uniquely has that property. Thus a definite description refers to an object *indirectly*, by expressing a property that determines its referent.

The negative part of the theory of direct reference says that the meanings of proper names and indexicals are not descriptive in this sense. The meanings of these expressions are not properties; their references are not determined by associated properties, in the way that the references of definite descriptions are; instead, these expressions refer *directly*. The positive part of the theory asserts that the meaning of a proper name or indexical is simply its referent.

One can accept the negative part of the theory without accepting the positive part; indeed, sometimes the term 'theory of direct reference' is used to label just the negative part.

The above description of the theory needs some technical refinements, which are discussed below.

HISTORY

The term 'direct reference' was introduced by David Kaplan (1989) in the 1970s, but the theory is almost certainly ancient in origin. John Stuart Mill was perhaps the first modern philosopher to argue that proper names are directly referring (Mill, 1843); thus, the theory of direct reference for proper names is sometimes called 'Millianism'. Gottlob Frege (1893) and Bertrand Russell (1905) formulated seemingly powerful arguments against the view, and presented alternative theories which were widely accepted in the late nineteenth and early twentieth centuries. Ruth Barcan Marcus revived the theory of direct reference in the 1960s (Marcus, 1961); and from the early 1970s, Saul Kripke (1980), Keith Donnellan (1972), David Kaplan (1989), and John Perry (1977) presented arguments for the theory that persuaded many philosophers of its viability. (*See* **Reference, Theories of**)

Frege's and Russell's arguments against direct reference for proper names turn on certain difficulties which are now commonly called *Frege's Puzzles*. The first puzzle is the *puzzle of informative identities*. If the meaning of a proper name is just its referent, then two proper names that refer to the same thing, such as 'Mark Twain' and 'Samuel Clemens', have the same meaning. It seems to follow that the two identity sentences below have the same meaning:

Mark Twain is Mark Twain. (1)

Mark Twain is Samuel Clemens. (2)

But these sentences seem to differ in meaning: sentence 1 seems to be uninformative, whereas sentence 2 seems to be highly informative.

The second puzzle is the *puzzle of apparent reference to nonexistents*. If the meaning of a proper name is its referent, then names that refer to nothing, such as 'Pegasus', should be meaningless. Therefore, sentences in which 'Pegasus' appears, such as 'Pegasus flies' and 'Pegasus does not exist', should also be meaningless. But these sentences seem to be meaningful; in fact, the second seems to be true.

The third puzzle is the *puzzle of substitution of coreferring names*. If the meaning of a proper name is just its referent, and we have two names that refer to the same thing, then we should be able to substitute one name for the other in any sentence without changing the meaning of the sentence. Moreover, the substitution should not change the truth-value of the sentence (that is, the sentence should not change from true to false or vice versa). But consider the following two sentences:

John believes that Mark Twain wrote *Huckleberry Finn*. (3)

John believes that Samuel Clemens wrote *Huckleberry Finn*. (4)

It seems possible for sentence 3 to be true while sentence 4 is false. So substitution of coreferring names can change the truth-value of a sentence.

Russell concluded from these puzzles that the meaning of a proper name is not its referent, but is instead descriptive in nature; ordinary proper names are just 'abbreviations' for definite descriptions that speakers have in mind when they use names (Russell, 1905). We can determine the description that a speaker associates with a name *N* by asking the speaker 'Who is *N*?' Thus, for some speakers, the meaning of 'Mark Twain' might be identical with the meaning of 'the author of *Huckleberry Finn*', while that of 'Samuel Clemens' might be 'the person who lives next door'. These descriptions express different properties, and so differ in meaning. Thus sentences 1 and 2 also differ in meaning, and one can be informative while the other is not. 'Pegasus' might mean the same as 'the flying horse'; this description is meaningful even if it fails to refer, so sentences containing it can be meaningful. Sentences 3 and 4 differ in meaning because the names 'Mark Twain' and 'Samuel Clemens' differ in meaning; they differ in truth-value because they attribute different beliefs to John.

Like Russell, Frege also attributed descriptive meanings, which he called 'senses', to proper names (Frege, 1893). Frege's solutions to the puzzles are similar to Russell's. There are differences between Frege and Russell which we ignore here. (*See* **Sense and Reference**)

ARGUMENTS FOR DIRECT REFERENCE

Many of the arguments in favor of direct reference are simply arguments against description theories of proper names.

According to Russell, the meaning for a speaker of a name *N* is the same as that of the definite description that the speaker would provide when asked 'Who is *N*?' Suppose that we ask Fred 'Who is Mark Twain?' and he answers 'The author of *Huckleberry Finn*.' Now consider the following two sentences:

If there is exactly one author of *Huckleberry Finn*, then Mark Twain is the author of *Huckleberry Finn*. (5)

If there is exactly one author of *Huckleberry Finn*, then the author of *Huckleberry Finn* is the author of *Huckleberry Finn*. (6)

The only difference between them is that sentence 6 contains the definite description 'the author of *Huckleberry Finn*' where sentence 5 contains the name 'Mark Twain'. Thus, according to Russell, sentences 5 and 6 should have the same meaning for Fred. But Kripke (1980) and Donnellan (1972) present several reasons for thinking that these sentences do not have the same meaning for Fred.

Kripke points out that sentence 6 expresses a necessary truth. So if sentence 5 means the same thing as sentence 6, then sentence 5 should also express a necessary truth. But sentence 5 clearly does not express a necessary truth; after all, Mark Twain might have died before he wrote *Huckleberry Finn* (as Fred would admit). A similar point would hold for just about any definite description that Fred might provide.

The next argument comes from both Donnellan and Kripke. Imagine that *Huckleberry Finn* was written by Samuel Clemens's cousin, Clyde Clemens, who died shortly after completing it; Twain/Clemens stole the manuscript, passed it off as his own, and since then has been commonly thought to be the author. Now if this is the case, then sentence 5 is false, even as spoken by Fred. But the description theory entails that it is true, for according to that theory, sentence 5 (in Fred's mouth) means the same as sentence 6, which is obviously still true in our imaginary situation. Furthermore, according to the description theory, the name 'Mark Twain' (in Fred's mouth) refers in this situation to the referent of the description 'the author of *Huckleberry Finn*', which is Clyde. But this is incorrect.

Finally, Kripke notes that Russell's objections take for granted that every speaker who uses the name 'Mark Twain' can formulate a definite description that 'identifies' the referent of the name. But this assumption is incorrect. When asked 'Who is Mark Twain?', some people may answer 'A writer'. This description does not determine a unique referent for the name; yet many of these people may be unable to produce a better identifying description.

Kaplan (1989) and Perry (1977) present similar arguments against description theories of indexicals. Consider the hypothesis that the descriptive meaning of an utterance of 'you' is the same as that of an utterance of 'the person I am addressing now' by the same speaker. Then utterances of the following two sentences by the same speaker should mean the same thing:

If I am addressing exactly one person now, then you are the person I am addressing now. (7)

If I am addressing exactly one person now, then the person I am addressing now is the person I am addressing now. (8)

However, sentence 8 expresses a necessary truth, whereas sentence 7 does not. To see that sentence 7 does not, imagine that I utter it while addressing Mary. What I say is true, but it could have been false; for, after all, I could have addressed someone other than Mary, or no one at all.

Examples involving indexicals, like sentence 7, show that we need to distinguish between two different sorts of meaning. Suppose that two different people utter sentence 7 while addressing different people. There is a sense in which the two utterances of 'you' share a meaning, which we may call the *linguistic meaning* of 'you'. But there is another sense in which those utterances differ in meaning: let's say that the two utterances of 'you', and the two utterances of sentence 7 as a whole, differ in *content*. Call the content of an utterance of a full sentence a *proposition*. The above arguments show that the contents of utterances of indexicals are not descriptive. Perhaps the linguistic meanings of indexicals are in some way descriptive, but that is an open question. (*See* **Philosophy of Language; Meaning**)

It is now possible to state the theory of direct reference more accurately. According to the negative part, the content of an utterance of a proper name or indexical is not descriptive (is not a property). According to the positive part, the content of an utterance of such an expression is simply the referent of the utterance; moreover, an utterance of a sentence containing such an expression expresses a *singular proposition*: a proposition containing the referent of the expression as a constituent.

PROBLEMS WITH DIRECT REFERENCE

The most important problems for the theory of direct reference are those posed by Frege's Puzzles. Advocates of direct reference have proposed solutions to these puzzles, but whether their solutions are successful is a subject of much current debate.

In response to the puzzle of informative identities, Nathan Salmon (1986), Scott Soames (1987), and other direct reference theorists say that sentences 1 and 2 express the same singular proposition, and, in that sense, mean the same thing. But they say that it is possible for a person to grasp that proposition in distinct ways. Different direct reference theorists have different views about what these ways of grasping propositions are. Some think that they are *mental representations*, and so hold that sentences 1 and 2, and the names 'Mark Twain' and 'Samuel Clemens', are connected with different mental representations in most speakers. This may explain why sentences 1 and 2 seem to differ in informativeness, and why many speakers incorrectly think that they differ in meaning. (*See* **Propositional Attitudes**)

In response to the puzzle of apparent reference to nonexistents, Salmon (1998) claims that names like 'Pegasus' refer to mythical or fictional objects that are created by our story-telling activities.

In response to the puzzle of substitution of coreferring names, Salmon (1986) and Soames (1987) claim that sentences 3 and 4 express the same proposition, and must have the same truth-value. But utterances of them differ in what they suggest about John: sentence 3 correctly suggests that John would assent to 'Mark Twain wrote *Huckleberry Finn*', whereas sentence 4 incorrectly suggests that he would assent to 'Samuel Clemens wrote *Huckleberry Finn*'. This might mislead some speakers into thinking that the sentences themselves differ in truth-value.

DIRECT REFERENCE AND COGNITIVE SCIENCE

The theory of direct reference implies that ordinary speakers are less reliable at judging differences in meaning than some cognitive scientists may think. For example, most speakers would judge that sentences 1 and 2 differ in meaning; but according to the theory of direct reference, such speakers are mistaken.

The theory also implies that two agents who believe and desire the same propositions may nonetheless behave quite differently. Suppose that John assents to 'Mark Twain is the author of *Huckleberry Finn*' and dissents from 'Samuel Clemens is the author of *Huckleberry Finn*'; suppose Mary assents to both. Now according to Salmon and Soames, John also believes that Clemens is the author of *Huckleberry Finn*, just like Mary (whatever he might say to the contrary). But they might behave differently in similar situations. Suppose, for example, that both assent to 'I want Twain to autograph my copy of *Huckleberry Finn*', and both are told 'Clemens is in the next room.' Then Mary will go next door to get her copy of *Huckleberry Finn* autographed, but John will not. The reason they will behave differently is that John does not believe the proposition in the same *way* as Mary: he believes it in a 'Twain' way, but not in a 'Clemens' way, whereas Mary believes it in both ways. Examples like this lead some philosophers to think that, if the theory of direct reference is correct, then ordinary belief attributions do not provide explanations of behavior of the sort we ordinarily expect; full explanations need to mention the ways in which propositions are grasped.

References

Donnellan K (1972) Proper names and identifying descriptions. In: Davidson D and Harman G (eds) *Semantics of Natural Language*, pp. 356–379. New York, NY: Humanities Press.

Frege G (1893) Über Sinn und Bedeutung. *Zeitschrift für Philosophie und Philosophische Kritik* **100**: 25–50. [In German. Translated as: Frege G (1952) On sense and reference. In: Geach P and Black M (eds and trans) *Translations From Philosophical Writings*, pp. 56–78. Oxford, UK: Blackwell.]

Kaplan D (1989) Demonstratives. In: Almog J, Perry J and Wettstein H (eds) *Themes From Kaplan*, pp. 481–614. Oxford, UK: Oxford University Press.

Kripke S (1980) *Naming and Necessity*. Cambridge, MA: Harvard University Press.
Marcus R (1961) Modalities and intensional languages. *Synthese* **13**: 303–322.
Mill JS (1843) *A System of Logic*. London, UK: Longman.
Perry J (1977) Frege on demonstratives. *Philosophical Review* **86**: 474–497.
Russell B (1905) On denoting. *Mind* **14**: 479–493.
Salmon N (1986) *Frege's Puzzle*. Cambridge, MA: MIT Press.
Salmon N (1998) Nonexistence. *Nous* **32**: 277–319.
Soames S (1987) Direct reference, propositional attitudes, and semantic content. *Philosophical Topics* **15**: 47–87.

Further Reading

Salmon N (1989) Reference and information content: names and descriptions. In: Gabbay D and Guenthner F (eds) *Handbook of Philosophical Logic*, vol. IV, pp. 409–461. Dordrecht, The Netherlands: Reidel.
Devitt M (1989) Against direct reference. *Midwest Studies in Philosophy* **14**: 206–240.

Discourse Processing

Intermediate article

Barbara Di Eugenio, University of Illinois, Chicago, Illinois, USA

CONTENTS

The theory of discourse processing concerns the computational processes underlying the interpretation and production of text encompassing more than one sentence – i.e., discourse. Discourse is generally taken to be written, and often, but not always, monologic.

INTRODUCTION

Two phenomena are considered intrinsic to discourse processing: the interpretation and production of phrases and utterances whose meaning depends on the discourse context; and the fact that a sequence of two or more utterances almost always conveys a meaning that is more than the sum of the meanings of the individual utterances. Consider the following example:

> As soon as they got to the beach, Karin jumped into the water. She was so hot from the long drive. (1)

Example 1 illustrates the issues most closely associated with the above phenomena: respectively, reference resolution and production, and text coherence.

Reference resolution concerns the interpretation of those noun phrases speakers use to refer to what are called discourse entities; i.e. entities in the model of the discourse – for example, *Karin*, *she*, *the long drive*. Reference resolution is closely related to the notion of processing of anaphora. This article will discuss the converse problem of referential expression generation, namely, how to choose a specific referential expression among all those that can potentially be used to refer to a discourse entity. (*See* **Anaphora, Processing of**)

It is difficult to define text coherence exactly. We could define it as the quality of a text that is 'tied' together in just the right way. It is text that can be readily comprehended by the hearer, apparently without effort; at the same time, it is text where relations between different sentences are not so explicit as to make it uninteresting. Example 1 is coherent; however, consider the following:

> As soon as they got to the beach, Karin jumped into the water. She hates ice cream. (2)

Example 2 sounds incoherent: it is likely that the hearer will wonder about the connection between hating ice cream and jumping into the water. The hearer may proceed to invent scenarios in which the text makes sense: for example, Karin had ice cream on the way to the beach, it gave her a stomach ache, and her way to deal with stomach aches is to swim. Scenario building of this sort is an exercise precisely in accounting for text coherence; i.e., in explaining text in terms of sentence connections to one another. However, this does not mean that every possible link between the sentences should be explicit. Expanding Example 1 results in a more tedious text, not a clearer one:

> As soon as they got to the beach, Karin jumped into the water. She was so hot from the long drive, so she wanted to cool down. Because the temperature of the sea is generally much lower than that of the air, going for a swim accomplished her goal. (3)

Thus, text coherence appears to obey Grice's maxims of quantity. (*See* **Implicature**)

Text coherence encompasses more than appropriate connections between individual sentences. Discourse appears to have a hierarchical structure: sentences are part of segments, which in turn are part of superordinate segments. Informally, a segment can be defined as a group of locally coherent utterances (see below for a more formal definition). Consider the following discourse:

(a) Georgia called Jeffrey on the phone.
  (b) She wanted to wish him happy birthday.
  (c) She also asked him if she could borrow his tent.
    (d) She had bought a tent herself a few months back.
    (e) However, it got torn on her summer hikes.
(f) After the phone call, she went out for a jog. (4)

Discourse 4 is about Georgia's activities. Intuitively, we recognize that sentences (a) through (e) form a subsegment $S_{a\text{–}e}$ of the whole discourse, as they pertain to Georgia's phone call to Jeffrey. In turn, sentences (c), (d), and (e) form the subsegment $S_{c\text{–}e}$ of $S_{a\text{–}e}$ that concerns the request for the tent; and sentences (d) and (e) form subsegment $S_{d\text{–}e}$ of $S_{a\text{–}e}$, because together they provide a justification for the request in sentence (c).

We will now look at how different researchers account for text coherence in both its manifestations, connections between sentences and discourse segmentation, and how coherent discourses can be interpreted and generated.

THEORIES OF DISCOURSE STRUCTURE

Two theories of discourse structure came to the fore in the mid-1980s, and these are still the most prominent: Grosz and Sidner's (1986) and Mann and Thompson's (1988).

Grosz and Sidner's theory (GST) sees discourse structure as the surface manifestation of the relationships among elements of the intentional structure underlying the discourse. The intentional structure is comprised of the intentions that a speaker brings to the discourse. There will be a primary intention, the discourse purpose (DP), which is, the intention that underlies engaging in that particular discourse. Further, a discourse segment purpose (DSP) is associated to each discourse segment, which is fully individuated by the corresponding DSP. Each DSP specifies how the specific segment contributes to achieving the overall DP. A plausible DP underlying the whole of discourse 4 is *Tell hearer about Georgia's activities*. The DSP associated to the subsegment $S_{d\text{–}e}$ could be something like *Explain to hearer why Georgia needs to borrow Jeffrey's tent*. GST does not specify which intentions can count as DPs or DSPs, except by noting that they are meant to be recognized (cf. Grice's notion of utterance-level intentions). DSPs can be related to one another only via two relationships: dominance and satisfaction-precedence. DSP_1 dominates DSP_2 if DSP_2 is intended to provide part of the satisfaction of DSP_1. DSP_1 satisfaction-precedes DSP_2 if DSP_1 must be satisfied before DSP_2. Grosz and Sidner argue that the intentional structure of the discourse is also intertwined with the attentional state, the set of entities that are salient at any point in the discourse. Attentional state is modeled by a set of focus spaces, which are associated to discourse segments, and contain the entities salient within the corresponding discourse segments. The processing of focus spaces is modeled via a stack. Shifts of attentional state that are local to a discourse segment are outside the scope of GST, but are accounted for by centering theory (Grosz *et al.*, 1995; Walker *et al.*, 1998).

Mann and Thompson (1988) propose 'rhetorical structure theory' (RST) as a descriptive framework that identifies hierarchical structure in text. RST is based on relations between two non-overlapping text spans, the 'nucleus' and the 'satellite'. The nucleus is the central member of the pair; the satellite is peripheral. Relations include an effect and constraints on nucleus and satellites. The relations defined by Mann and Thompson include 'elaboration', 'enablement', 'evidence', and 'contrast'; comparable inventories of discourse relations have been proposed by a number of other researchers (e.g. Hobbs, 1979; Lascarides and Asher, 1993). For example, the evidence relation has as effect that the belief of the hearer in the nucleus is increased, and a constraint that the hearer will find the satellite believable. In Mann and Thompson's view, an analyst will first identify the minimal units of the analysis, which they assume to be clauses. Then, the analyst will start applying relation schemas to adjacent text spans, which are minimal units or, recursively, constituents of relations. In the end there will be one relation schema encompassing text spans that cover the whole text.

From these brief descriptions, we can see that GST mainly accounts for the segmentation aspect of discourse coherence, but does not address how individual sentences are linked to one another by domain or rhetorical relations. As Grosz and Sidner believe that the intentions underlying discourse are too diverse, they argue that it would impossible to enumerate the intentions that can serve as DSPs; hence, they conclude that enumerating a fixed set of relations, as in RST, is wrong. On the other hand, RST accounts both for individual relations between individual sentences and for hierarchical segmentation of the discourse. The latter is a side effect of how the RST analysis is conducted. Note that an analysis of a discourse according to GST generally results in fewer and shallower segments than an RST analysis.

Grosz and Sidner present their theory as a computational account of discourse processing, but they do not provide much insight into the underlying computational processes, except by proposing that the attentional state is modeled as a stack. Mann and Thompson do not make any computational claims; however, their theory has been widely used in computational linguistics. The question thus arises, which processing paradigm is most appropriate for each theory. GST lends itself to a top-down model of discourse processing: the hearer recognizes the DP, and then recursively the subordinate DSPs. RST, on the other hand, lends itself more directly to a bottom-up interpretation of discourse.

A synthesis of GST and RST has been proposed (Moser and Moore, 1996). The synthesis is based on the observation that the dominance relation between intentions in GST closely corresponds to the 'nucleus versus satellite' distinction between text spans in RST.

INTERPRETATION OF DISCOURSE

Discourse interpretation consists of the computational inferences that compute the extended meaning of discourse. We can divide the approaches into two main groups: logical approaches that compute domain and rhetorical relations between sentences in written texts (Lascarides and Asher, 1993), and plan inference approaches that compute the speech acts performed by participants in a dialogue (Perrault and Allen, 1980; Litman and Allen, 1990; Carberry and Lambert, 1999). Plan inference approaches have been applied mainly to dialogue; nevertheless, they are considered part of discourse processing. Because of its inherent difficulty, not many researchers have tried to compute discourse segmentation as proposed in GST, but see Lochbaum (1998) for such an attempt.

Traditionally, approaches to inferring relations between sentences make use of some type of logical inferencing, such as a variant on non-monotonic logic or abduction. We will briefly look at approaches based on abduction (Hobbs *et al.*, 1993). Abduction is an unsound inference rule that reasons from an effect to a potential cause: for example, 'the alarm went off, so there is a burglar in the house'. Clearly, there may be other reasons why the alarm went off (e.g. the landlady forgot to switch it off). Abduction is a useful reasoning mechanism because it tries to find the best explanation for a fact. As far as discourse coherence is concerned, an abductive approach tries to find the most plausible coherence relation linking two utterances, on the basis of rhetorical, domain, and world knowledge. An abductive approach will build a full explanation that supports a specific coherence relation. For example, to establish a cause relation between the two sentences in example 2, an abductive approach would build an explanation, expressed in first-order predicate logic, akin to the one in example 3. The problem abduction has to face is how to choose the most plausible explanation among many possible ones. One can adopt heuristics, such as choosing the explanation that uses the fewest assumptions, or compute the probability of each explanation and choose the most likely one. Both approaches have serious flaws: the former, that even plausible heuristics can fail fairly often; the latter, that it is unclear over which space of events to compute those probabilities.

The computational approaches just discussed are not explicitly based on cognitive findings on text comprehension. Nevertheless, questions addressed by cognitive scientists and psycholinguists have affected computational models. Relevant issues include: inference control (i.e. which of the many possible inferences are made at comprehension, and which later, during recall); and how the connectedness of sentences affects reading times and the accuracy of recall. For example, it has been found that sentences that have a close causal connection are read faster and result in better content recall (Myers *et al.*, 1987).

Plan Inference

The plan inference approach to discourse has been applied mainly to dialogues, although applications to monologic discourse that describes one or more agents' actions have also been attempted. It originated at the end of the 1970s (Perrault and Allen, 1980), with the goal of providing an interpretation for indirect requests such as *I need to be in Boston on the 20th in the afternoon* (directed to a travel agent), or *The next train to Brighton* (directed to a clerk at the ticket booth). It rests on three components: the notion of speech acts from pragmatics; a theory of belief, desire, and intentions from computational linguistics, which in turn owes much to the philosophy of action; and planning models from artificial intelligence.

Every utterance counts as an action performed by the speaker (a speech act), such as asking or promising (Austin, 1962). (We are oversimplifying here. In reality there are three acts associated with each utterance: locutionary, illocutionary, and perlocutionary. The term 'speech act' has come to refer mostly to the illocutionary act.) Utterances can perform speech acts directly, as in example (a) below, or indirectly, as in example (b):

(a) Please find me a flight that arrives in Boston on the 20th in the afternoon.
(b) I need to be in Boston on the 20th in the afternoon. (5)

To explain how a statement such as example 5(b) can count as a request, proponents of the 'inferential' approach (Searle, 1975) contend that indirect speech acts concern felicity conditions on the corresponding direct act. For example, a request such as example 5(a) is felicitous under the assumption that the speaker wants to fly to Boston on that specific date and time. Example 5(b) then works because it explicitly states the speaker's mental attitude, once the hearer has recognized that the literal meaning of it is inappropriate and must be 'repaired' by some inference. (*See* **Pragmatics, Formal**)

Planning is a computational technique from artificial intelligence that, given a goal s_g to achieve, builds a plan, a partially ordered sequence of actions whose execution will bring the agent from the initial state s_0 to s_g. Often the plan is built as a tree, whose leaves are the actions to be executed; the internal nodes represent actions at a higher level, which decompose into lower-level actions. For example, if an agent has the goal *Attend conference in Washington* and lives in Chicago, the agent may build a plan that includes taking a flight from Chicago to Washington; in turn, to achieve taking the flight, the agent will need to buy a ticket, drive to the airport, and board the plane. Planners build plans on the basis of action operators, which must include: preconditions, the conditions that need to hold for the action to be executable; effects, what becomes true after performing the action; and body, a decomposition into a partially ordered set of sub-actions whose execution will result in the execution of the action.

In the plan inference approach, speech acts are modeled as action operators from planning. However, the logical language in which the operators are expressed is augmented with mental attitudes such as knowledge, beliefs, and desire. For instance, a formalization of `Request` (S, H, α) will include as a precondition that the speaker S wants the hearer H to perform action α (one of the felicity conditions on requests), and as an effect that H wants to perform α. Such a formalization can be used to build the interpretation of an indirect speech act via plan inference rules that work backwards from the utterance to its interpretation. One such rule is: if γ is a precondition of action α and H believes S to want γ, then it is plausible that H believes S to want α. Note that the representation can also be used by a regular planner to produce a speech act, starting from a communicative goal to be achieved.

The plan inference approach has been extensively used in dialogue modeling. The original model has been extended in various ways, such as by introducing different levels of inferred plans (e.g., the discourse plan and the domain plan) that the speaker is pursuing (Litman and Allen, 1990; Carberry and Lambert, 1999).

Approaches to discourse based on abduction, non-monotonic logic or plan inferencing, while elegant, suffer from brittleness. One missing domain axiom may cause the model to fail as it is not able to find a complete explanation. Thus, many implemented systems, instead of just using a logical approach, use information that can easily be derived from the surface form of the utterance, such as intonation, connectives, idiomatic expressions, and lexical associations between words (Reithinger and Maier, 1995; Qu *et al.*, 1997; Samuel *et al.*, 1998). These cues to the phenomenon of interest are derived from linguistic and corpus analysis (see below).

GENERATION OF DISCOURSE

From a computational point of view, discourse generation concerns the production of coherent, extended text. Whereas discourse processing is seen as the last stage in language interpretation, after parsing and semantic analysis, it is the first stage for language production. Computationally, language generation starts from a nonlinguistic representation of information that we can consider parceled into messages to be conveyed. The first task to be performed is discourse planning; i.e. imposing ordering and structure over the set of messages to be conveyed. This is followed by sentence planning and linearization, including sentence aggregation (grouping the elements of the discourse plan together into sentences) and the choice of referential terms to individuate the entities of interest. The final step is linguistic realization proper, namely, applying the rules of grammar in order to produce a text that is syntactically and morphologically correct.

Here, we concentrate on discourse planning, and on referring expression generation.

Planning and Linearization

There are two main approaches used to generate a coherent discourse: planning, and schemata.

The discourse planner is given a communicative goal to achieve such as `Intend` S (`Intend` H α). Communicative goals represent the speaker's intentions to affect the beliefs or goals of the hearer. The planner will build a plan consisting of rhetorical actions to achieve the given communicative goal. For example, to achieve `Intend` S (`Intend` H α), S may look for a β such that S expects H to want β, and then utter β as motivation for α (motivation is an RST relation):

> Come to the party on Saturday. I will make your favorite deviled eggs. (6)

The connection between discourse planning and the theories of discourse structure discussed earlier has mainly been achieved through RST. RST relations are recast in the terms of planning operators (Moore and Paris, 1993). The planner posts a high-level communicative goal such as `Intend` S (`Intend` H α) in terms of the effect ε the text should have on the reader. The planner will then search for an RST operator whose effect unifies with ε, and post the subgoals that correspond to constraints on the nucleus and satellite of that rhetorical relation. These subgoals are recursively expanded until the planner reaches the leaves of the rhetorical structure tree, which are expressible as simple clauses.

Schemata are an alternative approach to using a planner. Schemata represent common patterns that texts in a specific domain or genre follow (McKeown, 1985). A schema specifies how a particular discourse plan should be built using other schemata or messages, and the discourse relations that hold between different components of the discourse plan. Although schemata are not generally developed following a planning model, they can be considered as compilations of discourse plans produced by a planning system. As a mechanism for generation, schemata are less flexible, but easier to develop, than a fully-fledged discourse planner. For example, because schemata lack information on the intentions of the speaker, they cannot be used if the system needs to replan, for example if the explanation of a certain p is not understood by the hearer and the discourse planner needs to build a different explanation for p (Moore and Paris, 1993).

Note that a discourse plan, whether built by a planner or as a schema instantiation, does not encode decisions regarding how the leaves should be parceled into individual sentences, or how these sentences should be connected. For instance, the two sentences in example 6 could be linked in a variety of different ways, both paratactically and hypotactically. For example:

> Come to the party on Saturday if you don't want to miss your favorite deviled eggs. (7)

There are also more subtle decisions that need to be made. In example 6 the adjective *favorite* in the second clause may well be derived from a full proposition in the discourse plan, such as *You like the deviled eggs I make a lot*. This is why many researchers consider lexicalization as part of sentence planning. Lexicalization is concerned with choosing words to express concepts and relations.

The solutions proposed in the literature for sentence planning and linearization are diverse, but some general paradigms are beginning to emerge.

Establishment of Referential Terms

The task of generating referring expressions is to select words or phrases to identify discourse entities. The choice of referring expressions greatly affects the readability of a text. Compare the two texts below, the second of which always uses the nominal expression *Bill Gates*.

When a Stanford University professor asked for volunteers to have their heads scanned, Bill Gates was the first to volunteer.
The billionaire CEO of Microsoft Corporation sat patiently while a laser scanner orbited his head several times. A short time later, a 3-D image of Gates's head floated on a screen. (8)

When a Stanford University professor asked for volunteers to have their heads scanned, Bill Gates was the first to volunteer.
Bill Gates sat patiently while a laser scanner orbited Bill Gates's head several times. A short time later, a 3-D image of Bill Gates's head floated on a screen. (9)

In text 9, the repeated use of the proper name *Bill Gates* makes the text sound clumsy. The much more fluent text 8 makes use of different forms of proper names (*Bill Gates* or simply *Gates*), pronouns (*his*), and complex definite referring expressions (*the billionaire CEO of Microsoft Corporation*).

The problem of generating referring expressions can be subdivided into two tasks:

- Initial introduction, or how to perform the initial reference to a discourse entity.
- Subsequent references. This includes choosing between a pronoun and a definite description: if the latter is chosen, then the question is which features of the entity in question to include in the description.

The initial introduction and the choice of pronoun or definite description are generally performed by algorithms based on the 'given–new' distinction (Prince, 1981) or on centering (Grosz *et al.*, 1995; Walker *et al.*, 1998) or on a combination of the two. (*See* **Computability and Computational Complexity**)

Regarding the choice of appropriate definite descriptions, early approaches (Appelt, 1985) took a full planning approach to generating referring expressions. This means that in principle they could generate any description that satisfies a given communicative goal. As this approach was computationally inefficient, later approaches, most notably Dale's (1992), focused on the restricted problem of building a 'distinguishing description'. A distinguishing description is true only of the entity being described and of no others among the currently salient discourse entities. These algorithms generally aim at finding the minimal distinguishing description. However, even computing a minimal distinguishing description is an inherently hard computational task. Dale and Reiter (1995) showed that it is NP-hard by reducing it to a set cover problem. Moreover, humans do not produce minimal distinguishing descriptions, either because humans also face computational limitations, or because they intend to achieve other goals besides identification (Jordan, 2000). In text 8, the complex noun phrase *the billionaire CEO of Microsoft Corporation* may be used to introduce information that the hearer is not expected to know, or, more likely in this case, to remind the hearer of Gates's position. Algorithms used today try to strike a balance between conciseness of the definite description and reproducing human behavior, as observed in corpus analysis.

EMPIRICAL APPROACHES TO DISCOURSE

In the 1990s there was a shift in focus towards a rigorous empirical foundation for discourse processing work. The general methodology that has emerged (Walker and Moore, 1997) comprises the following components:

- Development and evaluation of coding schemes. Coding schemes are used to annotate language corpora for features deemed likely to affect the phenomena under study (e.g. correlates of discourse segments, minimality of referential expressions with respect to providing distinguishing descriptions). A necessary condition for a coding scheme to be useful is that it be reliable, namely, that two or more independent coders can use that coding scheme to annotate the same text in a 'similar enough' way. Much research has thus been devoted to measures of inter-coder agreement (Carletta, 1996; Di Eugenio, 2000).
- Extraction of information from the annotated corpus. Researchers use either statistical measures or data mining tools on the annotated features (Di Eugenio *et al.*, 1997; Poesio and Vieira, 1998; Samuel *et al.*, 1998; Jordan, 2000). The purpose is to verify hypotheses (e.g. the hypothesis that in real-world situations speakers use minimal distinguishing descriptions), and to find linguistic correlates of higher-level phenomena, such as intonation patterns and adverbs for discourse segmentation.
- Development of computational frameworks based on the information extracted from the corpus. For example, the result of an annotation for referring expressions is used to inform algorithms to generate referring expressions (Poesio and Vieira, 1998; Jordan, 2000).
- Evaluation. The computational models developed either theoretically or on the basis of corpus analysis need to be evaluated. This has motivated much interest in evaluation methodologies for computational models and implemented systems (e.g. Walker *et al.*, 1997). However, it is still too early to report specific results that determine which techniques, models, or systems

are the most promising. Systematic evaluations have only recently become the norm, and there is no standard test-bed of problems and phenomena that can be used to make comparisons across systems and techniques.

Acknowledgment

This work was partially supported by grant N00014-00-1-0640 from the Office of Naval Research, Cognitive, Neural and Biomolecular S&T Division.

References

Appelt D (1985) Planning English referring expressions. *Artificial Intelligence* **26**(1): 1–33. [Reprinted in: Grosz, Sparck Jones and Webber (eds) (1986) *Readings in Natural Language Processing*. Santa Monica, CA: Morgan Kaufmann.]

Austin JL (1962) *How to Do Things With Words*. Oxford, UK: Oxford University Press.

Carberry S and Lambert L (1999) A process model for recognizing communicative acts and modeling negotiation subdialogues. *Computational Linguistics* **25**: 1–53.

Carletta J (1996) Assessing agreement on classification tasks: the Kappa statistic. *Computational Linguistics* **22**: 249–254.

Dale R (1992) *Generating Referring Expressions*. Cambridge, MA: MIT Press.

Dale R and Reiter E (1995) Computational interpretations of the Gricean maxims in the generation of referring expressions. *Cognitive Science* **18**: 233–263.

Di Eugenio B (2000) On the usage of Kappa to evaluate agreement on coding tasks. In: Gavrilidou M, Carayannis G, Markantonatou S *et al.* (eds) *LREC2000, Proceedings of the Second International Conference on Language Resources and Evaluation*, pp. 441–444. Athens, Greece: National Technical University of Athens Press.

Di Eugenio B, Moore JD and Paolucci M (1997) Learning features that predict cue usage. In: *ACL-EACL97, Proceedings of the 35th Annual Meeting of the Association for Computational Linguistics*, pp. 80–87. San Francisco, CA: Morgan Kaufmann.

Grosz B, Joshi A and Weinstein S (1995) Centering: a framework for modeling the local coherence of discourse. *Computational Linguistics* **21**: 203–225.

Grosz B and Sidner C (1986) Attention, intentions, and the structure of discourse. *Computational Linguistics* **12**: 175–204.

Hobbs JR (1979) Coherence and co-reference. *Cognitive Science* **3**(1): 67–82.

Hobbs J, Stickel M, Appelt D and Martin P (1993) Interpretation as abduction. *Artificial Intelligence* **63**(1–2): 69–142.

Jordan PW (2000) *Intentional Influences on Object Redescriptions in Dialogue: Evidence from an Empirical Study*. PhD thesis, University of Pittsburgh.

Lascarides A and Asher N (1993) Temporal interpretation, discourse relations, and commonsense entailment. *Linguistics and Philosophy* **16**: 437–493.

Litman D and Allen J (1990) Discourse processing and commonsense plans. In: Cohen P, Morgan J and Pollack M (eds) *Intentions in Communication*, pp. 365–388. Cambridge, MA: MIT Press.

Lochbaum KE (1998) A collaborative planning model of intentional structure. *Computational Linguistics* **24**: 525–572.

Mann WC and Thompson S (1988) Rhetorical structure theory: toward a functional theory of text organization. *Text* **8**(3): 243–281.

McKeown KR (1985) *Text Generation: Using Discourse Strategies and Focus Constraints to Generate Natural Language Text*. Cambridge, UK: Cambridge University Press.

Moore JD and Paris CL (1993) Planning text for advisory dialogues: capturing intentional and rhetorical information. *Computational Linguistics* **19**: 651–695.

Moser M and Moore JD (1996) Towards a synthesis of two accounts of discourse structure. *Computational Linguistics* **22**: 409–419.

Myers JL, Shinjo M and Duffy SA (1987) Degree of causal relatedness and memory. *Journal of Verbal Learning and Verbal Behavior* **26**: 453–465.

Perrault R and Allen J (1980) A plan-based analysis of indirect speech-acts. *American Journal of Computational Linguistics* **6**: 167–182.

Poesio M and Vieira R (1998) A corpus-based investigation of definite description use. *Computational Linguistics* **24**: 183–216.

Prince E (1981) Toward a taxonomy of given–new information. In: Cole P (ed.) *Radical Pragmatics*, pp. 223–255. New York, NY: Academic Press.

Qu Y, Di Eugenio B, Lavie A, Levin L and Rosé CP (1997) Minimizing cumulative error in discourse context. In: Maier E, Mast M and LuperFoy S (eds) *Dialogue Processing in Spoken Language Systems*, pp. 171–182. Heidelberg, Germany: Springer-Verlag.

Reithinger N and Maier E (1995) Utilizing statistical dialogue act processing in Verbmobil. In: *ACL95, Proceedings of the 33rd Annual Meeting of the Association for Computational Linguistics*, pp. 116–121. Cambridge, MA: MIT Press.

Samuel K, Carberry S and Vijay-Shanker K (1998) Dialogue act tagging with transformation-based learning. In: *ACL/COLING 98, Proceedings of the 36th Annual Meeting of the Association for Computational Linguistics*, pp. 1150–1156. San Francisco, CA: Morgan Kaufmann.

Searle JR (1975) Indirect speech acts. In: Cole P and Morgan JL (eds) *Syntax and Semantics*, vol. III, *Speech Acts*, pp. 59–82. San Diego, CA: Academic Press. [Reprinted in: Davis S (ed.) (1991) *Pragmatics: A Reader*. New York, NY: Oxford University Press.]

Walker M, Joshi A and Prince E (eds) (1998) *Centering Theory in Discourse*. Oxford, UK: Oxford University Press.

Walker MA, Litman DJ, Kamm CA and Abella A (1997) PARADISE: a framework for evaluating spoken dialogue agents. In: *ACL-EACL97, Proceedings of the 35th Annual Meeting of the Association for Computational Linguistics*, pp. 271–280. San Francisco, CA: Morgan Kaufmann.

Walker MA and Moore JD (1997) Empirical studies in discourse. *Computational Linguistics* **23**: 1–12.

Further Reading

Allen J (1995) *Natural Language Understanding*, 2nd edn. Menlo Park, CA: Benjamin/Cummings.

Cohen PR, Morgan J and Pollack ME (eds) (1990) *Intentions in Communication*. Cambridge, MA: MIT Press.

Grosz B and Sidner C (1986) Attention, intentions, and the structure of discourse. *Computational Linguistics* **12**: 175–204.

Jurafsky D and Martin JH (2000) *Speech and Language Processing*. Englewood Cliffs, NJ: Prentice-Hall.

Levinson S (1983) *Pragmatics*. Cambridge, UK: Cambridge University Press.

Mann WC and Thompson S (1988) Rhetorical structure theory: toward a functional theory of text organization. *Text* **8**(3): 243–281.

Reiter E and Dale R (2000) *Building Applied Natural Language Generation Systems*. Cambridge, UK: Cambridge University Press.

Walker MA, Joshi A and Prince E (eds) (1998) *Centering Theory in Discourse*. Oxford, UK: Oxford University Press.

Walker MA and Moore JD (1997) Empirical studies in discourse. *Computational Linguistics*: **23**: 1–12. [Special issue on empirical studies in discourse.]

Webber BL (1999) Computational aspects of discourse and dialogue. In: Schiffrin D, Tannen D and Hamilton H (eds) *The Handbook of Discourse Analysis*. Oxford, UK: Blackwell.

Disfluencies in Spoken Language

Intermediate article

Jean E Fox Tree, University of California, Santa Cruz, California, USA

CONTENTS

Speech disfluencies are phenomena that cause a break in the smooth flow of talk, affecting speaking and understanding. Everyone who talks produces disfluencies. Several decades of research have provided a good understanding of the types of disfluencies people produce, and a beginning understanding of why disfluencies occur and how they impact addressees.

CATEGORIES OF DISFLUENCIES

Disfluencies are any of a group of phenomena that cause a break in the smooth flow of spoken talk. Three main categories of disfluencies are: (1) pauses; (2) *um*s and *uh*s; and (3) repetitions, replacements, and restarts, although there are others. Disfluencies are normal nonfluent speech, as opposed to dysfluencies, or abnormal nonfluent speech, such as stuttering (Wingate, 1984).

A pause is a stretch of speech that is heard as silence. Pauses are usually identified as times when the speaker is not saying anything, but they may also be heard when there is no actual silence in the speech, but rather a slowing down of tempo. Traditionally, only silences over a quarter of a second long were considered meaningful breaks in talk (Goldman-Eisler, 1968), but some have argued that pauses over a tenth of a second long should form the lower cutoff point (Hieke *et al.*, 1983). Most pauses in spontaneous talk are under 1 second long (Jefferson, 1989); in fact, pauses over 3 seconds have defined conversational lapses (McLaughlin and Cody, 1982). Pauses have been further categorized by their position in the clause, such as whether they are within or between sentences, or their purpose, such as whether or not they are produced for rhetorical effect. Although only a subset of pauses may be disfluent pauses, it can be hard to determine whether a pause is a disfluency or not.

Ums and *uhs* describe a group of sounds that sound, in English, like /um/ and /uh/, with some variation in the shape of the vowel (/em/ and /eh/, for example). They are sometimes referred to as fillers or filled pauses. These labels descend from historical comparisons between *unfilled pauses* (pauses) and *filled pauses* (*ums* and *uhs*; Maclay and Osgood, 1959); *ums* and *uhs* were thought to be ways to fill silence to show that a speaker intends to continue speaking (Cook, 1971). But this theoretical position has not held up over time. To avoid the implication that *ums* and *uhs* are equivalent to or alternative versions of silent pauses, some researchers identify them only as *ums* and *uhs*, or use the label *interjection*. Instead of being equivalent to pauses, *ums* and *uhs* indicate the lengths of upcoming pauses, with *ums* indicating major pauses and *uhs* minor pauses (Clark, 1994; Clark and Wasow, 1998; Smith and Clark, 1993).

Repetitions, replacements, and restarts are stretches of speech where people have: (1) stopped the smooth flow of speech; (2) possibly uttered a pause, *um* or *uh*, or other words such as *I mean* or *you know*; and (3) resumed their talk (Clark, 1996). In repetitions, words are repeated exactly in the resumption, as in 'of her of her daughter'. In replacements, some words are repeated but some are changed, as in 'there were a lot of tricks that the um tricks and toys that the ant could play with'. In restarts, no words are repeated, as in 'what would you-can I help you?'. Although people can detect problems that need correcting after hearing themselves start to say something wrong (Levelt, 1989), people can also choose in advance when they will suspend their speech, detecting problems while speaking and suspending their speech when they have the continuation ready (Blackmer and Mitton, 1991; Fox Tree and Clark, 1997). People also choose how to resume fluent talk, resulting in various resumptions.

The part of speech that is stopped is sometimes called the *reparandum*, and the part of speech that is resumed is sometimes called the *repair* (Levelt, 1983). The term 'repair' has also been used to refer to repetitions, replacements, and restarts as a group along with similar phenomena (Fox Tree and Clark, 1997). But the term 'repair' can imply the revising of something said earlier, which is not always the case for these disfluencies. For example, repetitions can be viewed as early commitments to speaking at particular moments with subsequent restorations of continuity in the resumption (Clark and Wasow, 1998), instead of as second occurrences' revising first occurrences, without changes.

The three types of disfluencies discussed here – pauses, *ums* and *uhs*, and repetitions, replacements, and restarts – are interrelated. Pauses can predict upcoming repetitions, replacements, and restarts. *Ums* and *uhs* indicate the lengths of upcoming pauses. And information between the suspension and resumption of repetitions, replacements, and restarts can contain pauses or *ums* and *uhs*.

DISFLUENCIES IN LANGUAGE PRODUCTION

Since at least the 1960s, disfluencies have been seen as windows into the speech production process. They could be the auditory remains of a problem in turning thoughts into words, or they could be the normal result of speakers' planning their talk. Hypotheses about disfluencies' aetiologies or purposes were arrived at by analyzing when they occurred.

Hypotheses about pauses were that they were epiphenomena of a general need for more processing time to produce talk (Levelt, 1989), or the result of more specific effort at lexical access (Goldman-Eisler, 1968; Maclay and Osgood, 1959; Martin and Strange, 1968), syntactic formulation (Brotherton, 1979; Clark and Wasow, 1998; Duez, 1982; Ferreira, 1991; Maclay and Osgood, 1959), or phonological encoding (Ferreira, 1991). Pauses were also thought to be used more purposefully for rhetorical effect (Duez, 1982; Kowal *et al.*, 1985), such as making people appear sincere (Maclay and Osgood, 1959). Pause placement also influenced hypotheses about the order of speech production processes, such as that syntactic formulation precedes lexical access (Maclay and Osgood, 1959).

Similar hypotheses were made about *ums* and *uhs*. Without distinguishing between proposals that they are symptoms or signals, *ums* and *uhs* have been thought to foreshadow: (1) general speech production difficulty (Brotherton, 1979; Reynolds and Paivio, 1968) or specific difficulty, such as upcoming delays (Clark, 1994) or error avoidance (Jefferson, 1974); (2) particular kinds of words, such as difficult to produce or unpredictable words (Brotherton, 1979; Tannenbaum *et al.*, 1965) or words with more competitors (Schachter *et al.*, 1991); (3) the major chunks of talk in a discourse (Swerts, 1998); (4) difficulty in planning what one wants to say and how to say it syntactically (Maclay and Osgood, 1959; Martin and Strange, 1968; Reynolds and Paivio, 1968); (5) speakers' desires to maintain control of the floor in a conversation (Maclay and Osgood, 1959); and (6) speakers' desires to show awareness of upcoming delays, to avoid being cast in a negative light by a silent pause (Smith and Clark, 1993).

Repetitions, replacements, and restarts in speech production come about because of a variety of problems, including conceptualizing ideas, formulating sentences, selecting words, or articulating utterances (Levelt, 1989). Different types of production trouble may yield different kinds of recovery (Tannenbaum *et al.*, 1965). One explanation for the reason repetitions, replacements, and restarts look the way they do is that speakers follow rules for making them well formed (Levelt, 1983, 1989). They are well formed if there is a way of converting the suspended talk and the resumption into a co-ordination; for example, because 'there were a lot of tricks that the- um tricks and toys that the ant could play with' could be filled out to the well-formed sentence 'there were a lot of tricks that the [ant could do, and] um tricks and toys that the ant could play with', the replacement without the bracketed talk is well-formed.

DISFLUENCIES IN LANGUAGE COMPREHENSION

Fewer researchers have explored the role of disfluencies in speech comprehension.

Disfluencies can be difficult to detect in talk, although listeners can detect them with effort (Martin, 1967; Martin and Strange, 1968). Detection of pauses may depend on where in the clause they fall; one study found that within-clause pauses can be detected if they are over 200 ms, but between-clause pauses need 500 ms to 1000 ms for detection (Boomer and Dittmann, 1962). Detection of *ums* and *uhs* varies depending on whether listeners are paying attention to what speakers are saying or how they are saying it (Christenfeld, 1995). Detection of repetitions and restarts takes place after the smooth flow of speech has stopped (Lickley and Bard, 1998).

None the less, effects of disfluencies on comprehension have been measured. There are generally two different measurement techniques, those that involve offline tasks (measuring comprehension after speech has been heard) and those that involve online tasks (measuring comprehension while speech is being heard).

In offline tasks, disfluencies have influenced what listeners think about a speaker's personality. For example, pauses can make a conversationalist appear less adept (McLaughlin and Cody, 1982), and also have implications for interpretations of what the speaker does or does not know (Brennan and Williams, 1995). Saying *um* can make people who know the answer to a question appear less sure of their answer, or give the appearance that someone who doesn't know the answer really does know it. *Ums* can also make people appear more relaxed compared to speech with the *ums* replaced by pauses, although pauses make people appear less relaxed than no pauses (Christenfeld, 1995). Offline tasks have also demonstrated that pauses can aid in syntactic disambiguation (Price *et al.*, 1991), and, if placed at syntactic boundaries, can improve recall of the gist of sentences (Reich, 1980). *Ums* and *uhs* can provide turn-ending or turn-continuation information, depending on whether they fall at grammatical or ungrammatical points (Cook and Lalljee, 1970).

In online tasks, disfluencies have been shown to produce a variety of effects. *Uhs* speed up the recognition of upcoming words in sentences but *ums* don't, a result that can be attributed to their differing roles in anticipating the lengths of upcoming pauses (Fox Tree, 2001). Attention may be heightened after hearing an *uh* in anticipation of the short upcoming pause and continuation, but it may not be after *um* because of the indeterminacy of the upcoming delay. Repetitions do not negatively affect recognition of subsequent words (Fox Tree, 1995), as would be expected if repetitions are a solution to a fluency problem as opposed to an error (Clark and Wasow, 1998). But certain kinds of restarts, those altering information mid-sentence, do slow recognition (Fox Tree, 1995). Restarts are only costly when listeners need to store information about one part of the discourse record while making the correction. Information between the suspension and resumption can help listeners follow the speaker successfully (Fox Tree and Schrock, 1999).

References

Blackmer ER and Mitton JL (1991) Theories of monitoring and the timing of repairs in spontaneous speech. *Cognition* **39**: 173–194.

Boomer DS and Dittmann AT (1962) Hesitation pauses and juncture pauses in speech. *Language and Speech* **5**: 215–220.

Brennan SE and Williams W (1995) The feeling of another's knowing: prosody and filled pauses as cues to listeners about the metacognitive states of speakers. *Journal of Memory and Language* **34**: 383–398.

Brotherton P (1979) Speaking and not speaking: processes for translating ideas into speech. In: Siegman AW and Feldstein S (eds) *Of Speech and Time*. New York: Wiley.

Christenfeld N (1995) Does it hurt to say um? *Journal of Nonverbal Behavior* **19**(3): 171–186.

Clark HH (1994) Managing problems in speaking. *Speech Communication* **15**: 243–250.

Clark HH (1996) *Using Language*. New York: Cambridge University Press.

Clark HH and Wasow T (1998) Repeating words in spontaneous speech. *Cognitive Psychology* **37**: 201–242.

Cook M (1971) The incidence of filled pauses in relation to part of speech. *Language and Speech* **14**(2): 135–139.

Cook M and Lalljee M (1970) The interpretation of pauses by the listener. *British Journal of Social and Clinical Psychology* **9**: 375–376.

Duez D (1982) Silent and non-silent pauses in three speech styles. *Language and Speech* **25**(1): 11–28.

Ferreira F (1991) Effects of length and syntactic complexity on initiation times for prepared utterances. *Journal of Memory and Language* **30**: 210–233.

Fox Tree JE (1995) The effects of false starts and repetitions on the processing of subsequent words in spontaneous speech. *Journal of Memory and Language* **34**: 709–738.

Fox Tree JE (2001) Listeners' uses of um and uh in speech comprehension. *Memory and Cognition* **29**(2): 320–326.

Fox Tree JE and Clark HH (1997) Pronouncing 'the' as 'thee' to signal problems in speaking. *Cognition* **62**(2): 151–167.

Fox Tree JE and Schrock JC (1999) Discourse markers in spontaneous speech: oh what a difference an oh makes. *Journal of Memory and Language* **40**: 280–295.

Goldman-Eisler F (1968) *Psycholinguistics: Experiments in Spontaneous Speech*. New York: Academic Press.

Hieke A, Kowal S and O'Connell DC (1983) The trouble with 'articulatory' pauses. *Language and Speech* **26**(3): 203–214.

Jefferson G (1974) Error correction as an interactional resource. *Language in Society* **2**: 181–199.

Jefferson G (1989) Preliminary notes on a possible metric which provides for a 'standard maximum' silence of approximately one second in conversation. In: Roger D and Bull P (eds) *Conversation: An Interdisciplinary Perspective*. Philadelphia, PA: Multilingual Matters.

Kowal S, Bassett MR and O'Connell DC (1985) The spontaneity of media interviews. *Journal of Psycholinguistic Research* **14**(1): 1–18.

Levelt WJM (1983) Monitoring and self-repair in speech. *Cognition* **14**(1): 41–104.

Levelt WJM (1989) *Speaking: From Intention to Articulation*. Cambridge, MA: MIT Press.

Lickley RJ and Bard EG (1998) When can listeners detect disfluency in spontaneous speech? *Language and Speech* **41**(2): 203–226.

Maclay H and Osgood CE (1959) Hesitation phenomena in spontaneous English speech. *Word* **75**: 19–44.

Martin JG (1967) Hesitations in the speaker's production and listener's reproduction of utterances. *Journal of Verbal Learning and Verbal Behavior* **6**(6): 903–909.

Martin JG and Strange W (1968) The perception of hesitation in spontaneous speech. *Perception & Psychophysics* **3**(6): 427–438.

McLaughlin ML and Cody MJ (1982) Awkward silences: Behavioral antecedents and consequences of the conversational lapse. *Human Communication Research* **8**(4): 299–316.

Price PJ, Ostendorf M, Shattuck-Hufragel S and Fong C (1991) The use of prosody in syntactic disambiguation. *Journal of the Acoustical Society of America* **90**(6): 2956–2970.

Reich SS (1980) Significance of pauses for speech perception. *Journal of Psycholinguistic Research* **9**(4): 379–389.

Reynolds A and Paivio A (1968) Cognitive and emotional determinants of speech. *Canadian Journal of Psychology* **22**(3): 164–175.

Schachter S, Christenfeld N, Ravina B and Bilous F (1991) Speech disfluency and the structure of knowledge. *Journal of Personality and Social Psychology* **60**(3): 362–367.

Smith VL and Clark HH (1993) On the course of answering questions. *Journal of Memory and Language* **32**: 25–38.

Swerts M (1998) Filled pauses as markers of discourse structure. *Journal of Pragmatics* **30**: 485–496.

Tannenbaum PH, Williams F and Hillier CS (1965) Word predictability in the environments of hesitations. *Journal of Verbal Learning and Verbal Behavior* **4**: 134–140.

Wingate ME (1984) Fluency, disfluency, dysfluency, and stuttering. *Journal of Fluency Disorders* **17**: 163–168.

Further Reading

Clark HH (1996) *Using Language*. New York, NY: Cambridge University Press.

Fox Tree JE (2000) Coordinating spontaneous talk. In: Wheeldon LR (ed.) *Aspects of Language Production*, pp. 375–406. Philadelphia, PA: Psychology Press.

Levelt WJM (1989) *Speaking: From Intention to Articulation*. Cambridge, MA: MIT Press.

Nofsinger RE (1991) *Everyday Conversation*. Prospect Heights, IL: Waveland.

Disorders of Body Image

Introductory article

Giovanni Berlucchi, University of Verona, Verona, Italy
Salvatore M Aglioti, University of Rome, Rome, Italy

CONTENTS

Body image (or schema) is the complex of perceptions, beliefs, and representations about one's own body that is included in the notion of self. Amputations, cerebral lesions, and psychiatric disorders may induce dramatic alterations of this image.

INTRODUCTION

The ability to distinguish self from nonself is a hallmark of all living organisms. In its most elementary expression, in unicellular animals, the distinction is based on simple physicochemical reactions which serve nutritional and self-defence purposes. Multicellular organisms retain such simple reactions for the same purposes, but if they possess a complex brain (particularly a human brain), the self–nonself distinction takes up an entirely different meaning, insofar as it provides the basis for self-perception and personal identity.

Our brain is housed in a body that is at the same time the instrument of all brain-generated patterns of behavior, and the container of all sensory receptors that inform the brain about the external world and about the body. How can our brain distinguish sensory reports about the external world from those about our own body?

A seemingly logical hypothesis is that we know the external world through visual, auditory, and olfactory receptors, all of which respond to stimuli from objects remote from the body, while we know our body through the somatosensory receptors for touch, thermoception, nociception, and proprioception, all of which are acted upon by stimulants directly applied to the body itself. Accordingly, the physiologist Sherrington distinguished between exteroception, which informs the brain about the external world, particularly through the distance receptors, and interoception and proprioception, which inform the brain about states and changes of states in the body.

However, awareness of one's own body can be gained not only through interoception and proprioception, but also through distance receptors that monitor body postures and movements. Vision, for example, is a major source of information not only about the external world, but also about one's body parts that can be seen directly or through mirrors or other reflecting surfaces. Similarly, somatosensory receptors can supply the brain with direct information about the body, but can also be used to explore one's surroundings and to recognize objects by contact.

In short, all kinds of receptors and all sense organs can make their own special contributions to the separate representations of the body and the external world that the brain concurrently entertains.

SELF-CONSCIOUSNESS

Self-consciousness as awareness of one's own being refers to knowledge of one's own mind as well as to knowledge of one's own body. In contrast to purely mental views of self-consciousness, many current psychological and physiological approaches to the concept of self-consciousness emphasize the importance of the awareness of one's own body.

In the nineteenth century the psychologist William James stated that the nucleus of the self is always the bodily existence felt to be present at the time, and that the entire feeling of one's own mental activities is really a feeling of bodily activities, mainly in the head (motor adjustments of eyeballs, eyelids and eyebrows) and the throat (changes of breathing due to movements of the soft palate, posterior nares, glottis, and so on).

Modern analyses suggest that from earliest infancy, simultaneously perceived flows of multisensory information enable us to gain a realistic and accurate perception of the relations between our own body and our physical environment. This perception, which is perhaps the earliest form of self-knowledge, is complemented – again at an early age – by the perception of reciprocated relations between our own behavior and that of other people: that is, by the sense of the self as an agent and target of social interactions. Babies a few weeks old engage in elaborated social exchanges, loaded with affective meanings, with their mothers and other individuals, and vocalize in response to heard language. Between one and two years of age, interactions with adults lead children to start to build up organized beliefs and memories, and to think that they have traits, attributes, and values. It is around this age that children not only develop language, but also begin to recognize themselves in a mirror – an ability that may be regarded as an objective index of self-consciousness (Figure 1). Among mammals, only humans, chimpanzees, and orangutans are thought to be endowed with the ability of mirror self-recognition.

BODY SCHEMA, BODY IMAGE, AND CORPOREAL AWARENESS

In classical neurology, notions about the functional representation of the body in the brain were first inspired by observations about postural regulation and perception. Even a minor movement of a single body part entails a widespread adjustment of the postural tone affecting most other body parts, suggesting that the brain keeps a moment-to-moment record of the postural state through the entire musculature. This usually unconscious record surfaces to consciousness when a person is asked to report a postural change imposed on the whole body or a part of it by an examiner.

The term 'body schema' was originally coined to denote this current internal model of one's own postural state, serving as the basis for the appreciation of subsequent postural changes. The term has later been extended to refer to a variety of normal performances which are indicative of a general awareness of the body, whether conscious or unconscious. In addition to the ability to appreciate active and passive postural changes and movements, such performances include the abilities to localize tactile stimuli, to move, name and point to specified body parts, and in general to map sensory inputs and motor outputs onto an orderly topographical model of the external anatomy of the human body.

All these abilities suggest the existence of a mental construct, termed 'corporeal or body awareness', which comprises the sense-impressions, perceptions, memories, and ideas about the dynamic organization of one's own body and its relation to other bodies. In current terminology sensory-perceptual components of corporeal awareness are preferentially named 'body schema', while conceptual and imaginative components are preferentially named 'body image'. This distinction, however, is not easy to make in practice, and the two terms are often used as synonyms. The term 'body image' is also employed to refer to the evaluative and emotional judgment, either self-appreciative or

Figure 1. Children begin to recognize themselves in a mirror at between one and two years of age. Before that, as shown by the behavior of this 7-month-old girl in front of the mirror, children tend to interact with their image as if it were another child.

self-critical, that one has of one's own body. Severe distortions of the body image are now thought to underlie psychogenic eating disorders, particularly the syndrome of anorexia nervosa in which people feel grossly fat even if they are extremely underweight.

The Extended Body Schema: Enlargement of the Body's Boundaries

Noncorporeal objects bearing some functional relation of contiguity to the body, such as clothes, ornaments and tools, may come to be felt as parts of the body itself. When one drives a nail into a wall with a hammer, the perceived end of the arm wielding the hammer is not the hand but the head of the hammer itself. This enlargement of the body's boundaries can be accounted for by the same mechanisms that are involved in the representation of the body *per se*. The execution of a voluntary movement of a limb under direct view activates a coherent multimodal representation of that limb in the body schema, based on the congruence between the internal knowledge of the moving command and the visual and proprioceptive feedbacks from the moving limb. An object held by the limb and moving jointly with it takes part in generating such feedbacks and is thus incorporated into the body schema.

A putative neuronal mechanism for including a noncorporeal object into awareness of the body has been found in the anterior parietal cortex of the monkey brain, where there exist neurons that respond to somatosensory and visual stimuli arising from the monkey's hands. If the monkey retrieves food with its hand, the visual receptive fields of these neurons are limited to that hand; but if the retrieval is helped by a rake, the visual receptive fields expand to include both hand and tool, as if they were a unified body part.

Vision and the Body Schema

Inanimate objects can be incorporated into the body schema even without visual feedback, as exemplified by the stick used by blind people for assistance in walking; but the importance of vision is evident from the fact that blindness, especially if congenital, can conspicuously distort the mental representation of the body.

Vision is also crucial for detecting the correspondence between a part of one's own body and the matching part of another person's body, as well as for the imitation of gestures and other movements. A strong innate tendency to imitate sounds and motor acts sets apart humans from other primates, and is probably a prerequisite both for social communication and for the self–nonself distinction. Hours or even minutes after delivery, babies can imitate orofacial and head movements performed by adult models in front of them. This deceivingly simple performance indicates that babies are able to identify a movement of a specific bodily part of the adult model, and to produce a similar movement in the corresponding part of their own anatomy.

This early capacity for visual imitation of elementary actions has suggested that humans are born equipped with a rudimentary body schema which during maturation undergoes a gradual refinement as a consequence of orderly interactions between visual, tactile, proprioceptive, and vestibular inputs. The latter inputs provide the sensory data for the appreciation of head position in the gravitational field, and for the detection of head acceleration and deceleration during linear and circular movements. In adults, the persistence of a systematic reciprocal relation between the visual perception of another person's body part and the representation of one's own corresponding body part is suggested by the finding that visual discrimination of postural changes in the arms of another person is facilitated during movements of the observer's arms but not legs, and vice versa.

Brain Regions Related to the Body Schema

The premotor cortex of the monkey contains neurons which become active when the monkey either performs a goal-directed movement or views a corresponding movement made by another monkey or human. These neurons, if present in the human brain, may provide a simple mechanism for the imitation of actions as well as for the detection of a correspondence between one's own body parts and the matching parts of another individual's body.

Analysis of cortical activation in normal observers during imitation of finger movements has suggested that a region of the left frontal lobe, corresponding to Broca's area, encodes the general goal shared by observed and imitated movements, whereas the precise kinesthetic representation of the same movements is encoded by specific regions of the right posterior parietal lobe, which may also distinguish between observed and self-produced movements.

Anatomically and physiologically the body is represented in a topographic fashion in somatosensory maps in the anterior parietal cortex, corresponding to the sensory homunculus, and in motor maps in the posterior frontal lobe, corresponding to the motor homunculus, but the relations between these cortical homunculi and corporeal awareness is far from simple. According to Melzack, corporeal awareness relies upon a large neural network where the somatosensory cortex, the parietal lobe, and the insular cortex play crucial and different roles, as suggested by studies of functional brain imaging, brain stimulations, and brain lesions.

Functional brain imaging has shown that a posterior parietal system (superior parietal cortex, intraparietal sulcus, and the adjacent, most rostral part of the inferior parietal lobule) is involved in mental transformations of body in space. The insular cortex is also involved in body awareness, particularly in relation to emotional aspects of this. Insular lesions can cause somatic hallucinations, and electrical stimulation near the insula induces illusions of body position changes and feelings of being outside one's body. Lesions of the cortical sensory homunculus, or of the cortical motor homunculus, induce tactile, proprioceptive, and motor deficits, but there is no evidence that the body parts numbed or paralyzed by such lesions are eliminated from the mental representation of the whole body.

Body awareness may be altered in the context of a diffuse cognitive impairment involving general mental functions such as attention, memory or language; but can some kind of brain damage selectively impair body awareness in the absence of major deficits in other cognitive domains? A positive answer to this question is provided by the striking alterations or mutilations of the mental representation of the body that can be observed after lesions of the cerebral hemispheres, especially the right cerebral hemisphere and the posterior parietal lobe.

Cerebral Lesions and Body Awareness

Patients with lesions of the right posterior parietal lobe may show a neglect (failure to attend to stimuli) of the left half of their body or parts of it. In many cases such symptoms occur in the setting of a general neglect of the left hemispace, both personal and extrapersonal, insofar as patients do not respond to (or mislocalize) all kinds of stimuli coming from their left side. In these cases, neglect of the left half of the body is best attributed to an overall impairment of spatial attention or space representation rather than to a selective disruption of the body schema. In other cases, however, at least some disturbances may be so pronounced or selective as to hint at a specific alteration of body awareness in the form of a *hemisomatagnosia* – that is, a defective or absent knowledge of half of the body.

Patients with right posterior parietal lesions may fail to detect tactile stimuli to the left side of the body even in the absence of basic sensory impairments, or may refer such stimuli to a corresponding position on the right side. Such mislocalization is called *allesthesia* or, in the case of stimuli to the hand, *allochiria*. Patients may even fail to groom or dress the left half of the body, or leave the left half of the face unshaved.

A pathologic attitude towards conspicuous motor and sensory deficits caused by a right hemisphere lesion on the left half of the body may manifest itself in various ways. Patients may appear unaware of their severe deficits to the point of vehemently denying any impairment at all, or they may admit the existence of deficits but show a lack of concern regarding them, or, on the contrary, exhibit hatred for the affected limbs.

The most severe manifestation of neglect for the left body is *somatoparaphrenia*, a condition in which otherwise rational and objective patients exhibit feelings that a body part does not belong to them, and express a resolute denial of ownership of such part. The neglected or disowned body parts are expunged from the mental representation of the body, so that in order to account for the material existence of these parts the patients resort to improbable rationalizations and confabulations: for example, they may claim that their disowned arm belongs to the examiner or to the previous occupant of the hospital bed. Noncorporeal objects previously associated with a disowned hand and included in the body schema, such as a ring, are similarly disowned, although ownership of such objects is immediately acknowledged when they are removed from the disowned hand (Figure 2). Somatoparaphrenia following posterior parietal lesions may also manifest as a distinct and irrepressible feeling of the existence of supernumerary limbs on the affected body half, perhaps suggesting that different aspects of body awareness are localized in different components of the parietal cortex.

Neglect of one side of the body is most commonly observed following lesions of the right posterior parietal lobe, but may also occur following frontal, cingulate or insular lesions, as well as following subcortical lesions in thalamus and basal ganglia, again on the right. Reports of similar symptoms on the right side of the body following left hemisphere lesions are extremely rare. There is,

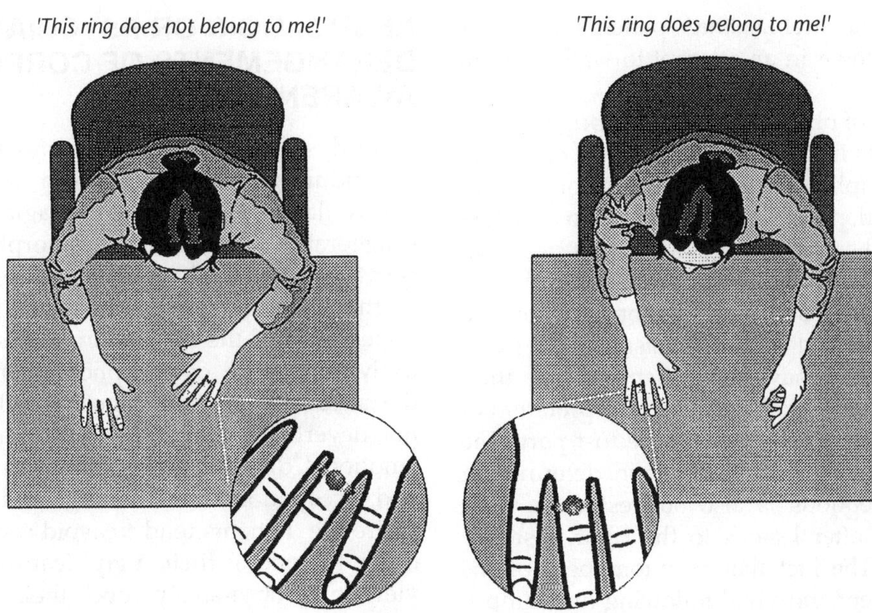

'This ring does not belong to me!'
'This ring does belong to me!'

Figure 2. A patient with a right hemisphere lesion exhibited a strong denial of ownership of the left side of the body. She also denied ownership of a ring that she had worn on the left hand as long as the ring was on that hand, but she promptly recognized the ring as her own when it was moved to the right hand. (See Aglioti, Smania, Manfredi and Berlucchi (1996) Disownership of left hand and objects related to it in a right-brain-damaged patient. *NeuroReport* **8**: 293–296).

however, a rare alteration of body awareness, *autotopagnosia*, which seems to depend on a left parietal lesion. Patients with this condition are unable to point to parts of their own or other people's body on verbal command, a disability that cannot be imputed to mental deterioration or language incomprehension because patients can carry out successfully verbal commands unrelated to the body, such as 'touch the pedal of a bicycle'. These patients also have difficulties in describing the spatial relations between body parts: for example, they may say that the mouth is between the nose and the eyes.

Often the difference in efficiency between responses to body-related commands and responses to body-unrelated commands is relative rather than absolute, and it can be argued that at least some cases of autotopagnosia suffer from a general disability to analyze a whole into parts, though such disability is maximally evident in relation to the body.

Autotopagnosia may occur in the setting of the Gerstmann syndrome, also associated with left parietal lesions, which is characterized by dysgraphia (writing disturbances), dyscalculia (calculation difficulties), right–left disorientation, and finger agnosia, the latter being a specific inability to identify and differentiate the fingers of both hands. The relation between autotopagnosia and finger agnosia is a complex one, because patients classified as autotopagnosic may show no finger agnosia, and patients with finger agnosia may perform normally in response to verbal commands targeted to other parts of the body.

The contrast between the symptoms of deranged body awareness associated with left parietal lesions and those associated with right parietal lesions has prompted the following simplistic but effective hypothesis about the relation between the parietal lobes and body awareness: the left parietal lobe houses a conceptual representation of the body, strongly based on a linguistic mediation, whereas the right parietal lobe houses a spatial, nonverbal representation of the body.

Phantom Limb Phenomena

Many people who have undergone amputation report rich and vivid perceptions originating from the amputated body part, which often occur in the form of excruciating pain precisely referred to that part. These 'phantom' perceptions are most common following limb amputation, though they also occur after amputation of other body parts such as the breast in women or the penis in men. Moreover, people with a congenital absence of one or more limbs have been reported to experience phantom perceptions from the limbs that they have never possessed. Even though phantom perceptions are in some sense illusory, they are usually so distinct and lifelike that patients can, for example, fall down in an attempt to walk with an amputated foot.

The analysis of phantom sensations provides important insights into the mechanisms of corporeal awareness. Peripheral activation of sensory nerves at the amputation scar can contribute to such experience, but there is clear evidence that phantom phenomena have major cerebral causes.

It has been proposed that phantom phenomena are primarily caused by the persisting activity of brain centers that have been deprived of their normal inputs, and by the brain's interpretation of this activity as originating from the lost part. The relevance of cerebral components in determining phantom perceptions is also suggested by their disappearance after lesions to the right posterior parietal lobe. The fact that phantom perceptions are most frequent and vivid following limb amputation is probably due to the functional relevance and the extensive cerebral representation of these body parts. A phantom limb may be perceived as identical in shape to the former real limb, thus suggesting that structures in the central nervous system are committed to the representation of that body part. Furthermore, the persistence of phantom limbs over decades after an amputation attests to a certain degree of stability of somatic representations. This strong tendency of the brain to maintain an intact representation of a lost body part, even a massive one, is perhaps an index of the preservation of the integrity of the self.

Changes in phantom sensations may reflect adjustments or reorganizational changes in the neural substrates representing the lost body part. Phantom limbs, for example, particularly when painful, may shrink in such a way that the hand is perceived as attached directly to the shoulder (the telescoping phenomenon). Complex dynamic aspects of the body schema are also revealed by the recent evidence in limb or breast amputees that vivid phantom sensations can arise as a result of tactile stimulations applied to body regions distant from the amputation line. Sensations in the phantom hand, for example, can be elicited by tactile stimuli delivered to the lower face on the side of the amputation. Like the concurrent veridical facial sensations, the evoked phantom sensations may convey precise information about different features of the facial stimuli. Given the representational contiguity of face and hand, phantom hand sensations from facial stimulation are probably caused by an appropriation of the original cerebral representation of the lost hand by sensory afferents to the adjacent face representation.

NEUROLOGIC OR PSYCHIATRIC DERANGEMENTS OF CORPOREAL AWARENESS

Body dysmorphic disorder, or *dysmorphophobia*, is an enduring excessive concern with a selected bodily flaw which is totally imaginary or grossly exaggerated. Targets for dysmorphophobic concerns may be the shape of the nose, the thickness of the hair, the size of the penis or breast, the appearance of the facial skin, and so forth. Some individuals with dysmorphophobia are aware of the absurdity of their concerns, while many are not; nevertheless, all of them suffer from an intense emotional distress which can disrupt their social and occupational functioning, and indeed their entire life. Patients tend to avoid social contacts in order to conceal their 'ugly' feature from others' view, may obsessively check their appearance in mirrors or, on the contrary, be morbidly afraid of seeing themselves in mirrors or photos. They may seek unnecessary surgical corrections, which invariably fail to eliminate their emotional problem, and may even be driven to such extreme decisions as physically injuring the offensive body part or committing suicide.

Dysmorphophobia differs from anorexia nervosa insofar as people affected by the former condition are concerned with a single feature of their bodily appearance, while in the latter condition it is the overall shape or size of the body that is at the center of the pathologic preoccupation. Moreover, the physical appearance and eating habits of people with dysmorphophobia are normal, whereas those of people with anorexia nervosa are not. Dysmorphophobia usually also differs from obsessive–compulsive disorder, because all patients with the latter typically recognize the absurdity of their disturbing thoughts, which they can at least temporarily hold in check with ritualistic behaviors that are of no help to patients with dysmorphophobia. The modest therapeutic success with serotonin reuptake inhibitors obtained in a few cases of obsessive–compulsive disorder or dysmorphophobia suggests that malfunctions of brain activity regulation by serotonin, a major central neurotransmitter, may underlie both disorders, but no definite notion about these putative pathogenetic mechanisms is yet available.

Body-centered delusions such as underestimation of the size of bodily parts are often observed in major psychiatric illnesses like schizophrenia, where such symptoms are more frequently related to the left body side, and depression, where they are more frequently related to the right body side.

Also, hypochondriac patients tend to refer more frequently to the right side of the body when expressing their complaints (e.g., a pain in an arm). Other deformations of the body image may be experienced during epileptic or migraine attacks, as reported by the British neurologist Macdonald Critchley. For example, a patient with a left parietal meningioma suffered from recurrent attacks of migraine during which the right side would feel bigger and stronger, as if there were a sharp line down the middle. The left side, however, would remain 'calm, cool and collected, while the right side would be tense, anxious, agitated, and highly strung'. These side differences may perhaps be pathologic expressions of the asymmetric functioning of the cerebral hemispheres, but no decisive evidence to support this possibility has been offered.

In heautoscopy (autoscopy) individuals experience a veritable visual hallucination of themselves. A celebrated case is that of the Swedish naturalist Linnaeus who, while examining plants and flowers in his garden, would sometimes see at a little distance his alter ego performing the same actions. A persistent feeling of living outside one's own body characterizes the depersonalization syndrome, while patients with the Cotard syndrome are affected by delusions that their body does not exist, suggesting a specific disorder of corporeal and/or egocentric space awareness.

All the disturbances of the body image described above must have an immediate neural basis, but so far knowledge about the brain activities involved in these conditions is discouragingly small. Even more obscure are the mechanisms by which cultural and social factors act on the brain to produce derangements of the body image. Deep-seated psychological problems related to family and occupational conditions are suspected to be involved in anorexia nervosa, which typically occurs in middle-class young women and shows a high incidence in professional models and ballet dancers. The pathologic attitude toward the body and the related drastic cutdown in eating may be triggered by an exacerbation of a culturally imposed view of thinness as a supreme hallmark of beauty. The endocrine manifestations of anorexia nervosa leave no doubt that the hypothalamus and the adenohypophysis are malfunctioning in this condition, but it is far from clear whether the hypothalamic dysfunction constitutes the primary neural problem or is secondary to starvation and weight loss. Perhaps, in a not too distant future, approaches based on functional brain imaging and other modern neurotechnologies will allow an understanding of the internally generated or externally triggered patterns of cerebral activity which result in this and other distortions of the body image.

CONCLUSION

Studies in nonhuman and human primates indicate that the brain generates separate representations of the body and external world. The term 'body image' (here used as a synonym of body schema) indicates the complex of beliefs, memories, and knowledge about one's own and others' anatomy. This complex mental construct, on which the concept of the self is based, is the product of a continuous interaction between central neural systems dedicated to the representation of the body and peripheral inputs including somatic, vestibular, and visual signals. The ways in which the human body is perceived and represented may change dramatically as a consequence of amputations and cerebral lesions, the effects of which can help localize the neural systems specialized in representing the body. The neural bases of body-related disorders in neurologic and psychiatric diseases are probably due to dysfunctions of these systems as well, but knowledge in this area is still preliminary, though potentially open to improvement thanks to modern approaches to the study of the brain.

Further Reading

Berlucchi G and Aglioti S (1997) The body in the brain: neural bases of corporeal awareness. *Trends in Neurosciences* **20**(12): 560–564.

Bermudez JL, Marcel A and Eilan N (eds) (1995) *The Body and the Self*. Cambridge, MA: MIT Press.

Critchley M (1979) *The Divine Banquet of the Brain*. New York, NY: Raven Press.

Head H and Holmes G (1911) Sensory disturbances from cerebral lesions. *Brain* **34**: 102–254.

Iriki A, Tanaka M and Iwamura Y (1996) Coding of modified body schema during tool use by macaque postcentral neurones. *NeuroReport* **7**: 2325–2330.

James W (1890) *The Principles of Psychology* [reprinted 1950]. New York, NY: Dover.

Meltzoff AN (1990) Towards a developmental cognitive science: the implications of cross-modal matching and imitation for the development of representation and memory in infancy. *Annals of the NY Academy of Science* **608**: 1–31.

Melzack R (1992) Phantom limbs. *Scientific American* **266**(4): 90–96.

Rizzolatti G, Fadiga L, Fogassi L and Gallese V (1999) Resonance behaviors and mirror neurons. *Archives Italiennes de Biologie* **137**: 85–100.

Snodgrass JG and Thompson RL (eds) (1997) *The Self Across Psychology*. New York, NY: New York Academy of Sciences Press.

Dissociation Methodology

Intermediate article

Mark G Packard, Yale University, New Haven, Connecticut, USA

CONTENTS

Dissociation methodology is an experimental design tool used to examine functional dichotomies and independence between and within various domains of psychological function.

INTRODUCTION

The brain is capable of perceiving electrical messages transmitted by the optic nerve as visual information, and those by the auditory nerve as sound information. The nineteenth-century German physiologist Johannes Müller (1801–1858) proposed the 'doctrine of specific nerve energies' to account for these observations. This hypothesis essentially holds that independent neural pathways determine the perceived quality of sensory information, and provides a basis for understanding the empirical data indicating that deaf individuals can possess normal eyesight and blind individuals can possess normal hearing. These findings also provide evidence of a dissociation of the neural circuitry underlying these two sensory capabilities, at least at some fundamental level of psychological processing.

In the psychological sciences, dissociation methodology has been used extensively as an experimental design tool to investigate possible functional dichotomies that might shape understanding of fundamental laws in various domains of cognition and behavior, such as language, memory, and emotion. An early example of the use of this methodology is provided by the discovery of the different roles played by Broca's area and Wernicke's area in language expression and reception, respectively.

In the field of cognitive neuroscience, scientists often discover evidence of a functional dichotomy at the neuroanatomical level, and speculate as to the differences in operating principles that best relate this anatomical (neural level) dissociation to behavior or cognition (psychological level).

Several functional dichotomies have been proposed in different areas of psychological research, and each has influenced theoretical debate in these areas. These include, for example, dissociations between object recognition and localization visual pathways, explicit versus implicit memory, conscious versus unconscious perception in 'blindsight', or effortful versus automatic attention. The development of each of these hypotheses, and several others on their respective topics, relies on dissociation methodology.

SINGLE VERSUS DOUBLE DISSOCIATIONS

A description of the types of dissociations that may be observed in a particular experiment can be readily understood using the following scenario. Consider a black box with two levers (X and Y) protruding from the sides, and containing a latched top and a sliding front drawer. When lever X is pulled, the top of the box flips open (box operation A). However, pulling lever X does not effect the operation of the box drawer. In attempting to elucidate the functions of lever X, this manipulation has produced a single dissociation: pulling lever X effects box operation A but not B. Next, when lever Y is pulled, the drawer in the front of the box slides open (box operation B). However, pulling lever Y does not effect the operation of the box top. These manipulations have produced a double dissociation, in which pulling lever X influences box function A but not B, and pulling lever Y influences box function B but not A. If the black box used in this example represents a mammalian brain, and the two levers represent different brain structures, then one might discover that selective damage to lever X impairs box function A but not B, and selective damage to lever Y impairs box function B but not A.

The double dissociation methodology used above illustrates a fundamental approach for examining the hypothesis that independent neural and/or psychological functions exist. Hans-Lukas Teuber (1955) originally coined the term 'double dissociation', and provided early arguments for the importance of this pattern of results in supporting the hypothesis of functional independence between neural structures. Teuber noted that neuropsychological studies employing dissociation methodology often reveal only a single dissociation (e.g., damage to structure X impairs performance of A but not B). Although it is tempting to conclude that A and B are functionally independent based on a single dissociation, it is also possible that their operation may be arranged in a hierarchical fashion. Clearly, if the primary goal of dissociation methodology is to provide information concerning the functional independence of various psychological and/or neural processes, then a double dissociation is necessary for the strongest version of such a hypothesis to be advanced.

Within the broader field of cognitive science, examples of double dissociations exist from research in traditional neuropsychology (the study of individuals who have suffered brain damage through injury or disease), more recent studies employing brain imaging techniques and interference tasks in normal humans, as well as research in nonhuman experimental animals. Research on the neurobiology of learning and memory can be considered an exemplar of the historical use of dissociation methodology in the psychological sciences, as this experimental tool has been used effectively across four decades of research in this area. Of course, the logic behind dissociation methodology and the use of this approach are not unique to memory research.

DISSOCIATION METHODOLOGY

Examples from Human Neuropsychology

A double dissociation of the roles of two cortical regions in facial identity illustrates the usefulness of dissociation methodology for providing information about the organization of a particular psychological function. Patients with bilateral damage of the ventromedial frontal cortex are able to recognize the identity of familiar faces in a normal fashion, but are unable to generate a discriminatory skin conductance response (SCR) to the same faces. In contrast, patients with bilateral occipitotemporal cortical damage display impaired identity recognition yet can generate discriminatory SCRs to familiar faces. These findings have been interpreted to suggest that a functional independence exists between the neural circuitry that processes somatic-based 'valence' and nonsomatic-based 'factual' information in this process. Note that interpretation of the psychological operating principles that might account for the anatomical double dissociation observed in any behavioral study ultimately involves debate over theoretical constructs such as emotion and memory, and often relies to some degree on *a priori* assumptions that are made at the time the behavioral tasks are developed. Nonetheless, an empirical double dissociation on task performance in humans with brain damage compromising different neuroanatomical structures provides the necessary evidence for postulating the existence of functionally independent systems.

Neuropsychological studies have also revealed double dissociations in performance of brain-damaged humans on pairs of memory tasks; patients with limbic-diencephalic damage acquire a probabilistic learning task normally, but memory for the training episode as assessed in an interview questionnaire is severely impaired. In contrast, patients with Parkinson disease are severely impaired in acquisition of the probabilistic task, yet demonstrate normal memory for the training episode. These findings have been interpreted as evidence of a double dissociation between the role of limbic-diencephalic and neostriatal brain regions in declarative memory and habit learning, respectively (Knowlton *et al.*, 1996). Again, separate theoretical accounts of the critical differences in the psychological operating principles that underlie the dissociations observed in performance of different learning and memory tasks have been offered and debated, and several 'dual-memory' theories describing different sets of principles exist. However, it is the use of dissociation methodology that has ultimately driven the development and refinement of these theories.

Imaging the Living Human Brain

The advent of technologies for imaging the living human brain has provided an extraordinary avenue for the use of dissociation methodology to elucidate the organization of various psychological functions. Neuroimaging studies have revealed double dissociations in the patterns of brain activity associated with performance of different tasks. For example, functional magnetic resonance imaging has been used to demonstrate a double dissociation in levels of activation of the left perisylvian cortex and the dorsolateral prefrontal cortex in working memory processes. Based on these findings, the former structure has been proposed to mediate working memory storage processes, and the latter to mediate executive control processes contributing to accurate working memory (Postle *et al.*, 1999). Other brain imaging research using injections of radiolabeled water in combination with positron emission tomography to measure regional cerebral blood flow has revealed dissociable roles of the ventral and dorsal human extrastriate cortex in object and spatial visual processing, respectively, consistent with a prominent dissociation in these visual pathways or processes originally developed in research with nonhuman primates (Haxby *et al.*, 1991).

Research in Experimental Animals

Research in experimental animals (e.g. nonhuman primates and rats), in which brain manipulations can be performed in a localized manner, provides further examples of empirical double dissociations. In monkeys, lesions of the medial temporal lobe impair performance of a delayed nonmatch to sample task, but do not affect acquisition of a concurrent visual discrimination (Mishkin and Petri, 1984). In contrast, lesions of the ventrocaudal neostriatum impair concurrent discrimination learning, but do not affect performance of delayed nonmatch to sample behavior (Fernandez-Ruiz *et al.*, 2001). In rats, a double dissociation between the roles of the hippocampal system and caudate nucleus in acquisition of spatial and visual discrimination tasks, respectively, has been observed following lesions of these two structures (Packard *et al.*, 1989).

The most compelling experimental design using dissociation methodology is one in which as many domains of psychological processes as possible can be equated across different tasks. An important feature of the pairs of tasks used in the rat study is that they each involved similar motor (maze running), sensory (use of visual cues), and motivational (appetitive) characteristics, and were hypothesized to differ primarily in mnemonic requirements.

The hypothesis of functional independence between brain areas in these memory tasks is strengthened by converging evidence from manipulations that affect neurochemical mechanisms. For example, localized intracerebral injections of drugs affecting various neurotransmitter systems have been used to doubly dissociate the roles of different brain regions in memory consolidation. Unlike traditional lesion techniques such as the use of irreversible or reversible lesions in which dissociations are revealed as an impairment in performance, drug injections have been used in these same brain regions to produce double dissociations using memory-enhancing agents (Packard and White, 1991).

CONCLUSION

Dissociation methodology has been used extensively to gather evidence supporting the possible existence of functional dichotomies in various psychological domains. The examples of double dissociations provided in this brief article illustrate the use of this methodology in developing theories of memory organization in the mammalian brain. Historically, the use of dissociation methodology has not been limited to this research topic, and important examples exist of dissociations in tasks measuring perceptual, attentional, and motivational processes, as well as complex language and numerical capabilities in humans.

An important goal of dissociation methodology is to provide evidence of functional dichotomies. However, as dissociation methodology has gained widespread use in psychological science, several scientists have stressed that such conclusions must be made cautiously, and others have raised reasonable concern over the use of dissociation methodology to develop a potentially endless taxonomic classification of various psychological functions. As considered earlier, in view of hierarchical relationships a single dissociation can never be interpreted as providing unequivocal evidence of functional independence. Moreover, even a double dissociation may not be an entirely sufficient condition for offering such a conclusion (Weiskrantz, 1968). For example, the observation of an empirical double dissociation of the effects of damage to different brain structures (X and Y) on performance in two different memory tasks (A and B) could conceivably be influenced by manipulation of a particular parametric setting such as task demand (e.g. delay interval or memory load; Olton, 1989). In this case, rather than posit a strong functional independence between the roles of brain structures X and Y in task performance, one might postulate that a third factor influencing performance on both tasks may be interacting differentially with each structure. Analyses of parametric settings in experiments using dissociation methodology are likely to be of particular importance in the psychological sciences, and may provide constraints on the number of fundamental dichotomies that are ultimately deemed dissociable at a functional level.

As dissociation methodology continues to find widespread use in brain imaging research, it should be noted that there are caveats in interpretation of data obtained in studies measuring localized increases in brain activity. For example, in some imaging studies activation of particular brain regions has been observed in tasks for which there is neuropsychological data indicating that damage to these same brain areas does not impair task performance. Thus, dissociations between the roles of different brain structures in behavior may be revealed in a neuropsychological study when brain damage is present, but not observed when imaging techniques are applied to the intact human brain. Conclusions about the necessary versus sufficient roles of different brain structures implicated in a dissociation study using neuroimaging techniques may be difficult, and converging evidence from neuropsychological testing in individuals with brain damage is likely to be important for interpretation of the ultimate functional significance of the dissociations observed. A similar caution is necessary in interpreting dissociations at the functional level using electrophysiological methodology, or more recently developed molecular science techniques that use gene expression as for a marker of nerve cell activation (e.g. c-*fos* activation). Nonetheless, if the functional dichotomies proposed to underlie the double dissociations observed at the neural and/or psychological level prove predictive across experimental settings using converging techniques, then the promise of dissociation methodology will continue to be the possible discovery of fundamental scientific laws.

References

Fernandez-Ruiz J, Wang J, Aigner TG and Mishkin M (2001) Visual habit formation in monkeys with neurotoxic lesions of the ventrocaudal neostriatum. *Proceedings of the National Academy of Sciences of the USA* **98**: 4196–4201.

Haxby JV, Grady CL, Horwitz B *et al.* (1991) Dissociation of object and spatial visual processing pathways in human extrastriate cortex. *Proceedings of the National Academy of Sciences of the USA* **88**: 1621–1625.

Knowlton BJ, Mangels JA and Squire LR (1996) A neostriatal habit learning system in humans. *Science* **273**: 1399–1402.

Mishkin M and Petri HL (1984) Memories and habits: some implications for the analysis of learning and retention. In: Squire LR and Butters N (eds) *Neuropsychology of Memory*, pp. 287–296. New York, NY: Guilford.

Olton DS (1989) Inferring psychological dissociations from experimental dissociations: the temporal context of episodic memory. In: Roediger HL and Craik FIM (eds) *Varieties of Memory and Consciousness: Essays in Honor of Endel Tulving*, pp. 161–174. Mahwah, NJ: Lawrence Erlbaum.

Packard MG and White NM (1991) Dissociation of hippocampus and caudate nucleus memory systems by pasttraining intracerebral injection of dopamine agonists. *Behavioral Neuroscience* **105**: 73–84.

Packard MG, Hirsh R and White NM (1989) Differential effects of fornix and caudate nucleus lesions on two radial maze tasks: evidence for multiple memory systems. *Journal of Neuroscience* **9**: 1465–1472.

Postle BR, Berger JS and D'Esposito M (1999) Functional neuroanatomical double dissociation of mnemonic and executive control processes contributing to working memory performance. *Proceedings of the National Academy of Sciences of the USA* **96**: 12959–12964.

Teuber HL (1955) Physiological psychology. *Annual Review of Psychology* **6**: 267–296.

Weiskrantz L (1968) Some traps and pontifications. In: Weiskrantz L (ed.) *Analysis of Behavioral Change*, pp. 415–429. New York, NY: Harper & Row.

Further Reading

Crowder RG (1989) Modularity and dissociations in memory systems. In: Roediger HL and Craik FIM (eds) *Varieties of Memory and Consciousness: Essays in Honor of Endel Tulving*, pp. 271–294. Mahwah, NJ: Lawrence Erlbaum.

Dunn JC and Kirsner K (1988) Discovering functionally independent mental processes: the principle of reversed association. *Psychological Review* **95**: 91–101.

Schallice T (1988) *From Neuropsychology to Mental Structure*. Cambridge, UK: Cambridge University Press.

Weiskrantz L (1991) Dissociations and associates in neuropsychology. In: Lister RG and Weingarter HJ (eds) *Perspectives on Cognitive Neuroscience*, pp. 157–164. New York, NY: Oxford University Press.

Distinctive Feature Theory

Introductory article

Elizabeth Hume-O'Haire, Ohio State University, Columbus, Ohio, USA
Stephen Winters, Ohio State University, Columbus, Ohio, USA

CONTENTS

The term 'distinctive features' is used in phonology to refer to the minimal units of sound that serve to distinguish the meaning of one word from another within a language. Distinctive features generally correspond to a specific articulatory or acoustic property of sound.

INTRODUCTION

Distinctive features are the universal set of cognitive properties associated with the speech sounds that are used in language. Distinctive features determine the contrasts which may exist between speech sounds, account for the ways in which these sounds may change, or *alternate*, and define the sets of sounds – known as natural classes – which may behave similarly in language. This limited set of features enables linguists to make powerful predictions about the kinds of sound structures that are expected to exist in the languages of the world. As universal properties of speech sounds, distinctive features also provide key insights into the cognitive organization of sound in language. The fact that distinctive features have such explanatory power makes their discovery one of the most important advances in linguistic science during the twentieth century.

THE FUNCTIONS OF DISTINCTIVE FEATURES

Contrast

The notion of contrast in language has been important for as long as linguists have been aware that the relationship between sound and meaning is arbitrary. For example, there is no logical connection between the sounds used to express the word 'tree' and the actual meaning of the word 'tree'. It does not matter, therefore, what particular sounds a language chooses to express the meaning of a word like 'tree'; what matters is that the sounds combined to form this word are distinct from those that combine to denote a word with a different meaning, such as 'true'. In other words, the sounds of language are meaningful only inasmuch as they *contrast* with one another.

Linguists originally conceived of contrast as a relationship that held between two entire sounds, such as between the [b] and [p] in a pair of words like 'bit' and 'pit'. Further investigation revealed, however, that such contrasts were always based on certain aspects of the articulation (i.e. the pronunciation) or the acoustics (i.e. the physical characteristics of sound waves) of the individual sounds. A [b] and a [p] differ, for instance, only in that the articulation of [b] involves the low-frequency vibration of the vocal cords (denoted by the distinctive feature [voice]) while the articulation of [p] does not. Likewise, the sound [s] contrasts with the sound [θ] (which is represented in English by *th*, as in 'with') in that, acoustically, [s] is louder and has more energy concentrated in higher frequencies, while [θ] is quieter and has acoustic energy spread across a broad range of frequencies. This acoustic distinction is represented with the feature [strident]. Any two sounds can differ meaningfully only in terms of such 'distinctive features' of their articulations or acoustics. Furthermore, linguists have observed that there are only a limited number of such features that can make meaningful distinctions. This insight has the important implication that the number and kind of meaningful sound distinctions in any language is limited by the set of available distinctive features. The fact that language is limited in this way – and is not completely arbitrary – provides significant evidence for the universal cognitive structures the mind imposes on language.

Natural Classes

Distinctive features further organize language by defining groups of sounds which may exhibit similar sound patterns. A *natural class* of sounds in a language consists of those sounds which share certain distinctive features to the exclusion of all other sounds in the language. Such natural classes of sounds often pattern together in similar ways. For example, the labio-velar sound [w], as in 'wit', cannot follow a specific group of sounds in English; [w] may follow [d] or [k] sounds, as in 'dwell' or 'quell', but it may never follow sounds like [b], [f], or [m]. That is, there are no words in English like 'bwell', 'fwell', or 'mwell'. The group of sounds which [w] cannot follow in English is collectively known as the natural class of *labial* consonants. The distinctive feature [labial] characterizes sounds that are articulated with the lips; as such, [labial] generalizes over the more specific phonetic categories of *bilabial* (pronounced with two lips, as in [b] or [p]) and *labio-dental* (pronounced with both the lips and the teeth, as in [f] or [v]). Though such finer distinctions could be made, phonological systems ignore them. The fact that distinctive features may draw broader distinctions than objectively exist in speech production or transmission provides further evidence for their essentially cognitive status. The observation that it is the sound [w] that fails to occur before a [labial] consonant also follows from the view that sounds defined by a common feature are predicted to pattern together, such as in the [labial] cooccurrence restriction; labio-velar [w] also belongs to the natural class of [labial] sounds.

Alternations

Speech sounds may also *alternate* with one another under certain conditions. The voiceless [s] sound, which may designate plural formations in English, alternates with a voiced [z] sound whenever it follows any of the natural class of voiced sounds at the end of a word which it pluralizes. Voiced [z] is always found after the voiced [b] in 'cabs', for instance, while the voiceless [s] always follows the voiceless [p] in 'caps'. The alternation here is not simply between the sounds [s] and [z] but between the distinctive feature of voicing which distinguishes the two. In fact, all sound alternations in language may be defined in terms of distinctive features. Furthermore, since distinctive features define natural classes of sounds, an entire group of sounds (as opposed to just one) is predicted to condition or undergo the same alternation. Thus, the pronunciation of the English plural ending as voiceless [s] occurs after all words ending in a voiceless sound, for example, [p, t, k, f]. Characterizing alternations in terms of distinctive features therefore not only captures the observation that speech sounds do not alternate arbitrarily but also provides linguists with a powerful predictive device.

THE NATURE OF DISTINCTIVE FEATURES

Phonetic Grounding

Although the primary motivation for the existence of a feature comes from considerations of language sound patterning as noted above, features are defined in the phonetic terms of articulations and acoustics. For example, the common articulatory property of sounds defined by the feature [coronal] for both consonants and vowels is a constriction involving the front or mid-portion of the tongue. Speakers use such articulations to produce sounds like [ʃ] (usually represented in English with the spelling *sh*), as in 'shirt', and [i], as in 'magazine'. Acoustically, the vocal tract configuration that this articulation creates results in a sound produced with a greater concentration of energy among the higher frequencies used in speech sounds.

Table 1 provides a list of some commonly used distinctive features along with their articulatory definitions.

Feature Values

A feature's ability to minimally contrast two sounds with each other is generally represented by listing that feature with either of two opposite values, indicated before a given feature by a plus or minus symbol. For example, the property of voicing which distinguishes [b] from [p] in English is defined by the feature values [+voice] and [−voice], respectively. The characterization of a sound as [+voice] indicates that it bears the property of vocal cord vibration, while the value [−voice] defines sounds which lack vocal cord vibration. Thus, it is possible to refer to natural classes defined by each of these two feature values.

There are certain features, however, for which the negative value never defines a natural class of sounds. In such cases, these features contrast by virtue of either their presence or absence in a sound; there is no negative feature value, as there is for voicing, which can be referred to in the formal description of the sound system. Such features are considered to be unary-valued, or privative. Place

Table 1. Some distinctive features and their articulatory definitions

[sonorant]	Sounds which are produced without an obstruction which would impede the flow of air that has passed through the vocal cords in voicing. Sonorant sounds include vowels, nasals (e.g. [n, m]), liquids (e.g. [l, r]), and glides (e.g. [j, w]).
[voice]	Sounds produced with vocal cord vibration, e.g. [b, z, n, r, i, w].
[nasal]	Sounds produced with a lowered velum, which allows air to pass through the nose, as in nasal consonants and nasalized vowels.
[continuant]	Sounds produced in such a way that air flows continuously through the vocal tract as in, among others, vowels and fricatives, e.g. [f, s].
[lateral]	Sounds produced by lowering the mid-section of the tongue at the sides, allowing air to flow out near the molars, e.g. [l].
[strident]	Sounds produced in such a way as to create a turbulent noise, e.g. [f, v, s, z].
[labial]	Sounds produced with a constriction formed by the lower lip, e.g. [p, b, f, v, m, o, u].
[coronal]	Sounds produced with a constriction formed by the front or mid-portion of the tongue, e.g. [t, z, ʃ, i, e].
[dorsal]	Sounds produced with a constriction formed by the back of the tongue, or dorsum, e.g. [k, g, u, o].
[high]	Sounds produced by raising the body of the tongue towards the roof of the mouth, e.g. [i, u].
[low]	Sounds produced by lowering the body of the tongue away from the roof of the mouth, e.g. [a].

of articulation features ([labial], [coronal], [dorsal]) are typically considered to be unary-valued, for example. Since it is impossible to refer to sounds defined by the lack of a given property, this approach makes the strong prediction that sounds defined by the absence of a given feature, for example nonlabial sounds, will not function as a natural class in sound systems.

FEATURE GEOMETRY

In the earliest versions of distinctive feature theory, all of the features that comprised a speech sound were considered to have the same status – no features were more closely related to each other than to any others. In more recent years, however, linguists have begun to recognize that there is internal organization of the features within a speech sound. The fact that certain features commonly pattern together motivates the idea that there are 'natural groupings' of features into higher-level functional units. For example, in many languages the features [labial], [coronal], and [dorsal] function together as a unit. In English, for instance, the nasal sound in a prefix such as 'en' typically takes on the place of articulation value of a following consonant, a process commonly referred to as *place assimilation*. Thus, the nasal is pronounced as coronal [n] before coronal sounds (as in 'encircle', 'endanger'), as labial [m] before labial sounds ('empower', 'embitter'), and as a dorsal nasal [ŋ] ('ng') before dorsal sounds ('encamp', 'encourage'). Notice that all three place features are involved and, furthermore, it is *only* the place of articulation value that is adopted from the following consonant. The three place features are thus patterning together as a functional unit.

An influential model of feature representation known as *feature geometry* incorporates the insight that certain features regularly function together by grouping these sets of features into constituents. Evidence from processes such as nasal place assimilation suggests that the place features [labial, coronal, dorsal] are grouped together into a single Place unit, as shown in (1). Since sounds are defined in terms of more than just place features, the representation in (1) constitutes only one of potentially many feature groupings that, combined together, define the universal geometry of speech sounds.

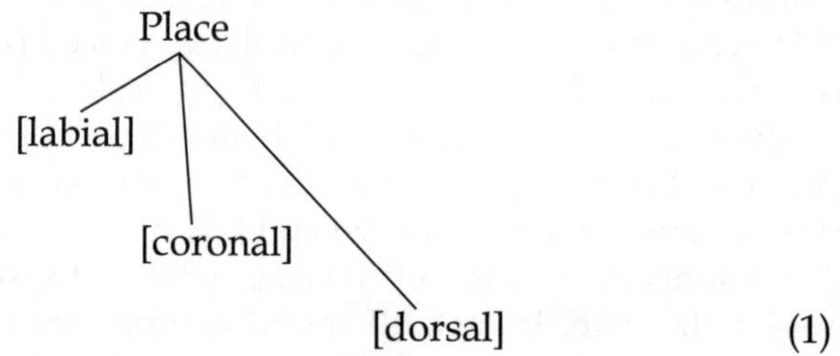

(1)

A crucial assumption embodied in this type of model is that phonological processes are described most simply by reference to a single element – either a feature (e.g. [labial], [voice]) or the head of a constituent (e.g. Place). As a result, the model makes strong predictions about the way in which features (and hence, speech sounds) function in language. For example, a phonological process may be described by reference to an entire constituent, such as Place, as would be the case for the process of nasal place assimilation in English described above. Or, a phonological process may be described in terms of any of the features individually. As seen above in the discussion of English plural formation, for example, the phonological process resulting in the alternation between voiceless [s] and voiced [z] can be described simply by reference to the feature [voice]. However, since processes are defined by reference to a single element, the model rules out phonological processes that would be described by a subset of features headed by a constituent. Given the feature grouping in (1), for instance, the sounds characterized by the features [coronal] and [labial] should never pattern together to the exclusion of those with the feature [dorsal]. Thus, no language should have a nasal assimilation rule in which the nasal consonant acquired the place feature of a following [coronal] or [labial] consonant, but not that of a [dorsal] consonant.

FEATURES AND MARKEDNESS

Linguists have long observed that, across languages, certain speech sounds are more common than others. The most common sounds are considered *unmarked* while progressively rarer sounds are considered more and more *marked*. Interestingly, there are certain groups of marked segments which never exist in a language unless another, less marked group of segments does as well. For instance, a language will never have voiced stops unless it also has voiceless stops, or nasal vowels without also having oral vowels. Since the existence of these marked natural classes always implies the existence of the less marked class, the relationship that holds between them is called an *implicational law*. Distinctive feature theory elegantly captures such cross-linguistic relationships between natural classes; for example, the implicational law that holds between voiced and voiceless stops can be characterized as: [+voice, −continuant] → [−voice, −continuant]. Informally stated, this rule claims that any language with voiced stops such as [b, d, g] must also have voiceless stops such as [p, t, k].

Universal patterns of markedness have also been claimed to correspond to the order in which children acquire contrasts (and thereby features) during language acquisition. Children generally acquire less marked sounds before acquiring the contrasts that define distinct, but more marked sounds in the system; extremely marked sounds may sometimes never be acquired at all. An apt example of the latter is the English *r* sound whose markedness stems in part from its articulatory complexity: it is commonly produced with multiple articulations, including lip rounding, retroflexion (curling or bunching of the front of the tongue), and pharyngeal constriction (a narrowing of the passageway between the root of the tongue and the back of the pharynx). A possible featural description of the sound would thus need to include three place features: [labial], [coronal], and [pharyngeal]. Conversely, the less marked [t] sound is specified for only a single place feature, [coronal]. The inherent difficulty of acquiring more marked sounds, such as English *r*, means that these sounds are not as likely to be passed on to succeeding generations as less marked sounds are. It only follows, therefore, that less marked sounds will appear in more languages than marked ones will.

CONCLUSION

Analyzing speech sounds in terms of fundamental properties known as distinctive features accounts for a wide variety of phenomena in language sound systems, including contrast, the grouping of sounds together into natural classes, and the alternations of sounds in various contexts. Along with this evidence, regular patterns in markedness, language acquisition, and historical change strongly attest to the psychological reality of features and the contrasts they define in speech sound systems. Understanding the nature of distinctive features and the various functions they have therefore provides crucial insights into the cognitive organization of sound in language. (*See* **Phonology**)

Further Reading

Anderson SR (1985) Roman Jakobson and the theory of distinctive features. *Phonology in the Twentieth Century*, chap. 5. Chicago and London: University of Chicago Press.

Chomsky N and Halle M (1968) The phonetic framework. *The Sound Pattern of English*, chap. 7. New York: Harper & Row.

Clements GN and Hume E (1995) The internal organization of speech sounds. In: Goldsmith J (ed.) *Handbook of Phonological Theory*, pp. 245–306. Oxford, UK and Cambridge, MA: Blackwell.

Halle M (1983) On distinctive features and their articulatory implementation. *Natural Language and Linguistic Theory* **1**: 91–105.

Jakobson R, Fant G and Halle M (1952) *Preliminaries to Speech Analysis*. Cambridge, MA: MIT Press.
Kenstowicz M (1994) *Phonology in Generative Grammar*, chaps 1 (The sounds of speech), 4 (The phonetic foundations of phonology), and 9 (Feature geometry, underspecification and constraints). Oxford, UK and Cambridge, MA: Blackwell.
McCarthy JJ (1994) The phonetics and phonology of Semitic pharyngeals. In: Keating P (ed.) *Phonological Structure and Phonetic Form: Papers from Laboratory Phonology III*, pp. 191–233. Cambridge, UK: Cambridge University Press.

Distributed Representations

Intermediate article

Tony Plate, Black Mesa Capital, Santa Fe, New Mexico, USA

CONTENTS

Distributed representations are a way of representing information in a pattern of activation over a set of neurons, in which each concept is represented by activation over multiple neurons, and each neuron participates in the representation of multiple concepts.

THE IMPORTANCE OF REPRESENTATION

The way information is represented has a large impact on the type of operations that are easy or practical to perform with it. Researchers working on neural models of cognitive tasks have taken representational issues especially seriously. It is a great challenge to work out how to effectively utilize the potentially vast computational power of the human brain, in the face of the unreliability and slowness of individual neurons. It does seem clear that any neural computational scheme that performs at near-human levels must make use of neural representations that make it possible to perform relatively high-level tasks in just a few 'steps' of neural computation. There is simply not enough time for many steps of computation in the time that people take to act when having a conversation, playing sports, etc.

PROPERTIES OF DISTRIBUTED REPRESENTATIONS

In distributed representations, concepts are represented by patterns of activity over a collection of neurons. This contrasts with local representations, in which each neuron represents a single concept, and each concept is represented by a single neuron. Researchers generally accept that a neural representation with the following two properties is a distributed representation (e.g., Hinton *et al.*, 1986):

- Each concept (e.g., an entity, token, or value) is represented by more than one neuron (by a pattern of neural activity in which more than one neuron is active).
- Each neuron participates in the representation of more than one concept.

Another equivalent property is that in a distributed representation one cannot interpret the meaning of activity on a single neuron in isolation: the meaning of activity on any particular neuron is dependent on the activity in other neurons (Thorpe, 1995).

The distinction between local and distributed representations is not always as clear as it might initially seem (see van Gelder, 1991, for discussion). Is a standard eight-bit binary encoding for numbers between zero and 255 a local or a distributed code? At the level of numbers each 'concept' (number) is represented by multiple 'neurons' (bits), and each neuron participates in representing many concepts. However, this encoding can also be viewed as a local representation of powers of two: 1, 2, 4, 8, etc. This is an example of where a representation is distributed at one level of interpretation but in which individual neurons represent finer-grained features or 'micro-features' in a localist fashion.

Similarity and Generalization

The similarity of two patterns is very important in distributed representations. Similarity of two patterns is usually computed as their dot-product or cosine. The cosine is preferable when total neural activity differs significantly across patterns.

Neural networks typically respond in a similar manner to similar inputs. Distributed representations are generally designed to take advantage of this; inputs that should result in similar responses are represented by similar activation patterns, and inputs that should result in different responses are represented by quite different activation patterns. One sees the same principle at work when a network develops distributed representations for itself, as in Hinton's (1986) family-tree learning network. Note that domain similarity can depend upon the task to be performed: two inputs that can be treated as the same for one task may need to be treated as different for another task. At the most basic level, learning is about determining which similarities and differences between inputs are and are not important for a particular task. A network that is either provided with, or that can learn, a good set of features to represent its input will often be able to generalize well from limited data.

Superposition, Multiple Concepts, Interference, and Ghosting

Using distributed representations, multiple concepts can be represented at the same time on the same set of neurons by superimposing their patterns together. Mathematically, superposition means some sort of addition, possibly followed by a thresholding or a normalizing operation.

How can we tell whether a particular pattern **x** is part of a superposition of patterns, denoted by **y**? The simplest, and most commonly used way is to check whether the similarity between **x** and **y**, exceeds some predetermined threshold. Another technique is discussed later in this article.

As more patterns are superimposed together, it can become difficult to tell whether or not a particular pattern is part of the superposition. When patterns appear in the result of a superposition, but were in fact not part of the superposition, this is known as interference or ghosting. The number of patterns that can be superimposed before ghosting becomes a problem depends on several aspects of the representation: the number of neurons (more neurons means less ghosting); the number of distinct patterns (more patterns means more ghosting); the degree of noise tolerance we wish the representation to have (higher noise tolerance means more ghosting); and the density of patterns.

Density and Sparse Representations

The total level of activity in a pattern is referred to as the sparseness, or alternatively, the density, of the representation. For a binary representation, this is the fraction of neurons that are active. For a continuous representation the Euclidean length of vectors is often used as the density. Patterns chosen to represent concepts in a model are often restricted to have the same density. Density of patterns in a representation can be tuned to optimize the properties of the representation, for example minimizing ghosting and maximizing capacity.

Sparse binary distributed representations have the attractive property that they can be superimposed using the binary-OR rule with little ghosting until the number of superimposed patterns becomes large. Sparse representations are also of interest because neurophysiological evidence suggests that representations used in the brain are quite sparse.

Advantages of Distributed Representations

Distributed representations are often held to have many advantages compared to symbolic (e.g., Lisp data structures) and local representations. The most common and important are as follows:

1. *Representational efficiency.* Distributed representations form a more efficient code than localist representations, provided that only a few concepts are to be represented at once. A localist representation using n neurons can represent just n different entities. A distributed representation using n binary neurons can represent up to 2^n different entities (using all possible patterns of zeros and ones).
2. *Mapping efficiency.* A microfeature-based distributed representation often allows a simple mapping (that uses few connections or weights) to solve a task. For example, suppose we wish to classify 100 different colored shapes as to whether or not they are yellow. Using a localist representation, this would require a connection from each neuron representing a yellow shape to the output neuron. Using a feature-based distributed representation, all that is required is a single connection from the neuron encoding 'yellow' to the output neuron. In general, a mapping that can be encoded in relatively few weights operating on a feature-based distributed representation can be learned from a relatively small training sample and will generalize correctly to examples not in the training set.

3. *Continuity* (in the mathematical sense). Representing concepts in continuous vector spaces allows powerful gradient-based learning techniques such as back propagation to be applied to many problems, including ones that might otherwise be seen as discrete symbolic problems.
4. *Soft capacity limits and graceful degradation*. Distributed representations typically have soft limits on how many concepts can be represented simultaneously before ghosting or interference becomes a serious problem. Also, the performance of neural networks using distributed representations tends to degrade gracefully in response to damage to the network or noise added to activations. Many researchers find these properties compellingly similar to performance observed in people.

On the other hand, local representations are far simpler to understand, implement, interpret, and work with. If distributed representations do not provide significant advantages for a particular application, it may be more appropriate to use a local representation. Page (2000) argues forcefully that this is the case in many cognitive modeling applications.

Major Areas of Research in Distributed Representations

The major areas of research in distributed representations are: (1) techniques for representing data more complex than simple tokens, such as data with compositional structure, continuous data, probability distributions; (2) properties of representational schemes, such as capacity, scaling, reconstruction accuracy; and (3) techniques for learning distributed representations.

CODING PROBLEMS AND TECHNIQUES

The Binding Problem

The problem of keeping track of which features or components belong to which objects is known as the 'binding problem'. Consider trying to represent the presence of several colored shapes. A simple local representation could have one set of features for color and another set of units for shape. A red circle would be represented by activity on the red unit and on the circle unit (Figure 1(a)). However, when we try to represent two colored objects simultaneously, we encounter a binding problem: does red + blue + circle + triangle mean red circle and blue triangle or blue circle and red triangle (Figure 1(b))? The binding problem can arise with local or distributed representations when features or components are represented independently and can belong to different objects.

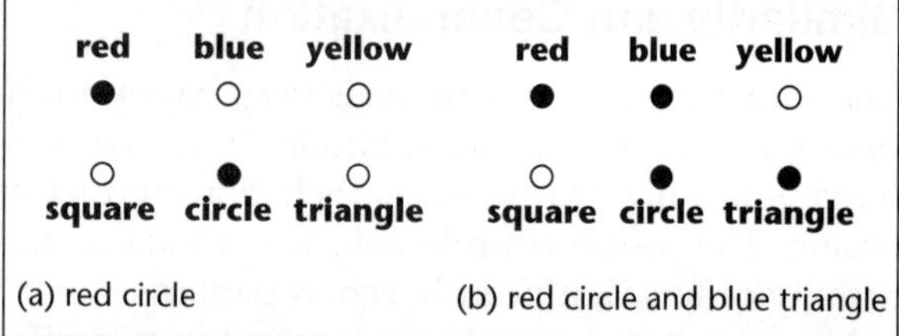

Figure 1. The binding problem: representations of multiple objects on independent feature sets can lose information about which features belong to which objects.

Conjunctive Coding

Conjunctive codes are a general approach to solving the binding problem in neural representations. A simple local conjunctive code has one neuron for every possible combination (conjunction) of a single value from each feature set. For example, to represent colored shapes with 10 possible colors and five possible shapes we would use 50 units. Figure 2 shows an example of this with just three colors and three shapes. This technique can also be used when the value on each feature dimension is encoded with a distributed representation: form the outer product of the patterns for each feature dimension. This outer-product operation underlies such well known associative memory schemes as the Willshaw net (Willshaw, 1989) and the Hopfield net (Hopfield, 1982).

This simple kind of conjunctive code is an example of how a neural code for a composite object can be computed in a systematic manner from the codes for its constituent parts or features. This is an important aspect of Hinton's influential idea of reduced representations, discussed later in this article.

There are two serious problems that occur with simple outer-product conjunctive codes. The first is inefficiency: resource requirements grow exponentially with the number of feature classes. With k feature classes (e.g., color, shape, size, etc.) each having n possible values, the total number of neurons required for a complete conjunctive code is n^k. The informational efficiency of such a code is very poor if only a handful of objects are to be represented simultaneously. Another way of looking at this is that the same set of neurons could represent far more information about each object, or about more objects, if a more efficient

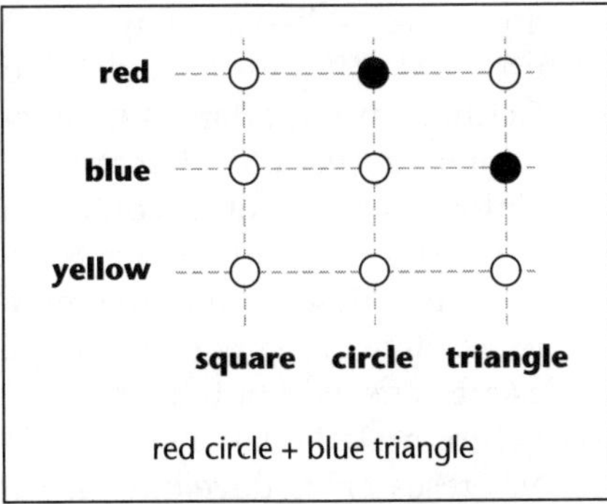

Figure 2. A conjunctive code representing two objects simultaneously.

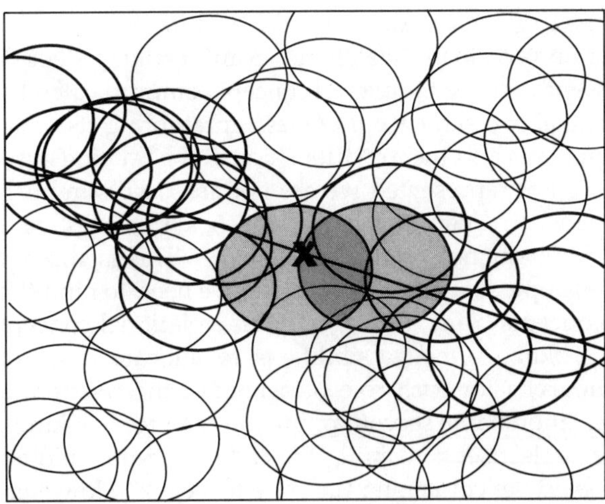

Figure 3. Coarse coding of two-dimensional positions. Each neuron is represented by an area showing its receptive field. Neurons that respond to the point X have gray receptive fields. Neurons that respond to some point along the solid line are shown with bolder receptive fields.

code were used. The second problem is that the conjunctive code can hide what makes objects similar, which places high resource demands on mapping and can make learning difficult. If the independent features of the input are useful for solving the problem or constructing the mapping, learning and mapping are simpler when input to a network represents each feature dimension independently rather than in a conjunctive manner.

Coarse Coding

One of the apparent paradoxes of neural codes is that a stimulus or entity can be represented more accurately by a collection of neurons with broad or coarse response functions than by a collection of neurons with more finely-tuned respone functions. This applies to the representation of both discrete and continuous stimuli or entities.

For representing continuous data, such as a position in space (with two or more dimensions), this means that one can increase the overall accuracy of coding schemes by decreasing the accuracy with which individual neurons represent a data point (i.e., by making the neurons code the data more coarsely). Consider a simple representation for a point in space, where each of n neurons has a randomly chosen centre (in input space), and responds to a data point within a radius r of its centre (its *receptive field*). A two-dimensional version of this is shown in Figure 3. Hinton *et al.* (1986) show that for neurons representing regions in a k-dimensional space, the inaccuracy of a distributed representation like this is proportional to $1/r^{k-1}$. For example, in a six-dimensional space, doubling the radius of each receptive field reduces the inaccuracy of the representation by a factor of 32 on each dimension.

When we look at the system from an information-theoretic viewpoint, it is not surprising that coarsening the receptive fields increases rather than decreases overall accuracy. With very small receptive fields, each neuron is active for only a small fraction of possible stimuli and thus each neuron carries very little information. With larger receptive fields, each neuron is active for a larger fraction of possible stimuli and thus carry more information.

Coarse coding can also be applied to representing discrete entities. The principle is the same: a single neuron is active for a moderately large proportion of possible stimuli, and different neurons cover dissimilar subsets of stimuli. For example, a distributed representation for animals based on semantic features might have one unit responding to animals that are small-to-medium in size AND moderate-to-high in ferocity; another unit might respond to animals that are medium-to-large in size AND moderate-to-high in ferocity, etc.

REPRESENTATIONS FOR INFORMATION WITH COMPLEX STRUCTURE

Many cognitive tasks involve understanding and processing information that has a very complex structure. The concepts communicated by human language almost always have a compositional nature, and quite often, the composition is hierarchical. In the sentence 'Joan watched Sam cook the eggs', there are a number of different concepts involved: 'Joan', 'watching', 'Sam', 'cooking', 'eggs'. Representing the overall concept communicated by the above sentence involves representing its components and their relationships. If the relationships are not represented we are not able to distinguish between 'Joan watched Sam cook the eggs' and 'Sam watched Joan cook the eggs'. This is another example of a binding problem; we need to bind the concepts 'Sam' and 'eggs' to the relational concept 'cooking'. One possibility is to allocate a set of neurons for each role or aspect of the relation: a relation name set containing the code for 'cooking', a subject set containing the code for 'Sam', and an object set containing the code for 'eggs'. However, there is also a hierarchical, or recursive, aspect to the sentence 'Joan watched Sam cook the eggs'. Once we have constructed an appropriate representation for 'Sam cooking eggs' we then need to bind it and 'Joan' to the relational concept 'watching'. An approach utilizing a fixed-size set of neurons for each role prevents any neat solution for representing the recursive aspects of the whole problem because the number of neurons in the representation of a relation is larger than the number of neurons allocated to represent the filler of a role.

The compositional and recursive nature of concepts is not limited to obviously linguistic tasks, but is widespread in human cognition; visual scene understanding, and analogy recall and matching are two examples. Although representing hierarchical compositional structures in the activations and/or weights of a neural network is not easy, it is important as it could eventually allow the application of powerful neural-network learning techniques to difficult problems such as language understanding and acquisition.

Reduced Descriptions

Hinton (1990) introduced the idea of 'reduced description' as a way of representing hierarchical compositional structure in fixed-size distributed representations. The basic idea is that a relation can be represented in two different ways: (1) as an expanded representation in which the fillers of each role are represented on separate groups of neurons (e.g., a relation with two arguments and a relation name could use three groups of n neurons); and (2) as a compressed or reduced description over just n neurons. Since the reduced description for a relation occupies just n neurons, it can be used as a filler in some other relation, which in turn could have a reduced description, and so on, allowing arbitrary levels of nesting. A reduced representation behaves like a pointer in symbolic data structures in that it gives a way of referring to another structure. An important difference is that unlike a pointer, a reduced description should carry some information about its contents that can be accessed without expanding the reduced description into a full description. This allows processing to be sensitive to components nested within reduced descriptions without having to unpack multiple levels of nested relations.

In the 1990s researchers developed a number of concrete neural schemes for implementing reduced descriptions. Many use some form of conjunctive role-filler bindings in which both the roles and fillers of a relation are represented with distributed patterns. Many of these schemes differ mainly in the binding operation they use; it turns out that there are many alternatives to a straight outer-product for forming a conjunctive code of two patterns.

RAAMs

Pollack (1990) used a bottleneck auto-encoder network to learn reduced descriptions for relations, which he called 'recursive auto-associative memory' (RAAM). A full, expanded relation was represented across the input units of the network: each filler on one group of input units, and possibly the relation name on a final group of input units. The hidden layer was the same size as one of the groups of input units and was intended to contain a reduced description of the full relation. The network learned, using back propagation, to compress the full relation down to a reduced description, which could then be expanded back out to a full relation. During learning the network had to discover simultaneously how to create and how to decode reduced descriptions for relations. Pollack showed that a network like this could reliably learn to represent hierarchically structured relations that were several levels deep.

Tensor Product Representations

Smolensky (1990) proposed a tensor-product formalism for representing recursive structure. This extends the outer-product role-filler binding operation to higher dimensions. For example, suppose a role and a filler are each represented by a pattern over a line of n neurons. Then their outer-product binding is a pattern over a square of n^2 neurons. This can be bound with another role vector (again a pattern over n neurons) by forming the tensor product, which in this situation is a pattern over a cube of n^3 neurons. This can be taken to arbitrarily deep levels of nesting. However, the number of neurons required increases exponentially with the depth of conceptual nesting.

Holographic Reduced Representations

Holographic reduced representations (HRRs) (Plate, 1995), are a role-filler binding scheme for recursive compositional structure based on conjunctive coding implemented using circular convolution. In HRRs, roles, fillers, labels, and entire relations are all represented as patterns over n neurons. Circular convolution is defined as $\mathbf{z} = \mathbf{x} \otimes \mathbf{y}$, where $z_i = \sum_{k=0}^{n-1} x_k y_{(i-k)\bmod n}$ (where n is the length of the vectors). Circular convolution is a conjunctive code that keeps dimensionality constant: given role and filler patterns over n neurons each, their circular convolution is also a pattern over n neurons. This makes it simple to build recursive structures. The encoding of a relation is the superposition of role-filler bindings (and possibly a relation label) and thus is also a pattern over n neurons and is easily used as a filler in another relation. For example, a reduced representation for the sentence 'Joan watched Sam cooking eggs' can be constructed as follows:

$$\mathbf{S_1} = \mathbf{cook} + \mathbf{cook}_{\mathrm{agt}} \otimes \mathbf{Sam} + \mathbf{cook}_{\mathrm{obj}} \otimes \mathbf{eggs} \tag{1}$$

$$\mathbf{S_2} = \mathbf{watch} + \mathbf{watch}_{\mathrm{agt}} \otimes \mathbf{Joan} + \mathbf{watch}_{\mathrm{obj}} \otimes \mathbf{S_1} \tag{2}$$

where all the variables are patterns over n neurons (**cook** and **watch** are relation labels; **Sam, eggs**, etc. are patterns representing entities; and $\mathbf{cook}_{\mathrm{agt}}$, etc. are patterns representing the roles of the relations). (*See* **Convolution-based Memory Models**)

A filler of a role in a relation in a reduced representation may be decoded by convolving with the approximate inverse of the role pattern. The approximate inverse of **x**, denoted by $\mathbf{x}^{\mathrm{T}}$, is a simple permutation of elements: $x_i^{\mathrm{T}} = x_{-i \bmod n}$, and has the property that $\mathbf{x}^{\mathrm{T}} \otimes (\mathbf{x} \otimes \mathbf{y}) \approx \mathbf{y}$.

For example, to recover the filler of the cook-agent role in $\mathbf{S_1}$, we compute $\mathbf{S_1} \otimes \mathbf{cook}_{\mathrm{agt}}^{\mathrm{T}}$, which results in the vector **Sam** + **noise**, because

$$\begin{aligned}\mathbf{cook}_{\mathrm{agt}}^{\mathrm{T}} \otimes \mathbf{S_1} &= \mathbf{cook}_{\mathrm{agt}}^{\mathrm{T}} \otimes (\mathbf{cook} + \mathbf{cook}_{\mathrm{agt}} \\ &\quad \otimes \mathbf{Sam} + \mathbf{cook}_{\mathrm{obj}} \otimes \mathbf{eggs}) \\ &= \mathbf{cook}_{\mathrm{agt}}^{\mathrm{T}} \otimes \mathbf{cook} + \mathbf{cook}_{\mathrm{agt}}^{\mathrm{T}} \\ &\quad \otimes \mathbf{cook}_{\mathrm{agt}} \otimes \mathbf{Sam} + \mathbf{cook}_{\mathrm{agt}}^{\mathrm{T}} \\ &\quad \otimes \mathbf{cook}_{\mathrm{obj}} \otimes \mathbf{eggs} \\ &= \mathbf{noise} + \mathbf{Sam} + \mathbf{noise}\end{aligned}$$

In order to recognize **Sam** + **noise** as the pattern **Sam**, we must pass it through an autoassociative clean-up memory that can take **Sam** + **noise** as input and return the pattern **Sam**. All potential decoding targets, such as lower-level patterns like **cook**, $\mathbf{cook}_{\mathrm{agt}}$, **Sam**, and higher-level patterns representing chunks, such as $\mathbf{S_1}$ and $\mathbf{S_2}$, must be stored in this long-term clean-up memory. A clean-up memory is also necessary to identify when a decoding target is not present. Without a clean-up memory it would be impossible to tell whether or not there was anything bound to $\mathbf{watch}_{\mathrm{agt}}$ in $\mathbf{S_1}$ because $\mathbf{S_1} \otimes \mathbf{watch}_{\mathrm{agt}}^{\mathrm{T}}$ is a pattern with similar statistical properties to any other pattern. With a clean-up memory containing all potential decoding targets, $\mathbf{S_1} \otimes \mathbf{watch}_{\mathrm{agt}}^{\mathrm{T}}$ can be identified as noise because it almost certainly will not be similar to anything in the clean-up memory.

Binary Spatter Codes

Kanerva's (1996) 'binary spatter code' is a scheme for encoding complex compositional structures in binary distributed representations. Binary spatter codes use binary vectors as patterns, element-wise exclusive-OR for encoding and decoding bindings, and a thresholded sum for superposition. They have similar properties to HRRs.

Holistic Processing

A major reason for interest in distributed representations of a complex structure is the potential for performing structure-sensitive processing without having to unpack hierarchical structures.

Determining similarity is one of the simplest types of processing. Plate (2000) shows that the dot-product of HRRs reflects both superficial similarity (similarity of components) and structural similarity (similarity of structural arrangement of components). The dot-product of HRRs composed of similar entities is higher if the entities are arranged in an analogical (isomorphic) structure. This means that HRRs can be used for fast (but approximate) detection of structural similarity. HRR computations can also be used to rapidly but imperfectly identify corresponding entities in isomorphisms (Plate, 2000; Eliasmith and Thagard, 2001).

Various authors have demonstrated that a variety of structure-sensitive manipulations can be performed on distributed representations without unpacking them. Pollack (1990) trained a feedforward network to transform reduced descriptions for propositions like (LOVED X Y) to ones for

(LOVED Y X) where the reduced descriptions were found by a RAAM. Chalmers (1990) trained a feedforward network to transform reduced descriptions of simple passive sentences to reduced descriptions of active sentences, where the reduced descriptions were found by a RAAM. Niklasson and van Gelder (1994) trained a feedforward network to do material conditional inference, and its reverse, on reduced descriptions found by a RAAM. Legendre *et al.* (1991) showed how tensor product representations for active sentences could be transformed to ones for passive sentences (and vice-versa) by a pre-calculated linear transform. Neumann (2001) trained networks to perform holistic transformations on a variety of representations, including RAAMs, HRRs, and binary spatter codes.

INTERPRETATION OF DISTRIBUTED REPRESENTATIONS

It is straightforward to tell what is represented in a model that uses symbolic or local neural representations. This is not the case with models that use distributed representations. It is important to note that for the purposes of processing the information present in a distributed representation, it is usually NOT necessary to identify what is represented in terms easily understandable to people. Indeed, the whole point of having a distributed representation is that it makes further processing simpler than performing the same computations on a more easily interpretable representation. Interpretation of a distributed representation is typically necessary either when outputs need to be computed in a readily interpretable form or when a person wants to gain insight into the internal workings of a model.

There are two quite different senses in which a distributed representation can be interpreted: (1) determine which items are represented in a pattern of activation, where patterns for individual items are known; and (2) determine what features a network has learned to use to represent a set of items.

Algorithms for Determining Which Items are Represented

A simple and common algorithm for determining the items present in a distributed pattern of activation is to compute the dot-product of each item with the pattern of activation. A feedforward neural network with a localist output representation performs this computation in its final layer of weights. Often it is useful to apply a thresholding or a winner-take-all function to the outputs to cut down the noise.

It is also possible to take an inferential approach to identifying the items present in a distributed code. The idea is to find the best explanation for the observed pattern of activities, in terms of the items that could be present. Zemel *et al.* (1998) do this in analyzing the potential of sparse distributed representations for representing probability distributions. This requires three pieces of knowledge, which can be combined using Bayes' rule: (1) the set of all possible probability distributions that could be represented, and the respective prior probability of each probability distribution (prior to knowledge of current activities); (2) the currently observed pattern of activity; and (3) for each possible probability distribution, the probability that it would generate the currently observed pattern of activity. In Zemel *et al.*'s approach, the probability distribution represented by a pattern of activity is the one with the maximum a-posteriori (MAP) probability (the posterior probability of a distribution is the product of its prior probability and the probability that it would have generated the currently observed pattern of activity).

Understanding Learned Representations

Principal components analysis is often used to gain understanding into a learned distributed representation. Elman (1991) trained a network to predict the next word in sentences generated from a simple recursive English-like language. The predictions made by the network indicated that it had managed to learn something about the recursive nature of the language. Elman used principal components analysis to gain some understanding of how the network represented state (i.e., its memory of what it had seen so far); he showed that a particular phrase caused a similar shape of trajectory through the principal component space independent of its level of embedding, and that level of embedding determined the position of the trajectory.

CONCLUSION

Distributed representations offer powerful representational principles that can be used in neural network approaches to learning and to modeling human cognitive performance. Research continues on three main aspects of distributed representation: schemes for representing data; properties of representational schemes; and ways of learning distributed representations. The ability to learn and use distributed representations of concepts, and to compose complex data structures out of these representations, offers the potential of building general-purpose computers that combine the power of symbolic manipulation with the robustness, learning, and generalization ability of neural networks.

References

Chalmers DJ (1990) Syntactic transformations on distributed representations. *Connection Science* **2**(1–2): 53–62.

Eliasmith E and Thagard P (2001) Integrating structure and meaning: a distributed model of analogical mapping. *Cognitive Science* **25**: 245–286.

Elman JL (1991) Distributed representations, simple recurrent networks, and grammatical structure. *Machine Learning* **7**: 194–220.

van Gelder T (1991) What is the 'D' in 'PDP'? An overview of the concept of distribution. In: Stich S, Rumelhart D and Ramsey W (eds) *Philosophy and Connectionist Theory*. Hillsdale, NJ: Lawrence Erlbaum.

Hinton GE (1986) Learning distributed representations of concepts. *Proceedings of the Eighth Annual Conference of the Cognitive Sciences*, pp. 1–12. Hillsdale, NJ: Lawrence Erlbaum.

Hinton GE (1990) Mapping part–whole hierarchies into connectionist networks. *Artificial Intelligence* **46**: 47–75.

Hinton GE, McClelland JL and Rumelhart DE (1986) Distributed representations. In: Rumelhart DE and McClelland JL and the PDP research group (eds) *Parallel Distributed Processing: Explorations in the Microstructure of Cognition*, vol. 1, pp. 77–109. Cambridge, MA: MIT Press.

Hopfield J (1982) Neural networks and physical systems with emergent collective computational abilities. *Proceedings of the National Academy of Sciences of the USA* **79**: 2554–2558.

Kanerva P (1996) Binary spatter-coding of ordered k-tuples. In: von der Malsburg C, von Seelen W, Vorbruggen J and Sendhoff B (eds) *Artificial Neural Networks – ICANN Proceedings*, vol. 1112 of Lecture Notes in Computer Science, pp. 869–873. Berlin, Germany: Springer.

Legendre G, Miyata Y and Smolensky P (1991) Distributed recursive structure processing. In Touretzky DS and Lippman R (eds) *Advances in Neural Information Processing Systems 3*, pp. 591–597. San Mateo, CA: Morgan Kaufmann.

Neumann J (2001) *Holistic Processing of Hierarchical Structures in Connectionist Networks*. PhD thesis, University of Edinburgh. [http://www.cogsci.ed.ac.uk/~jne/holistic_trafo/thesis.pdf.]

Niklasson LF and van Gelder T (1994) Can Connectionist Models Exhibit Non-Classical Structure Sensitivity? *Proceedings of the Sixteenth Annual Conference of The Cognitive Science Society*, pp. 664–669. Hillsdale, NJ: Lawrence Erlbaum.

Page (2000) Connectionist modeling in psychology: a localist manifesto. *Behavioral and Brain Sciences* **23**: 443–512.

Plate TA (1995) Holographic reduced representations. *IEEE Transactions on Neural Networks* **6**(3): 623–641.

Plate TA (2000) Structured operations with vector representations. *Expert Systems: The International Journal of Knowledge Engineering and Neural Networks: Special Issue on Connectionist Symbol Processing* **17**(1): 29–40.

Pollack JB (1990) Recursive distributed representations. *Artificial Intelligence* **46**(1–2): 77–105.

Smolensky P (1990) Tensor product variable binding and the representation of symbolic structures in connectionist systems. *Artificial Intelligence* **46**(1–2): 159–216.

Thorpe S (1995) Localized versus distributed representations. In: Arbib MA (ed.) *The Handbook of Brain Theory and Neural Networks*. Cambridge, MA: MIT Press.

Willshaw D (1989) Holography, associative memory, and inductive generalization. In: Hinton GE and Anderson JA (eds) *Parallel Models of Associative Memory* (updated edition), pp. 99–127. Hillsdale, NJ: Lawrence Erlbaum.

Zemel R, Dayan P and Pouget A (1998) Probabilistic Interpretation of Population Codes. *Neural Computation* **10**(2): 403–430.

Further Reading

Baldi P and Hornik K (1989) Neural networks and principal component analysis: learning from examples without local minima. *Neural Networks* **2**: 53–58.

Baum EB, Moody J and Wilczek F (1988) Internal representations for associative memory. *Biological Cybernetics* **59**: 217–228.

Bourlard H and Kamp Y (1988) Auto-association by multilayer perceptrons and singular value decomposition. *Biological Cybernetics* **59**: 291–294.

Deerwester S, Dumais ST, Furnas GW, Landauer TK and Harshman R (1990) Indexing by latent semantic analysis. *Journal of the American Society For Information Science* **41**: 391–407.

Halford G, Wilson WH and Phillips S (1998) Processing capacity defined by relational complexity: implications for comparative, developmental, and cognitive psychology. *Behavioral and Brain Sciences* **21**(6): 803–831.

Hinton GE, Dayan P, Frey BJ and Neal R (1995) The wake–sleep algorithm for unsupervised neural networks. *Science* **268**: 1158–1161.

Hinton GE and Ghahramani Z (1997) Generative models for discovering sparse distributed representations. *Philosophical Transactions of the Royal Society of London* **352**: 1177–1190.

Hummel JE and Holyoak KJ (1997) Distributed representations of structure: a theory of analogical access and mapping. *Psychological Review* **104**(3): 427–466.

LeCun Y, Boser B, Denker JS *et al.* (1989) Backpropagation applied to handwritten zip code recognition. *Neural Computation* **1**(4): 541–551.

Olshausen BA and Field DJ (1996) Emergence of simple-cell receptive field properties by learning a sparse code for natural images. *Nature* **381**: 607–609.

Rachkovskij DA (2001) Representation and processing of structures with binary sparse distributed codes. *IEEE Transactions on Knowledge and Data Engineering* **13**(2): 261–276.

Rumelhart DE and McClelland JL (1986) On learning the past tenses of English verbs. In: McClelland JL, Rumelhart DE and the PDP research group (eds) *Parallel Distributed Processing: Explorations in the Microstructure of Cognition*, vol. 2, pp. 216–271. Cambridge, MA: MIT Press.

Zemel RS and Hinton GE (1995) Learning population codes by minimizing description length. *Neural Computation* **7**(3): 549–564.

Down Syndrome

Introductory article

Ira T Lott, University of California, Irvine, California, USA

CONTENTS

Down syndrome is a chromosomal disorder characterized by specific dysmorphic features, organ malformations, and variable but often severe learning difficulties.

INTRODUCTION

Down syndrome (DS) is the most common known form of developmental disability. First described by Langdon Down in the *London Hospital Reports* in 1886, the disorder was given scientific focus by Lejeune who identified the chromosomal abnormality in 1960. Despite the mental and physical handicap invariably associated with the disorder, people with Down syndrome have been prominently represented in many areas of daily life, including sports and the entertainment industry.

INCIDENCE AND GENETIC BASIS

Down syndrome (DS) is a genetic disorder that occurs once in every 600–1000 live births and affects all races and both sexes equally. The presence of additional chromosomal material has been shown to be responsible for the condition. The vast majority of individuals with DS have received the extra chromosomal material from an error involving the separation of chromosomes during the cell divisional process of meiosis. During normal meiosis an originator cell containing two copies of each of the 23 chromosomes is divided into daughter cells, each containing a single copy of each chromosome. In DS an error in this process, nondisjunction, results in a sperm or egg cell containing two copies of chromosome 21 instead of the usual single copy. When this sperm or egg cell pairs at fertilization with the other normal complement containing a single copy of chromosome 21, the resulting fertilized egg has three copies of chromosome 21 instead of the usual two – hence the alternative name for the syndrome, 'trisomy 21'. Nondisjunction is much more likely to occur in maternal than paternal meiosis. The actual cause of this error is unknown but it is more likely to occur in older women.

In about 4% of cases the extra copy of chromosome 21 is the result of the attachment of chromosome 21 onto another chromosome. This type of DS is called 'translocation DS' and unlike the disjunction error it occurs equally in males or females and is not age-dependent. Interestingly, one cannot determine by appearance or physical examination whether the condition is a result of a nondisjunction or a translocation. The third mechanism discovered for DS involves a genetic error in which there is a coexistence of normal and trisomic cell lines. This type of DS is called 'mosaic'. It has been argued that people with the mosaic form of

DS have a less severe expression of the disorder since not all of the cells are trisomic. However, identifying mosaic DS based on its expression is difficult for even the most experienced clinicians.

It is not clear how the extra genetic material in DS accounts for the spectrum of clinical abnormalities in the disorder. The imbalance in genetic material appears to cause a pertubation in the timing of developmental sequences which eventually results in the clinical expression – phenotype – of the condition. Some individuals diagnosed with DS have only a tiny portion of chromosome 21 existing in triplicate. This small portion is enough to cause the phenotypic expression common to DS, and has thus been identified as the Down syndrome critical region.

CLINICAL SIGNS AND SYMPTOMS

Every physical feature of DS can be seen in a small percentage of the normal population. It is only when the features are combined into a single individual that one has the appearance or the phenotype of DS. Although Langdon Down became confused about the cause of DS, he was the first to describe the physical characteristics of the disorder. The middle portion of the face is flat. There is an upward slant of the eye sockets (or palpebral fissures). The irises are speckled (Brushfield spots). The hands and feet are short and broad. There is an incurving of the fifth finger (clinodactyly) and there is a separation between the first and second toe. The patterns of the skin creases in the palm impart a particular appearance to the hand of a person with DS, and the major crease across the palm harkens back to an earlier evolutionary appearance (simian crease). The tongue is thickened and this is one of the factors contributing to a lack of clarity in the speech of people with DS.

People with DS are subject to certain illnesses involving organ systems other than the brain. There is a high incidence of thyroid gland dysfunction, which can produce thyroid hormone excess or deficiency (the latter is more common). The immune cells of the body react abnormally in DS and this causes a predilection to infections such as pneumonia and hepatitis. The heart is malformed in DS about 50% of the time and the most common lesion is a defect between the atrium and the ventricle. In some cases surgical repair of the lesion is required, but in many instances the effect on cardiac function is minimal and further intervention is not necessary. The immune problem that underlies the tendency to infection may also contribute to an increased incidence of leukemia in DS. Most cases of leukemia in DS occur in children less than 10 years old.

The ear canals in DS are small and misshapen. As a result, children with DS have frequent ear infections and run the risk of hearing loss along with other complications. A hearing loss that would be a mild handicap for a normal child can be devastating to a school-age child with DS who is struggling against the cognitive disorder associated with the condition. Tongue size and enlargement of the tonsils may provoke periodic cessation of breathing during sleep in DS (sleep apnea) and result in fatigue as well as poor mental performance the next day.

Whether people with DS age prematurely has not been decided with certainty. Skin laxity, adult-onset diabetes, cataracts, and orthopedic difficulties would argue for a precocious biological aging. However, people with DS seem to have a low incidence of atherosclerotic heart disease and hypertension – factors that distinguish them from aging people in the general population. The tendency to Alzheimer disease is a special circumstance (see below).

COURSE OF DS

The infant with DS is often in a vulnerable state. Muscle tone is lax and motor development is often delayed. Infection always looms as a possibility. Congenital heart disease, immune dysfunction, thyroid abnormalities, and the tendency to leukemia are always present. Given so many potential obstacles, it is reassuring to see that so many children with DS grow up and thrive. Particular attention to people with DS is required across the life span, and a universal set of guidelines for preventive intervention is increasingly being applied within the medical profession.

NEURAL CORRELATES

The experienced neuropathologist can often make a diagnosis of Down syndrome by the appearance of the brain. Certain neuroimaging procedures afford the same opportunity. In DS, the brain is slightly underweight, foreshortened front to back, and has a steep decline to the posterior regions.

Muscle tone is abnormally low in infants with DS and this is a sentinel sign of the disorder at birth. The ligaments are lax and this provides a problem in stability for younger children with DS. More worrisome is the ligamentous laxity that occurs in the bones at the top of the spinal column. This atlantoaxial junction is subject to more movement in DS than seen in the general population and occasionally the spinal cord will be compressed by the vertebrae (atlantoaxial dislocation). For this reason, people with DS should have an X-ray examination of the spinal area before participating in certain sports.

Although a cognitive deficit is universally seen in DS, there is a broad range of potential abilities in the disorder. This variation suggests that we do not yet understand the real potential of people with this condition or how to account for the spread of abilities. Social abilities often allow the child and adult to function in a more adaptive manner than would be predicted on the basis of tests measuring general intelligence.

One area of challenge for people with DS about which there is universal agreement is the linguistic system. The typical person with DS has more difficulties with language than would be predicted on the basis of overall cognitive functioning. Verbal memory is a particular deficit. People with DS can not only recall fewer digits in a span than can people in the general population, but they also can recall fewer digits than intelligence-matched control subjects who do not have DS. The rate of acquiring new vocabulary is slower in DS than in children of similar mental age.

Epileptic seizures may occur at two separate points in the life span of an individual with DS. In infancy, seizures often take the form of sudden jerking movements called 'infantile spasms'. In later life the onset of seizures may herald the decline associated with Alzheimer disease. Since the early description of DS, personality characteristics have included a strong power of imitation, good sense of humor, and a tendency towards obstinacy. As a rule, people with DS appear to be outgoing, affectionate, with social quotients exceeding intelligence quotient (IQ) measures between 4 and 17 years of age. Older age in DS often brings bouts of depression.

LINKS WITH ALZHEIMER DISEASE

Alzheimer disease (AD) is the most commonly recognized form of dementia and is manifested by a progressive and ultimately fatal deterioration in brain function and capacity. From the late 1940s, researchers have found a curious association between DS and AD. Almost every brain examined from individuals with DS over the age of 40 years shows microscopic signs of the disorder. Often the primitive lesions of AD can be seen in brain tissue from children and young adults with DS. Despite the ubiquity of the microscopic signs of AD, not every person with DS develops clinical symptoms of the disorder. Prevalence rates vary widely, but average about 25%. The early symptoms of AD include change in personality, loss of ability to carry out complex daily skills, and (as mentioned above) the onset of epileptic seizures. As the decline progresses, awareness of the environment is lost and the ability to acquire recent memories ceases. In DS the disease may run a fatal course within 5 years. One of the factors imparting a predisposition to AD in DS relates to a gene located on chromosome 21 which is triplicated in DS. This gene is responsible for producing an overabundance of the peptide amyloid. Amyloid deposition in brain appears to have a key role in the pathogenesis of AD. (*See* **Alzheimer Disease**)

TREATMENT AND GENETIC COUNSELING

There is no medical cure for DS but there are many opportunities to prevent complications in the disorder and to provide a nurturing environment for the realization of full potential for people with this syndrome. A signal advance in modern society has been the provision of opportunities for people with DS (and other forms of learning difficulties) to live and work within the community. No longer are people with DS sent to institutions to live their lives apart from society. With increasing vigilance among medical professionals and the general willingness to intervene on behalf of the child with DS, the life span for people with DS has been greatly extended.

Genetic counseling for DS shares many of the same principles that are applied to other genetic conditions. The genetic counselor's role includes:

- interpreting recurrence risk for families where a DS birth has occurred
- assisting other medical professionals in explaining the diagnosis of DS, patient management, and treatment options
- facilitating psychological assessments where indicated for both patient and family members
- providing information to families about support groups.

The recurrence risk for DS is dependent on the genetic form of the disorder. The disclosure of full information on genetic risks often involves 'cytogenetic' testing of the patient and possibly family members. The recurrence risk to a couple with the classical form of trisomy 21 is about 1%, but there is a strong maternal age dependency which may increase this risk and requires explanation for couples engaging in family planning. In the translocation form of DS, the risk is dependent on whether the translocation was inherited by a parent or occurred for the first time in that individual child. A parent who can transmit a chromosome 21 translocation is referred to as a 'balanced translocation carrier'. Such a carrier parent is asymptomatic and has the normal chromosomal complement, except that part or all of one copy of chromosome 21 is attached to another chromosome. In this case the genetic risk varies widely and is dependent upon the chromosome to which the translocated segment of 21 is joined. The genetic risk for the mosaic form of DS depends on the percentage of mosaic cells in the individual as well as the percentage of trisomic cells in the gonadal tissues. Counseling for these rare forms of DS is complicated and always requires the services of a professional genetic counselor.

For families who wish to know whether the fetus being carried has DS, several chemical compounds have been identified in maternal blood that have a predictive value for diagnosis in up to 70% of cases. These are considered risk factors but are not diagnostic by themselves. More precise diagnostic data may be obtained prenatally by examining the amniotic fluid and determining the actual chromosomal complement of fetal cells. Maternal blood screening is best done at 15–17 weeks of gestation, whereas examination of amniotic fluid (amniocentesis) is generally done between 14 and 18 weeks of gestation. A newer procedure at around 10 weeks of gestation avoids amniocentesis and obtains fetal cells for genetic testing by sampling part of the placenta (chorionic villus biopsy). The complexity of prenatal diagnosis requires a close working relationship between professionals in obstetrics and in genetics. Many clinics specializing in these integrated services are available in modern medical systems.

Further Reading

Berg HM, Karlinsky H and Holland AJ (eds) (1993) *Alzheimer Disease, Down Syndrome, and Their Relationship*. Oxford, UK: Oxford Medical.

Beck MN (1962) *Expecting Adam: A True Story of Birth, Rebirth, and Everyday Magic*. New York, NY: Times Books.

Cicchetti D and Beeghly M (eds) (1990) *Children with Down Syndrome: A Developmental Perspective*. Cambridge, UK: Cambridge University Press.

Epstein CJ (1986) *The Consequences of Chromosome Imbalance: Principles, Mechanisms and Models*. Cambridge, UK: Cambridge University Press.

Hassold TJ and Patterson D (eds) (1999) *Down Syndrome: A Promising Future, Together*. New York, NY: Wiley-Liss.

Lott IT and McCoy EE (1992) *Down Syndrome: Advances in Medical Care*. New York, NY: Wiley-Liss.

Miller JF, Leddy M and Leavitt LA (1999) *Improving the Communication of People with Down Syndrome*. Baltimore, MD: Brooks.

Rondal JA (ed.) (1996) *Down's Syndrome: Psychological, Psychobiological and Socioeducational Perspectives*. San Diego, CA: Singular.

Selikowitz M (1990) *Down Syndrome: The Facts*. Oxford, UK: Oxford University Press.

Stratford B (1989) *Down's Syndrome: Past, Present and Future*. London, UK: Penguin.

Dreaming

See **Consciousness, Sleep, and Dreaming; Sleep and Dreaming**

Dualism

Intermediate article

David Robb, Davidson College, Davidson, North Carolina, USA

CONTENTS

Dualism is the view that the mental and the physical are distinct and mutually irreducible.

WHAT IS DUALISM?

As a candidate solution to the mind–body problem, dualism is opposed to the two main forms of monism: *materialism*, which reduces the mental to some part or aspect of the physical, and *idealism*, which reduces the physical to some part or aspect of the mental. Dualists regard monists of either sort as making the same mistake: conflating domains that are distinct and autonomous.

Dualism is a metaphysical doctrine, a view about the ultimate nature of reality. One often finds it packaged with other philosophical views, but it need not stand or fall with these. For example, while certain versions of dualism are congenial to theism, there is nothing in dualism *per se* that requires the existence of God, or that is committed to any particular religious tradition. Nor does dualism entail the controversial epistemological doctrines that often accompany it. These include the transparency of the mental – according to which the nature of the mind and its states is fully revealed in introspection – and the incorrigibility of first-person beliefs – according to which I am the final authority on the contents of my own mind. Dualism is compatible with the acceptance or rejection of either of these epistemological doctrines.

Even when isolated from these controversial views, dualism – at least in its most radical forms – is largely rejected by the cognitive science community, where materialistic monism predominates. Nevertheless, dualism has never completely disappeared from the intellectual scene, and it is enjoying something of a renaissance. In fact, as we will see, a moderate form of dualism is today widely accepted among cognitive scientists.

VARIETIES OF DUALISM

So far I have referred broadly to 'the mental' and 'the physical', but when dualists distinguish these realms, they normally have entities of a certain sort in mind. Here dualism divides into several related versions: there are dualisms of substances, properties, events, processes, facts, sentences, and more. This article will discuss only substance and property dualism, though much of the discussion applies to other versions as well.

Substance Dualism

A *substance* is a thing or object, something that has properties and persists through time. A substance dualist conceives of the mind as such an entity, one that, moreover, is distinct from the biological body or any part of it, such as the brain.

The most radical form of substance dualism is *Cartesian dualism*, named after its most famous modern defender, René Descartes. Although it has been out of fashion for sometime, Cartesian dualism has a number of contemporary defenders (e.g. Foster (1991)). A Cartesian dualist says that the mind is a nonphysical 'soul' or 'spirit', a substance entirely lacking in physical properties, including spatial location. As Descartes articulated the view, the mind's intrinsic properties are exclusively mental: they are modes (expressions) of consciousness, the essence of minds. By contrast, material substances (bodies) have only physical properties, modes of spatial extension, the essence of bodies.

This difference of essence entails an ontological independence: one kind of substance could exist without the other. Nevertheless, our minds are in fact 'embodied', albeit contingently and only temporarily. Descartes' own views on the nature of this embodiment are obscure; but the relation is often understood causally: for my mind to be embodied in a particular material substance is for it to have a privileged causal connection with that substance; only my mind can directly (by means of volitions) affect this biological body, and only this body can directly (by means of sensation) affect my mind.

For the Cartesian dualist, then, mind–body dependence is merely causal. But there are forms of substance dualism in which this dependence is much stronger. *Non-Cartesian substance dualists*, while insisting on the strict numerical distinctness of mind and body, allow that the mind has physical properties. Indeed, they claim, its physical properties are just those of the body or brain, for mind and body spatially coincide. E. J. Lowe is a contemporary defender of this more moderate form of substance dualism. He argues that the mind (or self) is a psychological substance, distinct from the body but sharing many of its physical properties (Lowe, 1996). Here the mind–body relation is analogous to the relation between a statue and the lump of clay composing it. Mind and body spatially coincide, but because they differ with respect to certain properties, they are distinct. In a similar vein, those who believe that the mind 'emerges' from the activities of the brain, and those who think that the brain 'constitutes' the mind, would also count as non-Cartesian substance dualists.

Property Dualism

Unlike substance dualists, property dualists need not postulate immaterial substances. They are willing to grant that the mind is nothing more than a complex physical thing. But they still insist that this substance's mental properties are not physical.

In its strongest form, property dualism says that mental properties are fundamentally different from – though they may be lawfully connected to – physical properties. We might call this *radical property dualism*. It is embraced by, for example, Chalmers (1996), at least about qualia. Chalmers believes that the qualitative, categorical features of our conscious experiences are different in kind from any physical properties, which he takes to be exclusively dispositional and structural. *Emergentists* also fall into this category: they take mental properties to be features of physical systems that have achieved a certain level of complexity. Such properties are at best only nomologically or causally dependent on the physical substrate from which they emerge.

Like substance dualism, property dualism exists in a more moderate form, according to which mental properties are not physical, yet always (perhaps necessarily) are realized in or supervene on the physical. The idea here is that mental properties are instantiated at a higher, more abstract level than their physical counterparts. In ascribing a mental property to someone, we abstract away from the physical details that realize, or implement, the mental property in that particular person. For example, in certain versions of functionalism, for a system S to have a mental property M is just for it to have some physical property P_s that within S plays the causal role definitive of M. Since the same M-defining role can be filled by different physical properties in different systems, we cannot identify M with P_s or with any other physical property. Yet there is a clear sense in which, within S, M is determined by, and in fact is 'nothing over and above', its physical realizer P_s. This realization relation is clearly much stronger than the causal or nomic psychophysical relations allowed by radical property dualism. Indeed, the relation here is so intimate that most adherents of moderate property dualism consider themselves materialists. This view is commonly called *nonreductive materialism*, but it is important to remember that it is, strictly speaking, a form of property dualism, since it denies that mental properties are physical.

ARGUMENTS FOR DUALISM

Although dualists have at times appealed to empirical considerations to support their view, the strongest dualist arguments, and those receiving the most philosophical attention, have been *a priori*. In the typical argument, some state of affairs incompatible with materialism is claimed to be clearly conceivable, and so possible. The dualist then moves from this possible state of affairs to a conclusion about the nature of the mental in the actual world.

Arguments for Substance Dualism

The most famous such argument occurs in Descartes' *Sixth Meditation*. A contemporary version proceeds along these lines (here we follow Descartes and other dualists in assuming that I am the same as my mind):

> I can clearly conceive of my existing in the absence of anything physical. (1)

That is, when I consider the matter carefully and rationally, there seems to be no contradiction in the idea of my existing in an entirely immaterial world. From this it follows that:

It's possible for me to exist the absence of anything physical. (2)

(Since what is clearly conceivable is at least possible.) But now note that:

If I am a physical thing, then I am essentially a physical thing. (3)

That is, any physical substance is physical by nature, and so could not exist as an immaterial thing. But then it follows that:

I am not a physical thing. (4)

From the claim of mere possibility in step 2 we have reached a conclusion about what is actually the case, via the linking premise 3.

In spite of its enormous influence among dualists, this argument has been challenged at almost every step. One line of objection faults the inference from step 1 to step 2. Even if I can clearly conceive of myself existing in the absence of anything physical, at most this shows that for all I know (about my nature) it's possible for me to exist in such a state. It doesn't imply that I really can exist in this way. That is, statement 1 at best establishes the epistemic possibility of my existing in an immaterial world, not its genuine, metaphysical possibility. Another line of objection challenges step 3: perhaps I am a physical thing, but not essentially so. Modal intuitions may differ here, but this objection at least puts the burden on the Cartesian to explain why a physical thing cannot be only contingently physical. Yet another line of objection is that the above argument, even if sound, merely establishes the more moderate, non-Cartesian form of substance dualism. That is, perhaps the argument only shows that I am not identical with any physical substance, not that my nature (and all my properties) are nonphysical. This is a delicate question: it turns in part on whether the non-Cartesian can allow for the possibility of disembodied existence. If so, then step 2 would not seem to be strong enough to support a Cartesian reading of the conclusion. (For further discussion of Descartes' argument, see Yablo, 1990.)

In any case, if a substance dualist wants merely to establish the non-Cartesian version, Descartes' modal argument is not required. Since moderate substance dualists claim only that mind and body are numerically distinct, they need only find some property possessed by one but not the other. Since the brain (say) constitutes the mind, the two substances share many of their properties, such as size, location, and so on. But mind and brain do seem to have different persistence conditions: certain kinds of neurophysiological damage, for example, would destroy the mind, but the brain (as a biological substance) would still exist. This difference in properties, the argument goes, entails a numerical difference – the mind is not the brain (or any other part of the body) – but it stops short of Cartesian dualism.

Arguments for Property Dualism

Philosophical arguments for property dualism, at least in its radical form, have also relied heavily on conceivability arguments. One thought experiment features super-neuroscientists able to examine a functioning brain down to the finest physical detail. They would never, it seems, find anything remotely resembling a thought or an experience. So mental properties are fundamentally different from anything physical. (For quite different versions of this argument, see G. W. Leibniz' *Monadology* and Jackson, 1982.) But a materialist might offer the following explanation for why the scientists fail to 'find' the mind: the mind and its contents can be accessed from two perspectives, the introspective, first-person perspective and the observational, third-person perspective. The scientists are in fact encountering mental properties as they observe the brain, but these mental properties aren't recognized as such because they're being accessed from an unfamiliar perspective. The resulting view would be a 'dualism of perspectives', but a materialism of the mind and its properties.

Another thought experiment marshalled in support of radical property dualism features the *zombie*, a being who, in spite of sharing all of the physical properties of a conscious being, is wholly lacking in conscious states (Chalmers, 1996). Such a being seems to be conceivable, yet if it is so much as possible, then our phenomenal properties (an important class of mental properties) are not the same as any of our physical properties. This argument is subject to some of the same criticisms that apply to Descartes. For example, are zombies really possible, or is their apparent conceivability explained by the limitations of our current knowledge? Just as further investigation may reveal, contra Descartes, that my disembodied existence is impossible, so we may eventually learn that zombies are impossible. At this point, however, the materialist can only offer the hope of such a discovery.

Property dualism in its more moderate form – nonreductive materialism – requires nothing as exotic as zombies. Here an appeal to multiple realizability is thought to suffice (Fodor, 1981). Any given mental property can be, and in fact is, realized by a variety of different physical properties in different species, different individuals of the same species, and even at different times in the same individual. Given that the same mental property is realized by many different physical properties, it cannot be identified with any one of them; hence mental properties are distinct from, though realized in, physical properties. This argument is often used to support certain versions of functionalism against the psychophysical identity theory. (For an identity theorist's response to the argument, see Kim, 1993.)

PROBLEMS FOR DUALISM

However one evaluates the preceding arguments, dualism faces a number of serious obstacles. These range from empirical objections, to the effect that dualism cannot be integrated into cognitive science, to philosophical and conceptual objections. The former are discussed in the final section of this article. Here I discuss the latter.

Charges of Incoherence and Vacuity

One of the reasons why substance dualism fell out of favour is that it seemed incoherent to a number of philosophers. Ryle (1949), for example, accused substance dualists of an egregious conceptual error, that of placing the mind in the wrong ontological category: having a mind isn't a matter of being (or being related to) a substance, immaterial or otherwise. Rather, talk of 'the mind' is merely talk about a set of capacities. Ryle thought that these capacities were exclusively behavioral, but one needn't be a behaviorist to appreciate his insight. A functionalist, for example, may also claim that talk about 'mind' doesn't refer to a kind of substance, but is just a way of describing the causal organization of an organism's mental and behavioral states.

Ryle's objection applies to any form of substance dualism, but there is another charge of incoherence directed specifically at the Cartesian variety: if minds are not located in space, there seems to be no way to individuate them. It is at least logically possible for there to be two qualitatively identical minds, that is, minds which share all of the same intrinsic properties. But by virtue of what, then, are they two minds and not one? Two qualitatively identical material substances can be distinguished by their different spatial locations, but there is no such medium to individuate Cartesian minds. Coupled with the (controversial) doctrine that there can be no entity without clear conditions of individuation, this objection could render Cartesian dualism incoherent. (See Hoffman and Rosenkrantz, 1991.)

Even if substance dualism is coherent, some have objected that as a proposed solution to the mind–body problem, dualism in any of its forms is devoid of positive theoretical content. It seems that dualists can tell us only that the mental is not physical; they can't say anything informative about the intrinsic nature of the mental. This charge of theoretical vacuity is most threatening to radical forms of dualism. Moderate forms – non-Cartesian substance dualism and nonreductive materialism – allow that minds or their mental properties are realized in the physical world, and the nature of this realization might contain enough resources for moderate dualists to say something positive about the mental. But can Cartesian or radical property dualism say anything more about the intrinsic nature of the mind? Perhaps the most obvious option here is also the most promising: discover the positive theoretical content of dualism by appealing to the direct introspective access we have to our own minds and their contents (Foster, 1991). Those who insist on drawing theoretical content only from objective, third-person sources will balk at this move, but the dualist will reply that first-person data cannot be ignored by any systematic study of the mind, at least any study that aspires to completeness.

The Problem of Interaction

The most serious problem for dualism arises from a seemingly undeniable fact: the mental and the physical causally interact. Such interaction is two-way: for example, my decision to get a drink causes me to walk to the refrigerator (mind-to-body causation); and putting my hand on a hot stove causes me to feel pain (body-to-mind causation). Dualists have had trouble integrating these common-sense facts about mind–body interaction into their ontology.

This problem is perhaps most acute for Cartesians, and again the mind's lack of spatial location is the source of the problem. There seems to be no way for a Cartesian mind to causally link to a location in space. Interacting bodies link by spatial contact; yet this mechanical account of the causal nexus cannot work for Cartesian mind–body interaction, since Cartesian minds cannot come into spatial contact with anything. Similar arguments apply in a more recent version of the problem of interaction, the 'pairing problem' (Kim, 2001). Imagine two qualitatively identical minds M_1 and M_2, and the bodies with which they (allegedly) causally interact, B_1 and B_2. What makes it the case that M_1 is causally paired with B_1, not B_2, and M_2 paired with B_2, not B_1? If these minds were located in space, we might appeal to their different spatial relations to B_1 and B_2 to resolve this question. But this resource is not available to the Cartesian. (For a Cartesian response to the pairing problem, see (Foster, 1991).)

One might think that the problem of interaction doesn't arise for property dualists, who after all can allow that the mind is a physical substance, causally related to the physical world. Yet property dualists face their own problems of causality, for we still would like mental properties to be causally relevant to bodily behavior. We wish, that is, not merely for minds (or mental events) to cause behavior; we wish them to cause behavior by virtue of their mental properties. And it has seemed to many opponents of property dualism that nonphysical properties cannot have a causal effect on the physical world. One much-discussed version of this objection appeals to the *causal closure* of the physical: the causal history of a physical event includes only physical events and their physical properties. At no point, that is, can a nonphysical event or property break into the network of physical causes. Causal closure raises immediate worries for both versions of property dualism: even for nonreductive materialism, since it is not clear that nonphysical properties, even if they are realized in the physical, can be causally relevant without violating closure (Kim, 1993).

Some property dualists cheerfully accept these consequences, thereby embracing *epiphenomenalism* (Jackson, 1982). Here mental properties are demoted to mere 'nomological danglers', properties caused by what goes on in the physical world, but not themselves having any causal efficacy in return. The mind's operations are, therefore, somewhat like the display on a computer's monitor: the changing patterns of colors reflect the internal operations of the computer without having any reciprocal effect on these operations. Epiphenomenalism is attractive for a number of reasons – for example, it allows us to embrace both the irreducibility of mental properties and the causal closure of the physical – yet it too has some obstacles to overcome. Besides being at odds with common sense, epiphenomenalism seems to rob us of an important kind of knowledge: our knowledge of our own mental states. If my qualia, for example, are causally impotent, how could I ever know about their existence and character? Yet first-person knowledge about conscious states is normally thought to be the most secure knowledge we can have. Either epiphenomenalists must bite the bullet and deny that we have such knowledge, or they have to explain how we can know about features of our minds from which we are causally isolated.

DUALISM AND COGNITIVE SCIENCE

What relevance might dualism have for contemporary cognitive science? Cartesian dualism is no longer considered seriously by most cognitive scientists, in spite of its renewed attention from metaphysicians. If cognitive scientists are substance dualists at all, they typically embrace only the non-Cartesian variety: the mind, if it can even rightly be called a substance, is constituted by the body or brain. In spite of this near-consensus, however, it is worth noting that there is nothing in cognitive science *per se* forcing one to reject Cartesian dualism. Most, if not all, theories in cognitive science are neutral with respect to the ultimate nature of the mind. Even psychological theories that 'locate' mental capacities or processes in the brain could be interpreted by a Cartesian dualist as merely revealing their 'neural correlates'. A scientific study of the mind requires systematic (lawful) correlations between mental states and the empirical states of the body and brain. But whether these states are one and the same, and indeed whether the brain is the mind at all, are questions outside the domain of such a study.

However, while empirical theories about the mind are compatible with Cartesian dualism, the progress of cognitive science is making the view less attractive than it was in the seventeenth century. The more cognitive scientists learn about the capacities of physical systems such as the brain, and the more they learn about the dependencies between mental and neural functioning, the less attractive Descartes' theory has become (Damasio, 1994).

In contrast to substance dualism, property dualism is becoming more popular. Indeed, nonreductive materialism is the dominant view in cognitive science, where it is almost taken for granted that mental phenomena exist at a higher, more abstract level than physical phenomena, even if the former are realized in the latter. Some philosophers question whether the hierarchy of 'levels' cognitive scientists speak of really requires a distinct class of mental properties, but this is how such talk is often understood. And while property dualism in its radical form is still a minority view, it has been earning more respect, especially among theorists of consciousness. Chalmers (1996), for example, believes that nonphysical qualia can be successfully integrated into a science of the mind. He believes that while no physical theory could ever fully capture the nature of our conscious states, cognitive science can fruitfully investigate the laws connecting consciousness with physical systems such as the brain.

Finally, perhaps the most interesting consequences of dualism for cognitive science are not so much metaphysical as methodological. If dualism in any of its forms is true, then any study of the mind is in an important sense an autonomous discipline: theorizing within cognitive science can proceed by and large independently of the physical sciences. As cognitive scientists formulate psychological laws and explanations, they need not await the perfection of, say, particle physics. Among some dualists, this autonomy takes a rather strong form. Descartes thought that the mind was wholly outside the domain of the emerging mechanistic science of his day. And much more recently Davidson (1980) has argued that the form of explanation in psychology is of a radically different sort than that in the physical sciences. Some nonreductive materialists soften this line on autonomy, allowing psychological explanations to be of the same sort as those in the physical sciences. They might even grant that the lower-level physical sciences can illuminate, and put important constraints on, higher-level theorizing about the mind. Yet because, in their view, mental theorizing occurs at a higher level, it will still retain a degree of autonomy.

References

Chalmers DJ (1996) The Conscious Mind. New York, NY: Oxford University Press.

Damasio A (1994) Descartes' Error. New York, NY: G. P. Putnam.

Davidson D (1980) Essays on Actions and Events. Oxford: Clarendon Press.

Fodor JA (1981) Special sciences. In: *Representations*. Cambridge, MA: MIT Press.

Foster J (1991) The Immaterial Self. London: Routledge.

Hoffman J and Rosenkrantz G (1991) Are souls unintelligible? *Philosophical Perspectives* **5**: 183–212.

Jackson F (1982) Epiphenomenal qualia. *Philosophical Quarterly* **32**: 127–136.

Kim J (1993) *Supervenience and Mind*. Cambridge, UK: Cambridge University Press.

Kim J (2001) Lonely souls. In: Corcoran K (ed.) *Soul, Body, and Survival*. Ithaca, NY: Cornell University Press.

Lowe EJ (1996) *Subjects of Experience*. Cambridge, UK: Cambridge University Press.

Ryle G (1949 [Reprinted 1984]) *The Concept of Mind*. Chicago, IL: University of Chicago Press.

Yablo S (1990) The real distinction between mind and body. *Canadian Journal of Philosophy*, Supplement **16**: 149–201.

Further Reading

Baker LR (1993) Metaphysics and mental causation. In: Heil J and Mele A (eds) *Mental Causation*. Oxford: Clarendon Press.

Hasker W (1999) *The Emergent Self*. Ithaca, NY: Cornell University Press.

Hill CS (1997) Imaginability, conceivability, possibility and the mind–body problem. *Philosophical Studies* **87**: 61–85.

Kim J (1999) Making sense of emergence. *Philosophical Studies* **95**: 3–36.

Levine J (2000) *Purple Haze: The Puzzle of Consciousness*. New York, NY: Oxford University Press.

Lowe EJ (2000) *An Introduction to the Philosophy of Mind*. Cambridge, UK: Cambridge University Press.

Merricks T (1994) A new objection to *a priori* arguments for dualism. *American Philosophical Quarterly* **31**: 81–85.

Shoemaker S (1984) *Identity, Cause, and Mind*. Cambridge, UK: Cambridge University Press.

Zimmerman D (1991) Two Cartesian arguments for the simplicity of the soul. *American Philosophical Quarterly* **28**: 217–226.

Dynamic Decision Makers, Classification of Types of

Intermediate article

Daniel Houser, University of Arizona, Tucson, Arizona, USA

CONTENTS

A dynamic decision problem is one that requires a sequence of decisions in a setting where pay-offs and alternatives available for later decisions depend on earlier choices. Analysis of data from dynamic behavioral experiments can shed light on the nature of the different types of dynamic decision makers in the population.

INTRODUCTION

Types

Suppose a general orders his troops into battle against an approaching enemy and valiantly leads the charge himself. One of his soldiers takes advantage of the momentary confusion surrounding the order by slipping away from the battlefield to an area of relative safety. One reasonable explanation for the different decisions made by the general and soldier is straightforward: they are different 'types' just in that they face different incentives. Victory in battle means lasting glory and honor for the general, while the soldier might receive little but the chance to fight again another day. It is less straightforward, however, to explain why one soldier flees while his comrades, who are in the same situation, rush forward with the general into combat. As a casual description, we might also say that 'observationally' identical soldiers who make different decisions are different 'types'.

In order to decide whether to flee or fight, soldiers must solve a dynamic decision problem. It is dynamic because their decision affects in a nontrivial way the alternatives available to them at later times. For example, the soldier who fights might be in a position to save the life of a wounded comrade on the battlefield. The soldier who flees might save his own life, but at the risk of being punished for desertion. Their eventual decision rests on idiosyncratic characteristics such as preferences and subjective assessments of battlefield risks.

Game theory defines people who have different preferences as different 'types'. Unfortunately, preferences are not observable. In practice, it is more useful to define people as different types of dynamic decision makers if, like the soldiers, they make different decisions in observationally identical dynamic situations.

Interest in classifying and characterizing dynamic decision makers has grown as the importance of accounting for type heterogeneity in dynamic economic models has become apparent. This importance stems from the fact that most of dynamic economics has as its final goal policy analysis. That is, the goal is to predict how different sorts of incentive structures (e.g. the tax system) affect dynamic decisions (e.g. work and educational choices). As the example of the soldiers makes clear, not everybody responds to the same incentives in the same way. Economists increasingly recognize that analyses, which assume that firms and societies can be described as collectives and modeled as though they were single agents, can often lead to very misleading conclusions and policy recommendations (Furubotn and Richter, 2000). Models that take account of type heterogeneity have the potential to improve policy analysis substantially.

Decision Rules

Economists use so-called 'decision rules' to describe the way a person's actions depend on personal information. 'Information' here should be thought of broadly as everything a person knows (including demographic variables) that could be relevant to a decision. The soldiers above, for example, can be thought of as having the choice either to fight or flee. When presented with the information that the enemy is approaching, the decision rules of some soldiers generate the decision to flee, while those of others generated the decision to fight. In general, people in observationally identical situations who make different decisions are viewed as using different decision rules, and a person's 'type' is defined by the decision rule they use. Increasingly, research is directed towards identifying, at least within narrow contexts, the number and nature of different decision rules, or types, that exist in the population.

Differences in decision rules can arise from differences in preferences. This is a good explanation in many situations, particularly when trying to explain idiosyncratic differences in tastes for, say, coffee and tea. On the other hand, differences along dimensions such as propensities to cooperate, which are well documented and involve higher-order cognitive processing, seem less naturally attributable to preferences. There is some evidence that differences in decision rules associated with higher-order functions are due to different cognitive algorithms employed to determine actions (e.g. McCabe *et al.*, 2001). Such differences are analogous to the difference between human and computer decision making. When a human plays chess against a computer, both parties have the same objective and information, yet their decision rules differ because they use different algorithms to determine their moves.

Expectations

A higher-order task of particular importance to dynamic decisions is expectation formation, because all dynamic decision problems require some sort of forward-looking behavior. Economists have been using the 'rational expectations' assumption to model forward-looking behavior for decades. However, numerous studies in economics and psychology suggest that expectations are not formed rationally. Moreover, it is straightforward to show that different expectation formation mechanisms lead to different dynamic decision rules.

An important, and often-replicated, finding in the literature on static decision making is that, except in very simple cases, people do not assess objective probabilities correctly (e.g. Camerer, 1995). Since probability assessment is cognitively difficult, it is presumably accomplished with idiosyncratic heuristics. Moreover, because probability assessment is fundamental to expectation formation, it seems likely that expectations are formed with idiosyncratic and imperfect heuristics. Although research in this area is still in its early stages, it seems plausible that heterogeneity in expectation formation underlies much of the idiosyncratic variation in dynamic decision rules.

TYPE ELICITATION

Stopping Experiments

Dynamic decision problems (DDPs) faced by individuals provide perhaps the simplest interesting environment in which to study dynamic decision rule heterogeneity. In these environments an individual makes several decisions sequentially, and the decisions made early in the problem affect the nature of the decision task later in the problem. For example, a person might first decide whether to bicycle or walk to work, and then decide on the route to follow. This is a DDP, because the set of candidate routes depends on the outcome of the first decision. (Economists contrast DDPs with sequential 'static' decision problems, where one makes a series of unrelated decisions.) It is important to understand how people actually solve DDPs, particularly when the dynamic nature of the problem involves deciding between different pay-offs at different times (so-called 'intertemporal' decision problems), because many actual consumption, savings, and labor supply decisions must be made within this context.

'Stopping problems', a widely studied class of DDPs, have proved useful tools in classifying and characterizing dynamic decision rules. In a simple stopping experiment, subjects receive payment offers sequentially from the experimenter until they accept one, at which time the experiment ends. Many variants of this basic design have been studied. For example, subjects might have to pay for offers; they might not know the distribution from which offers are generated; they might be able to accept previously rejected offers; and they might not know the exact amount of the offer, but only whether it is higher or lower than other offers. An advantage of this framework is that theoretical predictions about behavior under various decision rules are straightforward to derive. Different decision rules often imply different stopping points. Hence, observing stopping times allows one to draw simple and compelling inferences about the sorts of decision rules that are used in the population.

Analysis of stopping experiment data, using techniques such as that of El-Gamal and Grether (1995) discussed below, show that there is great heterogeneity in the ways subjects solve experimental stopping problems. There is little evidence to suggest that people decide in ways that are consistent with rational expectations. Instead, subjects seem to use sophisticated, nonstationary reservation pay-off heuristics. This means that subjects stop as soon as their pay-off is sufficiently high, where 'sufficiently' depends, for example, on the number of times they have had to search and on whether they are paying search costs.

There is a small set of reservation pay-off decision rules into which most subjects' behavior seems to fall. Moreover, there are two features that most of these rules share. Firstly, subjects who use them tend to stop searching somewhat earlier than a rational expectations searcher would. Secondly, these heuristics work well in the sense that subjects who use them earn only slightly less on average (often about one percent) than they would have if they had followed the rational expectations rule. Since the rational expectations rule is cognitively very complex to implement, there may be a sense in which using reservation pay-off heuristics is in fact optimal. For a survey of results from the experimental stopping experiment literature, see Cox and Oaxaca (1996).

The Voluntary Contribution Mechanism

Types can also be discerned in game environments where multiple subjects interact and make strategic decisions that affect each other's pay-off. The voluntary contribution mechanism (VCM) is an important example of such a game. There are N players, and player n has endowment w_n. Player n contributes g_n to the public good and leaves the remainder in a private account. The total contribution to the public good is $G = \sum_n g_n$. The interesting feature of the VCM is that the return on investment in the public account differs from that on investment in the private account. Without loss of generality we can suppose that the return to each player on the total investment in the public account is given by r while the return on the private account is set to unity. This means that the pay-off function for player n is

$$\Pi(g_n, \hat{g}_n) = (w_n - g_n) + rG \qquad (1)$$

where $\hat{g}_n$ represents the vector of contributions of everyone except player n. Provided that $r < 1$ it is easy to see that, given any arrangement of contributions by the other subjects, each player maximizes his or her individual pay-off by 'free-riding', or contributing zero to the public good. Hence, free-riding is a dominant Nash equilibrium strategy. But if $rN > 1$ then it is Pareto optimal for each player to contribute everything to the public good, and this strategy Pareto-dominates free-riding. The parameter r is the marginal per-capital return (MPCR). When designing VCM experiments, the MPCR and the number of subjects are usually chosen to exploit the tension between free-riding and Pareto optimality.

Experimental research with the VCM, has generated many widely-replicated results, including clear evidence of decision rule heterogeneity (for a survey see Ledyard, 1995). In particular, there is usually a subset of subjects, 'free-riders', whose decision is to contribute very little to the public good in every round, and another subset, 'cooperators', who systematically contribute a large fraction of their endowment to the public good. A third subset uses 'reciprocal' rules, trying to match others' contributions.

The presence of reciprocal decision rules suggests that group dynamics might be influenced by the type composition of groups. For example, the presence of players who contribute little or nothing to the public good could lead to decreasing aggregate contributions over time if reciprocators attempt to match free-riders' small contributions. Hence, a feedback system that is sensitive to the proportion of each type within the group could be created, and could affect the extent to which a group is able to sustain cooperation.

Recent research has found that type composition seems to have important effects on group dynamics (e.g. Gunthorsdottir *et al.*, 2001). In particular, the number of free-riders in a group influences that group's path over the course of a game. Without free-riders, groups are capable of sustaining high levels of contribution to the public good; while the presence of free-riders often pushes groups towards successively lower levels of contributions.

TYPES AND PERSONALITY SURVEYS

The ability to learn about a person's behavioral type from a personality survey would be useful, since the dependence of group outcomes on type composition implies that knowledge of types could be used to design groups (such as school classes) efficiently. However, whether behavioral types broadly and systematically correlate with personality surveys is an open question, and experiments have generated widely conflicting results. Nevertheless, some personality variables seem to correlate with propensities to cooperate in experiments. Personality dimensions displaying this correlation include Machiavellianism, self-monitoring, and three of the 'big five' personality traits.

Machiavellianism

Inspired by the writings of Niccolo Machiavelli (1469–1527), and first developed by Christie and Geis (1970), the Machiavellianism (or Mach) scale measures the extent to which a person agrees that the end sanctifies the means. People who score highly on the Mach scale tend to be manipulative, opportunistic, and rational. Low Machs tend to be more emotional and less likely to depart from social norms in order to pursue their own self-interest. While high Machs tend to be competitive and exploitative, low Machs are usually more willing to cooperate (Gunnthorsdottir, McCabe and Smith, 2002).

Self-monitoring

The 'self-monitoring' scale is an measure of the dependence of an individual's behavior on the social context. High self-monitors work to create the impression needed to obtain their social goals, while low self-monitors are less concerned about the impression they make. High self-monitors have been found to be more likely to cooperate, particularly in experiments where repeated interactions with the same counterpart are possible.

The Big Five

The 'big five' personality traits are extraversion, agreeableness, conscientiousness, neuroticism, and openness. Among these, extraversion and agreeableness seem to be positively correlated with cooperativeness, while neuroticism seems to be negatively correlated. The relation of the other two traits with cooperativeness is not clear.

Many other personality variables, including self-esteem and locus of control, have been studied in relation to cooperation, but without clear results. For further discussion on the connection between types and personality surveys, see Kurzban and Houser (2002).

STATISTICAL METHODS FOR TYPE CLASSIFICATION

Sophisticated statistical procedures are not usually required to determine whether subjects in an experiment behave according to a particular decision rule. Intuitively, all that is required is to compare actual decisions with those that would arise under a hypothesized behavior. Although the details depend on the experimental design, formal procedures to accomplish this sort of comparison are typically straightforward. A more difficult task, and one that typically requires sophisticated statistics, is to determine how people actually make decisions in a given dynamic environment.

Attempting to characterize actual decision making requires, at least, allowing for multiple decision rules to be used in the population. The task is then to determine the number of decision rules, and to assign each subject to a decision rule. Broadly, there are two ways in which this can be done. We will briefly summarize the two approaches, and then discuss in greater detail an instance of each of them.

One approach, exemplified by a procedure suggested by El-Gamal and Grether (1995), requires one to specify in advance a set of candidate decision rules. A statistical procedure is then used to choose a 'best' subset of these rules. Finally, each subject is assigned to one of the rules in the subset. An advantage of this sort of procedure is that it is relatively straightforward to implement. However, unless it is feasible to include all of the rules that subjects might possibly use in the prespecified superset, a potential drawback is that the right rules might not be included. Misspecification could mask underlying commonalities in subjects' play.

An alternative approach, exemplified by a method suggested by Houser *et al.* (2001), requires no assumptions about the number of decision rules used in the population, the nature of each decision rule, or the assignment of subjects to decision rules. This approach requires cluster analysis: subjects are clustered according to commonalities in their behavior.

The goal of both these approaches is to put each subject into a behavioral category. El-Gamal and Grether require one to specify the categories in advance, while Houser *et al.* allow the categories to be determined by the data. Of course, it may not be easy to assign behavioral labels to groups that follow statistically similar decision rules.

The Classification Procedure of El-Gamal and Grether

Suppose one has data from a behavioral laboratory experiment where each of N subjects makes T decisions. Let C^K denote the prespecified set of K heuristics (i.e., decision rules) that subjects might use to make these decisions, and let $c \in C^K$ denote a particular heuristic. The idea is to determine, for each subject, the number of decisions consistent with each possible heuristic, and then assign that subject to the heuristic that best fits his or her behavior.

To implement the procedure one assumes that each subject follows exactly one of the heuristics in C^K. In practise, of course, a subject's behavior may not be perfectly consistent with any of the heuristics in C^K. El-Gamal and Grether (1995) circumvent this problem by assuming that subjects follow their heuristics with error.

Heuristics are chosen that specify a subject's decision uniquely from his or her state. A subject's 'state', which generally changes after each decision, is a vector that summarizes all of the person's decision-relevant information. Let x_t^c be an indicator variable that takes the value one if the subject's tth decision agrees with heuristic c and takes the value zero otherwise. Assume that the decisions are made independently with common error rate ε.

Let $(x_{n1}, \ldots, x_{nT})$ be a vector denoting subject n's actions, and let $(x_{n1}^c, \ldots, x_{nT}^c)$ be a vector of zeros and ones that summarizes the consistency of the subject's choices with c. That is, assume that $x_{nj}^c = 1$ if decision rule c predicts decision x_{nj}, and $x_{nj}^c = 0$ otherwise. Then set $X_n^c = \sum_t x_{nt}^c$. The likelihood function for the subjects' actions is then:

$$f^c(x_{n1}, \ldots, x_{nT}) = (1 - \varepsilon/2)^{X_n^c} (\varepsilon/2)^{T - X_n^c} \tag{2}$$

It is natural to assign each subject to the heuristic from the candidate set that maximizes his or her likelihood function.

This model can be 'overfit': including a large number of heuristics in C^K would allow the statistical model to fit a sample arbitrarily well. Overfitting usually leads to results with little external validity. To avoid overfitting, El-Gamal and Grether suggest penalizing the likelihood for each additional heuristic that is included in the set of candidate heuristics. Let C^k denote a subset of $k \leq K$ decision rules. El-Gamal and Grether argue that a reasonable penalized log-likelihood is obtained by forming the Bayesian posterior that arises under the following priors: (1) the probability that the population includes exactly k heuristics is $1/2^k$; (2) all possible k-tuples of heuristics in any C^k are equally likely (each with probability $1/K^k$); (3) all allocations of heuristics to subjects are equally likely (each with probability $1/k^N$); (4) all error rates (between zero and one) are equally likely; and do not depend on the number of rules used in the population or on the way those rules are assigned. This generates the following penalized log-likelihood function:

$$\log\left(\prod_n \max_{c_n \in C^k} f^{c_n}(x_{n1}, \ldots, x_{nT})\right) - k \log 2 - N \log k - k \log K \tag{3}$$

Determination of the population of heuristics as well as the assignment of subjects to heuristics is accomplished by simply maximizing the above expression over the set of all possible k-tuples that can be formed from the set of K decision rules.

The Classification Procedure of Houser, Keane, and McCabe

Suppose that subjects solve a 15-period DDP. At each period, subjects choose either A or B, each of which results in a nonnegative monetary reward. Pay-offs are stochastic. The realizations of the random variables for period t occur before the decision at t is made, and the realizations of the random variables for period $t+1$ occur after the decision at t. Each subject's total pay-off is the sum of the rewards earned over the 15 periods. Subjects have complete information regarding the stochastic link between their current choices and future pay-offs, but the link is complicated and it is difficult to determine the decision rule that maximizes expected total pay-offs.

The goal is to learn about the dynamic decision rules that subjects actually use when solving this difficult problem. Houser *et al.* (2001) begin by assuming that subjects are rational in a weak sense. In particular, a subject will choose alternative A in period t if and only if, in period t, the value the person places on choosing A is greater than the value he or she places on choosing B. Because the problem is dynamic, the values that subjects place on A and B depend both on the immediate reward to each choice and on the way subjects believe that choice would influence their future pay-offs. Houser *et al.* assume that alternative valuations are additively separable into a 'present' component, which captures immediate rewards (in this case the immediate monetary pay-off), and a 'future' component, which captures any benefits expected to accrue in subsequent periods as a result of that choice (in this case future monetary pay-offs).

Since the present pay-off structure is known for each agent, differences in behavior result only from differences in the future component. Hence, all differences in decision rules between subjects can be captured by differences in the future component. Houser *et al.* propose clustering subjects into groups that seem to have similar future components, while simultaneously drawing inferences about the future components' forms. In this way, they avoid the need to prespecify the nature of the decision rules used by the subjects.

Drawing on Geweke and Keane (1999), Houser *et al.* model the unobserved future component of each alternative's value as a parametric function of the subject's information set I_{nt}. The information set can include anything the researcher believes is relevant to the subject when making his or her decision, such as choice and pay-off histories. Then, the value that subject n assigns to alternative $j \in \{A, B\}$ in period t, V_{njt} (I_{nt}), assuming that the person uses decision rule k, can be written

$$V_{njt}(I_{nt}|k) = w_{njt} + F(I_{n,t+1}|I_{nt}, j, \pi_k, \varsigma_{njtk}) \quad (4)$$

$$I_{n,t+1} = H(I_{nt}, j) \quad (5)$$

Here, w_{njt} is the known immediate pay-off associated with alternative j. $F(\cdot)$ represents the future component. It depends on the alternative j and information set I_{nt} and is characterized by a finite vector of parameters π_k, whose values determine the nature of decision rule k, and a random variable ς_{njtk} that accounts for idiosyncratic errors subjects make when attempting to implement decision rule k. (The researcher must specify the distribution of the idiosyncratic errors.) The function $H(\cdot)$ is the information set's (possibly stochastic) law of motion. It provides the dynamic link between current information, actions and future information. Note that it does not vary with the decision rule.

We denote the choice in period t of subject n following decision rule k with information I_{nt} by

$$d_k(I_{nt}) = \begin{cases} A & \text{if } Z_{nt}(I_{nt}|k) > 0 \\ B & \text{otherwise} \end{cases} \quad \text{for all } k \in K, \quad (6)$$

where $Z_{nt}(I_{nt}|k) = V_{nAt}(I_{nt}|k) - V_{nBt}$ $(I_{nt}|k)$.

The goal is to draw inferences about the parameters $\pi_k (k \in K)$, and about the probability with which each subject uses each decision rule. To this end, Houser *et al.* construct the likelihood function associated with this framework. This requires knowing the probability, conditional on a subject's information set, that he or she will choose A or B.

The probability that subject n using decision rule k chooses alternative A at period t, given that the person has information I_{nt}, is given by

$$\begin{aligned} P(d_k(I_{nt}) = A) &= P(V_{nAt}(I_{nt}) > V_{nBt}(I_{nt})) \\ &= P(w_{nAt} - w_{nBt} + f(I_{nt}|\pi_k) > 0) \quad (7) \end{aligned}$$

where $f(\cdot)$ is a stochastic function that represents the differenced future components $F(I_{n,t+1}|I_{nt}, A, \pi_k, \varsigma_{nAtk}) - F(I_{n,t+1}|I_{nt}, B, \pi_k, \varsigma_{nBtk})$. The conditional probability that B is chosen is one minus the conditional probability that A is chosen.

Knowing the conditional choice probabilities, it is straightforward to construct the likelihood function needed to draw inferences about the different decision rules used in the population, and the probability with which each subject uses each rule. Under the distributional assumptions made by Houser *et al.* the likelihood function corresponds to a mixture of normals probit model. Unfortunately, this likelihood can be computationally difficult to maximize. Further discussion of this point (and estimation strategies) can be found in Houser *et al.* (2001) and Geweke and Keane (1999).

CONCLUSION

Economists say that people who make different decisions in observationally identical situations are different 'types'. Decision rules form the link between a person's situation and decisions, and it is natural to define a person's type by the decision rule he or she uses. Investigating the nature of the various decision rules at use in the population is important, because the effects of incentives on behavior depend on the decision rules that incentives act upon.

Many economists are particularly interested in the decision rules people use in dynamic environments. Experimental studies have found that a small number of decision rules seem to explain most observed behavior in very narrow contexts, and that these rules do not usually include the rational expectations rule. Further research is needed to determine the nature and number of decision rules in the population, the relationship between decision rules used in different contexts, and consequent implications for individual and group outcomes and incentive structures. Such research may employ sophisticated statistical procedures that group people according to common behavioral patterns. These patterns may be either specified in advance or discerned directly from experimental data.

References

Camerer C (1995) Individual decision making. In: Kagel J and Roth A (eds) *The Handbook of Experimental Economics*. Princeton, NJ: Princeton University Press.

Christie R and Geis F (1970) *Studies in Machiavellianism*. New York, NY: Academic Press.

Cox JC and Oaxaca R (1996) Testing job search models: the laboratory approach. *Research in Labour Economics* **15**: 171–207. Greenwich, CT and London: JAI Press.

El-Gamal M and Grether D (1995) Are people Bayesian? Uncovering behavioral strategies. *Journal of the American Statistical Association* **90**: 1137–1145.

Furubotn EG and Richter R (2000) *Institutions and Economic Theory: The Contribution of the New Institutional Economics*. Ann Arbor, MI: University of Michigan Press.

Geweke J and Keane M (1999) Bayesian inference for dynamic discrete choice models without the need for dynamic programming. In: Mariano, Schuermann and Weeks (eds) *Simulation Based Inference and Econometrics: Methods and Applications*. Cambridge, UK: Cambridge University Press.

Gunthorsdottir A, Houser D, McCabe K and Ameden H (2001) Disposition, history and contributions in public goods experiments. [Working paper, University of Arizona.]

Gunnthorsdottir A, McCabe K and Smith V (2002) Using the Machiavellianism instrument to predict trustworthiness in a bargaining game. *Journal of Economic Psychology* **23**: 49–66.

Houser D, Keane M and McCabe K (2001) How do people actually solve dynamic decision problems? [Working paper, University of Arizona.]

Kurzban R and Houser D (2002) Individual differences in cooperation in a circular public goods game. *European Journal of Personality* **15**: 37–52.

Ledyard J (1995) Public goods: a survey of experimental research. In: Kagel J and Roth A (eds) *The Handbook of Experimental Economics*. Princeton, NJ: Princeton University Press.

McCabe K, Houser D, Ryan L, Smith V and Trouard T (2001) A functional imaging study of cooperation in two-person reciprocal exchange. *Proceedings of the National Academy of Science* **98**: 11832–11835.

Further Reading

Andreoni J (1995) Cooperation in public goods experiments: kindness or confusion? *American Economic Review* **85**: 891–904.

Geweke J, Houser D and Keane M (2001) Simulation based inference for dynamic multinomial choice models. In: Baltagi B (ed.) *Companion to Theoretical Econometrics*. Oxford, UK: Blackwell.

Geweke J and Keane M (1997) Mixture of normals probit models. [Federal Reserve Bank of Minneapolis staff report 237.]

Geweke J and Keane M (2001) Computationally intensive methods for integration in econometrics. In: Heckman J and Leamer E (eds) *Handbook of Econometrics*, vol. V. North Holland.

Gilks WR, Richardson S and Spiegelhalter DJ (1996) *Markov Chain Monte Carlo in Practice*. London, UK: Chapman & Hall.

Krusell P and Smith AA (1995) Rules of thumb in macroeconomic equilibrium: a quantitative analysis. *Journal of Economic Dynamics and Control* **20**: 527–558.

Lettau M and Uhlig H (1999) Rules of thumb versus dynamic programming. *American Economic Review* **89**: 148–174.

McLachlan GJ and Basford KE (1988) *Mixture Models: Inference and Applications to Clustering*. New York, NY: Marcel Dekker.

Sargent TJ (1987) *Dynamic Macroeconomic Theory*. Cambridge, MA: Harvard University Press.

Stokey NL and Lucas RE (1989) *Recursive Methods in Economic Dynamics*. Cambridge, MA: Harvard University Press.

Dynamical Systems Hypothesis in Cognitive Science

Intermediate article

Robert F Port, Indiana University, Bloomington, Indiana, USA

CONTENTS

The dynamical systems hypothesis in cognitive science identifies various research paradigms applying the mathematics of dynamical systems to understanding cognitive function.

OVERVIEW

The dynamical approach to cognition is allied with and partly inspired by research in neural science since the 1950s, in which dynamical equations have been found to provide excellent models for the behavior of single neurons (Hodgkin and Huxley, 1952). It also takes inspiration from work on gross motor activity by the limbs (e.g. Bernstein, 1967; Fel'dman, 1966). In the early 1950s, Ashby made the startling proposal that all of cognition might be accounted for with dynamical system models (Ashby, 1952), but little work followed directly from his speculation because of a lack of appropriate mathematical methods and computational tools to implement practical models. More recently, the connectionist movement (Rumelhart and McClelland, 1986) has provided insights and mathematical implementations of perception and learning, for example, that have helped revive interest in dynamical modeling.

The dynamical approach to cognition is also closely related to ideas about the embodiment of mind and the environmental situatedness of human cognition, since it emphasizes connections between behavior in neural and cognitive processes, on the one hand, and physiological and environmental events, on the other. The most important such connection is the dimension of time shared by all of these domains. This permits real-time coupling between domains, whereby the dynamics of one system influence the timing of events in another. Humans often couple many systems together: for example, when dancing to music, one's auditory perception system is coupled with environmental sound, and the gross motor system with both audition and musical sounds. Because of this connection between the world, the body and cognition, the method of differential equations is applicable to events at all levels of analysis over a wide range of timescales. This approach emphasizes change over time of relevant system variables. (*See* **Embodiment**; **Dynamical Systems, Philosophical Issues about**)

MATHEMATICAL CONTEXT

The mathematical models employed in dynamical systems research derive from many sources in biology and physics. Two schemas will be discussed out here. The first is the neural network idea, partially inspired by the remarkable equations of Hodgkin and Huxley (1952) which account for many neuronal phenomena in terms of the dynamics of cell membranes. Hodgkin and Huxley proposed a set of differential equations describing the flow of sodium and potassium ions through the axonal membrane during the passage of an action potential down the axon. These equations, which apply with slight modification to all neurons, led to attempts to account for whole cells (rather than patches of membrane) in terms of their likelihood of firing given various excitatory and inhibitory inputs. Interesting circuits of neuron-like units were constructed and simulated on computers. The Hodgkin–Huxley equations inspired many psychological models, such as those of Grossberg (1982, 1986), the connectionist network models (Rumelhart and McClelland, 1986), and models of neural oscillations (Kopell, 1995). (*See* **Hodgkin–Huxley**)

In this schema, it is hypothesized that each cell or cell group in a network follows an equation like

$$\frac{dA}{dT} = -\gamma A(t) + \delta(aE(t) - bI(t) + cS(t)) + k \quad (1)$$

indicating that the rate of change of activation (i.e., likelihood of firing) A at time t depends on the decay γ of A and a term representing inputs from other cells that are either excitatory, $E(t)$ (tending to increase the likelihood of firing), or inhibitory, $-I(t)$ (tending to decrease the likelihood of firing). For some units there may be an external physical stimulus, $S(t)$. A nonlinear function, $\delta(x)$, encourages all-or-none firing behavior, and the bias term k adjusts the value of the firing threshold. An equation of this general form can describe any neuron. Networks of units like these can exhibit a wide variety of behaviors, including many specific patterns of activity associated with animal nervous systems. (*See* **Neurons, Computation in**)

A second schema inspiring the dynamical approach to cognition is the classical equation for a simple oscillator like a pendulum. Indeed, it is obvious that arms and legs have many of the properties of pendulums. Pendular motion is a reasonable prototype for many limb motions. A similar system (lacking the complication of arc-shaped motion) is that of a mass and spring. It is described by the equation

$$m\frac{d^2x}{dt^2} + d\frac{dx}{dt} + k(x - x_0) = 0 \quad (2)$$

which specifies simple harmonic motion in terms of the mass m, the damping d, the spring's stiffness k, and the neutral position x_0 of the mass. Fel'dman (1966) used heavily damped harmonic motion to model a simple reach with the arm. If the neutral position x_0 (the attractor position when damped) can be externally set to the intended target position (for example, by adjusting the stiffness in springs representing flexor and extensor muscles), then movements from arbitrary distances and directions towards the target can occur – simply by allowing the neuromuscular system for the arm to settle to its fixed point, x_0. A number of experimental observations – for example, reaching maximum velocity in the middle of the gesture, higher maximum velocity for longer movements, automatic correction for an external perturbation, and the naturalness and ease of oscillatory motions at various rates – can be accounted for with a model using a mass and a spring with controllable stiffness, rest length and damping. (*See* **Motor Control and Learning**)

In the most general terms, a dynamical system may be defined as a set of quantitative variables (e.g., distances, activations, rates of change) that change simultaneously in real time due to influences on each other. These mutual influences can be described by differential or difference equations (van Gelder and Port, 1995). Newton's equations of motion for physical bodies were among the earliest dynamical models. Until the 1950s, the analysis of dynamical models was restricted to linear systems (such as eqns 1 and 2) containing no more than two or three variables. Since the 1970s, mathematical developments, simulations by digital computer programs and computer graphics have revolutionized modeling possibilities, and practical methods for studying nonlinear systems with many variables are now possible (Strogatz, 1994).

PERCEPTUAL MODELS

Dynamical models seem particularly appropriate to account for motor control and for perceptual recognition. In particular, there is a large body of research on temporal aspects of perception. (*See* **Perception: Overview**; **Perceptual Learning**; **Reaction Time**)

One well-known example of a dynamical model for general perception is the adaptive resonance theory (ART) model of Grossberg (1995). This neural network is defined by a series of differential equations, similar to eqn 1 above, describing how the activation of any given node is increased or decreased by stimulus inputs, excitation and inhibition from other nodes, and intrinsic decay. This depends on weights (represented as matrices for a, b and c in eqn 1) which are modified by previous successful perceptual events (simulating learning from experience). The model can discover the low-level features that are most useful for distinguishing frequent patterns in its stimulus environment (using unsupervised learning) and identify specific high-level patterns even from noisy or incomplete inputs. It can also reassign resources whenever a new significant pattern appears in its environment, without forgetting earlier patterns. Notions like 'successful perception' and 'significant pattern' are provided with mathematical specifications that drive the system toward greater 'understanding' of its environment.

To recognize an object such as a letter of the alphabet from visual input, the signal from a spatial retina-like system excites low-level 'feature' nodes. The pattern of activated features here feeds excitation through weighted connections to a higher set of 'identification' nodes. These nodes compete through mutual inhibition to identify the pattern. The best matching unit quickly wins by suppressing all its competitors. When the match is good enough, a 'resonance loop' is established between some sensory feature units and a particular classification unit. Only at this point is successful (and, according to Grossberg, conscious) identification achieved. This perceptual model is dynamic because it depends on differential equations that increase or decrease the activation of nodes in the network at various rates. Grossberg's research group has shown that variants of this model can account in a fairly natural way for many phenomena of visual perception, including those involving backward masking and reaction time. (*See* **Perceptual Systems: The Visual Model**)

HIGH-LEVEL MODELS

Dynamical models have also been applied to higher-level cognitive phenomena. Grossberg and colleagues have extended the ART model with mechanisms such as 'masking fields', so that, for example, the model can recognize words from temporally arriving auditory input. Several time-sensitive phenomena of speech perception can be successfully modeled in this way (Grossberg, 1986).

Models of human decision-making have traditionally applied the theory of expected utility, whereby evaluation of the relative advantages and disadvantages of each choice is made at a single point in time. But Townsend and Busemeyer (1995) have developed a 'decision field theory' that not only accounts for the likelihood of each eventual choice, but also accounts for many time-varying aspects of decision making, such as 'approach–avoidance' or vacillatory effects, and the fact that some decisions need more time than others. (*See* **Decision-making; Choice Selection**)

Some phenomena that at first glance seem to depend on high-level reasoning skills may in fact reflect more low-level properties of cognition. One startling result of this kind is in the 'A-not-B problem'. Infants (9 to 12 months old) will sometimes reach to grab a hidden object; yet when the object is moved to a new location, they often reach to the first location again. This puzzle was interpreted by Piaget (1954) as demonstrating a lack of the concept of 'object permanence', that is, that children have an inadequate understanding of objects, thinking that they somehow intrinsically belong to the place where they are first observed. However, Thelen *et al.* (2001) demonstrated a dynamical model for control of reaching that predicted sensitivity to a variety of temporal variables in a way that is supported by experimental tests. Thus what seems at first to be a property of abstract, high-level, static representations may turn out to result from less abstract time-sensitive processes, which are naturally modeled using dynamical equations. (*See* **Object Concept, Development of**)

RELATION TO SITUATED COGNITION AND CONNECTIONISM

From the perspective of 'situated cognition', the world, the body and the cognitive functions of the brain can all be analyzed using the same conceptual tools. This is important because it greatly simplifies our understanding of the mapping between these domains, and is readily interpreted as an illustration of the biological adaptation of the body and brain to the environment on short-term and long-term time scales. (*See* **Perception, Direct**)

Connectionist models are discrete dynamical systems, as are the learning algorithms used with them. But not all connectionist models study phenomena occurring in continuous time. Neural network models are frequently used to study time-varying phenomena, but other dynamical methods that do not employ connectionist networks are also available. The development of connectionist modeling since the 1980s has helped to move the field in the direction of the dynamical approach, but connectionist models are not always good illustrations of the dynamical hypothesis of cognition. (*See* **Language, Connectionist and Symbolic Representations of**)

CONTRAST WITH TRADITIONAL APPROACHES

The most widespread conceptualization of the mechanism of human cognition proposes that cognition resembles computational processes, like deductive reasoning or long division, by using symbolic representations (of objects and events in the world) that are manipulated by cognitive operations, modeling time only as serial order. These operations reorder or replace symbols, and draw deductions from them (Haugeland, 1985). The computational approach has been articulated as the 'physical symbol system hypothesis' (Newell and Simon, 1976). The theoretical framework of modern linguistics (Chomsky, 1963, 1965 and Chomsky and Halle, 1967) also falls squarely within this tradition since it views sentence generation and interpretation as a serially ordered process of manipulating word-like symbols (such as 'table' or 'go'), abstract syntactic symbols (such as 'noun phrase' or 'sentence') and letter-like symbols representing minimal speech sounds (such as /t/, /a/ or features like 'voiceless' or 'labial') in discrete time. The computational approach, applied to skills like the perceptual recognition of letters of the alphabet and sounds, or recognizing a person's distinctive gait, or the motor control that produces actions like reaching, walking or pronouncing a word, hypothesizes that essentially all processes of cognition are computational operations that manipulate digital representations in discrete time. The mathematics of such systems is based on the algebra of strings and graphs of symbol tokens. Chomsky's work on the foundations of such abstract algebras (Chomsky, 1963) served as the theoretical foundation for computer science as well as modern linguistic theory. (*See* **Representation, Philosophical Issues about; Computation, Philosophical Issues about; Symbol Systems; Syntax**)

It should be noted that the dynamical systems hypothesis for cognition is in no way incompatible with serially ordered operations on discrete symbols. However, proponents of the dynamical systems approach typically deny that most cognition can be satisfactorily understood in computational terms. They propose that any explanation of human symbolic processing must sooner or later include an account of its implementation in continuous time. The dynamical approach points out the inadequacy of assuming that a 'symbol processing mechanism' is available to human cognition, as a computer happens to be available to a programmer. In the dynamical framework, the discrete time of computational models is replaced with continuous time; first and second time derivatives are meaningful at each instant; and critical time points are specified by the environment or the body rather than by a discrete-time device jumping from one time point to the next. (*See* **Symbol Systems**)

STRENGTHS AND WEAKNESSES OF DYNAMICAL MODELS

Dynamical modeling offers many important advantages over traditional computational cognition. First, the biological implausibility of digital, discrete-time models remains a problem. How and where in the brain might there be a device that would behave like a computer chip, clicking along infallibly performing operations on digital units? One answer that has often been put forward is 'we don't really know how the brain works, anyway, so this hypothesis is as plausible as any other' (Chomsky, 1965 and 2000). Such an answer does not seem as reasonable today as it did in the 1960s. Certainly neurophysiological function exhibits many forms of discreteness; but this fact does not justify the postulation of whatever kind of units and operations would be useful for a digital model of cognition. (*See* **Computation, Philosophical Issues about; Categorical Perception**)

Secondly, temporal data can, by means of dynamical models, be incorporated directly into cognitive models. Phenomena such as processing time (e.g., reaction time, recognition time, response time), and temporal structure in motor behavior (e.g., reaching, speech production, locomotion, dance), and in stimulation (e.g., speech and music perception, interpersonal coordination while watching a tennis match), can be linked together if critical, events spanning several domains can be predicted in time. (*See* **Perception: The Ecological Approach; Perception, Direct**)

The language of dynamical systems provides a conceptual vocabulary that permits unification of cognitive processes in the brain with physiological processes in our bodily periphery and with environmental events external to the organism. Unification of processes across these fuzzy and partly artificial boundaries makes possible a truly embodied and situated understanding of all types of human behavior. Discrete-time models are always forced to draw a boundary somewhere to separate the discrete-time, digital aspects of cognition from continuous-time physiology (as articulated in Chomsky's, 1965 distinction between 'competence' and 'performance'). (*See* **Neuropsychological Development; Performance and Competence**)

Thirdly, cognitive development and 'run-time' processing can now be integrated, since learning and perceptual and motor behavior are governed by similar processes even if on different timescales. Symbolic or computational models were forced to treat learning and development as separate processes unrelated to motor and perceptual activity.

Fourthly, trumping the advantages given above, dynamical models include discrete-time, digital models as a special case. The converse is not true: the sampling of continuous events permits discrete simulation of continuous functions, but the simulation itself remains discrete, and only models a continuous function to an accuracy dependent on its sampling rate (Port *et al.*, 1995). Thus, any actual digital computer is also a dynamical system with real voltage values in continuous time that are discretized by an independent clock. Of course, computer scientists prefer not to regard them as continuous-valued dynamical systems (because it is much simpler to exploit their digital properties), but computer engineers have no choice. Hardware engineers have learned to constrain computer dynamics to be governed reliably by powerful attractors for each binary cell, ensuring that each bit settles into either one of two states before the next clock tick.

These strengths of dynamical modeling are of great importance to our understanding of human and animal cognition. But there are several weaknesses of dynamical modelling. First, the mathematics of dynamical models are more inscrutable and less developed than the mathematics of digital models. It is much more difficult to construct actual models, except for carefully constrained simple cases.

Second, during some cognitive phenomena (for example, performing long division, or designing an algorithm, and possibly some processes in the use of language) humans appear to rely on ordered operations on discrete symbols. Although dynamical models are capable of exhibiting digital behavior, how a neurally plausible model could perform these tasks remains a puzzle. It seems that computational models are simpler and more direct, even if they remain inherently insufficient.

DISCRETE VERSUS CONTINUOUS REPRESENTATIONS

Intuitively one of the major strengths of the traditional computational approach to cognition has been the seeming clarity of the traditional notion of a cognitive representation. Since cognition is conceived as functioning somewhat like a program in LISP, the representations are constructed from parts that resemble LISP 'atoms' and 's-expressions'. (*See* **Symbol Systems**; **Computation, Philosophical Issues about**)

A representation is a distinct data structure that has semantic content (with respect to the world outside or inside the cognitive system). Representations can be moved around or transformed as needed. Such tokens have an undeniable resemblance to words and phrases in natural language (Fodor, 1975). Thus, if one considers making a sandwich from bread and the ham in the refrigerator, one can imagine employing cognitive tokens standing for bread, the refrigerator, and so on. Thinking about sandwich assembly might be cognitively modeled using representations of sandwich components. Similarly, constructing the past tense of 'walk' can be modeled by concatenating the representation of 'walk' with the representation of '-ed'. However, this view runs into difficulties when we try to imagine thinking about actually slicing the bread or spreading the mayonnaise. How could discrete, word-like representations be deployed to yield successful slicing of bread? If this is instead to be handled by a nonrepresentational system (such as a dynamical one), then how could we combine these two seemingly incompatible types of systems? (*See* **Representation, Philosophical Issues about**; **Language of Thought**)

In the 1980s, the development of connectionist models, employing networks of interconnected nodes, provided the first alternative to the view of representations as context-invariant, manipulable tokens. In connectionist models, the result of a process of identification (of, say, an alphabetic character or a human face) is only a temporary pattern of activations across a particular set of nodes (modeling cells or cell groups), not something resembling a context-free object. The possibility of representation in this more flexible form led to the notion of distributed representations, where no apparent 'object' can be found to do the work of representing, but only a particular pattern distributed over a set of nodes that are also used for many other patterns. Connectionists emphasized that such a representation would not seem to be a good candidate for a *symbol token*, as conceived in the formalist or computational tradition, yet can still function as a representation for many of the same purposes.

The development of dynamical models of perception and motor tasks has led to further extension of the notion of the representational function to include time-varying trajectories, limit cycles, coupled limit cycles, and attractors towards which the system state may tend. From the dynamical viewpoint, static, computational representations will play a far more limited role in cognition. Indeed, some researchers have denied that static representations are ever needed for modeling any cognitive behavior (Brooks, 1997).

References

Ashby R (1952) *Design for a Brain*. London, UK: Chapman-Hall.

Bernstein N (1967) *The Control and Regulation of Movements*. London, UK: Pergamon Press.

Brooks R (1997) Intelligence without representation. In: Haugeland J (ed.) *Mind Design II*, pp. 395–420. Cambridge, MA: MIT Press.

Chomsky N (1963) Formal properties of grammars. In: Luce RD, Bush RR and Galanter E (eds) *Handbook of Mathematical Psychology*, vol. II, pp. 323–418. New York, NY: Wiley.

Chomsky N (1965) *Aspects of the Theory of Syntax*. Cambridge, MA: MIT Press.

Chomsky N (2000) Linguistics and brain science. In: Marantz A, Miyashita Y and O'Neil W (eds) *Image, Language and Brain*, pp. 13–28. Cambridge, MA: MIT Press.

Chomsky N and Halle M (1967) *The Sound Pattern of English*: Harper and Row.

Fel'dman AG (1966) Functional tuning of the nervous system with control of movement or maintenance of a steady posture – III. Mechanographic analysis of the execution by man of the simplest motor tasks. *Biophysics* **11**: 766–775.

Fodor J (1975) *The Language of Thought*. Cambridge, MA: Harvard University Press.

van Gelder T and Port RF (1995) It's about time: overview of the dynamical approach to cognition. In: Port RF and van Gelder T (eds) *Mind as Motion: Explorations in the Dynamics of Cognition*, pp. 1–43. Cambridge, MA: MIT Press.

Grossberg S (1982) Studies of mind and brain: neural principles of learning, perception, development, cognition, and motor control. Norwell, MA: Kluwer.

Grossberg S (1986) The adaptive self-organization of serial order in behavior: speech, language and motor control. In: Schwab NE and Nusbaum H (eds) *Pattern Recognition by Humans and Machines: Speech Perception*, pp. 187–294. Orlando, FL: Academic Press.

Grossberg S (1995) Neural dynamics of motion perception, recognition, learning and spatial cognition. In: Port RF and van Gelder T (eds) *Mind as Motion: Explorations in the Dynamics of Cognition*, pp. 449–490. Cambridge, MA: MIT Press.

Haugeland J (1985) *Artificial Intelligence: The Very Idea*. Cambridge, MA: MIT Press.

Hodgkin AL and Huxley AF (1952) A quantitative description of membrane current and its application to conduction and excitation in nerve. *Journal of Physiology* **117**: 500–544.

Kopell N (1995) Chains of coupled oscillators. In: Arbib M (ed.) *Handbook of Brain Theory and Neural Networks*, pp. 178–183. Cambridge, MA: MIT Press.

Newell A and Simon H (1976) Computer science and empirical inquiry. *Communications of the ACM* **19**: 113–126.

Piaget J (1954) *The Construction of Reality in the Child*. New York, NY: Basic Books.

Port RF, Cummins F and McAuley JD (1995) Naive time, temporal patterns and human audition. In: Port RF and van Gelder T (eds) *Mind as Motion: Explorations in the Dynamics of Cognition*, pp. 339–371. Cambridge, MA: MIT Press.

Rumelhart D and McChelland J (1986) *Parallel Distributed Processing*, vols 1 and 2. Bradford Books, MIT Press.

Strogatz SH (1994) *Nonlinear Dynamics and Chaos With Applications to Physics, Biology, Chemistry, and Engineering*. Reading, MA: Addison-Wesley.

Thelen E, Schöner G, Scheier C and Smith LB (2001) The dynamics of embodiment: a field theory of infant perseverative reaching. *Behavioral and Brain Sciences* **24**: 1–34.

Townsend J and Busemeyer J (1995) Dynamic representation of decision making. In: Port RF and van Gelder T (eds) *Mind as Motion: Explorations in the Dynamics of Cognition*, pp. 101–120. Cambridge, MA: MIT Press.

Further Reading

Abraham RH and Shaw CD (1982) *Dynamics: The Geometry of Behavior*, vol. I. Santa Cruz, CA: Ariel Press.

Clark A (1997) *Being There: Putting the Brain, Body and World Together Again*. Cambridge, MA: MIT Press.

Haugeland J (1985) *Artificial Intelligence: The Very Idea*. Cambridge, MA: MIT Press.

Kelso JAS (1995) *Dynamic Patterns: The Self-Organization of Brain and Behavior*. Cambridge, MA: MIT Press.

Port RF and van Gelder T (1995) *Mind as Motion: Explorations in the Dynamics of Cognition*. Cambridge, MA: MIT Press.

Thelen E and Smith LB (1994) *A Dynamical Systems Approach to the Development of Cognition and Action*. Cambridge, MA: MIT Press.

Dynamical Systems, Philosophical Issues about

Intermediate article

James Garson, University of Houston, Houston, Texas, USA

CONTENTS

Dynamical systems theory provides an alternative to the dominant paradigm in cognitive science that claims human intelligence results from digital computation. The dynamical account of cognition avoids symbolic representation, and stresses the importance of modeling human interactions with the external environment through time.

INTRODUCTION

Dynamical systems theory (DST) has provided a new paradigm for understanding complex systems. The concepts and methods developed to describe nonlinear dynamics have rapidly spread from physics, to chemistry, biology, neurology, ecology, geography, economics, and political science. It is natural to think that the same methods might contribute to cognitive science. The dynamical systems hypothesis (DSH) proposes that the methods and results of DST provide genuine new insights into the nature and explanation of cognition – insights that are missed by the traditional approach in cognitive science, which is to view human intellect as the product of symbolic computation. The differences between the symbolic and dynamical viewpoints raise a number of important philosophical issues. These include the role of symbolic representations, the importance of interaction with the environment, and the way in which time is treated in cognitive models.

WHAT ARE DYNAMICAL SYSTEMS?

Dynamical systems are models of the phenomena of nature that employ the methods of DST. DST is not so much a theory as a body of conceptual and mathematical tools that provide insight into the complex behavior that is generated in nonlinear systems, where solving the equations describing how properties (variables) change through time can be difficult or impossible. DST has developed concepts and graphical techniques for describing the qualitative nature of such systems. Where mathematical solutions for the equations that describe the changes in the variables are unavailable, progress can still be made by simulating a system's behavior on computers and by applying general knowledge drawn from the study of related systems. A main focus of this work is to identify a range of stable and unstable behaviors that the system takes on in responding to forces that affect it. (*See* **Dynamical Systems: Mathematics**)

Phase Space

The concept of phase space (or state space) is a foundational notion in DST. The phase space for a system has a separate dimension devoted to each variable in its equations. For a simple pendulum, for example, the phase space might have two dimensions, one for the position (x) and one for the velocity (v) of the pendulum bob. A phase space describing the brain's neural activity, on the other hand, might include a separate dimension for the firing frequency of each of over 100 billion neurons.

A point in a two-dimensional space indicates a value for each of the two dimensions. For example, the point (x, v) in the phase space for a pendulum might indicate that the position of its bob is x meters away from vertical, and that its velocity is v meters per second. If we give a pendulum bob a push, it will swing back and forth. This oscillating behavior corresponds to a moving point in phase space, with the x and v values of the point changing as the pendulum bob takes on new positions and velocities. The path this point takes is called the *phase trajectory*. Assuming no friction, the trajectory of a pendulum forms a closed loop. If there is friction, it forms a curve that spirals towards the origin ($x = 0$, $v = 0$), indicating that the pendulum bob is ultimately motionless in a vertical position.

Attractors

Most dynamical systems can be described by 'attractors', which are points, lines or higher dimensional surfaces in the phase space that represent stable motions of the system. A system that is disturbed so that its phase trajectory is moved away from an attractor will soon return to it. There are several fundamentally different kinds of attractors. In the case of a stable attractor, the phase trajectory evolves towards a point in the phase space. The resting position of the pendulum under friction is a good example, for here the system has reached a point of stable equilibrium. A second kind of trajectory is called a limit cycle. In this case, the pathway starting from initial conditions settles into a repeating sequence. For example, the friction-free pendulum traces a limit cycle. A third kind of attractor is the torus, a donut-shaped surface. A phase trajectory on a torus is a curve that may never loop back on itself. 'Strange attractors' are the signature of chaos. These attractors are very complex, and tend to 'fill' large regions of their phase space. Chaotic systems are practically impossible to predict because the slightest deviation in the values of their variables will be quickly magnified into large differences in outcome.

Chaos and Complexity

DST is sometimes called 'chaos theory', but the use of the world 'chaos' is misleading for two reasons. Firstly, it suggests randomness or disorder, yet many chaotic dynamical systems spontaneously create structure. Secondly, chaos is only one of several types of behavior studied by DST. It is not even accurate to refer to a system as chaotic, since chaotic behavior may come and go depending on differences in the system's parameters. Although chaos is not a central feature in most dynamical models of cognition, the possibility of chaotic and near-chaotic behavior has interesting implications.

Dynamical systems described by even very simple equations generally have complex trajectories. Such trajectories display a kind of richness of behavior that defies accurate characterization by any digital program or set of rules small enough to be actually written out or understood by humans. (Although the equations of the system might be easy to state, only an impractically large and fast digital computer could accurately predict the behavior determined by those equations.) The existence of such chaotic and near-chaotic complex behavior in dynamical systems is an important consideration in understanding the contribution of DST to cognitive science. (*See* **Real-valued Computation and Complexity**)

Self-organization and the Emergence of Order

In nonlinear dynamical systems, an activity that is seemingly disordered at the lowest level can still produce coherent and stable large-scale structures. Chaotic models of Jupiter's famous red spot suggest that such structures can persist autonomously for a long time, and that they do not need any special mechanism to create or maintain them (other than the natural behavior of the system as a whole). Contrary to our intuition that forms of order must always cancel out in a highly energetic and unguided system such as the atmosphere of a gigantic planet, DST shows that structures such as the red spot emerge naturally from the dynamics of gases on the surface of a heavy rotating sphere. It has been proposed that cognition may depend on the same kind of emergent self-organization. (*See* **Emergence**)

WHAT IS THE DYNAMICAL SYSTEMS HYPOTHESIS?

The dynamical systems hypothesis is that cognition is a dynamical system and so is best understood using the concepts and methods of DST. The DSH is often contrasted with the hypothesis that cognition should be modeled on the symbolic operation of a digital computer. The conflict between the two hypotheses centers around the concepts and methods of explanation they advocate. Symbolic modelers attempt to model human knowledge by postulating the existence of symbolic representations in the brain that record information about the external world. They also assume the existence of programs or sets of rules that the brain uses to transform these symbolic data into new forms. These program-guided transformations are thought to account for cognitive processes such as conceptualizing, planning, reasoning, decision making, and eventually action. (*See* **Symbolic versus Subsymbolic; Symbol Systems**)

Dynamicists, on the other hand, view the brain as a dynamical system which is continually affected by interaction with its environment. Dynamical representation is not symbolic, but is understood via the concepts that DST uses to understand structures in phase space, including equation parameters, trajectories or their parts, and attractors.

Levels of Description and the DSH

A common misconception is that the DSH demands that cognitive explanation be carried out in terms of variables for brain features such as the firing rates of neurons, and the strength of the connections between them. Such connectionist models do fall within the DSH. However, dynamical models can also be constructed at higher levels of description. For example, Townsend and Busemeyer (1995) have applied DST to characterize decision making. The variables of their models are clearly cognitive, for they measure the gains or losses expected for the various factors (or dimensions) that are relevant to the decision. (*See* **Connectionism**)

The Role of Representation in the DSH

One of the most important differences between symbolist and DSH approaches concerns the roles that representations play in the two paradigms. The symbolist theory assumes that cognition is correctly described as the execution of programs or subroutines, which are defined over representations. Here representations are directly implicated in a causal explanation of cognitive performance. On the DSH approach, the analogs of programs are the equations that govern the evolution of a system. These equations are not ordinarily defined over representations, but instead over the variables for the system's properties. From the DSH point of view, representations are not essential to an explanation of the mechanisms of cognition, since what matters is the way in which system variables evolve according to the system's equations of motion. Although dynamical explanation may mention representations, these are conceived of as emergent aspects of system activity.

An Illustration of the Difference between Symbolist and DSH Models

An illustration may help to explain the difference between symbolical and dynamical conceptions of representation. Consider two different ways of constructing a thermostat that controls a furnace. The 'symbolic' thermostat would collect temperature information about the outside world with sensors, and convert that information into numbers that are stored in a memory M of a small computer. A program running on that computer includes instructions such as 'if the number in M is smaller than 20 then turn on the furnace burner'. On the other hand, the 'dynamical' thermostat would consist of a device that connects the furnace to a bimetallic strip in such a way that as air temperature falls, the strip bends, thereby opening a valve that sends more fuel to the furnace's burner.

In the symbolic thermostat, digital representations of the temperature in the room interact with program commands to control the furnace. It would be inappropriate, however, to characterize the dynamical thermostat on the computational model, since no computation over data representing the world is performed. A dynamical model would provide equations describing the interactions between variables for room temperature and valve position. Although one might claim that the amount of bend in the bimetallic strip carries information about (or even represents) the temperature of the room, a satisfactory explanation of how the system works can be made without treating the strip as explicitly representational. The state of the strip is not symbolic data to be processed by a program; it is a part of the mechanism that directly ensures that room temperature and valve opening interact in the right way. Here, in contrast to the symbolist thermostat, we lack any meaningful distinction between the data and the procedures that operate on the data.

Time and Interaction in the DSH

The example of the dynamical and symbolic thermostats helps illustrate other features that distinguish the DSH from the symbolist hypothesis. The DSH promises to provide a richer framework for understanding the evolution of a system through time. The dynamical thermostat can be modeled with a system of equations that specify the relationships between room temperature, the furnace valve position, and its effect on room temperature. These equations specify the rates at which the variables change. A feedback system of this kind displays a rich variety of behavior including oscillations and overshoot when the feedback is too strong or delayed. A symbolist model does not have the well-developed temporal concepts that would explain such dynamic behavior: it does not explicitly treat time as a graded or real-valued quantity. The rate at which quantities change is largely ignored. When time is mentioned at all, it is in description of the sequence of steps a computer performs as it follows its program. Since time is treated as a sequence of discrete moments, rather than a continuous quantity, symbolist models have difficulty capturing interactions with the environment that depend on how quickly system variables change.

A second difference is that the symbolist has a tendency to view the system as isolated from the environment, being driven by its representations of the world rather than by the world itself. Dynamicists will claim that a better way of understanding cognition is to model the interaction with the world directly.

ARGUMENTS FOR THE DYNAMICAL SYSTEMS HYPOTHESIS

The DSH is in its infancy, and there is not yet conclusive evidence in its favor. However, the available evidence suggests that the DSH is well worth exploring. Firstly, the DSH has a record of success across a wide range of cognitive abilities, including perception, sensorimotor activity, language, attention, decision making, and development (Port and van Gelder, 1995). Secondly, as explained above, the treatment of time and interaction with the environment is richer and more natural in the DSH. Two further arguments for the DSH are worth discussing in detail.

Problems in the Symbolist Paradigm

Certain problems have emerged for the symbolist paradigm. Research programs in artificial intelligence that attempt to symbolically approximate human thought have met with difficulties. The complexity of symbolic programs increases rapidly as rules are elaborated to handle an endless stream of discovered exceptions. Disenchantment with symbolic methods, and interest in alternatives, has arisen from a belief that symbolic methods have failed.

According to the DSH, this failure is a symptom of the fact that the dynamics of the brain embody such complex information processing that even a marginally accurate symbolic approximation would require impossibly complex programs. Although cognition is clearly ordered in many respects, it may be that this order can be expressed in compact form only as 'soft laws' (Horgan and Tienson, 1990). Soft laws incorporate ineliminable exceptions (*'ceteris paribus'* clauses) that reflect the subtlety and flexibility found in human thought. If cognitive regularities are truly soft in this way, then attempts to characterize cognition in the form of programs that embody strict rules are bound to fail.

Even defenders of the symbolist paradigm have voiced doubts about whether symbolist methods can capture higher-level cognition. Given the persistent failure of the symbolist paradigm to present a plausible theory of the flexibility of thought and language, there is every reason to seek an alternative.

The Power of Dynamical Processing

One source of evidence for the DSH comes from the study of dynamical computation (Kauffman, 1993, chapters 5–6). Crutchfield and Young (1990) have examined the computational complexity of high-level system behavior over a range of underlying system dynamics. When a dynamical system exhibits short limit cycles, its computational powers are weak. At the interface between long limit cycles and chaos, the power increases to a maximum, while deeper in the chaotic realm, computational power wanes. Kauffman (1993, p. 221) modeled the evolution of dynamical systems whose survival depends on the ability to solve a simple task. His results lead to a similar conclusion: higher fitness corresponds to complex dynamical behavior at the 'edge of chaos'. Furthermore, he gives evidence that systems evolve more quickly when they operate in this realm.

Work on the superiority of analog over digital computation underscores the potential information processing advantages of dynamical models. Blum *et al.* (1989) have shown that analog computation provides a vast increase in the range of processing available to the brain, and Siegelmann and Sontag (1994) have shown the superiority of analog computation in models of noisy neural networks. Such evidence is far from conclusive, but it suggests that highly complex dynamical processing may account for the intelligent adaptability and avoidance of overly rigid or stereotypical behavior that characterize human cognitive performance.

PROBLEMS WITH THE DYNAMICAL SYSTEMS HYPOTHESIS

Since the DSH challenges an established paradigm, it is not surprising that it faces strong criticism. Some of the main objections are described below.

Problems in Dynamical Modeling

Modeling cognition with DST is not easy. Defining the relevant variables and the dynamical equations that govern them is a daunting task that too often depends on guesswork. Some tools available in

DST allow the researcher to investigate dynamical models even when the actual equations of motion are not known in detail. However, a fundamental problem remains. Even when one has the luck to produce a model that approximates fairly well the behavior to be explained, it can be difficult to determine whether that success is genuine or the result of 'fudge factors' written into the equations.

The 'Wrong Level' Objection

It has been claimed that the DSH attempts to explain cognition at the wrong level. Perhaps DST is useful in giving an account of the behavior of the brain in physical, chemical, or neurological terms, but this does not help us make progress in cognitive science. What is needed is a theory that explains human action in terms of perceptions, memories, beliefs, desires, reasons, plans, and goals. Although some aspects of perception and motor control may be illuminated by DST, the understanding of the central processes that govern human intelligence would seem to require a very different vocabulary from what appears to be available to DST.

One response to this objection has been outlined already, namely that DST is not limited to models framed at the neurological level. Furthermore, work on central processes within DST is in its infancy. Only time can tell whether DST models of a full range of higher cognitive abilities will be successful, either by employing variables for concepts already available in cognitive psychology, or by providing the tools that will allow us to construct more fruitful cognitive-level concepts.

The Systematicity Objection

One of the most widely discussed criticisms of the DSH appeals to the systematicity of higher-level cognitive abilities such as language and reasoning. The systematicity of language means that the ability to produce and understand some sentences is intrinsically connected to the ability to produce and understand others of related form. For example, no one with a command of English who understands 'John loves Mary' can fail to understand 'Mary loves John'.

The systematicity objection proposes that such facts can be explained only by assuming that cognition operates over symbolic strings (such as 'John loves Mary') composed of constituents ('John', 'loves', and 'Mary') that can be combined in different ways. A speaker of English computes the meaning of a string from the meanings of its constituents. If this is so, then understanding 'Mary loves John' can be accounted for as another instance of the same process. Some have claimed that no alternative solution can work, so that symbolic processing of representations with symbolic constituents is required to explain cognition.

It is indeed difficult to explain systematicity without adopting a model that explicitly employs symbolic constituents. Research in DST (van Gelder and Port, 1994) offers some hope that dynamical systems can create the needed combinatorial structures without implementing symbolist structures. However, a convincing response to the systematicity challenge will require more research.

PHILOSOPHICAL ISSUES ABOUT THE DYNAMICAL SYSTEMS HYPOTHESIS

The mission of cognitive science is to explain the highly flexible structure embodied in human intelligence. The DSH requires us to reconsider both the nature of that structure and the methods we use to explain it.

Representation

The DSH undermines common presuppositions about representation in the brain. Symbolists tend to think that having representations of the world is a prerequisite for having knowledge about it. The DSH stresses the idea that intelligence is the result of an ongoing interaction with the world that may not require explicit storage of information. Symbolists tend to think of representations as static entities waiting to be processed by programs. Dynamicists tend to view representation as the product of the system's dispositions to interact with the world as reflected in its attractor 'landscape'. The notion of a phase space in DST provides a framework for understanding the relationships between different representational states. Since representational features are defined in terms of a high-dimensional space, the notion of the distance or similarity between representations is well defined. Within this framework, the features of representations that account for their cognitive roles can be understood in terms of DST concepts that help characterize the similarities and differences to be found in phase space. (*See* **Language, Connectionist and Symbolic Representations of**)

Soft Laws

We tend to expect a science to provide us with the laws of its domain; so that cognitive science should discover the laws of human thought. But the DSH may require us to revise our understanding of the nature of laws, and hence of the goals of cognitive science. It has been widely noticed that the laws of psychology resist formulation. Some have concluded from this that the study of cognition can never meet the requirements of science; but the DSH suggests that we may need to revise our conception of laws in cognitive science. Soft laws incorporate *'ceteris paribus'* clauses, conceding the existence of exceptions. The DST perspective suggests that such laws may be preferable to hard laws in psychology and other sciences.

The Nature of Explanation in the DSH

The DSH also invites a new conception of explanation in cognitive science. To illustrate this point, let us contrast symbolist and dynamical explanations for how the leopard gets its spots. The symbolist assumes that the genetic code specifies a template for leopard spot patterns that guides the way spots are formed in leopard skin during development. The fetus incorporates a program that reads this template and fixes the colors that fetal skin cells will express. On this paradigm, explaining how the leopard gets its spots amounts to locating the template and the mechanism that reads it. The dynamicist's explanation is very different. The chemistry of fetal leopard skin can be modeled by a nonlinear dynamical system that spontaneously creates spots in leopard-like patterns. So, in a sense, the spot shapes have no explanation apart from the observation that the right dynamical systems make those patterns naturally.

This dynamical explanation may appear vacuous. The symbolist explanation accounts for the mechanism that forces fetal leopard skin to form the right pattern of spots. However, if the dynamicist is right, the project to find a template and program is bound to fail. The lesson of dynamical systems theory is that very complex patterns can emerge spontaneously from the seemingly random interaction of simple units such as skin cells. Patterns may emerge without symbolic guidance.

The same moral may apply to cognition. From the DSH perspective, intelligence is the product of self-organized structures that arise naturally. Searching for programs and data that guide the brain is like searching for a force that keeps objects moving at a constant velocity. No explanation for how cognitive systems behave is necessary beyond merely pointing out that the dynamical systems responsible spontaneously create that behavior.

THE DYNAMICAL SYSTEMS HYPOTHESIS IN COGNITIVE SCIENCE

The relationship between the DSH and alternative paradigms in cognitive science is complex.

Connectionism and the DSH

Connectionism (rather than DST) has so far presented the most popular alternative to the symbolist approach. Connectionists attempt to construct models of interconnected neurons in the brain that are capable of cognitive tasks. While some connectionists accept the thesis that the brain relies on symbolic processing, radical connectionists seek to show that neural networks can produce cognition using non-symbolic methods. Thus both dynamicists and radical connectionists are opposed to symbolism. Furthermore, neural networks are examples of dynamical systems, and some connectionists employ concepts from DST to help understand their behavior.

Symbolism and the DSH

It can be shown that a digital computer can be treated in DST as a special case of a dynamical system. Furthermore, dynamicists typically use digital computers to investigate the behavior of their models, so it may appear that everything they discover also has a symbolist account.

What is Unique about the DSH

However, the fact that symbolist and connectionist models can be treated as dynamical systems does not imply that the DSH is compatible with either paradigm. The main feature that distinguishes the three accounts is found not in *what* they model but in *how* they model. The DSH proposes that the concepts and methods of DST are what is needed to make progress in cognitive science. Symbolism employs entirely different methods and concepts. While some connectionists use tools drawn from DST to obtain insight into the functioning of their neural networks, they differ from dynamicists in their vision of cognition. The DSH views the mind as a structure in constant resonance with changes in the environment. It seeks an explanation for cognition in the complex and often self-organized structures that emerge from systems of nonlinear equations that determine how variables change value smoothly through time. Only time will tell which of these views is the right one or whether more than one of them may be needed to give a full account of the richness of the human mind.

References

Blum L, Shub S and Smale S (1989) On a theory of computation and complexity over the real numbers: NP-completeness, recursive functions and universal machines. *Bulletin of the American Mathematical Society* **21**(1): 1–46.

Crutchfield J and Young K (1990) Computation at the onset of chaos. In: Zurek W (ed.) *Complexity, Entropy, and Physics of Information*, pp. 223–269. Redwood City, CA: Addison-Wesley.

van Gelder T and Port R (1994) Beyond symbolic: prolegomena to a kama-sutra of compositionality. In: Honavaar V and Uhr L (eds) *Artificial Intelligence and Neural Networks: Steps Towards a Principled Integration*. pp. 107–125. New York, NY: Academic Press.

Horgan T and Tienson J (1990) Soft laws. *Midwest Studies in Philosophy* **15**: 256–279.

Kauffman S (1993) *The Origins of Order*. New York, NY: Oxford University Press.

Port R and van Gelder T (1995) *Mind as Motion: Explorations in the Dynamics of Cognition*. Cambridge, MA: MIT Press.

Siegelmann H and Sontag E (1994) Analog computation via neural networks. *Theoretical Computer Science* **131**: 331–360.

Townsend J and Busemeyer J (1995) Dynamic representation of decision making. In: Port R and van Gelder T (eds) *Mind as Motion: Explorations in the Dynamics of Cognition*, pp. 101–120. Cambridge, MA: MIT Press.

Further Reading

Clark A (1997) *Being There: Putting Brain Body and World Together Again*. Cambridge, MA: MIT Press.

Elman J (1995) Language as a dynamical system. In: Port R and van Gelder T (eds) *Mind as Motion: Explorations in the Dynamics of Cognition*, pp. 195–225. Cambridge, MA: MIT Press.

Gleick J (1987) *Chaos: Making the New Science*. New York, NY: Viking.

Gregson R (1995) *Computation, Dynamics and Cognition*. New York, NY: Oxford University Press.

Horgan T and Tienson J (1996) *Connectionism and the Philosophy of Psychology*. Cambridge, MA: MIT Press.

Kelso J (1995) *Dynamic Patterns: The Self Organization of Brain and Behavior*. Cambridge, MA: MIT Press.

Langton C (1991) Computation at the edge of chaos. In: Forrest S (ed.) *Emergent Computation*, pp. 12–37. Cambridge, MA: MIT Press.

Murray J (1988) How the leopard gets its spots. *Scientific American* **258**(3): 80–87.

Port R and van Gelder T (1995) *Mind as Motion: Explorations in the Dynamics of Cognition*. Cambridge, MA: MIT Press.

Skarda C and Freeman W (1987) How brains make chaos in order to make sense of the world. *Behavioral and Brain Sciences* **10**: 161–195.

Dynamical Systems: Mathematics

Intermediate article

Jerome R Busemeyer, University of Indiana, Bloomington, Indiana, USA

CONTENTS

Dynamical systems are mathematical models for describing how the state of a biological or an artificial system changes over time.

INTRODUCTION

A critical factor facilitating the 'cognitive revolution' of the early 1960s was the capability of formulating rigorous (computational or mathematical) models of how the inputs and outputs of mental processing systems change over time. The earliest approach was to view the mind as a dynamical cybernetic feedback system (Miller, Gallanter and Pribram, 1960). But this was abandoned in favour of another approach, which was to view the mind as a rule-based symbol processor (Newell and Simon, 1972). Most recently, developments in neural and connectionist networks have revived interest in a dynamical systems approach. More broadly, dynamical systems theory has been adopted by a wide range of fields in cognitive science, including the study of perceptual motor behavior, child development, speech and language, and artificial intelligence. There are excellent presentations of mathematical dynamical systems theory in Beltrami (1987), Luenberger (1979), Padulo and Arbib (1974), and Strogatz (1994).

ELEMENTS OF DYNAMICAL SYSTEMS

Generally speaking, a *dynamical system* is composed of three parts. The first part is the *state*, which is a representation of all the information about the system at a particular moment in time. As an example, the state of a computer can be summarized by a long list of the binary bits of information stored in the registers and memory banks at a given moment. The state of a brain model can be represented as a large-dimensional vector of positive real numbers, representing all the neural activations at a given moment. In general, the notation $X(t) = [x_1(t), \dots x_n(t)]$ will be used to denote the state of a system at time t.

The second part is the *state space*. This is a set that contains all of the possible states to which a system can be assigned. The state space of a computer is the set of all of the possible configurations for the n-element binary-valued list representing its state (a set of size 2^n). The state space of a brain model is the set of points contained in the positive region of the n-dimensional Cartesian vector space IR^n. The symbol Ω is used to denote the state space of a dynamical system, and $X(t) \in \Omega$.

The third part is the *state transition function*, which is used to update and change the state from one moment to another. For example, the state transition function of a computer is defined by the production rules that change the bits of information from the state at one step $X(t)$ to the state at the next step $X(t+1)$. The state transition function for a brain model is a continuous function of time that maps the brain state $X(t)$ at time t to another state $X(t+h)$ at a later moment. The symbol $\boldsymbol{T}$ is used to denote the state transition function that maps an initial state $X(t)$ into a new state $X(t+h)$:

$$X(t+h) = \boldsymbol{T}(X(t), t, t+h) \tag{1}$$

Whenever the state transition function is assumed to be a differentiable function of time, then we can define the *local generator* as the time derivative

$$\frac{\mathrm{d}\boldsymbol{T}}{\mathrm{d}t} = \lim_{h \to 0} \frac{X(t+h) - X(t)}{h} = \boldsymbol{f}(X(t), t) \tag{2}$$

Given an initial starting state $X(0)$, the local generator is used to generate a *trajectory* $X(t)$ for all $t > 0$. Provided that the local generator satisfies certain smoothness properties, a local generator is guaranteed to produce a unique trajectory from any given starting state. The objective of dynamical systems analysis is to understand all the possible trajectories produced by a local generator.

EXAMPLES OF DYNAMICAL SYSTEMS

Before describing the analysis of dynamical systems in more detail, it will be helpful to describe some well-studied examples.

Logistic Growth Model

Consider the following simple dynamical model of growth. To be concrete, suppose we wish to model a student's probability of performing a task correctly as a function of training (e.g., successfully playing a piece of classical music on the piano). Let $p(t)$ be the probability of correctly performing the task, and assume that this preference changes from one time point t to a later time point $t+h$ according to the following simple difference equation:

$$\frac{p(t+h) - p(t)}{h} = \alpha p(t) \times (1 - p(t)) \tag{3}$$

This model can be understood as a product of two parts. The second part, $1 - p(t)$, represents the amount that remains to be learned. The first part, $\alpha p(t)$, can be interpreted as the learning rate, which is an increasing function of the amount already learned. Note that as the probability approaches zero or one, then the change approaches zero; in effect, the probability remains bounded between zero and one.

This difference equation can be reformulated as a differential equation by allowing the time interval h to approach zero in the limit:

$$\frac{\mathrm{d}p(t)}{\mathrm{d}t} = \lim_{h \to 0} \frac{p(t+h) - p(t)}{h} = \alpha p(t)(1 - p(t)) \tag{4}$$

By separating the variables and integrating, we can solve this differential equation to yield the solution (Braun, 1975)

$$p(t) = (1 + c\mathrm{e}^{-\alpha t})^{-1} \tag{5}$$

where the constant c depends on the initial state, $p(0)$. Figure 1 is a time series plot that illustrates

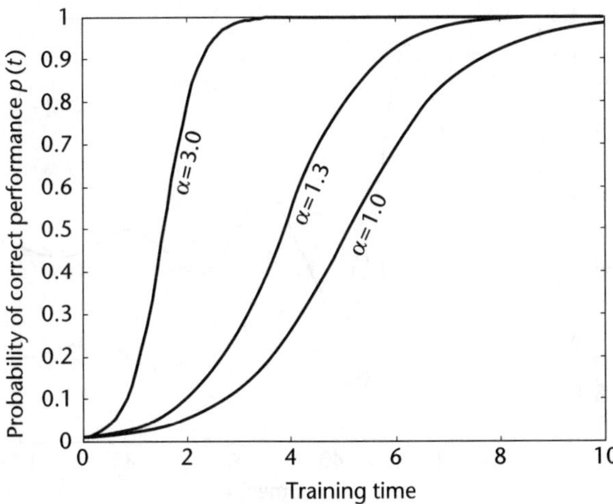

Figure 1. A time series plot showing the trajectories produced by the logistic growth model. The horizontal axis represents training time, and the vertical axis represents the probability of correct task performance.

performance over time (with $c = 100$ so that $p(0)$ is very close to zero). The model produces a family of smooth S-shaped curves that gradually rise from the initial level and approach the line of stable equilibrium $p = 1$. The growth rate, α, determines the steepness of the curve.

Linear System Model

Next consider the problem of developing a simple dynamical model of motivation to perform a task (Atkinson and Birch, 1970; Townsend and Busemeyer, 1989). Suppose a student is trying to decide how much effort to invest in a task over time, for example, the amount of time to allocate towards achieving an athletic, academic, or social goal. Let $q(t)$ be the amount of effort actually expended at time t, and let $p(t)$ be the student's strength of motivation for doing the task at time t.

First we assume that the rate of change in effort is directly influenced by the motivation state at a given time. But due to fatigue, the rate of change in effort also decreases in proportion to the current effort. These assumptions are incorporated into the following model:

$$\frac{\mathrm{d}q(t)}{\mathrm{d}t} = p(t) - \alpha q(t) \tag{6}$$

Secondly, we assume that the rate of change in motivation is determined by two factors. One is the valence difference between the anticipated rewards for success and punishments for failure, denoted by $v(t)$. Another is a satiation effect, which causes the motivation to decrease in proportion to the current effort. The following model is used to represent these assumptions:

$$\frac{\mathrm{d}p(t)}{\mathrm{d}t} = v(t) - \beta q(t) \tag{7}$$

Given the initial states, $p(0)$ and $q(0)$, and given the valence differences, $v(t)$, $t \geq 0$, eqns 6 and 7 define a simple dynamical system that can be used to model the behavior of a student's effort on a task over time. The valence $v(t)$, is called the *input* or *forcing term* of the system, and the coefficients α and β are called the *parameters* of the system.

Methods for solving *coupled systems* of linear differential equations, such as eqns 6 and 7, are covered in standard texts on ordinary differential equations (see e.g. Braun, 1975). In the special case when $p(0) = q(0) = 0$ and the input valence is constant ($v(t) = v$), the solution for effort is

$$q(t) = \frac{v}{\beta}\left(1 - \mathrm{e}^{-\frac{1}{2}\alpha t}(\sin\theta t + \frac{\alpha}{2}\cos\theta t)\right) \tag{8}$$

where $\theta^2 = \beta - (\frac{\alpha}{2})^2$. (This can be checked by differentiating eqn 8 with respect to t and showing that it satisfies both eqns 6 and 7 and the initial conditions.)

The time series plot in figure 2 illustrates the behavior predicted by the model (with the parameters $\alpha = 0.25$, $\beta = 1.00$ and $v = 0.50$). The horizontal axis shows the time elapsed on the task, and the vertical axis shows the effort spent. The curve oscillates up and down (between zero and one), approaching the asymptotic level of effort $q = 0.50$.

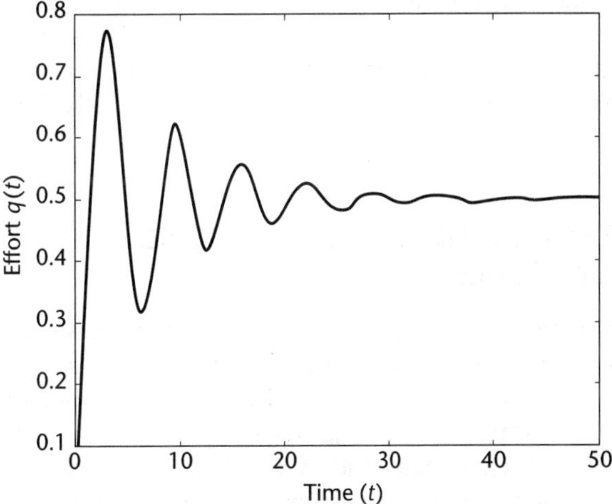

Figure 2. A time series plot showing the trajectory produced by the simple dynamic model of effort. The horizontal axis represents time elapsed on the task, and the vertical axis represents the amount of effort as a function of time.

Predator–Prey Model

Suppose a researcher is trying to decide how much effort to spend on generating and testing ideas. Let x_1 be the number of candidate ideas generated for testing, and let x_2 be the number of ideas tested. Note that an idea cannot be tested until it has been generated; and once it has been tested it is eliminated from the candidate pool. Consider the following simple nonlinear model of this process:

$$\frac{dx_1}{dt} = \alpha x_1 - \beta x_1 x_2 - \gamma x_1^2 \quad (9)$$

$$\frac{dx_2}{dt} = \phi x_1 x_2 - \lambda x_2 \quad (10)$$

In eqn 9, the coefficient α allows candidate ideas to grow over time, the coefficient β reflects depletion of candidate ideas that are generated and tested, and the coefficient γ represents interference between ideas. In eqn 10, the coefficient ϕ represents the increase in testing produced by a successful test and the coefficient λ provides for a fatigue effect from testing ideas. All of these coefficients are assumed to be positive.

Eqns 9 and 10 form what is called a predator–prey model. In this example, the idea-testing process is playing the role of the predator, which is preying on the idea-generating process. When dealing with nonlinear differential equations such as this, the most practical method for finding solutions is to use numerical integration routines, which are available in mathematical programming languages such as Matlab® or Mathematica®.

Figure 3 shows a time series plot for the ideas generated and tested (with $\alpha = 0.2$, $\beta = 0.3$, $\gamma = 0.1$, $\phi = 0.6$, $\lambda = 0.5$, and $x_1(0) = x_2(0) = 0.1$). As seen in the figure, generated ideas must build up first, and testing ideas dominates later. Eventually, both processes approach a steady state.

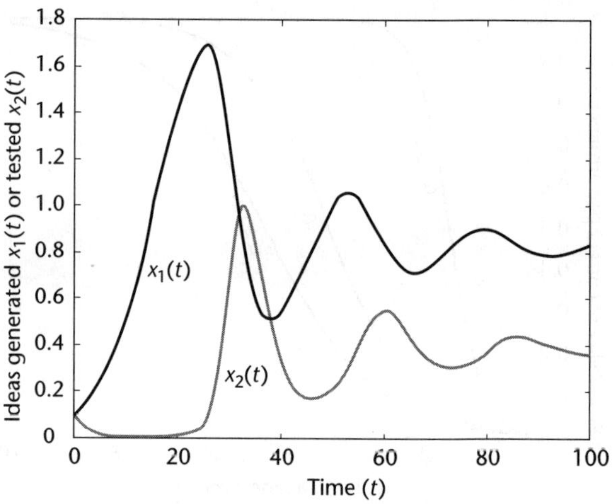

Figure 3. A time series plot showing the trajectories produced by the predator–prey model of idea generation and testing. The horizontal axis represents time elapsed on the task, and the vertical axis represents the number of ideas generated and tested, where generation leads testing.

PROPERTIES OF DYNAMICAL SYSTEMS

Dynamical systems describe the general laws that systems obey. Various special cases of the general law can be derived simply by changing either the initial state $X(0)$, or the system parameters. For a given initial state and a fixed set of parameter values, a dynamical system generates a unique trajectory or path through the state space as a function of time (like that shown in figure 1). The primary goal of dynamical systems theory is to develop analytic methods for studying all the possible trajectories produced by a dynamical system, and to understand how these trajectories change as a function of the initial state and system parameters.

Phase Portraits and Attractors

Figure 4 illustrates these ideas using the simple linear system model described above. The four panels shown in the figure are called *phase portraits.* The two-dimensional plane within each portrait represents the state space of the system. The four different portraits were computed from eqns 6 and 7 by varying the parameter α in eqn 6. The remaining parameters of the model were fixed at $\beta = 1$ and $v = 1$.

Each arrow inside each phase portrait is a velocity vector representing the directional rate of change in motivation and effort that occurs at a particular state of the system. In other words, the head of an arrow indicates where the state will move in the next instant, given the current state indicated by tail of the arrow. Thus the arrows indicate the flow of the dynamical system. The smooth curve following the flow within each portrait indicates the trajectory produced by the system with initial state $p(0) = q(0) = 0$. Different trajectories would result from different choices of initial state. In fact, each initial state defines a unique trajectory, and therefore the trajectories never cross (provided that the local generator satisfies certain smoothness conditions).

The four panels illustrate how the dynamical properties of the model depend on the parameter

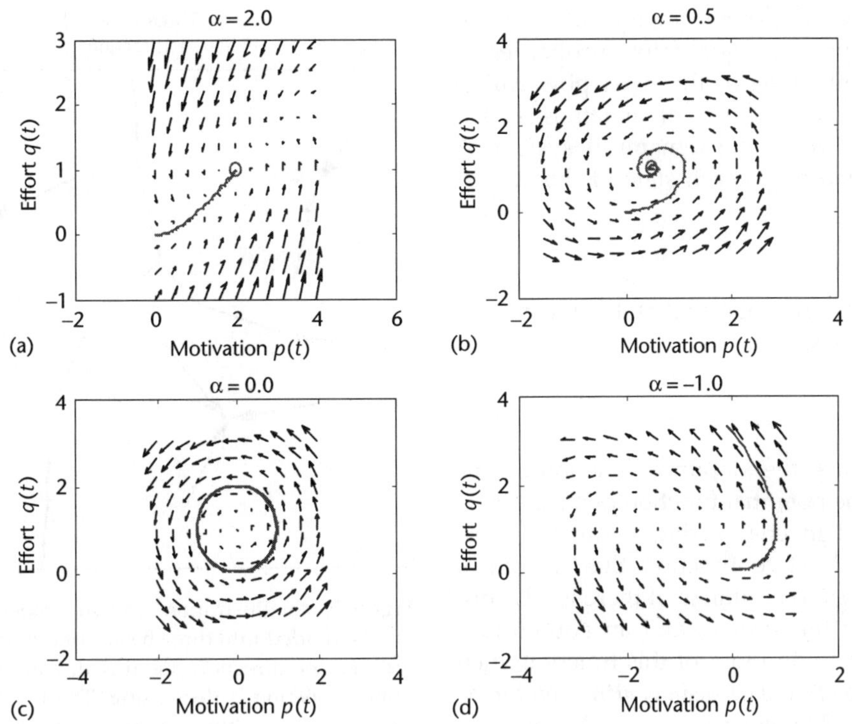

Figure 4. A display of four different phase portraits, one for each setting of the parameter α in Eqns 6 and 7. The horizontal axis within each portrait represents the preference state, and the vertical axis represents the level of effort. Each arrow is a vector indicating the direction and rate of change in the system at a particular point in the state space. The smooth curve within each portrait shows the trajectory produced by setting the initial state equal to zero for preference and effort.

α. The top left panel shows all of the arrows flowing towards a stable equilibrium point located at state $[p, q] = [2, 1]$. In this case, each trajectory moves steadily towards the equilibrium without any oscillation. The top right panel shows the arrows spiralling in towards a stable equilibrium point located at $[p, q] = [0.5, 1]$. The bottom left panel shows the arrows flowing in a circular manner around a central point located at $[p, q] = [0, 1]$. In this case, each trajectory oscillates indefinitely, like a clock. Finally, the bottom right panel shows the arrows spiralling away from an unstable equilibrium point located at $[p, q] = [-1, 1]$. In this case, the system shoots off towards infinity.

All four of these phase portraits contain a special state called an *equilibrium point*. But the nature of the equilibrium varies from one to another. An equilibrium point $\boldsymbol{X}^*$ has the special property that the local generator is zero at this point:

$$f(\boldsymbol{X}^*) = 0 \tag{11}$$

Thus no change occurs when the system is in this state. The equilibrium points for the top left and top right portraits are called *stable* equilibrium points or *attractors* because the system eventually tends towards these equilibrium points whenever the state of the system starts within a close proximity of these points. The equilibrium point for the bottom right panel is an *unstable* equilibrium point or *repellor* because if the system is placed an arbitrarily small distance away from that point, it eventually drifts further away (For rigorous definitions, see any of the references cited at the end of the Introduction.)

Stability Analysis

The model defined by eqns 6 and 7 is an example of a linear system. A linear system enjoys the special property of allowing only a single equilibrium point (provided that the system is nonsingular). The stability of equilibrium points for linear systems can be easily determined by checking the eigenvalues of the linear equations (see (Braun, 1975)). Nonlinear systems, however, allow multiple equilibrium points, of which some may be stable and others unstable, and a more general method of *stability analysis* is required to determine their properties.

There are several general mathematical techniques for studying the qualitative properties of equilibrium points. One of the most powerful is based on the construction of what is called a *Liapunov function* for the dynamical system. A Liapunov function maps each state of the system to a real number

$$V: \Omega \rightarrow \mathbb{R} \tag{12}$$

and it has the special property that its time derivative never increases:

$$\frac{\mathrm{d}V}{\mathrm{d}t} \leq 0 \tag{13}$$

for all t. In physics, the Liapunov function can be interpreted as the potential function of a conservative system; and in engineering it can be interpreted as the objective function that a control system is designed to minimize. If there is a Liapunov function for the system, and an equilibrium point X^* is a local minimum of this function, then X^* is a stable attractor. The *basin of attraction* for X^* is the largest possible region for which X^* serves as the attractor. Thus, once the system enters the basin of attraction for X^*, then it never leaves, and it converges towards the attractor. When a Liapunov function is defined over the entire state space, then the state space can be partitioned into a collection of attraction basins with a single stable attractor located within each basin (see figure 5).

To illustrate the idea of a Liapunov function, consider the predator–prey model described earlier. Figure 6(a) shows the phase portrait for this model.

This nonlinear model has three equilibrium points – $[0, 0]$, $[\frac{\alpha}{\gamma}, 0]$ and $\left[\frac{\lambda}{\phi}, \frac{\phi\alpha - \lambda\gamma}{\beta\phi}\right]$ – but only the last one is stable. For convenience, define $[x_1{}^*, x_2{}^*] = \left[\frac{\lambda}{\phi}, \frac{\phi\alpha - \lambda\gamma}{\beta\phi}\right]$. (In the case of figure 6 $[x_1{}^*, x_2{}^*] = [0.83, 0.39]$.)

It can be shown that the Liapunov function for this example is

$$V(x_1, x_2) = \lambda\left(\frac{x_1}{x_1{}^*} - \ln\frac{x_1}{x_1{}^*}\right) + \alpha\left(\frac{x_2}{x_2{}^*} - \ln\frac{x_2}{x_2{}^*}\right) \tag{14}$$

The time derivative of this function is

$$\frac{\mathrm{d}V(x_1, x_2)}{\mathrm{d}t} = \frac{-\lambda^2\gamma}{\phi}\left(1 - \frac{x_1}{x_1{}^*}\right)^2 \leq 0 \tag{15}$$

which is nonincreasing for all positive values of the state variables. The partial derivative of V with respect to the state variable is zero at $[x_1{}^*, x_2{}^*]$, so this point is a local minimum of V; hence it is a stable attractor for all positive values of the state variables.

Three basins of attraction

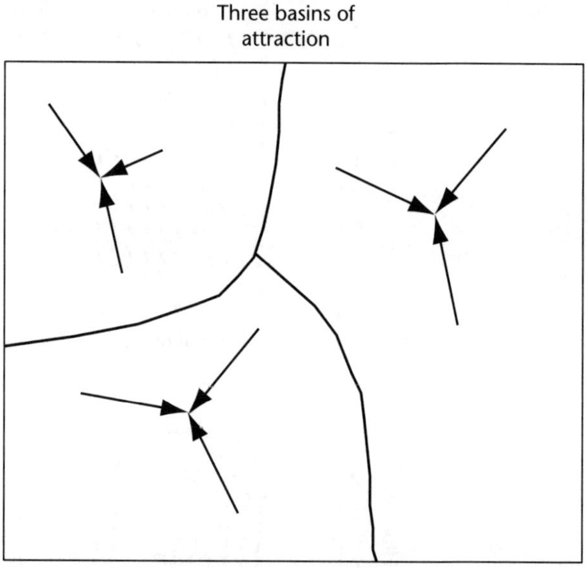

Figure 5. An illustration of a two dimensional state space that is divided into three basins of attraction. The arrows indicate the direction pointing downhill, where the Liapunov function is decreasing. The meeting point of the three arrows within each basin represents the stable attractor at the local medium. The curves indicate the boundaries that separate each basin.

Figure 6(b) shows the surface of the Liapunov function over the state space (using the same parameters). The surface has a minimum at the equilibrium point $[x_1{}^*, x_2{}^*] = [0.83, 0.39]$. This point is the stable attractor associated with the basin of attraction inside the positive region of the state space.

Bifurcation Analysis and Catastrophe Theory

A *bifurcation* is said to occur if the equilibrium points undergo qualitative changes as a result of small continuous changes in the model parameters. The parameter values at which these bifurcations occur are called bifurcation points. To illustrate the idea of bifurcation, consider the following example of a slightly more complex version of the predator–prey model:

$$\frac{\mathrm{d}x_1}{\mathrm{d}t} = (0.4)x_1 - (0.04)x_1{}^2 - (0.2)\frac{x_1 x_2}{1 + x_1} \tag{16}$$

$$\frac{\mathrm{d}x_2}{\mathrm{d}t} = \alpha x_2\left(1 - (0.5)\frac{x_1}{x_2}\right) \tag{17}$$

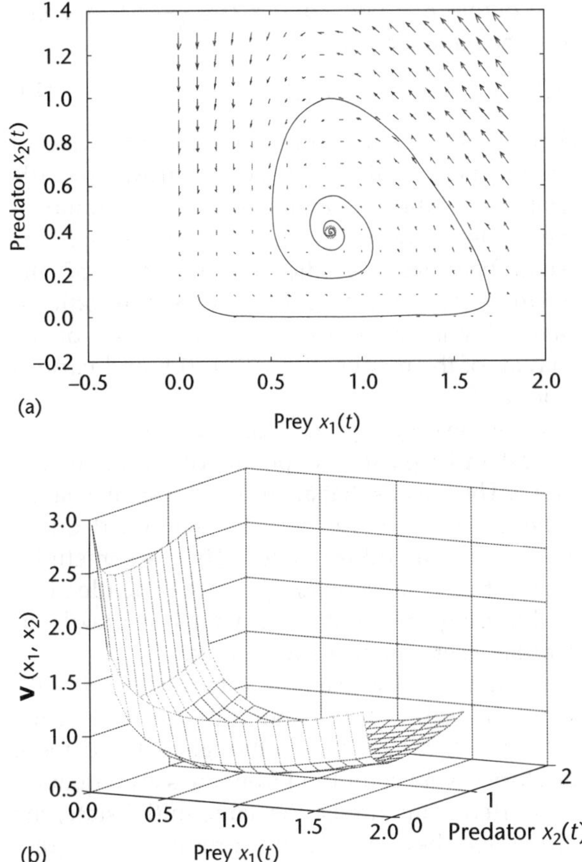

Figure 6. (a) An illustration of the phase portrait produced by the predator–prey model of effort for generating and testing ideas. (b) An illustration of the Liapunov function corresponding to this model. In (b), the plane represents the state space of the model, and the surface on the vertical axis above the plane represents the value of the Liapunov function for each point in the state space.

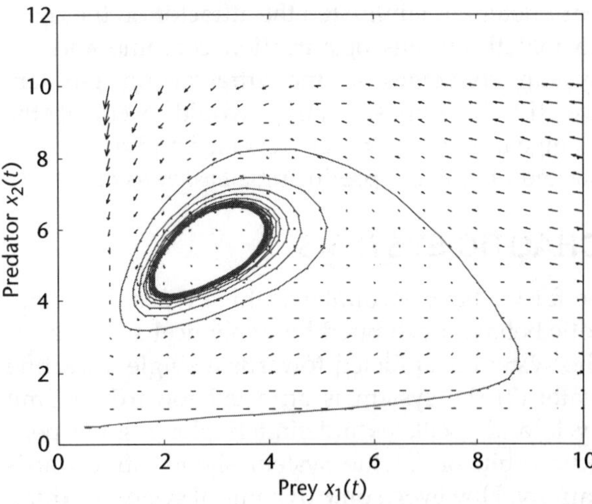

Figure 7. Phase portrait and trajectory produced by a modified predator–prey model that exhibits limit cycle behavior.

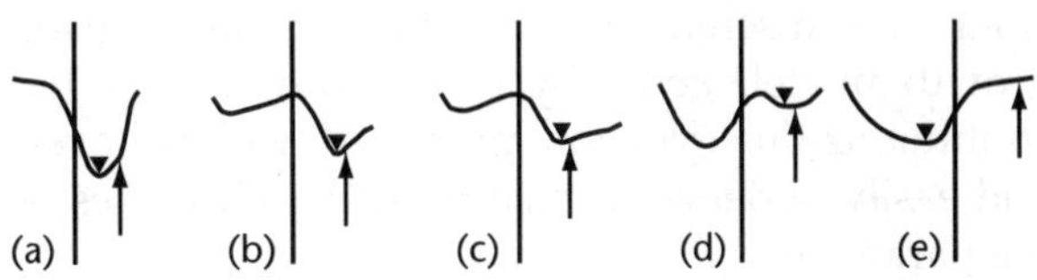

Figure 8. Five hypothetical Liapunov functions produced by small changes in a parameter of a dynamic system. The upward pointing arrow indicates the starting position. The solid triangle indicates the final equilibrium state. Note that the equilibrium makes a sudden jump into a new basin of attraction in the last case.

This example has only one free parameter, α, which is the focus of this bifurcation analysis. For this model, the location of the positive-valued equilibrium point $[x_1{}^*, x_2{}^*] = [2.7, 5.4]$ is independent of the parameter α (see (Beltrami, 1987)). Also, for large values of α, this equilibrium point is a stable attractor. The trajectory of the model looks very much like that shown in figure 6(a).

As the parameter α decreases, a qualitative change in the dynamics appears. The equilibrium point changes from an attractor to a repellor. A new behavior appears in which the asymptotic trajectory is attracted towards a *limit cycle* or asymptotic periodic orbit. In this case, the attractor is the set of points of the limit cycle. Figure 7 shows the phase portrait for this case, and the trajectory produced when the system is started at the initial state $[x_1(0), x_2(0)] = [0.5, 0.5]$ and $\alpha = 0.1$. This is an example of what is known as a *Hopf bifurcation*.

Catastrophe theory (Zeeman, 1977) is concerned with bifucations that result in discontinuous jumps in stable equilibrium points. Figure 8 illustrates the idea of a catastrophe. In this figure, each of the five curves represent a Liapunov function defined over a one-dimensional state space. The variations between the curves are produced by making small changes in a single parameter of the dynamical system. The upward-pointing arrow indicates the starting position of the system. The first curve, on the far left, exhibits an attractor on the right to which the system converges in the limit. Following a birfucation, the second, third and fourth curves exhibit two attractors, on the left and right, and a repellor in the middle. However, since the initial state lies inside the right basin, the system continues to converge to the attractor on the right. Finally, for the last curve, the change in parameter has eliminated the attractor on the right, so that the basins of attraction combine and the system converges to the attractor on the left. Exactly the opposite jump would occur if the change in parameter were reversed and the system started on the left, producing a *hysteresis* effect.

CHAOTIC SYSTEMS

So far we have encountered three types of asymptotic behavior exhibited by dynamical systems: (1) the system is attracted towards a single attracting state; (2) the system is attracted towards a limit cycle and oscillates indefinitely along some periodic orbit; or (3) the system shoots off towards infinity. However, some dynamical systems exhibit aperiodic behavior that does not fall into any of these three categories. This new category of *strange attractors* is exhibited by *chaotic dynamical systems*.

Chaotic behavior can arise from what appear to be very simple dynamical models. Let us reconsider the discrete-time version of the logistic growth model, given by eqn 3. (Recall that the continuous-time version produced well-behaved and easily understood trajectories for all values of the parameter α).

Setting $h = 1$ for the discrete-time model yields:

$$p(t+1) - p(t) = \alpha p(t)(1 - p(t)) \qquad (18)$$

If we define $\beta = 1 + \alpha$, and $x(t) = \frac{\alpha}{1+\alpha} p(t)$, we can write the equation as.

$$x(t+1) = \beta x(t)(1 - x(t)) \qquad (19)$$

When $\beta < 1$, the system described by eqn 19 decays to zero; when $1 < \beta < 3$, the system grows towards a stable equilibrium point, as in the continuous-time model. However, as β increases above 3, the system becomes periodic, and for $\beta > 3.57$ the system breaks down and becomes aperiodic or chaotic. Figure 9 shows a time series plot of the behavior of the model with $x(0) = 0.1$, and various values of β.

One of the defining features of a chaotic dynamical system is sensitive dependence on initial conditions. This means that an arbitrarily small change in the initial state is eventually magnified into a large change in future states. This is sometimes referred to as the *butterfly effect*, after the idea that the fluttering of a butterfly's wings in Brazil can eventually set off a tornado in Texas.

For example, if we set $\beta = 4$, then $x(0) = 0.1000$ yields $x(10\,000) = 0.2098$, but $x(0) = 0.1001$ yields $x(10\,000) = 0.9819$. Thus, although both of these trajectories were computed from the same equation and started from almost the same initial state, the trajectories they eventually produce are quite different.

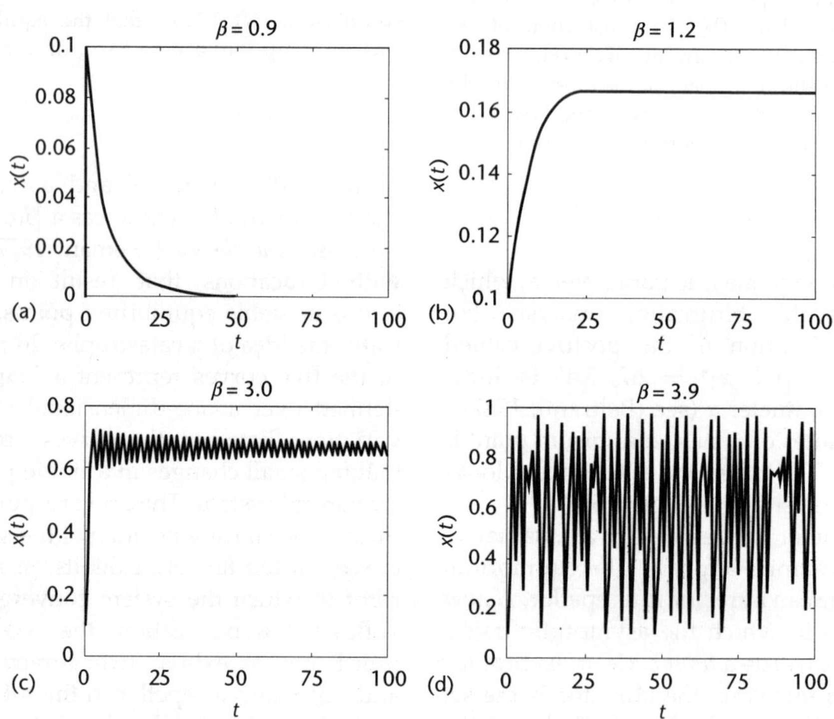

Figure 9. A display of four different time series plots produced by changing the parameter β in the discrete-time logistic growth model.

A rigorous method for identifying chaotic dynamical systems is based on an index, called the Liapunov index, λ:

$$\lambda = \lim_{n\to\infty} \frac{1}{n} \sum_{i=0}^{n-1} \ln|f'(x_i)| \tag{20}$$

The function f in eqn 20 refers to the local generator for a univariate discrete dynamical system. For example, for the logistic model f is defined by the right-hand side of eqn 19, and $f'(x) = \beta(1 - 2x)$ in this case. This index provides a measurement of sensitivity to initial conditions. A dynamical system is chaotic when the Liapunov index is positive ($\lambda > 0$).

Figure 10 shows the Liapunov index plotted as a function of β for the logistic model (using $x(0) = 0.10$; however, the pattern does not depend on this starting position). Note that the index is never positive until $\beta > 3.57$, at which point the model becomes chaotic. It is also interesting to note that the system occasionally returns to periodic behavior at a few higher values of β.

The discrete-time logistic model is the simplest example of a chaotic dynamical system. However, chaotic behavior is not limited to discrete-time systems. Many (and more complex) examples of continuous-time chaotic systems have also been studied.

Chaotic behavior may seem to be an 'undesirable' property for a dynamical system. But these models do provide an alternative to stochastic dynamical models for describing the unpredictability and variability manifested by many complex biological systems.

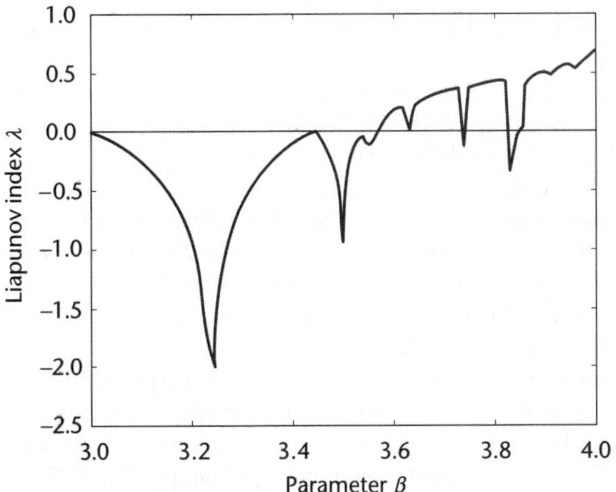

Figure 10. A plot of the Liapunov index as a function of the parameter β for the discrete-time logistic growth model.

TYPES OF DYNAMICAL SYSTEMS

Cognitive scientists have employed many different types of dynamical systems, ranging from production rule models of computers (Newell and Simon, 1972) to artificial neural network models of the brain (Grossberg, 1982; Rumelhart and McClelland, 1986). These models differ according to some basic characteristics as described below.

Discrete Versus Continuous State Spaces

A computer system has a *discrete* state space, which contains a countable number of possible states. The state space for the brain model is an example of a *continuous* vector space, which contains an uncountable number of states. The latter is also endowed with additional properties, including scalar multiplication of states ($aX \in \Omega$ for any real number a and state X), addition of states ($X_1 + X_2 \in \Omega$ for any X_1 and $X_2 \in \Omega$), and distances between states ($\| X_1 - X_2 \|$). Dynamical systems theory usually assumes that state spaces are vector spaces.

Discrete Versus Continuous Time Indices

The computer model is an example of a discrete-time system, in which a countable set of time points is indexed by the set of natural numbers $\{0, 1, 2, 3, \ldots\}$. The brain model is an example of a continuous-time system, in which an uncountable set of time points is indexed by the set of nonnegative real numbers $[0, \infty]$. Originally dynamical systems theory was concerned only with continuous-time systems, but now both types are studied in parallel.

Linear Versus Nonlinear Systems

The computer model is an example of a discrete nonlinear system in which the production rules produce jumps from state to state. Some early neural models used continuous linear state transition functions, but more recent neural models use continuous nonlinear transition functions. In general, a dynamical system is linear if the local generator f satisfies the following superposition property for two arbitrary states X_1 and X_2 and scales a and b:

$$f(aX_1 + bX_2, t) = af(X_1, t) + bf(X_2, t) \tag{21}$$

In this case, the local generator can be written as a linear combination of the state variables:

$$f_j(X(t),t) = \sum_{i=1}^{n} \alpha_i(t)x_i(t) \quad (22)$$

The coefficients used to define the linear combination are called the system parameters. The model of effort defined in eqns 6 and 7 is an example of a linear model. The logistic and predator–prey models are both examples of nonlinear models.

Time-invariant Versus Time-varying Systems

The computer model is an example of a time-invariant system, in which the state transition function does not change over time. A brain model, on the other hand, may require a time-varying system to allow for growth, development and aging. In general, the system is time-invariant if the local generator is independent of the time index:

$$f(X,t) = f(X) \quad (23)$$

One way to redefine a time-varying system as a time-invariant system is to add an extra state variable and set it equal to the time index. For example, the model of effort defined in eqns 6 and 7 contains a possibly time-varying input, $v(t)$. However, this two-dimensional time-varying system $[p, q]$ can be transformed into a three-dimensional time-invariant system $[p, q, x_3]$ by defining a third state variable, $x_3 = t$. The new three-dimensional system can then be described by three equations:

$$\frac{dq}{dt} = p - \alpha q \quad (24)$$

$$\frac{dp}{dt} = v(x_3) - \beta q \quad (25)$$

$$\frac{dx_3}{dt} = 1 \quad (26)$$

Deterministic Versus Stochastic Systems

The computer model is an example of a deterministic system: if we know the exact state of the system at time t then we can predict the state of the system at time $t+1$. Our simple model of effort (eqns 6 and 7) was also formulated as a deterministic system. However, models of the brain must account for the inherent unpredictability of human behavior. In the past, this was usually accomplished by allowing a subset of the variables in the state vector to be stochastic, or by allowing the initial state vector to be a random vector. Recently, deterministic but chaotic dynamical systems have been explored as alternatives to stochastic dynamical systems.

Bhattacharya and Waymire (1990) provide an introduction to stochastic dynamical systems theory.

SUMMARY

Dynamical systems theory was originally developed to solve problems arising in physics and engineering. Now cognitive scientists are making use of this approach, especially in applications of connectionist and neural models of cognition. Most applications are much more complex than the examples given in this article. Nevertheless, cognitive scientists have successfully applied these ideas to various substantive areas including pattern recognition, motor behavior, cognitive development, learning, thinking, and decision-making. There is little doubt that dynamical systems theory has much to contribute to cognitive science.

References

Atkinson JW and Birch D (1970) *The Dynamics of Action*. New York, NY: Wiley.

Bhattacharya RN and Waymire EC (1990) *Stochastic Processes With Applications*. New York, NY: Wiley.

Beltrami E (1987) *Mathematics for Dynamic Modeling*. New York, NY: Academic Press.

Braun M (1975) *Differential Equations and Their Applications*. New York, NY: Springer.

Grossberg S (1982) *Studies of Mind and Brain*. Reidel.

Luenberger DG (1979) *Introduction to Dynamic Systems*. New York, NY: Wiley.

Miller G, Galanter E and Pribram KH (1960) *Plans and the Structure of Behavior*. New York, NY: Holt, Rinehart, Winston.

Newell A and Simon HA (1972) *Human Problem Solving*. Englewood Cliffs, NJ: Prentice Hall.

Padulo L and Arbib MA (1974) *System Theory*. Philadelphia, PA: Saunders.

Rumelhart DE and McClelland JL (1986) *Parallel Distributed Processing: Explorations in the Microstructure of Cognition*, vol. I. Cambridge, MA: MIT Press.

Strogatz SH (1994) *Nonlinear Dynamics and Chaos*. Reading, MA: Addison-Wesley.

Townsend JT and Busemeyer JR (1989) Approach-avoidance: return to dynamic decision behavior. In: Izawa C (ed.) *Current Issues in Cognitive Processes: The Flowerree Symposium on Cognition*, pp. 107–133. Hillsdale, NJ: Erlbaum.

Zeeman EC (1977) *Catastrophe Theory: Selected Papers 1972–1977*. Reading, MA: Addison-Wesley.

Further Reading

Anderson JA (1997) *Introduction to Neural Networks*. Cambridge, MA: MIT Press.

Gleick J (1987) *Chaos: Making a New Science*. New York, NY: Viking Press.

Golden RM (1996) *Mathematical Methods for Neural Network Design and Analysis*. Cambridge, MA: MIT Press.

Grossberg S (1988) *Neural Networks and Natural Intelligence*. Cambridge, MA: MIT Press.

Haykin S (1994) *Neural Networks*. New York, NY: Macmillan.

Port RF and Van Gelder T (1995) *Mind As Motion*. Cambridge, MA: MIT Press.

Dyslexia

Introductory article

Max Coltheart, Macquarie University, Sydney, New South Wales, Australia

CONTENTS

Dyslexia is a specific impairment in the ability to read; it may be acquired (impaired reading caused by brain damage in a previously literate person) or developmental (failure ever to have learnt to read adequately).

INTRODUCTION

When we learn to read, we build up a system in our mind that is capable of turning the printed word into a pronunciation (that is how we read aloud) or a meaning (that is how we understand text). This is an information processing system: it takes in information in one form (the visual appearance of printed words) and gives out information in a different form (the pronunciation or the meaning of the printed words). In some people, however, this system no longer works, even though it used to; and other people never manage to learn the system properly in the first place. Both kinds of difficulty with reading are referred to as 'dyslexia': the first is acquired dyslexia and the second is developmental dyslexia.

There are many people who learnt to read perfectly but who then, because of brain damage (due to a stroke, for example, or an injury to the head), lost some of their ability to read. Such people may still be able to perform normally other mental activities such as remembering, or recognizing people and objects; sometimes they are also normal at speaking, writing, and spelling – but they can no longer read adequately. The brain damage has specifically impaired the mental information processing system which had been used for the job of reading. This is acquired dyslexia.

Other people never manage to learn to read properly – even people who are normal in intelligence and who had no difficulty as children in learning how to carry out other mental activities such as remembering, recognizing people and objects, and arithmetic calculations. This is developmental dyslexia.

If we are to understand these abnormalities of reading, we first need to understand normal reading. We need to know what the mental information processing system used for reading is like, how it works, and how reading is learnt. Experimental psychologists discovered a great deal about this in the last quarter of the twentieth century.

Suppose normal readers are asked to read aloud some nonsense word such as 'bloof'. All will be able to do so. This fact is simple yet instructive. It tells us that reading aloud does not depend solely on being able to look up, in a kind of mental dictionary, what pronunciation had been learnt for the string of letters. If that were so, no one would be able to read 'bloof' aloud, since no pronunciation could have been previously learned for that particular letter string, because the reader would never have seen it before. How do normal readers read aloud such unfamiliar material? This can be achieved because part of what normal readers have learnt is knowledge about the rules relating letters to sounds; they have learnt how 'b' is pronounced, how 'l' is pronounced, how 'oo' is pronounced, and how 'f' is pronounced. These rules apply even to letter strings never before seen; so they can be used for reading aloud such unfamiliar material.

Could this system of rules be all that is needed for adequate reading aloud? No, because there are many words in English that disobey these rules. The rule for how to pronounce 'oo', used to read 'bloof' aloud, is disobeyed by the real words 'blood' and 'good'. Despite this, normal readers will read 'blood' and 'good' aloud correctly if asked to. How do they do this? It can be achieved because these are words these readers have seen before, and this has allowed them to learn the correct pronunciations (and meanings) of the words and store these in a kind of mental dictionary.

ACQUIRED DYSLEXIA

The mental information processing system used for reading thus contains two different procedures for reading aloud – one a system of rules specifying the relationships of letters to sounds; the other a 'dictionary' procedure that allows the reader to retrieve information previously learnt about familiar words. To read aloud unfamiliar material requires the first procedure. To read aloud words that disobey the rules requires the second procedure.

This mental information processing system must be located somewhere in the brain. If different parts of this system are located in different parts of the brain, theoretically brain damage could harm the rule system without affecting the dictionary look-up system. What would the reading of such a person be like? Any familiar word could still be read aloud correctly, since the dictionary look-up system allows this, but unfamiliar letter strings such as 'bloof' could no longer be read aloud. This kind of acquired dyslexia does occur: it is known as 'phonological' dyslexia, and the first case was reported in 1979; numerous other cases have since been reported.

Suppose instead that brain damage harmed the dictionary look-up system without affecting the rule system. The affected person would still be able to read unfamiliar letter strings such as 'bloof', and real words that obey the rules (such as 'bloom') could still be read aloud. However, words that disobey the rules would be affected; if 'blood' could not be looked up in the dictionary, the rules would have to be used to read it aloud, and that will give an error with the 'oo' part of the word. This kind of acquired dyslexia is known as 'surface' dyslexia; the first case was reported in 1973, and numerous other cases have since been reported.

Several other types of acquired dyslexia have been discovered. For example, if it is true that words disobeying the rules are read aloud by looking them up in a mental dictionary, perhaps this involves reading them via information about the meanings of the words: the route is from print to meaning and then from meaning to speech. If so, anyone in whom brain damage has affected knowledge of word meaning would be impaired at reading words that disobey the rules. This turns out not to be so. In Alzheimer disease and other forms of dementia, knowledge about what words mean will sooner or later be lost. In some unfortunate people to whom this has happened, reading aloud of words, even words that disobey the rules, can still be normal. The person might no longer have any idea what the word 'blood' means, yet would still be able to read it aloud correctly. So here we have a form of acquired dyslexia in which what is impaired is reading *comprehension*, with reading aloud being unaffected.

In yet another well-documented type of acquired dyslexia, known as 'deep' dyslexia, the brain-damaged reader will read a word as some other word similar in meaning; the word 'canary' might be read as 'parrot', or the word 'wrist' as 'watch'. Reading by people with deep dyslexia shows a number of other symptoms too. Words that have a concrete meaning ('leopard', 'cigar') are much more likely to be read aloud correctly than words that are abstract in meaning ('idea', 'character'). The small grammatical words of the language, even though they are generally very short and very common ('the', 'and', 'if', 'but'), are rarely read aloud correctly. Unfamiliar letter strings such as 'bloof' cannot be read aloud at all.

Finally, there is 'letter-by-letter reading', so called because sufferers from this kind of acquired dyslexia typically spell out aloud words they are trying to read, letter by letter, rather than being able to recognize the word immediately as a whole. If they manage to name each letter in the word correctly, then they can generally say the whole word correctly. Despite this severe difficulty in reading, such people can be perfectly normal at writing and spelling.

The discovery of these distinct patterns of acquired dyslexia tells us not only that the mental reading system consists of numerous different components, but also that these different components are located in different parts of the brain; if that were not so, brain damage could not affect one part of the reading system while leaving other parts unaffected.

Only a limited amount is so far known about exactly which parts of the brain are damaged in these different kinds of acquired dyslexia. There is persuasive evidence that the limited reading achievable by people with deep dyslexia depends upon the use of reading mechanisms in the right hemisphere of the brain which may play little or no part in normal reading. Letter-by-letter reading also involves the right hemisphere, since these patients typically have damage to the visual areas of the left hemisphere (so that visual identification of letters cannot occur in the left hemisphere), and also damage to one part of the corpus callosum (which is responsible for transmitting information between the two hemispheres of the brain). The effect of this damage is that after letters are identified in the right hemisphere their transmission to the reading system in the left hemisphere is abnormally slow and inefficient.

When knowledge about word meanings is affected in Alzheimer disease and other forms of dementia (in patients with impaired text comprehension but intact ability to read aloud) this is due to progressive deterioration of the temporal lobes of the brain, especially the left temporal lobe. Virtually nothing is currently known about which brain regions are specifically impaired in surface dyslexia or phonological dyslexia.

One might assume that, since in acquired dyslexia some part of the brain that is needed for reading is permanently damaged, there could be no effective treatment to improve the reading of such people; but this turns out not to be so. Previously normal readers who no longer recognize the word 'blood', and so can only try to read it using the rules, can learn to recognize it again; that is, surface dyslexia can respond to treatment if this is designed appropriately. Phonological dyslexia, deep dyslexia, and letter-by-letter reading have also been shown to be remediable by appropriate treatment.

DEVELOPMENTAL DYSLEXIA

Turning now to a consideration of difficulties in learning to read – developmental dyslexia – a natural question to ask is the following. If the different components of the reading system can be separately impaired by various forms of brain damage to produce different kinds of acquired dyslexia, is the particular component of the reading system causing the difficulty in learning to read different in different children? If that were so, there would be different kinds of developmental dyslexia, just as there are different kinds of acquired dyslexia. This turns out to be the case.

Some children with developmental dyslexia have a particular problem in learning or using the system of rules specifying the relationships of letters to sounds. This is known as 'developmental phonological dyslexia'. Other children with developmental dyslexia have a particular problem learning or using the dictionary look-up system needed for fluently recognizing words as wholes – this is developmental surface dyslexia.

It is clear that developmental dyslexia is partly a genetic condition involving abnormalities on at least two specific chromosomes, namely chromosome 6 and chromosome 15. It may be the case that the abnormality of chromosome 6 is characteristic of developmental phonological dyslexia and is associated with some general difficulty on processing speech sounds; and it may also be the case that the abnormality on chromosome 15 is characteristic of developmental surface dyslexia.

Since there are genetic influences at work here, one might wonder whether it is possible to remedy these developmental dyslexias. However, it does not follow from the fact that a disorder has a genetic basis that it cannot be treated; and in fact research has shown that both of these kinds of developmental dyslexia respond well to appropriate treatment.

Many children with developmental dyslexia exhibit the symptoms of both of these kinds of developmental dyslexia: they are poor at both learning or using the system of rules specifying the relationships of letters to sounds and also at learning or using the dictionary look-up system needed for fluently recognizing words as wholes. The most plausible explanation for the prevalence of this commonly occurring 'mixed' developmental dyslexia is that each of these reading procedures assists the child in learning the other. A child who is, say, 8 years old will have an auditory vocabulary of perhaps 10 000 words, but a small vocabulary of words that can be recognized in print. Such a child will therefore constantly come across printed words which cannot be recognized in print but which would be recognized if heard. If the child could apply rules relating letters and sounds to such words, sounding out the words would allow them to be recognized. This is a method by which children could teach themselves to recognize whole words; but it is unavailable to the child who is poor at using such rules. So being poor at the rule-based reading route will make it difficult to build up the dictionary look-up reading route.

The reverse is probably also true. How do children learn what these rules are? One way might be by reflecting on the spellings and pronunciations of words they already know, and working out from these what the rules must be. If so, children who have few words in their mental dictionary will have an impoverished database upon which to reflect, and that will limit what these children can learn about letter–sound rules.

CONCLUSION

Great advances have been made in our understanding of developmental and acquired dyslexia. The crucial step was the recognition that the system we use to read is a mental information processing system that contains a number of different processing components, including a letter recognition system, a mental dictionary storing the spellings, meanings and pronunciations of the words the reader knows, and a system of rules specifying the relationship between letters and sounds. Studies of people with acquired dyslexia have shown that brain damage can impair particular components of the reading system while not affecting others. Studies of people with developmental dyslexia have shown that such people can have difficulty learning particular components of the reading system while being able to learn others well; and such specific difficulties in learning to read are associated with specific genetic abnormalities.

Further Reading

Castles A and Coltheart M (1993) Varieties of developmental dyslexia. *Cognition* **47**: 149–180.

Castles A, Datta H, Gayan J and Olson RK (1999) Varieties of developmental reading disorder: genetic and environmental influences. *Journal of Experimental Child Psychology* **72**: 73–94.

Coltheart M and Jackson N (1998) Defining dyslexia. *Child Psychology and Psychiatry Reviews* **3**: 12–16.

Habib M (2000) The neurological basis of developmental dyslexia: an overview and working hypothesis. *Brain* **123**: 2373–2399.

Jackson N and Coltheart M (2001) *Routes to Reading Success and Failure*. Philadelphia, PA: Psychology Press/Taylor & Francis.

Marshall JC (1989) The description and interpretation of acquired and developmental reading disorders. In: Galaburda AM (ed.) *From Reading to Neurons: Issues in The Biology of Language and Cognition*, pp. 69–86. Cambridge, MA: MIT Press.

Seymour PHK (1990) Developmental dyslexia. In: Eysenck MW (ed.) *Cognitive Psychology: An International Review*, pp. 135–196. New York, NY: John Wiley.

Early Experience and Cognitive Organization

Introductory article

Barbara Landau, Johns Hopkins University, Baltimore, Maryland, USA

CONTENTS

There is much support for the idea that cognitive development is specialized, with different aspects of knowledge developing under different mechanisms and timelines. Many aspects of mature knowledge can be traced to innate origins. At the same time, variation in early experience due to genetic defects, blindness or deafness can lead to significant reorganization of the developing brain. Thus human cognitive development is constrained by innate potential, but is also affected by experience during early development.

INTRODUCTION

One of the great mysteries of human development is how our knowledge comes to be organized in the way in which it is. Two competing views have quite different answers to this. The nativist view holds that we are born with specific capacities to organize the world in the way that we do. This view can be traced to the thinking of philosophers such as Immanuel Kant and René Descartes. In contrast, the empiricist view holds that experience after birth molds our knowledge. This view is traceable to philosophers such as John Locke and David Hume. The nativist view is supported by the fact that different aspects of our knowledge have quite different forms, and these develop quite early in life without any formal tutoring. Yet some variations in early experience can have profound effects on brain organization, suggesting that there is some flexibility in the way in which the brain and mind organize the world. This article will consider two types of change in early experience. One is due to changes in an individual's genetic endowment, and the other is due to changes in an individual's sensory and perceptual experience.

SPECIALIZATION IN COGNITIVE DOMAINS

When considering the relative roles of innate potential and experience, it is important to understand that different aspects of knowledge have very different types of organization – that is, they are 'specialized' in the sense that the principles which govern one domain of knowledge may be irrelevant or inapplicable to the next domain. For example, our capacity for language allows us to learn language early in life, and to fluently produce and understand even quite complex sentences throughout life. This capacity reflects our knowledge of language, which is a complex, rule-governed system that allows us to produce and comprehend an infinite number of sentences. In contrast, our capacity to move around the world and find our way, to remember the spatial locations of objects, and to read maps, is part of a different system of knowledge. This system allows us to learn the spatial relationships between places in space, and to understand how to get from one place to another, even if we are travelling along novel routes. This is part of our spatial knowledge system, but it is irrelevant to our knowledge of language. Other cognitive domains are specialized as well. For example, our knowledge of number, causality, time and social relationships is governed in all of these cases by rules and representations that are distinctly different from each other.

Examining development under conditions of varying genetic endowment and/or different types of sensory and perceptual experience can shed light on the question of how cognitive specialization emerges. The nativist view would suggest that there are strong biological constraints on the types of knowledge that humans can develop, and that changes in genetic endowment might have specific, targeted effects on cognition. At the same time, the empiricist view would suggest that changes in sensory or perceptual endowment – such as congenital blindness or deafness – might lead to changes in the way in which the brain's organization supports cognition. Evidence from recent research supports aspects of both views.

WILLIAMS SYNDROME: A GENETIC DISORDER THAT RESULTS IN SELECTIVE COGNITIVE IMPAIRMENT

Recently, cognitive researchers have become interested in Williams syndrome, a relatively rare genetic syndrome (1 in 15 000 live births) which is caused by a microdeletion of material (approximately 20 genes) on chromosome 7. Individuals with Williams syndrome show distinctive physical characteristics and an unusual cognitive profile. In general, these individuals are moderately mentally retarded, with an average IQ of around 60 (where 100 is the average for the general population). However, their retardation is not 'across the board'. Rather, individuals with Williams syndrome show severe impairments in their understanding of spatial relationships, but considerable strength in language ability. Their severe spatial impairments are manifested in a number of ways. One of the clearest examples is that they have extreme difficulty in copying simple figures. Figure 1 shows an example of drawings produced by two 11-year-olds with Williams syndrome, and one drawing produced by a normally developing child who was younger, but who had the same 'mental age' on an intelligence test, and had a roughly average IQ. It can be seen that the individuals with Williams syndrome could easily duplicate the colors in the model, but could not copy the spatial relationships shown there, resulting in a very distorted copy. In contrast, the normally developing child could reproduce both the colors and the spatial organization. The spatial impairment in people with Williams syndrome contrasts strikingly with their strong language capacities, which are demonstrated by a fluent, articulate

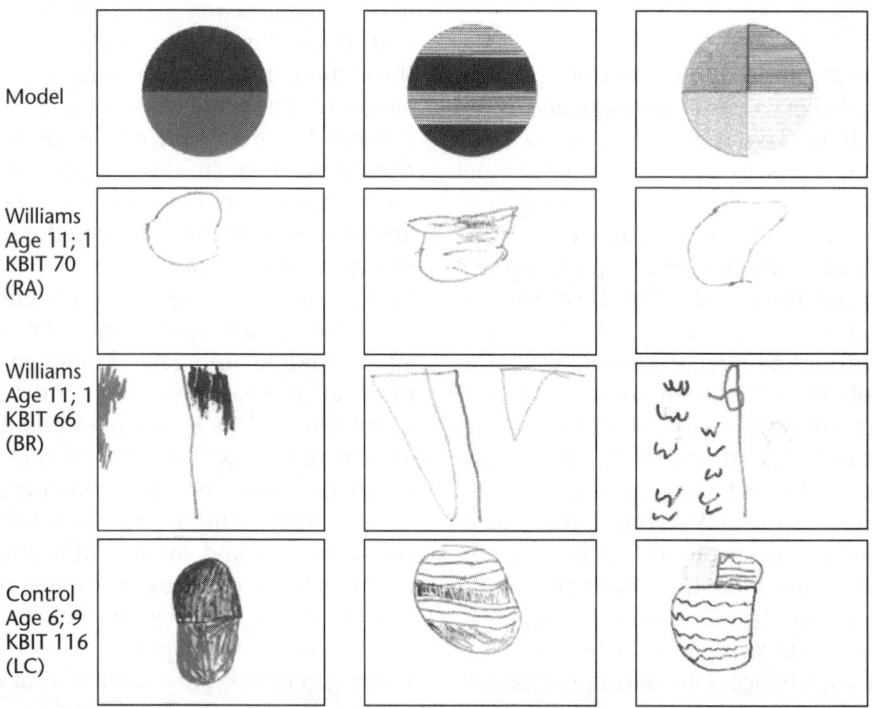

Figure 1. [*Figure is also reproduced in color section.*] Copies made by two 11-year-old individuals with Williams syndrome and a normally developing 6-year-old who was matched for mental age. Each model contains between two and four colors, and these are correctly copied by the children with Williams syndrome as well as by controls. However, there is considerable impairment of the spatial organization of Williams syndrome copies, resulting in unrecognizable configurations. Ages are stated in years and months. KBIT scores are from the Kaufman Brief Intelligence Test, and represent approximate IQ equivalents.

conversational style, relatively high scores on vocabulary tests, and the capacity to produce and understand sentences that are grammatically complex. Individuals with Williams syndrome also tend to be musical, and some people with this syndrome are highly skilled in musical activities such as singing and playing musical instruments. The striking impairment of spatial capacities, coupled with the fact that language appears to be strongly preserved in Williams syndrome, supports the idea that cognitive specialization might ultimately be linked to aspects of genetic endowment.

Researchers are currently asking many questions about the emergence of cognitive specialization in this syndrome. For example, it may be that the spatial impairment is more apparent in some aspects of the spatial knowledge system than in others. Some authors have conjectured that the profile of strengths and weaknesses within spatial cognition may be linked to somewhat unusual organization of the brain in Williams syndrome. One possibility is that the spatial impairment may be limited to certain specific spatial functions that are performed by the brain. The most obvious examples of spatial impairment in people with Williams syndrome are shown when these individuals are asked to copy an existing pattern, either by arranging a set of blocks to copy a pattern, or by drawing a copy of a model, as in Figure 1. However, people with Williams syndrome do not appear to show impaired performance on other spatial tasks. For example, they do not appear to have difficulties in recognizing faces or identifying pictures of objects, even when the objects are presented from very unusual viewpoints, as in Figure 2. In order to identify objects under these types of conditions, it would seem to be necessary to recognize both the parts of the objects and their spatial relationships. Thus this aspect of spatial capacity might be relatively spared in Williams syndrome, even though other aspects of spatial capacity are clearly impaired. Evidence from cases of normal adults who then sustain brain damage (e.g. through strokes) suggests that object recognition and identification may be carried by the brain's 'ventral' stream. The relative strength of object recognition in Williams syndrome may indicate selective sparing of functions carried by this stream. As another example, individuals with Williams syndrome tend to have some difficulty in planning visual–motor acts, such as posting a letter through a mail slot. Yet they have less difficulty when they must simply view the slot and compare its orientation to another sample slot. Again, some authors have conjectured that the relative difficulty in acting on objects, compared with simply perceiving and matching them, could be due to selective impairment of the brain's 'dorsal' stream. Finally, individuals with Williams syndrome perform better than normal children matched for mental age on tests of 'biological motion' perception, in which they perceive a moving animate figure that is shown only by a set of individual dots. Overall, the evidence suggests that there is specialization in the breakdown of spatial cognition in Williams syndrome. This and other hypotheses are currently being actively tested by researchers.

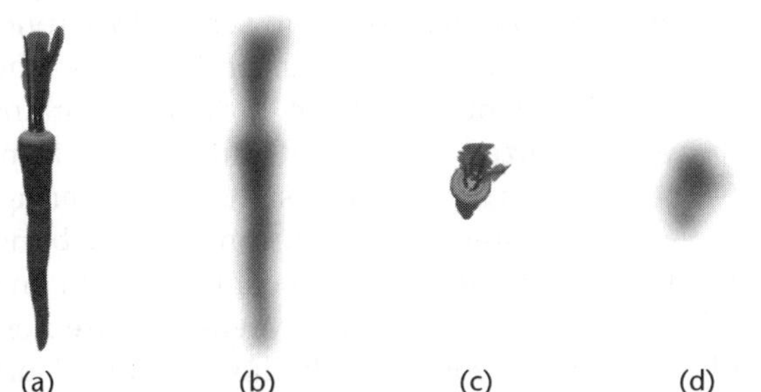

Figure 2. [*Figure is also reproduced in color section.*] These objects are easily recognized and named by children with Williams syndrome, despite the fact that they have profound spatial impairments. The objects are all the same carrot, but vary in whether they are common viewpoints (a and b) or unusual ones (c and d), and also in whether they are clear images (a and c) or blurred ones (b and d).

DOWN SYNDROME: A GENETIC DISORDER THAT SHOWS A MORE GENERAL DECLINE IN COGNITION

In comparison with Williams syndrome, Down syndrome has a very different cognitive profile, again suggesting that cognitive development may be partially under the control of genetic endowment. Down syndrome occurs in roughly 1 in 1000 births, and in the majority of cases it is associated with an additional copy of chromosome 21 (resulting in three rather than two chromosomes at that locus – hence the name 'trisomy 21'). Individuals with Down syndrome tend to have distinctive physical characteristics, although these are different from those in Williams syndrome. Furthermore, there are some gross anatomical differences between the brains of individuals with Down syndrome and those of individuals with Williams syndrome. Finally, the cognitive profile of the two groups is quite different. Individuals with Down syndrome are moderately retarded, but their cognitive profile is relatively even. That is, they are impaired in various cognitive domains to roughly the same extent, and they do not show the striking profile of strengths and weaknesses that is characteristic of individuals with Williams syndrome.

A number of studies have directly compared the spatial and linguistic capacities of individuals with Down and Williams syndromes. The results of spatial studies show that although both groups are impaired relative to normally developing individuals, people with Williams syndrome perform even more poorly on some spatial tasks than individuals with Down syndrome who have the same IQ. In addition, they may use different types of solutions for the same spatial problems. For example, several reports indicate that individuals with Down syndrome can replicate more global spatial aspects of a model, whereas individuals with Williams syndrome are more likely to replicate accurately the more local aspects of a model – the smaller elements that are put together to make up the overall pattern. It is possible that these different patterns of spatial impairment in Down versus Williams syndrome are due to different types of brain organization and impairment.

The results of language studies show that although neither group performs at a level commensurate with its chronological age (i.e. they are both impaired to some extent), individuals with Williams syndrome outstrip those with Down syndrome who have the same IQ. Moreover, when individuals with Williams syndrome are tested on grammatically complex structures, they tend to do well compared with individuals with Down syndrome. The spontaneous speech of individuals with Down syndrome does not exhibit a large amount of grammatical complexity. For example, longer sentences are primarily produced by stringing together simple phrases with conjunctors such as 'and'. In contrast, individuals with Williams syndrome produce longer sentences that exhibit a considerable degree of grammatical complexity, including relative clauses (e.g. 'The boy that was looking at the man jumped over the fence'). The fact that these can be fluently produced by individuals who are retarded points to the possibility that language capacity may develop rather normally in the case of Williams syndrome. In the case of Down syndrome, the general dampening of linguistic capacity is roughly commensurate with the overall retardation of these individuals. The combination of cases indicates that changes in genetic endowment may have highly specific and targeted consequences for cognitive development.

CONGENITAL BLINDNESS AND DEAFNESS

Sensory deficits such as congenital blindness and deafness may result in reorganization of the brain. There is considerable evidence that knowledge in a variety of domains can develop normally in individuals who are blind or deaf from birth. For example, congenitally blind individuals show the capacity to understand locations in space, to get from one place to another by independent locomotion, and to read haptic or Braille maps. As another example, congenitally deaf individuals acquire manual (signed) languages effortlessly, just as hearing individuals effortlessly acquire spoken languages. In these cases and others, the knowledge that develops is organized in the same manner as in individuals who are sighted or hearing. This indicates that cognitive organization can emerge guided by principles that do not depend on these particular types of sensory or perceptual experience (i.e. seeing or hearing).

However, there is also evidence of flexibility in the way in which the brain accomplishes these cognitive functions. This flexibility is evident from studies of brain activity following sensory or perceptual deprivation. A number of studies have shown that changes in early experience can have a major impact on the organization of the brain's cortex. If an individual is deprived of input from sensory receptors, the cortical areas that are normally activated by those receptors are 'taken over' by other types of input. For example, in adults who have had a limb amputated, the cortical area that used to represent that limb (i.e. be activated by it) now becomes responsive to stimuli from other areas of the body that would normally activate neighboring areas of the cortex. In effect, the areas of cortex that have been 'abandoned' by the amputated limb are commandeered to serve other functions. Other research shows that individuals who have been blind or deaf from birth have brains that are organized somewhat differently from those of individuals who are sighted or hearing. For example, in some research, congenitally blind individuals who perform tasks using the skin and hands show brain activation in areas that are normally 'designated' for vision. In this case, the visual areas are commandeered for touch, and hence the area of cortex dedicated to touch is expanded in these individuals. Other research has shown that congenitally deaf individuals show unusual brain activation patterns when they pay attention to stimuli in their peripheral visual field – that is, outside the central view. Additional evidence shows that these individuals may be more accurate at detecting these stimuli than hearing individuals. In essence, congenital auditory deprivation appears to lead to a reorganization of the brain in which there are changes in the ways in which visual stimuli are processed.

SUMMARY

Mature knowledge reflects the development of a set of highly complex, differentiated and specialized systems, including language, spatial knowledge, number, causality and other domains. These specialized systems are partly under the control of genetic endowment and partly under the control of variations in experience. The results of genetic deficits are highly specific, with some syndromes targeting specific knowledge systems and other syndromes having more general effects. The results of massive variation in experience, such as congenital blindness or deafness, are some degree of reorganization in the brain, indicating plasticity during early development.

Further Reading

Gallistel CR, Brown A, Carey S, Gelman R and Keil F (1991) Lessons from animal learning for the study of cognitive development. In: Carey S and Gelman R (eds) *The Epigenesis of Mind: Essays on Biology and Cognition*, pp. 3–36. Hillsdale, NJ: Erlbaum.

Gazzaniga MS, Ivry RB and Mangun GR (2000) *Cognitive Neuroscience: The Biology of the Mind*. New York: Norton.

Johnson M (1997) *Developmental Cognitive Neuroscience*. Cambridge, MA: Blackwell.

Jordan H, Reiss J, Hoffman JE and Landau B (2001) Intact perception of biological motion in the face of profound spatial deficits: Williams syndrome. *Psychological Science* **13**(2): 162–167.

Mervis C, Morris CA, Bertrand J and Robinson B (1999) Williams syndrome: findings from an integrated program of research. In: Tager-Flusberg H (ed.) *Neurodevelopmental Disorders: Contributions to a New Framework From the Cognitive Neurosciences*, pp. 65–110. Cambridge, MA: MIT Press.

Uecker A, Mangan PA, Obrzut JE and Nadel L (1993) Down syndrome in neurobiological perspective: an emphasis on spatial cognition. *Journal of Clinical Child Psychology* **22**: 266–276.

Ebbinghaus, Hermann

Introductory article

Robert R Hoffman, Institute for Human and Machine Cognition, University of West Florida, Pensacola, Florida, USA

Michael Bamberg, Clark University, Worcester, Massachusetts, USA

CONTENTS

Hermann Ebbinghaus (1850–1909) was a German psychologist whose books, research, and ideas had a great effect on early psychological theory. He is often credited with founding the experimental psychology of the 'higher mental processes'.

INTRODUCTION

Hermann Ebbinghaus was born on 23 January 1850 in the industrial town of Barmen, in the Rhine Province of the kingdom of Prussia. He studied classics, languages, and philosophy, and completed his doctoral dissertation at the University of Bonn in 1873. After working for some years as a tutor, he happened to read about the new research on psychophysics, and became inspired to study the 'higher mental processes'. In 1878, Ebbinghaus began formal experiments on memory, conducted in his home. A monograph on the work was published in 1885. Within a year he was promoted to a salaried professorship at the Friedrich Wilhelm University in Berlin. Journals that published psychological research were beginning to spring up everywhere and Germany needed a general journal. Ebbinghaus helped establish the *Zeitschrift für Psychologie und Physiologie der Sinnesorgane* (Journal for the Psychology and Physiology of the Sense Organs) and served as its editor for 22 years. In 1893 Ebbinghaus took a professorship at Breslau University in the Prussian province of Silesia.

Ebbinghaus's psychology textbook appeared in 1897, and was the most popular and widely used general psychology text for many years.

Ebbinghaus was known as an eloquent lecturer and excellent teacher. He was a man with vision, a champion of the view that psychology should be emancipated from philosophy, and the higher mental processes studied experimentally.

Ebbinghaus died of pneumonia in 1909. At that year's psychology conference at Clark University (to which Ebbinghaus had been invited), Cornell University psychologist Edward B. Titchener began with a eulogy: 'As I approach the topic of this lecture, what is uppermost in my mind is a sense of irreparable loss. When the cable brought the bad news, last February, that Ebbinghaus was dead ... the feeling that took precedence even of personal sorrow was the wonder of what experimental psychology would do without him.'

EBBINGHAUS'S RESEARCH

The monograph *On Memory* has three aspects: the experiments themselves, a discussion of statistical analyses of data, and some theorizing. Much of the theorizing concerns the strength and vividness of associations and the search for 'mathematical rules for mental events'. Ebbinghaus's discussion of the basic statistical methods was so clear and exact that many psychologists had their students read Ebbinghaus's book just for its discussion of statistics. Each data point that entered into his analyses was an average of the learning times (or average number of repetitions needed to reach a learning criterion) over a large number of lists. The averages were used to compute a distribution of means, and results were then described in terms of standard errors: the percentage of cases falling under a given area of the distribution.

The monograph reported 19 studies conducted in the years 1879–1880 and 1883–1884. To conduct the research Ebbinghaus first prepared a pool of 'all possible syllables' – 2300 in all (quite a few of them were words in German, English or French). A few examples are heim, beis, ship, dush, noir, noch, dach, wash, born, for, zuch, dauch, shok, hal, dauf, fich, theif, hatim, shish, and rur. Pacing himself with a metronome, and reading the lists aloud with a poetic meter, he proceeded to memorize lists of syllables. Using a set of buttons on a string, he was able to keep track of the number of repetitions he needed in order to learn a list to the point that he could give one perfect recitation. The experiments were an ambitious project and required great effort.

Ebbinghaus led a ritualistic, almost monastic life during these experiments, learning and recalling lists every day for months on end, dozens of experiments and replications of experiments, each involving multiple trials and hour after hour of data collection and careful record-keeping. Imagine learning 84 600 syllables in 6600 lists, taking more than 830 h! Although the number of list repetitions involved for every experiment cannot be determined for some of the studies on the basis of what Ebbinghaus said in his monograph, for experiment 2 alone Ebbinghaus engaged in 189 501 repetitions of lists.

The first two experiments had the goal of showing that the variability of the average learning times over a large number of lists was within limits that would be scientifically acceptable. He emphasized that the standard errors he obtained compared favorably with the precision of measurement in the physical and biological sciences (e.g. measurements of the speed of neural conduction, or measurements of the mechanical equivalent of heat). In fact, his 'probable errors' of about 7% were more precise than the physical measurements and very close to those for the biological measurements.

The list of Ebbinghaus's findings includes many of the basic phenomena that are discussed to this day in books on the psychology of memory. His research demonstrated the viability of the method of savings or 'ease of relearning' as a means of measuring the strength of association. He demonstrated the effects of fatigue and time of day on retention; the effect of list length on the number of repetitions it takes to learn material; the 'decay of memory' as a function of the delay between acquisition and memory test (with delays spanning hours, days, weeks and even years). Ebbinghaus demonstrated the effect of 'distributed versus massed' practice. He demonstrated what came to be called the 'serial position' effect (i.e. better memory for material that falls near the beginning and near the ending of a list). Ebbinghaus also measured what would come to be called the short-term memory span – 'the number of syllables which I can repeat without error after a single reading is about seven. One can, with a certain justification, look upon this number as a measure of the ideas of this sort which I can grasp in a single unitary conscious act.'

Textbooks on general and cognitive psychology preserve a myth about Ebbinghaus, which is that he conducted experiments in which he memorized 'nonsense' syllables. It was not the syllables that were nonsense – in the examples given above, the first 10 are all meaningful in one or another of the languages Ebbinghaus knew – it was the task of learning a list of semantically unconnected items that Ebbinghaus refers to as involving an 'impression of nonsense'. Indeed, the term he preferred for his lists, *Vorstellungsreihen* (literally, 'presentation series'), could just as well be translated as 'image series', and Ebbinghaus discussed at some length how the strength and vividness of memory images should be related to the effort taken in learning them.

Another contradiction to the myth is that Ebbinghaus did not begin his studies by attempting to memorize lists of syllables. Instead, he began with a task more familiar to teachers – and to the pupils whom Ebbinghaus taught – the memorization of poetry. His preliminary trials with poetry showed that the material was learned too quickly. He found no need for multiple repetitions (meaning that he could not obtain enough data about trials and time to criterion in order to generate statistically reliable laws) and he was also concerned that the material could not be systematically and quantitatively varied (lists of numbers did not afford enough variety either); hence his eventual choice of syllable lists. However, his research did not end with the memorization of syllable lists. Ebbinghaus memorized stanzas from Byron's *Don Juan* in order to address the question of the role of meaningfulness in the associative process. Over a period of 4 days he conducted seven separate tests, each test involving the learning of six stanzas. Each test took about 20 min and involved about eight repetitions of each stanza. Given that each stanza consisted of about 80 syllables, he could compute that meaningfulness resulted in a large advantage: about one-tenth the effort in terms of the number of repetitions needed to achieve one perfect recitation. Most important to Ebbinghaus was the fact that the findings with the poetry confirmed the findings for the syllables: general laws were in operation.

INTELLIGENCE TESTING

The school board of Breslau had commissioned Ebbinghaus to generate mental tests that could be used to determine the best distribution of study hours for schoolchildren. He invented the completion method to see how well children could perceive relationships, combine information, and arrive at correct conclusions. In the task, students would have to fill in the missing letters in sentences such as 'WH__ WILLY ___ TWO _____ OLD, HE _____ __ _ RED FARM_____'. This type of task is still used in modern intelligence and aptitude tests. Along with digit memory and a rapid calculation task, the results showed a clear effect of age and individual differences. However, only the results from the method of combinations showed a relation to the children's grades. Ebbinghaus's work on intelligence testing was thus some of the very first research on this topic. According to Woodworth, the completion method was probably a better test of intelligence than any other method available at the time. Alfred Binet was working on mental testing at the time, and Binet was encouraged by Ebbinghaus' studies of school children. The original Binet–Simon scale included Ebbinghaus's method of relearning of lists (of words) as well as the sentence completion task.

IMPACT ON PSYCHOLOGY

Ebbinghaus's monograph received mixed reviews when it was published, but American psychologist William James praised the work, pointing out the author's 'heroic efforts'. To James, 'this particular series of experiments [was] the entering wedge of a new method of incalculable reach' (p. 199). Once Ebbinghaus's work became known in the USA, other psychologists began conducting studies of learning. Ebbinghaus became a model of the experimental psychologist, whose theoretical speculations were brief and cautious but whose research was rigorous in its method and its use of statistics. He provided a model for the use of experimental logic, including the testing of alternative hypotheses by setting up experimental situations where rival hypotheses would make differing predictions, and also a sensitivity to what are today called 'experimenter bias effects' (especially important to Ebbinghaus because he was his own subject). Finally, he provided experimental psychology with a model for preparing a research report: the now-traditional ordering of introduction, methods, results, and discussion sections.

Further Reading

Boring EG (1929) *A History of Experimental Psychology.* New York, NY: Appleton-Century-Crofts.

Ebbinghaus H (1885) *Über das Gedächtnis* ('On Memory') Leipzig: Duncker & Humblot. Translated by Ruger H and Busenius C (1913) New York, NY: Columbia University Teacher's College.

Fechner GT (1860) *Elemente der Psychophysik* ('Elements of Psychophysics'). Leipzig, Germany: Breithaus & Hartel.

Herrmann DJ and Chaffin R (1987) Memory before Ebbinghaus. In: Gorfein DS and Hoffman RR (eds) *Memory and Learning: The Ebbinghaus Centennial Conference*, pp. 35–56. Hillsdale, NJ: Lawrence Erlbaum.

Hoffman RR, Bringmann W, Bamberg M and Klein R (1987) Some historical observations on Ebbinghaus. In: Gorfein DS and Hoffman RR (eds), *Memory and Learning: The Ebbinghaus Centennial Conference*, pp. 57–76. Hillsdale, NJ: Lawrence Erlbaum.
Hothersall D (1984) *A History of Psychology*. New York, NY: Random House.
Jaensch ER (1909) Hermann Ebbinghaus. *Zeitschrift für Psychologie* **51**: 3–8.
James W (1885) Experiments in memory. *Science* **6**: 198–199.
Peterson J (1925) *Early Conceptions and Tests of Intelligence*. Chicago, IL: World Book Co.
Stigler SM (1978) Some forgotten work on memory. *Journal of Experimental Psychology: Human Learning and Memory* **4**: 1–4.
Titchener EB (1910) The past decade in experimental psychology. *American Journal of Psychology* **21**: 404–421.
Woodworth RS (1909) Hermann Ebbinghaus. *Journal of Philosophy and Scientific Methods* **6**: 253–256.
Woodworth RS (1938) *Experimental Psychology*. New York, NY: Holt.

Economics Experiments, Learning in

Advanced article

Jacob K Goeree, University of Amsterdam, Amsterdam, The Netherlands
Charles A Holt, University of Virginia, Charlottesville, Virginia, USA

CONTENTS

Models of learning in economics are used to explain how people use information about past prices and other signals to make good decisions. These models can be tested with economics experiments in which decisions are made in a sequence of rounds or trading periods.

INTRODUCTION

The main focus of economic analysis is on equilibrium steady states, e.g. on prices determined by the intersection of supply and demand. The preoccupation with equilibrium is perhaps due to the fact that many markets operate for protracted periods of time under fairly stationary conditions. The awareness that there may be multiple equilibria, some of which are bad for all concerned, has raised interest in why behavior might converge to one equilibrium and not to another. As a result, there is renewed interest among economists in mathematical models of learning that were studied extensively by psychologists in the 1960s. This article will describe two of those models, 'reinforcement learning' and Bayesian 'belief learning'. These models and their generalizations will be discussed in the context of a binary prediction task, which may generate the kind of behavior that is known in the psychology literature as 'probability matching'.

We will then use these learning models to analyze behavior in an economic market where firms choose prices. Markets and games are more complex than individual decision tasks, in that people's choices affect others' beliefs. One role of learning models in such situations is to provide an explanation of the dynamic paths of prices, which can shed light on the nature of adjustment towards equilibrium. The equilibrium is characterized by an unchanging (steady-state) distribution of beliefs across individuals, which we call a 'stochastic learning equilibrium'.

TYPES OF LEARNING MODELS

We will introduce the basic learning models in the context of a binary prediction task that has been familiar to psychologists since the 1950s. This task is of special interest because humans are thought to be slow learners in this context. The typical setup involves two lights, each with a corresponding lever (or computer key). A signal light indicates that a decision can be made, and then one of the levers is pressed. Finally, one of the lights is illuminated. Animal subjects like rats and chicks are reinforced with food pellets when the prediction is correct. Human subjects are sometimes told to 'do your best' to predict accurately or to 'maximize the number of correct choices'. In other studies, humans are paid small amounts, typically pennies, for correct choices, and penalties may be deducted for incorrect choices.

The general result seems to be that humans are subject to 'probability matching', predicting each event with a frequency that approximately matches the frequency with which it actually occurs. For example, if the left light is illuminated three-fourths of the time, then subjects would come to learn this by experience and would tend to predict 'left' three-fourths of the time.

This behavior is not rational. A consistent prediction of the more likely event will be correct three-fourths of the time. Matching behavior will only generate a correct prediction with a probability of $\frac{3}{4}\times\frac{3}{4}+\frac{1}{4}\times\frac{1}{4}$ (the first term corresponds to predicting the more likely event with probability $\frac{3}{4}$ and being correct with this prediction three-fourths of the time, and similarly for the second term). Thus, the probability of being correct under probability matching is $\frac{5}{8}<\frac{3}{4}$.

In a recent summary of the probability-matching literature, the psychologist Fantino (1998, pp. 360–361) concludes: 'Human subjects do not behave optimally. Instead they match the proportion of their choices to the probability of reinforcement.... This behavior is perplexing given that non-humans are quite adept at optimal behavior in this situation.' For example, Mackintosh (1996) conducted probability-matching experiments with chicks and rats, and the choice frequencies were well above the probability-matching predictions in most treatments.

The resolution of this paradox may be found in the work of Sidney Siegel, who is perhaps the psychologist who has had the largest impact on experiments in economics. His early work from the 1960s exhibits a high standard of careful reporting and procedures, appropriate statistical techniques, and the use of financial incentives where appropriate. His work on probability matching is a good example. In one experiment, 36 male Penn State students were allowed to make predictions for 100 trials, and then 12 of these students were brought back on a later day to make predictions in 200 more trials (Siegel *et al.*, 1964). The proportions of predictions for the more likely event are shown in Figure 1, in which each point is an average over 20 trials.

The 12 subjects in the 'no pay' treatment were simply told to 'do your best' to predict which light bulb would be illuminated. These prediction rates begin at about 0.5, as would be expected in early trials with no information about which event is more likely. Notice that the proportion of predictions for the more likely event converges to 0.75, as predicted by probability matching, with very close matching from about trial 100.

In the 'pay/loss' treatment, 12 participants received 5¢ for each correct prediction, and they lost 5¢ for each incorrect prediction. The rate seems to converge to about 0.9.

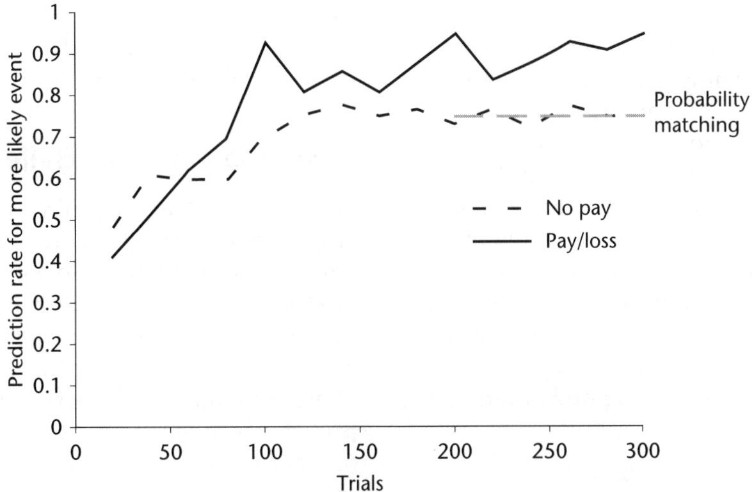

Figure 1. Prediction frequencies for an event that occurs with probability $\frac{3}{4}$ (Siegel *et al.*, 1964). Each point plotted represents an average over 20 trials. The 'no pay' group were simply told to 'do your best'. The 'pay/loss' group received 5¢ for each correct prediction and forfeited 5¢ for each incorrect prediction.

A third 'pay' treatment offered a 5¢ reward for a correct prediction but no loss for an incorrect prediction. The results (not shown) are in between those of the other two treatments, and clearly above 0.75.

Clearly, then, incentives matter. Probability matching is not observed with incentives in this context.

Siegel's findings suggest a resolution to the paradox that rats are smarter than humans in binary prediction tasks. You cannot tell a rat to 'do your best': incentives such as food pellets must be used. Consequently, the choice proportions are closer to those observed with financially motivated human subjects. In a recent survey of over 50 years of probability-matching experiments, Vulkan (2000) separates those studies that used real incentives from those that did not, and he concluded that probability matching is generally not observed with real pay-offs. However, humans can still be surprisingly slow learners in this simple setting. For this reason, probability-matching data are particularly interesting for valuating alternative learning theories.

Reinforcement Learning

One prominent theory of learning associates changes in behavior to the reinforcements actually received. Initially, when no reinforcements have been received, it is natural to assume that the choice probabilities for each decision are equal to one-half. Suppose that in the experiment there is a reinforcement of x for each correct prediction, and nothing otherwise. So if one predicts event L and is correct, then the probability of choosing L should increase. The extent of the behavioral change is assumed to depend on the size of the reinforcement. As the total earnings received for a particular decision increase, the probability of making that decision is assumed to increase. Suppose that event L has been predicted N_L times and that the predictions have sometimes been correct and sometimes not. Then the total earnings for predicting L, denoted by e_L, would be less than xN_L. Similarly, let e_R be the total earnings from the correct R predictions. One way to model the effect of total earnings for each decision on choice probabilities is to choose L or R with the following probabilities:

$$P(\text{choose L}) = \frac{\alpha + e_L}{2\alpha + e_L + e_R} \quad (1)$$

$$P(\text{choose R}) = \frac{\alpha + e_R}{2\alpha + e_L + e_R} \quad (2)$$

The parameter α determines how quickly learning responds to the reinforcements. Note that, as additional reinforcements are received, they are added into the relevant numerator, and to both denominators, to ensure that the probabilities sum to 1.

This kind of model might explain some aspects of behavior in probability-matching experiments with financial incentives. The choice probabilities would be equal initially, but a prediction of the more likely event will be correct 75% of the time, and the resulting asymmetries in reinforcement would tend to raise the prediction probability for that event, and the total earnings for this event would tend to be much larger than for the other event. If L is the more likely event, then e_L would be growing faster, so that e_R/e_L would tend to get smaller. Thus the probability of choosing L would tend to converge to 1.

This learning model can be simulated by using past accumulated earnings to compute choice probabilities. Then a random-number generator determines the actual choices. To make our data comparable with Siegel's data, we simulate decisions of a cohort of 1000 individuals for 300 periods, and calculate the 20-period choice averages for the more likely event.

The simulations were done for $\alpha = 5$ and $x = 1$. The value of α was chosen to create some initial inertia in behavior, which will tend to disappear after 40 or 50 periods. Setting α equal to 5 is analogous to having had each decision reinforced five times initially. Figure 2 shows simulated choice averages together with Siegel's original data. The simulated data are smoother, and start somewhat higher to start with, but the general pattern and final tendencies are similar. Erev and Roth (1998) have used reinforcement learning to explain behavior in simple matrix games.

A Simple Model of Belief Learning

With reinforcement learning, beliefs are not explicitly modeled. An alternative approach that is more natural to economists is to model learning in terms of (Bayesian) updating of beliefs. Given the symmetry of Siegel's experimental setup, a person's initial beliefs ought to be that each event is equally likely, but the first observation should raise the probability associated with the event that was just observed. Moreover, the probability of event L should be an increasing function of the number N_L of times that this event has been observed, and a decreasing function of the number N_R of times that event R has been observed. Let N be the total

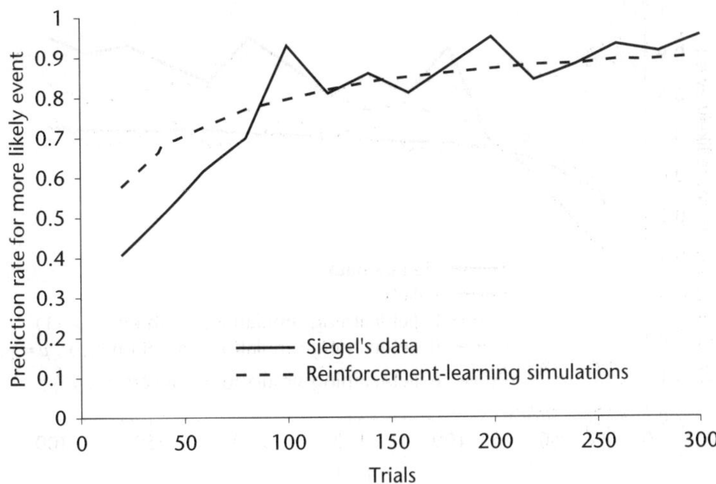

Figure 2. Data for Siegel's probability-matching experiment ('pay/loss' condition), with reinforcement-learning simulation data superimposed ($\alpha = 5$).

number of observations to date. Then a standard belief-learning model is:

$$P(\mathrm{L}) = \frac{\beta + N_{\mathrm{L}}}{2\beta + N} \quad (3)$$

$$P(\mathrm{R}) = \frac{\beta + N_{\mathrm{R}}}{2\beta + N} \quad (4)$$

where $N = N_{\mathrm{L}} + N_{\mathrm{R}}$. Note that β determines how quickly the probabilities respond to the new information; a large value of β will keep these probabilities close to $\frac{1}{2}$. These formulae for calculating probabilities can be derived from Bayesian statistical principles (DeGroot, 1970, p. 160). In the early periods, the totals N_{L} and N_{R} might sometimes switch in terms of which one is higher, but the more likely event will soon dominate, and therefore $P(\mathrm{L})$ will be greater than $\frac{1}{2}$.

The beliefs determine the expected pay-offs (or utilities) for each decision, which in turn determine the decisions made. In theory, the decision with the highest expected pay-off should be selected with certainty. The prediction of the belief-learning model is, therefore, that all people will eventually predict the more likely event every time.

In an experiment, however, some randomness in decision-making might be observed if the expected pay-offs for the two decisions are similar. This randomness may be due to changes in emotions, calculation errors, selective forgetting of past experience, etc. Following Luce (1959), we introduce some 'noise' via a probabilistic choice model, where decision probabilities are positively but not perfectly related to expected pay-offs. Let π_{L} and π_{R} denote the expected pay-offs from choosing 'left' and 'right' respectively. Luce provided a set of axioms under which the choice probability is calculated as

$$P(\text{choose L}) = \frac{(\pi_{\mathrm{L}})^{1/\mu}}{(\pi_{\mathrm{L}})^{1/\mu} + (\pi_{\mathrm{R}})^{1/\mu}} \quad (5)$$

$$P(\text{choose R}) = \frac{(\pi_{\mathrm{R}})^{1/\mu}}{(\pi_{\mathrm{L}})^{1/\mu} + (\pi_{\mathrm{R}})^{1/\mu}} \quad (6)$$

The parameter μ is an 'error' parameter. It determines the sensitivity of choice probabilities to differences in expected pay-offs. In the limit when μ tends to zero, the decision with the higher expected pay-off is selected with probability 1. In the other extreme as μ gets large, behavior is random and independent of pay-offs.

In the probability-matching experiment, the expected pay-off of choosing 'left' is the reward (of 1, say) times the probability of 'left' that represents the person's beliefs. Thus the expected pay-off of 'left' is $P(\mathrm{L})$, and similarly the expected pay-off of 'right' is $P(\mathrm{R})$. $P(\mathrm{L})$ is greater than one-half if 'left' is more likely, and the error parameter μ determines how close the choice probability for the more likely event is to 1.

Figure 3 shows a simulation of the belief-learning model for $\beta = 20$. The thin solid line represents the average of the belief probabilities for the 12 simulated subjects. Notice that beliefs start close to one-half and converge to the true probability of the more likely event $\left(\frac{3}{4}\right)$. These beliefs determine expected pay-offs, and hence choice probabilities, via equations 5 and 6.

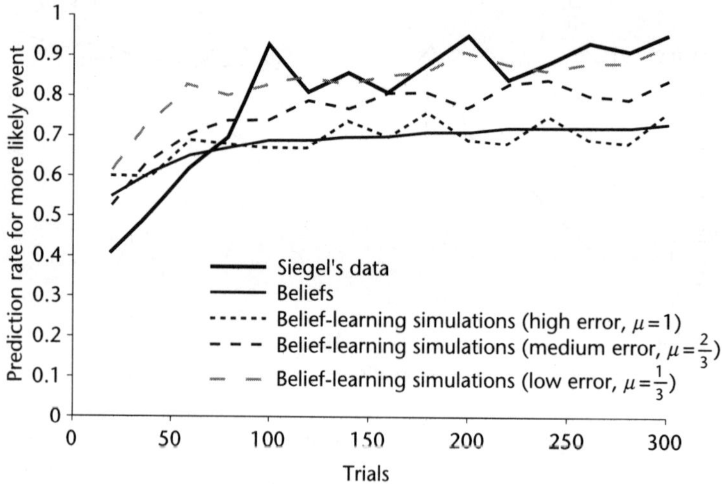

Figure 3. Data for Siegel's probability-matching experiment ('pay/loss' condition), with belief-learning simulation data superimposed ($\beta = 20$). The error parameter μ ranges from 1 (high error) to $\frac{1}{3}$ (low error).

The dashed lines show the simulated average choice frequencies for three different levels of the error parameter. With high error ($\mu = 1$), the choice frequencies are close to the belief line, which would correspond to probability matching. This result is to be expected, since expected pay-offs are equal to belief probabilities. The denominator on the right-hand side of equations 5 and 6 is 1 when $\mu = 1$, and hence the probability of choosing 'left' equals π_L, which is equal to the belief probability.

As the error is reduced, the simulated choice frequencies move upwards towards the optimal level of 1. The top line, with $\mu = \frac{1}{3}$, converges to the level of about 0.9, which is close to the choice frequency observed by Siegel.

The simulations in Figure 3 were done for a cohort of size 12, to be comparable with Siegel's experiment. This allows us to see the degree of variation in the simulated data with a small group. In order to predict the average over a large number of individuals, we ran the simulation 1000 times, and the average proportions of choices for the more likely event were: 0.76 for $\mu = 1$, 0.80 for $\mu = 0.67$, and 0.87 for $\mu = 0.33$.

Generalizations

Both of the learning models discussed above are somewhat simple, and this is part of their appeal. The reinforcement model builds in some randomness in behavior, and has the appealing feature that incentives matter. But it has less of a cognitive element. There is no reinforcement for decisions not made. For example, suppose that a person chooses L three times in a row (by chance) and is wrong each time. Since no reinforcement is received, the choice probabilities stay at 0.5 even after three incorrect predictions. This seems unreasonable. People do learn something in the absence of previous reinforcement, since they realize that making a good decision may result in higher earnings in the next round. Camerer and Ho (1999) have developed a generalization of reinforcement learning that contains some elements of belief learning. Roughly speaking, outcomes that are observed receive partial reinforcement even if nothing is earned.

These learning models can be enriched in other ways to obtain better predictions of behavior. For example, the sums of event observations in the belief-learning model weigh each observation equally. It may be reasonable to allow for 'forgetting' in some contexts, so that the observation of an event in the most recent trial may carry more weight than something observed a long time ago. This can be done by replacing sums with weighted sums. For example, if event L was observed three times, N_L in equation 3 would be 3, which can be thought of as $1 + 1 + 1$. If the most recent observation (listed on the right in this sum) is twice as prominent as the one before it, then the prior event would get a weight of one-half, and the one before that would get a weight of one-fourth, and so on. This type of 'recency' effect will be discussed below in the context of an interactive market game.

Finally, 'Luce's probabilistic-choice rule' (equations 5 and 6) is often replaced with the 'logit rule':

$$P(\text{choose L}) = \frac{\exp(\pi_L/\mu)}{\exp(\pi_L/\mu) + \exp(\pi_R/\mu)} \qquad (7)$$

$$P(\text{choose R}) = \frac{\exp(\pi_R/\mu)}{\exp(\pi_L/\mu) + \exp(\pi_R/\mu)} \quad (8)$$

where μ is an error parameter as before. The Luce and logit rules are often similar in effect, and both are commonly used. The logit probabilities are unchanged when all pay-offs are increased by a constant, and the Luce probabilities are unchanged when all pay-offs are multiplied by a positive constant.

LEARNING AND PRICE DYNAMICS IN A MARKET GAME

We use a simple price competition example from Capra *et al.* (2002) to illustrate the effects of learning in an interactive setting. Consider a market game in which firms 1 and 2 simultaneously choose prices p_1 and p_2 in the range [60, 160] (units are cents). Demand is assumed to be a fixed total quantity ('perfectly inelastic'). The sales quantity of the firm with the lower price $p_{\min}$ is normalized to be one, so the low-price firm earns an amount equal to its price. The high-price firm sells a 'residual' amount R, which is less than 1. The amount by which this residual is less than 1 indicates the degree of buyer responsiveness to price. The high-price firm has to match the lower price in order to make any sales, but some sales are lost because of the initially higher price. We assume that the high-price firm only earns $Rp_{\min}$, where $R<1$. In the event of a tie, the $1+R$ sales units are shared equally, so that each seller earns $(1+R)\,p_{\min}$.

As long as the high-price firm obtains less than half the market ($R<1$), the Nash equilibrium prediction is for both firms to set the lowest possible price of 60. To see this, note that at any common price, each firm has an incentive to undercut the other by a small amount to increase its market share. Therefore, the unique Nash equilibrium involves both firms charging the lowest possible price. The harsh competitive nature of the Nash prediction seems to go against the simple economic intuition that the degree of buyer inertia will affect the behavior of firms. When $R=0.8$, say, the loss from having the higher price is relatively small, and firms should be more likely to set prices above 60 when there is a small chance that rivals will do the same. Indeed, in the extreme case when $R=1$ it becomes a dominant strategy for both firms to choose the highest possible price of 160. While a standard Nash analysis predicts no change as long as $R<1$ (and then an abrupt change when $R=1$), it seems plausible that prices will gradually rise with R.

We ran an experiment based on this market game, using six cohorts of 10 subjects each. Each group of 10 subjects was randomly paired, with new partners in each of 10 periods. A period began with all subjects selecting a price in the interval [60, 160]. After these prices were recorded, subjects were matched, and each person was informed about the other's price choice. Pay-offs were calculated as described above: the low-price firm earned an amount equal to its price, and the high-price firm earned R times the lowest price. Three sessions were conducted with $R=0.2$ and three with $R=0.8$. Figure 4 shows the period-by-period average price choices. The upper solid line shows the average prices when buyers were relatively unresponsive ($R=0.8$), and the lower solid line shows average prices when buyers were relatively responsive ($R=0.2$). Recall that the Nash equilibrium was 60 for both treatments, as shown by the horizontal dashed line at 60. As intuition suggests, changes in the buyers' responsiveness has a large effect on price, even though the Nash equilibrium remains unchanged.

Notice that prices start high and stay high in the $R=0.8$ treatment, while prices decline before leveling off in the $R=0.2$ treatment. Standard economic models cannot explain either the levels or the patterns of adjustment. Our approach is to consider a naive learning model in which players use observations of rivals' past prices to update their beliefs about others' future actions. In turn, the expected pay-offs based on these beliefs determine players' choice probabilities via a logit rule. This model was used to simulate behavior in the experiment.

To obtain a tractable model, the price range [60, 160] is divided into 101 one-cent categories. Players assign weights to each category and use observations of their rivals' choices to update these weights as follows: each period, all weights are 'discounted' by a factor ρ and the discounted weight of the observed category is increased by 1. In other words, the weight w of an observed category is updated as $w \to \rho\, w+1$, whereas the other weights are updated as $w \to \rho w$. The belief probabilities in each period are obtained by dividing the weight of each category by the sum of all the weights. Hence, the learning parameter ρ determines the importance of new observations relative to previous information. Since the most recent observation gets a weight of 1, a lower value of ρ reduces the importance of prior history and increases recency effects.

Generally ρ will be between 0 and 1. When $\rho=0$, the observations prior to the most recent one are ignored, and the model is one of best response to the previously observed price (Cournot dynamics).

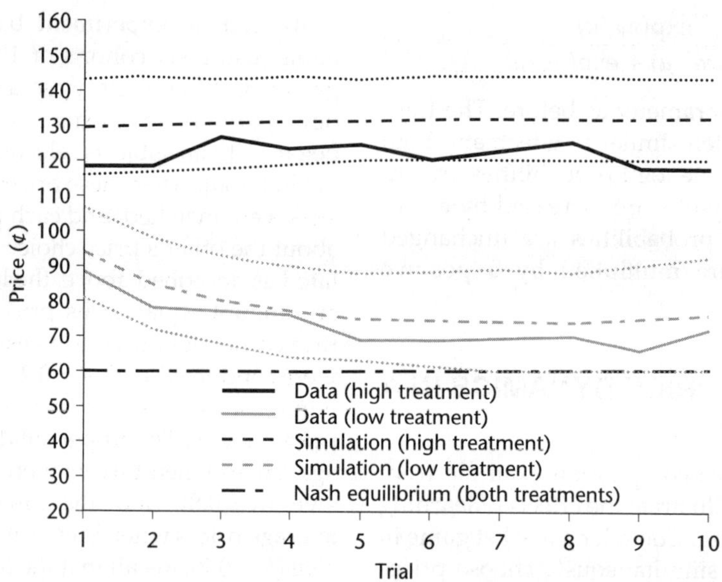

Figure 4. Data and simulations (plus or minus two standard deviations indicated by dotted lines) for a market game. In the 'high' treatment, buyers were relatively unresponsive to differences in price; in the 'low' treatment, buyers were relatively responsive. The simulations are based on a simple belief-learning model using a logit rule to determine probabilities.

At the other extreme, when $\rho = 1$, the model reduces to 'fictitious play', in which each observation is given equal weight, regardless of the number of periods that have elapsed since. For intermediate values of ρ, the weight given to past observations declines geometrically over time.

The expected pay-off for player i choosing a price in category j is denoted by $\pi_i^e(j|\rho)$. This determines player i's decision probabilities via the logit rule

$$P_i(j|p) = \frac{\exp(\pi_i^e(j|\rho)/\mu)}{\sum_{k=1}^{101} \exp(\pi_i^e(k|\rho)/\mu)}, \; j = 1, \ldots, 101 \qquad (9)$$

Choice probabilities and expected pay-offs depend on the learning parameter. In this dynamic model, beliefs, and hence choices, depend on the history of what has been observed up to that point. Since individual histories are realizations of a stochastic process, the predictions of this model will be stochastic and can be analyzed with simulation techniques.

The structure of the computer simulation program matches that of the experiment reported below: for each session, or 'run', there are 10 simulated subjects who are randomly matched in a sequence of 10 periods. We specify initial prior beliefs for each subject to be uniform on the integers in the set [60, 160]. These priors determine expected pay-offs for each price, which in turn determine the choice probabilities via the logit rule in equation 9. The simulation begins by determining each simulated player's actual price choice for period 1 as a draw from the logit probabilistic response to the pay-offs for priors that are uniform on [60, 160]. The simulated players are randomly divided into five pairs, and each player 'sees' the other's actual price choice. These price observations are used to update players' beliefs using the naive learning rule explained above, with a learning parameter $\rho = 0.72$ (which was estimated from the data). The updated beliefs, which become the priors for period 2, will not all be the same if the simulated subjects encountered different price choices in period 1. The process is repeated, with the period-2 priors determining expected pay-offs, which in turn determine the logit choice probabilities, and hence the observed price realizations for that period. The whole process is repeated for 10 periods.

Figure 4 shows the sequences of average prices obtained from 1000 simulations, together with dotted lines indicating two standard deviations of the average. These simulation results predict that average prices decline in the $R = 0.2$ treatment and stay the same in the $R = 0.8$ treatment, as observed in the data. Thus, the learning model explains the salient features of the experimental data: both the directions of adjustment and the steady-state levels.

STOCHASTIC LEARNING EQUILIBRIUM

Next we consider what the learning model implies about the long-run steady-state distribution of price decisions. In particular, will learning generate a price distribution that corresponds to some equilibrium?

At any point in time, different people will have different experiences or histories. These differences may be due to the randomness in individuals' decisions or to randomness in the random matching. For each person, the history of what they have seen will determine a probability distribution over their decisions. This mapping of histories to decision probabilities may be direct, as in reinforcement learning. Alternatively, histories may generate beliefs, which in turn produce decisions via a probabilistic choice rule. The decisions made are then appended to the existing histories, forming new histories. Because of the randomness in decision-making, there will be a probability distribution over all possible histories.

In a steady state of the learning model, the probability distribution over histories remains unchanged over time. The 'stochastic learning equilibrium' is defined as the steady-state probability distribution over histories. This formulation is general and includes many learning models as special cases. Goeree and Holt (2002) show that this equilibrium always exists when there is a finite number of decisions and players have finite (but possibly long) memories.

For example, consider the extreme case where a person can only remember the two most recent observations in the probability-matching experiment. There are four possible remembered histories: LL, LR, RL, and RR, with exogenously determined probabilities of $\frac{3}{4} \times \frac{3}{4}$, $\frac{3}{4} \times \frac{1}{4}$, $\frac{1}{4} \times \frac{3}{4}$, and $\frac{1}{4} \times \frac{1}{4}$, respectively. A stochastic learning equilibrium in this context would be a vector of transition probabilities between these states. The formulation of this model in terms of histories (instead of single-period choice distributions) allows the possibility of dynamic effects such as cycles and endogenous learning rules. The focus on histories (sequences of vectors of players' decisions) also facilitates the proof that a stochastic learning exists under fairly general conditions.

Given a specific learning rule, it is possible to determine the stochastic learning equilibrium. To illustrate, consider the market price game under two extreme conditions, fictitious play ($\rho = 1$) and Cournot best response ($\rho = 0$). Since there is no 'forgetfulness' in fictitious play, any steady-state distribution of decisions will eventually be fully learned by all players, i.e. the empirical frequencies of price draws from the distribution will converge to that distribution. In this case, each player is making a logit probabilistic best response to the empirical distribution, and these best responses match the empirical distribution. Notice that all players must have identical beliefs in this equilibrium. This is known as a 'quantal response equilibrium' as defined by McKelvey and Palfrey (1995.)

When $\rho = 0$, a player's history is simply the most recent observation, and beliefs are necessarily different across players. These differences in individuals' beliefs add extra randomness into the steady state. Figure 5 illustrates these observations for the high-*R* treatment of the price-choice game. The solid line represents the stochastic learning equilibrium with an infinite memory ($\rho = 1$), and the dashed line represents the price distribution for the case of one-period memory ($\rho = 0$). Both of these distributions are hump-shaped, with means near the price average observed in the experiment. The implied distribution of price choices is flatter and more dispersed for the case of one-period memory, since beliefs are being moved around by recent observations, which introduces extra randomness. Both cases, however, capture the salient feature of the prices observed in the high-*R* treatment of the market experiment. In particular, price averages are more than twice as high as the unique Nash equilibrium prediction.

When maximum-likelihood techniques are used to estimate the learning parameter from the choices made by the human subjects, the resulting estimate ($\rho = 0.72$) is intermediate between the extreme cases shown in Figure 5, and the resulting steady-state price distribution will also be intermediate. In fact, the weights determined by powers of 0.72 decline very quickly, and the equilibrium price distribution is rather close to the flatter ($\rho = 0$) case, as can be confirmed with computer simulations. Simulations of individual cohorts of 10 subjects (not shown) show the same up-and-down patterns exhibited by comparably sized cohorts of humans. The simulation averages shown in Figure 4 track the main features of the human data: prices start high and stay high in one treatment, and they start high and decline towards the Nash prediction in the other. Thus, computer simulations of learning models can explain data patterns that are not predicted with standard equilibrium techniques. In fact, we ran the computer simulations before we ran the experiments with human subjects, using the learning and error parameter estimates from a previous experiment (Capra *et al.*,

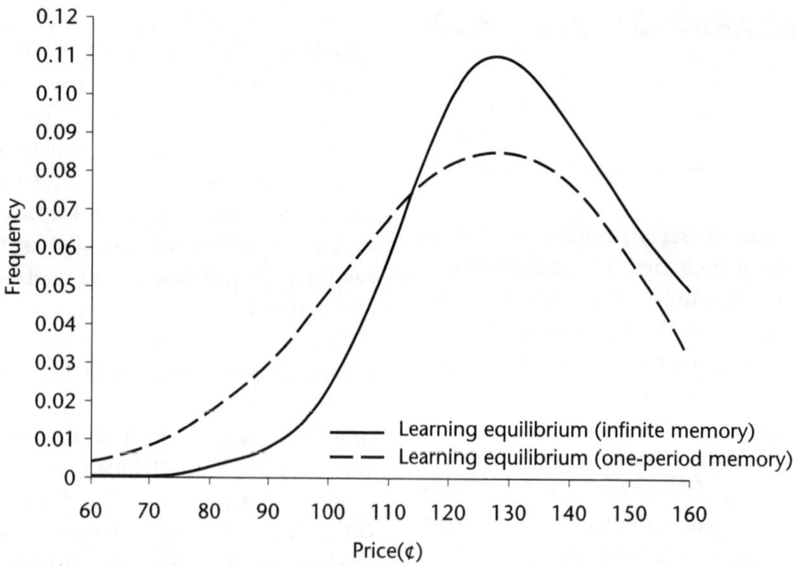

Figure 5. Stochastic learning equilibrium distribution of prices in the price-choice game for $R = 0.8$.

1999). The simulations helped us select two values of the treatment parameter R that would ensure that there would be a strong treatment effect that is not predicted by the Nash equilibrium.

SUMMARY

The learning models presented here were pioneered by Bush and Mosteller (1955), and the stochastic choice models were introduced by the mathematical psychologist Luce (1959) and others. These techniques no longer receive much attention in the psychology literature, where the main interest is in theories of learning, biases, and heuristics that have a richer cognitive content. Yet they have yielded important insights in explaining economics experiments where the anonymity and repetitiveness of market interactions dominate. The incorporation of insights from the literature on heuristics and biases may also prove to be valuable in the future.

The belief- and reinforcement-based learning models depend on past history in a somewhat mechanical manner. In contrast, some laboratory experiments provide situations in which learning seems to proceed in response to cognitive insights. For example, if one subject observes that another has earned more money, the first person may decide to try to imitate the other. Consider a market in which subjects choose 'production quantities' simultaneously, with the advance knowledge that the price at which all production is sold will be a decreasing function of the total quantity produced. Since all output is sold at the same price, the person with the highest quantity will have the highest sales revenue. To the extent that high revenues translate into high profits, the sellers with low quantities may be tempted to imitate the high-output strategies of those who have higher earnings. In this context, the process of imitating high-production sellers can cause the total production to be high. The implications of imitation learning have been studied in a series of recent economics experiments. Offerman and Sonnemans (1998), for example, find evidence that learning is induced to some extent both by imitation of others and by one's own experience.

When some people are following predictable learning patterns, it may be advantageous to try to manipulate others' beliefs via 'strategic teaching'. For example, a dominant seller may punish new entrants by expanding production quantity and thereby driving prices down. This behavior may be intended to 'teach' potential rivals not to enter. This is an important area for future research, and it is complicated by the fact that the person doing the teaching should have a mental model of the others' learning processes.

In some economic situations, learning may occur as a sudden realization that some different decision will provide higher earnings or will avoid losses. A first step in the study of this type of learning may be to measure biological indicators of mental activity for economic tasks that may involve sharp changes in behavior or attempts to anticipate others' decisions (McCabe *et al.*, 2000).

References

Bush R and Mosteller F (1955) *Stochastic Models for Learning*. New York, NY: Wiley.

Camerer C and Ho T-H (1999) Experience weighted attraction learning in normal-form games. *Econometrica* **67**: 827–874.

Capra CM, Goeree JK, Gomez R and Holt CA (1999) Anomalous behavior in a traveler's dilemma? *American Economic Review* **89**(3): 678–690.

Capra CM, Goeree JK, Gomez R and Holt CA (2002) Learning and noisy equilibrium behavior in an experimental study of imperfect price competition. *International Economic Review* **43**(3): 613–636.

DeGroot MH (1970) *Optimal Statistical Decisions*. New York, NY: McGraw-Hill.

Erev I and Roth AE (1998) Predicting how people play games: reinforcement learning in experimental games with unique, mixed strategy equilibria. *American Economic Review* **88**(4): 848–881.

Fantino E (1998) Behavior analysis and decision making. *Journal of the Experimental Analysis of Behavior* **69**: 355–364.

Goeree JK and Holt CA (2002) Stochastic learning equilibrium. Working paper, University of Virginia. [Presented at the Economic Science Association Meetings in New York City, June 2000.]

Luce D (1959) *Individual Choice Behavior*. New York, NY: Wiley.

Mackintosh NJ (1969) Comparative psychology of serial reversal and probability learning: rats, birds, and fish. In: Gilbert R and Sutherland NS (eds) *Animal Discrimination Learning*, pp. 137–167. London: Academic Press.

McCabe K, Coricelli G, Houser D, Ryan L and Smith VL (2000) Other minds in the brain: a functional imaging study of 'theory of mind' in two-person exchange. Working paper, University of Arizona.

McKelvey RD and Palfrey TR (1995) Quantal response equilibria for normal form games. *Games and Economic Behavior* **10**: 6–38.

Offerman T and Sonnemans J (1998) Learning by experience and learning by imitating successful others. *Journal of Economic Behavior and Organization* **34**(4): 559–575.

Siegel S, Siegel A and Andrews J (1964) *Choice, Strategy, and Utility*. New York, NY: McGraw-Hill.

Vulkan N (2000) An economist's perspective on probability matching. *Journal of Economic Surveys* **14**(1): 101–118.

Further Reading

Chen Y and Tang FF (1998) Learning and incentive compatible mechanisms for public goods provision: an experimental study. *Journal of Political Economy* **106**: 633–662.

Cooper DJ, Garvin S and Kagel JH (1994) Adaptive learning vs. equilibrium refinements in an entry limit pricing game. *RAND Journal of Economics* **106**(3): 662–683.

Fudenberg D and Levine DK (1998) *Learning in Games*. Cambridge, MA: MIT Press.

Goeree JK and Holt CA (1999) Stochastic game theory: for playing games, not just for doing theory. *Proceedings of the National Academy of Sciences* **96**: 10564–10567.

Economics, Experimental Methods in

Advanced article

Vernon L Smith, George Mason University, Arlington, Virginia, USA

CONTENTS

Experimental economics uses laboratory methodology to examine motivated human behavior and its interpretation in small group interactive games, and in bidding, auctioning, and market institutions. Subjects earn cash payments depending upon their joint interactive decisions, and the rules governing their interactions.

INTRODUCTION

It is useful to distinguish three complex self-ordering systems: the internal order of the mind (Hayek, 1952); the external order of social exchange between minds (McCabe and Smith, 2001); and the extended order of cooperation through markets and other cultural institutions (Hayek, 1988). Our concern here is with the first two.

We focus on social exchange because it was the cooperative behaviors registered in two-person anonymous interaction that first alerted experimental economists to a significant class of phenomena that violate certain static equilibrium concepts in game theory. Game theory is about strategic interdependent choice when the pay-off benefit to each of two or more people depends jointly on their decisions. These refutations generated alternative interpretations of that theory, and motivated questions directly concerned with the internal order of the mind. They are now leading to the study of the neural correlates of human decision-making in two-person strategic interactions.

Why do we study anonymous interactions? First, our theoretical model of single-play games assumes strangers without a history or a future, and anonymity provides the required control for testing this theory. Also, it is well documented that the effects of face-to-face interaction hide more subtle procedural effects in yielding cooperative outcomes (Hoffman and Spitzer, 1985). As will be illustrated in the experiments reported below, the anonymity condition provides great scope for exploring the natural human instinct for social exchange, and how it is affected by context, reward, and procedure.

Why Should Context Matter?

Context matters because all memory involves relationships and is associative. For example, priming experiments use cues to improve retrieval from memory because of associations between the cue and the stimulus. People perform better at completing word fragments (filling in missing letters) on words they have observed beforehand in lists, even if they are not told that the words appeared in the earlier list. Some priming effects are almost equally strong whether the interval between the original list and the test is a matter of hours or days. Furthermore, being able to state that a word was seen before does not correlate with improved completion performance. How one perceives a current task depends upon unconscious cues to past experience that are triggered by the context of the task. Two decision tasks with the same underlying logical structure may lead to different responses because they are embedded in different contexts and invoke different memory experiences (Gazzaniga *et al.*, 1998, pp. 258–261). This is because of the fundamental, though nonintuitive, nature of perception.

In the early 1950s Hayek articulated certain principles of perception, which are consistent with current neuroscientific understanding:

1. It is incorrect to suppose that experience is formed from the receipt of sensory impulses reflecting unchanging attributes of external objects in the physical environment. Rather, the process by which we learn about the external environment involves a relationship between current conditions and our past experience of similar conditions (Hayek, 1952, p. 165).
2. Categories are formed by the mind according to the relative frequency with which current perception and memory (past perceptions) concur (p. 64). What are

stored in memory are external stimuli modified by processing systems whose organization is conditioned by past experience of stimuli. All perception is produced by memory.

3. This leads to a 'constant dynamic interaction between perception and memory, which explains the ... identity of processing and representational networks of the cortex that modern evidence indicates' (Fuster, 1999, p. 89). 'Although devoid of mathematical elaboration, Hayek's model clearly contains most of the elements of those later network models of associative memory ... [and] comes closer, in some respects, to being neurophysiologically verifiable than those models developed 50 to 60 years after his.' (Fuster, 1999, pp. 88–89).

Hayek's model is incomplete, and did not influence the research it anticipated, but it captures the idea that perception is self-organized, created from abstract function combined with experience. This is relevant to the question of why context is important in the experiments reported below.

MENTAL MODULES AND EVOLUTIONARY PSYCHOLOGY

Evolutionary psychologists argue that the mind consists of circuitry, or interactive modules, that are specialized for vision, for language learning, for socialization, and for a host of other functions (Cosmides and Tooby, 1992). Language and socialization, which are of recent evolutionary origin, are hypothesized to have evolved in the two to three million years during which humans subsisted as hunter–gatherers. It is in this evolutionary environment of our ancestors (EEA) that humans developed mechanisms of social exchange in which assistance, meat, favors, information and other services and valuables were traded across time. This is evident in extant hunter–gatherer societies (e.g. the Ache of Paraguay) in which the product of the hunt is widely shared within the tribe as well as within the nuclear and extended family. In a world without refrigeration and only rare success in hunting, this made sense: if I am lucky in the hunt today, I share the meat with others; and tomorrow, when I fail to make a kill, you share your kill with me and with others. In contrast, the products of gathering – fruit, nuts, roots – depend more on effort than on luck, are much more predictable from day to day, and are shared only in the nuclear family where effort can be closely monitored. Traditions of sharing across time provide gains from exchange that support limited forms of specialization: women and children do the gathering; adult men do the hunting; older men make tools, advise in the hunt, and assist in gathering. Such patterns (subject to numerous variations) are common in tribal communities.

But delayed exchange across time based on reciprocity is hazardous. Favors cannot be retracted, and you might systematically fail to return mine. Without money – a recent invention not available in the EEA – it is adaptive to develop some skill in making judgments about who can or cannot be trusted. This puts a premium on 'mindreading', the ability to infer mental states from the words and actions of others. The minimal mental equipment required is a 'cheater-detector module' for social exchange. The results of experiments designed by Cosmides (1985) are consistent with the hypothesis that the human mind is attuned to detecting cheaters on perceived social contracts. With the development of language, our instincts for cheater detection were enhanced by gossip: comparing notes to determine those with good reputations for returning favors. Gossip, like language and reciprocity, is a human universal, an activity pursued in all human communities. None of this mental equipment was the product of our reason: rather, it was the unconscious product of the biological and cultural evolution that distinguished us from other primates.

Evolutionary psychologists see an inevitable tension between who we are (based on what we have inherited from the EEA) and the demands made on us by the world since the agricultural revolution 10,000 years ago. One account of this tension was articulated by Hayek: 'Part of our present difficulty is that we must constantly adjust our lives, our thoughts and our emotions, in order to live simultaneously within different kinds of orders according to different rules. If we were to apply the unmodified, uncurbed rules (a caring intervention to do visible good) ... of the small band or troop, or of, say, our families, to ... our wider civilization (the extended order of the market), as our instincts ... often make us wish to do, *we would destroy it*. Yet if we were always to apply the rules of the extended order (action in the self-interest within competitive markets) to our more intimate groupings, *we would crush them*. So we must learn to live in two sorts of world at once.' (Hayek, 1988, p. 18).

This observation raises questions about game theory, which postulates that the players are strictly self-interested and that this condition is common knowledge to all the players. How is action in the self-interest affected by whether the anonymous players are in an *n*-person market or a two-person interactive game? How do the players come to have 'common knowledge'? Does the procedural and instructional context of a two-person game affect cooperation by influencing how the players perceive the game? It is most natural to investigate such questions in experimental environments where monetary pay-offs, context, and interaction procedures can be controlled.

EXPERIMENTAL PROCEDURES

The experiments reported below show that context is important in the decision behavior we observe. This is to be expected, given what is known about the autobiographical character of memory and the interaction between current and past experience in creating memory. Below are reported behavioral results in two-person sequential-move game trees in which each pair plays once and only once through the move sequence defined by the tree, and the game is completely known to the subjects. However, the instructions for the experiments do not (except in systematic treatments) use words like 'game', 'play', 'player', 'opponent', or 'partner'; rather, reference is made to the 'decision tree', 'decision maker' 1 (DM1) or 2 (DM2), 'your counterpart', and other terms designed to provide a baseline context.

Your experience as a subject in a typical experiment might be as follows. You have been recruited to participate in an economics experiment for which you will be paid \$5 (or more, in some cases) for arriving on schedule, plus the amount in cash that you earn from your decisions, to be paid to you at the end. You arrive, sign in, receive \$5, and are assigned to a computer terminal in a large room with 40 stations. There are 11 other people, well spaced throughout the room. Each station is a partially enclosed booth, making it very easy to maintain your privacy. After everyone has arrived you log in to the experiment as directed on your screen. You read through the instructions for the experiment at your own pace, respond to the questions, and learn that in this experiment you are matched anonymously with another person in the room, whose identity you will never know, and vice versa. This does not mean that you know nothing about that person: it may seem evident that he or she is another 'like' person – for example, an undergraduate – with whom you may feel more or less of an in-group identity. Obviously, you bring with you a host of past experiences and impressions that you are likely to apply to the experiment.

Ultimatum Game Experiments

Consider the following simple two-stage two-person game. A fixed sum of money m is provided by the experimenter (e.g. m might be 10 one-dollar bills, or 10 ten-dollar bills). Player 1 moves first, proposing that a portion $x \leq m$ of the money be offered to player 2, player 1 retaining $m - x$. The offer is a 'take it or leave it' ultimatum. Player 2 then responds by either accepting the offer, in which case the experimenter pays $m - x$ to player 1 and x to player 2, or rejecting the offer, in which case each player receives 0.

Now consider four different instructional–procedural contexts in which an ultimatum game with this underlying abstract structure is played. In each case, imagine that you are the first mover (player 1 in the above abstract form). (See Hoffman *et al.* (2000) for instructional details, and for references to the literature and origins of the ultimatum game.)

Context 1: 'divide \$10'

In the first context, the instructions state that you and your anonymous counterpart have been 'provisionally allocated \$10'. Your task is to 'divide' the \$10 using the following procedure. You have been randomly assigned to the role of first mover. You (as person A) are asked to complete boxes (4) and (5) of the proposal form shown in Figure 1. The form then goes to your counterpart (person B) who checks 'Accept' or 'Reject'.

In this version, the \$10 consists of 10 one-dollar bills. In another version, there is \$100 (10 ten-dollar bills) to be divided.

Context 2: 'contest entitlement'

In the second context, each of the 12 people in the room takes the same general-knowledge quiz (10 questions). The results are used to determine the positions of persons A and B in each pairing. Your score is the number of questions answered

(1) Identification number..	#A
(2) Paired with..	#B
(3) Amount to divide...	\$10
(4) Person B receives	
(5) Person A receives (3)–(4)	
(6) Accept ☐	Reject ☐

Figure 1. Proposal form for an ultimatum game experiment using the 'divide \$10' context. You (as person A) are asked to complete boxes (4) and (5); the form then goes to your counterpart (person B) who checks 'Accept' or 'Reject'.

correctly, with ties broken in favor of the person who finished the quiz fastest. The scores are ranked from 1 (highest) to 12 (lowest). Those ranked from 1 to 6 will have 'earned' the right to be person A; the other six will be person B.

Context 3: 'exchange'

In the third context, person A is a 'seller' and B is a 'buyer'. A table lists the profit of the seller and of the buyer for each possible price ($0, $1, $2, ..., $10) charged by the seller if the buyer chooses to buy. The profit of the seller is equal to the price chosen; the profit of the buyer is $10 minus the price. The profit of each is zero if the buyer refuses to buy at the price chosen by the seller.

Context 4: 'contest–exchange'

The fourth context combines the second and third: 'sellers' are selected by a general-knowledge quiz. In one version the total amount is 10 one-dollar bills; in another, it is 10 ten-dollar bills.

Results of Ultimatum Game Experiments

The game-theoretic concept of sub-game perfect equilibrium (SPE) yields the same prediction in all of these versions of the ultimatum game (Selten, 1975): player 1 offers the minimum positive unit of account ($1 if m = $10, $10 if m = $100), and player 2 accepts the offer. The analysis assumes that each player is self-interested in the sense of always choosing the largest of two pay-offs for himself or herself; that this condition is common knowledge for the two players; and that player 1 applies backward induction to the decision problem faced by player 2, conditional on player 1's offer. Thus player 1 should reason that any positive pay-off is better than zero for player 2, and therefore, player 1 need only offer the minimum positive amount.

But there are other models of decision for games like the ultimatum. A problem with the above analysis is that, perhaps depending on context, the ultimatum interaction may be interpreted as a social exchange between any two anonymously matched players who normally read intentions into the actions of others (Baron-Cohen, 1995). Suppose that the ultimatum game is perceived as a social contract in which player 2 has a ('fair claim') entitlement to more than the minimum unit of account; then an offer of less than the perceived entitlement (say, only $1, or perhaps even $2 or $3) may be rejected by some players 2. Player 1, reading this potential mental state of player 2 (e.g. by imagining what he or she would do in the same circumstance), might then offer substantially more than $1 to ensure acceptance.

Observe that in context 1, the original $10 is allocated imprecisely to both players, and does not clearly belong to either person A or B. Further, a common interpretation of the word 'divide' involves the separation of some divisible quantity into equal parts. Moreover, in western culture the use of a lottery or other random device is recognized as a standard mechanism for 'fair' or equal treatment. Hence, the instructions can be interpreted as suggesting that the experimenter is engaged in a 'fair' treatment of the subjects. This can serve as a strong, albeit unconscious, cue that the subjects ought to be 'fair' in their treatment of each other.

By contrast, context 2 deliberately introduces a contest procedure, before the game itself, in which those who score the highest earn the right to be person A, and those who score the lowest will be person B. In this treatment, nothing is said about who has been initially allocated the money, and the word 'divide' is not used. Rather, person A must choose how much person B is to receive, and person B must choose to accept or reject the proposal. Consequently, the instructions may cue some norm of 'just desert' based on test performance.

In context 3, the abstract ultimatum game is embedded in a transaction between a buyer and a seller. In such exchanges, buyers (in western culture) do not normally question the right of the seller to move first by quoting a price, nor that of the buyer to respond with a decision to buy or not to buy.

Context 4 combines the implicit 'property right' norm of a seller with an explicit mechanism whereby subjects 'earn' the privilege of being the seller in a contest whose outcome provides the same opportunity for all participants, depending on their general knowledge. This treatment introduces the 'equal opportunity' norm, as opposed to 'equality of outcome'.

Table 1 summarizes the results from two different studies of ultimatum game bargaining with stakes of either 10 one-dollar or 10 ten-dollar bills, where the number of pairs of players varies from 23 to 27. Note that 'divide' with random entitlement corresponds to context 1; 'divide' with earned entitlement to context 2; 'exchange' with random entitlement to context 3; and 'exchange' with earned entitlement to context 4.

Comparing 'divide $10' with 'divide $100' under random entitlement, we observe a trivial difference

Table 1. Mean percentage offered in ultimatum games, by context treatment. Data from Hoffman *et al.* (1996) and (1994)

		$10 stakes 'Divide'	*$100 stakes* 'Exchange'	'Divide'	'Exchange'
Random entitlement	Mean offer	43.7%	37.1%	44.4%	(n/a)
	N	24	24	27	(n/a)
	Rejection rate[a]	8.3%	8.3%	3.7%	(n/a)
Earned entitlement	Mean offer	36.2%	30.8%	(n/a)	27.8%
	N	24	24	(n/a)	23
	Rejection rate[a]	0	12.5%	(n/a)	21.7%

[a]Percentage of the *N* pairs in which the second player rejects the offer of the first.

in the amount offered between the low stakes (43.7%) and the high stakes (44.4%). There is no significant difference in the rate at which offers are rejected (8.3% and 3.7% respectively).

When 'exchange' is combined with an earned entitlement, the increase in stakes seems to lower the offer percentage from 30.8% for $10 stakes to 27.8% for $100 stakes, but this difference is within the normal range of sampling error using different groups of subjects and is not significant. Surprisingly, this minuscule decline in the mean offer correlates with an increase in the rejection rate from 12.5% to 21.7%. In the high-stake game, three out of four subject players 1 offering $10 are rejected, and one offer of $30 is rejected. As we shall see below in other games, this behavior is associated with a strong human propensity to incur personal cost to punish those who are perceived as cheaters, even under strict anonymity (as in Cosmides, 1985).

Comparing the 'divide' and 'exchange' conditions with random entitlement and $10 stakes, the offer percentage declines from 43.7% to 37.1%, and comparing the 'divide' conditions with random and earned entitlement and $10 stakes the offer percentage declines from 43.7% to 36.2%. Both reductions are statistically significant. Even more significant is the reduction from 43.7% to 30.8% with the 'exchange' condition and earned entitlement. Moreover, in all four of these contexts the rejection rate is null or modest (0 to 12.5%).

The small proportion of offers rejected (except when the stakes are $100 in the 'contest–exchange' context, where the mean offers decline to 27.8%) indicates that players 1 generally read their counterparts well and offer a sufficient amount to avoid being rejected. The one exception shows that trying to push back the boundary, even if it seems justified by the higher stakes, may provoke rejections.

One obvious conclusion from these data is that the effect of context on behavior cannot be ignored in the ultimatum game: the percentage offered varies by over a third between the highest (44%) to the lowest (28%) measured values. Studies of cross-cultural variation in ultimatum offers show a comparable variation. Thus, a comparison of two hunter–gatherer and five modern cultures reveals a variation from a maximum of 48% (Los Angeles subjects) to a minimum of 26% (Machiguenga subjects from Peru) (Heinrich, 2000). These comparisons attempted to control for instructional differences across different languages, but of course this is inherently problematic in that one cannot be sure that the translations, or the procedures for handling the subjects, completely control for context across cultures. Nor can it be assumed that the pay-offs are subjectively comparable across currencies.

The instructional comparisons also call into question the extent to which one can define what is meant by 'unbiased' instructions. Some results may be robust with respect to instructional changes, but this can only be established empirically, since we know little about the sources of behavioral variation due to context. Indeed, unless such robustness is established no claims can be made concerning the relative 'neutrality' of instructions, and the extent to which differences in behavior can be attributed to differences between cultures.

Because of the nature of perception and memory, we should expect context to be an important factor. In the ultimatum game, the variation of observed results with systematic instructional changes designed to alter context shows clearly that context can and does matter.

Trust Games

Ultimatum games have been studied extensively, but because of their simplicity they leave unanswered many questions about what underlies the behavior manifest in them. For example, one cannot vary independently the cost of player 2's rejection of player 1's offer. The game is constant-sum, and is inherently confrontational: neither player can take action that increases the total gains from the transaction, and therefore the interpretation of the game as an exchange is limited.

We turn therefore to a somewhat richer class of two-person extensive-form trust games in which the return to equilibrium play, cooperation, defection, and the prospect of costly punishment of defection can be studied in a richer parameter space than that of the ultimatum game.

Figure 2(a) shows a trust game tree. Play starts at the top, node x_1, with player 1. Player 1 can move right; this stops the game, yielding \$7 to player 1 and \$14 to player 2. Alternatively, player 1 can move down, in which case player 2 has to choose a move at node x_2. If player 2 moves right, each player gets \$8. If player 2 moves down, player 1 can then move right at node x_3, yielding \$10 for each, or down, yielding \$12 for player 1 and \$6 for player 2.

The SPE is \$8 for each player. This is because at node x_1 player 1 can look ahead (use 'backward induction') to see that if play reaches node x_3 player 1 will want to move down. But player 2, also using backward induction, will see that at node x_2 player 2 should move right. Since this yields a higher pay-off to player 1, at node x_1 player 1 should move down.

The SPE outcome would prevail by the logic of self-interested players who always choose dominant strategies. There are other behavioral possibilities, however, depending on whether other preferences or perceptions of the interaction are applied.

If player 1 has other-regarding preferences (altruism, or utility from the other's pay-off), and is willing to incur some cost to greatly increase the pay-off to player 2, player 1 may move right at x_1. That way, at a cost of \$1, player 1 can increase player 2's pay-off by \$6, compared with the SPE. Thus, player 1 need have only a modest preference for an increase in player 2's welfare in order to move right.

At x_2, player 2 may move down, signaling to player 1 that such a move enables both to profit, provided that at x_3 player 1 cooperates by reciprocating player 2's favor. Alternatively, at x_3 player 1 can defect, by choosing the dominant strategy and moving down.

Figure 2(b) shows the tree for a punishment version of the trust game shown in Figure 2(a). The trees are identical except that at node x_3, player 1 chooses between the cooperation pay-off and passing back to player 2 at node x_4. Now player 2 decides whether to accept the defection or, at a cost, punish player 1 for the defection. By backward induction, the SPE is the same in the punishment version. The cooperation outcome can be justified (as a Nash equilibrium) only if the threat of punishment by player 2 at node x_4 is credible. But under the anonymity conditions, with no capacity to communicate, such a threat is not credible.

The outcome frequencies for the trust game ($N =$ 26 pairs), and for the trust–punishment game ($N = 29$ pairs) are summarized in Figure 3.

In neither game is there a single case of a player 1 choosing the altruism outcome: all choose to pass to player 2, seeking a higher pay-off for themselves, and being content to give player 2 a much smaller pay-off than would be achieved by altruism.

The sub-game perfect equilibrium is chosen by 54% of the pairs in the trust game, and 55% in the trust–punishment game. Thus, there is no significant difference in behavior between the two games in terms of the frequency with which players 2 offer to cooperate by passing to players 1 at x_2.

There is, however, a considerable difference in the response of players 1 to the offer to cooperate: only 50% cooperate in the trust game, while 85% cooperate when facing the prospect of punishment for defection.

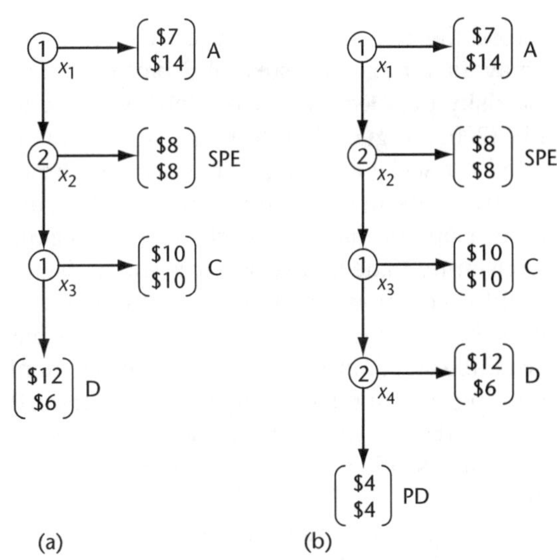

Figure 2. Trust game trees, (a) without punishment and (b) with punishment. At each terminal node the pay-off to player 1 is shown above the pay-off to player 2.

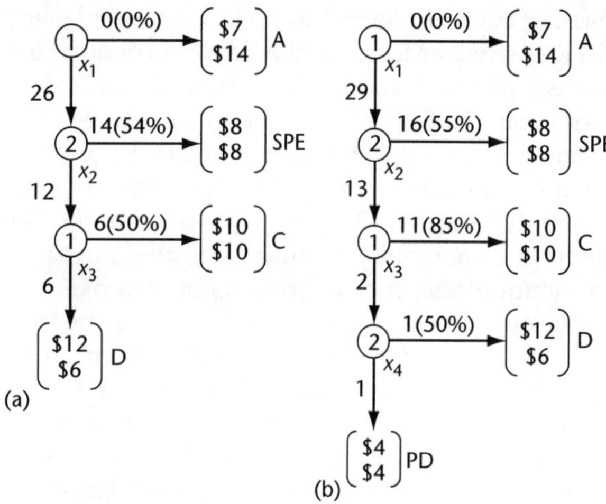

Figure 3. Experimental outcomes for the (a) trust and (b) trust–punishment games shown in Figure 2. A total of 26 subject pairs took part in the trust game and 29 in the trust–punishment game. The figures beside the arrows indicate the number of pairs following each route through the game tree and the percentage moving right at each decision node. The data are from McCabe *et al.* (2000). Source data for larger trees have been trimmed to eliminate rare outcomes, with commensurate reduction in sample size (from 30 to 29 in the punishment version).

Across the two games, why do nearly half of the players 1 forgo the sure SPE payoff in favor of the risky prospect of cooperation? McCabe and Smith (2001) argue that humans are eminently adapted for social exchange, or reciprocity among the individuals that constitute the small groups that form our primary networks of relationships. We constantly trade favors, services and assistance, with little conscious awareness of these trading relationships that are so much a part of our humanity. McCabe and Smith postulate an implicit mental accounting system for keeping track of trustworthy trading partners. This accounting system is part of the framework of our friendships and social connections.

Reciprocity is a human universal, characteristic of all cultures, as is the use of a spoken language. Like language, the form of reciprocity varies across cultures, but its common functionality is to produce gains from exchange. Smith (1998) argues that reciprocity in the family, extended family, and tribe is what ultimately led to the extended order of cooperation through market trade. He postulates that this proclivity for reciprocal social exchange is so natural and instinctive that it survives even in interactions between anonymously paired subjects in the two-person extensive-form games described above.

This interpretation has been reinforced by many other extensive-form game-tree experiments. Thus, in the game shown in Figure 4, player 1 chooses between the SPE, $10 for each, and passing to player 2 who chooses between pay-offs of $15 and $25 (for players 1 and 2 respectively) and pay-offs of $0 and $40. The move frequencies for 24 pairs of undergraduates are shown on the tree. Very similar outcomes prevail with a group of graduate students trained in economics and game theory (McCabe and Smith, 1999).

Effects of Context, Repetition, and Opportunity Costs in Trust Games

The sensitivity of cooperative behavior in trust games to the procedural, instructional, and opportunity cost context of the experiment has been demonstrated by several treatment manipulations.

'Partners' versus 'opponents'

Consider two treatment variations on the trust game of Figure 2(a): wherever the word 'counterpart' is used in the instructions to refer to the other decision maker, substitute the word 'partner' in one treatment and 'opponent' in the other (Burnham *et al.*, 2000). Subjects (156 pairs in total) were recruited in either 'small' groups of 12 in a session or 'large' groups of 24 in a session. In all sessions, half of the subjects (6 or 12) were randomly assigned to each of the two instructional conditions;

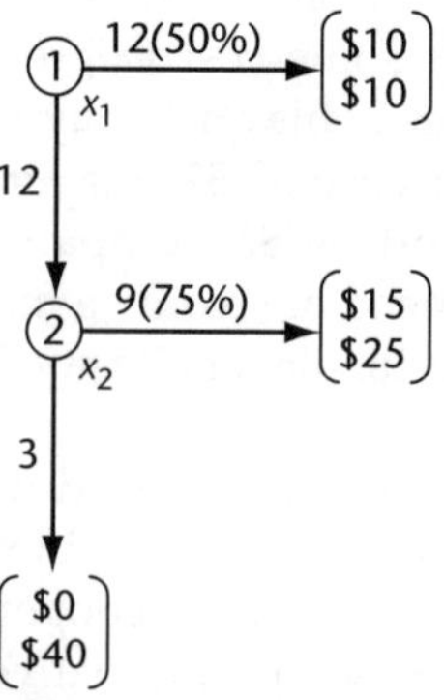

Figure 4. Experimental outcomes for a simple trust game (McCabe and Smith, 1999). A total of 24 subject pairs took part in the game. The figures beside the arrows indicate the number of pairs following each route though the game tree and the percentage moving right at each decision node.

each person was randomly paired with another and assigned to the player 1 or player 2 role; and the two experiments were run simultaneously in the same room. Neither group was informed that the other was reading slightly different instructions. Thus, the experimental design consisted of two group sizes, 12 or 24, and two instructional conditions, 'partner' and 'opponent'.

Each session began with a single play of the trust game. The subjects were then paid and informed that they would also participate in a second experiment. This second experiment used the same instructions except that the game was repeated for 10 periods of play. On each period of play, each person was matched with a new person, then randomly assigned the role of player 1 or 2. Each repetition was therefore between paired strangers. This is called 'repeat single' play. Repeat single play is like single play except that the subjects acquire experience under procedures that control for reputation formation across successive interactions.

It was found that 'partners' are more trusting (players 2 move down at node x_2) and more trustworthy (players 1 move right at x_3) than 'opponents'. (In the first single-play game, however, no difference was observed in the frequency of trust between the two treatments, but 68% of the 'partner' players 1 cooperated following an offer of cooperation, while only 33% of the 'opponent' players did.)

Over time (single play followed by 10 repeat single plays), with 'partners' trust increases through the first five plays then declines, while with 'opponents' trust steadily declines. Trustworthiness declines over time for 'partners', and remains low for 'opponents'. Hence, 'partners' learn to defect, but 'opponents' defect from the beginning.

Pairs who interact in groups of size 24 are less trusting than those in groups of size 12.

These results provide further support for the hypothesis, based on cortical memory theory, that context should matter. In this case, a simple two-level variation on the language used to describe the other person in each trial is sufficient to yield statistically significant differences in trust and trustworthiness.

Repeat single with and without punishment

The tendency for cooperation eventually to decline as play is repeated with distinct 'partners' is already suggested by Figure 3(a): of the 12 players 1 arriving at node x_3, half reciprocate and half defect. Hence, it is not profitable to offer cooperation in the trust game, and repetition with strangers is likely to cause a decline in cooperation, both offered and reciprocated, across time.

In the trust–punishment game in Figure 3(b), however, 85% reciprocate at node x_3, and only 15% defect, of which half are punished. Hence, it is profitable to offer cooperation, and it is not profitable to defect. This suggests that in repetition, using the repeat-single protocol, cooperation might not diminish. In fact, this is the case: when defection can be punished, the conditional probability of reciprocal cooperation by players 1 actually increases modestly across 15 periods of play (McCabe *et al.*, 2000).

Opportunity cost

An important implication of reciprocity theory (the value of option 'y' given up by choosing 'x') is that when person A chooses to forgo the SPE outcome and offer the cooperative option to person B, the pay-off alternatives should be such that person B sees clearly that person A is incurring an 'opportunity cost' – forgoing a smaller pay-off in an attempt to allow both persons to achieve larger pay-offs. There should be a cost incurred in order to gain from exchange. Failing this condition, the basis for an exchange, or reciprocation, is compromised: person B would be less likely to read clearly the intentions of person A, and person A will anticipate that an unclear message would be being conveyed.

Thus, in Figure 2, if instead of \$8 the SPE is \$10 for each player – identical to the 'cooperative' outcome – the outcome frequency results should change dramatically. This has been tested for the trust–punishment game tree in Figure 2(b) (McCabe *et al.*, 2002). The effect is to increase the frequency of the SPE outcome to about 95%. Thus, players have no difficulty concluding that the attempt to cooperate by player 2 at node x_2 is risky, and will not be chosen unless there is a compensating potential gain.

Another test of reciprocity is to contrast two versions of a game with the structure of Figure 4. Version 1 is like that in Figure 4, with different pay-offs but qualitatively the same outcomes. In version 2, player 1 has no option to move right. The prediction is that version 1 will yield more cooperative outcomes then version 2. In fact, defection is twice as frequent in version 2 as it is in version 1. The interpretation is that if nothing was given up by player 1 – the move did not deliberately forgo the pay-offs achievable at the SPE – then player 1's move does not constitute an 'offer'; so nothing need be reciprocated.

THEORY OF MIND AND ITS NEURAL CORRELATES

Experimental tests of non-cooperative equilibrium theory using anonymously paired subjects in two-person games consistently show that people do cooperate. Almost all subjects in the ultimatum game offer amounts in excess of the equilibrium predictions, and when they do offer equilibrium amounts their counterparts almost always reject the offer. Similarly, in trust games, up to half of subjects offer to cooperate at the risk of defection; and in varying degrees, depending on context, their counterparts cooperate at a cost to themselves. These data cannot be explained simply in terms of preferences – a utility for the other's payoff – nor can they be dismissed by the argument that the subjects are too unsophisticated or inadequately motivated.

A more satisfactory model is based on reciprocity and the human ability to communicate intentions through actions. This ability to invoke shared-attention, intention-detector, and 'mindreading' mechanisms in the brain is relevant to other observations of behavior in people impaired by frontal lobe damage and by autism.

Autism (whose genetic basis is indicated by its greater incidence in siblings and in identical twins) is characterized by 'mind blindness', a severe deficit in one's innate awareness of mental phenomena in other people. Children with autism fail developmentally to use pointing gestures to request objects or otherwise call the attention of others to items of joint interest. In contrast, blind children at age 3 are aware of what 'seeing' is in others, and will say 'see what I have'. At about age 3 or 4, normal children become aware of beliefs in others, and understand that others can hold false beliefs. Thus, shown a candy box, and asked what it contains, normal children will say that it contains candy. Upon opening the box, the child sees that pencils have replaced the candy. The child is then asked what the next child who comes in the room will think is in the box. Normal children will reply 'candy', whereas the majority of autistic children will say 'pencils' (Baron-Cohen, 1995).

Studies of autism, and of certain forms of brain damage from accidents or surgery, support hypotheses that particular regions of the brain have circuitry devoted to 'mindreading', an innate capacity for unconscious awareness of what others think or believe. Brain imaging studies of third-party false beliefs in story comprehension tasks have found activation in Broadman's area 8 (medial prefrontal cortex), and in other supporting regions such as the orbital frontal cortex (Fletcher *et al.*, 1995). This role of Broadman's area 8 has been specifically corroborated by functional magnetic resonance imaging of subjects playing trust and trust–punishment games like those presented above (McCabe *et al.*, 2001). These studies compare subjects' decision making when playing a human counterpart and when playing computer strategies with fixed known probabilities of moving 'left' or 'right'. Activation is significantly greater in the mindreading areas when playing a human than when playing a computer.

Thus, independent strands of research into the internal order of the mind and the external order of social exchange appear to be converging in support of the hypothesis that humans are so well adapted to personal exchange that reciprocity survives even in anonymous interactions.

References

Baron-Cohen S (1995) *Mindblindness: An Essay on Autism and Theory of Mind*. Cambridge, MA: MIT Press.

Burnham T, McCabe K and Smith VL (2000) Friend-or-foe intentionality priming in an extensive form trust game. *Journal of Economic Behavior and Organization* **43**: 57–73.

Cosmides L (1985) The logic of social exchange. *Cognition* **31**: 187–276.

Cosmides L and Tooby J (1992) Cognitive adaptations for social exchange. In: Cosmides L and Tooby J (eds) *The Adapted Mind*, pp. 163–228. New York, NY: Oxford University Press.

Fletcher P, Happe F, Frith U *et al.* (1995) Other minds in the brain: a functional imaging study of 'theory of mind' in story comprehension. *Cognition* **57**: 109–128.

Fuster J (1999) *Memory in the Cerebral Cortex*. Cambridge, MA: MIT Press.

Gazzaniga M, Ivry R and Mangun G (1998) *Cognitive Neuroscience*. New York, NY: Norton.

Hayek F (1952) *The Sensory Order*. Chicago, IL: University of Chicago Press.

Hayek F (1988) *The Fatal Conceit*. Chicago, IL: University of Chicago Press.

Heinrich J (2000) Does culture matter in economic behavior? *American Economic Review* **90**(4): 973–979.

Hoffman E (2000). In: Smith (2000), pp. 79–90.

Hoffman E and Spitzer M (1985) Entitlements, rights and fairness. *Journal of Legal Studies* **14**: 259–297.

Hoffman E, McCabe K and Smith VL (1996) On expectations and the monetary stakes in ultimatum games. *International Journal of Game Theory* **25**(3): 289–301. [Reprinted in Smith (2000).]

Hoffman E, McCabe K, Shachat K and Smith VL (1994) Preferences, property rights, and anonymity in bargaining games. *Games and Economic Behavior* **7**: 346–380. [Reprinted in Smith (2000).]

McCabe K and Smith VL (1999) A comparison of naïve and sophisticated subject behavior with game theoretic predictions. *Proceedings of the National Academy of Sciences of the USA* **97**: 3777–3781.

McCabe K and Smith VL (2001) Goodwill accounting and the process of exchange. In: Gigerenzer G and Selten R (eds) *Bounded Rationality: the Adaptive Toolbox*, pp. 319–340. Cambridge, MA: MIT Press.

McCabe K, Houser D, Ran L, Smith VL and Trouard T (2001) A functional imaging study of cooperation in two-person reciprocal exchange in process. *Proceedings of the National Academy of Sciences of the USA* **98**: 11832–11835.

McCabe K, Rassenti SJ and Smith VL (1996) Game theory and reciprocity in some extensive form experimental games. *Proceedings of the National Academy of Sciences of the USA* **93**: 13421–13428. [Reprinted in Smith (2000).]

McCabe K, Rigdon M and Smith VL (2002) Positive reciprocity and intentions in trust games. *Journal of Economic Behaviour and Organization* (in press).

Selten R (1975) Re-examination of the perfectness concept for equilibrium points in extensive games. *International Journal of Game Theory* **4**: 25–55.

Smith VL (1998) The two faces of Adam Smith. *Southern Economic Journal* **65**: 1–19.

Smith VL (2000) *Bargaining and Market Behavior*. Cambridge, UK: Cambridge, University Press.

Education, Learning in

Introductory article

Stanton Wortham, University of Pennsylvania, Philadelphia, Pennsylvania, USA

CONTENTS

Learning takes place in many settings, but educational institutions foster both breadth and depth of learning. Different types of teaching make very different assumptions about what learning is.

INTRODUCTION

Theories of learning have been applied most often in educational institutions. The relationship between cognitive science and education has benefited both scientists and practitioners. Scientists have used educational settings to develop and test their theories, and practitioners have used new knowledge about learning to design more effective education.

Broadly conceived, education is the process of continuing the human species. All humans are born immature, without the knowledge and skills they will need to function – without language, without knowing how to use complex tools, and so on. The species continues because adults communicate knowledge and skills to the next generation. This intergenerational transfer allows future generations to build on prior accomplishments.

Thus all humans teach. Whether they realize it or not, all teachers act as if some theory of learning is true. Particular ways of teaching make assumptions about what learning is. Furthermore, theories of learning themselves rest on conceptions of human nature. Different accounts of how people learn assume different things about what people are essentially like.

This article describes three broad theories of learning – together with the conceptions of human nature underlying these theories – and the types of educational practice that have been built on these theories. The article has two purposes. First, it is important to recognize the theories of learning and conceptions of human nature that underlie various types of schooling. The article describes how typical teacher and student behavior makes assumptions about how learning happens. Second, as theories of learning have developed, we have learned that earlier theories were too simple. The article describes how more complex accounts of learning and human nature are needed to guide educational practice.

BEHAVIOR

Theories of learning that focus on behavior are called 'behaviorist'. Behaviorists argue that humans should not consider themselves special. Copernicus showed that the earth was not the center of the solar system, and Darwin showed that humans were not qualitatively different from animals. Behaviorists further puncture our sense of superiority, arguing that humans do not have free will to act as they choose. 'A person does not act upon the world', B. F. Skinner said, 'the world acts upon him.' On this theory, the environment shapes people's behavior through reinforcement. Just as Darwin showed that organisms appear designed by a creator to fit their niche, even though adaptation is in fact a result of random variation and natural selection over time, behaviorists show that humans appear to reflect and choose their actions, while in fact their behavior has been shaped by reinforcement.

To learn, then, is to change one's behavior in response to reinforcement. This account of learning contains three central elements: behaviors by the organism, conditions present in the environment, and consequences that follow from various behaviors. People, like other animals, will generate various behaviors in a new situation. Some of these behaviors will result in positive consequences, while others will not. People learn to respond more often with behaviors that were reinforced positively in a given situation.

Behaviorist Education

On a behaviorist account, teaching is the systematic shaping of a student's behavior. The teacher has control and students are raw material to be shaped. Teachers arrange reinforcements so that students come to behave as teachers want them to. Scientists have successfully taught pigeons to play ping pong, for instance, by designing a long series of intermediate skills that lead from natural pigeon behavior to ping pong. They reinforce the pigeons for performing each of these intermediate skills, in turn, until the pigeons produce the target behavior. Similarly, teachers of human students should define the target behaviors, design a path of intermediate behaviors from what students can already do, then reinforce students at each step until they produce the target behavior.

Behaviorists have designed 'teaching machines' that dispense rewards as students accomplish prespecified tasks. One famous picture shows a small boy playing a piano, with a candy dispenser on top. Although these pictures now look outdated, many practices in today's schools presuppose a behaviorist account of learning. Discipline systems almost always rely on rewards and punishments to shape students' behavior. Grades are used as reinforcers. And many classroom practices, from worksheets to testing, involve teachers rewarding students for producing desired behavior.

Research in cognitive science from the second half of the twentieth century has shown that behaviorism is not an adequate theory of learning. People often act because they value activities intrinsically, not for external reinforcement. As described in the next section, people also develop complex representations of the world and reflect on their actions in a way that behaviorists denied. Why, then, do students and teachers so often act as if behaviorism were true?

Because it works. If you have control over effective reinforcers, you can shape people's behavior. Behaviorism is not false. It is true, but it is not the whole truth. Under certain circumstances, people do learn just like animals. The question is whether we should create more circumstances that encourage people to learn in this way. Cognitive scientists claim that we should not, because humans have the potential to learn in nonbehaviorist ways, and because students can develop deeper knowledge when encouraged to learn differently.

MIND

Theories of learning that focus on mental representations are called 'cognitivist'. Cognitive approaches to learning see humans as actively making sense of the environment. People develop mental models of the world and act on the basis of these models, not simply in response to reinforcements. When people encounter a new situation, they assimilate it to their own pre-existing models of the world. Learning involves expanding those mental models, in order to make them more accurate.

This account of learning distinguishes between genuine understanding and merely producing the right behavior. People often just parrot the right answer without understanding it, just as pigeons can play ping pong without understanding what they are doing. True learning involves a deeper grasp of the subject matter, such that people's mental models line up with the world. Furthermore, people cannot be forced to learn. True learning requires a change in people's internal models, and learners must change these models themselves.

Cognitive scientists have described various structures and processes that underlie learning. There seem to be some universal constraints, which presuppose people to certain broad types of mental models. Particular domains of knowledge are also organized in distinct ways, to facilitate learning. And individuals sometimes vary in the types of structures that they operate most effectively with.

For instance, there are different learning styles – some people learn most effectively through verbal explanations, while others learn more effectively through visual diagrams, and so on.

Cognitivist Education

From this perspective, learners need to develop deeper understandings, not just produce the right behaviors. Deeper understandings cannot be imposed on students, because they must construct their own mental models. So teachers do not shape students, nor do they deliver correct answers. Teachers should develop educational environments that push students to broaden and deepen their own models, thus opening up areas of the world that students have not thought about. After teachers have set up rich educational environments, ones that contain puzzles designed to provoke students to reflect, then they must allow students to explore. Teachers can challenge students, by pointing out contradictions in their beliefs, but students themselves must recognize the puzzles and work to solve them. Teachers can explain, but if students can only repeat a teacher's explanation then they have not truly learned. Students themselves must integrate new experience with their own developing mental models.

Assessment is a bigger challenge for cognitivist educators than for behaviorist ones. Behaviorists pre-define the educational objective, and they assess whether students produce the desired behavior. Genuine cognitive learning, in contrast, takes place internally. Teachers can infer about students' understandings, but they do not want to encourage rote learning by using simple tests. Instead of assessing whether students get the right answers, cognitivist educators try to assess underlying thought processes by examining how students reached certain answers.

Cognitivist theories of learning are more widely accepted than behaviorist ones. Nonetheless, there is less cognitivist teaching than behaviorist teaching in our schools. This happens partly because cognitivist education is difficult for both teachers and students. Because they are responsible for students' learning, it is hard for teachers to let students pursue their own ideas much of the time. Students also find it easier to write down what the teacher says, instead of developing their own accounts. This sort of resistance can be overcome, and many teachers do successfully encourage students to develop their own deeper understandings. But behaviorist practice has been harder to overcome than behaviorist theory.

SOCIETY

Theories of learning that go beyond mental representations to include social practices are called 'social cognitivist'. Cognitivist learners are autonomous, developing models themselves to make sense of the environment. Recent theories present the learner, instead, as a participant in social activities. Learning, on this account, is a transformation of participation in activity, not primarily the creation of mental models. Instead of simply developing their own representations, people become increasingly competent participants in the intellectual lives of those around them.

From this point of view, people learn as they more competently use tools to facilitate thought and action. Adults incorporate learners into their activity by teaching them how to use certain cognitive tools. Some of these tools are mental, such as mnemonic devices. Others are objects, such as maps. But learners do not have to construct them alone, because these tools have already been developed and can be borrowed from others.

Any theory of learning presupposes a 'unit of analysis'. This is the smallest unit that preserves essential behavior of the whole. In order to study the behavior of water, for example, one must understand the molecular level. Studying hydrogen and oxygen atoms separately will not allow one fully to understand the behavior of water. Similarly, one cannot fully understand learning solely by studying individuals' mental representations. Individual cognitions are essential, as hydrogen atoms are essential to water, but learning itself depends on a larger unit: a social activity, which includes individuals' mental representations, various cognitive tools, and others' knowledge and skill, all of which together allow learning.

Unlike behaviorists, and like cognitivists, social cognitivists describe how cognitive structures and processes mediate between the environment and people's actions. But social cognitivists emphasize that these mediating structures go beyond individuals' mental models to include tools and other aspects of social activities. Although some activities (such as conventional tests) do require individuals to think in isolation with limited tools, a full account of learning must analyze social activities in addition to mental representations.

Social Cognitivist Education

In a social cognitivist approach, both teacher and student are active. Instead of relying primarily on students' own exploration and model-building, the teacher acts as a competent practitioner of the activity being taught and brings tools for students to use. Teachers guide students as they begin to participate in the activity. This guidance allows students to do tasks that they would not be able to perform on their own. Students act like apprentices, at first doing minor parts of the task while observing others, then taking on increasing responsibility.

Teachers should design more naturalistic or 'authentic' activities for students to participate in, where the goal is competent participation in real activity. Many medical schools, for instance, now use 'problem-based learning' – in which groups of beginning students are given real, complex cases and asked to diagnose the problem. They must consult more expert practitioners, do research on relevant topics, and develop alternative diagnoses to present in class. Students thus learn how to participate in the practice of medical diagnosis, and they learn the relevant facts along the way.

From this perspective, testing is unnatural. If students must learn to participate competently in real activities, teachers should not test whether they can solve problems by themselves out of context. And because learning most often involves participating with others to accomplish a task, students should not be tested alone. Students should instead be asked to exhibit their mastery by participating competently in naturalistic activities.

Like behaviorism, pure cognitivism is only partly true. Just as people are often manipulated by reinforcements, people often rely primarily on their own mental models to understand the environment. But if our educational goal is to help young people build on the knowledge and skills that have been developed by previous generations, we should treat them neither as animals to be shaped nor as lone thinkers. We must help them grow into and expand the activities that make us human. This will require educational practices based on more complex accounts of learning.

Further Reading

Anderson J, Reder L and Simon H (1996) Situated learning and education. *Educational Researcher* **25**: 5–11.

Duckworth E (1987) *'The Having of Wonderful Ideas' and Other Essays on Teaching and Learning*. New York, NY: Teachers College Press.

Engeström Y, Miettinen R and Puramäki R (1999) *Perspectives on Activity Theory*. Cambridge, UK: Cambridge University Press.

Gardner H (1999) *Intelligence Reframed*. New York, NY: Basic Books.

Greeno J (1997) On claims that answer the wrong questions. *Educational Researcher* **26**: 5–17.

Hutchins E (1995) *Cognition in the Wild*. Cambridge, MA: MIT Press.

Lave J and Wenger E (1991) *Situated Learning*. Cambridge, UK: Cambridge University Press.

McGilly K (1994) *Classroom Lessons*. Cambridge, MA: MIT Press.

Piaget J (1967) *Six Psychological Studies*. New York, NY: Random House.

Renninger K (1998) Developmental psychology and instruction. In: Siegel I and Renninger K (eds) *Child Psychology in Practice*, pp. 211–274. New York, NY: John Wiley.

Rogoff B, Turkanis C and Bartlett L (2001) *Learning Together*. Oxford, UK: Oxford University Press.

Schwartz B (1985) *The Battle for Human Nature*. New York, NY: WW Norton.

Skinner BF (1968) *The Technology of Teaching*. New York, NY: Appleton-Century-Crofts.

Vygotsky L (1978) *Mind in Society*. Cambridge, MA: Harvard University Press.

Wertsch J (1998) *Mind as Action*. Oxford, UK: Oxford University Press.

Electroencephalography (EEG)

Introductory article

Terence W Picton, Rotman Research Institute, Toronto, Ontario, Canada
Ali Mazaheri, Rotman Research Institute, Toronto, Ontario, Canada

CONTENTS

Electroencephalography is a measurement of the brain's electrical activity. It provides information about the timing of intracerebral processes which can be used in conjunction with anatomical information derived from hemodynamic studies to learn about events in the human brain.

INTRODUCTION

Electroencephalography (EEG) is the recording of the electrical activity of the brain. When the neurons of the brain process information, they do so by changing the flow of electrical currents across their membranes. These changing currents, particularly those caused by the synaptic excitation and inhibition of cortical neurons, generate electric fields that can be recorded using small electrodes attached to the surface of the scalp. The potentials between different electrodes are amplified and displayed as they fluctuate over time.

The human EEG was first recorded in the 1920s by a German psychiatrist named Hans Berger. Since then, it has been used widely to investigate normal and abnormal brain function.

The EEG is characteristically recorded simultaneously from multiple locations on the scalp. The amplitude of the EEG is very small – usually several tens of microvolts. The recordings use special amplifiers that record the differences in potential between two electrodes. These differential amplifiers cancel out other large electrical activities, such as line noise, and allow us to see the small EEG signals. Two types of recordings are possible: the electrical activity can be recorded from each location relative to a common reference, or the activity can be recorded from each electrode relative to an adjacent electrode. The latter technique effectively records the slope or change of the scalp potentials over space rather than their absolute value relative to a single reference. Figure 1 shows the EEG signals recorded from ten scalp electrodes. The recordings are all relative to a linked-mastoid reference. The signals in Figure 1 are plotted according to the usual EEG convention that negativity at the first electrode relative to the second is shown as an upward deflection.

Electroencephalographic recordings fluctuate in time and are often 'rhythmic' in the sense that they alternate regularly. The most prominent rhythm in the EEG is the alpha (α) rhythm, which has a frequency of 8–13 cycles per second (Hz) and is recorded mainly over the posterior regions of the scalp close to the regions of the brain that process visual information. When the eyes are open the α rhythm is very small and when the eyes are closed it becomes large.

During a rhythm the neurons tend to fire synchronously, and their fields overlap and add to each other to cause the large scalp potentials. Interactions between cortical and thalamic neurons can cause the cortical neurons to fire periodically and thus generate rhythmic scalp potentials. When the visual areas are activated by real or imagined information, the neurons fire independently and their fields tend to cancel each other out. The EEG is then said to be 'desynchronized'. The transition between the α rhythm and the desynchronized EEG occurring with eye opening is shown in Figure 1.

As well as activity generated in the brain, the electrodes also pick up electrical potentials from other sources in the head. The eyes are the most prominent of these sources. Large potentials occur in the anterior regions of the scalp as the eyes or eyelids move. In Figure 1, the anterior regions become more negative as the eyes open, and there is a brief positive deflection during a blink. The scalp muscles also generate activity that is picked up in the EEG. This activity is characteristically faster than the activity arising from within the brain.

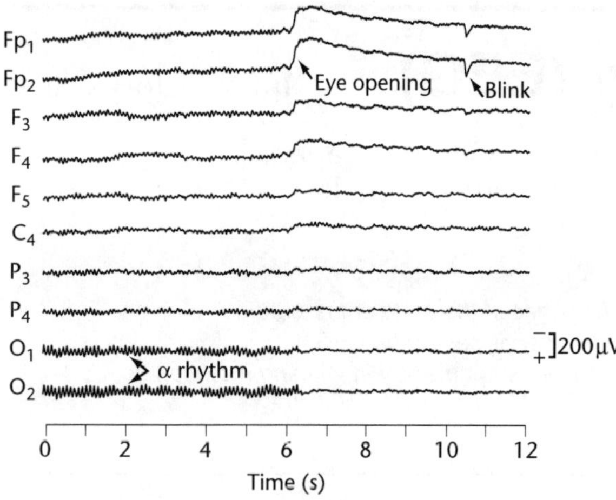

Figure 1. Human electroencephalogram (EEG). This 12 s recording was taken from a normal young woman. The EEG was recorded from ten scalp electrodes, with the activity at each electrode measured relative to a linked-mastoid reference. Negativity at the scalp electrode relative to the reference is plotted upwards. The electrodes are named according to their location on the scalp: Fp, frontopolar (forehead); F, frontal; C, central; P, parietal; O, occipital. Odd-numbered electrodes are on the left and even-numbered electrodes are on the right, and the number varies with the distance from the midline. At the beginning of the recording, the eyes were closed; halfway through the recording the woman opened her eyes. Movements of the eyes and eyelids when the eyes opened were recorded as a large negative wave in the Fp electrodes. While her eyes were open, the woman blinked; this was recorded as a brief positive wave in the Fp electrodes. While the eyes were closed there was a sustained rhythmic oscillation at 10 Hz in the posterior electrodes (O_1 and O_2) – the alpha rhythm.

FREQUENCY ANALYSIS

Electroencephalographic signals are usually plotted as changes in voltage over time, as in Figure 1. Because of the rhythmic nature of the signals, it is often informative to plot the signals according to the frequencies they contain. This change from the time domain to the frequency domain is accomplished using a Fourier transform. The frequency spectrum of the EEG signal then shows various peaks that denote the particular rhythms in the EEG. The frequency representation of the EEG recorded from the right occipital electrode at the back of the head (labeled O_2 in Figure 1) is shown in Figure 2. The α rhythm stands out as a peak in the spectrum recorded when the eyes are closed.

The spectrum of frequencies present in the EEG is broad. As well as the (α) activity recorded at frequencies of 8–13 Hz, the EEG also contains slower activity – theta (θ) activity at 4–7 Hz and delta (δ) activity at 0–3 Hz – and faster activity – beta (β) activity at 14–25 Hz and gamma (γ) activity at 25–50 Hz.

TEMPORAL ANALYSIS OF THE EEG SPECTRUM

The way in which the frequencies of the EEG change over time can show the changing state of the brain. Figure 3 shows the desynchronization of the α rhythm as the eyes open. The upper tracing shows the time-domain EEG recorded from the O_2 region of the scalp. The middle tracing shows the way the spectrum changes over time (EEG spectrograph). Spectra similar to those shown in Figure 2 are plotted with the frequency on the *y* axis and the amplitude of the activity demonstrated using the color scale (*z* axis). Whereas the two spectra in Figure 2 combine activity over several seconds, the multiple spectra plotted in Figure 3 are computed about 40 times a second. The α activity, which is present with the eyes closed and then goes away with eye-opening, shows up as the prominent dark red line at a frequency of 10 Hz. The lower tracing shows the amplitude of activity in the α frequency band over time. This follows the α rhythm over the time course portrayed in Figure 1.

The α rhythm can attenuate even when the eyes are closed. This can occur when the person is using the visual areas of the brain in processes such as problem-solving or imagining visual information. Sometimes the changes in the rhythms following an event (such as a stimulus) are small and can be measured only if the spectral changes are averaged over multiple trials.

SPATIAL ANALYSIS

The EEG signals are characterized by their distribution over the scalp as well as by the way they change in time. The EEG is spatiotemporal in nature. The visually reactive α rhythm is usually distinguished from other EEG activity by its posterior scalp location. Figure 4 shows the scalp topography of the α activity recorded in Figure 1. The α rhythm is typically slightly larger over the right posterior scalp than over the left.

Other rhythmic activities are recorded from different regions of the scalp. The mu rhythm has a similar frequency to the α rhythm but is located over the sensorimotor regions of the cortex and reacts to somatosensory input, motor activity or the thought of motor activity in the contralateral

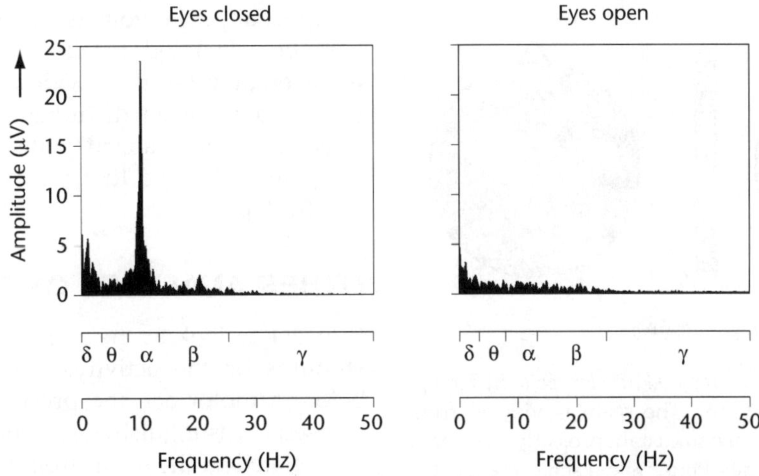

Figure 2. Frequency analysis of the electroencephalogram. The frequency content of the EEG signal at the O_2 electrode (lowest tracing in Figure 1) is shown. The spectrum on the left shows the amplitude of activity at the different frequencies when the eyes are closed (from 1 s to 5 s in Figure 1) and the spectrum on the right shows the pattern when the eyes are open (from 7 s to 11 s). The eyes-closed spectrum shows a prominent peak at 10 Hz.

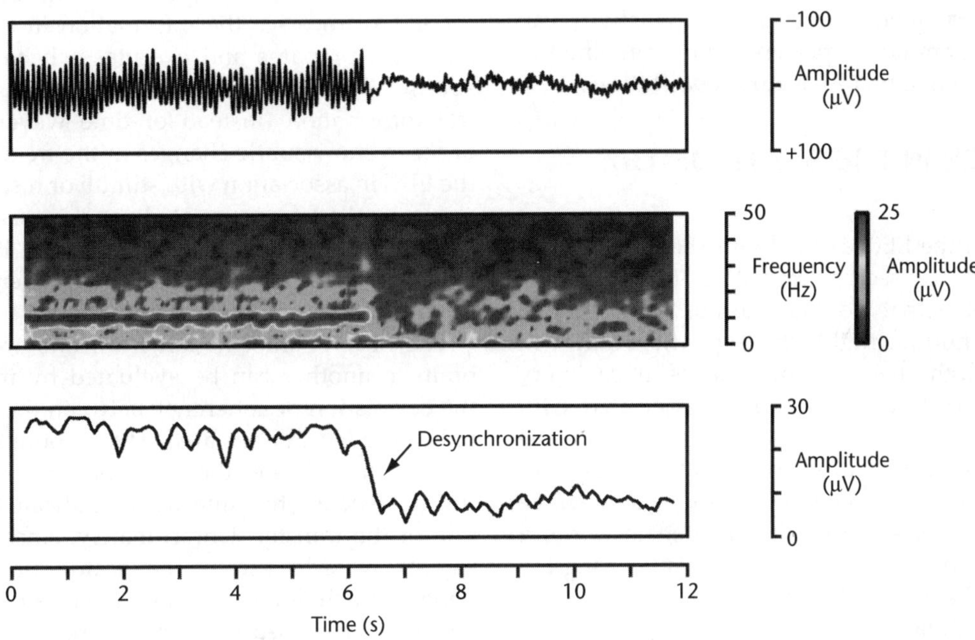

Figure 3. [*Figure is also reproduced in color section.*] Electroencephalographic changes over time. The upper tracing shows the EEG signal recorded from the O_2 electrode in Figure 1. The middle panel shows the spectrogram. This plots the changes in the spectrum over time: the frequency (on the x axis in Figure 2) is plotted vertically (y axis) and the amplitude of the activity is plotted on the z axis as color. A dark red line at 10 Hz represents the α rhythm. This line is clearly recognizable when the eyes are closed and disappears when the eyes are open. The lower tracing represents the amount of activity present in the 8–13 Hz frequency band as it changes over time. When the eyes open, this activity decreases (event-related desynchronization).

hand. Its name comes from the 'm' shape of the waves.

The γ rhythms of the brain have been studied in relation to memory and perception. These rapid rhythms may serve to bind separate aspects of a perceived object by synchronizing relevant neurons from different regions of the brain. The rhythms may also act as a signature for that

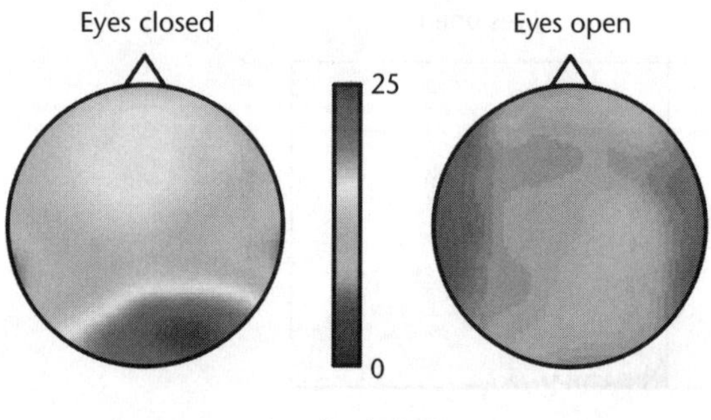

Figure 4. [*Figure is also reproduced in color section.*] Topography of the alpha rhythm. The scalp is viewed from above using an azimuthal equidistant projection centered at the vertex. The outside of the circle reaches the level of the ear canal. The α activity when the eyes are closed is maximally recorded from the posterior regions of the scalp, and is slightly greater on the right than on the left. These maps were based on activity from 65 scalp electrodes (10 of which are shown in Figure 1).

bound information. Bursts of EEG γ activity can occur as a stimulus is perceived or maintained in short-term memory. (*See* **Neural Oscillations**)

CHANGES IN THE STATE OF THE BRAIN

Changes in the EEG as the brain changes its state are most clearly seen during sleep. Many specific patterns of activity define various sleep stages. During a normal night of sleep, the brain will cycle through these various stages once every 90 min or so. In one stage, often associated with vivid visual dreams, there are many rapid eye movements (REM). Another stage of sleep is associated with widespread δ activity – slow-wave sleep. Bursts of activity with frequencies around 14 Hz – sleep spindles – are prominent during the transition between REM sleep and δ sleep. (*See* **Sleep and Dreaming**)

The EEG is used extensively to assess the abnormal brain states that occur with neurological disorders. An abnormal decrease of brain activity is usually associated with slow EEG waves. These can occur in localized regions of the scalp over areas of focal brain damage, or can be more widespread in cases of generalized brain dysfunction. After extensive brain damage there may be no electrical activity recorded from the brain – a state known as 'brain death'. An abnormal excitability of the neurons in the brain, which can occur in epilepsy, is usually associated with high-frequency EEG activity. The most common abnormality is a brief, sharp change in voltage called a spike. Spikes may be generated focally in one region of the brain or may be more widespread in association with a generalized disturbance of consciousness. A common generalized pattern is a combination of spike and wave that recurs at a rate of $3\,s^{-1}$. (*See* **Epilepsy**)

OTHER ANALYSES OF THE EEG

In order to look at the response of the brain to a stimulus or the activity in the brain occurring before a motor act, the process of averaging can be used to distinguish the cerebral activity specifically related to the event from the other activity of the EEG: the stimulus or the act is repeated and the EEG signals related to each occurrence of the event are averaged together. The specific activity – 'event-related potential' – remains the same during averaging, whereas the unrelated EEG activity tends to cancel itself out. The event-related potentials can show what is happening in the brain when a person processes the information in a stimulus and then prepares and executes a behavioral response. Averaging can also be used to assess spectral information (instead of time waveforms) in order to show subtle changes in the frequencies of the EEG in association with stimuli or responses – a process called 'event-related synchronization and desynchronization.' (*See* **Event-related Potentials and Mental Chronometry; Auditory Event-related Potentials; Visual Evoked Potentials**)

The flow of information from one area of the brain to another can be evaluated by measuring the correlation or coherency between the EEG signals recorded in each area. The amount of correlation can indicate the amount of information transferred, and the time lag can indicate the direction of the transfer. Unfortunately, since the EEG signals recorded from the scalp spread quite widely, much of the correlation between these signals may be caused by current spread in the scalp rather than information spread in the brain. Techniques that calculate the radial currents or extrapolate back from the scalp to the cortical surface can make these studies more precise. It would be even better to determine the actual intracerebral generators for the scalp-recorded activity and to correlate the waveforms generated in these sources. Although there is no unique solution to the problem of calculating the intracerebral sources for electrical signals recorded at the scalp surface, constraints imposed on the solutions from other imaging techniques, which provide anatomy and cortical activation patterns, can lead to sensible solutions.

RELATIONS TO OTHER BRAIN IMAGING METHODS

As well as generating electric fields in the extracellular fluid, the passage of currents through neuronal membranes generates magnetic fields. A magnetoencephalogram (MEG) can therefore be recorded from the surface of the scalp using specialized techniques. These signals are similar to EEG signals since they derive from the same currents in the brain; they differ, however, in terms of which neurons contribute to the scalp-recorded signals and how these signals spread to the recording sensors.

Both the EEG and the MEG are 'functional' tests. They differ from simple structural measurements of brain anatomy by measuring the activity of the brain as it is working. The EEG is therefore related to other tests of brain function that measure cerebral blood flow, such as positron emission tomography or functional magnetic resonance imaging. One difference between these hemodynamic measurements and the electromagnetic measurements is in the timing. The electromagnetic activity changes simultaneously with the neuronal activity, whereas the blood flow changes after a delay. However, blood flow measurements are much more accurate in terms of their anatomical localization. Activation patterns from hemodynamic studies may provide localization for the temporal changes in the EEG – either event-related potentials or event-related desynchronizations. (*See* **Neuroimaging**)

Further Reading

Aminoff MJ (1999) *Electrodiagnosis in Clinical Neurology*, 4th edn. New York, NY: Churchill Livingstone.

Dale A, Liu AK, Fischl BR *et al.* (2000) Dynamic statistical parametric mapping: combining fMRI and MEG for high resolution imaging of cortical activity. *Neuron* **26**: 55–67.

Gevins AS, Le J, Brickett P, Reutter B and Desmond J (1991) Seeing through the skull: advanced EEGs use MRIs to accurately measure cortical activity from the scalp. *Brain Topography* **4**: 125–131.

Gloor P (1969) *Hans Berger on the Electroencephalogram of Man (Electroencephalography and Clinical Neurophysiology*: supplement 28). Amsterdam, Netherlands: Elsevier.

McFarland DJ, Miner LA, Vaughan TM and Wolpaw JR (2000) Mu and beta rhythm topographies during motor imagery and actual movements. *Brain Topography* **12**: 177–186.

Niedermeyer E (1997) Alpha rhythms as physiological and abnormal phenomena. *International Journal of Psychophysiology* **26**: 31–49.

Niedermeyer E and Lopes da Silva F (1998) *Electroencephalography: Basic Principles, Clinical Applications and Related Fields*, 4th edn. Philadelphia, PA: Lippincott, Williams & Wilkins.

Nunez PL (1995) *Neocortical Dynamics and Human EEG Rhythms*. Oxford, UK: Oxford University Press.

Nunez PL, Silberstein RB, Shi Z *et al.* (1999) EEG coherency II: experimental comparisons of multiple measures. *Clinical Neurophysiology* **110**: 469–486.

Pfurtscheller G and Lopesda Silva F (1999) Event-related EEG/MEG synchronization and desynchronization: basic principles. *Clinical Neurophysiology* **110**: 1842–1857.

Singer W (2000) Response synchronization: a universal coding strategy for the definition of relations. In: Gazzaniga MS (ed.) *The New Cognitive Neuroscience*, pp. 325–338. Cambridge, MA: MIT Press.

Steriade M (1998) Cellular substrates of brain rhythms. In: Niedermeyer E and Lopes da Silva F (eds) *Electroencephalography: Basic Principles, Clinical Applications and Related Fields*, 4th edn, pp. 28–75. Philadelphia, PA: Lippincott, Williams & Wilkins.

Tallon-Baudry C and Bertrand O (1999) Oscillatory gamma activity in humans and its role in object representation. *Trends in Cognitive Sciences* **3**: 151–162.

Wong PKH (1991) *Introduction to Brain Topography*. New York, NY: Plenum.

Eliminativism

Intermediate article

William Ramsey, University of Notre Dame, Indiana, USA

CONTENTS

Eliminativism is the radical thesis that the common-sense conception of the mind is deeply mistaken and that certain mental states posited by it do not exist.

WHAT IS ELIMINATIVISM?

Eliminativism, or eliminative materialism as it is often called, is the thesis that our common-sense conception of the mind is deeply mistaken and that our ordinary notions of mental states will not belong to a scientifically respectable account of cognitive phenomena. Put more simply, it is the view that certain mental states, such as beliefs and desires, do not exist. Eliminativism is one of the most radical and controversial philosophical positions ever held. In many respects, it goes beyond other forms of scepticism since it challenges deep assumptions about the workings of our own minds. In his *Meditations*, Descartes offered a famously extreme form of doubt, questioning many of our beliefs about ourselves and the world. Nevertheless, he insisted that we can at least know with certainty that our minds contain thoughts and beliefs. Eliminativism claims not only that it is possible to doubt the existence of such mental states, but that it is correct to do so.

Eliminativism involves two central and controversial claims. It is worth examining each of these in some detail. For the sake of brevity, much of this article will focus upon our notion of belief, since it figures so prominently in discussions of eliminativism. Many of the arguments concerning belief are thought to generalize to other mental notions as well, especially other propositional attitudes.

Folk Psychology and the Theory Theory

The first claim of eliminativism is that we employ a theoretical framework to explain and predict intelligent behavior. This position is commonly called the 'theory theory', since it maintains that our 'folk' or 'common-sense' psychology is similar to other folk theories that we use to understand a range of different phenomena. As with most theories, folk psychology is assumed to consist of both universal generalizations (or laws) and theoretical posits, with the former capturing the counterfactual regularities found among the latter. For example, a generalization of folk psychology might be something like:

> If someone has the desire for X and
> the belief that the best way to get X is by
> doing Y, then (barring certain conditions)
> that person will tend to do Y. (1)

According to the theory theory, these generalizations function like the laws of scientific theories, though they are learned and used in a far more informal manner.

The theory theory maintains that the posits of folk psychology are the mental states that figure in our common-sense psychological explanations. As theoretical posits, these states account for observable effects (like statements and behavior), but are not themselves directly observed. If we concentrate on our folk notion of belief, the theory theory claims that common sense assigns two sorts of properties to these states. First, there are various causal properties. We assume beliefs are caused in certain circumstances, interact with desires and other cognitive states in certain ways, and bring about certain kinds of behavior. These causal roles help define and distinguish beliefs from other types of mental states. Second, beliefs are essentially about a wide range of different states of affairs. This inherit 'aboutness', or 'intentionality' (also sometimes called 'meaning', 'content', or 'semantic properties'), is commonly regarded as a unique feature of the mind. While conventional signs and linguistic symbols are meaningful, their meaning is derivative, stemming from the stipulations and interpretations of thinking creatures. Only beliefs and other mental representations have what is called 'original' or 'intrinsic' intentionality: a type of meaning not derived from other sources.

Eliminative Theory Change

The second claim of eliminativism is that folk psychology is severely mistaken about the actual nature of the mind. Eliminativists argue that the laws and posits of folk psychology radically misdescribe our minds; consequently, they denote nothing that is real. To understand eliminativism, it is important to note the difference between two types of theory change: reductive or retentive theory replacement, on the one hand, and eliminative, or ontologically radical, theory replacement, on the other. In the case of the former, the posits of a rejected theory – be it a folk theory or a former scientific theory – find a new home, perhaps with some modifications, in the superseding theory. For example, the notion of a planet survived the transition from Aristotelian physics to Newtonian physics, though it underwent some changes as Aristotle's notion was clearly mistaken in many respects. Despite problems with the Aristotelian framework, we did not conclude that it was so wrong that the notion of planet has no referent.

This can be contrasted with eliminative theory change, where a theoretical posit of an abandoned theory so misses the mark that we conclude that the notion fails to capture anything real. For example, early explanations of strange behavior posited the existence of malevolent spirits that were thought to possess the souls of unlucky individuals. As the theory of demonology was replaced by more sophisticated accounts of mental and neurological disorders, the notion of a supernatural demon was abandoned altogether. Since there is nothing in the new accounts with which demons can be reasonably identified, we have eliminated demons from our contemporary ontology.

Eliminativists predict that this sort of eliminative change will happen with the theoretical posits of folk psychology. The claim is not that mental states, such as beliefs, currently exist but will somehow be abolished as our scientific understanding of the mind grows. The claim is that mental states such as beliefs have never existed; hence, it is the concept of a belief that will be eventually eliminated from our mental taxonomy. There simply are no such things that have the causal and semantic properties we attribute to beliefs.

Eliminativism shares assumptions with both physicalism and dualism. Eliminativists agree with the physicalist claim that the actual causes of behavior are ultimately physical events, taking place inside the brain. However, eliminativists also agree with dualists that the mental states posited by common-sense psychology are not to be identified with anything inside the brain. Of course, whereas dualists hold this view because they think mental states are nonphysical, eliminativists hold this view because they think these states do not exist.

ARGUMENTS FOR ELIMINATIVISM

As one might expect, arguments for eliminativism typically take the form of arguments against folk psychology. By and large, these arguments fall into two groups. The first group involves general theoretical considerations that are relevant to the evaluation of any theory. The second group focuses upon problematic aspects of folk psychology itself.

Most of the arguments against folk psychology based upon theoretical considerations can be found in the writings of Paul and Patricia Churchland, perhaps the two strongest defenders of eliminativism. One such argument starts with the premise that a promising theory should offer a fertile research programme with considerable explanatory power and range. It then notes that folk psychology appears stagnant, making no real progress throughout much of history. Worse yet, there are many cognitive phenomena that folk psychology does not even begin to explain. Important aspects of the mind such as the nature of consciousness, the oddities of mental illness and the actual mechanisms of learning are all left unexplained by folk psychology. These and similar considerations suggest that folk psychology is ripe for replacement. Another argument looks at the track record of folk theories in general and notes how improbable it would be for this one to turn out true. Folk physics, folk biology, folk epidemiology, and so on, all proved to be radically mistaken. Folk psychology concerns a subject that is far more complex and difficult than these others. Hence, it seems implausible that just this one is right (Churchland, 1981).

In response to these theoretical arguments, many argue that folk psychology actually has stimulated a number of fruitful research programs in scientific psychology. Moreover, some of these do help explain a wide variety of cognitive phenomena not directly explained by folk psychology. Furthermore, it is one thing to claim that a folk theory is incomplete or unproductive, but another thing to claim that it is radically false (Horgan and Woodward, 1985). Defenders of folk psychology claim that these theoretical considerations cannot possibly outweigh the evidence provided by the everyday, ordinary experience of our own minds, which seems to support the reality of beliefs.

Regarding this last point, eliminativists often warn that we should be deeply suspicious of introspective 'evidence' about the inner workings of the mind. Since all forms of observation – including introspection – are, to some degree, 'theory-laden', our folk-theoretical framework may play a larger role in forming the content of our experienced inner lives than actual mental processes. Indeed, there is considerable empirical evidence that we lack reliable access to the actual workings and nature of cognitive processes. Thus, our inner observations of certain mental states may be like the observations of those who claimed to be able to 'observe' demonic spirits.

The second group of arguments looks at features that are unique to folk-psychological posits and challenges the likelihood that these features will be accommodated by a scientific account of the mind. The most widely discussed features are those associated with the alleged linguistic nature of beliefs and other propositional attitudes. Common sense appears to treat these states as having a form similar to public language sentences, with a compositional structure and syntax. For instance, the belief that John loves Mary appears to be composed of the concepts (or mental 'words') 'John', 'love', and 'Mary', and it differs from the belief that Mary loves John by virtue of something analogous to syntactic arrangement. Along with this quasilinguistic structure, beliefs resemble public sentences in another way: they have semantic properties. Beliefs, like sentences, are individuated by virtue of what they are about.

Both of these quasilinguistic features of propositional attitudes – their alleged sentential structure and their semantic properties – have motivated arguments in favour of eliminativism. With regard to the former, some writers have emphasized the apparent mismatch between the sentential form of propositional attitudes and the actual neurological structures of the brain. Whereas the former involves discrete symbols and a combinatorial syntax, the latter involves action potentials, spiking frequencies and spreading activation. It is hard to see where in the brain we are going to find sentence-like structures (Churchland, 1986).

Of course, this argument depends upon a certain view of folk psychology, namely, that it treats beliefs as having a sentential structure. Those who reject that claim, as many have, will not find the argument compelling. Even for those who find this reading of folk psychology plausible, there is a further difficulty regarding the relevance of neuroscience for determining the status of folk psychology. Many have insisted that just as the physical circuitry of a computer is the wrong place to look for computational symbol structures, so too, the neurological wiring of the brain is the wrong place to look for structures that might qualify as beliefs. Instead, the status of folk posits should be decided at a higher, more abstract level of analysis than the neurophysical. For many, the computer model of the mind – where the mind is regarded as the brain's program, abstracted from the neurological details – provides a more appropriate level at which to seek analogs to the posits of folk psychology.

The second type of argument against beliefs – focusing upon their intentional or semantic properties – concludes that folk-psychological concepts are inappropriate for any scientific theory of the mind, including a computational one. The main proponent of this argument has been Stephen Stich (1983). Stich claims that folk psychology individuates beliefs by virtue of their semantic properties; however, there are several reasons for thinking that a semantic taxonomy is ill-suited to scientific psychology. Firstly, because such a taxonomy depends upon matters outside the head, it will individuate mental states differently from other taxonomies, such as those based upon causal roles. Secondly, Stich has argued that the semantic content of a belief is ascribed on the basis of judgements of similarity. Consequently, these ascriptions are vague, and fail with subjects who are too dissimilar from ourselves, such as the very young and the mentally ill. Rather than adopt a folk-psychological ontology, Stich argues a scientific psychology should employ a syntactic taxonomy: one that individuates mental states by appeal to their purely non-semantic properties.

As Stich himself notes, even if folk posits are inappropriate for a scientific psychology, it need not follow that they do not exist. After all, there are plenty of entities that are vaguely defined or demarcated in ways that are inappropriate for science, but are nevertheless real. If our best scientific account posited states that share many features with beliefs, such as similar causal roles, then even if the two taxonomies differed in certain cases, we may still regard folk psychology as, in some sense, vindicated.

To generate a more robust form of eliminativism, we would need to show that there is nothing in our scientific psychology that shares the central properties we attribute to beliefs. In a more recent paper, Ramsey, Stich and Garon (1990) argue that certain connectionist models of memory and inference could serve as the basis for this stronger eliminativist claim. These connectionist models store information in a holistic, or distributed, manner; thus, there are no causally discrete, semantically evaluable data structures that represent specific propositions. It is not just that these models lack the sentential, compositional representations assumed in more traditional (or 'language of thought') models. Rather, in these networks there are no causally distinct structures that stand for anything specific. Consequently, there do not appear to be any structures in these models that might serve as candidates for the reduction of propositional attitudes. If these connectionist models should prove accurate, they argue, it will show that there are no such things as beliefs and propositional memories. This argument assumes that in a distributed network, it is impossible to specify which bits of information are causally responsible for various acts of cognition. Some have responded by insisting that, with highly sophisticated forms of analysis, it is possible to pick out casually relevant pieces of stored information.

While most arguments for eliminativism focus upon propositional attitudes, it should be noted that one philosopher, Daniel Dennett (1988), has endorsed an eliminativist stance towards a very different class of mental states: those commonly referred to as 'qualia'. Qualia are mental states picked out by virtue of their intrinsic qualitative or phenomenal character, like pain states. Such states seem intuitively to have a number of special features, such as being purely private, immediately perceived, and intrinsically subjective. Dennett discusses several cases – both actual and imaginary – to expose ways in which these ordinary intuitions about qualia are inconsistent. In suggesting that our concepts of qualia are deeply confused, he attempts to cast doubt on the reality of these states as they are commonly understood by folk psychology.

ARGUMENTS AGAINST ELIMINATIVISM

As might be expected of a theory that challenges our fundamental understanding of ourselves, eliminativism has been subjected to a wide range of criticisms. Here we will discuss four such criticisms which have enjoyed particular prominence.

Incoherence of Eliminativism

The first criticism claims that eliminativism is incoherent or in some way self-refuting. The charge is that eliminativism itself presupposes the existence of the very thing it claims not to exist, namely, belief states. A typical way this charge is made is to insist on some cognitive capacity or theoretical principle that is somehow required by eliminativism and yet requires the existence of beliefs. One popular example is the capacity to make an assertion, and that to assert something one must believe it. Hence, for eliminativism to be asserted as a thesis, the eliminativist must believe that it is true. But if the eliminativist has such a belief, then there are beliefs, and eliminativism is thereby proved false (Baker, 1987).

However, it is important to note that the eliminativist thesis itself – that there are no beliefs – is not conceptually incoherent. The criticism is not that eliminativism is self-contradictory, but that the eliminativist is doing something that goes against his or her own thesis. In the above example, this act was the making of an assertion: we are required, it is claimed, to believe what we assert. But this is exactly the sort of folk-psychological statement that an eliminativist would deny: a central tenet of the eliminativist position is that various capacities that we think involve beliefs actually don't. Thus the critic has merely begged the question, since eliminativists reject the idea that various acts (such as asserting a thesis or formulating a theory) require the existence of beliefs.

Misrepresentation of Folk Psychology

The next criticism of eliminativism challenges the way it characterizes folk psychology. This criticism has two distinct origins. The first origin is, at least partly, in the writings of Wittgenstein (1953) and Ryle (1949). According to one version of this view, common sense does not treat mental states, such as beliefs, as distinct inner causes of behavior. Instead, they are treated as dispositional states, which are used to adopt a certain heuristic 'stance' towards rational agents. Hence, ordinary talk about mental states should be interpreted as talk about abstract things that, although real, are not candidates for straightforward reductions to discrete neurophysical states. They serve in many non-explanatory endeavors – for example, they allow behavior to be interpreted as rational – without assigning any specific inner causal structure to the mind. Consequently, discoveries about the actual structure of the mind are viewed as irrelevant to the status of folk mental states (Dennett, 1987).

The second origin of skepticism regarding the theory theory is in contemporary cognitive science, and stems from a different model of common-sense psychological explanation and prediction. This view, known as the 'simulation theory', maintains that we predict and explain behavior not by using a theory, but by simulating what we would do in a comparable situation. According to this account, we predict an agent's behavior by taking our own information processing mechanisms 'off-line' and giving them data about the agent's background and circumstances including possible beliefs and desires. We then (subconsciously) use our own decision-making mechanisms to generate output which can thereby serve as predictions (and, in other circumstances, explanations) of the agent's behavior. If this account is correct, then no theory of the mind is used to explain and predict behavior. And if there is no theory of the mind, then there can be no false theory of the mind (Gordon, 1986).

In reply to these objections, defenders of the theory theory have turned to empirical work that supports their position. For example, the developmental psychologists Henry Wellman and Alison Gopnik have argued that in explaining and predicting behavior, children go through phases of development roughly analogous to the phases one would go through when acquiring a theory. Moreover, it turns out that children come to ascribe beliefs to themselves in the same way they ascribe beliefs to others. These considerations lend at least some support to the idea that our notion of belief is employed as the posit of a folk theory rather than as input to a simulation model (Gopnik and Wellman, 1992).

Success of Folk Psychology

Even if we allow that folk psychology is a theory, a third argument against eliminativism is that folk psychology seems to offer a fairly accurate account of cognitive processes. Apart from the way introspection appears to reveal beliefs and desires, we also use folk psychology successfully to predict and control the behavior of others. If folk psychology is so wrong then why are we so good at dealing with one another when we use it? The success of folk psychology suggests an 'inference to the best explanation' argument against eliminativism: the best explanation for the success we enjoy in explaining and predicting each other's behavior is that folk psychology is roughly true.

The eliminativist response to this argument is that any theory – especially one as near and dear to us as folk psychology – may appear successful even when it radically misrepresents reality. As philosophers of science often point out, when we are under the influence of a theory we discount the anomalies with which the theory struggles as insignificant, and attribute more power to the theory than it deserves. The degree to which folk psychology misdescribes mental phenomena may not be fully apparent until we have an alternative theory in hand.

Stich's Criticism of Eliminativism

The final argument against eliminativism comes from the recent writings of a former supporter, Stephen Stich. Stich's argument is complex, but we can outline it here. We noted above that eliminativism is committed to the thesis that the theoretical posits of folk psychology fail to refer to anything. But what exactly does such a claim amount to? Stich argues that this question is far more difficult than eliminativists have assumed. For example, we might think that reference failure occurs as the result of a certain mismatch between reality and the theory in which the posit is embedded. But how much mismatch is necessary before we can say that a given posit doesn't exist? For a variety of reasons, Stich suggests that this question has no definite answer. Consequently, the question of whether there really are such things as beliefs has no definite answer either, contrary to the eliminativist view. Of course, this seems to raise problems not just for the eliminativist, but for the folk-psychological reductionist as well. But Stich has presented compelling reasons for thinking that the usual terms of the debate over eliminativism need to be reconsidered (Stich, 1996).

ELIMINATIVISM AND COGNITIVE SCIENCE

Research in cognitive science has had an important influence on nearly all of the major debates concerning eliminativism. Arguments for eliminativism typically depend upon specific claims about the nature of common-sense psychology. Cognitive research is critical for determining the truth of these claims; and empirical work on our conception of the mind, including our introspective access to our own mental states, has played a central role in debates over the plausibility of the theory theory. As this research continues, we should expect to gain a deeper understanding of the nature of folk psychology.

Eliminativism requires that the transition from folk psychology to scientific psychology be of a certain type, namely, eliminative rather than reductive. This would require psychological theories that are hostile to the posits of common-sense psychology. Until recently, few psychological theories have been of this nature. The traditional computational paradigm which has dominated cognitive science makes explicit use of data structures that are naturally regarded as the scientific counterparts to folk mental states. However, if Ramsey, Stich and Garon are right, connectionist models may, for the first time, provide us with a plausible account of cognition that supports the denial of belief-like states. The future of eliminativism may depend upon which of these two major accounts of cognition proves correct.

The influence eliminativism has had on cognitive science research is less clear. Because so many theories in cognitive science presuppose the reality of folk-psychological states, both as explanatory posits and as phenomena to be explained, there is no doubt that a commitment to eliminativism would require a radical shift in cognitive science research. Of course, without a widely accepted, well-confirmed theory of cognition, we are not yet in a position to judge the status of folk psychology. Indeed, given the amount of speculative forecasting built into many eliminativist arguments, it is reasonable to ask why cognitive scientists should even consider it. The answer is that eliminativism reminds us to reconsider the constraints we intuitively impose on potential cognitive theories, and not to reject out of hand any psychological theory that failed to incorporate folk-psychological states. Thus, eliminativism is not so much a threat to cognitive science, as a view that potentially admits a range of new theoretical possibilities.

References

Baker L (1987) *Saving Belief*. Princeton, NJ: Princeton University Press.

Churchland PM (1981) Eliminative materialism and the propositional attitudes. *Journal of Philosophy* **78**: 67–90.

Churchland PS (1986) *Neurophilosophy: Toward a Unified Science of the Mind/Brain*. Cambridge, MA: MIT Press.

Dennett D (1987) *The Intentional Stance*. Cambridge, MA: MIT Press.

Dennett D (1988) Quining qualia. In: Marcel A and Bisiach E (eds) *Consciousness in Contemporary Science*, pp. 42–77. New York, NY: Oxford University Press.

Gopnik A and Wellman H (1992) Why the child's theory of mind really *is* a theory. *Mind and Language* **7**: 145–171.

Gordon R (1986) Folk psychology as simulation. *Mind and Language* **1**: 158–171.

Horgan T and Woodward J (1985) Folk psychology is here to stay. *Philosophical Review* **94**: 197–226.

Ramsey W, Stich S and Garon J (1990) Connectionism, eliminativism and the future of folk psychology. *Philosophical Perspectives* **4**: 499–533.

Ryle G (1949) *The Concept of Mind*. London: Hutchison.

Stich S (1983) *From Folk Psychology to Cognitive Science*. Cambridge, MA: MIT Press.

Stich S (1996) *Deconstructing the Mind*. New York, NY: Oxford University Press.

Wittgenstein L (1953) *Philosophical Investigations*. Oxford: Oxford University Press.

Further Reading

Carruthers P and Smith PK (1996) *Theories of Theories of Mind*. Cambridge, UK: Cambridge University Press.

Christensen SM and Turner DR (1993) *Folk Psychology and the Philosophy of Mind*. Hillsdale, NJ: Erlbaum.

Dennett D (1991) *Two contrasts: folk craft versus folk science, and belief versus opinion*. In: Greenwood J (ed.) *The Future of Folk Psychology*. New York, NY: Cambridge Univeristy Press.

Feyerabend P (1963) Materialism and the mind–body problem. *Review of Metaphysics* **17**: 49–66.

Fodor J (1987) *Psychosemantics*. Cambridge, MA: MIT Press.

Forster M and Saidel E (1994) Connectionism and the fate of folk psychology. *Philosophical Psychology* **7**: 437–452.

Goldman A (1992) In defense of the simulation theory. *Mind and Language* **7**: 104–119.

Greenwood J (1991) *The Future of Folk Psychology*. Cambridge, UK: Cambridge University Press.

Horgan T and Graham G (1990) In defense of southern fundamentalism. *Philosophical Studies* **62**: 107–134.

Rorty R (1970) In defense of eliminative materialism. *Review of Metaphysics* **24**: 112–121.

Sellars W (1956) Empiricism and the philosophy of mind. In: Feigl H and Scriven M (eds) *The Foundations of Science and the Concepts of of Psychology and Psychoanalysis*, pp. 253–329. Minneapolis, MN: University of Minnesota Press.

Smolensky P (1988) On the proper treatment of connectionism. *Behavioral and Brain Sciences* **11**: 1–74.

Wellman H (1990) *The Child's Theory of Mind*. Cambridge, MA: MIT Press.

Wilkes K (1993) The relationship between scientific and common sense psychology. In: Christensen S and Turner D (eds) *Folk Psychology and the Philosophy of Mind*, pp. 144–187. Hillsdale, NJ: Erlbaum.

Ellipsis

Intermediate article

Robert C May, University of California, Irvine, California, USA

CONTENTS

A linguistic ellipsis, most generally expressed, is a truncated or partial linguistic form. It is a linguistic form in which constituents normally occurring in a sentence are superficially absent, licenced by structurally present prior antecedents.

INTRODUCTION

A linguistic ellipsis, most generally expressed, is a truncated or partial linguistic form. This partiality is measured relative to a complete sentence; an elliptical sentence is one in which some of the constituent parts of a 'full' sentence are missing. For example, in answer to the question 'Who went to the store', one may answer 'Max went to the store'. In most contexts, however, speakers would avoid such prolixity and instead would employ an elliptical form: 'Max went', 'Max did', or even just 'Max' would suffice as answers to the question. The reason that they would suffice appears to be quite obvious; it is because they mean just what 'Max went to the store' means, except they express this more economically by leaving off at least part of what can otherwise be gleaned from the initial question. From such simple examples we can already observe a fundamental property of ellipsis that we would expect to be captured under any account of the relation between elliptical and non-elliptical forms: meaning is constant under ellipsis. Specifying this constancy is not as straightforward as it may initially appear, however, and trying to capture what it amounts to has been an on-going issue in linguistic theory. It has been an issue of particular importance because the accounts of elliptical phenomena have been used to bring empirical weight to fundamental claims about linguistic theory, including the extension of the notion of linguistic identity, the relation of syntax and semantics, and the abstractness of grammar.

To frame our discussion, we note the most well-known types of elliptical constructions that have been studied in the literature:

VP-ellipsis: Max went to the store, and Oscar did, too
gapping: Max went to the store, and Oscar to the arcade
sluicing: Max went to the store, but Oscar wondered why
stripping: Max saw Sally at the store, and Oscar, too
pseudo-gapping: Max loves Jane, and Harry does Sally
N′-ellipsis: Max's father went to the store, but Oscar's went to the arcade

Each of the these types of ellipsis have idiosyncratic properties. For example, with gapping there is a well-known correlation of the direction of gapping and word order, and with sluicing the restriction that the complementizer of a sluiced clause must be interrogative; these are among observations originating with Ross (1969, 1970). Our purpose here is not, however, to survey the differences between these various elliptical phenomena but to explore two fundamental properties that they all have in common: (1) ellipsis is of a syntactic constituent; (2) an antecedent occurrence of the elided constituent governs the ellipsis. For example, in VP-ellipsis the elided material is a verb phrase, and the ellipsis is licensed in the presence of a fully lexicalized antecedent: in the case above, 'went to the store'. The primary issue for ellipsis is how to properly characterize (1) and (2); what sorts of linguistic description are called for to capture these fundamentally grammatical aspects of ellipsis in their full generality? A number of subsidiary issues are implied by answers to the primary questions, including the nature of the grammatical mechanisms that account for the absence of the lexical material: is it a deletion of underlyingly present syntactic elements; or are these elements absent even at the underlying level? In what follows we will outline some of the main approaches that have been developed to these issues, focusing primarily on the case of VP-ellipsis, largely because the relevant theoretical issues have been most clearly and widely discussed in this context.

VP-ELLIPSIS

Syntactic Reduction: The Transformational Theory

The initial approaches to ellipsis within contemporary linguistic theory attempted to account for the semantic constancy alluded to above by reducing it to syntactic constancy. The idea here is quite intuitive: if an elliptical form can be seen as a syntactic repetition of a corresponding non-elliptical form, i.e. as simply two occurrences of the same syntactic form, then it would follow immediately that they have the same meaning. The first systematic investigations along these lines is found in the work of J. R. Ross in the late 1960s (Ross, 1967, 1969). The approach plays itself out very naturally with respect to the deep structure/surface structure distinction: elliptical and non-elliptical forms have the same deep structure but different surface structures, the difference arising from the application of transformations that delete syntactic structure to give the elliptical forms. That (1) and (2) fall together was taken as a natural consequence of assuming that the transformational rules involved delete syntactic constituents and do so under identity with an antecedent constituent. So, for example, because there is an antecedent occurrence of the verb phrase 'went to the store', 'Oscar did, too' can be derived from 'Max went to the store' by VP-deletion.

During this period in the development of generative grammar, it was generally assumed that deletion transformations were constrained by a general theoretical condition that required the 'recoverability' of deletions; cf. Peters and Ritchie (1973) for discussion of the formal importance of this condition for the theory of transformations. Informally put, a deletion transformation satisfies this condition if it is possible to reconstruct the deleted material from within the structural context that the rule applies. Deletions that apply under identity obviously meet this criterion. For example, VP-deletion would appear to meet the condition because within the overall sentence 'Max went to the store, and Oscar did, too' we can reconstruct the deleted VP as a copy of the antecedent VP. However, two problems were noticed that indicated quite clearly that such deletion rules fail to meet the recoverability criterion.

The first problem is that the domains in which transformational rules apply are not the same as the domains in which deleted constituents can match up with their antecedents. While the domain of transformational rules is the sentence, the antecedent/deletion pairing is not restricted to this domain but transcends sentential boundaries. So, in our example of VP-deletion, we could replace the conjunction with a full stop, turning one sentence into two: (1) 'Max went to the store' and (2) 'Oscar did, too'. Moreover, we could imagine discourses in which other sentences would be interpolated between them or, even more telling, discourses in which the two sentences were uttered by different speakers. VP-deletion, it would appear, is not a rule of *sentence* grammar, but of *discourse* grammar.

Notice that broadening the applicability of syntactic deletion operations to discourse does not impact on what would seem a more fundamental aspect of recoverability, the reconstructive aspect of deletion given by the syntactic identity of the deleted constituent and its antecedent. The following observations, initially made by Ross (1967) do, however, and herein lies the second problem. Consider the following sentence:

Max saw his mother, and Oscar did, too (1)

Clearly, there is a reading of (1) that entails that Max and Oscar saw one and the same person; they each saw Max's mother. This interpretation could easily be obtained on the transformational view by taking the deep structure of (1) to be roughly (1′):

Max saw Max's mother, and Oscar saw Max's mother (1′)

(1) can derived from (1′) first by applying VP-deletion, applicable given the identity of the verb phrases, to be followed by pronominalization in the first clause. The construal of (1) characterized in this way is not, however, the only construal available of this sentence. It can also be understood in a manner comparable to (1″):

Max saw Max's mother, and Oscar saw Oscar's mother (1″)

The problem is that although (1″) ought to be a possible underlying source for (1), it is not a structure to which VP-deletion can apply, since the verb phrases are not identical. Insofar as (1) can be derived from an underlying form like (1″), it will be a nonrecoverable deletion.

Semantic Non-Reduction: The Property Theory

The ambiguity shown by (1) between a 'strict identity' reading (1′) and a 'sloppy identity' reading (1″)

thus scotches a transformational account of the sort envisaged as deletion under identity. But what went wrong? The answer that emerged is that the underlying problem lies with the initial presumption, that semantic constancy can be reduced to syntactic constancy. Deletions, in some sense, must still be recoverable, not least because knowing what has been deleted is necessary for understanding an elliptical sentence. But what is to be recovered is not syntactic information *per se*, but information about the logico/semantic roles played by syntactic expressions; it is in the identity conditions applicable to these roles that we are to look for the conditions that govern deletion. The semantic constancy of ellipsis, in this view, is not to be reduced to something else, but is to have a direct semantic analysis.

The nonreductionist approach found its fullest hearing in the work of Ivan Sag (1976) and Edwin Williams (1977) in the mid-1970s. The central observation animating this approach is that a sentence such as 'Max saw his mother' is ambiguous, depending upon whether the verb phrase 'saw his mother' expresses the property of seeing Max's mother or the property of seeing one's own mother. These two properties are not unrelated; the latter is an abstraction of the former, being general where the former is particular. This ambiguity, however, is masked in simple sentences because 'Max saw his mother' has the same truth conditions under either interpretation of the verb phrase. It becomes unmasked in elliptical contexts: depending on which property the ellipsis is taken as being identical with, different interpretations are obtained. Thus, if in 'Max saw his mother, and Oscar did, too', the ellipsis is understood as the property 'saw Max's mother' the strict reading ensues; if the ellipsis is understood as 'saw his own mother', the sloppy reading follows. In either case note that what is elided is identical with the antecedent; they each express the same property. In this view, the constancy of ellipsis is thus a matter of property identity, and this is an inherently semantic notion.

These ideas found a natural representation by assuming that the logical representation of natural language incorporate aspects of a λ-calculus. These sorts of logistic systems (first introduced by Church), incorporate an operation that abstracts properties from propositions. This operation, known as λ-abstraction, derives from the proposition expressed by 'Max read *Moby Dick*' the property: λ*x* (*x* read Moby Dick). This is to be parsed as a λ-operator binding a variable in the following open sentence; it is interpreted as a characteristic function, taking an individual as argument and returning a truth value. If we supply an argument for this function, represented by placing it before the λ-expression: Max, λ*x* (*x* read Moby Dick), we can return to our original proposition by the inverse operation, λ-conversion; we effect this by placing the argument in the place of the variable in the open sentence, and erasing the λ-operator.

In their analysis of ellipsis, Sag and Williams make two basic assumptions about the semantics of natural language. First, following a suggestion of Barbara Partee (1975), they assume that verb phrases are interpreted as λ-expressions, so that the logical form of 'Max read Moby Dick' would be represented as immediately above. Second, they assume that anaphoric pronouns can be interpreted as either constants or variables. Taking these together, it follows that 'Max saw his mother' is representationally ambiguous between the following logical forms: 'Max, λ*x* (*x* saw Max's mother)' and 'Max, λ*x* (*x* read *x*'s mother)'. In the first form, the pronoun is represented as a constant specifying its anaphoric reference; in the latter, the pronoun occurs as a bound variable. Thus far these representations are only distinguished formally; via λ-conversion both convert to the same proposition. The distinction becomes more than this, however, when a third assumption comes into play: VP-ellipsis requires identity of λ-expressions. This gives two representations for 'Max saw his mother, and Oscar did, too':

Max, λ*x* (*x* saw Max's mother) & Oscar, λ*y* (*y* saw Max's mother) (2)

Max, λ*x* (*x* saw *x*'s mother) & Oscar, λ*y* (*y* saw *y*'s mother) (3)

The first representation is of the *strict* reading; the second clause means Oscar saw Max's mother. The second representation is of the *sloppy* reading; in it, the second clause means that Oscar saw his own mother; that is, Oscar saw Oscar's mother.

The part of the identity conditions in λ-expressions relevant to strict and sloppy identity is known as the alphabetic variance condition. This condition breaks down into two subcases. The first applies to λ-expressions if there are only bound occurrences of variables; in this case, the λ-expressions must be exactly the same up to alphabetic values of the variables. Thus, in (3) for example, the λ-expression on the left is nondistinct from that on the right because they are alphabetic variants, differing only in that where '*x*' occurs on the left, '*y*' occurs on the right. That is, it matters not that we have '*x*' on the right and '*y*' on the left so long as the pattern of binding remains unaltered. When this is changed, the condition is not satisfied; a consequence of this is that a sloppy reading is unavailable in 'Max saw his mother, and Oscar believes Jane did, too'; i.e. the right-hand clause cannot mean that Oscar believes that Jane saw Oscar's mother. This is because in the following representation, which would represent this reading, there are no λ-expressions that are alphabetic variants: 'Max, λx (x saw x's mother)' and 'Oscar, λz (z believes Jane, λy (y saw z's mother))'. In particular, 'λx (x saw x's mother)' and 'λy (y saw z's mother)' are not alphabetic variants because 'z' is free within the λ-expression while the corresponding occurrence of 'x' is bound. Not all free occurrences of variables within λ-expressions are illicit however; this is the effect of the second case of the condition that permits λ-expressions to be alphabetic variants only if the free variables are all bound by the same operator. This is what we find in the representation of 'Max saw everyone before Oscar did':

$\forall x$ (Max, λy (y saw x) before Oscar, λz (z saw x))

Although within each λ-expression there is a parallel occurrence of 'x' free, they are both bound by the universal quantifier, and hence are alphabetic variants.

The success of this account in overcoming the problems that plagued the prior transformational approach extends beyond the account of strict and sloppy identity. Because the notion of property identity on which the account depends is semantic, unlike in the syntactic account, which was limited by the structural extent of structural descriptions of transformational rules, there is nothing comparable that inherently restricts the context in which the identical properties may occur. Therefore, in the absence of some external constraint, the antecedent of an ellipsis may occur in positions quite detached in discourse from the ellipsis itself; the sentence in which the antecedent is expressed neither needs to be adjacent to the elliptical sentence, nor need it be uttered by the same speaker. All that is required is that the antecedent of the ellipsis be sufficiently salient in the surrounding context.

Property theory and the syntax of ellipsis

The property theory approach initiated by Sag and Williams has been highly influential in the study of ellipsis, and there have been any number of variations of this view. One important source of these variations arises from a changed perspective regarding what is the main syntactic issue raised by elliptical constructions. In the transformational deletion analysis, the concern was over what we may call the 'generation' problem: what are the syntactic operations that produce elliptical structures? But on the property theory, which assigns the explanatory role to semantics for the matters that were so troublesome *vis-à-vis* recoverability for the prior account, the focus is shifted to the mapping problem: how are syntactic structures translated into semantic structures? In particular, how are verb phrases translated into λ-expressions that satisfy the identity conditions (alphabetic variance)? Understanding the mapping problem in this way places a constraint on its solutions; however syntactic derivation is to be effected (i.e. whatever the solution to the generation problem is), it must be such that it allows for systematic translation. There are two broad approaches to the mapping problem falling within this constraint that differ in their views of the need to attribute syntactic structure to the ellipsis in order to generate the property it expresses.

The first, the rich syntax view, assumes that elliptical structures retain a syntactic relation to forms in which all constituents are structurally present. Thus, whatever procedures translate 'Oscar read *Moby Dick*' also translate 'Oscar did, too', because at the input to the translation, the latter has the same structure as the former. The rich syntax view thus calls for the 'reconstruction' of syntactic information in order to obtain a property that then serves as the basis of comparison for identifying salient antecedent properties in the context of an ellipsis. This is the view Sag and Williams take, although they differ on the derivational direction of this relation. Sag takes the elliptical structure to be derived from the non-elliptical by deletion, while Williams reverses the direction of the derivation, the non-elliptical arising from the elliptical by syntactic copying. The alternative, the poor syntax view, sees no need for such a syntactic relation; in this view, at all stages of derivation the elided phrase is missing, or if not actually missing is structurally noncomplex; i.e. an empty category with no internal constituents (cf. Hardt, 1993). Since there is no verb phrase, there is also no translation to a property, however, and hence, in contrast to the rich theory, there is no basis of comparison for determining the antecedent of an ellipsis. But, according to this view, there is in fact no need to derive this, for there is an independent model to draw upon, the anaphoric resolution of pronouns. In both the ellipsis/antecedent and the pronoun/antecedent relations, a possibly complex antecedent for a syntactically simple element is determined by the conditions on salience in context. Thus, the relation of ellipsis to antecedent in 'Max read *Moby Dick*, and Oscar did, too' is comparable to the relation of pronoun to antecedent in discourse (e.g. 'Herman Melville wrote *Billy Budd*. He is more famous, however, for *Moby Dick*') or even more directly to VP-anaphors like 'so' and 'it' (e.g. 'Max read *Moby Dick*, and so did Oscar') for which elided verb phrases are, in this view, the covert analogues. In the poor syntax view, then, there is no reconstruction as in the rich view; information relevant to the resolution of the ellipsis is only that which is found in the antecedent.

Problems with the property theory

In subsequent research a number of problems have emerged with the property theory. One case, initially observed by Shalom Lappin (1984) calls into question the validity of the second clause of the identity condition in λ-expressions on the basis of examples such as (4):

I know which book Max read, and which book Oscar didn't (4)

Recall that the restriction that λ-expressions are alphabetic variants only if the free variables are all bound by the same operator is what accounted for 'Max saw everyone before Oscar did' as discussed above. But in allowing for this case, (4) should be disallowed, for here the λ-expression corresponding to the elided phrase – λz (z read x) – contains a free variable bound by a different operator (i.e. the *wh*-phrase in the second clause) than that which binds the free variable in the λ-expression corresponding to the antecedent phrase.

A second sort of case, derived from initial observations of Schiebe (1971) and Dahl (1974), is among a series of cases most extensively discussed in joint research by Robert Fiengo and Robert May (Fiengo and May, 1994), who label them the 'eliminative puzzles of ellipsis'. It does not pertain directly to the identity condition but to a systematic overgeneration problem in the translation of pronouns. This case, which Fiengo and May called the 'many-pronouns puzzle', arises when the number of pronouns is increased beyond the one found in the standard examples used to illustrate strict and sloppy identity:

Max said he saw his mother, and Oscar did, too (5)

Here, the expectation is that there should be a four-way ambiguity. This is because whether a pronoun is translated as a constant or a variable is independent of how any other pronoun is translated, predicting, for n-many pronouns, 2^n readings. Thus in (5) the pronouns could be (i) both variables, (ii) both constants, (iii) the first one a variable, the second a constant, or (iv), vice versa, the first a constant, the second a variable. Given the correspondence of variables with the sloppy reading, and constants with the strict reading, (i) and (ii) will result in 'across-the-board' sloppy and strict readings, respectively, while (iii) and (iv) will give readings mixed between sloppy and strict. The problem is that in (5) we observe only three, not four, readings: precluded is the reading corresponding to (iv). The ellipsis in (5) cannot be glossed as '… and Oscar said Max saw Oscar's mother'. What we actually observe in this case, as well as in those of increasing complexity, are only $n+1$ readings; readings do not grow exponentially.

Syntactic Reduction Redux: The Dependency Theory

What the 'many-pronouns puzzle' indicates is that the assumption of translational independence embedded in the property theory's account of anaphora is incorrect, suggesting instead that what is involved is some dependence relation between the pronouns. The most highly developed theory that seeks to capture these dependencies is the dependency theory developed by Fiengo and May (1994). Central to the dependency theory picture is that dependencies must have a syntactic characterization, and in establishing this result, Fiengo and May turn away from the nonreductive account of ellipsis embedded in the property theory, and return to a reductive syntactic approach. The dependency theory assumes that a sentence such as 'Max saw his mother' is structurally ambiguous, the ambiguity being attributed to a distinction in the representation of anaphoric pronouns that indicates whether the pronoun is formally dependent on its antecedent or not. By Fiengo and May's conventions, grammatically anaphoric pronouns are represented by co-indexing, with those that are formally dependent on their antecedent marked by 'β', those that are not by 'α'. 'Max saw his mother' thus has the following pair of representations: 'Max$_1$ saw his$_1^\alpha$ mother' and 'Max$_1$ saw his$_1^\beta$ mother'. The dependency that a β-marked pronoun enters into is specified via a structural description of the sequence of categories that lies between the pronoun and its antecedent. So for the latter structure, this would be the dependency: $\langle(Max, his), 1, \langle NP, V, NP\rangle\rangle$, where the first member of the triple is the elements of the dependency, the (unique) antecedent and the dependent pronouns, the second the index of the elements in the dependency, and the third the string of categories that links the co-indexed elements together.

The dependency theory is applied to ellipsis by holding that the identity conditions that allow for ellipsis are satisfied in the following two sorts of structure, where, given the co-indexings, the first represents the strict reading and the second represents the sloppy reading.

Max$_1$ saw his$_1^\alpha$ mother, and Oscar$_2$ *saw his$_1^\alpha$ mother*

Max$_1$ saw his$_1^\beta$ mother, and Oscar$_2$ *saw his$_2^\beta$ mother*

While it is apparent that in the first structure the antecedent and elided verb phrases are simple syntactic copies, it is not in the second, for the pronouns are different in the two verb phrases. This discrepancy is reconciled by allowing for an identity condition such that dependencies are the same so long as there is the same sequence of categories, regardless of the index; dependencies that stand in this relation – same pattern, different index – Fiengo and May call '*i*-copies'. The dependencies in which the pronouns in the latter structure occur meet this criterion: since they are *i*-copies, they are sufficiently alike to allow for ellipsis even though the pronouns are syntactically distinct. On the other hand, where there are no dependencies to be calculated – where the pronouns are marked α, not β – there is no alternative but for the index of the pronoun to be unchanged.

Sloppy identity on the dependency theory view is thus the re-creation of an antecedent structural pattern of anaphora; strict identity, on the other hand, is the re-creation of the anaphora itself. Either may be extended to more complex structures, but when they are certain limitations arise. Thus, recall the 'many-pronouns puzzle' that swirled around examples such as (5) above – 'Max said he saw his mother, and Oscar did, too'. For this case, there are four possible combinations of indices for the antecedent; of these, only three give rise to well-formed elliptical structures:

Max$_1$ said he$_1^\alpha$ saw his$_1^\alpha$ mother, and Oscar$_2$ *said he$_1^\alpha$ saw his$_1^\alpha$ mother*

Max$_1$ said he$_1^\beta$ saw his$_1^\beta$ mother, and Oscar$_2$ *said he$_2^\beta$ saw his$_2^\beta$ mother*

Max$_1$ said he$_1^\beta$ saw his$_1^\alpha$ mother, and Oscar$_2$ *said he$_2^\beta$ saw his$_1^\alpha$ mother*

*Max$_1$ said he$_1^\alpha$ saw his$_1^\beta$ mother, and Oscar$_2$ *said he$_1^\alpha$ saw his$_2^\beta$ mother*

The first case is the across-the-board strict reading; since both pronouns are α, they have the same index in the antecedent and the ellipsis. The second case is across-the-board sloppy: the two pronouns are in a dependency, with 'Oscar' as the antecedent in the same way structurally as the pronouns in the prior clause are in a dependency with 'Max'. The third structure represents the mixed reading: the first pronoun is β, and thus may be in a dependency. The second pronoun, however, is α so it must be strict, i.e. co-indexed with 'Max', not 'Oscar'. The final case is the one that is excluded. This is because the dependency that reaches from the pronoun 'his' to 'Oscar' as antecedent is not structurally identical to any dependency in the prior clause. Insofar as there is a dependency in that clause, it must be to the closer possible antecedent, the pronoun 'he', but this is not structurally parallel to the dependency in the clause with the ellipsis that has a greater syntactic extent.

Antecedent-Contained Deletion

Thus far, the structural context of ellipsis we have considered has been that of a discourse; i.e. the ellipsis and its antecedent have each occurred in independent sentences. While the examples we have examined are ones in which these sentences are conjoined, this is not essential, for in all the examples 'and' could be replaced by a full stop. The one exception was 'Max saw everyone before Oscar did', but notice here that the ellipsis is contained in an adjunct, not a subordinate, clause, so that even this case falls under the generalization that the ellipsis and its antecedent are syntactically independent. The generalization, however, appears to clearly fail in the following case:

Dulles suspected everyone that Angleton did (6)

Sentence (6) is well formed, and is naturally understood to mean the same thing as its unelided counterpart, 'Dulles suspected everyone that Angleton suspected'. But in this case the elided VP is not independent of its antecedent – rather it is contained within it – and hence the name for this construction, 'antecedent-contained deletion'. Notice that not only does antecedent-contained deletion appear to run counter to the generalization, but it does so in a particularly curious way. We understand the antecedent of the elided VP to be that VP headed by the verb 'suspected'. But if we plug that VP into the ellipsis, it will again contain the ellipsis: 'Dulles suspected everyone that Angleton suspected everyone that Angleton did'.

Additional iterations of the process will continually give structures that still contain an ellipsis. Given this vicious regression, it is unclear how we are to establish the relation of ellipsis and antecedent.

The critical insight here was provided by Sag (1976). Sag observed that if we attend to the logical form of (6), that at the appropriate level of syntactic description it falls under the antecedence generalization. This is because (6) contains a quantified phrase, and this must be scoped out; if we assume that the logical structure of (6) is roughly as follows: '∀*x*: Angleton *suspected x* (Dulles suspected *x*)', we then need only observe that here the ellipsis (filled in and indicated by italics) is no longer contained within the antecedent. (This logical form is comparable to the logical form that would be assigned to the fully lexicalized counterpart of (6).) Thus, the significance of antecedent-contained deletion, given the generalization regarding the relation of ellipsis and antecedent, is that the notion of structure relevant to this generalization must be sufficiently abstract so as to represent the logical form of sentences. In May (1985) it is argued that this structure is a form of syntactic structure, a result of the syntactic rule QR, which gives (7) as a representation of (6):

[everyone that Angleton did [Dulles suspected *t*]] (7)

In this structure the ellipsis is no longer contained within the antecedent, and so we can now plug in the antecedent VP without any regress:

[everyone that Angleton *suspected t* [Dulles suspected *t*]] (7′)

May's account of antecedent-contained deletion has been one of the main arguments that has been cited in support of the view that there is a syntactic level – *LF* – that represents the logical structure of natural language; see May (1985, chap. 1), as well as Hornstein (1995), for a contrary view.

VP-ANAPHORA

Previously, we briefly mentioned the phenomena of VP-anaphora, citing examples like 'Max hit Oscar, and Harry did it, too' or 'Max hit Oscar, and Harry did so, too'. These cases are closely akin to VP-ellipsis except that they have an anaphoric element – 'so' or 'it' – rather than an ellipsis. VP-anaphora is distributionally more restricted than VP-ellipsis. For example, with stative verbs, VP-ellipsis is possible, but not the 'it' form of VP-anaphora, and the 'so' form is marginal; compare 'Max knows French, and Oscar does, too' with 'Max knows French, and Oscar does it, too' and 'Max knows French, and Oscar does so, too'. However, what is more relevant to the present discussion is that with VP-anaphora we can find the same ambiguity of strict and sloppy identity that we observed with VP-ellipsis; compare 'Max hit his mother, and Oscar did, too' with (8):

Max hit his mother, and Oscar did it, too (8)

As before, the second clause can be taken to mean that Oscar hit Max's mother (strict) or that Oscar hit Oscar's mother (sloppy).

Examples like (8) pose issues as to how we are to understand ellipsis. If VP-anaphora and VP-ellipsis display uniform behavior, it would seem natural to posit a uniform analysis. One way to do this would be to take ellipses as anaphoric elements, silent counterparts, if you will, of 'it' or 'so'. (For one account along these lines, couched in a variant of the property theory, see Hardt (1993).) In this view, in which VP-ellipsis is reduced to VP-anaphora, that ellipsis involves some sort of syntactic reconstruction is effectively denied. The alternative would be to run the reduction in the opposite direction by maintaining that at an appropriately abstract syntactic level VP-anaphora, like VP-ellipsis, is syntactically complex, and that the overt pronominal elements are but superficial syntactic reflexes. In either account, the goal would be to isolate a level of representation in which VP-anaphora and VP-ellipsis are structurally nondistinct in order to account for their common behavior.

This search for a common analysis, however, needs to be weighed against ways in which VP-anaphora and VP-ellipsis do not cluster but diverge in properties. One sort of divergence has already been mentioned: namely, the distributional differences. Another can be gleaned from examples of antecedent-contained deletion, as in (9):

Dulles talked to everyone that Angleton did (9)

If VP-ellipsis were just a variant of VP-anaphora, then we would expect that the VP-anaphora version of (9) would also be grammatical. But, as was observed by Fiengo and May (1994), it is not:

*Dulles talked to everyone that Angleton did it (9′)

This case indicates (along with the distributional facts cited above) that there are substantial differences between VP-anaphora and VP-ellipsis. It remains an open research question how observations like these can be integrated with those about strict and sloppy identity; until then it remains equivocal whether VP-ellipsis and VP-anaphora are a unified phenomenon.

SUMMARY

Returning to our initial observation, we have seen that capturing the basic intuition of semantic constancy under ellipsis devolves to the issue of the proper way to state the identity conditions that govern ellipsis. That ellipsis in general requires a notion of identity was the initial insight of the transformational account; its flaw, in a sense, was that the notion of identity it had available was not sufficiently abstract. What followed were attempts to find the right locus of abstractness for ellipsis. The property theory argues that the appropriate identity conditions are semantic and does not fundamentally question the relative lack of abstractness of the underlying syntax. The dependency theory, in contrast, argues for a more abstract notion of syntax by refining the criteria for identity of occurrences of syntactic categories. But regardless of how matters turn out, it is clear that in seeking to fix the identity conditions for ellipsis, issues fundamental to our conceptions of linguistic description have been raised. Phenomena relevant to these issues extend beyond what we have been able to consider here. Among them are further interactions with anaphora, such as the 'vehicle change' effect noticed by Fiengo and May (1994): the observation that 'Mary loves John, and he thinks that Sally does, too' is interpreted as comparable to 'Mary loves John, and he thinks that Sally loves him, too', not 'Mary loves John, and he thinks that Sally loves John, too'. Moreover, we have left aside discussion of the conditions on discourse that allow an ellipsis to be resolved, including conditions on discourse coherence with respect to antecedents that may occur in sentences at some degree of removal in the discourse (cf. Kehler, 2000), and on the abstractness of discourse: e.g. consider when Butch says to Sundance at the edge of the cliff before jumping: 'I will if you will', indicating that the antecedent need not even be uttered (cf. Chao, 1987). We have tried, however, to highlight some of the core issues that have animated the discussion and that have made understanding elliptical phenomena of continued interest within linguistic theory.

References

Chao W (1987) *On Ellipsis*. PhD dissertation. University of Massachusetts, Amherst, MA.

Dahl Ö (1974) How to open a sentence: abstraction in natural language, *Logical Grammar Reports*, no. 12. Göteberg, Sweden: University of Göteberg.

Fiengo R and May R (1994) *Indices and Identity*. Cambridge, MA: MIT Press.

Hardt D (1993) *VP Ellipsis: Form, Meaning, and Processing*. PhD dissertation, University of Pennsylvania.

Hornstein N (1995) *Logical Form*. Oxford, UK: Blackwell.

Kehler A (2000) Coherence and the resolution of ellipsis. *Linguistics and Philosophy* **23**: 533–575.

Lappin S (1984) VP-anaphora, quantifier scope, and logical form. *Linguistic Analysis* **13**: 273–315.

May R (1985) *Logical Form: Its Structure and Derivation*. Cambridge, MA: MIT Press.

Partee BH (1975) Deletion and variable binding. In: Keenan E (ed.) *Formal Semantics of Natural Language*, pp. 16–34. Cambridge, UK: Cambridge University Press.

Peters PS and Ritchie RW (1973) On the generative capacity of transformational grammars. *Information Sciences* **6**: 49–83

Ross JR (1967) *Constraints on Variables in Syntax*. PhD dissertation. Massachusetts Institute of Technology, Cambridge, MA.

Ross JR (1969) Guess who? In: Binnick RI, Davison A, Green GM and Morgan JL (eds) *Papers from the Fifth Regional Meeting of the Chicago Linguistic Society*, pp. 252–286. Chicago, IL: Department of Linguistics, University of Chicago.

Ross JR (1970) Gapping and the order of constituents. In: Bierwisch M and Heidolph K (eds) *Progress in Linguistics*. The Hague, Netherlands: Mouton.

Sag I (1976) *Deletion and Logical Form*. PhD dissertation. Massachusetts Institute of Technology, Cambridge, MA.

Schiebe T (1971) Zum Problem der grammatisch relevanten Identität. In: Kiefer F and Ruwet N (eds) *Generative Grammar in Europe*. Dordrecht, Netherlands: Reidel.

Williams E (1977) Discourse and logical form. *Linguistic Inquiry* **8**: 101–139.

Further Reading

Berman S and Hestvik A (eds) (1992) *Proceedings of the Stuttgart Ellipsis Workshop*. Heidelberg: Arbeitspapiere des Sonderforschungsbereichs 340, Bericht Nr. 29, IBM Germany.

Chung S, Ladusaw WA and McCloskey J (1995) Sluicing and logical form. *Natural Language Semantics* **3**: 239–282.

Dahl Ö (1973) On so-called 'Sloppy' identity. *Synthese* **26**: 81–112.

Dalrymple M, Shieber S and Perreira F (1991) Ellipsis and higher-order unification. *Linguistics and Philosophy* **14**: 399–452.

Hankamer J and Sag I (1984) Deep and surface anaphora. *Linguistic Inquiry* **7**: 391–426.

Kennedy C (1997) Antecedent contained deletion and the syntax of quantification. *Linguistic Inquiry* **28**: 662–688.

Lappin S (1996) The interpretation of ellipsis. In: Lappin S (ed.) *The Handbook of Contemporary Semantic Theory*, pp. 145–175. Oxford, UK: Blackwell.

Lappin S and Benmamoun EA (eds) (1999) *Fragments: Studies in Ellipsis and Gapping*. Oxford, UK: Oxford University Press.

Lobeck A (1995) *Ellipsis: Functional Heads, Licensing, and Identification*. New York, NY: Oxford University Press.

Merchant J (2001) *The Syntax of Silence – Sluicing, Islands, and the Theory of Ellipsis*. Oxford, UK: Oxford University Press.

Rooth M (1992) Ellipsis redundancy and reduction redundancy. In: Berman S and Hestvik A (eds) *Proceedings of the Stuttgart Ellipsis Workshop*. Heidelberg: Arbeitspapiere des Sonderforschungsbereichs 340, Bericht Nr. 29, IBM Germany.

Schwabe K and Zhang N (2000) *Ellipsis in Conjunction*. Tübingen, Germany: Max Niemeyer Verlag.

Steedman M (1990) Gapping and constituent coordination. *Linguistics and Philosophy* **13**: 207–264.

Zoerner CE (1995) *Coordination: The Syntax of &P*. PhD dissertation. University of California, Irvine, CA.

Embodiment

Intermediate article

Ronald Chrisley, University of Birmingham, Birmingham, UK
Tom Ziemke, University of Skövde, Skövde, Sweden

CONTENTS

An understanding of how cognition is realized or instantiated in a physical system, especially a body, may require or be required by an account of a system's embedding in its environment, its dynamical properties, its (especially phylogenetic) history and (especially biological) function, and its nonrepresentational or noncomputational properties.

INTRODUCTION

In recent years a number of researchers in cognitive science and artificial intelligence (AI) have criticized many traditional approaches to modeling, building and understanding cognitive systems as not placing sufficient emphasis on the body or physical realization of such systems. Non-embodied approaches to cognitive science typically involve some or all of the following features, to a greater or lesser extent:

- The belief that cognition is computation, and thus can be understood in an implementation-independent way, allowing cognitive science to proceed independently of biology and neuroscience.
- A search for general-purpose cognitive abilities, not relativized to any particular (biological, sensorimotor, physical) context or need.
- A method of analysis, modeling and design that for the most part ignores temporal aspects of cognition, in that it focuses on behaviors (e.g. chess playing) that are evaluated in terms of 'getting the right answer' rather than exhibiting a particular dynamic profile, and sees cognition as a module that mediates between the deliverances of a causally prior perceptual module and the inputs to an autonomous action system.

In contrast, embodied approaches to cognition typically involve some or all of the following features, again to varying extents:

- Acknowledgment of the role that the body and its sensorimotor processes can and do play in cognition. Some aspects of the system that would, on the traditional view, be considered mere matters of implementation, are instead taken to be crucial components.
- Understanding of cognition in the context of its (especially evolutionary) biological function: to support the activities of the body.
- A view of cognition as a real-time, situated activity, typically inseparable from and often fully interwoven with perception and action.

'Embodied' cognitive science or artificial intelligence, then, refers to a range of loosely affiliated philosophies, explanatory frameworks and design methodologies that strive to redress a perceived neglect of the body in cognitive science.

Since the mid-1980s there has been a rapid increase in interest in embodied cognition (and use of the term 'embodiment'), but there are many aspects of cognitive science and artificial intelligence research conducted in the 1960s and 1970s that take embodiment into account.

ISSUES CONCERNING EMBODIMENT

The issue of embodiment is closely related to, though distinct from, several other issues of recent interest in cognitive science.

Embeddedness

Recognizing the role of the body in cognition facilitates an approach that sees cognition as partly constituted by, or in terms of its relationship to, the environment (Clark, 1997). Thus, embodied cognitive science is naturally related to investigations of externalism (the belief that mental states are partially constituted by states external to the cognizer), situatedness (the importance, in cognitive activity, of a cognizer's location in and relations to the environment), offloading (using aspects of the environment, such as numerals when doing long division, to reduce cognitive load), scaffolding (the assistance a developing infant gains from, e.g. interacting with adults who already have the ability being acquired), and interactivity (cognitive phenomena, such as turn-taking, which depend crucially on the dynamics of interaction between a cognizer and an object or other cognizer).

Noncomputationalism and Nonrepresentationalism

Once one acknowledges the presence of the body, it is possible to use bodily properties, dynamics and configurations to explain some abilities, behaviors and phenomena that previously were thought to require explanation in computational or representational terms (Varela *et al.*, 1991). However, it is still unclear whether such explanations which advert to bodily states are themselves noncomputational and nonrepresentational, or whether they instead invoke a new form of computation and/or representation. It certainly seems that an embodied approach need not be anticomputationalist or antirepresentationalist (Clark, 1997).

Dynamics

Many researchers who have taken an embodied approach have found it useful to turn away from discontinuous, nontemporal, logic-based formalisms and instead use the continuous mathematics of change offered by dynamical systems theory as a way to characterize and explain cognitive phenomena (Port and van Gelder, 1995). Again, while there may be natural affinities between embodied cognitive science and these tools, it is certainly possible to have one without the other.

Biology

Perhaps the most obvious connection between the mind and the body is the brain: surely an important part of understanding the mind is to understand the neurophysiology underlying cognition (Churchland, 1986). But there are other connections with biology as well. For example, some researchers have maintained that one can best understand natural cognitive systems in terms of the biological function and purposes that cognitive faculties served in the phylogenetic development of their bodies (Millikan, 1984). This requires understanding not only bodies, past and present, but the evolutionary niches of such bodies as well. An important constraint, then, on models of human cognition will be whether the proposed architecture is an evolvable one: whether it is the kind of architecture that could have been reached through a process of natural selection, given the conditions known to have been in place in our natural history (Sloman, 2001).

VARIETIES OF EMBODIMENT

There are several dimensions of variation in the views and theories of embodiment currently being considered in cognitive science.

Criteria for Embodiment

Perhaps the most important dimension of variation concerns what criteria must be met for something to be an embodied cognizer: what is to count as a body? One can, partly following Ziemke (2001), distinguish a range of views on this question, from the least to the most constraining.

Physical realization

According to one view, to be embodied is just to be realized in some physical substrate. All work in cognitive science is about embodied systems in this sense: even a virtual web agent must be realized in some physical facts at any given time. Only purely mental entities or spirits would lack this kind of embodiment. Thus, even traditional cognitive science is not as disembodied as some have claimed. For example, the influential notion of a physical symbol system (Newell and Simon, 1976) explicitly acknowledges the requirement that a cognitive system be embodied in this (weakest) sense.

Physical embodiment

Physical embodiment requires that the realizing physical system be a coherent, integral system, that to some degree persists over time. This would rule out virtual web agents, the physical realization of which can be radically distributed over the entire planet, but could still include any conventional robot. A tension arises here: if, as some theorists have argued, human cognition extends into the tools and other physical environmental states we exploit, then the localized, biological body is less relevant to cognitive science than the extended, constantly changing, distributed physical system that at any given time includes parts of our environment as well as parts of our bodies. Thus, active externalism (Clark and Chalmers, 1999) may be incompatible with strong notions of embodiment.

Organismoid embodiment

According to another view, the localized physical realization of the system must share some (possibly superficial) characteristics with the bodies of natural organisms, in terms of form or sensorimotor capacities, but need not be alive in any sense. The most prominent, and perhaps the most complex, examples of organismoid embodiment are humanoid robots such as Cog (Brooks and Stein, 1994) and Kismet (Breazeal and Scassellati, 2000). Work with these robots is based on the view that research in AI and cognitive robotics, in order to be able to investigate human-level cognition, has to deal with systems that have bodies which, although artificial and possibly non-living, have at least some human-like characteristics. For example, Kismet is a humanoid robot that learns how to visually track objects. To do this, a human trainer must move objects of an appropriate size at an appropriate speed at an appropriate distance in front of Kismet's eyes. If the human trainer moves the tracked object too close to Kismet, Kismet responds by raising its eyebrows in a manner which in humans indicates a startle response. This naturally causes the human trainer to startle in return, which prompts a change in the training parameters of speed and distance. Thus, Kismet's organismoid embodiment, in the form of eyebrows and facial expressions, is an integral part of the human–robot training dynamic which should tend towards homeostasis.

Organismal embodiment

The strongest criterion of embodiment is that the body must not only be organism-like, but actually organismal and alive (e.g. Sharkey and Ziemke, 1998). Of course, this raises the question of what is required for something to be alive. Various answers to this question have been proposed, including the ability to metabolize, reproduce, regenerate, or grow; autonomy; and autopoiesis (e.g. Maturana and Varela, 1980; von Uexküll, 1928).

Other Dimensions of Variation

Another dimension of variation is the extent of the domain of cognition that requires an embodied approach. For example, one can ask whether reference to the body is required only for giving an account of low-level, sensorimotor aspects of human cognition, or if it is required for all forms of human cognition, including reasoning, mathematical thought, and language use (cf. the distinction between 'material' and 'full' embodiment in Nuñez (1999)).

Theorists might also disagree on how radical the effect of taking an embodied approach will be on the concepts, theories and methods of cognitive science. Clark (1999) distinguishes between simple and radical embodiment. In simple embodiment, the framework of traditional cognitive science is retained, and facts about embodiment are treated as mere constraints on theories of, for example, 'inner organization and processing'. Radical embodiment is more ambitious, and 'treats such facts as *profoundly altering the subject matter and theoretical framework of cognitive science*' (p. 348, emphasis in original). Radical embodiment is advocated by, for example, Lakoff and Johnson (1999). These authors claim that a central finding of cognitive science is that the mind is inherently embodied, and that this, together with the other two central findings (that thought is largely unconscious, and that abstract concepts are largely metaphorical), force us to reject not just the Western philosophy of mind, but most or all of Western philosophy, including especially Anglo-American analytic philosophy but also including postmodern philosophy. This throws everything into question: the nature of truth, meaning, time, space, language, rationality, and especially the self.

Finally, one can distinguish between epistemological and metaphysical approaches to an embodied cognitive science. An epistemological approach maintains that concepts concerning the body will be required to understand and explain cognition (even if the cognitive system itself is disembodied); a metaphysical approach maintains either that cognitive processes must be realized in a body or that AI should proceed by making embodied robots, but makes no claim as to whether embodiment must be adverted to in explanations of cognitive activity. For example, research involving the robot Shakey (Nilsson, 1984) was metaphysically embodied, but since its design and explanation focused primarily on Shakey's functional, computational aspects (namely, deliberation and planning) and not on Shakey's embodiment, this research was not epistemologically embodied. Conversely, computer simulations are by their very nature not (metaphysically) embodied, in most senses of 'embodiment', yet many researchers use (epistemologically) embodied simulations to model the crucial role a cognizer's body plays in its activity.

PHILOSOPHICAL CONCEPTIONS OF EMBODIMENT

Before the twentieth century, the most influential view of mind in Western thought was dualistic: the mind was regarded as composed of a separate, extensionless, nonphysical substance. This view led to many insoluble problems, both philosophical and empirical. For example, how do the mental and physical realms interact? How can we scientifically investigate something that is not in the physical world?

Behaviorism rejected dualism, and thereby opened up the possibility of scientific enquiry into an embodied mind, but it left little room for an understanding of the processes underlying much of mentality: it addressed little of what would be called 'cognition' today. Also, it actively avoided explaining or mentioning experience or consciousness (as did many later cognitivists). More promising steps towards an embodied understanding of mind were taken in the first part of the twentieth century: the relevance of von Uexküll's notion of the body has already been mentioned, but there were several other notable thinkers.

For example, in the 1920s Heidegger (1962) developed a phenomenology that understood human activity not as the result of the manipulation of context-free representations of objects, but as the contextualized experience of the body–environment system. Explicit, decontextualized representation of a hammer as an independent object occurs only when there is some kind of breakdown in the system (e.g. the hammer is too heavy). Dreyfus (1992) elaborated this and other aspects of Heidegger's philosophy into a critique of early, disembodied AI work, calling, for example, for systems that react to the particularities of the current perceptual/action situation rather than ones that attempt to create general-purpose, long-term plans. Although heartened by some aspects of neural or connectionist approaches to understanding the mind, Dreyfus provisionally concludes that, like their symbolic predecessors, connectionist models of cognition suffer from their lack of embodiment, in two ways. Firstly, by not being embedded in a real world with actual bodily concerns, most sophisticated connectionist learning must rely on the intervention of a human teacher, which prohibits the connectionist system from developing its own, genuine, meaningful attitude to the world. Secondly, he speculates that connectionist systems will never generalize in a way that we can recognize as being intelligent and meaningful until they have a form of life sufficiently similar to ours – which requires at least a body of some sort, perhaps even a humanoid one.

In the 1930s, Vygotsky (1978) claimed that language is an inherently socially situated activity, and that one can only understand a child's acquisition of language by recognizing this social context. It follows that inasmuch as social activity is embodied, the development and deployment of linguistic faculties will have to be understood as embodied as well.

While placing less emphasis on social embedding, Piaget (1954) was more explicit about the role of the body in the development of cognitive abilities. For example, his accounts typically made essential reference to the notion of a circular reaction, which in its primary form is the repetition of an activity in which the body starts in one configuration, goes through a series of intermediate stages, and eventually arrives at the starting configuration again. Thus, the kinds of abilities that an organism may acquire depend critically on what circular reactions are possible given that organism's body.

In the 1940s, Merleau-Ponty (1962) made the body central to his phenomenology of mind. For example, he claimed that we have intentions that we do not choose to have, by virtue of our bodies being the way they are. Furthermore, the way we perceive an object is determined by the modes of interaction that our bodies, given the nature of the object, allow. (This idea was a precursor of the notion of affordances (Gibson, 1979).) Even more strongly, Merleau-Ponty saw the body as the necessary medium for our having a world at all, with the nature of the activities of the body determining the nature of what could be experienced in our world. Further, the body could be augmented with tools to further develop the elements of our lived world.

Despite the emphasis that these thinkers placed on the body for understanding the mind and behavior, and partly because of their context outside the Anglo-American tradition, it is only recently that the notion of embodiment has had a significant influence on mainstream cognitive science and AI. Instead, these fields were, from their inception in the mid-1950s, dominated by the computer metaphor for mind (not surprising, perhaps, since the notion of computation had for centuries been a concept of the (human) mental activity of symbol and number manipulation). In particular, in the absence of any suggestion as to how they could constrain empirical investigation and modeling, philosophies of embodiment seemed metaphorical or even unintelligible to many cognitive and AI researchers at that time.

COGNITIVE SCIENCE AND EMBODIMENT

The fields of cognitive science and artificial intelligence have played a central role in the development of an embodied concept of mind and cognition. It was empirical work in cognitive science and artificial intelligence that allowed the development of a more robust and precise notion of embodied cognition to develop.

In particular, work involving mobile robots, at MIT Artifical Intelligence Laboratory and elsewhere, helped to establish the principles and concepts of embodiment and situatedness as the basis of a new approach to artificial intelligence and (later) an embodied cognitive science. For example, Brooks (1991) and his colleagues were able to get robots to perform tasks in the real world in real time that previously could only be done slowly and inflexibly, if at all. They did this by building robot bodies and robot controllers based on a design called 'subsumption architecture'. Rather than trying to graft a domain- and body-independent planning system onto a perceiving and acting robot, the subsumption architecture approach starts with an initial layer of simple perception–action mediation that implements some low-level behavior (e.g. obstacle avoidance). New behaviors (e.g. exploration) are added by adding further layers, which also mediate between perception and action in a simple way, but inhibit ('subsume') the lower layers when necessary to achieve the desired behavior. In such an architecture, there is no central locus of control, no separate planner, and no central model of the world that all processes must write to and read from in order to act appropriately. Communications between processes are not complex symbolic structures, but numerical values. What computation there is in the architecture is distributed, asynchronous, and non-hierarchical.

From an orthodox computationalist perspective, these design features have their disadvantages, but the designs of Brooks and his colleagues exploit the physical properties of the robots to overcome, or bypass, these limitations. For example, although internal communication between processes is limited, the world itself is often used as a medium for communication between the different layers and mechanisms. Much of what traditional thinking would say is required to perform a task is shown to be unnecessary if one takes advantage of regularities and information provided by the body–environment interaction.

The resulting empirical advances in robot engineering served as a springboard for a development and refinement of the notions and philosophies of embodiment. However, the failure of these approaches to quickly scale up to 'higher-level' aspects of cognition has led many to question the ability of the embodied approach to account for conceptual, abstract reasoning and representation. Kirsh (1991) correctly realizes that the issue is not one of representation per se; Brooks concedes that representation is required for some aspects of cognition. The question instead concerns having concepts: 'the ability to find an invariance across a range of concepts', as Kirsh puts it. As we move up a scale of accounting for ever more sophisticated cognitive activities, at what point must we stop limiting ourselves to designs that are tied to the particularities of the body, and begin to use designs that deploy concepts? Brooks (according to Kirsh) says 'almost never'; Kirsh disagrees, saying that much of not only reasoning and abstract thought but even perception and action must be understood in conceptual terms. Perhaps on a strict, rarefied notion of what concepts are, Kirsh is right: the explanation of concept-involving cognition can or must often go beyond what is provided by the body. But perhaps our very concept of concept needs revising (cf. Lakoff and Johnson, 1999); if we can understand how even full concept possession can be the result of being embodied in a particular way, then perhaps embodied robotics can serve as the model for far more of cognition than mere insect-like behavior.

The joint influence of embodied artificial intelligence research and philosophies and concepts of embodiment has prompted researchers to look for and formulate new forms of explanation for natural cognitive phenomena. For example, Thelen and Smith (1994) give an embodied explanation of the development of walking in infants. Rather than attempt to explain changes in gait as the result of changes in plans, rules or representations, Thelen and Smith give an elegant account that emphasizes changes in bodily factors such as the mass of the infant's leg. Such factors are then related to one another in a dynamical-systems framework, the phase transitions of which are used to explain the stage-like developments in infant walking behavior.

While the relevance of embodiment to research in mobile robotics, as discussed earlier, is obvious, some have claimed that concepts of embodiment are required for us to understand non-robotic artificial computational systems as well. Smith (1996) has argued that we will only be able truly to understand what is going on in ordinary desktop computers when we understand how the embodiment (in the sense of being located in space and time, having mass, and so on) of a computer enables it to achieve, for example, various forms of self-reference and even abstract mathematical reference. For example, it is the physically embodied 'two-ness' of a list `L=(a, b)` stored in computer memory that makes it possible for the computer to evaluate expressions such as `length(L)`.

Concepts of embodiment may also be necessary for us to theorize about the representational states of cognitive systems. The traditional means of specifying the content of, say, a belief state, is to provide a 'that' clause: a natural-language sentence that carries the same content as the belief that is being specified, for example, 'the child believes that the toy is within reach'. There are good reasons to believe that many representational contents, such as those of animals, infants, and sub-personal states, cannot be expressed in the conceptual framework of public language. To specify such contents for the purpose of a cognitive-scientific explanation, then, one may have to make essential reference to the body, and in particular the sensorimotor capabilities, of the system being explained (Cussins, 1990; Chrisley, 1995).

References

Breazeal C and Scassellati B (2000) Infant-like social interactions between a robot and a human caretaker. *Adaptive Behavior* **8**(1): 49–74.

Brooks RA (1991) Intelligence without representation. *Artificial Intelligence* **47**: 139–159.

Brooks RA and Stein LA (1994) Building brains for bodies. *Autonomous Robots* **1**: 7–25.

Chrisley R (1995) Taking embodiment seriously: nonconceptual content and robotics. In: Ford K, Glymour C and Hayes P (eds) *Android Epistemology*, pp. 141–166. Cambridge, MA: AAAI/MIT Press.

Churchland PS (1986) *Neurophilosophy: Toward a Unified Science of the Mind–Brain*. Cambridge, MA: MIT Press.

Clark A (1997) *Being There: Putting Brain, Body and World Together Again*. Cambridge, MA: MIT Press.

Clark A (1999) An embodied cognitive science? *Trends in Cognitive Science* **3**(9): 345–351.

Clark A and Chalmers D (1998) The extended mind. *Analysis* **58**: 10–23.

Cussins A (1990) The conectionist construction of concepts. In: Boden M (ed.) *The Philosophy of Artificial Intelligence*, pp. 368–440. Oxford, UK: Oxford University Press.

Dreyfus H (1992) *What Computers Still Can't Do*. Cambridge, MA: MIT Press.

Gibson JJ (1979) *The Ecological Approach to Visual Perception*. Boston, MA: Houghton Mifflin.

Heidegger M (1962) *Being and Time*. New York, NY: Harper and Row. [First published in 1927 as *Sein und Zeit*. Tübingen, Germany.]

Kirsh D (1991) Today the earwig, tomorrow man? *Artificial Intelligence* **47**: 161–184.

Lakoff G and Johnson M (1999) *Philosophy in the Flesh: The Embodied Mind and its Challenge to Western Thought*. New York, NY: Basic Books.

Maturana HR and Varela FJ (1980) *Autopoiesis and Cognition: The Realization of the Living*. Dordrecht, Netherlands: Reidel.

Merleau-Ponty M (1962) *Phenomenology of Perception*. London, UK: Routledge and Kegan Paul. [First published in 1945 as *Phénoménologie de la Perception*. Paris, France: Gallimard.]

Millikan R (1984) *Language, Thought and Other Biological Categories*. Cambridge, MA: MIT Press.

Newell A and Simon H (1976) Computer science as empirical enquiry: symbols and search. *Communications of the Association for Computing Machinery* **19**: 105–132.

Nilsson N (1984) *Shakey the Robot*. SRI Technical Note 323. Menlo Park, CA: SRI International.

Nuñez R (1999) Could the future taste purple? *Journal of Consciousness Studies* **6**(11–12): 41–60.

Piaget J (1954) *The Construction of Reality in the Child*. New York, NY: Basic Books. [First published in 1937 as *La Construction du Réel Chez l'Enfant*. Neuchâtel, Switzerland: Delachaux et Niestlé.]

Port R and van Gelder T (eds) (1995) *Mind As Motion: Explorations in the Dynamics of Cognition*. Cambridge, MA: MIT Press.

Sharkey NE and Ziemke T (1998) A consideration of the biological and psychological foundations of autonomous robotics. *Connection Science* **10**(3–4): 361–391.

Sloman A (2001) Evolvable biologically plausible visual architectures. In: Cootes T and Taylor C (eds) *The Proceedings of the British Machine Vision Conference, Manchester, September 2001*, pp. 313–322. Manchester, UK: BMVC Press.

Smith B (1996) *On the Origin of Objects*. Cambridge, MA: MIT Press.

Thelen E and Smith L (1994) A dynamic systems approach to the development of cognition and action. Cambridge, MA: MIT Press.

von Uexküll J (1928) *Theoretische Biologie*. Berlin: Springer.

Varela FJ, Thompson E and Rosch E (1991) *The Embodied Mind: Cognitive Science and Human Experience*. Cambridge, MA: MIT Press.

Vygotsky LS (1978) *Mind in Society: The Development of Higher Psychological Processes*. Cambridge, MA: Harvard University Press. [First published 1934 in Russian.]

Ziemke T (2001) Are robots embodied? In: Balkenius C, Zlatev J, Kozima H, Dautenhahn K and Breazeal C (eds) *Proceedings of the First International Workshop on Epigenetic Robotics: Modeling Cognitive Development in Robotic Systems*, pp. 75–83. Lund, Sweden: Lund University Cognitive Studies.

Further Reading

Brooks RA (1990) Elephants don't play chess. *Robotics and Autonomous Systems* **6**(1–2): 1–16.

Chrisley R (2000) *Artificial Intelligence: Critical Concepts*. London, UK: Routledge.

Hutchins E (1995) *Cognition in the Wild*. Cambridge, MA: MIT Press.

Pfeifer R and Scheier C (1999) *Understanding Intelligence*. Cambridge, MA: MIT Press.

Sheets-Johnstone M (1999) *The Primacy of Movement*. Amsterdam: John Benjamins.

von Uexküll J (1982) The theory of meaning. *Semiotica* **42**(1): 25–82.

Ziemke T (2001) The construction of 'reality' in the robot: constructivist perspectives on situated artificial intelligence and adaptive robotics. *Foundations of Science* **6**(1): 163–233.

Emergence

Intermediate article

Achim Stephan, University of Osnabrück, Osnabrück, Germany

CONTENTS

In ordinary language, to 'emerge' means to 'appear' or 'come into view'; but the technical use of the term is associated with features such as novelty, irreducibility, and unpredictability. The basic idea of emergence is that as systems become increasingly complex during evolution, some of them may exhibit novel properties that are neither predictable nor explainable on the basis of the laws governing the behavior of the systems' parts. Thus, complex wholes can come to have properties that are not reducible to the properties and relations of their constituents.

INTRODUCTION

During the 1990s, the term 'emergence' became widely used in such different fields as the philosophy of mind, self-organization, creativity, artificial life, dynamical systems, and connectionism. The term, however, is not used in a uniform way. It can imply novelty, unpredictability, irreducibility, and the unintended arising of systemic properties, particularly in artificial systems. Thus, it is rather controversial what the criteria are by which 'genuine' emergent phenomena should be distinguished from non-emergent phenomena. Some of the suggested criteria are very demanding, so that few, if any, properties would count as emergent. Others are inflationary, in that they count many, if not all, system properties as emergent. First of all, therefore, one should be clear about the various types of emergence. (*See* **Philosophy of Mind**; **Self-organizing Systems**; **Creativity**; **Artificial Life**; **Dynamical Systems, Philosophical Issues about**; **Connectionism**)

VARIETIES OF EMERGENTISM

Three theories among the different varieties of emergentism deserve particular interest: synchronic emergentism, diachronic emergentism, and a form of weak emergentism. In synchronic emergentism, the relationship between a system's properties and its microstructure (i.e. the arrangement and properties of the system's parts) is at the center of interest. A property of a system is taken to be emergent if it is irreducible, i.e. if it is not reducible to the arrangement and properties of the system's parts. In contrast, diachronic emergentism is mainly interested in predictability of novel properties. Those properties are taken to be emergent that could not have been predicted, in principle, before their first instantiation. Both of these stronger versions of emergentism are based on a common weak theory from which they can be developed by adding further theses.

Weak Emergentism

Physical monism

The first thesis of current theories of emergence – the thesis of physical monism – concerns the nature of systems that have emergent properties or structures. It says that the bearers of emergent features consist of physical entities only. Thus, all substance-dualistic positions are rejected; for they base properties such as being alive or having cognitive states on supernatural bearers, such as an entelechy or a *res cogitans* respectively. (*See* **Dualism; Descartes, René**)

Physical monism is the thesis that entities existing or coming into being in the universe consist solely of physical components. Likewise, properties, dispositions, behaviors, or structures classified as emergent are instantiated by systems consisting exclusively of physical entities.

Systemic properties

While the first thesis places emergent properties and structures within the framework of a physicalistic naturalism, the second thesis – the thesis of systemic properties – delimits the types of properties that are possible candidates for emergents. It is based on the idea that the general properties of a complex system fall into two classes: those that some of the system's parts also have, and those that none of the system's parts has. These latter properties are called systemic or collective properties.

The second thesis is that emergent properties are systemic properties. A property of a system is systemic if and only if the system possesses it but no proper part of the system possesses it.

Both artificial and natural systems with systemic properties exist. Those who would deny their existence would have to claim that all of a system's properties are instantiated already by some of the system's parts. Countless examples refute such a claim, e.g. it is among the properties of a leopard to run, but no part of it (head, heart, nor any cell assembly) can run; and it is among the properties of a connectionist network to recognize patterns, but no single part of it (unit, etc.) has this property.

Synchronic determination

The third thesis specifies the type of relationship that holds between a system's microstructure and its emergent properties. Namely, a system's properties and dispositions to behave depend nomologically on its microstructure, that is to say, on the properties and arrangement of its parts. There can be no difference in a system's systemic properties without some difference in the properties or arrangement of its parts.

Anyone who denies the thesis of synchronic determination has either to admit properties of a system that are not bound to the properties and arrangement of its parts, or to suppose that some other factors, in this case non-natural factors, are responsible for the different dispositions of systems that are identical in their microstructure. One may have to admit, for example, that there may exist objects that have the same parts in the same arrangement as diamonds, but that lack the diamond's hardness. This seems implausible. Equally implausible is the idea that there may exist two physically identical organisms, one viable and the other not. In the case of mental phenomena, opinions may be more divided; but one thing seems to be clear: anyone who believes, for example, that two physically identical creatures could be such that one is colorblind while the other is not, is not a physicalist.

Weak emergentism as sketched so far specifies the minimal criteria for emergent properties. It is the common base for all stronger theories of emergence. Moreover – and this is a reason for distinguishing it as a theory in its own right – it is held not only by some philosophers (e.g. Bunge, 1977), but also by some cognitive scientists (e.g. Varela *et al.*, 1991; Rumelhart and McClelland, 1986) in exactly its weak form. Weak emergentism is compatible with current reductionist approaches; and some champions of weak emergentism cite this compatibility as one of its merits compared with stronger versions of emergentism.

Synchronic Emergentism

The essential theses of the two more ambitious theories of emergence are the theses of irreducibility (synchronic emergentism) and of unpredictability (diachronic emergentism). These are closely connected: irreducible systemic properties are *eo ipso* unpredictable before their first appearance. Hence, synchronically emergent properties are also diachronically emergent, but not conversely.

A systemic property is irreducible if it cannot be explained reductively. For a reductive explanation to be successful several conditions must be met: the property to be reduced must be functionally construable or reconstruable; it must be shown that the specified functional role is filled by the system's parts and their mutual interactions; and the behavior of the system's parts must follow from the behavior they show in isolation or in simpler systems than the system in question. (It is an open question whether or not properties exist that demand a construction or reconstruction other than being functional.) If all these conditions are met, the behavior of the system's parts in other contexts reveals what systemic properties the actual system has. (*See* **Reduction**; **Functionalism**)

Since these three conditions are independent of each other, there are three different ways in which systemic properties may be irreducible. Namely, a systemic property is irreducible if: it is not functionally construable (or reconstruable); if it cannot be shown that the interactions between the system's parts fill the systemic property's specified functional role; or if the specific behavior of the system's components, over which the systemic property supervenes, does not follow from the component's behavior in isolation or in simpler configurations. (*See* **Supervenience**)

Thus, we have to distinguish three different types of irreducibility of systemic properties. Their consequences are also different. If a property is irreducible due to the irreducibility of its bearer's parts' behavior we seem to have an instance of 'downward causation'. For, if the parts' behavior is not reducible to their arrangement and the behavior they show in simpler systems, then there seems to exist some 'downward' causal influence, from the system itself or from its structure, on the behavior of the system's parts.

Such 'downward causation' would not violate the principle of the causal closure of the physical domain. We would just have to accept additional types of causal influences within the physical domain besides the known types of mutual interactions.

Likewise, if it cannot be shown that the interactions between the system's parts fill the specified functional role, it seems that the systemic property has causal powers that the microstructure does not have; hence in this case too there would be some downward causal influence.

In contrast, the occurrence of properties that are not functionally construable does not imply any kind of downward causation. Systems with properties that admit of no functional analysis need not be constituted in such a way that their components' behavior is irreducible. Nor is it implied that the system's structure has a downward causal influence on the system's parts. Thus there is no reason to assume that properties that cannot be analyzed themselves exert a causal influence on the system's parts. Rather, the question is how properties that cannot be functionally analyzed might have any causal role to play at all. And if one cannot see how they might play a causal role, then, it seems, such properties must be epiphenomena. (*See* **Epiphenomenalism**; **Mental Causation**)

Diachronic Emergentism

All diachronic theories of emergence are based on a thesis about the occurrence of genuine novelties in evolution. This thesis excludes all preformationist positions. According to this thesis, in the course of evolution exemplifications of genuine novelties occur again and again. Existing building blocks develop new configurations; new structures are formed that constitute new entities with new properties and behaviors.

However, the thesis of novelty does not by itself turn a weak theory of emergence into a strong one, since reductive physicalism remains compatible with such a variant of emergentism. Only the addition of the thesis of unpredictability, in principle, will lead to stronger forms of diachronic emergentism.

The first occurrence of a systemic property can be unpredictable for at least two different reasons. Firstly, it is unpredictable, in principle, if it is irreducible. This does not mean, however, that further occurrences of the property might not be predicted adequately. Secondly, it can be unpredictable because the microstructure of the system that exemplifies the property for the first time in evolution is unpredictable. Since in the first case the criteria for being unpredictable are identical with those for being irreducible, this notion of unpredictability will offer no theoretical gains beyond those afforded by the notion of irreducibility. Let us focus, therefore, on the second case: unpredictability of structure.

The structure of a newly formed system can itself be unpredictable for two reasons. If the universe is indeterministic, then its novel structures will be unpredictable. However, from an emergentist perspective, it is of no interest if a new structure's appearance is unpredictable only as a result of its indeterminacy – most emergentists claim that the development of new structures is governed by deterministic laws.

Nevertheless, deterministic formings of new structures can be unpredictable in principle if they are governed by laws of deterministic chaos. Against that claim one might argue that a Laplacean calculator could predict even chaotic processes correctly. Whether or not this 'actually' could be the case is not yet settled. It depends mainly on the question of what kind of information we allow such a creature to have. At least we can be sure that creatures of our mental capacities do not have these forecasting abilities, and thus, we can legitimately suppose that where chaos exists, structures exist that are unpredictable in principle.

The thesis of structure unpredictability is that the rise of a novel structure is unpredictable in principle if its formation is governed by laws of deterministic chaos. Likewise, any property that is instantiated by the novel structure is unpredictable in principle.

Although diachronic emergentism with the thesis of structure unpredictability implies the unpredictability of all properties instantiated by systems that emerge from chaotic processes, it does not thereby imply their irreducibility. The unpredictability, in principle, of systemic properties is entirely compatible with their being reducible to the microstructure of the system that instantiates them.

Synopsis

Figure 1 shows the logical relationships that hold between the different versions of emergentism.

Weak diachronic emergentism results from weak emergentism by adding a temporal dimension in the form of the thesis of novelty. Both versions are compatible with reductive physicalism. Weak theories of emergence are used today mainly in cognitive science, particularly for characterization of systemic properties of connectionist networks, and in theories of self-organization. Synchronic emergentism results from weak emergentism by adding the thesis of irreducibility. This version of emergentism is important for the philosophy of mind, particularly for debating nonreductive physicalism and qualia. It is not compatible with reductive physicalism. Strong diachronic emergentism differs from synchronic emergentism because of the temporal dimension in the thesis of novelty. Structure emergentism is entirely independent of synchronic emergentism. It results from weak diachronic emergentism by adding the thesis of structure unpredictability. Although structure emergentism emphasizes the boundaries of prediction within physicalistic approaches, it is compatible with reductive physicalism, and so it is weaker than synchronic emergentism. Theories of deterministic chaos can be considered as a type of structure emergentism. This perspective is important for evolutionary research. (*See* **Qualia**)

HISTORY

Although some hints of emergentist thinking can be found in the works of Empedocles, Epicurus, and Galen, the proper development of emergentism began in the mid nineteenth century in Britain. George Henry Lewes (1875) introduced the term 'emergent' into philosophy, to distinguish 'emergent' from 'resultant' effects. Here Lewes picked up on John Stuart Mill's distinction between 'homogeneous' and 'heterogeneous' effects: joint effects of causes are called heterogeneous (or emergent) if they are not the 'sum' of their separate effects; otherwise they are called homogeneous (Mill, 1974). Mill's distinction between 'ultimate' and

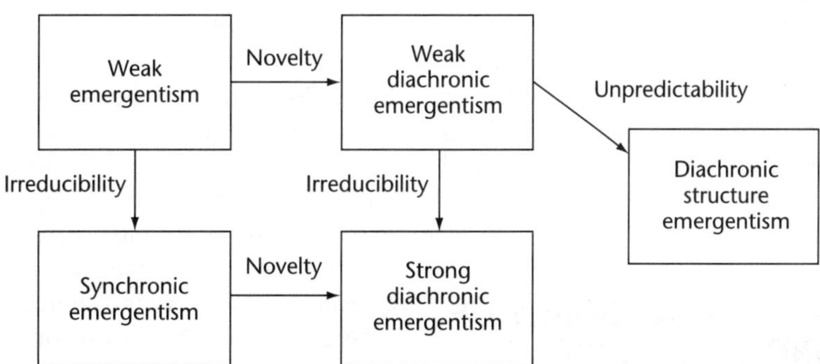

Figure 1. The logical relationships between the varieties of emergentism. Each arrow represents the addition of a thesis to a weaker theory.

'derivative' laws was also of great importance for the development of emergentist ideas. Some decades later, C. D. Broad oriented himself by Mill's distinctions and his subsequent considerations about the limits to explanation of psychophysical laws.

In the 1920s, theories of emergence began to attract greater philosophical and scientific interest. In rapid sequence the major works of British and American emergentism appeared: in 1920 Samuel Alexander's *Space, Time, and Deity*, in 1922 Roy Wood Sellars's *Evolutionary Naturalism*, in 1923 Conwy Lloyd Morgan's *Emergent Evolution*, and in 1925 Charles Dunbar Broad's *The Mind and its Place in Nature*.

Most of these philosophers' theories of emergence are reactions to the debate on the nature of life. While vitalists like Hans Driesch and Henri Bergson claimed, for the explanation of vital processes, the existence of supernatural entities such as an 'entelechy' or an *élan vital*, biological mechanists were trying to reduce all phenomena of life to physical and chemical processes without residue. Both positions seem to have implausible consequences: substance-dualistic approaches violate the principle of the causal closure of the physical domain, and it is hard to square them with evolutionary cosmologies; while mechanism does not seem to capture genuine organic and mental processes adequately. The emergentists steered a middle course. They denied both substance-dualistic and reductionist theories: they were nonreductive naturalists.

In the following decades, theories of emergence were much discussed. However, the criticism by Hempel, Oppenheim, and Nagel seemed to put an early end to emergentism, for their analysis led to an uninteresting concept of emergence as meaning nothing but: 'considering all theories we know of, we cannot explain why system *S* has property *P*' (Hempel and Oppenheim, 1948; Nagel, 1961).

With the decline of positivism, interest in metaphysical questions returned. It is the unsettled question about the nature of mental states that has helped emergentism to return to the philosophy of mind. The concept of emergence has also gained ground in such fields as self-organization, artificial life, the philosophy of science, and cognitive science.

EMERGENCE IN THE PHILOSOPHY OF MIND AND COGNITIVE SCIENCE

In different fields of philosophy and cognitive science the idea of 'emergence' has different roles. Thus, within the philosophy of mind, and particularly within the debate about qualia, there is a need for a strong notion of emergence; while within the fields of connectionism and artificial life, weaker notions of emergence suffice.

Emergentism as Nonreductive Physicalism

Within the philosophy of mind, emergentism is the most recent form of what has been called 'nonreductive physicalism' since the 1970s: a doctrine that in one way or other has tried to establish a compromise between physicalist reductionism and various sorts of dualism. First, physicalistic functionalism was seen as a species of nonreductivism because of its violation of biconditional bridge laws of the Nagelian type by its acceptance of multiply realizable mental properties. Subsequently, psychophysical supervenience was thought to be a theory of mind that is essentially both physicalistic and nonreductive. (*See* **Materialism**; **Dualism**; **Functionalism**; **Reduction**; **Supervenience**; **Multiple Realizability**)

Careful analyses, however, particularly by Jaegwon Kim (1993, 1998), revealed that both positions fall short of being what they are supposed to be. Physicalistic functionalism turned out to be reductionistic (it guarantees reductive explanations); and psychophysical supervenience, even in its strong form, turned out to be too weak to establish any specific theory of mind at all. Since even such diverse positions on the mind–body problem as reductive type physicalism and epiphenomenalism entail psychophysical supervenience, theories of supervenience fail to guarantee nonreductivism. In fact, it is synchronic emergentism that comprises the tenets originally associated with supervenience: property covariation and the dependence of supervenient properties on their subvenient bases are captured by the third thesis (synchronic determination) of weak emergentism; irreducibility, of course, is captured by the fourth tenet which is specific to synchronic emergentism. (*See* **Mind–Body Problem**; **Epiphenomenalism**)

Since weak emergentism (like mind–body supervenience) is compatible with both reductionism and nonreductionism, strong emergentism seems to be the only adequate representative of nonreductivism in recent philosophy of mind. An interesting question, however, is whether or not synchronic emergentism really is physicalism. Some philosophers maintain that such a position should be characterized as a kind of dualism, namely property dualism. However, insofar as psychophysical supervenience is regarded as defining minimal physicalism (Kim, 1998), synchronic emergentism can be seen as physicalism, too, and thus be treated as a genuine instance of nonreductive physicalism.

Qualia Emergentism

A case in point for the idea that nonreductive physicalism might be an adequate answer is the problem of phenomenal consciousness. Chalmers, Jackson, and Levine, among others, have argued in various ways that qualitative mental phenomena are not reducible to physical or functional states. If their arguments are sound, they imply strong emergentist positions. Most interesting and powerful seem to be Chalmers's argument for the 'hard problem' of consciousness and Levine's 'explanatory gap' argument. (*See* **Consciousness, Philosophical Issues about**; **Knowledge Argument, The**; **Explanatory Gap**)

According to Levine (1993) and Chalmers (1996), reductive explanations require two stages. The first stage involves the *a priori* process of working the concept of the property to be reduced 'into shape' for reduction by identifying the causal or functional role for which we are seeking the underlying mechanisms. The second stage involves the empirical work of discovering just what those underlying mechanisms are.

Since our concepts of phenomenal qualities do not seem to represent – at least in terms of their psychological contents – causal roles, a failure, in principle, of the first task seems to be unavoidable. Thus, to the extent that there is an element in our concept of qualitative character that is not captured by features of its causal role, qualia are irreducible emergent properties.

Emergence in Connectionism

Connectionism gives rise to emergentist considerations in several ways: trained networks show cognitive features such as 'rule following', 'schema formation', or 'pattern recognition', that their parts do not have. Thus, the systemic properties that a network acquires are weakly emergent. However, they are not irreducible: they are fully deducible from the network's structure, the properties of its units (their activation formulae), and the properties of their links (the distribution of weights, and the formulae for changing the weights). Thus, systemic properties of connectionist networks are not synchronically emergent.

Connectionists often make use of the word 'emergent' in its ordinary sense, sometimes intermingled with a more technical usage. For example, Rumelhart and McClelland (1986) say that a network's high-level properties 'emerge' from low-level interactions: rules and schemata come into being by themselves without being explicitly fed into the system. However, Rumelhart and McClelland do not thereby subscribe to emergentism. They mainly try to differentiate their position from traditional representationalism, accordingly to which all rules and schemata have to be fed in explicitly. (*See* **Representation, Philosophical Issues about**; **Implicit and Explicit Representation**)

Connectionist networks develop their specific distribution of link weights (their soft structures) in a somewhat evolution-like process. Again, this is not a case of genuine emergence. Since the changes of weights are calculable exactly if we know the initial magnitudes, we should not speak of structure emergence in connectionism. On the other hand, regarding their soft structures, networks show great plasticity, compared with other objects. Chemical compounds, for example, have no freedom to change their internal structure: the diamond's property of being hard is always manifest; it does not emerge.

Emergence of Creativity

The notion of emergence is also of interest in the field of creative cognition. There we seem to face a paradox: how is it ever possible to form a truly creative idea? If we could predict it, it would be determined and not creative. If we could not in principle anticipate it, how could we produce such an idea at all? Some psychologists assume that the cognitive structures involved in creative thinking have emergent properties that could be discovered when those structures are explored, at least some of which could not have been anticipated in advance. This seems like a postulation of unpredictable cognitive features as a result of structure emergence. (*See* **Creativity**)

Emergentism and Artificial Life

Within the field of artificial life, the notion of emergence is central. It refers to adaptive features of artificial systems that result both from 'clever' interactions of many simple components and from couplings between agents and their environments (including other agents). However, the term 'emergence' is not used in its strong sense here, since all phenomena studied in artificial life are reductively explainable, at least in principle. Rather, emergence is associated with behavior that is not centrally controlled and that cannot be reduced to the behavior of single components within hierarchical systems, but is seen as the result of the interactions of multiple simple components or as the outcome of the overall dynamics of the agent and its environment. Thus, the notion of emergence used in artificial life is close to that used in connectionism. (*See* **Artificial Life**)

ARGUMENTS FOR AND AGAINST EMERGENCE

Clearly weak emergent properties exist: indeed, one might ask why such properties should be called 'emergent' at all, and not just 'systemic'. Furthermore, since there are chaotic processes of structure formation, structure emergence exists too. Thus, what is really in question is synchronic emergence. What we need is an argument for the existence of properties that are not and will not be reductively explainable. Many natural scientists deny the existence of such properties, since they do not know of any properties that could not be reductively explained, at least in principle. Without exception, all systemic properties studied in the natural sciences are functionally construable, their functional roles are always filled by the interactions of their systems' parts, and the behavior of the parts of any system always seems to follow from their behavior in simpler systems. Therefore, some critics question whether it is useful to develop the notion of synchronic emergence at all. But, even if it should turn out that all systemic properties studied in the natural sciences are reductively explainable, it is useful to have the strong notion of synchronic emergence. More than any other notion, it can be used to clearly formulate nonreductive positions concerning the mind–body problem.

Whether or not synchronically emergent properties actually exist does not seem to depend on empirical, but rather on conceptual grounds. Among others, Broad, Levine, and Chalmers have argued forcefully that properties such as qualia are not functionally analyzable. If they are right, then phenomenal qualities may be emergent properties in the strong sense.

If mental properties such as qualia are emergent in the strong sense, then new problems arise. Some philosophers have claimed that irreducible properties necessarily exert downward causation. In the case of mental phenomena, however, this would conflict with the principle of the causal closure of the physical domain.

However, as we have seen above, properties that are irreducible for conceptual reasons do not imply downward causation. Rather, they give rise to another objection: how can properties that escape reconstruction via their causal role play any causal role at all? The reply to this objection mainly depends on our concept of causation. If we think that supervenient causation suffices for causation, then irreducible emergent properties can be causally efficacious. If we think that supervenient causation does not suffice, then irreducible emergent properties do not seem able to play any causal role. But these are still open questions (Kim, 1998; Stephan, 1997). (*See* **Mental Causation**)

References

Bunge M (1977) Emergence and the mind. *Neuroscience* **2**: 501–509.

Chalmers DJ (1996) *The Conscious Mind*. Oxford, UK: Oxford University Press.

Hempel CG and Oppenheim P (1948) Studies in the logic of explanation. *Philosophy of Science* **15**: 135–175.

Kim J (1993) *Supervenience and Mind*. Cambridge, UK: Cambridge University Press.

Kim J (1998) *Mind in a Physical World: An Essay on the Mind–Body Problem and Mental Causation*. Cambridge, MA: MIT Press.

Levine J (1993) On leaving out what it's like. In: Davies M and Humphreys GW (eds) *Consciousness*, pp. 121–136. Oxford, UK: Blackwell.

Lewes GH (1875) *Problems of Life and Mind*, vol. II. London, UK: Kegan Paul.

Mill JS (1974) *A System of Logic: Ratiocinative and Inductive*. Toronto, Canada: University of Toronto Press [First published 1843.]

Nagel E (1961) *The Structure of Science*. New York, NY: Routledge and Kegan Paul.

Rumelhart DE and McClelland JL (1986) PDP models and general issues in cognitive science. In: Rumelhart DE, McClelland JL and the PDP Research Group (eds) *Parallel Distributed Processing: Explorations in the Microstructure of Cognition*, vol. I, pp. 110–146. Cambridge, MA: MIT Press.

Stephan A (1997) Armchair arguments against emergentism. *Erkenntnis* **46**: 305–314.

Varela FJ, Thompson E and Rosch E (1991) *The Embodied Mind: Cognitive Science and Human Experience*. Cambridge, MA: MIT Press.

Further Reading

Beckermann A, Flohr H and Kim J (eds) (1992) *Emergence or Reduction? Essays on the Prospects of Nonreductive Physicalism*. Berlin and New York, NY: de Gruyter.

Bedau MA (1997) Weak emergence. *Philosophical Perspectives: Mind, Causation, and World* **11**: 375–399.

Clark A (1996) Happy couplings: emergence and explanatory interlock. In: Boden M (ed.) *The Philosophy of Artificial Life*, pp. 262–281. Oxford, UK: Oxford University Press.

Finke RA, Ward TB and Smith SM (1992) *Creative Cognition*. Cambridge, MA: MIT Press.
Holland JH (2000) *Emergence: From Chaos to Order*. Oxford, UK: Oxford University Press.
Humphreys P (1997) How properties emerge. *Philosophy of Science* **64**: 1–17.
Kim J (1999) Making sense of emergence. *Philosophical Studies* **95**: 3–36.
O'Connor T (1994) Emergent properties. *American Philosophical Quarterly* **31**: 92–104.
Stephan A (1999) *Emergenz. Von der Unvorhersagbarkeit zur Selbstorganisation*. Dresden and Munich, Germany: Dresden University Press. [In German.]
Wimsatt W (1997) Aggregativity: reductive heuristics for finding emergence. *Philosophy of Science* **64**: S372–S384.

Emotion

Introductory article

Paula M Niedenthal, National Centre for Scientific Research, Blaise Pascal University, Clermont-Ferrand, France

CONTENTS

Emotions are sets of processes involved in an organism's response to significant, goal-relevant life events. Such processes include expressive behavior, cognitive appraisals, physiological arousal, action tendencies and subjective feelings.

INTRODUCTION

It is difficult to imagine life without emotions. We would feel no joy at successfully accomplishing a task, no sadness at failing an examination, no anger when we witnessed displays of prejudice and discrimination. We would not feel ashamed upon insulting another individual in social interaction. Nevertheless, emotion as a field of study remained firmly ensconced in the realms of philosophy and literature for many centuries. Now of central interest to psychology, the cognitive sciences and the neurosciences, emotion is finally also a topic of empirical investigation. Neuropsychological and psychological investigation shows that it is indeed hard to imagine life without emotions, because emotions are essential for human functioning. Social interaction, decision-making and judgment would be very poor without the capacity to experience emotion. Normal and pathological emotion states are determined and controlled by almost all of the systems of the human body, and emotional states in turn influence the cognitive functions of attention, perception, categorization, memory and judgment.

EMOTIONS AS MULTICOMPONENT PROCESSES

A precise definition of emotion eludes scientists. This is because no single event or process constitutes an emotion. Emotions are sets of processes that involve different components including subjective feelings, but also expressive motor action, cognitive appraisals, physiological arousal, and tendencies to take particular actions. If you see a bear in the forest, a favorite example of the nineteenth-century psychologist William James, you might experience strong physiological arousal, have an urge to run, open your eyes and mouth wide, and feel something that you label as fear. In this example the components are quite coherent and conform to a common emotional experience. However, the different components of emotion can be decoupled. In different situations and across different cultures, social norms influence the expression and experience of emotion. For example, an adult in an industrialized Western country might feel like laughing at a funeral if a funny joke about a priest or a rabbi suddenly comes to mind; however, the person would probably suppress any laughter, avoid smiling, and display some degree of sadness. Such norms are one force that can decouple the multiple components of emotion. In this example the decoupled components are subjective experience and expressive behavior. (*See* **Cultural Psychology**)

Differentiation of the Component Processes: Are There Discrete Emotions?

Although scientists agree that emotions have many components, debate continues about the extent to which there are unique patterns or levels of each component process that correspond to discrete (basic or fundamental) emotions such as sadness, joy or fear.

Facial expression

The component of initial interest and debate was that of facial expression of emotion. Charles Darwin was an early proponent of the idea that facial expressions of emotions serve adaptive functions – he called them 'serviceable habits' – and have thus evolved as hardwired expressions of discrete subjective states. For example, disgust, Darwin suggested, is associated with a gesture that represents the expulsion of food from the mouth and avoidance of the intake of an odor through the nose, precisely because the function of disgust is to prevent individuals from ingesting dangerous substances. From this evolutionary perspective, facial expressions of emotion should be both distinctive and universal. Indeed, studies of people of different cultures, even those not exposed to Western media, blind children, and infants, suggest that there exist universal displays and recognition of the expressions of at least joy, sadness, fear, anger, disgust, and perhaps surprise. Thus, such facial gestures may be based in innate neural motor programs. However, such a conclusion is still widely debated because each existing demonstration of universality can be criticized on technical grounds. (*See* **Face Perception, Psychology of; Face Perception, Neural Basis of**)

Vocalizations

Somewhat less controversial is the study of vocal expression of emotion, also called emotional prosody. Research supports the idea that vocal expression is biologically based and that there is evolutionary continuity of vocal emotion expression. For example, data from studies conducted by behavioral biologists suggest that there are significant similarities in the vocal expression and communication of emotional states across species: angry states are typically expressed by loud, harsh vocalizations, while fear and anxiety are typically associated with high-pitched, shrill vocalizations. Furthermore, human perceivers show wide agreement about the emotions communicated by utterances characterized by specific patterns of physical parameters such as fundamental frequency and pitch.

Autonomic activity

William James, as well as a Danish contemporary named Lange, originally proposed the idea that there are distinctive patterns of autonomic arousal that correspond to discrete emotions. Although this peripheralist position was attacked for over a century, both by scientists who believed that arousal was nonspecific, and by those who believed that the autonomic nervous system responds too slowly to subserve discrete emotional states, research suggests that some physiological parameters do differentiate some pairs of emotion. Sadness and fear differ in their patterns of heart rate, for example. Other studies show that fear is characterized by increases in heart rate, contractility of the heart musculature and respiration rate, which together mobilize a flight response, while anger is characterized by a rise in diastolic blood pressure and in peripheral resistance, which together mobilize a fight response. (*See* **Autonomic Nervous System**)

Subjective experience

Despite intuition to the contrary, some emotion theorists do not believe that emotional states are differentiated as discrete categories (happiness, sadness, disgust) of subjective experience. Rather, they believe that the conscious experience of emotion can be accurately characterized by two underlying psychological dimensions. Early in the last century these dimensions were referred to as pleasantness–unpleasantness and excitement–depression, although now the dimensions are often termed 'valence' (positive–negative) and 'activation' (high–low). In the dimensional view, for instance, the emotion that is called fear can be described psychologically as a state that is negative in valence and high in activation. One answer to the question of whether subjective experience is differentiated in terms of a few dimensions or whether it is categorical, is to say that it depends on the person. Research suggests that some individuals do indeed experience their emotional states in terms of a few simple dimensions ('I feel good') while others make finer, categorical distinctions among states ('I feel joyful'; 'I feel proud').

THEORIES OF EMOTION

Theories of emotion are precise statements about the causes of an emotion, the order in which emotional processes unfold, and how the different components of emotion interact. Although within each category several different versions can be discerned, there are currently two major categories of emotion theory. These are evolutionary theories and cognitive appraisal theories. Depending upon which approach is adopted by a scientist, the methods for producing and measuring an emotion in the laboratory, the types of emotions under scrutiny, the way in emotions are thought to be elicited and regulated and the application of the research findings may be quite different.

Evolutionary Theories

The evolutionary approach to emotion is based in part on the thinking of Charles Darwin and begins with the assumption that emotions are biologically based and functional. In considering the evolution of emotion, Darwin focused largely on his idea that facial expressions of emotion are remnants of serviceable habits. In addition, he argued for an adaptive signaling function of facial expression: that is, facial expressions allow members of the same species to know the subjective experience of the expresser, and therefore the emotional significance of the situation, as well as the expresser's likely actions. In a more general way, the evolutionary perspective assumes that emotions motivate adaptive action, such as the tendency to flee when fearful and the tendency to fight when angry. The problems of adaptation thought to be associated with specific emotions are finding and consuming food and drink, locating shelter, seeking support from other members of the same species, being social, satisfying curiosity, appraising sexual partners, nurturing offspring, and escaping dangerous situations. From a biological perspective, then, emotions are responses that have evolved to motivate individuals to successfully pass on their genes to offspring. The three types of evidence cited in support of an evolutionary approach to emotion are: (1) blind children produce recognizable facial expressions of the basic emotions; (2) human facial expressions such as smiling have homologs in chimpanzee expressions; and (3) there are cross-cultural similarities in the antecedents of emotion. (*See* **Aggression and Defense, Neurohormonal Mechanisms of; Cultural Differences in Causal Attribution; Evolutionary Psychology: Theoretical Foundations**)

The implications of this approach are that there are biologically based systems that subserve emotions, in a discrete way. There also exist biologically relevant 'signal stimuli' that recruit specific emotional reactions and adaptive response tendencies in situations that are significant to the person for phylogenetic or ontogenetic reasons. At the behavioral level, emotional phenomena are displayed as action or motor tendencies. They thus recruit metabolic processes related to arousal and behavioral energetics. Such processes can be subjected to scientific investigation through psychophysiological indices such as cardiovascular and electrodermal responses. (*See* **Emotion, Neural Basis of**)

Cognitive Appraisal Theories

If the evolutionary approach links emotions to biological adaptation in the distant past, appraisal theories of emotion link emotions to 'higher level', cognitive processes of evaluation of meaning, causal attribution, and assessment of coping capabilities. The main principle of appraisal theories is that emotions are elicited and differentiated by individuals' evaluation of the significance or meaning of an object or event for themselves and their current goals on a number of dimensions or criteria. A classic demonstration of the role of appraisal in differentiating (although not eliciting) emotion was a study conducted by psychologists Schachter and Singer in the 1960s. These psychologists proposed that individuals sometimes experience physiological arousal that does not have a known cause. The arousal motivates individuals to explain the cause and nature of the arousal, which as consequence gives rise to a discrete emotional state. In their study the researchers injected participants with adrenaline (epinephrine), which enhances arousal, or with saline, a placebo which causes no physiological change. They then informed some of the participants injected with adrenaline to expect an increase in arousal; other participants were left uninformed about these effects. Later, when the adrenaline had taken effect for those who had received it, all participants were placed in a situation in which they interacted with a confederate of the researcher who acted in either a euphoric or an angry manner. Subsequent assessment of participants' emotions showed that those who had received adrenaline and were uninformed about the accompanying arousal – but not those participants who received placebo or who had an expectation of arousal – reported feeling the same emotions as those expressed by the confederate. That is, participants who experienced unexplained arousal sought an explanation in their environment and this appraisal determined their precise emotional state.

It is no longer widely believed that emotions are the products of unexplained arousal shaped by reference to events in the surrounding environment. However, modern cognitive appraisal theories do hold that discrete emotions result from processes of evaluation of significant events and of attributions of the causes of those events. In particular, whether unconsciously or consciously, individuals assess the degree to which events facilitate or impede their goals, whether they are pleasant or unpleasant, whether they are controllable or not, and whether they are novel or familiar. Depending upon the result of this appraisal, a discrete emotion is evoked. Tables 1 and 2 summarize examples of appraisal theories and the profiles that in theory give rise to certain emotions.

The implications of appraisal theories are that emotions do not unfold in a hardwired way in response to certain situations or objects, but that the emotional significance of the events and objects depends upon the goals and the coping capacities of each individual. Thus, appraisal theory can comfortably predict that one person will respond to the same stimulus with fear while a second will respond with anger. Emotions are differentiated and can be associated with different physiological processes and facial expressions in this view, but the antecedent of the emotion – the specific profile of appraisal – determines which discrete emotion is experienced.

COGNITIVE REPRESENTATION OF EMOTION

Individuals do not only experience emotions, they also verbally label emotions, represent their ideas of what emotions are as concepts, and can reflect on and remember their emotional feelings.

Emotion Labels

Subjective experience of emotion is largely revealed to scientists through emotion words. In the English language there are between 500 and 2000 words that refer more or less to emotions (e.g. shame), affective or valenced states (e.g. tranquility), and cognitive–affective states (e.g. vengeance). In Malay there are about 230 emotion words, and in Ifaluk there are only about 50. Clearly, then, although there is enormous cultural variation in the number of words individuals use to talk about emotions, they talk about emotions in all cultures. The basic categories of words, despite important differences, show general consistency across culture. (*See* **Cultural Psychology**)

The dimensional analysis of the meaning of emotion words usually reveals the same bipolar

Table 1. Comparison of the appraisal criteria postulated by different theorists

Scherer	*Frijda*	*Roseman*	*Smith/Ellsworth*
Novelty	Change		Attentional activity
• Suddenness			
• Familiarity	Familiarity		
• Predictability			
Intrinsic pleasantness	Valence		Pleasantness
Goal significance			
• Concern relevance	Focality	Appetitive/aversive motives	Importance
• Outcome probability	Certainty	Certainty	Certainty
• Expectation	Presence		
• Conduciveness	Open/closed	Motive consistency	Perceived obstacle/ Anticipated effort
• Urgency	Urgency		
Coping potential			
• Cause: agent	Intent/self–other	Agency	Human agency
• Cause: motive			
• Control	Modifiability	Control potential	Situational control
• Power	Controllability		
• Adjustment			
Compatibility standards			
• External	Value relevance		Legitimacy
• Internal			

Table 2. Examples of theoretically postulated appraisal profiles for different emotions

Stimulus evaluation checks	*Anger/rage*	*Fear/panic*	*Sadness*
Novelty			
• Suddenness	High	High	Low
• Familiarity	Low	Open	Low
• Predictability	Low	Low	Open
Intrinsic pleasantness	Open	Open	Open
Goal significance			
• Concern relevance	Order	Body	Open
• Outcome probability	Very high	High	Very high
• Expectation	Dissonant	Dissonant	Open
• Conduciveness	Obstruct	Obstruct	Obstruct
• Urgency	High	Very high	Low
Coping potential			
• Cause: agent	Other	Other/nature	Open
• Cause: motive	Intent	Open	Chance/neg
• Control	High	Open	Very low
• Power	High	Very low	Very low
• Adjustment	High	Low	Medium
Compatibility with standards			
• External	Low	Open	Open
• Internal	Low	Open	Open

Open-different appraisal results are compatible with the respective emotion.

structure as does an analysis of individuals' ratings of their subjective emotional states. That is, the meaning of the labels are organized around the two dimensions of valence and activation or arousal. More specifically, a circular representation of emotion words, a circumplex structure, is typically observed. Figure 1 illustrates this structure, showing the position of the basic or discrete emotions as well as other affective states. The circumplex shows that individuals make fine-grained distinctions among emotion words in their reference to pleasantness and level of arousal.

Emotion Prototypes

The categories that emotion labels refer to share many properties in common with categories of concrete objects such as animals, vehicles and furniture, as well as more abstract concepts such as personality disorders, negotiation strategies and scientific theories. First, the defining characteristics of the categories vary in their diagnosticity, or probability of being present in any given episode of an emotion. For instance, individuals know that the experience of anger might almost always include the feeling of the face growing hot or the muscles being tense, but only sometimes the action of striking out. Second, membership in the categories is a matter of degree, and is not 'all or none'. Thus, a good example of feeling fear is a situation in which you see a bear and run away, heart pounding and mouth open to scream. But being on a carnival ride or jumping off a high diving board, with all the physiological and subjective experiences these entail, may be less good examples of fear. The best examples of each emotion are the central or prototypic examples of the categories, while the others are considered borderline or less prototypic examples. Finally, each category of emotion contains a script for the ordering and patterning of events involved in that emotion. Individuals are able to describe the antecedent events, the behavioral reactions to those events and the physiological and expressive responses involved in a way that preserves the temporal and spatial structure of a prototypic emotion episode. (*See* **Conceptual Representations in Psychology**; **Schemas in Psychology**)

Representation of Emotional States

Emotional feelings and experiences themselves are also represented in memory. Individuals can remember what emotions feel like and even reproduce aspects of the feelings if they retrieve memories of past episodes. It has been proposed that emotional states, perhaps the discrete emotions, are represented as informational units in an associative network in memory. In such associative network models the informational units are

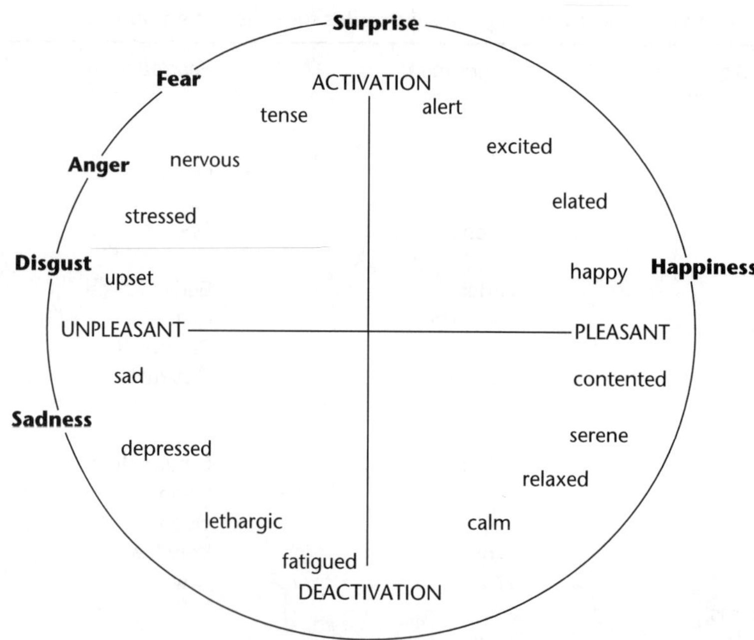

Figure 1. The circumplex structure of affective meaning that results from the multidimensional scaling of judgments of similarity between emotion words. The outer circle shows where several prototypical emotions fall. From Feldman Barrett L and Russell JA (1998) Independence and bipolarity in the structure of affect. *Journal of Personality and Social Psychology* **74**: 967–984.

proposed to be linked to memories of the experience of the emotion and to memories for times when these emotions were induced. The existence of mental associations between a representation of the feeling of emotion and memories and perhaps the semantic categories of knowledge about the emotion has important consequences for understanding the processing of information during emotional states. (*See* **Memory Models**; **Spreading-activation Networks**)

EMOTION–COGNITION INTERACTION

Emotional states, and indeed emotional traits such as depression and anxiety, influence the manner in which individuals process information. Such states and traits can affect the allocation of attention, the efficiency of perceptual encoding, the retrieval of memories from long-term memory, the learning of new information, decision-making and judgment, as well as the organization of conceptual material in memory. (*See* **Affective Disorders: Depression and Mania; Depression**)

Emotion and the Content of Cognition

The associative network model makes a specific prediction concerning the influence of emotion in information processing, which is reminiscent of expressions such as, 'She was so happy that she saw the world through rose-colored glasses.' According to the model, emotional states activate the unit of information representing that state in long-term memory. Through passive diffusion of activation, other information that has been associated with or causal to that same emotional state becomes activated too, and then influences the ease and efficiency with which new information is perceived and stored. The consequence of such activation is the facilitated processing of emotion-congruent information. For example, in a happy state an individual will allocate attention to and rapidly encode smiling faces, people he or she likes, and words or phrases with positive meanings. One study also showed that individuals in happy states who were watching a happy expression gradually disappear from a face of another person, perceived the expression to still convey happiness for a significantly longer time – even when almost completely neutral – than did individuals in a sad state. This means that the happy individuals were particularly efficient at detecting signs of a happy expression that was ambiguously neutral. Other research has shown that, while in a happy state, individuals are also better able to learn and retrieve information associated with happiness. Finally, the judgments of individuals in happy states tend to be optimistic rather than pessimistic in nature. In a sense, emotions mobilize the entire cognitive system to process information that is relevant to that state. (*See* **Priming**)

Emotion and Information-processing Strategies

Emotions also influence the actual cognitive processes that are employed in judgment and decision-making: that is, the structure of those processes. When individuals are in positive emotional states they process information less deeply, in a simple, heuristic way. This does not mean that they cannot make careful judgments or that they think poorly, only that they tend to use short cuts for solving problems and may sacrifice attention to detail. The same appears also to be true for individuals who are in states of anger. On the other hand, sadness is associated with a more careful, less schematic type of processing. Individuals who are sad are likely to scrutinize information and to engage in more systematic processing of that information. Research has shown, for example, that individuals in happy and angry states are more likely to base their judgments of the defendant in a hypothetical legal trial on racial stereotypes, than are individuals in sad or neutral states, who based their judgments more on the presented legal evidence.

There are a number of reasons why emotions are associated with more or less systematic styles of information processing. First, different emotions have different information values to the individuals experiencing them: happiness tends to indicate that the environment is safe and that all is going well; sadness indicates that there is a problem to be solved and that all is not going well; and anger suggests that fast action must be taken to solve a problem. Thus, happiness and anger have different signal values that nevertheless may result in the same tendency to process rapidly and in less detail. Sadness signals that careful attention to details of the environment is required. Second, happy states exert a high degree of cognitive load. Because happy states are associated with more information in memory – probably because individuals are on average happy more often than they are sad, angry or fearful – many different ideas come to mind during those states. The rush of associations can inhibit the ability to process additional incoming information in detail. In contrast, sad states, which are associated with less information stored in memory, at least among nondepressed individuals, tend not to trigger as many associations and tend therefore not to overload the system. Consequently individuals in sad states seem to have more capacity to process information in a systematic way. Third, different emotions have different motivational properties. Individuals who are feeling happy or angry may not be motivated to distract themselves from or otherwise change those states by processing new information in a systematic way. They protect their emotional states from change, albeit for different reasons. Individuals in nonclinical sad states, on the other hand, are often motivated to distract themselves from the sadness or engage in other mood regulation strategies. They are thus more likely to systematically process incoming information.

Emotion and Conceptual Organization

Finally, emotional states influence the way information is organized in memory, and therefore the content and structure of concepts. Natural categories in the environment are revealed by perceptual similarity among their members: it is easy to learn the category of birds because most examples of that category (most birds) have physical properties in common. Concepts, the representation of categories in memory, are therefore grounded in perceptual similarity. They are also organized theories that individuals possess about the origins and functioning of objects and events. However, during emotional states concepts may be organized somewhat differently. Rather than representing categories of things that belong together because they look alike or conform to a common theory of cause and effect, concepts may temporarily represent groups of objects and events that have elicited the same emotional state. For example, individuals are more likely during emotional states, compared with neutral states, to see emotional equivalence among events and objects, and group them together as being 'the same kind of thing', if they evoked the same emotion. (*See* **Conceptual Representations in Psychology**; **Concept Learning and Categorization: Models**)

REGULATION AND SUPPRESSION OF EMOTION

Emotion regulation includes processes that individuals use to influence whether they have an emotion, which emotions they experience and do not experience, the conditions under which they experience emotion, and how they express such states. Although Western cultures seem ambivalent about the need to regulate emotions, as conveyed in the conflicting expressions 'He who keeps a cool head prevails' versus 'Let your feelings be your guide', it is clear that individuals in all cultures attempt to control, alter, augment and suppress at least some emotions. Anthropologists and psychologists have observed specific social and cultural rules for doing so.

Emotion regulatory strategies are important to understand because they are fundamental to personality functioning and adjustment, and to psychological and physical health. Major depressive disorder, for example, is characterized by a deficit in experience of positive emotion and a surplus of processing negative emotions and negative information. Chronic hostility and anger inhibition are associated with hypertension and coronary heart disease. Emotion inhibition or suppression may enhance minor illnesses and accelerate the progression of major illness such as cancer.

Processes of Emotion Regulation

Given the multicomponent nature of emotions, it is not surprising that there are many different processes involved in the regulation of emotional states, and that these processes operate more or less on physiological, expressive, and experiential aspects of the emotion.

Antecedent-focused regulation involves regulation strategies that are mobilized in the service of avoiding or enhancing emotions in an anticipatory way. Potentially emotion-arousing situations or stimuli can be avoided or approached, aspects of the situations can be selectively attended to or ignored, and, as suggested by appraisal theories of emotions, situations can be (re)evaluated in ways that augment or diminish the probability of specific emotions.

Response-focused regulation involves attempts to alter emotional states once they have been elicited. Representative strategies include the suppression or enhancement of expressive and behavioral components of the emotion, and distraction of conscious attention away from the subjective experience of the emotional state to other events internal or external to the individual, such as distracting thoughts and stimuli, respectively. An example of the regulation of an expressive component of emotion as a means to regulate emotional state is the voluntary manipulation of facial expression of emotion. This is the deployment of the adage for people who are depressed to 'Put on a happy face', for example, in order to feel happier. Putting on a happy face may indeed regulate emotional experience through a number of different mechanisms. First, it may alter the social environment such that more positive experiences are possible. In addition, self-perceptual processes – noticing that one is smiling – may lead individuals to believe that indeed they are happy and to realize this belief. Finally, several facial feedback theories of emotion also hold that facial expression of emotions actually feed back through physiological systems to influence subjective state. Specifically, the strategic use of facial musculature may alter other neurochemical processes involved in emotion.

Effects of Different Regulation Strategies

Emotion regulation strategies are differentially effective and have different costs and benefits over time. Of particular interest has been comparison of reappraisal of negative situations compared with the suppression of negative expressive behavior and feelings. Research has generally demonstrated that reappraisal can diminish the subjective experience and expressive aspects of an emotion. Of course, reappraisal of some negative situations is quite difficult in the first place. Suppression of emotional expression and behavior may often (or even always) be possible. However, suppression is associated with increases in sympathetic nervous system activation, and thus can be a physiologically costly strategy. It is probably for this reason that individuals who habitually suppress negative emotions tend also to suffer both minor and more permanent health consequences.

Further Reading

Eich E, Kihlstrom JF, Bower GH, Forgas JP and Niedenthal PM (2000) *Cognition and Emotion*. Oxford: Oxford University Press.

Ekman P and Davidson RJ (eds) (1994) *The Nature of Emotion: Fundamental Questions*. Oxford: Oxford University Press.

Ekman P and Friesen WV (1971) Constants across cultures in the face and emotion. *Journal of Personality and Social Psychology* **17**: 124–129.

Feldman Barrett L and Russell JA (1998) Independence and bipolarity in the structure of affect. *Journal of Personality and Social Psychology* **74**: 967–984.

Forgas JP (2000) *Feeling and Thinking: The Role of Affect in Social Cognition*. Cambridge, UK: Cambridge University Press.

Frijda NH (1986) *The Emotions*. Cambridge, UK: Cambridge University Press.

Gross JJ (1998) The emerging field of emotion regulation: an integrative review. *Review of General Psychology* **2**: 1–29.

Izard CE (1991) *The Psychology of Emotions*. New York: Plenum Press.

Lane RD and Nadel L (eds) (2000) *The Cognitive Neuroscience of Emotion*. Oxford: Oxford University Press.

Lewis M and Haviland JM (1993) *Handbook of Emotion*. New York: Guilford Press.

Niedenthal PM, Halberstadt JB and Innes-Ker AK (1999) Emotional response categorization. *Psychological Review* **106**: 337–361.

Öhman A (1987) The psychophysiology of emotion: an evolutionary-cognitive perspective. *Advances in Psychophysiology* **2**: 79–127.

Ortony A, Clore GL and Foss MA (1987) The referential structure of the affective lexicon. *Cognitive Science* **11**: 361–384.

Russell JA (1991) Culture and the categorization of emotions. *Psychological Bulletin* **110**: 426–450.

Scherer KR (1999) Appraisal theory. In: Dalgleish T and Power M (eds) *Handbook of Cognition and Emotion*. New York: John Wiley.

Emotion, Neural Basis of

Introductory article

Jeannine V Morrone-Strupinsky, University of Arizona, Arizona, USA
Richard D Lane, University of Arizona, Arizona, USA

CONTENTS

Emotion is information about the extent to which goals are being met in interaction with the environment.

INTRODUCTION

Emotion is information about the extent to which goals are being met in interaction with the environment. Emotion is a mechanism that emerged in evolution to solve the problem of attributing value to classes of stimuli based on experience, and serves as a means of adapting rapidly to critical stimuli in the environment. It can be viewed as an extension of more rudimentary or reflex-based behavioral systems. For example, mating behavior that is reflexive in reptiles is elaborated and extended in mammals and contributes to the formation of social bonds.

Emotion provides a mechanism for rapidly resetting the organism in response to environmental contingencies. This resetting occurs cognitively (e.g. in attention and memory systems), physiologically (e.g. preparing for exertion, diminishing functions not needed during a crisis, such as digestion), and behaviorally (e.g. shifting the propensity for approach or avoidance behavior). Emotional information can be conveyed internally (experiential) or externally (expressive, for instance gestures or facial expressions), the latter reflecting the inherent social element in emotion, that of signaling to or communicating with other animals.

Thus, emotion involves evaluating the motivational significance of a stimulus and subsequently implementing motivated behavior. The function of emotion is manifold. In response to some change in the environment, emotions (1) interrupt behavior and focus attention on particular elements in one's surroundings, or on one's internal sensations or appraisals; (2) physiologically prepare the organism for alternative action; (3) serve as a means of communication among conspecifics; and (4) help label or 'tag' memories for significance, which can guide approach/avoidance tendencies in related situations in the future, as well as trigger memories of previous experiences that may affect progression of the current behavioral agenda.

WHAT INFLUENCES THE DEVELOPMENT OF EMOTIONS?

There has been spirited discussion about whether emotions are rooted in nature (i.e. genetically hard-wired neural circuits) or nurture (learning via experience). Indeed, it has been demonstrated that we possess certain rudimentary emotional functions at birth. However, these basic functions are then subjected to learning and environmental experiences of many kinds. For instance, Rene Spitz's observations of infants raised in hospitals illustrate that social development is disrupted if infants are deprived of appropriate emotional responses from the environment. More than a third of the infants died, and of those who survived, most experienced developmental delays. The effect of the environment is not well understood, but it is likely to be an important source of the heterogeneity observed between people in their emotional behavior. For instance, research shows that offspring of depressed caregivers are at increased risk of maladaptive development and emotional difficulties.

THE STRUCTURE OF EMOTION

Precisely how emotion is organized in the brain remains an unresolved issue. There is debate over whether different emotions should be considered as discrete entities, or viewed from a dimensional perspective. Proponents of the discrete or basic emotions model contend that there are unique circuits for particular emotions, such as fear, anger, joy, and disgust. Darwin suggested that distinct facial expressions reflected distinct neural circuits for each emotion. This is supported by evidence of the generation and recognition of prototypical facial expressions of emotion in all known cultures. Furthermore, specific neural circuits have been identified in animals by Panksepp for basic emotional states such as rage, fear, joy, and sorrow.

The dimensional approach postulates that emotion is founded on separable motivational systems, such as approach and avoidance (appetitive versus aversive), and can be subdivided into components of valence (unpleasantness–pleasantness) and arousal (unaroused–aroused). Dimensional models of valence are further bifurcated into bivariate and bipolar models; in the former positive and negative emotions are conceptualized as separate dimensions, in the latter they are viewed as opposite ends of the same continuum. Dimensional models can incorporate discrete emotions models, but discrete emotions models cannot encompass dimensional models. For example, discrete emotions can be mapped onto the plane defined by the arousal dimension on the vertical axis and the valence dimension on the horizontal axis.

Substantial evidence from psychometric and psychophysiologic studies in healthy people supports a dimensional perspective. For instance, Watson, Tellegen and colleagues have found that self-report of mood is organized along positive and negative affective dimensions. In addition, Lang and colleagues have demonstrated the validity of the dimensional perspective using emotional stimuli in different modalities (pictures and sounds) to evaluate both subjective emotional and psychophysiological response. Patterns of psychophysiological responses are consistent across positive or negative valence of stimuli, but there are no unique physiological signatures for discrete emotions. Furthermore, a cognitive neuroscientific perspective suggests that the brain mediates various cognitive functions by combining component processes from different brain regions (such as the visual system) rather than dedicating specific neurons or specific regions to one particular function, such as recognition of a particular face. Indeed, attempts to define unique circuitry for discrete emotions in human imaging studies have failed to yield nonoverlapping activation patterns. For instance, Lane and colleagues found that the discrete emotions of happiness, sadness, and disgust were each associated with increases in activity in the thalamus and medial prefrontal cortex. These three emotions were also associated with activation of anterior and posterior temporal structures, particularly when induced by film. Recalled sadness was associated with increased activation in the anterior insula. Happiness was distinguished from sadness by greater activity in the vicinity of ventral mesial frontal cortex. From these findings Lane and colleagues concluded that there are both common and unique components of the neural networks mediating discrete emotions.

A function as complex as emotion is likely to be mediated by the coordinated effort of a variety of subsystems. Much is known about which structures contribute to the cognitive elaboration of emotion. This process requires the conscious awareness of emotions in order to understand the origin and meaning of one's emotions, which can subsequently be incorporated into thought and behavior. However, it is not yet known which brain structures should definitely be included or excluded in this network, and our understanding of how the system works as a whole to mediate the variety of emotions that are observed behaviorally is limited.

HISTORICAL THEORIES OF THE NEURAL BASES OF EMOTION

Study of the neural bases of emotion was neglected throughout most of the nineteenth and twentieth centuries because the experience of emotion was considered too subjective and vague to study scientifically. Furthermore, behaviorist and cognitive approaches predominated in psychology during the second half of the twentieth century.

Visceral feedback theories of emotion are useful from a historical perspective and remain influential in contemporary work on emotion. Their main contribution is the realization that the nature of emotion is a combination of brain and bodily states. Although the brain probably has the more prominent role in the generation of emotions, the experiential aspect of emotions is influenced to a large degree by bodily state and the transmission of this information to the brain.

There are three main visceral feedback theories of emotion. The James–Lange theory of emotion states that the experience of emotion follows bodily feedback: that is, perceiving an emotional stimulus generates visceral sensations or autonomic arousal, which then mediates the experience of particular emotions. Put another way, emotion is simply the perception of visceral sensations, and each discrete emotion has a unique pattern of physiological arousal. The Cannon–Bard theory focuses on the central nervous system (which includes the brain and the spinal cord), particularly on the thalamus, a structure responsible for gating sensory input. Perceiving an emotional stimulus simultaneously generates an emotional state and induces bodily arousal via the sympathetic nervous system. The pattern of physiological arousal does not correspond with specific emotions. According to the Schachter–Singer theory, the experience of emotion is a combination of generalized physiological arousal and cognitive appraisal. Arousal occurs first in the chain of events, and one's appraisal of this inner state 'labels' the specific emotion being experienced. Thus, the Schachter–Singer theory agrees with the James–Lange theory that physiological arousal is necessary for emotion, but it differs from James–Lange theory in that it argues that each discrete emotion does not have a unique pattern of physiological arousal.

The three models therefore address 'bottom up' and 'top down' mechanisms that determine what the emotion is and how it is experienced. It is difficult to identify clearly a unique autonomic signature for basic emotions. To the extent that autonomic differences are observed between emotions, they are subtle and difficult to detect. (*See* **Autonomic Nervous System**)

Several neuroanatomical theories were proposed in the first half of the twentieth century as the basis for emotional function. They are not generally accepted today in their original forms, but elements of these theories hold some validity and served as the basis for later elaborations. The Papez circuit was the first specific neuroanatomical theory of the subjective experience of emotion. It identified a region of the brain for which the function was not known and attributed to it a function (emotion) that did not have a neural basis at the time. The Papez neural circuit of emotion detailed the flow of information from the sensory nuclei of the thalamus to the sensory cortex and on to the cingulate cortex, and from the thalamus to the mamillary bodies of the hypothalamus. It did not include the amygdala and orbitofrontal cortex, regions that are well supported today as integral to emotional responses, but did include the hippocampus and cingulate cortex; the latter probably participates in emotional experience. MacLean's theory of the triune brain proposed that the generation of emotions occurs in an evolutionarily old region of the brain of mammals ('paleomammalian'), which he deemed the limbic system or the 'visceral' brain. This region is located between the 'reptilian' brain, which controls basic and reflex motor functions, and the 'neomammalian' brain, which mediates higher cognitions. Like visceral feedback theories, the triune brain theory suggested that the subjective experience of emotion stems from the confluence of sensory input from the external world and internal visceral sensations in the limbic system. The concept of the limbic system is a subject of intense debate, as it has not been proved whether all of the structures included in the concept behave as a system in relation to emotional function, and the number of structures to be included is still undecided. The term persists, however, because a better alternative has not yet been found.

EMPIRICAL STUDIES OF EMOTION

The neural substrates of emotion have received considerable attention from researchers such as Damasio, LeDoux, Panksepp, and Rolls. Their work has largely been based on experimental findings in animals and observations in people with brain lesions. More recently functional brain imaging techniques such as positron emission tomography and functional magnetic resonance imaging have been used to study the neural substrates of emotion in healthy volunteers and, to a lesser extent, in clinical populations. Figure 1 illustrates the main brain structures that contribute to emotional function.

Studies of Emotion in Animals

Understanding of the neural bases of emotion has been informed substantially by experimental studies in animals. Lesion studies by Cannon and Bard led to the conclusion that the hypothalamus is essential to organized emotional responses. Since the hypothalamus regulates the autonomic nervous system, the Cannon–Bard theory proposed that bodily responses during emotional states are controlled by higher centers via the function of the hypothalamus. Concurrently, the subjective experience of emotion relies on input from the hypothalamus to the cortex, and thus emotional responses in decorticate animals were labeled as 'sham rage'.

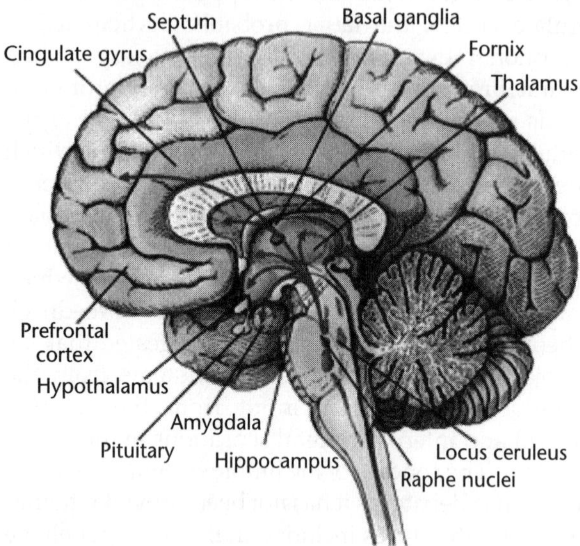

Figure 1. Brain structures that mediate emotion. In the brainstem, the locus ceruleus and raphe nucleus are the source of noradrenergic and serotonergic efferents to widespread regions of the cortex. These diffuse transmitter systems modulate neural activity related to emotion and other functions. The thalamus participates in evaluating the emotional significance of stimuli and organizing emotional behavior. The hypothalamus controls visceral functions. The amygdala and hippocampus mediate emotional conditioning and memory for emotional stimuli and their contextual surrounds, respectively. The basal ganglia facilitate the transition from motivation to action. A paralimbic structure, the cingulate gyrus, regulates attention and may contribute to the conscious experience of emotion. The prefrontal cortex coordinates functions throughout the emotion network, integrates information from the external world, and plays a major part in emotion regulation.

In the 1930s Klüver and Bucy demonstrated that extensive lesions of the anterior temporal lobe of rhesus monkeys produced profound behavioral changes, including psychic blindness (the tendency to approach animate or inanimate objects without hesitation or fear), excessive oral tendencies, hypermetamorphosis (the tendency to react and attend to all visual stimuli), decreased fear and aggression, and hypersexuality. Weiskrantz demonstrated in 1956 that the behavioral abnormalities of the Klüver–Bucy syndrome could be produced by lesions of the amygdala alone.

More recent animal work has delineated neural circuits for specific emotions. From this work, a consensus has developed that the amygdala is essential to emotion. In a programmatic line of research, LeDoux has mapped out the circuitry underlying fear. His work on auditory fear conditioning has demonstrated that there are separate subcortical and cortical pathways that contribute to the fear response. The subcortical pathway, or 'low road', enables quick responses to potentially dangerous stimuli, by the transmission of information from the sensory thalamus directly to the amygdala. The cortical pathway, or 'high road', responds more slowly, since information is passed from the sensory thalamus to the cortex and then to the amygdala. However, it more accurately represents the nature of the stimulus. Rolls has argued that LeDoux's 'low road' probably does not apply to most complex stimuli, but instead applies only to simple stimuli, such as tones – or perhaps evolutionarily prepared stimuli, such as sight of a snake. More complex stimuli require cortical processing before information about them is transferred to the orbitofrontal cortex and amygdala for the assignment of reinforcement value. The orbitofrontal cortex modulates amygdala activity and can suppress or override stimulus–reinforcement associations established in the amygdala. (*See* **Amygdala**)

Lesions of the amygdala prevent fear conditioning, which suggests that the amygdala is a primary structure involved in emotional function. Current concepts of the amygdala view it as a threat detector. Whalen theorizes that the amygdala monitors the environment for stimuli that signal an increased probability of threat. Although they indicate the presence of threat in the environment, fearful cues may be ambiguous in terms of the source of the threat. As a result, ambiguous stimuli that convey fear encourage processing of the contexts in which they are present, because they derive their meaning from their immediate surroundings. Lesions to the amygdala interfere with the ability to learn about both specific cues and contextual stimuli. In this vein, Whalen suggests that the amygdala facilitates activity in brain regions that encode contextual stimuli, such as the hippocampus. Indeed, hippocampal damage prevents the conditioning or association between contextual cues present with discrete stimuli and emotional states.

The ventral striatum, particularly the nucleus accumbens, has a primary role in motivational circuitry and reward-related behavior. The nucleus accumbens integrates inputs from the prefrontal cortex, amygdala, and the ventral tegmental area (which provides dopamine input). Dopaminergic input to the nucleus accumbens may play an important part in the reinforcing effects of substances of abuse.

Studies of Emotion in Humans

Studies in humans, particularly of people with brain damage, point towards a basis of emotional function similar to that in nonhuman animals. The most famous patient in the study of emotion, Phineas Gage, sustained damage to the frontal lobe, including the ventromedial prefrontal cortex, when a tamping iron pierced his skull in 1848. He changed from being a moral, upstanding man and a hard worker to spouting profanity, showing little regard for social conventions, and behaving irresponsibly. In general, damage to the lower middle portion of the frontal lobe (as in Phineas Gage) is associated with impulsive, disinhibited behavior, consistent with its modulatory influence on subcortical structures, including the amygdala. In contrast, damage to the upper portion of the frontal lobe is associated with amotivational behavior (i.e. failure to initiate behavior) consistent with the likely role of the dorsomedial frontal cortex in planning and executing goal-directed behavior.

Damasio has theorized that feedback from the body contributes to emotional experience. People with ventromedial frontal lobe damage lack physiological signs of emotional response, such as the skin conductance response (a measure of arousal), but they do have a conscious body state representation. As demonstrated by Bechara and colleagues, these patients also exhibit poor decision-making in gambling tasks in which risk-taking and reward assessment is involved. In particular, people without brain damage show electrodermal responses to risky choices when participating in the gambling task, prior to conscious awareness of how the game works, whereas people with ventromedial frontal lobe damage do not show such physiological responses and never learn the strategy of the gambling task. There is evidence that other brain regions, including the amygdala, and somatosensory cortices (S1, S2, and insula) are part of this neural system underlying the generation of somatic feedback. The insula has a particularly important role in detecting visceral sensations. Consequently, Damasio suggests that somatic feedback or 'markers' provide important information about the current state of the body to the brain, and influence subsequent decision-making.

Patients with bilateral amygdala damage but intact hippocampi show deficits in recognizing certain facial expressions of emotion, particularly fear and anger, although facial identification is unaffected. Patients with hippocampal damage but intact amygdalas display unconscious emotional conditioning, but have no memory for the conditioning events.

Studies of people with epilepsy have provided insight into the neural basis of emotion particularly negative emotions, which possibly are related to the adaptive significance and greater elaboration of varieties of negative emotions. Psychomotor epilepsy is often rooted in abnormal electrical activity in the temporal lobe. Prior to seizure activity patients report an 'aura', which is characterized by negative emotions such as fear, anger, and dejection, as well as by positive emotions such as affection and ecstasy. During psychomotor epileptic seizures, which often begin in the amygdala, patients report experiencing fearful feelings.

Functional imaging studies are beginning to reveal the neural substrates of emotional disorders. Activity in paralimbic structures, which include the anterior cingulate, orbitofrontal cortex, insula, and temporal pole, appears to differ in people with depression and certain anxiety disorders compared with controls. These brain structures are intermediate between limbic structures and neocortex in their phylogenetic origin, and are involved in coordination of memory and higher-level emotional functions. Mayberg, for example, proposes that major depression is associated with dysfunction of specific subregions of the anterior cingulate cortex, and that recovery from depression is associated with a reversal of some of these effects. A similar pattern of dysfunction has been observed in individuals with post-traumatic stress disorder (PTSD). In response to trauma-related stimuli, the amygdala and certain paralimbic structures such as the orbitofrontal cortex are more activated in the brains of individuals with PTSD. Owing to learning related to the trauma, individuals with PTSD may have a lower threshold in the detection of fearful stimuli by the amygdala. Importantly, the anterior cingulate was found to not be activated in individuals with PTSD in response to trauma-related stimuli. This may be related to the numbing of emotional experience that many PTSD patients report. (*See* **Anxiety Disorders**; **Affective Disorders: Depression and Mania**)

Thus, these regions subserve a variety of functions, including the integration of cognitive and emotional functions via infusion of cognitive functions with emotional significance. This could help to explain, for example, how cognitive functions may be altered in the context of depression or anxiety. Notably, there is overlap in the neural circuitry implicated in emotional disorders and in substance abuse. Breiter and colleagues demonstrated in people addicted to cocaine that this drug activates a number of limbic and paralimbic structures, including the nucleus accumbens, caudate, putamen, insula and cingulate, and deactivates the amygdala and temporal pole.

Emotion Measurement in Humans

A variety of methods are used to measure emotional function in humans. Subjective methods include self-report (by questionnaire, structured or open-ended interview, response to emotion induction or other experimental tasks) and behavioral observation. Psychophysiological data indirectly tap into neural function through peripheral indices such as muscle activity (electromyography) and autonomic nervous system activity, measured by electrodermal or skin conductance activity, heart rate, pupillary dilation, and the startle reflex (eyeblink). Psychophysiological measures are aligned more closely with dimensional models of emotion than with the discrete emotion theory because there is insufficient evidence for a specific profile of psychophysiological measures for particular emotions. For instance, activity of the corrugator muscle, which lies under the eyebrow and creates a furrowed brow, is associated with negative emotion, whereas activity of the zygomatic muscle, which lies under the cheek and raises the corners of the mouth during smiling, is associated with positive emotion. Arousal is positively correlated with both electrodermal activity and pupillary dilation, pleasantness is related to both heart rate increases and decreased startle response, and unpleasantness is related to pupillary dilation and an enhanced startle reflex.

Neuroimaging methods used in the study of emotional function include electroencephalography (EEG), single photon emission computed tomography (SPECT), positron emission tomography (PET), and functional magnetic resonance imaging (fMRI). In several PET and fMRI studies of induced emotion, both pleasant and unpleasant pictures activated the thalamus and medial prefrontal cortex more than neutral pictures did. Most studies point towards brainstem structures, the amygdala, hypothalamus, basal forebrain, ventromedial prefrontal cortex, cingulate cortex, and orbitofrontal cortex as active in neural circuits of emotion in a variety of emotional states. (*See* **Neuroimaging**)

On the basis of neuroimaging evidence, particularly that obtained with EEG, Davidson has theorized that there is hemispheric asymmetry or lateralization of emotions. People with greater left prefrontal cortical activation report greater positive and less negative affect and respond more strongly to positive compared with negative film clips than people who exhibit greater right frontal activation, who show the opposite pattern. Thus, left prefrontal activity is proposed to be related to approach or appetitive tendencies and right prefrontal activity is related to avoidance or aversive tendencies. Some studies, however, do not support this model. Issues to be resolved include the state versus trait nature of these hemispheric associations, and the time course and contexts in which this pattern is observed.

There are dissociations or loose coupling among the different components of emotion. Consequently, there is some question as to whether emotion represents a unified concept. Thus, one view holds that although data acquired using different methods is correlated, it should not be assumed that the findings all relate to a single causative influence or neural basis of emotion. From the same data, others conclude that there is a physiologic system mediating emotion, and this system was designed to have loosely coupled components that allow for optimal flexibility. For instance, it is adaptive in certain high-arousal situations to dampen one's subjective emotional responses. As more becomes known about the neural bases of emotion, these debates will move towards resolution.

CONSCIOUS AND UNCONSCIOUS EXPERIENCE OF EMOTION

It is now well established that people can display emotional behavior in the absence of concomitant emotional experience. Emotional responding can be elicited unconsciously, which allows for rapid responding in dangerous situations (fearful or angry stimuli both indicate danger), even before the stimulus is consciously perceived, conferring an evolutionary advantage. In his work on priming,

Ohman used a backward masking procedure in which the fearful stimulus is presented for a very brief period, followed by a neutral stimulus of longer duration. Ohman demonstrated that fear-relevant stimuli are more resistant to extinction than fear-irrelevant stimuli, and people show autonomic responses to conditioned angry faces without conscious recognition of the stimuli. Concordantly, Whalen found bilateral amygdala activation and deactivation during unconscious processing of fearful and happy faces, respectively. In addition, the sublenticular substantia innominata, which reciprocally connects the amygdala and the hypothalamus, was activated during unconscious processing of both happy and fearful faces. Morris, Ohman, and Dolan extended this finding in a PET study examining neural activity during the conscious and unconscious processing of aversively conditioned angry faces. The main findings were that the right amygdala was activated during unconscious processing and the left amygdala was activated during conscious processing of the conditioned faces. These findings are consistent with the thesis that unconscious processing of emotional stimuli occurs primarily at the subcortical level, where more basic, evolutionarily old functions are executed.

These findings are to be contrasted with the activation of paralimbic structures observed in studies of the conscious experience of emotion. LeDoux argues from animal research that emotional experience is represented in working memory. Human neuroimaging studies have demonstrated that the rostral anterior cingulate cortex and the medial prefrontal cortex participate in the representation of emotional experience, and lesions to these areas produce blunting of emotional experience. The conscious experience of emotion permits planning and flexibility of response, which would benefit long-term survival. It is likely that the structures needed for the conscious experience of emotion are not emotion-specific, which could explain the variability observed across individuals in the extent to which emotion is consciously experienced.

EMOTION AND MEMORY

Cahill and McGaugh have demonstrated in animals that the amygdala is involved in the formation of enhanced declarative memory for emotionally arousing events. The amygdala is not a site of long-term memory storage, but serves to influence memory storage processes in other brain regions, such as the hippocampus, striatum, and neocortex. Human studies show that administration of a drug that blocks the effects of noradrenaline (norepinephrine) decreases the enhancement of memory by emotional arousal. Imaging studies in humans demonstrate that activity in the amygdala while viewing emotional films correlates highly with memory of the films. Studies of emotion-dependent memory in healthy volunteers suggest that episodic memory, which reflects conscious awareness of one's personal experiences, appears to be organized along emotional lines. In other words, it is easier for one to recall emotionally significant memories if one is in a similar mood to the original experience.

INDIVIDUAL DIFFERENCES IN EMOTION

There is abundant research on variability across individuals in how they evaluate emotional stimuli, respond to emotional stimuli, and experience subjective feelings of emotion. From the perspective of the neural bases of emotion, pharmacological and neuroimaging work is quite informative.

For example, Depue and colleagues have demonstrated in humans that hormonal and affective response to a drug that stimulates the neurotransmitter dopamine through blockade of the reuptake transporter is associated with the personality trait of extraversion. Extraversion reflects the tendency to experience positive emotions based on sensitivity to rewarding stimuli, such as enjoying and valuing close interpersonal bonds, pursuing leadership roles, behaving assertively, and experiencing a subjective sense of potency in accomplishing goals. (*See* **Neurotransmitters**)

In an MRI study of personality influences on brain reactivity to emotional stimuli, Canli and colleagues found that extraversion was correlated with brain reactivity to positive picture stimuli in both cortical (frontal, temporal, including the cingulate gyrus) and subcortical (amygdala, caudate, putamen) regions, while neuroticism, or the tendency to experience anxiety, was correlated with brain reactivity to negative pictures in left frontal and temporal cortical regions. Extraversion was uncorrelated with response to negative pictures, relative to positive pictures, and neuroticism was uncorrelated with response to positive pictures, relative to negative pictures. Importantly, Canli and colleagues propose that studies of individual differences in the patterns of neural activity during emotional states provide a potential explanation for inconsistent findings across studies of emotion.

It is typically assumed that patterns of neural activity during emotional states are activated similarly across individuals, but evidence suggests that there is significant variability in the neural bases of emotion.

CONCLUSION

Much has been learned about the neural substrates of emotion from animal studies, observations in patients, and functional brain imaging studies in healthy volunteers and clinical populations. The fundamental constituents of the neural circuitry mediating emotion are being defined with increasing precision, and the neural basis of the interactions between emotion and cognitive functions is increasingly being understood. However, the precise mechanisms by which the neural networks mediating emotions work to orchestrate the range of normal and abnormal emotional responses in humans remain to be determined.

Further Reading

Damasio A (1994) *Descartes' Error: Emotion, Reason, and the Human Brain*. New York, NY: Grosset/Putnam.

Damasio A (1999) *The Feeling of What Happens: Body and Emotion in the Making of Consciousness*. New York, NY: Harcourt.

Davidson RJ and Irwin W (1999) The functional neuroanatomy of emotion and affective style. *Trends in Cognitive Sciences* **3**: 11–21.

Johnston VS (1999) *Why We Feel: The Science of Human Emotions*. Reading, MA: Perseus Books.

Lane RD, Nadel L, Ahern GL *et al*. (2000) *Cognitive Neuroscience of Emotion*. New York, NY: Oxford University Press.

Le Doux J (1996) *The Emotional Brain: The Mysterious Underpinnings of Emotional Life*. New York, NY: Simon & Schuster.

Le Doux J (2000) Emotion circuits in the brain. *Annual Review of Neuroscience* **23**: 155–184.

Lewis M and Haviland-Jones JM (2000) *Handbook of Emotions*. New York, NY: Guilford Press.

Panksepp J (1998) *Affective Neuroscience: The Foundations of Human and Animal Emotions*. New York, NY: Oxford University Press.

Rolls E (1999) *The Brain and Emotion*. New York, NY: Oxford University Press.

Emotion, Philosophical Issues about

Intermediate article

Paul E Griffiths, University of Pittsburgh, Pittsburgh, Pennsylvania, USA

CONTENTS

Philosophers have discussed the nature of emotions with particular reference to whether emotions are feelings, whether they are, or involve, cognitive states such as beliefs, whether they are human universals, whether the category of emotion is a unitary one, and the relationship between emotion and rationality.

CENTRAL PHILOSOPHICAL ISSUES ABOUT EMOTION

Feeling and Cognition

Until the twentieth century it was generally believed that emotions are feelings: subjective states of experience. Darwin carried out extensive empirical investigations of the physiological and behavioral components of emotion but interpreted these as merely external manifestations of the emotions themselves. He regarded emotions as closely akin to bodily sensations such as hunger or pain. Feelings also played a central role in the other important theory of the closing years of the nineteenth century, the James/Lange theory of emotion causation. William James provoked a long and productive tradition of research with his suggestion that emotion feelings are caused by the bodily changes associated with emotion, rather than the reverse. Nevertheless, he took the final stage in this causal sequence, the occurrence of emotion feelings, to be the central feature of an emotion considered as a *psychological* phenomenon.

Predictably, the rise of behaviorism led to the decline of the feeling theory of emotion. Behaviorists in psychology had no difficulty in fitting emotional reactions into their theoretical framework. John B. Watson suggested that all adult emotional reactions were conditioned responses based on three unconditioned reactions present in infants that he termed fear, rage, and pleasure. Behaviorist philosophers such as Gilbert Ryle attempted to analyze the meanings of sentences about emotion so as to eliminate any essential reference to subjective states of experience. The 'cognitive revolution' of the 1960s did not rehabilitate the feeling theory. Philosophers assimilated the new turn in the sciences of the mind by taking existing behavioral definitions of mental states and treating them as implicit definitions of underlying mental states which cause behavior. An emotion is an internal state that mediates causally between sensory inputs and behavioral outputs in a characteristic way. Crudely, anger is an internal state that takes slights as inputs and yields aggression as output. No reference to the quality of feeling associated with anger is required.

The consensus that emerged in the philosophy of emotion in the early 1960s and persists to the present day is that emotions are defined by the cognitions they involve. Later theorists have increasingly allowed feelings a role in emotion, but never one that determines the identity of the emotion. Instead, the role of emotion feelings is to add the 'heat' to 'hot cognition'. A leading contemporary philosopher of emotion, Patricia Greenspan, concludes that emotions are feelings of comfort or discomfort directed towards an evaluative thought about an external (or imaginary) stimulus (Greenspan, 1988). It is the evaluative thought that defines the emotion. Different negative emotions, such as anger and fear, are differentiated only by the evaluative thoughts they involve. Philosophers have generally held it to be a conceptual truth that emotions derive their identities from the thoughts associated with them, and so have not seen empirical results showing the differentiation of states of bodily arousal in different emotions as relevant to their research. Philosophers have, however, cited work on the 'cognitive labeling' of states of arousal as evidence that empirical findings converge on the same conclusion as their conceptual analyses. Many cite a famous 1962 study in which subjects were induced to describe the effects of identical adrenaline injections as either euphoria or anger (Schachter and Singer, 1962). The experimental design is quite complex, but in essence, some subjects were induced to discount the injection as a cause of their aroused state and these subjects described their state of arousal as an emotion appropriate to whichever cues the experimenters provided them with. (*See* **Cognitive Science: Philosophical Issues**)

Universality of Emotion

Emotions are widely believed to be a critical feature of moral agency, and are even more widely believed to be a critical part of aesthetic response. The claim that all healthy people display, recognize, and respond to the same emotions has been used to support the view that moral and aesthetic judgments can have universal validity. Conversely, if human emotions are as diverse as the concepts embodied in different languages, and if humans can understand the expressive repertoire only of their own cultural group, this would seem to support cultural relativism about ethics and aesthetics.

Until the 1970s there was a fairly solid consensus, based on anthropological fieldwork, that the emotions vary widely across cultures. The work of Paul Ekman and his collaborators has produced an equally widespread consensus that certain 'basic emotions' are found in all human cultures (Ekman, 1972). Ekman revived Darwin's experimental work on human facial expressions of emotion and demonstrated that a range of Western facial expressions of emotion could be reliably classified by members of a visually isolated, non-Western culture, and vice versa. He also reconfirmed many of Darwin's claims about the specific muscles used in these pancultural expressions. Other investigators demonstrated the early emergence of these expressions in human infants and established homologies with facial expressions in non-human primates. The widely accepted 'basic emotions' are fear, anger, surprise, sadness, joy, disgust, and perhaps contempt. (Each emotion term in this list refers to an operationally characterized, brief, involuntary response rather than to the full range of cases commonly referred to by the term.)

Cultural relativism about emotions was revived in the 1980s as part of a broader interest in the social construction of mental phenomena. This led to the first real involvement by analytic philosophers in the debate over universality, since the new arguments for social constructionism were as much conceptual as empirical. One influential argument starts from the widely accepted idea that an emotion involves a cognitive evaluation of the stimulus.

In that case, it is argued, cultural differences in how stimuli are represented will lead to cultural differences in emotion. If two cultures think differently about danger, then, since fear involves an evaluation of a stimulus as dangerous, fear in these two cultures will be a different emotion. Adherents of the basic emotions view are unimpressed by this argument since they define emotions by their behavioral and physiological characteristics and allow that there is a great deal of variation in what triggers the same emotion in different cultures.

Social constructionists also define the domain of emotion in a way that makes basic emotions research less relevant. The six or seven basic emotions seem to require minimal cognitive evaluation of the stimulus. Social constructionists often refuse to regard these physiological responses as emotions in themselves, reserving that term for the broader cognitive state of a person involved in a social situation in which he or she might be described as, for example, angry or jealous.

It is thus unclear whether the debate between the constructionists and their universalist opponents is more than merely semantic. One side has a preference for tractable, reductive explanations, even if these are of limited scope, and the other is concerned that science may neglect the social and cultural aspects of human emotion. (*See* **Social Cognition**)

Is Emotion a Natural Kind?

The neuroscientist Antonio Damasio recently defined an emotion as 'a specifically caused transition of the organism state' (Damasio, 1999, p. 282). Confronted by similar definitions, Alan Fridlund has remarked: 'Here, the logical question is what *isn't* emotion. Emotion has, in fact, replaced Bergson's *elan vital* and Freud's *libido* as the energetic basis of all human life' (Fridlund, 1994, p. 185). For many theorists emotion has indeed become synonymous with the whole affective life of the organism and perhaps with motivation itself. Damasio is well aware of this situation and is self-consciously using a familiar term for his own purposes in order to facilitate communication in what he sees as a period of conceptual upheaval (Damasio, 1999, p. 341). He does indeed have a very broad conception of emotion, to the extent that he takes it as axiomatic that a person is always in some emotional state or other.

Paul Griffiths has argued that the scientific investigation of the domain of affective phenomena has been hindered by a continued belief that 'the emotions' are a unitary kind of psychological state (Griffiths, 1997). Science aims to group phenomena into 'natural kinds': categories about which there are many, reliable generalizations to be discovered. The domain of emotion is so large that it is unlikely that all the psychological states in that domain form a natural kind. Hence there will be few if any reliable generalizations about emotion or, in other words, no theory of emotion in general. Scientific progress would be served by dividing up the domain and investigating groups of phenomena that are likely to form natural kinds, as has occurred in research into memory. Basic emotions theorists may well be investigating one such natural kind and social constructionists about emotion may be examining another – perhaps a certain kind of transient social role. New, more specific concepts will be required to replace the emotion concept, and a central role for philosophers of emotion is to facilitate this kind of conceptual revision.

Most philosophers of emotion see no serious problem with the category of emotion, although they admit that it is vague and covers a diverse range of phenomena. Their concern is with the proper analysis of the concept associated with the word 'emotion' in everyday language. Their analyses of the emotion concept are in reasonable agreement with those produced by psychologists studying the use of the term 'emotion' in Western cultures. There are clear paradigms of emotion, such as love, happiness, anger, fear, and sadness, and most philosophers define emotion so as to include these. Their definitions disagree over the same cases that produce disagreement between subjects in empirical studies, cases such as pride, hope, lust, pain, and hunger. Philosophical definitions include features that psychologists have argued are part of the prototype of the emotion concept in Western culture. Emotions are directed onto external states of affairs, are relatively short-lived, and have an evaluative aspect to them, such that their objects are judged to be either attractive or aversive. Most definitions also provide a role for emotion feelings.

Hence philosophers, like ordinary speakers, can achieve a reasonable level of agreement about what counts as an emotion, as opposed to a mood, a desire, or an intention. Whether the psychological states grouped together in this way form a single, productive object of scientific investigation, and whether other cultures conceptualize emotion in the same way, remains to be seen.

PHILOSOPHICAL VIEWS AND THEORIES OF EMOTION

Propositional Attitude Theories of Emotion

Since the early 1960s the cognitive or propositional attitude school has dominated the philosophy of emotion (Deigh, 1994). The basic commitments of this school are twofold. First, emotions are differentiated from one another by the cognitive states that they involve, as discussed above. Second, the cognitive states involved in emotion can be understood in terms of a *propositional attitude* theory of mental content. Mental states are attitudes, such as belief, desire, hope, and intention, to propositions. The nature of propositions is the subject of complex debate, but for present purposes we can treat them as representations of states of affairs.

The simplest propositional attitude theory is the judgmentalist theory, which identifies emotions with evaluative judgments. A person is angry if he or she has the attitude of belief to the proposition 'I have been wronged'. Other prominent varieties of propositional attitude theory are belief/desire theories, hybrid feeling theories, and 'seeing as' theories.

Belief/desire theories analyse emotions as combinations of beliefs and desires. For example, hope is the belief that some state of affairs is possible and the desire that it be actual.

Hybrid feeling theories, like that of Greenspan discussed above, analyze emotions as combinations of propositional attitudes and feelings. The feeling component is used to differentiate cold cognition from hot (emotional) cognition and in some theories to distinguish positive from negative emotions. The specific identity of the emotion is given by the propositional attitude component.

Finally, 'seeing as' theories have become increasingly popular. These theories cope with various anecdotal objections to earlier propositional attitude theories by noting that a subject's beliefs and desires about an object are not sufficient to constitute an emotion unless the subject 'sees' the object in the right way. A typical anecdote involves a mountain climber who is said to retain the same beliefs and desires as she fluctuates between seeing a climb as terrifying and as exhilarating. Earlier versions of this approach were inclined to treat 'seeing as' as a primitive, following some aspects of the later work of Wittgenstein. Contemporary versions analyze 'seeing as' in terms of attentional phenomena in cognition. Emotions are then biases in cognition that direct attention at some sources of information rather than others or lead to a higher weighting for one consideration than another, and thus lead to actions that would not have eventuated in the absence of the emotion.

The theories just outlined are primarily intended as analyses of the concept of emotion. They are assessed for their ability to correctly predict the author's intuitions about whether an emotion, or some specific emotion, occurs in an imaginary scenario. Some authors draw extensively on literature for these scenarios, others draw on more or less actual cases from the psychoanalytic literature. Some see their work as assisting the scientific investigation of emotion by more clearly defining its subject matter. Others see their work as complementary and parallel to scientific psychology. Philosophical psychology is often distressingly unclear about the identity of its subject population. It is conventional in the literature to refer to 'our emotions' and to 'commonsense' views about emotion, but it is unclear to what extent these locutions are to be read as limiting the claims made to the author's own community.

The Rationality of Emotions

The prime concern of the propositional attitude school in the philosophy of emotion has been with whether emotions are rational. The feeling theory of emotion is condemned for placing emotions outside the realm of rational evaluation. This is seen as part of a wider and invidious tendency to separate the realm of the moral from the realm of the rational. The attempt to bring these realms together is conceived by some authors as a proposal for the reform of moral discourse and by others as an attempt to do justice to one strain of 'our' everyday practice.

The simplest judgmentalist theory brings emotions back into the domain of reason by identifying them with beliefs. An emotion is rational if the beliefs composing it are justified by the evidence available to the subject. More complex propositional attitude theories give more complex accounts of the rationality of emotions. Belief/desire theories face the difficulty that formal accounts of rationality, such as decision theory, are confined to evaluating the suitability of means to ends and take the ends (desires) as given. So these theories must provide an account of what it is rational to desire. Hybrid feeling theories can evaluate the rationality of having one emotion rather than another, since the identity of an emotion is determined by its propositional attitude component. Whether the state is an emotion in the first place, however, relies on the feeling component, and so hybrid feeling theories must give some account of when it is rational to take one's cognition hot rather than cold. An extensive literature canvasses solutions to these and other difficulties with the project of rationally evaluating emotions. 'Seeing as' theories face their own difficulties, but also have new resources to bring to bear on the rationality question. The cognitive biases that constitute emotions can be evaluated for their heuristic value in generating true belief, successful action, and so forth, and judged rational if they are successful in these respects.

IMPACT OF COGNITIVE SCIENCE ON ISSUES ABOUT EMOTION

Evolutionary Psychology and the Universality of Emotion

Two contemporary schools of evolutionary psychology provide arguments for diametrically opposite views on the universality of emotion. John Tooby and Leda Cosmides urge the application of their well-known blueprint for the evolutionary study of the mind to the domain of emotion (Tooby and Cosmides, 1990). The mind is a collection of highly specialized, domain-specific cognitive devices, or modules, each adapted to a specific ecological problem in our evolutionary past. Existing work in basic emotions research is easily assimilated to this model, and these evolutionists see the six or seven pancultural responses confirmed to date as the first of a much larger number yet to be uncovered. A particularly confident prediction is that there will be a specific module for sexual jealousy. The psychology of emotion in modern subjects is conceived as the result of these many evolved, pancultural modules interacting with the environments found in different societies.

In stark contrast to these ideas, evolutionary arguments are used to support a tradition of emotion research that denies that emotions come in discrete types and emphasizes cultural variation in the psychology of emotion. *Transactional* theories of emotion see emotions as acts of social communication: 'nonverbal strategies of identity realignment and relationship reconfiguration which do not easily translate into the official idea of reasoned argument and information exchange' (Parkinson, 1995, p. 295). A central tenet of this approach is that emotional behaviors do not express emotions. Rather than being expressed in social interactions, an emotion actually *is* a particular kind of social interaction and emotions thus defined do not stand in a one-to-one relation to underlying psychological processes. The empirical research supporting the transactional view is aimed at showing the effects of social context both on the production of emotional behavior given a particular underlying cognitive state and on whether a given behavior is regarded as expressing an emotion. Transactional approaches naturally tend to be associated with the view that there is extensive cultural variation in emotion, since the forms of social interaction and the functions of emotions within those interactions will differ from society to society. The fundamental mechanisms underlying this variety, however, may be pancultural and subject to evolutionary explanation.

According to Alan Fridlund, the transactional view of emotion is strongly supported by the theories of animal communication found in contemporary sociobiology and behavioral ecology (Fridlund, 1994). Like most authors who have discussed the evolution of emotional expression, Fridlund treats these expressions as conveying information about an animal's motivational state. Organisms who transparently express the internal states that guide their future actions would be unlikely to succeed in evolutionary competition. Instead, he argues, selection would bring any existing signs of emotion under increasing voluntary control, enabling organisms to control the flow of information to their own advantage. Fridlund suggests that an evolutionary psychology of emotion will naturally interpret emotional behaviors not as expressions but as signals produced to manipulate the behavior of others. If we can infer other people's emotions when they would prefer us not to, then this must be the outcome of an 'arms race' between organisms seeking to predict the behavior of others and organisms seeking to manipulate their expectations. Fridlund's argument is certainly in line with the fundamental orientation of the game-theoretic literature on animal communication. However, evolutionary theory is notorious for its inability to predict the course of evolutionary change and it would be a mistake to give this theoretical argument much weight in comparison to empirical studies of the reliability, or lack thereof, with which people recognize one another's emotions. (*See* **Game Theory**)

The obvious objection to a transactional theory of emotion is that it cannot explain asocial emotions, such as fear of an asocial stimulus in a solitary subject. Fridlund has labored to demonstrate experimentally that audience effects play a cognitive role even in asocial emotions: 'What would people think if they saw me?' More fundamentally, however, transactional theorists, like the social constructionists before them, are prepared to re-define the domain of emotion in order to capture what they take to be psychological phenomena of a single natural kind and to exclude phenomena that are of a different kind (Parkinson, 1995, p. 303).

The Frame Problem and the Resurgence of the Feeling Theory

Recent work in cognitive neuroscience has shed new light on the relationship between emotion and cognition and has led to a revival of the feeling theory of emotion. Antonio Damasio has argued that effective reasoning is dependent on the capacity to experience emotion. Patients with bilateral lesions to the prefrontal cortex show both reduced emotionality and a diminished ability to allocate cognitive resources in such a way as to solve real-world problems. They do not, however, have deficits in abstract reasoning ability. Damasio interprets these findings as showing that emotion plays an essential role in labeling both data and goals for their relevance to the task in hand (Damasio, 1994).

Damasio's proposal must be regarded as highly tentative, and faces a number of difficulties (Rolls, 1999); but it has aroused considerable interest amongst cognitive scientists who have seen in 'affective computing' a possible solution to the frame problem: the problem of choosing all and only the relevant data without assessing all the available data for possible relevance (Picard, 1997). Damasio's theory bears a resemblance to some of the philosophical 'seeing as' theories which identify emotions with heuristic biases in cognition. In contrast to those theories, however, Damasio sees emotions themselves as feelings. If emotions function cognitively, then his proposal would be that cognitive priorities are assigned by calculating what is most relevant and important. This would not be a solution to the frame problem, but an instance of that problem. Damasio avoids this trap by using emotion *feelings* to prioritize cognition. He describes a class of 'primary emotions' that bear a strong affinity to Ekman's basic emotions. Damasio envisages emotional development as a process in which the feelings associated with the basic emotions become attached to particular cognitive states, giving rise to cognition/feeling composites that he labels 'secondary emotions'.

Damasio has so far given only a suggestive outline of his theory and it remains to be seen whether this sketch can be developed into a workable model of cognitive processes. Attempts to expand on Damasio's ideas to date resemble traditional behavior conditioning, with thoughts taking the place of behaviors and emotion feelings acting as reinforcers. The limitations of conditioning models as explanations of complex cognitive performances are well known.

Neurological Support for Twin Pathway Models of Emotion

Another important recent development is Joseph LeDoux's detailed mapping of the neural pathways involved in fear conditioning (LeDoux, 1996). Information about the stimulus activates many aspects of emotional response via a fast, 'low road' through subcortical structures, amongst which the amygdala is particularly important. A slower, 'high road' activates cortical structures and is essential for longer-term, planned, and often conscious responses to the same stimulus.

These findings are consistent with Ekman's proposal that an 'automatic appraisal mechanism' is associated with the basic emotions and operates independently of the formation of conscious or reportable judgments about the stimulus situation. LeDoux's findings also help to explain the experimental phenomenon of 'affective primacy', in which emotional associations with stimuli can be conditioned independently from paradigmatically cognitive responses to the same stimuli, such as recognition or recall.

Twin-pathway models suggest that at least for certain basic emotions the idea that an emotion involves a cognitive evaluation of the stimulus needs to be replaced with the idea that it involves two evaluations, which can conflict and which have complementary but independent cognitive functions. Twin-pathway models also provide some support for the many evolutionary accounts that see the basic emotions as 'quick and dirty' solutions to common survival problems.

CONCLUSION

Perhaps the most pressing philosophical question about emotion is the relationship between the philosophical psychology of emotion and the sciences of the mind. Even the most apparently purely philosophical issue, the rationality of emotion, is now the subject of simultaneous scientific study. A closer examination of the work of philosophers who claim to be concerned only with conceptual issues reveals that few of them fail to take some note of relevant empirical findings. Conversely, the exciting recent work of Antonio Damasio has led him to confront the traditional philosophical questions of the nature of mental representation and of conscious experience. These questions are arguably as much conceptual puzzles as empirical questions. The future of the field seems inevitably to be one of closer interdisciplinary cooperation.

References

Damasio AR (1994) *Descartes' Error: Emotion, Reason and the Human Brain*. New York, NY: Grosset/Putnam.

Damasio AR (1999) *The Feeling of What Happens: Body and Emotion in the Making of Consciousness*. New York, NY: Harcourt Brace.

Deigh J (1994) Cognitivism in the theory of emotions. *Ethics* **104**: 824–854.

Ekman P (1972) *Emotions in the Human Face*. New York, NY: Pergamon Press.

Fridlund A (1994) *Human Facial Expression: An Evolutionary View*. San Diego, CA: Academic Press.

Greenspan P (1988) *Emotions and Reasons: An Inquiry into Emotional Justification*. New York, NY: Routledge.

Griffiths PE (1997) *What Emotions Really Are: The Problem of Psychological Categories*. Chicago, IL: University of Chicago Press.

LeDoux J (1996) *The Emotional Brain: The Mysterious Underpinnings of Emotional Life*. New York, NY: Simon & Schuster.

Parkinson B (1995) *Ideas and Realities of Emotion*. London, UK: Routledge.

Picard R (1997) *Affective Computing*. Cambridge, MA: MIT Press.

Rolls ET (1999) *The Brain and Emotion*. Oxford, UK: Oxford University Press.

Schachter S and Singer JE (1962) Cognitive, social and physiological determinants of emotional state. *Psychological Review* **69**: 379–399.

Tooby J and Cosmides L (1990) The past explains the present: emotional adaptations and the structure of ancestral environments. *Ethology and Sociobiology* **11**: 375–424.

Further Reading

Calhoun C and Solomon RC (eds) (1984) *What is an Emotion? Classic Readings in Philosophical Psychology*. New York, NY: Oxford University Press.

Darwin C (1872) *The Expression of the Emotions in Man and Animals*. New York, NY: Philosophical Library.

De Sousa R (1991) *The Rationality of Emotion*. Cambridge, MA: MIT Press.

Ekman P (ed.) (1973) *Darwin and Facial Expression: A Century of Research in Review*. New York, NY: Academic Press.

Ekman P and Davidson RJ (eds) (1994) *The Nature of Emotion: Fundamental Questions*. New York, NY: Oxford University Press.

Harré R (ed.) (1986) *The Social Construction of the Emotions*. London, UK: Oxford University Press.

Panksepp J (1998) *Affective Neuroscience: The Foundations of Human and Animal Emotions*. New York, NY: Oxford University Press.

Encoding and Retrieval, Neural Basis of

Intermediate article

Lars Nyberg, Umeå University, Umeå, Sweden

CONTENTS

Cognitive theories describe encoding and retrieval as two distinct but interacting memory processes. This way of characterizing encoding and retrieval processes converges with analyses of their neural basis in that encoding and retrieval seem to engage specific as well as common brain areas.

INTRODUCTION

Encoding and retrieval are two fundamental memory processes. Encoding refers to the acquisition of information into memory, and retrieval to the use of previously encoded information. Retrieval can be *implicit* in the sense that previously encoded information affects subsequent behavior even though no conscious attempt is made to retrieve that information. Here, focus is on *episodic* memory retrieval as measured by tasks such as recall and recognition. Such tasks require *explicit* retrieval of previously encoded information in the sense that one has to think back to a previous study episode and retrieve information that was acquired at that particular time and place. Episodic information can be verbal or nonverbal; it can involve sensations of smell, taste, and touch; and it can represent emotional states.

THEORIES OF ENCODING AND RETRIEVAL AND THEIR INTERACTION

The division of the learning/memory process into encoding and retrieval (and an intermediate storage stage) became a major issue during the 1960s. Studies by Tulving and colleagues set the stage for subsequent empirical and theoretical work on retrieval processes. In a very influential paper (Tulving and Pearlstone, 1966), it was demonstrated that if encoding (and storage) conditions are held constant, the amount of information retrieved depends on the retrieval conditions.

During the 1970s, a considerable amount of research was focused on encoding. Much of this work was inspired by the *levels-of-processing* framework (Craik and Lockhart, 1972). In this framework, encoding was described in terms of various processing levels, and deeper processing was suggested to lead to better retention than shallow processing. It became clear that intention *per se* is not crucial for effective encoding but rather the way information is processed.

Another important topic of research during the 1970s concerned the interplay between encoding and retrieval processes. One set of findings gave rise to the *encoding specificity hypothesis* (Thomson and Tulving, 1970), by showing that retrieval cues are effective to the extent that they overlap with the encoded information. Another set of findings qualified predictions from the levels-of-processing framework by showing that deeper processing at encoding does not always lead to superior retention. Rather, what is optimal processing at encoding is dependent on the retrieval conditions. These findings formed the basis for the *transfer appropriate processing* principle (Morris *et al.*, 1977). The importance of interplay between encoding and retrieval processes is also salient in other theoretical accounts, such as Kolers' procedural viewpoint (Kolers, 1973).

Encoding–retrieval interplay has a central role in contemporary accounts of dissociations between measures of retrieval, most notably dissociations between explicit and implicit retrieval. Based on a classification of retrieval measures as 'conceptually-driven' (dependent on the encoded meaning of information) or 'data-driven' (dependent on the perceptual match between encoding and retrieval) (Jacoby, 1983), Roediger and colleagues have put forward a transfer-appropriate account of such dissociations (Roediger *et al.*, 1989).

CAN ENCODING AND RETRIEVAL BE DISTURBED SEPARATELY? LESION AND PHARMACOLOGICAL EVIDENCE

A fundamental question in the study of the neural basis of encoding and retrieval is whether encoding and retrieval can be selectively affected by brain damage. Here it should be noted that it is difficult to isolate the effects of brain damage to encoding or retrieval processes. If a lesion is made/has occurred prior to encoding and an effect is noted on subsequent retrieval performance, it is not clear whether the lesion affected encoding or retrieval processes. If a lesion occurs after encoding and prior to retrieval, it is still difficult to conclude that it affected retrieval processes. This is because there is evidence that consolidation of information in memory goes on for some time after the initial encoding. Moreover, it is possible that the lesion affected sites for memory storage rather than sites involved in the retrieval of stored information. With these caveats in mind, some suggestions regarding process-specific effects of brain damage will be considered.

Organic amnesia is characterized by dysfunctional episodic memory. This syndrome can result from lesions in the medial temporal lobes, diencephalon, or the basal forebrain. Collectively, these sites have been referred to as the 'expanded' limbic system (Markowitsch, 2000). This system, or various combinations of its constituent parts, has been associated with the transfer of incoming registered information for long-term memory storage. It has been proposed that the limbic structures can be regarded as 'bottleneck structures' in the sense that bilateral damage to any of them will lead to anterograde amnesia (Markowitsch, 2000).

Limbic regions may also play a role in retrieval. Several studies indicate that damage to limbic structures impairs retrieval of recent episodes, but it is a matter of debate whether retrieval of remote events is also affected. Moreover, specific regions in the infero-lateral prefrontal cortex and in the temporo-polar cortex (especially in the right hemisphere) seem to be critical for retrieval. Lesions to this regional combination result in retrograde amnesia (Markowitsch, 2000).

In a recent study (Rossi *et al.*, 2001), repetitive transcranial magnetic stimulation was used to transiently interfere with prefrontal brain activity during encoding and retrieval of pictures. It was found that the left dorsolateral prefrontal cortex was crucial for encoding, whereas the right dorsolateral prefrontal cortex was crucial for retrieval. This study provides strong evidence that encoding and retrieval can be disturbed separately, and highlights the role of prefrontal regions in this regard (cf. the section below on differences in functional brain activity).

Psychopharmacological studies also provide some hints that encoding and retrieval processes can be disturbed separately. A much-studied class of drugs is benzodiazepines. These drugs facilitate the transmission of gamma-aminobutyric acid (GABA), which is the major inhibitory neurotransmitter in the brain. Administration of benzodiazepine drugs before the encoding of new information leads to poor retention. By contrast, if the drug is administered at the test phase, performance is not affected (Curran, 2000). A similar effect has been observed for scopolamine, which acts as a cholinergic blocker. The observed pattern of effects suggests that these drugs impair encoding processes rather than retrieval processes, possibly by affecting the ability to form item–item or item–context associations (Curran, 2000).

ARE THERE SEPARATE NEURAL SYSTEMS FOR ENCODING AND RETRIEVAL?

The results from lesion and pharmacological studies indicate that brain damage or drug administration can have selective effects on encoding and retrieval processes. These patterns of results provide tentative support for the existence of separate neural systems for encoding and retrieval. Additional evidence comes from functional neuroimaging studies. In these studies the neural correlates of encoding and retrieval can be studied separately, and hence it can be explored whether there are distinct encoding and retrieval systems in the brain. Two main classes of functional neuroimaging techniques are hemodynamic methods and electromagnetic methods. This discussion is limited to hemodynamic registrations by positron emission tomography (PET) and functional magnetic resonance imaging (fMRI).

FUNCTIONAL NEUROIMAGING STUDIES

PET and fMRI rely on the changes in cerebral blood flow that accompany neural activity. To identify changes in brain activity that are associated with a process of interest, it is common to measure activity in at least two conditions (experimental and control condition). These conditions are carefully matched in all respects, except for the process of interest. By subtracting out brain activity associated with the

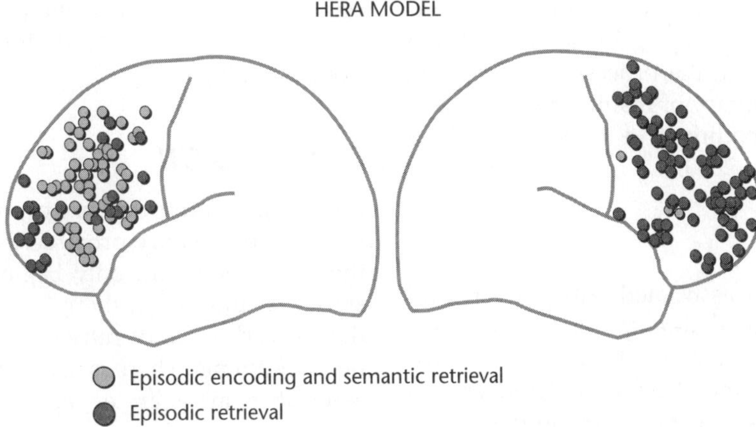

Figure 1. Differential activation of prefrontal cortex during encoding and retrieval of episodic information. Prefrontal activations from published PET and fMRI studies up to December 1999 are plotted on lateral brain outlines. Courtesy of Roberto Cabeza.

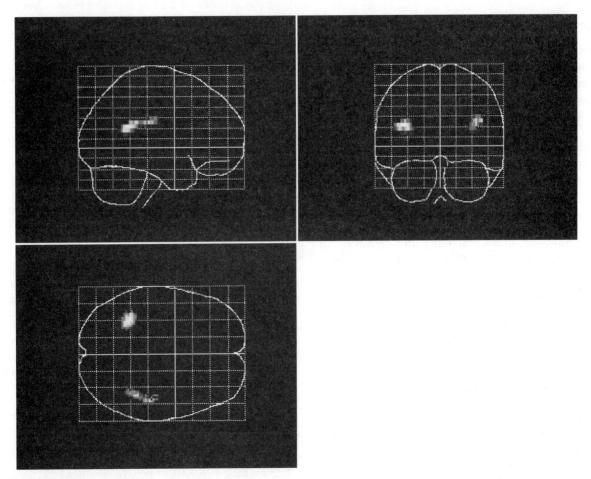

Figure 2. [*Figure is also reproduced in color section.*] Overlap in parietal activation during encoding and retrieval of spatial information. Activations are outlined in sagittal (top left), coronal (top right), and horizontal (bottom left) transparent brain maps. Reprinted with permission from 'Conjunction analysis of cortical activations common to encoding and retrieval' (Persson and Nyberg, 2000).

control condition from that associated with the experimental condition, it is possible to isolate brain regions associated with the process of interest. In what follows, a summary of results from PET and fMRI studies of encoding and retrieval is given (Cabeza and Nyberg, 2000).

Encoding

Episodic encoding is associated with the left prefrontal cortex and medial-temporal lobe regions. Left prefrontal activation has been observed for intentional encoding and also for incidental encoding (e.g. comparisons of deep and shallow semantic tasks). Medial-temporal activation during encoding has been related to novelty detection. There is some evidence that medial-temporal regions interact with material-specific regions during encoding (e.g. occipito-temporal regions during picture encoding), and that the laterality of encoding-related activity is modulated by material.

Retrieval

Retrieval is strongly associated with right prefrontal and medial parietal activation. Increased activity has also frequently been observed in the medial-temporal cortex and in the cerebellum. Activity in some regions is higher during successful than unsuccessful retrieval, whereas other regions' activity tends to be unaffected by level of retrieval or increase during more demanding (and less successful) retrieval conditions.

Differences and Similarities

As indicated above, encoding and retrieval have been associated with different regions of the frontal cortex, with encoding being associated with left prefrontal regions and retrieval with right prefrontal regions. This asymmetric involvement of prefrontal regions during encoding and retrieval is captured by the HERA (Hemispheric Encoding/Retrieval Asymmetry) model shown in Figure 1 (Nyberg *et al.*, 1996).

In addition to encoding–retrieval differences, several studies have reported overlap in activation patterns for encoding and retrieval (Nyberg, 2002). One example is that encoding and retrieval of spatial locations activate overlapping regions in the 'where-pathway' as shown in Figure 2 (Persson and Nyberg, 2000). Such overlap suggests that formation and recovery of memory representations engage the same brain regions. This possibility is in line with the view that memories are represented in a distributed fashion in/near regions that are involved during initial perception/encoding.

CONCLUSION

Encoding and retrieval are two distinct, but highly interactive, memory processes. The neural bases of these processes are only beginning to be understood. At this stage, there is converging evidence that encoding and retrieval engage specialized neural systems. There is also evidence that specific association areas are engaged during encoding of specific information as well as during subsequent retrieval of the same information. Areas where encoding and retrieval processes meet in the brain may represent storage sites.

References

Cabeza R and Nyberg L (2000) Imaging cognition II: an empirical review of 275 PET and fMRI studies. *Journal of Cognitive Neuroscience* **12**: 1–47.

Craik FIM and Lockhart RS (1972) Levels of processing: a framework for memory research. *Journal of Verbal Learning and Verbal Behavior* **11**: 671–684.

Curran HV (2000) Psychopharmacological perspectives on memory. In: Tulving E and Craik FIM (eds) *The Oxford Handbook of Memory*, pp. 539–554. New York, NY: Oxford University Press.

Jacoby LL (1983) Remembering the data: analyzing interactive processes in reading. *Journal of Verbal Learning and Verbal Behavior* **22**: 485–508.

Kolers PA (1973) Remembering operations. *Memory & Cognition* **1**: 347–355.

Markowitsch HJ (2000) Neuroanatomy of memory. In: Tulving E and Craik FIM (eds) *The Oxford Handbook of Memory*, pp. 465–484. New York, NY: Oxford University Press.

Morris CD, Bransford JD and Franks JJ (1977) Levels of processing versus transfer appropriate processing. *Journal of Verbal Learning and Verbal Behavior* **16**: 519–533.

Nyberg L (2002) Where encoding and retrieval meet in the brain. In: Squire LR and Schacter DL (eds) *Neuropsychology of Memory*, pp. 193–203. New York, NY: Guilford Press.

Nyberg L, Cabeza R and Tulving E (1996) PET studies of encoding and retrieval: the HERA model. *Psychonomic Bulletin & Review* **3**: 135–148.

Persson J and Nyberg L (2000) Conjunction analysis of cortical activations common to encoding and retrieval. *Microscopy Research Techniques* **51**: 39–44.

Roediger HL III, Weldon MS and Challis BH (1989) Explaining dissociations between implicit and explicit measures of retention: a processing account. In: Roediger HL III and Craik FIM (eds) *Varieties of Memory and Consciousness: Essays in Honour of Endel Tulving*, pp. 3–41. Hillsdale, NJ: Erlbaum.

Rossi S, Cappa SF, Babiloni C *et al.* (2001) Prefrontal cortex in long-term memory: an 'interference' approach using magnetic stimulation. *Nature Neuroscience* **4**: 948–952.

Thomson DM and Tulving E (1970) Associative encoding and retrieval: weak and strong cues. *Journal of Experimental Psychology* **86**: 255–262.

Tulving E and Pearlstone Z (1966) Availability versus accessibility of information in memory for words. *Journal of Verbal Learning and Verbal Behavior* **5**: 381–391.

Further Reading

Dolan RJ and Fletcher PC (1997) Dissociating prefrontal and hippocampal function in episodic memory encoding. *Nature* **388**: 582–585.

Kelley WM, Miezin FM, McDermott KB *et al.* (1998) Hemispheric specialization in human dorsal frontal cortex and medial temporal lobe for verbal and nonverbal memory encoding. *Neuron* **20**: 927–936.

Lepage M, Ghaffar O, Nyberg L and Tulving E (2000) Prefrontal cortex and episodic memory retrieval mode. *Proceedings of the National Academy of Sciences USA* **97**: 506–511.

Mayes AR (1995) Memory and amnesia. *Behavioural Brain Research* **66**: 29–36.

Nadel L and Moscovitch M (1997) Memory consolidation, retrograde amnesia and the hippocampal complex. *Current Opinion in Neurobiology* **7**: 217–227.

Nyberg L, McIntosh AR, Houle S, Nilsson L-G and Tulving E (1996) Activation of medial temporal structures during episodic memory retrieval. *Nature* **380**: 715–717.

Rugg MD and Allan K (2000) Event-related potential studies of memory. In: Tulving E and Craik FIM (eds) *The Oxford Handbook of Memory*, pp. 521–537. New York, NY: Oxford University Press.

Schacter DL and Wagner AD (1999) Medial temporal lobe activations in fMRI and PET studies of episodic encoding and retrieval. *Hippocampus* **9**: 7–24.

Squire LR (1992) Memory and the hippocampus: a synthesis from findings with rats, monkeys, and humans. *Psychological Review* **99**: 195–231.

Wagner AD, Schacter DL, Rotte M *et al.* (1998) Building memories: remembering and forgetting of verbal experiences as predicted by brain activity. *Science* **281**: 1188–1191.

Environmental Psychology

Introductory article

Stephen Kaplan, University of Michigan, Ann Arbor, Michigan, USA

CONTENTS

In the context of evolution and environment, human information processing has many connections with preference and motivation, with far-reaching implications for the possibilities of enhancing human effectiveness.

INTRODUCTION

If we think of an environment as a pattern of information, then environmental psychology and cognitive science have much to learn from each other. Computer scientists have provided a useful concept, the 'software environment', which connects these disciplines. When dealing with a new software program one faces a pattern of information that is only a small piece of the total collection of information that makes up the program. Yet to be comfortable with the program and make good use of its capabilities one has to build some sort of mental map of the whole program. This example gives a fairly good idea of what constitutes an environment from an informational perspective.

The environment that early humans faced was not, of course, a software environment, but a physical environment: if they had not been able to master that environment, their future development as a species would not have been possible. To understand environmental psychology from a cognitive point of view it is necessary to consider these two quite different environments: the one in which humans evolved, and the one that humans now inhabit.

THE ENVIRONMENT AS A SOURCE OF INFORMATION

The ancestors of the human species were originally tree dwellers. As latecomers to the savanna, humans adopted a life at the ecological margins, scavenging where they could, taking advantage of opportunities wherever possible, and getting away from predators before it was too late. The story of human evolution is the story of how a creature with limited physical assets managed to survive by its wits, by developing powerful and ingenious ways of dealing with environmental information.

Information Processing for Survival

The typical early human was home-based yet hunted a vast territory (probably about 250 square kilometers), in which one could easily get lost, so that a good memory for spatial information was essential. In addition to spatial knowledge, several other information-processing capabilities would have been essential to a creature surviving at the ecological margins: recognition, prediction, evaluation, and action.

Quick recognition of important objects in the environment, along with a capacity to discern the essence of what is happening, to get to the gist of things, would have been imperative. Quick recognition of objects is no small challenge (one that is still very difficult for computers). Interpretation of a scene is even more difficult, as it involves both the set of important objects and their spatial configuration.

Still, quick recognition is not sufficient. It is necessary not only to understand the present situation, but to anticipate what might happen next. Thus, prediction is a vital aspect of information processing.

Even recognition and prediction are not enough. One must evaluate whether the current situation and what might come next are likely to have good or bad consequences. And one must do so quickly, so that one can act before one is overtaken by events. Reviewing many similar events that have happened before, and pondering whether or not one would want them to be repeated, are often unaffordable luxuries. There must be a way to translate many different experiences into a quick and simple mandate for action.

The Cognitive Map

How can an animal with only a modest brain store so much spatial knowledge economically? The 'cognitive map' (some sort of durable mental model of the environment) provides a possible explanation. The capacity to recognize, think about and remember objects has been studied extensively. Considerable evidence suggests that experience leads to internal representations in the human mind, which stand for things in the world. These mental units make possible quick and confident recognition of objects despite the enormous variability in how they are experienced.

There is also evidence that people tend to associate things that occur near to one another in time; in other words, they readily form mental sequences that connect the representations of things that were experienced at about the same time. It is useful to think of these sequences as composed of 'nodes' (the representations of the objects) and 'paths' (the associations between them).

Sequences are necessary to cognitive maps, but not sufficient. What is needed is some way to integrate them into a coherent model of the environment. Suppose, for example, that there is a particular internal representation that stands for a landmark, such as the campus bell tower. Any sequence one experiences during one's walks on campus that include the bell tower will lead to the activation of this representation. Thus various routes will share this common node. Imagine, for example, that one is new on campus and only knows two routes: the 'weekday' pattern, from dorm to bell tower to classroom building, and the 'weekend' pattern, which includes the sequence from stadium to bell tower to cafeteria. With the knowledge of these two sequences, even if one has never traversed it, one also knows the route from dorm to bell tower to cafeteria. The cognitive map, therefore, is not a long list of separate sequences but a network of interlocking paths whose integration is due to their common elements.

Such an integration of nodes and sequences provides an economical way to store relevant information while ignoring much unnecessary detail. This cognitive map or mental model would have been a great help in avoiding getting lost, but it accomplishes much more than that. The capacity to form networks about the structure of the environment is likely to have served our distant ancestors even before the development of language. For example, early hunters, because of their limited weapons, had to attack potential prey from close proximity. In addition to spatial knowledge, they therefore needed considerable knowledge of prey species, including the capacity to recognize important signs and to interpret and predict behavior. The cognitive map is well suited to these challenges as well. The capacity to anticipate possible futures, and to recognize distinctive configurations and what they lead to, is called 'lookahead'. It is what enables people to use mental models to try out possible alternatives and ascertain possible consequences before making a decision. Given its antiquity, as well as its flexibility and generality, the cognitive map may in fact constitute a general-purpose way of storing knowledge that underlies much of what modern humans know and do.

INFLUENCES OF ENVIRONMENT ON AFFECT AND COGNITION

The importance of the cognitive map to the survival of early humans points to another issue, often neglected in cognitive science. Imagine an individual who finds being frequently confused or lost a satisfactory state of affairs. One can be reasonably sure that such an individual could not have been our ancestor. Indeed, although little note is taken of this fact in most current research and theory, modern humans, presumably like our ancestors, have strong negative feelings about being confused and, by contrast, find being knowledgeable and competent most enjoyable. Far from being totally separate domains, cognition and emotion (or affect, as it is sometimes called) are necessarily intimately linked. Strong feelings apply not only to the adequacy of our knowledge: they are also associated with the environments in which we feel competent or confused.

Preference and the Variables that Predict It

Experts trained to design landscapes consider form, line, color, and texture to be the essential elements of beauty in the landscape. However useful these concepts may be to the designer, an important component of what people like in the landscape is, of course, content. Trees, water, and smooth places to walk play an important role. There are, however, other properties of the preferred landscape that are about information rather than content. Two categories of informational factors have been identified: understanding (making sense of a setting and expecting that one could venture into it without getting lost), and exploration (having much to look at and the possibility of learning more as one ventures into the setting).

In a sense these two informational categories are in opposition to each other. What one already understands, while providing safety, may not offer opportunities for new learning; and a place that offers new learning opportunities might also be a place where one can get lost. Not surprisingly, environments where both learning and staying oriented are possible are strongly preferred.

Together, the often competing forces of understanding and exploration provide a solution to an important problem. As a knowledge-based organism, the human must be pulled towards opportunities for learning, for acquiring new information. Thus, when one is in a highly familiar environment the support for understanding is likely to be high, so one's choice of direction will be influenced by the path that offers greater opportunity to learn. At the same time, it is important that the organism not be pulled too far from what it understands. If its survival depends on information, it is dangerous to move too deeply into unfamiliar territory where it is not clear what to do or how to interpret events. The strong preference for exploration and understanding together leads the organism to seek new information without getting too far from what it knows and can handle.

Affect and the Environment

Pleasure and pain are often triggered by actual situations. The lookahead capability of cognitive maps, however, also enables us to experience pleasure and pain with respect to events that have not yet happened. For example, if one dislikes public speaking, having to make a presentation two weeks from now can create great pain. In this case the painful event is a prediction, not a reality. Similarly, the expectation of a holiday can give one pleasure long before one departs.

What makes the human reaction to environments associated with pain and pleasure particularly interesting is the subtlety of the environmental dimensions that are coded for pleasure or pain. People readily respond with strong negative reactions when their abilities to understand or explore the environment are blocked, when they are made to look foolish or experience harm to their sense of self, or when they feel incompetent or helpless. Human competence depends on the capacity to process information effectively. An environment that makes this difficult therefore lowers competence and, in turn, safety. Thus, it is adaptive to experience pain in such environments, leading one to quickly leave the situation and avoid it in the future. Correspondingly, environments that support people's information-processing needs enhance competence and confidence, and are experienced as pleasurable.

STRESS AND MENTAL FATIGUE

While pleasure and pain are powerful forces for moving individuals away from 'bad' environments and towards 'good' ones, they are less helpful for dealing with information within an environment. The environment is highly complex. Most of what is in one's field of view at any one time is of little or no relevance to one's purposes. To function effectively in a vast sea of stimulation, selection is essential. Humans have a variety of mechanisms for selection; one of the most important of these is attention.

Of the several kinds of attention, two are particularly important to understanding the human–environment relationship. 'Fascination' refers to a kind of attention that is automatic. A fascinating stimulus is one that is hard not to attend to. In evolutionary times, what was fascinating was what was important for human purposes, so that the problem of what to attend to was often solved automatically and effortlessly. In the modern world, many factors, such as advertising and the media in general, have utilized our capacity for fascination to so great a degree that, rather than supporting one's purposes, it often serves the purposes of others. A fascinating stimulus that is irrelevant to one's purposes is a distraction. When distractions are many, it is necessary to employ a different kind of attention, called 'directed attention', to maintain one's focus on what is important.

As we have seen, recognition depends on a process whereby patterns in the environment activate corresponding mental structures. There is, however, a limit to how many such structures can be active at once (5 +/− 2, assuming that the patterns are well learned). If what uses up this precious limited capacity is just whatever is biggest or brightest or most fascinating in the environment, what goes on in the mind will be determined by the environment and not by one's purposes. An all-too-frequent consequence is that one's mental activity will be dominated by someone else's purposes. Even when environmental patterns do support a purpose that an individual is committed to, that support is not necessarily enduring. A change in environmental configuration might present a new opportunity that fits some other purpose better. Thus, if one has no way of fending off distraction, environmental variability can undermine the persistence that is essential to effective goal-directed behavior. Some means of escaping from environmental control is necessary. Directed attention provides the mechanism for achieving this. This achievement, however, comes at a cost: directed attention requires effort and is subject to fatigue. Not surprisingly, the very contexts that require its use – resisting distraction, making sense of confusing situations, maintaining one's focus when what is important is not particularly interesting – are the very contexts that lead to such mental fatigue.

As fatigue increases one may eventually become aware of one's declining effectiveness, and this can cause stress. Stress can itself serve as a severe distraction, resulting in heavy demands on directed attention. Thus, although stress and fatigue are distinct, they often occur at the same time.

Environments that lead to mental fatigue or stress reduce confidence and competence while simultaneously increasing the sense of urgency and concern for self-preservation. One would hardly expect people to be helpful or responsive under such circumstances. In other words, environments that undermine competence and confidence will be unlikely to bring out the best in people.

USING THE ENVIRONMENT TO IMPROVE EFFECTIVENESS

The task of influencing one's own behavior is often achieved more readily by an indirect approach than by a direct one. Since people are responsive to environmental factors, arranging the environment to support appropriate behavior can be a particularly effective strategy. In addition, it does not call upon willpower, which depends on the same resource necessary for directed attention.

The attentional resource is perhaps the most important of the cognitive factors that one can do something about. Since mental fatigue is an ever-present threat and a serious handicap, ways to keep it under control and to recover from it are important interventions. Recent discoveries in environmental psychology are relevant here. Research on 'restorative environments' points to the importance of spending time in settings where fascination is readily available and where directed attention is little needed. Natural environments have been shown to serve this function particularly well. A related strategy is to make the everyday environment less costly. Such strategies as reducing distractions and avoiding noisy settings can protect directed attention from unnecessary demand. The focus on the attentional resource, which recasts the way one looks at one's environment, can have beneficial effects.

Both at the personal level and on a larger scale, strategies for achieving these benefits call for many small experiments to determine what works. In the individual case, 'know thyself' here takes on a more tangible meaning, as well as a greater urgency. The solutions, however, must go beyond the individual. People who are stressed and fatigued are not ideal neighbors. They contribute little to safety, stability, and civility. The pervasiveness of unreasonable behavior is one of the plagues of the modern world. Environments more responsive to human needs might help to temper this unfortunate trend.

Further Reading

Chown E, Kaplan S and Kortenkamp D (1995) Prototypes, location, and associative networks (PLAN): towards a unified theory of cognitive mapping. *Cognitive Science* **19**: 1–51.

Cimprich B (1993) Development of an intervention to restore attention in cancer patients. *Cancer Nursing* **16**: 83–92.

Clark A (1989) *Microcognition: Philosophy, Cognitive Science, and Parallel Distributed Processing*. Cambridge, MA: MIT Press.

James W (1892) *Psychology: The Briefer Course*. New York, NY: Holt.

Kaplan R, Kaplan S and Ryan RL (1998) *With People in Mind: Design and Management of Everyday Nature*. Washington, DC: Island Press.

Kaplan S (2001) Meditation, restoration and the management of mental fatigue. *Environment and Behavior* **33**: 480–506.

Kaplan S and Kaplan R (eds) (1978) *Humanscape: Environments for People*. Belmont, CA: Duxbury. [Republished by Ulrich's, Ann Arbor, MI, 1982.]

Kaplan S and Kaplan R (1982) *Cognition and Environment: Functioning in an Uncertain World*. New York, NY: Praeger. [Republished by Ulrich's, Ann Arbor, MI, 1989.]

Kearney AR and Kaplan S (1997) Toward a methodology for the measurement of the knowledge structures of ordinary people: The Conceptual Content Cognitive Map (3CM). *Environment and Behavior* **29**: 579–617.

Pfeiffer JE (1978) *The Emergence of Man*. New York, NY: Harper & Row.

Event Related Potentials

See **Auditory Event-related Potentials**

Evolution

See **Human Brain, Evolution of the**

Explicit Representation

See **Implicit and Explicit Representation**